D1560133

GƟRDIAN®

Mechanical Costs with RSMeans data

Joseph Kelble, Senior Editor

2020
43rd annual edition

Chief Data Officer
Noam Reininger

Vice President, Data
Tim Duggan

Principal Engineer
Bob Mewis *(1, 4)*

Contributing Editors
Brian Adams *(21, 22)*
Paul Cowan
Christopher Babbitt
Sam Babbitt
Stephen Bell
Michelle Curran
Antonio D'Aulerio *(26, 27, 28, 48)*

Matthew Doheny *(8, 9, 10)*
John Gomes *(13, 41)*
Derrick Hale, PE *(2, 31, 32, 33, 34, 35, 44, 46)*
Barry Hutchinson
Joseph Kelble *(14, 23, 25)*
Scott Keller *(3, 5)*
Charles Kibbee
Gerard Lafond, PE
Thomas Lane *(6, 7)*
Thomas Lyons
Jake MacDonald *(11, 12)*
John Melin, P.E.
Elisa Mello
Matthew Sorrentino

Kevin Souza
David Yazbek

Production Manager
Debbie Panarelli

Production
Jonathan Forgit
Sharon Larsen
Sheryl Rose
Janice Thalin

Data Quality Manager
Joseph Ingargiola

Innovation
Ray Diwakar
Kedar Gaikwad
Srini Narla

Cover Design
Blaire Collins

Data Analytics
David Byars
Ellen D'amico
Thomas Hauger
Cameron Jagoe
Matthew Kelliher-Gibson
Renee Rudicil

Numbers in italics are the divisional responsibilities for each editor. Please contact the designated editor directly with any questions.

RSMeans data from Gordian
Construction Publishers & Consultants
1099 Hingham Street, Suite 201
Rockland, MA 02370
United States of America
1.800.448.8182
RSMeans.com

Copyright 2019 by The Gordian Group Inc.
All rights reserved.
Cover photo © iStock.com/Freezingtime

Printed in the United States of America
ISSN 1524-3702
ISBN 978-1-950656-13-4

0020
$319.99 per copy (in United States)
Price is subject to change without prior notice.

Related Data and Services

Our engineers recommend the following products and services to complement *Mechanical Costs with RSMeans data*:

Annual Cost Data Books
2020 Assemblies Costs with RSMeans data
2020 Square Foot Costs with RSMeans data

Reference Books
Estimating Building Costs
RSMeans Estimating Handbook
Green Building: Project Planning & Estimating
How to Estimate with RSMeans data
Plan Reading & Material Takeoff
Project Scheduling & Management for Construction
Universal Design Ideas for Style, Comfort & Safety

Seminars and In-House Training
Unit Price Estimating
Training for our online estimating solution
Practical Project Management for Construction Professionals
Scheduling with MSProject for Construction Professionals
Mechanical & Electrical Estimating

RSMeans data
For access to the latest cost data, an intuitive search, and an easy-to-use estimate builder, take advantage of the time savings available from our online application.

To learn more visit: **RSMeans.com/online**

Enterprise Solutions
Building owners, facility managers, building product manufacturers, and attorneys across the public and private sectors engage with RSMeans data Enterprise to solve unique challenges where trusted construction cost data is critical.

To learn more visit: **RSMeans.com/Enterprise**

Custom Built Data Sets
Building and Space Models: Quickly plan construction costs across multiple locations based on geography, project size, building system component, product options, and other variables for precise budgeting and cost control.

Predictive Analytics: Accurately plan future builds with custom graphical interactive dashboards, negotiate future costs of tenant build-outs, and identify and compare national account pricing.

Consulting
Building Product Manufacturing Analytics: Validate your claims and assist with new product launches.

Third-Party Legal Resources: Used in cases of construction cost or estimate disputes, construction product failure vs. installation failure, eminent domain, class action construction product liability, and more.

API
For resellers or internal application integration, RSMeans data is offered via API. Deliver Unit, Assembly, and Square Foot Model data within your interface. To learn more about how you can provide your customers with the latest in localized construction cost data visit: **RSMeans.com/API**

Table of Contents

Foreword

The Value of RSMeans data from Gordian

Since 1942, RSMeans data has been the industry-standard materials, labor, and equipment cost information database for contractors, facility owners and managers, architects, engineers, and anyone else that requires the latest localized construction cost information. More than 75 years later, the objective remains the same: to provide facility and construction professionals with the most current and comprehensive construction cost database possible.

With the constant influx of new construction methods and materials, in addition to ever-changing labor and material costs, last year's cost data is not reliable for today's designs, estimates, or budgets. Gordian's cost engineers apply real-world construction experience to identify and quantify new building products and methodologies, adjust productivity rates, and adjust costs to local market conditions across the nation. This adds up to more than 22,000 hours in cost research annually. This unparalleled construction cost expertise is why so many facility and construction professionals rely on RSMeans data year over year.

About Gordian

Gordian originated in the spirit of innovation and a strong commitment to helping clients reach and exceed their construction goals. In 1982, Gordian's chairman and founder, Harry H. Mellon, created Job Order Contracting while serving as chief engineer at the Supreme Headquarters Allied Powers Europe. Job Order Contracting is a unique indefinite delivery/indefinite quantity (IDIQ) process, which enables facility owners to complete a substantial number of repair, maintenance, and construction projects with a single, competitively awarded contract. Realizing facility and infrastructure owners across various industries could greatly benefit from the time and cost saving advantages of this innovative construction procurement solution, he established Gordian in 1990.

Continuing the commitment to providing the most relevant and accurate facility and construction data, software, and expertise in the industry, Gordian enhanced the fortitude of its data with the acquisition of RSMeans in 2014. And in an effort to expand its facility management capabilities, Gordian acquired Sightlines, the leading provider of facilities benchmarking data and analysis, in 2015.

Our Offerings

Gordian is the leader in facility and construction cost data, software, and expertise for all phases of the building life cycle. From planning to design, procurement, construction, and operations, Gordian's solutions help clients maximize efficiency, optimize cost savings, and increase building quality with its highly specialized data engineers, software, and unique proprietary data sets.

Our Commitment

At Gordian, we do more than talk about the quality of our data and the usefulness of its application. We stand behind all of our RSMeans data—from historical cost indexes to construction materials and techniques—to craft current costs and predict future trends. If you have any questions about our products or services, please call us toll-free at 800.448.8182 or visit our website at gordian.com.

MasterFormat® 2016/ MasterFormat® 2018 Comparison Table

This table compares the 2016 edition of the Construction Specifications Institute's MasterFormat® to the expanded 2018 edition. For your convenience, all revised 2016 numbers and titles are listed along with the corresponding 2018 numbers and titles.

CSI 2016 MF ID	CSI 2016 MF Description	CSI 2018 MF ID	CSI 2018 MF Description
015632	Temporary Security	015733	Temporary Security
019308	Facility Maintenance Equipment	019308	Facilities Maintenance, Equipment
024200	Removal and Salvage of Construction Materials	024200	Removal and Diversion of Construction Materials
026600	Landfill Construction and Storage	026600	Landfills
040130	Unit Masonry Cleaning	04012052	Cleaning Masonry
068010	Composite Decking	067300	Composite Decking
072127	Reflective Insulation	072153	Reflective Insulation
072610	Above-Grade Vapor Retarders	072613	Above-Grade Vapor Retarders
074473	Metal Faced Panels	074433	Metal Faced Panels
075430	Ketone Ethylene Ester Roofing	075416	Ketone Ethylene Ester Roofing
077280	Vents	077280	Vent Options
081410	Wood Doors	081410	Doors, Wood
083410	Special Function Doors	083410	Specialized Function Doors
087125	Weatherstripping	087125	Door Weatherstripping
087530	Weatherstripping	087530	Window Weatherstripping
096223	Bamboo Flooring	096436	Bamboo Flooring
096720	Epoxy-Marble Chip Flooring	096716	Epoxy-Marble Chip Flooring
099103	Paint Restoration	090190	Maintenance of Painting and Coating
102833	Laundry Accessories	102823	Laundry Accessories
114700	Ice Machines	114700	Ice Making Machines
117610	Operating Room Equipment	117610	Equipment for Operating Rooms
117710	Radiology Equipment	117710	Equipment for Radiology
122310	Wood Interior Shutters	122313	Wood Interior Shutters
123580	Commercial Kitchen Casework	123539	Commercial Kitchen Casework
124636	Desk Accessories	124113	Desk Accessories
125273	Multiple Seating	126000	Multiple Seating
141210	Dumbwaiters	141000	Dumbwaiters
211113	Facility Water Distribution Piping	211113	Facility Fire Suppression Piping
233715	Louvers	233715	Air Outlets and Inlets, HVAC Louvers
260580	Wiring Connections	260583	Wiring Connections
270110	Operation and Maintenance of Communications Systems	270110	Operation and Maintenance of Communication Systems
272123	Data Communications Switches and Hubs	272129	Data Communications Switches and Hubs
283149	Carbon-Monoxide Detection Sensors	284611	Carbon-Monoxide Detection Sensors
284621	Fire Alarm	284620	Fire Alarm
316233	Drilled Micropiles	316333	Drilled Micropiles
323420	Fabricated Pedestrian Bridges	323413	Fabricated Pedestrian Bridges
333633	Utility Septic Tank Drainage Field	333633	Ground-Level AWWA D110 Type III Prestressed Conc. Wastewater Storage Tanks
337543	Shunt Reactors	337253	Shunt Reactors
350100	Operation and Maint. of Waterway & Marine Construction	350100	Operation and Maintenance of Waterway and Marine Construction

How the Cost Data Is Built: An Overview

Unit Prices*

All cost data have been divided into 50 divisions according to the MasterFormat® system of classification and numbering.

Assemblies*

The cost data in this section have been organized in an "Assemblies" format. These assemblies are the functional elements of a building and are arranged according to the 7 elements of the UNIFORMAT II classification system. For a complete explanation of a typical "Assembly", see "RSMeans data: Assemblies—How They Work."

Residential Models*

Model buildings for four classes of construction—economy, average, custom, and luxury—are developed and shown with complete costs per square foot.

Commercial/Industrial/ Institutional Models*

This section contains complete costs for 77 typical model buildings expressed as costs per square foot.

Green Commercial/Industrial/ Institutional Models*

This section contains complete costs for 25 green model buildings expressed as costs per square foot.

References*

This section includes information on Equipment Rental Costs, Crew Listings, Historical Cost Indexes, City Cost Indexes, Location Factors, Reference Tables, and Change Orders, as well as a listing of abbreviations.

- **Equipment Rental Costs:** Included are the average costs to rent and operate hundreds of pieces of construction equipment.
- **Crew Listings:** This section lists all the crews referenced in the cost data. A crew is composed of more than one trade classification and/or the addition of power equipment to any trade classification. Power equipment is included in the cost of the crew. Costs are shown both with bare labor rates and with the installing contractor's overhead and profit added. For each, the total crew cost per eight-hour day and the composite cost per labor-hour are listed.

Unit Cost data

Assembly Cost data

Square Foot Models

- **Historical Cost Indexes:** These indexes provide you with data to adjust construction costs over time.
- **City Cost Indexes:** All costs in this data set are U.S. national averages. Costs vary by region. You can adjust for this by CSI Division to over 730 cities in 900+ 3-digit zip codes throughout the U.S. and Canada by using this data.
- **Location Factors:** You can adjust total project costs to over 730 cities in 900+ 3-digit zip codes throughout the U.S. and Canada by using the weighted number, which applies across all divisions.
- **Reference Tables:** At the beginning of selected major classifications in the Unit Prices are reference numbers indicators. These numbers refer you to related information in the Reference Section. In this section, you'll find reference tables, explanations, and estimating information that support how we develop the unit price data, technical data, and estimating procedures.
- **Change Orders:** This section includes information on the factors that influence the pricing of change orders.

- **Abbreviations:** A listing of abbreviations used throughout this information, along with the terms they represent, is included.

Index (printed versions only)

A comprehensive listing of all terms and subjects will help you quickly find what you need when you are not sure where it occurs in MasterFormat®.

Conclusion

This information is designed to be as comprehensive and easy to use as possible.

The Construction Specifications Institute (CSI) and Construction Specifications Canada (CSC) have produced the 2018 edition of MasterFormat®, a system of titles and numbers used extensively to organize construction information.

All unit prices in the RSMeans cost data are now arranged in the 50-division MasterFormat® 2018 system.

---

* Not all information is available in all data sets

Note: The material prices in RSMeans cost data are "contractor's prices." They are the prices that contractors can expect to pay at the lumberyards, suppliers'/distributors' warehouses, etc. Small orders of specialty items would be higher than the costs shown, while very large orders, such as truckload lots, would be less. The variation would depend on the size, timing, and negotiating power of the contractor. The labor costs are primarily for new construction or major renovation rather than repairs or minor alterations. With reasonable exercise of judgment, the figures can be used for any building work.

Estimating with RSMeans data: Unit Prices

Following these steps will allow you to complete an accurate estimate using RSMeans data: Unit Prices.

1. Scope Out the Project

- Think through the project and identify the CSI divisions needed in your estimate.
- Identify the individual work tasks that will need to be covered in your estimate.
- The Unit Price data have been divided into 50 divisions according to CSI MasterFormat® 2018.
- In printed versions, the Unit Price Section Table of Contents on page 1 may also be helpful when scoping out your project.
- Experienced estimators find it helpful to begin with Division 2 and continue through completion. Division 1 can be estimated after the full project scope is known.

2. Quantify

- Determine the number of units required for each work task that you identified.
- Experienced estimators include an allowance for waste in their quantities. (Waste is not included in our Unit Price line items unless otherwise stated.)

3. Price the Quantities

- Use the search tools available to locate individual Unit Price line items for your estimate.
- Reference Numbers indicated within a Unit Price section refer to additional information that you may find useful.
- The crew indicates who is performing the work for that task. Crew codes are expanded in the Crew Listings in the Reference Section to include all trades and equipment that comprise the crew.
- The Daily Output is the amount of work the crew is expected to complete in one day.
- The Labor-Hours value is the amount of time it will take for the crew to install one unit of work.
- The abbreviated Unit designation indicates the unit of measure upon which the crew, productivity, and prices are based.
- Bare Costs are shown for materials, labor, and equipment needed to complete the Unit Price line item. Bare costs do not include waste, project overhead, payroll insurance, payroll taxes, main office overhead, or profit.
- The Total Incl O&P cost is the billing rate or invoice amount of the installing contractor or subcontractor who performs the work for the Unit Price line item.

4. Multiply

- Multiply the total number of units needed for your project by the Total Incl O&P cost for each Unit Price line item.
- Be careful that your take off unit of measure matches the unit of measure in the Unit column.
- The price you calculate is an estimate for a completed item of work.
- Keep scoping individual tasks, determining the number of units required for those tasks, matching each task with individual Unit Price line items, and multiplying quantities by Total Incl O&P costs.
- An estimate completed in this manner is priced as if a subcontractor, or set of subcontractors, is performing the work. The estimate does not yet include Project Overhead or Estimate Summary components such as general contractor markups on subcontracted work, general contractor office overhead and profit, contingency, and location factors.

5. Project Overhead

- Include project overhead items from Division 1–General Requirements.
- These items are needed to make the job run. They are typically, but not always, provided by the general contractor. Items include, but are not limited to, field personnel, insurance, performance bond, permits, testing, temporary utilities, field office and storage facilities, temporary scaffolding and platforms, equipment mobilization and demobilization, temporary roads and sidewalks, winter protection, temporary barricades and fencing, temporary security, temporary signs, field engineering and layout, final cleaning, and commissioning.
- Each item should be quantified and matched to individual Unit Price line items in Division 1, then priced and added to your estimate.
- An alternate method of estimating project overhead costs is to apply a percentage of the total project cost—usually 5% to 15% with an average of 10% (see General Conditions).
- Include other project related expenses in your estimate such as:
  - Rented equipment not itemized in the Crew Listings
  - Rubbish handling throughout the project (see section 02 41 19.19)

6. Estimate Summary

- Include sales tax as required by laws of your state or county.
- Include the general contractor's markup on self-performed work, usually 5% to 15% with an average of 10%.
- Include the general contractor's markup on subcontracted work, usually 5% to 15% with an average of 10%.
- Include the general contractor's main office overhead and profit:
  - RSMeans data provides general guidelines on the general contractor's main office overhead (see section 01 31 13.60 and Reference Number R013113-50).
  - Markups will depend on the size of the general contractor's operations, projected annual revenue, the level of risk, and the level of competition in the local area and for this project in particular.
- Include a contingency, usually 3% to 5%, if appropriate.
- Adjust your estimate to the project's location by using the City Cost Indexes or the Location Factors in the Reference Section:
  - Look at the rules in "How to Use the City Cost Indexes" to see how to apply the Indexes for your location.
  - When the proper Index or Factor has been identified for the project's location, convert it to a multiplier by dividing it by 100, then multiply that multiplier by your estimated total cost. The original estimated total cost will now be adjusted up or down from the national average to a total that is appropriate for your location.

Editors' Note:

We urge you to spend time reading and understanding the supporting material. An accurate estimate requires experience, knowledge, and careful calculation. The more you know about how we at RSMeans developed the data, the more accurate your estimate will be. In addition, it is important to take into consideration the reference material such as Equipment Listings, Crew Listings, City Cost Indexes, Location Factors, and Reference Tables.

How to Use the Cost Data: The Details

What's Behind the Numbers? The Development of Cost Data

RSMeans data engineers continually monitor developments in the construction industry in order to ensure reliable, thorough, and up-to-date cost information. While overall construction costs may vary relative to general economic conditions, price fluctuations within the industry are dependent upon many factors. Individual price variations may, in fact, be opposite to overall economic trends. Therefore, costs are constantly tracked and complete updates are performed yearly. Also, new items are frequently added in response to changes in materials and methods.

Costs in U.S. Dollars

All costs represent U.S. national averages and are given in U.S. dollars. The City Cost Index (CCI) with RSMeans data can be used to adjust costs to a particular location. The CCI for Canada can be used to adjust U.S. national averages to local costs in Canadian dollars. No exchange rate conversion is necessary because it has already been factored in.

[G] The processes or products identified by the green symbol in our publications have been determined to be environmentally responsible and/or resource-efficient solely by RSMeans data engineering staff. The inclusion of the green symbol does not represent compliance with any specific industry association or standard.

Material Costs

RSMeans data engineers contact manufacturers, dealers, distributors, and contractors all across the U.S. and Canada to determine national average material costs. If you have access to current material costs for your specific location, you may wish to make adjustments to reflect differences from the national average. Included within material costs are fasteners for a normal installation. RSMeans data engineers use manufacturers' recommendations, written specifications, and/or standard construction practices for the sizing and spacing of fasteners. Adjustments to material costs may be required for your specific application or location. The manufacturer's warranty is assumed. Extended warranties are not included in the material costs. **Material costs do not include sales tax.**

Labor Costs

Labor costs are based upon a mathematical average of trade-specific wages in 30 major U.S. cities. The type of wage (union, open shop, or residential) is identified on the inside back cover of printed publications or selected by the estimator when using the electronic products. Markups for the wages can also be found on the inside back cover of printed publications and/or under the labor references found in the electronic products.

- If wage rates in your area vary from those used, or if rate increases are expected within a given year, labor costs should be adjusted accordingly.

Labor costs reflect productivity based on actual working conditions. In addition to actual installation, these figures include time spent during a normal weekday on tasks, such as material receiving and handling, mobilization at the site, site movement, breaks, and cleanup.

Productivity data is developed over an extended period so as not to be influenced by abnormal variations and reflects a typical average.

Equipment Costs

Equipment costs include not only rental but also operating costs for equipment under normal use. The operating costs include parts and labor for routine servicing, such as the repair and replacement of pumps, filters, and worn lines. Normal operating expendables, such as fuel, lubricants, tires, and electricity (where applicable), are also included. Extraordinary operating expendables with highly variable wear patterns, such as diamond bits and blades, are excluded. These costs are included under materials. Equipment rental rates are obtained from industry sources throughout North America—contractors, suppliers, dealers, manufacturers, and distributors.

Rental rates can also be treated as reimbursement costs for contractor-owned equipment. Owned equipment costs include depreciation, loan payments, interest, taxes, insurance, storage, and major repairs.

Equipment costs do not include operators' wages.

Equipment Cost/Day—The cost of equipment required for each crew is included in the Crew Listings in the Reference Section (small tools that are considered essential everyday tools are not listed out separately). The Crew Listings itemize specialized tools and heavy equipment along with labor trades. The daily cost of itemized equipment included in a crew is based on dividing the weekly bare rental rate by 5 (number of working days per week), then adding the hourly operating cost times 8 (the number of hours per day). This Equipment Cost/Day is shown in the last column of the Equipment Rental Costs in the Reference Section.

Mobilization, Demobilization—The cost to move construction equipment from an equipment yard or rental company to the job site and back again is not included in equipment costs. Mobilization (to the site) and demobilization (from the site) costs can be found in the Unit Price Section. If a piece of equipment is already at the job site, it is not appropriate to utilize mobilization or demobilization costs again in an estimate.

Overhead and Profit

Total Cost including O&P for the installing contractor is shown in the last column of the Unit Price and/or Assemblies. This figure is the sum of the bare material cost plus 10% for profit, the bare labor cost plus total overhead and profit, and the bare equipment cost plus 10% for profit. Details for the calculation of overhead and profit on labor are shown on the inside back cover of the printed product and in the Reference Section of the electronic product.

General Conditions

Cost data in this data set are presented in two ways: Bare Costs and Total Cost including O&P (Overhead and Profit). General Conditions, or General Requirements, of the contract should also be added to the Total Cost including O&P when applicable. Costs for General Conditions are listed in Division 1 of the Unit Price Section and in the Reference Section.

General Conditions for the installing contractor may range from 0% to 10% of the Total Cost including O&P. For the general or prime contractor, costs for General Conditions may range from 5% to 15% of the Total Cost including O&P, with a figure of 10% as the most typical allowance. If applicable, the Assemblies and Models sections use costs that include the installing contractor's overhead and profit (O&P).

Factors Affecting Costs

Costs can vary depending upon a number of variables. Here's a listing of some factors that affect costs and points to consider.

Quality—The prices for materials and the workmanship upon which productivity is based represent sound construction work. They are also in line with industry standard and manufacturer specifications and are frequently used by federal, state, and local governments.

Overtime—We have made no allowance for overtime. If you anticipate premium time or work beyond normal working hours, be sure to make an appropriate adjustment to your labor costs.

Productivity—The productivity, daily output, and labor-hour figures for each line item are based on an eight-hour work day in daylight hours in moderate temperatures and up to a 14' working height unless otherwise indicated. For work that extends beyond normal work hours or is performed under adverse conditions, productivity may decrease.

Size of Project—The size, scope of work, and type of construction project will have a significant impact on cost. Economies of scale can reduce costs for large projects. Unit costs can often run higher for small projects.

Location—Material prices are for metropolitan areas. However, in dense urban areas, traffic and site storage limitations may increase costs. Beyond a 20-mile radius of metropolitan areas, extra trucking or transportation charges may also increase the material costs slightly. On the other hand, lower wage rates may be in effect. Be sure to consider both of these factors when preparing an estimate, particularly if the job site is located in a central city or remote rural location. In addition, highly specialized subcontract items may require travel and per-diem expenses for mechanics.

Other Factors—

- season of year
- contractor management
- weather conditions
- local union restrictions
- building code requirements
- availability of:
  - adequate energy
  - skilled labor
  - building materials
- owner's special requirements/restrictions
- safety requirements
- environmental considerations
- access

Unpredictable Factors—General business conditions influence "in-place" costs of all items. Substitute materials and construction methods may have to be employed. These may affect the installed cost and/or life cycle costs. Such factors may be difficult to evaluate and cannot necessarily be predicted on the basis of the job's location in a particular section of the country. Thus, where these factors apply, you may find significant but unavoidable cost variations for which you will have to apply a measure of judgment to your estimate.

Rounding of Costs

In printed publications only, all unit prices in excess of $5.00 have been rounded to make them easier to use and still maintain adequate precision of the results.

How Subcontracted Items Affect Costs

A considerable portion of all large construction jobs is usually subcontracted. In fact, the percentage done by subcontractors is constantly increasing and may run over 90%. Since the workers employed by these companies do nothing else but install their particular products, they soon become experts in that line. As a result, installation by these firms is accomplished so efficiently that the total in-place cost, even with the general contractor's overhead and profit, is no more, and often less, than if the principal contractor had handled the installation. Companies that deal with construction specialties are anxious to have their products perform well and, consequently, the installation will be the best possible.

Contingencies

The allowance for contingencies generally provides for unforeseen construction difficulties. On alterations or repair jobs, 20% is not too much. If drawings are final and only field contingencies are being considered, 2% or 3% is probably sufficient and often nothing needs to be added. Contractually, changes in plans will be covered by extras. The contractor should consider inflationary price trends and possible material shortages during the course of the job. These escalation factors are dependent upon both economic conditions and the anticipated time between the estimate and actual construction. If drawings are not complete or approved, or a budget cost is wanted, it is wise to add 5% to 10%. Contingencies, then, are a matter of judgment.

Important Estimating Considerations

The productivity, or daily output, of each craftsman or crew assumes a well-managed job where tradesmen with the proper tools and equipment, along with the appropriate construction materials, are present. Included are daily set-up and cleanup time, break time, and plan layout time. Unless otherwise indicated, time for material movement on site (for items that can be transported by hand) of up to 200' into the building and to the first or second floor is also included. If material has to be transported by other means, over greater distances, or to higher floors, an additional allowance should be considered by the estimator.

While horizontal movement is typically a sole function of distances, vertical transport introduces other variables that can significantly impact productivity. In an occupied building, the use of elevators (assuming access, size, and required protective measures are acceptable) must be understood at the time of the estimate. For new construction, hoist wait and cycle times can easily be 15 minutes and may result in scheduled access extending beyond the normal work day. Finally, all vertical transport will impose strict weight limits likely to preclude the use of any motorized material handling.

The productivity, or daily output, also assumes installation that meets manufacturer/designer/standard specifications. A time allowance for quality control checks, minor adjustments, and any task required to ensure proper function or operation is also included. For items that require connections to services, time is included for positioning, leveling, securing the unit, and making all the necessary connections (and start up where applicable) to ensure a complete installation. Estimating of the services themselves (electrical, plumbing, water, steam, hydraulics, dust collection, etc.) is separate.

In some cases, the estimator must consider the use of a crane and an appropriate crew for the installation of large or heavy items. For those situations where a crane is not included in the assigned crew and as part of the line item cost,

then equipment rental costs, mobilization and demobilization costs, and operator and support personnel costs must be considered.

Labor-Hours

The labor-hours expressed in this publication are derived by dividing the total daily labor-hours for the crew by the daily output. Based on average installation time and the assumptions listed above, the labor-hours include: direct labor, indirect labor, and nonproductive time. A typical day for a craftsman might include but is not limited to:

- Direct Work
  - ☐ Measuring and layout
  - ☐ Preparing materials
  - ☐ Actual installation
  - ☐ Quality assurance/quality control
- Indirect Work
  - ☐ Reading plans or specifications
  - ☐ Preparing space
  - ☐ Receiving materials
  - ☐ Material movement
  - ☐ Giving or receiving instruction
  - ☐ Miscellaneous
- Non-Work
  - ☐ Chatting
  - ☐ Personal issues
  - ☐ Breaks
  - ☐ Interruptions (i.e., sickness, weather, material or equipment shortages, etc.)

If any of the items for a typical day do not apply to the particular work or project situation, the estimator should make any necessary adjustments.

Final Checklist

Estimating can be a straightforward process provided you remember the basics. Here's a checklist of some of the steps you should remember to complete before finalizing your estimate.

Did you remember to:

- factor in the City Cost Index for your locale?
- take into consideration which items have been marked up and by how much?
- mark up the entire estimate sufficiently for your purposes?
- read the background information on techniques and technical matters that could impact your project time span and cost?
- include all components of your project in the final estimate?
- double check your figures for accuracy?
- call RSMeans data engineers if you have any questions about your estimate or the data you've used? Remember, Gordian stands behind all of our products, including our extensive RSMeans data solutions. If you have any questions about your estimate, about the costs you've used from our data, or even about the technical aspects of the job that may affect your estimate, feel free to call the Gordian RSMeans editors at 1.800.448.8182.

Unit Price Section

Table of Contents

1

RSMeans data: Unit Prices— How They Work

All RSMeans data: Unit Prices are organized in the same way.

03 30 Cast-In-Place Concrete

03 30 53 – Miscellaneous Cast-In-Place Concrete

03 30 53.40 Concrete In Place	Crew	Daily Output	Labor-Hours	Unit	Material	2020 Bare Costs Labor	Equipment	Total	Total Incl O&P
0010 **CONCRETE IN PLACE**									
0020 Including forms (4 uses), Grade 60 rebar, concrete (Portland cement									
0050 Type I), placement and finishing unless otherwise indicated									
0500 Chimney foundations (5000 psi), over 5 C.Y.	C-14C	32.22	3.476	C.Y.	178	175	.83	353.83	460
0510 (3500 psi), under 5 C.Y.	"	23.71	4.724	"	208	238	1.13	447.13	585
3540 Equipment pad (3000 psi), 3' x 3' x 6" thick	C-14H	45	1.067	Ea.	50.50	55	.60	106.10	139
3550 4' x 4' x 6" thick		30	1.600		78	82.50	.90	161.40	210
3560 5' x 5' x 8" thick		18	2.667		138	138	1.49	277.49	360
3570 6' x 6' x 8" thick		14	3.429		190	177	1.92	368.92	475
3580 8' x 8' x 10" thick		8	6		395	310	3.36	708.36	905
3590 10' x 10' x 12" thick		5	9.600		695	495	5.40	1,195.40	1,500
3800 Footings (3000 psi), spread under 1 C.Y.	C-14C	28	4	C.Y.	203	201	.96	404.96	525
3825 1 C.Y. to 5 C.Y.		43	2.605		240	131	.63	371.63	465
3850 Over 5 C.Y.		75	1.493		220	75	.36	295.36	355

It is important to understand the structure of RSMeans data: Unit Prices so that you can find information easily and use it correctly.

---

① Line Numbers

Line Numbers consist of 12 characters, which identify a unique location in the database for each task. The first 6 or 8 digits conform to the Construction Specifications Institute MasterFormat® 2018. The remainder of the digits are a further breakdown in order to arrange items in understandable groups of similar tasks. Line numbers are consistent across all of our publications, so a line number in any of our products will always refer to the same item of work.

② Descriptions

Descriptions are shown in a hierarchical structure to make them readable. In order to read a complete description, read up through the indents to the top of the section. Include everything that is above and to the left that is not contradicted by information below. For instance, the complete description for line 03 30 53.40 3550 is "Concrete in place, including forms (4 uses), Grade 60 rebar, concrete (Portland cement Type 1), placement and finishing unless otherwise indicated; Equipment pad (3000 psi), 4' × 4' × 6" thick."

③ RSMeans data

When using **RSMeans data**, it is important to read through an entire section to ensure that you use the data that most closely matches your work. Note that sometimes there is additional information shown in the section that may improve your price. There are frequently lines that further describe, add to, or adjust data for specific situations.

④ Reference Information

Gordian's RSMeans engineers have created **reference** information to assist you in your estimate. **If** there is information that applies to a section, it will be indicated at the start of the section. The Reference Section is located in the back of the data set.

⑤ Crews

Crews include labor and/or equipment necessary to accomplish each task. In this case, Crew C-14H is used. Gordian's RSMeans staff selects a crew to represent the workers and equipment that are typically used for that task. In this case, Crew C-14H consists of one carpenter foreman (outside), two carpenters, one rodman, one laborer, one cement finisher, and one gas engine vibrator. Details of all crews can be found in the Reference Section.

Crews - Standard

Crew No.	Bare Costs		Incl. Subs O & P		Cost Per Labor-Hour	
Crew C-14H	Hr.	Daily	Hr.	Daily	Bare Costs	Incl. O&P
1 Carpenter Foreman (outside)	$55.15	$441.20	$82.85	$662.80	$51.65	$77.36
2 Carpenters	53.15	850.40	79.85	1277.60		
1 Rodman (reinf.)	56.40	451.20	84.90	679.20		
1 Laborer	42.10	336.80	63.25	506.00		
1 Cement Finisher	49.95	399.60	73.45	587.60		
1 Gas Engine Vibrator		26.85		29.54	.56	.62
48 L.H., Daily Totals		$2506.05		$3742.74	$52.21	$77.97

6 Daily Output

The **Daily Output** is the amount of work that the crew can do in a normal 8-hour workday, including mobilization, layout, movement of materials, and cleanup. In this case, crew C-14H can install thirty 4' × 4' × 6" thick concrete pads in a day. Daily output is variable and based on many factors, including the size of the job, location, and environmental conditions. RSMeans data represents work done in daylight (or adequate lighting) and temperate conditions.

7 Labor-Hours

The figure in the **Labor-Hours** column is the amount of labor required to perform one unit of work—in this case the amount of labor required to construct one 4' × 4' equipment pad. This figure is calculated by dividing the number of hours of labor in the crew by the daily output (48 labor-hours divided by 30 pads = 1.6 hours of labor per pad). Multiply 1.6 times 60 to see the value in minutes: 60 × 1.6 = 96

minutes. Note: the labor-hour figure is not dependent on the crew size. A change in crew size will result in a corresponding change in daily output, but the labor-hours per unit of work will not change.

8 Unit of Measure

All RSMeans data: Unit Prices include the typical **Unit of Measure** used for estimating that item. For concrete-in-place the typical unit is cubic yards (C.Y.) or each (Ea.). For installing broadloom carpet it is square yard and for gypsum board it is square foot. The estimator needs to take special care that the unit in the data matches the unit in the take-off. Unit conversions may be found in the Reference Section.

9 Bare Costs

Bare Costs are the costs of materials, labor, and equipment that the installing contractor pays. They represent the cost, in U.S. dollars, for one unit of work. They do not include any markups for profit or labor burden.

10 Bare Total

The **Total column** represents the total bare cost for the installing contractor in U.S. dollars. In this case, the sum of $78 for material + $82.50 for labor + $.90 for equipment is $161.40.

11 Total Incl O&P

The **Total Incl O&P column** is the total cost, including overhead and profit, that the installing contractor will charge the customer. This represents the cost of materials plus 10% profit, the cost of labor plus labor burden and 10% profit, and the cost of equipment plus 10% profit. It does not include the general contractor's overhead and profit. Note: See the inside back cover of the printed product or the Reference Section of the electronic product for details on how the labor burden is calculated.

National Average

*The RSMeans data in our print publications represent a "national average" cost. This data should be modified to the project location using the **City Cost Indexes** or **Location Factors** tables found in the Reference Section. Use the Location Factors to adjust estimate totals if the project covers multiple trades. Use the City Cost Indexes (CCI) for single trade*

projects or projects where a more detailed analysis is required. All figures in the two tables are derived from the same research. The last row of data in the CCI—the weighted average—is the same as the numbers reported for each location in the location factor table.

RSMeans data: Unit Prices—
How They Work (Continued)

Project Name: Pre-Engineered Steel Building				Architect: As Shown				
Location:	**Anywhere, USA**						01/01/20	STD
Line Number	**Description**	**Qty**	**Unit**	**Material**	**Labor**	**Equipment**	**SubContract**	**Estimate Total**
03 30 53.40 3940	Strip footing, 12" x 24", reinforced	15	C.Y.	$2,565.00	$1,770.00	$8.40	$0.00	
03 30 53.40 3950	Strip footing, 12" x 36", reinforced	34	C.Y.	$5,610.00	$3,196.00	$15.30	$0.00	
03 11 13.65 3000	Concrete slab edge forms	500	L.F.	$165.00	$1,345.00	$0.00	$0.00	
03 22 11.10 0200	Welded wire fabric reinforcing	150	C.S.F.	$2,872.50	$4,350.00	$0.00	$0.00	
03 31 13.35 0300	Ready mix concrete, 4000 psi for slab on grade	278	C.Y.	$35,306.00	$0.00	$0.00	$0.00	
03 31 13.70 4300	Place, strike off & consolidate concrete slab	278	C.Y.	$0.00	$5,309.80	$136.22	$0.00	
03 35 13.30 0250	Machine float & trowel concrete slab	15,000	S.F.	$0.00	$9,900.00	$750.00	$0.00	
03 15 16.20 0140	Cut control joints in concrete slab	950	L.F.	$47.50	$418.00	$57.00	$0.00	
03 39 23.13 0300	Sprayed concrete curing membrane	150	C.S.F.	$1,867.50	$1,065.00	$0.00	$0.00	
Division 03	**Subtotal**			**$48,433.50**	**$27,353.80**	**$966.92**	**$0.00**	**$76,754.22**
08 36 13.10 2650	Manual 10' x 10' steel sectional overhead door	8	Ea.	$11,000.00	$3,760.00	$0.00	$0.00	
08 36 13.10 2860	Insulation and steel back panel for OH door	800	S.F.	$4,000.00	$0.00	$0.00	$0.00	
Division 08	**Subtotal**			**$15,000.00**	**$3,760.00**	**$0.00**	**$0.00**	**$18,760.00**
13 34 19.50 1100	Pre-Engineered Steel Building, 100' x 150' x 24'	15,000	SF Flr.	$0.00	$0.00	$0.00	$367,500.00	
13 34 19.50 6050	Framing for PESB door opening, 3' x 7'	4	Opng.	$0.00	$0.00	$0.00	$2,320.00	
13 34 19.50 6100	Framing for PESB door opening, 10' x 10'	8	Opng.	$0.00	$0.00	$0.00	$9,600.00	
13 34 19.50 6200	Framing for PESB window opening, 4' x 3'	6	Opng.	$0.00	$0.00	$0.00	$3,450.00	
13 34 19.50 5750	PESB door, 3' x 7', single leaf	4	Opng.	$2,920.00	$736.00	$0.00	$0.00	
13 34 19.50 7750	PESB sliding window, 4' x 3' with screen	6	Opng.	$2,940.00	$630.00	$67.80	$0.00	
13 34 19.50 6550	PESB gutter, eave type, 26 ga., painted	300	L.F.	$2,415.00	$864.00	$0.00	$0.00	
13 34 19.50 8650	PESB roof vent, 12" wide x 10' long	15	Ea.	$570.00	$3,465.00	$0.00	$0.00	
13 34 19.50 6900	PESB insulation, vinyl faced, 4" thick	27,400	S.F.	$11,782.00	$10,138.00	$0.00	$0.00	
Division 13	**Subtotal**			**$20,627.00**	**$15,833.00**	**$67.80**	**$382,870.00**	**$419,397.80**
			Subtotal	$84,060.50	$46,946.80	$1,034.72	$382,870.00	$514,912.02
Division 01	**General Requirements @ 7%**			5,884.24	3,286.28	72.43	26,800.90	
	❷		Estimate Subtotal	$89,944.74	$50,233.08	$1,107.15	$409,670.90	$514,912.02
	❸		Sales Tax @ 5%	4,497.24		55.36	10,241.77	
			Subtotal A	94,441.97	50,233.08	1,162.51	419,912.67	
	❹		GC O & P	9,444.20	25,769.57	116.25	41,991.27	
			Subtotal B	103,886.17	76,002.64	1,278.76	461,903.94	$643,071.51
	❺		Contingency @ 5%					32,153.58
			Subtotal C					$675,225.09
	❻		Bond @ $12/1000 +10% O&P					8,912.97
			Subtotal D					$684,138.06
	❼		Location Adjustment Factor		115.50			106,041.40
			Grand Total					**$790,179.46**

This estimate is based on an interactive spreadsheet. You are free to download it and adjust it to your methodology.
A copy of this spreadsheet is available at **RSMeans.com/2020books.**

Sample Estimate

This sample demonstrates the elements of an estimate, including a tally of the RSMeans data lines and a summary of the markups on a contractor's work to arrive at a total cost to the owner. The Location Factor with RSMeans data is added at the bottom of the estimate to adjust the cost of the work to a specific location.

---

① Work Performed

The body of the estimate shows the RSMeans data selected, including the line number, a brief description of each item, its take-off unit and quantity, and the bare costs of materials, labor, and equipment. This estimate also includes a column titled "SubContract." This data is taken from the column "Total Incl O&P" and represents the total that a subcontractor would charge a general contractor for the work, including the sub's markup for overhead and profit.

② Division 1, General Requirements

This is the first division numerically but the last division estimated. Division 1 includes project-wide needs provided by the general contractor. These requirements vary by project but may include temporary facilities and utilities, security, testing, project cleanup, etc. For small projects a percentage can be used—typically between 5% and 15% of project cost. For large projects the costs may be itemized and priced individually.

③ Sales Tax

If the work is subject to state or local sales taxes, the amount must be added to the estimate. Sales tax may be added to material costs, equipment costs, and subcontracted work. In this case, sales tax was added in all three categories. It was assumed that approximately half the subcontracted work would be material cost, so the tax was applied to 50% of the subcontract total.

④ GC O&P

This entry represents the general contractor's markup on material, labor, equipment, and subcontractor costs. Our standard markup on materials, equipment, and subcontracted work is 10%. In this estimate, the markup on the labor performed by the GC's workers uses "Skilled Workers Average" shown in Column F on the table "Installing Contractor's Overhead & Profit," which can be found on the inside back cover of the printed product or in the Reference Section of the electronic product.

⑤ Contingency

A factor for contingency may be added to any estimate to represent the cost of unknowns that may occur between the time that the estimate is performed and the time the project is constructed. The amount of the allowance will depend on the stage of design at which the estimate is done and the contractor's assessment of the risk involved. Refer to section 01 21 16.50 for contingency allowances.

⑥ Bonds

Bond costs should be added to the estimate. The figures here represent a typical performance bond, ensuring the owner that if the general contractor does not complete the obligations in the construction contract the bonding company will pay the cost for completion of the work.

⑦ Location Adjustment

Published prices are based on national average costs. If necessary, adjust the total cost of the project using a location factor from the "Location Factor" table or the "City Cost Index" table. Use location factors if the work is general, covering multiple trades. If the work is by a single trade (e.g., masonry) use the more specific data found in the "City Cost Indexes."

Estimating Tips
01 20 00 Price and Payment Procedures

- Allowances that should be added to estimates to cover contingencies and job conditions that are not included in the national average material and labor costs are shown in Section 01 21.

- When estimating historic preservation projects (depending on the condition of the existing structure and the owner's requirements), a 15–20% contingency or allowance is recommended, regardless of the stage of the drawings.

01 30 00 Administrative Requirements

- Before determining a final cost estimate, it is good practice to review all the items listed in Subdivisions 01 31 and 01 32 to make final adjustments for items that may need customizing to specific job conditions.

- Requirements for initial and periodic submittals can represent a significant cost to the General Requirements of a job. Thoroughly check the submittal specifications when estimating a project to determine any costs that should be included.

01 40 00 Quality Requirements

- All projects will require some degree of quality control. This cost is not included in the unit cost of construction listed in each division. Depending upon the terms of the contract, the various costs of inspection and testing can be the responsibility of either the owner or the contractor. Be sure to include the required costs in your estimate.

01 50 00 Temporary Facilities and Controls

- Barricades, access roads, safety nets, scaffolding, security, and many more requirements for the execution of a safe project are elements of direct cost. These costs can easily be overlooked when preparing an estimate. When looking through the major classifications of this subdivision, determine which items apply to each division in your estimate.

- Construction equipment rental costs can be found in the Reference Section in Section 01 54 33. Operators' wages are not included in equipment rental costs.

- Equipment mobilization and demobilization costs are not included in equipment rental costs and must be considered separately.

- The cost of small tools provided by the installing contractor for his workers is covered in the "Overhead" column on the "Installing Contractor's Overhead and Profit" table that lists labor trades, base rates, and markups. Therefore, it is included in the "Total Incl. O&P" cost of any unit price line item.

01 70 00 Execution and Closeout Requirements

- When preparing an estimate, thoroughly read the specifications to determine the requirements for Contract Closeout. Final cleaning, record documentation, operation and maintenance data, warranties and bonds, and spare parts and maintenance materials can all be elements of cost for the completion of a contract. Do not overlook these in your estimate.

Reference Numbers

Reference numbers are shown at the beginning of some major classifications. These numbers refer to related items in the Reference Section. The reference information may be an estimating procedure, an alternate pricing method, or technical information.

Note: Not all subdivisions listed here necessarily appear. ■

Same Data. Simplified.

Enjoy the convenience and efficiency of accessing your costs anywhere:

- **Skip the multiplier** by setting your location
- **Quickly search,** edit, favorite and share costs
- **Stay on top of price changes** with automatic updates

Discover more at rsmeans.com/online

01 11 Summary of Work

01 11 31 – Professional Consultants

01 11 31.10 Architectural Fees		Crew	Daily Output	Labor-Hours	Unit	Material	2020 Bare Costs Labor	Equipment	Total	Total Incl O&P
0010	**ARCHITECTURAL FEES** R011110-10									
2000	For "Greening" of building **G**				Project				3%	3%

01 11 31.30 Engineering Fees										
0010	**ENGINEERING FEES** R011110-30									
0020	Educational planning consultant, minimum				Project				.50%	.50%
0100	Maximum				"				2.50%	2.50%
0200	Electrical, minimum				Contrct				4.10%	4.10%
0300	Maximum								10.10%	10.10%
0400	Elevator & conveying systems, minimum								2.50%	2.50%
0500	Maximum								5%	5%
0600	Food service & kitchen equipment, minimum								8%	8%
0700	Maximum								12%	12%
1000	Mechanical (plumbing & HVAC), minimum								4.10%	4.10%
1100	Maximum								10.10%	10.10%

01 21 Allowances

01 21 16 – Contingency Allowances

01 21 16.50 Contingencies		Crew	Daily Output	Labor-Hours	Unit	Material	2020 Bare Costs Labor	Equipment	Total	Total Incl O&P
0010	**CONTINGENCIES**, Add to estimate									
0020	Conceptual stage				Project				20%	20%
0050	Schematic stage								15%	15%
0100	Preliminary working drawing stage (Design Dev.)								10%	10%
0150	Final working drawing stage								3%	3%

01 21 53 – Factors Allowance

01 21 53.50 Factors		Crew	Daily Output	Labor-Hours	Unit	Material	2020 Bare Costs Labor	Equipment	Total	Total Incl O&P
0010	**FACTORS** Cost adjustments R012153-10									
0100	Add to construction costs for particular job requirements									
0500	Cut & patch to match existing construction, add, minimum				Costs	2%	3%			
0550	Maximum					5%	9%			
0800	Dust protection, add, minimum					1%	2%			
0850	Maximum					4%	11%			
1100	Equipment usage curtailment, add, minimum					1%	1%			
1150	Maximum					3%	10%			
1400	Material handling & storage limitation, add, minimum					1%	1%			
1450	Maximum					6%	7%			
1700	Protection of existing work, add, minimum					2%	2%			
1750	Maximum					5%	7%			
2000	Shift work requirements, add, minimum						5%			
2050	Maximum						30%			
2300	Temporary shoring and bracing, add, minimum					2%	5%			
2350	Maximum					5%	12%			

01 21 53.60 Security Factors										
0010	**SECURITY FACTORS** R012153-60									
0100	Additional costs due to security requirements									
0110	Daily search of personnel, supplies, equipment and vehicles									
0120	Physical search, inventory and doc of assets, at entry				Costs	30%				
0130	At entry and exit					50%				
0140	Physical search, at entry					6.25%				
0150	At entry and exit					12.50%				
0160	Electronic scan search, at entry					2%				

8

01 21 Allowances

01 21 53 – Factors Allowance

01 21 53.60 Security Factors	Crew	Daily Output	Labor-Hours	Unit	Material	2020 Bare Costs Labor	Equipment	Total	Total Incl O&P
0170	At entry and exit				Costs		4%			
0180	Visual inspection only, at entry						.25%			
0190	At entry and exit						.50%			
0200	ID card or display sticker only, at entry						.12%			
0210	At entry and exit				▼		.25%			
0220	Day 1 as described below, then visual only for up to 5 day job duration									
0230	Physical search, inventory and doc of assets, at entry				Costs		5%			
0240	At entry and exit						10%			
0250	Physical search, at entry						1.25%			
0260	At entry and exit						2.50%			
0270	Electronic scan search, at entry						.42%			
0280	At entry and exit				▼		.83%			
0290	Day 1 as described below, then visual only for 6-10 day job duration									
0300	Physical search, inventory and doc of assets, at entry				Costs		2.50%			
0310	At entry and exit						5%			
0320	Physical search, at entry						.63%			
0330	At entry and exit						1.25%			
0340	Electronic scan search, at entry						.21%			
0350	At entry and exit				▼		.42%			
0360	Day 1 as described below, then visual only for 11-20 day job duration									
0370	Physical search, inventory and doc of assets, at entry				Costs		1.25%			
0380	At entry and exit						2.50%			
0390	Physical search, at entry						.31%			
0400	At entry and exit						.63%			
0410	Electronic scan search, at entry						.10%			
0420	At entry and exit				▼		.21%			
0430	Beyond 20 days, costs are negligible									
0440	Escort required to be with tradesperson during work effort				Costs		6.25%			

01 21 55 – Job Conditions Allowance

01 21 55.50 Job Conditions

		Crew	Daily Output	Labor-Hours	Unit	Material	Labor	Equipment	Total	Total Incl O&P
0010	**JOB CONDITIONS** Modifications to applicable									
0020	cost summaries									
0100	Economic conditions, favorable, deduct				Project				2%	2%
0200	Unfavorable, add								5%	5%
0300	Hoisting conditions, favorable, deduct								2%	2%
0400	Unfavorable, add								5%	5%
0700	Labor availability, surplus, deduct								1%	1%
0800	Shortage, add								10%	10%
0900	Material storage area, available, deduct								1%	1%
1000	Not available, add								2%	2%
1100	Subcontractor availability, surplus, deduct								5%	5%
1200	Shortage, add								12%	12%
1300	Work space, available, deduct								2%	2%
1400	Not available, add				▼				5%	5%

01 21 57 – Overtime Allowance

01 21 57.50 Overtime

		Crew	Daily Output	Labor-Hours	Unit	Material	Labor	Equipment	Total	Total Incl O&P
0010	**OVERTIME** for early completion of projects or where R012909-90									
0020	labor shortages exist, add to usual labor, up to				Costs		100%			

01 21 Allowances

01 21 63 – Taxes

01 21 63.10 Taxes		Crew	Daily Output	Labor-Hours	Unit	Material	2020 Bare Costs Labor	2020 Bare Costs Equipment	Total	Total Incl O&P
0010	**TAXES** R012909-80									
0020	Sales tax, State, average				%	5.08%				
0050	Maximum R012909-85					7.50%				
0200	Social Security, on first $118,500 of wages						7.65%			
0300	Unemployment, combined Federal and State, minimum R012909-86						.60%			
0350	Average						9.60%			
0400	Maximum						12%			

01 31 Project Management and Coordination

01 31 13 – Project Coordination

01 31 13.20 Field Personnel

		Crew	Daily Output	Labor-Hours	Unit	Material	Labor	Equipment	Total	Total Incl O&P
0010	**FIELD PERSONNEL**									
0020	Clerk, average				Week		495		495	750
0100	Field engineer, junior engineer						1,241		1,241	1,877
0120	Engineer						1,825		1,825	2,775
0140	Senior engineer						2,400		2,400	3,625
0160	General purpose laborer, average	1 Clab	.20	40			1,675		1,675	2,525
0180	Project manager, minimum						2,175		2,175	3,300
0200	Average						2,500		2,500	3,800
0220	Maximum						2,850		2,850	4,325
0240	Superintendent, minimum						2,125		2,125	3,225
0260	Average						2,325		2,325	3,525
0280	Maximum						2,650		2,650	4,025
0290	Timekeeper, average						1,350		1,350	2,050

01 31 13.30 Insurance

		Crew	Daily Output	Labor-Hours	Unit	Material	Labor	Equipment	Total	Total Incl O&P
0010	**INSURANCE** R013113-40									
0020	Builders risk, standard, minimum				Job				.24%	.24%
0050	Maximum R013113-50								.64%	.64%
0200	All-risk type, minimum								.25%	.25%
0250	Maximum R013113-60								.62%	.62%
0400	Contractor's equipment floater, minimum				Value				.50%	.50%
0450	Maximum				"				1.50%	1.50%
0600	Public liability, average				Job				2.02%	2.02%
0800	Workers' compensation & employer's liability, average									
0850	by trade, carpentry, general				Payroll		11.97%			
1000	Electrical						4.91%			
1150	Insulation						10.07%			
1450	Plumbing						5.77%			
1550	Sheet metal work (HVAC)						7.56%			

01 31 13.50 General Contractor's Mark-Up

		Crew	Daily Output	Labor-Hours	Unit	Material	Labor	Equipment	Total	Total Incl O&P
0010	**GENERAL CONTRACTOR'S MARK-UP** on Change Orders									
0200	Extra work, by subcontractors, add				%				10%	10%
0250	By General Contractor, add								15%	15%
0400	Omitted work, by subcontractors, deduct all but								5%	5%
0450	By General Contractor, deduct all but								7.50%	7.50%
0600	Overtime work, by subcontractors, add								15%	15%
0650	By General Contractor, add								10%	10%

01 31 Project Management and Coordination

01 31 13 – Project Coordination

01 31 13.80 Overhead and Profit

	Crew	Daily Output	Labor-Hours	Unit	Material	2020 Bare Costs Labor	Equipment	Total	Total Incl O&P
0010 **OVERHEAD & PROFIT** Allowance to add to items in this									
0020 book that do not include Subs O&P, average				%				25%	
0100 Allowance to add to items in this book that R013113-55									
0110 do include Subs O&P, minimum				%				5%	5%
0150 Average								10%	10%
0200 Maximum								15%	15%
0300 Typical, by size of project, under $100,000								30%	
0350 $500,000 project								25%	
0400 $2,000,000 project								20%	
0450 Over $10,000,000 project								15%	

01 31 13.90 Performance Bond

	Crew	Daily Output	Labor-Hours	Unit	Material	2020 Bare Costs Labor	Equipment	Total	Total Incl O&P
0010 **PERFORMANCE BOND** R013113-80									
0020 For buildings, minimum				Job				.60%	.60%
0100 Maximum				"				2.50%	2.50%

01 31 14 – Facilities Services Coordination

01 31 14.20 Lock Out/Tag Out

	Crew	Daily Output	Labor-Hours	Unit	Material	2020 Bare Costs Labor	Equipment	Total	Total Incl O&P
0010 **LOCK OUT/TAG OUT**									
0020 Miniature circuit breaker lock out device	1 Elec	220	.036	Ea.	24	2.23		26.23	30
0030 Miniature pin circuit breaker lock out device		220	.036		19.90	2.23		22.13	25.50
0040 Single circuit breaker lock out device		220	.036		21	2.23		23.23	26.50
0050 Multi-pole circuit breaker lock out device (15 to 225 Amp)		210	.038		19.90	2.34		22.24	25.50
0060 Large 3 pole circuit breaker lock out device (over 225 Amp)		210	.038		21	2.34		23.34	27
0080 Square D I-Line circuit breaker lock out device		210	.038		33	2.34		35.34	39.50
0090 Lock out disconnect switch, 30 to 100 Amp		330	.024		11	1.49		12.49	14.30
0100 100 to 400 Amp		330	.024		11	1.49		12.49	14.30
0110 Over 400 Amp		330	.024		11	1.49		12.49	14.30
0120 Lock out hasp for multiple lockout tags		200	.040		5.60	2.45		8.05	9.80
0130 Electrical cord plug lock out device		220	.036		10.50	2.23		12.73	14.85
0140 Electrical plug prong lock out device (3-wire grounding plug)		220	.036		6.95	2.23		9.18	10.95
0150 Wall switch lock out		200	.040		25	2.45		27.45	31.50
0160 Fire alarm pull station lock out	1 Stpi	200	.040		18.75	2.62		21.37	24.50
0170 Sprinkler valve tamper and flow switch lock out device	1 Skwk	220	.036		16.60	1.99		18.59	21.50
0180 Lock out sign		330	.024		18.90	1.33		20.23	23
0190 Lock out tag		440	.018		5	1		6	7

01 32 Construction Progress Documentation

01 32 33 – Photographic Documentation

01 32 33.50 Photographs

	Crew	Daily Output	Labor-Hours	Unit	Material	2020 Bare Costs Labor	Equipment	Total	Total Incl O&P
0010 **PHOTOGRAPHS**									
0020 8" x 10", 4 shots, 2 prints ea., std. mounting				Set	545			545	600
0100 Hinged linen mounts					550			550	605
0200 8" x 10", 4 shots, 2 prints each, in color					535			535	590
0300 For I.D. slugs, add to all above					5.10			5.10	5.60
1500 Time lapse equipment, camera and projector, buy				Ea.	2,800			2,800	3,100
1550 Rent per month				"	1,325			1,325	1,450
1700 Cameraman and processing, black & white				Day	1,300			1,300	1,425
1720 Color				"	1,500			1,500	1,650

For customer support on your Mechanical Costs with RSMeans Data, call 800.448.8182.

11

01 41 Regulatory Requirements

01 41 26 – Permit Requirements

01 41 26.50 Permits	Crew	Daily Output	Labor-Hours	Unit	Material	2020 Bare Costs Labor	Equipment	Total	Total Incl O&P
0010 **PERMITS**									
0020 Rule of thumb, most cities, minimum				Job				.50%	.50%
0100 Maximum				"				2%	2%

01 45 Quality Control

01 45 23 – Testing and Inspecting Services

01 45 23.50 Testing	Crew	Daily Output	Labor-Hours	Unit	Material	2020 Bare Costs Labor	Equipment	Total	Total Incl O&P
0010 **TESTING** and Inspecting Services									
5820 Non-destructive metal testing, dye penetrant				Day				556	612
5840 Magnetic particle								556	612
5860 Radiography								556	612
5880 Ultrasonic				↓				556	612
9000 Thermographic testing, for bldg envelope heat loss, average 2,000 S.F. [G]				Ea.				500	500

01 51 Temporary Utilities

01 51 13 – Temporary Electricity

01 51 13.80 Temporary Utilities	Crew	Daily Output	Labor-Hours	Unit	Material	2020 Bare Costs Labor	Equipment	Total	Total Incl O&P
0010 **TEMPORARY UTILITIES**									
0350 Lighting, lamps, wiring, outlets, 40,000 S.F. building, 8 strings	1 Elec	34	.235	CSF Flr	5.85	14.45		20.30	28
0360 16 strings	"	17	.471		11.75	29		40.75	56
0400 Power for temp lighting only, 6.6 KWH, per month								.92	1.01
0430 11.8 KWH, per month								1.65	1.82
0450 23.6 KWH, per month								3.30	3.63
0600 Power for job duration incl. elevator, etc., minimum								53	58
0650 Maximum				↓				110	121
0675 Temporary cooling				Ea.	1,025			1,025	1,125

01 52 Construction Facilities

01 52 13 – Field Offices and Sheds

01 52 13.20 Office and Storage Space	Crew	Daily Output	Labor-Hours	Unit	Material	2020 Bare Costs Labor	Equipment	Total	Total Incl O&P
0010 **OFFICE AND STORAGE SPACE**									
0020 Office trailer, furnished, no hookups, 20' x 8', buy	2 Skwk	1	16	Ea.	9,525	880		10,405	11,800
0250 Rent per month					195			195	214
0300 32' x 8', buy	2 Skwk	.70	22.857		15,300	1,250		16,550	18,700
0350 Rent per month					245			245	269
0400 50' x 10', buy	2 Skwk	.60	26.667		30,300	1,475		31,775	35,500
0450 Rent per month					355			355	390
0500 50' x 12', buy	2 Skwk	.50	32		25,800	1,750		27,550	31,100
0550 Rent per month					460			460	505
0700 For air conditioning, rent per month, add				↓	51.50			51.50	56.50
0800 For delivery, add per mile				Mile	12.20			12.20	13.40
0900 Bunk house trailer, 8' x 40' duplex dorm with kitchen, no hookups, buy	2 Carp	1	16	Ea.	89,000	850		89,850	99,000
0910 9 man with kitchen and bath, no hookups, buy		1	16		91,000	850		91,850	101,500
0920 18 man sleeper with bath, no hookups, buy		1	16	↓	98,000	850		98,850	109,500
1000 Portable buildings, prefab, on skids, economy, 8' x 8'		265	.060	S.F.	26.50	3.21		29.71	34
1100 Deluxe, 8' x 12'		150	.107	"	29.50	5.65		35.15	40.50
1200 Storage boxes, 20' x 8', buy	2 Skwk	1.80	8.889	Ea.	3,150	490		3,640	4,200

12

01 52 Construction Facilities

01 52 13 – Field Offices and Sheds

01 52 13.20 Office and Storage Space

		Crew	Daily Output	Labor-Hours	Unit	Material	2020 Bare Costs Labor	Equipment	Total	Total Incl O&P
1250	Rent per month				Ea.	86			86	94.50
1300	40' x 8', buy	2 Skwk	1.40	11.429		3,800	625		4,425	5,125
1350	Rent per month					126			126	139

01 52 13.40 Field Office Expense

		Crew	Daily Output	Labor-Hours	Unit	Material	Labor	Equipment	Total	Total Incl O&P
0010	**FIELD OFFICE EXPENSE**									
0100	Office equipment rental average				Month	209			209	230
0120	Office supplies, average				"	87			87	96
0125	Office trailer rental, see Section 01 52 13.20									
0140	Telephone bill; avg. bill/month incl. long dist.				Month	87			87	96
0160	Lights & HVAC				"	163			163	179

01 54 Construction Aids

01 54 16 – Temporary Hoists

01 54 16.50 Weekly Forklift Crew

		Crew	Daily Output	Labor-Hours	Unit	Material	Labor	Equipment	Total	Total Incl O&P
0010	**WEEKLY FORKLIFT CREW**									
0100	All-terrain forklift, 45' lift, 35' reach, 9000 lb. capacity	A-3P	.20	40	Week	2,125	1,875	4,000	5,225	

01 54 19 – Temporary Cranes

01 54 19.50 Daily Crane Crews

		Crew	Daily Output	Labor-Hours	Unit	Material	Labor	Equipment	Total	Total Incl O&P
0010	**DAILY CRANE CREWS** for small jobs, portal to portal	R015433-15								
0100	12-ton truck-mounted hydraulic crane	A-3H	1	8	Day		475	725	1,200	1,500
0200	25-ton	A-3I	1	8			475	800	1,275	1,600
0300	40-ton	A-3J	1	8			475	1,275	1,750	2,100
0400	55-ton	A-3K	1	16			880	1,500	2,380	2,950
0500	80-ton	A-3L	1	16			880	2,250	3,130	3,775
0900	If crane is needed on a Saturday, Sunday or Holiday									
0910	At time-and-a-half, add				Day		50%			
0920	At double time, add				"		100%			

01 54 19.60 Monthly Tower Crane Crew

		Crew	Daily Output	Labor-Hours	Unit	Material	Labor	Equipment	Total	Total Incl O&P
0010	**MONTHLY TOWER CRANE CREW**, excludes concrete footing									
0100	Static tower crane, 130' high, 106' jib, 6200 lb. capacity	A-3N	.05	176	Month	10,400	37,300	47,700	56,500	

01 54 23 – Temporary Scaffolding and Platforms

01 54 23.70 Scaffolding

		Crew	Daily Output	Labor-Hours	Unit	Material	Labor	Equipment	Total	Total Incl O&P
0010	**SCAFFOLDING**	R015423-10								
0015	Steel tube, regular, no plank, labor only to erect & dismantle									
0090	Building exterior, wall face, 1 to 5 stories, 6'-4" x 5' frames	3 Carp	8	3	C.S.F.		159		159	240
0200	6 to 12 stories	4 Carp	8	4			213		213	320
0301	13 to 20 stories	5 Clab	8	5			211		211	315
0460	Building interior, wall face area, up to 16' high	3 Carp	12	2			106		106	160
0560	16' to 40' high		10	2.400			128		128	192
0800	Building interior floor area, up to 30' high		150	.160	C.C.F.		8.50		8.50	12.80
0900	Over 30' high	4 Carp	160	.200	"		10.65		10.65	15.95
0906	Complete system for face of walls, no plank, material only rent/mo				C.S.F.	34			34	37.50
0908	Interior spaces, no plank, material only rent/mo				C.C.F.	3.88			3.88	4.27
0910	Steel tubular, heavy duty shoring, buy									
0920	Frames 5' high 2' wide				Ea.	89.50			89.50	98.50
0925	5' high 4' wide					106			106	116
0930	6' high 2' wide					107			107	118
0935	6' high 4' wide					115			115	126
0940	Accessories									

01 54 23.70 Scaffolding	Crew	Daily Output	Labor-Hours	Unit	Material	2020 Bare Costs Labor	Equipment	Total	Total Incl O&P
0945	Cross braces				Ea.	17.05			17.05	18.80
0950	U-head, 8" x 8"					20.50			20.50	22.50
0955	J-head, 4" x 8"					14.90			14.90	16.40
0960	Base plate, 8" x 8"					16.05			16.05	17.65
0965	Leveling jack				↓	35			35	38
1000	Steel tubular, regular, buy									
1100	Frames 3' high 5' wide				Ea.	92.50			92.50	102
1150	5' high 5' wide					108			108	119
1200	6'-4" high 5' wide					90.50			90.50	99.50
1350	7'-6" high 6' wide					151			151	166
1500	Accessories, cross braces					18.85			18.85	20.50
1550	Guardrail post					23.50			23.50	25.50
1600	Guardrail 7' section					8.20			8.20	9.05
1650	Screw jacks & plates					26			26	28.50
1700	Sidearm brackets					23.50			23.50	26
1750	8" casters					37.50			37.50	41
1800	Plank 2" x 10" x 16'-0"					64.50			64.50	71
1900	Stairway section					292			292	320
1910	Stairway starter bar					32.50			32.50	35.50
1920	Stairway inside handrail					54.50			54.50	60
1930	Stairway outside handrail					87.50			87.50	96
1940	Walk-thru frame guardrail				↓	42			42	46.50
2000	Steel tubular, regular, rent/mo.									
2100	Frames 3' high 5' wide				Ea.	4.50			4.50	4.95
2150	5' high 5' wide					4.50			4.50	4.95
2200	6'-4" high 5' wide					5.70			5.70	6.30
2250	7'-6" high 6' wide					9.90			9.90	10.90
2500	Accessories, cross braces					.90			.90	.99
2550	Guardrail post					.90			.90	.99
2600	Guardrail 7' section					.90			.90	.99
2650	Screw jacks & plates					1.80			1.80	1.98
2700	Sidearm brackets					1.80			1.80	1.98
2750	8" casters					7.20			7.20	7.90
2800	Outrigger for rolling tower					2.70			2.70	2.97
2850	Plank 2" x 10" x 16'-0"					9.95			9.95	10.90
2900	Stairway section					31.50			31.50	35
2940	Walk-thru frame guardrail				↓	2.25			2.25	2.48
3000	Steel tubular, heavy duty shoring, rent/mo.									
3250	5' high 2' & 4' wide				Ea.	8.45			8.45	9.30
3300	6' high 2' & 4' wide					8.45			8.45	9.30
3500	Accessories, cross braces					.90			.90	.99
3600	U-head, 8" x 8"					2.50			2.50	2.75
3650	J-head, 4" x 8"					2.50			2.50	2.75
3700	Base plate, 8" x 8"					.90			.90	.99
3750	Leveling jack					2.47			2.47	2.72
5700	Planks, 2" x 10" x 16'-0", labor only to erect & remove to 50' H	3 Carp	72	.333	↓		17.70		17.70	26.50
5800	Over 50' high	4 Carp	80	.400			21.50		21.50	32
6000	Heavy duty shoring for elevated slab forms to 8'-2" high, floor area									
6100	Labor only to erect & dismantle	4 Carp	16	2	C.S.F.		106		106	160
6110	Materials only, rent/mo.				"	43.50			43.50	48
6500	To 14'-8" high									
6600	Labor only to erect & dismantle	4 Carp	10	3.200	C.S.F.		170		170	256
6610	Materials only, rent/mo				"	63.50			63.50	70

01 54 23 – Temporary Scaffolding and Platforms

01 54 23.75 Scaffolding Specialties	Crew	Daily Output	Labor-Hours	Unit	Material	2020 Bare Costs Labor	Equipment	Total	Total Incl O&P
0010 **SCAFFOLDING SPECIALTIES**									
1200 Sidewalk bridge, heavy duty steel posts & beams, including									
1210 parapet protection & waterproofing (material cost is rent/month)									
1220 8' to 10' wide, 2 posts	3 Carp	15	1.600	L.F.	46	85		131	179
1230 3 posts	"	10	2.400	"	71	128		199	270
1500 Sidewalk bridge using tubular steel scaffold frames including									
1510 planking (material cost is rent/month)	3 Carp	45	.533	L.F.	8.40	28.50		36.90	52
1600 For 2 uses per month, deduct from all above					50%				
1700 For 1 use every 2 months, add to all above					100%				
1900 Catwalks, 20" wide, no guardrails, 7' span, buy				Ea.	153			153	168
2000 10' span, buy					213			213	234
3720 Putlog, standard, 8' span, with hangers, buy					80			80	88
3730 Rent per month					16.30			16.30	17.95
3750 12' span, buy					101			101	111
3755 Rent per month					20.50			20.50	22.50
3760 Trussed type, 16' span, buy					262			262	288
3770 Rent per month					24.50			24.50	27
3790 22' span, buy					284			284	315
3795 Rent per month					32.50			32.50	35.50
3800 Rolling ladders with handrails, 30" wide, buy, 2 step					320			320	355
4000 7 step					1,125			1,125	1,250
4050 10 step					1,625			1,625	1,775
4100 Rolling towers, buy, 5' wide, 7' long, 10' high					1,375			1,375	1,500
4200 For additional 5' high sections, to buy					255			255	280
4300 Complete incl. wheels, railings, outriggers,									
4350 21' high, to buy				Ea.	2,325			2,325	2,550
4400 Rent/month = 5% of purchase cost				"	116			116	128

01 54 36 – Equipment Mobilization

01 54 36.50 Mobilization

	Crew	Daily Output	Labor-Hours	Unit	Material	2020 Bare Costs Labor	Equipment	Total	Total Incl O&P
0010 **MOBILIZATION** (Use line item again for demobilization) R015436-50									
0015 Up to 25 mi. haul dist. (50 mi. RT for mob/demob crew)									
1200 Small equipment, placed in rear of, or towed by pickup truck	A-3A	4	2	Ea.		106	44	150	207
1300 Equipment hauled on 3-ton capacity towed trailer	A-3Q	2.67	3			159	92.50	251.50	340
1400 20-ton capacity	B-34U	2	8			410	221	631	855
1500 40-ton capacity	B-34N	2	8			425	340	765	1,000
1600 50-ton capacity	B-34V	1	24			1,300	985	2,285	3,000
1700 Crane, truck-mounted, up to 75 ton (driver only)	1 Eqhv	4	2			118		118	177
1800 Over 75 ton (with chase vehicle)	A-3E	2.50	6.400			345	70.50	415.50	600
2400 Crane, large lattice boom, requiring assembly	B-34W	.50	144			7,425	7,025	14,450	18,800
2500 For each additional 5 miles haul distance, add						10%	10%		
3000 For large pieces of equipment, allow for assembly/knockdown									
3100 For mob/demob of micro-tunneling equip, see Section 33 05 07.36									

For customer support on your Mechanical Costs with RSMeans Data, call 800.448.8182.

15

01 55 Vehicular Access and Parking

01 55 23 – Temporary Roads

01 55 23.50 Roads and Sidewalks	Crew	Daily Output	Labor-Hours	Unit	Material	2020 Bare Costs Labor	Equipment	Total	Total Incl O&P
0010	**ROADS AND SIDEWALKS** Temporary									
0050	Roads, gravel fill, no surfacing, 4" gravel depth	B-14	715	.067	S.Y.	3.38	2.97	.30	6.65	8.50
0100	8" gravel depth	"	615	.078	"	6.75	3.45	.35	10.55	13.05
1000	Ramp, 3/4" plywood on 2" x 6" joists, 16" OC	2 Carp	300	.053	S.F.	1.62	2.83		4.45	6.05
1100	On 2" x 10" joists, 16" OC	"	275	.058	"	2.25	3.09		5.34	7.10

01 56 Temporary Barriers and Enclosures

01 56 13 – Temporary Air Barriers

01 56 13.60 Tarpaulins

01 56 13.60 Tarpaulins	Crew	Daily Output	Labor-Hours	Unit	Material	2020 Bare Costs Labor	Equipment	Total	Total Incl O&P
0010	**TARPAULINS**									
0020	Cotton duck, 10-13.13 oz./S.Y., 6' x 8'				S.F.	.75			.75	.83
0050	30' x 30'					.63			.63	.69
0100	Polyvinyl coated nylon, 14-18 oz., minimum					1.49			1.49	1.64
0150	Maximum					1.49			1.49	1.64
0200	Reinforced polyethylene 3 mils thick, white					.05			.05	.06
0300	4 mils thick, white, clear or black					.16			.16	.18
0400	5.5 mils thick, clear					.19			.19	.21
0500	White, fire retardant					.60			.60	.66
0600	12 mils, oil resistant, fire retardant					.50			.50	.55
0700	8.5 mils, black					.26			.26	.29
0710	Woven polyethylene, 6 mils thick					.08			.08	.09
0730	Polyester reinforced w/integral fastening system, 11 mils thick					.20			.20	.22
0740	Polyethylene, reflective, 23 mils thick					1.34			1.34	1.47

01 56 13.90 Winter Protection

01 56 13.90 Winter Protection	Crew	Daily Output	Labor-Hours	Unit	Material	2020 Bare Costs Labor	Equipment	Total	Total Incl O&P
0010	**WINTER PROTECTION**									
0100	Framing to close openings	2 Clab	500	.032	S.F.	.51	1.35		1.86	2.58
0200	Tarpaulins hung over scaffolding, 8 uses, not incl. scaffolding		1500	.011		.25	.45		.70	.95
0250	Tarpaulin polyester reinf. w/integral fastening system, 11 mils thick		1600	.010		.22	.42		.64	.87
0300	Prefab fiberglass panels, steel frame, 8 uses		1200	.013		2.78	.56		3.34	3.90

01 56 16 – Temporary Dust Barriers

01 56 16.10 Dust Barriers, Temporary

01 56 16.10 Dust Barriers, Temporary	Crew	Daily Output	Labor-Hours	Unit	Material	2020 Bare Costs Labor	Equipment	Total	Total Incl O&P
0010	**DUST BARRIERS, TEMPORARY**, erect and dismantle									
0020	Spring loaded telescoping pole & head, to 12', erect and dismantle	1 Clab	240	.033	Ea.		1.40		1.40	2.11
0025	Cost per day (based upon 250 days)				Day	.22			.22	.24
0030	To 21', erect and dismantle	1 Clab	240	.033	Ea.		1.40		1.40	2.11
0035	Cost per day (based upon 250 days)				Day	.72			.72	.79
0040	Accessories, caution tape reel, erect and dismantle	1 Clab	480	.017	Ea.		.70		.70	1.05
0045	Cost per day (based upon 250 days)				Day	.33			.33	.37
0060	Foam rail and connector, erect and dismantle	1 Clab	240	.033	Ea.		1.40		1.40	2.11
0065	Cost per day (based upon 250 days)				Day	.10			.10	.11
0070	Caution tape	1 Clab	384	.021	C.L.F.	2.49	.88		3.37	4.06
0080	Zipper, standard duty		60	.133	Ea.	10.55	5.60		16.15	20
0090	Heavy duty		48	.167	"	10.65	7		17.65	22.50
0100	Polyethylene sheet, 4 mil		37	.216	Sq.	2.62	9.10		11.72	16.60
0110	6 mil		37	.216	"	4	9.10		13.10	18.10
1000	Dust partition, 6 mil polyethylene, 1" x 3" frame	2 Carp	2000	.008	S.F.	.33	.43		.76	1
1080	2" x 4" frame	"	2000	.008	"	.39	.43		.82	1.07
1085	Negative air machine, 1800 CFM				Ea.	890			890	975
1090	Adhesive strip application, 2" width	1 Clab	192	.042	C.L.F.	6.75	1.75		8.50	10.10
4000	Dust & infectious control partition, adj. to 10' high, obscured, 4' panel	2 Carp	90	.178	Ea.	580	9.45		589.45	655

01 56 Temporary Barriers and Enclosures

01 56 16 – Temporary Dust Barriers

01 56 16.10 Dust Barriers, Temporary

		Crew	Daily Output	Labor-Hours	Unit	Material	2020 Bare Costs Labor	2020 Bare Costs Equipment	Total	Total Incl O&P
4010	3' panel	2 Carp	90	.178	Ea.	550	9.45		559.45	620
4020	2' panel	↓	90	.178		430	9.45		439.45	490
4030	1' panel	1 Carp	90	.089		285	4.72		289.72	320
4040	6" panel	"	90	.089		265	4.72		269.72	299
4050	2' panel with HEPA filtered discharge port	2 Carp	90	.178		500	9.45		509.45	565
4060	3' panel with 32" door		90	.178		895	9.45		904.45	1,000
4070	4' panel with 36" door		90	.178		995	9.45		1,004.45	1,125
4080	4'-6" panel with 44" door		90	.178		1,200	9.45		1,209.45	1,350
4090	Hinged corner		80	.200		185	10.65		195.65	220
4100	Outside corner		80	.200		150	10.65		160.65	181
4110	T post		80	.200		150	10.65		160.65	181
4120	Accessories, ceiling grid clip		360	.044		7.45	2.36		9.81	11.75
4130	Panel locking clip		360	.044		5.15	2.36		7.51	9.20
4140	Panel joint closure strip		360	.044		8	2.36		10.36	12.35
4150	Screw jack	↓	360	.044		6.65	2.36		9.01	10.85
4160	Digital pressure difference guage					275			275	305
4180	Combination lockset	1 Carp	13	.615	↓	200	32.50		232.50	269
4185	Sealant tape, 2" wide	1 Clab	192	.042	C.L.F.	6.75	1.75		8.50	10.10
4190	System in place, including door and accessories									
4200	Based upon 25 uses	2 Carp	51	.314	L.F.	9.85	16.65		26.50	36
4210	Based upon 50 uses		51	.314		4.94	16.65		21.59	30.50
4230	Based upon 100 uses	↓	51	.314	↓	2.47	16.65		19.12	27.50

01 56 23 – Temporary Barricades

01 56 23.10 Barricades

		Crew	Daily Output	Labor-Hours	Unit	Material	2020 Bare Costs Labor	2020 Bare Costs Equipment	Total	Total Incl O&P
0010	**BARRICADES**									
0020	5' high, 3 rail @ 2" x 8", fixed	2 Carp	20	.800	L.F.	6.75	42.50		49.25	71.50
0150	Movable		30	.533		5.60	28.50		34.10	48.50
1000	Guardrail, wooden, 3' high, 1" x 6" on 2" x 4" posts		200	.080		1.44	4.25		5.69	8
1100	2" x 6" on 4" x 4" posts	↓	165	.097		2.80	5.15		7.95	10.85
1200	Portable metal with base pads, buy					14.45			14.45	15.90
1250	Typical installation, assume 10 reuses	2 Carp	600	.027	↓	2.31	1.42		3.73	4.67
1300	Barricade tape, polyethylene, 7 mil, 3" wide x 300' long roll	1 Clab	128	.063	Ea.	7.50	2.63		10.13	12.20
3000	Detour signs, set up and remove									
3010	Reflective aluminum, MUTCD, 24" x 24", post mounted	1 Clab	20	.400	Ea.	2.80	16.85		19.65	28.50
4000	Roof edge portable barrier stands and warning flags, 50 uses	1 Rohe	9100	.001	L.F.	.07	.03		.10	.12
4010	100 uses	"	9100	.001	"	.03	.03		.06	.09

01 56 26 – Temporary Fencing

01 56 26.50 Temporary Fencing

		Crew	Daily Output	Labor-Hours	Unit	Material	2020 Bare Costs Labor	2020 Bare Costs Equipment	Total	Total Incl O&P
0010	**TEMPORARY FENCING**									
0020	Chain link, 11 ga., 4' high	2 Clab	400	.040	L.F.	2	1.68		3.68	4.73
0100	6' high		300	.053		4.96	2.25		7.21	8.80
0200	Rented chain link, 6' high, to 1000' (up to 12 mo.)		400	.040		3.37	1.68		5.05	6.25
0250	Over 1000' (up to 12 mo.)	↓	300	.053		4.14	2.25		6.39	7.90
0350	Plywood, painted, 2" x 4" frame, 4' high	A-4	135	.178		6.90	8.95		15.85	21
0400	4" x 4" frame, 8' high	"	110	.218		13.45	10.95		24.40	31
0500	Wire mesh on 4" x 4" posts, 4' high	2 Carp	100	.160		11.15	8.50		19.65	25
0550	8' high	"	80	.200	↓	16.60	10.65		27.25	34

For customer support on your Mechanical Costs with RSMeans Data, call 800.448.8182.

17

01 56 Temporary Barriers and Enclosures

01 56 29 – Temporary Protective Walkways

01 56 29.50 Protection	Crew	Daily Output	Labor-Hours	Unit	Material	2020 Bare Costs Labor	2020 Bare Costs Equipment	Total	Total Incl O&P
0010 **PROTECTION**									
0020 Stair tread, 2" x 12" planks, 1 use	1 Carp	75	.107	Tread	5.15	5.65		10.80	14.20
0100 Exterior plywood, 1/2" thick, 1 use		65	.123		2.05	6.55		8.60	12.10
0200 3/4" thick, 1 use	↓	60	.133	↓	2.84	7.10		9.94	13.75
2200 Sidewalks, 2" x 12" planks, 2 uses		350	.023	S.F.	.86	1.22		2.08	2.78
2300 Exterior plywood, 2 uses, 1/2" thick		750	.011		.34	.57		.91	1.23
2400 5/8" thick		650	.012		.38	.65		1.03	1.40
2500 3/4" thick	↓	600	.013	↓	.47	.71		1.18	1.58

01 58 Project Identification

01 58 13 – Temporary Project Signage

01 58 13.50 Signs	Crew	Daily Output	Labor-Hours	Unit	Material	2020 Bare Costs Labor	2020 Bare Costs Equipment	Total	Total Incl O&P
0010 **SIGNS**									
0020 High intensity reflectorized, no posts, buy				Ea.	27			27	29.50

01 66 Product Storage and Handling Requirements

01 66 19 – Material Handling

01 66 19.10 Material Handling	Crew	Daily Output	Labor-Hours	Unit	Material	2020 Bare Costs Labor	2020 Bare Costs Equipment	Total	Total Incl O&P
0010 **MATERIAL HANDLING**									
0020 Above 2nd story, via stairs, per C.Y. of material per floor	2 Clab	145	.110	C.Y.		4.65		4.65	7
0030 Via elevator, per C.Y. of material	↓	240	.067			2.81		2.81	4.22
0050 Distances greater than 200', per C.Y. of material per each addl 200'	↓	300	.053	↓		2.25		2.25	3.37

01 74 Cleaning and Waste Management

01 74 13 – Progress Cleaning

01 74 13.20 Cleaning Up	Crew	Daily Output	Labor-Hours	Unit	Material	2020 Bare Costs Labor	2020 Bare Costs Equipment	Total	Total Incl O&P
0010 **CLEANING UP**									
0020 After job completion, allow, minimum				Job				.30%	.30%
0040 Maximum				"				1%	1%
0050 Cleanup of floor area, continuous, per day, during const.	A-5	24	.750	M.S.F.	2.47	32	2.04	36.51	53
0100 Final by GC at end of job	"	11.50	1.565	"	2.62	67	4.26	73.88	108

01 91 Commissioning

01 91 13 – General Commissioning Requirements

01 91 13.50 Building Commissioning	Crew	Daily Output	Labor-Hours	Unit	Material	2020 Bare Costs Labor	2020 Bare Costs Equipment	Total	Total Incl O&P
0010 **BUILDING COMMISSIONING**									
0100 Systems operation and verification during turnover				%				.25%	.25%
0150 Including all systems subcontractors								.50%	.50%
0200 Systems design assistance, operation, verification and training								.50%	.50%
0250 Including all systems subcontractors				↓				1%	1%

01 93 Facility Maintenance

01 93 13 – Facility Maintenance Procedures

01 93 13.15 Mechanical Facilities Maintenance		Crew	Daily Output	Labor-Hours	Unit	Material	2020 Bare Costs Labor	2020 Bare Costs Equipment	Total	Total Incl O&P
0010	**MECHANICAL FACILITIES MAINTENANCE**									
0100	Air conditioning system maintenance									
0130	Belt, replace	1 Stpi	15	.533	Ea.		35		35	52.50
0170	Fan, clean		16	.500			33		33	49
0180	Filter, remove, clean, replace		12	.667			43.50		43.50	65.50
0190	Flexible coupling alignment, inspect		40	.200			13.10		13.10	19.60
0200	Gas leak, locate and repair		4	2			131		131	196
0250	Pump packing gland, remove and replace		11	.727			47.50		47.50	71.50
0270	Tighten		32	.250			16.40		16.40	24.50
0290	Pump, disassemble and assemble		4	2			131		131	196
0300	Air pressure regulator, disassemble, clean, assemble	1 Skwk	4	2			110		110	166
0310	Repair or replace part	"	6	1.333			73		73	111
0320	Purging system	1 Stpi	16	.500			33		33	49
0400	Compressor, air, remove or install fan wheel	1 Skwk	20	.400			22		22	33
0410	Disassemble or assemble 2 cylinder, 2 stage		4	2			110		110	166
0420	4 cylinder, 4 stage		1	8			440		440	665
0430	Repair or replace part		2	4			219		219	330
0700	Demolition, for mech. demolition see Section 23 05 05.10 or 22 05 05.10									
0800	Ductwork, clean									
0810	Rectangular									
0820	6" G	1 Shee	187.50	.043	L.F.		2.66		2.66	4.03
0830	8" G		140.63	.057			3.54		3.54	5.40
0840	10" G		112.50	.071			4.43		4.43	6.70
0850	12" G		93.75	.085			5.30		5.30	8.05
0860	14" G		80.36	.100			6.20		6.20	9.40
0870	16" G		70.31	.114			7.10		7.10	10.75
0900	Round									
0910	4" G	1 Shee	358.10	.022	L.F.		1.39		1.39	2.11
0920	6" G		238.73	.034			2.09		2.09	3.17
0930	8" G		179.05	.045			2.78		2.78	4.22
0940	10" G		143.24	.056			3.48		3.48	5.30
0950	12" G		119.37	.067			4.18		4.18	6.35
0960	16" G		89.52	.089			5.55		5.55	8.45
1000	Expansion joint, not screwed, install or remove	1 Stpi	3	2.667	Ea.		175		175	262
1010	Repack	"	6	1.333	"		87.50		87.50	131
1200	Fire protection equipment									
1220	Fire hydrant, replace	Q-1	3	5.333	Ea.		310		310	465
1230	Service, lubricate, inspect, flush, clean	1 Plum	7	1.143			73.50		73.50	110
1240	Test	"	11	.727			47		47	70
1310	Inspect valves, pressure, nozzle	1 Spri	4	2	System		127		127	190
1800	Plumbing fixtures, for installation see Section 22 41 00									
1801	Plumbing fixtures									
1850	Open drain with toilet auger	1 Plum	16	.500	Ea.		32		32	48
1870	Plaster trap, clean	"	6	1.333			86		86	129
1881	Clean commode	1 Clab	20	.400		.16	16.85		17.01	25.50
1882	Clean commode seat		48	.167		.08	7		7.08	10.65
1884	Clean double sink		20	.400		.16	16.85		17.01	25.50
1886	Clean bathtub		12	.667		.32	28		28.32	42.50
1888	Clean fiberglass tub/shower		8	1		.40	42		42.40	64
1889	Clean faucet set		32	.250		.06	10.55		10.61	15.85
1891	Clean shower head		80	.100		.01	4.21		4.22	6.35
1893	Clean water heater		16	.500		.57	21		21.57	32
1900	Relief valve, test and adjust	1 Stpi	20	.400			26		26	39

For customer support on your Mechanical Costs with RSMeans Data, call 800.448.8182.

19

01 93 13 – Facility Maintenance Procedures

01 93 13.15 Mechanical Facilities Maintenance		Crew	Daily Output	Labor-Hours	Unit	Material	2020 Bare Costs		Total	Total Incl O&P
							Labor	Equipment		
1910	Clean pump, heater, or motor for whirlpool	1 Clab	16	.500	Ea.	.08	21		21.08	31.50
1920	Clean thermal cover for whirlpool	"	16	.500		.81	21		21.81	32.50
2000	Repair or replace, steam trap	1 Stpi	8	1			65.50		65.50	98
2020	Y-type or bell strainer		6	1.333			87.50		87.50	131
2040	Water trap or vacuum breaker, screwed joints	▼	13	.615	▼		40.50		40.50	60.50
2100	Steam specialties, clean									
2120	Air separator with automatic trap, 1" fittings	1 Stpi	12	.667	Ea.		43.50		43.50	65.50
2130	Bucket trap, 2" pipe		7	1.143			75		75	112
2150	Drip leg, 2" fitting		45	.178			11.65		11.65	17.45
2200	Thermodynamic trap, 1" fittings		50	.160			10.50		10.50	15.70
2210	Thermostatic		65	.123			8.05		8.05	12.05
2240	Screen and seat in Y-type strainer, plug type		25	.320			21		21	31.50
2242	Screen and seat in Y-type strainer, flange type		12	.667			43.50		43.50	65.50
2500	Valve, replace broken handwheel	▼	24	.333	▼		22		22	32.50
3000	Valve, overhaul, regulator, relief, flushometer, mixing									
3040	Cold water, gas	1 Stpi	5	1.600	Ea.		105		105	157
3050	Hot water, steam		3	2.667			175		175	262
3080	Globe, gate, check up to 4" cold water, gas		10	.800			52.50		52.50	78.50
3090	Hot water, steam		5	1.600			105		105	157
3100	Over 4" ID hot or cold line	▼	1.40	5.714			375		375	560
3120	Remove and replace, gate, globe or check up to 4"	Q-5	6	2.667			157		157	235
3130	Over 4"	"	2	8			470		470	705
3150	Repack up to 4"	1 Stpi	13	.615			40.50		40.50	60.50
3160	Over 4"	"	4	2	▼		131		131	196

Estimating Tips

02 30 00 Subsurface Investigation

In preparing estimates on structures involving earthwork or foundations, all information concerning soil characteristics should be obtained. Look particularly for hazardous waste, evidence of prior dumping of debris, and previous stream beds.

02 40 00 Demolition and Structure Moving

The costs shown for selective demolition do not include rubbish handling or disposal. These items should be estimated separately using RSMeans data or other sources.

- Historic preservation often requires that the contractor remove materials from the existing structure, rehab them, and replace them. The estimator must be aware of any related measures and precautions that must be taken when doing selective demolition and cutting and patching. Requirements may include special handling and storage, as well as security.

- In addition to Subdivision 02 41 00, you can find selective demolition items in each division. Example: Roofing demolition is in Division 7.
- Absent of any other specific reference, an approximate demolish-in-place cost can be obtained by halving the new-install labor cost. To remove for reuse, allow the entire new-install labor figure.

02 40 00 Building Deconstruction

This section provides costs for the careful dismantling and recycling of most low-rise building materials.

02 50 00 Containment of Hazardous Waste

This section addresses on-site hazardous waste disposal costs.

02 80 00 Hazardous Material Disposal/Remediation

This subdivision includes information on hazardous waste handling, asbestos remediation, lead remediation, and mold remediation. See reference numbers

R028213-20 and R028319-60 for further guidance in using these unit price lines.

02 90 00 Monitoring Chemical Sampling, Testing Analysis

This section provides costs for on-site sampling and testing hazardous waste.

Reference Numbers

Reference numbers are shown at the beginning of some major classifications. These numbers refer to related items in the Reference Section. The reference information may be an estimating procedure, an alternate pricing method, or technical information.

Note: Not all subdivisions listed here necessarily appear. ∎

Same Data. Simplified.

Enjoy the convenience and efficiency of accessing your costs anywhere:

- **Skip the multiplier** by setting your location
- **Quickly search,** edit, favorite and share costs
- **Stay on top of price changes** with automatic updates

Discover more at rsmeans.com/online

02 41 Demolition

02 41 13 – Selective Site Demolition

02 41 13.17 Demolish, Remove Pavement and Curb

		Crew	Daily Output	Labor-Hours	Unit	Material	2020 Bare Costs Labor	2020 Bare Costs Equipment	Total	Total Incl O&P
0010	**DEMOLISH, REMOVE PAVEMENT AND CURB** R024119-10									
5010	Pavement removal, bituminous roads, up to 3" thick	B-38	690	.058	S.Y.		2.76	1.80	4.56	6.10
5050	4"-6" thick		420	.095			4.53	2.96	7.49	10.05
5100	Bituminous driveways		640	.063			2.98	1.94	4.92	6.60
5200	Concrete to 6" thick, hydraulic hammer, mesh reinforced		255	.157			7.45	4.87	12.32	16.55
5300	Rod reinforced		200	.200	↓		9.50	6.20	15.70	21
5400	Concrete, 7"-24" thick, plain		33	1.212	C.Y.		57.50	37.50	95	128
5500	Reinforced	↓	24	1.667	"		79.50	51.50	131	176
5600	With hand held air equipment, bituminous, to 6" thick	B-39	1900	.025	S.F.		1.12	.19	1.31	1.89
5700	Concrete to 6" thick, no reinforcing		1600	.030			1.33	.22	1.55	2.24
5800	Mesh reinforced		1400	.034			1.52	.26	1.78	2.56
5900	Rod reinforced	↓	765	.063	↓		2.78	.47	3.25	4.68

02 41 13.23 Utility Line Removal

		Crew	Daily Output	Labor-Hours	Unit	Material	2020 Bare Costs Labor	2020 Bare Costs Equipment	Total	Total Incl O&P
0010	**UTILITY LINE REMOVAL**									
0015	No hauling, abandon catch basin or manhole	B-6	7	3.429	Ea.		157	30.50	187.50	269
0020	Remove existing catch basin or manhole, masonry		4	6			274	53.50	327.50	470
0030	Catch basin or manhole frames and covers, stored		13	1.846			84.50	16.45	100.95	145
0040	Remove and reset	↓	7	3.429			157	30.50	187.50	269
0900	Hydrants, fire, remove only	B-21A	5	8			420	85.50	505.50	720
0950	Remove and reset	"	2	20	↓		1,050	214	1,264	1,800

02 41 13.30 Minor Site Demolition

		Crew	Daily Output	Labor-Hours	Unit	Material	2020 Bare Costs Labor	2020 Bare Costs Equipment	Total	Total Incl O&P
0010	**MINOR SITE DEMOLITION** R024119-10									
4000	Sidewalk removal, bituminous, 2" thick	B-6	350	.069	S.Y.		3.14	.61	3.75	5.35
4010	2-1/2" thick		325	.074			3.38	.66	4.04	5.75
4050	Brick, set in mortar		185	.130			5.95	1.16	7.11	10.15
4100	Concrete, plain, 4"		160	.150			6.85	1.34	8.19	11.75
4110	Plain, 5"		140	.171			7.85	1.53	9.38	13.45
4120	Plain, 6"		120	.200			9.15	1.78	10.93	15.65
4200	Mesh reinforced, concrete, 4"		150	.160			7.30	1.43	8.73	12.50
4210	5" thick		131	.183			8.40	1.63	10.03	14.35
4220	6" thick	↓	112	.214	↓		9.80	1.91	11.71	16.80

02 41 19 – Selective Demolition

02 41 19.13 Selective Building Demolition

0010	**SELECTIVE BUILDING DEMOLITION**									
0020	Costs related to selective demolition of specific building components									
0025	are included under Common Work Results (XX 05)									
0030	in the component's appropriate division.									

02 41 19.19 Selective Demolition

		Crew	Daily Output	Labor-Hours	Unit	Material	2020 Bare Costs Labor	2020 Bare Costs Equipment	Total	Total Incl O&P
0010	**SELECTIVE DEMOLITION**, Rubbish Handling R024119-10									
0020	The following are to be added to the demolition prices									
0600	Dumpster, weekly rental, 1 dump/week, 6 C.Y. capacity (2 tons)				Week	415			415	455
0700	10 C.Y. capacity (3 tons)					480			480	530
0725	20 C.Y. capacity (5 tons) R024119-20					565			565	625
0800	30 C.Y. capacity (7 tons)					730			730	800
0840	40 C.Y. capacity (10 tons)				↓	775			775	850
2000	Load, haul, dump and return, 0'-50' haul, hand carried	2 Clab	24	.667	C.Y.		28		28	42
2005	Wheeled		37	.432			18.20		18.20	27.50
2040	0'-100' haul, hand carried		16.50	.970			41		41	61.50
2045	Wheeled		25	.640			27		27	40.50
2050	Forklift	A-3R	25	.320			16.95	10.25	27.20	37
2080	Haul and return, add per each extra 100' haul, hand carried	2 Clab	35.50	.451	↓		18.95		18.95	28.50

02 41 Demolition

02 41 19 – Selective Demolition

02 41 19.19 Selective Demolition

		Crew	Daily Output	Labor-Hours	Unit	Material	2020 Bare Costs Labor	2020 Bare Costs Equipment	Total	Total Incl O&P
2085	Wheeled	2 Clab	54	.296	C.Y.		12.45		12.45	18.75
2120	For travel in elevators, up to 10 floors, add		140	.114			4.81		4.81	7.25
2130	0'-50' haul, incl. up to 5 riser stairs, hand carried		23	.696			29.50		29.50	44
2135	Wheeled		35	.457			19.25		19.25	29
2140	6-10 riser stairs, hand carried		22	.727			30.50		30.50	46
2145	Wheeled		34	.471			19.80		19.80	30
2150	11-20 riser stairs, hand carried		20	.800			33.50		33.50	50.50
2155	Wheeled		31	.516			21.50		21.50	32.50
2160	21-40 riser stairs, hand carried		16	1			42		42	63.50
2165	Wheeled		24	.667			28		28	42
2170	0'-100' haul, incl. 5 riser stairs, hand carried		15	1.067			45		45	67.50
2175	Wheeled		23	.696			29.50		29.50	44
2180	6-10 riser stairs, hand carried		14	1.143			48		48	72.50
2185	Wheeled		21	.762			32		32	48
2190	11-20 riser stairs, hand carried		12	1.333			56		56	84.50
2195	Wheeled		18	.889			37.50		37.50	56
2200	21-40 riser stairs, hand carried		8	2			84		84	127
2205	Wheeled		12	1.333			56		56	84.50
2210	Haul and return, add per each extra 100' haul, hand carried		35.50	.451			18.95		18.95	28.50
2215	Wheeled		54	.296			12.45		12.45	18.75
2220	For each additional flight of stairs, up to 5 risers, add		550	.029	Flight		1.22		1.22	1.84
2225	6-10 risers, add		275	.058			2.45		2.45	3.68
2230	11-20 risers, add		138	.116			4.88		4.88	7.35
2235	21-40 risers, add		69	.232			9.75		9.75	14.65
3000	Loading & trucking, including 2 mile haul, chute loaded	B-16	45	.711	C.Y.		31.50	12.75	44.25	61.50
3040	Hand loading truck, 50' haul	"	48	.667			29.50	11.95	41.45	57.50
3080	Machine loading truck	B-17	120	.267			12.40	5.35	17.75	24.50
5000	Haul, per mile, up to 8 C.Y. truck	B-34B	1165	.007			.34	.49	.83	1.04
5100	Over 8 C.Y. truck	"	1550	.005			.25	.37	.62	.79

02 41 19.20 Selective Demolition, Dump Charges

		Crew	Daily Output	Labor-Hours	Unit	Material	2020 Bare Costs Labor	2020 Bare Costs Equipment	Total	Total Incl O&P
0010	**SELECTIVE DEMOLITION, DUMP CHARGES** R024119-10									
0020	Dump charges, typical urban city, tipping fees only									
0100	Building construction materials				Ton	74			74	81
0200	Trees, brush, lumber					63			63	69.50
0300	Rubbish only					63			63	69.50
0500	Reclamation station, usual charge					74			74	81

02 41 19.27 Selective Demolition, Torch Cutting

		Crew	Daily Output	Labor-Hours	Unit	Material	2020 Bare Costs Labor	2020 Bare Costs Equipment	Total	Total Incl O&P
0010	**SELECTIVE DEMOLITION, TORCH CUTTING** R024119-10									
0020	Steel, 1" thick plate	E-25	333	.024	L.F.	.90	1.43	.04	2.37	3.27
0040	1" diameter bar	"	600	.013	Ea.	.15	.79	.02	.96	1.43
1000	Oxygen lance cutting, reinforced concrete walls									
1040	12"-16" thick walls	1 Clab	10	.800	L.F.		33.50		33.50	50.50
1080	24" thick walls	"	6	1.333	"		56		56	84.50

For customer support on your Mechanical Costs with RSMeans Data, call 800.448.8182.

23

02 65 Underground Storage Tank Removal

02 65 10 – Underground Tank and Contaminated Soil Removal

02 65 10.30 Removal of Underground Storage Tanks		Crew	Daily Output	Labor-Hours	Unit	Material	2020 Bare Costs Labor	Equipment	Total	Total Incl O&P
0010	**REMOVAL OF UNDERGROUND STORAGE TANKS** R026510-20									
0011	Petroleum storage tanks, non-leaking									
0100	Excavate & load onto trailer									
0110	3,000 gal. to 5,000 gal. tank	G	B-14	4	12	Ea.		530	53.50	583.50	855
0120	6,000 gal. to 8,000 gal. tank	G	B-3A	3	13.333			600	229	829	1,150
0130	9,000 gal. to 12,000 gal. tank	G	"	2	20	↓		900	345	1,245	1,725
0190	Known leaking tank, add					%				100%	100%
0200	Remove sludge, water and remaining product from bottom									
0201	of tank with vacuum truck									
0300	3,000 gal. to 5,000 gal. tank	G	A-13	5	1.600	Ea.		85	148	233	289
0310	6,000 gal. to 8,000 gal. tank	G		4	2			106	184	290	360
0320	9,000 gal. to 12,000 gal. tank	G	↓	3	2.667	↓		141	246	387	480
0390	Dispose of sludge off-site, average					Gal.				6.25	6.80
0400	Insert inert solid CO_2 "dry ice" into tank									
0401	For cleaning/transporting tanks (1.5 lb./100 gal. cap)	G	1 Clab	500	.016	Lb.	1.22	.67		1.89	2.35
1020	Haul tank to certified salvage dump, 100 miles round trip									
1023	3,000 gal. to 5,000 gal. tank					Ea.				760	830
1026	6,000 gal. to 8,000 gal. tank									880	960
1029	9,000 gal. to 12,000 gal. tank					↓				1,050	1,150
1100	Disposal of contaminated soil to landfill									
1110	Minimum					C.Y.				145	160
1111	Maximum					"				400	440
1120	Disposal of contaminated soil to									
1121	bituminous concrete batch plant									
1130	Minimum					C.Y.				80	88
1131	Maximum					"				115	125
2010	Decontamination of soil on site incl poly tarp on top/bottom									
2011	Soil containment berm and chemical treatment									
2020	Minimum	G	B-11C	100	.160	C.Y.	7.65	7.90	2.14	17.69	22.50
2021	Maximum	G	"	100	.160	↓	9.90	7.90	2.14	19.94	25
2050	Disposal of decontaminated soil, minimum									135	150
2055	Maximum					↓				400	440

02 81 Transportation and Disposal of Hazardous Materials

02 81 20 – Hazardous Waste Handling

02 81 20.10 Hazardous Waste Cleanup/Pickup/Disposal

		Crew	Daily Output	Labor-Hours	Unit	Material	Labor	Equipment	Total	Total Incl O&P
0010	**HAZARDOUS WASTE CLEANUP/PICKUP/DISPOSAL**									
0100	For contractor rental equipment, i.e., dozer,									
0110	Front end loader, dump truck, etc., see 01 54 33 Reference Section									
1000	Solid pickup									
1100	55 gal. drums					Ea.				240	265
1120	Bulk material, minimum					Ton				190	210
1130	Maximum					"				595	655
1200	Transportation to disposal site									
1220	Truckload = 80 drums or 25 C.Y. or 18 tons									
1260	Minimum					Mile				3.95	4.45
1270	Maximum					"				7.25	7.98
3000	Liquid pickup, vacuum truck, stainless steel tank									
3100	Minimum charge, 4 hours									
3110	1 compartment, 2200 gallon					Hr.				140	155
3120	2 compartment, 5000 gallon					"				200	225

02 81 Transportation and Disposal of Hazardous Materials

02 81 20 – Hazardous Waste Handling

02 81 20.10 Hazardous Waste Cleanup/Pickup/Disposal	Crew	Daily Output	Labor-Hours	Unit	Material	2020 Bare Costs Labor	Equipment	Total	Total Incl O&P
3400	Transportation in 6900 gallon bulk truck				Mile				7.95	8.75
3410	In teflon lined truck				"				10.20	11.25
5000	Heavy sludge or dry vacuumable material				Hr.				140	155
6000	Dumpsite disposal charge, minimum				Ton				140	155
6020	Maximum				"				415	455

02 82 Asbestos Remediation

02 82 13 – Asbestos Abatement

02 82 13.39 Asbestos Remediation Plans and Methods

		Crew	Daily Output	Labor-Hours	Unit	Material	Labor	Equipment	Total	Total Incl O&P
0010	**ASBESTOS REMEDIATION PLANS AND METHODS**									
0100	Building Survey-Commercial Building				Ea.				2,200	2,400
0200	Asbestos Abatement Remediation Plan				"				1,350	1,475

02 82 13.41 Asbestos Abatement Equipment

		Crew	Daily Output	Labor-Hours	Unit	Material	Labor	Equipment	Total	Total Incl O&P
0010	**ASBESTOS ABATEMENT EQUIPMENT** R028213-20									
0011	Equipment and supplies, buy									
0200	Air filtration device, 2000 CFM				Ea.	950			950	1,050
0250	Large volume air sampling pump, minimum					315			315	350
0260	Maximum					345			345	380
0300	Airless sprayer unit, 2 gun					2,625			2,625	2,875
0350	Light stand, 500 watt					41			41	45
0400	Personal respirators									
0410	Negative pressure, 1/2 face, dual operation, minimum				Ea.	30			30	33
0420	Maximum					34			34	37.50
0450	P.A.P.R., full face, minimum					780			780	855
0460	Maximum					1,225			1,225	1,325
0470	Supplied air, full face, including air line, minimum					289			289	320
0480	Maximum					500			500	550
0500	Personnel sampling pump					237			237	261
1500	Power panel, 20 unit, including GFI					475			475	525
1600	Shower unit, including pump and filters					1,125			1,125	1,250
1700	Supplied air system (type C)					3,700			3,700	4,075
1750	Vacuum cleaner, HEPA, 16 gal., stainless steel, wet/dry					1,275			1,275	1,400
1760	55 gallon					1,475			1,475	1,600
1800	Vacuum loader, 9-18 ton/hr.					99,000			99,000	109,000
1900	Water atomizer unit, including 55 gal. drum					305			305	340
2000	Worker protection, whole body, foot, head cover & gloves, plastic					7.45			7.45	8.20
2500	Respirator, single use					30			30	33
2550	Cartridge for respirator					5.15			5.15	5.65
2570	Glove bag, 7 mil, 50" x 64"					8.75			8.75	9.60
2580	10 mil, 44" x 60"					7.90			7.90	8.65
2590	6 mil, 44" x 60"					4.83			4.83	5.30
6000	Disposable polyethylene bags, 6 mil, 3 C.F.					.77			.77	.85
6300	Disposable fiber drums, 3 C.F.					18.90			18.90	21
6400	Pressure sensitive caution labels, 3" x 5"					3.08			3.08	3.39
6450	11" x 17"					7.75			7.75	8.55
6500	Negative air machine, 1800 CFM					890			890	975

02 82 13.42 Preparation of Asbestos Containment Area

		Crew	Daily Output	Labor-Hours	Unit	Material	Labor	Equipment	Total	Total Incl O&P
0010	**PREPARATION OF ASBESTOS CONTAINMENT AREA** R028213-20									
0100	Pre-cleaning, HEPA vacuum and wet wipe, flat surfaces	A-9	12000	.005	S.F.	.01	.31		.32	.49
0200	Protect carpeted area, 2 layers 6 mil poly on 3/4" plywood	"	1000	.064		2.10	3.76		5.86	8.05

For customer support on your Mechanical Costs with RSMeans Data, call 800.448.8182.

25

02 82 Asbestos Remediation

02 82 13 – Asbestos Abatement

02 82 13.42 Preparation of Asbestos Containment Area

	Crew	Daily Output	Labor-Hours	Unit	Material	2020 Bare Costs Labor	Equipment	Total	Total Incl O&P
0300 Separation barrier, 2" x 4" @ 16", 1/2" plywood ea. side, 8' high	2 Carp	400	.040	S.F.	3.08	2.13		5.21	6.60
0310 12' high		320	.050		3.02	2.66		5.68	7.30
0320 16' high		200	.080		2.99	4.25		7.24	9.70
0400 Personnel decontam. chamber, 2" x 4" @ 16", 3/4" ply ea. side		280	.057		3.55	3.04		6.59	8.45
0450 Waste decontam. chamber, 2" x 4" studs @ 16", 3/4" ply ea. side		360	.044		3.55	2.36		5.91	7.45
0500 Cover surfaces with polyethylene sheeting									
0501 Including glue and tape									
0550 Floors, each layer, 6 mil	A-9	8000	.008	S.F.	.04	.47		.51	.77
0551 4 mil		9000	.007		.03	.42		.45	.67
0560 Walls, each layer, 6 mil		6000	.011		.04	.63		.67	1.01
0561 4 mil		7000	.009		.03	.54		.57	.85
0570 For heights above 14', add						20%			
0575 For heights above 20', add						30%			
0580 For fire retardant poly, add					100%				
0590 For large open areas, deduct					10%	20%			
0600 Seal floor penetrations with foam firestop to 36 sq. in.	2 Carp	200	.080	Ea.	13.90	4.25		18.15	21.50
0610 36 sq. in. to 72 sq. in.		125	.128		28	6.80		34.80	40.50
0615 72 sq. in. to 144 sq. in.		80	.200		55.50	10.65		66.15	77
0620 Wall penetrations, to 36 sq. in.		180	.089		13.90	4.72		18.62	22.50
0630 36 sq. in. to 72 sq. in.		100	.160		28	8.50		36.50	43.50
0640 72 sq. in. to 144 sq. in.		60	.267		55.50	14.15		69.65	82.50
0800 Caulk seams with latex	1 Carp	230	.035	L.F.	.17	1.85		2.02	2.97
0900 Set up neg. air machine, 1-2k CFM/25 M.C.F. volume	1 Asbe	4.30	1.860	Ea.		109		109	168
0950 Set up and remove portable shower unit	2 Asbe	4	4	"		235		235	360

02 82 13.43 Bulk Asbestos Removal

	Crew	Daily Output	Labor-Hours	Unit	Material	2020 Bare Costs Labor	Equipment	Total	Total Incl O&P
0010 **BULK ASBESTOS REMOVAL**									
0020 Includes disposable tools and 2 suits and 1 respirator filter/day/worker									
0200 Boiler insulation	A-9	480	.133	S.F.	.40	7.85		8.25	12.45
0210 With metal lath, add				%				50%	50%
0300 Boiler breeching or flue insulation	A-9	520	.123	S.F.	.31	7.25		7.56	11.45
0310 For active boiler, add				%				100%	100%
0400 Duct or AHU insulation	A-10B	440	.073	S.F.	.18	4.28		4.46	6.75
0500 Duct vibration isolation joints, up to 24 sq. in. duct	A-9	56	1.143	Ea.	2.87	67		69.87	106
0520 25 sq. in. to 48 sq. in. duct		48	1.333		3.35	78.50		81.85	124
0530 49 sq. in. to 76 sq. in. duct		40	1.600		4.01	94		98.01	148
0600 Pipe insulation, air cell type, up to 4" diameter pipe		900	.071	L.F.	.18	4.17		4.35	6.60
0610 4" to 8" diameter pipe		800	.080		.20	4.70		4.90	7.40
0620 10" to 12" diameter pipe		700	.091		.23	5.35		5.58	8.50
0630 14" to 16" diameter pipe		550	.116		.29	6.85		7.14	10.80
0650 Over 16" diameter pipe		650	.098	S.F.	.25	5.80		6.05	9.15
0700 With glove bag up to 3" diameter pipe		200	.320	L.F.	9.40	18.80		28.20	39.50
1000 Pipe fitting insulation up to 4" diameter pipe		320	.200	Ea.	.50	11.75		12.25	18.60
1100 6" to 8" diameter pipe		304	.211		.53	12.35		12.88	19.60
1110 10" to 12" diameter pipe		192	.333		.84	19.55		20.39	31
1120 14" to 16" diameter pipe		128	.500		1.25	29.50		30.75	46.50
1130 Over 16" diameter pipe		176	.364	S.F.	.91	21.50		22.41	34
1200 With glove bag, up to 8" diameter pipe		75	.853	L.F.	6.55	50		56.55	84
2000 Scrape foam fireproofing from flat surface		2400	.027	S.F.	.07	1.57		1.64	2.47
2100 Irregular surfaces		1200	.053		.13	3.13		3.26	4.96
3000 Remove cementitious material from flat surface		1800	.036		.09	2.09		2.18	3.31
3100 Irregular surface		1000	.064		.11	3.76		3.87	5.90
6000 Remove contaminated soil from crawl space by hand		400	.160	C.F.	.40	9.40		9.80	14.85

02 82 13 – Asbestos Abatement

02 82 13.43 Bulk Asbestos Removal		Crew	Daily Output	Labor-Hours	Unit	Material	2020 Bare Costs Labor	Equipment	Total	Total Incl O&P
6100	With large production vacuum loader	A-12	700	.091	C.F.	.23	5.35	1.05	6.63	9.65
7000	Radiator backing, not including radiator removal	A-9	1200	.053	S.F.	.13	3.13		3.26	4.96
9000	For type B (supplied air) respirator equipment, add				%				10%	10%

02 82 13.44 Demolition In Asbestos Contaminated Area		Crew	Daily Output	Labor-Hours	Unit	Material	2020 Bare Costs Labor	Equipment	Total	Total Incl O&P
0010	**DEMOLITION IN ASBESTOS CONTAMINATED AREA**									
0200	Ceiling, including suspension system, plaster and lath	A-9	2100	.030	S.F.	.08	1.79		1.87	2.83
0210	Finished plaster, leaving wire lath		585	.109		.27	6.40		6.67	10.15
0220	Suspended acoustical tile		3500	.018		.05	1.07		1.12	1.70
0230	Concealed tile grid system		3000	.021		.05	1.25		1.30	1.98
0240	Metal pan grid system		1500	.043		.11	2.51		2.62	3.97
0250	Gypsum board		2500	.026		.06	1.50		1.56	2.38
0260	Lighting fixtures up to 2' x 4'		72	.889	Ea.	2.23	52		54.23	82.50
0400	Partitions, non load bearing									
0410	Plaster, lath, and studs	A-9	690	.093	S.F.	.90	5.45		6.35	9.35
0450	Gypsum board and studs	"	1390	.046	"	.12	2.70		2.82	4.28
9000	For type B (supplied air) respirator equipment, add				%				10%	10%

02 82 13.45 OSHA Testing		Crew	Daily Output	Labor-Hours	Unit	Material	2020 Bare Costs Labor	Equipment	Total	Total Incl O&P
0010	**OSHA TESTING**									
0100	Certified technician, minimum				Day		340		340	340
0110	Maximum						670		670	670
0121	Industrial hygienist, minimum						385		385	385
0130	Maximum						755		755	755
0200	Asbestos sampling and PCM analysis, NIOSH 7400, minimum	1 Asbe	8	1	Ea.	15.90	58.50		74.40	108
0210	Maximum		4	2		27	117		144	210
1000	Cleaned area samples		8	1		203	58.50		261.50	315
1100	PCM air sample analysis, NIOSH 7400, minimum		8	1		17.70	58.50		76.20	110
1110	Maximum		4	2		2.30	117		119.30	183
1200	TEM air sample analysis, NIOSH 7402, minimum								80	106
1210	Maximum								360	450

02 82 13.46 Decontamination of Asbestos Containment Area		Crew	Daily Output	Labor-Hours	Unit	Material	2020 Bare Costs Labor	Equipment	Total	Total Incl O&P
0010	**DECONTAMINATION OF ASBESTOS CONTAINMENT AREA**									
0100	Spray exposed substrate with surfactant (bridging)									
0200	Flat surfaces	A-9	6000	.011	S.F.	.41	.63		1.04	1.41
0250	Irregular surfaces		4000	.016	"	.45	.94		1.39	1.94
0300	Pipes, beams, and columns		2000	.032	L.F.	.60	1.88		2.48	3.54
1000	Spray encapsulate polyethylene sheeting		8000	.008	S.F.	.41	.47		.88	1.17
1100	Roll down polyethylene sheeting		8000	.008	"		.47		.47	.72
1500	Bag polyethylene sheeting		400	.160	Ea.	1.01	9.40		10.41	15.50
2000	Fine clean exposed substrate, with nylon brush		2400	.027	S.F.		1.57		1.57	2.40
2500	Wet wipe substrate		4800	.013			.78		.78	1.20
2600	Vacuum surfaces, fine brush		6400	.010			.59		.59	.90
3000	Structural demolition									
3100	Wood stud walls	A-9	2800	.023	S.F.		1.34		1.34	2.06
3500	Window manifolds, not incl. window replacement		4200	.015			.89		.89	1.37
3600	Plywood carpet protection		2000	.032			1.88		1.88	2.88
4000	Remove custom decontamination facility	A-10A	8	3	Ea.	15.15	176		191.15	288
4100	Remove portable decontamination facility	3 Asbe	12	2	"	14.05	117		131.05	195
5000	HEPA vacuum, shampoo carpeting	A-9	4800	.013	S.F.	.11	.78		.89	1.32
9000	Final cleaning of protected surfaces	A-10A	8000	.003	"		.18		.18	.27

For customer support on your Mechanical Costs with RSMeans Data, call 800.448.8182.

27

02 82 Asbestos Remediation

02 82 13 – Asbestos Abatement

02 82 13.47 Asbestos Waste Pkg., Handling, and Disp.

02 82 13.47 Asbestos Waste Pkg., Handling, and Disp.		Crew	Daily Output	Labor-Hours	Unit	Material	2020 Bare Costs Labor	Equipment	Total	Total Incl O&P
0010	**ASBESTOS WASTE PACKAGING, HANDLING, AND DISPOSAL**									
0100	Collect and bag bulk material, 3 C.F. bags, by hand	A-9	400	.160	Ea.	.77	9.40		10.17	15.25
0200	Large production vacuum loader	A-12	880	.073		.99	4.27	.84	6.10	8.55
1000	Double bag and decontaminate	A-9	960	.067		.77	3.91		4.68	6.85
2000	Containerize bagged material in drums, per 3 C.F. drum	"	800	.080		18.90	4.70		23.60	28
3000	Cart bags 50' to dumpster	2 Asbe	400	.040			2.35		2.35	3.60
5000	Disposal charges, not including haul, minimum				C.Y.				61	67
5020	Maximum				"				355	395
9000	For type B (supplied air) respirator equipment, add				%				10%	10%

02 82 13.48 Asbestos Encapsulation With Sealants

02 82 13.48		Crew	Daily Output	Labor-Hours	Unit	Material	2020 Bare Costs Labor	Equipment	Total	Total Incl O&P
0010	**ASBESTOS ENCAPSULATION WITH SEALANTS**									
0100	Ceilings and walls, minimum	A-9	21000	.003	S.F.	.42	.18		.60	.73
0110	Maximum		10600	.006	"	.52	.35		.87	1.11
0300	Pipes to 12" diameter including minor repairs, minimum		800	.080	L.F.	.57	4.70		5.27	7.85
0310	Maximum		400	.160	"	1.27	9.40		10.67	15.80

02 87 Biohazard Remediation

02 87 13 – Mold Remediation

02 87 13.16 Mold Remediation Preparation and Containment

02 87 13.16		Crew	Daily Output	Labor-Hours	Unit	Material	2020 Bare Costs Labor	Equipment	Total	Total Incl O&P
0010	**MOLD REMEDIATION PREPARATION AND CONTAINMENT**									
6010	Preparation of mold containment area R028213-20									
6100	Pre-cleaning, HEPA vacuum and wet wipe, flat surfaces	A-9	12000	.005	S.F.	.01	.31		.32	.49
6300	Separation barrier, 2" x 4" @ 16", 1/2" plywood ea. side, 8' high	2 Carp	400	.040		3.70	2.13		5.83	7.25
6310	12' high		320	.050		3.52	2.66		6.18	7.85
6320	16' high		200	.080		2.40	4.25		6.65	9.05
6400	Personnel decontam. chamber, 2" x 4" @ 16", 3/4" ply ea. side		280	.057		4.47	3.04		7.51	9.50
6450	Waste decontam. chamber, 2" x 4" studs @ 16", 3/4" ply each side		360	.044		4.54	2.36		6.90	8.55
6500	Cover surfaces with polyethylene sheeting									
6501	Including glue and tape									
6550	Floors, each layer, 6 mil	A-9	8000	.008	S.F.	.04	.47		.51	.77
6551	4 mil		9000	.007		.03	.42		.45	.67
6560	Walls, each layer, 6 mil		6000	.011		.04	.63		.67	1.01
6561	4 mil		7000	.009		.03	.54		.57	.85
6570	For heights above 14', add						20%			
6575	For heights above 20', add						30%			
6580	For fire retardant poly, add					100%				
6590	For large open areas, deduct					10%	20%			
6600	Seal floor penetrations with foam firestop to 36 sq. in.	2 Carp	200	.080	Ea.	13.90	4.25		18.15	21.50
6610	36 sq. in. to 72 sq. in.		125	.128		28	6.80		34.80	40.50
6615	72 sq. in. to 144 sq. in.		80	.200		55.50	10.65		66.15	77
6620	Wall penetrations, to 36 sq. in.		180	.089		13.90	4.72		18.62	22.50
6630	36 sq. in. to 72 sq. in.		100	.160		28	8.50		36.50	43.50
6640	72 sq. in. to 144 sq. in.		60	.267		55.50	14.15		69.65	82.50
6800	Caulk seams with latex caulk	1 Carp	230	.035	L.F.	.17	1.85		2.02	2.97
6900	Set up neg. air machine, 1-2k CFM/25 M.C.F. volume	1 Asbe	4.30	1.860	Ea.		109		109	168

02 87 13.33 Removal and Disposal of Materials With Mold

02 87 13.33		Crew	Daily Output	Labor-Hours	Unit	Material	2020 Bare Costs Labor	Equipment	Total	Total Incl O&P
0010	**REMOVAL AND DISPOSAL OF MATERIALS WITH MOLD**									
0015	Demolition in mold contaminated area									
0200	Ceiling, including suspension system, plaster and lath	A-9	2100	.030	S.F.	.08	1.79		1.87	2.83
0210	Finished plaster, leaving wire lath		585	.109		.27	6.40		6.67	10.15

02 87 Biohazard Remediation

02 87 13 – Mold Remediation

02 87 13.33 Removal and Disposal of Materials With Mold	Crew	Daily Output	Labor-Hours	Unit	Material	2020 Bare Costs Labor	Equipment	Total	Total Incl O&P
0220	Suspended acoustical tile	A-9	3500	.018	S.F.	.05	1.07		1.12	1.70
0230	Concealed tile grid system		3000	.021		.05	1.25		1.30	1.98
0240	Metal pan grid system		1500	.043		.11	2.51		2.62	3.97
0250	Gypsum board		2500	.026		.06	1.50		1.56	2.38
0255	Plywood		2500	.026	▼	.06	1.50		1.56	2.38
0260	Lighting fixtures up to 2' x 4'	▼	72	.889	Ea.	2.23	52		54.23	82.50
0400	Partitions, non load bearing									
0410	Plaster, lath, and studs	A-9	690	.093	S.F.	.90	5.45		6.35	9.35
0450	Gypsum board and studs		1390	.046		.12	2.70		2.82	4.28
0465	Carpet & pad		1390	.046	▼	.12	2.70		2.82	4.28
0600	Pipe insulation, air cell type, up to 4" diameter pipe		900	.071	L.F.	.18	4.17		4.35	6.60
0610	4" to 8" diameter pipe		800	.080		.20	4.70		4.90	7.40
0620	10" to 12" diameter pipe		700	.091		.23	5.35		5.58	8.50
0630	14" to 16" diameter pipe		550	.116	▼	.29	6.85		7.14	10.80
0650	Over 16" diameter pipe	▼	650	.098	S.F.	.25	5.80		6.05	9.15
9000	For type B (supplied air) respirator equipment, add				%				10%	10%

Division Notes

		CREW	DAILY OUTPUT	LABOR-HOURS	UNIT	BARE COSTS				TOTAL INCL O&P
						MAT.	LABOR	EQUIP.	TOTAL	

Estimating Tips
General
- Carefully check all the plans and specifications. Concrete often appears on drawings other than structural drawings, including mechanical and electrical drawings for equipment pads. The cost of cutting and patching is often difficult to estimate. See Subdivision 03 81 for Concrete Cutting, Subdivision 02 41 19.16 for Cutout Demolition, Subdivision 03 05 05.10 for Concrete Demolition, and Subdivision 02 41 19.19 for Rubbish Handling (handling, loading, and hauling of debris).
- Always obtain concrete prices from suppliers near the job site. A volume discount can often be negotiated, depending upon competition in the area. Remember to add for waste, particularly for slabs and footings on grade.

03 10 00 Concrete Forming and Accessories
- A primary cost for concrete construction is forming. Most jobs today are constructed with prefabricated forms. The selection of the forms best suited for the job and the total square feet of forms required for efficient concrete forming and placing are key elements in estimating concrete construction. Enough forms must be available for erection to make efficient use of the concrete placing equipment and crew.
- Concrete accessories for forming and placing depend upon the systems used. Study the plans and specifications to ensure that all special accessory requirements have been included in the cost estimate, such as anchor bolts, inserts, and hangers.
- Included within costs for forms-in-place are all necessary bracing and shoring.

03 20 00 Concrete Reinforcing
- Ascertain that the reinforcing steel supplier has included all accessories, cutting, bending, and an allowance for lapping, splicing, and waste. A good rule of thumb is 10% for lapping, splicing, and waste. Also, 10% waste should be allowed for welded wire fabric.
- The unit price items in the subdivisions for Reinforcing In Place, Glass Fiber Reinforcing, and Welded Wire Fabric include the labor to install accessories such as beam and slab bolsters, high chairs, and bar ties and tie wire. The material cost for these accessories is not included; they may be obtained from the Accessories Subdivisions.

03 30 00 Cast-In-Place Concrete
- When estimating structural concrete, pay particular attention to requirements for concrete additives, curing methods, and surface treatments. Special consideration for climate, hot or cold, must be included in your estimate. Be sure to include requirements for concrete placing equipment and concrete finishing.
- For accurate concrete estimating, the estimator must consider each of the following major components individually: forms, reinforcing steel, ready-mix concrete, placement of the concrete, and finishing of the top surface. For faster estimating, Subdivision 03 30 53.40 for Concrete-In-Place can be used; here, various items of concrete work are presented that include the costs of all five major components (unless specifically stated otherwise).

03 40 00 Precast Concrete
03 50 00 Cast Decks and Underlayment
- The cost of hauling precast concrete structural members is often an important factor. For this reason, it is important to get a quote from the nearest supplier. It may become economically feasible to set up precasting beds on the site if the hauling costs are prohibitive.

Reference Numbers
Reference numbers are shown at the beginning of some major classifications. These numbers refer to related items in the Reference Section. The reference information may be an estimating procedure, an alternate pricing method, or technical information.

Note: Not all subdivisions listed here necessarily appear. ∎

Same Data. Simplified.
Enjoy the convenience and efficiency of accessing your costs anywhere:
- **Skip the multiplier** by setting your location
- **Quickly search,** edit, favorite and share costs
- **Stay on top of price changes** with automatic updates

Discover more at rsmeans.com/online

03 01 Maintenance of Concrete

03 01 30 – Maintenance of Cast-In-Place Concrete

03 01 30.62 Concrete Patching

	03 01 30.62 Concrete Patching	Crew	Daily Output	Labor-Hours	Unit	Material	2020 Bare Costs Labor	Equipment	Total	Total Incl O&P
0010	**CONCRETE PATCHING**									
0100	Floors, 1/4" thick, small areas, regular grout	1 Cefi	170	.047	S.F.	1.61	2.35		3.96	5.25
0150	Epoxy grout	"	100	.080	"	9.90	4		13.90	16.80
2000	Walls, including chipping, cleaning and epoxy grout									
2100	1/4" deep	1 Cefi	65	.123	S.F.	7.80	6.15		13.95	17.65
2150	1/2" deep		50	.160		15.60	8		23.60	29
2200	3/4" deep		40	.200		23.50	10		33.50	40

03 11 Concrete Forming

03 11 13 – Structural Cast-In-Place Concrete Forming

03 11 13.40 Forms In Place, Equipment Foundations

	03 11 13.40 Forms In Place, Equipment Foundations	Crew	Daily Output	Labor-Hours	Unit	Material	2020 Bare Costs Labor	Equipment	Total	Total Incl O&P
0010	**FORMS IN PLACE, EQUIPMENT FOUNDATIONS**									
0020	1 use	C-2	160	.300	SFCA	2.95	15.50		18.45	27
0050	2 use		190	.253		1.62	13.05		14.67	21.50
0100	3 use		200	.240		1.18	12.40		13.58	19.90
0150	4 use		205	.234		.96	12.10		13.06	19.20

03 11 13.45 Forms In Place, Footings

	03 11 13.45 Forms In Place, Footings	Crew	Daily Output	Labor-Hours	Unit	Material	2020 Bare Costs Labor	Equipment	Total	Total Incl O&P
0010	**FORMS IN PLACE, FOOTINGS**									
0020	Continuous wall, plywood, 1 use	C-1	375	.085	SFCA	6.85	4.30		11.15	14
0050	2 use		440	.073		3.77	3.66		7.43	9.65
0100	3 use		470	.068		2.74	3.43		6.17	8.15
0150	4 use		485	.066		2.23	3.32		5.55	7.45
5000	Spread footings, job-built lumber, 1 use		305	.105		2.28	5.30		7.58	10.45
5050	2 use		371	.086		1.27	4.35		5.62	7.95
5100	3 use		401	.080		.91	4.02		4.93	7.05
5150	4 use		414	.077		.74	3.89		4.63	6.65

03 11 13.65 Forms In Place, Slab On Grade

	03 11 13.65 Forms In Place, Slab On Grade	Crew	Daily Output	Labor-Hours	Unit	Material	2020 Bare Costs Labor	Equipment	Total	Total Incl O&P
0010	**FORMS IN PLACE, SLAB ON GRADE**									
3000	Edge forms, wood, 4 use, on grade, to 6" high	C-1	600	.053	L.F.	.33	2.69		3.02	4.40
6000	Trench forms in floor, wood, 1 use		160	.200	SFCA	2.05	10.10		12.15	17.40
6050	2 use		175	.183		1.13	9.20		10.33	15.10
6100	3 use		180	.178		.82	8.95		9.77	14.35
6150	4 use		185	.173		.67	8.70		9.37	13.85

03 15 Concrete Accessories

03 15 05 – Concrete Forming Accessories

03 15 05.75 Sleeves and Chases

	03 15 05.75 Sleeves and Chases	Crew	Daily Output	Labor-Hours	Unit	Material	2020 Bare Costs Labor	Equipment	Total	Total Incl O&P
0010	**SLEEVES AND CHASES**									
0100	Plastic, 1 use, 12" long, 2" diameter	1 Carp	100	.080	Ea.	2.32	4.25		6.57	8.95
0150	4" diameter		90	.089		4.97	4.72		9.69	12.55
0200	6" diameter		75	.107		10.85	5.65		16.50	20.50
0250	12" diameter		60	.133		28.50	7.10		35.60	41.50
5000	Sheet metal, 2" diameter	G	100	.080		1.54	4.25		5.79	8.10
5100	4" diameter	G	90	.089		1.93	4.72		6.65	9.20
5150	6" diameter	G	75	.107		2.03	5.65		7.68	10.75
5200	12" diameter	G	60	.133		4.24	7.10		11.34	15.30
6000	Steel pipe, 2" diameter	G	100	.080		3.26	4.25		7.51	10
6100	4" diameter	G	90	.089		17.20	4.72		21.92	26
6150	6" diameter	G	75	.107		36.50	5.65		42.15	48.50

03 15 Concrete Accessories

03 15 05 – Concrete Forming Accessories

03 15 05.75 Sleeves and Chases		Crew	Daily Output	Labor-Hours	Unit	Material	2020 Bare Costs Labor	Equipment	Total	Total Incl O&P
6200	12" diameter	G 1 Carp	60	.133	Ea.	85	7.10		92.10	104

03 15 16 – Concrete Construction Joints

03 15 16.20 Control Joints, Saw Cut

		Crew	Daily Output	Labor-Hours	Unit	Material	Labor	Equipment	Total	Total Incl O&P
0010	**CONTROL JOINTS, SAW CUT**									
0100	Sawcut control joints in green concrete									
0120	1" depth	C-27	2000	.008	L.F.	.03	.40	.06	.49	.69
0140	1-1/2" depth		1800	.009		.05	.44	.06	.55	.78
0160	2" depth	↓	1600	.010	↓	.07	.50	.07	.64	.88
0180	Sawcut joint reservoir in cured concrete									
0182	3/8" wide x 3/4" deep, with single saw blade	C-27	1000	.016	L.F.	.05	.80	.11	.96	1.36
0184	1/2" wide x 1" deep, with double saw blades		900	.018		.10	.89	.12	1.11	1.56
0186	3/4" wide x 1-1/2" deep, with double saw blades	↓	800	.020		.20	1	.14	1.34	1.84
0190	Water blast joint to wash away laitance, 2 passes	C-29	2500	.003			.13	.04	.17	.24
0200	Air blast joint to blow out debris and air dry, 2 passes	C-28	2000	.004	↓		.20	.02	.22	.31
0300	For backer rod, see Section 07 91 23.10									
0342	For joint sealant, see Section 07 92 13.20									

03 15 19 – Cast-In Concrete Anchors

03 15 19.10 Anchor Bolts

		Crew	Daily Output	Labor-Hours	Unit	Material	Labor	Equipment	Total	Total Incl O&P
0010	**ANCHOR BOLTS**									
0015	Made from recycled materials									
0025	Single bolts installed in fresh concrete, no templates									
0030	Hooked w/nut and washer, 1/2" diameter, 8" long	G 1 Carp	132	.061	Ea.	1.62	3.22		4.84	6.60
0040	12" long	G	131	.061		1.80	3.25		5.05	6.85
0050	5/8" diameter, 8" long	G	129	.062		3.39	3.30		6.69	8.70
0060	12" long	G	127	.063		4.18	3.35		7.53	9.65
0070	3/4" diameter, 8" long	G	127	.063		4.18	3.35		7.53	9.65
0080	12" long	G ↓	125	.064	↓	5.20	3.40		8.60	10.85
0090	2-bolt pattern, including job-built 2-hole template, per set									
0100	J-type, incl. hex nut & washer, 1/2" diameter x 6" long	G 1 Carp	21	.381	Set	5.60	20.50		26.10	36.50
0110	12" long	G	21	.381		6.35	20.50		26.85	37.50
0120	18" long	G	21	.381		7.40	20.50		27.90	38.50
0130	3/4" diameter x 8" long	G	20	.400		11.10	21.50		32.60	44
0140	12" long	G	20	.400		13.15	21.50		34.65	46.50
0150	18" long	G	20	.400		16.30	21.50		37.80	50
0160	1" diameter x 12" long	G	19	.421		26.50	22.50		49	62.50
0170	18" long	G	19	.421		31	22.50		53.50	68
0180	24" long	G	19	.421		37.50	22.50		60	74.50
0190	36" long	G	18	.444		50.50	23.50		74	91
0200	1-1/2" diameter x 18" long	G	17	.471		47.50	25		72.50	89.50
0210	24" long	G	16	.500		56	26.50		82.50	102
0300	L-type, incl. hex nut & washer, 3/4" diameter x 12" long	G	20	.400		18.20	21.50		39.70	52
0310	18" long	G	20	.400		22.50	21.50		44	57
0320	24" long	G	20	.400		27	21.50		48.50	61.50
0330	30" long	G	20	.400		33.50	21.50		55	69
0340	36" long	G	20	.400		38	21.50		59.50	73.50
0350	1" diameter x 12" long	G	19	.421		24.50	22.50		47	60.50
0360	18" long	G	19	.421		29.50	22.50		52	66
0370	24" long	G	19	.421		36	22.50		58.50	73
0380	30" long	G	19	.421		42	22.50		64.50	80
0390	36" long	G	18	.444		47.50	23.50		71	88
0400	42" long	G	18	.444		57.50	23.50		81	98.50
0410	48" long	G	18	.444		64	23.50		87.50	106

For customer support on your Mechanical Costs with RSMeans Data, call 800.448.8182.

33

03 15 19.10 Anchor Bolts		Crew	Daily Output	Labor-Hours	Unit	Material	2020 Bare Costs Labor	Equipment	Total	Total Incl O&P
0420	1-1/4" diameter x 18" long	1 Carp	18	.444	Set	43.50	23.50		67	83.50
0430	24" long		18	.444		51.50	23.50		75	92
0440	30" long		17	.471		59	25		84	103
0450	36" long		17	.471		66.50	25		91.50	111
0460	42" long	2 Carp	32	.500		75	26.50		101.50	123
0470	48" long		32	.500		85	26.50		111.50	134
0480	54" long		31	.516		100	27.50		127.50	151
0490	60" long		31	.516		109	27.50		136.50	161
0500	1-1/2" diameter x 18" long		33	.485		47	26		73	90
0510	24" long		32	.500		54	26.50		80.50	99.50
0520	30" long		31	.516		61	27.50		88.50	108
0530	36" long		30	.533		69.50	28.50		98	119
0540	42" long		30	.533		79	28.50		107.50	129
0550	48" long		29	.552		88	29.50		117.50	141
0560	54" long		28	.571		107	30.50		137.50	163
0570	60" long		28	.571		117	30.50		147.50	174
0580	1-3/4" diameter x 18" long		31	.516		71.50	27.50		99	120
0590	24" long		30	.533		83.50	28.50		112	135
0600	30" long		29	.552		96.50	29.50		126	150
0610	36" long		28	.571		110	30.50		140.50	167
0620	42" long		27	.593		123	31.50		154.50	184
0630	48" long		26	.615		135	32.50		167.50	198
0640	54" long		26	.615		167	32.50		199.50	233
0650	60" long		25	.640		180	34		214	249
0660	2" diameter x 24" long		27	.593		135	31.50		166.50	197
0670	30" long		27	.593		152	31.50		183.50	215
0680	36" long		26	.615		167	32.50		199.50	232
0690	42" long		25	.640		185	34		219	255
0700	48" long		24	.667		213	35.50		248.50	287
0710	54" long		23	.696		253	37		290	335
0720	60" long		23	.696		272	37		309	355
0730	66" long		22	.727		291	38.50		329.50	380
0740	72" long		21	.762		320	40.50		360.50	410
1000	4-bolt pattern, including job-built 4-hole template, per set									
1100	J-type, incl. hex nut & washer, 1/2" diameter x 6" long	1 Carp	19	.421	Set	8.50	22.50		31	43
1110	12" long		19	.421		9.95	22.50		32.45	44.50
1120	18" long		18	.444		12.10	23.50		35.60	49
1130	3/4" diameter x 8" long		17	.471		19.45	25		44.45	59
1140	12" long		17	.471		23.50	25		48.50	63.50
1150	18" long		17	.471		30	25		55	70.50
1160	1" diameter x 12" long		16	.500		50	26.50		76.50	95
1170	18" long		15	.533		59.50	28.50		88	108
1180	24" long		15	.533		72	28.50		100.50	122
1190	36" long		15	.533		98	28.50		126.50	151
1200	1-1/2" diameter x 18" long		13	.615		92	32.50		124.50	150
1210	24" long		12	.667		109	35.50		144.50	173
1300	L-type, incl. hex nut & washer, 3/4" diameter x 12" long		17	.471		33.50	25		58.50	74.50
1310	18" long		17	.471		42.50	25		67.50	84
1320	24" long		17	.471		51	25		76	93.50
1330	30" long		16	.500		64	26.50		90.50	111
1340	36" long		16	.500		73	26.50		99.50	120
1350	1" diameter x 12" long		16	.500		46	26.50		72.50	90.50
1360	18" long		15	.533		56.50	28.50		85	105

03 15 Concrete Accessories

03 15 19 – Cast-In Concrete Anchors

03 15 19.10 Anchor Bolts

			Crew	Daily Output	Labor-Hours	Unit	Material	2020 Bare Costs Labor	2020 Bare Costs Equipment	Total	Total Incl O&P
1370	24" long	G	1 Carp	15	.533	Set	69.50	28.50		98	119
1380	30" long	G		15	.533		81.50	28.50		110	132
1390	36" long	G		15	.533		92.50	28.50		121	145
1400	42" long	G		14	.571		112	30.50		142.50	169
1410	48" long	G		14	.571		125	30.50		155.50	184
1420	1-1/4" diameter x 18" long	G		14	.571		84.50	30.50		115	139
1430	24" long	G		14	.571		100	30.50		130.50	156
1440	30" long	G		13	.615		115	32.50		147.50	176
1450	36" long	G		13	.615		131	32.50		163.50	193
1460	42" long	G	2 Carp	25	.640		147	34		181	213
1470	48" long	G		24	.667		168	35.50		203.50	237
1480	54" long	G		23	.696		197	37		234	273
1490	60" long	G		23	.696		216	37		253	294
1500	1-1/2" diameter x 18" long	G		25	.640		91	34		125	151
1510	24" long	G		24	.667		105	35.50		140.50	169
1520	30" long	G		23	.696		119	37		156	187
1530	36" long	G		22	.727		136	38.50		174.50	208
1540	42" long	G		22	.727		155	38.50		193.50	228
1550	48" long	G		21	.762		173	40.50		213.50	252
1560	54" long	G		20	.800		211	42.50		253.50	296
1570	60" long	G		20	.800		231	42.50		273.50	320
1580	1-3/4" diameter x 18" long	G		22	.727		140	38.50		178.50	212
1590	24" long	G		21	.762		164	40.50		204.50	242
1600	30" long	G		21	.762		191	40.50		231.50	271
1610	36" long	G		20	.800		217	42.50		259.50	305
1620	42" long	G		19	.842		244	45		289	335
1630	48" long	G		18	.889		267	47		314	365
1640	54" long	G		18	.889		330	47		377	435
1650	60" long	G		17	.941		360	50		410	470
1660	2" diameter x 24" long	G		19	.842		267	45		312	360
1670	30" long	G		18	.889		300	47		347	400
1680	36" long	G		18	.889		330	47		377	435
1690	42" long	G		17	.941		370	50		420	480
1700	48" long	G		16	1		425	53		478	545
1710	54" long	G		15	1.067		505	56.50		561.50	640
1720	60" long	G		15	1.067		540	56.50		596.50	680
1730	66" long	G		14	1.143		580	60.50		640.50	725
1740	72" long	G		14	1.143		635	60.50		695.50	785
1990	For galvanized, add					Ea.	75%				

03 15 19.45 Machinery Anchors

			Crew	Daily Output	Labor-Hours	Unit	Material	2020 Bare Costs Labor	2020 Bare Costs Equipment	Total	Total Incl O&P
0010	**MACHINERY ANCHORS**, heavy duty, incl. sleeve, floating base nut,										
0020	lower stud & coupling nut, fiber plug, connecting stud, washer & nut.										
0030	For flush mounted embedment in poured concrete heavy equip. pads.										
0200	Stud & bolt, 1/2" diameter	G	E-16	40	.400	Ea.	61.50	23.50	3.68	88.68	109
0300	5/8" diameter	G		35	.457		67	26.50	4.20	97.70	120
0500	3/4" diameter	G		30	.533		79.50	31	4.90	115.40	142
0600	7/8" diameter	G		25	.640		93.50	37.50	5.90	136.90	168
0800	1" diameter	G		20	.800		102	47	7.35	156.35	194
0900	1-1/4" diameter	G		15	1.067		133	62.50	9.80	205.30	256

For customer support on your Mechanical Costs with RSMeans Data, call 800.448.8182.

35

03 21 Reinforcement Bars

03 21 11 – Plain Steel Reinforcement Bars

03 21 11.60 Reinforcing In Place

		Crew	Daily Output	Labor-Hours	Unit	Material	2020 Bare Costs Labor	Equipment	Total	Total Incl O&P
0010	**REINFORCING IN PLACE**, 50-60 ton lots, A615 Grade 60									
0020	Includes labor, but not material cost, to install accessories									
0030	Made from recycled materials									
0502	Footings, #4 to #7 [G]	4 Rodm	4200	.008	Lb.	.56	.43		.99	1.27
0552	#8 to #18 [G]		7200	.004		.56	.25		.81	1
0602	Slab on grade, #3 to #7 [G]	↓	4200	.008	↓	.56	.43		.99	1.27
0900	For other than 50-60 ton lots									
1000	Under 10 ton job, #3 to #7, add					25%	10%			
1010	#8 to #18, add					20%	10%			
1050	10-50 ton job, #3 to #7, add					10%				
1060	#8 to #18, add					5%				
1100	60-100 ton job, #3 to #7, deduct					5%				
1110	#8 to #18, deduct					10%				
1150	Over 100 ton job, #3 to #7, deduct					10%				
1160	#8 to #18, deduct					15%				

03 22 Fabric and Grid Reinforcing

03 22 11 – Plain Welded Wire Fabric Reinforcing

03 22 11.10 Plain Welded Wire Fabric

		Crew	Daily Output	Labor-Hours	Unit	Material	Labor	Equipment	Total	Total Incl O&P
0010	**PLAIN WELDED WIRE FABRIC** ASTM A185									
0020	Includes labor, but not material cost, to install accessories									
0030	Made from recycled materials									
0050	Sheets									
0100	6 x 6 - W1.4 x W1.4 (10 x 10) 21 lb./C.S.F. [G]	2 Rodm	35	.457	C.S.F.	16.25	26		42.25	57

03 22 13 – Galvanized Welded Wire Fabric Reinforcing

03 22 13.10 Galvanized Welded Wire Fabric

		Crew	Daily Output	Labor-Hours	Unit	Material	Labor	Equipment	Total	Total Incl O&P
0010	**GALVANIZED WELDED WIRE FABRIC**									
0100	Add to plain welded wire pricing for galvanized welded wire				Lb.	.26			.26	.28

03 22 16 – Epoxy-Coated Welded Wire Fabric Reinforcing

03 22 16.10 Epoxy-Coated Welded Wire Fabric

		Crew	Daily Output	Labor-Hours	Unit	Material	Labor	Equipment	Total	Total Incl O&P
0010	**EPOXY-COATED WELDED WIRE FABRIC**									
0100	Add to plain welded wire pricing for epoxy-coated welded wire				Lb.	.54			.54	.60

03 30 Cast-In-Place Concrete

03 30 53 – Miscellaneous Cast-In-Place Concrete

03 30 53.40 Concrete In Place

		Crew	Daily Output	Labor-Hours	Unit	Material	Labor	Equipment	Total	Total Incl O&P
0010	**CONCRETE IN PLACE**									
0020	Including forms (4 uses), Grade 60 rebar, concrete (Portland cement									
0050	Type I), placement and finishing unless otherwise indicated									
0500	Chimney foundations (5000 psi), over 5 C.Y.	C-14C	32.22	3.476	C.Y.	178	175	.83	353.83	460
0510	(3500 psi), under 5 C.Y.	"	23.71	4.724	"	208	238	1.13	447.13	585
3540	Equipment pad (3000 psi), 3' x 3' x 6" thick	C-14H	45	1.067	Ea.	50.50	55	.60	106.10	139
3550	4' x 4' x 6" thick		30	1.600		78	82.50	.90	161.40	210
3560	5' x 5' x 8" thick		18	2.667		138	138	1.49	277.49	360
3570	6' x 6' x 8" thick		14	3.429		190	177	1.92	368.92	475
3580	8' x 8' x 10" thick		8	6		395	310	3.36	708.36	905
3590	10' x 10' x 12" thick	↓	5	9.600	↓	695	495	5.40	1,195.40	1,500
3800	Footings (3000 psi), spread under 1 C.Y.	C-14C	28	4	C.Y.	203	201	.96	404.96	525

03 30 Cast-In-Place Concrete

03 30 53 – Miscellaneous Cast-In-Place Concrete

03 30 53.40 Concrete In Place

		Crew	Daily Output	Labor-Hours	Unit	Material	2020 Bare Costs		Total	Total Incl O&P
							Labor	Equipment		
3825	1 C.Y. to 5 C.Y.	C-14C	43	2.605	C.Y.	240	131	.63	371.63	465
3850	Over 5 C.Y.	↓	75	1.493		220	75	.36	295.36	355
3900	Footings, strip (3000 psi), 18" x 9", unreinforced	C-14L	40	2.400		154	118	.67	272.67	350
3920	18" x 9", reinforced	C-14C	35	3.200		181	161	.77	342.77	440
3925	20" x 10", unreinforced	C-14L	45	2.133		150	105	.60	255.60	325
3930	20" x 10", reinforced	C-14C	40	2.800		172	141	.67	313.67	400
3935	24" x 12", unreinforced	C-14L	55	1.745		148	86	.49	234.49	293
3940	24" x 12", reinforced	C-14C	48	2.333		171	118	.56	289.56	365
3945	36" x 12", unreinforced	C-14L	70	1.371		144	67.50	.38	211.88	261
3950	36" x 12", reinforced	C-14C	60	1.867		165	94	.45	259.45	320
4000	Foundation mat (3000 psi), under 10 C.Y.	↓	38.67	2.896		246	146	.70	392.70	490
4050	Over 20 C.Y.	↓	56.40	1.986		217	100	.48	317.48	390
4650	Slab on grade (3500 psi), not including finish, 4" thick	C-14E	60.75	1.449		147	74	.45	221.45	273
4700	6" thick	"	92	.957	↓	141	49	.30	190.30	230
4701	Thickened slab edge (3500 psi), for slab on grade poured									
4702	monolithically with slab; depth is in addition to slab thickness;									
4703	formed vertical outside edge, earthen bottom and inside slope									
4705	8" deep x 8" wide bottom, unreinforced	C-14L	2190	.044	L.F.	4.07	2.16	.01	6.24	7.75
4710	8" x 8", reinforced	C-14C	1670	.067		6.70	3.38	.02	10.10	12.45
4715	12" deep x 12" wide bottom, unreinforced	C-14L	1800	.053		8.30	2.63	.01	10.94	13.10
4720	12" x 12", reinforced	C-14C	1310	.086		13.15	4.31	.02	17.48	21
4725	16" deep x 16" wide bottom, unreinforced	C-14L	1440	.067		14	3.29	.02	17.31	20.50
4730	16" x 16", reinforced	C-14C	1120	.100		19.75	5.05	.02	24.82	29.50
4735	20" deep x 20" wide bottom, unreinforced	C-14L	1150	.083		21	4.12	.02	25.14	29.50
4740	20" x 20", reinforced	C-14C	920	.122		28.50	6.15	.03	34.68	40.50
4745	24" deep x 24" wide bottom, unreinforced	C-14L	930	.103		30	5.10	.03	35.13	40.50
4750	24" x 24", reinforced	C-14C	740	.151	↓	39.50	7.60	.04	47.14	55
4751	Slab on grade (3500 psi), incl. troweled finish, not incl. forms									
4760	or reinforcing, over 10,000 S.F., 4" thick	C-14F	3425	.021	S.F.	1.62	1	.01	2.63	3.27
4820	6" thick		3350	.021		2.36	1.02	.01	3.39	4.12
4840	8" thick		3184	.023		3.24	1.08	.01	4.33	5.15
4900	12" thick		2734	.026		4.85	1.25	.01	6.11	7.20
4950	15" thick	↓	2505	.029	↓	6.10	1.37	.01	7.48	8.75

03 31 Structural Concrete

03 31 13 – Heavyweight Structural Concrete

03 31 13.35 Heavyweight Concrete, Ready Mix

		Crew	Daily Output	Labor-Hours	Unit	Material	2020 Bare Costs		Total	Total Incl O&P
							Labor	Equipment		
0010	**HEAVYWEIGHT CONCRETE, READY MIX**, delivered									
0012	Includes local aggregate, sand, Portland cement (Type I) and water									
0015	Excludes all additives and treatments									
0020	2000 psi				C.Y.	107			107	117
0100	2500 psi					110			110	121
0150	3000 psi					129			129	142
0200	3500 psi					124			124	137
0300	4000 psi					127			127	140
1000	For high early strength (Portland cement Type III), add					10%				
1300	For winter concrete (hot water), add					5.35			5.35	5.90
1410	For mid-range water reducer, add					4.18			4.18	4.60
1420	For high-range water reducer/superplasticizer, add					6.40			6.40	7
1430	For retarder, add					3.45			3.45	3.80
1440	For non-Chloride accelerator, add					6.80			6.80	7.45

For customer support on your Mechanical Costs with RSMeans Data, call 800.448.8182.

37

03 31 Structural Concrete

03 31 13 – Heavyweight Structural Concrete

03 31 13.35 Heavyweight Concrete, Ready Mix	Crew	Daily Output	Labor-Hours	Unit	Material	2020 Bare Costs Labor	Equipment	Total	Total Incl O&P
1450	For Chloride accelerator, per 1%, add				C.Y.	4.10			4.10	4.51
1460	For fiber reinforcing, synthetic (1 lb./C.Y.), add					8.10			8.10	8.90
1500	For Saturday delivery, add				↓	9.20			9.20	10.10
1510	For truck holding/waiting time past 1st hour per load, add				Hr.	109			109	120
1520	For short load (less than 4 C.Y.), add per load				Ea.	85			85	93.50
2000	For all lightweight aggregate, add				C.Y.	45%				

03 31 13.70 Placing Concrete

	03 31 13.70 Placing Concrete	Crew	Daily Output	Labor-Hours	Unit	Material	2020 Bare Costs Labor	Equipment	Total	Total Incl O&P
0010	**PLACING CONCRETE**									
0020	Includes labor and equipment to place, level (strike off) and consolidate									
1200	Duct bank, direct chute	C-6	155	.310	C.Y.		13.55	.35	13.90	21
1900	Footings, continuous, shallow, direct chute	"	120	.400			17.50	.45	17.95	26.50
1950	Pumped	C-20	150	.427			19.25	3.09	22.34	32.50
2000	With crane and bucket	C-7	90	.800			36.50	12	48.50	68
2100	Footings, continuous, deep, direct chute	C-6	140	.343			15	.38	15.38	23
2150	Pumped	C-20	160	.400			18.05	2.90	20.95	30
2200	With crane and bucket	C-7	110	.655			30	9.80	39.80	56
2400	Footings, spread, under 1 C.Y., direct chute	C-6	55	.873			38	.98	38.98	58
2450	Pumped	C-20	65	.985			44.50	7.15	51.65	74.50
2500	With crane and bucket	C-7	45	1.600			73	24	97	137
2600	Over 5 C.Y., direct chute	C-6	120	.400			17.50	.45	17.95	26.50
2650	Pumped	C-20	150	.427			19.25	3.09	22.34	32.50
2700	With crane and bucket	C-7	100	.720			33	10.80	43.80	61.50
2900	Foundation mats, over 20 C.Y., direct chute	C-6	350	.137			6	.15	6.15	9.15
2950	Pumped	C-20	400	.160			7.25	1.16	8.41	12.10
3000	With crane and bucket	C-7	300	.240	↓		11	3.60	14.60	20.50

03 35 Concrete Finishing

03 35 13 – High-Tolerance Concrete Floor Finishing

	03 35 13.30 Finishing Floors, High Tolerance	Crew	Daily Output	Labor-Hours	Unit	Material	2020 Bare Costs Labor	Equipment	Total	Total Incl O&P
0010	**FINISHING FLOORS, HIGH TOLERANCE**									
0012	Finishing of fresh concrete flatwork requires that concrete									
0013	first be placed, struck off & consolidated									
0015	Basic finishing for various unspecified flatwork									
0100	Bull float only	C-10	4000	.006	S.F.		.28		.28	.42
0125	Bull float & manual float		2000	.012			.57		.57	.84
0150	Bull float, manual float & broom finish, w/edging & joints		1850	.013			.61		.61	.91
0200	Bull float, manual float & manual steel trowel	↓	1265	.019	↓		.90		.90	1.33
0210	For specified Random Access Floors in ACI Classes 1, 2, 3 and 4 to achieve									
0215	Composite Overall Floor Flatness and Levelness values up to FF35/FL25									
0250	Bull float, machine float & machine trowel (walk-behind)	C-10C	1715	.014	S.F.		.66	.05	.71	1.04
0300	Power screed, bull float, machine float & trowel (walk-behind)	C-10D	2400	.010			.47	.07	.54	.77
0350	Power screed, bull float, machine float & trowel (ride-on)	C-10E	4000	.006	↓		.28	.06	.34	.49
0352	For specified Random Access Floors in ACI Classes 5, 6, 7 and 8 to achieve									
0354	Composite Overall Floor Flatness and Levelness values up to FF50/FL50									
0356	Add for two-dimensional restraightening after power float	C-10	6000	.004	S.F.		.19		.19	.28
0358	For specified Random or Defined Access Floors in ACI Class 9 to achieve									
0360	Composite Overall Floor Flatness and Levelness values up to FF100/FL100									
0362	Add for two-dimensional restraightening after bull float & power float	C-10	3000	.008	S.F.		.38		.38	.56
0364	For specified Superflat Defined Access Floors in ACI Class 9 to achieve									
0366	Minimum Floor Flatness and Levelness values of FF100/FL100									
0368	Add for 2-dim'l restraightening after bull float, power float, power trowel	C-10	2000	.012	S.F.		.57		.57	.84

03 54 Cast Underlayment

03 54 16 – Hydraulic Cement Underlayment

03 54 16.50 Cement Underlayment	Crew	Daily Output	Labor-Hours	Unit	Material	2020 Bare Costs Labor	Equipment	Total	Total Incl O&P
0010 **CEMENT UNDERLAYMENT**									
2510 Underlayment, P.C. based, self-leveling, 4100 psi, pumped, 1/4" thick	C-8	20000	.003	S.F.	1.75	.13	.02	1.90	2.15
2520 1/2" thick		19000	.003		3.50	.14	.02	3.66	4.09
2530 3/4" thick		18000	.003		5.25	.15	.02	5.42	6.05
2540 1" thick		17000	.003		7	.15	.02	7.17	7.95
2550 1-1/2" thick		15000	.004		10.50	.17	.03	10.70	11.85
2560 Hand placed, 1/2" thick	C-18	450	.020		3.50	.85	.28	4.63	5.45
2610 Topping, P.C. based, self-leveling, 6100 psi, pumped, 1/4" thick	C-8	20000	.003		2.09	.13	.02	2.24	2.52
2620 1/2" thick		19000	.003		4.18	.14	.02	4.34	4.83
2630 3/4" thick		18000	.003		6.25	.15	.02	6.42	7.15
2660 1" thick		17000	.003		8.35	.15	.02	8.52	9.45
2670 1-1/2" thick		15000	.004		12.55	.17	.03	12.75	14.10
2680 Hand placed, 1/2" thick	C-18	450	.020		4.18	.85	.28	5.31	6.20

03 63 Epoxy Grouting

03 63 05 – Grouting of Dowels and Fasteners

03 63 05.10 Epoxy Only

	Crew	Daily Output	Labor-Hours	Unit	Material	2020 Bare Costs Labor	Equipment	Total	Total Incl O&P
0010 **EPOXY ONLY**									
1500 Chemical anchoring, epoxy cartridge, excludes layout, drilling, fastener									
1530 For fastener 3/4" diam. x 6" embedment	2 Skwk	72	.222	Ea.	4.78	12.20		16.98	23.50
1535 1" diam. x 8" embedment		66	.242		7.15	13.30		20.45	28
1540 1-1/4" diam. x 10" embedment		60	.267		14.35	14.65		29	38
1545 1-3/4" diam. x 12" embedment		54	.296		24	16.25		40.25	51
1550 14" embedment		48	.333		28.50	18.30		46.80	59
1555 2" diam. x 12" embedment		42	.381		38	21		59	73.50
1560 18" embedment		32	.500		48	27.50		75.50	94

03 82 Concrete Boring

03 82 13 – Concrete Core Drilling

03 82 13.10 Core Drilling

	Crew	Daily Output	Labor-Hours	Unit	Material	2020 Bare Costs Labor	Equipment	Total	Total Incl O&P
0010 **CORE DRILLING**									
0015 Includes bit cost, layout and set-up time									
0020 Reinforced concrete slab, up to 6" thick									
0100 1" diameter core	B-89A	17	.941	Ea.	.20	45.50	6.70	52.40	76.50
0150 For each additional inch of slab thickness in same hole, add		1440	.011		.03	.54	.08	.65	.94
0200 2" diameter core		16.50	.970		.26	47	6.90	54.16	79
0250 For each additional inch of slab thickness in same hole, add		1080	.015		.04	.72	.11	.87	1.25
0300 3" diameter core		16	1		.40	48.50	7.15	56.05	81.50
0350 For each additional inch of slab thickness in same hole, add		720	.022		.07	1.08	.16	1.31	1.86
0500 4" diameter core		15	1.067		.45	51.50	7.60	59.55	87
0550 For each additional inch of slab thickness in same hole, add		480	.033		.07	1.62	.24	1.93	2.78
0700 6" diameter core		14	1.143		.65	55.50	8.15	64.30	93
0750 For each additional inch of slab thickness in same hole, add		360	.044		.11	2.15	.32	2.58	3.72
0900 8" diameter core		13	1.231		1.07	59.50	8.80	69.37	101
0950 For each additional inch of slab thickness in same hole, add		288	.056		.18	2.69	.40	3.27	4.70
1100 10" diameter core		12	1.333		1.20	64.50	9.50	75.20	109
1150 For each additional inch of slab thickness in same hole, add		240	.067		.20	3.23	.48	3.91	5.60
1300 12" diameter core		11	1.455		1.84	70.50	10.40	82.74	119
1350 For each additional inch of slab thickness in same hole, add		206	.078		.31	3.77	.55	4.63	6.65

03 82 Concrete Boring

03 82 13 – Concrete Core Drilling

03 82 13.10 Core Drilling	Crew	Daily Output	Labor-Hours	Unit	Material	2020 Bare Costs Labor	2020 Bare Costs Equipment	Total	Total Incl O&P
1500	14" diameter core	B-89A	10	1.600	Ea.	1.84	77.50	11.40	90.74	132
1550	For each additional inch of slab thickness in same hole, add		180	.089		.31	4.31	.63	5.25	7.55
1700	18" diameter core		9	1.778		3.06	86	12.70	101.76	147
1750	For each additional inch of slab thickness in same hole, add		144	.111		.51	5.40	.79	6.70	9.55
1754	24" diameter core		8	2		4.15	97	14.30	115.45	166
1756	For each additional inch of slab thickness in same hole, add		120	.133		.69	6.45	.95	8.09	11.55
1760	For horizontal holes, add to above	▼		▼			20%	20%		
1770	Prestressed hollow core plank, 8" thick									
1780	1" diameter core	B-89A	17.50	.914	Ea.	.27	44.50	6.55	51.32	74.50
1790	For each additional inch of plank thickness in same hole, add		3840	.004		.03	.20	.03	.26	.37
1794	2" diameter core		17.25	.928		.35	45	6.60	51.95	75.50
1796	For each additional inch of plank thickness in same hole, add		2880	.006		.04	.27	.04	.35	.50
1800	3" diameter core		17	.941		.53	45.50	6.70	52.73	77
1810	For each additional inch of plank thickness in same hole, add		1920	.008		.07	.40	.06	.53	.75
1820	4" diameter core		16.50	.970		.60	47	6.90	54.50	79.50
1830	For each additional inch of plank thickness in same hole, add		1280	.013		.07	.61	.09	.77	1.09
1840	6" diameter core		15.50	1.032		.87	50	7.35	58.22	84.50
1850	For each additional inch of plank thickness in same hole, add		960	.017		.11	.81	.12	1.04	1.47
1860	8" diameter core		15	1.067		1.43	51.50	7.60	60.53	88
1870	For each additional inch of plank thickness in same hole, add		768	.021		.18	1.01	.15	1.34	1.88
1880	10" diameter core		14	1.143		1.60	55.50	8.15	65.25	94.50
1890	For each additional inch of plank thickness in same hole, add		640	.025		.20	1.21	.18	1.59	2.25
1900	12" diameter core		13.50	1.185		2.45	57.50	8.45	68.40	98.50
1910	For each additional inch of plank thickness in same hole, add	▼	548	.029	▼	.31	1.42	.21	1.94	2.70

03 82 16 – Concrete Drilling

03 82 16.10 Concrete Impact Drilling

		Crew	Daily Output	Labor-Hours	Unit	Material	Labor	Equipment	Total	Total Incl O&P
0010	**CONCRETE IMPACT DRILLING**									
0020	Includes bit cost, layout and set-up time, no anchors									
0050	Up to 4" deep in concrete/brick floors/walls									
0100	Holes, 1/4" diameter	1 Carp	75	.107	Ea.	.07	5.65		5.72	8.60
0150	For each additional inch of depth in same hole, add		430	.019		.02	.99		1.01	1.51
0200	3/8" diameter		63	.127		.05	6.75		6.80	10.20
0250	For each additional inch of depth in same hole, add		340	.024		.01	1.25		1.26	1.89
0300	1/2" diameter		50	.160		.05	8.50		8.55	12.85
0350	For each additional inch of depth in same hole, add		250	.032		.01	1.70		1.71	2.57
0400	5/8" diameter		48	.167		.10	8.85		8.95	13.40
0450	For each additional inch of depth in same hole, add		240	.033		.02	1.77		1.79	2.69
0500	3/4" diameter		45	.178		.11	9.45		9.56	14.30
0550	For each additional inch of depth in same hole, add		220	.036		.03	1.93		1.96	2.93
0600	7/8" diameter		43	.186		.17	9.90		10.07	15.05
0650	For each additional inch of depth in same hole, add		210	.038		.04	2.03		2.07	3.09
0700	1" diameter		40	.200		.18	10.65		10.83	16.15
0750	For each additional inch of depth in same hole, add		190	.042		.04	2.24		2.28	3.41
0800	1-1/4" diameter		38	.211		.33	11.20		11.53	17.15
0850	For each additional inch of depth in same hole, add		180	.044		.08	2.36		2.44	3.64
0900	1-1/2" diameter		35	.229		.43	12.15		12.58	18.70
0950	For each additional inch of depth in same hole, add	▼	165	.048	▼	.11	2.58		2.69	3.99
1000	For ceiling installations, add						40%			

Estimating Tips
04 05 00 Common Work Results for Masonry

- The terms mortar and grout are often used interchangeably—and incorrectly. Mortar is used to bed masonry units, seal the entry of air and moisture, provide architectural appearance, and allow for size variations in the units. Grout is used primarily in reinforced masonry construction and to bond the masonry to the reinforcing steel. Common mortar types are M (2500 psi), S (1800 psi), N (750 psi), and O (350 psi), and they conform to ASTM C270. Grout is either fine or coarse and conforms to ASTM C476, and in-place strengths generally exceed 2500 psi. Mortar and grout are different components of masonry construction and are placed by entirely different methods. An estimator should be aware of their unique uses and costs.

- Mortar is included in all assembled masonry line items. The mortar cost, part of the assembled masonry material cost, includes all ingredients, all labor, and all equipment required. Please see reference number R040513-10.

- Waste, specifically the loss/droppings of mortar and the breakage of brick and block, is included in all unit cost lines that include mortar and masonry units in this division. A factor of 25% is added for mortar and 3% for brick and concrete masonry units.

- Scaffolding or staging is not included in any of the Division 4 costs. Refer to Subdivision 01 54 23 for scaffolding and staging costs.

04 20 00 Unit Masonry

- The most common types of unit masonry are brick and concrete masonry. The major classifications of brick are building brick (ASTM C62), facing brick (ASTM C216), glazed brick, fire brick, and pavers. Many varieties of texture and appearance can exist within these classifications, and the estimator would be wise to check local custom and availability within the project area. For repair and remodeling jobs, matching the existing brick may be the most important criteria.

- Brick and concrete block are priced by the piece and then converted into a price per square foot of wall. Openings less than two square feet are generally ignored by the estimator because any savings in units used are offset by the cutting and trimming required.

- It is often difficult and expensive to find and purchase small lots of historic brick. Costs can vary widely. Many design issues affect costs, selection of mortar mix, and repairs or replacement of masonry materials. Cleaning techniques must be reflected in the estimate.

- All masonry walls, whether interior or exterior, require bracing. The cost of bracing walls during construction should be included by the estimator, and this bracing must remain in place until permanent bracing is complete. Permanent bracing of masonry walls is accomplished by masonry itself, in the form of pilasters or abutting wall corners, or by anchoring the walls to the structural frame. Accessories in the form of anchors, anchor slots, and ties are used, but their supply and installation can be by different trades. For instance, anchor slots on spandrel beams and columns are supplied and welded in place by the steel fabricator, but the ties from the slots into the masonry are installed by the bricklayer. Regardless of the installation method, the estimator must be certain that these accessories are accounted for in pricing.

Reference Numbers

Reference numbers are shown at the beginning of some major classifications. These numbers refer to related items in the Reference Section. The reference information may be an estimating procedure, an alternate pricing method, or technical information.

Note: Not all subdivisions listed here necessarily appear. ■

04 05 Common Work Results for Masonry

04 05 23 – Masonry Accessories

04 05 23.19 Masonry Cavity Drainage, Weepholes, and Vents	Crew	Daily Output	Labor-Hours	Unit	Material	2020 Bare Costs Labor	Equipment	Total	Total Incl O&P
0010 **MASONRY CAVITY DRAINAGE, WEEPHOLES, AND VENTS**									
0020 Extruded aluminum, 4" deep, 2-3/8" x 8-1/8"	1 Bric	30	.267	Ea.	41	13.90		54.90	66
0050 5" x 8-1/8"		25	.320		53	16.65		69.65	84
0100 2-1/4" x 25"		25	.320		89	16.65		105.65	124
0150 5" x 16-1/2"		22	.364		61.50	18.95		80.45	96.50
0175 5" x 24"		22	.364		93	18.95		111.95	131
0200 6" x 16-1/2"		22	.364		99.50	18.95		118.45	138
0250 7-3/4" x 16-1/2"		20	.400		92.50	21		113.50	134
0400 For baked enamel finish, add					35%				
0500 For cast aluminum, painted, add					60%				
1000 Stainless steel ventilators, 6" x 6"	1 Bric	25	.320		230	16.65		246.65	279
1050 8" x 8"		24	.333		257	17.35		274.35	310
1100 12" x 12"		23	.348		289	18.10		307.10	350
1150 12" x 6"		24	.333		279	17.35		296.35	330
1200 Foundation block vent, galv., 1-1/4" thk, 8" high, 16" long, no damper		30	.267		17.45	13.90		31.35	40
1250 For damper, add					3.34			3.34	3.67
1450 Drainage and ventilation fabric	2 Bric	1450	.011	S.F.	.82	.57		1.39	1.77

04 21 Clay Unit Masonry

04 21 13 – Brick Masonry

04 21 13.15 Chimney

	Crew	Daily Output	Labor-Hours	Unit	Material	2020 Bare Costs Labor	Equipment	Total	Total Incl O&P
0010 **CHIMNEY**, excludes foundation, scaffolding, grout and reinforcing									
0100 Brick, 16" x 16", 8" flue	D-1	18.20	.879	V.L.F.	26.50	41		67.50	91.50
0150 16" x 20" with one 8" x 12" flue		16	1		42.50	47		89.50	118
0200 16" x 24" with two 8" x 8" flues		14	1.143		63	53.50		116.50	151
0250 20" x 20" with one 12" x 12" flue		13.70	1.168		49.50	55		104.50	138
0300 20" x 24" with two 8" x 12" flues		12	1.333		71	62.50		133.50	173
0350 20" x 32" with two 12" x 12" flues		10	1.600		87	75		162	210

Estimating Tips

05 05 00 Common Work Results for Metals

- Nuts, bolts, washers, connection angles, and plates can add a significant amount to both the tonnage of a structural steel job and the estimated cost. As a rule of thumb, add 10% to the total weight to account for these accessories.

- Type 2 steel construction, commonly referred to as "simple construction," consists generally of field-bolted connections with lateral bracing supplied by other elements of the building, such as masonry walls or x-bracing. The estimator should be aware, however, that shop connections may be accomplished by welding or bolting. The method may be particular to the fabrication shop and may have an impact on the estimated cost.

05 10 00 Structural Steel

- Steel items can be obtained from two sources: a fabrication shop or a metals service center. Fabrication shops can fabricate items under more controlled conditions than crews in the field can. They are also more efficient and can produce items more economically. Metal service centers serve as a source of long mill shapes to both fabrication shops and contractors.

- Most line items in this structural steel subdivision, and most items in 05 50 00 Metal Fabrications, are indicated as being shop fabricated. The bare material cost for these shop fabricated items is the "Invoice Cost" from the shop and includes the mill base price of steel plus mill extras, transportation to the shop, shop drawings and detailing where warranted, shop fabrication and handling, sandblasting and a shop coat of primer paint, all necessary structural bolts, and delivery to the job site. The bare labor cost and bare equipment cost for these shop fabricated items are for field installation or erection.

- Line items in Subdivision 05 12 23.40 Lightweight Framing, and other items scattered in Division 5, are indicated as being field fabricated. The bare material cost for these field fabricated items is the "Invoice Cost" from the metals service center and includes the mill base price of steel plus mill extras, transportation to the metals service center, material handling, and delivery of long lengths of mill shapes to the job site. Material costs for structural bolts and welding rods should be added to the estimate. The bare labor cost and bare equipment cost for these items are for both field fabrication and field installation or erection, and include time for cutting, welding, and drilling in the fabricated metal items. Drilling into concrete and fasteners to fasten field fabricated items to other work is not included and should be added to the estimate.

05 20 00 Steel Joist Framing

- In any given project the total weight of open web steel joists is determined by the loads to be supported and the design. However, economies can be realized in minimizing the amount of labor used to place the joists. This is done by maximizing the joist spacing and therefore minimizing the number of joists required to be installed on the job. Certain spacings and locations may be required by the design, but in other cases maximizing the spacing and keeping it as uniform as possible will keep the costs down.

05 30 00 Steel Decking

- The takeoff and estimating of a metal deck involve more than the area of the floor or roof and the type of deck specified or shown on the drawings. Many different sizes and types of openings may exist. Small openings for individual pipes or conduits may be drilled after the floor/roof is installed, but larger openings may require special deck lengths as well as reinforcing or structural support. The estimator should determine who will be supplying this reinforcing. Additionally, some deck terminations are part of the deck package, such as screed angles and pour stops, and others will be part of the steel contract, such as angles attached to structural members and cast-in-place angles and plates. The estimator must ensure that all pieces are accounted for in the complete estimate.

05 50 00 Metal Fabrications

- The most economical steel stairs are those that use common materials, standard details, and most importantly, a uniform and relatively simple method of field assembly. Commonly available A36/A992 channels and plates are very good choices for the main stringers of the stairs, as are angles and tees for the carrier members. Risers and treads are usually made by specialty shops, and it is most economical to use a typical detail in as many places as possible. The stairs should be pre-assembled and shipped directly to the site. The field connections should be simple and straightforward enough to be accomplished efficiently, and with minimum equipment and labor.

Reference Numbers

Reference numbers are shown at the beginning of some major classifications. These numbers refer to related items in the Reference Section. The reference information may be an estimating procedure, an alternate pricing method, or technical information.

Note: Not all subdivisions listed here necessarily appear. ∎

Same Data. Simplified.

Enjoy the convenience and efficiency of accessing your costs anywhere:

- **Skip the multiplier** by setting your location
- **Quickly search,** edit, favorite and share costs
- **Stay on top of price changes** with automatic updates

Discover more at rsmeans.com/online

05 05 19 – Post-Installed Concrete Anchors

05 05 19.10 Chemical Anchors		Crew	Daily Output	Labor-Hours	Unit	Material	2020 Bare Costs Labor	Equipment	Total	Total Incl O&P
0010	**CHEMICAL ANCHORS**									
0020	Includes layout & drilling									
1430	Chemical anchor, w/rod & epoxy cartridge, 3/4" diameter x 9-1/2" long	B-89A	27	.593	Ea.	10.15	28.50	4.23	42.88	59.50
1435	1" diameter x 11-3/4" long		24	.667		20.50	32.50	4.76	57.76	76.50
1440	1-1/4" diameter x 14" long		21	.762		39	37	5.45	81.45	105
1445	1-3/4" diameter x 15" long		20	.800		68.50	39	5.70	113.20	140
1450	18" long		17	.941		82.50	45.50	6.70	134.70	167
1455	2" diameter x 18" long		16	1		113	48.50	7.15	168.65	206
1460	24" long		15	1.067		148	51.50	7.60	207.10	249

05 05 19.20 Expansion Anchors			Crew	Daily Output	Labor-Hours	Unit	Material	Labor	Equipment	Total	Total Incl O&P
0010	**EXPANSION ANCHORS**										
0100	Anchors for concrete, brick or stone, no layout and drilling										
0200	Expansion shields, zinc, 1/4" diameter, 1-5/16" long, single	G	1 Carp	90	.089	Ea.	.46	4.72		5.18	7.60
0300	1-3/8" long, double	G		85	.094		.66	5		5.66	8.25
0400	3/8" diameter, 1-1/2" long, single	G		85	.094		.71	5		5.71	8.30
0500	2" long, double	G		80	.100		1.20	5.30		6.50	9.30
0600	1/2" diameter, 2-1/16" long, single	G		80	.100		1.28	5.30		6.58	9.40
0700	2-1/2" long, double	G		75	.107		2.17	5.65		7.82	10.90
0800	5/8" diameter, 2-5/8" long, single	G		75	.107		2.22	5.65		7.87	10.95
0900	2-3/4" long, double	G		70	.114		3.64	6.05		9.69	13.15
1000	3/4" diameter, 2-3/4" long, single	G		70	.114		3.66	6.05		9.71	13.20
1100	3-15/16" long, double	G		65	.123		5.65	6.55		12.20	16.10
2100	Hollow wall anchors for gypsum wall board, plaster or tile										
2300	1/8" diameter, short	G	1 Carp	160	.050	Ea.	.31	2.66		2.97	4.33
2400	Long	G		150	.053		.32	2.83		3.15	4.61
2500	3/16" diameter, short	G		150	.053		.58	2.83		3.41	4.90
2600	Long	G		140	.057		.75	3.04		3.79	5.40
2700	1/4" diameter, short	G		140	.057		.65	3.04		3.69	5.30
2800	Long	G		130	.062		1.03	3.27		4.30	6.05
3000	Toggle bolts, bright steel, 1/8" diameter, 2" long	G		85	.094		.21	5		5.21	7.75
3100	4" long	G		80	.100		.26	5.30		5.56	8.30
3200	3/16" diameter, 3" long	G		80	.100		.30	5.30		5.60	8.35
3300	6" long	G		75	.107		.49	5.65		6.14	9.05
3400	1/4" diameter, 3" long	G		75	.107		.44	5.65		6.09	9
3500	6" long	G		70	.114		.57	6.05		6.62	9.80
3600	3/8" diameter, 3" long	G		70	.114		.88	6.05		6.93	10.10
3700	6" long	G		60	.133		1.58	7.10		8.68	12.40
3800	1/2" diameter, 4" long	G		60	.133		1.86	7.10		8.96	12.70
3900	6" long	G		50	.160		2.37	8.50		10.87	15.40
5000	Screw anchors for concrete, masonry,										
5100	stone & tile, no layout or drilling included										
5700	Lag screw shields, 1/4" diameter, short	G	1 Carp	90	.089	Ea.	.47	4.72		5.19	7.60
5800	Long	G		85	.094		.56	5		5.56	8.10
5900	3/8" diameter, short	G		85	.094		.74	5		5.74	8.30
6000	Long	G		80	.100		1.05	5.30		6.35	9.15
6100	1/2" diameter, short	G		80	.100		.98	5.30		6.28	9.10
6200	Long	G		75	.107		1.52	5.65		7.17	10.15
6300	5/8" diameter, short	G		70	.114		1.54	6.05		7.59	10.85
6400	Long	G		65	.123		2.21	6.55		8.76	12.30
6600	Lead, #6 & #8, 3/4" long	G		260	.031		.18	1.64		1.82	2.66
6700	#10 - #14, 1-1/2" long	G		200	.040		.48	2.13		2.61	3.72
6800	#16 & #18, 1-1/2" long	G		160	.050		.42	2.66		3.08	4.45

05 05 Common Work Results for Metals

05 05 19 – Post-Installed Concrete Anchors

05 05 19.20 Expansion Anchors

		Crew	Daily Output	Labor-Hours	Unit	Material	2020 Bare Costs Labor	Equipment	Total	Total Incl O&P
6900	Plastic, #6 & #8, 3/4" long	1 Carp	260	.031	Ea.	.05	1.64		1.69	2.52
7000	#8 & #10, 7/8" long		240	.033		.05	1.77		1.82	2.72
7100	#10 & #12, 1" long		220	.036		.07	1.93		2	2.98
7200	#14 & #16, 1-1/2" long		160	.050		.07	2.66		2.73	4.07
8000	Wedge anchors, not including layout or drilling									
8050	Carbon steel, 1/4" diameter, 1-3/4" long [G]	1 Carp	150	.053	Ea.	.62	2.83		3.45	4.94
8100	3-1/4" long [G]		140	.057		.82	3.04		3.86	5.45
8150	3/8" diameter, 2-1/4" long [G]		145	.055		.44	2.93		3.37	4.90
8200	5" long [G]		140	.057		.78	3.04		3.82	5.40
8250	1/2" diameter, 2-3/4" long [G]		140	.057		.99	3.04		4.03	5.65
8300	7" long [G]		125	.064		1.70	3.40		5.10	6.95
8350	5/8" diameter, 3-1/2" long [G]		130	.062		2.05	3.27		5.32	7.15
8400	8-1/2" long [G]		115	.070		4.36	3.70		8.06	10.35
8450	3/4" diameter, 4-1/4" long [G]		115	.070		2.97	3.70		6.67	8.80
8500	10" long [G]		95	.084		6.75	4.48		11.23	14.15
8550	1" diameter, 6" long [G]		100	.080		8.80	4.25		13.05	16.05
8575	9" long [G]		85	.094		11.40	5		16.40	20
8600	12" long [G]		75	.107		12.30	5.65		17.95	22
8650	1-1/4" diameter, 9" long [G]		70	.114		23.50	6.05		29.55	34.50
8700	12" long [G]		60	.133		30	7.10		37.10	43.50
8750	For type 303 stainless steel, add					350%				
8800	For type 316 stainless steel, add					450%				
8950	Self-drilling concrete screw, hex washer head, 3/16" diam. x 1-3/4" long [G]	1 Carp	300	.027	Ea.	.18	1.42		1.60	2.33
8960	2-1/4" long [G]		250	.032		.24	1.70		1.94	2.82
8970	Phillips flat head, 3/16" diam. x 1-3/4" long [G]		300	.027		.19	1.42		1.61	2.34
8980	2-1/4" long [G]		250	.032		.23	1.70		1.93	2.81

05 05 21 – Fastening Methods for Metal

05 05 21.15 Drilling Steel

		Crew	Daily Output	Labor-Hours	Unit	Material	2020 Bare Costs Labor	Equipment	Total	Total Incl O&P
0010	**DRILLING STEEL**									
1910	Drilling & layout for steel, up to 1/4" deep, no anchor									
1920	Holes, 1/4" diameter	1 Sswk	112	.071	Ea.	.10	4.12		4.22	6.55
1925	For each additional 1/4" depth, add		336	.024		.10	1.37		1.47	2.26
1930	3/8" diameter		104	.077		.08	4.43		4.51	7.05
1935	For each additional 1/4" depth, add		312	.026		.08	1.48		1.56	2.41
1940	1/2" diameter		96	.083		.08	4.80		4.88	7.65
1945	For each additional 1/4" depth, add		288	.028		.08	1.60		1.68	2.60
1950	5/8" diameter		88	.091		.16	5.25		5.41	8.35
1955	For each additional 1/4" depth, add		264	.030		.16	1.75		1.91	2.91
1960	3/4" diameter		80	.100		.17	5.75		5.92	9.25
1965	For each additional 1/4" depth, add		240	.033		.17	1.92		2.09	3.19
1970	7/8" diameter		72	.111		.24	6.40		6.64	10.30
1975	For each additional 1/4" depth, add		216	.037		.24	2.14		2.38	3.62
1980	1" diameter		64	.125		.26	7.20		7.46	11.60
1985	For each additional 1/4" depth, add		192	.042		.26	2.40		2.66	4.05
1990	For drilling up, add						40%			

05 05 21.90 Welding Steel

		Crew	Daily Output	Labor-Hours	Unit	Material	2020 Bare Costs Labor	Equipment	Total	Total Incl O&P
0010	**WELDING STEEL**, Structural									
0020	Field welding, 1/8" E6011, cost per welder, no operating engineer	E-14	8	1	Hr.	5.55	59.50	18.40	83.45	120
0200	With 1/2 operating engineer	E-13	8	1.500		5.55	86	18.40	109.95	159
0300	With 1 operating engineer	E-12	8	2		5.55	112	18.40	135.95	198
0500	With no operating engineer, 2# weld rod per ton	E-14	8	1	Ton	5.55	59.50	18.40	83.45	120
0600	8# E6011 per ton	"	2	4		22	238	73.50	333.50	480

For customer support on your Mechanical Costs with RSMeans Data, call 800.448.8182.

45

05 05 Common Work Results for Metals

05 05 21 – Fastening Methods for Metal

05 05 21.90 Welding Steel

		Crew	Daily Output	Labor-Hours	Unit	Material	2020 Bare Costs Labor	2020 Bare Costs Equipment	Total	Total Incl O&P
0800	With one operating engineer per welder, 2# E6011 per ton	E-12	8	2	Ton	5.55	112	18.40	135.95	198
0900	8# E6011 per ton	"	2	8	↓	22	450	73.50	545.50	795
1200	Continuous fillet, down welding									
1300	Single pass, 1/8" thick, 0.1#/L.F.	E-14	150	.053	L.F.	.28	3.17	.98	4.43	6.35
1400	3/16" thick, 0.2#/L.F.		75	.107		.55	6.35	1.96	8.86	12.70
1500	1/4" thick, 0.3#/L.F.		50	.160		.83	9.50	2.94	13.27	19.05
1610	5/16" thick, 0.4#/L.F.		38	.211		1.11	12.50	3.87	17.48	25
1800	3 passes, 3/8" thick, 0.5#/L.F.		30	.267		1.38	15.85	4.90	22.13	32
2010	4 passes, 1/2" thick, 0.7#/L.F.		22	.364		1.93	21.50	6.70	30.13	43.50
2200	5 to 6 passes, 3/4" thick, 1.3#/L.F.		12	.667		3.59	39.50	12.25	55.34	79.50
2400	8 to 11 passes, 1" thick, 2.4#/L.F.	▼	6	1.333		6.65	79.50	24.50	110.65	158
2600	For vertical joint welding, add						20%			
2700	Overhead joint welding, add						300%			
2900	For semi-automatic welding, obstructed joints, deduct						5%			
3000	Exposed joints, deduct				↓		15%			
4000	Cleaning and welding plates, bars, or rods									
4010	to existing beams, columns, or trusses	E-14	12	.667	L.F.	1.38	39.50	12.25	53.13	77

05 05 23 – Metal Fastenings

05 05 23.30 Lag Screws

			Crew	Daily Output	Labor-Hours	Unit	Material	2020 Bare Costs Labor	2020 Bare Costs Equipment	Total	Total Incl O&P
0010	**LAG SCREWS**										
0020	Steel, 1/4" diameter, 2" long	G	1 Carp	200	.040	Ea.	.10	2.13		2.23	3.30
0100	3/8" diameter, 3" long	G		150	.053		.31	2.83		3.14	4.60
0200	1/2" diameter, 3" long	G		130	.062		.72	3.27		3.99	5.70
0300	5/8" diameter, 3" long	G	↓	120	.067	↓	1.26	3.54		4.80	6.70

05 05 23.35 Machine Screws

			Crew	Daily Output	Labor-Hours	Unit	Material	2020 Bare Costs Labor	2020 Bare Costs Equipment	Total	Total Incl O&P
0010	**MACHINE SCREWS**										
0020	Steel, round head, #8 x 1" long	G	1 Carp	4.80	1.667	C	3.98	88.50		92.48	137
0110	#8 x 2" long	G		2.40	3.333		8.80	177		185.80	276
0200	#10 x 1" long	G		4	2		4.89	106		110.89	165
0300	#10 x 2" long	G	↓	2	4	↓	9.80	213		222.80	330

05 05 23.50 Powder Actuated Tools and Fasteners

			Crew	Daily Output	Labor-Hours	Unit	Material	2020 Bare Costs Labor	2020 Bare Costs Equipment	Total	Total Incl O&P
0010	**POWDER ACTUATED TOOLS & FASTENERS**										
0020	Stud driver, .22 caliber, single shot					Ea.	159			159	175
0100	.27 caliber, semi automatic, strip					"	460			460	505
0300	Powder load, single shot, .22 cal, power level 2, brown					C	6.25			6.25	6.85
0400	Strip, .27 cal, power level 4, red						10.30			10.30	11.35
0600	Drive pin, .300 x 3/4" long	G	1 Carp	4.80	1.667		4.48	88.50		92.98	138
0700	.300 x 3" long with washer	G	"	4	2	↓	11.95	106		117.95	173

05 05 23.55 Rivets

			Crew	Daily Output	Labor-Hours	Unit	Material	2020 Bare Costs Labor	2020 Bare Costs Equipment	Total	Total Incl O&P
0010	**RIVETS**										
0100	Aluminum rivet & mandrel, 1/2" grip length x 1/8" diameter	G	1 Carp	4.80	1.667	C	8.15	88.50		96.65	142
0200	3/16" diameter	G		4	2		11.65	106		117.65	173
0300	Aluminum rivet, steel mandrel, 1/8" diameter	G		4.80	1.667		10.30	88.50		98.80	144
0400	3/16" diameter	G		4	2		18.35	106		124.35	180
0500	Copper rivet, steel mandrel, 1/8" diameter	G		4.80	1.667		10.30	88.50		98.80	144
1200	Steel rivet and mandrel, 1/8" diameter	G		4.80	1.667		8.15	88.50		96.65	142
1300	3/16" diameter	G	↓	4	2	↓	11	106		117	172
1400	Hand riveting tool, standard					Ea.	78			78	86
1500	Deluxe						415			415	460
1600	Power riveting tool, standard						545			545	600
1700	Deluxe					↓	1,575			1,575	1,725

05 05 Common Work Results for Metals

05 05 23 – Metal Fastenings

05 05 23.70 Structural Blind Bolts		Crew	Daily Output	Labor-Hours	Unit	Material	2020 Bare Costs Labor	Equipment	Total	Total Incl O&P
0010	**STRUCTURAL BLIND BOLTS**									
0100	1/4" diameter x 1/4" grip	G	1 Sswk	240	.033	Ea.	1.84	1.92		3.76	5.05
0150	1/2" grip	G		216	.037		1.89	2.14		4.03	5.45
0200	3/8" diameter x 1/2" grip	G		232	.034		3.50	1.99		5.49	6.95
0250	3/4" grip	G		208	.038		3.14	2.22		5.36	6.95
0300	1/2" diameter x 1/2" grip	G		224	.036		6.25	2.06		8.31	10.15
0350	3/4" grip	G		200	.040		9.10	2.31		11.41	13.60
0400	5/8" diameter x 3/4" grip	G		216	.037		11.25	2.14		13.39	15.75
0450	1" grip	G	↓	192	.042	↓	15.70	2.40		18.10	21

05 12 Structural Steel Framing

05 12 23 – Structural Steel for Buildings

05 12 23.40 Lightweight Framing

			Crew	Daily Output	Labor-Hours	Unit	Material	2020 Bare Costs Labor	Equipment	Total	Total Incl O&P
0010	**LIGHTWEIGHT FRAMING**										
0015	Made from recycled materials										
0400	Angle framing, field fabricated, 4" and larger	G	E-3	440	.055	Lb.	.83	3.18	.33	4.34	6.25
0450	Less than 4" angles	G		265	.091		.86	5.30	.56	6.72	9.85
0600	Channel framing, field fabricated, 8" and larger	G		500	.048		.86	2.80	.29	3.95	5.65
0650	Less than 8" channels	G	↓	335	.072		.86	4.17	.44	5.47	8
1000	Continuous slotted channel framing system, shop fab, simple framing	G	2 Sswk	2400	.007		4.45	.38		4.83	5.50
1200	Complex framing	G	"	1600	.010		5.05	.58		5.63	6.45
1250	Plate & bar stock for reinforcing beams and trusses	G					1.58			1.58	1.74
1300	Cross bracing, rods, shop fabricated, 3/4" diameter	G	E-3	700	.034		1.72	2	.21	3.93	5.25
1310	7/8" diameter	G		850	.028		1.72	1.65	.17	3.54	4.67
1320	1" diameter	G		1000	.024		1.72	1.40	.15	3.27	4.25
1330	Angle, 5" x 5" x 3/8"	G		2800	.009		1.72	.50	.05	2.27	2.74
1350	Hanging lintels, shop fabricated	G	↓	850	.028		1.72	1.65	.17	3.54	4.67
1380	Roof frames, shop fabricated, 3'-0" square, 5' span	G	E-2	4200	.013		1.72	.76	.40	2.88	3.52
1400	Tie rod, not upset, 1-1/2" to 4" diameter, with turnbuckle	G	2 Sswk	800	.020		1.87	1.15		3.02	3.86
1420	No turnbuckle	G		700	.023		1.80	1.32		3.12	4.05
1500	Upset, 1-3/4" to 4" diameter, with turnbuckle	G		800	.020		1.87	1.15		3.02	3.86
1520	No turnbuckle	G	↓	700	.023	↓	1.80	1.32		3.12	4.05

05 12 23.60 Pipe Support Framing

			Crew	Daily Output	Labor-Hours	Unit	Material	2020 Bare Costs Labor	Equipment	Total	Total Incl O&P
0010	**PIPE SUPPORT FRAMING**										
0020	Under 10#/L.F., shop fabricated	G	E-4	3900	.008	Lb.	1.93	.48	.04	2.45	2.91
0200	10.1 to 15#/L.F.	G		4300	.007		1.90	.43	.03	2.36	2.81
0400	15.1 to 20#/L.F.	G		4800	.007		1.87	.39	.03	2.29	2.69
0600	Over 20#/L.F.	G	↓	5400	.006	↓	1.84	.34	.03	2.21	2.59

For customer support on your Mechanical Costs with RSMeans Data, call 800.448.8182.

47

Division Notes

	CREW	DAILY OUTPUT	LABOR-HOURS	UNIT	BARE COSTS				TOTAL INCL O&P
					MAT.	LABOR	EQUIP.	TOTAL	

Estimating Tips
06 05 00 Common Work Results for Wood, Plastics, and Composites

- Common to any wood-framed structure are the accessory connector items such as screws, nails, adhesives, hangers, connector plates, straps, angles, and hold-downs. For typical wood-framed buildings, such as residential projects, the aggregate total for these items can be significant, especially in areas where seismic loading is a concern. For floor and wall framing, the material cost is based on 10 to 25 lbs. of accessory connectors per MBF. Hold-downs, hangers, and other connectors should be taken off by the piece.

  Included with material costs are fasteners for a normal installation. Gordian's RSMeans engineers use manufacturers' recommendations, written specifications, and/or standard construction practice for the sizing and spacing of fasteners. Prices for various fasteners are shown for informational purposes only. Adjustments should be made if unusual fastening conditions exist.

06 10 00 Carpentry

- Lumber is a traded commodity and therefore sensitive to supply and demand in the marketplace. Even with "budgetary" estimating of wood-framed projects, it is advisable to call local suppliers for the latest market pricing.

- The common quantity unit for wood-framed projects is "thousand board feet" (MBF). A board foot is a volume of wood—1" x 1' x 1' or 144 cubic inches. Board-foot quantities are generally calculated using nominal material dimensions—dressed sizes are ignored. Board foot per lineal foot of any stick of lumber can be calculated by dividing the nominal cross-sectional area by 12. As an example, 2,000 lineal feet of 2 x 12 equates to 4 MBF by dividing the nominal area, 2 x 12, by 12, which equals 2, and multiplying that by 2,000 to give 4,000 board feet. This simple rule applies to all nominal dimensioned lumber.

- Waste is an issue of concern at the quantity takeoff for any area of construction. Framing lumber is sold in even foot lengths, i.e., 8', 10', 12', 14', 16', and depending on spans, wall heights, and the grade of lumber, waste is inevitable. A rule of thumb for lumber waste is 5–10% depending on material quality and the complexity of the framing.

- Wood in various forms and shapes is used in many projects, even where the main structural framing is steel, concrete, or masonry. Plywood as a back-up partition material and 2x boards used as blocking and cant strips around roof edges are two common examples. The estimator should ensure that the costs of all wood materials are included in the final estimate.

06 20 00 Finish Carpentry

- It is necessary to consider the grade of workmanship when estimating labor costs for erecting millwork and an interior finish. In practice, there are three grades: premium, custom, and economy. The RSMeans daily output for base and case moldings is in the range of 200 to 250 L.F. per carpenter per day. This is appropriate for most average custom-grade projects. For premium projects, an adjustment to productivity of 25–50% should be made, depending on the complexity of the job.

Reference Numbers
Reference numbers are shown at the beginning of some major classifications. These numbers refer to related items in the Reference Section. The reference information may be an estimating procedure, an alternate pricing method, or technical information.

Note: Not all subdivisions listed here necessarily appear. ■

---

06 16 36 – Wood Panel Product Sheathing

06 16 36.10 Sheathing	Crew	Daily Output	Labor-Hours	Unit	Material	2020 Bare Costs Labor	Equipment	Total	Total Incl O&P
0010 **SHEATHING**									
0012 Plywood on roofs, CDX									
0030 5/16" thick	2 Carp	1600	.010	S.F.	.60	.53		1.13	1.45
0035 Pneumatic nailed		1952	.008		.60	.44		1.04	1.30
0050 3/8" thick		1525	.010		.53	.56		1.09	1.42
0055 Pneumatic nailed		1860	.009		.53	.46		.99	1.27
0100 1/2" thick		1400	.011		.68	.61		1.29	1.66
0105 Pneumatic nailed		1708	.009		.68	.50		1.18	1.50
0200 5/8" thick		1300	.012		.77	.65		1.42	1.83
0205 Pneumatic nailed		1586	.010		.77	.54		1.31	1.66
0300 3/4" thick		1200	.013		.95	.71		1.66	2.10
0305 Pneumatic nailed		1464	.011		.95	.58		1.53	1.91
0500 Plywood on walls, with exterior CDX, 3/8" thick		1200	.013		.53	.71		1.24	1.64
0505 Pneumatic nailed		1488	.011		.53	.57		1.10	1.44
0600 1/2" thick		1125	.014		.68	.76		1.44	1.89
0605 Pneumatic nailed		1395	.011		.68	.61		1.29	1.67
0700 5/8" thick		1050	.015		.77	.81		1.58	2.07
0705 Pneumatic nailed		1302	.012		.77	.65		1.42	1.83
0800 3/4" thick		975	.016		.95	.87		1.82	2.35
0805 Pneumatic nailed		1209	.013		.95	.70		1.65	2.10

Estimating Tips
07 10 00 Dampproofing and Waterproofing

- Be sure of the job specifications before pricing this subdivision. The difference in cost between waterproofing and dampproofing can be great. Waterproofing will hold back standing water. Dampproofing prevents the transmission of water vapor. Also included in this section are vapor retarding membranes.

07 20 00 Thermal Protection

- Insulation and fireproofing products are measured by area, thickness, volume, or R-value. Specifications may give only what the specific R-value should be in a certain situation. The estimator may need to choose the type of insulation to meet that R-value.

07 30 00 Steep Slope Roofing
07 40 00 Roofing and Siding Panels

- Many roofing and siding products are bought and sold by the square. One square is equal to an area that measures 100 square feet.

  This simple change in unit of measure could create a large error if the estimator is not observant. Accessories necessary for a complete installation must be figured into any calculations for both material and labor.

07 50 00 Membrane Roofing
07 60 00 Flashing and Sheet Metal
07 70 00 Roofing and Wall Specialties and Accessories

- The items in these subdivisions compose a roofing system. No one component completes the installation, and all must be estimated. Built-up or single-ply membrane roofing systems are made up of many products and installation trades. Wood blocking at roof perimeters or penetrations, parapet coverings, reglets, roof drains, gutters, downspouts, sheet metal flashing, skylights, smoke vents, and roof hatches all need to be considered along with the roofing material. Several different installation trades will need to work together on the roofing system. Inherent difficulties in the scheduling and coordination of various trades must be accounted for when estimating labor costs.

07 90 00 Joint Protection

- To complete the weather-tight shell, the sealants and caulkings must be estimated. Where different materials meet—at expansion joints, at flashing penetrations, and at hundreds of other locations throughout a construction project—caulking and sealants provide another line of defense against water penetration. Often, an entire system is based on the proper location and placement of caulking or sealants. The detailed drawings that are included as part of a set of architectural plans show typical locations for these materials. When caulking or sealants are shown at typical locations, this means the estimator must include them for all the locations where this detail is applicable. Be careful to keep different types of sealants separate, and remember to consider backer rods and primers if necessary.

Reference Numbers

Reference numbers are shown at the beginning of some major classifications. These numbers refer to related items in the Reference Section. The reference information may be an estimating procedure, an alternate pricing method, or technical information.

Note: Not all subdivisions listed here necessarily appear. ∎

Same Data. Simplified.

Enjoy the convenience and efficiency of accessing your costs anywhere:

- **Skip the multiplier** by setting your location
- **Quickly search,** edit, favorite and share costs
- **Stay on top of price changes** with automatic updates

Discover more at rsmeans.com/online

07 65 10 – Sheet Metal Flashing

07 65 10.10 Sheet Metal Flashing and Counter Flashing	Crew	Daily Output	Labor-Hours	Unit	Material	2020 Bare Costs Labor	Equipment	Total	Total Incl O&P
0010 **SHEET METAL FLASHING AND COUNTER FLASHING**									
0011 Including up to 4 bends									
0020 Aluminum, mill finish, .013" thick	1 Rofc	145	.055	S.F.	1.01	2.55		3.56	5.30
0030 .016" thick		145	.055		1.15	2.55		3.70	5.50
0060 .019" thick		145	.055		1.50	2.55		4.05	5.85
0100 .032" thick		145	.055		1.38	2.55		3.93	5.75
0200 .040" thick		145	.055		2.33	2.55		4.88	6.75
0300 .050" thick		145	.055		3.22	2.55		5.77	7.75
0400 Painted finish, add					.34			.34	.37
1600 Copper, 16 oz. sheets, under 1000 lb.	1 Rofc	115	.070		8.15	3.21		11.36	14.30
1700 Over 4000 lb.		155	.052		8.15	2.38		10.53	12.95
1900 20 oz. sheets, under 1000 lb.		110	.073		10.45	3.36		13.81	17.05
2000 Over 4000 lb.		145	.055		9.95	2.55		12.50	15.15
2200 24 oz. sheets, under 1000 lb.		105	.076		15.30	3.52		18.82	22.50
2300 Over 4000 lb.		135	.059		14.55	2.74		17.29	20.50
2500 32 oz. sheets, under 1000 lb.		100	.080		21	3.70		24.70	29
2600 Over 4000 lb.		130	.062		19.75	2.84		22.59	26
5800 Lead, 2.5 lb./S.F., up to 12" wide		135	.059		6.10	2.74		8.84	11.20
5900 Over 12" wide		135	.059		4.05	2.74		6.79	9
8900 Stainless steel sheets, 32 ga.		155	.052		3.51	2.38		5.89	7.80
9000 28 ga.		155	.052		5	2.38		7.38	9.45
9100 26 ga.		155	.052		5.10	2.38		7.48	9.55
9200 24 ga.		155	.052		5.85	2.38		8.23	10.35
9400 Terne coated stainless steel, .015" thick, 28 ga.		155	.052		8.20	2.38		10.58	12.95
9500 .018" thick, 26 ga.		155	.052		9.15	2.38		11.53	14
9600 Zinc and copper alloy (brass), .020" thick		155	.052		10.10	2.38		12.48	15.05
9700 .027" thick		155	.052		12.25	2.38		14.63	17.45
9800 .032" thick		155	.052		15.45	2.38		17.83	21
9900 .040" thick		155	.052		20.50	2.38		22.88	26.50

07 65 13 – Laminated Sheet Flashing

07 65 13.10 Laminated Sheet Flashing

	Crew	Daily Output	Labor-Hours	Unit	Material	2020 Bare Costs Labor	Equipment	Total	Total Incl O&P
0010 **LAMINATED SHEET FLASHING**, Including up to 4 bends									
4300 Copper-clad stainless steel, .015" thick, under 500 lb.	1 Rofc	115	.070	S.F.	7.05	3.21		10.26	13.05
4400 Over 2000 lb.		155	.052		7.05	2.38		9.43	11.70
4600 .018" thick, under 500 lb.		100	.080		7.95	3.70		11.65	14.85
4700 Over 2000 lb.		145	.055		8.10	2.55		10.65	13.10
8550 Shower pan, 3 ply copper and fabric, 3 oz.		155	.052		4.12	2.38		6.50	8.45
8600 7 oz.		155	.052		5.45	2.38		7.83	9.90
9300 Stainless steel, paperbacked 2 sides, .005" thick		330	.024		4.04	1.12		5.16	6.30

07 65 19 – Plastic Sheet Flashing

07 65 19.10 Plastic Sheet Flashing and Counter Flashing

	Crew	Daily Output	Labor-Hours	Unit	Material	2020 Bare Costs Labor	Equipment	Total	Total Incl O&P
0010 **PLASTIC SHEET FLASHING AND COUNTER FLASHING**									
7300 Polyvinyl chloride, black, 10 mil	1 Rofc	285	.028	S.F.	.27	1.30		1.57	2.44
7400 20 mil		285	.028		.26	1.30		1.56	2.43
7600 30 mil		285	.028		.33	1.30		1.63	2.50
7700 60 mil		285	.028		.88	1.30		2.18	3.11
7900 Black or white for exposed roofs, 60 mil		285	.028		1.11	1.30		2.41	3.36
8060 PVC tape, 5" x 45 mils, for joint covers, 100 L.F./roll				Ea.	183			183	201
8850 Polyvinyl chloride, 30 mil	1 Rofc	160	.050	S.F.	1.47	2.31		3.78	5.45

07 71 Roof Specialties

07 71 16 – Manufactured Counterflashing Systems

07 71 16.20 Pitch Pockets, Variable Sizes		Crew	Daily Output	Labor-Hours	Unit	Material	2020 Bare Costs Labor	Equipment	Total	Total Incl O&P
0010	**PITCH POCKETS, VARIABLE SIZES**									
0100	Adjustable, 4" to 7", welded corners, 4" deep	1 Rofc	48	.167	Ea.	21.50	7.70		29.20	36.50
0200	Side extenders, 6"	"	240	.033	"	3.34	1.54		4.88	6.20

07 71 23 – Manufactured Gutters and Downspouts

07 71 23.10 Downspouts

		Crew	Daily Output	Labor-Hours	Unit	Material	2020 Bare Costs Labor	Equipment	Total	Total Incl O&P
0010	**DOWNSPOUTS**									
0020	Aluminum, embossed, .020" thick, 2" x 3"	1 Shee	190	.042	L.F.	.95	2.62		3.57	5.05
0100	Enameled		190	.042		1.37	2.62		3.99	5.50
4800	Steel, galvanized, round, corrugated, 2" or 3" diameter, 28 ga.		190	.042		2.30	2.62		4.92	6.50
4900	4" diameter, 28 ga.		145	.055		2.19	3.44		5.63	7.60
5100	5" diameter, 26 ga.		130	.062		3.86	3.83		7.69	10.05
5400	6" diameter, 28 ga.		105	.076		3.72	4.75		8.47	11.30
5500	26 ga.		105	.076		4.08	4.75		8.83	11.70

07 72 Roof Accessories

07 72 26 – Ridge Vents

07 72 26.10 Ridge Vents and Accessories

		Crew	Daily Output	Labor-Hours	Unit	Material	2020 Bare Costs Labor	Equipment	Total	Total Incl O&P
0010	**RIDGE VENTS AND ACCESSORIES**									
2300	Ridge vent strip, mill finish	1 Shee	155	.052	L.F.	3.95	3.22		7.17	9.25

07 72 53 – Snow Guards

07 72 53.10 Snow Guard Options

		Crew	Daily Output	Labor-Hours	Unit	Material	2020 Bare Costs Labor	Equipment	Total	Total Incl O&P
0010	**SNOW GUARD OPTIONS**									
0100	Slate & asphalt shingle roofs, fastened with nails	1 Rofc	160	.050	Ea.	12.45	2.31		14.76	17.50
0200	Standing seam metal roofs, fastened with set screws		48	.167		17.75	7.70		25.45	32.50
0300	Surface mount for metal roofs, fastened with solder		48	.167		7.75	7.70		15.45	21.50
0400	Double rail pipe type, including pipe		130	.062	L.F.	35	2.84		37.84	43

07 76 Roof Pavers

07 76 16 – Roof Decking Pavers

07 76 16.10 Roof Pavers and Supports

		Crew	Daily Output	Labor-Hours	Unit	Material	2020 Bare Costs Labor	Equipment	Total	Total Incl O&P
0010	**ROOF PAVERS AND SUPPORTS**									
1000	Roof decking pavers, concrete blocks, 2" thick, natural	1 Clab	115	.070	S.F.	3.58	2.93		6.51	8.35
1100	Colors		115	.070	"	3.71	2.93		6.64	8.50
1200	Support pedestal, bottom cap		960	.008	Ea.	2.66	.35		3.01	3.46
1300	Top cap		960	.008		4.88	.35		5.23	5.90
1400	Leveling shims, 1/16"		1920	.004		1.22	.18		1.40	1.60
1500	1/8"		1920	.004		1.22	.18		1.40	1.60
1600	Buffer pad		960	.008		2.54	.35		2.89	3.32
1700	PVC legs (4" SDR 35)		2880	.003	Inch	.14	.12		.26	.33
2000	Alternate pricing method, system in place		101	.079	S.F.	7.20	3.33		10.53	12.90

For customer support on your Mechanical Costs with RSMeans Data, call 800.448.8182.

53

07 84 13 – Penetration Firestopping

07 84 13.10 Firestopping		Crew	Daily Output	Labor-Hours	Unit	Material	2020 Bare Costs Labor	Equipment	Total	Total Incl O&P
0010	**FIRESTOPPING**	R078413-30								
0100	Metallic piping, non insulated									
0110	Through walls, 2" diameter	1 Carp	16	.500	Ea.	3.67	26.50		30.17	44
0120	4" diameter		14	.571		8.20	30.50		38.70	54.50
0130	6" diameter		12	.667		12.60	35.50		48.10	67
0140	12" diameter		10	.800		25	42.50		67.50	91.50
0150	Through floors, 2" diameter		32	.250		1.86	13.30		15.16	22
0160	4" diameter		28	.286		3.28	15.20		18.48	26.50
0170	6" diameter		24	.333		6.45	17.70		24.15	33.50
0180	12" diameter		20	.400		12.80	21.50		34.30	46
0190	Metallic piping, insulated									
0200	Through walls, 2" diameter	1 Carp	16	.500	Ea.	8.60	26.50		35.10	49.50
0210	4" diameter		14	.571		12.55	30.50		43.05	59.50
0220	6" diameter		12	.667		16.75	35.50		52.25	71.50
0230	12" diameter		10	.800		28.50	42.50		71	95.50
0240	Through floors, 2" diameter		32	.250		3.27	13.30		16.57	23.50
0250	4" diameter		28	.286		6.30	15.20		21.50	30
0260	6" diameter		24	.333		8.70	17.70		26.40	36
0270	12" diameter		20	.400		12.95	21.50		34.45	46
0280	Non metallic piping, non insulated									
0290	Through walls, 2" diameter	1 Carp	12	.667	Ea.	52.50	35.50		88	111
0300	4" diameter		10	.800		101	42.50		143.50	175
0310	6" diameter		8	1		186	53		239	285
0330	Through floors, 2" diameter		16	.500		36.50	26.50		63	80
0340	4" diameter		6	1.333		62	71		133	175
0350	6" diameter		6	1.333		135	71		206	255
0370	Ductwork, insulated & non insulated, round									
0380	Through walls, 6" diameter	1 Carp	12	.667	Ea.	16.55	35.50		52.05	71
0390	12" diameter		10	.800		28.50	42.50		71	95.50
0400	18" diameter		8	1		33.50	53		86.50	117
0410	Through floors, 6" diameter		16	.500		8.50	26.50		35	49.50
0420	12" diameter		14	.571		14.60	30.50		45.10	61.50
0430	18" diameter		12	.667		17.60	35.50		53.10	72.50
0440	Ductwork, insulated & non insulated, rectangular									
0450	With stiffener/closure angle, through walls, 6" x 12"	1 Carp	8	1	Ea.	21	53		74	103
0460	12" x 24"		6	1.333		36	71		107	146
0470	24" x 48"		4	2		84	106		190	252
0480	With stiffener/closure angle, through floors, 6" x 12"		10	.800		10.40	42.50		52.90	75.50
0490	12" x 24"		8	1		20	53		73	102
0500	24" x 48"		6	1.333		42	71		113	152
0510	Multi trade openings									
0520	Through walls, 6" x 12"	1 Carp	2	4	Ea.	52	213		265	380
0530	12" x 24"	"	1	8		189	425		614	850
0540	24" x 48"	2 Carp	1	16		670	850		1,520	2,025
0550	48" x 96"	"	.75	21.333		2,550	1,125		3,675	4,500
0560	Through floors, 6" x 12"	1 Carp	2	4		44.50	213		257.50	370
0570	12" x 24"	"	1	8		69.50	425		494.50	715
0580	24" x 48"	2 Carp	.75	21.333		154	1,125		1,279	1,875
0590	48" x 96"	"	.50	32		365	1,700		2,065	2,950
0600	Structural penetrations, through walls									
0610	Steel beams, W8 x 10	1 Carp	8	1	Ea.	75	53		128	163
0620	W12 x 14		6	1.333		109	71		180	225
0630	W21 x 44		5	1.600		154	85		239	297

07 84 Firestopping

07 84 13 – Penetration Firestopping

	07 84 13.10 Firestopping	Crew	Daily Output	Labor-Hours	Unit	Material	2020 Bare Costs Labor	Equipment	Total	Total Incl O&P
0640	W36 x 135	1 Carp	3	2.667	Ea.	229	142		371	465
0650	Bar joists, 18" deep		6	1.333		42	71		113	153
0660	24" deep		6	1.333		59	71		130	171
0670	36" deep		5	1.600		88.50	85		173.50	226
0680	48" deep	↓	4	2	↓	120	106		226	292
0690	Construction joints, floor slab at exterior wall									
0700	Precast, brick, block or drywall exterior									
0710	2" wide joint	1 Carp	125	.064	L.F.	10.30	3.40		13.70	16.45
0720	4" wide joint	"	75	.107	"	16.60	5.65		22.25	27
0730	Metal panel, glass or curtain wall exterior									
0740	2" wide joint	1 Carp	40	.200	L.F.	12.45	10.65		23.10	29.50
0750	4" wide joint	"	25	.320	"	18.65	17		35.65	46
0760	Floor slab to drywall partition									
0770	Flat joint	1 Carp	100	.080	L.F.	14.55	4.25		18.80	22.50
0780	Fluted joint		50	.160		16.35	8.50		24.85	31
0790	Etched fluted joint	↓	75	.107	↓	20.50	5.65		26.15	31
0800	Floor slab to concrete/masonry partition									
0810	Flat joint	1 Carp	75	.107	L.F.	15.45	5.65		21.10	25.50
0820	Fluted joint	"	50	.160	"	17.65	8.50		26.15	32.50
0830	Concrete/CMU wall joints									
0840	1" wide	1 Carp	100	.080	L.F.	20.50	4.25		24.75	29
0850	2" wide		75	.107		30	5.65		35.65	41.50
0860	4" wide	↓	50	.160	↓	36	8.50		44.50	52.50
0870	Concrete/CMU floor joints									
0880	1" wide	1 Carp	200	.040	L.F.	20.50	2.13		22.63	26
0890	2" wide		150	.053		25	2.83		27.83	32
0900	4" wide	↓	100	.080	↓	36.50	4.25		40.75	46.50

07 91 Preformed Joint Seals

07 91 13 – Compression Seals

07 91 13.10 Compression Seals

		Crew	Daily Output	Labor-Hours	Unit	Material	2020 Bare Costs Labor	Equipment	Total	Total Incl O&P
0010	**COMPRESSION SEALS**									
4900	O-ring type cord, 1/4"	1 Bric	472	.017	L.F.	.40	.88		1.28	1.78
4910	1/2"		440	.018		1.01	.95		1.96	2.54
4920	3/4"		424	.019		2	.98		2.98	3.69
4930	1"		408	.020		3.93	1.02		4.95	5.85
4940	1-1/4"		384	.021		9.40	1.08		10.48	12
4950	1-1/2"		368	.022		11.50	1.13		12.63	14.35
4960	1-3/4"		352	.023		13.50	1.18		14.68	16.65
4970	2"	↓	344	.023	↓	22.50	1.21		23.71	27

07 91 16 – Joint Gaskets

07 91 16.10 Joint Gaskets

		Crew	Daily Output	Labor-Hours	Unit	Material	2020 Bare Costs Labor	Equipment	Total	Total Incl O&P
0010	**JOINT GASKETS**									
4400	Joint gaskets, neoprene, closed cell w/adh, 1/8" x 3/8"	1 Bric	240	.033	L.F.	.36	1.73		2.09	3.03
4500	1/4" x 3/4"		215	.037		.66	1.94		2.60	3.67
4700	1/2" x 1"		200	.040		1.71	2.08		3.79	5.05
4800	3/4" x 1-1/2"	↓	165	.048	↓	1.85	2.52		4.37	5.85

07 91 23 – Backer Rods

07 91 23.10 Backer Rods		Crew	Daily Output	Labor-Hours	Unit	Material	2020 Bare Costs Labor	Equipment	Total	Total Incl O&P
0010	**BACKER RODS**									
0030	Backer rod, polyethylene, 1/4" diameter	1 Bric	4.60	1.739	C.L.F.	2.78	90.50		93.28	140
0050	1/2" diameter		4.60	1.739		4.10	90.50		94.60	142
0070	3/4" diameter		4.60	1.739		6.70	90.50		97.20	144
0090	1" diameter		4.60	1.739		14.80	90.50		105.30	153

07 91 26 – Joint Fillers

07 91 26.10 Joint Fillers

07 91 26.10 Joint Fillers		Crew	Daily Output	Labor-Hours	Unit	Material	2020 Bare Costs Labor	Equipment	Total	Total Incl O&P
0010	**JOINT FILLERS**									
4360	Butyl rubber filler, 1/4" x 1/4"	1 Bric	290	.028	L.F.	.23	1.44		1.67	2.43
4365	1/2" x 1/2"		250	.032		.92	1.67		2.59	3.53
4370	1/2" x 3/4"		210	.038		1.38	1.98		3.36	4.52
4375	3/4" x 3/4"		230	.035		2.06	1.81		3.87	5
4380	1" x 1"		180	.044		2.75	2.31		5.06	6.55
4390	For coloring, add					12%				
4980	Polyethylene joint backing, 1/4" x 2"	1 Bric	2.08	3.846	C.L.F.	14.25	200		214.25	320
4990	1/4" x 6"		1.28	6.250	"	32.50	325		357.50	530
5600	Silicone, room temp vulcanizing foam seal, 1/4" x 1/2"		1312	.006	L.F.	.46	.32		.78	.98
5610	1/2" x 1/2"		656	.012		.92	.64		1.56	1.97
5620	1/2" x 3/4"		442	.018		1.37	.94		2.31	2.94
5630	3/4" x 3/4"		328	.024		2.06	1.27		3.33	4.19
5640	1/8" x 1"		1312	.006		.46	.32		.78	.98
5650	1/8" x 3"		442	.018		1.37	.94		2.31	2.94
5670	1/4" x 3"		295	.027		2.75	1.41		4.16	5.15
5680	1/4" x 6"		148	.054		5.50	2.81		8.31	10.30
5690	1/2" x 6"		82	.098		11	5.10		16.10	19.80
5700	1/2" x 9"		52.50	.152		16.50	7.95		24.45	30
5710	1/2" x 12"		33	.242		22	12.60		34.60	43

07 92 Joint Sealants

07 92 13 – Elastomeric Joint Sealants

07 92 13.20 Caulking and Sealant Options

07 92 13.20 Caulking and Sealant Options		Crew	Daily Output	Labor-Hours	Unit	Material	2020 Bare Costs Labor	Equipment	Total	Total Incl O&P
0010	**CAULKING AND SEALANT OPTIONS**									
0050	Latex acrylic based, bulk				Gal.	31			31	34
0055	Bulk in place 1/4" x 1/4" bead	1 Bric	300	.027	L.F.	.10	1.39		1.49	2.21
0060	1/4" x 3/8"		294	.027		.16	1.42		1.58	2.33
0065	1/4" x 1/2"		288	.028		.22	1.45		1.67	2.43
0075	3/8" x 3/8"		284	.028		.24	1.47		1.71	2.49
0080	3/8" x 1/2"		280	.029		.32	1.49		1.81	2.61
0085	3/8" x 5/8"		276	.029		.40	1.51		1.91	2.74
0095	3/8" x 3/4"		272	.029		.48	1.53		2.01	2.85
0100	1/2" x 1/2"		275	.029		.43	1.51		1.94	2.78
0105	1/2" x 5/8"		269	.030		.54	1.55		2.09	2.94
0110	1/2" x 3/4"		263	.030		.65	1.58		2.23	3.11
0115	1/2" x 7/8"		256	.031		.76	1.63		2.39	3.30
0120	1/2" x 1"		250	.032		.87	1.67		2.54	3.47
0125	3/4" x 3/4"		244	.033		.97	1.71		2.68	3.66
0130	3/4" x 1"		225	.036		1.30	1.85		3.15	4.24
0135	1" x 1"		200	.040		1.73	2.08		3.81	5.05
0190	Cartridges				Gal.	33.50			33.50	37
0200	11 fl. oz. cartridge				Ea.	2.89			2.89	3.18

For customer support on your Mechanical Costs with RSMeans Data, call 800.448.8182.

07 92 Joint Sealants

07 92 13 – Elastomeric Joint Sealants

07 92 13.20 Caulking and Sealant Options		Crew	Daily Output	Labor-Hours	Unit	Material	2020 Bare Costs Labor	Equipment	Total	Total Incl O&P
0500	1/4" x 1/2"	1 Bric	288	.028	L.F.	.24	1.45		1.69	2.45
0600	1/2" x 1/2"		275	.029		.47	1.51		1.98	2.82
0800	3/4" x 3/4"		244	.033		1.06	1.71		2.77	3.76
0900	3/4" x 1"		225	.036		1.42	1.85		3.27	4.37
1000	1" x 1"		200	.040		1.77	2.08		3.85	5.10
1400	Butyl based, bulk				Gal.	43.50			43.50	48
1500	Cartridges				"	43.50			43.50	48
1700	1/4" x 1/2", 154 L.F./gal.	1 Bric	288	.028	L.F.	.28	1.45		1.73	2.50
1800	1/2" x 1/2", 77 L.F./gal.	"	275	.029	"	.57	1.51		2.08	2.92
2300	Polysulfide compounds, 1 component, bulk				Gal.	88.50			88.50	97.50
2600	1 or 2 component, in place, 1/4" x 1/4", 308 L.F./gal.	1 Bric	300	.027	L.F.	.29	1.39		1.68	2.42
2700	1/2" x 1/4", 154 L.F./gal.		288	.028		.57	1.45		2.02	2.82
2900	3/4" x 3/8", 68 L.F./gal.		272	.029		1.30	1.53		2.83	3.75
3000	1" x 1/2", 38 L.F./gal.		250	.032		2.33	1.67		4	5.10
3200	Polyurethane, 1 or 2 component				Gal.	54			54	59.50
3300	Cartridges				"	67.50			67.50	74.50
3500	Bulk, in place, 1/4" x 1/4"	1 Bric	300	.027	L.F.	.18	1.39		1.57	2.29
3655	1/2" x 1/4"		288	.028		.35	1.45		1.80	2.58
3800	3/4" x 3/8"		272	.029		.80	1.53		2.33	3.19
3900	1" x 1/2"		250	.032		1.40	1.67		3.07	4.06
4100	Silicone rubber, bulk				Gal.	68.50			68.50	75.50
4200	Cartridges				"	51.50			51.50	56.50

07 92 16 – Rigid Joint Sealants

07 92 16.10 Rigid Joint Sealants

		Crew	Daily Output	Labor-Hours	Unit	Material	Labor	Equipment	Total	Total Incl O&P
0010	**RIGID JOINT SEALANTS**									
5800	Tapes, sealant, PVC foam adhesive, 1/16" x 1/4"				C.L.F.	9.05			9.05	10
5900	1/16" x 1/2"					9.30			9.30	10.25
5950	1/16" x 1"					16.20			16.20	17.85
6000	1/8" x 1/2"					9.20			9.20	10.10

07 92 19 – Acoustical Joint Sealants

07 92 19.10 Acoustical Sealant

		Crew	Daily Output	Labor-Hours	Unit	Material	Labor	Equipment	Total	Total Incl O&P
0010	**ACOUSTICAL SEALANT**									
0020	Acoustical sealant, elastomeric, cartridges				Ea.	8.65			8.65	9.55
0025	In place, 1/4" x 1/4"	1 Bric	300	.027	L.F.	.35	1.39		1.74	2.49
0030	1/4" x 1/2"		288	.028		.71	1.45		2.16	2.97
0035	1/2" x 1/2"		275	.029		1.41	1.51		2.92	3.86
0040	1/2" x 3/4"		263	.030		2.12	1.58		3.70	4.73
0045	3/4" x 3/4"		244	.033		3.18	1.71		4.89	6.10
0050	1" x 1"		200	.040		5.65	2.08		7.73	9.40

For customer support on your Mechanical Costs with RSMeans Data, call 800.448.8182.

57

Division Notes

		CREW	DAILY OUTPUT	LABOR-HOURS	UNIT	BARE COSTS				TOTAL INCL O&P
						MAT.	LABOR	EQUIP.	TOTAL	

Estimating Tips
08 10 00 Doors and Frames

All exterior doors should be addressed for their energy conservation (insulation and seals).

- Most metal doors and frames look alike, but there may be significant differences among them. When estimating these items, be sure to choose the line item that most closely compares to the specification or door schedule requirements regarding:
  - □ type of metal
  - □ metal gauge
  - □ door core material
  - □ fire rating
  - □ finish
- Wood and plastic doors vary considerably in price. The primary determinant is the veneer material. Lauan, birch, and oak are the most common veneers. Other variables include the following:
  - □ hollow or solid core
  - □ fire rating
  - □ flush or raised panel
  - □ finish
- Door pricing includes bore for cylindrical locksets and mortise for hinges.

08 30 00 Specialty Doors and Frames

- There are many varieties of special doors, and they are usually priced per each. Add frames, hardware, or operators required for a complete installation.

08 40 00 Entrances, Storefronts, and Curtain Walls

- Glazed curtain walls consist of the metal tube framing and the glazing material. The cost data in this subdivision is presented for the metal tube framing alone or the composite wall. If your estimate requires a detailed takeoff of the framing, be sure to add the glazing cost and any tints.

08 50 00 Windows

- Steel windows are unglazed and aluminum can be glazed or unglazed. Some metal windows are priced without glass. Refer to 08 80 00 Glazing for glass pricing. The grade C indicates commercial grade windows, usually ASTM C-35.
- All wood windows and vinyl are priced preglazed. The glazing is insulating glass. Add the cost of screens and grills if required and not already included.

08 70 00 Hardware

- Hardware costs add considerably to the cost of a door. The most efficient method to determine the hardware requirements for a project is to review the door and hardware schedule together. One type of door may have different hardware, depending on the door usage.
- Door hinges are priced by the pair, with most doors requiring 1-1/2 pairs per door. The hinge prices do not include installation labor because it is included in door installation.

Hinges are classified according to the frequency of use, base material, and finish.

08 80 00 Glazing

- Different openings require different types of glass. The most common types are:
  - □ float
  - □ tempered
  - □ insulating
  - □ impact-resistant
  - □ ballistic-resistant
- Most exterior windows are glazed with insulating glass. Entrance doors and window walls, where the glass is less than 18" from the floor, are generally glazed with tempered glass. Interior windows and some residential windows are glazed with float glass.
- Coastal communities require the use of impact-resistant glass, dependent on wind speed.
- The insulation or 'u' value is a strong consideration, along with solar heat gain, to determine total energy efficiency.

Reference Numbers

Reference numbers are shown at the beginning of some major classifications. These numbers refer to related items in the Reference Section. The reference information may be an estimating procedure, an alternate pricing method, or technical information.

Note: Not all subdivisions listed here necessarily appear. ∎

---

08 34 Special Function Doors

08 34 13 – Cold Storage Doors

08 34 13.10 Doors for Cold Area Storage	Crew	Daily Output	Labor-Hours	Unit	Material	2020 Bare Costs Labor	2020 Bare Costs Equipment	Total	Total Incl O&P
0010 **DOORS FOR COLD AREA STORAGE**									
0020 Single, 20 ga. galvanized steel									
0300 Horizontal sliding, 5' x 7', manual operation, 3.5" thick	2 Carp	2	8	Ea.	3,150	425		3,575	4,100
0400 4" thick		2	8		3,650	425		4,075	4,650
0500 6" thick		2	8		3,400	425		3,825	4,400
0800 5' x 7', power operation, 2" thick		1.90	8.421		5,675	450		6,125	6,925
0900 4" thick		1.90	8.421		5,675	450		6,125	6,925
1000 6" thick		1.90	8.421		6,475	450		6,925	7,775
1300 9' x 10', manual operation, 2" insulation		1.70	9.412		4,450	500		4,950	5,650
1400 4" insulation		1.70	9.412		4,825	500		5,325	6,050
1500 6" insulation		1.70	9.412		5,675	500		6,175	7,000
1800 Power operation, 2" insulation		1.60	10		7,825	530		8,355	9,400
1900 4" insulation		1.60	10		7,975	530		8,505	9,575
2000 6" insulation	↓	1.70	9.412	↓	8,225	500		8,725	9,775
2300 For stainless steel face, add					25%				
3000 Hinged, lightweight, 3' x 7'-0", 2" thick	2 Carp	2	8	Ea.	1,300	425		1,725	2,075
3050 4" thick		1.90	8.421		1,625	450		2,075	2,450
3300 Polymer doors, 3' x 7'-0"		1.90	8.421		1,350	450		1,800	2,150
3350 6" thick		1.40	11.429		2,375	605		2,980	3,550
3600 Stainless steel, 3' x 7'-0", 4" thick		1.90	8.421		1,725	450		2,175	2,575
3650 6" thick		1.40	11.429		2,875	605		3,480	4,075
3900 Painted, 3' x 7'-0", 4" thick		1.90	8.421		1,350	450		1,800	2,150
3950 6" thick	↓	1.40	11.429	↓	2,325	605		2,930	3,500
5000 Bi-parting, electric operated									
5010 6' x 8' opening, galv. faces, 4" thick for cooler	2 Carp	.80	20	Opng.	7,350	1,075		8,425	9,700
5050 For freezer, 4" thick		.80	20		8,100	1,075		9,175	10,500
5300 For door buck framing and door protection, add		2.50	6.400		680	340		1,020	1,250
6000 Galvanized batten door, galvanized hinges, 4' x 7'		2	8		1,875	425		2,300	2,700
6050 6' x 8'		1.80	8.889		2,475	470		2,945	3,425
6500 Fire door, 3 hr., 6' x 8', single slide		.80	20		8,725	1,075		9,800	11,200
6550 Double, bi-parting	↓	.70	22.857	↓	11,300	1,225		12,525	14,300

08 91 Louvers

08 91 16 – Operable Wall Louvers

08 91 16.10 Movable Blade Louvers	Crew	Daily Output	Labor-Hours	Unit	Material	2020 Bare Costs Labor	2020 Bare Costs Equipment	Total	Total Incl O&P
0010 **MOVABLE BLADE LOUVERS**									
0100 PVC, commercial grade, 12" x 12"	1 Shee	20	.400	Ea.	82.50	25		107.50	129
0110 16" x 16"		14	.571		96	35.50		131.50	160
0120 18" x 18"		14	.571		98.50	35.50		134	163
0130 24" x 24"		14	.571		141	35.50		176.50	209
0140 30" x 30"		12	.667		170	41.50		211.50	249
0150 36" x 36"		12	.667		200	41.50		241.50	283
0300 Stainless steel, commercial grade, 12" x 12"		20	.400		230	25		255	291
0310 16" x 16"		14	.571		310	35.50		345.50	395
0320 18" x 18"		14	.571		310	35.50		345.50	395
0330 20" x 20"		14	.571		360	35.50		395.50	450
0340 24" x 24"		14	.571		425	35.50		460.50	525
0350 30" x 30"		12	.667		475	41.50		516.50	590
0360 36" x 36"	↓	12	.667	↓	580	41.50		621.50	700

08 91 Louvers

08 91 19 – Fixed Louvers

08 91 19.10 Aluminum Louvers	Crew	Daily Output	Labor-Hours	Unit	Material	2020 Bare Costs Labor	Equipment	Total	Total Incl O&P
0010 **ALUMINUM LOUVERS**									
0020 Aluminum with screen, residential, 8" x 8"	1 Carp	38	.211	Ea.	21.50	11.20		32.70	40.50
0100 12" x 12"		38	.211		17.85	11.20		29.05	36.50
0200 12" x 18"		35	.229		22.50	12.15		34.65	43
0250 14" x 24"		30	.267		32.50	14.15		46.65	57
0300 18" x 24"		27	.296		33.50	15.75		49.25	60.50
0500 24" x 30"		24	.333		70.50	17.70		88.20	104
0700 Triangle, adjustable, small		20	.400		73.50	21.50		95	113
0800 Large	▼	15	.533	▼	94	28.50		122.50	146
1200 Extruded aluminum, see Section 23 37 15.40									
2100 Midget, aluminum, 3/4" deep, 1" diameter	1 Carp	85	.094	Ea.	.98	5		5.98	8.60
2150 3" diameter		60	.133		3.22	7.10		10.32	14.20
2200 4" diameter		50	.160		6	8.50		14.50	19.40
2250 6" diameter	▼	30	.267		3.87	14.15		18.02	26
3000 PVC, commercial grade, 12" x 12"	1 Shee	20	.400		97	25		122	145
3010 12" x 18"		20	.400		109	25		134	158
3020 12" x 24"		20	.400		147	25		172	199
3030 14" x 24"		20	.400		153	25		178	207
3100 Aluminum, commercial grade, 12" x 12"		20	.400		170	25		195	225
3110 24" x 24"		16	.500		234	31		265	305
3120 24" x 36"		16	.500		335	31		366	415
3130 36" x 24"		16	.500		345	31		376	430
3140 36" x 36"		14	.571		430	35.50		465.50	530
3150 36" x 48"		10	.800		465	50		515	585
3160 48" x 36"		10	.800		500	50		550	625
3170 48" x 48"		10	.800		455	50		505	575
3180 60" x 48"		8	1		515	62.50		577.50	665
3190 60" x 60"	▼	8	1	▼	750	62.50		812.50	920

08 91 19.20 Steel Louvers

	Crew	Daily Output	Labor-Hours	Unit	Material	Labor	Equipment	Total	Total Incl O&P
0010 **STEEL LOUVERS**									
3300 Galvanized steel, fixed blades, commercial grade, 18" x 18"	1 Shee	20	.400	Ea.	163	25		188	217
3310 24" x 24"		20	.400		218	25		243	278
3320 24" x 36"		16	.500		284	31		315	360
3330 36" x 24"		16	.500		291	31		322	370
3340 36" x 36"		14	.571		355	35.50		390.50	450
3350 36" x 48"		10	.800		420	50		470	540
3360 48" x 36"		10	.800		410	50		460	525
3370 48" x 48"		10	.800		510	50		560	635
3380 60" x 48"		10	.800		560	50		610	690
3390 60" x 60"	▼	10	.800	▼	670	50		720	815

08 95 Vents

08 95 13 – Soffit Vents

08 95 13.10 Wall Louvers	Crew	Daily Output	Labor-Hours	Unit	Material	2020 Bare Costs Labor	Equipment	Total	Total Incl O&P
0010 **WALL LOUVERS**									
2340 Baked enamel finish	1 Carp	200	.040	L.F.	4.81	2.13		6.94	8.50
2400 Under eaves vent, aluminum, mill finish, 16" x 4"		48	.167	Ea.	2.01	8.85		10.86	15.50
2500 16" x 8"		48	.167	"	2.85	8.85		11.70	16.45

Estimating Tips
General
- Room Finish Schedule: A complete set of plans should contain a room finish schedule. If one is not available, it would be well worth the time and effort to obtain one.

09 20 00 Plaster and Gypsum Board
- Lath is estimated by the square yard plus a 5% allowance for waste. Furring, channels, and accessories are measured by the linear foot. An extra foot should be allowed for each accessory miter or stop.
- Plaster is also estimated by the square yard. Deductions for openings vary by preference, from zero deduction to 50% of all openings over 2 feet in width. The estimator should allow one extra square foot for each linear foot of horizontal interior or exterior angle located below the ceiling level. Also, double the areas of small radius work.
- Drywall accessories, studs, track, and acoustical caulking are all measured by the linear foot. Drywall taping is figured by the square foot. Gypsum wallboard is estimated by the square foot. No material deductions should be made for door or window openings under 32 S.F.

09 60 00 Flooring
- Tile and terrazzo areas are taken off on a square foot basis. Trim and base materials are measured by the linear foot. Accent tiles are listed per each. Two basic methods of installation are used. Mud set is approximately 30% more expensive than thin set.

The cost of grout is included with tile unit price lines unless otherwise noted. In terrazzo work, be sure to include the linear footage of embedded decorative strips, grounds, machine rubbing, and power cleanup.
- Wood flooring is available in strip, parquet, or block configuration. The latter two types are set in adhesives with quantities estimated by the square foot. The laying pattern will influence labor costs and material waste. In addition to the material and labor for laying wood floors, the estimator must make allowances for sanding and finishing these areas, unless the flooring is prefinished.
- Sheet flooring is measured by the square yard. Roll widths vary, so consideration should be given to use the most economical width, as waste must be figured into the total quantity. Consider also the installation methods available—direct glue down or stretched. Direct glue-down installation is assumed with sheet carpet unit price lines unless otherwise noted.

09 70 00 Wall Finishes
- Wall coverings are estimated by the square foot. The area to be covered is measured—length by height of the wall above the baseboards—to calculate the square footage of each wall. This figure is divided by the number of square feet in the single roll which is being used. Deduct, in full, the areas of openings such as doors and windows. Where a pattern match is required allow 25–30% waste.

09 80 00 Acoustic Treatment
- Acoustical systems fall into several categories. The takeoff of these materials should be by the square foot of area with a 5% allowance for waste. Do not forget about scaffolding, if applicable, when estimating these systems.

09 90 00 Painting and Coating
- New line items created for cut-ins with reference diagram.
- A major portion of the work in painting involves surface preparation. Be sure to include cleaning, sanding, filling, and masking costs in the estimate.
- Protection of adjacent surfaces is not included in painting costs. When considering the method of paint application, an important factor is the amount of protection and masking required. These must be estimated separately and may be the determining factor in choosing the method of application.

Reference Numbers
Reference numbers are shown at the beginning of some major classifications. These numbers refer to related items in the Reference Section. The reference information may be an estimating procedure, an alternate pricing method, or technical information.

Note: Not all subdivisions listed here necessarily appear. ■

Same Data. Simplified.

Enjoy the convenience and efficiency of accessing your costs anywhere:

- **Skip the multiplier** by setting your location
- **Quickly search,** edit, favorite and share costs
- **Stay on top of price changes** with automatic updates

Discover more at rsmeans.com/online

09 22 Supports for Plaster and Gypsum Board

09 22 03 – Fastening Methods for Finishes

09 22 03.20 Drilling Plaster/Drywall	Crew	Daily Output	Labor-Hours	Unit	Material	2020 Bare Costs Labor	Equipment	Total	Total Incl O&P
0010 **DRILLING PLASTER/DRYWALL**									
1100 Drilling & layout for drywall/plaster walls, up to 1" deep, no anchor									
1200 Holes, 1/4" diameter	1 Carp	150	.053	Ea.	.01	2.83		2.84	4.27
1300 3/8" diameter		140	.057		.01	3.04		3.05	4.57
1400 1/2" diameter		130	.062		.01	3.27		3.28	4.92
1500 3/4" diameter		120	.067		.01	3.54		3.55	5.30
1600 1" diameter		110	.073		.02	3.87		3.89	5.80
1700 1-1/4" diameter		100	.080		.04	4.25		4.29	6.45
1800 1-1/2" diameter	▼	90	.089		.05	4.72		4.77	7.15
1900 For ceiling installations, add				▼		40%			

09 69 Access Flooring

09 69 13 – Rigid-Grid Access Flooring

09 69 13.10 Access Floors	Crew	Daily Output	Labor-Hours	Unit	Material	2020 Bare Costs Labor	Equipment	Total	Total Incl O&P
0010 **ACCESS FLOORS**									
0015 Access floor pkg. including panel, pedestal, & stringers 1500 lbs. load									
0100 Package pricing, conc. fill panels, no fin., 6" ht.	4 Carp	750	.043	S.F.	18.45	2.27		20.72	24
0105 12" ht.		750	.043		18.55	2.27		20.82	24
0110 18" ht.		750	.043		19.20	2.27		21.47	24.50
0115 24" ht.		750	.043		19.05	2.27		21.32	24.50
0120 Package pricing, steel panels, no fin., 6" ht.		750	.043		15.15	2.27		17.42	20
0125 12" ht.		750	.043		20	2.27		22.27	25.50
0130 18" ht.		750	.043		20.50	2.27		22.77	26
0135 24" ht.		750	.043		20.50	2.27		22.77	26
0140 Package pricing, wood core panels, no fin., 6" ht.		750	.043		13.65	2.27		15.92	18.40
0145 12" ht.		750	.043		13.90	2.27		16.17	18.65
0150 18" ht.		750	.043		13.90	2.27		16.17	18.65
0155 24" ht.		750	.043		14.15	2.27		16.42	18.95
0160 Pkg. pricing, conc. fill pnls, no stringers, no fin., 6" ht, 1250 lbs. load		700	.046		11.05	2.43		13.48	15.80
0165 12" ht.		700	.046		10.25	2.43		12.68	14.90
0170 18" ht.		700	.046		11.20	2.43		13.63	15.95
0175 24" ht.	▼	700	.046		12.15	2.43		14.58	17.05
0250 Panels, 2' x 2' conc. fill, no fin.	2 Carp	500	.032		23	1.70		24.70	27.50
0255 With 1/8" high pressure laminate		500	.032		57.50	1.70		59.20	66
0260 Metal panels with 1/8" high pressure laminate		500	.032		68	1.70		69.70	77.50
0265 Wood core panels with 1/8" high pressure laminate		500	.032		44.50	1.70		46.20	51.50
0400 Aluminum panels, no fin.	▼	500	.032		35	1.70		36.70	41
0600 For carpet covering, add					9.25			9.25	10.15
0700 For vinyl floor covering, add					9.90			9.90	10.90
0900 For high pressure laminate covering, add					8.15			8.15	8.95
0910 For snap on stringer system, add	2 Carp	1000	.016	▼	1.60	.85		2.45	3.04
1000 Machine cutouts after initial installation	1 Carp	50	.160	Ea.	20.50	8.50		29	35.50
1050 Pedestals, 6" to 12"	2 Carp	85	.188		9.10	10		19.10	25
1100 Air conditioning grilles, 4" x 12"	1 Carp	17	.471		72.50	25		97.50	117
1150 4" x 18"	"	14	.571	▼	99	30.50		129.50	155
1200 Approach ramps, steel	2 Carp	60	.267	S.F.	28	14.15		42.15	52.50
1300 Aluminum	"	40	.400	"	34.50	21.50		56	70
1500 Handrail, 2 rail, aluminum	1 Carp	15	.533	L.F.	126	28.50		154.50	181

09 91 23.52 Miscellaneous, Interior	Crew	Daily Output	Labor-Hours	Unit	Material	2020 Bare Costs Labor	Equipment	Total	Total Incl O&P
0010 **MISCELLANEOUS, INTERIOR** R099100-10									
3800 Grilles, per side, oil base, primer coat, brushwork	1 Pord	520	.015	S.F.	.14	.68		.82	1.17
3850 Spray		1140	.007		.15	.31		.46	.62
3880 Paint 1 coat, brushwork		520	.015		.25	.68		.93	1.30
3900 Spray		1140	.007		.28	.31		.59	.77
3920 Paint 2 coats, brushwork		325	.025		.49	1.09		1.58	2.17
3940 Spray		650	.012		.56	.55		1.11	1.43
3950 Prime & paint 1 coat		325	.025		.39	1.09		1.48	2.06
3960 Prime & paint 2 coats		270	.030		.39	1.32		1.71	2.38
4500 Louvers, 1 side, primer, brushwork		524	.015		.09	.68		.77	1.11
4520 Paint 1 coat, brushwork		520	.015		.11	.68		.79	1.14
4530 Spray		1140	.007		.12	.31		.43	.60
4540 Paint 2 coats, brushwork		325	.025		.22	1.09		1.31	1.87
4550 Spray		650	.012		.24	.55		.79	1.07
4560 Paint 3 coats, brushwork		270	.030		.32	1.32		1.64	2.31
4570 Spray		500	.016		.36	.71		1.07	1.45
5000 Pipe, 1"-4" diameter, primer or sealer coat, oil base, brushwork	2 Pord	1250	.013	L.F.	.09	.57		.66	.95
5100 Spray		2165	.007		.09	.33		.42	.59
5200 Paint 1 coat, brushwork		1250	.013		.13	.57		.70	.99
5300 Spray		2165	.007		.11	.33		.44	.62
5350 Paint 2 coats, brushwork		775	.021		.23	.92		1.15	1.61
5400 Spray		1240	.013		.25	.57		.82	1.13
5420 Paint 3 coats, brushwork		775	.021		.34	.92		1.26	1.73
5450 5"-8" diameter, primer or sealer coat, brushwork		620	.026		.18	1.15		1.33	1.90
5500 Spray		1085	.015		.30	.65		.95	1.30
5550 Paint 1 coat, brushwork		620	.026		.35	1.15		1.50	2.08
5600 Spray		1085	.015		.38	.65		1.03	1.39
5650 Paint 2 coats, brushwork		385	.042		.45	1.85		2.30	3.25
5700 Spray		620	.026		.51	1.15		1.66	2.26
5720 Paint 3 coats, brushwork		385	.042		.68	1.85		2.53	3.49
5750 9"-12" diameter, primer or sealer coat, brushwork		415	.039		.28	1.71		1.99	2.85
5800 Spray		725	.022		.41	.98		1.39	1.91
5850 Paint 1 coat, brushwork		415	.039		.35	1.71		2.06	2.94
6000 Spray		725	.022		.39	.98		1.37	1.89
6200 Paint 2 coats, brushwork		260	.062		.68	2.73		3.41	4.81
6250 Spray		415	.039		.76	1.71		2.47	3.38
6270 Paint 3 coats, brushwork		260	.062		1.01	2.73		3.74	5.15
6300 13"-16" diameter, primer or sealer coat, brushwork		310	.052		.37	2.29		2.66	3.81
6350 Spray		540	.030		.41	1.32		1.73	2.41
6400 Paint 1 coat, brushwork		310	.052		.47	2.29		2.76	3.93
6450 Spray		540	.030		.52	1.32		1.84	2.53
6500 Paint 2 coats, brushwork		195	.082		.91	3.64		4.55	6.40
6550 Spray		310	.052		1.01	2.29		3.30	4.52
6600 Radiators, per side, primer, brushwork	1 Pord	520	.015	S.F.	.09	.68		.77	1.12
6620 Paint, 1 coat		520	.015		.08	.68		.76	1.11
6640 2 coats		340	.024		.22	1.04		1.26	1.79
6660 3 coats		283	.028		.32	1.26		1.58	2.22

For customer support on your Mechanical Costs with RSMeans Data, call 800.448.8182.

65

Division Notes

	CREW	DAILY OUTPUT	LABOR-HOURS	UNIT	BARE COSTS				TOTAL INCL O&P
					MAT.	LABOR	EQUIP.	TOTAL	

Estimating Tips
General

- The items in this division are usually priced per square foot or each.
- Many items in Division 10 require some type of support system or special anchors that are not usually furnished with the item. The required anchors must be added to the estimate in the appropriate division.
- Some items in Division 10, such as lockers, may require assembly before installation. Verify the amount of assembly required. Assembly can often exceed installation time.

10 20 00 Interior Specialties

- Support angles and blocking are not included in the installation of toilet compartments, shower/dressing compartments, or cubicles. Appropriate line items from Division 5 or 6 may need to be added to support the installations.
- Toilet partitions are priced by the stall. A stall consists of a side wall, pilaster, and door with hardware. Toilet tissue holders and grab bars are extra.
- The required acoustical rating of a folding partition can have a significant impact on costs. Verify the sound transmission coefficient rating of the panel priced against the specification requirements.

- Grab bar installation does not include supplemental blocking or backing to support the required load. When grab bars are installed at an existing facility, provisions must be made to attach the grab bars to a solid structure.

Reference Numbers

Reference numbers are shown at the beginning of some major classifications. These numbers refer to related items in the Reference Section. The reference information may be an estimating procedure, an alternate pricing method, or technical information.

Note: Not all subdivisions listed here necessarily appear. ■

---

10 31 Manufactured Fireplaces

10 31 13 – Manufactured Fireplace Chimneys

10 31 13.10 Fireplace Chimneys

		Crew	Daily Output	Labor-Hours	Unit	Material	2020 Bare Costs Labor	Equipment	Total	Total Incl O&P
0010	**FIREPLACE CHIMNEYS**									
0500	Chimney dbl. wall, all stainless, over 8'-6", 7" diam., add to fireplace	1 Carp	33	.242	V.L.F.	94	12.90		106.90	122
0600	10" diameter, add to fireplace		32	.250		103	13.30		116.30	133
0700	12" diameter, add to fireplace		31	.258		181	13.70		194.70	220
0800	14" diameter, add to fireplace		30	.267	↓	216	14.15		230.15	260
1000	Simulated brick chimney top, 4' high, 16" x 16"		10	.800	Ea.	515	42.50		557.50	630
1100	24" x 24"	↓	7	1.143	"	575	60.50		635.50	720

10 31 23 – Prefabricated Fireplaces

10 31 23.10 Fireplace, Prefabricated

		Crew	Daily Output	Labor-Hours	Unit	Material	2020 Bare Costs Labor	Equipment	Total	Total Incl O&P
0010	**FIREPLACE, PREFABRICATED**, free standing or wall hung									
0100	With hood & screen, painted	1 Carp	1.30	6.154	Ea.	1,650	325		1,975	2,300
0150	Average		1	8		1,775	425		2,200	2,600
0200	Stainless steel		.90	8.889	↓	3,300	470		3,770	4,325
1500	Simulated logs, gas fired, 40,000 BTU, 2' long, manual safety pilot		7	1.143	Set	560	60.50		620.50	705
1600	Adjustable flame remote pilot		6	1.333		1,325	71		1,396	1,550
1700	Electric, 1,500 BTU, 1'-6" long, incandescent flame		7	1.143		265	60.50		325.50	385
1800	1,500 BTU, LED flame	↓	6	1.333	↓	345	71		416	480

10 35 Stoves

10 35 13 – Heating Stoves

10 35 13.10 Wood Burning Stoves

		Crew	Daily Output	Labor-Hours	Unit	Material	2020 Bare Costs Labor	Equipment	Total	Total Incl O&P
0010	**WOOD BURNING STOVES**									
0015	Cast iron, less than 1,500 S.F.	2 Carp	1.30	12.308	Ea.	1,500	655		2,155	2,625
0020	1,500 to 2,000 S.F.		1	16		2,550	850		3,400	4,075
0030	greater than 2,000 S.F.	↓	.80	20		2,925	1,075		4,000	4,825
0050	For gas log lighter, add				↓	49.50			49.50	54.50

Estimating Tips
General

- The items in this division are usually priced per square foot or each. Many of these items are purchased by the owner for installation by the contractor. Check the specifications for responsibilities and include time for receiving, storage, installation, and mechanical and electrical hookups in the appropriate divisions.

- Many items in Division 11 require some type of support system that is not usually furnished with the item. Examples of these systems include blocking for the attachment of casework and support angles for ceiling-hung projection screens. The required blocking or supports must be added to the estimate in the appropriate division.

- Some items in Division 11 may require assembly or electrical hookups. Verify the amount of assembly required or the need for a hard electrical connection and add the appropriate costs.

Reference Numbers

Reference numbers are shown at the beginning of some major classifications. These numbers refer to related items in the Reference Section. The reference information may be an estimating procedure, an alternate pricing method, or technical information.

Same Data. Simplified.

Enjoy the convenience and efficiency of accessing your costs anywhere:

- **Skip the multiplier** by setting your location
- **Quickly search,** edit, favorite and share costs
- **Stay on top of price changes** with automatic updates

Discover more at rsmeans.com/online

11 11 Vehicle Service Equipment

11 11 13 – Compressed-Air Vehicle Service Equipment

11 11 13.10 Compressed Air Equipment	Crew	Daily Output	Labor-Hours	Unit	Material	2020 Bare Costs Labor	Equipment	Total	Total Incl O&P
0010 **COMPRESSED AIR EQUIPMENT**									
0030 Compressors, electric, 1-1/2 HP, standard controls	L-4	1.50	16	Ea.	505	800		1,305	1,750
0550 Dual controls		1.50	16		1,050	800		1,850	2,350
0600 5 HP, 115/230 volt, standard controls		1	24		1,725	1,200		2,925	3,725
0650 Dual controls	↓	1	24	↓	3,225	1,200		4,425	5,375

11 11 19 – Vehicle Lubrication Equipment

11 11 19.10 Lubrication Equipment

11 11 19.10 Lubrication Equipment	Crew	Daily Output	Labor-Hours	Unit	Material	2020 Bare Costs Labor	Equipment	Total	Total Incl O&P
0010 **LUBRICATION EQUIPMENT**									
3000 Lube equipment, 3 reel type, with pumps, not including piping	L-4	.50	48	Set	10,400	2,400		12,800	15,100
3700 Pump lubrication, pneumatic, not incl. air compressor									
3710 Oil/gear lube	Q-1	9.60	1.667	Ea.	890	96.50		986.50	1,125
3720 Grease	"	9.60	1.667	"	1,200	96.50		1,296.50	1,475

11 11 33 – Vehicle Spray Painting Equipment

11 11 33.10 Spray Painting Equipment

11 11 33.10 Spray Painting Equipment	Crew	Daily Output	Labor-Hours	Unit	Material	2020 Bare Costs Labor	Equipment	Total	Total Incl O&P
0010 **SPRAY PAINTING EQUIPMENT**									
4000 Spray painting booth, 26' long, complete	L-4	.40	60	Ea.	11,600	3,000		14,600	17,200

11 21 Retail and Service Equipment

11 21 73 – Commercial Laundry and Dry Cleaning Equipment

11 21 73.16 Drying and Conditioning Equipment

11 21 73.16 Drying and Conditioning Equipment	Crew	Daily Output	Labor-Hours	Unit	Material	2020 Bare Costs Labor	Equipment	Total	Total Incl O&P
0010 **DRYING AND CONDITIONING EQUIPMENT**									
0100 Dryers, not including rough-in									
1500 Industrial, 30 lb. capacity	1 Plum	2	4	Ea.	3,350	258		3,608	4,050
1600 50 lb. capacity	"	1.70	4.706	"	4,550	305		4,855	5,475

11 21 73.26 Commercial Washers and Extractors

11 21 73.26 Commercial Washers and Extractors	Crew	Daily Output	Labor-Hours	Unit	Material	2020 Bare Costs Labor	Equipment	Total	Total Incl O&P
0010 **COMMERCIAL WASHERS AND EXTRACTORS**, not including rough-in									
6000 Combination washer/extractor, 20 lb. capacity	L-6	1.50	8	Ea.	6,725	505		7,230	8,150
6100 30 lb. capacity		.80	15		10,300	950		11,250	12,700
6200 50 lb. capacity		.68	17.647		12,900	1,125		14,025	15,900
6300 75 lb. capacity		.30	40		19,200	2,525		21,725	24,900
6350 125 lb. capacity	↓	.16	75	↓	29,900	4,750		34,650	40,000

11 21 73.33 Coin-Operated Laundry Equipment

11 21 73.33 Coin-Operated Laundry Equipment	Crew	Daily Output	Labor-Hours	Unit	Material	2020 Bare Costs Labor	Equipment	Total	Total Incl O&P
0010 **COIN-OPERATED LAUNDRY EQUIPMENT**									
0990 Dryer, gas fired									
1000 Commercial, 30 lb. capacity, coin operated, single	1 Plum	3	2.667	Ea.	3,475	172		3,647	4,075
1100 Double stacked	"	2	4	"	8,275	258		8,533	9,475
5290 Clothes washer									
5300 Commercial, coin operated, average	1 Plum	3	2.667	Ea.	1,350	172		1,522	1,750

11 21 83 – Photo Processing Equipment

11 21 83.13 Darkroom Equipment

11 21 83.13 Darkroom Equipment	Crew	Daily Output	Labor-Hours	Unit	Material	2020 Bare Costs Labor	Equipment	Total	Total Incl O&P
0010 **DARKROOM EQUIPMENT**									
0020 Developing sink, 5" deep, 24" x 48"	Q-1	2	8	Ea.	500	465		965	1,250
0050 48" x 52"		1.70	9.412		1,250	545		1,795	2,200
0200 10" deep, 24" x 48"		1.70	9.412		1,625	545		2,170	2,625
0250 24" x 108"	↓	1.50	10.667	↓	3,875	620		4,495	5,175

11 30 Residential Equipment

11 30 13 – Residential Appliances

11 30 13.15 Cooking Equipment	Crew	Daily Output	Labor-Hours	Unit	Material	2020 Bare Costs Labor	Equipment	Total	Total Incl O&P
0010 **COOKING EQUIPMENT**									
0020 Cooking range, 30" free standing, 1 oven, minimum	2 Clab	10	1.600	Ea.	465	67.50		532.50	610
0050 Maximum		4	4		2,200	168		2,368	2,675
0150 2 oven, minimum		10	1.600		1,050	67.50		1,117.50	1,275
0200 Maximum		10	1.600		3,425	67.50		3,492.50	3,850

11 30 13.16 Refrigeration Equipment

	Crew	Daily Output	Labor-Hours	Unit	Material	2020 Bare Costs Labor	Equipment	Total	Total Incl O&P
0010 **REFRIGERATION EQUIPMENT**									
5200 Icemaker, automatic, 20 lbs./day	1 Plum	7	1.143	Ea.	1,375	73.50		1,448.50	1,600
5350 51 lbs./day	"	2	4	"	1,400	258		1,658	1,925

11 30 13.17 Kitchen Cleaning Equipment

	Crew	Daily Output	Labor-Hours	Unit	Material	2020 Bare Costs Labor	Equipment	Total	Total Incl O&P
0010 **KITCHEN CLEANING EQUIPMENT**									
2750 Dishwasher, built-in, 2 cycles, minimum	L-1	4	4	Ea.	310	252		562	715
2800 Maximum		2	8		445	505		950	1,250
2950 4 or more cycles, minimum		4	4		400	252		652	815
2960 Average		4	4		530	252		782	960
3000 Maximum		2	8		1,850	505		2,355	2,775

11 30 13.18 Waste Disposal Equipment

	Crew	Daily Output	Labor-Hours	Unit	Material	2020 Bare Costs Labor	Equipment	Total	Total Incl O&P
0010 **WASTE DISPOSAL EQUIPMENT**									
3300 Garbage disposal, sink type, minimum	L-1	10	1.600	Ea.	111	101		212	272
3350 Maximum	"	10	1.600	"	216	101		317	385

11 30 13.19 Kitchen Ventilation Equipment

	Crew	Daily Output	Labor-Hours	Unit	Material	2020 Bare Costs Labor	Equipment	Total	Total Incl O&P
0010 **KITCHEN VENTILATION EQUIPMENT**									
4150 Hood for range, 2 speed, vented, 30" wide, minimum	L-3	5	3.200	Ea.	107	184		291	395
4200 Maximum		3	5.333		1,000	305		1,305	1,550
4300 42" wide, minimum		5	3.200		151	184		335	440
4330 Custom		5	3.200		1,800	184		1,984	2,250
4350 Maximum		3	5.333		2,175	305		2,480	2,850
4500 For ventless hood, 2 speed, add					19.05			19.05	21
4650 For vented 1 speed, deduct from maximum					74			74	81.50

11 30 13.24 Washers

	Crew	Daily Output	Labor-Hours	Unit	Material	2020 Bare Costs Labor	Equipment	Total	Total Incl O&P
0010 **WASHERS**									
5000 Residential, 4 cycle, average	1 Plum	3	2.667	Ea.	990	172		1,162	1,350
6650 Washing machine, automatic, minimum		3	2.667		560	172		732	870
6700 Maximum		1	8		1,400	515		1,915	2,325

11 30 13.25 Dryers

	Crew	Daily Output	Labor-Hours	Unit	Material	2020 Bare Costs Labor	Equipment	Total	Total Incl O&P
0010 **DRYERS**									
0500 Gas fired residential, 16 lb. capacity, average	1 Plum	3	2.667	Ea.	720	172		892	1,050
7450 Vent kits for dryers	1 Carp	10	.800	"	48	42.50		90.50	117

11 30 15 – Miscellaneous Residential Appliances

11 30 15.13 Sump Pumps

	Crew	Daily Output	Labor-Hours	Unit	Material	2020 Bare Costs Labor	Equipment	Total	Total Incl O&P
0010 **SUMP PUMPS**									
6400 Cellar drainer, pedestal, 1/3 HP, molded PVC base	1 Plum	3	2.667	Ea.	141	172		313	410
6450 Solid brass	"	2	4	"	243	258		501	650
6460 Sump pump, see also Section 22 14 29.16									

11 30 15.23 Water Heaters

	Crew	Daily Output	Labor-Hours	Unit	Material	2020 Bare Costs Labor	Equipment	Total	Total Incl O&P
0010 **WATER HEATERS**									
6900 Electric, glass lined, 30 gallon, minimum	L-1	5	3.200	Ea.	970	201		1,171	1,375
6950 Maximum		3	5.333		1,350	335		1,685	1,975
7100 80 gallon, minimum		2	8		1,950	505		2,455	2,900
7150 Maximum		1	16		2,725	1,000		3,725	4,500

For customer support on your Mechanical Costs with RSMeans Data, call 800.448.8182.

71

11 30 Residential Equipment

11 30 15 – Miscellaneous Residential Appliances

11 30 15.23 Water Heaters

		Crew	Daily Output	Labor-Hours	Unit	Material	2020 Bare Costs Labor	2020 Bare Costs Equipment	Total	Total Incl O&P
7180	Gas, glass lined, 30 gallon, minimum	2 Plum	5	3.200	Ea.	1,900	206		2,106	2,375
7220	Maximum		3	5.333		2,625	345		2,970	3,425
7260	50 gallon, minimum		2.50	6.400		1,750	410		2,160	2,550
7300	Maximum	↓	1.50	10.667	↓	2,450	685		3,135	3,725

11 30 15.43 Air Quality

		Crew	Daily Output	Labor-Hours	Unit	Material	2020 Bare Costs Labor	2020 Bare Costs Equipment	Total	Total Incl O&P
0010	**AIR QUALITY**									
2450	Dehumidifier, portable, automatic, 15 pint	1 Elec	4	2	Ea.	208	123		331	410
2550	40 pint		3.75	2.133		247	131		378	465
3550	Heater, electric, built-in, 1250 watt, ceiling type, minimum		4	2		125	123		248	320
3600	Maximum		3	2.667		184	164		348	445
3700	Wall type, minimum		4	2		221	123		344	425
3750	Maximum		3	2.667		198	164		362	460
3900	1500 watt wall type, with blower		4	2		186	123		309	385
3950	3000 watt	↓	3	2.667		515	164		679	810
4850	Humidifier, portable, 8 gallons/day					158			158	174
5000	15 gallons/day				↓	207			207	227

11 32 Unit Kitchens

11 32 13 – Metal Unit Kitchens

11 32 13.10 Commercial Unit Kitchens

		Crew	Daily Output	Labor-Hours	Unit	Material	2020 Bare Costs Labor	2020 Bare Costs Equipment	Total	Total Incl O&P
0010	**COMMERCIAL UNIT KITCHENS**									
1500	Combination range, refrigerator and sink, 30" wide, minimum	L-1	2	8	Ea.	710	505		1,215	1,525
1550	Maximum		1	16		1,100	1,000		2,100	2,700
1570	60" wide, average		1.40	11.429		1,225	720		1,945	2,425
1590	72" wide, average		1.20	13.333		2,000	840		2,840	3,450
1600	Office model, 48" wide		2	8		1,425	505		1,930	2,325
1620	Refrigerator and sink only	↓	2.40	6.667	↓	2,625	420		3,045	3,525
1640	Combination range, refrigerator, sink, microwave									
1660	Oven and ice maker	L-1	.80	20	Ea.	4,900	1,250		6,150	7,250

11 41 Foodservice Storage Equipment

11 41 13 – Refrigerated Food Storage Cases

11 41 13.10 Refrigerated Food Cases

		Crew	Daily Output	Labor-Hours	Unit	Material	2020 Bare Costs Labor	2020 Bare Costs Equipment	Total	Total Incl O&P
0010	**REFRIGERATED FOOD CASES**									
0030	Dairy, multi-deck, 12' long	Q-5	3	5.333	Ea.	11,500	315		11,815	13,100
0100	For rear sliding doors, add					2,050			2,050	2,250
0200	Delicatessen case, service deli, 12' long, single deck	Q-5	3.90	4.103		8,750	242		8,992	9,975
0300	Multi-deck, 18 S.F. shelf display		3	5.333		8,375	315		8,690	9,700
0400	Freezer, self-contained, chest-type, 30 C.F.		3.90	4.103		4,800	242		5,042	5,625
0500	Glass door, upright, 78 C.F.		3.30	4.848		9,375	286		9,661	10,700
0600	Frozen food, chest type, 12' long		3.30	4.848		7,575	286		7,861	8,750
0700	Glass door, reach-in, 5 door		3	5.333		10,400	315		10,715	11,900
0800	Island case, 12' long, single deck		3.30	4.848		7,900	286		8,186	9,100
0900	Multi-deck		3	5.333		10,100	315		10,415	11,600
1000	Meat case, 12' long, single deck		3.30	4.848		7,875	286		8,161	9,075
1050	Multi-deck		3.10	5.161		10,200	305		10,505	11,700
1100	Produce, 12' long, single deck		3.30	4.848		6,475	286		6,761	7,550
1200	Multi-deck	↓	3.10	5.161		8,550	305		8,855	9,850

11 41 Foodservice Storage Equipment

11 41 13 – Refrigerated Food Storage Cases

11 41 13.20 Refrigerated Food Storage Equipment	Crew	Daily Output	Labor-Hours	Unit	Material	2020 Bare Costs Labor	Equipment	Total	Total Incl O&P
0010 **REFRIGERATED FOOD STORAGE EQUIPMENT**									
2350 Cooler, reach-in, beverage, 6' long	Q-1	6	2.667	Ea.	3,125	155		3,280	3,650
4300 Freezers, reach-in, 44 C.F.		4	4		4,425	232		4,657	5,225
4500 68 C.F.		3	5.333		5,575	310		5,885	6,600

11 44 Food Cooking Equipment

11 44 13 – Commercial Ranges

11 44 13.10 Cooking Equipment

	Crew	Daily Output	Labor-Hours	Unit	Material	2020 Bare Costs Labor	Equipment	Total	Total Incl O&P
0010 **COOKING EQUIPMENT**									
0020 Bake oven, gas, one section	Q-1	8	2	Ea.	5,675	116		5,791	6,425
0300 Two sections		7	2.286		9,375	133		9,508	10,500
0600 Three sections		6	2.667		12,400	155		12,555	13,800
0900 Electric convection, single deck	L-7	4	7		5,575	360		5,935	6,650
6350 Kettle, w/steam jacket, tilting, w/positive lock, SS, 20 gallons		7	4		8,650	205		8,855	9,825
6600 60 gallons		6	4.667		17,100	239		17,339	19,200

11 46 Food Dispensing Equipment

11 46 83 – Ice Machines

11 46 83.10 Commercial Ice Equipment

	Crew	Daily Output	Labor-Hours	Unit	Material	2020 Bare Costs Labor	Equipment	Total	Total Incl O&P
0010 **COMMERCIAL ICE EQUIPMENT**									
5800 Ice cube maker, 50 lbs./day	Q-1	6	2.667	Ea.	1,800	155		1,955	2,200
6050 500 lbs./day	"	4	4	"	2,600	232		2,832	3,200

11 48 Foodservice Cleaning and Disposal Equipment

11 48 13 – Commercial Dishwashers

11 48 13.10 Dishwashers

		Crew	Daily Output	Labor-Hours	Unit	Material	2020 Bare Costs Labor	Equipment	Total	Total Incl O&P
0010 **DISHWASHERS**										
2700 Dishwasher, commercial, rack type										
2720 10 to 12 racks/hour		Q-1	3.20	5	Ea.	3,450	290		3,740	4,225
2730 Energy star rated, 35 to 40 racks/hour	G		1.30	12.308		5,500	715		6,215	7,125
2740 50 to 60 racks/hour	G		1.30	12.308		10,600	715		11,315	12,700
2800 Automatic, 190 to 230 racks/hour		L-6	.35	34.286		14,600	2,175		16,775	19,300
2820 235 to 275 racks/hour			.25	48		28,000	3,050		31,050	35,400
2840 8,750 to 12,500 dishes/hour			.10	120		47,500	7,600		55,100	64,000

11 53 Laboratory Equipment

11 53 13 – Laboratory Fume Hoods

11 53 13.13 Recirculating Laboratory Fume Hoods

	Crew	Daily Output	Labor-Hours	Unit	Material	2020 Bare Costs Labor	Equipment	Total	Total Incl O&P
0010 **RECIRCULATING LABORATORY FUME HOODS**									
0600 Fume hood, with countertop & base, not including HVAC									
0610 Simple, minimum	2 Carp	5.40	2.963	L.F.	580	157		737	875
0620 Complex, including fixtures		2.40	6.667		1,125	355		1,480	1,775
0630 Special, maximum		1.70	9.412		1,325	500		1,825	2,200
0670 Service fixtures, average				Ea.	365			365	405
0680 For sink assembly with hot and cold water, add	1 Plum	1.40	5.714	"	780	370		1,150	1,400

11 53 Laboratory Equipment

11 53 19 – Laboratory Sterilizers

11 53 19.13 Sterilizers

		Crew	Daily Output	Labor-Hours	Unit	Material	2020 Bare Costs Labor	Equipment	Total	Total Incl O&P
0010	**STERILIZERS**									
0700	Glassware washer, undercounter, minimum	L-1	1.80	8.889	Ea.	6,850	560		7,410	8,350
0710	Maximum	"	1	16		15,200	1,000		16,200	18,200
1850	Utensil washer-sanitizer	1 Plum	2	4	↓	8,725	258		8,983	9,975

11 53 33 – Emergency Safety Appliances

11 53 33.13 Emergency Equipment

		Crew	Daily Output	Labor-Hours	Unit	Material	2020 Bare Costs Labor	Equipment	Total	Total Incl O&P
0010	**EMERGENCY EQUIPMENT**									
1400	Safety equipment, eye wash, hand held				Ea.	400			400	440
1450	Deluge shower				"	840			840	920

11 53 43 – Service Fittings and Accessories

11 53 43.13 Fittings

		Crew	Daily Output	Labor-Hours	Unit	Material	2020 Bare Costs Labor	Equipment	Total	Total Incl O&P
0010	**FITTINGS**									
1600	Sink, one piece plastic, flask wash, hose, free standing	1 Plum	1.60	5	Ea.	2,075	320		2,395	2,750
1610	Epoxy resin sink, 25" x 16" x 10"	"	2	4	"	227	258		485	635
8000	Alternate pricing method: as percent of lab furniture									
8050	Installation, not incl. plumbing & duct work				% Furn.				22%	22%
8100	Plumbing, final connections, simple system								10%	10%
8110	Moderately complex system								15%	15%
8120	Complex system								20%	20%
8150	Electrical, simple system								10%	10%
8160	Moderately complex system								20%	20%
8170	Complex system				↓				35%	35%

11 71 Medical Sterilizing Equipment

11 71 10 – Medical Sterilizers & Distillers

11 71 10.10 Sterilizers and Distillers

		Crew	Daily Output	Labor-Hours	Unit	Material	2020 Bare Costs Labor	Equipment	Total	Total Incl O&P
0010	**STERILIZERS AND DISTILLERS**									
0700	Distiller, water, steam heated, 50 gal. capacity	1 Plum	1.40	5.714	Ea.	25,500	370		25,870	28,700
5600	Sterilizers, floor loading, 26" x 62" x 42", single door, steam					155,000			155,000	170,500
5650	Double door, steam					239,000			239,000	262,500
5800	General purpose, 20" x 20" x 38", single door					12,500			12,500	13,700
6000	Portable, counter top, steam, minimum					2,900			2,900	3,175
6020	Maximum					5,075			5,075	5,575
6050	Portable, counter top, gas, 17" x 15" x 32-1/2"					45,400			45,400	49,900
6150	Manual washer/sterilizer, 16" x 16" x 26"	1 Plum	2	4	↓	62,000	258		62,258	69,000
6200	Steam generators, electric 10 kW to 180 kW, freestanding									
6250	Minimum	1 Elec	3	2.667	Ea.	12,200	164		12,364	13,600
6300	Maximum	"	.70	11.429		22,700	700		23,400	26,100
8200	Bed pan washer-sanitizer	1 Plum	2	4	↓	9,775	258		10,033	11,100

11 73 Patient Care Equipment

11 73 10 – Patient Treatment Equipment

11 73 10.10 Treatment Equipment	Crew	Daily Output	Labor-Hours	Unit	Material	2020 Bare Costs Labor	Equipment	Total	Total Incl O&P
0010 **TREATMENT EQUIPMENT**									
1800 Heat therapy unit, humidified, 26" x 78" x 28"				Ea.	4,250			4,250	4,675
8400 Whirlpool bath, mobile, sst, 18" x 24" x 60"					6,550			6,550	7,200
8450 Fixed, incl. mixing valves	1 Plum	2	4		4,825	258		5,083	5,700

11 74 Dental Equipment

11 74 10 – Dental Office Equipment

11 74 10.10 Diagnostic and Treatment Equipment

	Crew	Daily Output	Labor-Hours	Unit	Material	2020 Bare Costs Labor	Equipment	Total	Total Incl O&P
0010 **DIAGNOSTIC AND TREATMENT EQUIPMENT**									
0020 Central suction system, minimum	1 Plum	1.20	6.667	Ea.	1,175	430		1,605	1,925
0100 Maximum	"	.90	8.889		6,425	575		7,000	7,925
0600 Chair, electric or hydraulic, minimum	1 Skwk	.50	16		2,625	880		3,505	4,200
0700 Maximum		.25	32		3,975	1,750		5,725	7,025
2000 Light, ceiling mounted, minimum		8	1		875	55		930	1,050
2100 Maximum		8	1		2,050	55		2,105	2,325
2200 Unit light, minimum	2 Skwk	5.33	3.002		775	165		940	1,100
2210 Maximum		5.33	3.002		1,475	165		1,640	1,850
2220 Track light, minimum		3.20	5		1,725	274		1,999	2,325
2230 Maximum		3.20	5		2,300	274		2,574	2,950
2300 Sterilizers, steam portable, minimum					1,875			1,875	2,050
2350 Maximum					6,350			6,350	7,000
2600 Steam, institutional					2,650			2,650	2,900
2650 Dry heat, electric, portable, 3 trays					1,375			1,375	1,500

11 78 Mortuary Equipment

11 78 13 – Mortuary Refrigerators

11 78 13.10 Mortuary and Autopsy Equipment

	Crew	Daily Output	Labor-Hours	Unit	Material	2020 Bare Costs Labor	Equipment	Total	Total Incl O&P
0010 **MORTUARY AND AUTOPSY EQUIPMENT**									
0015 Autopsy table, standard	1 Plum	1	8	Ea.	10,000	515		10,515	11,800
0020 Deluxe	"	.60	13.333		16,400	860		17,260	19,300
3200 Mortuary refrigerator, end operated, 2 capacity					9,625			9,625	10,600
3300 6 capacity					16,300			16,300	17,900

11 78 16 – Crematorium Equipment

11 78 16.10 Crematory

	Crew	Daily Output	Labor-Hours	Unit	Material	2020 Bare Costs Labor	Equipment	Total	Total Incl O&P
0010 **CREMATORY**									
1500 Crematory, not including building, 1 place	Q-3	.20	160	Ea.	80,500	9,825		90,325	103,000
1750 2 place	"	.10	320	"	115,000	19,600		134,600	156,000

For customer support on your Mechanical Costs with RSMeans Data, call 800.448.8182.

75

11 81 Facility Maintenance Equipment

11 81 19 – Vacuum Cleaning Systems

11 81 19.10 Vacuum Cleaning	Crew	Daily Output	Labor-Hours	Unit	Material	2020 Bare Costs Labor	Equipment	Total	Total Incl O&P
0010 **VACUUM CLEANING**									
0020 Central, 3 inlet, residential	1 Skwk	.90	8.889	Total	1,325	490		1,815	2,200
0200 Commercial		.70	11.429		1,350	625		1,975	2,450
0400 5 inlet system, residential		.50	16		2,025	880		2,905	3,550
0600 7 inlet system, commercial		.40	20		2,600	1,100		3,700	4,500
0800 9 inlet system, residential		.30	26.667		4,225	1,475		5,700	6,850

11 82 Facility Solid Waste Handling Equipment

11 82 19 – Packaged Incinerators

11 82 19.10 Packaged Gas Fired Incinerators

		Crew	Daily Output	Labor-Hours	Unit	Material	2020 Bare Costs Labor	Equipment	Total	Total Incl O&P
0010	**PACKAGED GAS FIRED INCINERATORS**									
4400	Incinerator, gas, not incl. chimney, elec. or pipe, 50 lbs./hr., minimum	Q-3	.80	40	Ea.	42,000	2,450		44,450	49,900
4420	Maximum		.70	45.714		43,500	2,800		46,300	52,000
4440	200 lbs./hr., minimum (batch type)		.60	53.333		72,000	3,275		75,275	84,000
4460	Maximum (with feeder)		.50	64		84,500	3,925		88,425	99,000
4480	400 lbs./hr., minimum (batch type)		.30	107		88,000	6,550		94,550	106,500
4500	Maximum (with feeder)		.25	128		95,500	7,850		103,350	117,000
4520	800 lbs./hr., with feeder, minimum		.20	160		122,000	9,825		131,825	149,000
4540	Maximum		.17	188		211,000	11,500		222,500	249,500
4560	1,200 lbs./hr., with feeder, minimum		.15	213		161,500	13,100		174,600	197,000
4580	Maximum		.11	291		204,500	17,800		222,300	251,000
4600	2,000 lbs./hr., with feeder, minimum		.10	320		417,500	19,600		437,100	488,500
4620	Maximum		.05	640		653,500	39,300		692,800	777,500
4700	For heat recovery system, add, minimum		.25	128		101,500	7,850		109,350	124,000
4710	Add, maximum		.11	291		266,500	17,800		284,300	319,500
4720	For automatic ash conveyer, add		.50	64		41,900	3,925		45,825	52,000
4750	Large municipal incinerators, incl. stack, minimum		.25	128	Ton/day	21,700	7,850		29,550	35,700
4850	Maximum		.10	320	"	58,000	19,600		77,600	93,000

11 82 26 – Facility Waste Compactors

11 82 26.10 Compactors

		Crew	Daily Output	Labor-Hours	Unit	Material	2020 Bare Costs Labor	Equipment	Total	Total Incl O&P
0010	**COMPACTORS**									
0020	Compactors, 115 volt, 250 lbs./hr., chute fed	L-4	1	24	Ea.	12,600	1,200		13,800	15,600
0100	Hand fed		2.40	10		17,200	500		17,700	19,700
0300	Multi-bag, 230 volt, 600 lbs./hr., chute fed		1	24		16,800	1,200		18,000	20,300
0400	Hand fed		1	24		17,200	1,200		18,400	20,700
0500	Containerized, hand fed, 2 to 6 C.Y. containers, 250 lbs./hr.		1	24		16,700	1,200		17,900	20,200
0550	For chute fed, add per floor		1	24		1,500	1,200		2,700	3,475
1000	Heavy duty industrial compactor, 0.5 C.Y. capacity		1	24		11,200	1,200		12,400	14,100
1050	1.0 C.Y. capacity		1	24		17,100	1,200		18,300	20,600
1100	3.0 C.Y. capacity		.50	48		29,200	2,400		31,600	35,700
1150	5.0 C.Y. capacity		.50	48		36,900	2,400		39,300	44,200
1200	Combination shredder/compactor (5,000 lbs./hr.)		.50	48		71,500	2,400		73,900	82,000
1400	For handling hazardous waste materials, 55 gallon drum packer, std.					21,700			21,700	23,900
1410	55 gallon drum packer w/HEPA filter					26,400			26,400	29,100
1420	55 gallon drum packer w/charcoal & HEPA filter					35,200			35,200	38,800
1430	All of the above made explosion proof, add					1,725			1,725	1,900
5800	Shredder, industrial, minimum					29,900			29,900	32,900
5850	Maximum					160,000			160,000	175,500
5900	Baler, industrial, minimum					12,000			12,000	13,200
5950	Maximum					697,500			697,500	767,000

Estimating Tips

General

- The items and systems in this division are usually estimated, purchased, supplied, and installed as a unit by one or more subcontractors. The estimator must ensure that all parties are operating from the same set of specifications and assumptions, and that all necessary items are estimated and will be provided. Many times the complex items and systems are covered, but the more common ones, such as excavation or a crane, are overlooked for the very reason that everyone assumes nobody could miss them. The estimator should be the central focus and be able to ensure that all systems are complete.

- It is important to consider factors such as site conditions, weather, shape and size of building, as well as labor availability as they may impact the overall cost of erecting special structures and systems included in this division.

- Another area where problems can develop in this division is at the interface between systems.

The estimator must ensure, for instance, that anchor bolts, nuts, and washers are estimated and included for the air-supported structures and pre-engineered buildings to be bolted to their foundations. Utility supply is a common area where essential items or pieces of equipment can be missed or overlooked because each subcontractor may feel it is another's responsibility. The estimator should also be aware of certain items which may be supplied as part of a package but installed by others, and ensure that the installing contractor's estimate includes the cost of installation. Conversely, the estimator must also ensure that items are not costed by two different subcontractors, resulting in an inflated overall estimate.

13 30 00 Special Structures

- The foundations and floor slab, as well as rough mechanical and electrical, should be estimated, as this work is required for the assembly and erection of the structure. Generally, as noted in the data set, the pre-engineered building comes as a shell. Pricing is based on the size and structural design parameters stated in the reference section. Additional features, such as windows and doors with their related structural framing, must also be included by the estimator. Here again, the estimator must have a clear understanding of the scope of each portion of the work and all the necessary interfaces.

Reference Numbers

Reference numbers are shown at the beginning of some major classifications. These numbers refer to related items in the Reference Section. The reference information may be an estimating procedure, an alternate pricing method, or technical information.

Note: Not all subdivisions listed here necessarily appear. ■

Same Data. Simplified.

Enjoy the convenience and efficiency of accessing your costs anywhere:

- **Skip the multiplier** by setting your location
- **Quickly search,** edit, favorite and share costs
- **Stay on top of price changes** with automatic updates

Discover more at rsmeans.com/online

13 11 Swimming Pools

13 11 13 – Below-Grade Swimming Pools

13 11 13.50 Swimming Pools

		Crew	Daily Output	Labor-Hours	Unit	Material	2020 Bare Costs Labor	Equipment	Total	Total Incl O&P
0010	**SWIMMING POOLS** Residential in-ground, vinyl lined									
0020	Concrete sides, w/equip, sand bottom	B-52	300	.187	SF Surf	28.50	9.10	1.95	39.55	47
0100	Metal or polystyrene sides	B-14	410	.117		24	5.20	.52	29.72	35
0200	Add for vermiculite bottom				↓	1.82			1.82	2
0500	Gunite bottom and sides, white plaster finish									
0600	12' x 30' pool	B-52	145	.386	SF Surf	53	18.80	4.03	75.83	91
0720	16' x 32' pool		155	.361		48	17.60	3.77	69.37	83
0750	20' x 40' pool	↓	250	.224	↓	42.50	10.90	2.34	55.74	66
0810	Concrete bottom and sides, tile finish									
0820	12' x 30' pool	B-52	80	.700	SF Surf	53.50	34	7.30	94.80	118
0830	16' x 32' pool		95	.589		44.50	28.50	6.15	79.15	99
0840	20' x 40' pool	↓	130	.431	↓	35.50	21	4.50	61	75.50
1100	Motel, gunite with plaster finish, incl. medium									
1150	capacity filtration & chlorination	B-52	115	.487	SF Surf	65.50	23.50	5.10	94.10	113
1200	Municipal, gunite with plaster finish, incl. high									
1250	capacity filtration & chlorination	B-52	100	.560	SF Surf	84.50	27.50	5.85	117.85	140
1350	Add for formed gutters				L.F.	124			124	137
1360	Add for stainless steel gutters				"	370			370	405
1700	Filtration and deck equipment only, as % of total				Total				20%	20%
1800	Automatic vacuum, hand tools, etc., 20' x 40' pool				SF Pool				.56	.62
1900	5,000 S.F. pool				"				.11	.12
3000	Painting pools, preparation + 3 coats, 20' x 40' pool, epoxy	2 Pord	.33	48.485	Total	1,450	2,150		3,600	4,800
3100	Rubber base paint, 18 gallons	"	.33	48.485		1,250	2,150		3,400	4,575
3500	42' x 82' pool, 75 gallons, epoxy paint	3 Pord	.14	171		6,150	7,600		13,750	18,100
3600	Rubber base paint	"	.14	171	↓	5,150	7,600		12,750	17,000

13 11 23 – On-Grade Swimming Pools

13 11 23.50 Swimming Pools

		Crew	Daily Output	Labor-Hours	Unit	Material	2020 Bare Costs Labor	Equipment	Total	Total Incl O&P
0010	**SWIMMING POOLS** Residential above ground, steel construction									
0100	Round, 15' diam.	B-80A	3	8	Ea.	895	335	273	1,503	1,800
0120	18' diam.		2.50	9.600		1,025	405	330	1,760	2,100
0140	21' diam.		2	12		1,175	505	410	2,090	2,500
0160	24' diam.		1.80	13.333		1,350	560	455	2,365	2,850
0180	27' diam.		1.50	16		1,550	675	545	2,770	3,325
0200	30' diam.		1	24		1,575	1,000	820	3,395	4,175
0220	Oval, 12' x 24'		2.30	10.435		1,700	440	355	2,495	2,900
0240	15' x 30'		1.80	13.333		1,925	560	455	2,940	3,475
0260	18' x 33'	↓	1	24	↓	2,150	1,000	820	3,970	4,775

13 11 46 – Swimming Pool Accessories

13 11 46.50 Swimming Pool Equipment

		Crew	Daily Output	Labor-Hours	Unit	Material	2020 Bare Costs Labor	Equipment	Total	Total Incl O&P
0010	**SWIMMING POOL EQUIPMENT**									
0020	Diving stand, stainless steel, 3 meter	2 Carp	.40	40	Ea.	18,000	2,125		20,125	23,000
0300	1 meter	"	2.70	5.926		10,900	315		11,215	12,500
2100	Lights, underwater, 12 volt, with transformer, 300 watt	1 Elec	1	8		365	490		855	1,125
2200	110 volt, 500 watt, standard		1	8		291	490		781	1,050
2400	Low water cutoff type	↓	1	8	↓	310	490		800	1,075
2800	Heaters, see Section 23 52 28.10									

13 18 Ice Rinks

13 18 13 – Ice Rink Floor Systems

13 18 13.50 Ice Skating		Crew	Daily Output	Labor-Hours	Unit	Material	2020 Bare Costs Labor	Equipment	Total	Total Incl O&P
0010	**ICE SKATING** Equipment incl. refrigeration, plumbing & cooling									
0020	coils & concrete slab, 85' x 200' rink									
0300	55° system, 5 mos., 100 ton				Total	591,000			591,000	650,000
0700	90° system, 12 mos., 135 ton				"	683,000			683,000	751,500
1200	Subsoil heating system (recycled from compressor), 85' x 200'	Q-7	.27	119	Ea.	41,400	7,400		48,800	56,500
1300	Subsoil insulation, 2 lb. polystyrene with vapor barrier, 85' x 200'	2 Carp	.14	114	"	30,900	6,075		36,975	43,100

13 18 16 – Ice Rink Dasher Boards

13 18 16.50 Ice Rink Dasher Boards		Crew	Daily Output	Labor-Hours	Unit	Material	2020 Bare Costs Labor	Equipment	Total	Total Incl O&P
0010	**ICE RINK DASHER BOARDS**									
1000	Dasher boards, 1/2" H.D. polyethylene faced steel frame, 3' acrylic									
1020	screen at sides, 5' acrylic ends, 85' x 200'	F-5	.06	533	Ea.	144,500	28,600		173,100	202,000
1100	Fiberglass & aluminum construction, same sides and ends	"	.06	533	"	162,000	28,600		190,600	221,000

13 21 Controlled Environment Rooms

13 21 13 – Clean Rooms

13 21 13.50 Clean Room Components		Crew	Daily Output	Labor-Hours	Unit	Material	2020 Bare Costs Labor	Equipment	Total	Total Incl O&P
0010	**CLEAN ROOM COMPONENTS**									
1100	Clean room, soft wall, 12' x 12', Class 100	1 Carp	.18	44.444	Ea.	16,500	2,350		18,850	21,700
1110	Class 1,000		.18	44.444		16,800	2,350		19,150	22,100
1120	Class 10,000		.21	38.095		15,100	2,025		17,125	19,700
1130	Class 100,000		.21	38.095		12,800	2,025		14,825	17,200
2800	Ceiling grid support, slotted channel struts 4'-0" OC, ea. way				S.F.				5.90	6.50
3000	Ceiling panel, vinyl coated foil on mineral substrate									
3020	Sealed, non-perforated				S.F.				3.72	4.13
4000	Ceiling panel seal, silicone sealant, 150 L.F./gal.	1 Carp	150	.053	L.F.	.30	2.83		3.13	4.59
4100	Two sided adhesive tape	"	240	.033	"	.13	1.77		1.90	2.80
4200	Clips, one per panel				Ea.	1			1	1.10
6000	HEPA filter, 2' x 4', 99.97% eff., 3" dp beveled frame (silicone seal)					360			360	395
6040	6" deep skirted frame (channel seal)					500			500	550
6100	99.99% efficient, 3" deep beveled frame (silicone seal)					440			440	485
6140	6" deep skirted frame (channel seal)					485			485	530
6200	99.999% efficient, 3" deep beveled frame (silicone seal)					525			525	580
6240	6" deep skirted frame (channel seal)					565			565	620
7000	Wall panel systems, including channel strut framing									
7020	Polyester coated aluminum, particle board				S.F.				18.20	18.20
7100	Porcelain coated aluminum, particle board								32	32
7400	Wall panel support, slotted channel struts, to 12' high								16.35	16.35

13 21 26 – Cold Storage Rooms

13 21 26.50 Refrigeration		Crew	Daily Output	Labor-Hours	Unit	Material	2020 Bare Costs Labor	Equipment	Total	Total Incl O&P
0010	**REFRIGERATION**									
0020	Curbs, 12" high, 4" thick, concrete	2 Carp	58	.276	L.F.	3.30	14.65		17.95	25.50
1000	Doors, see Section 08 34 13.10									
2400	Finishes, 2 coat Portland cement plaster, 1/2" thick	1 Plas	48	.167	S.F.	.80	8.10		8.90	13
2500	For galvanized reinforcing mesh, add	1 Lath	335	.024		.61	1.24		1.85	2.49
2700	3/16" thick latex cement	1 Plas	88	.091		1.50	4.42		5.92	8.25
2900	For glass cloth reinforced ceilings, add	"	450	.018		.64	.86		1.50	1.99
3100	Fiberglass panels, 1/8" thick	1 Carp	149.45	.054		2.08	2.85		4.93	6.55
3200	Polystyrene, plastic finish ceiling, 1" thick		274	.029		2.84	1.55		4.39	5.45
3400	2" thick		274	.029		3.19	1.55		4.74	5.85
3500	4" thick		219	.037		4.05	1.94		5.99	7.40

13 21 26.50 Refrigeration	Crew	Daily Output	Labor-Hours	Unit	Material	2020 Bare Costs Labor	Equipment	Total	Total Incl O&P
3800	Floors, concrete, 4" thick	1 Cefi	93	.086	S.F.	1.67	4.30		5.97	8.15
3900	6" thick	"	85	.094	↓	2.62	4.70		7.32	9.80
4000	Insulation, 1" to 6" thick, cork				B.F.	1.50			1.50	1.65
4100	Urethane					.62			.62	.68
4300	Polystyrene, regular					.62			.62	.68
4400	Bead board					.29			.29	.32
4600	Installation of above, add per layer	2 Carp	657.60	.024	S.F.	.44	1.29		1.73	2.42
4700	Wall and ceiling juncture		298.90	.054	L.F.	1.41	2.85		4.26	5.80
4900	Partitions, galvanized sandwich panels, 4" thick, stock		219.20	.073	S.F.	10.10	3.88		13.98	16.95
5000	Aluminum or fiberglass	↓	219.20	.073	"	6.45	3.88		10.33	12.95
5200	Prefab walk-in, 7'-6" high, aluminum, incl. refrigeration, door & floor									
5210	not incl. partitions, 6' x 6'	2 Carp	54.80	.292	SF Flr.	105	15.50		120.50	140
5500	10' x 10'		82.20	.195		85	10.35		95.35	109
5700	12' x 14'		109.60	.146		76	7.75		83.75	95
5800	12' x 20'	↓	109.60	.146		115	7.75		122.75	139
6100	For 8'-6" high, add					5%				
6300	Rule of thumb for complete units, w/o doors & refrigeration, cooler	2 Carp	146	.110		166	5.80		171.80	192
6400	Freezer		109.60	.146	↓	113	7.75		120.75	136
6600	Shelving, plated or galvanized, steel wire type		360	.044	SF Hor.	8.35	2.36		10.71	12.70
6700	Slat shelf type		375	.043		17.65	2.27		19.92	23
6900	For stainless steel shelving, add	↓			↓	300%				
7000	Vapor barrier, on wood walls	2 Carp	1644	.010	S.F.	.19	.52		.71	.99
7200	On masonry walls	"	1315	.012	"	.33	.65		.98	1.33
7500	For air curtain doors, see Section 23 34 33.10									

13 24 Special Activity Rooms
13 24 16 – Saunas

13 24 16.50 Saunas and Heaters	Crew	Daily Output	Labor-Hours	Unit	Material	2020 Bare Costs Labor	Equipment	Total	Total Incl O&P
0010	**SAUNAS AND HEATERS**									
0020	Prefabricated, incl. heater & controls, 7' high, 6' x 4', C/C	L-7	2.20	12.727	Ea.	5,250	650		5,900	6,750
0050	6' x 4', C/P		2	14		4,325	715		5,040	5,850
0400	6' x 5', C/C		2	14		5,300	715		6,015	6,900
0450	6' x 5', C/P		2	14		5,225	715		5,940	6,825
0600	6' x 6', C/C		1.80	15.556		6,950	795		7,745	8,850
0650	6' x 6', C/P		1.80	15.556		5,525	795		6,320	7,275
0800	6' x 9', C/C		1.60	17.500		7,325	895		8,220	9,400
0850	6' x 9', C/P		1.60	17.500		6,500	895		7,395	8,500
1000	8' x 12', C/C		1.10	25.455		11,500	1,300		12,800	14,700
1050	8' x 12', C/P		1.10	25.455		8,575	1,300		9,875	11,400
1200	8' x 8', C/C		1.40	20		9,600	1,025		10,625	12,100
1250	8' x 8', C/P		1.40	20		7,200	1,025		8,225	9,450
1400	8' x 10', C/C		1.20	23.333		8,925	1,200		10,125	11,600
1450	8' x 10', C/P		1.20	23.333		8,375	1,200		9,575	11,000
1600	10' x 12', C/C		1	28		13,500	1,425		14,925	17,000
1650	10' x 12', C/P	↓	1	28		13,500	1,425		14,925	17,100
2500	Heaters only (incl. above), wall mounted, to 200 C.F.					1,025			1,025	1,125
2750	To 300 C.F.					1,150			1,150	1,250
3000	Floor standing, to 720 C.F., 10,000 watts, w/controls	1 Elec	3	2.667		2,575	164		2,739	3,100
3250	To 1,000 C.F., 16,000 watts	"	3	2.667	↓	3,725	164		3,889	4,350

13 24 26.50 Steam Baths and Components	Crew	Daily Output	Labor-Hours	Unit	Material	2020 Bare Costs Labor	Equipment	Total	Total Incl O&P
0010 **STEAM BATHS AND COMPONENTS**									
0020 Heater, timer & head, single, to 140 C.F.	1 Plum	1.20	6.667	Ea.	2,350	430		2,780	3,250
0500 To 300 C.F.		1.10	7.273		2,700	470		3,170	3,650
1000 Commercial size, with blow-down assembly, to 800 C.F.		.90	8.889		5,800	575		6,375	7,250
1500 To 2,500 C.F.		.80	10		8,200	645		8,845	10,000
2000 Multiple, motels, apts., 2 baths, w/blow-down assm., 500 C.F.	Q-1	1.30	12.308		6,650	715		7,365	8,400
2500 4 baths	"	.70	22.857		10,500	1,325		11,825	13,500
2700 Conversion unit for residential tub, including door					4,050			4,050	4,450

	Crew	Daily Output	Labor-Hours	Unit	Material	Labor	Equipment	Total	Total Incl O&P
0010 **COMFORT STATIONS** Prefab., stock, w/doors, windows & fixt.									
0100 Not incl. interior finish or electrical									
0300 Mobile, on steel frame, 2 unit				S.F.	190			190	209
0350 7 unit					320			320	350
0400 Permanent, including concrete slab, 2 unit	B-12J	50	.320		251	16.20	16.95	284.15	320
0500 6 unit	"	43	.372		201	18.85	19.70	239.55	271
0600 Alternate pricing method, mobile, 2 fixture				Fixture	6,675			6,675	7,350
0650 7 fixture					11,600			11,600	12,800
0700 Permanent, 2 unit	B-12J	.70	22.857		20,700	1,150	1,200	23,050	25,900
0750 6 unit	"	.50	32		17,800	1,625	1,700	21,125	23,900

13 34 23.16 Fabricated Control Booths

	Crew	Daily Output	Labor-Hours	Unit	Material	Labor	Equipment	Total	Total Incl O&P
0010 **FABRICATED CONTROL BOOTHS**									
0100 Guard House, prefab conc. w/bullet resistant doors & windows, roof & wiring									
0110 8' x 8', Level III	L-10	1	24	Ea.	54,500	1,400	470	56,370	62,500
0120 8' x 8', Level IV	"	1	24	"	72,000	1,400	470	73,870	81,500

	Crew	Daily Output	Labor-Hours	Unit	Material	Labor	Equipment	Total	Total Incl O&P
0010 **CATHODIC PROTECTION FOR UNDERGROUND STORAGE TANKS**									
1000 Anodes, magnesium type, 9 #	R-15	18.50	2.595	Ea.	38.50	156	15.05	209.55	291
1010 17 #		13	3.692		81.50	222	21.50	325	445
1020 32 #		10	4.800		123	288	28	439	595
1030 48 #		7.20	6.667		160	400	38.50	598.50	815
1100 Graphite type w/epoxy cap, 3" x 60" (32 #)	R-22	8.40	4.438		148	249		397	535
1110 4" x 80" (68 #)		6	6.213		246	350		596	790
1120 6" x 72" (80 #)		5.20	7.169		1,525	405		1,930	2,275
1130 6" x 36" (45 #)		9.60	3.883		770	218		988	1,175
2000 Rectifiers, silicon type, air cooled, 28 V/10 A	R-19	3.50	5.714		2,325	350		2,675	3,075
2010 20 V/20 A		3.50	5.714		2,325	350		2,675	3,100
2100 Oil immersed, 28 V/10 A		3	6.667		2,875	410		3,285	3,775
2110 20 V/20 A		3	6.667		3,250	410		3,660	4,175
3000 Anode backfill, coke breeze	R-22	3850	.010	Lb.	.30	.54		.84	1.14
4000 Cable, HMWPE, No. 8		2.40	15.533	M.L.F.	605	870		1,475	1,975
4010 No. 6		2.40	15.533		820	870		1,690	2,200
4020 No. 4		2.40	15.533		1,225	870		2,095	2,650

For customer support on your Mechanical Costs with RSMeans Data, call 800.448.8182.

81

13 47 Facility Protection

13 47 13 – Cathodic Protection

13 47 13.16 Cathodic Prot. for Underground Storage Tanks		Crew	Daily Output	Labor-Hours	Unit	Material	2020 Bare Costs Labor	Equipment	Total	Total Incl O&P
4030	No. 2	R-22	2.40	15.533	M.L.F.	1,900	870		2,770	3,400
4040	No. 1		2.20	16.945		2,575	950		3,525	4,275
4050	No. 1/0		2.20	16.945		3,350	950		4,300	5,100
4060	No. 2/0		2.20	16.945		5,250	950		6,200	7,200
4070	No. 4/0	↓	2	18.640	↓	7,225	1,050		8,275	9,500
5000	Test station, 7 terminal box, flush curb type w/lockable cover	R-19	12	1.667	Ea.	79.50	102		181.50	241
5010	Reference cell, 2" diam. PVC conduit, cplg., plug, set flush	"	4.00	4.167	"	158	256		414	555

Estimating Tips
General

- Many products in Division 14 will require some type of support or blocking for installation not included with the item itself. Examples are supports for conveyors or tube systems, attachment points for lifts, and footings for hoists or cranes. Add these supports in the appropriate division.

14 10 00 Dumbwaiters
14 20 00 Elevators

- Dumbwaiters and elevators are estimated and purchased in a method similar to buying a car. The manufacturer has a base unit with standard features. Added to this base unit price will be whatever options the owner or specifications require. Increased load capacity, additional vertical travel, additional stops, higher speed, and cab finish options are items to be considered. When developing an estimate for dumbwaiters and elevators, remember that some items needed by the installers may have to be included as part of the general contract.

Examples are:
- □ shaftway
- □ rail support brackets
- □ machine room
- □ electrical supply
- □ sill angles
- □ electrical connections
- □ pits
- □ roof penthouses
- □ pit ladders

Check the job specifications and drawings before pricing.

- Installation of elevators and handicapped lifts in historic structures can require significant additional costs. The associated structural requirements may involve cutting into and repairing finishes, moldings, flooring, etc. The estimator must account for these special conditions.

14 30 00 Escalators and Moving Walks

- Escalators and moving walks are specialty items installed by specialty contractors. There are numerous options associated with these items. For specific options, contact a manufacturer or contractor. In a method similar to estimating dumbwaiters and elevators, you should verify the extent of general contract work and add items as necessary.

14 40 00 Lifts
14 90 00 Other Conveying Equipment

- Products such as correspondence lifts, chutes, and pneumatic tube systems, as well as other items specified in this subdivision, may require trained installers. The general contractor might not have any choice as to who will perform the installation or when it will be performed. Long lead times are often required for these products, making early decisions in scheduling necessary.

Reference Numbers

Reference numbers are shown at the beginning of some major classifications. These numbers refer to related items in the Reference Section. The reference information may be an estimating procedure, an alternate pricing method, or technical information.

Note: Not all subdivisions listed here necessarily appear. ■

Same Data. Simplified.

Enjoy the convenience and efficiency of accessing your costs anywhere:

- **Skip the multiplier** by setting your location
- **Quickly search**, edit, favorite and share costs
- **Stay on top of price changes** with automatic updates

Discover more at rsmeans.com/online

14 91 Facility Chutes

14 91 33 – Laundry and Linen Chutes

14 91 33.10 Chutes

		Crew	Daily Output	Labor-Hours	Unit	Material	2020 Bare Costs Labor	Equipment	Total	Total Incl O&P
0011	**CHUTES**, linen, trash or refuse									
0050	Aluminized steel, 16 ga., 18" diameter	2 Shee	3.50	4.571	Floor	1,800	285		2,085	2,425
0100	24" diameter		3.20	5		2,000	310		2,310	2,675
0200	30" diameter		3	5.333		2,375	330		2,705	3,100
0300	36" diameter		2.80	5.714		2,725	355		3,080	3,550
0400	Galvanized steel, 16 ga., 18" diameter		3.50	4.571		1,075	285		1,360	1,625
0500	24" diameter		3.20	5		1,225	310		1,535	1,825
0600	30" diameter		3	5.333		1,375	330		1,705	2,000
0700	36" diameter		2.80	5.714		1,625	355		1,980	2,325
0800	Stainless steel, 18" diameter		3.50	4.571		3,375	285		3,660	4,150
0900	24" diameter		3.20	5		3,250	310		3,560	4,050
1000	30" diameter		3	5.333		3,975	330		4,305	4,875
1005	36" diameter		2.80	5.714	▼	4,200	355		4,555	5,175
1200	Linen chute bottom collector, aluminized steel		4	4	Ea.	1,525	249		1,774	2,050
1300	Stainless steel		4	4		1,950	249		2,199	2,525
1500	Refuse, bottom hopper, aluminized steel, 18" diameter		3	5.333		1,150	330		1,480	1,750
1600	24" diameter		3	5.333		1,350	330		1,680	1,975
1800	36" diameter	▼	3	5.333	▼	2,600	330		2,930	3,375

14 91 82 – Trash Chutes

14 91 82.10 Trash Chutes and Accessories

		Crew	Daily Output	Labor-Hours	Unit	Material	2020 Bare Costs Labor	Equipment	Total	Total Incl O&P
0010	**TRASH CHUTES AND ACCESSORIES**									
2900	Package chutes, spiral type, minimum	2 Shee	4.50	3.556	Floor	2,575	222		2,797	3,175
3000	Maximum	"	1.50	10.667	"	6,750	665		7,415	8,425

14 92 Pneumatic Tube Systems

14 92 10 – Conventional, Automatic and Computer Controlled Pneumatic Tube Systems

14 92 10.10 Pneumatic Tube Systems

		Crew	Daily Output	Labor-Hours	Unit	Material	2020 Bare Costs Labor	Equipment	Total	Total Incl O&P
0010	**PNEUMATIC TUBE SYSTEMS**									
0020	100' long, single tube, 2 stations, stock									
0100	3" diameter	2 Stpi	.12	133	Total	3,675	8,750		12,425	17,200
0300	4" diameter	"	.09	178	"	4,600	11,700		16,300	22,500
0400	Twin tube, two stations or more, conventional system									
0600	2-1/2" round	2 Stpi	62.50	.256	L.F.	43	16.80		59.80	72
0700	3" round		46	.348		40	23		63	78
0900	4" round		49.60	.323		50.50	21		71.50	87
1000	4" x 7" oval		37.60	.426	▼	99.50	28		127.50	151
1050	Add for blower		2	8	System	5,200	525		5,725	6,500
1110	Plus for each round station, add		7.50	2.133	Ea.	1,375	140		1,515	1,700
1150	Plus for each oval station, add		7.50	2.133	"	1,375	140		1,515	1,700
1200	Alternate pricing method: base cost, economy model		.75	21.333	Total	5,850	1,400		7,250	8,525
1300	Custom model		.25	64	"	11,700	4,200		15,900	19,100
1500	Plus total system length, add, for economy model		93.40	.171	L.F.	8.25	11.25		19.50	26
1600	For custom model		37.60	.426	"	25	28		53	69
1800	Completely automatic system, 4" round, 15 to 50 stations		.29	55.172	Station	22,600	3,625		26,225	30,200
2200	51 to 144 stations		.32	50		15,600	3,275		18,875	22,100
2400	6" round or 4" x 7" oval, 15 to 50 stations		.24	66.667		27,600	4,375		31,975	37,000
2800	51 to 144 stations	▼	.23	69.565	▼	21,100	4,550		25,650	30,000

Estimating Tips
22 10 00 Plumbing Piping and Pumps

This subdivision is primarily basic pipe and related materials. The pipe may be used by any of the mechanical disciplines, i.e., plumbing, fire protection, heating, and air conditioning.

Note: CPVC plastic piping approved for fire protection is located in 21 11 13.

- The labor adjustment factors listed in Subdivision 22 01 02.20 apply throughout Divisions 21, 22, and 23. CAUTION: the correct percentage may vary for the same items. For example, the percentage add for the basic pipe installation should be based on the maximum height that the installer must install for that particular section. If the pipe is to be located 14' above the floor but it is suspended on threaded rod from beams, the bottom flange of which is 18' high (4' rods), then the height is actually 18' and the add is 20%. The pipe cover, however, does not have to go above the 14' and so the add should be 10%.

- Most pipe is priced first as straight pipe with a joint (coupling, weld, etc.) every 10' and a hanger usually every 10'. There are exceptions with hanger spacing such as for cast iron pipe (5') and plastic pipe (3 per 10'). Following each type of pipe there are several lines listing sizes and the amount to be subtracted to delete couplings and hangers. This is for pipe that is to be buried or supported together on trapeze hangers. The reason that the couplings are deleted is that these runs are usually long, and frequently longer lengths of pipe are used. By deleting the couplings, the estimator is expected to look up and add back the correct reduced number of couplings.

- When preparing an estimate, it may be necessary to approximate the fittings. Fittings usually run between 25% and 50% of the cost of the pipe. The lower percentage is for simpler runs, and the higher number is for complex areas, such as mechanical rooms.

- For historic restoration projects, the systems must be as invisible as possible, and pathways must be sought for pipes, conduit, and ductwork. While installations in accessible spaces (such as basements and attics) are relatively straightforward to estimate, labor costs may be more difficult to determine when delivery systems must be concealed.

22 40 00 Plumbing Fixtures

- Plumbing fixture costs usually require two lines: the fixture itself and its "rough-in, supply, and waste."

- In the Assemblies Section (Plumbing D2010) for the desired fixture, the System Components Group at the center of the page shows the fixture on the first line. The rest of the list (fittings, pipe, tubing, etc.) will total up to what we refer to in the Unit Price section as "Rough-in, supply, waste, and vent." Note that for most fixtures we allow a nominal 5' of tubing to reach from the fixture to a main or riser.

- Remember that gas- and oil-fired units need venting.

Reference Numbers

Reference numbers are shown at the beginning of some major classifications. These numbers refer to related items in the Reference Section. The reference information may be an estimating procedure, an alternate pricing method, or technical information.

Note: Not all subdivisions listed here necessarily appear. ■

Same Data. Simplified.

Enjoy the convenience and efficiency of accessing your costs anywhere:

- **Skip the multiplier** by setting your location
- **Quickly search,** edit, favorite and share costs
- **Stay on top of price changes** with automatic updates

Discover more at rsmeans.com/online

22 01 Operation and Maintenance of Plumbing

22 01 02 – Labor Adjustments

22 01 02.10 Boilers, General		Crew	Daily Output	Labor-Hours	Unit	Material	2020 Bare Costs Labor	Equipment	Total	Total Incl O&P
0010	**BOILERS, GENERAL**, Prices do not include flue piping, elec. wiring,									
0020	gas or oil piping, boiler base, pad, or tankless unless noted									
0100	Boiler H.P.: 10 KW = 34 lb./steam/hr. = 33,475 BTU/hr.									
0150	To convert SFR to BTU rating: Hot water, 150 x SFR;									
0160	Forced hot water, 180 x SFR; steam, 240 x SFR									

22 01 02.20 Labor Adjustment Factors

		Crew	Daily Output	Labor-Hours	Unit	Material	2020 Bare Costs Labor	Equipment	Total	Total Incl O&P
0010	**LABOR ADJUSTMENT FACTORS** (For Div. 21, 22 and 23) R220102-20									
0100	Labor factors: The below are reasonable suggestions, but									
0110	each project must be evaluated for its own peculiarities, and									
0120	the adjustments be increased or decreased depending on the									
0130	severity of the special conditions.									
1000	Add to labor for elevated installation (Above floor level)									
1080	10' to 14.5' high R221113-70						10%			
1100	15' to 19.5' high						20%			
1120	20' to 24.5' high						25%			
1140	25' to 29.5' high						35%			
1160	30' to 34.5' high						40%			
1180	35' to 39.5' high						50%			
1200	40' and higher						55%			
2000	Add to labor for crawl space									
2100	3' high						40%			
2140	4' high						30%			
3000	Add to labor for multi-story building									
3010	For new construction (No elevator available)									
3100	Add for floors 3 thru 10						5%			
3110	Add for floors 11 thru 15						10%			
3120	Add for floors 16 thru 20						15%			
3130	Add for floors 21 thru 30						20%			
3140	Add for floors 31 and up						30%			
3170	For existing structure (Elevator available)									
3180	Add for work on floor 3 and above						2%			
4000	Add to labor for working in existing occupied buildings									
4100	Hospital						35%			
4140	Office building						25%			
4180	School						20%			
4220	Factory or warehouse						15%			
4260	Multi dwelling						15%			
5000	Add to labor, miscellaneous									
5100	Cramped shaft						35%			
5140	Congested area						15%			
5180	Excessive heat or cold						30%			
9000	Labor factors: The above are reasonable suggestions, but									
9010	each project should be evaluated for its own peculiarities.									
9100	Other factors to be considered are:									
9140	Movement of material and equipment through finished areas									
9180	Equipment room									
9220	Attic space									
9260	No service road									
9300	Poor unloading/storage area									
9340	Congested site area/heavy traffic									

22 05 Common Work Results for Plumbing

22 05 05 – Selective Demolition for Plumbing

22 05 05.10 Plumbing Demolition		Crew	Daily Output	Labor-Hours	Unit	Material	2020 Bare Costs Labor	Equipment	Total	Total Incl O&P
0010	**PLUMBING DEMOLITION** R220105-10									
0400	Air compressor, up thru 2 HP	Q-1	10	1.600	Ea.		93		93	139
0410	3 HP thru 7-1/2 HP R024119-10		5.60	2.857			166		166	248
0420	10 HP thru 15 HP		1.40	11.429			665		665	990
0430	20 HP thru 30 HP	Q-2	1.30	18.462			1,100		1,100	1,650
0500	Backflow preventer, up thru 2" diameter	1 Plum	17	.471			30.50		30.50	45.50
0510	2-1/2" thru 3" diameter	Q-1	10	1.600			93		93	139
0520	4" thru 6" diameter	"	5	3.200			186		186	278
0530	8" thru 10" diameter	Q-2	3	8			480		480	720
1900	Piping fittings, single connection, up thru 1-1/2" diameter	1 Plum	30	.267			17.20		17.20	25.50
1910	2" thru 4" diameter		14	.571			37		37	55
1980	Pipe hanger/support removal		80	.100			6.45		6.45	9.65
2000	Piping, metal, up thru 1-1/2" diameter		200	.040	L.F.		2.58		2.58	3.86
2050	2" thru 3-1/2" diameter		150	.053			3.44		3.44	5.15
2100	4" thru 6" diameter	2 Plum	100	.160			10.30		10.30	15.45
2150	8" thru 14" diameter	"	60	.267			17.20		17.20	25.50
2153	16" thru 20" diameter	Q-18	70	.343			21	1.52	22.52	33
2155	24" thru 26" diameter		55	.436			26.50	1.93	28.43	42
2156	30" thru 36" diameter		40	.600			36.50	2.66	39.16	58
2160	Plastic pipe with fittings, up thru 1-1/2" diameter	1 Plum	250	.032			2.06		2.06	3.09
2162	2" thru 3" diameter	"	200	.040			2.58		2.58	3.86
2164	4" thru 6" diameter	Q-1	200	.080			4.64		4.64	6.95
2166	8" thru 14" diameter		150	.107			6.20		6.20	9.25
2168	16" diameter		100	.160			9.30		9.30	13.90
2180	Pumps, all fractional horse-power		12	1.333	Ea.		77.50		77.50	116
2184	1 HP thru 5 HP		6	2.667			155		155	231
2186	7-1/2 HP thru 15 HP		2.50	6.400			370		370	555
2188	20 HP thru 25 HP	Q-2	4	6			360		360	540
2190	30 HP thru 60 HP		.80	30			1,800		1,800	2,700
2192	75 HP thru 100 HP		.60	40			2,400		2,400	3,600
2194	150 HP		.50	48			2,875		2,875	4,325
2230	Temperature maintenance cable	1 Plum	1200	.007	L.F.		.43		.43	.64
3100	Tanks, water heaters and liquid containers									
3110	Up thru 45 gallons	Q-1	22	.727	Ea.		42		42	63
3120	50 thru 120 gallons		14	1.143			66.50		66.50	99
3130	130 thru 240 gallons		7.60	2.105			122		122	183
3140	250 thru 500 gallons		5.40	2.963			172		172	257
3150	600 thru 1,000 gallons	Q-2	1.60	15			900		900	1,350
3160	1,100 thru 2,000 gallons		.70	34.286			2,050		2,050	3,075
3170	2,100 thru 4,000 gallons		.50	48			2,875		2,875	4,325
9100	Valve, metal valves or strainers and similar, up thru 1-1/2" diameter	1 Stpi	28	.286			18.75		18.75	28
9110	2" thru 3" diameter	Q-1	11	1.455			84.50		84.50	126
9120	4" thru 6" diameter	"	8	2			116		116	174
9130	8" thru 14" diameter	Q-2	8	3			180		180	270
9140	16" thru 20" diameter		2	12			720		720	1,075
9150	24" diameter		1.20	20			1,200		1,200	1,800
9200	Valve, plastic, up thru 1-1/2" diameter	1 Plum	42	.190			12.30		12.30	18.35
9210	2" thru 3" diameter		15	.533			34.50		34.50	51.50
9220	4" thru 6" diameter		12	.667			43		43	64.50
9300	Vent flashing and caps		55	.145			9.35		9.35	14.05
9350	Water filter, commercial, 1" thru 1-1/2"	Q-1	2	8			465		465	695
9360	2" thru 2-1/2"	"	1.60	10			580		580	870
9400	Water heaters									

For customer support on your Mechanical Costs with RSMeans Data, call 800.448.8182.

87

22 05 Common Work Results for Plumbing

22 05 05 – Selective Demolition for Plumbing

22 05 05.10 Plumbing Demolition

		Crew	Daily Output	Labor-Hours	Unit	Material	2020 Bare Costs Labor	2020 Bare Costs Equipment	Total	Total Incl O&P
9410	Up thru 245 GPH	Q-1	2.40	6.667	Ea.		385		385	580
9420	250 thru 756 GPH		1.60	10			580		580	870
9430	775 thru 1,640 GPH	↓	.80	20			1,150		1,150	1,725
9440	1,650 thru 4,000 GPH	Q-2	.50	48			2,875		2,875	4,325
9470	Water softener	Q-1	2	8	↓		465		465	695

22 05 23 – General-Duty Valves for Plumbing Piping

22 05 23.10 Valves, Brass

		Crew	Daily Output	Labor-Hours	Unit	Material	2020 Bare Costs Labor	2020 Bare Costs Equipment	Total	Total Incl O&P
0010	**VALVES, BRASS**									
0032	For motorized valves, see Section 23 09 53.10									
0500	Gas cocks, threaded									
0510	1/4"	1 Plum	26	.308	Ea.	15.80	19.85		35.65	47
0520	3/8"		24	.333		15.80	21.50		37.30	49.50
0530	1/2"		24	.333		13.40	21.50		34.90	47
0540	3/4"		22	.364		18.10	23.50		41.60	55
0550	1"		19	.421		34	27		61	78
0560	1-1/4"		15	.533		57	34.50		91.50	114
0570	1-1/2"		13	.615		85.50	39.50		125	154
0580	2"	↓	11	.727	↓	111	47		158	192
0672	For larger sizes use lubricated plug valve, Section 23 05 23.70									

22 05 23.20 Valves, Bronze

		Crew	Daily Output	Labor-Hours	Unit	Material	2020 Bare Costs Labor	2020 Bare Costs Equipment	Total	Total Incl O&P
0010	**VALVES, BRONZE** R220523-90									
1020	Angle, 150 lb., rising stem, threaded									
1030	1/8"	1 Plum	24	.333	Ea.	142	21.50		163.50	188
1040	1/4"		24	.333		162	21.50		183.50	210
1050	3/8"		24	.333		165	21.50		186.50	213
1060	1/2"		22	.364		169	23.50		192.50	221
1070	3/4"		20	.400		230	26		256	291
1080	1"		19	.421		315	27		342	385
1090	1-1/4"		15	.533		355	34.50		389.50	445
1100	1-1/2"	↓	13	.615	↓	525	39.50		564.50	635
1102	Soldered same price as threaded									
1110	2"	1 Plum	11	.727	Ea.	845	47		892	1,000
1300	Ball									
1304	Soldered									
1312	3/8"	1 Plum	21	.381	Ea.	19.70	24.50		44.20	58
1316	1/2"		18	.444		18.70	28.50		47.20	63.50
1320	3/4"		17	.471		33	30.50		63.50	82
1324	1"		15	.533		44	34.50		78.50	100
1328	1-1/4"		13	.615		46	39.50		85.50	110
1332	1-1/2"		11	.727		86.50	47		133.50	166
1336	2"		9	.889		92	57.50		149.50	187
1340	2-1/2"		7	1.143		485	73.50		558.50	645
1344	3"	↓	5	1.600	↓	570	103		673	780
1350	Single union end									
1358	3/8"	1 Plum	21	.381	Ea.	26.50	24.50		51	65.50
1362	1/2"		18	.444		29.50	28.50		58	75.50
1366	3/4"		17	.471		56	30.50		86.50	108
1370	1"		15	.533		65.50	34.50		100	124
1374	1-1/4"		13	.615		102	39.50		141.50	172
1378	1-1/2"		11	.727		145	47		192	230
1382	2"	↓	9	.889	↓	194	57.50		251.50	299
1398	Threaded, 150 psi									

22 05 23.20 Valves, Bronze	Crew	Daily Output	Labor-Hours	Unit	Material	2020 Bare Costs Labor	Equipment	Total	Total Incl O&P
1400	1/4"	1 Plum	24	.333	Ea.	19.55	21.50		41.05	53.50
1430	3/8"		24	.333		18.25	21.50		39.75	52
1450	1/2"		22	.364		18	23.50		41.50	55
1460	3/4"		20	.400		35.50	26		61.50	77.50
1470	1"		19	.421		32	27		59	75.50
1480	1-1/4"		15	.533		50	34.50		84.50	107
1490	1-1/2"		13	.615		75.50	39.50		115	143
1500	2"		11	.727		83.50	47		130.50	162
1510	2-1/2"		9	.889		264	57.50		321.50	375
1520	3"		8	1		425	64.50		489.50	560
1522	Solder the same price as threaded									
1600	Butterfly, 175 psi, full port, solder or threaded ends									
1610	Stainless steel disc and stem									
1620	1/4"	1 Plum	24	.333	Ea.	24	21.50		45.50	58.50
1630	3/8"		24	.333		16.85	21.50		38.35	50.50
1640	1/2"		22	.364		21	23.50		44.50	58.50
1650	3/4"		20	.400		29	26		55	70.50
1660	1"		19	.421		36	27		63	80
1670	1-1/4"		15	.533		69	34.50		103.50	128
1680	1-1/2"		13	.615		73.50	39.50		113	141
1690	2"		11	.727		93	47		140	173
1750	Check, swing, class 150, regrinding disc, threaded									
1800	1/8"	1 Plum	24	.333	Ea.	82.50	21.50		104	123
1830	1/4"		24	.333		76.50	21.50		98	117
1840	3/8"		24	.333		71	21.50		92.50	111
1850	1/2"		24	.333		77	21.50		98.50	117
1860	3/4"		20	.400		94	26		120	142
1870	1"		19	.421		178	27		205	237
1880	1-1/4"		15	.533		210	34.50		244.50	283
1890	1-1/2"		13	.615		300	39.50		339.50	390
1900	2"		11	.727		380	47		427	490
1910	2-1/2"	Q-1	15	1.067		925	62		987	1,125
1920	3"	"	13	1.231		1,275	71.50		1,346.50	1,500
2000	For 200 lb., add					5%	10%			
2040	For 300 lb., add					15%	15%			
2060	Check swing, 300 lb., lead free unless noted, sweat, 3/8" size	1 Plum	24	.333	Ea.	96	21.50		117.50	138
2070	1/2"		24	.333		96	21.50		117.50	138
2080	3/4"		20	.400		130	26		156	182
2090	1"		19	.421		191	27		218	251
2100	1-1/4"		15	.533		269	34.50		303.50	350
2110	1-1/2"		13	.615		315	39.50		354.50	410
2120	2"		11	.727		465	47		512	580
2130	2-1/2", not lead free	Q-1	15	1.067		750	62		812	920
2140	3", not lead free	"	13	1.231		950	71.50		1,021.50	1,150
2350	Check, lift, class 150, horizontal composition disc, threaded									
2430	1/4"	1 Plum	24	.333	Ea.	162	21.50		183.50	211
2440	3/8"		24	.333		206	21.50		227.50	259
2450	1/2"		24	.333		179	21.50		200.50	229
2460	3/4"		20	.400		219	26		245	280
2470	1"		19	.421		320	27		347	390
2480	1-1/4"		15	.533		420	34.50		454.50	510
2490	1-1/2"		13	.615		500	39.50		539.50	610
2500	2"		11	.727		845	47		892	1,000

For customer support on your Mechanical Costs with RSMeans Data, call 800.448.8182.

89

22 05 23.20 Valves, Bronze		Crew	Daily Output	Labor-Hours	Unit	Material	2020 Bare Costs Labor	Equipment	Total	Total Incl O&P
2850	Gate, N.R.S., soldered, 125 psi									
2900	3/8"	1 Plum	24	.333	Ea.	75.50	21.50		97	115
2920	1/2"		24	.333		60.50	21.50		82	98.50
2940	3/4"		20	.400		74	26		100	120
2950	1"		19	.421		78.50	27		105.50	127
2960	1 1/4"		15	.533		152	34.50		186.50	219
2970	1-1/2"		13	.615		184	39.50		223.50	263
2980	2"		11	.727		213	47		260	305
2990	2-1/2"	Q-1	15	1.067		515	62		577	660
3000	3"	"	13	1.231		570	71.50		641.50	735
3350	Threaded, class 150									
3410	1/4"	1 Plum	24	.333	Ea.	81	21.50		102.50	121
3420	3/8"		24	.333		98.50	21.50		120	140
3430	1/2"		24	.333		78.50	21.50		100	119
3440	3/4"		20	.400		104	26		130	153
3450	1"		19	.421		134	27		161	189
3460	1-1/4"		15	.533		137	34.50		171.50	203
3470	1-1/2"		13	.615		269	39.50		308.50	355
3480	2"		11	.727		234	47		281	325
3490	2-1/2"	Q-1	15	1.067		700	62		762	865
3500	3"	"	13	1.231		865	71.50		936.50	1,050
3600	Gate, flanged, 150 lb.									
3610	1"	1 Plum	7	1.143	Ea.	1,175	73.50		1,248.50	1,375
3620	1-1/2"		6	1.333		1,450	86		1,536	1,725
3630	2"		5	1.600		2,500	103		2,603	2,900
3634	2-1/2"	Q-1	5	3.200		1,975	186		2,161	2,450
3640	3"	"	4.50	3.556		3,850	206		4,056	4,550
3850	Rising stem, soldered, 300 psi									
3900	3/8"	1 Plum	24	.333	Ea.	155	21.50		176.50	203
3920	1/2"		24	.333		157	21.50		178.50	205
3940	3/4"		20	.400		169	26		195	225
3950	1"		19	.421		229	27		256	293
3960	1-1/4"		15	.533		315	34.50		349.50	400
3970	1-1/2"		13	.615		385	39.50		424.50	485
3980	2"		11	.727		615	47		662	750
3990	2-1/2"	Q-1	15	1.067		1,325	62		1,387	1,550
4000	3"	"	13	1.231		2,025	71.50		2,096.50	2,350
4250	Threaded, class 150									
4310	1/4"	1 Plum	24	.333	Ea.	90	21.50		111.50	131
4320	3/8"		24	.333		90	21.50		111.50	131
4330	1/2"		24	.333		69.50	21.50		91	109
4340	3/4"		20	.400		81.50	26		107.50	128
4350	1"		19	.421		125	27		152	179
4360	1-1/4"		15	.533		132	34.50		166.50	197
4370	1-1/2"		13	.615		167	39.50		206.50	243
4380	2"		11	.727		224	47		271	315
4390	2-1/2"	Q-1	15	1.067		670	62		732	835
4400	3"	"	13	1.231		935	71.50		1,006.50	1,125
4500	For 300 psi, threaded, add					100%	15%			
4540	For chain operated type, add					15%				
4850	Globe, class 150, rising stem, threaded									
4920	1/4"	1 Plum	24	.333	Ea.	123	21.50		144.50	168
4940	3/8"		24	.333		121	21.50		142.50	166

22 05 23 – General-Duty Valves for Plumbing Piping

22 05 23.20 Valves, Bronze		Crew	Daily Output	Labor-Hours	Unit	Material	2020 Bare Costs Labor	Equipment	Total	Total Incl O&P
4950	1/2"	1 Plum	24	.333	Ea.	106	21.50		127.50	149
4960	3/4"		20	.400		172	26		198	228
4970	1"		19	.421		239	27		266	305
4980	1-1/4"		15	.533		292	34.50		326.50	370
4990	1-1/2"		13	.615		460	39.50		499.50	565
5000	2"		11	.727		690	47		737	830
5010	2-1/2"	Q-1	15	1.067		1,475	62		1,537	1,725
5020	3"	"	13	1.231		2,125	71.50		2,196.50	2,425
5120	For 300 lb. threaded, add					50%	15%			
5130	Globe, 300 lb., sweat, 3/8" size	1 Plum	24	.333	Ea.	134	21.50		155.50	180
5140	1/2"		24	.333		130	21.50		151.50	175
5150	3/4"		20	.400		187	26		213	245
5160	1"		19	.421		297	27		324	365
5170	1-1/4"		15	.533		365	34.50		399.50	450
5180	1-1/2"		13	.615		445	39.50		484.50	550
5190	2"		11	.727		1,050	47		1,097	1,225
5200	2-1/2"	Q-1	15	1.067		1,625	62		1,687	1,900
5210	3"	"	13	1.231		2,100	71.50		2,171.50	2,400
5600	Relief, pressure & temperature, self-closing, ASME, threaded									
5640	3/4"	1 Plum	28	.286	Ea.	257	18.40		275.40	310
5650	1"		24	.333		425	21.50		446.50	495
5660	1-1/4"		20	.400		805	26		831	925
5670	1-1/2"		18	.444		1,550	28.50		1,578.50	1,750
5680	2"		16	.500		1,675	32		1,707	1,900
5950	Pressure, poppet type, threaded									
6000	1/2"	1 Plum	30	.267	Ea.	90	17.20		107.20	125
6040	3/4"	"	28	.286	"	104	18.40		122.40	143
6400	Pressure, water, ASME, threaded									
6440	3/4"	1 Plum	28	.286	Ea.	119	18.40		137.40	159
6450	1"		24	.333		340	21.50		361.50	405
6460	1-1/4"		20	.400		485	26		511	575
6470	1-1/2"		18	.444		750	28.50		778.50	865
6480	2"		16	.500		1,075	32		1,107	1,250
6490	2-1/2"		15	.533		4,400	34.50		4,434.50	4,900
6900	Reducing, water pressure									
6920	300 psi to 25-75 psi, threaded or sweat									
6940	1/2"	1 Plum	24	.333	Ea.	470	21.50		491.50	550
6950	3/4"		20	.400		550	26		576	645
6960	1"		19	.421		850	27		877	975
6970	1-1/4"		15	.533		1,475	34.50		1,509.50	1,675
6980	1-1/2"		13	.615		2,225	39.50		2,264.50	2,500
6990	2"		11	.727		3,350	47		3,397	3,750
7100	For built-in by-pass or 10-35 psi, add					53			53	58.50
7700	High capacity, 250 psi to 25-75 psi, threaded									
7740	1/2"	1 Plum	24	.333	Ea.	890	21.50		911.50	1,000
7780	3/4"		20	.400		780	26		806	895
7790	1"		19	.421		1,050	27		1,077	1,225
7800	1-1/4"		15	.533		1,825	34.50		1,859.50	2,075
7810	1-1/2"		13	.615		2,675	39.50		2,714.50	3,000
7820	2"		11	.727		3,900	47		3,947	4,375
7830	2-1/2"		9	.889		5,575	57.50		5,632.50	6,200
7840	3"		8	1		6,575	64.50		6,639.50	7,350
7850	3" flanged (iron body)	Q-1	10	1.600		6,575	93		6,668	7,375

22 05 23.20 Valves, Bronze	Crew	Daily Output	Labor-Hours	Unit	Material	2020 Bare Costs Labor	Equipment	Total	Total Incl O&P
7860	4" flanged (iron body)	Q-1	8	2	Ea.	8,450	116		8,566	9,475
7920	For higher pressure, add					25%				
8000	Silent check, bronze trim									
8010	Compact wafer type, for 125 or 150 lb. flanges									
8020	1-1/2"	1 Plum	11	.727	Ea.	490	47		537	610
8021	2"	"	9	.889		520	57.50		577.50	655
8022	2-1/2"	Q-1	9	1.778		545	103		648	755
8023	3"		8	2		605	116		721	845
8024	4"		5	3.200		1,050	186		1,236	1,425
8025	5"	Q-2	6	4		1,000	241		1,241	1,450
8026	6"		5	4.800		1,200	289		1,489	1,750
8027	8"		4.50	5.333		2,150	320		2,470	2,825
8028	10"		4	6		5,850	360		6,210	7,000
8029	12"		3	8		13,300	480		13,780	15,300
8050	For 250 or 300 lb. flanges, thru 6" no change									
8051	For 8" and 10", add				Ea.	40%	10%			
8060	Full flange wafer type, 150 lb.									
8061	1"	1 Plum	14	.571	Ea.	320	37		357	405
8062	1-1/4"		12	.667		335	43		378	435
8063	1-1/2"		11	.727		395	47		442	505
8064	2"		9	.889		535	57.50		592.50	670
8065	2-1/2"	Q-1	9	1.778		600	103		703	815
8066	3"		8	2		665	116		781	905
8067	4"		5	3.200		980	186		1,166	1,350
8068	5"	Q-2	6	4		1,275	241		1,516	1,750
8069	6"	"	5	4.800		1,675	289		1,964	2,275
8080	For 300 lb., add					40%	10%			
8100	Globe type, 150 lb.									
8110	2"	1 Plum	9	.889	Ea.	620	57.50		677.50	765
8111	2-1/2"	Q-1	9	1.778		765	103		868	1,000
8112	3"		8	2		915	116		1,031	1,175
8113	4"		5	3.200		1,275	186		1,461	1,675
8114	5"	Q-2	6	4		1,550	241		1,791	2,050
8115	6"	"	5	4.800		2,150	289		2,439	2,800
8130	For 300 lb., add					20%	10%			
8140	Screwed end type, 250 lb.									
8141	1/2"	1 Plum	24	.333	Ea.	56	21.50		77.50	93.50
8142	3/4"		20	.400		56	26		82	100
8143	1"		19	.421		63.50	27		90.50	111
8144	1-1/4"		15	.533		88	34.50		122.50	148
8145	1-1/2"		13	.615		98.50	39.50		138	169
8146	2"		11	.727		133	47		180	216
8350	Tempering, water, sweat connections									
8400	1/2"	1 Plum	24	.333	Ea.	118	21.50		139.50	162
8440	3/4"	"	20	.400	"	167	26		193	222
8650	Threaded connections									
8700	1/2"	1 Plum	24	.333	Ea.	162	21.50		183.50	210
8740	3/4"		20	.400		1,050	26		1,076	1,225
8750	1"		19	.421		1,200	27		1,227	1,350
8760	1-1/4"		15	.533		1,975	34.50		2,009.50	2,225
8770	1-1/2"		13	.615		2,025	39.50		2,064.50	2,275
8780	2"		11	.727		3,000	47		3,047	3,375

22 05 23.60 Valves, Plastic		Crew	Daily Output	Labor-Hours	Unit	Material	2020 Bare Costs Labor	Equipment	Total	Total Incl O&P
0010	**VALVES, PLASTIC** R220523-90									
1100	Angle, PVC, threaded									
1110	1/4"	1 Plum	26	.308	Ea.	60	19.85		79.85	95.50
1120	1/2"		26	.308		81.50	19.85		101.35	119
1130	3/4"		25	.320		96.50	20.50		117	137
1140	1"		23	.348		117	22.50		139.50	163
1150	Ball, PVC, socket or threaded, true union									
1230	1/2"	1 Plum	26	.308	Ea.	39	19.85		58.85	72.50
1240	3/4"		25	.320		50	20.50		70.50	86
1250	1"		23	.348		55.50	22.50		78	94.50
1260	1-1/4"		21	.381		94	24.50		118.50	140
1270	1-1/2"		20	.400		86	26		112	134
1280	2"		17	.471		136	30.50		166.50	195
1290	2-1/2"	Q-1	26	.615		236	35.50		271.50	315
1300	3"		24	.667		272	38.50		310.50	355
1310	4"		20	.800		455	46.50		501.50	575
1360	For PVC, flanged, add					100%	15%			
1450	Double union 1/2"	1 Plum	26	.308		37.50	19.85		57.35	70.50
1460	3/4"		25	.320		43.50	20.50		64	79
1470	1"		23	.348		57	22.50		79.50	96.50
1480	1-1/4"		21	.381		72.50	24.50		97	116
1490	1-1/2"		20	.400		95	26		121	143
1500	2"		17	.471		109	30.50		139.50	166
1650	CPVC, socket or threaded, single union									
1700	1/2"	1 Plum	26	.308	Ea.	61.50	19.85		81.35	97
1720	3/4"		25	.320		81	20.50		101.50	120
1730	1"		23	.348		92.50	22.50		115	136
1750	1-1/4"		21	.381		144	24.50		168.50	196
1760	1-1/2"		20	.400		146	26		172	200
1770	2"		17	.471		202	30.50		232.50	268
1780	3"	Q-1	24	.667		755	38.50		793.50	890
1840	For CPVC, flanged, add					65%	15%			
1880	For true union, socket or threaded, add					50%	5%			
2050	Polypropylene, threaded									
2100	1/4"	1 Plum	26	.308	Ea.	44	19.85		63.85	78
2120	3/8"		26	.308		45.50	19.85		65.35	79.50
2130	1/2"		26	.308		39.50	19.85		59.35	73
2140	3/4"		25	.320		49	20.50		69.50	85
2150	1"		23	.348		55.50	22.50		78	94.50
2160	1-1/4"		21	.381		74	24.50		98.50	118
2170	1-1/2"		20	.400		91.50	26		117.50	140
2180	2"		17	.471		123	30.50		153.50	182
2190	3"	Q-1	24	.667		300	38.50		338.50	390
2200	4"	"	20	.800		500	46.50		546.50	620
2550	PVC, three way, socket or threaded									
2600	1/2"	1 Plum	26	.308	Ea.	57	19.85		76.85	92
2640	3/4"		25	.320		67.50	20.50		88	106
2650	1"		23	.348		81.50	22.50		104	123
2660	1-1/2"		20	.400		117	26		143	168
2670	2"		17	.471		155	30.50		185.50	217
2680	3"	Q-1	24	.667		465	38.50		503.50	570
2740	For flanged, add					60%	15%			

22 05 23.60 Valves, Plastic	Crew	Daily Output	Labor-Hours	Unit	Material	2020 Bare Costs Labor	Equipment	Total	Total Incl O&P
3150 Ball check, PVC, socket or threaded									
3200 1/4"	1 Plum	26	.308	Ea.	48	19.85		67.85	82.50
3220 3/8"		26	.308		48	19.85		67.85	82.50
3240 1/2"		26	.308		46.50	19.85		66.35	80.50
3250 3/4"		25	.320		52	20.50		72.50	88
3260 1"		23	.348		65	22.50		87.50	105
3270 1-1/4"		21	.381		124	24.50		148.50	173
3280 1-1/2"		20	.400		109	26		135	159
3290 2"		17	.471		149	30.50		179.50	209
3310 3"	Q-1	24	.667		415	38.50		453.50	515
3320 4"	"	20	.800		585	46.50		631.50	710
3360 For PVC, flanged, add					50%	15%			
3750 CPVC, socket or threaded									
3800 1/2"	1 Plum	26	.308	Ea.	66.50	19.85		86.35	103
3840 3/4"		25	.320		85.50	20.50		106	126
3850 1"		23	.348		104	22.50		126.50	148
3860 1-1/2"		20	.400		175	26		201	231
3870 2"		17	.471		227	30.50		257.50	295
3880 3"	Q-1	24	.667		700	38.50		738.50	830
3920 4"	"	20	.800		1,000	46.50		1,046.50	1,175
3930 For CPVC, flanged, add					40%	15%			
4340 Polypropylene, threaded									
4360 1/2"	1 Plum	26	.308	Ea.	49.50	19.85		69.35	84
4400 3/4"		25	.320		69	20.50		89.50	107
4440 1"		23	.348		73	22.50		95.50	114
4450 1-1/2"		20	.400		141	26		167	194
4460 2"		17	.471		179	30.50		209.50	243
4500 For polypropylene flanged, add					200%	15%			
4850 Foot valve, PVC, socket or threaded									
4900 1/2"	1 Plum	34	.235	Ea.	62	15.15		77.15	91
4930 3/4"		32	.250		70	16.10		86.10	101
4940 1"		28	.286		91.50	18.40		109.90	128
4950 1-1/4"		27	.296		173	19.10		192.10	219
4960 1-1/2"		26	.308		175	19.85		194.85	223
4970 2"		24	.333		202	21.50		223.50	254
4980 3"		20	.400		480	26		506	570
4990 4"		18	.444		845	28.50		873.50	975
5000 For flanged, add					25%	10%			
5050 CPVC, socket or threaded									
5060 1/2"	1 Plum	34	.235	Ea.	77.50	15.15		92.65	108
5070 3/4"		32	.250		99.50	16.10		115.60	134
5080 1"		28	.286		122	18.40		140.40	162
5090 1-1/4"		27	.296		194	19.10		213.10	242
5100 1-1/2"		26	.308		194	19.85		213.85	243
5110 2"		24	.333		243	21.50		264.50	299
5120 3"		20	.400		495	26		521	585
5130 4"		18	.444		895	28.50		923.50	1,025
5140 For flanged, add					25%	10%			
5280 Needle valve, PVC, threaded									
5300 1/4"	1 Plum	26	.308	Ea.	66	19.85		85.85	102
5340 3/8"		26	.308		76.50	19.85		96.35	114
5360 1/2"		26	.308		76	19.85		95.85	113
5380 For polypropylene, add					10%				

22 05 Common Work Results for Plumbing

22 05 23 – General-Duty Valves for Plumbing Piping

22 05 23.60 Valves, Plastic

		Crew	Daily Output	Labor-Hours	Unit	Material	2020 Bare Costs Labor	Equipment	Total	Total Incl O&P
5800	Y check, PVC, socket or threaded									
5820	1/2"	1 Plum	26	.308	Ea.	68	19.85		87.85	105
5840	3/4"		25	.320		74	20.50		94.50	112
5850	1"		23	.348		82	22.50		104.50	124
5860	1-1/4"		21	.381		129	24.50		153.50	179
5870	1-1/2"		20	.400		141	26		167	194
5880	2"		17	.471		175	30.50		205.50	238
5890	2-1/2"		15	.533		370	34.50		404.50	460
5900	3"	Q-1	24	.667		365	38.50		403.50	460
5910	4"	"	20	.800		610	46.50		656.50	740
5960	For PVC flanged, add					45%	15%			
6350	Y sediment strainer, PVC, socket or threaded									
6400	1/2"	1 Plum	26	.308	Ea.	56.50	19.85		76.35	91.50
6440	3/4"		24	.333		65.50	21.50		87	105
6450	1"		23	.348		67.50	22.50		90	108
6460	1-1/4"		21	.381		113	24.50		137.50	162
6470	1-1/2"		20	.400		133	26		159	185
6480	2"		17	.471		142	30.50		172.50	202
6490	2-1/2"		15	.533		345	34.50		379.50	430
6500	3"	Q-1	24	.667		340	38.50		378.50	435
6510	4"	"	20	.800		580	46.50		626.50	705
6560	For PVC, flanged, add					55%	15%			

22 05 29 – Hangers and Supports for Plumbing Piping and Equipment

22 05 29.10 Hangers & Supp. for Plumb'g/HVAC Pipe/Equip.

		Crew	Daily Output	Labor-Hours	Unit	Material	2020 Bare Costs Labor	Equipment	Total	Total Incl O&P
0010	**HANGERS AND SUPPORTS FOR PLUMB'G/HVAC PIPE/EQUIP.**									
0011	TYPE numbers per MSS-SP58									
0050	Brackets									
0060	Beam side or wall, malleable iron, TYPE 34									
0070	3/8" threaded rod size	1 Plum	48	.167	Ea.	5.10	10.75		15.85	21.50
0080	1/2" threaded rod size		48	.167		4	10.75		14.75	20.50
0090	5/8" threaded rod size		48	.167		11.70	10.75		22.45	29
0100	3/4" threaded rod size		48	.167		21.50	10.75		32.25	39.50
0110	7/8" threaded rod size		48	.167		13.95	10.75		24.70	31.50
0120	For concrete installation, add						30%			
0150	Wall, welded steel, medium, TYPE 32									
0160	0 size, 12" wide, 18" deep	1 Plum	34	.235	Ea.	183	15.15		198.15	225
0170	1 size, 18" wide, 24" deep		34	.235		218	15.15		233.15	263
0180	2 size, 24" wide, 30" deep		34	.235		289	15.15		304.15	340
0200	Beam attachment, welded, TYPE 22									
0202	3/8"	Q-15	80	.200	Ea.	12	11.60	1.33	24.93	32
0203	1/2"		76	.211		13.15	12.20	1.40	26.75	34
0204	5/8"		72	.222		13.55	12.90	1.48	27.93	36
0205	3/4"		68	.235		14.05	13.65	1.56	29.26	37.50
0206	7/8"		64	.250		19.30	14.50	1.66	35.46	44.50
0207	1"		56	.286		41.50	16.55	1.90	59.95	72.50
0300	Clamps									
0310	C-clamp, for mounting on steel beam flange, w/locknut, TYPE 23									
0320	3/8" threaded rod size	1 Plum	160	.050	Ea.	3.38	3.22		6.60	8.55
0330	1/2" threaded rod size		160	.050		4.76	3.22		7.98	10.05
0340	5/8" threaded rod size		160	.050		5.30	3.22		8.52	10.60
0350	3/4" threaded rod size		160	.050		6.85	3.22		10.07	12.30
0352	7/8" threaded rod size		140	.057		17.30	3.68		20.98	24.50

For customer support on your Mechanical Costs with RSMeans Data, call 800.448.8182.

22 05 29 – Hangers and Supports for Plumbing Piping and Equipment

22 05 29.10 Hangers & Supp. for Plumb'g/HVAC Pipe/Equip.	Crew	Daily Output	Labor-Hours	Unit	Material	2020 Bare Costs Labor	Equipment	Total	Total Incl O&P
0400 High temperature to 1050°F, alloy steel									
0410 4" pipe size	Q-1	106	.151	Ea.	28	8.75		36.75	43.50
0420 6" pipe size		106	.151		43	8.75		51.75	60
0430 8" pipe size		97	.165		48	9.55		57.55	67.50
0440 10" pipe size		84	.190		81.50	11.05		92.55	106
0450 12" pipe size		72	.222		82	12.90		94.90	110
0460 14" pipe size		64	.250		253	14.50		267.50	300
0470 16" pipe size		56	.286		269	16.55		285.55	320
0480 Beam clamp, flange type, TYPE 25									
0482 For 3/8" bolt	1 Plum	48	.167	Ea.	4.85	10.75		15.60	21.50
0483 For 1/2" bolt		44	.182		6.15	11.70		17.85	24.50
0484 For 5/8" bolt		40	.200		6.10	12.90		19	26
0485 For 3/4" bolt		36	.222		6.10	14.30		20.40	28
0486 For 1" bolt		32	.250		6.15	16.10		22.25	31
0500 I-beam, for mounting on bottom flange, strap iron, TYPE 21									
0530 4" flange size	1 Plum	93	.086	Ea.	6.75	5.55		12.30	15.70
0540 5" flange size		92	.087		7.40	5.60		13	16.55
0550 6" flange size		90	.089		8.40	5.75		14.15	17.80
0560 7" flange size		88	.091		9.35	5.85		15.20	19.05
0570 8" flange size		86	.093		10.05	6		16.05	20
0600 One hole, vertical mounting, malleable iron									
0610 1/2" pipe size	1 Plum	160	.050	Ea.	1.27	3.22		4.49	6.20
0620 3/4" pipe size		145	.055		1.35	3.56		4.91	6.80
0630 1" pipe size		136	.059		1.46	3.79		5.25	7.25
0640 1-1/4" pipe size		128	.063		2.56	4.03		6.59	8.85
0650 1-1/2" pipe size		120	.067		2.96	4.30		7.26	9.70
0660 2" pipe size		112	.071		4.27	4.60		8.87	11.60
0670 2-1/2" pipe size		104	.077		8.45	4.96		13.41	16.65
0680 3" pipe size		96	.083		9.85	5.35		15.20	18.85
0690 3-1/2" pipe size		90	.089		9.55	5.75		15.30	19.05
0700 4" pipe size		84	.095		13.50	6.15		19.65	24
0750 Riser or extension pipe, carbon steel, TYPE 8									
0756 1/2" pipe size	1 Plum	52	.154	Ea.	2.51	9.90		12.41	17.60
0760 3/4" pipe size		48	.167		3.48	10.75		14.23	19.95
0770 1" pipe size		47	.170		4.04	10.95		14.99	21
0780 1-1/4" pipe size		46	.174		5.45	11.20		16.65	23
0790 1-1/2" pipe size		45	.178		5.65	11.45		17.10	23.50
0800 2" pipe size		43	.186		5.65	12		17.65	24
0810 2-1/2" pipe size		41	.195		6.65	12.60		19.25	26
0820 3" pipe size		40	.200		6.70	12.90		19.60	26.50
0830 3-1/2" pipe size		39	.205		6.05	13.20		19.25	26.50
0840 4" pipe size		38	.211		9.20	13.55		22.75	30.50
0850 5" pipe size		37	.216		14.30	13.95		28.25	37
0860 6" pipe size		36	.222		16.30	14.30		30.60	39.50
0870 8" pipe size		34	.235		30	15.15		45.15	55.50
0880 10" pipe size		32	.250		30.50	16.10		46.60	57.50
0890 12" pipe size		28	.286		43.50	18.40		61.90	75.50
0900 For plastic coating 3/4" to 4", add					190%				
0910 For copper plating 3/4" to 4", add					58%				
0950 Two piece, complete, carbon steel, medium weight, TYPE 4									
0960 1/2" pipe size	Q-1	137	.117	Ea.	2.67	6.75		9.42	13.10
0970 3/4" pipe size		134	.119		2.73	6.95		9.68	13.35
0980 1" pipe size		132	.121		2.51	7.05		9.56	13.25

22 05 29.10 Hangers & Supp. for Plumb'g/HVAC Pipe/Equip.	Crew	Daily Output	Labor-Hours	Unit	Material	2020 Bare Costs Labor	Equipment	Total	Total Incl O&P
0990	1-1/4" pipe size	Q-1	130	.123	Ea.	3.30	7.15		10.45	14.35
1000	1-1/2" pipe size		126	.127		4.40	7.35		11.75	15.85
1010	2" pipe size		124	.129		5.20	7.50		12.70	16.90
1020	2-1/2" pipe size		120	.133		5.60	7.75		13.35	17.70
1030	3" pipe size		117	.137		5.25	7.95		13.20	17.65
1040	3-1/2" pipe size		114	.140		11.80	8.15		19.95	25
1050	4" pipe size		110	.145		9	8.45		17.45	22.50
1060	5" pipe size		106	.151		17	8.75		25.75	32
1070	6" pipe size		104	.154		22	8.90		30.90	37.50
1080	8" pipe size		100	.160		21.50	9.30		30.80	38
1090	10" pipe size		96	.167		51.50	9.65		61.15	71
1100	12" pipe size		89	.180		77.50	10.45		87.95	101
1110	14" pipe size		82	.195		87	11.30		98.30	112
1120	16" pipe size		68	.235		136	13.65		149.65	171
1130	For galvanized, add					45%				
1150	Insert, concrete									
1160	Wedge type, carbon steel body, malleable iron nut, galvanized									
1170	1/4" threaded rod size	1 Plum	96	.083	Ea.	8.40	5.35		13.75	17.30
1180	3/8" threaded rod size		96	.083		20.50	5.35		25.85	30.50
1190	1/2" threaded rod size		96	.083		39	5.35		44.35	51
1200	5/8" threaded rod size		96	.083		6.55	5.35		11.90	15.30
1210	3/4" threaded rod size		96	.083		12.30	5.35		17.65	21.50
1220	7/8" threaded rod size		96	.083		12.30	5.35		17.65	21.50
1250	Pipe guide sized for insulation									
1260	No. 1, 1" pipe size, 1" thick insulation	1 Stpi	26	.308	Ea.	143	20		163	187
1270	No. 2, 1-1/4"-2" pipe size, 1" thick insulation		23	.348		169	23		192	220
1280	No. 3, 1-1/4"-2" pipe size, 1-1/2" thick insulation		21	.381		169	25		194	224
1290	No. 4, 2-1/2"-3-1/2" pipe size, 1-1/2" thick insulation		18	.444		183	29		212	245
1300	No. 5, 4"-5" pipe size, 1-1/2" thick insulation		16	.500		197	33		230	265
1310	No. 6, 5"-6" pipe size, 2" thick insulation	Q-5	21	.762		226	45		271	315
1320	No. 7, 8" pipe size, 2" thick insulation		16	1		292	59		351	410
1330	No. 8, 10" pipe size, 2" thick insulation		12	1.333		435	78.50		513.50	600
1340	No. 9, 12" pipe size, 2" thick insulation	Q-6	17	1.412		435	86.50		521.50	610
1350	No. 10, 12"-14" pipe size, 2-1/2" thick insulation		16	1.500		520	92		612	705
1360	No. 11, 16" pipe size, 2-1/2" thick insulation		10.50	2.286		520	140		660	780
1370	No. 12, 16"-18" pipe size, 3" thick insulation		9	2.667		740	163		903	1,050
1380	No. 13, 20" pipe size, 3" thick insulation		7.50	3.200		740	196		936	1,100
1390	No. 14, 24" pipe size, 3" thick insulation		7	3.429		1,025	210		1,235	1,475
1400	Bands									
1410	Adjustable band, carbon steel, for non-insulated pipe, TYPE 7									
1420	1/2" pipe size	Q-1	142	.113	Ea.	.38	6.55		6.93	10.20
1430	3/4" pipe size		140	.114		.38	6.65		7.03	10.30
1440	1" pipe size		137	.117		.38	6.75		7.13	10.55
1450	1-1/4" pipe size		134	.119		.46	6.95		7.41	10.85
1460	1-1/2" pipe size		131	.122		.41	7.10		7.51	11.05
1470	2" pipe size		129	.124		.41	7.20		7.61	11.20
1480	2-1/2" pipe size		125	.128		.74	7.40		8.14	11.90
1490	3" pipe size		122	.131		.79	7.60		8.39	12.25
1500	3-1/2" pipe size		119	.134		.95	7.80		8.75	12.70
1510	4" pipe size		114	.140		1.29	8.15		9.44	13.60
1520	5" pipe size		110	.145		2.23	8.45		10.68	15.10
1530	6" pipe size		108	.148		2.68	8.60		11.28	15.80
1540	8" pipe size		104	.154		4.08	8.90		12.98	17.85

22 05 29 – Hangers and Supports for Plumbing Piping and Equipment

22 05 29.10 Hangers & Supp. for Plumb'g/HVAC Pipe/Equip.	Crew	Daily Output	Labor-Hours	Unit	Material	2020 Bare Costs Labor	Equipment	Total	Total Incl O&P
1550	For copper plated, add					50%			
1560	For galvanized, add					30%			
1570	For plastic coating, add					30%			
1600	Adjusting nut malleable iron, steel band, TYPE 9								
1610	1/2" pipe size, galvanized band	Q-1	137	.117	Ea.	5.75	6.75		12.50	16.50
1620	3/4" pipe size, galvanized band		135	.119		5.70	6.85		12.55	16.60
1630	1" pipe size, galvanized band		132	.121		5.40	7.05		12.45	16.45
1640	1-1/4" pipe size, galvanized band		129	.124		5.85	7.20		13.05	17.15
1650	1-1/2" pipe size, galvanized band		126	.127		5.60	7.35		12.95	17.15
1660	2" pipe size, galvanized band		124	.129		6.35	7.50		13.85	18.15
1670	2-1/2" pipe size, galvanized band		120	.133		9.95	7.75		17.70	22.50
1680	3" pipe size, galvanized band		117	.137		10.40	7.95		18.35	23.50
1690	3-1/2" pipe size, galvanized band		114	.140		32	8.15		40.15	47
1700	4" pipe size, cadmium plated band		110	.145		28	8.45		36.45	43.50
1740	For plastic coated band, add					35%			
1750	For completely copper coated, add					45%			
1800	Clevis, adjustable, carbon steel, for non-insulated pipe, TYPE 1								
1810	1/2" pipe size	Q-1	137	.117	Ea.	1.12	6.75		7.87	11.40
1820	3/4" pipe size		135	.119		1.19	6.85		8.04	11.60
1830	1" pipe size		132	.121		1.51	7.05		8.56	12.15
1840	1-1/4" pipe size		129	.124		1.68	7.20		8.88	12.60
1850	1-1/2" pipe size		126	.127		1.15	7.35		8.50	12.25
1860	2" pipe size		124	.129		2.15	7.50		9.65	13.55
1870	2-1/2" pipe size		120	.133		3.55	7.75		11.30	15.45
1880	3" pipe size		117	.137		4.22	7.95		12.17	16.50
1890	3-1/2" pipe size		114	.140		4.50	8.15		12.65	17.15
1900	4" pipe size		110	.145		3.15	8.45		11.60	16.10
1910	5" pipe size		106	.151		4.27	8.75		13.02	17.80
1920	6" pipe size		104	.154		5.40	8.90		14.30	19.30
1930	8" pipe size		100	.160		10.60	9.30		19.90	25.50
1940	10" pipe size		96	.167		17.05	9.65		26.70	33
1950	12" pipe size		89	.180		22	10.45		32.45	40
1960	14" pipe size		82	.195		36	11.30		47.30	57
1970	16" pipe size		68	.235		70.50	13.65		84.15	98
1971	18" pipe size		54	.296		90	17.20		107.20	125
1972	20" pipe size		38	.421		139	24.50		163.50	190
1980	For galvanized, add					66%			
1990	For copper plated 1/2" to 4", add					77%			
2000	For light weight 1/2" to 4", deduct					13%			
2010	Insulated pipe type, 3/4" to 12" pipe, add					180%			
2020	Insulated pipe type, chrome-moly U-strap, add					530%			
2250	Split ring, malleable iron, for non-insulated pipe, TYPE 11								
2260	1/2" pipe size	Q-1	137	.117	Ea.	5.05	6.75		11.80	15.70
2270	3/4" pipe size		135	.119		5.15	6.85		12	16
2280	1" pipe size		132	.121		5.45	7.05		12.50	16.50
2290	1-1/4" pipe size		129	.124		6.80	7.20		14	18.20
2300	1-1/2" pipe size		126	.127		8.25	7.35		15.60	20
2310	2" pipe size		124	.129		9.35	7.50		16.85	21.50
2320	2-1/2" pipe size		120	.133		13.45	7.75		21.20	26.50
2330	3" pipe size		117	.137		16.85	7.95		24.80	30.50
2340	3-1/2" pipe size		114	.140		17.25	8.15		25.40	31
2350	4" pipe size		110	.145		17.55	8.45		26	32
2360	5" pipe size		106	.151		23.50	8.75		32.25	38.50

22 05 29.10 Hangers & Supp. for Plumb'g/HVAC Pipe/Equip.	Crew	Daily Output	Labor-Hours	Unit	Material	2020 Bare Costs Labor	Equipment	Total	Total Incl O&P
2370	6" pipe size	Q-1	104	.154	Ea.	52	8.90		60.90	71
2380	8" pipe size	↓	100	.160	↓	78	9.30		87.30	99.50
2390	For copper plated, add					8%				
2400	Channels, steel, 3/4" x 1-1/2"	1 Plum	80	.100	L.F.	2.98	6.45		9.43	12.95
2404	1-1/2" x 1-1/2"		70	.114		3.68	7.35		11.03	15.05
2408	1-7/8" x 1-1/2"		60	.133		21	8.60		29.60	36.50
2412	3" x 1-1/2"		50	.160	↓	19.85	10.30		30.15	37.50
2416	Hangers, trapeze channel support, 12 ga. 1-1/2" x 1-1/2", 12" wide, steel		8.80	.909	Ea.	26	58.50		84.50	116
2418	18" wide, steel		8.30	.964		27.50	62		89.50	124
2430	Spring nuts, long, 1/4"		120	.067		.65	4.30		4.95	7.15
2432	3/8"		100	.080		.75	5.15		5.90	8.55
2434	1/2"		80	.100		.83	6.45		7.28	10.55
2436	5/8"		80	.100		3.71	6.45		10.16	13.75
2438	3/4"		75	.107		6.20	6.85		13.05	17.15
2440	Spring nuts, short, 1/4"		120	.067		1.36	4.30		5.66	7.95
2442	3/8"		100	.080		1.70	5.15		6.85	9.55
2444	1/2"	↓	80	.100	↓	1.81	6.45		8.26	11.65
2500	Washer, flat steel									
2502	3/8"	1 Plum	240	.033	Ea.	.08	2.15		2.23	3.30
2503	1/2"		220	.036		.21	2.34		2.55	3.74
2504	5/8"		200	.040		.35	2.58		2.93	4.25
2505	3/4"		180	.044		.61	2.86		3.47	4.96
2506	7/8"		160	.050		.43	3.22		3.65	5.30
2507	1"		140	.057		1.38	3.68		5.06	7
2508	1-1/4"	↓	120	.067	↓	1.52	4.30		5.82	8.10
2520	Nut, steel, hex									
2522	3/8"	1 Plum	200	.040	Ea.	.22	2.58		2.80	4.10
2523	1/2"		180	.044		.53	2.86		3.39	4.87
2524	5/8"		160	.050		.91	3.22		4.13	5.80
2525	3/4"		140	.057		1.42	3.68		5.10	7.05
2526	7/8"		120	.067		2.09	4.30		6.39	8.75
2527	1"		100	.080		3.15	5.15		8.30	11.15
2528	1-1/4"	↓	80	.100	↓	8.10	6.45		14.55	18.60
2532	Turnbuckle, TYPE 13									
2534	3/8"	1 Plum	80	.100	Ea.	5.10	6.45		11.55	15.25
2535	1/2"		72	.111		5.65	7.15		12.80	16.95
2536	5/8"		64	.125		10.05	8.05		18.10	23
2537	3/4"		56	.143		12.80	9.20		22	28
2538	7/8"		48	.167		33.50	10.75		44.25	53
2539	1"		40	.200		45.50	12.90		58.40	69.50
2540	1-1/4"	↓	32	.250	↓	84	16.10		100.10	117
2650	Rods, carbon steel									
2660	Continuous thread									
2670	1/4" thread size	1 Plum	144	.056	L.F.	2.43	3.58		6.01	8
2680	3/8" thread size		144	.056		2.59	3.58		6.17	8.20
2690	1/2" thread size		144	.056		4.07	3.58		7.65	9.85
2700	5/8" thread size		144	.056		5.75	3.58		9.33	11.70
2710	3/4" thread size		144	.056		10.60	3.58		14.18	17.05
2720	7/8" thread size	↓	144	.056		13.35	3.58		16.93	20
2721	1" thread size	Q-1	160	.100		22.50	5.80		28.30	33.50
2722	1-1/8" thread size	"	120	.133		24.50	7.75		32.25	38.50
2725	1/4" thread size, bright finish	1 Plum	144	.056		1.57	3.58		5.15	7.10
2726	1/2" thread size, bright finish	"	144	.056	↓	5.10	3.58		8.68	10.95

22 05 29.10 Hangers & Supp. for Plumb'g/HVAC Pipe/Equip.	Crew	Daily Output	Labor-Hours	Unit	Material	2020 Bare Costs Labor	2020 Bare Costs Equipment	Total	Total Incl O&P
2730	For galvanized, add					40%				
2820	Rod couplings									
2821	3/8"	1 Plum	60	.133	Ea.	1.74	8.60		10.34	14.75
2822	1/2"		54	.148		2.43	9.55		11.98	16.95
2823	5/8"		48	.167		3.09	10.75		13.84	19.50
2824	3/4"		44	.182		4.34	11.70		16.04	22.50
2825	7/8"		40	.200		7.35	12.90		20.25	27.50
2826	1"		34	.235		9.70	15.15		24.85	33
2827	1-1/8"	▼	30	.267	▼	16.40	17.20		33.60	43.50
2860	Pipe hanger assy, adj. clevis, saddle, rod, clamp, insul. allowance									
2864	1/2" pipe size	Q-5	35	.457	Ea.	26	27		53	69
2866	3/4" pipe size		34.80	.460		27.50	27		54.50	70.50
2868	1" pipe size		34.60	.462		27	27.50		54.50	70.50
2869	1-1/4" pipe size		34.30	.466		27	27.50		54.50	71
2870	1-1/2" pipe size		33.90	.472		27.50	28		55.50	71.50
2872	2" pipe size		33.30	.480		30	28.50		58.50	75.50
2874	2-1/2" pipe size		32.30	.495		33.50	29		62.50	80
2876	3" pipe size		31.20	.513		36	30.50		66.50	85
2880	4" pipe size		30.70	.521		38.50	31		69.50	88.50
2884	6" pipe size		29.80	.537		49.50	31.50		81	102
2888	8" pipe size		28	.571		67.50	33.50		101	125
2892	10" pipe size		25.20	.635		74.50	37.50		112	138
2896	12" pipe size	▼	23.20	.690	▼	117	40.50		157.50	189
2900	Rolls									
2910	Adjustable yoke, carbon steel with CI roll, TYPE 43									
2918	2" pipe size	Q-1	140	.114	Ea.	11.70	6.65		18.35	23
2920	2-1/2" pipe size		137	.117		12.40	6.75		19.15	24
2930	3" pipe size		131	.122		13.25	7.10		20.35	25
2940	3-1/2" pipe size		124	.129		18.45	7.50		25.95	31.50
2950	4" pipe size		117	.137		18.15	7.95		26.10	32
2960	5" pipe size		110	.145		30	8.45		38.45	45.50
2970	6" pipe size		104	.154		40.50	8.90		49.40	58
2980	8" pipe size		96	.167		58	9.65		67.65	78.50
2990	10" pipe size		80	.200		74.50	11.60		86.10	99.50
3000	12" pipe size		68	.235		108	13.65		121.65	140
3010	14" pipe size		56	.286		185	16.55		201.55	229
3020	16" pipe size	▼	48	.333	▼	224	19.35		243.35	275
3050	Chair, carbon steel with CI roll									
3060	2" pipe size	1 Plum	68	.118	Ea.	14.65	7.60		22.25	27.50
3070	2-1/2" pipe size		65	.123		15.75	7.95		23.70	29
3080	3" pipe size		62	.129		16.80	8.30		25.10	31
3090	3-1/2" pipe size		60	.133		20.50	8.60		29.10	35.50
3100	4" pipe size		58	.138		19.90	8.90		28.80	35.50
3110	5" pipe size		56	.143		23	9.20		32.20	39.50
3120	6" pipe size		53	.151		31	9.75		40.75	48.50
3130	8" pipe size		50	.160		42	10.30		52.30	61.50
3140	10" pipe size		48	.167		56.50	10.75		67.25	78
3150	12" pipe size	▼	46	.174		86	11.20		97.20	111
3170	Single pipe roll (see line 2650 for rods), TYPE 41, 1" pipe size	Q-1	137	.117		9.85	6.75		16.60	21
3180	1-1/4" pipe size		131	.122		10.45	7.10		17.55	22
3190	1-1/2" pipe size		129	.124		10.45	7.20		17.65	22
3200	2" pipe size		124	.129		10.45	7.50		17.95	22.50
3210	2-1/2" pipe size	▼	118	.136		11.15	7.85		19	24

22 05 29 – Hangers and Supports for Plumbing Piping and Equipment

22 05 29.10 Hangers & Supp. for Plumb'g/HVAC Pipe/Equip.	Crew	Daily Output	Labor-Hours	Unit	Material	2020 Bare Costs Labor	2020 Bare Costs Equipment	Total	Total Incl O&P
3220	3" pipe size	Q-1	115	.139	Ea.	11.75	8.05		19.80	25
3230	3-1/2" pipe size		113	.142		12.80	8.20		21	26.50
3240	4" pipe size		112	.143		13	8.30		21.30	26.50
3250	5" pipe size		110	.145		15.10	8.45		23.55	29.50
3260	6" pipe size		101	.158		30	9.20		39.20	47
3270	8" pipe size		90	.178		32.50	10.30		42.80	51
3280	10" pipe size		80	.200		42	11.60		53.60	63.50
3290	12" pipe size		68	.235		68.50	13.65		82.15	95.50
3291	14" pipe size		56	.286		104	16.55		120.55	139
3292	16" pipe size		48	.333		135	19.35		154.35	177
3293	18" pipe size		40	.400		229	23		252	287
3294	20" pipe size		35	.457		189	26.50		215.50	248
3296	24" pipe size		30	.533		320	31		351	395
3297	30" pipe size		25	.640		585	37		622	695
3298	36" pipe size	▼	20	.800	▼	620	46.50		666.50	750
3300	Saddles (add vertical pipe riser, usually 3" diameter)									
3310	Pipe support, complete, adjust., CI saddle, TYPE 36									
3320	2-1/2" pipe size	1 Plum	96	.083	Ea.	111	5.35		116.35	130
3330	3" pipe size		88	.091		113	5.85		118.85	133
3340	3-1/2" pipe size		79	.101		111	6.55		117.55	132
3350	4" pipe size		68	.118		160	7.60		167.60	187
3360	5" pipe size		64	.125		169	8.05		177.05	198
3370	6" pipe size		59	.136		165	8.75		173.75	194
3380	8" pipe size		53	.151		170	9.75		179.75	202
3390	10" pipe size		50	.160		198	10.30		208.30	232
3400	12" pipe size	▼	48	.167	▼	209	10.75		219.75	246
3450	For standard pipe support, one piece, CI, deduct					34%				
3460	For stanchion support, CI with steel yoke, deduct					60%				
3550	Insulation shield 1" thick, 1/2" pipe size, TYPE 40	1 Asbe	100	.080	Ea.	3.37	4.69		8.06	10.90
3560	3/4" pipe size		100	.080		4.06	4.69		8.75	11.65
3570	1" pipe size		98	.082		4.39	4.79		9.18	12.20
3580	1-1/4" pipe size		98	.082		4.68	4.79		9.47	12.50
3590	1-1/2" pipe size		96	.083		4.35	4.89		9.24	12.30
3600	2" pipe size		96	.083		4.53	4.89		9.42	12.50
3610	2-1/2" pipe size		94	.085		4.67	4.99		9.66	12.80
3620	3" pipe size		94	.085		5.55	4.99		10.54	13.75
3630	2" thick, 3-1/2" pipe size		92	.087		5.95	5.10		11.05	14.35
3640	4" pipe size		92	.087		9.85	5.10		14.95	18.70
3650	5" pipe size		90	.089		12.05	5.20		17.25	21.50
3660	6" pipe size		90	.089		13.85	5.20		19.05	23.50
3670	8" pipe size		88	.091		19.65	5.35		25	29.50
3680	10" pipe size		88	.091		27.50	5.35		32.85	38.50
3690	12" pipe size		86	.093		29.50	5.45		34.95	41
3700	14" pipe size		86	.093		57.50	5.45		62.95	71.50
3710	16" pipe size		84	.095		56.50	5.60		62.10	71
3720	18" pipe size		84	.095		59	5.60		64.60	73
3730	20" pipe size		82	.098		64.50	5.70		70.20	80
3732	24" pipe size	▼	80	.100	▼	76	5.85		81.85	93
3750	Covering protection saddle, TYPE 39									
3760	1" covering size									
3770	3/4" pipe size	1 Plum	68	.118	Ea.	5.60	7.60		13.20	17.50
3780	1" pipe size		68	.118		5.60	7.60		13.20	17.50
3790	1-1/4" pipe size	▼	68	.118		5.60	7.60		13.20	17.50

22 05 29 – Hangers and Supports for Plumbing Piping and Equipment

22 05 29.10 Hangers & Supp. for Plumb'g/HVAC Pipe/Equip.	Crew	Daily Output	Labor-Hours	Unit	Material	2020 Bare Costs Labor	Equipment	Total	Total Incl O&P
3800	1-1/2" pipe size	1 Plum	66	.121	Ea.	6	7.80		13.80	18.30
3810	2" pipe size		66	.121		6	7.80		13.80	18.30
3820	2-1/2" pipe size		64	.125		6.05	8.05		14.10	18.75
3830	3" pipe size		64	.125		8	8.05		16.05	21
3840	3-1/2" pipe size		62	.129		9.55	8.30		17.85	23
3850	4" pipe size		62	.129		8.85	8.30		17.15	22
3860	5" pipe size		60	.133		9.65	8.60		18.25	23.50
3870	6" pipe size		60	.133		11.55	8.60		20.15	25.50
3900	1-1/2" covering size									
3910	3/4" pipe size	1 Plum	68	.118	Ea.	7.20	7.60		14.80	19.25
3920	1" pipe size		68	.118		7.15	7.60		14.75	19.25
3930	1-1/4" pipe size		68	.118		7.20	7.60		14.80	19.25
3940	1-1/2" pipe size		66	.121		6.70	7.80		14.50	19.10
3950	2" pipe size		66	.121		7.25	7.80		15.05	19.65
3960	2-1/2" pipe size		64	.125		8.70	8.05		16.75	21.50
3970	3" pipe size		64	.125		8.70	8.05		16.75	21.50
3980	3-1/2" pipe size		62	.129		9	8.30		17.30	22.50
3990	4" pipe size		62	.129		9.15	8.30		17.45	22.50
4000	5" pipe size		60	.133		9.90	8.60		18.50	24
4010	6" pipe size		60	.133		12.45	8.60		21.05	26.50
4020	8" pipe size		58	.138		15.60	8.90		24.50	30.50
4022	10" pipe size		56	.143		15.60	9.20		24.80	31
4024	12" pipe size		54	.148		37.50	9.55		47.05	56
4028	2" covering size									
4029	2-1/2" pipe size	1 Plum	62	.129	Ea.	9.50	8.30		17.80	23
4032	3" pipe size		60	.133		10.40	8.60		19	24.50
4033	4" pipe size		58	.138		10.40	8.90		19.30	25
4034	6" pipe size		56	.143		14.90	9.20		24.10	30
4035	8" pipe size		54	.148		17.70	9.55		27.25	34
4080	10" pipe size		58	.138		20	8.90		28.90	35.50
4090	12" pipe size		56	.143		42	9.20		51.20	60.50
4100	14" pipe size		56	.143		43	9.20		52.20	61.50
4110	16" pipe size		54	.148		57	9.55		66.55	77.50
4120	18" pipe size		54	.148		61	9.55		70.55	82
4130	20" pipe size		52	.154		68	9.90		77.90	90
4150	24" pipe size		50	.160		79.50	10.30		89.80	103
4160	30" pipe size		48	.167		87.50	10.75		98.25	112
4180	36" pipe size		45	.178		101	11.45		112.45	128
4186	2-1/2" covering size									
4187	3" pipe size	1 Plum	58	.138	Ea.	10.60	8.90		19.50	25
4188	4" pipe size		56	.143		11.55	9.20		20.75	26.50
4189	6" pipe size		52	.154		17.65	9.90		27.55	34.50
4190	8" pipe size		48	.167		19.70	10.75		30.45	37.50
4191	10" pipe size		44	.182		21.50	11.70		33.20	41.50
4192	12" pipe size		40	.200		50	12.90		62.90	74.50
4193	14" pipe size		36	.222		50	14.30		64.30	76.50
4194	16" pipe size		32	.250		60	16.10		76.10	90
4195	18" pipe size		28	.286		67	18.40		85.40	101
4200	Sockets									
4210	Rod end, malleable iron, TYPE 16									
4220	1/4" thread size	1 Plum	240	.033	Ea.	1.64	2.15		3.79	5
4230	3/8" thread size		240	.033		1.67	2.15		3.82	5.05
4240	1/2" thread size		230	.035		2.20	2.24		4.44	5.75

22 05 29.10 Hangers & Supp. for Plumb'g/HVAC Pipe/Equip.	Crew	Daily Output	Labor-Hours	Unit	Material	2020 Bare Costs Labor	2020 Bare Costs Equipment	Total	Total Incl O&P
4250	5/8" thread size	1 Plum	225	.036	Ea.	4.51	2.29		6.80	8.40
4260	3/4" thread size		220	.036		5.90	2.34		8.24	10
4270	7/8" thread size		210	.038		8.65	2.46		11.11	13.20
4290	Strap, 1/2" pipe size, TYPE 26	Q-1	142	.113		1.94	6.55		8.49	11.95
4300	3/4" pipe size		140	.114		1.97	6.65		8.62	12.05
4310	1" pipe size		137	.117		2.77	6.75		9.52	13.20
4320	1-1/4" pipe size		134	.119		2.84	6.95		9.79	13.45
4330	1-1/2" pipe size		131	.122		3.08	7.10		10.18	14
4340	2" pipe size		129	.124		3.13	7.20		10.33	14.20
4350	2-1/2" pipe size		125	.128		4.91	7.40		12.31	16.50
4360	3" pipe size		122	.131		6.25	7.60		13.85	18.25
4370	3-1/2" pipe size		119	.134		7	7.80		14.80	19.35
4380	4" pipe size		114	.140		7.35	8.15		15.50	20.50
4400	U-bolt, carbon steel									
4410	Standard, with nuts, TYPE 42									
4420	1/2" pipe size	1 Plum	160	.050	Ea.	1.41	3.22		4.63	6.35
4430	3/4" pipe size		158	.051		1.22	3.26		4.48	6.20
4450	1" pipe size		152	.053		1.25	3.39		4.64	6.50
4460	1-1/4" pipe size		148	.054		1.57	3.48		5.05	6.95
4470	1-1/2" pipe size		143	.056		1.67	3.61		5.28	7.25
4480	2" pipe size		139	.058		1.81	3.71		5.52	7.55
4490	2-1/2" pipe size		134	.060		2.96	3.85		6.81	9
4500	3" pipe size		128	.063		3.31	4.03		7.34	9.70
4510	3-1/2" pipe size		122	.066		4.29	4.23		8.52	11
4520	4" pipe size		117	.068		4.42	4.41		8.83	11.45
4530	5" pipe size		114	.070		4.39	4.52		8.91	11.60
4540	6" pipe size		111	.072		7.35	4.64		11.99	15.05
4550	8" pipe size		109	.073		8.85	4.73		13.58	16.80
4560	10" pipe size		107	.075		15.15	4.82		19.97	24
4570	12" pipe size		104	.077		21	4.96		25.96	30.50
4580	For plastic coating on 1/2" thru 6" size, add					150%				
4700	U-hook, carbon steel, requires mounting screws or bolts									
4710	3/4" thru 2" pipe size									
4720	6" long	1 Plum	96	.083	Ea.	1.16	5.35		6.51	9.35
4730	8" long		96	.083		1.46	5.35		6.81	9.65
4740	10" long		96	.083		1.59	5.35		6.94	9.80
4750	12" long		96	.083		1.75	5.35		7.10	10
4760	For copper plated, add					50%				
7000	Roof supports									
7006	Duct									
7010	Rectangular, open, 12" off roof									
7020	To 18" wide	Q-9	26	.615	Ea.	163	34.50		197.50	232
7030	To 24" wide	"	22	.727		190	41		231	271
7040	To 36" wide	Q-10	30	.800		217	46.50		263.50	310
7050	To 48" wide		28	.857		244	50		294	345
7060	To 60" wide		24	1		271	58		329	385
7100	Equipment									
7120	Equipment support	Q-5	20	.800	Ea.	71	47		118	149
7300	Pipe									
7310	Roller type									
7320	Up to 2-1/2" diam. pipe									
7324	3-1/2" off roof	Q-5	24	.667	Ea.	22.50	39.50		62	83.50
7326	Up to 10" off roof	"	20	.800	"	28	47		75	101

22 05 29.10 Hangers & Supp. for Plumb'g/HVAC Pipe/Equip.	Crew	Daily Output	Labor-Hours	Unit	Material	2020 Bare Costs Labor	Equipment	Total	Total Incl O&P
7340	2-1/2" to 3-1/2" diam. pipe								
7342	Up to 16" off roof	Q-5	18	.889	Ea.	49.50	52.50		102	133
7360	4" to 5" diam. pipe								
7362	Up to 12" off roof	Q-5	16	1	Ea.	70.50	59		129.50	166
7400	Strut/channel type								
7410	Up to 2-1/2" diam. pipo								
7424	3-1/2" off roof	Q-5	24	.667	Ea.	20	39.50		59.50	81
7426	Up to 10" off roof	"	20	.800	"	25.50	47		72.50	98.50
7440	Strut and roller type								
7452	2-1/2" to 3-1/2" diam. pipe								
7454	Up to 16" off roof	Q-5	18	.889	Ea.	71	52.50		123.50	157
7460	Strut and hanger type								
7470	Up to 3" diam. pipe								
7474	Up to 8" off roof	Q-5	19	.842	Ea.	45	49.50		94.50	124
8000	Pipe clamp, plastic, 1/2" CTS	1 Plum	80	.100		.20	6.45		6.65	9.85
8010	3/4" CTS		73	.110		.24	7.05		7.29	10.80
8020	1" CTS		68	.118		.54	7.60		8.14	11.95
8080	Economy clamp, 1/4" CTS		175	.046		.05	2.95		3	4.47
8090	3/8" CTS		168	.048		.05	3.07		3.12	4.65
8100	1/2" CTS		160	.050		.05	3.22		3.27	4.88
8110	3/4" CTS		145	.055		.05	3.56		3.61	5.35
8200	Half clamp, 1/2" CTS		80	.100		.07	6.45		6.52	9.75
8210	3/4" CTS		73	.110		.10	7.05		7.15	10.65
8300	Suspension clamp, 1/2" CTS		80	.100		.24	6.45		6.69	9.90
8310	3/4" CTS		73	.110		.22	7.05		7.27	10.80
8320	1" CTS		68	.118		.53	7.60		8.13	11.95
8400	Insulator, 1/2" CTS		80	.100		.36	6.45		6.81	10.05
8410	3/4" CTS		73	.110		.37	7.05		7.42	10.95
8420	1" CTS		68	.118		.39	7.60		7.99	11.80
8500	J hook clamp with nail, 1/2" CTS		240	.033		.09	2.15		2.24	3.31
8501	3/4" CTS	▼	240	.033	▼	.11	2.15		2.26	3.33
8800	Wire cable support system								
8810	Cable with hook terminal and locking device								
8830	2 mm (.079") diam. cable (100 lb. cap.)								
8840	1 m (3.3') length, with hook	1 Shee	96	.083	Ea.	3.90	5.20		9.10	12.15
8850	2 m (6.6') length, with hook		84	.095		4.53	5.95		10.48	14
8860	3 m (9.9') length, with hook	▼	72	.111		4.47	6.90		11.37	15.40
8870	5 m (16.4') length, with hook	Q-9	60	.267		5.55	14.95		20.50	28.50
8880	10 m (32.8') length, with hook	"	30	.533	▼	8.05	30		38.05	54.50
8900	3 mm (.118") diam. cable (200 lb. cap.)								
8910	1 m (3.3') length, with hook	1 Shee	96	.083	Ea.	9.80	5.20		15	18.65
8920	2 m (6.6') length, with hook		84	.095		10.85	5.95		16.80	21
8930	3 m (9.9') length, with hook	▼	72	.111		11.80	6.90		18.70	23.50
8940	5 m (16.4') length, with hook	Q-9	60	.267		10.40	14.95		25.35	34
8950	10 m (32.8') length, with hook	"	30	.533	▼	13.90	30		43.90	61
9000	Cable system accessories								
9010	Anchor bolt, 3/8", with nut	1 Shee	140	.057	Ea.	1.13	3.56		4.69	6.65
9020	Air duct corner protector		160	.050		1.51	3.12		4.63	6.40
9030	Air duct support attachment	▼	140	.057	▼	1.82	3.56		5.38	7.40
9040	Flange clip, hammer-on style								
9044	For flange thickness 3/32"-9/64", 160 lb. cap.	1 Shee	180	.044	Ea.	.44	2.77		3.21	4.68
9048	For flange thickness 1/8"-1/4", 200 lb. cap.		160	.050		.32	3.12		3.44	5.10
9052	For flange thickness 5/16"-1/2", 200 lb. cap.	▼	150	.053	▼	.65	3.32		3.97	5.75

22 05 29 – Hangers and Supports for Plumbing Piping and Equipment

22 05 29.10 Hangers & Supp. for Plumb'g/HVAC Pipe/Equip.	Crew	Daily Output	Labor-Hours	Unit	Material	2020 Bare Costs Labor	Equipment	Total	Total Incl O&P
9056	For flange thickness 9/16"-3/4", 200 lb. cap.	1 Shee	140	.057	Ea.	.89	3.56		4.45	6.40
9060	Wire insulation protection tube	↓	180	.044	L.F.	.54	2.77		3.31	4.79
9070	Wire cutter				Ea.	44			44	48.50

22 05 33 – Heat Tracing for Plumbing Piping

22 05 33.20 Temperature Maintenance Cable

		Crew	Daily Output	Labor-Hours	Unit	Material	2020 Bare Costs Labor	Equipment	Total	Total Incl O&P
0010	**TEMPERATURE MAINTENANCE CABLE**									
0040	Components									
0080	Heating cable									
0100	208 V									
0150	140°F	Q-1	1060.80	.015	L.F.	10.15	.87		11.02	12.50
0200	120 V									
0220	125°F	Q-1	1060.80	.015	L.F.	9.50	.87		10.37	11.75
0300	Power kit w/1 end seal	1 Elec	48.80	.164	Ea.	136	10.05		146.05	165
0310	Splice kit		35.50	.225		161	13.85		174.85	198
0320	End seal		160	.050		12.75	3.07		15.82	18.55
0330	Tee kit w/1 end seal		26.80	.299		166	18.30		184.30	210
0340	Powered splice w/2 end seals		20	.400		118	24.50		142.50	167
0350	Powered tee kit w/3 end seals		18	.444		152	27.50		179.50	208
0360	Cross kit w/2 end seals	↓	18.60	.430	↓	156	26.50		182.50	212
0500	Recommended thickness of fiberglass insulation									
0510	Pipe size									
0520	1/2" to 1" use 1" insulation									
0530	1-1/4" to 2" use 1-1/2" insulation									
0540	2-1/2" to 6" use 2" insulation									
0560	NOTE: For pipe sizes 1-1/4" and smaller use 1/4" larger diameter									
0570	insulation to allow room for installation over cable.									

22 05 48 – Vibration and Seismic Controls for Plumbing Piping and Equipment

22 05 48.40 Vibration Absorbers

		Crew	Daily Output	Labor-Hours	Unit	Material	2020 Bare Costs Labor	Equipment	Total	Total Incl O&P
0010	**VIBRATION ABSORBERS**									
0100	Hangers, neoprene flex									
0200	10-120 lb. capacity				Ea.	24			24	26.50
0220	75-550 lb. capacity					42			42	46
0240	250-1,100 lb. capacity					78			78	86
0260	1,000-4,000 lb. capacity					144			144	158
0500	Spring flex, 60 lb. capacity					35.50			35.50	39.50
0520	450 lb. capacity					64.50			64.50	71
0540	900 lb. capacity					81			81	89
0560	1,100-1,300 lb. capacity				↓	78			78	86
0600	Rubber in shear									
0610	45-340 lb., up to 1/2" rod size	1 Stpi	22	.364	Ea.	23	24		47	61
0620	130-700 lb., up to 3/4" rod size		20	.400		50	26		76	94
0630	50-1,000 lb., up to 3/4" rod size	↓	18	.444		66	29		95	116
1000	Mounts, neoprene, 45-380 lb. capacity					17.35			17.35	19.10
1020	250-1,100 lb. capacity					63			63	69
1040	1,000-4,000 lb. capacity					116			116	127
1100	Spring flex, 60 lb. capacity					79			79	86.50
1120	165 lb. capacity					82			82	90
1140	260 lb. capacity					80.50			80.50	89
1160	450 lb. capacity					115			115	127
1180	600 lb. capacity					100			100	110
1200	750 lb. capacity					157			157	173
1220	900 lb. capacity					152			152	167

22 05 Common Work Results for Plumbing

22 05 48 – Vibration and Seismic Controls for Plumbing Piping and Equipment

22 05 48.40 Vibration Absorbers

	22 05 48.40 Vibration Absorbers	Crew	Daily Output	Labor-Hours	Unit	Material	2020 Bare Costs Labor	Equipment	Total	Total Incl O&P
1240	1,100 lb. capacity				Ea.	175			175	192
1260	1,300 lb. capacity					168			168	184
1280	1,500 lb. capacity					178			178	195
1300	1,800 lb. capacity					238			238	262
1320	2,200 lb. capacity					299			299	330
1340	2,600 lb. capacity					330			330	365
1399	Spring type									
1400	2 piece									
1410	50-1,000 lb.	1 Stpi	12	.667	Ea.	210	43.50		253.50	297
1420	1,100-1,600 lb.	"	12	.667	"	226	43.50		269.50	315
1500	Double spring open									
1510	150-450 lb.	1 Stpi	24	.333	Ea.	144	22		166	191
1520	500-1,000 lb.		24	.333		144	22		166	191
1530	1,100-1,600 lb.		24	.333		144	22		166	191
1540	1,700-2,400 lb.		24	.333		156	22		178	205
1550	2,500-3,400 lb.		24	.333		266	22		288	325
2000	Pads, cork rib, 18" x 18" x 1", 10-50 psi					196			196	215
2020	18" x 36" x 1", 10-50 psi					390			390	430
2100	Shear flexible pads, 18" x 18" x 3/8", 20-70 psi					78.50			78.50	86.50
2120	18" x 36" x 3/8", 20-70 psi					195			195	215
2150	Laminated neoprene and cork									
2160	1" thick	1 Stpi	16	.500	S.F.	87.50	33		120.50	145
2200	Neoprene elastomer isolation bearing pad									
2230	464 psi, 5/8" x 19-11/16" x 39-3/8"	1 Stpi	16	.500	Ea.	345	33		378	430
3000	Note overlap in capacities due to deflections									

22 05 53 – Identification for Plumbing Piping and Equipment

22 05 53.10 Piping System Identification Labels

	22 05 53.10 Piping System Identification Labels	Crew	Daily Output	Labor-Hours	Unit	Material	2020 Bare Costs Labor	Equipment	Total	Total Incl O&P
0010	**PIPING SYSTEM IDENTIFICATION LABELS**									
0100	Indicate contents and flow direction									
0106	Pipe markers									
0110	Plastic snap around									
0114	1/2" pipe	1 Plum	80	.100	Ea.	6.90	6.45		13.35	17.25
0116	3/4" pipe		80	.100		7.25	6.45		13.70	17.60
0118	1" pipe		80	.100		7.25	6.45		13.70	17.60
0120	2" pipe		75	.107		8.60	6.85		15.45	19.75
0122	3" pipe		70	.114		11.25	7.35		18.60	23.50
0124	4" pipe		60	.133		11.55	8.60		20.15	25.50
0126	6" pipe		60	.133		12.15	8.60		20.75	26.50
0128	8" pipe		56	.143		17.10	9.20		26.30	32.50
0130	10" pipe		56	.143		18.80	9.20		28	34.50
0200	Over 10" pipe size		50	.160		22	10.30		32.30	39.50
1110	Self adhesive									
1114	1" pipe	1 Plum	80	.100	Ea.	3.29	6.45		9.74	13.25
1116	2" pipe		75	.107		3.46	6.85		10.31	14.10
1118	3" pipe		70	.114		5.30	7.35		12.65	16.85
1120	4" pipe		60	.133		5.30	8.60		13.90	18.65
1122	6" pipe		60	.133		5.30	8.60		13.90	18.70
1124	8" pipe		56	.143		7.60	9.20		16.80	22
1126	10" pipe		56	.143		7.60	9.20		16.80	22
1200	Over 10" pipe size		50	.160		10.65	10.30		20.95	27
2000	Valve tags									
2010	Numbered plus identifying legend									

22 05 Common Work Results for Plumbing

22 05 53 – Identification for Plumbing Piping and Equipment

22 05 53.10 Piping System Identification Labels		Crew	Daily Output	Labor-Hours	Unit	Material	2020 Bare Costs Labor	Equipment	Total	Total Incl O&P
2100	Brass, 2" diameter	1 Plum	40	.200	Ea.	3.16	12.90		16.06	23
2200	Plastic, 1-1/2" diameter	"	40	.200	"	2.99	12.90		15.89	22.50

22 07 Plumbing Insulation

22 07 16 – Plumbing Equipment Insulation

22 07 16.10 Insulation for Plumbing Equipment

			Crew	Daily Output	Labor-Hours	Unit	Material	Labor	Equipment	Total	Total Incl O&P
0010	**INSULATION FOR PLUMBING EQUIPMENT**										
2900	Domestic water heater wrap kit										
2920	1-1/2" with vinyl jacket, 20 to 60 gal.	G	1 Plum	8	1	Ea.	16.70	64.50		81.20	115

22 07 19 – Plumbing Piping Insulation

22 07 19.10 Piping Insulation

			Crew	Daily Output	Labor-Hours	Unit	Material	Labor	Equipment	Total	Total Incl O&P
0010	**PIPING INSULATION**										
0110	Insulation req'd. is based on the surface size/area to be covered										
0230	Insulated protectors (ADA)										
0235	For exposed piping under sinks or lavatories										
0240	Vinyl coated foam, velcro tabs										
0245	P Trap, 1-1/4" or 1-1/2"		1 Plum	32	.250	Ea.	16.95	16.10		33.05	42.50
0260	Valve and supply cover										
0265	1/2", 3/8", and 7/16" pipe size		1 Plum	32	.250	Ea.	16.40	16.10		32.50	42
0280	Tailpiece offset (wheelchair)										
0285	1-1/4" pipe size		1 Plum	32	.250	Ea.	13.15	16.10		29.25	38.50
0600	Pipe covering (price copper tube one size less than IPS)										
1000	Mineral wool										
1010	Preformed, 1200°F, plain										
1014	1" wall										
1016	1/2" iron pipe size	G	Q-14	230	.070	L.F.	1.92	3.67		5.59	7.75
1018	3/4" iron pipe size	G		220	.073		1.95	3.84		5.79	8.05
1022	1" iron pipe size	G		210	.076		2.05	4.02		6.07	8.40
1024	1-1/4" iron pipe size	G		205	.078		2.11	4.12		6.23	8.60
1026	1-1/2" iron pipe size	G		205	.078		1.99	4.12		6.11	8.50
1028	2" iron pipe size	G		200	.080		2.69	4.22		6.91	9.45
1030	2-1/2" iron pipe size	G		190	.084		2.76	4.44		7.20	9.85
1032	3" iron pipe size	G		180	.089		2.96	4.69		7.65	10.45
1034	4" iron pipe size	G		150	.107		3.64	5.65		9.29	12.65
1036	5" iron pipe size	G		140	.114		4.12	6.05		10.17	13.80
1038	6" iron pipe size	G		120	.133		4.40	7.05		11.45	15.65
1040	7" iron pipe size	G		110	.145		5.15	7.70		12.85	17.45
1042	8" iron pipe size	G		100	.160		7.10	8.45		15.55	21
1044	9" iron pipe size	G		90	.178		7.45	9.40		16.85	22.50
1046	10" iron pipe size	G		90	.178		8.45	9.40		17.85	23.50
1050	1-1/2" wall										
1052	1/2" iron pipe size	G	Q-14	225	.071	L.F.	3.27	3.75		7.02	9.35
1054	3/4" iron pipe size	G		215	.074		3.36	3.93		7.29	9.75
1056	1" iron pipe size	G		205	.078		3.40	4.12		7.52	10.05
1058	1-1/4" iron pipe size	G		200	.080		3.47	4.22		7.69	10.30
1060	1-1/2" iron pipe size	G		200	.080		3.64	4.22		7.86	10.50
1062	2" iron pipe size	G		190	.084		3.86	4.44		8.30	11.05
1064	2-1/2" iron pipe size	G		180	.089		4.20	4.69		8.89	11.80
1066	3" iron pipe size	G		165	.097		4.38	5.10		9.48	12.65
1068	4" iron pipe size	G		140	.114		5.25	6.05		11.30	15.05

22 07 19 – Plumbing Piping Insulation

22 07 19.10 Piping Insulation		Crew	Daily Output	Labor-Hours	Unit	Material	2020 Bare Costs Labor	Equipment	Total	Total Incl O&P
1070	5" iron pipe size	G Q-14	130	.123	L.F.	5.60	6.50		12.10	16.10
1072	6" iron pipe size	G	110	.145		5.80	7.70		13.50	18.15
1074	7" iron pipe size	G	100	.160		6.85	8.45		15.30	20.50
1076	8" iron pipe size	G	90	.178		7.60	9.40		17	23
1078	9" iron pipe size	G	85	.188		8.75	9.95		10.70	25
1080	10" iron pipe size	G	80	.200		9.60	10.55		20.15	27
1082	12" iron pipe size	G	75	.213		10.95	11.25		22.20	29.50
1084	14" iron pipe size	G	70	.229		12.55	12.05		24.60	32.50
1086	16" iron pipe size	G	65	.246		13.80	13		26.80	35
1088	18" iron pipe size	G	60	.267		16.05	14.05		30.10	39
1090	20" iron pipe size	G	55	.291		17.15	15.35		32.50	42.50
1092	22" iron pipe size	G	50	.320		19.60	16.90		36.50	47.50
1094	24" iron pipe size	G	45	.356		21	18.75		39.75	52
1100	2" wall									
1102	1/2" iron pipe size	G Q-14	220	.073	L.F.	4.23	3.84		8.07	10.55
1104	3/4" iron pipe size	G	210	.076		4.39	4.02		8.41	11
1106	1" iron pipe size	G	200	.080		4.65	4.22		8.87	11.60
1108	1-1/4" iron pipe size	G	190	.084		4.99	4.44		9.43	12.30
1110	1-1/2" iron pipe size	G	190	.084		5.20	4.44		9.64	12.50
1112	2" iron pipe size	G	180	.089		5.45	4.69		10.14	13.20
1114	2-1/2" iron pipe size	G	170	.094		6.15	4.97		11.12	14.45
1116	3" iron pipe size	G	160	.100		6.60	5.30		11.90	15.35
1118	4" iron pipe size	G	130	.123		7.70	6.50		14.20	18.40
1120	5" iron pipe size	G	120	.133		8.55	7.05		15.60	20
1122	6" iron pipe size	G	100	.160		8.85	8.45		17.30	22.50
1124	8" iron pipe size	G	80	.200		10.75	10.55		21.30	28
1126	10" iron pipe size	G	70	.229		13.15	12.05		25.20	33
1128	12" iron pipe size	G	65	.246		14.60	13		27.60	36
1130	14" iron pipe size	G	60	.267		16.30	14.05		30.35	39.50
1132	16" iron pipe size	G	55	.291		18.45	15.35		33.80	44
1134	18" iron pipe size	G	50	.320		19.95	16.90		36.85	48
1136	20" iron pipe size	G	45	.356		23	18.75		41.75	54.50
1138	22" iron pipe size	G	45	.356		25	18.75		43.75	56.50
1140	24" iron pipe size	G	40	.400		26.50	21		47.50	61.50
1150	4" wall									
1152	1-1/4" iron pipe size	G Q-14	170	.094	L.F.	13.45	4.97		18.42	22.50
1154	1-1/2" iron pipe size	G	165	.097		14	5.10		19.10	23.50
1156	2" iron pipe size	G	155	.103		14.25	5.45		19.70	24
1158	4" iron pipe size	G	105	.152		18	8.05		26.05	32
1160	6" iron pipe size	G	75	.213		21	11.25		32.25	41
1162	8" iron pipe size	G	60	.267		24	14.05		38.05	48
1164	10" iron pipe size	G	50	.320		29	16.90		45.90	58
1166	12" iron pipe size	G	45	.356		32	18.75		50.75	64
1168	14" iron pipe size	G	40	.400		34	21		55	70
1170	16" iron pipe size	G	35	.457		38.50	24		62.50	79.50
1172	18" iron pipe size	G	32	.500		41.50	26.50		68	86
1174	20" iron pipe size	G	30	.533		44.50	28		72.50	92
1176	22" iron pipe size	G	28	.571		50	30		80	102
1178	24" iron pipe size	G	26	.615		53	32.50		85.50	109
4280	Cellular glass, closed cell foam, all service jacket, sealant,									
4281	working temp. (-450°F to +900°F), 0 water vapor transmission									
4284	1" wall									
4286	1/2" iron pipe size	G Q-14	120	.133	L.F.	9.40	7.05		16.45	21

22 07 19 – Plumbing Piping Insulation

22 07 19.10 Piping Insulation		Crew	Daily Output	Labor-Hours	Unit	Material	2020 Bare Costs Labor	Equipment	Total	Total Incl O&P
4300	1-1/2" wall									
4301	1" iron pipe size	G Q-14	105	.152	L.F.	10.40	8.05		18.45	24
4304	2-1/2" iron pipe size	G	90	.178		13.70	9.40		23.10	29.50
4306	3" iron pipe size	G	85	.188		17.50	9.95		27.45	34.50
4308	4" iron pipe size	G	70	.229		24.50	12.05		36.55	45.50
4310	5" iron pipe size	G	65	.246		25.50	13		38.50	48
4320	2" wall									
4322	1" iron pipe size	G Q-14	100	.160	L.F.	16.45	8.45		24.90	31
4324	2-1/2" iron pipe size	G	85	.188		21	9.95		30.95	39
4326	3" iron pipe size	G	80	.200		21	10.55		31.55	39.50
4328	4" iron pipe size	G	65	.246		29	13		42	52
4330	5" iron pipe size	G	60	.267		31.50	14.05		45.55	56
4332	6" iron pipe size	G	50	.320		36.50	16.90		53.40	66
4336	8" iron pipe size	G	40	.400		42	21		63	78.50
4338	10" iron pipe size	G	35	.457		45	24		69	86.50
4350	2-1/2" wall									
4360	12" iron pipe size	G Q-14	32	.500	L.F.	75	26.50		101.50	123
4362	14" iron pipe size	G "	28	.571	"	82.50	30		112.50	138
4370	3" wall									
4378	6" iron pipe size	G Q-14	48	.333	L.F.	55	17.60		72.60	87.50
4380	8" iron pipe size	G	38	.421		60	22		82	100
4382	10" iron pipe size	G	33	.485		69	25.50		94.50	115
4384	16" iron pipe size	G	25	.640		98.50	34		132.50	161
4386	18" iron pipe size	G	22	.727		127	38.50		165.50	199
4388	20" iron pipe size	G	20	.800		135	42		177	214
4400	3-1/2" wall									
4412	12" iron pipe size	G Q-14	27	.593	L.F.	79.50	31.50		111	136
4414	14" iron pipe size	G "	25	.640	"	96.50	34		130.50	158
4430	4" wall									
4446	16" iron pipe size	G Q-14	22	.727	L.F.	92.50	38.50		131	161
4448	18" iron pipe size	G	20	.800		118	42		160	195
4450	20" iron pipe size	G	18	.889		135	47		182	220
4480	Fittings, average with fabric and mastic									
4484	1" wall									
4486	1/2" iron pipe size	G	1 Asbe	40	.200	Ea.	7.40	11.75		19.15	26
4500	1-1/2" wall									
4502	1" iron pipe size	G	1 Asbe	38	.211	Ea.	8.15	12.35		20.50	28
4504	2-1/2" iron pipe size	G	32	.250		18.40	14.65		33.05	42.50
4506	3" iron pipe size	G	30	.267		18.20	15.65		33.85	44
4508	4" iron pipe size	G	28	.286		19.30	16.75		36.05	46.50
4510	5" iron pipe size	G	24	.333		25	19.55		44.55	57.50
4520	2" wall									
4522	1" iron pipe size	G	1 Asbe	36	.222	Ea.	13.10	13.05		26.15	34.50
4524	2-1/2" iron pipe size	G	30	.267		20	15.65		35.65	46
4526	3" iron pipe size	G	28	.286		22.50	16.75		39.25	50.50
4528	4" iron pipe size	G	24	.333		25.50	19.55		45.05	58
4530	5" iron pipe size	G	22	.364		42.50	21.50		64	80
4532	6" iron pipe size	G	20	.400		51.50	23.50		75	93
4536	8" iron pipe size	G	12	.667		89	39		128	158
4538	10" iron pipe size	G	8	1		102	58.50		160.50	202
4550	2-1/2" wall									
4560	12" iron pipe size	G	1 Asbe	6	1.333	Ea.	189	78		267	330
4562	14" iron pipe size	G "	4	2	"	207	117		324	410

For customer support on your Mechanical Costs with RSMeans Data, call 800.448.8182.

109

22 07 19.10 Piping Insulation		Crew	Daily Output	Labor-Hours	Unit	Material	2020 Bare Costs Labor	Equipment	Total	Total Incl O&P
4570	3" wall									
4578	6" iron pipe size	G	1 Asbe	16	.500	Ea.	63.50	29.50		93	115
4580	8" iron pipe size	G		10	.800		94.50	47		141.50	176
4582	10" iron pipe size	G	↓	6	1.333	↓	128	78		206	261
4900	Calcium silicate, with 8 oz. canvas cover										
5100	1" wall, 1/2" iron pipe size	G	Q 14	170	.094	L.F.	4.29	4.97		9.26	12.35
5130	3/4" iron pipe size	G		170	.094		4.32	4.97		9.29	12.40
5140	1" iron pipe size	G		170	.094		4.23	4.97		9.20	12.30
5150	1-1/4" iron pipe size	G		165	.097		4.29	5.10		9.39	12.55
5160	1-1/2" iron pipe size	G		165	.097		4.34	5.10		9.44	12.60
5170	2" iron pipe size	G		160	.100		5	5.30		10.30	13.60
5180	2-1/2" iron pipe size	G		160	.100		5.30	5.30		10.60	13.95
5190	3" iron pipe size	G		150	.107		5.80	5.65		11.45	15
5200	4" iron pipe size	G		140	.114		7.15	6.05		13.20	17.10
5210	5" iron pipe size	G		135	.119		7.55	6.25		13.80	17.90
5220	6" iron pipe size	G		130	.123		8	6.50		14.50	18.75
5280	1-1/2" wall, 1/2" iron pipe size	G		150	.107		4.65	5.65		10.30	13.75
5310	3/4" iron pipe size	G		150	.107		4.74	5.65		10.39	13.85
5320	1" iron pipe size	G		150	.107		5.15	5.65		10.80	14.30
5330	1-1/4" iron pipe size	G		145	.110		5.55	5.80		11.35	15.05
5340	1-1/2" iron pipe size	G		145	.110		5.95	5.80		11.75	15.50
5350	2" iron pipe size	G		140	.114		6.55	6.05		12.60	16.45
5360	2-1/2" iron pipe size	G		140	.114		7.15	6.05		13.20	17.10
5370	3" iron pipe size	G		135	.119		7.50	6.25		13.75	17.85
5380	4" iron pipe size	G		125	.128		8.70	6.75		15.45	19.90
5390	5" iron pipe size	G		120	.133		9.80	7.05		16.85	21.50
5400	6" iron pipe size	G		110	.145		10.10	7.70		17.80	23
5402	8" iron pipe size	G		95	.168		13.25	8.90		22.15	28
5404	10" iron pipe size	G		85	.188		17.30	9.95		27.25	34.50
5406	12" iron pipe size	G		80	.200		21	10.55		31.55	39
5408	14" iron pipe size	G		75	.213		23	11.25		34.25	43
5410	16" iron pipe size	G		70	.229		26	12.05		38.05	47
5412	18" iron pipe size	G		65	.246		28.50	13		41.50	51.50
5460	2" wall, 1/2" iron pipe size	G		135	.119		7.20	6.25		13.45	17.50
5490	3/4" iron pipe size	G		135	.119		7.55	6.25		13.80	17.90
5500	1" iron pipe size	G		135	.119		7.95	6.25		14.20	18.35
5510	1-1/4" iron pipe size	G		130	.123		8.50	6.50		15	19.30
5520	1-1/2" iron pipe size	G		130	.123		8.90	6.50		15.40	19.70
5530	2" iron pipe size	G		125	.128		9.40	6.75		16.15	20.50
5540	2-1/2" iron pipe size	G		125	.128		11.10	6.75		17.85	22.50
5550	3" iron pipe size	G		120	.133		11.20	7.05		18.25	23
5560	4" iron pipe size	G		115	.139		12.95	7.35		20.30	25.50
5570	5" iron pipe size	G		110	.145		14.75	7.70		22.45	28
5580	6" iron pipe size	G		105	.152		16.15	8.05		24.20	30
5581	8" iron pipe size	G		95	.168		19.25	8.90		28.15	34.50
5582	10" iron pipe size	G		85	.188		24	9.95		33.95	42
5583	12" iron pipe size	G		80	.200		26.50	10.55		37.05	45.50
5584	14" iron pipe size	G		75	.213		29.50	11.25		40.75	50
5585	16" iron pipe size	G		70	.229		32.50	12.05		44.55	54
5586	18" iron pipe size	G		65	.246		35.50	13		48.50	59
5587	3" wall, 1-1/4" iron pipe size	G		100	.160		13.15	8.45		21.60	27.50
5588	1-1/2" iron pipe size	G		100	.160		13.25	8.45		21.70	27.50
5589	2" iron pipe size	G	↓	95	.168	↓	13.75	8.90		22.65	29

22 07 19 – Plumbing Piping Insulation

22 07 19.10 Piping Insulation		Crew	Daily Output	Labor-Hours	Unit	Material	2020 Bare Costs Labor	Equipment	Total	Total Incl O&P
5590	2-1/2" iron pipe size	G Q-14	95	.168	L.F.	16	8.90		24.90	31.50
5591	3" iron pipe size	G	90	.178		16.20	9.40		25.60	32.50
5592	4" iron pipe size	G	85	.188		20.50	9.95		30.45	38
5593	6" iron pipe size	G	80	.200		25	10.55		35.55	43.50
5594	8" iron pipe size	G	75	.213		29.50	11.25		40.75	50
5595	10" iron pipe size	G	65	.246		35.50	13		48.50	59.50
5596	12" iron pipe size	G	60	.267		39.50	14.05		53.55	65
5597	14" iron pipe size	G	55	.291		44	15.35		59.35	72
5598	16" iron pipe size	G	50	.320		49	16.90		65.90	79.50
5599	18" iron pipe size	G	45	.356		53.50	18.75		72.25	88
5600	Calcium silicate, no cover									
5720	1" wall, 1/2" iron pipe size	G Q-14	180	.089	L.F.	3.91	4.69		8.60	11.50
5740	3/4" iron pipe size	G	180	.089		3.91	4.69		8.60	11.50
5750	1" iron pipe size	G	180	.089		3.74	4.69		8.43	11.30
5760	1-1/4" iron pipe size	G	175	.091		3.82	4.83		8.65	11.60
5770	1-1/2" iron pipe size	G	175	.091		3.85	4.83		8.68	11.65
5780	2" iron pipe size	G	170	.094		4.42	4.97		9.39	12.50
5790	2-1/2" iron pipe size	G	170	.094		4.69	4.97		9.66	12.80
5800	3" iron pipe size	G	160	.100		5.05	5.30		10.35	13.65
5810	4" iron pipe size	G	150	.107		6.30	5.65		11.95	15.55
5820	5" iron pipe size	G	145	.110		6.55	5.80		12.35	16.15
5830	6" iron pipe size	G	140	.114		6.90	6.05		12.95	16.80
5900	1-1/2" wall, 1/2" iron pipe size	G	160	.100		4.14	5.30		9.44	12.65
5920	3/4" iron pipe size	G	160	.100		4.20	5.30		9.50	12.70
5930	1" iron pipe size	G	160	.100		4.57	5.30		9.87	13.15
5940	1-1/4" iron pipe size	G	155	.103		4.94	5.45		10.39	13.80
5950	1-1/2" iron pipe size	G	155	.103		5.30	5.45		10.75	14.20
5960	2" iron pipe size	G	150	.107		5.85	5.65		11.50	15.10
5970	2-1/2" iron pipe size	G	150	.107		6.40	5.65		12.05	15.65
5980	3" iron pipe size	G	145	.110		6.65	5.80		12.45	16.30
5990	4" iron pipe size	G	135	.119		7.70	6.25		13.95	18.10
6000	5" iron pipe size	G	130	.123		8.70	6.50		15.20	19.50
6010	6" iron pipe size	G	120	.133		8.90	7.05		15.95	20.50
6020	7" iron pipe size	G	115	.139		10.45	7.35		17.80	22.50
6030	8" iron pipe size	G	105	.152		11.75	8.05		19.80	25.50
6040	9" iron pipe size	G	100	.160		14.10	8.45		22.55	28.50
6050	10" iron pipe size	G	95	.168		15.55	8.90		24.45	31
6060	12" iron pipe size	G	90	.178		18.50	9.40		27.90	35
6070	14" iron pipe size	G	85	.188		21	9.95		30.95	38.50
6080	16" iron pipe size	G	80	.200		23.50	10.55		34.05	41.50
6090	18" iron pipe size	G	75	.213		26	11.25		37.25	46
6120	2" wall, 1/2" iron pipe size	G	145	.110		6.55	5.80		12.35	16.15
6140	3/4" iron pipe size	G	145	.110		6.90	5.80		12.70	16.50
6150	1" iron pipe size	G	145	.110		7.25	5.80		13.05	16.95
6160	1-1/4" iron pipe size	G	140	.114		7.75	6.05		13.80	17.80
6170	1-1/2" iron pipe size	G	140	.114		8.10	6.05		14.15	18.15
6180	2" iron pipe size	G	135	.119		8.50	6.25		14.75	18.95
6190	2-1/2" iron pipe size	G	135	.119		10.25	6.25		16.50	21
6200	3" iron pipe size	G	130	.123		10.30	6.50		16.80	21.50
6210	4" iron pipe size	G	125	.128		11.90	6.75		18.65	23.50
6220	5" iron pipe size	G	120	.133		13.55	7.05		20.60	26
6230	6" iron pipe size	G	115	.139		14.80	7.35		22.15	27.50
6240	7" iron pipe size	G	110	.145		16.05	7.70		23.75	29.50

For customer support on your Mechanical Costs with RSMeans Data, call 800.448.8182.

111

22 07 19 – Plumbing Piping Insulation

22 07 19.10 Piping Insulation		Crew	Daily Output	Labor-Hours	Unit	Material	2020 Bare Costs Labor	Equipment	Total	Total Incl O&P
6250	8" iron pipe size	G Q-14	105	.152	L.F.	18.05	8.05		26.10	32.50
6260	9" iron pipe size	G	100	.160		20	8.45		28.45	35
6270	10" iron pipe size	G	95	.168		22	8.90		30.90	38
6280	12" iron pipe size	G	90	.178		24.50	9.40		33.90	41.50
6290	14" iron pipe size	G	85	.188		27	9.95		36.95	45.50
6300	16" iron pipe size	G	80	.200		30	10.55		40.55	49
6310	18" iron pipe size	G	75	.213		32.50	11.25		43.75	53.50
6320	20" iron pipe size	G	65	.246		40	13		53	64
6330	22" iron pipe size	G	60	.267		44.50	14.05		58.55	70.50
6340	24" iron pipe size	G	55	.291		45.50	15.35		60.85	73.50
6360	3" wall, 1/2" iron pipe size	G	115	.139		11.85	7.35		19.20	24.50
6380	3/4" iron pipe size	G	115	.139		11.90	7.35		19.25	24.50
6390	1" iron pipe size	G	115	.139		12	7.35		19.35	24.50
6400	1-1/4" iron pipe size	G	110	.145		12.20	7.70		19.90	25
6410	1-1/2" iron pipe size	G	110	.145		12.20	7.70		19.90	25
6420	2" iron pipe size	G	105	.152		12.65	8.05		20.70	26.50
6430	2-1/2" iron pipe size	G	105	.152		14.90	8.05		22.95	28.50
6440	3" iron pipe size	G	100	.160		15	8.45		23.45	29.50
6450	4" iron pipe size	G	95	.168		19.15	8.90		28.05	34.50
6460	5" iron pipe size	G	90	.178		21	9.40		30.40	37.50
6470	6" iron pipe size	G	90	.178		23.50	9.40		32.90	40
6480	7" iron pipe size	G	85	.188		26	9.95		35.95	44
6490	8" iron pipe size	G	85	.188		28	9.95		37.95	46
6500	9" iron pipe size	G	80	.200		31.50	10.55		42.05	50.50
6510	10" iron pipe size	G	75	.213		33.50	11.25		44.75	54.50
6520	12" iron pipe size	G	70	.229		37	12.05		49.05	59.50
6530	14" iron pipe size	G	65	.246		41.50	13		54.50	66
6540	16" iron pipe size	G	60	.267		46	14.05		60.05	72
6550	18" iron pipe size	G	55	.291		50.50	15.35		65.85	79
6560	20" iron pipe size	G	50	.320		60	16.90		76.90	92
6570	22" iron pipe size	G	45	.356		65	18.75		83.75	101
6580	24" iron pipe size	G	40	.400		70	21		91	110
6600	Fiberglass, with all service jacket									
6640	1/2" wall, 1/2" iron pipe size	G Q-14	250	.064	L.F.	.75	3.38		4.13	6.05
6660	3/4" iron pipe size	G	240	.067		.86	3.52		4.38	6.35
6670	1" iron pipe size	G	230	.070		.89	3.67		4.56	6.65
6680	1-1/4" iron pipe size	G	220	.073		.95	3.84		4.79	6.95
6690	1-1/2" iron pipe size	G	220	.073		1.08	3.84		4.92	7.10
6700	2" iron pipe size	G	210	.076		1.16	4.02		5.18	7.45
6710	2-1/2" iron pipe size	G	200	.080		1.21	4.22		5.43	7.85
6840	1" wall, 1/2" iron pipe size	G	240	.067		.92	3.52		4.44	6.40
6860	3/4" iron pipe size	G	230	.070		.98	3.67		4.65	6.75
6870	1" iron pipe size	G	220	.073		1.08	3.84		4.92	7.10
6880	1-1/4" iron pipe size	G	210	.076		1.16	4.02		5.18	7.45
6890	1-1/2" iron pipe size	G	210	.076		1.25	4.02		5.27	7.55
6900	2" iron pipe size	G	200	.080		1.78	4.22		6	8.45
6910	2-1/2" iron pipe size	G	190	.084		1.79	4.44		6.23	8.75
6920	3" iron pipe size	G	180	.089		1.94	4.69		6.63	9.35
6930	3-1/2" iron pipe size	G	170	.094		2.32	4.97		7.29	10.20
6940	4" iron pipe size	G	150	.107		2.58	5.65		8.23	11.50
6950	5" iron pipe size	G	140	.114		2.84	6.05		8.89	12.35
6960	6" iron pipe size	G	120	.133		3.07	7.05		10.12	14.20
6970	7" iron pipe size	G	110	.145		3.65	7.70		11.35	15.80

22 07 19.10 Piping Insulation		Crew	Daily Output	Labor-Hours	Unit	Material	2020 Bare Costs Labor	Equipment	Total	Total Incl O&P
6980	8" iron pipe size	G	Q-14	100	.160	L.F.	5	8.45		13.45	18.45
6990	9" iron pipe size	G		90	.178		5.30	9.40		14.70	20
7000	10" iron pipe size	G		90	.178		5.35	9.40		14.75	20.50
7010	12" iron pipe size	G		80	.200		5.85	10.55		16.40	22.50
7020	14" iron pipe size	G		80	.200		7.05	10.55		17.60	24
7030	16" iron pipe size	G		70	.229		9	12.05		21.05	28.50
7040	18" iron pipe size	G		70	.229		10	12.05		22.05	29.50
7050	20" iron pipe size	G		60	.267		11.15	14.05		25.20	34
7060	24" iron pipe size	G		60	.267		13.60	14.05		27.65	36.50
7080	1-1/2" wall, 1/2" iron pipe size	G		230	.070		2.06	3.67		5.73	7.90
7100	3/4" iron pipe size	G		220	.073		2.06	3.84		5.90	8.15
7110	1" iron pipe size	G		210	.076		2.21	4.02		6.23	8.60
7120	1-1/4" iron pipe size	G		200	.080		2.40	4.22		6.62	9.15
7130	1-1/2" iron pipe size	G		200	.080		2.81	4.22		7.03	9.60
7140	2" iron pipe size	G		190	.084		2.79	4.44		7.23	9.85
7150	2-1/2" iron pipe size	G		180	.089		2.98	4.69		7.67	10.50
7160	3" iron pipe size	G		170	.094		3.12	4.97		8.09	11.10
7170	3-1/2" iron pipe size	G		160	.100		3.42	5.30		8.72	11.85
7180	4" iron pipe size	G		140	.114		3.95	6.05		10	13.60
7190	5" iron pipe size	G		130	.123		3.97	6.50		10.47	14.30
7200	6" iron pipe size	G		110	.145		4.20	7.70		11.90	16.40
7210	7" iron pipe size	G		100	.160		4.66	8.45		13.11	18.10
7220	8" iron pipe size	G		90	.178		5.90	9.40		15.30	21
7230	9" iron pipe size	G		85	.188		6.10	9.95		16.05	22
7240	10" iron pipe size	G		80	.200		6.35	10.55		16.90	23
7250	12" iron pipe size	G		75	.213		7.20	11.25		18.45	25
7260	14" iron pipe size	G		70	.229		8.65	12.05		20.70	28
7270	16" iron pipe size	G		65	.246		11.30	13		24.30	32.50
7280	18" iron pipe size	G		60	.267		12.65	14.05		26.70	35.50
7290	20" iron pipe size	G		55	.291		12.95	15.35		28.30	37.50
7300	24" iron pipe size	G		50	.320		15.85	16.90		32.75	43.50
7320	2" wall, 1/2" iron pipe size	G		220	.073		3.20	3.84		7.04	9.40
7340	3/4" iron pipe size	G		210	.076		3.30	4.02		7.32	9.80
7350	1" iron pipe size	G		200	.080		3.52	4.22		7.74	10.35
7360	1-1/4" iron pipe size	G		190	.084		3.71	4.44		8.15	10.90
7370	1-1/2" iron pipe size	G		190	.084		4.47	4.44		8.91	11.70
7380	2" iron pipe size	G		180	.089		4.08	4.69		8.77	11.70
7390	2-1/2" iron pipe size	G		170	.094		4.39	4.97		9.36	12.50
7400	3" iron pipe size	G		160	.100		4.67	5.30		9.97	13.25
7410	3-1/2" iron pipe size	G		150	.107		5.05	5.65		10.70	14.20
7420	4" iron pipe size	G		130	.123		6.25	6.50		12.75	16.85
7430	5" iron pipe size	G		120	.133		6.20	7.05		13.25	17.65
7440	6" iron pipe size	G		100	.160		6.90	8.45		15.35	20.50
7450	7" iron pipe size	G		90	.178		7.30	9.40		16.70	22.50
7460	8" iron pipe size	G		80	.200		7.80	10.55		18.35	25
7470	9" iron pipe size	G		75	.213		8.55	11.25		19.80	26.50
7480	10" iron pipe size	G		70	.229		9.30	12.05		21.35	29
7490	12" iron pipe size	G		65	.246		10.45	13		23.45	31.50
7500	14" iron pipe size	G		60	.267		13.60	14.05		27.65	36.50
7510	16" iron pipe size	G		55	.291		14.90	15.35		30.25	40
7520	18" iron pipe size	G		50	.320		16.60	16.90		33.50	44.50
7530	20" iron pipe size	G		45	.356		18.90	18.75		37.65	50
7540	24" iron pipe size	G		40	.400		20	21		41	54.50

For customer support on your Mechanical Costs with RSMeans Data, call 800.448.8182.

113

22 07 19.10 Piping Insulation		Crew	Daily Output	Labor-Hours	Unit	Material	2020 Bare Costs Labor	Equipment	Total	Total Incl O&P
7560	2-1/2" wall, 1/2" iron pipe size	G	Q-14	210	.076	L.F.	3.79	4.02		7.81	10.30
7562	3/4" iron pipe size	G		200	.080		3.96	4.22		8.18	10.85
7564	1" iron pipe size	G		190	.084		4.13	4.44		8.57	11.35
7566	1-1/4" iron pipe size	G		185	.086		4.28	4.56		8.84	11.70
7568	1-1/2" iron pipe size	C		180	.089		4.50	4.69		9.19	12.15
7570	2" iron pipe size	G		170	.094		4.72	4.97		9.69	12.85
7572	2-1/2" iron pipe size	G		160	.100		5.45	5.30		10.75	14.10
7574	3" iron pipe size	G		150	.107		5.70	5.65		11.35	14.95
7576	3-1/2" iron pipe size	G		140	.114		6.25	6.05		12.30	16.10
7578	4" iron pipe size	G		120	.133		6.55	7.05		13.60	18.05
7580	5" iron pipe size	G		110	.145		5.65	7.70		13.35	18.05
7582	6" iron pipe size	G		90	.178		9.45	9.40		18.85	25
7584	7" iron pipe size	G		80	.200		9.50	10.55		20.05	26.50
7586	8" iron pipe size	G		70	.229		10.05	12.05		22.10	29.50
7588	9" iron pipe size	G		65	.246		11	13		24	32
7590	10" iron pipe size	G		60	.267		12	14.05		26.05	34.50
7592	12" iron pipe size	G		55	.291		14.55	15.35		29.90	39.50
7594	14" iron pipe size	G		50	.320		17.15	16.90		34.05	45
7596	16" iron pipe size	G		45	.356		19.70	18.75		38.45	50.50
7598	18" iron pipe size	G		40	.400		21.50	21		42.50	56
7602	24" iron pipe size	G		30	.533		28	28		56	74
7620	3" wall, 1/2" iron pipe size	G		200	.080		4.85	4.22		9.07	11.85
7622	3/4" iron pipe size	G		190	.084		5.15	4.44		9.59	12.45
7624	1" iron pipe size	G		180	.089		5.45	4.69		10.14	13.20
7626	1-1/4" iron pipe size	G		175	.091		5.55	4.83		10.38	13.50
7628	1-1/2" iron pipe size	G		170	.094		5.80	4.97		10.77	14.05
7630	2" iron pipe size	G		160	.100		6.25	5.30		11.55	15
7632	2-1/2" iron pipe size	G		150	.107		6.50	5.65		12.15	15.80
7634	3" iron pipe size	G		140	.114		6.95	6.05		13	16.85
7636	3-1/2" iron pipe size	G		130	.123		7.65	6.50		14.15	18.40
7638	4" iron pipe size	G		110	.145		8.20	7.70		15.90	21
7640	5" iron pipe size	G		100	.160		9.30	8.45		17.75	23
7642	6" iron pipe size	G		80	.200		9.95	10.55		20.50	27
7644	7" iron pipe size	G		70	.229		11.40	12.05		23.45	31
7646	8" iron pipe size	G		60	.267		12.40	14.05		26.45	35
7648	9" iron pipe size	G		55	.291		13.35	15.35		28.70	38
7650	10" iron pipe size	G		50	.320		14.35	16.90		31.25	42
7652	12" iron pipe size	G		45	.356		17.90	18.75		36.65	48.50
7654	14" iron pipe size	G		40	.400		21	21		42	55.50
7656	16" iron pipe size	G		35	.457		23.50	24		47.50	63
7658	18" iron pipe size	G		32	.500		25	26.50		51.50	68
7660	20" iron pipe size	G		30	.533		27.50	28		55.50	73
7662	24" iron pipe size	G		28	.571		34.50	30		64.50	84.50
7664	26" iron pipe size	G		26	.615		38.50	32.50		71	92
7666	30" iron pipe size	G		24	.667		48	35		83	107
7800	For fiberglass with standard canvas jacket, deduct						5%			
7802	For fittings, add 3 L.F. for each fitting									
7804	plus 4 L.F. for each flange of the fitting									
7820	Polyethylene tubing flexible closed cell foam, UV resistant									
7828	Standard temperature (-90°F to +212°F)									
7830	3/8" wall, 1/8" iron pipe size	G	1 Asbe	130	.062	L.F.	.33	3.61		3.94	5.90
7831	1/4" iron pipe size	G		130	.062		.34	3.61		3.95	5.90
7832	3/8" iron pipe size	G		130	.062		.33	3.61		3.94	5.90

22 07 19 – Plumbing Piping Insulation

22 07 19.10 Piping Insulation		Crew	Daily Output	Labor-Hours	Unit	Material	2020 Bare Costs Labor	Equipment	Total	Total Incl O&P
7833	1/2" iron pipe size	G	1 Asbe	126	.063	L.F.	.43	3.72		4.15	6.15
7834	3/4" iron pipe size	G		122	.066		.45	3.85		4.30	6.40
7835	1" iron pipe size	G		120	.067		.55	3.91		4.46	6.60
7836	1-1/4" iron pipe size	G		118	.068		.70	3.98		4.68	6.85
7837	1-1/2" iron pipe size	G		118	.068		.76	3.98		4.74	6.95
7838	2" iron pipe size	G		116	.069		1.12	4.05		5.17	7.45
7839	2-1/2" iron pipe size	G		114	.070		1.39	4.12		5.51	7.85
7840	3" iron pipe size	G		112	.071		2.26	4.19		6.45	8.95
7842	1/2" wall, 1/8" iron pipe size	G		120	.067		.42	3.91		4.33	6.45
7843	1/4" iron pipe size	G		120	.067		.52	3.91		4.43	6.55
7844	3/8" iron pipe size	G		120	.067		.55	3.91		4.46	6.60
7845	1/2" iron pipe size	G		118	.068		.63	3.98		4.61	6.80
7846	3/4" iron pipe size	G		116	.069		.64	4.05		4.69	6.90
7847	1" iron pipe size	G		114	.070		.71	4.12		4.83	7.10
7848	1-1/4" iron pipe size	G		112	.071		.82	4.19		5.01	7.35
7849	1-1/2" iron pipe size	G		110	.073		.95	4.27		5.22	7.60
7850	2" iron pipe size	G		108	.074		1.61	4.34		5.95	8.40
7851	2-1/2" iron pipe size	G		106	.075		2.06	4.43		6.49	9.05
7852	3" iron pipe size	G		104	.077		2.68	4.51		7.19	9.90
7853	3-1/2" iron pipe size	G		102	.078		3.01	4.60		7.61	10.35
7854	4" iron pipe size	G		100	.080		3.59	4.69		8.28	11.15
7855	3/4" wall, 1/8" iron pipe size	G		110	.073		.63	4.27		4.90	7.25
7856	1/4" iron pipe size	G		110	.073		.68	4.27		4.95	7.30
7857	3/8" iron pipe size	G		108	.074		.90	4.34		5.24	7.65
7858	1/2" iron pipe size	G		106	.075		1.07	4.43		5.50	8
7859	3/4" iron pipe size	G		104	.077		1.27	4.51		5.78	8.35
7860	1" iron pipe size	G		102	.078		1.49	4.60		6.09	8.70
7861	1-1/4" iron pipe size	G		100	.080		1.76	4.69		6.45	9.15
7862	1-1/2" iron pipe size	G		100	.080		1.89	4.69		6.58	9.30
7863	2" iron pipe size	G		98	.082		2.87	4.79		7.66	10.50
7864	2-1/2" iron pipe size	G		96	.083		3.56	4.89		8.45	11.40
7865	3" iron pipe size	G		94	.085		3.81	4.99		8.80	11.85
7866	3-1/2" iron pipe size	G		92	.087		5.10	5.10		10.20	13.45
7867	4" iron pipe size	G		90	.089		5.70	5.20		10.90	14.25
7868	1" wall, 1/4" iron pipe size	G		100	.080		1.56	4.69		6.25	8.90
7869	3/8" iron pipe size	G		98	.082		1.51	4.79		6.30	9
7870	1/2" iron pipe size	G		96	.083		1.97	4.89		6.86	9.65
7871	3/4" iron pipe size	G		94	.085		2.26	4.99		7.25	10.15
7872	1" iron pipe size	G		92	.087		2.76	5.10		7.86	10.90
7873	1-1/4" iron pipe size	G		90	.089		3.13	5.20		8.33	11.45
7874	1-1/2" iron pipe size	G		90	.089		3.46	5.20		8.66	11.80
7875	2" iron pipe size	G		88	.091		4.64	5.35		9.99	13.30
7876	2-1/2" iron pipe size	G		86	.093		6.30	5.45		11.75	15.30
7877	3" iron pipe size	G		84	.095		6.90	5.60		12.50	16.20
7878	Contact cement, quart can	G				Ea.	13.20			13.20	14.55
7879	Rubber tubing, flexible closed cell foam										
7880	3/8" wall, 1/4" iron pipe size	G	1 Asbe	120	.067	L.F.	.33	3.91		4.24	6.35
7900	3/8" iron pipe size	G		120	.067		.46	3.91		4.37	6.50
7910	1/2" iron pipe size	G		115	.070		.50	4.08		4.58	6.80
7920	3/4" iron pipe size	G		115	.070		.57	4.08		4.65	6.90
7930	1" iron pipe size	G		110	.073		.60	4.27		4.87	7.20
7940	1-1/4" iron pipe size	G		110	.073		.68	4.27		4.95	7.30
7950	1-1/2" iron pipe size	G		110	.073		.79	4.27		5.06	7.40

22 07 19 – Plumbing Piping Insulation

22 07 19.10 Piping Insulation		Crew	Daily Output	Labor-Hours	Unit	Material	2020 Bare Costs Labor	Equipment	Total	Total Incl O&P
8100	1/2" wall, 1/4" iron pipe size	G	1 Asbe	90	.089	L.F.	.99	5.20		6.19	9.10
8120	3/8" iron pipe size	G		90	.089		1.05	5.20		6.25	9.15
8130	1/2" iron pipe size	G		89	.090		1.06	5.25		6.31	9.25
8140	3/4" iron pipe size	G		89	.090		1.22	5.25		6.47	9.45
8150	1" iron pipe size	G		88	.091		.86	5.35		6.21	9.15
8160	1-1/4" iron pipe size	G		87	.092		1.21	5.40		6.61	9.65
8170	1-1/2" iron pipe size	G		87	.092		1.89	5.40		7.29	10.40
8180	2" iron pipe size	G		86	.093		2.38	5.45		7.83	11
8190	2-1/2" iron pipe size	G		86	.093		2.60	5.45		8.05	11.25
8200	3" iron pipe size	G		85	.094		2.78	5.50		8.28	11.55
8210	3-1/2" iron pipe size	G		85	.094		3.81	5.50		9.31	12.70
8220	4" iron pipe size	G		80	.100		4.12	5.85		9.97	13.55
8230	5" iron pipe size	G		80	.100		5.60	5.85		11.45	15.15
8240	6" iron pipe size	G		75	.107		5.60	6.25		11.85	15.75
8300	3/4" wall, 1/4" iron pipe size	G		90	.089		.94	5.20		6.14	9.05
8320	3/8" iron pipe size	G		90	.089		1.10	5.20		6.30	9.20
8330	1/2" iron pipe size	G		89	.090		1.13	5.25		6.38	9.35
8340	3/4" iron pipe size	G		89	.090		2.13	5.25		7.38	10.45
8350	1" iron pipe size	G		88	.091		2.10	5.35		7.45	10.50
8360	1-1/4" iron pipe size	G		87	.092		2.46	5.40		7.86	11
8370	1-1/2" iron pipe size	G		87	.092		3.07	5.40		8.47	11.70
8380	2" iron pipe size	G		86	.093		4.27	5.45		9.72	13.10
8390	2-1/2" iron pipe size	G		86	.093		4.89	5.45		10.34	13.80
8400	3" iron pipe size	G		85	.094		5.50	5.50		11	14.55
8410	3-1/2" iron pipe size	G		85	.094		5.20	5.50		10.70	14.25
8420	4" iron pipe size	G		80	.100		6.80	5.85		12.65	16.50
8430	5" iron pipe size	G		80	.100		8.15	5.85		14	17.95
8440	6" iron pipe size	G		80	.100		9.90	5.85		15.75	19.90
8444	1" wall, 1/2" iron pipe size	G		86	.093		3.18	5.45		8.63	11.90
8445	3/4" iron pipe size	G		84	.095		3.90	5.60		9.50	12.90
8446	1" iron pipe size	G		84	.095		3.49	5.60		9.09	12.45
8447	1-1/4" iron pipe size	G		82	.098		3.83	5.70		9.53	13
8448	1-1/2" iron pipe size	G		82	.098		6.15	5.70		11.85	15.55
8449	2" iron pipe size	G		80	.100		7.55	5.85		13.40	17.30
8450	2-1/2" iron pipe size	G		80	.100		8.45	5.85		14.30	18.30
8456	Rubber insulation tape, 1/8" x 2" x 30'	G				Ea.	23			23	25
8460	Polyolefin tubing, flexible closed cell foam, UV stabilized, work										
8462	temp. -165°F to +210°F, 0 water vapor transmission										
8464	3/8" wall, 1/8" iron pipe size	G	1 Asbe	140	.057	L.F.	.49	3.35		3.84	5.70
8466	1/4" iron pipe size	G		140	.057		.51	3.35		3.86	5.70
8468	3/8" iron pipe size	G		140	.057		.57	3.35		3.92	5.80
8470	1/2" iron pipe size	G		136	.059		.63	3.45		4.08	6
8472	3/4" iron pipe size	G		132	.061		.63	3.55		4.18	6.15
8474	1" iron pipe size	G		130	.062		.84	3.61		4.45	6.45
8476	1-1/4" iron pipe size	G		128	.063		1.03	3.67		4.70	6.80
8478	1-1/2" iron pipe size	G		128	.063		1.26	3.67		4.93	7.05
8480	2" iron pipe size	G		126	.063		1.52	3.72		5.24	7.35
8482	2-1/2" iron pipe size	G		123	.065		2.21	3.81		6.02	8.30
8484	3" iron pipe size	G		121	.066		2.46	3.88		6.34	8.65
8486	4" iron pipe size	G		118	.068		5.05	3.98		9.03	11.65
8500	1/2" wall, 1/8" iron pipe size	G		130	.062		.72	3.61		4.33	6.35
8502	1/4" iron pipe size	G		130	.062		.77	3.61		4.38	6.40
8504	3/8" iron pipe size	G		130	.062		.83	3.61		4.44	6.45

22 07 Plumbing Insulation

22 07 19 – Plumbing Piping Insulation

22 07 19.10 Piping Insulation

		Crew	Daily Output	Labor-Hours	Unit	Material	2020 Bare Costs Labor	Equipment	Total	Total Incl O&P
8506	1/2" iron pipe size G	1 Asbe	128	.063	L.F.	.89	3.67		4.56	6.65
8508	3/4" iron pipe size G		126	.063		1.05	3.72		4.77	6.85
8510	1" iron pipe size G		123	.065		1.19	3.81		5	7.15
8512	1-1/4" iron pipe size G		121	.066		1.41	3.88		5.29	7.50
8514	1-1/2" iron pipe size G		119	.067		1.69	3.94		5.63	7.90
8516	2" iron pipe size G		117	.068		2.14	4.01		6.15	8.50
8518	2-1/2" iron pipe size G		114	.070		2.90	4.12		7.02	9.50
8520	3" iron pipe size G		112	.071		3.77	4.19		7.96	10.60
8522	4" iron pipe size G		110	.073		5.20	4.27		9.47	12.30
8534	3/4" wall, 1/8" iron pipe size G		120	.067		1.10	3.91		5.01	7.20
8536	1/4" iron pipe size G		120	.067		1.16	3.91		5.07	7.30
8538	3/8" iron pipe size G		117	.068		1.33	4.01		5.34	7.60
8540	1/2" iron pipe size G		114	.070		1.47	4.12		5.59	7.90
8542	3/4" iron pipe size G		112	.071		1.59	4.19		5.78	8.20
8544	1" iron pipe size G		110	.073		2.24	4.27		6.51	9
8546	1-1/4" iron pipe size G		108	.074		2.95	4.34		7.29	9.90
8548	1-1/2" iron pipe size G		108	.074		3.50	4.34		7.84	10.50
8550	2" iron pipe size G		106	.075		4.42	4.43		8.85	11.65
8552	2-1/2" iron pipe size G		104	.077		5.55	4.51		10.06	13.10
8554	3" iron pipe size G		102	.078		7.20	4.60		11.80	15
8556	4" iron pipe size G		100	.080		9.15	4.69		13.84	17.25
8570	1" wall, 1/8" iron pipe size G		110	.073		1.98	4.27		6.25	8.75
8572	1/4" iron pipe size G		108	.074		2.37	4.34		6.71	9.25
8574	3/8" iron pipe size G		106	.075		2.08	4.43		6.51	9.10
8576	1/2" iron pipe size G		104	.077		2.12	4.51		6.63	9.30
8578	3/4" iron pipe size G		102	.078		2.82	4.60		7.42	10.15
8580	1" iron pipe size G		100	.080		3.16	4.69		7.85	10.70
8582	1-1/4" iron pipe size G		97	.082		3.79	4.84		8.63	11.60
8584	1-1/2" iron pipe size G		97	.082		4.32	4.84		9.16	12.20
8586	2" iron pipe size G		95	.084		5.60	4.94		10.54	13.75
8588	2-1/2" iron pipe size G		93	.086		7.95	5.05		13	16.50
8590	3" iron pipe size G		91	.088		9.70	5.15		14.85	18.60
8606	Contact adhesive (R-320) G				Qt.	19.40			19.40	21.50
8608	Contact adhesive (R-320) G				Gal.	77.50			77.50	85.50
8610	NOTE: Preslit/preglued vs. unslit, same price									

22 07 19.30 Piping Insulation Protective Jacketing, PVC

		Crew	Daily Output	Labor-Hours	Unit	Material	2020 Bare Costs Labor	Equipment	Total	Total Incl O&P
0010	**PIPING INSULATION PROTECTIVE JACKETING, PVC**									
0100	PVC, white, 48" lengths cut from roll goods									
0120	20 mil thick									
0140	Size based on OD of insulation									
0150	1-1/2" ID	Q-14	270	.059	L.F.	.29	3.13		3.42	5.10
0152	2" ID		260	.062		.39	3.25		3.64	5.40
0154	2-1/2" ID		250	.064		.45	3.38		3.83	5.70
0156	3" ID		240	.067		.53	3.52		4.05	6
0158	3-1/2" ID		230	.070		.61	3.67		4.28	6.30
0160	4" ID		220	.073		.68	3.84		4.52	6.65
0162	4-1/2" ID		210	.076		.77	4.02		4.79	7
0164	5" ID		200	.080		.84	4.22		5.06	7.40
0166	5-1/2" ID		190	.084		.92	4.44		5.36	7.80
0168	6" ID		180	.089		1	4.69		5.69	8.30
0170	6-1/2" ID		175	.091		1.08	4.83		5.91	8.60
0172	7" ID		170	.094		1.15	4.97		6.12	8.90

22 07 19 – Plumbing Piping Insulation

22 07 19.30 Piping Insulation Protective Jacketing, PVC		Crew	Daily Output	Labor-Hours	Unit	Material	2020 Bare Costs Labor	Equipment	Total	Total Incl O&P
0174	7-1/2" ID	Q-14	164	.098	L.F.	1.25	5.15		6.40	9.30
0176	8" ID		161	.099		1.33	5.25		6.58	9.50
0178	8-1/2" ID		158	.101		1.40	5.35		6.75	9.75
0180	9" ID		155	.103		1.48	5.45		6.93	10
0182	9-1/2" ID		152	.105		1.56	5.55		7.11	10.25
0184	10" ID		149	.107		1.63	5.65		7.28	10.50
0186	10-1/2" ID		146	.110		1.72	5.80		7.52	10.80
0188	11" ID		143	.112		1.80	5.90		7.70	11.05
0190	11-1/2" ID		140	.114		1.87	6.05		7.92	11.30
0192	12" ID		137	.117		1.95	6.15		8.10	11.60
0194	12-1/2" ID		134	.119		2.03	6.30		8.33	11.95
0195	13" ID		132	.121		2.11	6.40		8.51	12.10
0196	13-1/2" ID		132	.121		2.19	6.40		8.59	12.20
0198	14" ID		130	.123		2.27	6.50		8.77	12.45
0200	15" ID		128	.125		2.42	6.60		9.02	12.80
0202	16" ID		126	.127		2.58	6.70		9.28	13.15
0204	17" ID		124	.129		2.74	6.80		9.54	13.45
0206	18" ID		122	.131		2.90	6.90		9.80	13.85
0208	19" ID		120	.133		3.05	7.05		10.10	14.15
0210	20" ID		118	.136		3.24	7.15		10.39	14.55
0212	21" ID		116	.138		3.37	7.30		10.67	14.90
0214	22" ID		114	.140		3.54	7.40		10.94	15.25
0216	23" ID		112	.143		3.70	7.55		11.25	15.65
0218	24" ID		110	.145		3.86	7.70		11.56	16.05
0220	25" ID		108	.148		4.02	7.80		11.82	16.40
0222	26" ID		106	.151		4.18	7.95		12.13	16.85
0224	27" ID		104	.154		4.34	8.10		12.44	17.20
0226	28" ID		102	.157		4.49	8.30		12.79	17.65
0228	29" ID		100	.160		4.66	8.45		13.11	18.10
0230	30" ID	↓	98	.163	↓	4.82	8.60		13.42	18.55
0300	For colors, add				Ea.	10%				
1000	30 mil thick									
1010	Size based on OD of insulation									
1020	2" ID	Q-14	260	.062	L.F.	.55	3.25		3.80	5.60
1022	2-1/2" ID		250	.064		.67	3.38		4.05	5.95
1024	3" ID		240	.067		.80	3.52		4.32	6.30
1026	3-1/2" ID		230	.070		.91	3.67		4.58	6.65
1028	4" ID		220	.073		1.02	3.84		4.86	7
1030	4-1/2" ID		210	.076		1.14	4.02		5.16	7.40
1032	5" ID		200	.080		1.26	4.22		5.48	7.90
1034	5-1/2" ID		190	.084		1.38	4.44		5.82	8.30
1036	6" ID		180	.089		1.49	4.69		6.18	8.85
1038	6-1/2" ID		175	.091		1.61	4.83		6.44	9.15
1040	7" ID		170	.094		1.73	4.97		6.70	9.55
1042	7-1/2" ID		164	.098		1.86	5.15		7.01	9.95
1044	8" ID		161	.099		1.97	5.25		7.22	10.20
1046	8-1/2" ID		158	.101		2.09	5.35		7.44	10.50
1048	9" ID		155	.103		2.21	5.45		7.66	10.80
1050	9-1/2" ID		152	.105		2.33	5.55		7.88	11.10
1052	10" ID		149	.107		2.44	5.65		8.09	11.40
1054	10-1/2" ID		146	.110		2.56	5.80		8.36	11.70
1056	11" ID		143	.112		2.69	5.90		8.59	12
1058	11-1/2" ID	↓	140	.114	↓	2.80	6.05		8.85	12.35

22 07 19.30 Piping Insulation Protective Jacketing, PVC		Crew	Daily Output	Labor-Hours	Unit	Material	2020 Bare Costs Labor	Equipment	Total	Total Incl O&P
1060	12" ID	Q-14	137	.117	L.F.	2.91	6.15		9.06	12.65
1062	12-1/2" ID		134	.119		3.04	6.30		9.34	13.05
1063	13" ID		132	.121		3.16	6.40		9.56	13.30
1064	13-1/2" ID		132	.121		3.28	6.40		9.68	13.40
1066	14" ID		130	.123		3.39	6.50		9.89	13.70
1068	15" ID		128	.125		3.63	6.60		10.23	14.15
1070	16" ID		126	.127		3.87	6.70		10.57	14.55
1072	17" ID		124	.129		4.11	6.80		10.91	14.95
1074	18" ID		122	.131		4.34	6.90		11.24	15.40
1076	19" ID		120	.133		4.58	7.05		11.63	15.85
1078	20" ID		118	.136		4.80	7.15		11.95	16.30
1080	21" ID		116	.138		5.05	7.30		12.35	16.75
1082	22" ID		114	.140		5.25	7.40		12.65	17.15
1084	23" ID		112	.143		5.55	7.55		13.10	17.70
1086	24" ID		110	.145		5.75	7.70		13.45	18.15
1088	25" ID		108	.148		6	7.80		13.80	18.60
1090	26" ID		106	.151		6.25	7.95		14.20	19.10
1092	27" ID		104	.154		6.50	8.10		14.60	19.60
1094	28" ID		102	.157		6.70	8.30		15	20
1096	29" ID		100	.160		6.95	8.45		15.40	20.50
1098	30" ID		98	.163		7.20	8.60		15.80	21
1300	For colors, add				Ea.	10%				
2000	PVC, white, fitting covers									
2020	Fiberglass insulation inserts included with sizes 1-3/4" thru 9-3/4"									
2030	Size is based on OD of insulation									
2040	90° elbow fitting									
2060	1-3/4"	Q-14	135	.119	Ea.	.55	6.25		6.80	10.20
2062	2"		130	.123		.68	6.50		7.18	10.70
2064	2-1/4"		128	.125		.79	6.60		7.39	11
2068	2-1/2"		126	.127		.84	6.70		7.54	11.20
2070	2-3/4"		123	.130		.99	6.85		7.84	11.65
2072	3"		120	.133		.96	7.05		8.01	11.85
2074	3-3/8"		116	.138		1.13	7.30		8.43	12.45
2076	3-3/4"		113	.142		1.21	7.45		8.66	12.80
2078	4-1/8"		110	.145		1.60	7.70		9.30	13.55
2080	4-3/4"		105	.152		1.93	8.05		9.98	14.45
2082	5-1/4"		100	.160		2.25	8.45		10.70	15.45
2084	5-3/4"		95	.168		2.84	8.90		11.74	16.75
2086	6-1/4"		90	.178		4.73	9.40		14.13	19.60
2088	6-3/4"		87	.184		4.99	9.70		14.69	20.50
2090	7-1/4"		85	.188		6.25	9.95		16.20	22
2092	7-3/4"		83	.193		6.60	10.15		16.75	23
2094	8-3/4"		80	.200		8.45	10.55		19	25.50
2096	9-3/4"		77	.208		11.30	10.95		22.25	29.50
2098	10-7/8"		74	.216		12.60	11.40		24	31.50
2100	11-7/8"		71	.225		14.45	11.90		26.35	34
2102	12-7/8"		68	.235		19.95	12.40		32.35	41
2104	14-1/8"		66	.242		21	12.80		33.80	42.50
2106	15-1/8"		64	.250		22.50	13.20		35.70	45
2108	16-1/8"		63	.254		24.50	13.40		37.90	47.50
2110	17-1/8"		62	.258		27	13.60		40.60	51
2112	18-1/8"		61	.262		36.50	13.85		50.35	61.50
2114	19-1/8"		60	.267		47.50	14.05		61.55	73.50

22 07 19.30 Piping Insulation Protective Jacketing, PVC		Crew	Daily Output	Labor-Hours	Unit	Material	2020 Bare Costs Labor	Equipment	Total	Total Incl O&P
2116	20-1/8"	Q-14	59	.271	Ea.	61	14.30		75.30	89
2200	45° elbow fitting									
2220	1-3/4" thru 9-3/4" same price as 90° elbow fitting									
2320	10-7/8"	Q-14	74	.216	Ea.	12.60	11.40		24	31.50
2322	11-7/8"		71	.225		13.85	11.90		25.75	33.50
2324	12-7/8"		68	.235		15.50	12.40		27.90	36
2326	14-1/8"		66	.242		17.65	12.80		30.45	39
2328	15-1/8"		64	.250		19	13.20		32.20	41.50
2330	16-1/8"		63	.254		21.50	13.40		34.90	44.50
2332	17-1/8"		62	.258		24.50	13.60		38.10	48
2334	18-1/8"		61	.262		30	13.85		43.85	54.50
2336	19-1/8"		60	.267		40.50	14.05		54.55	66.50
2338	20-1/8"		59	.271		46	14.30		60.30	72.50
2400	Tee fitting									
2410	1-3/4"	Q-14	96	.167	Ea.	1.04	8.80		9.84	14.65
2412	2"		94	.170		1.18	9		10.18	15.10
2414	2-1/4"		91	.176		1.27	9.30		10.57	15.65
2416	2-1/2"		88	.182		1.39	9.60		10.99	16.30
2418	2-3/4"		85	.188		1.53	9.95		11.48	16.95
2420	3"		82	.195		1.66	10.30		11.96	17.65
2422	3-3/8"		79	.203		1.91	10.70		12.61	18.50
2424	3-3/4"		76	.211		2.37	11.10		13.47	19.65
2426	4-1/8"		73	.219		2.57	11.55		14.12	20.50
2428	4-3/4"		70	.229		3.19	12.05		15.24	22
2430	5-1/4"		67	.239		3.84	12.60		16.44	23.50
2432	5-3/4"		63	.254		5.10	13.40		18.50	26
2434	6-1/4"		60	.267		6.70	14.05		20.75	29
2436	6-3/4"		59	.271		8.30	14.30		22.60	31
2438	7-1/4"		57	.281		13.40	14.80		28.20	38
2440	7-3/4"		54	.296		14.70	15.65		30.35	40
2442	8-3/4"		52	.308		17.85	16.25		34.10	44.50
2444	9-3/4"		50	.320		21	16.90		37.90	49
2446	10-7/8"		48	.333		21.50	17.60		39.10	50.50
2448	11-7/8"		47	.340		23.50	17.95		41.45	53.50
2450	12-7/8"		46	.348		26	18.35		44.35	56.50
2452	14-1/8"		45	.356		28.50	18.75		47.25	60.50
2454	15-1/8"		44	.364		31	19.20		50.20	63.50
2456	16-1/8"		43	.372		33	19.65		52.65	66.50
2458	17-1/8"		42	.381		35.50	20		55.50	70
2460	18-1/8"		41	.390		39	20.50		59.50	74.50
2462	19-1/8"		40	.400		42.50	21		63.50	79.50
2464	20-1/8"		39	.410		47	21.50		68.50	85
4000	Mechanical grooved fitting cover, including insert									
4020	90° elbow fitting									
4030	3/4" & 1"	Q-14	140	.114	Ea.	5.05	6.05		11.10	14.80
4040	1-1/4" & 1-1/2"		135	.119		6.20	6.25		12.45	16.40
4042	2"		130	.123		8.60	6.50		15.10	19.45
4044	2-1/2"		125	.128		9.60	6.75		16.35	21
4046	3"		120	.133		10.75	7.05		17.80	22.50
4048	3-1/2"		115	.139		12.45	7.35		19.80	25
4050	4"		110	.145		13.85	7.70		21.55	27
4052	5"		100	.160		17.25	8.45		25.70	32
4054	6"		90	.178		25.50	9.40		34.90	42.50

For customer support on your Mechanical Costs with RSMeans Data, call 800.448.8182.

22 07 19.30 Piping Insulation Protective Jacketing, PVC

		Crew	Daily Output	Labor-Hours	Unit	Material	2020 Bare Costs Labor	Equipment	Total	Total Incl O&P
4056	8"	Q-14	80	.200	Ea.	27.50	10.55		38.05	46.50
4058	10"		75	.213		35.50	11.25		46.75	56.50
4060	12"		68	.235		51	12.40		63.40	75
4062	14"		65	.246		61	13		74	87.50
4064	16"		63	.254		83	13.40		96.40	112
4066	18"	▼	61	.262	▼	114	13.85		127.85	148
4100	45° elbow fitting									
4120	3/4" & 1"	Q-14	140	.114	Ea.	4.53	6.05		10.58	14.25
4130	1-1/4" & 1-1/2"		135	.119		5.60	6.25		11.85	15.75
4140	2"		130	.123		8.40	6.50		14.90	19.20
4142	2-1/2"		125	.128		9.35	6.75		16.10	20.50
4144	3"		120	.133		9.60	7.05		16.65	21.50
4146	3-1/2"		115	.139		12.15	7.35		19.50	24.50
4148	4"		110	.145		15.45	7.70		23.15	29
4150	5"		100	.160		16.75	8.45		25.20	31.50
4152	6"		90	.178		23.50	9.40		32.90	40.50
4154	8"		80	.200		24.50	10.55		35.05	43
4156	10"		75	.213		31.50	11.25		42.75	52.50
4158	12"		68	.235		47.50	12.40		59.90	71
4160	14"		65	.246		64	13		77	90.50
4162	16"		63	.254		78.50	13.40		91.90	107
4164	18"	▼	61	.262	▼	101	13.85		114.85	133
4200	Tee fitting									
4220	3/4" & 1"	Q-14	93	.172	Ea.	6.60	9.10		15.70	21
4230	1-1/4" & 1-1/2"		90	.178		8.10	9.40		17.50	23.50
4240	2"		87	.184		13.55	9.70		23.25	30
4242	2-1/2"		84	.190		15.10	10.05		25.15	32
4244	3"		80	.200		16.20	10.55		26.75	34
4246	3-1/2"		77	.208		17.35	10.95		28.30	36
4248	4"		73	.219		21.50	11.55		33.05	42
4250	5"		67	.239		23.50	12.60		36.10	45.50
4252	6"		60	.267		37	14.05		51.05	62
4254	8"		54	.296		38.50	15.65		54.15	66.50
4256	10"		50	.320		46.50	16.90		63.40	77
4258	12"		46	.348		72	18.35		90.35	108
4260	14"		43	.372		101	19.65		120.65	141
4262	16"		42	.381		105	20		125	147
4264	18"	▼	41	.390	▼	127	20.50		147.50	171

22 07 19.40 Pipe Insulation Protective Jacketing, Aluminum

		Crew	Daily Output	Labor-Hours	Unit	Material	2020 Bare Costs Labor	Equipment	Total	Total Incl O&P
0010	**PIPE INSULATION PROTECTIVE JACKETING, ALUMINUM**									
0100	Metal roll jacketing									
0120	Aluminum with polykraft moisture barrier									
0140	Smooth, based on OD of insulation, .016" thick									
0180	1/2" ID	Q-14	220	.073	L.F.	.29	3.84		4.13	6.20
0190	3/4" ID		215	.074		.37	3.93		4.30	6.45
0200	1" ID		210	.076		.45	4.02		4.47	6.65
0210	1-1/4" ID		205	.078		.54	4.12		4.66	6.90
0220	1-1/2" ID		202	.079		.62	4.18		4.80	7.10
0230	1-3/4" ID		199	.080		.79	4.24		5.03	7.35
0240	2" ID		195	.082		.88	4.33		5.21	7.60
0250	2-1/4" ID		191	.084		.98	4.42		5.40	7.90
0260	2-1/2" ID	▼	187	.086		.97	4.52		5.49	8

For customer support on your Mechanical Costs with RSMeans Data, call 800.448.8182.

121

22 07 19.40 Pipe Insulation Protective Jacketing, Aluminum		Crew	Daily Output	Labor-Hours	Unit	Material	2020 Bare Costs Labor	Equipment	Total	Total Incl O&P
0270	2-3/4" ID	Q-14	184	.087	L.F.	1.06	4.59		5.65	8.20
0280	3" ID		180	.089		1.27	4.69		5.96	8.60
0290	3-1/4" ID		176	.091		1.37	4.80		6.17	8.85
0300	3-1/2" ID		172	.093		1.33	4.91		6.24	9
0310	3-3/4" ID		169	.095		1.41	5		6.41	9.20
0320	4" ID		165	.097		1.50	5.10		6.60	9.50
0330	4-1/4" ID		161	.099		1.74	5.25		6.99	9.95
0340	4-1/2" ID		157	.102		1.68	5.40		7.08	10.10
0350	4-3/4" ID		154	.104		1.93	5.50		7.43	10.50
0360	5" ID		150	.107		2.03	5.65		7.68	10.90
0370	5-1/4" ID		146	.110		2.12	5.80		7.92	11.25
0380	5-1/2" ID		143	.112		2.02	5.90		7.92	11.25
0390	5-3/4" ID		139	.115		2.33	6.10		8.43	11.90
0400	6" ID		135	.119		2.42	6.25		8.67	12.25
0410	6-1/4" ID		133	.120		2.51	6.35		8.86	12.50
0420	6-1/2" ID		131	.122		2.37	6.45		8.82	12.50
0430	7" ID		128	.125		2.52	6.60		9.12	12.90
0440	7-1/4" ID		125	.128		2.61	6.75		9.36	13.20
0450	7-1/2" ID		123	.130		2.69	6.85		9.54	13.50
0460	8" ID		121	.132		2.86	7		9.86	13.85
0470	8-1/2" ID		119	.134		3.03	7.10		10.13	14.25
0480	9" ID		116	.138		3.20	7.30		10.50	14.70
0490	9-1/2" ID		114	.140		3.38	7.40		10.78	15.05
0500	10" ID		112	.143		3.55	7.55		11.10	15.50
0510	10-1/2" ID		110	.145		4.09	7.70		11.79	16.30
0520	11" ID		107	.150		3.88	7.90		11.78	16.35
0530	11-1/2" ID		105	.152		4.06	8.05		12.11	16.80
0540	12" ID		103	.155		4.24	8.20		12.44	17.25
0550	12-1/2" ID		100	.160		4.41	8.45		12.86	17.80
0560	13" ID		99	.162		4.58	8.55		13.13	18.15
0570	14" ID		98	.163		4.93	8.60		13.53	18.65
0580	15" ID		96	.167		5.25	8.80		14.05	19.30
0590	16" ID		95	.168		5.60	8.90		14.50	19.80
0600	17" ID		93	.172		5.95	9.10		15.05	20.50
0610	18" ID		92	.174		6.20	9.20		15.40	21
0620	19" ID		90	.178		6.65	9.40		16.05	21.50
0630	20" ID		89	.180		7	9.50		16.50	22.50
0640	21" ID		87	.184		7.35	9.70		17.05	23
0650	22" ID		86	.186		7.65	9.80		17.45	23.50
0660	23" ID		84	.190		8	10.05		18.05	24.50
0670	24" ID	▼	83	.193		8.35	10.15		18.50	25
0710	For smooth .020" thick, add					27%	10%			
0720	For smooth .024" thick, add					52%	20%			
0730	For smooth .032" thick, add					104%	33%			
0800	For stucco embossed, add					1%				
0820	For corrugated, add				▼	2.50%				
0900	White aluminum with polysurlyn moisture barrier									
0910	Smooth, % is an add to polykraft lines of same thickness									
0940	For smooth .016" thick, add				L.F.	35%				
0960	For smooth .024" thick, add				"	22%				
1000	Aluminum fitting covers									
1010	Size is based on OD of insulation									
1020	90° LR elbow, 2 piece									

22 07 19 – Plumbing Piping Insulation

22 07 19.40 Pipe Insulation Protective Jacketing, Aluminum		Crew	Daily Output	Labor-Hours	Unit	Material	2020 Bare Costs Labor	Equipment	Total	Total Incl O&P
1100	1-1/2"	Q-14	140	.114	Ea.	4.37	6.05		10.42	14.05
1110	1-3/4"		135	.119		4.37	6.25		10.62	14.40
1120	2"		130	.123		5.50	6.50		12	16
1130	2-1/4"		128	.125		5.50	6.60		12.10	16.20
1140	2-1/2"		126	.127		5.50	6.70		12.20	16.35
1150	2-3/4"		123	.130		5.50	6.85		12.35	16.60
1160	3"		120	.133		5.95	7.05		13	17.35
1170	3-1/4"		117	.137		5.95	7.20		13.15	17.65
1180	3-1/2"		115	.139		7.05	7.35		14.40	19
1190	3-3/4"		113	.142		7.25	7.45		14.70	19.40
1200	4"		110	.145		7.55	7.70		15.25	20
1210	4-1/4"		108	.148		7.80	7.80		15.60	20.50
1220	4-1/2"		106	.151		7.80	7.95		15.75	21
1230	4-3/4"		104	.154		7.80	8.10		15.90	21
1240	5"		102	.157		8.75	8.30		17.05	22.50
1250	5-1/4"		100	.160		10.10	8.45		18.55	24
1260	5-1/2"		97	.165		12.60	8.70		21.30	27
1270	5-3/4"		95	.168		12.60	8.90		21.50	27.50
1280	6"		92	.174		10.75	9.20		19.95	26
1290	6-1/4"		90	.178		10.75	9.40		20.15	26
1300	6-1/2"		87	.184		12.95	9.70		22.65	29
1310	7"		85	.188		15.70	9.95		25.65	32.50
1320	7-1/4"		84	.190		15.70	10.05		25.75	33
1330	7-1/2"		83	.193		23	10.15		33.15	41
1340	8"		82	.195		17.20	10.30		27.50	35
1350	8-1/2"		80	.200		32	10.55		42.55	51
1360	9"		78	.205		32	10.85		42.85	51.50
1370	9-1/2"		77	.208		23.50	10.95		34.45	43
1380	10"		76	.211		23.50	11.10		34.60	43
1390	10-1/2"		75	.213		24	11.25		35.25	44
1400	11"		74	.216		24	11.40		35.40	44
1410	11-1/2"		72	.222		27	11.75		38.75	48
1420	12"		71	.225		27	11.90		38.90	48.50
1430	12-1/2"		69	.232		48.50	12.25		60.75	72
1440	13"		68	.235		48.50	12.40		60.90	72
1450	14"		66	.242		65	12.80		77.80	91
1460	15"		64	.250		68	13.20		81.20	95
1470	16"		63	.254		74	13.40		87.40	102
2000	45° elbow, 2 piece									
2010	2-1/2"	Q-14	126	.127	Ea.	4.54	6.70		11.24	15.30
2020	2-3/4"		123	.130		4.54	6.85		11.39	15.55
2030	3"		120	.133		5.15	7.05		12.20	16.45
2040	3-1/4"		117	.137		5.15	7.20		12.35	16.75
2050	3-1/2"		115	.139		5.85	7.35		13.20	17.70
2060	3-3/4"		113	.142		5.85	7.45		13.30	17.90
2070	4"		110	.145		5.95	7.70		13.65	18.35
2080	4-1/4"		108	.148		6.80	7.80		14.60	19.45
2090	4-1/2"		106	.151		6.80	7.95		14.75	19.70
2100	4-3/4"		104	.154		6.80	8.10		14.90	19.90
2110	5"		102	.157		7.95	8.30		16.25	21.50
2120	5-1/4"		100	.160		7.95	8.45		16.40	21.50
2130	5-1/2"		97	.165		8.35	8.70		17.05	22.50
2140	6"		92	.174		8.35	9.20		17.55	23.50

For customer support on your Mechanical Costs with RSMeans Data, call 800.448.8182.

123

22 07 Plumbing Insulation

22 07 19 – Plumbing Piping Insulation

22 07 19.40 Pipe Insulation Protective Jacketing, Aluminum	Crew	Daily Output	Labor-Hours	Unit	Material	2020 Bare Costs Labor	Equipment	Total	Total Incl O&P
2150	6-1/2"	Q-14	87	.184	Ea.	11.55	9.70		21.25	27.50
2160	7"		85	.188		11.55	9.95		21.50	28
2170	7-1/2"		83	.193		11.65	10.15		21.80	28.50
2100	8"		82	.195		11.65	10.30		21.95	28.50
2190	8-1/2"		80	.200		14.50	10.55		25.05	32
2200	9"		78	.205		14.50	10.85		25.35	32.50
2210	9-1/2"		77	.208		20	10.95		30.95	39
2220	10"		76	.211		20	11.10		31.10	39
2230	10-1/2"		75	.213		18.80	11.25		30.05	38
2240	11"		74	.216		18.80	11.40		30.20	38
2250	11-1/2"		72	.222		22.50	11.75		34.25	43
2260	12"		71	.225		22.50	11.90		34.40	43.50
2270	13"		68	.235		26.50	12.40		38.90	48
2280	14"		66	.242		32.50	12.80		45.30	55.50
2290	15"		64	.250		50	13.20		63.20	75.50
2300	16"		63	.254		54.50	13.40		67.90	80.50
2310	17"		62	.258		53	13.60		66.60	79.50
2320	18"		61	.262		60.50	13.85		74.35	88
2330	19"		60	.267		74	14.05		88.05	103
2340	20"		59	.271		71.50	14.30		85.80	101
2350	21"		58	.276		77	14.55		91.55	108
3000	Tee, 4 piece									
3010	2-1/2"	Q-14	88	.182	Ea.	29.50	9.60		39.10	47.50
3020	2-3/4"		86	.186		29.50	9.80		39.30	47.50
3030	3"		84	.190		33.50	10.05		43.55	52.50
3040	3-1/4"		82	.195		33.50	10.30		43.80	53
3050	3-1/2"		80	.200		35	10.55		45.55	54.50
3060	4"		78	.205		35.50	10.85		46.35	55.50
3070	4-1/4"		76	.211		37	11.10		48.10	58
3080	4-1/2"		74	.216		37	11.40		48.40	58.50
3090	4-3/4"		72	.222		37	11.75		48.75	59
3100	5"		70	.229		38.50	12.05		50.55	61
3110	5-1/4"		68	.235		38.50	12.40		50.90	61.50
3120	5-1/2"		66	.242		40.50	12.80		53.30	64
3130	6"		64	.250		40.50	13.20		53.70	65
3140	6-1/2"		60	.267		44.50	14.05		58.55	70
3150	7"		58	.276		44.50	14.55		59.05	71
3160	7-1/2"		56	.286		49	15.10		64.10	77
3170	8"		54	.296		49	15.65		64.65	78
3180	8-1/2"		52	.308		50.50	16.25		66.75	80.50
3190	9"		50	.320		50.50	16.90		67.40	81.50
3200	9-1/2"		49	.327		37.50	17.25		54.75	67.50
3210	10"		48	.333		37.50	17.60		55.10	68
3220	10-1/2"		47	.340		40	17.95		57.95	71.50
3230	11"		46	.348		40	18.35		58.35	72
3240	11-1/2"		45	.356		42	18.75		60.75	75.50
3250	12"		44	.364		42	19.20		61.20	76
3260	13"		43	.372		44	19.65		63.65	78.50
3270	14"		42	.381		46.50	20		66.50	82
3280	15"		41	.390		50	20.50		70.50	86.50
3290	16"		40	.400		51.50	21		72.50	89
3300	17"		39	.410		59	21.50		80.50	98
3310	18"		38	.421		61.50	22		83.50	102

22 07 Plumbing Insulation

22 07 19 – Plumbing Piping Insulation

22 07 19.40 Pipe Insulation Protective Jacketing, Aluminum		Crew	Daily Output	Labor-Hours	Unit	Material	2020 Bare Costs Labor	Equipment	Total	Total Incl O&P
3320	19"	Q-14	37	.432	Ea.	68	23		91	110
3330	20"		36	.444		69.50	23.50		93	113
3340	22"		35	.457		88.50	24		112.50	134
3350	23"		34	.471		90.50	25		115.50	138
3360	24"		31	.516		93	27		120	144

22 07 19.50 Pipe Insulation Protective Jacketing, St. Stl.

		Crew	Daily Output	Labor-Hours	Unit	Material	2020 Bare Costs Labor	Equipment	Total	Total Incl O&P
0010	**PIPE INSULATION PROTECTIVE JACKETING, STAINLESS STEEL**									
0100	Metal roll jacketing									
0120	Type 304 with moisture barrier									
0140	Smooth, based on OD of insulation, .010" thick									
0260	2-1/2" ID	Q-14	250	.064	L.F.	3.10	3.38		6.48	8.60
0270	2-3/4" ID		245	.065		3.37	3.45		6.82	9
0280	3" ID		240	.067		3.64	3.52		7.16	9.40
0290	3-1/4" ID		235	.068		3.95	3.59		7.54	9.85
0300	3-1/2" ID		230	.070		4.19	3.67		7.86	10.25
0310	3-3/4" ID		225	.071		4.50	3.75		8.25	10.70
0320	4" ID		220	.073		4.75	3.84		8.59	11.15
0330	4-1/4" ID		215	.074		5	3.93		8.93	11.55
0340	4-1/2" ID		210	.076		5.30	4.02		9.32	12
0350	5" ID		200	.080		5.85	4.22		10.07	12.95
0360	5-1/2" ID		190	.084		6.40	4.44		10.84	13.85
0370	6" ID		180	.089		6.95	4.69		11.64	14.85
0380	6-1/2" ID		175	.091		7.50	4.83		12.33	15.65
0390	7" ID		170	.094		8.05	4.97		13.02	16.50
0400	7-1/2" ID		164	.098		8.60	5.15		13.75	17.35
0410	8" ID		161	.099		9.15	5.25		14.40	18.10
0420	8-1/2" ID		158	.101		9.70	5.35		15.05	18.85
0430	9" ID		155	.103		10.25	5.45		15.70	19.65
0440	9-1/2" ID		152	.105		10.80	5.55		16.35	20.50
0450	10" ID		149	.107		11.35	5.65		17	21
0460	10-1/2" ID		146	.110		11.90	5.80		17.70	22
0470	11" ID		143	.112		12.45	5.90		18.35	23
0480	12" ID		137	.117		13.55	6.15		19.70	24.50
0490	13" ID		132	.121		14.65	6.40		21.05	26
0500	14" ID		130	.123		15.75	6.50		22.25	27.50
0700	For smooth .016" thick, add					45%	33%			
1000	Stainless steel, Type 316, fitting covers									
1010	Size is based on OD of insulation									
1020	90° LR elbow, 2 piece									
1100	1-1/2"	Q-14	126	.127	Ea.	12.90	6.70		19.60	24.50
1110	2-3/4"		123	.130		13.20	6.85		20.05	25
1120	3"		120	.133		13.80	7.05		20.85	26
1130	3-1/4"		117	.137		13.80	7.20		21	26.50
1140	3-1/2"		115	.139		14.50	7.35		21.85	27
1150	3-3/4"		113	.142		15.35	7.45		22.80	28.50
1160	4"		110	.145		16.85	7.70		24.55	30.50
1170	4-1/4"		108	.148		22.50	7.80		30.30	36.50
1180	4-1/2"		106	.151		22.50	7.95		30.45	37
1190	5"		102	.157		23	8.30		31.30	37.50
1200	5-1/2"		97	.165		34	8.70		42.70	51
1210	6"		92	.174		37.50	9.20		46.70	55
1220	6-1/2"		87	.184		51.50	9.70		61.20	72

For customer support on your Mechanical Costs with RSMeans Data, call 800.448.8182.

125

22 07 Plumbing Insulation

22 07 19 – Plumbing Piping Insulation

22 07 19.50 Pipe Insulation Protective Jacketing, St. Stl.	Crew	Daily Output	Labor-Hours	Unit	Material	2020 Bare Costs Labor	Equipment	Total	Total Incl O&P
1230	7"	Q-14	85	.188	Ea.	51.50	9.95		61.45	72.50
1240	7-1/2"		83	.193		60	10.15		70.15	82
1250	8"		80	.200		60	10.55		70.55	82.50
1260	8 1/2"		80	.200		62.50	10.55		73.05	84.50
1270	9"		78	.205		95	10.85		105.85	121
1280	9-1/2"		77	.208		93	10.95		103.95	119
1290	10"		76	.211		93	11.10		104.10	119
1300	10-1/2"		75	.213		111	11.25		122.25	140
1310	11"		74	.216		106	11.40		117.40	135
1320	12"		71	.225		121	11.90		132.90	151
1330	13"		68	.235		167	12.40		179.40	203
1340	14"	▼	66	.242	▼	168	12.80		180.80	205
2000	45° elbow, 2 piece									
2010	2-1/2"	Q-14	126	.127	Ea.	11	6.70		17.70	22.50
2020	2-3/4"		123	.130		11	6.85		17.85	22.50
2030	3"		120	.133		11.85	7.05		18.90	24
2040	3-1/4"		117	.137		11.85	7.20		19.05	24
2050	3-1/2"		115	.139		12	7.35		19.35	24.50
2060	3-3/4"		113	.142		12	7.45		19.45	24.50
2070	4"		110	.145		15.40	7.70		23.10	29
2080	4-1/4"		108	.148		21.50	7.80		29.30	36
2090	4-1/2"		106	.151		21.50	7.95		29.45	36.50
2100	4-3/4"		104	.154		21.50	8.10		29.60	36.50
2110	5"		102	.157		22	8.30		30.30	36.50
2120	5-1/2"		97	.165		22	8.70		30.70	37.50
2130	6"		92	.174		26	9.20		35.20	42.50
2140	6-1/2"		87	.184		26	9.70		35.70	43.50
2150	7"		85	.188		44	9.95		53.95	64
2160	7-1/2"		83	.193		45	10.15		55.15	65
2170	8"		82	.195		45	10.30		55.30	65.50
2180	8-1/2"		80	.200		52.50	10.55		63.05	74
2190	9"		78	.205		52.50	10.85		63.35	74.50
2200	9-1/2"		77	.208		62	10.95		72.95	85
2210	10"		76	.211		62	11.10		73.10	85
2220	10-1/2"		75	.213		74.50	11.25		85.75	99.50
2230	11"		74	.216		74.50	11.40		85.90	99.50
2240	12"		71	.225		81	11.90		92.90	107
2250	13"	▼	68	.235	▼	95	12.40		107.40	124

22 11 Facility Water Distribution

22 11 13 – Facility Water Distribution Piping

22 11 13.14 Pipe, Brass

		Crew	Daily Output	Labor-Hours	Unit	Material	2020 Bare Costs Labor	Equipment	Total	Total Incl O&P
0010	**PIPE, BRASS**, Plain end									
0900	Field threaded, coupling & clevis hanger assembly 10' OC									
0920	Regular weight									
1120	1/2" diameter	1 Plum	48	.167	L.F.	8.30	10.75		19.05	25
1140	3/4" diameter		46	.174		10.95	11.20		22.15	29
1160	1" diameter	▼	43	.186		7.15	12		19.15	26
1180	1-1/4" diameter	Q-1	72	.222		23.50	12.90		36.40	45.50
1200	1-1/2" diameter		65	.246		28	14.30		42.30	52
1220	2" diameter	▼	53	.302		18.85	17.50		36.35	47

22 11 Facility Water Distribution

22 11 13 – Facility Water Distribution Piping

22 11 13.14 Pipe, Brass

		Crew	Daily Output	Labor-Hours	Unit	Material	2020 Bare Costs Labor	Equipment	Total	Total Incl O&P
1240	2-1/2" diameter	Q-1	41	.390	L.F.	32.50	22.50		55	70
1260	3" diameter	↓	31	.516		43.50	30		73.50	93
1300	4" diameter	Q-2	37	.649	↓	151	39		190	225
1930	To delete coupling & hanger, subtract									
1940	1/2" diam.					40%	46%			
1950	3/4" diam. to 1-1/2" diam.					39%	39%			
1960	2" diam. to 4" diam.					48%	35%			

22 11 13.16 Pipe Fittings, Brass

		Crew	Daily Output	Labor-Hours	Unit	Material	2020 Bare Costs Labor	Equipment	Total	Total Incl O&P
0010	**PIPE FITTINGS, BRASS**, Rough bronze, threaded, lead free.									
1000	Standard wt., 90° elbow									
1040	1/8"	1 Plum	13	.615	Ea.	22.50	39.50		62	84
1060	1/4"		13	.615		22.50	39.50		62	84
1080	3/8"		13	.615		22.50	39.50		62	84
1100	1/2"		12	.667		22.50	43		65.50	89
1120	3/4"		11	.727		30	47		77	103
1140	1"	↓	10	.800		48.50	51.50		100	131
1160	1-1/4"	Q-1	17	.941		78.50	54.50		133	168
1180	1-1/2"		16	1		97.50	58		155.50	194
1200	2"		14	1.143		157	66.50		223.50	272
1220	2-1/2"		11	1.455		380	84.50		464.50	545
1240	3"	↓	8	2		580	116		696	815
1260	4"	Q-2	11	2.182		1,175	131		1,306	1,500
1280	5"		8	3		2,800	180		2,980	3,350
1300	6"	↓	7	3.429		4,800	206		5,006	5,575
1500	45° elbow, 1/8"	1 Plum	13	.615		27.50	39.50		67	90
1540	1/4"		13	.615		27.50	39.50		67	90
1560	3/8"		13	.615		27.50	39.50		67	90
1580	1/2"		12	.667		27.50	43		70.50	95
1600	3/4"		11	.727		39	47		86	113
1620	1"	↓	10	.800		66.50	51.50		118	151
1640	1-1/4"	Q-1	17	.941		106	54.50		160.50	198
1660	1-1/2"		16	1		133	58		191	234
1680	2"		14	1.143		215	66.50		281.50	335
1700	2-1/2"		11	1.455		410	84.50		494.50	580
1720	3"	↓	8	2		545	116		661	775
1740	4"	Q-2	11	2.182		1,225	131		1,356	1,550
1760	5"		8	3		2,275	180		2,455	2,775
1780	6"	↓	7	3.429		3,150	206		3,356	3,775
2000	Tee, 1/8"	1 Plum	9	.889		26.50	57.50		84	115
2040	1/4"		9	.889		26.50	57.50		84	115
2060	3/8"		9	.889		26.50	57.50		84	115
2080	1/2"		8	1		26.50	64.50		91	126
2100	3/4"		7	1.143		37.50	73.50		111	151
2120	1"	↓	6	1.333		67.50	86		153.50	203
2140	1-1/4"	Q-1	10	1.600		116	93		209	267
2160	1-1/2"		9	1.778		131	103		234	298
2180	2"		8	2		217	116		333	415
2200	2-1/2"		7	2.286		515	133		648	770
2220	3"	↓	5	3.200		790	186		976	1,150
2240	4"	Q-2	7	3.429		1,950	206		2,156	2,450
2260	5"		5	4.800		3,350	289		3,639	4,100
2280	6"	↓	4	6		5,525	360		5,885	6,625

22 11 13 – Facility Water Distribution Piping

22 11 13.16 Pipe Fittings, Brass

		Crew	Daily Output	Labor-Hours	Unit	Material	2020 Bare Costs Labor	2020 Bare Costs Equipment	Total	Total Incl O&P
2500	Coupling, 1/8"	1 Plum	26	.308	Ea.	18.70	19.85		38.55	50
2540	1/4"		22	.364		18.70	23.50		42.20	55.50
2560	3/8"		18	.444		18.70	28.50		47.20	63.50
2580	1/2"		15	.533		18.80	34.50		53.30	72
2600	3/4"		14	.571		26.50	37		63.50	84
2620	1"		13	.615		45	39.50		84.50	109
2640	1-1/4"	Q-1	22	.727		75	42		117	146
2660	1-1/2"		20	.800		97.50	46.50		144	177
2680	2"		18	.889		161	51.50		212.50	254
2700	2-1/2"		14	1.143		275	66.50		341.50	405
2720	3"		10	1.600		380	93		473	560
2740	4"	Q-2	12	2		780	120		900	1,025
2760	5"		10	2.400		1,625	144		1,769	2,000
2780	6"		9	2.667		2,325	160		2,485	2,825
3000	Union, 125 lb.									
3020	1/8"	1 Plum	12	.667	Ea.	41	43		84	110
3040	1/4"		12	.667		43	43		86	112
3060	3/8"		12	.667		43	43		86	112
3080	1/2"		11	.727		43	47		90	118
3100	3/4"		10	.800		59.50	51.50		111	143
3120	1"		9	.889		91.50	57.50		149	187
3140	1-1/4"	Q-1	16	1		132	58		190	233
3160	1-1/2"		15	1.067		158	62		220	266
3180	2"		13	1.231		208	71.50		279.50	335
3200	2-1/2"		10	1.600		725	93		818	940
3220	3"		7	2.286		970	133		1,103	1,275
3240	4"	Q-2	10	2.400		2,800	144		2,944	3,325
3320	For 250 lb. (navy pattern), add					100%				

22 11 13.23 Pipe/Tube, Copper

		Crew	Daily Output	Labor-Hours	Unit	Material	2020 Bare Costs Labor	2020 Bare Costs Equipment	Total	Total Incl O&P
0010	**PIPE/TUBE, COPPER**, Solder joints R221113-50									
0100	Solder									
0120	Solder, lead free, roll				Lb.	14.25			14.25	15.70
1000	Type K tubing, couplings & clevis hanger assemblies 10' OC									
1100	1/4" diameter	1 Plum	84	.095	L.F.	4.25	6.15		10.40	13.85
1120	3/8" diameter		82	.098		4.76	6.30		11.06	14.65
1140	1/2" diameter		78	.103		5.30	6.60		11.90	15.70
1160	5/8" diameter		77	.104		6.75	6.70		13.45	17.40
1180	3/4" diameter		74	.108		8.65	6.95		15.60	19.95
1200	1" diameter		66	.121		12.80	7.80		20.60	26
1220	1-1/4" diameter		56	.143		15.30	9.20		24.50	30.50
1240	1-1/2" diameter		50	.160		18.60	10.30		28.90	36
1260	2" diameter		40	.200		27	12.90		39.90	49.50
1280	2-1/2" diameter	Q-1	60	.267		44	15.45		59.45	71.50
1300	3" diameter		54	.296		58	17.20		75.20	89
1320	3-1/2" diameter		42	.381		80	22		102	121
1330	4" diameter		38	.421		97.50	24.50		122	144
1340	5" diameter		32	.500		140	29		169	198
1360	6" diameter	Q-2	38	.632		207	38		245	284
1380	8" diameter	"	34	.706		360	42.50		402.50	465
1390	For other than full hard temper, add					13%				
1440	For silver solder, add						15%			
1800	For medical clean (oxygen class), add					12%				

22 11 13 – Facility Water Distribution Piping

22 11 13.23 Pipe/Tube, Copper		Crew	Daily Output	Labor-Hours	Unit	Material	2020 Bare Costs Labor	Equipment	Total	Total Incl O&P
1950	To delete cplgs. & hngrs., 1/4"-1" pipe, subtract					27%	60%			
1960	1-1/4"-3" pipe, subtract					14%	52%			
1970	3-1/2"-5" pipe, subtract					10%	60%			
1980	6"-8" pipe, subtract					19%	53%			
2000	Type L tubing, couplings & clevis hanger assemblies 10' OC									
2100	1/4" diameter	1 Plum	88	.091	L.F.	2.67	5.85		8.52	11.70
2120	3/8" diameter		84	.095		3.34	6.15		9.49	12.85
2140	1/2" diameter		81	.099		3.68	6.35		10.03	13.60
2160	5/8" diameter		79	.101		5.75	6.55		12.30	16.05
2180	3/4" diameter		76	.105		4.71	6.80		11.51	15.35
2200	1" diameter		68	.118		7.60	7.60		15.20	19.70
2220	1-1/4" diameter		58	.138		12	8.90		20.90	26.50
2240	1-1/2" diameter		52	.154		11.60	9.90		21.50	27.50
2260	2" diameter		42	.190		18.85	12.30		31.15	39.50
2280	2-1/2" diameter	Q-1	62	.258		30	14.95		44.95	55.50
2300	3" diameter		56	.286		46.50	16.55		63.05	76
2320	3-1/2" diameter		43	.372		60	21.50		81.50	98.50
2340	4" diameter		39	.410		65.50	24		89.50	108
2360	5" diameter		34	.471		148	27.50		175.50	203
2380	6" diameter	Q-2	40	.600		151	36		187	220
2400	8" diameter	"	36	.667		250	40		290	335
2410	For other than full hard temper, add					21%				
2590	For silver solder, add						15%			
2900	For medical clean (oxygen class), add					12%				
2940	To delete cplgs. & hngrs., 1/4"-1" pipe, subtract					37%	63%			
2960	1-1/4"-3" pipe, subtract					12%	53%			
2970	3-1/2"-5" pipe, subtract					12%	63%			
2980	6"-8" pipe, subtract					24%	55%			
3000	Type M tubing, couplings & clevis hanger assemblies 10' OC									
3100	1/4" diameter	1 Plum	90	.089	L.F.	3.72	5.75		9.47	~ 12.65
3120	3/8" diameter		87	.092		4.09	5.95		10.04	13.35
3140	1/2" diameter		84	.095		3.56	6.15		9.71	13.10
3160	5/8" diameter		81	.099		4.94	6.35		11.29	15
3180	3/4" diameter		78	.103		5.05	6.60		11.65	15.45
3200	1" diameter		70	.114		8.50	7.35		15.85	20.50
3220	1-1/4" diameter		60	.133		12.30	8.60		20.90	26.50
3240	1-1/2" diameter		54	.148		14.45	9.55		24	30
3260	2" diameter		44	.182		21.50	11.70		33.20	41
3280	2-1/2" diameter	Q-1	64	.250		31.50	14.50		46	56.50
3300	3" diameter		58	.276		39.50	16		55.50	67.50
3320	3-1/2" diameter		45	.356		58.50	20.50		79	95
3340	4" diameter		40	.400		75	23		98	117
3360	5" diameter		36	.444		138	26		164	191
3370	6" diameter	Q-2	42	.571		197	34.50		231.50	269
3380	8" diameter	"	38	.632		325	38		363	415
3440	For silver solder, add						15%			
3960	To delete cplgs. & hngrs., 1/4"-1" pipe, subtract					35%	65%			
3970	1-1/4"-3" pipe, subtract					19%	56%			
3980	3-1/2"-5" pipe, subtract					13%	65%			
3990	6"-8" pipe, subtract					28%	58%			
4000	Type DWV tubing, couplings & clevis hanger assemblies 10' OC									
4100	1-1/4" diameter	1 Plum	60	.133	L.F.	12.45	8.60		21.05	26.50
4120	1-1/2" diameter		54	.148		12.60	9.55		22.15	28

For customer support on your Mechanical Costs with RSMeans Data, call 800.448.8182.

129

22 11 13.23 Pipe/Tube, Copper		Crew	Daily Output	Labor-Hours	Unit	Material	2020 Bare Costs Labor	Equipment	Total	Total Incl O&P
4140	2" diameter	1 Plum	44	.182	L.F.	18.30	11.70		30	37.50
4160	3" diameter	Q-1	58	.276		29	16		45	56
4180	4" diameter		40	.400		60.50	23		83.50	101
4200	5" diameter		36	.444		110	26		136	160
4220	6" diameter	Q-2	42	.571		161	34.50		195.50	229
4240	8" diameter	"	38	.632		510	38		548	620
4730	To delete cplgs. & hngrs., 1-1/4"-2" pipe, subtract					16%	53%			
4740	3"-4" pipe, subtract					13%	60%			
4750	5"-8" pipe, subtract					23%	58%			
5200	ACR tubing, type L, hard temper, cleaned and									
5220	capped, no couplings or hangers									
5240	3/8" OD				L.F.	1.76			1.76	1.94
5250	1/2" OD					2.64			2.64	2.90
5260	5/8" OD					3.30			3.30	3.63
5270	3/4" OD					4.48			4.48	4.93
5280	7/8" OD					5.05			5.05	5.55
5290	1-1/8" OD					7.20			7.20	7.90
5300	1-3/8" OD					9.65			9.65	10.60
5310	1-5/8" OD					12.60			12.60	13.85
5320	2-1/8" OD					20.50			20.50	22.50
5330	2-5/8" OD					27.50			27.50	30.50
5340	3-1/8" OD					34			34	37.50
5350	3-5/8" OD					63.50			63.50	70
5360	4-1/8" OD					65			65	72
5380	ACR tubing, type L, hard, cleaned and capped									
5381	No couplings or hangers									
5384	3/8"	1 Stpi	160	.050	L.F.	1.76	3.28		5.04	6.85
5385	1/2"		160	.050		2.64	3.28		5.92	7.80
5386	5/8"		160	.050		3.30	3.28		6.58	8.55
5387	3/4"		130	.062		4.48	4.03		8.51	11
5388	7/8"		130	.062		5.05	4.03		9.08	11.60
5389	1-1/8"		115	.070		7.20	4.56		11.76	14.70
5390	1-3/8"		100	.080		9.65	5.25		14.90	18.45
5391	1-5/8"		90	.089		12.60	5.85		18.45	22.50
5392	2-1/8"		80	.100		20.50	6.55		27.05	32.50
5393	2-5/8"	Q-5	125	.128		27.50	7.55		35.05	42
5394	3-1/8"		105	.152		34	9		43	51
5395	4-1/8"		95	.168		65	9.95		74.95	87
5800	Refrigeration tubing, dryseal, 50' coils									
5840	1/8" OD				Coil	34			34	37.50
5850	3/16" OD					46			46	50.50
5860	1/4" OD					46.50			46.50	51
5870	5/16" OD					71			71	78
5880	3/8" OD					64.50			64.50	71
5890	1/2" OD					93			93	102
5900	5/8" OD					124			124	136
5910	3/4" OD					145			145	159
5920	7/8" OD					147			147	162
5930	1-1/8" OD					330			330	365
5940	1-3/8" OD					555			555	615
5950	1-5/8" OD					710			710	780
9400	Sub assemblies used in assembly systems									
9410	Chilled water unit, coil connections per unit under 10 ton	Q-5	.80	20	System	1,200	1,175		2,375	3,100

22 11 13 – Facility Water Distribution Piping

22 11 13.23 Pipe/Tube, Copper

		Crew	Daily Output	Labor-Hours	Unit	Material	2020 Bare Costs Labor	Equipment	Total	Total Incl O&P
9420	Chilled water unit, coil connections per unit 10 ton and up	Q-5	1	16	System	2,175	945		3,120	3,825
9430	Chilled water dist. piping per ton, less than 61 ton systems	↓	26	.615		23	36.50		59.50	79.50
9440	Chilled water dist. piping per ton, 61 through 120 ton systems	Q-6	31	.774		53	47.50		100.50	130
9450	Chilled water dist. piping/ton, 135 ton systems and up	Q-8	25.40	1.260		71	78.50	4.20	153.70	201
9510	Refrigerant piping/ton of cooling for remote condensers	Q-5	2	8		375	470		845	1,125
9520	Refrigerant piping per ton up to 10 ton w/remote condensing unit		2.40	6.667		201	395		596	810
9530	Refrigerant piping per ton, 20 ton w/remote condensing unit		2	8		245	470		715	975
9540	Refrigerant piping per ton, 40 ton w/remote condensing unit	↓	1.90	8.421		330	495		825	1,100
9550	Refrigerant piping per ton, 75-80 ton w/remote condensing unit	Q-6	2.40	10		515	610		1,125	1,475
9560	Refrigerant piping per ton, 100 ton w/remote condensing unit	"	2.20	10.909	↓	710	665		1,375	1,775

22 11 13.25 Pipe/Tube Fittings, Copper

		Crew	Daily Output	Labor-Hours	Unit	Material	2020 Bare Costs Labor	Equipment	Total	Total Incl O&P
0010	**PIPE/TUBE FITTINGS, COPPER**, Wrought unless otherwise noted									
0020	For silver solder, add						15%			
0040	Solder joints, copper x copper									
0070	90° elbow, 1/4"	1 Plum	22	.364	Ea.	3.88	23.50		27.38	39.50
0090	3/8"		22	.364		4.15	23.50		27.65	39.50
0100	1/2"		20	.400		1.26	26		27.26	40
0110	5/8"		19	.421		2.77	27		29.77	43.50
0120	3/4"		19	.421		2.68	27		29.68	43.50
0130	1"		16	.500		7.05	32		39.05	56
0140	1-1/4"		15	.533		11.65	34.50		46.15	64.50
0150	1-1/2"		13	.615		16.10	39.50		55.60	77
0160	2"	↓	11	.727		29.50	47		76.50	103
0170	2-1/2"	Q-1	13	1.231		51	71.50		122.50	163
0180	3"		11	1.455		86.50	84.50		171	221
0190	3-1/2"		10	1.600		281	93		374	450
0200	4"		9	1.778		236	103		339	415
0210	5"	↓	6	2.667		775	155		930	1,075
0220	6"	Q-2	9	2.667		1,025	160		1,185	1,375
0230	8"	"	8	3		3,800	180		3,980	4,475
0250	45° elbow, 1/4"	1 Plum	22	.364		7.75	23.50		31.25	43.50
0270	3/8"		22	.364		6.80	23.50		30.30	42.50
0280	1/2"		20	.400		2.75	26		28.75	41.50
0290	5/8"		19	.421		11.45	27		38.45	53
0300	3/4"		19	.421		4.46	27		31.46	45.50
0310	1"		16	.500		11.20	32		43.20	60.50
0320	1-1/4"		15	.533		16	34.50		50.50	69
0330	1-1/2"		13	.615		18.20	39.50		57.70	79.50
0340	2"	↓	11	.727		30.50	47		77.50	104
0350	2-1/2"	Q-1	13	1.231		49.50	71.50		121	162
0360	3"		13	1.231		88	71.50		159.50	204
0370	3-1/2"		10	1.600		122	93		215	273
0380	4"		9	1.778		173	103		276	345
0390	5"	↓	6	2.667		635	155		790	930
0400	6"	Q-2	9	2.667		995	160		1,155	1,350
0410	8"	"	8	3		4,300	180		4,480	5,000
0450	Tee, 1/4"	1 Plum	14	.571		8.45	37		45.45	64.50
0470	3/8"		14	.571		6.85	37		43.85	62.50
0480	1/2"		13	.615		2.44	39.50		41.94	62
0490	5/8"		12	.667		16.25	43		59.25	82.50
0500	3/4"		12	.667		6.05	43		49.05	71
0510	1"	↓	10	.800	↓	17.05	51.50		68.55	96

22 11 13.25 Pipe/Tube Fittings, Copper		Crew	Daily Output	Labor-Hours	Unit	Material	2020 Bare Costs Labor	Equipment	Total	Total Incl O&P
0520	1-1/4"	1 Plum	9	.889	Ea.	24.50	57.50		82	113
0530	1-1/2"		8	1		36.50	64.50		101	137
0540	2"	▼	7	1.143		58	73.50		131.50	174
0550	2-1/2"	Q-1	8	2		111	116		227	296
0560	3"		7	2.286		158	133		291	370
0570	3-1/2"		6	2.667		475	155		630	750
0580	4"		5	3.200		365	186		551	680
0590	5"	▼	4	4		1,075	232		1,307	1,550
0600	6"	Q-2	6	4		1,475	241		1,716	1,975
0610	8"	"	5	4.800		5,875	289		6,164	6,900
0612	Tee, reducing on the outlet, 1/4"	1 Plum	15	.533		17.05	34.50		51.55	70.50
0613	3/8"		15	.533		15.55	34.50		50.05	68.50
0614	1/2"		14	.571		15.60	37		52.60	72
0615	5/8"		13	.615		30.50	39.50		70	93
0616	3/4"		12	.667		8	43		51	73.50
0617	1"		11	.727		29	47		76	102
0618	1-1/4"		10	.800		32.50	51.50		84	113
0619	1-1/2"		9	.889		32	57.50		89.50	121
0620	2"	▼	8	1		64.50	64.50		129	168
0621	2-1/2"	Q-1	9	1.778		137	103		240	305
0622	3"		8	2		163	116		279	355
0623	4"		6	2.667		299	155		454	560
0624	5"	▼	5	3.200		1,725	186		1,911	2,175
0625	6"	Q-2	7	3.429		2,150	206		2,356	2,675
0626	8"	"	6	4		10,800	241		11,041	12,300
0630	Tee, reducing on the run, 1/4"	1 Plum	15	.533		23.50	34.50		58	77.50
0631	3/8"		15	.533		31	34.50		65.50	85.50
0632	1/2"		14	.571		20.50	37		57.50	77.50
0633	5/8"		13	.615		29	39.50		68.50	91.50
0634	3/4"		12	.667		21	43		64	87.50
0635	1"		11	.727		26	47		73	98.50
0636	1-1/4"		10	.800		41	51.50		92.50	122
0637	1-1/2"		9	.889		72	57.50		129.50	165
0638	2"	▼	8	1		82.50	64.50		147	188
0639	2-1/2"	Q-1	9	1.778		177	103		280	350
0640	3"		8	2		320	116		436	530
0641	4"		6	2.667		585	155		740	875
0642	5"	▼	5	3.200		1,950	186		2,136	2,425
0643	6"	Q-2	7	3.429		2,950	206		3,156	3,550
0644	8"	"	6	4		10,200	241		10,441	11,600
0650	Coupling, 1/4"	1 Plum	24	.333		1.07	21.50		22.57	33
0670	3/8"		24	.333		1.53	21.50		23.03	33.50
0680	1/2"		22	.364		.96	23.50		24.46	36
0690	5/8"		21	.381		4.23	24.50		28.73	41
0700	3/4"		21	.381		2.71	24.50		27.21	39.50
0710	1"		18	.444		5.30	28.50		33.80	49
0715	1-1/4"		17	.471		7.80	30.50		38.30	54
0716	1-1/2"		15	.533		10.80	34.50		45.30	63.50
0718	2"	▼	13	.615		15.30	39.50		54.80	76.50
0721	2-1/2"	Q-1	15	1.067		40.50	62		102.50	137
0722	3"		13	1.231		47	71.50		118.50	159
0724	3-1/2"		8	2		124	116		240	310
0726	4"	▼	7	2.286		154	133		287	370

132

22 11 Facility Water Distribution

22 11 13 – Facility Water Distribution Piping

22 11 13.25 Pipe/Tube Fittings, Copper		Crew	Daily Output	Labor-Hours	Unit	Material	2020 Bare Costs Labor	Equipment	Total	Total Incl O&P
0728	5"	Q-1	6	2.667	Ea.	228	155		383	480
0731	6"	Q-2	8	3		355	180		535	660
0732	8"	"	7	3.429		1,275	206		1,481	1,700
0741	Coupling, reducing, concentric									
0743	1/2"	1 Plum	23	.348	Ea.	2.85	22.50		25.35	36.50
0745	3/4"		21.50	.372		5.65	24		29.65	42
0747	1"		19.50	.410		8.80	26.50		35.30	49
0748	1-1/4"		18	.444		12.15	28.50		40.65	56.50
0749	1-1/2"		16	.500		17.05	32		49.05	67
0751	2"		14	.571		25.50	37		62.50	83
0752	2-1/2"		13	.615		56	39.50		95.50	121
0753	3"	Q-1	14	1.143		44.50	66.50		111	148
0755	4"	"	8	2		133	116		249	320
0757	5"	Q-2	7.50	3.200		755	192		947	1,125
0759	6"		7	3.429		1,200	206		1,406	1,625
0761	8"		6.50	3.692		3,825	222		4,047	4,525
0771	Cap, sweat									
0773	1/2"	1 Plum	40	.200	Ea.	1.27	12.90		14.17	20.50
0775	3/4"		38	.211		2.36	13.55		15.91	23
0777	1"		32	.250		5.50	16.10		21.60	30
0778	1-1/4"		29	.276		6.95	17.80		24.75	34
0779	1-1/2"		26	.308		12.80	19.85		32.65	43.50
0781	2"		22	.364		19.70	23.50		43.20	56.50
0791	Flange, sweat									
0793	3"	Q-1	22	.727	Ea.	275	42		317	370
0795	4"		18	.889		415	51.50		466.50	530
0797	5"		12	1.333		720	77.50		797.50	905
0799	6"	Q-2	18	1.333		750	80		830	945
0801	8"	"	16	1.500		1,275	90		1,365	1,525
0850	Unions, 1/4"	1 Plum	21	.381		41	24.50		65.50	82
0870	3/8"		21	.381		41.50	24.50		66	82
0880	1/2"		19	.421		22.50	27		49.50	65
0890	5/8"		18	.444		89	28.50		117.50	141
0900	3/4"		18	.444		23	28.50		51.50	68
0910	1"		15	.533		47.50	34.50		82	104
0920	1-1/4"		14	.571		78.50	37		115.50	142
0930	1-1/2"		12	.667		118	43		161	195
0940	2"		10	.800		152	51.50		203.50	245
0950	2-1/2"	Q-1	12	1.333		385	77.50		462.50	540
0960	3"	"	10	1.600		1,000	93		1,093	1,250
0980	Adapter, copper x male IPS, 1/4"	1 Plum	20	.400		13.55	26		39.55	53.50
0990	3/8"		20	.400		8.50	26		34.50	48
1000	1/2"		18	.444		3.64	28.50		32.14	47
1010	3/4"		17	.471		6.10	30.50		36.60	52
1020	1"		15	.533		15.55	34.50		50.05	68.50
1030	1-1/4"		13	.615		26.50	39.50		66	88.50
1040	1-1/2"		12	.667		26.50	43		69.50	93.50
1050	2"		11	.727		44.50	47		91.50	119
1060	2-1/2"	Q-1	10.50	1.524		182	88.50		270.50	330
1070	3"		10	1.600		226	93		319	390
1080	3-1/2"		9	1.778		219	103		322	395
1090	4"		8	2		247	116		363	445
1200	5", cast		6	2.667		2,200	155		2,355	2,650

22 11 Facility Water Distribution

22 11 13 – Facility Water Distribution Piping

22 11 13.25 Pipe/Tube Fittings, Copper		Crew	Daily Output	Labor-Hours	Unit	Material	2020 Bare Costs Labor	Equipment	Total	Total Incl O&P
1210	6", cast	Q-2	8.50	2.824	Ea.	2,475	170		2,645	2,975
1214	Adapter, copper x female IPS									
1216	1/2"	1 Plum	18	.444	Ea.	5.75	28.50		34.25	49.50
1218	3/4"		17	.471		7.90	30.50		38.40	54
1220	1"		15	.533		18.15	34.50		52.65	71.50
1221	1-1/4"		13	.615		26	39.50		65.50	88
1222	1-1/2"		12	.667		41.50	43		84.50	110
1224	2"		11	.727		56	47		103	132
1250	Cross, 1/2"		10	.800		28.50	51.50		80	109
1260	3/4"		9.50	.842		55	54.50		109.50	142
1270	1"		8	1		93.50	64.50		158	200
1280	1-1/4"		7.50	1.067		135	69		204	251
1290	1-1/2"		6.50	1.231		192	79.50		271.50	330
1300	2"	▼	5.50	1.455		365	94		459	540
1310	2-1/2"	Q-1	6.50	2.462		835	143		978	1,125
1320	3"	"	5.50	2.909	▼	710	169		879	1,025
1500	Tee fitting, mechanically formed (Type 1, 'branch sizes up to 2 in.')									
1520	1/2" run size, 3/8" to 1/2" branch size	1 Plum	80	.100	Ea.		6.45		6.45	9.65
1530	3/4" run size, 3/8" to 3/4" branch size		60	.133			8.60		8.60	12.85
1540	1" run size, 3/8" to 1" branch size		54	.148			9.55		9.55	14.30
1550	1-1/4" run size, 3/8" to 1-1/4" branch size		48	.167			10.75		10.75	16.10
1560	1-1/2" run size, 3/8" to 1-1/2" branch size		40	.200			12.90		12.90	19.30
1570	2" run size, 3/8" to 2" branch size		35	.229			14.75		14.75	22
1580	2-1/2" run size, 1/2" to 2" branch size		32	.250			16.10		16.10	24
1590	3" run size, 1" to 2" branch size		26	.308			19.85		19.85	29.50
1600	4" run size, 1" to 2" branch size	▼	24	.333	▼		21.50		21.50	32
1640	Tee fitting, mechanically formed (Type 2, branches 2-1/2" thru 4")									
1650	2-1/2" run size, 2-1/2" branch size	1 Plum	12.50	.640	Ea.		41.50		41.50	61.50
1660	3" run size, 2-1/2" to 3" branch size		12	.667			43		43	64.50
1670	3-1/2" run size, 2-1/2" to 3-1/2" branch size		11	.727			47		47	70
1680	4" run size, 2-1/2" to 4" branch size		10.50	.762			49		49	73.50
1698	5" run size, 2" to 4" branch size		9.50	.842			54.50		54.50	81
1700	6" run size, 2" to 4" branch size		8.50	.941			60.50		60.50	91
1710	8" run size, 2" to 4" branch size	▼	7	1.143	▼		73.50		73.50	110
1800	ACR fittings, OD size									
1802	Tee, straight									
1808	5/8"	1 Stpi	12	.667	Ea.	2.96	43.50		46.46	69
1810	3/4"		12	.667		22	43.50		65.50	89.50
1812	7/8"		10	.800		7.15	52.50		59.65	86.50
1813	1"		10	.800		60	52.50		112.50	145
1814	1-1/8"		10	.800		25.50	52.50		78	107
1816	1-3/8"		9	.889		29.50	58.50		88	120
1818	1-5/8"		8	1		53	65.50		118.50	156
1820	2-1/8"	▼	7	1.143		71	75		146	191
1822	2-5/8"	Q-5	8	2		153	118		271	345
1824	3-1/8"		7	2.286		220	135		355	445
1826	4-1/8"	▼	5	3.200	▼	415	189		604	745
1830	90° elbow									
1836	5/8"	1 Stpi	19	.421	Ea.	6.10	27.50		33.60	48
1838	3/4"		19	.421		11.05	27.50		38.55	53.50
1840	7/8"		16	.500		11	33		44	61
1842	1-1/8"		16	.500		14.70	33		47.70	65
1844	1-3/8"	▼	15	.533	▼	14.20	35		49.20	68

For customer support on your Mechanical Costs with RSMeans Data, call 800.448.8182.

22 11 13.25 Pipe/Tube Fittings, Copper		Crew	Daily Output	Labor-Hours	Unit	Material	2020 Bare Costs Labor	Equipment	Total	Total Incl O&P
1846	1-5/8"	1 Stpi	13	.615	Ea.	22	40.50		62.50	85
1848	2-1/8"	↓	11	.727		40.50	47.50		88	116
1850	2-5/8"	Q-5	13	1.231		86	72.50		158.50	204
1852	3-1/8"		11	1.455		104	86		190	243
1854	4-1/8"	↓	9	1.778	↓	228	105		333	410
1860	Coupling									
1866	5/8"	1 Stpi	21	.381	Ea.	1.27	25		26.27	39
1868	3/4"		21	.381		3.46	25		28.46	41.50
1870	7/8"		18	.444		2.49	29		31.49	46
1871	1"		18	.444		8.55	29		37.55	53
1872	1-1/8"		18	.444		5.25	29		34.25	49.50
1874	1-3/8"		17	.471		9.20	31		40.20	56
1876	1-5/8"		15	.533		12.15	35		47.15	66
1878	2-1/8"	↓	13	.615		20.50	40.50		61	83
1880	2-5/8"	Q-5	15	1.067		39.50	63		102.50	138
1882	3-1/8"		13	1.231		53.50	72.50		126	168
1884	4-1/8"	↓	7	2.286	↓	118	135		253	330
2000	DWV, solder joints, copper x copper									
2030	90° elbow, 1-1/4"	1 Plum	13	.615	Ea.	18.50	39.50		58	80
2050	1-1/2"		12	.667		24.50	43		67.50	91.50
2070	2"	↓	10	.800		46	51.50		97.50	128
2090	3"	Q-1	10	1.600		85.50	93		178.50	233
2100	4"	"	9	1.778		545	103		648	755
2150	45° elbow, 1-1/4"	1 Plum	13	.615		15.15	39.50		54.65	76
2170	1-1/2"		12	.667		14.05	43		57.05	80
2180	2"	↓	10	.800		29	51.50		80.50	109
2190	3"	Q-1	10	1.600		59	93		152	204
2200	4"	"	9	1.778		94	103		197	258
2250	Tee, sanitary, 1-1/4"	1 Plum	9	.889		28.50	57.50		86	117
2270	1-1/2"		8	1		35.50	64.50		100	136
2290	2"	↓	7	1.143		55	73.50		128.50	171
2310	3"	Q-1	7	2.286		213	133		346	435
2330	4"	"	6	2.667		520	155		675	805
2400	Coupling, 1-1/4"	1 Plum	14	.571		7.75	37		44.75	63.50
2420	1-1/2"		13	.615		9.60	39.50		49.10	70
2440	2"	↓	11	.727		13.35	47		60.35	84.50
2460	3"	Q-1	11	1.455		30.50	84.50		115	160
2480	4"	"	10	1.600	↓	68	93		161	214
2602	Traps, see Section 22 13 16.60									
3500	Compression joint fittings									
3510	As used for plumbing and oil burner work									
3520	Fitting price includes nuts and sleeves									
3540	Sleeve, 1/8"				Ea.	.15			.15	.17
3550	3/16"					.15			.15	.17
3560	1/4"					.05			.05	.06
3570	5/16"					.15			.15	.17
3580	3/8"					.30			.30	.33
3600	1/2"					.41			.41	.45
3620	Nut, 1/8"					.20			.20	.22
3630	3/16"					.23			.23	.25
3640	1/4"					.21			.21	.23
3650	5/16"					.34			.34	.37
3660	3/8"				↓	.36			.36	.40

22 11 13.25 Pipe/Tube Fittings, Copper		Crew	Daily Output	Labor-Hours	Unit	Material	2020 Bare Costs Labor	Equipment	Total	Total Incl O&P
3670	1/2"				Ea.	.64			.64	.70
3710	Union, 1/8"	1 Plum	26	.308		1.40	19.85		21.25	31
3720	3/16"		24	.333		1.33	21.50		22.83	33.50
3730	1/4"		24	.333		1.80	21.50		23.30	34
3740	5/16"		23	.348		2.30	22.50		24.80	36
3750	3/8"		22	.364		2.04	23.50		25.54	37
3760	1/2"		22	.364		3.21	23.50		26.71	38.50
3780	5/8"		21	.381		2.94	24.50		27.44	39.50
3820	Union tee, 1/8"		17	.471		5.30	30.50		35.80	51.50
3830	3/16"		16	.500		3.06	32		35.06	51.50
3840	1/4"		15	.533		3.49	34.50		37.99	55.50
3850	5/16"		15	.533		3.89	34.50		38.39	56
3860	3/8"		15	.533		3.58	34.50		38.08	55.50
3870	1/2"		15	.533		7.85	34.50		42.35	60
3910	Union elbow, 1/4"		24	.333		1.94	21.50		23.44	34
3920	5/16"		23	.348		2.73	22.50		25.23	36.50
3930	3/8"		22	.364		3.70	23.50		27.20	39
3940	1/2"		22	.364		5.55	23.50		29.05	41
3980	Female connector, 1/8"		26	.308		.97	19.85		20.82	30.50
4000	3/16" x 1/8"		24	.333		1.37	21.50		22.87	33.50
4010	1/4" x 1/8"		24	.333		1.53	21.50		23.03	33.50
4020	1/4"		24	.333		2.06	21.50		23.56	34.50
4030	3/8" x 1/4"		22	.364		2.06	23.50		25.56	37.50
4040	1/2" x 3/8"		22	.364		3.27	23.50		26.77	38.50
4050	5/8" x 1/2"		21	.381		3.85	24.50		28.35	40.50
4090	Male connector, 1/8"		26	.308		1.41	19.85		21.26	31
4100	3/16" x 1/8"		24	.333		.90	21.50		22.40	33
4110	1/4" x 1/8"		24	.333		1.19	21.50		22.69	33.50
4120	1/4"		24	.333		1.19	21.50		22.69	33.50
4130	5/16" x 1/8"		23	.348		1.41	22.50		23.91	35
4140	5/16" x 1/4"		23	.348		1.24	22.50		23.74	35
4150	3/8" x 1/8"		22	.364		1.67	23.50		25.17	37
4160	3/8" x 1/4"		22	.364		1.91	23.50		25.41	37
4170	3/8"		22	.364		1.92	23.50		25.42	37
4180	3/8" x 1/2"		22	.364		2.61	23.50		26.11	38
4190	1/2" x 3/8"		22	.364		2.55	23.50		26.05	38
4200	1/2"		22	.364		2.85	23.50		26.35	38
4210	5/8" x 1/2"		21	.381		3.30	24.50		27.80	40
4240	Male elbow, 1/8"		26	.308		2.30	19.85		22.15	32
4250	3/16" x 1/8"		24	.333		1.97	21.50		23.47	34
4260	1/4" x 1/8"		24	.333		1.91	21.50		23.41	34
4270	1/4"		24	.333		2.17	21.50		23.67	34.50
4280	3/8" x 1/4"		22	.364		2.76	23.50		26.26	38
4290	3/8"		22	.364		4.73	23.50		28.23	40
4300	1/2" x 1/4"		22	.364		4.58	23.50		28.08	40
4310	1/2" x 3/8"		22	.364		7.20	23.50		30.70	43
4340	Female elbow, 1/8"		26	.308		3.86	19.85		23.71	34
4350	1/4" x 1/8"		24	.333		2.81	21.50		24.31	35
4360	1/4"		24	.333		3.68	21.50		25.18	36
4370	3/8" x 1/4"		22	.364		3.62	23.50		27.12	39
4380	1/2" x 3/8"		22	.364		4.84	23.50		28.34	40.50
4390	1/2"		22	.364		6.50	23.50		30	42
4420	Male run tee, 1/4" x 1/8"		15	.533		3.43	34.50		37.93	55.50

22 11 13 – Facility Water Distribution Piping

22 11 13.25 Pipe/Tube Fittings, Copper		Crew	Daily Output	Labor-Hours	Unit	Material	2020 Bare Costs Labor	Equipment	Total	Total Incl O&P
4430	5/16" x 1/8"	1 Plum	15	.533	Ea.	5.15	34.50		39.65	57
4440	3/8" x 1/4"		15	.533		5.30	34.50		39.80	57.50
4480	Male branch tee, 1/4" x 1/8"		15	.533		3.98	34.50		38.48	56
4490	1/4"		15	.533		4.28	34.50		38.78	56
4500	3/8" x 1/4"		15	.533		3.90	34.50		38.40	56
4510	1/2" x 3/8"		15	.533		6.05	34.50		40.55	58
4520	1/2"	▼	15	.533	▼	9.70	34.50		44.20	62
4800	Flare joint fittings									
4810	Refrigeration fittings									
4820	Flare joint nuts and labor not incl. in price. Add 1 nut per jnt.									
4830	90° elbow, 1/4"				Ea.	2.51			2.51	2.76
4840	3/8"					2.07			2.07	2.28
4850	1/2"					2.66			2.66	2.93
4860	5/8"					3.66			3.66	4.03
4870	3/4"					11.90			11.90	13.10
5030	Tee, 1/4"					2.20			2.20	2.42
5040	5/16"					2.58			2.58	2.84
5050	3/8"					2.43			2.43	2.67
5060	1/2"					3.67			3.67	4.04
5070	5/8"					5.30			5.30	5.80
5080	3/4"					6.75			6.75	7.45
5140	Union, 3/16"					.91			.91	1
5150	1/4"					.69			.69	.76
5160	5/16"					1.27			1.27	1.40
5170	3/8"					.96			.96	1.06
5180	1/2"					1.43			1.43	1.57
5190	5/8"					2.29			2.29	2.52
5200	3/4"					7.80			7.80	8.60
5260	Long flare nut, 3/16"	1 Stpi	42	.190		.40	12.50		12.90	19.15
5270	1/4"		41	.195		.72	12.80		13.52	19.95
5280	5/16"		40	.200		1.30	13.10		14.40	21
5290	3/8"		39	.205		2.06	13.45		15.51	22.50
5300	1/2"		38	.211		2.39	13.80		16.19	23
5310	5/8"		37	.216		4.09	14.15		18.24	25.50
5320	3/4"		34	.235		16.20	15.40		31.60	41
5380	Short flare nut, 3/16"		42	.190		4.11	12.50		16.61	23
5390	1/4"		41	.195		.50	12.80		13.30	19.70
5400	5/16"		40	.200		.60	13.10		13.70	20.50
5410	3/8"		39	.205		.69	13.45		14.14	21
5420	1/2"		38	.211		1.18	13.80		14.98	22
5430	5/8"		36	.222		1.65	14.55		16.20	24
5440	3/4"	▼	34	.235		5.75	15.40		21.15	29.50
5500	90° elbow flare by MIPS, 1/4"					2.10			2.10	2.31
5510	3/8"					2.04			2.04	2.24
5520	1/2"					3.68			3.68	4.05
5530	5/8"					6.95			6.95	7.65
5540	3/4"					12.30			12.30	13.55
5600	Flare by FIPS, 1/4"					3.78			3.78	4.16
5610	3/8"					3.41			3.41	3.75
5620	1/2"					4.83			4.83	5.30
5670	Flare by sweat, 1/4"					2.29			2.29	2.52
5680	3/8"					5.05			5.05	5.60
5690	1/2"				▼	6.25			6.25	6.85

22 11 13.25 Pipe/Tube Fittings, Copper		Crew	Daily Output	Labor-Hours	Unit	Material	2020 Bare Costs Labor	Equipment	Total	Total Incl O&P
5700	5/8"				Ea.	16.50			16.50	18.15
5760	Tee flare by IPS, 1/4"					3.14			3.14	3.45
5770	3/8"					4.92			4.92	5.40
5780	1/2"					8.30			8.30	9.10
5790	5/8"					6.45			6.45	7.10
5850	Connector, 1/4"					.74			.74	.81
5860	3/8"					1.10			1.10	1.21
5870	1/2"					1.85			1.85	2.04
5880	5/8"					3.77			3.77	4.15
5890	3/4"					5.65			5.65	6.20
5950	Seal cap, 1/4"					.48			.48	.53
5960	3/8"					.95			.95	1.05
5970	1/2"					.97			.97	1.07
5980	5/8"					2.38			2.38	2.62
5990	3/4"				▼	6.05			6.05	6.65
6000	Water service fittings									
6010	Flare joints nut and labor are included in the fitting price.									
6020	90° elbow, C x C, 3/8"	1 Plum	19	.421	Ea.	57	27		84	103
6030	1/2"		18	.444		70.50	28.50		99	121
6040	3/4"		16	.500		88	32		120	145
6050	1"		15	.533		148	34.50		182.50	215
6080	2"		10	.800		615	51.50		666.50	755
6090	90° elbow, C x MPT, 3/8"		19	.421		43	27		70	88
6100	1/2"		18	.444		53.50	28.50		82	102
6110	3/4"		16	.500		62	32		94	116
6120	1"		15	.533		170	34.50		204.50	239
6130	1-1/4"		13	.615		375	39.50		414.50	475
6140	1-1/2"		12	.667		335	43		378	430
6150	2"		10	.800		305	51.50		356.50	410
6160	90° elbow, C x FPT, 3/8"		19	.421		46.50	27		73.50	91.50
6170	1/2"		18	.444		57.50	28.50		86	106
6180	3/4"		16	.500		71	32		103	126
6190	1"		15	.533		164	34.50		198.50	232
6200	1-1/4"		13	.615		124	39.50		163.50	196
6210	1-1/2"		12	.667		475	43		518	590
6220	2"		10	.800		685	51.50		736.50	830
6230	Tee, C x C x C, 3/8"		13	.615		90	39.50		129.50	159
6240	1/2"		12	.667		103	43		146	178
6250	3/4"		11	.727		125	47		172	207
6260	1"		10	.800		187	51.50		238.50	282
6330	Tube nut, C x nut seat, 3/8"		40	.200		22	12.90		34.90	43.50
6340	1/2"		38	.211		22	13.55		35.55	44.50
6350	3/4"		34	.235		18.65	15.15		33.80	43
6360	1"		32	.250		38	16.10		54.10	65.50
6380	Coupling, C x C, 3/8"		19	.421		55.50	27		82.50	102
6390	1/2"		18	.444		63	28.50		91.50	113
6400	3/4"		16	.500		80	32		112	136
6410	1"		15	.533		148	34.50		182.50	214
6420	1-1/4"		12	.667		239	43		282	330
6430	1-1/2"		12	.667		360	43		403	465
6440	2"		10	.800		545	51.50		596.50	675
6450	Adapter, C x FPT, 3/8"		19	.421		42	27		69	86.50
6460	1/2"		18	.444		47.50	28.50		76	95.50

22 11 13 – Facility Water Distribution Piping

22 11 13.25 Pipe/Tube Fittings, Copper

		Crew	Daily Output	Labor-Hours	Unit	Material	2020 Bare Costs Labor	2020 Bare Costs Equipment	Total	Total Incl O&P
6470	3/4"	1 Plum	16	.500	Ea.	57.50	32		89.50	112
6480	1"		15	.533		123	34.50		157.50	187
6490	1-1/4"		13	.615		108	39.50		147.50	179
6500	1-1/2"		12	.667		261	43		304	350
6510	2"		10	.800		297	51.50		348.50	400
6520	Adapter, C x MPT, 3/8"		19	.421		34.50	27		61.50	78.50
6530	1/2"		18	.444		42.50	28.50		71	90
6540	3/4"		16	.500		59	32		91	113
6550	1"		15	.533		104	34.50		138.50	166
6560	1-1/4"		13	.615		227	39.50		266.50	310
6570	1-1/2"		12	.667		261	43		304	355
6580	2"		10	.800		315	51.50		366.50	420
6992	Tube connector fittings, See Section 22 11 13.76 for plastic ftng.									
7000	Insert type brass/copper, 100 psi @ 180°F, CTS									
7010	Adapter MPT 3/8" x 1/2" CTS	1 Plum	29	.276	Ea.	2.90	17.80		20.70	29.50
7020	1/2" x 1/2"		26	.308		3	19.85		22.85	33
7030	3/4" x 1/2"		26	.308		3.82	19.85		23.67	33.50
7040	3/4" x 3/4"		25	.320		4.52	20.50		25.02	36
7050	Adapter CTS 1/2" x 1/2" sweat		24	.333		3.82	21.50		25.32	36
7060	3/4" x 3/4" sweat		22	.364		1.43	23.50		24.93	36.50
7070	Coupler center set 3/8" CTS		25	.320		1.43	20.50		21.93	32.50
7080	1/2" CTS		23	.348		3.67	22.50		26.17	37.50
7090	3/4" CTS		22	.364		1.47	23.50		24.97	36.50
7100	Elbow 90°, copper 3/8"		25	.320		2.98	20.50		23.48	34.50
7110	1/2" CTS		23	.348		2	22.50		24.50	35.50
7120	3/4" CTS		22	.364		2.63	23.50		26.13	38
7130	Tee copper 3/8" CTS		17	.471		3.79	30.50		34.29	49.50
7140	1/2" CTS		15	.533		2.67	34.50		37.17	54.50
7150	3/4" CTS		14	.571		4.16	37		41.16	59.50
7160	3/8" x 3/8" x 1/2"		16	.500		4.21	32		36.21	52.50
7170	1/2" x 3/8" x 1/2"		15	.533		2.84	34.50		37.34	54.50
7180	3/4" x 1/2" x 3/4"		14	.571		3.91	37		40.91	59.50

22 11 13.27 Pipe/Tube, Grooved Joint for Copper

		Crew	Daily Output	Labor-Hours	Unit	Material	2020 Bare Costs Labor	2020 Bare Costs Equipment	Total	Total Incl O&P
0010	**PIPE/TUBE, GROOVED JOINT FOR COPPER**									
4000	Fittings: coupling material required at joints not incl. in fitting price.									
4001	Add 1 selected coupling, material only, per joint for installed price.									
4010	Coupling, rigid style									
4018	2" diameter	1 Plum	50	.160	Ea.	30.50	10.30		40.80	49
4020	2-1/2" diameter	Q-1	80	.200		34	11.60		45.60	55
4022	3" diameter		67	.239		38.50	13.85		52.35	62.50
4024	4" diameter		50	.320		57.50	18.55		76.05	91.50
4026	5" diameter		40	.400		96	23		119	141
4028	6" diameter	Q-2	50	.480		127	29		156	183
4100	Elbow, 90° or 45°									
4108	2" diameter	1 Plum	25	.320	Ea.	55.50	20.50		76	92
4110	2-1/2" diameter	Q-1	40	.400		60	23		83	101
4112	3" diameter		33	.485		84	28		112	135
4114	4" diameter		25	.640		193	37		230	268
4116	5" diameter		20	.800		545	46.50		591.50	670
4118	6" diameter	Q-2	25	.960		870	57.50		927.50	1,050
4200	Tee									
4208	2" diameter	1 Plum	17	.471	Ea.	91	30.50		121.50	146

For customer support on your Mechanical Costs with RSMeans Data, call 800.448.8182.

139

22 11 13 – Facility Water Distribution Piping

22 11 13.27 Pipe/Tube, Grooved Joint for Copper

		Crew	Daily Output	Labor-Hours	Unit	Material	2020 Bare Costs Labor	2020 Bare Costs Equipment	Total	Total Incl O&P
4210	2-1/2" diameter	Q-1	27	.593	Ea.	96.50	34.50		131	158
4212	3" diameter		22	.727		144	42		186	221
4214	4" diameter		17	.941		315	54.50		369.50	425
4216	5" diameter		13	1.231		880	71.50		951.50	1,075
4218	6" diameter	Q-2	17	1.412		1,075	85		1,160	1,325
4300	Reducer, concentric									
4310	3" x 2-1/2" diameter	Q-1	35	.457	Ea.	79.50	26.50		106	127
4312	4" x 2-1/2" diameter		32	.500		163	29		192	223
4314	4" x 3" diameter		29	.552		163	32		195	227
4316	5" x 3" diameter		25	.640		455	37		492	555
4318	5" x 4" diameter		22	.727		455	42		497	565
4320	6" x 3" diameter	Q-2	28	.857		495	51.50		546.50	620
4322	6" x 4" diameter		26	.923		495	55.50		550.50	630
4324	6" x 5" diameter		24	1		495	60		555	635
4350	Flange, w/groove gasket									
4351	ANSI class 125 and 150									
4355	2" diameter	1 Plum	23	.348	Ea.	229	22.50		251.50	286
4356	2-1/2" diameter	Q-1	37	.432		239	25		264	300
4358	3" diameter		31	.516		250	30		280	320
4360	4" diameter		23	.696		274	40.50		314.50	360
4362	5" diameter		19	.842		370	49		419	480
4364	6" diameter	Q-2	23	1.043		400	63		463	535

22 11 13.29 Pipe, Fittings and Valves, Copper, Pressed-Joint

		Crew	Daily Output	Labor-Hours	Unit	Material	2020 Bare Costs Labor	2020 Bare Costs Equipment	Total	Total Incl O&P
0010	**PIPE, FITTINGS AND VALVES, COPPER, PRESSED-JOINT**									
0040	Pipe/tube includes coupling & clevis type hanger assy's, 10' OC									
0120	Type K									
0130	1/2" diameter	1 Plum	78	.103	L.F.	5.50	6.60		12.10	15.95
0134	3/4" diameter		74	.108		8.85	6.95		15.80	20
0138	1" diameter		66	.121		13.25	7.80		21.05	26.50
0142	1-1/4" diameter		56	.143		15.75	9.20		24.95	31
0146	1-1/2" diameter		50	.160		19.75	10.30		30.05	37.50
0150	2" diameter		40	.200		28.50	12.90		41.40	51
0154	2-1/2" diameter	Q-1	60	.267		49	15.45		64.45	76.50
0158	3" diameter		54	.296		64.50	17.20		81.70	96.50
0162	4" diameter		38	.421		98	24.50		122.50	145
0180	To delete cplgs. & hngrs., 1/2" pipe, subtract					19%	48%			
0184	3/4"-2" pipe, subtract					14%	46%			
0186	2-1/2"-4" pipe, subtract					24%	34%			
0220	Type L									
0230	1/2" diameter	1 Plum	81	.099	L.F.	3.90	6.35		10.25	13.85
0234	3/4" diameter		76	.105		4.93	6.80		11.73	15.55
0238	1" diameter		68	.118		8.05	7.60		15.65	20
0242	1-1/4" diameter		58	.138		12.45	8.90		21.35	27
0246	1-1/2" diameter		52	.154		12.80	9.90		22.70	29
0250	2" diameter		42	.190		20	12.30		32.30	40.50
0254	2-1/2" diameter	Q-1	62	.258		35	14.95		49.95	61
0258	3" diameter		56	.286		53	16.55		69.55	83.50
0262	4" diameter		39	.410		66	24		90	108
0280	To delete cplgs. & hngrs., 1/2" pipe, subtract					21%	52%			
0284	3/4"-2" pipe, subtract					17%	46%			
0286	2-1/2"-4" pipe, subtract					23%	35%			
0320	Type M									

22 11 Facility Water Distribution

22 11 13 – Facility Water Distribution Piping

22 11 13.29 Pipe, Fittings and Valves, Copper, Pressed-Joint	Crew	Daily Output	Labor-Hours	Unit	Material	2020 Bare Costs Labor	Equipment	Total	Total Incl O&P
0330	1/2" diameter	1 Plum	84	.095	L.F.	3.78	6.15		9.93	13.35
0334	3/4" diameter		78	.103		5.25	6.60		11.85	15.70
0338	1" diameter		70	.114		8.95	7.35		16.30	21
0342	1-1/4" diameter		60	.133		12.75	8.60		21.35	27
0346	1-1/2" diameter		54	.148		15.65	9.55		25.20	31.50
0350	2" diameter		44	.182		22.50	11.70		34.20	42.50
0354	2-1/2" diameter	Q-1	64	.250		36.50	14.50		51	61.50
0358	3" diameter		58	.276		46.50	16		62.50	75
0362	4" diameter		40	.400		75.50	23		98.50	118
0380	To delete cplgs. & hngrs., 1/2" pipe, subtract					32%	49%			
0384	3/4"-2" pipe, subtract					21%	46%			
0386	2-1/2"-4" pipe, subtract					25%	36%			
1600	Fittings									
1610	Press joints, copper x copper									
1620	Note: Reducing fittings show most expensive size combination.									
1800	90° elbow, 1/2"	1 Plum	36.60	.219	Ea.	3.58	14.10		17.68	25
1810	3/4"		27.50	.291		5.80	18.75		24.55	34.50
1820	1"		25.90	.309		11.60	19.90		31.50	43
1830	1-1/4"		20.90	.383		23	24.50		47.50	62.50
1840	1-1/2"		18.30	.437		44	28		72	90.50
1850	2"		15.70	.510		62	33		95	117
1860	2-1/2"	Q-1	25.90	.618		179	36		215	250
1870	3"		22	.727		225	42		267	310
1880	4"		16.30	.982		278	57		335	390
2000	45° elbow, 1/2"	1 Plum	36.60	.219		4.20	14.10		18.30	25.50
2010	3/4"		27.50	.291		4.95	18.75		23.70	33.50
2020	1"		25.90	.309		15.80	19.90		35.70	47.50
2030	1-1/4"		20.90	.383		23	24.50		47.50	62
2040	1-1/2"		18.30	.437		36.50	28		64.50	82.50
2050	2"		15.70	.510		51	33		84	105
2060	2-1/2"	Q-1	25.90	.618		120	36		156	186
2070	3"		22	.727		169	42		211	248
2080	4"		16.30	.982		238	57		295	345
2200	Tee, 1/2"	1 Plum	27.50	.291		5.45	18.75		24.20	34
2210	3/4"		20.70	.386		9.40	25		34.40	48
2220	1"		19.40	.412		17.15	26.50		43.65	59
2230	1-1/4"		15.70	.510		29.50	33		62.50	81.50
2240	1-1/2"		13.80	.580		56.50	37.50		94	118
2250	2"		11.80	.678		69.50	43.50		113	142
2260	2-1/2"	Q-1	19.40	.825		224	48		272	320
2270	3"		16.50	.970		276	56		332	390
2280	4"		12.20	1.311		395	76		471	545
2400	Tee, reducing on the outlet									
2410	3/4"	1 Plum	20.70	.386	Ea.	8.15	25		33.15	46.50
2420	1"		19.40	.412		19.95	26.50		46.45	62
2430	1-1/4"		15.70	.510		29	33		62	81
2440	1-1/2"		13.80	.580		61	37.50		98.50	123
2450	2"		11.80	.678		96	43.50		139.50	172
2460	2-1/2"	Q-1	19.40	.825		300	48		348	400
2470	3"		16.50	.970		360	56		416	480
2480	4"		12.20	1.311		450	76		526	610
2600	Tee, reducing on the run									
2610	3/4"	1 Plum	20.70	.386	Ea.	16	25		41	55

For customer support on your Mechanical Costs with RSMeans Data, call 800.448.8182.

141

22 11 13.29 Pipe, Fittings and Valves, Copper, Pressed-Joint		Crew	Daily Output	Labor-Hours	Unit	Material	2020 Bare Costs Labor	Equipment	Total	Total Incl O&P
2620	1"	1 Plum	19.40	.412	Ea.	30.50	26.50		57	73.50
2630	1-1/4"		15.70	.510		61	33		94	116
2640	1-1/2"		13.80	.580		93.50	37.50		131	159
2650	2"		11.80	.678		102	43.50		145.50	179
2660	2-1/2"	Q-1	19.40	.825		335	48		383	440
2670	3"		16.50	.970		400	56		456	525
2680	4"		12.20	1.311		525	76		601	690
2800	Coupling, 1/2"	1 Plum	36.60	.219		3.16	14.10		17.26	24.50
2810	3/4"		27.50	.291		4.88	18.75		23.63	33.50
2820	1"		25.90	.309		9.85	19.90		29.75	41
2830	1-1/4"		20.90	.383		12.30	24.50		36.80	50.50
2840	1-1/2"		18.30	.437		22.50	28		50.50	67
2850	2"		15.70	.510		29	33		62	80.50
2860	2-1/2"	Q-1	25.90	.618		89	36		125	152
2870	3"		22	.727		113	42		155	188
2880	4"		16.30	.982		160	57		217	261
3000	Union, 1/2"	1 Plum	36.60	.219		28	14.10		42.10	51.50
3010	3/4"		27.50	.291		35	18.75		53.75	66.50
3020	1"		25.90	.309		56.50	19.90		76.40	92
3030	1-1/4"		20.90	.383		82.50	24.50		107	128
3040	1-1/2"		18.30	.437		108	28		136	161
3050	2"		15.70	.510		174	33		207	240
3200	Adapter, tube to MPT									
3210	1/2"	1 Plum	15.10	.530	Ea.	4.10	34		38.10	55.50
3220	3/4"		13.60	.588		7.60	38		45.60	65
3230	1"		11.60	.690		13.70	44.50		58.20	81.50
3240	1-1/4"		9.90	.808		29.50	52		81.50	111
3250	1-1/2"		9	.889		41.50	57.50		99	131
3260	2"		7.90	1.013		80	65.50		145.50	186
3270	2-1/2"	Q-1	13.40	1.194		181	69.50		250.50	305
3280	3"		10.60	1.509		228	87.50		315.50	380
3290	4"		7.70	2.078		277	121		398	485
3400	Adapter, tube to FPT									
3410	1/2"	1 Plum	15.10	.530	Ea.	5.05	34		39.05	56.50
3420	3/4"		13.60	.588		8	38		46	65.50
3430	1"		11.60	.690		15.15	44.50		59.65	83
3440	1-1/4"		9.90	.808		34.50	52		86.50	116
3450	1-1/2"		9	.889		49	57.50		106.50	140
3460	2"		7.90	1.013		82.50	65.50		148	189
3470	2-1/2"	Q-1	13.40	1.194		203	69.50		272.50	325
3480	3"		10.60	1.509		335	87.50		422.50	495
3490	4"		7.70	2.078		430	121		551	650
3600	Flange									
3620	1"	1 Plum	36.20	.221	Ea.	147	14.25		161.25	183
3630	1-1/4"		29.30	.273		210	17.60		227.60	258
3640	1-1/2"		25.60	.313		232	20		252	285
3650	2"		22	.364		261	23.50		284.50	320
3660	2-1/2"	Q-1	36.20	.442		273	25.50		298.50	340
3670	3"		30.80	.519		330	30		360	410
3680	4"		22.80	.702		370	40.50		410.50	465
3800	Cap, 1/2"	1 Plum	53.10	.151		6.85	9.70		16.55	22
3810	3/4"		39.80	.201		11.50	12.95		24.45	32
3820	1"		37.50	.213		17.75	13.75		31.50	40

22 11 Facility Water Distribution

22 11 13 – Facility Water Distribution Piping

22 11 13.29 Pipe, Fittings and Valves, Copper, Pressed-Joint		Crew	Daily Output	Labor-Hours	Unit	Material	2020 Bare Costs Labor	Equipment	Total	Total Incl O&P
3830	1-1/4"	1 Plum	30.40	.263	Ea.	21	16.95		37.95	48.50
3840	1-1/2"		26.60	.301		32.50	19.40		51.90	65
3850	2"		22.80	.351		39.50	22.50		62	77.50
3860	2-1/2"	Q-1	37.50	.427		124	25		149	174
3870	3"		31.90	.502		157	29		186	217
3880	4"		23.60	.678		192	39.50		231.50	270
4000	Reducer									
4010	3/4"	1 Plum	27.50	.291	Ea.	14.45	18.75		33.20	44
4020	1"		25.90	.309		26.50	19.90		46.40	59
4030	1-1/4"		20.90	.383		39	24.50		63.50	80
4040	1-1/2"		18.30	.437		46.50	28		74.50	93.50
4050	2"		15.70	.510		64.50	33		97.50	120
4060	2-1/2"	Q-1	25.90	.618		179	36		215	251
4070	3"		22	.727		242	42		284	330
4080	4"		16.30	.982		310	57		367	430
4100	Stub out, 1/2"	1 Plum	50	.160		9.65	10.30		19.95	26
4110	3/4"		35	.229		14.90	14.75		29.65	38.50
4120	1"		30	.267		20.50	17.20		37.70	48
6000	Valves									
6200	Ball valve									
6210	1/2"	1 Plum	25.60	.313	Ea.	40	20		60	73.50
6220	3/4"		19.20	.417		53	27		80	98
6230	1"		18.10	.442		63.50	28.50		92	113
6240	1-1/4"		14.70	.544		103	35		138	167
6250	1-1/2"		12.80	.625		254	40.50		294.50	340
6260	2"		11	.727		365	47		412	475
6400	Check valve									
6410	1/2"	1 Plum	30	.267	Ea.	34.50	17.20		51.70	63.50
6420	3/4"		22.50	.356		44.50	23		67.50	83.50
6430	1"		21.20	.377		51	24.50		75.50	93
6440	1-1/4"		17.20	.465		74.50	30		104.50	127
6450	1-1/2"		15	.533		105	34.50		139.50	168
6460	2"		12.90	.620		191	40		231	270
6600	Butterfly valve, lug type									
6660	2-1/2"	Q-1	9	1.778	Ea.	163	103		266	335
6670	3"		8	2		200	116		316	395
6680	4"		5	3.200		250	186		436	555

22 11 13.44 Pipe, Steel

		Crew	Daily Output	Labor-Hours	Unit	Material	2020 Bare Costs Labor	Equipment	Total	Total Incl O&P
0010	**PIPE, STEEL**		R221113-50							
0020	All pipe sizes are to Spec. A-53 unless noted otherwise		R221113-70							
0032	Schedule 10, see Line 22 11 13.48 0500									
0050	Schedule 40, threaded, with couplings, and clevis hanger									
0060	assemblies sized for covering, 10' OC									
0540	Black, 1/4" diameter	1 Plum	66	.121	L.F.	6.15	7.80		13.95	18.50
0550	3/8" diameter		65	.123		6.85	7.95		14.80	19.40
0560	1/2" diameter		63	.127		4.03	8.20		12.23	16.70
0570	3/4" diameter		61	.131		4.40	8.45		12.85	17.50
0580	1" diameter		53	.151		7.80	9.75		17.55	23
0590	1-1/4" diameter	Q-1	89	.180		8.20	10.45		18.65	24.50
0600	1-1/2" diameter		80	.200		8.75	11.60		20.35	27
0610	2" diameter		64	.250		6.90	14.50		21.40	29
0620	2-1/2" diameter		50	.320		11.40	18.55		29.95	40.50

For customer support on your Mechanical Costs with RSMeans Data, call 800.448.8182.

143

22 11 13.44 Pipe, Steel		Crew	Daily Output	Labor-Hours	Unit	Material	2020 Bare Costs Labor	Equipment	Total	Total Incl O&P
0630	3" diameter	Q-1	43	.372	L.F.	14.35	21.50		35.85	48.50
0640	3-1/2" diameter		40	.400		21	23		44	58
0650	4" diameter	↓	36	.444	↓	23	26		49	64
0809	A-106, gr. A/B, seamless w/cplgs. & clevis hanger assemblies									
0011	1/4" diameter	1 Plum	66	.121	L.F.	9.15	7.80		16.95	22
0812	3/8" diameter		65	.123		9.20	7.95		17.15	22
0813	1/2" diameter		63	.127		9.40	8.20		17.60	22.50
0814	3/4" diameter		61	.131		12.30	8.45		20.75	26
0815	1" diameter	↓	53	.151		11.30	9.75		21.05	27
0816	1-1/4" diameter	Q-1	89	.180		13.20	10.45		23.65	30
0817	1-1/2" diameter		80	.200		19.55	11.60		31.15	39
0819	2" diameter		64	.250		15.70	14.50		30.20	39
0821	2-1/2" diameter		50	.320		21	18.55		39.55	51
0822	3" diameter		43	.372		26.50	21.50		48	61.50
0823	4" diameter	↓	36	.444	↓	41	26		67	83.50
1220	To delete coupling & hanger, subtract									
1230	1/4" diam. to 3/4" diam.					31%	56%			
1240	1" diam. to 1-1/2" diam.					23%	51%			
1250	2" diam. to 4" diam.					23%	41%			
1280	All pipe sizes are to Spec. A-53 unless noted otherwise									
1281	Schedule 40, threaded, with couplings and clevis hanger									
1282	assemblies sized for covering, 10' OC									
1290	Galvanized, 1/4" diameter	1 Plum	66	.121	L.F.	8.10	7.80		15.90	20.50
1300	3/8" diameter		65	.123		8.35	7.95		16.30	21
1310	1/2" diameter		63	.127		4.10	8.20		12.30	16.75
1320	3/4" diameter		61	.131		5	8.45		13.45	18.15
1330	1" diameter	↓	53	.151		7.90	9.75		17.65	23.50
1340	1-1/4" diameter	Q-1	89	.180		8.85	10.45		19.30	25.50
1350	1-1/2" diameter		80	.200		9.45	11.60		21.05	28
1360	2" diameter		64	.250		7.90	14.50		22.40	30
1370	2-1/2" diameter		50	.320		13.15	18.55		31.70	42.50
1380	3" diameter		43	.372		16	21.50		37.50	50
1390	3-1/2" diameter		40	.400		23	23		46	59.50
1400	4" diameter	↓	36	.444	↓	25.50	26		51.50	66.50
1750	To delete coupling & hanger, subtract									
1760	1/4" diam. to 3/4" diam.					31%	56%			
1770	1" diam. to 1-1/2" diam.					23%	51%			
1780	2" diam. to 4" diam.					23%	41%			
2000	Welded, sch. 40, on yoke & roll hanger assy's, sized for covering, 10' OC									
2040	Black, 1" diameter	Q-15	93	.172	L.F.	7.15	10	1.14	18.29	24
2050	1-1/4" diameter		84	.190		8.20	11.05	1.27	20.52	27
2060	1-1/2" diameter		76	.211		8.65	12.20	1.40	22.25	29.50
2070	2" diameter		61	.262		6.60	15.20	1.74	23.54	32
2080	2-1/2" diameter		47	.340		11.80	19.75	2.26	33.81	45
2090	3" diameter		43	.372		16.65	21.50	2.47	40.62	53.50
2100	3-1/2" diameter		39	.410		17.45	24	2.73	44.18	57.50
2110	4" diameter		37	.432		24	25	2.88	51.88	67
2120	5" diameter	↓	32	.500		37	29	3.33	69.33	87.50
2130	6" diameter	Q-16	36	.667		45	40	2.95	87.95	113
2140	8" diameter		29	.828		71.50	50	3.67	125.17	157
2150	10" diameter		24	1		90	60	4.43	154.43	194
2160	12" diameter		19	1.263		107	76	5.60	188.60	237
2170	14" diameter (two rod roll type hanger for 14" diam. and up)	↓	15	1.600		104	96	7.10	207.10	267

22 11 13 – Facility Water Distribution Piping

22 11 13.44 Pipe, Steel		Crew	Daily Output	Labor- Hours	Unit	Material	2020 Bare Costs Labor	Equipment	Total	Total Incl O&P
2180	16" diameter (two rod roll type hanger)	Q-16	13	1.846	L.F.	172	111	8.20	291.20	365
2190	18" diameter (two rod roll type hanger)		11	2.182		147	131	9.65	287.65	370
2200	20" diameter (two rod roll type hanger)		9	2.667		156	160	11.80	327.80	425
2220	24" diameter (two rod roll type hanger)	↓	8	3	↓	203	180	13.30	396.30	510
2345	Sch. 40, A-53, gr. A/B, ERW, welded w/hngrs.									
2346	2" diameter	Q-15	61	.262	L.F.	10.20	15.20	1.74	27.14	36
2347	2-1/2" diameter		47	.340		12.40	19.75	2.26	34.41	45.50
2348	3" diameter		43	.372		14.15	21.50	2.47	38.12	51
2349	4" diameter	↓	38	.421		20.50	24.50	2.80	47.80	62
2350	6" diameter	Q-16	37	.649		30.50	39	2.87	72.37	95.50
2351	8" diameter		29	.828		73	50	3.67	126.67	159
2352	10" diameter	↓	24	1	↓	93.50	60	4.43	157.93	198
2560	To delete hanger, subtract									
2570	1" diam. to 1-1/2" diam.					15%	34%			
2580	2" diam. to 3-1/2" diam.					9%	21%			
2590	4" diam. to 12" diam.					5%	12%			
2596	14" diam. to 24" diam.					3%	10%			
2640	Galvanized, 1" diameter	Q-15	93	.172	L.F.	7.85	10	1.14	18.99	25
2650	1-1/4" diameter		84	.190		9	11.05	1.27	21.32	28
2660	1-1/2" diameter		76	.211		9.55	12.20	1.40	23.15	30.50
2670	2" diameter		61	.262		7.75	15.20	1.74	24.69	33.50
2680	2-1/2" diameter		47	.340		10.80	19.75	2.26	32.81	44
2690	3" diameter		43	.372		14.45	21.50	2.47	38.42	51
2700	3-1/2" diameter		39	.410		18.90	24	2.73	45.63	59.50
2710	4" diameter		37	.432		21	25	2.88	48.88	63.50
2720	5" diameter	↓	32	.500		26.50	29	3.33	58.83	76
2730	6" diameter	Q-16	36	.667		32.50	40	2.95	75.45	99
2740	8" diameter		29	.828		45	50	3.67	98.67	128
2750	10" diameter		24	1		90	60	4.43	154.43	194
2760	12" diameter	↓	19	1.263	↓	117	76	5.60	198.60	248
3160	To delete hanger, subtract									
3170	1" diam. to 1-1/2" diam.					30%	34%			
3180	2" diam. to 3-1/2" diam.					19%	21%			
3190	4" diam. to 12" diam.					10%	12%			
3250	Flanged, 150 lb. weld neck, on yoke & roll hangers									
3260	sized for covering, 10' OC									
3290	Black, 1" diameter	Q-15	70	.229	L.F.	10.60	13.25	1.52	25.37	33
3300	1-1/4" diameter		64	.250		11.65	14.50	1.66	27.81	36
3310	1-1/2" diameter		58	.276		12.10	16	1.83	29.93	39.50
3320	2" diameter		45	.356		15.25	20.50	2.36	38.11	50.50
3330	2-1/2" diameter		36	.444		23	26	2.96	51.96	67.50
3340	3" diameter		32	.500		28.50	29	3.33	60.83	78.50
3350	3-1/2" diameter		29	.552		31	32	3.67	66.67	86
3360	4" diameter		26	.615		33	35.50	4.09	72.59	94.50
3370	5" diameter	↓	21	.762		51.50	44	5.05	100.55	129
3380	6" diameter	Q-16	25	.960		62.50	57.50	4.25	124.25	160
3390	8" diameter		19	1.263		101	76	5.60	182.60	231
3400	10" diameter		16	1.500		141	90	6.65	237.65	297
3410	12" diameter	↓	14	1.714		175	103	7.60	285.60	355
3470	For 300 lb. flanges, add					63%				
3480	For 600 lb. flanges, add				↓	310%				
3960	To delete flanges & hanger, subtract									
3970	1" diam. to 2" diam.					76%	65%			

22 11 13.44 Pipe, Steel	Crew	Daily Output	Labor-Hours	Unit	Material	2020 Bare Costs Labor	Equipment	Total	Total Incl O&P
3980	2-1/2" diam. to 4" diam.					62%	59%		
3990	5" diam. to 12" diam.					60%	46%		
4040	Galvanized, 1" diameter	Q-15	70	.229	L.F.	11.25	13.25	1.52	26.02	34
4050	1-1/4" diameter		64	.250		12.45	14.50	1.66	28.61	37
4060	1-1/2" diameter		58	.276		12.95	16	1.83	30.78	40.50
4070	2" diameter		45	.356		14.65	20.50	2.36	37.51	49.50
4080	2-1/2" diameter		36	.444		22.50	26	2.96	51.46	67
4090	3" diameter		32	.500		26	29	3.33	58.33	75.50
4100	3-1/2" diameter		29	.552		32	32	3.67	67.67	87.50
4110	4" diameter		26	.615		30	35.50	4.09	69.59	91
4120	5" diameter		21	.762		41	44	5.05	90.05	117
4130	6" diameter	Q-16	25	.960		50	57.50	4.25	111.75	146
4140	8" diameter		19	1.263		63.50	76	5.60	145.10	190
4150	10" diameter		16	1.500		127	90	6.65	223.65	282
4160	12" diameter		14	1.714		150	103	7.60	260.60	325
4220	For 300 lb. flanges, add					65%			
4240	For 600 lb. flanges, add					350%			
4660	To delete flanges & hanger, subtract								
4670	1" diam. to 2" diam.					73%	65%		
4680	2-1/2" diam. to 4" diam.					58%	59%		
4690	5" diam. to 12" diam.					43%	46%		
4750	Schedule 80, threaded, with couplings, and clevis hanger assemblies								
4760	sized for covering, 10' OC								
4790	Black, 1/4" diameter	1 Plum	54	.148	L.F.	7.60	9.55		17.15	22.50
4800	3/8" diameter		53	.151		9.70	9.75		19.45	25
4810	1/2" diameter		52	.154		4.87	9.90		14.77	20
4820	3/4" diameter		50	.160		6.45	10.30		16.75	22.50
4830	1" diameter		45	.178		7.35	11.45		18.80	25.50
4840	1-1/4" diameter	Q-1	75	.213		9.40	12.35		21.75	29
4850	1-1/2" diameter		69	.232		10.35	13.45		23.80	31.50
4860	2" diameter		56	.286		13.20	16.55		29.75	39.50
4870	2-1/2" diameter		44	.364		22	21		43	55.50
4880	3" diameter		38	.421		27	24.50		51.50	66.50
4890	3-1/2" diameter		35	.457		32.50	26.50		59	75.50
4900	4" diameter		32	.500		36	29		65	83.50
5061	A-106, gr. A/B seamless with cplgs. & clevis hanger assemblies, 1/4" diam.	1 Plum	63	.127		7	8.20		15.20	19.95
5062	3/8" diameter		62	.129		8	8.30		16.30	21.50
5063	1/2" diameter		61	.131		6.65	8.45		15.10	19.95
5064	3/4" diameter		57	.140		7.65	9.05		16.70	22
5065	1" diameter		51	.157		8.90	10.10		19	25
5066	1-1/4" diameter	Q-1	85	.188		12.40	10.90		23.30	30
5067	1-1/2" diameter		77	.208		16.95	12.05		29	36.50
5071	2" diameter		61	.262		15.60	15.20		30.80	40
5072	2-1/2" diameter		48	.333		18.35	19.35		37.70	49
5073	3" diameter		41	.390		24	22.50		46.50	60.50
5074	4" diameter		35	.457		34.50	26.50		61	77.50
5430	To delete coupling & hanger, subtract								
5440	1/4" diam. to 1/2" diam.					31%	54%		
5450	3/4" diam. to 1-1/2" diam.					28%	49%		
5460	2" diam. to 4" diam.					21%	40%		
5510	Galvanized, 1/4" diameter	1 Plum	54	.148	L.F.	9.90	9.55		19.45	25
5520	3/8" diameter		53	.151		10.70	9.75		20.45	26.50
5530	1/2" diameter		52	.154		5.65	9.90		15.55	21

22 11 13 – Facility Water Distribution Piping

22 11 13.44 Pipe, Steel	Crew	Daily Output	Labor-Hours	Unit	Material	2020 Bare Costs Labor	Equipment	Total	Total Incl O&P
5540	3/4" diameter	1 Plum	50	.160	L.F.	7.10	10.30		17.40	23.50
5550	1" diameter	↓	45	.178		8.25	11.45		19.70	26.50
5560	1-1/4" diameter	Q-1	75	.213		10.70	12.35		23.05	30.50
5570	1-1/2" diameter		69	.232		11.90	13.45		25.35	33
5580	2" diameter		56	.286		15.45	16.55		32	42
5590	2-1/2" diameter		44	.364		25	21		46	59
5600	3" diameter		38	.421		31	24.50		55.50	70.50
5610	3-1/2" diameter		35	.457		38	26.50		64.50	81.50
5620	4" diameter	↓	32	.500	↓	39.50	29		68.50	87
5930	To delete coupling & hanger, subtract									
5940	1/4" diam. to 1/2" diam.					31%	54%			
5950	3/4" diam. to 1-1/2" diam.					28%	49%			
5960	2" diam. to 4" diam.					21%	40%			
6000	Welded, on yoke & roller hangers									
6010	sized for covering, 10' OC									
6040	Black, 1" diameter	Q-15	85	.188	L.F.	7.15	10.90	1.25	19.30	25.50
6050	1-1/4" diameter		79	.203		9	11.75	1.35	22.10	29
6060	1-1/2" diameter		72	.222		9.95	12.90	1.48	24.33	32
6070	2" diameter		57	.281		12	16.30	1.87	30.17	40
6080	2-1/2" diameter		44	.364		20.50	21	2.42	43.92	56.50
6090	3" diameter		40	.400		25	23	2.66	50.66	65
6100	3-1/2" diameter		34	.471		30	27.50	3.13	60.63	77.50
6110	4" diameter		33	.485		32	28	3.22	63.22	81
6120	5" diameter, A-106B	↓	26	.615		57.50	35.50	4.09	97.09	122
6130	6" diameter, A-106B	Q-16	30	.800		67.50	48	3.54	119.04	150
6140	8" diameter, A-106B		25	.960		103	57.50	4.25	164.75	205
6150	10" diameter, A-106B		20	1.200		153	72	5.30	230.30	282
6160	12" diameter, A-106B	↓	15	1.600	↓	345	96	7.10	448.10	530
6540	To delete hanger, subtract									
6550	1" diam. to 1-1/2" diam.					30%	14%			
6560	2" diam. to 3" diam.					23%	9%			
6570	3-1/2" diam. to 5" diam.					12%	6%			
6580	6" diam. to 12" diam.					10%	4%			
6610	Galvanized, 1" diameter	Q-15	85	.188	L.F.	8.05	10.90	1.25	20.20	26.50
6650	1-1/4" diameter		79	.203		10.30	11.75	1.35	23.40	30.50
6660	1-1/2" diameter		72	.222		11.50	12.90	1.48	25.88	33.50
6670	2" diameter		57	.281		14.20	16.30	1.87	32.37	42
6680	2-1/2" diameter		44	.364		23.50	21	2.42	46.92	60
6690	3" diameter		40	.400		29	23	2.66	54.66	69.50
6700	3-1/2" diameter		34	.471		35	27.50	3.13	65.63	83
6710	4" diameter		33	.485		35.50	28	3.22	66.72	84.50
6720	5" diameter, A-106B	↓	26	.615		42	35.50	4.09	81.59	104
6730	6" diameter, A-106B	Q-16	30	.800		59	48	3.54	110.54	141
6740	8" diameter, A-106B		25	.960		109	57.50	4.25	170.75	211
6750	10" diameter, A-106B		20	1.200		249	72	5.30	326.30	390
6760	12" diameter, A-106B	↓	15	1.600	↓	565	96	7.10	668.10	770
7150	To delete hanger, subtract									
7160	1" diam. to 1-1/2" diam.					26%	14%			
7170	2" diam. to 3" diam.					16%	9%			
7180	3-1/2" diam. to 5" diam.					10%	6%			
7190	6" diam. to 12" diam.					7%	4%			
7250	Flanged, 300 lb. weld neck, on yoke & roll hangers									
7260	sized for covering, 10' OC									

22 11 Facility Water Distribution

22 11 13 – Facility Water Distribution Piping

22 11 13.44 Pipe, Steel	Crew	Daily Output	Labor-Hours	Unit	Material	2020 Bare Costs Labor	Equipment	Total	Total Incl O&P
7290	Black, 1" diameter	Q-15	66	.242	L.F.	14.40	14.05	1.61	30.06	38.50
7300	1-1/4" diameter		61	.262		16.25	15.20	1.74	33.19	43
7310	1-1/2" diameter		54	.296		17.20	17.20	1.97	36.37	46.50
7320	2" diameter		47	.381		22	22	2.53	46.53	60
7330	2-1/2" diameter		33	.485		31.50	28	3.22	62.72	80
7340	3" diameter		29	.552		34.50	32	3.67	70.17	90
7350	3-1/2" diameter		24	.667		47	38.50	4.43	89.93	114
7360	4" diameter		23	.696		49	40.50	4.63	94.13	120
7370	5" diameter		19	.842		82.50	49	5.60	137.10	170
7380	6" diameter	Q-16	23	1.043		92.50	63	4.62	160.12	201
7390	8" diameter		17	1.412		149	85	6.25	240.25	298
7400	10" diameter		14	1.714		241	103	7.60	351.60	425
7410	12" diameter		12	2		440	120	8.85	568.85	675
7470	For 600 lb. flanges, add					100%				
7940	To delete flanges & hanger, subtract									
7950	1" diam. to 1-1/2" diam.					75%	66%			
7960	2" diam. to 3" diam.					62%	60%			
7970	3-1/2" diam. to 5" diam.					54%	66%			
7980	6" diam. to 12" diam.					55%	62%			
8040	Galvanized, 1" diameter	Q-15	66	.242	L.F.	15.30	14.05	1.61	30.96	39.50
8050	1-1/4" diameter		61	.262		17.55	15.20	1.74	34.49	44
8060	1-1/2" diameter		54	.296		18.75	17.20	1.97	37.92	48
8070	2" diameter		42	.381		24	22	2.53	48.53	62.50
8080	2-1/2" diameter		33	.485		34.50	28	3.22	65.72	83.50
8090	3" diameter		29	.552		38.50	32	3.67	74.17	94.50
8100	3-1/2" diameter		24	.667		52.50	38.50	4.43	95.43	120
8110	4" diameter		23	.696		52.50	40.50	4.63	97.63	123
8120	5" diameter, A-106B		19	.842		67	49	5.60	121.60	153
8130	6" diameter, A-106B	Q-16	23	1.043		84	63	4.62	151.62	192
8140	8" diameter, A-106B		17	1.412		155	85	6.25	246.25	305
8150	10" diameter, A-106B		14	1.714		335	103	7.60	445.60	530
8160	12" diameter, A-106B		12	2		660	120	8.85	788.85	920
8240	For 600 lb. flanges, add					100%				
8300	To delete flanges & hangers, subtract									
8310	1" diam. to 1-1/2" diam.					72%	66%			
8320	2" diam. to 3" diam.					59%	60%			
8330	3-1/2" diam. to 5" diam.					51%	66%			
8340	6" diam. to 12" diam.					49%	62%			
9000	Threading pipe labor, one end, all schedules through 80									
9010	1/4" through 3/4" pipe size	1 Plum	80	.100	Ea.		6.45		6.45	9.65
9020	1" through 2" pipe size		73	.110			7.05		7.05	10.55
9030	2-1/2" pipe size		53	.151			9.75		9.75	14.55
9040	3" pipe size		50	.160			10.30		10.30	15.45
9050	3-1/2" pipe size	Q-1	89	.180			10.45		10.45	15.60
9060	4" pipe size		73	.219			12.70		12.70	19
9070	5" pipe size		53	.302			17.50		17.50	26
9080	6" pipe size		46	.348			20		20	30
9090	8" pipe size		29	.552			32		32	48
9100	10" pipe size		21	.762			44		44	66
9110	12" pipe size		13	1.231			71.50		71.50	107
9120	Cutting pipe labor, one cut									
9124	Shop fabrication, machine cut									
9126	Schedule 40, straight pipe									

22 11 13.44 Pipe, Steel		Crew	Daily Output	Labor-Hours	Unit	Material	2020 Bare Costs Labor	Equipment	Total	Total Incl O&P
9128	2" pipe size or less	1 Stpi	62	.129	Ea.		8.45		8.45	12.65
9130	2-1/2" pipe size		56	.143			9.35		9.35	14
9132	3" pipe size		42	.190			12.50		12.50	18.70
9134	4" pipe size		31	.258			16.90		16.90	25.50
9136	5" pipe size		26	.308			20		20	30
9138	6" pipe size		19	.421			27.50		27.50	41.50
9140	8" pipe size		14	.571			37.50		37.50	56
9142	10" pipe size		10	.800			52.50		52.50	78.50
9144	12" pipe size		7	1.143			75		75	112
9146	14" pipe size	Q-5	10.50	1.524			90		90	135
9148	16" pipe size		8.60	1.860			110		110	164
9150	18" pipe size		7	2.286			135		135	202
9152	20" pipe size		5.80	2.759			163		163	244
9154	24" pipe size		4	4			236		236	355
9160	Schedule 80, straight pipe									
9164	2" pipe size or less	1 Stpi	42	.190	Ea.		12.50		12.50	18.70
9166	2-1/2" pipe size		37	.216			14.15		14.15	21
9168	3" pipe size		31	.258			16.90		16.90	25.50
9170	4" pipe size		23	.348			23		23	34
9172	5" pipe size		18	.444			29		29	43.50
9174	6" pipe size		14.60	.548			36		36	54
9176	8" pipe size		10	.800			52.50		52.50	78.50
9178	10" pipe size	Q-5	14	1.143			67.50		67.50	101
9180	12" pipe size	"	10	1.600			94.50		94.50	141
9200	Welding labor per joint									
9210	Schedule 40									
9230	1/2" pipe size	Q-15	32	.500	Ea.		29	3.33	32.33	47
9240	3/4" pipe size		27	.593			34.50	3.94	38.44	56
9250	1" pipe size		23	.696			40.50	4.63	45.13	65.50
9260	1-1/4" pipe size		20	.800			46.50	5.30	51.80	75.50
9270	1-1/2" pipe size		19	.842			49	5.60	54.60	79
9280	2" pipe size		16	1			58	6.65	64.65	94.50
9290	2-1/2" pipe size		13	1.231			71.50	8.20	79.70	116
9300	3" pipe size		12	1.333			77.50	8.85	86.35	126
9310	4" pipe size		10	1.600			93	10.65	103.65	151
9320	5" pipe size		9	1.778			103	11.80	114.80	167
9330	6" pipe size		8	2			116	13.30	129.30	189
9340	8" pipe size		5	3.200			186	21.50	207.50	300
9350	10" pipe size		4	4			232	26.50	258.50	375
9360	12" pipe size		3	5.333			310	35.50	345.50	505
9370	14" pipe size		2.60	6.154			355	41	396	580
9380	16" pipe size		2.20	7.273			420	48.50	468.50	685
9390	18" pipe size		2	8			465	53	518	755
9400	20" pipe size		1.80	8.889			515	59	574	835
9410	22" pipe size		1.70	9.412			545	62.50	607.50	885
9420	24" pipe size		1.50	10.667			620	71	691	1,000
9450	Schedule 80									
9460	1/2" pipe size	Q-15	27	.593	Ea.		34.50	3.94	38.44	56
9470	3/4" pipe size		23	.696			40.50	4.63	45.13	65.50
9480	1" pipe size		20	.800			46.50	5.30	51.80	75.50
9490	1-1/4" pipe size		19	.842			49	5.60	54.60	79
9500	1-1/2" pipe size		18	.889			51.50	5.90	57.40	83.50
9510	2" pipe size		15	1.067			62	7.10	69.10	100

22 11 13.44 Pipe, Steel

		Crew	Daily Output	Labor-Hours	Unit	Material	2020 Bare Costs Labor	Equipment	Total	Total Incl O&P
9520	2-1/2" pipe size	Q-15	12	1.333	Ea.		77.50	8.85	86.35	126
9530	3" pipe size		11	1.455			84.50	9.65	94.15	137
9540	4" pipe size		8	2			116	13.30	129.30	189
9550	5" pipe size		6	2.667			155	17.75	172.75	251
9560	6" pipe size		5	3.200			186	21.50	207.50	300
9570	8" pipe size		4	4			232	26.50	258.50	375
9580	10" pipe size		3	5.333			310	35.50	345.50	505
9590	12" pipe size		2	8			465	53	518	755
9600	14" pipe size	Q-16	2.60	9.231			555	41	596	875
9610	16" pipe size		2.30	10.435			630	46	676	990
9620	18" pipe size		2	12			720	53	773	1,125
9630	20" pipe size		1.80	13.333			800	59	859	1,275
9640	22" pipe size		1.60	15			900	66.50	966.50	1,425
9650	24" pipe size		1.50	16			960	71	1,031	1,525

22 11 13.45 Pipe Fittings, Steel, Threaded

		Crew	Daily Output	Labor-Hours	Unit	Material	2020 Bare Costs Labor	Equipment	Total	Total Incl O&P
0010	**PIPE FITTINGS, STEEL, THREADED**									
0020	Cast iron									
0040	Standard weight, black									
0060	90° elbow, straight									
0070	1/4"	1 Plum	16	.500	Ea.	11.95	32		43.95	61
0080	3/8"		16	.500		17.30	32		49.30	67
0090	1/2"		15	.533		7.60	34.50		42.10	60
0100	3/4"		14	.571		7.90	37		44.90	63.50
0110	1"		13	.615		9.35	39.50		48.85	70
0120	1-1/4"	Q-1	22	.727		13.25	42		55.25	77.50
0130	1-1/2"		20	.800		18.35	46.50		64.85	89.50
0140	2"		18	.889		28.50	51.50		80	109
0150	2-1/2"		14	1.143		68.50	66.50		135	175
0160	3"		10	1.600		113	93		206	263
0170	3-1/2"		8	2		305	116		421	510
0180	4"		6	2.667		209	155		364	460
0250	45° elbow, straight									
0260	1/4"	1 Plum	16	.500	Ea.	14.75	32		46.75	64.50
0270	3/8"		16	.500		15.85	32		47.85	65.50
0280	1/2"		15	.533		11.60	34.50		46.10	64.50
0300	3/4"		14	.571		11.65	37		48.65	68
0320	1"		13	.615		13.70	39.50		53.20	74.50
0330	1-1/4"	Q-1	22	.727		18.40	42		60.40	83
0340	1-1/2"		20	.800		30.50	46.50		77	103
0350	2"		18	.889		35	51.50		86.50	116
0360	2-1/2"		14	1.143		91.50	66.50		158	200
0370	3"		10	1.600		145	93		238	298
0380	3-1/2"		8	2		340	116		456	550
0400	4"		6	2.667		300	155		455	560
0500	Tee, straight									
0510	1/4"	1 Plum	10	.800	Ea.	18.70	51.50		70.20	97.50
0520	3/8"		10	.800		18.20	51.50		69.70	97
0530	1/2"		9	.889		11.80	57.50		69.30	98.50
0540	3/4"		9	.889		13.75	57.50		71.25	101
0550	1"		8	1		12.25	64.50		76.75	110
0560	1-1/4"	Q-1	14	1.143		22.50	66.50		89	124
0570	1-1/2"		13	1.231		29	71.50		100.50	139

22 11 13.45 Pipe Fittings, Steel, Threaded

		Crew	Daily Output	Labor-Hours	Unit	Material	2020 Bare Costs Labor	Equipment	Total	Total Incl O&P
0580	2"	Q-1	11	1.455	Ea.	40.50	84.50		125	171
0590	2-1/2"		9	1.778		105	103		208	270
0600	3"		6	2.667		161	155		316	410
0610	3-1/2"		5	3.200		325	186		511	640
0620	4"		4	4		315	232		547	690
0660	Tee, reducing, run or outlet									
0661	1/2"	1 Plum	9	.889	Ea.	29	57.50		86.50	118
0662	3/4"		9	.889		21	57.50		78.50	109
0663	1"		8	1		28.50	64.50		93	128
0664	1-1/4"	Q-1	14	1.143		40	66.50		106.50	143
0665	1-1/2"		13	1.231		41	71.50		112.50	152
0666	2"		11	1.455		51.50	84.50		136	183
0667	2-1/2"		9	1.778		119	103		222	284
0668	3"		6	2.667		208	155		363	460
0669	3-1/2"		5	3.200		515	186		701	845
0670	4"		4	4		455	232		687	845
0674	Reducer, concentric									
0675	3/4"	1 Plum	18	.444	Ea.	21	28.50		49.50	66.50
0676	1"	"	15	.533		14.35	34.50		48.85	67.50
0677	1-1/4"	Q-1	26	.615		37	35.50		72.50	94
0678	1-1/2"		24	.667		57.50	38.50		96	121
0679	2"		21	.762		71.50	44		115.50	145
0680	2-1/2"		18	.889		101	51.50		152.50	189
0681	3"		14	1.143		162	66.50		228.50	277
0682	3-1/2"		12	1.333		320	77.50		397.50	465
0683	4"		10	1.600		320	93		413	490
0687	Reducer, eccentric									
0688	3/4"	1 Plum	16	.500	Ea.	44.50	32		76.50	97
0689	1"	"	14	.571		48	37		85	108
0690	1-1/4"	Q-1	25	.640		73.50	37		110.50	137
0691	1-1/2"		22	.727		97	42		139	169
0692	2"		20	.800		135	46.50		181.50	218
0693	2-1/2"		16	1		198	58		256	305
0694	3"		12	1.333		300	77.50		377.50	445
0695	3-1/2"		10	1.600		455	93		548	640
0696	4"		9	1.778		585	103		688	800
0700	Standard weight, galvanized cast iron									
0720	90° elbow, straight									
0730	1/4"	1 Plum	16	.500	Ea.	21.50	32		53.50	71.50
0740	3/8"		16	.500		21.50	32		53.50	71.50
0750	1/2"		15	.533		22	34.50		56.50	75.50
0760	3/4"		14	.571		21	37		58	78
0770	1"		13	.615		28	39.50		67.50	90
0780	1-1/4"	Q-1	22	.727		37.50	42		79.50	104
0790	1-1/2"		20	.800		59.50	46.50		106	135
0800	2"		18	.889		76	51.50		127.50	161
0810	2-1/2"		14	1.143		156	66.50		222.50	270
0820	3"		10	1.600		237	93		330	400
0830	3-1/2"		8	2		380	116		496	590
0840	4"		6	2.667		435	155		590	710
0900	45° elbow, straight									
0910	1/4"	1 Plum	16	.500	Ea.	23	32		55	73.50
0920	3/8"		16	.500		20	32		52	70

For customer support on your Mechanical Costs with RSMeans Data, call 800.448.8182.

151

22 11 13.45 Pipe Fittings, Steel, Threaded	Crew	Daily Output	Labor-Hours	Unit	Material	2020 Bare Costs Labor	Equipment	Total	Total Incl O&P
0930	1/2"	1 Plum	15	.533	Ea.	21	34.50		55.50	75
0940	3/4"		14	.571		24.50	37		61.50	81.50
0950	1"	↓	13	.615		34	39.50		73.50	97
0960	1-1/4"	Q-1	22	.727		52	42		94	120
0970	1-1/2"		20	.800		69	46.50		115.50	146
0980	2"		18	.889		96.50	51.50		148	183
0990	2-1/2"		14	1.143		136	66.50		202.50	249
1000	3"		10	1.600		305	93		398	475
1010	3-1/2"		8	2		525	116		641	755
1020	4"	↓	6	2.667	↓	475	155		630	755
1100	Tee, straight									
1110	1/4"	1 Plum	10	.800	Ea.	20	51.50		71.50	99
1120	3/8"		10	.800		22.50	51.50		74	102
1130	1/2"		9	.889		26.50	57.50		84	115
1140	3/4"		9	.889		34.50	57.50		92	124
1150	1"	↓	8	1		32.50	64.50		97	133
1160	1-1/4"	Q-1	14	1.143		57	66.50		123.50	162
1170	1-1/2"		13	1.231		75.50	71.50		147	190
1180	2"		11	1.455		94	84.50		178.50	229
1190	2-1/2"		9	1.778		191	103		294	365
1200	3"		6	2.667		455	155		610	730
1210	3-1/2"		5	3.200		535	186		721	870
1220	4"	↓	4	4	↓	590	232		822	995
1300	Extra heavy weight, black									
1310	Couplings, steel straight									
1320	1/4"	1 Plum	19	.421	Ea.	5.55	27		32.55	46.50
1330	3/8"		19	.421		6.05	27		33.05	47
1340	1/2"		19	.421		8.20	27		35.20	49.50
1350	3/4"		18	.444		8.75	28.50		37.25	52.50
1360	1"	↓	15	.533		11.15	34.50		45.65	64
1370	1-1/4"	Q-1	26	.615		17.85	35.50		53.35	73
1380	1-1/2"		24	.667		17.85	38.50		56.35	77.50
1390	2"		21	.762		27	44		71	96
1400	2-1/2"		18	.889		40.50	51.50		92	122
1410	3"		14	1.143		48	66.50		114.50	152
1420	3-1/2"		12	1.333		64.50	77.50		142	187
1430	4"	↓	10	1.600	↓	76	93		169	223
1510	90° elbow, straight									
1520	1/2"	1 Plum	15	.533	Ea.	39	34.50		73.50	94.50
1530	3/4"		14	.571		39.50	37		76.50	98.50
1540	1"	↓	13	.615		48	39.50		87.50	113
1550	1-1/4"	Q-1	22	.727		71.50	42		113.50	142
1560	1-1/2"		20	.800		88.50	46.50		135	167
1580	2"		18	.889		109	51.50		160.50	197
1590	2-1/2"		14	1.143		266	66.50		332.50	390
1600	3"		10	1.600		360	93		453	535
1610	4"	↓	6	2.667	↓	785	155		940	1,100
1650	45° elbow, straight									
1660	1/2"	1 Plum	15	.533	Ea.	55.50	34.50		90	113
1670	3/4"		14	.571		53.50	37		90.50	114
1680	1"	↓	13	.615		64.50	39.50		104	131
1690	1-1/4"	Q-1	22	.727		106	42		148	180
1700	1-1/2"	↓	20	.800	↓	117	46.50		163.50	199

22 11 13.45 Pipe Fittings, Steel, Threaded		Crew	Daily Output	Labor-Hours	Unit	Material	2020 Bare Costs Labor	Equipment	Total	Total Incl O&P
1710	2"	Q-1	18	.889	Ea.	166	51.50		217.50	260
1720	2-1/2"	▼	14	1.143	▼	287	66.50		353.50	415
1800	Tee, straight									
1810	1/2"	1 Plum	9	.889	Ea.	61	57.50		118.50	153
1820	3/4"		9	.889		61	57.50		118.50	153
1830	1"	▼	8	1		73.50	64.50		138	178
1840	1-1/4"	Q-1	14	1.143		110	66.50		176.50	220
1850	1-1/2"		13	1.231		141	71.50		212.50	262
1860	2"		11	1.455		175	84.50		259.50	320
1870	2-1/2"		9	1.778		370	103		473	560
1880	3"		6	2.667		525	155		680	805
1890	4"	▼	4	4	▼	1,025	232		1,257	1,475
4000	Standard weight, black									
4010	Couplings, steel straight, merchants									
4030	1/4"	1 Plum	19	.421	Ea.	1.48	27		28.48	42
4040	3/8"		19	.421		1.79	27		28.79	42.50
4050	1/2"		19	.421		1.91	27		28.91	42.50
4060	3/4"		18	.444		2.42	28.50		30.92	45.50
4070	1"	▼	15	.533		3.39	34.50		37.89	55
4080	1-1/4"	Q-1	26	.615		4.32	35.50		39.82	58.50
4090	1-1/2"		24	.667		5.50	38.50		44	64
4100	2"		21	.762		7.85	44		51.85	74.50
4110	2-1/2"		18	.889		25.50	51.50		77	105
4120	3"		14	1.143		36	66.50		102.50	139
4130	3-1/2"		12	1.333		64	77.50		141.50	186
4140	4"	▼	10	1.600		64	93		157	209
4166	Plug, 1/4"	1 Plum	38	.211		3.37	13.55		16.92	24
4167	3/8"		38	.211		3.03	13.55		16.58	24
4168	1/2"		38	.211		3.37	13.55		16.92	24
4169	3/4"		32	.250		8.90	16.10		25	34
4170	1"	▼	30	.267		9.50	17.20		26.70	36
4171	1-1/4"	Q-1	52	.308		10.90	17.85		28.75	38.50
4172	1-1/2"		48	.333		15.50	19.35		34.85	46
4173	2"		42	.381		20	22		42	55
4176	2-1/2"		36	.444		29.50	26		55.50	71
4180	4"	▼	20	.800	▼	62.50	46.50		109	138
4200	Standard weight, galvanized									
4210	Couplings, steel straight, merchants									
4230	1/4"	1 Plum	19	.421	Ea.	1.72	27		28.72	42.50
4240	3/8"		19	.421		2.20	27		29.20	43
4250	1/2"		19	.421		2.33	27		29.33	43
4260	3/4"		18	.444		2.91	28.50		31.41	46
4270	1"	▼	15	.533		4.07	34.50		38.57	56
4280	1-1/4"	Q-1	26	.615		5.20	35.50		40.70	59.50
4290	1-1/2"		24	.667		6.45	38.50		44.95	65
4300	2"		21	.762		9.65	44		53.65	76.50
4310	2-1/2"		18	.889		27	51.50		78.50	107
4320	3"		14	1.143		36	66.50		102.50	139
4330	3-1/2"		12	1.333		73.50	77.50		151	197
4340	4"	▼	10	1.600	▼	73.50	93		166.50	220
4370	Plug, galvanized, square head									
4374	1/2"	1 Plum	38	.211	Ea.	8.75	13.55		22.30	30
4375	3/4"	▼	32	.250	▼	8.15	16.10		24.25	33

22 11 13.45 Pipe Fittings, Steel, Threaded	Crew	Daily Output	Labor-Hours	Unit	Material	2020 Bare Costs Labor	Equipment	Total	Total Incl O&P
4376	1"	1 Plum	30	.267	Ea.	8.15	17.20		25.35	34.50
4377	1-1/4"	Q-1	52	.308		13.60	17.85		31.45	41.50
4378	1-1/2"		48	.333		18.35	19.35		37.70	49
4379	2"		42	.381		23	22		45	58
4380	2-1/2"		36	.444		48	26		74	91.50
4381	3"		28	.571		61.50	33		94.50	118
4382	4"		20	.800		169	46.50		215.50	256
4385	8"		20	.800		127	46.50		173.50	210
4700	Nipple, black									
4710	1/2" x 4" long	1 Plum	19	.421	Ea.	3.40	27		30.40	44
4712	3/4" x 4" long		18	.444		4.11	28.50		32.61	47.50
4714	1" x 4" long		15	.533		5.70	34.50		40.20	58
4716	1-1/4" x 4" long	Q-1	26	.615		7.15	35.50		42.65	61.50
4718	1-1/2" x 4" long		24	.667		8.40	38.50		46.90	67.50
4720	2" x 4" long		21	.762		11.75	44		55.75	79
4722	2-1/2" x 4" long		18	.889		32	51.50		83.50	112
4724	3" x 4" long		14	1.143		40.50	66.50		107	144
4726	4" x 4" long		10	1.600		48.50	93		141.50	193
4800	Nipple, galvanized									
4810	1/2" x 4" long	1 Plum	19	.421	Ea.	4.17	27		31.17	45
4812	3/4" x 4" long		18	.444		5.15	28.50		33.65	48.50
4814	1" x 4" long		15	.533		6.90	34.50		41.40	59
4816	1-1/4" x 4" long	Q-1	26	.615		8.50	35.50		44	63
4818	1-1/2" x 4" long		24	.667		10.75	38.50		49.25	70
4820	2" x 4" long		21	.762		13.65	44		57.65	81
4822	2-1/2" x 4" long		18	.889		36	51.50		87.50	117
4824	3" x 4" long		14	1.143		47.50	66.50		114	152
4826	4" x 4" long		10	1.600		64	93		157	210
5000	Malleable iron, 150 lb.									
5020	Black									
5040	90° elbow, straight									
5060	1/4"	1 Plum	16	.500	Ea.	5.90	32		37.90	54.50
5070	3/8"		16	.500		5.90	32		37.90	54.50
5080	1/2"		15	.533		4	34.50		38.50	56
5090	3/4"		14	.571		4.12	37		41.12	59.50
5100	1"		13	.615		8.35	39.50		47.85	68.50
5110	1-1/4"	Q-1	22	.727		13.85	42		55.85	78.50
5120	1-1/2"		20	.800		14.60	46.50		61.10	85.50
5130	2"		18	.889		25.50	51.50		77	105
5140	2-1/2"		14	1.143		71	66.50		137.50	178
5150	3"		10	1.600		104	93		197	254
5160	3-1/2"		8	2		287	116		403	490
5170	4"		6	2.667		196	155		351	445
5250	45° elbow, straight									
5270	1/4"	1 Plum	16	.500	Ea.	8.85	32		40.85	58
5280	3/8"		16	.500		8.85	32		40.85	58
5290	1/2"		15	.533		6.75	34.50		41.25	59
5300	3/4"		14	.571		8.35	37		45.35	64
5310	1"		13	.615		10.50	39.50		50	71
5320	1-1/4"	Q-1	22	.727		18.55	42		60.55	83.50
5330	1-1/2"		20	.800		23	46.50		69.50	95
5340	2"		18	.889		34.50	51.50		86	115
5350	2-1/2"		14	1.143		100	66.50		166.50	209

22 11 13.45 Pipe Fittings, Steel, Threaded		Crew	Daily Output	Labor-Hours	Unit	Material	2020 Bare Costs Labor	2020 Bare Costs Equipment	Total	Total Incl O&P
5360	3"	Q-1	10	1.600	Ea.	131	93		224	283
5370	3-1/2"		8	2		261	116		377	460
5380	4"	↓	6	2.667	↓	256	155		411	510
5450	Tee, straight									
5470	1/4"	1 Plum	10	.800	Ea.	8.55	51.50		60.05	86.50
5480	3/8"		10	.800		8.55	51.50		60.05	86.50
5490	1/2"		9	.889		5.45	57.50		62.95	91.50
5500	3/4"		9	.889		7.85	57.50		65.35	94
5510	1"	↓	8	1		13.40	64.50		77.90	111
5520	1-1/4"	Q-1	14	1.143		21.50	66.50		88	123
5530	1-1/2"		13	1.231		27	71.50		98.50	137
5540	2"		11	1.455		46	84.50		130.50	177
5550	2-1/2"		9	1.778		99	103		202	263
5560	3"		6	2.667		146	155		301	390
5570	3-1/2"		5	3.200		340	186		526	655
5580	4"	↓	4	4	↓	355	232		587	735
5601	Tee, reducing, on outlet									
5602	1/2"	1 Plum	9	.889	Ea.	12.75	57.50		70.25	99.50
5603	3/4"		9	.889		13.15	57.50		70.65	100
5604	1"	↓	8	1		17.30	64.50		81.80	116
5605	1-1/4"	Q-1	14	1.143		37	66.50		103.50	140
5606	1-1/2"		13	1.231		38	71.50		109.50	149
5607	2"		11	1.455		54	84.50		138.50	186
5608	2-1/2"		9	1.778		136	103		239	305
5609	3"		6	2.667		205	155		360	455
5610	3-1/2"		5	3.200		490	186		676	815
5611	4"	↓	4	4	↓	415	232		647	800
5650	Coupling									
5670	1/4"	1 Plum	19	.421	Ea.	7.30	27		34.30	48.50
5680	3/8"		19	.421		7.30	27		34.30	48.50
5690	1/2"		19	.421		5.65	27		32.65	46.50
5700	3/4"		18	.444		6.60	28.50		35.10	50.50
5710	1"	↓	15	.533		9.90	34.50		44.40	62.50
5720	1-1/4"	Q-1	26	.615		11.95	35.50		47.45	66.50
5730	1-1/2"		24	.667		15.80	38.50		54.30	75.50
5740	2"		21	.762		23.50	44		67.50	91.50
5750	2-1/2"		18	.889		64.50	51.50		116	148
5760	3"		14	1.143		87.50	66.50		154	195
5770	3-1/2"		12	1.333		161	77.50		238.50	293
5780	4"	↓	10	1.600		176	93		269	330
5840	Reducer, concentric, 1/4"	1 Plum	19	.421		7.70	27		34.70	49
5850	3/8"		19	.421		8.40	27		35.40	50
5860	1/2"		19	.421		7	27		34	48
5870	3/4"		16	.500		8	32		40	57
5880	1"	↓	15	.533		12.65	34.50		47.15	65.50
5890	1-1/4"	Q-1	26	.615		15.65	35.50		51.15	71
5900	1-1/2"		24	.667		21	38.50		59.50	81
5910	2"		21	.762		30	44		74	99
5911	2-1/2"		18	.889		68.50	51.50		120	152
5912	3"		14	1.143		95.50	66.50		162	204
5913	3-1/2"		12	1.333		310	77.50		387.50	455
5914	4"	↓	10	1.600		230	93		323	390
5981	Bushing, 1/4"	1 Plum	19	.421	↓	1.09	27		28.09	41.50

22 11 13.45 Pipe Fittings, Steel, Threaded	Crew	Daily Output	Labor-Hours	Unit	Material	2020 Bare Costs Labor	Equipment	Total	Total Incl O&P
5982	3/8"	1 Plum	19	.421	Ea.	5.70	27		32.70	47
5983	1/2"		19	.421		6	27		33	47
5984	3/4"		16	.500		6.95	32		38.95	55.50
5985	1"		15	.533		10.70	34.50		45.20	63.50
5986	1-1/4"	Q-1	26	.615		13.25	35.50		48.75	68
5987	1-1/2"		24	.667		11.45	38.50		49.95	70.50
5988	2"		21	.762		14.35	44		58.35	82
5989	Cap, 1/4"	1 Plum	38	.211		6.85	13.55		20.40	28
5991	3/8"		38	.211		6.45	13.55		20	27.50
5992	1/2"		38	.211		4.13	13.55		17.68	25
5993	3/4"		32	.250		5.60	16.10		21.70	30
5994	1"		30	.267		6.75	17.20		23.95	33
5995	1-1/4"	Q-1	52	.308		8.90	17.85		26.75	36.50
5996	1-1/2"		48	.333		12.25	19.35		31.60	42.50
5997	2"		42	.381		17.85	22		39.85	52.50
6000	For galvanized elbows, tees, and couplings, add					20%				
6058	For galvanized reducers, caps and bushings, add					20%				
6100	90° elbow, galvanized, 150 lb., reducing									
6110	3/4" x 1/2"	1 Plum	15.40	.519	Ea.	11.70	33.50		45.20	63
6112	1" x 3/4"		14	.571		15.05	37		52.05	71.50
6114	1" x 1/2"		14.50	.552		15.95	35.50		51.45	70.50
6116	1-1/4" x 1"	Q-1	24.20	.661		25.50	38.50		64	85.50
6118	1-1/4" x 3/4"		25.40	.630		30.50	36.50		67	88
6120	1-1/4" x 1/2"		26.20	.611		32	35.50		67.50	88.50
6122	1-1/2" x 1-1/4"		21.60	.741		40.50	43		83.50	109
6124	1-1/2" x 1"		23.50	.681		40.50	39.50		80	104
6126	1-1/2" x 3/4"		24.60	.650		40.50	37.50		78	101
6128	2" x 1-1/2"		20.50	.780		47	45.50		92.50	120
6130	2" x 1-1/4"		21	.762		54	44		98	126
6132	2" x 1"		22.80	.702		56	40.50		96.50	123
6134	2" x 3/4"		23.90	.669		57	39		96	121
6136	2-1/2" x 2"		12.30	1.301		160	75.50		235.50	289
6138	2-1/2" x 1-1/2"		12.50	1.280		178	74		252	305
6140	3" x 2-1/2"		8.60	1.860		290	108		398	480
6142	3" x 2"		11.80	1.356		252	78.50		330.50	395
6144	4" x 3"		8.20	1.951		655	113		768	895
6160	90° elbow, black, 150 lb., reducing									
6170	1" x 3/4"	1 Plum	14	.571	Ea.	10.25	37		47.25	66.50
6174	1-1/2" x 1"	Q-1	23.50	.681		24.50	39.50		64	86
6178	1-1/2" x 3/4"		24.60	.650		28	37.50		65.50	87
6182	2" x 1-1/2"		20.50	.780		35.50	45.50		81	107
6186	2" x 1"		22.80	.702		40	40.50		80.50	105
6190	2" x 3/4"		23.90	.669		42.50	39		81.50	105
6194	2-1/2" x 2"		12.30	1.301		97	75.50		172.50	219
7000	Union, with brass seat									
7010	1/4"	1 Plum	15	.533	Ea.	26.50	34.50		61	80.50
7020	3/8"		15	.533		21	34.50		55.50	75
7030	1/2"		14	.571		19.15	37		56.15	76
7040	3/4"		13	.615		18.95	39.50		58.45	80.50
7050	1"		12	.667		24.50	43		67.50	91.50
7060	1-1/4"	Q-1	21	.762		41	44		85	112
7070	1-1/2"		19	.842		51.50	49		100.50	130
7080	2"		17	.941		59.50	54.50		114	147

22 11 13.45 Pipe Fittings, Steel, Threaded	Crew	Daily Output	Labor-Hours	Unit	Material	2020 Bare Costs Labor	Equipment	Total	Total Incl O&P
7090 2-1/2"	Q-1	13	1.231	Ea.	176	71.50		247.50	300
7100 3"	↓	9	1.778	↓	184	103		287	355
7120 Union, galvanized									
7124 1/2"	1 Plum	14	.571	Ea.	25.50	37		62.50	83
7125 3/4"		13	.615		29	39.50		68.50	91.50
7126 1"	↓	12	.667		38.50	43		81.50	107
7127 1-1/4"	Q-1	21	.762		55.50	44		99.50	127
7128 1-1/2"		19	.842		67	49		116	147
7129 2"		17	.941		77	54.50		131.50	167
7130 2-1/2"		13	1.231		268	71.50		339.50	400
7131 3"	↓	9	1.778	↓	375	103		478	565
7500 Malleable iron, 300 lb.									
7520 Black									
7540 90° elbow, straight, 1/4"	1 Plum	16	.500	Ea.	20.50	32		52.50	70.50
7560 3/8"		16	.500		18.35	32		50.35	68
7570 1/2"		15	.533		24	34.50		58.50	77.50
7580 3/4"		14	.571		27	37		64	84.50
7590 1"	↓	13	.615		34.50	39.50		74	97.50
7600 1-1/4"	Q-1	22	.727		50	42		92	118
7610 1-1/2"		20	.800		59	46.50		105.50	135
7620 2"		18	.889		85	51.50		136.50	171
7630 2-1/2"		14	1.143		220	66.50		286.50	340
7640 3"		10	1.600		253	93		346	415
7650 4"	↓	6	2.667		595	155		750	885
7700 45° elbow, straight, 1/4"	1 Plum	16	.500		30.50	32		62.50	82
7720 3/8"		16	.500		30.50	32		62.50	81.50
7730 1/2"		15	.533		34	34.50		68.50	89
7740 3/4"		14	.571		38	37		75	97
7750 1"	↓	13	.615		42	39.50		81.50	106
7760 1-1/4"	Q-1	22	.727		67.50	42		109.50	137
7770 1-1/2"		20	.800		88	46.50		134.50	166
7780 2"		18	.889		133	51.50		184.50	223
7790 2-1/2"		14	1.143		297	66.50		363.50	425
7800 3"		10	1.600		390	93		483	570
7810 4"	↓	6	2.667		890	155		1,045	1,200
7850 Tee, straight, 1/4"	1 Plum	10	.800		26	51.50		77.50	106
7870 3/8"		10	.800		27.50	51.50		79	108
7880 1/2"		9	.889		35	57.50		92.50	124
7890 3/4"		9	.889		38	57.50		95.50	127
7900 1"	↓	8	1		45.50	64.50		110	147
7910 1-1/4"	Q-1	14	1.143		65	66.50		131.50	171
7920 1-1/2"		13	1.231		76	71.50		147.50	191
7930 2"		11	1.455		112	84.50		196.50	250
7940 2-1/2"		9	1.778		273	103		376	455
7950 3"		6	2.667		390	155		545	660
7960 4"	↓	4	4		1,100	232		1,332	1,575
8050 Couplings, straight, 1/4"	1 Plum	19	.421		18.60	27		45.60	61
8070 3/8"		19	.421		18.60	27		45.60	61
8080 1/2"		19	.421		21	27		48	63.50
8090 3/4"		18	.444		24	28.50		52.50	69.50
8100 1"	↓	15	.533		27.50	34.50		62	81.50
8110 1-1/4"	Q-1	26	.615		32.50	35.50		68	89.50
8120 1-1/2"	↓	24	.667	↓	48.50	38.50		87	112

For customer support on your Mechanical Costs with RSMeans Data, call 800.448.8182.

157

22 11 13 – Facility Water Distribution Piping

22 11 13.45 Pipe Fittings, Steel, Threaded	Crew	Daily Output	Labor-Hours	Unit	Material	2020 Bare Costs Labor	Equipment	Total	Total Incl O&P
8130	2"	Q-1	21	.762	Ea.	70	44		114	143
8140	2-1/2"		18	.889		129	51.50		180.50	219
8150	3"		14	1.143		186	66.50		252.50	305
8160	4"		10	1.600		355	93		448	530
8162	6"	▼	10	1.600	▼	355	93		448	530
8200	Galvanized									
8220	90° elbow, straight, 1/4"	1 Plum	16	.500	Ea.	37	32		69	88.50
8222	3/8"		16	.500		37.50	32		69.50	89.50
8224	1/2"		15	.533		44.50	34.50		79	101
8226	3/4"		14	.571		50	37		87	110
8228	1"	▼	13	.615		64	39.50		103.50	130
8230	1-1/4"	Q-1	22	.727		101	42		143	174
8232	1-1/2"		20	.800		109	46.50		155.50	190
8234	2"		18	.889		184	51.50		235.50	279
8236	2-1/2"		14	1.143		410	66.50		476.50	555
8238	3"		10	1.600		485	93		578	675
8240	4"	▼	6	2.667	▼	1,450	155		1,605	1,825
8280	45° elbow, straight									
8282	1/2"	1 Plum	15	.533	Ea.	66.50	34.50		101	125
8284	3/4"		14	.571		77	37		114	140
8286	1"	▼	13	.615		84.50	39.50		124	152
8288	1-1/4"	Q-1	22	.727		131	42		173	207
8290	1-1/2"		20	.800		169	46.50		215.50	256
8292	2"	▼	18	.889		234	51.50		285.50	335
8310	Tee, straight, 1/4"	1 Plum	10	.800		52.50	51.50		104	135
8312	3/8"		10	.800		53.50	51.50		105	136
8314	1/2"		9	.889		67	57.50		124.50	160
8316	3/4"		9	.889		74	57.50		131.50	167
8318	1"	▼	8	1		91	64.50		155.50	197
8320	1-1/4"	Q-1	14	1.143		130	66.50		196.50	242
8322	1-1/2"		13	1.231		135	71.50		206.50	256
8324	2"		11	1.455		216	84.50		300.50	365
8326	2-1/2"		9	1.778		685	103		788	905
8328	3"		6	2.667		775	155		930	1,075
8330	4"	▼	4	4		2,000	232		2,232	2,550
8380	Couplings, straight, 1/4"	1 Plum	19	.421		24.50	27		51.50	67
8382	3/8"		19	.421		35	27		62	79
8384	1/2"		19	.421		36	27		63	80
8386	3/4"		18	.444		40.50	28.50		69	87.50
8388	1"	▼	15	.533		53.50	34.50		88	111
8390	1-1/4"	Q-1	26	.615		69.50	35.50		105	130
8392	1-1/2"		24	.667		98.50	38.50		137	166
8394	2"		21	.762		119	44		163	197
8396	2-1/2"		18	.889		315	51.50		366.50	420
8398	3"		14	1.143		375	66.50		441.50	510
8399	4"	▼	10	1.600	▼	570	93		663	765
8529	Black									
8530	Reducer, concentric, 1/4"	1 Plum	19	.421	Ea.	25	27		52	68
8531	3/8"		19	.421		24.50	27		51.50	67.50
8532	1/2"		17	.471		29.50	30.50		60	78
8533	3/4"		16	.500		34.50	32		66.50	85.50
8534	1"	▼	15	.533		54.50	34.50		89	112
8535	1-1/4"	Q-1	26	.615	▼	74.50	35.50		110	136

22 11 Facility Water Distribution

22 11 13 – Facility Water Distribution Piping

22 11 13.45 Pipe Fittings, Steel, Threaded		Crew	Daily Output	Labor-Hours	Unit	Material	2020 Bare Costs Labor	Equipment	Total	Total Incl O&P
8536	1-1/2"	Q-1	24	.667	Ea.	76	38.50		114.50	142
8537	2"	↓	21	.762		100	44		144	177
8550	Cap, 1/4"	1 Plum	38	.211		19.15	13.55		32.70	41.50
8551	3/8"		38	.211		19.15	13.55		32.70	41.50
8552	1/2"		34	.235		18.95	15.15		34.10	43.50
8553	3/4"		32	.250		24	16.10		40.10	50.50
8554	1"	↓	30	.267		31.50	17.20		48.70	60
8555	1-1/4"	Q-1	52	.308		35	17.85		52.85	65
8556	1-1/2"		48	.333		51	19.35		70.35	85
8557	2"	↓	42	.381		70.50	22		92.50	111
8570	Plug, 1/4"	1 Plum	38	.211		3.68	13.55		17.23	24.50
8571	3/8"		38	.211		3.50	13.55		17.05	24.50
8572	1/2"		34	.235		3.59	15.15		18.74	26.50
8573	3/4"		32	.250		4.48	16.10		20.58	29
8574	1"	↓	30	.267		6.85	17.20		24.05	33
8575	1-1/4"	Q-1	52	.308		14.05	17.85		31.90	42
8576	1-1/2"		48	.333		16.15	19.35		35.50	47
8577	2"	↓	42	.381		25.50	22		47.50	61
9500	Union with brass seat, 1/4"	1 Plum	15	.533		45	34.50		79.50	101
9530	3/8"		15	.533		45.50	34.50		80	102
9540	1/2"		14	.571		36.50	37		73.50	95.50
9550	3/4"		13	.615		40.50	39.50		80	104
9560	1"	↓	12	.667		53	43		96	123
9570	1-1/4"	Q-1	21	.762		86.50	44		130.50	161
9580	1-1/2"		19	.842		90.50	49		139.50	173
9590	2"		17	.941		112	54.50		166.50	205
9600	2-1/2"		13	1.231		360	71.50		431.50	505
9610	3"		9	1.778		470	103		573	670
9620	4"	↓	5	3.200		1,475	186		1,661	1,900
9630	Union, all iron, 1/4"	1 Plum	15	.533		69.50	34.50		104	128
9650	3/8"		15	.533		69.50	34.50		104	128
9660	1/2"		14	.571		73	37		110	136
9670	3/4"		13	.615		78.50	39.50		118	146
9680	1"	↓	12	.667		96	43		139	171
9690	1-1/4"	Q-1	21	.762		148	44		192	229
9700	1-1/2"		19	.842		180	49		229	271
9710	2"		17	.941		228	54.50		282.50	335
9720	2-1/2"		13	1.231		405	71.50		476.50	550
9730	3"	↓	9	1.778		770	103		873	1,000
9750	For galvanized unions, add				↓	15%				
9757	Forged steel, 3000 lb.									
9758	Black									
9760	90° elbow, 1/4"	1 Plum	16	.500	Ea.	19.25	32		51.25	69
9761	3/8"		16	.500		19.25	32		51.25	69
9762	1/2"		15	.533		15	34.50		49.50	68
9763	3/4"		14	.571		18.45	37		55.45	75.50
9764	1"	↓	13	.615		29.50	39.50		69	92
9765	1-1/4"	Q-1	22	.727		52.50	42		94.50	121
9766	1-1/2"		20	.800		67.50	46.50		114	144
9767	2"	↓	18	.889		82	51.50		133.50	167
9780	45° elbow, 1/4"	1 Plum	16	.500		24.50	32		56.50	75
9781	3/8"		16	.500		24.50	32		56.50	75
9782	1/2"		15	.533		24	34.50		58.50	78

For customer support on your Mechanical Costs with RSMeans Data, call 800.448.8182.

159

22 11 13.45 Pipe Fittings, Steel, Threaded	Crew	Daily Output	Labor-Hours	Unit	Material	2020 Bare Costs Labor	Equipment	Total	Total Incl O&P
9783	3/4"	1 Plum	14	.571	Ea.	28	37		65	85.50
9784	1"		13	.615		38	39.50		77.50	102
9785	1-1/4"	Q-1	22	.727		52.50	42		94.50	121
9786	1-1/2"		20	.800		75.50	46.50		122	153
9787	2"		18	.889		104	51.50		155.50	191
9800	Tee, 1/4"	1 Plum	10	.800		23.50	51.50		75	103
9801	3/8"		10	.800		23.50	51.50		75	103
9802	1/2"		9	.889		21	57.50		78.50	109
9803	3/4"		9	.889		28.50	57.50		86	117
9804	1"		8	1		37	64.50		101.50	138
9805	1-1/4"	Q-1	14	1.143		73	66.50		139.50	179
9806	1-1/2"		13	1.231		85	71.50		156.50	201
9807	2"		11	1.455		99.50	84.50		184	235
9820	Reducer, concentric, 1/4"	1 Plum	19	.421		12.85	27		39.85	54.50
9821	3/8"		19	.421		13.40	27		40.40	55.50
9822	1/2"		17	.471		13.40	30.50		43.90	60.50
9823	3/4"		16	.500		15.90	32		47.90	65.50
9824	1"		15	.533		20.50	34.50		55	74.50
9825	1-1/4"	Q-1	26	.615		36	35.50		71.50	93
9826	1-1/2"		24	.667		39	38.50		77.50	101
9827	2"		21	.762		57	44		101	129
9840	Cap, 1/4"	1 Plum	38	.211		8.10	13.55		21.65	29.50
9841	3/8"		38	.211		7.75	13.55		21.30	29
9842	1/2"		34	.235		7.55	15.15		22.70	31
9843	3/4"		32	.250		10.50	16.10		26.60	35.50
9844	1"		30	.267		16.10	17.20		33.30	43
9845	1-1/4"	Q-1	52	.308		24.50	17.85		42.35	53
9846	1-1/2"		48	.333		29	19.35		48.35	60.50
9847	2"		42	.381		42	22		64	79
9860	Plug, 1/4"	1 Plum	38	.211		4.09	13.55		17.64	25
9861	3/8"		38	.211		3.72	13.55		17.27	24.50
9862	1/2"		34	.235		3.58	15.15		18.73	26.50
9863	3/4"		32	.250		5.55	16.10		21.65	30
9864	1"		30	.267		8.50	17.20		25.70	35
9865	1-1/4"	Q-1	52	.308		16.40	17.85		34.25	44.50
9866	1-1/2"		48	.333		18.65	19.35		38	49.50
9867	2"		42	.381		29.50	22		51.50	65.50
9880	Union, bronze seat, 1/4"	1 Plum	15	.533		65	34.50		99.50	123
9881	3/8"		15	.533		65	34.50		99.50	123
9882	1/2"		14	.571		62.50	37		99.50	124
9883	3/4"		13	.615		84.50	39.50		124	153
9884	1"		12	.667		94.50	43		137.50	169
9885	1-1/4"	Q-1	21	.762		173	44		217	256
9886	1-1/2"		19	.842		189	49		238	281
9887	2"		17	.941		224	54.50		278.50	330
9900	Coupling, 1/4"	1 Plum	19	.421		7.60	27		34.60	49
9901	3/8"		19	.421		7.60	27		34.60	49
9902	1/2"		17	.471		6.20	30.50		36.70	52.50
9903	3/4"		16	.500		8	32		40	57
9904	1"		15	.533		13.90	34.50		48.40	67
9905	1-1/4"	Q-1	26	.615		23.50	35.50		59	79
9906	1-1/2"		24	.667		30	38.50		68.50	91
9907	2"		21	.762		40	44		84	111

22 11 13.47 Pipe Fittings, Steel	Crew	Daily Output	Labor-Hours	Unit	Material	2020 Bare Costs Labor	Equipment	Total	Total Incl O&P
0010 **PIPE FITTINGS, STEEL**, flanged, welded & special									
0020 Flanged joints, CI, standard weight, black. One gasket & bolt									
0040 set, mat'l only, required at each joint, not included (see line 0620)									
0060 90° elbow, straight, 1-1/2" pipe size	Q-1	14	1.143	Ea.	690	66.50		756.50	860
0080 2" pipe size		13	1.231		400	71.50		471.50	545
0090 2-1/2" pipe size		12	1.333		430	77.50		507.50	590
0100 3" pipe size		11	1.455		360	84.50		444.50	520
0110 4" pipe size		8	2		445	116		561	665
0120 5" pipe size	▼	7	2.286		1,050	133		1,183	1,350
0130 6" pipe size	Q-2	9	2.667		695	160		855	1,000
0140 8" pipe size		8	3		1,200	180		1,380	1,600
0150 10" pipe size		7	3.429		2,625	206		2,831	3,200
0160 12" pipe size	▼	6	4	▼	5,300	241		5,541	6,175
0171 90° elbow, reducing									
0172 2-1/2" by 2" pipe size	Q-1	12	1.333	Ea.	1,350	77.50		1,427.50	1,600
0173 3" by 2-1/2" pipe size		11	1.455		1,325	84.50		1,409.50	1,575
0174 4" by 3" pipe size		8	2		985	116		1,101	1,250
0175 5" by 3" pipe size	▼	7	2.286		2,175	133		2,308	2,575
0176 6" by 4" pipe size	Q-2	9	2.667		1,225	160		1,385	1,600
0177 8" by 6" pipe size		8	3		1,775	180		1,955	2,250
0178 10" by 8" pipe size		7	3.429		3,400	206		3,606	4,050
0179 12" by 10" pipe size	▼	6	4		6,525	241		6,766	7,525
0200 45° elbow, straight, 1-1/2" pipe size	Q-1	14	1.143		840	66.50		906.50	1,025
0220 2" pipe size		13	1.231		575	71.50		646.50	740
0230 2-1/2" pipe size		12	1.333		615	77.50		692.50	790
0240 3" pipe size		11	1.455		600	84.50		684.50	785
0250 4" pipe size		8	2		675	116		791	915
0260 5" pipe size	▼	7	2.286		1,625	133		1,758	1,975
0270 6" pipe size	Q-2	9	2.667		1,100	160		1,260	1,450
0280 8" pipe size		8	3		1,600	180		1,780	2,050
0290 10" pipe size		7	3.429		3,400	206		3,606	4,050
0300 12" pipe size	▼	6	4	▼	5,200	241		5,441	6,075
0310 Cross, straight									
0311 2-1/2" pipe size	Q-1	6	2.667	Ea.	1,275	155		1,430	1,625
0312 3" pipe size		5	3.200		1,325	186		1,511	1,725
0313 4" pipe size		4	4		1,725	232		1,957	2,250
0314 5" pipe size	▼	3	5.333		3,950	310		4,260	4,800
0315 6" pipe size	Q-2	5	4.800		3,625	289		3,914	4,425
0316 8" pipe size		4	6		5,875	360		6,235	7,000
0317 10" pipe size		3	8		6,975	480		7,455	8,400
0318 12" pipe size	▼	2	12		11,500	720		12,220	13,700
0350 Tee, straight, 1-1/2" pipe size	Q-1	10	1.600		795	93		888	1,025
0370 2" pipe size		9	1.778		435	103		538	635
0380 2-1/2" pipe size		8	2		635	116		751	875
0390 3" pipe size		7	2.286		445	133		578	690
0400 4" pipe size		5	3.200		675	186		861	1,025
0410 5" pipe size	▼	4	4		1,825	232		2,057	2,350
0420 6" pipe size	Q-2	6	4		985	241		1,226	1,425
0430 8" pipe size		5	4.800		1,675	289		1,964	2,275
0440 10" pipe size		4	6		4,525	360		4,885	5,525
0450 12" pipe size	▼	3	8	▼	7,100	480		7,580	8,525
0459 Tee, reducing on outlet									

22 11 Facility Water Distribution

22 11 13 – Facility Water Distribution Piping

22 11 13.47 Pipe Fittings, Steel		Crew	Daily Output	Labor-Hours	Unit	Material	2020 Bare Costs Labor	Equipment	Total	Total Incl O&P
0460	2-1/2" by 2" pipe size	Q-1	8	2	Ea.	1,375	116		1,491	1,700
0461	3" by 2-1/2" pipe size		7	2.286		1,325	133		1,458	1,650
0462	4" by 3" pipe size		5	3.200		1,400	186		1,586	1,800
0463	5" by 4" pipe size		4	4		3,100	232		3,332	3,775
0464	6" by 4" pipe size	Q-2	6	4		1,350	241		1,591	1,825
0465	8" by 6" pipe size		5	4.800		1,125	289		1,414	1,675
0466	10" by 8" pipe size		4	6		4,775	360		5,135	5,800
0467	12" by 10" pipe size		3	8		8,325	480		8,805	9,875
0476	Reducer, concentric									
0477	3" by 2-1/2"	Q-1	12	1.333	Ea.	800	77.50		877.50	995
0478	4" by 3"		9	1.778		895	103		998	1,150
0479	5" by 4"		8	2		1,375	116		1,491	1,700
0480	6" by 4"	Q-2	10	2.400		1,200	144		1,344	1,550
0481	8" by 6"		9	2.667		1,475	160		1,635	1,875
0482	10" by 8"		8	3		3,100	180		3,280	3,700
0483	12" by 10"		7	3.429		5,375	206		5,581	6,225
0492	Reducer, eccentric									
0493	4" by 3"	Q-1	8	2	Ea.	1,450	116		1,566	1,775
0494	5" by 4"	"	7	2.286		2,275	133		2,408	2,700
0495	6" by 4"	Q-2	9	2.667		1,375	160		1,535	1,775
0496	8" by 6"		8	3		1,750	180		1,930	2,200
0497	10" by 8"		7	3.429		4,425	206		4,631	5,175
0498	12" by 10"		6	4		6,050	241		6,291	7,000
0500	For galvanized elbows and tees, add					100%				
0520	For extra heavy weight elbows and tees, add					140%				
0620	Gasket and bolt set, 150 lb., 1/2" pipe size	1 Plum	20	.400		3.83	26		29.83	42.50
0622	3/4" pipe size		19	.421		4.13	27		31.13	45
0624	1" pipe size		18	.444		4.62	28.50		33.12	48
0626	1-1/4" pipe size		17	.471		4.36	30.50		34.86	50.50
0628	1-1/2" pipe size		15	.533		4.38	34.50		38.88	56.50
0630	2" pipe size		13	.615		8.90	39.50		48.40	69.50
0640	2-1/2" pipe size		12	.667		8.45	43		51.45	74
0650	3" pipe size		11	.727		9.10	47		56.10	80
0660	3-1/2" pipe size		9	.889		20	57.50		77.50	108
0670	4" pipe size		8	1		17.90	64.50		82.40	116
0680	5" pipe size		7	1.143		27	73.50		100.50	140
0690	6" pipe size		6	1.333		28.50	86		114.50	160
0700	8" pipe size		5	1.600		31	103		134	188
0710	10" pipe size		4.50	1.778		57	115		172	234
0720	12" pipe size		4.20	1.905		59	123		182	249
0730	14" pipe size		4	2		54.50	129		183.50	253
0740	16" pipe size		3	2.667		65.50	172		237.50	330
0750	18" pipe size		2.70	2.963		127	191		318	425
0760	20" pipe size		2.30	3.478		206	224		430	560
0780	24" pipe size		1.90	4.211		257	271		528	690
0790	26" pipe size		1.60	5		350	320		670	865
0810	30" pipe size		1.40	5.714		680	370		1,050	1,300
0830	36" pipe size		1.10	7.273		1,275	470		1,745	2,100
0850	For 300 lb. gasket set, add					40%				
2000	Flanged unions, 125 lb., black, 1/2" pipe size	1 Plum	17	.471	Ea.	103	30.50		133.50	159
2040	3/4" pipe size		17	.471		143	30.50		173.50	203
2050	1" pipe size		16	.500		138	32		170	200
2060	1-1/4" pipe size	Q-1	28	.571		164	33		197	230

162

22 11 13.47 Pipe Fittings, Steel		Crew	Daily Output	Labor-Hours	Unit	Material	2020 Bare Costs Labor	Equipment	Total	Total Incl O&P
2070	1-1/2" pipe size	Q-1	27	.593	Ea.	151	34.50		185.50	218
2080	2" pipe size		26	.615		178	35.50		213.50	250
2090	2-1/2" pipe size		24	.667		241	38.50		279.50	325
2100	3" pipe size		22	.727		273	42		315	365
2110	3-1/2" pipe size		18	.889		460	51.50		511.50	585
2120	4" pipe size		16	1		370	58		428	495
2130	5" pipe size		14	1.143		865	66.50		931.50	1,050
2140	6" pipe size	Q-2	19	1.263		810	76		886	1,000
2150	8" pipe size	"	16	1.500		1,875	90		1,965	2,175
2200	For galvanized unions, add					150%				
2290	Threaded flange									
2300	Cast iron									
2310	Black, 125 lb., per flange									
2320	1" pipe size	1 Plum	27	.296	Ea.	53	19.10		72.10	87
2330	1-1/4" pipe size	Q-1	44	.364		69.50	21		90.50	108
2340	1-1/2" pipe size		40	.400		59	23		82	99
2350	2" pipe size		36	.444		64.50	26		90.50	109
2360	2-1/2" pipe size		28	.571		75	33		108	132
2370	3" pipe size		20	.800		97	46.50		143.50	177
2380	3-1/2" pipe size		16	1		138	58		196	239
2390	4" pipe size		12	1.333		119	77.50		196.50	247
2400	5" pipe size		10	1.600		184	93		277	340
2410	6" pipe size	Q-2	14	1.714		209	103		312	385
2420	8" pipe size		12	2		330	120		450	540
2430	10" pipe size		10	2.400		585	144		729	855
2440	12" pipe size		8	3		1,325	180		1,505	1,725
2460	For galvanized flanges, add					95%				
2490	Blind flange									
2492	Cast iron									
2494	Black, 125 lb., per flange									
2496	1" pipe size	1 Plum	27	.296	Ea.	92.50	19.10		111.60	131
2500	1-1/2" pipe size	Q-1	40	.400		100	23		123	145
2502	2" pipe size		36	.444		117	26		143	168
2504	2-1/2" pipe size		28	.571		125	33		158	187
2506	3" pipe size		20	.800		155	46.50		201.50	240
2508	4" pipe size		12	1.333		198	77.50		275.50	335
2510	5" pipe size		10	1.600		320	93		413	495
2512	6" pipe size	Q-2	14	1.714		350	103		453	540
2514	8" pipe size		12	2		550	120		670	785
2516	10" pipe size		10	2.400		815	144		959	1,100
2518	12" pipe size		8	3		1,575	180		1,755	2,000
2520	For galvanized flanges, add					80%				
2570	Threaded flange									
2580	Forged steel									
2590	Black, 150 lb., per flange									
2600	1/2" pipe size	1 Plum	30	.267	Ea.	27.50	17.20		44.70	56
2610	3/4" pipe size		28	.286		27.50	18.40		45.90	58
2620	1" pipe size		27	.296		27.50	19.10		46.60	59
2630	1-1/4" pipe size	Q-1	44	.364		27.50	21		48.50	62
2640	1-1/2" pipe size		40	.400		27.50	23		50.50	65
2650	2" pipe size		36	.444		31	26		57	72.50
2660	2-1/2" pipe size		28	.571		42.50	33		75.50	96.50
2670	3" pipe size		20	.800		43	46.50		89.50	117

For customer support on your Mechanical Costs with RSMeans Data, call 800.448.8182.

163

22 11 13.47 Pipe Fittings, Steel		Crew	Daily Output	Labor-Hours	Unit	Material	2020 Bare Costs Labor	Equipment	Total	Total Incl O&P
2690	4" pipe size	Q-1	12	1.333	Ea.	48.50	77.50		126	169
2700	5" pipe size	↓	10	1.600		79.50	93		172.50	227
2710	6" pipe size	Q-2	14	1.714		85	103		188	248
2720	8" pipe size		12	2		149	120		269	345
2730	10" pipe size	↓	10	2.400	↓	266	144		410	510
2860	Black, 300 lb., per flange									
2870	1/2" pipe size	1 Plum	30	.267	Ea.	30.50	17.20		47.70	59
2880	3/4" pipe size		28	.286		30.50	18.40		48.90	61
2890	1" pipe size	↓	27	.296		30.50	19.10		49.60	62
2900	1-1/4" pipe size	Q-1	44	.364		30.50	21		51.50	65
2910	1-1/2" pipe size		40	.400		30.50	23		53.50	68
2920	2" pipe size		36	.444		40	26		66	82.50
2930	2-1/2" pipe size		28	.571		49.50	33		82.50	104
2940	3" pipe size		20	.800		50.50	46.50		97	125
2960	4" pipe size	↓	12	1.333		72.50	77.50		150	196
2970	6" pipe size	Q-2	14	1.714	↓	157	103		260	325
3000	Weld joint, butt, carbon steel, standard weight									
3040	90° elbow, long radius									
3050	1/2" pipe size	Q-15	16	1	Ea.	44	58	6.65	108.65	143
3060	3/4" pipe size		16	1		44	58	6.65	108.65	143
3070	1" pipe size		16	1		26.50	58	6.65	91.15	123
3080	1-1/4" pipe size		14	1.143		26.50	66.50	7.60	100.60	136
3090	1-1/2" pipe size		13	1.231		26.50	71.50	8.20	106.20	145
3100	2" pipe size		10	1.600		22.50	93	10.65	126.15	175
3110	2-1/2" pipe size		8	2		35	116	13.30	164.30	227
3120	3" pipe size		7	2.286		30	133	15.20	178.20	248
3130	4" pipe size		5	3.200		56.50	186	21.50	264	365
3136	5" pipe size	↓	4	4		112	232	26.50	370.50	500
3140	6" pipe size	Q-16	5	4.800		126	289	21.50	436.50	590
3150	8" pipe size		3.75	6.400		237	385	28.50	650.50	865
3160	10" pipe size		3	8		470	480	35.50	985.50	1,275
3170	12" pipe size		2.50	9.600		665	575	42.50	1,282.50	1,650
3180	14" pipe size		2	12		1,025	720	53	1,798	2,250
3190	16" pipe size		1.50	16		1,400	960	71	2,431	3,075
3191	18" pipe size		1.25	19.200		1,750	1,150	85	2,985	3,750
3192	20" pipe size		1.15	20.870		2,425	1,250	92.50	3,767.50	4,625
3194	24" pipe size	↓	1.02	23.529	↓	3,625	1,425	104	5,154	6,250
3200	45° elbow, long									
3210	1/2" pipe size	Q-15	16	1	Ea.	65.50	58	6.65	130.15	166
3220	3/4" pipe size		16	1		65.50	58	6.65	130.15	166
3230	1" pipe size		16	1		23	58	6.65	87.65	120
3240	1-1/4" pipe size		14	1.143		23	66.50	7.60	97.10	133
3250	1-1/2" pipe size		13	1.231		23	71.50	8.20	102.70	142
3260	2" pipe size		10	1.600		23	93	10.65	126.65	176
3270	2-1/2" pipe size		8	2		28	116	13.30	157.30	220
3280	3" pipe size		7	2.286		29	133	15.20	177.20	247
3290	4" pipe size		5	3.200		52	186	21.50	259.50	360
3296	5" pipe size	↓	4	4		79	232	26.50	337.50	460
3300	6" pipe size	Q-16	5	4.800		103	289	21.50	413.50	565
3310	8" pipe size		3.75	6.400		171	385	28.50	584.50	795
3320	10" pipe size		3	8		340	480	35.50	855.50	1,125
3330	12" pipe size		2.50	9.600		480	575	42.50	1,097.50	1,450
3340	14" pipe size		2	12		650	720	53	1,423	1,850

22 11 13.47 Pipe Fittings, Steel		Crew	Daily Output	Labor-Hours	Unit	Material	2020 Bare Costs Labor	Equipment	Total	Total Incl O&P
3341	16" pipe size	Q-16	1.50	16	Ea.	1,150	960	71	2,181	2,800
3342	18" pipe size		1.25	19.200		1,650	1,150	85	2,885	3,625
3343	20" pipe size		1.15	20.870		1,700	1,250	92.50	3,042.50	3,850
3345	24" pipe size		1.05	22.857		2,525	1,375	101	4,001	4,925
3346	26" pipe size		.85	28.235		2,825	1,700	125	4,650	5,825
3347	30" pipe size		.45	53.333		3,125	3,200	236	6,561	8,475
3349	36" pipe size		.38	63.158		3,425	3,800	280	7,505	9,750
3350	Tee, straight									
3352	For reducing tees and concentrics see starting line 4600									
3360	1/2" pipe size	Q-15	10	1.600	Ea.	116	93	10.65	219.65	279
3370	3/4" pipe size		10	1.600		116	93	10.65	219.65	279
3380	1" pipe size		10	1.600		57.50	93	10.65	161.15	214
3390	1-1/4" pipe size		9	1.778		72	103	11.80	186.80	246
3400	1-1/2" pipe size		8	2		72	116	13.30	201.30	268
3410	2" pipe size		6	2.667		57.50	155	17.75	230.25	315
3420	2-1/2" pipe size		5	3.200		79.50	186	21.50	287	390
3430	3" pipe size		4	4		88.50	232	26.50	347	470
3440	4" pipe size		3	5.333		124	310	35.50	469.50	640
3446	5" pipe size		2.50	6.400		205	370	42.50	617.50	825
3450	6" pipe size	Q-16	3	8		213	480	35.50	728.50	995
3460	8" pipe size		2.50	9.600		370	575	42.50	987.50	1,325
3470	10" pipe size		2	12		730	720	53	1,503	1,950
3480	12" pipe size		1.60	15		1,025	900	66.50	1,991.50	2,550
3481	14" pipe size		1.30	18.462		1,800	1,100	82	2,982	3,725
3482	16" pipe size		1	24		2,000	1,450	106	3,556	4,475
3483	18" pipe size		.80	30		3,175	1,800	133	5,108	6,350
3484	20" pipe size		.75	32		5,000	1,925	142	7,067	8,525
3486	24" pipe size		.70	34.286		6,450	2,050	152	8,652	10,300
3487	26" pipe size		.55	43.636		7,075	2,625	193	9,893	11,900
3488	30" pipe size		.30	80		7,750	4,800	355	12,905	16,100
3490	36" pipe size		.25	96		8,550	5,775	425	14,750	18,500
3491	Eccentric reducer, 1-1/2" pipe size	Q-15	14	1.143		44.50	66.50	7.60	118.60	156
3492	2" pipe size		11	1.455		58	84.50	9.65	152.15	200
3493	2-1/2" pipe size		9	1.778		63.50	103	11.80	178.30	237
3494	3" pipe size		8	2		83	116	13.30	212.30	280
3495	4" pipe size		6	2.667		166	155	17.75	338.75	435
3496	6" pipe size	Q-16	5	4.800		223	289	21.50	533.50	700
3497	8" pipe size		4	6		258	360	26.50	644.50	855
3498	10" pipe size		3	8		440	480	35.50	955.50	1,250
3499	12" pipe size		2.50	9.600		680	575	42.50	1,297.50	1,650
3501	Cap, 1-1/2" pipe size	Q-15	28	.571		20.50	33	3.80	57.30	76
3502	2" pipe size		22	.727		22.50	42	4.84	69.34	93
3503	2-1/2" pipe size		18	.889		23.50	51.50	5.90	80.90	109
3504	3" pipe size		16	1		23.50	58	6.65	88.15	120
3505	4" pipe size		12	1.333		34	77.50	8.85	120.35	163
3506	6" pipe size	Q-16	10	2.400		58.50	144	10.65	213.15	292
3507	8" pipe size		8	3		89	180	13.30	282.30	380
3508	10" pipe size		6	4		160	241	17.70	418.70	555
3509	12" pipe size		5	4.800		247	289	21.50	557.50	725
3511	14" pipe size		4	6		480	360	26.50	866.50	1,100
3512	16" pipe size		4	6		535	360	26.50	921.50	1,150
3513	18" pipe size		3	8		515	480	35.50	1,030.50	1,325
3517	Weld joint, butt, carbon steel, extra strong									

22 11 13.47 Pipe Fittings, Steel	Crew	Daily Output	Labor-Hours	Unit	Material	2020 Bare Costs Labor	Equipment	Total	Total Incl O&P
3519	90° elbow, long								
3520	1/2" pipe size	Q-15	13	1.231	Ea.	58.50	71.50	8.20	138.20	181
3530	3/4" pipe size		12	1.333		58.50	77.50	8.85	144.85	190
3540	1" pipe size		11	1.455		28.50	84.50	9.65	122.65	168
3550	1-1/4" pipe size		10	1.600		28.50	93	10.65	132.15	182
3560	1-1/2" pipe size		9	1.778		28.50	103	11.80	143.30	199
3570	2" pipe size		8	2		29	116	13.30	158.30	221
3580	2-1/2" pipe size		7	2.286		40.50	133	15.20	188.70	260
3590	3" pipe size		6	2.667		52	155	17.75	224.75	310
3600	4" pipe size		4	4		86	232	26.50	344.50	470
3606	5" pipe size	▼	3.50	4.571		206	265	30.50	501.50	655
3610	6" pipe size	Q-16	4.50	5.333		217	320	23.50	560.50	745
3620	8" pipe size		3.50	6.857		410	410	30.50	850.50	1,100
3630	10" pipe size		2.50	9.600		870	575	42.50	1,487.50	1,875
3640	12" pipe size	▼	2.25	10.667	▼	1,075	640	47.50	1,762.50	2,175
3650	45° elbow, long								
3660	1/2" pipe size	Q-15	13	1.231	Ea.	65	71.50	8.20	144.70	188
3670	3/4" pipe size		12	1.333		65	77.50	8.85	151.35	197
3680	1" pipe size		11	1.455		30.50	84.50	9.65	124.65	170
3690	1-1/4" pipe size		10	1.600		30.50	93	10.65	134.15	184
3700	1-1/2" pipe size		9	1.778		30.50	103	11.80	145.30	201
3710	2" pipe size		8	2		30.50	116	13.30	159.80	222
3720	2-1/2" pipe size		7	2.286		67.50	133	15.20	215.70	289
3730	3" pipe size		6	2.667		39	155	17.75	211.75	294
3740	4" pipe size		4	4		61.50	232	26.50	320	440
3746	5" pipe size	▼	3.50	4.571		146	265	30.50	441.50	590
3750	6" pipe size	Q-16	4.50	5.333		169	320	23.50	512.50	690
3760	8" pipe size		3.50	6.857		291	410	30.50	731.50	970
3770	10" pipe size		2.50	9.600		560	575	42.50	1,177.50	1,525
3780	12" pipe size	▼	2.25	10.667	▼	830	640	47.50	1,517.50	1,925
3800	Tee, straight								
3810	1/2" pipe size	Q-15	9	1.778	Ea.	143	103	11.80	257.80	325
3820	3/4" pipe size		8.50	1.882		137	109	12.50	258.50	330
3830	1" pipe size		8	2		54.50	116	13.30	183.80	248
3840	1-1/4" pipe size		7	2.286		54	133	15.20	202.20	274
3850	1-1/2" pipe size		6	2.667		54.50	155	17.75	227.25	310
3860	2" pipe size		5	3.200		61	186	21.50	268.50	370
3870	2-1/2" pipe size		4	4		102	232	26.50	360.50	485
3880	3" pipe size		3.50	4.571		128	265	30.50	423.50	570
3890	4" pipe size		2.50	6.400		153	370	42.50	565.50	770
3896	5" pipe size	▼	2.25	7.111		390	410	47.50	847.50	1,100
3900	6" pipe size	Q-16	2.25	10.667		300	640	47.50	987.50	1,350
3910	8" pipe size		2	12		575	720	53	1,348	1,775
3920	10" pipe size		1.75	13.714		865	825	61	1,751	2,250
3930	12" pipe size	▼	1.50	16		1,275	960	71	2,306	2,925
4000	Eccentric reducer, 1-1/2" pipe size	Q-15	10	1.600		20.50	93	10.65	124.15	174
4010	2" pipe size		9	1.778		56.50	103	11.80	171.30	230
4020	2-1/2" pipe size		8	2		82.50	116	13.30	211.80	279
4030	3" pipe size		7	2.286		69	133	15.20	217.20	291
4040	4" pipe size		5	3.200		111	186	21.50	318.50	425
4046	5" pipe size	▼	4.70	3.404		300	197	22.50	519.50	650
4050	6" pipe size	Q-16	4.50	5.333		305	320	23.50	648.50	840
4060	8" pipe size	▼	3.50	6.857	▼	405	410	30.50	845.50	1,100

22 11 13 – Facility Water Distribution Piping

22 11 13.47 Pipe Fittings, Steel		Crew	Daily Output	Labor-Hours	Unit	Material	2020 Bare Costs Labor	Equipment	Total	Total Incl O&P
4070	10" pipe size	Q-16	2.50	9.600	Ea.	690	575	42.50	1,307.50	1,675
4080	12" pipe size		2.25	10.667		880	640	47.50	1,567.50	1,975
4090	14" pipe size		2.10	11.429		1,650	685	50.50	2,385.50	2,875
4100	16" pipe size		1.90	12.632		2,125	760	56	2,941	3,500
4151	Cap, 1-1/2" pipe size	Q-15	24	.667		25.50	38.50	4.43	68.43	91
4152	2" pipe size		18	.889		22.50	51.50	5.90	79.90	108
4153	2-1/2" pipe size		16	1		30.50	58	6.65	95.15	128
4154	3" pipe size		14	1.143		34	66.50	7.60	108.10	145
4155	4" pipe size		10	1.600		45.50	93	10.65	149.15	201
4156	6" pipe size	Q-16	9	2.667		95.50	160	11.80	267.30	360
4157	8" pipe size		7	3.429		143	206	15.20	364.20	485
4158	10" pipe size		5	4.800		217	289	21.50	527.50	695
4159	12" pipe size		4	6		294	360	26.50	680.50	895
4190	Weld fittings, reducing, standard weight									
4200	Welding ring w/spacer pins, 2" pipe size				Ea.	2.23			2.23	2.45
4210	2-1/2" pipe size					2.16			2.16	2.38
4220	3" pipe size					2.25			2.25	2.48
4230	4" pipe size					2.81			2.81	3.09
4236	5" pipe size					3.46			3.46	3.81
4240	6" pipe size					3.29			3.29	3.62
4250	8" pipe size					3.65			3.65	4.02
4260	10" pipe size					4.22			4.22	4.64
4270	12" pipe size					4.93			4.93	5.40
4280	14" pipe size					5.65			5.65	6.25
4290	16" pipe size					6.65			6.65	7.30
4300	18" pipe size					7.35			7.35	8.05
4310	20" pipe size					9.25			9.25	10.20
4330	24" pipe size					12.85			12.85	14.10
4340	26" pipe size					15.85			15.85	17.40
4350	30" pipe size					18.40			18.40	20.50
4370	36" pipe size					23			23	25.50
4600	Tee, reducing on outlet									
4601	2-1/2" x 2" pipe size	Q-15	5	3.200	Ea.	106	186	21.50	313.50	420
4602	3" x 2-1/2" pipe size		4	4		151	232	26.50	409.50	540
4604	4" x 3" pipe size		3	5.333		122	310	35.50	467.50	640
4605	5" x 4" pipe size		2.50	6.400		340	370	42.50	752.50	975
4606	6" x 5" pipe size	Q-16	3	8		475	480	35.50	990.50	1,275
4607	8" x 6" pipe size		2.50	9.600		585	575	42.50	1,202.50	1,550
4608	10" x 8" pipe size		2	12		840	720	53	1,613	2,050
4609	12" x 10" pipe size		1.60	15		1,325	900	66.50	2,291.50	2,875
4610	16" x 12" pipe size		1.50	16		2,475	960	71	3,506	4,250
4611	14" x 12" pipe size		1.52	15.789		2,400	950	70	3,420	4,125
4618	Reducer, concentric									
4619	2-1/2" by 2" pipe size	Q-15	10	1.600	Ea.	42	93	10.65	145.65	197
4620	3" by 2-1/2" pipe size		9	1.778		34.50	103	11.80	149.30	205
4621	3-1/2" by 3" pipe size		8	2		112	116	13.30	241.30	310
4622	4" by 2-1/2" pipe size		7	2.286		57.50	133	15.20	205.70	278
4623	5" by 3" pipe size		7	2.286		144	133	15.20	292.20	375
4624	6" by 4" pipe size	Q-16	6	4		120	241	17.70	378.70	510
4625	8" by 6" pipe size		5	4.800		122	289	21.50	432.50	590
4626	10" by 8" pipe size		4	6		254	360	26.50	640.50	850
4627	12" by 10" pipe size		3	8		292	480	35.50	807.50	1,075
4660	Reducer, eccentric									

For customer support on your Mechanical Costs with RSMeans Data, call 800.448.8182.

167

22 11 13.47 Pipe Fittings, Steel		Crew	Daily Output	Labor-Hours	Unit	Material	2020 Bare Costs Labor	Equipment	Total	Total Incl O&P
4662	3" x 2" pipe size	Q-15	8	2	Ea.	75	116	13.30	204.30	271
4664	4" x 3" pipe size		6	2.667		91	155	17.75	263.75	350
4666	4" x 2" pipe size	↓	6	2.667		111	155	17.75	283.75	375
4670	6" x 4" pipe size	Q-16	5	4.800		188	289	21.50	498.50	660
4672	6" x 3" pipe size		5	4.800		295	289	21.50	605.50	780
4676	8" x 6" pipe size		4	6		240	360	26.50	626.50	835
4678	8" x 4" pipe size		4	6		440	360	26.50	826.50	1,050
4682	10" x 8" pipe size		3	8		315	480	35.50	830.50	1,100
4684	10" x 6" pipe size		3	8		430	480	35.50	945.50	1,225
4688	12" x 10" pipe size		2.50	9.600		490	575	42.50	1,107.50	1,450
4690	12" x 8" pipe size		2.50	9.600		770	575	42.50	1,387.50	1,750
4691	14" x 12" pipe size		2.20	10.909		860	655	48.50	1,563.50	1,975
4693	16" x 14" pipe size		1.80	13.333		1,025	800	59	1,884	2,400
4694	16" x 12" pipe size		2	12		1,750	720	53	2,523	3,050
4696	18" x 16" pipe size	↓	1.60	15	↓	1,075	900	66.50	2,041.50	2,625
5000	Weld joint, socket, forged steel, 3000 lb., schedule 40 pipe									
5010	90° elbow, straight									
5020	1/4" pipe size	Q-15	22	.727	Ea.	29.50	42	4.84	76.34	101
5030	3/8" pipe size		22	.727		29.50	42	4.84	76.34	101
5040	1/2" pipe size		20	.800		16.45	46.50	5.30	68.25	93.50
5050	3/4" pipe size		20	.800		13.65	46.50	5.30	65.45	90.50
5060	1" pipe size		20	.800		22	46.50	5.30	73.80	99.50
5070	1-1/4" pipe size		18	.889		42.50	51.50	5.90	99.90	130
5080	1-1/2" pipe size		16	1		41.50	58	6.65	106.15	140
5090	2" pipe size		12	1.333		62	77.50	8.85	148.35	194
5100	2-1/2" pipe size		10	1.600		206	93	10.65	309.65	380
5110	3" pipe size		8	2		355	116	13.30	484.30	580
5120	4" pipe size	↓	6	2.667	↓	925	155	17.75	1,097.75	1,275
5130	45° elbow, straight									
5134	1/4" pipe size	Q-15	22	.727	Ea.	29.50	42	4.84	76.34	101
5135	3/8" pipe size		22	.727		29.50	42	4.84	76.34	101
5136	1/2" pipe size		20	.800		22	46.50	5.30	73.80	99.50
5137	3/4" pipe size		20	.800		25.50	46.50	5.30	77.30	103
5140	1" pipe size		20	.800		33.50	46.50	5.30	85.30	112
5150	1-1/4" pipe size		18	.889		46	51.50	5.90	103.40	134
5160	1-1/2" pipe size		16	1		56	58	6.65	120.65	156
5170	2" pipe size		12	1.333		90.50	77.50	8.85	176.85	225
5180	2-1/2" pipe size		10	1.600		238	93	10.65	341.65	410
5190	3" pipe size		8	2		395	116	13.30	524.30	625
5200	4" pipe size	↓	6	2.667	↓	775	155	17.75	947.75	1,100
5250	Tee, straight									
5254	1/4" pipe size	Q-15	15	1.067	Ea.	32.50	62	7.10	101.60	136
5255	3/8" pipe size		15	1.067		32.50	62	7.10	101.60	136
5256	1/2" pipe size		13	1.231		20.50	71.50	8.20	100.20	139
5257	3/4" pipe size		13	1.231		25	71.50	8.20	104.70	144
5260	1" pipe size		13	1.231		34	71.50	8.20	113.70	154
5270	1-1/4" pipe size		12	1.333		52.50	77.50	8.85	138.85	184
5280	1-1/2" pipe size		11	1.455		68.50	84.50	9.65	162.65	212
5290	2" pipe size		8	2		100	116	13.30	229.30	299
5300	2-1/2" pipe size		6	2.667		246	155	17.75	418.75	520
5310	3" pipe size		5	3.200		590	186	21.50	797.50	945
5320	4" pipe size	↓	4	4	↓	1,100	232	26.50	1,358.50	1,600
5350	For reducing sizes, add					60%				

22 11 Facility Water Distribution

22 11 13 – Facility Water Distribution Piping

22 11 13.47 Pipe Fittings, Steel	Crew	Daily Output	Labor-Hours	Unit	Material	2020 Bare Costs Labor	Equipment	Total	Total Incl O&P
5450	Couplings									
5451	1/4" pipe size	Q-15	23	.696	Ea.	19.25	40.50	4.63	64.38	86.50
5452	3/8" pipe size		23	.696		19.25	40.50	4.63	64.38	86.50
5453	1/2" pipe size		21	.762		8.75	44	5.05	57.80	81
5454	3/4" pipe size		21	.762		11.30	44	5.05	60.35	84
5460	1" pipe size		20	.800		12.45	46.50	5.30	64.25	89
5470	1-1/4" pipe size		20	.800		22	46.50	5.30	73.80	100
5480	1-1/2" pipe size		18	.889		24.50	51.50	5.90	81.90	111
5490	2" pipe size		14	1.143		39	66.50	7.60	113.10	150
5500	2-1/2" pipe size		12	1.333		87.50	77.50	8.85	173.85	222
5510	3" pipe size		9	1.778		196	103	11.80	310.80	385
5520	4" pipe size		7	2.286		295	133	15.20	443.20	540
5570	Union, 1/4" pipe size		21	.762		40.50	44	5.05	89.55	116
5571	3/8" pipe size		21	.762		40.50	44	5.05	89.55	116
5572	1/2" pipe size		19	.842		34	49	5.60	88.60	117
5573	3/4" pipe size		19	.842		39.50	49	5.60	94.10	123
5574	1" pipe size		19	.842		51	49	5.60	105.60	136
5575	1-1/4" pipe size		17	.941		82	54.50	6.25	142.75	179
5576	1-1/2" pipe size		15	1.067		89.50	62	7.10	158.60	199
5577	2" pipe size		11	1.455		109	84.50	9.65	203.15	257
5600	Reducer, 1/4" pipe size		23	.696		45.50	40.50	4.63	90.63	116
5601	3/8" pipe size		23	.696		48.50	40.50	4.63	93.63	119
5602	1/2" pipe size		21	.762		31	44	5.05	80.05	106
5603	3/4" pipe size		21	.762		31	44	5.05	80.05	106
5604	1" pipe size		21	.762		36.50	44	5.05	85.55	112
5605	1-1/4" pipe size		19	.842		55	49	5.60	109.60	140
5607	1-1/2" pipe size		17	.941		53	54.50	6.25	113.75	146
5608	2" pipe size		13	1.231		58.50	71.50	8.20	138.20	181
5612	Cap, 1/4" pipe size		46	.348		18.10	20	2.31	40.41	52.50
5613	3/8" pipe size		46	.348		18.10	20	2.31	40.41	52.50
5614	1/2" pipe size		42	.381		10.95	22	2.53	35.48	48
5615	3/4" pipe size		42	.381		12.75	22	2.53	37.28	50
5616	1" pipe size		42	.381		19.35	22	2.53	43.88	57.50
5617	1-1/4" pipe size		38	.421		23	24.50	2.80	50.30	64.50
5618	1-1/2" pipe size		34	.471		32	27.50	3.13	62.63	79.50
5619	2" pipe size		26	.615		49	35.50	4.09	88.59	112
5630	T-O-L, 1/4" pipe size, nozzle		23	.696		8.25	40.50	4.63	53.38	74.50
5631	3/8" pipe size, nozzle		23	.696		8.40	40.50	4.63	53.53	75
5632	1/2" pipe size, nozzle		22	.727		8.25	42	4.84	55.09	77.50
5633	3/4" pipe size, nozzle		21	.762		9.50	44	5.05	58.55	82
5634	1" pipe size, nozzle		20	.800		11.10	46.50	5.30	62.90	87.50
5635	1-1/4" pipe size, nozzle		18	.889		15.80	51.50	5.90	73.20	101
5636	1-1/2" pipe size, nozzle		16	1		17.55	58	6.65	82.20	114
5637	2" pipe size, nozzle		12	1.333		19.15	77.50	8.85	105.50	147
5638	2-1/2" pipe size, nozzle		10	1.600		67	93	10.65	170.65	225
5639	4" pipe size, nozzle		6	2.667		151	155	17.75	323.75	415
5640	W-O-L, 1/4" pipe size, nozzle		23	.696		17.60	40.50	4.63	62.73	85
5641	3/8" pipe size, nozzle		23	.696		16.65	40.50	4.63	61.78	84
5642	1/2" pipe size, nozzle		22	.727		17.10	42	4.84	63.94	87
5643	3/4" pipe size, nozzle		21	.762		18.05	44	5.05	67.10	91.50
5644	1" pipe size, nozzle		20	.800		18.85	46.50	5.30	70.65	96
5645	1-1/4" pipe size, nozzle		18	.889		21.50	51.50	5.90	78.90	108
5646	1-1/2" pipe size, nozzle		16	1		22.50	58	6.65	87.15	119

For customer support on your Mechanical Costs with RSMeans Data, call 800.448.8182.

169

22 11 13 – Facility Water Distribution Piping

22 11 13.47 Pipe Fittings, Steel		Crew	Daily Output	Labor-Hours	Unit	Material	2020 Bare Costs Labor	Equipment	Total	Total Incl O&P
5647	2" pipe size, nozzle	Q-15	12	1.333	Ea.	22.50	77.50	8.85	108.85	151
5648	2-1/2" pipe size, nozzle		10	1.600		51.50	93	10.65	155.15	208
5649	3" pipe size, nozzle		8	2		56.50	116	13.30	185.80	251
5650	4" pipe size, nozzle		6	2.667		71.50	155	17.75	244.25	330
5651	5" pipe size, nozzle		5	3.200		176	186	21.50	383.50	495
5652	6" pipe size, nozzle		4	4		199	232	26.50	457.50	595
5653	8" pipe size, nozzle		3	5.333		380	310	35.50	725.50	920
5654	10" pipe size, nozzle		2.60	6.154		525	355	41	921	1,150
5655	12" pipe size, nozzle		2.20	7.273		1,000	420	48.50	1,468.50	1,775
5674	S-O-L, 1/4" pipe size, outlet		23	.696		9.80	40.50	4.63	54.93	76.50
5675	3/8" pipe size, outlet		23	.696		9.80	40.50	4.63	54.93	76.50
5676	1/2" pipe size, outlet		22	.727		10.25	42	4.84	57.09	79.50
5677	3/4" pipe size, outlet		21	.762		10.40	44	5.05	59.45	83
5678	1" pipe size, outlet		20	.800		11.50	46.50	5.30	63.30	88
5679	1-1/4" pipe size, outlet		18	.889		19.25	51.50	5.90	76.65	105
5680	1-1/2" pipe size, outlet		16	1		19.25	58	6.65	83.90	115
5681	2" pipe size, outlet		12	1.333		19.75	77.50	8.85	106.10	147
6000	Weld-on flange, forged steel									
6020	Slip-on, 150 lb. flange (welded front and back)									
6050	1/2" pipe size	Q-15	18	.889	Ea.	19.85	51.50	5.90	77.25	106
6060	3/4" pipe size		18	.889		19.85	51.50	5.90	77.25	106
6070	1" pipe size		17	.941		19.85	54.50	6.25	80.60	110
6080	1-1/4" pipe size		16	1		19.85	58	6.65	84.50	116
6090	1-1/2" pipe size		15	1.067		19.85	62	7.10	88.95	122
6100	2" pipe size		12	1.333		26	77.50	8.85	112.35	155
6110	2-1/2" pipe size		10	1.600		38.50	93	10.65	142.15	193
6120	3" pipe size		9	1.778		32	103	11.80	146.80	203
6130	3-1/2" pipe size		7	2.286		34	133	15.20	182.20	252
6140	4" pipe size		6	2.667		33	155	17.75	205.75	287
6150	5" pipe size		5	3.200		67.50	186	21.50	275	375
6160	6" pipe size	Q-16	6	4		54	241	17.70	312.70	440
6170	8" pipe size		5	4.800		94.50	289	21.50	405	560
6180	10" pipe size		4	6		166	360	26.50	552.50	755
6190	12" pipe size		3	8		249	480	35.50	764.50	1,025
6191	14" pipe size		2.50	9.600		280	575	42.50	897.50	1,225
6192	16" pipe size		1.80	13.333		495	800	59	1,354	1,800
6200	300 lb. flange									
6210	1/2" pipe size	Q-15	17	.941	Ea.	24.50	54.50	6.25	85.25	115
6220	3/4" pipe size		17	.941		24.50	54.50	6.25	85.25	115
6230	1" pipe size		16	1		24.50	58	6.65	89.15	121
6240	1-1/4" pipe size		13	1.231		24.50	71.50	8.20	104.20	143
6250	1-1/2" pipe size		12	1.333		24.50	77.50	8.85	110.85	153
6260	2" pipe size		11	1.455		36.50	84.50	9.65	130.65	177
6270	2-1/2" pipe size		9	1.778		36.50	103	11.80	151.30	207
6280	3" pipe size		7	2.286		38	133	15.20	186.20	257
6290	4" pipe size		6	2.667		57.50	155	17.75	230.25	315
6300	5" pipe size		4	4		111	232	26.50	369.50	495
6310	6" pipe size	Q-16	5	4.800		108	289	21.50	418.50	575
6320	8" pipe size		4	6		184	360	26.50	570.50	770
6330	10" pipe size		3.40	7.059		325	425	31.50	781.50	1,025
6340	12" pipe size		2.80	8.571		340	515	38	893	1,175
6400	Welding neck, 150 lb. flange									
6410	1/2" pipe size	Q-15	40	.400	Ea.	25.50	23	2.66	51.16	65.50

22 11 13.47 Pipe Fittings, Steel		Crew	Daily Output	Labor-Hours	Unit	Material	2020 Bare Costs Labor	Equipment	Total	Total Incl O&P
6420	3/4" pipe size	Q-15	36	.444	Ea.	25.50	26	2.96	54.46	70
6430	1" pipe size		32	.500		25.50	29	3.33	57.83	75
6440	1-1/4" pipe size		29	.552		25.50	32	3.67	61.17	80
6450	1-1/2" pipe size		26	.615		25.50	35.50	4.09	65.09	86
6460	2" pipe size		20	.800		34	46.50	5.30	85.80	113
6470	2-1/2" pipe size		16	1		37	58	6.65	101.65	135
6480	3" pipe size		14	1.143		41	66.50	7.60	115.10	152
6500	4" pipe size		10	1.600		42	93	10.65	145.65	197
6510	5" pipe size		8	2		66.50	116	13.30	195.80	262
6520	6" pipe size	Q-16	10	2.400		63.50	144	10.65	218.15	298
6530	8" pipe size		7	3.429		118	206	15.20	339.20	455
6540	10" pipe size		6	4		188	241	17.70	446.70	585
6550	12" pipe size		5	4.800		296	289	21.50	606.50	780
6551	14" pipe size		4.50	5.333		440	320	23.50	783.50	990
6552	16" pipe size		3	8		595	480	35.50	1,110.50	1,425
6553	18" pipe size		2.50	9.600		800	575	42.50	1,417.50	1,800
6554	20" pipe size		2.30	10.435		970	630	46	1,646	2,075
6556	24" pipe size		2	12		1,300	720	53	2,073	2,575
6557	26" pipe size		1.70	14.118		1,425	850	62.50	2,337.50	2,925
6558	30" pipe size		.90	26.667		1,625	1,600	118	3,343	4,325
6559	36" pipe size		.75	32		1,875	1,925	142	3,942	5,100
6560	300 lb. flange									
6570	1/2" pipe size	Q-15	36	.444	Ea.	34.50	26	2.96	63.46	80
6580	3/4" pipe size		34	.471		34.50	27.50	3.13	65.13	82.50
6590	1" pipe size		30	.533		34.50	31	3.55	69.05	88.50
6600	1-1/4" pipe size		28	.571		34.50	33	3.80	71.30	91.50
6610	1-1/2" pipe size		24	.667		34.50	38.50	4.43	77.43	101
6620	2" pipe size		18	.889		45.50	51.50	5.90	102.90	134
6630	2-1/2" pipe size		14	1.143		52	66.50	7.60	126.10	164
6640	3" pipe size		12	1.333		44.50	77.50	8.85	130.85	175
6650	4" pipe size		8	2		76	116	13.30	205.30	272
6660	5" pipe size		7	2.286		111	133	15.20	259.20	335
6670	6" pipe size	Q-16	9	2.667		113	160	11.80	284.80	380
6680	8" pipe size		6	4		216	241	17.70	474.70	620
6690	10" pipe size		5	4.800		415	289	21.50	725.50	910
6700	12" pipe size		4	6		450	360	26.50	836.50	1,075
6710	14" pipe size		3.50	6.857		975	410	30.50	1,415.50	1,725
6720	16" pipe size		2	12		1,150	720	53	1,923	2,400
7740	Plain ends for plain end pipe, mechanically coupled									
7750	Cplg. & labor required at joints not included, add 1 per									
7760	joint for installed price, see line 9180									
7770	Malleable iron, painted, unless noted otherwise									
7800	90° elbow, 1"				Ea.	154			154	170
7810	1-1/2"					182			182	201
7820	2"					271			271	299
7830	2-1/2"					320			320	350
7840	3"					330			330	365
7860	4"					375			375	410
7870	5" welded steel					430			430	470
7880	6"					540			540	595
7890	8" welded steel					1,025			1,025	1,125
7900	10" welded steel					1,300			1,300	1,425
7910	12" welded steel					1,450			1,450	1,600

22 11 13.47 Pipe Fittings, Steel		Crew	Daily Output	Labor-Hours	Unit	Material	2020 Bare Costs Labor	Equipment	Total	Total Incl O&P
7970	45° elbow, 1"				Ea.	92.50			92.50	102
7980	1-1/2"					137			137	151
7990	2"					288			288	315
8000	2-1/2"					288			288	315
8010	3"					330			330	365
8030	4"					345			345	380
8040	5" welded steel					430			430	470
8050	6"					490			490	540
8060	8"					575			575	635
8070	10" welded steel					595			595	655
8080	12" welded steel					1,075			1,075	1,175
8140	Tee, straight 1"					172			172	189
8150	1-1/2"					222			222	244
8160	2"					222			222	244
8170	2-1/2"					289			289	320
8180	3"					445			445	490
8200	4"					635			635	700
8210	5" welded steel					870			870	955
8220	6"					755			755	830
8230	8" welded steel					1,100			1,100	1,200
8240	10" welded steel					1,725			1,725	1,900
8250	12" welded steel					2,000			2,000	2,200
8340	Segmentally welded steel, painted									
8390	Wye 2"				Ea.	251			251	276
8400	2-1/2"					251			251	276
8410	3"					300			300	330
8430	4"					390			390	430
8440	5"					570			570	625
8450	6"					695			695	760
8460	8"					1,025			1,025	1,125
8470	10"					1,675			1,675	1,850
8480	12"					1,725			1,725	1,900
8540	Wye, lateral 2"					299			299	330
8550	2-1/2"					345			345	380
8560	3"					410			410	450
8580	4"					560			560	620
8590	5"					955			955	1,050
8600	6"					985			985	1,075
8610	8"					1,650			1,650	1,825
8620	10"					2,425			2,425	2,675
8630	12"					3,025			3,025	3,325
8690	Cross, 2"					181			181	199
8700	2-1/2"					181			181	199
8710	3"					216			216	237
8730	4"					490			490	540
8740	5"					560			560	615
8750	6"					710			710	780
8760	8"					1,100			1,100	1,225
8770	10"					1,425			1,425	1,575
8780	12"					1,825			1,825	2,025
8800	Tees, reducing 2" x 1"					235			235	259
8810	2" x 1-1/2"					235			235	259
8820	3" x 1"					217			217	238

22 11 13 – Facility Water Distribution Piping

22 11 13.47 Pipe Fittings, Steel		Crew	Daily Output	Labor-Hours	Unit	Material	2020 Bare Costs Labor	Equipment	Total	Total Incl O&P
8830	3" x 1-1/2"				Ea.	298			298	330
8840	3" x 2"					216			216	237
8850	4" x 1"					315			315	350
8860	4" x 1-1/2"					315			315	350
8870	4" x 2"					470			470	515
8880	4" x 2-1/2"					485			485	530
8890	4" x 3"					485			485	530
8900	6" x 2"					400			400	440
8910	6" x 3"					298			298	330
8920	6" x 4"					380			380	420
8930	8" x 2"					575			575	635
8940	8" x 3"					625			625	690
8950	8" x 4"					670			670	735
8960	8" x 5"					470			470	515
8970	8" x 6"					640			640	705
8980	10" x 4"					530			530	585
8990	10" x 6"					545			545	600
9000	10" x 8"					850			850	940
9010	12" x 6"					1,325			1,325	1,450
9020	12" x 8"					1,050			1,050	1,150
9030	12" x 10"				▼	1,025			1,025	1,150
9080	Adapter nipples 3" long									
9090	1"				Ea.	27			27	29.50
9100	1-1/2"					27			27	29.50
9110	2"					27			27	29.50
9120	2-1/2"					31			31	34.50
9130	3"					38			38	41.50
9140	4"					61.50			61.50	67.50
9150	6"				▼	157			157	172
9180	Coupling, mechanical, plain end pipe to plain end pipe or fitting									
9190	1"	Q-1	29	.552	Ea.	93.50	32		125.50	151
9200	1-1/2"		28	.571		93.50	33		126.50	153
9210	2"		27	.593		93.50	34.50		128	155
9220	2-1/2"		26	.615		93.50	35.50		129	157
9230	3"		25	.640		136	37		173	206
9240	3-1/2"		24	.667		157	38.50		195.50	231
9250	4"	▼	22	.727		157	42		199	236
9260	5"	Q-2	28	.857		223	51.50		274.50	325
9270	6"		24	1		273	60		333	390
9280	8"		19	1.263		480	76		556	645
9290	10"		16	1.500		625	90		715	820
9300	12"	▼	12	2	▼	785	120		905	1,050
9310	Outlets for precut holes through pipe wall									
9331	Strapless type, with gasket									
9332	4" to 8" pipe x 1/2"	1 Plum	13	.615	Ea.	102	39.50		141.50	172
9333	4" to 8" pipe x 3/4"		13	.615		109	39.50		148.50	180
9334	10" pipe and larger x 1/2"		11	.727		102	47		149	182
9335	10" pipe and larger x 3/4"	▼	11	.727	▼	109	47		156	190
9341	Thermometer wells with gasket									
9342	4" to 8" pipe, 6" stem	1 Plum	14	.571	Ea.	153	37		190	223
9343	8" pipe and larger, 6" stem	"	13	.615	"	153	39.50		192.50	228
9800	Mech. tee, cast iron, grooved or threaded, w/nuts and bolts									
9940	For galvanized fittings for plain end pipe, add					20%				

22 11 13.48 Pipe, Fittings and Valves, Steel, Grooved-Joint		Crew	Daily Output	Labor-Hours	Unit	Material	2020 Bare Costs Labor	Equipment	Total	Total Incl O&P
0010	**PIPE, FITTINGS AND VALVES, STEEL, GROOVED-JOINT**									
0012	Fittings are ductile iron. Steel fittings noted.									
0020	Pipe includes coupling & clevis type hanger assemblies, 10' OC									
0500	Schedule 10, black									
0550	2" diameter	1 Plum	43	.186	L.F.	9.40	12		21.40	28.50
0560	2-1/2" diameter	Q-1	61	.262		13.25	15.20		28.45	37.50
0570	3" diameter		55	.291		14.65	16.85		31.50	41.50
0580	3-1/2" diameter		53	.302		18.40	17.50		35.90	46.50
0590	4" diameter		49	.327		15.15	18.95		34.10	45
0600	5" diameter		40	.400		18.75	23		41.75	55
0610	6" diameter	Q-2	46	.522		24.50	31.50		56	74
0620	8" diameter	"	41	.585		36.50	35		71.50	92.50
0700	To delete couplings & hangers, subtract									
0710	2" diam. to 5" diam.					25%	20%			
0720	6" diam. to 8" diam.					27%	15%			
1000	Schedule 40, black									
1040	3/4" diameter	1 Plum	71	.113	L.F.	6.55	7.25		13.80	18.05
1050	1" diameter		63	.127		6.45	8.20		14.65	19.35
1060	1-1/4" diameter		58	.138		7.70	8.90		16.60	22
1070	1-1/2" diameter		51	.157		8.40	10.10		18.50	24.50
1080	2" diameter 90/lr + 25% F.H = 112		40	.200		9.70	12.90		22.60	30
1090	2-1/2" diameter 32	Q-1	57	.281		16.50	16.30		32.80	42.50
1100	3" diameter 36		50	.320		22	18.55 -6		40.55	52
1110	4" diameter 40		45	.356		26	20.50 -6		46.50	59.50
1120	5" diameter 48		37	.432		41	25		66	82.50
1130	6" diameter 64	Q-2	42	.571		52.50	34.50		87	109
1140	8" diameter 73		37	.649		84.50	39		123.50	152
1150	10" diameter 87		31	.774		114	46.50		160.50	195
1160	12" diameter 100		27	.889		128	53.50		181.50	220
1170	14" diameter		20	1.200		138	72		210	260
1180	16" diameter		17	1.412		215	85		300	365
1190	18" diameter		14	1.714		207	103		310	380
1200	20" diameter		12	2		252	120		372	455
1210	24" diameter		10	2.400		284	144		428	525
1740	To delete coupling & hanger, subtract									
1750	3/4" diam. to 2" diam.					65%	27%			
1760	2-1/2" diam. to 5" diam.					41%	18%			
1770	6" diam. to 12" diam.					31%	13%			
1780	14" diam. to 24" diam.					35%	10%			
1800	Galvanized									
1840	3/4" diameter	1 Plum	71	.113	L.F.	7.15	7.25		14.40	18.70
1850	1" diameter		63	.127		7.10	8.20		15.30	20
1860	1-1/4" diameter		58	.138		8.50	8.90		17.40	22.50
1870	1-1/2" diameter		51	.157		9.25	10.10		19.35	25.50
1880	2" diameter		40	.200		10.80	12.90		23.70	31
1890	2-1/2" diameter	Q-1	57	.281		16.70	16.30		33	43
1900	3" diameter		50	.320		19.65	18.55		38.20	49.50
1910	4" diameter		45	.356		23	20.50		43.50	56.50
1920	5" diameter		37	.432		30.50	25		55.50	71
1930	6" diameter	Q-2	42	.571		39.50	34.50		74	95
1940	8" diameter		37	.649		58.50	39		97.50	123
1950	10" diameter		31	.774		114	46.50		160.50	195

22 11 13 – Facility Water Distribution Piping

22 11 13.48 Pipe, Fittings and Valves, Steel, Grooved-Joint		Crew	Daily Output	Labor-Hours	Unit	Material	2020 Bare Costs Labor	Equipment	Total	Total Incl O&P
1960	12" diameter	Q-2	27	.889	L.F.	137	53.50		190.50	231
2540	To delete coupling & hanger, subtract									
2550	3/4" diam. to 2" diam.					36%	27%			
2560	2-1/2" diam. to 5" diam.					19%	18%			
2570	6" diam. to 12" diam.					14%	13%			
2600	Schedule 80, black									
2610	3/4" diameter	1 Plum	65	.123	L.F.	8	7.95		15.95	20.50
2650	1" diameter		61	.131		8.50	8.45		16.95	22
2660	1-1/4" diameter		55	.145		10.55	9.35		19.90	25.50
2670	1-1/2" diameter		49	.163		11.75	10.50		22.25	28.50
2680	2" diameter		38	.211		14.20	13.55		27.75	36
2690	2-1/2" diameter	Q-1	54	.296		22	17.20		39.20	49.50
2700	3" diameter		48	.333		27	19.35		46.35	59
2710	4" diameter		44	.364		35.50	21		56.50	70.50
2720	5" diameter		35	.457		63	26.50		89.50	109
2730	6" diameter	Q-2	40	.600		72.50	36		108.50	134
2740	8" diameter		35	.686		113	41.50		154.50	187
2750	10" diameter		29	.828		169	50		219	260
2760	12" diameter		24	1		365	60		425	490
3240	To delete coupling & hanger, subtract									
3250	3/4" diam. to 2" diam.					30%	25%			
3260	2-1/2" diam. to 5" diam.					14%	17%			
3270	6" diam. to 12" diam.					12%	12%			
3300	Galvanized									
3310	3/4" diameter	1 Plum	65	.123	L.F.	8.60	7.95		16.55	21.50
3350	1" diameter		61	.131		9.40	8.45		17.85	23
3360	1-1/4" diameter		55	.145		11.85	9.35		21.20	27
3370	1-1/2" diameter		46	.174		13.30	11.20		24.50	31.50
3380	2" diameter		38	.211		17	13.55		30.55	39
3390	2-1/2" diameter	Q-1	54	.296		26	17.20		43.20	54
3400	3" diameter		48	.333		32	19.35		51.35	64
3410	4" diameter		44	.364		39.50	21		60.50	75
3420	5" diameter		35	.457		47	26.50		73.50	91
3430	6" diameter	Q-2	40	.600		65	36		101	126
3440	8" diameter		35	.686		120	41.50		161.50	194
3450	10" diameter		29	.828		273	50		323	375
3460	12" diameter		24	1		590	60		650	740
3920	To delete coupling & hanger, subtract									
3930	3/4" diam. to 2" diam.					30%	25%			
3940	2-1/2" diam. to 5" diam.					15%	17%			
3950	6" diam. to 12" diam.					11%	12%			
3990	Fittings: coupling material required at joints not incl. in fitting price.									
3994	Add 1 selected coupling, material only, per joint for installed price.									
4000	Elbow, 90° or 45°, painted									
4030	3/4" diameter	1 Plum	50	.160	Ea.	82.50	10.30		92.80	106
4040	1" diameter		50	.160		44.50	10.30		54.80	64.50
4050	1-1/4" diameter		40	.200		44.50	12.90		57.40	68.50
4060	1-1/2" diameter		33	.242		44.50	15.60		60.10	72.50
4070	2" diameter		25	.320		44.50	20.50		65	80
4080	2-1/2" diameter	Q-1	40	.400		44.50	23		67.50	83.50
4090	3" diameter		33	.485		78	28		106	128
4100	4" diameter		25	.640		84.50	37		121.50	149
4110	5" diameter		20	.800		201	46.50		247.50	291

For customer support on your Mechanical Costs with RSMeans Data, call 800.448.8182.

175

22 11 13.48 Pipe, Fittings and Valves, Steel, Grooved-Joint	Crew	Daily Output	Labor-Hours	Unit	Material	2020 Bare Costs Labor	Equipment	Total	Total Incl O&P
4120	6" diameter	Q-2	25	.960	Ea.	236	57.50		293.50	345
4130	8" diameter		21	1.143		490	68.50		558.50	645
4140	10" diameter		18	1.333		900	80		980	1,100
4150	12" diameter		15	1.600		975	96		1,071	1,225
4170	14" diameter		12	2		1,025	120		1,145	1,300
4180	16" diameter		11	2.182		1,125	131		1,256	1,425
4190	18" diameter	Q-3	14	2.286		1,700	140		1,840	2,050
4200	20" diameter		12	2.667		2,225	164		2,389	2,700
4210	24" diameter		10	3.200		3,225	196		3,421	3,850
4250	For galvanized elbows, add					26%				
4690	Tee, painted									
4700	3/4" diameter	1 Plum	38	.211	Ea.	88.50	13.55		102.05	118
4740	1" diameter		33	.242		68.50	15.60		84.10	98.50
4750	1-1/4" diameter		27	.296		68.50	19.10		87.60	104
4760	1-1/2" diameter		22	.364		68.50	23.50		92	110
4770	2" diameter		17	.471		68.50	30.50		99	121
4780	2-1/2" diameter	Q-1	27	.593		68.50	34.50		103	127
4790	3" diameter		22	.727		93.50	42		135.50	166
4800	4" diameter		17	.941		142	54.50		196.50	238
4810	5" diameter		13	1.231		330	71.50		401.50	470
4820	6" diameter	Q-2	17	1.412		380	85		465	545
4830	8" diameter		14	1.714		835	103		938	1,075
4840	10" diameter		12	2		1,050	120		1,170	1,350
4850	12" diameter		10	2.400		1,375	144		1,519	1,750
4851	14" diameter		9	2.667		1,375	160		1,535	1,750
4852	16" diameter		8	3		1,550	180		1,730	1,975
4853	18" diameter	Q-3	10	3.200		1,925	196		2,121	2,425
4854	20" diameter		9	3.556		2,775	218		2,993	3,375
4855	24" diameter		8	4		4,225	245		4,470	5,025
4900	For galvanized tees, add					24%				
4906	Couplings, rigid style, painted									
4908	1" diameter	1 Plum	100	.080	Ea.	34.50	5.15		39.65	45.50
4909	1-1/4" diameter		100	.080		32	5.15		37.15	42.50
4910	1-1/2" diameter		67	.119		34.50	7.70		42.20	49.50
4912	2" diameter		50	.160		43.50	10.30		53.80	63.50
4914	2-1/2" diameter	Q-1	80	.200		49.50	11.60		61.10	72
4916	3" diameter		67	.239		57.50	13.85		71.35	83.50
4918	4" diameter		50	.320		79.50	18.55		98.05	116
4920	5" diameter		40	.400		102	23		125	148
4922	6" diameter	Q-2	50	.480		135	29		164	192
4924	8" diameter		42	.571		213	34.50		247.50	286
4926	10" diameter		35	.686		410	41.50		451.50	510
4928	12" diameter		32	.750		460	45		505	575
4930	14" diameter		24	1		495	60		555	635
4931	16" diameter		20	1.200		450	72		522	605
4932	18" diameter		18	1.333		530	80		610	705
4933	20" diameter		16	1.500		725	90		815	930
4934	24" diameter		13	1.846		930	111		1,041	1,200
4940	Flexible, standard, painted									
4950	3/4" diameter	1 Plum	100	.080	Ea.	25	5.15		30.15	35
4960	1" diameter		100	.080		25	5.15		30.15	35
4970	1-1/4" diameter		80	.100		32.50	6.45		38.95	45
4980	1-1/2" diameter		67	.119		35	7.70		42.70	50

22 11 Facility Water Distribution

22 11 13 – Facility Water Distribution Piping

22 11 13.48 Pipe, Fittings and Valves, Steel, Grooved-Joint		Crew	Daily Output	Labor-Hours	Unit	Material	2020 Bare Costs Labor	Equipment	Total	Total Incl O&P
4990	2" diameter	1 Plum	50	.160	Ea.	38	10.30		48.30	57
5000	2-1/2" diameter	Q-1	80	.200		44	11.60		55.60	66
5010	3" diameter		67	.239		48.50	13.85		62.35	74
5020	3-1/2" diameter		57	.281		69.50	16.30		85.80	101
5030	4" diameter		50	.320		70	18.55		88.55	105
5040	5" diameter		40	.400		105	23		128	151
5050	6" diameter	Q-2	50	.480		125	29		154	180
5070	8" diameter		42	.571		202	34.50		236.50	274
5090	10" diameter		35	.686		330	41.50		371.50	420
5110	12" diameter		32	.750		375	45		420	480
5120	14" diameter		24	1		530	60		590	670
5130	16" diameter		20	1.200		695	72		767	870
5140	18" diameter		18	1.333		810	80		890	1,000
5150	20" diameter		16	1.500		1,275	90		1,365	1,525
5160	24" diameter		13	1.846		1,400	111		1,511	1,700
5176	Lightweight style, painted									
5178	1-1/2" diameter	1 Plum	67	.119	Ea.	31	7.70		38.70	46
5180	2" diameter	"	50	.160		37.50	10.30		47.80	57
5182	2-1/2" diameter	Q-1	80	.200		44	11.60		55.60	65.50
5184	3" diameter		67	.239		48.50	13.85		62.35	74
5186	3-1/2" diameter		57	.281		59.50	16.30		75.80	89.50
5188	4" diameter		50	.320		70	18.55		88.55	105
5190	5" diameter		40	.400		104	23		127	150
5192	6" diameter	Q-2	50	.480		124	29		153	179
5194	8" diameter		42	.571		200	34.50		234.50	272
5196	10" diameter		35	.686		410	41.50		451.50	510
5198	12" diameter		32	.750		455	45		500	570
5200	For galvanized couplings, add					33%				
5220	Tee, reducing, painted									
5225	2" x 1-1/2" diameter	Q-1	38	.421	Ea.	144	24.50		168.50	195
5226	2-1/2" x 2" diameter		28	.571		144	33		177	208
5227	3" x 2-1/2" diameter		23	.696		127	40.50		167.50	201
5228	4" x 3" diameter		18	.889		171	51.50		222.50	265
5229	5" x 4" diameter		15	1.067		365	62		427	495
5230	6" x 4" diameter	Q-2	18	1.333		400	80		480	560
5231	8" x 6" diameter		15	1.600		835	96		931	1,075
5232	10" x 8" diameter		13	1.846		920	111		1,031	1,200
5233	12" x 10" diameter		11	2.182		1,125	131		1,256	1,450
5234	14" x 12" diameter		10	2.400		1,475	144		1,619	1,825
5235	16" x 12" diameter		9	2.667		1,225	160		1,385	1,600
5236	18" x 12" diameter	Q-3	12	2.667		1,475	164		1,639	1,875
5237	18" x 16" diameter		11	2.909		1,875	178		2,053	2,325
5238	20" x 16" diameter		10	3.200		2,450	196		2,646	3,000
5239	24" x 20" diameter		9	3.556		3,875	218		4,093	4,575
5240	Reducer, concentric, painted									
5241	2-1/2" x 2" diameter	Q-1	43	.372	Ea.	52	21.50		73.50	89.50
5242	3" x 2-1/2" diameter		35	.457		62.50	26.50		89	108
5243	4" x 3" diameter		29	.552		75	32		107	131
5244	5" x 4" diameter		22	.727		103	42		145	177
5245	6" x 4" diameter	Q-2	26	.923		120	55.50		175.50	215
5246	8" x 6" diameter		23	1.043		310	63		373	435
5247	10" x 8" diameter		20	1.200		605	72		677	775
5248	12" x 10" diameter		16	1.500		1,075	90		1,165	1,325

For customer support on your Mechanical Costs with RSMeans Data, call 800.448.8182.

177

22 11 13 – Facility Water Distribution Piping

22 11 13.48 Pipe, Fittings and Valves, Steel, Grooved-Joint		Crew	Daily Output	Labor-Hours	Unit	Material	2020 Bare Costs Labor	Equipment	Total	Total Incl O&P
5255	Eccentric, painted									
5256	2-1/2" x 2" diameter	Q-1	42	.381	Ea.	107	22		129	151
5257	3" x 2-1/2" diameter		34	.471		123	27.50		150.50	176
5258	4" x 3" diameter		28	.571		149	33		182	214
5259	5" x 4" diameter		21	.762		202	44		246	288
5260	6" x 4" diameter	Q-2	25	.960		235	57.50		292.50	345
5261	8" x 6" diameter		22	1.091		475	65.50		540.50	620
5262	10" x 8" diameter		19	1.263		1,250	76		1,326	1,500
5263	12" x 10" diameter		15	1.600		1,725	96		1,821	2,050
5270	Coupling, reducing, painted									
5272	2" x 1-1/2" diameter	1 Plum	52	.154	Ea.	50	9.90		59.90	70
5274	2-1/2" x 2" diameter	Q-1	82	.195		64.50	11.30		75.80	87.50
5276	3" x 2" diameter		69	.232		73	13.45		86.45	100
5278	4" x 2" diameter		52	.308		115	17.85		132.85	154
5280	5" x 4" diameter		42	.381		130	22		152	176
5282	6" x 4" diameter	Q-2	52	.462		195	28		223	257
5284	8" x 6" diameter	"	44	.545		291	33		324	370
5290	Outlet coupling, painted									
5294	1-1/2" x 1" pipe size	1 Plum	65	.123	Ea.	66	7.95		73.95	84.50
5296	2" x 1" pipe size	"	48	.167		67	10.75		77.75	90
5298	2-1/2" x 1" pipe size	Q-1	78	.205		103	11.90		114.90	131
5300	2-1/2" x 1-1/4" pipe size		70	.229		116	13.25		129.25	147
5302	3" x 1" pipe size		65	.246		131	14.30		145.30	166
5304	4" x 3/4" pipe size		48	.333		145	19.35		164.35	188
5306	4" x 1-1/2" pipe size		46	.348		205	20		225	256
5308	6" x 1-1/2" pipe size	Q-2	44	.545		290	33		323	370
5320	Outlet, strap-on T, painted									
5324	Threaded female branch									
5330	2" pipe x 1-1/2" branch size	1 Plum	50	.160	Ea.	47	10.30		57.30	67.50
5334	2-1/2" pipe x 1-1/2" branch size	Q-1	80	.200		57.50	11.60		69.10	81
5338	3" pipe x 1-1/2" branch size		67	.239		59.50	13.85		73.35	86
5342	3" pipe x 2" branch size		59	.271		68.50	15.75		84.25	99
5346	4" pipe x 2" branch size		50	.320		72	18.55		90.55	107
5350	4" pipe x 2-1/2" branch size		47	.340		75	19.75		94.75	112
5354	4" pipe x 3" branch size		43	.372		80	21.50		101.50	121
5358	5" pipe x 3" branch size		40	.400		98.50	23		121.50	143
5362	6" pipe x 2" branch size	Q-2	50	.480		90	29		119	142
5366	6" pipe x 3" branch size		47	.511		103	30.50		133.50	159
5370	6" pipe x 4" branch size		44	.545		114	33		147	175
5374	8" pipe x 2-1/2" branch size		42	.571		157	34.50		191.50	225
5378	8" pipe x 4" branch size		38	.632		170	38		208	244
5390	Grooved branch									
5394	2" pipe x 1-1/4" branch size	1 Plum	78	.103	Ea.	47	6.60		53.60	62
5398	2" pipe x 1-1/2" branch size	"	50	.160		47	10.30		57.30	67.50
5402	2-1/2" pipe x 1-1/2" branch size	Q-1	80	.200		57.50	11.60		69.10	81
5406	3" pipe x 1-1/2" branch size		67	.239		59.50	13.85		73.35	86
5410	3" pipe x 2" branch size		59	.271		68.50	15.75		84.25	99
5414	4" pipe x 2" branch size		50	.320		72	18.55		90.55	107
5418	4" pipe x 2-1/2" branch size		47	.340		75	19.75		94.75	112
5422	4" pipe x 3" branch size		43	.372		80	21.50		101.50	121
5426	5" pipe x 3" branch size		40	.400		98.50	23		121.50	143
5430	6" pipe x 2" branch size	Q-2	50	.480		90	29		119	142
5434	6" pipe x 3" branch size		47	.511		103	30.50		133.50	159

22 11 13 – Facility Water Distribution Piping

22 11 13.48 Pipe, Fittings and Valves, Steel, Grooved-Joint	Crew	Daily Output	Labor-Hours	Unit	Material	2020 Bare Costs Labor	Equipment	Total	Total Incl O&P
5438	6" pipe x 4" branch size	Q-2	44	.545	Ea.	114	33		147	175
5442	8" pipe x 2-1/2" branch size		42	.571		157	34.50		191.50	225
5446	8" pipe x 4" branch size		38	.632		170	38		208	244
5750	Flange, w/groove gasket, black steel									
5754	See Line 22 11 13.47 0620 for gasket & bolt set									
5760	ANSI class 125 and 150, painted									
5780	2" pipe size	1 Plum	23	.348	Ea.	145	22.50		167.50	194
5790	2-1/2" pipe size	Q-1	37	.432		181	25		206	237
5800	3" pipe size		31	.516		195	30		225	259
5820	4" pipe size		23	.696		260	40.50		300.50	345
5830	5" pipe size		19	.842		300	49		349	405
5840	6" pipe size	Q-2	23	1.043		330	63		393	455
5850	8" pipe size		17	1.412		370	85		455	535
5860	10" pipe size		14	1.714		585	103		688	800
5870	12" pipe size		12	2		765	120		885	1,025
5880	14" pipe size		10	2.400		1,400	144		1,544	1,775
5890	16" pipe size		9	2.667		1,625	160		1,785	2,050
5900	18" pipe size		6	4		2,000	241		2,241	2,550
5910	20" pipe size		5	4.800		2,425	289		2,714	3,075
5920	24" pipe size		4.50	5.333		3,100	320		3,420	3,875
5940	ANSI class 350, painted									
5946	2" pipe size	1 Plum	23	.348	Ea.	182	22.50		204.50	234
5948	2-1/2" pipe size	Q-1	37	.432		210	25		235	269
5950	3" pipe size		31	.516		287	30		317	360
5952	4" pipe size		23	.696		380	40.50		420.50	480
5954	5" pipe size		19	.842		435	49		484	555
5956	6" pipe size	Q-2	23	1.043		505	63		568	650
5958	8" pipe size		17	1.412		580	85		665	760
5960	10" pipe size		14	1.714		920	103		1,023	1,175
5962	12" pipe size		12	2		985	120		1,105	1,250
6100	Cross, painted									
6110	2" diameter	1 Plum	12.50	.640	Ea.	122	41.50		163.50	196
6112	2-1/2" diameter	Q-1	20	.800		122	46.50		168.50	204
6114	3" diameter		15.50	1.032		216	60		276	330
6116	4" diameter		12.50	1.280		360	74		434	505
6118	6" diameter	Q-2	12.50	1.920		935	115		1,050	1,200
6120	8" diameter		10.50	2.286		1,225	137		1,362	1,550
6122	10" diameter		9	2.667		2,025	160		2,185	2,475
6124	12" diameter		7.50	3.200		2,925	192		3,117	3,525
7400	Suction diffuser									
7402	Grooved end inlet x flanged outlet									
7410	3" x 3"	Q-1	27	.593	Ea.	1,175	34.50		1,209.50	1,350
7412	4" x 4"		19	.842		1,600	49		1,649	1,825
7414	5" x 5"		14	1.143		1,850	66.50		1,916.50	2,150
7416	6" x 6"	Q-2	20	1.200		2,350	72		2,422	2,675
7418	8" x 8"		15	1.600		4,375	96		4,471	4,950
7420	10" x 10"		12	2		5,925	120		6,045	6,700
7422	12" x 12"		9	2.667		9,775	160		9,935	10,900
7424	14" x 14"		7	3.429		9,525	206		9,731	10,800
7426	16" x 14"		6	4		9,950	241		10,191	11,300
7500	Strainer, tee type, painted									
7506	2" pipe size	1 Plum	21	.381	Ea.	785	24.50		809.50	895
7508	2-1/2" pipe size	Q-1	30	.533		825	31		856	950

22 11 13.48 Pipe, Fittings and Valves, Steel, Grooved-Joint	Crew	Daily Output	Labor-Hours	Unit	Material	2020 Bare Costs Labor	Equipment	Total	Total Incl O&P
7510	3" pipe size	Q-1	28	.571	Ea.	925	33		958	1,075
7512	4" pipe size		20	.800		1,050	46.50		1,096.50	1,225
7514	5" pipe size	▼	15	1.067		1,525	62		1,587	1,775
7516	6" pipe size	Q-2	23	1.043		1,650	63		1,713	1,900
7518	8" pipe size		16	1.500		2,525	90		2,615	2,900
7520	10" pipe size		13	1.846		3,700	111		3,811	4,250
7522	12" pipe size		10	2.400		4,750	144		4,894	5,450
7524	14" pipe size		8	3		12,900	180		13,080	14,500
7526	16" pipe size	▼	7	3.429	▼	16,200	206		16,406	18,100
7570	Expansion joint, max. 3" travel									
7572	2" diameter	1 Plum	24	.333	Ea.	990	21.50		1,011.50	1,125
7574	3" diameter	Q-1	31	.516		1,125	30		1,155	1,300
7576	4" diameter	"	23	.696		1,500	40.50		1,540.50	1,700
7578	6" diameter	Q-2	38	.632	▼	2,325	38		2,363	2,600
7790	Valves: coupling material required at joints not incl. in valve price.									
7794	Add 1 selected coupling, material only, per joint for installed price.									
7800	Ball valve w/handle, carbon steel trim									
7810	1-1/2" pipe size	1 Plum	31	.258	Ea.	187	16.65		203.65	231
7812	2" pipe size	"	24	.333		178	21.50		199.50	228
7814	2-1/2" pipe size	Q-1	38	.421		450	24.50		474.50	530
7816	3" pipe size		31	.516		730	30		760	845
7818	4" pipe size	▼	23	.696		1,125	40.50		1,165.50	1,300
7820	6" pipe size	Q-2	24	1	▼	3,475	60		3,535	3,925
7830	With gear operator									
7834	2-1/2" pipe size	Q-1	38	.421	Ea.	910	24.50		934.50	1,025
7836	3" pipe size		31	.516		1,300	30		1,330	1,475
7838	4" pipe size	▼	23	.696		1,650	40.50		1,690.50	1,850
7840	6" pipe size	Q-2	24	1	▼	3,875	60		3,935	4,375
7870	Check valve									
7874	2-1/2" pipe size	Q-1	38	.421	Ea.	375	24.50		399.50	450
7876	3" pipe size		31	.516		440	30		470	530
7878	4" pipe size		23	.696		465	40.50		505.50	575
7880	5" pipe size	▼	18	.889		775	51.50		826.50	930
7882	6" pipe size	Q-2	24	1		920	60		980	1,100
7884	8" pipe size		19	1.263		1,250	76		1,326	1,500
7886	10" pipe size		16	1.500		3,375	90		3,465	3,850
7888	12" pipe size	▼	13	1.846	▼	4,000	111		4,111	4,575
7900	Plug valve, balancing, w/lever operator									
7906	3" pipe size	Q-1	31	.516	Ea.	925	30		955	1,075
7908	4" pipe size	"	23	.696		1,150	40.50		1,190.50	1,300
7909	6" pipe size	Q-2	24	1	▼	1,750	60		1,810	2,025
7916	With gear operator									
7920	3" pipe size	Q-1	31	.516	Ea.	1,750	30		1,780	1,975
7922	4" pipe size	"	23	.696		1,825	40.50		1,865.50	2,050
7924	6" pipe size	Q-2	24	1		2,450	60		2,510	2,800
7926	8" pipe size		19	1.263		3,775	76		3,851	4,275
7928	10" pipe size		16	1.500		4,775	90		4,865	5,375
7930	12" pipe size	▼	13	1.846	▼	7,275	111		7,386	8,175
8000	Butterfly valve, 2 position handle, with standard trim									
8010	1-1/2" pipe size	1 Plum	30	.267	Ea.	350	17.20		367.20	410
8020	2" pipe size	"	23	.348		350	22.50		372.50	420
8030	3" pipe size	Q-1	30	.533		505	31		536	600
8050	4" pipe size	"	22	.727	▼	555	42		597	675

22 11 Facility Water Distribution

22 11 13 – Facility Water Distribution Piping

22 11 13.48 Pipe, Fittings and Valves, Steel, Grooved-Joint	Crew	Daily Output	Labor-Hours	Unit	Material	2020 Bare Costs Labor	Equipment	Total	Total Incl O&P
8070	6" pipe size	Q-2	23	1.043	Ea.	1,125	63		1,188	1,325
8080	8" pipe size		18	1.333		1,450	80		1,530	1,725
8090	10" pipe size	↓	15	1.600	↓	2,000	96		2,096	2,350
8200	With stainless steel trim									
8240	1-1/2" pipe size	1 Plum	30	.267	Ea.	445	17.20		462.20	515
8250	2" pipe size	"	23	.348		445	22.50		467.50	525
8270	3" pipe size	Q-1	30	.533		600	31		631	705
8280	4" pipe size	"	22	.727		650	42		692	780
8300	6" pipe size	Q-2	23	1.043		1,200	63		1,263	1,425
8310	8" pipe size		18	1.333		3,000	80		3,080	3,425
8320	10" pipe size		15	1.600		4,800	96		4,896	5,425
8322	12" pipe size		12	2		5,750	120		5,870	6,500
8324	14" pipe size		10	2.400		7,375	144		7,519	8,325
8326	16" pipe size	↓	9	2.667		12,500	160		12,660	13,900
8328	18" pipe size	Q-3	12	2.667		13,100	164		13,264	14,600
8330	20" pipe size		10	3.200		16,100	196		16,296	18,000
8332	24" pipe size	↓	9	3.556	↓	21,600	218		21,818	24,000
8336	Note: sizes 8" up w/manual gear operator									
9000	Cut one groove, labor									
9010	3/4" pipe size	Q-1	152	.105	Ea.		6.10		6.10	9.15
9020	1" pipe size		140	.114			6.65		6.65	9.90
9030	1-1/4" pipe size		124	.129			7.50		7.50	11.20
9040	1-1/2" pipe size		114	.140			8.15		8.15	12.20
9050	2" pipe size		104	.154			8.90		8.90	13.35
9060	2-1/2" pipe size		96	.167			9.65		9.65	14.45
9070	3" pipe size		88	.182			10.55		10.55	15.80
9080	3-1/2" pipe size		83	.193			11.20		11.20	16.75
9090	4" pipe size		78	.205			11.90		11.90	17.80
9100	5" pipe size		72	.222			12.90		12.90	19.30
9110	6" pipe size		70	.229			13.25		13.25	19.85
9120	8" pipe size		54	.296			17.20		17.20	25.50
9130	10" pipe size		38	.421			24.50		24.50	36.50
9140	12" pipe size		30	.533			31		31	46.50
9150	14" pipe size		20	.800			46.50		46.50	69.50
9160	16" pipe size		19	.842			49		49	73
9170	18" pipe size		18	.889			51.50		51.50	77
9180	20" pipe size		17	.941			54.50		54.50	81.50
9190	24" pipe size	↓	15	1.067	↓		62		62	92.50
9210	Roll one groove									
9220	3/4" pipe size	Q-1	266	.060	Ea.		3.49		3.49	5.20
9230	1" pipe size		228	.070			4.07		4.07	6.10
9240	1-1/4" pipe size		200	.080			4.64		4.64	6.95
9250	1-1/2" pipe size		178	.090			5.20		5.20	7.80
9260	2" pipe size		116	.138			8		8	11.95
9270	2-1/2" pipe size		110	.145			8.45		8.45	12.65
9280	3" pipe size		100	.160			9.30		9.30	13.90
9290	3-1/2" pipe size		94	.170			9.85		9.85	14.75
9300	4" pipe size		86	.186			10.80		10.80	16.15
9310	5" pipe size		84	.190			11.05		11.05	16.55
9320	6" pipe size		80	.200			11.60		11.60	17.35
9330	8" pipe size		66	.242			14.05		14.05	21
9340	10" pipe size		58	.276			16		16	24
9350	12" pipe size	↓	46	.348			20		20	30

22 11 13 – Facility Water Distribution Piping

22 11 13.48 Pipe, Fittings and Valves, Steel, Grooved-Joint	Crew	Daily Output	Labor-Hours	Unit	Material	2020 Bare Costs Labor	2020 Bare Costs Equipment	Total	Total Incl O&P
9360	14" pipe size	Q-1	30	.533	Ea.		31		31	46.50
9370	16" pipe size		28	.571			33		33	49.50
9380	18" pipe size		27	.593			34.50		34.50	51.50
9390	20" pipe size		25	.640			37		37	55.50
9400	24" pipe size	▼	23	.696	▼		40.50		40.50	60.50

22 11 13.60 Tubing, Stainless Steel

		Crew	Daily Output	Labor-Hours	Unit	Material	2020 Bare Costs Labor	2020 Bare Costs Equipment	Total	Total Incl O&P
0010	**TUBING, STAINLESS STEEL**									
5000	Tubing									
5010	Type 304, no joints, no hangers									
5020	.035 wall									
5021	1/4"	1 Plum	160	.050	L.F.	6.70	3.22		9.92	12.20
5022	3/8"		160	.050		6.25	3.22		9.47	11.70
5023	1/2"		160	.050		10.60	3.22		13.82	16.45
5024	5/8"		160	.050		10.25	3.22		13.47	16.05
5025	3/4"		133	.060		10.80	3.88		14.68	17.65
5026	7/8"		133	.060		11.05	3.88		14.93	17.95
5027	1"	▼	114	.070	▼	14.65	4.52		19.17	23
5040	.049 wall									
5041	1/4"	1 Plum	160	.050	L.F.	7.05	3.22		10.27	12.60
5042	3/8"		160	.050		10.60	3.22		13.82	16.45
5043	1/2"		160	.050		10.45	3.22		13.67	16.25
5044	5/8"		160	.050		11.70	3.22		14.92	17.65
5045	3/4"		133	.060		12.40	3.88		16.28	19.45
5046	7/8"		133	.060		10.80	3.88		14.68	17.65
5047	1"	▼	114	.070	▼	12	4.52		16.52	19.95
5060	.065 wall									
5061	1/4"	1 Plum	160	.050	L.F.	7.60	3.22		10.82	13.20
5062	3/8"		160	.050		10.60	3.22		13.82	16.45
5063	1/2"		160	.050		10.15	3.22		13.37	15.95
5064	5/8"		160	.050		13	3.22		16.22	19.10
5065	3/4"		133	.060		14.55	3.88		18.43	22
5066	7/8"		133	.060		13.20	3.88		17.08	20.50
5067	1"	▼	114	.070	▼	12.55	4.52		17.07	20.50
5210	Type 316									
5220	.035 wall									
5221	1/4"	1 Plum	160	.050	L.F.	6.10	3.22		9.32	11.50
5222	3/8"		160	.050		6.30	3.22		9.52	11.75
5223	1/2"		160	.050		9.45	3.22		12.67	15.20
5224	5/8"		160	.050		9.80	3.22		13.02	15.60
5225	3/4"		133	.060		12.15	3.88		16.03	19.15
5226	7/8"		133	.060		18.10	3.88		21.98	25.50
5227	1"	▼	114	.070		16.30	4.52		20.82	24.50
5240	.049 wall									
5241	1/4"	1 Plum	160	.050	L.F.	5.30	3.22		8.52	10.65
5242	3/8"		160	.050		9.80	3.22		13.02	15.55
5243	1/2"		160	.050		10.35	3.22		13.57	16.15
5244	5/8"		160	.050		11.25	3.22		14.47	17.15
5245	3/4"		133	.060		12.20	3.88		16.08	19.20
5246	7/8"		133	.060		20.50	3.88		24.38	28.50
5247	1"	▼	114	.070	▼	14.55	4.52		19.07	23
5260	.065 wall									
5261	1/4"	1 Plum	160	.050	L.F.	7.20	3.22		10.42	12.75

22 11 Facility Water Distribution

22 11 13 – Facility Water Distribution Piping

22 11 13.60 Tubing, Stainless Steel		Crew	Daily Output	Labor-Hours	Unit	Material	2020 Bare Costs Labor	Equipment	Total	Total Incl O&P
5262	3/8"	1 Plum	160	.050	L.F.	10.35	3.22		13.57	16.20
5263	1/2"		160	.050		10.05	3.22		13.27	15.85
5264	5/8"		160	.050		12.10	3.22		15.32	18.10
5265	3/4"		133	.060		14.05	3.88		17.93	21.50
5266	7/8"		133	.060		21	3.88		24.88	29
5267	1"		114	.070		18.25	4.52		22.77	27

22 11 13.61 Tubing Fittings, Stainless Steel

		Crew	Daily Output	Labor-Hours	Unit	Material	2020 Bare Costs Labor	Equipment	Total	Total Incl O&P
0010	**TUBING FITTINGS, STAINLESS STEEL**									
8200	Tube fittings, compression type									
8202	Type 316									
8204	90° elbow									
8206	1/4"	1 Plum	24	.333	Ea.	17.30	21.50		38.80	51
8207	3/8"		22	.364		21.50	23.50		45	58.50
8208	1/2"		22	.364		35	23.50		58.50	73.50
8209	5/8"		21	.381		39.50	24.50		64	80
8210	3/4"		21	.381		62	24.50		86.50	105
8211	7/8"		20	.400		95	26		121	144
8212	1"		20	.400		118	26		144	169
8220	Union tee									
8222	1/4"	1 Plum	15	.533	Ea.	24.50	34.50		59	78.50
8224	3/8"		15	.533		31.50	34.50		66	86
8225	1/2"		15	.533		48.50	34.50		83	105
8226	5/8"		14	.571		54.50	37		91.50	115
8227	3/4"		14	.571		73	37		110	135
8228	7/8"		13	.615		141	39.50		180.50	215
8229	1"		13	.615		156	39.50		195.50	232
8234	Union									
8236	1/4"	1 Plum	24	.333	Ea.	12	21.50		33.50	45
8237	3/8"		22	.364		17.10	23.50		40.60	54
8238	1/2"		22	.364		25.50	23.50		49	63
8239	5/8"		21	.381		33	24.50		57.50	73
8240	3/4"		21	.381		41.50	24.50		66	82
8241	7/8"		20	.400		67.50	26		93.50	113
8242	1"		20	.400		71.50	26		97.50	117
8250	Male connector									
8252	1/4" x 1/4"	1 Plum	24	.333	Ea.	7.70	21.50		29.20	40.50
8253	3/8" x 3/8"		22	.364		12	23.50		35.50	48
8254	1/2" x 1/2"		22	.364		17.75	23.50		41.25	54.50
8256	3/4" x 3/4"		21	.381		27	24.50		51.50	66.50
8258	1" x 1"		20	.400		47.50	26		73.50	90.50

22 11 13.64 Pipe, Stainless Steel

		Crew	Daily Output	Labor-Hours	Unit	Material	2020 Bare Costs Labor	Equipment	Total	Total Incl O&P
0010	**PIPE, STAINLESS STEEL**									
0020	Welded, with clevis type hanger assemblies, 10' OC									
0500	Schedule 5, type 304									
0540	1/2" diameter	Q-15	128	.125	L.F.	13.70	7.25	.83	21.78	27
0550	3/4" diameter		116	.138		18.70	8	.92	27.62	33.50
0560	1" diameter		103	.155		17.05	9	1.03	27.08	33.50
0570	1-1/4" diameter		93	.172		20.50	10	1.14	31.64	39
0580	1-1/2" diameter		85	.188		23	10.90	1.25	35.15	43
0590	2" diameter		69	.232		33.50	13.45	1.54	48.49	58.50
0600	2-1/2" diameter		53	.302		47.50	17.50	2.01	67.01	80.50
0610	3" diameter		48	.333		73.50	19.35	2.22	95.07	112

22 11 13.64 Pipe, Stainless Steel		Crew	Daily Output	Labor-Hours	Unit	Material	2020 Bare Costs Labor	2020 Bare Costs Equipment	Total	Total Incl O&P
0620	4" diameter	Q-15	44	.364	L.F.	127	21	2.42	150.42	173
0630	5" diameter	↓	36	.444		154	26	2.96	182.96	211
0640	6" diameter	Q-16	42	.571		320	34.50	2.53	357.03	405
0650	8" diameter		34	.706		490	42.50	3.13	535.63	605
0660	10" diameter		26	.923		300	55.50	4.09	359.59	420
0670	12" diameter	↓	21	1.143	↓	395	68.50	5.05	468.55	545
0700	To delete hangers, subtract									
0710	1/2" diam. to 1-1/2" diam.					8%	19%			
0720	2" diam. to 5" diam.					4%	9%			
0730	6" diam. to 12" diam.					3%	4%			
0750	For small quantities, add				L.F.	10%				
1250	Schedule 5, type 316									
1290	1/2" diameter	Q-15	128	.125	L.F.	13.05	7.25	.83	21.13	26
1300	3/4" diameter		116	.138		21	8	.92	29.92	36.50
1310	1" diameter		103	.155		21.50	9	1.03	31.53	38.50
1320	1-1/4" diameter		93	.172		25	10	1.14	36.14	43.50
1330	1-1/2" diameter		85	.188		44	10.90	1.25	56.15	66
1340	2" diameter		69	.232		59	13.45	1.54	73.99	86.50
1350	2-1/2" diameter		53	.302		91.50	17.50	2.01	111.01	129
1360	3" diameter		48	.333		115	19.35	2.22	136.57	158
1370	4" diameter		44	.364		139	21	2.42	162.42	187
1380	5" diameter	↓	36	.444		159	26	2.96	187.96	217
1390	6" diameter	Q-16	42	.571		190	34.50	2.53	227.03	263
1400	8" diameter		34	.706		286	42.50	3.13	331.63	380
1410	10" diameter		26	.923		335	55.50	4.09	394.59	455
1420	12" diameter	↓	21	1.143		430	68.50	5.05	503.55	585
1490	For small quantities, add				↓	10%				
1940	To delete hanger, subtract									
1950	1/2" diam. to 1-1/2" diam.					5%	19%			
1960	2" diam. to 5" diam.					3%	9%			
1970	6" diam. to 12" diam.					2%	4%			
2000	Schedule 10, type 304									
2040	1/4" diameter	Q-15	131	.122	L.F.	9.10	7.10	.81	17.01	21.50
2050	3/8" diameter		128	.125		8.15	7.25	.83	16.23	20.50
2060	1/2" diameter		125	.128		13.95	7.40	.85	22.20	27.50
2070	3/4" diameter		113	.142		16.40	8.20	.94	25.54	31.50
2080	1" diameter		100	.160		15.30	9.30	1.06	25.66	32
2090	1-1/4" diameter		91	.176		25.50	10.20	1.17	36.87	45
2100	1-1/2" diameter		83	.193		22	11.20	1.28	34.48	42
2110	2" diameter		67	.239		23.50	13.85	1.59	38.94	48.50
2120	2-1/2" diameter		51	.314		42	18.20	2.09	62.29	75.50
2130	3" diameter		46	.348		50.50	20	2.31	72.81	88
2140	4" diameter		42	.381		44	22	2.53	68.53	84.50
2150	5" diameter	↓	35	.457		66.50	26.50	3.04	96.04	116
2160	6" diameter	Q-16	40	.600		60	36	2.66	98.66	123
2170	8" diameter		33	.727		102	44	3.22	149.22	181
2180	10" diameter		25	.960		140	57.50	4.25	201.75	245
2190	12" diameter	↓	21	1.143		174	68.50	5.05	247.55	300
2250	For small quantities, add				↓	10%				
2650	To delete hanger, subtract									
2660	1/4" diam. to 3/4" diam.					9%	22%			
2670	1" diam. to 2" diam.					4%	15%			
2680	2-1/2" diam. to 5" diam.					3%	8%			

22 11 13.64 Pipe, Stainless Steel	Crew	Daily Output	Labor-Hours	Unit	Material	2020 Bare Costs Labor	Equipment	Total	Total Incl O&P
2690	6" diam. to 12" diam.					3%	4%			
2750	Schedule 10, type 316									
2790	1/4" diameter	Q-15	131	.122	L.F.	7.15	7.10	.81	15.06	19.35
2800	3/8" diameter		128	.125		8.70	7.25	.83	16.78	21.50
2810	1/2" diameter		125	.128		11.10	7.40	.85	19.35	24
2820	3/4" diameter		113	.142		20.50	8.20	.94	29.64	36
2830	1" diameter		100	.160		23.50	9.30	1.06	33.86	41
2840	1-1/4" diameter		91	.176		22.50	10.20	1.17	33.87	41.50
2850	1-1/2" diameter		83	.193		30.50	11.20	1.28	42.98	51.50
2860	2" diameter		67	.239		32	13.85	1.59	47.44	57.50
2870	2-1/2" diameter		51	.314		39	18.20	2.09	59.29	72.50
2880	3" diameter		46	.348		59	20	2.31	81.31	97.50
2890	4" diameter		42	.381		59	22	2.53	83.53	101
2900	5" diameter	▼	35	.457		56.50	26.50	3.04	86.04	105
2910	6" diameter	Q-16	40	.600		91	36	2.66	129.66	157
2920	8" diameter		33	.727		133	44	3.22	180.22	215
2930	10" diameter		25	.960		174	57.50	4.25	235.75	283
2940	12" diameter	▼	21	1.143		218	68.50	5.05	291.55	350
2990	For small quantities, add				▼	10%				
3430	To delete hanger, subtract									
3440	1/4" diam. to 3/4" diam.					6%	22%			
3450	1" diam. to 2" diam.					3%	15%			
3460	2-1/2" diam. to 5" diam.					2%	8%			
3470	6" diam. to 12" diam.					2%	4%			
3500	Threaded, couplings and clevis hanger assemblies, 10' OC									
3520	Schedule 40, type 304									
3540	1/4" diameter	1 Plum	54	.148	L.F.	10.20	9.55		19.75	25.50
3550	3/8" diameter		53	.151		11.15	9.75		20.90	27
3560	1/2" diameter		52	.154		12.40	9.90		22.30	28.50
3570	3/4" diameter		51	.157		20.50	10.10		30.60	37.50
3580	1" diameter	▼	45	.178		24.50	11.45		35.95	44
3590	1-1/4" diameter	Q-1	76	.211		39	12.20		51.20	61.50
3600	1-1/2" diameter		69	.232		40	13.45		53.45	64.50
3610	2" diameter		57	.281		52	16.30		68.30	82
3620	2-1/2" diameter		44	.364		98	21		119	140
3630	3" diameter	▼	38	.421		113	24.50		137.50	162
3640	4" diameter	Q-2	51	.471		131	28.50		159.50	187
3740	For small quantities, add				▼	10%				
4200	To delete couplings & hangers, subtract									
4210	1/4" diam. to 3/4" diam.					15%	56%			
4220	1" diam. to 2" diam.					18%	49%			
4230	2-1/2" diam. to 4" diam.					34%	40%			
4250	Schedule 40, type 316									
4290	1/4" diameter	1 Plum	54	.148	L.F.	23.50	9.55		33.05	40
4300	3/8" diameter		53	.151		25.50	9.75		35.25	42.50
4310	1/2" diameter		52	.154		28.50	9.90		38.40	46.50
4320	3/4" diameter		51	.157		32.50	10.10		42.60	50.50
4330	1" diameter	▼	45	.178		45	11.45		56.45	66.50
4340	1-1/4" diameter	Q-1	76	.211		74	12.20		86.20	99.50
4350	1-1/2" diameter		69	.232		67.50	13.45		80.95	94
4360	2" diameter		57	.281		90.50	16.30		106.80	124
4370	2-1/2" diameter		44	.364		130	21		151	175
4380	3" diameter	▼	38	.421	▼	152	24.50		176.50	204

22 11 13 - Facility Water Distribution Piping

22 11 13.64 Pipe, Stainless Steel	Crew	Daily Output	Labor-Hours	Unit	Material	2020 Bare Costs Labor	Equipment	Total	Total Incl O&P
4390	4" diameter	Q-2	51	.471	L.F.	187	28.50		215.50	249
4490	For small quantities, add					10%				
4900	To delete couplings & hangers, subtract									
4910	1/4" diam. to 3/4" diam.					12%	56%			
4920	1" diam. to 2" diam.					14%	49%			
4930	2-1/2" diam. to 4" diam.					27%	40%			
5000	Schedule 80, type 304									
5040	1/4" diameter	1 Plum	53	.151	L.F.	21	9.75		30.75	37.50
5050	3/8" diameter		52	.154		37	9.90		46.90	56
5060	1/2" diameter		51	.157		31.50	10.10		41.60	49.50
5070	3/4" diameter		48	.167		38	10.75		48.75	57.50
5080	1" diameter		43	.186		55	12		67	78.50
5090	1-1/4" diameter	Q-1	73	.219		59	12.70		71.70	83.50
5100	1-1/2" diameter		67	.239		67.50	13.85		81.35	95
5110	2" diameter		54	.296		108	17.20		125.20	145
5190	For small quantities, add					10%				
5700	To delete couplings & hangers, subtract									
5710	1/4" diam. to 3/4" diam.					10%	53%			
5720	1" diam. to 2" diam.					14%	47%			
5750	Schedule 80, type 316									
5790	1/4" diameter	1 Plum	53	.151	L.F.	42.50	9.75		52.25	61.50
5800	3/8" diameter		52	.154		42.50	9.90		52.40	61.50
5810	1/2" diameter		51	.157		54.50	10.10		64.60	75
5820	3/4" diameter		48	.167		54	10.75		64.75	75.50
5830	1" diameter		43	.186		51.50	12		63.50	75
5840	1-1/4" diameter	Q-1	73	.219		77.50	12.70		90.20	104
5850	1-1/2" diameter		67	.239		104	13.85		117.85	135
5860	2" diameter		54	.296		117	17.20		134.20	154
5950	For small quantities, add					10%				
7000	To delete couplings & hangers, subtract									
7010	1/4" diam. to 3/4" diam.					9%	53%			
7020	1" diam. to 2" diam.					14%	47%			
8000	Weld joints with clevis type hanger assemblies, 10' OC									
8010	Schedule 40, type 304									
8050	1/8" pipe size	Q-15	126	.127	L.F.	7.55	7.35	.84	15.74	20
8060	1/4" pipe size		125	.128		8.85	7.40	.85	17.10	22
8070	3/8" pipe size		122	.131		9.55	7.60	.87	18.02	23
8080	1/2" pipe size		118	.136		10.25	7.85	.90	19	24
8090	3/4" pipe size		109	.147		17.70	8.50	.98	27.18	33.50
8100	1" pipe size		95	.168		19.80	9.75	1.12	30.67	38
8110	1-1/4" pipe size		86	.186		30.50	10.80	1.24	42.54	51
8120	1-1/2" pipe size		78	.205		32	11.90	1.36	45.26	54.50
8130	2" pipe size		62	.258		38.50	14.95	1.72	55.17	67
8140	2-1/2" pipe size		49	.327		66	18.95	2.17	87.12	103
8150	3" pipe size		44	.364		69.50	21	2.42	92.92	111
8160	3-1/2" pipe size		44	.364		88	21	2.42	111.42	131
8170	4" pipe size		39	.410		71	24	2.73	97.73	117
8180	5" pipe size		32	.500		121	29	3.33	153.33	180
8190	6" pipe size	Q-16	37	.649		160	39	2.87	201.87	238
8191	6" pipe size		37	.649		157	39	2.87	198.87	234
8200	8" pipe size		29	.828		164	50	3.67	217.67	260
8210	10" pipe size		24	1		273	60	4.43	337.43	395
8220	12" pipe size		20	1.200		515	72	5.30	592.30	680

22 11 13.64 Pipe, Stainless Steel	Crew	Daily Output	Labor-Hours	Unit	Material	2020 Bare Costs Labor	Equipment	Total	Total Incl O&P
8300	Schedule 40, type 316								
8310	1/8" pipe size	Q-15	126	.127	L.F.	20	7.35	.84	28.19	34
8320	1/4" pipe size		125	.128		22	7.40	.85	30.25	36
8330	3/8" pipe size		122	.131		23.50	7.60	.87	31.97	38.50
8340	1/2" pipe size		118	.136		26	7.85	.90	34.75	41
8350	3/4" pipe size		109	.147		29	8.50	.98	38.48	46
8360	1" pipe size		95	.168		39.50	9.75	1.12	50.37	59.50
8370	1-1/4" pipe size		86	.186		63	10.80	1.24	75.04	87
8380	1-1/2" pipe size		78	.205		57.50	11.90	1.36	70.76	83
8390	2" pipe size		62	.258		74	14.95	1.72	90.67	106
8400	2-1/2" pipe size		49	.327		91.50	18.95	2.17	112.62	132
8410	3" pipe size		44	.364		99.50	21	2.42	122.92	143
8420	3-1/2" pipe size		44	.364		109	21	2.42	132.42	154
8430	4" pipe size		39	.410		114	24	2.73	140.73	165
8440	5" pipe size	▼	32	.500		180	29	3.33	212.33	245
8450	6" pipe size	Q-16	37	.649		263	39	2.87	304.87	350
8460	8" pipe size		29	.828		271	50	3.67	324.67	375
8470	10" pipe size		24	1		335	60	4.43	399.43	465
8480	12" pipe size	▼	20	1.200	▼	455	72	5.30	532.30	615
8500	Schedule 80, type 304								
8510	1/4" pipe size	Q-15	110	.145	L.F.	19.40	8.45	.97	28.82	35
8520	3/8" pipe size		109	.147		35	8.50	.98	44.48	53
8530	1/2" pipe size		106	.151		29.50	8.75	1	39.25	46.50
8540	3/4" pipe size		96	.167		35	9.65	1.11	45.76	54
8550	1" pipe size		87	.184		50.50	10.65	1.22	62.37	73
8560	1-1/4" pipe size		81	.198		49.50	11.45	1.31	62.26	73
8570	1-1/2" pipe size		74	.216		55	12.55	1.44	68.99	80.50
8580	2" pipe size		58	.276		93.50	16	1.83	111.33	129
8590	2-1/2" pipe size		46	.348		104	20	2.31	126.31	147
8600	3" pipe size		41	.390		123	22.50	2.60	148.10	172
8610	4" pipe size	▼	33	.485		154	28	3.22	185.22	215
8630	6" pipe size	Q-16	30	.800	▼	279	48	3.54	330.54	380
8640	Schedule 80, type 316								
8650	1/4" pipe size	Q-15	110	.145	L.F.	40.50	8.45	.97	49.92	58
8660	3/8" pipe size		109	.147		40.50	8.50	.98	49.98	58.50
8670	1/2" pipe size		106	.151		52	8.75	1	61.75	71.50
8680	3/4" pipe size		96	.167		51	9.65	1.11	61.76	71.50
8690	1" pipe size		87	.184		45.50	10.65	1.22	57.37	67.50
8700	1-1/4" pipe size		81	.198		63.50	11.45	1.31	76.26	88
8710	1-1/2" pipe size		74	.216		87	12.55	1.44	100.99	116
8720	2" pipe size		58	.276		94	16	1.83	111.83	129
8730	2-1/2" pipe size		46	.348		95.50	20	2.31	117.81	138
8740	3" pipe size		41	.390		148	22.50	2.60	173.10	200
8760	4" pipe size	▼	33	.485	▼	195	28	3.22	226.22	260
8770	6" pipe size	Q-16	30	.800	▼	340	48	3.54	391.54	450
9100	Threading pipe labor, sst, one end, schedules 40 & 80								
9110	1/4" through 3/4" pipe size	1 Plum	61.50	.130	Ea.		8.40		8.40	12.55
9120	1" through 2" pipe size		55.90	.143			9.20		9.20	13.80
9130	2-1/2" pipe size		41.50	.193			12.40		12.40	18.60
9140	3" pipe size		38.50	.208			13.40		13.40	20
9150	3-1/2" pipe size	Q-1	68.40	.234			13.55		13.55	20.50
9160	4" pipe size		73	.219			12.70		12.70	19
9170	5" pipe size	▼	40.70	.393	▼		23		23	34

22 11 Facility Water Distribution

22 11 13 - Facility Water Distribution Piping

22 11 13.64 Pipe, Stainless Steel		Crew	Daily Output	Labor-Hours	Unit	Material	2020 Bare Costs Labor	Equipment	Total	Total Incl O&P
9180	6" pipe size	Q-1	35.40	.452	Ea.		26		26	39
9190	8" pipe size		22.30	.717			41.50		41.50	62.50
9200	10" pipe size		16.10	.994			57.50		57.50	86.50
9210	12" pipe size		12.30	1.301			75.50		75.50	113
9250	Welding labor per joint for stainless steel									
9260	Schedule 5 and 10									
9270	1/4" pipe size	Q-15	36	.444	Ea.		26	2.96	28.96	42
9280	3/8" pipe size		35	.457			26.50	3.04	29.54	43
9290	1/2" pipe size		35	.457			26.50	3.04	29.54	43
9300	3/4" pipe size		28	.571			33	3.80	36.80	53.50
9310	1" pipe size		25	.640			37	4.26	41.26	60
9320	1-1/4" pipe size		22	.727			42	4.84	46.84	68.50
9330	1-1/2" pipe size		21	.762			44	5.05	49.05	71.50
9340	2" pipe size		18	.889			51.50	5.90	57.40	83.50
9350	2-1/2" pipe size		12	1.333			77.50	8.85	86.35	126
9360	3" pipe size		9.73	1.644			95.50	10.95	106.45	155
9370	4" pipe size		7.37	2.171			126	14.45	140.45	204
9380	5" pipe size		6.15	2.602			151	17.30	168.30	245
9390	6" pipe size		5.71	2.802			163	18.65	181.65	264
9400	8" pipe size		3.69	4.336			251	29	280	405
9410	10" pipe size		2.91	5.498			320	36.50	356.50	515
9420	12" pipe size		2.31	6.926			400	46	446	650
9500	Schedule 40									
9510	1/4" pipe size	Q-15	28	.571	Ea.		33	3.80	36.80	53.50
9520	3/8" pipe size		27	.593			34.50	3.94	38.44	56
9530	1/2" pipe size		25.40	.630			36.50	4.19	40.69	59
9540	3/4" pipe size		22.22	.720			42	4.79	46.79	68
9550	1" pipe size		20.25	.790			46	5.25	51.25	74.50
9560	1-1/4" pipe size		18.82	.850			49.50	5.65	55.15	80
9570	1-1/2" pipe size		17.78	.900			52	6	58	84.50
9580	2" pipe size		15.09	1.060			61.50	7.05	68.55	100
9590	2-1/2" pipe size		7.96	2.010			117	13.35	130.35	189
9600	3" pipe size		6.43	2.488			144	16.55	160.55	234
9610	4" pipe size		4.88	3.279			190	22	212	310
9620	5" pipe size		4.26	3.756			218	25	243	355
9630	6" pipe size		3.77	4.244			246	28	274	400
9640	8" pipe size		2.44	6.557			380	43.50	423.50	620
9650	10" pipe size		1.92	8.333			485	55.50	540.50	785
9660	12" pipe size		1.52	10.526			610	70	680	990
9750	Schedule 80									
9760	1/4" pipe size	Q-15	21.55	.742	Ea.		43	4.94	47.94	70
9770	3/8" pipe size		20.75	.771			44.50	5.15	49.65	72.50
9780	1/2" pipe size		19.54	.819			47.50	5.45	52.95	77
9790	3/4" pipe size		17.09	.936			54.50	6.25	60.75	88.50
9800	1" pipe size		15.58	1.027			59.50	6.85	66.35	96.50
9810	1-1/4" pipe size		14.48	1.105			64	7.35	71.35	104
9820	1-1/2" pipe size		13.68	1.170			68	7.80	75.80	111
9830	2" pipe size		11.61	1.378			80	9.15	89.15	130
9840	2-1/2" pipe size		6.12	2.614			152	17.40	169.40	246
9850	3" pipe size		4.94	3.239			188	21.50	209.50	305
9860	4" pipe size		3.75	4.267			247	28.50	275.50	400
9870	5" pipe size		3.27	4.893			284	32.50	316.50	460
9880	6" pipe size		2.90	5.517			320	36.50	356.50	520

For customer support on your Mechanical Costs with RSMeans Data, call 800.448.8182.

22 11 13.64 Pipe, Stainless Steel

		Crew	Daily Output	Labor-Hours	Unit	Material	2020 Bare Costs Labor	2020 Bare Costs Equipment	Total	Total Incl O&P
9890	8" pipe size	Q-15	1.87	8.556	Ea.		495	57	552	810
9900	10" pipe size		1.48	10.811			625	72	697	1,025
9910	12" pipe size		1.17	13.675			795	91	886	1,275
9920	Schedule 160, 1/2" pipe size		17	.941			54.50	6.25	60.75	88.50
9930	3/4" pipe size		14.81	1.080			62.50	7.20	69.70	102
9940	1" pipe size		13.50	1.185			68.50	7.90	76.40	112
9950	1-1/4" pipe size		12.55	1.275			74	8.50	82.50	120
9960	1-1/2" pipe size		11.85	1.350			78.50	9	87.50	127
9970	2" pipe size		10	1.600			93	10.65	103.65	151
9980	3" pipe size		4.28	3.738			217	25	242	355
9990	4" pipe size		3.25	4.923			286	32.50	318.50	460

22 11 13.66 Pipe Fittings, Stainless Steel

		Crew	Daily Output	Labor-Hours	Unit	Material	2020 Bare Costs Labor	2020 Bare Costs Equipment	Total	Total Incl O&P
0010	**PIPE FITTINGS, STAINLESS STEEL**									
0100	Butt weld joint, schedule 5, type 304									
0120	90° elbow, long									
0140	1/2"	Q-15	17.50	.914	Ea.	18.05	53	6.10	77.15	106
0150	3/4"		14	1.143		18.05	66.50	7.60	92.15	127
0160	1"		12.50	1.280		19.05	74	8.50	101.55	141
0170	1-1/4"		11	1.455		26	84.50	9.65	120.15	165
0180	1-1/2"		10.50	1.524		22	88.50	10.15	120.65	168
0190	2"		9	1.778		26	103	11.80	140.80	196
0200	2-1/2"		6	2.667		60	155	17.75	232.75	315
0210	3"		4.86	3.292		56	191	22	269	370
0220	3-1/2"		4.27	3.747		165	217	25	407	535
0230	4"		3.69	4.336		94	251	29	374	510
0240	5"		3.08	5.195		350	300	34.50	684.50	875
0250	6"	Q-16	4.29	5.594		271	335	25	631	830
0260	8"		2.76	8.696		570	525	38.50	1,133.50	1,450
0270	10"		2.18	11.009		880	660	49	1,589	2,025
0280	12"		1.73	13.873		1,250	835	61.50	2,146.50	2,700
0320	For schedule 5, type 316, add					30%				
0600	45° elbow, long									
0620	1/2"	Q-15	17.50	.914	Ea.	18.05	53	6.10	77.15	106
0630	3/4"		14	1.143		18.05	66.50	7.60	92.15	127
0640	1"		12.50	1.280		19.05	74	8.50	101.55	141
0650	1-1/4"		11	1.455		26	84.50	9.65	120.15	165
0660	1-1/2"		10.50	1.524		22	88.50	10.15	120.65	168
0670	2"		9	1.778		26	103	11.80	140.80	196
0680	2-1/2"		6	2.667		60	155	17.75	232.75	315
0690	3"		4.86	3.292		45	191	22	258	360
0700	3-1/2"		4.27	3.747		165	217	25	407	535
0710	4"		3.69	4.336		76	251	29	356	490
0720	5"		3.08	5.195		281	300	34.50	615.50	800
0730	6"	Q-16	4.29	5.594		190	335	25	550	745
0740	8"		2.76	8.696		400	525	38.50	963.50	1,275
0750	10"		2.18	11.009		700	660	49	1,409	1,825
0760	12"		1.73	13.873		875	835	61.50	1,771.50	2,275
0800	For schedule 5, type 316, add					25%				
1100	Tee, straight									
1130	1/2"	Q-15	11.66	1.372	Ea.	54	79.50	9.15	142.65	189
1140	3/4"		9.33	1.715		54	99.50	11.40	164.90	221
1150	1"		8.33	1.921		57	111	12.75	180.75	244

For customer support on your Mechanical Costs with RSMeans Data, call 800.448.8182.

189

22 11 13.66 Pipe Fittings, Stainless Steel		Crew	Daily Output	Labor-Hours	Unit	Material	2020 Bare Costs Labor	Equipment	Total	Total Incl O&P
1160	1-1/4"	Q-15	7.33	2.183	Ea.	46	127	14.50	187.50	256
1170	1-1/2"		7	2.286		45	133	15.20	193.20	264
1180	2"		6	2.667		47	155	17.75	219.75	305
1190	2-1/2"		4	4		113	232	26.50	371.50	500
1200	3"		3.24	4.938		86	286	33	405	560
1210	3-1/2"		2.85	5.614		237	325	37.50	599.50	785
1220	4"		2.46	6.504		128	375	43.50	546.50	755
1230	5"		2	8		405	465	53	923	1,200
1240	6"	Q-16	2.85	8.421		320	505	37.50	862.50	1,150
1250	8"		1.84	13.043		680	785	58	1,523	2,000
1260	10"		1.45	16.552		1,100	995	73.50	2,168.50	2,800
1270	12"		1.15	20.870		1,525	1,250	92.50	2,867.50	3,650
1320	For schedule 5, type 316, add					25%				
2000	Butt weld joint, schedule 10, type 304									
2020	90° elbow, long									
2040	1/2"	Q-15	17	.941	Ea.	16.55	54.50	6.25	77.30	107
2050	3/4"		14	1.143		12.75	66.50	7.60	86.85	121
2060	1"		12.50	1.280		14.65	74	8.50	97.15	137
2070	1-1/4"		11	1.455		24	84.50	9.65	118.15	163
2080	1-1/2"		10.50	1.524		17	88.50	10.15	115.65	162
2090	2"		9	1.778		20	103	11.80	134.80	189
2100	2-1/2"		6	2.667		55	155	17.75	227.75	310
2110	3"		4.86	3.292		51.50	191	22	264.50	365
2120	3-1/2"		4.27	3.747		152	217	25	394	520
2130	4"		3.69	4.336		72.50	251	29	352.50	485
2140	5"		3.08	5.195		320	300	34.50	654.50	845
2150	6"	Q-16	4.29	5.594		208	335	25	568	760
2160	8"		2.76	8.696		440	525	38.50	1,003.50	1,325
2170	10"		2.18	11.009		680	660	49	1,389	1,800
2180	12"		1.73	13.873		965	835	61.50	1,861.50	2,375
2500	45° elbow, long									
2520	1/2"	Q-15	17.50	.914	Ea.	16.55	53	6.10	75.65	104
2530	3/4"		14	1.143		16.55	66.50	7.60	90.65	126
2540	1"		12.50	1.280		14.65	74	8.50	97.15	137
2550	1-1/4"		11	1.455		24	84.50	9.65	118.15	163
2560	1-1/2"		10.50	1.524		17	88.50	10.15	115.65	162
2570	2"		9	1.778		20	103	11.80	134.80	189
2580	2-1/2"		6	2.667		55	155	17.75	227.75	310
2590	3"		4.86	3.292		41.50	191	22	254.50	355
2600	3-1/2"		4.27	3.747		152	217	25	394	520
2610	4"		3.69	4.336		58.50	251	29	338.50	470
2620	5"		3.08	5.195		257	300	34.50	591.50	770
2630	6"	Q-16	4.29	5.594		147	335	25	507	695
2640	8"		2.76	8.696		310	525	38.50	873.50	1,175
2650	10"		2.18	11.009		540	660	49	1,249	1,650
2660	12"		1.73	13.873		675	835	61.50	1,571.50	2,075
2670	Reducer, concentric									
2674	1" x 3/4"	Q-15	13.25	1.208	Ea.	31	70	8.05	109.05	148
2676	2" x 1-1/2"	"	9.75	1.641		25.50	95	10.90	131.40	182
2678	6" x 4"	Q-16	4.91	4.888		76.50	294	21.50	392	550
2680	Caps									
2682	1"	Q-15	25	.640	Ea.	32	37	4.26	73.26	95
2684	1-1/2"		21	.762		38.50	44	5.05	87.55	114

22 11 13.66 Pipe Fittings, Stainless Steel		Crew	Daily Output	Labor-Hours	Unit	Material	2020 Bare Costs Labor	Equipment	Total	Total Incl O&P
2685	2"	Q-15	18	.889	Ea.	36	51.50	5.90	93.40	123
2686	4"	↓	7.38	2.168		57.50	126	14.40	197.90	267
2687	6"	Q-16	8.58	2.797	↓	92	168	12.40	272.40	365
3000	Tee, straight									
3030	1/2"	Q-15	11.66	1.372	Ea.	49.50	79.50	9.15	138.15	184
3040	3/4"		9.33	1.715		49.50	99.50	11.40	160.40	216
3050	1"		8.33	1.921		44	111	12.75	167.75	230
3060	1-1/4"		7.33	2.183		55.50	127	14.50	197	266
3070	1-1/2"		7	2.286		47	133	15.20	195.20	267
3080	2"		6	2.667		49.50	155	17.75	222.25	305
3090	2-1/2"		4	4		63	232	26.50	321.50	445
3100	3"		3.24	4.938		82	286	33	401	555
3110	3-1/2"		2.85	5.614		121	325	37.50	483.50	660
3120	4"		2.46	6.504		99	375	43.50	517.50	720
3130	5"	↓	2	8		375	465	53	893	1,175
3140	6"	Q-16	2.85	8.421		247	505	37.50	789.50	1,075
3150	8"		1.84	13.043		525	785	58	1,368	1,825
3151	10"		1.45	16.552		850	995	73.50	1,918.50	2,525
3152	12"	↓	1.15	20.870		1,175	1,250	92.50	2,517.50	3,275
3154	For schedule 10, type 316, add				↓	25%				
3281	Butt weld joint, schedule 40, type 304									
3284	90° elbow, long, 1/2"	Q-15	12.70	1.260	Ea.	19.30	73	8.40	100.70	139
3288	3/4"		11.10	1.441		19.30	83.50	9.60	112.40	157
3289	1"		10.13	1.579		20	91.50	10.50	122	171
3290	1-1/4"		9.40	1.702		26.50	98.50	11.30	136.30	190
3300	1-1/2"		8.89	1.800		21	104	11.95	136.95	192
3310	2"		7.55	2.119		30.50	123	14.10	167.60	233
3320	2-1/2"		3.98	4.020		58	233	26.50	317.50	445
3330	3"		3.21	4.984		74.50	289	33	396.50	555
3340	3-1/2"		2.83	5.654		276	330	37.50	643.50	835
3350	4"		2.44	6.557		129	380	43.50	552.50	760
3360	5"	↓	2.13	7.512		430	435	50	915	1,175
3370	6"	Q-16	2.83	8.481		375	510	37.50	922.50	1,225
3380	8"		1.83	13.115		635	790	58	1,483	1,925
3390	10"		1.44	16.667		1,325	1,000	74	2,399	3,050
3400	12"	↓	1.14	21.053	↓	1,700	1,275	93.50	3,068.50	3,875
3410	For schedule 40, type 316, add					25%				
3460	45° elbow, long, 1/2"	Q-15	12.70	1.260	Ea.	19.30	73	8.40	100.70	139
3470	3/4"		11.10	1.441		19.30	83.50	9.60	112.40	157
3480	1"		10.13	1.579		20	91.50	10.50	122	171
3490	1-1/4"		9.40	1.702		26.50	98.50	11.30	136.30	190
3500	1-1/2"		8.89	1.800		21	104	11.95	136.95	192
3510	2"		7.55	2.119		30.50	123	14.10	167.60	233
3520	2-1/2"		3.98	4.020		58	233	26.50	317.50	445
3530	3"		3.21	4.984		59	289	33	381	535
3540	3-1/2"		2.83	5.654		276	330	37.50	643.50	835
3550	4"		2.44	6.557		92	380	43.50	515.50	720
3560	5"	↓	2.13	7.512		305	435	50	790	1,050
3570	6"	Q-16	2.83	8.481		266	510	37.50	813.50	1,100
3580	8"		1.83	13.115		445	790	58	1,293	1,725
3590	10"		1.44	16.667		935	1,000	74	2,009	2,600
3600	12"	↓	1.14	21.053		1,200	1,275	93.50	2,568.50	3,300
3610	For schedule 40, type 316, add					25%				

For customer support on your Mechanical Costs with RSMeans Data, call 800.448.8182.

191

22 11 13 – Facility Water Distribution Piping

22 11 13.66 Pipe Fittings, Stainless Steel	Crew	Daily Output	Labor-Hours	Unit	Material	2020 Bare Costs Labor	Equipment	Total	Total Incl O&P
3660	Tee, straight 1/2"	Q-15	8.46	1.891	Ea.	49.50	110	12.60	172.10	232
3670	3/4"		7.40	2.162		49.50	125	14.40	188.90	258
3680	1"		6.74	2.374		52.50	138	15.80	206.30	281
3690	1-1/4"		6.27	2.552		122	148	16.95	286.95	375
3700	1-1/2"		5.92	2.703		56	157	17.95	230.95	315
3710	2"		5.03	3.181		83.50	184	21	288.50	390
3720	2-1/2"		2.65	6.038		106	350	40	496	685
3730	3"		2.14	7.477		107	435	49.50	591.50	820
3740	3-1/2"		1.88	8.511		201	495	56.50	752.50	1,025
3750	4"		1.62	9.877		201	575	65.50	841.50	1,150
3760	5"	▼	1.42	11.268		465	655	75	1,195	1,575
3770	6"	Q-16	1.88	12.766		375	770	56.50	1,201.50	1,625
3780	8"		1.22	19.672		755	1,175	87	2,017	2,700
3790	10"		.96	25		1,475	1,500	111	3,086	4,000
3800	12"	▼	.76	31.579	▼	1,950	1,900	140	3,990	5,150
3810	For schedule 40, type 316, add					25%				
3820	Tee, reducing on outlet, 3/4" x 1/2"	Q-15	7.73	2.070	Ea.	57	120	13.75	190.75	258
3822	1" x 1/2"		7.24	2.210		71.50	128	14.70	214.20	287
3824	1" x 3/4"		6.96	2.299		66	133	15.30	214.30	289
3826	1-1/4" x 1"		6.43	2.488		162	144	16.55	322.55	410
3828	1-1/2" x 1/2"		6.58	2.432		97.50	141	16.15	254.65	335
3830	1-1/2" x 3/4"		6.35	2.520		91	146	16.75	253.75	335
3832	1-1/2" x 1"		6.18	2.589		70	150	17.20	237.20	320
3834	2" x 1"		5.50	2.909		125	169	19.35	313.35	410
3836	2" x 1-1/2"		5.30	3.019		105	175	20	300	400
3838	2-1/2" x 2"		3.15	5.079		158	295	34	487	650
3840	3" x 1-1/2"		2.72	5.882		160	340	39	539	730
3842	3" x 2"		2.65	6.038		133	350	40	523	715
3844	4" x 2"		2.10	7.619		335	440	50.50	825.50	1,075
3846	4" x 3"		1.77	9.040		242	525	60	827	1,125
3848	5" x 4"	▼	1.48	10.811		560	625	72	1,257	1,625
3850	6" x 3"	Q-16	2.19	10.959		615	660	48.50	1,323.50	1,725
3852	6" x 4"		2.04	11.765		535	710	52	1,297	1,700
3854	8" x 4"		1.46	16.438		1,250	990	73	2,313	2,925
3856	10" x 8"		.69	34.783		2,100	2,100	154	4,354	5,600
3858	12" x 10"	▼	.55	43.636		2,800	2,625	193	5,618	7,225
3950	Reducer, concentric, 3/4" x 1/2"	Q-15	11.85	1.350		36.50	78.50	9	124	167
3952	1" x 3/4"		10.60	1.509		38.50	87.50	10.05	136.05	185
3954	1-1/4" x 3/4"		10.19	1.570		94.50	91	10.45	195.95	252
3956	1-1/4" x 1"		9.76	1.639		48	95	10.90	153.90	207
3958	1-1/2" x 3/4"		9.88	1.619		80	94	10.75	184.75	241
3960	1-1/2" x 1"		9.47	1.690		64.50	98	11.25	173.75	230
3962	2" x 1"		8.65	1.850		32.50	107	12.30	151.80	211
3964	2" x 1-1/2"		8.16	1.961		32	114	13.05	159.05	220
3966	2-1/2" x 1"		5.71	2.802		139	163	18.65	320.65	415
3968	2-1/2" x 2"		5.21	3.071		70.50	178	20.50	269	370
3970	3" x 1"		4.88	3.279		95.50	190	22	307.50	415
3972	3" x 1-1/2"		4.72	3.390		48.50	197	22.50	268	375
3974	3" x 2"		4.51	3.548		46	206	23.50	275.50	385
3976	4" x 2"		3.69	4.336		65	251	29	345	480
3978	4" x 3"		2.77	5.776		48.50	335	38.50	422	595
3980	5" x 3"		2.56	6.250		290	365	41.50	696.50	910
3982	5" x 4"	▼	2.27	7.048	▼	234	410	47	691	920

For customer support on your Mechanical Costs with RSMeans Data, call 800.448.8182.

22 11 13 – Facility Water Distribution Piping

22 11 13.66 Pipe Fittings, Stainless Steel	Crew	Daily Output	Labor-Hours	Unit	Material	2020 Bare Costs Labor	Equipment	Total	Total Incl O&P
3984	6" x 3"	Q-16	3.57	6.723	Ea.	150	405	30	585	805
3986	6" x 4"		3.19	7.524		123	455	33.50	611.50	845
3988	8" x 4"		2.44	9.836		325	590	43.50	958.50	1,300
3990	8" x 6"		2.22	10.811		242	650	48	940	1,300
3992	10" x 6"		1.91	12.565		450	755	55.50	1,260.50	1,675
3994	10" x 8"		1.61	14.907		375	895	66	1,336	1,850
3995	12" x 6"		1.63	14.724		825	885	65	1,775	2,300
3996	12" x 8"		1.41	17.021		640	1,025	75.50	1,740.50	2,300
3997	12" x 10"		1.27	18.898		450	1,125	83.50	1,658.50	2,275
4000	Socket weld joint, 3,000 lb., type 304									
4100	90° elbow									
4140	1/4"	Q-15	13.47	1.188	Ea.	45	69	7.90	121.90	161
4150	3/8"		12.97	1.234		59	71.50	8.20	138.70	181
4160	1/2"		12.21	1.310		63	76	8.70	147.70	193
4170	3/4"		10.68	1.498		74	87	9.95	170.95	222
4180	1"		9.74	1.643		111	95.50	10.90	217.40	278
4190	1-1/4"		9.05	1.768		194	103	11.75	308.75	380
4200	1-1/2"		8.55	1.871		236	109	12.45	357.45	435
4210	2"		7.26	2.204		380	128	14.65	522.65	625
4300	45° elbow									
4340	1/4"	Q-15	13.47	1.188	Ea.	85	69	7.90	161.90	205
4350	3/8"		12.97	1.234		85	71.50	8.20	164.70	210
4360	1/2"		12.21	1.310		85	76	8.70	169.70	217
4370	3/4"		10.68	1.498		96.50	87	9.95	193.45	247
4380	1"		9.74	1.643		140	95.50	10.90	246.40	310
4390	1-1/4"		9.05	1.768		230	103	11.75	344.75	420
4400	1-1/2"		8.55	1.871		231	109	12.45	352.45	430
4410	2"		7.26	2.204		420	128	14.65	562.65	665
4500	Tee									
4540	1/4"	Q-15	8.97	1.784	Ea.	60	103	11.85	174.85	234
4550	3/8"		8.64	1.852		72	107	12.30	191.30	254
4560	1/2"		8.13	1.968		88	114	13.10	215.10	282
4570	3/4"		7.12	2.247		102	130	14.95	246.95	325
4580	1"		6.48	2.469		137	143	16.40	296.40	385
4590	1-1/4"		6.03	2.653		243	154	17.65	414.65	515
4600	1-1/2"		5.69	2.812		350	163	18.70	531.70	650
4610	2"		4.83	3.313		530	192	22	744	895
5000	Socket weld joint, 3,000 lb., type 316									
5100	90° elbow									
5140	1/4"	Q-15	13.47	1.188	Ea.	55.50	69	7.90	132.40	173
5150	3/8"		12.97	1.234		64.50	71.50	8.20	144.20	187
5160	1/2"		12.21	1.310		78	76	8.70	162.70	209
5170	3/4"		10.68	1.498		103	87	9.95	199.95	254
5180	1"		9.74	1.643		146	95.50	10.90	252.40	315
5190	1-1/4"		9.05	1.768		260	103	11.75	374.75	450
5200	1-1/2"		8.55	1.871		293	109	12.45	414.45	500
5210	2"		7.26	2.204		500	128	14.65	642.65	755
5300	45° elbow									
5340	1/4"	Q-15	13.47	1.188	Ea.	110	69	7.90	186.90	233
5350	3/8"		12.97	1.234		110	71.50	8.20	189.70	237
5360	1/2"		12.21	1.310		112	76	8.70	196.70	247
5370	3/4"		10.68	1.498		124	87	9.95	220.95	277
5380	1"		9.74	1.643		187	95.50	10.90	293.40	360

For customer support on your Mechanical Costs with RSMeans Data, call 800.448.8182.

193

22 11 13.66 Pipe Fittings, Stainless Steel	Crew	Daily Output	Labor-Hours	Unit	Material	2020 Bare Costs Labor	Equipment	Total	Total Incl O&P
5390	1-1/4"	Q-15	9.05	1.768	Ea.	267	103	11.75	381.75	460
5400	1-1/2"		8.55	1.871		300	109	12.45	421.45	505
5410	2"		7.26	2.204		450	128	14.65	592.65	695
5500	Tee									
5540	1/4"	Q-15	8.97	1.784	Ea.	75	103	11.85	189.85	251
5550	3/8"		8.64	1.852		91.50	107	12.30	210.80	276
5560	1/2"		8.13	1.968		102	114	13.10	229.10	297
5570	3/4"		7.12	2.247		126	130	14.95	270.95	350
5580	1"		6.48	2.469		193	143	16.40	352.40	445
5590	1-1/4"		6.03	2.653		305	154	17.65	476.65	590
5600	1-1/2"		5.69	2.812		430	163	18.70	611.70	740
5610	2"		4.83	3.313		690	192	22	904	1,075
5700	For socket weld joint, 6,000 lb., type 304 and 316, add					100%				
6000	Threaded companion flange									
6010	Stainless steel, 150 lb., type 304									
6020	1/2" diam.	1 Plum	30	.267	Ea.	46	17.20		63.20	76
6030	3/4" diam.		28	.286		51	18.40		69.40	83.50
6040	1" diam.		27	.296		56	19.10		75.10	90
6050	1-1/4" diam.	Q-1	44	.364		72.50	21		93.50	111
6060	1-1/2" diam.		40	.400		72.50	23		95.50	114
6070	2" diam.		36	.444		94.50	26		120.50	143
6080	2-1/2" diam.		28	.571		132	33		165	196
6090	3" diam.		20	.800		139	46.50		185.50	222
6110	4" diam.		12	1.333		189	77.50		266.50	325
6130	6" diam.	Q-2	14	1.714		340	103		443	530
6140	8" diam.	"	12	2		655	120		775	900
6150	For type 316, add					40%				
6260	Weld flanges, stainless steel, type 304									
6270	Slip on, 150 lb. (welded, front and back)									
6280	1/2" diam.	Q-15	12.70	1.260	Ea.	41	73	8.40	122.40	163
6290	3/4" diam.		11.11	1.440		42	83.50	9.60	135.10	182
6300	1" diam.		10.13	1.579		46.50	91.50	10.50	148.50	200
6310	1-1/4" diam.		9.41	1.700		62.50	98.50	11.30	172.30	229
6320	1-1/2" diam.		8.89	1.800		62.50	104	11.95	178.45	238
6330	2" diam.		7.55	2.119		81	123	14.10	218.10	289
6340	2-1/2" diam.		3.98	4.020		113	233	26.50	372.50	505
6350	3" diam.		3.21	4.984		122	289	33	444	605
6370	4" diam.		2.44	6.557		166	380	43.50	589.50	800
6390	6" diam.	Q-16	1.89	12.698		252	765	56.50	1,073.50	1,500
6400	8" diam.	"	1.22	19.672		475	1,175	87	1,737	2,400
6410	For type 316, add					40%				
6530	Weld neck 150 lb.									
6540	1/2" diam.	Q-15	25.40	.630	Ea.	21.50	36.50	4.19	62.19	83
6550	3/4" diam.		22.22	.720		26	42	4.79	72.79	96.50
6560	1" diam.		20.25	.790		29.50	46	5.25	80.75	106
6570	1-1/4" diam.		18.82	.850		41.50	49.50	5.65	96.65	126
6580	1-1/2" diam.		17.78	.900		41.50	52	6	99.50	130
6590	2" diam.		15.09	1.060		46.50	61.50	7.05	115.05	151
6600	2-1/2" diam.		7.96	2.010		75.50	117	13.35	205.85	272
6610	3" diam.		6.43	2.488		76	144	16.55	236.55	320
6630	4" diam.		4.88	3.279		110	190	22	322	430
6640	5" diam.		4.26	3.756		139	218	25	382	505
6650	6" diam.	Q-16	5.66	4.240		163	255	18.80	436.80	580

22 11 13.66 Pipe Fittings, Stainless Steel		Crew	Daily Output	Labor-Hours	Unit	Material	2020 Bare Costs Labor	Equipment	Total	Total Incl O&P
6652	8" diam.	Q-16	3.65	6.575	Ea.	286	395	29	710	935
6654	10" diam.		2.88	8.333		400	500	37	937	1,225
6656	12" diam.		2.28	10.526		740	635	46.50	1,421.50	1,825
6670	For type 316, add					23%				
7000	Threaded joint, 150 lb., type 304									
7030	90° elbow									
7040	1/8"	1 Plum	13	.615	Ea.	29.50	39.50		69	92
7050	1/4"		13	.615		29.50	39.50		69	92
7070	3/8"		13	.615		34.50	39.50		74	97.50
7080	1/2"		12	.667		30.50	43		73.50	98
7090	3/4"		11	.727		36.50	47		83.50	111
7100	1"		10	.800		50.50	51.50		102	133
7110	1-1/4"	Q-1	17	.941		78.50	54.50		133	168
7120	1-1/2"		16	1		90	58		148	186
7130	2"		14	1.143		130	66.50		196.50	242
7140	2-1/2"		11	1.455		315	84.50		399.50	475
7150	3"		8	2		460	116		576	680
7160	4"	Q-2	11	2.182		775	131		906	1,050
7180	45° elbow									
7190	1/8"	1 Plum	13	.615	Ea.	43	39.50		82.50	107
7200	1/4"		13	.615		43	39.50		82.50	107
7210	3/8"		13	.615		43.50	39.50		83	108
7220	1/2"		12	.667		44	43		87	113
7230	3/4"		11	.727		48	47		95	123
7240	1"		10	.800		55.50	51.50		107	138
7250	1-1/4"	Q-1	17	.941		77.50	54.50		132	167
7260	1-1/2"		16	1		100	58		158	197
7270	2"		14	1.143		143	66.50		209.50	257
7280	2-1/2"		11	1.455		445	84.50		529.50	615
7290	3"		8	2		655	116		771	895
7300	4"	Q-2	11	2.182		1,175	131		1,306	1,500
7320	Tee, straight									
7330	1/8"	1 Plum	9	.889	Ea.	45.50	57.50		103	136
7340	1/4"		9	.889		45.50	57.50		103	136
7350	3/8"		9	.889		48.50	57.50		106	139
7360	1/2"		8	1		46	64.50		110.50	148
7370	3/4"		7	1.143		49.50	73.50		123	165
7380	1"		6.50	1.231		64	79.50		143.50	190
7390	1-1/4"	Q-1	11	1.455		107	84.50		191.50	244
7400	1-1/2"		10	1.600		138	93		231	291
7410	2"		9	1.778		171	103		274	340
7420	2-1/2"		7	2.286		455	133		588	700
7430	3"		5	3.200		690	186		876	1,050
7440	4"	Q-2	7	3.429		1,700	206		1,906	2,175
7460	Coupling, straight									
7470	1/8"	1 Plum	19	.421	Ea.	10.85	27		37.85	52.50
7480	1/4"		19	.421		13.35	27		40.35	55
7490	3/8"		19	.421		16	27		43	58
7500	1/2"		19	.421		21.50	27		48.50	64
7510	3/4"		18	.444		28.50	28.50		57	74
7520	1"		15	.533		45.50	34.50		80	102
7530	1-1/4"	Q-1	26	.615		88.50	35.50		124	151
7540	1-1/2"		24	.667		82	38.50		120.50	149

For customer support on your Mechanical Costs with RSMeans Data, call 800.448.8182.

195

22 11 13 – Facility Water Distribution Piping

22 11 13.66 Pipe Fittings, Stainless Steel		Crew	Daily Output	Labor-Hours	Unit	Material	2020 Bare Costs Labor	Equipment	Total	Total Incl O&P
7550	2"	Q-1	21	.762	Ea.	137	44		181	217
7560	2-1/2"		18	.889		320	51.50		371.50	425
7570	3"		14	1.143		435	66.50		501.50	580
7580	4"	Q-2	16	1.500		605	90		695	800
7600	Reducer, concentric, 1/2"	1 Plum	12	.667		23	43		66	89.50
7610	3/4"		11	.727		29.50	47		76.50	103
7612	1"		10	.800		49	51.50		100.50	131
7614	1-1/4"	Q-1	17	.941		104	54.50		158.50	196
7616	1-1/2"		16	1		114	58		172	212
7618	2"		14	1.143		177	66.50		243.50	294
7620	2-1/2"		11	1.455		470	84.50		554.50	640
7622	3"		8	2		520	116		636	750
7624	4"	Q-2	11	2.182		865	131		996	1,150
7710	Union									
7720	1/8"	1 Plum	12	.667	Ea.	53.50	43		96.50	124
7730	1/4"		12	.667		53.50	43		96.50	124
7740	3/8"		12	.667		63	43		106	134
7750	1/2"		11	.727		78	47		125	156
7760	3/4"		10	.800		107	51.50		158.50	194
7770	1"		9	.889		155	57.50		212.50	256
7780	1-1/4"	Q-1	16	1		385	58		443	510
7790	1-1/2"		15	1.067		415	62		477	550
7800	2"		13	1.231		525	71.50		596.50	680
7810	2-1/2"		10	1.600		1,000	93		1,093	1,250
7820	3"		7	2.286		1,325	133		1,458	1,675
7830	4"	Q-2	10	2.400		1,825	144		1,969	2,250
7838	Caps									
7840	1/2"	1 Plum	24	.333	Ea.	14.25	21.50		35.75	47.50
7841	3/4"		22	.364		20.50	23.50		44	57.50
7842	1"		20	.400		33	26		59	75
7843	1-1/2"	Q-1	32	.500		83.50	29		112.50	136
7844	2"	"	28	.571		106	33		139	166
7845	4"	Q-2	22	1.091		415	65.50		480.50	555
7850	For 150 lb., type 316, add					25%				
8750	Threaded joint, 2,000 lb., type 304									
8770	90° elbow									
8780	1/8"	1 Plum	13	.615	Ea.	33.50	39.50		73	96
8790	1/4"		13	.615		33.50	39.50		73	96
8800	3/8"		13	.615		41	39.50		80.50	105
8810	1/2"		12	.667		56	43		99	126
8820	3/4"		11	.727		64.50	47		111.50	141
8830	1"		10	.800		86	51.50		137.50	172
8840	1-1/4"	Q-1	17	.941		140	54.50		194.50	236
8850	1-1/2"		16	1		224	58		282	335
8860	2"		14	1.143		315	66.50		381.50	445
8880	45° elbow									
8890	1/8"	1 Plum	13	.615	Ea.	69.50	39.50		109	136
8900	1/4"		13	.615		69.50	39.50		109	136
8910	3/8"		13	.615		87	39.50		126.50	155
8920	1/2"		12	.667		87	43		130	160
8930	3/4"		11	.727		96	47		143	175
8940	1"		10	.800		114	51.50		165.50	203
8950	1-1/4"	Q-1	17	.941		199	54.50		253.50	300

22 11 13.66 Pipe Fittings, Stainless Steel	Crew	Daily Output	Labor-Hours	Unit	Material	2020 Bare Costs Labor	Equipment	Total	Total Incl O&P
8960	1-1/2"	Q-1	16	1	Ea.	267	58		325	380
8970	2"		14	1.143		310	66.50		376.50	440
8990	Tee, straight									
9000	1/8"	1 Plum	9	.889	Ea.	43	57.50		100.50	133
9010	1/4"		9	.889		43	57.50		100.50	133
9020	3/8"		9	.889		55.50	57.50		113	147
9030	1/2"		8	1		71	64.50		135.50	175
9040	3/4"		7	1.143		90.50	73.50		164	210
9050	1"		6.50	1.231		113	79.50		192.50	243
9060	1-1/4"	Q-1	11	1.455		195	84.50		279.50	340
9070	1-1/2"		10	1.600		282	93		375	450
9080	2"		9	1.778		435	103		538	630
9100	For couplings and unions use 3,000 lb., type 304									
9120	2,000 lb., type 316									
9130	90° elbow									
9140	1/8"	1 Plum	13	.615	Ea.	41.50	39.50		81	105
9150	1/4"		13	.615		41.50	39.50		81	105
9160	3/8"		13	.615		48.50	39.50		88	113
9170	1/2"		12	.667		63	43		106	134
9180	3/4"		11	.727		77	47		124	155
9190	1"		10	.800		116	51.50		167.50	205
9200	1-1/4"	Q-1	17	.941		208	54.50		262.50	310
9210	1-1/2"		16	1		239	58		297	350
9220	2"		14	1.143		405	66.50		471.50	545
9240	45° elbow									
9250	1/8"	1 Plum	13	.615	Ea.	88	39.50		127.50	156
9260	1/4"		13	.615		88	39.50		127.50	156
9270	3/8"		13	.615		88	39.50		127.50	156
9280	1/2"		12	.667		88	43		131	161
9300	3/4"		11	.727		99	47		146	179
9310	1"		10	.800		150	51.50		201.50	242
9320	1-1/4"	Q-1	17	.941		214	54.50		268.50	315
9330	1-1/2"		16	1		293	58		351	405
9340	2"		14	1.143		360	66.50		426.50	495
9360	Tee, straight									
9370	1/8"	1 Plum	9	.889	Ea.	53	57.50		110.50	144
9380	1/4"		9	.889		53	57.50		110.50	144
9390	3/8"		9	.889		64.50	57.50		122	156
9400	1/2"		8	1		80.50	64.50		145	185
9410	3/4"		7	1.143		99.50	73.50		173	219
9420	1"		6.50	1.231		152	79.50		231.50	286
9430	1-1/4"	Q-1	11	1.455		246	84.50		330.50	395
9440	1-1/2"		10	1.600		355	93		448	530
9450	2"		9	1.778		565	103		668	775
9470	For couplings and unions use 3,000 lb., type 316									
9490	3,000 lb., type 304									
9510	Coupling									
9520	1/8"	1 Plum	19	.421	Ea.	13.15	27		40.15	55
9530	1/4"		19	.421		14.30	27		41.30	56.50
9540	3/8"		19	.421		18.40	27		45.40	61
9550	1/2"		19	.421		21.50	27		48.50	64
9560	3/4"		18	.444		30	28.50		58.50	75.50
9570	1"		15	.533		44	34.50		78.50	99.50

22 11 13 – Facility Water Distribution Piping

22 11 13.66 Pipe Fittings, Stainless Steel

		Crew	Daily Output	Labor-Hours	Unit	Material	2020 Bare Costs Labor	2020 Bare Costs Equipment	Total	Total Incl O&P
9580	1-1/4"	Q-1	26	.615	Ea.	93	35.50		128.50	156
9590	1-1/2"		24	.667		129	38.50		167.50	200
9600	2"		21	.762		147	44		191	228
9620	Union									
9630	1/8"	1 Plum	12	.667	Ea.	101	43		144	176
9640	1/4"		12	.667		110	43		153	186
9650	3/8"		12	.667		112	43		155	188
9660	1/2"		11	.727		89.50	47		136.50	169
9670	3/4"		10	.800		138	51.50		189.50	228
9680	1"		9	.889		212	57.50		269.50	320
9690	1-1/4"	Q-1	16	1		390	58		448	515
9700	1-1/2"		15	1.067		440	62		502	580
9710	2"		13	1.231		590	71.50		661.50	755
9730	3,000 lb., type 316									
9750	Coupling									
9770	1/8"	1 Plum	19	.421	Ea.	14.80	27		41.80	57
9780	1/4"		19	.421		19.55	27		46.55	62
9790	3/8"		19	.421		21	27		48	63.50
9800	1/2"		19	.421		23.50	27		50.50	66
9810	3/4"		18	.444		33.50	28.50		62	79.50
9820	1"		15	.533		62.50	34.50		97	121
9830	1-1/4"	Q-1	26	.615		140	35.50		175.50	208
9840	1-1/2"		24	.667		163	38.50		201.50	238
9850	2"		21	.762		226	44		270	315
9870	Union									
9880	1/8"	1 Plum	12	.667	Ea.	115	43		158	192
9890	1/4"		12	.667		115	43		158	192
9900	3/8"		12	.667		133	43		176	212
9910	1/2"		11	.727		134	47		181	217
9920	3/4"		10	.800		174	51.50		225.50	269
9930	1"		9	.889		275	57.50		332.50	385
9940	1-1/4"	Q-1	16	1		480	58		538	615
9950	1-1/2"		15	1.067		595	62		657	750
9960	2"		13	1.231		755	71.50		826.50	935

22 11 13.74 Pipe, Plastic

		Crew	Daily Output	Labor-Hours	Unit	Material	2020 Bare Costs Labor	2020 Bare Costs Equipment	Total	Total Incl O&P
0010	**PIPE, PLASTIC**									
0020	Fiberglass reinforced, couplings 10' OC, clevis hanger assy's, 3 per 10'									
0080	General service									
0120	2" diameter	Q-1	59	.271	L.F.	17.50	15.75		33.25	43
0140	3" diameter		52	.308		27.50	17.85		45.35	57
0150	4" diameter		48	.333		23	19.35		42.35	54.50
0160	6" diameter		39	.410		43	24		67	83
0170	8" diameter	Q-2	49	.490		67.50	29.50		97	118
0180	10" diameter		41	.585		95.50	35		130.50	158
0190	12" diameter		36	.667		125	40		165	198
0600	PVC, high impact/pressure, cplgs. 10' OC, clevis hanger assy's, 3 per 10'									
1020	Schedule 80									
1070	1/2" diameter	1 Plum	50	.160	L.F.	6.85	10.30		17.15	23
1080	3/4" diameter		47	.170		7.95	10.95		18.90	25
1090	1" diameter		43	.186		13.25	12		25.25	32.50
1100	1-1/4" diameter		39	.205		15.40	13.20		28.60	37
1110	1-1/2" diameter		34	.235		16.75	15.15		31.90	41

22 11 13.74 Pipe, Plastic		Crew	Daily Output	Labor- Hours	Unit	Material	2020 Bare Costs Labor	Equipment	Total	Total Incl O&P
1120	2" diameter	Q-1	55	.291	L.F.	20.50	16.85		37.35	48
1140	3" diameter		50	.320		40.50	18.55		59.05	73
1150	4" diameter		46	.348		43	20		63	77
1170	6" diameter		38	.421		66.50	24.50		91	110
1730	To delete coupling & hangers, subtract									
1740	1/2" diam.					62%	80%			
1750	3/4" diam. to 1-1/4" diam.					58%	73%			
1760	1-1/2" diam. to 6" diam.					40%	57%			
1800	PVC, couplings 10' OC, clevis hanger assemblies, 3 per 10'									
1820	Schedule 40									
1860	1/2" diameter	1 Plum	54	.148	L.F.	4.90	9.55		14.45	19.70
1870	3/4" diameter		51	.157		5.40	10.10		15.50	21
1880	1" diameter		46	.174		9.45	11.20		20.65	27
1890	1-1/4" diameter		42	.190		9.95	12.30		22.25	29.50
1900	1-1/2" diameter		36	.222		10.25	14.30		24.55	33
1910	2" diameter	Q-1	59	.271		12.15	15.75		27.90	37
1920	2-1/2" diameter		56	.286		20	16.55		36.55	47.50
1930	3" diameter		53	.302		22	17.50		39.50	50.50
1940	4" diameter		48	.333		14.80	19.35		34.15	45.50
1950	5" diameter		43	.372		23.50	21.50		45	58.50
1960	6" diameter		39	.410		30.50	24		54.50	69
1970	8" diameter	Q-2	48	.500		37	30		67	86
1980	10" diameter		43	.558		77	33.50		110.50	135
1990	12" diameter		42	.571		93	34.50		127.50	154
2000	14" diameter		31	.774		151	46.50		197.50	236
2010	16" diameter		23	1.043		220	63		283	335
2340	To delete coupling & hangers, subtract									
2360	1/2" diam. to 1-1/4" diam.					65%	74%			
2370	1-1/2" diam. to 6" diam.					44%	57%			
2380	8" diam. to 12" diam.					41%	53%			
2390	14" diam. to 16" diam.					48%	45%			
2420	Schedule 80									
2440	1/4" diameter	1 Plum	58	.138	L.F.	4.46	8.90		13.36	18.20
2450	3/8" diameter		55	.145		4.46	9.35		13.81	18.95
2460	1/2" diameter		50	.160		5.15	10.30		15.45	21
2470	3/4" diameter		47	.170		5.55	10.95		16.50	22.50
2480	1" diameter		43	.186		9.90	12		21.90	29
2490	1-1/4" diameter		39	.205		10.45	13.20		23.65	31.50
2500	1-1/2" diameter		34	.235		10.65	15.15		25.80	34.50
2510	2" diameter	Q-1	55	.291		11.55	16.85		28.40	38
2520	2-1/2" diameter		52	.308		19.55	17.85		37.40	48
2530	3" diameter		50	.320		24	18.55		42.55	54.50
2540	4" diameter		46	.348		19.30	20		39.30	51
2550	5" diameter		42	.381		29.50	22		51.50	65.50
2560	6" diameter		38	.421		37	24.50		61.50	77
2570	8" diameter	Q-2	47	.511		47	30.50		77.50	97.50
2580	10" diameter		42	.571		110	34.50		144.50	173
2590	12" diameter		38	.632		134	38		172	204
2830	To delete coupling & hangers, subtract									
2840	1/4" diam. to 1/2" diam.					66%	80%			
2850	3/4" diam. to 1-1/4" diam.					61%	73%			
2860	1-1/2" diam. to 6" diam.					41%	57%			
2870	8" diam. to 12" diam.					31%	50%			

22 11 13.74 Pipe, Plastic		Crew	Daily Output	Labor-Hours	Unit	Material	2020 Bare Costs Labor	Equipment	Total	Total Incl O&P
2900	Schedule 120									
2910	1/2" diameter	1 Plum	50	.160	L.F.	5.60	10.30		15.90	21.50
2950	3/4" diameter		47	.170		6.20	10.95		17.15	23
2960	1" diameter		43	.186		10.60	12		22.60	29.50
2970	1-1/4" diameter		39	.205		11.90	13.20		25.10	33
2980	1-1/2" diameter	↓	33	.242		12.25	15.60		27.85	37
2990	2" diameter	Q-1	54	.296		14.35	17.20		31.55	41.50
3000	2-1/2" diameter		52	.308		25	17.85		42.85	54.50
3010	3" diameter		49	.327		36	18.95		54.95	68
3020	4" diameter		45	.356		25.50	20.50		46	59
3030	6" diameter	↓	37	.432	↓	63	25		88	107
3240	To delete coupling & hangers, subtract									
3250	1/2" diam. to 1-1/4" diam.					52%	74%			
3260	1-1/2" diam. to 4" diam.					30%	57%			
3270	6" diam.					17%	50%			
3300	PVC, pressure, couplings 10' OC, clevis hanger assy's, 3 per 10'									
3310	SDR 26, 160 psi									
3350	1-1/4" diameter	1 Plum	42	.190	L.F.	9.85	12.30		22.15	29
3360	1-1/2" diameter	"	36	.222		10.60	14.30		24.90	33
3370	2" diameter	Q-1	59	.271		11.05	15.75		26.80	35.50
3380	2-1/2" diameter		56	.286		19.35	16.55		35.90	46.50
3390	3" diameter		53	.302		22	17.50		39.50	50.50
3400	4" diameter		48	.333		16.35	19.35		35.70	47
3420	6" diameter	↓	39	.410		33.50	24		57.50	72.50
3430	8" diameter	Q-2	48	.500	↓	49	30		79	98.50
3660	To delete coupling & clevis hanger assy's, subtract									
3670	1-1/4" diam.					63%	68%			
3680	1-1/2" diam. to 4" diam.					48%	57%			
3690	6" diam. to 8" diam.					60%	54%			
3720	SDR 21, 200 psi, 1/2" diameter	1 Plum	54	.148	L.F.	4.80	9.55		14.35	19.60
3740	3/4" diameter		51	.157		5.05	10.10		15.15	20.50
3750	1" diameter		46	.174		8.90	11.20		20.10	26.50
3760	1-1/4" diameter		42	.190		10	12.30		22.30	29.50
3770	1-1/2" diameter	↓	36	.222		10.55	14.30		24.85	33
3780	2" diameter	Q-1	59	.271		10.45	15.75		26.20	35
3790	2-1/2" diameter		56	.286		19.85	16.55		36.40	47
3800	3" diameter		53	.302		21	17.50		38.50	49
3810	4" diameter		48	.333		16.95	19.35		36.30	47.50
3830	6" diameter	↓	39	.410		29.50	24		53.50	68
3840	8" diameter	Q-2	48	.500	↓	48	30		78	97.50
4000	To delete coupling & hangers, subtract									
4010	1/2" diam. to 3/4" diam.					71%	77%			
4020	1" diam. to 1-1/4" diam.					63%	70%			
4030	1-1/2" diam. to 6" diam.					44%	57%			
4040	8" diam.					46%	54%			
4100	DWV type, schedule 40, couplings 10' OC, clevis hanger assy's, 3 per 10'									
4210	ABS, schedule 40, foam core type									
4212	Plain end black									
4214	1-1/2" diameter	1 Plum	39	.205	L.F.	8.60	13.20		21.80	29.50
4216	2" diameter	Q-1	62	.258		9.25	14.95		24.20	32.50
4218	3" diameter		56	.286		17.60	16.55		34.15	44.50
4220	4" diameter		51	.314		9.80	18.20		28	38
4222	6" diameter	↓	42	.381	↓	21.50	22		43.50	56.50

22 11 13.74 Pipe, Plastic

		Crew	Daily Output	Labor-Hours	Unit	Material	2020 Bare Costs Labor	2020 Bare Costs Equipment	Total	Total Incl O&P
4240	To delete coupling & hangers, subtract									
4244	1-1/2" diam. to 6" diam.					43%	48%			
4400	PVC									
4410	1-1/4" diameter	1 Plum	42	.190	L.F.	9.10	12.30		21.40	28.50
4420	1-1/2" diameter	"	36	.222		8.25	14.30		22.55	30.50
4460	2" diameter	Q-1	59	.271		9.15	15.75		24.90	33.50
4470	3" diameter		53	.302		17.45	17.50		34.95	45
4480	4" diameter		48	.333		19.20	19.35		38.55	50
4490	6" diameter		39	.410		18.60	24		42.60	56
4500	8" diameter	Q-2	48	.500		25.50	30		55.50	73
4510	To delete coupling & hangers, subtract									
4520	1-1/4" diam. to 1-1/2" diam.					48%	60%			
4530	2" diam. to 8" diam.					42%	54%			
4532	to delete hangers, 2" diam. to 8" diam.	Q-1	50	.320	L.F.	5.50	18.55		24.05	34
4550	PVC, schedule 40, foam core type									
4552	Plain end, white									
4554	1-1/2" diameter	1 Plum	39	.205	L.F.	8.40	13.20		21.60	29
4556	2" diameter	Q-1	62	.258		8.95	14.95		23.90	32.50
4558	3" diameter		56	.286		17.05	16.55		33.60	44
4560	4" diameter		51	.314		18.55	18.20		36.75	47.50
4562	6" diameter		42	.381		17.25	22		39.25	52
4564	8" diameter	Q-2	51	.471		23.50	28.50		52	68
4568	10" diameter		48	.500		27.50	30		57.50	75
4570	12" diameter		46	.522		30.50	31.50		62	80.50
4580	To delete coupling & hangers, subtract									
4582	1-1/2" diam. to 2" diam.					58%	54%			
4584	3" diam. to 12" diam.					46%	42%			
4800	PVC, clear pipe, cplgs. 10' OC, clevis hanger assy's 3 per 10', Sched. 40									
4840	1/4" diameter	1 Plum	59	.136	L.F.	4.93	8.75		13.68	18.50
4850	3/8" diameter		56	.143		5.05	9.20		14.25	19.35
4860	1/2" diameter		54	.148		5.90	9.55		15.45	21
4870	3/4" diameter		51	.157		6.60	10.10		16.70	22.50
4880	1" diameter		46	.174		11.60	11.20		22.80	29.50
4890	1-1/4" diameter		42	.190		13.10	12.30		25.40	33
4900	1-1/2" diameter		36	.222		13.95	14.30		28.25	37
4910	2" diameter	Q-1	59	.271		16.45	15.75		32.20	41.50
4920	2-1/2" diameter		56	.286		28	16.55		44.55	56
4930	3" diameter		53	.302		32.50	17.50		50	62
4940	3-1/2" diameter		50	.320		40	18.55		58.55	72
4950	4" diameter		48	.333		28.50	19.35		47.85	60
5250	To delete coupling & hangers, subtract									
5260	1/4" diam. to 3/8" diam.					60%	81%			
5270	1/2" diam. to 3/4" diam.					41%	77%			
5280	1" diam. to 1-1/2" diam.					26%	67%			
5290	2" diam. to 4" diam.					16%	58%			
5300	CPVC, socket joint, couplings 10' OC, clevis hanger assemblies, 3 per 10'									
5302	Schedule 40									
5304	1/2" diameter	1 Plum	54	.148	L.F.	5.80	9.55		15.35	20.50
5305	3/4" diameter		51	.157		6.95	10.10		17.05	23
5306	1" diameter		46	.174		11.35	11.20		22.55	29.50
5307	1-1/4" diameter		42	.190		12.85	12.30		25.15	32.50
5308	1-1/2" diameter		36	.222		12.45	14.30		26.75	35
5309	2" diameter	Q-1	59	.271		15.70	15.75		31.45	41

For customer support on your Mechanical Costs with RSMeans Data, call 800.448.8182.

201

22 11 Facility Water Distribution

22 11 13 – Facility Water Distribution Piping

22 11 13.74 Pipe, Plastic		Crew	Daily Output	Labor-Hours	Unit	Material	2020 Bare Costs Labor	Equipment	Total	Total Incl O&P
5310	2-1/2" diameter	Q-1	56	.286	L.F.	28	16.55		44.55	56
5311	3" diameter		53	.302		32	17.50		49.50	61
5312	4" diameter		48	.333		32.50	19.35		51.85	65
5314	6" diameter		43	.372		55.50	21.50		77	93.50
5318	To delete coupling & hangers, subtract									
5319	1/2" diam. to 3/4" diam.					37%	77%			
5320	1" diam. to 1-1/4" diam.					27%	70%			
5321	1-1/2" diam. to 3" diam.					21%	57%			
5322	4" diam. to 6" diam.					16%	57%			
5324	Schedule 80									
5325	1/2" diameter	1 Plum	50	.160	L.F.	6.10	10.30		16.40	22
5326	3/4" diameter		47	.170		6.95	10.95		17.90	24
5327	1" diameter		43	.186		12	12		24	31
5328	1-1/4" diameter		39	.205		13.55	13.20		26.75	34.50
5329	1-1/2" diameter		34	.235		14.70	15.15		29.85	38.50
5330	2" diameter	Q-1	55	.291		17.30	16.85		34.15	44.50
5331	2-1/2" diameter		52	.308		30.50	17.85		48.35	60
5332	3" diameter		50	.320		32.50	18.55		51.05	63.50
5333	4" diameter		46	.348		31.50	20		51.50	64.50
5334	6" diameter		38	.421		61.50	24.50		86	104
5335	8" diameter	Q-2	47	.511		170	30.50		200.50	233
5339	To delete couplings & hangers, subtract									
5340	1/2" diam. to 3/4" diam.					44%	77%			
5341	1" diam. to 1-1/4" diam.					32%	71%			
5342	1-1/2" diam. to 4" diam.					25%	58%			
5343	6" diam. to 8" diam.					20%	53%			
5360	CPVC, threaded, couplings 10' OC, clevis hanger assemblies, 3 per 10'									
5380	Schedule 40									
5460	1/2" diameter	1 Plum	54	.148	L.F.	6.65	9.55		16.20	21.50
5470	3/4" diameter		51	.157		8.45	10.10		18.55	24.50
5480	1" diameter		46	.174		12.90	11.20		24.10	31
5490	1-1/4" diameter		42	.190		14.05	12.30		26.35	34
5500	1-1/2" diameter		36	.222		13.45	14.30		27.75	36.50
5510	2" diameter	Q-1	59	.271		16.90	15.75		32.65	42
5520	2-1/2" diameter		56	.286		29.50	16.55		46.05	57.50
5530	3" diameter		53	.302		33.50	17.50		51	63
5540	4" diameter		48	.333		40	19.35		59.35	73.50
5550	6" diameter		43	.372		59	21.50		80.50	97.50
5730	To delete coupling & hangers, subtract									
5740	1/2" diam. to 3/4" diam.					37%	77%			
5750	1" diam. to 1-1/4" diam.					27%	70%			
5760	1-1/2" diam. to 3" diam.					21%	57%			
5770	4" diam. to 6" diam.					16%	57%			
5800	Schedule 80									
5860	1/2" diameter	1 Plum	50	.160	L.F.	6.95	10.30		17.25	23
5870	3/4" diameter		47	.170		8.45	10.95		19.40	25.50
5880	1" diameter		43	.186		13.55	12		25.55	33
5890	1-1/4" diameter		39	.205		14.75	13.20		27.95	36
5900	1-1/2" diameter		34	.235		15.70	15.15		30.85	40
5910	2" diameter	Q-1	55	.291		18.55	16.85		35.40	46
5920	2-1/2" diameter		52	.308		31.50	17.85		49.35	61.50
5930	3" diameter		50	.320		34.50	18.55		53.05	66
5940	4" diameter		46	.348		39	20		59	73

22 11 13 – Facility Water Distribution Piping

22 11 13.74 Pipe, Plastic		Crew	Daily Output	Labor-Hours	Unit	Material	2020 Bare Costs Labor	Equipment	Total	Total Incl O&P
5950	6" diameter	Q-1	38	.421	L.F.	65	24.50		89.50	108
5960	8" diameter	Q-2	47	.511	↓	168	30.50		198.50	231
6060	To delete couplings & hangers, subtract									
6070	1/2" diam. to 3/4" diam.					44%	77%			
6080	1" diam. to 1-1/4" diam.					32%	71%			
6090	1-1/2" diam. to 4" diam.					25%	58%			
6100	6" diam. to 8" diam.					20%	53%			
6240	CTS, 1/2" diameter	1 Plum	54	.148	L.F.	4.80	9.55		14.35	19.60
6250	3/4" diameter		51	.157		10.05	10.10		20.15	26
6260	1" diameter		46	.174		13.85	11.20		25.05	32
6270	1-1/4" diameter		42	.190		18.65	12.30		30.95	39
6280	1-1/2" diameter	↓	36	.222		22	14.30		36.30	46
6290	2" diameter	Q-1	59	.271	↓	33	15.75		48.75	59.50
6370	To delete coupling & hangers, subtract									
6380	1/2" diam.					51%	79%			
6390	3/4" diam.					40%	76%			
6392	1" thru 2" diam.					72%	68%			
7280	PEX, flexible, no couplings or hangers									
7282	Note: For labor costs add 25% to the couplings and fittings labor total.									
7285	For fittings see section 23 83 16.10 7000									
7300	Non-barrier type, hot/cold tubing rolls									
7310	1/4" diameter x 100'				L.F.	.56			.56	.62
7350	3/8" diameter x 100'					.55			.55	.61
7360	1/2" diameter x 100'					.76			.76	.84
7370	1/2" diameter x 500'					.74			.74	.81
7380	1/2" diameter x 1000'					.72			.72	.79
7400	3/4" diameter x 100'					1.04			1.04	1.14
7410	3/4" diameter x 500'					1.17			1.17	1.29
7420	3/4" diameter x 1000'					1.17			1.17	1.29
7460	1" diameter x 100'					2.02			2.02	2.22
7470	1" diameter x 300'					2.02			2.02	2.22
7480	1" diameter x 500'					2.03			2.03	2.23
7500	1-1/4" diameter x 100'					3.44			3.44	3.78
7510	1-1/4" diameter x 300'					3.44			3.44	3.78
7540	1-1/2" diameter x 100'					4.71			4.71	5.20
7550	1-1/2" diameter x 300'				↓	4.72			4.72	5.20
7596	Most sizes available in red or blue									
7700	Non-barrier type, hot/cold tubing straight lengths									
7710	1/2" diameter x 20'				L.F.	.68			.68	.75
7750	3/4" diameter x 20'					1.20			1.20	1.32
7760	1" diameter x 20'					2.04			2.04	2.24
7770	1-1/4" diameter x 20'					3.65			3.65	4.02
7780	1-1/2" diameter x 20'					4.73			4.73	5.20
7790	2" diameter				↓	9.30			9.30	10.25
7796	Most sizes available in red or blue									
9000	Polypropylene pipe									
9002	For fusion weld fittings and accessories see line 22 11 13.76 9400									
9004	Note: sizes 1/2" thru 4" use socket fusion									
9005	Sizes 6" thru 10" use butt fusion									
9010	SDR 7.4 (domestic hot water piping)									
9011	Enhanced to minimize thermal expansion and high temperature life									
9016	13' lengths, size is ID, includes joints 13' OC and hangers 3 per 10'									
9020	3/8" diameter	1 Plum	53	.151	L.F.	3.20	9.75		12.95	18.05

22 11 13.74 Pipe, Plastic

		Crew	Daily Output	Labor-Hours	Unit	Material	2020 Bare Costs Labor	Equipment	Total	Total Incl O&P
9022	1/2" diameter	1 Plum	52	.154	L.F.	3.67	9.90		13.57	18.90
9024	3/4" diameter		50	.160		4.25	10.30		14.55	20
9026	1" diameter		45	.178		5.40	11.45		16.85	23
9028	1-1/4" diameter		40	.200		7.90	12.90		20.80	28
9030	1-1/2" diameter		35	.229		10.45	14.75		25.20	33.50
9032	2" diameter	Q-1	58	.276		14.50	16		30.50	40
9034	2-1/2" diameter		55	.291		19.55	16.85		36.40	47
9036	3" diameter		52	.308		25.50	17.85		43.35	54.50
9038	3-1/2" diameter		49	.327		35	18.95		53.95	66.50
9040	4" diameter		46	.348		37	20		57	71
9042	6" diameter		39	.410		50	24		74	90.50
9044	8" diameter	Q-2	48	.500		75	30		105	128
9046	10" diameter	"	43	.558		118	33.50		151.50	180
9050	To delete joint & hangers, subtract									
9052	3/8" diam. to 1" diam.					45%	65%			
9054	1-1/4" diam. to 4" diam.					15%	45%			
9056	6" diam. to 10" diam.					5%	24%			
9060	SDR 11 (domestic cold water piping)									
9062	13' lengths, size is ID, includes joints 13' OC and hangers 3 per 10'									
9064	1/2" diameter	1 Plum	57	.140	L.F.	3.33	9.05		12.38	17.20
9066	3/4" diameter		54	.148		3.70	9.55		13.25	18.35
9068	1" diameter		49	.163		4.43	10.50		14.93	20.50
9070	1-1/4" diameter		45	.178		6.45	11.45		17.90	24.50
9072	1-1/2" diameter		40	.200		7.95	12.90		20.85	28
9074	2" diameter	Q-1	62	.258		10.95	14.95		25.90	34.50
9076	2-1/2" diameter		59	.271		14.05	15.75		29.80	39
9078	3" diameter		56	.286		18.65	16.55		35.20	45.50
9080	3-1/2" diameter		53	.302		26.50	17.50		44	55
9082	4" diameter		50	.320		29.50	18.55		48.05	60.50
9084	6" diameter		47	.340		34.50	19.75		54.25	67
9086	8" diameter	Q-2	51	.471		52.50	28.50		81	100
9088	10" diameter	"	46	.522		78.50	31.50		110	134
9090	To delete joint & hangers, subtract									
9092	1/2" diam. to 1" diam.					45%	65%			
9094	1-1/4" diam. to 4" diam.					15%	45%			
9096	6" diam. to 10" diam.					5%	24%			

22 11 13.76 Pipe Fittings, Plastic

		Crew	Daily Output	Labor-Hours	Unit	Material	2020 Bare Costs Labor	Equipment	Total	Total Incl O&P
0010	**PIPE FITTINGS, PLASTIC**									
0030	Epoxy resin, fiberglass reinforced, general service									
0100	3"	Q-1	20.80	.769	Ea.	118	44.50		162.50	196
0110	4"		16.50	.970		121	56		177	217
0120	6"		10.10	1.584		234	92		326	395
0130	8"	Q-2	9.30	2.581		430	155		585	705
0140	10"		8.50	2.824		540	170		710	850
0150	12"		7.60	3.158		775	190		965	1,150
0170	Elbow, 90°, flanged									
0172	2"	Q-1	23	.696	Ea.	253	40.50		293.50	340
0173	3"		16	1		290	58		348	405
0174	4"		13	1.231		380	71.50		451.50	520
0176	6"		8	2		685	116		801	930
0177	8"	Q-2	9	2.667		1,250	160		1,410	1,625
0178	10"		7	3.429		1,700	206		1,906	2,175

22 11 13.76 Pipe Fittings, Plastic	Crew	Daily Output	Labor-Hours	Unit	Material	2020 Bare Costs Labor	Equipment	Total	Total Incl O&P
0179 12"	Q-2	5	4.800	Ea.	2,275	289		2,564	2,925
0186 Elbow, 45°, flanged									
0188 2"	Q-1	23	.696	Ea.	253	40.50		293.50	340
0189 3"		16	1		290	58		348	405
0190 4"		13	1.231		380	71.50		451.50	520
0192 6"		8	2		685	116		801	930
0193 8"	Q-2	9	2.667		1,250	160		1,410	1,625
0194 10"		7	3.429		1,750	206		1,956	2,225
0195 12"		5	4.800		2,375	289		2,664	3,050
0352 Tee, flanged									
0354 2"	Q-1	17	.941	Ea.	345	54.50		399.50	455
0355 3"		10	1.600		455	93		548	640
0356 4"		8	2		510	116		626	735
0358 6"		5	3.200		875	186		1,061	1,250
0359 8"	Q-2	6	4		1,650	241		1,891	2,175
0360 10"		5	4.800		2,400	289		2,689	3,075
0361 12"		4	6		3,325	360		3,685	4,200
0365 Wye, flanged									
0367 2"	Q-1	17	.941	Ea.	680	54.50		734.50	825
0368 3"		10	1.600		935	93		1,028	1,175
0369 4"		8	2		1,250	116		1,366	1,550
0371 6"		5	3.200		1,775	186		1,961	2,225
0372 8"	Q-2	6	4		2,875	241		3,116	3,525
0373 10"		5	4.800		4,475	289		4,764	5,350
0374 12"		4	6		6,500	360		6,860	7,700
0380 Couplings									
0410 2"	Q-1	33.10	.483	Ea.	27	28		55	71.50
0420 3"		20.80	.769		31.50	44.50		76	102
0430 4"		16.50	.970		43.50	56		99.50	132
0440 6"		10.10	1.584		101	92		193	249
0450 8"	Q-2	9.30	2.581		171	155		326	420
0460 10"		8.50	2.824		233	170		403	510
0470 12"		7.60	3.158		335	190		525	655
0473 High corrosion resistant couplings, add					30%				
0474 Reducer, concentric, flanged									
0475 2" x 1-1/2"	Q-1	30	.533	Ea.	660	31		691	770
0476 3" x 2"		24	.667		715	38.50		753.50	850
0477 4" x 3"		19	.842		805	49		854	960
0479 6" x 4"		15	1.067		805	62		867	980
0480 8" x 6"	Q-2	16	1.500		1,325	90		1,415	1,575
0481 10" x 8"		13	1.846		1,225	111		1,336	1,525
0482 12" x 10"		11	2.182		1,900	131		2,031	2,300
0486 Adapter, bell x male or female									
0488 2"	Q-1	28	.571	Ea.	33.50	33		66.50	86
0489 3"		20	.800		50.50	46.50		97	125
0491 4"		17	.941		59	54.50		113.50	147
0492 6"		12	1.333		138	77.50		215.50	268
0493 8"	Q-2	15	1.600		198	96		294	360
0494 10"	"	11	2.182		285	131		416	510
0528 Flange									
0532 2"	Q-1	46	.348	Ea.	43.50	20		63.50	78
0533 3"		32	.500		61	29		90	111
0534 4"		26	.615		77.50	35.50		113	139

For customer support on your Mechanical Costs with RSMeans Data, call 800.448.8182.

205

22 11 13 – Facility Water Distribution Piping

22 11 13.76 Pipe Fittings, Plastic	Crew	Daily Output	Labor-Hours	Unit	Material	2020 Bare Costs Labor	Equipment	Total	Total Incl O&P
0536	6"	Q-1	16	1	Ea.	147	58		205	249
0537	8"	Q-2	18	1.333		225	80		305	370
0538	10"		14	1.714		320	103		423	505
0539	12"		10	2.400		400	144		544	655
2100	PVC schedule 80, socket joint									
2110	90° elbow, 1/2"	1 Plum	30.30	.264	Ea.	2.67	17		19.67	28.50
2130	3/4"		26	.308		3.42	19.85		23.27	33.50
2140	1"		22.70	.352		5.55	22.50		28.05	40
2150	1-1/4"		20.20	.396		7.35	25.50		32.85	46
2160	1-1/2"		18.20	.440		7.90	28.50		36.40	51
2170	2"	Q-1	33.10	.483		9.55	28		37.55	52.50
2180	3"		20.80	.769		25	44.50		69.50	94.50
2190	4"		16.50	.970		38	56		94	126
2200	6"		10.10	1.584		109	92		201	257
2210	8"	Q-2	9.30	2.581		299	155		454	560
2250	45° elbow, 1/2"	1 Plum	30.30	.264		5.05	17		22.05	31
2270	3/4"		26	.308		7.70	19.85		27.55	38
2280	1"		22.70	.352		11.55	22.50		34.05	46.50
2290	1-1/4"		20.20	.396		14.60	25.50		40.10	54
2300	1-1/2"		18.20	.440		17.35	28.50		45.85	61.50
2310	2"	Q-1	33.10	.483		22.50	28		50.50	67
2320	3"		20.80	.769		57	44.50		101.50	130
2330	4"		16.50	.970		104	56		160	198
2340	6"		10.10	1.584		130	92		222	282
2350	8"	Q-2	9.30	2.581		283	155		438	540
2400	Tee, 1/2"	1 Plum	20.20	.396		7.55	25.50		33.05	46.50
2420	3/4"		17.30	.462		7.90	30		37.90	53
2430	1"		15.20	.526		9.90	34		43.90	62
2440	1-1/4"		13.50	.593		27	38		65	86.50
2450	1-1/2"		12.10	.661		27	42.50		69.50	94
2460	2"	Q-1	20	.800		34	46.50		80.50	107
2470	3"		13.90	1.151		46	67		113	151
2480	4"		11	1.455		53.50	84.50		138	185
2490	6"		6.70	2.388		183	139		322	410
2500	8"	Q-2	6.20	3.871		420	233		653	810
2510	Flange, socket, 150 lb., 1/2"	1 Plum	55.60	.144		14.55	9.25		23.80	30
2514	3/4"		47.60	.168		15.55	10.85		26.40	33.50
2518	1"		41.70	.192		17.35	12.35		29.70	37.50
2522	1-1/2"		33.30	.240		18.30	15.50		33.80	43
2526	2"	Q-1	60.60	.264		24.50	15.30		39.80	49.50
2530	4"		30.30	.528		52.50	30.50		83	104
2534	6"		18.50	.865		82.50	50		132.50	166
2538	8"	Q-2	17.10	1.404		147	84.50		231.50	288
2550	Coupling, 1/2"	1 Plum	30.30	.264		5.10	17		22.10	31
2570	3/4"		26	.308		7.55	19.85		27.40	38
2580	1"		22.70	.352		7.45	22.50		29.95	42
2590	1-1/4"		20.20	.396		10.15	25.50		35.65	49
2600	1-1/2"		18.20	.440		11.40	28.50		39.90	55
2610	2"	Q-1	33.10	.483		12.25	28		40.25	55.50
2620	3"		20.80	.769		33.50	44.50		78	104
2630	4"		16.50	.970		43.50	56		99.50	132
2640	6"		10.10	1.584		93	92		185	241
2650	8"	Q-2	9.30	2.581		122	155		277	365

22 11 Facility Water Distribution

22 11 13 – Facility Water Distribution Piping

22 11 13.76 Pipe Fittings, Plastic		Crew	Daily Output	Labor-Hours	Unit	Material	2020 Bare Costs Labor	Equipment	Total	Total Incl O&P
2660	10"	Q-2	8.50	2.824	Ea.	420	170		590	715
2670	12"	↓	7.60	3.158	↓	485	190		675	815
2700	PVC (white), schedule 40, socket joints									
2760	90° elbow, 1/2"	1 Plum	33.30	.240	Ea.	.52	15.50		16.02	23.50
2770	3/4"		28.60	.280		.59	18.05		18.64	27.50
2780	1"		25	.320		1.05	20.50		21.55	32
2790	1-1/4"		22.20	.360		1.83	23		24.83	37
2800	1-1/2"	↓	20	.400		1.99	26		27.99	40.50
2810	2"	Q-1	36.40	.440		3.11	25.50		28.61	41.50
2820	2-1/2"		26.70	.599		9.60	35		44.60	62.50
2830	3"		22.90	.699		11.35	40.50		51.85	73
2840	4"		18.20	.879		20.50	51		71.50	99
2850	5"		12.10	1.322		52	76.50		128.50	173
2860	6"	↓	11.10	1.441		64.50	83.50		148	196
2870	8"	Q-2	10.30	2.330		166	140		306	395
2980	45° elbow, 1/2"	1 Plum	33.30	.240		.86	15.50		16.36	24
2990	3/4"		28.60	.280		1.34	18.05		19.39	28.50
3000	1"		25	.320		1.60	20.50		22.10	33
3010	1-1/4"		22.20	.360		2.23	23		25.23	37.50
3020	1-1/2"	↓	20	.400		2.80	26		28.80	41.50
3030	2"	Q-1	36.40	.440		3.65	25.50		29.15	42
3040	2-1/2"		26.70	.599		9.50	35		44.50	62.50
3050	3"		22.90	.699		14.65	40.50		55.15	76.50
3060	4"		18.20	.879		26.50	51		77.50	106
3070	5"		12.10	1.322		52	76.50		128.50	173
3080	6"	↓	11.10	1.441		65.50	83.50		149	197
3090	8"	Q-2	10.30	2.330		157	140		297	385
3180	Tee, 1/2"	1 Plum	22.20	.360		.65	23		23.65	35.50
3190	3/4"		19	.421		.75	27		27.75	41.50
3200	1"		16.70	.479		1.39	31		32.39	47.50
3210	1-1/4"		14.80	.541		2.16	35		37.16	54.50
3220	1-1/2"	↓	13.30	.602		2.64	39		41.64	61
3230	2"	Q-1	24.20	.661		3.84	38.50		42.34	61.50
3240	2-1/2"		17.80	.899		12.60	52		64.60	92
3250	3"		15.20	1.053		17.15	61		78.15	110
3260	4"		12.10	1.322		30	76.50		106.50	148
3270	5"		8.10	1.975		72.50	115		187.50	251
3280	6"	↓	7.40	2.162		101	125		226	300
3290	8"	Q-2	6.80	3.529		235	212		447	580
3380	Coupling, 1/2"	1 Plum	33.30	.240		.34	15.50		15.84	23.50
3390	3/4"		28.60	.280		.47	18.05		18.52	27.50
3400	1"		25	.320		.83	20.50		21.33	32
3410	1-1/4"		22.20	.360		1.13	23		24.13	36
3420	1-1/2"	↓	20	.400		1.22	26		27.22	40
3430	2"	Q-1	36.40	.440		1.86	25.50		27.36	40
3440	2-1/2"		26.70	.599		4.11	35		39.11	56.50
3450	3"		22.90	.699		7.70	40.50		48.20	69
3460	4"		18.20	.879		9.60	51		60.60	87
3470	5"		12.10	1.322		16.95	76.50		93.45	134
3480	6"	↓	11.10	1.441		30.50	83.50		114	159
3490	8"	Q-2	10.30	2.330		54.50	140		194.50	270
3600	Cap, schedule 40, PVC socket 1/2"	1 Plum	60.60	.132		.47	8.50		8.97	13.25
3610	3/4"	↓	51.90	.154	↓	.55	9.95		10.50	15.45

22 11 13.76 Pipe Fittings, Plastic		Crew	Daily Output	Labor-Hours	Unit	Material	2020 Bare Costs Labor	Equipment	Total	Total Incl O&P
3620	1"	1 Plum	45.50	.176	Ea.	.86	11.35		12.21	17.90
3630	1-1/4"		40.40	.198		1.21	12.75		13.96	20.50
3640	1-1/2"	↓	36.40	.220		1.34	14.15		15.49	22.50
3650	2"	Q-1	66.10	.242		1.60	14.05		15.65	23
3660	2-1/2"		48.50	.330		5.05	19.15		24.20	34
3670	3"		41.60	.385		5.55	22.50		28.05	39.50
3680	4"	↓	33.10	.483		12.65	28		40.65	56
3690	6"	↓	20.20	.792		30.50	46		76.50	103
3700	8"	Q-2	18.60	1.290	↓	76.50	77.50		154	200
3710	Reducing insert, schedule 40, socket weld									
3712	3/4"	1 Plum	31.50	.254	Ea.	.53	16.35		16.88	25
3713	1"		27.50	.291		.97	18.75		19.72	29
3715	1-1/2"	↓	22	.364		1.37	23.50		24.87	36.50
3716	2"	Q-1	40	.400		2.27	23		25.27	37
3717	4"		20	.800		12	46.50		58.50	82.50
3718	6"	↓	12.20	1.311		30	76		106	147
3719	8"	Q-2	11.30	2.124	↓	110	128		238	310
3730	Reducing insert, socket weld x female/male thread									
3732	1/2"	1 Plum	38.30	.209	Ea.	2.37	13.45		15.82	22.50
3733	3/4"		32.90	.243		1.47	15.65		17.12	25
3734	1"		28.80	.278		2.06	17.90		19.96	29.50
3736	1-1/2"	↓	23	.348		3.70	22.50		26.20	37.50
3737	2"	Q-1	41.90	.382		3.96	22		25.96	37.50
3738	4"	"	20.90	.766	↓	39	44.50		83.50	110
3742	Male adapter, socket weld x male thread									
3744	1/2"	1 Plum	38.30	.209	Ea.	.45	13.45		13.90	20.50
3745	3/4"		32.90	.243		.51	15.65		16.16	24
3746	1"		28.80	.278		.91	17.90		18.81	28
3748	1-1/2"	↓	23	.348		1.58	22.50		24.08	35
3749	2"	Q-1	41.90	.382		1.94	22		23.94	35
3750	4"	"	20.90	.766	↓	10.75	44.50		55.25	78.50
3754	Female adapter, socket weld x female thread									
3756	1/2"	1 Plum	38.30	.209	Ea.	.57	13.45		14.02	20.50
3757	3/4"		32.90	.243		.73	15.65		16.38	24.50
3758	1"		28.80	.278		.84	17.90		18.74	28
3760	1-1/2"	↓	23	.348		1.49	22.50		23.99	35
3761	2"	Q-1	41.90	.382		2	22		24	35
3762	4"	"	20.90	.766	↓	11.25	44.50		55.75	79
3800	PVC, schedule 80, socket joints									
3810	Reducing insert									
3812	3/4"	1 Plum	28.60	.280	Ea.	1.47	18.05		19.52	28.50
3813	1"		25	.320		4.21	20.50		24.71	35.50
3815	1-1/2"	↓	20	.400		9.85	26		35.85	49.50
3816	2"	Q-1	36.40	.440		14	25.50		39.50	53.50
3817	4"		18.20	.879		53.50	51		104.50	135
3818	6"	↓	11.10	1.441		64	83.50		147.50	196
3819	8"	Q-2	10.20	2.353	↓	430	142		572	680
3830	Reducing insert, socket weld x female/male thread									
3832	1/2"	1 Plum	34.80	.230	Ea.	10.35	14.80		25.15	33.50
3833	3/4"		29.90	.268		6.15	17.25		23.40	33
3834	1"		26.10	.307		9.65	19.75		29.40	40
3836	1-1/2"	↓	20.90	.383		12.15	24.50		36.65	50.50
3837	2"	Q-1	38	.421		17.75	24.50		42.25	56

22 11 13.76 Pipe Fittings, Plastic		Crew	Daily Output	Labor-Hours	Unit	Material	2020 Bare Costs Labor	Equipment	Total	Total Incl O&P
3838	4"	Q-1	19	.842	Ea.	86	49		135	168
3844	Adapter, male socket x male thread									
3846	1/2"	1 Plum	34.80	.230	Ea.	3.89	14.80		18.69	26.50
3847	3/4"		29.90	.268		4.29	17.25		21.54	30.50
3848	1"		26.10	.307		7.40	19.75		27.15	37.50
3850	1-1/2"		20.90	.383		12.45	24.50		36.95	50.50
3851	2"	Q-1	38	.421		18.05	24.50		42.55	56.50
3852	4"	"	19	.842		40.50	49		89.50	118
3860	Adapter, female socket x female thread									
3862	1/2"	1 Plum	34.80	.230	Ea.	4.66	14.80		19.46	27
3863	3/4"		29.90	.268		6.95	17.25		24.20	33.50
3864	1"		26.10	.307		10.20	19.75		29.95	41
3866	1-1/2"		20.90	.383		20.50	24.50		45	59.50
3867	2"	Q-1	38	.421		35.50	24.50		60	75.50
3868	4"	"	19	.842		108	49		157	192
3872	Union, socket joints									
3874	1/2"	1 Plum	25.80	.310	Ea.	8.80	20		28.80	39.50
3875	3/4"		22.10	.362		11.15	23.50		34.65	47.50
3876	1"		19.30	.415		12.70	26.50		39.20	54
3878	1-1/2"		15.50	.516		28.50	33.50		62	81.50
3879	2"	Q-1	28.10	.569		39	33		72	92
3888	Cap									
3890	1/2"	1 Plum	54.50	.147	Ea.	5.40	9.45		14.85	20
3891	3/4"		46.70	.171		4.34	11.05		15.39	21.50
3892	1"		41	.195		8.40	12.60		21	28
3894	1-1/2"		32.80	.244		12.45	15.70		28.15	37
3895	2"	Q-1	59.50	.269		29	15.60		44.60	55.50
3896	4"		30	.533		75.50	31		106.50	130
3897	6"		18.20	.879		218	51		269	315
3898	8"	Q-2	16.70	1.437		280	86.50		366.50	440
4500	DWV, ABS, non pressure, socket joints									
4540	1/4 bend, 1-1/4"	1 Plum	20.20	.396	Ea.	4.48	25.50		29.98	43
4560	1-1/2"	"	18.20	.440		3.44	28.50		31.94	46.50
4570	2"	Q-1	33.10	.483		5.30	28		33.30	48
4580	3"		20.80	.769		13.45	44.50		57.95	82
4590	4"		16.50	.970		27.50	56		83.50	115
4600	6"		10.10	1.584		118	92		210	267
4650	1/8 bend, same as 1/4 bend									
4800	Tee, sanitary									
4820	1-1/4"	1 Plum	13.50	.593	Ea.	5.95	38		43.95	63.50
4830	1-1/2"	"	12.10	.661		5.15	42.50		47.65	69.50
4840	2"	Q-1	20	.800		7.90	46.50		54.40	78
4850	3"		13.90	1.151		21.50	67		88.50	124
4860	4"		11	1.455		38.50	84.50		123	169
4862	Tee, sanitary, reducing, 2" x 1-1/2"		22	.727		7	42		49	70.50
4864	3" x 2"		15.30	1.046		13.40	60.50		73.90	106
4868	4" x 3"		12.10	1.322		37.50	76.50		114	157
4870	Combination Y and 1/8 bend									
4872	1-1/2"	1 Plum	12.10	.661	Ea.	12.25	42.50		54.75	77.50
4874	2"	Q-1	20	.800		13.70	46.50		60.20	84.50
4876	3"		13.90	1.151		32	67		99	136
4878	4"		11	1.455		62	84.50		146.50	194
4880	3" x 1-1/2"		15.50	1.032		32.50	60		92.50	126

For customer support on your Mechanical Costs with RSMeans Data, call 800.448.8182.

209

22 11 13 – Facility Water Distribution Piping

22 11 13.76 Pipe Fittings, Plastic		Crew	Daily Output	Labor-Hours	Unit	Material	2020 Bare Costs Labor	Equipment	Total	Total Incl O&P
4882	4" x 3"	Q-1	12.10	1.322	Ea.	49.50	76.50		126	170
4900	Wye, 1-1/4"	1 Plum	13.50	.593		6.85	38		44.85	64.50
4902	1-1/2"	"	12.10	.661		7.85	42.50		50.35	72.50
4904	2"	Q-1	20	.800		10.20	46.50		56.70	80.50
4906	3"		13.90	1.151		24	67		91	127
4908	4"		11	1.455		49	84.50		133.50	180
4910	6"		6.70	2.388		146	139		285	370
4918	3" x 1-1/2"		15.50	1.032		19.40	60		79.40	111
4920	4" x 3"		12.10	1.322		38.50	76.50		115	158
4922	6" x 4"	▼	6.90	2.319		116	134		250	330
4930	Double wye, 1-1/2"	1 Plum	9.10	.879		25	56.50		81.50	113
4932	2"	Q-1	16.60	.964		30	56		86	117
4934	3"		10.40	1.538		70.50	89		159.50	212
4936	4"		8.25	1.939		138	112		250	320
4940	2" x 1-1/2"		16.80	.952		27.50	55		82.50	113
4942	3" x 2"		10.60	1.509		49	87.50		136.50	185
4944	4" x 3"		8.45	1.893		109	110		219	284
4946	6" x 4"		7.25	2.207		156	128		284	365
4950	Reducer bushing, 2" x 1-1/2"		36.40	.440		2.72	25.50		28.22	41
4952	3" x 1-1/2"		27.30	.586		11.95	34		45.95	64
4954	4" x 2"		18.20	.879		22.50	51		73.50	101
4956	6" x 4"	▼	11.10	1.441		62.50	83.50		146	194
4960	Couplings, 1-1/2"	1 Plum	18.20	.440		1.66	28.50		30.16	44.50
4962	2"	Q-1	33.10	.483		2.22	28		30.22	44.50
4963	3"		20.80	.769		6.30	44.50		50.80	74
4964	4"		16.50	.970		11.35	56		67.35	96.50
4966	6"		10.10	1.584		47.50	92		139.50	190
4970	2" x 1-1/2"		33.30	.480		4.80	28		32.80	47
4972	3" x 1-1/2"		21	.762		13.85	44		57.85	81.50
4974	4" x 3"	▼	16.70	.958		21	55.50		76.50	107
4978	Closet flange, 4"	1 Plum	32	.250		11.60	16.10		27.70	37
4980	4" x 3"	"	34	.235	▼	14.10	15.15		29.25	38
5000	DWV, PVC, schedule 40, socket joints									
5040	1/4 bend, 1-1/4"	1 Plum	20.20	.396	Ea.	8.80	25.50		34.30	47.50
5060	1-1/2"	"	18.20	.440		2.48	28.50		30.98	45
5070	2"	Q-1	33.10	.483		3.91	28		31.91	46.50
5080	3"		20.80	.769		11.65	44.50		56.15	80
5090	4"		16.50	.970		23	56		79	109
5100	6"	▼	10.10	1.584		79.50	92		171.50	226
5105	8"	Q-2	9.30	2.581		122	155		277	365
5106	10"	"	8.50	2.824		365	170		535	655
5110	1/4 bend, long sweep, 1-1/2"	1 Plum	18.20	.440		5.85	28.50		34.35	49
5112	2"	Q-1	33.10	.483		6.50	28		34.50	49
5114	3"		20.80	.769		15.05	44.50		59.55	83.50
5116	4"	▼	16.50	.970		28.50	56		84.50	116
5150	1/8 bend, 1-1/4"	1 Plum	20.20	.396		6.05	25.50		31.55	44.50
5170	1-1/2"	"	18.20	.440		2.46	28.50		30.96	45
5180	2"	Q-1	33.10	.483		3.64	28		31.64	46
5190	3"		20.80	.769		10.45	44.50		54.95	78.50
5200	4"		16.50	.970		18.95	56		74.95	105
5210	6"	▼	10.10	1.584		71.50	92		163.50	217
5215	8"	Q-2	9.30	2.581		117	155		272	360
5216	10"	▼	8.50	2.824	▼	238	170		408	515

22 11 13.76 Pipe Fittings, Plastic		Crew	Daily Output	Labor-Hours	Unit	Material	2020 Bare Costs Labor	Equipment	Total	Total Incl O&P
5217	12"	Q-2	7.60	3.158	Ea.	294	190		484	610
5250	Tee, sanitary 1-1/4"	1 Plum	13.50	.593		9.45	38		47.45	67.50
5254	1-1/2"	"	12.10	.661		4.41	42.50		46.91	69
5255	2"	Q-1	20	.800		6.50	46.50		53	76.50
5256	3"		13.90	1.151		17.05	67		84.05	119
5257	4"		11	1.455		31	84.50		115.50	161
5259	6"		6.70	2.388		126	139		265	345
5261	8"	Q-2	6.20	3.871		293	233		526	675
5276	Tee, sanitary, reducing									
5281	2" x 1-1/2" x 1-1/2"	Q-1	23	.696	Ea.	5.70	40.50		46.20	67
5282	2" x 1-1/2" x 2"		22	.727		6.95	42		48.95	70.50
5283	2" x 2" x 1-1/2"		22	.727		5.75	42		47.75	69.50
5284	3" x 3" x 1-1/2"		15.50	1.032		12.45	60		72.45	103
5285	3" x 3" x 2"		15.30	1.046		12.85	60.50		73.35	105
5286	4" x 4" x 1-1/2"		12.30	1.301		33	75.50		108.50	149
5287	4" x 4" x 2"		12.20	1.311		27	76		103	144
5288	4" x 4" x 3"		12.10	1.322		36.50	76.50		113	156
5291	6" x 6" x 4"		6.90	2.319		122	134		256	335
5294	Tee, double sanitary									
5295	1-1/2"	1 Plum	9.10	.879	Ea.	9.80	56.50		66.30	96
5296	2"	Q-1	16.60	.964		13.20	56		69.20	98
5297	3"		10.40	1.538		37	89		126	175
5298	4"		8.25	1.939		59.50	112		171.50	234
5303	Wye, reducing									
5304	2" x 1-1/2" x 1-1/2"	Q-1	23	.696	Ea.	11.15	40.50		51.65	73
5305	2" x 2" x 1-1/2"		22	.727		9.75	42		51.75	73.50
5306	3" x 3" x 2"		15.30	1.046		31.50	60.50		92	126
5307	4" x 4" x 2"		12.20	1.311		23.50	76		99.50	140
5309	4" x 4" x 3"		12.10	1.322		31.50	76.50		108	150
5314	Combination Y & 1/8 bend, 1-1/2"	1 Plum	12.10	.661		10.75	42.50		53.25	76
5315	2"	Q-1	20	.800		12.90	46.50		59.40	83.50
5317	3"		13.90	1.151		29	67		96	132
5318	4"		11	1.455		58	84.50		142.50	190
5319	6"		6.70	2.388		189	139		328	415
5320	8"	Q-2	6.20	3.871		287	233		520	665
5321	10"		5.70	4.211		975	253		1,228	1,450
5322	12"		5.10	4.706		1,225	283		1,508	1,775
5324	Combination Y & 1/8 bend, reducing									
5325	2" x 2" x 1-1/2"	Q-1	22	.727	Ea.	15.15	42		57.15	79.50
5327	3" x 3" x 1-1/2"		15.50	1.032		27	60		87	119
5328	3" x 3" x 2"		15.30	1.046		19.95	60.50		80.45	113
5329	4" x 4" x 2"		12.20	1.311		30.50	76		106.50	148
5331	Wye, 1-1/4"	1 Plum	13.50	.593		12.10	38		50.10	70.50
5332	1-1/2"	"	12.10	.661		8.20	42.50		50.70	73
5333	2"	Q-1	20	.800		7.90	46.50		54.40	78
5334	3"		13.90	1.151		21.50	67		88.50	124
5335	4"		11	1.455		39	84.50		123.50	169
5336	6"		6.70	2.388		114	139		253	335
5337	8"	Q-2	6.20	3.871		217	233		450	590
5338	10"		5.70	4.211		475	253		728	905
5339	12"		5.10	4.706		770	283		1,053	1,275
5341	2" x 1-1/2"	Q-1	22	.727		9.85	42		51.85	74
5342	3" x 1-1/2"		15.50	1.032		14.60	60		74.60	106

For customer support on your Mechanical Costs with RSMeans Data, call 800.448.8182.

211

22 11 13.76 Pipe Fittings, Plastic	Crew	Daily Output	Labor-Hours	Unit	Material	2020 Bare Costs Labor	Equipment	Total	Total Incl O&P
5343	4" x 3"	Q-1	12.10	1.322	Ea.	32	76.50		108.50	150
5344	6" x 4"		6.90	2.319		87	134		221	297
5345	8" x 6"	Q-2	6.40	3.750		188	226		414	545
5347	Double wye, 1-1/2"	1 Plum	9.10	.879		18.25	56.50		74.75	105
5348	2"	Q-1	16.60	.964		20.50	56		76.50	106
5349	3"		10.40	1.538		42.50	89		131.50	181
5350	4"		8.25	1.939		85.50	112		197.50	263
5353	Double wye, reducing									
5354	2" x 2" x 1-1/2" x 1-1/2"	Q-1	16.80	.952	Ea.	18.60	55		73.60	103
5355	3" x 3" x 2" x 2"		10.60	1.509		31.50	87.50		119	166
5356	4" x 4" x 3" x 3"		8.45	1.893		68	110		178	239
5357	6" x 6" x 4" x 4"		7.25	2.207		239	128		367	455
5374	Coupling, 1-1/4"	1 Plum	20.20	.396		5.70	25.50		31.20	44.50
5376	1-1/2"	"	18.20	.440		1.19	28.50		29.69	44
5378	2"	Q-1	33.10	.483		1.62	28		29.62	44
5380	3"		20.80	.769		5.65	44.50		50.15	73
5390	4"		16.50	.970		9.65	56		65.65	94.50
5400	6"		10.10	1.584		31.50	92		123.50	173
5402	8"	Q-2	9.30	2.581		57	155		212	295
5404	2" x 1-1/2"	Q-1	33.30	.480		3.63	28		31.63	45.50
5406	3" x 1-1/2"		21	.762		10.65	44		54.65	78
5408	4" x 3"		16.70	.958		17.35	55.50		72.85	102
5410	Reducer bushing, 2" x 1-1/4"		36.50	.438		3.38	25.50		28.88	41.50
5411	2" x 1-1/2"		36.40	.440		2.07	25.50		27.57	40.50
5412	3" x 1-1/2"		27.30	.586		10.10	34		44.10	62
5413	3" x 2"		27.10	.590		5.25	34		39.25	57.50
5414	4" x 2"		18.20	.879		17.35	51		68.35	95.50
5415	4" x 3"		16.70	.958		8.90	55.50		64.40	93
5416	6" x 4"		11.10	1.441		45	83.50		128.50	175
5418	8" x 6"	Q-2	10.20	2.353		89	142		231	310
5425	Closet flange 4"	Q-1	32	.500		13.70	29		42.70	58.50
5426	4" x 3"	"	34	.471		13.50	27.50		41	56
5450	Solvent cement for PVC, industrial grade, per quart				Qt.	31			31	34
5500	CPVC, Schedule 80, threaded joints									
5540	90° elbow, 1/4"	1 Plum	32	.250	Ea.	13	16.10		29.10	38.50
5560	1/2"		30.30	.264		7.55	17		24.55	34
5570	3/4"		26	.308		11.30	19.85		31.15	42
5580	1"		22.70	.352		15.85	22.50		38.35	51.50
5590	1-1/4"		20.20	.396		30.50	25.50		56	71.50
5600	1-1/2"		18.20	.440		33	28.50		61.50	78.50
5610	2"	Q-1	33.10	.483		44	28		72	90.50
5620	2-1/2"		24.20	.661		137	38.50		175.50	208
5630	3"		20.80	.769		147	44.50		191.50	228
5640	4"		16.50	.970		230	56		286	335
5650	6"		10.10	1.584		267	92		359	430
5660	45° elbow same as 90° elbow									
5700	Tee, 1/4"	1 Plum	22	.364	Ea.	25.50	23.50		49	63
5702	1/2"		20.20	.396		25.50	25.50		51	66
5704	3/4"		17.30	.462		36.50	30		66.50	84.50
5706	1"		15.20	.526		39.50	34		73.50	94
5708	1-1/4"		13.50	.593		39.50	38		77.50	101
5710	1-1/2"		12.10	.661		41.50	42.50		84	110
5712	2"	Q-1	20	.800		46	46.50		92.50	120

212

For customer support on your Mechanical Costs with RSMeans Data, call 800.448.8182.

22 11 13.76 Pipe Fittings, Plastic	Crew	Daily Output	Labor-Hours	Unit	Material	2020 Bare Costs Labor	Equipment	Total	Total Incl O&P
5714	2-1/2"	Q-1	16.20	.988	Ea.	225	57.50		282.50	335
5716	3"		13.90	1.151		263	67		330	390
5718	4"		11	1.455		625	84.50		709.50	810
5720	6"		6.70	2.388		710	139		849	985
5730	Coupling, 1/4"	1 Plum	32	.250		16.65	16.10		32.75	42.50
5732	1/2"		30.30	.264		13.75	17		30.75	40.50
5734	3/4"		26	.308		22	19.85		41.85	54
5736	1"		22.70	.352		25	22.50		47.50	61.50
5738	1-1/4"		20.20	.396		26.50	25.50		52	67
5740	1-1/2"		18.20	.440		28.50	28.50		57	74
5742	2"	Q-1	33.10	.483		33.50	28		61.50	79
5744	2-1/2"		24.20	.661		60.50	38.50		99	124
5746	3"		20.80	.769		70	44.50		114.50	145
5748	4"		16.50	.970		143	56		199	241
5750	6"		10.10	1.584		196	92		288	355
5752	8"	Q-2	9.30	2.581		410	155		565	680
5900	CPVC, Schedule 80, socket joints									
5904	90° elbow, 1/4"	1 Plum	32	.250	Ea.	12.45	16.10		28.55	37.50
5906	1/2"		30.30	.264		4.88	17		21.88	31
5908	3/4"		26	.308		6.25	19.85		26.10	36.50
5910	1"		22.70	.352		9.90	22.50		32.40	45
5912	1-1/4"		20.20	.396		21.50	25.50		47	61.50
5914	1-1/2"		18.20	.440		24	28.50		52.50	68.50
5916	2"	Q-1	33.10	.483		29	28		57	73.50
5918	2-1/2"		24.20	.661		66.50	38.50		105	131
5920	3"		20.80	.769		75	44.50		119.50	150
5922	4"		16.50	.970		135	56		191	233
5924	6"		10.10	1.584		272	92		364	435
5926	8"		9.30	1.720		665	100		765	880
5930	45° elbow, 1/4"	1 Plum	32	.250		18.50	16.10		34.60	44.50
5932	1/2"		30.30	.264		5.95	17		22.95	32
5934	3/4"		26	.308		7.80	19.85		27.65	38
5936	1"		22.70	.352		13.70	22.50		36.20	49
5938	1-1/4"		20.20	.396		27	25.50		52.50	67.50
5940	1-1/2"		18.20	.440		27.50	28.50		56	73
5942	2"	Q-1	33.10	.483		27.50	28		55.50	72
5944	2-1/2"		24.20	.661		63.50	38.50		102	127
5946	3"		20.80	.769		81.50	44.50		126	157
5948	4"		16.50	.970		98.50	56		154.50	192
5950	6"		10.10	1.584		310	92		402	480
5952	8"		9.30	1.720		715	100		815	935
5960	Tee, 1/4"	1 Plum	22	.364		11.45	23.50		34.95	47.50
5962	1/2"		20.20	.396		11.45	25.50		36.95	50.50
5964	3/4"		17.30	.462		11.65	30		41.65	57.50
5966	1"		15.20	.526		14.25	34		48.25	66.50
5968	1-1/4"		13.50	.593		30	38		68	90
5970	1-1/2"		12.10	.661		34.50	42.50		77	102
5972	2"	Q-1	20	.800		34	46.50		80.50	107
5974	2-1/2"		16.20	.988		97.50	57.50		155	193
5976	3"		13.90	1.151		97.50	67		164.50	207
5978	4"		11	1.455		130	84.50		214.50	269
5980	6"		6.70	2.388		340	139		479	575
5982	8"	Q-2	6.20	3.871		965	233		1,198	1,400

22 11 13.76 Pipe Fittings, Plastic	Crew	Daily Output	Labor-Hours	Unit	Material	2020 Bare Costs Labor	Equipment	Total	Total Incl O&P
5990	Coupling, 1/4"	1 Plum	32	.250	Ea.	13.25	16.10		29.35	38.50
5992	1/2"		30.30	.264		5.15	17		22.15	31
5994	3/4"		26	.308		7.20	19.85		27.05	37.50
5996	1"		22.70	.352		9.70	22.50		32.20	44.50
5998	1-1/4"		20.20	.396		14.55	25.50		40.05	54
6000	1-1/2"	▼	18.20	.440		18.25	28.50		46.75	62.50
6002	2"	Q-1	33.10	.483		21.50	28		49.50	65.50
6004	2-1/2"		24.20	.661		47.50	38.50		86	110
6006	3"		20.80	.769		51.50	44.50		96	124
6008	4"		16.50	.970		67.50	56		123.50	158
6010	6"	▼	10.10	1.584		159	92		251	310
6012	8"	Q-2	9.30	2.581	▼	430	155		585	700
6200	CTS, 100 psi at 180°F, hot and cold water									
6230	90° elbow, 1/2"	1 Plum	20	.400	Ea.	.32	26		26.32	39
6250	3/4"		19	.421		.52	27		27.52	41
6251	1"		16	.500		1.65	32		33.65	50
6252	1-1/4"		15	.533		3.04	34.50		37.54	55
6253	1-1/2"	▼	14	.571		5.50	37		42.50	61
6254	2"	Q-1	23	.696		10.55	40.50		51.05	72
6260	45° elbow, 1/2"	1 Plum	20	.400		.40	26		26.40	39
6280	3/4"		19	.421		.69	27		27.69	41.50
6281	1"		16	.500		1.87	32		33.87	50
6282	1-1/4"		15	.533		3.75	34.50		38.25	55.50
6283	1-1/2"	▼	14	.571		5.45	37		42.45	61
6284	2"	Q-1	23	.696		12.40	40.50		52.90	74
6290	Tee, 1/2"	1 Plum	13	.615		.40	39.50		39.90	60
6310	3/4"		12	.667		.75	43		43.75	65.50
6311	1"		11	.727		3.80	47		50.80	74
6312	1-1/4"		10	.800		5.85	51.50		57.35	83.50
6313	1-1/2"	▼	10	.800		7.65	51.50		59.15	85.50
6314	2"	Q-1	17	.941		12.35	54.50		66.85	95
6320	Coupling, 1/2"	1 Plum	22	.364		.26	23.50		23.76	35.50
6340	3/4"		21	.381		.34	24.50		24.84	37
6341	1"		18	.444		1.55	28.50		30.05	44.50
6342	1-1/4"		17	.471		2.01	30.50		32.51	47.50
6343	1-1/2"	▼	16	.500		3.07	32		35.07	51.50
6344	2"	Q-1	28	.571	▼	5.40	33		38.40	55.50
6360	Solvent cement for CPVC, commercial grade, per quart				Qt.	51.50			51.50	56.50
7340	PVC flange, slip-on, Sch 80 std., 1/2"	1 Plum	22	.364	Ea.	14.65	23.50		38.15	51
7350	3/4"		21	.381		15.65	24.50		40.15	54
7360	1"		18	.444		17.45	28.50		45.95	62
7370	1-1/4"		17	.471		17.95	30.50		48.45	65.50
7380	1-1/2"	▼	16	.500		18.35	32		50.35	68
7390	2"	Q-1	26	.615		24.50	35.50		60	80.50
7400	2-1/2"		24	.667		37.50	38.50		76	99.50
7410	3"		18	.889		41.50	51.50		93	123
7420	4"		15	1.067		52.50	62		114.50	151
7430	6"	▼	10	1.600		83	93		176	230
7440	8"	Q-2	11	2.182		148	131		279	360
7550	Union, schedule 40, socket joints, 1/2"	1 Plum	19	.421		5.20	27		32.20	46.50
7560	3/4"		18	.444		5.80	28.50		34.30	49.50
7570	1"		15	.533		6	34.50		40.50	58
7580	1-1/4"	▼	14	.571	▼	17.85	37		54.85	74.50

22 11 13.76 Pipe Fittings, Plastic	Crew	Daily Output	Labor-Hours	Unit	Material	2020 Bare Costs Labor	Equipment	Total	Total Incl O&P
7590	1-1/2"	1 Plum	13	.615	Ea.	20	39.50		59.50	81.50
7600	2"	Q-1	20	.800	▼	27	46.50		73.50	99.50
7992	Polybutyl/polyethyl pipe, for copper fittings see Line 22 11 13.25 7000									
8000	Compression type, PVC, 160 psi cold water									
8010	Coupling, 3/4" CTS	1 Plum	21	.381	Ea.	4.27	24.50		28.77	41
8020	1" CTS		18	.444		6	28.50		34.50	49.50
8030	1-1/4" CTS		17	.471		8.35	30.50		38.85	54.50
8040	1-1/2" CTS		16	.500		10	32		42	59
8050	2" CTS		15	.533		15.85	34.50		50.35	69
8060	Female adapter, 3/4" FPT x 3/4" CTS		23	.348		6.35	22.50		28.85	40.50
8070	3/4" FPT x 1" CTS		21	.381		7.35	24.50		31.85	44.50
8080	1" FPT x 1" CTS		20	.400		7.35	26		33.35	46.50
8090	1-1/4" FPT x 1-1/4" CTS		18	.444		9.60	28.50		38.10	53.50
8100	1-1/2" FPT x 1-1/2" CTS		16	.500		10.95	32		42.95	60
8110	2" FPT x 2" CTS		13	.615		16.05	39.50		55.55	77
8130	Male adapter, 3/4" MPT x 3/4" CTS		23	.348		5.30	22.50		27.80	39.50
8140	3/4" MPT x 1" CTS		21	.381		6.25	24.50		30.75	43.50
8150	1" MPT x 1" CTS		20	.400		6.25	26		32.25	45.50
8160	1-1/4" MPT x 1-1/4" CTS		18	.444		8.45	28.50		36.95	52.50
8170	1-1/2" MPT x 1-1/2" CTS		16	.500		10.15	32		42.15	59
8180	2" MPT x 2" CTS		13	.615		13.15	39.50		52.65	74
8200	Spigot adapter, 3/4" IPS x 3/4" CTS		23	.348		2.22	22.50		24.72	36
8210	3/4" IPS x 1" CTS		21	.381		2.71	24.50		27.21	39.50
8220	1" IPS x 1" CTS		20	.400		2.70	26		28.70	41.50
8230	1-1/4" IPS x 1-1/4" CTS		18	.444		4.07	28.50		32.57	47.50
8240	1-1/2" IPS x 1-1/2" CTS		16	.500		4.26	32		36.26	52.50
8250	2" IPS x 2" CTS	▼	13	.615	▼	5.30	39.50		44.80	65.50
8270	Price includes insert stiffeners									
8280	250 psi is same price as 160 psi									
8300	Insert type, nylon, 160 & 250 psi, cold water									
8310	Clamp ring stainless steel, 3/4" IPS	1 Plum	115	.070	Ea.	2.78	4.48		7.26	9.75
8320	1" IPS		107	.075		3.07	4.82		7.89	10.60
8330	1-1/4" IPS		101	.079		2.85	5.10		7.95	10.80
8340	1-1/2" IPS		95	.084		3.95	5.45		9.40	12.45
8350	2" IPS		85	.094		4.74	6.05		10.79	14.30
8370	Coupling, 3/4" IPS		22	.364		.95	23.50		24.45	36
8380	1" IPS		19	.421		.99	27		27.99	41.50
8390	1-1/4" IPS		18	.444		1.47	28.50		29.97	44.50
8400	1-1/2" IPS		17	.471		1.73	30.50		32.23	47.50
8410	2" IPS		16	.500		3.36	32		35.36	51.50
8430	Elbow, 90°, 3/4" IPS		22	.364		1.89	23.50		25.39	37
8440	1" IPS		19	.421		2.10	27		29.10	43
8450	1-1/4" IPS		18	.444		2.35	28.50		30.85	45.50
8460	1-1/2" IPS		17	.471		2.77	30.50		33.27	48.50
8470	2" IPS		16	.500		3.86	32		35.86	52.50
8490	Male adapter, 3/4" IPS x 3/4" MPT		25	.320		.95	20.50		21.45	32
8500	1" IPS x 1" MPT		21	.381		.98	24.50		25.48	37.50
8510	1-1/4" IPS x 1-1/4" MPT		20	.400		1.55	26		27.55	40
8520	1-1/2" IPS x 1-1/2" MPT		18	.444		1.73	28.50		30.23	45
8530	2" IPS x 2" MPT		15	.533		3.33	34.50		37.83	55
8550	Tee, 3/4" IPS		14	.571		1.83	37		38.83	57
8560	1" IPS		13	.615		2.40	39.50		41.90	62
8570	1-1/4" IPS	▼	12	.667	▼	3.73	43		46.73	68.50

22 11 13.76 Pipe Fittings, Plastic		Crew	Daily Output	Labor-Hours	Unit	Material	2020 Bare Costs Labor	Equipment	Total	Total Incl O&P
8580	1-1/2" IPS	1 Plum	11	.727	Ea.	4.24	47		51.24	74.50
8590	2" IPS	↓	10	.800	↓	8.35	51.50		59.85	86
8610	Insert type, PVC, 100 psi @ 180°F, hot & cold water									
8620	Coupler, male, 3/8" CTS x 3/8" MPT	1 Plum	29	.276	Ea.	.76	17.80		18.56	27.50
8630	3/8" CTS x 1/2" MPT		28	.286		.76	18.40		19.16	28.50
8640	1/2" CTS x 1/2" MPT		27	.296		.78	19.10		19.88	29.50
8650	1/2" CTS x 3/4" MPT		26	.308		2.46	19.85		22.31	32
8660	3/4" CTS x 1/2" MPT		25	.320		2.22	20.50		22.72	33.50
8670	3/4" CTS x 3/4" MPT		25	.320		.93	20.50		21.43	32
8700	Coupling, 3/8" CTS x 1/2" CTS		25	.320		4.44	20.50		24.94	36
8710	1/2" CTS		23	.348		5.65	22.50		28.15	40
8730	3/4" CTS		22	.364		12.35	23.50		35.85	48.50
8750	Elbow 90°, 3/8" CTS		25	.320		4.15	20.50		24.65	35.50
8760	1/2" CTS		23	.348		5.10	22.50		27.60	39
8770	3/4" CTS		22	.364		7.65	23.50		31.15	43.50
8800	Rings, crimp, copper, 3/8" CTS		120	.067		.18	4.30		4.48	6.65
8810	1/2" CTS		117	.068		.20	4.41		4.61	6.80
8820	3/4" CTS		115	.070		.26	4.48		4.74	7
8850	Reducer tee, bronze, 3/8" x 3/8" x 1/2" CTS		17	.471		5.50	30.50		36	51.50
8860	1/2" x 1/2" x 3/4" CTS		15	.533		7.40	34.50		41.90	59.50
8870	3/4" x 1/2" x 1/2" CTS		14	.571		7.40	37		44.40	63
8890	3/4" x 3/4" x 1/2" CTS		14	.571		7.50	37		44.50	63.50
8900	1" x 1/2" x 1/2" CTS		14	.571		12.40	37		49.40	68.50
8930	Tee, 3/8" CTS		17	.471		3.24	30.50		33.74	49
8940	1/2" CTS		15	.533		2.92	34.50		37.42	54.50
8950	3/4" CTS	↓	14	.571	↓	3.61	37		40.61	59
8960	Copper rings included in fitting price									
9000	Flare type, assembled, acetal, hot & cold water									
9010	Coupling, 1/4" & 3/8" CTS	1 Plum	24	.333	Ea.	3.55	21.50		25.05	36
9020	1/2" CTS		22	.364		4.07	23.50		27.57	39.50
9030	3/4" CTS		21	.381		6	24.50		30.50	43
9040	1" CTS		18	.444		7.65	28.50		36.15	51.50
9050	Elbow 90°, 1/4" CTS		26	.308		3.94	19.85		23.79	34
9060	3/8" CTS		24	.333		4.23	21.50		25.73	36.50
9070	1/2" CTS		22	.364		5	23.50		28.50	40.50
9080	3/4" CTS		21	.381		7.65	24.50		32.15	45
9090	1" CTS		18	.444		9.90	28.50		38.40	54
9110	Tee, 1/4" CTS		16	.500		4.33	32		36.33	53
9114	3/8" CTS		15	.533		4.42	34.50		38.92	56.50
9120	1/2" CTS		14	.571		5.55	37		42.55	61
9130	3/4" CTS		13	.615		8.70	39.50		48.20	69
9140	1" CTS	↓	12	.667	↓	11.70	43		54.70	77.50
9400	Polypropylene, fittings and accessories									
9404	Fittings fusion welded, sizes are ID									
9408	Note: sizes 1/2" thru 4" use socket fusion									
9410	Sizes 6" thru 10" use butt fusion									
9416	Coupling									
9420	3/8"	1 Plum	39	.205	Ea.	.92	13.20		14.12	21
9422	1/2"		37.40	.214		1.23	13.80		15.03	22
9424	3/4"		35.40	.226		1.36	14.55		15.91	23.50
9426	1"		29.70	.269		1.79	17.35		19.14	28
9428	1-1/4"		27.60	.290		2.14	18.70		20.84	30.50
9430	1-1/2"	↓	24.80	.323		4.51	21		25.51	36

22 11 13.76 Pipe Fittings, Plastic

		Crew	Daily Output	Labor-Hours	Unit	Material	2020 Bare Costs Labor	Equipment	Total	Total Incl O&P
9432	2"	Q-1	43	.372	Ea.	9	21.50		30.50	42.50
9434	2-1/2"		35.60	.449		10.10	26		36.10	50
9436	3"		30.90	.518		22	30		52	69.50
9438	3-1/2"		27.80	.576		36	33.50		69.50	89.50
9440	4"		25	.640		47.50	37		84.50	108
9442	Reducing coupling, female to female									
9446	2" to 1-1/2"	Q-1	49	.327	Ea.	13	18.95		31.95	43
9448	2-1/2" to 2"		41.20	.388		14.25	22.50		36.75	49
9450	3" to 2-1/2"		33.10	.483		19.25	28		47.25	63
9470	Reducing bushing, female to female									
9472	1/2" to 3/8"	1 Plum	38.20	.209	Ea.	1.23	13.50		14.73	21.50
9474	3/4" to 3/8" or 1/2"		36.50	.219		1.36	14.15		15.51	22.50
9476	1" to 3/4" or 1/2"		33.10	.242		1.81	15.60		17.41	25.50
9478	1-1/4" to 3/4" or 1"		28.70	.279		2.79	17.95		20.74	30
9480	1-1/2" to 1/2" thru 1-1/4"		26.20	.305		4.62	19.70		24.32	34.50
9482	2" to 1/2" thru 1-1/2"	Q-1	43	.372		9.25	21.50		30.75	42.50
9484	2-1/2" to 1/2" thru 2"		41.20	.388		10.35	22.50		32.85	45
9486	3" to 1-1/2" thru 2-1/2"		33.10	.483		23	28		51	67.50
9488	3-1/2" to 2" thru 3"		29.20	.548		37	32		69	88
9490	4" to 2-1/2" thru 3-1/2"		26.20	.611		57.50	35.50		93	117
9491	6" to 4" SDR 7.4		16.50	.970		69	56		125	160
9492	6" to 4" SDR 11		16.50	.970		69	56		125	160
9493	8" to 6" SDR 7.4		10.10	1.584		102	92		194	250
9494	8" to 6" SDR 11		10.10	1.584		69.50	92		161.50	215
9495	10" to 8" SDR 7.4		7.80	2.051		139	119		258	330
9496	10" to 8" SDR 11		7.80	2.051		100	119		219	288
9500	90° elbow									
9504	3/8"	1 Plum	39	.205	Ea.	1.26	13.20		14.46	21
9506	1/2"		37.40	.214		1.31	13.80		15.11	22
9508	3/4"		35.40	.226		1.68	14.55		16.23	24
9510	1"		29.70	.269		2.42	17.35		19.77	28.50
9512	1-1/4"		27.60	.290		3.73	18.70		22.43	32
9514	1-1/2"		24.80	.323		8	21		29	40
9516	2"	Q-1	43	.372		12.30	21.50		33.80	46
9518	2-1/2"		35.60	.449		27.50	26		53.50	69
9520	3"		30.90	.518		45.50	30		75.50	95
9522	3-1/2"		27.80	.576		64.50	33.50		98	121
9524	4"		25	.640		99.50	37		136.50	165
9526	6" SDR 7.4		5.55	2.883		113	167		280	375
9528	6" SDR 11		5.55	2.883		89	167		256	350
9530	8" SDR 7.4	Q-2	8.10	2.963		295	178		473	590
9532	8" SDR 11		8.10	2.963		229	178		407	520
9534	10" SDR 7.4		7.50	3.200		390	192		582	720
9536	10" SDR 11		7.50	3.200		365	192		557	695
9551	45° elbow									
9554	3/8"	1 Plum	39	.205	Ea.	1.26	13.20		14.46	21
9556	1/2"		37.40	.214		1.31	13.80		15.11	22
9558	3/4"		35.40	.226		1.68	14.55		16.23	24
9564	1"		29.70	.269		2.42	17.35		19.77	28.50
9566	1-1/4"		27.60	.290		3.73	18.70		22.43	32
9568	1-1/2"		24.80	.323		8	21		29	40
9570	2"	Q-1	43	.372		12.15	21.50		33.65	46
9572	2-1/2"		35.60	.449		27	26		53	68.50

For customer support on your Mechanical Costs with RSMeans Data, call 800.448.8182.

217

22 11 13.76 Pipe Fittings, Plastic		Crew	Daily Output	Labor-Hours	Unit	Material	2020 Bare Costs Labor	Equipment	Total	Total Incl O&P
9574	3"	Q-1	30.90	.518	Ea.	50	30		80	100
9576	3-1/2"		27.80	.576		71	33.50		104.50	128
9578	4"		25	.640		109	37		146	176
9580	6" SDR 7.4		5.55	2.883		126	167		293	390
9582	6" SDR 11		5.55	2.883		104	167		271	365
9584	8" SDR 7.4	Q-2	8.10	2.963		274	178		452	565
9586	8" SDR 11		8.10	2.963		239	178		417	530
9588	10" SDR 7.4		7.50	3.200		440	192		632	775
9590	10" SDR 11		7.50	3.200		365	192		557	695
9600	Tee									
9604	3/8"	1 Plum	26	.308	Ea.	1.68	19.85		21.53	31.50
9606	1/2"		24.90	.321		2.31	20.50		22.81	33.50
9608	3/4"		23.70	.338		2.42	22		24.42	35
9610	1"		19.90	.402		3.05	26		29.05	42.50
9612	1-1/4"		18.50	.432		4.69	28		32.69	46.50
9614	1-1/2"		16.60	.482		13.40	31		44.40	61.50
9616	2"	Q-1	26.80	.597		18.25	34.50		52.75	72
9618	2-1/2"		23.80	.672		30	39		69	91.50
9620	3"		20.60	.777		60	45		105	134
9622	3-1/2"		18.40	.870		94	50.50		144.50	179
9624	4"		16.70	.958		110	55.50		165.50	204
9626	6" SDR 7.4		3.70	4.324		137	251		388	525
9628	6" SDR 11		3.70	4.324		167	251		418	560
9630	8" SDR 7.4	Q-2	5.40	4.444		375	267		642	815
9632	8" SDR 11		5.40	4.444		330	267		597	765
9634	10" SDR 7.4		5	4.800		645	289		934	1,150
9636	10" SDR 11		5	4.800		560	289		849	1,050
9638	For reducing tee use same tee price									
9660	End cap									
9662	3/8"	1 Plum	78	.103	Ea.	1.80	6.60		8.40	11.90
9664	1/2"		74.60	.107		1.90	6.90		8.80	12.45
9666	3/4"		71.40	.112		2.42	7.20		9.62	13.45
9668	1"		59.50	.134		2.92	8.65		11.57	16.15
9670	1-1/4"		55.60	.144		4.62	9.25		13.87	19
9672	1-1/2"		49.50	.162		6.35	10.40		16.75	22.50
9674	2"	Q-1	80	.200		10.70	11.60		22.30	29
9676	2-1/2"		71.40	.224		15.45	13		28.45	36.50
9678	3"		61.70	.259		35	15.05		50.05	61
9680	3-1/2"		55.20	.290		42	16.80		58.80	71
9682	4"		50	.320		64	18.55		82.55	98.50
9684	6" SDR 7.4		26.50	.604		88	35		123	149
9686	6" SDR 11		26.50	.604		71	35		106	131
9688	8" SDR 7.4	Q-2	16.30	1.472		88	88.50		176.50	230
9690	8" SDR 11		16.30	1.472		77.50	88.50		166	218
9692	10" SDR 7.4		14.90	1.611		132	97		229	290
9694	10" SDR 11		14.90	1.611		94	97		191	248
9800	Accessories and tools									
9802	Pipe clamps for suspension, not including rod or beam clamp									
9804	3/8"	1 Plum	74	.108	Ea.	2.25	6.95		9.20	12.95
9805	1/2"		70	.114		2.83	7.35		10.18	14.10
9806	3/4"		68	.118		2.69	7.60		10.29	14.30
9807	1"		66	.121		3.55	7.80		11.35	15.60
9808	1-1/4"		64	.125		3.64	8.05		11.69	16.05

For customer support on your Mechanical Costs with RSMeans Data, call 800.448.8182.

22 11 Facility Water Distribution

22 11 13 – Facility Water Distribution Piping

22 11 13.76 Pipe Fittings, Plastic		Crew	Daily Output	Labor-Hours	Unit	Material	2020 Bare Costs Labor	Equipment	Total	Total Incl O&P
9809	1-1/2"	1 Plum	62	.129	Ea.	3.98	8.30		12.28	16.85
9810	2"	Q-1	110	.145		4.94	8.45		13.39	18.10
9811	2-1/2"		104	.154		6.35	8.90		15.25	20.50
9812	3"		98	.163		6.85	9.45		16.30	21.50
9813	3-1/2"		92	.174		7.50	10.10		17.60	23.50
9814	4"		86	.186		8.40	10.80		19.20	25.50
9815	6"		70	.229		10.20	13.25		23.45	31
9816	8"	Q-2	100	.240		36.50	14.45		50.95	62
9817	10"	"	94	.255		42	15.35		57.35	69
9820	Pipe cutter									
9822	For 3/8" thru 1-1/4"				Ea.	113			113	124
9824	For 1-1/2" thru 4"				"	296			296	325
9826	Note: Pipes may be cut with standard									
9827	iron saw with blades for plastic.									
9982	For plastic hangers see Line 22 05 29.10 8000									
9986	For copper/brass fittings see Line 22 11 13.25 7000									

22 11 13.78 Pipe, High Density Polyethylene Plastic (HDPE)

		Crew	Daily Output	Labor-Hours	Unit	Material	2020 Bare Costs Labor	Equipment	Total	Total Incl O&P
0010	**PIPE, HIGH DENSITY POLYETHYLENE PLASTIC (HDPE)**									
0020	Not incl. hangers, trenching, backfill, hoisting or digging equipment.									
0030	Standard length is 40', add a weld for each joint									
0035	For HDPE weld machine see 015433401685 in equipment rental									
0040	Single wall									
0050	Straight									
0054	1" diameter DR 11				L.F.	.89			.89	.98
0058	1-1/2" diameter DR 11					1.14			1.14	1.25
0062	2" diameter DR 11					1.89			1.89	2.08
0066	3" diameter DR 11					2.29			2.29	2.52
0070	3" diameter DR 17					1.83			1.83	2.01
0074	4" diameter DR 11					3.83			3.83	4.21
0078	4" diameter DR 17					3.83			3.83	4.21
0082	6" diameter DR 11					9.50			9.50	10.45
0086	6" diameter DR 17					6.30			6.30	6.90
0090	8" diameter DR 11					15.80			15.80	17.40
0094	8" diameter DR 26					7.40			7.40	8.15
0098	10" diameter DR 11					25			25	27
0102	10" diameter DR 26					11.45			11.45	12.60
0106	12" diameter DR 11					36			36	40
0110	12" diameter DR 26					17.15			17.15	18.85
0114	16" diameter DR 11					55			55	61
0118	16" diameter DR 26					25			25	27
0122	18" diameter DR 11					70.50			70.50	77.50
0126	18" diameter DR 26					32.50			32.50	35.50
0130	20" diameter DR 11					85.50			85.50	94.50
0134	20" diameter DR 26					38			38	42
0138	22" diameter DR 11					105			105	115
0142	22" diameter DR 26					47.50			47.50	52.50
0146	24" diameter DR 11					124			124	136
0150	24" diameter DR 26					55			55	61
0154	28" diameter DR 17					114			114	126
0158	28" diameter DR 26					76			76	84
0162	30" diameter DR 21					107			107	117
0166	30" diameter DR 26					87.50			87.50	96.50

For customer support on your Mechanical Costs with RSMeans Data, call 800.448.8182.

219

22 11 13.78 Pipe, High Density Polyethylene Plastic (HDPE)	Crew	Daily Output	Labor-Hours	Unit	Material	2020 Bare Costs Labor	Equipment	Total	Total Incl O&P
0170	36" diameter DR 26				L.F.	87.50			87.50	96.50
0174	42" diameter DR 26					170			170	186
0178	48" diameter DR 26					223			223	245
0182	54" diameter DR 26				↓	280			280	310
0300	90° elbow									
0304	1" diameter DR 11				Ea.	5.50			5.50	6.05
0308	1-1/2" diameter DR 11					7.30			7.30	8.05
0312	2" diameter DR 11					7.30			7.30	8.05
0316	3" diameter DR 11					14.60			14.60	16.05
0320	3" diameter DR 17					14.60			14.60	16.05
0324	4" diameter DR 11					20.50			20.50	22.50
0328	4" diameter DR 17					20.50			20.50	22.50
0332	6" diameter DR 11					46.50			46.50	51.50
0336	6" diameter DR 17					46.50			46.50	51.50
0340	8" diameter DR 11					115			115	127
0344	8" diameter DR 26					102			102	112
0348	10" diameter DR 11					430			430	475
0352	10" diameter DR 26					400			400	440
0356	12" diameter DR 11					455			455	500
0360	12" diameter DR 26					410			410	450
0364	16" diameter DR 11					540			540	590
0368	16" diameter DR 26					530			530	585
0372	18" diameter DR 11					675			675	745
0376	18" diameter DR 26					635			635	700
0380	20" diameter DR 11					795			795	875
0384	20" diameter DR 26					770			770	845
0388	22" diameter DR 11					820			820	905
0392	22" diameter DR 26					795			795	875
0396	24" diameter DR 11					930			930	1,025
0400	24" diameter DR 26					900			900	990
0404	28" diameter DR 17					1,125			1,125	1,250
0408	28" diameter DR 26					1,050			1,050	1,175
0412	30" diameter DR 17					1,600			1,600	1,750
0416	30" diameter DR 26					1,450			1,450	1,600
0420	36" diameter DR 26					1,850			1,850	2,050
0424	42" diameter DR 26					2,400			2,400	2,625
0428	48" diameter DR 26					2,775			2,775	3,075
0432	54" diameter DR 26				▼	6,625			6,625	7,300
0500	45° elbow									
0512	2" diameter DR 11				Ea.	5.85			5.85	6.45
0516	3" diameter DR 11					14.60			14.60	16.05
0520	3" diameter DR 17					14.60			14.60	16.05
0524	4" diameter DR 11					20.50			20.50	22.50
0528	4" diameter DR 17					20.50			20.50	22.50
0532	6" diameter DR 11					46.50			46.50	51.50
0536	6" diameter DR 17					46.50			46.50	51.50
0540	8" diameter DR 11					115			115	127
0544	8" diameter DR 26					67			67	74
0548	10" diameter DR 11					430			430	475
0552	10" diameter DR 26					400			400	440
0556	12" diameter DR 11					455			455	500
0560	12" diameter DR 26					410			410	450
0564	16" diameter DR 11				↓	239			239	263

22 11 13 – Facility Water Distribution Piping

22 11 13.78 Pipe, High Density Polyethylene Plastic (HDPE)	Crew	Daily Output	Labor-Hours	Unit	Material	2020 Bare Costs Labor	Equipment	Total	Total Incl O&P
0568	16" diameter DR 26				Ea.	226			226	248
0572	18" diameter DR 11					253			253	279
0576	18" diameter DR 26					232			232	255
0580	20" diameter DR 11					385			385	425
0584	20" diameter DR 26					365			365	400
0588	22" diameter DR 11					530			530	585
0592	22" diameter DR 26					505			505	555
0596	24" diameter DR 11					650			650	715
0600	24" diameter DR 26					625			625	685
0604	28" diameter DR 17					740			740	810
0608	28" diameter DR 26					715			715	790
0612	30" diameter DR 17					915			915	1,000
0616	30" diameter DR 26					890			890	980
0620	36" diameter DR 26					1,125			1,125	1,250
0624	42" diameter DR 26					1,450			1,450	1,600
0628	48" diameter DR 26					1,575			1,575	1,725
0632	54" diameter DR 26				▼	2,100			2,100	2,325
0700	Tee									
0704	1" diameter DR 11				Ea.	7.60			7.60	8.40
0708	1-1/2" diameter DR 11					10.70			10.70	11.75
0712	2" diameter DR 11					9.15			9.15	10.10
0716	3" diameter DR 11					16.75			16.75	18.45
0720	3" diameter DR 17					16.75			16.75	18.45
0724	4" diameter DR 11					24.50			24.50	27
0728	4" diameter DR 17					24.50			24.50	27
0732	6" diameter DR 11					61			61	67
0736	6" diameter DR 17					61			61	67
0740	8" diameter DR 11					151			151	166
0744	8" diameter DR 17					151			151	166
0748	10" diameter DR 11					450			450	495
0752	10" diameter DR 17					450			450	495
0756	12" diameter DR 11					600			600	660
0760	12" diameter DR 17					600			600	660
0764	16" diameter DR 11					315			315	350
0768	16" diameter DR 17					261			261	287
0772	18" diameter DR 11					445			445	490
0776	18" diameter DR 17					365			365	405
0780	20" diameter DR 11					545			545	600
0784	20" diameter DR 17					445			445	490
0788	22" diameter DR 11					700			700	770
0792	22" diameter DR 17					555			555	610
0796	24" diameter DR 11					870			870	955
0800	24" diameter DR 17					730			730	805
0804	28" diameter DR 17					1,425			1,425	1,550
0812	30" diameter DR 17					1,625			1,625	1,775
0820	36" diameter DR 17					2,675			2,675	2,950
0824	42" diameter DR 26					2,975			2,975	3,275
0828	48" diameter DR 26				▼	3,200			3,200	3,525
1000	Flange adptr, w/back-up ring and 1/2 cost of plated bolt set									
1004	1" diameter DR 11				Ea.	29			29	32
1008	1-1/2" diameter DR 11					29			29	32
1012	2" diameter DR 11					18.30			18.30	20
1016	3" diameter DR 11				▼	21.50			21.50	23.50

For customer support on your Mechanical Costs with RSMeans Data, call 800.448.8182.

221

22 11 13 – Facility Water Distribution Piping

22 11 13.78 Pipe, High Density Polyethylene Plastic (HDPE)	Crew	Daily Output	Labor-Hours	Unit	Material	2020 Bare Costs Labor	Equipment	Total	Total Incl O&P
1020	3" diameter DR 17				Ea.	21.50			21.50	23.50
1024	4" diameter DR 11					29			29	32
1028	4" diameter DR 17					29			29	32
1032	6" diameter DR 11					41			41	45.50
1036	6" diameter DR 17					41			41	45.50
1040	8" diameter DR 11					59.50			59.50	65.50
1044	8" diameter DR 26					59.50			59.50	65.50
1048	10" diameter DR 11					94.50			94.50	104
1052	10" diameter DR 26					94.50			94.50	104
1056	12" diameter DR 11					139			139	153
1060	12" diameter DR 26					139			139	153
1064	16" diameter DR 11					297			297	325
1068	16" diameter DR 26					297			297	325
1072	18" diameter DR 11					385			385	425
1076	18" diameter DR 26					385			385	425
1080	20" diameter DR 11					535			535	590
1084	20" diameter DR 26					535			535	590
1088	22" diameter DR 11					580			580	640
1092	22" diameter DR 26					580			580	640
1096	24" diameter DR 17					630			630	695
1100	24" diameter DR 32.5					630			630	695
1104	28" diameter DR 15.5					855			855	940
1108	28" diameter DR 32.5					855			855	940
1112	30" diameter DR 11					990			990	1,100
1116	30" diameter DR 21					990			990	1,100
1120	36" diameter DR 26					1,075			1,075	1,200
1124	42" diameter DR 26					1,225			1,225	1,350
1128	48" diameter DR 26					1,500			1,500	1,650
1132	54" diameter DR 26				▼	1,825			1,825	2,025
1200	Reducer									
1208	2" x 1-1/2" diameter DR 11				Ea.	8.75			8.75	9.65
1212	3" x 2" diameter DR 11					8.75			8.75	9.65
1216	4" x 2" diameter DR 11					10.20			10.20	11.25
1220	4" x 3" diameter DR 11					13.15			13.15	14.45
1224	6" x 4" diameter DR 11					30.50			30.50	33.50
1228	8" x 6" diameter DR 11					46.50			46.50	51.50
1232	10" x 8" diameter DR 11					80.50			80.50	88.50
1236	12" x 8" diameter DR 11					131			131	144
1240	12" x 10" diameter DR 11					105			105	116
1244	14" x 12" diameter DR 11					117			117	128
1248	16" x 14" diameter DR 11					149			149	164
1252	18" x 16" diameter DR 11					181			181	199
1256	20" x 18" diameter DR 11					360			360	395
1260	22" x 20" diameter DR 11					440			440	485
1264	24" x 22" diameter DR 11					495			495	545
1268	26" x 24" diameter DR 11					585			585	640
1272	28" x 24" diameter DR 11					745			745	820
1276	32" x 28" diameter DR 17					965			965	1,050
1280	36" x 32" diameter DR 17				▼	1,325			1,325	1,450
4000	Welding labor per joint, not including welding machine									
4010	Pipe joint size (cost based on thickest wall for each diam.)									
4030	1" pipe size	4 Skwk	273	.117	Ea.		6.45		6.45	9.70
4040	1-1/2" pipe size		175	.183			10.05		10.05	15.15

22 11 Facility Water Distribution

22 11 13 – Facility Water Distribution Piping

22 11 13.78 Pipe, High Density Polyethylene Plastic (HDPE)	Crew	Daily Output	Labor-Hours	Unit	Material	2020 Bare Costs Labor	Equipment	Total	Total Incl O&P
4050	2" pipe size	4 Skwk	128	.250	Ea.		13.70		13.70	20.50
4060	3" pipe size		100	.320			17.55		17.55	26.50
4070	4" pipe size	↓	77	.416			23		23	34.50
4080	6" pipe size	5 Skwk	63	.635			35		35	52.50
4090	8" pipe size		48	.833			45.50		45.50	69
4100	10" pipe size	↓	40	1			55		55	83
4110	12" pipe size	6 Skwk	41	1.171			64		64	97
4120	16" pipe size		34	1.412			77.50		77.50	117
4130	18" pipe size	↓	32	1.500			82.50		82.50	124
4140	20" pipe size	8 Skwk	37	1.730			95		95	143
4150	22" pipe size		35	1.829			100		100	152
4160	24" pipe size		34	1.882			103		103	156
4170	28" pipe size		33	1.939			106		106	161
4180	30" pipe size		32	2			110		110	166
4190	36" pipe size	↓	31	2.065			113		113	171
4200	42" pipe size	↓	30	2.133			117		117	177
4210	48" pipe size	9 Skwk	33	2.182			120		120	181
4220	54" pipe size	"	31	2.323	↓		127		127	193
5000	Dual wall contained pipe									
5040	Straight									
5054	1" DR 11 x 3" DR 11				L.F.	5.35			5.35	5.90
5058	1" DR 11 x 4" DR 11					5.65			5.65	6.25
5062	1-1/2" DR 11 x 4" DR 17					6.25			6.25	6.90
5066	2" DR 11 x 4" DR 17					6.60			6.60	7.25
5070	2" DR 11 x 6" DR 17					10			10	11
5074	3" DR 11 x 6" DR 17					11.20			11.20	12.30
5078	3" DR 11 x 6" DR 26					9.55			9.55	10.50
5086	3" DR 17 x 8" DR 17					12.15			12.15	13.40
5090	4" DR 11 x 8" DR 17					16.85			16.85	18.55
5094	4" DR 17 x 8" DR 26					13.20			13.20	14.50
5098	6" DR 11 x 10" DR 17					26			26	29
5102	6" DR 17 x 10" DR 26					17.80			17.80	19.60
5106	6" DR 26 x 10" DR 26					16.70			16.70	18.35
5110	8" DR 17 x 12" DR 26					25.50			25.50	28.50
5114	8" DR 26 x 12" DR 32.5					23.50			23.50	26
5118	10" DR 17 x 14" DR 26					38.50			38.50	42
5122	10" DR 17 x 16" DR 26					27.50			27.50	30.50
5126	10" DR 26 x 16" DR 26					36.50			36.50	40
5130	12" DR 26 x 16" DR 26					38.50			38.50	42.50
5134	12" DR 17 x 18" DR 26					53.50			53.50	58.50
5138	12" DR 26 x 18" DR 26					50.50			50.50	56
5142	14" DR 26 x 20" DR 32.5					51.50			51.50	57
5146	16" DR 26 x 22" DR 32.5					57			57	63
5150	18" DR 26 x 24" DR 32.5					71			71	78.50
5154	20" DR 32.5 x 28" DR 32.5					80.50			80.50	88.50
5158	22" DR 32.5 x 30" DR 32.5					80.50			80.50	88.50
5162	24" DR 32.5 x 32" DR 32.5					83.50			83.50	92
5166	36" DR 32.5 x 42" DR 32.5				↓	97.50			97.50	107
5300	Force transfer coupling									
5354	1" DR 11 x 3" DR 11				Ea.	188			188	207
5358	1" DR 11 x 4" DR 17					221			221	243
5362	1-1/2" DR 11 x 4" DR 17					212			212	233
5366	2" DR 11 x 4" DR 17				↓	212			212	233

22 11 13.78 Pipe, High Density Polyethylene Plastic (HDPE)	Crew	Daily Output	Labor-Hours	Unit	Material	2020 Bare Costs Labor	Equipment	Total	Total Incl O&P
5370	2" DR 11 x 6" DR 17				Ea.	365			365	400
5374	3" DR 11 x 6" DR 17					390			390	425
5378	3" DR 11 x 6" DR 26					390			390	425
5382	3" DR 11 x 8" DR 11					390			390	430
5386	3" DR 11 x 8" DR 17					425			425	465
5390	4" DR 11 x 8" DR 17					435			435	475
5394	4" DR 17 x 8" DR 26					440			440	480
5398	6" DR 11 x 10" DR 17					520			520	570
5402	6" DR 17 x 10" DR 26					495			495	545
5406	6" DR 26 x 10" DR 26					675			675	745
5410	8" DR 17 x 12" DR 26					720			720	790
5414	8" DR 26 x 12" DR 32.5					830			830	910
5418	10" DR 17 x 14" DR 26					980			980	1,075
5422	10" DR 17 x 16" DR 26					990			990	1,100
5426	10" DR 26 x 16" DR 26					1,100			1,100	1,200
5430	12" DR 26 x 16" DR 26					1,350			1,350	1,475
5434	12" DR 17 x 18" DR 26					1,350			1,350	1,475
5438	12" DR 26 x 18" DR 26					1,475			1,475	1,625
5442	14" DR 26 x 20" DR 32.5					1,650			1,650	1,800
5446	16" DR 26 x 22" DR 32.5					1,825			1,825	2,000
5450	18" DR 26 x 24" DR 32.5					2,000			2,000	2,200
5454	20" DR 32.5 x 28" DR 32.5					2,200			2,200	2,425
5458	22" DR 32.5 x 30" DR 32.5					2,425			2,425	2,675
5462	24" DR 32.5 x 32" DR 32.5					2,675			2,675	2,950
5466	36" DR 32.5 x 42" DR 32.5					2,875			2,875	3,175
5600	90° elbow									
5654	1" DR 11 x 3" DR 11				Ea.	206			206	226
5658	1" DR 11 x 4" DR 17					243			243	267
5662	1-1/2" DR 11 x 4" DR 17					250			250	275
5666	2" DR 11 x 4" DR 17					244			244	269
5670	2" DR 11 x 6" DR 17					355			355	390
5674	3" DR 11 x 6" DR 17					380			380	420
5678	3" DR 17 x 6" DR 26					390			390	425
5682	3" DR 17 x 8" DR 11					520			520	575
5686	3" DR 17 x 8" DR 17					555			555	610
5690	4" DR 11 x 8" DR 17					470			470	520
5694	4" DR 17 x 8" DR 26					450			450	495
5698	6" DR 11 x 10" DR 17					820			820	905
5702	6" DR 17 x 10" DR 26					835			835	920
5706	6" DR 26 x 10" DR 26					815			815	900
5710	8" DR 17 x 12" DR 26					1,100			1,100	1,200
5714	8" DR 26 x 12" DR 32.5					905			905	995
5718	10" DR 17 x 14" DR 26					1,075			1,075	1,175
5722	10" DR 17 x 16" DR 26					1,050			1,050	1,150
5726	10" DR 26 x 16" DR 26					1,025			1,025	1,150
5730	12" DR 26 x 16" DR 26					1,050			1,050	1,175
5734	12" DR 17 x 18" DR 26					1,150			1,150	1,275
5738	12" DR 26 x 18" DR 26					1,175			1,175	1,300
5742	14" DR 26 x 20" DR 32.5					1,275			1,275	1,400
5746	16" DR 26 x 22" DR 32.5					1,325			1,325	1,450
5750	18" DR 26 x 24" DR 32.5					1,375			1,375	1,500
5754	20" DR 32.5 x 28" DR 32.5					1,425			1,425	1,550
5758	22" DR 32.5 x 30" DR 32.5					1,475			1,475	1,600

22 11 13.78 Pipe, High Density Polyethylene Plastic (HDPE)	Crew	Daily Output	Labor-Hours	Unit	Material	2020 Bare Costs Labor	Equipment	Total	Total Incl O&P
5762	24" DR 32.5 x 32" DR 32.5				Ea.	1,525			1,525	1,675
5766	36" DR 32.5 x 42" DR 32.5				↓	1,575			1,575	1,725
5800	45° elbow									
5804	1" DR 11 x 3" DR 11				Ea.	222			222	244
5808	1" DR 11 x 4" DR 17					195			195	215
5812	1-1/2" DR 11 x 4" DR 17					199			199	219
5816	2" DR 11 x 4" DR 17					190			190	209
5820	2" DR 11 x 6" DR 17					272			272	299
5824	3" DR 11 x 6" DR 17					292			292	320
5828	3" DR 17 x 6" DR 26					295			295	325
5832	3" DR 17 x 8" DR 11					375			375	410
5836	3" DR 17 x 8" DR 17					420			420	465
5840	4" DR 11 x 8" DR 17					345			345	380
5844	4" DR 17 x 8" DR 26					350			350	385
5848	6" DR 11 x 10" DR 17					545			545	600
5852	6" DR 17 x 10" DR 26					555			555	610
5856	6" DR 26 x 10" DR 26					625			625	685
5860	8" DR 17 x 12" DR 26					740			740	815
5864	8" DR 26 x 12" DR 32.5					695			695	765
5868	10" DR 17 x 14" DR 26					785			785	860
5872	10" DR 17 x 16" DR 26					755			755	830
5876	10" DR 26 x 16" DR 26					800			800	880
5880	12" DR 26 x 16" DR 26					720			720	795
5884	12" DR 17 x 18" DR 26					1,025			1,025	1,125
5888	12" DR 26 x 18" DR 26					910			910	1,000
5892	14" DR 26 x 20" DR 32.5					990			990	1,100
5896	16" DR 26 x 22" DR 32.5					1,025			1,025	1,125
5900	18" DR 26 x 24" DR 32.5					1,100			1,100	1,200
5904	20" DR 32.5 x 28" DR 32.5					1,075			1,075	1,200
5908	22" DR 32.5 x 30" DR 32.5					1,125			1,125	1,225
5912	24" DR 32.5 x 32" DR 32.5					1,150			1,150	1,275
5916	36" DR 32.5 x 42" DR 32.5				↓	1,200			1,200	1,325
6000	Access port with 4" riser									
6050	1" DR 11 x 4" DR 17				Ea.	284			284	310
6054	1-1/2" DR 11 x 4" DR 17					286			286	315
6058	2" DR 11 x 6" DR 17					350			350	385
6062	3" DR 11 x 6" DR 17					355			355	390
6066	3" DR 17 x 6" DR 26					345			345	380
6070	3" DR 17 x 8" DR 11					355			355	390
6074	3" DR 17 x 8" DR 17					370			370	405
6078	4" DR 11 x 8" DR 17					380			380	415
6082	4" DR 17 x 8" DR 26					360			360	400
6086	6" DR 11 x 10" DR 17					440			440	485
6090	6" DR 17 x 10" DR 26					405			405	445
6094	6" DR 26 x 10" DR 26					425			425	465
6098	8" DR 17 x 12" DR 26					455			455	500
6102	8" DR 26 x 12" DR 32.5				↓	445			445	490
6200	End termination with vent plug									
6204	1" DR 11 x 3" DR 11				Ea.	116			116	128
6208	1" DR 11 x 4" DR 17					141			141	155
6212	1-1/2" DR 11 x 4" DR 17					141			141	155
6216	2" DR 11 x 4" DR 17					141			141	155
6220	2" DR 11 x 6" DR 17				↓	286			286	315

For customer support on your Mechanical Costs with RSMeans Data, call 800.448.8182.

225

22 11 13.78 Pipe, High Density Polyethylene Plastic (HDPE)		Crew	Daily Output	Labor-Hours	Unit	Material	2020 Bare Costs Labor	Equipment	Total	Total Incl O&P
6224	3" DR 11 x 6" DR 17				Ea.	286			286	315
6228	3" DR 17 x 6" DR 26					286			286	315
6232	3" DR 17 x 8" DR 11					276			276	305
6236	3" DR 17 x 8" DR 17					310			310	340
6240	4" DR 11 x 8" DR 17					370			370	405
6244	4" DR 17 x 8" DR 26					345			345	380
6248	6" DR 11 x 10" DR 17					400			400	440
6252	6" DR 17 x 10" DR 26					370			370	410
6256	6" DR 26 x 10" DR 26					540			540	595
6260	8" DR 17 x 12" DR 26					565			565	625
6264	8" DR 26 x 12" DR 32.5					680			680	745
6268	10" DR 17 x 14" DR 26					900			900	990
6272	10" DR 17 x 16" DR 26					850			850	935
6276	10" DR 26 x 16" DR 26					955			955	1,050
6280	12" DR 26 x 16" DR 26					1,050			1,050	1,175
6284	12" DR 17 x 18" DR 26					1,225			1,225	1,350
6288	12" DR 26 x 18" DR 26					1,350			1,350	1,475
6292	14" DR 26 x 20" DR 32.5					1,525			1,525	1,675
6296	16" DR 26 x 22" DR 32.5					1,700			1,700	1,850
6300	18" DR 26 x 24" DR 32.5					1,900			1,900	2,100
6304	20" DR 32.5 x 28" DR 32.5					2,125			2,125	2,325
6308	22" DR 32.5 x 30" DR 32.5					2,375			2,375	2,600
6312	24" DR 32.5 x 32" DR 32.5					2,650			2,650	2,925
6316	36" DR 32.5 x 42" DR 32.5				▼	2,925			2,925	3,225
6600	Tee									
6604	1" DR 11 x 3" DR 11				Ea.	239			239	263
6608	1" DR 11 x 4" DR 17					260			260	286
6612	1-1/2" DR 11 x 4" DR 17					269			269	296
6616	2" DR 11 x 4" DR 17					244			244	268
6620	2" DR 11 x 6" DR 17					390			390	430
6624	3" DR 11 x 6" DR 17					460			460	505
6628	3" DR 17 x 6" DR 26					435			435	480
6632	3" DR 17 x 8" DR 11					535			535	590
6636	3" DR 17 x 8" DR 17					580			580	640
6640	4" DR 11 x 8" DR 17					700			700	770
6644	4" DR 17 x 8" DR 26					700			700	770
6648	6" DR 11 x 10" DR 17					910			910	1,000
6652	6" DR 17 x 10" DR 26					915			915	1,000
6656	6" DR 26 x 10" DR 26					1,125			1,125	1,225
6660	8" DR 17 x 12" DR 26					1,125			1,125	1,225
6664	8" DR 26 x 12" DR 32.5					1,450			1,450	1,600
6668	10" DR 17 x 14" DR 26					1,675			1,675	1,850
6672	10" DR 17 x 16" DR 26					1,875			1,875	2,050
6676	10" DR 26 x 16" DR 26					1,875			1,875	2,050
6680	12" DR 26 x 16" DR 26					2,125			2,125	2,350
6684	12" DR 17 x 18" DR 26					2,125			2,125	2,350
6688	12" DR 26 x 18" DR 26					2,425			2,425	2,650
6692	14" DR 26 x 20" DR 32.5					2,725			2,725	3,000
6696	16" DR 26 x 22" DR 32.5					3,075			3,075	3,400
6700	18" DR 26 x 24" DR 32.5					3,500			3,500	3,850
6704	20" DR 32.5 x 28" DR 32.5					3,950			3,950	4,350
6708	22" DR 32.5 x 30" DR 32.5					4,475			4,475	4,925
6712	24" DR 32.5 x 32" DR 32.5					5,075			5,075	5,575

22 11 Facility Water Distribution

22 11 13 – Facility Water Distribution Piping

22 11 13.78 Pipe, High Density Polyethylene Plastic (HDPE)	Crew	Daily Output	Labor-Hours	Unit	Material	2020 Bare Costs Labor	Equipment	Total	Total Incl O&P
6716	36" DR 32.5 x 42" DR 32.5				Ea.	5,825			5,825	6,400
6800	Wye									
6816	2" DR 11 x 4" DR 17				Ea.	595			595	655
6820	2" DR 11 x 6" DR 17					735			735	810
6824	3" DR 11 x 6" DR 17					780			780	855
6828	3" DR 17 x 6" DR 26					780			780	855
6832	3" DR 17 x 8" DR 11					830			830	910
6836	3" DR 17 x 8" DR 17					900			900	990
6840	4" DR 11 x 8" DR 17					990			990	1,100
6844	4" DR 17 x 8" DR 26					625			625	685
6848	6" DR 11 x 10" DR 17					1,525			1,525	1,675
6852	6" DR 17 x 10" DR 26					950			950	1,050
6856	6" DR 26 x 10" DR 26					1,325			1,325	1,450
6860	8" DR 17 x 12" DR 26					1,850			1,850	2,025
6864	8" DR 26 x 12" DR 32.5					1,525			1,525	1,675
6868	10" DR 17 x 14" DR 26					1,625			1,625	1,800
6872	10" DR 17 x 16" DR 26					1,750			1,750	1,925
6876	10" DR 26 x 16" DR 26					1,900			1,900	2,100
6880	12" DR 26 x 16" DR 26					2,050			2,050	2,250
6884	12" DR 17 x 18" DR 26					2,200			2,200	2,425
6888	12" DR 26 x 18" DR 26					2,400			2,400	2,625
6892	14" DR 26 x 20" DR 32.5					2,550			2,550	2,825
6896	16" DR 26 x 22" DR 32.5					2,700			2,700	2,975
6900	18" DR 26 x 24" DR 32.5					3,100			3,100	3,400
6904	20" DR 32.5 x 28" DR 32.5					3,225			3,225	3,550
6908	22" DR 32.5 x 30" DR 32.5					3,500			3,500	3,850
6912	24" DR 32.5 x 32" DR 32.5					3,775			3,775	4,150
9000	Welding labor per joint, not including welding machine									
9010	Pipe joint size, outer pipe (cost based on the thickest walls)									
9020	Straight pipe									
9050	3" pipe size	4 Skwk	96	.333	Ea.		18.30		18.30	27.50
9060	4" pipe size	"	77	.416			23		23	34.50
9070	6" pipe size	5 Skwk	60	.667			36.50		36.50	55.50
9080	8" pipe size	"	40	1			55		55	83
9090	10" pipe size	6 Skwk	41	1.171			64		64	97
9100	12" pipe size		39	1.231			67.50		67.50	102
9110	14" pipe size		38	1.263			69.50		69.50	105
9120	16" pipe size		35	1.371			75		75	114
9130	18" pipe size	8 Skwk	45	1.422			78		78	118
9140	20" pipe size		42	1.524			83.50		83.50	126
9150	22" pipe size		40	1.600			88		88	133
9160	24" pipe size		38	1.684			92.50		92.50	140
9170	28" pipe size		37	1.730			95		95	143
9180	30" pipe size		36	1.778			97.50		97.50	147
9190	32" pipe size		35	1.829			100		100	152
9200	42" pipe size		32	2			110		110	166

22 11 19 – Domestic Water Piping Specialties

22 11 19.10 Flexible Connectors

0010	**FLEXIBLE CONNECTORS**, Corrugated, 5/8" OD, 3/4" ID									
0050	Gas, seamless brass, steel fittings									
0200	12" long	1 Plum	36	.222	Ea.	6.80	14.30		21.10	29
0220	18" long		36	.222		8.55	14.30		22.85	31

For customer support on your Mechanical Costs with RSMeans Data, call 800.448.8182.

227

22 11 19.10 Flexible Connectors

		Crew	Daily Output	Labor-Hours	Unit	Material	2020 Bare Costs Labor	Equipment	Total	Total Incl O&P
0240	24" long	1 Plum	34	.235	Ea.	10	15.15		25.15	33.50
0260	30" long		34	.235		12.45	15.15		27.60	36
0280	36" long		32	.250		11.90	16.10		28	37
0320	48" long		30	.267		15.10	17.20		32.30	42
0340	60" long		30	.267		17.95	17.20		35.15	45.50
0360	72" long		30	.267		21	17.20		38.20	48.50
2000	Water, copper tubing, dielectric separators									
2100	12" long	1 Plum	36	.222	Ea.	8.20	14.30		22.50	30.50
2220	15" long		36	.222		9.15	14.30		23.45	31.50
2240	18" long		36	.222		9.10	14.30		23.40	31.50
2260	24" long		34	.235		12.20	15.15		27.35	36

22 11 19.14 Flexible Metal Hose

		Crew	Daily Output	Labor-Hours	Unit	Material	2020 Bare Costs Labor	Equipment	Total	Total Incl O&P
0010	**FLEXIBLE METAL HOSE**, Connectors, standard lengths									
0100	Bronze braided, bronze ends									
0120	3/8" diameter x 12"	1 Stpi	26	.308	Ea.	28.50	20		48.50	61
0140	1/2" diameter x 12"		24	.333		29	22		51	64.50
0160	3/4" diameter x 12"		20	.400		39.50	26		65.50	82.50
0180	1" diameter x 18"		19	.421		51	27.50		78.50	97.50
0200	1-1/2" diameter x 18"		13	.615		57	40.50		97.50	123
0220	2" diameter x 18"		11	.727		93.50	47.50		141	175
1000	Carbon steel ends									
1020	1/4" diameter x 12"	1 Stpi	28	.286	Ea.	36	18.75		54.75	67.50
1040	3/8" diameter x 12"		26	.308		37	20		57	70.50
1060	1/2" diameter x 12"		24	.333		32	22		54	68
1080	1/2" diameter x 24"		24	.333		61.50	22		83.50	101
1120	3/4" diameter x 12"		20	.400		47	26		73	90.50
1140	3/4" diameter x 24"		20	.400		75	26		101	122
1160	3/4" diameter x 36"		20	.400		82	26		108	130
1180	1" diameter x 18"		19	.421		65.50	27.50		93	114
1200	1" diameter x 30"		19	.421		69.50	27.50		97	118
1220	1" diameter x 36"		19	.421		104	27.50		131.50	156
1240	1-1/4" diameter x 18"		15	.533		88.50	35		123.50	150
1260	1-1/4" diameter x 36"		15	.533		141	35		176	208
1280	1-1/2" diameter x 18"		13	.615		121	40.50		161.50	194
1300	1-1/2" diameter x 36"		13	.615		145	40.50		185.50	221
1320	2" diameter x 24"		11	.727		182	47.50		229.50	272
1340	2" diameter x 36"		11	.727		183	47.50		230.50	274
1360	2-1/2" diameter x 24"		9	.889		350	58.50		408.50	470
1380	2-1/2" diameter x 36"		9	.889		126	58.50		184.50	226
1400	3" diameter x 24"		7	1.143		465	75		540	620
1420	3" diameter x 36"		7	1.143		1,025	75		1,100	1,250
2000	Carbon steel braid, carbon steel solid ends									
2100	1/2" diameter x 12"	1 Stpi	24	.333	Ea.	50.50	22		72.50	88.50
2120	3/4" diameter x 12"		20	.400		78.50	26		104.50	126
2140	1" diameter x 12"		19	.421		112	27.50		139.50	165
2160	1-1/4" diameter x 12"		15	.533		55	35		90	113
2180	1-1/2" diameter x 12"		13	.615		60.50	40.50		101	127
3000	Stainless steel braid, welded on carbon steel ends									
3100	1/2" diameter x 12"	1 Stpi	24	.333	Ea.	74.50	22		96.50	115
3120	3/4" diameter x 12"		20	.400		90	26		116	138
3140	3/4" diameter x 24"		20	.400		103	26		129	152
3160	3/4" diameter x 36"		20	.400		119	26		145	170

22 11 19 – Domestic Water Piping Specialties

22 11 19.14 Flexible Metal Hose		Crew	Daily Output	Labor-Hours	Unit	Material	2020 Bare Costs Labor	Equipment	Total	Total Incl O&P
3180	1" diameter x 12"	1 Stpi	19	.421	Ea.	108	27.50		135.50	161
3200	1" diameter x 24"		19	.421		115	27.50		142.50	168
3220	1" diameter x 36"		19	.421		147	27.50		174.50	204
3240	1-1/4" diameter x 12"		15	.533		151	35		186	219
3260	1-1/4" diameter x 24"		15	.533		178	35		213	249
3280	1-1/4" diameter x 36"		15	.533		200	35		235	273
3300	1-1/2" diameter x 12"		13	.615		62.50	40.50		103	130
3320	1-1/2" diameter x 24"		13	.615		186	40.50		226.50	266
3340	1-1/2" diameter x 36"	▼	13	.615	▼	226	40.50		266.50	310
3400	Metal stainless steel braid, over corrugated stainless steel, flanged ends									
3410	150 psi									
3420	1/2" diameter x 12"	1 Stpi	24	.333	Ea.	91	22		113	133
3430	1" diameter x 12"		20	.400		136	26		162	189
3440	1-1/2" diameter x 12"		15	.533		146	35		181	213
3450	2-1/2" diameter x 9"		12	.667		64.50	43.50		108	137
3460	3" diameter x 9"		9	.889		32	58.50		90.50	122
3470	4" diameter x 9"		7	1.143		41	75		116	157
3480	4" diameter x 30"		5	1.600		410	105		515	605
3490	4" diameter x 36"		4.80	1.667		430	109		539	635
3500	6" diameter x 11"		5	1.600		75	105		180	239
3510	6" diameter x 36"		3.80	2.105		525	138		663	780
3520	8" diameter x 12"		4	2		296	131		427	520
3530	10" diameter x 13"	▼	3	2.667		213	175		388	495
3540	12" diameter x 14"	Q-5	4	4	▼	335	236		571	725
6000	Molded rubber with helical wire reinforcement									
6010	150 psi									
6020	1-1/2" diameter x 12"	1 Stpi	15	.533	Ea.	31	35		66	86.50
6030	2" diameter x 12"		12	.667		78	43.50		121.50	152
6040	3" diameter x 12"		8	1		81.50	65.50		147	188
6050	4" diameter x 12"		6	1.333		103	87.50		190.50	245
6060	6" diameter x 18"		4	2		155	131		286	365
6070	8" diameter x 24"		3	2.667		217	175		392	500
6080	10" diameter x 24"	▼	2	4		263	262		525	680
6090	12" diameter x 24"	Q-5	3	5.333	▼	300	315		615	800
7000	Molded teflon with stainless steel flanges									
7010	150 psi									
7020	2-1/2" diameter x 3-3/16"	Q-1	7.80	2.051	Ea.	725	119		844	975
7030	3" diameter x 3-5/8"		6.50	2.462		550	143		693	820
7040	4" diameter x 3-5/8"		5	3.200		700	186		886	1,050
7050	6" diameter x 4"		4.30	3.721		975	216		1,191	1,400
7060	8" diameter x 6"	▼	3.80	4.211	▼	1,550	244		1,794	2,075

22 11 19.18 Mixing Valve

		Crew	Daily Output	Labor-Hours	Unit	Material	Labor	Equipment	Total	Total Incl O&P
0010	**MIXING VALVE**, Automatic, water tempering.									
0040	1/2" size	1 Stpi	19	.421	Ea.	625	27.50		652.50	730
0050	3/4" size		18	.444		660	29		689	770
0100	1" size		16	.500		930	33		963	1,075
0120	1-1/4" size		13	.615		1,250	40.50		1,290.50	1,425
0140	1-1/2" size		10	.800		1,525	52.50		1,577.50	1,750
0160	2" size		8	1		1,900	65.50		1,965.50	2,200
0170	2-1/2" size		6	1.333		1,900	87.50		1,987.50	2,225
0180	3" size		4	2		4,550	131		4,681	5,200
0190	4" size	▼	3	2.667	▼	4,550	175		4,725	5,250

22 11 Facility Water Distribution

22 11 19 – Domestic Water Piping Specialties

22 11 19.22 Pressure Reducing Valve	Crew	Daily Output	Labor-Hours	Unit	Material	2020 Bare Costs Labor	Equipment	Total	Total Incl O&P
0010 **PRESSURE REDUCING VALVE**, Steam, pilot operated.									
0100 Threaded, iron body									
0200 1-1/2" size	1 Stpi	8	1	Ea.	2,125	65.50		2,190.50	2,425
0220 2" size	"	5	1.600	"	2,450	105		2,555	2,850
1000 Flanged, iron body, 125 lb. flanges									
1020 2" size	1 Stpi	8	1	Ea.	3,400	65.50		3,465.50	3,850
1040 2-1/2" size	"	4	2		2,525	131		2,656	2,975
1060 3" size	Q-5	4.50	3.556		3,225	210		3,435	3,875
1080 4" size	"	3	5.333		4,050	315		4,365	4,925
1500 For 250 lb. flanges, add					5%				

22 11 19.26 Pressure Regulators

	Crew	Daily Output	Labor-Hours	Unit	Material	Labor	Equipment	Total	Total Incl O&P
0010 **PRESSURE REGULATORS**									
0100 Gas appliance regulators									
0106 Main burner and pilot applications									
0108 Rubber seat poppet type									
0109 1/8" pipe size	1 Stpi	24	.333	Ea.	19.65	22		41.65	54
0110 1/4" pipe size		24	.333		23.50	22		45.50	58
0112 3/8" pipe size		24	.333		24	22		46	58.50
0113 1/2" pipe size		24	.333		26.50	22		48.50	61.50
0114 3/4" pipe size		20	.400		31	26		57	73
0122 Lever action type									
0123 3/8" pipe size	1 Stpi	24	.333	Ea.	30	22		52	65.50
0124 1/2" pipe size		24	.333		30	22		52	65.50
0125 3/4" pipe size		20	.400		58	26		84	103
0126 1" pipe size		19	.421		58	27.50		85.50	106
0132 Double diaphragm type									
0133 3/8" pipe size	1 Stpi	24	.333	Ea.	46	22		68	83.50
0134 1/2" pipe size		24	.333		59.50	22		81.50	98
0135 3/4" pipe size		20	.400		96	26		122	144
0136 1" pipe size		19	.421		110	27.50		137.50	163
0137 1-1/4" pipe size		15	.533		395	35		430	490
0138 1-1/2" pipe size		13	.615		725	40.50		765.50	860
0139 2" pipe size		11	.727		725	47.50		772.50	870
0140 2-1/2" pipe size	Q-5	15	1.067		1,400	63		1,463	1,650
0141 3" pipe size		13	1.231		1,400	72.50		1,472.50	1,650
0142 4" pipe size (flanged)		8	2		2,500	118		2,618	2,925
0160 Main burner only									
0162 Straight-thru-flow design									
0163 1/2" pipe size	1 Stpi	24	.333	Ea.	48.50	22		70.50	86
0164 3/4" pipe size		20	.400		66.50	26		92.50	112
0165 1" pipe size		19	.421		88	27.50		115.50	138
0166 1-1/4" pipe size		15	.533		88	35		123	149
0200 Oil, light, hot water, ordinary steam, threaded									
0220 Bronze body, 1/4" size	1 Stpi	24	.333	Ea.	197	22		219	250
0230 3/8" size		24	.333		190	22		212	242
0240 1/2" size		24	.333		233	22		255	290
0250 3/4" size		20	.400		278	26		304	345
0260 1" size		19	.421		430	27.50		457.50	510
0270 1-1/4" size		15	.533		575	35		610	685
0280 1-1/2" size		13	.615		595	40.50		635.50	715
0290 2" size		11	.727		1,250	47.50		1,297.50	1,450
0320 Iron body, 1/4" size		24	.333		128	22		150	173

22 11 Facility Water Distribution

22 11 19 – Domestic Water Piping Specialties

22 11 19.26 Pressure Regulators		Crew	Daily Output	Labor-Hours	Unit	Material	2020 Bare Costs Labor	Equipment	Total	Total Incl O&P
0330	3/8" size	1 Stpi	24	.333	Ea.	172	22		194	223
0340	1/2" size		24	.333		157	22		179	206
0350	3/4" size		20	.400		191	26		217	249
0360	1" size		19	.421		247	27.50		274.50	315
0370	1-1/4" size		15	.533		345	35		380	435
0380	1-1/2" size		13	.615		380	40.50		420.50	480
0390	2" size		11	.727		550	47.50		597.50	675
0500	Oil, heavy, viscous fluids, threaded									
0520	Bronze body, 3/8" size	1 Stpi	24	.333	Ea.	300	22		322	365
0530	1/2" size		24	.333		365	22		387	435
0540	3/4" size		20	.400		430	26		456	515
0550	1" size		19	.421		505	27.50		532.50	600
0560	1-1/4" size		15	.533		760	35		795	890
0570	1-1/2" size		13	.615		870	40.50		910.50	1,025
0600	Iron body, 3/8" size		24	.333		239	22		261	296
0620	1/2" size		24	.333		290	22		312	355
0630	3/4" size		20	.400		330	26		356	400
0640	1" size		19	.421		365	27.50		392.50	445
0650	1-1/4" size		15	.533		560	35		595	675
0660	1-1/2" size		13	.615		570	40.50		610.50	685
0800	Process steam, wet or super heated, monel trim, threaded									
0820	Bronze body, 1/4" size	1 Stpi	24	.333	Ea.	625	22		647	725
0830	3/8" size		24	.333		625	22		647	725
0840	1/2" size		24	.333		720	22		742	825
0850	3/4" size		20	.400		845	26		871	970
0860	1" size		19	.421		1,100	27.50		1,127.50	1,250
0870	1-1/4" size		15	.533		1,350	35		1,385	1,550
0880	1-1/2" size		13	.615		1,625	40.50		1,665.50	1,850
0920	Iron body, max 125 PSIG press out, 1/4" size		24	.333		820	22		842	935
0930	3/8" size		24	.333		845	22		867	965
0940	1/2" size		24	.333		1,050	22		1,072	1,175
0950	3/4" size		20	.400		1,275	26		1,301	1,450
0960	1" size		19	.421		1,625	27.50		1,652.50	1,825
0970	1-1/4" size		15	.533		1,950	35		1,985	2,200
0980	1-1/2" size		13	.615		2,350	40.50		2,390.50	2,625
3000	Steam, high capacity, bronze body, stainless steel trim									
3020	Threaded, 1/2" diameter	1 Stpi	24	.333	Ea.	3,000	22		3,022	3,325
3030	3/4" diameter		24	.333		3,050	22		3,072	3,400
3040	1" diameter		19	.421		3,425	27.50		3,452.50	3,800
3060	1-1/4" diameter		15	.533		3,475	35		3,510	3,875
3080	1-1/2" diameter		13	.615		4,325	40.50		4,365.50	4,800
3100	2" diameter		11	.727		5,275	47.50		5,322.50	5,900
3120	2-1/2" diameter	Q-5	12	1.333		6,575	78.50		6,653.50	7,350
3140	3" diameter	"	11	1.455		7,450	86		7,536	8,325
3500	Flanged connection, iron body, 125 lb. W.S.P.									
3520	3" diameter	Q-5	11	1.455	Ea.	8,275	86		8,361	9,225
3540	4" diameter	"	5	3.200	"	10,400	189		10,589	11,800

For customer support on your Mechanical Costs with RSMeans Data, call 800.448.8182.

231

22 11 19 – Domestic Water Piping Specialties

22 11 19.26 Pressure Regulators	Crew	Daily Output	Labor-Hours	Unit	Material	2020 Bare Costs Labor	Equipment	Total	Total Incl O&P
9002	For water pressure regulators, see Section 22 05 23.20								

22 11 19.30 Pressure and Temperature Safety Plug

		Crew	Daily Output	Labor-Hours	Unit	Material	2020 Bare Costs Labor	Equipment	Total	Total Incl O&P
0010	**PRESSURE & TEMPERATURE SAFETY PLUG**									
1000	3/4" external thread, 3/8" diam. element									
1020	Carbon steel									
1050	7-1/2" insertion	1 Stpi	32	.250	Ea.	50.50	16.40		66.90	80
1120	304 stainless steel									
1150	7-1/2" insertion	1 Stpi	32	.250	Ea.	48	16.40		64.40	77
1220	316 stainless steel									
1250	7-1/2" insertion	1 Stpi	32	.250	Ea.	58.50	16.40		74.90	88.50

22 11 19.32 Pressure and Temperature Measurement Plug

		Crew	Daily Output	Labor-Hours	Unit	Material	2020 Bare Costs Labor	Equipment	Total	Total Incl O&P
0010	**PRESSURE & TEMPERATURE MEASUREMENT PLUG**									
0020	A permanent access port for insertion of									
0030	a pressure or temperature measuring probe									
0100	Plug, brass									
0110	1/4" MNPT, 1-1/2" long	1 Stpi	32	.250	Ea.	6.55	16.40		22.95	31.50
0120	3" long		31	.258		12.20	16.90		29.10	39
0140	1/2" MNPT, 1-1/2" long		30	.267		9	17.50		26.50	36
0150	3" long		29	.276		14.80	18.10		32.90	43.50
0200	Pressure gauge probe adapter									
0210	1/8" diameter, 1-1/2" probe				Ea.	21			21	23.50
0220	3" probe				"	35			35	38.50
0300	Temperature gauge, 5" stem									
0310	Analog				Ea.	10.90			10.90	12
0330	Digital				"	29.50			29.50	32.50
0400	Pressure gauge, compound									
0410	1/4" MNPT				Ea.	32.50			32.50	35.50
0500	Pressure and temperature test kit									
0510	Contains 2 thermometers and 2 pressure gauges									
0520	Kit				Ea.	355			355	390

22 11 19.34 Sleeves and Escutcheons

		Crew	Daily Output	Labor-Hours	Unit	Material	2020 Bare Costs Labor	Equipment	Total	Total Incl O&P
0010	**SLEEVES & ESCUTCHEONS**									
0100	Pipe sleeve									
0110	Steel, w/water stop, 12" long, with link seal									
0120	2" diam. for 1/2" carrier pipe	1 Plum	8.40	.952	Ea.	64.50	61.50		126	163
0130	2-1/2" diam. for 3/4" carrier pipe		8	1		73	64.50		137.50	177
0140	2-1/2" diam. for 1" carrier pipe		8	1		69	64.50		133.50	173
0150	3" diam. for 1-1/4" carrier pipe		7.20	1.111		89	71.50		160.50	205
0160	3-1/2" diam. for 1-1/2" carrier pipe		6.80	1.176		97	76		173	219
0170	4" diam. for 2" carrier pipe		6	1.333		98	86		184	237
0180	4" diam. for 2-1/2" carrier pipe		6	1.333		98	86		184	237
0190	5" diam. for 3" carrier pipe		5.40	1.481		115	95.50		210.50	270
0200	6" diam. for 4" carrier pipe		4.80	1.667		138	107		245	310
0210	10" diam. for 6" carrier pipe	Q-1	8	2		271	116		387	470
0220	12" diam. for 8" carrier pipe		7.20	2.222		370	129		499	605
0230	14" diam. for 10" carrier pipe		6.40	2.500		385	145		530	635
0240	16" diam. for 12" carrier pipe		5.80	2.759		425	160		585	710
0250	18" diam. for 14" carrier pipe		5.20	3.077		780	178		958	1,125
0260	24" diam. for 18" carrier pipe		4	4		1,125	232		1,357	1,600
0270	24" diam. for 20" carrier pipe		4	4		975	232		1,207	1,425
0280	30" diam. for 24" carrier pipe		3.20	5		1,400	290		1,690	1,950
0500	Wall sleeve									

22 11 Facility Water Distribution

22 11 19 – Domestic Water Piping Specialties

22 11 19.34 Sleeves and Escutcheons

		Crew	Daily Output	Labor-Hours	Unit	Material	2020 Bare Costs Labor	Equipment	Total	Total Incl O&P
0510	Ductile iron with rubber gasket seal									
0520	3"	1 Plum	8.40	.952	Ea.	213	61.50		274.50	325
0530	4"		7.20	1.111		227	71.50		298.50	355
0540	6"		6	1.333		276	86		362	435
0550	8"		4	2		1,225	129		1,354	1,550
0560	10"		3	2.667		1,500	172		1,672	1,900
0570	12"		2.40	3.333		1,750	215		1,965	2,250
5000	Escutcheon									
5100	Split ring, pipe									
5110	Chrome plated									
5120	1/2"	1 Plum	160	.050	Ea.	1.10	3.22		4.32	6.05
5130	3/4"		160	.050		1.29	3.22		4.51	6.25
5140	1"		135	.059		1.43	3.82		5.25	7.25
5150	1-1/2"		115	.070		1.86	4.48		6.34	8.75
5160	2"		100	.080		2.16	5.15		7.31	10.10
5170	4"		80	.100		5	6.45		11.45	15.15
5180	6"		68	.118		7.90	7.60		15.50	20
5400	Shallow flange type									
5410	Chrome plated steel									
5420	1/2" CTS	1 Plum	180	.044	Ea.	.33	2.86		3.19	4.65
5430	3/4" CTS		180	.044		.37	2.86		3.23	4.70
5440	1/2" IPS		180	.044		.40	2.86		3.26	4.73
5450	3/4" IPS		180	.044		.44	2.86		3.30	4.77
5460	1" IPS		175	.046		.48	2.95		3.43	4.94
5470	1-1/2" IPS		170	.047		.81	3.03		3.84	5.45
5480	2" IPS		160	.050		.99	3.22		4.21	5.90

22 11 19.38 Water Supply Meters

		Crew	Daily Output	Labor-Hours	Unit	Material	2020 Bare Costs Labor	Equipment	Total	Total Incl O&P
0010	**WATER SUPPLY METERS**									
1000	Detector, serves dual systems such as fire and domestic or									
1020	process water, wide range cap., UL and FM approved									
1100	3" mainline x 2" by-pass, 400 GPM	Q-1	3.60	4.444	Ea.	8,025	258		8,283	9,200
1140	4" mainline x 2" by-pass, 700 GPM	"	2.50	6.400		8,025	370		8,395	9,375
1180	6" mainline x 3" by-pass, 1,600 GPM	Q-2	2.60	9.231		12,300	555		12,855	14,300
1220	8" mainline x 4" by-pass, 2,800 GPM		2.10	11.429		18,200	685		18,885	21,000
1260	10" mainline x 6" by-pass, 4,400 GPM		2	12		26,000	720		26,720	29,700
1300	10" x 12" mainlines x 6" by-pass, 5,400 GPM		1.70	14.118		35,300	850		36,150	40,100
2000	Domestic/commercial, bronze									
2020	Threaded									
2060	5/8" diameter, to 20 GPM	1 Plum	16	.500	Ea.	54	32		86	108
2080	3/4" diameter, to 30 GPM		14	.571		98.50	37		135.50	163
2100	1" diameter, to 50 GPM		12	.667		149	43		192	229
2300	Threaded/flanged									
2340	1-1/2" diameter, to 100 GPM	1 Plum	8	1	Ea.	365	64.50		429.50	495
2360	2" diameter, to 160 GPM	"	6	1.333	"	495	86		581	675
2600	Flanged, compound									
2640	3" diameter, 320 GPM	Q-1	3	5.333	Ea.	2,650	310		2,960	3,400
2660	4" diameter, to 500 GPM		1.50	10.667		4,250	620		4,870	5,600
2680	6" diameter, to 1,000 GPM		1	16		6,850	930		7,780	8,950
2700	8" diameter, to 1,800 GPM		.80	20		10,700	1,150		11,850	13,500
7000	Turbine									
7260	Flanged									
7300	2" diameter, to 160 GPM	1 Plum	7	1.143	Ea.	615	73.50		688.50	785

For customer support on your Mechanical Costs with RSMeans Data, call 800.448.8182.

233

22 11 Facility Water Distribution

22 11 19 – Domestic Water Piping Specialties

22 11 19.38 Water Supply Meters

		Crew	Daily Output	Labor-Hours	Unit	Material	2020 Bare Costs Labor	Equipment	Total	Total Incl O&P
7320	3" diameter, to 450 GPM	Q-1	3.60	4.444	Ea.	1,100	258		1,358	1,575
7340	4" diameter, to 650 GPM	"	2.50	6.400		1,825	370		2,195	2,550
7360	6" diameter, to 1,800 GPM	Q-2	2.60	9.231		3,000	555		3,555	4,125
7380	8" diameter, to 2,500 GPM		2.10	11.429		5,150	685		5,835	6,700
7400	10" diameter, to 5,500 GPM	↓	1.70	14.118	↓	6,925	850		7,775	8,900

22 11 19.42 Backflow Preventers

		Crew	Daily Output	Labor-Hours	Unit	Material	2020 Bare Costs Labor	Equipment	Total	Total Incl O&P
0010	**BACKFLOW PREVENTERS**, Includes valves									
0020	and four test cocks, corrosion resistant, automatic operation									
1000	Double check principle									
1010	Threaded, with ball valves									
1020	3/4" pipe size	1 Plum	16	.500	Ea.	241	32		273	315
1030	1" pipe size		14	.571		253	37		290	335
1040	1-1/2" pipe size		10	.800		620	51.50		671.50	760
1050	2" pipe size	↓	7	1.143	↓	665	73.50		738.50	845
1080	Threaded, with gate valves									
1100	3/4" pipe size	1 Plum	16	.500	Ea.	1,150	32		1,182	1,300
1120	1" pipe size		14	.571		1,150	37		1,187	1,325
1140	1-1/2" pipe size		10	.800		1,225	51.50		1,276.50	1,425
1160	2" pipe size	↓	7	1.143	↓	1,825	73.50		1,898.50	2,100
1200	Flanged, valves are gate									
1210	3" pipe size	Q-1	4.50	3.556	Ea.	2,975	206		3,181	3,575
1220	4" pipe size	"	3	5.333		3,000	310		3,310	3,775
1230	6" pipe size	Q-2	3	8		4,675	480		5,155	5,850
1240	8" pipe size		2	12		7,975	720		8,695	9,850
1250	10" pipe size	↓	1	24	↓	12,000	1,450		13,450	15,400
1300	Flanged, valves are OS&Y									
1370	1" pipe size	1 Plum	5	1.600	Ea.	1,275	103		1,378	1,550
1374	1-1/2" pipe size		5	1.600		1,350	103		1,453	1,625
1378	2" pipe size	↓	4.80	1.667		1,950	107		2,057	2,300
1380	3" pipe size	Q-1	4.50	3.556		2,850	206		3,056	3,450
1400	4" pipe size	"	3	5.333		4,675	310		4,985	5,600
1420	6" pipe size	Q-2	3	8		6,700	480		7,180	8,100
1430	8" pipe size	"	2	12	↓	14,000	720		14,720	16,500
4000	Reduced pressure principle									
4100	Threaded, bronze, valves are ball									
4120	3/4" pipe size	1 Plum	16	.500	Ea.	530	32		562	635
4140	1" pipe size		14	.571		560	37		597	670
4150	1-1/4" pipe size		12	.667		1,125	43		1,168	1,300
4160	1-1/2" pipe size		10	.800		1,150	51.50		1,201.50	1,350
4180	2" pipe size	↓	7	1.143	↓	1,325	73.50		1,398.50	1,575
5000	Flanged, bronze, valves are OS&Y									
5060	2-1/2" pipe size	Q-1	5	3.200	Ea.	5,475	186		5,661	6,300
5080	3" pipe size		4.50	3.556		6,600	206		6,806	7,550
5100	4" pipe size	↓	3	5.333		7,725	310		8,035	8,975
5120	6" pipe size	Q-2	3	8	↓	12,000	480		12,480	13,900
5200	Flanged, iron, valves are gate									
5210	2-1/2" pipe size	Q-1	5	3.200	Ea.	3,200	186		3,386	3,775
5220	3" pipe size		4.50	3.556		3,300	206		3,506	3,950
5230	4" pipe size	↓	3	5.333		3,775	310		4,085	4,650
5240	6" pipe size	Q-2	3	8		5,675	480		6,155	6,975
5250	8" pipe size		2	12		9,575	720		10,295	11,600
5260	10" pipe size	↓	1	24	↓	16,200	1,450		17,650	20,100

22 11 Facility Water Distribution

22 11 19 – Domestic Water Piping Specialties

22 11 19.42 Backflow Preventers

		Crew	Daily Output	Labor-Hours	Unit	Material	2020 Bare Costs Labor	2020 Bare Costs Equipment	Total	Total Incl O&P
5600	Flanged, iron, valves are OS&Y									
5660	2-1/2" pipe size	Q-1	5	3.200	Ea.	3,300	186		3,486	3,900
5680	3" pipe size		4.50	3.556		3,275	206		3,481	3,900
5700	4" pipe size		3	5.333		4,725	310		5,035	5,675
5720	6" pipe size	Q-2	3	8		5,825	480		6,305	7,150
5740	8" pipe size		2	12		10,300	720		11,020	12,400
5760	10" pipe size		1	24		19,800	1,450		21,250	24,000

22 11 19.50 Vacuum Breakers

		Crew	Daily Output	Labor-Hours	Unit	Material	2020 Bare Costs Labor	2020 Bare Costs Equipment	Total	Total Incl O&P
0010	**VACUUM BREAKERS**									
0013	See also backflow preventers Section 22 11 19.42									
1000	Anti-siphon continuous pressure type									
1010	Max. 150 psi - 210°F									
1020	Bronze body									
1030	1/2" size	1 Stpi	24	.333	Ea.	192	22		214	244
1040	3/4" size		20	.400		192	26		218	250
1050	1" size		19	.421		197	27.50		224.50	258
1060	1-1/4" size		15	.533		390	35		425	485
1070	1-1/2" size		13	.615		475	40.50		515.50	585
1080	2" size		11	.727		490	47.50		537.50	610
1200	Max. 125 psi with atmospheric vent									
1210	Brass, in-line construction									
1220	1/4" size	1 Stpi	24	.333	Ea.	143	22		165	190
1230	3/8" size	"	24	.333		143	22		165	190
1260	For polished chrome finish, add					13%				
2000	Anti-siphon, non-continuous pressure type									
2010	Hot or cold water 125 psi - 210°F									
2020	Bronze body									
2030	1/4" size	1 Stpi	24	.333	Ea.	88.50	22		110.50	130
2040	3/8" size		24	.333		88.50	22		110.50	130
2050	1/2" size		24	.333		99	22		121	142
2060	3/4" size		20	.400		119	26		145	169
2070	1" size		19	.421		183	27.50		210.50	244
2080	1-1/4" size		15	.533		320	35		355	410
2090	1-1/2" size		13	.615		350	40.50		390.50	445
2100	2" size		11	.727		585	47.50		632.50	715
2110	2-1/2" size		8	1		1,675	65.50		1,740.50	1,950
2120	3" size		6	1.333		2,225	87.50		2,312.50	2,575
2150	For polished chrome finish, add					50%				
3000	Air gap fitting									
3020	1/2" NPT size	1 Plum	19	.421	Ea.	61.50	27		88.50	108
3030	1" NPT size	"	15	.533		70.50	34.50		105	129
3040	2" NPT size	Q-1	21	.762		138	44		182	218
3050	3" NPT size		14	1.143		273	66.50		339.50	400
3060	4" NPT size		10	1.600		273	93		366	440

22 11 19.54 Water Hammer Arresters/Shock Absorbers

		Crew	Daily Output	Labor-Hours	Unit	Material	2020 Bare Costs Labor	2020 Bare Costs Equipment	Total	Total Incl O&P
0010	**WATER HAMMER ARRESTERS/SHOCK ABSORBERS**									
0490	Copper									
0500	3/4" male IPS for 1 to 11 fixtures	1 Plum	12	.667	Ea.	31	43		74	98.50
0600	1" male IPS for 12 to 32 fixtures		8	1		51.50	64.50		116	153
0700	1-1/4" male IPS for 33 to 60 fixtures		8	1		50.50	64.50		115	153
0800	1-1/2" male IPS for 61 to 113 fixtures		8	1		73	64.50		137.50	177
0900	2" male IPS for 114 to 154 fixtures		8	1		108	64.50		172.50	215

22 11 Facility Water Distribution

22 11 19 – Domestic Water Piping Specialties

22 11 19.54 Water Hammer Arresters/Shock Absorbers		Crew	Daily Output	Labor-Hours	Unit	Material	2020 Bare Costs Labor	Equipment	Total	Total Incl O&P
1000	2-1/2" male IPS for 155 to 330 fixtures	1 Plum	4	2	Ea.	330	129		459	560
4000	Bellows type									
4010	3/4" FNPT, to 11 fixture units	1 Plum	10	.800	Ea.	355	51.50		406.50	465
4020	1" FNPT, to 32 fixture units		8.80	.909		725	58.50		783.50	885
4030	1" FNPT, to 60 fixture units		8.80	.909		1,075	58.50		1,133.50	1,300
4040	1" FNPT, to 113 fixture units		8.80	.909		2,725	58.50		2,783.50	3,075
4050	1" FNPT, to 154 fixture units		6.60	1.212		3,075	78		3,153	3,500
4060	1-1/2" FNPT, to 300 fixture units		5	1.600		3,775	103		3,878	4,300

22 11 19.64 Hydrants

		Crew	Daily Output	Labor-Hours	Unit	Material	2020 Bare Costs Labor	Equipment	Total	Total Incl O&P
0010	**HYDRANTS**									
0050	Wall type, moderate climate, bronze, encased									
0200	3/4" IPS connection	1 Plum	16	.500	Ea.	1,025	32		1,057	1,175
0300	1" IPS connection		14	.571		1,125	37		1,162	1,300
0500	Anti-siphon type, 3/4" connection		16	.500		815	32		847	945
1000	Non-freeze, bronze, exposed									
1100	3/4" IPS connection, 4" to 9" thick wall	1 Plum	14	.571	Ea.	470	37		507	570
1120	10" to 14" thick wall		12	.667		445	43		488	555
1140	15" to 19" thick wall		12	.667		550	43		593	670
1160	20" to 24" thick wall		10	.800		830	51.50		881.50	985
1200	For 1" IPS connection, add					15%	10%			
1240	For 3/4" adapter type vacuum breaker, add				Ea.	83			83	91.50
1280	For anti-siphon type, add				"	176			176	194
2000	Non-freeze bronze, encased, anti-siphon type									
2100	3/4" IPS connection, 5" to 9" thick wall	1 Plum	14	.571	Ea.	1,275	37		1,312	1,450
2120	10" to 14" thick wall		12	.667		1,625	43		1,668	1,875
2140	15" to 19" thick wall		12	.667		1,725	43		1,768	1,975
2160	20" to 24" thick wall		10	.800		1,800	51.50		1,851.50	2,050
2200	For 1" IPS connection, add					10%	10%			
3000	Ground box type, bronze frame, 3/4" IPS connection									
3080	Non-freeze, all bronze, polished face, set flush									
3100	2' depth of bury	1 Plum	8	1	Ea.	1,325	64.50		1,389.50	1,550
3120	3' depth of bury		8	1		1,425	64.50		1,489.50	1,675
3140	4' depth of bury		8	1		1,525	64.50		1,589.50	1,775
3160	5' depth of bury		7	1.143		1,625	73.50		1,698.50	1,875
3180	6' depth of bury		7	1.143		1,700	73.50		1,773.50	1,950
3200	7' depth of bury		6	1.333		1,675	86		1,761	1,975
3220	8' depth of bury		5	1.600		1,775	103		1,878	2,100
3240	9' depth of bury		4	2		1,875	129		2,004	2,275
3260	10' depth of bury		4	2		1,975	129		2,104	2,350
3400	For 1" IPS connection, add					15%	10%			
3450	For 1-1/4" IPS connection, add					325%	14%			
3500	For 1-1/2" IPS connection, add					370%	18%			
3550	For 2" IPS connection, add					445%	24%			
3600	For tapped drain port in box, add					107			107	118
4000	Non-freeze, CI body, bronze frame & scoriated cover									
4010	with hose storage									
4100	2' depth of bury	1 Plum	7	1.143	Ea.	2,200	73.50		2,273.50	2,525
4120	3' depth of bury		7	1.143		2,300	73.50		2,373.50	2,625
4140	4' depth of bury		7	1.143		2,375	73.50		2,448.50	2,700
4160	5' depth of bury		6.50	1.231		2,425	79.50		2,504.50	2,775
4180	6' depth of bury		6	1.333		2,450	86		2,536	2,825
4200	7' depth of bury		5.50	1.455		2,550	94		2,644	2,950

22 11 Facility Water Distribution

22 11 19 – Domestic Water Piping Specialties

22 11 19.64 Hydrants		Crew	Daily Output	Labor-Hours	Unit	2020 Bare Costs				Total Incl O&P
						Material	Labor	Equipment	Total	
4220	8' depth of bury	1 Plum	5	1.600	Ea.	2,625	103		2,728	3,025
4240	9' depth of bury		4.50	1.778		2,825	115		2,940	3,300
4260	10' depth of bury		4	2		2,825	129		2,954	3,300
4280	For 1" IPS connection, add					505			505	555
4300	For tapped drain port in box, add					113			113	124
5000	Moderate climate, all bronze, polished face									
5020	and scoriated cover, set flush									
5100	3/4" IPS connection	1 Plum	16	.500	Ea.	830	32		862	965
5120	1" IPS connection	"	14	.571		2,000	37		2,037	2,275
5200	For tapped drain port in box, add					113			113	124
6000	Ground post type, all non-freeze, all bronze, aluminum casing									
6010	guard, exposed head, 3/4" IPS connection									
6100	2' depth of bury	1 Plum	8	1	Ea.	1,275	64.50		1,339.50	1,500
6120	3' depth of bury		8	1		1,300	64.50		1,364.50	1,550
6140	4' depth of bury		8	1		1,475	64.50		1,539.50	1,725
6160	5' depth of bury		7	1.143		1,450	73.50		1,523.50	1,700
6180	6' depth of bury		7	1.143		1,775	73.50		1,848.50	2,050
6200	7' depth of bury		6	1.333		1,675	86		1,761	1,950
6220	8' depth of bury		5	1.600		1,775	103		1,878	2,100
6240	9' depth of bury		4	2		1,875	129		2,004	2,275
6260	10' depth of bury		4	2		2,000	129		2,129	2,375
6300	For 1" IPS connection, add					40%	10%			
6350	For 1-1/4" IPS connection, add					140%	14%			
6400	For 1-1/2" IPS connection, add					225%	18%			
6450	For 2" IPS connection, add					315%	24%			

22 11 23 – Domestic Water Pumps

22 11 23.10 General Utility Pumps		Crew	Daily Output	Labor-Hours	Unit	Material	Labor	Equipment	Total	Total Incl O&P
0010	**GENERAL UTILITY PUMPS**									
2000	Single stage									
3000	Double suction,									
3140	50 HP, 5" D x 6" S	Q-2	.33	72.727	Ea.	9,900	4,375		14,275	17,500
3180	60 HP, 6" D x 8" S	Q-3	.30	107		17,500	6,550		24,050	29,100
3190	75 HP, to 2,500 GPM		.28	114		16,500	7,000		23,500	28,600
3220	100 HP, to 3,000 GPM		.26	123		20,300	7,550		27,850	33,600
3240	150 HP, to 4,000 GPM		.24	133		23,600	8,175		31,775	38,200
4000	Centrifugal, end suction, mounted on base									
4010	Horizontal mounted, with drip proof motor, rated @ 100' head									
4020	Vertical split case, single stage									
4040	100 GPM, 5 HP, 1-1/2" discharge	Q-1	1.70	9.412	Ea.	4,725	545		5,270	6,000
4050	200 GPM, 10 HP, 2" discharge		1.30	12.308		4,925	715		5,640	6,500
4060	250 GPM, 10 HP, 3" discharge		1.28	12.500		5,500	725		6,225	7,125
4070	300 GPM, 15 HP, 2" discharge	Q-2	1.56	15.385		6,475	925		7,400	8,475
4080	500 GPM, 20 HP, 4" discharge		1.44	16.667		7,725	1,000		8,725	10,000
4090	750 GPM, 30 HP, 4" discharge		1.20	20		8,325	1,200		9,525	11,000
4100	1,050 GPM, 40 HP, 5" discharge		1	24		9,900	1,450		11,350	13,100
4110	1,500 GPM, 60 HP, 6" discharge		.60	40		17,500	2,400		19,900	22,900
4120	2,000 GPM, 75 HP, 6" discharge		.50	48		18,900	2,875		21,775	25,100
4130	3,000 GPM, 100 HP, 8" discharge		.40	60		22,500	3,600		26,100	30,200
4200	Horizontal split case, single stage									
4210	100 GPM, 7.5 HP, 1-1/2" discharge	Q-1	1.70	9.412	Ea.	5,250	545		5,795	6,600
4220	250 GPM, 15 HP, 2-1/2" discharge	"	1.30	12.308		7,800	715		8,515	9,650
4230	500 GPM, 20 HP, 4" discharge	Q-2	1.60	15		10,600	900		11,500	13,100

For customer support on your Mechanical Costs with RSMeans Data, call 800.448.8182.

237

22 11 23 – Domestic Water Pumps

22 11 23.10 General Utility Pumps		Crew	Daily Output	Labor-Hours	Unit	Material	2020 Bare Costs Labor	Equipment	Total	Total Incl O&P
4240	750 GPM, 25 HP, 5" discharge	Q-2	1.54	15.584	Ea.	10,700	935		11,635	13,100
4250	1,000 GPM, 40 HP, 5" discharge	↓	1.20	20		13,400	1,200		14,600	16,500
4260	1,500 GPM, 50 HP, 6" discharge	Q-3	1.42	22.535		17,000	1,375		18,375	20,800
4270	2,000 GPM, 75 HP, 8" discharge		1.14	28.070		21,900	1,725		23,625	26,700
4280	3,000 GPM, 100 HP, 10" discharge		.96	33.333		22,800	2,050		24,850	28,100
4290	3,500 GPM, 150 HP, 10" discharge		.86	37.209		37,700	2,275		39,975	44,900
4300	4,000 GPM, 200 HP, 10" discharge	↓	.66	48.485	↓	27,800	2,975		30,775	35,100
4330	Horizontal split case, two stage, 500' head									
4340	100 GPM, 40 HP, 1-1/2" discharge	Q-2	1.70	14.118	Ea.	17,800	850		18,650	20,900
4350	200 GPM, 50 HP, 1-1/2" discharge	"	1.44	16.667		19,600	1,000		20,600	23,000
4360	300 GPM, 75 HP, 2" discharge	Q-3	1.57	20.382		25,600	1,250		26,850	30,000
4370	400 GPM, 100 HP, 3" discharge		1.14	28.070		31,100	1,725		32,825	36,800
4380	800 GPM, 200 HP, 4" discharge	↓	.86	37.209	↓	41,800	2,275		44,075	49,400
5000	Centrifugal, in-line									
5006	Vertical mount, iron body, 125 lb. flgd, 1,800 RPM TEFC mtr									
5010	Single stage									
5012	.5 HP, 1-1/2" suction & discharge	Q-1	3.20	5	Ea.	1,525	290		1,815	2,100
5014	.75 HP, 2" suction & discharge		2.80	5.714		2,950	330		3,280	3,750
5015	1 HP, 3" suction & discharge		2.50	6.400		3,250	370		3,620	4,125
5016	1.5 HP, 3" suction & discharge		2.30	6.957		3,200	405		3,605	4,125
5018	2 HP, 4" suction & discharge		2.10	7.619		4,450	440		4,890	5,550
5020	3 HP, 5" suction & discharge		1.90	8.421		4,900	490		5,390	6,100
5030	5 HP, 6" suction & discharge		1.60	10		5,550	580		6,130	6,975
5040	7.5 HP, 6" suction & discharge	↓	1.30	12.308		6,400	715		7,115	8,125
5050	10 HP, 8" suction & discharge	Q-2	1.80	13.333		8,500	800		9,300	10,600
5060	15 HP, 8" suction & discharge		1.70	14.118		10,000	850		10,850	12,300
5064	20 HP, 8" suction & discharge		1.60	15		12,300	900		13,200	14,900
5070	30 HP, 8" suction & discharge		1.50	16		14,700	960		15,660	17,700
5080	40 HP, 8" suction & discharge		1.40	17.143		17,700	1,025		18,725	21,100
5090	50 HP, 8" suction & discharge		1.30	18.462		18,000	1,100		19,100	21,500
5094	60 HP, 8" suction & discharge	↓	.90	26.667		20,100	1,600		21,700	24,500
5100	75 HP, 8" suction & discharge	Q-3	.60	53.333		21,600	3,275		24,875	28,700
5110	100 HP, 8" suction & discharge	"	.50	64	↓	24,600	3,925		28,525	32,900

22 11 23.11 Miscellaneous Pumps

		Crew	Daily Output	Labor-Hours	Unit	Material	2020 Bare Costs Labor	Equipment	Total	Total Incl O&P
0010	**MISCELLANEOUS PUMPS**									
0020	Water pump, portable, gasoline powered									
0100	170 GPH, 2" discharge	Q-1	11	1.455	Ea.	780	84.50		864.50	980
0110	343 GPH, 3" discharge		10.50	1.524		955	88.50		1,043.50	1,175
0120	608 GPH, 4" discharge	↓	10	1.600	↓	1,550	93		1,643	1,875
0500	Pump, propylene body, housing and impeller									
0510	22 GPM, 1/3 HP, 40' HD	1 Plum	5	1.600	Ea.	920	103		1,023	1,150
0520	33 GPM, 1/2 HP, 40' HD	Q-1	5	3.200		840	186		1,026	1,200
0530	53 GPM, 3/4 HP, 40' HD	"	4	4	↓	1,025	232		1,257	1,475
0600	Rotary pump, CI									
0614	282 GPH, 3/4 HP, 1" discharge	Q-1	4.50	3.556	Ea.	1,200	206		1,406	1,625
0618	277 GPH, 1 HP, 1" discharge		4	4		1,200	232		1,432	1,675
0624	1,100 GPH, 1.5 HP, 1-1/4" discharge		3.60	4.444		2,975	258		3,233	3,650
0628	1,900 GPH, 2 HP, 1-1/4" discharge	↓	3.20	5	↓	4,200	290		4,490	5,050
1000	Turbine pump, CI									
1010	50 GPM, 2 HP, 3" discharge	Q-1	.80	20	Ea.	4,075	1,150		5,225	6,200
1020	100 GPM, 3 HP, 4" discharge	Q-2	.96	25		6,750	1,500		8,250	9,675
1030	250 GPM, 15 HP, 6" discharge	"	.94	25.532		8,350	1,525		9,875	11,500

22 11 Facility Water Distribution

22 11 23 – Domestic Water Pumps

22 11 23.11 Miscellaneous Pumps

22 11 23.11 Miscellaneous Pumps		Crew	Daily Output	Labor-Hours	Unit	Material	2020 Bare Costs		Total	Total Incl O&P
							Labor	Equipment		
1040	500 GPM, 25 HP, 6" discharge	Q-3	1.22	26.230	Ea.	8,450	1,600		10,050	11,700
1050	1,000 GPM, 50 HP, 8" discharge		1.14	28.070		11,600	1,725		13,325	15,400
1060	2,000 GPM, 100 HP, 10" discharge		1	32		13,800	1,975		15,775	18,200
1070	3,000 GPM, 150 HP, 10" discharge		.80	40		22,200	2,450		24,650	28,100
1080	4,000 GPM, 200 HP, 12" discharge		.70	45.714		24,300	2,800		27,100	30,900
1090	6,000 GPM, 300 HP, 14" discharge		.60	53.333		37,000	3,275		40,275	45,600
1100	10,000 GPM, 300 HP, 18" discharge	▼	.58	55.172	▼	42,800	3,375		46,175	52,000
2000	Centrifugal stainless steel pumps									
2100	100 GPM, 100' TDH, 3 HP	Q-1	1.80	8.889	Ea.	12,100	515		12,615	14,100
2130	250 GPM, 100' TDH, 10 HP	"	1.28	12.500		17,000	725		17,725	19,700
2160	500 GPM, 100' TDH, 20 HP	Q-2	1.44	16.667	▼	19,400	1,000		20,400	22,800
2200	Vertical turbine stainless steel pumps									
2220	100 GPM, 100' TDH, 7.5 HP	Q-2	.95	25.263	Ea.	68,000	1,525		69,525	77,000
2240	250 GPM, 100' TDH, 15 HP		.94	25.532		75,000	1,525		76,525	85,000
2260	500 GPM, 100' TDH, 20 HP	▼	.93	25.806		82,500	1,550		84,050	93,000
2280	750 GPM, 100' TDH, 30 HP	Q-3	1.19	26.891		89,500	1,650		91,150	101,000
2300	1,000 GPM, 100' TDH, 40 HP	"	1.16	27.586		97,000	1,700		98,700	109,000
2320	100 GPM, 200' TDH, 15 HP	Q-2	.94	25.532		104,000	1,525		105,525	117,000
2340	250 GPM, 200' TDH, 25 HP	Q-3	1.22	26.230		111,500	1,600		113,100	125,000
2360	500 GPM, 200' TDH, 40 HP		1.16	27.586		118,500	1,700		120,200	133,000
2380	750 GPM, 200' TDH, 60 HP		1.13	28.319		120,000	1,725		121,725	134,500
2400	1,000 GPM, 200' TDH, 75 HP	▼	1.12	28.571	▼	123,500	1,750		125,250	138,000

22 11 23.13 Domestic-Water Packaged Booster Pumps

		Crew	Daily Output	Labor-Hours	Unit	Material	Labor	Equipment	Total	Total Incl O&P
0010	**DOMESTIC-WATER PACKAGED BOOSTER PUMPS**									
0200	Pump system, with diaphragm tank, control, press. switch									
0300	1 HP pump	Q-1	1.30	12.308	Ea.	7,500	715		8,215	9,325
0400	1-1/2 HP pump		1.25	12.800		7,575	740		8,315	9,450
0420	2 HP pump		1.20	13.333		7,300	775		8,075	9,175
0440	3 HP pump	▼	1.10	14.545		7,900	845		8,745	9,950
0460	5 HP pump	Q-2	1.50	16		8,200	960		9,160	10,500
0480	7-1/2 HP pump		1.42	16.901		9,725	1,025		10,750	12,200
0500	10 HP pump	▼	1.34	17.910	▼	10,100	1,075		11,175	12,800
2000	Pump system, variable speed, base, controls, starter									
2010	Duplex, 100' head									
2020	400 GPM, 7-1/2 HP, 4" discharge	Q-2	.70	34.286	Ea.	44,200	2,050		46,250	52,000
2025	Triplex, 100' head									
2030	1,000 GPM, 15 HP, 6" discharge	Q-2	.50	48	Ea.	49,200	2,875		52,075	58,500
2040	1,700 GPM, 30 HP, 6" discharge	"	.30	80	"	72,000	4,800		76,800	86,000

22 12 Facility Potable-Water Storage Tanks

22 12 21 – Facility Underground Potable-Water Storage Tanks

22 12 21.13 Fiberglass, Undrgrnd Pot.-Water Storage Tanks

		Crew	Daily Output	Labor-Hours	Unit	Material	Labor	Equipment	Total	Total Incl O&P
0010	**FIBERGLASS, UNDERGROUND POTABLE-WATER STORAGE TANKS**									
0020	Excludes excavation, backfill & piping									
0030	Single wall									
2000	600 gallon capacity	B-21B	3.75	10.667	Ea.	4,125	490	126	4,741	5,425
2010	1,000 gallon capacity		3.50	11.429		5,275	525	135	5,935	6,725
2020	2,000 gallon capacity		3.25	12.308		7,500	565	145	8,210	9,250
2030	4,000 gallon capacity		3	13.333		10,100	610	157	10,867	12,200
2040	6,000 gallon capacity		2.65	15.094		11,400	695	178	12,273	13,800
2050	8,000 gallon capacity	▼	2.30	17.391		13,900	800	205	14,905	16,700

For customer support on your Mechanical Costs with RSMeans Data, call 800.448.8182.

239

22 12 Facility Potable-Water Storage Tanks

22 12 21 – Facility Underground Potable-Water Storage Tanks

22 12 21.13 Fiberglass, Undrgrnd Pot.-Water Storage Tanks	Crew	Daily Output	Labor-Hours	Unit	Material	2020 Bare Costs Labor	Equipment	Total	Total Incl O&P
2060	10,000 gallon capacity	B-21B	2	20	Ea.	15,400	920	236	16,556	18,600
2070	12,000 gallon capacity		1.50	26.667		21,600	1,225	315	23,140	25,900
2080	15,000 gallon capacity		1	40		24,500	1,825	470	26,795	30,300
2090	20,000 gallon capacity		.75	53.333		31,800	2,450	630	34,880	39,400
2100	25,000 gallon capacity		.50	80		47,400	3,675	940	52,015	58,500
2110	30,000 gallon capacity		.35	114		57,000	5,250	1,350	63,600	72,000
2120	40,000 gallon capacity		.30	133		80,000	6,125	1,575	87,700	99,000

22 12 23 – Facility Indoor Potable-Water Storage Tanks

22 12 23.13 Facility Steel, Indoor Pot.-Water Storage Tanks

		Crew	Daily Output	Labor-Hours	Unit	Material	2020 Bare Costs Labor	Equipment	Total	Total Incl O&P
0010	**FACILITY STEEL, INDOOR POT.-WATER STORAGE TANKS**									
2000	Galvanized steel, 15 gal., 14" diam. x 26" LOA	1 Plum	12	.667	Ea.	1,500	43		1,543	1,725
2060	30 gal., 14" diam. x 49" LOA		11	.727		1,775	47		1,822	2,025
2080	80 gal., 20" diam. x 64" LOA		9	.889		2,800	57.50		2,857.50	3,150
2100	135 gal., 24" diam. x 75" LOA		6	1.333		4,025	86		4,111	4,550
2120	240 gal., 30" diam. x 86" LOA		4	2		7,850	129		7,979	8,825
2140	300 gal., 36" diam. x 76" LOA		3	2.667		11,200	172		11,372	12,600
2160	400 gal., 36" diam. x 100" LOA	Q-1	4	4		13,700	232		13,932	15,400
2180	500 gal., 36" diam. x 126" LOA	"	3	5.333		17,000	310		17,310	19,200
3000	Glass lined, P.E., 80 gal., 20" diam. x 60" LOA	1 Plum	9	.889		4,050	57.50		4,107.50	4,525
3060	140 gal., 24" diam. x 75" LOA		6	1.333		5,575	86		5,661	6,250
3080	200 gal., 30" diam. x 71" LOA		4	2		8,050	129		8,179	9,050
3100	350 gal., 36" diam. x 86" LOA		3	2.667		9,250	172		9,422	10,500
3120	450 gal., 42" diam. x 79" LOA	Q-1	4	4		13,300	232		13,532	15,000
3140	600 gal., 48" diam. x 81" LOA		3	5.333		15,900	310		16,210	18,000
3160	750 gal., 48" diam. x 105" LOA		3	5.333		17,800	310		18,110	20,100
3180	900 gal., 54" diam. x 95" LOA		2.50	6.400		25,500	370		25,870	28,700
3200	1,500 gal., 54" diam. x 153" LOA		2	8		32,200	465		32,665	36,100
3220	1,800 gal., 54" diam. x 181" LOA		1.50	10.667		35,700	620		36,320	40,200
3240	2,000 gal., 60" diam. x 165" LOA		1	16		41,000	930		41,930	46,500
3260	3,500 gal., 72" diam. x 201" LOA	Q-2	1.50	16		51,000	960		51,960	57,500

22 13 Facility Sanitary Sewerage

22 13 16 – Sanitary Waste and Vent Piping

22 13 16.40 Pipe Fittings, Cast Iron for Drainage

		Crew	Daily Output	Labor-Hours	Unit	Material	2020 Bare Costs Labor	Equipment	Total	Total Incl O&P
0010	**PIPE FITTINGS, CAST IRON FOR DRAINAGE**, Special									
1000	Drip pan elbow (safety valve discharge elbow)									
1010	Cast iron, threaded inlet									
1014	2-1/2"	Q-1	8	2	Ea.	1,600	116		1,716	1,950
1015	3"		6.40	2.500		1,725	145		1,870	2,125
1017	4"		4.80	3.333		2,325	193		2,518	2,850
1018	Cast iron, flanged inlet									
1019	6"	Q-2	3.60	6.667	Ea.	2,525	400		2,925	3,375
1020	8"	"	2.60	9.231	"	2,675	555		3,230	3,775

22 13 16.60 Traps

		Crew	Daily Output	Labor-Hours	Unit	Material	2020 Bare Costs Labor	Equipment	Total	Total Incl O&P
0010	**TRAPS**									
4700	Copper, drainage, drum trap									
4800	3" x 5" solid, 1-1/2" pipe size	1 Plum	16	.500	Ea.	156	32		188	220
4840	3" x 6" swivel, 1-1/2" pipe size	"	16	.500	"	315	32		347	395
5100	P trap, standard pattern									
5200	1-1/4" pipe size	1 Plum	18	.444	Ea.	102	28.50		130.50	156

22 13 16 – Sanitary Waste and Vent Piping

22 13 16.60 Traps		Crew	Daily Output	Labor-Hours	Unit	Material	2020 Bare Costs Labor	Equipment	Total	Total Incl O&P
5240	1-1/2" pipe size	1 Plum	17	.471	Ea.	111	30.50		141.50	169
5260	2" pipe size		15	.533		184	34.50		218.50	254
5280	3" pipe size	↓	11	.727	↓	525	47		572	650
5340	With cleanout, swivel joint and slip joint									
5360	1-1/4" pipe size	1 Plum	18	.444	Ea.	124	28.50		152.50	179
5400	1-1/2" pipe size		17	.471		199	30.50		229.50	265
5420	2" pipe size	↓	15	.533	↓	176	34.50		210.50	246
6710	ABS DWV P trap, solvent weld joint									
6720	1-1/2" pipe size	1 Plum	18	.444	Ea.	10.50	28.50		39	54.50
6722	2" pipe size		17	.471		14.05	30.50		44.55	61
6724	3" pipe size		15	.533		55.50	34.50		90	113
6726	4" pipe size	↓	14	.571	↓	114	37		151	181
6732	PVC DWV P trap, solvent weld joint									
6733	1-1/2" pipe size	1 Plum	18	.444	Ea.	8.95	28.50		37.45	53
6734	2" pipe size		17	.471		11.15	30.50		41.65	58
6735	3" pipe size		15	.533		38	34.50		72.50	93
6736	4" pipe size		14	.571		85	37		122	149
6760	PP DWV, dilution trap, 1-1/2" pipe size		16	.500		274	32		306	350
6770	P trap, 1-1/2" pipe size		17	.471		68	30.50		98.50	121
6780	2" pipe size		16	.500		101	32		133	159
6790	3" pipe size		14	.571		186	37		223	259
6800	4" pipe size		13	.615		310	39.50		349.50	405
6830	S trap, 1-1/2" pipe size		16	.500		55.50	32		87.50	109
6840	2" pipe size		15	.533		83	34.50		117.50	143
6850	Universal trap, 1-1/2" pipe size		14	.571		112	37		149	178
6860	PVC DWV hub x hub, basin trap, 1-1/4" pipe size		18	.444		54	28.50		82.50	102
6870	Sink P trap, 1-1/2" pipe size		18	.444		14.45	28.50		42.95	59
6880	Tubular S trap, 1-1/2" pipe size	↓	17	.471	↓	26.50	30.50		57	74.50
6890	PVC sch. 40 DWV, drum trap									
6900	1-1/2" pipe size	1 Plum	16	.500	Ea.	37	32		69	88.50
6910	P trap, 1-1/2" pipe size		18	.444		8.80	28.50		37.30	52.50
6920	2" pipe size		17	.471		11.85	30.50		42.35	58.50
6930	3" pipe size		15	.533		40.50	34.50		75	96.50
6940	4" pipe size		14	.571		91.50	37		128.50	156
6950	P trap w/clean out, 1-1/2" pipe size		18	.444		14.90	28.50		43.40	59.50
6960	2" pipe size		17	.471		24.50	30.50		55	72.50
6970	P trap adjustable, 1-1/2" pipe size		17	.471		11.60	30.50		42.10	58.50
6980	P trap adj. w/union & cleanout, 1-1/2" pipe size	↓	16	.500	↓	47.50	32		79.50	101

22 13 16.80 Vent Flashing and Caps

		Crew	Daily Output	Labor-Hours	Unit	Material	2020 Bare Costs Labor	Equipment	Total	Total Incl O&P
0010	**VENT FLASHING AND CAPS**									
0120	Vent caps									
0140	Cast iron									
0180	2-1/2" to 3-5/8" pipe	1 Plum	21	.381	Ea.	52	24.50		76.50	94
0190	4" to 4-1/8" pipe	"	19	.421	"	75	27		102	124
0900	Vent flashing									
1000	Aluminum with lead ring									
1020	1-1/4" pipe	1 Plum	20	.400	Ea.	5.95	26		31.95	45
1030	1-1/2" pipe		20	.400		5.75	26		31.75	45
1040	2" pipe		18	.444		5.65	28.50		34.15	49
1050	3" pipe		17	.471		6.25	30.50		36.75	52.50
1060	4" pipe	↓	16	.500	↓	7.55	32		39.55	56.50
1350	Copper with neoprene ring									

For customer support on your Mechanical Costs with RSMeans Data, call 800.448.8182.

241

22 13 Facility Sanitary Sewerage

22 13 16 – Sanitary Waste and Vent Piping

22 13 16.80 Vent Flashing and Caps

		Crew	Daily Output	Labor-Hours	Unit	Material	2020 Bare Costs Labor	Equipment	Total	Total Incl O&P
1400	1-1/4" pipe	1 Plum	20	.400	Ea.	74	26		100	120
1430	1-1/2" pipe		20	.400		74	26		100	120
1440	2" pipe		18	.444		74	28.50		102.50	125
1450	3" pipe		17	.471		89.50	30.50		120	144
1460	4" pipe		16	.500		89.50	32		121.50	147
2000	Galvanized with neoprene ring									
2020	1-1/4" pipe	1 Plum	20	.400	Ea.	16.40	26		42.40	56.50
2030	1-1/2" pipe		20	.400		17.45	26		43.45	57.50
2040	2" pipe		18	.444		16	28.50		44.50	60.50
2050	3" pipe		17	.471		19.30	30.50		49.80	66.50
2060	4" pipe		16	.500		17.65	32		49.65	67.50
2980	Neoprene, one piece									
3000	1-1/4" pipe	1 Plum	24	.333	Ea.	3.29	21.50		24.79	35.50
3030	1-1/2" pipe		24	.333		3.35	21.50		24.85	35.50
3040	2" pipe		23	.348		4.79	22.50		27.29	39
3050	3" pipe		21	.381		7	24.50		31.50	44
3060	4" pipe		20	.400		10.45	26		36.45	50
4000	Lead, 4#, 8" skirt, vent through roof									
4100	2" pipe	1 Plum	18	.444	Ea.	43	28.50		71.50	90.50
4110	3" pipe		17	.471		48.50	30.50		79	99
4120	4" pipe		16	.500		56	32		88	110
4130	6" pipe		14	.571		84	37		121	148

22 13 26 – Sanitary Waste Separators

22 13 26.10 Separators

		Crew	Daily Output	Labor-Hours	Unit	Material	2020 Bare Costs Labor	Equipment	Total	Total Incl O&P
0010	**SEPARATORS**, Entrainment eliminator, steel body, 150 PSIG									
0100	1/4" size	1 Stpi	24	.333	Ea.	276	22		298	340
0120	1/2" size		24	.333		284	22		306	350
0140	3/4" size		20	.400		284	26		310	350
0160	1" size		19	.421		310	27.50		337.50	380
0180	1-1/4" size		15	.533		325	35		360	415
0200	1-1/2" size		13	.615		360	40.50		400.50	455
0220	2" size		11	.727		400	47.50		447.50	510
0240	2-1/2" size	Q-5	15	1.067		1,725	63		1,788	2,000
0260	3" size		13	1.231		1,900	72.50		1,972.50	2,200
0280	4" size		10	1.600		2,375	94.50		2,469.50	2,775
0300	5" size		6	2.667		2,800	157		2,957	3,300
0320	6" size		3	5.333		3,050	315		3,365	3,850
0340	8" size	Q-6	4.40	5.455		3,525	335		3,860	4,375
0360	10" size	"	4	6		5,125	365		5,490	6,200
1000	For 300 PSIG, add					15%				

22 13 29 – Sanitary Sewerage Pumps

22 13 29.13 Wet-Pit-Mounted, Vertical Sewerage Pumps

		Crew	Daily Output	Labor-Hours	Unit	Material	2020 Bare Costs Labor	Equipment	Total	Total Incl O&P
0010	**WET-PIT-MOUNTED, VERTICAL SEWERAGE PUMPS**									
0020	Controls incl. alarm/disconnect panel w/wire. Excavation not included									
0260	Simplex, 9 GPM at 60 PSIG, 91 gal. tank				Ea.	3,525			3,525	3,900
0300	Unit with manway, 26" ID, 18" high					4,150			4,150	4,575
0340	26" ID, 36" high					4,075			4,075	4,500
0380	43" ID, 4' high					4,250			4,250	4,675
0600	Simplex, 9 GPM at 60 PSIG, 150 gal. tank, indoor					4,275			4,275	4,700
0700	Unit with manway, 26" ID, 36" high					4,975			4,975	5,475
0740	26" ID, 4' high					5,025			5,025	5,525
2000	Duplex, 18 GPM at 60 PSIG, 150 gal. tank, indoor					8,175			8,175	9,000

22 13 Facility Sanitary Sewerage

22 13 29 – Sanitary Sewerage Pumps

22 13 29.13 Wet-Pit-Mounted, Vertical Sewerage Pumps	Crew	Daily Output	Labor-Hours	Unit	Material	2020 Bare Costs Labor	Equipment	Total	Total Incl O&P
2060 Unit with manway, 43" ID, 4' high				Ea.	8,975			8,975	9,875
2400 For core only				▼	2,050			2,050	2,250
3000 Indoor residential type installation									
3020 Simplex, 9 GPM at 60 PSIG, 91 gal. HDPE tank				Ea.	3,625			3,625	3,975

22 13 29.14 Sewage Ejector Pumps

	Crew	Daily Output	Labor-Hours	Unit	Material	2020 Bare Costs Labor	Equipment	Total	Total Incl O&P
0010 **SEWAGE EJECTOR PUMPS**, With operating and level controls									
0100 Simplex system incl. tank, cover, pump 15' head									
0500 37 gal. PE tank, 12 GPM, 1/2 HP, 2" discharge	Q-1	3.20	5	Ea.	505	290		795	990
0510 3" discharge		3.10	5.161		550	299		849	1,050
0530 87 GPM, .7 HP, 2" discharge		3.20	5		770	290		1,060	1,275
0540 3" discharge		3.10	5.161		835	299		1,134	1,375
0600 45 gal. coated stl. tank, 12 GPM, 1/2 HP, 2" discharge		3	5.333		900	310		1,210	1,450
0610 3" discharge		2.90	5.517		935	320		1,255	1,500
0630 87 GPM, .7 HP, 2" discharge		3	5.333		1,150	310		1,460	1,750
0640 3" discharge		2.90	5.517		1,225	320		1,545	1,825
0660 134 GPM, 1 HP, 2" discharge		2.80	5.714		1,250	330		1,580	1,875
0680 3" discharge		2.70	5.926		1,325	345		1,670	1,975
0700 70 gal. PE tank, 12 GPM, 1/2 HP, 2" discharge		2.60	6.154		970	355		1,325	1,600
0710 3" discharge		2.40	6.667		1,050	385		1,435	1,725
0730 87 GPM, .7 HP, 2" discharge		2.50	6.400		1,275	370		1,645	1,975
0740 3" discharge		2.30	6.957		1,325	405		1,730	2,075
0760 134 GPM, 1 HP, 2" discharge		2.20	7.273		1,375	420		1,795	2,125
0770 3" discharge		2	8		1,450	465		1,915	2,300
0800 75 gal. coated stl. tank, 12 GPM, 1/2 HP, 2" discharge		2.40	6.667		1,075	385		1,460	1,750
0810 3" discharge		2.20	7.273		1,125	420		1,545	1,850
0830 87 GPM, .7 HP, 2" discharge		2.30	6.957		1,350	405		1,755	2,100
0840 3" discharge		2.10	7.619		1,425	440		1,865	2,225
0860 134 GPM, 1 HP, 2" discharge		2	8		1,450	465		1,915	2,300
0880 3" discharge	▼	1.80	8.889	▼	1,525	515		2,040	2,450
1040 Duplex system incl. tank, covers, pumps									
1060 110 gal. fiberglass tank, 24 GPM, 1/2 HP, 2" discharge	Q-1	1.60	10	Ea.	1,950	580		2,530	3,025
1080 3" discharge		1.40	11.429		2,050	665		2,715	3,250
1100 174 GPM, .7 HP, 2" discharge		1.50	10.667		2,525	620		3,145	3,700
1120 3" discharge		1.30	12.308		2,625	715		3,340	3,950
1140 268 GPM, 1 HP, 2" discharge		1.20	13.333		2,725	775		3,500	4,150
1160 3" discharge	▼	1	16		2,825	930		3,755	4,500
1260 135 gal. coated stl. tank, 24 GPM, 1/2 HP, 2" discharge	Q-2	1.70	14.118		2,000	850		2,850	3,475
2000 3" discharge		1.60	15		2,125	900		3,025	3,700
2640 174 GPM, .7 HP, 2" discharge		1.60	15		2,625	900		3,525	4,225
2660 3" discharge		1.50	16		2,775	960		3,735	4,500
2700 268 GPM, 1 HP, 2" discharge		1.30	18.462		2,850	1,100		3,950	4,775
3040 3" discharge		1.10	21.818		3,025	1,300		4,325	5,300
3060 275 gal. coated stl. tank, 24 GPM, 1/2 HP, 2" discharge		1.50	16		2,500	960		3,460	4,200
3080 3" discharge		1.40	17.143		2,550	1,025		3,575	4,350
3100 174 GPM, .7 HP, 2" discharge		1.40	17.143		3,225	1,025		4,250	5,100
3120 3" discharge		1.30	18.462		3,475	1,100		4,575	5,475
3140 268 GPM, 1 HP, 2" discharge		1.10	21.818		3,550	1,300		4,850	5,875
3160 3" discharge	▼	.90	26.667	▼	3,750	1,600		5,350	6,500
3260 Pump system accessories, add									
3300 Alarm horn and lights, 115 V mercury switch	Q-1	8	2	Ea.	102	116		218	286
3340 Switch, mag. contactor, alarm bell, light, 3 level control		5	3.200		495	186		681	825
3380 Alternator, mercury switch activated	▼	4	4	▼	895	232		1,127	1,325

For customer support on your Mechanical Costs with RSMeans Data, call 800.448.8182.

243

22 14 Facility Storm Drainage

22 14 29 – Sump Pumps

22 14 29.13 Wet-Pit-Mounted, Vertical Sump Pumps

	Crew	Daily Output	Labor-Hours	Unit	Material	2020 Bare Costs Labor	Equipment	Total	Total Incl O&P
0010 **WET-PIT-MOUNTED, VERTICAL SUMP PUMPS**									
0400 Molded PVC base, 21 GPM at 15' head, 1/3 HP	1 Plum	5	1.600	Ea.	141	103		244	310
0800 Iron base, 21 GPM at 15' head, 1/3 HP		5	1.600		155	103		258	325
1200 Solid brass, 21 GPM at 15' head, 1/3 HP		5	1.600		243	103		346	420
2000 Sump pump, single stage									
2010 25 GPM, 1 HP, 1-1/2" discharge	Q-1	1.80	8.889	Ea.	3,825	515		4,340	5,000
2020 75 GPM, 1-1/2 HP, 2" discharge		1.50	10.667		4,050	620		4,670	5,375
2030 100 GPM, 2 HP, 2-1/2" discharge		1.30	12.308		4,125	715		4,840	5,625
2040 150 GPM, 3 HP, 3" discharge		1.10	14.545		4,125	845		4,970	5,825
2050 200 GPM, 3 HP, 3" discharge		1	16		4,375	930		5,305	6,225
2060 300 GPM, 10 HP, 4" discharge	Q-2	1.20	20		4,725	1,200		5,925	7,000
2070 500 GPM, 15 HP, 5" discharge		1.10	21.818		5,375	1,300		6,675	7,875
2080 800 GPM, 20 HP, 6" discharge		1	24		6,350	1,450		7,800	9,125
2090 1,000 GPM, 30 HP, 6" discharge		.85	28.235		6,975	1,700		8,675	10,200
2100 1,600 GPM, 50 HP, 8" discharge		.72	33.333		10,900	2,000		12,900	15,000
2110 2,000 GPM, 60 HP, 8" discharge	Q-3	.85	37.647		11,100	2,300		13,400	15,700
2202 For general purpose float switch, copper coated float, add	Q-1	5	3.200		107	186		293	395

22 14 29.16 Submersible Sump Pumps

	Crew	Daily Output	Labor-Hours	Unit	Material	2020 Bare Costs Labor	Equipment	Total	Total Incl O&P
0010 **SUBMERSIBLE SUMP PUMPS**									
1000 Elevator sump pumps, automatic									
1010 Complete systems, pump, oil detector, controls and alarm									
1020 1-1/2" discharge, does not include the sump pit/tank									
1040 1/3 HP, 115 V	1 Plum	4.40	1.818	Ea.	1,925	117		2,042	2,275
1050 1/2 HP, 115 V		4	2		2,025	129		2,154	2,425
1060 1/2 HP, 230 V		4	2		2,000	129		2,129	2,400
1070 3/4 HP, 115 V		3.60	2.222		2,100	143		2,243	2,550
1080 3/4 HP, 230 V		3.60	2.222		2,075	143		2,218	2,500
1100 Sump pump only									
1110 1/3 HP, 115 V	1 Plum	6.40	1.250	Ea.	179	80.50		259.50	320
1120 1/2 HP, 115 V		5.80	1.379		254	89		343	415
1130 1/2 HP, 230 V		5.80	1.379		294	89		383	460
1140 3/4 HP, 115 V		5.40	1.481		340	95.50		435.50	520
1150 3/4 HP, 230 V		5.40	1.481		380	95.50		475.50	565
1200 Oil detector, control and alarm only									
1210 115 V	1 Plum	8	1	Ea.	1,775	64.50		1,839.50	2,050
1220 230 V	"	8	1	"	1,775	64.50		1,839.50	2,050
7000 Sump pump, automatic									
7100 Plastic, 1-1/4" discharge, 1/4 HP	1 Plum	6.40	1.250	Ea.	168	80.50		248.50	305
7140 1/3 HP		6	1.333		224	86		310	375
7160 1/2 HP		5.40	1.481		245	95.50		340.50	415
7180 1-1/2" discharge, 1/2 HP		5.20	1.538		325	99		424	505
7500 Cast iron, 1-1/4" discharge, 1/4 HP		6	1.333		225	86		311	375
7540 1/3 HP		6	1.333		264	86		350	420
7560 1/2 HP		5	1.600		320	103		423	505

22 14 53 – Rainwater Storage Tanks

22 14 53.13 Fiberglass, Rainwater Storage Tank

	Crew	Daily Output	Labor-Hours	Unit	Material	2020 Bare Costs Labor	Equipment	Total	Total Incl O&P
0010 **FIBERGLASS, RAINWATER STORAGE TANK**									
2000 600 gallon	B-21B	3.75	10.667	Ea.	4,125	490	126	4,741	5,425
2010 1,000 gallon		3.50	11.429		5,275	525	135	5,935	6,725
2020 2,000 gallon		3.25	12.308		7,500	565	145	8,210	9,250
2030 4,000 gallon		3	13.333		10,100	610	157	10,867	12,200
2040 6,000 gallon		2.65	15.094		11,400	695	178	12,273	13,800

244

For customer support on your Mechanical Costs with RSMeans Data, call 800.448.8182.

22 14 Facility Storm Drainage

22 14 53 – Rainwater Storage Tanks

22 14 53.13 Fiberglass, Rainwater Storage Tank		Crew	Daily Output	Labor-Hours	Unit	Material	2020 Bare Costs Labor	Equipment	Total	Total Incl O&P
2050	8,000 gallon	B-21B	2.30	17.391	Ea.	13,900	800	205	14,905	16,700
2060	10,000 gallon		2	20		15,400	920	236	16,556	18,600
2070	12,000 gallon		1.50	26.667		21,600	1,225	315	23,140	25,900
2080	15,000 gallon		1	40		24,500	1,825	470	26,795	30,300
2090	20,000 gallon		.75	53.333		31,800	2,450	630	34,880	39,400
2100	25,000 gallon		.50	80		47,400	3,675	940	52,015	58,500
2110	30,000 gallon		.35	114		57,000	5,250	1,350	63,600	72,000
2120	40,000 gallon		.30	133		80,000	6,125	1,575	87,700	99,000

22 15 General Service Compressed-Air Systems

22 15 13 – General Service Compressed-Air Piping

22 15 13.10 Compressor Accessories

		Crew	Daily Output	Labor-Hours	Unit	Material	2020 Bare Costs Labor	Equipment	Total	Total Incl O&P
0010	**COMPRESSOR ACCESSORIES**									
1700	Refrigerated air dryers with ambient air filters									
1710	10 CFM	Q-5	8	2	Ea.	785	118		903	1,050
1715	15 CFM		7.30	2.192		1,075	129		1,204	1,375
1716	20 CFM		6.90	2.319		2,150	137		2,287	2,575
1720	25 CFM		6.60	2.424		1,575	143		1,718	1,950
1730	50 CFM		6.20	2.581		1,875	152		2,027	2,300
1740	75 CFM		5.80	2.759		1,925	163		2,088	2,350
1750	100 CFM		5.60	2.857		3,100	169		3,269	3,650
1751	125 CFM		5.40	2.963		3,300	175		3,475	3,875
1752	175 CFM		5	3.200		3,575	189		3,764	4,200
1753	200 CFM		4.80	3.333		6,000	197		6,197	6,900
1754	300 CFM		4.40	3.636		6,900	215		7,115	7,925
1755	400 CFM		4	4		8,825	236		9,061	10,100
1756	500 CFM		3.80	4.211		9,200	248		9,448	10,500
1757	600 CFM		3.40	4.706		11,700	278		11,978	13,300
1758	800 CFM		3	5.333		19,600	315		19,915	22,000
1759	1,000 CFM		2.60	6.154		20,800	365		21,165	23,300
3460	Air filter, regulator, lubricator combination									
3470	Flush mount									
3480	Adjustable range 0-140 psi									
3500	1/8" NPT, 34 SCFM	1 Stpi	17	.471	Ea.	107	31		138	163
3510	1/4" NPT, 61 SCFM		17	.471		192	31		223	257
3520	3/8" NPT, 85 SCFM		16	.500		196	33		229	265
3530	1/2" NPT, 150 SCFM		15	.533		210	35		245	284
3540	3/4" NPT, 171 SCFM		14	.571		260	37.50		297.50	340
3550	1" NPT, 150 SCFM		13	.615		415	40.50		455.50	515
4000	Couplers, air line, sleeve type									
4010	Female, connection size NPT									
4020	1/4"	1 Stpi	38	.211	Ea.	7.80	13.80		21.60	29
4030	3/8"		36	.222		12.60	14.55		27.15	36
4040	1/2"		35	.229		20.50	15		35.50	45
4050	3/4"		34	.235		22.50	15.40		37.90	48
4100	Male									
4110	1/4"	1 Stpi	38	.211	Ea.	7.95	13.80		21.75	29.50
4120	3/8"		36	.222		11.25	14.55		25.80	34.50
4130	1/2"		35	.229		19.90	15		34.90	44.50
4140	3/4"		34	.235		22.50	15.40		37.90	47.50
4150	Coupler, combined male and female halves									

For customer support on your Mechanical Costs with RSMeans Data, call 800.448.8182.

245

22 15 General Service Compressed-Air Systems

22 15 13 – General Service Compressed-Air Piping

22 15 13.10 Compressor Accessories	Crew	Daily Output	Labor-Hours	Unit	Material	2020 Bare Costs Labor	Equipment	Total	Total Incl O&P
4160	1/2"	1 Stpi	17	.471	Ea.	40.50	31		71.50	90.50
4170	3/4"	"	15	.533	"	45	35		80	102

22 15 19 – General Service Packaged Air Compressors and Receivers

22 15 19.10 Air Compressors

		Crew	Daily Output	Labor-Hours	Unit	Material	2020 Bare Costs Labor	Equipment	Total	Total Incl O&P
0010	**AIR COMPRESSORS**									
5250	Air, reciprocating air cooled, splash lubricated, tank mounted									
5300	Single stage, 1 phase, 140 psi									
5303	1/2 HP, 17 gal. tank	1 Stpi	3	2.667	Ea.	2,000	175		2,175	2,450
5305	3/4 HP, 30 gal. tank		2.60	3.077		1,700	202		1,902	2,175
5307	1 HP, 30 gal. tank		2.20	3.636		2,075	238		2,313	2,650
5309	2 HP, 30 gal. tank	Q-5	4	4		2,575	236		2,811	3,200
5310	3 HP, 30 gal. tank		3.60	4.444		2,700	262		2,962	3,375
5314	3 HP, 60 gal. tank		3.50	4.571		3,175	270		3,445	3,875
5320	5 HP, 60 gal. tank		3.20	5		3,375	295		3,670	4,175
5330	5 HP, 80 gal. tank		3	5.333		3,325	315		3,640	4,125
5340	7.5 HP, 80 gal. tank		2.60	6.154		4,550	365		4,915	5,550
5600	2 stage pkg., 3 phase									
5650	6 CFM at 125 psi, 1-1/2 HP, 60 gal. tank	Q-5	3	5.333	Ea.	2,575	315		2,890	3,325
5670	10.9 CFM at 125 psi, 3 HP, 80 gal. tank		1.50	10.667		3,300	630		3,930	4,600
5680	38.7 CFM at 125 psi, 10 HP, 120 gal. tank		.60	26.667		6,275	1,575		7,850	9,250
5690	105 CFM at 125 psi, 25 HP, 250 gal. tank	Q-6	.60	40		16,300	2,450		18,750	21,600
5800	With single stage pump									
5850	8.3 CFM at 125 psi, 2 HP, 80 gal. tank	Q-6	3.50	6.857	Ea.	3,175	420		3,595	4,125
5860	38.7 CFM at 125 psi, 10 HP, 120 gal. tank	"	.90	26.667	"	5,800	1,625		7,425	8,825
6000	Reciprocating, 2 stage, tank mtd, 3 ph, cap rated @ 175 PSIG									
6050	Pressure lubricated, hvy. duty, 9.7 CFM, 3 HP, 120 gal. tank	Q-5	1.30	12.308	Ea.	4,975	725		5,700	6,550
6054	5 CFM, 1-1/2 HP, 80 gal. tank		2.80	5.714		3,100	335		3,435	3,900
6056	6.4 CFM, 2 HP, 80 gal. tank		2	8		3,175	470		3,645	4,200
6058	8.1 CFM, 3 HP, 80 gal. tank		1.70	9.412		3,400	555		3,955	4,550
6059	14.8 CFM, 5 HP, 80 gal. tank		1	16		3,425	945		4,370	5,175
6060	16.5 CFM, 5 HP, 120 gal. tank		1	16		5,625	945		6,570	7,600
6063	13 CFM, 6 HP, 80 gal. tank		.90	17.778		5,225	1,050		6,275	7,325
6066	19.8 CFM, 7.5 HP, 80 gal. tank		.80	20		5,150	1,175		6,325	7,425
6070	25.8 CFM, 7-1/2 HP, 120 gal. tank		.80	20		5,400	1,175		6,575	7,700
6078	34.8 CFM, 10 HP, 80 gal. tank		.70	22.857		7,800	1,350		9,150	10,600
6080	34.8 CFM, 10 HP, 120 gal. tank		.60	26.667		8,325	1,575		9,900	11,500
6090	53.7 CFM, 15 HP, 120 gal. tank	Q-6	.80	30		8,750	1,825		10,575	12,400
6100	76.7 CFM, 20 HP, 120 gal. tank		.70	34.286		12,600	2,100		14,700	17,000
6104	76.7 CFM, 20 HP, 240 gal. tank		.68	35.294		12,200	2,150		14,350	16,600
6110	90.1 CFM, 25 HP, 120 gal. tank		.63	38.095		12,700	2,325		15,025	17,500
6120	101 CFM, 30 HP, 120 gal. tank		.57	42.105		11,300	2,575		13,875	16,300
6130	101 CFM, 30 HP, 250 gal. tank		.52	46.154		12,800	2,825		15,625	18,300
6200	Oil-less, 13.6 CFM, 5 HP, 120 gal. tank	Q-5	.88	18.182		13,700	1,075		14,775	16,600
6210	13.6 CFM, 5 HP, 250 gal. tank		.80	20		14,600	1,175		15,775	17,900
6220	18.2 CFM, 7.5 HP, 120 gal. tank		.73	21.918		13,600	1,300		14,900	16,900
6230	18.2 CFM, 7.5 HP, 250 gal. tank		.67	23.881		14,600	1,400		16,000	18,200
6250	30.5 CFM, 10 HP, 120 gal. tank		.57	28.070		16,100	1,650		17,750	20,200
6260	30.5 CFM, 10 HP, 250 gal. tank		.53	30.189		17,000	1,775		18,775	21,500
6270	41.3 CFM, 15 HP, 120 gal. tank	Q-6	.70	34.286		17,400	2,100		19,500	22,400
6280	41.3 CFM, 15 HP, 250 gal. tank	"	.67	35.821		18,500	2,200		20,700	23,600

22 31 Domestic Water Softeners

22 31 13 – Residential Domestic Water Softeners

22 31 13.10 Residential Water Softeners	Crew	Daily Output	Labor-Hours	Unit	Material	2020 Bare Costs Labor	Equipment	Total	Total Incl O&P
0010 **RESIDENTIAL WATER SOFTENERS**									
7350 Water softener, automatic, to 30 grains per gallon	2 Plum	5	3.200	Ea.	395	206		601	745
7400 To 100 grains per gallon	"	4	4	"	940	258		1,198	1,400

22 35 Domestic Water Heat Exchangers

22 35 30 – Water Heating by Steam

22 35 30.10 Water Heating Transfer Package

	Crew	Daily Output	Labor-Hours	Unit	Material	2020 Bare Costs Labor	Equipment	Total	Total Incl O&P
0010 **WATER HEATING TRANSFER PACKAGE**, Complete controls,									
0020 expansion tank, converter, air separator									
1000 Hot water, 180°F enter, 200°F leaving, 15# steam									
1010 One pump system, 28 GPM	Q-6	.75	32	Ea.	21,800	1,950		23,750	26,900
1020 35 GPM		.70	34.286		24,700	2,100		26,800	30,400
1040 55 GPM		.65	36.923		28,000	2,250		30,250	34,200
1060 130 GPM		.55	43.636		35,500	2,675		38,175	43,100
1080 255 GPM		.40	60		46,600	3,675		50,275	56,500
1100 550 GPM		.30	80		67,000	4,900		71,900	81,500
1120 800 GPM		.25	96		74,000	5,875		79,875	90,500
1220 Two pump system, 28 GPM		.70	34.286		29,700	2,100		31,800	35,800
1240 35 GPM		.65	36.923		36,100	2,250		38,350	43,100
1260 55 GPM		.60	40		35,900	2,450		38,350	43,100
1280 130 GPM		.50	48		49,400	2,925		52,325	59,000
1300 255 GPM		.35	68.571		65,000	4,200		69,200	78,000
1320 550 GPM		.25	96		79,500	5,875		85,375	96,000
1340 800 GPM		.20	120		102,500	7,350		109,850	123,500

22 41 Residential Plumbing Fixtures

22 41 39 – Residential Faucets, Supplies and Trim

22 41 39.10 Faucets and Fittings

	Crew	Daily Output	Labor-Hours	Unit	Material	2020 Bare Costs Labor	Equipment	Total	Total Incl O&P
0010 **FAUCETS AND FITTINGS**									
5000 Sillcock, compact, brass, IPS or copper to hose	1 Plum	24	.333	Ea.	12.05	21.50		33.55	45.50
6000 Stop and waste valves, bronze									
6100 Angle, solder end 1/2"	1 Plum	24	.333	Ea.	28	21.50		49.50	63
6110 3/4"		20	.400		36	26		62	78
6300 Straightway, solder end 3/8"		24	.333		20.50	21.50		42	54.50
6310 1/2"		24	.333		20.50	21.50		42	54.50
6320 3/4"		20	.400		21	26		47	61.50
6410 Straightway, threaded 1/2"		24	.333		22	21.50		43.50	56
6420 3/4"		20	.400		24.50	26		50.50	65.50
6430 1"		19	.421		27	27		54	70

For customer support on your Mechanical Costs with RSMeans Data, call 800.448.8182.

247

22 51 Swimming Pool Plumbing Systems

22 51 19 – Swimming Pool Water Treatment Equipment

22 51 19.50 Swimming Pool Filtration Equipment		Crew	Daily Output	Labor-Hours	Unit	Material	2020 Bare Costs Labor	Equipment	Total	Total Incl O&P
0010	**SWIMMING POOL FILTRATION EQUIPMENT**									
0900	Filter system, sand or diatomite type, incl. pump, 6,000 gal./hr.	2 Plum	1.80	8.889	Total	2,375	575		2,950	3,475
1020	Add for chlorination system, 800 S.F. pool		3	5.333	Ea.	235	345		580	775
1040	5,000 S.F. pool		3	5.333	"	1,950	345		2,295	2,650

22 52 Fountain Plumbing Systems

22 52 16 – Fountain Pumps

22 52 16.10 Fountain Water Pumps

		Crew	Daily Output	Labor-Hours	Unit	Material	2020 Bare Costs Labor	Equipment	Total	Total Incl O&P
0010	**FOUNTAIN WATER PUMPS**									
0100	Pump w/controls									
0200	Single phase, 100' cord, 1/2 HP pump	2 Skwk	4.40	3.636	Ea.	1,325	199		1,524	1,750
0300	3/4 HP pump		4.30	3.721		1,375	204		1,579	1,825
0400	1 HP pump		4.20	3.810		1,925	209		2,134	2,450
0500	1-1/2 HP pump		4.10	3.902		2,575	214		2,789	3,175
0600	2 HP pump		4	4		4,350	219		4,569	5,100
0700	Three phase, 200' cord, 5 HP pump		3.90	4.103		6,250	225		6,475	7,225
0800	7-1/2 HP pump		3.80	4.211		12,600	231		12,831	14,200
0900	10 HP pump		3.70	4.324		15,200	237		15,437	17,200
1000	15 HP pump		3.60	4.444		21,500	244		21,744	24,000
2000	DESIGN NOTE: Use two horsepower per surface acre.									

22 52 33 – Fountain Ancillary

22 52 33.10 Fountain Miscellaneous

		Crew	Daily Output	Labor-Hours	Unit	Material	2020 Bare Costs Labor	Equipment	Total	Total Incl O&P
0010	**FOUNTAIN MISCELLANEOUS**									
1300	Lights w/mounting kits, 200 watt	2 Skwk	18	.889	Ea.	1,200	49		1,249	1,400
1400	300 watt		18	.889		1,350	49		1,399	1,575
1500	500 watt		18	.889		1,525	49		1,574	1,750
1600	Color blender		12	1.333		600	73		673	770

22 62 Vacuum Systems for Laboratory and Healthcare Facilities

22 62 19 – Vacuum Equipment for Laboratory and Healthcare Facilities

22 62 19.70 Healthcare Vacuum Equipment

		Crew	Daily Output	Labor-Hours	Unit	Material	2020 Bare Costs Labor	Equipment	Total	Total Incl O&P
0010	**HEALTHCARE VACUUM EQUIPMENT**									
0300	Dental oral									
0310	Duplex									
0330	165 SCFM with 77 gal. separator	Q-2	1.30	18.462	Ea.	52,500	1,100		53,600	59,000
1100	Vacuum system									
1110	Vacuum outlet alarm panel	1 Plum	3.20	2.500	Ea.	1,100	161		1,261	1,450
2000	Medical, with receiver									
2100	Rotary vane type, lubricated, with controls									
2110	Simplex									
2120	1.5 HP, 80 gal. tank	Q-1	8	2	Ea.	6,650	116		6,766	7,500
2130	2 HP, 80 gal. tank		7	2.286		6,900	133		7,033	7,775
2140	3 HP, 80 gal. tank		6.60	2.424		7,275	141		7,416	8,200
2150	5 HP, 80 gal. tank		6	2.667		7,925	155		8,080	8,950
2200	Duplex									
2210	1 HP, 80 gal. tank	Q-1	8.40	1.905	Ea.	10,600	110		10,710	11,900
2220	1.5 HP, 80 gal. tank		7.80	2.051		10,900	119		11,019	12,200
2230	2 HP, 80 gal. tank		6.80	2.353		11,800	136		11,936	13,200
2240	3 HP, 120 gal. tank		6	2.667		12,800	155		12,955	14,300

22 62 Vacuum Systems for Laboratory and Healthcare Facilities

22 62 19 – Vacuum Equipment for Laboratory and Healthcare Facilities

22 62 19.70 Healthcare Vacuum Equipment

22 62 19.70 Healthcare Vacuum Equipment	Crew	Daily Output	Labor-Hours	Unit	Material	2020 Bare Costs Labor	Equipment	Total	Total Incl O&P
2250	5 HP, 120 gal. tank	Q-1	5.40	2.963	Ea.	14,600	172		14,772	16,300
2260	7.5 HP, 200 gal. tank	Q-2	7	3.429		20,900	206		21,106	23,200
2270	10 HP, 200 gal. tank		6.40	3.750		24,100	226		24,326	26,800
2280	15 HP, 200 gal. tank		5.80	4.138		43,100	249		43,349	47,800
2290	20 HP, 200 gal. tank		5	4.800		49,100	289		49,389	54,500
2300	25 HP, 200 gal. tank		4	6		57,000	360		57,360	63,500
2400	Triplex									
2410	7.5 HP, 200 gal. tank	Q-2	6.40	3.750	Ea.	33,100	226		33,326	36,700
2420	10 HP, 200 gal. tank		5.70	4.211		38,500	253		38,753	42,800
2430	15 HP, 200 gal. tank		4.90	4.898		65,000	295		65,295	72,000
2440	20 HP, 200 gal. tank		4	6		74,500	360		74,860	82,000
2450	25 HP, 200 gal. tank		3.70	6.486		86,000	390		86,390	95,000
2500	Quadruplex									
2510	7.5 HP, 200 gal. tank	Q-2	5.30	4.528	Ea.	41,200	272		41,472	45,700
2520	10 HP, 200 gal. tank		4.70	5.106		47,500	305		47,805	53,000
2530	15 HP, 200 gal. tank		4	6		86,000	360		86,360	95,000
2540	20 HP, 200 gal. tank		3.60	6.667		98,000	400		98,400	108,000
2550	25 HP, 200 gal. tank		3.20	7.500		113,500	450		113,950	125,500
4000	Liquid ring type, water sealed, with controls									
4200	Duplex									
4210	1.5 HP, 120 gal. tank	Q-1	6	2.667	Ea.	21,000	155		21,155	23,300
4220	3 HP, 120 gal. tank	"	5.50	2.909		21,900	169		22,069	24,400
4230	4 HP, 120 gal. tank	Q-2	6.80	3.529		23,400	212		23,612	26,000
4240	5 HP, 120 gal. tank		6.50	3.692		24,700	222		24,922	27,500
4250	7.5 HP, 200 gal. tank		6.20	3.871		30,000	233		30,233	33,400
4260	10 HP, 200 gal. tank		5.80	4.138		37,500	249		37,749	41,600
4270	15 HP, 200 gal. tank		5.10	4.706		45,400	283		45,683	50,500
4280	20 HP, 200 gal. tank		4.60	5.217		58,500	315		58,815	65,000
4290	30 HP, 200 gal. tank		4	6		70,000	360		70,360	77,500

22 63 Gas Systems for Laboratory and Healthcare Facilities

22 63 13 – Gas Piping for Laboratory and Healthcare Facilities

22 63 13.70 Healthcare Gas Piping

22 63 13.70 Healthcare Gas Piping	Crew	Daily Output	Labor-Hours	Unit	Material	2020 Bare Costs Labor	Equipment	Total	Total Incl O&P
0010	**HEALTHCARE GAS PIPING**									
1000	Nitrogen or oxygen system									
1010	Cylinder manifold									
1020	5 cylinder	1 Plum	.80	10	Ea.	5,900	645		6,545	7,450
1026	10 cylinder	"	.40	20		7,100	1,300		8,400	9,725
1050	Nitrogen generator, 30 LPM	Q-5	.40	40		28,900	2,350		31,250	35,300
3000	Outlets and valves									
3010	Recessed, wall mounted									
3012	Single outlet	1 Plum	3.20	2.500	Ea.	70.50	161		231.50	320
3100	Ceiling outlet									
3190	Zone valve with box									
3210	2" valve size	1 Plum	3.20	2.500	Ea.	415	161		576	695
4000	Alarm panel, medical gases and vacuum									
4010	Alarm panel	1 Plum	3.20	2.500	Ea.	1,100	161		1,261	1,450

For customer support on your Mechanical Costs with RSMeans Data, call 800.448.8182.

249

Division Notes

	CREW	DAILY OUTPUT	LABOR-HOURS	UNIT	BARE COSTS				TOTAL INCL O&P
					MAT.	LABOR	EQUIP.	TOTAL	

Estimating Tips

The labor adjustment factors listed in Subdivision 22 01 02.20 also apply to Division 23.

23 10 00 Facility Fuel Systems

- The prices in this subdivision for above- and below-ground storage tanks do not include foundations or hold-down slabs, unless noted. The estimator should refer to Divisions 3 and 31 for foundation system pricing. In addition to the foundations, required tank accessories, such as tank gauges, leak detection devices, and additional manholes and piping, must be added to the tank prices.

23 50 00 Central Heating Equipment

- When estimating the cost of an HVAC system, check to see who is responsible for providing and installing the temperature control system. It is possible to overlook controls, assuming that they would be included in the electrical estimate.
- When looking up a boiler, be careful on specified capacity. Some manufacturers rate their products on output while others use input.
- Include HVAC insulation for pipe, boiler, and duct (wrap and liner).
- Be careful when looking up mechanical items to get the correct pressure rating and connection type (thread, weld, flange).

23 70 00 Central HVAC Equipment

- Combination heating and cooling units are sized by the air conditioning requirements. (See Reference No. R236000-20 for the preliminary sizing guide.)
- A ton of air conditioning is nominally 400 CFM.
- Rectangular duct is taken off by the linear foot for each size, but its cost is usually estimated by the pound. Remember that SMACNA standards now base duct on internal pressure.
- Prefabricated duct is estimated and purchased like pipe: straight sections and fittings.
- Note that cranes or other lifting equipment are not included on any lines in Division 23. For example, if a crane is required to lift a heavy piece of pipe into place high above a gym floor, or to put a rooftop unit on the roof of a four-story building, etc., it must be added. Due to the potential for extreme variation—from nothing additional required to a major crane or helicopter—we feel that including a nominal amount for "lifting contingency" would be useless and detract from the accuracy of the estimate. When using equipment rental cost data from RSMeans, do not forget to include the cost of the operator(s).

Reference Numbers

Reference numbers are shown at the beginning of some major classifications. These numbers refer to related items in the Reference Section. The reference information may be an estimating procedure, an alternate pricing method, or technical information.

Note: Not all subdivisions listed here necessarily appear. ■

Same Data. Simplified.

Enjoy the convenience and efficiency of accessing your costs anywhere:

- **Skip the multiplier** by setting your location
- **Quickly search,** edit, favorite and share costs
- **Stay on top of price changes** with automatic updates

Discover more at rsmeans.com/online

---

23 05 02 – HVAC General

23 05 02.10 Air Conditioning, General		Crew	Daily Output	Labor-Hours	Unit	Material	2020 Bare Costs Labor	Equipment	Total	Total Incl O&P
0010	**AIR CONDITIONING, GENERAL** Prices are for standard efficiencies (SEER 13)									
0020	for upgrade to SEER 14 add					10%				

23 05 05 – Selective Demolition for HVAC

23 05 05.10 HVAC Demolition

		Crew	Daily Output	Labor-Hours	Unit	Material	Labor	Equipment	Total	Total Incl O&P
0010	**HVAC DEMOLITION** R220105-10									
0100	Air conditioner, split unit, 3 ton	Q-5	2	8	Ea.		470		470	705
0150	Package unit, 3 ton R024119-10	Q-6	3	8			490		490	735
0190	Rooftop, self contained, up to 5 ton	1 Plum	1.20	6.667	▼		430		430	645
0250	Air curtain	Q-9	20	.800	L.F.		45		45	68
0254	Air filters, up thru 16,000 CFM		20	.800	Ea.		45		45	68
0256	20,000 thru 60,000 CFM	▼	16	1			56		56	85
0297	Boiler blowdown	Q-5	8	2	▼		118		118	177
0298	Boilers									
0300	Electric, up thru 148 kW	Q-19	2	12	Ea.		715		715	1,075
0310	150 thru 518 kW	"	1	24			1,425		1,425	2,150
0320	550 thru 2,000 kW	Q-21	.40	80			4,900		4,900	7,325
0330	2,070 kW and up	"	.30	107			6,525		6,525	9,750
0340	Gas and/or oil, up thru 150 MBH	Q-7	2.20	14.545			910		910	1,350
0350	160 thru 2,000 MBH		.80	40			2,500		2,500	3,725
0360	2,100 thru 4,500 MBH		.50	64			4,000		4,000	5,975
0370	4,600 thru 7,000 MBH		.30	107			6,650		6,650	9,950
0380	7,100 thru 12,000 MBH		.16	200			12,500		12,500	18,700
0390	12,200 thru 25,000 MBH	▼	.12	267			16,600		16,600	24,900
0400	Central station air handler unit, up thru 15 ton	Q-5	1.60	10			590		590	885
0410	17.5 thru 30 ton	"	.80	20	▼		1,175		1,175	1,775
0430	Computer room unit									
0434	Air cooled split, up thru 10 ton	Q-5	.67	23.881	Ea.		1,400		1,400	2,100
0436	12 thru 23 ton		.53	30.189			1,775		1,775	2,675
0440	Chilled water, up thru 10 ton		1.30	12.308			725		725	1,075
0444	12 thru 23 ton		1	16			945		945	1,425
0450	Glycol system, up thru 10 ton		.53	30.189			1,775		1,775	2,675
0454	12 thru 23 ton		.40	40			2,350		2,350	3,525
0460	Water cooled, not including condenser, up thru 10 ton		.80	20			1,175		1,175	1,775
0464	12 thru 23 ton		.60	26.667			1,575		1,575	2,350
0600	Condenser, up thru 50 ton	▼	1	16			945		945	1,425
0610	51 thru 100 ton	Q-6	.90	26.667			1,625		1,625	2,450
0620	101 thru 1,000 ton	"	.70	34.286			2,100		2,100	3,150
0660	Condensing unit, up thru 10 ton	Q-5	1.25	12.800			755		755	1,125
0670	11 thru 50 ton	"	.40	40			2,350		2,350	3,525
0680	60 thru 100 ton	Q-6	.30	80			4,900		4,900	7,325
0700	Cooling tower, up thru 400 ton		.80	30			1,825		1,825	2,750
0710	450 thru 600 ton		.53	45.283			2,775		2,775	4,150
0720	700 thru 1,300 ton	▼	.40	60			3,675		3,675	5,500
0780	Dehumidifier, up thru 155 lb./hr.	Q-1	8	2			116		116	174
0790	240 lb./hr. and up	"	2	8	▼		465		465	695
1560	Ductwork									
1570	Metal, steel, sst, fabricated	Q-9	1000	.016	Lb.		.90		.90	1.36
1580	Aluminum, fabricated		485	.033	"		1.85		1.85	2.81
1590	Spiral, prefabricated		400	.040	L.F.		2.24		2.24	3.40
1600	Fiberglass, prefabricated		400	.040			2.24		2.24	3.40
1610	Flex, prefabricated		500	.032			1.79		1.79	2.72
1620	Glass fiber reinforced plastic, prefabricated	▼	280	.057	▼		3.20		3.20	4.86

23 05 05.10 HVAC Demolition		Crew	Daily Output	Labor-Hours	Unit	Material	2020 Bare Costs Labor	2020 Bare Costs Equipment	Total	Total Incl O&P
1630	Diffusers, registers or grills, up thru 20" max dimension	1 Shee	50	.160	Ea.		9.95		9.95	15.10
1640	21 thru 36" max dimension		36	.222			13.85		13.85	21
1650	Above 36" max dimension		30	.267			16.60		16.60	25
1700	Evaporator, up thru 12,000 BTUH	Q-5	5.30	3.019			178		178	267
1710	12,500 thru 30,000 BTUH	"	2.70	5.926			350		350	525
1720	31,000 BTUH and up	Q-6	1.50	16			980		980	1,475
1730	Evaporative cooler, up thru 5 HP	Q-9	2.70	5.926			330		330	505
1740	10 thru 30 HP	"	.67	23.881			1,350		1,350	2,025
1750	Exhaust systems									
1760	Exhaust components	1 Shee	8	1	System		62.50		62.50	94.50
1770	Weld fume hoods	"	20	.400	Ea.		25		25	38
2120	Fans, up thru 1 HP or 2,000 CFM	Q-9	8	2			112		112	170
2124	1-1/2 thru 10 HP or 20,000 CFM		5.30	3.019			169		169	257
2128	15 thru 30 HP or above 20,000 CFM		4	4			224		224	340
2150	Fan coil air conditioner, chilled water, up thru 7.5 ton	Q-5	14	1.143			67.50		67.50	101
2154	Direct expansion, up thru 10 ton		8	2			118		118	177
2158	11 thru 30 ton		2	8			470		470	705
2170	Flue shutter damper	Q-9	8	2			112		112	170
2200	Furnace, electric	Q-20	2	10			570		570	865
2300	Gas or oil, under 120 MBH	Q-9	4	4			224		224	340
2340	Over 120 MBH	"	3	5.333			299		299	455
2730	Heating and ventilating unit	Q-5	2.70	5.926			350		350	525
2740	Heater, electric, wall, baseboard and quartz	1 Elec	10	.800			49		49	73
2750	Heater, electric, unit, cabinet, fan and convector	"	8	1			61.50		61.50	91.50
2760	Heat exchanger, shell and tube type	Q-5	1.60	10			590		590	885
2770	Plate type	Q-6	.60	40			2,450		2,450	3,675
2810	Heat pump									
2820	Air source, split, 4 thru 10 ton	Q-5	.90	17.778	Ea.		1,050		1,050	1,575
2830	15 thru 25 ton	Q-6	.80	30			1,825		1,825	2,750
2850	Single package, up thru 12 ton	Q-5	1	16			945		945	1,425
2860	Water source, up thru 15 ton	"	.90	17.778			1,050		1,050	1,575
2870	20 thru 50 ton	Q-6	.80	30			1,825		1,825	2,750
2910	Heat recovery package, up thru 20,000 CFM	Q-5	2	8			470		470	705
2920	25,000 CFM and up		1.20	13.333			785		785	1,175
2930	Heat transfer package, up thru 130 GPM		.80	20			1,175		1,175	1,775
2934	255 thru 800 GPM		.42	38.095			2,250		2,250	3,375
2940	Humidifier		10.60	1.509			89		89	133
2961	Hydronic unit heaters, up thru 200 MBH		14	1.143			67.50		67.50	101
2962	Above 200 MBH		8	2			118		118	177
2964	Valance units		32	.500			29.50		29.50	44
2966	Radiant floor heating									
2967	System valves, controls, manifolds	Q-5	16	1	Ea.		59		59	88.50
2968	Per room distribution		8	2			118		118	177
2970	Hydronic heating, baseboard radiation		16	1			59		59	88.50
2976	Convectors and free standing radiators		18	.889			52.50		52.50	78.50
2980	Induced draft fan, up thru 1 HP	Q-9	4.60	3.478			195		195	296
2984	1-1/2 HP thru 7-1/2 HP	"	2.20	7.273			410		410	620
2988	Infrared unit	Q-5	16	1			59		59	88.50
2992	Louvers	1 Shee	46	.174	S.F.		10.85		10.85	16.45
3000	Mechanical equipment, light items. Unit is weight, not cooling.	Q-5	.90	17.778	Ton		1,050		1,050	1,575
3600	Heavy items		1.10	14.545	"		860		860	1,275
3720	Make-up air unit, up thru 6,000 CFM		3	5.333	Ea.		315		315	470
3730	6,500 thru 30,000 CFM		1.60	10			590		590	885

Handwritten annotations: use 50/u (row 1640); use 800/u (row 1710); use 400/u (row 1730); use 300/u (row 2124); use 200/u (row 2961); use 150/u (row 2970); use 25/SF (row 2992)

23 05 05 – Selective Demolition for HVAC

23 05 05.10 HVAC Demolition		Crew	Daily Output	Labor-Hours	Unit	Material	2020 Bare Costs Labor	Equipment	Total	Total Incl O&P
3740	35,000 thru 75,000 CFM	Q-6	1	24	Ea.		1,475		1,475	2,200
3800	Mixing boxes, constant and VAV	Q-9	18	.889			50		50	75.50
4000	Packaged terminal air conditioner, up thru 18,000 BTUH	Q-5	8	2			118		118	177
4010	24,000 thru 48,000 BTUH	"	2.80	5.714	▼		335		335	505
5000	Refrigerant compressor, reciprocating or scroll									
5010	Up thru 5 ton	1 Stpi	6	1.333	Ea.		87.50		87.50	131
5020	5.08 thru 10 ton	Q-5	6	2.667			157		157	235
5030	15 thru 50 ton	"	3	5.333			315		315	470
5040	60 thru 130 ton	Q-6	2.80	8.571	▼		525		525	785
5090	Remove refrigerant from system	1 Stpi	40	.200	Lb.		13.10		13.10	19.60
5100	Rooftop air conditioner, up thru 10 ton	Q-5	1.40	11.429	Ea.		675		675	1,000
5110	12 thru 40 ton	Q-6	1	24			1,475		1,475	2,200
5120	50 thru 140 ton		.50	48			2,925		2,925	4,400
5130	150 thru 300 ton	▼	.30	80			4,900		4,900	7,325
6000	Self contained single package air conditioner, up thru 10 ton	Q-5	1.60	10			590		590	885
6010	15 thru 60 ton	Q-6	1.20	20			1,225		1,225	1,825
6100	Space heaters, up thru 200 MBH	Q-5	10	1.600			94.50		94.50	141
6110	Over 200 MBH		5	3.200			189		189	283
6200	Split ductless, both sections	▼	8	2			118		118	177
6300	Steam condensate meter	1 Stpi	11	.727			47.50		47.50	71.50
6600	Thru-the-wall air conditioner	L-2	8	2	▼		93		93	141
7000	Vent chimney, prefabricated, up thru 12" diameter	Q-9	94	.170	V.L.F.		9.55		9.55	14.50
7010	14" thru 36" diameter		40	.400			22.50		22.50	34
7020	38" thru 48" diameter	▼	32	.500			28		28	42.50
7030	54" thru 60" diameter	Q-10	14	1.714	▼		99.50		99.50	151
7400	Ventilators, up thru 14" neck diameter	Q-9	58	.276	Ea.		15.45		15.45	23.50
7410	16" thru 50" neck diameter		40	.400			22.50		22.50	34
7450	Relief vent, up thru 24" x 96"		22	.727			41		41	62
7460	48" x 60" thru 96" x 144"	▼	10	1.600			89.50		89.50	136
8000	Water chiller up thru 10 ton	Q-5	2.50	6.400			380		380	565
8010	15 thru 100 ton	Q-6	.48	50			3,050		3,050	4,575
8020	110 thru 500 ton	Q-7	.29	110			6,875		6,875	10,300
8030	600 thru 1000 ton		.23	139			8,675		8,675	13,000
8040	1100 ton and up	▼	.20	160			9,975		9,975	14,900
8400	Window air conditioner	1 Carp	16	.500	▼		26.50		26.50	40
8401	HVAC demo, water chiller, 1100 ton and up	"	12	.667	Ton		35.50		35.50	53

23 05 23 – General-Duty Valves for HVAC Piping

23 05 23.20 Valves, Bronze/Brass

		Crew	Daily Output	Labor-Hours	Unit	Material	2020 Bare Costs Labor	Equipment	Total	Total Incl O&P
0010	**VALVES, BRONZE/BRASS**									
0020	Brass									
1300	Ball combination valves, shut-off and union									
1310	Solder, with strainer, drain and PT ports									
1320	1/2"	1 Stpi	17	.471	Ea.	70	31		101	123
1330	3/4"		16	.500		76.50	33		109.50	134
1340	1"		14	.571		104	37.50		141.50	171
1350	1-1/4"		12	.667		133	43.50		176.50	212
1360	1-1/2"		10	.800		195	52.50		247.50	294
1370	2"	▼	8	1	▼	245	65.50		310.50	370
1410	Threaded, with strainer, drain and PT ports									
1420	1/2"	1 Stpi	20	.400	Ea.	70	26		96	116
1430	3/4"		18	.444		76.50	29		105.50	128
1440	1"	▼	17	.471	▼	104	31		135	161

23 05 23 – General-Duty Valves for HVAC Piping

23 05 23.20 Valves, Bronze/Brass		Crew	Daily Output	Labor-Hours	Unit	Material	2020 Bare Costs Labor	Equipment	Total	Total Incl O&P
1450	1-1/4"	1 Stpi	12	.667	Ea.	133	43.50		176.50	212
1460	1-1/2"		11	.727		195	47.50		242.50	287
1470	2"		9	.889		245	58.50		303.50	355

23 05 23.30 Valves, Iron Body

		Crew	Daily Output	Labor-Hours	Unit	Material	2020 Bare Costs Labor	Equipment	Total	Total Incl O&P
0010	**VALVES, IRON BODY**　　R220523-90									
0022	For grooved joint, see Section 22 11 13.48									
0100	Angle, 125 lb.									
0110	Flanged									
0116	2"	1 Plum	5	1.600	Ea.	1,525	103		1,628	1,825
0118	4"	Q-1	3	5.333		2,600	310		2,910	3,350
0120	6"	Q-2	3	8		5,700	480		6,180	7,000
0122	8"	"	2.50	9.600		8,150	575		8,725	9,825
0560	Butterfly, lug type, pneumatic operator, 2" size	1 Stpi	14	.571		680	37.50		717.50	800
0570	3"	Q-1	8	2		720	116		836	970
0580	4"	"	5	3.200		775	186		961	1,125
0590	6"	Q-2	5	4.800		1,050	289		1,339	1,575
0600	8"		4.50	5.333		1,225	320		1,545	1,825
0610	10"		4	6		1,575	360		1,935	2,275
0620	12"		3	8		1,875	480		2,355	2,775
0630	14"		2.30	10.435		2,575	630		3,205	3,775
0640	18"		1.50	16		4,400	960		5,360	6,275
0650	20"		1	24		5,625	1,450		7,075	8,350
0790	Butterfly, lug type, gear operated, 2" size, 200 lb. except noted	1 Plum	14	.571		370	37		407	460
0800	2-1/2"	Q-1	9	1.778		370	103		473	565
0810	3"		8	2		390	116		506	605
0820	4"		5	3.200		455	186		641	7
0830	5"	Q-2	5	4.800		690	289		979	1,
0840	6"		5	4.800		625	289		914	1
0850	8"		4.50	5.333		815	320		1,135	
0860	10"		4	6		1,375	360		1,735	2,050
0870	12"		3	8		1,550	480		2,030	2,450
0880	14", 150 lb.		2.30	10.435		2,500	630		3,130	3,700
0890	16", 150 lb.		1.75	13.714		3,825	825		4,650	5,425
0900	18", 150 lb.		1.50	16		4,625	960		5,585	6,550
0910	20", 150 lb.		1	24		6,200	1,450		7,650	8,975
0930	24", 150 lb.		.75	32		10,500	1,925		12,425	14,500
1020	Butterfly, wafer type, gear actuator, 200 lb.									
1030	2"	1 Plum	14	.571	Ea.	118	37		155	185
1040	2-1/2"	Q-1	9	1.778		125	103		228	291
1050	3"		8	2		124	116		240	310
1060	4"		5	3.200		139	186		325	430
1070	5"	Q-2	5	4.800		157	289		446	600
1080	6"		5	4.800		177	289		466	625
1090	8"		4.50	5.333		220	320		540	720
1100	10"		4	6		295	360		655	865
1110	12"		3	8		685	480		1,165	1,475
1200	Wafer type, lever actuator, 200 lb.									
1220	2"	1 Plum	14	.571	Ea.	203	37		240	278
1230	2-1/2"	Q-1	9	1.778		206	103		309	380
1240	3"		8	2		258	116		374	455
1250	4"		5	3.200		265	186		451	570
1260	5"	Q-2	5	4.800		395	289		684	865

For customer support on your Mechanical Costs with RSMeans Data, call 800.448.8182.

255

23 05 23.30 Valves, Iron Body		Crew	Daily Output	Labor-Hours	Unit	Material	2020 Bare Costs Labor	Equipment	Total	Total Incl O&P
1270	6"	Q-2	5	4.800	Ea.	445	289		734	920
1280	8"		4.50	5.333		700	320		1,020	1,250
1290	10"		4	6		740	360		1,100	1,350
1300	12"		3	8		1,250	480		1,730	2,100
1650	Gate, 125 lb., N.R.S.									
2150	Flanged									
2200	2"	1 Plum	5	1.600	Ea.	610	103		713	830
2240	2-1/2"	Q-1	5	3.200		630	186		816	970
2260	3"		4.50	3.556		710	206		916	1,100
2280	4"		3	5.333		1,000	310		1,310	1,575
2290	5"	Q-2	3.40	7.059		1,725	425		2,150	2,525
2300	6"		3	8		1,725	480		2,205	2,625
2320	8"		2.50	9.600		2,950	575		3,525	4,125
2340	10"		2.20	10.909		5,175	655		5,830	6,675
2360	12"		1.70	14.118		7,125	850		7,975	9,100
2370	14"		1.30	18.462		10,700	1,100		11,800	13,500
2380	16"		1	24		14,800	1,450		16,250	18,500
2420	For 250 lb. flanged, add					200%	10%			
3550	OS&Y, 125 lb., flanged									
3600	2"	1 Plum	5	1.600	Ea.	400	103		503	595
3640	2-1/2"	Q-1	5	3.200		430	186		616	750
3660	3"		4.50	3.556		460	206		666	820
3670	3-1/2"		3	5.333		770	310		1,080	1,300
3680	4"		3	5.333		685	310		995	1,225
3690	5"	Q-2	3.40	7.059		1,075	425		1,500	1,825
3700	6"		3	8		1,075	480		1,555	1,925
3720	8"		2.50	9.600		1,925	575		2,500	3,000
3740	10"		2.20	10.909		3,525	655		4,180	4,850
3760	12"		1.70	14.118		4,800	850		5,650	6,550
3770	14"		1.30	18.462		10,400	1,100		11,500	13,100
3780	16"		1	24		14,800	1,450		16,250	18,500
3790	18"		.80	30		24,200	1,800		26,000	29,300
3800	20"		.60	40		25,300	2,400		27,700	31,400
3830	24"		.50	48		38,000	2,875		40,875	46,100
3900	For 175 lb., flanged, add					200%	10%			
4350	Globe, OS&Y									
4540	Class 125, flanged									
4550	2"	1 Plum	5	1.600	Ea.	935	103		1,038	1,175
4560	2-1/2"	Q-1	5	3.200		940	186		1,126	1,300
4570	3"		4.50	3.556		1,150	206		1,356	1,550
4580	4"		3	5.333		1,625	310		1,935	2,275
4590	5"	Q-2	3.40	7.059		2,975	425		3,400	3,900
4600	6"		3	8		2,975	480		3,455	4,000
4610	8"		2.50	9.600		5,850	575		6,425	7,300
4612	10"		2.20	10.909		11,500	655		12,155	13,700
4614	12"		1.70	14.118		11,000	850		11,850	13,400
5040	Class 250, flanged									
5050	2"	1 Plum	4.50	1.778	Ea.	1,500	115		1,615	1,825
5060	2-1/2"	Q-1	4.50	3.556		1,950	206		2,156	2,450
5070	3"		4	4		2,025	232		2,257	2,575
5080	4"		2.70	5.926		2,950	345		3,295	3,775
5090	5"	Q-2	3	8		6,225	480		6,705	7,575
5100	6"		2.70	8.889		6,475	535		7,010	7,925

For customer support on your Mechanical Costs with RSMeans Data, call 800.448.8182.

23 05 23.30 Valves, Iron Body		Crew	Daily Output	Labor-Hours	Unit	Material	2020 Bare Costs Labor	Equipment	Total	Total Incl O&P
5110	8"	Q-2	2.20	10.909	Ea.	8,975	655		9,630	10,900
5120	10"		2	12		12,900	720		13,620	15,300
5130	12"	↓	1.60	15		24,200	900		25,100	28,100
5240	Valve sprocket rim w/chain, for 2" valve	1 Stpi	30	.267		185	17.50		202.50	229
5250	2-1/2" valve		27	.296		180	19.40		199.40	227
5260	3-1/2" valve		25	.320		194	21		215	245
5270	6" valve		20	.400		192	26		218	251
5280	8" valve		18	.444		225	29		254	291
5290	12" valve		16	.500		237	33		270	310
5300	16" valve		12	.667		258	43.50		301.50	350
5310	20" valve		10	.800		258	52.50		310.50	365
5320	36" valve	↓	8	1	↓	405	65.50		470.50	545
5450	Swing check, 125 lb., threaded									
5500	2"	1 Plum	11	.727	Ea.	575	47		622	700
5540	2-1/2"	Q-1	15	1.067		710	62		772	875
5550	3"		13	1.231		770	71.50		841.50	955
5560	4"	↓	10	1.600	↓	1,225	93		1,318	1,500
5950	Flanged									
5994	1"	1 Plum	7	1.143	Ea.	289	73.50		362.50	430
5998	1-1/2"		6	1.333		475	86		561	655
6000	2"	↓	5	1.600		510	103		613	720
6040	2-1/2"	Q-1	5	3.200		365	186		551	680
6050	3"		4.50	3.556		490	206		696	850
6060	4"	↓	3	5.333		785	310		1,095	1,325
6070	6"	Q-2	3	8		1,325	480		1,805	2,175
6080	8"		2.50	9.600		2,525	575		3,100	3,650
6090	10"		2.20	10.909		4,275	655		4,930	5,675
6100	12"		1.70	14.118		7,000	850		7,850	8,975
6102	14"		1.55	15.484		13,400	930		14,330	16,200
6104	16"		1.40	17.143		17,000	1,025		18,025	20,300
6110	18"		1.30	18.462		25,900	1,100		27,000	30,200
6112	20"		1	24		49,000	1,450		50,450	56,000
6114	24"	↓	.75	32	↓	66,000	1,925		67,925	75,500
6160	For 250 lb. flanged, add					200%	20%			
6600	Silent check, bronze trim									
6610	Compact wafer type, for 125 or 150 lb. flanges									
6630	1-1/2"	1 Plum	11	.727	Ea.	162	47		209	248
6640	2"	"	9	.889		198	57.50		255.50	305
6650	2-1/2"	Q-1	9	1.778		215	103		318	390
6660	3"		8	2		236	116		352	435
6670	4"	↓	5	3.200		290	186		476	600
6680	5"	Q-2	6	4		390	241		631	785
6690	6"		6	4		530	241		771	945
6700	8"		4.50	5.333		885	320		1,205	1,450
6710	10"		4	6		1,550	360		1,910	2,250
6720	12"	↓	3	8	↓	2,875	480		3,355	3,875
6740	For 250 or 300 lb. flanges, thru 6" no change									
6741	For 8" and 10", add				Ea.	11%	10%			
6750	Twin disc									
6752	2"	1 Plum	9	.889	Ea.	425	57.50		482.50	550
6754	4"	Q-1	5	3.200		705	186		891	1,050
6756	6"	Q-2	5	4.800		1,050	289		1,339	1,575
6758	8"		4.50	5.333	↓	1,600	320		1,920	2,250

23 05 23.30 Valves, Iron Body

		Crew	Daily Output	Labor-Hours	Unit	Material	2020 Bare Costs Labor	2020 Bare Costs Equipment	Total	Total Incl O&P
6760	10"	Q-2	4	6	Ea.	2,575	360		2,935	3,375
6762	12"		3	8		3,325	480		3,805	4,400
6764	18"		1.50	16		14,400	960		15,360	17,300
6766	24"		.75	32		20,300	1,925		22,225	25,200
6800	Full flange type, 150 lb.									
6810	1"	1 Plum	14	.571	Ea.	184	37		221	257
6811	1-1/4"		12	.667		194	43		237	278
6812	1-1/2"		11	.727		225	47		272	315
6813	2"		9	.889		257	57.50		314.50	370
6814	2-1/2"	Q-1	9	1.778		305	103		408	495
6815	3"		8	2		335	116		451	540
6816	4"		5	3.200		400	186		586	720
6817	5"	Q-2	6	4		630	241		871	1,050
6818	6"		5	4.800		830	289		1,119	1,350
6819	8"		4.50	5.333		1,425	320		1,745	2,050
6820	10"		4	6		1,775	360		2,135	2,500
6840	For 250 lb., add					30%	10%			
6900	Globe type, 125 lb.									
6910	2"	1 Plum	9	.889	Ea.	365	57.50		422.50	490
6911	2-1/2"	Q-1	9	1.778		490	103		593	695
6912	3"		8	2		440	116		556	655
6913	4"		5	3.200		770	186		956	1,125
6914	5"	Q-2	6	4		1,100	241		1,341	1,550
6915	6"		5	4.800		1,350	289		1,639	1,900
6916	8"		4.50	5.333		2,450	320		2,770	3,175
6917	10"		4	6		3,100	360		3,460	3,950
6918	12"		3	8		5,325	480		5,805	6,600
6919	14"		2.30	10.435		7,000	630		7,630	8,650
6920	16"		1.75	13.714		7,650	825		8,475	9,650
6921	18"		1.50	16		11,400	960		12,360	14,000
6922	20"		1	24		13,800	1,450		15,250	17,400
6923	24"		.75	32		15,100	1,925		17,025	19,500
6940	For 250 lb., add					40%	10%			
6980	Screwed end type, 125 lb.									
6981	1"	1 Plum	19	.421	Ea.	97.50	27		124.50	148
6982	1-1/4"		15	.533		114	34.50		148.50	177
6983	1-1/2"		13	.615		139	39.50		178.50	213
6984	2"		11	.727		190	47		237	279

23 05 23.50 Valves, Multipurpose

		Crew	Daily Output	Labor-Hours	Unit	Material	2020 Bare Costs Labor	2020 Bare Costs Equipment	Total	Total Incl O&P
0010	**VALVES, MULTIPURPOSE**									
0100	Functions as a shut off, balancing, check & metering valve									
1000	Cast iron body									
1010	Threaded									
1020	1-1/2" size	1 Stpi	11	.727	Ea.	470	47.50		517.50	590
1030	2" size	"	8	1		570	65.50		635.50	730
1040	2-1/2" size	Q-5	5	3.200		620	189		809	965
1200	Flanged									
1210	3" size	Q-5	4.20	3.810	Ea.	710	225		935	1,125
1220	4" size	"	3	5.333		1,450	315		1,765	2,050
1230	5" size	Q-6	3.80	6.316		1,700	385		2,085	2,450
1240	6" size		3	8		2,125	490		2,615	3,075
1250	8" size		2.50	9.600		3,675	585		4,260	4,925

23 05 Common Work Results for HVAC

23 05 23 – General-Duty Valves for HVAC Piping

23 05 23.50 Valves, Multipurpose

		Crew	Daily Output	Labor-Hours	Unit	Material	2020 Bare Costs Labor	Equipment	Total	Total Incl O&P
1260	10" size	Q-6	2.20	10.909	Ea.	5,275	665		5,940	6,800
1270	12" size		2.10	11.429		9,400	700		10,100	11,400
1280	14" size	↓	2	12	↓	14,100	735		14,835	16,600

23 05 23.70 Valves, Semi-Steel

		Crew	Daily Output	Labor-Hours	Unit	Material	2020 Bare Costs Labor	Equipment	Total	Total Incl O&P
0010	**VALVES, SEMI-STEEL** R220523-90									
1020	Lubricated plug valve, threaded, 200 psi									
1030	1/2"	1 Plum	18	.444	Ea.	119	28.50		147.50	174
1040	3/4"		16	.500		125	32		157	186
1050	1"		14	.571		135	37		172	203
1060	1-1/4"		12	.667		159	43		202	240
1070	1-1/2"		11	.727		154	47		201	240
1080	2"	↓	8	1		208	64.50		272.50	325
1090	2-1/2"	Q-1	5	3.200		299	186		485	610
1100	3"	"	4.50	3.556	↓	425	206		631	780
6990	Flanged, 200 psi									
7000	2"	1 Plum	8	1	Ea.	277	64.50		341.50	400
7010	2-1/2"	Q-1	5	3.200		390	186		576	710
7020	3"		4.50	3.556		470	206		676	830
7030	4"		3	5.333		585	310		895	1,100
7036	5"	↓	2.50	6.400		1,325	370		1,695	2,000
7040	6"	Q-2	3	8		1,675	480		2,155	2,550
7050	8"		2.50	9.600		2,225	575		2,800	3,325
7060	10"		2.20	10.909		3,650	655		4,305	5,000
7070	12"	↓	1.70	14.118	↓	5,625	850		6,475	7,475

23 05 23.80 Valves, Steel

		Crew	Daily Output	Labor-Hours	Unit	Material	2020 Bare Costs Labor	Equipment	Total	Total Incl O&P
0010	**VALVES, STEEL** R220523-90									
0800	Cast									
1350	Check valve, swing type, 150 lb., flanged									
1370	1"	1 Plum	10	.800	Ea.	360	51.50		411.50	475
1400	2"	"	8	1		765	64.50		829.50	935
1440	2-1/2"	Q-1	5	3.200		885	186		1,071	1,250
1450	3"		4.50	3.556		900	206		1,106	1,300
1460	4"	↓	3	5.333		1,100	310		1,410	1,700
1470	6"	Q-2	3	8		1,725	480		2,205	2,625
1480	8"		2.50	9.600		2,925	575		3,500	4,075
1490	10"		2.20	10.909		5,100	655		5,755	6,575
1500	12"		1.70	14.118		6,275	850		7,125	8,175
1510	14"		1.30	18.462		9,350	1,100		10,450	12,000
1520	16"	↓	1	24		10,700	1,450		12,150	14,000
1540	For 300 lb., flanged, add						50%	15%		
1548	For 600 lb., flanged, add						110%	20%		
1571	300 lb., 2"	1 Plum	7.40	1.081		680	69.50		749.50	855
1572	2-1/2"	Q-1	4.20	3.810		1,025	221		1,246	1,450
1573	3"		4	4		1,025	232		1,257	1,475
1574	4"	↓	2.80	5.714		1,400	330		1,730	2,050
1575	6"	Q-2	2.90	8.276		2,700	500		3,200	3,725
1576	8"		2.40	10		4,000	600		4,600	5,300
1577	10"		2.10	11.429		6,000	685		6,685	7,600
1578	12"		1.60	15		8,675	900		9,575	10,900
1579	14"		1.20	20		12,900	1,200		14,100	16,000
1581	16"	↓	.90	26.667	↓	16,400	1,600		18,000	20,500
1950	Gate valve, 150 lb., flanged									

For customer support on your Mechanical Costs with RSMeans Data, call 800.448.8182.

259

23 05 23.80 Valves, Steel		Crew	Daily Output	Labor-Hours	Unit	Material	2020 Bare Costs Labor	Equipment	Total	Total Incl O&P
2000	2"	1 Plum	8	1	Ea.	645	64.50		709.50	800
2040	2-1/2"	Q-1	5	3.200		1,075	186		1,261	1,450
2050	3"		4.50	3.556		910	206		1,116	1,300
2060	4"		3	5.333		1,150	310		1,460	1,725
2070	6"	Q-2	3	8		1,825	480		2,305	2,725
2080	8"		2.50	9.600		2,800	575		3,375	3,950
2090	10"		2.20	10.909		4,175	655		4,830	5,575
2100	12"		1.70	14.118		5,700	850		6,550	7,550
2110	14"		1.30	18.462		9,200	1,100		10,300	11,800
2120	16"		1	24		11,900	1,450		13,350	15,300
2130	18"		.80	30		15,600	1,800		17,400	19,800
2140	20"		.60	40		18,100	2,400		20,500	23,500
2650	300 lb., flanged									
2700	2"	1 Plum	7.40	1.081	Ea.	1,025	69.50		1,094.50	1,225
2740	2-1/2"	Q-1	4.20	3.810		1,425	221		1,646	1,875
2750	3"		4	4		1,200	232		1,432	1,675
2760	4"		2.80	5.714		1,700	330		2,030	2,350
2770	6"	Q-2	2.90	8.276		2,950	500		3,450	4,000
2780	8"		2.40	10		4,625	600		5,225	5,975
2790	10"		2.10	11.429		6,000	685		6,685	7,625
2800	12"		1.60	15		8,650	900		9,550	10,900
2810	14"		1.20	20		17,000	1,200		18,200	20,600
2820	16"		.90	26.667		21,400	1,600		23,000	25,900
2830	18"		.70	34.286		30,500	2,050		32,550	36,700
2840	20"		.50	48		34,200	2,875		37,075	42,000
3650	Globe valve, 150 lb., flanged									
3700	2"	1 Plum	8	1	Ea.	950	64.50		1,014.50	1,150
3740	2-1/2"	Q-1	5	3.200		1,200	186		1,386	1,600
3750	3"		4.50	3.556		1,200	206		1,406	1,625
3760	4"		3	5.333		1,750	310		2,060	2,400
3770	6"	Q-2	3	8		2,750	480		3,230	3,750
3780	8"		2.50	9.600		5,175	575		5,750	6,550
3790	10"		2.20	10.909		9,475	655		10,130	11,400
3800	12"		1.70	14.118		10,700	850		11,550	13,100
4080	300 lb., flanged									
4100	2"	1 Plum	7.40	1.081	Ea.	1,275	69.50		1,344.50	1,500
4140	2-1/2"	Q-1	4.20	3.810		1,725	221		1,946	2,225
4150	3"		4	4		1,725	232		1,957	2,250
4160	4"		2.80	5.714		2,375	330		2,705	3,125
4170	6"	Q-2	2.90	8.276		4,300	500		4,800	5,500
4180	8"		2.40	10		6,175	600		6,775	7,700
4190	10"		2.10	11.429		14,000	685		14,685	16,400
4200	12"		1.60	15		16,900	900		17,800	19,900
4680	600 lb., flanged									
4700	2"	1 Plum	7	1.143	Ea.	1,775	73.50		1,848.50	2,050
4740	2-1/2"	Q-1	4	4		2,850	232		3,082	3,475
4750	3"		3.60	4.444		2,875	258		3,133	3,525
4760	4"		2.50	6.400		4,375	370		4,745	5,350
4770	6"	Q-2	2.60	9.231		9,250	555		9,805	11,000
4780	8"	"	2.10	11.429		12,900	685		13,585	15,200
4800	Silent check, 316 S.S. trim									
4810	Full flange type, 150 lb.									
4811	1"	1 Plum	14	.571	Ea.	640	37		677	760

23 05 23 – General-Duty Valves for HVAC Piping

23 05 23.80 Valves, Steel		Crew	Daily Output	Labor-Hours	Unit	Material	2020 Bare Costs Labor	Equipment	Total	Total Incl O&P
4812	1-1/4"	1 Plum	12	.667	Ea.	640	43		683	770
4813	1-1/2"		11	.727		705	47		752	845
4814	2"		9	.889		840	57.50		897.50	1,000
4815	2-1/2"	Q-1	9	1.778		1,025	103		1,128	1,275
4816	3"		8	2		1,175	116		1,291	1,475
4817	4"		5	3.200		1,625	186		1,811	2,050
4818	5"	Q-2	6	4		2,025	241		2,266	2,600
4819	6"		5	4.800		2,600	289		2,889	3,275
4820	8"		4.50	5.333		3,800	320		4,120	4,650
4821	10"		4	6		5,350	360		5,710	6,450
4840	For 300 lb., add					20%	10%			
4860	For 600 lb., add					50%	15%			
4900	Globe type, 150 lb.									
4910	2"	1 Plum	9	.889	Ea.	995	57.50		1,052.50	1,175
4911	2-1/2"	Q-1	9	1.778		1,200	103		1,303	1,475
4912	3"		8	2		1,325	116		1,441	1,650
4913	4"		5	3.200		1,775	186		1,961	2,225
4914	5"	Q-2	6	4		2,250	241		2,491	2,825
4915	6"		5	4.800		2,750	289		3,039	3,450
4916	8"		4.50	5.333		3,875	320		4,195	4,750
4917	10"		4	6		6,725	360		7,085	7,950
4918	12"		3	8		8,250	480		8,730	9,800
4919	14"		2.30	10.435		10,900	630		11,530	12,900
4920	16"		1.75	13.714		15,300	825		16,125	18,000
4921	18"		1.50	16		16,000	960		16,960	19,100
4940	For 300 lb., add					30%	10%			
4960	For 600 lb., add					60%	15%			
5150	Forged									
5340	Ball valve, 1,500 psi, threaded, 1/4" size	1 Plum	24	.333	Ea.	75.50	21.50		97	115
5350	3/8"		24	.333		74	21.50		95.50	114
5360	1/2"		24	.333		103	21.50		124.50	146
5370	3/4"		20	.400		121	26		147	172
5380	1"		19	.421		154	27		181	210
5390	1-1/4"		15	.533		203	34.50		237.50	275
5400	1-1/2"		13	.615		250	39.50		289.50	335
5410	2"		11	.727		355	47		402	460
5460	Ball valve, 800 lb., socket weld, 1/4" size	Q-15	19	.842		75.50	49	5.60	130.10	162
5470	3/8"		19	.842		75.50	49	5.60	130.10	162
5480	1/2"		19	.842		94	49	5.60	148.60	182
5490	3/4"		19	.842		125	49	5.60	179.60	216
5500	1"		15	1.067		156	62	7.10	225.10	271
5510	1-1/4"		13	1.231		211	71.50	8.20	290.70	350
5520	1-1/2"		11	1.455		267	84.50	9.65	361.15	430
5530	2"		8.50	1.882		350	109	12.50	471.50	560
5550	Ball valve, 150 lb., flanged									
5560	4"	Q-1	3	5.333	Ea.	1,775	310		2,085	2,425
5570	6"	Q-2	3	8		2,425	480		2,905	3,400
5580	8"	"	2.50	9.600		5,725	575		6,300	7,175
5650	Check valve, class 800, horizontal									
5651	Socket									
5652	1/4"	Q-15	19	.842	Ea.	86	49	5.60	140.60	174
5654	3/8"		19	.842		86	49	5.60	140.60	174
5656	1/2"		19	.842		86	49	5.60	140.60	174

23 05 23.80 Valves, Steel		Crew	Daily Output	Labor-Hours	Unit	Material	2020 Bare Costs Labor	Equipment	Total	Total Incl O&P
5658	3/4"	Q-15	19	.842	Ea.	91.50	49	5.60	146.10	180
5660	1"		15	1.067		108	62	7.10	177.10	219
5662	1-1/4"		13	1.231		212	71.50	8.20	291.70	350
5664	1-1/2"		11	1.455		212	84.50	9.65	306.15	370
5666	2"		8.50	1.882		297	109	12.50	418.50	500
5698	Threaded									
5700	1/4"	1 Plum	24	.333	Ea.	86	21.50		107.50	127
5720	3/8"		24	.333		86	21.50		107.50	127
5730	1/2"		24	.333		86	21.50		107.50	127
5740	3/4"		20	.400		92	26		118	140
5750	1"		19	.421		108	27		135	160
5760	1-1/4"		15	.533		212	34.50		246.50	285
5770	1-1/2"		13	.615		212	39.50		251.50	293
5780	2"		11	.727		297	47		344	395
5840	For class 150, flanged, add					100%	15%			
5860	For class 300, flanged, add					120%	20%			
6100	Gate, class 800, OS&Y, socket									
6102	3/8"	Q-15	19	.842	Ea.	56.50	49	5.60	111.10	141
6103	1/2"		19	.842		56.50	49	5.60	111.10	141
6104	3/4"		19	.842		62	49	5.60	116.60	148
6105	1"		15	1.067		75	62	7.10	144.10	183
6106	1-1/4"		13	1.231		142	71.50	8.20	221.70	272
6107	1-1/2"		11	1.455		142	84.50	9.65	236.15	293
6108	2"		8.50	1.882		188	109	12.50	309.50	385
6118	Threaded									
6120	3/8"	1 Plum	24	.333	Ea.	56.50	21.50		78	94
6130	1/2"		24	.333		56.50	21.50		78	94
6140	3/4"		20	.400		62	26		88	107
6150	1"		19	.421		75	27		102	123
6160	1-1/4"		15	.533		142	34.50		176.50	208
6170	1-1/2"		13	.615		142	39.50		181.50	216
6180	2"		11	.727		188	47		235	277
6260	For OS&Y, flanged, add					100%	20%			
6700	Globe, OS&Y, class 800, socket									
6710	1/4"	Q-15	19	.842	Ea.	87	49	5.60	141.60	175
6720	3/8"		19	.842		87	49	5.60	141.60	175
6730	1/2"		19	.842		87	49	5.60	141.60	175
6740	3/4"		19	.842		100	49	5.60	154.60	189
6750	1"		15	1.067		130	62	7.10	199.10	243
6760	1-1/4"		13	1.231		255	71.50	8.20	334.70	395
6770	1-1/2"		11	1.455		255	84.50	9.65	349.15	415
6780	2"		8.50	1.882		325	109	12.50	446.50	535
6860	For OS&Y, flanged, add					300%	20%			
6880	Threaded									
6882	1/4"	1 Plum	24	.333	Ea.	87	21.50		108.50	128
6884	3/8"		24	.333		87	21.50		108.50	128
6886	1/2"		24	.333		81	21.50		102.50	121
6888	3/4"		20	.400		100	26		126	149
6890	1"		19	.421		130	27		157	184
6892	1-1/4"		15	.533		255	34.50		289.50	330
6894	1-1/2"		13	.615		255	39.50		294.50	340
6896	2"		11	.727		325	47		372	430

23 05 23.90 Valves, Stainless Steel		Crew	Daily Output	Labor-Hours	Unit	Material	2020 Bare Costs Labor	Equipment	Total	Total Incl O&P
0010	**VALVES, STAINLESS STEEL** R220523-90									
1610	Ball, threaded 1/4"	1 Stpi	24	.333	Ea.	47.50	22		69.50	84.50
1620	3/8"		24	.333		53.50	22		75.50	91
1630	1/2"		22	.364		47.50	24		71.50	87.50
1640	3/4"		20	.400		88.50	26		114.50	136
1650	1"		19	.421		96	27.50		123.50	147
1660	1-1/4"		15	.533		193	35		228	266
1670	1-1/2"		13	.615		240	40.50		280.50	325
1680	2"		11	.727		291	47.50		338.50	390
1700	Check, 200 lb., threaded									
1710	1/4"	1 Plum	24	.333	Ea.	145	21.50		166.50	191
1720	1/2"		22	.364		145	23.50		168.50	194
1730	3/4"		20	.400		149	26		175	203
1750	1"		19	.421		194	27		221	255
1760	1-1/2"		13	.615		365	39.50		404.50	460
1770	2"		11	.727		620	47		667	750
1800	150 lb., flanged									
1810	2-1/2"	Q-1	5	3.200	Ea.	1,925	186		2,111	2,400
1820	3"		4.50	3.556		1,975	206		2,181	2,475
1830	4"		3	5.333		2,950	310		3,260	3,700
1840	6"	Q-2	3	8		5,200	480		5,680	6,450
1850	8"	"	2.50	9.600		10,600	575		11,175	12,600
2100	Gate, OS&Y, 150 lb., flanged									
2120	1/2"	1 Plum	18	.444	Ea.	515	28.50		543.50	610
2140	3/4"		16	.500		495	32		527	595
2150	1"		14	.571		625	37		662	745
2160	1-1/2"		11	.727		1,200	47		1,247	1,400
2170	2"		8	1		1,425	64.50		1,489.50	1,675
2180	2-1/2"	Q-1	5	3.200		1,800	186		1,986	2,275
2190	3"		4.50	3.556		1,950	206		2,156	2,450
2200	4"		3	5.333		2,675	310		2,985	3,425
2205	5"		2.80	5.714		5,125	330		5,455	6,150
2210	6"	Q-2	3	8		5,000	480		5,480	6,225
2220	8"		2.50	9.600		8,675	575		9,250	10,400
2230	10"		2.30	10.435		15,200	630		15,830	17,600
2240	12"		1.90	12.632		20,200	760		20,960	23,400
2260	For 300 lb., flanged, add					120%	15%			
2600	600 lb., flanged									
2620	1/2"	1 Plum	16	.500	Ea.	184	32		216	250
2640	3/4"		14	.571		199	37		236	274
2650	1"		12	.667		239	43		282	330
2660	1-1/2"		10	.800		380	51.50		431.50	495
2670	2"		7	1.143		525	73.50		598.50	690
2680	2-1/2"	Q-1	4	4		6,625	232		6,857	7,650
2690	3"	"	3.60	4.444		6,625	258		6,883	7,675
3100	Globe, OS&Y, 150 lb., flanged									
3120	1/2"	1 Plum	18	.444	Ea.	495	28.50		523.50	585
3140	3/4"		16	.500		535	32		567	640
3150	1"		14	.571		700	37		737	825
3160	1-1/2"		11	.727		1,100	47		1,147	1,275
3170	2"		8	1		1,425	64.50		1,489.50	1,675
3180	2-1/2"	Q-1	5	3.200		3,075	186		3,261	3,650

For customer support on your Mechanical Costs with RSMeans Data, call 800.448.8182.

263

23 05 23 – General-Duty Valves for HVAC Piping

23 05 23.90 Valves, Stainless Steel

		Crew	Daily Output	Labor-Hours	Unit	Material	2020 Bare Costs Labor	2020 Bare Costs Equipment	Total	Total Incl O&P
3190	3"	Q-1	4.50	3.556	Ea.	3,075	206		3,281	3,675
3200	4"	↓	3	5.333		4,925	310		5,235	5,875
3210	6"	Q-2	3	8	↓	8,275	480		8,755	9,825
5000	Silent check, 316 S.S. body and trim									
5010	Compact wafer type, 300 lb.									
5020	1"	1 Plum	14	.571	Ea.	1,175	37		1,212	1,350
5021	1-1/4"		12	.667		1,275	43		1,318	1,475
5022	1-1/2"		11	.727		1,325	47		1,372	1,525
5023	2"	↓	9	.889		1,500	57.50		1,557.50	1,750
5024	2-1/2"	Q-1	9	1.778		2,075	103		2,178	2,425
5025	3"	"	8	2	↓	2,400	116		2,516	2,800
5100	Full flange wafer type, 150 lb.									
5110	1"	1 Plum	14	.571	Ea.	830	37		867	970
5111	1-1/4"		12	.667		895	43		938	1,050
5112	1-1/2"		11	.727		1,050	47		1,097	1,225
5113	2"	↓	9	.889		1,325	57.50		1,382.50	1,550
5114	2-1/2"	Q-1	9	1.778		1,625	103		1,728	1,925
5115	3"		8	2		1,800	116		1,916	2,150
5116	4"	↓	5	3.200		2,675	186		2,861	3,200
5117	5"	Q-2	6	4		3,100	241		3,341	3,750
5118	6"		5	4.800		3,875	289		4,164	4,700
5119	8"		4.50	5.333		5,100	320		5,420	6,075
5120	10"		4	6	↓	10,000	360		10,360	11,500
5200	Globe type, 300 lb.									
5210	2"	1 Plum	9	.889	Ea.	1,450	57.50		1,507.50	1,675
5211	2-1/2"	Q-1	9	1.778		1,800	103		1,903	2,125
5212	3"		8	2		2,200	116		2,316	2,600
5213	4"	↓	5	3.200		3,075	186		3,261	3,650
5214	5"	Q-2	6	4		4,450	241		4,691	5,250
5215	6"		5	4.800		4,850	289		5,139	5,750
5216	8"		4.50	5.333		6,850	320		7,170	8,025
5217	10"		4	6	↓	8,700	360		9,060	10,100
5300	Screwed end type, 300 lb.									
5310	1/2"	1 Plum	24	.333	Ea.	88.50	21.50		110	130
5311	3/4"		20	.400		109	26		135	158
5312	1"		19	.421		125	27		152	179
5313	1-1/4"		15	.533		134	34.50		168.50	200
5314	1-1/2"		13	.615		175	39.50		214.50	253
5315	2"	↓	11	.727	↓	217	47		264	310

23 05 93 – Testing, Adjusting, and Balancing for HVAC

23 05 93.50 Piping, Testing

		Crew	Daily Output	Labor-Hours	Unit	Material	2020 Bare Costs Labor	2020 Bare Costs Equipment	Total	Total Incl O&P
0010	**PIPING, TESTING**									
0100	Nondestructive testing									
0120	1" - 4" pipe									
0140	0-250 L.F.	1 Stpi	1.33	6.015	Ea.		395		395	590
0160	250-500 L.F.	"	.80	10			655		655	980
0180	500-1000 L.F.	Q-5	1.14	14.035			830		830	1,250
0200	1000-2000 L.F.		.80	20			1,175		1,175	1,775
0320	0-250 L.F.		1	16			945		945	1,425
0340	250-500 L.F.		.73	21.918			1,300		1,300	1,925
0360	500-1000 L.F.		.53	30.189			1,775		1,775	2,675
0380	1000-2000 L.F.	↓	.38	42.105	↓		2,475		2,475	3,725

23 05 Common Work Results for HVAC

23 05 93 – Testing, Adjusting, and Balancing for HVAC

23 05 93.50 Piping, Testing		Crew	Daily Output	Labor-Hours	Unit	Material	2020 Bare Costs Labor	Equipment	Total	Total Incl O&P
1000	Pneumatic pressure test, includes soaping joints									
1120	1" - 4" pipe									
1140	0-250 L.F.	Q-5	2.67	5.993	Ea.	12.30	355		367.30	545
1160	250-500 L.F.		1.33	12.030		24.50	710		734.50	1,075
1180	500-1000 L.F.		.80	20		37	1,175		1,212	1,825
1200	1000-2000 L.F.		.50	32		49	1,900		1,949	2,875
1300	6" - 10" pipe									
1320	0-250 L.F.	Q-5	1.33	12.030	Ea.	12.30	710		722.30	1,075
1340	250-500 L.F.		.67	23.881		24.50	1,400		1,424.50	2,125
1360	500-1000 L.F.		.40	40		49	2,350		2,399	3,575
1380	1000-2000 L.F.		.25	64		61.50	3,775		3,836.50	5,725
2000	X-Ray of welds									
2110	2" diam.	1 Stpi	8	1	Ea.	17.25	65.50		82.75	117
2120	3" diam.		8	1		17.25	65.50		82.75	117
2130	4" diam.		8	1		26	65.50		91.50	127
2140	6" diam.		8	1		26	65.50		91.50	127
2150	8" diam.		6.60	1.212		26	79.50		105.50	148
2160	10" diam.		6	1.333		34.50	87.50		122	169
3000	Liquid penetration of welds									
3110	2" diam.	1 Stpi	14	.571	Ea.	13.05	37.50		50.55	70.50
3120	3" diam.		13.60	.588		13.05	38.50		51.55	72
3130	4" diam.		13.40	.597		13.05	39		52.05	73
3140	6" diam.		13.20	.606		13.05	39.50		52.55	74
3150	8" diam.		13	.615		19.60	40.50		60.10	82
3160	10" diam.		12.80	.625		19.60	41		60.60	83

23 07 HVAC Insulation

23 07 13 – Duct Insulation

23 07 13.10 Duct Thermal Insulation

			Crew	Daily Output	Labor-Hours	Unit	Material	Labor	Equipment	Total	Total Incl O&P
0010	**DUCT THERMAL INSULATION**										
0110	Insulation req'd. is based on the surface size/area to be covered										
3000	Ductwork										
3020	Blanket type, fiberglass, flexible										
3030	Fire rated for grease and hazardous exhaust ducts										
3060	1-1/2" thick		Q-14	84	.190	S.F.	4.49	10.05		14.54	20.50
3090	Fire rated for plenums										
3100	1/2" x 24" x 25'		Q-14	1.94	8.247	Roll	177	435		612	865
3110	1/2" x 24" x 25'			98	.163	S.F.	3.54	8.60		12.14	17.15
3120	1/2" x 48" x 25'			1.04	15.385	Roll	350	810		1,160	1,625
3126	1/2" x 48" x 25'			104	.154	S.F.	3.49	8.10		11.59	16.30
3140	FSK vapor barrier wrap, .75 lb. density										
3160	1" thick	G	Q-14	350	.046	S.F.	.22	2.41		2.63	3.94
3170	1-1/2" thick	G		320	.050		.27	2.64		2.91	4.35
3180	2" thick	G		300	.053		.33	2.81		3.14	4.68
3190	3" thick	G		260	.062		.51	3.25		3.76	5.55
3200	4" thick	G		242	.066		.69	3.49		4.18	6.10
3210	Vinyl jacket, same as FSK										
3280	Unfaced, 1 lb. density										
3310	1" thick	G	Q-14	360	.044	S.F.	.28	2.35		2.63	3.91
3320	1-1/2" thick	G		330	.048		.38	2.56		2.94	4.35
3330	2" thick	G		310	.052		.50	2.72		3.22	4.73

For customer support on your Mechanical Costs with RSMeans Data, call 800.448.8182.

265

23 07 HVAC Insulation

23 07 13 – Duct Insulation

23 07 13.10 Duct Thermal Insulation

		Crew	Daily Output	Labor-Hours	Unit	Material	2020 Bare Costs Labor	Equipment	Total	Total Incl O&P
3400	FSK facing, 1 lb. density									
3420	1-1/2" thick	G	Q-14	310	.052	S.F.	.38	2.72		3.10	4.60
3430	2" thick	G	"	300	.053	"	.51	2.81		3.32	4.88
3450	FSK facing, 1.5 lb. density									
3470	1-1/2" thick	G	Q-14	300	.053	S.F.	.54	2.81		3.35	4.91
3480	2" thick	G	"	290	.055	"	.66	2.91		3.57	5.20
3730	Sheet insulation									
3760	Polyethylene foam, closed cell, UV resistant									
3770	Standard temperature (-90°F to +212°F)									
3771	1/4" thick	G	Q-14	450	.036	S.F.	1.83	1.88		3.71	4.89
3772	3/8" thick	G		440	.036		2.61	1.92		4.53	5.80
3773	1/2" thick	G		420	.038		3.21	2.01		5.22	6.60
3774	3/4" thick	G		400	.040		4.58	2.11		6.69	8.30
3775	1" thick	G		380	.042		6.20	2.22		8.42	10.20
3776	1-1/2" thick	G		360	.044		9.80	2.35		12.15	14.35
3777	2" thick	G		340	.047		12.95	2.48		15.43	18.05
3778	2-1/2" thick	G		320	.050		16.60	2.64		19.24	22.50
3779	Adhesive (see line 7878)									
3780	Foam, rubber									
3782	1" thick	G	1 Stpi	50	.160	S.F.	3.05	10.50		13.55	19.05
3795	Finishes									
3800	Stainless steel woven mesh		Q-14	100	.160	S.F.	.99	8.45		9.44	14.05
3810	For .010" stainless steel, add			160	.100		3.26	5.30		8.56	11.70
3820	18 oz. fiberglass cloth, pasted on			170	.094		1.02	4.97		5.99	8.75
3900	8 oz. canvas, pasted on			180	.089		.30	4.69		4.99	7.55
3940	For .016" aluminum jacket, add			200	.080		1.03	4.22		5.25	7.65
7000	Board insulation									
7020	Mineral wool, 1200° F									
7022	6 lb. density, plain									
7024	1" thick	G	Q-14	370	.043	S.F.	.34	2.28		2.62	3.87
7026	1-1/2" thick	G		350	.046		.51	2.41		2.92	4.26
7028	2" thick	G		330	.048		.68	2.56		3.24	4.68
7030	3" thick	G		300	.053		.84	2.81		3.65	5.25
7032	4" thick	G		280	.057		1.37	3.02		4.39	6.15
7038	8 lb. density, plain									
7040	1" thick	G	Q-14	360	.044	S.F.	.42	2.35		2.77	4.06
7042	1-1/2" thick	G		340	.047		.61	2.48		3.09	4.48
7044	2" thick	G		320	.050		.83	2.64		3.47	4.96
7046	3" thick	G		290	.055		1.25	2.91		4.16	5.85
7048	4" thick	G		270	.059		1.62	3.13		4.75	6.60
7060	10 lb. density, plain									
7062	1" thick	G	Q-14	350	.046	S.F.	.58	2.41		2.99	4.34
7064	1-1/2" thick	G		330	.048		.87	2.56		3.43	4.89
7066	2" thick	G		310	.052		1.16	2.72		3.88	5.45
7068	3" thick	G		280	.057		1.75	3.02		4.77	6.55
7070	4" thick	G		260	.062		2.22	3.25		5.47	7.45
7878	Contact cement, quart can					Ea.	13.20			13.20	14.55

23 07 16 – HVAC Equipment Insulation

23 07 16.10 HVAC Equipment Thermal Insulation

		Crew	Daily Output	Labor-Hours	Unit	Material	2020 Bare Costs Labor	Equipment	Total	Total Incl O&P
0010	**HVAC EQUIPMENT THERMAL INSULATION**									
0110	Insulation req'd. is based on the surface size/area to be covered									
1000	Boiler, 1-1/2" calcium silicate only	G	Q-14	110	.145	S.F.	4.47	7.70		12.17	16.70

23 07 16.10 HVAC Equipment Thermal Insulation		Crew	Daily Output	Labor-Hours	Unit	Material	2020 Bare Costs Labor	Equipment	Total	Total Incl O&P
1020	Plus 2" fiberglass	G Q-14	80	.200	S.F.	6.05	10.55		16.60	23
2000	Breeching, 2" calcium silicate									
2020	Rectangular	G Q-14	42	.381	S.F.	8.65	20		28.65	40.50
2040	Round	G "	38.70	.413	"	9	22		31	43.50
2300	Calcium silicate block, +200°F to +1,200°F									
2310	On irregular surfaces, valves and fittings									
2340	1" thick	G Q-14	30	.533	S.F.	3.90	28		31.90	47.50
2360	1-1/2" thick	G	25	.640		4.40	34		38.40	57
2380	2" thick	G	22	.727		4.19	38.50		42.69	63.50
2400	3" thick	G ↓	18	.889	↓	6.45	47		53.45	79
2410	On plane surfaces									
2420	1" thick	G Q-14	126	.127	S.F.	3.90	6.70		10.60	14.60
2430	1-1/2" thick	G	120	.133		4.40	7.05		11.45	15.65
2440	2" thick	G	100	.160		4.19	8.45		12.64	17.55
2450	3" thick	G ↓	70	.229	↓	6.45	12.05		18.50	25.50

23 09 13.60 Water Level Controls		Crew	Daily Output	Labor-Hours	Unit	Material	2020 Bare Costs Labor	Equipment	Total	Total Incl O&P
0010	**WATER LEVEL CONTROLS**									
1000	Electric water feeder	1 Stpi	12	.667	Ea.	375	43.50		418.50	480
2000	Feeder cut-off combination									
2100	Steam system up to 5,000 sq. ft.	1 Stpi	12	.667	Ea.	710	43.50		753.50	845
2200	Steam system above 5,000 sq. ft.		12	.667		1,000	43.50		1,043.50	1,175
2300	Steam and hot water, high pressure	↓	10	.800	↓	1,200	52.50		1,252.50	1,375
3000	Low water cut-off for hot water boiler, 50 psi maximum									
3100	1" top & bottom equalizing pipes, manual reset	1 Stpi	14	.571	Ea.	460	37.50		497.50	560
3200	1" top & bottom equalizing pipes		14	.571		420	37.50		457.50	515
3300	2-1/2" side connection for nipple-to-boiler	↓	14	.571	↓	415	37.50		452.50	510
4000	Low water cut-off for low pressure steam with quick hook-up ftgs.									
4100	For installation in gauge glass tappings	1 Stpi	16	.500	Ea.	340	33		373	425
4200	Built-in type, 2-1/2" tap, 3-1/8" insertion		16	.500		271	33		304	345
4300	Built-in type, 2-1/2" tap, 1-3/4" insertion		16	.500		298	33		331	375
4400	Side connection to 2-1/2" tapping		16	.500		320	33		353	400
5000	Pump control, low water cut-off and alarm switch	↓	14	.571	↓	865	37.50		902.50	1,000
9000	Water gauges, complete									
9010	Rough brass, wheel type									
9020	125 psi at 350°F									
9030	3/8" pipe size	1 Stpi	11	.727	Ea.	74.50	47.50		122	153
9040	1/2" pipe size	"	10	.800	"	79	52.50		131.50	166
9060	200 psi at 400°F									
9070	3/8" pipe size	1 Stpi	11	.727	Ea.	93	47.50		140.50	174
9080	1/2" pipe size		10	.800		91.50	52.50		144	180
9090	3/4" pipe size	↓	9	.889	↓	119	58.50		177.50	217
9130	Rough brass, chain lever type									
9140	250 psi at 400°F									
9200	Polished brass, wheel type									
9210	200 psi at 400°F									
9220	3/8" pipe size	1 Stpi	11	.727	Ea.	177	47.50		224.50	266
9230	1/2" pipe size		10	.800		179	52.50		231.50	276
9240	3/4" pipe size	↓	9	.889	↓	189	58.50		247.50	295

23 09 13 – Instrumentation and Control Devices for HVAC

23 09 13.60 Water Level Controls		Crew	Daily Output	Labor-Hours	Unit	Material	2020 Bare Costs Labor	Equipment	Total	Total Incl O&P
9260	Polished brass, chain lever type									
9270	250 psi at 400°F									
9280	1/2" pipe size	1 Stpi	10	.800	Ea.	259	52.50		311.50	365
9290	3/4" pipe size	"	9	.889	"	281	58.50		339.50	395
9400	Bronze, high pressure, ASME									
9410	1/2" pipe size	1 Stpi	10	.800	Ea.	320	52.50		372.50	435
9420	3/4" pipe size	"	9	.889	"	345	58.50		403.50	460
9460	Chain lever type									
9470	1/2" pipe size	1 Stpi	10	.800	Ea.	415	52.50		467.50	535
9480	3/4" pipe size	"	9	.889	"	460	58.50		518.50	590
9500	316 stainless steel, high pressure, ASME									
9510	500 psi at 450°F									
9520	1/2" pipe size	1 Stpi	10	.800	Ea.	930	52.50		982.50	1,100
9530	3/4" pipe size	"	9	.889	"	900	58.50		958.50	1,075

23 09 43 – Pneumatic Control System for HVAC

23 09 43.10 Pneumatic Control Systems

		Crew	Daily Output	Labor-Hours	Unit	Material	2020 Bare Costs Labor	Equipment	Total	Total Incl O&P
0010	**PNEUMATIC CONTROL SYSTEMS** R230500-10									
0011	Including a nominal 50' of tubing. Add control panelboard if req'd.									
0100	Heating and ventilating, split system									
0200	Mixed air control, economizer cycle, panel readout, tubing									
0220	Up to 10 tons [G]	Q-19	.68	35.294	Ea.	4,600	2,100		6,700	8,225
0240	For 10 to 20 tons [G]		.63	37.915		4,950	2,275		7,225	8,825
0260	For over 20 tons [G]		.58	41.096		5,375	2,450		7,825	9,575
0270	Enthalpy cycle, up to 10 tons		.50	48.387		5,100	2,900		8,000	9,950
0280	For 10 to 20 tons		.46	52.174		5,500	3,125		8,625	10,700
0290	For over 20 tons		.42	56.604		5,750	3,375		9,125	11,400
0300	Heating coil, hot water, 3 way valve,									
0320	Freezestat, limit control on discharge, readout	Q-5	.69	23.088	Ea.	3,425	1,350		4,775	5,800
0500	Cooling coil, chilled water, room									
0520	Thermostat, 3 way valve	Q-5	2	8	Ea.	1,475	470		1,945	2,325
0600	Cooling tower, fan cycle, damper control,									
0620	Control system including water readout in/out at panel	Q-19	.67	35.821	Ea.	6,050	2,150		8,200	9,875
1000	Unit ventilator, day/night operation,									
1100	freezestat, ASHRAE, cycle 2	Q-19	.91	26.374	Ea.	3,250	1,575		4,825	5,925
2000	Compensated hot water from boiler, valve control,									
2100	readout and reset at panel, up to 60 GPM	Q-19	.55	43.956	Ea.	6,300	2,625		8,925	10,900
2120	For 120 GPM		.51	47.059		6,475	2,825		9,300	11,300
2140	For 240 GPM		.49	49.180		7,050	2,950		10,000	12,200
3000	Boiler room combustion air, damper to 5 S.F., controls		1.37	17.582		3,025	1,050		4,075	4,900
3500	Fan coil, heating and cooling valves, 4 pipe control system		3	8		1,375	480		1,855	2,225
3600	Heat exchanger system controls		.86	27.907		2,850	1,675		4,525	5,625
3900	Multizone control (one per zone), includes thermostat, damper									
3910	motor and reset of discharge temperature	Q-5	.51	31.373	Ea.	2,925	1,850		4,775	6,000
4000	Pneumatic thermostat, including controlling room radiator valve	"	2.43	6.593		880	390		1,270	1,550
4040	Program energy saving optimizer [G]	Q-19	1.21	19.786		7,575	1,175		8,750	10,100
4060	Pump control system	"	3	8		1,350	480		1,830	2,200
4080	Reheat coil control system, not incl. coil	Q-5	2.43	6.593		1,150	390		1,540	1,825
4500	Air supply for pneumatic control system									
4600	Tank mounted duplex compressor, starter, alternator,									
4620	piping, dryer, PRV station and filter									
4630	1/2 HP	Q-19	.68	35.139	Ea.	11,300	2,100		13,400	15,600
4640	3/4 HP		.64	37.383		11,400	2,225		13,625	15,900

23 09 Instrumentation and Control for HVAC

23 09 43 – Pneumatic Control System for HVAC

23 09 43.10 Pneumatic Control Systems

23 09 43.10 Pneumatic Control Systems		Crew	Daily Output	Labor-Hours	Unit	Material	2020 Bare Costs Labor	2020 Bare Costs Equipment	Total	Total Incl O&P
4650	1 HP	Q-19	.61	39.539	Ea.	12,900	2,375		15,275	17,700
4660	1-1/2 HP		.58	41.739		13,300	2,500		15,800	18,300
4680	3 HP		.55	43.956		18,100	2,625		20,725	23,800
4690	5 HP		.42	57.143		32,800	3,425		36,225	41,200
4800	Main air supply, includes 3/8" copper main and labor	Q-5	1.82	8.791	C.L.F.	380	520		900	1,200
7000	Static pressure control for air handling unit, includes pressure									
7010	sensor, receiver controller, readout and damper motors	Q-19	.64	37.383	Ea.	9,000	2,225		11,225	13,300
8600	VAV boxes, incl. thermostat, damper motor, reheat coil & tubing	Q-5	1.46	10.989	"	1,350	650		2,000	2,475

23 09 53 – Pneumatic and Electric Control System for HVAC

23 09 53.10 Control Components

23 09 53.10 Control Components		Crew	Daily Output	Labor-Hours	Unit	Material	2020 Bare Costs Labor	2020 Bare Costs Equipment	Total	Total Incl O&P
0010	**CONTROL COMPONENTS**									
0500	Aquastats									
0508	Immersion type									
0510	High/low limit, breaks contact w/temp rise	1 Stpi	4	2	Ea.	385	131		516	620
0514	Sequencing, break 2 switches w/temp rise		4	2		310	131		441	535
0518	Circulating, makes contact w/temp rise		4	2		280	131		411	505
0600	Carbon monoxide detector system									
0606	Panel	1 Stpi	4	2	Ea.	960	131		1,091	1,250
0610	Sensor		7.30	1.096		645	72		717	820
0680	Controller for VAV box, includes actuator		7.30	1.096		281	72		353	420
0700	Controller, receiver									
0730	Pneumatic, panel mount, single input	1 Plum	8	1	Ea.	460	64.50		524.50	600
0740	With conversion mounting bracket		8	1		460	64.50		524.50	600
0750	Dual input, with control point adjustment		7	1.143		635	73.50		708.50	810
0850	Electric, single snap switch	1 Elec	4	2		485	123		608	720
0860	Dual snap switches		3	2.667		655	164		819	965
0870	Humidity controller		8	1		273	61.50		334.50	390
0880	Load limiting controller		8	1		1,000	61.50		1,061.50	1,225
0890	Temperature controller		8	1		570	61.50		631.50	715
0900	Control panel readout									
0910	Panel up to 12 indicators	1 Stpi	1.20	6.667	Ea.	310	435		745	995
0914	Panel up to 24 indicators		.86	9.302		315	610		925	1,250
0918	Panel up to 48 indicators		.50	16		760	1,050		1,810	2,400
1000	Enthalpy control, boiler water temperature control									
1010	governed by outdoor temperature, with timer	1 Elec	3	2.667	Ea.	355	164		519	635
1300	Energy control/monitor									
1320	BTU computer meter and controller	1 Stpi	1	8	Ea.	1,075	525		1,600	1,950
1600	Flow meters									
1610	Gas	1 Stpi	4	2	Ea.	112	131		243	320
1620	Liquid	"	4	2	"	161	131		292	375
1640	Freezestat									
1644	20' sensing element, adjustable	1 Stpi	3.50	2.286	Ea.	93	150		243	325
2000	Gauges, pressure or vacuum									
2100	2" diameter dial	1 Stpi	32	.250	Ea.	10.55	16.40		26.95	36
2200	2-1/2" diameter dial		32	.250		12.80	16.40		29.20	38.50
2300	3-1/2" diameter dial		32	.250		18.15	16.40		34.55	44.50
2400	4-1/2" diameter dial		32	.250		20	16.40		36.40	46.50
2700	Flanged iron case, black ring									
2800	3-1/2" diameter dial	1 Stpi	32	.250	Ea.	101	16.40		117.40	137
2900	4-1/2" diameter dial		32	.250		105	16.40		121.40	140
3000	6" diameter dial		32	.250		167	16.40		183.40	209
3010	Steel case, 0-300 psi									

For customer support on your Mechanical Costs with RSMeans Data, call 800.448.8182.

269

23 09 53.10 Control Components	Crew	Daily Output	Labor-Hours	Unit	Material	2020 Bare Costs Labor	Equipment	Total	Total Incl O&P
3012	2" diameter dial	1 Stpi	16	.500	Ea.	8.60	33		41.60	58.50
3014	4" diameter dial	"	16	.500	"	23	33		56	74.50
3020	Aluminum case, 0-300 psi									
3022	3-1/2" diameter dial	1 Stpi	16	.500	Ea.	12.60	33		45.60	63
3024	4-1/2" diameter dial		16	.500		122	33		155	183
3026	6" diameter dial		16	.500		193	33		226	262
3028	8-1/2" diameter dial		16	.500		335	33		368	415
3030	Brass case, 0-300 psi									
3032	2" diameter dial	1 Stpi	16	.500	Ea.	49	33		82	103
3034	4-1/2" diameter dial	"	16	.500	"	104	33		137	163
3040	Steel case, high pressure, 0-10,000 psi									
3042	4-1/2" diameter dial	1 Stpi	16	.500	Ea.	315	33		348	395
3044	6-1/2" diameter dial		16	.500		325	33		358	405
3046	8-1/2" diameter dial		16	.500		430	33		463	525
3080	Pressure gauge, differential, magnehelic									
3084	0-2" W.C., with air filter kit	1 Stpi	6	1.333	Ea.	84.50	87.50		172	224
3300	For compound pressure-vacuum, add					18%				
3350	Humidistat									
3360	Pneumatic operation									
3361	Room humidistat, direct acting	1 Stpi	12	.667	Ea.	375	43.50		418.50	480
3362	Room humidistat, reverse acting		12	.667		375	43.50		418.50	480
3363	Room humidity transmitter		17	.471		405	31		436	490
3364	Duct mounted controller		12	.667		440	43.50		483.50	550
3365	Duct mounted transmitter		12	.667		400	43.50		443.50	505
3366	Humidity indicator, 3-1/2"		28	.286		132	18.75		150.75	173
3390	Electric operated	1 Shee	8	1		50	62.50		112.50	150
3400	Relays									
3430	Pneumatic/electric	1 Plum	16	.500	Ea.	310	32		342	390
3440	Pneumatic proportioning		8	1		280	64.50		344.50	405
3450	Pneumatic switching		12	.667		151	43		194	232
3460	Selector, 3 point		6	1.333		108	86		194	248
3470	Pneumatic time delay		8	1		345	64.50		409.50	475
3500	Sensor, air operated									
3520	Humidity	1 Plum	16	.500	Ea.	440	32		472	535
3540	Pressure		16	.500		66	32		98	121
3560	Temperature		12	.667		162	43		205	244
3600	Electric operated									
3620	Humidity	1 Elec	8	1	Ea.	284	61.50		345.50	400
3650	Pressure		8	1		325	61.50		386.50	450
3680	Temperature		10	.800		121	49		170	206
3700	Switches									
3710	Minimum position									
3720	Electrical	1 Stpi	4	2	Ea.	21	131		152	219
3730	Pneumatic	"	4	2	"	191	131		322	405
4000	Thermometers									
4100	Dial type, 3-1/2" diameter, vapor type, union connection	1 Stpi	32	.250	Ea.	226	16.40		242.40	274
4120	Liquid type, union connection		32	.250		435	16.40		451.40	505
4130	Remote reading, 15' capillary		32	.250		235	16.40		251.40	283
4500	Stem type, 6-1/2" case, 2" stem, 1/2" NPT		32	.250		60	16.40		76.40	90.50
4520	4" stem, 1/2" NPT		32	.250		73.50	16.40		89.90	106
4600	9" case, 3-1/2" stem, 3/4" NPT		28	.286		93.50	18.75		112.25	131
4620	6" stem, 3/4" NPT		28	.286		111	18.75		129.75	151
4640	8" stem, 3/4" NPT		28	.286		187	18.75		205.75	234

23 09 53 – Pneumatic and Electric Control System for HVAC

23 09 53.10 Control Components	Crew	Daily Output	Labor-Hours	Unit	Material	2020 Bare Costs Labor	Equipment	Total	Total Incl O&P
4660	12" stem, 1" NPT	1 Stpi	26	.308	Ea.	164	20		184	211
4670	Bi-metal, dial type, steel case brass stem									
4672	2" dial, 4" - 9" stem	1 Stpi	16	.500	Ea.	40	33		73	93
4673	2-1/2" dial, 4" - 9" stem		16	.500		40	33		73	93
4674	3-1/2" dial, 4" - 9" stem	↓	16	.500	↓	49	33		82	103
4680	Mercury filled, industrial, union connection type									
4682	Angle stem, 7" scale	1 Stpi	16	.500	Ea.	61	33		94	116
4683	9" scale		16	.500		167	33		200	233
4684	12" scale		16	.500		214	33		247	285
4686	Straight stem, 7" scale		16	.500		163	33		196	228
4687	9" scale		16	.500		145	33		178	209
4688	12" scale	↓	16	.500	↓	171	33		204	237
4690	Mercury filled, industrial, separable socket type, with well									
4692	Angle stem, with socket, 7" scale	1 Stpi	16	.500	Ea.	198	33		231	267
4693	9" scale		16	.500		197	33		230	266
4694	12" scale		16	.500		234	33		267	305
4696	Straight stem, with socket, 7" scale		16	.500		154	33		187	218
4697	9" scale		16	.500		158	33		191	223
4698	12" scale	↓	16	.500	↓	189	33		222	257
5000	Thermostats									
5030	Manual	1 Stpi	8	1	Ea.	42.50	65.50		108	145
5040	1 set back, electric, timed G		8	1		77.50	65.50		143	184
5050	2 set back, electric, timed G		8	1		257	65.50		322.50	380
5100	Locking cover	↓	20	.400		20	26		46	61
5200	24 hour, automatic, clock G	1 Shee	8	1		204	62.50		266.50	320
5220	Electric, low voltage, 2 wire	1 Elec	13	.615		58	38		96	120
5230	3 wire		10	.800		58	49		107	137
5236	Heating/cooling, low voltage, with clock	↓	8	1	↓	227	61.50		288.50	340
5240	Pneumatic									
5250	Single temp., single pressure	1 Stpi	8	1	Ea.	213	65.50		278.50	335
5251	Dual pressure		8	1		365	65.50		430.50	500
5252	Dual temp., dual pressure		8	1		284	65.50		349.50	410
5253	Reverse acting w/averaging element		8	1		261	65.50		326.50	385
5254	Heating-cooling w/deadband		8	1		475	65.50		540.50	625
5255	Integral w/piston top valve actuator		8	1		231	65.50		296.50	350
5256	Dual temp., dual pressure		8	1		236	65.50		301.50	355
5257	Low limit, 8' averaging element		8	1		183	65.50		248.50	299
5258	Room single temp. proportional		8	1		75	65.50		140.50	181
5260	Dual temp, direct acting for VAV fan		8	1		272	65.50		337.50	395
5262	Capillary tube type, 20', 30°F to 100°F	↓	8	1	↓	320	65.50		385.50	455
5300	Transmitter, pneumatic									
5320	Temperature averaging element	Q-1	8	2	Ea.	151	116		267	340
5350	Pressure differential	1 Plum	7	1.143		1,275	73.50		1,348.50	1,500
5370	Humidity, duct		8	1		400	64.50		464.50	535
5380	Room		12	.667		405	43		448	510
5390	Temperature, with averaging element	↓	6	1.333	↓	175	86		261	320
5500	Timer/time clocks									
5510	7 day, 12 hr. battery	1 Stpi	2.60	3.077	Ea.	222	202		424	545
5600	Recorders									
5610	Hydrograph humidity, wall mtd., aluminum case, grad. chart									
5612	8"	1 Stpi	1.60	5	Ea.	920	330		1,250	1,525
5614	10"	"	1.60	5	"	1,100	330		1,430	1,700
5620	Thermo-hydrograph, wall mtd., grad. chart, 5' SS probe									

23 09 53 – Pneumatic and Electric Control System for HVAC

23 09 53.10 Control Components	Crew	Daily Output	Labor-Hours	Unit	Material	2020 Bare Costs Labor	Equipment	Total	Total Incl O&P
5622	8"	1 Stpi	1.60	5	Ea.	1,100	330		1,430	1,700
5624	10"	"	1.60	5	"	1,600	330		1,930	2,250
5630	Time of operation, wall mtd., 1 pen cap, 24 hr. clock									
5632	6"	1 Stpi	1.60	5	Ea.	435	330		765	970
5634	8"	"	1.60	5	"	670	330		1,000	1,225
5640	Pressure & vacuum, bourdon tube type									
5642	Flush mount, 1 pen									
5644	4"	1 Stpi	1.60	5	Ea.	495	330		825	1,025
5645	6"		1.60	5		355	330		685	880
5646	8"		1.60	5		700	330		1,030	1,250
5648	10"		1.60	5		1,400	330		1,730	2,050
6000	Valves, motorized zone									
6100	Sweat connections, 1/2" C x C	1 Stpi	20	.400	Ea.	136	26		162	188
6110	3/4" C x C		20	.400		138	26		164	191
6120	1" C x C		19	.421		185	27.50		212.50	245
6140	1/2" C x C, with end switch, 2 wire		20	.400		137	26		163	190
6150	3/4" C x C, with end switch, 2 wire		20	.400		118	26		144	169
6160	1" C x C, with end switch, 2 wire		19	.421		179	27.50		206.50	238
7090	Valves, motor controlled, including actuator									
7100	Electric motor actuated									
7200	Brass, two way, screwed									
7210	1/2" pipe size	L-6	36	.333	Ea.	385	21		406	455
7220	3/4" pipe size		30	.400		625	25.50		650.50	725
7230	1" pipe size		28	.429		620	27		647	720
7240	1-1/2" pipe size		19	.632		960	40		1,000	1,100
7250	2" pipe size		16	.750		1,250	47.50		1,297.50	1,450
7350	Brass, three way, screwed									
7360	1/2" pipe size	L-6	33	.364	Ea.	340	23		363	410
7370	3/4" pipe size		27	.444		395	28		423	475
7380	1" pipe size		25.50	.471		410	30		440	495
7384	1-1/4" pipe size		21	.571		450	36		486	550
7390	1-1/2" pipe size		17	.706		495	45		540	610
7400	2" pipe size		14	.857		780	54.50		834.50	935
7550	Iron body, two way, flanged									
7560	2-1/2" pipe size	L-6	4	3	Ea.	1,175	190		1,365	1,550
7570	3" pipe size		3	4		1,200	254		1,454	1,700
7580	4" pipe size		2	6		2,175	380		2,555	2,950
7850	Iron body, three way, flanged									
7860	2-1/2" pipe size	L-6	3	4	Ea.	1,175	254		1,429	1,675
7870	3" pipe size		2.50	4.800		1,325	305		1,630	1,925
7880	4" pipe size		2	6		1,700	380		2,080	2,450
8000	Pneumatic, air operated									
8050	Brass, two way, screwed									
8060	1/2" pipe size, class 250	1 Plum	24	.333	Ea.	174	21.50		195.50	223
8070	3/4" pipe size, class 250		20	.400		208	26		234	268
8080	1" pipe size, class 250		19	.421		243	27		270	310
8090	1-1/4" pipe size, class 125		15	.533		300	34.50		334.50	380
8100	1-1/2" pipe size, class 125		13	.615		385	39.50		424.50	485
8110	2" pipe size, class 125		11	.727		445	47		492	560
8180	Brass, three way, screwed									
8190	1/2" pipe size, class 250	1 Plum	22	.364	Ea.	184	23.50		207.50	237
8200	3/4" pipe size, class 250		18	.444		209	28.50		237.50	273
8210	1" pipe size, class 250		17	.471		244	30.50		274.50	315

23 09 53.10 Control Components	Crew	Daily Output	Labor-Hours	Unit	Material	2020 Bare Costs Labor	Equipment	Total	Total Incl O&P
8214	1-1/4" pipe size, class 250	1 Plum	14	.571	Ea.	330	37		367	415
8220	1-1/2" pipe size, class 125		11	.727		390	47		437	495
8230	2" pipe size, class 125		9	.889		455	57.50		512.50	585
8450	Iron body, two way, flanged									
8510	2-1/2" pipe size, class 125	Q-1	5	3.200	Ea.	930	186		1,116	1,300
8520	3" pipe size, class 125		4.50	3.556		1,025	206		1,231	1,425
8530	4" pipe size, class 125		3	5.333		1,775	310		2,085	2,425
8540	5" pipe size, class 125	Q-2	3.40	7.059		3,875	425		4,300	4,900
8550	6" pipe size, class 125	"	3	8		3,300	480		3,780	4,350
8560	Iron body, three way, flanged									
8570	2-1/2" pipe size, class 125	Q-1	4.50	3.556	Ea.	1,000	206		1,206	1,400
8580	3" pipe size, class 125		4	4		1,200	232		1,432	1,650
8590	4" pipe size, class 125		2.50	6.400		2,350	370		2,720	3,125
8600	6" pipe size, class 125	Q-2	3	8		3,550	480		4,030	4,650
9005	Pneumatic system misc. components									
9010	Adding/subtracting repeater	1 Stpi	20	.400	Ea.	167	26		193	223
9020	Adjustable ratio network		16	.500		158	33		191	222
9030	Comparator		20	.400		400	26		426	485
9040	Cumulator									
9041	Air switching	1 Stpi	20	.400	Ea.	151	26		177	206
9042	Averaging		20	.400		216	26		242	276
9043	2:1 ratio		20	.400		375	26		401	450
9044	Sequencing		20	.400		96.50	26		122.50	145
9045	Two-position		16	.500		189	33		222	257
9046	Two-position pilot		16	.500		360	33		393	445
9050	Damper actuator									
9051	For smoke control	1 Stpi	8	1	Ea.	315	65.50		380.50	445
9052	For ventilation	"	8	1	"	243	65.50		308.50	365
9053	Series duplex pedestal mounted									
9054	For inlet vanes on fans, compressors	1 Stpi	6	1.333	Ea.	1,925	87.50		2,012.50	2,225
9060	Enthalpy logic center		16	.500		375	33		408	460
9070	High/low pressure selector		22	.364		61	24		85	103
9080	Signal limiter		20	.400		96.50	26		122.50	145
9081	Transmitter		28	.286		37	18.75		55.75	68.50
9090	Optimal start									
9092	Mass temperature transmitter	1 Stpi	18	.444	Ea.	145	29		174	203
9100	Step controller with positioner, time delay restrictor									
9110	recycler air valve, switches and cam settings									
9112	With cabinet									
9113	6 points	1 Stpi	4	2	Ea.	1,150	131		1,281	1,475
9114	6 points, 6 manual sequences		2	4		1,000	262		1,262	1,500
9115	8 points		3	2.667		1,250	175		1,425	1,625
9200	Pressure controller and switches									
9210	High static pressure limit	1 Stpi	8	1	Ea.	288	65.50		353.50	415
9220	Pressure transmitter		8	1		228	65.50		293.50	350
9230	Differential pressure transmitter		8	1		895	65.50		960.50	1,075
9240	Static pressure transmitter		8	1		1,075	65.50		1,140.50	1,300
9250	Proportional-only control, single input		6	1.333		460	87.50		547.50	635
9260	Proportional plus integral, single input		6	1.333		650	87.50		737.50	845
9270	Proportional-only, dual input		6	1.333		635	87.50		722.50	830
9280	Proportional plus integral, dual input		6	1.333		795	87.50		882.50	1,000
9281	Time delay for above proportional units		16	.500		187	33		220	255
9290	Proportional/2 position controller units		6	1.333		465	87.50		552.50	640

For customer support on your Mechanical Costs with RSMeans Data, call 800.448.8182.

273

23 09 53.10 Control Components	Crew	Daily Output	Labor-Hours	Unit	Material	2020 Bare Costs Labor	Equipment	Total	Total Incl O&P
9300	Differential pressure control direct/reverse	1 Stpi	6	1.333	Ea.	475	87.50		562.50	655
9310	Air pressure reducing valve									
9311	1/8" size	1 Stpi	18	.444	Ea.	44	29		73	91.50
9315	Precision valve, 1/4" size		17	.471		246	31		277	315
9320	Air flow controller		12	.667		178	43.50		221.50	262
9330	Booster relay volume amplifier		18	.444		104	29		133	158
9340	Series restrictor, straight		60	.133		5.30	8.75		14.05	18.90
9341	T-fitting		40	.200		7.75	13.10		20.85	28
9342	In-line adjustable		48	.167		34.50	10.95		45.45	54.50
9350	Diode tee		40	.200		17.80	13.10		30.90	39
9351	Restrictor tee		40	.200		17.80	13.10		30.90	39
9360	Pneumatic gradual switch		8	1		298	65.50		363.50	430
9361	Selector switch		8	1		108	65.50		173.50	217
9370	Electro-pneumatic motor driven servo		10	.800		1,150	52.50		1,202.50	1,350
9390	Fan control switch and mounting base		16	.500		107	33		140	167
9400	Circulating pump sequencer	↓	14	.571	↓	1,250	37.50		1,287.50	1,425
9410	Pneumatic tubing, fittings and accessories									
9414	Tubing, urethane									
9415	1/8" OD x 1/16" ID	1 Stpi	120	.067	L.F.	.70	4.37		5.07	7.30
9416	1/4" OD x 1/8" ID		115	.070		.82	4.56		5.38	7.70
9417	5/32" OD x 3/32" ID	↓	110	.073	↓	.46	4.77		5.23	7.65
9420	Coupling, straight									
9422	Barb x barb									
9423	1/4" x 5/32"	1 Stpi	160	.050	Ea.	.80	3.28		4.08	5.80
9424	1/4" x 1/4"		158	.051		.64	3.32		3.96	5.65
9425	3/8" x 3/8"		154	.052		.67	3.41		4.08	5.85
9426	3/8" x 1/4"		150	.053		.54	3.50		4.04	5.85
9427	1/2" x 1/4"		148	.054		5.15	3.54		8.69	10.95
9428	1/2" x 3/8"		144	.056		1.02	3.64		4.66	6.55
9429	1/2" x 1/2"	↓	140	.057	↓	.69	3.75		4.44	6.35
9440	Tube x tube									
9441	1/4" x 1/4"	1 Stpi	100	.080	Ea.	2.17	5.25		7.42	10.25
9442	3/8" x 1/4"		96	.083		4.89	5.45		10.34	13.55
9443	3/8" x 3/8"		92	.087		2.69	5.70		8.39	11.50
9444	1/2" x 3/8"		88	.091		5.25	5.95		11.20	14.65
9445	1/2" x 1/2"	↓	84	.095	↓	4.93	6.25		11.18	14.75
9450	Elbow coupling									
9452	Barb x barb									
9454	1/4" x 1/4"	1 Stpi	158	.051	Ea.	1.37	3.32		4.69	6.50
9455	3/8" x 3/8"		154	.052		3.72	3.41		7.13	9.20
9456	1/2" x 1/2"	↓	140	.057	↓	5.70	3.75		9.45	11.85
9460	Tube x tube									
9462	1/2" x 1/2"	1 Stpi	84	.095	Ea.	8.35	6.25		14.60	18.55
9470	Tee coupling									
9472	Barb x barb x barb									
9474	1/4" x 1/4" x 5/32"	1 Stpi	108	.074	Ea.	1.75	4.86		6.61	9.20
9475	1/4" x 1/4" x 1/4"		104	.077		1.14	5.05		6.19	8.80
9476	3/8" x 3/8" x 5/32"		102	.078		3.04	5.15		8.19	11.05
9477	3/8" x 3/8" x 1/4"		99	.081		1.75	5.30		7.05	9.90
9478	3/8" x 3/8" x 3/8"		98	.082		5.10	5.35		10.45	13.60
9479	1/2" x 1/2" x 1/4"		96	.083		4.61	5.45		10.06	13.20
9480	1/2" x 1/2" x 3/8"		95	.084		2.61	5.50		8.11	11.10
9481	1/2" x 1/2" x 1/2"	↓	92	.087	↓	4.43	5.70		10.13	13.40

23 09 53.10 Control Components	Crew	Daily Output	Labor-Hours	Unit	Material	2020 Bare Costs Labor	Equipment	Total	Total Incl O&P
9484	Tube x tube x tube									
9485	1/4" x 1/4" x 1/4"	1 Stpi	66	.121	Ea.	4.63	7.95		12.58	17
9486	3/8" x 1/4" x 1/4"		64	.125		9.85	8.20		18.05	23
9487	3/8" x 3/8" x 1/4"		61	.131		6.25	8.60		14.85	19.70
9488	3/8" x 3/8" x 3/8"		60	.133		6.55	8.75		15.30	20.50
9489	1/2" x 1/2" x 3/8"		58	.138		15	9.05		24.05	30
9490	1/2" x 1/2" x 1/2"		56	.143		10.75	9.35		20.10	26
9492	Needle valve									
9494	Tube x tube									
9495	1/4" x 1/4"	1 Stpi	86	.093	Ea.	4.46	6.10		10.56	14.05
9496	3/8" x 3/8"	"	82	.098	"	9.45	6.40		15.85	19.95
9600	Electronic system misc. components									
9610	Electric motor damper actuator	1 Elec	8	1	Ea.	515	61.50		576.50	660
9620	Damper position indicator		10	.800		123	49		172	209
9630	Pneumatic-electronic transducer		12	.667		60.50	41		101.50	128
9700	Modulating step controller									
9701	For 2-5 steps	1 Elec	5.30	1.509	Ea.	880	92.50		972.50	1,100
9702	6-10 steps		3.20	2.500		272	153		425	525
9703	11-15 steps		2.70	2.963		550	182		732	875
9704	16-20 steps		1.60	5		820	305		1,125	1,350
9710	Staging network									
9740	Accessory unit regulator									
9750	Accessory power supply	1 Elec	18	.444	Ea.	228	27.50		255.50	292
9760	Step down transformer	"	16	.500	"	187	30.50		217.50	252

23 11 13.10 Fuel Oil Specialties	Crew	Daily Output	Labor-Hours	Unit	Material	2020 Bare Costs Labor	Equipment	Total	Total Incl O&P
0010	**FUEL OIL SPECIALTIES**									
0020	Foot valve, single poppet, metal to metal construction									
0040	Bevel seat, 1/2" diameter	1 Stpi	20	.400	Ea.	80.50	26		106.50	128
0060	3/4" diameter		18	.444		113	29		142	168
0080	1" diameter		16	.500		113	33		146	173
0100	1-1/4" diameter		15	.533		165	35		200	235
0120	1-1/2" diameter		13	.615		209	40.50		249.50	290
0140	2" diameter		11	.727		260	47.50		307.50	360
0160	Foot valve, double poppet, metal to metal construction									
0164	1" diameter	1 Stpi	15	.533	Ea.	169	35		204	238
0166	1-1/2" diameter	"	12	.667	"	284	43.50		327.50	380
0400	Fuel fill box, flush type									
0408	Nonlocking, watertight									
0410	1-1/2" diameter	1 Stpi	12	.667	Ea.	23	43.50		66.50	90.50
0440	2" diameter		10	.800		18.80	52.50		71.30	99
0450	2-1/2" diameter		9	.889		66	58.50		124.50	160
0460	3" diameter		7	1.143		54	75		129	171
0470	4" diameter		5	1.600		77	105		182	242
0500	Locking inner cover									
0510	2" diameter	1 Stpi	8	1	Ea.	83	65.50		148.50	190
0520	2-1/2" diameter		7	1.143		111	75		186	234
0530	3" diameter		5	1.600		119	105		224	288
0540	4" diameter		4	2		154	131		285	365

23 11 13.10 Fuel Oil Specialties		Crew	Daily Output	Labor-Hours	Unit	Material	2020 Bare Costs Labor	Equipment	Total	Total Incl O&P
0600	Fuel system components									
0620	Spill container	1 Stpi	4	2	Ea.	1,000	131		1,131	1,325
0640	Fill adapter, 4", straight drop		8	1		60	65.50		125.50	164
0680	Fill cap, 4"		30	.267		42.50	17.50		60	72.50
0700	Extractor fitting, 4" x 1-1/2"		8	1		545	65.50		610.50	700
0740	Vapor hose adapter, 4"	▼	8	1		128	65.50		193.50	238
0760	Wood gage stick, 10'					13.25			13.25	14.60
1000	Oil filters, 3/8" IPT, 20 gal. per hour	1 Stpi	20	.400		43	26		69	86.50
1020	32 gal. per hour		18	.444		58.50	29		87.50	108
1040	40 gal. per hour		16	.500		56.50	33		89.50	112
1060	50 gal. per hour	▼	14	.571	▼	61.50	37.50		99	124
2000	Remote tank gauging system, self contained									
2100	Single tank kit/8 sensors inputs w/printer	1 Stpi	2.50	3.200	Ea.	4,225	210		4,435	4,975
2120	Two tank kit/8 sensors inputs w/printer		2	4		5,750	262		6,012	6,725
3000	Valve, ball check, globe type, 3/8" diameter		24	.333		13.05	22		35.05	47
3500	Fusible, 3/8" diameter		24	.333		14.70	22		36.70	48.50
3600	1/2" diameter		24	.333		40	22		62	76.50
3610	3/4" diameter		20	.400		79	26		105	126
3620	1" diameter		19	.421		229	27.50		256.50	294
4000	Nonfusible, 3/8" diameter	▼	24	.333	▼	27	22		49	62.50
4500	Shutoff, gate type, lever handle, spring-fusible kit									
4520	1/4" diameter	1 Stpi	14	.571	Ea.	44	37.50		81.50	104
4540	3/8" diameter		12	.667		42.50	43.50		86	113
4560	1/2" diameter		10	.800		60.50	52.50		113	145
4570	3/4" diameter	▼	8	1	▼	113	65.50		178.50	222
4580	Lever handle, requires weight and fusible kit									
4600	1" diameter	1 Stpi	9	.889	Ea.	400	58.50		458.50	525
4620	1-1/4" diameter		8	1		545	65.50		610.50	700
4640	1-1/2" diameter		7	1.143		455	75		530	610
4660	2" diameter	▼	6	1.333		905	87.50		992.50	1,125
4680	For fusible link, weight and braided wire, add					5%				
5000	Vent alarm, whistling signal					38			38	42
5500	Vent protector/breather, 1-1/4" diameter	1 Stpi	32	.250		22.50	16.40		38.90	49
5520	1-1/2" diameter		32	.250		26	16.40		42.40	53
5540	2" diameter		32	.250		53.50	16.40		69.90	83.50
5560	3" diameter		28	.286		75	18.75		93.75	111
5580	4" diameter		24	.333		91.50	22		113.50	134
5600	Dust cap, breather, 2"	▼	40	.200	▼	12.05	13.10		25.15	33
8000	Fuel oil and tank heaters									
8020	Electric, capacity rated at 230 volts									
8040	Immersion element in steel manifold									
8060	96 GPH at 50°F rise	Q-5	6.40	2.500	Ea.	1,350	148		1,498	1,700
8070	128 GPH at 50°F rise		6.20	2.581		1,350	152		1,502	1,700
8080	160 GPH at 50°F rise		5.90	2.712		1,550	160		1,710	1,950
8090	192 GPH at 50°F rise		5.50	2.909		1,700	172		1,872	2,100
8100	240 GPH at 50°F rise		5.10	3.137		1,950	185		2,135	2,425
8110	288 GPH at 50°F rise		4.60	3.478		2,125	205		2,330	2,625
8120	384 GPH at 50°F rise		3.10	5.161		2,475	305		2,780	3,175
8130	480 GPH at 50°F rise		2.30	6.957		3,125	410		3,535	4,050
8140	576 GPH at 50°F rise	▼	2.10	7.619		3,125	450		3,575	4,100
8300	Suction stub, immersion type									
8320	75" long, 750 watts	1 Stpi	14	.571	Ea.	890	37.50		927.50	1,025
8330	99" long, 2000 watts	▼	12	.667	▼	1,275	43.50		1,318.50	1,500

23 11 13 – Facility Fuel-Oil Piping

23 11 13.10 Fuel Oil Specialties

		Crew	Daily Output	Labor-Hours	Unit	Material	2020 Bare Costs Labor	Equipment	Total	Total Incl O&P
8340	123" long, 3000 watts	1 Stpi	10	.800	Ea.	1,375	52.50		1,427.50	1,600
8660	Steam, cross flow, rated at 5 PSIG									
8680	42 GPH	Q-5	7	2.286	Ea.	2,050	135		2,185	2,450
8690	73 GPH		6.70	2.388		2,475	141		2,616	2,900
8700	112 GPH		6.20	2.581		2,475	152		2,627	2,950
8710	158 GPH		5.80	2.759		2,700	163		2,863	3,225
8720	187 GPH		4	4		3,825	236		4,061	4,550
8730	270 GPH		3.60	4.444		4,000	262		4,262	4,800
8740	365 GPH	Q-6	4.90	4.898		4,850	300		5,150	5,775
8750	635 GPH		3.70	6.486		5,925	395		6,320	7,125
8760	845 GPH		2.50	9.600		8,950	585		9,535	10,700
8770	1420 GPH		1.60	15		14,600	920		15,520	17,400
8780	2100 GPH		1.10	21.818		20,000	1,325		21,325	24,000

23 11 23 – Facility Natural-Gas Piping

23 11 23.10 Gas Meters

		Crew	Daily Output	Labor-Hours	Unit	Material	2020 Bare Costs Labor	Equipment	Total	Total Incl O&P
0010	**GAS METERS**									
4000	Residential									
4010	Gas meter, residential, 3/4" pipe size	1 Plum	14	.571	Ea.	305	37		342	390
4020	Gas meter, residential, 1" pipe size		12	.667		290	43		333	385
4030	Gas meter, residential, 1-1/4" pipe size		10	.800		254	51.50		305.50	355

23 11 23.20 Gas Piping, Flexible (Csst)

		Crew	Daily Output	Labor-Hours	Unit	Material	2020 Bare Costs Labor	Equipment	Total	Total Incl O&P
0010	**GAS PIPING, FLEXIBLE (CSST)**									
0100	Tubing with lightning protection									
0110	3/8"	1 Stpi	65	.123	L.F.	3.22	8.05		11.27	15.60
0120	1/2"		62	.129		3.62	8.45		12.07	16.65
0130	3/4"		60	.133		4.71	8.75		13.46	18.30
0140	1"		55	.145		7.25	9.55		16.80	22
0150	1-1/4"		50	.160		8.40	10.50		18.90	25
0160	1-1/2"		45	.178		14.75	11.65		26.40	33.50
0170	2"		40	.200		21	13.10		34.10	43
0200	Tubing for underground/underslab burial									
0210	3/8"	1 Stpi	65	.123	L.F.	4.48	8.05		12.53	17
0220	1/2"		62	.129		5.10	8.45		13.55	18.30
0230	3/4"		60	.133		6.50	8.75		15.25	20.50
0240	1"		55	.145		8.75	9.55		18.30	24
0250	1-1/4"		50	.160		11.75	10.50		22.25	28.50
0260	1-1/2"		45	.178		22	11.65		33.65	41.50
0270	2"		40	.200		26	13.10		39.10	48.50
3000	Fittings									
3010	Straight									
3100	Tube to NPT									
3110	3/8"	1 Stpi	29	.276	Ea.	13.35	18.10		31.45	41.50
3120	1/2"		27	.296		14.35	19.40		33.75	45
3130	3/4"		25	.320		19.60	21		40.60	53
3140	1"		23	.348		30.50	23		53.50	67.50
3150	1-1/4"		20	.400		68	26		94	114
3160	1-1/2"		17	.471		138	31		169	198
3170	2"		15	.533		235	35		270	310
3200	Coupling									
3210	3/8"	1 Stpi	29	.276	Ea.	23	18.10		41.10	52.50
3220	1/2"		27	.296		26.50	19.40		45.90	58
3230	3/4"		25	.320		36.50	21		57.50	71.50

For customer support on your Mechanical Costs with RSMeans Data, call 800.448.8182.

277

23 11 Facility Fuel Piping

23 11 23 – Facility Natural-Gas Piping

23 11 23.20 Gas Piping, Flexible (Csst)	Crew	Daily Output	Labor-Hours	Unit	Material	2020 Bare Costs Labor	Equipment	Total	Total Incl O&P
3240	1"	1 Stpi	23	.348	Ea.	56.50	23		79.50	96
3250	1-1/4"		20	.400		125	26		151	177
3260	1-1/2"		17	.471		262	31		293	335
3270	2"		15	.533		445	35		480	545
3300	Flange fitting									
3310	3/8"	1 Stpi	25	.320	Ea.	20.50	21		41.50	54
3320	1/2"		22	.364		19.50	24		43.50	57
3330	3/4"		19	.421		25.50	27.50		53	69.50
3340	1"		16	.500		35	33		68	87.50
3350	1-1/4"		12	.667		81	43.50		124.50	155
3400	90° flange valve									
3410	3/8"	1 Stpi	25	.320	Ea.	39	21		60	74.50
3420	1/2"		22	.364		39.50	24		63.50	79
3430	3/4"		19	.421		48.50	27.50		76	95
4000	Tee									
4120	1/2"	1 Stpi	20.50	.390	Ea.	41.50	25.50		67	84.50
4130	3/4"		19	.421		55.50	27.50		83	103
4140	1"		17.50	.457		99.50	30		129.50	155
5000	Reducing									
5110	Tube to NPT									
5120	3/4" to 1/2" NPT	1 Stpi	26	.308	Ea.	21	20		41	53
5130	1" to 3/4" NPT	"	24	.333	"	32.50	22		54.50	68.50
5200	Reducing tee									
5210	1/2" x 3/8" x 3/8"	1 Stpi	21	.381	Ea.	55	25		80	98
5220	3/4" x 1/2" x 1/2"		20	.400		51.50	26		77.50	96
5230	1" x 3/4" x 1/2"		18	.444		85.50	29		114.50	138
5240	1-1/4" x 1-1/4" x 1"		15.60	.513		228	33.50		261.50	300
5250	1-1/2" x 1-1/2" x 1-1/4"		13.30	.602		350	39.50		389.50	445
5260	2" x 2" x 1-1/2"		11.80	.678		525	44.50		569.50	640
5300	Manifold with four ports and mounting bracket									
5302	Labor to mount manifold does not include making pipe									
5304	connections which are included in fitting labor.									
5310	3/4" x 1/2" x 1/2" (4)	1 Stpi	76	.105	Ea.	42	6.90		48.90	57
5330	1-1/4" x 1" x 3/4" (4)		72	.111		60.50	7.30		67.80	77.50
5350	2" x 1-1/2" x 1" (4)		68	.118		76	7.70		83.70	95
5600	Protective striker plate									
5610	Quarter plate, 3" x 2"	1 Stpi	88	.091	Ea.	.93	5.95		6.88	9.90
5620	Half plate, 3" x 7"		82	.098		2.30	6.40		8.70	12.10
5630	Full plate, 3" x 12"		78	.103		3.92	6.70		10.62	14.35

23 12 Facility Fuel Pumps

23 12 13 – Facility Fuel-Oil Pumps

23 12 13.10 Pump and Motor Sets

		Crew	Daily Output	Labor-Hours	Unit	Material	2020 Bare Costs Labor	Equipment	Total	Total Incl O&P
0010	**PUMP AND MOTOR SETS**									
1810	Light fuel and diesel oils									
1820	20 GPH, 1/3 HP	Q-5	6	2.667	Ea.	1,600	157		1,757	2,000
1830	27 GPH, 1/3 HP		6	2.667		1,600	157		1,757	2,000
1840	80 GPH, 1/3 HP		5	3.200		1,600	189		1,789	2,025
1850	145 GPH, 1/2 HP		4	4		1,700	236		1,936	2,225
1860	277 GPH, 1 HP		4	4		1,900	236		2,136	2,425
1870	700 GPH, 1-1/2 HP		3	5.333		4,400	315		4,715	5,325

23 12 Facility Fuel Pumps

23 12 13 – Facility Fuel-Oil Pumps

23 12 13.10 Pump and Motor Sets	Crew	Daily Output	Labor-Hours	Unit	Material	2020 Bare Costs Labor	Equipment	Total	Total Incl O&P
1880	1000 GPH, 2 HP	Q-5	3	5.333	Ea.	5,000	315		5,315	5,975
1890	1800 GPH, 5 HP	↓	1.80	8.889	↓	8,100	525		8,625	9,675

23 13 Facility Fuel-Storage Tanks

23 13 13 – Facility Underground Fuel-Oil, Storage Tanks

23 13 13.09 Single-Wall Steel Fuel-Oil Tanks

		Crew	Daily Output	Labor-Hours	Unit	Material	2020 Bare Costs Labor	Equipment	Total	Total Incl O&P
0010	**SINGLE-WALL STEEL FUEL-OIL TANKS**									
5000	Tanks, steel ugnd., sti-p3, not incl. hold-down bars									
5500	Excavation, pad, pumps and piping not included									
5510	Single wall, 500 gallon capacity, 7 ga. shell	Q-5	2.70	5.926	Ea.	1,875	350		2,225	2,600
5520	1,000 gallon capacity, 7 ga. shell	"	2.50	6.400		2,925	380		3,305	3,800
5530	2,000 gallon capacity, 1/4" thick shell	Q-7	4.60	6.957		3,725	435		4,160	4,750
5535	2,500 gallon capacity, 7 ga. shell	Q-5	3	5.333		5,675	315		5,990	6,700
5540	5,000 gallon capacity, 1/4" thick shell	Q-7	3.20	10		13,400	625		14,025	15,600
5560	10,000 gallon capacity, 1/4" thick shell		2	16		10,800	1,000		11,800	13,400
5580	15,000 gallon capacity, 5/16" thick shell		1.70	18.824		11,200	1,175		12,375	14,100
5600	20,000 gallon capacity, 5/16" thick shell		1.50	21.333		22,600	1,325		23,925	26,900
5610	25,000 gallon capacity, 3/8" thick shell		1.30	24.615		29,600	1,525		31,125	34,900
5620	30,000 gallon capacity, 3/8" thick shell		1.10	29.091		35,000	1,825		36,825	41,200
5630	40,000 gallon capacity, 3/8" thick shell		.90	35.556		44,500	2,225		46,725	52,500
5640	50,000 gallon capacity, 3/8" thick shell	↓	.80	40	↓	49,800	2,500		52,300	58,000

23 13 13.13 Dbl-Wall Steel, Undrgrnd Fuel-Oil, Stor. Tanks

		Crew	Daily Output	Labor-Hours	Unit	Material	2020 Bare Costs Labor	Equipment	Total	Total Incl O&P
0010	**DOUBLE-WALL STEEL, UNDERGROUND FUEL-OIL, STORAGE TANKS**									
6200	Steel, underground, 360°, double wall, UL listed,									
6210	with sti-P3 corrosion protection,									
6220	(dielectric coating, cathodic protection, electrical									
6230	isolation) 30 year warranty,									
6240	not incl. manholes or hold-downs.									
6250	500 gallon capacity	Q-5	2.40	6.667	Ea.	3,325	395		3,720	4,250
6260	1,000 gallon capacity	"	2.25	7.111		3,675	420		4,095	4,650
6270	2,000 gallon capacity	Q-7	4.16	7.692		4,850	480		5,330	6,050
6280	3,000 gallon capacity		3.90	8.205		6,950	510		7,460	8,425
6290	4,000 gallon capacity		3.64	8.791		8,200	550		8,750	9,850
6300	5,000 gallon capacity		2.91	10.997		8,800	685		9,485	10,700
6310	6,000 gallon capacity		2.42	13.223		11,600	825		12,425	14,000
6320	8,000 gallon capacity		2.08	15.385		11,200	960		12,160	13,700
6330	10,000 gallon capacity		1.82	17.582		13,900	1,100		15,000	17,000
6340	12,000 gallon capacity		1.70	18.824		16,600	1,175		17,775	20,100
6350	15,000 gallon capacity		1.33	24.060		20,700	1,500		22,200	25,100
6360	20,000 gallon capacity		1.33	24.060		34,200	1,500		35,700	40,000
6370	25,000 gallon capacity		1.16	27.586		38,100	1,725		39,825	44,500
6380	30,000 gallon capacity		1.03	31.068		37,800	1,950		39,750	44,500
6390	40,000 gallon capacity		.80	40	↓	115,000	2,500		117,500	129,500
6395	50,000 gallon capacity		.73	43.836	↓	139,500	2,725		142,225	157,500
6400	For hold-downs 500-2,000 gal., add		16	2	Set	226	125		351	435
6410	For hold-downs 3,000-6,000 gal., add		12	2.667		385	166		551	670
6420	For hold-downs 8,000-12,000 gal., add		11	2.909		455	182		637	770
6430	For hold-downs 15,000 gal., add		9	3.556		675	222		897	1,075
6440	For hold-downs 20,000 gal., add		8	4		780	250		1,030	1,225
6450	For hold-downs 20,000 gal. plus, add	↓	6	5.333	↓	1,000	335		1,335	1,600
6500	For manways, add				Ea.	1,050			1,050	1,175

For customer support on your Mechanical Costs with RSMeans Data, call 800.448.8182.

279

23 13 Facility Fuel-Storage Tanks

23 13 13 – Facility Underground Fuel-Oil, Storage Tanks

23 13 13.13 Dbl-Wall Steel, Undrgrnd Fuel-Oil, Stor. Tanks		Crew	Daily Output	Labor-Hours	Unit	Material	2020 Bare Costs Labor	Equipment	Total	Total Incl O&P
6600	In place with hold-downs									
6652	550 gallon capacity	Q-5	1.84	8.696	Ea.	3,550	515		4,065	4,675
0500	Horizontal propane tank, 500 gal., above ground	"	5	3.200	"	8,975	189		9,164	10,200

23 13 13.23 Glass-Fiber-Reinfcd-Plastic, Fuel-Oil, Storage

		Crew	Daily Output	Labor-Hours	Unit	Material	2020 Bare Costs Labor	Equipment	Total	Total Incl O&P
0010	**GLASS-FIBER-REINFCD-PLASTIC, UNDERGRND. FUEL-OIL, STORAGE**									
0210	Fiberglass, underground, single wall, UL listed, not including									
0220	manway or hold-down strap									
0225	550 gallon capacity	Q-5	2.67	5.993	Ea.	4,075	355		4,430	5,025
0230	1,000 gallon capacity	"	2.46	6.504		5,275	385		5,660	6,375
0240	2,000 gallon capacity	Q-7	4.57	7.002		8,275	435		8,710	9,750
0245	3,000 gallon capacity		3.90	8.205		8,400	510		8,910	10,000
0250	4,000 gallon capacity		3.55	9.014		11,100	560		11,660	13,000
0255	5,000 gallon capacity		3.20	10		10,400	625		11,025	12,300
0260	6,000 gallon capacity		2.67	11.985		11,800	750		12,550	14,100
0270	8,000 gallon capacity		2.29	13.974		14,900	870		15,770	17,700
0280	10,000 gallon capacity		2	16		15,600	1,000		16,600	18,700
0282	12,000 gallon capacity		1.88	17.021		20,900	1,050		21,950	24,600
0284	15,000 gallon capacity		1.68	19.048		27,900	1,200		29,100	32,500
0290	20,000 gallon capacity		1.45	22.069		33,200	1,375		34,575	38,600
0300	25,000 gallon capacity		1.28	25		58,500	1,550		60,050	67,000
0320	30,000 gallon capacity		1.14	28.070		104,500	1,750		106,250	117,500
0340	40,000 gallon capacity		.89	35.955		114,000	2,250		116,250	129,000
0360	48,000 gallon capacity		.81	39.506		177,000	2,475		179,475	198,000
0500	For manway, fittings and hold-downs, add					20%	15%			
0600	For manways, add					2,525			2,525	2,775
1000	For helical heating coil, add	Q-5	2.50	6.400		5,450	380		5,830	6,575
1020	Fiberglass, underground, double wall, UL listed									
1030	includes manways, not incl. hold-down straps									
1040	600 gallon capacity	Q-5	2.42	6.612	Ea.	9,100	390		9,490	10,600
1050	1,000 gallon capacity	"	2.25	7.111		12,400	420		12,820	14,300
1060	2,500 gallon capacity	Q-7	4.16	7.692		17,300	480		17,780	19,700
1070	3,000 gallon capacity		3.90	8.205		19,400	510		19,910	22,100
1080	4,000 gallon capacity		3.64	8.791		19,600	550		20,150	22,300
1090	6,000 gallon capacity		2.42	13.223		26,000	825		26,825	29,800
1100	8,000 gallon capacity		2.08	15.385		33,000	960		33,960	37,700
1110	10,000 gallon capacity		1.82	17.582		38,800	1,100		39,900	44,300
1120	12,000 gallon capacity		1.70	18.824		48,300	1,175		49,475	55,000
1122	15,000 gallon capacity		1.52	21.053		67,000	1,325		68,325	76,000
1124	20,000 gallon capacity		1.33	24.060		79,500	1,500		81,000	90,000
1126	25,000 gallon capacity		1.16	27.586		92,000	1,725		93,725	103,500
1128	30,000 gallon capacity		1.03	31.068		110,500	1,950		112,450	124,500
1140	For hold-down straps, add					2%	10%			
1150	For hold-downs 500-4,000 gal., add	Q-7	16	2	Set	505	125		630	740
1160	For hold-downs 5,000-15,000 gal., add		8	4		1,000	250		1,250	1,475
1170	For hold-downs 20,000 gal., add		5.33	6.004		1,500	375		1,875	2,200
1180	For hold-downs 25,000 gal., add		4	8		2,025	500		2,525	2,975
1190	For hold-downs 30,000 gal., add		2.60	12.308		3,025	770		3,795	4,475
2210	Fiberglass, underground, single wall, UL listed, including									
2220	hold-down straps, no manways									
2225	550 gallon capacity	Q-5	2	8	Ea.	4,600	470		5,070	5,750
2230	1,000 gallon capacity	"	1.88	8.511		5,775	500		6,275	7,100
2240	2,000 gallon capacity	Q-7	3.55	9.014		8,775	560		9,335	10,500

23 13 Facility Fuel-Storage Tanks

23 13 13 – Facility Underground Fuel-Oil, Storage Tanks

23 13 13.23 Glass-Fiber-Reinfcd-Plastic, Fuel-Oil, Storage

		Crew	Daily Output	Labor-Hours	Unit	Material	2020 Bare Costs Labor	Equipment	Total	Total Incl O&P
2250	4,000 gallon capacity	Q-7	2.90	11.034	Ea.	11,600	690		12,290	13,800
2260	6,000 gallon capacity		2	16		12,800	1,000		13,800	15,600
2270	8,000 gallon capacity		1.78	17.978		15,900	1,125		17,025	19,200
2280	10,000 gallon capacity		1.60	20		16,600	1,250		17,850	20,200
2282	12,000 gallon capacity		1.52	21.053		21,900	1,325		23,225	26,100
2284	15,000 gallon capacity		1.39	23.022		28,900	1,425		30,325	34,000
2290	20,000 gallon capacity		1.14	28.070		34,700	1,750		36,450	40,800
2300	25,000 gallon capacity		.96	33.333		60,500	2,075		62,575	69,500
2320	30,000 gallon capacity	▼	.80	40	▼	107,500	2,500		110,000	121,500
3020	Fiberglass, underground, double wall, UL listed									
3030	includes manways and hold-down straps									
3040	600 gallon capacity	Q-5	1.86	8.602	Ea.	9,600	510		10,110	11,400
3050	1,000 gallon capacity	"	1.70	9.412		12,900	555		13,455	15,000
3060	2,500 gallon capacity	Q-7	3.29	9.726		17,800	605		18,405	20,500
3070	3,000 gallon capacity		3.13	10.224		19,900	640		20,540	22,800
3080	4,000 gallon capacity		2.93	10.922		20,100	680		20,780	23,100
3090	6,000 gallon capacity		1.86	17.204		27,000	1,075		28,075	31,300
3100	8,000 gallon capacity		1.65	19.394		34,000	1,200		35,200	39,200
3110	10,000 gallon capacity		1.48	21.622		39,800	1,350		41,150	45,800
3120	12,000 gallon capacity		1.40	22.857		49,300	1,425		50,725	56,000
3122	15,000 gallon capacity		1.28	25		68,000	1,550		69,550	77,500
3124	20,000 gallon capacity		1.06	30.189		81,000	1,875		82,875	92,000
3126	25,000 gallon capacity		.90	35.556		94,000	2,225		96,225	107,000
3128	30,000 gallon capacity	▼	.74	43.243	▼	113,500	2,700		116,200	129,000

23 13 23 – Facility Aboveground Fuel-Oil, Storage Tanks

23 13 23.16 Horizontal, Stl, Abvgrd Fuel-Oil, Storage Tanks

		Crew	Daily Output	Labor-Hours	Unit	Material	2020 Bare Costs Labor	Equipment	Total	Total Incl O&P
0010	**HORIZONTAL, STEEL, ABOVEGROUND FUEL-OIL, STORAGE TANKS**									
3000	Steel, storage, aboveground, including cradles, coating,									
3020	fittings, not including foundation, pumps or piping									
3040	Single wall, 275 gallon	Q-5	5	3.200	Ea.	510	189		699	845
3060	550 gallon	"	2.70	5.926		4,400	350		4,750	5,375
3080	1,000 gallon	Q-7	5	6.400		7,175	400		7,575	8,475
3100	1,500 gallon		4.75	6.737		10,100	420		10,520	11,700
3120	2,000 gallon		4.60	6.957		13,500	435		13,935	15,600
3140	5,000 gallon		3.20	10		21,400	625		22,025	24,500
3150	10,000 gallon		2	16		40,800	1,000		41,800	46,400
3160	15,000 gallon		1.70	18.824		57,000	1,175		58,175	65,000
3170	20,000 gallon		1.45	22.069		74,000	1,375		75,375	83,500
3180	25,000 gallon		1.30	24.615		84,500	1,525		86,025	95,500
3190	30,000 gallon	▼	1.10	29.091		104,000	1,825		105,825	117,000
3320	Double wall, 500 gallon capacity	Q-5	2.40	6.667		2,000	395		2,395	2,800
3330	2,000 gallon capacity	Q-7	4.15	7.711		6,650	480		7,130	8,050
3340	4,000 gallon capacity		3.60	8.889		14,500	555		15,055	16,700
3350	6,000 gallon capacity		2.40	13.333		16,400	830		17,230	19,400
3360	8,000 gallon capacity		2	16		19,500	1,000		20,500	23,000
3370	10,000 gallon capacity		1.80	17.778		30,500	1,100		31,600	35,200
3380	15,000 gallon capacity		1.50	21.333		41,100	1,325		42,425	47,200
3390	20,000 gallon capacity		1.30	24.615		48,400	1,525		49,925	55,500
3400	25,000 gallon capacity		1.15	27.826		60,000	1,725		61,725	68,500
3410	30,000 gallon capacity	▼	1	32		67,000	2,000		69,000	76,500

For customer support on your Mechanical Costs with RSMeans Data, call 800.448.8182.

281

23 13 Facility Fuel-Storage Tanks

23 13 23 – Facility Aboveground Fuel-Oil, Storage Tanks

23 13 23.26 Horizontal, Conc., Abvgrd Fuel-Oil, Stor. Tanks	Crew	Daily Output	Labor-Hours	Unit	Material	2020 Bare Costs Labor	Equipment	Total	Total Incl O&P
0010	**HORIZONTAL, CONCRETE, ABOVEGROUND FUEL-OIL, STORAGE TANKS**								
0050	Concrete, storage, aboveground, including pad & pump								
0100	500 gallon	F-3	2	20	Ea.	10,500	1,075	236	11,811	13,400
0200	1,000 gallon	"	2	20		14,700	1,075	236	16,011	18,000
0300	2,000 gallon	F-4	2	24		18,900	1,300	490	20,690	23,300
0400	4,000 gallon		2	24		24,100	1,300	490	25,890	29,000
0500	8,000 gallon		2	24		37,700	1,300	490	39,490	44,000
0600	12,000 gallon	↓	2	24	↓	50,500	1,300	490	52,290	58,000

23 21 Hydronic Piping and Pumps

23 21 20 – Hydronic HVAC Piping Specialties

23 21 20.10 Air Control

		Crew	Daily Output	Labor-Hours	Unit	Material	2020 Bare Costs Labor	Equipment	Total	Total Incl O&P
0010	**AIR CONTROL**									
0030	Air separator, with strainer									
0040	2" diameter	Q-5	6	2.667	Ea.	1,575	157		1,732	1,950
0080	2-1/2" diameter		5	3.200		1,750	189		1,939	2,200
0100	3" diameter		4	4		2,700	236		2,936	3,325
0120	4" diameter	↓	3	5.333		3,900	315		4,215	4,750
0130	5" diameter	Q-6	3.60	6.667		4,950	410		5,360	6,050
0140	6" diameter		3.40	7.059		5,950	430		6,380	7,200
0160	8" diameter		3	8		8,875	490		9,365	10,500
0180	10" diameter		2.20	10.909		13,700	665		14,365	16,100
0200	12" diameter		1.70	14.118		22,800	865		23,665	26,400
0210	14" diameter		1.30	18.462		27,200	1,125		28,325	31,600
0220	16" diameter		1	24		40,600	1,475		42,075	46,800
0230	18" diameter		.80	30		53,500	1,825		55,325	61,500
0240	20" diameter	↓	.60	40	↓	59,500	2,450		61,950	69,000
0300	Without strainer									
0310	2" diameter	Q-5	6	2.667	Ea.	970	157		1,127	1,300
0320	2-1/2" diameter		5	3.200		1,150	189		1,339	1,550
0330	3" diameter		4	4		1,625	236		1,861	2,125
0340	4" diameter		3	5.333		2,550	315		2,865	3,300
0350	5" diameter	↓	2.40	6.667		3,450	395		3,845	4,400
0360	6" diameter	Q-6	3.40	7.059		3,950	430		4,380	5,000
0370	8" diameter		3	8		5,450	490		5,940	6,725
0380	10" diameter		2.20	10.909		8,025	665		8,690	9,825
0390	12" diameter		1.70	14.118		12,300	865		13,165	14,900
0400	14" diameter		1.30	18.462		18,500	1,125		19,625	22,000
0410	16" diameter		1	24		25,800	1,475		27,275	30,600
0420	18" diameter		.80	30		33,700	1,825		35,525	39,800
0430	20" diameter	↓	.60	40	↓	40,400	2,450		42,850	48,100
1000	Micro-bubble separator for total air removal, closed loop system									
1010	Requires bladder type tank in system.									
1020	Water (hot or chilled) or glycol system									
1030	Threaded									
1040	3/4" diameter	1 Stpi	20	.400	Ea.	89.50	26		115.50	138
1050	1" diameter		19	.421		104	27.50		131.50	156
1060	1-1/4" diameter		16	.500		139	33		172	202
1070	1-1/2" diameter		13	.615		182	40.50		222.50	261
1080	2" diameter	↓	11	.727	↓	1,225	47.50		1,272.50	1,425

23 21 Hydronic Piping and Pumps

23 21 20 – Hydronic HVAC Piping Specialties

23 21 20.10 Air Control

		Crew	Daily Output	Labor-Hours	Unit	Material	2020 Bare Costs Labor	2020 Bare Costs Equipment	Total	Total Incl O&P
1090	2-1/2" diameter	Q-5	15	1.067	Ea.	1,475	63		1,538	1,725
1100	3" diameter		13	1.231		1,975	72.50		2,047.50	2,275
1110	4" diameter	▼	10	1.600	▼	2,175	94.50		2,269.50	2,550
1230	Flanged									
1250	2" diameter	1 Stpi	8	1	Ea.	1,700	65.50		1,765.50	1,975
1260	2-1/2" diameter	Q-5	5	3.200		1,850	189		2,039	2,300
1270	3" diameter		4.50	3.556		2,500	210		2,710	3,075
1280	4" diameter	▼	3	5.333		2,700	315		3,015	3,450
1290	5" diameter	Q-6	3.40	7.059		4,075	430		4,505	5,150
1300	6" diameter	"	3	8	▼	5,450	490		5,940	6,725
1400	Larger sizes available									
1590	With extended tank dirt catcher									
1600	Flanged									
1620	2" diameter	1 Stpi	7.40	1.081	Ea.	3,850	71		3,921	4,325
1630	2-1/2" diameter	Q-5	4.20	3.810		4,175	225		4,400	4,925
1640	3" diameter		4	4		5,575	236		5,811	6,475
1650	4" diameter	▼	2.80	5.714		6,175	335		6,510	7,275
1660	5" diameter	Q-6	3.10	7.742		10,200	475		10,675	11,900
1670	6" diameter	"	2.90	8.276	▼	12,600	505		13,105	14,600
1800	With drain/dismantle/cleaning access flange									
1810	Flanged									
1820	2" diameter	Q-5	4.60	3.478	Ea.	3,125	205		3,330	3,750
1830	2-1/2" diameter		4	4		3,375	236		3,611	4,075
1840	3" diameter		3.60	4.444		4,550	262		4,812	5,425
1850	4" diameter	▼	2.50	6.400		5,225	380		5,605	6,300
1860	5" diameter	Q-6	2.80	8.571		8,700	525		9,225	10,400
1870	6" diameter	"	2.60	9.231	▼	11,200	565		11,765	13,100
2000	Boiler air fitting, separator									
2010	1-1/4"	Q-5	22	.727	Ea.	267	43		310	355
2020	1-1/2"		20	.800		515	47		562	640
2021	2"	▼	18	.889	▼	890	52.50		942.50	1,050
2400	Compression tank air fitting									
2410	For tanks 9" to 24" diameter	1 Stpi	15	.533	Ea.	100	35		135	163
2420	For tanks 100 gallon or larger	"	12	.667	"	385	43.50		428.50	490

23 21 20.14 Air Purging Scoop

		Crew	Daily Output	Labor-Hours	Unit	Material	2020 Bare Costs Labor	2020 Bare Costs Equipment	Total	Total Incl O&P
0010	**AIR PURGING SCOOP**, With tappings.									
0020	For air vent and expansion tank connection									
0100	1" pipe size, threaded	1 Stpi	19	.421	Ea.	29.50	27.50		57	73.50
0110	1-1/4" pipe size, threaded		15	.533		29.50	35		64.50	84.50
0120	1-1/2" pipe size, threaded		13	.615		64.50	40.50		105	132
0130	2" pipe size, threaded	▼	11	.727		74	47.50		121.50	153
0140	2-1/2" pipe size, threaded	Q-5	15	1.067		160	63		223	270
0150	3" pipe size, threaded		13	1.231		208	72.50		280.50	340
0160	4" 150 lb. flanges	▼	3	5.333	▼	465	315		780	980

23 21 20.18 Automatic Air Vent

		Crew	Daily Output	Labor-Hours	Unit	Material	2020 Bare Costs Labor	2020 Bare Costs Equipment	Total	Total Incl O&P
0010	**AUTOMATIC AIR VENT**									
0020	Cast iron body, stainless steel internals, float type									
0060	1/2" NPT inlet, 300 psi	1 Stpi	12	.667	Ea.	180	43.50		223.50	264
0140	3/4" NPT inlet, 300 psi		12	.667		180	43.50		223.50	264
0180	1/2" NPT inlet, 250 psi		10	.800		375	52.50		427.50	495
0220	3/4" NPT inlet, 250 psi		10	.800		375	52.50		427.50	495
0260	1" NPT inlet, 250 psi	▼	10	.800	▼	550	52.50		602.50	685

For customer support on your Mechanical Costs with RSMeans Data, call 800.448.8182.

283

23 21 Hydronic Piping and Pumps

23 21 20 – Hydronic HVAC Piping Specialties

23 21 20.18 Automatic Air Vent

		Crew	Daily Output	Labor-Hours	Unit	Material	2020 Bare Costs Labor	Equipment	Total	Total Incl O&P
0340	1-1/2" NPT inlet, 250 psi	Q-5	12	1.333	Ea.	1,175	78.50		1,253.50	1,400
0380	2" NPT inlet, 250 psi	"	12	1.333		1,175	78.50		1,253.50	1,400
0600	Forged steel body, stainless steel internals, float type									
0640	1/2" NPT inlet, 750 psi	1 Stpi	12	.667	Ea.	1,250	43.50		1,293.50	1,450
0680	3/4" NPT inlet, 750 psi		12	.667		1,250	43.50		1,293.50	1,450
0760	3/4" NPT inlet, 1,000 psi		10	.800		1,875	52.50		1,927.50	2,125
0800	1" NPT inlet, 1,000 psi	Q-5	12	1.333		1,875	78.50		1,953.50	2,175
0880	1-1/2" NPT inlet, 1,000 psi		10	1.600		5,250	94.50		5,344.50	5,925
0920	2" NPT inlet, 1,000 psi		10	1.600		5,250	94.50		5,344.50	5,925
1100	Formed steel body, noncorrosive									
1110	1/8" NPT inlet, 150 psi	1 Stpi	32	.250	Ea.	15.70	16.40		32.10	42
1120	1/4" NPT inlet, 150 psi		32	.250		53.50	16.40		69.90	83.50
1130	3/4" NPT inlet, 150 psi		32	.250		53.50	16.40		69.90	83.50
1300	Chrome plated brass, automatic/manual, for radiators									
1310	1/8" NPT inlet, nickel plated brass	1 Stpi	32	.250	Ea.	9.25	16.40		25.65	34.50

23 21 20.22 Circuit Sensor

		Crew	Daily Output	Labor-Hours	Unit	Material	2020 Bare Costs Labor	Equipment	Total	Total Incl O&P
0010	**CIRCUIT SENSOR**, Flow meter									
0020	Metering stations									
0040	Wafer orifice insert type									
0060	2-1/2" pipe size	Q-5	12	1.333	Ea.	300	78.50		378.50	450
0100	3" pipe size		11	1.455		340	86		426	505
0140	4" pipe size		8	2		395	118		513	610
0180	5" pipe size		7.30	2.192		495	129		624	740
0220	6" pipe size		6.40	2.500		590	148		738	870
0260	8" pipe size	Q-6	5.30	4.528		815	277		1,092	1,325
0280	10" pipe size		4.60	5.217		945	320		1,265	1,525
0360	12" pipe size		4.20	5.714		1,525	350		1,875	2,200

23 21 20.26 Circuit Setter

		Crew	Daily Output	Labor-Hours	Unit	Material	2020 Bare Costs Labor	Equipment	Total	Total Incl O&P
0010	**CIRCUIT SETTER**, Balance valve									
0012	Bronze body, soldered									
0013	1/2" size	1 Stpi	24	.333	Ea.	81	22		103	122
0014	3/4" size	"	20	.400	"	86	26		112	134
0018	Threaded									
0019	1/2" pipe size	1 Stpi	22	.364	Ea.	84.50	24		108.50	129
0020	3/4" pipe size		20	.400		90	26		116	138
0040	1" pipe size		18	.444		117	29		146	172
0050	1-1/4" pipe size		15	.533		171	35		206	242
0060	1-1/2" pipe size		12	.667		202	43.50		245.50	288
0080	2" pipe size		10	.800		290	52.50		342.50	400
0100	2-1/2" pipe size	Q-5	15	1.067		655	63		718	815
0120	3" pipe size	"	10	1.600		920	94.50		1,014.50	1,175
0130	Cast iron body, flanged									
0136	3" pipe size, flanged	Q-5	4	4	Ea.	925	236		1,161	1,375
0140	4" pipe size	"	3	5.333		1,375	315		1,690	2,000
0200	For differential meter, accurate to 1%, add					915			915	1,000
0300	For differential meter 400 psi/200°F continuous, add					2,500			2,500	2,750

23 21 20.30 Cocks, Drains and Specialties

		Crew	Daily Output	Labor-Hours	Unit	Material	2020 Bare Costs Labor	Equipment	Total	Total Incl O&P
0010	**COCKS, DRAINS AND SPECIALTIES**									
1000	Boiler drain									
1010	Pipe thread to hose									
1020	Bronze									
1030	1/2" size	1 Stpi	36	.222	Ea.	15.25	14.55		29.80	39

23 21 20.30 Cocks, Drains and Specialties		Crew	Daily Output	Labor-Hours	Unit	Material	2020 Bare Costs Labor	Equipment	Total	Total Incl O&P
1040	3/4" size	1 Stpi	34	.235	Ea.	16.50	15.40		31.90	41
1100	Solder to hose									
1110	Bronze									
1120	1/2" size	1 Stpi	46	.174	Ea.	15.25	11.40		26.65	34
1130	3/4" size	"	44	.182	"	16.60	11.90		28.50	36
1600	With built-in vacuum breaker									
1610	1/2" I.P. or solder	1 Stpi	36	.222	Ea.	38	14.55		52.55	63.50
1630	With tamper proof vacuum breaker									
1640	1/2" I.P. or solder	1 Stpi	36	.222	Ea.	51.50	14.55		66.05	79
1650	3/4" I.P. or solder	"	34	.235	"	56.50	15.40		71.90	85
3000	Cocks									
3010	Air, lever or tee handle									
3020	Bronze, single thread									
3030	1/8" size	1 Stpi	52	.154	Ea.	12.90	10.10		23	29.50
3040	1/4" size		46	.174		13.40	11.40		24.80	32
3050	3/8" size		40	.200		14	13.10		27.10	35
3060	1/2" size	↓	36	.222	↓	16.50	14.55		31.05	40
3100	Bronze, double thread									
3110	1/8" size	1 Stpi	26	.308	Ea.	16.20	20		36.20	48
3120	1/4" size		22	.364		16.85	24		40.85	54
3130	3/8" size		18	.444		17.85	29		46.85	63
3140	1/2" size	↓	15	.533	↓	22	35		57	76.50
4000	Steam									
4010	Bronze									
4020	1/8" size	1 Stpi	26	.308	Ea.	24	20		44	56.50
4030	1/4" size		22	.364		25.50	24		49.50	63.50
4040	3/8" size		18	.444		27	29		56	73
4050	1/2" size		15	.533		27	35		62	82
4060	3/4" size		14	.571		33.50	37.50		71	93
4070	1" size		13	.615		48.50	40.50		89	114
4080	1-1/4" size		11	.727		66.50	47.50		114	145
4090	1-1/2" size	↓	10	.800		102	52.50		154.50	191
4100	2" size		9	.889	↓	145	58.50		203.50	247
4300	Bronze, 3 way									
4320	1/4" size	1 Stpi	15	.533	Ea.	52.50	35		87.50	110
4330	3/8" size		12	.667		52.50	43.50		96	123
4340	1/2" size		10	.800		46.50	52.50		99	130
4350	3/4" size		9.50	.842		80	55		135	171
4360	1" size		8.50	.941		96.50	61.50		158	199
4370	1-1/4" size		7.50	1.067		144	70		214	264
4380	1-1/2" size		6.50	1.231		168	80.50		248.50	305
4390	2" size	↓	6	1.333	↓	242	87.50		329.50	395
4500	Gauge cock, brass									
4510	1/4" FPT	1 Stpi	24	.333	Ea.	13.60	22		35.60	47.50
4512	1/4" MPT	"	24	.333	"	17.10	22		39.10	51.50
4600	Pigtail, steam syphon									
4604	1/4"	1 Stpi	24	.333	Ea.	24.50	22		46.50	59.50
4650	Snubber valve									
4654	1/4"	1 Stpi	22	.364	Ea.	10.25	24		34.25	47
4660	Nipple, black steel									
4664	1/4" x 3"	1 Stpi	37	.216	Ea.	3.51	14.15		17.66	25

For customer support on your Mechanical Costs with RSMeans Data, call 800.448.8182.

285

23 21 20.34 Dielectric Unions	Crew	Daily Output	Labor-Hours	Unit	Material	2020 Bare Costs Labor	Equipment	Total	Total Incl O&P
0010 **DIELECTRIC UNIONS**, Standard gaskets for water and air									
0020 250 psi maximum pressure									
0280 Female IPT to sweat, straight									
0300 1/2" pipe size	1 Plum	24	.333	Ea.	8.40	21.50		29.90	41.50
0340 3/4" pipe size		20	.400		9.40	26		35.40	49
0360 1" pipe size		19	.421		12.65	27		39.65	54.50
0380 1-1/4" pipe size		15	.533		17.35	34.50		51.85	70.50
0400 1-1/2" pipe size		13	.615		29.50	39.50		69	92
0420 2" pipe size		11	.727		40	47		87	114
0580 Female IPT to brass pipe thread, straight									
0600 1/2" pipe size	1 Plum	24	.333	Ea.	18.10	21.50		39.60	52
0640 3/4" pipe size		20	.400		19.85	26		45.85	60.50
0660 1" pipe size		19	.421		37	27		64	81
0680 1-1/4" pipe size		15	.533		37.50	34.50		72	92.50
0700 1-1/2" pipe size		13	.615		54.50	39.50		94	120
0720 2" pipe size		11	.727		110	47		157	191
0780 Female IPT to female IPT, straight									
0800 1/2" pipe size	1 Plum	24	.333	Ea.	15.90	21.50		37.40	49.50
0840 3/4" pipe size		20	.400		18.25	26		44.25	58.50
0860 1" pipe size		19	.421		24	27		51	66.50
0880 1-1/4" pipe size		15	.533		31.50	34.50		66	86.50
0900 1-1/2" pipe size		13	.615		63.50	39.50		103	129
0920 2" pipe size		11	.727		94	47		141	173
2000 175 psi maximum pressure									
2180 Female IPT to sweat									
2240 2" pipe size	1 Plum	9	.889	Ea.	188	57.50		245.50	292
2260 2-1/2" pipe size	Q-1	15	1.067		222	62		284	335
2280 3" pipe size		14	1.143		305	66.50		371.50	435
2300 4" pipe size		11	1.455		805	84.50		889.50	1,000
2480 Female IPT to brass pipe									
2500 1-1/2" pipe size	1 Plum	11	.727	Ea.	218	47		265	310
2540 2" pipe size	"	9	.889		229	57.50		286.50	340
2560 2-1/2" pipe size	Q-1	15	1.067		320	62		382	445
2580 3" pipe size		14	1.143		330	66.50		396.50	460
2600 4" pipe size		11	1.455		470	84.50		554.50	645

23 21 20.38 Expansion Couplings

23 21 20.38 Expansion Couplings	Crew	Daily Output	Labor-Hours	Unit	Material	Labor	Equipment	Total	Total Incl O&P
0010 **EXPANSION COUPLINGS**, Hydronic									
0100 Copper to copper, sweat									
1000 Baseboard riser fitting, 5" stub by coupling 12" long									
1020 1/2" diameter	1 Stpi	24	.333	Ea.	23.50	22		45.50	58.50
1040 3/4" diameter		20	.400		34.50	26		60.50	77
1060 1" diameter		19	.421		46.50	27.50		74	92.50
1080 1-1/4" diameter		15	.533		61	35		96	120
1180 9" stub by tubing 8" long									
1200 1/2" diameter	1 Stpi	24	.333	Ea.	21	22		43	55.50
1220 3/4" diameter		20	.400		28	26		54	70
1240 1" diameter		19	.421		39	27.50		66.50	84
1260 1-1/4" diameter		15	.533		58	35		93	117

23 21 Hydronic Piping and Pumps

23 21 20 – Hydronic HVAC Piping Specialties

23 21 20.42 Expansion Joints	Crew	Daily Output	Labor-Hours	Unit	Material	2020 Bare Costs Labor	Equipment	Total	Total Incl O&P
0010 **EXPANSION JOINTS**									
0100 Bellows type, neoprene cover, flanged spool									
0140 6" face to face, 1-1/4" diameter	1 Stpi	11	.727	Ea.	264	47.50		311.50	360
0160 1-1/2" diameter	"	10.60	.755		264	49.50		313.50	365
0180 2" diameter	Q-5	13.30	1.203		267	71		338	400
0190 2-1/2" diameter		12.40	1.290		276	76		352	420
0200 3" diameter		11.40	1.404		310	83		393	465
0210 4" diameter		8.40	1.905		335	112		447	540
0220 5" diameter		7.60	2.105		410	124		534	635
0230 6" diameter		6.80	2.353		420	139		559	675
0240 8" diameter		5.40	2.963		490	175		665	800
0250 10" diameter		5	3.200		675	189		864	1,025
0260 12" diameter		4.60	3.478		775	205		980	1,150
0480 10" face to face, 2" diameter		13	1.231		385	72.50		457.50	535
0500 2-1/2" diameter		12	1.333		405	78.50		483.50	565
0520 3" diameter		11	1.455		410	86		496	585
0540 4" diameter		8	2		470	118		588	690
0560 5" diameter		7	2.286		560	135		695	815
0580 6" diameter		6	2.667		580	157		737	870
0600 8" diameter		5	3.200		690	189		879	1,050
0620 10" diameter		4.60	3.478		760	205		965	1,150
0640 12" diameter		4	4		945	236		1,181	1,400
0660 14" diameter		3.80	4.211		1,175	248		1,423	1,650
0680 16" diameter		2.90	5.517		1,350	325		1,675	1,950
0700 18" diameter		2.50	6.400		1,525	380		1,905	2,250
0720 20" diameter		2.10	7.619		1,600	450		2,050	2,425
0740 24" diameter		1.80	8.889		1,875	525		2,400	2,825
0760 26" diameter		1.40	11.429		2,075	675		2,750	3,300
0780 30" diameter		1.20	13.333		2,325	785		3,110	3,750
0800 36" diameter	▼	1	16	▼	2,875	945		3,820	4,600
1000 Bellows with internal sleeves and external covers									
1010 Stainless steel, 150 lb.									
1020 With male threads									
1040 3/4" diameter	1 Stpi	20	.400	Ea.	122	26		148	173
1050 1" diameter		19	.421		131	27.50		158.50	186
1060 1-1/4" diameter		16	.500		133	33		166	195
1070 1-1/2" diameter		13	.615		163	40.50		203.50	240
1080 2" diameter	▼	11	.727	▼	175	47.50		222.50	265
1110 With flanged ends									
1120 1-1/4" diameter	1 Stpi	12	.667	Ea.	130	43.50		173.50	209
1130 1-1/2" diameter		11	.727		148	47.50		195.50	235
1140 2" diameter	▼	9	.889		201	58.50		259.50	310
1150 3" diameter	Q-5	8	2		390	118		508	600
1160 4" diameter	"	5	3.200		1,025	189		1,214	1,400
1170 5" diameter	Q-6	6	4		1,225	245		1,470	1,700
1180 6" diameter		5	4.800		1,475	294		1,769	2,075
1190 8" diameter		4.50	5.333		2,250	325		2,575	2,975
1200 10" diameter		4	6		2,850	365		3,215	3,675
1210 12" diameter	▼	3.60	6.667	▼	3,400	410		3,810	4,350

For customer support on your Mechanical Costs with RSMeans Data, call 800.448.8182.

287

23 21 20.46 Expansion Tanks	Crew	Daily Output	Labor-Hours	Unit	Material	2020 Bare Costs Labor	Equipment	Total	Total Incl O&P
0010	**EXPANSION TANKS**								
1400	Plastic, corrosion resistant, see Plumbing Costs								
1507	Underground fuel-oil storage tanks, see Section 23 13 13								
1512	Tank leak detection systems, see Section 28 33 33.50								
2000	Steel, liquid expansion, ASME, painted, 15 gallon capacity	Q-5	17	.941	Ea.	855	55.50		910.50	1,025
2020	24 gallon capacity		14	1.143		760	67.50		827.50	935
2040	30 gallon capacity		12	1.333		985	78.50		1,063.50	1,200
2060	40 gallon capacity		10	1.600		1,150	94.50		1,244.50	1,425
2080	60 gallon capacity		8	2		1,375	118		1,493	1,700
2100	80 gallon capacity		7	2.286		1,475	135		1,610	1,825
2120	100 gallon capacity		6	2.667		2,000	157		2,157	2,425
2130	120 gallon capacity		5	3.200		2,150	189		2,339	2,625
2140	135 gallon capacity		4.50	3.556		2,225	210		2,435	2,775
2150	175 gallon capacity		4	4		3,525	236		3,761	4,225
2160	220 gallon capacity		3.60	4.444		3,975	262		4,237	4,775
2170	240 gallon capacity		3.30	4.848		4,150	286		4,436	4,975
2180	305 gallon capacity		3	5.333		5,850	315		6,165	6,900
2190	400 gallon capacity	▼	2.80	5.714	▼	7,175	335		7,510	8,400
2360	Galvanized								
2370	15 gallon capacity	Q-5	17	.941	Ea.	1,275	55.50		1,330.50	1,475
2380	24 gallon capacity		14	1.143		1,550	67.50		1,617.50	1,825
2390	30 gallon capacity		12	1.333		1,600	78.50		1,678.50	1,900
2400	40 gallon capacity		10	1.600		1,875	94.50		1,969.50	2,200
2410	60 gallon capacity		8	2		2,175	118		2,293	2,575
2420	80 gallon capacity		7	2.286		2,725	135		2,860	3,200
2430	100 gallon capacity		6	2.667		3,550	157		3,707	4,125
2440	120 gallon capacity		5	3.200		3,375	189		3,564	4,000
2450	135 gallon capacity		4.50	3.556		4,000	210		4,210	4,725
2460	175 gallon capacity		4	4		4,900	236		5,136	5,750
2470	220 gallon capacity		3.60	4.444		6,725	262		6,987	7,800
2480	240 gallon capacity		3.30	4.848		7,050	286		7,336	8,175
2490	305 gallon capacity		3	5.333		8,425	315		8,740	9,750
2500	400 gallon capacity		2.80	5.714		12,600	335		12,935	14,300
3000	Steel ASME expansion, rubber diaphragm, 19 gal. cap. accept.		12	1.333		3,200	78.50		3,278.50	3,650
3020	31 gallon capacity		8	2		3,575	118		3,693	4,125
3040	61 gallon capacity		6	2.667		5,225	157		5,382	5,975
3060	79 gallon capacity		5	3.200		5,150	189		5,339	5,925
3080	119 gallon capacity		4	4		5,425	236		5,661	6,300
3100	158 gallon capacity		3.80	4.211		7,525	248		7,773	8,650
3120	211 gallon capacity		3.30	4.848		8,700	286		8,986	10,000
3140	317 gallon capacity		2.80	5.714		11,400	335		11,735	13,000
3160	422 gallon capacity		2.60	6.154		16,800	365		17,165	19,000
3180	528 gallon capacity	▼	2.40	6.667	▼	18,500	395		18,895	20,900

23 21 20.50 Float Valves

		Crew	Daily Output	Labor-Hours	Unit	Material	Labor	Equipment	Total	Total Incl O&P
0010	**FLOAT VALVES**									
0020	With ball and bracket									
0030	Single seat, threaded									
0040	Brass body									
0050	1/2"	1 Stpi	11	.727	Ea.	108	47.50		155.50	191
0060	3/4"		9	.889		119	58.50		177.50	218
0070	1"		7	1.143		166	75		241	295
0080	1-1/2"	▼	4.50	1.778	▼	238	117		355	435

23 21 Hydronic Piping and Pumps

23 21 20 – Hydronic HVAC Piping Specialties

23 21 20.50 Float Valves		Crew	Daily Output	Labor-Hours	Unit	Material	2020 Bare Costs Labor	Equipment	Total	Total Incl O&P
0090	2"	1 Stpi	3.60	2.222	Ea.	235	146		381	475
0300	For condensate receivers, CI, in-line mount									
0320	1" inlet	1 Stpi	7	1.143	Ea.	154	75		229	281
0360	For condensate receiver, CI, external float, flanged tank mount									
0370	3/4" inlet	1 Stpi	5	1.600	Ea.	117	105		222	285

23 21 20.54 Flow Check Control

		Crew	Daily Output	Labor-Hours	Unit	Material	2020 Bare Costs Labor	Equipment	Total	Total Incl O&P
0010	**FLOW CHECK CONTROL**									
0100	Bronze body, soldered									
0110	3/4" size	1 Stpi	20	.400	Ea.	77	26		103	124
0120	1" size	"	19	.421	"	93	27.50		120.50	144
0200	Cast iron body, threaded									
0210	3/4" size	1 Stpi	20	.400	Ea.	59.50	26		85.50	105
0220	1" size		19	.421		67	27.50		94.50	116
0230	1-1/4" size		15	.533		81.50	35		116.50	142
0240	1-1/2" size		13	.615		125	40.50		165.50	198
0250	2" size		11	.727		179	47.50		226.50	269
0300	Flanged inlet, threaded outlet									
0310	2-1/2" size	1 Stpi	8	1	Ea.	360	65.50		425.50	495
0320	3" size	Q-5	9	1.778		435	105		540	635
0330	4" size	"	7	2.286		835	135		970	1,125

23 21 20.58 Hydronic Heating Control Valves

		Crew	Daily Output	Labor-Hours	Unit	Material	2020 Bare Costs Labor	Equipment	Total	Total Incl O&P
0010	**HYDRONIC HEATING CONTROL VALVES**									
0050	Hot water, nonelectric, thermostatic									
0100	Radiator supply, 1/2" diameter	1 Stpi	24	.333	Ea.	78.50	22		100.50	119
0120	3/4" diameter		20	.400		74.50	26		100.50	121
0140	1" diameter		19	.421		88	27.50		115.50	139
0160	1-1/4" diameter		15	.533		108	35		143	172
0500	For low pressure steam, add					25%				
1000	Manual, radiator supply									
1010	1/2" pipe size, angle union	1 Stpi	24	.333	Ea.	71	22		93	111
1020	3/4" pipe size, angle union		20	.400		89.50	26		115.50	138
1030	1" pipe size, angle union		19	.421		114	27.50		141.50	167
1100	Radiator, balancing, straight, sweat connections									
1110	1/2" pipe size	1 Stpi	24	.333	Ea.	25	22		47	60
1120	3/4" pipe size		20	.400		34.50	26		60.50	77
1130	1" pipe size		19	.421		56.50	27.50		84	104
1200	Steam, radiator, supply									
1210	1/2" pipe size, angle union	1 Stpi	24	.333	Ea.	67	22		89	107
1220	3/4" pipe size, angle union		20	.400		73.50	26		99.50	120
1230	1" pipe size, angle union		19	.421		84	27.50		111.50	134
1240	1-1/4" pipe size, angle union		15	.533		109	35		144	173
8000	System balancing and shut-off									
8020	Butterfly, quarter turn, calibrated, threaded or solder									
8040	Bronze, -30°F to +350°F, pressure to 175 psi									
8060	1/2" size	1 Stpi	22	.364	Ea.	21	24		45	59
8070	3/4" size		20	.400		29	26		55	71
8080	1" size		19	.421		36	27.50		63.50	81
8090	1-1/4" size		15	.533		69	35		104	129
8100	1-1/2" size		13	.615		73.50	40.50		114	142
8110	2" size		11	.727		93	47.50		140.50	175

For customer support on your Mechanical Costs with RSMeans Data, call 800.448.8182.

289

23 21 20.62 Liquid Drainers	Crew	Daily Output	Labor-Hours	Unit	Material	2020 Bare Costs Labor	Equipment	Total	Total Incl O&P
0010 **LIQUID DRAINERS**									
0100 Guided lever type									
0110 Cast iron body, threaded									
0120 1/2" pipe size	1 Stpi	22	.364	Ea.	139	24		163	188
0130 3/4" pipe size		20	.400		360	26		386	435
0140 1" pipe size		19	.421		580	27.50		607.50	680
0150 1-1/2" pipe size		11	.727		1,075	47.50		1,122.50	1,250
0160 2" pipe size		8	1		1,075	65.50		1,140.50	1,275
0200 Forged steel body, threaded									
0210 1/2" pipe size	1 Stpi	18	.444	Ea.	1,175	29		1,204	1,350
0220 3/4" pipe size		16	.500		1,175	33		1,208	1,350
0230 1" pipe size		14	.571		1,775	37.50		1,812.50	2,000
0240 1-1/2" pipe size	Q-5	10	1.600		6,050	94.50		6,144.50	6,825
0250 2" pipe size	"	8	2		6,050	118		6,168	6,850
0300 Socket weld									
0310 1/2" pipe size	Q-17	16	1	Ea.	1,575	59	6.65	1,640.65	1,825
0320 3/4" pipe size		14	1.143		1,575	67.50	7.60	1,650.10	1,825
0330 1" pipe size		12	1.333		2,325	78.50	8.85	2,412.35	2,675
0340 1-1/2" pipe size		9	1.778		6,250	105	11.80	6,366.80	7,050
0350 2" pipe size		7	2.286		6,150	135	15.20	6,300.20	7,000
0400 Flanged									
0410 1/2" pipe size	1 Stpi	12	.667	Ea.	3,225	43.50		3,268.50	3,625
0420 3/4" pipe size		10	.800		3,225	52.50		3,277.50	3,625
0430 1" pipe size		7	1.143		4,075	75		4,150	4,575
0440 1-1/2" pipe size	Q-5	9	1.778		8,350	105		8,455	9,325
0450 2" pipe size	"	7	2.286		8,350	135		8,485	9,375
1000 Fixed lever and snap action type									
1100 Cast iron body, threaded									
1110 1/2" pipe size	1 Stpi	22	.364	Ea.	214	24		238	272
1120 3/4" pipe size		20	.400		565	26		591	660
1130 1" pipe size		19	.421		565	27.50		592.50	660
1200 Forged steel body, threaded									
1210 1/2" pipe size	1 Stpi	18	.444	Ea.	1,500	29		1,529	1,700
1220 3/4" pipe size		16	.500		1,500	33		1,533	1,700
1230 1" pipe size		14	.571		3,625	37.50		3,662.50	4,050
1240 1-1/4" pipe size		12	.667		3,475	43.50		3,518.50	3,900
2000 High pressure high leverage type									
2100 Forged steel body, threaded									
2110 1/2" pipe size	1 Stpi	18	.444	Ea.	2,775	29		2,804	3,100
2120 3/4" pipe size		16	.500		2,875	33		2,908	3,225
2130 1" pipe size		14	.571		4,500	37.50		4,537.50	5,000
2140 1-1/4" pipe size		12	.667		4,825	43.50		4,868.50	5,400
2150 1-1/2" pipe size	Q-5	10	1.600		6,825	94.50		6,919.50	7,675
2160 2" pipe size	"	8	2		6,825	118		6,943	7,700
2200 Socket weld									
2210 1/2" pipe size	Q-17	16	1	Ea.	3,125	59	6.65	3,190.65	3,525
2220 3/4" pipe size		14	1.143		3,000	67.50	7.60	3,075.10	3,400
2230 1" pipe size		12	1.333		4,875	78.50	8.85	4,962.35	5,475
2240 1-1/4" pipe size		11	1.455		5,075	86	9.65	5,170.65	5,725
2250 1-1/2" pipe size		9	1.778		7,200	105	11.80	7,316.80	8,075
2260 2" pipe size		7	2.286		7,200	135	15.20	7,350.20	8,125
2300 Flanged									

23 21 Hydronic Piping and Pumps

23 21 20 – Hydronic HVAC Piping Specialties

23 21 20.62 Liquid Drainers		Crew	Daily Output	Labor-Hours	Unit	Material	2020 Bare Costs Labor	Equipment	Total	Total Incl O&P
2310	1/2" pipe size	1 Stpi	12	.667	Ea.	4,000	43.50		4,043.50	4,475
2320	3/4" pipe size		10	.800		4,175	52.50		4,227.50	4,675
2330	1" pipe size		7	1.143		6,325	75		6,400	7,075
2340	1-1/4" pipe size		6	1.333		6,325	87.50		6,412.50	7,100
2350	1-1/2" pipe size	Q-5	9	1.778		8,650	105		8,755	9,675
2360	2" pipe size	"	7	2.286		8,650	135		8,785	9,725
3000	Guided lever dual gravity type									
3100	Cast iron body, threaded									
3110	1/2" pipe size	1 Stpi	22	.364	Ea.	525	24		549	615
3120	3/4" pipe size		20	.400		505	26		531	600
3130	1" pipe size		19	.421		735	27.50		762.50	850
3140	1-1/2" pipe size		11	.727		1,425	47.50		1,472.50	1,650
3150	2" pipe size		8	1		1,500	65.50		1,565.50	1,750
3200	Forged steel body, threaded									
3210	1/2" pipe size	1 Stpi	18	.444	Ea.	1,500	29		1,529	1,700
3220	3/4" pipe size		16	.500		1,500	33		1,533	1,700
3230	1" pipe size		14	.571		2,250	37.50		2,287.50	2,525
3240	1-1/2" pipe size	Q-5	10	1.600		5,975	94.50		6,069.50	6,725
3250	2" pipe size	"	8	2		5,975	118		6,093	6,750
3300	Socket weld									
3310	1/2" pipe size	Q-17	16	1	Ea.	1,700	59	6.65	1,765.65	1,975
3320	3/4" pipe size		14	1.143		1,700	67.50	7.60	1,775.10	1,975
3330	1" pipe size		12	1.333		2,475	78.50	8.85	2,562.35	2,850
3340	1-1/2" pipe size		9	1.778		6,450	105	11.80	6,566.80	7,275
3350	2" pipe size		7	2.286		6,200	135	15.20	6,350.20	7,025
3400	Flanged									
3410	1/2" pipe size	1 Stpi	12	.667	Ea.	2,750	43.50		2,793.50	3,100
3420	3/4" pipe size		10	.800		2,750	52.50		2,802.50	3,100
3430	1" pipe size		7	1.143		3,750	75		3,825	4,225
3440	1-1/2" pipe size	Q-5	9	1.778		7,650	105		7,755	8,550
3450	2" pipe size	"	7	2.286		7,650	135		7,785	8,600

23 21 20.66 Monoflow Tee Fitting

		Crew	Daily Output	Labor-Hours	Unit	Material	2020 Bare Costs Labor	Equipment	Total	Total Incl O&P
0010	**MONOFLOW TEE FITTING**									
1100	For one pipe hydronic, supply and return									
1110	Copper, soldered									
1120	3/4" x 1/2" size	1 Stpi	13	.615	Ea.	21	40.50		61.50	83.50
1130	1" x 1/2" size		12	.667		21	43.50		64.50	88.50
1140	1" x 3/4" size		11	.727		21	47.50		68.50	94.50
1150	1-1/4" x 1/2" size		11	.727		21.50	47.50		69	95
1160	1-1/4" x 3/4" size		10	.800		21.50	52.50		74	102
1170	1-1/2" x 3/4" size		10	.800		31.50	52.50		84	113
1180	1-1/2" x 1" size		9	.889		31.50	58.50		90	122
1190	2" x 3/4" size		9	.889		46	58.50		104.50	138
1200	2" x 1" size		8	1		46	65.50		111.50	149

23 21 20.70 Steam Traps

		Crew	Daily Output	Labor-Hours	Unit	Material	2020 Bare Costs Labor	Equipment	Total	Total Incl O&P
0010	**STEAM TRAPS**									
0030	Cast iron body, threaded									
0040	Inverted bucket									
0050	1/2" pipe size	1 Stpi	12	.667	Ea.	170	43.50		213.50	253
0070	3/4" pipe size		10	.800		280	52.50		332.50	390
0100	1" pipe size		9	.889		430	58.50		488.50	560
0120	1-1/4" pipe size		8	1		650	65.50		715.50	815

For customer support on your Mechanical Costs with RSMeans Data, call 800.448.8182.

291

23 21 Hydronic Piping and Pumps

23 21 20 – Hydronic HVAC Piping Specialties

23 21 20.70 Steam Traps		Crew	Daily Output	Labor-Hours	Unit	Material	2020 Bare Costs Labor	Equipment	Total	Total Incl O&P
0130	1-1/2" pipe size	1 Stpi	7	1.143	Ea.	820	75		895	1,000
0140	2" pipe size	↓	6	1.333	↓	1,375	87.50		1,462.50	1,625
0200	With thermic vent & check valve									
0202	1/2" pipe size	1 Stpi	12	.667	Ea.	220	43.50		263.50	310
0204	3/4" pipe size		10	.800		283	52.50		335.50	390
0206	1" pipe size		9	.889		495	58.50		553.50	630
0208	1-1/4" pipe size		8	1		755	65.50		820.50	930
0210	1-1/2" pipe size		7	1.143		970	75		1,045	1,175
0212	2" pipe size	↓	6	1.333	↓	1,475	87.50		1,562.50	1,750
1000	Float & thermostatic, 15 psi									
1010	3/4" pipe size	1 Stpi	16	.500	Ea.	168	33		201	233
1020	1" pipe size		15	.533		193	35		228	265
1030	1-1/4" pipe size		13	.615		239	40.50		279.50	325
1040	1-1/2" pipe size		9	.889		360	58.50		418.50	480
1060	2" pipe size	↓	6	1.333	↓	1,000	87.50		1,087.50	1,225
1100	For vacuum breaker, add				↓	74.50			74.50	81.50
1110	Differential condensate controller									
1120	With union and manual metering valve									
1130	1/2" pipe size	1 Stpi	24	.333	Ea.	740	22		762	850
1140	3/4" pipe size		20	.400		875	26		901	1,000
1150	1" pipe size		19	.421		1,175	27.50		1,202.50	1,350
1160	1-1/4" pipe size		15	.533		1,575	35		1,610	1,800
1170	1-1/2" pipe size		13	.615		1,900	40.50		1,940.50	2,150
1180	2" pipe size	↓	11	.727	↓	2,475	47.50		2,522.50	2,800
1290	Brass body, threaded									
1300	Thermostatic, angle union, 25 psi									
1310	1/2" pipe size	1 Stpi	24	.333	Ea.	70	22		92	110
1320	3/4" pipe size		20	.400		107	26		133	157
1330	1" pipe size		19	.421		182	27.50		209.50	243
1990	Carbon steel body, threaded									
2000	Controlled disc, integral strainer & blow down									
2010	Threaded									
2020	3/8" pipe size	1 Stpi	24	.333	Ea.	335	22		357	405
2030	1/2" pipe size		24	.333		335	22		357	405
2040	3/4" pipe size		20	.400		450	26		476	535
2050	1" pipe size	↓	19	.421	↓	575	27.50		602.50	675
2100	Socket weld									
2110	3/8" pipe size	Q-17	17	.941	Ea.	400	55.50	6.25	461.75	530
2120	1/2" pipe size		16	1		400	59	6.65	465.65	535
2130	3/4" pipe size		14	1.143		520	67.50	7.60	595.10	680
2140	1" pipe size	↓	12	1.333	↓	650	78.50	8.85	737.35	845
2170	Flanged, 600 lb. ASA									
2180	3/8" pipe size	1 Stpi	12	.667	Ea.	900	43.50		943.50	1,050
2190	1/2" pipe size		12	.667		900	43.50		943.50	1,050
2200	3/4" pipe size		10	.800		1,075	52.50		1,127.50	1,250
2210	1" pipe size	↓	7	1.143	↓	1,250	75		1,325	1,475
5000	Forged steel body									
5010	Inverted bucket									
5020	Threaded									
5030	1/2" pipe size	1 Stpi	12	.667	Ea.	605	43.50		648.50	730
5040	3/4" pipe size		10	.800		1,050	52.50		1,102.50	1,225
5050	1" pipe size		9	.889		1,575	58.50		1,633.50	1,825
5060	1-1/4" pipe size	↓	8	1		2,300	65.50		2,365.50	2,625

23 21 20.70 Steam Traps		Crew	Daily Output	Labor-Hours	Unit	Material	2020 Bare Costs Labor	Equipment	Total	Total Incl O&P
5070	1-1/2" pipe size	1 Stpi	7	1.143	Ea.	3,250	75		3,325	3,675
5080	2" pipe size	↓	6	1.333	↓	5,725	87.50		5,812.50	6,400
5100	Socket weld									
5110	1/2" pipe size	Q-17	16	1	Ea.	720	59	6.65	785.65	890
5120	3/4" pipe size		14	1.143		1,150	67.50	7.60	1,225.10	1,375
5130	1" pipe size		12	1.333		1,675	78.50	8.85	1,762.35	1,975
5140	1-1/4" pipe size		10	1.600		2,400	94.50	10.65	2,505.15	2,775
5150	1-1/2" pipe size		9	1.778		3,325	105	11.80	3,441.80	3,825
5160	2" pipe size	↓	7	2.286	↓	5,775	135	15.20	5,925.20	6,575
5200	Flanged									
5210	1/2" pipe size	1 Stpi	12	.667	Ea.	1,100	43.50		1,143.50	1,300
5220	3/4" pipe size		10	.800		1,825	52.50		1,877.50	2,075
5230	1" pipe size		7	1.143		2,475	75		2,550	2,825
5240	1-1/4" pipe size	↓	6	1.333		3,375	87.50		3,462.50	3,825
5250	1-1/2" pipe size	Q-5	9	1.778		4,300	105		4,405	4,875
5260	2" pipe size	"	7	2.286	↓	6,750	135		6,885	7,625
6000	Stainless steel body									
6010	Inverted bucket									
6020	Threaded									
6030	1/2" pipe size	1 Stpi	12	.667	Ea.	203	43.50		246.50	289
6040	3/4" pipe size		10	.800		266	52.50		318.50	370
6050	1" pipe size	↓	9	.889	↓	945	58.50		1,003.50	1,125
6100	Threaded with thermic vent									
6110	1/2" pipe size	1 Stpi	12	.667	Ea.	195	43.50		238.50	281
6120	3/4" pipe size		10	.800		258	52.50		310.50	360
6130	1" pipe size	↓	9	.889	↓	925	58.50		983.50	1,100
6200	Threaded with check valve									
6210	1/2" pipe size	1 Stpi	12	.667	Ea.	285	43.50		328.50	380
6220	3/4" pipe size	"	10	.800	"	395	52.50		447.50	515
6300	Socket weld									
6310	1/2" pipe size	Q-17	16	1	Ea.	232	59	6.65	297.65	350
6320	3/4" pipe size	"	14	1.143	"	282	67.50	7.60	357.10	420

23 21 20.74 Strainers, Basket Type

		Crew	Daily Output	Labor-Hours	Unit	Material	2020 Bare Costs Labor	Equipment	Total	Total Incl O&P
0010	**STRAINERS, BASKET TYPE,** Perforated stainless steel basket									
0100	Brass or monel available									
2000	Simplex style									
2300	Bronze body									
2320	Screwed, 3/8" pipe size	1 Stpi	22	.364	Ea.	191	24		215	246
2340	1/2" pipe size		20	.400		199	26		225	258
2360	3/4" pipe size		17	.471		305	31		336	380
2380	1" pipe size		15	.533		380	35		415	475
2400	1-1/4" pipe size		13	.615		415	40.50		455.50	520
2420	1-1/2" pipe size		12	.667		425	43.50		468.50	530
2440	2" pipe size	↓	10	.800		625	52.50		677.50	765
2460	2-1/2" pipe size	Q-5	15	1.067		1,050	63		1,113	1,250
2480	3" pipe size	"	14	1.143		1,550	67.50		1,617.50	1,800
2600	Flanged, 2" pipe size	1 Stpi	6	1.333		860	87.50		947.50	1,075
2620	2-1/2" pipe size	Q-5	4.50	3.556		1,500	210		1,710	1,975
2640	3" pipe size		3.50	4.571		1,700	270		1,970	2,250
2660	4" pipe size	↓	3	5.333		2,575	315		2,890	3,300
2680	5" pipe size	Q-6	3.40	7.059		4,100	430		4,530	5,175
2700	6" pipe size		3	8		5,000	490		5,490	6,225

For customer support on your Mechanical Costs with RSMeans Data, call 800.448.8182.

293

23 21 20.74 Strainers, Basket Type	Crew	Daily Output	Labor-Hours	Unit	Material	2020 Bare Costs Labor	Equipment	Total	Total Incl O&P
2710	8" pipe size	Q-6	2.50	9.600	Ea.	8,150	585		8,735	9,850
3600	Iron body									
3700	Screwed, 3/8" pipe size	1 Stpi	22	.364	Ea.	165	24		189	217
3720	1/2" pipe size		20	.400		170	26		196	226
3740	3/4" pipe size		17	.471		217	31		248	284
3760	1" pipe size		15	.533		221	35		256	296
3780	1-1/4" pipe size		13	.615		289	40.50		329.50	380
3800	1-1/2" pipe size		12	.667		320	43.50		363.50	415
3820	2" pipe size		10	.800		380	52.50		432.50	500
3840	2-1/2" pipe size	Q-5	15	1.067		510	63		573	655
3860	3" pipe size	"	14	1.143		615	67.50		682.50	780
4000	Flanged, 2" pipe size	1 Stpi	6	1.333		560	87.50		647.50	745
4020	2-1/2" pipe size	Q-5	4.50	3.556		755	210		965	1,150
4040	3" pipe size		3.50	4.571		790	270		1,060	1,275
4060	4" pipe size		3	5.333		1,225	315		1,540	1,825
4080	5" pipe size	Q-6	3.40	7.059		1,825	430		2,255	2,675
4100	6" pipe size		3	8		2,350	490		2,840	3,300
4120	8" pipe size		2.50	9.600		4,225	585		4,810	5,525
4140	10" pipe size		2.20	10.909		9,300	665		9,965	11,200
6000	Cast steel body									
6400	Screwed, 1" pipe size	1 Stpi	15	.533	Ea.	360	35		395	450
6410	1-1/4" pipe size		13	.615		540	40.50		580.50	650
6420	1-1/2" pipe size		12	.667		560	43.50		603.50	685
6440	2" pipe size		10	.800		740	52.50		792.50	895
6460	2-1/2" pipe size	Q-5	15	1.067		1,075	63		1,138	1,275
6480	3" pipe size	"	14	1.143		1,375	67.50		1,442.50	1,600
6560	Flanged, 2" pipe size	1 Stpi	6	1.333		1,450	87.50		1,537.50	1,725
6580	2-1/2" pipe size	Q-5	4.50	3.556		2,150	210		2,360	2,700
6600	3" pipe size		3.50	4.571		2,450	270		2,720	3,100
6620	4" pipe size		3	5.333		2,600	315		2,915	3,325
6640	6" pipe size	Q-6	3	8		4,750	490		5,240	5,950
6660	8" pipe size	"	2.50	9.600		9,625	585		10,210	11,500
7000	Stainless steel body									
7200	Screwed, 1" pipe size	1 Stpi	15	.533	Ea.	465	35		500	570
7210	1-1/4" pipe size		13	.615		725	40.50		765.50	860
7220	1-1/2" pipe size		12	.667		725	43.50		768.50	865
7240	2" pipe size		10	.800		1,075	52.50		1,127.50	1,275
7260	2-1/2" pipe size	Q-5	15	1.067		1,525	63		1,588	1,775
7280	3" pipe size	"	14	1.143		2,100	67.50		2,167.50	2,425
7400	Flanged, 2" pipe size	1 Stpi	6	1.333		1,650	87.50		1,737.50	1,950
7420	2-1/2" pipe size	Q-5	4.50	3.556		3,225	210		3,435	3,850
7440	3" pipe size		3.50	4.571		3,300	270		3,570	4,025
7460	4" pipe size		3	5.333		5,175	315		5,490	6,175
7480	6" pipe size	Q-6	3	8		10,900	490		11,390	12,700
7500	8" pipe size	"	2.50	9.600		15,300	585		15,885	17,800
8100	Duplex style									
8200	Bronze body									
8240	Screwed, 3/4" pipe size	1 Stpi	16	.500	Ea.	1,325	33		1,358	1,525
8260	1" pipe size		14	.571		1,325	37.50		1,362.50	1,525
8280	1-1/4" pipe size		12	.667		2,700	43.50		2,743.50	3,050
8300	1-1/2" pipe size		11	.727		2,700	47.50		2,747.50	3,050
8320	2" pipe size		9	.889		4,225	58.50		4,283.50	4,725
8340	2-1/2" pipe size	Q-5	14	1.143		5,450	67.50		5,517.50	6,100

For customer support on your Mechanical Costs with RSMeans Data, call 800.448.8182.

23 21 Hydronic Piping and Pumps

23 21 20 – Hydronic HVAC Piping Specialties

23 21 20.74 Strainers, Basket Type

		Crew	Daily Output	Labor-Hours	Unit	Material	2020 Bare Costs Labor	Equipment	Total	Total Incl O&P
8420	Flanged, 2" pipe size	1 Stpi	6	1.333	Ea.	5,225	87.50		5,312.50	5,875
8440	2-1/2" pipe size	Q-5	4.50	3.556		7,225	210		7,435	8,275
8460	3" pipe size		3.50	4.571		7,875	270		8,145	9,050
8480	4" pipe size	▼	3	5.333		11,500	315		11,815	13,200
8500	5" pipe size	Q-6	3.40	7.059		25,900	430		26,330	29,000
8520	6" pipe size	"	3	8	▼	25,100	490		25,590	28,400
8700	Iron body									
8740	Screwed, 3/4" pipe size	1 Stpi	16	.500	Ea.	1,675	33		1,708	1,900
8760	1" pipe size		14	.571		1,675	37.50		1,712.50	1,900
8780	1-1/4" pipe size		12	.667		1,850	43.50		1,893.50	2,100
8800	1-1/2" pipe size		11	.727		1,850	47.50		1,897.50	2,100
8820	2" pipe size	▼	9	.889		3,100	58.50		3,158.50	3,500
8840	2-1/2" pipe size	Q-5	14	1.143		3,425	67.50		3,492.50	3,875
9000	Flanged, 2" pipe size	1 Stpi	6	1.333		2,825	87.50		2,912.50	3,225
9020	2-1/2" pipe size	Q-5	4.50	3.556		3,550	210		3,760	4,225
9040	3" pipe size		3.50	4.571		3,875	270		4,145	4,650
9060	4" pipe size	▼	3	5.333		6,525	315		6,840	7,650
9080	5" pipe size	Q-6	3.40	7.059		14,300	430		14,730	16,300
9100	6" pipe size		3	8		14,300	490		14,790	16,400
9120	8" pipe size		2.50	9.600		25,100	585		25,685	28,500
9140	10" pipe size		2.20	10.909		40,700	665		41,365	45,800
9160	12" pipe size		1.70	14.118		44,800	865		45,665	50,500
9170	14" pipe size		1.40	17.143		52,500	1,050		53,550	59,500
9180	16" pipe size	▼	1	24	▼	64,500	1,475		65,975	73,000
9300	Cast steel body									
9340	Screwed, 1" pipe size	1 Stpi	14	.571	Ea.	2,125	37.50		2,162.50	2,400
9360	1-1/2" pipe size		11	.727		3,425	47.50		3,472.50	3,850
9380	2" pipe size		9	.889		4,550	58.50		4,608.50	5,075
9460	Flanged, 2" pipe size	▼	6	1.333		3,950	87.50		4,037.50	4,450
9480	2-1/2" pipe size	Q-5	4.50	3.556		5,775	210		5,985	6,700
9500	3" pipe size		3.50	4.571		6,300	270		6,570	7,325
9520	4" pipe size	▼	3	5.333		7,775	315		8,090	9,025
9540	6" pipe size	Q-6	3	8		14,600	490		15,090	16,700
9560	8" pipe size	"	2.50	9.600	▼	33,400	585		33,985	37,600
9700	Stainless steel body									
9740	Screwed, 1" pipe size	1 Stpi	14	.571	Ea.	3,250	37.50		3,287.50	3,625
9760	1-1/2" pipe size		11	.727		4,875	47.50		4,922.50	5,425
9780	2" pipe size		9	.889		7,175	58.50		7,233.50	7,975
9860	Flanged, 2" pipe size	▼	6	1.333		7,400	87.50		7,487.50	8,275
9880	2-1/2" pipe size	Q-5	4.50	3.556		12,400	210		12,610	14,000
9900	3" pipe size		3.50	4.571		13,500	270		13,770	15,300
9920	4" pipe size	▼	3	5.333		17,600	315		17,915	19,900
9940	6" pipe size	Q-6	3	8		26,900	490		27,390	30,300
9960	8" pipe size	"	2.50	9.600		69,000	585		69,585	77,000

23 21 20.76 Strainers, Y Type, Bronze Body

		Crew	Daily Output	Labor-Hours	Unit	Material	2020 Bare Costs Labor	Equipment	Total	Total Incl O&P
0010	**STRAINERS, Y TYPE, BRONZE BODY**									
0050	Screwed, 125 lb., 1/4" pipe size	1 Stpi	24	.333	Ea.	44	22		66	81
0070	3/8" pipe size		24	.333		45.50	22		67.50	83
0100	1/2" pipe size		20	.400		45.50	26		71.50	89
0120	3/4" pipe size		19	.421		57	27.50		84.50	105
0140	1" pipe size		17	.471		66.50	31		97.50	120
0150	1-1/4" pipe size	▼	15	.533		134	35		169	201

For customer support on your Mechanical Costs with RSMeans Data, call 800.448.8182.

295

23 21 20.76 Strainers, Y Type, Bronze Body

		Crew	Daily Output	Labor-Hours	Unit	Material	2020 Bare Costs Labor	Equipment	Total	Total Incl O&P
0160	1-1/2" pipe size	1 Stpi	14	.571	Ea.	144	37.50		181.50	214
0180	2" pipe size		13	.615		190	40.50		230.50	270
0182	3" pipe size		12	.667		1,200	43.50		1,243.50	1,400
0200	300 lb., 2-1/2" pipe size	Q-5	17	.941		865	55.50		920.50	1,025
0220	3" pipe size		16	1		1,325	59		1,384	1,575
0240	4" pipe size		15	1.067		2,425	63		2,488	2,775
0500	For 300 lb. rating 1/4" thru 2", add					15%				
1000	Flanged, 150 lb., 1-1/2" pipe size	1 Stpi	11	.727	Ea.	545	47.50		592.50	670
1020	2" pipe size	"	8	1		630	65.50		695.50	790
1030	2-1/2" pipe size	Q-5	5	3.200		1,125	189		1,314	1,525
1040	3" pipe size		4.50	3.556		1,375	210		1,585	1,850
1060	4" pipe size		3	5.333		2,125	315		2,440	2,825
1080	5" pipe size	Q-6	3.40	7.059		3,025	430		3,455	3,975
1100	6" pipe size		3	8		4,100	490		4,590	5,250
1106	8" pipe size		2.60	9.231		4,450	565		5,015	5,750
1500	For 300 lb. rating, add					40%				

23 21 20.78 Strainers, Y Type, Iron Body

		Crew	Daily Output	Labor-Hours	Unit	Material	2020 Bare Costs Labor	Equipment	Total	Total Incl O&P
0010	**STRAINERS, Y TYPE, IRON BODY**									
0050	Screwed, 250 lb., 1/4" pipe size	1 Stpi	20	.400	Ea.	20	26		46	61
0070	3/8" pipe size		20	.400		20	26		46	61
0100	1/2" pipe size		20	.400		20	26		46	61
0120	3/4" pipe size		18	.444		23.50	29		52.50	69
0140	1" pipe size		16	.500		27.50	33		60.50	79
0150	1-1/4" pipe size		15	.533		43.50	35		78.50	101
0160	1-1/2" pipe size		12	.667		55.50	43.50		99	127
0180	2" pipe size		8	1		83	65.50		148.50	189
0200	2-1/2" pipe size	Q-5	12	1.333		288	78.50		366.50	435
0220	3" pipe size		11	1.455		310	86		396	475
0240	4" pipe size		5	3.200		530	189		719	865
0500	For galvanized body, add					50%				
1000	Flanged, 125 lb., 1-1/2" pipe size	1 Stpi	11	.727	Ea.	167	47.50		214.50	256
1020	2" pipe size	"	8	1		177	65.50		242.50	293
1030	2-1/2" pipe size	Q-5	5	3.200		161	189		350	460
1040	3" pipe size		4.50	3.556		188	210		398	520
1060	4" pipe size		3	5.333		375	315		690	885
1080	5" pipe size	Q-6	3.40	7.059		405	430		835	1,100
1100	6" pipe size		3	8		675	490		1,165	1,475
1120	8" pipe size		2.50	9.600		1,300	585		1,885	2,300
1140	10" pipe size		2	12		2,300	735		3,035	3,625
1160	12" pipe size		1.70	14.118		3,525	865		4,390	5,175
1170	14" pipe size		1.30	18.462		6,500	1,125		7,625	8,850
1180	16" pipe size		1	24		9,225	1,475		10,700	12,400
1500	For 250 lb. rating, add					20%				
2000	For galvanized body, add					50%				
2500	For steel body, add					40%				

23 21 20.80 Suction Diffusers

		Crew	Daily Output	Labor-Hours	Unit	Material	2020 Bare Costs Labor	Equipment	Total	Total Incl O&P
0010	**SUCTION DIFFUSERS**									
0100	Cast iron body with integral straightening vanes, strainer									
1000	Flanged									
1010	2" inlet, 1-1/2" pump side	1 Stpi	6	1.333	Ea.	415	87.50		502.50	585
1020	2" pump side	"	5	1.600		490	105		595	695
1030	3" inlet, 2" pump side	Q-5	6.50	2.462		520	145		665	790

296

23 21 20 – Hydronic HVAC Piping Specialties

23 21 20.80 Suction Diffusers		Crew	Daily Output	Labor-Hours	Unit	Material	2020 Bare Costs Labor	Equipment	Total	Total Incl O&P
1040	2-1/2" pump side	Q-5	5.30	3.019	Ea.	760	178		938	1,100
1050	3" pump side		4.50	3.556		770	210		980	1,150
1060	4" inlet, 3" pump side		3.50	4.571		880	270		1,150	1,375
1070	4" pump side	▼	3	5.333		1,025	315		1,340	1,600
1080	5" inlet, 4" pump side	Q-6	3.80	6.316		1,150	385		1,535	1,850
1090	5" pump side		3.40	7.059		1,375	430		1,805	2,175
1100	6" inlet, 4" pump side		3.30	7.273		1,200	445		1,645	2,000
1110	5" pump side		3.10	7.742		1,450	475		1,925	2,300
1120	6" pump side		3	8		1,525	490		2,015	2,400
1130	8" inlet, 6" pump side		2.70	8.889		1,725	545		2,270	2,725
1140	8" pump side		2.50	9.600		2,975	585		3,560	4,150
1150	10" inlet, 6" pump side		2.40	10		3,475	610		4,085	4,750
1160	8" pump side		2.30	10.435		3,450	640		4,090	4,750
1170	10" pump side		2.20	10.909		4,975	665		5,640	6,475
1180	12" inlet, 8" pump side		2	12		5,250	735		5,985	6,875
1190	10" pump side		1.80	13.333		5,900	815		6,715	7,725
1200	12" pump side		1.70	14.118		6,525	865		7,390	8,500
1210	14" inlet, 12" pump side		1.50	16		7,875	980		8,855	10,200
1220	14" pump side	▼	1.30	18.462	▼	9,575	1,125		10,700	12,200

23 21 20.84 Thermoflo Indicator

		Crew	Daily Output	Labor-Hours	Unit	Material	Labor	Equipment	Total	Total Incl O&P
0010	**THERMOFLO INDICATOR**, For balancing									
1000	Sweat connections, 1-1/4" pipe size	1 Stpi	12	.667	Ea.	790	43.50		833.50	935
1020	1-1/2" pipe size		10	.800		805	52.50		857.50	965
1040	2" pipe size		8	1		845	65.50		910.50	1,025
1060	2-1/2" pipe size	▼	7	1.143		1,275	75		1,350	1,525
2000	Flange connections, 3" pipe size	Q-5	5	3.200		1,550	189		1,739	1,975
2020	4" pipe size		4	4		1,850	236		2,086	2,375
2030	5" pipe size		3.50	4.571		2,325	270		2,595	2,950
2040	6" pipe size		3	5.333		2,475	315		2,790	3,200
2060	8" pipe size	▼	2	8	▼	3,025	470		3,495	4,025

23 21 20.88 Venturi Flow

		Crew	Daily Output	Labor-Hours	Unit	Material	Labor	Equipment	Total	Total Incl O&P
0010	**VENTURI FLOW**, Measuring device									
0050	1/2" diameter	1 Stpi	24	.333	Ea.	310	22		332	375
0100	3/4" diameter		20	.400		284	26		310	355
0120	1" diameter		19	.421		305	27.50		332.50	375
0140	1-1/4" diameter		15	.533		380	35		415	470
0160	1-1/2" diameter		13	.615		380	40.50		420.50	475
0180	2" diameter	▼	11	.727		405	47.50		452.50	515
0200	2-1/2" diameter	Q-5	16	1		555	59		614	700
0220	3" diameter		14	1.143		570	67.50		637.50	730
0240	4" diameter		11	1.455		855	86		941	1,075
0260	5" diameter	Q-6	4	6		1,125	365		1,490	1,775
0280	6" diameter		3.50	6.857		1,250	420		1,670	2,000
0300	8" diameter		3	8		1,600	490		2,090	2,475
0320	10" diameter		2	12		3,775	735		4,510	5,250
0330	12" diameter		1.80	13.333		5,225	815		6,040	6,975
0340	14" diameter		1.60	15		5,925	920		6,845	7,900
0350	16" diameter	▼	1.40	17.143		6,775	1,050		7,825	9,025
0500	For meter, add				▼	2,425			2,425	2,650

For customer support on your Mechanical Costs with RSMeans Data, call 800.448.8182.

297

23 21 20 – Hydronic HVAC Piping Specialties

23 21 20.94 Weld End Ball Joints		Crew	Daily Output	Labor-Hours	Unit	Material	2020 Bare Costs Labor	Equipment	Total	Total Incl O&P
0010	**WELD END BALL JOINTS**, Steel									
0050	2-1/2" diameter	Q-17	13	1.231	Ea.	840	72.50	8.20	920.70	1,050
0100	3" diameter		12	1.333		990	78.50	8.85	1,077.35	1,225
0120	4" diameter	↓	11	1.455		1,450	86	9.65	1,545.65	1,750
0140	5" diameter	Q-18	14	1.714		2,000	105	7.60	2,112.60	2,375
0160	6" diameter		12	2		2,300	122	8.85	2,430.85	2,725
0180	8" diameter		9	2.667		3,350	163	11.80	3,524.80	3,950
0200	10" diameter		8	3		4,900	184	13.30	5,097.30	5,675
0220	12" diameter	↓	6	4	↓	6,850	245	17.70	7,112.70	7,900

23 21 20.98 Zone Valves

		Crew	Daily Output	Labor-Hours	Unit	Material	2020 Bare Costs Labor	Equipment	Total	Total Incl O&P
0010	**ZONE VALVES**									
1000	Bronze body									
1010	2 way, 125 psi									
1020	Standard, 65' pump head									
1030	1/2" soldered	1 Stpi	24	.333	Ea.	144	22		166	191
1040	3/4" soldered		20	.400		144	26		170	197
1050	1" soldered		19	.421		155	27.50		182.50	212
1060	1-1/4" soldered	↓	15	.533	↓	181	35		216	252
1200	High head, 125' pump head									
1210	1/2" soldered	1 Stpi	24	.333	Ea.	208	22		230	262
1220	3/4" soldered		20	.400		216	26		242	276
1230	1" soldered	↓	19	.421	↓	235	27.50		262.50	300
1300	Geothermal, 150' pump head									
1310	3/4" soldered	1 Stpi	20	.400	Ea.	226	26		252	288
1320	1" soldered		19	.421		262	27.50		289.50	330
1330	3/4" threaded	↓	20	.400	↓	229	26		255	291
3000	3 way, 125 psi									
3200	Bypass, 65' pump head									
3210	1/2" soldered	1 Stpi	24	.333	Ea.	212	22		234	267
3220	3/4" soldered		20	.400		224	26		250	286
3230	1" soldered		19	.421		255	27.50		282.50	325
9000	Transformer, for up to 3 zone valves	↓	20	.400	↓	31.50	26		57.50	73.50

23 21 23 – Hydronic Pumps

23 21 23.13 In-Line Centrifugal Hydronic Pumps

		Crew	Daily Output	Labor-Hours	Unit	Material	2020 Bare Costs Labor	Equipment	Total	Total Incl O&P
0010	**IN-LINE CENTRIFUGAL HYDRONIC PUMPS**									
0600	Bronze, sweat connections, 1/40 HP, in line									
0640	3/4" size	Q-1	16	1	Ea.	275	58		333	385
1000	Flange connection, 3/4" to 1-1/2" size									
1040	1/12 HP	Q-1	6	2.667	Ea.	710	155		865	1,025
1060	1/8 HP		6	2.667		1,250	155		1,405	1,600
1100	1/3 HP		6	2.667		1,400	155		1,555	1,750
1140	2" size, 1/6 HP		5	3.200		1,750	186		1,936	2,200
1180	2-1/2" size, 1/4 HP		5	3.200		2,200	186		2,386	2,700
1220	3" size, 1/4 HP		4	4		2,400	232		2,632	2,975
1260	1/3 HP		4	4		2,825	232		3,057	3,475
1300	1/2 HP		4	4		2,950	232		3,182	3,575
1340	3/4 HP		4	4		3,000	232		3,232	3,650
1380	1 HP	↓	4	4	↓	4,775	232		5,007	5,600
2000	Cast iron, flange connection									
2040	3/4" to 1-1/2" size, in line, 1/12 HP	Q-1	6	2.667	Ea.	470	155		625	750
2060	1/8 HP		6	2.667	↓	780	155		935	1,100

298

For customer support on your Mechanical Costs with RSMeans Data, call 800.448.8182.

23 21 Hydronic Piping and Pumps

23 21 23 – Hydronic Pumps

23 21 23.13 In-Line Centrifugal Hydronic Pumps

		Crew	Daily Output	Labor-Hours	Unit	Material	2020 Bare Costs Labor	Equipment	Total	Total Incl O&P
2100	1/3 HP	Q-1	6	2.667	Ea.	870	155		1,025	1,175
2140	2" size, 1/6 HP		5	3.200		955	186		1,141	1,325
2180	2-1/2" size, 1/4 HP		5	3.200		1,150	186		1,336	1,525
2220	3" size, 1/4 HP		4	4		1,150	232		1,382	1,625
2260	1/3 HP		4	4		1,525	232		1,757	2,025
2300	1/2 HP		4	4		1,575	232		1,807	2,075
2340	3/4 HP		4	4		1,825	232		2,057	2,350
2380	1 HP	▼	4	4	▼	2,625	232		2,857	3,225
2600	For nonferrous impeller, add					3%				
3000	High head, bronze impeller									
3030	1-1/2" size, 1/2 HP	Q-1	5	3.200	Ea.	1,400	186		1,586	1,825
3040	1-1/2" size, 3/4 HP		5	3.200		1,625	186		1,811	2,075
3050	2" size, 1 HP		4	4		1,975	232		2,207	2,525
3090	2" size, 1-1/2 HP	▼	4	4	▼	2,400	232		2,632	2,975
4000	Close coupled, end suction, bronze impeller									
4040	1-1/2" size, 1-1/2 HP, to 40 GPM	Q-1	3	5.333	Ea.	2,500	310		2,810	3,225
4090	2" size, 2 HP, to 50 GPM		3	5.333		3,025	310		3,335	3,800
4100	2" size, 3 HP, to 90 GPM		2.30	6.957		3,075	405		3,480	3,975
4190	2-1/2" size, 3 HP, to 150 GPM		2	8		3,350	465		3,815	4,400
4300	3" size, 5 HP, to 225 GPM		1.80	8.889		3,900	515		4,415	5,075
4410	3" size, 10 HP, to 350 GPM		1.60	10		6,100	580		6,680	7,575
4420	4" size, 7-1/2 HP, to 350 GPM	▼	1.60	10		6,250	580		6,830	7,750
4520	4" size, 10 HP, to 600 GPM	Q-2	1.70	14.118		6,475	850		7,325	8,400
4530	5" size, 15 HP, to 1,000 GPM		1.70	14.118		6,500	850		7,350	8,425
4610	5" size, 20 HP, to 1,350 GPM		1.50	16		6,925	960		7,885	9,050
4620	5" size, 25 HP, to 1,550 GPM	▼	1.50	16	▼	8,325	960		9,285	10,600
5000	Base mounted, bronze impeller, coupling guard									
5040	1-1/2" size, 1-1/2 HP, to 40 GPM	Q-1	2.30	6.957	Ea.	7,300	405		7,705	8,625
5090	2" size, 2 HP, to 50 GPM		2.30	6.957		8,175	405		8,580	9,600
5100	2" size, 3 HP, to 90 GPM		2	8		9,000	465		9,465	10,600
5190	2-1/2" size, 3 HP, to 150 GPM		1.80	8.889		10,400	515		10,915	12,200
5300	3" size, 5 HP, to 225 GPM		1.60	10		11,300	580		11,880	13,300
5410	4" size, 5 HP, to 350 GPM		1.50	10.667		13,100	620		13,720	15,300
5420	4" size, 7-1/2 HP, to 350 GPM	▼	1.50	10.667		14,100	620		14,720	16,400
5520	5" size, 10 HP, to 600 GPM	Q-2	1.60	15		19,400	900		20,300	22,700
5530	5" size, 15 HP, to 1,000 GPM		1.60	15		20,900	900		21,800	24,400
5610	6" size, 20 HP, to 1,350 GPM		1.40	17.143		22,600	1,025		23,625	26,500
5620	6" size, 25 HP, to 1,550 GPM	▼	1.40	17.143	▼	26,600	1,025		27,625	30,900
5800	The above pump capacities are based on 1,800 RPM,									
5810	at a 60' head. Increasing the RPM									
5820	or decreasing the head will increase the GPM.									

23 21 29 – Automatic Condensate Pump Units

23 21 29.10 Condensate Removal Pump System

			Crew	Daily Output	Labor-Hours	Unit	Material	Labor	Equipment	Total	Total Incl O&P
0010	**CONDENSATE REMOVAL PUMP SYSTEM**										
0020	Pump with 1 gal. ABS tank										
0100	115 V										
0120	1/50 HP, 200 GPH	G	1 Stpi	12	.667	Ea.	188	43.50		231.50	272
0140	1/18 HP, 270 GPH	G		10	.800		186	52.50		238.50	284
0160	1/5 HP, 450 GPH	G	▼	8	1	▼	425	65.50		490.50	570
0200	230 V										
0240	1/18 HP, 270 GPH		1 Stpi	10	.800	Ea.	208	52.50		260.50	310
0260	1/5 HP, 450 GPH	G	"	8	1	"	495	65.50		560.50	645

For customer support on your Mechanical Costs with RSMeans Data, call 800.448.8182.

299

23 22 Steam and Condensate Piping and Pumps

23 22 13 – Steam and Condensate Heating Piping

23 22 13.23 Aboveground Steam and Condensate Piping	Crew	Daily Output	Labor-Hours	Unit	Material	2020 Bare Costs Labor	Equipment	Total	Total Incl O&P
0010 **ABOVEGROUND STEAM AND CONDENSATE HEATING PIPING**									
0020 Condensate meter									
0100 500 lb. per hour	1 Stpi	14	.571	Ea.	4,300	37.50		4,337.50	4,775
0140 1500 lb. per hour		7	1.143		4,825	75		4,900	5,425
0160 3000 lb. per hour	↓	5	1.600		5,675	105		5,780	6,375
0200 12,000 lb. per hour	Q-5	3.50	4.571	↓	6,475	270		6,745	7,500

23 22 23 – Steam Condensate Pumps

23 22 23.10 Condensate Return System

	Crew	Daily Output	Labor-Hours	Unit	Material	2020 Bare Costs Labor	Equipment	Total	Total Incl O&P
0010 **CONDENSATE RETURN SYSTEM**									
2000 Simplex									
2010 With pump, motor, CI receiver, float switch									
2020 3/4 HP, 15 GPM	Q-1	1.80	8.889	Ea.	7,175	515		7,690	8,650
2100 Duplex									
2110 With 2 pumps and motors, CI receiver, float switch, alternator									
2120 3/4 HP, 15 GPM, 15 gal. CI receiver _640⁻/gpm_	Q-1	1.40	11.429	Ea.	7,825	665		8,490	9,600
2130 1 HP, 25 GPM _460⁻/gpm_		1.20	13.333		9,400	775		10,175	11,500
2140 1-1/2 HP, 45 GPM _302⁻/gpm_		1	16		11,100	930		12,030	13,600
2150 1-1/2 HP, 60 GPM _250⁻/gpm_	↓	1	16	↓	12,200	930		13,130	14,900

23 23 Refrigerant Piping

23 23 13 – Refrigerant Piping Valves

23 23 13.10 Valves

	Crew	Daily Output	Labor-Hours	Unit	Material	2020 Bare Costs Labor	Equipment	Total	Total Incl O&P
0010 **VALVES**									
8100 Check valve, soldered									
8110 5/8"	1 Stpi	36	.222	Ea.	94	14.55		108.55	126
8114 7/8"		26	.308		114	20		134	156
8118 1-1/8"		18	.444		129	29		158	186
8122 1-3/8"		14	.571		200	37.50		237.50	276
8126 1-5/8"		13	.615		250	40.50		290.50	335
8130 2-1/8"		12	.667		285	43.50		328.50	380
8134 2-5/8"	Q-5	22	.727		405	43		448	515
8138 3-1/8"	"	20	.800	↓	540	47		587	665
8500 Refrigeration valve, packless, soldered									
8510 1/2"	1 Stpi	38	.211	Ea.	47	13.80		60.80	72.50
8514 5/8"		36	.222		48.50	14.55		63.05	75.50
8518 7/8"	↓	26	.308	↓	183	20		203	231
8520 Packed, soldered									
8522 1-1/8"	1 Stpi	18	.444	Ea.	237	29		266	305
8526 1-3/8"		14	.571		385	37.50		422.50	480
8530 1-5/8"		13	.615		370	40.50		410.50	465
8534 2-1/8"	↓	12	.667		835	43.50		878.50	985
8538 2-5/8"	Q-5	22	.727		1,250	43		1,293	1,450
8542 3-1/8"		20	.800		1,250	47		1,297	1,450
8546 4-1/8"	↓	18	.889	↓	2,075	52.50		2,127.50	2,350
8600 Solenoid valve, flange/solder									
8610 1/2"	1 Stpi	38	.211	Ea.	261	13.80		274.80	310
8614 5/8"		36	.222		355	14.55		369.55	415
8618 3/4"		30	.267		455	17.50		472.50	525
8622 7/8"		26	.308		535	20		555	620
8626 1-1/8"	↓	18	.444		645	29		674	755

23 23 Refrigerant Piping

23 23 13 – Refrigerant Piping Valves

23 23 13.10 Valves		Crew	Daily Output	Labor-Hours	Unit	Material	2020 Bare Costs Labor	Equipment	Total	Total Incl O&P
8630	1-3/8"	1 Stpi	14	.571	Ea.	780	37.50		817.50	910
8634	1-5/8"		13	.615		1,100	40.50		1,140.50	1,250
8638	2-1/8"	↓	12	.667	↓	1,200	43.50		1,243.50	1,400
8800	Thermostatic valve, flange/solder									
8810	1/2-3 ton, 3/8" x 5/8"	1 Stpi	9	.889	Ea.	127	58.50		185.50	226
8814	4-5 ton, 1/2" x 7/8"		7	1.143		151	75		226	278
8818	6-8 ton, 5/8" x 7/8"		5	1.600		280	105		385	465
8822	7-12 ton, 7/8" x 1-1/8"		4	2		330	131		461	560
8826	15-20 ton, 7/8" x 1-3/8"	↓	3.20	2.500	↓	288	164		452	560

23 23 16 – Refrigerant Piping Specialties

23 23 16.10 Refrigerant Piping Component Specialties

		Crew	Daily Output	Labor-Hours	Unit	Material	2020 Bare Costs Labor	Equipment	Total	Total Incl O&P
0010	**REFRIGERANT PIPING COMPONENT SPECIALTIES**									
0600	Accumulator									
0610	3/4"	1 Stpi	8.80	.909	Ea.	52.50	59.50		112	147
0614	7/8"		6.40	1.250		67.50	82		149.50	198
0618	1-1/8"		4.80	1.667		125	109		234	300
0622	1-3/8"		4	2		123	131		254	330
0626	1-5/8"		3.20	2.500		129	164		293	385
0630	2-1/8"		2.40	3.333		345	219		564	705
0700	Condensate drip/drain pan, electric, 10" x 9" x 2-3/4"	↓	24	.333	↓	106	22		128	150
1000	Filter dryer									
1010	Replaceable core type, solder									
1020	1/2"	1 Stpi	20	.400	Ea.	69	26		95	115
1030	5/8"		19	.421		68	27.50		95.50	117
1040	7/8"		18	.444		38	29		67	85.50
1050	1-1/8"		15	.533		101	35		136	164
1060	1-3/8"		14	.571		157	37.50		194.50	229
1070	1-5/8"		12	.667		157	43.50		200.50	239
1080	2-1/8"		10	.800		460	52.50		512.50	585
1090	2-5/8"		9	.889		685	58.50		743.50	835
1100	3-1/8"	↓	8	1	↓	845	65.50		910.50	1,025
1200	Sealed in-line, solder									
1210	1/4"	1 Stpi	22	.364	Ea.	30.50	24		54.50	69
1220	3/8"		21	.381		25.50	25		50.50	65.50
1230	1/2"		20	.400		30.50	26		56.50	72.50
1260	5/8"		19	.421		41.50	27.50		69	87
1270	7/8"		18	.444		45.50	29		74.50	93.50
1290	1-1/8"	↓	15	.533	↓	69	35		104	128
4000	P-trap, suction line, solder									
4010	5/8"	1 Stpi	19	.421	Ea.	49	27.50		76.50	95
4020	3/4"		19	.421		69	27.50		96.50	118
4030	7/8"		18	.444		48.50	29		77.50	97
4040	1-1/8"		15	.533		79	35		114	140
4050	1-3/8"		14	.571		132	37.50		169.50	201
4060	1-5/8"		12	.667		247	43.50		290.50	340
4070	2-1/8"	↓	10	.800	↓	330	52.50		382.50	445
5000	Sightglass									
5010	Moisture and liquid indicator, solder									
5020	1/4"	1 Stpi	22	.364	Ea.	18.25	24		42.25	55.50
5030	3/8"		21	.381		18.30	25		43.30	57.50
5040	1/2"		20	.400		23	26		49	64.50
5050	5/8"	↓	19	.421		24	27.50		51.50	68

For customer support on your Mechanical Costs with RSMeans Data, call 800.448.8182.

301

23 23 Refrigerant Piping

23 23 16 – Refrigerant Piping Specialties

23 23 16.10 Refrigerant Piping Component Specialties		Crew	Daily Output	Labor-Hours	Unit	Material	2020 Bare Costs Labor	Equipment	Total	Total Incl O&P
5060	7/8"	1 Stpi	18	.444	Ea.	35.50	29		64.50	82.50
5070	1-1/8"		15	.533		38.50	35		73.50	94.50
5080	1-3/8"		14	.571		76	37.50		113.50	140
5090	1-5/8"		12	.667		85.50	43.50		129	160
5100	2-1/8"		10	.800		89.50	52.50		142	177
7400	Vacuum pump set									
7410	Two stage, high vacuum continuous duty	1 Stpi	8	1	Ea.	4,000	65.50		4,065.50	4,475

23 23 16.16 Refrigerant Line Sets

		Crew	Daily Output	Labor-Hours	Unit	Material	2020 Bare Costs Labor	Equipment	Total	Total Incl O&P
0010	**REFRIGERANT LINE SETS**, Standard									
0100	Copper tube									
0110	1/2" insulation, both tubes									
0120	Combination 1/4" and 1/2" tubes									
0130	10' set	Q-5	42	.381	Ea.	48.50	22.50		71	87
0135	15' set		42	.381		70.50	22.50		93	111
0140	20' set		40	.400		76	23.50		99.50	119
0150	30' set		37	.432		107	25.50		132.50	156
0160	40' set		35	.457		134	27		161	188
0170	50' set		32	.500		166	29.50		195.50	227
0180	100' set		22	.727		360	43		403	460
0300	Combination 1/4" and 3/4" tubes									
0310	10' set	Q-5	40	.400	Ea.	54.50	23.50		78	95.50
0320	20' set		38	.421		96.50	25		121.50	143
0330	30' set		35	.457		144	27		171	199
0340	40' set		33	.485		188	28.50		216.50	250
0350	50' set		30	.533		237	31.50		268.50	310
0380	100' set		20	.800		540	47		587	665
0500	Combination 3/8" & 3/4" tubes									
0510	10' set	Q-5	28	.571	Ea.	62.50	33.50		96	120
0520	20' set		36	.444		97.50	26		123.50	146
0530	30' set		34	.471		132	28		160	188
0540	40' set		31	.516		174	30.50		204.50	237
0550	50' set		28	.571		204	33.50		237.50	276
0580	100' set		18	.889		605	52.50		657.50	745
0700	Combination 3/8" & 1-1/8" tubes									
0710	10' set	Q-5	36	.444	Ea.	109	26		135	159
0720	20' set		33	.485		183	28.50		211.50	244
0730	30' set		31	.516		246	30.50		276.50	315
0740	40' set		28	.571		370	33.50		403.50	455
0750	50' set		26	.615		355	36.50		391.50	450
0900	Combination 1/2" & 3/4" tubes									
0910	10' set	Q-5	37	.432	Ea.	66	25.50		91.50	111
0920	20' set		35	.457		117	27		144	170
0930	30' set		33	.485		176	28.50		204.50	237
0940	40' set		30	.533		232	31.50		263.50	300
0950	50' set		27	.593		292	35		327	375
0980	100' set		17	.941		660	55.50		715.50	810
2100	Combination 1/2" & 1-1/8" tubes									
2110	10' set	Q-5	35	.457	Ea.	113	27		140	165
2120	20' set		31	.516		193	30.50		223.50	258
2130	30' set		29	.552		290	32.50		322.50	370
2140	40' set		25	.640		390	38		428	485
2150	50' set		14	1.143		475	67.50		542.50	625

23 23 Refrigerant Piping

23 23 16 - Refrigerant Piping Specialties

23 23 16.16 Refrigerant Line Sets	Crew	Daily Output	Labor-Hours	Unit	Material	2020 Bare Costs Labor	2020 Bare Costs Equipment	Total	Total Incl O&P
2300 For 1" thick insulation add					30%	15%			
3000 Refrigerant line sets, min-split, flared									
3100 Combination 1/4" & 3/8" tubes									
3120 15' set	Q-5	41	.390	Ea.	69.50	23		92.50	111
3140 25' set		38.50	.416		98.50	24.50		123	145
3160 35' set		37	.432		124	25.50		149.50	175
3180 50' set		30	.533		169	31.50		200.50	233
3200 Combination 1/4" & 1/2" tubes									
3220 15' set	Q-5	41	.390	Ea.	74.50	23		97.50	117
3240 25' set		38.50	.416		101	24.50		125.50	148
3260 35' set		37	.432		130	25.50		155.50	181
3280 50' set		30	.533		176	31.50		207.50	240

23 23 23 - Refrigerants

23 23 23.10 Anti-Freeze

0010 **ANTI-FREEZE**, Inhibited									
0900 Ethylene glycol concentrated									
1000 55 gallon drums, small quantities				Gal.	12.90			12.90	14.20
1200 Large quantities.					11.80			11.80	12.95
2000 Propylene glycol, for solar heat, small quantities					23			23	25.50
2100 Large quantities					9.90			9.90	10.90

23 23 23.20 Refrigerant

0010 **REFRIGERANT**									
4420 R-22, 30 lb. disposable cylinder				Lb.	22			22	24.50
4428 R-134A, 30 lb. disposable cylinder					14.20			14.20	15.60
4434 R-407C, 30 lb. disposable cylinder					10.65			10.65	11.70
4440 R-408A, 25 lb. disposable cylinder					26			26	29
4450 R-410A, 25 lb. disposable cylinder					11.80			11.80	12.95
4470 R-507, 25 lb. disposable cylinder					7.05			7.05	7.75

23 31 HVAC Ducts and Casings

23 31 13 - Metal Ducts

23 31 13.13 Rectangular Metal Ducts

0010 **RECTANGULAR METAL DUCTS**									
0020 Fabricated rectangular, includes fittings, joints, supports,									
0021 allowance for flexible connections and field sketches.									
0030 Does not include "as-built dwgs." or insulation. R233100-30									
0031 NOTE: Fabrication and installation are combined									
0040 as LABOR cost. Approx. 25% fittings assumed. R233100-40									
0042 Fabrication/Inst. is to commercial quality standards									
0043 (SMACNA or equiv.) for structure, sealing, leak testing, etc. R233100-50									
0050 Add to labor for elevated installation									
0051 of fabricated ductwork									
0052 10' to 15' high						6%			
0053 15' to 20' high						12%			
0054 20' to 25' high						15%			
0055 25' to 30' high						21%			
0056 30' to 35' high						24%			
0057 35' to 40' high						30%			
0058 Over 40' high						33%			
0072 For duct insulation see Line 23 07 13.10 3000									

23 31 13 – Metal Ducts

23 31 13.13 Rectangular Metal Ducts

		Crew	Daily Output	Labor-Hours	Unit	Material	2020 Bare Costs Labor	Equipment	Total	Total Incl O&P
0100	Aluminum, alloy 3003-H14, under 100 lb.	Q-10	75	.320	Lb.	3.20	18.60		21.80	31.50
0110	100 to 500 lb.		80	.300		1.89	17.45		19.34	28.50
0120	500 to 1,000 lb.		95	.253		1.91	14.70		16.61	24.50
0140	1,000 to 2,000 lb.		120	.200		1.86	11.65		13.51	19.70
0150	2,000 to 5,000 lb.		130	.185		1.80	10.75		12.55	18.30
0160	Over 5,000 lb.		145	.166		1.86	9.65		11.51	16.65
0500	Galvanized steel, under 200 lb.		235	.102		.60	5.95		6.55	9.65
0520	200 to 500 lb.		245	.098		.57	5.70		6.27	9.30
0540	500 to 1,000 lb.		255	.094		.54	5.45		5.99	8.90
0560	1,000 to 2,000 lb.		265	.091		.55	5.25		5.80	8.60
0570	2,000 to 5,000 lb.		275	.087		.56	5.05		5.61	8.30
0580	Over 5,000 lb.	▼	285	.084	▼	.55	4.90		5.45	8.05
0600	For large quantities special prices available from supplier									
1000	Stainless steel, type 304, under 100 lb.	Q-10	165	.145	Lb.	4.78	8.45		13.23	18.10
1020	100 to 500 lb.		175	.137		3.88	7.95		11.83	16.35
1030	500 to 1,000 lb.		190	.126		2.77	7.35		10.12	14.20
1040	1,000 to 2,000 lb.		200	.120		2.46	7		9.46	13.30
1050	2,000 to 5,000 lb.		225	.107		2.17	6.20		8.37	11.80
1060	Over 5,000 lb.	▼	235	.102	▼	2.67	5.95		8.62	11.95
1080	Note: Minimum order cost exceeds per lb. cost for min. wt.									
1100	For medium pressure ductwork, add				Lb.		15%			
1200	For high pressure ductwork, add						40%			
1210	For welded ductwork, add						85%			
1220	For 30% fittings, add						11%			
1224	For 40% fittings, add						34%			
1228	For 50% fittings, add						56%			
1232	For 60% fittings, add						79%			
1236	For 70% fittings, add						101%			
1240	For 80% fittings, add						124%			
1244	For 90% fittings, add						147%			
1248	For 100% fittings, add				▼		169%			
1252	Note: Fittings add includes time for detailing and installation.									
1276	Duct tape, 2" x 60 yd roll				Ea.	6.75			6.75	7.45
1277	Duct sealer	1 Shee	304.72	.026	L.F.	6.75	1.64		8.39	9.95

23 31 13.16 Round and Flat-Oval Spiral Ducts

		Crew	Daily Output	Labor-Hours	Unit	Material	2020 Bare Costs Labor	Equipment	Total	Total Incl O&P
0010	**ROUND AND FLAT-OVAL SPIRAL DUCTS**									
0020	Fabricated round and flat oval spiral, includes hangers,									
0021	supports and field sketches.									
1280	Add to labor for elevated installation									
1282	of prefabricated (purchased) ductwork									
1283	10' to 15' high						10%			
1284	15' to 20' high						20%			
1285	20' to 25' high						25%			
1286	25' to 30' high						35%			
1287	30' to 35' high						40%			
1288	35' to 40' high						50%			
1289	Over 40' high						55%			
5400	Spiral preformed, steel, galv., straight lengths, max 10" spwg.									
5410	4" diameter, 26 ga.	Q-9	360	.044	L.F.	1.93	2.49		4.42	5.90
5416	5" diameter, 26 ga.		320	.050		1.98	2.80		4.78	6.45
5420	6" diameter, 26 ga.		280	.057		2.03	3.20		5.23	7.10
5425	7" diameter, 26 ga.		240	.067		2.27	3.74		6.01	8.15

23 31 HVAC Ducts and Casings

23 31 13 – Metal Ducts

23 31 13.16 Round and Flat-Oval Spiral Ducts	Crew	Daily Output	Labor-Hours	Unit	Material	2020 Bare Costs Labor	Equipment	Total	Total Incl O&P
5430	8" diameter, 26 ga.	Q-9	200	.080	L.F.	2.90	4.49		7.39	10
5440	10" diameter, 26 ga.		160	.100		3.54	5.60		9.14	12.40
5450	12" diameter, 26 ga.		120	.133		4.24	7.50		11.74	16
5460	14" diameter, 26 ga.		80	.200		4.63	11.20		15.83	22
5480	16" diameter, 24 ga.		60	.267		6.40	14.95		21.35	29.50
5490	18" diameter, 24 ga.		50	.320		7.75	17.95		25.70	35.50
5500	20" diameter, 24 ga.	Q-10	65	.369		8.60	21.50		30.10	42
5510	22" diameter, 24 ga.		60	.400		9.45	23.50		32.95	46
5520	24" diameter, 24 ga.		55	.436		10.30	25.50		35.80	50
5540	30" diameter, 22 ga.		45	.533		13.90	31		44.90	62.50
5600	36" diameter, 22 ga.		40	.600		19.10	35		54.10	74
5800	Connector, 4" diameter	Q-9	100	.160	Ea.	3.31	8.95		12.26	17.25
5810	5" diameter		94	.170		3.55	9.55		13.10	18.40
5820	6" diameter		88	.182		3.72	10.20		13.92	19.55
5840	8" diameter		78	.205		4.47	11.50		15.97	22.50
5860	10" diameter		70	.229		4.48	12.80		17.28	24.50
5880	12" diameter		50	.320		5.60	17.95		23.55	33
5900	14" diameter		44	.364		5.45	20.50		25.95	37
5920	16" diameter		40	.400		6	22.50		28.50	40.50
5930	18" diameter		37	.432		6.55	24.50		31.05	44
5940	20" diameter		34	.471		8.15	26.50		34.65	49
5950	22" diameter		31	.516		8.95	29		37.95	54
5960	24" diameter		28	.571		9.50	32		41.50	59
5980	30" diameter		22	.727		15.25	41		56.25	79
6000	36" diameter		18	.889		17.45	50		67.45	94.50
6300	Elbow, 45°, 4" diameter		60	.267		6.40	14.95		21.35	29.50
6310	5" diameter		52	.308		6.75	17.25		24	33.50
6320	6" diameter		44	.364		6.90	20.50		27.40	38.50
6340	8" diameter		28	.571		8.15	32		40.15	57.50
6360	10" diameter		18	.889		8.45	50		58.45	85
6380	12" diameter		13	1.231		9.50	69		78.50	115
6400	14" diameter		11	1.455		13.15	81.50		94.65	138
6420	16" diameter		10	1.600		21	89.50		110.50	159
6430	18" diameter		9.60	1.667		52	93.50		145.50	199
6440	20" diameter	Q-10	14	1.714		58	99.50		157.50	215
6450	22" diameter		13	1.846		74	107		181	245
6460	24" diameter		12	2		77	116		193	261
6480	30" diameter		9	2.667		121	155		276	370
6500	36" diameter		7	3.429		149	199		348	465
6600	Elbow, 90°, 4" diameter	Q-9	60	.267		4.61	14.95		19.56	27.50
6610	5" diameter		52	.308		4.87	17.25		22.12	31.50
6620	6" diameter		44	.364		5	20.50		25.50	36.50
6625	7" diameter		36	.444		5.40	25		30.40	44
6630	8" diameter		28	.571		6.15	32		38.15	55.50
6640	10" diameter		18	.889		7.65	50		57.65	84
6650	12" diameter		13	1.231		9.55	69		78.55	116
6660	14" diameter		11	1.455		16.45	81.50		97.95	142
6670	16" diameter		10	1.600		17.05	89.50		106.55	155
6676	18" diameter		9.60	1.667		69.50	93.50		163	219
6680	20" diameter	Q-10	14	1.714		77	99.50		176.50	236
6684	22" diameter		13	1.846		72.50	107		179.50	243
6690	24" diameter		12	2		83	116		199	268
6700	30" diameter		9	2.667		168	155		323	420

For customer support on your Mechanical Costs with RSMeans Data, call 800.448.8182.

305

23 31 13.16 Round and Flat-Oval Spiral Ducts	Crew	Daily Output	Labor-Hours	Unit	Material	2020 Bare Costs Labor	Equipment	Total	Total Incl O&P
6710	36" diameter	Q-10	7	3.429	Ea.	223	199		422	545
6720	End cap									
6722	4"	Q-9	110	.145	Ea.	4.86	8.15		13.01	17.70
6724	6"		94	.170		4.86	9.55		14.41	19.85
6725	8"		88	.182		6.25	10.20		16.45	22.50
6726	10"		78	.205		7.60	11.50		19.10	26
6727	12"		70	.229		8.35	12.80		21.15	28.50
6728	14"		50	.320		11.70	17.95		29.65	40
6729	16"		44	.364		13.25	20.50		33.75	45.50
6730	18"		40	.400		17.75	22.50		40.25	53.50
6731	20"		37	.432		35.50	24.50		60	76.50
6732	22"		34	.471		40	26.50		66.50	84
6733	24"		30	.533		37	30		67	86
6800	Reducing coupling, 6" x 4"		46	.348		17.40	19.50		36.90	48.50
6820	8" x 6"		40	.400		20	22.50		42.50	56
6840	10" x 8"		32	.500		22	28		50	66.50
6860	12" x 10"		24	.667		22	37.50		59.50	81
6880	14" x 12"		20	.800		21	45		66	91
6900	16" x 14"		18	.889		26.50	50		76.50	105
6920	18" x 16"	▼	16	1		38.50	56		94.50	128
6940	20" x 18"	Q-10	24	1		49.50	58		107.50	143
6950	22" x 20"		23	1.043		49.50	60.50		110	146
6960	24" x 22"		22	1.091		56	63.50		119.50	158
6980	30" x 28"		18	1.333		111	77.50		188.50	240
7000	36" x 34"	▼	16	1.500		153	87		240	300
7100	Tee, 90°, 4" diameter	Q-9	40	.400		13.70	22.50		36.20	49
7110	5" diameter		34.60	.462		14.20	26		40.20	55
7120	6" diameter		30	.533		14.85	30		44.85	62
7130	8" diameter		19	.842		18.70	47		65.70	92
7140	10" diameter		12	1.333		28	75		103	144
7150	12" diameter		9	1.778		33.50	99.50		133	188
7160	14" diameter		7	2.286		45	128		173	243
7170	16" diameter		6.70	2.388		72.50	134		206.50	283
7176	18" diameter	▼	6	2.667		106	150		256	345
7180	20" diameter	Q-10	8.70	2.759		126	160		286	380
7190	24" diameter		7	3.429		188	199		387	505
7200	30" diameter		6	4		252	233		485	630
7210	36" diameter	▼	5	4.800	▼	320	279		599	780
7250	Saddle tap									
7254	4" diameter on 6" diameter	Q-9	36	.444	Ea.	20.50	25		45.50	60.50
7256	5" diameter on 7" diameter		35.80	.447		22.50	25		47.50	62.50
7258	6" diameter on 8" diameter		35.40	.452		22.50	25.50		48	63
7260	6" diameter on 10" diameter		35.20	.455		22.50	25.50		48	63
7262	6" diameter on 12" diameter		35	.457		22.50	25.50		48	63.50
7264	7" diameter on 9" diameter		35.20	.455		25.50	25.50		51	66.50
7266	8" diameter on 10" diameter		35	.457		25.50	25.50		51	67
7268	8" diameter on 12" diameter		34.80	.460		25.50	26		51.50	67
7270	8" diameter on 14" diameter		34.50	.464		26	26		52	68
7272	10" diameter on 12" diameter		34.20	.468		28.50	26		54.50	71
7274	10" diameter on 14" diameter		33.80	.473		28.50	26.50		55	71.50
7276	10" diameter on 16" diameter		33.40	.479		28.50	27		55.50	71.50
7278	10" diameter on 18" diameter		33	.485		31	27		58	75.50
7280	12" diameter on 14" diameter		32.60	.491	▼	34.50	27.50		62	79.50

23 31 HVAC Ducts and Casings

23 31 13 – Metal Ducts

23 31 13.16 Round and Flat-Oval Spiral Ducts

		Crew	Daily Output	Labor-Hours	Unit	Material	2020 Bare Costs Labor	Equipment	Total	Total Incl O&P
7282	12" diameter on 18" diameter	Q-9	32.20	.497	Ea.	36	28		64	82.50
7284	14" diameter on 16" diameter		31.60	.506		51	28.50		79.50	99.50
7286	14" diameter on 20" diameter		31	.516		51	29		80	101
7288	16" diameter on 20" diameter		30	.533		69	30		99	122
7290	18" diameter on 20" diameter		28	.571		85	32		117	142
7292	20" diameter on 22" diameter		26	.615		110	34.50		144.50	174
7294	22" diameter on 24" diameter	↓	24	.667	↓	122	37.50		159.50	191
7400	Steel, PVC coated both sides, straight lengths									
7410	4" diameter, 26 ga.	Q-9	360	.044	L.F.	4.43	2.49		6.92	8.65
7416	5" diameter, 26 ga.		240	.067		4.65	3.74		8.39	10.75
7420	6" diameter, 26 ga.		280	.057		4.42	3.20		7.62	9.70
7440	8" diameter, 26 ga.		200	.080		5.95	4.49		10.44	13.35
7460	10" diameter, 26 ga.		160	.100		7.50	5.60		13.10	16.75
7480	12" diameter, 26 ga.		120	.133		8.45	7.50		15.95	20.50
7500	14" diameter, 26 ga.		80	.200		10.05	11.20		21.25	28
7520	16" diameter, 24 ga.		60	.267		12.35	14.95		27.30	36
7540	18" diameter, 24 ga.	↓	45	.356		13.85	19.95		33.80	45.50
7560	20" diameter, 24 ga.	Q-10	65	.369		15.45	21.50		36.95	49.50
7580	24" diameter, 24 ga.		55	.436		18.60	25.50		44.10	59
7600	30" diameter, 22 ga.		45	.533		28	31		59	77.50
7620	36" diameter, 22 ga.	↓	40	.600	↓	33.50	35		68.50	90
7890	Connector, 4" diameter	Q-9	100	.160	Ea.	2.71	8.95		11.66	16.60
7896	5" diameter		83	.193		3.02	10.80		13.82	19.70
7900	6" diameter		88	.182		3.43	10.20		13.63	19.20
7920	8" diameter		78	.205		4.06	11.50		15.56	22
7940	10" diameter		70	.229		4.71	12.80		17.51	24.50
7960	12" diameter		50	.320		5.55	17.95		23.50	33
7980	14" diameter		44	.364		5.90	20.50		26.40	37.50
8000	16" diameter		40	.400		6.95	22.50		29.45	41.50
8020	18" diameter		37	.432		7.75	24.50		32.25	45.50
8040	20" diameter		34	.471		9.35	26.50		35.85	50.50
8060	24" diameter		28	.571		11.55	32		43.55	61
8080	30" diameter		22	.727		15.40	41		56.40	79
8100	36" diameter		18	.889		18.80	50		68.80	96
8390	Elbow, 45°, 4" diameter		60	.267		26	14.95		40.95	51
8396	5" diameter		36	.444		25	25		50	65.50
8400	6" diameter		44	.364		25	20.50		45.50	58.50
8420	8" diameter		28	.571		23	32		55	73.50
8440	10" diameter		18	.889		23.50	50		73.50	102
8460	12" diameter		13	1.231		24	69		93	131
8480	14" diameter		11	1.455		24.50	81.50		106	151
8500	16" diameter		10	1.600		23.50	89.50		113	162
8520	18" diameter	↓	9	1.778		26.50	99.50		126	180
8540	20" diameter	Q-10	13	1.846		83.50	107		190.50	255
8560	24" diameter		11	2.182		104	127		231	305
8580	30" diameter		9	2.667		172	155		327	425
8600	36" diameter	↓	7	3.429		202	199		401	525
8800	Elbow, 90°, 4" diameter	Q-9	60	.267		27	14.95		41.95	52
8804	5" diameter		52	.308		26.50	17.25		43.75	55
8806	6" diameter		44	.364		26.50	20.50		47	60
8808	8" diameter		28	.571		25.50	32		57.50	76.50
8810	10" diameter		18	.889		25.50	50		75.50	104
8812	12" diameter	↓	13	1.231		26	69		95	134

For customer support on your Mechanical Costs with RSMeans Data, call 800.448.8182.

307

23 31 HVAC Ducts and Casings

23 31 13 - Metal Ducts

23 31 13.16 Round and Flat-Oval Spiral Ducts

		Crew	Daily Output	Labor-Hours	Unit	Material	2020 Bare Costs Labor	Equipment	Total	Total Incl O&P
8814	14" diameter	Q-9	11	1.455	Ea.	29.50	81.50		111	157
8816	16" diameter		10	1.600		30.50	89.50		120	170
8818	18" diameter	↓	9.60	1.667		37.50	93.50		131	184
8820	20" diameter	Q-10	14	1.714		47.50	99.50		147	203
8822	24" diameter		12	2		109	116		225	296
8824	30" diameter		9	2.667		213	155		368	470
8826	36" diameter	↓	7	3.429	↓	335	199		534	670
9000	Reducing coupling									
9040	6" x 4"	Q-9	46	.348	Ea.	19.40	19.50		38.90	51
9060	8" x 6"		40	.400		21	22.50		43.50	57
9080	10" x 8"		32	.500		22.50	28		50.50	67.50
9100	12" x 10"		24	.667		23.50	37.50		61	82
9120	14" x 12"		20	.800		28.50	45		73.50	99.50
9140	16" x 14"		18	.889		40	50		90	120
9160	18" x 16"	↓	16	1		50	56		106	140
9180	20" x 18"	Q-10	24	1		57	58		115	151
9200	24" x 22"		22	1.091		76.50	63.50		140	181
9220	30" x 28"		18	1.333		118	77.50		195.50	248
9240	36" x 34"	↓	16	1.500	↓	171	87		258	320
9800	Steel, stainless, straight lengths									
9810	3" diameter, 26 ga.	Q-9	400	.040	L.F.	15.25	2.24		17.49	20
9820	4" diameter, 26 ga.		360	.044		17.70	2.49		20.19	23
9830	5" diameter, 26 ga.		320	.050		19.20	2.80		22	25.50
9840	6" diameter, 26 ga.		280	.057		19.35	3.20		22.55	26.50
9850	7" diameter, 26 ga.		240	.067		35.50	3.74		39.24	44.50
9860	8" diameter, 26 ga.		200	.080		26.50	4.49		30.99	36
9870	9" diameter, 26 ga.		180	.089		38.50	4.99		43.49	50
9880	10" diameter, 26 ga.		160	.100		50.50	5.60		56.10	64
9890	12" diameter, 26 ga.		120	.133		72	7.50		79.50	90.50
9900	14" diameter, 26 ga.	↓	80	.200	↓	75	11.20		86.20	99.50

23 31 13.17 Round Grease Duct

		Crew	Daily Output	Labor-Hours	Unit	Material	2020 Bare Costs Labor	Equipment	Total	Total Incl O&P
0010	**ROUND GREASE DUCT**									
1280	Add to labor for elevated installation									
1282	of prefabricated (purchased) ductwork									
1283	10' to 15' high						10%			
1284	15' to 20' high						20%			
1285	20' to 25' high						25%			
1286	25' to 30' high						35%			
1287	30' to 35' high						40%			
1288	35' to 40' high						50%			
1289	Over 40' high						55%			
4020	Zero clearance, 2 hr. fire rated, UL listed, 3" double wall									
4030	304 interior, aluminized exterior, all sizes are ID									
4032	For 316 interior and aluminized exterior, add						6%			
4034	For 304 interior and 304 exterior, add						21%			
4036	For 316 interior and 316 exterior, add						40%			
4040	Straight									
4052	6"	Q-9	34.20	.468	L.F.	77	26		103	125
4054	8"		29.64	.540		88.50	30.50		119	144
4056	10"		27.36	.585		99	33		132	159
4058	12"		25.10	.637		113	36		149	178
4060	14"	↓	23.94	.668	↓	128	37.50		165.50	198

23 31 13.17 Round Grease Duct	Crew	Daily Output	Labor-Hours	Unit	Material	2020 Bare Costs Labor	Equipment	Total	Total Incl O&P
4062	16"	Q-9	22.80	.702	L.F.	139	39.50		178.50	212
4064	18"	↓	21.66	.739		163	41.50		204.50	243
4066	20"	Q-10	20.52	1.170		186	68		254	305
4068	22"		19.38	1.238		210	72		282	340
4070	24"		18.24	1.316		238	76.50		314.50	380
4072	26"		17.67	1.358		255	79		334	400
4074	28"		17.10	1.404		272	81.50		353.50	425
4076	30"		15.96	1.504		287	87.50		374.50	450
4078	32"		15.39	1.559		310	90.50		400.50	480
4080	36"	↓	14.25	1.684	↓	345	98		443	530
4100	Adjustable section, 30" long									
4104	6"	Q-9	17.10	.936	Ea.	234	52.50		286.50	340
4106	8"		14.82	1.080		287	60.50		347.50	405
4108	10"		13.68	1.170		325	65.50		390.50	455
4110	12"		12.54	1.276		360	71.50		431.50	505
4112	14"		11.97	1.337		405	75		480	560
4114	16"		11.40	1.404		465	78.50		543.50	630
4116	18"	↓	10.83	1.477		515	83		598	695
4118	20"	Q-10	10.26	2.339		605	136		741	870
4120	22"		9.69	2.477		680	144		824	970
4122	24"		9.12	2.632		775	153		928	1,075
4124	26"		8.83	2.718		830	158		988	1,150
4126	28"		8.55	2.807		860	163		1,023	1,200
4128	30"		8.26	2.906		945	169		1,114	1,300
4130	32"		7.98	3.008		1,000	175		1,175	1,375
4132	36"	↓	6.84	3.509	↓	1,125	204		1,329	1,525
4140	Tee, 90°, grease									
4144	6"	Q-9	13.68	1.170	Ea.	415	65.50		480.50	555
4146	8"		12.54	1.276		455	71.50		526.50	610
4148	10"		11.97	1.337		505	75		580	670
4150	12"		11.40	1.404		585	78.50		663.50	765
4152	14"		10.26	1.559		665	87.50		752.50	870
4154	16"		9.12	1.754		735	98.50		833.50	960
4156	18"		7.98	2.005		865	112		977	1,125
4158	20"		9.69	1.651		1,000	92.50		1,092.50	1,250
4160	22"		8.26	1.937		1,125	109		1,234	1,400
4162	24"		6.84	2.339		1,225	131		1,356	1,550
4164	26"	↓	6.55	2.443		1,350	137		1,487	1,700
4166	28"	Q-10	6.27	3.828		1,500	223		1,723	2,000
4168	30"		5.98	4.013		1,775	233		2,008	2,325
4170	32"		5.70	4.211		2,100	245		2,345	2,675
4172	36"	↓	5.13	4.678	↓	2,325	272		2,597	2,975
4180	Cleanout tee cap									
4184	6"	Q-9	21.09	.759	Ea.	84.50	42.50		127	158
4186	8"		19.38	.826		89	46.50		135.50	168
4188	10"		18.24	.877		96	49		145	180
4190	12"		17.10	.936		103	52.50		155.50	193
4192	14"		15.96	1.003		108	56		164	205
4194	16"		14.25	1.123		120	63		183	228
4196	18"	↓	13.68	1.170		127	65.50		192.50	240
4198	20"	Q-10	15.39	1.559		141	90.50		231.50	293
4200	22"		12.54	1.914		152	111		263	335
4202	24"	↓	11.97	2.005	↓	166	117		283	360

For customer support on your Mechanical Costs with RSMeans Data, call 800.448.8182.

309

23 31 HVAC Ducts and Casings

23 31 13 – Metal Ducts

23 31 13.17 Round Grease Duct		Crew	Daily Output	Labor-Hours	Unit	Material	2020 Bare Costs Labor	Equipment	Total	Total Incl O&P
4204	26"	Q-10	11.40	2.105	Ea.	181	122		303	385
4206	28"		10.83	2.216		197	129		326	410
4208	30"		10.26	2.339		214	136		350	440
4210	32"		9.69	2.477		234	144		378	475
4212	36"		8.55	2.807		274	163		437	550
4220	Elbow, 45°									
4224	6"	Q-9	17.10	.936	Ea.	288	52.50		340.50	395
4226	8"		14.82	1.080		325	60.50		385.50	445
4228	10"		13.68	1.170		370	65.50		435.50	505
4230	12"		12.54	1.276		420	71.50		491.50	570
4232	14"		11.97	1.337		470	75		545	635
4234	16"		11.40	1.404		535	78.50		613.50	705
4236	18"		10.83	1.477		605	83		688	790
4238	20"	Q-10	10.26	2.339		685	136		821	955
4240	22"		9.69	2.477		765	144		909	1,050
4242	24"		9.12	2.632		875	153		1,028	1,200
4244	26"		8.83	2.718		925	158		1,083	1,275
4246	28"		8.55	2.807		1,100	163		1,263	1,475
4248	30"		8.26	2.906		1,100	169		1,269	1,475
4250	32"		7.98	3.008		1,325	175		1,500	1,725
4252	36"		6.84	3.509		1,775	204		1,979	2,250
4260	Elbow, 90°									
4264	6"	Q-9	17.10	.936	Ea.	565	52.50		617.50	700
4266	8"		14.82	1.080		635	60.50		695.50	790
4268	10"		13.68	1.170		775	65.50		840.50	950
4270	12"		12.54	1.276		875	71.50		946.50	1,075
4272	14"		11.97	1.337		940	75		1,015	1,150
4274	16"		11.40	1.404		1,050	78.50		1,128.50	1,300
4276	18"		10.83	1.477		1,275	83		1,358	1,525
4278	20"	Q-10	10.26	2.339		1,350	136		1,486	1,700
4280	22"		9.69	2.477		1,525	144		1,669	1,900
4282	24"		9.12	2.632		1,700	153		1,853	2,100
4284	26"		8.83	2.718		1,825	158		1,983	2,275
4286	28"		8.55	2.807		1,925	163		2,088	2,350
4288	30"		8.26	2.906		2,225	169		2,394	2,700
4290	32"		7.98	3.008		2,575	175		2,750	3,125
4292	36"		6.84	3.509		3,475	204		3,679	4,125
4300	Support strap									
4304	6"	Q-9	25.50	.627	Ea.	98	35		133	162
4306	8"		22	.727		114	41		155	188
4308	10"		20.50	.780		118	44		162	196
4310	12"		18.80	.851		126	47.50		173.50	212
4312	14"		17.90	.894		136	50		186	225
4314	16"		17	.941		137	53		190	230
4316	18"		16.20	.988		139	55.50		194.50	237
4318	20"	Q-10	15.40	1.558		154	90.50		244.50	305
4320	22"	"	14.50	1.655		170	96.50		266.50	335
4340	Plate support assembly									
4344	6"	Q-9	14.82	1.080	Ea.	141	60.50		201.50	247
4346	8"		12.54	1.276		166	71.50		237.50	292
4348	10"		11.40	1.404		178	78.50		256.50	315
4350	12"		10.26	1.559		190	87.50		277.50	340
4352	14"		9.69	1.651		223	92.50		315.50	385

For customer support on your Mechanical Costs with RSMeans Data, call 800.448.8182.

23 31 13.17 Round Grease Duct		Crew	Daily Output	Labor-Hours	Unit	Material	2020 Bare Costs Labor	2020 Bare Costs Equipment	Total	Total Incl O&P
4354	16"	Q-9	9.12	1.754	Ea.	234	98.50		332.50	405
4356	18"	▼	8.55	1.871		248	105		353	430
4358	20"	Q-10	9.12	2.632		260	153		413	520
4360	22"		8.55	2.807		266	163		429	540
4362	24"		7.98	3.008		269	175		444	560
4364	26"		7.69	3.121		300	181		481	605
4366	28"		7.41	3.239		325	188		513	645
4368	30"		7.12	3.371		365	196		561	695
4370	32"		6.84	3.509		405	204		609	760
4372	36"	▼	5.70	4.211	▼	485	245		730	905
4380	Wall guide assembly									
4384	6"	Q-9	19.10	.838	Ea.	183	47		230	274
4386	8"		16.50	.970		209	54.50		263.50	315
4388	10"		15.30	1.046		221	58.50		279.50	330
4390	12"		14.10	1.135		231	63.50		294.50	350
4392	14"		13.40	1.194		250	67		317	375
4394	16"		12.75	1.255		253	70.50		323.50	385
4396	18"	▼	12.10	1.322		256	74		330	395
4398	20"	Q-10	11.50	2.087		286	121		407	500
4400	22"		10.90	2.202		310	128		438	535
4402	24"		10.20	2.353		345	137		482	590
4404	26"		9.90	2.424		380	141		521	635
4406	28"		9.60	2.500		395	145		540	655
4408	30"		9.30	2.581		450	150		600	725
4410	32"		9	2.667		480	155		635	760
4412	36"	▼	7.60	3.158	▼	525	184		709	855
4420	Hood transition, flanged or unflanged									
4424	6"	Q-9	15.60	1.026	Ea.	78.50	57.50		136	174
4426	8"		13.50	1.185		89.50	66.50		156	200
4428	10"		12.50	1.280		99.50	72		171.50	218
4430	12"		11.40	1.404		119	78.50		197.50	250
4432	14"		10.90	1.468		128	82.50		210.50	266
4434	16"		10.40	1.538		138	86.50		224.50	283
4436	18"	▼	9.80	1.633		158	91.50		249.50	315
4438	20"	Q-10	9.40	2.553		167	148		315	410
4440	22"		8.80	2.727		181	159		340	440
4442	24"		8.30	2.892		203	168		371	480
4444	26"		8.10	2.963		208	172		380	490
4446	28"		7.80	3.077		221	179		400	515
4448	30"		7.50	3.200		240	186		426	545
4450	32"		7.30	3.288		251	191		442	565
4452	36"	▼	6.20	3.871	▼	279	225		504	645
4460	Fan adapter									
4464	6"	Q-9	15	1.067	Ea.	227	60		287	340
4466	8"		13	1.231		257	69		326	390
4468	10"		12	1.333		281	75		356	425
4470	12"		11	1.455		315	81.50		396.50	470
4472	14"		10.50	1.524		340	85.50		425.50	505
4474	16"		10	1.600		370	89.50		459.50	545
4476	18"	▼	9.50	1.684		405	94.50		499.50	590
4478	20"	Q-10	9	2.667		430	155		585	705
4480	22"		8.50	2.824		460	164		624	755
4482	24"	▼	8	3		485	174		659	800

For customer support on your Mechanical Costs with RSMeans Data, call 800.448.8182.

311

23 31 HVAC Ducts and Casings

23 31 13 – Metal Ducts

23 31 13.17 Round Grease Duct

		Crew	Daily Output	Labor-Hours	Unit	Material	2020 Bare Costs Labor	2020 Bare Costs Equipment	Total	Total Incl O&P
4484	26"	Q-10	7.70	3.117	Ea.	515	181		696	840
4486	28"		7.50	3.200		540	186		726	875
4488	30"		7.20	3.333		570	194		764	925
4490	32"		7	3.429		600	199		799	960
4492	36"		6	4		665	233		898	1,100
4500	Tapered increaser/reducer									
4530	6" x 10" diam.	Q-9	13.40	1.194	Ea.	290	67		357	420
4540	6" x 20" diam.		10.10	1.584		293	89		382	460
4550	8" x 10" diam.		13.20	1.212		350	68		418	490
4560	8" x 20" diam.		9.90	1.616		350	90.50		440.50	520
4570	10" x 20" diam.		10.60	1.509		410	84.50		494.50	580
4580	10" x 28" diam.	Q-10	8.60	2.791		410	162		572	695
4590	12" x 20" diam.		8.80	2.727		470	159		629	755
4600	14" x 28" diam.		8.40	2.857		535	166		701	840
4610	16" x 20" diam.		8.60	2.791		595	162		757	895
4620	16" x 30" diam.		8.30	2.892		590	168		758	905
4630	18" x 24" diam.		8.80	2.727		645	159		804	950
4640	18" x 30" diam.		8.10	2.963		650	172		822	975
4650	20" x 24" diam.		8.30	2.892		705	168		873	1,025
4660	20" x 32" diam.		8	3		705	174		879	1,050
4670	24" x 26" diam.		8	3		820	174		994	1,175
4680	24" x 36" diam.		7.30	3.288		830	191		1,021	1,200
4690	28" x 30" diam.		7.90	3.038		945	177		1,122	1,325
4700	28" x 36" diam.		7.10	3.380		945	197		1,142	1,350
4710	30" x 36" diam.		6.90	3.478		1,000	202		1,202	1,400
4730	Many intermediate standard sizes of tapered fittings are available									

23 31 13.19 Metal Duct Fittings

		Crew	Daily Output	Labor-Hours	Unit	Material	2020 Bare Costs Labor	2020 Bare Costs Equipment	Total	Total Incl O&P
0010	**METAL DUCT FITTINGS**									
0350	Duct collar, spin-in type									
0352	Without damper									
0354	Round									
0356	4"	1 Shee	32	.250	Ea.	3.82	15.60		19.42	27.50
0357	5"		30	.267		4.22	16.60		20.82	29.50
0358	6"		28	.286		5.60	17.80		23.40	33
0359	7"		26	.308		4.76	19.15		23.91	34.50
0360	8"		24	.333		4.76	21		25.76	37
0361	9"		22	.364		5.20	22.50		27.70	40.50
0362	10"		20	.400		5.25	25		30.25	44
0363	12"		17	.471		7.75	29.50		37.25	53
2000	Fabrics for flexible connections, with metal edge		100	.080	L.F.	3.89	4.98		8.87	11.85
2100	Without metal edge		160	.050	"	3.53	3.12		6.65	8.60

23 31 16 – Nonmetal Ducts

23 31 16.13 Fibrous-Glass Ducts

		Crew	Daily Output	Labor-Hours	Unit	Material	2020 Bare Costs Labor	2020 Bare Costs Equipment	Total	Total Incl O&P
0010	**FIBROUS-GLASS DUCTS**									
1280	Add to labor for elevated installation									
1282	of prefabricated (purchased) ductwork									
1283	10' to 15' high						10%			
1284	15' to 20' high						20%			
1285	20' to 25' high						25%			
1286	25' to 30' high						35%			
1287	30' to 35' high						40%			
1288	35' to 40' high						50%			

23 31 HVAC Ducts and Casings

23 31 16 – Nonmetal Ducts

23 31 16.13 Fibrous-Glass Ducts

		Crew	Daily Output	Labor-Hours	Unit	Material	2020 Bare Costs Labor	Equipment	Total	Total Incl O&P
1289	Over 40' high						55%			
3490	Rigid fiberglass duct board, foil reinf. kraft facing									
3500	Rectangular, 1" thick, alum. faced, (FRK), std. weight	Q-10	350	.069	SF Surf	.89	3.99		4.88	7.05

23 31 16.16 Thermoset Fiberglass-Reinforced Plastic Ducts

		Crew	Daily Output	Labor-Hours	Unit	Material	2020 Bare Costs Labor	Equipment	Total	Total Incl O&P
0010	**THERMOSET FIBERGLASS-REINFORCED PLASTIC DUCTS**									
1280	Add to labor for elevated installation									
1282	of prefabricated (purchased) ductwork									
1283	10' to 15' high						10%			
1284	15' to 20' high						20%			
1285	20' to 25' high						25%			
1286	25' to 30' high						35%			
1287	30' to 35' high						40%			
1288	35' to 40' high						50%			
1289	Over 40' high						55%			
3550	Rigid fiberglass reinforced plastic, FM approved									
3552	for acid fume and smoke exhaust system, nonflammable									
3554	Straight, 4" diameter	Q-9	200	.080	L.F.	10.35	4.49		14.84	18.20
3555	6" diameter		146	.110		13.60	6.15		19.75	24.50
3556	8" diameter		106	.151		16.45	8.45		24.90	31
3557	10" diameter		85	.188		19.75	10.55		30.30	37.50
3558	12" diameter		68	.235		23	13.20		36.20	45.50
3559	14" diameter		50	.320		34	17.95		51.95	64.50
3560	16" diameter		40	.400		38	22.50		60.50	76
3561	18" diameter		31	.516		42.50	29		71.50	90.50
3562	20" diameter	Q-10	44	.545		46.50	31.50		78	99.50
3563	22" diameter		40	.600		50.50	35		85.50	109
3564	24" diameter		37	.649		54.50	37.50		92	117
3565	26" diameter		34	.706		58.50	41		99.50	127
3566	28" diameter		28	.857		63	50		113	145
3567	30" diameter		24	1		67	58		125	162
3568	32" diameter		22.60	1.062		71	62		133	172
3569	34" diameter		21.60	1.111		75	64.50		139.50	181
3570	36" diameter		20.60	1.165		79.50	68		147.50	191
3571	38" diameter		19.80	1.212		122	70.50		192.50	241
3572	42" diameter		18.30	1.311		137	76.50		213.50	266
3573	46" diameter		17	1.412		151	82		233	291
3574	50" diameter		15.90	1.509		165	88		253	315
3575	54" diameter		14.90	1.611		181	93.50		274.50	340
3576	60" diameter		13.60	1.765		207	103		310	385
3577	64" diameter		12.80	1.875		305	109		414	500
3578	68" diameter		12.20	1.967		320	114		434	530
3579	72" diameter		11.90	2.017		345	117		462	560
3584	Note: joints are cemented with									
3586	fiberglass resin, included in material cost.									
3590	Elbow, 90°, 4" diameter	Q-9	20.30	.788	Ea.	50	44		94	122
3591	6" diameter		13.80	1.159		70	65		135	176
3592	8" diameter		10	1.600		77	89.50		166.50	221
3593	10" diameter		7.80	2.051		101	115		216	285
3594	12" diameter		6.50	2.462		128	138		266	350
3595	14" diameter		5.50	2.909		157	163		320	420
3596	16" diameter		4.80	3.333		187	187		374	490
3597	18" diameter		4.40	3.636		268	204		472	605

For customer support on your Mechanical Costs with RSMeans Data, call 800.448.8182.

313

23 31 HVAC Ducts and Casings

23 31 16 – Nonmetal Ducts

	23 31 16.16 Thermoset Fiberglass-Reinforced Plastic Ducts	Crew	Daily Output	Labor-Hours	Unit	Material	2020 Bare Costs Labor	Equipment	Total	Total Incl O&P
3598	20" diameter	Q-10	5.90	4.068	Ea.	282	237		519	670
3599	22" diameter		5.40	4.444		345	258		603	770
3600	24" diameter		4.90	4.898		400	285		685	870
3601	26" diameter		4.60	5.217		405	305		710	905
3602	28" diameter		4.20	5.714		505	330		835	1,050
3603	30" diameter		3.90	6.154		565	360		925	1,175
3604	32" diameter		3.70	6.486		625	375		1,000	1,250
3605	34" diameter		3.45	6.957		735	405		1,140	1,425
3606	36" diameter		3.20	7.500		845	435		1,280	1,600
3607	38" diameter		3.07	7.818		1,200	455		1,655	2,025
3608	42" diameter		2.70	8.889		1,450	515		1,965	2,375
3609	46" diameter		2.54	9.449		1,675	550		2,225	2,675
3610	50" diameter		2.33	10.300		1,950	600		2,550	3,050
3611	54" diameter		2.14	11.215		2,225	650		2,875	3,450
3612	60" diameter		1.94	12.371		2,975	720		3,695	4,375
3613	64" diameter		1.82	13.187		4,325	765		5,090	5,950
3614	68" diameter		1.71	14.035		4,800	815		5,615	6,550
3615	72" diameter		1.62	14.815		5,325	860		6,185	7,150
3626	Elbow, 45°, 4" diameter	Q-9	22.30	.717		48	40		88	114
3627	6" diameter		15.20	1.053		66.50	59		125.50	163
3628	8" diameter		11	1.455		70.50	81.50		152	202
3629	10" diameter		8.60	1.860		74.50	104		178.50	240
3630	12" diameter		7.20	2.222		82	125		207	279
3631	14" diameter		6.10	2.623		100	147		247	335
3632	16" diameter		5.28	3.030		119	170		289	390
3633	18" diameter		4.84	3.306		140	185		325	435
3634	20" diameter	Q-10	6.50	3.692		162	215		377	505
3635	22" diameter		5.94	4.040		220	235		455	595
3636	24" diameter		5.40	4.444		251	258		509	665
3637	26" diameter		5.06	4.743		285	276		561	735
3638	28" diameter		4.62	5.195		310	300		610	800
3639	30" diameter		4.30	5.581		355	325		680	880
3640	32" diameter		4.07	5.897		395	345		740	955
3641	34" diameter		3.80	6.316		465	365		830	1,075
3642	36" diameter		3.52	6.818		540	395		935	1,200
3643	38" diameter		3.38	7.101		755	415		1,170	1,450
3644	42" diameter		2.97	8.081		905	470		1,375	1,700
3645	46" diameter		2.80	8.571		1,050	500		1,550	1,925
3646	50" diameter		2.56	9.375		1,225	545		1,770	2,175
3647	54" diameter		2.35	10.213		1,400	595		1,995	2,425
3648	60" diameter		2.13	11.268		1,875	655		2,530	3,075
3649	64" diameter		2	12		2,725	700		3,425	4,050
3650	68" diameter		1.88	12.766		3,000	740		3,740	4,425
3651	72" diameter		1.78	13.483		3,325	785		4,110	4,850
3660	Tee, 90°, 4" diameter	Q-9	14.06	1.138		27.50	64		91.50	127
3661	6" diameter		9.52	1.681		42	94.50		136.50	189
3662	8" diameter		6.98	2.292		57.50	129		186.50	258
3663	10" diameter		5.45	2.936		74.50	165		239.50	330
3664	12" diameter		4.60	3.478		93.50	195		288.50	400
3665	14" diameter		3.86	4.145		136	232		368	505
3666	16" diameter		3.40	4.706		162	264		426	580
3667	18" diameter		3.05	5.246		192	294		486	655
3668	20" diameter	Q-10	4.12	5.825		224	340		564	760

314

For customer support on your Mechanical Costs with RSMeans Data, call 800.448.8182.

23 31 16.16 Thermoset Fiberglass-Reinforced Plastic Ducts		Crew	Daily Output	Labor-Hours	Unit	Material	2020 Bare Costs Labor	Equipment	Total	Total Incl O&P
3669	22" diameter	Q-10	3.78	6.349	Ea.	256	370		626	840
3670	24" diameter		3.46	6.936		289	405		694	930
3671	26" diameter		3.20	7.500		325	435		760	1,025
3672	28" diameter		2.96	8.108		365	470		835	1,125
3673	30" diameter		2.75	8.727		410	505		915	1,225
3674	32" diameter		2.54	9.449		450	550		1,000	1,325
3675	34" diameter		2.40	10		495	580		1,075	1,425
3676	36" diameter		2.27	10.573		545	615		1,160	1,525
3677	38" diameter		2.15	11.163		820	650		1,470	1,875
3678	42" diameter		1.91	12.565		985	730		1,715	2,175
3679	46" diameter		1.78	13.483		1,150	785		1,935	2,475
3680	50" diameter		1.63	14.724		1,350	855		2,205	2,775
3681	54" diameter		1.50	16		1,575	930		2,505	3,125
3682	60" diameter		1.35	17.778		1,950	1,025		2,975	3,725
3683	64" diameter		1.27	18.898		2,925	1,100		4,025	4,900
3684	68" diameter		1.20	20		3,275	1,175		4,450	5,375
3685	72" diameter	▼	1.13	21.239		3,650	1,225		4,875	5,900
3690	For Y @ 45°, add					22%				
3700	Blast gate, 4" diameter	Q-9	22.30	.717		133	40		173	207
3701	6" diameter		15.20	1.053		150	59		209	255
3702	8" diameter		11	1.455		202	81.50		283.50	345
3703	10" diameter		8.60	1.860		231	104		335	410
3704	12" diameter		7.20	2.222		282	125		407	500
3705	14" diameter		6.10	2.623		315	147		462	570
3706	16" diameter		5.28	3.030		345	170		515	640
3707	18" diameter	▼	4.84	3.306		350	185		535	665
3708	20" diameter	Q-10	6.50	3.692		410	215		625	775
3709	22" diameter		5.94	4.040		445	235		680	840
3710	24" diameter		5.40	4.444		475	258		733	915
3711	26" diameter		5.06	4.743		510	276		786	980
3712	28" diameter		4.62	5.195		550	300		850	1,075
3713	30" diameter		4.30	5.581		580	325		905	1,125
3714	32" diameter		4.07	5.897		635	345		980	1,225
3715	34" diameter		3.80	6.316		695	365		1,060	1,325
3716	36" diameter		3.52	6.818		745	395		1,140	1,425
3717	38" diameter		3.38	7.101		1,000	415		1,415	1,725
3718	42" diameter		2.97	8.081		1,150	470		1,620	2,000
3719	46" diameter		2.80	8.571		1,325	500		1,825	2,200
3720	50" diameter		2.56	9.375		1,525	545		2,070	2,500
3721	54" diameter		2.35	10.213		1,700	595		2,295	2,775
3722	60" diameter		2.13	11.268		2,000	655		2,655	3,200
3723	64" diameter		2	12		2,225	700		2,925	3,475
3724	68" diameter		1.88	12.766		2,450	740		3,190	3,825
3725	72" diameter	▼	1.78	13.483		2,700	785		3,485	4,175
3740	Reducers, 4" diameter	Q-9	24.50	.653		33	36.50		69.50	92
3741	6" diameter		16.70	.958		49.50	53.50		103	136
3742	8" diameter		12	1.333		67	75		142	187
3743	10" diameter		9.50	1.684		83	94.50		177.50	234
3744	12" diameter		7.90	2.025		99	114		213	281
3745	14" diameter		6.70	2.388		115	134		249	330
3746	16" diameter		5.80	2.759		132	155		287	380
3747	18" diameter	▼	5.30	3.019		149	169		318	420
3748	20" diameter	Q-10	7.20	3.333		166	194		360	475

For customer support on your Mechanical Costs with RSMeans Data, call 800.448.8182.

315

23 31 16 – Nonmetal Ducts

23 31 16.16 Thermoset Fiberglass-Reinforced Plastic Ducts

		Crew	Daily Output	Labor-Hours	Unit	Material	2020 Bare Costs Labor	Equipment	Total	Total Incl O&P
3749	22" diameter	Q-10	6.50	3.692	Ea.	267	215		482	620
3750	24" diameter		5.90	4.068		289	237		526	680
3751	26" diameter		5.60	4.286		315	249		564	725
3752	28" diameter		5.10	4.706		340	274		614	790
3753	30" diameter		4.70	5.106		365	297		662	850
3754	32" diameter		4.50	5.333		390	310		700	900
3755	34" diameter		4.20	5.714		415	330		745	960
3756	36" diameter		3.90	6.154		440	360		800	1,025
3757	38" diameter		3.70	6.486		665	375		1,040	1,300
3758	42" diameter		3.30	7.273		730	425		1,155	1,450
3759	46" diameter		3.10	7.742		805	450		1,255	1,575
3760	50" diameter		2.80	8.571		870	500		1,370	1,725
3761	54" diameter		2.60	9.231		945	535		1,480	1,850
3762	60" diameter		2.34	10.256		1,050	595		1,645	2,075
3763	64" diameter		2.20	10.909		1,625	635		2,260	2,750
3764	68" diameter		2.06	11.650		1,725	675		2,400	2,925
3765	72" diameter		1.96	12.245		1,850	710		2,560	3,125
3780	Flange, per each, 4" diameter	Q-9	40.60	.394		80.50	22		102.50	122
3781	6" diameter		27.70	.578		94	32.50		126.50	152
3782	8" diameter		20.50	.780		117	44		161	196
3783	10" diameter		16.30	.982		141	55		196	239
3784	12" diameter		13.90	1.151		162	64.50		226.50	276
3785	14" diameter		11.30	1.416		187	79.50		266.50	325
3786	16" diameter		9.70	1.649		210	92.50		302.50	370
3787	18" diameter		8.30	1.928		233	108		341	420
3788	20" diameter	Q-10	11.30	2.124		256	124		380	470
3789	22" diameter		10.40	2.308		279	134		413	510
3790	24" diameter		9.50	2.526		305	147		452	560
3791	26" diameter		8.80	2.727		325	159		484	600
3792	28" diameter		8.20	2.927		350	170		520	645
3793	30" diameter		7.60	3.158		370	184		554	690
3794	32" diameter		7.20	3.333		395	194		589	730
3795	34" diameter		6.80	3.529		420	205		625	770
3796	36" diameter		6.40	3.750		445	218		663	815
3797	38" diameter		6.10	3.934		465	229		694	860
3798	42" diameter		5.50	4.364		510	254		764	945
3799	46" diameter		5	4.800		560	279		839	1,050
3800	50" diameter		4.60	5.217		605	305		910	1,125
3801	54" diameter		4.20	5.714		650	330		980	1,225
3802	60" diameter		3.80	6.316		720	365		1,085	1,350
3803	64" diameter		3.60	6.667		765	390		1,155	1,425
3804	68" diameter		3.40	7.059		815	410		1,225	1,525
3805	72" diameter		3.20	7.500		860	435		1,295	1,600

23 31 16.19 PVC Ducts

		Crew	Daily Output	Labor-Hours	Unit	Material	2020 Bare Costs Labor	Equipment	Total	Total Incl O&P
0010	**PVC DUCTS**									
4000	Rigid plastic, corrosive fume resistant PVC									
4020	Straight, 6" diameter	Q-9	220	.073	L.F.	14.15	4.08		18.23	22
4040	8" diameter		160	.100		20.50	5.60		26.10	31
4060	10" diameter		120	.133		26	7.50		33.50	40
4070	12" diameter		100	.160		30	8.95		38.95	46.50
4080	14" diameter		70	.229		37.50	12.80		50.30	61
4090	16" diameter		62	.258		43.50	14.45		57.95	70

23 31 HVAC Ducts and Casings

23 31 16 – Nonmetal Ducts

23 31 16.19 PVC Ducts

		Crew	Daily Output	Labor-Hours	Unit	Material	2020 Bare Costs Labor	Equipment	Total	Total Incl O&P
4100	18" diameter	Q-9	58	.276	L.F.	60	15.45		75.45	89.50
4110	20" diameter	Q-10	75	.320		54.50	18.60		73.10	88
4130	24" diameter	"	55	.436		74.50	25.50		100	121
4250	Coupling, 6" diameter	Q-9	88	.182	Ea.	29	10.20		39.20	47.50
4270	8" diameter		78	.205		38	11.50		49.50	59.50
4290	10" diameter		70	.229		43.50	12.80		56.30	67
4300	12" diameter		55	.291		47	16.30		63.30	76.50
4310	14" diameter		44	.364		60.50	20.50		81	97.50
4320	16" diameter		40	.400		64	22.50		86.50	105
4330	18" diameter		38	.421		83	23.50		106.50	128
4340	20" diameter	Q-10	55	.436		98.50	25.50		124	147
4360	24" diameter	"	45	.533		125	31		156	184
4470	Elbow, 90°, 6" diameter	Q-9	44	.364		132	20.50		152.50	177
4490	8" diameter		28	.571		148	32		180	211
4510	10" diameter		18	.889		185	50		235	279
4520	12" diameter		15	1.067		179	60		239	288
4530	14" diameter		11	1.455		248	81.50		329.50	395
4540	16" diameter		10	1.600		315	89.50		404.50	480
4550	18" diameter	Q-10	15	1.600		380	93		473	560
4560	20" diameter		14	1.714		640	99.50		739.50	850
4580	24" diameter		12	2		785	116		901	1,025
4750	Elbow 45°, use 90° and deduct					25%				

23 31 16.21 Polypropylene Ducts

		Crew	Daily Output	Labor-Hours	Unit	Material	2020 Bare Costs Labor	Equipment	Total	Total Incl O&P
0010	**POLYPROPYLENE DUCTS**									
0020	Rigid plastic, corrosive fume resistant, flame retardant, PP									
0110	Straight, 4" diameter	Q-9	200	.080	L.F.	7.95	4.49		12.44	15.55
0120	6" diameter		146	.110		12.15	6.15		18.30	22.50
0140	8" diameter		106	.151		15.20	8.45		23.65	29.50
0160	10" diameter		85	.188		22.50	10.55		33.05	40.50
0170	12" diameter		68	.235		40.50	13.20		53.70	64.50
0180	14" diameter		50	.320		45.50	17.95		63.45	77.50
0190	16" diameter		40	.400		61	22.50		83.50	102
0200	18" diameter		31	.516		98	29		127	152
0210	20" diameter	Q-10	44	.545		112	31.50		143.50	171
0220	22" diameter		40	.600		126	35		161	191
0230	24" diameter		37	.649		174	37.50		211.50	248
0250	28" diameter		28	.857		234	50		284	335
0270	32" diameter		24	1		264	58		322	380
0400	Coupling									
0410	4" diameter	Q-9	23	.696	Ea.	22.50	39		61.50	83.50
0420	6" diameter		16	1		27	56		83	115
0430	8" diameter		11.40	1.404		31.50	78.50		110	154
0440	10" diameter		9	1.778		42	99.50		141.50	198
0450	12" diameter		8	2		49.50	112		161.50	225
0460	14" diameter		6.60	2.424		82.50	136		218.50	297
0470	16" diameter		5.70	2.807		88.50	157		245.50	335
0480	18" diameter		5.20	3.077		117	173		290	390
0490	20" diameter	Q-10	6.80	3.529		128	205		333	450
0500	22" diameter		6.20	3.871		345	225		570	720
0510	24" diameter		5.80	4.138		435	241		676	840
0520	28" diameter		5.50	4.364		390	254		644	815
0530	32" diameter		5.20	4.615		425	268		693	875

For customer support on your Mechanical Costs with RSMeans Data, call 800.448.8182.

317

23 31 HVAC Ducts and Casings

23 31 16 - Nonmetal Ducts

23 31 16.21 Polypropylene Ducts		Crew	Daily Output	Labor-Hours	Unit	Material	2020 Bare Costs Labor	Equipment	Total	Total Incl O&P
0800	Elbow, 90°									
0810	4" diameter	Q-9	22	.727	Ea.	34.50	41		75.50	100
0820	6" diameter		15.40	1.039		48.50	58.50		107	142
0830	8" diameter		11	1.455		59	81.50		140.50	189
0840	10" diameter		8.60	1.860		85.50	104		189.50	252
0850	12" diameter		7.70	2.078		220	117		337	420
0860	14" diameter		6.40	2.500		254	140		394	495
0870	16" diameter	▼	5.40	2.963		271	166		437	550
0880	18" diameter	Q-10	7.50	3.200		790	186		976	1,150
0890	20" diameter		6.60	3.636		790	211		1,001	1,200
0896	22" diameter		6.10	3.934		1,075	229		1,304	1,525
0900	24" diameter		5.60	4.286		1,275	249		1,524	1,775
0910	28" diameter		5.40	4.444		2,175	258		2,433	2,800
0920	32" diameter	▼	5.20	4.615	▼	2,850	268		3,118	3,550
1000	Elbow, 45°									
1020	4" diameter	Q-9	22	.727	Ea.	26	41		67	90.50
1030	6" diameter		15.40	1.039		41.50	58.50		100	134
1040	8" diameter		11	1.455		47	81.50		128.50	176
1050	10" diameter		8.60	1.860		63	104		167	228
1060	12" diameter		7.70	2.078		170	117		287	365
1070	14" diameter		6.40	2.500		191	140		331	425
1080	16" diameter	▼	5.40	2.963		205	166		371	475
1090	18" diameter	Q-10	7.50	3.200		475	186		661	800
1100	20" diameter		6.60	3.636		580	211		791	955
1106	22" diameter		6.10	3.934		815	229		1,044	1,250
1110	24" diameter		5.60	4.286		920	249		1,169	1,375
1120	28" diameter		5.40	4.444		1,450	258		1,708	2,000
1130	32" diameter	▼	5.20	4.615	▼	1,800	268		2,068	2,400
1200	Tee									
1210	4" diameter	Q-9	16.50	.970	Ea.	151	54.50		205.50	250
1220	6" diameter		11.60	1.379		154	77.50		231.50	286
1230	8" diameter		8.30	1.928		191	108		299	375
1240	10" diameter		6.50	2.462		270	138		408	505
1250	12" diameter		5.80	2.759		375	155		530	650
1260	14" diameter		4.80	3.333		475	187		662	805
1270	16" diameter	▼	4.05	3.951		545	222		767	935
1280	18" diameter	Q-10	5.60	4.286		685	249		934	1,125
1290	20" diameter		4.90	4.898		750	285		1,035	1,250
1300	22" diameter		4.40	5.455		925	315		1,240	1,500
1310	24" diameter		4.20	5.714		1,675	330		2,005	2,325
1320	28" diameter		4.10	5.854		2,250	340		2,590	3,000
1330	32" diameter	▼	4	6	▼	2,775	350		3,125	3,575

23 31 16.22 Duct Board

23 31 16.22 Duct Board		Crew	Daily Output	Labor-Hours	Unit	Material	2020 Bare Costs Labor	Equipment	Total	Total Incl O&P
0010	**DUCT BOARD**									
3800	Rigid, resin bonded fibrous glass, FSK									
3810	Temperature, bacteria and fungi resistant									
3820	1" thick	Q-14	150	.107	S.F.	1.68	5.65		7.33	10.50
3830	1-1/2" thick		130	.123		2.22	6.50		8.72	12.40
3840	2" thick	▼	120	.133	▼	3.82	7.05		10.87	15

For customer support on your Mechanical Costs with RSMeans Data, call 800.448.8182.

23 33 13.13 Volume-Control Dampers	Crew	Daily Output	Labor-Hours	Unit	Material	2020 Bare Costs Labor	Equipment	Total	Total Incl O&P
0010 **VOLUME-CONTROL DAMPERS**									
5990 Multi-blade dampers, opposed blade, 8" x 6"	1 Shee	24	.333	Ea.	32	21		53	66.50
5994 8" x 8"		22	.364		34	22.50		56.50	72
5996 10" x 10"		21	.381		33.50	23.50		57	73
6000 12" x 12"		21	.381		46	23.50		69.50	86.50
6020 12" x 18"		18	.444		61.50	27.50		89	110
6030 14" x 10"		20	.400		45	25		70	87.50
6031 14" x 14"		17	.471		44.50	29.50		74	93.50
6032 16" x 10"		18	.444		48	27.50		75.50	94.50
6033 16" x 12"		17	.471		44.50	29.50		74	93.50
6035 16" x 16"		16	.500		55.50	31		86.50	109
6036 18" x 14"		16	.500		55.50	31		86.50	109
6037 18" x 16"		15	.533		75	33		108	133
6038 18" x 18"		15	.533		80	33		113	139
6040 18" x 24"		12	.667		104	41.50		145.50	178
6060 18" x 28"		10	.800		117	50		167	204
6068 20" x 6"		18	.444		42.50	27.50		70	88.50
6069 20" x 8"		16	.500		48	31		79	100
6070 20" x 16"		14	.571		80	35.50		115.50	142
6071 20" x 18"		13	.615		88	38.50		126.50	155
6072 20" x 20"		13	.615		79	38.50		117.50	145
6073 22" x 6"		20	.400		43.50	25		68.50	85.50
6074 22" x 18"		14	.571		83	35.50		118.50	146
6075 22" x 20"		12	.667		102	41.50		143.50	175
6076 24" x 16"		11	.727		94.50	45.50		140	173
6077 22" x 22"		10	.800		114	50		164	202
6078 24" x 20"		8	1		114	62.50		176.50	221
6080 24" x 24"		8	1		121	62.50		183.50	228
6100 24" x 28"		6	1.333		141	83		224	281
6110 26" x 26"	↓	6	1.333		143	83		226	283
6120 28" x 28"	Q-9	11	1.455		134	81.50		215.50	272
6130 30" x 18"		10	1.600		109	89.50		198.50	256
6132 30" x 24"		7	2.286		154	128		282	365
6133 30" x 30"		6.60	2.424		216	136		352	445
6135 32" x 32"		6.40	2.500		236	140		376	470
6150 34" x 34"		6	2.667		253	150		403	505
6151 36" x 12"		10	1.600		106	89.50		195.50	253
6152 36" x 16"		8	2		172	112		284	360
6156 36" x 32"		6.30	2.540		253	142		395	495
6157 36" x 34"		6.20	2.581		267	145		412	510
6158 36" x 36"		6	2.667		340	150		490	595
6160 44" x 28"		5.80	2.759		286	155		441	550
6180 48" x 36"		5.60	2.857		400	160		560	685
6200 56" x 36"		5.40	2.963		470	166		636	765
6220 60" x 36"		5.20	3.077		505	173		678	815
6240 60" x 44"	↓	5	3.200	↓	585	179		764	910
7500 Variable volume modulating motorized damper, incl. elect. mtr.									
7504 8" x 6"	1 Shee	15	.533	Ea.	177	33		210	245
7506 10" x 6"		14	.571		177	35.50		212.50	248
7510 10" x 10"		13	.615		184	38.50		222.50	261
7520 12" x 12"		12	.667		190	41.50		231.50	272
7522 12" x 16"		11	.727		191	45.50		236.50	279
7524 16" x 10"	↓	12	.667	↓	188	41.50		229.50	269

For customer support on your Mechanical Costs with RSMeans Data, call 800.448.8182.

319

23 33 13.13 Volume-Control Dampers

		Crew	Daily Output	Labor-Hours	Unit	Material	Labor	Equipment	Total	Total Incl O&P
7526	16" x 14"	1 Shee	10	.800	Ea.	197	50		247	292
7528	16" x 18"		9	.889		203	55.50		258.50	305
7540	18" x 12"		10	.800		197	50		247	293
7542	18" x 18"		8	1		206	62.50		268.50	320
7544	20" x 14"		8	1		213	62.50		275.50	330
7546	20" x 18"		7	1.143		220	71		291	350
7560	24" x 12"		8	1		215	62.50		277.50	330
7562	24" x 18"		7	1.143		232	71		303	365
7568	28" x 10"		7	1.143		215	71		286	345
7580	28" x 16"		6	1.333		215	83		298	360
7590	30" x 14"		5	1.600		221	99.50		320.50	395
7600	30" x 18"		4	2		296	125		421	515
7610	30" x 24"		3.80	2.105		395	131		526	635
7690	48" x 48"	Q-9	6	2.667		1,400	150		1,550	1,750
7694	6' x 14' w/2 motors	"	3	5.333		4,450	299		4,749	5,325
7700	For thermostat, add	1 Shee	8	1		48.50	62.50		111	148
7800	For transformer 40 VA capacity, add	"	16	.500		29.50	31		60.50	80
8000	Multi-blade dampers, parallel blade									
8100	8" x 8"	1 Shee	24	.333	Ea.	119	21		140	163
8120	12" x 8"		22	.364		119	22.50		141.50	166
8140	16" x 10"		20	.400		152	25		177	206
8160	18" x 12"		18	.444		168	27.50		195.50	226
8180	22" x 12"		15	.533		176	33		209	245
8200	24" x 16"		11	.727		194	45.50		239.50	283
8220	28" x 16"		10	.800		218	50		268	315
8240	30" x 16"		8	1		225	62.50		287.50	340
8260	30" x 18"		7	1.143		270	71		341	405
8400	Round damper, butterfly, vol. control w/lever lock reg.									
8410	6" diam.	1 Shee	22	.364	Ea.	33.50	22.50		56	71
8412	7" diam.		21	.381		39.50	23.50		63	79.50
8414	8" diam.		20	.400		42	25		67	84
8416	9" diam.		19	.421		42	26		68	86
8418	10" diam.		18	.444		45	27.50		72.50	91.50
8420	12" diam.		16	.500		46	31		77	98
8422	14" diam.		14	.571		54.50	35.50		90	114
8424	16" diam.		13	.615		58	38.50		96.50	122
8426	18" diam.		12	.667		71.50	41.50		113	142
8428	20" diam.		11	.727		99	45.50		144.50	178
8430	24" diam.		10	.800		123	50		173	211
8432	30" diam.		9	.889		172	55.50		227.50	273
8434	36" diam.		8	1		229	62.50		291.50	345
8500	Round motor operated damper									
8510	6" diam.	1 Shee	22	.364	Ea.	91	22.50		113.50	135
8512	7" diam.		21	.381		98.50	23.50		122	144
8514	8" diam.		20	.400		101	25		126	149
8516	10" diam.		18	.444		112	27.50		139.50	166
8518	12" diam.		16	.500		118	31		149	178
8520	14" diam.		14	.571		134	35.50		169.50	202

23 33 13.16 Fire Dampers

		Crew	Daily Output	Labor-Hours	Unit	Material	Labor	Equipment	Total	Total Incl O&P
0010	**FIRE DAMPERS**									
3000	Fire damper, curtain type, 1-1/2 hr. rated, vertical, 6" x 6"	1 Shee	24	.333	Ea.	35	21		56	70
3020	8" x 6"		22	.364		35	22.50		57.50	73

23 33 13.16 Fire Dampers		Crew	Daily Output	Labor-Hours	Unit	Material	2020 Bare Costs Labor	Equipment	Total	Total Incl O&P
3021	8" x 8"	1 Shee	22	.364	Ea.	30	22.50		52.50	67.50
3030	10" x 10"		22	.364		35	22.50		57.50	73
3040	12" x 6"		22	.364		35	22.50		57.50	73
3044	12" x 12"		20	.400		35	25		60	76.50
3060	20" x 6"		18	.444		39.50	27.50		67	85.50
3080	12" x 8"		22	.364		29.50	22.50		52	67
3100	24" x 8"		16	.500		44.50	31		75.50	96.50
3120	12" x 10"		21	.381		35	23.50		58.50	74.50
3140	24" x 10"		15	.533		62	33		95	119
3160	36" x 10"		12	.667		80.50	41.50		122	152
3180	16" x 12"		20	.400		53.50	25		78.50	96.50
3200	24" x 12"		13	.615		50	38.50		88.50	113
3220	48" x 12"		10	.800		78	50		128	162
3238	14" x 14"		19	.421		56	26		82	102
3240	16" x 14"		18	.444		59	27.50		86.50	107
3260	24" x 14"		12	.667		71	41.50		112.50	141
3280	30" x 14"		11	.727		78.50	45.50		124	155
3298	16" x 16"		18	.444		49	27.50		76.50	96
3300	18" x 16"		17	.471		67	29.50		96.50	119
3320	24" x 16"		11	.727		56.50	45.50		102	131
3340	36" x 16"		10	.800		70.50	50		120.50	153
3356	18" x 18"		16	.500		69	31		100	124
3360	24" x 18"		10	.800		84	50		134	168
3380	48" x 18"		8	1		121	62.50		183.50	229
3398	20" x 20"		10	.800		58	50		108	139
3400	24" x 20"		8	1		66	62.50		128.50	167
3420	36" x 20"		7	1.143		79	71		150	195
3440	24" x 22"		7	1.143		90	71		161	207
3460	30" x 22"		6	1.333		99	83		182	235
3478	24" x 24"		8	1		60	62.50		122.50	161
3480	26" x 24"		7	1.143		88.50	71		159.50	206
3484	32" x 24"	Q-9	13	1.231		82	69		151	195
3500	48" x 24"		12	1.333		106	75		181	230
3520	28" x 26"		13	1.231		99	69		168	214
3540	30" x 28"		12	1.333		110	75		185	234
3560	48" x 30"		11	1.455		160	81.50		241.50	300
3580	48" x 36"		10	1.600		166	89.50		255.50	320
3590	32" x 40"		10	1.600		128	89.50		217.50	277
3600	44" x 44"		10	1.600		175	89.50		264.50	330
3620	48" x 48"		8	2		196	112		308	385
3700	UL label included in above									
3800	For horizontal operation, add				Ea.	20%				
3900	For cap for blades out of air stream, add					20%				
4000	For 10" 22 ga., UL approved sleeve, add					35%				
4100	For 10" 22 ga., UL sleeve, 100% free area, add					65%				
4200	For oversize openings group dampers									
4210	Fire damper, round, Type A, 1-1/2 hour									
4214	4" diam.	1 Shee	20	.400	Ea.	55.50	25		80.50	99
4216	5" diam.		16	.500		53	31		84	106
4218	6" diam.		14.30	.559		53.50	35		88.50	112
4220	7" diam.		13.30	.602		60.50	37.50		98	124
4222	8" diam.		11.40	.702		60.50	43.50		104	133
4224	9" diam.		10.80	.741		58.50	46		104.50	135

Handwritten annotations: near 3320 "2 x 1.33 = 2.6⁻ SF", "131 ÷ 2.6⁻ = 49/SF"; near 3478 "40/SF"; near 3620 "25⁻/SF"

23 33 13 – Dampers

23 33 13.16 Fire Dampers		Crew	Daily Output	Labor-Hours	Unit	Material	2020 Bare Costs		Total	Total Incl O&P
							Labor	Equipment		
4226	10" diam.	1 Shee	10	.800	Ea.	58.50	50		108.50	140
4228	12" diam.		9.50	.842		63	52.50		115.50	149
4230	14" diam.		8.20	.976		85.50	61		146.50	186
4232	16" diam.		7.60	1.053		101	65.50		166.50	211
4234	18" diam.		7.20	1.111		121	69		190	238
4236	20" diam.		6.40	1.250		137	78		215	268
4238	22" diam.		6.10	1.311		166	81.50		247.50	305
4240	24" diam.		5.70	1.404		173	87.50		260.50	325
4242	26" diam.		5.50	1.455		196	90.50		286.50	350
4244	28" diam.		5.40	1.481		184	92.50		276.50	340
4246	30" diam.	▼	5.30	1.509	▼	215	94		309	380
4502	Power open-spring close, sleeve and actuator motor mounted, 1-1/2 hour									
4506	6" x 6"	1 Shee	24	.333	Ea.	82.50	21		103.50	122
4510	8" x 8"		22	.364		291	22.50		313.50	355
4520	16" x 8"		20	.400		360	25		385	435
4540	18" x 8"		18	.444		365	27.50		392.50	445
4560	20" x 8"		16	.500		365	31		396	450
4580	10" x 10"		21	.381		350	23.50		373.50	420
4600	24" x 10"		15	.533		415	33		448	505
4620	30" x 10"		12	.667		515	41.50		556.50	630
4640	12" x 12"		20	.400		400	25		425	480
4660	18" x 12"		18	.444		375	27.50		402.50	455
4680	24" x 12"		13	.615		425	38.50		463.50	530
4700	30" x 12"		11	.727		420	45.50		465.50	535
4720	14" x 14"		17	.471		430	29.50		459.50	515
4740	16" x 14"		18	.444		430	27.50		457.50	510
4760	20" x 14"		14	.571		410	35.50		445.50	505
4780	24" x 14"		12	.667		435	41.50		476.50	545
4800	30" x 14"		10	.800		430	50		480	545
4820	16" x 16"		16	.500		440	31		471	535
4840	20" x 16"		14	.571		430	35.50		465.50	530
4860	24" x 16"		11	.727		435	45.50		480.50	550
4880	30" x 16"		8	1		475	62.50		537.50	615
4900	18" x 18"		15	.533		465	33		498	560
5000	24" x 18"		10	.800		465	50		515	590
5020	36" x 18"		7	1.143		480	71		551	635
5040	20" x 20"		13	.615		460	38.50		498.50	565
5060	24" x 20"		8	1		465	62.50		527.50	605
5080	30" x 20"		8	1		485	62.50		547.50	625
5100	36" x 20"		7	1.143		565	71		636	735
5120	24" x 24"		8	1		505	62.50		567.50	650
5130	30" x 24"	▼	7.60	1.053		520	65.50		585.50	670
5140	36" x 24"	Q-9	12	1.333		530	75		605	700
5141	36" x 30"		10	1.600		625	89.50		714.50	820
5142	40" x 36"		8	2		1,125	112		1,237	1,425
5143	48" x 48"	▼	4	4		2,025	224		2,249	2,575
5150	Damper operator motor, 24 or 120 volt	1 Shee	16	.500	▼	495	31		526	595

23 33 13.25 Exhaust Vent Damper

		Crew	Daily Output	Labor-Hours	Unit	Material	Labor	Equipment	Total	Total Incl O&P
0010	**EXHAUST VENT DAMPER**, Auto, OB with elect. actuator.									
1110	6" x 6"	1 Shee	24	.333	Ea.	97.50	21		118.50	139
1114	8" x 6"		23	.348		99	21.50		120.50	142
1116	8" x 8"	▼	22	.364	▼	102	22.50		124.50	147

23 33 Air Duct Accessories

23 33 13 – Dampers

23 33 13.25 Exhaust Vent Damper

		Crew	Daily Output	Labor-Hours	Unit	Material	2020 Bare Costs Labor	2020 Bare Costs Equipment	Total	Total Incl O&P
1118	10" x 6"	1 Shee	23	.348	Ea.	102	21.50		123.50	145
1120	10" x 10"		22	.364		111	22.50		133.50	157
1122	12" x 6"		22	.364		106	22.50		128.50	152
1124	12" x 10"		21	.381		115	23.50		138.50	163
1126	12" x 12"		20	.400		112	25		137	161
1128	14" x 6"		21	.381		105	23.50		128.50	152
1130	14" x 10"		19	.421		119	26		145	171
1132	16" x 6"		21	.381		106	23.50		129.50	153
1134	16" x 8"		20	.400		118	25		143	168
1136	18" x 6"		20	.400		114	25		139	164
1138	18" x 8"		19	.421		119	26		145	171

23 33 13.28 Splitter Damper Assembly

		Crew	Daily Output	Labor-Hours	Unit	Material	2020 Bare Costs Labor	2020 Bare Costs Equipment	Total	Total Incl O&P
0010	**SPLITTER DAMPER ASSEMBLY**									
7000	Self locking, 1' rod	1 Shee	24	.333	Ea.	29.50	21		50.50	64
7020	3' rod		22	.364		40.50	22.50		63	79
7040	4' rod		20	.400		45	25		70	87.50
7060	6' rod		18	.444		54.50	27.50		82	102

23 33 13.32 Relief Damper

		Crew	Daily Output	Labor-Hours	Unit	Material	2020 Bare Costs Labor	2020 Bare Costs Equipment	Total	Total Incl O&P
0010	**RELIEF DAMPER**, Electronic bypass with tight seal									
8310	8" x 6"	1 Shee	22	.364	Ea.	265	22.50		287.50	325
8314	10" x 6"		22	.364		265	22.50		287.50	325
8318	10" x 10"		21	.381		272	23.50		295.50	335
8322	12" x 12"		20	.400		283	25		308	350
8326	12" x 16"		19	.421		292	26		318	360
8330	16" x 10"		20	.400		305	25		330	380
8334	16" x 14"		18	.444		310	27.50		337.50	380
8338	16" x 18"		17	.471		340	29.50		369.50	420
8342	18" x 12"		18	.444		340	27.50		367.50	415
8346	18" x 18"		15	.533		305	33		338	385
8350	20" x 14"		14	.571		320	35.50		355.50	410
8354	20" x 18"		13	.615		325	38.50		363.50	420
8358	24" x 12"		13	.615		330	38.50		368.50	420
8362	24" x 18"		10	.800		325	50		375	435
8363	24" x 24"		10	.800		355	50		405	465
8364	24" x 36"		9	.889		420	55.50		475.50	550
8365	24" x 48"		6	1.333		470	83		553	645
8366	28" x 10"		13	.615		231	38.50		269.50	310
8370	28" x 16"		10	.800		360	50		410	475
8374	30" x 14"		11	.727		242	45.50		287.50	335
8378	30" x 18"		8	1		370	62.50		432.50	500
8382	30" x 24"		6	1.333		390	83		473	550
8390	46" x 36"		4	2		520	125		645	765
8394	48" x 48"		3	2.667		570	166		736	880
8396	54" x 36"		2	4		600	249		849	1,050

23 33 19 – Duct Silencers

23 33 19.10 Duct Silencers

		Crew	Daily Output	Labor-Hours	Unit	Material	2020 Bare Costs Labor	2020 Bare Costs Equipment	Total	Total Incl O&P
0010	**DUCT SILENCERS**									
9000	Silencers, noise control for air flow, duct				MCFM	87			87	96
9004	Duct sound trap, packaged									
9010	12" x 18" x 36", 2,000 CFM	1 Shee	12	.667	Ea.	695	41.50		736.50	825
9011	24" x 18" x 36", 9,000 CFM	Q-9	80	.200		775	11.20		786.20	870
9012	24" x 24" x 36", 9,000 CFM		60	.267		1,175	14.95		1,189.95	1,300

For customer support on your Mechanical Costs with RSMeans Data, call 800.448.8182.

323

23 33 Air Duct Accessories

23 33 19 – Duct Silencers

23 33 19.10 Duct Silencers

		Crew	Daily Output	Labor-Hours	Unit	Material	2020 Bare Costs Labor	Equipment	Total	Total Incl O&P
9013	24" x 30" x 36", 9,000 CFM	Q-9	50	.320	Ea.	1,250	17.95		1,267.95	1,400
9014	24" x 36" x 36", 9,000 CFM		40	.400		1,550	22.50		1,572.50	1,725
9015	24" x 48" x 36", 9,000 CFM		28	.571		2,025	32		2,057	2,275
9018	36" x 18" x 36", 6,300 CFM		42	.381		1,550	21.50		1,571.50	1,725
9019	36" x 36" x 36", 6,300 CFM		21	.762		3,100	42.50		3,142.50	3,500
9020	36" x 48" x 36", 6,300 CFM		14	1.143		4,025	64		4,089	4,550
9021	36" x 60" x 36", 6,300 CFM		13	1.231		5,125	69		5,194	5,725
9022	48" x 48" x 36", 6,300 CFM		12	1.333		5,425	75		5,500	6,100
9023	24" x 24" x 60", 6,000 CFM		3.60	4.444		1,775	249		2,024	2,350
9026	24" x 30" x 60", 7,000 CFM		3.50	4.571		1,975	256		2,231	2,575
9030	24" x 36" x 60", 8,000 CFM		3.40	4.706		2,325	264		2,589	2,950
9034	24" x 48" x 60", 11,000 CFM		3.30	4.848		1,900	272		2,172	2,475
9038	36" x 18" x 60", 6,300 CFM		3.50	4.571		1,525	256		1,781	2,075
9042	36" x 36" x 60", 12,000 CFM		3.30	4.848		2,500	272		2,772	3,150
9046	36" x 48" x 60", 17,000 CFM		3.10	5.161		3,125	289		3,414	3,900
9050	36" x 60" x 60", 20,000 CFM		2.80	5.714		3,925	320		4,245	4,775
9054	48" x 48" x 60", 23,000 CFM		2.60	6.154		4,400	345		4,745	5,375
9056	48" x 60" x 60", 23,000 CFM		2.10	7.619		5,525	425		5,950	6,725

23 33 23 – Turning Vanes

23 33 23.13 Air Turning Vanes

		Crew	Daily Output	Labor-Hours	Unit	Material	2020 Bare Costs Labor	Equipment	Total	Total Incl O&P
0010	**AIR TURNING VANES**									
9400	Turning vane components									
9410	Turning vane rail	1 Shee	160	.050	L.F.	.84	3.12		3.96	5.65
9420	Double thick, factory fab. vane		300	.027		1.26	1.66		2.92	3.91
9428	12" high set		170	.047		1.92	2.93		4.85	6.55
9432	14" high set		160	.050		2.24	3.12		5.36	7.20
9434	16" high set		150	.053		2.56	3.32		5.88	7.85
9436	18" high set		144	.056		2.88	3.46		6.34	8.40
9438	20" high set		138	.058		3.20	3.61		6.81	9
9440	22" high set		130	.062		3.52	3.83		7.35	9.65
9442	24" high set		124	.065		3.84	4.02		7.86	10.30
9444	26" high set		116	.069		4.16	4.30		8.46	11.10
9446	30" high set		112	.071		4.80	4.45		9.25	12.05

23 33 33 – Duct-Mounting Access Doors

23 33 33.13 Duct Access Doors

		Crew	Daily Output	Labor-Hours	Unit	Material	2020 Bare Costs Labor	Equipment	Total	Total Incl O&P
0010	**DUCT ACCESS DOORS**									
1000	Duct access door, insulated, 6" x 6"	1 Shee	14	.571	Ea.	22.50	35.50		58	79
1020	10" x 10"		11	.727		23.50	45.50		69	94
1040	12" x 12"		10	.800		28	50		78	106
1050	12" x 18"		9	.889		48	55.50		103.50	137
1060	16" x 12"		9	.889		39	55.50		94.50	127
1070	18" x 18"		8	1		40	62.50		102.50	139
1074	24" x 18"		8	1		69	62.50		131.50	171
1080	24" x 24"		8	1		60	62.50		122.50	161

23 33 46 – Flexible Ducts

23 33 46.10 Flexible Air Ducts

		Crew	Daily Output	Labor-Hours	Unit	Material	2020 Bare Costs Labor	Equipment	Total	Total Incl O&P
0010	**FLEXIBLE AIR DUCTS**									
1280	Add to labor for elevated installation									
1282	of prefabricated (purchased) ductwork									
1283	10' to 15' high						10%			
1284	15' to 20' high						20%			

23 33 46 – Flexible Ducts

23 33 46.10 Flexible Air Ducts		Crew	Daily Output	Labor-Hours	Unit	Material	2020 Bare Costs Labor	Equipment	Total	Total Incl O&P
1285	20' to 25' high						25%			
1286	25' to 30' high						35%			
1287	30' to 35' high						40%			
1288	35' to 40' high						50%			
1289	Over 40' high						55%			
1300	Flexible, coated fiberglass fabric on corr. resist. metal helix									
1400	pressure to 12" (WG) UL-181									
1500	Noninsulated, 3" diameter		Q-9	400	.040	L.F.	1.36	2.24		3.60	4.90
1520	4" diameter			360	.044		1.45	2.49		3.94	5.40
1540	5" diameter			320	.050		1.57	2.80		4.37	6
1560	6" diameter			280	.057		1.85	3.20		5.05	6.90
1580	7" diameter			240	.067		2.05	3.74		5.79	7.90
1600	8" diameter			200	.080		2.32	4.49		6.81	9.35
1620	9" diameter			180	.089		2.51	4.99		7.50	10.30
1640	10" diameter			160	.100		2.97	5.60		8.57	11.75
1660	12" diameter			120	.133		3.50	7.50		11	15.20
1680	14" diameter			80	.200		4.27	11.20		15.47	21.50
1700	16" diameter			60	.267		6.15	14.95		21.10	29.50
1800	For plastic cable tie, add					Ea.	.61			.61	.67
1900	Insulated, 1" thick, PE jacket, 3" diameter	G	Q-9	380	.042	L.F.	2.68	2.36		5.04	6.55
1910	4" diameter	G		340	.047		3.26	2.64		5.90	7.60
1920	5" diameter	G		300	.053		3.42	2.99		6.41	8.30
1940	6" diameter	G		260	.062		3.61	3.45		7.06	9.20
1960	7" diameter	G		220	.073		4.16	4.08		8.24	10.80
1980	8" diameter	G		180	.089		4.04	4.99		9.03	12
2000	9" diameter	G		160	.100		4.74	5.60		10.34	13.70
2020	10" diameter	G		140	.114		5.15	6.40		11.55	15.40
2040	12" diameter	G		100	.160		6.05	8.95		15	20.50
2060	14" diameter	G		80	.200		7.40	11.20		18.60	25
2080	16" diameter	G		60	.267		10.55	14.95		25.50	34
2100	18" diameter	G		45	.356		11.90	19.95		31.85	43
2120	20" diameter	G	Q-10	65	.369		10.90	21.50		32.40	44.50
2500	Insulated, heavy duty, coated fiberglass fabric									
2520	4" diameter	G	Q-9	340	.047	L.F.	4.40	2.64		7.04	8.85
2540	5" diameter	G		300	.053		5.25	2.99		8.24	10.30
2560	6" diameter	G		260	.062		5.20	3.45		8.65	11
2580	7" diameter	G		220	.073		6.55	4.08		10.63	13.40
2600	8" diameter	G		180	.089		6.50	4.99		11.49	14.70
2620	9" diameter	G		160	.100		8.10	5.60		13.70	17.40
2640	10" diameter	G		140	.114		8	6.40		14.40	18.50
2660	12" diameter	G		100	.160		9.45	8.95		18.40	24
2680	14" diameter	G		80	.200		11.40	11.20		22.60	29.50
2700	16" diameter	G		60	.267		15.25	14.95		30.20	39.50
2720	18" diameter	G		45	.356		17.15	19.95		37.10	49
2800	Flexible, aluminum, pressure to 12" (WG) UL-181									
2820	Non-insulated									
2830	3" diameter		Q-9	400	.040	L.F.	1.34	2.24		3.58	4.87
2831	4" diameter			360	.044		1.55	2.49		4.04	5.50
2832	5" diameter			320	.050		2.04	2.80		4.84	6.50
2833	6" diameter			280	.057		2.60	3.20		5.80	7.70
2834	7" diameter			240	.067		3.13	3.74		6.87	9.10
2835	8" diameter			200	.080		3.26	4.49		7.75	10.40
2836	9" diameter			180	.089		4.18	4.99		9.17	12.15

23 33 46.10 Flexible Air Ducts		Crew	Daily Output	Labor-Hours	Unit	Material	2020 Bare Costs Labor	Equipment	Total	Total Incl O&P
2837	10" diameter	Q-9	160	.100	L.F.	4.96	5.60		10.56	13.95
2838	12" diameter		120	.133		6.05	7.50		13.55	18
2839	14" diameter		80	.200		7.45	11.20		18.65	25
2840	15" diameter		70	.229		7.85	12.80		20.65	28
2841	16" diameter		60	.267		11.90	14.95		26.85	35.50
2842	18" diameter		45	.356		11.85	19.95		31.80	43
2843	20" diameter	Q-10	65	.369		15.20	21.50		36.70	49
2880	Insulated, 1" thick with 3/4 lb., PE jacket									
2890	3" diameter G	Q-9	380	.042	L.F.	3.14	2.36		5.50	7.05
2891	4" diameter G		340	.047		4.27	2.64		6.91	8.70
2892	5" diameter G		300	.053		4.80	2.99		7.79	9.85
2893	6" diameter G		260	.062		5.65	3.45		9.10	11.50
2894	7" diameter G		220	.073		6.75	4.08		10.83	13.65
2895	8" diameter G		180	.089		6.85	4.99		11.84	15.10
2896	9" diameter G		160	.100		8.05	5.60		13.65	17.35
2897	10" diameter G		140	.114		9.50	6.40		15.90	20
2898	12" diameter G		100	.160		11.20	8.95		20.15	26
2899	14" diameter G		80	.200		12.60	11.20		23.80	31
2900	15" diameter G		70	.229		12.55	12.80		25.35	33.50
2901	16" diameter G		60	.267		17.95	14.95		32.90	42
2902	18" diameter G		45	.356		21	19.95		40.95	53
3000	Transitions, 3" deep, square to round									
3010	6" x 6" to round	Q-9	44	.364	Ea.	12.15	20.50		32.65	44.50
3014	6" x 12" to round		42	.381		22	21.50		43.50	56.50
3018	8" x 8" to round		40	.400		14.35	22.50		36.85	50
3022	9" x 9" to round		38	.421		14.35	23.50		37.85	52
3026	9" x 12" to round		37	.432		28	24.50		52.50	68
3030	9" x 15" to round		36	.444		33.50	25		58.50	74.50
3034	10" x 10" to round		34	.471		15.75	26.50		42.25	57.50
3038	10" x 22" to round		32	.500		40	28		68	86.50
3042	12" x 12" to round		30	.533		17	30		47	64
3046	12" x 18" to round		28	.571		40	32		72	92.50
3050	12" x 24" to round		26	.615		40	34.50		74.50	96.50
3054	15" x 15" to round		25	.640		23	36		59	80
3058	15" x 18" to round		24	.667		43.50	37.50		81	104
3062	18" x 18" to round		22	.727		33	41		74	98
3066	21" x 21" to round		21	.762		32.50	42.50		75	101
3070	22" x 22" to round		20	.800		36	45		81	108
3074	24" x 24" to round		18	.889		37.50	50		87.50	117
5000	Flexible, aluminum, acoustical, pressure to 2" (WG), NFPA-90A									
5010	Fiberglass insulation 1-1/2" thick, 1/2 lb. density									
5020	Polyethylene jacket, UL approved									
5026	5" diameter G	Q-9	300	.053	L.F.	5.30	2.99		8.29	10.40
5030	6" diameter G		260	.062		6.40	3.45		9.85	12.25
5034	7" diameter G		220	.073		7.55	4.08		11.63	14.50
5038	8" diameter G		180	.089		7.80	4.99		12.79	16.10
5042	9" diameter G		160	.100		9.10	5.60		14.70	18.50
5046	10" diameter G		140	.114		10.65	6.40		17.05	21.50
5050	12" diameter G		100	.160		12.65	8.95		21.60	27.50
5054	14" diameter G		80	.200		14.60	11.20		25.80	33
5058	16" diameter G		60	.267		20.50	14.95		35.45	45
5072	Hospital grade, PE jacket, UL approved									
5076	5" diameter G	Q-9	300	.053	L.F.	5.75	2.99		8.74	10.85

23 33 46 – Flexible Ducts

23 33 46.10 Flexible Air Ducts

		Crew	Daily Output	Labor-Hours	Unit	Material	2020 Bare Costs Labor	2020 Bare Costs Equipment	Total	Total Incl O&P
5080	6" diameter	G Q-9	260	.062	L.F.	6.90	3.45		10.35	12.85
5084	7" diameter	G	220	.073		8.15	4.08		12.23	15.15
5088	8" diameter	G	180	.089		8.40	4.99		13.39	16.80
5092	9" diameter	G	160	.100		9.75	5.60		15.35	19.25
5096	10" diameter	G	140	.114		11.50	6.40		17.90	22.50
5100	12" diameter	G	100	.160		13.65	8.95		22.60	28.50
5104	14" diameter	G	80	.200		15.75	11.20		26.95	34.50
5108	16" diameter	G	60	.267		22.50	14.95		37.45	47
5140	Flexible, aluminum, silencer, pressure to 12" (WG), NFPA-90A									
5144	Fiberglass insulation 1-1/2" thick, 1/2 lb. density									
5148	Aluminum outer shell, UL approved									
5152	5" diameter	G Q-9	290	.055	L.F.	14.80	3.09		17.89	21
5156	6" diameter	G	250	.064		13.65	3.59		17.24	20.50
5160	7" diameter	G	210	.076		20.50	4.27		24.77	29
5164	8" diameter	G	170	.094		20.50	5.30		25.80	30.50
5168	9" diameter	G	150	.107		21	6		27	32.50
5172	10" diameter	G	130	.123		21	6.90		27.90	33.50
5176	12" diameter	G	90	.178		25	9.95		34.95	42.50
5180	14" diameter	G	70	.229		30.50	12.80		43.30	53
5184	16" diameter	G	50	.320		39	17.95		56.95	69.50
5200	Hospital grade, aluminum shell, UL approved									
5204	5" diameter	G Q-9	280	.057	L.F.	15.40	3.20		18.60	22
5208	6" diameter	G	240	.067		14.25	3.74		17.99	21.50
5212	7" diameter	G	200	.080		21	4.49		25.49	30
5216	8" diameter	G	160	.100		21	5.60		26.60	31.50
5220	9" diameter	G	140	.114		22	6.40		28.40	34
5224	10" diameter	G	130	.123		22	6.90		28.90	34.50
5228	12" diameter	G	80	.200		26	11.20		37.20	45.50
5232	14" diameter	G	60	.267		32	14.95		46.95	57.50
5236	16" diameter	G	40	.400		40.50	22.50		63	78.50

23 33 53 – Duct Liners

23 33 53.10 Duct Liner Board

		Crew	Daily Output	Labor-Hours	Unit	Material	2020 Bare Costs Labor	2020 Bare Costs Equipment	Total	Total Incl O&P
0010	**DUCT LINER BOARD**									
3340	Board type fiberglass liner, FSK, 1-1/2 lb. density									
3344	1" thick	G Q-14	150	.107	S.F.	1.06	5.65		6.71	9.80
3345	1-1/2" thick	G	130	.123		1.16	6.50		7.66	11.25
3346	2" thick	G	120	.133		1.34	7.05		8.39	12.25
3348	3" thick	G	110	.145		1.75	7.70		9.45	13.75
3350	4" thick	G	100	.160		2.13	8.45		10.58	15.30
3356	3 lb. density, 1" thick	G	150	.107		1.34	5.65		6.99	10.10
3358	1-1/2" thick	G	130	.123		1.49	6.50		7.99	11.60
3360	2" thick	G	120	.133		2.07	7.05		9.12	13.10
3362	2-1/2" thick	G	110	.145		2.44	7.70		10.14	14.50
3364	3" thick	G	100	.160		2.80	8.45		11.25	16.05
3366	4" thick	G	90	.178		3.50	9.40		12.90	18.25
3370	6 lb. density, 1" thick	G	140	.114		1.90	6.05		7.95	11.35
3374	1-1/2" thick	G	120	.133		2.54	7.05		9.59	13.60
3378	2" thick	G	100	.160		3.20	8.45		11.65	16.45
3490	Board type, fiberglass liner, 3 lb. density									
3680	No finish									
3700	1" thick	G Q-14	170	.094	S.F.	.73	4.97		5.70	8.45
3710	1-1/2" thick	G	140	.114		1.10	6.05		7.15	10.45

For customer support on your Mechanical Costs with RSMeans Data, call 800.448.8182.

327

23 33 Air Duct Accessories

23 33 53 – Duct Liners

23 33 53.10 Duct Liner Board		Crew	Daily Output	Labor-Hours	Unit	Material	2020 Bare Costs Labor	2020 Bare Costs Equipment	Total	Total Incl O&P
3720	2" thick	G Q-14	130	.123	S.F.	1.49	6.50		7.99	11.60
3940	Board type, non-fibrous foam									
3950	Temperature, bacteria and fungi resistant									
3960	1" thick	G Q-14	150	.107	S.F.	2.76	5.65		8.41	11.70
3970	1-1/2" thick	G	130	.123		4.89	6.50		11.39	15.35
3980	2" thick	G	120	.133		4.13	7.05		11.18	15.35

23 34 HVAC Fans

23 34 13 – Axial HVAC Fans

23 34 13.10 Axial Flow HVAC Fans

		Crew	Daily Output	Labor-Hours	Unit	Material	2020 Bare Costs Labor	2020 Bare Costs Equipment	Total	Total Incl O&P
0010	**AXIAL FLOW HVAC FANS**									
0020	Air conditioning and process air handling									
0500	Axial flow, constant speed									
0505	Direct drive, 1/8" S.P.									
0510	12", 1,060 CFM, 1/6 HP	Q-20	3	6.667	Ea.	815	380		1,195	1,475
0514	12", 2,095 CFM, 1/2 HP		3	6.667		730	380		1,110	1,375
0518	16", 2,490 CFM, 1/3 HP		2.80	7.143		725	410		1,135	1,400
0522	20", 4,130 CFM, 3/4 HP		2.60	7.692		1,150	440		1,590	1,925
0526	22", 4,700 CFM, 3/4 HP		2.60	7.692		1,675	440		2,115	2,525
0530	24", 5,850 CFM, 1 HP		2.50	8		1,775	455		2,230	2,650
0534	24", 7,925 CFM, 1-1/2 HP		2.40	8.333		1,750	475		2,225	2,650
0538	30", 10,640 CFM, 2 HP		2.20	9.091		2,800	520		3,320	3,850
0542	30", 14,765 CFM, 2-1/2 HP		2.10	9.524		3,075	545		3,620	4,200
0546	36", 16,780 CFM, 2 HP		2.10	9.524		3,325	545		3,870	4,475
0550	36", 22,920 CFM, 5 HP	▼	1.80	11.111	▼	3,700	635		4,335	5,025
0560	Belt drive, 1/8" S.P.									
0562	15", 2,800 CFM, 1/3 HP	Q-20	3.20	6.250	Ea.	1,350	355		1,705	2,025
0564	15", 3,400 CFM, 1/2 HP		3	6.667		1,375	380		1,755	2,075
0568	18", 3,280 CFM, 1/3 HP		2.80	7.143		1,300	410		1,710	2,075
0572	18", 3,900 CFM, 1/2 HP		2.80	7.143		1,475	410		1,885	2,250
0576	18", 5,250 CFM, 1 HP		2.70	7.407		1,450	425		1,875	2,250
0584	24", 6,430 CFM, 1 HP		2.50	8		2,150	455		2,605	3,050
0588	24", 8,860 CFM, 2 HP		2.40	8.333		2,275	475		2,750	3,225
0592	30", 9,250 CFM, 1 HP		2.20	9.091		2,775	520		3,295	3,825
0596	30", 16,900 CFM, 5 HP		2	10		2,875	570		3,445	4,025
0604	36", 14,475 CFM, 2 HP		2.30	8.696		3,225	495		3,720	4,300
0608	36", 20,080 CFM, 5 HP		1.80	11.111		3,450	635		4,085	4,750
0616	42", 29,000 CFM, 7-1/2 HP		1.40	14.286		4,325	815		5,140	6,000
1500	Vaneaxial, low pressure, 2,000 CFM, 1/2 HP		3.60	5.556		2,400	315		2,715	3,100
1520	4,000 CFM, 1 HP		3.20	6.250		2,500	355		2,855	3,300
1540	8,000 CFM, 2 HP		2.80	7.143		3,550	410		3,960	4,525
1560	16,000 CFM, 5 HP	▼	2.40	8.333	▼	5,150	475		5,625	6,400

23 34 14 – Blower HVAC Fans

23 34 14.10 Blower Type HVAC Fans

		Crew	Daily Output	Labor-Hours	Unit	Material	2020 Bare Costs Labor	2020 Bare Costs Equipment	Total	Total Incl O&P
0010	**BLOWER TYPE HVAC FANS**									
2000	Blowers, direct drive with motor, complete									
2020	1,045 CFM @ 0.5" S.P., 1/5 HP	Q-20	18	1.111	Ea.	350	63.50		413.50	480
2040	1,385 CFM @ 0.5" S.P., 1/4 HP		18	1.111		365	63.50		428.50	495
2060	1,640 CFM @ 0.5" S.P., 1/3 HP		18	1.111		370	63.50		433.50	505
2080	1,760 CFM @ 0.5" S.P., 1/2 HP	▼	18	1.111	▼	360	63.50		423.50	495

23 34 HVAC Fans

23 34 14 – Blower HVAC Fans

23 34 14.10 Blower Type HVAC Fans

	Crew	Daily Output	Labor-Hours	Unit	Material	2020 Bare Costs Labor	Equipment	Total	Total Incl O&P
2090	4 speed								
2100	1,164 to 1,739 CFM @ 0.5" S.P., 1/3 HP	Q-20	16	1.250	Ea.	335	71.50		406.50	480
2120	1,467 to 2,218 CFM @ 1.0" S.P., 3/4 HP	"	14	1.429	"	385	81.50		466.50	545
2500	Ceiling fan, right angle, extra quiet, 0.10" S.P.									
2520	95 CFM	Q-20	20	1	Ea.	305	57		362	420
2540	210 CFM		19	1.053		360	60		420	485
2560	385 CFM		18	1.111		455	63.50		518.50	600
2580	885 CFM		16	1.250		900	71.50		971.50	1,100
2600	1,650 CFM		13	1.538		1,250	88		1,338	1,500
2620	2,960 CFM		11	1.818		1,675	104		1,779	1,975
2640	For wall or roof cap, add	1 Shee	16	.500		305	31		336	385
2660	For straight thru fan, add					10%				
2680	For speed control switch, add	1 Elec	16	.500		166	30.50		196.50	229
7500	Utility set, steel construction, pedestal, 1/4" S.P.									
7520	Direct drive, 150 CFM, 1/8 HP	Q-20	6.40	3.125	Ea.	975	179		1,154	1,350
7540	485 CFM, 1/6 HP		5.80	3.448		1,100	197		1,297	1,500
7560	1,950 CFM, 1/2 HP		4.80	4.167		1,425	238		1,663	1,925
7580	2,410 CFM, 3/4 HP		4.40	4.545		2,650	260		2,910	3,325
7600	3,328 CFM, 1-1/2 HP		3	6.667		2,625	380		3,005	3,475
7680	V-belt drive, drive cover, 3 phase									
7700	800 CFM, 1/4 HP	Q-20	6	3.333	Ea.	1,050	190		1,240	1,475
7720	1,300 CFM, 1/3 HP		5	4		1,125	229		1,354	1,575
7740	2,000 CFM, 1 HP		4.60	4.348		1,325	248		1,573	1,825
7760	2,900 CFM, 3/4 HP		4.20	4.762		1,775	272		2,047	2,350
7780	3,600 CFM, 3/4 HP		4	5		2,200	286		2,486	2,825
7800	4,800 CFM, 1 HP		3.50	5.714		2,575	325		2,900	3,350
7820	6,700 CFM, 1-1/2 HP		3	6.667		3,175	380		3,555	4,075
7830	7,500 CFM, 2 HP		2.50	8		4,325	455		4,780	5,450
7840	11,000 CFM, 3 HP		2	10		5,775	570		6,345	7,225
7860	13,000 CFM, 3 HP		1.60	12.500		5,875	715		6,590	7,525
7880	15,000 CFM, 5 HP		1	20		6,075	1,150		7,225	8,400
7900	17,000 CFM, 7-1/2 HP		.80	25		6,500	1,425		7,925	9,300
7920	20,000 CFM, 7-1/2 HP		.80	25		7,750	1,425		9,175	10,700

23 34 16 – Centrifugal HVAC Fans

23 34 16.10 Centrifugal Type HVAC Fans

	Crew	Daily Output	Labor-Hours	Unit	Material	2020 Bare Costs Labor	Equipment	Total	Total Incl O&P
0010	**CENTRIFUGAL TYPE HVAC FANS**									
0200	In-line centrifugal, supply/exhaust booster									
0220	aluminum wheel/hub, disconnect switch, 1/4" S.P.									
0240	500 CFM, 10" diameter connection	Q-20	3	6.667	Ea.	1,625	380		2,005	2,350
0260	1,380 CFM, 12" diameter connection		2	10		1,750	570		2,320	2,800
0280	1,520 CFM, 16" diameter connection		2	10		1,700	570		2,270	2,750
0300	2,560 CFM, 18" diameter connection		1	20		1,850	1,150		3,000	3,775
0320	3,480 CFM, 20" diameter connection		.80	25		2,175	1,425		3,600	4,550
0326	5,080 CFM, 20" diameter connection		.75	26.667		2,375	1,525		3,900	4,900
0340	7,500 CFM, 22" diameter connection		.70	28.571		2,950	1,625		4,575	5,725
0350	10,000 CFM, 27" diameter connection		.65	30.769		3,450	1,750		5,200	6,450
3500	Centrifugal, airfoil, motor and drive, complete									
3520	1,000 CFM, 1/2 HP	Q-20	2.50	8	Ea.	2,100	455		2,555	3,000
3540	2,000 CFM, 1 HP		2	10		2,250	570		2,820	3,375
3560	4,000 CFM, 3 HP		1.80	11.111		3,000	635		3,635	4,250
3580	8,000 CFM, 7-1/2 HP		1.40	14.286		4,525	815		5,340	6,200
3600	12,000 CFM, 10 HP		1	20		5,875	1,150		7,025	8,175

For customer support on your Mechanical Costs with RSMeans Data, call 800.448.8182.

329

23 34 16 – Centrifugal HVAC Fans

23 34 16.10 Centrifugal Type HVAC Fans	Crew	Daily Output	Labor-Hours	Unit	Material	2020 Bare Costs Labor	Equipment	Total	Total Incl O&P
4000 Single width, belt drive, not incl. motor, capacities									
4020 at 2,000 fpm, 2.5" S.P. for indicated motor									
4040 6,900 CFM, 5 HP	Q-9	2.40	6.667	Ea.	4,225	375		4,600	5,225
4060 10,340 CFM, 7-1/2 HP		2.20	7.273		5,900	410		6,310	7,100
4080 15,320 CFM, 10 HP		2	8		6,425	450		6,875	7,725
4100 22,780 CFM, 15 HP		1.80	8.889		10,300	500		10,800	12,200
4120 33,840 CFM, 20 HP		1.60	10		13,700	560		14,260	16,000
4140 41,400 CFM, 25 HP		1.40	11.429		16,200	640		16,840	18,800
4160 50,100 CFM, 30 HP		.80	20		22,400	1,125		23,525	26,300
4200 Double width wheel, 12,420 CFM, 7-1/2 HP		2.20	7.273		5,600	410		6,010	6,775
4220 18,620 CFM, 15 HP		2	8		8,675	450		9,125	10,200
4240 27,580 CFM, 20 HP		1.80	8.889		9,800	500		10,300	11,600
4260 40,980 CFM, 25 HP		1.50	10.667		15,300	600		15,900	17,800
4280 60,920 CFM, 40 HP		1	16		22,400	895		23,295	26,000
4300 74,520 CFM, 50 HP		.80	20		24,500	1,125		25,625	28,600
4320 90,160 CFM, 50 HP		.70	22.857		36,200	1,275		37,475	41,900
4340 110,300 CFM, 60 HP		.50	32		49,000	1,800		50,800	56,500
4360 134,960 CFM, 75 HP		.40	40		65,000	2,250		67,250	75,000
5000 Utility set, centrifugal, V belt drive, motor									
5020 1/4" S.P., 1,200 CFM, 1/4 HP	Q-20	6	3.333	Ea.	1,925	190		2,115	2,425
5040 1,520 CFM, 1/3 HP		5	4		2,575	229		2,804	3,175
5060 1,850 CFM, 1/2 HP		4	5		2,550	286		2,836	3,250
5080 2,180 CFM, 3/4 HP		3	6.667		2,950	380		3,330	3,825
5100 1/2" S.P., 3,600 CFM, 1 HP		2	10		3,100	570		3,670	4,275
5120 4,250 CFM, 1-1/2 HP		1.60	12.500		3,800	715		4,515	5,250
5140 4,800 CFM, 2 HP		1.40	14.286		4,625	815		5,440	6,325
5160 6,920 CFM, 5 HP		1.30	15.385		5,575	880		6,455	7,475
5180 7,700 CFM, 7-1/2 HP		1.20	16.667		6,775	950		7,725	8,900
5200 For explosion proof motor, add					15%				
5300 Fume exhauster without hose and intake nozzle									
5310 630 CFM, 1-1/2 HP	Q-20	2	10	Ea.	880	570		1,450	1,825
5320 Hose extension kit 5'	"	20	1	"	35.50	57		92.50	126
5500 Fans, industrial exhauster, for air which may contain granular matl.									
5520 1,000 CFM, 1-1/2 HP	Q-20	2.50	8	Ea.	3,300	455		3,755	4,350
5540 2,000 CFM, 3 HP		2	10		4,000	570		4,570	5,275
5560 4,000 CFM, 7-1/2 HP		1.80	11.111		5,750	635		6,385	7,275
5580 8,000 CFM, 15 HP		1.40	14.286		7,425	815		8,240	9,400
5600 12,000 CFM, 30 HP		1	20		11,500	1,150		12,650	14,300
7000 Roof exhauster, centrifugal, aluminum housing, 12" galvanized									
7020 curb, bird screen, back draft damper, 1/4" S.P.									
7100 Direct drive, 320 CFM, 11" sq. damper	Q-20	7	2.857	Ea.	785	163		948	1,100
7120 600 CFM, 11" sq. damper		6	3.333		880	190		1,070	1,250
7140 815 CFM, 13" sq. damper		5	4		1,050	229		1,279	1,525
7160 1,450 CFM, 13" sq. damper		4.20	4.762		1,750	272		2,022	2,325
7180 2,050 CFM, 16" sq. damper		4	5		1,900	286		2,186	2,500
7200 V-belt drive, 1,650 CFM, 12" sq. damper		6	3.333		1,425	190		1,615	1,875
7220 2,750 CFM, 21" sq. damper		5	4		1,800	229		2,029	2,350
7230 3,500 CFM, 21" sq. damper		4.50	4.444		2,025	254		2,279	2,625
7240 4,910 CFM, 23" sq. damper		4	5		2,500	286		2,786	3,175
7260 8,525 CFM, 28" sq. damper		3	6.667		3,250	380		3,630	4,150
7280 13,760 CFM, 35" sq. damper		2	10		4,500	570		5,070	5,825
7300 20,558 CFM, 43" sq. damper		1	20		8,950	1,150		10,100	11,600
7320 For 2 speed winding, add					15%				

23 34 HVAC Fans

23 34 16 – Centrifugal HVAC Fans

23 34 16.10 Centrifugal Type HVAC Fans

		Crew	Daily Output	Labor-Hours	Unit	Material	2020 Bare Costs Labor	2020 Bare Costs Equipment	Total	Total Incl O&P
7340	For explosion proof motor, add				Ea.	745			745	820
7360	For belt driven, top discharge, add			↓		15%				
7400	Roof mounted kitchen exhaust, aluminum, centrifugal									
7410	Direct drive, 1/4" S.P., 2 speed, temp to 200°F									
7414	1/3 HP, 2,038 CFM	Q-20	7	2.857	Ea.	635	163		798	945
7416	1/2 HP, 2,393 CFM		6	3.333		625	190		815	975
7418	3/4 HP, 4,329 CFM	↓	5	4	↓	960	229		1,189	1,400
7424	Belt drive, 1/4" S.P., temp to 250°F									
7426	3/4 HP, 2,757 CFM	Q-20	5	4	Ea.	980	229		1,209	1,425
7428	1-1/2 HP, 4,913 CFM		4	5		1,425	286		1,711	1,975
7430	1-1/2 HP, 5,798 CFM	↓	4	5	↓	1,450	286		1,736	2,025
7450	Upblast, propeller, w/backdraft damper, bird screen & curb									
7454	30,300 CFM @ 3/8" S.P., 5 HP	Q-20	1.60	12.500	Ea.	5,075	715		5,790	6,650
7458	36,000 CFM @ 1/2" S.P., 15 HP	"	1.60	12.500	"	7,200	715		7,915	9,000
8500	Wall exhausters, centrifugal, auto damper, 1/8" S.P.									
8520	Direct drive, 610 CFM, 1/20 HP	Q-20	14	1.429	Ea.	425	81.50		506.50	595
8540	796 CFM, 1/12 HP		13	1.538		1,075	88		1,163	1,300
8560	822 CFM, 1/6 HP		12	1.667		1,075	95		1,170	1,325
8580	1,320 CFM, 1/4 HP		12	1.667		1,375	95		1,470	1,650
8600	1,756 CFM, 1/4 HP		11	1.818		1,275	104		1,379	1,550
8620	1,983 CFM, 1/4 HP		10	2		1,300	114		1,414	1,600
8640	2,900 CFM, 1/2 HP		9	2.222		1,375	127		1,502	1,725
8660	3,307 CFM, 3/4 HP		8	2.500		1,500	143		1,643	1,875
8670	5,940 CFM, 1 HP	↓	7	2.857	↓	1,675	163		1,838	2,100
9500	V-belt drive, 3 phase									
9520	2,800 CFM, 1/4 HP	Q-20	9	2.222	Ea.	1,925	127		2,052	2,325
9540	3,740 CFM, 1/2 HP		8	2.500		2,000	143		2,143	2,425
9560	4,400 CFM, 3/4 HP		7	2.857		2,025	163		2,188	2,475
9580	5,700 CFM, 1-1/2 HP	↓	6	3.333	↓	2,125	190		2,315	2,625

23 34 23 – HVAC Power Ventilators

23 34 23.10 HVAC Power Circulators and Ventilators

			Crew	Daily Output	Labor-Hours	Unit	Material	2020 Bare Costs Labor	2020 Bare Costs Equipment	Total	Total Incl O&P
0010	**HVAC POWER CIRCULATORS AND VENTILATORS**										
3000	Paddle blade air circulator, 3 speed switch										
3020	42", 5,000 CFM high, 3,000 CFM low	G	1 Elec	2.40	3.333	Ea.	105	205		310	420
3040	52", 6,500 CFM high, 4,000 CFM low	G	"	2.20	3.636	"	144	223		367	490
3100	For antique white motor, same cost										
3200	For brass plated motor, same cost										
3300	For light adaptor kit, add	G				Ea.	40			40	44
3310	Industrial grade, reversible, 4 blade										
3312	5,500 CFM		Q-20	5	4	Ea.	139	229		368	500
3314	7,600 CFM			5	4		165	229		394	525
3316	21,015 CFM		↓	4	5	↓	193	286		479	640
4000	High volume, low speed (HVLS), paddle blade air circulator										
4010	Variable speed, reversible, 1 HP motor, motor control panel,										
4020	motor drive cable, control cable with remote, and safety cable.										
4140	8' diameter		Q-5	1.50	10.667	Ea.	4,925	630		5,555	6,350
4150	10' diameter			1.45	11.034		4,975	650		5,625	6,425
4160	12' diameter			1.40	11.429		5,175	675		5,850	6,700
4170	14' diameter			1.35	11.852		5,275	700		5,975	6,850
4180	16' diameter			1.30	12.308		5,325	725		6,050	6,925
4190	18' diameter			1.25	12.800		5,350	755		6,105	7,025
4200	20' diameter		↓	1.20	13.333		5,725	785		6,510	7,475

23 34 23.10 HVAC Power Circulators and Ventilators	Crew	Daily Output	Labor-Hours	Unit	Material	2020 Bare Costs Labor	2020 Bare Costs Equipment	Total	Total Incl O&P
4210 24' diameter	Q-5	1	16	Ea.	5,850	945		6,795	7,850
6000 Propeller exhaust, wall shutter									
6020 Direct drive, one speed, 0.075" S.P.									
6100 653 CFM, 1/30 HP	Q-20	10	2	Ea.	236	114		350	435
6120 1,033 CFM, 1/20 HP		9	2.222		325	127		452	545
6140 1,323 CFM, 1/15 HP		8	2.500		390	143		533	645
6160 2,444 CFM, 1/4 HP		7	2.857		425	163		588	715
6300 V-belt drive, 3 phase									
6320 6,175 CFM, 3/4 HP	Q-20	5	4	Ea.	1,200	229		1,429	1,650
6340 7,500 CFM, 3/4 HP		5	4		1,275	229		1,504	1,750
6360 10,100 CFM, 1 HP		4.50	4.444		1,500	254		1,754	2,050
6380 14,300 CFM, 1-1/2 HP		4	5		1,775	286		2,061	2,375
6400 19,800 CFM, 2 HP		3	6.667		2,400	380		2,780	3,225
6420 26,250 CFM, 3 HP		2.60	7.692		2,425	440		2,865	3,325
6440 38,500 CFM, 5 HP		2.20	9.091		2,975	520		3,495	4,050
6460 46,000 CFM, 7-1/2 HP		2	10		3,200	570		3,770	4,400
6480 51,500 CFM, 10 HP		1.80	11.111		3,175	635		3,810	4,425
6490 V-belt drive, 115 V, residential, whole house									
6500 Ceiling-wall, 5,200 CFM, 1/4 HP, 30" x 30"	1 Shee	6	1.333	Ea.	605	83		688	795
6510 7,500 CFM, 1/3 HP, 36" x 36"		5	1.600		635	99.50		734.50	850
6520 10,500 CFM, 1/3 HP, 42" x 42"		5	1.600		665	99.50		764.50	880
6530 13,200 CFM, 1/3 HP, 48" x 48"		4	2		735	125		860	1,000
6540 15,445 CFM, 1/2 HP, 48" x 48"		4	2		765	125		890	1,025
6550 17,025 CFM, 1/2 HP, 54" x 54"		4	2		1,475	125		1,600	1,825
6560 For two speed motor, add					20%				
6570 Shutter, automatic, ceiling/wall									
6580 30" x 30"	1 Shee	8	1	Ea.	198	62.50		260.50	310
6590 36" x 36"		8	1		226	62.50		288.50	345
6600 42" x 42"		8	1		282	62.50		344.50	405
6610 48" x 48"		7	1.143		305	71		376	445
6620 54" x 54"		6	1.333		405	83		488	570
6630 Timer, shut off, to 12 hour		20	.400		65	25		90	110
6650 Residential, bath exhaust, grille, back draft damper									
6660 50 CFM	Q-20	24	.833	Ea.	61	47.50		108.50	139
6670 110 CFM		22	.909		116	52		168	207
6672 180 CFM		22	.909		152	52		204	246
6673 210 CFM		22	.909		143	52		195	237
6674 260 CFM		22	.909		245	52		297	350
6675 300 CFM		22	.909		177	52		229	273
6680 Light combination, squirrel cage, 100 watt, 70 CFM		24	.833		123	47.50		170.50	207
6700 Light/heater combination, ceiling mounted									
6710 70 CFM, 1,450 watt	Q-20	24	.833	Ea.	139	47.50		186.50	225
6800 Heater combination, recessed, 70 CFM		24	.833		83.50	47.50		131	164
6820 With 2 infrared bulbs		23	.870		100	49.50		149.50	185
6840 Wall mount, 170 CFM		22	.909		281	52		333	390
6846 Ceiling mount, 180 CFM		22	.909		284	52		336	390
6900 Kitchen exhaust, grille, complete, 160 CFM		22	.909		104	52		156	194
6910 180 CFM		20	1		97	57		154	194
6920 270 CFM		18	1.111		214	63.50		277.50	330
6930 350 CFM		16	1.250		146	71.50		217.50	269
6940 Residential roof jacks and wall caps									
6944 Wall cap with back draft damper									
6946 3" & 4" diam. round duct	1 Shee	11	.727	Ea.	26.50	45.50		72	97.50

23 34 23 – HVAC Power Ventilators

23 34 23.10 HVAC Power Circulators and Ventilators	Crew	Daily Output	Labor-Hours	Unit	Material	2020 Bare Costs Labor	Equipment	Total	Total Incl O&P
6948	6" diam. round duct	1 Shee	11	.727	Ea.	76	45.50		121.50	153
6958	Roof jack with bird screen and back draft damper									
6960	3" & 4" diam. round duct	1 Shee	11	.727	Ea.	18.85	45.50		64.35	89.50
6962	3-1/4" x 10" rectangular duct	"	10	.800	"	35	50		85	114
6980	Transition									
6982	3-1/4" x 10" to 6" diam. round	1 Shee	20	.400	Ea.	36	25		61	77.50
8020	Attic, roof type									
8030	Aluminum dome, damper & curb									
8080	12" diameter, 1,000 CFM (gravity)	1 Elec	10	.800	Ea.	640	49		689	780
8090	16" diameter, 1,500 CFM (gravity)		9	.889		775	54.50		829.50	930
8100	20" diameter, 2,500 CFM (gravity)		8	1		950	61.50		1,011.50	1,150
8110	26" diameter, 4,000 CFM (gravity)		7	1.143		1,150	70		1,220	1,375
8120	32" diameter, 6,500 CFM (gravity)		6	1.333		1,575	82		1,657	1,850
8130	38" diameter, 8,000 CFM (gravity)		5	1.600		2,350	98		2,448	2,725
8140	50" diameter, 13,000 CFM (gravity)	↓	4	2	↓	3,400	123		3,523	3,900
8160	Plastic, ABS dome									
8180	1,050 CFM	1 Elec	14	.571	Ea.	189	35		224	259
8200	1,600 CFM	"	12	.667	"	282	41		323	370
8240	Attic, wall type, with shutter, one speed									
8250	12" diameter, 1,000 CFM	1 Elec	14	.571	Ea.	430	35		465	525
8260	14" diameter, 1,500 CFM		12	.667		465	41		506	575
8270	16" diameter, 2,000 CFM	↓	9	.889	↓	530	54.50		584.50	660
8290	Whole house, wall type, with shutter, one speed									
8300	30" diameter, 4,800 CFM	1 Elec	7	1.143	Ea.	1,125	70		1,195	1,350
8310	36" diameter, 7,000 CFM		6	1.333		1,225	82		1,307	1,475
8320	42" diameter, 10,000 CFM		5	1.600		1,375	98		1,473	1,675
8330	48" diameter, 16,000 CFM	↓	4	2		1,700	123		1,823	2,050
8340	For two speed, add				↓	103			103	113
8350	Whole house, lay-down type, with shutter, one speed									
8360	30" diameter, 4,500 CFM	1 Elec	8	1	Ea.	1,200	61.50		1,261.50	1,425
8370	36" diameter, 6,500 CFM		7	1.143		1,300	70		1,370	1,525
8380	42" diameter, 9,000 CFM		6	1.333		1,425	82		1,507	1,675
8390	48" diameter, 12,000 CFM		5	1.600		1,600	98		1,698	1,925
8440	For two speed, add					77.50			77.50	85
8450	For 12 hour timer switch, add	1 Elec	32	.250	↓	77.50	15.35		92.85	108

23 34 33 – Air Curtains

23 34 33.10 Air Barrier Curtains

		Crew	Daily Output	Labor-Hours	Unit	Material	2020 Bare Costs Labor	Equipment	Total	Total Incl O&P
0010	**AIR BARRIER CURTAINS**, Incl. motor starters, transformers,									
0050	and door switches									
2450	Conveyor openings or service windows									
3000	Service window, 5' high x 25" wide	2 Shee	5	3.200	Ea.	345	199		544	680
3100	Environmental separation									
3110	Door heights up to 8', low profile, super quiet									
3120	Unheated, variable speed									
3130	36" wide	2 Shee	4	4	Ea.	765	249		1,014	1,225
3134	42" wide		3.80	4.211		780	262		1,042	1,250
3138	48" wide		3.60	4.444		810	277		1,087	1,300
3142	60" wide	↓	3.40	4.706		865	293		1,158	1,400
3146	72" wide	Q-3	4.60	6.957		1,050	425		1,475	1,800
3150	96" wide		4.40	7.273		1,625	445		2,070	2,450
3154	120" wide		4.20	7.619		1,600	465		2,065	2,450
3158	144" wide	↓	4	8	↓	2,000	490		2,490	2,925

23 34 33 – Air Curtains

23 34 33.10 Air Barrier Curtains		Crew	Daily Output	Labor-Hours	Unit	Material	2020 Bare Costs Labor	Equipment	Total	Total Incl O&P
3200	Door heights up to 10'									
3210	Unheated									
3230	36" wide	2 Shee	3.80	4.211	Ea.	735	262		997	1,200
3234	42" wide		3.60	4.444		795	277		1,072	1,300
3238	48" wide		3.40	4.706		780	293		1,073	1,300
3242	60" wide	▼	3.20	5		1,100	310		1,410	1,700
3246	72" wide	Q-3	4.40	7.273		1,325	445		1,770	2,150
3250	96" wide		4.20	7.619		1,575	465		2,040	2,425
3254	120" wide		4	8		2,250	490		2,740	3,200
3258	144" wide	▼	3.80	8.421	▼	2,275	515		2,790	3,275
3300	Door heights up to 12'									
3310	Unheated									
3334	42" wide	2 Shee	3.40	4.706	Ea.	1,125	293		1,418	1,675
3338	48" wide		3.20	5		1,125	310		1,435	1,725
3342	60" wide	▼	3	5.333		1,150	330		1,480	1,750
3346	72" wide	Q-3	4.20	7.619		1,925	465		2,390	2,800
3350	96" wide		4	8		2,225	490		2,715	3,175
3354	120" wide		3.80	8.421		2,525	515		3,040	3,550
3358	144" wide	▼	3.60	8.889	▼	3,650	545		4,195	4,825
3400	Door heights up to 16'									
3410	Unheated									
3438	48" wide	2 Shee	3	5.333	Ea.	1,475	330		1,805	2,125
3442	60" wide	"	2.80	5.714		1,550	355		1,905	2,250
3446	72" wide	Q-3	3.80	8.421		2,700	515		3,215	3,750
3450	96" wide		3.60	8.889		3,175	545		3,720	4,300
3454	120" wide		3.40	9.412		3,625	575		4,200	4,875
3458	144" wide	▼	3.20	10	▼	3,625	615		4,240	4,925
3470	Heated, electric									
3474	48" wide	2 Shee	2.90	5.517	Ea.	2,500	345		2,845	3,275
3478	60" wide	"	2.70	5.926		2,550	370		2,920	3,350
3482	72" wide	Q-3	3.70	8.649		4,275	530		4,805	5,500
3486	96" wide		3.50	9.143		4,150	560		4,710	5,425
3490	120" wide		3.30	9.697		4,725	595		5,320	6,100
3494	144" wide	▼	3.10	10.323	▼	5,550	635		6,185	7,050

23 34 39 – Ceiling Fans

		Crew	Daily Output	Labor-Hours	Unit	Material	2020 Bare Costs Labor	Equipment	Total	Total Incl O&P
2301	Ceiling fan, Powerfoil, 8' diameter	R-1A	2.40	6.667	Ea.	8,650	370		9,020	10,100
2302	10'		2.40	6.667		8,750	370		9,120	10,200
2303	12'		2.40	6.667		8,850	370		9,220	10,300
2304	14'		2.40	6.667		8,950	370		9,320	10,400
2305	16'		2	8		9,050	440		9,490	10,600
2306	18'		2	8		9,150	440		9,590	10,800
2307	20'		2	8		9,250	440		9,690	10,900
2308	24'	▼	2	8	▼	9,450	440		9,890	11,100
2400	Powerfoil with Hybrid Airfoil									
2401	12'	R-1A	2	8	Ea.	9,350	440		9,790	11,000
2402	14'		2	8		9,450	440		9,890	11,100
2403	16'		1.75	9.143		9,550	505		10,055	11,300
2404	18'		1.75	9.143		9,650	505		10,155	11,400
2405	20'		1.75	9.143		9,750	505		10,255	11,500
2406	24'	▼	1.75	9.143	▼	9,975	505		10,480	11,800

23 35 16.10 Engine Exhaust Removal Systems	Crew	Daily Output	Labor-Hours	Unit	Material	2020 Bare Costs Labor	2020 Bare Costs Equipment	Total	Total Incl O&P
0010 **ENGINE EXHAUST REMOVAL SYSTEMS** D3090-320									
0500　Engine exhaust, garage, in-floor system									
0510　　Single tube outlet assemblies									
0520　　　For transite pipe ducting, self-storing tube									
0530　　　　3" tubing adapter plate	1 Shee	16	.500	Ea.	262	31		293	335
0540　　　　4" tubing adapter plate		16	.500		266	31		297	340
0550　　　　5" tubing adapter plate	↓	16	.500	↓	264	31		295	340
0600　　　For vitrified tile ducting									
0610　　　　3" tubing adapter plate, self-storing tube	1 Shee	16	.500	Ea.	260	31		291	335
0620　　　　4" tubing adapter plate, self-storing tube		16	.500		265	31		296	340
0660　　　　5" tubing adapter plate, self-storing tube	↓	16	.500	↓	265	31		296	340
0800　　Two tube outlet assemblies									
0810　　　For transite pipe ducting, self-storing tube									
0820　　　　3" tubing, dual exhaust adapter plate	1 Shee	16	.500	Ea.	267	31		298	340
0850　　　For vitrified tile ducting									
0860　　　　3" tubing, dual exhaust, self-storing tube	1 Shee	16	.500	Ea.	300	31		331	380
0870　　　　3" tubing, double outlet, non-storing tubes	"	16	.500	"	300	31		331	380
0900　Accessories for metal tubing (overhead systems also)									
0910　　Adapters, for metal tubing end									
0920　　　3" tail pipe type				Ea.	51			51	56.50
0930　　　4" tail pipe type					56			56	62
0940　　　5" tail pipe type					57			57	63
0990　　　5" diesel stack type					283			283	310
1000　　　6" diesel stack type				↓	335			335	370
1100　　Bullnose (guide) required for in-floor assemblies									
1110　　　3" tubing size				Ea.	30.50			30.50	33.50
1120　　　4" tubing size					31.50			31.50	35
1130　　　5" tubing size				↓	33.50			33.50	36.50
1150　　Plain rings, for tubing end									
1160　　　3" tubing size				Ea.	23			23	25
1170　　　4" tubing size				"	38.50			38.50	42.50
1200　　Tubing, galvanized, flexible (for overhead systems also)									
1210　　　3" ID				L.F.	10.80			10.80	11.85
1220　　　4" ID					13.20			13.20	14.55
1230　　　5" ID					15.55			15.55	17.15
1240　　　6" ID				↓	18			18	19.80
1250　　Stainless steel, flexible (for overhead system also)									
1260　　　3" ID				L.F.	24			24	26.50
1270　　　4" ID					33			33	36
1280　　　5" ID					37.50			37.50	41
1290　　　6" ID				↓	43.50			43.50	47.50
1500　Engine exhaust, garage, overhead components, for neoprene tubing									
1510　　Alternate metal tubing & accessories see above									
1550　　Adapters, for neoprene tubing end									
1560　　　3" tail pipe, adjustable, neoprene				Ea.	58.50			58.50	64
1570　　　3" tail pipe, heavy wall neoprene					66			66	72.50
1580　　　4" tail pipe, heavy wall neoprene					103			103	113
1590　　　5" tail pipe, heavy wall neoprene				↓	117			117	129
1650　　Connectors, tubing									
1660　　　3" interior, aluminum				Ea.	23			23	25.50
1670　　　4" interior, aluminum					38.50			38.50	42.50
1710　　　5" interior, neoprene					59			59	64.50
1750　　　3" spiralock, neoprene				↓	23			23	25

For customer support on your Mechanical Costs with RSMeans Data, call 800.448.8182.

335

23 35 Special Exhaust Systems

23 35 16 – Engine Exhaust Systems

23 35 16.10 Engine Exhaust Removal Systems		Crew	Daily Output	Labor-Hours	Unit	Material	2020 Bare Costs Labor	Equipment	Total	Total Incl O&P
1760	4" spiralock, neoprene				Ea.	38.50			38.50	42.50
1780	Y for 3" ID tubing, neoprene, dual exhaust					194			194	213
1790	Y for 4" ID tubing, aluminum, dual exhaust					191			191	210
1850	Elbows, aluminum, splice into tubing for strap									
1860	3" neoprene tubing size				Ea.	54			54	59.50
1870	4" neoprene tubing size					36			36	40
1900	Flange assemblies, connect tubing to overhead duct					69.50			69.50	76.50
2000	Hardware and accessories									
2020	Cable, galvanized, 1/8" diameter				L.F.	.47			.47	.52
2040	Cleat, tie down cable or rope				Ea.	5.90			5.90	6.50
2060	Pulley					8.15			8.15	9
2080	Pulley hook, universal					5.90			5.90	6.50
2100	Rope, nylon, 1/4" diameter				L.F.	.41			.41	.45
2120	Winch, 1" diameter				Ea.	119			119	131
2150	Lifting strap, mounts on neoprene									
2160	3" tubing size				Ea.	29.50			29.50	32.50
2170	4" tubing size					29.50			29.50	32.50
2180	5" tubing size					29.50			29.50	32.50
2190	6" tubing size					29.50			29.50	32.50
2200	Tubing, neoprene, 11' lengths									
2210	3" ID				L.F.	12.25			12.25	13.50
2220	4" ID					16.90			16.90	18.55
2230	5" ID					29			29	32
2500	Engine exhaust, thru-door outlet									
2510	3" tube size	1 Carp	16	.500	Ea.	64	26.50		90.50	111
2530	4" tube size	"	16	.500	"	69.50	26.50		96	117
3000	Tubing, exhaust, flex hose, with									
3010	coupler, damper and tail pipe adapter									
3020	Neoprene									
3040	3" x 20'	1 Shee	6	1.333	Ea.	360	83		443	525
3050	4" x 15'		5.40	1.481		420	92.50		512.50	605
3060	4" x 20'		5	1.600		505	99.50		604.50	705
3070	5" x 15'		4.40	1.818		645	113		758	880
3100	Galvanized									
3110	3" x 20'	1 Shee	6	1.333	Ea.	315	83		398	475
3120	4" x 17'		5.60	1.429		350	89		439	515
3130	4" x 20'		5	1.600		385	99.50		484.50	575
3140	5" x 17'		4.60	1.739		415	108		523	620
8000	Blower, for tailpipe exhaust system									
8010	Direct drive									
8012	495 CFM, 1/3 HP	Q-9	5	3.200	Ea.	2,550	179		2,729	3,075
8014	1,445 CFM, 3/4 HP		4	4		3,175	224		3,399	3,850
8016	1,840 CFM, 1-1/2 HP		3	5.333		3,800	299		4,099	4,625
8030	Beltdrive									
8032	1,400 CFM, 1 HP	Q-9	4	4	Ea.	3,175	224		3,399	3,850
8034	2,023 CFM, 1-1/2 HP		3	5.333		4,425	299		4,724	5,325
8036	2,750 CFM, 2 HP		2	8		5,225	450		5,675	6,425
8038	4,400 CFM, 3 HP		1.80	8.889		7,700	500		8,200	9,225
8040	7,060 CFM, 5 HP		1.40	11.429		10,900	640		11,540	13,000

23 35 Special Exhaust Systems

23 35 43 – Welding Fume Elimination Systems

23 35 43.10 Welding Fume Elimination System Components

	Crew	Daily Output	Labor-Hours	Unit	Material	2020 Bare Costs Labor	Equipment	Total	Total Incl O&P
0010 **WELDING FUME ELIMINATION SYSTEM COMPONENTS**									
7500 Welding fume elimination accessories for garage exhaust systems									
7600 Cut off (blast gate)									
7610 3" tubing size, 3" x 6" opening	1 Shee	24	.333	Ea.	27.50	21		48.50	62
7620 4" tubing size, 4" x 8" opening		24	.333		28	21		49	62.50
7630 5" tubing size, 5" x 10" opening		24	.333		32	21		53	67
7640 6" tubing size		24	.333		33.50	21		54.50	68.50
7650 8" tubing size	↓	24	.333	↓	43.50	21		64.50	79
7700 Hoods, magnetic, with handle & screen									
7710 3" tubing size, 3" x 6" opening	1 Shee	24	.333	Ea.	94.50	21		115.50	136
7720 4" tubing size, 4" x 8" opening		24	.333		102	21		123	145
7730 5" tubing size, 5" x 10" opening	↓	24	.333	↓	102	21		123	145

23 36 Air Terminal Units

23 36 13 – Constant-Air-Volume Units

23 36 13.10 Constant Volume Mixing Boxes

	Crew	Daily Output	Labor-Hours	Unit	Material	2020 Bare Costs Labor	Equipment	Total	Total Incl O&P
0010 **CONSTANT VOLUME MIXING BOXES**									
5180 Mixing box, includes electric or pneumatic motor									
5192 Recommend use with silencer, see Line 23 33 19.10 0010									
5200 Constant volume, 150 to 270 CFM	Q-9	12	1.333	Ea.	1,200	75		1,275	1,450
5210 270 to 600 CFM		11	1.455		1,150	81.50		1,231.50	1,400
5230 550 to 1,000 CFM		9	1.778		1,150	99.50		1,249.50	1,425
5240 1,000 to 1,600 CFM		8	2		1,250	112		1,362	1,575
5250 1,300 to 1,900 CFM		6	2.667		1,300	150		1,450	1,650
5260 550 to 2,640 CFM		5.60	2.857		1,425	160		1,585	1,800
5270 650 to 3,120 CFM	↓	5.20	3.077		1,425	173		1,598	1,825

23 36 16 – Variable-Air-Volume Units

23 36 16.10 Variable Volume Mixing Boxes

	Crew	Daily Output	Labor-Hours	Unit	Material	2020 Bare Costs Labor	Equipment	Total	Total Incl O&P
0010 **VARIABLE VOLUME MIXING BOXES**									
5180 Mixing box, includes electric or pneumatic motor									
5192 Recommend use with attenuator, see Line 23 33 19.10 0010									
5500 VAV cool only, pneumatic, pressure independent 300 to 600 CFM	Q-9	11	1.455	Ea.	815	81.50		896.50	1,025
5510 500 to 1,000 CFM		9	1.778		835	99.50		934.50	1,075
5520 800 to 1,600 CFM		9	1.778		865	99.50		964.50	1,100
5530 1,100 to 2,000 CFM		8	2		885	112		997	1,150
5540 1,500 to 3,000 CFM		7	2.286		940	128		1,068	1,225
5550 2,000 to 4,000 CFM	↓	6	2.667	↓	950	150		1,100	1,275
5600 VAV, HW coils, damper, actuator and thermostat									
5610 200 CFM	Q-9	11	1.455	Ea.	1,275	81.50		1,356.50	1,525
5620 400 CFM		10	1.600		1,275	89.50		1,364.50	1,525
5630 600 CFM		10	1.600		1,275	89.50		1,364.50	1,525
5640 800 CFM		8	2		1,325	112		1,437	1,625
5650 1,000 CFM		8	2		1,325	112		1,437	1,625
5660 1,250 CFM		6	2.667		1,450	150		1,600	1,800
5670 1,500 CFM		6	2.667		1,450	150		1,600	1,800
5680 2,000 CFM		4	4		1,675	224		1,899	2,175
5684 3,000 CFM	↓	3.80	4.211	↓	1,825	236		2,061	2,350
5700 VAV cool only, fan powered, damper, actuator, thermostat									
5710 200 CFM	Q-9	10	1.600	Ea.	1,725	89.50		1,814.50	2,025
5720 400 CFM	↓	9	1.778	↓	1,975	99.50		2,074.50	2,300

For customer support on your Mechanical Costs with RSMeans Data, call 800.448.8182.

337

23 36 16 – Variable-Air-Volume Units

23 36 16.10 Variable Volume Mixing Boxes		Crew	Daily Output	Labor-Hours	Unit	Material	2020 Bare Costs Labor	2020 Bare Costs Equipment	Total	Total Incl O&P
5730	600 CFM	Q-9	9	1.778	Ea.	1,975	99.50		2,074.50	2,300
5740	800 CFM		7	2.286		2,100	128		2,228	2,525
5750	1,000 CFM		7	2.286		2,100	128		2,228	2,525
5760	1,250 CFM		5	3.200		2,275	179		2,454	2,775
5770	1,500 CFM		5	3.200		2,275	179		2,454	2,775
5780	2,000 CFM	▼	4	4	▼	2,500	224		2,724	3,100
5800	VAV fan powr'd, with HW coils, dampers, actuators, thermostat									
5810	200 CFM	Q-9	10	1.600	Ea.	2,275	89.50		2,364.50	2,625
5820	400 CFM		9	1.778		2,350	99.50		2,449.50	2,750
5830	600 CFM		9	1.778		2,350	99.50		2,449.50	2,750
5840	800 CFM		7	2.286		2,500	128		2,628	2,950
5850	1,000 CFM		7	2.286		2,500	128		2,628	2,950
5860	1,250 CFM		5	3.200		2,825	179		3,004	3,400
5870	1,500 CFM		5	3.200		2,825	179		3,004	3,400
5880	2,000 CFM	▼	3.50	4.571	▼	3,050	256		3,306	3,750

23 37 Air Outlets and Inlets

23 37 13 – Diffusers, Registers, and Grilles

23 37 13.10 Diffusers

			Crew	Daily Output	Labor-Hours	Unit	Material	2020 Bare Costs Labor	2020 Bare Costs Equipment	Total	Total Incl O&P
0010	**DIFFUSERS**, Aluminum, opposed blade damper unless noted										
0100	Ceiling, linear, also for sidewall										
0120	2" wide	R233100-30	1 Shee	32	.250	L.F.	21	15.60		36.60	46.50
0140	3" wide			30	.267		24	16.60		40.60	51.50
0160	4" wide	R233700-60		26	.308		24	19.15		43.15	55.50
0180	6" wide			24	.333		30	21		51	64.50
0200	8" wide			22	.364		40.50	22.50		63	79
0220	10" wide			20	.400		39.50	25		64.50	81.50
0240	12" wide		▼	18	.444	▼	50.50	27.50		78	97.50
0260	For floor or sill application, add						15%				
0500	Perforated, 24" x 24" lay-in panel size, 6" x 6"		1 Shee	16	.500	Ea.	166	31		197	230
0520	8" x 8"			15	.533		181	33		214	250
0530	9" x 9"			14	.571		177	35.50		212.50	248
0540	10" x 10"			14	.571		177	35.50		212.50	249
0560	12" x 12"			12	.667		184	41.50		225.50	266
0580	15" x 15"			11	.727		196	45.50		241.50	285
0590	16" x 16"			11	.727		207	45.50		252.50	297
0600	18" x 18"			10	.800		222	50		272	320
0610	20" x 20"			10	.800		239	50		289	340
0620	24" x 24"			9	.889		284	55.50		339.50	395
1000	Rectangular, 1 to 4 way blow, 6" x 6"			16	.500		43.50	31		74.50	95.50
1010	8" x 8"			15	.533		63	33		96	120
1014	9" x 9"			15	.533		53.50	33		86.50	110
1016	10" x 10"			15	.533		83.50	33		116.50	143
1020	12" x 6"			15	.533		78	33		111	137
1040	12" x 9"			14	.571		84.50	35.50		120	147
1060	12" x 12"			12	.667		76	41.50		117.50	147
1070	14" x 6"			13	.615		85	38.50		123.50	152
1074	14" x 14"			12	.667		133	41.50		174.50	210
1080	18" x 12"			11	.727		150	45.50		195.50	234
1120	15" x 15"			10	.800		94	50		144	180
1140	18" x 15"		▼	9	.889	▼	172	55.50		227.50	273

23 37 13.10 Diffusers		Crew	Daily Output	Labor-Hours	Unit	Material	2020 Bare Costs Labor	Equipment	Total	Total Incl O&P
1150	18" x 18"	1 Shee	9	.889	Ea.	129	55.50		184.50	225
1160	21" x 21"		8	1		232	62.50		294.50	350
1170	24" x 12"		10	.800		177	50		227	270
1180	24" x 24"		7	1.143		298	71		369	440
1500	Round, butterfly damper, steel, diffuser size, 6" diameter		18	.444		13.80	27.50		41.30	57
1520	8" diameter		16	.500		14.70	31		45.70	63.50
1540	10" diameter		14	.571		18.55	35.50		54.05	74.50
1560	12" diameter		12	.667		25	41.50		66.50	90.50
1580	14" diameter		10	.800		31.50	50		81.50	111
1600	18" diameter		9	.889		77.50	55.50		133	169
1700	Round, steel, adjustable core, annular segmented damper, diffuser size									
1704	6" diameter	1 Shee	18	.444	Ea.	82.50	27.50		110	133
1708	8" diameter		16	.500		98	31		129	156
1712	10" diameter		14	.571		116	35.50		151.50	182
1716	12" diameter		12	.667		143	41.50		184.50	220
1720	14" diameter		10	.800		175	50		225	269
1724	18" diameter		9	.889		320	55.50		375.50	440
1728	20" diameter		9	.889		370	55.50		425.50	490
1732	24" diameter		8	1		360	62.50		422.50	490
1736	30" diameter		7	1.143		880	71		951	1,075
1740	36" diameter		6	1.333		1,075	83		1,158	1,300
2000	T-bar mounting, 24" x 24" lay-in frame, 6" x 6"		16	.500		73.50	31		104.50	129
2020	8" x 8"		14	.571		73.50	35.50		109	135
2040	12" x 12"		12	.667		90	41.50		131.50	162
2060	16" x 16"		11	.727		125	45.50		170.50	206
2080	18" x 18"		10	.800		123	50		173	211
2500	Combination supply and return									
2520	21" x 21" supply, 15" x 15" return	Q-9	10	1.600	Ea.	175	89.50		264.50	330
2540	24" x 24" supply, 18" x 18" return		9.50	1.684		231	94.50		325.50	395
2560	27" x 27" supply, 18" x 18" return		9	1.778		255	99.50		354.50	430
2580	30" x 30" supply, 21" x 21" return		8.50	1.882		345	106		451	540
2600	33" x 33" supply, 21" x 21" return		8	2		375	112		487	580
2620	36" x 36" supply, 24" x 24" return		7.50	2.133		440	120		560	665
3000	Baseboard, white enameled steel									
3100	18" long	1 Shee	20	.400	Ea.	7.65	25		32.65	46.50
3120	24" long		18	.444		15.55	27.50		43.05	59
3140	48" long		16	.500		28.50	31		59.50	79
3400	For matching return, deduct					10%				
4000	Floor, steel, adjustable pattern									
4100	2" x 10"	1 Shee	34	.235	Ea.	10.70	14.65		25.35	34
4120	2" x 12"		32	.250		11.75	15.60		27.35	36.50
4140	2" x 14"		30	.267		13.30	16.60		29.90	39.50
4200	4" x 10"		28	.286		11.95	17.80		29.75	40
4220	4" x 12"		26	.308		13.20	19.15		32.35	43.50
4240	4" x 14"		25	.320		15.45	19.95		35.40	47
4260	6" x 10"		26	.308		18.65	19.15		37.80	49.50
4280	6" x 12"		24	.333		18.85	21		39.85	52
4300	6" x 14"		22	.364		20.50	22.50		43	57
4500	Linear bar diffuser in wall, sill or ceiling									
4510	2" wide	1 Shee	32	.250	L.F.	65.50	15.60		81.10	95.50
4512	3" wide		30	.267		70.50	16.60		87.10	103
4514	4" wide		26	.308		87.50	19.15		106.65	125
4516	6" wide		24	.333		108	21		129	151

23 37 13.10 Diffusers

		Crew	Daily Output	Labor-Hours	Unit	Material	2020 Bare Costs Labor	Equipment	Total	Total Incl O&P
4520	10" wide	1 Shee	20	.400	L.F.	144	25		169	196
4522	12" wide	↓	18	.444	↓	161	27.50		188.50	219
5000	Sidewall, aluminum, 3 way dispersion									
5100	8" x 4"	1 Shee	24	.333	Ea.	6.60	21		27.60	39
5110	8" x 6"		24	.333		7.10	21		28.10	39.50
5120	10" x 4"		22	.364		6.75	22.50		29.25	42
5130	10" x 6"		22	.364		7.20	22.50		29.70	42.50
5140	10" x 8"		20	.400		10.20	25		35.20	49
5160	12" x 4"		18	.444		7.50	27.50		35	50.50
5170	12" x 6"		17	.471		8	29.50		37.50	53.50
5180	12" x 8"		16	.500		10.75	31		41.75	59.50
5200	14" x 4"		15	.533		8.35	33		41.35	59.50
5220	14" x 6"		14	.571		9.20	35.50		44.70	64
5240	14" x 8"		13	.615		12	38.50		50.50	71
5260	16" x 6"	↓	12.50	.640		13.15	40		53.15	75
6000	For steel diffusers instead of aluminum, deduct				↓	10%				

23 37 13.30 Grilles

		Crew	Daily Output	Labor-Hours	Unit	Material	2020 Bare Costs Labor	Equipment	Total	Total Incl O&P
0010	**GRILLES**									
0020	Aluminum, unless noted otherwise									
0100	Air supply, single deflection, adjustable									
0120	8" x 4"	1 Shee	30	.267	Ea.	12.90	16.60		29.50	39
0140	8" x 8"		28	.286		15.65	17.80		33.45	44
0160	10" x 4"		24	.333		14.25	21		35.25	47
0180	10" x 10"		23	.348		18	21.50		39.50	53
0200	12" x 6"		23	.348		11.70	21.50		33.20	46
0220	12" x 12"		22	.364		20.50	22.50		43	57
0230	14" x 6"		23	.348		14.45	21.50		35.95	49
0240	14" x 8"		23	.348		17.85	21.50		39.35	52.50
0260	14" x 14"		22	.364		27	22.50		49.50	64.50
0270	18" x 8"		23	.348		26	21.50		47.50	61.50
0280	18" x 10"		23	.348		25.50	21.50		47	61
0300	18" x 18"		21	.381		35.50	23.50		59	75
0320	20" x 12"		22	.364		21	22.50		43.50	57.50
0340	20" x 20"		21	.381		44.50	23.50		68	85
0350	24" x 8"		20	.400		19.85	25		44.85	60
0360	24" x 14"		18	.444		36	27.50		63.50	82
0380	24" x 24"		15	.533		49.50	33		82.50	105
0400	30" x 8"		20	.400		41	25		66	83
0420	30" x 10"		19	.421		27	26		53	69.50
0440	30" x 12"		18	.444		29.50	27.50		57	74.50
0460	30" x 16"		17	.471		44	29.50		73.50	93
0480	30" x 18"		17	.471		58.50	29.50		88	109
0500	30" x 30"		14	.571		90	35.50		125.50	153
0520	36" x 12"		17	.471		50	29.50		79.50	99.50
0540	36" x 16"		16	.500		54	31		85	107
0560	36" x 18"		15	.533		70.50	33		103.50	128
0570	36" x 20"		15	.533		66.50	33		99.50	124
0580	36" x 24"		14	.571		100	35.50		135.50	164
0600	36" x 28"		13	.615		88.50	38.50		127	156
0620	36" x 30"		12	.667		135	41.50		176.50	211
0640	36" x 32"		12	.667		109	41.50		150.50	183
0660	36" x 34"	↓	11	.727		139	45.50		184.50	222

23 37 13.30 Grilles		Crew	Daily Output	Labor-Hours	Unit	Material	2020 Bare Costs Labor	Equipment	Total	Total Incl O&P
0680	36" x 36"	1 Shee	11	.727	Ea.	175	45.50		220.50	261
0700	For double deflecting, add					70%				
1000	Air return, steel, 6" x 6"	1 Shee	26	.308		23.50	19.15		42.65	55
1020	10" x 6"		24	.333		23.50	21		44.50	57.50
1040	14" x 6"		23	.348		28	21.50		49.50	64
1060	10" x 8"		23	.348		28	21.50		49.50	64
1080	16" x 8"		22	.364		35	22.50		57.50	73
1100	12" x 12"		22	.364		33.50	22.50		56	71
1120	24" x 12"		18	.444		43.50	27.50		71	90
1140	30" x 12"		16	.500		56.50	31		87.50	110
1160	14" x 14"		22	.364		38	22.50		60.50	76.50
1180	16" x 16"		22	.364		42	22.50		64.50	81
1200	18" x 18"		21	.381		48.50	23.50		72	89.50
1220	24" x 18"		16	.500		54.50	31		85.50	108
1240	36" x 18"		15	.533		84.50	33		117.50	144
1260	24" x 24"		15	.533		65.50	33		98.50	123
1280	36" x 24"		14	.571		93	35.50		128.50	157
1300	48" x 24"		12	.667		156	41.50		197.50	235
1320	48" x 30"		11	.727		203	45.50		248.50	292
1340	36" x 36"		13	.615		123	38.50		161.50	193
1360	48" x 36"		11	.727		255	45.50		300.50	350
1380	48" x 48"		8	1		279	62.50		341.50	400
2000	Door grilles, 12" x 12"		22	.364		72	22.50		94.50	114
2020	18" x 12"		22	.364		83	22.50		105.50	126
2040	24" x 12"		18	.444		93.50	27.50		121	145
2060	18" x 18"		18	.444		115	27.50		142.50	169
2080	24" x 18"		16	.500		131	31		162	192
2100	24" x 24"		15	.533		185	33		218	254
3000	Filter grille with filter, 12" x 12"		24	.333		65	21		86	103
3020	18" x 12"		20	.400		83	25		108	130
3040	24" x 18"		18	.444		105	27.50		132.50	157
3060	24" x 24"		16	.500		126	31		157	187
3080	30" x 24"		14	.571		140	35.50		175.50	208
3100	30" x 30"		13	.615		165	38.50		203.50	240
3950	Eggcrate, framed, 6" x 6" opening		26	.308		16.95	19.15		36.10	47.50
3954	8" x 8" opening		24	.333		23.50	21		44.50	57.50
3960	10" x 10" opening		23	.348		25	21.50		46.50	60.50
3970	12" x 12" opening		22	.364		29.50	22.50		52	67
3980	14" x 14" opening		22	.364		36	22.50		58.50	74
3984	16" x 16" opening		21	.381		44	23.50		67.50	84
3990	18" x 18" opening		21	.381		50.50	23.50		74	92
4020	22" x 22" opening		17	.471		59.50	29.50		89	110
4040	24" x 24" opening		15	.533		76.50	33		109.50	135
4044	28" x 28" opening		14	.571		114	35.50		149.50	179
4048	36" x 36" opening		12	.667		130	41.50		171.50	206
4050	48" x 24" opening		12	.667		165	41.50		206.50	244
4060	Eggcrate, lay-in, T-bar system									
4070	48" x 24" sheet	1 Shee	40	.200	Ea.	77.50	12.45		89.95	104
5000	Transfer grille, vision proof, 8" x 4"		30	.267		15.50	16.60		32.10	42
5020	8" x 6"		28	.286		23	17.80		40.80	52.50
5040	8" x 8"		26	.308		25.50	19.15		44.65	57
5060	10" x 6"		24	.333		24	21		45	58
5080	10" x 10"		23	.348		24.50	21.50		46	60

23 37 13.30 Grilles		Crew	Daily Output	Labor-Hours	Unit	Material	2020 Bare Costs Labor	2020 Bare Costs Equipment	Total	Total Incl O&P
5090	12" x 6"	1 Shee	23	.348	Ea.	20	21.50		41.50	55
5100	12" x 10"		22	.364		26.50	22.50		49	64
5110	12" x 12"		22	.364		29	22.50		51.50	66.50
5120	14" x 10"		22	.364		36	22.50		58.50	74
5140	16" x 8"		21	.381		36	23.50		59.50	75.50
5160	18" x 12"		20	.400		38	25		63	79.50
5170	18" x 18"		21	.381		53.50	23.50		77	94.50
5180	20" x 12"		20	.400		41	25		66	83
5200	20" x 20"		19	.421		64.50	26		90.50	111
5220	24" x 12"		17	.471		44	29.50		73.50	92.50
5240	24" x 24"		16	.500		88.50	31		119.50	145
5260	30" x 6"		15	.533		41	33		74	95.50
5280	30" x 8"		15	.533		46	33		79	101
5300	30" x 12"		14	.571		59.50	35.50		95	120
5320	30" x 16"		13	.615		70.50	38.50		109	136
5340	30" x 20"		12	.667		89	41.50		130.50	161
5360	30" x 24"		11	.727		110	45.50		155.50	190
5380	30" x 30"	▼	10	.800		140	50		190	230
6000	For steel grilles instead of aluminum in above, deduct				▼	10%				
6200	Plastic, eggcrate, lay-in, T-bar system									
6210	48" x 24" sheet	1 Shee	50	.160	Ea.	22	9.95		31.95	39
6250	Steel door louver									
6270	With fire link, steel only									
6350	12" x 18"	1 Shee	22	.364	Ea.	225	22.50		247.50	282
6410	12" x 24"		21	.381		250	23.50		273.50	310
6470	18" x 18"		20	.400		250	25		275	315
6530	18" x 24"		19	.421		288	26		314	355
6660	24" x 24"	▼	15	.533	▼	291	33		324	370

23 37 13.60 Registers		Crew	Daily Output	Labor-Hours	Unit	Material	2020 Bare Costs Labor	2020 Bare Costs Equipment	Total	Total Incl O&P
0010	**REGISTERS**									
0980	Air supply									
1000	Ceiling/wall, O.B. damper, anodized aluminum									
1010	One or two way deflection, adj. curved face bars									
1014	6" x 6"	1 Shee	24	.333	Ea.	28	21		49	62
1020	8" x 4"		26	.308		30.50	19.15		49.65	62.50
1040	8" x 8"		24	.333		35	21		56	70
1060	10" x 6"		20	.400		21.50	25		46.50	61.50
1080	10" x 10"		19	.421		28.50	26		54.50	71.50
1100	12" x 6"		19	.421		32.50	26		58.50	75.50
1120	12" x 12"		18	.444		37	27.50		64.50	82.50
1140	14" x 8"		17	.471		22	29.50		51.50	69
1160	14" x 14"		18	.444		39.50	27.50		67	85.50
1170	16" x 16"		17	.471		43.50	29.50		73	92.50
1180	18" x 8"		18	.444		49.50	27.50		77	96.50
1200	18" x 18"		17	.471		90.50	29.50		120	144
1220	20" x 4"		19	.421		20	26		46	62
1240	20" x 6"		18	.444		30.50	27.50		58	75.50
1260	20" x 8"		18	.444		50	27.50		77.50	97
1280	20" x 20"		17	.471		110	29.50		139.50	166
1290	22" x 22"		15	.533		126	33		159	190
1300	24" x 4"		17	.471		25	29.50		54.50	72
1320	24" x 6"	▼	16	.500	▼	56	31		87	110

23 37 13 – Diffusers, Registers, and Grilles

23 37 13.60 Registers		Crew	Daily Output	Labor-Hours	Unit	Material	2020 Bare Costs Labor	Equipment	Total	Total Incl O&P
1340	24" x 8"	1 Shee	13	.615	Ea.	61.50	38.50		100	126
1350	24" x 18"		12	.667		113	41.50		154.50	187
1360	24" x 24"		11	.727		112	45.50		157.50	193
1380	30" x 4"		16	.500		31	31		62	81.50
1400	30" x 6"		15	.533		70.50	33		103.50	128
1420	30" x 8"		14	.571		80.50	35.50		116	143
1440	30" x 24"		12	.667		200	41.50		241.50	283
1460	30" x 30"	▼	10	.800	▼	256	50		306	360
1504	4 way deflection, adjustable curved face bars									
1510	6" x 6"	1 Shee	26	.308	Ea.	33.50	19.15		52.65	66
1514	8" x 8"		24	.333		42	21		63	77.50
1518	10" x 10"		19	.421		34.50	26		60.50	78
1522	12" x 6"		19	.421		39	26		65	83
1526	12" x 12"		18	.444		44.50	27.50		72	91
1530	14" x 14"		18	.444		47	27.50		74.50	94
1534	16" x 16"		17	.471		52.50	29.50		82	102
1538	18" x 18"		17	.471		108	29.50		137.50	164
1542	22" x 22"	▼	16	.500	▼	152	31		183	215
1980	One way deflection, adj. vert. or horiz. face bars									
1990	6" x 6"	1 Shee	26	.308	Ea.	24.50	19.15		43.65	56
2000	8" x 4"		26	.308		23	19.15		42.15	54.50
2020	8" x 8"		24	.333		27.50	21		48.50	61.50
2040	10" x 6"		20	.400		23.50	25		48.50	63.50
2060	10" x 10"		19	.421		27	26		53	70
2080	12" x 6"		19	.421		25	26		51	67.50
2100	12" x 12"		18	.444		33	27.50		60.50	78
2120	14" x 8"		17	.471		30.50	29.50		60	78
2140	14" x 12"		18	.444		53	27.50		80.50	101
2160	14" x 14"		18	.444		60	27.50		87.50	108
2180	16" x 6"		18	.444		30.50	27.50		58	75.50
2200	16" x 12"		18	.444		58.50	27.50		86	107
2220	16" x 16"		17	.471		62.50	29.50		92	113
2240	18" x 8"		18	.444		51.50	27.50		79	98.50
2260	18" x 12"		17	.471		42.50	29.50		72	91.50
2280	18" x 18"		16	.500		91	31		122	148
2300	20" x 10"		18	.444		61	27.50		88.50	109
2320	20" x 16"		16	.500		85	31		116	141
2340	20" x 20"		16	.500		111	31		142	170
2360	24" x 12"		15	.533		77	33		110	135
2380	24" x 16"		14	.571		103	35.50		138.50	167
2400	24" x 20"		12	.667		128	41.50		169.50	204
2420	24" x 24"		10	.800		121	50		171	209
2440	30" x 12"		13	.615		99.50	38.50		138	168
2460	30" x 16"		13	.615		90.50	38.50		129	158
2480	30" x 24"		11	.727		143	45.50		188.50	227
2500	30" x 30"		9	.889		259	55.50		314.50	370
2520	36" x 12"		11	.727		124	45.50		169.50	205
2540	36" x 24"		10	.800		250	50		300	350
2560	36" x 36"	▼	8	1		420	62.50		482.50	555
2600	For 2 way deflect., adj. vert. or horiz. face bars, add					40%				
2700	Above registers in steel instead of aluminum, deduct				▼	10%				
3000	Baseboard, hand adj. damper, enameled steel									
3012	8" x 6"	1 Shee	26	.308	Ea.	8.10	19.15		27.25	38

For customer support on your Mechanical Costs with RSMeans Data, call 800.448.8182.

343

23 37 Air Outlets and Inlets

23 37 13 – Diffusers, Registers, and Grilles

23 37 13.60 Registers		Crew	Daily Output	Labor-Hours	Unit	Material	2020 Bare Costs Labor	Equipment	Total	Total Incl O&P
3020	10" x 6"	1 Shee	24	.333	Ea.	8.80	21		29.80	41
3040	12" x 5"		23	.348		7.70	21.50		29.20	41.50
3060	12" x 6"		23	.348		11.45	21.50		32.95	45.50
3080	12" x 8"		22	.364		13.85	22.50		36.35	50
3100	14" x 6"	▼	20	.400	▼	10.40	25		35.40	49.50
4000	Floor, toe operated damper, enameled steel									
4020	4" x 8"	1 Shee	32	.250	Ea.	14.65	15.60		30.25	39.50
4040	4" x 12"		26	.308		17.20	19.15		36.35	48
4060	6" x 8"		28	.286		15.70	17.80		33.50	44.50
4080	6" x 14"		22	.364		19.90	22.50		42.40	56.50
4100	8" x 10"		22	.364		17.90	22.50		40.40	54
4120	8" x 16"		20	.400		15.80	25		40.80	55.50
4140	10" x 10"		20	.400		21.50	25		46.50	61.50
4160	10" x 16"		18	.444		26.50	27.50		54	71
4180	12" x 12"		18	.444		26.50	27.50		54	71
4200	12" x 24"		16	.500		30	31		61	80.50
4220	14" x 14"		16	.500		56.50	31		87.50	110
4240	14" x 20"	▼	15	.533	▼	53	33		86	109
4300	Spiral pipe supply register									
4310	Steel, with air scoop									
4320	4" x 12", for 8" thru 13" diameter duct	1 Shee	25	.320	Ea.	84	19.95		103.95	123
4330	4" x 18", for 8" thru 13" diameter duct		18	.444		99.50	27.50		127	151
4340	6" x 12", for 14" thru 21" diameter duct		19	.421		91	26		117	140
4350	6" x 16", for 14" thru 21" diameter duct		18	.444		102	27.50		129.50	154
4360	6" x 20", for 14" thru 21" diameter duct		17	.471		109	29.50		138.50	165
4370	6" x 24", for 14" thru 21" diameter duct		16	.500		132	31		163	193
4380	8" x 16", for 22" thru 31" diameter duct		19	.421		104	26		130	154
4390	8" x 18", for 22" thru 31" diameter duct		18	.444		112	27.50		139.50	165
4400	8" x 24", for 22" thru 31" diameter duct	▼	15	.533	▼	140	33		173	205
4980	Air return									
5000	Ceiling or wall, fixed 45° face blades									
5010	Adjustable O.B. damper, anodized aluminum									
5020	4" x 8"	1 Shee	26	.308	Ea.	19.95	19.15		39.10	51
5040	6" x 8"		24	.333		26	21		47	60
5060	6" x 10"		19	.421		24.50	26		50.50	67
5070	6" x 12"		19	.421		30	26		56	73.50
5080	6" x 16"		18	.444		30.50	27.50		58	76
5100	8" x 10"		19	.421		30.50	26		56.50	73.50
5120	8" x 12"		16	.500		33.50	31		64.50	84.50
5140	10" x 10"		18	.444		28	27.50		55.50	72.50
5160	10" x 16"		17	.471		52	29.50		81.50	102
5170	12" x 12"		18	.444		33	27.50		60.50	78.50
5180	12" x 18"		18	.444		43	27.50		70.50	89.50
5184	12" x 24"		14	.571		74.50	35.50		110	136
5200	12" x 30"		12	.667		68.50	41.50		110	139
5210	12" x 36"		10	.800		88.50	50		138.50	173
5220	16" x 16"		17	.471		74	29.50		103.50	126
5240	18" x 18"		16	.500		62	31		93	116
5250	18" x 24"		13	.615		81	38.50		119.50	147
5254	18" x 30"		11	.727		100	45.50		145.50	179
5260	18" x 36"		10	.800		171	50		221	264
5270	20" x 20"		10	.800		78.50	50		128.50	162
5274	20" x 24"	▼	12	.667		87.50	41.50		129	159

23 37 Air Outlets and Inlets

23 37 13 – Diffusers, Registers, and Grilles

23 37 13.60 Registers

		Crew	Daily Output	Labor-Hours	Unit	Material	2020 Bare Costs Labor	Equipment	Total	Total Incl O&P
5280	24" x 24"	1 Shee	11	.727	Ea.	113	45.50		158.50	193
5290	24" x 30"		10	.800		135	50		185	224
5300	24" x 36"		8	1		179	62.50		241.50	292
5320	24" x 48"		6	1.333		230	83		313	380
5340	30" x 30"	↓	6.50	1.231		177	76.50		253.50	310
5344	30" x 120"	Q-9	6	2.667		1,075	150		1,225	1,400
5360	36" x 36"	1 Shee	6	1.333		288	83		371	440
5370	40" x 32"	"	6	1.333	↓	272	83		355	425
5400	Ceiling or wall, removable-reversible core, single deflection									
5402	Adjustable O.B. damper, radiused frame with border, aluminum									
5410	8" x 4"	1 Shee	26	.308	Ea.	22	19.15		41.15	53
5414	8" x 6"		24	.333		22	21		43	55.50
5418	10" x 6"		19	.421		27.50	26		53.50	70.50
5422	10" x 10"		18	.444		33	27.50		60.50	78.50
5426	12" x 6"		19	.421		31	26		57	74
5430	12" x 12"		18	.444		40.50	27.50		68	86.50
5434	16" x 16"		16	.500		46.50	31		77.50	99
5438	18" x 18"		17	.471		64.50	29.50		94	116
5442	20" x 20"		16	.500		86.50	31		117.50	143
5446	24" x 12"		15	.533		53.50	33		86.50	110
5450	24" x 18"		13	.615		86.50	38.50		125	153
5454	24" x 24"		11	.727		102	45.50		147.50	182
5458	30" x 12"		13	.615		64.50	38.50		103	129
5462	30" x 18"		12.50	.640		93.50	40		133.50	164
5466	30" x 24"		11	.727		218	45.50		263.50	310
5470	30" x 30"		9	.889		240	55.50		295.50	350
5474	36" x 12"		11	.727		80	45.50		125.50	157
5478	36" x 18"		10.50	.762		112	47.50		159.50	195
5482	36" x 24"		10	.800		232	50		282	330
5486	36" x 30"		9	.889		252	55.50		307.50	360
5490	36" x 36"	↓	8	1	↓	273	62.50		335.50	395
6000	For steel construction instead of aluminum, deduct					10%				

23 37 15 – Air Outlets and Inlets, HVAC Louvers

23 37 15.40 HVAC Louvers

		Crew	Daily Output	Labor-Hours	Unit	Material	2020 Bare Costs Labor	Equipment	Total	Total Incl O&P
0010	**HVAC LOUVERS**									
0100	Aluminum, extruded, with screen, mill finish									
1002	Brick vent, see also Section 04 05 23.19									
1100	Standard, 4" deep, 8" wide, 5" high	1 Shee	24	.333	Ea.	37	21		58	72.50
1200	Modular, 4" deep, 7-3/4" wide, 5" high		24	.333		38.50	21		59.50	74
1300	Speed brick, 4" deep, 11-5/8" wide, 3-7/8" high		24	.333		38.50	21		59.50	74
1400	Fuel oil brick, 4" deep, 8" wide, 5" high		24	.333	↓	66.50	21		87.50	105
2000	Cooling tower and mechanical equip., screens, light weight		40	.200	S.F.	16.65	12.45		29.10	37.50
2020	Standard weight		35	.229		44	14.25		58.25	70
2500	Dual combination, automatic, intake or exhaust		20	.400		60.50	25		85.50	105
2520	Manual operation		20	.400		45	25		70	87.50
2540	Electric or pneumatic operation		20	.400	↓	45	25		70	87.50
2560	Motor, for electric or pneumatic	↓	14	.571	Ea.	520	35.50		555.50	625
3000	Fixed blade, continuous line									
3100	Mullion type, stormproof	1 Shee	28	.286	S.F.	45	17.80		62.80	76.50
3200	Stormproof		28	.286		45	17.80		62.80	76.50
3300	Vertical line	↓	28	.286	↓	53.50	17.80		71.30	85.50
3500	For damper to use with above, add					50%	30%			

For customer support on your Mechanical Costs with RSMeans Data, call 800.448.8182.

345

23 37 Air Outlets and Inlets

23 37 15 – Air Outlets and Inlets, HVAC Louvers

23 37 15.40 HVAC Louvers	Crew	Daily Output	Labor-Hours	Unit	Material	2020 Bare Costs Labor	Equipment	Total	Total Incl O&P
3520	Motor, for damper, electric or pneumatic	1 Shee	14	.571	Ea.	520	35.50		555.50	625
4000	Operating, 45°, manual, electric or pneumatic		24	.333	S.F.	60	21		81	97.50
4100	Motor, for electric or pneumatic		14	.571	Ea.	520	35.50		555.50	625
4200	Penthouse, roof		56	.143	S.F.	26.50	8.90		35.40	43
4300	Walls		40	.200		62.50	12.45		74.95	88
5000	Thinline, under 4" thick, fixed blade	↓	40	.200	↓	26	12.45		38.45	47.50
5010	Finishes, applied by mfr. at additional cost, available in colors									
5020	Prime coat only, add				S.F.	3.57			3.57	3.93
5040	Baked enamel finish coating, add					6.60			6.60	7.25
5060	Anodized finish, add					7.15			7.15	7.85
5080	Duranodic finish, add					12.95			12.95	14.25
5100	Fluoropolymer finish coating, add				↓	20.50			20.50	22.50
9000	Stainless steel, fixed blade, continuous line									
9010	Hospital grade									
9110	20 ga.	1 Shee	22	.364	S.F.	43	22.50		65.50	82
9980	For small orders (under 10 pieces), add				"	25%				

23 37 23 – HVAC Gravity Ventilators

23 37 23.10 HVAC Gravity Air Ventilators

		Crew	Daily Output	Labor-Hours	Unit	Material	2020 Bare Costs Labor	Equipment	Total	Total Incl O&P
0010	**HVAC GRAVITY AIR VENTILATORS**, Includes base									
1280	Rotary ventilators, wind driven, galvanized									
1300	4" neck diameter	Q-9	20	.800	Ea.	52	45		97	125
1320	5" neck diameter		18	.889		55	50		105	136
1340	6" neck diameter		16	1		52.50	56		108.50	143
1360	8" neck diameter		14	1.143		61	64		125	164
1380	10" neck diameter		12	1.333		67	75		142	187
1400	12" neck diameter		10	1.600		73	89.50		162.50	216
1420	14" neck diameter		10	1.600		124	89.50		213.50	272
1440	16" neck diameter		9	1.778		144	99.50		243.50	310
1460	18" neck diameter		9	1.778		174	99.50		273.50	345
1480	20" neck diameter		8	2		209	112		321	400
1500	24" neck diameter		8	2		320	112		432	525
1520	30" neck diameter		7	2.286		375	128		503	610
1540	36" neck diameter	↓	6	2.667	↓	605	150		755	890
1600	For aluminum, add					30%				
2000	Stationary, gravity, syphon, galvanized									
2100	3" neck diameter, 40 CFM	Q-9	24	.667	Ea.	30.50	37.50		68	90.50
2120	4" neck diameter, 50 CFM		20	.800		32.50	45		77.50	104
2140	5" neck diameter, 58 CFM		18	.889		35.50	50		85.50	115
2160	6" neck diameter, 66 CFM		16	1		37	56		93	126
2180	7" neck diameter, 86 CFM		15	1.067		50	60		110	146
2200	8" neck diameter, 110 CFM		14	1.143		51	64		115	153
2220	10" neck diameter, 140 CFM		12	1.333		68.50	75		143.50	189
2240	12" neck diameter, 160 CFM		10	1.600		85.50	89.50		175	230
2260	14" neck diameter, 250 CFM		10	1.600		136	89.50		225.50	286
2280	16" neck diameter, 380 CFM		9	1.778		175	99.50		274.50	345
2300	18" neck diameter, 500 CFM		9	1.778		197	99.50		296.50	365
2320	20" neck diameter, 625 CFM		8	2		235	112		347	430
2340	24" neck diameter, 900 CFM		8	2		290	112		402	490
2360	30" neck diameter, 1,375 CFM		7	2.286		405	128		533	640
2380	36" neck diameter, 2,000 CFM		6	2.667		405	150		555	670
2400	42" neck diameter, 3,000 CFM		4	4		605	224		829	1,000
2410	48" neck diameter, 4,000 CFM	↓	3	5.333		965	299		1,264	1,525

23 37 23 – HVAC Gravity Ventilators

23 37 23.10 HVAC Gravity Air Ventilators	Crew	Daily Output	Labor-Hours	Unit	Material	2020 Bare Costs Labor	Equipment	Total	Total Incl O&P
2500	For aluminum, add				Ea.	30%			
2520	For stainless steel, add					60%			
2530	For copper, add					100%			
3000	Rotating chimney cap, galvanized, 4" neck diameter	Q-9	20	.800		21	45		66	91.50
3020	5" neck diameter		18	.889		21	50		71	99
3040	6" neck diameter		16	1		21	56		77	109
3060	7" neck diameter		15	1.067		27.50	60		87.50	121
3080	8" neck diameter		14	1.143		27.50	64		91.50	127
3100	10" neck diameter		12	1.333		36	75		111	153
3600	Stationary chimney rain cap, galvanized, 3" neck diameter		24	.667		3.50	37.50		41	60.50
3620	4" neck diameter		20	.800		3.83	45		48.83	72
3640	6" neck diameter		16	1		4.55	56		60.55	90
3680	8" neck diameter		14	1.143		5.70	64		69.70	103
3700	10" neck diameter		12	1.333		6.90	75		81.90	121
3720	12" neck diameter		10	1.600		10.15	89.50		99.65	147
3740	14" neck diameter		10	1.600		16.10	89.50		105.60	154
3760	16" neck diameter		9	1.778		16.95	99.50		116.45	170
3770	18" neck diameter		9	1.778		27	99.50		126.50	181
3780	20" neck diameter		8	2		36	112		148	210
3790	24" neck diameter		8	2		45.50	112		157.50	220
4200	Stationary mushroom, aluminum, 16" orifice diameter		10	1.600		720	89.50		809.50	925
4220	26" orifice diameter		6.15	2.602		1,050	146		1,196	1,400
4230	30" orifice diameter		5.71	2.802		1,550	157		1,707	1,950
4240	38" orifice diameter		5	3.200		2,225	179		2,404	2,725
4250	42" orifice diameter		4.70	3.404		2,950	191		3,141	3,550
4260	50" orifice diameter		4.44	3.604		3,500	202		3,702	4,150
5000	Relief vent								
5500	Rectangular, aluminum, galvanized curb								
5510	intake/exhaust, 0.033" SP								
5580	500 CFM, 12" x 12"	Q-9	8.60	1.860	Ea.	790	104		894	1,025
5600	600 CFM, 12" x 16"		8	2		880	112		992	1,150
5620	750 CFM, 12" x 20"		7.20	2.222		945	125		1,070	1,225
5640	1,000 CFM, 12" x 24"		6.60	2.424		985	136		1,121	1,275
5660	1,500 CFM, 12" x 36"		5.80	2.759		1,325	155		1,480	1,700
5680	3,000 CFM, 20" x 42"		4	4		1,725	224		1,949	2,250
5700	5,000 CFM, 20" x 72"		3	5.333		2,500	299		2,799	3,200
5720	6,000 CFM, 24" x 72"		2.30	6.957		2,700	390		3,090	3,575
5740	10,000 CFM, 48" x 60"		1.80	8.889		3,425	500		3,925	4,500
5760	12,000 CFM, 48" x 72"		1.60	10		4,025	560		4,585	5,275
5780	13,750 CFM, 60" x 66"		1.40	11.429		4,450	640		5,090	5,875
5800	15,000 CFM, 60" x 72"		1.30	12.308		4,775	690		5,465	6,300
5820	18,000 CFM, 72" x 72"		1.20	13.333		5,400	750		6,150	7,075
5880	Size is throat area, volume is at 500 fpm								
7000	Note: sizes based on exhaust. Intake, with 0.125" SP								
7100	loss, approximately twice listed capacity.								

For customer support on your Mechanical Costs with RSMeans Data, call 800.448.8182.

347

23 38 Ventilation Hoods

23 38 13 - Commercial-Kitchen Hoods

23 38 13.10 Hood and Ventilation Equipment

		Crew	Daily Output	Labor-Hours	Unit	Material	2020 Bare Costs Labor	Equipment	Total	Total Incl O&P
0010	**HOOD AND VENTILATION EQUIPMENT**									
2970	Exhaust hood, sst, gutter on all sides, 4' x 4' x 2'	1 Carp	1.80	4.444	Ea.	4,075	236		4,311	4,825
2980	4' x 4' x 7'	"	1.60	5		6,750	266		7,016	7,825
7800	Vent hood, wall canopy with fire protection, 30"	L-3A	9	1.333		390	76.50		466.50	545
7810	Without fire protection, 36"		10	1.200		430	69		499	575
7820	Island canopy with fire protection, 30"		7	1.714		760	98.50		858.50	985
7830	Without fire protection, 36"		8	1.500		785	86.50		871.50	990
7840	Back shelf with fire protection, 30"		11	1.091		410	63		473	545
7850	Without fire protection, black, 36"		12	1		450	57.50		507.50	585
7852	Without fire protection, stainless steel, 36"		12	1		470	57.50		527.50	600
7860	Range hood & CO_2 system, 30"	1 Carp	2.50	3.200		4,325	170		4,495	5,000
7950	Hood fire protection system, electric stove	Q-1	3	5.333		2,025	310		2,335	2,700
7952	Hood fire protection system, gas stove	"	3	5.333		2,350	310		2,660	3,050

23 38 13.20 Kitchen Ventilation

		Crew	Daily Output	Labor-Hours	Unit	Material	2020 Bare Costs Labor	Equipment	Total	Total Incl O&P
0010	**KITCHEN VENTILATION**, Commercial									
6000	Exhaust hoods and fans									
6100	Centrifugal grease extraction									
6110	Water wash type									
6120	Per foot of length	Q-10	7.10	3.380	L.F.	1,925	197		2,122	2,400
6200	Non water wash type									
6210	Per foot of length	Q-10	13.40	1.791	L.F.	1,400	104		1,504	1,700
6400	Island style ventilator									
6410	Water wash type									
6420	Per foot of length	Q-10	7.10	3.380	L.F.	2,550	197		2,747	3,100
6460	Non water wash type									
6470	Per foot of length	Q-10	13.40	1.791	L.F.	870	104		974	1,125

23 41 Particulate Air Filtration

23 41 13 - Panel Air Filters

23 41 13.10 Panel Type Air Filters

		Crew	Daily Output	Labor-Hours	Unit	Material	2020 Bare Costs Labor	Equipment	Total	Total Incl O&P
0010	**PANEL TYPE AIR FILTERS**									
2950	Mechanical media filtration units									
3000	High efficiency type, with frame, non-supported G				MCFM	41			41	45
3100	Supported type G				"	52.50			52.50	58
5500	Throwaway glass or paper media type, 12" x 36" x 1"				Ea.	2.48			2.48	2.73

23 41 16 - Renewable-Media Air Filters

23 41 16.10 Disposable Media Air Filters

		Crew	Daily Output	Labor-Hours	Unit	Material	2020 Bare Costs Labor	Equipment	Total	Total Incl O&P
0010	**DISPOSABLE MEDIA AIR FILTERS**									
5000	Renewable disposable roll				C.S.F.	5.25			5.25	5.75
5800	Filter, bag type									
5810	90-95% efficiency									
5820	24" x 12" x 29", 0.75-1.25 MCFM	1 Shee	1.90	4.211	Ea.	60.50	262		322.50	465
5830	24" x 24" x 29", 1.5-2.5 MCFM	"	1.90	4.211	"	97.50	262		359.50	505
5850	80-85% efficiency									
5860	24" x 12" x 29", 0.75-1.25 MCFM	1 Shee	1.90	4.211	Ea.	49.50	262		311.50	455
5870	24" x 24" x 29", 1.5-2.5 MCFM	"	1.90	4.211	"	91	262		353	500

23 41 Particulate Air Filtration

23 41 19 – Washable Air Filters

23 41 19.10 Permanent Air Filters

	Crew	Daily Output	Labor-Hours	Unit	Material	2020 Bare Costs Labor	Equipment	Total	Total Incl O&P
0010 **PERMANENT AIR FILTERS**									
4500 Permanent washable	G			MCFM	25			25	27.50

23 41 23 – Extended Surface Filters

23 41 23.10 Expanded Surface Filters

	Crew	Daily Output	Labor-Hours	Unit	Material	2020 Bare Costs Labor	Equipment	Total	Total Incl O&P
0010 **EXPANDED SURFACE FILTERS**									
4000 Medium efficiency, extended surface	G			MCFM	6.65			6.65	7.35

23 41 33 – High-Efficiency Particulate Filtration

23 41 33.10 HEPA Filters

	Crew	Daily Output	Labor-Hours	Unit	Material	2020 Bare Costs Labor	Equipment	Total	Total Incl O&P
0010 **HEPA FILTERS**									
6000 HEPA filter complete w/particle board,									
6010 kraft paper frame, separator material									
6020 95% DOP efficiency									
6030 12" x 12" x 6", 150 CFM	1 Shee	3.70	2.162	Ea.	44.50	135		179.50	253
6034 24" x 12" x 6", 375 CFM		1.90	4.211		74	262		336	480
6038 24" x 18" x 6", 450 CFM		1.90	4.211		91	262		353	500
6042 24" x 24" x 6", 700 CFM		1.90	4.211		102	262		364	515
6046 12" x 12" x 12", 250 CFM		3.70	2.162		75	135		210	287
6050 24" x 12" x 12", 500 CFM		1.90	4.211		106	262		368	515
6054 24" x 18" x 12", 875 CFM		1.90	4.211		148	262		410	565
6058 24" x 24" x 12", 1,000 CFM		1.90	4.211		159	262		421	575
6100 99% DOP efficiency									
6110 12" x 12" x 6", 150 CFM	1 Shee	3.70	2.162	Ea.	46.50	135		181.50	255
6114 24" x 12" x 6", 325 CFM		1.90	4.211		77	262		339	485
6118 24" x 18" x 6", 550 CFM		1.90	4.211		94.50	262		356.50	505
6122 24" x 24" x 6", 775 CFM		1.90	4.211		106	262		368	515
6124 24" x 48" x 6", 775 CFM		1.30	6.154		236	385		621	840
6126 12" x 12" x 12", 250 CFM		3.60	2.222		78	138		216	296
6130 24" x 12" x 12", 500 CFM		1.90	4.211		110	262		372	520
6134 24" x 18" x 12", 775 CFM		1.90	4.211		154	262		416	570
6138 24" x 24" x 12", 1,100 CFM		1.90	4.211		166	262		428	580
6500 HEPA filter housing, 14 ga. galv. sheet metal									
6510 12" x 12" x 6"	1 Shee	2.50	3.200	Ea.	670	199		869	1,025
6514 24" x 12" x 6"		2	4		745	249		994	1,200
6518 12" x 12" x 12"		2.40	3.333		670	208		878	1,050
6522 14" x 12" x 12"		2.30	3.478		670	217		887	1,075
6526 24" x 18" x 6"		1.90	4.211		950	262		1,212	1,450
6530 24" x 24" x 6"		1.80	4.444		865	277		1,142	1,375
6534 24" x 48" x 6"		1.70	4.706		1,150	293		1,443	1,725
6538 24" x 72" x 6"		1.60	5		1,475	310		1,785	2,100
6542 24" x 18" x 12"		1.80	4.444		950	277		1,227	1,475
6546 24" x 24" x 12"		1.70	4.706		1,075	293		1,368	1,625
6550 24" x 48" x 12"		1.60	5		1,250	310		1,560	1,850
6554 24" x 72" x 12"		1.50	5.333		1,550	330		1,880	2,200
6558 48" x 48" x 6"	Q-9	2.80	5.714		1,700	320		2,020	2,350
6562 48" x 72" x 6"		2.60	6.154		2,050	345		2,395	2,775
6566 48" x 96" x 6"		2.40	6.667		2,600	375		2,975	3,425
6570 48" x 48" x 12"		2.70	5.926		1,700	330		2,030	2,375
6574 48" x 72" x 12"		2.50	6.400		2,050	360		2,410	2,800
6578 48" x 96" x 12"		2.30	6.957		2,600	390		2,990	3,450
6582 114" x 72" x 12"		2	8		3,600	450		4,050	4,625

For customer support on your Mechanical Costs with RSMeans Data, call 800.448.8182.

349

23 42 Gas-Phase Air Filtration

23 42 13 – Activated-Carbon Air Filtration

23 42 13.10 Charcoal Type Air Filtration	Crew	Daily Output	Labor-Hours	Unit	Material	2020 Bare Costs Labor	Equipment	Total	Total Incl O&P
0010 **CHARCOAL TYPE AIR FILTRATION**									
0050 Activated charcoal type, full flow				MCFM	650			650	715
0060 Full flow, impregnated media 12" deep					225			225	248
0070 HEPA filter & frame for field erection					410			410	450
0080 HEPA filter-diffuser, ceiling install.					340			340	375

23 43 Electronic Air Cleaners

23 43 13 – Washable Electronic Air Cleaners

23 43 13.10 Electronic Air Cleaners

	Crew	Daily Output	Labor-Hours	Unit	Material	2020 Bare Costs Labor	Equipment	Total	Total Incl O&P
0010 **ELECTRONIC AIR CLEANERS**									
2000 Electronic air cleaner, duct mounted									
2150 1,000 CFM	1 Shee	4	2	Ea.	425	125		550	660
2200 1,200 CFM		3.80	2.105		570	131		701	825
2250 1,400 CFM		3.60	2.222		600	138		738	870
2260 2,000 CFM		3.20	2.500		635	156		791	935

23 51 Breechings, Chimneys, and Stacks

23 51 13 – Draft Control Devices

23 51 13.13 Draft-Induction Fans

	Crew	Daily Output	Labor-Hours	Unit	Material	2020 Bare Costs Labor	Equipment	Total	Total Incl O&P
0010 **DRAFT-INDUCTION FANS**									
1000 Breeching installation									
1800 Hot gas, 600°F, variable pitch pulley and motor									
1840 6" diam. inlet, 1/4 HP, 1 phase, 400 CFM	Q-9	6	2.667	Ea.	1,725	150		1,875	2,100
1860 8" diam. inlet, 1/4 HP, 1 phase, 1,120 CFM		4	4		2,150	224		2,374	2,700
1870 9" diam. inlet, 3/4 HP, 1 phase, 1,440 CFM		3.60	4.444		2,650	249		2,899	3,275
1880 10" diam. inlet, 3/4 HP, 1 phase, 2,000 CFM		3.30	4.848		2,625	272		2,897	3,300
1900 12" diam. inlet, 3/4 HP, 3 phase, 2,960 CFM		3	5.333		2,925	299		3,224	3,675
1910 14" diam. inlet, 1 HP, 3 phase, 4,160 CFM		2.60	6.154		2,900	345		3,245	3,700
1920 16" diam. inlet, 2 HP, 3 phase, 5,500 CFM		2.30	6.957		3,200	390		3,590	4,125
1950 20" diam. inlet, 3 HP, 3 phase, 9,760 CFM		1.50	10.667		4,675	600		5,275	6,050
1960 22" diam. inlet, 5 HP, 3 phase, 13,360 CFM		1	16		8,500	895		9,395	10,700
1980 24" diam. inlet, 7-1/2 HP, 3 phase, 17,760 CFM		.80	20		9,625	1,125		10,750	12,300
2300 For multi-blade damper at fan inlet, add					20%				
3600 Chimney-top installation									
3700 6" size	1 Shee	8	1	Ea.	1,425	62.50		1,487.50	1,650
3740 8" size		7	1.143		1,425	71		1,496	1,675
3750 10" size		6.50	1.231		2,075	76.50		2,151.50	2,400
3780 13" size		6	1.333		1,925	83		2,008	2,225
3880 For speed control switch, add					138			138	152
3920 For thermal fan control, add					138			138	152
5500 Flue blade style damper for draft control,									
5510 locking quadrant blade									
5550 8" size	Q-9	8	2	Ea.	470	112		582	690
5560 9" size		7.50	2.133		475	120		595	705
5570 10" size		7	2.286		490	128		618	730
5580 12" size		6.50	2.462		500	138		638	760
5590 14" size		6	2.667		490	150		640	765
5600 16" size		5.50	2.909		505	163		668	805
5610 18" size		5	3.200		530	179		709	855

23 51 Breechings, Chimneys, and Stacks

23 51 13 – Draft Control Devices

23 51 13.13 Draft-Induction Fans

		Crew	Daily Output	Labor-Hours	Unit	Material	2020 Bare Costs Labor	Equipment	Total	Total Incl O&P
5620	20" size	Q-9	4.50	3.556	Ea.	565	199		764	920
5630	22" size		4	4		570	224		794	965
5640	24" size		3.50	4.571		600	256		856	1,050
5650	27" size		3	5.333		630	299		929	1,150
5660	30" size		2.50	6.400		715	360		1,075	1,325
5670	32" size		2	8		750	450		1,200	1,500
5680	36" size	↓	1.50	10.667	↓	830	600		1,430	1,825

23 51 13.16 Vent Dampers

		Crew	Daily Output	Labor-Hours	Unit	Material	2020 Bare Costs Labor	Equipment	Total	Total Incl O&P
0010	**VENT DAMPERS**									
5000	Vent damper, bi-metal, gas, 3" diameter	Q-9	24	.667	Ea.	71	37.50		108.50	135
5010	4" diameter		24	.667		70	37.50		107.50	134
5020	5" diameter		23	.696		71	39		110	137
5030	6" diameter		22	.727		70	41		111	139
5040	7" diameter		21	.762		72.50	42.50		115	145
5050	8" diameter	↓	20	.800	↓	77.50	45		122.50	153

23 51 13.19 Barometric Dampers

		Crew	Daily Output	Labor-Hours	Unit	Material	2020 Bare Costs Labor	Equipment	Total	Total Incl O&P
0010	**BAROMETRIC DAMPERS**									
1000	Barometric, gas fired system only, 6" size for 5" and 6" pipes	1 Shee	20	.400	Ea.	105	25		130	154
1020	7" size, for 6" and 7" pipes		19	.421		115	26		141	166
1040	8" size, for 7" and 8" pipes		18	.444		148	27.50		175.50	205
1060	9" size, for 8" and 9" pipes	↓	16	.500	↓	164	31		195	228
2000	All fuel, oil, oil/gas, coal									
2020	10" for 9" and 10" pipes	1 Shee	15	.533	Ea.	255	33		288	330
2040	12" for 11" and 12" pipes		15	.533		335	33		368	415
2060	14" for 13" and 14" pipes		14	.571		435	35.50		470.50	530
2080	16" for 15" and 16" pipes		13	.615		605	38.50		643.50	725
2100	18" for 17" and 18" pipes		12	.667		805	41.50		846.50	950
2120	20" for 19" and 21" pipes	↓	10	.800		965	50		1,015	1,125
2140	24" for 22" and 25" pipes	Q-9	12	1.333		1,175	75		1,250	1,425
2160	28" for 26" and 30" pipes		10	1.600		1,475	89.50		1,564.50	1,725
2180	32" for 31" and 34" pipes	↓	8	2		1,875	112		1,987	2,225
3260	For thermal switch for above, add	1 Shee	24	.333	↓	105	21		126	148

23 51 23 – Gas Vents

23 51 23.10 Gas Chimney Vents

		Crew	Daily Output	Labor-Hours	Unit	Material	2020 Bare Costs Labor	Equipment	Total	Total Incl O&P
0010	**GAS CHIMNEY VENTS**, Prefab metal, UL listed									
0020	Gas, double wall, galvanized steel									
0080	3" diameter	Q-9	72	.222	V.L.F.	7.60	12.45		20.05	27.50
0100	4" diameter		68	.235		9.35	13.20		22.55	30.50
0120	5" diameter		64	.250		10.60	14		24.60	33
0140	6" diameter		60	.267		12.65	14.95		27.60	36.50
0160	7" diameter		56	.286		24.50	16		40.50	51.50
0180	8" diameter		52	.308		26	17.25		43.25	54.50
0200	10" diameter		48	.333		49.50	18.70		68.20	83
0220	12" diameter		44	.364		58.50	20.50		79	95.50
0240	14" diameter		42	.381		97.50	21.50		119	140
0260	16" diameter		40	.400		140	22.50		162.50	188
0280	18" diameter	↓	38	.421		173	23.50		196.50	226
0300	20" diameter	Q-10	36	.667		205	39		244	285
0320	22" diameter		34	.706		261	41		302	350
0340	24" diameter	↓	32	.750	↓	320	43.50		363.50	415
0600	For 4", 5" and 6" oval, add					50%				
0650	Gas, double wall, galvanized steel, fittings									

For customer support on your Mechanical Costs with RSMeans Data, call 800.448.8182.

351

23 51 23.10 Gas Chimney Vents		Crew	Daily Output	Labor-Hours	Unit	Material	2020 Bare Costs Labor	Equipment	Total	Total Incl O&P
0660	Elbow 45°, 3" diameter	Q-9	36	.444	Ea.	13.45	25		38.45	53
0670	4" diameter		34	.471		16.45	26.50		42.95	58
0680	5" diameter		32	.500		19.05	28		47.05	63.50
0690	6" diameter		30	.533		23.50	30		53.50	71.50
0700	7" diameter		28	.571		38	32		70	90.50
0710	8" diameter		26	.615		50	34.50		84.50	108
0720	10" diameter		24	.667		106	37.50		143.50	174
0730	12" diameter		22	.727		113	41		154	186
0740	14" diameter		21	.762		178	42.50		220.50	261
0750	16" diameter		20	.800		231	45		276	320
0760	18" diameter	↓	19	.842		305	47		352	405
0770	20" diameter	Q-10	18	1.333		340	77.50		417.50	490
0780	22" diameter		17	1.412		545	82		627	725
0790	24" diameter	↓	16	1.500	↓	695	87		782	895
0916	Adjustable length									
0918	3" diameter, to 12"	Q-9	36	.444	Ea.	16.95	25		41.95	56.50
0920	4" diameter, to 12"		34	.471		19.70	26.50		46.20	61.50
0924	6" diameter, to 12"		30	.533		25.50	30		55.50	73.50
0928	8" diameter, to 12"		26	.615		47.50	34.50		82	105
0930	10" diameter, to 18"		24	.667		148	37.50		185.50	219
0932	12" diameter, to 18"		22	.727		150	41		191	227
0936	16" diameter, to 18"		20	.800		300	45		345	400
0938	18" diameter, to 18"	↓	19	.842		340	47		387	445
0944	24" diameter, to 18"	Q-10	16	1.500		655	87		742	850
0950	Elbow 90°, adjustable, 3" diameter	Q-9	36	.444		23	25		48	63.50
0960	4" diameter		34	.471		27	26.50		53.50	69.50
0970	5" diameter		32	.500		32.50	28		60.50	78.50
0980	6" diameter		30	.533		39	30		69	88.50
0990	7" diameter		28	.571		67	32		99	123
1010	8" diameter		26	.615		68.50	34.50		103	128
1020	Wall thimble, 4 to 7" adjustable, 3" diameter		36	.444		14.40	25		39.40	54
1022	4" diameter		34	.471		16.20	26.50		42.70	58
1024	5" diameter		32	.500		19.40	28		47.40	64
1026	6" diameter		30	.533		20	30		50	67.50
1028	7" diameter		28	.571		41.50	32		73.50	94.50
1030	8" diameter		26	.615		51	34.50		85.50	109
1040	Roof flashing, 3" diameter		36	.444		7.85	25		32.85	46.50
1050	4" diameter		34	.471		9.15	26.50		35.65	50
1060	5" diameter		32	.500		27.50	28		55.50	73
1070	6" diameter		30	.533		22	30		52	69.50
1080	7" diameter		28	.571		29	32		61	80
1090	8" diameter		26	.615		31	34.50		65.50	86.50
1100	10" diameter		24	.667		41.50	37.50		79	102
1110	12" diameter		22	.727		58	41		99	126
1120	14" diameter		20	.800		132	45		177	213
1130	16" diameter		18	.889		158	50		208	250
1140	18" diameter	↓	16	1		222	56		278	330
1150	20" diameter	Q-10	18	1.333		288	77.50		365.50	435
1160	22" diameter		14	1.714		360	99.50		459.50	545
1170	24" diameter		12	2		415	116		531	630
1200	Tee, 3" diameter	Q-9	27	.593		35.50	33		68.50	89.50
1210	4" diameter		26	.615		38	34.50		72.50	94
1220	5" diameter	↓	25	.640	↓	40	36		76	98.50

For customer support on your Mechanical Costs with RSMeans Data, call 800.448.8182.

23 51 23.10 Gas Chimney Vents		Crew	Daily Output	Labor-Hours	Unit	Material	2020 Bare Costs Labor	Equipment	Total	Total Incl O&P
1230	6" diameter	Q-9	24	.667	Ea.	45.50	37.50		83	107
1240	7" diameter		23	.696		65	39		104	131
1250	8" diameter		22	.727		72.50	41		113.50	142
1260	10" diameter		21	.762		191	42.50		233.50	275
1270	12" diameter		20	.800		195	45		240	283
1280	14" diameter		18	.889		350	50		400	460
1290	16" diameter		16	1		520	56		576	655
1300	18" diameter		14	1.143		625	64		689	780
1310	20" diameter	Q-10	17	1.412		855	82		937	1,075
1320	22" diameter		13	1.846		1,100	107		1,207	1,375
1330	24" diameter		12	2		1,275	116		1,391	1,575
1460	Tee cap, 3" diameter	Q-9	45	.356		2.38	19.95		22.33	32.50
1470	4" diameter		42	.381		2.57	21.50		24.07	35.50
1480	5" diameter		40	.400		3.31	22.50		25.81	37.50
1490	6" diameter		37	.432		4.71	24.50		29.21	42
1500	7" diameter		35	.457		8.65	25.50		34.15	48.50
1510	8" diameter		34	.471		8.65	26.50		35.15	49.50
1520	10" diameter		32	.500		69.50	28		97.50	119
1530	12" diameter		30	.533		69	30		99	122
1540	14" diameter		28	.571		72.50	32		104.50	128
1550	16" diameter		25	.640		74	36		110	136
1560	18" diameter		24	.667		87.50	37.50		125	153
1570	20" diameter	Q-10	27	.889		97.50	51.50		149	186
1580	22" diameter		22	1.091		171	63.50		234.50	284
1590	24" diameter		21	1.143		200	66.50		266.50	320
1750	Top, 3" diameter	Q-9	46	.348		17.75	19.50		37.25	49
1760	4" diameter		44	.364		18.65	20.50		39.15	51.50
1770	5" diameter		42	.381		17.95	21.50		39.45	52.50
1780	6" diameter		40	.400		23.50	22.50		46	60
1790	7" diameter		38	.421		51.50	23.50		75	92.50
1800	8" diameter		36	.444		63	25		88	108
1810	10" diameter		34	.471		103	26.50		129.50	154
1820	12" diameter		32	.500		129	28		157	185
1830	14" diameter		30	.533		205	30		235	271
1840	16" diameter		28	.571		260	32		292	335
1850	18" diameter		26	.615		360	34.50		394.50	450
1860	20" diameter	Q-10	28	.857		555	50		605	685
1870	22" diameter		22	1.091		830	63.50		893.50	1,000
1880	24" diameter		20	1.200		1,050	70		1,120	1,250
1900	Gas, double wall, galvanized steel, oval									
1904	4" x 1'	Q-9	68	.235	V.L.F.	18.60	13.20		31.80	40.50
1906	5" x 1'		64	.250		92.50	14		106.50	124
1908	5"/6" x 1'		60	.267		46.50	14.95		61.45	73.50
1910	Oval fittings									
1912	Adjustable length									
1914	4" diameter to 12" long	Q-9	34	.471	Ea.	19.35	26.50		45.85	61.50
1916	5" diameter to 12" long		32	.500		21.50	28		49.50	66
1918	5"/6" diameter to 12" long		30	.533		26	30		56	74
1920	Elbow 45°									
1922	4"	Q-9	34	.471	Ea.	31.50	26.50		58	74.50
1924	5"		32	.500		62.50	28		90.50	111
1926	5"/6"		30	.533		64	30		94	116
1930	Elbow 45°, flat									

For customer support on your Mechanical Costs with RSMeans Data, call 800.448.8182.

353

23 51 23 – Gas Vents

23 51 23.10 Gas Chimney Vents		Crew	Daily Output	Labor-Hours	Unit	Material	2020 Bare Costs Labor	Equipment	Total	Total Incl O&P
1932	4"	Q-9	34	.471	Ea.	31	26.50		57.50	74
1934	5"		32	.500		62.50	28		90.50	111
1936	5"/6"		30	.533		65	30		95	117
1940	Top									
1942	4"	Q-9	44	.364	Ea.	37	20.50		57.50	71.50
1944	5"		42	.381		35.50	21.50		57	71.50
1946	5"/6"		40	.400		46	22.50		68.50	85
1950	Adjustable flashing									
1952	4"	Q-9	34	.471	Ea.	13.25	26.50		39.75	54.50
1954	5"		32	.500		39	28		67	85
1956	5"/6"		30	.533		40	30		70	89.50
1960	Tee									
1962	4"	Q-9	26	.615	Ea.	46	34.50		80.50	103
1964	5"		25	.640		95.50	36		131.50	160
1966	5"/6"		24	.667		97	37.50		134.50	163
1970	Tee with short snout									
1972	4"	Q-9	26	.615	Ea.	49	34.50		83.50	107

23 51 26 – All-Fuel Vent Chimneys

23 51 26.10 All-Fuel Vent Chimneys, Press. Tight, Dbl. Wall

		Crew	Daily Output	Labor-Hours	Unit	Material	2020 Bare Costs Labor	Equipment	Total	Total Incl O&P
0010	**ALL-FUEL VENT CHIMNEYS, PRESSURE TIGHT, DOUBLE WALL**									
3200	All fuel, pressure tight, double wall, 1" insulation, UL listed, 1,400°F.									
3210	304 stainless steel liner, aluminized steel outer jacket									
3220	6" diameter	Q-9	60	.267	L.F.	60.50	14.95		75.45	89
3221	8" diameter		52	.308		65	17.25		82.25	97.50
3222	10" diameter		48	.333		72.50	18.70		91.20	109
3223	12" diameter		44	.364		70.50	20.50		91	109
3224	14" diameter		42	.381		90.50	21.50		112	132
3225	16" diameter		40	.400		106	22.50		128.50	151
3226	18" diameter		38	.421		114	23.50		137.50	162
3227	20" diameter	Q-10	36	.667		136	39		175	209
3228	24" diameter		32	.750		174	43.50		217.50	257
3229	28" diameter		30	.800		200	46.50		246.50	291
3230	32" diameter		27	.889		227	51.50		278.50	330
3231	36" diameter		25	.960		251	56		307	360
3232	42" diameter		22	1.091		285	63.50		348.50	410
3233	48" diameter		19	1.263		335	73.50		408.50	480
3260	For 316 stainless steel liner, add					30%				
3280	All fuel, pressure tight, double wall fittings									
3284	304 stainless steel inner, aluminized steel jacket									
3288	Adjustable 20"/29" section									
3292	6" diameter	Q-9	30	.533	Ea.	204	30		234	270
3293	8" diameter		26	.615		218	34.50		252.50	293
3294	10" diameter		24	.667		245	37.50		282.50	325
3295	12" diameter		22	.727		279	41		320	365
3296	14" diameter		21	.762		295	42.50		337.50	390
3297	16" diameter		20	.800		345	45		390	450
3298	18" diameter		19	.842		395	47		442	505
3299	20" diameter	Q-10	18	1.333		430	77.50		507.50	595
3300	24" diameter		16	1.500		555	87		642	740
3301	28" diameter		15	1.600		615	93		708	815
3302	32" diameter		14	1.714		715	99.50		814.50	940
3303	36" diameter		12	2		840	116		956	1,100

23 51 Breechings, Chimneys, and Stacks

23 51 26 – All-Fuel Vent Chimneys

23 51 26.10 All-Fuel Vent Chimneys, Press. Tight, Dbl. Wall		Crew	Daily Output	Labor-Hours	Unit	Material	2020 Bare Costs Labor	Equipment	Total	Total Incl O&P
3304	42" diameter	Q-10	11	2.182	Ea.	960	127		1,087	1,250
3305	48" diameter	↓	10	2.400	↓	1,100	140		1,240	1,400
3350	Elbow 90° fixed									
3354	6" diameter	Q-9	30	.533	Ea.	420	30		450	510
3355	8" diameter		26	.615		475	34.50		509.50	575
3356	10" diameter		24	.667		540	37.50		577.50	645
3357	12" diameter		22	.727		610	41		651	730
3358	14" diameter		21	.762		690	42.50		732.50	825
3359	16" diameter		20	.800		780	45		825	930
3360	18" diameter	↓	19	.842		880	47		927	1,050
3361	20" diameter	Q-10	18	1.333		995	77.50		1,072.50	1,225
3362	24" diameter		16	1.500		1,275	87		1,362	1,525
3363	28" diameter		15	1.600		1,450	93		1,543	1,750
3364	32" diameter		14	1.714		1,975	99.50		2,074.50	2,300
3365	36" diameter		12	2		2,600	116		2,716	3,050
3366	42" diameter		11	2.182		3,125	127		3,252	3,650
3367	48" diameter	↓	10	2.400	↓	3,525	140		3,665	4,075
3380	For 316 stainless steel liner, add					30%				
3400	Elbow 45°									
3404	6" diameter	Q-9	30	.533	Ea.	212	30		242	279
3405	8" diameter		26	.615		238	34.50		272.50	315
3406	10" diameter		24	.667		271	37.50		308.50	355
3407	12" diameter		22	.727		310	41		351	400
3408	14" diameter		21	.762		350	42.50		392.50	450
3409	16" diameter		20	.800		395	45		440	500
3410	18" diameter	↓	19	.842		445	47		492	560
3411	20" diameter	Q-10	18	1.333		515	77.50		592.50	690
3412	24" diameter		16	1.500		660	87		747	855
3413	28" diameter		15	1.600		740	93		833	950
3414	32" diameter		14	1.714		995	99.50		1,094.50	1,250
3415	36" diameter		12	2		1,325	116		1,441	1,650
3416	42" diameter		11	2.182		1,550	127		1,677	1,900
3417	48" diameter	↓	10	2.400		1,750	140		1,890	2,125
3430	For 316 stainless steel liner, add				↓	30%				
3450	Tee 90°									
3454	6" diameter	Q-9	24	.667	Ea.	252	37.50		289.50	335
3455	8" diameter		22	.727		275	41		316	360
3456	10" diameter		21	.762		310	42.50		352.50	405
3457	12" diameter		20	.800		360	45		405	465
3458	14" diameter		18	.889		405	50		455	525
3459	16" diameter		16	1		455	56		511	590
3460	18" diameter	↓	14	1.143		535	64		599	685
3461	20" diameter	Q-10	17	1.412		610	82		692	795
3462	24" diameter		12	2		760	116		876	1,000
3463	28" diameter		11	2.182		920	127		1,047	1,225
3464	32" diameter		10	2.400		1,275	140		1,415	1,600
3465	36" diameter		9	2.667		1,350	155		1,505	1,725
3466	42" diameter		6	4		1,675	233		1,908	2,175
3467	48" diameter	↓	5	4.800		1,925	279		2,204	2,525
3480	For tee cap, add					35%	20%			
3500	For 316 stainless steel liner, add				↓	30%				
3520	Plate support, galvanized									
3524	6" diameter	Q-9	26	.615	Ea.	125	34.50		159.50	191

For customer support on your Mechanical Costs with RSMeans Data, call 800.448.8182.

355

23 51 26 – All-Fuel Vent Chimneys

23 51 26.10 All-Fuel Vent Chimneys, Press. Tight, Dbl. Wall		Crew	Daily Output	Labor-Hours	Unit	Material	2020 Bare Costs Labor	Equipment	Total	Total Incl O&P
3525	8" diameter	Q-9	22	.727	Ea.	147	41		188	223
3526	10" diameter		20	.800		160	45		205	244
3527	12" diameter		18	.889		168	50		218	261
3528	14" diameter		17	.941		198	53		251	298
3529	16" diameter		16	1		209	56		265	315
3530	18" diameter		15	1.067		220	60		280	335
3531	20" diameter	Q-10	16	1.500		231	87		318	385
3532	24" diameter		14	1.714		240	99.50		339.50	415
3533	28" diameter		13	1.846		295	107		402	490
3534	32" diameter		12	2		360	116		476	570
3535	36" diameter		10	2.400		445	140		585	700
3536	42" diameter		9	2.667		510	155		665	795
3537	48" diameter		8	3		585	174		759	910
3570	Bellows, lined									
3574	6" diameter	Q-9	30	.533	Ea.	1,350	30		1,380	1,525
3575	8" diameter		26	.615		1,400	34.50		1,434.50	1,600
3576	10" diameter		24	.667		1,425	37.50		1,462.50	1,625
3577	12" diameter		22	.727		1,475	41		1,516	1,675
3578	14" diameter		21	.762		1,525	42.50		1,567.50	1,750
3579	16" diameter		20	.800		1,575	45		1,620	1,825
3580	18" diameter		19	.842		1,650	47		1,697	1,875
3581	20" diameter	Q-10	18	1.333		1,675	77.50		1,752.50	1,975
3582	24" diameter		16	1.500		1,975	87		2,062	2,300
3583	28" diameter		15	1.600		2,100	93		2,193	2,450
3584	32" diameter		14	1.714		2,350	99.50		2,449.50	2,725
3585	36" diameter		12	2		2,575	116		2,691	3,000
3586	42" diameter		11	2.182		2,850	127		2,977	3,325
3587	48" diameter		10	2.400		3,350	140		3,490	3,875
3590	For all 316 stainless steel construction, add					55%				
3600	Ventilated roof thimble, 304 stainless steel									
3620	6" diameter	Q-9	26	.615	Ea.	264	34.50		298.50	345
3624	8" diameter		22	.727		259	41		300	345
3625	10" diameter		20	.800		272	45		317	365
3626	12" diameter		18	.889		282	50		332	385
3627	14" diameter		17	.941		305	53		358	415
3628	16" diameter		16	1		325	56		381	445
3629	18" diameter		15	1.067		350	60		410	475
3630	20" diameter	Q-10	16	1.500		370	87		457	540
3631	24" diameter		14	1.714		410	99.50		509.50	605
3632	28" diameter		13	1.846		435	107		542	640
3633	32" diameter		12	2		470	116		586	690
3634	36" diameter		10	2.400		505	140		645	765
3635	42" diameter		9	2.667		565	155		720	855
3636	48" diameter		8	3		620	174		794	950
3650	For 316 stainless steel, add					30%				
3670	Exit cone, 316 stainless steel only									
3674	6" diameter	Q-9	46	.348	Ea.	199	19.50		218.50	249
3675	8" diameter		42	.381		204	21.50		225.50	258
3676	10" diameter		40	.400		210	22.50		232.50	265
3677	12" diameter		38	.421		230	23.50		253.50	289
3678	14" diameter		37	.432		239	24.50		263.50	300
3679	16" diameter		36	.444		294	25		319	365
3680	18" diameter		35	.457		320	25.50		345.50	395

23 51 Breechings, Chimneys, and Stacks

23 51 26 – All-Fuel Vent Chimneys

23 51 26.10 All-Fuel Vent Chimneys, Press. Tight, Dbl. Wall

		Crew	Daily Output	Labor-Hours	Unit	Material	2020 Bare Costs Labor	Equipment	Total	Total Incl O&P
3681	20" diameter	Q-10	28	.857	Ea.	385	50		435	500
3682	24" diameter		26	.923		510	53.50		563.50	640
3683	28" diameter		25	.960		600	56		656	750
3684	32" diameter		24	1		745	58		803	910
3685	36" diameter		22	1.091		905	63.50		968.50	1,100
3686	42" diameter		21	1.143		1,050	66.50		1,116.50	1,250
3687	48" diameter		20	1.200		1,200	70		1,270	1,425
3720	Roof guide, 304 stainless steel									
3724	6" diameter	Q-9	25	.640	Ea.	93	36		129	157
3725	8" diameter		21	.762		108	42.50		150.50	184
3726	10" diameter		19	.842		119	47		166	202
3727	12" diameter		17	.941		123	53		176	215
3728	14" diameter		16	1		143	56		199	242
3729	16" diameter		15	1.067		151	60		211	257
3730	18" diameter		14	1.143		160	64		224	274
3731	20" diameter	Q-10	15	1.600		169	93		262	325
3732	24" diameter		13	1.846		176	107		283	355
3733	28" diameter		12	2		215	116		331	410
3734	32" diameter		11	2.182		264	127		391	480
3735	36" diameter		9	2.667		315	155		470	580
3736	42" diameter		8	3		370	174		544	670
3737	48" diameter		7	3.429		420	199		619	765
3750	For 316 stainless steel, add					30%				
3770	Rain cap with bird screen									
3774	6" diameter	Q-9	46	.348	Ea.	305	19.50		324.50	365
3775	8" diameter		42	.381		350	21.50		371.50	420
3776	10" diameter		40	.400		410	22.50		432.50	485
3777	12" diameter		38	.421		475	23.50		498.50	555
3778	14" diameter		37	.432		545	24.50		569.50	635
3779	16" diameter		36	.444		620	25		645	720
3780	18" diameter		35	.457		705	25.50		730.50	815
3781	20" diameter	Q-10	28	.857		800	50		850	955
3782	24" diameter		26	.923		965	53.50		1,018.50	1,125
3783	28" diameter		25	.960		1,150	56		1,206	1,325
3784	32" diameter		24	1		1,300	58		1,358	1,525
3785	36" diameter		22	1.091		1,525	63.50		1,588.50	1,800
3786	42" diameter		21	1.143		1,825	66.50		1,891.50	2,100
3787	48" diameter		20	1.200		2,075	70		2,145	2,375

23 51 26.30 All-Fuel Vent Chimneys, Double Wall, St. Stl.

		Crew	Daily Output	Labor-Hours	Unit	Material	2020 Bare Costs Labor	Equipment	Total	Total Incl O&P
0010	**ALL-FUEL VENT CHIMNEYS, DOUBLE WALL, STAINLESS STEEL**									
7780	All fuel, pressure tight, double wall, 4" insulation, UL listed, 1,400°F.									
7790	304 stainless steel liner, aluminized steel outer jacket									
7800	6" diameter	Q-9	60	.267	V.L.F.	64.50	14.95		79.45	93.50
7804	8" diameter		52	.308		77	17.25		94.25	111
7806	10" diameter		48	.333		87.50	18.70		106.20	125
7808	12" diameter		44	.364		101	20.50		121.50	142
7810	14" diameter		42	.381		113	21.50		134.50	157
7880	For 316 stainless steel liner, add				L.F.	30%				
8000	All fuel, double wall, stainless steel fittings									
8010	Roof support, 6" diameter	Q-9	30	.533	Ea.	120	30		150	178
8030	8" diameter		26	.615		136	34.50		170.50	202
8040	10" diameter		24	.667		151	37.50		188.50	223

23 51 Breechings, Chimneys, and Stacks

23 51 26 – All-Fuel Vent Chimneys

23 51 26.30 All-Fuel Vent Chimneys, Double Wall, St. Stl.		Crew	Daily Output	Labor-Hours	Unit	Material	2020 Bare Costs Labor	Equipment	Total	Total Incl O&P
8050	12" diameter	Q-9	22	.727	Ea.	157	41		198	234
8060	14" diameter		21	.762		167	42.50		209.50	248
8100	Elbow 45°, 6" diameter		30	.533		252	30		282	325
8140	8" diameter		26	.615		286	34.50		320.50	370
8160	10" diameter		24	.667		330	37.50		367.50	420
8180	12" diameter		22	.727		375	41		416	470
8200	14" diameter		21	.762		410	42.50		452.50	515
8300	Insulated tee, 6" diameter		30	.533		296	30		326	370
8360	8" diameter		26	.615		330	34.50		364.50	420
8380	10" diameter		24	.667		350	37.50		387.50	440
8400	12" diameter		22	.727		420	41		461	520
8420	14" diameter		20	.800		490	45		535	610
8500	Boot tee, 6" diameter		28	.571		610	32		642	720
8520	8" diameter		24	.667		670	37.50		707.50	790
8530	10" diameter		22	.727		770	41		811	905
8540	12" diameter		20	.800		905	45		950	1,075
8550	14" diameter		18	.889		945	50		995	1,125
8600	Rain cap with bird screen, 6" diameter		30	.533		297	30		327	370
8640	8" diameter		26	.615		350	34.50		384.50	440
8660	10" diameter		24	.667		410	37.50		447.50	505
8680	12" diameter		22	.727		470	41		511	580
8700	14" diameter		21	.762		535	42.50		577.50	655
8800	Flat roof flashing, 6" diameter		30	.533		111	30		141	168
8840	8" diameter		26	.615		121	34.50		155.50	186
8860	10" diameter		24	.667		132	37.50		169.50	202
8880	12" diameter		22	.727		145	41		186	222
8900	14" diameter		21	.762		147	42.50		189.50	227

23 52 Heating Boilers

23 52 13 – Electric Boilers

23 52 13.10 Electric Boilers, ASME

			Crew	Daily Output	Labor-Hours	Unit	Material	2020 Bare Costs Labor	Equipment	Total	Total Incl O&P
0010	**ELECTRIC BOILERS, ASME**, Standard controls and trim	R235000-30									
1000	Steam, 6 KW, 20.5 MBH		Q-19	1.20	20	Ea.	4,450	1,200		5,650	6,650
1040	9 KW, 30.7 MBH	D3020-102		1.20	20		4,400	1,200		5,600	6,625
1060	18 KW, 61.4 MBH			1.20	20		4,525	1,200		5,725	6,750
1080	24 KW, 81.8 MBH	D3020-104		1.10	21.818		5,250	1,300		6,550	7,725
1120	36 KW, 123 MBH			1.10	21.818		5,800	1,300		7,100	8,350
1160	60 KW, 205 MBH			1	24		7,250	1,425		8,675	10,200
1220	112 KW, 382 MBH			.75	32		9,725	1,925		11,650	13,600
1240	148 KW, 505 MBH			.65	36.923		10,500	2,200		12,700	14,800
1260	168 KW, 573 MBH			.60	40		20,600	2,400		23,000	26,200
1280	222 KW, 758 MBH			.55	43.636		24,000	2,600		26,600	30,300
1300	296 KW, 1,010 MBH			.45	53.333		25,300	3,200		28,500	32,600
1320	300 KW, 1,023 MBH			.40	60		26,600	3,575		30,175	34,600
1340	370 KW, 1,263 MBH			.35	68.571		29,400	4,100		33,500	38,500
1360	444 KW, 1,515 MBH			.30	80		31,600	4,775		36,375	42,000
1380	518 KW, 1,768 MBH		Q-21	.36	88.889		32,900	5,450		38,350	44,300
1400	592 KW, 2,020 MBH			.34	94.118		36,000	5,775		41,775	48,200
1420	666 KW, 2,273 MBH			.32	100		37,500	6,125		43,625	50,500
1460	740 KW, 2,526 MBH			.28	114		39,000	7,000		46,000	53,500
1480	814 KW, 2,778 MBH			.25	128		40,100	7,825		47,925	56,000

For customer support on your Mechanical Costs with RSMeans Data, call 800.448.8182.

23 52 Heating Boilers

23 52 13 – Electric Boilers

23 52 13.10 Electric Boilers, ASME

		Crew	Daily Output	Labor-Hours	Unit	Material	2020 Bare Costs Labor	Equipment	Total	Total Incl O&P
1500	962 KW, 3,283 MBH	Q-21	.22	145	Ea.	45,700	8,900		54,600	64,000
1520	1,036 KW, 3,536 MBH		.20	160		48,600	9,800		58,400	68,000
1540	1,110 KW, 3,788 MBH		.19	168		50,500	10,300		60,800	71,000
1560	2,070 KW, 7,063 MBH		.18	178		69,500	10,900		80,400	93,000
1580	2,250 KW, 7,677 MBH		.17	188		81,500	11,500		93,000	106,500
1600	2,340 KW, 7,984 MBH		.16	200		88,000	12,200		100,200	115,000
2000	Hot water, 7.5 KW, 25.6 MBH	Q-19	1.30	18.462		5,425	1,100		6,525	7,600
2020	15 KW, 51.2 MBH		1.30	18.462		5,450	1,100		6,550	7,650
2040	30 KW, 102 MBH		1.20	20		5,850	1,200		7,050	8,200
2060	45 KW, 164 MBH		1.20	20		5,700	1,200		6,900	8,025
2070	60 KW, 205 MBH		1.20	20		5,800	1,200		7,000	8,175
2080	75 KW, 256 MBH		1.10	21.818		6,150	1,300		7,450	8,700
2100	90 KW, 307 MBH		1.10	21.818		6,125	1,300		7,425	8,700
2120	105 KW, 358 MBH		1	24		6,950	1,425		8,375	9,775
2140	120 KW, 410 MBH		.90	26.667		6,875	1,600		8,475	9,950
2160	135 KW, 461 MBH		.75	32		7,550	1,925		9,475	11,200
2180	150 KW, 512 MBH		.65	36.923		7,925	2,200		10,125	12,000
2200	165 KW, 563 MBH		.60	40		8,275	2,400		10,675	12,700
2220	296 KW, 1,010 MBH		.55	43.636		15,300	2,600		17,900	20,800
2280	370 KW, 1,263 MBH		.40	60		17,000	3,575		20,575	24,100
2300	444 KW, 1,515 MBH		.35	68.571		20,600	4,100		24,700	28,800
2340	518 KW, 1,768 MBH	Q-21	.44	72.727		22,900	4,450		27,350	31,800
2360	592 KW, 2,020 MBH		.43	74.419		24,900	4,550		29,450	34,200
2400	666 KW, 2,273 MBH		.40	80		27,100	4,900		32,000	37,100
2420	740 KW, 2,526 MBH		.39	82.051		27,100	5,025		32,125	37,300
2440	814 KW, 2,778 MBH		.38	84.211		28,900	5,150		34,050	39,500
2460	888 KW, 3,031 MBH		.37	86.486		31,400	5,300		36,700	42,500
2480	962 KW, 3,283 MBH		.36	88.889		33,700	5,450		39,150	45,100
2500	1,036 KW, 3,536 MBH		.34	94.118		34,700	5,775		40,475	46,800
2520	1,110 KW, 3,788 MBH		.33	96.970		38,000	5,925		43,925	50,500
2540	1,440 KW, 4,915 MBH		.32	100		42,400	6,125		48,525	56,000
2560	1,560 KW, 5,323 MBH		.31	103		48,100	6,325		54,425	62,500
2580	1,680 KW, 5,733 MBH		.30	107		51,500	6,525		58,025	66,500
2600	1,800 KW, 6,143 MBH		.29	110		55,000	6,750		61,750	70,500
2620	1,980 KW, 6,757 MBH		.28	114		58,000	7,000		65,000	74,000
2640	2,100 KW, 7,167 MBH		.27	119		69,000	7,250		76,250	86,500
2660	2,220 KW, 7,576 MBH		.26	123		66,500	7,525		74,025	85,000
2680	2,400 KW, 8,191 MBH		.25	128		69,500	7,825		77,325	88,000
2700	2,610 KW, 8,905 MBH		.24	133		74,000	8,175		82,175	93,500
2720	2,790 KW, 9,519 MBH		.23	139		77,500	8,525		86,025	98,000
2740	2,970 KW, 10,133 MBH		.21	152		79,500	9,325		88,825	101,000
2760	3,150 KW, 10,748 MBH		.19	168		84,500	10,300		94,800	108,000
2780	3,240 KW, 11,055 MBH		.18	178		86,000	10,900		96,900	111,500
2800	3,420 KW, 11,669 MBH		.17	188		92,000	11,500		103,500	118,500
2820	3,600 KW, 12,283 MBH		.16	200		96,500	12,200		108,700	124,500

23 52 16 – Condensing Boilers

23 52 16.24 Condensing Boilers

			Crew	Daily Output	Labor-Hours	Unit	Material	Labor	Equipment	Total	Total Incl O&P
0010	**CONDENSING BOILERS**, Cast iron, high efficiency										
0020	Packaged with standard controls, circulator and trim										
0030	Intermittent (spark) pilot, natural or LP gas										
0040	Hot water, DOE MBH output (AFUE %)										
0100	42 MBH (84.0%)	G	Q-5	1.80	8.889	Ea.	1,625	525		2,150	2,575

For customer support on your Mechanical Costs with RSMeans Data, call 800.448.8182.

359

23 52 Heating Boilers

23 52 16 - Condensing Boilers

23 52 16.24 Condensing Boilers

			Crew	Daily Output	Labor-Hours	Unit	Material	2020 Bare Costs Labor	Equipment	Total	Total Incl O&P
0120	57 MBH (84.3%)	G	Q-5	1.60	10	Ea.	1,775	590		2,365	2,825
0140	85 MBH (84.0%)	G		1.40	11.429		1,975	675		2,650	3,175
0160	112 MBH (83.7%)	G		1.20	13.333		2,225	785		3,010	3,625
0180	140 MBH (83.3%)	G	Q-6	1.60	15		2,500	920		3,420	4,125
0200	167 MBH (83.0%)	G		1.40	17.143		2,825	1,050		3,875	4,700
0220	194 MBH (82.7%)	G		1.20	20		3,125	1,225		4,350	5,250

23 52 19 - Pulse Combustion Boilers

23 52 19.20 Pulse Type Combustion Boilers

			Crew	Daily Output	Labor-Hours	Unit	Material	2020 Bare Costs Labor	Equipment	Total	Total Incl O&P
0010	**PULSE TYPE COMBUSTION BOILERS**, High efficiency										
7990	Special feature gas fired boilers										
8000	Pulse combustion, standard controls/trim										
8010	Hot water, DOE MBH output (AFUE %)										
8030	71 MBH (95.2%)		Q-5	1.60	10	Ea.	3,975	590		4,565	5,225
8050	94 MBH (95.3%)	G		1.40	11.429		4,250	675		4,925	5,675
8080	139 MBH (95.6%)	G		1.20	13.333		4,950	785		5,735	6,625
8090	207 MBH (95.4%)			1.16	13.793		5,600	815		6,415	7,400
8120	270 MBH (96.4%)			1.12	14.286		7,750	845		8,595	9,775
8130	365 MBH (91.7%)			1.07	14.953		8,750	880		9,630	11,000

23 52 23 - Cast-Iron Boilers

23 52 23.20 Gas-Fired Boilers

			Crew	Daily Output	Labor-Hours	Unit	Material	2020 Bare Costs Labor	Equipment	Total	Total Incl O&P
0010	**GAS-FIRED BOILERS**, Natural or propane, standard controls, packaged	R235000-10									
1000	Cast iron, with insulated jacket										
2000	Steam, gross output, 81 MBH	R235000-20	Q-7	1.40	22.857	Ea.	2,750	1,425		4,175	5,150
2020	102 MBH			1.30	24.615		2,625	1,525		4,150	5,200
2040	122 MBH	R235000-30		1	32		3,250	2,000		5,250	6,575
2060	163 MBH			.90	35.556		3,400	2,225		5,625	7,075
2080	203 MBH			.90	35.556		3,950	2,225		6,175	7,675
2100	240 MBH			.85	37.647		4,025	2,350		6,375	7,950
2120	280 MBH			.80	40		4,525	2,500		7,025	8,700
2140	320 MBH			.70	45.714		5,200	2,850		8,050	9,975
2160	360 MBH			.63	51.200		5,475	3,200		8,675	10,800
2180	400 MBH			.56	56.838		5,825	3,550		9,375	11,700
2200	440 MBH			.51	62.500		6,250	3,900		10,150	12,700
2220	544 MBH			.45	71.588		9,000	4,475		13,475	16,600
2240	765 MBH			.43	74.419		11,300	4,650		15,950	19,400
2260	892 MBH			.38	84.211		13,300	5,250		18,550	22,600
2280	1,275 MBH			.34	94.118		16,700	5,875		22,575	27,200
2300	1,530 MBH			.32	100		19,000	6,250		25,250	30,300
2320	1,875 MBH			.30	107		25,200	6,650		31,850	37,800
2340	2,170 MBH			.26	122		27,000	7,625		34,625	41,100
2360	2,675 MBH			.20	163		28,800	10,200		39,000	46,900
2380	3,060 MBH			.19	172		30,000	10,700		40,700	49,100
2400	3,570 MBH			.18	182		35,800	11,300		47,100	56,500
2420	4,207 MBH			.16	205		39,200	12,800		52,000	62,500
2440	4,720 MBH			.15	208		63,500	13,000		76,500	89,500
2460	5,660 MBH			.15	221		71,500	13,800		85,300	99,500
2480	6,100 MBH			.13	246		86,500	15,400		101,900	118,000
2500	6,390 MBH			.12	267		90,500	16,600		107,100	124,500
2520	6,680 MBH			.11	291		92,500	18,200		110,700	129,000
2540	6,970 MBH			.10	320		97,000	20,000		117,000	136,500
3000	Hot water, gross output, 80 MBH			1.46	21.918		1,975	1,375		3,350	4,225
3020	100 MBH			1.35	23.704		2,575	1,475		4,050	5,050

23 52 23 – Cast-Iron Boilers

23 52 23.20 Gas-Fired Boilers		Crew	Daily Output	Labor-Hours	Unit	Material	2020 Bare Costs Labor	Equipment	Total	Total Incl O&P
3040	122 MBH	Q-7	1.10	29.091	Ea.	2,825	1,825		4,650	5,825
3060	163 MBH		1	32		3,275	2,000		5,275	6,600
3080	203 MBH		1	32		3,950	2,000		5,950	7,350
3100	240 MBH		.95	33.684		3,825	2,100		5,925	7,375
3120	280 MBH		.90	35.556		4,375	2,225		6,600	8,150
3140	320 MBH		.80	40		4,275	2,500		6,775	8,425
3160	360 MBH		.71	45.070		4,900	2,800		7,700	9,575
3180	400 MBH		.64	50		5,175	3,125		8,300	10,400
3200	440 MBH		.58	54.983		5,550	3,425		8,975	11,200
3220	544 MBH		.51	62.992		8,600	3,925		12,525	15,400
3240	765 MBH		.46	70.022		10,800	4,375		15,175	18,500
3260	1,088 MBH		.40	80		13,900	5,000		18,900	22,800
3280	1,275 MBH		.36	89.888		16,300	5,600		21,900	26,400
3300	1,530 MBH		.31	105		18,700	6,550		25,250	30,300
3320	2,000 MBH		.26	125		23,300	7,800		31,100	37,300
3340	2,312 MBH		.22	148		27,600	9,250		36,850	44,200
3360	2,856 MBH		.20	160		30,100	9,975		40,075	48,000
3380	3,264 MBH		.18	180		31,900	11,200		43,100	52,000
3400	3,996 MBH		.16	195		34,900	12,200		47,100	56,500
3420	4,488 MBH		.15	211		38,000	13,100		51,100	61,500
3440	4,720 MBH		.15	221		69,000	13,800		82,800	96,500
3460	5,520 MBH		.14	229		100,000	14,300		114,300	132,000
3480	6,100 MBH		.13	250		139,500	15,600		155,100	177,000
3500	6,390 MBH		.11	286		117,000	17,800		134,800	155,500
3520	6,680 MBH		.10	311		122,000	19,400		141,400	163,000
3540	6,970 MBH	▼	.09	360		128,000	22,400		150,400	174,500
7000	For tankless water heater, add					10%				
7050	For additional zone valves up to 312 MBH, add				▼	199			199	219

23 52 23.30 Gas/Oil Fired Boilers

		Crew	Daily Output	Labor-Hours	Unit	Material	Labor	Equipment	Total	Total Incl O&P
0010	**GAS/OIL FIRED BOILERS**, Combination with burners and controls, packaged									
1000	Cast iron with insulated jacket									
2000	Steam, gross output, 720 MBH	Q-7	.43	74.074	Ea.	15,200	4,625		19,825	23,600
2020	810 MBH		.38	83.990		15,200	5,250		20,450	24,600
2040	1,084 MBH		.34	93.023		17,500	5,800		23,300	28,000
2060	1,360 MBH		.33	98.160		20,000	6,125		26,125	31,200
2080	1,600 MBH		.30	107		20,700	6,675		27,375	32,800
2100	2,040 MBH		.25	130		24,100	8,125		32,225	38,600
2120	2,450 MBH		.21	156		25,400	9,750		35,150	42,500
2140	2,700 MBH		.19	166		26,800	10,300		37,100	45,000
2160	3,000 MBH		.18	176		29,000	11,000		40,000	48,300
2180	3,270 MBH		.17	184		34,000	11,500		45,500	54,500
2200	3,770 MBH		.17	192		69,500	12,000		81,500	94,500
2220	4,070 MBH		.16	200		73,500	12,500		86,000	99,000
2240	4,650 MBH		.15	211		77,500	13,100		90,600	104,500
2260	5,230 MBH		.14	224		85,500	14,000		99,500	115,000
2280	5,520 MBH		.14	235		93,500	14,700		108,200	124,500
2300	5,810 MBH		.13	248		95,000	15,500		110,500	127,500
2320	6,100 MBH		.12	260		97,000	16,200		113,200	131,000
2340	6,390 MBH		.11	296		100,000	18,500		118,500	137,500
2360	6,680 MBH		.10	320		99,500	20,000		119,500	139,500
2380	6,970 MBH	▼	.09	372	▼	106,000	23,200		129,200	151,000
2900	Hot water, gross output									

For customer support on your Mechanical Costs with RSMeans Data, call 800.448.8182.

361

23 52 23 – Cast-Iron Boilers

23 52 23.30 Gas/Oil Fired Boilers

		Crew	Daily Output	Labor-Hours	Unit	Material	2020 Bare Costs Labor	Equipment	Total	Total Incl O&P
2910	200 MBH 80⁻/mbh	Q-6	.62	39.024	Ea.	11,400	2,400		13,800	16,100
2920	300 MBH		.49	49.080		11,400	3,000		14,400	17,000
2930	400 MBH		.41	57.971		13,300	3,550		16,850	20,000
2940	500 MBH		.36	67.039		14,300	4,100		18,400	22,000
3000	584 MBH 38⁻/mbh	Q-7	.44	72.072		14,100	4,500		18,600	22,300
3020	876 MBH		.41	79.012		21,000	4,925		25,925	30,500
3040	1,168 MBH 35⁻/mbh		.31	104		28,500	6,475		34,975	41,000
3060	1,460 MBH		.28	113		33,200	7,050		40,250	47,100
3080	2,044 MBH 27⁻/mbh		.26	122		39,900	7,625		47,525	55,500
3100	2,628 MBH		.21	150		47,800	9,375		57,175	66,500
3120	3,210 MBH		.18	175		52,500	10,900		63,400	74,000
3140	3,796 MBH		.17	186		58,500	11,600		70,100	82,000
3160	4,088 MBH 22⁻/mbh		.16	195		66,500	12,200		78,700	91,500
3180	4,672 MBH		.16	204		72,500	12,700		85,200	99,000
3200	5,256 MBH		.15	218		82,500	13,600		96,100	111,000
3220	6,000 MBH, 179 BHP 25⁻/mbh		.13	256		116,500	16,000		132,500	152,500
3240	7,130 MBH, 213 BHP		.08	386		122,000	24,100		146,100	170,500
3260	9,800 MBH, 286 BHP		.06	533		140,500	33,300		173,800	204,500
3280	10,900 MBH, 325.6 BHP		.05	593		170,500	37,000		207,500	243,000
3290	12,200 MBH, 364.5 BHP		.05	667		186,000	41,600		227,600	267,500
3300	13,500 MBH, 403.3 BHP		.04	727		206,000	45,400		251,400	294,500

23 52 23.40 Oil-Fired Boilers

		Crew	Daily Output	Labor-Hours	Unit	Material	2020 Bare Costs Labor	Equipment	Total	Total Incl O&P
0010	**OIL-FIRED BOILERS**, Standard controls, flame retention burner, packaged									
1000	Cast iron, with insulated flush jacket									
2000	Steam, gross output, 109 MBH	Q-7	1.20	26.667	Ea.	2,375	1,675		4,050	5,100
2020	144 MBH		1.10	29.091		2,650	1,825		4,475	5,650
2040	173 MBH		1	32		3,000	2,000		5,000	6,300
2060	207 MBH		.90	35.556		3,225	2,225		5,450	6,850
2080	236 MBH		.85	37.647		3,600	2,350		5,950	7,500
2100	300 MBH		.70	45.714		4,600	2,850		7,450	9,325
2120	480 MBH		.50	64		6,550	4,000		10,550	13,200
2140	665 MBH		.45	71.111		8,500	4,425		12,925	16,000
2160	794 MBH		.41	78.049		9,450	4,875		14,325	17,700
2180	1,084 MBH		.38	85.106		10,800	5,300		16,100	19,900
2200	1,360 MBH		.33	98.160		13,300	6,125		19,425	23,800
2220	1,600 MBH		.26	122		17,700	7,625		25,325	30,900
2240	2,175 MBH		.24	134		17,400	8,350		25,750	31,600
2260	2,480 MBH		.21	156		20,300	9,750		30,050	36,900
2280	3,000 MBH		.19	170		23,400	10,600		34,000	41,600
2300	3,550 MBH		.17	187		27,800	11,700		39,500	48,100
2320	3,820 MBH		.16	200		37,200	12,500		49,700	59,500
2340	4,360 MBH		.15	215		40,800	13,400		54,200	65,000
2360	4,940 MBH		.14	225		68,000	14,100		82,100	95,500
2380	5,520 MBH		.14	235		80,000	14,700		94,700	110,000
2400	6,100 MBH		.13	256		91,500	16,000		107,500	124,500
2420	6,390 MBH		.11	291		96,500	18,200		114,700	133,000
2440	6,680 MBH		.10	314		99,000	19,600		118,600	138,500
2460	6,970 MBH		.09	364		102,500	22,700		125,200	147,000
3000	Hot water, same price as steam									
4000	For tankless coil in smaller sizes, add				Ea.	15%				

23 52 Heating Boilers

23 52 23 – Cast-Iron Boilers

23 52 23.60 Solid-Fuel Boilers

		Crew	Daily Output	Labor-Hours	Unit	Material	2020 Bare Costs Labor	2020 Bare Costs Equipment	Total	Total Incl O&P
0010	**SOLID-FUEL BOILERS**									
3000	Stoker fired (coal/wood/biomass) cast iron with flush jacket and									
3400	insulation, steam or water, gross output, 1,280 MBH	Q-6	.36	66.667	Ea.	173,500	4,075		177,575	197,000
3420	1,460 MBH		.30	80		180,000	4,900		184,900	205,500
3440	1,640 MBH		.28	85.714		193,000	5,250		198,250	220,500
3460	1,820 MBH		.26	92.308		209,000	5,650		214,650	238,000
3480	2,000 MBH		.25	96		225,500	5,875		231,375	257,000
3500	2,360 MBH		.23	104		226,500	6,375		232,875	258,500
3540	2,725 MBH	▼	.20	120		234,000	7,350		241,350	268,500
3800	2,950 MBH	Q-7	.16	200		247,500	12,500		260,000	291,000
3820	3,210 MBH		.15	213		261,000	13,300		274,300	307,500
3840	3,480 MBH		.14	229		263,000	14,300		277,300	311,000
3860	3,745 MBH		.14	229		261,000	14,300		275,300	308,500
3880	4,000 MBH		.13	246		263,000	15,400		278,400	312,500
3900	4,200 MBH		.13	246		265,500	15,400		280,900	315,000
3920	4,400 MBH		.12	267		267,500	16,600		284,100	319,500
3940	4,600 MBH	▼	.12	267	▼	271,000	16,600		287,600	323,000

23 52 26 – Steel Boilers

23 52 26.40 Oil-Fired Boilers

		Crew	Daily Output	Labor-Hours	Unit	Material	2020 Bare Costs Labor	2020 Bare Costs Equipment	Total	Total Incl O&P
0010	**OIL-FIRED BOILERS**, Standard controls, flame retention burner									
5000	Steel, with insulated flush jacket									
7000	Hot water, gross output, 103 MBH	Q-6	1.60	15	Ea.	2,000	920		2,920	3,575
7020	122 MBH		1.45	16.506		2,150	1,000		3,150	3,875
7040	137 MBH		1.36	17.595		2,325	1,075		3,400	4,150
7060	168 MBH		1.30	18.405		2,275	1,125		3,400	4,200
7080	225 MBH		1.22	19.704		4,000	1,200		5,200	6,200
7100	315 MBH		.96	25.105		5,075	1,525		6,600	7,875
7120	420 MBH		.70	34.483		5,925	2,100		8,025	9,675
7140	525 MBH		.57	42.403		7,650	2,600		10,250	12,300
7180	735 MBH		.48	50.104		8,125	3,075		11,200	13,500
7220	1,050 MBH		.37	65.753		21,300	4,025		25,325	29,400
7280	2,310 MBH		.21	115		28,200	7,025		35,225	41,500
7320	3,150 MBH	▼	.13	185	▼	45,400	11,300		56,700	67,000
7340	For tankless coil in steam or hot water, add					7%				

23 52 26.70 Packaged Water Tube Boilers

		Crew	Daily Output	Labor-Hours	Unit	Material	2020 Bare Costs Labor	2020 Bare Costs Equipment	Total	Total Incl O&P
0010	**PACKAGED WATER TUBE BOILERS**									
2000	Packaged water tube, #2 oil, steam or hot water, gross output									
2010	200 MBH	Q-6	1.50	16	Ea.	4,775	980		5,755	6,725
2014	275 MBH		1.40	17.143		5,550	1,050		6,600	7,675
2018	360 MBH		1.10	21.818		6,825	1,325		8,150	9,525
2022	520 MBH		.65	36.923		8,250	2,250		10,500	12,500
2026	600 MBH		.60	40		11,300	2,450		13,750	16,200
2030	720 MBH		.55	43.636		12,200	2,675		14,875	17,400
2034	960 MBH	▼	.48	50		14,400	3,050		17,450	20,400
2040	1,200 MBH	Q-7	.50	64		28,300	4,000		32,300	37,200
2044	1,440 MBH		.45	71.111		34,100	4,425		38,525	44,200
2060	1,600 MBH		.40	80		40,100	5,000		45,100	51,500
2068	1,920 MBH		.35	91.429		43,500	5,700		49,200	56,500
2072	2,160 MBH		.33	96.970		48,200	6,050		54,250	62,000
2080	2,400 MBH		.30	107		54,000	6,650		60,650	69,500
2100	3,200 MBH	▼	.25	128	▼	61,500	7,975		69,475	80,000

For customer support on your Mechanical Costs with RSMeans Data, call 800.448.8182.

363

23 52 26 – Steel Boilers

23 52 26.70 Packaged Water Tube Boilers	Crew	Daily Output	Labor-Hours	Unit	Material	2020 Bare Costs Labor	Equipment	Total	Total Incl O&P
2120	4,800 MBH	Q-7	.20	160	Ea.	80,500	9,975		90,475	103,500
2200	Gas fired									
2204	200 MBH	Q-6	1.50	16	Ea.	8,625	980		9,605	11,000
2208	275 MBH		1.40	17.143		9,025	1,050		10,075	11,500
2212	360 MBH		1.10	21.818		9,300	1,325		10,625	12,200
2216	520 MBH		.65	36.923		8,375	2,250		10,625	12,600
2220	600 MBH		.60	40		10,700	2,450		13,150	15,500
2224	720 MBH		.55	43.636		10,300	2,675		12,975	15,300
2228	960 MBH		.48	50		18,700	3,050		21,750	25,200
2232	1,220 MBH	Q-7	.50	64		19,700	4,000		23,700	27,600
2236	1,440 MBH		.45	71.111		20,600	4,425		25,025	29,400
2240	1,680 MBH		.40	80		32,700	5,000		37,700	43,500
2244	1,920 MBH		.35	91.429		29,800	5,700		35,500	41,400
2248	2,160 MBH		.33	96.970		33,500	6,050		39,550	45,900
2252	2,400 MBH		.30	107		37,600	6,650		44,250	51,500

23 52 28 – Swimming Pool Boilers

23 52 28.10 Swimming Pool Heaters

		Crew	Daily Output	Labor-Hours	Unit	Material	2020 Bare Costs Labor	Equipment	Total	Total Incl O&P
0010	**SWIMMING POOL HEATERS**, Not including wiring, external									
0020	piping, base or pad									
0160	Gas fired, input, 155 MBH	Q-6	1.50	16	Ea.	1,850	980		2,830	3,500
0200	199 MBH		1	24		1,975	1,475		3,450	4,375
0220	250 MBH		.70	34.286		2,175	2,100		4,275	5,525
0240	300 MBH		.60	40		2,275	2,450		4,725	6,175
0260	399 MBH		.50	48		2,550	2,925		5,475	7,200
0280	500 MBH		.40	60		8,500	3,675		12,175	14,900
0300	650 MBH		.35	68.571		9,050	4,200		13,250	16,200
0320	750 MBH		.33	72.727		9,875	4,450		14,325	17,600
0360	990 MBH		.22	109		13,300	6,675		19,975	24,600
0370	1,260 MBH		.21	114		17,700	7,000		24,700	30,000
0380	1,440 MBH		.19	126		19,000	7,725		26,725	32,500
0400	1,800 MBH		.14	171		21,000	10,500		31,500	38,800
0410	2,070 MBH		.13	185		24,800	11,300		36,100	44,200
2000	Electric, 12 KW, 4,800 gallon pool	Q-19	3	8		2,200	480		2,680	3,150
2020	15 KW, 7,200 gallon pool		2.80	8.571		2,225	510		2,735	3,225
2040	24 KW, 9,600 gallon pool		2.40	10		2,575	600		3,175	3,750
2060	30 KW, 12,000 gallon pool		2	12		2,650	715		3,365	4,000
2080	36 KW, 14,400 gallon pool		1.60	15		3,075	895		3,970	4,725
2100	57 KW, 24,000 gallon pool		1.20	20		3,875	1,200		5,075	6,025
9000	To select pool heater: 12 BTUH x S.F. pool area									
9010	X temperature differential = required output									
9050	For electric, KW = gallons x 2.5 divided by 1,000									
9100	For family home type pool, double the									
9110	Rated gallon capacity = 1/2°F rise per hour									

23 52 39 – Fire-Tube Boilers

23 52 39.13 Scotch Marine Boilers

		Crew	Daily Output	Labor-Hours	Unit	Material	2020 Bare Costs Labor	Equipment	Total	Total Incl O&P
0010	**SCOTCH MARINE BOILERS**									
1000	Packaged fire tube, #2 oil, gross output									
1006	15 psi steam									
1010	1,005 MBH, 30 HP	Q-6	.33	72.072	Ea.	63,000	4,400		67,400	76,000
1014	1,675 MBH, 50 HP	"	.26	93.023		71,000	5,700		76,700	86,500
1020	3,348 MBH, 100 HP	Q-7	.21	152		105,500	9,500		115,000	130,000
1030	5,025 MBH, 150 HP		.17	193		133,000	12,000		145,000	164,000

23 52 39.13 Scotch Marine Boilers		Crew	Daily Output	Labor-Hours	Unit	Material	2020 Bare Costs Labor	Equipment	Total	Total Incl O&P
1040	6,696 MBH, 200 HP	Q-7	.14	224	Ea.	157,500	14,000		171,500	194,000
1050	8,375 MBH, 250 HP		.14	232		154,000	14,500		168,500	191,000
1060	10,044 MBH, 300 HP		.13	252		181,500	15,700		197,200	223,000
1080	16,740 MBH, 500 HP		.08	381		246,500	23,800		270,300	307,000
1100	23,435 MBH, 700 HP		.07	485		302,000	30,300		332,300	378,000
1102	26,780 MBH, 800 HP		.07	485		330,000	30,300		360,300	408,500
1104	30,125 MBH, 900 HP		.06	500		602,000	31,200		633,200	708,500
1106	33,475 MBH, 1,000 HP		.06	525		637,000	32,700		669,700	749,500
1107	36,825 MBH, 1,100 HP		.06	552		683,500	34,400		717,900	803,500
1108	40,170 MBH, 1,200 HP		.06	582		717,000	36,300		753,300	843,000
1110	46,865 MBH, 1,400 HP		.05	604		755,000	37,700		792,700	887,000
1112	53,560 MBH, 1,600 HP		.05	627		820,500	39,200		859,700	961,000
1114	60,250 MBH, 1,800 HP		.05	667		847,500	41,600		889,100	995,000
1116	66,950 MBH, 2,000 HP		.05	696		897,500	43,400		940,900	1,052,500
1118	73,650 MBH, 2,200 HP		.04	744		970,000	46,400		1,016,400	1,136,000
1120	To fire #6, add		.83	38.554		18,300	2,400		20,700	23,700
1140	To fire #6, and gas, add		.42	76.190		26,900	4,750		31,650	36,700
1141	1,675 MBH, 50 HP	Q-6	.26	93.023		93,500	5,700		99,200	111,500
1142	6,696 MBH, 200 HP	Q-7	.14	224		162,500	14,000		176,500	200,000
1143	10,044 MBH, 300 HP		.13	252		200,000	15,700		215,700	243,500
1144	16,740 MBH, 500 HP		.08	381		262,500	23,800		286,300	324,000
1145	23,435 MBH, 700 HP		.07	485		316,000	30,300		346,300	393,000
1146	30,125 MBH, 900 HP		.06	500		555,000	31,200		586,200	657,000
1160	For high pressure, add					29,900			29,900	32,800
1180	For duplex package feed system									
1200	To 3,348 MBH boiler, add	Q-7	.54	59.259	Ea.	14,500	3,700		18,200	21,400
1220	To 6,696 MBH boiler, add		.41	78.049		17,100	4,875		21,975	26,200
1240	To 10,044 MBH boiler, add		.38	84.211		19,700	5,250		24,950	29,500
1260	To 16,740 MBH boiler, add		.28	114		22,400	7,125		29,525	35,400
1280	To 23,435 MBH boiler, add		.25	128		26,500	7,975		34,475	41,100
1300	150 psi steam									
1310	1,005 MBH, 30 HP	Q-6	.33	72.072	Ea.	61,000	4,400		65,400	74,000
1320	1,675 MBH, 50 HP	"	.26	93.023		80,000	5,700		85,700	96,500
1330	3,350 MBH, 100 HP	Q-7	.21	152		126,500	9,500		136,000	153,500
1340	5,025 MBH, 150 HP		.17	193		161,500	12,000		173,500	195,500
1350	6,700 MBH, 200 HP		.14	224		152,500	14,000		166,500	188,500
1360	10,050 MBH, 300 HP		.13	252		227,500	15,700		243,200	273,500
1400	Packaged Scotch Marine, #6 oil									
1410	15 psi steam									
1420	1,675 MBH, 50 HP	Q-6	.25	97.166	Ea.	97,000	5,950		102,950	116,000
1430	3,350 MBH, 100 HP	Q-7	.20	159		113,000	9,925		122,925	139,500
1440	5,025 MBH, 150 HP		.16	203		138,500	12,600		151,100	171,000
1450	6,700 MBH, 200 HP		.14	234		158,000	14,600		172,600	196,000
1460	8,375 MBH, 250 HP		.13	244		203,000	15,200		218,200	246,500
1470	10,050 MBH, 300 HP		.12	262		228,500	16,400		244,900	276,000
1480	13,400 MBH, 400 HP		.09	352		280,500	21,900		302,400	341,500
1500	150 psi steam									
1510	1,675 MBH, 50 HP	Q-6	.25	97.166	Ea.	115,500	5,950		121,450	136,000
1520	3,350 MBH, 100 HP	Q-7	.20	159		145,000	9,925		154,925	174,500
1530	5,025 MBH, 150 HP		.16	203		167,000	12,600		179,600	203,000
1540	6,700 MBH, 200 HP		.14	234		196,500	14,600		211,100	238,000
1550	8,375 MBH, 250 HP		.13	244		214,500	15,200		229,700	259,000
1560	10,050 MUBH, 300 HP		.12	262		239,000	16,400		255,400	287,500

23 52 Heating Boilers

23 52 39 – Fire-Tube Boilers

23 52 39.13 Scotch Marine Boilers	Crew	Daily Output	Labor- Hours	Unit	Material	2020 Bare Costs Labor	Equipment	Total	Total Incl O&P
1570	13,400 MBH, 400 HP	Q-7	.09	352	Ea.	277,000	21,900		298,900	338,000

23 52 84 – Boiler Blowdown

23 52 84.10 Boiler Blowdown Systems

0010	**BOILER BLOWDOWN SYSTEMS**									
1010	Boiler blowdown, auto/manual to 2,000 MBH	Q-5	3.75	4.267	Ea.	6,975	252		7,227	8,050
1020	7,300 MBH	"	3	5.333	"	7,525	315		7,840	8,775

23 52 88 – Burners

23 52 88.10 Replacement Type Burners

0010	**REPLACEMENT TYPE BURNERS**									
0990	Residential, conversion, gas fired, LP or natural									
1000	Gun type, atmospheric input 50 to 225 MBH	Q-1	2.50	6.400	Ea.	1,025	370		1,395	1,675
1020	100 to 400 MBH		2	8		1,725	465		2,190	2,600
1040	300 to 1,000 MBH	↓	1.70	9.412	↓	5,125	545		5,670	6,475
2000	Commercial and industrial, gas/oil, input									
2050	400 MBH	Q-1	1.50	10.667	Ea.	4,400	620		5,020	5,775
2090	670 MBH		1.40	11.429		4,425	665		5,090	5,875
2140	1,155 MBH		1.30	12.308		4,925	715		5,640	6,500
2200	1,800 MBH		1.20	13.333		4,950	775		5,725	6,600
2260	3,000 MBH		1.10	14.545		5,125	845		5,970	6,925
2320	4,100 MBH	↓	1	16	↓	6,850	930		7,780	8,950
3000	Flame retention oil fired assembly, input									
3020	.50 to 2.25 GPH	Q-1	2.40	6.667	Ea.	360	385		745	975
3040	2.0 to 5.0 GPH		2	8		335	465		800	1,075
3060	3.0 to 7.0 GPH		1.80	8.889		625	515		1,140	1,450
3080	6.0 to 12.0 GPH	↓	1.60	10		1,000	580		1,580	1,975
4600	Gas safety, shut off valve, 3/4" threaded	1 Stpi	20	.400		190	26		216	247
4610	1" threaded		19	.421		183	27.50		210.50	243
4620	1-1/4" threaded		15	.533		206	35		241	280
4630	1-1/2" threaded		13	.615		223	40.50		263.50	305
4640	2" threaded	↓	11	.727		250	47.50		297.50	345
4650	2-1/2" threaded	Q-1	15	1.067		286	62		348	410
4660	3" threaded		13	1.231		395	71.50		466.50	540
4670	4" flanged	↓	3	5.333		2,800	310		3,110	3,550
4680	6" flanged	Q-2	3	8	↓	6,250	480		6,730	7,600

23 53 Heating Boiler Feedwater Equipment

23 53 24 – Shot Chemical Feeder

23 53 24.10 Chemical Feeder

0010	**CHEMICAL FEEDER**									
0200	Shot chem feeder, by pass, in line mount, 300 PSIG, 1.7 gallon	2 Stpi	1.80	8.889	Ea.	279	585		864	1,175
0230	12 gallon		.93	17.204		510	1,125		1,635	2,250
0400	300 lb., ASME, 5 gallon		1.24	12.903		315	845		1,160	1,625
0420	10 gallon	↓	.93	17.204	↓	470	1,125		1,595	2,225

23 54 Furnaces

23 54 13 – Electric-Resistance Furnaces

23 54 13.10 Electric Furnaces

		Crew	Daily Output	Labor-Hours	Unit	Material	2020 Bare Costs Labor	Equipment	Total	Total Incl O&P
0010	**ELECTRIC FURNACES**, Hot air, blowers, std. controls									
0011	not including gas, oil or flue piping									
1000	Electric, UL listed									
1070	10.2 MBH	Q-20	4.40	4.545	Ea.	350	260		610	775
1080	17.1 MBH		4.60	4.348		430	248		678	845
1100	34.1 MBH		4.40	4.545		535	260		795	980
1120	51.6 MBH		4.20	4.762		590	272		862	1,050
1140	68.3 MBH		4	5		610	286		896	1,100
1160	85.3 MBH	▼	3.80	5.263	▼	650	300		950	1,175

23 54 16 – Fuel-Fired Furnaces

23 54 16.13 Gas-Fired Furnaces

		Crew	Daily Output	Labor-Hours	Unit	Material	2020 Bare Costs Labor	Equipment	Total	Total Incl O&P
0010	**GAS-FIRED FURNACES**									
3000	Gas, AGA certified, upflow, direct drive models									
3020	45 MBH input	Q-9	4	4	Ea.	700	224		924	1,100
3040	60 MBH input		3.80	4.211		730	236		966	1,175
3060	75 MBH input		3.60	4.444		835	249		1,084	1,300
3100	100 MBH input		3.20	5		825	280		1,105	1,325
3120	125 MBH input		3	5.333		855	299		1,154	1,400
3130	150 MBH input		2.80	5.714		1,750	320		2,070	2,400
3140	200 MBH input		2.60	6.154		3,525	345		3,870	4,400
3160	300 MBH input		2.30	6.957		4,125	390		4,515	5,125
3180	400 MBH input	▼	2	8	▼	5,500	450		5,950	6,725
3200	Gas fired wall furnace									
3204	Horizontal flow									
3208	7.7 MBH	Q-9	7	2.286	Ea.	840	128		968	1,125
3212	14 MBH		6.50	2.462		850	138		988	1,150
3216	24 MBH		5	3.200		825	179		1,004	1,175
3220	49 MBH		4	4		1,075	224		1,299	1,550
3224	65 MBH	▼	3.60	4.444	▼	1,175	249		1,424	1,675
3240	Up-flow									
3244	7.7 MBH	Q-9	7	2.286	Ea.	765	128		893	1,025
3248	14 MBH		6.50	2.462		765	138		903	1,050
3252	24 MBH		5	3.200		815	179		994	1,175
3254	35 MBH		4.50	3.556		855	199		1,054	1,250
3256	49 MBH		4	4		1,650	224		1,874	2,175
3260	65 MBH	▼	3.60	4.444	▼	1,475	249		1,724	2,000
3500	Gas furnace									
3510	Up-flow									
3514	1,200 CFM, 45 MBH	Q-9	4	4	Ea.	1,175	224		1,399	1,625
3518	75 MBH		3.60	4.444		1,250	249		1,499	1,750
3522	100 MBH		3.20	5		1,400	280		1,680	1,975
3526	125 MBH		3	5.333		1,425	299		1,724	2,000
3530	150 MBH		2.80	5.714		1,600	320		1,920	2,225
3540	2,000 CFM, 75 MBH		3.60	4.444		1,375	249		1,624	1,900
3544	100 MBH		3.20	5		1,550	280		1,830	2,150
3548	120 MBH	▼	3	5.333	▼	1,650	299		1,949	2,250
3560	Horizontal flow									
3564	1,200 CFM, 45 MBH	Q-9	4	4	Ea.	1,325	224		1,549	1,800
3568	60 MBH		3.80	4.211		1,400	236		1,636	1,875
3572	70 MBH		3.60	4.444		1,425	249		1,674	1,950
3576	2,000 CFM, 90 MBH		3.30	4.848		1,575	272		1,847	2,125
3580	110 MBH	▼	3.10	5.161	▼	1,825	289		2,114	2,475

For customer support on your Mechanical Costs with RSMeans Data, call 800.448.8182.

367

23 54 16.14 Condensing Furnaces

		Crew	Daily Output	Labor-Hours	Unit	Material	2020 Bare Costs Labor	Equipment	Total	Total Incl O&P
0010	**CONDENSING FURNACES**, High efficiency									
0020	Oil fired, packaged, complete									
0030	Upflow									
0040	Output @ 95% A.F.U.E.									
0100	49 MBH @ 1,000 CFM	Q-9	3.70	4.324	Ea.	7,375	243		7,618	8,475
0110	73.5 MBH @ 2,000 CFM		3.60	4.444		7,825	249		8,074	9,000
0120	96 MBH @ 2,000 CFM		3.40	4.706		7,825	264		8,089	9,025
0130	115.6 MBH @ 2,000 CFM		3.40	4.706		7,825	264		8,089	9,025
0140	147 MBH @ 2,000 CFM		3.30	4.848		18,400	272		18,672	20,700
0150	192 MBH @ 4,000 CFM		2.60	6.154		19,000	345		19,345	21,400
0170	231.5 MBH @ 4,000 CFM		2.30	6.957		19,000	390		19,390	21,500
0260	For variable speed motor, add					785			785	865
0270	Note: Also available in horizontal, counterflow and lowboy configurations.									

23 54 16.16 Oil-Fired Furnaces

		Crew	Daily Output	Labor-Hours	Unit	Material	2020 Bare Costs Labor	Equipment	Total	Total Incl O&P
0010	**OIL-FIRED FURNACES**									
6000	Oil, UL listed, atomizing gun type burner									
6020	56 MBH output	Q-9	3.60	4.444	Ea.	3,275	249		3,524	3,975
6030	84 MBH output		3.50	4.571		3,400	256		3,656	4,150
6040	95 MBH output		3.40	4.706		3,400	264		3,664	4,125
6060	134 MBH output		3.20	5		3,325	280		3,605	4,100
6080	151 MBH output		3	5.333		3,700	299		3,999	4,525
6100	200 MBH input		2.60	6.154		4,175	345		4,520	5,125
6120	300 MBH input		2.30	6.957		4,375	390		4,765	5,400
6140	400 MBH input		2	8		5,000	450		5,450	6,175
6200	Up flow									
6210	105 MBH	Q-9	3.40	4.706	Ea.	2,125	264		2,389	2,725
6220	140 MBH		3.30	4.848		2,775	272		3,047	3,450
6230	168 MBH		2.90	5.517		5,100	310		5,410	6,100
6260	Lo-boy style									
6270	105 MBH	Q-9	3.40	4.706	Ea.	3,325	264		3,589	4,050
6280	140 MBH		3.30	4.848		3,600	272		3,872	4,350
6290	189 MBH		2.80	5.714		5,350	320		5,670	6,350
6320	Down flow									
6330	105 MBH	Q-9	3.40	4.706	Ea.	3,325	264		3,589	4,075
6340	140 MBH		3.30	4.848		3,600	272		3,872	4,350
6350	168 MBH		2.90	5.517		5,975	310		6,285	7,050
6380	Horizontal									
6390	105 MBH	Q-9	3.40	4.706	Ea.	2,200	264		2,464	2,825
6400	140 MBH		3.30	4.848		3,600	272		3,872	4,350
6410	189 MBH		2.80	5.714		5,250	320		5,570	6,250

23 54 16.21 Solid Fuel-Fired Furnaces

			Crew	Daily Output	Labor-Hours	Unit	Material	2020 Bare Costs Labor	Equipment	Total	Total Incl O&P
0010	**SOLID FUEL-FIRED FURNACES**										
6020	Wood fired furnaces										
6030	Includes hot water coil, thermostat, and auto draft control										
6040	24" long firebox	G	Q-9	4	4	Ea.	4,500	224		4,724	5,300
6050	30" long firebox	G		3.60	4.444		5,225	249		5,474	6,125
6060	With fireplace glass doors	G		3.20	5		6,750	280		7,030	7,850
6200	Wood/oil fired furnaces, includes two thermostats										
6210	Includes hot water coil and auto draft control										
6240	24" long firebox	G	Q-9	3.40	4.706	Ea.	5,525	264		5,789	6,475
6250	30" long firebox	G		3	5.333		6,975	299		7,274	8,125
6260	With fireplace glass doors	G		2.80	5.714		7,800	320		8,120	9,050

For customer support on your Mechanical Costs with RSMeans Data, call 800.448.8182.

23 54 Furnaces

23 54 16 – Fuel-Fired Furnaces

23 54 16.21 Solid Fuel-Fired Furnaces

		Crew	Daily Output	Labor-Hours	Unit	Material	2020 Bare Costs Labor	2020 Bare Costs Equipment	Total	Total Incl O&P
6400	Wood/gas fired furnaces, includes two thermostats									
6410	Includes hot water coil and auto draft control									
6440	24" long firebox [G]	Q-9	2.80	5.714	Ea.	6,100	320		6,420	7,200
6450	30" long firebox [G]		2.40	6.667		6,975	375		7,350	8,250
6460	With fireplace glass doors [G]		2	8		8,725	450		9,175	10,300
6600	Wood/oil/gas fired furnaces, optional accessories									
6610	Hot air plenum	Q-9	16	1	Ea.	120	56		176	217
6620	Safety heat dump		24	.667		104	37.50		141.50	171
6630	Auto air intake		18	.889		161	50		211	254
6640	Cold air return package		14	1.143		175	64		239	289
6650	Wood fork					56.50			56.50	62
6700	Wood fired outdoor furnace									
6740	24" long firebox [G]	Q-9	3.80	4.211	Ea.	4,200	236		4,436	4,975
6760	Wood fired outdoor furnace, optional accessories									
6770	Chimney section, stainless steel, 6" ID x 3' long [G]	Q-9	36	.444	Ea.	210	25		235	269
6780	Chimney cap, stainless steel [G]	"	40	.400	"	127	22.50		149.50	174
6800	Wood fired hot water furnace									
6820	Includes 200 gal. hot water storage, thermostat, and auto draft control									
6840	30" long firebox [G]	Q-9	2.10	7.619	Ea.	9,325	425		9,750	11,000
6850	Water to air heat exchanger									
6870	Includes mounting kit and blower relay									
6880	140 MBH, 18.75" W x 18.75" L	Q-9	7.50	2.133	Ea.	385	120		505	605
6890	200 MBH, 24" W x 24" L	"	7	2.286	"	550	128		678	800
6900	Water to water heat exchanger									
6940	100 MBH	Q-9	6.50	2.462	Ea.	390	138		528	640
6960	290 MBH	"	6	2.667	"	560	150		710	840
7000	Optional accessories									
7010	Large volume circulation pump (2 included)	Q-9	14	1.143	Ea.	253	64		317	375
7020	Air bleed fittings (package)		24	.667		54	37.50		91.50	116
7030	Domestic water preheater		6	2.667		229	150		379	480
7040	Smoke pipe kit		4	4		96	224		320	445

23 54 24 – Furnace Components for Cooling

23 54 24.10 Furnace Components and Combinations

		Crew	Daily Output	Labor-Hours	Unit	Material	2020 Bare Costs Labor	2020 Bare Costs Equipment	Total	Total Incl O&P
0010	**FURNACE COMPONENTS AND COMBINATIONS**									
0080	Coils, A.C. evaporator, for gas or oil furnaces									
0090	Add-on, with holding charge									
0100	Upflow									
0120	1-1/2 ton cooling	Q-5	4	4	Ea.	244	236		480	625
0130	2 ton cooling		3.70	4.324		305	255		560	715
0140	3 ton cooling		3.30	4.848		430	286		716	900
0150	4 ton cooling		3	5.333		570	315		885	1,100
0160	5 ton cooling		2.70	5.926		615	350		965	1,200
0300	Downflow									
0330	2-1/2 ton cooling	Q-5	3	5.333	Ea.	365	315		680	870
0340	3-1/2 ton cooling		2.60	6.154		495	365		860	1,100
0350	5 ton cooling		2.20	7.273		510	430		940	1,200
0600	Horizontal									
0630	2 ton cooling	Q-5	3.90	4.103	Ea.	440	242		682	845
0640	3 ton cooling		3.50	4.571		450	270		720	900
0650	4 ton cooling		3.20	5		540	295		835	1,025
0660	5 ton cooling		2.90	5.517		540	325		865	1,075
2000	Cased evaporator coils for air handlers									

For customer support on your Mechanical Costs with RSMeans Data, call 800.448.8182.

369

23 54 Furnaces

23 54 24 – Furnace Components for Cooling

23 54 24.10 Furnace Components and Combinations

		Crew	Daily Output	Labor-Hours	Unit	Material	2020 Bare Costs Labor	2020 Bare Costs Equipment	Total	Total Incl O&P
2100	1-1/2 ton cooling	Q-5	4.40	3.636	Ea.	310	215		525	665
2110	2 ton cooling		4.10	3.902		375	230		605	755
2120	2-1/2 ton cooling		3.90	4.103		400	242		642	800
2130	3 ton cooling		3.70	4.324		470	255		725	895
2140	3-1/2 ton cooling		3.50	4.571		600	270		870	1,075
2150	4 ton cooling		3.20	5		630	295		925	1,125
2160	5 ton cooling	▼	2.90	5.517	▼	625	325		950	1,175
3010	Air handler, modular									
3100	With cased evaporator cooling coil									
3120	1-1/2 ton cooling	Q-5	3.80	4.211	Ea.	1,050	248		1,298	1,525
3130	2 ton cooling		3.50	4.571		1,100	270		1,370	1,625
3140	2-1/2 ton cooling		3.30	4.848		1,175	286		1,461	1,725
3150	3 ton cooling		3.10	5.161		1,450	305		1,755	2,050
3160	3-1/2 ton cooling		2.90	5.517		1,700	325		2,025	2,350
3170	4 ton cooling		2.50	6.400		1,650	380		2,030	2,400
3180	5 ton cooling	▼	2.10	7.619	▼	2,125	450		2,575	3,000
3500	With no cooling coil									
3520	1-1/2 ton coil size	Q-5	12	1.333	Ea.	1,025	78.50		1,103.50	1,250
3530	2 ton coil size		10	1.600		1,100	94.50		1,194.50	1,350
3540	2-1/2 ton coil size		10	1.600		1,075	94.50		1,169.50	1,325
3554	3 ton coil size		9	1.778		1,075	105		1,180	1,325
3560	3-1/2 ton coil size		9	1.778		1,150	105		1,255	1,400
3570	4 ton coil size		8.50	1.882		1,275	111		1,386	1,575
3580	5 ton coil size	▼	8	2	▼	1,575	118		1,693	1,925
4000	With heater									
4120	5 kW, 17.1 MBH	Q-5	16	1	Ea.	485	59		544	625
4130	7.5 kW, 25.6 MBH		15.60	1.026		830	60.50		890.50	1,000
4140	10 kW, 34.2 MBH		15.20	1.053		1,225	62		1,287	1,450
4150	12.5 kW, 42.7 MBH		14.80	1.081		1,375	64		1,439	1,625
4160	15 kW, 51.2 MBH		14.40	1.111		1,825	65.50		1,890.50	2,100
4170	25 kW, 85.4 MBH		14	1.143		2,500	67.50		2,567.50	2,850
4180	30 kW, 102 MBH	▼	13	1.231		2,900	72.50		2,972.50	3,275

23 55 Fuel-Fired Heaters

23 55 13 – Fuel-Fired Duct Heaters

23 55 13.16 Gas-Fired Duct Heaters

		Crew	Daily Output	Labor-Hours	Unit	Material	2020 Bare Costs Labor	2020 Bare Costs Equipment	Total	Total Incl O&P
0010	**GAS-FIRED DUCT HEATERS**, Includes burner, controls, stainless steel									
0020	heat exchanger. Gas fired, electric ignition									
0030	Indoor installation									
0080	100 MBH output	Q-5	5	3.200	Ea.	2,975	189		3,164	3,550
0100	120 MBH output		4	4		3,275	236		3,511	3,950
0130	200 MBH output		2.70	5.926		4,225	350		4,575	5,150
0140	240 MBH output		2.30	6.957		5,250	410		5,660	6,400
0160	280 MBH output		2	8		4,775	470		5,245	5,950
0180	320 MBH output	▼	1.60	10		5,225	590		5,815	6,625
0300	For powered venter and adapter, add				▼	535			535	585
0502	For required flue pipe, see Section 23 51 23.10									
1000	Outdoor installation, with power venter									
1020	75 MBH output	Q-5	4	4	Ea.	3,550	236		3,786	4,250
1040	94 MBH output		4	4		3,775	236		4,011	4,500
1060	120 MBH output	▼	4	4	▼	4,075	236		4,311	4,850

For customer support on your Mechanical Costs with RSMeans Data, call 800.448.8182.

23 55 Fuel-Fired Heaters

23 55 13 – Fuel-Fired Duct Heaters

23 55 13.16 Gas-Fired Duct Heaters

		Crew	Daily Output	Labor-Hours	Unit	Material	2020 Bare Costs Labor	2020 Bare Costs Equipment	Total	Total Incl O&P
1080	157 MBH output	Q-5	3.50	4.571	Ea.	4,275	270		4,545	5,100
1100	187 MBH output		3	5.333		5,500	315		5,815	6,525
1120	225 MBH output		2.50	6.400		5,725	380		6,105	6,875
1140	300 MBH output		1.80	8.889		6,375	525		6,900	7,775
1160	375 MBH output		1.60	10		6,800	590		7,390	8,350
1180	450 MBH output		1.40	11.429		8,300	675		8,975	10,100
1200	600 MBH output	↓	1	16		10,500	945		11,445	12,900
1300	Aluminized exchanger, subtract					15%				
1500	For two stage gas valve, add				↓	870			870	955

23 55 23 – Gas-Fired Radiant Heaters

23 55 23.10 Infrared Type Heating Units

		Crew	Daily Output	Labor-Hours	Unit	Material	2020 Bare Costs Labor	2020 Bare Costs Equipment	Total	Total Incl O&P
0010	**INFRARED TYPE HEATING UNITS**									
0020	Gas fired, unvented, electric ignition, 100% shutoff.									
0030	Piping and wiring not included									
0100	Input, 30 MBH	Q-5	6	2.667	Ea.	1,050	157		1,207	1,400
0120	45 MBH		5	3.200		995	189		1,184	1,375
0140	50 MBH		4.50	3.556		1,000	210		1,210	1,425
0160	60 MBH		4	4		1,025	236		1,261	1,475
0180	75 MBH		3	5.333		1,025	315		1,340	1,600
0200	90 MBH		2.50	6.400		1,050	380		1,430	1,725
0220	105 MBH		2	8		1,100	470		1,570	1,925
0240	120 MBH	↓	2	8	↓	1,625	470		2,095	2,500
1000	Gas fired, vented, electric ignition, tubular									
1020	Piping and wiring not included, 20' to 80' lengths									
1030	Single stage, input, 60 MBH	Q-6	4.50	5.333	Ea.	1,550	325		1,875	2,200
1040	80 MBH		3.90	6.154		1,550	375		1,925	2,275
1050	100 MBH		3.40	7.059		1,550	430		1,980	2,350
1060	125 MBH		2.90	8.276		1,525	505		2,030	2,425
1070	150 MBH		2.70	8.889		1,525	545		2,070	2,500
1080	170 MBH		2.50	9.600		1,525	585		2,110	2,550
1090	200 MBH	↓	2.20	10.909	↓	1,725	665		2,390	2,900
1100	Note: Final pricing may vary due to									
1110	tube length and configuration package selected									
1130	Two stage, input, 60 MBH high, 45 MBH low	Q-6	4.50	5.333	Ea.	1,900	325		2,225	2,575
1140	80 MBH high, 60 MBH low		3.90	6.154		1,900	375		2,275	2,650
1150	100 MBH high, 65 MBH low		3.40	7.059		1,900	430		2,330	2,725
1160	125 MBH high, 95 MBH low		2.90	8.276		1,925	505		2,430	2,850
1170	150 MBH high, 100 MBH low		2.70	8.889		1,925	545		2,470	2,925
1180	170 MBH high, 125 MBH low		2.50	9.600		2,125	585		2,710	3,225
1190	200 MBH high, 150 MBH low	↓	2.20	10.909	↓	2,125	665		2,790	3,350
1220	Note: Final pricing may vary due to									
1230	tube length and configuration package selected									

23 55 33 – Fuel-Fired Unit Heaters

23 55 33.13 Oil-Fired Unit Heaters

		Crew	Daily Output	Labor-Hours	Unit	Material	2020 Bare Costs Labor	2020 Bare Costs Equipment	Total	Total Incl O&P
0010	**OIL-FIRED UNIT HEATERS**, Cabinet, grilles, fan, ctrl., burner, no piping									
6000	Oil fired, suspension mounted, 94 MBH output	Q-5	4	4	Ea.	4,875	236		5,111	5,700
6040	140 MBH output		3	5.333		5,200	315		5,515	6,200
6060	184 MBH output	↓	3	5.333	↓	5,600	315		5,915	6,625

For customer support on your Mechanical Costs with RSMeans Data, call 800.448.8182.

371

23 55 33.16 Gas-Fired Unit Heaters	Crew	Daily Output	Labor-Hours	Unit	Material	2020 Bare Costs Labor	Equipment	Total	Total Incl O&P
0010 **GAS-FIRED UNIT HEATERS**, Cabinet, grilles, fan, ctrls., burner, no piping									
0022 thermostat, no piping. For flue see Section 23 51 23.10									
1000 Gas fired, floor mounted									
1100 60 MBH output	Q-5	10	1.600	Ea.	765	94.50		859.50	985
1120 80 MBH output		9	1.778		775	105		880	1,000
1140 100 MBH output		8	2		845	118		963	1,100
1160 120 MBH output		7	2.286		970	135		1,105	1,275
1180 180 MBH output		6	2.667		1,200	157		1,357	1,550
1500 Rooftop mounted, power vent, stainless steel exchanger									
1520 75 MBH input	Q-6	4	6	Ea.	5,925	365		6,290	7,050
1540 100 MBH input		3.60	6.667		6,175	410		6,585	7,400
1560 125 MBH input		3.30	7.273		6,450	445		6,895	7,775
1580 150 MBH input		3	8		6,950	490		7,440	8,350
1600 175 MBH input		2.60	9.231		7,125	565		7,690	8,675
1620 225 MBH input		2.30	10.435		8,100	640		8,740	9,875
1640 300 MBH input		1.90	12.632		9,350	775		10,125	11,500
1660 350 MBH input		1.40	17.143		10,000	1,050		11,050	12,600
1680 450 MBH input		1.20	20		10,700	1,225		11,925	13,600
1720 700 MBH input		.80	30		13,700	1,825		15,525	17,900
1760 1,200 MBH input		.30	80		13,800	4,900		18,700	22,500
1900 For aluminized steel exchanger, subtract					10%				
2000 Suspension mounted, propeller fan, 20 MBH output	Q-5	8.50	1.882		1,725	111		1,836	2,075
2020 40 MBH output		7.50	2.133		1,650	126		1,776	2,025
2040 60 MBH output		7	2.286		1,850	135		1,985	2,225
2060 80 MBH output		6	2.667		2,000	157		2,157	2,425
2080 100 MBH output		5.50	2.909		2,150	172		2,322	2,600
2100 130 MBH output		5	3.200		2,375	189		2,564	2,875
2120 140 MBH output		4.50	3.556		2,575	210		2,785	3,175
2140 160 MBH output		4	4		2,600	236		2,836	3,200
2160 180 MBH output		3.50	4.571		3,150	270		3,420	3,875
2180 200 MBH output		3	5.333		3,100	315		3,415	3,875
2200 240 MBH output		2.70	5.926		3,400	350		3,750	4,275
2220 280 MBH output		2.30	6.957		3,775	410		4,185	4,775
2240 320 MBH output		2	8		4,400	470		4,870	5,550
2500 For powered venter and adapter, add					400			400	440
3000 Suspension mounted, blower type, 40 MBH output	Q-5	6.80	2.353		810	139		949	1,100
3020 60 MBH output		6.60	2.424		935	143		1,078	1,250
3040 84 MBH output		5.80	2.759		875	163		1,038	1,200
3060 104 MBH output		5.20	3.077		945	182		1,127	1,325
3080 140 MBH output		4.30	3.721		1,075	220		1,295	1,500
3100 180 MBH output		3.30	4.848		1,500	286		1,786	2,075
3120 240 MBH output		2.50	6.400		1,675	380		2,055	2,400
3140 280 MBH output		2	8		1,475	470		1,945	2,325
4000 Suspension mounted, sealed combustion system,									
4020 Aluminized steel exchanger, powered vent									
4040 100 MBH output	Q-5	5	3.200	Ea.	3,125	189		3,314	3,700
4060 120 MBH output		4.70	3.404		3,725	201		3,926	4,400
4080 160 MBH output		3.70	4.324		4,350	255		4,605	5,150
4100 200 MBH output		2.90	5.517		5,250	325		5,575	6,250
4120 240 MBH output		2.50	6.400		5,700	380		6,080	6,850
4140 320 MBH output		1.70	9.412		6,825	555		7,380	8,350
5000 Wall furnace, 17.5 MBH output		6	2.667		710	157		867	1,025

23 55 Fuel-Fired Heaters

23 55 33 – Fuel-Fired Unit Heaters

23 55 33.16 Gas-Fired Unit Heaters		Crew	Daily Output	Labor-Hours	Unit	Material	2020 Bare Costs Labor	Equipment	Total	Total Incl O&P
5020	24 MBH output	Q-5	5	3.200	Ea.	805	189		994	1,175
5040	35 MBH output	↓	4	4	↓	1,150	236		1,386	1,600

23 56 Solar Energy Heating Equipment

23 56 16 – Packaged Solar Heating Equipment

23 56 16.40 Solar Heating Systems

			Crew	Daily Output	Labor-Hours	Unit	Material	2020 Bare Costs Labor	Equipment	Total	Total Incl O&P
0010	**SOLAR HEATING SYSTEMS**	R235616-60									
0020	System/package prices, not including connecting										
0030	pipe, insulation, or special heating/plumbing fixtures										
0152	For solar ultraviolet pipe insulation see Section 22 07 19.10										
0500	Hot water, standard package, low temperature										
0540	1 collector, circulator, fittings, 65 gal. tank	G	Q-1	.50	32	Ea.	4,400	1,850		6,250	7,600
0580	2 collectors, circulator, fittings, 120 gal. tank	G		.40	40		5,750	2,325		8,075	9,800
0620	3 collectors, circulator, fittings, 120 gal. tank	G	↓	.34	47.059	↓	7,850	2,725		10,575	12,700
0700	Medium temperature package										
0720	1 collector, circulator, fittings, 80 gal. tank	G	Q-1	.50	32	Ea.	5,450	1,850		7,300	8,750
0740	2 collectors, circulator, fittings, 120 gal. tank	G		.40	40		7,400	2,325		9,725	11,600
0780	3 collectors, circulator, fittings, 120 gal. tank	G	↓	.30	53.333		8,250	3,100		11,350	13,700
0980	For each additional 120 gal. tank, add	G				↓	1,775			1,775	1,975

23 56 19 – Solar Heating Components

23 56 19.50 Solar Heating Ancillary

			Crew	Daily Output	Labor-Hours	Unit	Material	2020 Bare Costs Labor	Equipment	Total	Total Incl O&P
0010	**SOLAR HEATING ANCILLARY**										
2300	Circulators, air										
2310	Blowers										
2330	100-300 S.F. system, 1/10 HP	G	Q-9	16	1	Ea.	266	56		322	380
2340	300-500 S.F. system, 1/5 HP	G		15	1.067		355	60		415	480
2350	Two speed, 100-300 S.F., 1/10 HP	G		14	1.143		147	64		211	259
2400	Reversible fan, 20" diameter, 2 speed	G		18	.889		116	50		166	204
2550	Booster fan 6" diameter, 120 CFM	G		16	1		38	56		94	127
2570	6" diameter, 225 CFM	G		16	1		46.50	56		102.50	136
2580	8" diameter, 150 CFM	G		16	1		42.50	56		98.50	132
2590	8" diameter, 310 CFM	G		14	1.143		65.50	64		129.50	169
2600	8" diameter, 425 CFM	G		14	1.143		73.50	64		137.50	178
2650	Rheostat	G		32	.500		15.95	28		43.95	60
2660	Shutter/damper	G		12	1.333		58.50	75		133.50	178
2670	Shutter motor	G	↓	16	1		146	56		202	246
2800	Circulators, liquid, 1/25 HP, 5.3 GPM	G	Q-1	14	1.143		205	66.50		271.50	325
2820	1/20 HP, 17 GPM	G		12	1.333		284	77.50		361.50	425
2850	1/20 HP, 17 GPM, stainless steel	G		12	1.333		262	77.50		339.50	405
2870	1/12 HP, 30 GPM	G	↓	10	1.600	↓	360	93		453	535
3000	Collector panels, air with aluminum absorber plate										
3010	Wall or roof mount										
3040	Flat black, plastic glazing										
3080	4' x 8'	G	Q-9	6	2.667	Ea.	685	150		835	980
3100	4' x 10'	G		5	3.200	"	840	179		1,019	1,200
3200	Flush roof mount, 10' to 16' x 22" wide	G		96	.167	L.F.	165	9.35		174.35	195
3210	Manifold, by L.F. width of collectors	G	↓	160	.100	"	161	5.60		166.60	187
3300	Collector panels, liquid with copper absorber plate										
3320	Black chrome, tempered glass glazing										
3330	Alum. frame, 4' x 8', 5/32" single glazing	G	Q-1	9.50	1.684	Ea.	1,075	97.50		1,172.50	1,325

For customer support on your Mechanical Costs with RSMeans Data, call 800.448.8182.

373

23 56 19.50 Solar Heating Ancillary		Crew	Daily Output	Labor-Hours	Unit	Material	2020 Bare Costs Labor	2020 Bare Costs Equipment	Total	Total Incl O&P
3390	Alum. frame, 4' x 10', 5/32" single glazing	G Q-1	6	2.667	Ea.	1,225	155		1,380	1,575
3450	Flat black, alum. frame, 3.5' x 7.5'	G	9	1.778		815	103		918	1,050
3500	4' x 8'	G	5.50	2.909		1,025	169		1,194	1,375
3520	4' x 10'	G	10	1.600		1,225	93		1,318	1,500
3540	4' x 12.5'	G	5	3.200		1,300	186		1,486	1,700
3550	Liquid with fin tube absorber plate									
3560	Alum. frame 4' x 8' tempered glass	G Q-1	10	1.600	Ea.	605	93		698	805
3580	Liquid with vacuum tubes, 4' x 6'-10"	G	9	1.778		915	103		1,018	1,150
3600	Liquid, full wetted, plastic, alum. frame, 4' x 10'	G	5	3.200		335	186		521	650
3650	Collector panel mounting, flat roof or ground rack	G	7	2.286		253	133		386	475
3670	Roof clamps	G	70	.229	Set	3.27	13.25		16.52	23.50
3700	Roof strap, teflon	G 1 Plum	205	.039	L.F.	25	2.51		27.51	31.50
3900	Differential controller with two sensors									
3930	Thermostat, hard wired	G 1 Plum	8	1	Ea.	112	64.50		176.50	220
3950	Line cord and receptacle	G	12	.667		154	43		197	234
4050	Pool valve system	G	2.50	3.200		182	206		388	510
4070	With 12 VAC actuator	G	2	4		335	258		593	750
4080	Pool pump system, 2" pipe size	G	6	1.333		205	86		291	355
4100	Five station with digital read-out	G	3	2.667		271	172		443	555
4150	Sensors									
4200	Brass plug, 1/2" MPT	G 1 Plum	32	.250	Ea.	16.70	16.10		32.80	42.50
4210	Brass plug, reversed	G	32	.250		26.50	16.10		42.60	53
4220	Freeze prevention	G	32	.250		25	16.10		41.10	51.50
4240	Screw attached	G	32	.250		10.05	16.10		26.15	35
4250	Brass, immersion	G	32	.250		15.50	16.10		31.60	41
4300	Heat exchanger									
4315	includes coil, blower, circulator									
4316	and controller for DHW and space hot air									
4330	Fluid to air coil, up flow, 45 MBH	G Q-1	4	4	Ea.	330	232		562	710
4380	70 MBH	G	3.50	4.571		375	265		640	805
4400	80 MBH	G	3	5.333		495	310		805	1,000
4580	Fluid to fluid package includes two circulating pumps									
4590	expansion tank, check valve, relief valve									
4600	controller, high temperature cutoff and sensors	G Q-1	2.50	6.400	Ea.	810	370		1,180	1,450
4650	Heat transfer fluid									
4700	Propylene glycol, inhibited anti-freeze	G 1 Plum	28	.286	Gal.	23	18.40		41.40	53
4800	Solar storage tanks, knocked down									
4810	Air, galvanized steel clad, double wall, 4" fiberglass insulation									
5120	45 mil reinforced polypropylene lining,									
5140	4' high, 4' x 4' = 64 C.F./450 gallons	G Q-9	2	8	Ea.	3,700	450		4,150	4,750
5150	4' x 8' = 128 C.F./900 gallons	G	1.50	10.667		5,875	600		6,475	7,375
5160	4' x 12' = 190 C.F./1,300 gallons	G	1.30	12.308		7,425	690		8,115	9,225
5170	8' x 8' = 250 C.F./1,700 gallons	G	1	16		7,700	895		8,595	9,825
5190	6'-3" high, 7' x 7' = 306 C.F./2,000 gallons	G Q-10	1.20	20		13,900	1,175		15,075	17,100
5200	7' x 10'-6" = 459 C.F./3,000 gallons	G	.80	30		17,300	1,750		19,050	21,700
5210	7' x 14' = 613 C.F./4,000 gallons	G	.60	40		22,400	2,325		24,725	28,100
5220	10'-6" x 10'-6" = 689 C.F./4,500 gallons	G	.50	48		20,800	2,800		23,600	27,000
5230	10'-6" x 14' = 919 C.F./6,000 gallons	G	.40	60		24,200	3,500		27,700	32,000
5240	14' x 14' = 1,225 C.F./8,000 gallons	G Q-11	.40	80		29,000	4,750		33,750	39,100
5250	14' x 17'-6" = 1,531 C.F./10,000 gallons	G	.30	107		31,400	6,325		37,725	44,200
5260	17'-6" x 17'-6" = 1,914 C.F./12,500 gallons	G	.25	128		35,500	7,600		43,100	50,500
5270	17'-6" x 21' = 2,297 C.F./15,000 gallons	G	.20	160		40,300	9,500		49,800	59,000
5280	21' x 21' = 2,756 C.F./18,000 gallons	G	.18	178		43,100	10,500		53,600	63,500

23 56 Solar Energy Heating Equipment

23 56 19 – Solar Heating Components

23 56 19.50 Solar Heating Ancillary		Crew	Daily Output	Labor-Hours	Unit	Material	2020 Bare Costs Labor	Equipment	Total	Total Incl O&P
5290	30 mil reinforced Hypalon lining, add					.02%				
7000	Solar control valves and vents									
7050	Air purger, 1" pipe size	G	1 Plum	12	.667	Ea.	46.50	43		89.50	116
7070	Air eliminator, automatic 3/4" size	G		32	.250		31	16.10		47.10	58
7090	Air vent, automatic, 1/8" fitting	G		32	.250		17.90	16.10		34	43.50
7100	Manual, 1/8" NPT	G		32	.250		2.69	16.10		18.79	27
7120	Backflow preventer, 1/2" pipe size	G		16	.500		90.50	32		122.50	148
7130	3/4" pipe size	G		16	.500		89	32		121	146
7150	Balancing valve, 3/4" pipe size	G		20	.400		59.50	26		85.50	104
7180	Draindown valve, 1/2" copper tube	G		9	.889		219	57.50		276.50	325
7200	Flow control valve, 1/2" pipe size	G		22	.364		140	23.50		163.50	189
7220	Expansion tank, up to 5 gal.	G		32	.250		70.50	16.10		86.60	102
7250	Hydronic controller (aquastat)	G		8	1		217	64.50		281.50	335
7400	Pressure gauge, 2" dial	G		32	.250		25	16.10		41.10	51.50
7450	Relief valve, temp. and pressure 3/4" pipe size	G	▼	30	.267	▼	25.50	17.20		42.70	53.50
7500	Solenoid valve, normally closed										
7520	Brass, 3/4" NPT, 24V	G	1 Plum	9	.889	Ea.	137	57.50		194.50	236
7530	1" NPT, 24V	G		9	.889		224	57.50		281.50	335
7750	Vacuum relief valve, 3/4" pipe size	G	▼	32	.250	▼	31	16.10		47.10	58
7800	Thermometers										
7820	Digital temperature monitoring, 4 locations	G	1 Plum	2.50	3.200	Ea.	138	206		344	460
7900	Upright, 1/2" NPT	G		8	1		39.50	64.50		104	140
7970	Remote probe, 2" dial	G		8	1		34.50	64.50		99	135
7990	Stem, 2" dial, 9" stem	G	▼	16	.500	▼	22.50	32		54.50	73
8250	Water storage tank with heat exchanger and electric element										
8270	66 gal. with 2" x 2 lb. density insulation	G	1 Plum	1.60	5	Ea.	2,150	320		2,470	2,850
8300	80 gal. with 2" x 2 lb. density insulation	G		1.60	5		1,950	320		2,270	2,625
8380	120 gal. with 2" x 2 lb. density insulation	G		1.40	5.714		1,900	370		2,270	2,650
8400	120 gal. with 2" x 2 lb. density insul., 40 S.F. heat coil	G	▼	1.40	5.714	▼	2,825	370		3,195	3,675
8500	Water storage module, plastic										
8600	Tubular, 12" diameter, 4' high	G	1 Carp	48	.167	Ea.	99	8.85		107.85	122
8610	12" diameter, 8' high	G		40	.200		157	10.65		167.65	188
8620	18" diameter, 5' high	G		38	.211		176	11.20		187.20	210
8630	18" diameter, 10' high	G	▼	32	.250		263	13.30		276.30	310
8640	58" diameter, 5' high	G	2 Carp	32	.500		815	26.50		841.50	935
8650	Cap, 12" diameter	G					20			20	22
8660	18" diameter	G				▼	25.50			25.50	28

23 57 Heat Exchangers for HVAC

23 57 16 – Steam-to-Water Heat Exchangers

23 57 16.10 Shell/Tube Type Steam-to-Water Heat Exch.

0010	SHELL AND TUBE TYPE STEAM-TO-WATER HEAT EXCHANGERS										
0016	Shell & tube type, 2 or 4 pass, 3/4" OD copper tubes,										
0020	C.I. heads, C.I. tube sheet, steel shell										
0100	Hot water 40°F to 180°F, by steam at 10 psi										
0120	8 GPM 402/gpm	Q-5	6	2.667	Ea.	2,750	157		2,907	3,250
0140	10 GPM 480/		5	3.200		4,125	189		4,314	4,825
0160	40 GPM 134/		4	4		6,400	236		6,636	7,375
0180	64 GPM 170/		2	8		9,350	470		9,820	11,000
0200	96 GPM 160/	▼	1	16		12,500	945		13,445	15,200
0220	120 GPM	Q-6	1.50	16		16,000	980		16,980	19,100

For customer support on your Mechanical Costs with RSMeans Data, call 800.448.8182.

375

23 57 Heat Exchangers for HVAC

23 57 16 – Steam-to-Water Heat Exchangers

23 57 16.10 Shell/Tube Type Steam-to-Water Heat Exch.	Crew	Daily Output	Labor-Hours	Unit	Material	2020 Bare Costs Labor	Equipment	Total	Total Incl O&P
0240	168 GPM *145/gpm*	Q-6	1	24	Ea.	20,100	1,475		21,575	24,400
0260	240 GPM *156/"*		.80	30		31,500	1,825		33,325	37,400
0300	600 GPM *132/"*	↓	.70	34.286	↓	69,500	2,100		71,600	79,500
0500	For bronze head and tube sheet, add					50%				

23 57 19 – Liquid-to-Liquid Heat Exchangers

23 57 19.13 Plate-Type, Liquid-to-Liquid Heat Exchangers

		Crew	Daily Output	Labor-Hours	Unit	Material	Labor	Equipment	Total	Total Incl O&P
0010	PLATE-TYPE, LIQUID-TO-LIQUID HEAT EXCHANGERS									
3000	Plate type,									
3100	400 GPM *134/gpm*	Q-6	.80	30	Ea.	45,700	1,825		47,525	53,500
3120	800 GPM *114/"*	"	.50	48		79,000	2,925		81,925	91,000
3140	1,200 GPM *115/"*	Q-7	.34	94.118		117,000	5,875		122,875	138,000
3160	1,800 GPM *162/"*	"	.24	133	↓	155,500	8,325		163,825	183,500

23 57 19.16 Shell-Type, Liquid-to-Liquid Heat Exchangers

		Crew	Daily Output	Labor-Hours	Unit	Material	Labor	Equipment	Total	Total Incl O&P
0010	SHELL-TYPE, LIQUID-TO-LIQUID HEAT EXCHANGERS									
1000	Hot water 40°F to 140°F, by water at 200°F									
1020	7 GPM *535/gpm*	Q-5	6	2.667	Ea.	3,225	157		3,382	3,750
1040	16 GPM *330/"*		5	3.200		4,550	189		4,739	5,275
1060	34 GPM *233/"*		4	4		6,875	236		7,111	7,925
1080	55 GPM *204/"*		3	5.333		9,750	315		10,065	11,200
1100	74 GPM *197/"*		1.50	10.667		12,500	630		13,130	14,600
1120	86 GPM *230/"*	↓	1.40	11.429		16,700	675		17,375	19,400
1140	112 GPM *214/"*	Q-6	2	12		20,800	735		21,535	24,000
1160	126 GPM *235/"*		1.80	13.333		25,900	815		26,715	29,700
1180	152 GPM *254/"*	↓	1	24	↓	33,100	1,475		34,575	38,600

23 61 Refrigerant Compressors

23 61 13 – Centrifugal Refrigerant Compressors

23 61 13.10 Centrifugal Compressors

		Crew	Daily Output	Labor-Hours	Unit	Material	Labor	Equipment	Total	Total Incl O&P
0010	CENTRIFUGAL COMPRESSORS									
0100	Refrigeration, centrifugal including switches & protective devices									
0210	200 ton	Q-7	.23	137	Ea.	34,000	8,575		42,575	50,000
0220	300 ton		.17	187		36,300	11,700		48,000	57,500
0230	400 ton		.14	227		40,600	14,200		54,800	66,000
0240	500 ton		.14	237		40,900	14,800		55,700	67,000
0250	600 ton		.13	248		47,600	15,500		63,100	75,500
0260	700 ton		.12	260		48,800	16,200		65,000	78,000
0270	800 ton		.12	271		61,000	16,900		77,900	92,500
0280	900 ton		.11	288		69,000	18,000		87,000	102,500
0290	1,000 ton		.11	296		78,500	18,500		97,000	113,500
0300	1,100 ton		.10	308		81,000	19,200		100,200	118,000
0310	1,200 ton		.10	317		83,000	19,800		102,800	121,000
0320	1,300 ton		.10	327		81,500	20,400		101,900	120,500
0330	1,400 ton		.09	340		86,000	21,200		107,200	126,500
0340	1,500 ton	↓	.09	352	↓	90,000	21,900		111,900	132,000

23 61 15 – Rotary Refrigerant Compressors

23 61 15.10 Rotary Compressors

		Crew	Daily Output	Labor-Hours	Unit	Material	Labor	Equipment	Total	Total Incl O&P
0010	ROTARY COMPRESSORS									
0100	Refrigeration, hermetic, switches and protective devices									
0210	1.25 ton	1 Stpi	3	2.667	Ea.	251	175		426	540
0220	1.42 ton	↓	3	2.667	↓	254	175		429	540

23 61 Refrigerant Compressors

23 61 15 – Rotary Refrigerant Compressors

23 61 15.10 Rotary Compressors

23 61 15.10 Rotary Compressors	Crew	Daily Output	Labor-Hours	Unit	Material	2020 Bare Costs Labor	2020 Bare Costs Equipment	Total	Total Incl O&P
0230	1.68 ton	1 Stpi	2.80	2.857	Ea.	276	187		463	585
0240	2.00 ton		2.60	3.077		284	202		486	615
0250	2.37 ton		2.50	3.200		320	210		530	670
0260	2.67 ton		2.40	3.333		335	219		554	695
0270	3.53 ton		2.30	3.478		370	228		598	745
0280	4.43 ton		2.20	3.636		470	238		708	870
0290	5.08 ton		2.10	3.810		520	250		770	945
0300	5.22 ton		2	4		730	262		992	1,200
0310	7.31 ton	Q-5	3	5.333		1,175	315		1,490	1,750
0320	9.95 ton		2.60	6.154		1,375	365		1,740	2,050
0330	11.6 ton		2.50	6.400		1,750	380		2,130	2,500
0340	15.25 ton		2.30	6.957		2,425	410		2,835	3,275
0350	17.7 ton		2.10	7.619		2,575	450		3,025	3,525

23 61 16 – Reciprocating Refrigerant Compressors

23 61 16.10 Reciprocating Compressors

23 61 16.10		Crew	Daily Output	Labor-Hours	Unit	Material	2020 Bare Costs Labor	2020 Bare Costs Equipment	Total	Total Incl O&P
0010	**RECIPROCATING COMPRESSORS**									
0990	Refrigeration, recip. hermetic, switches & protective devices									
1000	10 ton	Q-5	1	16	Ea.	15,200	945		16,145	18,100
1100	20 ton	Q-6	.72	33.333		23,700	2,050		25,750	29,200
1200	30 ton		.64	37.500		28,700	2,300		31,000	35,000
1300	40 ton		.44	54.545		33,000	3,325		36,325	41,300
1400	50 ton		.20	120		34,300	7,350		41,650	48,800
1500	75 ton	Q-7	.27	119		38,000	7,400		45,400	53,000
1600	130 ton	"	.21	152		43,400	9,500		52,900	62,000

23 61 19 – Scroll Refrigerant Compressors

23 61 19.10 Scroll Compressors

23 61 19.10		Crew	Daily Output	Labor-Hours	Unit	Material	2020 Bare Costs Labor	2020 Bare Costs Equipment	Total	Total Incl O&P
0010	**SCROLL COMPRESSORS**									
1800	Refrigeration, scroll type									
1810	1.5 ton	1 Stpi	2.70	2.963	Ea.	935	194		1,129	1,325
1820	2.5 ton		2.50	3.200		930	210		1,140	1,350
1830	2.75 ton		2.35	3.404		760	223		983	1,175
1840	3 ton		2.30	3.478		980	228		1,208	1,425
1850	3.5 ton		2.25	3.556		1,000	233		1,233	1,450
1860	4 ton		2.20	3.636		1,200	238		1,438	1,675
1870	5 ton		2.10	3.810		1,275	250		1,525	1,775
1880	6 ton	Q-5	3.70	4.324		1,450	255		1,705	1,975
1900	7.5 ton		3.20	5		2,625	295		2,920	3,350
1910	10 ton		3.10	5.161		2,800	305		3,105	3,525
1920	13 ton		2.90	5.517		2,850	325		3,175	3,625
1960	25 ton		2.20	7.273		5,175	430		5,605	6,350

For customer support on your Mechanical Costs with RSMeans Data, call 800.448.8182.

377

23 62 Packaged Compressor and Condenser Units

23 62 13 – Packaged Air-Cooled Refrigerant Compressor and Condenser Units

23 62 13.10 Packaged Air-Cooled Refrig. Condensing Units	Crew	Daily Output	Labor-Hours	Unit	Material	2020 Bare Costs Labor	Equipment	Total	Total Incl O&P
0010 **PACKAGED AIR-COOLED REFRIGERANT CONDENSING UNITS**									
0020 Condensing unit									
0030 Air cooled, compressor, standard controls									
0050 1.5 ton	Q-5	2.50	6.400	Ea.	1,025	380		1,405	1,700
0100 2 ton		2.10	7.619		1,050	450		1,500	1,825
0200 2.5 ton		1.70	9.412		1,225	555		1,780	2,175
0300 3 ton		1.30	12.308		1,250	725		1,975	2,450
0350 3.5 ton		1.10	14.545		1,525	860		2,385	2,950
0400 4 ton		.90	17.778		1,700	1,050		2,750	3,450
0500 5 ton		.60	26.667		2,025	1,575		3,600	4,575
0550 7.5 ton		.55	29.091		3,350	1,725		5,075	6,250
0560 8.5 ton		.53	30.189		3,475	1,775		5,250	6,500
0600 10 ton		.50	32		3,525	1,900		5,425	6,700
0620 12.5 ton		.48	33.333		4,475	1,975		6,450	7,875
0650 15 ton	↓	.40	40		7,850	2,350		10,200	12,200
0700 20 ton	Q-6	.40	60		9,350	3,675		13,025	15,800
0720 25 ton		.35	68.571		14,300	4,200		18,500	22,100
0750 30 ton		.30	80		16,600	4,900		21,500	25,500
0800 40 ton		.20	120		20,100	7,350		27,450	33,200
0840 50 ton		.18	133		21,700	8,150		29,850	36,000
0860 60 ton		.16	150		23,000	9,175		32,175	39,000
0900 70 ton		.14	171		28,700	10,500		39,200	47,300
1000 80 ton		.12	200		31,900	12,200		44,100	53,500
1010 90 ton		.09	267		34,700	16,300		51,000	62,500
1100 100 ton	↓	.09	267	↓	40,300	16,300		56,600	68,500

23 62 23 – Packaged Water-Cooled Refrigerant Compressor and Condenser Units

23 62 23.10 Pckgd. Water-Cooled Refrigerant Condns. Units

	Crew	Daily Output	Labor-Hours	Unit	Material	Labor	Equipment	Total	Total Incl O&P
0010 **PACKAGED WATER-COOLED REFRIGERANT CONDENSING UNITS**									
2000 Water cooled, compressor, heat exchanger, controls									
2100 5 ton	Q-5	.70	22.857	Ea.	16,600	1,350		17,950	20,300
2110 10 ton		.60	26.667		21,600	1,575		23,175	26,100
2200 15 ton	↓	.50	32		29,700	1,900		31,600	35,400
2300 20 ton	Q-6	.40	60		33,300	3,675		36,975	42,200
2310 30 ton		.30	80		32,600	4,900		37,500	43,200
2400 40 ton	↓	.20	120	↓	35,200	7,350		42,550	49,700

23 63 Refrigerant Condensers

23 63 13 – Air-Cooled Refrigerant Condensers

23 63 13.10 Air-Cooled Refrig. Condensers

	Crew	Daily Output	Labor-Hours	Unit	Material	Labor	Equipment	Total	Total Incl O&P
0010 **AIR-COOLED REFRIG. CONDENSERS**									
0080 Air cooled, belt drive, propeller fan									
0220 45 ton	Q-6	.70	34.286	Ea.	10,500	2,100		12,600	14,700
0240 50 ton		.69	34.985		11,200	2,150		13,350	15,600
0260 54 ton		.64	37.795		12,200	2,300		14,500	16,900
0280 59 ton		.58	41.308		13,400	2,525		15,925	18,500
0300 65 ton		.53	45.541		15,900	2,775		18,675	21,700
0320 73 ton		.47	51.173		17,500	3,125		20,625	23,900
0340 81 ton		.42	56.738		18,900	3,475		22,375	26,000
0360 86 ton		.40	60.302		19,900	3,700		23,600	27,400
0380 88 ton	↓	.39	61.697		21,400	3,775		25,175	29,200

23 63 Refrigerant Condensers

23 63 13 – Air-Cooled Refrigerant Condensers

23 63 13.10 Air-Cooled Refrig. Condensers		Crew	Daily Output	Labor-Hours	Unit	Material	2020 Bare Costs Labor	Equipment	Total	Total Incl O&P
0400	101 ton 335/T	Q-7	.45	70.640	Ea.	24,500	4,400		28,900	33,500
0500	159 ton 250-/T		.31	103		26,300	6,425		32,725	38,600
0600	228 ton		.22	148		54,500	9,250		63,750	74,000
0650	320 ton		.16	206		52,500	12,900		65,400	77,500
0660	475 ton 320/T		.11	296		113,500	18,500		132,000	152,500
1500	May be specified single or multi-circuit									
1550	Air cooled, direct drive, propeller fan									
1590	1 ton 2375/T	Q-5	3.80	4.211	Ea.	1,825	248		2,073	2,375
1600	1-1/2 ton		3.60	4.444		2,225	262		2,487	2,850
1620	2 ton		3.20	5		2,400	295		2,695	3,075
1630	3 ton		2.40	6.667		2,675	395		3,070	3,550
1640	5 ton 1260/T		2	8		5,100	470		5,570	6,300
1650	8 ton		1.80	8.889		6,150	525		6,675	7,550
1660	10 ton 920/T		1.40	11.429		7,475	675		8,150	9,225
1670	12 ton		1.30	12.308		8,450	725		9,175	10,400
1680	14 ton		1.20	13.333		9,650	785		10,435	11,800
1690	16 ton 825-/T		1.10	14.545		10,800	860		11,660	13,200
1700	21 ton		1	16		12,200	945		13,145	14,800
1720	26 ton		.84	19.002		13,700	1,125		14,825	16,800
1740	30 ton 690/T		.70	22.792		17,100	1,350		18,450	20,800
1760	41 ton	Q-6	.77	31.008		18,800	1,900		20,700	23,600
1780	52 ton 630/T		.66	36.419		26,800	2,225		29,025	32,800
1800	63 ton		.55	44.037		29,600	2,700		32,300	36,600
1820	76 ton		.45	52.980		33,600	3,250		36,850	41,900
1840	86 ton		.40	60		41,500	3,675		45,175	51,000
1860	97 ton		.35	67.989		47,300	4,150		51,450	58,000
1880	105 ton 638/T	Q-7	.44	73.563		54,500	4,600		59,100	67,000
1890	118 ton		.39	82.687		58,500	5,150		63,650	72,000
1900	126 ton		.36	88.154		68,500	5,500		74,000	83,000
1910	136 ton		.34	95.238		76,500	5,950		82,450	93,000
1920	142 ton 672-/T		.32	99.379		78,500	6,200		84,700	95,500

23 63 30 – Lubricants

23 63 30.10 Lubricant Oils

		Crew	Daily Output	Labor-Hours	Unit	Material	2020 Bare Costs Labor	Equipment	Total	Total Incl O&P
0010	**LUBRICANT OILS**									
8000	Oils									
8100	Lubricating									
8120	Oil, lubricating				Oz.	.32			.32	.35
8500	Refrigeration									
8520	Oil, refrigeration				Gal.	48			48	53
8525	Oil, refrigeration				Qt.	12			12	13.20

23 63 33 – Evaporative Refrigerant Condensers

23 63 33.10 Evaporative Condensers

		Crew	Daily Output	Labor-Hours	Unit	Material	2020 Bare Costs Labor	Equipment	Total	Total Incl O&P
0010	**EVAPORATIVE CONDENSERS**									
3400	Evaporative, copper coil, pump, fan motor									
3440	10 ton	Q-5	.54	29.630	Ea.	12,600	1,750		14,350	16,400
3460	15 ton		.50	32		12,700	1,900		14,600	16,800
3480	20 ton		.47	34.043		13,000	2,000		15,000	17,300
3500	25 ton		.45	35.556		10,200	2,100		12,300	14,400
3520	30 ton		.42	38.095		10,700	2,250		12,950	15,200
3540	40 ton	Q-6	.49	48.980		11,500	3,000		14,500	17,100
3560	50 ton		.39	61.538		13,900	3,775		17,675	20,900
3580	65 ton		.35	68.571		14,200	4,200		18,400	21,900

For customer support on your Mechanical Costs with RSMeans Data, call 800.448.8182.

379

23 63 Refrigerant Condensers

23 63 33 – Evaporative Refrigerant Condensers

23 63 33.10 Evaporative Condensers		Crew	Daily Output	Labor-Hours	Unit	Material	2020 Bare Costs Labor	Equipment	Total	Total Incl O&P
3600	80 ton	Q-6	.33	72.727	Ea.	15,700	4,450		20,150	24,000
3620	90 ton	▼	.29	82.759		17,200	5,075		22,275	26,500
3640	100 ton	Q-7	.36	88.889		19,600	5,550		25,150	29,900
3660	110 ton		.33	96.970		25,600	6,050		31,650	37,200
3680	125 ton		.30	107		27,700	6,650		34,350	40,400
3700	135 ton		.28	114		29,300	7,125		36,425	43,000
3720	150 ton		.25	128		31,900	7,975		39,875	47,100
3740	165 ton		.23	139		33,000	8,675		41,675	49,300
3760	185 ton	▼	.22	145		37,900	9,075		46,975	55,500
3860	For fan damper control, add	Q-5	2	8	▼	890	470		1,360	1,675

23 64 Packaged Water Chillers

23 64 13 – Absorption Water Chillers

23 64 13.13 Direct-Fired Absorption Water Chillers

		Crew	Daily Output	Labor-Hours	Unit	Material	2020 Bare Costs Labor	Equipment	Total	Total Incl O&P
0010	**DIRECT-FIRED ABSORPTION WATER CHILLERS**									
3000	Gas fired, air cooled									
3220	5 ton	Q-5	.60	26.667	Ea.	8,850	1,575		10,425	12,100
3270	10 ton	"	.40	40	"	23,200	2,350		25,550	29,000
4000	Water cooled, duplex									
4130	100 ton	Q-7	.13	246	Ea.	144,000	15,400		159,400	181,500
4140	200 ton		.11	283		188,000	17,700		205,700	233,500
4150	300 ton		.11	299		232,500	18,700		251,200	284,000
4160	400 ton		.10	317		302,500	19,800		322,300	362,000
4170	500 ton		.10	330		381,000	20,600		401,600	450,000
4180	600 ton		.09	340		450,000	21,200		471,200	527,000
4190	700 ton		.09	360		499,000	22,400		521,400	582,000
4200	800 ton		.08	381		602,000	23,800		625,800	698,000
4210	900 ton		.08	400		698,000	25,000		723,000	805,000
4220	1,000 ton	▼	.08	421	▼	775,500	26,300		801,800	892,500

23 64 13.16 Indirect-Fired Absorption Water Chillers

		Crew	Daily Output	Labor-Hours	Unit	Material	2020 Bare Costs Labor	Equipment	Total	Total Incl O&P
0010	**INDIRECT-FIRED ABSORPTION WATER CHILLERS**									
0020	Steam or hot water, water cooled									
0050	100 ton	Q-7	.13	240	Ea.	132,500	15,000		147,500	168,500
0100	148 ton		.12	258		245,500	16,100		261,600	294,000
0200	200 ton		.12	276		273,000	17,200		290,200	326,500
0240	250 ton		.11	291		299,000	18,200		317,200	356,000
0300	354 ton		.11	305		375,500	19,000		394,500	441,500
0400	420 ton		.10	323		414,500	20,200		434,700	486,000
0500	665 ton		.09	340		576,000	21,200		597,200	665,500
0600	750 ton		.09	364		670,000	22,700		692,700	771,000
0700	850 ton		.08	386		718,500	24,100		742,600	826,000
0800	955 ton		.08	410		768,000	25,600		793,600	883,500
0900	1,125 ton		.08	421		887,000	26,300		913,300	1,015,000
1000	1,250 ton		.07	444		949,500	27,700		977,200	1,085,500
1100	1,465 ton		.07	464		1,130,500	28,900		1,159,400	1,287,000
1200	1,660 ton	▼	.07	478	▼	1,330,000	29,800		1,359,800	1,507,500
2000	For two stage unit, add					80%	25%			

23 64 16 – Centrifugal Water Chillers

23 64 16.10 Centrifugal Type Water Chillers	Crew	Daily Output	Labor- Hours	Unit	Material	2020 Bare Costs Labor	Equipment	Total	Total Incl O&P
0010 **CENTRIFUGAL TYPE WATER CHILLERS**, With standard controls									
0020 Centrifugal liquid chiller, water cooled									
0030 not including water tower									
0100 2,000 ton (twin 1,000 ton units)	Q-7	.07	478	Ea.	808,500	29,800		838,300	934,000
0274 Centrifugal, packaged unit, water cooled, not incl. tower									
0278 200 ton	Q-7	.19	172	Ea.	131,000	10,700		141,700	160,500
0279 300 ton		.14	234		152,500	14,600		167,100	190,000
0280 400 ton		.11	283		150,500	17,700		168,200	192,000
0282 450 ton		.11	291		181,500	18,200		199,700	227,000
0290 500 ton		.11	296		184,500	18,500		203,000	230,500
0292 550 ton		.11	305		231,000	19,000		250,000	282,500
0300 600 ton		.10	311		248,500	19,400		267,900	302,000
0302 650 ton		.10	320		263,500	20,000		283,500	319,500
0304 700 ton		.10	327		283,000	20,400		303,400	342,000
0306 750 ton		.10	333		303,000	20,800		323,800	364,000
0310 800 ton		.09	340		323,000	21,200		344,200	387,000
0312 850 ton		.09	352		343,000	21,900		364,900	410,500
0316 900 ton		.09	360		364,000	22,400		386,400	434,000
0318 950 ton		.09	364		384,000	22,700		406,700	456,500
0320 1,000 ton		.09	372		404,500	23,200		427,700	479,500
0324 1,100 ton		.08	386		368,000	24,100		392,100	440,500
0330 1,200 ton		.08	395		441,500	24,700		466,200	522,500
0340 1,300 ton		.08	410		438,500	25,600		464,100	521,000
0350 1,400 ton		.08	421		515,000	26,300		541,300	606,000
0360 1,500 ton		.08	427		544,500	26,600		571,100	638,500
0370 1,600 ton		.07	438		579,500	27,400		606,900	678,500
0380 1,700 ton		.07	451		616,500	28,100		644,600	720,000
0390 1,800 ton		.07	457		653,000	28,500		681,500	761,000
0400 1,900 ton		.07	471		688,500	29,400		717,900	801,000
0430 2,000 ton		.07	492		726,500	30,700		757,200	845,000
0440 2,500 ton	↓	.06	525	↓	910,000	32,700		942,700	1,050,000

23 64 19 – Reciprocating Water Chillers

23 64 19.10 Reciprocating Type Water Chillers

	Crew	Daily Output	Labor- Hours	Unit	Material	Labor	Equipment	Total	Total Incl O&P
0010 **RECIPROCATING TYPE WATER CHILLERS**, With standard controls									
0494 Water chillers, integral air cooled condenser									
0538 60 ton cooling	Q-7	.28	115	Ea.	45,200	7,175		52,375	60,500
0552 80 ton cooling		.26	123		56,500	7,675		64,175	73,500
0554 90 ton cooling		.25	126		60,000	7,850		67,850	78,000
0640 150 ton cooling		.23	138		99,500	8,600		108,100	122,500
0654 190 ton cooling		.22	144		122,000	9,000		131,000	147,500
0660 210 ton cooling		.22	148		123,500	9,250		132,750	150,000
0664 275 ton cooling		.21	156		154,500	9,750		164,250	184,500
0666 300 ton cooling		.20	160		167,500	9,975		177,475	199,500
0668 330 ton cooling		.20	164		197,500	10,200		207,700	233,000
0670 360 ton cooling		.19	168		218,500	10,500		229,000	256,000
0672 390 ton cooling	↓	.19	173	↓	229,500	10,800		240,300	268,500
0980 Water cooled, multiple compressor, semi-hermetic, tower not incl.									
1090 45 ton cooling	Q-7	.29	111	Ea.	27,200	6,925		34,125	40,300
1130 75 ton cooling		.23	139		45,100	8,675		53,775	62,500
1140 85 ton cooling		.21	152		52,000	9,475		61,475	71,000
1150 95 ton cooling		.19	165		51,000	10,300		61,300	71,500
1160 100 ton cooling	↓	.18	180	↓	49,800	11,200		61,000	72,000

23 64 Packaged Water Chillers

23 64 19 – Reciprocating Water Chillers

23 64 19.10 Reciprocating Type Water Chillers	Crew	Daily Output	Labor-Hours	Unit	Material	2020 Bare Costs Labor	2020 Bare Costs Equipment	Total	Total Incl O&P
1170	115 ton cooling	Q-7	.17	190	Ea.	53,500	11,900		65,400	77,000
1180	125 ton cooling		.16	196		54,000	12,300		66,300	78,000
1200	145 ton cooling		.16	203		61,500	12,600		74,100	86,500
1210	155 ton cooling		.15	211		64,500	13,100		77,600	90,500
1300	Water cooled, single compressor, semi-hermetic, tower not incl.									
1320	40 ton cooling	Q-7	.30	108	Ea.	17,200	6,750		23,950	29,000
1340	50 ton cooling		.28	114		21,400	7,100		28,500	34,100
1360	60 ton cooling		.25	126		23,800	7,850		31,650	37,900
4000	Packaged chiller, remote air cooled condensers not incl.									
4020	15 ton cooling	Q-7	.30	108	Ea.	20,500	6,750		27,250	32,600
4030	20 ton cooling		.28	116		21,200	7,225		28,425	34,200
4040	25 ton cooling		.25	126		26,700	7,850		34,550	41,100
4050	35 ton cooling		.24	134		32,700	8,350		41,050	48,500
4060	40 ton cooling		.22	144		29,000	9,000		38,000	45,400
4066	45 ton cooling		.21	150		26,300	9,325		35,625	42,900
4070	50 ton cooling		.21	154		31,300	9,600		40,900	48,800
4080	60 ton cooling		.20	164		32,900	10,200		43,100	51,500
4090	75 ton cooling		.18	174		62,000	10,900		72,900	84,000
4100	85 ton cooling		.17	184		63,500	11,500		75,000	86,500
4110	95 ton cooling		.17	194		54,000	12,100		66,100	77,000
4120	105 ton cooling		.16	204		69,500	12,700		82,200	95,000
4130	115 ton cooling		.15	213		61,000	13,300		74,300	87,000
4140	125 ton cooling		.14	224		64,000	14,000		78,000	91,000
4150	145 ton cooling		.14	234		73,500	14,600		88,100	103,000

23 64 23 – Scroll Water Chillers

23 64 23.10 Scroll Water Chillers

		Crew	Daily Output	Labor-Hours	Unit	Material	Labor	Equipment	Total	Total Incl O&P
0010	**SCROLL WATER CHILLERS**, With standard controls									
0480	Packaged w/integral air cooled condenser									
0482	10 ton cooling	Q-7	.34	94.118	Ea.	15,200	5,875		21,075	25,500
0490	15 ton cooling		.37	86.486		18,300	5,400		23,700	28,300
0500	20 ton cooling		.34	94.118		23,300	5,875		29,175	34,500
0510	25 ton cooling		.34	94.118		27,800	5,875		33,675	39,400
0515	30 ton cooling		.31	102		29,800	6,350		36,150	42,300
0517	35 ton cooling		.31	105		32,600	6,550		39,150	45,700
0520	40 ton cooling		.30	108		34,800	6,750		41,550	48,400
0528	45 ton cooling		.29	110		38,300	6,850		45,150	52,500
0536	50 ton cooling		.28	113		40,700	7,050		47,750	55,500
0540	70 ton cooling		.27	119		55,000	7,400		62,400	71,500
0590	100 ton cooling		.25	128		72,500	7,975		80,475	92,000
0600	110 ton cooling		.24	133		74,500	8,325		82,825	94,500
0630	130 ton cooling		.24	133		88,500	8,325		96,825	110,000
0670	170 ton cooling		.23	139		111,000	8,675		119,675	135,000
0671	210 ton cooling		.22	145		123,500	9,075		132,575	149,500
0675	250 ton cooling		.21	152		143,500	9,500		153,000	171,500
0676	275 ton cooling		.21	152		154,500	9,500		164,000	184,000
0677	300 ton cooling		.20	160		167,500	9,975		177,475	199,500
0678	330 ton cooling		.20	160		197,500	9,975		207,475	232,500
0679	390 ton cooling		.19	168		229,500	10,500		240,000	268,000
0680	Scroll water cooled, single compressor, hermetic, tower not incl.									
0700	2 ton cooling	Q-5	.57	28.070	Ea.	3,475	1,650		5,125	6,300
0710	5 ton cooling		.57	28.070		4,150	1,650		5,800	7,050
0720	6 ton cooling		.42	38.005		5,250	2,250		7,500	9,125

23 64 Packaged Water Chillers

23 64 23 – Scroll Water Chillers

23 64 23.10 Scroll Water Chillers

23 64 23.10 Scroll Water Chillers		Crew	Daily Output	Labor-Hours	Unit	Material	2020 Bare Costs Labor	Equipment	Total	Total Incl O&P
0740	8 ton cooling	Q-5	.31	52.117	Ea.	5,725	3,075		8,800	10,900
0760	10 ton cooling	Q-6	.36	67.039		6,575	4,100		10,675	13,400
0780	15 ton cooling	"	.36	66.667		10,300	4,075		14,375	17,400
0800	20 ton cooling	Q-7	.38	83.990		11,800	5,250		17,050	20,800
0820	30 ton cooling	"	.33	96.096		13,200	6,000		19,200	23,500
0900	Scroll water cooled, multiple compressor, hermetic, tower not incl.									
0910	15 ton cooling	Q-6	.36	66.667	Ea.	20,500	4,075		24,575	28,600
0930	30 ton cooling	Q-7	.36	88.889		26,700	5,550		32,250	37,600
0935	35 ton cooling		.31	103		32,700	6,450		39,150	45,700
0940	40 ton cooling		.30	107		29,000	6,650		35,650	41,900
0950	50 ton cooling		.28	114		31,300	7,125		38,425	45,100
0960	60 ton cooling		.25	128		32,900	7,975		40,875	48,200
0962	75 ton cooling		.23	139		62,000	8,675		70,675	81,000
0964	85 ton cooling		.23	139		63,500	8,675		72,175	82,500
0970	100 ton cooling		.18	178		69,500	11,100		80,600	92,500
0980	115 ton cooling		.17	188		61,000	11,700		72,700	84,500
0990	125 ton cooling		.16	200		64,000	12,500		76,500	88,500
0995	145 ton cooling		.16	200		73,500	12,500		86,000	99,500

23 64 26 – Rotary-Screw Water Chillers

23 64 26.10 Rotary-Screw Type Water Chillers

23 64 26.10 Rotary-Screw Type Water Chillers		Crew	Daily Output	Labor-Hours	Unit	Material	2020 Bare Costs Labor	Equipment	Total	Total Incl O&P
0010	**ROTARY-SCREW TYPE WATER CHILLERS**, With standard controls									
0110	Screw, liquid chiller, air cooled, insulated evaporator									
0120	130 ton	Q-7	.14	229	Ea.	98,500	14,300		112,800	130,000
0124	160 ton		.13	246		125,000	15,400		140,400	160,500
0128	180 ton		.13	250		140,500	15,600		156,100	178,000
0132	210 ton		.12	258		147,000	16,100		163,100	185,500
0136	270 ton		.12	267		177,000	16,600		193,600	219,500
0140	320 ton		.12	276		221,500	17,200		238,700	269,500
0200	Packaged unit, water cooled, not incl. tower									
0210	80 ton	Q-7	.14	224	Ea.	48,500	14,000		62,500	74,500
0220	100 ton		.14	230		59,000	14,400		73,400	86,500
0230	150 ton		.13	241		78,000	15,000		93,000	108,000
0240	200 ton		.13	252		83,500	15,700		99,200	115,500
0250	250 ton		.12	260		108,500	16,200		124,700	144,000
0260	300 ton		.12	267		130,000	16,600		146,600	168,000
0270	350 ton		.12	276		153,000	17,200		170,200	194,500
1450	Water cooled, tower not included									
1560	135 ton cooling, screw compressors	Q-7	.14	235	Ea.	59,000	14,700		73,700	87,000
1580	150 ton cooling, screw compressors		.13	241		68,500	15,000		83,500	98,000
1620	200 ton cooling, screw compressors		.13	250		94,500	15,600		110,100	127,500
1660	291 ton cooling, screw compressors		.12	260		98,500	16,200		114,700	133,000

23 64 33 – Modular Water Chillers

23 64 33.10 Direct Expansion Type Water Chillers

23 64 33.10 Direct Expansion Type Water Chillers		Crew	Daily Output	Labor-Hours	Unit	Material	2020 Bare Costs Labor	Equipment	Total	Total Incl O&P
0010	**DIRECT EXPANSION TYPE WATER CHILLERS**, With standard controls									
8000	Direct expansion, shell and tube type, for built up systems									
8020	1 ton	Q-5	2	8	Ea.	8,125	470		8,595	9,625
8030	5 ton		1.90	8.421		13,500	495		13,995	15,500
8040	10 ton		1.70	9.412		16,700	555		17,255	19,200
8050	20 ton		1.50	10.667		19,100	630		19,730	21,900
8060	30 ton		1	16		24,300	945		25,245	28,100
8070	50 ton		.90	17.778		39,000	1,050		40,050	44,500
8080	100 ton	Q-6	.90	26.667		69,000	1,625		70,625	78,500

For customer support on your Mechanical Costs with RSMeans Data, call 800.448.8182.

383

23 65 Cooling Towers

23 65 13 – Forced-Draft Cooling Towers

23 65 13.10 Forced-Draft Type Cooling Towers		Crew	Daily Output	Labor-Hours	Unit	Material	2020 Bare Costs Labor	Equipment	Total	Total Incl O&P
0010	**FORCED-DRAFT TYPE COOLING TOWERS**, Packaged units D3030–310									
0070	Galvanized steel									
0080	Induced draft, crossflow									
0100	Vertical, belt drive, 61 tons	Q-6	90	.267	TonAC	203	16.30		219.30	248
0150	100 ton		100	.240		195	14.70		209.70	237
0200	115 ton		109	.220		170	13.45		183.45	207
0250	131 ton		120	.200		204	12.25		216.25	242
0260	162 ton		132	.182		165	11.10		176.10	198
1000	For higher capacities, use multiples									
1500	Induced air, double flow									
1900	Vertical, gear drive, 167 ton	Q-6	126	.190	TonAC	160	11.65		171.65	193
2000	297 ton		129	.186		98.50	11.40		109.90	126
2100	582 ton		132	.182		55	11.10		66.10	77
2150	849 ton		142	.169		74	10.35		84.35	96.50
2200	1,016 ton		150	.160		74.50	9.80		84.30	96.50
2500	Blow through, centrifugal type									
2510	50 ton	Q-6	2.28	10.526	Ea.	13,500	645		14,145	15,800
2520	75 ton		1.52	15.789		17,400	965		18,365	20,600
2524	100 ton		1.22	19.672		21,200	1,200		22,400	25,100
2528	125 ton	Q-7	1.31	24.427		23,600	1,525		25,125	28,300
2532	200 ton		.82	39.216		33,000	2,450		35,450	40,000
2536	250 ton		.65	49.231		38,600	3,075		41,675	47,100
2540	300 ton		.54	59.259		44,400	3,700		48,100	54,500
2544	350 ton		.47	68.085		50,000	4,250		54,250	61,500
2548	400 ton		.41	78.049		52,500	4,875		57,375	65,000
2552	450 ton		.36	88.154		63,000	5,500		68,500	77,000
2556	500 ton		.32	100		67,000	6,250		73,250	83,500
2560	550 ton		.30	108		72,000	6,725		78,725	89,500
2564	600 ton		.27	118		75,500	7,350		82,850	94,000
2568	650 ton		.25	127		84,000	7,950		91,950	104,500
2572	700 ton		.23	137		98,500	8,575		107,075	121,500
2576	750 ton		.22	147		106,000	9,150		115,150	130,000
2580	800 ton		.21	152		114,500	9,500		124,000	140,000
2584	850 ton		.20	157		122,000	9,800		131,800	148,500
2588	900 ton		.20	162		129,000	10,100		139,100	157,000
2592	950 ton		.19	170		136,000	10,600		146,600	166,000
2596	1,000 ton		.18	181		140,500	11,300		151,800	171,500
2700	Axial fan, induced draft									
2710	50 ton	Q-6	2.28	10.526	Ea.	11,800	645		12,445	14,000
2720	75 ton		1.52	15.789		13,900	965		14,865	16,800
2724	100 ton		1.22	19.672		14,700	1,200		15,900	18,000
2728	125 ton	Q-7	1.31	24.427		18,100	1,525		19,625	22,300
2732	200 ton		.82	39.216		30,800	2,450		33,250	37,600
2736	250 ton		.65	49.231		34,700	3,075		37,775	42,800
2740	300 ton		.54	59.259		40,500	3,700		44,200	50,000
2744	350 ton		.47	68.085		43,300	4,250		47,550	54,000
2748	400 ton		.41	78.049		49,800	4,875		54,675	62,000
2752	450 ton		.36	88.154		55,000	5,500		60,500	68,500
2756	500 ton		.32	100		58,500	6,250		64,750	74,000
2760	550 ton		.30	108		65,500	6,725		72,225	82,000
2764	600 ton		.27	118		69,00C	7,350		76,350	87,000
2768	650 ton		.25	127		75,500	7,950		83,450	95,000
2772	700 ton		.23	137		86,000	8,575		94,575	107,500

23 65 Cooling Towers

23 65 13 – Forced-Draft Cooling Towers

23 65 13.10 Forced-Draft Type Cooling Towers	Crew	Daily Output	Labor-Hours	Unit	Material	2020 Bare Costs Labor	Equipment	Total	Total Incl O&P
2776	750 ton	Q-7	.22	147	Ea.	89,500	9,150		98,650	111,500
2780	800 ton		.21	152		93,000	9,500		102,500	116,500
2784	850 ton		.20	157		94,000	9,800		103,800	117,500
2788	900 ton		.20	162		97,000	10,100		107,100	121,500
2792	950 ton		.19	170		102,000	10,600		112,600	128,500
2796	1,000 ton		.18	181		108,500	11,300		119,800	136,000
3000	For higher capacities, use multiples									
3500	For pumps and piping, add	Q-6	38	.632	TonAC	101	38.50		139.50	169
4000	For absorption systems, add				"	75%	75%			
4100	Cooling water chemical feeder	Q-5	3	5.333	Ea.	425	315		740	940
5000	Fiberglass tower on galvanized steel support structure									
5010	Draw thru									
5100	100 ton	Q-6	1.40	17.143	Ea.	14,000	1,050		15,050	17,000
5120	120 ton		1.20	20		16,700	1,225		17,925	20,200
5140	140 ton		1	24		18,100	1,475		19,575	22,100
5160	160 ton		.80	30		22,900	1,825		24,725	28,000
5180	180 ton		.65	36.923		26,400	2,250		28,650	32,500
5251	600 ton		.16	150		77,500	9,175		86,675	98,500
5252	1,000 ton		.10	250		129,000	15,300		144,300	164,500
5300	For stainless steel support structure, add					30%				
5360	For higher capacities, use multiples of each size									
6000	Stainless steel									
6010	Induced draft, crossflow, horizontal, belt drive									
6100	57 ton	Q-6	1.50	16	Ea.	29,900	980		30,880	34,400
6120	91 ton		.99	24.242		36,700	1,475		38,175	42,600
6140	111 ton		.43	55.814		49,600	3,425		53,025	59,500
6160	126 ton		.22	109		54,000	6,675		60,675	69,500
6170	Induced draft, crossflow, vertical, gear drive									
6172	167 ton	Q-6	.75	32	Ea.	61,500	1,950		63,450	70,500
6174	297 ton		.43	55.814		68,000	3,425		71,425	79,500
6176	582 ton		.23	104		107,000	6,375		113,375	127,000
6178	849 ton 210/ton		.17	141		151,000	8,625		159,625	179,500
6180	1,016 ton 216/Ton		.15	160		185,500	9,800		195,300	219,000

23 65 33 – Liquid Coolers

23 65 33.10 Liquid Cooler, Closed Circuit Type Cooling Towers

		Crew	Daily Output	Labor-Hours	Unit	Material	2020 Bare Costs Labor	Equipment	Total	Total Incl O&P
0010	**LIQUID COOLER, CLOSED CIRCUIT TYPE COOLING TOWERS**									
0070	Packaged options include: vibration switch, sump sweeper piping,									
0076	grooved coil connections, fan motor inverter capable,									
0080	304 st. stl. cold water basin, external platform w/ladder.									
0100	50 ton	Q-6	2	12	Ea.	36,500	735		37,235	41,200
0150	100 ton	"	1.13	21.239		73,000	1,300		74,300	82,500
0200	150 ton	Q-7	.98	32.653		109,500	2,050		111,550	123,500
0250	200 ton		.78	41.026		146,000	2,550		148,550	164,500
0300	250 ton		.63	50.794		182,500	3,175		185,675	205,500

For customer support on your Mechanical Costs with RSMeans Data, call 800.448.8182.

385

23 72 Air-to-Air Energy Recovery Equipment

23 72 13 – Heat-Wheel Air-to-Air Energy-Recovery Equipment

23 72 13.10 Heat-Wheel Air-to-Air Energy Recovery Equip.

		Crew	Daily Output	Labor-Hours	Unit	Material	2020 Bare Costs Labor	2020 Bare Costs Equipment	Total	Total Incl O&P
0010	**HEAT-WHEEL AIR-TO-AIR ENERGY RECOVERY EQUIPMENT**									
0100	Air to air									
4000	Enthalpy recovery wheel									
4010	1,000 max CFM ⚟G	Q-9	1.20	13.333	Ea.	7,675	750		8,425	9,575
4020	2,000 max CFM ⚟G		1	16		8,550	895		9,445	10,800
4030	4,000 max CFM ⚟G		.80	20		9,875	1,125		11,000	12,600
4040	6,000 max CFM ⚟G	↓	.70	22.857		11,500	1,275		12,775	14,700
4050	8,000 max CFM ⚟G	Q-10	1	24		13,400	1,400		14,800	16,900
4060	10,000 max CFM ⚟G		.90	26.667		16,000	1,550		17,550	20,100
4070	20,000 max CFM ⚟G		.80	30		27,700	1,750		29,450	33,100
4080	25,000 max CFM ⚟G		.70	34.286		35,600	2,000		37,600	42,100
4090	30,000 max CFM ⚟G		.50	48		39,500	2,800		42,300	47,600
4100	40,000 max CFM ⚟G		.45	53.333		52,000	3,100		55,100	61,500
4110	50,000 max CFM ⚟G	↓	.40	60	↓	60,000	3,500		63,500	71,500

23 72 16 – Heat-Pipe Air-To-Air Energy-Recovery Equipment

23 72 16.10 Heat Pipes

		Crew	Daily Output	Labor-Hours	Unit	Material	2020 Bare Costs Labor	2020 Bare Costs Equipment	Total	Total Incl O&P
0010	**HEAT PIPES**									
8000	Heat pipe type, glycol, 50% efficient									
8010	100 MBH, 1,700 CFM	1 Stpi	.80	10	Ea.	4,625	655		5,280	6,075
8020	160 MBH, 2,700 CFM		.60	13.333		6,225	875		7,100	8,150
8030	620 MBH, 4,000 CFM	↓	.40	20	↓	9,400	1,300		10,700	12,300

23 73 Indoor Central-Station Air-Handling Units

23 73 13 – Modular Indoor Central-Station Air-Handling Units

23 73 13.10 Air-Handling Units

		Crew	Daily Output	Labor-Hours	Unit	Material	2020 Bare Costs Labor	2020 Bare Costs Equipment	Total	Total Incl O&P
0010	**AIR-HANDLING UNITS**, Built-Up									
0100	With cooling/heating coil section, filters, mixing box									
0880	Single zone, horizontal/vertical									
0890	Constant volume									
0900	1,600 CFM	Q-5	1.20	13.333	Ea.	5,575	785		6,360	7,300
0906	2,000 CFM		1.10	14.545		7,800	860		8,660	9,850
0910	3,000 CFM		1	16		9,775	945		10,720	12,200
0916	4,000 CFM	↓	.95	16.842		12,800	995		13,795	15,600
0920	5,000 CFM	Q-6	1.40	17.143		15,600	1,050		16,650	18,700
0926	6,500 CFM		1.30	18.462		19,500	1,125		20,625	23,100
0930	7,500 CFM		1.20	20		24,900	1,225		26,125	29,200
0936	9,200 CFM		1.10	21.818		28,100	1,325		29,425	32,900
0940	11,500 CFM		1	24		32,400	1,475		33,875	37,800
0946	13,200 CFM		.90	26.667		33,400	1,625		35,025	39,300
0950	16,500 CFM		.80	30		48,700	1,825		50,525	56,500
0960	19,500 CFM		.70	34.286		57,000	2,100		59,100	65,500
0970	22,000 CFM		.60	40		64,500	2,450		66,950	74,000
0980	27,000 CFM		.50	48		65,500	2,925		68,425	77,000
0990	34,000 CFM		.40	60		83,500	3,675		87,175	97,500
1000	40,000 CFM		.30	80		100,000	4,900		104,900	117,500
1010	47,000 CFM	↓	.20	120	↓	110,000	7,350		117,350	132,000
2300	Multi-zone, horizontal or vertical, blow-thru fan									
2310	3,000 CFM	Q-5	1	16	Ea.	7,650	945		8,595	9,850
2314	4,000 CFM	"	.95	16.842		10,500	995		11,495	13,100
2318	5,000 CFM	Q-6	1.40	17.143		13,100	1,050		14,150	16,000

23 73 Indoor Central-Station Air-Handling Units

23 73 13 – Modular Indoor Central-Station Air-Handling Units

23 73 13.10 Air-Handling Units

23 73 13.10 Air-Handling Units		Crew	Daily Output	Labor-Hours	Unit	Material	2020 Bare Costs Labor	Equipment	Total	Total Incl O&P
2322	6,500 CFM	Q-6	1.30	18.462	Ea.	17,000	1,125		18,125	20,400
2326	7,500 CFM		1.20	20		18,800	1,225		20,025	22,500
2330	9,200 CFM		1.10	21.818		24,200	1,325		25,525	28,600
2334	11,500 CFM		1	24		26,600	1,475		28,075	31,500
2336	13,200 CFM		.90	26.667		30,500	1,625		32,125	36,000
2340	16,500 CFM		.80	30		38,000	1,825		39,825	44,600
2344	19,500 CFM		.70	34.286		44,900	2,100		47,000	52,500
2348	22,000 CFM		.60	40		50,500	2,450		52,950	59,000
2350	22,500 CFM		.55	43.636		51,500	2,675		54,175	61,000
2352	27,000 CFM		.50	48		54,500	2,925		57,425	64,000
2356	34,000 CFM		.40	60		68,500	3,675		72,175	81,000
2360	40,000 CFM		.30	80		74,500	4,900		79,400	89,500
2364	47,000 CFM	↓	.20	120	↓	94,500	7,350		101,850	115,000

23 73 13.20 Air-Handling Units, Packaged Indoor Type

23 73 13.20 Air-Handling Units, Packaged Indoor Type		Crew	Daily Output	Labor-Hours	Unit	Material	2020 Bare Costs Labor	Equipment	Total	Total Incl O&P
0010	**AIR-HANDLING UNITS, PACKAGED INDOOR TYPE**									
0040	Cooling coils may be chilled water or DX									
0050	Heating coils may be hot water, steam or electric									
1090	Constant volume									
1200	2,000 CFM	Q-5	1.10	14.545	Ea.	10,900	860		11,760	13,300
1400	5,000 CFM	Q-6	.80	30		17,700	1,825		19,525	22,200
1550	10,000 CFM		.54	44.444		26,000	2,725		28,725	32,700
1670	15,000 CFM		.38	63.158		53,000	3,875		56,875	64,500
1700	20,000 CFM		.31	77.419		67,500	4,725		72,225	81,500
1710	30,000 CFM	↓	.27	88.889	↓	92,000	5,450		97,450	109,000
2300	Variable air volume									
2330	2,000 CFM	Q-5	1	16	Ea.	15,600	945		16,545	18,600
2340	5,000 CFM	Q-6	.70	34.286		21,600	2,100		23,700	26,900
2350	10,000 CFM		.50	48		41,900	2,925		44,825	50,500
2360	15,000 CFM		.40	60		61,000	3,675		64,675	73,000
2370	20,000 CFM		.29	82.759		78,500	5,075		83,575	93,500
2380	30,000 CFM	↓	.20	120		94,500	7,350		101,850	115,000

23 73 39 – Indoor, Direct Gas-Fired Heating and Ventilating Units

23 73 39.10 Make-Up Air Unit

23 73 39.10 Make-Up Air Unit		Crew	Daily Output	Labor-Hours	Unit	Material	2020 Bare Costs Labor	Equipment	Total	Total Incl O&P
0010	**MAKE-UP AIR UNIT**									
0020	Indoor suspension, natural/LP gas, direct fired,									
0032	standard control. For flue see Section 23 51 23.10									
0040	70°F temperature rise, MBH is input									
0100	75 MBH input	Q-6	3.60	6.667	Ea.	5,225	410		5,635	6,325
0120	100 MBH input		3.40	7.059		5,400	430		5,830	6,600
0140	125 MBH input		3.20	7.500		5,750	460		6,210	7,000
0160	150 MBH input		3	8		6,325	490		6,815	7,675
0180	175 MBH input		2.80	8.571		6,475	525		7,000	7,900
0200	200 MBH input		2.60	9.231		6,825	565		7,390	8,375
0220	225 MBH input		2.40	10		7,075	610		7,685	8,725
0240	250 MBH input		2	12		8,175	735		8,910	10,100
0260	300 MBH input		1.90	12.632		8,375	775		9,150	10,400
0280	350 MBH input		1.75	13.714		10,200	840		11,040	12,500
0300	400 MBH input		1.60	15		16,500	920		17,420	19,600
0320	840 MBH input		1.20	20		26,700	1,225		27,925	31,200
0340	1,200 MBH input	↓	.88	27.273		36,800	1,675		38,475	43,000
0600	For discharge louver assembly, add					5%				
0700	For filters, add					10%				

For customer support on your Mechanical Costs with RSMeans Data, call 800.448.8182.

387

23 73 Indoor Central-Station Air-Handling Units

23 73 39 – Indoor, Direct Gas-Fired Heating and Ventilating Units

23 73 39.10 Make-Up Air Unit	Crew	Daily Output	Labor-Hours	Unit	Material	2020 Bare Costs Labor	Equipment	Total	Total Incl O&P
0800	For air shut-off damper section, add				Ea.	30%				

23 74 Packaged Outdoor HVAC Equipment

23 74 13 – Packaged, Outdoor, Central-Station Air-Handling Units

23 74 13.10 Packaged, Outdoor Type, Central-Station AHU

		Crew	Daily Output	Labor-Hours	Unit	Material	2020 Bare Costs Labor	Equipment	Total	Total Incl O&P
0010	**PACKAGED, OUTDOOR TYPE, CENTRAL-STATION AIR-HANDLING UNITS**									
2900	Cooling coils may be chilled water or DX									
2910	Heating coils may be hot water, steam or electric									
3000	Weathertight, ground or rooftop									
3010	Constant volume									
3100	2,000 CFM	Q-6	1	24	Ea.	23,900	1,475		25,375	28,500
3140	5,000 CFM		.79	30.380		29,100	1,850		30,950	34,900
3150	10,000 CFM		.52	46.154		63,500	2,825		66,325	74,000
3160	15,000 CFM		.41	58.537		92,000	3,575		95,575	106,500
3170	20,000 CFM		.30	80		127,000	4,900		131,900	147,500
3180	30,000 CFM		.24	100		145,500	6,125		151,625	169,500
3200	Variable air volume									
3210	2,000 CFM	Q-6	.96	25	Ea.	15,200	1,525		16,725	19,100
3240	5,000 CFM		.73	32.877		30,600	2,000		32,600	36,700
3250	10,000 CFM		.50	48		65,000	2,925		67,925	76,000
3260	15,000 CFM		.40	60		100,000	3,675		103,675	115,500
3270	20,000 CFM		.29	82.759		129,500	5,075		134,575	150,000
3280	30,000 CFM		.20	120		153,000	7,350		160,350	179,500

23 74 23 – Packaged, Outdoor, Heating-Only Makeup-Air Units

23 74 23.16 Packaged, Ind.-Fired, In/Outdoor

		Crew	Daily Output	Labor-Hours	Unit	Material	2020 Bare Costs Labor	Equipment	Total	Total Incl O&P
0010	**PACKAGED, IND.-FIRED, IN/OUTDOOR,** Heat-Only Makeup-Air Units									
1000	Rooftop unit, natural gas, gravity vent, S.S. exchanger									
1010	70°F temperature rise, MBH is input									
1020	250 MBH	Q-6	4	6	Ea.	16,000	365		16,365	18,300
1040	400 MBH		3.60	6.667		18,300	410		18,710	20,800
1060	550 MBH		3.30	7.273		21,200	445		21,645	24,100
1080	750 MBH		3	8		23,500	490		23,990	26,600
1100	1,000 MBH		2.60	9.231		26,300	565		26,865	29,700
1120	1,750 MBH		2.30	10.435		30,900	640		31,540	35,000
1140	2,500 MBH		1.90	12.632		45,700	775		46,475	51,500
1160	3,250 MBH		1.40	17.143		50,500	1,050		51,550	57,000
1180	4,000 MBH		1.20	20		68,000	1,225		69,225	77,000
1200	6,000 MBH		1	24		86,000	1,475		87,475	96,500
1600	For cleanable filters, add					5%				
1700	For electric modulating gas control, add					10%				

23 74 33 – Dedicated Outdoor-Air Units

23 74 33.10 Rooftop Air Conditioners

			Crew	Daily Output	Labor-Hours	Unit	Material	2020 Bare Costs Labor	Equipment	Total	Total Incl O&P
0010	**ROOFTOP AIR CONDITIONERS,** Standard controls, curb, economizer										
1000	Single zone, electric cool, gas heat										
1100	3 ton cooling, 60 MBH heating	D3050-150	Q-5	.70	22.857	Ea.	2,900	1,350		4,250	5,200
1120	4 ton cooling, 95 MBH heating	D3050-155		.61	26.403		3,750	1,550		5,300	6,450
1140	5 ton cooling, 112 MBH heating	R236000-10		.56	28.521		4,475	1,675		6,150	7,450
1145	6 ton cooling, 140 MBH heating	R236000-20		.52	30.769		5,125	1,825		6,950	8,375
1150	7.5 ton cooling, 170 MBH heating			.50	32.258		5,725	1,900		7,625	9,150
1156	8.5 ton cooling, 170 MBH heating			.46	34.783		7,200	2,050		9,250	11,000

23 74 33.10 Rooftop Air Conditioners	Crew	Daily Output	Labor-Hours	Unit	Material	2020 Bare Costs Labor	2020 Bare Costs Equipment	Total	Total Incl O&P
1160	10 ton cooling, 200 MBH heating	Q-6	.67	35.982	Ea.	9,675	2,200		11,875	13,900
1170	12.5 ton cooling, 230 MBH heating		.63	37.975		10,600	2,325		12,925	15,200
1180	15 ton cooling, 270 MBH heating		.57	42.032		13,300	2,575		15,875	18,500
1190	17.5 ton cooling, 330 MBH heating		.52	45.889		14,000	2,800		16,800	19,600
1200	20 ton cooling, 360 MBH heating	Q-7	.67	47.976		23,400	3,000		26,400	30,200
1210	25 ton cooling, 450 MBH heating		.56	57.554		34,800	3,600		38,400	43,700
1220	30 ton cooling, 540 MBH heating		.47	68.376		36,300	4,275		40,575	46,300
1240	40 ton cooling, 675 MBH heating		.35	91.168		44,400	5,700		50,100	57,500
1260	50 ton cooling, 810 MBH heating		.28	114		50,500	7,100		57,600	66,000
1265	60 ton cooling, 900 MBH heating		.23	137		60,000	8,525		68,525	79,000
1270	80 ton cooling, 1,000 MBH heating		.20	160		78,500	9,975		88,475	101,500
1275	90 ton cooling, 1,200 MBH heating		.17	188		109,500	11,700		121,200	138,000
1280	100 ton cooling, 1,350 MBH heating		.14	229		125,000	14,300		139,300	159,000
1300	Electric cool, electric heat									
1310	5 ton	Q-5	.63	25.600	Ea.	5,675	1,500		7,175	8,475
1314	10 ton	Q-6	.74	32.432		8,550	1,975		10,525	12,400
1318	15 ton	"	.64	37.795		12,900	2,300		15,200	17,700
1322	20 ton	Q-7	.74	43.185		17,100	2,700		19,800	22,900
1326	25 ton		.62	51.696		21,400	3,225		24,625	28,400
1330	30 ton		.52	61.657		25,800	3,850		29,650	34,100
1334	40 ton		.39	82.051		34,300	5,125		39,425	45,400
1338	50 ton		.31	103		40,700	6,400		47,100	54,500
1342	60 ton		.26	123		49,000	7,675		56,675	65,500
1346	75 ton		.21	154		61,000	9,600		70,600	82,000
2000	Multizone, electric cool, gas heat, economizer									
2100	15 ton cooling, 360 MBH heating	Q-7	.61	52.545	Ea.	66,500	3,275		69,775	78,000
2120	20 ton cooling, 360 MBH heating		.53	60.038		72,000	3,750		75,750	85,000
2140	25 ton cooling, 450 MBH heating		.45	71.910		87,500	4,475		91,975	102,500
2160	28 ton cooling, 450 MBH heating		.41	79.012		99,500	4,925		104,425	116,500
2180	30 ton cooling, 540 MBH heating		.37	85.562		110,500	5,350		115,850	129,500
2200	40 ton cooling, 540 MBH heating		.28	114		126,500	7,100		133,600	149,500
2210	50 ton cooling, 540 MBH heating		.23	142		161,000	8,875		169,875	190,500
2220	70 ton cooling, 1,500 MBH heating		.16	199		171,500	12,400		183,900	207,000
2240	80 ton cooling, 1,500 MBH heating		.14	229		195,500	14,300		209,800	237,000
2260	90 ton cooling, 1,500 MBH heating		.13	256		200,500	16,000		216,500	244,500
2280	105 ton cooling, 1,500 MBH heating		.11	291		223,000	18,200		241,200	272,500
2400	For hot water heat coil, deduct					5%				
2500	For steam heat coil, deduct					2%				
2600	For electric heat, deduct					3%	5%			
4000	Single zone, electric cool, gas heat, variable air volume									
4100	12.5 ton cooling, 230 MBH heating	Q-6	.55	43.716	Ea.	13,400	2,675		16,075	18,700
4120	18 ton cooling, 330 MBH heating	"	.46	52.747		17,400	3,225		20,625	24,000
4140	25 ton cooling, 450 MBH heating	Q-7	.48	66.116		39,200	4,125		43,325	49,300
4160	40 ton cooling, 675 MBH heating		.31	105		51,500	6,550		58,050	66,500
4180	60 ton cooling, 900 MBH heating		.20	158		74,000	9,825		83,825	96,000
4200	80 ton cooling, 1,000 MBH heating		.15	209		95,000	13,100		108,100	124,500
5000	Single zone, electric cool only									
5050	3 ton cooling	Q-5	.88	18.203	Ea.	2,750	1,075		3,825	4,625
5060	4 ton cooling		.76	21.108		3,575	1,250		4,825	5,800
5070	5 ton cooling		.70	22.792		3,650	1,350		5,000	6,050
5080	6 ton cooling		.65	24.502		4,350	1,450		5,800	6,950
5090	7.5 ton cooling		.62	25.806		6,425	1,525		7,950	9,350
5100	8.5 ton cooling		.59	26.981		7,025	1,600		8,625	10,100

For customer support on your Mechanical Costs with RSMeans Data, call 800.448.8182.

389

23 74 33.10 Rooftop Air Conditioners		Crew	Daily Output	Labor-Hours	Unit	Material	2020 Bare Costs Labor	Equipment	Total	Total Incl O&P
5110	10 ton cooling	Q-6	.83	28.812	Ea.	8,400	1,775		10,175	11,900
5120	12.5 ton cooling		.80	30		11,500	1,825		13,325	15,400
5130	15 ton cooling		.71	33.613		15,300	2,050		17,350	20,000
5140	17.5 ton cooling		.65	36.697		15,800	2,250		18,050	20,800
5150	20 ton cooling	Q-7	.83	38.415		22,400	2,400		24,800	28,200
5160	25 ton cooling		.70	45.977		22,700	2,875		25,575	29,300
5170	30 ton cooling		.59	54.701		24,500	3,425		27,925	32,100
5180	40 ton cooling		.44	73.059		33,000	4,550		37,550	43,100
5190	50 ton cooling		.38	84.211		38,500	5,250		43,750	50,500
5200	60 ton cooling		.28	114		50,000	7,125		57,125	65,500
5400	For low heat, add					7%				
5410	For high heat, add					10%				
6000	Single zone electric cooling with variable volume distribution									
6020	20 ton cooling	Q-7	.72	44.199	Ea.	23,600	2,750		26,350	30,100
6030	25 ton cooling		.61	52.893		26,000	3,300		29,300	33,600
6040	30 ton cooling		.51	62.868		30,100	3,925		34,025	39,000
6050	40 ton cooling		.38	83.990		34,700	5,250		39,950	46,100
6060	50 ton cooling		.31	105		37,100	6,550		43,650	50,500
6070	60 ton cooling		.25	126		49,700	7,850		57,550	66,500
6200	For low heat, add					7%				
6210	For high heat, add					10%				
7000	Multizone, cool/heat, variable volume distribution									
7100	50 ton cooling	Q-7	.20	164	Ea.	117,500	10,200		127,700	145,000
7110	70 ton cooling		.14	229		164,500	14,300		178,800	202,500
7120	90 ton cooling		.11	296		180,000	18,500		198,500	225,500
7130	105 ton cooling		.10	333		214,000	20,800		234,800	266,500
7140	120 ton cooling		.08	381		244,500	23,800		268,300	304,500
7150	140 ton cooling		.07	444		261,000	27,700		288,700	328,500
7300	Penthouse unit, cool/heat, variable volume distribution									
7400	150 ton cooling	Q-7	.10	311	Ea.	410,000	19,400		429,400	480,000
7410	170 ton cooling		.10	320		450,500	20,000		470,500	525,500
7420	200 ton cooling		.10	327		530,000	20,400		550,400	613,000
7430	225 ton cooling		.09	340		557,500	21,200		578,700	645,000
7440	250 ton cooling		.09	360		595,000	22,400		617,400	688,000
7450	270 ton cooling		.09	364		643,000	22,700		665,700	741,000
7460	300 ton cooling		.09	372		714,500	23,200		737,700	820,000
9400	Sub assemblies for assembly systems									
9410	Ductwork, per ton, rooftop 1 zone units	Q-9	.96	16.667	Ton	290	935		1,225	1,750
9420	Ductwork, per ton, rooftop multizone units		.50	32		375	1,800		2,175	3,150
9430	Ductwork, VAV cooling/ton, rooftop multizone, 1 zone/ton		.32	50		455	2,800		3,255	4,750
9440	Ductwork per ton, packaged water & air cooled units		1.26	12.698		61.50	710		771.50	1,150
9450	Ductwork per ton, split system remote condensing units		1.32	12.121		58	680		738	1,100
9500	Ductwork package for residential units, 1 ton		.80	20	Ea.	410	1,125		1,535	2,150
9520	2 ton		.40	40		825	2,250		3,075	4,300
9530	3 ton		.27	59.259		1,225	3,325		4,550	6,400
9540	4 ton	Q-10	.30	80		1,650	4,650		6,300	8,850
9550	5 ton		.23	104		2,050	6,075		8,125	11,500
9560	6 ton		.20	120		2,475	6,975		9,450	13,300
9570	7 ton		.17	141		2,875	8,200		11,075	15,700
9580	8 ton		.15	160		3,300	9,300		12,600	17,700

23 76 Evaporative Air-Cooling Equipment

23 76 13 – Direct Evaporative Air Coolers

23 76 13.10 Evaporative Coolers		Crew	Daily Output	Labor-Hours	Unit	Material	2020 Bare Costs Labor	Equipment	Total	Total Incl O&P
0010	**EVAPORATIVE COOLERS** (Swamp Coolers) ducted, not incl. duct.									
0100	Side discharge style, capacities at 0.25" S.P.									
0120	1,785 CFM, 1/3 HP, 115 V [G]	Q-9	5	3.200	Ea.	465	179		644	780
0140	2,740 CFM, 1/3 HP, 115 V [G]		4.50	3.556		565	199		764	920
0160	3,235 CFM, 1/2 HP, 115 V [G]		4	4		575	224		799	970
0180	3,615 CFM, 1/2 HP, 230 V [G]		3.60	4.444		695	249		944	1,150
0200	4,215 CFM, 3/4 HP, 230 V [G]		3.20	5		745	280		1,025	1,250
0220	5,255 CFM, 1 HP, 115/230 V [G]		3	5.333		1,150	299		1,449	1,700
0240	6,090 CFM, 1 HP, 230/460 V [G]		2.80	5.714		2,225	320		2,545	2,925
0260	8,300 CFM, 1-1/2 HP, 230/460 V [G]		2.60	6.154		1,950	345		2,295	2,675
0280	8,360 CFM, 1-1/2 HP, 230/460 V [G]		2.20	7.273		2,000	410		2,410	2,825
0300	9,725 CFM, 2 HP, 230/460 V [G]		1.80	8.889		2,200	500		2,700	3,175
0320	11,715 CFM, 3 HP, 230/460 V [G]		1.40	11.429		2,550	640		3,190	3,775
0340	14,410 CFM, 5 HP, 230/460 V [G]		1	16		3,000	895		3,895	4,650
0400	For two-speed motor, add					5%				
0500	For down discharge style, add					10%				

23 76 26 – Fan Coil Evaporators for Coolers/Freezers

23 76 26.10 Evaporators

		Crew	Daily Output	Labor-Hours	Unit	Material	2020 Bare Costs Labor	Equipment	Total	Total Incl O&P
0010	**EVAPORATORS**, DX coils, remote compressors not included.									
1000	Coolers, reach-in type, above freezing temperatures									
1300	Shallow depth, wall mount, 7 fins per inch, air defrost									
1310	600 BTUH, 8" fan, rust-proof core	Q-5	3.80	4.211	Ea.	710	248		958	1,150
1320	900 BTUH, 8" fan, rust-proof core		3.30	4.848		825	286		1,111	1,325
1330	1,200 BTUH, 8" fan, rust-proof core		3	5.333		1,050	315		1,365	1,625
1340	1,800 BTUH, 8" fan, rust-proof core		2.40	6.667		1,050	395		1,445	1,750
1350	2,500 BTUH, 10" fan, rust-proof core		2.20	7.273		1,100	430		1,530	1,875
1360	3,500 BTUH, 10" fan		2	8		1,175	470		1,645	1,975
1370	4,500 BTUH, 10" fan		1.90	8.421		1,225	495		1,720	2,100
1390	For pan drain, add					63.50			63.50	70
1600	Undercounter refrigerators, ceiling or wall mount,									
1610	8 fins per inch, air defrost, rust-proof core									
1630	800 BTUH, one 6" fan	Q-5	5	3.200	Ea.	287	189		476	600
1640	1,300 BTUH, two 6" fans		4	4		370	236		606	760
1650	1,700 BTUH, two 6" fans		3.60	4.444		415	262		677	845
2000	Coolers, reach-in and walk-in types, above freezing temperatures									
2600	Two-way discharge, ceiling mount, 150 to 4,100 CFM,									
2610	above 34°F applications, air defrost									
2630	900 BTUH, 7 fins per inch, 8" fan	Q-5	4	4	Ea.	565	236		801	980
2660	2,500 BTUH, 7 fins per inch, 10" fan		2.60	6.154		860	365		1,225	1,500
2690	5,500 BTUH, 7 fins per inch, 12" fan		1.70	9.412		1,425	555		1,980	2,400
2720	8,500 BTUH, 8 fins per inch, 16" fan		1.20	13.333		1,700	785		2,485	3,050
2750	15,000 BTUH, 7 fins per inch, 18" fan		1.10	14.545		2,650	860		3,510	4,175
2770	24,000 BTUH, 7 fins per inch, two 16" fans		1	16		3,950	945		4,895	5,775
2790	30,000 BTUH, 7 fins per inch, two 18" fans		.90	17.778		4,900	1,050		5,950	6,975
2850	Two-way discharge, low profile, ceiling mount,									
2860	8 fins per inch, 200 to 570 CFM, air defrost									
2880	800 BTUH, one 6" fan	Q-5	4	4	Ea.	340	236		576	730
2890	1,300 BTUH, two 6" fans		3.50	4.571		410	270		680	855
2900	1,800 BTUH, two 6" fans		3.30	4.848		495	286		781	975
2910	2,700 BTUH, three 6" fans		2.70	5.926		595	350		945	1,175
3000	Coolers, walk-in type, above freezing temperatures									
3300	General use, ceiling mount, 108 to 2,080 CFM									

For customer support on your Mechanical Costs with RSMeans Data, call 800.448.8182.

391

23 76 26.10 Evaporators		Crew	Daily Output	Labor-Hours	Unit	Material	2020 Bare Costs Labor	2020 Bare Costs Equipment	Total	Total Incl O&P
3320	600 BTUH, 7 fins per inch, 6" fan	Q-5	5.20	3.077	Ea.	345	182		527	650
3340	1,200 BTUH, 7 fins per inch, 8" fan		3.70	4.324		480	255		735	905
3360	1,800 BTUH, 7 fins per inch, 10" fan		2.70	5.926		670	350		1,020	1,275
3380	3,500 BTUH, 7 fins per inch, 12" fan		2.40	6.667		985	395		1,380	1,675
3400	5,500 BTUH, 8 fins per inch, 12" fan		2	8		1,100	470		1,570	1,900
3430	8,500 BTUH, 8 fins per inch, 16" fan		1.40	11.429		1,650	675		2,325	2,800
3460	15,000 BTUH, 7 fins per inch, two 16" fans	▼	1.10	14.545	▼	2,625	860		3,485	4,175
3640	Low velocity, high latent load, ceiling mount,									
3650	1,050 to 2,420 CFM, air defrost, 6 fins per inch									
3670	6,700 BTUH, two 10" fans	Q-5	1.30	12.308	Ea.	1,775	725		2,500	3,025
3680	10,000 BTUH, three 10" fans		1	16		2,250	945		3,195	3,900
3690	13,500 BTUH, three 10" fans		1	16		2,850	945		3,795	4,550
3700	18,000 BTUH, four 10" fans		1	16		3,550	945		4,495	5,325
3710	26,500 BTUH, four 10" fans	▼	.80	20		4,500	1,175		5,675	6,725
3730	For electric defrost, add				▼	17%				
5000	Freezers and coolers, reach-in type, above 34°F									
5030	to sub-freezing temperature range, low latent load,									
5050	air defrost, 7 fins per inch									
5070	1,200 BTUH, 8" fan, rustproof core	Q-5	4.40	3.636	Ea.	730	215		945	1,125
5080	1,500 BTUH, 8" fan, rustproof core		4.20	3.810		755	225		980	1,175
5090	1,800 BTUH, 8" fan, rustproof core		3.70	4.324		780	255		1,035	1,250
5100	2,500 BTUH, 10" fan, rustproof core		2.90	5.517		855	325		1,180	1,425
5110	3,500 BTUH, 10" fan, rustproof core		2.50	6.400		870	380		1,250	1,525
5120	4,500 BTUH, 12" fan, rustproof core	▼	2.10	7.619	▼	1,075	450		1,525	1,850
6000	Freezers and coolers, walk-in type									
6050	1,960 to 18,000 CFM, medium profile									
6060	Standard motor, 6 fins per inch, to -30°F, all aluminum									
6080	10,500 BTUH, air defrost, one 18" fan	Q-5	1.40	11.429	Ea.	1,675	675		2,350	2,825
6090	12,500 BTUH, air defrost, one 18" fan		1.40	11.429		1,800	675		2,475	2,975
6100	16,400 BTUH, air defrost, one 18" fan		1.30	12.308		1,875	725		2,600	3,150
6110	20,900 BTUH, air defrost, one 18" fan		1	16		2,200	945		3,145	3,850
6120	27,000 BTUH, air defrost, two 18" fans	▼	1	16		3,100	945		4,045	4,825
6130	32,900 BTUH, air defrost, two 20" fans	Q-6	1.30	18.462		3,450	1,125		4,575	5,475
6140	39,000 BTUH, air defrost, two 20" fans		1.25	19.200		3,975	1,175		5,150	6,125
6150	44,100 BTUH, air defrost, three 20" fans		1	24		4,600	1,475		6,075	7,275
6155	53,000 BTUH, air defrost, three 20" fans		1	24		4,850	1,475		6,325	7,550
6160	66,200 BTUH, air defrost, four 20" fans		1	24		5,100	1,475		6,575	7,800
6170	78,000 BTUH, air defrost, four 20" fans		.80	30		5,875	1,825		7,700	9,200
6180	88,200 BTUH, air defrost, four 24" fans		.60	40		6,325	2,450		8,775	10,700
6190	110,000 BTUH, air defrost, four 24" fans	▼	.50	48		7,650	2,925		10,575	12,800
6330	Hot gas defrost, standard motor units, add					35%				
6370	Electric defrost, 230V, standard motor, add					16%				
6410	460V, standard or high capacity motor, add				▼	25%				
6800	Eight fins per inch, increases BTUH 75%									
6810	Standard capacity, add				Ea.	5%				
6830	Hot gas & electric defrost not recommended									
7000	800 to 4,150 CFM, 12" fans									
7010	Four fins per inch									
7030	3,400 BTUH, 1 fan	Q-5	2.80	5.714	Ea.	560	335		895	1,125
7040	4,200 BTUH, 1 fan		2.40	6.667		675	395		1,070	1,325
7050	5,300 BTUH, 1 fan		1.90	8.421		785	495		1,280	1,600
7060	6,800 BTUH, 1 fan		1.70	9.412		900	555		1,455	1,825
7070	8,400 BTUH, 2 fans	▼	1.60	10	▼	805	590		1,395	1,775

23 76 Evaporative Air-Cooling Equipment

23 76 26 – Fan Coil Evaporators for Coolers/Freezers

23 76 26.10 Evaporators		Crew	Daily Output	Labor-Hours	Unit	Material	2020 Bare Costs Labor	Equipment	Total	Total Incl O&P
7080	10,500 BTUH, 2 fans	Q-5	1.40	11.429	Ea.	1,025	675		1,700	2,125
7090	13,000 BTUH, 3 fans		1.30	12.308		1,650	725		2,375	2,900
7100	17,000 BTUH, 4 fans		1.20	13.333		2,000	785		2,785	3,375
7110	21,500 BTUH, 5 fans	↓	1	16		2,300	945		3,245	3,950
7160	Six fins per inch increases BTUH 35%, add				↓	5%				
7180	Eight fins per inch increases BTUH 53%,									
7190	not recommended for below 35°F, add				Ea.	8%				
7210	For 12°F temperature differential, add					20%				
7230	For adjustable thermostat control, add				↓	131			131	144
7240	For hot gas defrost, except 8 fin models									
7250	on applications below 35°F, add				Ea.	58%				
7260	For electric defrost, except 8 fin models									
7270	on applications below 35°F, add				Ea.	36%				
8000	Freezers, pass-thru door uprights, walk-in storage									
8300	Low temperature, thin profile, electric defrost									
8310	138 to 1,800 CFM, 5 fins per inch									
8320	900 BTUH, one 6" fan	Q-5	3.30	4.848	Ea.	435	286		721	910
8340	1,500 BTUH, two 6" fans		2.20	7.273		655	430		1,085	1,350
8360	2,600 BTUH, three 6" fans		1.90	8.421		915	495		1,410	1,750
8370	3,300 BTUH, four 6" fans		1.70	9.412		1,025	555		1,580	1,950
8380	4,400 BTUH, five 6" fans		1.40	11.429		1,425	675		2,100	2,575
8400	7,000 BTUH, three 10" fans		1.30	12.308		1,800	725		2,525	3,050
8410	8,700 BTUH, four 10" fans	↓	1.10	14.545		2,250	860		3,110	3,750
8450	For air defrost, deduct				↓	271			271	298

23 81 Decentralized Unitary HVAC Equipment

23 81 13 – Packaged Terminal Air-Conditioners

23 81 13.10 Packaged Cabinet Type Air-Conditioners

		Crew	Daily Output	Labor-Hours	Unit	Material	2020 Bare Costs Labor	Equipment	Total	Total Incl O&P
0010	**PACKAGED CABINET TYPE AIR-CONDITIONERS**, Cabinet, wall sleeve,									
0100	louver, electric heat, thermostat, manual changeover; 208 V									
0200	6,000 BTUH cooling, 8,800 BTU heat	Q-5	6	2.667	Ea.	725	157		882	1,025
0220	9,000 BTUH cooling, 13,900 BTU heat		5	3.200		975	189		1,164	1,350
0240	12,000 BTUH cooling, 13,900 BTU heat		4	4		1,550	236		1,786	2,050
0260	15,000 BTUH cooling, 13,900 BTU heat		3	5.333		1,525	315		1,840	2,175
0280	18,000 BTUH cooling, 10 KW heat		2.40	6.667		2,700	395		3,095	3,575
0300	24,000 BTUH cooling, 10 KW heat		1.90	8.421		2,725	495		3,220	3,750
0320	30,000 BTUH cooling, 10 KW heat		1.40	11.429		2,900	675		3,575	4,200
0340	36,000 BTUH cooling, 10 KW heat		1.25	12.800		3,075	755		3,830	4,500
0360	42,000 BTUH cooling, 10 KW heat		1	16		3,225	945		4,170	4,975
0380	48,000 BTUH cooling, 10 KW heat	↓	.90	17.778		3,450	1,050		4,500	5,375
0500	For hot water coil, increase heat by 10%, add					5%	10%			
1000	For steam, increase heat output by 30%, add				↓	8%	10%			

23 81 19 – Self-Contained Air-Conditioners

23 81 19.10 Window Unit Air Conditioners

		Crew	Daily Output	Labor-Hours	Unit	Material	2020 Bare Costs Labor	Equipment	Total	Total Incl O&P
0010	**WINDOW UNIT AIR CONDITIONERS**									
4000	Portable/window, 15 amp, 125 V grounded receptacle required									
4060	5,000 BTUH	1 Carp	8	1	Ea.	305	53		358	420
4340	6,000 BTUH		8	1		239	53		292	345
4480	8,000 BTUH		6	1.333		385	71		456	530
4500	10,000 BTUH	↓	6	1.333		650	71		721	820

For customer support on your Mechanical Costs with RSMeans Data, call 800.448.8182.

393

23 81 Decentralized Unitary HVAC Equipment

23 81 19 – Self-Contained Air-Conditioners

23 81 19.10 Window Unit Air Conditioners	Crew	Daily Output	Labor-Hours	Unit	Material	2020 Bare Costs Labor	Equipment	Total	Total Incl O&P
4520	12,000 BTUH	L-2	8	2	Ea.	2,000	93		2,093	2,350
4600	Window/thru-the-wall, 15 amp, 230 V grounded receptacle required									
4780	18,000 BTUH	L-2	6	2.667	Ea.	715	124		839	975
4940	25,000 BTUH		4	4		1,075	186		1,261	1,475
4960	29,000 BTUH	↓	4	4	↓	1,050	186		1,236	1,425

23 81 19.20 Self-Contained Single Package

		Crew	Daily Output	Labor-Hours	Unit	Material	Labor	Equipment	Total	Total Incl O&P
0010	**SELF-CONTAINED SINGLE PACKAGE**									
0100	Air cooled, for free blow or duct, not incl. remote condenser									
0110	Constant volume									
0200	3 ton cooling	Q-5	1	16	Ea.	4,050	945		4,995	5,875
0210	4 ton cooling	"	.80	20		4,400	1,175		5,575	6,625
0220	5 ton cooling	Q-6	1.20	20		4,525	1,225		5,750	6,800
0230	7.5 ton cooling	"	.90	26.667		5,650	1,625		7,275	8,675
0240	10 ton cooling	Q-7	1	32		8,350	2,000		10,350	12,200
0250	15 ton cooling		.95	33.684		10,300	2,100		12,400	14,500
0260	20 ton cooling		.90	35.556		13,800	2,225		16,025	18,500
0270	25 ton cooling		.85	37.647		20,700	2,350		23,050	26,200
0280	30 ton cooling		.80	40		26,400	2,500		28,900	32,800
0300	40 ton cooling		.60	53.333		37,300	3,325		40,625	46,000
0320	50 ton cooling	↓	.50	64		50,500	4,000		54,500	62,000
0340	60 ton cooling	Q-8	.40	80	↓	61,000	5,000	266	66,266	75,000
0380	With hot water heat, variable air volume									
0390	10 ton cooling	Q-7	.96	33.333	Ea.	11,600	2,075		13,675	15,900
0400	20 ton cooling		.86	37.209		19,500	2,325		21,825	25,000
0410	30 ton cooling		.76	42.105		35,300	2,625		37,925	42,700
0420	40 ton cooling		.58	55.172		48,200	3,450		51,650	58,000
0430	50 ton cooling	↓	.48	66.667		65,500	4,150		69,650	78,000
0440	60 ton cooling	Q-8	.38	84.211	↓	80,500	5,250	280	86,030	96,500
0490	For duct mounting, no price change									
0500	For steam heating coils, add				Ea.	10%	10%			
0550	Packaged, with electric heat									
0552	Constant volume									
0560	5 ton cooling	Q-6	1.20	20	Ea.	8,300	1,225		9,525	11,000
0570	10 ton cooling	Q-7	1	32		14,200	2,000		16,200	18,600
0580	20 ton cooling		.90	35.556		22,000	2,225		24,225	27,500
0590	30 ton cooling		.80	40		40,300	2,500		42,800	48,000
0600	40 ton cooling		.60	53.333		52,000	3,325		55,325	62,500
0610	50 ton cooling	↓	.50	64	↓	71,500	4,000		75,500	84,500
0700	Variable air volume									
0710	10 ton cooling	Q-7	1	32	Ea.	16,600	2,000		18,600	21,200
0720	20 ton cooling		.90	35.556		26,400	2,225		28,625	32,300
0730	30 ton cooling		.80	40		46,500	2,500		49,000	54,500
0740	40 ton cooling		.60	53.333		59,500	3,325		62,825	70,500
0750	50 ton cooling	↓	.50	64		81,000	4,000		85,000	95,500
0760	60 ton cooling	Q-8	.40	80	↓	93,500	5,000	266	98,766	110,500
1000	Water cooled for free blow or duct, not including tower									
1010	Constant volume									
1100	3 ton cooling	Q-6	1	24	Ea.	3,750	1,475		5,225	6,325
1120	5 ton cooling		1	24		4,875	1,475		6,350	7,575
1130	7.5 ton cooling		.80	30		6,625	1,825		8,450	10,100
1140	10 ton cooling	Q-7	.90	35.556		8,725	2,225		10,950	12,900
1150	15 ton cooling	↓	.85	37.647		12,400	2,350		14,750	17,200

23 81 Decentralized Unitary HVAC Equipment

23 81 19 – Self-Contained Air-Conditioners

23 81 19.20 Self-Contained Single Package

		Crew	Daily Output	Labor-Hours	Unit	Material	2020 Bare Costs Labor	2020 Bare Costs Equipment	Total	Total Incl O&P
1160	20 ton cooling	Q-7	.80	40	Ea.	27,700	2,500		30,200	34,100
1170	25 ton cooling		.75	42.667		32,600	2,650		35,250	39,800
1180	30 ton cooling		.70	45.714		36,600	2,850		39,450	44,600
1200	40 ton cooling		.40	80		44,500	5,000		49,500	56,500
1220	50 ton cooling		.30	107		50,500	6,650		57,150	65,500
1240	60 ton cooling	Q-8	.30	107		58,000	6,650	355	65,005	74,500
1300	For hot water or steam heat coils, add					12%	10%			
2000	Water cooled with electric heat, not including tower									
2010	Constant volume									
2100	5 ton cooling	Q-6	1.20	20	Ea.	8,675	1,225		9,900	11,400
2120	10 ton cooling	Q-7	1	32		14,500	2,000		16,500	19,000
2130	15 ton cooling	"	.90	35.556		20,700	2,225		22,925	26,100
2400	Variable air volume									
2440	10 ton cooling	Q-7	1	32	Ea.	17,000	2,000		19,000	21,700
2450	15 ton cooling	"	.90	35.556	"	25,000	2,225		27,225	30,800
2600	Water cooled, hot water coils, not including tower									
2610	Variable air volume									
2640	10 ton cooling	Q-7	1	32	Ea.	12,000	2,000		14,000	16,300
2650	15 ton cooling	"	.80	40	"	18,000	2,500		20,500	23,500

23 81 23 – Computer-Room Air-Conditioners

23 81 23.10 Computer Room Units

		Crew	Daily Output	Labor-Hours	Unit	Material	2020 Bare Costs Labor	2020 Bare Costs Equipment	Total	Total Incl O&P
0010	**COMPUTER ROOM UNITS** D3050-185									
1000	Air cooled, includes remote condenser but not									
1020	interconnecting tubing or refrigerant									
1080	3 ton	Q-5	.50	32	Ea.	27,600	1,900		29,500	33,200
1120	5 ton		.45	35.556		29,500	2,100		31,600	35,600
1160	6 ton		.30	53.333		55,000	3,150		58,150	65,000
1200	8 ton		.27	59.259		55,000	3,500		58,500	65,500
1240	10 ton		.25	64		57,500	3,775		61,275	69,000
1280	15 ton		.22	72.727		63,500	4,300		67,800	76,000
1290	18 ton		.20	80		71,500	4,725		76,225	85,500
1300	20 ton	Q-6	.26	92.308		76,000	5,650		81,650	92,000
1320	22 ton		.24	100		77,000	6,125		83,125	93,500
1360	30 ton		.21	114		94,500	7,000		101,500	114,500
2200	Chilled water, for connection to									
2220	existing chiller system of adequate capacity									
2260	5 ton	Q-5	.74	21.622	Ea.	20,000	1,275		21,275	23,900
2280	6 ton		.52	30.769		19,800	1,825		21,625	24,500
2300	8 ton		.50	32		20,600	1,900		22,500	25,500
2320	10 ton		.49	32.653		20,700	1,925		22,625	25,700
2330	12 ton		.49	32.990		21,800	1,950		23,750	26,800
2360	15 ton		.48	33.333		22,600	1,975		24,575	27,900
2400	20 ton		.46	34.783		24,100	2,050		26,150	29,600
2440	25 ton	Q-6	.63	38.095		25,600	2,325		27,925	31,700
2460	30 ton		.57	42.105		27,800	2,575		30,375	34,500
2480	35 ton		.52	46.154		28,100	2,825		30,925	35,100
2500	45 ton		.46	52.174		29,700	3,200		32,900	37,400
2520	50 ton		.42	57.143		34,000	3,500		37,500	42,700
2540	60 ton		.38	63.158		36,200	3,875		40,075	45,600
2560	77 ton		.35	68.571		57,000	4,200		61,200	69,500
2580	95 ton		.32	75		66,500	4,600		71,100	80,500
4000	Glycol system, complete except for interconnecting tubing									

For customer support on your Mechanical Costs with RSMeans Data, call 800.448.8182.

395

23 81 Decentralized Unitary HVAC Equipment

23 81 23 – Computer-Room Air-Conditioners

23 81 23.10 Computer Room Units	Crew	Daily Output	Labor-Hours	Unit	Material	2020 Bare Costs Labor	Equipment	Total	Total Incl O&P
4060	3 ton	Q-5	.40	40	Ea.	34,700	2,350		37,050	41,600
4100	5 ton		.38	42.105		37,900	2,475		40,375	45,400
4120	6 ton		.25	64		53,500	3,775		57,275	64,500
4140	8 ton		.23	69.565		61,000	4,100		65,100	73,000
4160	10 ton		.21	76.190		65,000	4,500		69,500	78,000
4180	12 ton	↓	.19	84.211		65,500	4,975		70,475	79,500
4200	15 ton	Q-6	.26	92.308		80,500	5,650		86,150	97,000
4240	20 ton		.24	100		88,000	6,125		94,125	106,000
4280	22 ton		.23	104		92,500	6,375		98,875	111,500
4430	30 ton	↓	.22	109	↓	115,000	6,675		121,675	136,500
8000	Water cooled system, not including condenser,									
8020	water supply or cooling tower									
8060	3 ton	Q-5	.62	25.806	Ea.	28,100	1,525		29,625	33,200
8100	5 ton		.54	29.630		31,700	1,750		33,450	37,500
8120	6 ton		.35	45.714		40,600	2,700		43,300	48,700
8140	8 ton		.33	48.485		50,000	2,850		52,850	59,500
8160	10 ton		.31	51.613		52,000	3,050		55,050	61,500
8180	12 ton		.29	55.172		52,500	3,250		55,750	62,500
8200	15 ton	↓	.27	59.259		61,000	3,500		64,500	72,000
8240	20 ton	Q-6	.38	63.158		65,500	3,875		69,375	78,000
8280	22 ton		.35	68.571		68,500	4,200		72,700	82,000
8300	30 ton	↓	.30	80	↓	85,500	4,900		90,400	101,500

23 81 26 – Split-System Air-Conditioners

23 81 26.10 Split Ductless Systems	Crew	Daily Output	Labor-Hours	Unit	Material	2020 Bare Costs Labor	Equipment	Total	Total Incl O&P
0010	**SPLIT DUCTLESS SYSTEMS**									
0100	Cooling only, single zone									
0110	Wall mount									
0120	3/4 ton cooling	Q-5	2	8	Ea.	1,125	470		1,595	1,925
0130	1 ton cooling		1.80	8.889		1,250	525		1,775	2,150
0140	1-1/2 ton cooling		1.60	10		1,975	590		2,565	3,050
0150	2 ton cooling		1.40	11.429		2,275	675		2,950	3,500
0160	3 ton cooling	↓	1.20	13.333	↓	2,275	785		3,060	3,675
1000	Ceiling mount									
1020	2 ton cooling	Q-5	1.40	11.429	Ea.	2,000	675		2,675	3,200
1030	3 ton cooling	"	1.20	13.333	"	2,575	785		3,360	4,025
3000	Multizone									
3010	Wall mount									
3020	2 @ 3/4 ton cooling	Q-5	1.80	8.889	Ea.	3,425	525		3,950	4,550
5000	Cooling/Heating									
5010	Wall mount									
5110	1 ton cooling	Q-5	1.70	9.412	Ea.	1,300	555		1,855	2,250
5120	1-1/2 ton cooling	"	1.50	10.667	"	2,000	630		2,630	3,150
7000	Accessories for all split ductless systems									
7010	Add for ambient frost control	Q-5	8	2	Ea.	129	118		247	320
7020	Add for tube/wiring kit (line sets)									
7030	15' kit	Q-5	32	.500	Ea.	91	29.50		120.50	144
7036	25' kit		28	.571		158	33.50		191.50	224
7040	35' kit		24	.667		205	39.50		244.50	285
7050	50' kit	↓	20	.800	↓	206	47		253	297

23 81 Decentralized Unitary HVAC Equipment

23 81 29 – Variable Refrigerant Flow HVAC Systems

23 81 29.10 Heat Pump, Gas Driven	Crew	Daily Output	Labor-Hours	Unit	Material	2020 Bare Costs Labor	Equipment	Total	Total Incl O&P
0010 **HEAT PUMP, GAS DRIVEN**, Variable refrigerant volume (VRV) type									
0020 Not including interconnecting tubing or multi-zone controls									
1000 For indoor fan VRV type AHU see 23 82 19.40									
1010 Multi-zone split									
1020 Outdoor unit									
1100 8 tons cooling, for up to 17 zones	Q-5	1.30	12.308	Ea.	30,200	725		30,925	34,300
1110 Isolation rails		2.60	6.154	Pair	1,075	365		1,440	1,725
1160 15 tons cooling, for up to 33 zones		1	16	Ea.	38,600	945		39,545	43,800
1170 Isolation rails	▼	2.60	6.154	Pair	1,250	365		1,615	1,925
2000 Packaged unit									
2020 Outdoor unit									
2200 11 tons cooling	Q-5	1.30	12.308	Ea.	38,900	725		39,625	43,900
2210 Roof curb adapter	"	2.60	6.154		675	365		1,040	1,275
2220 Thermostat	1 Stpi	1.30	6.154	▼	405	405		810	1,050

23 81 43 – Air-Source Unitary Heat Pumps

23 81 43.10 Air-Source Heat Pumps	Crew	Daily Output	Labor-Hours	Unit	Material	2020 Bare Costs Labor	Equipment	Total	Total Incl O&P
0010 **AIR-SOURCE HEAT PUMPS**, Not including interconnecting tubing									
1000 Air to air, split system, not including curbs, pads, fan coil and ductwork									
1010 For curbs/pads see Section 23 91 10									
1012 Outside condensing unit only, for fan coil see Section 23 82 19.10									
1015 1.5 ton cooling, 7 MBH heat @ 0°F	Q-5	2.40	6.667	Ea.	1,625	395		2,020	2,400
1020 2 ton cooling, 8.5 MBH heat @ 0°F		2	8		1,775	470		2,245	2,650
1030 2.5 ton cooling, 10 MBH heat @ 0°F		1.60	10		1,950	590		2,540	3,025
1040 3 ton cooling, 13 MBH heat @ 0°F		1.20	13.333		2,150	785		2,935	3,525
1050 3.5 ton cooling, 18 MBH heat @ 0°F		1	16		2,275	945		3,220	3,925
1054 4 ton cooling, 24 MBH heat @ 0°F		.80	20		2,475	1,175		3,650	4,500
1060 5 ton cooling, 27 MBH heat @ 0°F		.50	32		2,700	1,900		4,600	5,800
1080 7.5 ton cooling, 33 MBH heat @ 0°F	▼	.45	35.556		3,975	2,100		6,075	7,525
1100 10 ton cooling, 50 MBH heat @ 0°F	Q-6	.64	37.500		6,575	2,300		8,875	10,700
1120 15 ton cooling, 64 MBH heat @ 0°F		.50	48		10,400	2,925		13,325	15,800
1130 20 ton cooling, 85 MBH heat @ 0°F		.35	68.571		18,600	4,200		22,800	26,800
1140 25 ton cooling, 119 MBH heat @ 0°F	▼	.25	96	▼	21,800	5,875		27,675	32,800
1500 Single package, not including curbs, pads, or plenums									
1502 0.5 ton cooling, supplementary heat included	Q-5	8	2	Ea.	3,550	118		3,668	4,075
1504 0.75 ton cooling, supplementary heat included		6	2.667		4,150	157		4,307	4,800
1506 1 ton cooling, supplementary heat included		4	4		3,425	236		3,661	4,125
1510 1.5 ton cooling, 5 MBH heat @ 0°F		1.55	10.323		3,375	610		3,985	4,625
1520 2 ton cooling, 6.5 MBH heat @ 0°F		1.50	10.667		3,325	630		3,955	4,625
1540 2.5 ton cooling, 8 MBH heat @ 0°F		1.40	11.429		3,300	675		3,975	4,625
1560 3 ton cooling, 10 MBH heat @ 0°F		1.20	13.333		3,775	785		4,560	5,325
1570 3.5 ton cooling, 11 MBH heat @ 0°F		1	16		4,125	945		5,070	5,975
1580 4 ton cooling, 13 MBH heat @ 0°F		.96	16.667		4,425	985		5,410	6,350
1620 5 ton cooling, 27 MBH heat @ 0°F		.65	24.615		5,125	1,450		6,575	7,825
1640 7.5 ton cooling, 35 MBH heat @ 0°F	▼	.40	40		7,900	2,350		10,250	12,200
1648 10 ton cooling, 45 MBH heat @ 0°F	Q-6	.40	60		10,600	3,675		14,275	17,200
1652 12 ton cooling, 50 MBH heat @ 0°F	"	.36	66.667	▼	11,300	4,075		15,375	18,500
1696 Supplementary electric heat coil incl., unless noted otherwise									
6000 Air to water, single package, excluding storage tank and ductwork									
6010 Includes circulating water pump, air duct connections, digital temperature									
6020 controller with remote tank temp. probe and sensor for storage tank.									
6040 Water heating - air cooling capacity									
6110 35.5 MBH heat water, 2.3 ton cool air	Q-5	1.60	10	Ea.	17,300	590		17,890	20,000

For customer support on your Mechanical Costs with RSMeans Data, call 800.448.8182.

397

23 81 Decentralized Unitary HVAC Equipment

23 81 43 – Air-Source Unitary Heat Pumps

23 81 43.10 Air-Source Heat Pumps	Crew	Daily Output	Labor-Hours	Unit	Material	2020 Bare Costs Labor	Equipment	Total	Total Incl O&P
6120	58 MBH heat water, 3.8 ton cool air	Q-5	1.10	14.545	Ea.	19,800	860		20,660	23,000
6130	76 MBH heat water, 4.9 ton cool air		.87	18.391		23,600	1,075		24,675	27,500
6140	98 MBH heat water, 6.5 ton cool air		.62	25.806		30,500	1,525		32,025	35,800
6150	113 MBH heat water, 7.4 ton cool air		.59	27.119		33,300	1,600		34,900	39,000
6160	142 MBH heat water, 9.2 ton cool air		.52	30.769		40,500	1,825		42,325	47,300
6170	171 MBH heat water, 11.1 ton cool air		.49	32.653		47,500	1,925		49,425	55,000

23 81 46 – Water-Source Unitary Heat Pumps

23 81 46.10 Water Source Heat Pumps

		Crew	Daily Output	Labor-Hours	Unit	Material	2020 Bare Costs Labor	Equipment	Total	Total Incl O&P
0010	**WATER SOURCE HEAT PUMPS**, Not incl. connecting tubing or water source									
2000	Water source to air, single package									
2100	1 ton cooling, 13 MBH heat @ 75°F	Q-5	2	8	Ea.	2,125	470		2,595	3,050
2120	1.5 ton cooling, 17 MBH heat @ 75°F		1.80	8.889		2,075	525		2,600	3,050
2140	2 ton cooling, 19 MBH heat @ 75°F		1.70	9.412		2,475	555		3,030	3,550
2160	2.5 ton cooling, 25 MBH heat @ 75°F		1.60	10		2,700	590		3,290	3,850
2180	3 ton cooling, 27 MBH heat @ 75°F		1.40	11.429		2,750	675		3,425	4,025
2190	3.5 ton cooling, 29 MBH heat @ 75°F		1.30	12.308		2,975	725		3,700	4,350
2200	4 ton cooling, 31 MBH heat @ 75°F		1.20	13.333		3,125	785		3,910	4,600
2220	5 ton cooling, 29 MBH heat @ 75°F		.90	17.778		3,400	1,050		4,450	5,300
2240	7.5 ton cooling, 35 MBH heat @ 75°F		.60	26.667		6,900	1,575		8,475	9,925
2250	8 ton cooling, 40 MBH heat @ 75°F		.58	27.586		7,000	1,625		8,625	10,100
2260	10 ton cooling, 50 MBH heat @ 75°F		.53	30.189		7,675	1,775		9,450	11,100
2280	15 ton cooling, 64 MBH heat @ 75°F	Q-6	.47	51.064		14,900	3,125		18,025	21,000
2300	20 ton cooling, 100 MBH heat @ 75°F		.41	58.537		16,200	3,575		19,775	23,200
2310	25 ton cooling, 100 MBH heat @ 75°F		.32	75		22,100	4,600		26,700	31,200
2320	30 ton cooling, 128 MBH heat @ 75°F		.24	102		24,200	6,250		30,450	36,000
2340	40 ton cooling, 200 MBH heat @ 75°F		.21	117		34,200	7,175		41,375	48,300
2360	50 ton cooling, 200 MBH heat @ 75°F		.15	160		38,600	9,800		48,400	57,000
3960	For supplementary heat coil, add					10%				
4000	For increase in capacity thru use									
4020	of solar collector, size boiler at 60%									

23 82 Convection Heating and Cooling Units

23 82 13 – Valance Heating and Cooling Units

23 82 13.16 Valance Units

		Crew	Daily Output	Labor-Hours	Unit	Material	2020 Bare Costs Labor	Equipment	Total	Total Incl O&P
0010	**VALANCE UNITS**									
6000	Valance units, complete with 1/2" cooling coil, enclosure									
6020	2 tube	Q-5	18	.889	L.F.	37	52.50		89.50	120
6040	3 tube		16	1		44.50	59		103.50	138
6060	4 tube		16	1		52	59		111	146
6080	5 tube		15	1.067		57.50	63		120.50	158
6100	6 tube		15	1.067		62.50	63		125.50	163
6120	8 tube		14	1.143		91	67.50		158.50	201
6200	For 3/4" cooling coil, add					10%				

23 82 16 – Air Coils

23 82 16.10 Flanged Coils

0010	**FLANGED COILS**
0100	Basic water, DX, or condenser coils
0110	Copper tubes, alum. fins, galv. end sheets
0112	H is finned height, L is finned length
0120	3/8" x 0.016" tubing, 0.0065 aluminum fins

For customer support on your Mechanical Costs with RSMeans Data, call 800.448.8182.

23 82 Convection Heating and Cooling Units

23 82 16 – Air Coils

23 82 16.10 Flanged Coils		Crew	Daily Output	Labor-Hours	Unit	Material	2020 Bare Costs Labor	Equipment	Total	Total Incl O&P
0130	2 row, 8 fins per inch									
0140	4" H x 12" L	Q-5	48	.333	Ea.	1,225	19.65		1,244.65	1,375
0150	4" H x 24" L		38.38	.417		1,275	24.50		1,299.50	1,425
0160	4" H x 48" L		19.25	.831		1,325	49		1,374	1,525
0170	4" H x 72" L		12.80	1.250		1,475	74		1,549	1,725
0180	6" H x 12" L		48	.333		1,200	19.65		1,219.65	1,350
0190	6" H x 24" L		25.60	.625		1,275	37		1,312	1,450
0200	6" H x 48" L		12.80	1.250		1,375	74		1,449	1,600
0210	6" H x 72" L		8.53	1.876		1,850	111		1,961	2,225
0220	10" H x 12" L		30.73	.521		1,300	30.50		1,330.50	1,475
0230	10" H x 24" L		15.33	1.044		1,400	61.50		1,461.50	1,625
0240	10" H x 48" L		7.69	2.081		1,500	123		1,623	1,825
0250	10" H x 72" L		5.12	3.125		1,650	184		1,834	2,075
0260	12" H x 12" L		25.60	.625		1,400	37		1,437	1,575
0270	12" H x 24" L		12.80	1.250		1,425	74		1,499	1,675
0280	12" H x 48" L		6.40	2.500		1,550	148		1,698	1,950
0290	12" H x 72" L		4.27	3.747		1,750	221		1,971	2,250
0300	16" H x 16" L		14.38	1.113		1,425	65.50		1,490.50	1,675
0310	16" H x 24" L		9.59	1.668		1,525	98.50		1,623.50	1,825
0320	16" H x 48" L		4.80	3.333		1,675	197		1,872	2,125
0330	16" H x 72" L		3.20	5		1,875	295		2,170	2,525
0340	20" H x 24" L		7.69	2.081		1,600	123		1,723	1,950
0350	20" H x 48" L		3.84	4.167		1,775	246		2,021	2,325
0360	20" H x 72" L		2.56	6.250		2,250	370		2,620	3,050
0370	24" H x 24" L		6.40	2.500		1,700	148		1,848	2,100
0380	24" H x 48" L		3.20	5		1,875	295		2,170	2,525
0390	24" H x 72" L		2.13	7.512		2,400	445		2,845	3,325
0400	30" H x 28" L		4.39	3.645		1,850	215		2,065	2,375
0410	30" H x 48" L		2.56	6.250		2,325	370		2,695	3,100
0420	30" H x 72" L		1.71	9.357		2,600	550		3,150	3,675
0430	30" H x 84" L		1.46	10.959		2,775	645		3,420	4,025
0440	34" H x 32" L		3.39	4.720		1,950	278		2,228	2,575
0450	34" H x 48" L		2.26	7.080		2,425	420		2,845	3,275
0460	34" H x 72" L		1.51	10.596		2,700	625		3,325	3,900
0470	34" H x 84" L	▼	1.29	12.403	▼	2,925	730		3,655	4,325
0600	For 10 fins per inch, add					1%				
0610	For 12 fins per inch, add					3%				
0620	For 4 row, add				Ea.	30%	50%			
0630	For 6 row, add					60%	100%			
0640	For 8 row, add				▼	80%	150%			
1100	1/2" x 0.017" tubing, 0.0065 aluminum fins									
1110	2 row, 8 fins per inch									
1120	5" H x 6" L	Q-5	40	.400	Ea.	1,225	23.50		1,248.50	1,375
1130	5" H x 20" L		37.10	.431		1,275	25.50		1,300.50	1,475
1140	5" H x 45" L		16.41	.975		1,325	57.50		1,382.50	1,525
1150	5" H x 90" L		8.21	1.949		1,825	115		1,940	2,175
1160	7.5" H x 5" L		38.50	.416		1,175	24.50		1,199.50	1,325
1170	7.5" H x 20" L		24.62	.650		1,275	38.50		1,313.50	1,450
1180	7.5" H x 45" L		10.94	1.463		1,400	86.50		1,486.50	1,675
1190	7.5" H x 90" L		5.46	2.930		2,025	173		2,198	2,475
1200	12.5" H x 10" L		29.49	.543		1,300	32		1,332	1,475
1210	12.5" H x 20" L		14.75	1.085		1,400	64		1,464	1,625
1220	12.5" H x 45" L		6.55	2.443		1,575	144		1,719	1,950

23 82 16.10 Flanged Coils		Crew	Daily Output	Labor-Hours	Unit	Material	2020 Bare Costs Labor	Equipment	Total	Total Incl O&P
1230	12.5" H x 90" L	Q-5	3.28	4.878	Ea.	2,125	288		2,413	2,775
1240	15" H x 15" L		16.41	.975		1,350	57.50		1,407.50	1,550
1250	15" H x 30" L		8.21	1.949		1,550	115		1,665	1,900
1260	15" H x 45" L		5.46	2.930		1,650	173		1,823	2,075
1270	15" H x 90" L		2.73	5.861		2,175	345		2,520	2,925
1280	20" H x 20" L		9.22	1.735		1,600	102		1,702	1,900
1290	20" H x 45" L		4.10	3.902		1,975	230		2,205	2,525
1300	20" H x 90" L		2.05	7.805		2,475	460		2,935	3,425
1310	25" H x 25" L		5.90	2.712		1,700	160		1,860	2,125
1320	25" H x 45" L		3.28	4.878		2,025	288		2,313	2,650
1330	25" H x 90" L		1.64	9.756		2,575	575		3,150	3,675
1340	30" H x 30" L		4.10	3.902		2,100	230		2,330	2,650
1350	30" H x 45" L		2.73	5.861		2,125	345		2,470	2,850
1360	30" H x 90" L		1.37	11.679		2,775	690		3,465	4,075
1370	35" H x 35" L		3.01	5.316		3,025	315		3,340	3,800
1380	35" H x 45" L		2.34	6.838		2,250	405		2,655	3,075
1390	35" H x 90" L		1.17	13.675		3,000	805		3,805	4,500
1400	42.5" H x 45" L		1.93	8.290		2,475	490		2,965	3,450
1410	42.5" H x 75" L		1.16	13.793		3,000	815		3,815	4,525
1420	42.5" H x 105" L		.83	19.277		3,550	1,125		4,675	5,600
1600	For 10 fins per inch, add					2%				
1610	For 12 fins per inch, add					5%				
1620	For 4 row, add					35%	50%			
1630	For 6 row, add					65%	100%			
1640	For 8 row, add					90%	150%			
2010	5/8" x 0.020" tubing, 0.0065 aluminum fins									
2020	2 row, 8 fins per inch									
2030	6" H x 6" L	Q-5	38	.421	Ea.	1,075	25		1,100	1,225
2040	6" H x 24" L		24.96	.641		1,150	38		1,188	1,325
2050	6" H x 48" L		12.48	1.282		1,225	75.50		1,300.50	1,475
2060	6" H x 96" L		6.24	2.564		1,675	151		1,826	2,050
2070	12" H x 12" L		24.96	.641		1,225	38		1,263	1,400
2080	12" H x 24" L		12.48	1.282		1,425	75.50		1,500.50	1,675
2090	12" H x 48" L		6.24	2.564		1,575	151		1,726	1,950
2100	12" H x 96" L		3.12	5.128		2,850	305		3,155	3,575
2110	18" H x 24" L		8.32	1.923		1,575	113		1,688	1,925
2120	18" H x 48" L		4.16	3.846		1,575	227		1,802	2,075
2130	18" H x 96" L		2.08	7.692		2,200	455		2,655	3,075
2140	24" H x 24" L		6.24	2.564		1,675	151		1,826	2,075
2150	24" H x 48" L		3.12	5.128		1,750	305		2,055	2,375
2160	24" H x 96" L		1.56	10.256		2,400	605		3,005	3,525
2170	30" H x 30" L		3.99	4.010		1,825	237		2,062	2,350
2180	30" H x 48" L		2.50	6.400		2,050	380		2,430	2,825
2190	30" H x 96" L		1.25	12.800		2,700	755		3,455	4,100
2200	30" H x 120" L		1	16		2,725	945		3,670	4,425
2210	36" H x 36" L		2.77	5.776		2,050	340		2,390	2,775
2220	36" H x 48" L		2.08	7.692		2,175	455		2,630	3,075
2230	36" H x 90" L		1.11	14.414		2,600	850		3,450	4,125
2240	36" H x 120" L		.83	19.277		2,950	1,125		4,075	4,925
2250	42" H x 48" L		1.78	8.989		2,325	530		2,855	3,350
2260	42" H x 90" L		.95	16.842		2,875	995		3,870	4,650
2270	42" H x 120" L		.71	22.535		3,925	1,325		5,250	6,325
2280	51" H x 48" L		1.47	10.884		2,575	640		3,215	3,775

23 82 Convection Heating and Cooling Units

23 82 16 – Air Coils

23 82 16.10 Flanged Coils		Crew	Daily Output	Labor-Hours	Unit	Material	2020 Bare Costs Labor	Equipment	Total	Total Incl O&P
2290	51" H x 96" L	Q-5	.73	21.918	Ea.	3,975	1,300		5,275	6,300
2300	51" H x 120" L	▼	.59	27.119	▼	4,000	1,600		5,600	6,800
2500	For 10 fins per inch, add					3%				
2510	For 12 fins per inch, add					6%				
2520	For 4 row, add				Ea.	40%	50%			
2530	For 6 row, add					80%	100%			
2540	For 8 row, add				▼	110%	150%			
3000	Hot water booster coils									
3010	Copper tubes, alum. fins, galv. end sheets									
3012	H is finned height, L is finned length									
3020	1/2" x 0.017" tubing, 0.0065 aluminum fins									
3030	1 row, 10 fins per inch									
3040	5" H x 10" L	Q-5	40	.400	Ea.	875	23.50		898.50	1,000
3050	5" H x 25" L		33.18	.482		950	28.50		978.50	1,100
3060	10" H x 10" L		39	.410		910	24		934	1,025
3070	10" H x 20" L		20.73	.772		980	45.50		1,025.50	1,150
3080	10" H x 30" L		13.83	1.157		1,025	68.50		1,093.50	1,225
3090	15" H x 15" L		18.44	.868		925	51		976	1,100
3110	15" H x 20" L		13.83	1.157		1,025	68.50		1,093.50	1,225
3120	15" H x 30" L		9.22	1.735		1,075	102		1,177	1,325
3130	20" H x 20" L		10.37	1.543		1,075	91		1,166	1,300
3140	20" H x 30" L		6.91	2.315		1,125	137		1,262	1,425
3150	20" H x 40" L		5.18	3.089		1,175	182		1,357	1,575
3160	25" H x 25" L		6.64	2.410		1,150	142		1,292	1,500
3170	25" H x 35" L		4.74	3.376		1,200	199		1,399	1,625
3180	25" H x 45" L		3.69	4.336		1,325	256		1,581	1,825
3190	30" H x 30" L		4.61	3.471		1,225	205		1,430	1,650
3200	30" H x 40" L		3.46	4.624		1,275	273		1,548	1,800
3210	30" H x 50" L	▼	2.76	5.797		1,350	340		1,690	2,000
3300	For 2 row, add				▼	20%	50%			
3400	1/2" x 0.020" tubing, 0.008 aluminum fins									
3410	2 row, 10 fins per inch									
3420	20" H x 27" L	Q-5	7.27	2.201	Ea.	1,275	130		1,405	1,600
3430	20" H x 36" L		5.44	2.941		1,350	174		1,524	1,725
3440	20" H x 45" L		4.35	3.678		1,325	217		1,542	1,800
3450	25" H x 45" L		3.49	4.585		1,700	270		1,970	2,275
3460	25" H x 54" L		2.90	5.517		1,900	325		2,225	2,550
3470	30" H x 54" L		2.42	6.612		2,025	390		2,415	2,800
3480	35" H x 54" L		2.07	7.729		2,175	455		2,630	3,050
3490	35" H x 63" L		1.78	8.989		2,175	530		2,705	3,200
3500	40" H x 72" L		1.36	11.765		2,600	695		3,295	3,900
3510	55" H x 63" L		1.13	14.159		2,900	835		3,735	4,425
3520	55" H x 72" L		.99	16.162		3,200	955		4,155	4,950
3530	55" H x 81" L		.88	18.182		3,475	1,075		4,550	5,425
3540	75" H x 81" L		.64	25		4,275	1,475		5,750	6,900
3550	75" H x 90" L		.58	27.586		4,350	1,625		5,975	7,225
3560	75" H x 108" L	▼	.48	33.333	▼	4,875	1,975		6,850	8,325
3800	5/8" x 0.020" tubing, 0.0065 aluminum fins									
3810	1 row, 10 fins per inch									
3820	6" H x 12" L	Q-5	38	.421	Ea.	795	25		820	910
3830	6" H x 24" L		28.80	.556		860	33		893	1,000
3840	6" H x 30" L		23.04	.694		795	41		836	935
3850	12" H x 12" L		28.80	.556		895	33		928	1,025

For customer support on your Mechanical Costs with RSMeans Data, call 800.448.8182.

401

23 82 Convection Heating and Cooling Units

23 82 16 – Air Coils

23 82 16.10 Flanged Coils	Crew	Daily Output	Labor-Hours	Unit	Material	2020 Bare Costs Labor	Equipment	Total	Total Incl O&P
3860	12" H x 24" L	Q-5	14.40	1.111	Ea.	800	65.50		865.50	980
3870	12" H x 30" L		11.52	1.389		940	82		1,022	1,150
3880	18" H x 18" L		12.80	1.250		1,100	74		1,174	1,300
3890	18" H x 24" L		9.60	1.667		1,125	98.50		1,223.50	1,400
3900	18" H x 36" L		6.40	2.500		1,175	148		1,323	1,525
3910	24" H x 24" L		7.20	2.222		1,175	131		1,306	1,500
3920	24" H x 30" L		4.80	3.333		1,225	197		1,422	1,650
3930	24" H x 36" L		5.05	3.168		1,225	187		1,412	1,625
3940	30" H x 30" L		4.61	3.471		1,250	205		1,455	1,675
3950	30" H x 36" L		3.85	4.159		1,300	245		1,545	1,800
3960	30" H x 42" L		3.29	4.863		1,300	287		1,587	1,875
3970	36" H x 36" L		3.20	5		1,325	295		1,620	1,925
3980	36" H x 42" L	▼	2.74	5.839		1,375	345		1,720	2,050
4100	For 2 row, add				▼	30%	50%			
4400	Steam and hot water heating coils									
4410	Copper tubes, alum. fins. galv. end sheets									
4412	H is finned height, L is finned length									
4420	5/8" tubing, to 175 PSIG working pressure									
4430	1 row, 8 fins per inch									
4440	6" H x 6" L	Q-5	38	.421	Ea.	1,275	25		1,300	1,425
4450	6" H x 12" L		37	.432		1,275	25.50		1,300.50	1,450
4460	6" H x 24" L		24.48	.654		1,375	38.50		1,413.50	1,550
4470	6" H x 36" L		16.32	.980		1,325	58		1,383	1,550
4480	6" H x 48" L		12.24	1.307		1,375	77		1,452	1,625
4490	6" H x 84" L		6.99	2.289		2,100	135		2,235	2,525
4500	6" H x 108" L		5.44	2.941		2,300	174		2,474	2,775
4510	12" H x 12" L		24.48	.654		1,425	38.50		1,463.50	1,600
4520	12" H x 24" L		12.24	1.307		1,450	77		1,527	1,725
4530	12" H x 36" L		8.16	1.961		1,600	116		1,716	1,950
4540	12" H x 48" L		6.12	2.614		1,575	154		1,729	1,975
4550	12" H x 84" L		3.50	4.571		2,425	270		2,695	3,050
4560	12" H x 108" L		2.72	5.882		2,975	345		3,320	3,800
4570	18" H x 18" L		10.88	1.471		1,550	87		1,637	1,825
4580	18" H x 24" L		8.16	1.961		1,600	116		1,716	1,925
4590	18" H x 48" L		4.08	3.922		1,875	231		2,106	2,400
4600	18" H x 84" L		2.33	6.867		3,025	405		3,430	3,925
4610	18" H x 108" L		1.81	8.840		3,300	520		3,820	4,400
4620	24" H x 24" L		6.12	2.614		1,875	154		2,029	2,275
4630	24" H x 48" L		3.06	5.229		2,175	310		2,485	2,850
4640	24" H x 84" L		1.75	9.143		3,350	540		3,890	4,475
4650	24" H x 108" L		1.36	11.765		4,025	695		4,720	5,475
4660	36" H x 36" L		2.72	5.882		2,425	345		2,770	3,200
4670	36" H x 48" L		2.04	7.843		2,650	465		3,115	3,600
4680	36" H x 84" L		1.17	13.675		4,525	805		5,330	6,175
4690	36" H x 108" L		.91	17.582		5,350	1,025		6,375	7,425
4700	48" H x 48" L		1.53	10.458		3,675	615		4,290	4,975
4710	48" H x 84" L		.87	18.391		5,350	1,075		6,425	7,500
4720	48" H x 108" L		.68	23.529		6,400	1,400		7,800	9,100
4730	48" H x 126" L	▼	.58	27.586	▼	6,750	1,625		8,375	9,850
4910	For 10 fins per inch, add					2%				
4920	For 12 fins per inch, add					5%				
4930	For 2 row, add					30%	50%			
4940	For 3 row, add					50%	100%			

23 82 16.10 Flanged Coils

		Daily Output	Labor-Hours	Unit	Material	2020 Bare Costs Labor	2020 Bare Costs Equipment	Total	Total Incl O&P
4950	For 4 row, add				80%	123%			
4960	For 230 PSIG heavy duty, add				20%	30%			
6000	Steam, heavy duty, 200 PSIG								
6010	Copper tubes, alum. fins, galv. end sheets								
6012	H is finned height, L is finned length								
6020	1" x 0.035" tubing, 0.010 aluminum fins								
6030	1 row, 8 fins per inch								
6040	12" H x 12" L	Q-5	24.48	.654	Ea.	1,425	38.50		1,463.50	1,625
6050	12" H x 24" L		12.24	1.307		1,725	77		1,802	2,025
6060	12" H x 36" L		8.16	1.961		1,850	116		1,966	2,200
6070	12" H x 48" L		6.12	2.614		1,900	154		2,054	2,300
6080	12" H x 84" L		3.50	4.571		2,475	270		2,745	3,125
6090	12" H x 108" L		2.72	5.882		3,150	345		3,495	4,000
6100	18" H x 24" L		8.16	1.961		1,900	116		2,016	2,250
6110	18" H x 36" L		5.44	2.941		2,125	174		2,299	2,575
6120	18" H x 48" L		4.08	3.922		2,125	231		2,356	2,675
6130	18" H x 84" L		2.33	6.867		3,625	405		4,030	4,575
6140	18" H x 108" L		1.81	8.840		4,200	520		4,720	5,400
6150	24" H x 24" L		6.12	2.614		2,275	154		2,429	2,725
6160	24" H x 36" L		4.08	3.922		2,400	231		2,631	2,975
6170	24" H x 48" L		3.06	5.229		2,700	310		3,010	3,400
6180	24" H x 84" L		1.75	9.143		4,000	540		4,540	5,200
6190	24" H x 108" L		1.36	11.765		4,975	695		5,670	6,525
6200	30" H x 36" L		3.26	4.908		3,225	290		3,515	3,975
6210	30" H x 48" L		2.45	6.531		3,575	385		3,960	4,500
6220	30" H x 84" L		1.40	11.429		4,900	675		5,575	6,375
6230	30" H x 108" L		1.09	14.679		5,825	865		6,690	7,700
6240	36" H x 36" L		2.72	5.882		3,600	345		3,945	4,500
6250	36" H x 48" L		2.04	7.843		3,775	465		4,240	4,850
6260	36" H x 84" L		1.17	13.675		5,525	805		6,330	7,300
6270	36" H x 108" L		.91	17.582		6,575	1,025		7,600	8,775
6280	36" H x 120" L		.82	19.512		7,025	1,150		8,175	9,450
6290	48" H x 48" L		1.53	10.458		5,175	615		5,790	6,600
6300	48" H x 84" L		.87	18.391		6,875	1,075		7,950	9,200
6310	48" H x 108" L		.68	23.529		8,225	1,400		9,625	11,100
6320	48" H x 120" L		.61	26.230		8,800	1,550		10,350	12,000
6400	For 10 fins per inch, add				4%				
6410	For 12 fins per inch, add				10%				

23 82 16.20 Duct Heaters

		Daily Output	Labor-Hours	Unit	Material	2020 Bare Costs Labor	2020 Bare Costs Equipment	Total	Total Incl O&P
0010	**DUCT HEATERS**, Electric, 480 V, 3 Ph.								
0020	Finned tubular insert, 500°F								
0100	8" wide x 6" high, 4.0 kW	Q-20	16	1.250	Ea.	825	71.50		896.50	1,025
0120	12" high, 8.0 kW		15	1.333		1,375	76		1,451	1,625
0140	18" high, 12.0 kW		14	1.429		1,925	81.50		2,006.50	2,225
0160	24" high, 16.0 kW		13	1.538		2,475	88		2,563	2,850
0180	30" high, 20.0 kW		12	1.667		3,025	95		3,120	3,475
0300	12" wide x 6" high, 6.7 kW		15	1.333		875	76		951	1,075
0320	12" high, 13.3 kW		14	1.429		1,425	81.50		1,506.50	1,675
0340	18" high, 20.0 kW		13	1.538		1,975	88		2,063	2,300
0360	24" high, 26.7 kW		12	1.667		2,550	95		2,645	2,950
0380	30" high, 33.3 kW		11	1.818		3,125	104		3,229	3,575
0500	18" wide x 6" high, 13.3 kW		14	1.429		965	81.50		1,046.50	1,200

For customer support on your Mechanical Costs with RSMeans Data, call 800.448.8182.

403

23 82 16.20 Duct Heaters		Crew	Daily Output	Labor-Hours	Unit	Material	2020 Bare Costs Labor	Equipment	Total	Total Incl O&P
0520	12" high, 26.7 kW	Q-20	13	1.538	Ea.	1,625	88		1,713	1,900
0540	18" high, 40.0 kW		12	1.667		2,150	95		2,245	2,525
0560	24" high, 53.3 kW		11	1.818		2,875	104		2,979	3,300
0580	30" high, 66.7 kW		10	2		3,575	114		3,689	4,125
0700	24" wide x 6" high, 17.8 kW		13	1.538		1,075	88		1,163	1,300
0720	12" high, 35.6 kW		12	1.667		1,750	95		1,845	2,075
0740	18" high, 53.3 kW		11	1.818		2,525	104		2,629	2,925
0760	24" high, 71.1 kW		10	2		3,175	114		3,289	3,675
0780	30" high, 88.9 kW		9	2.222		3,900	127		4,027	4,500
0900	30" wide x 6" high, 22.2 kW		12	1.667		1,125	95		1,220	1,400
0920	12" high, 44.4 kW		11	1.818		1,850	104		1,954	2,200
0940	18" high, 66.7 kW		10	2		2,675	114		2,789	3,125
0960	24" high, 88.9 kW		9	2.222		3,500	127		3,627	4,050
0980	30" high, 111.0 kW		8	2.500		4,125	143		4,268	4,775
1400	Note decreased kW available for									
1410	each duct size at same cost									
1420	See line 5000 for modifications and accessories									
2000	Finned tubular flange with insulated									
2020	terminal box, 500°F									
2100	12" wide x 36" high, 54 kW	Q-20	10	2	Ea.	4,075	114		4,189	4,650
2120	40" high, 60 kW		9	2.222		4,675	127		4,802	5,350
2200	24" wide x 36" high, 118.8 kW		9	2.222		5,175	127		5,302	5,875
2220	40" high, 132 kW		8	2.500		5,250	143		5,393	6,000
2400	36" wide x 8" high, 40 kW		11	1.818		2,100	104		2,204	2,475
2420	16" high, 80 kW		10	2		2,750	114		2,864	3,225
2440	24" high, 120 kW		9	2.222		3,625	127		3,752	4,175
2460	32" high, 160 kW		8	2.500		4,625	143		4,768	5,325
2480	36" high, 180 kW		7	2.857		5,700	163		5,863	6,525
2500	40" high, 200 kW		6	3.333		6,325	190		6,515	7,250
2600	40" wide x 8" high, 45 kW		11	1.818		2,150	104		2,254	2,525
2620	16" high, 90 kW		10	2		2,975	114		3,089	3,450
2640	24" high, 135 kW		9	2.222		3,775	127		3,902	4,350
2660	32" high, 180 kW		8	2.500		5,025	143		5,168	5,750
2680	36" high, 202.5 kW		7	2.857		5,800	163		5,963	6,625
2700	40" high, 225 kW		6	3.333		6,600	190		6,790	7,550
2800	48" wide x 8" high, 54.8 kW		10	2		2,275	114		2,389	2,675
2820	16" high, 109.8 kW		9	2.222		3,150	127		3,277	3,650
2840	24" high, 164.4 kW		8	2.500		4,000	143		4,143	4,625
2860	32" high, 219.2 kW		7	2.857		5,325	163		5,488	6,125
2880	36" high, 246.6 kW		6	3.333		6,125	190		6,315	7,050
2900	40" high, 274 kW		5	4		6,925	229		7,154	7,975
3000	56" wide x 8" high, 64 kW		9	2.222		2,575	127		2,702	3,050
3020	16" high, 128 kW		8	2.500		3,575	143		3,718	4,150
3040	24" high, 192 kW		7	2.857		4,325	163		4,488	5,025
3060	32" high, 256 kW		6	3.333		6,050	190		6,240	6,975
3080	36" high, 288 kW		5	4		6,950	229		7,179	8,000
3100	40" high, 320 kW		4	5		7,675	286		7,961	8,875
3200	64" wide x 8" high, 74 kW		8	2.500		2,650	143		2,793	3,150
3220	16" high, 148 kW		7	2.857		3,675	163		3,838	4,300
3240	24" high, 222 kW		6	3.333		4,675	190		4,865	5,450
3260	32" high, 296 kW		5	4		6,275	229		6,504	7,250
3280	36" high, 333 kW		4	5		7,475	286		7,761	8,650
3300	40" high, 370 kW		3	6.667		8,250	380		8,630	9,650

23 82 Convection Heating and Cooling Units

23 82 16 – Air Coils

23 82 16.20 Duct Heaters

		Crew	Daily Output	Labor-Hours	Unit	Material	2020 Bare Costs Labor	Equipment	Total	Total Incl O&P
3800	Note decreased kW available for									
3820	each duct size at same cost									
5000	Duct heater modifications and accessories									
5120	T.C.O. limit auto or manual reset	Q-20	42	.476	Ea.	135	27		162	190
5140	Thermostat		28	.714		580	41		621	695
5160	Overheat thermocouple (removable)		7	2.857		820	163		983	1,150
5180	Fan interlock relay		18	1.111		197	63.50		260.50	315
5200	Air flow switch		20	1		171	57		228	275
5220	Split terminal box cover	↓	100	.200	↓	56.50	11.45		67.95	79.50
8000	To obtain BTU multiply kW by 3413									

23 82 19 – Fan Coil Units

23 82 19.10 Fan Coil Air Conditioning

		Crew	Daily Output	Labor-Hours	Unit	Material	2020 Bare Costs Labor	Equipment	Total	Total Incl O&P
0010	**FAN COIL AIR CONDITIONING**									
0030	Fan coil AC, cabinet mounted, filters and controls									
0100	Chilled water, 1/2 ton cooling	Q-5	8	2	Ea.	575	118		693	810
0110	3/4 ton cooling		7	2.286		700	135		835	970
0120	1 ton cooling		6	2.667		820	157		977	1,125
0140	1-1/2 ton cooling		5.50	2.909		865	172		1,037	1,200
0150	2 ton cooling		5.25	3.048		1,175	180		1,355	1,550
0160	2-1/2 ton cooling		5	3.200		1,800	189		1,989	2,250
0180	3 ton cooling	↓	4	4	↓	1,975	236		2,211	2,525
0262	For hot water coil, add					40%	10%			
0300	Console, 2 pipe with electric heat									
0310	1/2 ton cooling	Q-5	8	2	Ea.	1,500	118		1,618	1,800
0315	3/4 ton cooling		7	2.286		1,625	135		1,760	1,975
0320	1 ton cooling		6	2.667		1,750	157		1,907	2,150
0330	1-1/2 ton cooling		5.50	2.909		2,150	172		2,322	2,625
0340	2 ton cooling		4.70	3.404		2,850	201		3,051	3,450
0345	2-1/2 ton cooling		4.70	3.404		4,125	201		4,326	4,850
0350	3 ton cooling		4	4		5,525	236		5,761	6,450
0360	4 ton cooling		4	4		5,750	236		5,986	6,700
0940	Direct expansion, for use w/air cooled condensing unit, 1-1/2 ton cooling		5	3.200		685	189		874	1,050
0950	2 ton cooling		4.80	3.333		670	197		867	1,025
0960	2-1/2 ton cooling		4.40	3.636		785	215		1,000	1,175
0970	3 ton cooling		3.80	4.211		960	248		1,208	1,425
0980	3-1/2 ton cooling		3.60	4.444		990	262		1,252	1,500
0990	4 ton cooling		3.40	4.706		995	278		1,273	1,525
1000	5 ton cooling		3	5.333		1,250	315		1,565	1,850
1020	7-1/2 ton cooling	↓	3	5.333		1,775	315		2,090	2,425
1040	10 ton cooling	Q-6	2.60	9.231		2,250	565		2,815	3,350
1042	12-1/2 ton cooling		2.30	10.435		2,875	640		3,515	4,100
1050	15 ton cooling		1.60	15		2,950	920		3,870	4,625
1060	20 ton cooling		.70	34.286		4,325	2,100		6,425	7,900
1070	25 ton cooling		.65	36.923		5,225	2,250		7,475	9,125
1080	30 ton cooling	↓	.60	40		5,675	2,450		8,125	9,925
1500	For hot water coil, add				↓	40%	10%			
1512	For condensing unit add see Section 23 62									
3000	Chilled water, horizontal unit, housing, 2 pipe, fan control, no valves									
3100	1/2 ton cooling	Q-5	8	2	Ea.	1,425	118		1,543	1,750
3110	1 ton cooling		6	2.667		1,675	157		1,832	2,075
3120	1-1/2 ton cooling		5.50	2.909		1,875	172		2,047	2,325
3130	2 ton cooling	↓	5.25	3.048		2,250	180		2,430	2,750

For customer support on your Mechanical Costs with RSMeans Data, call 800.448.8182.

405

23 82 19.10 Fan Coil Air Conditioning

		Crew	Daily Output	Labor-Hours	Unit	Material	2020 Bare Costs Labor	Equipment	Total	Total Incl O&P
3140	3 ton cooling	Q-5	4	4	Ea.	2,775	236		3,011	3,400
3150	3-1/2 ton cooling		3.80	4.211		3,075	248		3,323	3,750
3160	4 ton cooling		3.80	4.211		3,075	248		3,323	3,750
3170	5 ton cooling		3.40	4.706		3,550	278		3,828	4,325
3180	6 ton cooling		3.40	4.706		3,550	278		3,828	4,325
3190	7 ton cooling		2.90	5.517		3,875	325		4,200	4,725
3200	8 ton cooling		2.90	5.517		3,875	325		4,200	4,725
3210	10 ton cooling	Q-6	2.80	8.571		4,050	525		4,575	5,225
3212	12-1/2 ton cooling		2.50	9.600		3,350	585		3,935	4,550
3214	15 ton cooling		2.30	10.435		3,975	640		4,615	5,325
3216	20 ton cooling		2.10	11.429		4,575	700		5,275	6,100
3218	25 ton cooling		1.80	13.333		6,775	815		7,590	8,675
3220	30 ton cooling		1.60	15		8,325	920		9,245	10,500
3240	For hot water coil, add					40%	10%			
4000	With electric heat, 2 pipe									
4100	1/2 ton cooling	Q-5	8	2	Ea.	1,675	118		1,793	2,025
4105	3/4 ton cooling		7	2.286		1,875	135		2,010	2,250
4110	1 ton cooling		6	2.667		2,050	157		2,207	2,475
4120	1-1/2 ton cooling		5.50	2.909		2,275	172		2,447	2,750
4130	2 ton cooling		5.25	3.048		2,875	180		3,055	3,425
4135	2-1/2 ton cooling		4.80	3.333		3,975	197		4,172	4,675
4140	3 ton cooling		4	4		5,300	236		5,536	6,175
4150	3-1/2 ton cooling		3.80	4.211		6,250	248		6,498	7,250
4160	4 ton cooling		3.60	4.444		7,200	262		7,462	8,325
4170	5 ton cooling		3.40	4.706		8,725	278		9,003	10,000
4180	6 ton cooling		3.20	5		8,725	295		9,020	10,000
4190	7 ton cooling		2.90	5.517		8,975	325		9,300	10,400
4200	8 ton cooling		2.20	7.273		9,575	430		10,005	11,100
4210	10 ton cooling	Q-6	2.80	8.571		10,400	525		10,925	12,200

23 82 19.20 Heating and Ventilating Units

		Crew	Daily Output	Labor-Hours	Unit	Material	2020 Bare Costs Labor	Equipment	Total	Total Incl O&P
0010	**HEATING AND VENTILATING UNITS,** Classroom units									
0020	Includes filter, heating/cooling coils, standard controls									
0080	750 CFM, 2 tons cooling	Q-6	2	12	Ea.	4,750	735		5,485	6,350
0100	1,000 CFM, 2-1/2 tons cooling		1.60	15		5,550	920		6,470	7,475
0120	1,250 CFM, 3 tons cooling		1.40	17.143		5,725	1,050		6,775	7,875
0140	1,500 CFM, 4 tons cooling		.80	30		6,125	1,825		7,950	9,500
0160	2,000 CFM, 5 tons cooling		.50	48		7,200	2,925		10,125	12,300
0500	For electric heat, add					35%				
1000	For no cooling, deduct					25%	10%			

23 82 19.40 Fan Coil Air Conditioning

			Crew	Daily Output	Labor-Hours	Unit	Material	2020 Bare Costs Labor	Equipment	Total	Total Incl O&P
0010	**FAN COIL AIR CONDITIONING,** Variable refrigerant volume (VRV) type										
0020	Not including interconnecting tubing or multi-zone controls										
0030	For VRV condensing unit see section 23 81 29.10										
0050	Indoor type, ducted										
0100	Vertical concealed										
0130	1 ton cooling	G	Q-5	2.97	5.387	Ea.	2,650	320		2,970	3,375
0140	1.5 ton cooling	G		2.77	5.776		2,675	340		3,015	3,450
0150	2 ton cooling	G		2.70	5.926		2,900	350		3,250	3,700
0160	2.5 ton cooling	G		2.60	6.154		3,050	365		3,415	3,900
0170	3 ton cooling	G		2.50	6.400		3,100	380		3,480	3,975
0180	3.5 ton cooling	G		2.30	6.957		3,250	410		3,660	4,200
0190	4 ton cooling	G		2.19	7.306		3,275	430		3,705	4,250

23 82 19 – Fan Coil Units

23 82 19.40 Fan Coil Air Conditioning		Crew	Daily Output	Labor-Hours	Unit	Material	2020 Bare Costs Labor	Equipment	Total	Total Incl O&P
0200	4.5 ton cooling [G]	Q-5	2.08	7.692	Ea.	3,600	455		4,055	4,625
0230	Outside air connection possible									
1100	Ceiling concealed									
1130	0.6 ton cooling [G]	Q-5	2.60	6.154	Ea.	1,725	365		2,090	2,450
1140	0.75 ton cooling [G]		2.60	6.154		1,775	365		2,140	2,525
1150	1 ton cooling [G]		2.60	6.154		1,875	365		2,240	2,625
1152	Screening door [G]	1 Stpi	5.20	1.538		104	101		205	266
1160	1.5 ton cooling [G]	Q-5	2.60	6.154		1,950	365		2,315	2,700
1162	Screening door [G]	1 Stpi	5.20	1.538		121	101		222	284
1170	2 ton cooling [G]	Q-5	2.60	6.154		2,300	365		2,665	3,075
1172	Screening door [G]	1 Stpi	5.20	1.538		104	101		205	266
1180	2.5 ton cooling [G]	Q-5	2.60	6.154		2,650	365		3,015	3,475
1182	Screening door [G]	1 Stpi	5.20	1.538		181	101		282	350
1190	3 ton cooling [G]	Q-5	2.60	6.154		2,825	365		3,190	3,650
1192	Screening door [G]	1 Stpi	5.20	1.538		181	101		282	350
1200	4 ton cooling [G]	Q-5	2.30	6.957		2,675	410		3,085	3,575
1202	Screening door [G]	1 Stpi	5.20	1.538		181	101		282	350
1220	6 ton cooling [G]	Q-5	2.08	7.692		4,900	455		5,355	6,075
1230	8 ton cooling [G]	"	1.89	8.466		5,500	500		6,000	6,800
1260	Outside air connection possible									
4050	Indoor type, duct-free									
4100	Ceiling mounted cassette									
4130	0.75 ton cooling [G]	Q-5	3.46	4.624	Ea.	2,050	273		2,323	2,650
4140	1 ton cooling [G]		2.97	5.387		2,175	320		2,495	2,875
4150	1.5 ton cooling [G]		2.77	5.776		2,250	340		2,590	2,950
4160	2 ton cooling [G]		2.60	6.154		2,375	365		2,740	3,175
4170	2.5 ton cooling [G]		2.50	6.400		2,450	380		2,830	3,275
4180	3 ton cooling [G]		2.30	6.957		2,625	410		3,035	3,500
4190	4 ton cooling [G]		2.08	7.692		2,925	455		3,380	3,900
4198	For ceiling cassette decoration panel, add [G]	1 Stpi	5.20	1.538		310	101		411	495
4230	Outside air connection possible									
4500	Wall mounted									
4530	0.6 ton cooling [G]	Q-5	5.20	3.077	Ea.	1,050	182		1,232	1,425
4540	0.75 ton cooling [G]		5.20	3.077		1,100	182		1,282	1,475
4550	1 ton cooling [G]		4.16	3.846		1,250	227		1,477	1,725
4560	1.5 ton cooling [G]		3.77	4.244		1,375	250		1,625	1,900
4570	2 ton cooling [G]		3.46	4.624		1,500	273		1,773	2,050
4590	Condensate pump for the above air handlers [G]	1 Stpi	6.46	1.238		225	81		306	370
4700	Floor standing unit									
4730	1 ton cooling [G]	Q-5	2.60	6.154	Ea.	1,775	365		2,140	2,500
4740	1.5 ton cooling [G]		2.60	6.154		2,000	365		2,365	2,775
4750	2 ton cooling [G]		2.60	6.154		2,150	365		2,515	2,925
4800	Floor standing unit, concealed									
4830	1 ton cooling [G]	Q-5	2.60	6.154	Ea.	1,700	365		2,065	2,425
4840	1.5 ton cooling [G]		2.60	6.154		1,925	365		2,290	2,675
4850	2 ton cooling [G]		2.60	6.154		2,075	365		2,440	2,825
4880	Outside air connection possible									
8000	Accessories									
8100	Branch divergence pipe fitting									
8110	Capacity under 76 MBH	1 Stpi	2.50	3.200	Ea.	164	210		374	495
8120	76 MBH to 112 MBH		2.50	3.200		188	210		398	520
8130	112 MBH to 234 MBH		2.50	3.200		530	210		740	895
8200	Header pipe fitting									

For customer support on your Mechanical Costs with RSMeans Data, call 800.448.8182.

407

23 82 Convection Heating and Cooling Units

23 82 19 – Fan Coil Units

23 82 19.40 Fan Coil Air Conditioning	Crew	Daily Output	Labor-Hours	Unit	Material	2020 Bare Costs Labor	Equipment	Total	Total Incl O&P
8210 Max 4 branches									
8220 Capacity under 76 MBH	1 Stpi	1.70	4.706	Ea.	233	310		543	715
8260 Max 8 branches									
8270 76 MBH to 112 MBH	1 Stpi	1.70	4.706	Ea.	475	310		785	985
8280 112 MBH to 234 MBH	"	1.30	6.154	"	755	405		1,160	1,425

23 82 29 – Radiators

23 82 29.10 Hydronic Heating

	Crew	Daily Output	Labor-Hours	Unit	Material	Labor	Equipment	Total	Total Incl O&P
0010 **HYDRONIC HEATING**, Terminal units, not incl. main supply pipe									
1000 Radiation									
1100 Panel, baseboard, C.I., including supports, no covers	Q-5	46	.348	L.F.	49	20.50		69.50	84.50
3000 Radiators, cast iron									
3100 Free standing or wall hung, 6 tube, 25" high	Q-5	96	.167	Section	61	9.85		70.85	81.50
3150 4 tube, 25" high		96	.167		47.50	9.85		57.35	66.50
3200 4 tube, 19" high	↓	96	.167	↓	42	9.85		51.85	60.50
3250 Adj. brackets, 2 per wall radiator up to 30 sections	1 Stpi	32	.250	Ea.	66	16.40		82.40	97
3500 Recessed, 20" high x 5" deep, without grille	Q-5	60	.267	Section	51.50	15.75		67.25	80.50
3525 Free standing or wall hung, 30" high	"	60	.267		46.50	15.75		62.25	74.50
3600 For inlet grille, add				↓	6.40			6.40	7.05
9500 To convert SFR to BTU rating: Hot water, 150 x SFR									
9510 Forced hot water, 180 x SFR; steam, 240 x SFR									

23 82 33 – Convectors

23 82 33.10 Convector Units

	Crew	Daily Output	Labor-Hours	Unit	Material	Labor	Equipment	Total	Total Incl O&P
0010 **CONVECTOR UNITS**, Terminal units, not incl. main supply pipe									
2204 Convector, multifin, 2 pipe w/cabinet									
2210 17" H x 24" L	Q-5	10	1.600	Ea.	110	94.50		204.50	262
2214 17" H x 36" L		8.60	1.860		165	110		275	345
2218 17" H x 48" L		7.40	2.162		220	128		348	435
2222 21" H x 24" L		9	1.778		124	105		229	293
2226 21" H x 36" L		8.20	1.951		185	115		300	375
2228 21" H x 48" L	↓	6.80	2.353	↓	247	139		386	480
2240 For knob operated damper, add					140%				
2241 For metal trim strips, add	Q-5	64	.250	Ea.	14.25	14.75		29	37.50
2243 For snap-on inlet grille, add					10%	10%			
2245 For hinged access door, add	Q-5	64	.250	Ea.	40.50	14.75		55.25	66.50
2246 For air chamber, auto-venting, add	"	58	.276	"	8.75	16.30		25.05	34

23 82 36 – Finned-Tube Radiation Heaters

23 82 36.10 Finned Tube Radiation

	Crew	Daily Output	Labor-Hours	Unit	Material	Labor	Equipment	Total	Total Incl O&P
0010 **FINNED TUBE RADIATION**, Terminal units, not incl. main supply pipe									
1150 Fin tube, wall hung, 14" slope top cover, with damper									
1200 1-1/4" copper tube, 4-1/4" alum. fin	Q-5	38	.421	L.F.	52.50	25		77.50	95
1250 1-1/4" steel tube, 4-1/4" steel fin		36	.444		43	26		69	86.50
1255 2" steel tube, 4-1/4" steel fin	↓	32	.500	↓	47	29.50		76.50	95.50
1260 21", two tier, slope top, w/damper									
1262 1-1/4" copper tube, 4-1/4" alum. fin	Q-5	30	.533	L.F.	62	31.50		93.50	115
1264 1-1/4" steel tube, 4-1/4" steel fin		28	.571		51	33.50		84.50	107
1266 2" steel tube, 4-1/4" steel fin	↓	25	.640	↓	55	38		93	117
1270 10", flat top, with damper									
1272 1-1/4" copper tube, 4-1/4" alum. fin	Q-5	38	.421	L.F.	25	25		50	64.50
1274 1-1/4" steel tube, 4-1/4" steel fin		36	.444		42.50	26		68.50	86
1276 2" steel tube, 4-1/4" steel fin	↓	32	.500	↓	45.50	29.50		75	94
1280 17", two tier, flat top, w/damper									

23 82 Convection Heating and Cooling Units

23 82 36 – Finned-Tube Radiation Heaters

23 82 36.10 Finned Tube Radiation

		Crew	Daily Output	Labor-Hours	Unit	Material	2020 Bare Costs Labor	2020 Bare Costs Equipment	Total	Total Incl O&P
1282	1-1/4" copper tube, 4-1/4" alum. fin	Q-5	30	.533	L.F.	55	31.50		86.50	108
1284	1-1/4" steel tube, 4-1/4" steel fin		28	.571		51.50	33.50		85	107
1286	2" steel tube, 4-1/4" steel fin	↓	25	.640	↓	55	38		93	117
1301	Rough in wall hung, steel fin w/supply & balance valves	1 Stpi	.91	8.791	Ea.	345	575		920	1,250
1310	Baseboard, pkgd, 1/2" copper tube, alum. fin, 7" high	Q-5	60	.267	L.F.	11.95	15.75		27.70	36.50
1320	3/4" copper tube, alum. fin, 7" high		58	.276		8.05	16.30		24.35	33.50
1340	1" copper tube, alum. fin, 8-7/8" high		56	.286		21	16.85		37.85	48
1360	1-1/4" copper tube, alum. fin, 8-7/8" high		54	.296		31	17.50		48.50	60.50
1380	1-1/4" IPS steel tube with steel fins	↓	52	.308	↓	32.50	18.15		50.65	62.50
1381	Rough in baseboard panel & fin tube, supply & balance valves	1 Stpi	1.06	7.547	Ea.	279	495		774	1,050
1500	Note: fin tube may also require corners, caps, etc.									

23 82 39 – Unit Heaters

23 82 39.13 Cabinet Unit Heaters

		Crew	Daily Output	Labor-Hours	Unit	Material	2020 Bare Costs Labor	2020 Bare Costs Equipment	Total	Total Incl O&P
0010	**CABINET UNIT HEATERS**									
5400	Cabinet, horizontal, hot water, blower type									
5420	20 MBH	Q-5	10	1.600	Ea.	1,175	94.50		1,269.50	1,450
5430	60 MBH		7	2.286		2,075	135		2,210	2,475
5440	100 MBH		5	3.200		2,100	189		2,289	2,600
5450	120 MBH	↓	4	4	↓	2,125	236		2,361	2,675

23 82 39.16 Propeller Unit Heaters

		Crew	Daily Output	Labor-Hours	Unit	Material	2020 Bare Costs Labor	2020 Bare Costs Equipment	Total	Total Incl O&P
0010	**PROPELLER UNIT HEATERS**									
3950	Unit heaters, propeller, 115 V 2 psi steam, 60°F entering air									
4000	Horizontal, 12 MBH	Q-5	12	1.333	Ea.	415	78.50		493.50	580
4020	28.2 MBH		10	1.600		540	94.50		634.50	735
4040	36.5 MBH		8	2		625	118		743	860
4060	43.9 MBH		8	2		625	118		743	860
4080	56.5 MBH		7.50	2.133		690	126		816	950
4100	65.6 MBH		7	2.286		710	135		845	980
4120	87.6 MBH		6.50	2.462		750	145		895	1,050
4140	96.8 MBH		6	2.667		880	157		1,037	1,200
4160	133.3 MBH		5	3.200		995	189		1,184	1,375
4180	157.6 MBH		4	4		1,225	236		1,461	1,700
4200	197.7 MBH		3	5.333		1,400	315		1,715	2,000
4220	257.2 MBH		2.50	6.400		1,600	380		1,980	2,350
4240	286.9 MBH		2	8		1,825	470		2,295	2,700
4260	364 MBH		1.80	8.889		2,325	525		2,850	3,325
4270	404 MBH	↓	1.60	10	↓	2,375	590		2,965	3,500
4300	Vertical diffuser same price									
4310	Vertical flow, 40 MBH	Q-5	11	1.455	Ea.	640	86		726	835
4314	58.5 MBH		8	2		635	118		753	875
4318	92 MBH		7	2.286		790	135		925	1,075
4322	109.7 MBH		6	2.667		955	157		1,112	1,275
4326	131 MBH		4	4		955	236		1,191	1,400
4330	160 MBH		3	5.333		1,025	315		1,340	1,600
4334	194 MBH		2.20	7.273		1,175	430		1,605	1,950
4338	212 MBH		2.10	7.619		1,450	450		1,900	2,275
4342	247 MBH		1.96	8.163		1,675	480		2,155	2,575
4346	297 MBH	↓	1.80	8.889		1,900	525		2,425	2,875
4350	333 MBH	Q-6	1.90	12.632		2,000	775		2,775	3,350
4354	420 MBH (460 V)		1.80	13.333		2,525	815		3,340	4,000
4358	500 MBH (460 V)		1.71	14.035		3,125	860		3,985	4,725
4362	570 MBH (460 V)	↓	1.40	17.143	↓	4,075	1,050		5,125	6,050

For customer support on your Mechanical Costs with RSMeans Data, call 800.448.8182.

409

23 82 Convection Heating and Cooling Units

23 82 39 – Unit Heaters

23 82 39.16 Propeller Unit Heaters		Crew	Daily Output	Labor-Hours	Unit	Material	2020 Bare Costs Labor	Equipment	Total	Total Incl O&P
4366	620 MBH (460 V)	Q-6	1.30	18.462	Ea.	4,725	1,125		5,850	6,900
4370	960 MBH (460 V)	↓	1.10	21.818	↓	8,200	1,325		9,525	11,000
4400	Unit heaters, propeller, hot water, 115 V, 60 degree F entering air									
4404	Horizontal									
4408	6 MBH	Q-5	11	1.455	Ea.	415	86		501	590
4412	12 MBH		11	1.455		415	86		501	590
4416	16 MBH		10	1.600		470	94.50		564.50	655
4420	24 MBH		9	1.778		540	105		645	750
4424	29 MBH		8	2		630	118		748	870
4428	47 MBH		7	2.286		690	135		825	960
4432	63 MBH		6	2.667		755	157		912	1,075
4436	81 MBH		5.50	2.909		920	172		1,092	1,250
4440	90 MBH		5	3.200		995	189		1,184	1,375
4460	133 MBH		4	4		1,225	236		1,461	1,700
4464	139 MBH		3.50	4.571		1,400	270		1,670	1,925
4468	198 MBH		2.50	6.400		1,600	380		1,980	2,350
4472	224 MBH		2	8		1,800	470		2,270	2,675
4476	273 MBH	↓	1.50	10.667	↓	2,325	630		2,955	3,500
4500	Vertical									
4504	8 MBH	Q-5	13	1.231	Ea.	1,425	72.50		1,497.50	1,650
4508	12 MBH		13	1.231		1,425	72.50		1,497.50	1,650
4512	16 MBH		13	1.231		1,425	72.50		1,497.50	1,650
4516	30 MBH		12	1.333		1,425	78.50		1,503.50	1,675
4520	43 MBH		11	1.455		1,425	86		1,511	1,675
4524	57 MBH		8	2		1,575	118		1,693	1,900
4528	68 MBH		7	2.286		1,600	135		1,735	1,950
4540	105 MBH		6	2.667		1,450	157		1,607	1,825
4544	123 MBH		5	3.200		1,675	189		1,864	2,100
4550	140 MBH		5	3.200		1,675	189		1,864	2,125
4554	156 MBH		4	4		2,075	236		2,311	2,650
4560	210 MBH		4	4		2,400	236		2,636	3,000
4564	223 MBH		3	5.333		3,150	315		3,465	3,925
4568	257 MBH	↓	2	8	↓	3,150	470		3,620	4,150

23 83 Radiant Heating Units

23 83 16 – Radiant-Heating Hydronic Piping

23 83 16.10 Radiant Floor Heating

0010	**RADIANT FLOOR HEATING**									
0100	Tubing, PEX (cross-linked polyethylene)									
0110	Oxygen barrier type for systems with ferrous materials									
0120	1/2"	Q-5	800	.020	L.F.	1.05	1.18		2.23	2.93
0130	3/4"		535	.030		1.55	1.76		3.31	4.35
0140	1"	↓	400	.040	↓	2.32	2.36		4.68	6.10
0200	Non barrier type for ferrous free systems									
0210	1/2"	Q-5	800	.020	L.F.	.57	1.18		1.75	2.40
0220	3/4"		535	.030		1	1.76		2.76	3.74
0230	1"	↓	400	.040	↓	1.79	2.36		4.15	5.50
1000	Manifolds									
1110	Brass									
1120	With supply and return valves, flow meter, thermometer,									
1122	auto air vent and drain/fill valve.									

23 83 Radiant Heating Units

23 83 16 – Radiant-Heating Hydronic Piping

23 83 16.10 Radiant Floor Heating		Crew	Daily Output	Labor-Hours	Unit	Material	2020 Bare Costs Labor	Equipment	Total	Total Incl O&P
1130	1", 2 circuit	Q-5	14	1.143	Ea.	355	67.50		422.50	495
1140	1", 3 circuit		13.50	1.185		405	70		475	550
1150	1", 4 circuit		13	1.231		435	72.50		507.50	590
1154	1", 5 circuit		12.50	1.280		525	75.50		600.50	690
1158	1", 6 circuit		12	1.333		590	78.50		668.50	765
1162	1", 7 circuit		11.50	1.391		640	82		722	825
1166	1", 8 circuit		11	1.455		685	86		771	885
1172	1", 9 circuit		10.50	1.524		760	90		850	970
1174	1", 10 circuit		10	1.600		795	94.50		889.50	1,025
1178	1", 11 circuit		9.50	1.684		820	99.50		919.50	1,050
1182	1", 12 circuit	▼	9	1.778	▼	930	105		1,035	1,175
1610	Copper manifold header (cut to size)									
1620	1" header, 12 circuit 1/2" sweat outlets	Q-5	3.33	4.805	Ea.	117	283		400	555
1630	1-1/4" header, 12 circuit 1/2" sweat outlets		3.20	5		136	295		431	590
1640	1-1/4" header, 12 circuit 3/4" sweat outlets		3	5.333		143	315		458	625
1650	1-1/2" header, 12 circuit 3/4" sweat outlets		3.10	5.161		173	305		478	645
1660	2" header, 12 circuit 3/4" sweat outlets	▼	2.90	5.517	▼	257	325		582	770
3000	Valves									
3110	Thermostatic zone valve actuator with end switch	Q-5	40	.400	Ea.	51.50	23.50		75	92
3114	Thermostatic zone valve actuator	"	36	.444	"	98	26		124	147
3120	Motorized straight zone valve with operator complete									
3130	3/4"	Q-5	35	.457	Ea.	160	27		187	217
3140	1"		32	.500		174	29.50		203.50	235
3150	1-1/4"	▼	29.60	.541	▼	219	32		251	289
3500	4 way mixing valve, manual, brass									
3530	1"	Q-5	13.30	1.203	Ea.	230	71		301	360
3540	1-1/4"		11.40	1.404		249	83		332	395
3550	1-1/2"		11	1.455		315	86		401	475
3560	2"		10.60	1.509		455	89		544	635
3800	Mixing valve motor, 4 way for valves, 1" and 1-1/4"		34	.471		390	28		418	465
3810	Mixing valve motor, 4 way for valves, 1-1/2" and 2"	▼	30	.533	▼	435	31.50		466.50	525
5000	Radiant floor heating, zone control panel									
5120	4 zone actuator valve control, expandable	Q-5	20	.800	Ea.	163	47		210	251
5130	6 zone actuator valve control, expandable		18	.889		271	52.50		323.50	375
6070	Thermal track, straight panel for long continuous runs, 5.333 S.F.		40	.400		36	23.50		59.50	75.50
6080	Thermal track, utility panel, for direction reverse at run end, 5.333 S.F.		40	.400		36	23.50		59.50	75.50
6090	Combination panel, for direction reverse plus straight run, 5.333 S.F.	▼	40	.400	▼	36	23.50		59.50	75.50
7000	PEX tubing fittings									
7100	Compression type									
7116	Coupling									
7120	1/2" x 1/2"	1 Stpi	27	.296	Ea.	6.95	19.40		26.35	36.50
7124	3/4" x 3/4"	"	23	.348	"	16.35	23		39.35	52
7130	Adapter									
7132	1/2" x female sweat 1/2"	1 Stpi	27	.296	Ea.	4.51	19.40		23.91	34
7134	1/2" x female sweat 3/4"		26	.308		5.05	20		25.05	35.50
7136	5/8" x female sweat 3/4"	▼	24	.333		7.25	22		29.25	40.50
7140	Elbow									
7142	1/2" x female sweat 1/2"	1 Stpi	27	.296	Ea.	7.30	19.40		26.70	37
7144	1/2" x female sweat 3/4"		26	.308		8.55	20		28.55	39.50
7146	5/8" x female sweat 3/4"	▼	24	.333	▼	9.60	22		31.60	43
7200	Insert type									
7206	PEX x male NPT									
7210	1/2" x 1/2"	1 Stpi	29	.276	Ea.	2.94	18.10		21.04	30

For customer support on your Mechanical Costs with RSMeans Data, call 800.448.8182.

411

23 83 16 – Radiant-Heating Hydronic Piping

23 83 16.10 Radiant Floor Heating		Crew	Daily Output	Labor-Hours	Unit	Material	2020 Bare Costs Labor	Equipment	Total	Total Incl O&P
7220	3/4" x 3/4"	1 Stpi	27	.296	Ea.	4.35	19.40		23.75	34
7230	1" x 1"		26	.308		7.30	20		27.30	38
7300	PEX coupling									
7310	1/2" x 1/2"	1 Stpi	30	.267	Ea.	.64	17.50		18.14	26.50
7320	3/4" x 3/4"		29	.276		.81	18.10		18.91	28
7330	1" x 1"		28	.286		1.40	18.75		20.15	29.50
7400	PEX stainless crimp ring									
7410	1/2" x 1/2"	1 Stpi	86	.093	Ea.	.56	6.10		6.66	9.75
7420	3/4" x 3/4"		84	.095		.77	6.25		7.02	10.20
7430	1" x 1"		82	.098		1.11	6.40		7.51	10.75

23 83 33 – Electric Radiant Heaters

23 83 33.10 Electric Heating

		Crew	Daily Output	Labor-Hours	Unit	Material	2020 Bare Costs Labor	Equipment	Total	Total Incl O&P
0010	**ELECTRIC HEATING**, not incl. conduit or feed wiring									
1100	Rule of thumb: Baseboard units, including control	1 Elec	4.40	1.818	kW	116	112		228	294
1300	Baseboard heaters, 2' long, 350 watt		8	1	Ea.	29	61.50		90.50	123
1400	3' long, 750 watt		8	1		35	61.50		96.50	130
1600	4' long, 1,000 watt		6.70	1.194		40.50	73.50		114	154
1800	5' long, 935 watt		5.70	1.404		44	86		130	177
2000	6' long, 1,500 watt		5	1.600		57	98		155	209
2200	7' long, 1,310 watt		4.40	1.818		61	112		173	234
2400	8' long, 2,000 watt		4	2		70	123		193	260
2600	9' long, 1,680 watt		3.60	2.222		86.50	136		222.50	298
2800	10' long, 1,875 watt		3.30	2.424		162	149		311	400
2950	Wall heaters with fan, 120 to 277 volt									
3160	Recessed, residential, 750 watt	1 Elec	6	1.333	Ea.	94.50	82		176.50	226
3170	1,000 watt		6	1.333		94.50	82		176.50	226
3180	1,250 watt		5	1.600		107	98		205	264
3190	1,500 watt		4	2		107	123		230	300
3210	2,000 watt		4	2		99	123		222	292
3230	2,500 watt		3.50	2.286		236	140		376	470
3240	3,000 watt		3	2.667		365	164		529	650
3250	4,000 watt		2.70	2.963		370	182		552	675
3260	Commercial, 750 watt		6	1.333		211	82		293	355
3270	1,000 watt		6	1.333		211	82		293	355
3280	1,250 watt		5	1.600		211	98		309	380
3290	1,500 watt		4	2		195	123		318	395
3300	2,000 watt		4	2		211	123		334	415
3310	2,500 watt		3.50	2.286		211	140		351	440
3320	3,000 watt		3	2.667		425	164		589	715
3330	4,000 watt		2.70	2.963		500	182		682	820
3600	Thermostats, integral		16	.500		29	30.50		59.50	77.50
3800	Line voltage, 1 pole		8	1		16.30	61.50		77.80	109
3810	2 pole		8	1		23.50	61.50		85	118
4000	Heat trace system, 400 degree									
4020	115 V, 2.5 watts/L.F.	1 Elec	530	.015	L.F.	11	.93		11.93	13.50
4030	5 watts/L.F.		530	.015		11	.93		11.93	13.50
4050	10 watts/L.F.		530	.015		9.20	.93		10.13	11.50
4060	208 V, 5 watts/L.F.		530	.015		11	.93		11.93	13.50
4080	480 V, 8 watts/L.F.		530	.015		11	.93		11.93	13.50
4200	Heater raceway									
4260	Heat transfer cement									
4280	1 gallon				Ea.	64.50			64.50	71

23 83 Radiant Heating Units

23 83 33 – Electric Radiant Heaters

23 83 33.10 Electric Heating	Crew	Daily Output	Labor-Hours	Unit	Material	2020 Bare Costs Labor	Equipment	Total	Total Incl O&P
4300	5 gallon				Ea.	252			252	277
4320	Cable tie									
4340	3/4" pipe size	1 Elec	470	.017	Ea.	.02	1.04		1.06	1.57
4360	1" pipe size		444	.018		.02	1.11		1.13	1.67
4380	1-1/4" pipe size		400	.020		.02	1.23		1.25	1.85
4400	1-1/2" pipe size		355	.023		.02	1.38		1.40	2.08
4420	2" pipe size		320	.025		.05	1.53		1.58	2.34
4440	3" pipe size		160	.050		.05	3.07		3.12	4.63
4460	4" pipe size		100	.080		.06	4.91		4.97	7.35
4480	Thermostat NEMA 3R, 22 amp, 0-150 degree, 10' cap.		8	1		193	61.50		254.50	305
4500	Thermostat NEMA 4X, 25 amp, 40 degree, 5-1/2' cap.		7	1.143		248	70		318	375
4520	Thermostat NEMA 4X, 22 amp, 25-325 degree, 10' cap.		7	1.143		660	70		730	830
4540	Thermostat NEMA 4X, 22 amp, 15-140 degree		6	1.333		560	82		642	735
4580	Thermostat NEMA 4, 7, 9, 22 amp, 25-325 degree, 10' cap.		3.60	2.222		810	136		946	1,100
4600	Thermostat NEMA 4, 7, 9, 22 amp, 15-140 degree		3	2.667		815	164		979	1,150
4720	Fiberglass application tape, 36 yard roll		11	.727		75	44.50		119.50	149
5000	Radiant heating ceiling panels, 2' x 4', 500 watt		16	.500		405	30.50		435.50	490
5050	750 watt		16	.500		430	30.50		460.50	520
5200	For recessed plaster frame, add		32	.250		138	15.35		153.35	175
5300	Infrared quartz heaters, 120 volts, 1,000 watt		6.70	1.194		350	73.50		423.50	495
5350	1,500 watt		5	1.600		335	98		433	515
5400	240 volts, 1,500 watt		5	1.600		405	98		503	590
5450	2,000 watt		4	2		335	123		458	555
5500	3,000 watt		3	2.667		330	164		494	610
5550	4,000 watt		2.60	3.077		380	189		569	700
5570	Modulating control		.80	10		141	615		756	1,075
5600	Unit heaters, heavy duty, with fan & mounting bracket									
5650	Single phase, 208-240-277 volt, 3 kW	1 Elec	6	1.333	Ea.	705	82		787	895
5750	5 kW		5.50	1.455		800	89		889	1,025
5800	7 kW		5	1.600		1,375	98		1,473	1,650
5850	10 kW		4	2		1,100	123		1,223	1,375
5950	15 kW		3.80	2.105		1,850	129		1,979	2,225
6000	480 volt, 3 kW		6	1.333		470	82		552	635
6020	4 kW		5.80	1.379		700	84.50		784.50	895
6040	5 kW		5.50	1.455		710	89		799	915
6060	7 kW		5	1.600		1,150	98		1,248	1,425
6080	10 kW		4	2		1,100	123		1,223	1,400
6100	13 kW		3.80	2.105		1,950	129		2,079	2,350
6120	15 kW		3.70	2.162		2,150	133		2,283	2,575
6140	20 kW		3.50	2.286		3,075	140		3,215	3,600
6300	3 phase, 208-240 volt, 5 kW		5.50	1.455		575	89		664	770
6320	7 kW		5	1.600		940	98		1,038	1,175
6340	10 kW		4	2		805	123		928	1,075
6360	15 kW		3.70	2.162		1,600	133		1,733	1,950
6380	20 kW		3.50	2.286		2,500	140		2,640	2,950
6400	25 kW		3.30	2.424		3,725	149		3,874	4,325
6500	480 volt, 5 kW		5.50	1.455		650	89		739	850
6520	7 kW		5	1.600		930	98		1,028	1,175
6540	10 kW		4	2		860	123		983	1,125
6560	13 kW		3.80	2.105		1,650	129		1,779	2,025
6580	15 kW		3.70	2.162		1,475	133		1,608	1,825
6600	20 kW		3.50	2.286		2,325	140		2,465	2,750
6620	25 kW		3.30	2.424		3,450	149		3,599	4,025

For customer support on your Mechanical Costs with RSMeans Data, call 800.448.8182.

413

23 83 33.10 Electric Heating	Crew	Daily Output	Labor-Hours	Unit	Material	2020 Bare Costs Labor	2020 Bare Costs Equipment	Total	Total Incl O&P
6630	30 kW	1 Elec	3	2.667	Ea.	3,025	164		3,189	3,575
6640	40 kW		2	4		4,050	245		4,295	4,825
6650	50 kW	↓	1.60	5	↓	6,675	305		6,980	7,800
6800	Vertical discharge heaters, with fan									
6820	Single phase, 208-240-277 volt, 10 kW	1 Elec	4	2	Ea.	860	123		983	1,125
6840	15 kW		3.70	2.162		1,650	133		1,783	2,000
6900	3 phase, 208-240 volt, 10 kW		4	2		1,050	123		1,173	1,350
6920	15 kW		3.70	2.162		1,600	133		1,733	1,975
6940	20 kW		3.50	2.286		2,925	140		3,065	3,425
6960	25 kW		3.30	2.424		3,175	149		3,324	3,725
6980	30 kW		3	2.667		4,350	164		4,514	5,025
7000	40 kW		2	4		4,325	245		4,570	5,125
7020	50 kW		1.60	5		6,950	305		7,255	8,100
7100	480 volt, 10 kW		4	2		1,200	123		1,323	1,500
7120	15 kW		3.70	2.162		1,950	133		2,083	2,350
7140	20 kW		3.50	2.286		2,650	140		2,790	3,125
7160	25 kW		3.30	2.424		3,450	149		3,599	4,025
7180	30 kW		3	2.667		3,800	164		3,964	4,425
7200	40 kW	↓	2	4	↓	3,925	245		4,170	4,700
7900	Cabinet convector heaters, 240 volt, three phase,									
7920	3' long, 2,000 watt	1 Elec	5.30	1.509	Ea.	2,300	92.50		2,392.50	2,675
7940	3,000 watt		5.30	1.509		2,400	92.50		2,492.50	2,800
7960	4,000 watt		5.30	1.509		2,475	92.50		2,567.50	2,850
7980	6,000 watt		4.60	1.739		2,675	107		2,782	3,100
8000	8,000 watt		4.60	1.739		2,750	107		2,857	3,175
8020	4' long, 4,000 watt		4.60	1.739		2,500	107		2,607	2,900
8040	6,000 watt		4	2		2,600	123		2,723	3,025
8060	8,000 watt		4	2		2,675	123		2,798	3,100
8080	10,000 watt	↓	4	2	↓	2,700	123		2,823	3,150
8100	Available also in 208 or 277 volt									
8200	Cabinet unit heaters, 120 to 277 volt, 1 pole,									
8220	wall mounted, 2 kW	1 Elec	4.60	1.739	Ea.	2,075	107		2,182	2,450
8230	3 kW		4.60	1.739		2,075	107		2,182	2,425
8240	4 kW		4.40	1.818		2,150	112		2,262	2,550
8250	5 kW		4.40	1.818		2,275	112		2,387	2,675
8260	6 kW		4.20	1.905		2,250	117		2,367	2,650
8270	8 kW		4	2		2,300	123		2,423	2,700
8280	10 kW		3.80	2.105		2,350	129		2,479	2,775
8290	12 kW		3.50	2.286		2,575	140		2,715	3,050
8300	13.5 kW		2.90	2.759		3,650	169		3,819	4,275
8310	16 kW		2.70	2.963		3,025	182		3,207	3,625
8320	20 kW		2.30	3.478		3,825	213		4,038	4,525
8330	24 kW		1.90	4.211		4,000	258		4,258	4,800
8350	Recessed, 2 kW		4.40	1.818		2,275	112		2,387	2,675
8370	3 kW		4.40	1.818		2,325	112		2,437	2,725
8380	4 kW		4.20	1.905		2,200	117		2,317	2,600
8390	5 kW		4.20	1.905		2,300	117		2,417	2,700
8400	6 kW		4	2		2,350	123		2,473	2,750
8410	8 kW		3.80	2.105		2,525	129		2,654	3,000
8420	10 kW		3.50	2.286		3,150	140		3,290	3,675
8430	12 kW		2.90	2.759		3,175	169		3,344	3,750
8440	13.5 kW		2.70	2.963		3,050	182		3,232	3,625
8450	16 kW	↓	2.30	3.478		3,250	213		3,463	3,900

23 83 Radiant Heating Units

23 83 33 – Electric Radiant Heaters

23 83 33.10 Electric Heating		Crew	Daily Output	Labor-Hours	Unit	Material	2020 Bare Costs Labor	Equipment	Total	Total Incl O&P
8460	20 kW	1 Elec	1.90	4.211	Ea.	3,775	258		4,033	4,550
8470	24 kW		1.60	5		3,725	305		4,030	4,550
8490	Ceiling mounted, 2 kW		3.20	2.500		2,100	153		2,253	2,525
8510	3 kW		3.20	2.500		2,200	153		2,353	2,650
8520	4 kW		3	2.667		2,250	164		2,414	2,725
8530	5 kW		3	2.667		2,325	164		2,489	2,800
8540	6 kW		2.80	2.857		2,325	175		2,500	2,825
8550	8 kW		2.40	3.333		2,450	205		2,655	3,000
8560	10 kW		2.20	3.636		3,050	223		3,273	3,675
8570	12 kW		2	4		2,625	245		2,870	3,275
8580	13.5 kW		1.50	5.333		2,550	325		2,875	3,275
8590	16 kW		1.30	6.154		2,675	380		3,055	3,475
8600	20 kW		.90	8.889		3,700	545		4,245	4,875
8610	24 kW		.60	13.333		3,725	820		4,545	5,325
8630	208 to 480 V, 3 pole									
8650	Wall mounted, 2 kW	1 Elec	4.60	1.739	Ea.	2,025	107		2,132	2,400
8670	3 kW		4.60	1.739		2,375	107		2,482	2,775
8680	4 kW		4.40	1.818		2,425	112		2,537	2,850
8690	5 kW		4.40	1.818		2,500	112		2,612	2,925
8700	6 kW		4.20	1.905		2,525	117		2,642	2,950
8710	8 kW		4	2		2,625	123		2,748	3,050
8720	10 kW		3.80	2.105		2,775	129		2,904	3,250
8730	12 kW		3.50	2.286		2,825	140		2,965	3,300
8740	13.5 kW		2.90	2.759		2,825	169		2,994	3,350
8750	16 kW		2.70	2.963		2,925	182		3,107	3,500
8760	20 kW		2.30	3.478		4,475	213		4,688	5,250
8770	24 kW		1.90	4.211		4,200	258		4,458	4,975
8790	Recessed, 2 kW		4.40	1.818		2,250	112		2,362	2,650
8810	3 kW		4.40	1.818		2,300	112		2,412	2,725
8820	4 kW		4.20	1.905		2,400	117		2,517	2,825
8830	5 kW		4.20	1.905		2,425	117		2,542	2,850
8840	6 kW		4	2		2,350	123		2,473	2,750
8850	8 kW		3.80	2.105		2,425	129		2,554	2,875
8860	10 kW		3.50	2.286		2,450	140		2,590	2,900
8870	12 kW		2.90	2.759		2,750	169		2,919	3,250
8880	13.5 kW		2.70	2.963		2,375	182		2,557	2,900
8890	16 kW		2.30	3.478		5,175	213		5,388	6,025
8900	20 kW		1.90	4.211		4,300	258		4,558	5,100
8920	24 kW		1.60	5		4,350	305		4,655	5,250
8940	Ceiling mount, 2 kW		3.20	2.500		2,325	153		2,478	2,775
8950	3 kW		3.20	2.500		2,150	153		2,303	2,575
8960	4 kW		3	2.667		2,475	164		2,639	2,975
8970	5 kW		3	2.667		2,175	164		2,339	2,625
8980	6 kW		2.80	2.857		2,475	175		2,650	2,975
8990	8 kW		2.40	3.333		2,550	205		2,755	3,125
9000	10 kW		2.20	3.636		2,350	223		2,573	2,925
9020	13.5 kW		1.50	5.333		3,150	325		3,475	3,950
9030	16 kW		1.30	6.154		4,300	380		4,680	5,300
9040	20 kW		.90	8.889		4,300	545		4,845	5,525
9060	24 kW		.60	13.333		5,325	820		6,145	7,075
9230	13.5 kW, 40,956 BTU		2.20	3.636		2,075	223		2,298	2,600
9250	24 kW, 81,912 BTU		2	4		2,025	245		2,270	2,600

For customer support on your Mechanical Costs with RSMeans Data, call 800.448.8182.

415

23 84 Humidity Control Equipment

23 84 13 – Humidifiers

23 84 13.10 Humidifier Units	Crew	Daily Output	Labor-Hours	Unit	Material	2020 Bare Costs Labor	Equipment	Total	Total Incl O&P
0010 **HUMIDIFIER UNITS**									
0520 Steam, room or duct, filter, regulators, auto. controls, 220 V									
0540 11 lb./hr.	Q-5	6	2.667	Ea.	2,775	157		2,932	3,275
0560 22 lb./hr.		5	3.200		3,050	189		3,239	3,650
0580 33 lb./hr.		4	4		3,150	236		3,386	3,800
0600 50 lb./hr.		4	4		3,850	236		4,086	4,575
0620 100 lb./hr.		3	5.333		5,175	315		5,490	6,175
0640 150 lb./hr.		2.50	6.400		6,550	380		6,930	7,775
0660 200 lb./hr.	▼	2	8	▼	8,225	470		8,695	9,750
0700 With blower									
0720 11 lb./hr.	Q-5	5.50	2.909	Ea.	4,175	172		4,347	4,850
0740 22 lb./hr.		4.75	3.368		4,800	199		4,999	5,575
0760 33 lb./hr.		3.75	4.267		4,650	252		4,902	5,500
0780 50 lb./hr.		3.50	4.571		5,450	270		5,720	6,400
0800 100 lb./hr.		2.75	5.818		6,075	345		6,420	7,225
0820 150 lb./hr.		2	8		9,075	470		9,545	10,700
0840 200 lb./hr.	▼	1.50	10.667	▼	11,300	630		11,930	13,400
1000 Steam for duct installation, manifold with pneum. controls									
1010 3-191 lb./hr., 24" x 24"	Q-5	5	3.200	Ea.	2,350	189		2,539	2,850
1020 3-191 lb./hr., 48" x 48"		4.60	3.478		4,275	205		4,480	5,000
1030 65-334 lb./hr., 36" x 36"		4.80	3.333		4,175	197		4,372	4,900
1040 65-334 lb./hr., 72" x 72"		4	4		4,525	236		4,761	5,325
1050 100-690 lb./hr., 48" x 48"		4	4		5,875	236		6,111	6,800
1060 100-690 lb./hr., 84" x 84"		3.60	4.444		6,300	262		6,562	7,325
1070 220-2000 lb./hr., 72" x 72"		3.80	4.211		6,125	248		6,373	7,100
1080 220-2000 lb./hr., 144" x 144"	▼	3.40	4.706	▼	7,150	278		7,428	8,300
1090 For electrically controlled unit, add					805			805	885
1092 For additional manifold, add					25%	15%			
1200 Electric steam, self generating, with manifold									
1210 1-96 lb./hr., 12" x 72"	Q-5	3.60	4.444	Ea.	10,700	262		10,962	12,200
1220 20-168 lb./hr., 12" x 72"		3.60	4.444		16,700	262		16,962	18,700
1290 For room mounting, add	▼	16	1	▼	270	59		329	385
5000 Furnace type, wheel bypass									
5020 10 GPD	1 Stpi	4	2	Ea.	203	131		334	420
5040 14 GPD		3.80	2.105		211	138		349	440
5060 19 GPD	▼	3.60	2.222		305	146		451	560

23 84 16 – Mechanical Dehumidification Units

23 84 16.10 Dehumidifier Units

23 84 16.10 Dehumidifier Units	Crew	Daily Output	Labor-Hours	Unit	Material	Labor	Equipment	Total	Total Incl O&P
0010 **DEHUMIDIFIER UNITS**									
6000 Self contained with filters and standard controls									
6040 1.5 lb./hr., 50 CFM	1 Plum	8	1	Ea.	8,575	64.50		8,639.50	9,525
6060 3 lb./hr., 150 CFM	Q-1	12	1.333		9,200	77.50		9,277.50	10,200
6065 6 lb./hr., 150 CFM		9	1.778		15,000	103		15,103	16,700
6070 16 to 20 lb./hr., 600 CFM		5	3.200		28,300	186		28,486	31,500
6080 30 to 40 lb./hr., 1,125 CFM		4	4		44,100	232		44,332	48,800
6090 60 to 75 lb./hr., 2,250 CFM		3	5.333		57,500	310		57,810	64,000
6100 120 to 155 lb./hr., 4,500 CFM		2	8		102,000	465		102,465	112,500
6110 240 to 310 lb./hr., 9,000 CFM	▼	1.50	10.667		146,000	620		146,620	161,500
6120 400 to 515 lb./hr., 15,000 CFM	Q-2	1.60	15		183,000	900		183,900	202,500
6130 530 to 690 lb./hr., 20,000 CFM		1.40	17.143		198,000	1,025		199,025	219,500
6140 800 to 1030 lb./hr., 30,000 CFM		1.20	20		225,500	1,200		226,700	250,000
6150 1060 to 1375 lb./hr., 40,000 CFM	▼	1	24	▼	314,000	1,450		315,450	347,500

23 91 10 – Prefabricated Curbs, Pads and Stands

23 91 10.10 Prefabricated Pads and Stands	Crew	Daily Output	Labor-Hours	Unit	Material	2020 Bare Costs Labor	Equipment	Total	Total Incl O&P
0010 **PREFABRICATED PADS AND STANDS**									
6000 Pad, fiberglass reinforced concrete with polystyrene foam core									
6070 24" x 24"	1 Shee	16	.500	Ea.	26	31		57	76
6090 24" x 36"		12	.667		41.50	41.50		83	109
6110 24" x 42"	↓	8	1		47.50	62.50		110	147
6150 26" x 36"	Q-9	8	2		43	112		155	218
6170 28" x 38"	1 Shee	8	1		47	62.50		109.50	146
6190 30" x 30"		12	.667		43	41.50		84.50	111
6220 30" x 36"	↓	8	1		51	62.50		113.50	151
6240 30" x 40"	Q-9	8	2		56	112		168	232
6260 32" x 32"		7	2.286		48.50	128		176.50	247
6280 36" x 36"		8	2		59.50	112		171.50	236
6300 36" x 40"		7	2.286		64.50	128		192.50	265
6320 36" x 48"		7	2.286		79	128		207	281
6340 36" x 54"		6	2.667		96	150		246	330
6360 36" x 60" x 3"		7	2.286		109	128		237	315
6400 41" x 47" x 3"		6	2.667		101	150		251	340
6490 26" round		10	1.600		25.50	89.50		115	164
6550 30" round		9	1.778		32	99.50		131.50	186
6600 36" round	↓	8	2		62.50	112		174.50	239
6700 For 3" thick pad, add				↓	20%	5%			
8800 Stand, corrosion-resistant plastic									
8850 Heat pump, 6" high	1 Shee	32	.250	Ea.	5.35	15.60		20.95	29.50
8870 12" high	"	32	.250	"	13.60	15.60		29.20	38.50

For customer support on your Mechanical Costs with RSMeans Data, call 800.448.8182.

417

Division Notes

	CREW	DAILY OUTPUT	LABOR-HOURS	UNIT	BARE COSTS				TOTAL INCL O&P
					MAT.	LABOR	EQUIP.	TOTAL	

Estimating Tips

26 05 00 Common Work Results for Electrical

- Conduit should be taken off in three main categories—power distribution, branch power, and branch lighting—so the estimator can concentrate on systems and components, therefore making it easier to ensure all items have been accounted for.

- For cost modifications for elevated conduit installation, add the percentages to labor according to the height of installation and only to the quantities exceeding the different height levels, not to the total conduit quantities. Refer to subdivision 26 01 02.20 for labor adjustment factors.

- Remember that aluminum wiring of equal ampacity is larger in diameter than copper and may require larger conduit.

- If more than three wires at a time are being pulled, deduct percentages from the labor hours of that grouping of wires.

- When taking off grounding systems, identify separately the type and size of wire, and list each unique type of ground connection.

- The estimator should take the weights of materials into consideration when completing a takeoff. Topics to consider include: How will the materials be supported? What methods of support are available? How high will the support structure have to reach? Will the final support structure be able to withstand the total burden? Is the support material included or separate from the fixture, equipment, and material specified?

- Do not overlook the costs for equipment used in the installation. If scaffolding or highlifts are available in the field, contractors may use them in lieu of the proposed ladders and rolling staging.

26 20 00 Low-Voltage Electrical Transmission

- Supports and concrete pads may be shown on drawings for the larger equipment, or the support system may be only a piece of plywood for the back of a panelboard. In either case, they must be included in the costs.

26 40 00 Electrical and Cathodic Protection

- When taking off cathodic protection systems, identify the type and size of cable, and list each unique type of anode connection.

26 50 00 Lighting

- Fixtures should be taken off room by room using the fixture schedule, specifications, and the ceiling plan. For large concentrations of lighting fixtures in the same area, deduct the percentages from labor hours.

Reference Numbers

Reference numbers are shown at the beginning of some major classifications. These numbers refer to related items in the Reference Section. The reference information may be an estimating procedure, an alternate pricing method, or technical information.

Note: Not all subdivisions listed here necessarily appear. ■

Same Data. Simplified.

Enjoy the convenience and efficiency of accessing your costs anywhere:

- **Skip the multiplier** by setting your location
- **Quickly search,** edit, favorite and share costs
- **Stay on top of price changes** with automatic updates

Discover more at rsmeans.com/online

Note: "Powered in part by CINX™, based on liscensed proprietary information of Harrison Publishing House, Inc."

Note: The following companies, in part, have been used as a reference source for some of the material prices used in Division 26:

Electriflex

Trade Service

26 01 02.20 Labor Adjustment Factors	Crew	Daily Output	Labor-Hours	Unit	Material	2020 Bare Costs Labor	Equipment	Total	Total Incl O&P
0010 **LABOR ADJUSTMENT FACTORS** (For Div. 26, 27 and 28) R260519-90									
0100 Subtract from labor for Economy of Scale for Wire									
0110 4-5 wires						25%			
0120 6-10 wires						30%			
0130 11-15 wires						35%			
0140 over 15 wires						40%			
0150 Labor adjustment factors (For Div. 26, 27, 28 and 48)									
0200 Labor factors: The below are reasonable suggestions, but									
0210 each project must be evaluated for its own peculiarities, and									
0220 the adjustments be increased or decreased depending on the									
0230 severity of the special conditions.									
1000 Add to labor for elevated installation (above floor level)									
1010 10' to 14.5' high						10%			
1020 15' to 19.5' high						20%			
1030 20' to 24.5' high						25%			
1040 25' to 29.5' high						35%			
1050 30' to 34.5' high						40%			
1060 35' to 39.5' high						50%			
1070 40' and higher						55%			
2000 Add to labor for crawl space									
2010 3' high						40%			
2020 4' high						30%			
3000 Add to labor for multi-story building									
3100 For new construction (No elevator available)									
3110 Add for floors 3 thru 10						5%			
3120 Add for floors 11 thru 15						10%			
3130 Add for floors 16 thru 20						15%			
3140 Add for floors 21 thru 30						20%			
3150 Add for floors 31 and up						30%			
3200 For existing structure (Elevator available)									
3210 Add for work on floor 3 and above						2%			
4000 Add to labor for working in existing occupied buildings									
4010 Hospital						35%			
4020 Office building						25%			
4030 School						20%			
4040 Factory or warehouse						15%			
4050 Multi-dwelling						15%			
5000 Add to labor, miscellaneous									
5010 Cramped shaft						35%			
5020 Congested area						15%			
5030 Excessive heat or cold						30%			
6000 Labor factors: the above are reasonable suggestions, but									
6100 each project should be evaluated for its own peculiarities									
6200 Other factors to be considered are:									
6210 Movement of material and equipment through finished areas						10%			
6220 Equipment room min security direct access w/authorization						15%			
6230 Attic space						25%			
6240 No service road						25%			
6250 Poor unloading/storage area, no hydraulic lifts or jacks						20%			
6260 Congested site area/heavy traffic						20%			
7000 Correctional facilities (no compounding division 1 adjustment factors)									
7010 Minimum security w/facilities escort						30%			
7020 Medium security w/facilities and correctional officer escort						40%			

26 01 Operation and Maintenance of Electrical Systems

26 01 02 – Labor Adjustment

26 01 02.20 Labor Adjustment Factors	Crew	Daily Output	Labor-Hours	Unit	Material	2020 Bare Costs Labor	2020 Bare Costs Equipment	Total	Total Incl O&P
7030 Max security w/facilities & correctional officer escort (no inmate contact)						50%			

26 05 Common Work Results for Electrical

26 05 33 – Raceway and Boxes for Electrical Systems

26 05 33.95 Cutting and Drilling

	26 05 33.95 Cutting and Drilling	Crew	Daily Output	Labor-Hours	Unit	Material	2020 Bare Costs Labor	2020 Bare Costs Equipment	Total	Total Incl O&P
0010	**CUTTING AND DRILLING**									
0100	Hole drilling to 10' high, concrete wall									
0110	8" thick, 1/2" pipe size	R-31	12	.667	Ea.	.27	41	5.20	46.47	67
0120	3/4" pipe size		12	.667		.27	41	5.20	46.47	67
0130	1" pipe size		9.50	.842		.35	51.50	6.60	58.45	84.50
0140	1-1/4" pipe size		9.50	.842		.35	51.50	6.60	58.45	84.50
0150	1-1/2" pipe size		9.50	.842		.35	51.50	6.60	58.45	84.50
0160	2" pipe size		4.40	1.818		.53	112	14.20	126.73	182
0170	2-1/2" pipe size		4.40	1.818		.53	112	14.20	126.73	182
0180	3" pipe size		4.40	1.818		.53	112	14.20	126.73	182
0190	3-1/2" pipe size		3.30	2.424		.60	149	18.95	168.55	243
0200	4" pipe size		3.30	2.424		.60	149	18.95	168.55	243
0500	12" thick, 1/2" pipe size		9.40	.851		.40	52	6.65	59.05	85
0520	3/4" pipe size		9.40	.851		.40	52	6.65	59.05	85
0540	1" pipe size		7.30	1.096		.52	67	8.55	76.07	110
0560	1-1/4" pipe size		7.30	1.096		.52	67	8.55	76.07	110
0570	1-1/2" pipe size		7.30	1.096		.52	67	8.55	76.07	110
0580	2" pipe size		3.60	2.222		.80	136	17.35	154.15	223
0590	2-1/2" pipe size		3.60	2.222		.80	136	17.35	154.15	223
0600	3" pipe size		3.60	2.222		.80	136	17.35	154.15	223
0610	3-1/2" pipe size		2.80	2.857		.90	175	22.50	198.40	286
0630	4" pipe size		2.50	3.200		.90	196	25	221.90	320
0650	16" thick, 1/2" pipe size		7.60	1.053		.54	64.50	8.20	73.24	106
0670	3/4" pipe size		7	1.143		.54	70	8.95	79.49	114
0690	1" pipe size		6	1.333		.69	82	10.40	93.09	134
0710	1-1/4" pipe size		5.50	1.455		.69	89	11.35	101.04	146
0730	1-1/2" pipe size		5.50	1.455		.69	89	11.35	101.04	146
0750	2" pipe size		3	2.667		1.06	164	21	186.06	268
0770	2-1/2" pipe size		2.70	2.963		1.06	182	23	206.06	298
0790	3" pipe size		2.50	3.200		1.06	196	25	222.06	320
0810	3-1/2" pipe size		2.30	3.478		1.20	213	27	241.20	350
0830	4" pipe size		2	4		1.20	245	31	277.20	400
0850	20" thick, 1/2" pipe size		6.40	1.250		.67	76.50	9.75	86.92	125
0870	3/4" pipe size		6	1.333		.67	82	10.40	93.07	134
0890	1" pipe size		5	1.600		.86	98	12.50	111.36	161
0910	1-1/4" pipe size		4.80	1.667		.86	102	13	115.86	167
0930	1-1/2" pipe size		4.60	1.739		.86	107	13.60	121.46	175
0950	2" pipe size		2.70	2.963		1.33	182	23	206.33	298
0970	2-1/2" pipe size		2.40	3.333		1.33	205	26	232.33	335
0990	3" pipe size		2.20	3.636		1.33	223	28.50	252.83	360
1010	3-1/2" pipe size		2	4		1.50	245	31	277.50	400
1030	4" pipe size		1.70	4.706		1.50	289	37	327.50	470
1050	24" thick, 1/2" pipe size		5.50	1.455		.81	89	11.35	101.16	146
1070	3/4" pipe size		5.10	1.569		.81	96	12.25	109.06	157
1090	1" pipe size		4.30	1.860		1.04	114	14.55	129.59	187
1110	1-1/4" pipe size		4	2		1.04	123	15.60	139.64	201

For customer support on your Mechanical Costs with RSMeans Data, call 800.448.8182.

421

26 05 33.95 Cutting and Drilling	Crew	Daily Output	Labor-Hours	Unit	Material	2020 Bare Costs Labor	2020 Bare Costs Equipment	Total	Total Incl O&P
1130	1-1/2" pipe size	R-31	4	2	Ea.	1.04	123	15.60	139.64	201
1150	2" pipe size		2.40	3.333		1.59	205	26	232.59	335
1170	2-1/2" pipe size		2.20	3.636		1.59	223	28.50	253.09	365
1190	3" pipe size		2	4		1.59	245	31	277.59	400
1210	3-1/2" pipe size		1.80	4.444		1.80	273	34.50	309.30	445
1230	4" pipe size		1.50	5.333		1.80	325	41.50	368.30	535
1500	Brick wall, 8" thick, 1/2" pipe size		18	.444		.27	27.50	3.47	31.24	44.50
1520	3/4" pipe size		18	.444		.27	27.50	3.47	31.24	44.50
1540	1" pipe size		13.30	.602		.35	37	4.70	42.05	60.50
1560	1-1/4" pipe size		13.30	.602		.35	37	4.70	42.05	60.50
1580	1-1/2" pipe size		13.30	.602		.35	37	4.70	42.05	60.50
1600	2" pipe size		5.70	1.404		.53	86	10.95	97.48	141
1620	2-1/2" pipe size		5.70	1.404		.53	86	10.95	97.48	141
1640	3" pipe size		5.70	1.404		.53	86	10.95	97.48	141
1660	3-1/2" pipe size		4.40	1.818		.60	112	14.20	126.80	182
1680	4" pipe size		4	2		.60	123	15.60	139.20	201
1700	12" thick, 1/2" pipe size		14.50	.552		.40	34	4.31	38.71	55.50
1720	3/4" pipe size		14.50	.552		.40	34	4.31	38.71	55.50
1740	1" pipe size		11	.727		.52	44.50	5.70	50.72	73.50
1760	1-1/4" pipe size		11	.727		.52	44.50	5.70	50.72	73.50
1780	1-1/2" pipe size		11	.727		.52	44.50	5.70	50.72	73.50
1800	2" pipe size		5	1.600		.80	98	12.50	111.30	161
1820	2-1/2" pipe size		5	1.600		.80	98	12.50	111.30	161
1840	3" pipe size		5	1.600		.80	98	12.50	111.30	161
1860	3-1/2" pipe size		3.80	2.105		.90	129	16.45	146.35	211
1880	4" pipe size		3.30	2.424		.90	149	18.95	168.85	243
1900	16" thick, 1/2" pipe size		12.30	.650		.54	40	5.10	45.64	65.50
1920	3/4" pipe size		12.30	.650		.54	40	5.10	45.64	65.50
1940	1" pipe size		9.30	.860		.69	53	6.70	60.39	86.50
1960	1-1/4" pipe size		9.30	.860		.69	53	6.70	60.39	86.50
1980	1-1/2" pipe size		9.30	.860		.69	53	6.70	60.39	86.50
2000	2" pipe size		4.40	1.818		1.06	112	14.20	127.26	183
2010	2-1/2" pipe size		4.40	1.818		1.06	112	14.20	127.26	183
2030	3" pipe size		4.40	1.818		1.06	112	14.20	127.26	183
2050	3-1/2" pipe size		3.30	2.424		1.20	149	18.95	169.15	243
2070	4" pipe size		3	2.667		1.20	164	21	186.20	268
2090	20" thick, 1/2" pipe size		10.70	.748		.67	46	5.85	52.52	75.50
2110	3/4" pipe size		10.70	.748		.67	46	5.85	52.52	75.50
2130	1" pipe size		8	1		.86	61.50	7.80	70.16	101
2150	1-1/4" pipe size		8	1		.86	61.50	7.80	70.16	101
2170	1-1/2" pipe size		8	1		.86	61.50	7.80	70.16	101
2190	2" pipe size		4	2		1.33	123	15.60	139.93	202
2210	2-1/2" pipe size		4	2		1.33	123	15.60	139.93	202
2230	3" pipe size		4	2		1.33	123	15.60	139.93	202
2250	3-1/2" pipe size		3	2.667		1.50	164	21	186.50	269
2270	4" pipe size		2.70	2.963		1.50	182	23	206.50	298
2290	24" thick, 1/2" pipe size		9.40	.851		.81	52	6.65	59.46	85.50
2310	3/4" pipe size		9.40	.851		.81	52	6.65	59.46	85.50
2330	1" pipe size		7.10	1.127		1.04	69	8.80	78.84	114
2350	1-1/4" pipe size		7.10	1.127		1.04	69	8.80	78.84	114
2370	1-1/2" pipe size		7.10	1.127		1.04	69	8.80	78.84	114
2390	2" pipe size		3.60	2.222		1.59	136	17.35	154.94	224
2410	2-1/2" pipe size		3.60	2.222		1.59	136	17.35	154.94	224

26 05 Common Work Results for Electrical

26 05 33 – Raceway and Boxes for Electrical Systems

26 05 33.95 Cutting and Drilling

		Crew	Daily Output	Labor-Hours	Unit	Material	2020 Bare Costs Labor	Equipment	Total	Total Incl O&P
2430	3" pipe size	R-31	3.60	2.222	Ea.	1.59	136	17.35	154.94	224
2450	3-1/2" pipe size		2.80	2.857		1.80	175	22.50	199.30	287
2470	4" pipe size	▼	2.50	3.200	▼	1.80	196	25	222.80	320
3000	Knockouts to 8' high, metal boxes & enclosures									
3020	With hole saw, 1/2" pipe size	1 Elec	53	.151	Ea.		9.25		9.25	13.80
3040	3/4" pipe size		47	.170			10.45		10.45	15.55
3050	1" pipe size		40	.200			12.25		12.25	18.25
3060	1-1/4" pipe size		36	.222			13.65		13.65	20.50
3070	1-1/2" pipe size		32	.250			15.35		15.35	23
3080	2" pipe size		27	.296			18.20		18.20	27
3090	2-1/2" pipe size		20	.400			24.50		24.50	36.50
4010	3" pipe size		16	.500			30.50		30.50	45.50
4030	3-1/2" pipe size		13	.615			38		38	56
4050	4" pipe size	▼	11	.727	▼		44.50		44.50	66.50

26 05 83 – Wiring Connections

26 05 83.10 Motor Connections

		Crew	Daily Output	Labor-Hours	Unit	Material	2020 Bare Costs Labor	Equipment	Total	Total Incl O&P
0010	**MOTOR CONNECTIONS**									
0020	Flexible conduit and fittings, 115 volt, 1 phase, up to 1 HP motor	1 Elec	8	1	Ea.	5.90	61.50		67.40	98
0050	2 HP motor		6.50	1.231		10.40	75.50		85.90	123
0100	3 HP motor		5.50	1.455		9.20	89		98.20	143
0110	230 volt, 3 phase, 3 HP motor		6.78	1.180		7.15	72.50		79.65	116
0112	5 HP motor		5.47	1.463		6.05	89.50		95.55	141
0114	7-1/2 HP motor		4.61	1.735		9.35	106		115.35	169
0120	10 HP motor		4.20	1.905		17.40	117		134.40	193
0150	15 HP motor		3.30	2.424		17.40	149		166.40	240
0200	25 HP motor		2.70	2.963		29.50	182		211.50	305
0400	50 HP motor		2.20	3.636		54.50	223		277.50	390
0600	100 HP motor		1.50	5.333		121	325		446	620
1500	460 volt, 5 HP motor, 3 phase		8	1		6.35	61.50		67.85	98.50
1520	10 HP motor		8	1		6.35	61.50		67.85	98.50
1530	25 HP motor		6	1.333		11.25	82		93.25	134
1540	30 HP motor		6	1.333		11.25	82		93.25	134
1550	40 HP motor		5	1.600		16.70	98		114.70	164
1560	50 HP motor		5	1.600		22.50	98		120.50	171
1570	60 HP motor		3.80	2.105		25	129		154	220
1580	75 HP motor		3.50	2.286		31.50	140		171.50	244
1590	100 HP motor		2.50	3.200		53.50	196		249.50	350
1600	125 HP motor		2	4		58.50	245		303.50	430
1610	150 HP motor		1.80	4.444		60	273		333	470
1620	200 HP motor		1.50	5.333		99.50	325		424.50	595
2005	460 volt, 5 HP motor, 3 phase, w/sealtite		8	1		9.30	61.50		70.80	102
2010	10 HP motor		8	1		9.30	61.50		70.80	102
2015	25 HP motor		6	1.333		17	82		99	141
2020	30 HP motor		6	1.333		17	82		99	141
2025	40 HP motor		5	1.600		30	98		128	180
2030	50 HP motor		5	1.600		34.50	98		132.50	184
2035	60 HP motor		3.80	2.105		38.50	129		167.50	234
2040	75 HP motor		3.50	2.286		45	140		185	259
2045	100 HP motor		2.50	3.200		80	196		276	380
2055	150 HP motor		1.80	4.444		86.50	273		359.50	500
2060	200 HP motor	▼	1.50	5.333	▼	273	325		598	785

For customer support on your Mechanical Costs with RSMeans Data, call 800.448.8182.

423

26 05 Common Work Results for Electrical

26 05 90 – Residential Applications

26 05 90.10 Residential Wiring	Crew	Daily Output	Labor-Hours	Unit	Material	2020 Bare Costs Labor	Equipment	Total	Total Incl O&P
0010 RESIDENTIAL WIRING									
0020 20' avg. runs and #14/2 wiring incl. unless otherwise noted									
1000 Service & panel, includes 24' SE-AL cable, service eye, meter,									
1010 Socket, panel board, main bkr., ground rod, 15 or 20 amp									
1020 1-pole circuit breakers, and misc. hardware									
1100 100 amp, with 10 branch breakers	1 Elec	1.19	6.723	Ea.	335	410		745	980
1110 With PVC conduit and wire		.92	8.696		370	535		905	1,200
1120 With RGS conduit and wire		.73	10.959		570	670		1,240	1,625
1150 150 amp, with 14 branch breakers		1.03	7.767		805	475		1,280	1,600
1170 With PVC conduit and wire		.82	9.756		875	600		1,475	1,850
1180 With RGS conduit and wire		.67	11.940		1,200	735		1,935	2,425
1200 200 amp, with 18 branch breakers	2 Elec	1.80	8.889		1,025	545		1,570	1,925
1220 With PVC conduit and wire		1.46	10.959		1,100	670		1,770	2,200
1230 With RGS conduit and wire		1.24	12.903		1,500	790		2,290	2,825
1800 Lightning surge suppressor	1 Elec	32	.250		85	15.35		100.35	117
2000 Switch devices									
2100 Single pole, 15 amp, ivory, with a 1-gang box, cover plate,									
2110 Type NM (Romex) cable	1 Elec	17.10	.468	Ea.	15.55	28.50		44.05	59.50
2120 Type MC cable		14.30	.559		26	34.50		60.50	80
2130 EMT & wire		5.71	1.401		36	86		122	168
2150 3-way, #14/3, type NM cable		14.55	.550		10.10	33.50		43.60	61
2170 Type MC cable		12.31	.650		24.50	40		64.50	86.50
2180 EMT & wire		5	1.600		30.50	98		128.50	180
2200 4-way, #14/3, type NM cable		14.55	.550		18.20	33.50		51.70	70
2220 Type MC cable		12.31	.650		32.50	40		72.50	95.50
2230 EMT & wire		5	1.600		38.50	98		136.50	189
2250 S.P., 20 amp, #12/2, type NM cable		13.33	.600		12.30	37		49.30	68.50
2270 Type MC cable		11.43	.700		22	43		65	88
2280 EMT & wire		4.85	1.649		35	101		136	190
2290 S.P. rotary dimmer, 600 W, no wiring		17	.471		32	29		61	78
2300 S.P. rotary dimmer, 600 W, type NM cable		14.55	.550		35.50	33.50		69	89
2320 Type MC cable		12.31	.650		46	40		86	111
2330 EMT & wire		5	1.600		57.50	98		155.50	209
2350 3-way rotary dimmer, type NM cable		13.33	.600		24	37		61	81.50
2370 Type MC cable		11.43	.700		34.50	43		77.50	102
2380 EMT & wire		4.85	1.649		46	101		147	202
2400 Interval timer wall switch, 20 amp, 1-30 min., #12/2									
2410 Type NM cable	1 Elec	14.55	.550	Ea.	59	33.50		92.50	115
2420 Type MC cable		12.31	.650		64.50	40		104.50	131
2430 EMT & wire		5	1.600		81.50	98		179.50	236
2500 Decorator style									
2510 S.P., 15 amp, type NM cable	1 Elec	17.10	.468	Ea.	21	28.50		49.50	65.50
2520 Type MC cable		14.30	.559		32	34.50		66.50	86
2530 EMT & wire		5.71	1.401		41.50	86		127.50	174
2550 3-way, #14/3, type NM cable		14.55	.550		15.65	33.50		49.15	67.50
2570 Type MC cable		12.31	.650		30	40		70	92.50
2580 EMT & wire		5	1.600		36	98		134	186
2600 4-way, #14/3, type NM cable		14.55	.550		24	33.50		57.50	76
2620 Type MC cable		12.31	.650		38	40		78	102
2630 EMT & wire		5	1.600		44	98		142	195
2650 S.P., 20 amp, #12/2, type NM cable		13.33	.600		17.85	37		54.85	74.50
2670 Type MC cable		11.43	.700		27.50	43		70.50	94

424

26 05 90.10 Residential Wiring	Crew	Daily Output	Labor-Hours	Unit	Material	2020 Bare Costs Labor	Equipment	Total	Total Incl O&P
2680	EMT & wire	1 Elec	4.85	1.649	Ea.	40.50	101		141.50	196
2700	S.P., slide dimmer, type NM cable		17.10	.468		37.50	28.50		66	84
2720	Type MC cable		14.30	.559		48	34.50		82.50	104
2730	EMT & wire		5.71	1.401		59.50	86		145.50	194
2750	S.P., touch dimmer, type NM cable		17.10	.468		53	28.50		81.50	101
2770	Type MC cable		14.30	.559		64	34.50		98.50	121
2780	EMT & wire		5.71	1.401		75	86		161	211
2800	3-way touch dimmer, type NM cable		13.33	.600		49.50	37		86.50	110
2820	Type MC cable		11.43	.700		60.50	43		103.50	131
2830	EMT & wire		4.85	1.649		71.50	101		172.50	230
3000	Combination devices									
3100	S.P. switch/15 amp recpt., ivory, 1-gang box, plate									
3110	Type NM cable	1 Elec	11.43	.700	Ea.	22	43		65	88
3120	Type MC cable		10	.800		32.50	49		81.50	109
3130	EMT & wire		4.40	1.818		44	112		156	214
3150	S.P. switch/pilot light, type NM cable		11.43	.700		23.50	43		66.50	89.50
3170	Type MC cable		10	.800		34	49		83	111
3180	EMT & wire		4.43	1.806		45	111		156	215
3190	2-S.P. switches, 2-#14/2, no wiring		14	.571		13	35		48	66.50
3200	2-S.P. switches, 2-#14/2, type NM cables		10	.800		24	49		73	99.50
3220	Type MC cable		8.89	.900		38.50	55		93.50	125
3230	EMT & wire		4.10	1.951		46	120		166	229
3250	3-way switch/15 amp recpt., #14/3, type NM cable		10	.800		30	49		79	106
3270	Type MC cable		8.89	.900		44	55		99	131
3280	EMT & wire		4.10	1.951		50	120		170	233
3300	2-3 way switches, 2-#14/3, type NM cables		8.89	.900		38	55		93	124
3320	Type MC cable		8	1		60	61.50		121.50	158
3330	EMT & wire		4	2		56.50	123		179.50	246
3350	S.P. switch/20 amp recpt., #12/2, type NM cable		10	.800		40	49		89	117
3370	Type MC cable		8.89	.900		45.50	55		100.50	132
3380	EMT & wire		4.10	1.951		62.50	120		182.50	247
3400	Decorator style									
3410	S.P. switch/15 amp recpt., type NM cable	1 Elec	11.43	.700	Ea.	27.50	43		70.50	94
3420	Type MC cable		10	.800		38	49		87	115
3430	EMT & wire		4.40	1.818		49.50	112		161.50	221
3450	S.P. switch/pilot light, type NM cable		11.43	.700		29	43		72	95.50
3470	Type MC cable		10	.800		39.50	49		88.50	117
3480	EMT & wire		4.40	1.818		50.50	112		162.50	222
3500	2-S.P. switches, 2-#14/2, type NM cables		10	.800		29.50	49		78.50	106
3520	Type MC cable		8.89	.900		44.50	55		99.50	131
3530	EMT & wire		4.10	1.951		51.50	120		171.50	235
3550	3-way/15 amp recpt., #14/3, type NM cable		10	.800		35.50	49		84.50	112
3570	Type MC cable		8.89	.900		49.50	55		104.50	137
3580	EMT & wire		4.10	1.951		55.50	120		175.50	239
3650	2-3 way switches, 2-#14/3, type NM cables		8.89	.900		43.50	55		98.50	130
3670	Type MC cable		8	1		65.50	61.50		127	164
3680	EMT & wire		4	2		62	123		185	252
3700	S.P. switch/20 amp recpt., #12/2, type NM cable		10	.800		45.50	49		94.50	123
3720	Type MC cable		8.89	.900		51	55		106	138
3730	EMT & wire		4.10	1.951		68	120		188	253
4000	Receptacle devices									
4010	Duplex outlet, 15 amp recpt., ivory, 1-gang box, plate									
4015	Type NM cable	1 Elec	14.55	.550	Ea.	8.90	33.50		42.40	60

26 05 90.10 Residential Wiring	Crew	Daily Output	Labor-Hours	Unit	Material	2020 Bare Costs Labor	Equipment	Total	Total Incl O&P
4020	Type MC cable	1 Elec	12.31	.650	Ea.	19.55	40		59.55	81
4030	EMT & wire		5.33	1.501		29.50	92		121.50	170
4050	With #12/2, type NM cable		12.31	.650		10.50	40		50.50	71
4070	Type MC cable		10.67	.750		19.95	46		65.95	90.50
4080	EMT & wire		4.71	1.699		33	104		137	191
4100	20 amp recpt., #12/2, type NM cable		12.31	.650		19.50	40		59.50	81
4120	Type MC cable		10.67	.750		29	46		75	101
4130	EMT & wire	▼	4.71	1.699	▼	42	104		146	201
4140	For GFI see Section 26 05 90.10 line 4300 below									
4150	Decorator style, 15 amp recpt., type NM cable	1 Elec	14.55	.550	Ea.	14.45	33.50		47.95	66
4170	Type MC cable		12.31	.650		25	40		65	87
4180	EMT & wire		5.33	1.501		35	92		127	176
4200	With #12/2, type NM cable		12.31	.650		16.05	40		56.05	77
4220	Type MC cable		10.67	.750		25.50	46		71.50	96.50
4230	EMT & wire		4.71	1.699		38.50	104		142.50	198
4250	20 amp recpt., #12/2, type NM cable		12.31	.650		25	40		65	87
4270	Type MC cable		10.67	.750		34.50	46		80.50	107
4280	EMT & wire		4.71	1.699		47.50	104		151.50	208
4300	GFI, 15 amp recpt., type NM cable		12.31	.650		21	40		61	82.50
4320	Type MC cable		10.67	.750		31.50	46		77.50	103
4330	EMT & wire		4.71	1.699		41.50	104		145.50	201
4350	GFI with #12/2, type NM cable		10.67	.750		22.50	46		68.50	93
4370	Type MC cable		9.20	.870		32	53.50		85.50	115
4380	EMT & wire		4.21	1.900		45	117		162	224
4400	20 amp recpt., #12/2, type NM cable		10.67	.750		54	46		100	128
4420	Type MC cable		9.20	.870		63.50	53.50		117	150
4430	EMT & wire		4.21	1.900		76.50	117		193.50	258
4500	Weather-proof cover for above receptacles, add	▼	32	.250	▼	2.02	15.35		17.37	25
4550	Air conditioner outlet, 20 amp-240 volt recpt.									
4560	30' of #12/2, 2 pole circuit breaker									
4570	Type NM cable	1 Elec	10	.800	Ea.	61	49		110	140
4580	Type MC cable		9	.889		72	54.50		126.50	160
4590	EMT & wire		4	2		84	123		207	275
4600	Decorator style, type NM cable		10	.800		66	49		115	146
4620	Type MC cable		9	.889		77	54.50		131.50	166
4630	EMT & wire	▼	4	2	▼	88.50	123		211.50	281
4650	Dryer outlet, 30 amp-240 volt recpt., 20' of #10/3									
4660	2 pole circuit breaker									
4670	Type NM cable	1 Elec	6.41	1.248	Ea.	54.50	76.50		131	174
4680	Type MC cable		5.71	1.401		62.50	86		148.50	197
4690	EMT & wire	▼	3.48	2.299	▼	73	141		214	291
4700	Range outlet, 50 amp-240 volt recpt., 30' of #8/3									
4710	Type NM cable	1 Elec	4.21	1.900	Ea.	82.50	117		199.50	265
4720	Type MC cable		4	2		133	123		256	330
4730	EMT & wire		2.96	2.703		105	166		271	360
4750	Central vacuum outlet, type NM cable		6.40	1.250		58	76.50		134.50	178
4770	Type MC cable		5.71	1.401		70.50	86		156.50	206
4780	EMT & wire	▼	3.48	2.299	▼	88	141		229	305
4800	30 amp-110 volt locking recpt., #10/2 circ. bkr.									
4810	Type NM cable	1 Elec	6.20	1.290	Ea.	67	79		146	192
4820	Type MC cable		5.40	1.481		83	91		174	227
4830	EMT & wire	▼	3.20	2.500	▼	98	153		251	335
4900	Low voltage outlets									

26 05 90 – Residential Applications

26 05 90.10 Residential Wiring		Crew	Daily Output	Labor-Hours	Unit	Material	2020 Bare Costs Labor	Equipment	Total	Total Incl O&P
4910	Telephone recpt., 20' of 4/C phone wire	1 Elec	26	.308	Ea.	8.80	18.90		27.70	37.50
4920	TV recpt., 20' of RG59U coax wire, F type connector	"	16	.500	"	17.60	30.50		48.10	65
4950	Door bell chime, transformer, 2 buttons, 60' of bellwire									
4970	Economy model	1 Elec	11.50	.696	Ea.	57.50	42.50		100	127
4980	Custom model		11.50	.696		111	42.50		153.50	186
4990	Luxury model, 3 buttons		9.50	.842		188	51.50		239.50	284
6000	Lighting outlets									
6050	Wire only (for fixture), type NM cable	1 Elec	32	.250	Ea.	5.95	15.35		21.30	29.50
6070	Type MC cable		24	.333		11.30	20.50		31.80	43
6080	EMT & wire		10	.800		20.50	49		69.50	95.50
6100	Box (4"), and wire (for fixture), type NM cable		25	.320		15.05	19.65		34.70	45.50
6120	Type MC cable		20	.400		20.50	24.50		45	59
6130	EMT & wire		11	.727		29.50	44.50		74	99
6200	Fixtures (use with line 6050 or 6100 above)									
6210	Canopy style, economy grade	1 Elec	40	.200	Ea.	22	12.25		34.25	43
6220	Custom grade		40	.200		53	12.25		65.25	76.50
6250	Dining room chandelier, economy grade		19	.421		82	26		108	129
6260	Custom grade		19	.421		320	26		346	395
6270	Luxury grade		15	.533		1,300	32.50		1,332.50	1,475
6310	Kitchen fixture (fluorescent), economy grade		30	.267		72.50	16.35		88.85	104
6320	Custom grade		25	.320		147	19.65		166.65	191
6350	Outdoor, wall mounted, economy grade		30	.267		30.50	16.35		46.85	58
6360	Custom grade		30	.267		119	16.35		135.35	155
6370	Luxury grade		25	.320		247	19.65		266.65	300
6410	Outdoor PAR floodlights, 1 lamp, 150 watt		20	.400		28	24.50		52.50	67
6420	2 lamp, 150 watt each		20	.400		45	24.50		69.50	86
6430	For infrared security sensor, add		32	.250		95	15.35		110.35	127
6450	Outdoor, quartz-halogen, 300 watt flood		20	.400		40	24.50		64.50	81
6600	Recessed downlight, round, pre-wired, 50 or 75 watt trim		30	.267		68	16.35		84.35	99.50
6610	With shower light trim		30	.267		93.50	16.35		109.85	128
6620	With wall washer trim		28	.286		94	17.55		111.55	129
6630	With eye-ball trim		28	.286		85	17.55		102.55	120
6700	Porcelain lamp holder		40	.200		2.76	12.25		15.01	21.50
6710	With pull switch		40	.200		10.70	12.25		22.95	30
6750	Fluorescent strip, 2-20 watt tube, wrap around diffuser, 24"		24	.333		45.50	20.50		66	80.50
6770	2-34 watt tubes, 48"		20	.400		160	24.50		184.50	213
6800	Bathroom heat lamp, 1-250 watt		28	.286		33.50	17.55		51.05	63
6810	2-250 watt lamps		28	.286		64.50	17.55		82.05	96.50
6820	For timer switch, see Section 26 05 90.10 line 2400									
6900	Outdoor post lamp, incl. post, fixture, 35' of #14/2									
6910	Type NM cable	1 Elec	3.50	2.286	Ea.	325	140		465	565
6920	Photo-eye, add		27	.296		27	18.20		45.20	57
6950	Clock dial time switch, 24 hr., w/enclosure, type NM cable		11.43	.700		72.50	43		115.50	144
6970	Type MC cable		11	.727		83	44.50		127.50	158
6980	EMT & wire		4.85	1.649		93	101		194	253
7000	Alarm systems									
7050	Smoke detectors, box, #14/3, type NM cable	1 Elec	14.55	.550	Ea.	33.50	33.50		67	87
7070	Type MC cable		12.31	.650		44	40		84	108
7080	EMT & wire		5	1.600		50	98		148	201
7090	For relay output to security system, add					10.25			10.25	11.25
8000	Residential equipment									
8050	Disposal hook-up, incl. switch, outlet box, 3' of flex									
8060	20 amp-1 pole circ. bkr., and 25' of #12/2									

For customer support on your Mechanical Costs with RSMeans Data, call 800.448.8182.

427

26 05 90.10 Residential Wiring	Crew	Daily Output	Labor-Hours	Unit	Material	2020 Bare Costs Labor	Equipment	Total	Total Incl O&P
8070	Type NM cable	1 Elec	10	.800	Ea.	29	49		78	105
8080	Type MC cable		8	1		39	61.50		100.50	135
8090	EMT & wire		5	1.600		55	98		153	206
8100	Trash compactor or dishwasher hook-up, incl. outlet box,									
8110	3' of flex, 15 amp-1 pole circ. bkr., and 25' of #14/2									
8120	Type NM cable	1 Elec	10	.800	Ea.	16.05	49		65.05	90.50
8130	Type MC cable		8	1		28	61.50		89.50	123
8140	EMT & wire		5	1.600		41.50	98		139.50	192
8150	Hot water sink dispenser hook-up, use line 8100									
8200	Vent/exhaust fan hook-up, type NM cable	1 Elec	32	.250	Ea.	5.95	15.35		21.30	29.50
8220	Type MC cable		24	.333		11.30	20.50		31.80	43
8230	EMT & wire		10	.800		20.50	49		69.50	95.50
8250	Bathroom vent fan, 50 CFM (use with above hook-up)									
8260	Economy model	1 Elec	15	.533	Ea.	18.80	32.50		51.30	69
8270	Low noise model		15	.533		47.50	32.50		80	101
8280	Custom model		12	.667		117	41		158	190
8300	Bathroom or kitchen vent fan, 110 CFM									
8310	Economy model	1 Elec	15	.533	Ea.	66.50	32.50		99	122
8320	Low noise model	"	15	.533	"	96.50	32.50		129	155
8350	Paddle fan, variable speed (w/o lights)									
8360	Economy model (AC motor)	1 Elec	10	.800	Ea.	136	49		185	222
8362	With light kit		10	.800		176	49		225	266
8370	Custom model (AC motor)		10	.800		345	49		394	455
8372	With light kit		10	.800		385	49		434	500
8380	Luxury model (DC motor)		8	1		315	61.50		376.50	435
8382	With light kit		8	1		355	61.50		416.50	480
8390	Remote speed switch for above, add		12	.667		40.50	41		81.50	106
8500	Whole house exhaust fan, ceiling mount, 36", variable speed									
8510	Remote switch, incl. shutters, 20 amp-1 pole circ. bkr.									
8520	30' of #12/2, type NM cable	1 Elec	4	2	Ea.	1,375	123		1,498	1,700
8530	Type MC cable		3.50	2.286		1,400	140		1,540	1,750
8540	EMT & wire		3	2.667		1,425	164		1,589	1,800
8600	Whirlpool tub hook-up, incl. timer switch, outlet box									
8610	3' of flex, 20 amp-1 pole GFI circ. bkr.									
8620	30' of #12/2, type NM cable	1 Elec	5	1.600	Ea.	129	98		227	288
8630	Type MC cable		4.20	1.905		136	117		253	325
8640	EMT & wire		3.40	2.353		149	144		293	380
8650	Hot water heater hook-up, incl. 1-2 pole circ. bkr., box;									
8660	3' of flex, 20' of #10/2, type NM cable	1 Elec	5	1.600	Ea.	29	98		127	178
8670	Type MC cable		4.20	1.905		42	117		159	220
8680	EMT & wire		3.40	2.353		48	144		192	268
9000	Heating/air conditioning									
9050	Furnace/boiler hook-up, incl. firestat, local on-off switch									
9060	Emergency switch, and 40' of type NM cable	1 Elec	4	2	Ea.	57	123		180	246
9070	Type MC cable		3.50	2.286		72	140		212	288
9080	EMT & wire		1.50	5.333		93.50	325		418.50	590
9100	Air conditioner hook-up, incl. local 60 amp disc. switch									
9110	3' sealtite, 40 amp, 2 pole circuit breaker									
9130	40' of #8/2, type NM cable	1 Elec	3.50	2.286	Ea.	144	140		284	365
9140	Type MC cable		3	2.667		212	164		376	475
9150	EMT & wire		1.30	6.154		185	380		565	765
9200	Heat pump hook-up, 1-40 & 1-100 amp 2 pole circ. bkr.									
9210	Local disconnect switch, 3' sealtite									

26 05 Common Work Results for Electrical

26 05 90 – Residential Applications

26 05 90.10 Residential Wiring	Crew	Daily Output	Labor-Hours	Unit	Material	2020 Bare Costs Labor	Equipment	Total	Total Incl O&P
9220	40' of #8/2 & 30' of #3/2								
9230	Type NM cable	1 Elec	1.30	6.154	Ea.	520	380		900	1,125
9240	Type MC cable		1.08	7.407		550	455		1,005	1,275
9250	EMT & wire		.94	8.511		535	520		1,055	1,350
9500	Thermostat hook-up, using low voltage wire									
9520	Heating only, 25' of #18-3	1 Elec	24	.333	Ea.	6.75	20.50		27.25	38
9530	Heating/cooling, 25' of #18-4	"	20	.400	"	8.60	24.50		33.10	46

26 24 Switchboards and Panelboards

26 24 19 – Motor-Control Centers

26 24 19.40 Motor Starters and Controls

		Crew	Daily Output	Labor-Hours	Unit	Material	2020 Bare Costs Labor	Equipment	Total	Total Incl O&P
0010	**MOTOR STARTERS AND CONTROLS**									
0050	Magnetic, FVNR, with enclosure and heaters, 480 volt									
0080	2 HP, size 00	1 Elec	3.50	2.286	Ea.	196	140		336	425
0100	5 HP, size 0		2.30	3.478		345	213		558	700
0200	10 HP, size 1		1.60	5		262	305		567	745
0300	25 HP, size 2	2 Elec	2.20	7.273		490	445		935	1,200
0400	50 HP, size 3		1.80	8.889		800	545		1,345	1,700
0500	100 HP, size 4		1.20	13.333		1,775	820		2,595	3,175
0600	200 HP, size 5		.90	17.778		4,150	1,100		5,250	6,200
0610	400 HP, size 6		.80	20		18,300	1,225		19,525	21,900
0620	NEMA 7, 5 HP, size 0	1 Elec	1.60	5		1,375	305		1,680	1,950
0630	10 HP, size 1	"	1.10	7.273		1,425	445		1,870	2,250
0640	25 HP, size 2	2 Elec	1.80	8.889		2,300	545		2,845	3,325
0650	50 HP, size 3		1.20	13.333		3,450	820		4,270	5,025
0660	100 HP, size 4		.90	17.778		5,575	1,100		6,675	7,750
0670	200 HP, size 5		.50	32		13,300	1,975		15,275	17,500
0700	Combination, with motor circuit protectors, 5 HP, size 0	1 Elec	1.80	4.444		985	273		1,258	1,475
0800	10 HP, size 1	"	1.30	6.154		1,025	380		1,405	1,675
0900	25 HP, size 2	2 Elec	2	8		1,425	490		1,915	2,300
1000	50 HP, size 3		1.32	12.121		2,075	745		2,820	3,375
1200	100 HP, size 4		.80	20		4,450	1,225		5,675	6,725
1220	NEMA 7, 5 HP, size 0	1 Elec	1.30	6.154		3,550	380		3,930	4,475
1230	10 HP, size 1	"	1	8		3,650	490		4,140	4,725
1240	25 HP, size 2	2 Elec	1.32	12.121		4,875	745		5,620	6,450
1250	50 HP, size 3		.80	20		8,050	1,225		9,275	10,700
1260	100 HP, size 4		.60	26.667		12,500	1,625		14,125	16,200
1270	200 HP, size 5		.40	40		27,200	2,450		29,650	33,700
1400	Combination, with fused switch, 5 HP, size 0	1 Elec	1.80	4.444		590	273		863	1,050
1600	10 HP, size 1	"	1.30	6.154		630	380		1,010	1,250
1800	25 HP, size 2	2 Elec	2	8		1,025	490		1,515	1,850
2000	50 HP, size 3		1.32	12.121		1,725	745		2,470	3,000
2200	100 HP, size 4		.80	20		3,025	1,225		4,250	5,150
3500	Magnetic FVNR with NEMA 12, enclosure & heaters, 480 volt									
3600	5 HP, size 0	1 Elec	2.20	3.636	Ea.	237	223		460	590
3700	10 HP, size 1	"	1.50	5.333		355	325		680	875
3800	25 HP, size 2	2 Elec	2	8		665	490		1,155	1,475
3900	50 HP, size 3		1.60	10		1,025	615		1,640	2,050
4000	100 HP, size 4		1	16		2,450	980		3,430	4,150
4100	200 HP, size 5		.80	20		5,900	1,225		7,125	8,300
4200	Combination, with motor circuit protectors, 5 HP, size 0	1 Elec	1.70	4.706		785	289		1,074	1,300

For customer support on your Mechanical Costs with RSMeans Data, call 800.448.8182.

429

26 24 19 – Motor-Control Centers

26 24 19.40 Motor Starters and Controls

		Crew	Daily Output	Labor-Hours	Unit	Material	2020 Bare Costs Labor	2020 Bare Costs Equipment	Total	Total Incl O&P
4300	10 HP, size 1	1 Elec	1.20	6.667	Ea.	815	410		1,225	1,500
4400	25 HP, size 2	2 Elec	1.80	8.889		1,225	545		1,770	2,150
4500	50 HP, size 3		1.20	13.333		2,000	820		2,820	3,400
4600	100 HP, size 4	↓	.74	21.622		4,475	1,325		5,800	6,900
4700	Combination, with fused switch, 5 HP, size 0	1 Elec	1.70	4.706		745	289		1,034	1,250
4800	10 HP, size 1	"	1.20	6.667		775	410		1,185	1,450
4900	25 HP, size 2	2 Elec	1.80	8.889		1,175	545		1,720	2,100
5000	50 HP, size 3		1.20	13.333		1,900	820		2,720	3,300
5100	100 HP, size 4	↓	.74	21.622	↓	3,825	1,325		5,150	6,200
5200	Factory installed controls, adders to size 0 thru 5									
5300	Start-stop push button	1 Elec	32	.250	Ea.	49.50	15.35		64.85	77.50
5400	Hand-off-auto-selector switch		32	.250		49.50	15.35		64.85	77.50
5500	Pilot light		32	.250		93	15.35		108.35	125
5600	Start-stop-pilot		32	.250		143	15.35		158.35	180
5700	Auxiliary contact, NO or NC		32	.250		68	15.35		83.35	98
5800	NO-NC	↓	32	.250	↓	136	15.35		151.35	173

26 29 Low-Voltage Controllers

26 29 13 – Enclosed Controllers

26 29 13.20 Control Stations

		Crew	Daily Output	Labor-Hours	Unit	Material	2020 Bare Costs Labor	2020 Bare Costs Equipment	Total	Total Incl O&P
0010	**CONTROL STATIONS**									
0050	NEMA 1, heavy duty, stop/start	1 Elec	8	1	Ea.	153	61.50		214.50	260
0100	Stop/start, pilot light		6.20	1.290		208	79		287	345
0200	Hand/off/automatic		6.20	1.290		113	79		192	242
0400	Stop/start/reverse		5.30	1.509		206	92.50		298.50	365
0500	NEMA 7, heavy duty, stop/start		6	1.333		555	82		637	730
0600	Stop/start, pilot light	↓	4	2	↓	680	123		803	930

26 29 23 – Variable-Frequency Motor Controllers

26 29 23.10 Variable Frequency Drives/Adj. Frequency Drives

			Crew	Daily Output	Labor-Hours	Unit	Material	2020 Bare Costs Labor	2020 Bare Costs Equipment	Total	Total Incl O&P
0010	**VARIABLE FREQUENCY DRIVES/ADJ. FREQUENCY DRIVES**										
0100	Enclosed (NEMA 1), 460 volt, for 3 HP motor size	G	1 Elec	.80	10	Ea.	1,825	615		2,440	2,925
0110	5 HP motor size	G		.80	10		2,050	615		2,665	3,175
0120	7.5 HP motor size	G		.67	11.940		2,450	735		3,185	3,775
0130	10 HP motor size	G	↓	.67	11.940		2,775	735		3,510	4,150
0140	15 HP motor size	G	2 Elec	.89	17.978		3,450	1,100		4,550	5,450
0150	20 HP motor size	G		.89	17.978		4,350	1,100		5,450	6,425
0160	25 HP motor size	G		.67	23.881		5,125	1,475		6,600	7,825
0170	30 HP motor size	G		.67	23.881		6,250	1,475		7,725	9,050
0180	40 HP motor size	G		.67	23.881		7,225	1,475		8,700	10,100
0190	50 HP motor size	G	↓	.53	30.189		9,650	1,850		11,500	13,400
0200	60 HP motor size	G	R-3	.56	35.714		11,700	2,175	340	14,215	16,500
0210	75 HP motor size	G		.56	35.714		13,700	2,175	340	16,215	18,700
0220	100 HP motor size	G		.50	40		16,800	2,450	380	19,630	22,500
0230	125 HP motor size	G		.50	40		18,200	2,450	380	21,030	24,100
0240	150 HP motor size	G		.50	40		23,600	2,450	380	26,430	30,100
0250	200 HP motor size	G	↓	.42	47.619		30,600	2,900	450	33,950	38,500
1100	Custom-engineered, 460 volt, for 3 HP motor size	G	1 Elec	.56	14.286		3,425	875		4,300	5,075
1110	5 HP motor size	G		.56	14.286		3,450	875		4,325	5,100
1120	7.5 HP motor size	G		.47	17.021		3,100	1,050		4,150	4,950
1130	10 HP motor size	G	↓	.47	17.021	↓	3,250	1,050		4,300	5,150

26 29 23.10 Variable Frequency Drives/Adj. Frequency Drives		Crew	Daily Output	Labor-Hours	Unit	Material	2020 Bare Costs Labor	Equipment	Total	Total Incl O&P
1140	15 HP motor size	G	2 Elec	.62	25.806	Ea.	4,700	1,575		6,275	7,525
1150	20 HP motor size	G		.62	25.806		4,350	1,575		5,925	7,125
1160	25 HP motor size	G		.47	34.043		5,375	2,100		7,475	9,025
1170	30 HP motor size	G		.47	34.043		6,625	2,100		8,725	10,400
1180	40 HP motor size	G		.47	34.043		7,925	2,100		10,025	11,800
1190	50 HP motor size	G		.37	43.243		10,500	2,650		13,150	15,500
1200	60 HP motor size	G	R-3	.39	51.282		15,800	3,125	485	19,410	22,600
1210	75 HP motor size	G		.39	51.282		14,600	3,125	485	18,210	21,300
1220	100 HP motor size	G		.35	57.143		18,500	3,500	540	22,540	26,100
1230	125 HP motor size	G		.35	57.143		19,800	3,500	540	23,840	27,600
1240	150 HP motor size	G		.35	57.143		22,500	3,500	540	26,540	30,600
1250	200 HP motor size	G		.29	68.966		25,700	4,225	650	30,575	35,200
2000	For complex & special design systems to meet specific										
2010	requirements, obtain quote from vendor.										

For customer support on your Mechanical Costs with RSMeans Data, call 800.448.8182.

431

Division Notes

		CREW	DAILY OUTPUT	LABOR-HOURS	UNIT	BARE COSTS				TOTAL INCL O&P
						MAT.	LABOR	EQUIP.	TOTAL	

Estimating Tips

- When estimating material costs for electronic safety and security systems, it is always prudent to obtain manufacturers' quotations for equipment prices and special installation requirements that may affect the total cost.

- Fire alarm systems consist of control panels, annunciator panels, batteries with rack, charger, and fire alarm actuating and indicating devices. Some fire alarm systems include speakers, telephone lines, door closer controls, and other components. Be careful not to overlook the costs related to installation for these items. Also be aware of costs for integrated automation instrumentation and terminal devices, control equipment, control wiring, and programming. Insurance underwriters may have specific requirements for the type of materials to be installed or design requirements based on the hazard to be protected. Local jurisdictions may have requirements not covered by code. It is advisable to be aware of any special conditions.

- Security equipment includes items such as CCTV, access control, and other detection and identification systems to perform alert and alarm functions. Be sure to consider the costs related to installation for this security equipment, such as for integrated automation instrumentation and terminal devices, control equipment, control wiring, and programming.

Reference Numbers

Reference numbers are shown at the beginning of some major classifications. These numbers refer to related items in the Reference Section. The reference information may be an estimating procedure, an alternate pricing method, or technical information.

Same Data. Simplified.

Enjoy the convenience and efficiency of accessing your costs anywhere:

- **Skip the multiplier** by setting your location
- **Quickly search,** edit, favorite and share costs
- **Stay on top of price changes** with automatic updates

Discover more at rsmeans.com/online

Note: Trade Service, in part, has been used as a reference source for some of the material prices used in Division 28.

28 18 Security Access Detection Equipment

28 18 11 – Security Access Metal Detectors

28 18 11.13 Security Access Metal Detectors

		Crew	Daily Output	Labor-Hours	Unit	Material	2020 Bare Costs Labor	Equipment	Total	Total Incl O&P
0010	**SECURITY ACCESS METAL DETECTORS**									
0240	Metal detector, hand-held, wand type, unit only				Ea.	129			129	138
0250	Metal detector, walk through portal type, single zone	1 Elec	2	4		3,750	245		3,995	4,500
0260	Multi-zone	"	2	4		4,875	245		5,120	5,725

28 18 13 – Security Access X-Ray Equipment

28 18 13.16 Security Access X-Ray Equipment

		Crew	Daily Output	Labor-Hours	Unit	Material	2020 Bare Costs Labor	Equipment	Total	Total Incl O&P
0010	**SECURITY ACCESS X-RAY EQUIPMENT**									
0290	X-ray machine, desk top, for mail/small packages/letters	1 Elec	4	2	Ea.	3,425	123		3,548	3,950
0300	Conveyor type, incl. monitor		2	4		15,700	245		15,945	17,700
0310	Includes additional features		2	4		27,500	245		27,745	30,700
0320	X-ray machine, large unit, for airports, incl. monitor	2 Elec	1	16		39,400	980		40,380	44,800
0330	Full console	"	.50	32		67,500	1,975		69,475	77,000

28 18 15 – Security Access Explosive Detection Equipment

28 18 15.23 Security Access Explosive Detection Equipment

		Crew	Daily Output	Labor-Hours	Unit	Material	2020 Bare Costs Labor	Equipment	Total	Total Incl O&P
0010	**SECURITY ACCESS EXPLOSIVE DETECTION EQUIPMENT**									
0270	Explosives detector, walk through portal type	1 Elec	2	4	Ea.	48,800	245		49,045	54,000
0280	Hand-held, battery operated				"				25,500	28,100

28 23 Video Management System

28 23 23 – Video Surveillance Systems Infrastructure

28 23 23.50 Video Surveillance Equipments

		Crew	Daily Output	Labor-Hours	Unit	Material	2020 Bare Costs Labor	Equipment	Total	Total Incl O&P
0010	**VIDEO SURVEILLANCE EQUIPMENTS**									
0200	Video cameras, wireless, hidden in exit signs, clocks, etc., incl. receiver	1 Elec	3	2.667	Ea.	171	164		335	430
0210	Accessories for video recorder, single camera		3	2.667		195	164		359	460
0220	For multiple cameras		3	2.667		1,675	164		1,839	2,100
0230	Video cameras, wireless, for under vehicle searching, complete		2	4		15,200	245		15,445	17,100
0234	Master monitor station, 3 doors x 5 color monitor with tilt feature		2	4		985	245		1,230	1,450

28 31 Intrusion Detection

28 31 16 – Intrusion Detection Systems Infrastructure

28 31 16.50 Intrusion Detection

		Crew	Daily Output	Labor-Hours	Unit	Material	2020 Bare Costs Labor	Equipment	Total	Total Incl O&P
0010	**INTRUSION DETECTION**, not including wires & conduits									
0100	Burglar alarm, battery operated, mechanical trigger	1 Elec	4	2	Ea.	289	123		412	505
0200	Electrical trigger		4	2		345	123		468	565
0400	For outside key control, add		8	1		89.50	61.50		151	190
0600	For remote signaling circuitry, add		8	1		143	61.50		204.50	249
0800	Card reader, flush type, standard		2.70	2.963		725	182		907	1,075
1000	Multi-code		2.70	2.963		1,250	182		1,432	1,650

28 42 Gas Detection and Alarm

28 42 15 – Gas Detection Sensors

28 42 15.50 Tank Leak Detection Systems	Crew	Daily Output	Labor-Hours	Unit	Material	2020 Bare Costs Labor	Equipment	Total	Total Incl O&P
0010 **TANK LEAK DETECTION SYSTEMS** Liquid and vapor									
0100 For hydrocarbons and hazardous liquids/vapors									
0120 Controller, data acquisition, incl. printer, modem, RS232 port									
0140 24 channel, for use with all probes				Ea.	3,125			3,125	3,450
0160 9 channel, for external monitoring				"	1,225			1,225	1,350
0200 Probes									
0210 Well monitoring									
0220 Liquid phase detection				Ea.	725			725	800
0230 Hydrocarbon vapor, fixed position					870			870	960
0240 Hydrocarbon vapor, float mounted					615			615	675
0250 Both liquid and vapor hydrocarbon					645			645	710
0300 Secondary containment, liquid phase									
0310 Pipe trench/manway sump				Ea.	765			765	840
0320 Double wall pipe and manual sump					640			640	705
0330 Double wall fiberglass annular space					430			430	470
0340 Double wall steel tank annular space					282			282	310
0500 Accessories									
0510 Modem, non-dedicated phone line				Ea.	281			281	310
0600 Monitoring, internal									
0610 Automatic tank gauge, incl. overfill				Ea.	1,325			1,325	1,450
0620 Product line				"	1,375			1,375	1,525
0700 Monitoring, special									
0710 Cathodic protection				Ea.	695			695	765
0720 Annular space chemical monitor				"	945			945	1,050

28 46 Fire Detection and Alarm

28 46 11 – Fire Sensors and Detectors

28 46 11.21 Carbon-Monoxide Detection Sensors

	Crew	Daily Output	Labor-Hours	Unit	Material	2020 Bare Costs Labor	Equipment	Total	Total Incl O&P
0010 **CARBON-MONOXIDE DETECTION SENSORS**									
8400 Smoke and carbon monoxide alarm battery operated photoelectric low profile	1 Elec	24	.333	Ea.	51.50	20.50		72	87
8410 low profile photoelectric battery powered		24	.333		40	20.50		60.50	75
8420 photoelectric low profile sealed lithium		24	.333		70.50	20.50		91	108
8430 Photoelectric low profile sealed lithium smoke and CO with voice combo		24	.333		50.50	20.50		71	86
8500 Carbon monoxide sensor, wall mount 1Mod 1 relay output smoke & heat		24	.333		490	20.50		510.50	570
8700 Carbon monoxide detector, battery operated, wall mounted		16	.500		43	30.50		73.50	93
8710 Hardwired, wall and ceiling mounted		8	1		84.50	61.50		146	185
8720 Duct mounted		8	1		350	61.50		411.50	475

28 46 11.27 Other Sensors

	Crew	Daily Output	Labor-Hours	Unit	Material	2020 Bare Costs Labor	Equipment	Total	Total Incl O&P
0010 **OTHER SENSORS**									
5200 Smoke detector, ceiling type	1 Elec	6.20	1.290	Ea.	120	79		199	251
5240 Smoke detector, addressable type		6	1.333		224	82		306	370
5400 Duct type		3.20	2.500		295	153		448	555
5420 Duct addressable type		3.20	2.500		520	153		673	805

28 46 11.50 Fire and Heat Detectors

	Crew	Daily Output	Labor-Hours	Unit	Material	2020 Bare Costs Labor	Equipment	Total	Total Incl O&P
0010 **FIRE & HEAT DETECTORS**									
5000 Detector, rate of rise	1 Elec	8	1	Ea.	52	61.50		113.50	149
5100 Fixed temp fire alarm	"	7	1.143	"	42	70		112	151

For customer support on your Mechanical Costs with RSMeans Data, call 800.448.8182.

435

28 46 20.50 Alarm Panels and Devices	Crew	Daily Output	Labor-Hours	Unit	Material	2020 Bare Costs Labor	Equipment	Total	Total Incl O&P
0010 **ALARM PANELS AND DEVICES**, not including wires & conduits									
3600 4 zone	2 Elec	2	8	Ea.	410	490		900	1,175
3800 8 zone		1	16		875	980		1,855	2,425
3810 5 zone		1.50	10.667		700	655		1,355	1,750
3900 10 zone		1.25	12.800		1,025	785		1,810	2,325
4000 12 zone		.67	23.988		2,350	1,475		3,825	4,775
4020 Alarm device, tamper, flow	1 Elec	8	1		230	61.50		291.50	345
4050 Actuating device		8	1		340	61.50		401.50	460
4200 Battery and rack		4	2		460	123		583	695
4400 Automatic charger		8	1		525	61.50		586.50	670
4600 Signal bell		8	1		66.50	61.50		128	165
4610 Fire alarm signal bell 10" red 20-24 V P		8	1		154	61.50		215.50	261
4800 Trouble buzzer or manual station		8	1		88.50	61.50		150	189
5425 Duct smoke and heat detector 2 wire		8	1		125	61.50		186.50	230
5430 Fire alarm duct detector controller		3	2.667		254	164		418	525
5435 Fire alarm duct detector sensor kit		8	1		75	61.50		136.50	174
5440 Remote test station for smoke detector duct type		5.30	1.509		54.50	92.50		147	198
5460 Remote fire alarm indicator light		5.30	1.509		25	92.50		117.50	166
5600 Strobe and horn		5.30	1.509		137	92.50		229.50	289
5610 Strobe and horn (ADA type)		5.30	1.509		165	92.50		257.50	320
5620 Visual alarm (ADA type)		6.70	1.194		119	73.50		192.50	240
5800 electric bell		6.70	1.194		54	73.50		127.50	169
6000 Door holder, electro-magnetic		4	2		92	123		215	284
6200 Combination holder and closer		3.20	2.500		146	153		299	390
6600 Drill switch		8	1		400	61.50		461.50	530
6800 Master box		2.70	2.963		6,675	182		6,857	7,600
7000 Break glass station		8	1		58	61.50		119.50	155
7800 Remote annunciator, 8 zone lamp		1.80	4.444		212	273		485	640
8000 12 zone lamp	2 Elec	2.60	6.154		415	380		795	1,025
8200 16 zone lamp	"	2.20	7.273		370	445		815	1,075

Estimating Tips
31 05 00 Common Work Results for Earthwork

- Estimating the actual cost of performing earthwork requires careful consideration of the variables involved. This includes items such as type of soil, whether water will be encountered, dewatering, whether banks need bracing, disposal of excavated earth, and length of haul to fill or spoil sites, etc. If the project has large quantities of cut or fill, consider raising or lowering the site to reduce costs, while paying close attention to the effect on site drainage and utilities.

- If the project has large quantities of fill, creating a borrow pit on the site can significantly lower the costs.

- It is very important to consider what time of year the project is scheduled for completion. Bad weather can create large cost overruns from dewatering, site repair, and lost productivity from cold weather.

Reference Numbers

Reference numbers are shown at the beginning of some major classifications. These numbers refer to related items in the Reference Section. The reference information may be an estimating procedure, an alternate pricing method, or technical information.

Note: Not all subdivisions listed here necessarily appear. ■

Same Data. Simplified.

Enjoy the convenience and efficiency of accessing your costs anywhere:

- **Skip the multiplier** by setting your location
- **Quickly search,** edit, favorite and share costs
- **Stay on top of price changes** with automatic updates

Discover more at rsmeans.com/online

31 05 Common Work Results for Earthwork

31 05 23 – Cement and Concrete for Earthwork

31 05 23.30 Plant Mixed Bituminous Concrete	Crew	Daily Output	Labor-Hours	Unit	Material	2020 Bare Costs Labor	Equipment	Total	Total Incl O&P
0010 **PLANT MIXED BITUMINOUS CONCRETE**									
0020 Asphaltic concrete plant mix (145 lb./C.F.)				Ton	68			68	74.50
0040 Asphaltic concrete less than 300 tons add trucking costs									
0200 All weather patching mix, hot				Ton	66			66	73
0250 Cold patch					69.50			69.50	76.50
0300 Berm mix					68			68	75

31 23 Excavation and Fill

31 23 16 – Excavation

31 23 16.13 Excavating, Trench

		Crew	Daily Output	Labor-Hours	Unit	Material	2020 Bare Costs Labor	Equipment	Total	Total Incl O&P
0010	**EXCAVATING, TRENCH** G1030-805									
0011	Or continuous footing									
0020	Common earth with no sheeting or dewatering included									
0050	1' to 4' deep, 3/8 C.Y. excavator	B-11C	150	.107	B.C.Y.		5.25	1.43	6.68	9.45
0060	1/2 C.Y. excavator	B-11M	200	.080			3.95	1.16	5.11	7.20
0090	4' to 6' deep, 1/2 C.Y. excavator	"	200	.080			3.95	1.16	5.11	7.20
0100	5/8 C.Y. excavator	B-12Q	250	.064			3.24	2.39	5.63	7.50
0110	3/4 C.Y. excavator	B-12F	300	.053			2.70	2.31	5.01	6.60
0300	1/2 C.Y. excavator, truck mounted	B-12J	200	.080			4.05	4.23	8.28	10.70
0500	6' to 10' deep, 3/4 C.Y. excavator	B-12F	225	.071			3.60	3.09	6.69	8.80
0600	1 C.Y. excavator, truck mounted	B-12K	400	.040			2.03	2.43	4.46	5.70
0900	10' to 14' deep, 3/4 C.Y. excavator	B-12F	200	.080			4.05	3.47	7.52	9.85
1000	1-1/2 C.Y. excavator	B-12B	540	.030			1.50	1.27	2.77	3.65
1300	14' to 20' deep, 1 C.Y. excavator	B-12A	320	.050			2.53	2.49	5.02	6.50
1340	20' to 24' deep, 1 C.Y. excavator	"	288	.056			2.81	2.76	5.57	7.25
1352	4' to 6' deep, 1/2 C.Y. excavator w/trench box	B-13H	188	.085			4.31	5.15	9.46	12.10
1354	5/8 C.Y. excavator	"	235	.068			3.45	4.10	7.55	9.65
1356	3/4 C.Y. excavator	B-13G	282	.057			2.87	2.88	5.75	7.50
1362	6' to 10' deep, 3/4 C.Y. excavator w/trench box		212	.075			3.82	3.83	7.65	9.95
1374	10' to 14' deep, 3/4 C.Y. excavator w/trench box		188	.085			4.31	4.32	8.63	11.20
1376	1-1/2 C.Y. excavator	B-13E	508	.032			1.60	1.59	3.19	4.13
1381	14' to 20' deep, 1 C.Y. excavator w/trench box	B-13D	301	.053			2.69	3.03	5.72	7.35
1386	20' to 24' deep, 1 C.Y. excavator w/trench box	"	271	.059			2.99	3.37	6.36	8.20
1400	By hand with pick and shovel 2' to 6' deep, light soil	1 Clab	8	1			42		42	63.50
1500	Heavy soil	"	4	2			84		84	127
1700	For tamping backfilled trenches, air tamp, add	A-1G	100	.080	E.C.Y.		3.37	.54	3.91	5.65
1900	Vibrating plate, add	B-18	180	.133	"		5.70	.92	6.62	9.55
2100	Trim sides and bottom for concrete pours, common earth		1500	.016	S.F.		.68	.11	.79	1.15
2300	Hardpan		600	.040	"		1.71	.28	1.99	2.87
5020	Loam & sandy clay with no sheeting or dewatering included									
5050	1' to 4' deep, 3/8 C.Y. tractor loader/backhoe	B-11C	162	.099	B.C.Y.		4.88	1.32	6.20	8.75
5060	1/2 C.Y. excavator	B-11M	216	.074			3.66	1.08	4.74	6.70
5080	4' to 6' deep, 1/2 C.Y. excavator	"	216	.074			3.66	1.08	4.74	6.70
5090	5/8 C.Y. excavator	B-12Q	276	.058			2.94	2.17	5.11	6.80
5100	3/4 C.Y. excavator	B-12F	324	.049			2.50	2.14	4.64	6.10
5130	1/2 C.Y. excavator, truck mounted	B-12J	216	.074			3.75	3.92	7.67	9.90
5140	6' to 10' deep, 3/4 C.Y. excavator	B-12F	243	.066			3.33	2.86	6.19	8.15
5160	1 C.Y. excavator, truck mounted	B-12K	432	.037			1.88	2.25	4.13	5.30
5190	10' to 14' deep, 3/4 C.Y. excavator	B-12F	216	.074			3.75	3.21	6.96	9.15
5210	1-1/2 C.Y. excavator	B-12B	583	.027			1.39	1.18	2.57	3.38
5250	14' to 20' deep, 1 C.Y. excavator	B-12A	346	.046			2.34	2.30	4.64	6.05

31 23 16.13 Excavating, Trench

		Crew	Daily Output	Labor-Hours	Unit	Material	2020 Bare Costs Labor	Equipment	Total	Total Incl O&P
5300	20' to 24' deep, 1 C.Y. excavator	B-12A	311	.051	B.C.Y.		2.61	2.56	5.17	6.70
5352	4' to 6' deep, 1/2 C.Y. excavator w/trench box	B-13H	205	.078			3.95	4.70	8.65	11.05
5354	5/8 C.Y. excavator	"	257	.062			3.15	3.75	6.90	8.85
5356	3/4 C.Y. excavator	B-13G	308	.052			2.63	2.64	5.27	6.85
5362	6' to 10' deep, 3/4 C.Y. excavator w/trench box		231	.069			3.51	3.51	7.02	9.10
5370	10' to 14' deep, 3/4 C.Y. excavator w/trench box	↓	205	.078			3.95	3.96	7.91	10.25
5374	1-1/2 C.Y. excavator	B-13E	554	.029			1.46	1.45	2.91	3.79
5382	14' to 20' deep, 1 C.Y. excavator w/trench box	B-13D	329	.049			2.46	2.78	5.24	6.75
5392	20' to 24' deep, 1 C.Y. excavator w/trench box	"	295	.054	↓		2.75	3.10	5.85	7.50
6020	Sand & gravel with no sheeting or dewatering included									
6050	1' to 4' deep, 3/8 C.Y. excavator	B-11C	165	.097	B.C.Y.		4.79	1.30	6.09	8.65
6060	1/2 C.Y. excavator	B-11M	220	.073			3.60	1.06	4.66	6.55
6080	4' to 6' deep, 1/2 C.Y. excavator	"	220	.073			3.60	1.06	4.66	6.55
6090	5/8 C.Y. excavator	B-12Q	275	.058			2.95	2.18	5.13	6.80
6100	3/4 C.Y. excavator	B-12F	330	.048			2.46	2.10	4.56	6
6130	1/2 C.Y. excavator, truck mounted	B-12J	220	.073			3.68	3.85	7.53	9.75
6140	6' to 10' deep, 3/4 C.Y. excavator	B-12F	248	.065			3.27	2.80	6.07	8
6160	1 C.Y. excavator, truck mounted	B-12K	440	.036			1.84	2.21	4.05	5.20
6190	10' to 14' deep, 3/4 C.Y. excavator	B-12F	220	.073			3.68	3.16	6.84	8.95
6210	1-1/2 C.Y. excavator	B-12B	594	.027			1.36	1.16	2.52	3.31
6250	14' to 20' deep, 1 C.Y. excavator	B-12A	352	.045			2.30	2.26	4.56	5.95
6300	20' to 24' deep, 1 C.Y. excavator	"	317	.050			2.56	2.51	5.07	6.60
6352	4' to 6' deep, 1/2 C.Y. excavator w/trench box	B-13H	209	.077			3.88	4.61	8.49	10.90
6362	6' to 10' deep, 3/4 C.Y. excavator w/trench box	B-13G	236	.068			3.43	3.44	6.87	8.95
6370	10' to 14' deep, 3/4 C.Y. excavator w/trench box	"	209	.077			3.88	3.89	7.77	10.05
6374	1-1/2 C.Y. excavator	B-13E	564	.028			1.44	1.43	2.87	3.72
6382	14' to 20' deep, 1 C.Y. excavator w/trench box	B-13D	334	.048			2.43	2.73	5.16	6.65
6392	20' to 24' deep, 1 C.Y. excavator w/trench box	"	301	.053	↓		2.69	3.03	5.72	7.35
7020	Dense hard clay with no sheeting or dewatering included									
7050	1' to 4' deep, 3/8 C.Y. excavator	B-11C	132	.121	B.C.Y.		6	1.62	7.62	10.80
7060	1/2 C.Y. excavator	B-11M	176	.091			4.49	1.32	5.81	8.20
7080	4' to 6' deep, 1/2 C.Y. excavator	"	176	.091			4.49	1.32	5.81	8.20
7090	5/8 C.Y. excavator	B-12Q	220	.073			3.68	2.72	6.40	8.50
7100	3/4 C.Y. excavator	B-12F	264	.061			3.07	2.63	5.70	7.50
7130	1/2 C.Y. excavator, truck mounted	B-12J	176	.091			4.60	4.81	9.41	12.20
7140	6' to 10' deep, 3/4 C.Y. excavator	B-12F	198	.081			4.09	3.51	7.60	10
7160	1 C.Y. excavator, truck mounted	B-12K	352	.045			2.30	2.77	5.07	6.50
7190	10' to 14' deep, 3/4 C.Y. excavator	B-12F	176	.091			4.60	3.94	8.54	11.25
7210	1-1/2 C.Y. excavator	B-12B	475	.034			1.71	1.45	3.16	4.15
7250	14' to 20' deep, 1 C.Y. excavator	B-12A	282	.057			2.87	2.82	5.69	7.40
7300	20' to 24' deep, 1 C.Y. excavator	"	254	.063	↓		3.19	3.13	6.32	8.20

31 23 16.14 Excavating, Utility Trench

		Crew	Daily Output	Labor-Hours	Unit	Material	2020 Bare Costs Labor	Equipment	Total	Total Incl O&P
0010	**EXCAVATING, UTILITY TRENCH**									
0011	Common earth									
0050	Trenching with chain trencher, 12 HP, operator walking									
0100	4" wide trench, 12" deep	B-53	800	.010	L.F.		.53	.20	.73	1.01
0150	18" deep		750	.011			.57	.21	.78	1.08
0200	24" deep		700	.011			.61	.22	.83	1.16
0300	6" wide trench, 12" deep		650	.012			.65	.24	.89	1.25
0350	18" deep		600	.013			.71	.26	.97	1.35
0400	24" deep		550	.015			.77	.29	1.06	1.46
0450	36" deep	↓	450	.018	↓		.94	.35	1.29	1.79

For customer support on your Mechanical Costs with RSMeans Data, call 800.448.8182.

439

31 23 Excavation and Fill

31 23 16 – Excavation

31 23 16.14 Excavating, Utility Trench

		Crew	Daily Output	Labor-Hours	Unit	Material	2020 Bare Costs Labor	2020 Bare Costs Equipment	Total	Total Incl O&P
0600	8" wide trench, 12" deep	B-53	475	.017	L.F.		.89	.33	1.22	1.69
0650	18" deep		400	.020			1.06	.39	1.45	2.01
0700	24" deep		350	.023			1.21	.45	1.66	2.30
0750	36" deep	▼	300	.027	▼		1.41	.52	1.93	2.69
1000	Backfill by hand including compaction, add									
1050	4" wide trench, 12" deep	A-1G	800	.010	L.F.		.42	.07	.49	.70
1100	18" deep		530	.015			.64	.10	.74	1.06
1150	24" deep		400	.020			.84	.14	.98	1.42
1300	6" wide trench, 12" deep		540	.015			.62	.10	.72	1.05
1350	18" deep		405	.020			.83	.13	.96	1.40
1400	24" deep		270	.030			1.25	.20	1.45	2.09
1450	36" deep		180	.044			1.87	.30	2.17	3.14
1600	8" wide trench, 12" deep		400	.020			.84	.14	.98	1.42
1650	18" deep		265	.030			1.27	.20	1.47	2.13
1700	24" deep		200	.040			1.68	.27	1.95	2.83
1750	36" deep	▼	135	.059	▼		2.49	.40	2.89	4.19
2000	Chain trencher, 40 HP operator riding									
2050	6" wide trench and backfill, 12" deep	B-54	1200	.007	L.F.		.35	.34	.69	.90
2100	18" deep		1000	.008			.42	.40	.82	1.07
2150	24" deep		975	.008			.44	.41	.85	1.10
2200	36" deep		900	.009			.47	.45	.92	1.19
2250	48" deep		750	.011			.57	.54	1.11	1.44
2300	60" deep		650	.012			.65	.62	1.27	1.66
2400	8" wide trench and backfill, 12" deep		1000	.008			.42	.40	.82	1.07
2450	18" deep		950	.008			.45	.42	.87	1.14
2500	24" deep		900	.009			.47	.45	.92	1.19
2550	36" deep		800	.010			.53	.50	1.03	1.34
2600	48" deep		650	.012			.65	.62	1.27	1.66
2700	12" wide trench and backfill, 12" deep		975	.008			.44	.41	.85	1.10
2750	18" deep		860	.009			.49	.47	.96	1.25
2800	24" deep		800	.010			.53	.50	1.03	1.34
2850	36" deep		725	.011			.58	.55	1.13	1.48
3000	16" wide trench and backfill, 12" deep		835	.010			.51	.48	.99	1.29
3050	18" deep		750	.010			.57	.54	1.11	1.44
3100	24" deep	▼	700	.011	▼		.61	.57	1.18	1.54
3200	Compaction with vibratory plate, add								35%	35%
5100	Hand excavate and trim for pipe bells after trench excavation									
5200	8" pipe	1 Clab	155	.052	L.F.		2.17		2.17	3.26
5300	18" pipe	"	130	.062	"		2.59		2.59	3.89

31 23 16.16 Structural Excavation for Minor Structures

		Crew	Daily Output	Labor-Hours	Unit	Material	2020 Bare Costs Labor	2020 Bare Costs Equipment	Total	Total Incl O&P
0010	**STRUCTURAL EXCAVATION FOR MINOR STRUCTURES**									
0015	Hand, pits to 6' deep, sandy soil	1 Clab	8	1	B.C.Y.		42		42	63.50
0100	Heavy soil or clay		4	2			84		84	127
0300	Pits 6' to 12' deep, sandy soil		5	1.600			67.50		67.50	101
0500	Heavy soil or clay		3	2.667			112		112	169
0700	Pits 12' to 18' deep, sandy soil		4	2			84		84	127
0900	Heavy soil or clay	▼	2	4	▼		168		168	253
1500	For wet or muck hand excavation, add to above								50%	50%

31 23 16.30 Drilling and Blasting Rock

		Crew	Daily Output	Labor-Hours	Unit	Material	2020 Bare Costs Labor	2020 Bare Costs Equipment	Total	Total Incl O&P
0010	**DRILLING AND BLASTING ROCK**									
0020	Rock, open face, under 1,500 C.Y.	B-47	225	.107	B.C.Y.	4.54	4.95	6.90	16.39	20
0100	Over 1,500 C.Y.	▼	300	.080	▼	4.54	3.71	5.20	13.45	16.25

31 23 Excavation and Fill

31 23 16 - Excavation

31 23 16.30 Drilling and Blasting Rock	Crew	Daily Output	Labor-Hours	Unit	Material	2020 Bare Costs Labor	2020 Bare Costs Equipment	Total	Total Incl O&P
0200	Areas where blasting mats are required, under 1,500 C.Y.	B-47	175	.137	B.C.Y.	4.54	6.35	8.90	19.79	24.50
0250	Over 1,500 C.Y.		250	.096		4.54	4.45	6.20	15.19	18.55
2200	Trenches, up to 1,500 C.Y.		22	1.091		13.15	50.50	70.50	134.15	169
2300	Over 1,500 C.Y.		26	.923		13.15	43	60	116.15	145

31 23 19 - Dewatering

31 23 19.20 Dewatering Systems

31 23 19.20 Dewatering Systems	Crew	Daily Output	Labor-Hours	Unit	Material	2020 Bare Costs Labor	2020 Bare Costs Equipment	Total	Total Incl O&P
0010	**DEWATERING SYSTEMS**									
0020	Excavate drainage trench, 2' wide, 2' deep	B-11C	90	.178	C.Y.		8.80	2.38	11.18	15.75
0100	2' wide, 3' deep, with backhoe loader	"	135	.119			5.85	1.58	7.43	10.55
0200	Excavate sump pits by hand, light soil	1 Clab	7.10	1.127			47.50		47.50	71.50
0300	Heavy soil	"	3.50	2.286			96		96	145
0500	Pumping 8 hrs., attended 2 hrs./day, incl. 20 L.F.									
0550	of suction hose & 100 L.F. discharge hose									
0600	2" diaphragm pump used for 8 hrs.	B-10H	4	3	Day		156	24.50	180.50	260
0650	4" diaphragm pump used for 8 hrs.	B-10I	4	3			156	37	193	274
0800	8 hrs. attended, 2" diaphragm pump	B-10H	1	12			620	98.50	718.50	1,050
0900	3" centrifugal pump	B-10J	1	12			620	92	712	1,025
1000	4" diaphragm pump	B-10I	1	12			620	148	768	1,100
1100	6" centrifugal pump	B-10K	1	12			620	294	914	1,250
1300	CMP, incl. excavation 3' deep, 12" diameter	B-6	115	.209	L.F.	12.40	9.55	1.86	23.81	30
1400	18" diameter		100	.240	"	18.80	11	2.14	31.94	39.50
1600	Sump hole construction, incl. excavation and gravel, pit		1250	.019	C.F.	1.10	.88	.17	2.15	2.72
1700	With 12" gravel collar, 12" pipe, corrugated, 16 ga.		70	.343	L.F.	23	15.70	3.05	41.75	52.50
1800	15" pipe, corrugated, 16 ga.		55	.436		30	19.95	3.89	53.84	67.50
1900	18" pipe, corrugated, 16 ga.		50	.480		35	22	4.28	61.28	76
2000	24" pipe, corrugated, 14 ga.		40	.600		42	27.50	5.35	74.85	93
2200	Wood lining, up to 4' x 4', add		300	.080	SFCA	16.40	3.66	.71	20.77	24.50
9950	See Section 31 23 19.40 for wellpoints									
9960	See Section 31 23 19.30 for deep well systems									

31 23 19.30 Wells

31 23 19.30 Wells	Crew	Daily Output	Labor-Hours	Unit	Material	2020 Bare Costs Labor	2020 Bare Costs Equipment	Total	Total Incl O&P
0010	**WELLS**									
0011	For dewatering 10' to 20' deep, 2' diameter									
0020	with steel casing, minimum	B-6	165	.145	V.L.F.	37.50	6.65	1.30	45.45	53
0050	Average		98	.245		44.50	11.20	2.18	57.88	68
0100	Maximum		49	.490		48	22.50	4.36	74.86	91
0300	For dewatering pumps see 01 54 33 in Reference Section									
0500	For domestic water wells, see Section 33 21 13.10									

31 23 19.40 Wellpoints

31 23 19.40 Wellpoints	Crew	Daily Output	Labor-Hours	Unit	Material	2020 Bare Costs Labor	2020 Bare Costs Equipment	Total	Total Incl O&P
0010	**WELLPOINTS** R312319-90									
0011	For equipment rental, see 01 54 33 in Reference Section									
0100	Installation and removal of single stage system									
0110	Labor only, 0.75 labor-hours per L.F.	1 Clab	10.70	.748	LF Hdr		31.50		31.50	47.50
0200	2.0 labor-hours per L.F.	"	4	2	"		84		84	127
0400	Pump operation, 4 @ 6 hr. shifts									
0410	Per 24 hr. day	4 Eqlt	1.27	25.197	Day		1,325		1,325	2,000
0500	Per 168 hr. week, 160 hr. straight, 8 hr. double time		.18	178	Week		9,425		9,425	14,100
0550	Per 4.3 week month		.04	800	Month		42,400		42,400	63,500
0600	Complete installation, operation, equipment rental, fuel &									
0610	removal of system with 2" wellpoints 5' OC									
0700	100' long header, 6" diameter, first month	4 Eqlt	3.23	9.907	LF Hdr	160	525		685	960
0800	Thereafter, per month		4.13	7.748		128	410		538	755
1000	200' long header, 8" diameter, first month		6	5.333		154	283		437	590

31 23 19 – Dewatering

31 23 19.40 Wellpoints	Crew	Daily Output	Labor-Hours	Unit	Material	2020 Bare Costs Labor	Equipment	Total	Total Incl O&P
1100	Thereafter, per month	4 Eqlt	8.39	3.814	LF Hdr	72	202		274	380
1300	500' long header, 8" diameter, first month		10.63	3.010		56	160		216	300
1400	Thereafter, per month		20.91	1.530		40	81		121	165
1600	1,000' long header, 10" diameter, first month		11.62	2.754		48	146		194	271
1700	Thereafter, per month		41.81	.765		24	40.50		64.50	87
1900	Note: above figures include pumping 168 hrs. per week,									
1910	the pump operator, and one stand-by pump.									

31 23 23 – Fill

31 23 23.13 Backfill

		Crew	Daily Output	Labor-Hours	Unit	Material	2020 Bare Costs Labor	Equipment	Total	Total Incl O&P
0010	**BACKFILL** G1030–805									
0015	By hand, no compaction, light soil	1 Clab	14	.571	L.C.Y.		24		24	36
0100	Heavy soil		11	.727	"		30.50		30.50	46
0300	Compaction in 6" layers, hand tamp, add to above		20.60	.388	E.C.Y.		16.35		16.35	24.50
0400	Roller compaction operator walking, add	B-10A	100	.120			6.20	1.66	7.86	11.15
0500	Air tamp, add	B-9D	190	.211			8.95	1.67	10.62	15.30
0600	Vibrating plate, add	A-1D	60	.133			5.60	.53	6.13	9.05
0800	Compaction in 12" layers, hand tamp, add to above	1 Clab	34	.235			9.90		9.90	14.90
0900	Roller compaction operator walking, add	B-10A	150	.080			4.15	1.11	5.26	7.40
1000	Air tamp, add	B-9	285	.140			5.95	1.25	7.20	10.35
1100	Vibrating plate, add	A-1E	90	.089			3.74	1.84	5.58	7.60
1300	Dozer backfilling, bulk, up to 300' haul, no compaction	B-10B	1200	.010	L.C.Y.		.52	1.25	1.77	2.16
1400	Air tamped, add	B-11B	80	.200	E.C.Y.		9.50	4.97	14.47	19.70
1600	Compacting backfill, 6" to 12" lifts, vibrating roller	B-10C	800	.015			.78	2.52	3.30	3.94
1700	Sheepsfoot roller	B-10D	750	.016			.83	2.59	3.42	4.09
1900	Dozer backfilling, trench, up to 300' haul, no compaction	B-10B	900	.013	L.C.Y.		.69	1.67	2.36	2.88
2000	Air tamped, add	B-11B	80	.200	E.C.Y.		9.50	4.97	14.47	19.70
2200	Compacting backfill, 6" to 12" lifts, vibrating roller	B-10C	700	.017			.89	2.88	3.77	4.50
2300	Sheepsfoot roller	B-10D	650	.018			.96	2.99	3.95	4.71

31 23 23.16 Fill By Borrow and Utility Bedding

		Crew	Daily Output	Labor-Hours	Unit	Material	2020 Bare Costs Labor	Equipment	Total	Total Incl O&P
0010	**FILL BY BORROW AND UTILITY BEDDING**									
0049	Utility bedding, for pipe & conduit, not incl. compaction									
0050	Crushed or screened bank run gravel	B-6	150	.160	L.C.Y.	21	7.30	1.43	29.73	35.50
0100	Crushed stone 3/4" to 1/2"		150	.160		27.50	7.30	1.43	36.23	43
0200	Sand, dead or bank		150	.160		18.25	7.30	1.43	26.98	32.50
0500	Compacting bedding in trench	A-1D	90	.089	E.C.Y.		3.74	.35	4.09	6
0600	If material source exceeds 2 miles, add for extra mileage.									
0610	See Section 31 23 23.20 for hauling mileage add.									

31 23 23.17 General Fill

		Crew	Daily Output	Labor-Hours	Unit	Material	2020 Bare Costs Labor	Equipment	Total	Total Incl O&P
0010	**GENERAL FILL**									
0011	Spread dumped material, no compaction									
0020	By dozer	B-10B	1000	.012	L.C.Y.		.62	1.50	2.12	2.58
0100	By hand	1 Clab	12	.667	"		28		28	42
0500	Gravel fill, compacted, under floor slabs, 4" deep	B-37	10000	.005	S.F.	.49	.21	.03	.73	.89
0600	6" deep		8600	.006		.74	.25	.03	1.02	1.21
0700	9" deep		7200	.007		1.23	.30	.04	1.57	1.83
0800	12" deep		6000	.008		1.72	.35	.04	2.11	2.47
1000	Alternate pricing method, 4" deep		120	.400	E.C.Y.	37	17.70	2.13	56.83	69.50
1100	6" deep		160	.300		37	13.30	1.60	51.90	62
1200	9" deep		200	.240		37	10.60	1.28	48.88	58
1300	12" deep		220	.218		37	9.65	1.16	47.81	56.50
1400	Granular fill				L.C.Y.	24.50			24.50	27

31 23 Excavation and Fill

31 23 23 – Fill

31 23 23.20 Hauling	Crew	Daily Output	Labor-Hours	Unit	Material	2020 Bare Costs Labor	Equipment	Total	Total Incl O&P
0010 **HAULING**									
0011 Excavated or borrow, loose cubic yards									
0012 no loading equipment, including hauling, waiting, loading/dumping									
0013 time per cycle (wait, load, travel, unload or dump & return)									
0014 8 C.Y. truck, 15 MPH avg., cycle 0.5 miles, 10 min. wait/ld./uld.	B-34A	320	.025	L.C.Y.		1.22	1.34	2.56	3.30
0016 cycle 1 mile		272	.029			1.44	1.57	3.01	3.89
0018 cycle 2 miles		208	.038			1.88	2.06	3.94	5.10
0020 cycle 4 miles		144	.056			2.72	2.97	5.69	7.35
0022 cycle 6 miles		112	.071			3.50	3.82	7.32	9.45
0024 cycle 8 miles		88	.091			4.45	4.86	9.31	12
0026 20 MPH avg., cycle 0.5 mile		336	.024			1.17	1.27	2.44	3.15
0028 cycle 1 mile		296	.027			1.32	1.45	2.77	3.57
0030 cycle 2 miles		240	.033			1.63	1.78	3.41	4.40
0032 cycle 4 miles		176	.045			2.22	2.43	4.65	6
0034 cycle 6 miles		136	.059			2.88	3.15	6.03	7.75
0036 cycle 8 miles		112	.071			3.50	3.82	7.32	9.45
0044 25 MPH avg., cycle 4 miles		192	.042			2.04	2.23	4.27	5.50
0046 cycle 6 miles		160	.050			2.45	2.68	5.13	6.60
0048 cycle 8 miles		128	.063			3.06	3.34	6.40	8.25
0050 30 MPH avg., cycle 4 miles		216	.037			1.81	1.98	3.79	4.90
0052 cycle 6 miles		176	.045			2.22	2.43	4.65	6
0054 cycle 8 miles		144	.056			2.72	2.97	5.69	7.35
0114 15 MPH avg., cycle 0.5 mile, 15 min. wait/ld./uld.		224	.036			1.75	1.91	3.66	4.72
0116 cycle 1 mile		200	.040			1.96	2.14	4.10	5.30
0118 cycle 2 miles		168	.048			2.33	2.55	4.88	6.30
0120 cycle 4 miles		120	.067			3.26	3.57	6.83	8.80
0122 cycle 6 miles		96	.083			4.08	4.46	8.54	11
0124 cycle 8 miles		80	.100			4.90	5.35	10.25	13.25
0126 20 MPH avg., cycle 0.5 mile		232	.034			1.69	1.85	3.54	4.56
0128 cycle 1 mile		208	.038			1.88	2.06	3.94	5.10
0130 cycle 2 miles		184	.043			2.13	2.33	4.46	5.75
0132 cycle 4 miles		144	.056			2.72	2.97	5.69	7.35
0134 cycle 6 miles		112	.071			3.50	3.82	7.32	9.45
0136 cycle 8 miles		96	.083			4.08	4.46	8.54	11
0144 25 MPH avg., cycle 4 miles		152	.053			2.58	2.82	5.40	6.95
0146 cycle 6 miles		128	.063			3.06	3.34	6.40	8.25
0148 cycle 8 miles		112	.071			3.50	3.82	7.32	9.45
0150 30 MPH avg., cycle 4 miles		168	.048			2.33	2.55	4.88	6.30
0152 cycle 6 miles		144	.056			2.72	2.97	5.69	7.35
0154 cycle 8 miles		120	.067			3.26	3.57	6.83	8.80
0214 15 MPH avg., cycle 0.5 mile, 20 min. wait/ld./uld.		176	.045			2.22	2.43	4.65	6
0216 cycle 1 mile		160	.050			2.45	2.68	5.13	6.60
0218 cycle 2 miles		136	.059			2.88	3.15	6.03	7.75
0220 cycle 4 miles		104	.077			3.77	4.12	7.89	10.20
0222 cycle 6 miles		88	.091			4.45	4.86	9.31	12
0224 cycle 8 miles		72	.111			5.45	5.95	11.40	14.70
0226 20 MPH avg., cycle 0.5 mile		176	.045			2.22	2.43	4.65	6
0228 cycle 1 mile		168	.048			2.33	2.55	4.88	6.30
0230 cycle 2 miles		144	.056			2.72	2.97	5.69	7.35
0232 cycle 4 miles		120	.067			3.26	3.57	6.83	8.80
0234 cycle 6 miles		96	.083			4.08	4.46	8.54	11
0236 cycle 8 miles		88	.091			4.45	4.86	9.31	12

For customer support on your Mechanical Costs with RSMeans Data, call 800.448.8182.

443

31 23 23.20 Hauling		Crew	Daily Output	Labor-Hours	Unit	Material	2020 Bare Costs		Total	Total Incl O&P
							Labor	Equipment		
0244	25 MPH avg., cycle 4 miles	B-34A	128	.063	L.C.Y.		3.06	3.34	6.40	8.25
0246	cycle 6 miles		112	.071			3.50	3.82	7.32	9.45
0248	cycle 8 miles		96	.083			4.08	4.46	8.54	11
0250	30 MPH avg., cycle 4 miles		136	.059			2.88	3.15	6.03	7.75
0252	cycle 6 miles		120	.067			3.26	3.57	6.83	8.80
0254	cycle 8 miles		104	.077			3.77	4.12	7.89	10.20
0314	15 MPH avg., cycle 0.5 mile, 25 min. wait/ld./uld.		144	.056			2.72	2.97	5.69	7.35
0316	cycle 1 mile		128	.063			3.06	3.34	6.40	8.25
0318	cycle 2 miles		112	.071			3.50	3.82	7.32	9.45
0320	cycle 4 miles		96	.083			4.08	4.46	8.54	11
0322	cycle 6 miles		80	.100			4.90	5.35	10.25	13.25
0324	cycle 8 miles		64	.125			6.10	6.70	12.80	16.50
0326	20 MPH avg., cycle 0.5 mile		144	.056			2.72	2.97	5.69	7.35
0328	cycle 1 mile		136	.059			2.88	3.15	6.03	7.75
0330	cycle 2 miles		120	.067			3.26	3.57	6.83	8.80
0332	cycle 4 miles		104	.077			3.77	4.12	7.89	10.20
0334	cycle 6 miles		88	.091			4.45	4.86	9.31	12
0336	cycle 8 miles		80	.100			4.90	5.35	10.25	13.25
0344	25 MPH avg., cycle 4 miles		112	.071			3.50	3.82	7.32	9.45
0346	cycle 6 miles		96	.083			4.08	4.46	8.54	11
0348	cycle 8 miles		88	.091			4.45	4.86	9.31	12
0350	30 MPH avg., cycle 4 miles		112	.071			3.50	3.82	7.32	9.45
0352	cycle 6 miles		104	.077			3.77	4.12	7.89	10.20
0354	cycle 8 miles		96	.083			4.08	4.46	8.54	11
0414	15 MPH avg., cycle 0.5 mile, 30 min. wait/ld./uld.		120	.067			3.26	3.57	6.83	8.80
0416	cycle 1 mile		112	.071			3.50	3.82	7.32	9.45
0418	cycle 2 miles		96	.083			4.08	4.46	8.54	11
0420	cycle 4 miles		80	.100			4.90	5.35	10.25	13.25
0422	cycle 6 miles		72	.111			5.45	5.95	11.40	14.70
0424	cycle 8 miles		64	.125			6.10	6.70	12.80	16.50
0426	20 MPH avg., cycle 0.5 mile		120	.067			3.26	3.57	6.83	8.80
0428	cycle 1 mile		112	.071			3.50	3.82	7.32	9.45
0430	cycle 2 miles		104	.077			3.77	4.12	7.89	10.20
0432	cycle 4 miles		88	.091			4.45	4.86	9.31	12
0434	cycle 6 miles		80	.100			4.90	5.35	10.25	13.25
0436	cycle 8 miles		72	.111			5.45	5.95	11.40	14.70
0444	25 MPH avg., cycle 4 miles		96	.083			4.08	4.46	8.54	11
0446	cycle 6 miles		88	.091			4.45	4.86	9.31	12
0448	cycle 8 miles		80	.100			4.90	5.35	10.25	13.25
0450	30 MPH avg., cycle 4 miles		96	.083			4.08	4.46	8.54	11
0452	cycle 6 miles		88	.091			4.45	4.86	9.31	12
0454	cycle 8 miles		80	.100			4.90	5.35	10.25	13.25
0514	15 MPH avg., cycle 0.5 mile, 35 min. wait/ld./uld.		104	.077			3.77	4.12	7.89	10.20
0516	cycle 1 mile		96	.083			4.08	4.46	8.54	11
0518	cycle 2 miles		88	.091			4.45	4.86	9.31	12
0520	cycle 4 miles		72	.111			5.45	5.95	11.40	14.70
0522	cycle 6 miles		64	.125			6.10	6.70	12.80	16.50
0524	cycle 8 miles		56	.143			7	7.65	14.65	18.90
0526	20 MPH avg., cycle 0.5 mile		104	.077			3.77	4.12	7.89	10.20
0528	cycle 1 mile		96	.083			4.08	4.46	8.54	11
0530	cycle 2 miles		96	.083			4.08	4.46	8.54	11
0532	cycle 4 miles		80	.100			4.90	5.35	10.25	13.25
0534	cycle 6 miles		72	.111			5.45	5.95	11.40	14.70

31 23 23.20 Hauling

		Crew	Daily Output	Labor-Hours	Unit	Material	Labor	Equipment	Total	Total Incl O&P
0536	cycle 8 miles	B-34A	64	.125	L.C.Y.		6.10	6.70	12.80	16.50
0544	25 MPH avg., cycle 4 miles		88	.091			4.45	4.86	9.31	12
0546	cycle 6 miles		80	.100			4.90	5.35	10.25	13.25
0548	cycle 8 miles		72	.111			5.45	5.95	11.40	14.70
0550	30 MPH avg., cycle 4 miles		88	.091			4.45	4.86	9.31	12
0552	cycle 6 miles		80	.100			4.90	5.35	10.25	13.25
0554	cycle 8 miles		72	.111			5.45	5.95	11.40	14.70
1014	12 C.Y. truck, cycle 0.5 mile, 15 MPH avg., 15 min. wait/ld./uld.	B-34B	336	.024			1.17	1.71	2.88	3.63
1016	cycle 1 mile		300	.027			1.31	1.91	3.22	4.06
1018	cycle 2 miles		252	.032			1.55	2.27	3.82	4.83
1020	cycle 4 miles		180	.044			2.18	3.18	5.36	6.75
1022	cycle 6 miles		144	.056			2.72	3.98	6.70	8.45
1024	cycle 8 miles		120	.067			3.26	4.77	8.03	10.15
1025	cycle 10 miles		96	.083			4.08	5.95	10.03	12.65
1026	20 MPH avg., cycle 0.5 mile		348	.023			1.13	1.65	2.78	3.50
1028	cycle 1 mile		312	.026			1.26	1.84	3.10	3.90
1030	cycle 2 miles		276	.029			1.42	2.08	3.50	4.41
1032	cycle 4 miles		216	.037			1.81	2.65	4.46	5.65
1034	cycle 6 miles		168	.048			2.33	3.41	5.74	7.25
1036	cycle 8 miles		144	.056			2.72	3.98	6.70	8.45
1038	cycle 10 miles		120	.067			3.26	4.77	8.03	10.15
1040	25 MPH avg., cycle 4 miles		228	.035			1.72	2.51	4.23	5.35
1042	cycle 6 miles		192	.042			2.04	2.98	5.02	6.35
1044	cycle 8 miles		168	.048			2.33	3.41	5.74	7.25
1046	cycle 10 miles		144	.056			2.72	3.98	6.70	8.45
1050	30 MPH avg., cycle 4 miles		252	.032			1.55	2.27	3.82	4.83
1052	cycle 6 miles		216	.037			1.81	2.65	4.46	5.65
1054	cycle 8 miles		180	.044			2.18	3.18	5.36	6.75
1056	cycle 10 miles		156	.051			2.51	3.67	6.18	7.80
1060	35 MPH avg., cycle 4 miles		264	.030			1.48	2.17	3.65	4.61
1062	cycle 6 miles		228	.035			1.72	2.51	4.23	5.35
1064	cycle 8 miles		204	.039			1.92	2.81	4.73	5.95
1066	cycle 10 miles		180	.044			2.18	3.18	5.36	6.75
1068	cycle 20 miles		120	.067			3.26	4.77	8.03	10.15
1069	cycle 30 miles		84	.095			4.66	6.80	11.46	14.50
1070	cycle 40 miles		72	.111			5.45	7.95	13.40	16.90
1072	40 MPH avg., cycle 6 miles		240	.033			1.63	2.39	4.02	5.05
1074	cycle 8 miles		216	.037			1.81	2.65	4.46	5.65
1076	cycle 10 miles		192	.042			2.04	2.98	5.02	6.35
1078	cycle 20 miles		120	.067			3.26	4.77	8.03	10.15
1080	cycle 30 miles		96	.083			4.08	5.95	10.03	12.65
1082	cycle 40 miles		72	.111			5.45	7.95	13.40	16.90
1084	cycle 50 miles		60	.133			6.55	9.55	16.10	20.50
1094	45 MPH avg., cycle 8 miles		216	.037			1.81	2.65	4.46	5.65
1096	cycle 10 miles		204	.039			1.92	2.81	4.73	5.95
1098	cycle 20 miles		132	.061			2.97	4.34	7.31	9.20
1100	cycle 30 miles		108	.074			3.63	5.30	8.93	11.30
1102	cycle 40 miles		84	.095			4.66	6.80	11.46	14.50
1104	cycle 50 miles		72	.111			5.45	7.95	13.40	16.90
1106	50 MPH avg., cycle 10 miles		216	.037			1.81	2.65	4.46	5.65
1108	cycle 20 miles		144	.056			2.72	3.98	6.70	8.45
1110	cycle 30 miles		108	.074			3.63	5.30	8.93	11.30
1112	cycle 40 miles		84	.095			4.66	6.80	11.46	14.50

For customer support on your Mechanical Costs with RSMeans Data, call 800.448.8182.

445

31 23 23 – Fill

	31 23 23.20 Hauling	Crew	Daily Output	Labor-Hours	Unit	Material	2020 Bare Costs Labor	Equipment	Total	Total Incl O&P
1114	cycle 50 miles	B-34B	72	.111	L.C.Y.		5.45	7.95	13.40	16.90
1214	15 MPH avg., cycle 0.5 mile, 20 min. wait/ld./uld.		264	.030			1.48	2.17	3.65	4.61
1216	cycle 1 mile		240	.033			1.63	2.39	4.02	5.05
1218	cycle 2 miles		204	.039			1.92	2.81	4.73	5.95
1220	cycle 4 miles		156	.051			2.51	3.67	6.18	7.80
1222	cycle 6 miles		132	.061			2.97	4.34	7.31	9.20
1224	cycle 8 miles		108	.074			3.63	5.30	8.93	11.30
1225	cycle 10 miles		96	.083			4.08	5.95	10.03	12.65
1226	20 MPH avg., cycle 0.5 mile		264	.030			1.48	2.17	3.65	4.61
1228	cycle 1 mile		252	.032			1.55	2.27	3.82	4.83
1230	cycle 2 miles		216	.037			1.81	2.65	4.46	5.65
1232	cycle 4 miles		180	.044			2.18	3.18	5.36	6.75
1234	cycle 6 miles		144	.056			2.72	3.98	6.70	8.45
1236	cycle 8 miles		132	.061			2.97	4.34	7.31	9.20
1238	cycle 10 miles		108	.074			3.63	5.30	8.93	11.30
1240	25 MPH avg., cycle 4 miles		192	.042			2.04	2.98	5.02	6.35
1242	cycle 6 miles		168	.048			2.33	3.41	5.74	7.25
1244	cycle 8 miles		144	.056			2.72	3.98	6.70	8.45
1246	cycle 10 miles		132	.061			2.97	4.34	7.31	9.20
1250	30 MPH avg., cycle 4 miles		204	.039			1.92	2.81	4.73	5.95
1252	cycle 6 miles		180	.044			2.18	3.18	5.36	6.75
1254	cycle 8 miles		156	.051			2.51	3.67	6.18	7.80
1256	cycle 10 miles		144	.056			2.72	3.98	6.70	8.45
1260	35 MPH avg., cycle 4 miles		216	.037			1.81	2.65	4.46	5.65
1262	cycle 6 miles		192	.042			2.04	2.98	5.02	6.35
1264	cycle 8 miles		168	.048			2.33	3.41	5.74	7.25
1266	cycle 10 miles		156	.051			2.51	3.67	6.18	7.80
1268	cycle 20 miles		108	.074			3.63	5.30	8.93	11.30
1269	cycle 30 miles		72	.111			5.45	7.95	13.40	16.90
1270	cycle 40 miles		60	.133			6.55	9.55	16.10	20.50
1272	40 MPH avg., cycle 6 miles		192	.042			2.04	2.98	5.02	6.35
1274	cycle 8 miles		180	.044			2.18	3.18	5.36	6.75
1276	cycle 10 miles		156	.051			2.51	3.67	6.18	7.80
1278	cycle 20 miles		108	.074			3.63	5.30	8.93	11.30
1280	cycle 30 miles		84	.095			4.66	6.80	11.46	14.50
1282	cycle 40 miles		72	.111			5.45	7.95	13.40	16.90
1284	cycle 50 miles		60	.133			6.55	9.55	16.10	20.50
1294	45 MPH avg., cycle 8 miles		180	.044			2.18	3.18	5.36	6.75
1296	cycle 10 miles		168	.048			2.33	3.41	5.74	7.25
1298	cycle 20 miles		120	.067			3.26	4.77	8.03	10.15
1300	cycle 30 miles		96	.083			4.08	5.95	10.03	12.65
1302	cycle 40 miles		72	.111			5.45	7.95	13.40	16.90
1304	cycle 50 miles		60	.133			6.55	9.55	16.10	20.50
1306	50 MPH avg., cycle 10 miles		180	.044			2.18	3.18	5.36	6.75
1308	cycle 20 miles		132	.061			2.97	4.34	7.31	9.20
1310	cycle 30 miles		96	.083			4.08	5.95	10.03	12.65
1312	cycle 40 miles		84	.095			4.66	6.80	11.46	14.50
1314	cycle 50 miles		72	.111			5.45	7.95	13.40	16.90
1414	15 MPH avg., cycle 0.5 mile, 25 min. wait/ld./uld.		204	.039			1.92	2.81	4.73	5.95
1416	cycle 1 mile		192	.042			2.04	2.98	5.02	6.35
1418	cycle 2 miles		168	.048			2.33	3.41	5.74	7.25
1420	cycle 4 miles		132	.061			2.97	4.34	7.31	9.20
1422	cycle 6 miles		120	.067			3.26	4.77	8.03	10.15

31 23 23.20 Hauling		Crew	Daily Output	Labor-Hours	Unit	Material	2020 Bare Costs Labor	2020 Bare Costs Equipment	Total	Total Incl O&P
1424	cycle 8 miles	B-34B	96	.083	L.C.Y.		4.08	5.95	10.03	12.65
1425	cycle 10 miles		84	.095			4.66	6.80	11.46	14.50
1426	20 MPH avg., cycle 0.5 mile		216	.037			1.81	2.65	4.46	5.65
1428	cycle 1 mile		204	.039			1.92	2.81	4.73	5.95
1430	cycle 2 miles		180	.044			2.18	3.18	5.36	6.75
1432	cycle 4 miles		156	.051			2.51	3.67	6.18	7.80
1434	cycle 6 miles		132	.061			2.97	4.34	7.31	9.20
1436	cycle 8 miles		120	.067			3.26	4.77	8.03	10.15
1438	cycle 10 miles		96	.083			4.08	5.95	10.03	12.65
1440	25 MPH avg., cycle 4 miles		168	.048			2.33	3.41	5.74	7.25
1442	cycle 6 miles		144	.056			2.72	3.98	6.70	8.45
1444	cycle 8 miles		132	.061			2.97	4.34	7.31	9.20
1446	cycle 10 miles		108	.074			3.63	5.30	8.93	11.30
1450	30 MPH avg., cycle 4 miles		168	.048			2.33	3.41	5.74	7.25
1452	cycle 6 miles		156	.051			2.51	3.67	6.18	7.80
1454	cycle 8 miles		132	.061			2.97	4.34	7.31	9.20
1456	cycle 10 miles		120	.067			3.26	4.77	8.03	10.15
1460	35 MPH avg., cycle 4 miles		180	.044			2.18	3.18	5.36	6.75
1462	cycle 6 miles		156	.051			2.51	3.67	6.18	7.80
1464	cycle 8 miles		144	.056			2.72	3.98	6.70	8.45
1466	cycle 10 miles		132	.061			2.97	4.34	7.31	9.20
1468	cycle 20 miles		96	.083			4.08	5.95	10.03	12.65
1469	cycle 30 miles		72	.111			5.45	7.95	13.40	16.90
1470	cycle 40 miles		60	.133			6.55	9.55	16.10	20.50
1472	40 MPH avg., cycle 6 miles		168	.048			2.33	3.41	5.74	7.25
1474	cycle 8 miles		156	.051			2.51	3.67	6.18	7.80
1476	cycle 10 miles		144	.056			2.72	3.98	6.70	8.45
1478	cycle 20 miles		96	.083			4.08	5.95	10.03	12.65
1480	cycle 30 miles		84	.095			4.66	6.80	11.46	14.50
1482	cycle 40 miles		60	.133			6.55	9.55	16.10	20.50
1484	cycle 50 miles		60	.133			6.55	9.55	16.10	20.50
1494	45 MPH avg., cycle 8 miles		156	.051			2.51	3.67	6.18	7.80
1496	cycle 10 miles		144	.056			2.72	3.98	6.70	8.45
1498	cycle 20 miles		108	.074			3.63	5.30	8.93	11.30
1500	cycle 30 miles		84	.095			4.66	6.80	11.46	14.50
1502	cycle 40 miles		72	.111			5.45	7.95	13.40	16.90
1504	cycle 50 miles		60	.133			6.55	9.55	16.10	20.50
1506	50 MPH avg., cycle 10 miles		156	.051			2.51	3.67	6.18	7.80
1508	cycle 20 miles		120	.067			3.26	4.77	8.03	10.15
1510	cycle 30 miles		96	.083			4.08	5.95	10.03	12.65
1512	cycle 40 miles		72	.111			5.45	7.95	13.40	16.90
1514	cycle 50 miles		60	.133			6.55	9.55	16.10	20.50
1614	15 MPH avg., cycle 0.5 mile, 30 min. wait/ld./uld.		180	.044			2.18	3.18	5.36	6.75
1616	cycle 1 mile		168	.048			2.33	3.41	5.74	7.25
1618	cycle 2 miles		144	.056			2.72	3.98	6.70	8.45
1620	cycle 4 miles		120	.067			3.26	4.77	8.03	10.15
1622	cycle 6 miles		108	.074			3.63	5.30	8.93	11.30
1624	cycle 8 miles		84	.095			4.66	6.80	11.46	14.50
1625	cycle 10 miles		84	.095			4.66	6.80	11.46	14.50
1626	20 MPH avg., cycle 0.5 mile		180	.044			2.18	3.18	5.36	6.75
1628	cycle 1 mile		168	.048			2.33	3.41	5.74	7.25
1630	cycle 2 miles		156	.051			2.51	3.67	6.18	7.80
1632	cycle 4 miles		132	.061			2.97	4.34	7.31	9.20

For customer support on your Mechanical Costs with RSMeans Data, call 800.448.8182.

447

31 23 23.20 Hauling		Crew	Daily Output	Labor-Hours	Unit	Material	2020 Bare Costs Labor	2020 Bare Costs Equipment	Total	Total Incl O&P
1634	cycle 6 miles	B-34B	120	.067	L.C.Y.		3.26	4.77	8.03	10.15
1636	cycle 8 miles		108	.074			3.63	5.30	8.93	11.30
1638	cycle 10 miles		96	.083			4.08	5.95	10.03	12.65
1640	25 MPH avg., cycle 4 miles		144	.056			2.72	3.98	6.70	8.45
1642	cycle 6 miles		132	.061			2.97	4.34	7.31	9.20
1644	cycle 8 miles		108	.074			3.63	5.30	8.93	11.30
1646	cycle 10 miles		108	.074			3.63	5.30	8.93	11.30
1650	30 MPH avg., cycle 4 miles		144	.056			2.72	3.98	6.70	8.45
1652	cycle 6 miles		132	.061			2.97	4.34	7.31	9.20
1654	cycle 8 miles		120	.067			3.26	4.77	8.03	10.15
1656	cycle 10 miles		108	.074			3.63	5.30	8.93	11.30
1660	35 MPH avg., cycle 4 miles		156	.051			2.51	3.67	6.18	7.80
1662	cycle 6 miles		144	.056			2.72	3.98	6.70	8.45
1664	cycle 8 miles		132	.061			2.97	4.34	7.31	9.20
1666	cycle 10 miles		120	.067			3.26	4.77	8.03	10.15
1668	cycle 20 miles		84	.095			4.66	6.80	11.46	14.50
1669	cycle 30 miles		72	.111			5.45	7.95	13.40	16.90
1670	cycle 40 miles		60	.133			6.55	9.55	16.10	20.50
1672	40 MPH avg., cycle 6 miles		144	.056			2.72	3.98	6.70	8.45
1674	cycle 8 miles		132	.061			2.97	4.34	7.31	9.20
1676	cycle 10 miles		120	.067			3.26	4.77	8.03	10.15
1678	cycle 20 miles		96	.083			4.08	5.95	10.03	12.65
1680	cycle 30 miles		72	.111			5.45	7.95	13.40	16.90
1682	cycle 40 miles		60	.133			6.55	9.55	16.10	20.50
1684	cycle 50 miles		48	.167			8.15	11.95	20.10	25.50
1694	45 MPH avg., cycle 8 miles		144	.056			2.72	3.98	6.70	8.45
1696	cycle 10 miles		132	.061			2.97	4.34	7.31	9.20
1698	cycle 20 miles		96	.083			4.08	5.95	10.03	12.65
1700	cycle 30 miles		84	.095			4.66	6.80	11.46	14.50
1702	cycle 40 miles		60	.133			6.55	9.55	16.10	20.50
1704	cycle 50 miles		60	.133			6.55	9.55	16.10	20.50
1706	50 MPH avg., cycle 10 miles		132	.061			2.97	4.34	7.31	9.20
1708	cycle 20 miles		108	.074			3.63	5.30	8.93	11.30
1710	cycle 30 miles		84	.095			4.66	6.80	11.46	14.50
1712	cycle 40 miles		72	.111			5.45	7.95	13.40	16.90
1714	cycle 50 miles		60	.133			6.55	9.55	16.10	20.50
2000	Hauling, 8 C.Y. truck, small project cost per hour	B-34A	8	1	Hr.		49	53.50	102.50	133
2100	12 C.Y. truck	B-34B	8	1			49	71.50	120.50	153
2150	16.5 C.Y. truck	B-34C	8	1			49	78.50	127.50	160
2175	18 C.Y. 8 wheel truck	B-34I	8	1			49	89	138	172
2200	20 C.Y. truck	B-34D	8	1			49	80.50	129.50	162
2300	Grading at dump, or embankment if required, by dozer	B-10B	1000	.012	L.C.Y.		.62	1.50	2.12	2.58
2310	Spotter at fill or cut, if required	1 Clab	8	1	Hr.		42		42	63.50
9014	18 C.Y. truck, 8 wheels,15 min. wait/ld./uld.,15 MPH, cycle 0.5 mi.	B-34I	504	.016	L.C.Y.		.78	1.42	2.20	2.72
9016	cycle 1 mile		450	.018			.87	1.59	2.46	3.05
9018	cycle 2 miles		378	.021			1.04	1.89	2.93	3.63
9020	cycle 4 miles		270	.030			1.45	2.64	4.09	5.10
9022	cycle 6 miles		216	.037			1.81	3.31	5.12	6.35
9024	cycle 8 miles		180	.044			2.18	3.97	6.15	7.60
9025	cycle 10 miles		144	.056			2.72	4.96	7.68	9.55
9026	20 MPH avg., cycle 0.5 mile		522	.015			.75	1.37	2.12	2.62
9028	cycle 1 mile		468	.017			.84	1.52	2.36	2.93
9030	cycle 2 miles		414	.019			.95	1.72	2.67	3.32

For customer support on your Mechanical Costs with RSMeans Data, call 800.448.8182.

31 23 23.20 Hauling		Crew	Daily Output	Labor-Hours	Unit	Material	2020 Bare Costs Labor	2020 Bare Costs Equipment	Total	Total Incl O&P
9032	cycle 4 miles	B-34I	324	.025	L.C.Y.		1.21	2.20	3.41	4.23
9034	cycle 6 miles		252	.032			1.55	2.83	4.38	5.45
9036	cycle 8 miles		216	.037			1.81	3.31	5.12	6.35
9038	cycle 10 miles		180	.044			2.18	3.97	6.15	7.60
9040	25 MPH avg., cycle 4 miles		342	.023			1.14	2.09	3.23	4.02
9042	cycle 6 miles		288	.028			1.36	2.48	3.84	4.77
9044	cycle 8 miles		252	.032			1.55	2.83	4.38	5.45
9046	cycle 10 miles		216	.037			1.81	3.31	5.12	6.35
9050	30 MPH avg., cycle 4 miles		378	.021			1.04	1.89	2.93	3.63
9052	cycle 6 miles		324	.025			1.21	2.20	3.41	4.23
9054	cycle 8 miles		270	.030			1.45	2.64	4.09	5.10
9056	cycle 10 miles		234	.034			1.67	3.05	4.72	5.85
9060	35 MPH avg., cycle 4 miles		396	.020			.99	1.80	2.79	3.46
9062	cycle 6 miles		342	.023			1.14	2.09	3.23	4.02
9064	cycle 8 miles		288	.028			1.36	2.48	3.84	4.77
9066	cycle 10 miles		270	.030			1.45	2.64	4.09	5.10
9068	cycle 20 miles		162	.049			2.42	4.41	6.83	8.45
9070	cycle 30 miles		126	.063			3.11	5.65	8.76	10.90
9072	cycle 40 miles		90	.089			4.35	7.95	12.30	15.20
9074	40 MPH avg., cycle 6 miles		360	.022			1.09	1.98	3.07	3.81
9076	cycle 8 miles		324	.025			1.21	2.20	3.41	4.23
9078	cycle 10 miles		288	.028			1.36	2.48	3.84	4.77
9080	cycle 20 miles		180	.044			2.18	3.97	6.15	7.60
9082	cycle 30 miles		144	.056			2.72	4.96	7.68	9.55
9084	cycle 40 miles		108	.074			3.63	6.60	10.23	12.70
9086	cycle 50 miles		90	.089			4.35	7.95	12.30	15.20
9094	45 MPH avg., cycle 8 miles		324	.025			1.21	2.20	3.41	4.23
9096	cycle 10 miles		306	.026			1.28	2.33	3.61	4.49
9098	cycle 20 miles		198	.040			1.98	3.60	5.58	6.95
9100	cycle 30 miles		144	.056			2.72	4.96	7.68	9.55
9102	cycle 40 miles		126	.063			3.11	5.65	8.76	10.90
9104	cycle 50 miles		108	.074			3.63	6.60	10.23	12.70
9106	50 MPH avg., cycle 10 miles		324	.025			1.21	2.20	3.41	4.23
9108	cycle 20 miles		216	.037			1.81	3.31	5.12	6.35
9110	cycle 30 miles		162	.049			2.42	4.41	6.83	8.45
9112	cycle 40 miles		126	.063			3.11	5.65	8.76	10.90
9114	cycle 50 miles		108	.074			3.63	6.60	10.23	12.70
9214	20 min. wait/ld./uld.,15 MPH, cycle 0.5 mi.		396	.020			.99	1.80	2.79	3.46
9216	cycle 1 mile		360	.022			1.09	1.98	3.07	3.81
9218	cycle 2 miles		306	.026			1.28	2.33	3.61	4.49
9220	cycle 4 miles		234	.034			1.67	3.05	4.72	5.85
9222	cycle 6 miles		198	.040			1.98	3.60	5.58	6.95
9224	cycle 8 miles		162	.049			2.42	4.41	6.83	8.45
9225	cycle 10 miles		144	.056			2.72	4.96	7.68	9.55
9226	20 MPH avg., cycle 0.5 mile		396	.020			.99	1.80	2.79	3.46
9228	cycle 1 mile		378	.021			1.04	1.89	2.93	3.63
9230	cycle 2 miles		324	.025			1.21	2.20	3.41	4.23
9232	cycle 4 miles		270	.030			1.45	2.64	4.09	5.10
9234	cycle 6 miles		216	.037			1.81	3.31	5.12	6.35
9236	cycle 8 miles		198	.040			1.98	3.60	5.58	6.95
9238	cycle 10 miles		162	.049			2.42	4.41	6.83	8.45
9240	25 MPH avg., cycle 4 miles		288	.028			1.36	2.48	3.84	4.77
9242	cycle 6 miles		252	.032			1.55	2.83	4.38	5.45

31 23 Excavation and Fill

31 23 23 – Fill

31 23 23.20 Hauling		Crew	Daily Output	Labor-Hours	Unit	Material	2020 Bare Costs Labor	Equipment	Total	Total Incl O&P
9244	cycle 8 miles	B-34I	216	.037	L.C.Y.		1.81	3.31	5.12	6.35
9246	cycle 10 miles		198	.040			1.98	3.60	5.58	6.95
9250	30 MPH avg., cycle 4 miles		306	.026			1.28	2.33	3.61	4.49
9252	cycle 6 miles		270	.030			1.45	2.64	4.09	5.10
9254	cycle 8 miles		234	.034			1.67	3.05	4.72	5.85
9256	cycle 10 miles		216	.037			1.81	3.31	5.12	6.35
9260	35 MPH avg., cycle 4 miles		324	.025			1.21	2.20	3.41	4.23
9262	cycle 6 miles		288	.028			1.36	2.48	3.84	4.77
9264	cycle 8 miles		252	.032			1.55	2.83	4.38	5.45
9266	cycle 10 miles		234	.034			1.67	3.05	4.72	5.85
9268	cycle 20 miles		162	.049			2.42	4.41	6.83	8.45
9270	cycle 30 miles		108	.074			3.63	6.60	10.23	12.70
9272	cycle 40 miles		90	.089			4.35	7.95	12.30	15.20
9274	40 MPH avg., cycle 6 miles		288	.028			1.36	2.48	3.84	4.77
9276	cycle 8 miles		270	.030			1.45	2.64	4.09	5.10
9278	cycle 10 miles		234	.034			1.67	3.05	4.72	5.85
9280	cycle 20 miles		162	.049			2.42	4.41	6.83	8.45
9282	cycle 30 miles		126	.063			3.11	5.65	8.76	10.90
9284	cycle 40 miles		108	.074			3.63	6.60	10.23	12.70
9286	cycle 50 miles		90	.089			4.35	7.95	12.30	15.20
9294	45 MPH avg., cycle 8 miles		270	.030			1.45	2.64	4.09	5.10
9296	cycle 10 miles		252	.032			1.55	2.83	4.38	5.45
9298	cycle 20 miles		180	.044			2.18	3.97	6.15	7.60
9300	cycle 30 miles		144	.056			2.72	4.96	7.68	9.55
9302	cycle 40 miles		108	.074			3.63	6.60	10.23	12.70
9304	cycle 50 miles		90	.089			4.35	7.95	12.30	15.20
9306	50 MPH avg., cycle 10 miles		270	.030			1.45	2.64	4.09	5.10
9308	cycle 20 miles		198	.040			1.98	3.60	5.58	6.95
9310	cycle 30 miles		144	.056			2.72	4.96	7.68	9.55
9312	cycle 40 miles		126	.063			3.11	5.65	8.76	10.90
9314	cycle 50 miles		108	.074			3.63	6.60	10.23	12.70
9414	25 min. wait/ld./uld.,15 MPH, cycle 0.5 mi.		306	.026			1.28	2.33	3.61	4.49
9416	cycle 1 mile		288	.028			1.36	2.48	3.84	4.77
9418	cycle 2 miles		252	.032			1.55	2.83	4.38	5.45
9420	cycle 4 miles		198	.040			1.98	3.60	5.58	6.95
9422	cycle 6 miles		180	.044			2.18	3.97	6.15	7.60
9424	cycle 8 miles		144	.056			2.72	4.96	7.68	9.55
9425	cycle 10 miles		126	.063			3.11	5.65	8.76	10.90
9426	20 MPH avg., cycle 0.5 mile		324	.025			1.21	2.20	3.41	4.23
9428	cycle 1 mile		306	.026			1.28	2.33	3.61	4.49
9430	cycle 2 miles		270	.030			1.45	2.64	4.09	5.10
9432	cycle 4 miles		234	.034			1.67	3.05	4.72	5.85
9434	cycle 6 miles		198	.040			1.98	3.60	5.58	6.95
9436	cycle 8 miles		180	.044			2.18	3.97	6.15	7.60
9438	cycle 10 miles		144	.056			2.72	4.96	7.68	9.55
9440	25 MPH avg., cycle 4 miles		252	.032			1.55	2.83	4.38	5.45
9442	cycle 6 miles		216	.037			1.81	3.31	5.12	6.35
9444	cycle 8 miles		198	.040			1.98	3.60	5.58	6.95
9446	cycle 10 miles		180	.044			2.18	3.97	6.15	7.60
9450	30 MPH avg., cycle 4 miles		252	.032			1.55	2.83	4.38	5.45
9452	cycle 6 miles		234	.034			1.67	3.05	4.72	5.85
9454	cycle 8 miles		198	.040			1.98	3.60	5.58	6.95
9456	cycle 10 miles		180	.044			2.18	3.97	6.15	7.60

31 23 23.20 Hauling		Crew	Daily Output	Labor-Hours	Unit	Material	2020 Bare Costs Labor	Equipment	Total	Total Incl O&P
9460	35 MPH avg., cycle 4 miles	B-34I	270	.030	L.C.Y.		1.45	2.64	4.09	5.10
9462	cycle 6 miles		234	.034			1.67	3.05	4.72	5.85
9464	cycle 8 miles		216	.037			1.81	3.31	5.12	6.35
9466	cycle 10 miles		198	.040			1.98	3.60	5.58	6.95
9468	cycle 20 miles		144	.056			2.72	4.96	7.68	9.55
9470	cycle 30 miles		108	.074			3.63	6.60	10.23	12.70
9472	cycle 40 miles		90	.089			4.35	7.95	12.30	15.20
9474	40 MPH avg., cycle 6 miles		252	.032			1.55	2.83	4.38	5.45
9476	cycle 8 miles		234	.034			1.67	3.05	4.72	5.85
9478	cycle 10 miles		216	.037			1.81	3.31	5.12	6.35
9480	cycle 20 miles		144	.056			2.72	4.96	7.68	9.55
9482	cycle 30 miles		126	.063			3.11	5.65	8.76	10.90
9484	cycle 40 miles		90	.089			4.35	7.95	12.30	15.20
9486	cycle 50 miles		90	.089			4.35	7.95	12.30	15.20
9494	45 MPH avg., cycle 8 miles		234	.034			1.67	3.05	4.72	5.85
9496	cycle 10 miles		216	.037			1.81	3.31	5.12	6.35
9498	cycle 20 miles		162	.049			2.42	4.41	6.83	8.45
9500	cycle 30 miles		126	.063			3.11	5.65	8.76	10.90
9502	cycle 40 miles		108	.074			3.63	6.60	10.23	12.70
9504	cycle 50 miles		90	.089			4.35	7.95	12.30	15.20
9506	50 MPH avg., cycle 10 miles		234	.034			1.67	3.05	4.72	5.85
9508	cycle 20 miles		180	.044			2.18	3.97	6.15	7.60
9510	cycle 30 miles		144	.056			2.72	4.96	7.68	9.55
9512	cycle 40 miles		108	.074			3.63	6.60	10.23	12.70
9514	cycle 50 miles		90	.089			4.35	7.95	12.30	15.20
9614	30 min. wait/ld./uld.,15 MPH, cycle 0.5 mi.		270	.030			1.45	2.64	4.09	5.10
9616	cycle 1 mile		252	.032			1.55	2.83	4.38	5.45
9618	cycle 2 miles		216	.037			1.81	3.31	5.12	6.35
9620	cycle 4 miles		180	.044			2.18	3.97	6.15	7.60
9622	cycle 6 miles		162	.049			2.42	4.41	6.83	8.45
9624	cycle 8 miles		126	.063			3.11	5.65	8.76	10.90
9625	cycle 10 miles		126	.063			3.11	5.65	8.76	10.90
9626	20 MPH avg., cycle 0.5 mile		270	.030			1.45	2.64	4.09	5.10
9628	cycle 1 mile		252	.032			1.55	2.83	4.38	5.45
9630	cycle 2 miles		234	.034			1.67	3.05	4.72	5.85
9632	cycle 4 miles		198	.040			1.98	3.60	5.58	6.95
9634	cycle 6 miles		180	.044			2.18	3.97	6.15	7.60
9636	cycle 8 miles		162	.049			2.42	4.41	6.83	8.45
9638	cycle 10 miles		144	.056			2.72	4.96	7.68	9.55
9640	25 MPH avg., cycle 4 miles		216	.037			1.81	3.31	5.12	6.35
9642	cycle 6 miles		198	.040			1.98	3.60	5.58	6.95
9644	cycle 8 miles		180	.044			2.18	3.97	6.15	7.60
9646	cycle 10 miles		162	.049			2.42	4.41	6.83	8.45
9650	30 MPH avg., cycle 4 miles		216	.037			1.81	3.31	5.12	6.35
9652	cycle 6 miles		198	.040			1.98	3.60	5.58	6.95
9654	cycle 8 miles		180	.044			2.18	3.97	6.15	7.60
9656	cycle 10 miles		162	.049			2.42	4.41	6.83	8.45
9660	35 MPH avg., cycle 4 miles		234	.034			1.67	3.05	4.72	5.85
9662	cycle 6 miles		216	.037			1.81	3.31	5.12	6.35
9664	cycle 8 miles		198	.040			1.98	3.60	5.58	6.95
9666	cycle 10 miles		180	.044			2.18	3.97	6.15	7.60
9668	cycle 20 miles		126	.063			3.11	5.65	8.76	10.90
9670	cycle 30 miles		108	.074			3.63	6.60	10.23	12.70

31 23 23.20 Hauling

		Crew	Daily Output	Labor-Hours	Unit	Material	2020 Bare Costs Labor	Equipment	Total	Total Incl O&P
9672	cycle 40 miles	B-34I	90	.089	L.C.Y.		4.35	7.95	12.30	15.20
9674	40 MPH avg., cycle 6 miles		216	.037			1.81	3.31	5.12	6.35
9676	cycle 8 miles		198	.040			1.98	3.60	5.58	6.95
9678	cycle 10 miles		180	.044			2.18	3.97	6.15	7.60
9680	cycle 20 miles		144	.056			2.72	4.96	7.68	9.55
9682	cycle 30 miles		108	.074			3.63	6.60	10.23	12.70
9684	cycle 40 miles		90	.089			4.35	7.95	12.30	15.20
9686	cycle 50 miles		72	.111			5.45	9.90	15.35	19.05
9694	45 MPH avg., cycle 8 miles		216	.037			1.81	3.31	5.12	6.35
9696	cycle 10 miles		198	.040			1.98	3.60	5.58	6.95
9698	cycle 20 miles		144	.056			2.72	4.96	7.68	9.55
9700	cycle 30 miles		126	.063			3.11	5.65	8.76	10.90
9702	cycle 40 miles		108	.074			3.63	6.60	10.23	12.70
9704	cycle 50 miles		90	.089			4.35	7.95	12.30	15.20
9706	50 MPH avg., cycle 10 miles		198	.040			1.98	3.60	5.58	6.95
9708	cycle 20 miles		162	.049			2.42	4.41	6.83	8.45
9710	cycle 30 miles		126	.063			3.11	5.65	8.76	10.90
9712	cycle 40 miles		108	.074			3.63	6.60	10.23	12.70
9714	cycle 50 miles		90	.089			4.35	7.95	12.30	15.20

31 41 Shoring

31 41 16 – Sheet Piling

31 41 16.10 Sheet Piling Systems

			Crew	Daily Output	Labor-Hours	Unit	Material	2020 Bare Costs Labor	Equipment	Total	Total Incl O&P
0010	**SHEET PILING SYSTEMS**										
0020	Sheet piling, 50,000 psi steel, not incl. wales, 22 psf, left in place		B-40	10.81	5.920	Ton	1,875	325	320	2,520	2,900
0100	Drive, extract & salvage	R314116-45		6	10.667		525	590	580	1,695	2,100
0300	20' deep excavation, 27 psf, left in place			12.95	4.942		1,875	273	269	2,417	2,775
0400	Drive, extract & salvage			6.55	9.771		525	540	530	1,595	1,975
1200	15' deep excavation, 22 psf, left in place			983	.065	S.F.	21.50	3.60	3.54	28.64	33.50
1300	Drive, extract & salvage			545	.117		5.90	6.50	6.40	18.80	23.50
1500	20' deep excavation, 27 psf, left in place			960	.067		27.50	3.68	3.62	34.80	39.50
1600	Drive, extract & salvage			485	.132		7.65	7.30	7.15	22.10	27.50
2100	Rent steel sheet piling and wales, first month					Ton	335			335	370
2200	Per added month						33.50			33.50	37
2300	Rental piling left in place, add to rental						1,200			1,200	1,325
2500	Wales, connections & struts, 2/3 salvage						535			535	585
3900	Wood, solid sheeting, incl. wales, braces and spacers,	R314116-40									
3910	drive, extract & salvage, 8' deep excavation		B-31	330	.121	S.F.	2.06	5.40	.78	8.24	11.25
4000	10' deep, 50 S.F./hr. in & 150 S.F./hr. out			300	.133		2.12	5.95	.86	8.93	12.20
4100	12' deep, 45 S.F./hr. in & 135 S.F./hr. out			270	.148		2.18	6.60	.95	9.73	13.40
4200	14' deep, 42 S.F./hr. in & 126 S.F./hr. out			250	.160		2.24	7.15	1.03	10.42	14.35
4300	16' deep, 40 S.F./hr. in & 120 S.F./hr. out			240	.167		2.31	7.45	1.07	10.83	14.90
4400	18' deep, 38 S.F./hr. in & 114 S.F./hr. out			230	.174		2.39	7.80	1.12	11.31	15.55
4520	Left in place, 8' deep, 55 S.F./hr.			440	.091		3.70	4.06	.58	8.34	10.80
4540	10' deep, 50 S.F./hr.			400	.100		3.90	4.47	.64	9.01	11.70
4560	12' deep, 45 S.F./hr.			360	.111		4.11	4.97	.71	9.79	12.75
4565	14' deep, 42 S.F./hr.			335	.119		4.35	5.35	.77	10.47	13.65
4570	16' deep, 40 S.F./hr.			320	.125		4.63	5.60	.80	11.03	14.40
4700	Alternate pricing, left in place, 8' deep			1.76	22.727	M.B.F.	830	1,025	146	2,001	2,600
4800	Drive, extract and salvage, 8' deep			1.32	30.303	"	740	1,350	195	2,285	3,050

452

31 52 Cofferdams

31 52 16 - Timber Cofferdams

31 52 16.10 Cofferdams	Crew	Daily Output	Labor-Hours	Unit	Material	2020 Bare Costs Labor	Equipment	Total	Total Incl O&P
0010 **COFFERDAMS**									
0011 Incl. mobilization and temporary sheeting									
0080 Soldier beams & lagging H-piles with 3" wood sheeting									
0090 horizontal between piles, including removal of wales & braces									
0100 No hydrostatic head, 15' deep, 1 line of braces, minimum	B-50	545	.206	S.F.	8.60	10.75	4.72	24.07	31
0200 Maximum		495	.226		9.55	11.85	5.20	26.60	34.50
1300 No hydrostatic head, left in place, 15' deep, 1 line of braces, min.		635	.176		11.50	9.25	4.05	24.80	31
1400 Maximum		575	.195		12.30	10.20	4.47	26.97	34
2350 Lagging only, 3" thick wood between piles 8' OC, minimum	B-46	400	.120		1.91	5.80	.10	7.81	11.10
2370 Maximum		250	.192		2.87	9.30	.17	12.34	17.60
2400 Open sheeting no bracing, for trenches to 10' deep, min.		1736	.028		.86	1.34	.02	2.22	3.03
2450 Maximum		1510	.032		.96	1.54	.03	2.53	3.44
2500 Tie-back method, add to open sheeting, add, minimum								20%	20%
2550 Maximum								60%	60%
2700 Tie-backs only, based on tie-backs total length, minimum	B-46	86.80	.553	L.F.	17.25	27	.48	44.73	60.50
2750 Maximum		38.50	1.247	"	30.50	60.50	1.07	92.07	127
3500 Tie-backs only, typical average, 25' long		2	24	Ea.	760	1,175	20.50	1,955.50	2,625
3600 35' long		1.58	30.380	"	1,025	1,475	26	2,526	3,400
6000 See also Section 31 41 16.10									

For customer support on your Mechanical Costs with RSMeans Data, call 800.448.8182.

453

Division Notes

		CREW	DAILY OUTPUT	LABOR-HOURS	UNIT	BARE COSTS				TOTAL INCL O&P
						MAT.	LABOR	EQUIP.	TOTAL	

Estimating Tips

32 01 00 Operations and Maintenance of Exterior Improvements

- Recycling of asphalt pavement is becoming very popular and is an alternative to removal and replacement. It can be a good value engineering proposal if removed pavement can be recycled, either at the project site or at another site that is reasonably close to the project site. Sections on repair of flexible and rigid pavement are included.

32 10 00 Bases, Ballasts, and Paving

- When estimating paving, keep in mind the project schedule. Also note that prices for asphalt and concrete are generally higher in the cold seasons. Lines for pavement markings, including tactile warning systems and fence lines, are included.

32 90 00 Planting

- The timing of planting and guarantee specifications often dictate the costs for establishing tree and shrub growth and a stand of grass or ground cover. Establish the work performance schedule to coincide with the local planting season. Maintenance and growth guarantees can add 20–100% to the total landscaping cost and can be contractually cumbersome. The cost to replace trees and shrubs can be as high as 5% of the total cost, depending on the planting zone, soil conditions, and time of year.

Reference Numbers

Reference numbers are shown at the beginning of some major classifications. These numbers refer to related items in the Reference Section. The reference information may be an estimating procedure, an alternate pricing method, or technical information.

Note: Not all subdivisions listed here necessarily appear. ■

Same Data. Simplified.

Enjoy the convenience and efficiency of accessing your costs anywhere:

- **Skip the multiplier** by setting your location
- **Quickly search,** edit, favorite and share costs
- **Stay on top of price changes** with automatic updates

Discover more at rsmeans.com/online

32 12 Flexible Paving

32 12 16 – Asphalt Paving

32 12 16.13 Plant-Mix Asphalt Paving

		Crew	Daily Output	Labor-Hours	Unit	Material	2020 Bare Costs Labor	Equipment	Total	Total Incl O&P
0010	**PLANT-MIX ASPHALT PAVING**									
0020	For highways and large paved areas, exludes hauling									
0025	See Section 31 23 23.20 for hauling costs									
0080	Binder course, 1-1/2" thick	B-25	7725	.011	S.Y.	5.55	.53	.35	6.43	7.25
0120	2" thick		6345	.014		7.40	.64	.43	8.47	9.55
0130	2-1/2" thick		5620	.016		9.25	.72	.48	10.45	11.75
0160	3" thick		4905	.018		11.05	.83	.55	12.43	14.05
0170	3-1/2" thick		4520	.019		12.90	.90	.60	14.40	16.20
0200	4" thick		4140	.021		14.75	.98	.65	16.38	18.45
0300	Wearing course, 1" thick	B-25B	10575	.009		3.66	.43	.28	4.37	4.98
0340	1-1/2" thick		7725	.012		6.15	.59	.38	7.12	8.05
0380	2" thick		6345	.015		8.25	.71	.46	9.42	10.70
0420	2-1/2" thick		5480	.018		10.20	.83	.54	11.57	13.05
0460	3" thick		4900	.020		12.15	.92	.60	13.67	15.40
0470	3-1/2" thick		4520	.021		14.25	1	.65	15.90	17.90
0480	4" thick		4140	.023		16.30	1.09	.71	18.10	20.50
0500	Open graded friction course	B-25C	5000	.010		2.55	.45	.47	3.47	4
0800	Alternate method of figuring paving costs									
0810	Binder course, 1-1/2" thick	B-25	630	.140	Ton	68	6.45	4.29	78.74	89
0811	2" thick		690	.128		68	5.90	3.91	77.81	87.50
0812	3" thick		800	.110		68	5.10	3.37	76.47	86
0813	4" thick		900	.098		68	4.53	3	75.53	84.50
0850	Wearing course, 1" thick	B-25B	575	.167		73.50	7.85	5.10	86.45	98.50
0851	1-1/2" thick		630	.152		73.50	7.20	4.66	85.36	97
0852	2" thick		690	.139		73.50	6.55	4.26	84.31	95.50
0853	2-1/2" thick		765	.125		73.50	5.90	3.84	83.24	94
0854	3" thick		800	.120		73.50	5.65	3.67	82.82	93.50
1000	Pavement replacement over trench, 2" thick	B-37	90	.533	S.Y.	7.60	23.50	2.84	33.94	47
1050	4" thick		70	.686		15.10	30.50	3.65	49.25	66
1080	6" thick		55	.873		24	38.50	4.65	67.15	89.50

32 12 16.14 Asphaltic Concrete Paving

		Crew	Daily Output	Labor-Hours	Unit	Material	2020 Bare Costs Labor	Equipment	Total	Total Incl O&P
0011	**ASPHALTIC CONCRETE PAVING**, parking lots & driveways									
0015	No asphalt hauling included									
0018	Use 6.05 C.Y. per inch per M.S.F. for hauling									
0020	6" stone base, 2" binder course, 1" topping	B-25C	9000	.005	S.F.	1.92	.25	.26	2.43	2.78
0025	2" binder course, 2" topping		9000	.005		2.36	.25	.26	2.87	3.27
0030	3" binder course, 2" topping		9000	.005		2.78	.25	.26	3.29	3.73
0035	4" binder course, 2" topping		9000	.005		3.19	.25	.26	3.70	4.17
0040	1-1/2" binder course, 1" topping		9000	.005		1.72	.25	.26	2.23	2.56
0042	3" binder course, 1" topping		9000	.005		2.33	.25	.26	2.84	3.24
0045	3" binder course, 3" topping		9000	.005		3.23	.25	.26	3.74	4.22
0050	4" binder course, 3" topping		9000	.005		3.63	.25	.26	4.14	4.67
0055	4" binder course, 4" topping		9000	.005		4.08	.25	.26	4.59	5.15
0300	Binder course, 1-1/2" thick		35000	.001		.62	.06	.07	.75	.85
0400	2" thick		25000	.002		.80	.09	.09	.98	1.12
0500	3" thick		15000	.003		1.23	.15	.16	1.54	1.76
0600	4" thick		10800	.004		1.62	.21	.22	2.05	2.34
0800	Sand finish course, 3/4" thick		41000	.001		.31	.06	.06	.43	.48
0900	1" thick		34000	.001		.38	.07	.07	.52	.60
1000	Fill pot holes, hot mix, 2" thick	B-16	4200	.008		.82	.34	.14	1.30	1.56
1100	4" thick		3500	.009		1.20	.41	.16	1.77	2.11
1120	6" thick		3100	.010		1.61	.46	.18	2.25	2.66

32 12 Flexible Paving

32 12 16 – Asphalt Paving

32 12 16.14 Asphaltic Concrete Paving	Crew	Daily Output	Labor-Hours	Unit	Material	2020 Bare Costs Labor	Equipment	Total	Total Incl O&P
1140 Cold patch, 2" thick	B-51	3000	.016	S.F.	.87	.69	.07	1.63	2.07
1160 4" thick	↓	2700	.018	↓	1.66	.77	.07	2.50	3.07
1180 6" thick	↓	1900	.025	↓	2.58	1.09	.10	3.77	4.59

32 84 Planting Irrigation

32 84 23 – Underground Sprinklers

32 84 23.10 Sprinkler Irrigation System

		Crew	Daily Output	Labor-Hours	Unit	Material	2020 Bare Costs Labor	Equipment	Total	Total Incl O&P
0010	**SPRINKLER IRRIGATION SYSTEM**									
0011	For lawns									
0800	Residential system, custom, 1" supply	B-20	2000	.012	S.F.	.27	.56		.83	1.15
0900	1-1/2" supply	"	1800	.013	"	.42	.63		1.05	1.40

Division Notes

	CREW	DAILY OUTPUT	LABOR-HOURS	UNIT	BARE COSTS				TOTAL INCL O&P
					MAT.	LABOR	EQUIP.	TOTAL	

Estimating Tips
33 10 00 Water Utilities
33 30 00 Sanitary Sewerage Utilities
33 40 00 Storm Drainage Utilities

- Never assume that the water, sewer, and drainage lines will go in at the early stages of the project. Consider the site access needs before dividing the site in half with open trenches, loose pipe, and machinery obstructions. Always inspect the site to establish that the site drawings are complete. Check off all existing utilities on your drawings as you locate them. Be especially careful with underground utilities because appurtenances are sometimes buried during regrading or repaving operations. If you find any discrepancies, mark up the site plan for further research. Differing site conditions can be very costly if discovered later in the project.

- See also Section 33 01 00 for restoration of pipe where removal/replacement may be undesirable. Use of new types of piping materials can reduce the overall project cost. Owners/design engineers should consider the installing contractor as a valuable source of current information on utility products and local conditions that could lead to significant cost savings.

Reference Numbers
Reference numbers are shown at the beginning of some major classifications. These numbers refer to related items in the Reference Section. The reference information may be an estimating procedure, an alternate pricing method, or technical information.

Note: Not all subdivisions listed here necessarily appear. ■

Same Data. Simplified.

Enjoy the convenience and efficiency of accessing your costs anywhere:

- **Skip the multiplier** by setting your location
- **Quickly search,** edit, favorite and share costs
- **Stay on top of price changes** with automatic updates

Discover more at rsmeans.com/online

33 01 10.10 Corrosion Resistance	Crew	Daily Output	Labor-Hours	Unit	Material	2020 Bare Costs Labor	Equipment	Total	Total Incl O&P
0010 **CORROSION RESISTANCE**									
0012 Wrap & coat, add to pipe, 4" diameter				L.F.	2.38			2.38	2.62
0020 5" diameter					3.04			3.04	3.34
0040 6" diameter					3.67			3.67	4.04
0060 8" diameter					5			5	5.50
0080 10" diameter					6.30			6.30	6.95
0100 12" diameter					7.30			7.30	8
0120 14" diameter					8.35			8.35	9.20
0140 16" diameter					10.55			10.55	11.60
0160 18" diameter					11			11	12.10
0180 20" diameter					12.25			12.25	13.50
0200 24" diameter					14.75			14.75	16.20
0220 Small diameter pipe, 1" diameter, add					1.01			1.01	1.11
0240 2" diameter					1.28			1.28	1.41
0260 2-1/2" diameter					1.61			1.61	1.77
0280 3" diameter					1.93			1.93	2.12
0300 Fittings, field covered, add				S.F.	9.40			9.40	10.35
0500 Coating, bituminous, per diameter inch, 1 coat, add				L.F.	.21			.21	.23
0540 3 coat					.57			.57	.63
0560 Coal tar epoxy, per diameter inch, 1 coat, add					.21			.21	.23
0600 3 coat					.66			.66	.73
1000 Polyethylene H.D. extruded, 0.025" thk., 1/2" diameter, add					.10			.10	.11
1020 3/4" diameter					.15			.15	.17
1040 1" diameter					.19			.19	.21
1060 1-1/4" diameter					.25			.25	.28
1080 1-1/2" diameter					.28			.28	.31
1100 0.030" thk., 2" diameter					.39			.39	.43
1120 2-1/2" diameter					.50			.50	.55
1140 0.035" thk., 3" diameter					.56			.56	.62
1160 3-1/2" diameter					.65			.65	.72
1180 4" diameter					.81			.81	.89
1200 0.040" thk., 5" diameter					.93			.93	1.02
1220 6" diameter					1.14			1.14	1.25
1240 8" diameter					1.46			1.46	1.61
1260 10" diameter					1.79			1.79	1.97
1280 12" diameter					2.15			2.15	2.37
1300 0.060" thk., 14" diameter					2.76			2.76	3.04
1320 16" diameter					3.16			3.16	3.48
1340 18" diameter					3.30			3.30	3.63
1360 20" diameter					3.67			3.67	4.04
1380 Fittings, field wrapped, add				S.F.	4.30			4.30	4.73

33 01 10.20 Pipe Repair

	Crew	Daily Output	Labor-Hours	Unit	Material	2020 Bare Costs Labor	Equipment	Total	Total Incl O&P
0010 **PIPE REPAIR**									
0020 Not including excavation or backfill									
0100 Clamp, stainless steel, lightweight, for steel pipe									
0110 3" long, 1/2" diameter pipe	1 Plum	34	.235	Ea.	11.15	15.15		26.30	35
0120 3/4" diameter pipe		32	.250		14.10	16.10		30.20	39.50
0130 1" diameter pipe		30	.267		12.50	17.20		29.70	39.50
0140 1-1/4" diameter pipe		28	.286		13.15	18.40		31.55	42
0150 1-1/2" diameter pipe		26	.308		16.85	19.85		36.70	48
0160 2" diameter pipe		24	.333		18.15	21.50		39.65	52
0170 2-1/2" diameter pipe		23	.348		17.40	22.50		39.90	52.50

33 01 10 – Operation and Maintenance of Water Utilities

33 01 10.20 Pipe Repair		Crew	Daily Output	Labor-Hours	Unit	Material	2020 Bare Costs Labor	Equipment	Total	Total Incl O&P
0180	3" diameter pipe	1 Plum	22	.364	Ea.	24.50	23.50		48	62
0190	3-1/2" diameter pipe		21	.381		25	24.50		49.50	64
0200	4" diameter pipe	B-20	44	.545		23	25.50		48.50	63.50
0210	5" diameter pipe		42	.571		26.50	27		53.50	69.50
0220	6" diameter pipe		38	.632		29.50	29.50		59	77
0230	8" diameter pipe		30	.800		35.50	37.50		73	96
0240	10" diameter pipe		28	.857		120	40.50		160.50	193
0250	12" diameter pipe		24	1		131	47		178	216
0260	14" diameter pipe		22	1.091		166	51.50		217.50	261
0270	16" diameter pipe		20	1.200		179	56.50		235.50	282
0280	18" diameter pipe		18	1.333		206	62.50		268.50	320
0290	20" diameter pipe		16	1.500		234	70.50		304.50	365
0300	24" diameter pipe		14	1.714		238	80.50		318.50	385
0360	For 6" long, add					100%	40%			
0370	For 9" long, add					200%	100%			
0380	For 12" long, add					300%	150%			
0390	For 18" long, add					500%	200%			
0400	Pipe freezing for live repairs of systems 3/8" to 6"									
0410	Note: Pipe freezing can also be used to install a valve into a live system									
0420	Pipe freezing each side 3/8"	2 Skwk	8	2	Ea.	550	110		660	770
0425	Pipe freezing each side 3/8", second location same kit		8	2		32	110		142	202
0430	Pipe freezing each side 3/4"		8	2		515	110		625	735
0435	Pipe freezing each side 3/4", second location same kit		8	2		32	110		142	202
0440	Pipe freezing each side 1-1/2"		6	2.667		515	146		661	790
0445	Pipe freezing each side 1-1/2", second location same kit		6	2.667		32	146		178	257
0450	Pipe freezing each side 2"		6	2.667		945	146		1,091	1,275
0455	Pipe freezing each side 2", second location same kit		6	2.667		32	146		178	257
0460	Pipe freezing each side 2-1/2" to 3"		6	2.667		955	146		1,101	1,275
0465	Pipe freezing each side 2-1/2" to 3", second location same kit		6	2.667		64	146		210	292
0470	Pipe freezing each side 4"		4	4		1,575	219		1,794	2,050
0475	Pipe freezing each side 4", second location same kit		4	4		73	219		292	410
0480	Pipe freezing each side 5" to 6"		4	4		4,400	219		4,619	5,175
0485	Pipe freezing each side 5" to 6", second location same kit		4	4		218	219		437	570
0490	Pipe freezing extra 20 lb. CO_2 cylinders (3/8" to 2" - 1 ea, 3" - 2 ea)					231			231	254
0500	Pipe freezing extra 50 lb. CO_2 cylinders (4" - 2 ea, 5"-6" - 6 ea)					530			530	585
1000	Clamp, stainless steel, with threaded service tap									
1040	Full seal for iron, steel, PVC pipe									
1100	6" long, 2" diameter pipe	1 Plum	17	.471	Ea.	115	30.50		145.50	173
1110	2-1/2" diameter pipe		16	.500		114	32		146	174
1120	3" diameter pipe		15.60	.513		112	33		145	173
1130	3-1/2" diameter pipe		15	.533		135	34.50		169.50	200
1140	4" diameter pipe	B-20	32	.750		125	35.50		160.50	190
1150	6" diameter pipe		28	.857		164	40.50		204.50	241
1160	8" diameter pipe		21	1.143		176	53.50		229.50	275
1170	10" diameter pipe		20	1.200		232	56.50		288.50	340
1180	12" diameter pipe		17	1.412		266	66.50		332.50	390
1200	8" long, 2" diameter pipe	1 Plum	11.72	.683		117	44		161	195
1210	2-1/2" diameter pipe		11	.727		127	47		174	210
1220	3" diameter pipe		10.75	.744		125	48		173	210
1230	3-1/2" diameter pipe		10.34	.774		132	50		182	220
1240	4" diameter pipe	B-20	22	1.091		158	51.50		209.50	252
1250	6" diameter pipe		19.31	1.243		162	58.50		220.50	267
1260	8" diameter pipe		14.48	1.657		211	78		289	350

For customer support on your Mechanical Costs with RSMeans Data, call 800.448.8182.

461

33 01 10.20 Pipe Repair		Crew	Daily Output	Labor-Hours	Unit	Material	2020 Bare Costs Labor	Equipment	Total	Total Incl O&P
1270	10" diameter pipe	B-20	13.80	1.739	Ea.	243	82		325	390
1280	12" diameter pipe	↓	11.72	2.048		305	96.50		401.50	485
1300	12" long, 2" diameter pipe	1 Plum	9.44	.847		248	54.50		302.50	355
1310	2-1/2" diameter pipe		8.89	.900		204	58		262	310
1320	3" diameter pipe		8.67	.923		191	59.50		250.50	299
1330	3-1/2" diameter pipe	↓	8.33	.960		228	62		290	345
1340	4" diameter pipe	B-20	17.78	1.350		237	63.50		300.50	355
1350	6" diameter pipe		15.56	1.542		243	72.50		315.50	375
1360	8" diameter pipe		11.67	2.057		288	96.50		384.50	460
1370	10" diameter pipe		11.11	2.160		470	102		572	675
1380	12" diameter pipe	↓	9.44	2.542		475	120		595	705
1400	20" long, 2" diameter pipe	1 Plum	8.10	.988		231	63.50		294.50	350
1410	2-1/2" diameter pipe		7.62	1.050		272	67.50		339.50	400
1420	3" diameter pipe		7.43	1.077		345	69.50		414.50	485
1430	3-1/2" diameter pipe	↓	7.14	1.120		310	72		382	450
1440	4" diameter pipe	B-20	15.24	1.575		340	74		414	485
1450	6" diameter pipe		13.33	1.800		405	84.50		489.50	575
1460	8" diameter pipe		10	2.400		495	113		608	715
1470	10" diameter pipe		9.52	2.521		555	119		674	795
1480	12" diameter pipe	↓	8.10	2.963	↓	660	139		799	935
1600	Clamp, stainless steel, single section									
1640	Full seal for iron, steel, PVC pipe									
1700	6" long, 2" diameter pipe	1 Plum	17	.471	Ea.	91	30.50		121.50	146
1710	2-1/2" diameter pipe		16	.500		94	32		126	151
1720	3" diameter pipe		15.60	.513		86.50	33		119.50	145
1730	3-1/2" diameter pipe	↓	15	.533		113	34.50		147.50	176
1740	4" diameter pipe	B-20	32	.750		123	35.50		158.50	188
1750	6" diameter pipe		27	.889		152	42		194	231
1760	8" diameter pipe		21	1.143		161	53.50		214.50	258
1770	10" diameter pipe		20	1.200		227	56.50		283.50	335
1780	12" diameter pipe	↓	17	1.412		218	66.50		284.50	340
1800	8" long, 2" diameter pipe	1 Plum	11.72	.683		126	44		170	205
1805	2-1/2" diameter pipe		11.03	.725		129	46.50		175.50	212
1810	3" diameter pipe		10.76	.743		136	48		184	222
1815	3-1/2" diameter pipe	↓	10.34	.774		124	50		174	212
1820	4" diameter pipe	B-20	22.07	1.087		147	51		198	239
1825	6" diameter pipe		19.31	1.243		187	58.50		245.50	294
1830	8" diameter pipe		14.48	1.657		223	78		301	365
1835	10" diameter pipe		13.79	1.740		300	82		382	460
1840	12" diameter pipe	↓	11.72	2.048		335	96.50		431.50	515
1850	12" long, 2" diameter pipe	1 Plum	9.44	.847		190	54.50		244.50	290
1855	2-1/2" diameter pipe		8.89	.900		196	58		254	305
1860	3" diameter pipe		8.67	.923		215	59.50		274.50	325
1865	3-1/2" diameter pipe	↓	8.33	.960		197	62		259	310
1870	4" diameter pipe	B-20	17.78	1.350		224	63.50		287.50	345
1875	6" diameter pipe		15.56	1.542		305	72.50		377.50	445
1880	8" diameter pipe		11.67	2.057		330	96.50		426.50	510
1885	10" diameter pipe		11.11	2.160		450	102		552	645
1890	12" diameter pipe	↓	9.44	2.542		535	120		655	770
1900	20" long, 2" diameter pipe	1 Plum	8.10	.988		197	63.50		260.50	315
1905	2-1/2" diameter pipe		7.62	1.050		218	67.50		285.50	340
1910	3" diameter pipe		7.43	1.077		250	69.50		319.50	380
1915	3-1/2" diameter pipe	↓	7.14	1.120		320	72		392	465

33 01 10 - Operation and Maintenance of Water Utilities

33 01 10.20 Pipe Repair		Crew	Daily Output	Labor-Hours	Unit	Material	2020 Bare Costs Labor	Equipment	Total	Total Incl O&P
1920	4" diameter pipe	B-20	15.24	1.575	Ea.	425	74		499	575
1925	6" diameter pipe		13.33	1.800		465	84.50		549.50	640
1930	8" diameter pipe		10	2.400		555	113		668	780
1935	10" diameter pipe		9.52	2.521		700	119		819	950
1940	12" diameter pipe		8.10	2.963		830	139		969	1,125
2000	Clamp, stainless steel, two section									
2040	Full seal, for iron, steel, PVC pipe									
2100	8" long, 4" diameter pipe	B-20	24	1	Ea.	230	47		277	325
2110	6" diameter pipe		20	1.200		264	56.50		320.50	375
2120	8" diameter pipe		13	1.846		305	87		392	470
2130	10" diameter pipe		12	2		293	94		387	460
2140	12" diameter pipe		10	2.400		395	113		508	605
2200	10" long, 4" diameter pipe		16	1.500		300	70.50		370.50	435
2210	6" diameter pipe		13	1.846		340	87		427	505
2220	8" diameter pipe		9	2.667		380	125		505	605
2230	10" diameter pipe		8	3		490	141		631	750
2240	12" diameter pipe		7	3.429		565	161		726	870
2242	Clamp, stainless steel, three section									
2244	Full seal, for iron, steel, PVC pipe									
2250	10" long, 14" diameter pipe	B-20	6.40	3.750	Ea.	655	176		831	985
2260	16" diameter pipe		6	4		695	188		883	1,050
2270	18" diameter pipe		5	4.800		795	226		1,021	1,225
2280	20" diameter pipe		4.60	5.217		905	245		1,150	1,375
2290	24" diameter pipe		4	6		1,250	282		1,532	1,800
2320	For 12" long, add to 10"					15%	25%			
2330	For 20" long, add to 10"					70%	55%			
8000	For internal cleaning and inspection, see Section 33 01 30.11									
8100	For pipe testing, see Section 23 05 93.50									

33 01 30 - Operation and Maintenance of Sewer Utilities

33 01 30.11 Television Inspection of Sewers

	33 01 30.11 Television Inspection of Sewers	Crew	Daily Output	Labor-Hours	Unit	Material	2020 Bare Costs Labor	Equipment	Total	Total Incl O&P
0010	**TELEVISION INSPECTION OF SEWERS**									
0100	Pipe internal cleaning & inspection, cleaning, pressure pipe systems									
0120	Pig method, lengths 1000' to 10,000'									
0140	4" diameter thru 24" diameter, minimum				L.F.				3.60	4.14
0160	Maximum				"				18	21
6000	Sewage/sanitary systems									
6100	Power rodder with header & cutters									
6110	Mobilization charge, minimum				Total				695	800
6120	Mobilization charge, maximum				"				9,125	10,600
6140	Cleaning 4"-12" diameter				L.F.				6	6.60
6190	14"-24" diameter								8	8.80
6240	30" diameter								9.60	10.50
6250	36" diameter								12	13.20
6260	48" diameter								16	17.60
6270	60" diameter								24	26.50
6280	72" diameter								48	53
9000	Inspection, television camera with video									
9060	up to 500 linear feet				Total				715	820

For customer support on your Mechanical Costs with RSMeans Data, call 800.448.8182.

463

33 01 Operation and Maintenance of Utilities

33 01 30 – Operation and Maintenance of Sewer Utilities

33 01 30.23 Pipe Bursting

		Crew	Daily Output	Labor-Hours	Unit	Material	2020 Bare Costs Labor	Equipment	Total	Total Incl O&P
0010	**PIPE BURSTING**									
0011	300' runs, replace with HDPE pipe									
0020	Not including excavation, backfill, shoring, or dewatering									
0100	6" to 15" diameter, minimum				L.F.				200	220
0200	Maximum								550	605
0300	18" to 36" diameter, minimum								650	715
0400	Maximum								1,050	1,150
0500	Mobilize and demobilize, minimum				Job				3,000	3,300
0600	Maximum				"				32,100	35,400

33 01 30.74 Sliplining, Excludes Cleaning

		Crew	Daily Output	Labor-Hours	Unit	Material	2020 Bare Costs Labor	Equipment	Total	Total Incl O&P
0010	**SLIPLINING, excludes cleaning** and video inspection									
0020	Pipe relined with one pipe size smaller than original (4" for 6")									
0100	6" diameter, original size	B-6B	600	.080	L.F.	5.60	3.42	1.67	10.69	13.20
0150	8" diameter, original size		600	.080		9.60	3.42	1.67	14.69	17.55
0200	10" diameter, original size		600	.080		11.25	3.42	1.67	16.34	19.40
0250	12" diameter, original size		400	.120		16	5.15	2.50	23.65	28
0300	14" diameter, original size		400	.120		15.60	5.15	2.50	23.25	27.50
0350	16" diameter, original size	B-6C	300	.160		18.30	6.85	7.75	32.90	39
0400	18" diameter, original size	"	300	.160		28	6.85	7.75	42.60	50
1000	Pipe HDPE lining, make service line taps	B-6	4	6	Ea.	87	274	53.50	414.50	565

33 05 Common Work Results for Utilities

33 05 07 – Trenchless Installation of Utility Piping

33 05 07.23 Utility Boring and Jacking

		Crew	Daily Output	Labor-Hours	Unit	Material	2020 Bare Costs Labor	Equipment	Total	Total Incl O&P
0010	**UTILITY BORING AND JACKING**									
0011	Casing only, 100' minimum,									
0020	not incl. jacking pits or dewatering									
0100	Roadwork, 1/2" thick wall, 24" diameter casing	B-42	20	3.200	L.F.	122	152	52	326	420
0200	36" diameter		16	4		215	190	65	470	595
0300	48" diameter		15	4.267		300	203	69.50	572.50	710
0500	Railroad work, 24" diameter		15	4.267		122	203	69.50	394.50	515
0600	36" diameter		14	4.571		215	217	74.50	506.50	650
0700	48" diameter		12	5.333		300	253	87	640	805
0900	For ledge, add								20%	20%
1000	Small diameter boring, 3", sandy soil	B-82	900	.018		22	.85	.20	23.05	26
1040	Rocky soil	"	500	.032		22	1.52	.36	23.88	27
1100	Prepare jacking pits, incl. mobilization & demobilization, minimum				Ea.				3,225	3,700
1101	Maximum				"				22,000	25,500

33 05 07.36 Microtunneling

		Crew	Daily Output	Labor-Hours	Unit	Material	2020 Bare Costs Labor	Equipment	Total	Total Incl O&P
0010	**MICROTUNNELING**									
0011	Not including excavation, backfill, shoring,									
0020	or dewatering, average 50'/day, slurry method									
0100	24" to 48" outside diameter, minimum				L.F.				965	965
0110	Adverse conditions, add				"				500	500
1000	Rent microtunneling machine, average monthly lease				Month				97,500	107,000
1010	Operating technician				Day				630	705
1100	Mobilization and demobilization, minimum				Job				41,200	45,900
1110	Maximum				"				445,500	490,500

33 05 Common Work Results for Utilities

33 05 61 – Concrete Manholes

33 05 61.10 Storm Drainage Manholes, Frames and Covers	Crew	Daily Output	Labor-Hours	Unit	Material	2020 Bare Costs Labor	Equipment	Total	Total Incl O&P
0010 **STORM DRAINAGE MANHOLES, FRAMES & COVERS**									
0020 Excludes footing, excavation, backfill (See line items for frame & cover)									
0050 Brick, 4' inside diameter, 4' deep	D-1	1	16	Ea.	615	750		1,365	1,825
0100 6' deep		.70	22.857		870	1,075		1,945	2,575
0150 8' deep		.50	32		1,125	1,500		2,625	3,500
0200 For depths over 8', add		4	4	V.L.F.	95.50	188		283.50	390
0400 Concrete blocks (radial), 4' ID, 4' deep		1.50	10.667	Ea.	430	500		930	1,225
0500 6' deep		1	16		565	750		1,315	1,775
0600 8' deep		.70	22.857		705	1,075		1,780	2,400
0700 For depths over 8', add		5.50	2.909	V.L.F.	71.50	136		207.50	286
0800 Concrete, cast in place, 4' x 4', 8" thick, 4' deep	C-14H	2	24	Ea.	580	1,250	13.45	1,843.45	2,500
0900 6' deep		1.50	32		830	1,650	17.90	2,497.90	3,400
1000 8' deep		1	48		1,200	2,475	27	3,702	5,075
1100 For depths over 8', add		8	6	V.L.F.	133	310	3.36	446.36	615
1110 Precast, 4' ID, 4' deep	B-22	4.10	7.317	Ea.	880	360	69	1,309	1,600
1120 6' deep		3	10		1,025	495	94.50	1,614.50	1,975
1130 8' deep		2	15		1,200	740	142	2,082	2,600
1140 For depths over 8', add		16	1.875	V.L.F.	134	92.50	17.70	244.20	305
1150 5' ID, 4' deep	B-6	3	8	Ea.	1,725	365	71.50	2,161.50	2,525
1160 6' deep		2	12		2,075	550	107	2,732	3,225
1170 8' deep		1.50	16		2,625	730	143	3,498	4,125
1180 For depths over 8', add		12	2	V.L.F.	325	91.50	17.80	434.30	515
1190 6' ID, 4' deep		2	12	Ea.	2,600	550	107	3,257	3,800
1200 6' deep		1.50	16		2,975	730	143	3,848	4,525
1210 8' deep		1	24		3,675	1,100	214	4,989	5,925
1220 For depths over 8', add		8	3	V.L.F.	410	137	26.50	573.50	685
1250 Slab tops, precast, 8" thick									
1300 4' diameter manhole	B-6	8	3	Ea.	282	137	26.50	445.50	545
1400 5' diameter manhole		7.50	3.200		440	146	28.50	614.50	735
1500 6' diameter manhole		7	3.429		705	157	30.50	892.50	1,050
3800 Steps, heavyweight cast iron, 7" x 9"	1 Bric	40	.200		18.05	10.40		28.45	35.50
3900 8" x 9"		40	.200		21.50	10.40		31.90	40
3928 12" x 10-1/2"		40	.200		29.50	10.40		39.90	48.50
4000 Standard sizes, galvanized steel		40	.200		26.50	10.40		36.90	45
4100 Aluminum		40	.200		30	10.40		40.40	49
4150 Polyethylene		40	.200		32.50	10.40		42.90	52

33 05 97 – Identification and Signage for Utilities

33 05 97.05 Utility Connection

	Crew	Daily Output	Labor-Hours	Unit	Material	2020 Bare Costs Labor	Equipment	Total	Total Incl O&P
0010 **UTILITY CONNECTION**									
0020 Water, sanitary, stormwater, gas, single connection	B-14	1	48	Ea.	3,050	2,125	214	5,389	6,775
0030 Telecommunication	"	3	16	"	400	710	71	1,181	1,600

33 05 97.10 Utility Accessories

		Crew	Daily Output	Labor-Hours	Unit	Material	2020 Bare Costs Labor	Equipment	Total	Total Incl O&P
0010 **UTILITY ACCESSORIES**	G1030–805									
0400 Underground tape, detectable, reinforced, alum. foil core, 2"		1 Clab	150	.053	C.L.F.	9	2.25		11.25	13.25
0500 6"		"	140	.057	"	38	2.41		40.41	45

For customer support on your Mechanical Costs with RSMeans Data, call 800.448.8182.

465

33 11 13.10 Wells and Accessories	Crew	Daily Output	Labor-Hours	Unit	Material	2020 Bare Costs Labor	Equipment	Total	Total Incl O&P
0010 **WELLS & ACCESSORIES**									
0011 Domestic									
0100 Drilled, 4" to 6" diameter	B-23	120	.333	L.F.		14.15	13.25	27.40	36
0200 8" diameter	"	95.20	.420	"		17.85	16.70	34.55	45.50
0400 Gravel pack well, 40' deep, incl. gravel & casing, complete									
0500 24" diameter casing x 18" diameter screen	B-23	.13	308	Total	38,500	13,100	12,200	63,800	75,500
0600 36" diameter casing x 18" diameter screen		.12	333	"	40,800	14,200	13,300	68,300	81,000
0800 Observation wells, 1-1/4" riser pipe	↓	163	.245	V.L.F.	19.90	10.45	9.75	40.10	48.50
0900 For flush Buffalo roadway box, add	1 Skwk	16.60	.482	Ea.	53	26.50		79.50	98
1200 Test well, 2-1/2" diameter, up to 50' deep (15 to 50 GPM)	B-23	1.51	26.490	"	830	1,125	1,050	3,005	3,775
1300 Over 50' deep, add	"	121.80	.328	L.F.	22	13.95	13.05	49	60
1500 Pumps, installed in wells to 100' deep, 4" submersible									
1510 1/2 HP	Q-1	3.22	4.969	Ea.	660	288		948	1,150
1520 3/4 HP		2.66	6.015		795	350		1,145	1,400
1600 1 HP	↓	2.29	6.987		975	405		1,380	1,675
1700 1-1/2 HP	Q-22	1.60	10		2,150	580	294	3,024	3,575
1800 2 HP		1.33	12.030		1,800	700	355	2,855	3,450
1900 3 HP		1.14	14.035		2,425	815	415	3,655	4,350
2000 5 HP		1.14	14.035		2,850	815	415	4,080	4,800
2050 Remove and install motor only, 4 HP		1.14	14.035		1,450	815	415	2,680	3,275
3000 Pump, 6" submersible, 25' to 150' deep, 25 HP, 103 to 400 GPM		.89	17.978		9,325	1,050	530	10,905	12,400
3100 25' to 500' deep, 30 HP, 104 to 400 GPM	↓	.73	21.918	↓	11,000	1,275	645	12,920	14,700
8110 Well screen assembly, stainless steel, 2" diameter	B-23A	273	.088	L.F.	81.50	4.19	3.23	88.92	100
8120 3" diameter		253	.095		143	4.52	3.49	151.01	168
8130 4" diameter		200	.120		178	5.70	4.41	188.11	209
8140 5" diameter		168	.143		184	6.80	5.25	196.05	218
8150 6" diameter		126	.190		204	9.10	7	220.10	246
8160 8" diameter		98.50	.244		274	11.60	8.95	294.55	325
8170 10" diameter		73	.329		340	15.65	12.10	367.75	410
8180 12" diameter		62.50	.384		395	18.30	14.10	427.40	480
8190 14" diameter		54.30	.442		445	21	16.25	482.25	540
8200 16" diameter		48.30	.497		490	23.50	18.25	531.75	595
8210 18" diameter		39.20	.612		620	29	22.50	671.50	755
8220 20" diameter		31.20	.769		700	36.50	28.50	765	855
8230 24" diameter		23.80	1.008		855	48	37	940	1,050
8240 26" diameter		21	1.143		960	54.50	42	1,056.50	1,175
8244 Well casing or drop pipe, PVC, 1/2" diameter		550	.044		1.31	2.08	1.60	4.99	6.30
8245 3/4" diameter		550	.044		1.32	2.08	1.60	5	6.35
8246 1" diameter		550	.044		1.36	2.08	1.60	5.04	6.40
8247 1-1/4" diameter		520	.046		1.71	2.20	1.70	5.61	7.05
8248 1-1/2" diameter		490	.049		1.81	2.33	1.80	5.94	7.45
8249 1-3/4" diameter		380	.063		1.85	3.01	2.32	7.18	9.10
8250 2" diameter		280	.086		2.12	4.08	3.15	9.35	11.90
8252 3" diameter		260	.092		4.16	4.40	3.39	11.95	14.90
8254 4" diameter		205	.117		4.46	5.60	4.30	14.36	18
8255 5" diameter		170	.141		4.48	6.75	5.20	16.43	20.50
8256 6" diameter		130	.185		8.60	8.80	6.80	24.20	30
8258 8" diameter		100	.240		13	11.45	8.80	33.25	41
8260 10" diameter		73	.329		19.90	15.65	12.10	47.65	59
8262 12" diameter		62	.387		24	18.45	14.25	56.70	69
8300 Slotted PVC, 1-1/4" diameter		521	.046		2.89	2.20	1.69	6.78	8.35
8310 1-1/2" diameter		488	.049		3.27	2.34	1.81	7.42	9.10
8320 2" diameter	↓	273	.088	↓	4.09	4.19	3.23	11.51	14.35

33 11 13 – Potable Water Supply Wells

33 11 13.10 Wells and Accessories

		Crew	Daily Output	Labor-Hours	Unit	Material	2020 Bare Costs Labor	Equipment	Total	Total Incl O&P
8330	3" diameter	B-23A	253	.095	L.F.	6.75	4.52	3.49	14.76	18.05
8340	4" diameter		200	.120		8.35	5.70	4.41	18.46	22.50
8350	5" diameter		168	.143		15.05	6.80	5.25	27.10	32.50
8360	6" diameter		126	.190		16.30	9.10	7	32.40	39
8370	8" diameter		98.50	.244		24	11.60	8.95	44.55	53.50
8400	Artificial gravel pack, 2" screen, 6" casing	B-23B	174	.138		4.68	6.55	6.40	17.63	22
8405	8" casing		111	.216		6.30	10.30	10.05	26.65	33.50
8410	10" casing		74.50	.322		8.80	15.35	14.95	39.10	49
8415	12" casing		60	.400		14.05	19.05	18.60	51.70	64.50
8420	14" casing		50.20	.478		14.65	23	22	59.65	74.50
8425	16" casing		40.70	.590		18.55	28	27.50	74.05	92.50
8430	18" casing		36	.667		22.50	32	31	85.50	106
8435	20" casing		29.50	.814		26	39	38	103	128
8440	24" casing		25.70	.934		30	44.50	43.50	118	147
8445	26" casing		24.60	.976		33	46.50	45.50	125	156
8450	30" casing		20	1.200		38	57	55.50	150.50	189
8455	36" casing		16.40	1.463		40.50	69.50	68	178	225
8500	Develop well		8	3	Hr.	300	143	139	582	695
8550	Pump test well		8	3		96.50	143	139	378.50	475
8560	Standby well	B-23A	8	3		105	143	110	358	450
8570	Standby, drill rig		8	3			143	110	253	335
8580	Surface seal well, concrete filled		1	24	Ea.	1,025	1,150	880	3,055	3,850
8590	Well test pump, install & remove	B-23	1	40			1,700	1,600	3,300	4,300
8600	Well sterilization, chlorine	2 Clab	1	16		141	675		816	1,150
8610	Well water pressure switch	1 Clab	12	.667		98	28		126	150
8630	Well water pressure switch with manual reset	"	12	.667		130	28		158	185
9950	See Section 31 23 19.40 for wellpoints									
9960	See Section 31 23 19.30 for drainage wells									

33 11 13.20 Water Supply Wells, Pumps

		Crew	Daily Output	Labor-Hours	Unit	Material	2020 Bare Costs Labor	Equipment	Total	Total Incl O&P
0010	**WATER SUPPLY WELLS, PUMPS**									
0011	With pressure control									
1000	Deep well, jet, 42 gal. galvanized tank									
1040	3/4 HP	1 Plum	.80	10	Ea.	1,125	645		1,770	2,225
3000	Shallow well, jet, 30 gal. galvanized tank									
3040	1/2 HP	1 Plum	2	4	Ea.	915	258		1,173	1,375

33 14 Water Utility Transmission and Distribution

33 14 13 – Public Water Utility Distribution Piping

33 14 13.15 Water Supply, Ductile Iron Pipe

		Crew	Daily Output	Labor-Hours	Unit	Material	2020 Bare Costs Labor	Equipment	Total	Total Incl O&P
0010	**WATER SUPPLY, DUCTILE IRON PIPE** R331113-80									
0020	Not including excavation or backfill									
2000	Pipe, class 50 water piping, 18' lengths									
2020	Mechanical joint, 4" diameter	B-21A	200	.200	L.F.	42.50	10.45	2.14	55.09	64.50
2040	6" diameter		160	.250		51	13.05	2.68	66.73	78.50
2060	8" diameter		133.33	.300		53.50	15.70	3.21	72.41	86
2080	10" diameter		114.29	.350		71	18.30	3.75	93.05	110
2100	12" diameter		105.26	.380		90.50	19.85	4.07	114.42	134
2120	14" diameter		100	.400		105	21	4.28	130.28	152
2140	16" diameter		72.73	.550		112	29	5.90	146.90	173
2160	18" diameter		68.97	.580		129	30.50	6.20	165.70	194
2170	20" diameter		57.14	.700		132	36.50	7.50	176	208

For customer support on your Mechanical Costs with RSMeans Data, call 800.448.8182.

467

33 14 13.15 Water Supply, Ductile Iron Pipe		Crew	Daily Output	Labor-Hours	Unit	Material	2020 Bare Costs Labor	Equipment	Total	Total Incl O&P
2180	24" diameter	B-21A	47.06	.850	L.F.	168	44.50	9.10	221.60	262
3000	Push-on joint, 4" diameter		400	.100		23	5.25	1.07	29.32	34.50
3020	6" diameter		333.33	.120		23.50	6.25	1.29	31.04	37
3040	8" diameter		200	.200		32.50	10.45	2.14	45.09	54
3060	10" diameter		181.82	.220		50	11.50	2.36	63.86	75
3080	12" diameter		160	.250		53	13.05	2.68	68.73	80.50
3100	14" diameter		133.33	.300		53	15.70	3.21	71.91	85
3120	16" diameter		114.29	.350		55	18.30	3.75	77.05	92
3140	18" diameter		100	.400		61	21	4.28	86.28	103
3160	20" diameter		88.89	.450		63.50	23.50	4.82	91.82	110
3180	24" diameter	↓	76.92	.520	↓	86.50	27	5.55	119.05	142
8000	Piping, fittings, mechanical joint, AWWA C110									
8006	90° bend, 4" diameter	B-20A	16	2	Ea.	176	101		277	345
8020	6" diameter		12.80	2.500		256	126		382	470
8040	8" diameter	↓	10.67	2.999		500	152		652	775
8060	10" diameter	B-21A	11.43	3.500		730	183	37.50	950.50	1,125
8080	12" diameter		10.53	3.799		960	199	40.50	1,199.50	1,400
8100	14" diameter		10	4		1,375	209	43	1,627	1,875
8120	16" diameter		7.27	5.502		1,750	288	59	2,097	2,425
8140	18" diameter		6.90	5.797		2,400	305	62	2,767	3,175
8160	20" diameter		5.71	7.005		2,875	365	75	3,315	3,775
8180	24" diameter	↓	4.70	8.511		4,975	445	91	5,511	6,250
8200	Wye or tee, 4" diameter	B-20A	10.67	2.999		390	152		542	650
8220	6" diameter		8.53	3.751		585	190		775	925
8240	8" diameter	↓	7.11	4.501		890	228		1,118	1,325
8260	10" diameter	B-21A	7.62	5.249		1,175	274	56	1,505	1,775
8280	12" diameter		7.02	5.698		1,975	298	61	2,334	2,675
8300	14" diameter		6.67	5.997		2,425	315	64	2,804	3,225
8320	16" diameter		4.85	8.247		2,750	430	88.50	3,268.50	3,775
8340	18" diameter		4.60	8.696		4,500	455	93	5,048	5,725
8360	20" diameter		3.81	10.499		6,525	550	112	7,187	8,125
8380	24" diameter	↓	3.14	12.739		8,800	665	136	9,601	10,800
8398	45° bend, 4" diameter	B-20A	16	2		206	101		307	380
8400	6" diameter		12.80	2.500		295	126		421	515
8405	8" diameter	↓	10.67	2.999		430	152		582	700
8410	12" diameter	B-21A	10.53	3.799		875	199	40.50	1,114.50	1,300
8420	16" diameter		7.27	5.502		1,725	288	59	2,072	2,400
8430	20" diameter		5.71	7.005		2,450	365	75	2,890	3,325
8440	24" diameter	↓	4.70	8.511		3,450	445	91	3,986	4,575
8450	Decreaser, 6" x 4" diameter	B-20A	14.22	2.250		262	114		376	460
8460	8" x 6" diameter	"	11.64	2.749		350	139		489	595
8470	10" x 6" diameter	B-21A	13.33	3.001		425	157	32	614	740
8480	12" x 6" diameter		12.70	3.150		620	165	33.50	818.50	965
8490	16" x 6" diameter		10	4		1,075	209	43	1,327	1,525
8500	20" x 6" diameter	↓	8.42	4.751	↓	1,800	248	51	2,099	2,425
8552	For water utility valves see Section 33 14 19									
8700	Joint restraint, ductile iron mechanical joints									
8710	4" diameter	B-20A	32	1	Ea.	32.50	50.50		83	112
8720	6" diameter		25.60	1.250		41	63		104	140
8730	8" diameter		21.33	1.500		59	76		135	179
8740	10" diameter		18.28	1.751		102	88.50		190.50	246
8750	12" diameter		16.84	1.900		117	96		213	273
8760	14" diameter	↓	16	2	↓	177	101		278	345

33 14 13 - Public Water Utility Distribution Piping

33 14 13.15 Water Supply, Ductile Iron Pipe

		Crew	Daily Output	Labor-Hours	Unit	Material	2020 Bare Costs Labor	Equipment	Total	Total Incl O&P
8770	16" diameter	B-20A	11.64	2.749	Ea.	190	139		329	415
8780	18" diameter		11.03	2.901		267	147		414	515
8785	20" diameter		9.14	3.501		330	177		507	630
8790	24" diameter		7.53	4.250		450	215		665	815
9600	Steel sleeve with tap, 4" diameter	B-20	3	8		505	375		880	1,125
9620	6" diameter		2	12		555	565		1,120	1,475
9630	8" diameter		2	12		720	565		1,285	1,650

33 14 13.25 Water Supply, Polyvinyl Chloride Pipe

		Crew	Daily Output	Labor-Hours	Unit	Material	2020 Bare Costs Labor	Equipment	Total	Total Incl O&P
0010	**WATER SUPPLY, POLYVINYL CHLORIDE PIPE** R331113-80									
0020	Not including excavation or backfill, unless specified									
2100	PVC pipe, Class 150, 1-1/2" diameter	Q-1A	750	.013	L.F.	.61	.86		1.47	1.96
2120	2" diameter		686	.015		.88	.95		1.83	2.39
2140	2-1/2" diameter		500	.020		1.18	1.30		2.48	3.24
2160	3" diameter	B-20	430	.056		1.59	2.62		4.21	5.70
3010	AWWA C905, PR 100, DR 25									
3030	14" diameter	B-20A	213	.150	L.F.	14.20	7.60		21.80	27
3040	16" diameter		200	.160		17.30	8.10		25.40	31
3050	18" diameter		160	.200		27.50	10.10		37.60	45.50
3060	20" diameter		133	.241		30.50	12.15		42.65	52
3070	24" diameter		107	.299		45	15.10		60.10	72
3080	30" diameter		80	.400		74.50	20		94.50	113
3090	36" diameter		80	.400		115	20		135	158
3100	42" diameter		60	.533		153	27		180	209
3200	48" diameter		60	.533		200	27		227	261
4520	Pressure pipe Class 150, SDR 18, AWWA C900, 4" diameter		380	.084		3.03	4.26		7.29	9.75
4530	6" diameter		316	.101		5	5.10		10.10	13.15
4540	8" diameter		264	.121		8.60	6.15		14.75	18.65
4550	10" diameter		220	.145		12.65	7.35		20	25
4560	12" diameter		186	.172		17.75	8.70		26.45	32.50
8000	Fittings with rubber gasket									
8003	Class 150, DR 18									
8006	90° bend , 4" diameter	B-20	100	.240	Ea.	42.50	11.30		53.80	63.50
8020	6" diameter		90	.267		74	12.55		86.55	100
8040	8" diameter		80	.300		139	14.10		153.10	175
8060	10" diameter		50	.480		365	22.50		387.50	435
8080	12" diameter		30	.800		445	37.50		482.50	545
8100	Tee, 4" diameter		90	.267		63	12.55		75.55	88
8120	6" diameter		80	.300		139	14.10		153.10	175
8140	8" diameter		70	.343		199	16.10		215.10	243
8160	10" diameter		40	.600		660	28		688	770
8180	12" diameter		20	1.200		905	56.50		961.50	1,075
8200	45° bend, 4" diameter		100	.240		41.50	11.30		52.80	62.50
8220	6" diameter		90	.267		77	12.55		89.55	103
8240	8" diameter		50	.480		137	22.50		159.50	184
8260	10" diameter		50	.480		305	22.50		327.50	370
8280	12" diameter		30	.800		400	37.50		437.50	495
8300	Reducing tee 6" x 4"		100	.240		126	11.30		137.30	156
8320	8" x 6"		90	.267		202	12.55		214.55	241
8330	10" x 6"		90	.267		340	12.55		352.55	395
8340	10" x 8"		90	.267		365	12.55		377.55	420
8350	12" x 6"		90	.267		420	12.55		432.55	480
8360	12" x 8"		90	.267		440	12.55		452.55	505

For customer support on your Mechanical Costs with RSMeans Data, call 800.448.8182.

469

33 14 13.25 Water Supply, Polyvinyl Chloride Pipe

		Crew	Daily Output	Labor-Hours	Unit	Material	2020 Bare Costs Labor	Equipment	Total	Total Incl O&P
8400	Tapped service tee (threaded type) 6" x 6" x 3/4"	B-20	100	.240	Ea.	96	11.30		107.30	123
8430	6" x 6" x 1"		90	.267		96	12.55		108.55	125
8440	6" x 6" x 1-1/2"		90	.267		90.50	12.55		103.05	118
8450	6" x 6" x 2"		90	.267		96	12.55		108.55	125
8460	8" x 8" x 3/4"		90	.267		126	12.55		138.55	158
8470	8" x 8" x 1"		90	.267		126	12.55		138.55	158
8480	8" x 8" x 1-1/2"		90	.267		126	12.55		138.55	158
8490	8" x 8" x 2"		90	.267		126	12.55		138.55	158
8500	Repair coupling 4"		100	.240		23.50	11.30		34.80	43
8520	6" diameter		90	.267		36.50	12.55		49.05	59.50
8540	8" diameter		50	.480		90	22.50		112.50	133
8560	10" diameter		50	.480		175	22.50		197.50	227
8580	12" diameter		50	.480		265	22.50		287.50	325
8600	Plug end 4"		100	.240		21	11.30		32.30	40
8620	6" diameter		90	.267		43	12.55		55.55	66
8640	8" diameter		50	.480		63.50	22.50		86	104
8660	10" diameter		50	.480		97.50	22.50		120	141
8680	12" diameter	▼	50	.480	▼	120	22.50		142.50	166
8700	PVC pipe, joint restraint									
8710	4" diameter	B-20A	32	1	Ea.	46.50	50.50		97	128
8720	6" diameter		25.60	1.250		58	63		121	159
8730	8" diameter		21.33	1.500		84.50	76		160.50	207
8740	10" diameter		18.28	1.751		139	88.50		227.50	286
8750	12" diameter		16.84	1.900		146	96		242	305
8760	14" diameter		16	2		201	101		302	375
8770	16" diameter		11.64	2.749		270	139		409	505
8780	18" diameter		11.03	2.901		335	147		482	585
8785	20" diameter		9.14	3.501		445	177		622	755
8790	24" diameter	▼	7.53	4.250	▼	515	215		730	885

33 14 13.35 Water Supply, HDPE

		Crew	Daily Output	Labor-Hours	Unit	Material	2020 Bare Costs Labor	Equipment	Total	Total Incl O&P
0010	**WATER SUPPLY, HDPE**									
0011	Butt fusion joints, SDR 21 40' lengths not including excavation or backfill									
0100	4" diameter	B-22A	400	.100	L.F.	2.59	4.85	2.01	9.45	12.35
0200	6" diameter		380	.105		5.60	5.10	2.12	12.82	16.20
0300	8" diameter		320	.125		9.60	6.05	2.52	18.17	22.50
0400	10" diameter		300	.133		11.25	6.45	2.68	20.38	25
0500	12" diameter		260	.154		16	7.45	3.10	26.55	32
0600	14" diameter	B-22B	220	.182		15.60	8.80	6.60	31	37.50
0700	16" diameter		180	.222		18.30	10.75	8.10	37.15	45
0800	18" diameter		140	.286		28	13.85	10.40	52.25	63.50
0850	20" diameter		130	.308		34	14.90	11.20	60.10	72.50
0900	24" diameter	▼	100	.400	▼	55.50	19.40	14.55	89.45	106
1000	Fittings									
1100	Elbows, 90 degrees									
1200	4" diameter	B-22A	32	1.250	Ea.	17.25	60.50	25	102.75	138
1300	6" diameter		28	1.429		47.50	69	28.50	145	188
1400	8" diameter		24	1.667		113	81	33.50	227.50	282
1500	10" diameter		18	2.222		254	108	44.50	406.50	490
1600	12" diameter		12	3.333		297	162	67	526	640
1700	14" diameter	B-22B	9	4.444		630	215	162	1,007	1,200
1800	16" diameter		6	6.667		935	325	242	1,502	1,775
1900	18" diameter	▼	4	10	▼	1,050	485	365	1,900	2,275

33 14 Water Utility Transmission and Distribution

33 14 13 – Public Water Utility Distribution Piping

33 14 13.35 Water Supply, HDPE

		Crew	Daily Output	Labor-Hours	Unit	Material	2020 Bare Costs Labor	Equipment	Total	Total Incl O&P
2000	24" diameter	B-22B	3	13.333	Ea.	1,825	645	485	2,955	3,500
2100	Tees									
2200	4" diameter	B-22A	30	1.333	Ea.	22.50	64.50	27	114	152
2300	6" diameter		26	1.538		74	74.50	31	179.50	227
2400	8" diameter		22	1.818		130	88	36.50	254.50	315
2500	10" diameter		15	2.667		172	129	53.50	354.50	440
2600	12" diameter		10	4		400	194	80.50	674.50	820
2700	14" diameter	B-22B	8	5		475	242	182	899	1,075
2800	16" diameter		6	6.667		555	325	242	1,122	1,350
2900	18" diameter		4	10		670	485	365	1,520	1,875
3000	24" diameter		2	20		895	970	725	2,590	3,225
4100	Caps									
4110	4" diameter	B-22A	34	1.176	Ea.	15.40	57	23.50	95.90	128
4120	6" diameter		30	1.333		33.50	64.50	27	125	164
4130	8" diameter		26	1.538		57	74.50	31	162.50	209
4150	10" diameter		20	2		139	97	40	276	345
4160	12" diameter		14	2.857		168	138	57.50	363.50	455

33 14 17 – Site Water Utility Service Laterals

33 14 17.15 Tapping, Crosses and Sleeves

		Crew	Daily Output	Labor-Hours	Unit	Material	2020 Bare Costs Labor	Equipment	Total	Total Incl O&P
0010	**TAPPING, CROSSES AND SLEEVES**									
4000	Drill and tap pressurized main (labor only)									
4100	6" main, 1" to 2" service	Q-1	3	5.333	Ea.		310		310	465
4150	8" main, 1" to 2" service	"	2.75	5.818	"		335		335	505
4500	Tap and insert gate valve									
4600	8" main, 4" branch	B-21	3.20	8.750	Ea.		425	59	484	705
4650	6" branch		2.70	10.370			505	70	575	835
4700	10" main, 4" branch		2.70	10.370			505	70	575	835
4750	6" branch		2.35	11.915			580	80.50	660.50	965
4800	12" main, 6" branch		2.35	11.915			580	80.50	660.50	965
4850	8" branch		2.35	11.915			580	80.50	660.50	965
7020	Crosses, 4" x 4"		37	.757		1,450	37	5.10	1,492.10	1,625
7030	6" x 4"		25	1.120		1,650	54.50	7.55	1,712.05	1,925
7040	6" x 6"		25	1.120		1,725	54.50	7.55	1,787.05	2,000
7060	8" x 6"		21	1.333		2,050	65	9	2,124	2,350
7080	8" x 8"		21	1.333		2,175	65	9	2,249	2,500
7100	10" x 6"		21	1.333		4,150	65	9	4,224	4,675
7120	10" x 10"		21	1.333		4,475	65	9	4,549	5,025
7140	12" x 6"		18	1.556		4,125	76	10.50	4,211.50	4,675
7160	12" x 12"		18	1.556		5,300	76	10.50	5,386.50	5,950
7180	14" x 6"		16	1.750		10,200	85.50	11.80	10,297.30	11,300
7200	14" x 14"		16	1.750		10,800	85.50	11.80	10,897.30	12,000
7220	16" x 6"		14	2		10,900	97.50	13.50	11,011	12,200
7240	16" x 10"		14	2		11,100	97.50	13.50	11,211	12,400
7260	16" x 16"		14	2		11,700	97.50	13.50	11,811	13,000
7280	18" x 6"		10	2.800		16,400	137	18.90	16,555.90	18,200
7300	18" x 12"		10	2.800		16,500	137	18.90	16,655.90	18,400
7320	18" x 18"		10	2.800		17,100	137	18.90	17,255.90	19,000
7340	20" x 6"		8	3.500		13,200	171	23.50	13,394.50	14,900
7360	20" x 12"		8	3.500		14,200	171	23.50	14,394.50	15,900
7380	20" x 20"		8	3.500		21,000	171	23.50	21,194.50	23,400
7400	24" x 6"		6	4.667		17,100	228	31.50	17,359.50	19,200
7420	24" x 12"		6	4.667		17,400	228	31.50	17,659.50	19,600

For customer support on your Mechanical Costs with RSMeans Data, call 800.448.8182.

471

33 14 17.15 Tapping, Crosses and Sleeves	Crew	Daily Output	Labor-Hours	Unit	Material	2020 Bare Costs Labor	Equipment	Total	Total Incl O&P
7440	24" x 18"	B-21	6	4.667	Ea.	27,100	228	31.50	27,359.50	30,200
7460	24" x 24"		6	4.667		27,300	228	31.50	27,559.50	30,500
7600	Cut-in sleeves with rubber gaskets, 4"		18	1.556		560	76	10.50	646.50	740
7620	6"		12	2.333		705	114	15.75	834.75	965
7640	8"		10	2.800		895	137	18.90	1,050.90	1,200
7660	10"		10	2.800		1,250	137	18.90	1,405.90	1,600
7680	12"		9	3.111		1,800	152	21	1,973	2,225
7800	Cut-in valves with rubber gaskets, 4"		18	1.556		560	76	10.50	646.50	740
7820	6"		12	2.333		715	114	15.75	844.75	975
7840	8"		10	2.800		1,000	137	18.90	1,155.90	1,325
7860	10"		10	2.800		1,775	137	18.90	1,930.90	2,175
7880	12"		9	3.111		2,350	152	21	2,523	2,825
7900	Tapping valve 4", MJ, ductile iron		18	1.556		455	76	10.50	541.50	625
7920	6", MJ, ductile iron		12	2.333		620	114	15.75	749.75	870
8000	Sleeves with rubber gaskets, 4" x 4"		37	.757		1,050	37	5.10	1,092.10	1,200
8010	6" x 4"		25	1.120		985	54.50	7.55	1,047.05	1,175
8020	6" x 6"		25	1.120		1,300	54.50	7.55	1,362.05	1,525
8030	8" x 4"		21	1.333		1,025	65	9	1,099	1,225
8040	8" x 6"		21	1.333		910	65	9	984	1,100
8060	8" x 8"		21	1.333		1,350	65	9	1,424	1,575
8070	10" x 4"		21	1.333		830	65	9	904	1,025
8080	10" x 6"		21	1.333		1,025	65	9	1,099	1,225
8090	10" x 8"		21	1.333		1,525	65	9	1,599	1,800
8100	10" x 10"		21	1.333		2,225	65	9	2,299	2,550
8110	12" x 4"		18	1.556		1,375	76	10.50	1,461.50	1,625
8120	12" x 6"		18	1.556		1,425	76	10.50	1,511.50	1,675
8130	12" x 8"		18	1.556		1,500	76	10.50	1,586.50	1,775
8135	12" x 10"		18	1.556		2,600	76	10.50	2,686.50	2,975
8140	12" x 12"		18	1.556		2,725	76	10.50	2,811.50	3,125
8160	14" x 6"		16	1.750		1,600	85.50	11.80	1,697.30	1,900
8180	14" x 14"		16	1.750		2,575	85.50	11.80	2,672.30	3,000
8200	16" x 6"		14	2		1,875	97.50	13.50	1,986	2,225
8220	16" x 10"		14	2		3,750	97.50	13.50	3,861	4,275
8240	16" x 16"		14	2		9,700	97.50	13.50	9,811	10,900
8260	18" x 6"		10	2.800		1,150	137	18.90	1,305.90	1,500
8280	18" x 12"		10	2.800		2,925	137	18.90	3,080.90	3,425
8300	18" x 18"		10	2.800		8,650	137	18.90	8,805.90	9,750
8320	20" x 6"		8	3.500		2,425	171	23.50	2,619.50	2,950
8340	20" x 12"		8	3.500		4,750	171	23.50	4,944.50	5,500
8360	20" x 20"		8	3.500		10,600	171	23.50	10,794.50	11,900
8380	24" x 6"		6	4.667		2,325	228	31.50	2,584.50	2,925
8400	24" x 12"		6	4.667		6,350	228	31.50	6,609.50	7,350
8420	24" x 18"		6	4.667		11,300	228	31.50	11,559.50	12,800
8440	24" x 24"	↓	6	4.667		13,300	228	31.50	13,559.50	15,000
8800	Hydrant valve box, 6' long	B-20	20	1.200		230	56.50		286.50	340
8820	8' long		18	1.333		305	62.50		367.50	430
8830	Valve box w/lid 4' deep		14	1.714		96.50	80.50		177	227
8840	Valve box and large base w/lid	↓	14	1.714	↓	325	80.50		405.50	480

33 14 19.10 Valves	Crew	Daily Output	Labor-Hours	Unit	Material	2020 Bare Costs Labor	Equipment	Total	Total Incl O&P
0010 **VALVES**, water distribution									
0011 See Sections 22 05 23.20 and 22 05 23.60									
3000 Butterfly valves with boxes, cast iron, mechanical joint									
3100 4" diameter	B-6	6	4	Ea.	550	183	35.50	768.50	925
3140 6" diameter		6	4		725	183	35.50	943.50	1,100
3180 8" diameter		6	4		910	183	35.50	1,128.50	1,325
3300 10" diameter		6	4		1,475	183	35.50	1,693.50	1,925
3340 12" diameter		6	4		1,650	183	35.50	1,868.50	2,150
3400 14" diameter		4	6		2,825	274	53.50	3,152.50	3,575
3440 16" diameter		4	6		4,150	274	53.50	4,477.50	5,050
3460 18" diameter		4	6		4,950	274	53.50	5,277.50	5,925
3480 20" diameter		4	6		6,525	274	53.50	6,852.50	7,650
3500 24" diameter		4	6		10,800	274	53.50	11,127.50	12,400
3510 30" diameter		4	6		11,600	274	53.50	11,927.50	13,300
3520 36" diameter		4	6		14,400	274	53.50	14,727.50	16,300
3530 42" diameter		4	6		18,900	274	53.50	19,227.50	21,300
3540 48" diameter	▼	4	6	▼	24,300	274	53.50	24,627.50	27,200
3600 With lever operator									
3610 4" diameter	B-6	6	4	Ea.	455	183	35.50	673.50	815
3614 6" diameter		6	4		625	183	35.50	843.50	1,000
3616 8" diameter		6	4		815	183	35.50	1,033.50	1,200
3618 10" diameter		6	4		1,375	183	35.50	1,593.50	1,825
3620 12" diameter		6	4		1,550	183	35.50	1,768.50	2,050
3622 14" diameter		4	6		2,500	274	53.50	2,827.50	3,225
3624 16" diameter		4	6		3,825	274	53.50	4,152.50	4,675
3626 18" diameter		4	6		4,625	274	53.50	4,952.50	5,575
3628 20" diameter		4	6		6,200	274	53.50	6,527.50	7,300
3630 24" diameter	▼	4	6	▼	10,500	274	53.50	10,827.50	12,100
3700 Check valves, flanged									
3710 4" diameter	B-6	6	4	Ea.	785	183	35.50	1,003.50	1,175
3714 6" diameter		6	4		1,325	183	35.50	1,543.50	1,775
3716 8" diameter		6	4		2,525	183	35.50	2,743.50	3,100
3718 10" diameter		6	4		4,275	183	35.50	4,493.50	5,025
3720 12" diameter		6	4		7,000	183	35.50	7,218.50	8,025
3726 18" diameter		4	6		25,900	274	53.50	26,227.50	29,000
3730 24" diameter	▼	4	6	▼	66,000	274	53.50	66,327.50	73,000
3800 Gate valves, C.I., 125 psi, mechanical joint, w/boxes									
3810 4" diameter	B-6	6	4	Ea.	780	183	35.50	998.50	1,175
3814 6" diameter		6	4		1,175	183	35.50	1,393.50	1,625
3816 8" diameter		6	4		2,025	183	35.50	2,243.50	2,550
3818 10" diameter		6	4		3,625	183	35.50	3,843.50	4,300
3820 12" diameter		6	4		4,900	183	35.50	5,118.50	5,700
3822 14" diameter		4	6		10,700	274	53.50	11,027.50	12,300
3824 16" diameter		4	6		15,100	274	53.50	15,427.50	17,100
3826 18" diameter		4	6		24,500	274	53.50	24,827.50	27,500
3828 20" diameter		4	6		25,600	274	53.50	25,927.50	28,700
3830 24" diameter		4	6		38,300	274	53.50	38,627.50	42,700
3831 30" diameter		4	6		59,000	274	53.50	59,327.50	65,500
3832 36" diameter	▼	4	6	▼	86,000	274	53.50	86,327.50	95,000

For customer support on your Mechanical Costs with RSMeans Data, call 800.448.8182.

473

33 14 Water Utility Transmission and Distribution

33 14 19 – Valves and Hydrants for Water Utility Service

33 14 19.20 Valves

	33 14 19.20 Valves	Crew	Daily Output	Labor-Hours	Unit	Material	2020 Bare Costs Labor	Equipment	Total	Total Incl O&P
0010	**VALVES**									
0011	Special trim or use									
9000	Valves, gate valve, N.R.S. PIV with post, 4" diameter	B-6	6	4	Ea.	2,000	183	35.50	2,218.50	2,525
9020	6" diameter		6	4		2,400	183	35.50	2,618.50	2,975
9040	8" diameter		6	4		3,250	183	35.50	3,468.50	3,875
9060	10" diameter		6	4		4,850	183	35.50	5,068.50	5,650
9080	12" diameter		6	4		6,125	183	35.50	6,343.50	7,050
9100	14" diameter		6	4		11,700	183	35.50	11,918.50	13,200
9120	OS&Y, 4" diameter		6	4		920	183	35.50	1,138.50	1,325
9140	6" diameter		6	4		1,275	183	35.50	1,493.50	1,725
9160	8" diameter		6	4		1,650	183	35.50	1,868.50	2,150
9180	10" diameter		6	4		2,575	183	35.50	2,793.50	3,175
9200	12" diameter		6	4		4,025	183	35.50	4,243.50	4,750
9220	14" diameter		4	6		6,150	274	53.50	6,477.50	7,250
9400	Check valves, rubber disc, 2-1/2" diameter		6	4		475	183	35.50	693.50	840
9420	3" diameter		6	4		595	183	35.50	813.50	970
9440	4" diameter		6	4		785	183	35.50	1,003.50	1,175
9480	6" diameter		6	4		1,325	183	35.50	1,543.50	1,775
9500	8" diameter		6	4		2,525	183	35.50	2,743.50	3,100
9520	10" diameter		6	4		4,275	183	35.50	4,493.50	5,025
9540	12" diameter		6	4		7,000	183	35.50	7,218.50	8,025
9542	14" diameter		4	6		6,750	274	53.50	7,077.50	7,900
9700	Detector check valves, reducing, 4" diameter		6	4		1,125	183	35.50	1,343.50	1,575
9720	6" diameter		6	4		1,925	183	35.50	2,143.50	2,450
9740	8" diameter		6	4		2,875	183	35.50	3,093.50	3,500
9760	10" diameter		6	4		5,025	183	35.50	5,243.50	5,850
9800	Galvanized, 4" diameter		6	4		2,025	183	35.50	2,243.50	2,550
9820	6" diameter		6	4		2,700	183	35.50	2,918.50	3,275
9840	8" diameter		6	4		4,025	183	35.50	4,243.50	4,750
9860	10" diameter		6	4		5,425	183	35.50	5,643.50	6,275

33 16 Water Utility Storage Tanks

33 16 11 – Elevated Composite Water Storage Tanks

33 16 11.50 Elevated Water Storage Tanks

	33 16 11.50 Elevated Water Storage Tanks	Crew	Daily Output	Labor-Hours	Unit	Material	2020 Bare Costs Labor	Equipment	Total	Total Incl O&P
0010	**ELEVATED WATER STORAGE TANKS**									
0011	Not incl. pipe, pumps or foundation									
3000	Elevated water tanks, 100' to bottom capacity line, incl. painting									
3010	50,000 gallons				Ea.				185,000	204,000
3300	100,000 gallons								280,000	307,500
3400	250,000 gallons								751,500	826,500
3600	500,000 gallons								1,336,000	1,470,000
3700	750,000 gallons								1,622,000	1,783,500
3900	1,000,000 gallons								2,322,000	2,556,000

33 16 23 – Ground-Level Steel Water Storage Tanks

33 16 23.13 Steel Water Storage Tanks

	33 16 23.13 Steel Water Storage Tanks	Crew	Daily Output	Labor-Hours	Unit	Material	2020 Bare Costs Labor	Equipment	Total	Total Incl O&P
0010	**STEEL WATER STORAGE TANKS**									
0910	Steel, ground level, ht./diam. less than 1, not incl. fdn., 100,000 gallons				Ea.				202,000	244,500
1000	250,000 gallons								295,500	324,000
1200	500,000 gallons								417,000	458,500
1250	750,000 gallons								538,000	591,500

For customer support on your Mechanical Costs with RSMeans Data, call 800.448.8182.

33 16 Water Utility Storage Tanks

33 16 23 – Ground-Level Steel Water Storage Tanks

33 16 23.13 Steel Water Storage Tanks

33 16 23.13 Steel Water Storage Tanks	Crew	Daily Output	Labor-Hours	Unit	Material	2020 Bare Costs Labor	Equipment	Total	Total Incl O&P
1300	1,000,000 gallons				Ea.				558,000	725,500
1500	2,000,000 gallons								1,043,000	1,148,000
1600	4,000,000 gallons								2,121,000	2,333,000
1800	6,000,000 gallons								3,095,000	3,405,000
1850	8,000,000 gallons								4,068,000	4,475,000
1910	10,000,000 gallons								5,050,000	5,554,500
2100	Steel standpipes, ht./diam. more than 1,100' to overflow, no fdn.									
2200	500,000 gallons				Ea.				546,500	600,500
2400	750,000 gallons								722,500	794,500
2500	1,000,000 gallons								1,060,500	1,167,000
2700	1,500,000 gallons								1,749,000	1,923,000
2800	2,000,000 gallons								2,327,000	2,559,000

33 16 36 – Ground-Level Reinforced Concrete Water Storage Tanks

33 16 36.16 Prestressed Conc. Water Storage Tanks

33 16 36.16 Prestressed Conc. Water Storage Tanks	Crew	Daily Output	Labor-Hours	Unit	Material	2020 Bare Costs Labor	Equipment	Total	Total Incl O&P
0010	PRESTRESSED CONC. WATER STORAGE TANKS									
0020	Not including fdn., pipe or pumps, 250,000 gallons				Ea.				299,000	329,500
0100	500,000 gallons								487,000	536,000
0300	1,000,000 gallons								707,000	807,500
0400	2,000,000 gallons								1,072,000	1,179,000
0600	4,000,000 gallons								1,706,000	1,877,000
0700	6,000,000 gallons								2,266,000	2,493,000
0750	8,000,000 gallons								2,924,000	3,216,000
0800	10,000,000 gallons								3,533,000	3,886,000

33 16 56 – Ground-Level Plastic Water Storage Cisterns

33 16 56.23 Plastic-Coated Fabric Pillow Water Tanks

33 16 56.23 Plastic-Coated Fabric Pillow Water Tanks	Crew	Daily Output	Labor-Hours	Unit	Material	2020 Bare Costs Labor	Equipment	Total	Total Incl O&P
0010	PLASTIC-COATED FABRIC PILLOW WATER TANKS									
7000	Water tanks, vinyl coated fabric pillow tanks, freestanding, 5,000 gallons	4 Clab	4	8	Ea.	3,800	335		4,135	4,675
7100	Supporting embankment not included, 25,000 gallons	6 Clab	2	24		13,500	1,000		14,500	16,300
7200	50,000 gallons	8 Clab	1.50	42.667		20,500	1,800		22,300	25,300
7300	100,000 gallons	9 Clab	.90	80		37,700	3,375		41,075	46,500
7400	150,000 gallons		.50	144		44,400	6,050		50,450	58,000
7500	200,000 gallons		.40	180		84,500	7,575		92,075	104,500
7600	250,000 gallons		.30	240		89,000	10,100		99,100	113,000

33 31 Sanitary Sewerage Piping

33 31 11 – Public Sanitary Sewerage Gravity Piping

33 31 11.15 Sewage Collection, Concrete Pipe

		Crew	Daily Output	Labor-Hours	Unit	Material	2020 Bare Costs Labor	Equipment	Total	Total Incl O&P
0010	SEWAGE COLLECTION, CONCRETE PIPE									
0020	See Section 33 42 11.60 for sewage/drainage collection, concrete pipe									

33 31 11.25 Sewage Collection, Polyvinyl Chloride Pipe

		Crew	Daily Output	Labor-Hours	Unit	Material	2020 Bare Costs Labor	Equipment	Total	Total Incl O&P
0010	SEWAGE COLLECTION, POLYVINYL CHLORIDE PIPE									
0020	Not including excavation or backfill									
2000	20' lengths, SDR 35, B&S, 4" diameter	B-20	375	.064	L.F.	1.67	3.01		4.68	6.35
2040	6" diameter		350	.069		4.05	3.22		7.27	9.30
2080	13' lengths, SDR 35, B&S, 8" diameter		335	.072		6.70	3.37		10.07	12.40
2120	10" diameter	B-21	330	.085		11.30	4.14	.57	16.01	19.30
2160	12" diameter		320	.088		15.90	4.27	.59	20.76	24.50
2170	14" diameter		280	.100		19.45	4.88	.68	25.01	29.50
2200	15" diameter		240	.117		22	5.70	.79	28.49	33.50
2250	16" diameter		220	.127		29	6.20	.86	36.06	42

For customer support on your Mechanical Costs with RSMeans Data, call 800.448.8182.

475

33 31 Sanitary Sewerage Piping

33 31 11 – Public Sanitary Sewerage Gravity Piping

33 31 11.25 Sewage Collection, Polyvinyl Chloride Pipe	Crew	Daily Output	Labor-Hours	Unit	Material	2020 Bare Costs Labor	Equipment	Total	Total Incl O&P
4000	Piping, DWV PVC, no exc./bkfill., 10' L, Sch 40, 4" diameter	B-20	375	.064	L.F.	3.83	3.01		6.84	8.75
4010	6" diameter		350	.069		8.25	3.22		11.47	13.90
4020	8" diameter		335	.072		13.40	3.37		16.77	19.80

33 41 Subdrainage

33 41 16 – Subdrainage Piping

33 41 16.25 Piping, Subdrainage, Corrugated Metal

		Crew	Daily Output	Labor-Hours	Unit	Material	2020 Bare Costs Labor	Equipment	Total	Total Incl O&P
0010	PIPING, SUBDRAINAGE, CORRUGATED METAL									
0021	Not including excavation and backfill									
2010	Aluminum, perforated									
2020	6" diameter, 18 ga.	B-20	380	.063	L.F.	6.40	2.97		9.37	11.50
2200	8" diameter, 16 ga.	"	370	.065		8.80	3.05		11.85	14.30
2220	10" diameter, 16 ga.	B-21	360	.078		11	3.79	.53	15.32	18.40
2240	12" diameter, 16 ga.		285	.098		12.30	4.79	.66	17.75	21.50
2260	18" diameter, 16 ga.		205	.137		18.50	6.65	.92	26.07	31.50
3000	Uncoated galvanized, perforated									
3020	6" diameter, 18 ga.	B-20	380	.063	L.F.	6.65	2.97		9.62	11.75
3200	8" diameter, 16 ga.	"	370	.065		8.10	3.05		11.15	13.50
3220	10" diameter, 16 ga.	B-21	360	.078		8.60	3.79	.53	12.92	15.75
3240	12" diameter, 16 ga.		285	.098		9.55	4.79	.66	15	18.50
3260	18" diameter, 16 ga.		205	.137		14.65	6.65	.92	22.22	27
4000	Steel, perforated, asphalt coated									
4020	6" diameter, 18 ga.	B-20	380	.063	L.F.	6.35	2.97		9.32	11.45
4030	8" diameter, 18 ga.	"	370	.065		9	3.05		12.05	14.50
4040	10" diameter, 16 ga.	B-21	360	.078		10.05	3.79	.53	14.37	17.35
4050	12" diameter, 16 ga.		285	.098		11.30	4.79	.66	16.75	20.50
4060	18" diameter, 16 ga.		205	.137		17.80	6.65	.92	25.37	30.50

33 41 16.30 Piping, Subdrainage, Plastic

		Crew	Daily Output	Labor-Hours	Unit	Material	2020 Bare Costs Labor	Equipment	Total	Total Incl O&P
0010	PIPING, SUBDRAINAGE, PLASTIC									
0020	Not including excavation and backfill									
2100	Perforated PVC, 4" diameter	B-14	314	.153	L.F.	1.67	6.75	.68	9.10	12.75
2110	6" diameter		300	.160		4.05	7.10	.71	11.86	15.90
2120	8" diameter		290	.166		6.10	7.30	.74	14.14	18.50
2130	10" diameter		280	.171		10.45	7.60	.76	18.81	23.50
2140	12" diameter		270	.178		14.75	7.85	.79	23.39	29

33 42 Stormwater Conveyance

33 42 11 – Stormwater Gravity Piping

33 42 11.40 Piping, Storm Drainage, Corrugated Metal

		Crew	Daily Output	Labor-Hours	Unit	Material	2020 Bare Costs Labor	Equipment	Total	Total Incl O&P
0010	PIPING, STORM DRAINAGE, CORRUGATED METAL									
0020	Not including excavation or backfill									
2000	Corrugated metal pipe, galvanized									
2020	Bituminous coated with paved invert, 20' lengths									
2040	8" diameter, 16 ga.	B-14	330	.145	L.F.	8.35	6.45	.65	15.45	19.50
2060	10" diameter, 16 ga.		260	.185		8.65	8.15	.82	17.62	22.50
2080	12" diameter, 16 ga.		210	.229		12.40	10.10	1.02	23.52	30
2100	15" diameter, 16 ga.		200	.240		14.60	10.60	1.07	26.27	33
2120	18" diameter, 16 ga.		190	.253		18.80	11.20	1.12	31.12	38.50
2140	24" diameter, 14 ga.		160	.300		24.50	13.30	1.34	39.14	48.50

33 42 Stormwater Conveyance

33 42 11 – Stormwater Gravity Piping

33 42 11.40 Piping, Storm Drainage, Corrugated Metal

		Crew	Daily Output	Labor-Hours	Unit	Material	2020 Bare Costs Labor	Equipment	Total	Total Incl O&P
2160	30" diameter, 14 ga.	B-13	120	.467	L.F.	29	21.50	4.84	55.34	69.50
2180	36" diameter, 12 ga.		120	.467		35	21.50	4.84	61.34	76
2200	48" diameter, 12 ga.	↓	100	.560		46.50	26	5.80	78.30	96
2220	60" diameter, 10 ga.	B-13B	75	.747		80	34.50	13.05	127.55	153
2240	72" diameter, 8 ga.	"	45	1.244	↓	84.50	57.50	22	164	203
2500	Galvanized, uncoated, 20' lengths									
2520	8" diameter, 16 ga.	B-14	355	.135	L.F.	7.85	6	.60	14.45	18.25
2540	10" diameter, 16 ga.		280	.171		8.40	7.60	.76	16.76	21.50
2560	12" diameter, 16 ga.		220	.218		9.75	9.65	.97	20.37	26.50
2580	15" diameter, 16 ga.		220	.218		11.05	9.65	.97	21.67	28
2600	18" diameter, 16 ga.		205	.234		13.50	10.35	1.04	24.89	31.50
2620	24" diameter, 14 ga.	↓	175	.274		23	12.15	1.22	36.37	45
2640	30" diameter, 14 ga.	B-13	130	.431		32	19.85	4.47	56.32	70
2660	36" diameter, 12 ga.		130	.431		39	19.85	4.47	63.32	77
2680	48" diameter, 12 ga.	↓	110	.509		54	23.50	5.30	82.80	100
2690	60" diameter, 10 ga.	B-13B	78	.718	↓	84	33	12.55	129.55	155
2780	End sections, 8" diameter	B-14	35	1.371	Ea.	78.50	60.50	6.10	145.10	184
2785	10" diameter		35	1.371		87	60.50	6.10	153.60	193
2790	12" diameter		35	1.371		100	60.50	6.10	166.60	208
2800	18" diameter	↓	30	1.600		117	71	7.10	195.10	243
2810	24" diameter	B-13	25	2.240		197	103	23	323	395
2820	30" diameter		25	2.240		350	103	23	476	565
2825	36" diameter		20	2.800		470	129	29	628	740
2830	48" diameter	↓	10	5.600		940	258	58	1,256	1,475
2835	60" diameter	B-13B	5	11.200		1,725	515	196	2,436	2,900
2840	72" diameter	"	4	14	↓	2,175	645	245	3,065	3,625

33 42 11.60 Sewage/Drainage Collection, Concrete Pipe

		Crew	Daily Output	Labor-Hours	Unit	Material	2020 Bare Costs Labor	Equipment	Total	Total Incl O&P
0010	**SEWAGE/DRAINAGE COLLECTION, CONCRETE PIPE**									
0020	Not including excavation or backfill									
1000	Non-reinforced pipe, extra strength, B&S or T&G joints									
1010	6" diameter	B-14	265.04	.181	L.F.	7.80	8	.81	16.61	21.50
1020	8" diameter		224	.214		8.55	9.50	.95	19	24.50
1030	10" diameter		216	.222		9.50	9.85	.99	20.34	26.50
1040	12" diameter		200	.240		10.05	10.60	1.07	21.72	28
1050	15" diameter		180	.267		14.40	11.80	1.19	27.39	35
1060	18" diameter		144	.333		18.05	14.75	1.48	34.28	43.50
1070	21" diameter		112	.429		18.95	18.95	1.91	39.81	51.50
1080	24" diameter	↓	100	.480	↓	22	21	2.14	45.14	58.50
2000	Reinforced culvert, class 3, no gaskets									
2010	12" diameter	B-14	150	.320	L.F.	16.15	14.15	1.42	31.72	41
2020	15" diameter		150	.320		22	14.15	1.42	37.57	47
2030	18" diameter		132	.364		24.50	16.10	1.62	42.22	53
2035	21" diameter		120	.400		28.50	17.70	1.78	47.98	60
2040	24" diameter	↓	100	.480		28.50	21	2.14	51.64	66
2045	27" diameter	B-13	92	.609		47.50	28	6.30	81.80	101
2050	30" diameter		88	.636		52	29.50	6.60	88.10	108
2060	36" diameter		72	.778		83	36	8.05	127.05	153
2070	42" diameter	B-13B	68	.824		99.50	38	14.40	151.90	182
2080	48" diameter		64	.875		100	40.50	15.30	155.80	187
2085	54" diameter		56	1		147	46	17.50	210.50	249
2090	60" diameter		48	1.167		181	53.50	20.50	255	300
2100	72" diameter	↓	40	1.400	↓	273	64.50	24.50	362	425

For customer support on your Mechanical Costs with RSMeans Data, call 800.448.8182.

477

33 42 11 – Stormwater Gravity Piping

33 42 11.60 Sewage/Drainage Collection, Concrete Pipe	Crew	Daily Output	Labor-Hours	Unit	Material	2020 Bare Costs Labor	Equipment	Total	Total Incl O&P
2120	84" diameter	B-13B	32	1.750	L.F.	405	80.50	30.50	516	600
2140	96" diameter		24	2.333		485	107	41	633	740
2200	With gaskets, class 3, 12" diameter	B-21	168	.167		17.80	8.15	1.13	27.08	33
2220	15" diameter		160	.175		24	8.55	1.18	33.73	40.50
2230	18" diameter		152	.184		27	9	1.24	37.24	45
2235	21" diameter		152	.184		31.50	9	1.24	41.74	49.50
2240	24" diameter		136	.206		35.50	10.05	1.39	46.94	55.50
2260	30" diameter	B-13	88	.636		60	29.50	6.60	96.10	117
2270	36" diameter	"	72	.778		92.50	36	8.05	136.55	164
2290	48" diameter	B-13B	64	.875		114	40.50	15.30	169.80	202
2310	72" diameter	"	40	1.400		294	64.50	24.50	383	450
2330	Flared ends, 12" diameter	B-21	31	.903	Ea.	305	44	6.10	355.10	410
2340	15" diameter		25	1.120		350	54.50	7.55	412.05	475
2400	18" diameter		20	1.400		400	68.50	9.45	477.95	555
2420	24" diameter		14	2		455	97.50	13.50	566	665
2440	36" diameter	B-13	10	5.600		990	258	58	1,306	1,550
3080	Radius pipe, add to pipe prices, 12" to 60" diameter				L.F.	50%				
3090	Over 60" diameter, add				"	20%				
3500	Reinforced elliptical, 8' lengths, C507 class 3									
3520	14" x 23" inside, round equivalent 18" diameter	B-21	82	.341	L.F.	40.50	16.65	2.30	59.45	72
3530	24" x 38" inside, round equivalent 30" diameter	B-13	58	.966		69.50	44.50	10	124	154
3540	29" x 45" inside, round equivalent 36" diameter		52	1.077		106	49.50	11.15	166.65	204
3550	38" x 60" inside, round equivalent 48" diameter		38	1.474		171	68	15.30	254.30	305
3560	48" x 76" inside, round equivalent 60" diameter		26	2.154		215	99	22.50	336.50	410
3570	58" x 91" inside, round equivalent 72" diameter		22	2.545		330	117	26.50	473.50	565
3780	Concrete slotted pipe, class 4 mortar joint									
3800	12" diameter	B-21	168	.167	L.F.	33	8.15	1.13	42.28	49.50
3840	18" diameter	"	152	.184	"	37	9	1.24	47.24	55.50
3900	Concrete slotted pipe, Class 4 O-ring joint									
3940	12" diameter	B-21	168	.167	L.F.	30.50	8.15	1.13	39.78	47
3960	18" diameter	"	152	.184	"	32	9	1.24	42.24	50

33 42 13 – Stormwater Culverts

33 42 13.15 Oval Arch Culverts

		Crew	Daily Output	Labor-Hours	Unit	Material	Labor	Equipment	Total	Total Incl O&P
0010	**OVAL ARCH CULVERTS**									
3000	Corrugated galvanized or aluminum, coated & paved									
3020	17" x 13", 16 ga., 15" equivalent	B-14	200	.240	L.F.	14.80	10.60	1.07	26.47	33.50
3040	21" x 15", 16 ga., 18" equivalent		150	.320		19.25	14.15	1.42	34.82	44
3060	28" x 20", 14 ga., 24" equivalent		125	.384		23	17	1.71	41.71	52.50
3080	35" x 24", 14 ga., 30" equivalent		100	.480		30.50	21	2.14	53.64	68
3100	42" x 29", 12 ga., 36" equivalent	B-13	100	.560		36	26	5.80	67.80	84.50
3120	49" x 33", 12 ga., 42" equivalent		90	.622		43.50	28.50	6.45	78.45	98
3140	57" x 38", 12 ga., 48" equivalent		75	.747		56	34.50	7.75	98.25	122
3160	Steel, plain oval arch culverts, plain									
3180	17" x 13", 16 ga., 15" equivalent	B-14	225	.213	L.F.	13.25	9.45	.95	23.65	30
3200	21" x 15", 16 ga., 18" equivalent		175	.274		17.25	12.15	1.22	30.62	38.50
3220	28" x 20", 14 ga., 24" equivalent		150	.320		27	14.15	1.42	42.57	52.50
3240	35" x 24", 14 ga., 30" equivalent	B-13	108	.519		32	24	5.40	61.40	77
3260	42" x 29", 12 ga., 36" equivalent		108	.519		40	24	5.40	69.40	86
3280	49" x 33", 12 ga., 42" equivalent		92	.609		59.50	28	6.30	93.80	114
3300	57" x 38", 12 ga., 48" equivalent		75	.747		69	34.50	7.75	111.25	136
3320	End sections, 17" x 13"		22	2.545	Ea.	139	117	26.50	282.50	360
3340	42" x 29"		17	3.294	"	425	152	34	611	735

33 42 Stormwater Conveyance

33 42 13 – Stormwater Culverts

33 42 13.15 Oval Arch Culverts	Crew	Daily Output	Labor-Hours	Unit	Material	2020 Bare Costs Labor	Equipment	Total	Total Incl O&P
3360	Multi-plate arch, steel	B-20	1690	.014	Lb.	1.39	.67		2.06	2.54

33 42 33 – Stormwater Curbside Drains and Inlets

33 42 33.13 Catch Basins

		Crew	Daily Output	Labor-Hours	Unit	Material	2020 Bare Costs Labor	Equipment	Total	Total Incl O&P
0010	**CATCH BASINS**									
0011	Not including footing & excavation									
1600	Frames & grates, C.I., 24" square, 500 lb.	B-6	7.80	3.077	Ea.	375	141	27.50	543.50	650
1700	26" D shape, 600 lb.		7	3.429		720	157	30.50	907.50	1,050
1800	Light traffic, 18" diameter, 100 lb.		10	2.400		215	110	21.50	346.50	425
1900	24" diameter, 300 lb.		8.70	2.759		199	126	24.50	349.50	435
2000	36" diameter, 900 lb.		5.80	4.138		765	189	37	991	1,175
2100	Heavy traffic, 24" diameter, 400 lb.		7.80	3.077		295	141	27.50	463.50	565
2200	36" diameter, 1,150 lb.		3	8		935	365	71.50	1,371.50	1,650
2300	Mass. State standard, 26" diameter, 475 lb.		7	3.429		750	157	30.50	937.50	1,100
2400	30" diameter, 620 lb.		7	3.429		340	157	30.50	527.50	645
2500	Watertight, 24" diameter, 350 lb.		7.80	3.077		415	141	27.50	583.50	695
2600	26" diameter, 500 lb.		7	3.429		450	157	30.50	637.50	765
2700	32" diameter, 575 lb.		6	4		960	183	35.50	1,178.50	1,375
2800	3 piece cover & frame, 10" deep,									
2900	1,200 lb., for heavy equipment	B-6	3	8	Ea.	1,200	365	71.50	1,636.50	1,950
3000	Raised for paving 1-1/4" to 2" high									
3100	4 piece expansion ring									
3200	20" to 26" diameter	1 Clab	3	2.667	Ea.	213	112		325	405
3300	30" to 36" diameter	"	3	2.667	"	295	112		407	495
3320	Frames and covers, existing, raised for paving, 2", including									
3340	row of brick, concrete collar, up to 12" wide frame	B-6	18	1.333	Ea.	52	61	11.90	124.90	162
3360	20" to 26" wide frame		11	2.182		78.50	100	19.45	197.95	258
3380	30" to 36" wide frame		9	2.667		97	122	24	243	315
3400	Inverts, single channel brick	D-1	3	5.333		112	250		362	505
3500	Concrete		5	3.200		124	150		274	365
3600	Triple channel, brick		2	8		182	375		557	770
3700	Concrete		3	5.333		156	250		406	550

33 52 Hydrocarbon Transmission and Distribution

33 52 13 – Liquid Hydrocarbon Piping

33 52 13.16 Gasoline Piping

		Crew	Daily Output	Labor-Hours	Unit	Material	2020 Bare Costs Labor	Equipment	Total	Total Incl O&P
0010	**GASOLINE PIPING**									
0020	Primary containment pipe, fiberglass-reinforced									
0030	Plastic pipe 15' & 30' lengths									
0040	2" diameter	Q-6	425	.056	L.F.	7	3.45		10.45	12.85
0050	3" diameter		400	.060		10.65	3.67		14.32	17.20
0060	4" diameter		375	.064		13.85	3.92		17.77	21
0100	Fittings									
0110	Elbows, 90° & 45°, bell ends, 2"	Q-6	24	1	Ea.	40.50	61		101.50	137
0120	3" diameter		22	1.091		56	66.50		122.50	162
0130	4" diameter		20	1.200		70	73.50		143.50	188
0200	Tees, bell ends, 2"		21	1.143		57	70		127	168
0210	3" diameter		18	1.333		68	81.50		149.50	197
0220	4" diameter		15	1.600		81.50	98		179.50	237
0230	Flanges bell ends, 2"		24	1		31.50	61		92.50	126
0240	3" diameter		22	1.091		38.50	66.50		105	142

For customer support on your Mechanical Costs with RSMeans Data, call 800.448.8182.

479

33 52 13 – Liquid Hydrocarbon Piping

33 52 13.16 Gasoline Piping

		Crew	Daily Output	Labor-Hours	Unit	Material	2020 Bare Costs Labor	Equipment	Total	Total Incl O&P
0250	4" diameter	Q-6	20	1.200	Ea.	43.50	73.50		117	158
0260	Sleeve couplings, 2"		21	1.143		12.05	70		82.05	118
0270	3" diameter		18	1.333		18.05	81.50		99.55	142
0280	4" diameter		15	1.600		22.50	98		120.50	172
0290	Threaded adapters, 2"		21	1.143		19.25	70		89.25	126
0300	3" diameter		18	1.333		32	81.50		113.50	158
0310	4" diameter		15	1.600		35.50	98		133.50	186
0320	Reducers, 2"		27	.889		27	54.50		81.50	111
0330	3" diameter		22	1.091		27	66.50		93.50	130
0340	4" diameter	▼	20	1.200	▼	36	73.50		109.50	150
1010	Gas station product line for secondary containment (double wall)									
1100	Fiberglass reinforced plastic pipe 25' lengths									
1120	Pipe, plain end, 3" diameter	Q-6	375	.064	L.F.	29.50	3.92		33.42	38.50
1130	4" diameter		350	.069		33.50	4.20		37.70	43.50
1140	5" diameter		325	.074		35.50	4.52		40.02	46.50
1150	6" diameter	▼	300	.080	▼	39.50	4.89		44.39	51
1200	Fittings									
1230	Elbows, 90° & 45°, 3" diameter	Q-6	18	1.333	Ea.	131	81.50		212.50	266
1240	4" diameter		16	1.500		163	92		255	315
1250	5" diameter		14	1.714		170	105		275	345
1260	6" diameter		12	2		211	122		333	415
1270	Tees, 3" diameter		15	1.600		157	98		255	320
1280	4" diameter		12	2		184	122		306	385
1290	5" diameter		9	2.667		299	163		462	575
1300	6" diameter		6	4		340	245		585	740
1310	Couplings, 3" diameter		18	1.333		54.50	81.50		136	182
1320	4" diameter		16	1.500		117	92		209	266
1330	5" diameter		14	1.714		218	105		323	395
1340	6" diameter		12	2		320	122		442	535
1350	Cross-over nipples, 3" diameter		18	1.333		10.30	81.50		91.80	133
1360	4" diameter		16	1.500		12.55	92		104.55	151
1370	5" diameter		14	1.714		16.30	105		121.30	175
1380	6" diameter		12	2		17.55	122		139.55	202
1400	Telescoping, reducers, concentric 4" x 3"		18	1.333		44.50	81.50		126	171
1410	5" x 4"		17	1.412		94.50	86.50		181	233
1420	6" x 5"	▼	16	1.500	▼	237	92		329	400

33 52 16 – Gas Hydrocarbon Piping

33 52 16.20 Piping, Gas Service and Distribution, P.E.

		Crew	Daily Output	Labor-Hours	Unit	Material	2020 Bare Costs Labor	Equipment	Total	Total Incl O&P
0010	**PIPING, GAS SERVICE AND DISTRIBUTION, POLYETHYLENE**									
0020	Not including excavation or backfill									
1000	60 psi coils, compression coupling @ 100', 1/2" diameter, SDR 11	B-20A	608	.053	L.F.	.49	2.66		3.15	4.53
1010	1" diameter, SDR 11		544	.059		1.14	2.97		4.11	5.70
1040	1-1/4" diameter, SDR 11		544	.059		1.64	2.97		4.61	6.25
1100	2" diameter, SDR 11		488	.066		2.48	3.31		5.79	7.70
1160	3" diameter, SDR 11	▼	408	.078		6.20	3.96		10.16	12.75
1500	60 psi 40' joints with coupling, 3" diameter, SDR 11	B-21A	408	.098		7.35	5.15	1.05	13.55	16.95
1540	4" diameter, SDR 11		352	.114		13.35	5.95	1.22	20.52	25
1600	6" diameter, SDR 11		328	.122		35.50	6.40	1.31	43.21	50
1640	8" diameter, SDR 11	▼	272	.147	▼	54	7.70	1.58	63.28	72.50

33 61 13 – Underground Hydronic Energy Distribution

33 61 13.20 Pipe Conduit, Prefabricated/Preinsulated	Crew	Daily Output	Labor-Hours	Unit	Material	2020 Bare Costs Labor	Equipment	Total	Total Incl O&P
0010 **PIPE CONDUIT, PREFABRICATED/PREINSULATED**									
0020 Does not include trenching, fittings or crane.									
0300 For cathodic protection, add 12 to 14%									
0310 of total built-up price (casing plus service pipe)									
0580 Polyurethane insulated system, 250°F max. temp.									
0620 Black steel service pipe, standard wt., 1/2" insulation									
0660 3/4" diam. pipe size	Q-17	54	.296	L.F.	88.50	17.50	1.97	107.97	126
0670 1" diam. pipe size		50	.320		97.50	18.90	2.13	118.53	138
0680 1-1/4" diam. pipe size		47	.340		109	20	2.26	131.26	152
0690 1-1/2" diam. pipe size		45	.356		118	21	2.36	141.36	164
0700 2" diam. pipe size		42	.381		122	22.50	2.53	147.03	171
0710 2-1/2" diam. pipe size		34	.471		124	28	3.13	155.13	182
0720 3" diam. pipe size		28	.571		144	33.50	3.80	181.30	213
0730 4" diam. pipe size		22	.727		182	43	4.84	229.84	269
0740 5" diam. pipe size		18	.889		232	52.50	5.90	290.40	340
0750 6" diam. pipe size	Q-18	23	1.043		269	64	4.62	337.62	395
0760 8" diam. pipe size		19	1.263		395	77.50	5.60	478.10	555
0770 10" diam. pipe size		16	1.500		500	92	6.65	598.65	695
0780 12" diam. pipe size		13	1.846		625	113	8.20	746.20	865
0790 14" diam. pipe size		11	2.182		695	133	9.65	837.65	975
0800 16" diam. pipe size		10	2.400		800	147	10.65	957.65	1,100
0810 18" diam. pipe size		8	3		920	184	13.30	1,117.30	1,300
0820 20" diam. pipe size		7	3.429		1,025	210	15.20	1,250.20	1,450
0830 24" diam. pipe size		6	4		1,250	245	17.70	1,512.70	1,750
0900 For 1" thick insulation, add					10%				
0940 For 1-1/2" thick insulation, add					13%				
0980 For 2" thick insulation, add					20%				
1500 Gland seal for system, 3/4" diam. pipe size	Q-17	32	.500	Ea.	1,025	29.50	3.33	1,057.83	1,175
1510 1" diam. pipe size		32	.500		1,025	29.50	3.33	1,057.83	1,175
1540 1-1/4" diam. pipe size		30	.533		1,100	31.50	3.55	1,135.05	1,250
1550 1-1/2" diam. pipe size		30	.533		1,100	31.50	3.55	1,135.05	1,250
1560 2" diam. pipe size		28	.571		1,300	33.50	3.80	1,337.30	1,475
1570 2-1/2" diam. pipe size		26	.615		1,400	36.50	4.09	1,440.59	1,600
1580 3" diam. pipe size		24	.667		1,500	39.50	4.43	1,543.93	1,725
1590 4" diam. pipe size		22	.727		1,775	43	4.84	1,822.84	2,025
1600 5" diam. pipe size		19	.842		2,200	49.50	5.60	2,255.10	2,475
1610 6" diam. pipe size	Q-18	26	.923		2,325	56.50	4.09	2,385.59	2,650
1620 8" diam. pipe size		25	.960		2,700	58.50	4.25	2,762.75	3,075
1630 10" diam. pipe size		23	1.043		3,175	64	4.62	3,243.62	3,600
1640 12" diam. pipe size		21	1.143		3,500	70	5.05	3,575.05	3,950
1650 14" diam. pipe size		19	1.263		3,925	77.50	5.60	4,008.10	4,450
1660 16" diam. pipe size		18	1.333		4,650	81.50	5.90	4,737.40	5,225
1670 18" diam. pipe size		16	1.500		4,925	92	6.65	5,023.65	5,575
1680 20" diam. pipe size		14	1.714		5,600	105	7.60	5,712.60	6,325
1690 24" diam. pipe size		12	2		6,200	122	8.85	6,330.85	7,025
2000 Elbow, 45° for system									
2020 3/4" diam. pipe size	Q-17	14	1.143	Ea.	675	67.50	7.60	750.10	850
2040 1" diam. pipe size		13	1.231		690	72.50	8.20	770.70	880
2050 1-1/4" diam. pipe size		11	1.455		785	86	9.65	880.65	1,000
2060 1-1/2" diam. pipe size		9	1.778		790	105	11.80	906.80	1,050
2070 2" diam. pipe size		6	2.667		845	157	17.75	1,019.75	1,175
2080 2-1/2" diam. pipe size		4	4		930	236	26.50	1,192.50	1,400
2090 3" diam. pipe size		3.50	4.571		1,075	270	30.50	1,375.50	1,625

For customer support on your Mechanical Costs with RSMeans Data, call 800.448.8182.

481

33 61 Hydronic Energy Distribution

33 61 13 – Underground Hydronic Energy Distribution

33 61 13.20 Pipe Conduit, Prefabricated/Preinsulated

		Crew	Daily Output	Labor-Hours	Unit	Material	2020 Bare Costs Labor	Equipment	Total	Total Incl O&P
2100	4" diam. pipe size	Q-17	3	5.333	Ea.	1,250	315	35.50	1,600.50	1,875
2110	5" diam. pipe size	↓	2.80	5.714		1,600	335	38	1,973	2,325
2120	6" diam. pipe size	Q-18	4	6		1,825	365	26.50	2,216.50	2,575
2130	8" diam. pipe size		3	8		2,650	490	35.50	3,175.50	3,675
2140	10" diam. pipe size		2.40	10		3,250	610	44.50	3,904.50	4,550
2150	12" diam. pipe size		2	12		4,275	735	53	5,063	5,850
2160	14" diam. pipe size		1.80	13.333		5,325	815	59	6,199	7,150
2170	16" diam. pipe size		1.60	15		6,325	920	66.50	7,311.50	8,400
2180	18" diam. pipe size		1.30	18.462		7,925	1,125	82	9,132	10,500
2190	20" diam. pipe size		1	24		10,000	1,475	106	11,581	13,300
2200	24" diam. pipe size	↓	.70	34.286		12,600	2,100	152	14,852	17,100
2260	For elbow, 90°, add					25%				
2300	For tee, straight, add					85%	30%			
2340	For tee, reducing, add					170%	30%			
2380	For weldolet, straight, add				↓	50%				
2400	Polyurethane insulation, 1"									
2410	FRP carrier and casing									
2420	4"	Q-5	18	.889	L.F.	56.50	52.50		109	141
2422	6"	Q-6	23	1.043		98.50	64		162.50	204
2424	8"		19	1.263		160	77.50		237.50	293
2426	10"		16	1.500		216	92		308	375
2428	12"	↓	13	1.846	↓	282	113		395	480
2430	FRP carrier and PVC casing									
2440	4"	Q-5	18	.889	L.F.	23.50	52.50		76	105
2444	8"	Q-6	19	1.263		48.50	77.50		126	170
2446	10"	↓	16	1.500		71	92		163	215
2448	12"	↓	13	1.846	↓	97	113		210	275
2450	PVC carrier and casing									
2460	4"	Q-1	36	.444	L.F.	12.25	26		38.25	52
2462	6"	"	29	.552		17.05	32		49.05	67
2464	8"	Q-2	36	.667		23.50	40		63.50	86
2466	10"		32	.750		32.50	45		77.50	103
2468	12"	↓	31	.774	↓	34.50	46.50		81	108

33 63 Steam Energy Distribution

33 63 13 – Underground Steam and Condensate Distribution Piping

33 63 13.10 Calcium Silicate Insulated System

		Crew	Daily Output	Labor-Hours	Unit	Material	2020 Bare Costs Labor	Equipment	Total	Total Incl O&P
0010	**CALCIUM SILICATE INSULATED SYSTEM**									
0011	High temp. (1200 degrees F)									
2840	Steel casing with protective exterior coating									
2850	6-5/8" diameter	Q-18	52	.462	L.F.	130	28	2.04	160.04	188
2860	8-5/8" diameter		50	.480		141	29.50	2.13	172.63	201
2870	10-3/4" diameter		47	.511		165	31	2.26	198.26	231
2880	12-3/4" diameter		44	.545		180	33.50	2.42	215.92	251
2890	14" diameter		41	.585		203	36	2.59	241.59	280
2900	16" diameter		39	.615		218	37.50	2.73	258.23	300
2910	18" diameter		36	.667		240	41	2.95	283.95	330
2920	20" diameter		34	.706		271	43	3.13	317.13	365
2930	22" diameter		32	.750		375	46	3.32	424.32	480
2940	24" diameter		29	.828		430	50.50	3.67	484.17	550
2950	26" diameter	↓	26	.923	↓	490	56.50	4.09	550.59	630

33 63 13 – Underground Steam and Condensate Distribution Piping

33 63 13.10 Calcium Silicate Insulated System	Crew	Daily Output	Labor-Hours	Unit	Material	2020 Bare Costs Labor	Equipment	Total	Total Incl O&P
2960	28" diameter	Q-18	23	1.043	L.F.	605	64	4.62	673.62	765
2970	30" diameter		21	1.143		650	70	5.05	725.05	825
2980	32" diameter		19	1.263		730	77.50	5.60	813.10	920
2990	34" diameter		18	1.333		735	81.50	5.90	822.40	940
3000	36" diameter		16	1.500		790	92	6.65	888.65	1,025
3040	For multi-pipe casings, add					10%				
3060	For oversize casings, add					2%				
3400	Steel casing gland seal, single pipe									
3420	6-5/8" diameter	Q-18	25	.960	Ea.	1,975	58.50	4.25	2,037.75	2,275
3440	8-5/8" diameter		23	1.043		2,325	64	4.62	2,393.62	2,650
3450	10-3/4" diameter		21	1.143		2,575	70	5.05	2,650.05	2,950
3460	12-3/4" diameter		19	1.263		3,050	77.50	5.60	3,133.10	3,475
3470	14" diameter		17	1.412		3,375	86.50	6.25	3,467.75	3,850
3480	16" diameter		16	1.500		3,950	92	6.65	4,048.65	4,500
3490	18" diameter		15	1.600		4,400	98	7.10	4,505.10	4,975
3500	20" diameter		13	1.846		4,850	113	8.20	4,971.20	5,525
3510	22" diameter		12	2		5,450	122	8.85	5,580.85	6,200
3520	24" diameter		11	2.182		6,125	133	9.65	6,267.65	6,950
3530	26" diameter		10	2.400		7,000	147	10.65	7,157.65	7,925
3540	28" diameter		9.50	2.526		7,900	155	11.20	8,066.20	8,950
3550	30" diameter		9	2.667		8,050	163	11.80	8,224.80	9,100
3560	32" diameter		8.50	2.824		9,050	173	12.50	9,235.50	10,200
3570	34" diameter		8	3		9,850	184	13.30	10,047.30	11,100
3580	36" diameter		7	3.429		10,500	210	15.20	10,725.20	11,800
3620	For multi-pipe casings, add					5%				
4000	Steel casing anchors, single pipe									
4020	6-5/8" diameter	Q-18	8	3	Ea.	1,775	184	13.30	1,972.30	2,250
4040	8-5/8" diameter		7.50	3.200		1,850	196	14.20	2,060.20	2,325
4050	10-3/4" diameter		7	3.429		2,400	210	15.20	2,625.20	2,975
4060	12-3/4" diameter		6.50	3.692		2,575	226	16.35	2,817.35	3,200
4070	14" diameter		6	4		3,050	245	17.70	3,312.70	3,725
4080	16" diameter		5.50	4.364		3,525	267	19.35	3,811.35	4,300
4090	18" diameter		5	4.800		3,975	294	21.50	4,290.50	4,850
4100	20" diameter		4.50	5.333		4,400	325	23.50	4,748.50	5,350
4110	22" diameter		4	6		4,900	365	26.50	5,291.50	5,950
4120	24" diameter		3.50	6.857		5,325	420	30.50	5,775.50	6,525
4130	26" diameter		3	8		6,025	490	35.50	6,550.50	7,400
4140	28" diameter		2.50	9.600		6,625	585	42.50	7,252.50	8,225
4150	30" diameter		2	12		7,050	735	53	7,838	8,900
4160	32" diameter		1.50	16		8,400	980	71	9,451	10,800
4170	34" diameter		1	24		9,450	1,475	106	11,031	12,700
4180	36" diameter		1	24		10,300	1,475	106	11,881	13,600
4220	For multi-pipe, add					5%	20%			
4800	Steel casing elbow									
4820	6-5/8" diameter	Q-18	15	1.600	Ea.	2,525	98	7.10	2,630.10	2,925
4830	8-5/8" diameter		15	1.600		2,700	98	7.10	2,805.10	3,100
4850	10-3/4" diameter		14	1.714		3,250	105	7.60	3,362.60	3,750
4860	12-3/4" diameter		13	1.846		3,850	113	8.20	3,971.20	4,400
4870	14" diameter		12	2		4,125	122	8.85	4,255.85	4,750
4880	16" diameter		11	2.182		4,500	133	9.65	4,642.65	5,150
4890	18" diameter		10	2.400		5,150	147	10.65	5,307.65	5,875
4900	20" diameter		9	2.667		5,500	163	11.80	5,674.80	6,300
4910	22" diameter		8	3		5,900	184	13.30	6,097.30	6,800

For customer support on your Mechanical Costs with RSMeans Data, call 800.448.8182.

483

33 63 13 – Underground Steam and Condensate Distribution Piping

33 63 13.10 Calcium Silicate Insulated System	Crew	Daily Output	Labor-Hours	Unit	Material	2020 Bare Costs Labor	Equipment	Total	Total Incl O&P
4920	24" diameter	Q-18	7	3.429	Ea.	6,500	210	15.20	6,725.20	7,475
4930	26" diameter		6	4		7,125	245	17.70	7,387.70	8,225
4940	28" diameter		5	4.800		7,700	294	21.50	8,015.50	8,925
4950	30" diameter		4	6		7,725	365	26.50	8,116.50	9,075
4960	32" diameter		3	8		8,775	490	35.50	9,300.50	10,400
4970	34" diameter		2	12		9,650	735	53	10,438	11,800
4980	36" diameter	▼	2	12	▼	10,300	735	53	11,088	12,500
5500	Black steel service pipe, std. wt., 1" thick insulation									
5510	3/4" diameter pipe size	Q-17	54	.296	L.F.	79.50	17.50	1.97	98.97	115
5540	1" diameter pipe size		50	.320		81.50	18.90	2.13	102.53	121
5550	1-1/4" diameter pipe size		47	.340		104	20	2.26	126.26	146
5560	1-1/2" diameter pipe size		45	.356		109	21	2.36	132.36	153
5570	2" diameter pipe size		42	.381		118	22.50	2.53	143.03	166
5580	2-1/2" diameter pipe size		34	.471		124	28	3.13	155.13	181
5590	3" diameter pipe size		28	.571		108	33.50	3.80	145.30	174
5600	4" diameter pipe size		22	.727		121	43	4.84	168.84	202
5610	5" diameter pipe size	▼	18	.889		136	52.50	5.90	194.40	235
5620	6" diameter pipe size	Q-18	23	1.043	▼	147	64	4.62	215.62	262
6000	Black steel service pipe, std. wt., 1-1/2" thick insul.									
6010	3/4" diameter pipe size	Q-17	54	.296	L.F.	94	17.50	1.97	113.47	131
6040	1" diameter pipe size		50	.320		99	18.90	2.13	120.03	140
6050	1-1/4" diameter pipe size		47	.340		107	20	2.26	129.26	149
6060	1-1/2" diameter pipe size		45	.356		106	21	2.36	129.36	150
6070	2" diameter pipe size		42	.381		113	22.50	2.53	138.03	161
6080	2-1/2" diameter pipe size		34	.471		104	28	3.13	135.13	159
6090	3" diameter pipe size		28	.571		115	33.50	3.80	152.30	182
6100	4" diameter pipe size		22	.727		132	43	4.84	179.84	214
6110	5" diameter pipe size	▼	18	.889		157	52.50	5.90	215.40	258
6120	6" diameter pipe size	Q-18	23	1.043		164	64	4.62	232.62	281
6130	8" steam pipe size		19	1.263		208	77.50	5.60	291.10	350
6140	10" diameter pipe size		16	1.500		278	92	6.65	376.65	450
6150	12" diameter pipe size	▼	13	1.846		370	113	8.20	491.20	585
6190	For 2" thick insulation, add					15%				
6220	For 2-1/2" thick insulation, add					25%				
6260	For 3" thick insulation, add				▼	30%				
6800	Black steel service pipe, ex. hvy. wt., 1" thick insul.									
6820	3/4" diameter pipe size	Q-17	50	.320	L.F.	95	18.90	2.13	116.03	136
6840	1" diameter pipe size		47	.340		101	20	2.26	123.26	143
6850	1-1/4" diameter pipe size		44	.364		112	21.50	2.42	135.92	158
6860	1-1/2" diameter pipe size		42	.381		114	22.50	2.53	139.03	161
6870	2" diameter pipe size		40	.400		125	23.50	2.66	151.16	175
6880	2-1/2" diameter pipe size		31	.516		102	30.50	3.43	135.93	162
6890	3" diameter pipe size		27	.593		118	35	3.94	156.94	187
6900	4" diameter pipe size		21	.762		129	45	5.05	179.05	215
6910	5" diameter pipe size	▼	17	.941		156	55.50	6.25	217.75	262
6920	6" diameter pipe size	Q-18	22	1.091	▼	166	66.50	4.83	237.33	288
7400	Black steel service pipe, ex. hvy. wt., 1-1/2" thick insul.									
7420	3/4" diameter pipe size	Q-17	50	.320	L.F.	70	18.90	2.13	91.03	108
7440	1" diameter pipe size		47	.340		88.50	20	2.26	110.76	129
7450	1-1/4" diameter pipe size		44	.364		97.50	21.50	2.42	121.42	142
7460	1-1/2" diameter pipe size		42	.381		96	22.50	2.53	121.03	142
7470	2" diameter pipe size		40	.400		93	23.50	2.66	119.16	140
7480	2-1/2" diameter pipe size	▼	31	.516	▼	95.50	30.50	3.43	129.43	154

33 63 Steam Energy Distribution

33 63 13 – Underground Steam and Condensate Distribution Piping

33 63 13.10 Calcium Silicate Insulated System	Crew	Daily Output	Labor-Hours	Unit	Material	2020 Bare Costs Labor	Equipment	Total	Total Incl O&P
7490	3" diameter pipe size	Q-17	27	.593	L.F.	105	35	3.94	143.94	173
7500	4" diameter pipe size		21	.762		134	45	5.05	184.05	220
7510	5" diameter pipe size	▼	17	.941		185	55.50	6.25	246.75	293
7520	6" diameter pipe size	Q-18	22	1.091		197	66.50	4.83	268.33	320
7530	8" diameter pipe size		18	1.333		270	81.50	5.90	357.40	425
7540	10" diameter pipe size		15	1.600		300	98	7.10	405.10	490
7550	12" diameter pipe size	▼	13	1.846		360	113	8.20	481.20	580
7590	For 2" thick insulation, add					13%				
7640	For 2-1/2" thick insulation, add					18%				
7680	For 3" thick insulation, add				▼	24%				

33 63 13.20 Combined Steam Pipe With Condensate Return

		Crew	Daily Output	Labor-Hours	Unit	Material	2020 Bare Costs Labor	Equipment	Total	Total Incl O&P
0010	**COMBINED STEAM PIPE WITH CONDENSATE RETURN**									
9010	8" and 4" in 24" case	Q-18	100	.240	L.F.	415	14.70	1.06	430.76	480
9020	6" and 3" in 20" case		104	.231		405	14.10	1.02	420.12	465
9030	3" and 1-1/2" in 16" case		110	.218		305	13.35	.97	319.32	355
9040	2" and 1-1/4" in 12-3/4" case		114	.211		255	12.90	.93	268.83	300
9050	1-1/2" and 1-1/4" in 10-3/4" case	▼	116	.207	▼	266	12.65	.92	279.57	315
9100	Steam pipe only (no return)									
9110	6" in 18" case	Q-18	108	.222	L.F.	330	13.60	.98	344.58	385
9120	4" in 14" case		112	.214		255	13.10	.95	269.05	300
9130	3" in 14" case		114	.211		242	12.90	.93	255.83	286
9140	2-1/2" in 12-3/4" case		118	.203		225	12.45	.90	238.35	268
9150	2" in 12-3/4" case	▼	122	.197	▼	213	12.05	.87	225.92	253

For customer support on your Mechanical Costs with RSMeans Data, call 800.448.8182.

485

Division Notes

	CREW	DAILY OUTPUT	LABOR-HOURS	UNIT	BARE COSTS				TOTAL INCL O&P
					MAT.	LABOR	EQUIP.	TOTAL	

Estimating Tips

This section involves equipment and construction costs for air, noise, and odor pollution control systems. These systems may be interrelated and care must be taken that the complete systems are estimated. For example, air pollution equipment may include dust and air-entrained particles that have to be collected. The vacuum systems could be noisy, requiring silencers to reduce noise pollution, and the collected solids have to be disposed of to prevent solid pollution.

Reference Numbers

Reference numbers are shown at the beginning of some major classifications. These numbers refer to related items in the Reference Section. The reference information may be an estimating procedure, an alternate pricing method, or technical information.

Division 44

Pollution & Waste Control Equipment

Same Data. Simplified.

Enjoy the convenience and efficiency of accessing your costs anywhere:

- **Skip the multiplier** by setting your location
- **Quickly search,** edit, favorite and share costs
- **Stay on top of price changes** with automatic updates

Discover more at rsmeans.com/online

44 11 Particulate Control Equipment

44 11 16 - Fugitive Dust Barrier Systems

44 11 16.10 Dust Collection Systems	Crew	Daily Output	Labor-Hours	Unit	Material	2020 Bare Costs Labor	Equipment	Total	Total Incl O&P
0010 **DUST COLLECTION SYSTEMS** Commercial/Industrial									
0120 Central vacuum units									
0130 Includes stand, filters and motorized shaker									
0200 500 CFM, 10" inlet, 2 HP	Q-20	2.40	8.333	Ea.	4,850	475		5,325	6,050
0220 1,000 CFM, 10" inlet, 3 HP		2.20	9.091		5,100	520		5,620	6,375
0240 1,500 CFM, 10" inlet, 5 HP		2	10		5,775	570		6,345	7,250
0260 3,000 CFM, 13" inlet, 10 HP		1.50	13.333		16,100	760		16,860	18,900
0280 5,000 CFM, 16" inlet, 2 @ 10 HP	↓	1	20	↓	17,100	1,150		18,250	20,500
1000 Vacuum tubing, galvanized									
1100 2-1/8" OD, 16 ga.	Q-9	440	.036	L.F.	3.60	2.04		5.64	7.05
1110 2-1/2" OD, 16 ga.		420	.038		3.98	2.14		6.12	7.60
1120 3" OD, 16 ga.		400	.040		4.80	2.24		7.04	8.70
1130 3-1/2" OD, 16 ga.		380	.042		6.80	2.36		9.16	11.10
1140 4" OD, 16 ga.		360	.044		7.40	2.49		9.89	11.90
1150 5" OD, 14 ga.		320	.050		13.30	2.80		16.10	18.85
1160 6" OD, 14 ga.		280	.057		14.85	3.20		18.05	21
1170 8" OD, 14 ga.		200	.080		21.50	4.49		25.99	30.50
1180 10" OD, 12 ga.		160	.100		48.50	5.60		54.10	62
1190 12" OD, 12 ga.		120	.133		51.50	7.50		59	68
1200 14" OD, 12 ga.	↓	80	.200	↓	65	11.20		76.20	88.50
1940 Hose, flexible wire reinforced rubber									
1956 3" diam.	Q-9	400	.040	L.F.	6.90	2.24		9.14	10.95
1960 4" diam.		360	.044		8.10	2.49		10.59	12.70
1970 5" diam.		320	.050		10.05	2.80		12.85	15.30
1980 6" diam.	↓	280	.057	↓	10.70	3.20		13.90	16.60
2000 90° elbow, slip fit									
2110 2-1/8" diam.	Q-9	70	.229	Ea.	13.05	12.80		25.85	34
2120 2-1/2" diam.		65	.246		18.25	13.80		32.05	41
2130 3" diam.		60	.267		24.50	14.95		39.45	49.50
2140 3-1/2" diam.		55	.291		30.50	16.30		46.80	58
2150 4" diam.		50	.320		38	17.95		55.95	69
2160 5" diam.		45	.356		72	19.95		91.95	109
2170 6" diam.		40	.400		99	22.50		121.50	143
2180 8" diam.	↓	30	.533	↓	189	30		219	253
2400 45° elbow, slip fit									
2410 2-1/8" diam.	Q-9	70	.229	Ea.	11.45	12.80		24.25	32
2420 2-1/2" diam.		65	.246		16.85	13.80		30.65	39.50
2430 3" diam.		60	.267		20.50	14.95		35.45	45
2440 3-1/2" diam.		55	.291		25.50	16.30		41.80	52.50
2450 4" diam.		50	.320		33	17.95		50.95	63.50
2460 5" diam.		45	.356		56	19.95		75.95	91.50
2470 6" diam.		40	.400		75.50	22.50		98	117
2480 8" diam.	↓	35	.457	↓	148	25.50		173.50	202
2800 90° TY, slip fit thru 6" diam.									
2810 2-1/8" diam.	Q-9	42	.381	Ea.	21	21.50		42.50	56
2820 2-1/2" diam.		39	.410		27.50	23		50.50	65
2830 3" diam.		36	.444		36.50	25		61.50	78.50
2840 3-1/2" diam.		33	.485		46.50	27		73.50	92.50
2850 4" diam.		30	.533		45.50	30		75.50	95.50
2860 5" diam.		27	.593		74.50	33		107.50	133
2870 6" diam.	↓	24	.667		111	37.50		148.50	179
2880 8" diam., butt end					180			180	198
2890 10" diam., butt end			↓		565			565	625

44 11 Particulate Control Equipment

44 11 16 – Fugitive Dust Barrier Systems

44 11 16.10 Dust Collection Systems		Crew	Daily Output	Labor-Hours	Unit	Material	2020 Bare Costs Labor	Equipment	Total	Total Incl O&P
2900	12" diam., butt end				Ea.	565			565	625
2910	14" diam., butt end					1,025			1,025	1,125
2920	6" x 4" diam., butt end					127			127	140
2930	8" x 4" diam., butt end					241			241	265
2940	10" x 4" diam., butt end					315			315	350
2950	12" x 4" diam., butt end					360			360	395
3100	90° elbow, butt end, segmented									
3110	8" diam., butt end, segmented				Ea.	158			158	174
3120	10" diam., butt end, segmented					470			470	515
3130	12" diam., butt end, segmented					595			595	655
3140	14" diam., butt end, segmented					745			745	820
3200	45° elbow, butt end, segmented									
3210	8" diam., butt end, segmented				Ea.	148			148	163
3220	10" diam., butt end, segmented					340			340	375
3230	12" diam., butt end, segmented					345			345	380
3240	14" diam., butt end, segmented					640			640	705
3400	All butt end fittings require one coupling per joint.									
3410	Labor for fitting included with couplings.									
3460	Compression coupling, galvanized, neoprene gasket									
3470	2-1/8" diam.	Q-9	44	.364	Ea.	13.55	20.50		34.05	46
3480	2-1/2" diam.		44	.364		13.55	20.50		34.05	46
3490	3" diam.		38	.421		19.45	23.50		42.95	57.50
3500	3-1/2" diam.		35	.457		22	25.50		47.50	63
3510	4" diam.		33	.485		24	27		51	67
3520	5" diam.		29	.552		27	31		58	77
3530	6" diam.		26	.615		33	34.50		67.50	89
3540	8" diam.		22	.727		57.50	41		98.50	125
3550	10" diam.		20	.800		93.50	45		138.50	171
3560	12" diam.		18	.889		115	50		165	202
3570	14" diam.		16	1		150	56		206	250
3800	Air gate valves, galvanized									
3810	2-1/8" diam.	Q-9	30	.533	Ea.	122	30		152	180
3820	2-1/2" diam.		28	.571		133	32		165	195
3830	3" diam.		26	.615		141	34.50		175.50	209
3840	4" diam.		23	.696		164	39		203	239
3850	6" diam.		18	.889		228	50		278	325

44 11 96 – Dry Plate Electrostatic Precipitator Equipment

44 11 96.10 Electrostatic Precipitators

		Crew	Daily Output	Labor-Hours	Unit	Material	2020 Bare Costs Labor	Equipment	Total	Total Incl O&P
0010	**ELECTROSTATIC PRECIPITATORS**									
4100	Smoke pollution control									
4110	Exhaust air volume									
4120	1 to 2.4 MCFM	Q-10	2.15	11.163	MCFM	35,500	650		36,150	40,000
4130	2.4 to 4.8 MCFM		2.70	8.889		60,500	515		61,015	67,500
4140	4.8 to 7.2 MCFM		3.60	6.667		82,500	390		82,890	91,500
4150	7.2 to 12 MCFM		5.40	4.444		143,000	258		143,258	158,000

For customer support on your Mechanical Costs with RSMeans Data, call 800.448.8182.

489

Division Notes

		CREW	DAILY OUTPUT	LABOR-HOURS	UNIT	BARE COSTS				TOTAL INCL O&P
						MAT.	LABOR	EQUIP.	TOTAL	

Estimating Tips

This division contains information about water and wastewater equipment and systems, which was formerly located in Division 44. The main areas of focus are total wastewater treatment plants and components of wastewater treatment plants. Also included in this section are oil/water separators for wastewater treatment.

Reference Numbers

Reference numbers are shown at the beginning of some major classifications. These numbers refer to related items in the Reference Section. The reference information may be an estimating procedure, an alternate pricing method, or technical information.

46 23 Grit Removal And Handling Equipment

46 23 23 – Vortex Grit Removal Equipment

46 23 23.10 Rainwater Filters		Crew	Daily Output	Labor-Hours	Unit	Material	2020 Bare Costs Labor	Equipment	Total	Total Incl O&P
0010	**RAINWATER FILTERS**									
0100	42 gal./min	B-21	3.50	8	Ea.	40,600	390	54	41,044	45,300
0200	65 gal./min		3.50	8		40,700	390	54	41,144	45,300
0300	208 gal./min		3.50	8		40,700	390	54	41,144	45,400

46 25 Oil and Grease Separation and Removal Equipment

46 25 13 – Coalescing Oil-Water Separators

46 25 13.20 Oil/Water Separators

		Crew	Daily Output	Labor-Hours	Unit	Material	2020 Bare Costs Labor	Equipment	Total	Total Incl O&P
0010	**OIL/WATER SEPARATORS**									
0020	Underground, tank only									
0030	Excludes excavation, backfill & piping									
0100	200 GPM	B-21	3.50	8	Ea.	40,400	390	54	40,844	45,000
0110	400 GPM		3.25	8.615		85,500	420	58	85,978	94,500
0120	600 GPM		2.75	10.182		90,000	495	68.50	90,563.50	100,500
0130	800 GPM		2.50	11.200		108,000	545	75.50	108,620.50	119,500
0140	1,000 GPM		2	14		122,000	685	94.50	122,779.50	135,500
0150	1,200 GPM		1.50	18.667		134,000	910	126	135,036	148,500
0160	1,500 GPM		1	28		154,000	1,375	189	155,564	171,500

46 25 16 – API Oil-Water Separators

46 25 16.10 Oil/Water Separators

		Crew	Daily Output	Labor-Hours	Unit	Material	2020 Bare Costs Labor	Equipment	Total	Total Incl O&P
0010	**OIL/WATER SEPARATORS**									
0020	Complete system, not including excavation and backfill									
0100	Treated capacity 0.2 C.F./second	B-22	1	30	Ea.	9,225	1,475	284	10,984	12,600
0200	0.5 C.F./second	B-13	.75	74.667		11,800	3,450	775	16,025	19,000
0300	1.0 C.F./second		.60	93.333		17,000	4,300	970	22,270	26,200
0400	2.4 to 3.0 C.F./second		.30	187		24,800	8,600	1,925	35,325	42,300
0500	11 C.F./second		.17	329		29,100	15,200	3,425	47,725	58,500
0600	22 C.F./second	B-13	.10	560		48,500	25,800	5,800	80,100	98,500

Assemblies Section

Table of Contents

Table of Contents

RSMeans data: Assemblies—How They Work

Assemblies estimating provides a fast and reasonably accurate way to develop construction costs. An assembly is the grouping of individual work items—with appropriate quantities—to provide a cost for a major construction component in a convenient unit of measure.

An assemblies estimate is often used during early stages of design development to compare the cost impact of various design alternatives on total building cost.

Assemblies estimates are also used as an efficient tool to verify construction estimates.

Assemblies estimates do not require a completed design or detailed drawings. Instead, they are based on the general size of the structure and other known parameters of the project. The degree of accuracy of an assemblies estimate is generally within +/- 15%.

Most assemblies consist of three major elements: a graphic, the system components, and the cost data itself. The **Graphic** is a visual representation showing the typical appearance of the assembly

❶ Unique 12-character Identifier

Our assemblies are identified by a **unique 12-character identifier**. The assemblies are numbered using UNIFORMAT II, ASTM Standard E1557. The first 5 characters represent this system to Level 3. The last 7 characters represent further breakdown in order to arrange items in understandable groups of similar tasks. Line numbers are consistent across all of our publications, so a line number in any assemblies data set will always refer to the same item.

❷ Reference Box

Information is available in the Reference Section to assist the estimator with estimating procedures, alternate pricing methods, and additional technical information.

The **Reference Box** indicates the exact location of this information in the Reference Section. The "R" stands for "reference," and the remaining characters are the line numbers.

❸ Narrative Descriptions

Our assemblies descriptions appear in two formats: narrative and table. **Narrative descriptions** are shown in a hierarchical structure to make them readable. In order to read a complete description, read up through the indents to the top of the section. Include everything that is above and to the left that is not contradicted by information below.

D30 HVAC

D3010 Energy Supply

Basis for Heat Loss Estimate, Apartment Type Structures:

1. Masonry walls and flat roof are insulated. U factor is assumed at .08.
2. Window glass area taken as BOCA minimum, 1/10th of floor area. Double insulating glass with 1/4" air space, U = .65.
3. Infiltration = 0.3 C.F. per hour per S.F. of net wall.
4. Concrete floor loss is 2 BTUH per S.F.
5. Temperature difference taken as 70°F.
6. Ventilating or makeup air has not been included and must be added if desired. Air shafts are not used.

System Components	QUANTITY	UNIT	COST EACH MAT.	COST EACH INST.	COST EACH TOTAL
❶ SYSTEM D3010 510 1760					
HEATING SYSTEM, FIN TUBE RADIATION, FORCED HOT WATER					
1,000 S.F. AREA, 10,000 C.F. VOLUME					
Boiler, oil fired, CI, burner, ctrls/insul/breech/pipe/ftng/valves, 109 MBH	1.000	Ea.	4,550	4,375	8,925
Circulating pump, CI flange connection, 1/12 HP	1.000	Ea.	520	231	751
Expansion tank, painted steel, ASME 18 Gal capacity	1.000	Ea.	835	101	936
Storage tank, steel, above ground, 275 Gal capacity w/supports	1.000	Ea.	560	283	843
Copper tubing type L, solder joint, hanger 10' OC, 3/4" diam	100.000	L.F.	520	1,015	1,535
Radiation, 3/4" copper tube w/alum fin baseboard pkg, 7" high	30.000	L.F.	265.50	735	1,000.50
Pipe covering, calcium silicate w/cover, 1' wall, 3/4' diam	100.000	L.F.	472	765	1,237
TOTAL			7,722.50	7,505	15,227.50
COST PER S.F.			7.72	7.51	15.23

D3010 510	Apartment Building Heating - Fin Tube Radiation		COST PER S.F. MAT.	COST PER S.F. INST.	COST PER S.F. TOTAL
1740	Heating systems, fin tube radiation, forced hot water				
1760	1,000 S.F. area, 10,000 C.F. volume		7.73	7.50	15.23
1800	10,000 S.F. area, 100,000 C.F. volume		3.50	4.52	8.02
1840	20,000 S.F. area, 200,000 C.F. volume	R235000 -10 ❷	3.81	5.05	8.86
1880 ❸	30,000 S.F. area, 300,000 C.F. volume	R235000 -20	3.69	4.91	8.60
1890					

For supplemental customizable square foot estimating forms, visit: **RSMeans.com/2020books** in question. It is frequently accompanied by additional explanatory technical information describing the class of items. The **System Components** is a listing of the individual tasks that make up the assembly, including the quantity and unit of measure for each item, along with the cost of material and installation. The **Assemblies** data below lists prices for other similar systems with dimensional and/or size variations.

All of our assemblies costs represent the cost for the installing contractor. An allowance for profit has been added to all material, labor, and equipment rental costs. A markup for labor burdens, including workers' compensation, fixed overhead, and business overhead, is included with installation costs.

The information in RSMeans cost data represents a "national average" cost. This data should be modified to the project location using the **City Cost Indexes** or **Location Factors** tables found in the Reference Section.

D30 HVAC

D3010 Energy Supply

Basis for Heat Loss Estimate, Apartment Type Structures:

1. Masonry walls and flat roof are insulated. U factor is assumed at .08.
2. Window glass area taken as BOCA minimum, 1/10th of floor area. Double insulating glass with 1/4" air space, U = .65.
3. Infiltration = 0.3 C.F. per hour per S.F. of net wall.
4. Concrete floor loss is 2 BTUH per S.F.
5. Temperature difference taken as 70°F.
6. Ventilating or makeup air has not been included and must be added if desired. Air shafts are not used.

4 Unit of Measure

All RSMeans data: Assemblies include a typical **Unit of Measure** used for estimating that item. For instance, for continuous footings or foundation walls the unit is linear feet (L.F.). For spread footings the unit is each (Ea.). The estimator needs to take special care that the unit in the data matches the unit in the takeoff. Abbreviations and unit conversions can be found in the Reference Section.

5 System Components

System components are listed separately to detail what is included in the development of the total system price.

System Components	QUANTITY	UNIT	COST EACH MAT.	COST EACH INST.	COST EACH TOTAL
SYSTEM D3010 510 1760					
HEATING SYSTEM, FIN TUBE RADIATION, FORCED HOT WATER					
1,000 S.F. AREA, 10,000 C.F. VOLUME					
Boiler, oil fired, CI, burner, ctrls/insul/breech/pipe/fttng/valves, 109 MBH	1.000	Ea.	4,550	4,375	8,925
Circulating pump, CI flange connection, 1/12 HP	1.000	Ea.	520	231	751
Expansion tank, painted steel, ASME 18 Gal capacity	1.000	Ea.	835	101	936
Storage tank, steel, above ground, 275 Gal capacity w/supports	1.000	Ea.	560	283	843
Copper tubing type L, solder joint, hanger 10' OC, 3/4" diam	100.000	L.F.	520	1,015	1,535
Radiation, 3/4" copper tube w/alum fin baseboard pkg, 7" high	30.000	L.F.	265.50	735	1,000.50
Pipe covering, calcium silicate w/cover, 1' wall, 3/4' diam	100.000	L.F.	472	765	1,237
TOTAL			7,722.50	7,505	15,227.50
COST PER S.F.			7.72	7.51	15.23

D3010 510	Apartment Building Heating - Fin Tube Radiation		COST PER S.F. MAT.	COST PER S.F. INST.	COST PER S.F. TOTAL
1740	Heating systems, fin tube radiation, forced hot water				
1760	1,000 S.F. area, 10,000 C.F. volume		7.73	7.50	15.23
1800	10,000 S.F. area, 100,000 C.F. volume	R235000 -10	3.50	4.52	8.02
1840	20,000 S.F. area, 200,000 C.F. volume		3.81	5.05	8.86
1880	30,000 S.F. area, 300,000 C.F. volume	R235000 -20	3.69	4.91	8.60
1890					

RSMeans data: Assemblies— How They Work (Continued)

Sample Estimate

This sample demonstrates the elements of an estimate, including a tally of the RSMeans data lines. Published assemblies costs include all markups for labor burden and profit for the installing contractor. This estimate adds a summary of the markups applied by a general contractor on the installing contractor's work. These figures represent the total cost to the owner. The location factor with RSMeans data is applied at the bottom of the estimate to adjust the cost of the work to a specific location.

Project Name:	Interior Fit-out, ABC Office			
Location:	Anywhere, USA	Date: 1/1/2020		STD
Assembly Number	Description	Qty.	Unit	Subtotal
❶ C1010 124 1200	Wood partition, 2 x 4 @ 16" OC w/5/8" FR gypsum board	560	S.F.	$2,996.00
C1020 114 1800	Metal door & frame, flush hollow core, 3'-0" x 7'-0"	2	Ea.	$2,620.00
C3010 230 0080	Painting, brushwork, primer & 2 coats	1,120	S.F.	$1,467.20
C3020 410 0140	Carpet, tufted, nylon, roll goods, 12' wide, 26 oz	240	S.F.	$928.80
C3030 210 6000	Acoustic ceilings, 24" x 48" tile, tee grid suspension	200	S.F.	$1,326.00
D5020 125 0560	Receptacles incl plate, box, conduit, wire, 20 A duplex	8	Ea.	$2,428.00
D5020 125 0720	Light switch incl plate, box, conduit, wire, 20 A single pole	2	Ea.	$591.00
D5020 210 0560	Fluorescent fixtures, recess mounted, 20 per 1000 SF	200	S.F.	$2,320.00
	Assembly Subtotal			**$14,677.00**
	Sales Tax @ **❷**	5 %		$ 366.93
	General Requirements @ **❸**	7 %		$ 1,027.39
	Subtotal A			**$16,071.32**
	GC Overhead @ **❹**	5 %		$ 803.57
	Subtotal B			**$16,874.88**
	GC Profit @ **❺**	5 %		$ 843.74
	Subtotal C			**$17,718.62**
	Adjusted by Location Factor **❻**	115.5		$ 20,465.01
	Architects Fee @ **❼**	8 %		$ 1,637.20
	Contingency @ **❽**	15 %		$ 3,069.75
	Project Total Cost			**$ 25,171.96**

This estimate is based on an interactive spreadsheet. You are free to download it and adjust it to your methodology.
A copy of this spreadsheet is available at **RSMeans.com/2020books.**

1. Work Performed

The body of the estimate shows the RSMeans data selected, including line numbers, a brief description of each item, its takeoff quantity and unit, and the total installed cost, including the installing contractor's overhead and profit.

2. Sales Tax

If the work is subject to state or local sales taxes, the amount must be added to the estimate. In a conceptual estimate it can be assumed that one half of the total represents material costs. Therefore, apply the sales tax rate to 50% of the assembly subtotal.

3. General Requirements

This item covers project-wide needs provided by the general contractor. These items vary by project but may include temporary facilities and utilities, security, testing, project cleanup, etc. In assemblies estimates a percentage is used—typically between 5% and 15% of project cost.

4. General Contractor Overhead

This entry represents the general contractor's markup on all work to cover project administration costs.

5. General Contractor Profit

This entry represents the GC's profit on all work performed. The value included here can vary widely by project and is influenced by the GC's perception of the project's financial risk and market conditions.

6. Location Factor

RSMeans published data are based on national average costs. If necessary, adjust the total cost of the project using a location factor from the "Location Factor" table or the "City Cost Indexes" table found in the Reference Section. Use location factors if the work is general, covering the work of multiple trades. If the work is by a single trade (e.g., masonry) use the more specific data found in the City Cost Indexes.

To adjust costs by location factors, multiply the base cost by the factor and divide by 100.

7. Architect's Fee

If appropriate, add the design cost to the project estimate. These fees vary based on project complexity and size. Typical design and engineering fees can be found in the Reference Section.

8. Contingency

A factor for contingency may be added to any estimate to represent the cost of unknowns that may occur between the time that the estimate is performed and the time the project is constructed. The amount of the allowance will depend on the stage of design at which the estimate is done, as well as the contractor's assessment of the risk involved.

Same Data. Simplified.

Enjoy the convenience and efficiency of accessing your costs anywhere:

- **Skip the multiplier** by setting your location
- **Quickly search**, edit, favorite and share costs
- **Stay on top of price changes** with automatic updates

Discover more at rsmeans.com/online

A20 Basement Construction

A2020 Basement Walls

A2020 220	Subdrainage Piping	COST PER L.F.		
		MAT.	INST.	TOTAL
2000	Piping, excavation & backfill excluded, PVC, perforated			
2110	3" diameter	1.84	4.53	6.37
2130	4" diameter	1.84	4.53	6.37
2140	5" diameter	4.46	4.86	9.32
2150	6" diameter	4.46	4.86	9.32
3000	Metal alum. or steel, perforated asphalt coated			
3150	6" diameter	7.05	4.47	11.52
3160	8" diameter	9.70	4.59	14.29
3170	10" diameter	12.10	6.30	18.40
3180	12" diameter	13.55	7.95	21.50
3220	18" diameter	20.50	11	31.50
4000	Corrugated HDPE tubing			
4130	4" diameter	.83	.84	1.67
4150	6" diameter	2.30	1.12	3.42
4160	8" diameter	4.52	1.45	5.97
4180	12" diameter	7.55	5	12.55
4200	15" diameter	11.15	5.65	16.80
4220	18" diameter	13.70	8.20	21.90

In this closed-loop indirect collection system, fluid with a low freezing temperature, such as propylene glycol, transports heat from the collectors to water storage. The transfer fluid is contained in a closed-loop consisting of collectors, supply and return piping, and a remote heat exchanger. The heat exchanger transfers heat energy from the

fluid in the collector loop to potable water circulated in a storage loop. A typical two-or-three panel system contains 5 to 6 gallons of heat transfer fluid.

When the collectors become approximately 20°F warmer than the storage temperature, a controller activates the circulator on the collector and storage

loops. The circulators will move the fluid and potable water through the heat exchanger until heat collection no longer occurs. At that point, the system shuts down. Since the heat transfer medium is a fluid with a very low freezing temperature, there is no need for it to be drained from the system between periods of collection.

D2020 Domestic Water Distribution

System Components	QUANTITY	UNIT	COST EACH		
			MAT.	INST.	TOTAL
SYSTEM D2020 265 2760					
SOLAR, CLOSED LOOP, ADD-ON HOT WATER SYS., EXTERNAL HEAT EXCHANGER					
3/4" TUBING, TWO 3'X7' BLACK CHROME COLLECTORS					
A,B,G,L,K,M Heat exchanger fluid-fluid pkg incl 2 circulators, expansion tank,					
Check valve, relief valve, controller, hi temp cutoff, & 2 sensors	1.000	Ea.	890	555	1,445
C Thermometer, 2" dial	3.000	Ea.	75	144	219
D, T Fill & drain valve, brass, 3/4" connection	1.000	Ea.	13.25	32	45.25
E Air vent, manual, 1/8" fitting	2.000	Ea.	5.92	48	53.92
F Air purger	1.000	Ea.	51	64.50	115.50
H Strainer, Y type, bronze body, 3/4" IPS	1.000	Ea.	63	41.50	104.50
I Valve, gate, bronze, NRS, soldered 3/4" diam	6.000	Ea.	489	231	720
J Neoprene vent flashing	2.000	Ea.	23	77	100
N-1, N Relief valve temp & press, 150 psi 210°F self-closing 3/4" IPS	1.000	Ea.	28	25.50	53.50
O Pipe covering, urethane, ultraviolet cover, 1" wall 3/4" diam	20.000	L.F.	62	141	203
P Pipe covering, fiberglass, all service jacket, 1" wall, 3/4" diam	50.000	L.F.	54	282.50	336.50
Q Collector panel solar energy blk chrome on copper, 1/8" temp glass 3'x7'	2.000	Ea.	2,350	292	2,642
Roof clamps for solar energy collector panels	2.000	Set	7.20	39.70	46.90
R Valve, swing check, bronze, regrinding disc, 3/4" diam	2.000	Ea.	206	77	283
S Pressure gauge, 60 psi, 2" dial	1.000	Ea.	27.50	24	51.50
U Valve, water tempering, bronze, sweat connections, 3/4" diam	1.000	Ea.	183	38.50	221.50
W-2, V Tank water storage w/heating element, drain, relief valve, existing	1.000	Ea.			
Copper tubing type L, solder joint, hanger 10' OC 3/4" diam	20.000	L.F.	104	203	307
Copper tubing, type M, solder joint, hanger 10' OC 3/4" diam	70.000	L.F.	388.50	693	1,081.50
Sensor wire, #22-2 conductor multistranded	.500	C.L.F.	6.50	36.50	43
Solar energy heat transfer fluid, propylene glycol anti-freeze	6.000	Gal.	153	165	318
Wrought copper fittings & solder, 3/4" diam	76.000	Ea.	224.20	3,078	3,302.20
TOTAL			5,404.07	6,288.70	11,692.77

D2020 265	Solar, Closed Loop, Add-On Hot Water Systems		COST EACH		
			MAT.	INST.	TOTAL
2550	Solar, closed loop, add-on hot water system, external heat exchanger				
2570	3/8" tubing, 3 ea. 4' x 4'-4" vacuum tube collectors		5,950	5,875	11,825
2580	1/2" tubing, 4 ea. 4 x 4'-4" vacuum tube collectors	R235616 -60	6,600	6,375	12,975
2600	2 ea. 3'x7' black chrome collectors		4,950	6,000	10,950
2620	3 ea. 3'x7' black chrome collectors		6,125	6,175	12,300
2640	2 ea. 3'x7' flat black collectors		4,375	6,025	10,400
2660	3 ea. 3'x7' flat black collectors		5,275	6,200	11,475
2700	3/4" tubing, 3 ea. 3'x7' black chrome collectors		6,575	6,450	13,025
2720	3 ea. 3'x7' flat black absorber plate collectors		5,750	6,475	12,225
2740	2 ea. 4'x9' flat black w/plastic glazing collectors		5,300	6,500	11,800
2760	2 ea. 3'x7' black chrome collectors		5,400	6,300	11,700
2780	1" tubing, 4 ea 2'x9' plastic absorber & glazing collectors		6,650	7,300	13,950
2800	4 ea. 3'x7' black chrome absorber collectors		8,700	7,325	16,025
2820	4 ea. 3'x7' flat black absorber collectors		7,575	7,350	14,925

For customer support on your Mechanical Costs with RSMeans Data, call 800.448.8182.

505

In the drainback indirect-collection system, the heat transfer fluid is distilled water contained in a loop consisting of collectors, supply and return piping, and an unpressurized holding tank. A large heat exchanger containing incoming potable water is immersed in the holding tank. When a controller activates solar collection, the distilled water is pumped through the

collectors and heated and pumped back down to the holding tank. When the temperature differential between the water in the collectors and water in storage is such that collection no longer occurs, the pump turns off and gravity causes the distilled water in the collector loop to drain back to the holding tank. All the loop piping is pitched so that the water can drain out

of the collectors and piping and not freeze there. As hot water is needed in the home, incoming water first flows through the holding tank with the immersed heat exchanger and is warmed and then flows through a conventional heater for any supplemental heating that is necessary.

506

For customer support on your Mechanical Costs with RSMeans Data, call 800.448.8182.

D2020 Domestic Water Distribution

System Components	QUANTITY	UNIT	COST EACH		
			MAT.	INST.	TOTAL
SYSTEM D2020 270 2760					
SOLAR, DRAINBACK, ADD ON, HOT WATER, IMMERSED HEAT EXCHANGER					
3/4″ TUBING, THREE EA 3′X7′ BLACK CHROME COLLECTOR					
A, B Differential controller 2 sensors, thermostat, solar energy system	1.000	Ea.	169	64.50	233.50
C Thermometer 2″ dial	3.000	Ea.	75	144	219
D, T Fill & drain valve, brass, 3/4″ connection	1.000	Ea.	13.25	32	45.25
E-1 Automatic air vent 1/8″ fitting	1.000	Ea.	19.70	24	43.70
H Strainer, Y type, bronze body, 3/4″ IPS	1.000	Ea.	63	41.50	104.50
I Valve, gate, bronze, NRS, soldered 3/4″ diam	2.000	Ea.	163	77	240
J Neoprene vent flashing	2.000	Ea.	23	77	100
L Circulator, solar heated liquid, 1/20 HP	1.000	Ea.	310	116	426
N Relief valve temp. & press. 150 psi 210°F self-closing 3/4″ IPS	1.000	Ea.	28	25.50	53.50
O Pipe covering, urethane, ultraviolet cover, 1″ wall, 3/4″ diam	20.000	L.F.	62	141	203
P Pipe covering, fiberglass, all service jacket, 1″ wall, 3/4″ diam	50.000	L.F.	54	282.50	336.50
Q Collector panel solar energy blk chrome on copper, 1/8″ temp glas 3′x7′	3.000	Ea.	3,525	438	3,963
Roof clamps for solar energy collector panels	3.000	Set	10.80	59.55	70.35
R Valve, swing check, bronze, regrinding disc, 3/4″ diam	1.000	Ea.	103	38.50	141.50
U Valve, water tempering, bronze sweat connections, 3/4″ diam	1.000	Ea.	183	38.50	221.50
V Tank, water storage w/heating element, drain, relief valve, existing	1.000	Ea.			
W Tank, water storage immersed heat exchr elec 2″x1/2# insul 120 gal	1.000	Ea.	2,100	550	2,650
X Valve, globe, bronze, rising stem, 3/4″ diam, soldered	3.000	Ea.	567	115.50	682.50
Y Flow control valve	1.000	Ea.	154	35	189
Z Valve, ball, bronze, solder 3/4″ diam, solar loop flow control	1.000	Ea.	39	38.50	77.50
Copper tubing, type L, solder joint, hanger 10′ OC 3/4″ diam	20.000	L.F.	104	203	307
Copper tubing, type M, solder joint, hanger 10′ OC 3/4″ diam	70.000	L.F.	388.50	693	1,081.50
Sensor wire, #22-2 conductor, multistranded	.500	C.L.F.	6.50	36.50	43
Wrought copper fittings & solder, 3/4″ diam	76.000	Ea.	224.20	3,078	3,302.20
TOTAL			8,384.95	6,349.05	14,734

D2020 270	Solar, Drainback, Hot Water Systems		COST EACH		
			MAT.	INST.	TOTAL
2550	Solar, drainback, hot water, immersed heat exchanger	R235616 -60			
2560	3/8″ tubing, 3 ea. 4′ x 4′-4″ vacuum tube collectors		7,650	5,725	13,375
2580	1/2″ tubing, 4 ea 4′ x 4′-4″ vacuum tube collectors, 80 gal tank		8,300	6,200	14,500
2600	120 gal tank		8,250	6,275	14,525
2640	2 ea. 3′x7′ blk chrome collectors, 80 gal tank		6,625	5,825	12,450
2660	3 ea. 3′x7′ blk chrome collectors, 120 gal tank		7,750	6,075	13,825
2700	2 ea. 3′x7′ flat blk collectors, 120 gal tank		6,025	5,925	11,950
2720	3 ea. 3′x7′ flat blk collectors, 120 gal tank		6,925	6,100	13,025
2760	3/4″ tubing, 3 ea 3′x7′ black chrome collectors, 120 gal tank		8,375	6,350	14,725
2780	3 ea. 3′x7′ flat black absorber collectors, 120 gal tank		7,550	6,375	13,925
2800	2 ea. 4′x9′ flat blk w/plastic glazing collectors 120 gal tank		7,100	6,400	13,500
2840	1″ tubing, 4 ea. 2′x9′ plastic absorber & glazing collectors, 120 gal tank		8,550	7,200	15,750
2860	4 ea. 3′x7′ black chrome absorber collectors, 120 gal tank		10,600	7,225	17,825
2880	4 ea. 3′x7′ flat black absorber collectors, 120 gal tank		9,475	7,250	16,725

In the draindown direct-collection system, incoming domestic water is heated in the collectors. When the controller activates solar collection, domestic water is first heated as it flows through the collectors and is then pumped to storage. When conditions are no longer suitable for heat collection, the pump shuts off and the water in the loop drains down and out of the system by means of solenoid valves and properly pitched piping.

D2020 Domestic Water Distribution

System Components	QUANTITY	UNIT	COST EACH		
			MAT.	INST.	TOTAL
SYSTEM D2020 275 2760					
SOLAR, DRAINDOWN, HOT WATER, DIRECT COLLECTION					
3/4″ TUBING, THREE 3′X7′ BLACK CHROME COLLECTORS					
A, B Differential controller, 2 sensors, thermostat, solar energy system	1.000	Ea.	169	64.50	233.50
A-1 Solenoid valve, solar heating loop, brass, 3/4″ diam, 24 volts	3.000	Ea.	450	256.50	706.50
B-1 Solar energy sensor, freeze prevention	1.000	Ea.	27.50	24	51.50
C Thermometer, 2″ dial	3.000	Ea.	75	144	219
E-1 Vacuum relief valve, 3/4″ diam	1.000	Ea.	34	24	58
F-1 Air vent, automatic, 1/8″ fitting	1.000	Ea.	19.70	24	43.70
H Strainer, Y type, bronze body, 3/4″ IPS	1.000	Ea.	63	41.50	104.50
I Valve, gate, bronze, NRS, soldered, 3/4″ diam	2.000	Ea.	163	77	240
J Vent flashing neoprene	2.000	Ea.	23	77	100
K Circulator, solar heated liquid, 1/25 HP	1.000	Ea.	225	99	324
N Relief valve temp & press 150 psi 210°F self-closing 3/4″ IPS	1.000	Ea.	28	25.50	53.50
O Pipe covering, urethane, ultraviolet cover, 1″ wall, 3/4″ diam	20.000	L.F.	62	141	203
P Pipe covering, fiberglass, all service jacket, 1″ wall, 3/4″ diam	50.000	L.F.	54	282.50	336.50
Roof clamps for solar energy collector panels	3.000	Set	10.80	59.55	70.35
Q Collector panel solar energy blk chrome on copper, 1/8″ temp glass 3′x7′	3.000	Ea.	3,525	438	3,963
R Valve, swing check, bronze, regrinding disc, 3/4″ diam, soldered	2.000	Ea.	206	77	283
T Drain valve, brass, 3/4″ connection	2.000	Ea.	26.50	64	90.50
U Valve, water tempering, bronze, sweat connections, 3/4″ diam	1.000	Ea.	183	38.50	221.50
W-2, W Tank, water storage elec elem 2″x1/2# insul 120 gal	1.000	Ea.	2,100	550	2,650
X Valve, globe, bronze, rising stem, 3/4″ diam, soldered	1.000	Ea.	189	38.50	227.50
Copper tubing, type L, solder joints, hangers 10′ OC 3/4″ diam	20.000	L.F.	104	203	307
Copper tubing, type M, solder joints, hangers 10′ OC 3/4″ diam	70.000	L.F.	388.50	693	1,081.50
Sensor wire, #22-2 conductor, multistranded	.500	C.L.F.	6.50	36.50	43
Wrought copper fittings & solder, 3/4″ diam	76.000	Ea.	224.20	3,078	3,302.20
TOTAL			8,356.70	6,556.55	14,913.25

D2020 275	Solar, Draindown, Hot Water Systems		COST EACH		
			MAT.	INST.	TOTAL
2550	Solar, draindown, hot water				
2560	3/8″ tubing, 3 ea. 4′ x 4′-4″ vacuum tube collectors, 80 gal tank		7,400	6,250	13,650
2580	1/2″ tubing, 4 ea. 4′ x 4′-4″ vacuum tube collectors, 80 gal tank	R235616 -60	8,400	6,425	14,825
2600	120 gal tank		8,350	6,475	14,825
2640	2 ea. 3′x7′ black chrome collectors, 80 gal tank		6,750	6,050	12,800
2660	3 ea. 3′x7′ black chrome collectors, 120 gal tank		7,875	6,300	14,175
2700	2 ea. 3′x7′ flat black collectors, 120 gal tank		6,150	6,150	12,300
2720	3 ea. 3′x7′ flat black collectors, 120 gal tank		6,950	6,275	13,225
2760	3/4″ tubing, 3 ea. 3′x7′ black chrome collectors, 120 gal tank		8,350	6,550	14,900
2780	3 ea. 3′x7′ flat collectors, 120 gal tank		7,525	6,575	14,100
2800	2 ea. 4′x9′ flat black & plastic glazing collectors, 120 gal tank		7,075	6,600	13,675
2840	1″ tubing, 4 ea. 2′x9′ plastic absorber & glazing collectors, 120 gal tank		12,300	8,700	21,000
2860	4 ea. 3′x7′ black chrome absorber collectors, 120 gal tank		11,900	8,225	20,125
2880	4 ea. 3′x7′ flat black absorber collectors, 120 gal tank		9,700	7,450	17,150

In the recirculation system, (a direct-collection system), incoming domestic water is heated in the collectors. When the controller activates solar collection, domestic water is heated as it flows through the collectors and then it flows back to storage. When conditions are not suitable for heat collection, the pump shuts off and the flow of the water stops. In this type of system, water remains in the collector loop at all times. A "frost sensor" at the collector activates the circulation of warm water from storage through the collectors when protection from freezing is required.

D2020 Domestic Water Distribution

System Components	QUANTITY	UNIT	COST EACH		
			MAT.	INST.	TOTAL
SYSTEM D2020 280 2820					
SOLAR, RECIRCULATION, HOT WATER					
3/4" TUBING, TWO 3'X7' BLACK CHROME COLLECTORS					
A, B Differential controller 2 sensors, thermostat, for solar energy system	1.000	Ea.	169	64.50	233.50
A-1 Solenoid valve, solar heating loop, brass, 3/4" IPS, 24 volts	2.000	Ea.	300	171	471
B-1 Solar energy sensor freeze prevention	1.000	Ea.	27.50	24	51.50
C Thermometer, 2" dial	3.000	Ea.	75	144	219
D Drain valve, brass, 3/4" connection	1.000	Ea.	13.25	32	45.25
F-1 Air vent, automatic, 1/8" fitting	1.000	Ea.	19.70	24	43.70
H Strainer, Y type, bronze body, 3/4" IPS	1.000	Ea.	63	41.50	104.50
I Valve, gate, bronze, 125 lb, soldered 3/4" diam	3.000	Ea.	244.50	115.50	360
J Vent flashing, neoprene	2.000	Ea.	23	77	100
L Circulator, solar heated liquid, 1/20 HP	1.000	Ea.	310	116	426
N Relief valve, temp & press 150 psi 210° F self-closing 3/4" IPS	2.000	Ea.	56	51	107
O Pipe covering, urethane, ultraviolet cover, 1" wall, 3/4" diam	20.000	L.F.	62	141	203
P Pipe covering, fiberglass, all service jacket, 1" wall 3/4" diam	50.000	L.F.	54	282.50	336.50
Q Collector panel solar energy blk chrome on copper, 1/8" temp glass 3'x7'	2.000	Ea.	2,350	292	2,642
Roof clamps for solar energy collector panels	2.000	Set	7.20	39.70	46.90
R Valve, swing check, bronze, 125 lb, regrinding disc, soldered 3/4" diam	2.000	Ea.	206	77	283
U Valve, water tempering, bronze, sweat connections, 3/4" diam	1.000	Ea.	183	38.50	221.50
V Tank, water storage, w/heating element, drain, relief valve, existing	1.000	Ea.			
X Valve, globe, bronze, 125 lb, 3/4" diam	2.000	Ea.	378	77	455
Copper tubing, type L, solder joints, hangers 10' OC 3/4" diam	20.000	L.F.	104	203	307
Copper tubing, type M, solder joints, hangers 10' OC 3/4" diam	70.000	L.F.	388.50	693	1,081.50
Wrought copper fittings & solder, 3/4" diam	76.000	Ea.	224.20	3,078	3,302.20
Sensor wire, #22-2 conductor, multistranded	.500	C.L.F.	6.50	36.50	43
TOTAL			5,264.35	5,818.70	11,083.05

D2020 280	Solar, Recirculation, Domestic Hot Water Systems		COST EACH		
			MAT.	INST.	TOTAL
2550	Solar, recirculation, hot water				
2560	3/8" tubing, 3 ea. 4' x 4'-4" vacuum tube collectors		5,700	5,425	11,125
2580	1/2" tubing, 4 ea. 4' x 4'-4" vacuum tube collectors	R235616	6,350	5,900	12,250
2640	2 ea. 3'x7' black chrome collectors	-60	4,700	5,525	10,225
2660	3 ea. 3'x7' black chrome collectors		5,875	5,700	11,575
2700	2 ea. 3'x7' flat black collectors		4,150	5,550	9,700
2720	3 ea. 3'x7' flat black collectors		5,025	5,725	10,750
2760	3/4" tubing, 3 ea. 3'x7' black chrome collectors		6,450	5,975	12,425
2780	3 ea. 3'x7' flat black absorber plate collectors		5,600	6,000	11,600
2800	2 ea. 4'x9' flat black w/plastic glazing collectors		5,175	6,025	11,200
2820	2 ea. 3'x7' black chrome collectors		5,275	5,825	11,100
2840	1" tubing, 4 ea. 2'x9' black plastic absorber & glazing collectors		6,850	6,825	13,675
2860	4 ea. 3'x7' black chrome absorber collectors		8,875	6,850	15,725
2880	4 ea. 3'x7' flat black absorber collectors		7,750	6,875	14,625

The thermosyphon domestic hot water system, a direct collection system, operates under city water pressure and does not require pumps for system operation. An insulated water storage tank is located above the collectors. As the sun heats the collectors, warm water in them rises by means of natural convection; the colder water in the storage tank flows into the collectors by means of gravity. As long as the sun is shining the water continues to flow through the collectors and to become warmer.

To prevent freezing, the system must be drained or the collectors covered with an insulated lid when the temperature drops below 32°F.

D20 Plumbing

D2020 Domestic Water Distribution

System Components	QUANTITY	UNIT	COST EACH		
			MAT.	INST.	TOTAL
SYSTEM D2020 285 0960					
SOLAR, THERMOSYPHON, WATER HEATER					
3/4" TUBING, TWO 3'X7' BLACK CHROME COLLECTORS					
D-1 Framing lumber, fir, 2" x 6" x 8', tank cradle	.008	M.B.F.	6.04	8.20	14.24
F-1 Framing lumber, fir, 2" x 4" x 24', sleepers	.016	M.B.F.	12.48	24.80	37.28
I Valve, gate, bronze, 125 lb, soldered 1/2" diam	2.000	Ea.	133	64	197
J Vent flashing, neoprene	4.000	Ea.	46	154	200
O Pipe covering, urethane, ultraviolet cover, 1" wall, 1/2"diam	40.000	L.F.	93.20	278	371.20
P Pipe covering fiberglass all service jacket 1" wall 1/2" diam	160.000	L.F.	161.60	864	1,025.60
Q Collector panel solar, blk chrome on copper, 3/16" temp glass 3'-6" x 7.5'	2.000	Ea.	1,790	308	2,098
Y Flow control valve, globe, bronze, 125#, soldered, 1/2" diam	1.000	Ea.	117	32	149
U Valve, water tempering, bronze,sweat connections, 1/2" diam	1.000	Ea.	130	32	162
W Tank, water storage, solar energy system, 80 Gal, 2" x 1/2 lb insul	1.000	Ea.	2,150	480	2,630
Copper tubing type L, solder joints, hangers 10' OC 1/2" diam	150.000	L.F.	607.50	1,432.50	2,040
Copper tubing type M, solder joints, hangers 10' OC 1/2" diam	50.000	L.F.	196	460	656
Sensor wire, #22-2 gauge multistranded	.500	C.L.F.	6.50	36.50	43
Wrought copper fittings & solder, 1/2" diam	75.000	Ea.	104.25	2,887.50	2,991.75
TOTAL			5,553.57	7,061.50	12,615.07

D2020 285	Thermosyphon, Hot Water		COST EACH		
			MAT.	INST.	TOTAL
0960 0970	Solar, thermosyphon, hot water, two collector system	R235616 -60	5,550	7,050	12,600

513

ATTIC FLOOR

CWS

HWS

This domestic hot water pre-heat system includes heat exchanger with a circulating pump, blower, air-to-water, coil and controls, mounted in the upper collector manifold. Heat from the hot air coming out of the collectors is transferred through the heat exchanger. For each degree of DHW preheating gained, one degree less heating is needed from the fuel fired water heater. The system is simple, inexpensive to operate and can provide a substantial portion of DHW requirements for modest additional cost.

D2020 Domestic Water Distribution

System Components	QUANTITY	UNIT	COST EACH		
			MAT.	INST.	TOTAL
SYSTEM D2020 290 2560					
SOLAR, HOT WATER, AIR TO WATER HEAT EXCHANGE					
THREE COLLECTORS, OPTICAL BLACK ON ALUM., 7.5' x 3.5',80 GAL TANK					
A, B Differential controller, 2 sensors, thermostat, solar energy system	1.000	Ea.	169	64.50	233.50
C Thermometer, 2" dial	2.000	Ea.	50	96	146
C-1 Heat exchanger, air to fluid, up flow 70 MBH	1.000	Ea.	410	395	805
E Air vent, manual, for solar energy system 1/8" fitting	1.000	Ea.	2.96	24	26.96
F Air purger	1.000	Ea.	51	64.50	115.50
G Expansion tank	1.000	Ea.	77.50	24	101.50
H Strainer, Y type, bronze body, 1/2" IPS	1.000	Ea.	50	39	89
I Valve, gate, bronze, 125 lb, NRS, soldered 1/2" diam	5.000	Ea.	332.50	160	492.50
N Relief valve, temp & pressure solar 150 psi 210°F self-closing	1.000	Ea.	28	25.50	53.50
P Pipe covering, fiberglass, all service jacket, 1" wall, 1/2" diam	60.000	L.F.	60.60	324	384.60
Q Collector panel solar energy, air, black on alum plate, 7.5' x 3.5'	3.000	Ea.	2,685	462	3,147
B-1, R-1 Shutter damper for solar heater circulator	1.000	Ea.	64.50	113	177.50
B-1, R-1 Shutter motor for solar heater circulator	1.000	Ea.	161	85	246
R Backflow preventer, 1/2" pipe size	2.000	Ea.	199	96	295
T Drain valve, brass, 3/4" connection	1.000	Ea.	13.25	32	45.25
U Valve, water tempering, bronze, sweat connections, 1/2 " diam	1.000	Ea.	130	32	162
W Tank, water storage, solar energy system, 80 Gal, 2" x 2 lb insul	1.000	Ea.	2,150	480	2,630
V Tank, water storage, w/heating element, drain, relief valve, existing	1.000	System			
X Valve, globe, bronze, 125 lb, rising stem, 1/2" diam	1.000	Ea.	117	32	149
Copper tubing type M, solder joints, hangers 10' OC, 1/2" diam	50.000	L.F.	196	460	656
Copper tubing type L, solder joints, hangers 10' OC 1/2" diam	10.000	L.F.	40.50	95.50	136
Wrought copper fittings & solder, 1/2" diam	10.000	Ea.	13.90	385	398.90
Sensor wire, #22-2 conductor multistranded	.500	C.L.F.	6.50	36.50	43
Q-1, Q-2 Ductwork, fiberglass, aluminized jacket, 1-1/2" thick, 8" diam	32.000	S.F.	228.80	241.60	470.40
Q-3 Manifold for flush mount solar energy collector panels, air	6.000	L.F.	1,068	51	1,119
TOTAL			8,305.01	3,818.10	12,123.11

D2020 290	Solar, Hot Water, Air To Water Heat Exchange		COST EACH		
			MAT.	INST.	TOTAL
2550	Solar, hot water, air to water heat exchange				
2560	Three collectors, optical black on aluminum, 7.5' x 3.5', 80 Gal tank		8,300	3,825	12,125
2580	Four collectors, optical black on aluminum, 7.5' x 3.5', 80 Gal tank	R235616 -60	9,600	4,050	13,650
2600	Four collectors, optical black on aluminum, 7.5' x 3.5', 120 Gal tank		9,550	4,125	13,675

In this closed-loop indirect collection system, fluid with a low freezing temperature, such as propylene glycol, transports heat from the collectors to water storage. The transfer fluid is contained in a closed-loop consisting of collectors, supply and return piping, and a heat exchanger immersed in the storage tank. A typical two-or-three panel system contains 5 to 6 gallons of heat transfer fluid.

When the collectors become approximately 20°F warmer than the storage temperature, a controller activates the circulator. The circulator moves the fluid continuously through the collectors until the temperature difference between the collectors and storage is such that heat collection no longer occurs; at that point, the circulator shuts off. Since the heat transfer fluid has a very low freezing temperature, there is no need for it to be drained from the collectors between periods of collection.

D2020 Domestic Water Distribution

System Components	QUANTITY	UNIT	COST EACH		
			MAT.	INST.	TOTAL
SYSTEM D2020 295 2760					
SOLAR, CLOSED LOOP, HOT WATER SYSTEM, IMMERSED HEAT EXCHANGER					
3/4" TUBING, THREE 3' X 7' BLACK CHROME COLLECTORS					
A, B Differential controller, 2 sensors, thermostat, solar energy system	1.000	Ea.	169	64.50	233.50
C Thermometer 2" dial	3.000	Ea.	75	144	219
D, T Fill & drain valves, brass, 3/4" connection	3.000	Ea.	39.75	96	135.75
E Air vent, manual, 1/8" fitting	1.000	Ea.	2.96	24	26.96
F Air purger	1.000	Ea.	51	64.50	115.50
G Expansion tank	1.000	Ea.	77.50	24	101.50
I Valve, gate, bronze, NRS, soldered 3/4" diam	3.000	Ea.	244.50	115.50	360
J Neoprene vent flashing	2.000	Ea.	23	77	100
K Circulator, solar heated liquid, 1/25 HP	1.000	Ea.	225	99	324
N-1, N Relief valve, temp & press 150 psi 210°F self-closing 3/4" IPS	2.000	Ea.	56	51	107
O Pipe covering, urethane ultraviolet cover, 1" wall, 3/4" diam	20.000	L.F.	62	141	203
P Pipe covering, fiberglass, all service jacket, 1" wall, 3/4" diam	50.000	L.F.	54	282.50	336.50
Roof clamps for solar energy collector panel	3.000	Set	10.80	59.55	70.35
Q Collector panel solar blk chrome on copper, 1/8" temp glass, 3'x7'	3.000	Ea.	3,525	438	3,963
R-1 Valve, swing check, bronze, regrinding disc, 3/4" diam, soldered	1.000	Ea.	103	38.50	141.50
S Pressure gauge, 60 psi, 2-1/2" dial	1.000	Ea.	27.50	24	51.50
U Valve, water tempering, bronze, sweat connections, 3/4" diam	1.000	Ea.	183	38.50	221.50
W-2, W Tank, water storage immersed heat exchr elec elem 2"x2# insul 120 Gal	1.000	Ea.	2,100	550	2,650
X Valve, globe, bronze, rising stem, 3/4" diam, soldered	1.000	Ea.	189	38.50	227.50
Copper tubing type L, solder joint, hanger 10' OC 3/4" diam	20.000	L.F.	104	203	307
Copper tubing, type M, solder joint, hanger 10' OC 3/4" diam	70.000	L.F.	388.50	693	1,081.50
Sensor wire, #22-2 conductor multistranded	.500	C.L.F.	6.50	36.50	43
Solar energy heat transfer fluid, propylene glycol, anti-freeze	6.000	Gal.	153	165	318
Wrought copper fittings & solder, 3/4" diam	76.000	Ea.	224.20	3,078	3,302.20
TOTAL			8,094.21	6,545.55	14,639.76

D2020 295	Solar, Closed Loop, Hot Water Systems		COST EACH		
			MAT.	INST.	TOTAL
2550	Solar, closed loop, hot water system, immersed heat exchanger				
2560	3/8" tubing, 3 ea. 4' x 4'-4" vacuum tube collectors, 80 gal. tank		7,500	5,925	13,425
2580	1/2" tubing, 4 ea. 4' x 4'-4" vacuum tube collectors, 80 gal. tank		8,150	6,400	14,550
2600	120 gal. tank		8,100	6,475	14,575
2640	2 ea. 3'x7' black chrome collectors, 80 gal. tank	R235616 -60	6,575	6,125	12,700
2660	120 gal. tank		6,450	6,125	12,575
2700	2 ea. 3'x7' flat black collectors, 120 gal. tank		5,900	6,125	12,025
2720	3 ea. 3'x7' flat black collectors, 120 gal. tank		6,800	6,300	13,100
2760	3/4" tubing, 3 ea. 3'x7' black chrome collectors, 120 gal. tank		8,100	6,550	14,650
2780	3 ea. 3'x7' flat black collectors, 120 gal. tank		7,250	6,575	13,825
2800	2 ea. 4'x9' flat black w/plastic glazing collectors 120 gal. tank		6,825	6,600	13,425
2840	1" tubing, 4 ea. 2'x9' plastic absorber & glazing collectors 120 gal. tank		8,050	7,350	15,400
2860	4 ea. 3'x7' black chrome collectors, 120 gal. tank		10,100	7,375	17,475
2880	4 ea. 3'x7' flat black absorber collectors, 120 gal. tank		8,975	7,425	16,400

517

For customer support on your Mechanical Costs with RSMeans Data, call 800.448.8182.

Basis for Heat Loss Estimate, Apartment Type Structures:

1. Masonry walls and flat roof are insulated. U factor is assumed at .08.
2. Window glass area taken as BOCA minimum, 1/10th of floor area. Double insulating glass with 1/4″ air space, U = .65.
3. Infiltration = 0.3 C.F. per hour per S.F. of net wall.
4. Concrete floor loss is 2 BTUH per S.F.
5. Temperature difference taken as 70°F.
6. Ventilating or makeup air has not been included and must be added if desired. Air shafts are not used.

System Components	QUANTITY	UNIT	COST EACH MAT.	COST EACH INST.	COST EACH TOTAL
SYSTEM D3010 510 1760					
HEATING SYSTEM, FIN TUBE RADIATION, FORCED HOT WATER					
1,000 S.F. AREA, 10,000 C.F. VOLUME					
Boiler, oil fired, CI, burner, ctrls/insul/breech/pipe/ftng/valves, 109 MBH	1.000	Ea.	4,550	4,375	8,925
Circulating pump, CI flange connection, 1/12 HP	1.000	Ea.	520	231	751
Expansion tank, painted steel, ASME 18 Gal capacity	1.000	Ea.	835	101	936
Storage tank, steel, above ground, 275 Gal capacity w/supports	1.000	Ea.	560	283	843
Copper tubing type L, solder joint, hanger 10′ OC, 3/4″ diam	100.000	L.F.	520	1,015	1,535
Radiation, 3/4″ copper tube w/alum fin baseboard pkg, 7″ high	30.000	L.F.	265.50	735	1,000.50
Pipe covering, calcium silicate w/cover, 1′ wall, 3/4′ diam	100.000	L.F.	472	765	1,237
TOTAL			7,722.50	7,505	15,227.50
COST PER S.F.			7.72	7.51	15.23

D3010 510	Apartment Building Heating - Fin Tube Radiation		COST PER S.F. MAT.	COST PER S.F. INST.	COST PER S.F. TOTAL
1740	Heating systems, fin tube radiation, forced hot water				
1760	1,000 S.F. area, 10,000 C.F. volume		7.73	7.50	15.23
1800	10,000 S.F. area, 100,000 C.F. volume	R235000 -10	3.50	4.52	8.02
1840	20,000 S.F. area, 200,000 C.F. volume		3.81	5.05	8.86
1880	30,000 S.F. area, 300,000 C.F. volume	R235000 -20	3.69	4.91	8.60
1890					

D3010 Energy Supply

Fin Tube Radiator

Basis for Heat Loss Estimate, Factory or Commercial Type Structures:

1. Walls and flat roof are of lightly insulated concrete block or metal. U factor is assumed at .17.
2. Windows and doors are figured at 1-1/2 S.F. per linear foot of building perimeter. The U factor for flat single glass is 1.13.
3. Infiltration is approximately 5 C.F. per hour per S.F. of wall.
4. Concrete floor loss is 2 BTUH per S.F. of floor.
5. Temperature difference is assumed at 70°F.
6. Ventilation or makeup air has not been included and must be added if desired.

System Components	QUANTITY	UNIT	COST EACH MAT.	COST EACH INST.	COST EACH TOTAL
SYSTEM D3010 520 1960					
HEATING SYSTEM, FIN TUBE RADIATION, FORCED HOT WATER					
1,000 S.F. BLDG., ONE FLOOR					
Boiler, oil fired, CI, burner/ctrls/insul/breech/pipe/ftngs/valves, 109 MBH	1.000	Ea.	4,550	4,375	8,925
Expansion tank, painted steel, ASME 18 Gal capacity	1.000	Ea.	3,525	118	3,643
Storage tank, steel, above ground, 550 Gal capacity w/supports	1.000	Ea.	4,850	525	5,375
Circulating pump, CI flanged, 1/8 HP	1.000	Ea.	860	231	1,091
Pipe, steel, black, sch. 40, threaded, cplg & hngr 10'OC, 1-1/2" diam.	260.000	L.F.	2,496	4,511	7,007
Pipe covering, calcium silicate w/cover , 1" wall, 1-1/2" diam	260.000	L.F.	1,240.20	2,041	3,281.20
Radiation, steel 1-1/4" tube & 4-1/4" fin w/cover & damper, wall hung	42.000	L.F.	1,995	1,638	3,633
Rough in, steel fin tube radiation w/supply & balance valves	4.000	Set	1,520	3,440	4,960
TOTAL			21,036.20	16,879	37,915.20
COST PER S.F.			21.04	16.88	37.92

D3010 520	Commercial Building Heating - Fin Tube Radiation		COST PER S.F. MAT.	COST PER S.F. INST.	COST PER S.F. TOTAL
1940	Heating systems, fin tube radiation, forced hot water				
1960	1,000 S.F. bldg, one floor		21	16.90	37.90
2000	10,000 S.F., 100,000 C.F., total two floors	R235000 -10	5.40	6.45	11.85
2040	100,000 S.F., 1,000,000 C.F., total three floors		2.01	2.92	4.93
2080	1,000,000 S.F., 10,000,000 C.F., total five floors	R235000 -20	1.09	1.53	2.62
2090					

519

For customer support on your Mechanical Costs with RSMeans Data, call 800.448.8182.

D30 HVAC

D3010 Energy Supply

Unit Heater

Basis for Heat Loss Estimate, Factory or Commercial Type Structures:

1. Walls and flat roof are of lightly insulated concrete block or metal. U factor is assumed at .17.
2. Windows and doors are figured at 1-1/2 S.F. per linear foot of building perimeter. The U factor for flat single glass is 1.13.
3. Infiltration is approximately 5 C.F. per hour per S.F. of wall.
4. Concrete floor loss is 2 BTUH per S.F. of floor.
5. Temperature difference is assumed at 70°F.
6. Ventilation or makeup air has not been included and must be added if desired.

System Components	QUANTITY	UNIT	COST EACH		
			MAT.	INST.	TOTAL
SYSTEM D3010 530 1880					
HEATING SYSTEM, TERMINAL UNIT HEATERS, FORCED HOT WATER					
1,000 S.F. BLDG., ONE FLOOR					
Boiler oil fired, CI, burner/ctrls/insul/breech/pipe/ftngs/valves, 109 MBH	1.000	Ea.	4,550	4,375	8,925
Expansion tank, painted steel, ASME 18 Gal capacity	1.000	Ea.	3,525	118	3,643
Storage tank, steel, above ground, 550 Gal capacity w/supports	1.000	Ea.	4,850	525	5,375
Circulating pump, CI, flanged, 1/8 HP	1.000	Ea.	860	231	1,091
Pipe, steel, black, schedule 40, threaded, cplg & hngr 10' OC 1-1/2" diam	260.000	L.F.	2,496	4,511	7,007
Pipe covering, calcium silicate w/cover, 1" wall, 1-1/2" diam	260.000	L.F.	1,240.20	2,041	3,281.20
Unit heater, 1 speed propeller, horizontal, 200° EWT, 26.9 MBH	2.000	Ea.	1,390	354	1,744
Unit heater piping hookup with controls	2.000	Set	1,890	3,350	5,240
TOTAL			20,801.20	15,505	36,306.20
COST PER S.F.			20.80	15.51	36.31

D3010 530	Commercial Bldg. Heating - Terminal Unit Heaters		COST PER S.F.		
			MAT.	INST.	TOTAL
1860	Heating systems, terminal unit heaters, forced hot water				
1880	1,000 S.F. bldg., one floor		20.80	15.50	36.30
1920	10,000 S.F. bldg., 100,000 C.F. total two floors	R235000 -10	4.88	5.35	10.23
1960	100,000 S.F. bldg., 1,000,000 C.F. total three floors		2.06	2.69	4.75
2000	1,000,000 S.F. bldg., 10,000,000 C.F. total five floors	R235000 -20	1.43	1.72	3.15
2010					

Many styles of active solar energy systems exist. Those shown on the following page represent the majority of systems now being installed in different regions of the country.

The five active domestic hot water (DHW) systems which follow are typically specified as two or three panel systems with additional variations being the type of glazing and size of the storage tanks. Various combinations have been costed for the user's evaluation and comparison. The basic specifications from which the following systems were developed satisfy the construction detail requirements specified in the HUD Intermediate Minimum Property Standards (IMPS). If these standards are not complied with, the renewable energy system's costs could be significantly lower than shown.

To develop the system's specifications and costs it was necessary to make a number of assumptions about the systems. Certain systems are more appropriate to one climatic region than another or the systems may require modifications to be usable in particular locations. Specific instances in which the systems are not appropriate throughout the country as specified, or in which modification will be needed include the following:

- The freeze protection mechanisms provided in the DHW systems vary greatly. In harsh climates, where freeze is a major concern, a closed-loop indirect collection system may be more appropriate than a direct collection system.
- The thermosyphon water heater system described cannot be used when temperatures drop below 32°F.
- In warm climates it may be necessary to modify the systems installed to prevent overheating.

For each renewable resource (solar) system a schematic diagram and descriptive summary of the system is provided along with a list of all the components priced as part of the system. The costs were developed based on these specifications.

Considerations affecting costs which may increase or decrease beyond the estimates presented here include the following:

- Special structural qualities (allowance for earthquake, future expansion, high winds, and unusual spans or shapes);
- Isolated building site or rough terrain that would affect the transportation of personnel, material, or equipment;
- Unusual climatic conditions during the construction process;
- Substitution of other materials or system components for those used in the system specifications.

In this closed-loop indirect collection system, fluid with a low freezing temperature, propylene glycol, transports heat from the collectors to water storage. The transfer fluid is contained in a closed loop consisting of collectors, supply and return piping, and a heat exchanger

immersed in the storage tank.

When the collectors become approximately 20°F warmer than the storage temperature, the controller activates the circulator. The circulator moves the fluid continuously until the temperature difference between fluid in

the collectors and storage is such that the collection will no longer occur and then the circulator turns off. Since the heat transfer fluid has a very low freezing temperature, there is no need for it to be drained from the collectors between periods of collection.

D3010   Energy Supply

System Components	QUANTITY	UNIT	COST EACH		
			MAT.	INST.	TOTAL
SYSTEM D3010 650 2750					
SOLAR, CLOSED LOOP, SPACE/HOT WATER					
1″ TUBING, TEN 3′X7′ BLK CHROME ON COPPER ABSORBER COLLECTORS					
A, B Differential controller 2 sensors, thermostat, solar energy system	2.000	Ea.	338	129	467
C Thermometer, 2″ dial	10.000	Ea.	250	480	730
C-1 Heat exchanger, solar energy system, fluid to air, up flow, 80 MBH	1.000	Ea.	545	465	1,010
D, T Fill & drain valves, brass, 3/4″ connection	5.000	Ea.	66.25	160	226.25
D-1 Fan center	1.000	Ea.	131	162	293
E Air vent, manual, 1/8″ fitting	1.000	Ea.	2.96	24	26.96
E-2 Thermostat, 2 stage for sensing room temperature	1.000	Ea.	283	98	381
F Air purger	2.000	Ea.	102	129	231
F-1 Controller, liquid temperature, solar energy system	1.000	Ea.	129	154	283
G Expansion tank	2.000	Ea.	155	48	203
I Valve, gate, bronze, 125 lb, soldered, 1″ diam	5.000	Ea.	432.50	202.50	635
J Vent flashing, neoprene	2.000	Ea.	23	77	100
K Circulator, solar heated liquid, 1/25 HP	2.000	Ea.	450	198	648
L Circulator, solar heated liquid, 1/20 HP	1.000	Ea.	310	116	426
N Relief valve, temp & pressure 150 psi 210°F self-closing	4.000	Ea.	112	102	214
N-1 Relief valve, pressure poppet, bronze, 30 psi, 3/4″ IPS	3.000	Ea.	345	82.50	427.50
O Pipe covering, urethane, ultraviolet cover, 1″ wall, 1″ diam	50.000	L.F.	155	352.50	507.50
P Pipe covering, fiberglass, all service jacket, 1″ wall, 1″ diam	60.000	L.F.	64.80	339	403.80
Q Collector panel solar energy blk chrome on copper 1/8″ temp glass 3′x7′	10.000	Ea.	11,750	1,460	13,210
Roof clamps for solar energy collector panel	10.000	Set	36	198.50	234.50
R Valve, swing check, bronze, 125 lb, regrinding disc, 3/4″ & 1″ diam	4.000	Ea.	784	162	946
S Pressure gage, 0-60 psi, for solar energy system	2.000	Ea.	55	48	103
U Valve, water tempering, bronze, sweat connections, 3/4″ diam	1.000	Ea.	183	38.50	221.50
W-2, W-1, W Tank, water storage immersed heat xchr elec elem 2″x1/2# ins 120 gal	4.000	Ea.	8,400	2,200	10,600
X Valve, globe, bronze, 125 lb, rising stem, 1″ diam	3.000	Ea.	789	121.50	910.50
Y Valve, flow control	1.000	Ea.	154	35	189
Copper tubing, type M, solder joint, hanger 10′ OC 1″ diam	110.000	L.F.	1,028.50	1,210	2,238.50
Copper tubing, type L, solder joint, hanger 10′ OC 3/4″ diam	20.000	L.F.	104	203	307
Wrought copper fittings & solder, 3/4″ & 1″ diam	121.000	Ea.	937.75	5,808	6,745.75
Sensor, wire, #22-2 conductor, multistranded	.700	C.L.F.	9.10	51.10	60.20
Ductwork, galvanized steel, for heat exchanger	8.000	Lb.	5.28	72	77.28
Solar energy heat transfer fluid propylene glycol, anti-freeze	25.000	Gal.	637.50	687.50	1,325
TOTAL			28,767.64	15,613.60	44,381.24

D3010 650	Solar, Closed Loop, Space/Hot Water Systems		COST EACH		
			MAT.	INST.	TOTAL
2540	Solar, closed loop, space/hot water				
2550	1/2″ tubing, 12 ea. 4′x4′4″ vacuum tube collectors		26,600	14,600	41,200
2600	3/4″ tubing, 12 ea. 4′x4′4″ vacuum tube collectors	R235616	28,000	15,000	43,000
2650	10 ea. 3′ x 7′ black chrome absorber collectors	-60	27,700	14,600	42,300
2700	10 ea. 3′ x 7′ flat black absorber collectors		24,900	14,700	39,600
2750	1″ tubing, 10 ea. 3′ x 7′ black chrome absorber collectors		28,800	15,600	44,400
2800	10 ea. 3′ x 7′ flat black absorber collectors		26,100	15,700	41,800
2850	6 ea. 4′ x 9′ flat black w/plastic glazing collectors		23,900	15,600	39,500
2900	12 ea. 2′ x 9′ plastic absorber and glazing collectors		25,200	15,900	41,100

For customer support on your Mechanical Costs with RSMeans Data, call 800.448.8182.

523

This draindown pool system uses a differential thermostat similar to those used in solar domestic hot water and space heating applications. To heat the pool, the pool water passes through the conventional pump-filter loop and then flows through the collectors. When collection is not possible, or when the pool temperature is reached, all water drains from the solar loop back to the pool through the existing piping. The modes are controlled by solenoid valves or other automatic valves in conjunction with a vacuum breaker relief valve, which facilitates draindown.

D3010 Energy Supply

System Components	QUANTITY	UNIT	COST EACH		
			MAT.	INST.	TOTAL
SYSTEM D3010 660 2640					
SOLAR SWIMMING POOL HEATER, ROOF MOUNTED COLLECTORS					
TEN 4' X 10' FULLY WETTED UNGLAZED PLASTIC ABSORBERS					
A Differential thermostat/controller, 110V, adj pool pump system	1.000	Ea.	365	385	750
A-1 Solenoid valve, PVC, normally 1 open 1 closed (included)	2.000	Ea.			
B Sensor, thermistor type (included)	2.000	Ea.			
E-1 Valve, vacuum relief	1.000	Ea.	34	24	58
Q Collector panel, solar energy, plastic, liquid full wetted, 4' x 10'	10.000	Ea.	3,700	2,780	6,480
R Valve, ball check, PVC, socket, 1-1/2" diam	1.000	Ea.	120	38.50	158.50
Z Valve, ball, PVC, socket, 1-1/2" diam	3.000	Ea.	285	115.50	400.50
Pipe, PVC, sch 40, 1-1/2" diam	80.000	L.F.	904	1,720	2,624
Pipe fittings, PVC sch 40, socket joint, 1-1/2" diam	10.000	Ea.	21.90	385	406.90
Sensor wire, #22-2 conductor, multistranded	.500	C.L.F.	6.50	36.50	43
Roof clamps for solar energy collector panels	10.000	Set	36	198.50	234.50
Roof strap, teflon for solar energy collector panels	26.000	L.F.	715	97.76	812.76
TOTAL			6,187.40	5,780.76	11,968.16

D3010 660	Solar Swimming Pool Heater Systems		COST EACH		
			MAT.	INST.	TOTAL
2530	Solar swimming pool heater systems, roof mounted collectors				
2540	10 ea. 3'x7' black chrome absorber, 1/8" temp. glass		14,200	4,450	18,650
2560	10 ea. 4'x8' black chrome absorber, 3/16" temp. glass	R235616	16,000	5,300	21,300
2580	10 ea. 3'8"x6' flat black absorber, 3/16" temp. glass	-60	11,400	4,550	15,950
2600	10 ea. 4'x9' flat black absorber, plastic glazing		13,700	5,525	19,225
2620	10 ea. 2'x9' rubber absorber, plastic glazing		15,800	5,975	21,775
2640	10 ea. 4'x10' fully wetted unglazed plastic absorber		6,175	5,775	11,950
2660	Ground mounted collectors				
2680	10 ea. 3'x7' black chrome absorber, 1/8" temp. glass		14,300	4,975	19,275
2700	10 ea. 4'x8' black chrome absorber, 3/16" temp. glass		16,100	5,825	21,925
2720	10 ea. 3'8"x6' flat blk absorber, 3/16" temp. glass		11,500	5,050	16,550
2740	10 ea. 4'x9' flat blk absorber, plastic glazing		13,800	6,050	19,850
2760	10 ea. 2'x9' rubber absorber, plastic glazing		15,900	6,300	22,200
2780	10 ea. 4'x10' fully wetted unglazed plastic absorber		6,250	6,300	12,550

525

For customer support on your **Mechanical Costs** with RSMeans Data, call 800.448.8182.

The complete Solar Air Heating System provides maximum savings of conventional fuel with both space heating and year-round domestic hot water heating. It allows for the home air conditioning to operate simultaneously and independently from the solar domestic water heating in summer. The system's modes of operation are:

Mode 1: The building is heated directly from the collectors with air circulated by the Solar Air Mover.

Mode 2: When heat is not needed in the building, the dampers change within the air mover to circulate the air from the collectors to the rock storage bin.

Mode 3: When heat is not available from the collector array and is available in rock storage, the air mover draws heated air from rock storage and directs it into the building. When heat is not available from the collectors or the rock storage bin, the auxiliary heating unit will provide heat for the building. The size of the collector array is typically 25% the size of the main floor area.

D3010 Energy Supply

System Components	QUANTITY	UNIT	COST EACH		
			MAT.	INST.	TOTAL
SYSTEM D3010 675 1210					
SOLAR, SPACE/HOT WATER, AIR TO WATER HEAT EXCHANGE					
A, B Differential controller 2 sensors thermos., solar energy sys liquid loop	1.000	Ea.	169	64.50	233.50
A-1, B Differential controller 2 sensors 6 station solar energy sys air loop	1.000	Ea.	299	257	556
B-1 Solar energy sensor, freeze prevention	1.000	Ea.	27.50	24	51.50
C Thermometer for solar energy system, 2" dial	2.000	Ea.	50	96	146
C-1 Heat exchanger, solar energy system, air to fluid, up flow, 70 MBH	1.000	Ea.	410	395	805
D Drain valve, brass, 3/4" connection	2.000	Ea.	26.50	64	90.50
E Air vent, manual, for solar energy system 1/8" fitting	1.000	Ea.	2.96	24	26.96
E-1 Thermostat, 2 stage for sensing room temperature	1.000	Ea.	283	98	381
F Air purger	1.000	Ea.	51	64.50	115.50
G Expansion tank, for solar energy system	1.000	Ea.	77.50	24	101.50
I Valve, gate, bronze, 125 lb, NRS, soldered 3/4" diam	2.000	Ea.	163	77	240
K Circulator, solar heated liquid, 1/25 HP	1.000	Ea.	225	99	324
N Relief valve temp & press 150 psi 210°F self-closing, 3/4" IPS	1.000	Ea.	28	25.50	53.50
N-1 Relief valve, pressure, poppet, bronze, 30 psi, 3/4" IPS	1.000	Ea.	115	27.50	142.50
P Pipe covering, fiberglass, all service jacket, 1" wall, 3/4" diam	60.000	L.F.	64.80	339	403.80
Q-3 Manifold for flush mount solar energy collector panels	20.000	L.F.	3,560	170	3,730
Q Collector panel solar energy, air, black on alum. plate, 7.5' x 3.5'	10.000	Ea.	8,950	1,540	10,490
R Valve, swing check, bronze, 125 lb, regrinding disc, 3/4" diam	2.000	Ea.	206	77	283
S Pressure gage, 2" dial, for solar energy system	1.000	Ea.	27.50	24	51.50
U Valve, water tempering, bronze, sweat connections, 3/4" diam	1.000	Ea.	183	38.50	221.50
W-2, W Tank, water storage, solar, elec element 2"x1/2# insul, 80 Gal	1.000	Ea.	2,150	480	2,630
X Valve, globe, bronze, 125 lb, soldered, 3/4" diam	1.000	Ea.	189	38.50	227.50
Copper tubing type L, solder joints, hangers 10' OC 3/4" diam	10.000	L.F.	52	101.50	153.50
Copper tubing type M, solder joints, hangers 10' OC 3/4" diam	60.000	L.F.	333	594	927
Wrought copper fittings & solder, 3/4" diam	26.000	Ea.	76.70	1,053	1,129.70
Sensor wire, #22-2 conductor multistranded	1.200	C.L.F.	15.60	87.60	103.20
Q-1 Duct work, rigid fiberglass, rectangular	400.000	S.F.	392	2,420	2,812
Duct work, spiral preformed, steel, PVC coated both sides, 12" x 10"	8.000	Ea.	204	452	656
R-2 Shutter/damper for solar heater circulator	9.000	Ea.	580.50	1,017	1,597.50
K-1 Shutter motor for solar heater circulator blower	9.000	Ea.	1,449	765	2,214
K-1 Fan, solar energy heated air circulator, space & DHW system	1.000	Ea.	1,775	2,725	4,500
Z-1 Tank, solar energy air storage, 6'-3"H 7'x7' = 306 CF/2000 Gal	1.000	Ea.	15,300	1,775	17,075
Z-2 Crushed stone 1-1/2"	11.000	C.Y.	335.50	92.07	427.57
C-2 Thermometer, remote probe, 2" dial	1.000	Ea.	38	96.50	134.50
R-1 Solenoid valve	1.000	Ea.	150	85.50	235.50
V, R-4, R-3 Furnace, supply diffusers, return grilles, existing	1.000	Ea.	15,300	1,775	17,075
TOTAL			53,259.06	17,086.17	70,345.23

D3010 675	Air To Water Heat Exchange		COST EACH		
			MAT.	INST.	TOTAL
1210 1220	Solar, air to water heat exchange, for space/hot water heating	R235616 -60	53,500	17,100	70,600

For customer support on your Mechanical Costs with RSMeans Data, call 800.448.8182.

527

Small Electric Boiler
System Considerations:
1. Terminal units are fin tube baseboard radiation rated at 720 BTU/hr with 200° water temperature or 820 BTU/hr steam.
2. Primary use being for residential or smaller supplementary areas, the floor levels are based on 7-1/2' ceiling heights.
3. All distribution piping is copper for boilers through 205 MBH. All piping for larger systems is steel pipe.

Boiler   **Baseboard Radiation**

System Components	QUANTITY	UNIT	COST EACH		
			MAT.	INST.	TOTAL
SYSTEM D3020 102 1120					
SMALL HEATING SYSTEM, HYDRONIC, ELECTRIC BOILER					
1,480 S.F., 61 MBH, STEAM, 1 FLOOR					
Boiler, electric steam, std cntrls, trim, ftngs and valves, 18 KW, 61.4 MBH	1.000	Ea.	5,472.50	1,952.50	7,425
Copper tubing type L, solder joint, hanger 10'OC, 1-1/4" diam	160.000	L.F.	2,112	2,128	4,240
Radiation, 3/4" copper tube w/alum fin baseboard pkg 7" high	60.000	L.F.	531	1,470	2,001
Rough in baseboard panel or fin tube with valves & traps	10.000	Set	3,050	7,400	10,450
Pipe covering, calcium silicate w/cover, 1" wall 1-1/4" diam	160.000	L.F.	755.20	1,256	2,011.20
Low water cut-off, quick hookup, in gage glass tappings	1.000	Ea.	375	49	424
TOTAL			12,295.70	14,255.50	26,551.20
COST PER S.F.			8.31	9.63	17.94

D3020 102	Small Heating Systems, Hydronic, Electric Boilers		COST PER S.F.		
			MAT.	INST.	TOTAL
1100	Small heating systems, hydronic, electric boilers				
1120	Steam, 1 floor, 1480 S.F., 61 M.B.H.		8.31	9.66	17.97
1160	3,000 S.F., 123 M.B.H.	R235000 -10	6.40	8.45	14.85
1200	5,000 S.F., 205 M.B.H.		5.30	7.80	13.10
1240	2 floors, 12,400 S.F., 512 M.B.H.	R235000 -20	4.42	7.75	12.17
1280	3 floors, 24,800 S.F., 1023 M.B.H.		4.62	7.65	12.27
1320	34,750 S.F., 1,433 M.B.H.	R235000 -30	4.28	7.45	11.73
1360	Hot water, 1 floor, 1,000 S.F., 41 M.B.H.		14.25	5.35	19.60
1400	2,500 S.F., 103 M.B.H.		9.65	9.65	19.30
1440	2 floors, 4,850 S.F., 205 M.B.H.		8.90	11.55	20.45
1480	3 floors, 9,700 S.F., 410 M.B.H.		9.50	11.95	21.45

D3020 Heat Generating Systems

Boiler

Unit Heater

Large Electric Boiler System Considerations:

1. Terminal units are all unit heaters of the same size. Quantities are varied to accommodate total requirements.
2. All air is circulated through the heaters a minimum of three times per hour.
3. As the capacities are adequate for commercial use, floor levels are based on 10′ ceiling heights.
4. All distribution piping is black steel pipe.

System Components	QUANTITY	UNIT	COST EACH		
			MAT.	INST.	TOTAL
SYSTEM D3020 104 1240					
LARGE HEATING SYSTEM, HYDRONIC, ELECTRIC BOILER					
9,280 S.F., 135 KW, 461 MBH, 1 FLOOR					
Boiler, electric hot water, std ctrls, trim, ftngs, valves, 135 KW, 461 MBH	1.000	Ea.	12,450	4,275	16,725
Expansion tank, painted steel, 60 Gal capacity ASME	1.000	Ea.	5,750	235	5,985
Circulating pump, CI, close cpld, 50 GPM, 2 HP, 2″ pipe conn	1.000	Ea.	3,325	465	3,790
Unit heater, 1 speed propeller, horizontal, 200° EWT, 72.7 MBH	7.000	Ea.	7,000	1,799	8,799
Unit heater piping hookup with controls	7.000	Set	6,615	11,725	18,340
Pipe, steel, black, schedule 40, welded, 2-1/2″ diam	380.000	L.F.	4,940	12,156.20	17,096.20
Pipe covering, calcium silicate w/cover, 1″ wall, 2-1/2″ diam	380.000	L.F.	2,223	3,078	5,301
TOTAL			42,303	33,733.20	76,036.20
COST PER S.F.			4.56	3.64	8.20

D3020 104	Large Heating Systems, Hydronic, Electric Boilers		COST PER S.F.		
			MAT.	INST.	TOTAL
1230	Large heating systems, hydronic, electric boilers				
1240	9,280 S.F., 135 K.W., 461 M.B.H., 1 floor		4.56	3.63	8.19
1280	14,900 S.F., 240 K.W., 820 M.B.H., 2 floors		5.75	5.90	11.65
1320	18,600 S.F., 296 K.W., 1,010 M.B.H., 3 floors	R235000 -10	5.70	6.45	12.15
1360	26,100 S.F., 420 K.W., 1,432 M.B.H., 4 floors		5.60	6.25	11.85
1400	39,100 S.F., 666 K.W., 2,273 M.B.H., 4 floors	R235000 -20	4.82	5.25	10.07
1440	57,700 S.F., 900 K.W., 3,071 M.B.H., 5 floors		4.83	5.20	10.03
1480	111,700 S.F., 1,800 K.W., 6,148 M.B.H., 6 floors	R235000 -30	4.31	4.46	8.77
1520	149,000 S.F., 2,400 K.W., 8,191 M.B.H., 8 floors		4.27	4.45	8.72
1560	223,300 S.F., 3,600 K.W., 12,283 M.B.H., 14 floors		4.53	5.05	9.58

529

For customer support on your **Mechanical Costs** with RSMeans Data, call 800.448.8182.

D3020 Heat Generating Systems

Unit Heater

Fossil Fuel Boiler System Considerations:

1. Terminal units are horizontal unit heaters. Quantities are varied to accommodate total heat loss per building.
2. Unit heater selection was determined by their capacity to circulate the building volume a minimum of three times per hour in addition to the BTU output.

3. Systems shown are forced hot water. Steam boilers cost slightly more than hot water boilers. However, this is compensated for by the smaller size or fewer terminal units required with steam.
4. Floor levels are based on 10′ story heights.
5. MBH requirements are gross boiler output.

System Components	QUANTITY	UNIT	COST EACH MAT.	COST EACH INST.	COST EACH TOTAL
SYSTEM D3020 108 1280					
HEATING SYSTEM, HYDRONIC, FOSSIL FUEL, TERMINAL UNIT HEATERS					
CAST IRON BOILER, GAS, 80 MBH, 1,070 S.F. BUILDING					
Boiler, gas, hot water, CI, burner, controls, insulation, breeching, 80 MBH	1.000	Ea.	2,283.75	2,152.50	4,436.25
Pipe, steel, black, schedule 40, threaded, cplg & hngr 10′OC, 2″ diam	200.000	L.F.	1,520	4,300	5,820
Unit heater, 1 speed propeller, horizontal, 200° EWT, 72.7 MBH	2.000	Ea.	2,000	514	2,514
Unit heater piping hookup with controls	2.000	Set	1,890	3,350	5,240
Expansion tank, painted steel, ASME, 18 Gal capacity	1.000	Ea.	3,525	118	3,643
Circulating pump, CI, flange connection, 1/12 HP	1.000	Ea.	520	231	751
Pipe covering, calcium silicate w/cover, 1″ wall, 2″ diam	200.000	L.F.	1,100	1,620	2,720
TOTAL			12,838.75	12,285.50	25,124.25
COST PER S.F.			12	11.48	23.48

D3020 108	Heating Systems, Unit Heaters		COST PER S.F. MAT.	COST PER S.F. INST.	COST PER S.F. TOTAL
1260	Heating systems, hydronic, fossil fuel, terminal unit heaters,				
1280	Cast iron boiler, gas, 80 M.B.H., 1,070 S.F. bldg.		11.96	11.50	23.46
1320	163 M.B.H., 2,140 S.F. bldg.	R235000 -10	8.05	7.80	15.85
1360	544 M.B.H., 7,250 S.F. bldg.		5.80	5.55	11.35
1400	1,088 M.B.H., 14,500 S.F. bldg.	R235000 -20	4.95	5.30	10.25
1440	3,264 M.B.H., 43,500 S.F. bldg.		4.17	4.02	8.19
1480	5,032 M.B.H., 67,100 S.F. bldg.	R235000 -30	4.68	4.08	8.76
1520	Oil, 109 M.B.H., 1,420 S.F. bldg.		13.35	10.20	23.55
1560	235 M.B.H., 3,150 S.F. bldg.		7.85	7.40	15.25
1600	940 M.B.H., 12,500 S.F. bldg.		6.10	4.86	10.96
1640	1,600 M.B.H., 21,300 S.F. bldg.		5.35	4.66	10.01
1680	2,480 M.B.H., 33,100 S.F. bldg.		4.95	4.21	9.16
1720	3,350 M.B.H., 44,500 S.F. bldg.		4.79	4.30	9.09
1760	Coal, 148 M.B.H., 1,975 S.F. bldg.		98	6.65	104.65
1800	300 M.B.H., 4,000 S.F. bldg.		50.50	5.25	55.75
1840	2,360 M.B.H., 31,500 S.F. bldg.		11.90	4.26	16.16

Fin Tube Radiator

Fossil Fuel Boiler System Considerations:

1. Terminal units are commercial steel fin tube radiation. Quantities are varied to accommodate total heat loss per building.
2. Systems shown are forced hot water. Steam boilers cost slightly more than hot water boilers. However, this is compensated for by the smaller size or fewer terminal units required with steam.
3. Floor levels are based on 10' story heights.
4. MBH requirements are gross boiler output.

System Components	QUANTITY	UNIT	COST EACH		
			MAT.	INST.	TOTAL
SYSTEM D3020 110 3240					
HEATING SYSTEM, HYDRONIC, FOSSIL FUEL, FIN TUBE RADIATION					
CAST IRON BOILER, GAS, 80 MBH, 1,070 S.F. BLDG.					
Boiler, gas, hot water, CI, burner, controls, insulation, breeching, 80 MBH	1.000	Ea.	2,175	2,050	4,225
Pipe, steel, black, schedule 40, threaded, 2" diam	200.000	L.F.	1,520	4,300	5,820
Radiation, steel 1-1/4" tube & 4-1/4" fin w/cover & damper, wall hung	80.000	L.F.	3,800	3,120	6,920
Rough-in, wall hung steel fin radiation with supply & balance valves	8.000	Set	3,040	6,880	9,920
Expansion tank, painted steel, ASME, 18 Gal capacity	1.000	Ea.	3,525	118	3,643
Circulating pump, CI flanged, 1/12 HP	1.000	Ea.	520	231	751
Pipe covering, calcium silicate w/cover, 1" wall, 2" diam	200.000	L.F.	1,100	1,620	2,720
TOTAL			15,680	18,319	33,999
COST PER S.F.			14.65	17.12	31.77

D3020 110	Heating System, Fin Tube Radiation		COST PER S.F.		
			MAT.	INST.	TOTAL
3230	Heating systems, hydronic, fossil fuel, fin tube radiation				
3240	Cast iron boiler, gas, 80 MBH, 1,070 S.F. bldg.		14.67	17.10	31.77
3280	169 M.B.H., 2,140 S.F. bldg.	R235000 -10	9.35	10.90	20.25
3320	544 M.B.H., 7,250 S.F. bldg.		7.80	9.35	17.15
3360	1,088 M.B.H., 14,500 S.F. bldg.	R235000 -20	7.10	9.25	16.35
3400	3,264 M.B.H., 43,500 S.F. bldg.		6.50	8.15	14.65
3440	5,032 M.B.H., 67,100 S.F. bldg.	R235000 -30	7.05	8.25	15.30
3480	Oil, 109 M.B.H., 1,420 S.F. bldg.		18.30	18.80	37.10
3520	235 M.B.H., 3,150 S.F. bldg.		9.85	11.15	21
3560	940 M.B.H., 12,500 S.F. bldg.		8.30	8.85	17.15
3600	1,600 M.B.H., 21,300 S.F. bldg.		7.65	8.80	16.45
3640	2,480 M.B.H., 33,100 S.F. bldg.		7.30	8.40	15.70
3680	3,350 M.B.H., 44,500 S.F. bldg.		7.10	8.45	15.55
3720	Coal, 148 M.B.H., 1,975 S.F. bldg.		99.50	10.40	109.90
3760	300 M.B.H., 4,000 S.F. bldg.		52.50	8.95	61.45
3800	2,360 M.B.H., 31,500 S.F. bldg.		14.20	8.35	22.55
4080	Steel boiler, oil, 97 M.B.H., 1,300 S.F. bldg.		13.85	15.50	29.35
4120	315 M.B.H., 4,550 S.F. bldg.		6.85	7.85	14.70
4160	525 M.B.H., 7,000 S.F. bldg.		9.95	9.05	19
4200	1,050 M.B.H., 14,000 S.F. bldg.		9.10	9.05	18.15
4240	2,310 M.B.H., 30,800 S.F. bldg.		7.70	8.40	16.10
4280	3,150 M.B.H., 42,000 S.F. bldg.		7.75	8.55	16.30

531

For customer support on your Mechanical Costs with RSMeans Data, call 800.448.8182.

D3020   Heat Generating Systems

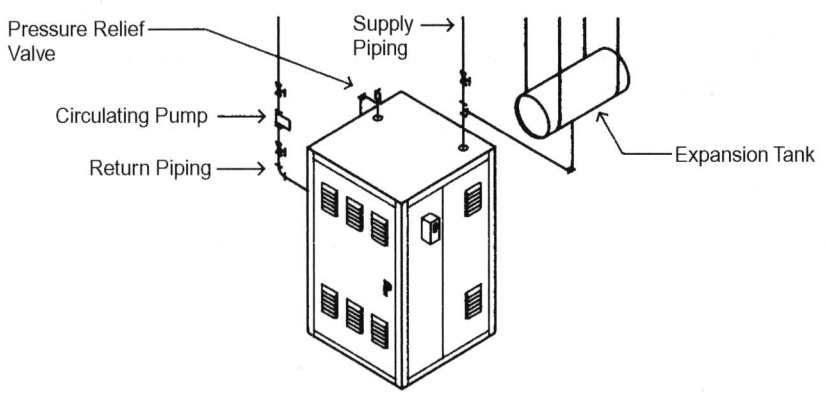

Electric Boiler, Hot Water

System Components	QUANTITY	UNIT	COST EACH		
			MAT.	INST.	TOTAL
SYSTEM D3020 126 1010					
BOILER, ELECTRIC, HOT WATER, 15 kW, 52 MBH					
Boilers, electric, ASME, hot water, 15 KW, 52 MBH	1.000	Ea.	6,000	1,650	7,650
Pipe, black steel, Sch 40, threaded, W/coupling & hangers, 10' OC, 3/4" dia	11.500	L.F.	55.66	145.48	201.14
Pipe, black steel, Sch 40, threaded, W/coupling & hangers, 10' OC, 1" dia	21.500	L.F.	183.83	312.83	496.66
Elbow, 90°, black, straight, 3/4" dia.	6.000	Ea.	27.18	330	357.18
Elbow, 90°, black, straight, 1" dia.	6.000	Ea.	55.20	357	412.20
Tee, black, straight, 3/4" dia.	1.000	Ea.	8.60	85.50	94.10
Tee, black, straight, 1" dia.	2.000	Ea.	29.50	193	222.50
Tee, black, reducing, 1" dia.	2.000	Ea.	38	193	231
Union, black with brass seat, 3/4" dia.	1.000	Ea.	21	59.50	80.50
Union, black with brass seat, 1" dia.	2.000	Ea.	54	129	183
Pipe nipples, 3/4" diam.	3.000	Ea.	7.26	18.98	26.24
Pipe nipples, 1" diam.	3.000	Ea.	12.83	21.83	34.66
Valves, bronze, gate, N.R.S., threaded, class 150, 3/4" size	2.000	Ea.	228	77	305
Valves, bronze, gate, N.R.S., threaded, class 150, 1" size	2.000	Ea.	296	81	377
Thermometer, stem type, 9" case, 8" stem, 3/4" NPT	1.000	Ea.	412	56	468
Tank, steel, liquid expansion, ASME, painted, 15 gallon capacity	1.000	Ea.	940	83	1,023
Pump, circulating, bronze, flange connection, 3/4" to 1-1/2" size, 1/8 HP	1.000	Ea.	1,375	231	1,606
Insulation, fiberglass pipe covering, 1" wall, 1" IPS	20.000	L.F.	23.80	118	141.80
TOTAL			9,767.86	4,142.12	13,909.98

D3020 126	Electric Boiler, Hot Water	COST EACH		
		MAT.	INST.	TOTAL
1010	Boiler, electric, hot water, 15 kW, 52 MBH	9,775	4,150	13,925
1020	30 kW, 103 MBH	10,200	4,350	14,550
1030	60 kW, 205 MBH	10,500	4,500	15,000
1040	120 kW, 410 MBH	12,400	5,400	17,800
1050	150 kW, 510 MBH	19,200	10,600	29,800
1060	210 kW, 716 MBH	21,800	11,600	33,400
1070	296 kW, 1010 MBH	29,800	11,900	41,700
1080	444 kW, 1515 MBH	35,800	16,100	51,900
1090	666 kW, 2273 MBH	43,400	17,400	60,800
1100	900 kW, 3071 MBH	55,000	20,200	75,200
1110	1320 kW, 4505 MBH	62,500	21,200	83,700
1120	2100 kW, 7167 MBH	97,500	25,600	123,100

D3020 Heat Generating Systems

D3020 126	Electric Boiler, Hot Water	COST EACH		
		MAT.	INST.	TOTAL
1130	2610 kW, 8905 MBH	113,500	30,900	144,400
1140	3600 kW, 12,283 MBH	138,000	37,000	175,000

D3020 128	Electric Boiler, Steam	COST EACH		
		MAT.	INST.	TOTAL
1010	Boiler, electric, steam, 18 kW, 61.4 MBH	6,450	3,975	10,425
1020	36 kW, 123 MBH	8,050	4,225	12,275
1030	60 kW, 205 MBH	20,300	8,550	28,850
1040	120 kW, 409 MBH	22,800	9,400	32,200
1050	150 kW, 512 MBH	26,700	14,400	41,100
1060	210 kW, 716 MBH	42,300	15,400	57,700
1070	300 kW, 1023 MBH	45,100	16,800	61,900
1080	510 kW, 1740 MBH	54,000	21,900	75,900
1090	720 kW, 2456 MBH	59,000	22,900	81,900
1100	1080 kW, 3685 MBH	65,000	27,300	92,300
1110	1260 kW, 4300 MBH	73,000	29,100	102,100
1120	1620 kW, 5527 MBH	77,000	32,700	109,700
1130	2070 kW, 7063 MBH	100,000	34,400	134,400
1140	2340 kW, 7984 MBH	127,500	39,100	166,600

533

For customer support on your Mechanical Costs with RSMeans Data, call 800.448.8182.

Cast Iron Boiler, Hot Water, Gas Fired

System Components	QUANTITY	UNIT	COST EACH		
			MAT.	INST.	TOTAL
SYSTEM D3020 130 1010					
BOILER, CAST IRON, GAS, HOT WATER, 100 MBH					
Boilers, gas fired, std controls, CI, insulated, HW, gross output 100 MBH	1.000	Ea.	2,825	2,225	5,050
Pipe, black steel, Sch 40, threaded, W/coupling & hangers, 10' OC, 3/4" dia	22.500	L.F.	108.90	284.63	393.53
Pipe, black steel, Sch 40, threaded, W/coupling & hangers, 10' OC, 1" dia.	21.500	L.F.	183.83	312.83	496.66
Elbow, 90°, black, straight, 3/4" dia.	9.000	Ea.	40.77	495	535.77
Elbow, 90°, black, straight, 1" dia.	6.000	Ea.	55.20	357	412.20
Tee, black, straight, 3/4" dia.	2.000	Ea.	17.20	171	188.20
Tee, black, straight, 1" dia.	2.000	Ea.	29.50	193	222.50
Tee, black, reducing, 1" dia.	2.000	Ea.	38	193	231
Pipe cap, black, 3/4" dia.	1.000	Ea.	6.15	24	30.15
Union, black with brass seat, 3/4" dia.	2.000	Ea.	42	119	161
Union, black with brass seat, 1" dia.	2.000	Ea.	54	129	183
Pipe nipples, black, 3/4" dia.	5.000	Ea.	12.10	31.63	43.73
Pipe nipples, black, 1" dia.	3.000	Ea.	12.83	21.83	34.66
Valves, bronze, gate, N.R.S., threaded, class 150, 3/4" size	2.000	Ea.	228	77	305
Valves, bronze, gate, N.R.S., threaded, class 150, 1" size	2.000	Ea.	296	81	377
Gas cock, brass, 3/4" size	1.000	Ea.	19.90	35	54.90
Thermometer, stem type, 9" case, 8" stem, 3/4" NPT	2.000	Ea.	412	56	468
Tank, steel, liquid expansion, ASME, painted, 15 gallon capacity	1.000	Ea.	940	83	1,023
Pump, circulating, bronze, flange connection, 3/4" to 1-1/2" size, 1/8 HP	1.000	Ea.	1,375	231	1,606
Vent chimney, all fuel, pressure tight, double wall, SS, 6" dia.	20.000	L.F.	1,330	450	1,780
Vent chimney, elbow, 90° fixed, 6" dia.	2.000	Ea.	930	91	1,021
Vent chimney, Tee, 6" dia.	2.000	Ea.	554	113	667
Vent chimney, ventilated roof thimble, 6" dia.	1.000	Ea.	290	52.50	342.50
Vent chimney, flat roof flashing, 6" dia.	1.000	Ea.	122	45.50	167.50
Vent chimney, stack cap, 6" diameter	1.000	Ea.	335	29.50	364.50
Insulation, fiberglass pipe covering, 1" wall, 1" IPS	20.000	L.F.	23.80	118	141.80
TOTAL			10,281.18	6,019.42	16,300.60

D3020 130	Boiler, Cast Iron, Hot Water, Gas	COST EACH		
		MAT.	INST.	TOTAL
1010	Boiler, cast iron, gas, hot water, 100 MBH	10,300	6,025	16,325
1020	200 MBH	12,000	6,875	18,875
1030	320 MBH	10,900	7,475	18,375
1040	440 MBH	15,100	9,650	24,750
1050	544 MBH	24,600	15,300	39,900
1060	765 MBH	29,600	16,400	46,000

D3020 Heat Generating Systems

D3020 130	Boiler, Cast Iron, Hot Water, Gas	COST EACH		
		MAT.	INST.	TOTAL
1070	1088 MBH	33,100	17,500	50,600
1080	1530 MBH	39,600	23,100	62,700
1090	2312 MBH	49,500	27,100	76,600
1100	2856 MBH	58,500	29,300	87,800
1110	3808 MBH	65,500	34,300	99,800
1120	4720 MBH	104,500	38,400	142,900
1130	6100 MBH	183,000	42,800	225,800
1140	6970 MBH	170,500	52,500	223,000

D3020 134	Boiler, Cast Iron, Steam, Gas	COST EACH		
		MAT.	INST.	TOTAL
1010	Boiler, cast iron, gas, steam, 100 MBH	8,125	5,825	13,950
1020	200 MBH	20,200	10,900	31,100
1030	320 MBH	21,800	12,300	34,100
1040	544 MBH	29,200	19,400	48,600
1050	765 MBH	32,900	20,200	53,100
1060	1275 MBH	39,000	22,200	61,200
1070	2675 MBH	55,500	32,300	87,800
1080	3570 MBH	67,000	36,300	103,300
1090	4720 MBH	97,500	38,700	136,200
1100	6100 MBH	125,500	45,000	170,500
1110	6970 MBH	137,500	52,500	190,000

D3020 136	Boiler, Cast Iron, Hot Water, Gas/Oil	COST EACH		
		MAT.	INST.	TOTAL
1010	Boiler, cast iron, gas & oil, hot water, 584 MBH	30,900	16,900	47,800
1020	876 MBH	41,000	18,200	59,200
1030	1168 MBH	49,200	20,500	69,700
1040	1460 MBH	55,500	24,700	80,200
1050	2044 MBH	63,000	25,500	88,500
1060	2628 MBH	80,000	30,900	110,900
1070	3210 MBH	85,000	33,200	118,200
1080	3796 MBH	92,000	34,300	126,300
1090	4672 MBH	107,500	35,900	143,400
1100	5256 MBH	133,000	43,400	176,400
1110	6000 MBH	158,000	44,100	202,100
1120	9800 MBH	197,500	75,500	273,000
1130	12,200 MBH	248,000	88,000	336,000
1140	13,500 MBH	269,500	93,500	363,000

D3020 138	Boiler, Cast Iron, Steam, Gas/Oil	COST EACH		
		MAT.	INST.	TOTAL
1010	Boiler, cast iron, gas & oil, steam, 810 MBH	37,500	21,900	59,400
1020	1360 MBH	42,900	23,400	66,300
1030	2040 MBH	50,500	30,000	80,500
1040	2700 MBH	53,500	33,400	86,900
1050	3270 MBH	65,000	37,400	102,400
1060	3770 MBH	104,000	38,100	142,100
1070	4650 MBH	116,000	43,200	159,200
1080	5520 MBH	133,500	45,500	179,000
1090	6100 MBH	137,500	47,800	185,300
1100	6970 MBH	147,500	58,000	205,500

For customer support on your Mechanical Costs with RSMeans Data, call 800.448.8182.

535

D3020 Heat Generating Systems

Base Mounted End-Suction Pump

System Components	QUANTITY	UNIT	COST EACH		
			MAT.	INST.	TOTAL
SYSTEM D3020 330 1010					
PUMP, BASE MTD. WITH MOTOR, END-SUCTION, 2-1/2″ SIZE, 3 HP, TO 150 GPM					
Pump, circulating, CI, base mounted, 2-1/2″ size, 3 HP, to 150 GPM	1.000	Ea.	11,400	770	12,170
Pipe, black steel, Sch. 40, on yoke & roll hangers, 10′ O.C., 2-1/2″ dia	12.000	L.F.	156	383.88	539.88
Elbow, 90°, weld joint, steel, 2-1/2″ pipe size	1.000	Ea.	38.50	188.65	227.15
Flange, weld neck, 150 LB, 2-1/2″ pipe size	8.000	Ea.	328	754.40	1,082.40
Valve, iron body, gate, 125 lb., N.R.S., flanged, 2-1/2″ size	1.000	Ea.	690	278	968
Strainer, Y type, iron body, flanged, 125 lb., 2-1/2″ pipe size	1.000	Ea.	177	283	460
Multipurpose valve, CI body, 2″ size	1.000	Ea.	630	98	728
Expansion joint, flanged spool, 6″ F to F, 2-1/2″ dia	2.000	Ea.	610	228	838
T-O-L, weld joint, socket, 1/4″ pipe size, nozzle	2.000	Ea.	18.20	131.20	149.40
T-O-L, weld joint, socket, 1/2″ pipe size, nozzle	2.000	Ea.	18.20	136.60	154.80
Control gauges, pressure or vacuum, 3-1/2″ diameter dial	2.000	Ea.	40	49	89
Pressure/temperature relief plug, 316 SS, 3/4″ OD, 7-1/2″ insertion	2.000	Ea.	128	49	177
Insulation, fiberglass pipe covering, 1-1/2″ wall, 2-1/2″ IPS	12.000	L.F.	39.36	86.40	125.76
Water flow	1.000	Ea.	1,400	1,200	2,600
Pump balancing	1.000	Ea.		345	345
TOTAL			15,673.26	4,981.13	20,654.39

D3020 330	Circulating Pump Systems, End Suction	COST EACH		
		MAT.	INST.	TOTAL
1010	Pump, base mtd with motor, end-suction, 2-1/2″ size, 3 HP, to 150 GPM *80″/gpm*	15,700	4,975	20,675
1020	3″ size, 5 HP, to 225 GPM *64″/*	19,100	5,625	24,725
1030	4″ size, 7-1/2 HP, to 350 GPM *48″/*	21,700	6,675	28,375
1040	5″ size, 15 HP, to 1000 GPM *42″/*	32,400	9,875	42,275
1050	6″ size, 25 HP, to 1550 GPM *22″/*	43,500	11,900	55,400

D3020 340	Circulating Pump Systems, Double Suction	COST EACH		
		MAT.	INST.	TOTAL
1010	Pump, base mtd w/motor, double suction, 6″ size, 50 HP, to 1200 GPM	25,100	16,900	42,000
1020	8″ size, 75 HP, to 2500 GPM	39,500	22,600	62,100
1030	8″ size, 100 HP, to 3000 GPM	53,500	25,200	78,700
1040	10″ size, 150 HP, to 4000 GPM	74,000	28,500	102,500

D3030 Cooling Generating Systems

Chilled Water Supply & Return Piping

Air Cooled Water Chiller Unit

Roof

Insulate

Return

Fan Coil Unit

Supply — Finish Ceiling

Design Assumptions: The chilled water, air cooled systems priced, utilize reciprocating hermetic compressors and propeller-type condenser fans. Piping with pumps and expansion tanks is included based on a two pipe system. No ducting is included and the fan-coil units are cooling only. Water treatment and balancing are not included. Chilled water piping is insulated. Area distribution is through the use of multiple fan coil units. Fewer but larger fan coil units with duct distribution would be approximately the same S.F. cost.

System Components	QUANTITY	UNIT	COST EACH		
			MAT.	INST.	TOTAL
SYSTEM D3030 110 1200					
PACKAGED CHILLER, AIR COOLED, WITH FAN COIL UNIT					
APARTMENT CORRIDORS, 3,000 S.F., 5.50 TON					
Fan coil air conditioning unit, cabinet mounted & filters chilled water	1.000	Ea.	3,986.78	650.72	4,637.50
Water chiller, air conditioning unit, air cooled	1.000	Ea.	7,067.50	2,420	9,487.50
Chilled water unit coil connections	1.000	Ea.	1,325	1,775	3,100
Chilled water distribution piping	440.000	L.F.	11,000	23,980	34,980
TOTAL			23,379.28	28,825.72	52,205
COST PER S.F.			7.79	9.61	17.40

*Cooling requirements would lead to choosing a water cooled unit.

D3030 110	Chilled Water, Air Cooled Condenser Systems	COST PER S.F.		
		MAT.	INST.	TOTAL
1180	Packaged chiller, air cooled, with fan coil unit			
1200	Apartment corridors, 3,000 S.F., 5.50 ton	7.80	9.60	17.40
1240	6,000 S.F., 11.00 ton	6.35	7.65	14
1280	10,000 S.F., 18.33 ton	5.40	5.80	11.20
1320	20,000 S.F., 36.66 ton	4.16	4.17	8.33
1360	40,000 S.F., 73.33 ton	5.35	4.18	9.53
1440	Banks and libraries, 3,000 S.F., 12.50 ton	11.45	11.15	22.60
1480	6,000 S.F., 25.00 ton	10.40	9.05	19.45
1520	10,000 S.F., 41.66 ton	8.05	6.30	14.35
1560	20,000 S.F., 83.33 ton	9	5.65	14.65
1600	40,000 S.F., 167 ton*			
1680	Bars and taverns, 3,000 S.F., 33.25 ton	19.55	13.05	32.60
1720	6,000 S.F., 66.50 ton	17.45	9.90	27.35
1760	10,000 S.F., 110.83 ton	13.15	2.98	16.13
1800	20,000 S.F., 220 ton*			
1840	40,000 S.F., 440 ton*			
1920	Bowling alleys, 3,000 S.F., 17.00 ton	14.30	12.50	26.80
1960	6,000 S.F., 34.00 ton	10.95	8.75	19.70
2000	10,000 S.F., 56.66 ton	9.40	6.25	15.65
2040	20,000 S.F., 113.33 ton	9.95	5.70	15.65
2080	40,000 S.F., 227 ton*			
2160	Department stores, 3,000 S.F., 8.75 ton	10.75	10.70	21.45
2200	6,000 S.F., 17.50 ton	8.05	8.20	16.25
2240	10,000 S.F., 29.17 ton	6.50	5.90	12.40
2280	20,000 S.F., 58.33 ton	5.40	4.30	9.70
2320	40,000 S.F., 116.66 ton	6.55	4.29	10.84

D3030 Cooling Generating Systems

D3030 110	Chilled Water, Air Cooled Condenser Systems	COST PER S.F.		
		MAT.	INST.	TOTAL
2400	Drug stores, 3,000 S.F., 20.00 ton	16.20	12.90	29.10
2440	6000 S.F., 40.00 ton	12.75	9.40	22.15
2480	10,000 S.F., 66.66 ton	13.70	8.05	21.75
2520	20,000 S.F., 133.33 ton	11.80	5.95	17.75
2560	40,000 S.F., 267 ton*			
2640	Factories, 2,000 S.F., 10.00 ton	10.15	10.75	20.90
2680	6,000 S.F., 20.00 ton	9.10	8.65	17.75
2720	10,000 S.F., 33.33 ton	7	6	13
2760	20,000 S.F., 66.66 ton	7.90	5.40	13.30
2800	40,000 S.F., 133.33 ton	7.05	4.40	11.45
2880	Food supermarkets, 3,000 S.F., 8.50 ton	10.60	10.65	21.25
2920	6,000 S.F., 17.00 ton	7.90	8.15	16.05
2960	10,000 S.F., 28.33 ton	6.15	5.75	11.90
3000	20,000 S.F., 56.66 ton	5.20	4.20	9.40
3040	40,000 S.F., 113.33 ton	6.25	4.30	10.55
3120	Medical centers, 3,000 S.F., 7.00 ton	9.60	10.35	19.95
3160	6,000 S.F., 14.00 ton	7.15	7.90	15.05
3200	10,000 S.F., 23.33 ton	6.30	6	12.30
3240	20,000 S.F., 46.66 ton	4.96	4.39	9.35
3280	40,000 S.F., 93.33 ton	5.85	4.23	10.08
3360	Offices, 3,000 S.F., 9.50 ton	9.65	10.55	20.20
3400	6,000 S.F., 19.00 ton	8.85	8.55	17.40
3440	10,000 S.F., 31.66 ton	6.80	5.95	12.75
3480	20,000 S.F., 63.33 ton	7.95	5.50	13.45
3520	40,000 S.F., 126.66 ton	7	4.44	11.44
3600	Restaurants, 3,000 S.F., 15.00 ton	12.80	11.60	24.40
3640	6,000 S.F., 30.00 ton	10.50	8.75	19.25
3680	10,000 S.F., 50.00 ton	9.10	6.35	15.45
3720	20,000 S.F., 100.00 ton	10.25	6	16.25
3760	40,000 S.F., 200 ton*			
3840	Schools and colleges, 3,000 S.F., 11.50 ton	10.95	11	21.95
3880	6,000 S.F., 23.00 ton	9.90	8.90	18.80
3920	10,000 S.F., 38.33 ton	7.65	6.20	13.85
3960	20,000 S.F., 76.66 ton	8.85	5.70	14.55
4000	40,000 S.F., 153 ton*			

D3030 Cooling Generating Systems

Reciprocating Package Chiller
Condenser Water
Cooling Tower
Chilled Water Supply & Return Piping
Cooling Tower Water Makeup
Roof Structure
Finish Ceiling
Insulate Return
Supply
Fan Coil Unit

General: Water cooled chillers are available in the same sizes as air cooled units. They are also available in larger capacities.

Design Assumptions: The chilled water systems with water cooled condenser include reciprocating hermetic compressors, water cooling tower, pumps, piping and expansion tanks and are based on a two pipe system. Chilled water piping is insulated. No ducts are included and fan-coil units are cooling only. Area distribution is through use of multiple fan coil units. Fewer but larger fan coil units with duct distribution would be approximately the same S.F. cost. Water treatment and balancing are not included.

System Components	QUANTITY	UNIT	COST EACH		
			MAT.	INST.	TOTAL
SYSTEM D3030 115 1320					
PACKAGED CHILLER, WATER COOLED, WITH FAN COIL UNIT					
APARTMENT CORRIDORS, 4,000 S.F., 7.33 TON					
Fan coil air conditioner unit, cabinet mounted & filters, chilled water	2.000	Ea.	5,313.53	867.27	6,180.80
Water chiller, water cooled, 1 compressor, hermetic scroll,	1.000	Ea.	5,770.80	4,213.60	9,984.40
Cooling tower, draw thru single flow, belt drive	1.000	Ea.	1,634.59	179.59	1,814.18
Cooling tower pumps & piping	1.000	System	813.63	425.14	1,238.77
Chilled water unit coil connections	2.000	Ea.	2,650	3,550	6,200
Chilled water distribution piping	520.000	L.F.	13,000	28,340	41,340
TOTAL			29,182.55	37,575.60	66,758.15
COST PER S.F.			7.30	9.39	16.69

*Cooling requirements would lead to choosing a water cooled unit.

D3030 115	Chilled Water, Cooling Tower Systems		COST PER S.F.		
			MAT.	INST.	TOTAL
1300	Packaged chiller, water cooled, with fan coil unit				
1320	Apartment corridors, 4,000 S.F., 7.33 ton		7.30	9.40	16.70
1360	6,000 S.F., 11.00 ton	R236000 -20	6.10	8.05	14.15
1400	10,000 S.F., 18.33 ton		6.05	6.05	12.10
1440	20,000 S.F., 26.66 ton		4.86	4.35	9.21
1480	40,000 S.F., 73.33 ton		6	4.34	10.34
1520	60,000 S.F., 110.00 ton		5.70	4.61	10.31
1600	Banks and libraries, 4,000 S.F., 16.66 ton		13	10.20	23.20
1640	6,000 S.F., 25.00 ton		11.35	8.95	20.30
1680	10,000 S.F., 41.66 ton		8.75	6.60	15.35
1720	20,000 S.F., 83.33 ton		10.30	5.95	16.25
1760	40,000 S.F., 166.66 ton		9.25	7.80	17.05
1800	60,000 S.F., 250.00 ton		9.50	8.20	17.70
1880	Bars and taverns, 4,000 S.F., 44.33 ton		20.50	12.60	33.10
1920	6,000 S.F., 66.50 ton		22	13.15	35.15
1960	10,000 S.F., 110.83 ton		19.75	10.05	29.80
2000	20,000 S.F., 221.66 ton		21	10.55	31.55
2040	40,000 S.F., 440 ton*				
2080	60,000 S.F., 660 ton*				
2160	Bowling alleys, 4,000 S.F., 22.66 ton		14.95	11.40	26.35
2200	6,000 S.F., 34.00 ton		13.30	9.60	22.90
2240	10,000 S.F., 56.66 ton		10.55	7.15	17.70
2280	20,000 S.F., 113.33 ton		11.15	6.50	17.65
2320	40,000 S.F., 226.66 ton		12.05	7.75	19.80

For customer support on your Mechanical Costs with RSMeans Data, call 800.448.8182.

539

D3030 Cooling Generating Systems

D3030 115	Chilled Water, Cooling Tower Systems	COST PER S.F.		
		MAT.	INST.	TOTAL
2360	60,000 S.F., 340 ton			
2440	Department stores, 4,000 S.F., 11.66 ton	8.10	10.10	18.20
2480	6,000 S.F., 17.50 ton	9.65	8.35	18
2520	10,000 S.F., 29.17 ton	7.30	6.05	13.35
2560	20,000 S.F., 58.33 ton	5.50	4.58	10.08
2600	40,000 S.F., 116.66 ton	6.80	4.65	11.45
2640	60,000 S.F., 175.00 ton	8.40	7.55	15.95
2720	Drug stores, 4,000 S.F., 26.66 ton	16.75	11.40	28.15
2760	6,000 S.F., 40.00 ton	13.90	9.75	23.65
2800	10,000 S.F., 66.66 ton	14.20	8.85	23.05
2840	20,000 S.F., 133.33 ton	11.80	6.90	18.70
2880	40,000 S.F., 266.67 ton	12.55	8.75	21.30
2920	60,000 S.F., 400 ton*			
3000	Factories, 4,000 S.F., 13.33 ton	11.25	9.70	20.95
3040	6,000 S.F., 20.00 ton	9.60	8.50	18.10
3080	10,000 S.F., 33.33 ton	7.95	6.25	14.20
3120	20,000 S.F., 66.66 ton	8.05	5.70	13.75
3160	40,000 S.F., 133.33 ton	7.05	4.86	11.91
3200	60,000 S.F., 200.00 ton	8.65	7.90	16.55
3280	Food supermarkets, 4,000 S.F., 11.33 ton	8	10.05	18.05
3320	6,000 S.F., 17.00 ton	8.55	8.30	16.85
3360	10,000 S.F., 28.33 ton	7.15	6	13.15
3400	20,000 S.F., 56.66 ton	5.65	4.59	10.24
3440	40,000 S.F., 113.33 ton	6.75	4.65	11.40
3480	60,000 S.F., 170.00 ton	8.35	7.55	15.90
3560	Medical centers, 4.000 S.F., 9.33 ton	7	9.25	16.25
3600	6,000 S.F., 14.00 ton	8.40	8	16.40
3640	10,000 S.F., 23.33 ton	6.80	5.95	12.75
3680	20,000 S.F., 46.66 ton	5.10	4.50	9.60
3720	40,000 S.F., 93.33 ton	6.55	4.54	11.09
3760	60,000 S.F., 140.00 ton	7.20	7.65	14.85
3840	Offices, 4,000 S.F., 12.66 ton	10.90	9.65	20.55
3880	6,000 S.F., 19.00 ton	9.55	8.75	18.30
3920	10,000 S.F., 31.66 ton	7.90	6.30	14.20
3960	20,000 S.F., 63.33 ton	7.95	5.75	13.70
4000	40,000 S.F., 126.66 ton	8.20	7.40	15.60
4040	60,000 S.F., 190.00 ton	8.35	7.80	16.15
4120	Restaurants, 4,000 S.F., 20.00 ton	13.45	10.60	24.05
4160	6,000 S.F., 30.00 ton	12.05	9	21.05
4200	10,000 S.F., 50.00 ton	9.80	6.85	16.65
4240	20,000 S.F., 100.00 ton	11.40	6.40	17.80
4280	40,000 S.F., 200.00 ton	9.65	7.50	17.15
4320	60,000 S.F., 300.00 ton	10.55	8.50	19.05
4400	Schools and colleges, 4,000 S.F., 15.33 ton	12.35	10	22.35
4440	6,000 S.F., 23.00 ton	10.75	8.80	19.55
4480	10,000 S.F., 38.33 ton	8.30	6.45	14.75
4520	20,000 S.F., 76.66 ton	9.90	5.90	15.80
4560	40,000 S.F., 153.33 ton	8.75	7.60	16.35
4600	60,000 S.F., 230.00 ton	8.80	7.90	16.70

D3030 Cooling Generating Systems

Chilled Water Piping

Reciprocating Chiller, Water Cooled

Condenser Water Piping

System Components	QUANTITY	UNIT	COST EACH		
			MAT.	INST.	TOTAL
SYSTEM D3030 130 1010					
CHILLER, RECIPROCATING, WATER COOLED, STD. CONTROLS, 60 TON					
Water chiller, water cooled, 60 ton cooling	1.000	Ea.	36,200	12,000	48,200
Pipe, black steel, welded, Sch. 40, on yoke & roll hgrs, 10' O.C., 4" dia	40.000	L.F.	1,060	1,626.80	2,686.80
Elbow, 90°, weld joint, butt, 4" pipe size	8.000	Ea.	500	2,412	2,912
Flange, welding neck, 150 lb., 4" pipe size	16.000	Ea.	744	2,411.20	3,155.20
Valves, iron body, butterfly, lug type, lever actuator, 4" size	4.000	Ea.	1,168	1,112	2,280
Expansion joints, bellows type, flange spool, 6" F to F, 4" dia	4.000	Ea.	1,480	672	2,152
T-O-L socket, weld joint, 1/4" pipe size, nozzle	4.000	Ea.	36.40	262.40	298.80
T-O-L socket, weld joint, 3/4" pipe size, nozzle	4.000	Ea.	41.80	286.40	328.20
Thermometers, stem type, 9" case, 8" stem, 3/4" NPT	4.000	Ea.	824	112	936
Gauges, pressure or vacuum, 3-1/2" diameter dial	4.000	Ea.	80	98	178
Insulation, fiberglass pipe covering, 1-1/2" wall, 4" IPS	20.000	L.F.	87	185	272
Chiller balancing	1.000	Ea.		665	665
TOTAL			42,221.20	21,842.80	64,064

D3030 130		Reciprocating Chiller, Water Cooled	COST EACH		
			MAT.	INST.	TOTAL
1010	Chiller, reciprocating, water cooled, std. controls, 60 Ton		42,200	21,800	64,000
1020	100 Ton		83,500	27,700	111,200
1030	150 Ton		84,500	35,700	120,200
1040	200 Ton		113,000	37,600	150,600

D3030 135		Reciprocating Chiller, Air Cooled	COST EACH		
			MAT.	INST.	TOTAL
1010	Chiller, reciprocating, air cooled, std. controls, 20 Ton		28,100	12,700	40,800
1020	30 Ton		35,200	13,400	48,600
1030	40 Ton		41,000	14,300	55,300

Chilled Water Piping

Centrifugal Chiller, Water Cooled, Hermetic

Condenser Water Piping

System Components	QUANTITY	UNIT	COST EACH		
			MAT.	INST.	TOTAL
SYSTEM D3030 140 1010					
CHILLER, CENTRIF., WATER COOLED, PKGD. HERMETIC, STD. CONTROLS, 200 TON					
Water chiller, cntfgl, water cooled, not incl. tower, 200 ton	1.000	Ea.	144,500	16,100	160,600
Pipe, black steel, welded, Sch. 40, on yoke & roll hgrs, 10' OC, 6" dia	40.000	L.F.	1,980	2,530	4,510
Elbow, 90°, weld joint, 6" pipe size	8.000	Ea.	1,104	3,628	4,732
Flange, welding neck, 150 lb., 6" pipe size	16.000	Ea.	1,120	3,643.20	4,763.20
Valves, iron body, butterfly, lug type, lever actuator, 6" size	4.000	Ea.	1,960	1,720	3,680
Expansion joints, bellows type, flange spool, 6" F to F, 6" dia	4.000	Ea.	1,860	832	2,692
T-O-L socket, weld jiont, 1/4" pipe size, nozzle	4.000	Ea.	36.40	262.40	298.80
T-O-L socket, weld joint, 3/4" pipe size, nozzle	4.000	Ea.	41.80	286.40	328.20
Thermometers, stem type, 9" case, 8" stem, 3/4" NPT	4.000	Ea.	824	112	936
Gauges, pressure or vacuum, 3-1/2" diameter dial	4.000	Ea.	80	98	178
Insulation, fiberglass pipe covering, 1-1/2" wall, 6" IPS	40.000	L.F.	184.80	472	656.80
Chiller balancing	1.000	Ea.		665	665
TOTAL			153,691	30,349	184,040

D3030 140		Centrifugal Chiller, Water Cooled	COST EACH		
			MAT.	INST.	TOTAL
1010	Chiller, centrif., water cooled, pkgd. hermetic, std controls, 200 Ton		153,500	30,400	183,900
1020		400 Ton	183,500	45,900	229,400
1030		1,000 Ton	486,500	68,000	554,500
1040		1,500 Ton	659,000	84,000	743,000

Chilled Water Piping
(Condenser Water Piping Similar)

Gas Absorption Chiller, Water Cooled

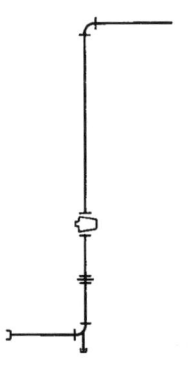

Gas Piping

System Components	QUANTITY	UNIT	COST EACH MAT.	COST EACH INST.	COST EACH TOTAL
SYSTEM D3030 145 1010					
CHILLER, GAS FIRED ABSORPTION, WATER COOLED, STD. CONTROLS, 800 TON					
Absorption water chiller, gas fired, water cooled, 800 ton	1.000	Ea.	662,500	35,600	698,100
Pipe, black steel, weld, Sch. 40, on yoke & roll hangers, 10' O.C., 10" dia	20.000	L.F.	1,980	1,897.60	3,877.60
Pipe, black steel, welded, Sch. 40, 14" dia., (two rod roll type hanger)	44.000	L.F.	5,060	6,679.20	11,739.20
Pipe, black steel, weld, Sch. 40, on yoke & roll hangers, 10' O.C., 4" dia	20.000	L.F.	530	813.40	1,343.40
Elbow, 90°, weld joint, 10" pipe size	4.000	Ea.	2,060	3,036	5,096
Elbow, 90°, weld joint, 14" pipe size	6.000	Ea.	6,750	6,801	13,551
Elbow, 90°, weld joint, 4" pipe size	3.000	Ea.	187.50	904.50	1,092
Flange, welding neck, 150 lb., 10" pipe size	8.000	Ea.	1,656	3,036	4,692
Flange, welding neck, 150 lb., 14" pipe size	14.000	Ea.	6,790	7,084	13,874
Flange, welding neck, 150 lb., 4" pipe size	3.000	Ea.	139.50	452.10	591.60
Valves, iron body, butterfly, lug type, gear operated, 10" size	2.000	Ea.	3,000	1,080	4,080
Valves, iron body, butterfly, lug type, gear operated, 14" size	5.000	Ea.	13,750	4,700	18,450
Valves, semi-steel, lubricated plug valve, flanged, 4" pipe size	1.000	Ea.	645	465	1,110
Expansion joints, bellows type, flange spool, 6" F to F, 10" dia	2.000	Ea.	1,490	566	2,056
Expansion joints, bellows type, flange spool, 10" F to F, 14" dia	2.000	Ea.	2,550	740	3,290
T-O-L socket, weld joint, 1/4" pipe size, nozzle	4.000	Ea.	36.40	262.40	298.80
T-O-L socket, weld joint, 3/4" pipe size, nozzle	4.000	Ea.	41.80	286.40	328.20
Thermometers, stem type, 9" case, 8" stem, 3/4" NPT	4.000	Ea.	824	112	936
Gauges, pressure or vacuum, 3-1/2" diameter dial	4.000	Ea.	80	98	178
Insulation, fiberglass pipe covering, 2" wall, 10" IPS	20.000	L.F.	205	370	575
Vent chimney, all fuel, pressure tight, double wall, SS, 12" dia	20.000	L.F.	1,550	620	2,170
Chiller balancing	1.000	Ea.		665	665
TOTAL			711,825.20	76,268.60	788,093.80

D3030 145	Gas Absorption Chiller, Water Cooled	COST EACH MAT.	COST EACH INST.	COST EACH TOTAL
1010	Chiller, gas fired absorption, water cooled, std. ctrls, 800 Ton	712,000	76,000	788,000
1020	1000 Ton	932,500	98,000	1,030,500

D3030 150	Steam Absorption Chiller, Water Cooled	COST EACH MAT.	COST EACH INST.	COST EACH TOTAL
1010	Chiller, steam absorption, water cooled, std. ctrls, 750 Ton	807,000	90,500	897,500
1020	955 Ton	940,500	108,000	1,048,500
1030	1465 Ton	1,361,500	123,500	1,485,000
1040	1660 Ton	1,637,500	140,500	1,778,000

D3030 Cooling Generating Systems

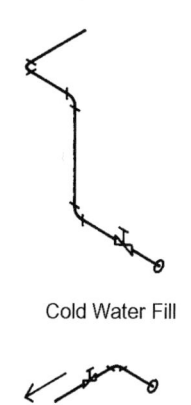

Cold Water Fill

Drain To Waste Piping

Cooling Tower

Condenser Water Piping

System Components	QUANTITY	UNIT	COST EACH		
430/T Less ATC			MAT.	INST.	TOTAL
SYSTEM D3030 310 1010					
COOLING TOWER, GALVANIZED STEEL, PACKAGED UNIT, DRAW THRU, 60 TON					
Cooling towers, draw thru, single flow, belt drive, 60 tons	1.000	Ea.	13,380	1,470	14,850
Pipe, black steel, welded, Sch. 40, on yoke & roll hangers, 10' OC, 4" dia.	40.000	L.F.	1,060	1,626.80	2,686.80
Tubing, copper, Type L, couplings & hangers 10' OC, 1-1/2" dia.	20.000	L.F.	255	297	552
Elbow, 90°, copper, cu x cu, 1-1/2" dia.	3.000	Ea.	53.10	178.50	231.60
Elbow, 90°, weld joint, 4" pipe size	6.000	Ea.	375	1,809	2,184
Flange, welding neck, 150 lb., 4" pipe size	6.000	Ea.	279	904.20	1,183.20
Valves, bronze, gate, N.R.S., soldered, 125 psi, 1-1/2" size	1.000	Ea.	203	59.50	262.50
Valves, iron body, butterfly, lug type, lever actuator, 4" size	2.000	Ea.	584	556	1,140
Electric heat trace system, 400 degree, 115v, 5 watts/L.F.	160.000	L.F.	1,936	220.80	2,156.80
Insulation, fiberglass pipe covering, , 1-1/2" wall, 4" IPS	40.000	L.F.	174	370	544
Insulation, fiberglass pipe covering, 1" wall, 1-1/2" IPS	20.000	L.F.	27.60	123	150.60
Cooling tower condenser control system	1.000	Ea.	6,675	3,200	9,875
Cooling tower balancing	1.000	Ea.		510	510
TOTAL			25,001.70	11,324.80	36,326.50 (≈26 000)

D3030 310	Galvanized Draw Through Cooling Tower	COST EACH		
		MAT.	INST.	TOTAL
1010	Cooling tower, galvanized steel, packaged unit, draw thru, 60 Ton	25,000	11,300	36,300
1020	110 Ton	33,600	13,300	46,900
1030	300 Ton	49,200	20,100	69,300
1040	600 Ton	60,500	30,000	90,500
1050	1000 Ton	121,500	44,200	165,700

D3030 320	Fiberglass Draw Through Cooling Tower	COST EACH		
		MAT.	INST.	TOTAL
1010	Cooling tower, fiberglass, packaged unit, draw thru, 100 Ton	27,000	11,400	38,400
1020	120 Ton	31,400	12,900	44,300
1030	140 Ton	36,400	17,200	53,600
1040	160 Ton	49,300	22,700	72,000
1050	180 Ton	69,000	32,900	101,900

D3030 330	Stainless Steel Draw Through Cooling Tower	COST EACH		
		MAT.	INST.	TOTAL
1010	Cooling tower, stainless steel, packaged unit, draw thru, 60 Ton	44,500	11,300	55,800
1020	110 Ton	53,500	13,300	66,800
1030	300 Ton	91,000	20,100	111,100
1040	600 Ton	141,500	29,500	171,000
1050	1000 Ton	244,000	44,200	288,200

D3040 Distribution Systems

Built-Up Air Handling Unit

System Components	QUANTITY	UNIT	COST EACH		
			MAT.	INST.	TOTAL
SYSTEM D3040 106 1010					
AHU, FIELD FAB., BLT. UP, COOL/HEAT COILS, FLTRS., CONST. VOL., 40,000 CFM					
AHU, Built-up, cool/heat coils, filters, mix box, constant volume	1.000	Ea.	110,000	7,325	117,325
Pipe, black steel, welded, Sch. 40, on yoke & roll hgrs, 10' OC, 3" dia.	26.000	L.F.	475.80	915.72	1,391.52
Pipe, black steel, welded, Sch. 40, on yoke & roll hgrs, 10' OC, 4" dia	20.000	L.F.	530	813.40	1,343.40
Elbow, 90°, weld joint, 3" pipe size	5.000	Ea.	165	1,073.75	1,238.75
Elbow, 90°, weld joint, 4" pipe size	4.000	Ea.	250	1,206	1,456
Tee, welded, reducing on outlet, 3" by 2-1/2" pipe	3.000	Ea.	498	1,123.50	1,621.50
Tee, welded, reducing on outlet, 4" by 3" pipe	2.000	Ea.	270	1,008	1,278
Reducer, eccentric, weld joint, 3" pipe size	4.000	Ea.	364	754.60	1,118.60
Reducer, eccentric, weld joint, 4" pipe size	4.000	Ea.	728	1,002	1,730
Flange, weld neck, 150 LB, 2-1/2" pipe size	3.000	Ea.	123	282.90	405.90
Flange, weld neck, 150 lb., 3" pipe size	14.000	Ea.	630	1,502.90	2,132.90
Flange, weld neck, 150 lb., 4" pipe size	10.000	Ea.	465	1,507	1,972
Valves, iron body, butterfly, lug type, lever actuator, 3" size	2.000	Ea.	566	348	914
Valves, iron body, butterfly, lug type, lever actuator, 4" size	2.000	Ea.	584	556	1,140
Strainer, Y type, iron body, flanged, 125 lb., 3" pipe size	1.000	Ea.	207	315	522
Strainer, Y type, iron body, flanged, 125 lb., 4" pipe size	1.000	Ea.	415	470	885
Valve, electric motor actuated, iron body, 3 way, flanged, 2-1/2" pipe	1.000	Ea.	1,300	380	1,680
Valve, electric motor actuated, iron body, two way, flanged, 3" pipe size	1.000	Ea.	1,325	380	1,705
Expansion joints, bellows type, flange spool, 6" F to F, 3" dia	2.000	Ea.	680	248	928
Expansion joints, bellows type, flange spool, 6" F to F, 4" dia	2.000	Ea.	740	336	1,076
T-O-L socket, weld joint, 1/4" pipe size, nozzle	4.000	Ea.	36.40	262.40	298.80
T-O-L socket, weld joint, 1/2" pipe size, nozzle	12.000	Ea.	109.20	819.60	928.80
Thermometers, stem type, 9" case, 8" stem, 3/4" NPT	4.000	Ea.	824	112	936
Pressure/temperature relief plug, 316 SS, 3/4"OD, 7-1/2" insertion	4.000	Ea.	256	98	354
Insulation, fiberglass pipe covering, 1-1/2" wall, 3" IPS	26.000	L.F.	89.18	198.90	288.08
Insulation, fiberglass pipe covering, 1-1/2" wall, 4" IPS	20.000	L.F.	87	185	272
Fan coil control system	1.000	Ea.	1,500	715	2,215
Re-heat coil balancing	2.000	Ea.		290	290
Air conditioner balancing	1.000	Ea.		910	910
TOTAL			123,217.58	25,138.67	148,356.25

D3040 106	Field Fabricated Air Handling Unit	COST EACH		
		MAT.	INST.	TOTAL
1010	AHU, Field fab, blt up, cool/heat coils, fltrs, const vol, 40,000 CFM	123,000	25,200	148,200
1020	60,000 CFM	130,500	38,800	169,300
1030	75,000 CFM	132,000	40,900	172,900

D30 HVAC

D3040 Distribution Systems

D3040 108	Field Fabricated VAV Air Handling Unit	COST EACH		
		MAT.	INST.	TOTAL
1010	AHU, Field fab, blt up, cool/heat coils, fltrs, VAV, 75,000 CFM	134,000	46,900	180,900
1020	100,000 CFM	214,500	62,000	276,500
1030	150,000 CFM	349,500	99,000	448,500

D3040 Distribution Systems

| Chilled Water Piping | Central Station Air Handling Unit | Hot Water Piping |

System Components	QUANTITY	UNIT	COST EACH		
			MAT.	INST.	TOTAL
SYSTEM D3040 110 1010					
AHU, CENTRAL STATION, COOL/HEAT COILS, CONST. VOL., FLTRS., 2,000 CFM					
Central station air handling unit, chilled water, 2,000 CFM	1.000	Ea.	12,000	1,275	13,275
Pipe, black steel, Sch 40, threaded, W/cplgs & hangers, 10' OC, 1" dia.	32.500	L.F.	277.88	472.88	750.76
Pipe, black steel, Sch 40, threaded, W/cplg & hangers, 10' OC, 1-1/4" dia.	26.000	L.F.	234	405.60	639.60
Elbow, 90°, black, straight, 1" dia.	4.000	Ea.	36.80	238	274.80
Elbow, 90°, black, straight, 1-1/4" dia.	4.000	Ea.	61	252	313
Tee, black, reducing, 1" dia.	8.000	Ea.	152	772	924
Tee, black, reducing, 1-1/4" dia.	8.000	Ea.	328	792	1,120
Reducer, concentric, black, 1" dia.	4.000	Ea.	55.60	206	261.60
Reducer, concentric, black, 1-1/4" dia.	4.000	Ea.	69	214	283
Union, black with brass seat, 1" dia.	4.000	Ea.	108	258	366
Union, black with brass seat, 1-1/4" dia.	2.000	Ea.	91	132	223
Pipe nipple, black, 1" dia	13.000	Ea.	55.58	94.58	150.16
Pipe nipple, black, 1-1/4" dia	12.000	Ea.	54	93.60	147.60
Valves, bronze, gate, N.R.S., threaded, class 150, 1" size	2.000	Ea.	296	81	377
Valves, bronze, gate, N.R.S., threaded, class 150, 1-1/4" size	2.000	Ea.	302	103	405
Strainer, Y type, bronze body, screwed, 150 lb., 1" pipe size	1.000	Ea.	73.50	46	119.50
Strainer, Y type, bronze body, screwed, 150 lb., 1-1/4" pipe size	1.000	Ea.	148	52.50	200.50
Valve, electric motor actuated, brass, 3 way, screwed, 3/4" pipe	1.000	Ea.	435	42	477
Valve, electric motor actuated, brass, 2 way, screwed, 1" pipe size	1.000	Ea.	680	40.50	720.50
Thermometers, stem type, 9" case, 8" stem, 3/4" NPT	4.000	Ea.	824	112	936
Pressure/temperature relief plug, 316 SS, 3/4"OD, 7-1/2" insertion	4.000	Ea.	256	98	354
Insulation, fiberglass pipe covering, 1" wall, 1" IPS	26.000	L.F.	30.94	153.40	184.34
Insulation, fiberglass pipe covering, 1" wall, 1-1/4" IPS	20.000	L.F.	25.60	123	148.60
Fan coil control system	1.000	Ea.	1,500	715	2,215
Re-heat coil balancing	2.000	Ea.		290	290
Multizone A/C balancing	1.000	Ea.		680	680
TOTAL			18,093.90	7,742.06	25,835.96

D3040 110	Central Station Air Handling Unit	COST EACH		
		MAT.	INST.	TOTAL
1010	AHU, Central station, cool/heat coils, constant vol, fltrs, 2,000 CFM	18,100	7,750	25,850
1020	5,000 CFM	27,100	9,975	37,075
1030	10,000 CFM	39,200	14,000	53,200
1040	15,000 CFM	72,500	19,100	91,600
1050	20,000 CFM	89,000	22,300	111,300

D3040 Distribution Systems

D3040 112	Central Station Air Handling Unit, VAV	COST EACH		
		MAT.	INST.	TOTAL
1010	AHU, Central station, cool/heat coils, VAV, fltrs, 5,000 CFM	31,400	10,400	41,800
1020	10,000 CFM	56,500	14,400	70,900
1030	15,000 CFM	81,500	18,800	100,300
1040	20,000 CFM	100,500	22,800	123,300
1050	30,000 CFM	118,500	26,200	144,700

D30 HVAC

D3040 Distribution Systems

Cooling Coil — Two-Way Valve Piping

Heating Coil — Three-Way Valve Piping

Exhaust Section
Mixing Box
Filter Section
Coil Section
Return Air Fan Section
Roof Mounting Curb
Supply Air Fan Section

Rooftop Air Handling Unit

System Components	QUANTITY	UNIT	COST EACH MAT.	COST EACH INST.	COST EACH TOTAL
SYSTEM D3040 114 1010					
AHU, ROOFTOP, COOL/HEAT COILS, FLTRS., CONST. VOL., 2,000 CFM					
AHU, Built-up, cool/heat coils, filter, mix box, rooftop, constant volume	1.000	Ea.	26,300	2,200	28,500
Pipe, black steel, Sch 40, threaded, W/cplgs & hangers, 10' OC, 1" dia.	32.500	L.F.	277.88	472.88	750.76
Pipe, black steel, Sch 40, threaded, W/cplg & hangers, 10' OC, 1-1/4" dia.	26.000	L.F.	234	405.60	639.60
Elbow, 90°, black, straight, 1" dia.	4.000	Ea.	36.80	238	274.80
Elbow, 90°, black, straight, 1-1/4" dia.	4.000	Ea.	61	252	313
Tee, black, reducing, 1" dia.	8.000	Ea.	152	772	924
Tee, black, reducing, 1-1/4" dia.	8.000	Ea.	328	792	1,120
Reducer, black, concentric, 1" dia.	4.000	Ea.	55.60	206	261.60
Reducer, black, concentric, 1-1/4" dia.	4.000	Ea.	69	214	283
Union, black with brass seat, 1" dia.	4.000	Ea.	108	258	366
Union, black with brass seat, 1-1/4" dia.	2.000	Ea.	91	132	223
Pipe nipples, 1" dia	13.000	Ea.	55.58	94.58	150.16
Pipe nipples, 1-1/4" dia	12.000	Ea.	54	93.60	147.60
Valves, bronze, gate, N.R.S., threaded, class 150, 1" size	2.000	Ea.	296	81	377
Valves, bronze, gate, N.R.S., threaded, class 150, 1-1/4" size	2.000	Ea.	302	103	405
Strainer, Y type, bronze body, screwed, 150 lb., 1" pipe size	1.000	Ea.	73.50	46	119.50
Strainer, Y type, bronze body, screwed, 150 lb., 1-1/4" pipe size	1.000	Ea.	148	52.50	200.50
Valve, electric motor actuated, brass, 3 way, screwed, 3/4" pipe	1.000	Ea.	435	42	477
Valve, electric motor actuated, brass, 2 way, screwed, 1" pipe size	1.000	Ea.	680	40.50	720.50
Thermometers, stem type, 9" case, 8" stem, 3/4" NPT	4.000	Ea.	824	112	936
Pressure/temperature relief plug, 316 SS, 3/4"OD, 7-1/2" insertion	4.000	Ea.	256	98	354
Insulation, fiberglass pipe covering, 1" wall, 1" IPS	26.000	L.F.	30.94	153.40	184.34
Insulation, fiberglass pipe covering, 1" wall, 1-1/4" IPS	20.000	L.F.	25.60	123	148.60
Insulation pipe covering .016" aluminum jacket finish, add	45.000	S.F.	50.85	292.50	343.35
Fan coil control system	1.000	Ea.	1,500	715	2,215
Re-heat coil balancing	2.000	Ea.		290	290
Rooftop unit heat/cool balancing	1.000	Ea.		530	530
TOTAL			32,444.75	8,809.56	41,254.31

D3040 114	Rooftop Air Handling Unit	COST EACH MAT.	COST EACH INST.	COST EACH TOTAL
1010	AHU, Rooftop, cool/heat coils, constant vol., fltrs, 2,000 CFM	32,400	8,800	41,200
1020	5,000 CFM	39,900	10,200	50,100

D30 HVAC

D3040 Distribution Systems

D3040 114	Rooftop Air Handling Unit	COST EACH		
		MAT.	INST.	TOTAL
1030	10,000 CFM	80,500	14,300	94,800
1040	15,000 CFM	115,000	19,000	134,000
1050	20,000 CFM	154,500	23,000	177,500

(handwritten: 948/cfm, 893/cfm, 885/cfm)

D3040 116	Rooftop Air Handling Unit, VAV	COST EACH		
		MAT.	INST.	TOTAL
1010	AHU, Rooftop, cool/heat coils, VAV, fltrs, 5,000 CFM	41,500	10,400	51,900
1020	10,000 CFM	82,000	14,500	96,500
1030	15,000 CFM	124,000	19,200	143,200
1040	20,000 CFM	157,000	23,200	180,200
1050	30,000 CFM	183,000	26,700	209,700

D3040 Distribution Systems

Galvanized Steel Duct

Fiberglass Duct Insulation

Flexible Fiberglass Duct

Horizontal Fan Coil Air Conditioning System

System Components	QUANTITY	UNIT	COST EACH		
			MAT.	INST.	TOTAL
SYSTEM D3040 124 1010					
FAN COIL A/C SYSTEM, HORIZONTAL, W/HOUSING, CONTROLS, 2 PIPE, 1/2 TON					
Fan coil A/C, horizontal housing, filters, chilled water, 1/2 ton cooling	1.000	Ea.	1,575	177	1,752
Pipe, black steel, Sch 40, threaded, W/cplgs & hangers, 10' OC, 3/4" dia.	21.500	L.F.	104.06	271.98	376.04
Elbow, 90°, black, straight, 3/4" dia.	6.000	Ea.	27.18	330	357.18
Tee, black, straight, 3/4" dia.	2.000	Ea.	17.20	171	188.20
Union, black with brass seat, 3/4" dia.	2.000	Ea.	42	119	161
Pipe nipples, 3/4" diam	3.000	Ea.	7.26	18.98	26.24
Valves, bronze, gate, N.R.S., threaded, class 150, 3/4" size	2.000	Ea.	228	77	305
Circuit setter, balance valve, bronze body, threaded, 3/4" pipe size	1.000	Ea.	99	39	138
Insulation, fiberglass pipe covering, 1" wall, 3/4" IPS	20.000	L.F.	21.60	113	134.60
Ductwork, 12" x 8" fabricated, galvanized steel, 12 LF	55.000	Lb.	36.30	495	531.30
Insulation, ductwork, blanket type, fiberglass, 1" thk, 1-1/2 LB density	40.000	S.F.	197.60	618	815.60
Diffusers, aluminum, OB damper, ceiling, perf, 24"x24" panel size, 6"x6"	2.000	Ea.	364	95	459
Ductwork, flexible, fiberglass fabric, insulated, 1"thk, PE jacket, 6" dia	16.000	L.F.	63.52	84	147.52
Round volume control damper 6" dia.	2.000	Ea.	73	69	142
Fan coil unit balancing	1.000	Ea.		98	98
Re-heat coil balancing	1.000	Ea.		145	145
Diffuser/register, high, balancing	2.000	Ea.		272	272
TOTAL			2,855.72	3,192.96	6,048.68

D3040 118	Fan Coil A/C Unit, Two Pipe	COST EACH		
		MAT.	INST.	TOTAL
1010	Fan coil A/C system, cabinet mounted, controls, 2 pipe, 1/2 Ton	1,175	1,450	2,625
1020	1 Ton	1,675	1,625	3,300
1030	1-1/2 Ton	1,725	1,650	3,375
1040	2 Ton	2,725	2,000	4,725
1050	3 Ton	3,625	2,100	5,725

D3040 120	Fan Coil A/C Unit, Two Pipe, Electric Heat	COST EACH		
		MAT.	INST.	TOTAL
1010	Fan coil A/C system, cabinet mntd, elect. ht, controls, 2 pipe, 1/2 Ton	2,175	1,450	3,625
1020	1 Ton	2,700	1,625	4,325
1030	1-1/2 Ton	3,150	1,650	4,800
1040	2 Ton	4,600	2,050	6,650
1050	3 Ton	7,550	2,100	9,650

D30 HVAC

D3040   Distribution Systems

D3040 122	Fan Coil A/C Unit, Four Pipe	COST EACH		
		MAT.	INST.	TOTAL
1010	Fan coil A/C system, cabinet mounted, controls, 4 pipe, 1/2 Ton	2,000	2,825	4,825
1020	1 Ton	2,600	3,000	5,600
1030	1-1/2 Ton	2,650	3,050	5,700
1040	2 Ton	3,275	3,150	6,425
1050	3 Ton	4,775	3,375	8,150

D3040 124	Fan Coil A/C, Horizontal, Duct Mount, 2 Pipe	COST EACH		
		MAT.	INST.	TOTAL
1010	Fan coil A/C system, horizontal w/housing, controls, 2 pipe, 1/2 Ton	2,850	3,200	6,050
1020	1 Ton	3,900	5,100	9,000
1030	1-1/2 Ton	4,825	6,825	11,650
1040	2 Ton	6,175	8,475	14,650
1050	3 Ton	7,950	11,800	19,750
1060	3-1/2 Ton	8,275	12,100	20,375
1070	4 Ton	8,700	14,100	22,800
1080	5 Ton	10,500	18,100	28,600
1090	6 Ton	10,700	19,300	30,000
1100	7 Ton	10,900	19,800	30,700
1110	8 Ton	10,900	20,100	31,000
1120	10 Ton	11,200	20,600	31,800

D3040 126	Fan Coil A/C, Horiz., Duct Mount, 2 Pipe, Elec. Ht.	COST EACH		
		MAT.	INST.	TOTAL
1010	Fan coil A/C system, horiz. hsng, elect. ht, ctrls, 2 pipe, 1/2 Ton	3,125	3,200	6,325
1020	1 Ton	4,300	5,100	9,400
1030	1-1/2 Ton	5,250	6,825	12,075
1040	2 Ton	6,850	8,475	15,325
1050	3 Ton	10,700	12,100	22,800
1060	3-1/2 Ton	11,800	12,100	23,900
1070	4 Ton	13,200	14,100	27,300
1080	5 Ton	16,200	18,100	34,300
1090	6 Ton	16,300	19,300	35,600
1100	7 Ton	17,000	20,000	37,000
1110	8 Ton	17,200	20,700	37,900
1120	10 Ton	18,200	19,800	38,000

D3040 128	Fan Coil A/C, Horizontal, Duct Mount, 4 Pipe	COST EACH		
		MAT.	INST.	TOTAL
1010	Fan coil A/C system, horiz. w/cabinet, controls, 4 pipe, 1/2 Ton	4,025	4,550	8,575
1020	1 Ton	5,200	6,600	11,800
1030	1-1/2 Ton	6,200	8,225	14,425
1040	2 Ton	7,225	9,625	16,850
1050	3 Ton	9,450	13,300	22,750
1060	3-1/2 Ton	11,100	16,800	27,900
1070	4 Ton	11,100	17,000	28,100
1080	5 Ton	13,600	23,900	37,500
1090	6 Ton	15,200	28,900	44,100
1100	7 Ton	17,400	32,600	50,000
1110	8 Ton	18,000	34,900	52,900
1120	10 Ton	21,000	42,100	63,100

D3040 Distribution Systems

Fiberglass Duct Insulation

Galvanized Steel Duct

Flexible Insulated Duct

VAV Terminal

Fan Powered VAV Terminal

System Components	QUANTITY	UNIT	COST EACH		
			MAT.	INST.	TOTAL
SYSTEM D3040 132 1010					
VAV TERMINAL, COOLING ONLY, WITH ACTUATOR/CONTROLS, 200 CFM					
Mixing box variable volume 300 to 600 CFM, cool only	1.000	Ea.	984.50	136.40	1,120.90
Ductwork, 12" x 8" fabricated, galvanized steel, 12 LF	55.000	Lb.	36.30	495	531.30
Insulation, ductwork, blanket type, fiberglass, 1" thk, 1-1/2 LB density	40.000	S.F.	197.60	618	815.60
Diffusers, aluminum, OB damper, ceiling, perf, 24"x24" panel size, 6"x6"	2.000	Ea.	364	95	459
Ductwork, flexible, fiberglass fabric, insulated, 1"thk, PE jacket, 6" dia	16.000	L.F.	63.52	84	147.52
Round volume control damper 6" dia.	2.000	Ea.	73"	69	142
VAV box balancing	1.000	Ea.		90.50	90.50
Diffuser/register, high, balancing	2.000	Ea.		272	272
TOTAL			1,718.92	1,859.90	3,578.82

D3040 132	VAV Terminal, Cooling Only	COST EACH		
		MAT.	INST.	TOTAL
1010	VAV Terminal, cooling only, with actuator/controls, 200 CFM	1,725	1,850	3,575
1020	400 CFM	2,250	3,575	5,825
1030	600 CFM	2,975	5,325	8,300
1040	800 CFM	3,275	6,725	10,000
1050	1000 CFM	3,750	7,750	11,500
1060	1250 CFM	4,500	10,100	14,600
1070	1500 CFM	4,950	12,100	17,050
1080	2000 CFM	6,475	17,100	23,575

D3040 134	VAV Terminal, Hot Water Reheat	COST EACH		
		MAT.	INST.	TOTAL
1010	VAV Terminal, cooling, HW reheat, with actuator/controls, 200 CFM	3,425	3,075	6,500
1020	400 CFM	3,975	4,650	8,625
1030	600 CFM	4,650	6,525	11,175
1040	800 CFM	5,000	7,950	12,950
1050	1000 CFM	5,450	8,975	14,425
1060	1250 CFM	6,325	11,400	17,725
1070	1500 CFM	7,050	13,700	20,750
1080	2000 CFM	8,750	18,700	27,450

D3040 136	Fan Powered VAV Terminal, Cooling Only	COST EACH		
		MAT.	INST.	TOTAL
1010	VAV Terminal, cooling, fan powrd, with actuator/controls, 200 CFM	2,625	1,850	4,475
1020	400 CFM	3,425	3,425	6,850

D3040 Distribution Systems

D3040 136	Fan Powered VAV Terminal, Cooling Only	COST EACH		
		MAT.	INST.	TOTAL
1030	600 CFM	4,100	5,325	9,425
1040	800 CFM	4,575	6,650	11,225
1050	1000 CFM	5,025	7,650	12,675
1060	1250 CFM	5,950	10,200	16,150
1070	1500 CFM	6,375	12,200	18,575
1080	2000 CFM	8,075	17,400	25,475

D3040 138	Fan Powered VAV Terminal, Hot Water Reheat	COST EACH		
		MAT.	INST.	TOTAL
1010	VAV Terminal, cool, HW reht, fan powrd, with actuator/ctrls, 200 CFM	4,525	3,050	7,575
1020	400 CFM	5,175	4,675	9,850
1030	600 CFM	5,850	6,525	12,375
1040	800 CFM	6,300	7,875	14,175
1050	1000 CFM	6,750	8,900	15,650
1060	1250 CFM	7,875	11,400	19,275
1070	1500 CFM	8,625	13,800	22,425
1080	2000 CFM	10,300	19,100	29,400

D30 HVAC

D3040 Distribution Systems

Axial Flow Centrifugal Fan
Belt Drive

Belt Drive Utility Set

Centrifugal Roof Exhaust Fan
Direct Drive

System Components	QUANTITY	UNIT	COST EACH		
			MAT.	INST.	TOTAL
SYSTEM D3040 220 1010					
FAN SYSTEM, IN-LINE CENTRIFUGAL, 500 CFM					
Fans, in-line centrifugal, supply/exhaust, 500 CFM, 10″ dia conn	1.000	Ea.	1,775	575	2,350
Louver, alum., w/screen, damper, mill fin, fxd blade, cont line, stormproof	4.050	S.F.	200.48	109.35	309.83
Motor for louver damper, electric or pneumatic	1.000	Ea.	570	54	624
Ductwork, 12″ x 8″ fabricated, galvanized steel, 60 LF	200.000	Lb.	163.80	2,249	2,412.80
Grilles, aluminum, air supply, single deflection, adjustable, 12″ x 6″	1.000	Ea.	38.70	99	137.70
Fire damper, curtain type, vertical, 12″ x 8″	1.000	Ea.	32.50	34.50	67
Duct access door, insulated, 10″ x 10″	1.000	Ea.	25.50	68.50	94
Duct access door, insulated, 16″ x 12″	1.000	Ea.	42.50	84	126.50
Pressure controller/switch, air flow controller	1.000	Ea.	196	65.50	261.50
Fan control switch and mounting base	1.000	Ea.	118	49	167
Fan balancing	1.000	Ea.		455	455
Diffuser/register balancing	3.000	Ea.		271.50	271.50
TOTAL			3,162.48	4,114.35	7,276.83

D3040 220	Centrifugal In Line Fan Systems	COST EACH		
		MAT.	INST.	TOTAL
1010	Fan system, in-line centrifugal, 500 CFM	3,150	4,125	7,275
1020	1300 CFM	3,725	7,125	10,850
1030	1500 CFM	3,950	8,225	12,175
1040	2500 CFM	5,925	25,100	31,025
1050	3500 CFM	7,400	33,900	41,300
1060	5000 CFM	9,425	50,500	59,925
1070	7500 CFM	11,300	51,500	62,800
1080	10,000 CFM	13,500	60,500	74,000

D3040 230	Utility Set Fan Systems	COST EACH		
		MAT.	INST.	TOTAL
1010	Utility fan set system, belt drive, 2000 CFM	2,875	13,300	16,175
1020	3500 CFM	4,825	23,500	28,325
1030	5000 CFM	6,025	32,700	38,725
1040	7500 CFM	8,550	37,900	46,450
1050	10,000 CFM	11,400	49,900	61,300
1060	15,000 CFM	11,800	51,000	62,800
1070	20,000 CFM	14,100	57,500	71,600

D30 HVAC

D3040 Distribution Systems

D3040 240	Roof Exhaust Fan Systems	COST EACH		
		MAT.	INST.	TOTAL
1010	Roof vent. system, centrifugal, alum., galv curb, BDD, 500 CFM	1,375	2,700	4,075
1020	800 CFM	1,900	5,200	7,100
1030	1500 CFM	2,800	7,700	10,500
1040	2750 CFM	3,800	17,500	21,300
1050	3500 CFM	4,650	22,500	27,150
1060	5000 CFM	6,350	37,100	43,450
1070	8500 CFM	8,225	47,700	55,925
1080	13,800 CFM	11,600	69,500	81,100

D3040 Distribution Systems

Commercial/Industrial Dust Collection System

System Components	QUANTITY	UNIT	COST EACH		
			MAT.	INST.	TOTAL
SYSTEM D3040 250 1010					
COMMERCIAL/INDUSTRIAL VACUUM DUST COLLECTION SYSTEM, 500 CFM					
Dust collection central vac unit inc. stand, filter & shaker, 500 CFM 2 HP	1.000	Ea.	5,325	720	6,045
Galvanized tubing, 16 ga., 4" OD	80.000	L.F.	648	302.40	950.40
Galvanized tubing, 90° ell, 4" dia.	3.000	Ea.	126	81	207
Galvanized tubing, 45° ell, 4" dia.	4.000	Ea.	146	108	254
Galvanized tubing, TY slip fit, 4" dia.	4.000	Ea.	200	182	382
Flexible rubber hose, 4" dia.	32.000	L.F.	284.80	120.96	405.76
Galvanized air gate valve, 4" dia.	4.000	Ea.	720	236	956
Galvanized compression coupling, 4" dia.	36.000	Ea.	936	1,476	2,412
Pipe hangers, steel, 4" pipe size	16.000	Ea.	496	202.40	698.40
TOTAL			8,881.80	3,428.76	12,310.56

D3040 250	Commercial/Industrial Vacuum Dust Collection	COST EACH		
		MAT.	INST.	TOTAL
1010	Commercial/industrial, vacuum dust collection system, 500 CFM	8,875	3,425	12,300
1020	1000 CFM	12,500	3,900	16,400
1030	1500 CFM	15,800	4,325	20,125
1040	3000 CFM	36,500	6,050	42,550
1050	5000 CFM	46,100	9,100	55,200

D3040 Distribution Systems

Commercial Kitchen Exhaust/Make-up Air Rooftop System

System Components	QUANTITY	UNIT	COST EACH MAT.	COST EACH INST.	COST EACH TOTAL
SYSTEM D3040 260 1010					
COMMERCIAL KITCHEN EXHAUST/MAKE-UP AIR SYSTEM, ROOFTOP, GAS, 2000 CFM					
Make-up air rooftop unit, gas, 750 MBH	1.000	Ea.	5,931.10	168.32	6,099.42
Pipe, black steel, Sch 40, thrded, nipples, W/cplngs/hngrs, 10' OC, 1" dia.	22.000	L.F.	188.10	320.10	508.20
Elbow, 90°, black steel, straight, 3/4" dia.	2.000	Ea.	9.06	110	119.06
Elbow, 90°, black steel, straight, 1" dia.	3.000	Ea.	27.60	178.50	206.10
Tee, black steel, reducing, 1" dia.	1.000	Ea.	19	96.50	115.50
Union, black with brass seat, 1" dia.	1.000	Ea.	27	64.50	91.50
Pipe nipples, black, 3/4" dia	2.000	Ea.	9.68	25.30	34.98
Pipe cap, malleable iron, black, 1" dia.	1.000	Ea.	7.45	25.50	32.95
Gas cock, brass, 1" size	1.000	Ea.	37.50	40.50	78
Ductwork, 20" x 14" fabricated, galv. steel, 30 LF	220.000	Lb.	129.80	1,826	1,955.80
Ductwork, 18" x 10" fabricated, galv. steel, 12 LF	72.000	Lb.	42.48	597.60	640.08
Ductwork, 30" x 10" fabricated, galv. steel, 30 LF	260.000	Lb.	153.40	2,158	2,311.40
Ductwork, 16" x 10" fabricated, galv. steel, 12 LF	70.000	Lb.	41.30	581	622.30
Kitchen ventilation, island style, water wash, Stainless Steel	5.000	L.F.	14,000	1,490	15,490
Control system, Make-up Air Unit/Exhauster	1.000	Ea.	3,225	2,775	6,000
Rooftop unit heat/cool balancing	1.000	Ea.		530	530
Roof fan balancing	1.000	Ea.		300	300
TOTAL			23,848.47	11,286.82	35,135.29

D3040 260	Kitchen Exhaust/Make-Up Air	COST EACH MAT.	COST EACH INST.	COST EACH TOTAL
1010	Commercial kitchen exhaust/make-up air system, rooftop, gas, 2000 CFM	23,800	11,300	35,100
1020	3000 CFM	35,700	15,700	51,400
1030	5000 CFM	41,500	18,600	60,100
1040	8000 CFM	61,000	27,500	88,500
1050	12,000 CFM	79,000	32,300	111,300
1060	16,000 CFM	97,500	49,500	147,000

D3040   Distribution Systems

Plate Heat Exchanger   Shell and Tube Heat Exchanger

System Components	QUANTITY	UNIT	COST EACH		
			MAT.	INST.	TOTAL
SYSTEM D3040 620 1010					
SHELL & TUBE HEAT EXCHANGER, 40 GPM					
Heat exchanger, 4 pass, 3/4″ O.D. copper tubes, by steam at 10 psi, 40 GPM	1.000	Ea.	7,025	355	7,380
Pipe, black steel, Sch 40, threaded, w/couplings & hangers, 10′ OC, 2″ dia.	50.500	L.F.	383.80	1,085.75	1,469.55
Elbow, 90°, straight, 2″ dia.	16.000	Ea.	448	1,232	1,680
Reducer, black steel, concentric, 2″ dia.	2.000	Ea.	66	132	198
Valves, bronze, gate, rising stem, threaded, class 150, 2″ size	6.000	Ea.	1,482	420	1,902
Valves, bronze, globe, class 150, rising stem, threaded, 2″ size	2.000	Ea.	1,520	140	1,660
Strainers, Y type, bronze body, screwed, 125 lb, 2″ pipe size	2.000	Ea.	418	121	539
Valve, electric motor actuated, brass, 2 way, screwed, 1-1/2″ pipe size	1.000	Ea.	1,050	60	1,110
Union, black with brass seat, 2″ dia.	8.000	Ea.	524	652	1,176
Tee, black, straight, 2″ dia.	9.000	Ea.	400.50	1,134	1,534.50
Tee, black, reducing run and outlet, 2″ dia.	4.000	Ea.	226	504	730
Pipe nipple, black, 2″ dia.	2.000	Ea.	3.80	10.75	14.55
Thermometers, stem type, 9″ case, 8″ stem, 3/4″ NPT	2.000	Ea.	412	56	468
Gauges, pressure or vacuum, 3-1/2″ diameter dial	2.000	Ea.	40	49	89
Insulation, fiberglass pipe covering, 1″ wall, 2″ IPS	40.000	L.F.	78.40	260	338.40
Heat exchanger control system	1.000	Ea.	3,125	2,500	5,625
Coil balancing	1.000	Ea.		145	145
TOTAL			17,202.50	8,856.50	26,059

D3040 610	Heat Exchanger, Plate Type	COST EACH		
		MAT.	INST.	TOTAL
1010	Plate heat exchanger, 400 GPM	66,500	18,100	84,600
1020	800 GPM	107,000	23,000	130,000
1030	1200 GPM	160,500	31,200	191,700
1040	1800 GPM	216,000	38,900	254,900

D3040 620	Heat Exchanger, Shell & Tube	COST EACH		
		MAT.	INST.	TOTAL
1010	Shell & tube heat exchanger, 40 GPM	17,200	8,850	26,050
1020	96 GPM	28,700	16,000	44,700
1030	240 GPM	54,000	21,700	75,700
1040	600 GPM	108,500	30,400	138,900

| Cabinet Unit Heater | Gas Unit Heater | Hydronic Unit Heater |

System Components			COST EACH		
	QUANTITY	UNIT	MAT.	INST.	TOTAL
SYSTEM D3050 120 1010					
SPACE HEATER, SUSPENDED, GAS FIRED, PROPELLER FAN, 20 MBH					
Space heater, propeller fan, 20 MBH output	1.000	Ea.	1,900	166	2,066
Pipe, black steel, Sch 40, threaded, w/coupling & hangers, 10' OC, 3/4" dia	21.000	L.F.	101.64	265.65	367.29
Elbow, 90°, black steel, straight, 1/2" dia.	3.000	Ea.	13.20	154.50	167.70
Elbow, 90°, black steel, straight, 3/4" dia.	3.000	Ea.	13.59	165	178.59
Tee, black steel, reducing, 3/4" dia.	1.000	Ea.	14.45	85.50	99.95
Union, black with brass seat, 3/4" dia.	1.000	Ea.	21	59.50	80.50
Pipe nipples, black, 1/2" dia	2.000	Ea.	8.86	24.50	33.36
Pipe nipples, black, 3/4" dia	2.000	Ea.	4.84	12.65	17.49
Pipe cap, black, 3/4" dia.	1.000	Ea.	6.15	24	30.15
Gas cock, brass, 3/4" size	1.000	Ea.	19.90	35	54.90
Thermostat, 1 set back, electric, timed	1.000	Ea.	85.50	98	183.50
Wiring, thermostat hook-up,25' of #18-3	1.000	Ea.	7.45	30.50	37.95
Vent chimney, prefab metal, U.L. listed, gas, double wall, galv. st, 4" dia	12.000	L.F.	123.60	240	363.60
Vent chimney, gas, double wall, galv. steel, elbow 90°, 4" dia.	2.000	Ea.	59	80	139
Vent chimney, gas, double wall, galv. steel, Tee, 4" dia.	1.000	Ea.	41.50	52.50	94
Vent chimney, gas, double wall, galv. steel, T cap, 4" dia.	1.000	Ea.	2.83	32.50	35.33
Vent chimney, gas, double wall, galv. steel, roof flashing, 4" dia.	1.000	Ea.	10.05	40	50.05
Vent chimney, gas, double wall, galv. steel, top, 4" dia.	1.000	Ea.	20.50	31	51.50
TOTAL			2,454.06	1,596.80	4,050.86

D3050 120	Unit Heaters, Gas	COST EACH		
		MAT.	INST.	TOTAL
1010	Space heater, suspended, gas fired, propeller fan, 20 MBH	2,450	1,600	4,050
1020	60 MBH	2,625	1,675	4,300
1030	100 MBH	3,000	1,750	4,750
1040	160 MBH	3,625	1,925	5,550
1050	200 MBH	4,525	2,150	6,675
1060	280 MBH	5,275	2,275	7,550
1070	320 MBH	6,950	2,600	9,550

D3050 130	Unit Heaters, Hydronic	COST EACH		
		MAT.	INST.	TOTAL
1010	Space heater, suspended, horiz. mount, HW, prop. fan, 20 MBH	2,475	1,700	4,175
1020	60 MBH	3,400	1,925	5,325

For customer support on your Mechanical Costs with RSMeans Data, call 800.448.8182.

D3050 Terminal & Package Units

D3050 130	Unit Heaters, Hydronic	COST EACH		
		MAT.	INST.	TOTAL
1030	100 MBH	3,875	2,100	5,975
1040	150 MBH	4,300	2,325	6,625
1050	200 MBH	5,250	2,600	7,850
1060	300 MBH	5,950	3,050	9,000

D3050 140	Cabinet Unit Heaters, Hydronic	COST EACH		
		MAT.	INST.	TOTAL
1010	Unit heater, cabinet type, horizontal blower, hot water, 20 MBH	3,100	1,675	4,775
1020	60 MBH	4,575	1,850	6,425
1030	100 MBH	4,850	2,025	6,875
1040	120 MBH	4,850	2,100	6,950

For customer support on your Mechanical Costs with RSMeans Data, call 800.448.8182.

Roof — Roof Top Unit

Return Duct

Return Grille

Supply Diff.

Insulated Supply Duct

Finish Ceiling

System Description: Rooftop single zone units are electric cooling and gas heat. Duct systems are low velocity, galvanized steel supply and return. Price variations between sizes are due to several factors. Jumps in the cost of the rooftop unit occur when the manufacturer shifts from the largest capacity unit on a small frame to the smallest capacity on the next larger

frame, or changes from one compressor to two. As the unit capacity increases for larger areas the duct distribution grows in proportion. For most applications there is a tradeoff point where it is less expensive and more efficient to utilize smaller units with short simple distribution systems. Larger units also require larger initial supply and return ducts which can create a space

problem. Supplemental heat may be desired in colder locations. The table below is based on one unit supplying the area listed. The 10,000 S.F. unit for bars and taverns is not listed because a nominal 110 ton unit would be required and this is above the normal single zone rooftop capacity.

System Components	QUANTITY	UNIT	COST EACH		
			MAT.	INST.	TOTAL
SYSTEM D3050 150 1280					
ROOFTOP, SINGLE ZONE, AIR CONDITIONER					
APARTMENT CORRIDORS, 500 S.F., .92 TON		S.F.			
Rooftop air conditioner, 1 zone, electric cool, standard controls, curb	1.000	Ea.	1,460.50	931.50	2,392
Ductwork package for rooftop single zone units	1.000	System	294.40	1,311	1,605.40
TOTAL			1,754.90	2,242.50	3,997.40
COST PER S.F.			3.51	4.49	8

*Size would suggest multiple units.

D3050 150	Rooftop Single Zone Unit Systems		COST PER S.F.		
			MAT.	INST.	TOTAL
1260	Rooftop, single zone, air conditioner				
1280	Apartment corridors, 500 S.F., .92 ton		3.50	4.50	8
1320	1,000 S.F., 1.83 ton	R236000	3.50	4.46	7.96
1360	1500 S.F., 2.75 ton	-20	2.48	3.67	6.15
1400	3,000 S.F., 5.50 ton		2.40	3.54	5.94
1440	5,000 S.F., 9.17 ton		2.53	3.22	5.75
1480	10,000 S.F., 18.33 ton		2.95	3.02	5.97
1560	Banks or libraries, 500 S.F., 2.08 ton		7.95	10.15	18.10
1600	1,000 S.F., 4.17 ton		5.65	8.35	14
1640	1,500 S.F., 6.25 ton		5.45	8.05	13.50
1680	3,000 S.F., 12.50 ton		5.75	7.30	13.05
1720	5,000 S.F., 20.80 ton		6.70	6.85	13.55
1760	10,000 S.F., 41.67 ton		6.40	6.85	13.25
1840	Bars and taverns, 500 S.F. 5.54 ton		12.70	13.50	26.20
1880	1,000 S.F., 11.08 ton		13.50	11.55	25.05
1920	1,500 S.F., 16.62 ton		12.55	10.75	23.30
1960	3,000 S.F., 33.25 ton		16.50	10.25	26.75
2000	5,000 S.F., 55.42 ton		14.05	10.25	24.30
2040	10,000 S.F., 110.83 ton*				
2080	Bowling alleys, 500 S.F., 2.83 ton		7.65	11.35	19
2120	1,000 S.F., 5.67 ton		7.40	10.95	18.35
2160	1,500 S.F., 8.50 ton		7.80	9.95	17.75
2200	3,000 S.F., 17.00 ton		7.30	9.55	16.85
2240	5,000 S.F., 28.33 ton		9.35	9.25	18.60
2280	10,000 S.F., 56.67 ton		8.10	9.30	17.40
2360	Department stores, 500 S.F., 1.46 ton		5.55	7.10	12.65

D3050 Terminal & Package Units

D3050 150	Rooftop Single Zone Unit Systems	COST PER S.F.		
		MAT.	INST.	TOTAL
2400	1,000 S.F., 2.92 ton	3.94	5.85	9.79
2440	1,500 S.F., 4.37 ton	3.80	5.60	9.40
2480	3,000 S.F., 8.75 ton	4.02	5.10	9.12
2520	5,000 S.F., 14.58 ton	3.77	4.91	8.68
2560	10,000 S.F., 29.17 ton	4.81	4.78	9.59
2640	Drug stores, 500 S.F., 3.33 ton	9	13.35	22.35
2680	1,000 S.F., 6.67 ton	8.70	12.85	21.55
2720	1,500 S.F., 10.00 ton	9.20	11.70	20.90
2760	3,000 S.F., 20.00 ton	10.70	11	21.70
2800	5,000 S.F., 33.33 ton	11	10.90	21.90
2840	10,000 S.F., 66.67 ton	9.55	10.90	20.45
2920	Factories, 500 S.F., 1.67 ton	6.35	8.10	14.45
2960	1,000 S.F., 3.33 ton	4.51	6.70	11.21
3000	1,500 S.F., 5.00 ton	4.35	6.45	10.80
3040	3,000 S.F., 10.00 ton	4.60	5.85	10.45
3080	5,000 S.F., 16.67 ton	4.31	5.60	9.91
3120	10,000 S.F., 33.33 ton	5.50	5.45	10.95
3200	Food supermarkets, 500 S.F., 1.42 ton	5.40	6.90	12.30
3240	1,000 S.F., 2.83 ton	3.80	5.65	9.45
3280	1,500 S.F., 4.25 ton	3.70	5.45	9.15
3320	3,000 S.F., 8.50 ton	3.91	4.98	8.89
3360	5,000 S.F., 14.17 ton	3.67	4.77	8.44
3400	10,000 S.F., 28.33 ton	4.68	4.64	9.32
3480	Medical centers, 500 S.F., 1.17 ton	4.45	5.70	10.15
3520	1,000 S.F., 2.33 ton	4.45	5.70	10.15
3560	1,500 S.F., 3.50 ton	3.16	4.69	7.85
3600	3,000 S.F., 7.00 ton	3.05	4.51	7.56
3640	5,000 S.F., 11.67 ton	3.22	4.10	7.32
3680	10,000 S.F., 23.33 ton	3.75	3.84	7.59
3760	Offices, 500 S.F., 1.58 ton	6.05	7.70	13.75
3800	1,000 S.F., 3.17 ton	4.28	6.35	10.63
3840	1,500 S.F., 4.75 ton	4.13	6.10	10.23
3880	3,000 S.F., 9.50 ton	4.37	5.55	9.92
3920	5,000 S.F., 15.83 ton	4.09	5.30	9.39
3960	10,000 S.F., 31.67 ton	5.20	5.20	10.40
4000	Restaurants, 500 S.F., 2.50 ton	9.55	12.20	21.75
4040	1,000 S.F., 5.00 ton	6.55	9.65	16.20
4080	1,500 S.F., 7.50 ton	6.90	8.80	15.70
4120	3,000 S.F., 15.00 ton	6.45	8.40	14.85
4160	5,000 S.F., 25.00 ton	8.05	8.25	16.30
4200	10,000 S.F., 50.00 ton	7.15	8.20	15.35
4240	Schools and colleges, 500 S.F., 1.92 ton	7.30	9.35	16.65
4280	1,000 S.F., 3.83 ton	5.20	7.70	12.90
4320	1,500 S.F., 5.75 ton	5	7.40	12.40
4360	3,000 S.F., 11.50 ton	5.30	6.75	12.05
4400	5,000 S.F., 19.17 ton	6.15	6.30	12.45
4440	10,000 S.F., 38.33 ton	5.90	6.30	12.20
5000				
9000	Components of ductwork packages for above systems, per ton of cooling:			
9010	Ductwork; galvanized steel, 120 pounds			
9020	Insulation; fiberglass 2" thick, FRK faced, 52 S.F.			
9030	Diffusers; aluminum, 24" x 12", one			
9040	Registers; aluminum, return with dampers, one			

D3050 Terminal & Package Units

System Description: Rooftop units are multizone with up to 12 zones, and include electric cooling, gas heat, thermostats, filters, supply and return fans complete. Duct systems are low velocity, galvanized steel supply and return with insulated supplies.

Multizone units cost more per ton of cooling than single zone. However, they offer flexibility where load conditions are varied due to heat generating areas or exposure to radiational heating. For example, perimeter offices on the "sunny side" may require cooling at the same time "shady side" or central offices may require heating. It is possible to accomplish similar results using duct heaters in branches of the single zone unit. However, heater location could be a problem and total system operating energy efficiency could be lower.

System Components			COST EACH		
	QUANTITY	UNIT	MAT.	INST.	TOTAL
SYSTEM D3050 155 1280					
ROOFTOP, MULTIZONE, AIR CONDITIONER					
APARTMENT CORRIDORS, 3,000 S.F., 5.50 TON					
Rooftop multizone unit, standard controls, curb	1.000	Ea.	32,120	2,156	34,276
Ductwork package for rooftop multizone units	1.000	System	2,282.50	14,987.50	17,270
TOTAL			34,402.50	17,143.50	51,546
COST PER S.F.			11.47	5.71	17.18

Note A: Small single zone unit recommended.
Note B: A combination of multizone units recommended.

D3050 155	Rooftop Multizone Unit Systems		COST PER S.F.		
			MAT.	INST.	TOTAL
1240	Rooftop, multizone, air conditioner				
1260	Apartment corridors, 1,500 S.F., 2.75 ton. See Note A.				
1280	3,000 S.F., 5.50 ton	R236000 -20	11.47	5.70	17.17
1320	10,000 S.F., 18.30 ton		8.05	5.50	13.55
1360	15,000 S.F., 27.50 ton		7.90	5.50	13.40
1400	20,000 S.F., 36.70 ton		8.20	5.50	13.70
1440	25,000 S.F., 45.80 ton		7.65	5.50	13.15
1520	Banks or libraries, 1,500 S.F., 6.25 ton		26	13	39
1560	3,000 S.F., 12.50 ton		22	12.70	34.70
1600	10,000 S.F., 41.67 ton		17.40	12.55	29.95
1640	15,000 S.F., 62.50 ton		12.95	12.45	25.40
1680	20,000 S.F., 83.33 ton		12.95	12.45	25.40
1720	25,000 S.F., 104.00 ton		11.45	12.40	23.85
1800	Bars and taverns, 1,500 S.F., 16.62 ton		56	18.70	74.70
1840	3,000 S.F., 33.24 ton		47	18.05	65.05
1880	10,000 S.F., 110.83 ton		28	17.95	45.95
1920	15,000 S.F., 165 ton, See Note B.				
1960	20,000 S.F., 220 ton, See Note B.				
2000	25,000 S.F., 275 ton, See Note B.				
2080	Bowling alleys, 1,500 S.F., 8.50 ton		35.50	17.65	53.15
2120	3,000 S.F., 17.00 ton		30	17.30	47.30
2160	10,000 S.F., 56.70 ton		23.50	17.05	40.55
2200	15,000 S.F., 85.00 ton		17.60	16.95	34.55
2240	20,000 S.F., 113.00 ton		15.55	16.85	32.40
2280	25,000 S.F., 140.00 ton see Note B.				
2360	Department stores, 1,500 S.F., 4.37 ton, See Note A.				

D3050 Terminal & Package Units

D3050 155	Rooftop Multizone Unit Systems	COST PER S.F.		
		MAT.	INST.	TOTAL
2400	3,000 S.F., 8.75 ton	18.25	9.10	27.35
2440	10,000 S.F., 29.17 ton	13	8.75	21.75
2480	15,000 S.F., 43.75 ton	12.15	8.80	20.95
2520	20,000 S.F., 58.33 ton	9.05	8.70	17.75
2560	25,000 S.F., 72.92 ton	9.05	8.75	17.80
2640	Drug stores, 1,500 S.F., 10.00 ton	41.50	21	62.50
2680	3,000 S.F., 20.00 ton	29.50	20	49.50
2720	10,000 S.F., 66.66 ton	20.50	19.95	40.45
2760	15,000 S.F., 100.00 ton	18.35	19.90	38.25
2800	20,000 S.F., 135 ton, See Note B.			
2840	25,000 S.F., 165 ton, See Note B.			
2920	Factories, 1,500 S.F., 5 ton, See Note A.			
2960	3,000 S.F., 10.00 ton	21	10.40	31.40
3000	10,000 S.F., 33.33 ton	14.90	9.95	24.85
3040	15,000 S.F., 50.00 ton	13.90	10.05	23.95
3080	20,000 S.F., 66.66 ton	10.35	9.95	20.30
3120	25,000 S.F., 83.33 ton	10.35	9.95	20.30
3200	Food supermarkets, 1,500 S.F., 4.25 ton, See Note A.			
3240	3,000 S.F., 8.50 ton	17.75	8.85	26.60
3280	10,000 S.F., 28.33 ton	12.20	8.45	20.65
3320	15,000 S.F., 42.50 ton	11.85	8.55	20.40
3360	20,000 S.F., 56.67 ton	8.80	8.45	17.25
3400	25,000 S.F., 70.83 ton	8.80	8.45	17.25
3480	Medical centers, 1,500 S.F., 3.5 ton, See Note A.			
3520	3,000 S.F., 7.00 ton	14.60	7.25	21.85
3560	10,000 S.F., 23.33 ton	9.95	7	16.95
3600	15,000 S.F., 35.00 ton	10.40	7	17.40
3640	20,000 S.F., 46.66 ton	9.75	7.05	16.80
3680	25,000 S.F., 58.33 ton	7.25	7	14.25
3760	Offices, 1,500 S.F., 4.75 ton, See Note A.			
3800	3,000 S.F., 9.50 ton	19.80	9.85	29.65
3840	10,000 S.F., 31.66 ton	14.15	9.45	23.60
3880	15,000 S.F., 47.50 ton	13.20	9.55	22.75
3920	20,000 S.F., 63.33 ton	9.85	9.45	19.30
3960	25,000 S.F., 79.16 ton	9.85	9.45	19.30
4000	Restaurants, 1,500 S.F., 7.50 ton	31.50	15.60	47.10
4040	3,000 S.F., 15.00 ton	26.50	15.25	41.75
4080	10,000 S.F., 50.00 ton	21	15.05	36.05
4120	15,000 S.F., 75.00 ton	15.55	14.95	30.50
4160	20,000 S.F., 100.00 ton	13.75	14.90	28.65
4200	25,000 S.F., 125 ton, See Note B.			
4240	Schools and colleges, 1,500 S.F., 5.75 ton	24	11.95	35.95
4280	3,000 S.F., 11.50 ton	20	11.70	31.70
4320	10,000 S.F., 38.33 ton	16	11.55	27.55
4360	15,000 S.F., 57.50 ton	11.90	11.45	23.35
4400	20,000 S.F.,76.66 ton	11.90	11.45	23.35
4440	25,000 S.F., 95.83 ton	11	11.45	22.45
4450				
9000	Components of ductwork packages for above systems, per ton of cooling:			
9010	Ductwork; galvanized steel, 240 pounds			
9020	Insulation; fiberglass, 2" thick, FRK faced, 104 S.F.			
9030	Diffusers; aluminum, 24" x 12", two			
9040	Registers; aluminum, return with dampers, one			

565

For customer support on your Mechanical Costs with RSMeans Data, call 800.448.8182.

D3050 Terminal & Package Units

System Description: Self-contained, single package water cooled units include cooling tower, pump, piping allowance. Systems for 1000 S.F. and up include duct and diffusers to provide for even distribution of air. Smaller units distribute air through a supply air plenum, which is integral with the unit.

Returns are not ducted and supplies are not insulated.

Hot water or steam heating coils are included but piping to boiler and the boiler itself is not included.

Where local codes or conditions permit single pass cooling for the smaller units, deduct 10%.

System Components	QUANTITY	UNIT	COST EACH		
			MAT.	INST.	TOTAL
SYSTEM D3050 160 1300					
SELF-CONTAINED, WATER COOLED UNIT					
APARTMENT CORRIDORS, 500 S.F., .92 TON					
Self-contained, water cooled, single package air conditioner unit	1.000	Ea.	1,386	739.20	2,125.20
Ductwork package for water or air cooled packaged units	1.000	System	62.10	989	1,051.10
Cooling tower, draw thru single flow, belt drive	1.000	Ea.	205.16	22.54	227.70
Cooling tower pumps & piping	1.000	System	102.12	53.36	155.48
TOTAL			1,755.38	1,804.10	3,559.48
COST PER S.F.			3.51	3.61	7.12

D3050 160	Self-contained, Water Cooled Unit Systems		COST PER S.F.		
			MAT.	INST.	TOTAL
1280	Self-contained, water cooled unit		3.50	3.62	7.12
1300	Apartment corridors, 500 S.F., .92 ton		3.50	3.60	7.10
1320	1,000 S.F., 1.83 ton	R236000	3.50	3.60	7.10
1360	3,000 S.F., 5.50 ton	-20	2.90	3.01	5.91
1400	5,000 S.F., 9.17 ton		2.67	2.79	5.46
1440	10,000 S.F., 18.33 ton		3.80	2.50	6.30
1520	Banks or libraries, 500 S.F., 2.08 ton		7.70	3.69	11.39
1560	1,000 S.F., 4.17 ton		6.60	6.85	13.45
1600	3,000 S.F., 12.50 ton		6.05	6.35	12.40
1640	5,000 S.F., 20.80 ton		8.65	5.65	14.30
1680	10,000 S.F., 41.66 ton		7.30	5.70	13
1760	Bars and taverns, 500 S.F., 5.54 ton		16.70	6.25	22.95
1800	1,000 S.F., 11.08 ton		15.70	10.90	26.60
1840	3,000 S.F., 33.25 ton		20.50	8.60	29.10
1880	5,000 S.F., 55.42 ton		17.05	9	26.05
1920	10,000 S.F., 110.00 ton		16.55	8.90	25.45
2000	Bowling alleys, 500 S.F., 2.83 ton		10.45	5.05	15.50
2040	1,000 S.F., 5.66 ton		8.95	9.30	18.25
2080	3,000 S.F., 17.00 ton		11.75	7.70	19.45
2120	5,000 S.F., 28.33 ton		10.65	7.45	18.10
2160	10,000 S.F., 56.66 ton		8.90	7.65	16.55
2200	Department stores, 500 S.F., 1.46 ton		5.40	2.59	7.99
2240	1,000 S.F., 2.92 ton		4.62	4.79	9.41
2280	3,000 S.F., 8.75 ton		4.25	4.45	8.70
2320	5,000 S.F., 14.58 ton		6.05	3.98	10.03
2360	10,000 S.F., 29.17 ton		5.50	3.84	9.34

D30 HVAC

D3050 Terminal & Package Units

D3050 160	Self-contained, Water Cooled Unit Systems	COST PER S.F.		
		MAT.	INST.	TOTAL
2440	Drug stores, 500 S.F., 3.33 ton	12.30	5.95	18.25
2480	1,000 S.F., 6.66 ton	10.55	10.95	21.50
2520	3,000 S.F., 20.00 ton	13.85	9.10	22.95
2560	5,000 S.F., 33.33 ton	14.50	8.95	23.45
2600	10,000 S.F., 66.66 ton	10.50	9	19.50
2680	Factories, 500 S.F., 1.66 ton	6.15	2.95	9.10
2720	1,000 S.F. 3.37 ton	5.35	5.55	10.90
2760	3,000 S.F., 10.00 ton	4.86	5.05	9.91
2800	5,000 S.F., 16.66 ton	6.90	4.53	11.43
2840	10,000 S.F., 33.33 ton	6.25	4.37	10.62
2920	Food supermarkets, 500 S.F., 1.42 ton	5.20	2.51	7.71
2960	1,000 S.F., 2.83 ton	5.40	5.55	10.95
3000	3,000 S.F., 8.50 ton	4.48	4.65	9.13
3040	5,000 S.F., 14.17 ton	4.12	4.32	8.44
3080	10,000 S.F., 28.33 ton	5.30	3.72	9.02
3160	Medical centers, 500 S.F., 1.17 ton	4.30	2.08	6.38
3200	1,000 S.F., 2.33 ton	4.47	4.59	9.06
3240	3,000 S.F., 7.00 ton	3.70	3.84	7.54
3280	5,000 S.F., 11.66 ton	3.40	3.56	6.96
3320	10,000 S.F., 23.33 ton	4.84	3.19	8.03
3400	Offices, 500 S.F., 1.58 ton	5.85	2.81	8.66
3440	1,000 S.F., 3.17 ton	6.05	6.20	12.25
3480	3,000 S.F., 9.50 ton	4.61	4.82	9.43
3520	5,000 S.F., 15.83 ton	6.55	4.31	10.86
3560	10,000 S.F., 31.67 ton	5.95	4.16	10.11
3640	Restaurants, 500 S.F., 2.50 ton	9.25	4.44	13.69
3680	1,000 S.F., 5.00 ton	7.95	8.20	16.15
3720	3,000 S.F., 15.00 ton	7.30	7.60	14.90
3760	5,000 S.F., 25.00 ton	10.40	6.80	17.20
3800	10,000 S.F., 50.00 ton	6.45	6.25	12.70
3880	Schools and colleges, 500 S.F., 1.92 ton	7.05	3.40	10.45
3920	1,000 S.F., 3.83 ton	6.05	6.30	12.35
3960	3,000 S.F., 11.50 ton	5.60	5.85	11.45
4000	5,000 S.F., 19.17 ton	7.95	5.20	13.15
4040	10,000 S.F., 38.33 ton	6.70	5.20	11.90
4050				
9000	Components of ductwork packages for above systems, per ton of cooling:			
9010	Ductwork; galvanized steel, 108 pounds			
9020	Diffusers, aluminum, 24" x 12", two			

D3050 Terminal & Package Units

System Description: Self-contained air cooled units with remote air cooled condenser and interconnecting tubing. Systems for 1000 S.F. and up include duct and diffusers. Smaller units distribute air directly.

Returns are not ducted and supplies are not insulated.

Potential savings may be realized by using a single zone rooftop system or through-the-wall unit, especially in the smaller capacities, if the application permits.

Hot water or steam heating coils are included but piping to boiler and the boiler itself is not included.

Condenserless models are available for 15% less where remote refrigerant source is available.

System Components	QUANTITY	UNIT	COST EACH MAT.	COST EACH INST.	COST EACH TOTAL
SYSTEM D3050 165 1320					
SELF-CONTAINED, AIR COOLED UNIT					
APARTMENT CORRIDORS, 500 S.F., .92 TON					
Air cooled, package unit	1.000	Ea.	1,513	484.50	1,997.50
Ductwork package for water or air cooled packaged units	1.000	System	62.10	989	1,051.10
Refrigerant piping	1.000	System	381.80	648.60	1,030.40
Air cooled condenser, direct drive, propeller fan	1.000	Ea.	914.50	182.90	1,097.40
TOTAL			2,871.40	2,305	5,176.40
COST PER S.F.			5.74	4.61	10.35

D3050 165	Self-contained, Air Cooled Unit Systems		COST PER S.F. MAT.	COST PER S.F. INST.	COST PER S.F. TOTAL
1300	Self-contained, air cooled unit				
1320	Apartment corridors, 500 S.F., .92 ton		5.75	4.60	10.35
1360	1,000 S.F., 1.83 ton	R236000	5.65	4.57	10.22
1400	3,000 S.F., 5.50 ton	-20	4.94	4.26	9.20
1440	5,000 S.F., 9.17 ton		4.25	4.05	8.30
1480	10,000 S.F., 18.33 ton		3.59	3.72	7.31
1560	Banks or libraries, 500 S.F., 2.08 ton		12.50	5.90	18.40
1600	1,000 S.F., 4.17 ton		11.20	9.65	20.85
1640	3,000 S.F., 12.50 ton		9.65	9.20	18.85
1680	5,000 S.F., 20.80 ton		8.25	8.45	16.70
1720	10,000 S.F., 41.66 ton		8.80	8.30	17.10
1800	Bars and taverns, 500 S.F., 5.54 ton		28	12.85	40.85
1840	1,000 S.F., 11.08 ton		25.50	18.55	44.05
1880	3,000 S.F., 33.25 ton		24	16.05	40.05
1920	5,000 S.F., 55.42 ton		24.50	16.05	40.55
1960	10,000 S.F., 110.00 ton		24.50	15.95	40.45
2040	Bowling alleys, 500 S.F., 2.83 ton		17.15	8.05	25.20
2080	1,000 S.F., 5.66 ton		15.25	13.15	28.40
2120	3,000 S.F., 17.00 ton		11.10	11.50	22.60
2160	5,000 S.F., 28.33 ton		12.35	11.25	23.60
2200	10,000 S.F., 56.66 ton		12.60	11.25	23.85
2240	Department stores, 500 S.F., 1.46 ton		8.85	4.15	13
2280	1,000 S.F., 2.92 ton		7.85	6.80	14.65
2320	3,000 S.F., 8.75 ton		6.75	6.45	13.20

D30 HVAC

D3050 Terminal & Package Units

D3050 165	Self-contained, Air Cooled Unit Systems	COST PER S.F.		
		MAT.	INST.	TOTAL
2360	5,000 S.F., 14.58 ton	6.75	6.45	13.20
2400	10,000 S.F., 29.17 ton	6.35	5.80	12.15
2480	Drug stores, 500 S.F., 3.33 ton	20	9.50	29.50
2520	1,000 S.F., 6.66 ton	17.95	15.45	33.40
2560	3,000 S.F., 20.00 ton	13.25	13.55	26.80
2600	5,000 S.F., 33.33 ton	14.50	13.25	27.75
2640	10,000 S.F., 66.66 ton	15	13.30	28.30
2720	Factories, 500 S.F., 1.66 ton	10.25	4.79	15.04
2760	1,000 S.F., 3.33 ton	9.05	7.75	16.80
2800	3,000 S.F., 10.00 ton	7.70	7.35	15.05
2840	5,000 S.F., 16.66 ton	6.50	6.75	13.25
2880	10,000 S.F., 33.33 ton	7.25	6.60	13.85
2960	Food supermarkets, 500 S.F., 1.42 ton	8.60	4.03	12.63
3000	1,000 S.F., 2.83 ton	8.70	7.05	15.75
3040	3,000 S.F., 8.50 ton	7.65	6.60	14.25
3080	5,000 S.F., 14.17 ton	6.55	6.25	12.80
3120	10,000 S.F., 28.33 ton	6.15	5.65	11.80
3200	Medical centers, 500 S.F., 1.17 ton	7.10	3.33	10.43
3240	1,000 S.F., 2.33 ton	7.25	5.85	13.10
3280	3,000 S.F., 7.00 ton	6.30	5.45	11.75
3320	5,000 S.F., 16.66 ton	5.40	5.15	10.55
3360	10,000 S.F., 23.33 ton	4.58	4.74	9.32
3440	Offices, 500 S.F., 1.58 ton	9.60	4.51	14.11
3480	1,000 S.F., 3.16 ton	9.85	7.95	17.80
3520	3,000 S.F., 9.50 ton	7.35	7	14.35
3560	5,000 S.F., 15.83 ton	6.30	6.45	12.75
3600	10,000 S.F., 31.66 ton	6.90	6.25	13.15
3680	Restaurants, 500 S.F., 2.50 ton	15.10	7.10	22.20
3720	1,000 S.F., 5.00 ton	13.50	11.65	25.15
3760	3,000 S.F., 15.00 ton	11.60	11.05	22.65
3800	5,000 S.F., 25.00 ton	10.85	9.95	20.80
3840	10,000 S.F., 50.00 ton	11.40	9.90	21.30
3920	Schools and colleges, 500 S.F., 1.92 ton	11.60	5.45	17.05
3960	1,000 S.F., 3.83 ton	10.40	8.90	19.30
4000	3,000 S.F., 11.50 ton	8.90	8.45	17.35
4040	5,000 S.F., 19.17 ton	7.50	7.80	15.30
4080	10,000 S.F., 38.33 ton	8.15	7.60	15.75
4090				
9000	Components of ductwork packages for above systems, per ton of cooling:			
9010	Ductwork; galvanized steel, 102 pounds			
9020	Diffusers; aluminum, 24" x 12", two			

D3050 Terminal & Package Units

Air Cooled Condensing Unit

Refrigerant Piping

Roof

Supply Duct

Fin. Ceiling

Return Grille **DX Air Handling Unit** **Supply Diffuser**

General: Split systems offer several important advantages which should be evaluated when a selection is to be made. They provide a greater degree of flexibility in component selection which permits an accurate match-up of the proper equipment size and type with the particular needs of the building. This allows for maximum use of modern energy saving concepts in heating and cooling. Outdoor installation of the air cooled condensing unit allows space savings in the building and also isolates the equipment operating sounds from building occupants.

Design Assumptions: The systems below are comprised of a direct expansion air handling unit and air cooled condensing unit with interconnecting copper tubing. Ducts and diffusers are also included for distribution of air. Systems are priced for cooling only. Heat can be added as desired either by putting hot water/steam coils into the air unit or into the duct supplying the particular area of need. Gas fired duct furnaces are also available. Refrigerant liquid line is insulated.

System Components	QUANTITY	UNIT	COST EACH		
			MAT.	INST.	TOTAL
SYSTEM D3050 170 1280					
SPLIT SYSTEM, AIR COOLED CONDENSING UNIT					
APARTMENT CORRIDORS, 1,000 S.F., 1.80 TON					
Fan coil AC unit, cabinet mntd & filters direct expansion air cool	1.000	Ea.	503.25	172.02	675.27
Ductwork package, for split system, remote condensing unit	1.000	System	58.10	937.88	995.98
Refrigeration piping	1.000	System	404.43	1,079.70	1,484.13
Condensing unit, air cooled, incls compressor & standard controls	1.000	Ea.	1,372.50	689.30	2,061.80
TOTAL			2,338.28	2,878.90	5,217.18
COST PER S.F.			2.34	2.88	5.22

*Cooling requirements would lead to choosing a water cooled unit.

D3050 170	Split Systems With Air Cooled Condensing Units		COST PER S.F.		
			MAT.	INST.	TOTAL
1260	Split system, air cooled condensing unit				
1280	Apartment corridors, 1,000 S.F., 1.83 ton		2.35	2.88	5.23
1320	2,000 S.F., 3.66 ton	R236000	1.82	2.91	4.73
1360	5,000 S.F., 9.17 ton	-20	1.70	3.63	5.33
1400	10,000 S.F., 18.33 ton		1.99	3.96	5.95
1440	20,000 S.F., 36.66 ton		2.25	4.04	6.29
1520	Banks and libraries, 1,000 S.F., 4.17 ton		4.15	6.65	10.80
1560	2,000 S.F., 8.33 ton		3.83	8.25	12.08
1600	5,000 S.F., 20.80 ton		4.51	9	13.51
1640	10,000 S.F., 41.66 ton		5.10	9.20	14.30
1680	20,000 S.F., 83.32 ton		6.20	9.55	15.75
1760	Bars and taverns, 1,000 S.F., 11.08 ton		8.65	13	21.65
1800	2,000 S.F., 22.16 ton		12.30	15.60	27.90
1840	5,000 S.F., 55.42 ton		11.15	14.60	25.75
1880	10,000 S.F., 110.84 ton		13.95	15.20	29.15
1920	20,000 S.F., 220 ton*				
2000	Bowling alleys, 1,000 S.F., 5.66 ton		5.70	12.35	18.05
2040	2,000 S.F., 11.33 ton		5.25	11.25	16.50
2080	5,000 S.F., 28.33 ton		6.15	12.25	18.40
2120	10,000 S.F., 56.66 ton		6.95	12.50	19.45
2160	20,000 S.F., 113.32 ton		9.55	13.50	23.05
2320	Department stores, 1,000 S.F., 2.92 ton		2.87	4.52	7.39
2360	2,000 S.F., 5.83 ton		2.93	6.35	9.28
2400	5,000 S.F., 14.58 ton		2.69	5.80	8.49

D30 HVAC

D3050 Terminal & Package Units

D3050 170	Split Systems With Air Cooled Condensing Units	COST PER S.F.		
		MAT.	INST.	TOTAL
2440	10,000 S.F., 29.17 ton	3.17	6.30	9.47
2480	20,000 S.F., 58.33 ton	3.56	6.40	9.96
2560	Drug stores, 1,000 S.F., 6.66 ton	6.70	14.50	21.20
2600	2,000 S.F., 13.32 ton	6.15	13.20	19.35
2640	5,000 S.F., 33.33 ton	8.15	14.70	22.85
2680	10,000 S.F., 66.66 ton	8.60	15.35	23.95
2720	20,000 S.F., 133.32 ton*			
2800	Factories, 1,000 S.F., 3.33 ton	3.30	5.15	8.45
2840	2,000 S.F., 6.66 ton	3.35	7.25	10.60
2880	5,000 S.F., 16.66 ton	3.62	7.20	10.82
2920	10,000 S.F., 33.33 ton	4.07	7.35	11.42
2960	20,000 S.F., 66.66 ton	4.28	7.65	11.93
3040	Food supermarkets, 1,000 S.F., 2.83 ton	2.80	4.40	7.20
3080	2,000 S.F., 5.66 ton	2.85	6.15	9
3120	5,000 S.F., 14.66 ton	2.62	5.60	8.22
3160	10,000 S.F., 28.33 ton	3.07	6.15	9.22
3200	20,000 S.F., 56.66 ton	3.45	6.25	9.70
3280	Medical centers, 1,000 S.F., 2.33 ton	2.49	3.57	6.06
3320	2,000 S.F., 4.66 ton	2.35	5.10	7.45
3360	5,000 S.F., 11.66 ton	2.15	4.63	6.78
3400	10,000 S.F., 23.33 ton	2.53	5.05	7.58
3440	20,000 S.F., 46.66 ton	2.85	5.15	8
3520	Offices, 1,000 S.F., 3.17 ton	3.12	4.92	8.04
3560	2,000 S.F., 6.33 ton	3.18	6.90	10.08
3600	5,000 S.F., 15.83 ton	2.92	6.30	9.22
3640	10,000 S.F., 31.66 ton	3.43	6.85	10.28
3680	20,000 S.F., 63.32 ton	4.07	7.30	11.37
3760	Restaurants, 1,000 S.F., 5.00 ton	5.05	10.90	15.95
3800	2,000 S.F., 10.00 ton	4.62	9.90	14.52
3840	5,000 S.F., 25.00 ton	5.45	10.85	16.30
3880	10,000 S.F., 50.00 ton	6.10	11.05	17.15
3920	20,000 S.F., 100.00 ton	8.40	11.95	20.35
4000	Schools and colleges, 1,000 S.F., 3.83 ton	3.82	6.10	9.92
4040	2,000 S.F., 7.66 ton	3.54	7.60	11.14
4080	5,000 S.F., 19.17 ton	4.15	8.30	12.45
4120	10,000 S.F., 38.33 ton	4.68	8.45	13.13
4160	20,000 S.F., 76.66 ton	4.94	8.80	13.74
5000				
9000	Components of ductwork packages for above systems, per ton of cooling:			
9010	Ductwork; galvanized steel, 102 pounds			
9020	Diffusers; aluminum, 24" x 12", two			

571

For customer support on your Mechanical Costs with RSMeans Data, call 800.448.8182.

D3050 Terminal & Package Units

Rooftop Air Conditioner Unit

System Components	QUANTITY	UNIT	COST EACH		
			MAT.	INST.	TOTAL
SYSTEM D3050 175 1010					
A/C, ROOFTOP, DX COOL, GAS HEAT, CURB, ECONOMIZER, FILTERS., 5 TON					
Roof top A/C, curb, economizer, sgl zone, elec cool, gas ht, 5 ton, 112 MBH	1.000	Ea.	4,925	2,525	7,450
Pipe, black steel, Sch 40, threaded, w/cplgs & hangers, 10' OC, 1" dia	21.000	L.F.	179.55	305.55	485.10
Elbow, 90°, black, straight, 3/4" dia.	3.000	Ea.	13.59	165	178.59
Elbow, 90°, black, straight, 1" dia.	3.000	Ea.	27.60	178.50	206.10
Tee, black, reducing, 1" dia.	1.000	Ea.	19	96.50	115.50
Union, black with brass seat, 1" dia.	1.000	Ea.	27	64.50	91.50
Pipe nipple, black, 3/4" dia	2.000	Ea.	9.68	25.30	34.98
Pipe nipple, black, 1" dia	2.000	Ea.	8.55	14.55	23.10
Cap, black, 1" dia.	1.000	Ea.	7.45	25.50	32.95
Gas cock, brass, 1" size	1.000	Ea.	37.50	40.50	78
Control system, pneumatic, Rooftop A/C unit	1.000	Ea.	3,225	2,775	6,000
Rooftop unit heat/cool balancing	1.000	Ea.		530	530
TOTAL			8,479.92	6,745.90	15,225.82

D3050 175	Rooftop Air Conditioner, Const. Volume	COST EACH		
		MAT.	INST.	TOTAL
1010	A/C, Rooftop, DX cool, gas heat, curb, economizer, fltrs, 5 Ton	8,475	6,750	15,225
1020	7-1/2 Ton	9,850	7,025	16,875
1030	12-1/2 Ton	15,400	7,700	23,100
1040	18 Ton	19,100	8,425	27,525
1050	25 Ton	42,000	9,600	51,600
1060	40 Ton	52,500	12,800	65,300

D3050 180	Rooftop Air Conditioner, Variable Air Volume	COST EACH		
		MAT.	INST.	TOTAL
1010	A/C, Rooftop, DX cool, gas heat, curb, ecmizr, fltrs, VAV, 12-1/2 Ton	18,400	8,225	26,625
1020	18 Ton	22,900	9,050	31,950
1030	25 Ton	46,800	10,400	57,200
1040	40 Ton	60,000	14,100	74,100
1050	60 Ton	85,500	19,200	104,700
1060	80 Ton	109,000	24,000	133,000

D30 HVAC

D3050 Terminal & Package Units

Computer rooms impose special requirements on air conditioning systems. A prime requirement is reliability, due to the potential monetary loss that could be incurred by a system failure. A second basic requirement is the tolerance of control with which temperature and humidity are regulated, and dust eliminated. As the air conditioning system reliability is so vital, the additional cost of reserve capacity and redundant components is often justified.

System Descriptions: Computer areas may be environmentally controlled by one of three methods as follows:

1. Self-contained Units
These are units built to higher standards of performance and reliability.

They usually contain alarms and controls to indicate component operation failure, filter change, etc. It should be remembered that these units in the room will occupy space that is relatively expensive to build and that all alterations and service of the equipment will also have to be accomplished within the computer area.

2. Decentralized Air Handling Units
In operation these are similar to the self-contained units except that their cooling capability comes from remotely located refrigeration equipment as refrigerant or chilled water. As no compressors or refrigerating equipment are required in the air units, they are smaller and require less service than

self-contained units. An added plus for this type of system occurs if some of the computer components themselves also require chilled water for cooling.

3. Central System Supply
Cooling is obtained from a central source which, since it is not located within the computer room, may have excess capacity and permit greater flexibility without interfering with the computer components. System performance criteria must still be met.

Note: The costs shown below do not include an allowance for ductwork or piping.

D3050 185	Computer Room Cooling Units	MAT.	INST.	TOTAL
0560	Computer room unit, air cooled, includes remote condenser			
0580	3 ton	30,400	2,825	33,225
0600	5 ton	32,400	3,150	35,550
0620	8 ton	60,500	5,225	65,725
0640	10 ton	63,500	5,650	69,150
0660	15 ton	69,500	6,425	75,925
0680	20 ton	84,500	9,150	93,650
0700	23 ton	104,000	10,500	114,500
0800	Chilled water, for connection to existing chiller system			
0820	5 ton	22,000	1,900	23,900
0840	8 ton	22,700	2,825	25,525
0860	10 ton	22,800	2,875	25,675
0880	15 ton	24,900	2,950	27,850
0900	20 ton	26,500	3,075	29,575
0920	23 ton	28,200	3,500	31,700
1000	Glycol system, complete except for interconnecting tubing			
1020	3 ton	38,100	3,525	41,625
1040	5 ton	41,700	3,725	45,425
1060	8 ton	67,000	6,150	73,150
1080	10 ton	71,500	6,725	78,225
1100	15 ton	88,500	8,450	96,950
1120	20 ton	97,000	9,150	106,150
1140	23 ton	102,000	9,550	111,550
1240	Water cooled, not including condenser water supply or cooling tower			
1260	3 ton	30,900	2,275	33,175
1280	5 ton	34,900	2,625	37,525
1300	8 ton	55,000	4,275	59,275
1320	15 ton	67,000	5,225	72,225
1340	20 ton	72,000	5,775	77,775
1360	23 ton	75,500	6,275	81,775

D3050 201	Packaged A/C, Elec. Ht., Const. Volume	MAT.	INST.	TOTAL
1010	A/C Packaged, DX, air cooled, electric heat, constant vol., 5 Ton	10,000	6,525	16,525
1020	10 Ton	16,700	8,850	25,550
1030	20 Ton	25,800	11,800	37,600
1040	30 Ton	46,000	13,000	59,000

For customer support on your Mechanical Costs with RSMeans Data, call 800.448.8182.

573

D3050 Terminal & Package Units

D3050 201	Packaged A/C, Elec. Ht., Const. Volume	COST EACH		
		MAT.	INST.	TOTAL
1050	40 Ton	59,500	14,700	74,200
1060	50 Ton	80,500	16,300	96,800

D3050 202	Packaged Air Conditioner, Elect. Heat, VAV	COST EACH		
		MAT.	INST.	TOTAL
1010	A/C Packaged, DX, air cooled, electric heat, VAV, 10 Ton	18,900	8,275	27,175
1020	20 Ton	29,900	10,900	40,800
1030	30 Ton	52,000	11,900	63,900
1040	40 Ton	66,500	13,600	80,100
1050	50 Ton	90,500	15,200	105,700
1060	60 Ton	104,000	17,800	121,800

D3050 203	Packaged A/C, Hot Wtr. Heat, Const. Volume	COST EACH		
		MAT.	INST.	TOTAL
1010	A/C Packaged, DX, air cooled, H/W heat, constant vol., 5 Ton	8,725	8,825	17,550
1020	10 Ton	13,500	11,500	25,000
1030	20 Ton	21,100	14,700	35,800
1040	30 Ton	38,400	18,900	57,300
1050	40 Ton	52,000	21,200	73,200
1060	50 Ton	69,500	22,900	92,400

D3050 204	Packaged A/C, Hot Wtr. Heat, VAV	COST EACH		
		MAT.	INST.	TOTAL
1010	A/C Packaged, DX, air cooled, H/W heat, VAV, 10 Ton	15,800	10,800	26,600
1020	20 Ton	25,600	14,100	39,700
1030	30 Ton	44,500	17,600	62,100
1040	40 Ton	59,000	19,100	78,100
1050	50 Ton	79,000	22,100	101,100
1060	60 Ton	95,500	24,900	120,400

D3050 Terminal & Package Units

Condenser Supply Water Piping

Self–Contained Air Conditioner Unit, Water Cooled

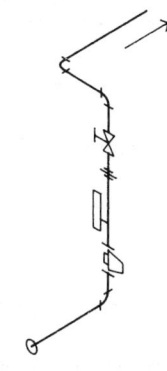

Condenser Return Water Piping

System Components	QUANTITY	UNIT	COST EACH		
			MAT.	INST.	TOTAL
SYSTEM D3050 210 1010					
A/C, SELF CONTAINED, SINGLE PKG., WATER COOLED, ELECT. HEAT, 5 TON					
Self-contained, water cooled, elect. heat, not inc tower, 5 ton, const vol	1.000	Ea.	9,525	1,825	11,350
Pipe, black steel, Sch 40, threaded, w/cplg & hangers, 10' OC, 1-1/4" dia.	21.500	L.F.	193.50	335.40	528.90
Elbow, 90°, black, straight, 1-1/4" dia.	6.000	Ea.	91.50	378	469.50
Tee, black, straight, 1-1/4" dia.	2.000	Ea.	48	198	246
Tee, black, reducing, 1-1/4" dia.	2.000	Ea.	82	198	280
Thermometers, stem type, 9" case, 8" stem, 3/4" NPT	2.000	Ea.	412	56	468
Union, black with brass seat, 1-1/4" dia.	2.000	Ea.	91	132	223
Pipe nipple, black, 1-1/4" dia	3.000	Ea.	13.50	23.40	36.90
Valves, bronze, gate, N.R.S., threaded, class 150, 1-1/4" size	2.000	Ea.	302	103	405
Circuit setter, bal valve, bronze body, threaded, 1-1/4" pipe size	1.000	Ea.	189	52.50	241.50
Insulation, fiberglass pipe covering, 1" wall, 1-1/4" IPS	20.000	L.F.	25.60	123	148.60
Control system, pneumatic, A/C Unit with heat	1.000	Ea.	3,225	2,775	6,000
Re-heat coil balancing	1.000	Ea.		145	145
Rooftop unit heat/cool balancing	1.000	Ea.		530	530
TOTAL			14,198.10	6,874.30	21,072.40

D3050 210	AC Unit, Package, Elec. Ht., Water Cooled	COST EACH		
		MAT.	INST.	TOTAL
1010	A/C, Self contained, single pkg., water cooled, elect. heat, 5 Ton	14,200	6,875	21,075
1020	10 Ton	21,200	8,475	29,675
1030	20 Ton	28,500	10,900	39,400
1040	30 Ton	50,500	11,800	62,300
1050	40 Ton	63,500	13,100	76,600
1060	50 Ton	85,000	15,700	100,700

D3050 215	AC Unit, Package, Elec. Ht., Water Cooled, VAV	COST EACH		
		MAT.	INST.	TOTAL
1010	A/C, Self contained, single pkg., water cooled, elect. ht, VAV, 10 Ton	23,900	8,475	32,375
1020	20 Ton	33,200	10,900	44,100
1030	30 Ton	57,000	11,800	68,800
1040	40 Ton	71,500	13,100	84,600
1050	50 Ton	96,000	15,700	111,700
1060	60 Ton	110,500	19,000	129,500

575

For customer support on your Mechanical Costs with RSMeans Data, call 800.448.8182.

D3050 Terminal & Package Units

D3050 220	AC Unit, Package, Hot Water Coil, Water Cooled	COST EACH		
		MAT.	INST.	TOTAL
1010	A/C, Self contn'd, single pkg., water cool, H/W ht, const. vol, 5 Ton	12,700	9,550	22,250
1020	10 Ton	17,500	11,400	28,900
1030	20 Ton	42,500	14,300	56,800
1040	30 Ton	55,500	18,200	73,700
1050	40 Ton	65,500	22,100	87,600
1060	50 Ton	74,500	27,200	101,700

D3050 225	AC Unit, Package, HW Coil, Water Cooled, VAV	COST EACH		
		MAT.	INST.	TOTAL
1010	A/C, Self contn'd, single pkg., water cool, H/W ht, VAV, 10 Ton	20,300	10,700	31,000
1020	20 Ton	28,900	14,000	42,900
1030	30 Ton	50,000	17,400	67,400
1040	40 Ton	64,500	19,000	83,500
1050	50 Ton	85,500	22,500	108,000
1060	60 Ton	102,000	24,400	126,400

D30 HVAC

D3050 Terminal & Package Units

Note: Water source units do not include an allowance for any type of water source or piping to it.

Heat Pump, Air/Air, Rooftop w/Economizer

System Components			COST EACH		
	QUANTITY	UNIT	MAT.	INST.	TOTAL
SYSTEM D3050 245 1010					
HEAT PUMP, ROOFTOP, AIR/AIR, CURB., ECONOMIZER., SUP. ELECT. HEAT, 10 TON					
Heat pump, air to air, rooftop, curb, economizer, elect ht, 10 ton cooling	1.000	Ea.	74,500	5,500	80,000
Ductwork, 32" x 12" fabricated, galvanized steel, 16 LF	176.000	Lb.	116.16	1,584	1,700.16
Insulation, ductwork, blanket type, fiberglass, 1" thk, 1-1/2 LB density	117.000	S.F.	577.98	1,807.65	2,385.63
Rooftop unit heat/cool balancing	1.000	Ea.		530	530
TOTAL			75,194.14	9,421.65	84,615.79

D3050 230	Heat Pump, Water Source, Central Station	COST EACH		
		MAT.	INST.	TOTAL
1010	Heat pump, central station, water source, constant vol., 5 Ton	5,175	3,850	9,025
1020	10 Ton	10,400	5,375	15,775
1030	20 Ton	20,300	10,100	30,400
1040	30 Ton	29,500	14,700	44,200
1050	40 Ton	40,500	16,000	56,500
1060	50 Ton	45,900	21,600	67,500

D3050 235	Heat Pump, Water Source, Console	COST EACH		
		MAT.	INST.	TOTAL
1020	Heat pump, console, water source, 1 Ton	3,125	2,100	5,225
1030	1-1/2 Ton	3,050	2,175	5,225
1040	2 Ton	4,175	2,575	6,750
1050	3 Ton	4,475	2,750	7,225
1060	3-1/2 Ton	4,725	2,825	7,550

D3050 240	Heat Pump, Horizontal, Duct Mounted	COST EACH		
		MAT.	INST.	TOTAL
1020	Heat pump, horizontal, ducted, water source, 1 Ton	3,650	4,500	8,150
1030	1-1/2 Ton	3,800	5,650	9,450
1040	2 Ton	4,950	6,750	11,700
1050	3 Ton	5,650	8,825	14,475
1060	3-1/2 Ton	5,900	9,125	15,025

D3050 Terminal & Package Units

D3050 245	Heat Pump, Rooftop, Air/Air	COST EACH		
		MAT.	INST.	TOTAL
1010	Heat pump, rooftop, air/air, curb, economizer, sup. elect heat, 10 Ton	75,000	9,425	84,425
1020	20 Ton	106,000	15,300	121,300
1030	30 Ton	145,000	21,100	166,100
1040	40 Ton	153,000	24,500	177,500
1050	50 Ton	188,000	27,400	215,400
1060	60 Ton	199,000	31,000	230,000

D3050 Terminal & Package Units

Thru—Wall Heat Pump   Thru—Wall Air Conditioner

System Components	QUANTITY	UNIT	COST EACH		
			MAT.	INST.	TOTAL
SYSTEM D3050 255 1010 **A/C UNIT, THRU-THE-WALL, SUPPLEMENTAL ELECT. HEAT, CABINET, LOUVER, 1/2 TON** A/C unit, thru-wall, electric heat., cabinet, louver, 1/2 ton	1.000	Ea.	800	235	1,035
TOTAL			800	235	1,035

D3050 255	**Thru-Wall A/C Unit**	COST EACH		
		MAT.	INST.	TOTAL
1010	A/C Unit, thru-the-wall, supplemental elect. heat, cabinet, louver, 1/2 Ton	800	235	1,035
1020	3/4 Ton	1,075	283	1,358
1030	1 Ton	1,700	355	2,055
1040	1-1/2 Ton	2,975	590	3,565
1050	2 Ton	3,000	745	3,745

D3050 260	**Thru-Wall Heat Pump**	COST EACH		
		MAT.	INST.	TOTAL
1010	Heat pump, thru-the-wall, cabinet, louver, 1/2 Ton	3,900	177	4,077
1020	3/4 Ton	4,575	235	4,810
1030	1 Ton	3,775	355	4,130
1040	Supplemental. elect. heat, 1-1/2 Ton	3,725	910	4,635
1050	2 Ton	3,675	940	4,615

Roof Exhaust Fan

10″ Dia. Fiberglass 90° Elbows →

10″ Dia. Fiberglass Reinforced Duct

Fume Hood →

Fume Hood Exhaust

System Components	QUANTITY	UNIT	COST EACH		
			MAT.	INST.	TOTAL
SYSTEM D3090 310 1010					
FUME HOOD EXHAUST SYSTEM, 3′ LONG, 1000 CFM					
Fume hood	3.000	L.F.	3,750	1,590	5,340
Ductwork, plastic, FM approved for acid fume & smoke, straight, 10″ dia	60.000	L.F.	1,290	960	2,250
Ductwork, plastic, FM approved for acid fume & smoke, elbow 90°, 10″ dia	3.000	Ea.	333	522	855
Fan, corrosive fume resistant plastic, 1,630 CFM, 1/2 HP	1.000	Ea.	7,600	430	8,030
Fan control switch & mounting base	1.000	Ea.	118	49	167
Roof fan balancing	1.000	Ea.		300	300
TOTAL			13,091	3,851	16,942

D3090 310	Fume Hood Exhaust Systems	COST EACH		
		MAT.	INST.	TOTAL
1010	Fume hood exhaust system, 3′ long, 1000 CFM	13,100	3,850	16,950
1020	4′ long, 2000 CFM	14,600	4,875	19,475
1030	6′ long, 3500 CFM	19,600	7,275	26,875
1040	6′ long, 5000 CFM	29,400	8,550	37,950
1050	10′ long, 8000 CFM	37,300	11,700	49,000

Cast Iron Garage Exhaust System

Dual Exhaust System

System Components	QUANTITY	UNIT	COST EACH		
			MAT.	INST.	TOTAL
SYSTEM D3090 320 1040					
GARAGE, EXHAUST, SINGLE 3″ EXHAUST OUTLET, CARS & LIGHT TRUCKS					
A Outlet top assy, for engine exhaust system with adapters and ftngs, 3″ diam	1.000	Ea.	286	47.50	333.50
F Bullnose (guide) for engine exhaust system, 3″ diam	1.000	Ea.	33.50		33.50
G Galvanized flexible tubing for engine exhaust system, 3″ diam	8.000	L.F.	94.80		94.80
H Adapter for metal tubing end of engine exhaust system, 3″ tail pipe	1.000	Ea.	56.50		56.50
J Pipe, sewer, cast iron, push-on joint, 8″ diam.	18.000	L.F.	1,404	855	2,259
Excavating utility trench, chain trencher, 8″ wide, 24″ deep	18.000	L.F.		41.40	41.40
Backfill utility trench by hand, incl. compaction, 8″ wide 24″ deep	18.000	L.F.		50.94	50.94
Stand for blower, concrete over polystyrene core, 6″ high	1.000	Ea.	5.90	23.50	29.40
AC&V duct spiral reducer 10″x8″	1.000	Ea.	24	42.50	66.50
AC&V duct spiral reducer 12″x10″	1.000	Ea.	24.50	56.50	81
AC&V duct spiral preformed 45° elbow, 8″ diam	1.000	Ea.	8.95	48.50	57.45
AC&V utility fan, belt drive, 3 phase, 2000 CFM, 1 HP	2.000	Ea.	2,900	750	3,650
Safety switch, heavy duty fused, 240V, 3 pole, 30 amp	1.000	Ea.	113	228	341
TOTAL			4,951.15	2,143.84	7,094.99

D3090 320	Garage Exhaust Systems	COST PER BAY		
		MAT.	INST.	TOTAL
1040	Garage, single 3″ exhaust outlet, cars & light trucks, one bay	4,950	2,150	7,100
1060	Additional bays up to seven bays	1,250	575	1,825
1500	4″ outlet, trucks, one bay	4,975	2,150	7,125
1520	Additional bays up to six bays	1,275	575	1,850
1600	5″ outlet, diesel trucks, one bay	5,250	2,150	7,400
1650	Additional single bays up to six bays	1,700	680	2,380
1700	Two adjoining bays	5,250	2,150	7,400
2000	Dual exhaust, 3″ outlets, pair of adjoining bays	6,250	2,675	8,925
2100	Additional pairs of adjoining bays	1,925	680	2,605

NOTES:
1) ALL HOUSEKEEPING PAD REINFORCING SHOULD TO ACI STANDARDS FOR MINIMUM AREA AND CONCRETE COVERAGE
2) HIGH CONCENTRATED LOADS OR POINT SUPPORTED EQUIPMENT REQUIRE SPECIAL TIE DOWN CONSIDERATIONS NOT COVERED BY THIS DETAIL
3) MINIMUM OF (4) HK PAD TO FLOOR ANCHOR BOLTS REQUIRED FOR EACH HOUSEKEEPING PAD.

System Components	QUANTITY	UNIT	COST EACH		
			MAT.	INST.	TOTAL
SYSTEM D3090 750 1000					
SEISMIC PAD FOR 20-38 SELF-CONTAINED UNIT					
Form equipment pad	28.000	L.F.	33.88	133	166.88
Reinforcing in hold down pad, #3 to #7	60.000	Lb.	37.20	39	76.20
Concrete 3000 PSI	.866	C.Y.	122.97		122.97
Place concrete pad	.866	C.Y.		32.41	32.41
Finish concrete equipment pad	47.000	S.F.		42.77	42.77
Drill for 3/8 inch slab anchors	10.000	Ea.	.50	101.50	102
Exisiting slab anchors 3/8 inch	10.000	Ea.	8.60	45.60	54.20
Equipment anchors 3/8 inch for plate/foot	8.000	Ea.	15.84	39.04	54.88
TOTAL			218.99	433.32	652.31

D3090 750	Seismic Concrete Equipment Pads	COST PER EACH		
		MAT.	INST.	TOTAL
1000	Seismic pad for 20-38 tons Self-contained AC unit without economizer .25g	219	435	654
1010	0.5 g seismic acceleration	219	435	654
1020	0.75 g seismic acceleration	219	435	654
1030	1.0 g seismic acceleration	219	435	654
1040	1.5 g seismic acceleration	223	495	718
1050	2.0 g seismic acceleration	238	680	918
1060	2.5 g seismic acceleration	248	855	1,103
1100	Seismic pad for 40-80 tons Self contained AC unit without economizer .25g	345	630	975
1110	0.5 g seismic acceleration	345	630	975
1120	0.75 g seismic acceleration	345	630	975
1130	1.0 g seismic acceleration	345	630	975
1140	1.5 g seismic acceleration	375	895	1,270
1150	2.0 g seismic acceleration	435	1,400	1,835
1160	2.5 g seismic acceleration	500	2,000	2,500

D30 HVAC

D3090 Other HVAC Systems/Equip

D3090 750	Seismic Concrete Equipment Pads	COST PER EACH		
		MAT.	INST.	TOTAL
1200	Seismic pad for 80-110 tons Self contained AC unit without economizer .25g	380	675	1,055
1210	0.5 g seismic acceleration	380	675	1,055
1220	0.75 g seismic acceleration	380	675	1,055
1230	1.0 g seismic acceleration	380	675	1,055
1240	1.5 g seismic acceleration	410	965	1,375
1250	2.0 g seismic acceleration	475	1,525	2,000
1260	2.5 g seismic acceleration	545	2,150	2,695
1990	Seismic pad for self-contained AC units for isolator/snubber bolts			
2000	Seismic pad for 20-38 tons Self contained AC unit without economizer .25g	235	475	710
2010	0.5 g seismic acceleration	235	475	710
2020	0.75 g seismic acceleration	235	475	710
2030	1.0 g seismic acceleration	235	475	710
2040	1.5 g seismic acceleration	238	535	773
2050	2.0 g seismic acceleration	257	730	987
2060	2.5 g seismic acceleration	268	905	1,173
2100	Seismic pad for 40-80 tons Self contained AC unit without economizer .25g	365	675	1,040
2110	0.5 g seismic acceleration	365	675	1,040
2120	0.75 g seismic acceleration	365	675	1,040
2130	1.0 g seismic acceleration	365	675	1,040
2140	1.5 g seismic acceleration	395	945	1,340
2150	2.0 g seismic acceleration	460	1,475	1,935
2160	2.5 g seismic acceleration	530	2,075	2,605
2200	Seismic pad for 80-110 tons Self contained AC unit without economizer .25g	400	725	1,125
2210	0.5 g seismic acceleration	400	725	1,125
2220	0.75 g seismic acceleration	400	725	1,125
2230	1.0 g seismic acceleration	400	725	1,125
2240	1.5 g seismic acceleration	430	1,000	1,430
2250	2.0 g seismic acceleration	505	1,575	2,080
2260	2.5 g seismic acceleration	580	2,225	2,805
2990	Seismic pad for self-contained AC economizer units for plate/foot attachment			
3000	Seismic pad for 20-38 tons Self contained AC unit economizer .25 g-2.0 g	146	251	397
3010	2.5 g seismic acceleration	155	330	485
3100	Seismic pad for 40-80 tons Self contained AC unit economizer .25-1.5 g	196	335	531
3110	2.0 g seismic acceleration	202	390	592
3120	2.5 g seismic acceleration	209	460	669
3190	Seismic pad for self-contained AC unit economizer for isolator/snubber bolts			
3200	Seismic pad for 20-38 tons Self contained AC unit economizer .25 g-2.0 g	152	266	418
3210	2.5 g seismic acceleration	164	355	519
3300	Seismic pad for 40-80 tons Self contained AC unit economizer .25-1.5 g	208	365	573
3310	2.0 g seismic acceleration	214	420	634
3320	2.5 g seismic acceleration	221	490	711

For customer support on your Mechanical Costs with RSMeans Data, call 800.448.8182.

Same Data. Simplified.

Enjoy the convenience and efficiency of accessing your costs anywhere:

- **Skip the multiplier** by setting your location
- **Quickly search,** edit, favorite and share costs
- **Stay on top of price changes** with automatic updates

Discover more at rsmeans.com/online

G1030   Site Earthwork

Trenching Systems are shown on a cost per linear foot basis. The systems include: excavation; backfill and removal of spoil; and compaction for various depths and trench bottom widths. The backfill has been reduced to accommodate a pipe of suitable diameter and bedding.

The slope for trench sides varies from none to 1:1.

The Expanded System Listing shows Trenching Systems that range from 2' to 12' in width. Depths range from 2' to 25'.

System Components		QUANTITY	UNIT	COST PER L.F.		
				EQUIP.	LABOR	TOTAL
SYSTEM G1030 805 1310						
TRENCHING, COMMON EARTH, NO SLOPE, 2' WIDE, 2' DP, 3/8 C.Y. BUCKET						
Excavation, trench, hyd. backhoe, track mtd., 3/8 C.Y. bucket		.148	B.C.Y.	.23	1.17	1.40
Backfill and load spoil, from stockpile		.153	L.C.Y.	.13	.36	.49
Compaction by vibrating plate, 6" lifts, 4 passes		.118	E.C.Y.	.03	.43	.46
Remove excess spoil, 8 C.Y. dump truck, 2 mile roundtrip		.040	L.C.Y.	.16	.20	.36
	TOTAL			.55	2.16	2.71

G1030 805	Trenching Common Earth	COST PER L.F.		
		EQUIP.	LABOR	TOTAL
1310	Trenching, common earth, no slope, 2' wide, 2' deep, 3/8 C.Y. bucket	.55	2.16	2.71
1320	3' deep, 3/8 C.Y. bucket	.77	3.22	3.99
1330	4' deep, 3/8 C.Y. bucket	.98	4.31	5.29
1340	6' deep, 3/8 C.Y. bucket	1.28	5.55	6.83
1350	8' deep, 1/2 C.Y. bucket	1.65	7.35	9
1360	10' deep, 1 C.Y. bucket	3.57	8.80	12.37
1400	4' wide, 2' deep, 3/8 C.Y. bucket	1.29	4.28	5.57
1410	3' deep, 3/8 C.Y. bucket	1.72	6.40	8.12
1420	4' deep, 1/2 C.Y. bucket	1.92	7.15	9.07
1430	6' deep, 1/2 C.Y. bucket	3.18	11.50	14.68
1440	8' deep, 1/2 C.Y. bucket	6.65	14.95	21.60
1450	10' deep, 1 C.Y. bucket	8.10	18.55	26.65
1460	12' deep, 1 C.Y. bucket	10.45	23.50	33.95
1470	15' deep, 1-1/2 C.Y. bucket	7.65	21	28.65
1480	18' deep, 2-1/2 C.Y. bucket	12.65	29.50	42.15
1520	6' wide, 6' deep, 5/8 C.Y. bucket w/trench box	9	17.30	26.30
1530	8' deep, 3/4 C.Y. bucket	11.65	22.50	34.15
1540	10' deep, 1 C.Y. bucket	11.35	23.50	34.85
1550	12' deep, 1-1/2 C.Y. bucket	10.30	25	35.30
1560	16' deep, 2-1/2 C.Y. bucket	16	31.50	47.50
1570	20' deep, 3-1/2 C.Y. bucket	20.50	37.50	58
1580	24' deep, 3-1/2 C.Y. bucket	24	45	69
1640	8' wide, 12' deep, 1-1/2 C.Y. bucket w/trench box	14.35	32	46.35
1650	15' deep, 1-1/2 C.Y. bucket	18.60	42	60.60
1660	18' deep, 2-1/2 C.Y. bucket	23	42	65
1680	24' deep, 3-1/2 C.Y. bucket	32.50	58.50	91
1730	10' wide, 20' deep, 3-1/2 C.Y. bucket w/trench box	26.50	55.50	82
1740	24' deep, 3-1/2 C.Y. bucket	39	66.50	105.50
1780	12' wide, 20' deep, 3-1/2 C.Y. bucket w/trench box	41.50	70.50	112
1790	25' deep, bucket	51.50	89.50	141
1800	1/2 to 1 slope, 2' wide, 2' deep, 3/8 C.Y. bucket	.77	3.22	3.99
1810	3' deep, 3/8 C.Y. bucket	1.25	5.65	6.90
1820	4' deep, 3/8 C.Y. bucket	1.84	8.60	10.44
1840	6' deep, 3/8 C.Y. bucket	3.02	13.95	16.97

G1030 Site Earthwork

G1030 805	Trenching Common Earth	COST PER L.F.		
		EQUIP.	LABOR	TOTAL
1860	8' deep, 1/2 C.Y. bucket	4.75	22	26.75
1880	10' deep, 1 C.Y. bucket	12.30	31	43.30
2300	4' wide, 2' deep, 3/8 C.Y. bucket	1.50	5.35	6.85
2310	3' deep, 3/8 C.Y. bucket	2.20	8.85	11.05
2320	4' deep, 1/2 C.Y. bucket	2.70	10.90	13.60
2340	6' deep, 1/2 C.Y. bucket	5.30	20.50	25.80
2360	8' deep, 1/2 C.Y. bucket	12.95	30	42.95
2380	10' deep, 1 C.Y. bucket	18	42.50	60.50
2400	12' deep, 1 C.Y. bucket	23.50	57.50	81
2430	15' deep, 1-1/2 C.Y. bucket	21.50	61.50	83
2460	18' deep, 2-1/2 C.Y. bucket	43.50	96	139.50
2840	6' wide, 6' deep, 5/8 C.Y. bucket w/trench box	13	25.50	38.50
2860	8' deep, 3/4 C.Y. bucket	18.90	38.50	57.40
2880	10' deep, 1 C.Y. bucket	15.45	39	54.45
2900	12' deep, 1-1/2 C.Y. bucket	19.60	51	70.60
2940	16' deep, 2-1/2 C.Y. bucket	36.50	75	111.50
2980	20' deep, 3-1/2 C.Y. bucket	51	101	152
3020	24' deep, 3-1/2 C.Y. bucket	72	138	210
3100	8' wide, 12' deep, 1-1/2 C.Y. bucket w/trench box	24	58	82
3120	15' deep, 1-1/2 C.Y. bucket	35	84.50	119.50
3140	18' deep, 2-1/2 C.Y. bucket	50.50	100	150.50
3180	24' deep, 3-1/2 C.Y. bucket	80.50	151	231.50
3270	10' wide, 20' deep, 3-1/2 C.Y. bucket w/trench box	52	116	168
3280	24' deep, 3-1/2 C.Y. bucket	89	165	254
3370	12' wide, 20' deep, 3-1/2 C.Y. bucket w/trench box	75	135	210
3380	25' deep, 3-1/2 C.Y. bucket	104	193	297
3500	1 to 1 slope, 2' wide, 2' deep, 3/8 C.Y. bucket	.99	4.31	5.30
3520	3' deep, 3/8 C.Y. bucket	3.14	9.85	12.99
3540	4' deep, 3/8 C.Y. bucket	2.70	12.95	15.65
3560	6' deep, 1/2 C.Y. bucket	3.02	13.90	16.92
3580	8' deep, 1/2 C.Y. bucket	5.90	28	33.90
3600	10' deep, 1 C.Y. bucket	21	53.50	74.50
3800	4' wide, 2' deep, 3/8 C.Y. bucket	1.72	6.45	8.17
3820	3' deep, 3/8 C.Y. bucket	2.69	11.25	13.94
3840	4' deep, 1/2 C.Y. bucket	3.47	14.60	18.07
3860	6' deep, 1/2 C.Y. bucket	7.50	29.50	37
3880	8' deep, 1/2 C.Y. bucket	19.30	45.50	64.80
3900	10' deep, 1 C.Y. bucket	28	66	94
3920	12' deep, 1 C.Y. bucket	41	95.50	136.50
3940	15' deep, 1-1/2 C.Y. bucket	35.50	102	137.50
3960	18' deep, 2-1/2 C.Y. bucket	59.50	132	191.50
4030	6' wide, 6' deep, 5/8 C.Y. bucket w/trench box	17.15	34.50	51.65
4040	8' deep, 3/4 C.Y. bucket	25	47.50	72.50
4050	10' deep, 1 C.Y. bucket	22.50	57.50	80
4060	12' deep, 1-1/2 C.Y. bucket	30	77.50	107.50
4070	16' deep, 2-1/2 C.Y. bucket	57.50	118	175.50
4080	20' deep, 3-1/2 C.Y. bucket	82.50	164	246.50
4090	24' deep, 3-1/2 C.Y. bucket	120	231	351
4500	8' wide, 12' deep, 1-1/2 C.Y. bucket w/trench box	34	84.50	118.50
4550	15' deep, 1-1/2 C.Y. bucket	51	127	178
4600	18' deep, 2-1/2 C.Y. bucket	76.50	155	231.50
4650	24' deep, 3-1/2 C.Y. bucket	129	244	373
4800	10' wide, 20' deep, 3-1/2 C.Y. bucket w/trench box	77.50	177	254.50
4850	24' deep, 3-1/2 C.Y. bucket	137	258	395
4950	12' wide, 20' deep, 3-1/2 C.Y. bucket w/trench box	108	199	307
4980	25' deep, 3-1/2 C.Y. bucket	155	291	446

G1030 Site Earthwork

Trenching Systems are shown on a cost per linear foot basis. The systems include: excavation; backfill and removal of spoil; and compaction for various depths and trench bottom widths. The backfill has been reduced to accommodate a pipe of suitable diameter and bedding.

The slope for trench sides varies from none to 1:1.

The Expanded System Listing shows Trenching Systems that range from 2' to 12' in width. Depths range from 2' to 25'.

System Components	QUANTITY	UNIT	COST PER L.F.		
			EQUIP.	LABOR	TOTAL
SYSTEM G1030 806 1310					
TRENCHING, LOAM & SANDY CLAY, NO SLOPE, 2' WIDE, 2' DP, 3/8 C.Y. BUCKET					
Excavation, trench, hyd. backhoe, track mtd., 3/8 C.Y. bucket	.148	B.C.Y.	.21	1.08	1.29
Backfill and load spoil, from stockpile	.165	L.C.Y.	.14	.39	.53
Compaction by vibrating plate 18" wide, 6" lifts, 4 passes	.118	E.C.Y.	.03	.43	.46
Remove excess spoil, 8 C.Y. dump truck, 2 mile roundtrip	.042	L.C.Y.	.16	.21	.37
TOTAL			.54	2.11	2.65

G1030 806	Trenching Loam & Sandy Clay	COST PER L.F.		
		EQUIP.	LABOR	TOTAL
1310	Trenching, loam & sandy clay, no slope, 2' wide, 2' deep, 3/8 C.Y. bucket	.54	2.11	2.65
1320	3' deep, 3/8 C.Y. bucket	.81	3.44	4.25
1330	4' deep, 3/8 C.Y. bucket	.97	4.20	5.15
1340	6' deep, 3/8 C.Y. bucket	1.80	5	6.80
1350	8' deep, 1/2 C.Y. bucket	2.35	6.60	8.95
1360	10' deep, 1 C.Y. bucket	2.63	7.05	9.70
1400	4' wide, 2' deep, 3/8 C.Y. bucket	1.33	4.19	5.50
1410	3' deep, 3/8 C.Y. bucket	1.74	6.30	8.05
1420	4' deep, 1/2 C.Y. bucket	1.94	7.05	9
1430	6' deep, 1/2 C.Y. bucket	4.26	10.40	14.65
1440	8' deep, 1/2 C.Y. bucket	6.50	14.75	21.50
1450	10' deep, 1 C.Y. bucket	6.25	15.05	21.50
1460	12' deep, 1 C.Y. bucket	7.85	18.75	26.50
1470	15' deep, 1-1/2 C.Y. bucket	7.95	22	30
1480	18' deep, 2-1/2 C.Y. bucket	10.75	24	35
1520	6' wide, 6' deep, 5/8 C.Y. bucket w/trench box	8.75	16.95	25.50
1530	8' deep, 3/4 C.Y. bucket	11.20	22	33
1540	10' deep, 1 C.Y. bucket	10.40	22.50	33
1550	12' deep, 1-1/2 C.Y. bucket	10.15	25	35
1560	16' deep, 2-1/2 C.Y. bucket	15.65	32	47.50
1570	20' deep, 3-1/2 C.Y. bucket	18.75	37.50	56.50
1580	24' deep, 3-1/2 C.Y. bucket	23.50	46	69.50
1640	8' wide, 12' deep, 1-1/4 C.Y. bucket w/trench box	14.30	32	46.50
1650	15' deep, 1-1/2 C.Y. bucket	17.50	41	58.50
1660	18' deep, 2-1/2 C.Y. bucket	23.50	46	69.50
1680	24' deep, 3-1/2 C.Y. bucket	32	59	91
1730	10' wide, 20' deep, 3-1/2 C.Y. bucket w/trench box	32	59.50	91.50
1740	24' deep, 3-1/2 C.Y. bucket	40	73.50	114
1780	12' wide, 20' deep, 3-1/2 C.Y. bucket w/trench box	39	70.50	110
1790	25' deep, 3-1/2 C.Y. bucket	50.50	91	142
1800	1/2:1 slope, 2' wide, 2' deep, 3/8 C.Y. bucket	.75	3.15	3.90
1810	3' deep, 3/8 C.Y. bucket	1.23	5.50	6.75
1820	4' deep, 3/8 C.Y. bucket	1.81	8.40	10.20
1840	6' deep, 3/8 C.Y. bucket	4.33	12.50	16.85

G1030 Site Earthwork

G1030 806	Trenching Loam & Sandy Clay	COST PER L.F.		
		EQUIP.	LABOR	TOTAL
1860	8' deep, 1/2 C.Y. bucket	6.85	20	27
1880	10' deep, 1 C.Y. bucket	9	25	34
2300	4' wide, 2' deep, 3/8 C.Y. bucket	1.54	5.25	6.80
2310	3' deep, 3/8 C.Y. bucket	2.23	8.65	10.90
2320	4' deep, 1/2 C.Y. bucket	2.71	10.75	13.45
2340	6' deep, 1/2 C.Y. bucket	7.20	18.45	25.50
2360	8' deep, 1/2 C.Y. bucket	12.60	30	42.50
2380	10' deep, 1 C.Y. bucket	13.75	34.50	48.50
2400	12' deep, 1 C.Y. bucket	23	57	80
2430	15' deep, 1-1/2 C.Y. bucket	22.50	63.50	86
2460	18' deep, 2-1/2 C.Y. bucket	42.50	97.50	140
2840	6' wide, 6' deep, 5/8 C.Y. bucket w/trench box	12.65	25.50	38
2860	8' deep, 3/4 C.Y. bucket	18.20	37.50	55.50
2880	10' deep, 1 C.Y. bucket	18.70	42.50	61
2900	12' deep, 1-1/2 C.Y. bucket	19.80	51.50	71.50
2940	16' deep, 2-1/2 C.Y. bucket	36	75.50	112
2980	20' deep, 3-1/2 C.Y. bucket	49.50	102	152
3020	24' deep, 3-1/2 C.Y. bucket	70	140	210
3100	8' wide, 12' deep, 1-1/2 C.Y. bucket w/trench box	24	58.50	82.50
3120	15' deep, 1-1/2 C.Y. bucket	32.50	82	115
3140	18' deep, 2-1/2 C.Y. bucket	49	101	150
3180	24' deep, 3-1/2 C.Y. bucket	78.50	153	232
3270	10' wide, 20' deep, 3-1/2 C.Y. bucket w/trench box	63	124	187
3280	24' deep, 3-1/2 C.Y. bucket	86.50	168	255
3320	12' wide, 20' deep, 3-1/2 C.Y. bucket w/trench box	69.50	135	205
3380	25' deep, 3-1/2 C.Y. bucket w/trench box	95	181	276
3500	1:1 slope, 2' wide, 2' deep, 3/8 C.Y. bucket	.97	4.20	5.15
3520	3' deep, 3/8 C.Y. bucket	1.71	7.90	9.60
3540	4' deep, 3/8 C.Y. bucket	2.66	12.60	15.25
3560	6' deep, 1/2 C.Y. bucket	4.33	12.50	16.85
3580	8' deep, 1/2 C.Y. bucket	11.35	33.50	45
3600	10' deep, 1 C.Y. bucket	15.35	43	58.50
3800	4' wide, 2' deep, 3/8 C.Y. bucket	1.74	6.30	8.05
3820	3' deep, 1/2 C.Y. bucket	2.71	11	13.70
3840	4' deep, 1/2 C.Y. bucket	3.48	14.40	17.90
3860	6' deep, 1/2 C.Y. bucket	10.15	26.50	36.50
3880	8' deep, 1/2 C.Y. bucket	18.75	45	64
3900	10' deep, 1 C.Y. bucket	21.50	54	75.50
3920	12' deep, 1 C.Y. bucket	30.50	76.50	107
3940	15' deep, 1-1/2 C.Y. bucket	37	105	142
3960	18' deep, 2-1/2 C.Y. bucket	58.50	134	193
4030	6' wide, 6' deep, 5/8 C.Y. bucket w/trench box	16.60	34.50	51
4040	8' deep, 3/4 C.Y. bucket	25	53	78
4050	10' deep, 1 C.Y. bucket	27	62	89
4060	12' deep, 1-1/2 C.Y. bucket	29.50	78	108
4070	16' deep, 2-1/2 C.Y. bucket	56	120	176
4080	20' deep, 3-1/2 C.Y. bucket	80.50	167	248
4090	24' deep, 3-1/2 C.Y. bucket	117	234	350
4500	8' wide, 12' deep, 1-1/4 C.Y. bucket w/trench box	33.50	85	119
4550	15' deep, 1-1/2 C.Y. bucket	48	123	171
4600	18' deep, 2-1/2 C.Y. bucket	75	157	232
4650	24' deep, 3-1/2 C.Y. bucket	125	248	375
4800	10' wide, 20' deep, 3-1/2 C.Y. bucket w/trench box	93.50	189	283
4850	24' deep, 3-1/2 C.Y. bucket	133	262	395
4950	12' wide, 20' deep, 3-1/2 C.Y. bucket w/trench box	100	200	300
4980	25' deep, 3-1/2 C.Y. bucket	151	295	445

589

For customer support on your Mechanical Costs with RSMeans Data, call 800.448.8182.

G1030 Site Earthwork

Trenching Systems are shown on a cost per linear foot basis. The systems include: excavation; backfill and removal of spoil; and compaction for various depths and trench bottom widths. The backfill has been reduced to accommodate a pipe of suitable diameter and bedding.

The slope for trench sides varies from none to 1:1.

The Expanded System Listing shows Trenching Systems that range from 2' to 12' in width. Depths range from 2' to 25'.

System Components			COST PER L.F.		
	QUANTITY	UNIT	EQUIP.	LABOR	TOTAL
SYSTEM G1030 807 1310					
TRENCHING, SAND & GRAVEL, NO SLOPE, 2' WIDE, 2' DEEP, 3/8 C.Y. BUCKET					
Excavation, trench, hyd. backhoe, track mtd., 3/8 C.Y. bucket	.148	B.C.Y.	.21	1.07	1.28
Backfill and load spoil, from stockpile	.140	L.C.Y.	.12	.33	.45
Compaction by vibrating plate 18" wide, 6" lifts, 4 passes	.118	E.C.Y.	.03	.43	.46
Remove excess spoil, 8 C.Y. dump truck, 2 mile roundtrip	.035	L.C.Y.	.14	.17	.31
TOTAL			.50	2	2.50

G1030 807	Trenching Sand & Gravel	COST PER L.F.		
		EQUIP.	LABOR	TOTAL
1310	Trenching, sand & gravel, no slope, 2' wide, 2' deep, 3/8 C.Y. bucket	.50	2	2.50
1320	3' deep, 3/8 C.Y. bucket	.75	3.32	4.07
1330	4' deep, 3/8 C.Y. bucket	.89	3.99	4.88
1340	6' deep, 3/8 C.Y. bucket	1.71	4.76	6.45
1350	8' deep, 1/2 C.Y. bucket	2.23	6.30	8.55
1360	10' deep, 1 C.Y. bucket	2.45	6.65	9.10
1400	4' wide, 2' deep, 3/8 C.Y. bucket	1.19	3.95	5.15
1410	3' deep, 3/8 C.Y. bucket	1.57	5.95	7.50
1420	4' deep, 1/2 C.Y. bucket	1.76	6.70	8.45
1430	6' deep, 1/2 C.Y. bucket	4.03	9.90	13.95
1440	8' deep, 1/2 C.Y. bucket	6.15	14	20
1450	10' deep, 1 C.Y. bucket	5.85	14.20	20
1460	12' deep, 1 C.Y. bucket	7.40	17.75	25
1470	15' deep, 1-1/2 C.Y. bucket	7.45	20.50	28
1480	18' deep, 2-1/2 C.Y. bucket	10.15	22.50	32.50
1520	6' wide, 6' deep, 5/8 C.Y. bucket w/trench box	8.25	16.05	24.50
1530	8' deep, 3/4 C.Y. bucket	10.65	21	31.50
1540	10' deep, 1 C.Y. bucket	9.80	21.50	31.50
1550	12' deep, 1-1/2 C.Y. bucket	9.50	24	33.50
1560	16' deep, 2 C.Y. bucket	14.90	30	45
1570	20' deep, 3-1/2 C.Y. bucket	17.75	35	53
1580	24' deep, 3-1/2 C.Y. bucket	22.50	43	65.50
1640	8' wide, 12' deep, 1-1/2 C.Y. bucket w/trench box	13.30	30	43.50
1650	15' deep, 1-1/2 C.Y. bucket	16.25	38.50	55
1660	18' deep, 2-1/2 C.Y. bucket	22.50	43	65.50
1680	24' deep, 3-1/2 C.Y. bucket	30	55.50	85.50
1730	10' wide, 20' deep, 3-1/2 C.Y. bucket w/trench box	30	55.50	85.50
1740	24' deep, 3-1/2 C.Y. bucket	37.50	68.50	106
1780	12' wide, 20' deep, 3-1/2 C.Y. bucket w/trench box	36.50	66	103
1790	25' deep, 3-1/2 C.Y. bucket	47	85	132
1800	1/2:1 slope, 2' wide, 2' deep, 3/8 C.Y. bucket	.70	2.99	3.69
1810	3' deep, 3/8 C.Y. bucket	1.14	5.25	6.40
1820	4' deep, 3/8 C.Y. bucket	1.68	8	9.70
1840	6' deep, 3/8 C.Y. bucket	4.12	11.95	16.05

G1030 Site Earthwork

G1030 807	Trenching Sand & Gravel	COST PER L.F.		
		EQUIP.	LABOR	TOTAL
1860	8' deep, 1/2 C.Y. bucket	6.50	19.10	25.50
1880	10' deep, 1 C.Y. bucket	8.40	23.50	32
2300	4' wide, 2' deep, 3/8 C.Y. bucket	1.38	4.96	6.35
2310	3' deep, 3/8 C.Y. bucket	2.01	8.20	10.20
2320	4' deep, 1/2 C.Y. bucket	2.46	10.15	12.60
2340	6' deep, 1/2 C.Y. bucket	6.85	17.65	24.50
2360	8' deep, 1/2 C.Y. bucket	11.95	28.50	40.50
2380	10' deep, 1 C.Y. bucket	12.95	32.50	45.50
2400	12' deep, 1 C.Y. bucket	22	54	76
2430	15' deep, 1-1/2 C.Y. bucket	21	60	81
2460	18' deep, 2-1/2 C.Y. bucket	40.50	91.50	132
2840	6' wide, 6' deep, 5/8 C.Y. bucket w/trench box	12	24.50	36.50
2860	8' deep, 3/4 C.Y. bucket	17.30	35.50	53
2880	10' deep, 1 C.Y. bucket	17.70	40	57.50
2900	12' deep, 1-1/2 C.Y. bucket	18.60	48.50	67
2940	16' deep, 2 C.Y. bucket	34	71.50	106
2980	20' deep, 3-1/2 C.Y. bucket	47	96	143
3020	24' deep, 3-1/2 C.Y. bucket	66.50	132	199
3100	8' wide, 12' deep, 1-1/4 C.Y. bucket w/trench box	22.50	55	77.50
3120	15' deep, 1-1/2 C.Y. bucket	30.50	77	108
3140	18' deep, 2-1/2 C.Y. bucket	46.50	95.50	142
3180	24' deep, 3-1/2 C.Y. bucket	74	144	218
3270	10' wide, 20' deep, 3-1/2 C.Y. bucket w/trench box	59	116	175
3280	24' deep, 3-1/2 C.Y. bucket	82	157	239
3370	12' wide, 20' deep, 3-1/2 C.Y. bucket w/trench box	66	129	195
3380	25' deep, 3-1/2 C.Y. bucket	95	181	276
3500	1:1 slope, 2' wide, 2' deep, 3/8 C.Y. bucket	1.81	5.05	6.85
3520	3' deep, 3/8 C.Y. bucket	1.57	7.50	9.05
3540	4' deep, 3/8 C.Y. bucket	2.46	12.05	14.50
3560	6' deep, 3/8 C.Y. bucket	4.13	11.95	16.10
3580	8' deep, 1/2 C.Y. bucket	10.80	32	43
3600	10' deep, 1 C.Y. bucket	14.35	40.50	55
3800	4' wide, 2' deep, 3/8 C.Y. bucket	1.58	5.95	7.55
3820	3' deep, 3/8 C.Y. bucket	2.47	10.50	12.95
3840	4' deep, 1/2 C.Y. bucket	3.17	13.65	16.80
3860	6' deep, 1/2 C.Y. bucket	9.70	25.50	35
3880	8' deep, 1/2 C.Y. bucket	17.80	42.50	60.50
3900	10' deep, 1 C.Y. bucket	20	51	71
3920	12' deep, 1 C.Y. bucket	29	72.50	102
3940	15' deep, 1-1/2 C.Y. bucket	35	99.50	135
3960	18' deep, 2-1/2 C.Y. bucket	55.50	126	182
4030	6' wide, 6' deep, 5/8 C.Y. bucket w/trench box	15.75	32.50	48.50
4040	8' deep, 3/4 C.Y. bucket	24	50.50	74.50
4050	10' deep, 1 C.Y. bucket	25.50	58.50	84
4060	12' deep, 1-1/2 C.Y. bucket	27.50	73.50	101
4070	16' deep, 2 C.Y. bucket	53.50	113	167
4080	20' deep, 3-1/2 C.Y. bucket	76	157	233
4090	24' deep, 3-1/2 C.Y. bucket	110	220	330
4500	8' wide, 12' deep, 1-1/2 C.Y. bucket w/trench box	31.50	80	112
4550	15' deep, 1-1/2 C.Y. bucket	44.50	116	161
4600	18' deep, 2-1/2 C.Y. bucket	71	148	219
4650	24' deep, 3-1/2 C.Y. bucket	118	233	350
4800	10' wide, 20' deep, 3-1/2 C.Y. bucket w/trench box	88.50	177	266
4850	24' deep, 3-1/2 C.Y. bucket	126	246	370
4950	12' wide, 20' deep, 3-1/2 C.Y. bucket w/trench box	94.50	187	282
4980	25' deep, 3-1/2 C.Y. bucket	143	277	420

G10 Site Preparation

G1030 Site Earthwork

The Pipe Bedding System is shown for various pipe diameters. Compacted bank sand is used for pipe bedding and to fill 12" over the pipe. No backfill is included. Various side slopes are shown to accommodate different soil conditions. Pipe sizes vary from 6" to 84" diameter.

System Components	QUANTITY	UNIT	COST PER L.F.		
			MAT.	INST.	TOTAL
SYSTEM G1030 815 1440					
PIPE BEDDING, SIDE SLOPE 0 TO 1, 1' WIDE, PIPE SIZE 6" DIAMETER					
Borrow, bank sand, 2 mile haul, machine spread	.086	C.Y.	1.71	.72	2.43
Compaction, vibrating plate	.086	C.Y.		.23	.23
TOTAL			1.71	.95	2.66

G1030 815	Pipe Bedding	COST PER L.F.		
		MAT.	INST.	TOTAL
1440	Pipe bedding, side slope 0 to 1, 1' wide, pipe size 6" diameter	1.71	.95	2.66
1460	2' wide, pipe size 8" diameter	3.70	2.05	5.75
1480	Pipe size 10" diameter	3.78	2.09	5.87
1500	Pipe size 12" diameter	3.86	2.14	6
1520	3' wide, pipe size 14" diameter	6.25	3.46	9.71
1540	Pipe size 15" diameter	6.30	3.50	9.80
1560	Pipe size 16" diameter	6.40	3.53	9.93
1580	Pipe size 18" diameter	6.50	3.60	10.10
1600	4' wide, pipe size 20" diameter	9.25	5.10	14.35
1620	Pipe size 21" diameter	9.30	5.15	14.45
1640	Pipe size 24" diameter	9.50	5.25	14.75
1660	Pipe size 30" diameter	9.70	5.35	15.05
1680	6' wide, pipe size 32" diameter	16.60	9.15	25.75
1700	Pipe size 36" diameter	17	9.40	26.40
1720	7' wide, pipe size 48" diameter	27	14.95	41.95
1740	8' wide, pipe size 60" diameter	33	18.20	51.20
1760	10' wide, pipe size 72" diameter	45.50	25.50	71
1780	12' wide, pipe size 84" diameter	60.50	33.50	94
2140	Side slope 1/2 to 1, 1' wide, pipe size 6" diameter	3.19	1.77	4.96
2160	2' wide, pipe size 8" diameter	5.40	3.01	8.41
2180	Pipe size 10" diameter	5.80	3.22	9.02
2200	Pipe size 12" diameter	6.15	3.41	9.56
2220	3' wide, pipe size 14" diameter	8.90	4.93	13.83
2240	Pipe size 15" diameter	9.10	5.05	14.15
2260	Pipe size 16" diameter	9.35	5.15	14.50
2280	Pipe size 18" diameter	9.80	5.45	15.25
2300	4' wide, pipe size 20" diameter	12.95	7.15	20.10
2320	Pipe size 21" diameter	13.25	7.35	20.60
2340	Pipe size 24" diameter	14.10	7.80	21.90
2360	Pipe size 30" diameter	15.60	8.65	24.25
2380	6' wide, pipe size 32" diameter	23	12.75	35.75
2400	Pipe size 36" diameter	24.50	13.55	38.05
2420	7' wide, pipe size 48" diameter	38	21	59
2440	8' wide, pipe size 60" diameter	48.50	27	75.50
2460	10' wide, pipe size 72" diameter	66.50	37	103.50
2480	12' wide, pipe size 84" diameter	87	48.50	135.50
2620	Side slope 1 to 1, 1' wide, pipe size 6" diameter	4.67	2.58	7.25
2640	2' wide, pipe size 8" diameter	7.20	3.98	11.18

G1030 Site Earthwork

G1030 815	Pipe Bedding	COST PER L.F.		
		MAT.	INST.	TOTAL
2660	Pipe size 10″ diameter	7.80	4.33	12.13
2680	Pipe size 12″ diameter	8.50	4.72	13.22
2700	3′ wide, pipe size 14″ diameter	11.55	6.40	17.95
2720	Pipe size 15″ diameter	11.90	6.60	18.50
2740	Pipe size 16″ diameter	12.30	6.80	19.10
2760	Pipe size 18″ diameter	13.15	7.25	20.40
2780	4′ wide, pipe size 20″ diameter	16.70	9.25	25.95
2800	Pipe size 21″ diameter	17.15	9.50	26.65
2820	Pipe size 24″ diameter	18.60	10.30	28.90
2840	Pipe size 30″ diameter	21.50	11.90	33.40
2860	6′ wide, pipe size 32″ diameter	29.50	16.30	45.80
2880	Pipe size 36″ diameter	32	17.70	49.70
2900	7′ wide, pipe size 48″ diameter	49.50	27.50	77
2920	8′ wide, pipe size 60″ diameter	64	35.50	99.50
2940	10′ wide, pipe size 72″ diameter	87.50	48.50	136
2960	12′ wide, pipe size 84″ diameter	114	63	177

G3010 Water Supply

G3010 110	Water Distribution Piping	COST PER L.F.		
		MAT.	INST.	TOTAL
2000	Piping, excav. & backfill excl., ductile iron class 250, mech. joint			
2130	4" diameter	46.50	18	64.50
2150	6" diameter	56	22.50	78.50
2160	8" diameter	59	27	86
2170	10" diameter	78	31.50	109.50
2180	12" diameter	99.50	34.50	134
2210	16" diameter	123	49.50	172.50
2220	18" diameter	142	52.50	194.50
3000	Tyton joint			
3130	4" diameter	25.50	9.05	34.55
3150	6" diameter	26	10.80	36.80
3160	8" diameter	36	18	54
3170	10" diameter	55	19.85	74.85
3180	12" diameter	58	22.50	80.50
3210	16" diameter	60.50	31.50	92
3220	18" diameter	67	36	103
3230	20" diameter	70	40.50	110.50
3250	24" diameter	95.50	46.50	142
4000	Copper tubing, type K			
4050	3/4" diameter	7.75	3.47	11.22
4060	1" diameter	10.65	4.34	14.99
4080	1-1/2" diameter	16.45	5.25	21.70
4090	2" diameter	25.50	6.05	31.55
4110	3" diameter	53	10.35	63.35
4130	4" diameter	88	14.60	102.60
4150	6" diameter	148	27	175
5000	Polyvinyl chloride class 160, S.D.R. 26			
5130	1-1/2" diameter	.67	1.29	1.96
5150	2" diameter	.97	1.42	2.39
5160	4" diameter	3.33	6.40	9.73
5170	6" diameter	5.50	7.65	13.15
5180	8" diameter	9.45	9.20	18.65
6000	Polyethylene 160 psi, S.D.R. 7			
6050	3/4" diameter	.44	1.85	2.29
6060	1" diameter	.64	2	2.64
6080	1-1/2" diameter	1.06	2.16	3.22
6090	2" diameter	1.58	2.66	4.24

G3020 110	Drainage & Sewage Piping	COST PER L.F.		
		MAT.	INST.	TOTAL
2000	Piping, excavation & backfill excluded, PVC, plain			
2130	4" diameter	1.84	4.53	6.37
2150	6" diameter	4.46	4.86	9.32
2160	8" diameter	7.35	5.05	12.40
2900	Box culvert, precast, 8' long			
3000	6' x 3'	292	35.50	327.50
3020	6' x 7'	350	40	390
3040	8' x 3'	385	44	429
3060	8' x 8'	455	50	505
3080	10' x 3'	445	45.50	490.50
3100	10' x 8'	665	62.50	727.50
3120	12' x 3'	910	50	960
3140	12' x 8'	1,025	74.50	1,099.50
4000	Concrete, nonreinforced			
4150	6" diameter	8.55	12.95	21.50
4160	8" diameter	9.40	15.30	24.70
4170	10" diameter	10.45	15.85	26.30
4180	12" diameter	11.05	17.15	28.20
4200	15" diameter	15.85	19	34.85
4220	18" diameter	19.85	23.50	43.35
4250	24" diameter	24	34.50	58.50
4400	Reinforced, no gasket			
4580	12" diameter	17.80	23	40.80
4600	15" diameter	24	23	47
4620	18" diameter	27	26	53
4650	24" diameter	31.50	34.50	66
4670	30" diameter	57	51.50	108.50
4680	36" diameter	91	62.50	153.50
4690	42" diameter	109	73	182
4700	48" diameter	110	77.50	187.50
4720	60" diameter	199	103	302
4730	72" diameter	300	124	424
4740	84" diameter	445	155	600
4800	With gasket			
4980	12" diameter	19.55	13.45	33
5000	15" diameter	26.50	14.15	40.65
5020	18" diameter	30	14.85	44.85
5050	24" diameter	39	16.65	55.65
5070	30" diameter	66	51.50	117.50
5080	36" diameter	102	62.50	164.50
5090	42" diameter	100	62	162
5100	48" diameter	125	77.50	202.50
5120	60" diameter	240	98.50	338.50
5130	72" diameter	325	124	449
5140	84" diameter	535	206	741
5700	Corrugated metal, alum. or galv. bit. coated			
5760	8" diameter	9.15	10.35	19.50
5770	10" diameter	9.55	13.15	22.70
5780	12" diameter	13.65	16.30	29.95
5800	15" diameter	16.05	17.15	33.20
5820	18" diameter	20.50	18.05	38.55
5850	24" diameter	27	21.50	48.50
5870	30" diameter	32	37.50	69.50
5880	36" diameter	38.50	37.50	76
5900	48" diameter	51	45	96
5920	60" diameter	87.50	66	153.50
5930	72" diameter	93	110	203
6000	Plain			

G3020 Sanitary Sewer

G3020 110	Drainage & Sewage Piping	COST PER L.F.		
		MAT.	INST.	TOTAL
6060	8" diameter	8.60	9.65	18.25
6070	10" diameter	9.25	12.25	21.50
6080	12" diameter	10.70	15.55	26.25
6100	15" diameter	12.20	15.55	27.75
6120	18" diameter	14.85	16.70	31.55
6140	24" diameter	25.50	19.55	45.05
6170	30" diameter	35.50	34.50	70
6180	36" diameter	42.50	34.50	77
6200	48" diameter	59	41	100
6220	60" diameter	92	63.50	155.50
6230	72" diameter	181	159	340
6300	Steel or alum. oval arch, coated & paved invert			
6400	15" equivalent diameter	16.30	17.15	33.45
6420	18" equivalent diameter	21	23	44
6450	24" equivalent diameter	25	27.50	52.50
6470	30" equivalent diameter	33.50	34.50	68
6480	36" equivalent diameter	39.50	45	84.50
6490	42" equivalent diameter	48	50	98
6500	48" equivalent diameter	62	60	122
6600	Plain			
6700	15" equivalent diameter	14.60	15.20	29.80
6720	18" equivalent diameter	19	19.55	38.55
6750	24" equivalent diameter	29.50	23	52.50
6770	30" equivalent diameter	35	42	77
6780	36" equivalent diameter	44	42	86
6790	42" equivalent diameter	65	49	114
6800	48" equivalent diameter	76	60	136
8000	Polyvinyl chloride SDR 35			
8130	4" diameter	1.84	4.53	6.37
8150	6" diameter	4.46	4.86	9.32
8160	8" diameter	7.35	5.05	12.40
8170	10" diameter	12.45	6.85	19.30
8180	12" diameter	17.50	7.05	24.55
8200	15" diameter	24	9.40	33.40

G3030 Storm Sewer

Manhole   Catch Basin

The Manhole and Catch Basin System includes: excavation with a backhoe; a formed concrete footing; frame and cover; cast iron steps and compacted backfill.

The Expanded System Listing shows manholes that have a 4', 5' and 6' inside diameter riser. Depths range from 4' to 14'. Construction material shown is either concrete, concrete block, precast concrete, or brick.

System Components	QUANTITY	UNIT	COST PER EACH		
			MAT.	INST.	TOTAL
SYSTEM G3030 210 1920					
MANHOLE/CATCH BASIN, BRICK, 4' I.D. RISER, 4' DEEP					
Excavation, hydraulic backhoe, 3/8 C.Y. bucket	14.815	B.C.Y.		130.22	130.22
Trim sides and bottom of excavation	64.000	S.F.		73.60	73.60
Forms in place, manhole base, 4 uses	20.000	SFCA	16.40	117	133.40
Reinforcing in place footings, #4 to #7	.019	Ton	23.75	24.70	48.45
Concrete, 3000 psi	.925	C.Y.	131.35		131.35
Place and vibrate concrete, footing, direct chute	.925	C.Y.		53.72	53.72
Catch basin or MH, brick, 4' ID, 4' deep	1.000	Ea.	675	1,150	1,825
Catch basin or MH steps; heavy galvanized cast iron	1.000	Ea.	19.85	15.80	35.65
Catch basin or MH frame and cover	1.000	Ea.	325	241	566
Fill, granular	12.954	L.C.Y.	349.76		349.76
Backfill, spread with wheeled front end loader	12.954	L.C.Y.		37.31	37.31
Backfill compaction, 12" lifts, air tamp	12.954	E.C.Y.		133.82	133.82
TOTAL			1,541.11	1,977.17	3,518.28

G3030 210	Manholes & Catch Basins	COST PER EACH		
		MAT.	INST.	TOTAL
1920	Manhole/catch basin, brick, 4' I.D. riser, 4' deep	1,550	1,975	3,525
1940	6' deep	2,150	2,725	4,875
1960	8' deep	2,875	3,750	6,625
1980	10' deep	3,375	4,650	8,025
3000	12' deep	4,200	5,075	9,275
3020	14' deep	5,175	7,100	12,275
3200	Block, 4' I.D. riser, 4' deep	1,325	1,575	2,900
3220	6' deep	1,825	2,250	4,075
3240	8' deep	2,425	3,100	5,525
3260	10' deep	2,875	3,850	6,725
3280	12' deep	3,650	4,900	8,550
3300	14' deep	4,550	5,975	10,525
4620	Concrete, cast-in-place, 4' I.D. riser, 4' deep	1,500	2,700	4,200
4640	6' deep	2,125	3,600	5,725
4660	8' deep	2,975	5,250	8,225
4680	10' deep	3,550	6,500	10,050
4700	12' deep	4,475	8,075	12,550
4720	14' deep	5,525	9,675	15,200
5820	Concrete, precast, 4' I.D. riser, 4' deep	1,825	1,450	3,275
5840	6' deep	2,325	1,950	4,275

G3030 Storm Sewer

G3030 210	Manholes & Catch Basins	COST PER EACH		
		MAT.	INST.	TOTAL
5860	8' deep	2,975	2,775	5,750
5880	10' deep	3,550	3,400	6,950
5900	12' deep	4,475	4,100	8,575
5920	14' deep	5,525	5,350	10,875
6000	5' I.D. riser, 4' deep	2,850	1,575	4,425
6020	6' deep	3,575	2,200	5,775
6040	8' deep	4,650	2,900	7,550
6060	10' deep	5,950	3,700	9,650
6080	12' deep	7,400	4,725	12,125
6100	14' deep	8,975	5,800	14,775
6200	6' I.D. riser, 4' deep	3,975	2,050	6,025
6220	6' deep	4,825	2,700	7,525
6240	8' deep	6,150	3,800	9,950
6260	10' deep	7,700	4,825	12,525
6280	12' deep	9,400	6,100	15,500
6300	14' deep	11,300	7,425	18,725

G30 Site Mechanical Utilities

G3060 Fuel Distribution

G3060 110	Gas Service Piping	COST PER L.F.		
		MAT.	INST.	TOTAL
2000	Piping, excavation & backfill excluded, polyethylene			
2070	1-1/4" diam, SDR 10	1.80	4.46	6.26
2090	2" diam, SDR 11	2.72	4.97	7.69
2110	3" diam, SDR 11.5	6.80	5.95	12.75
2130	4" diam, SDR 11	14.70	10.25	24.95
2150	6" diam, SDR 21	39	11	50
2160	8" diam, SDR 21	59.50	13.25	72.75
3000	Steel, schedule 40, plain end, tarred & wrapped			
3060	1" diameter	5.60	10.20	15.80
3090	2" diameter	8.75	10.90	19.65
3110	3" diameter	14.55	11.75	26.30
3130	4" diameter	18.90	18.30	37.20
3140	5" diameter	27.50	21	48.50
3150	6" diameter	33.50	26	59.50
3160	8" diameter	53	33.50	86.50

G3060 Fuel Distribution

Fiberglass Underground Fuel Storage Tank, Double Wall

System Components	QUANTITY	UNIT	COST EACH		
			MAT.	INST.	TOTAL
SYSTEM G3060 310 1010					
STORAGE TANK, FUEL, UNDERGROUND, DOUBLE WALL FIBERGLASS, 600 GAL.					
Excavating trench, no sheeting or dewatering, 6'-10 ' D, 3/4 CY hyd backhoe	18.000	Ea.		158.22	158.22
Corrosion resistance wrap & coat, small diam. pipe, 1" diam., add	120.000	Ea.	133.20		133.20
Fiberglass-reinforced elbows 90 & 45 degree, 2 inch	7.000	Ea.	315	640.50	955.50
Fiberglass-reinforced threaded adapters, 2 inch	2.000	Ea.	42	210	252
Remote tank gauging system, 30', 5" pointer travel	1.000	Ea.	6,325	390	6,715
Corrosion resistance wrap & coat, small diam. pipe. 4 " diam. add	30.000	Ea.	78.60		78.60
Pea gravel	9.000	Ea.	1,134	495	1,629
Reinforcing in hold down pad, #3 to #7	120.000	Ea.	74.40	78	152.40
Tank leak detector system, 8 channel, external monitoring	1.000	Ea.	1,350		1,350
Tubbing copper, type L, 3/4" diam.	120.000	Ea.	624	1,218	1,842
Elbow 90 degree ,wrought cu x cu 3/4 " diam.	4.000	Ea.	11.80	162	173.80
Pipe, black steel welded, sch 40, 3" diam.	30.000	Ea.	549	1,056.60	1,605.60
Elbow 90 degree, steel welded butt joint, 3 " diam.	3.000	Ea.	99	644.25	743.25
Fiberglass-reinforced plastic pipe, 2 inch	156.000	Ea.	1,201.20	803.40	2,004.60
Fiberglass-reinforced plastic pipe, 3 inch	36.000	Ea.	1,170	210.60	1,380.60
Fiberglass-reinforced elbows 90 & 45 degree, 3 inch	3.000	Ea.	432	366	798
Tank leak detection system, probes, well monitoring, liquid phase detection	2.000	Ea.	1,600		1,600
Double wall fiberglass annular space	1.000	Ea.	470		470
Flange, weld neck, 150 lb, 3 " pipe size	1.000	Ea.	45	107.35	152.35
Vent protector/breather, 2" diam.	1.000	Ea.	59	24.50	83.50
Corrosion resistance wrap & coat, small diam. pipe, 2" diam., add	36.000	Ea.	50.76		50.76
Concrete hold down pad, 6' x 5' x 8" thick	30.000	Ea.	112.50	50.70	163.20
600 gallon capacity, underground storage tank, double wall fiberglass	1.000	Ea.	10,000	585	10,585
Tank, for hold-downs 500-4,000 gal., add	1.000	Ea.	555	187	742
Foot valve, single poppet, 3/4" diam.	1.000	Ea.	124	43.50	167.50
Fuel fill box, locking inner cover, 3" diam.	1.000	Ea.	131	157	288
Fuel oil specialties, valve,ball chk, globe type fusible, 3/4" diam.	2.000	Ea.	174	78	252
TOTAL			26,860.46	7,665.62	34,526.08

G3060 310		Fiberglass Fuel Tank, Double Wall	COST EACH		
			MAT.	INST.	TOTAL
1010	Storage tank, fuel, underground, double wall fiberglass, 600 Gal.		26,900	7,675	34,575
1020		2500 Gal.	41,000	10,500	51,500
1030		4000 Gal.	49,000	13,000	62,000
1040		6000 Gal.	58,000	14,600	72,600

G30 Site Mechanical Utilities

G3060 Fuel Distribution

G3060 310	Fiberglass Fuel Tank, Double Wall	COST EACH		
		MAT.	INST.	TOTAL
1050	8000 Gal.	72,000	17,600	89,600
1060	10,000 Gal.	82,500	20,200	102,700
1070	15,000 Gal.	118,000	23,200	141,200
1080	20,000 Gal.	142,500	28,100	170,600
1090	25,000 Gal.	163,500	32,700	196,200
1100	30,000 Gal.	187,500	34,600	222,100

For customer support on your Mechanical Costs with RSMeans Data, call 800.448.8182.

G3060 Fuel Distribution

Remote Gauge —
3" Fill Box —
2" Vent Cap
Manway Extension
Return
Pea Gravel Backfill
Concrete Slab w/ Reinf. Steel
16" Manhole
Hold Down Rods and Lugs
Supply

Steel Underground Fuel Storage Tank, Double Wall

System Components	QUANTITY	UNIT	COST EACH		
			MAT.	INST.	TOTAL
SYSTEM G3060 320 1010					
STORAGE TANK, FUEL, UNDERGROUND, DOUBLE WALL STEEL, 500 GAL.					
Fiberglass-reinforced plastic pipe, 2 inch	156.000	Ea.	1,201.20	803.40	2,004.60
Fiberglass-reinforced plastic pipe, 3 inch	36.000	Ea.	1,170	210.60	1,380.60
Fiberglass-reinforced elbows 90 & 45 degree, 3 inch	3.000	Ea.	432	366	798
Tank leak detection system, probes, well monitoring, liquid phase detection	2.000	Ea.	1,600		1,600
Flange, weld neck, 150 lb, 3" pipe size	1.000	Ea.	45	107.35	152.35
Vent protector/breather, 2" diam.	1.000	Ea.	59	24.50	83.50
Concrete hold down pad, 6' x 5' x 8" thick	30.000	Ea.	112.50	50.70	163.20
Tanks, for hold-downs, 500-2,000 gal, add	1.000	Ea.	249	187	436
Double wall steel tank annular space	1.000	Ea.	310		310
Foot valve, single poppet, 3/4" diam.	1.000	Ea.	124	43.50	167.50
Fuel fill box, locking inner cover, 3" diam.	1.000	Ea.	131	157	288
Fuel oil specialties, valve, ball chk, globe type fusible, 3/4" diam.	2.000	Ea.	174	78	252
Pea gravel	9.000	Ea.	1,134	495	1,629
Reinforcing in hold down pad, #3 to #7	120.000	Ea.	74.40	78	152.40
Tank, for manways, add	1.000	Ea.	1,175		1,175
Tank leak detector system, 8 channel, external monitoring	1.000	Ea.	1,350		1,350
Tubbing copper, type L, 3/4" diam.	120.000	Ea.	624	1,218	1,842
Elbow 90 degree, wrought cu x cu 3/4" diam.	8.000	Ea.	23.60	324	347.60
Pipe, black steel welded, sch 40, 3" diam.	30.000	Ea.	549	1,056.60	1,605.60
Elbow 90 degree, steel welded butt joint, 3" diam.	3.000	Ea.	99	644.25	743.25
Excavating trench, no sheeting or dewatering, 6'-10' D, 3/4 CY hyd backhoe	18.000	Ea.		158.22	158.22
Fiberglass-reinforced elbows 90 & 45 degree, 2 inch	7.000	Ea.	315	640.50	955.50
Fiberglass-reinforced threaded adapters, 2 inch	2.000	Ea.	42	210	252
500 gallon capacity, underground storage tank, double wall steel	1.000	Ea.	3,650	590	4,240
Remote tank gauging system, 30', 5" pointer travel	1.000	Ea.	6,325	390	6,715
TOTAL			20,968.70	7,832.62	28,801.32

G3060 320	Steel Fuel Tank, Double Wall	COST EACH		
		MAT.	INST.	TOTAL
1010	Storage tank, fuel, underground, double wall steel, 500 Gal.	21,000	7,825	28,825
1020	2000 Gal.	26,800	10,200	37,000
1030	4000 Gal.	37,000	13,100	50,100
1040	6000 Gal.	42,200	14,500	56,700
1050	8000 Gal.	48,300	17,900	66,200
1060	10,000 Gal.	55,500	20,500	76,000

G30 Site Mechanical Utilities

G3060 Fuel Distribution

G3060 320	Steel Fuel Tank, Double Wall	COST EACH		
		MAT.	INST.	TOTAL
1070	15,000 Gal.	67,500	23,800	91,300
1080	20,000 Gal.	93,000	28,500	121,500
1090	25,000 Gal.	103,000	32,700	135,700
1100	30,000 Gal.	106,000	34,700	140,700
1110	40,000 Gal.	203,500	43,600	247,100
1120	50,000 Gal.	246,500	52,500	299,000

G3060 325	Steel Tank, Above Ground, Single Wall	COST EACH		
		MAT.	INST.	TOTAL
1010	Storage tank, fuel, above ground, single wall steel, 550 Gal.	15,400	5,600	21,000
1020	2000 Gal.	25,500	5,725	31,225
1030	5000 Gal.	36,100	8,750	44,850
1040	10,000 Gal.	57,500	9,300	66,800
1050	15,000 Gal.	75,500	9,550	85,050
1060	20,000 Gal.	94,000	9,850	103,850
1070	25,000 Gal.	105,500	10,100	115,600
1080	30,000 Gal.	127,000	10,600	137,600

G3060 330	Steel Tank, Above Ground, Double Wall	COST EACH		
		MAT.	INST.	TOTAL
1010	Storage tank, fuel, above ground, double wall steel, 500 Gal.	14,500	5,650	20,150
1020	2000 Gal.	19,600	5,800	25,400
1030	4000 Gal.	28,500	5,900	34,400
1040	6000 Gal.	32,600	9,050	41,650
1050	8000 Gal.	36,300	9,300	45,600
1060	10,000 Gal.	48,300	9,450	57,750
1070	15,000 Gal.	60,000	9,800	69,800
1080	20,000 Gal.	68,000	10,100	78,100
1090	25,000 Gal.	81,000	10,400	91,400
1100	30,000 Gal.	88,500	10,800	99,300

Reference Section

All the reference information is in one section, making it easy to find what you need to know and easy to use the data set on a daily basis. This section is visually identified by a vertical black bar on the page edges.

In this Reference Section, we've included Equipment Rental Costs, a listing of rental and operating costs; Crew Listings, a full listing of all crews and equipment and their costs; Historical Cost Indexes for cost comparisons over time; City Cost Indexes and Location Factors for adjusting costs to the region you are in; Reference Tables, where you will find explanations, estimating information and procedures, and technical data; Change Orders, information on pricing changes to contract documents; and an explanation of all the Abbreviations in the data set.

Table of Contents

Estimating Tips

- This section contains the average costs to rent and operate hundreds of pieces of construction equipment. This is useful information when one is estimating the time and material requirements of any particular operation in order to establish a unit or total cost. Bare equipment costs shown on a unit cost line include, not only rental, but also operating costs for equipment under normal use.

Rental Costs

- Equipment rental rates are obtained from the following industry sources throughout North America: contractors, suppliers, dealers, manufacturers, and distributors.

- Rental rates vary throughout the country, with larger cities generally having lower rates. Lease plans for new equipment are available for periods in excess of six months, with a percentage of payments applying toward purchase.

- Monthly rental rates vary from 2% to 5% of the purchase price of the equipment depending on the anticipated life of the equipment and its wearing parts.

- Weekly rental rates are about 1/3 of the monthly rates, and daily rental rates are about 1/3 of the weekly rate.

- Rental rates can also be treated as reimbursement costs for contractor-owned equipment. Owned equipment costs include depreciation, loan payments, interest, taxes, insurance, storage, and major repairs.

Operating Costs

- The operating costs include parts and labor for routine servicing, such as the repair and replacement of pumps, filters, and worn lines. Normal operating expendables, such as fuel, lubricants, tires, and electricity (where applicable), are also included.

- Extraordinary operating expendables with highly variable wear patterns, such as diamond bits and blades, are excluded. These costs can be found as material costs in the Unit Price section.

- The hourly operating costs listed do not include the operator's wages.

Equipment Cost/Day

- Any power equipment required by a crew is shown in the Crew Listings with a daily cost.

- This daily cost of equipment needed by a crew includes both the rental cost and the operating cost and is based on dividing the weekly rental rate by 5 (the number of working days in the week), then adding the hourly operating cost multiplied by 8 (the number of hours in a day). This "Equipment Cost/Day" is shown in the far right column of the Equipment Rental section.

- If equipment is needed for only one or two days, it is best to develop your own cost by including components for daily rent and hourly operating costs. This is important when the listed Crew for a task does not contain the equipment needed, such as a crane for lifting mechanical heating/cooling equipment up onto a roof.

- If the quantity of work is less than the crew's Daily Output shown for a Unit Price line item that includes a bare unit equipment cost, the recommendation is to estimate one day's rental cost and operating cost for equipment shown in the Crew Listing for that line item.

- Please note, in some cases the equipment description in the crew is followed by a time period in parenthesis. For example: (daily) or (monthly). In these cases the equipment cost/day is calculated by adding the rental cost per time period to the hourly operating cost multiplied by 8.

Mobilization, Demobilization Costs

- The cost to move construction equipment from an equipment yard or rental company to the job site and back again is not included in equipment rental costs listed in the Reference Section. It is also not included in the bare equipment cost of any Unit Price line item or in any equipment costs shown in the Crew Listings.

- Mobilization (to the site) and demobilization (from the site) costs can be found in the Unit Price section.

- If a piece of equipment is already at the job site, it is not appropriate to utilize mobilization or demobilization costs again in an estimate. ■

Same Data. Simplified.

Enjoy the convenience and efficiency of accessing your costs anywhere:

- **Skip the multiplier** by setting your location
- **Quickly search,** edit, favorite and share costs
- **Stay on top of price changes** with automatic updates

Discover more at rsmeans.com/online

01 54 33 | Equipment Rental

		UNIT	HOURLY OPER. COST	RENT PER DAY	RENT PER WEEK	RENT PER MONTH	EQUIPMENT COST/DAY	
10	0010	**CONCRETE EQUIPMENT RENTAL** without operators [R015433 -10]						**10**
	0200	Bucket, concrete lightweight, 1/2 C.Y.	Ea.	.87	38.50	115	345	30	
	0300	1 C.Y.		.98	63.50	190	570	45.80	
	0400	1-1/2 C.Y.		1.23	60	180	540	45.85	
	0500	2 C.Y.		1.34	73.50	220	660	54.70	
	0580	8 C.Y.		6.47	93.50	280	840	107.75	
	0600	Cart, concrete, self-propelled, operator walking, 10 C.F.		2.85	175	525	1,575	127.85	
	0700	Operator riding, 18 C.F.		4.81	192	575	1,725	153.45	
	0800	Conveyer for concrete, portable, gas, 16" wide, 26' long		10.61	160	480	1,450	180.85	
	0900	46' long		10.98	175	525	1,575	192.85	
	1000	56' long		11.15	192	575	1,725	204.15	
	1100	Core drill, electric, 2-1/2 H.P., 1" to 8" bit diameter		1.56	83.50	250	750	62.50	
	1150	11 H.P., 8" to 18" cores		5.38	119	356.32	1,075	114.30	
	1200	Finisher, concrete floor, gas, riding trowel, 96" wide		9.64	153	459.60	1,375	169	
	1300	Gas, walk-behind, 3 blade, 36" trowel		2.03	96	287.50	865	73.75	
	1400	4 blade, 48" trowel		3.06	104	312.50	940	87	
	1500	Float, hand-operated (Bull float), 48" wide		.08	12.35	37	111	8.05	
	1570	Curb builder, 14 H.P., gas, single screw		14.00	253	760	2,275	263.95	
	1590	Double screw		15.00	253	760	2,275	272	
	1600	Floor grinder, concrete and terrazzo, electric, 22" path		3.03	134	401.75	1,200	104.60	
	1700	Edger, concrete, electric, 7" path		1.18	57.50	172.50	520	43.95	
	1750	Vacuum pick-up system for floor grinders, wet/dry		1.61	102	305.71	915	74.05	
	1800	Mixer, powered, mortar and concrete, gas, 6 C.F., 18 H.P.		7.39	97	291	875	117.35	
	1900	10 C.F., 25 H.P.		8.97	114	342.50	1,025	140.30	
	2000	16 C.F.		9.33	144	432.50	1,300	161.15	
	2100	Concrete, stationary, tilt drum, 2 C.Y.		7.21	80	240	720	105.70	
	2120	Pump, concrete, truck mounted, 4" line, 80' boom		29.79	287	860	2,575	410.35	
	2140	5" line, 110' boom		37.34	287	860	2,575	470.75	
	2160	Mud jack, 50 C.F. per hr.		6.43	228	685	2,050	188.45	
	2180	225 C.F. per hr.		8.52	293	880	2,650	244.15	
	2190	Shotcrete pump rig, 12 C.Y./hr.		13.93	223	670	2,000	245.40	
	2200	35 C.Y./hr.		15.75	287	860	2,575	298	
	2600	Saw, concrete, manual, gas, 18 H.P.		5.52	112	337	1,000	111.55	
	2650	Self-propelled, gas, 30 H.P.		7.87	81	242.71	730	111.50	
	2675	V-groove crack chaser, manual, gas, 6 H.P.		1.64	100	300	900	73.10	
	2700	Vibrators, concrete, electric, 60 cycle, 2 H.P.		.47	73	218.50	655	47.45	
	2800	3 H.P.		.56	73.50	221	665	48.70	
	2900	Gas engine, 5 H.P.		1.54	16.85	50.61	152	22.45	
	3000	8 H.P.		2.08	17.05	51.12	153	26.85	
	3050	Vibrating screed, gas engine, 8 H.P.		2.80	88	263.50	790	75.10	
	3120	Concrete transit mixer, 6 x 4, 250 H.P., 8 C.Y., rear discharge		50.57	70	210	630	446.55	
	3200	Front discharge		58.71	135	405	1,225	550.65	
	3300	6 x 6, 285 H.P., 12 C.Y., rear discharge		57.97	150	450	1,350	553.80	
	3400	Front discharge		60.41	170	510	1,525	585.25	
20	0010	**EARTHWORK EQUIPMENT RENTAL** without operators [R015433 -10]							**20**
	0040	Aggregate spreader, push type, 8' to 12' wide	Ea.	2.59	75	225	675	65.75	
	0045	Tailgate type, 8' wide		2.54	63.50	190	570	58.30	
	0055	Earth auger, truck mounted, for fence & sign posts, utility poles		13.81	150	450	1,350	200.50	
	0060	For borings and monitoring wells		42.52	83.50	250	750	390.20	
	0070	Portable, trailer mounted		2.29	100	300	900	78.35	
	0075	Truck mounted, for caissons, water wells		85.14	150	450	1,350	771.10	
	0080	Horizontal boring machine, 12" to 36" diameter, 45 H.P.		22.70	104	312	935	244	
	0090	12" to 48" diameter, 65 H.P.		31.16	108	325	975	314.25	
	0095	Auger, for fence posts, gas engine, hand held		.45	84	251.50	755	53.85	
	0100	Excavator, diesel hydraulic, crawler mounted, 1/2 C.Y. cap.		21.66	465	1,394.28	4,175	452.10	
	0120	5/8 C.Y. capacity		28.95	610	1,833.22	5,500	598.25	
	0140	3/4 C.Y. capacity		32.56	725	2,168.88	6,500	694.25	
	0150	1 C.Y. capacity		41.09	780	2,333	7,000	795.30	

01 54 33 | Equipment Rental

		UNIT	HOURLY OPER. COST	RENT PER DAY	RENT PER WEEK	RENT PER MONTH	EQUIPMENT COST/DAY	
0200	1-1/2 C.Y. capacity	Ea.	48.44	500	1,500	4,500	687.55	20
0300	2 C.Y. capacity		56.41	835	2,500	7,500	951.30	
0320	2-1/2 C.Y. capacity		82.39	1,350	4,027.92	12,100	1,465	
0325	3-1/2 C.Y. capacity		119.76	2,000	6,000	18,000	2,158	
0330	4-1/2 C.Y. capacity		151.17	3,675	11,000	33,000	3,409	
0335	6 C.Y. capacity		191.81	3,225	9,650	29,000	3,464	
0340	7 C.Y. capacity		174.67	3,400	10,200	30,600	3,437	
0342	Excavator attachments, bucket thumbs		3.39	258	774.60	2,325	182.05	
0345	Grapples		3.13	222	666.16	2,000	158.30	
0346	Hydraulic hammer for boom mounting, 4000 ft lb.		13.44	890	2,670	8,000	641.50	
0347	5000 ft lb.		15.90	950	2,850	8,550	697.25	
0348	8000 ft lb.		23.47	1,200	3,600	10,800	907.75	
0349	12,000 ft lb.		25.64	1,100	3,333	10,000	871.75	
0350	Gradall type, truck mounted, 3 ton @ 15' radius, 5/8 C.Y.		43.31	835	2,500	7,500	846.50	
0370	1 C.Y. capacity		59.22	835	2,500	7,500	973.80	
0400	Backhoe-loader, 40 to 45 H.P., 5/8 C.Y. capacity		11.86	244	732.50	2,200	241.40	
0450	45 H.P. to 60 H.P., 3/4 C.Y. capacity		17.97	117	350	1,050	213.75	
0460	80 H.P., 1-1/4 C.Y. capacity		20.30	117	350	1,050	232.40	
0470	112 H.P., 1-1/2 C.Y. capacity		32.89	610	1,833.22	5,500	629.75	
0482	Backhoe-loader attachment, compactor, 20,000 lb.		6.42	155	464.76	1,400	144.30	
0485	Hydraulic hammer, 750 ft lb.		3.67	107	320.17	960	93.40	
0486	Hydraulic hammer, 1200 ft lb.		6.53	205	614.52	1,850	175.15	
0500	Brush chipper, gas engine, 6" cutter head, 35 H.P.		9.14	247	740	2,225	221.10	
0550	Diesel engine, 12" cutter head, 130 H.P.		23.60	340	1,020	3,050	392.80	
0600	15" cutter head, 165 H.P.		26.51	415	1,239.36	3,725	459.90	
0750	Bucket, clamshell, general purpose, 3/8 C.Y.		1.40	91.50	275	825	66.15	
0800	1/2 C.Y.		1.51	91.50	275	825	67.10	
0850	3/4 C.Y.		1.64	91.50	275	825	68.10	
0900	1 C.Y.		1.70	91.50	275	825	68.55	
0950	1-1/2 C.Y.		2.78	91.50	275	825	77.25	
1000	2 C.Y.		2.91	91.50	275	825	78.30	
1010	Bucket, dragline, medium duty, 1/2 C.Y.		.82	91.50	275	825	61.55	
1020	3/4 C.Y.		.78	91.50	275	825	61.25	
1030	1 C.Y.		.80	91.50	275	825	61.35	
1040	1-1/2 C.Y.		1.26	91.50	275	825	65.05	
1050	2 C.Y.		1.29	91.50	275	825	65.30	
1070	3 C.Y.		2.07	91.50	275	825	71.60	
1200	Compactor, manually guided 2-drum vibratory smooth roller, 7.5 H.P.		7.20	181	542.50	1,625	166.10	
1250	Rammer/tamper, gas, 8"		2.20	48	144.59	435	46.50	
1260	15"		2.62	55	165.25	495	54	
1300	Vibratory plate, gas, 18" plate, 3000 lb. blow		2.12	24.50	72.81	218	31.55	
1350	21" plate, 5000 lb. blow		2.61	241	722.50	2,175	165.35	
1370	Curb builder/extruder, 14 H.P., gas, single screw		13.99	253	760	2,275	263.95	
1390	Double screw		14.99	253	760	2,275	271.95	
1500	Disc harrow attachment, for tractor		.47	82.50	246.84	740	53.15	
1810	Feller buncher, shearing & accumulating trees, 100 H.P.		39.08	460	1,380	4,150	588.60	
1860	Grader, self-propelled, 25,000 lb.		33.25	1,100	3,333	10,000	932.60	
1910	30,000 lb.		32.76	1,325	4,000	12,000	1,062	
1920	40,000 lb.		51.73	1,550	4,667	14,000	1,347	
1930	55,000 lb.		66.73	1,775	5,333	16,000	1,600	
1950	Hammer, pavement breaker, self-propelled, diesel, 1000 to 1250 lb.		28.31	600	1,800	5,400	586.50	
2000	1300 to 1500 lb.		42.67	1,000	3,020.94	9,075	945.55	
2050	Pile driving hammer, steam or air, 4150 ft lb. @ 225 bpm		12.11	500	1,500	4,500	396.90	
2100	8750 ft lb. @ 145 bpm		14.30	700	2,100	6,300	534.45	
2150	15,000 ft lb. @ 60 bpm		14.63	835	2,500	7,500	617.05	
2200	24,450 ft lb. @ 111 bpm		15.64	965	2,900	8,700	705.15	
2250	Leads, 60' high for pile driving hammers up to 20,000 ft lb.		3.66	300	900	2,700	209.25	
2300	90' high for hammers over 20,000 ft lb.		5.43	540	1,620	4,850	367.45	

For customer support on your Mechanical Costs with RSMeans Data, call 800.448.8182.

609

		UNIT	HOURLY OPER. COST	RENT PER DAY	RENT PER WEEK	RENT PER MONTH	EQUIPMENT COST/DAY	
01 54 33	**Equipment Rental**							
2350	Diesel type hammer, 22,400 ft lb.	Ea.	17.76	490	1,471.74	4,425	436.45	20
2400	41,300 ft lb.		25.61	620	1,859.04	5,575	576.65	
2450	141,000 ft lb.		41.20	980	2,943.48	8,825	918.35	
2500	Vib. elec. hammer/extractor, 200 kW diesel generator, 34 H.P.		41.25	715	2,143.06	6,425	758.60	
2550	80 H.P.		72.81	1,025	3,098.40	9,300	1,202	
2600	150 H.P.		134.74	2,000	5,964.42	17,900	2,271	
2700	Hydro Excavator w/EXT boom 12 C.Y., 1200 gallons		37.70	1,600	4,800	14,400	1,262	
2800	Log chipper, up to 22" diameter, 600 H.P.		46.13	305	915	2,750	552	
2850	Logger, for skidding & stacking logs, 150 H.P.		43.40	930	2,785	8,350	904.20	
2860	Mulcher, diesel powered, trailer mounted		17.99	305	915	2,750	326.90	
2900	Rake, spring tooth, with tractor		14.67	370	1,110.26	3,325	339.45	
3000	Roller, vibratory, tandem, smooth drum, 20 H.P.		7.78	320	967	2,900	255.65	
3050	35 H.P.		10.10	260	779.76	2,350	236.75	
3100	Towed type vibratory compactor, smooth drum, 50 H.P.		25.20	520	1,566	4,700	514.75	
3150	Sheepsfoot, 50 H.P.		25.56	385	1,161.90	3,475	436.90	
3170	Landfill compactor, 220 H.P.		69.80	1,650	4,985	15,000	1,555	
3200	Pneumatic tire roller, 80 H.P.		12.88	405	1,213.54	3,650	345.75	
3250	120 H.P.		19.33	665	1,988.14	5,975	552.25	
3300	Sheepsfoot vibratory roller, 240 H.P.		62.02	1,425	4,260.30	12,800	1,348	
3320	340 H.P.		83.57	2,175	6,500	19,500	1,969	
3350	Smooth drum vibratory roller, 75 H.P.		23.27	655	1,962.32	5,875	578.60	
3400	125 H.P.		27.53	740	2,220.52	6,650	664.35	
3410	Rotary mower, brush, 60", with tractor		18.73	360	1,084.44	3,250	366.75	
3420	Rototiller, walk-behind, gas, 5 H.P.		2.13	60	180	540	53.05	
3422	8 H.P.		2.80	132	395	1,175	101.40	
3440	Scrapers, towed type, 7 C.Y. capacity		6.42	127	382.14	1,150	127.80	
3450	10 C.Y. capacity		7.18	170	511.24	1,525	159.70	
3500	15 C.Y. capacity		7.38	196	588.70	1,775	176.75	
3525	Self-propelled, single engine, 14 C.Y. capacity		132.89	2,225	6,660	20,000	2,395	
3550	Dual engine, 21 C.Y. capacity		140.95	2,500	7,500	22,500	2,628	
3600	31 C.Y. capacity		187.28	3,625	10,844.40	32,500	3,667	
3640	44 C.Y. capacity		231.98	4,650	13,942.80	41,800	4,644	
3650	Elevating type, single engine, 11 C.Y. capacity		61.68	1,075	3,200	9,600	1,133	
3700	22 C.Y. capacity		114.28	1,625	4,850	14,600	1,884	
3710	Screening plant, 110 H.P. w/5' x 10' screen		21.07	645	1,933	5,800	555.15	
3720	5' x 16' screen		26.60	1,325	4,000	12,000	1,013	
3850	Shovel, crawler-mounted, front-loading, 7 C.Y. capacity		218.00	3,925	11,773.92	35,300	4,099	
3855	12 C.Y. capacity		335.89	5,450	16,318.24	49,000	5,951	
3860	Shovel/backhoe bucket, 1/2 C.Y.		2.68	73	218.95	655	65.25	
3870	3/4 C.Y.		2.66	82	245.81	735	70.40	
3880	1 C.Y.		2.75	91	272.66	820	76.50	
3890	1-1/2 C.Y.		2.94	107	320.17	960	87.60	
3910	3 C.Y.		3.43	145	433.78	1,300	114.15	
3950	Stump chipper, 18" deep, 30 H.P.		6.88	232	697	2,100	194.45	
4110	Dozer, crawler, torque converter, diesel 80 H.P.		25.18	335	1,000	3,000	401.40	
4150	105 H.P.		34.23	600	1,800	5,400	633.85	
4200	140 H.P.		41.15	720	2,166	6,500	762.40	
4260	200 H.P.		62.97	1,675	5,000	15,000	1,504	
4310	300 H.P.		80.49	1,875	5,600	16,800	1,764	
4360	410 H.P.		106.42	3,200	9,630	28,900	2,777	
4370	500 H.P.		132.98	3,900	11,670	35,000	3,398	
4380	700 H.P.		229.47	5,475	16,421.52	49,300	5,120	
4400	Loader, crawler, torque conv., diesel, 1-1/2 C.Y., 80 H.P.		29.45	550	1,651	4,950	565.80	
4450	1-1/2 to 1-3/4 C.Y., 95 H.P.		30.18	695	2,091.42	6,275	659.75	
4510	1-3/4 to 2-1/4 C.Y., 130 H.P.		47.61	965	2,900	8,700	960.90	
4530	2-1/2 to 3-1/4 C.Y., 190 H.P.		57.61	1,175	3,540	10,600	1,169	
4560	3-1/2 to 5 C.Y., 275 H.P.		71.19	1,525	4,595.96	13,800	1,489	
4610	Front end loader, 4WD, articulated frame, diesel, 1 to 1-1/4 C.Y., 70 H.P.		16.58	282	846.90	2,550	302	

		01 54 33	Equipment Rental	UNIT	HOURLY OPER. COST	RENT PER DAY	RENT PER WEEK	RENT PER MONTH	EQUIPMENT COST/DAY	
20	4620	1-1/2 to 1-3/4 C.Y., 95 H.P.	Ea.	19.94	440	1,320	3,950	423.50	20
	4650	1-3/4 to 2 C.Y., 130 H.P.		21.00	395	1,187.72	3,575	405.55	
	4710	2-1/2 to 3-1/2 C.Y., 145 H.P.		29.44	780	2,333	7,000	702.10	
	4730	3 to 4-1/2 C.Y., 185 H.P.		31.99	890	2,667	8,000	789.35	
	4760	5-1/4 to 5-3/4 C.Y., 270 H.P.		53.03	890	2,666	8,000	957.50	
	4810	7 to 9 C.Y., 475 H.P.		90.90	2,550	7,667	23,000	2,261	
	4870	9 to 11 C.Y., 620 H.P.		131.52	2,700	8,107.48	24,300	2,674	
	4880	Skid-steer loader, wheeled, 10 C.F., 30 H.P. gas		9.54	169	506.07	1,525	177.55	
	4890	1 C.Y., 78 H.P., diesel		18.38	420	1,265.18	3,800	400.10	
	4892	Skid-steer attachment, auger		.74	145	433.50	1,300	92.65	
	4893	Backhoe		.74	122	366.64	1,100	79.25	
	4894	Broom		.70	140	420.25	1,250	89.70	
	4895	Forks		.15	32	96	288	20.45	
	4896	Grapple		.72	88.50	265.25	795	58.80	
	4897	Concrete hammer		1.05	183	550	1,650	118.40	
	4898	Tree spade		.60	103	309.84	930	66.75	
	4899	Trencher		.65	102	305	915	66.20	
	4900	Trencher, chain, boom type, gas, operator walking, 12 H.P.		4.16	206	618.25	1,850	156.95	
	4910	Operator riding, 40 H.P.		16.64	450	1,343.75	4,025	401.85	
	5000	Wheel type, diesel, 4' deep, 12" wide		68.50	965	2,891.84	8,675	1,126	
	5100	6' deep, 20" wide		87.32	1,050	3,127.50	9,375	1,324	
	5150	Chain type, diesel, 5' deep, 8" wide		16.25	360	1,084.44	3,250	346.90	
	5200	Diesel, 8' deep, 16" wide		89.39	1,925	5,783.68	17,400	1,872	
	5202	Rock trencher, wheel type, 6" wide x 18" deep		46.98	90	270	810	429.85	
	5206	Chain type, 18" wide x 7' deep		104.38	283	850	2,550	1,005	
	5210	Tree spade, self-propelled		13.65	230	690	2,075	247.20	
	5250	Truck, dump, 2-axle, 12 ton, 8 C.Y. payload, 220 H.P.		23.88	395	1,185	3,550	428.05	
	5300	Three axle dump, 16 ton, 12 C.Y. payload, 400 H.P.		44.50	360	1,084.44	3,250	572.90	
	5310	Four axle dump, 25 ton, 18 C.Y. payload, 450 H.P.		49.85	525	1,575.02	4,725	713.80	
	5350	Dump trailer only, rear dump, 16-1/2 C.Y.		5.73	151	454.43	1,375	136.70	
	5400	20 C.Y.		6.18	170	511.24	1,525	151.70	
	5450	Flatbed, single axle, 1-1/2 ton rating		19.00	73.50	221.02	665	196.15	
	5500	3 ton rating		23.05	1,050	3,180	9,550	820.40	
	5550	Off highway rear dump, 25 ton capacity		62.67	1,475	4,389.40	13,200	1,379	
	5600	35 ton capacity		66.90	665	2,000	6,000	935.20	
	5610	50 ton capacity		83.87	1,825	5,499.66	16,500	1,771	
	5620	65 ton capacity		89.56	2,000	5,990.24	18,000	1,915	
	5630	100 ton capacity		121.24	2,950	8,830.44	26,500	2,736	
	6000	Vibratory plow, 25 H.P., walking		6.77	300	900	2,700	234.20	
40	0010	**GENERAL EQUIPMENT RENTAL** without operators R015433 -10							40
	0020	Aerial lift, scissor type, to 20' high, 1200 lb. capacity, electric	Ea.	3.48	129	385.75	1,150	105	
	0030	To 30' high, 1200 lb. capacity		3.77	203	607.67	1,825	151.70	
	0040	Over 30' high, 1500 lb. capacity		5.13	243	727.50	2,175	186.60	
	0070	Articulating boom, to 45' high, 500 lb. capacity, diesel R015433 -15		9.92	250	750	2,250	229.35	
	0075	To 60' high, 500 lb. capacity		13.66	300	900	2,700	289.25	
	0080	To 80' high, 500 lb. capacity		16.05	900	2,702.25	8,100	668.85	
	0085	To 125' high, 500 lb. capacity		18.34	1,525	4,603.50	13,800	1,067	
	0100	Telescoping boom to 40' high, 500 lb. capacity, diesel		11.24	315	945	2,825	278.90	
	0105	To 45' high, 500 lb. capacity		12.51	320	965	2,900	293.05	
	0110	To 60' high, 500 lb. capacity		16.36	300	900	2,700	310.90	
	0115	To 80' high, 500 lb. capacity		21.27	355	1,067	3,200	383.55	
	0120	To 100' high, 500 lb. capacity		28.71	865	2,587.75	7,775	747.25	
	0125	To 120' high, 500 lb. capacity		29.16	1,450	4,348.50	13,000	1,103	
	0195	Air compressor, portable, 6.5 CFM, electric		.90	44.50	133	400	33.80	
	0196	Gasoline		.65	56	167.50	505	38.70	
	0200	Towed type, gas engine, 60 CFM		9.43	129	387.50	1,175	152.95	
	0300	160 CFM		10.47	198	595	1,775	202.80	

01 54 33 | Equipment Rental

		UNIT	HOURLY OPER. COST	RENT PER DAY	RENT PER WEEK	RENT PER MONTH	EQUIPMENT COST/DAY	
40	0400	Diesel engine, rotary screw, 250 CFM	Ea.	12.08	175	524	1,575	201.50	**40**
	0500	365 CFM		16.00	310	937	2,800	315.40	
	0550	450 CFM		19.95	277	832.25	2,500	326.05	
	0600	600 CFM		34.10	248	743.62	2,225	421.50	
	0700	750 CFM		34.62	435	1,306.50	3,925	538.25	
	0930	Air tools, breaker, pavement, 60 lb.		.57	81.50	245	735	53.50	
	0940	80 lb.		.56	81	242.50	730	53	
	0950	Drills, hand (jackhammer), 65 lb.		.67	68	203.50	610	46.05	
	0960	Track or wagon, swing boom, 4" drifter		54.66	1,025	3,104	9,300	1,058	
	0970	5" drifter		63.30	1,025	3,104	9,300	1,127	
	0975	Track mounted quarry drill, 6" diameter drill		101.91	1,900	5,665	17,000	1,948	
	0980	Dust control per drill		1.04	25	75.50	227	23.40	
	0990	Hammer, chipping, 12 lb.		.60	46	138	415	32.40	
	1000	Hose, air with couplings, 50' long, 3/4" diameter		.07	12	36	108	7.75	
	1100	1" diameter		.08	12.35	37	111	8	
	1200	1-1/2" diameter		.22	37.50	112.50	340	24.25	
	1300	2" diameter		.24	45	135	405	28.90	
	1400	2-1/2" diameter		.36	57.50	172.50	520	37.35	
	1410	3" diameter		.42	58.50	175	525	38.35	
	1450	Drill, steel, 7/8" x 2'		.08	12.90	38.73	116	8.40	
	1460	7/8" x 6'		.12	19.60	58.87	177	12.70	
	1520	Moil points		.03	7	21	63	4.40	
	1525	Pneumatic nailer w/accessories		.48	39.50	118	355	27.40	
	1530	Sheeting driver for 60 lb. breaker		.04	7.75	23.24	69.50	5	
	1540	For 90 lb. breaker		.13	10.50	31.50	94.50	7.35	
	1550	Spade, 25 lb.		.50	7.40	22.21	66.50	8.45	
	1560	Tamper, single, 35 lb.		.59	48.50	145.75	435	33.85	
	1570	Triple, 140 lb.		.89	61.50	184.87	555	44.05	
	1580	Wrenches, impact, air powered, up to 3/4" bolt		.43	49.50	148.25	445	33.05	
	1590	Up to 1-1/4" bolt		.58	79.50	238.50	715	52.30	
	1600	Barricades, barrels, reflectorized, 1 to 99 barrels		.03	4	12	36	2.65	
	1610	100 to 200 barrels		.02	4.41	13.22	39.50	2.85	
	1620	Barrels with flashers, 1 to 99 barrels		.03	6.40	19.16	57.50	4.10	
	1630	100 to 200 barrels		.03	5.10	15.34	46	3.30	
	1640	Barrels with steady burn type C lights		.05	8.45	25.30	76	5.45	
	1650	Illuminated board, trailer mounted, with generator		3.28	139	418.28	1,250	109.85	
	1670	Portable barricade, stock, with flashers, 1 to 6 units		.03	6.35	19.11	57.50	4.10	
	1680	25 to 50 units		.03	5.95	17.82	53.50	3.85	
	1685	Butt fusion machine, wheeled, 1.5 HP electric, 2" - 8" diameter pipe		2.63	225	675	2,025	156.05	
	1690	Tracked, 20 HP diesel, 4"-12" diameter pipe		11.23	560	1,685	5,050	426.85	
	1695	83 HP diesel, 8" - 24" diameter pipe		51.32	1,100	3,325	9,975	1,076	
	1700	Carts, brick, gas engine, 1000 lb. capacity		2.94	65	195	585	62.55	
	1800	1500 lb., 7-1/2' lift		2.92	69.50	208	625	64.95	
	1822	Dehumidifier, medium, 6 lb./hr., 150 CFM		1.19	76.50	229.28	690	55.35	
	1824	Large, 18 lb./hr., 600 CFM		2.19	585	1,750	5,250	367.55	
	1830	Distributor, asphalt, trailer mounted, 2000 gal., 38 H.P. diesel		10.99	355	1,058.62	3,175	299.65	
	1840	3000 gal., 38 H.P. diesel		12.87	380	1,136.08	3,400	330.15	
	1850	Drill, rotary hammer, electric		1.11	71.50	214	640	51.70	
	1860	Carbide bit, 1-1/2" diameter, add to electric rotary hammer		.03	41.50	125	375	25.25	
	1865	Rotary, crawler, 250 H.P.		135.77	2,300	6,868.12	20,600	2,460	
	1870	Emulsion sprayer, 65 gal., 5 H.P. gas engine		2.77	107	320.17	960	86.15	
	1880	200 gal., 5 H.P. engine		7.22	179	537.06	1,600	165.20	
	1900	Floor auto-scrubbing machine, walk-behind, 28" path		5.62	222	667	2,000	178.40	
	1930	Floodlight, mercury vapor, or quartz, on tripod, 1000 watt		.46	36.50	110	330	25.65	
	1940	2000 watt		.59	28	84.69	254	21.65	
	1950	Floodlights, trailer mounted with generator, 1 - 300 watt light		3.54	78.50	235.48	705	75.45	
	1960	2 - 1000 watt lights		4.49	87.50	262.33	785	88.35	
	2000	4 - 300 watt lights		4.24	100	299.51	900	93.85	

For customer support on your Mechanical Costs with RSMeans Data, call 800.448.8182.

01 54 33 | Equipment Rental

		UNIT	HOURLY OPER. COST	RENT PER DAY	RENT PER WEEK	RENT PER MONTH	EQUIPMENT COST/DAY	
40	2005	Foam spray rig, incl. box trailer, compressor, generator, proportioner	Ea.	25.46	535	1,600.84	4,800	523.85	40
	2015	Forklift, pneumatic tire, rough terr, straight mast, 5000 lb, 12' lift, gas		18.59	219	655.83	1,975	279.90	
	2025	8000 lb, 12' lift		22.68	360	1,084.44	3,250	398.35	
	2030	5000 lb, 12' lift, diesel		15.41	244	733.29	2,200	269.90	
	2035	8000 lb, 12' lift, diesel		16.70	277	831.40	2,500	299.90	
	2045	All terrain, telescoping boom, diesel, 5000 lb, 10' reach, 19' lift		17.20	233	700	2,100	277.60	
	2055	6600 lb, 29' reach, 42' lift		21.04	233	700	2,100	308.30	
	2065	10,000 lb, 31' reach, 45' lift		23.03	315	950	2,850	374.20	
	2070	Cushion tire, smooth floor, gas, 5000 lb capacity		8.23	247	741.50	2,225	214.10	
	2075	8000 lb capacity		11.33	275	826.25	2,475	255.90	
	2085	Diesel, 5000 lb capacity		7.73	210	629.25	1,900	187.65	
	2090	12,000 lb capacity		12.01	400	1,194.50	3,575	335	
	2095	20,000 lb capacity		17.20	660	1,980	5,950	533.60	
	2100	Generator, electric, gas engine, 1.5 kW to 3 kW		2.57	46.50	140	420	48.55	
	2200	5 kW		3.21	91	272.50	820	80.15	
	2300	10 kW		5.91	108	322.50	970	111.80	
	2400	25 kW		7.38	405	1,210	3,625	301.10	
	2500	Diesel engine, 20 kW		9.18	229	687.50	2,075	210.95	
	2600	50 kW		15.90	370	1,110	3,325	349.20	
	2700	100 kW		28.51	445	1,340.50	4,025	496.20	
	2800	250 kW		54.19	750	2,249.33	6,750	883.40	
	2850	Hammer, hydraulic, for mounting on boom, to 500 ft lb.		2.89	93.50	279.89	840	79.10	
	2860	1000 ft lb.		4.59	139	418.28	1,250	120.40	
	2900	Heaters, space, oil or electric, 50 MBH		1.46	46.50	140	420	39.65	
	3000	100 MBH		2.71	46.50	140	420	49.70	
	3100	300 MBH		7.90	135	405	1,225	144.20	
	3150	500 MBH		13.12	200	600	1,800	224.95	
	3200	Hose, water, suction with coupling, 20' long, 2" diameter		.02	5.65	17	51	3.55	
	3210	3" diameter		.03	14.15	42.50	128	8.75	
	3220	4" diameter		.03	28	84	252	17.05	
	3230	6" diameter		.11	40.50	121.50	365	25.20	
	3240	8" diameter		.27	53.50	160	480	34.15	
	3250	Discharge hose with coupling, 50' long, 2" diameter		.01	6.50	19.50	58.50	4	
	3260	3" diameter		.01	7.35	22	66	4.50	
	3270	4" diameter		.02	21	62.50	188	12.65	
	3280	6" diameter		.06	29	87	261	17.90	
	3290	8" diameter		.24	37.50	112.50	340	24.40	
	3295	Insulation blower		.83	117	350	1,050	76.65	
	3300	Ladders, extension type, 16' to 36' long		.18	41.50	125	375	26.45	
	3400	40' to 60' long		.64	120	360.50	1,075	77.20	
	3405	Lance for cutting concrete		2.20	65	195	585	56.60	
	3407	Lawn mower, rotary, 22", 5 H.P.		1.05	38.50	115	345	31.40	
	3408	48" self-propelled		2.89	138	415	1,250	106.10	
	3410	Level, electronic, automatic, with tripod and leveling rod		1.05	37.50	112	335	30.80	
	3430	Laser type, for pipe and sewer line and grade		2.17	117	350	1,050	87.35	
	3440	Rotating beam for interior control		.90	64	192.50	580	45.70	
	3460	Builder's optical transit, with tripod and rod		.10	37.50	112	335	23.20	
	3500	Light towers, towable, with diesel generator, 2000 watt		4.25	101	303.64	910	94.75	
	3600	4000 watt		4.50	165	495	1,475	135	
	3700	Mixer, powered, plaster and mortar, 6 C.F., 7 H.P.		2.05	83.50	250	750	66.40	
	3800	10 C.F., 9 H.P.		2.24	124	372.50	1,125	92.40	
	3850	Nailer, pneumatic		.48	33.50	100.18	300	23.85	
	3900	Paint sprayers complete, 8 CFM		.85	61.50	184.87	555	43.75	
	4000	17 CFM		1.60	110	330.50	990	78.85	
	4020	Pavers, bituminous, rubber tires, 8' wide, 50 H.P., diesel		31.93	570	1,704.12	5,100	596.25	
	4030	10' wide, 150 H.P.		95.62	1,950	5,835.32	17,500	1,932	
	4050	Crawler, 8' wide, 100 H.P., diesel		87.59	2,050	6,170.98	18,500	1,935	
	4060	10' wide, 150 H.P.		103.97	2,350	7,048.86	21,100	2,241	

		UNIT	HOURLY OPER. COST	RENT PER DAY	RENT PER WEEK	RENT PER MONTH	EQUIPMENT COST/DAY	
01 54 33 | Equipment Rental								
40	4070	Concrete paver, 12' to 24' wide, 250 H.P.	Ea.	87.62	1,675	5,060.72	15,200	1,713	**40**
	4080	Placer-spreader-trimmer, 24' wide, 300 H.P.		117.51	2,550	7,668.54	23,000	2,474	
	4100	Pump, centrifugal gas pump, 1-1/2" diam., 65 GPM		3.92	54	162.15	485	63.80	
	4200	2" diameter, 130 GPM		4.98	44	132.50	400	66.35	
	4300	3" diameter, 250 GPM		5.12	56	167.50	505	74.45	
	4400	6" diameter, 1500 GPM		22.24	91.50	275	825	232.90	
	4500	Submersible electric pump, 1-1/4" diameter, 55 GPM		.40	36	107.50	325	24.70	
	4600	1-1/2" diameter, 83 GPM		.44	43.50	130	390	29.55	
	4700	2" diameter, 120 GPM		1.64	61.50	185	555	50.15	
	4800	3" diameter, 300 GPM		3.03	109	327.50	985	89.75	
	4900	4" diameter, 560 GPM		14.75	61.50	185	555	155	
	5000	6" diameter, 1590 GPM		22.08	65	195	585	215.60	
	5100	Diaphragm pump, gas, single, 1-1/2" diameter		1.13	38.50	115	345	32	
	5200	2" diameter		3.98	91.50	275	825	86.80	
	5300	3" diameter		4.05	95	285	855	89.40	
	5400	Double, 4" diameter		6.03	95	285	855	105.25	
	5450	Pressure washer 5 GPM, 3000 psi		3.87	110	330	990	96.95	
	5460	7 GPM, 3000 psi		4.94	85	255	765	90.50	
	5470	High pressure water jet 10 ksi		39.55	720	2,160	6,475	748.40	
	5480	40 ksi		27.88	980	2,940	8,825	811.05	
	5500	Trash pump, self-priming, gas, 2" diameter		3.82	108	325	975	95.55	
	5600	Diesel, 4" diameter		6.68	162	485	1,450	150.40	
	5650	Diesel, 6" diameter		16.85	162	485	1,450	231.80	
	5655	Grout Pump		18.70	281	841.73	2,525	317.90	
	5700	Salamanders, L.P. gas fired, 100,000 BTU		2.88	57.50	172.50	520	57.55	
	5705	50,000 BTU		1.66	23.50	71	213	27.50	
	5720	Sandblaster, portable, open top, 3 C.F. capacity		.60	132	395	1,175	83.80	
	5730	6 C.F. capacity		1.00	132	395	1,175	87	
	5740	Accessories for above		.14	24	71.26	214	15.35	
	5750	Sander, floor		.77	73.50	220	660	50.15	
	5760	Edger		.52	35	105	315	25.15	
	5800	Saw, chain, gas engine, 18" long		1.75	63.50	190	570	52	
	5900	Hydraulic powered, 36" long		.78	58.50	175	525	41.25	
	5950	60" long		.78	65	195	585	45.25	
	6000	Masonry, table mounted, 14" diameter, 5 H.P.		1.32	76.50	230	690	56.55	
	6050	Portable cut-off, 8 H.P.		1.81	77.50	232.50	700	61	
	6100	Circular, hand held, electric, 7-1/4" diameter		.23	13.85	41.50	125	10.10	
	6200	12" diameter		.24	41	122.50	370	26.40	
	6250	Wall saw, w/hydraulic power, 10 H.P.		3.29	98.50	296	890	85.50	
	6275	Shot blaster, walk-behind, 20" wide		4.73	281	841.73	2,525	206.20	
	6280	Sidewalk broom, walk-behind		2.24	82.50	247.87	745	67.50	
	6300	Steam cleaner, 100 gallons per hour		3.34	82.50	247.87	745	76.30	
	6310	200 gallons per hour		4.33	100	299.51	900	94.55	
	6340	Tar Kettle/Pot, 400 gallons		16.48	127	380	1,150	207.85	
	6350	Torch, cutting, acetylene-oxygen, 150' hose, excludes gases		.45	15.30	45.96	138	12.80	
	6360	Hourly operating cost includes tips and gas		20.92	7	20.98	63	171.55	
	6410	Toilet, portable chemical		.13	23	69.20	208	14.90	
	6420	Recycle flush type		.16	28.50	85.72	257	18.45	
	6430	Toilet, fresh water flush, garden hose,		.19	34	102.25	305	22	
	6440	Hoisted, non-flush, for high rise		.16	28	83.66	251	18	
	6465	Tractor, farm with attachment		17.37	390	1,172.50	3,525	373.40	
	6480	Trailers, platform, flush deck, 2 axle, 3 ton capacity		1.69	94.50	284	850	70.30	
	6500	25 ton capacity		6.23	143	428.61	1,275	135.55	
	6600	40 ton capacity		8.04	203	609.35	1,825	186.20	
	6700	3 axle, 50 ton capacity		8.72	225	676.48	2,025	205.05	
	6800	75 ton capacity		11.08	300	898.54	2,700	268.30	
	6810	Trailer mounted cable reel for high voltage line work		5.89	28.50	85	255	64.10	
	6820	Trailer mounted cable tensioning rig		11.67	28.50	85	255	110.40	

01 54 33 | Equipment Rental

		UNIT	HOURLY OPER. COST	RENT PER DAY	RENT PER WEEK	RENT PER MONTH	EQUIPMENT COST/DAY	
40								**40**
6830	Cable pulling rig	Ea.	73.77	28.50	85	255	607.15	
6850	Portable cable/wire puller, 8000 lb max pulling capacity		3.70	120	360	1,075	101.60	
6900	Water tank trailer, engine driven discharge, 5000 gallons		7.16	158	475.09	1,425	152.25	
6925	10,000 gallons		9.75	215	645.50	1,925	207.10	
6950	Water truck, off highway, 6000 gallons		71.75	835	2,504.54	7,525	1,075	
7010	Tram car for high voltage line work, powered, 2 conductor		6.88	28.50	85	255	72.05	
7020	Transit (builder's level) with tripod		.10	17.55	52.67	158	11.30	
7030	Trench box, 3000 lb., 6' x 8'		.56	96.50	290.22	870	62.50	
7040	7200 lb., 6' x 20'		.72	187	560	1,675	117.75	
7050	8000 lb., 8' x 16'		1.08	186	557.71	1,675	120.15	
7060	9500 lb., 8' x 20'		1.20	232	697.14	2,100	149.05	
7065	11,000 lb., 8' x 24'		1.26	219	655.83	1,975	141.25	
7070	12,000 lb., 10' x 20'		1.49	263	790.09	2,375	169.95	
7100	Truck, pickup, 3/4 ton, 2 wheel drive		9.24	61.50	184.87	555	110.85	
7200	4 wheel drive		9.48	167	500	1,500	175.85	
7250	Crew carrier, 9 passenger		12.66	108	325	975	166.25	
7290	Flat bed truck, 20,000 lb. GVW		15.26	133	397.63	1,200	201.60	
7300	Tractor, 4 x 2, 220 H.P.		22.25	215	645.50	1,925	307.10	
7410	330 H.P.		32.33	294	883.04	2,650	435.25	
7500	6 x 4, 380 H.P.		36.09	340	1,022.47	3,075	493.25	
7600	450 H.P.		44.23	415	1,239.36	3,725	601.75	
7610	Tractor, with A frame, boom and winch, 225 H.P.		24.74	293	877.88	2,625	373.50	
7620	Vacuum truck, hazardous material, 2500 gallons		12.79	310	929.52	2,800	288.25	
7625	5,000 gallons		13.02	440	1,316.82	3,950	367.55	
7650	Vacuum, HEPA, 16 gallon, wet/dry		.85	122	365	1,100	79.80	
7655	55 gallon, wet/dry		.78	25.50	76.50	230	21.50	
7660	Water tank, portable		.73	160	480.25	1,450	101.90	
7690	Sewer/catch basin vacuum, 14 C.Y., 1500 gallons		17.31	665	1,988.14	5,975	536.15	
7700	Welder, electric, 200 amp		3.81	33.50	100	300	50.50	
7800	300 amp		5.55	103	310	930	106.40	
7900	Gas engine, 200 amp		8.95	58.50	175	525	106.55	
8000	300 amp		10.13	110	330	990	147	
8100	Wheelbarrow, any size		.06	11.15	33.50	101	7.20	
8200	Wrecking ball, 4000 lb.		2.50	60	180	540	56	
50	0010	**HIGHWAY EQUIPMENT RENTAL** without operators	R015433 -10						**50**
	0050	Asphalt batch plant, portable drum mixer, 100 ton/hr.	Ea.	88.41	1,550	4,621.78	13,900	1,632	
	0060	200 ton/hr.		101.99	1,650	4,931.62	14,800	1,802	
	0070	300 ton/hr.		119.86	1,925	5,783.68	17,400	2,116	
	0100	Backhoe attachment, long stick, up to 185 H.P., 10.5' long		.37	25.50	76.43	229	18.25	
	0140	Up to 250 H.P., 12' long		.41	28.50	85.72	257	20.45	
	0180	Over 250 H.P., 15' long		.56	39	116.71	350	27.85	
	0200	Special dipper arm, up to 100 H.P., 32' long		1.16	79.50	238.58	715	56.95	
	0240	Over 100 H.P., 33' long		1.44	100	299.51	900	71.45	
	0280	Catch basin/sewer cleaning truck, 3 ton, 9 C.Y., 1000 gal.		35.39	420	1,265.18	3,800	536.15	
	0300	Concrete batch plant, portable, electric, 200 C.Y./hr.		24.18	560	1,678.30	5,025	529.15	
	0520	Grader/dozer attachment, ripper/scarifier, rear mounted, up to 135 H.P.		3.15	63.50	190.04	570	63.20	
	0540	Up to 180 H.P.		4.13	95.50	287.12	860	90.50	
	0580	Up to 250 H.P.		5.85	153	459.60	1,375	138.75	
	0700	Pvmt. removal bucket, for hyd. excavator, up to 90 H.P.		2.16	58	174.54	525	52.20	
	0740	Up to 200 H.P.		2.31	74.50	223.08	670	63.05	
	0780	Over 200 H.P.		2.52	91	273.69	820	74.90	
	0900	Aggregate spreader, self-propelled, 187 H.P.		50.60	740	2,220.52	6,650	848.90	
	1000	Chemical spreader, 3 C.Y.		3.17	96.50	290	870	83.35	
	1900	Hammermill, traveling, 250 H.P.		67.35	515	1,550	4,650	848.80	
	2000	Horizontal borer, 3" diameter, 13 H.P. gas driven		5.42	232	695	2,075	182.35	
	2150	Horizontal directional drill, 20,000 lb. thrust, 78 H.P. diesel		27.58	530	1,590	4,775	538.65	
	2160	30,000 lb. thrust, 115 H.P.		33.90	615	1,850	5,550	641.20	
	2170	50,000 lb. thrust, 170 H.P.		48.60	710	2,135	6,400	815.80	

01 54 33 | Equipment Rental

		UNIT	HOURLY OPER. COST	RENT PER DAY	RENT PER WEEK	RENT PER MONTH	EQUIPMENT COST/DAY	
50	2190	Mud trailer for HDD, 1500 gallons, 175 H.P., gas	Ea.	25.50	175	525	1,575	309	50
	2200	Hydromulcher, diesel, 3000 gallon, for truck mounting		17.43	227	680	2,050	275.45	
	2300	Gas, 600 gallon		7.49	95	285	855	116.95	
	2400	Joint & crack cleaner, walk behind, 25 H.P.		3.16	45.50	136	410	52.45	
	2500	Filler, trailer mounted, 400 gallons, 20 H.P.		8.35	147	440	1,325	154.75	
	3000	Paint striper, self-propelled, 40 gallon, 22 H.P.		6.76	122	365	1,100	127.10	
	3100	120 gallon, 120 H.P.		19.23	380	1,140	3,425	381.80	
	3200	Post drivers, 6" I-Beam frame, for truck mounting		12.41	320	960	2,875	291.30	
	3400	Road sweeper, self-propelled, 8' wide, 90 H.P.		35.91	715	2,143.06	6,425	715.85	
	3450	Road sweeper, vacuum assisted, 4 C.Y., 220 gallons		58.28	670	2,013.96	6,050	869	
	4000	Road mixer, self-propelled, 130 H.P.		46.23	825	2,478.72	7,425	865.60	
	4100	310 H.P.		75.01	2,150	6,480.82	19,400	1,896	
	4220	Cold mix paver, incl. pug mill and bitumen tank, 165 H.P.		94.97	2,325	6,945.58	20,800	2,149	
	4240	Pavement brush, towed		3.43	100	299.51	900	87.30	
	4250	Paver, asphalt, wheel or crawler, 130 H.P., diesel		94.23	2,275	6,816.48	20,400	2,117	
	4300	Paver, road widener, gas, 1' to 6', 67 H.P.		46.66	975	2,917.66	8,750	956.80	
	4400	Diesel, 2' to 14', 88 H.P.		56.38	1,150	3,459.88	10,400	1,143	
	4600	Slipform pavers, curb and gutter, 2 track, 75 H.P.		57.83	1,250	3,769.72	11,300	1,217	
	4700	4 track, 165 H.P.		35.69	845	2,530.36	7,600	791.55	
	4800	Median barrier, 215 H.P.		58.43	1,350	4,027.92	12,100	1,273	
	4901	Trailer, low bed, 75 ton capacity		10.71	282	846.90	2,550	255.05	
	5000	Road planer, walk behind, 10" cutting width, 10 H.P.		2.45	243	730	2,200	165.65	
	5100	Self-propelled, 12" cutting width, 64 H.P.		8.26	190	570	1,700	180.05	
	5120	Traffic line remover, metal ball blaster, truck mounted, 115 H.P.		46.56	905	2,720	8,150	916.50	
	5140	Grinder, truck mounted, 115 H.P.		50.89	905	2,720	8,150	951.15	
	5160	Walk-behind, 11 H.P.		3.56	142	425	1,275	113.45	
	5200	Pavement profiler, 4' to 6' wide, 450 H.P.		216.58	1,275	3,800	11,400	2,493	
	5300	8' to 10' wide, 750 H.P.		331.58	1,325	3,975	11,900	3,448	
	5400	Roadway plate, steel, 1" x 8' x 20'		.09	61	182.50	550	37.20	
	5600	Stabilizer, self-propelled, 150 H.P.		41.14	1,025	3,100	9,300	949.10	
	5700	310 H.P.		76.18	1,300	3,900	11,700	1,389	
	5800	Striper, truck mounted, 120 gallon paint, 460 H.P.		48.74	340	1,015	3,050	592.95	
	5900	Thermal paint heating kettle, 115 gallons		7.71	61.50	185	555	98.65	
	6000	Tar kettle, 330 gallon, trailer mounted		12.27	96.50	290	870	156.20	
	7000	Tunnel locomotive, diesel, 8 to 12 ton		29.76	620	1,859.04	5,575	609.85	
	7005	Electric, 10 ton		29.25	705	2,117.24	6,350	657.40	
	7010	Muck cars, 1/2 C.Y. capacity		2.30	26.50	80.04	240	34.40	
	7020	1 C.Y. capacity		2.51	35	104.31	315	40.95	
	7030	2 C.Y. capacity		2.66	39	116.71	350	44.60	
	7040	Side dump, 2 C.Y. capacity		2.87	48	144.59	435	51.90	
	7050	3 C.Y. capacity		3.85	53	159.05	475	62.65	
	7060	5 C.Y. capacity		5.62	68.50	205.53	615	86.10	
	7100	Ventilating blower for tunnel, 7-1/2 H.P.		2.14	52.50	158.02	475	48.70	
	7110	10 H.P.		2.42	55	165.25	495	52.40	
	7120	20 H.P.		3.54	71.50	214.82	645	71.30	
	7140	40 H.P.		6.14	94.50	284.02	850	105.90	
	7160	60 H.P.		8.69	102	304.68	915	130.45	
	7175	75 H.P.		10.37	158	475.09	1,425	177.95	
	7180	200 H.P.		20.78	310	934.68	2,800	353.20	
	7800	Windrow loader, elevating		53.94	1,650	4,975	14,900	1,427	
60	0010	**LIFTING AND HOISTING EQUIPMENT RENTAL** without operators	R015433-10						60
	0150	Crane, flatbed mounted, 3 ton capacity	Ea.	14.41	201	604.19	1,825	236.10	
	0200	Crane, climbing, 106' jib, 6000 lb. capacity, 410 fpm	R312316-45	39.72	2,600	7,800	23,400	1,878	
	0300	101' jib, 10,250 lb. capacity, 270 fpm		46.43	2,275	6,800	20,400	1,731	
	0500	Tower, static, 130' high, 106' jib, 6200 lb. capacity at 400 fpm		45.16	2,250	6,715	20,100	1,704	
	0520	Mini crawler spider crane, up to 24" wide, 1990 lb. lifting capacity		12.50	550	1,652.48	4,950	430.50	
	0525	Up to 30" wide, 6450 lb. lifting capacity		14.52	655	1,962.32	5,875	508.65	
	0530	Up to 52" wide, 6680 lb. lifting capacity		23.10	800	2,401.26	7,200	665.05	

01 54 33 | Equipment Rental

		UNIT	HOURLY OPER. COST	RENT PER DAY	RENT PER WEEK	RENT PER MONTH	EQUIPMENT COST/DAY	
60	0535	Up to 55" wide, 8920 lb. lifting capacity	Ea.	25.79	885	2,659.46	7,975	738.25	60
	0540	Up to 66" wide, 13,350 lb. lifting capacity		34.92	1,375	4,131.20	12,400	1,106	
	0600	Crawler mounted, lattice boom, 1/2 C.Y., 15 tons at 12' radius		36.96	830	2,483	7,450	792.30	
	0700	3/4 C.Y., 20 tons at 12' radius		50.42	930	2,790	8,375	961.35	
	0800	1 C.Y., 25 tons at 12' radius		67.42	985	2,950	8,850	1,129	
	0900	1-1/2 C.Y., 40 tons at 12' radius		66.31	1,125	3,375	10,100	1,206	
	1000	2 C.Y., 50 tons at 12' radius		88.77	1,325	4,000	12,000	1,510	
	1100	3 C.Y., 75 tons at 12' radius		75.26	2,325	7,000	21,000	2,002	
	1200	100 ton capacity, 60' boom		85.91	2,675	8,000	24,000	2,287	
	1300	165 ton capacity, 60' boom		106.10	3,000	9,000	27,000	2,649	
	1400	200 ton capacity, 70' boom		138.21	3,825	11,500	34,500	3,406	
	1500	350 ton capacity, 80' boom		182.20	4,175	12,500	37,500	3,958	
	1600	Truck mounted, lattice boom, 6 x 4, 20 tons at 10' radius		39.76	1,950	5,850	17,600	1,488	
	1700	25 tons at 10' radius		42.73	2,325	7,000	21,000	1,742	
	1800	8 x 4, 30 tons at 10' radius		45.54	2,500	7,500	22,500	1,864	
	1900	40 tons at 12' radius		48.55	2,725	8,200	24,600	2,028	
	2000	60 tons at 15' radius		53.69	1,650	4,950	14,900	1,419	
	2050	82 tons at 15' radius		59.43	1,775	5,350	16,100	1,545	
	2100	90 tons at 15' radius		66.39	1,950	5,825	17,500	1,696	
	2200	115 tons at 15' radius		74.90	2,175	6,525	19,600	1,904	
	2300	150 tons at 18' radius		81.09	2,700	8,100	24,300	2,269	
	2350	165 tons at 18' radius		87.03	2,425	7,275	21,800	2,151	
	2400	Truck mounted, hydraulic, 12 ton capacity		29.50	390	1,175	3,525	471	
	2500	25 ton capacity		36.36	485	1,450	4,350	580.85	
	2550	33 ton capacity		50.67	900	2,700	8,100	945.35	
	2560	40 ton capacity		49.47	900	2,700	8,100	935.80	
	2600	55 ton capacity		53.78	915	2,750	8,250	980.20	
	2700	80 ton capacity		75.71	1,475	4,400	13,200	1,486	
	2720	100 ton capacity		74.96	1,550	4,675	14,000	1,535	
	2740	120 ton capacity		102.81	1,825	5,500	16,500	1,922	
	2760	150 ton capacity		109.92	2,050	6,125	18,400	2,104	
	2800	Self-propelled, 4 x 4, with telescoping boom, 5 ton		15.14	430	1,285	3,850	378.10	
	2900	12-1/2 ton capacity		21.42	430	1,285	3,850	428.30	
	3000	15 ton capacity		34.42	450	1,350	4,050	545.35	
	3050	20 ton capacity		24.02	650	1,950	5,850	582.20	
	3100	25 ton capacity		36.69	1,425	4,250	12,800	1,144	
	3150	40 ton capacity		44.90	660	1,975	5,925	754.20	
	3200	Derricks, guy, 20 ton capacity, 60' boom, 75' mast		22.74	1,425	4,250	12,800	1,032	
	3300	100' boom, 115' mast		36.04	2,000	6,000	18,000	1,488	
	3400	Stiffleg, 20 ton capacity, 70' boom, 37' mast		25.41	615	1,850	5,550	573.25	
	3500	100' boom, 47' mast		39.32	665	2,000	6,000	714.55	
	3550	Helicopter, small, lift to 1250 lb. maximum, w/pilot		99.14	2,150	6,435	19,300	2,080	
	3600	Hoists, chain type, overhead, manual, 3/4 ton		.14	10.25	30.70	92	7.30	
	3900	10 ton		.79	6.20	18.59	56	10	
	4000	Hoist and tower, 5000 lb. cap., portable electric, 40' high		5.12	142	426	1,275	126.20	
	4100	For each added 10' section, add		.12	31.50	95	285	19.95	
	4200	Hoist and single tubular tower, 5000 lb. electric, 100' high		6.96	105	315	945	118.65	
	4300	For each added 6'-6" section, add		.21	38.50	115	345	24.65	
	4400	Hoist and double tubular tower, 5000 lb., 100' high		7.57	105	315	945	123.60	
	4500	For each added 6'-6" section, add		.23	41.50	125	375	26.80	
	4550	Hoist and tower, mast type, 6000 lb., 100' high		8.24	94.50	284	850	122.70	
	4570	For each added 10' section, add		.13	31.50	95	285	20.05	
	4600	Hoist and tower, personnel, electric, 2000 lb., 100' @ 125 fpm		17.50	25	75	225	155	
	4700	3000 lb., 100' @ 200 fpm		20.02	25	75	225	175.15	
	4800	3000 lb., 150' @ 300 fpm		22.22	25	75	225	192.75	
	4900	4000 lb., 100' @ 300 fpm		22.98	25	75	225	198.85	
	5000	6000 lb., 100' @ 275 fpm		24.70	25	75	225	212.60	
	5100	For added heights up to 500', add	L.F.	.01	3.33	10	30	2.10	

01 54 33 | Equipment Rental

			UNIT	HOURLY OPER. COST	RENT PER DAY	RENT PER WEEK	RENT PER MONTH	EQUIPMENT COST/DAY	
60	5200	Jacks, hydraulic, 20 ton	Ea.	.05	19.65	59	177	12.20	60
	5500	100 ton		.40	26	78.50	236	18.90	
	6100	Jacks, hydraulic, climbing w/50' jackrods, control console, 30 ton cap.		2.17	31	93	279	35.90	
	6150	For each added 10' jackrod section, add		.05	5	15	45	3.40	
	6300	50 ton capacity		3.48	33.50	100	300	47.85	
	6350	For each added 10' jackrod section, add		.06	5	15	45	3.50	
	6500	125 ton capacity		9.10	51.50	155	465	103.85	
	6550	For each added 10' jackrod section, add		.61	5	15	45	7.90	
	6600	Cable jack, 10 ton capacity with 200' cable		1.82	35.50	107	320	35.95	
	6650	For each added 50' of cable, add		.22	15	45	135	10.75	
70	0010	**WELLPOINT EQUIPMENT RENTAL** without operators							70
	0020	Based on 2 months rental							
	0100	Combination jetting & wellpoint pump, 60 H.P. diesel	Ea.	15.67	298	895	2,675	304.35	
	0200	High pressure gas jet pump, 200 H.P., 300 psi	"	33.83	275	825	2,475	435.65	
	0300	Discharge pipe, 8" diameter	L.F.	.01	1.40	4.20	12.60	.90	
	0350	12" diameter		.01	2.07	6.20	18.60	1.35	
	0400	Header pipe, flows up to 150 GPM, 4" diameter		.01	.73	2.20	6.60	.50	
	0500	400 GPM, 6" diameter		.01	1.07	3.20	9.60	.70	
	0600	800 GPM, 8" diameter		.01	1.40	4.20	12.60	.95	
	0700	1500 GPM, 10" diameter		.01	1.73	5.20	15.60	1.15	
	0800	2500 GPM, 12" diameter		.03	2.07	6.20	18.60	1.45	
	0900	4500 GPM, 16" diameter		.03	2.40	7.20	21.50	1.70	
	0950	For quick coupling aluminum and plastic pipe, add		.03	9.35	28	84	5.85	
	1100	Wellpoint, 25' long, with fittings & riser pipe, 1-1/2" or 2" diameter	Ea.	.07	132	395	1,175	79.55	
	1200	Wellpoint pump, diesel powered, 4" suction, 20 H.P.		7.00	150	450	1,350	146	
	1300	6" suction, 30 H.P.		9.39	167	500	1,500	175.15	
	1400	8" suction, 40 H.P.		12.73	250	750	2,250	251.80	
	1500	10" suction, 75 H.P.		18.77	265	795	2,375	309.20	
	1600	12" suction, 100 H.P.		27.24	298	895	2,675	396.90	
	1700	12" suction, 175 H.P.		38.98	315	950	2,850	501.80	
80	0010	**MARINE EQUIPMENT RENTAL** without operators							80
	0200	Barge, 400 Ton, 30' wide x 90' long	Ea.	17.63	1,200	3,588.98	10,800	858.85	
	0240	800 Ton, 45' wide x 90' long		22.14	1,475	4,415.22	13,200	1,060	
	2000	Tugboat, diesel, 100 H.P.		29.57	238	712.63	2,150	379.10	
	2040	250 H.P.		57.41	430	1,291	3,875	717.50	
	2080	380 H.P.		124.99	1,300	3,873	11,600	1,774	
	3000	Small work boat, gas, 16-foot, 50 H.P.		11.35	48	143.56	430	119.50	
	4000	Large, diesel, 48-foot, 200 H.P.		74.68	1,375	4,105.38	12,300	1,418	

Crew A-1

Crew A-1	Bare Costs Hr.	Daily	Incl. Subs O&P Hr.	Daily	Cost Per Labor-Hour Bare Costs	Incl. O&P
1 Building Laborer	$42.10	$336.80	$63.25	$506.00	$42.10	$63.25
1 Concrete Saw, Gas Manual		111.55		122.71	13.94	15.34
8 L.H., Daily Totals		$448.35		$628.71	$56.04	$78.59

Crew A-1A

Crew A-1A	Hr.	Daily	Hr.	Daily	Bare Costs	Incl. O&P
1 Skilled Worker	$54.85	$438.80	$82.95	$663.60	$54.85	$82.95
1 Shot Blaster, 20"		206.20		226.82	25.77	28.35
8 L.H., Daily Totals		$645.00		$890.42	$80.63	$111.30

Crew A-1B

Crew A-1B	Hr.	Daily	Hr.	Daily	Bare Costs	Incl. O&P
1 Building Laborer	$42.10	$336.80	$63.25	$506.00	$42.10	$63.25
1 Concrete Saw		111.50		122.65	13.94	15.33
8 L.H., Daily Totals		$448.30		$628.65	$56.04	$78.58

Crew A-1C

Crew A-1C	Hr.	Daily	Hr.	Daily	Bare Costs	Incl. O&P
1 Building Laborer	$42.10	$336.80	$63.25	$506.00	$42.10	$63.25
1 Chain Saw, Gas, 18"		52.00		57.20	6.50	7.15
8 L.H., Daily Totals		$388.80		$563.20	$48.60	$70.40

Crew A-1D

Crew A-1D	Hr.	Daily	Hr.	Daily	Bare Costs	Incl. O&P
1 Building Laborer	$42.10	$336.80	$63.25	$506.00	$42.10	$63.25
1 Vibrating Plate, Gas, 18"		31.55		34.70	3.94	4.34
8 L.H., Daily Totals		$368.35		$540.71	$46.04	$67.59

Crew A-1E

Crew A-1E	Hr.	Daily	Hr.	Daily	Bare Costs	Incl. O&P
1 Building Laborer	$42.10	$336.80	$63.25	$506.00	$42.10	$63.25
1 Vibrating Plate, Gas, 21"		165.35		181.88	20.67	22.74
8 L.H., Daily Totals		$502.15		$687.88	$62.77	$85.99

Crew A-1F

Crew A-1F	Hr.	Daily	Hr.	Daily	Bare Costs	Incl. O&P
1 Building Laborer	$42.10	$336.80	$63.25	$506.00	$42.10	$63.25
1 Rammer/Tamper, Gas, 8"		46.50		51.15	5.81	6.39
8 L.H., Daily Totals		$383.30		$557.15	$47.91	$69.64

Crew A-1G

Crew A-1G	Hr.	Daily	Hr.	Daily	Bare Costs	Incl. O&P
1 Building Laborer	$42.10	$336.80	$63.25	$506.00	$42.10	$63.25
1 Rammer/Tamper, Gas, 15"		54.00		59.40	6.75	7.42
8 L.H., Daily Totals		$390.80		$565.40	$48.85	$70.67

Crew A-1H

Crew A-1H	Hr.	Daily	Hr.	Daily	Bare Costs	Incl. O&P
1 Building Laborer	$42.10	$336.80	$63.25	$506.00	$42.10	$63.25
1 Exterior Steam Cleaner		76.30		83.93	9.54	10.49
8 L.H., Daily Totals		$413.10		$589.93	$51.64	$73.74

Crew A-1J

Crew A-1J	Hr.	Daily	Hr.	Daily	Bare Costs	Incl. O&P
1 Building Laborer	$42.10	$336.80	$63.25	$506.00	$42.10	$63.25
1 Cultivator, Walk-Behind, 5 H.P.		53.05		58.35	6.63	7.29
8 L.H., Daily Totals		$389.85		$564.36	$48.73	$70.54

Crew A-1K

Crew A-1K	Hr.	Daily	Hr.	Daily	Bare Costs	Incl. O&P
1 Building Laborer	$42.10	$336.80	$63.25	$506.00	$42.10	$63.25
1 Cultivator, Walk-Behind, 8 H.P.		101.40		111.54	12.68	13.94
8 L.H., Daily Totals		$438.20		$617.54	$54.77	$77.19

Crew A-1M

Crew A-1M	Hr.	Daily	Hr.	Daily	Bare Costs	Incl. O&P
1 Building Laborer	$42.10	$336.80	$63.25	$506.00	$42.10	$63.25
1 Snow Blower, Walk-Behind		67.50		74.25	8.44	9.28
8 L.H., Daily Totals		$404.30		$580.25	$50.54	$72.53

Crew A-2

Crew A-2	Hr.	Daily	Hr.	Daily	Bare Costs	Incl. O&P
2 Laborers	$42.10	$673.60	$63.25	$1012.00	$43.82	$65.77
1 Truck Driver (light)	47.25	378.00	70.80	566.40		
1 Flatbed Truck, Gas, 1.5 Ton		196.15		215.76	8.17	8.99
24 L.H., Daily Totals		$1247.75		$1794.17	$51.99	$74.76

Crew A-2A

Crew A-2A	Hr.	Daily	Hr.	Daily	Bare Costs	Incl. O&P
2 Laborers	$42.10	$673.60	$63.25	$1012.00	$43.82	$65.77
1 Truck Driver (light)	47.25	378.00	70.80	566.40		
1 Flatbed Truck, Gas, 1.5 Ton		196.15		215.76		
1 Concrete Saw		111.50		122.65	12.82	14.10
24 L.H., Daily Totals		$1359.25		$1916.82	$56.64	$79.87

Crew A-2B

Crew A-2B	Hr.	Daily	Hr.	Daily	Bare Costs	Incl. O&P
1 Truck Driver (light)	$47.25	$378.00	$70.80	$566.40	$47.25	$70.80
1 Flatbed Truck, Gas, 1.5 Ton		196.15		215.76	24.52	26.97
8 L.H., Daily Totals		$574.15		$782.16	$71.77	$97.77

Crew A-3A

Crew A-3A	Hr.	Daily	Hr.	Daily	Bare Costs	Incl. O&P
1 Equip. Oper. (light)	$53.00	$424.00	$79.20	$633.60	$53.00	$79.20
1 Pickup Truck, 4x4, 3/4 Ton		175.85		193.44	21.98	24.18
8 L.H., Daily Totals		$599.85		$827.03	$74.98	$103.38

Crew A-3B

Crew A-3B	Hr.	Daily	Hr.	Daily	Bare Costs	Incl. O&P
1 Equip. Oper. (medium)	$56.75	$454.00	$84.85	$678.80	$52.85	$79.10
1 Truck Driver (heavy)	48.95	391.60	73.35	586.80		
1 Dump Truck, 12 C.Y., 400 H.P.		572.90		630.19		
1 F.E. Loader, W.M., 2.5 C.Y.		702.10		772.31	79.69	87.66
16 L.H., Daily Totals		$2120.60		$2668.10	$132.54	$166.76

Crew A-3C

Crew A-3C	Hr.	Daily	Hr.	Daily	Bare Costs	Incl. O&P
1 Equip. Oper. (light)	$53.00	$424.00	$79.20	$633.60	$53.00	$79.20
1 Loader, Skid Steer, 78 H.P.		400.10		440.11	50.01	55.01
8 L.H., Daily Totals		$824.10		$1073.71	$103.01	$134.21

Crew A-3D

Crew A-3D	Hr.	Daily	Hr.	Daily	Bare Costs	Incl. O&P
1 Truck Driver (light)	$47.25	$378.00	$70.80	$566.40	$47.25	$70.80
1 Pickup Truck, 4x4, 3/4 Ton		175.85		193.44		
1 Flatbed Trailer, 25 Ton		135.55		149.10	38.92	42.82
8 L.H., Daily Totals		$689.40		$908.94	$86.17	$113.62

Crew A-3E

Crew A-3E	Hr.	Daily	Hr.	Daily	Bare Costs	Incl. O&P
1 Equip. Oper. (crane)	$59.20	$473.60	$88.50	$708.00	$54.08	$80.92
1 Truck Driver (heavy)	48.95	391.60	73.35	586.80		
1 Pickup Truck, 4x4, 3/4 Ton		175.85		193.44	10.99	12.09
16 L.H., Daily Totals		$1041.05		$1488.23	$65.07	$93.01

Crew A-3F

Crew A-3F	Hr.	Daily	Hr.	Daily	Bare Costs	Incl. O&P
1 Equip. Oper. (crane)	$59.20	$473.60	$88.50	$708.00	$54.08	$80.92
1 Truck Driver (heavy)	48.95	391.60	73.35	586.80		
1 Pickup Truck, 4x4, 3/4 Ton		175.85		193.44		
1 Truck Tractor, 6x4, 380 H.P.		493.25		542.58		
1 Lowbed Trailer, 75 Ton		255.05		280.56	57.76	63.54
16 L.H., Daily Totals		$1789.35		$2311.36	$111.83	$144.46

Crews - Standard

Crew No.	Bare Costs Hr.	Daily	Incl. Subs O&P Hr.	Daily	Cost Per Labor-Hour Bare Costs	Incl. O&P
Crew A-3G	Hr.	Daily	Hr.	Daily	Bare Costs	Incl. O&P
1 Equip. Oper. (crane)	$59.20	$473.60	$88.50	$708.00	$54.08	$80.92
1 Truck Driver (heavy)	48.95	391.60	73.35	586.80		
1 Pickup Truck, 4x4, 3/4 Ton		175.85		193.44		
1 Truck Tractor, 6x4, 450 H.P.		601.75		661.92		
1 Lowbed Trailer, 75 Ton		255.05		280.56	64.54	70.99
16 L.H., Daily Totals		$1897.85		$2430.72	$118.62	$151.92

Crew A-3H	Hr.	Daily	Hr.	Daily	Bare Costs	Incl. O&P
1 Equip. Oper. (crane)	$59.20	$473.60	$88.50	$708.00	$59.20	$88.50
1 Hyd. Crane, 12 Ton (Daily)		724.85		797.34	90.61	99.67
8 L.H., Daily Totals		$1198.45		$1505.34	$149.81	$188.17

Crew A-3I	Hr.	Daily	Hr.	Daily	Bare Costs	Incl. O&P
1 Equip. Oper. (crane)	$59.20	$473.60	$88.50	$708.00	$59.20	$88.50
1 Hyd. Crane, 25 Ton (Daily)		801.40		881.54	100.18	110.19
8 L.H., Daily Totals		$1275.00		$1589.54	$159.38	$198.69

Crew A-3J	Hr.	Daily	Hr.	Daily	Bare Costs	Incl. O&P
1 Equip. Oper. (crane)	$59.20	$473.60	$88.50	$708.00	$59.20	$88.50
1 Hyd. Crane, 40 Ton (Daily)		1272.00		1399.20	159.00	174.90
8 L.H., Daily Totals		$1745.60		$2107.20	$218.20	$263.40

Crew A-3K	Hr.	Daily	Hr.	Daily	Bare Costs	Incl. O&P
1 Equip. Oper. (crane)	$59.20	$473.60	$88.50	$708.00	$54.88	$82.03
1 Equip. Oper. (oiler)	50.55	404.40	75.55	604.40		
1 Hyd. Crane, 55 Ton (Daily)		1362.00		1498.20		
1 P/U Truck, 3/4 Ton (Daily)		142.15		156.37	94.01	103.41
16 L.H., Daily Totals		$2382.15		$2966.97	$148.88	$185.44

Crew A-3L	Hr.	Daily	Hr.	Daily	Bare Costs	Incl. O&P
1 Equip. Oper. (crane)	$59.20	$473.60	$88.50	$708.00	$54.88	$82.03
1 Equip. Oper. (oiler)	50.55	404.40	75.55	604.40		
1 Hyd. Crane, 80 Ton (Daily)		2101.00		2311.10		
1 P/U Truck, 3/4 Ton (Daily)		142.15		156.37	140.20	154.22
16 L.H., Daily Totals		$3121.15		$3779.86	$195.07	$236.24

Crew A-3M	Hr.	Daily	Hr.	Daily	Bare Costs	Incl. O&P
1 Equip. Oper. (crane)	$59.20	$473.60	$88.50	$708.00	$54.88	$82.03
1 Equip. Oper. (oiler)	50.55	404.40	75.55	604.40		
1 Hyd. Crane, 100 Ton (Daily)		2227.00		2449.70		
1 P/U Truck, 3/4 Ton (Daily)		142.15		156.37	148.07	162.88
16 L.H., Daily Totals		$3247.15		$3918.47	$202.95	$244.90

Crew A-3N	Hr.	Daily	Hr.	Daily	Bare Costs	Incl. O&P
1 Equip. Oper. (crane)	$59.20	$473.60	$88.50	$708.00	$59.20	$88.50
1 Tower Crane (monthly)		1693.00		1862.30	211.63	232.79
8 L.H., Daily Totals		$2166.60		$2570.30	$270.82	$321.29

Crew A-3P	Hr.	Daily	Hr.	Daily	Bare Costs	Incl. O&P
1 Equip. Oper. (light)	$53.00	$424.00	$79.20	$633.60	$53.00	$79.20
1 A.T. Forklift, 31' reach, 45' lift		374.20		411.62	46.77	51.45
8 L.H., Daily Totals		$798.20		$1045.22	$99.78	$130.65

Crew A-3Q	Hr.	Daily	Hr.	Daily	Bare Costs	Incl. O&P
1 Equip. Oper. (light)	$53.00	$424.00	$79.20	$633.60	$53.00	$79.20
1 Pickup Truck, 4x4, 3/4 Ton		175.85		193.44		
1 Flatbed Trailer, 3 Ton		70.30		77.33	30.77	33.85
8 L.H., Daily Totals		$670.15		$904.37	$83.77	$113.05

Crew A-3R	Hr.	Daily	Hr.	Daily	Bare Costs	Incl. O&P
1 Equip. Oper. (light)	$53.00	$424.00	$79.20	$633.60	$53.00	$79.20
1 Forklift, Smooth Floor, 8,000 Lb.		255.90		281.49	31.99	35.19
8 L.H., Daily Totals		$679.90		$915.09	$84.99	$114.39

Crew A-4	Hr.	Daily	Hr.	Daily	Bare Costs	Incl. O&P
2 Carpenters	$53.15	$850.40	$79.85	$1277.60	$50.23	$75.25
1 Painter, Ordinary	44.40	355.20	66.05	528.40		
24 L.H., Daily Totals		$1205.60		$1806.00	$50.23	$75.25

Crew A-5	Hr.	Daily	Hr.	Daily	Bare Costs	Incl. O&P
2 Laborers	$42.10	$673.60	$63.25	$1012.00	$42.67	$64.09
.25 Truck Driver (light)	47.25	94.50	70.80	141.60		
.25 Flatbed Truck, Gas, 1.5 Ton		49.04		53.94	2.72	3.00
18 L.H., Daily Totals		$817.14		$1207.54	$45.40	$67.09

Crew A-6	Hr.	Daily	Hr.	Daily	Bare Costs	Incl. O&P
1 Instrument Man	$54.85	$438.80	$82.95	$663.60	$52.92	$79.67
1 Rodman/Chainman	51.00	408.00	76.40	611.20		
1 Level, Electronic		30.80		33.88	1.93	2.12
16 L.H., Daily Totals		$877.60		$1308.68	$54.85	$81.79

Crew A-7	Hr.	Daily	Hr.	Daily	Bare Costs	Incl. O&P
1 Chief of Party	$66.05	$528.40	$99.00	$792.00	$57.30	$86.12
1 Instrument Man	54.85	438.80	82.95	663.60		
1 Rodman/Chainman	51.00	408.00	76.40	611.20		
1 Level, Electronic		30.80		33.88	1.28	1.41
24 L.H., Daily Totals		$1406.00		$2100.68	$58.58	$87.53

Crew A-8	Hr.	Daily	Hr.	Daily	Bare Costs	Incl. O&P
1 Chief of Party	$66.05	$528.40	$99.00	$792.00	$55.73	$83.69
1 Instrument Man	54.85	438.80	82.95	663.60		
2 Rodmen/Chainmen	51.00	816.00	76.40	1222.40		
1 Level, Electronic		30.80		33.88	.96	1.06
32 L.H., Daily Totals		$1814.00		$2711.88	$56.69	$84.75

Crew A-9	Hr.	Daily	Hr.	Daily	Bare Costs	Incl. O&P
1 Asbestos Foreman	$59.15	$473.20	$90.80	$726.40	$58.71	$90.14
7 Asbestos Workers	58.65	3284.40	90.05	5042.80		
64 L.H., Daily Totals		$3757.60		$5769.20	$58.71	$90.14

Crew A-10A	Hr.	Daily	Hr.	Daily	Bare Costs	Incl. O&P
1 Asbestos Foreman	$59.15	$473.20	$90.80	$726.40	$58.82	$90.30
2 Asbestos Workers	58.65	938.40	90.05	1440.80		
24 L.H., Daily Totals		$1411.60		$2167.20	$58.82	$90.30

Crew A-10B	Hr.	Daily	Hr.	Daily	Bare Costs	Incl. O&P
1 Asbestos Foreman	$59.15	$473.20	$90.80	$726.40	$58.77	$90.24
3 Asbestos Workers	58.65	1407.60	90.05	2161.20		
32 L.H., Daily Totals		$1880.80		$2887.60	$58.77	$90.24

Crew A-10C	Hr.	Daily	Hr.	Daily	Bare Costs	Incl. O&P
3 Asbestos Workers	$58.65	$1407.60	$90.05	$2161.20	$58.65	$90.05
1 Flatbed Truck, Gas, 1.5 Ton		196.15		215.76	8.17	8.99
24 L.H., Daily Totals		$1603.75		$2376.97	$66.82	$99.04

For customer support on your Mechanical Costs with RSMeans Data, call 800.448.8182.

Crew A-10D

Crew A-10D	Hr.	Daily	Hr.	Daily	Bare Costs	Incl. O&P
2 Asbestos Workers	$58.65	$938.40	$90.05	$1440.80	$56.76	$86.04
1 Equip. Oper. (crane)	59.20	473.60	88.50	708.00		
1 Equip. Oper. (oiler)	50.55	404.40	75.55	604.40		
1 Hydraulic Crane, 33 Ton		945.35		1039.89	29.54	32.50
32 L.H., Daily Totals		$2761.75		$3793.09	$86.30	$118.53

Crew A-11

Crew A-11	Hr.	Daily	Hr.	Daily	Bare Costs	Incl. O&P
1 Asbestos Foreman	$59.15	$473.20	$90.80	$726.40	$58.71	$90.14
7 Asbestos Workers	58.65	3284.40	90.05	5042.80		
2 Chip. Hammers, 12 Lb., Elec.		64.80		71.28	1.01	1.11
64 L.H., Daily Totals		$3822.40		$5840.48	$59.73	$91.26

Crew A-12

Crew A-12	Hr.	Daily	Hr.	Daily	Bare Costs	Incl. O&P
1 Asbestos Foreman	$59.15	$473.20	$90.80	$726.40	$58.71	$90.14
7 Asbestos Workers	58.65	3284.40	90.05	5042.80		
1 Trk-Mtd Vac, 14 CY, 1500 Gal.		536.15		589.76		
1 Flatbed Truck, 20,000 GVW		201.60		221.76	11.53	12.68
64 L.H., Daily Totals		$4495.35		$6580.73	$70.24	$102.82

Crew A-13

Crew A-13	Hr.	Daily	Hr.	Daily	Bare Costs	Incl. O&P
1 Equip. Oper. (light)	$53.00	$424.00	$79.20	$633.60	$53.00	$79.20
1 Trk-Mtd Vac, 14 CY, 1500 Gal.		536.15		589.76		
1 Flatbed Truck, 20,000 GVW		201.60		221.76	92.22	101.44
8 L.H., Daily Totals		$1161.75		$1445.13	$145.22	$180.64

Crew B-1

Crew B-1	Hr.	Daily	Hr.	Daily	Bare Costs	Incl. O&P
1 Labor Foreman (outside)	$44.10	$352.80	$66.25	$530.00	$42.77	$64.25
2 Laborers	42.10	673.60	63.25	1012.00		
24 L.H., Daily Totals		$1026.40		$1542.00	$42.77	$64.25

Crew B-1A

Crew B-1A	Hr.	Daily	Hr.	Daily	Bare Costs	Incl. O&P
1 Labor Foreman (outside)	$44.10	$352.80	$66.25	$530.00	$42.77	$64.25
2 Laborers	42.10	673.60	63.25	1012.00		
2 Cutting Torches		25.60		28.16		
2 Sets of Gases		343.10		377.41	15.36	16.90
24 L.H., Daily Totals		$1395.10		$1947.57	$58.13	$81.15

Crew B-1B

Crew B-1B	Hr.	Daily	Hr.	Daily	Bare Costs	Incl. O&P
1 Labor Foreman (outside)	$44.10	$352.80	$66.25	$530.00	$46.88	$70.31
2 Laborers	42.10	673.60	63.25	1012.00		
1 Equip. Oper. (crane)	59.20	473.60	88.50	708.00		
2 Cutting Torches		25.60		28.16		
2 Sets of Gases		343.10		377.41		
1 Hyd. Crane, 12 Ton		471.00		518.10	26.24	28.86
32 L.H., Daily Totals		$2339.70		$3173.67	$73.12	$99.18

Crew B-1C

Crew B-1C	Hr.	Daily	Hr.	Daily	Bare Costs	Incl. O&P
1 Labor Foreman (outside)	$44.10	$352.80	$66.25	$530.00	$42.77	$64.25
2 Laborers	42.10	673.60	63.25	1012.00		
1 Telescoping Boom Lift, to 60'		310.90		341.99	12.95	14.25
24 L.H., Daily Totals		$1337.30		$1883.99	$55.72	$78.50

Crew B-1D

Crew B-1D	Hr.	Daily	Hr.	Daily	Bare Costs	Incl. O&P
2 Laborers	$42.10	$673.60	$63.25	$1012.00	$42.10	$63.25
1 Small Work Boat, Gas, 50 H.P.		119.50		131.45		
1 Pressure Washer, 7 GPM		90.50		99.55	13.13	14.44
16 L.H., Daily Totals		$883.60		$1243.00	$55.23	$77.69

Crew B-1E

Crew B-1E	Hr.	Daily	Hr.	Daily	Bare Costs	Incl. O&P
1 Labor Foreman (outside)	$44.10	$352.80	$66.25	$530.00	$42.60	$64.00
3 Laborers	42.10	1010.40	63.25	1518.00		
1 Work Boat, Diesel, 200 H.P.		1418.00		1559.80		
2 Pressure Washers, 7 GPM		181.00		199.10	49.97	54.97
32 L.H., Daily Totals		$2962.20		$3806.90	$92.57	$118.97

Crew B-1F

Crew B-1F	Hr.	Daily	Hr.	Daily	Bare Costs	Incl. O&P
2 Skilled Workers	$54.85	$877.60	$82.95	$1327.20	$50.60	$76.38
1 Laborer	42.10	336.80	63.25	506.00		
1 Small Work Boat, Gas, 50 H.P.		119.50		131.45		
1 Pressure Washer, 7 GPM		90.50		99.55	8.75	9.63
24 L.H., Daily Totals		$1424.40		$2064.20	$59.35	$86.01

Crew B-1G

Crew B-1G	Hr.	Daily	Hr.	Daily	Bare Costs	Incl. O&P
2 Laborers	$42.10	$673.60	$63.25	$1012.00	$42.10	$63.25
1 Small Work Boat, Gas, 50 H.P.		119.50		131.45	7.47	8.22
16 L.H., Daily Totals		$793.10		$1143.45	$49.57	$71.47

Crew B-1H

Crew B-1H	Hr.	Daily	Hr.	Daily	Bare Costs	Incl. O&P
2 Skilled Workers	$54.85	$877.60	$82.95	$1327.20	$50.60	$76.38
1 Laborer	42.10	336.80	63.25	506.00		
1 Small Work Boat, Gas, 50 H.P.		119.50		131.45	4.98	5.48
24 L.H., Daily Totals		$1333.90		$1964.65	$55.58	$81.86

Crew B-1J

Crew B-1J	Hr.	Daily	Hr.	Daily	Bare Costs	Incl. O&P
1 Labor Foreman (inside)	$42.60	$340.80	$64.00	$512.00	$42.35	$63.63
1 Laborer	42.10	336.80	63.25	506.00		
16 L.H., Daily Totals		$677.60		$1018.00	$42.35	$63.63

Crew B-1K

Crew B-1K	Hr.	Daily	Hr.	Daily	Bare Costs	Incl. O&P
1 Carpenter Foreman (inside)	$53.65	$429.20	$80.60	$644.80	$53.40	$80.22
1 Carpenter	53.15	425.20	79.85	638.80		
16 L.H., Daily Totals		$854.40		$1283.60	$53.40	$80.22

Crew B-2

Crew B-2	Hr.	Daily	Hr.	Daily	Bare Costs	Incl. O&P
1 Labor Foreman (outside)	$44.10	$352.80	$66.25	$530.00	$42.50	$63.85
4 Laborers	42.10	1347.20	63.25	2024.00		
40 L.H., Daily Totals		$1700.00		$2554.00	$42.50	$63.85

Crew B-2A

Crew B-2A	Hr.	Daily	Hr.	Daily	Bare Costs	Incl. O&P
1 Labor Foreman (outside)	$44.10	$352.80	$66.25	$530.00	$42.77	$64.25
2 Laborers	42.10	673.60	63.25	1012.00		
1 Telescoping Boom Lift, to 60'		310.90		341.99	12.95	14.25
24 L.H., Daily Totals		$1337.30		$1883.99	$55.72	$78.50

Crew B-3

Crew B-3	Hr.	Daily	Hr.	Daily	Bare Costs	Incl. O&P
1 Labor Foreman (outside)	$44.10	$352.80	$66.25	$530.00	$47.16	$70.72
2 Laborers	42.10	673.60	63.25	1012.00		
1 Equip. Oper. (medium)	56.75	454.00	84.85	678.80		
2 Truck Drivers (heavy)	48.95	783.20	73.35	1173.60		
1 Crawler Loader, 3 C.Y.		1169.00		1285.90		
2 Dump Trucks, 12 C.Y., 400 H.P.		1145.80		1260.38	48.23	53.05
48 L.H., Daily Totals		$4578.40		$5940.68	$95.38	$123.76

Crew B-3A

Crew B-3A	Hr.	Daily	Hr.	Daily	Bare Costs	Incl. O&P
4 Laborers	$42.10	$1347.20	$63.25	$2024.00	$45.03	$67.57
1 Equip. Oper. (medium)	56.75	454.00	84.85	678.80		
1 Hyd. Excavator, 1.5 C.Y.		687.55		756.30	17.19	18.91
40 L.H., Daily Totals		$2488.75		$3459.11	$62.22	$86.48

For customer support on your Mechanical Costs with RSMeans Data, call 800.448.8182.

621

Crews - Standard

Crew No.	Bare Costs Hr.	Daily	Incl. Subs O&P Hr.	Daily	Cost Per Labor-Hour Bare Costs	Incl. O&P
Crew B-3B	Hr.	Daily	Hr.	Daily	Bare Costs	Incl. O&P
2 Laborers	$42.10	$673.60	$63.25	$1012.00	$47.48	$71.17
1 Equip. Oper. (medium)	56.75	454.00	84.85	678.80		
1 Truck Driver (heavy)	48.95	391.60	73.35	586.80		
1 Backhoe Loader, 80 H.P.		232.40		255.64		
1 Dump Truck, 12 C.Y., 400 H.P.		572.90		630.19	25.17	27.68
32 L.H., Daily Totals		$2324.50		$3163.43	$72.64	$98.86
Crew B-3C	Hr.	Daily	Hr.	Daily	Bare Costs	Incl. O&P
3 Laborers	$42.10	$1010.40	$63.25	$1518.00	$45.76	$68.65
1 Equip. Oper. (medium)	56.75	454.00	84.85	678.80		
1 Crawler Loader, 4 C.Y.		1489.00		1637.90	46.53	51.18
32 L.H., Daily Totals		$2953.40		$3834.70	$92.29	$119.83
Crew B-4	Hr.	Daily	Hr.	Daily	Bare Costs	Incl. O&P
1 Labor Foreman (outside)	$44.10	$352.80	$66.25	$530.00	$43.58	$65.43
4 Laborers	42.10	1347.20	63.25	2024.00		
1 Truck Driver (heavy)	48.95	391.60	73.35	586.80		
1 Truck Tractor, 220 H.P.		307.10		337.81		
1 Flatbed Trailer, 40 Ton		186.20		204.82	10.28	11.30
48 L.H., Daily Totals		$2584.90		$3683.43	$53.85	$76.74
Crew B-5	Hr.	Daily	Hr.	Daily	Bare Costs	Incl. O&P
1 Labor Foreman (outside)	$44.10	$352.80	$66.25	$530.00	$46.57	$69.85
4 Laborers	42.10	1347.20	63.25	2024.00		
2 Equip. Oper. (medium)	56.75	908.00	84.85	1357.60		
1 Air Compressor, 250 cfm		201.50		221.65		
2 Breakers, Pavement, 60 lb.		107.00		117.70		
2 -50' Air Hoses, 1.5"		48.50		53.35		
1 Crawler Loader, 3 C.Y.		1169.00		1285.90	27.25	29.98
56 L.H., Daily Totals		$4134.00		$5590.20	$73.82	$99.83
Crew B-5A	Hr.	Daily	Hr.	Daily	Bare Costs	Incl. O&P
1 Labor Foreman (outside)	$44.10	$352.80	$66.25	$530.00	$46.76	$70.11
6 Laborers	42.10	2020.80	63.25	3036.00		
2 Equip. Oper. (medium)	56.75	908.00	84.85	1357.60		
1 Equip. Oper. (light)	53.00	424.00	79.20	633.60		
2 Truck Drivers (heavy)	48.95	783.20	73.35	1173.60		
1 Air Compressor, 365 cfm		315.40		346.94		
2 Breakers, Pavement, 60 lb.		107.00		117.70		
8 -50' Air Hoses, 1"		64.00		70.40		
2 Dump Trucks, 8 C.Y., 220 H.P.		856.10		941.71	13.98	15.38
96 L.H., Daily Totals		$5831.30		$8207.55	$60.74	$85.50
Crew B-5B	Hr.	Daily	Hr.	Daily	Bare Costs	Incl. O&P
1 Powderman	$54.85	$438.80	$82.95	$663.60	$52.53	$78.78
2 Equip. Oper. (medium)	56.75	908.00	84.85	1357.60		
3 Truck Drivers (heavy)	48.95	1174.80	73.35	1760.40		
1 F.E. Loader, W.M., 2.5 C.Y.		702.10		772.31		
3 Dump Trucks, 12 C.Y., 400 H.P.		1718.70		1890.57		
1 Air Compressor, 365 cfm		315.40		346.94	57.00	62.70
48 L.H., Daily Totals		$5257.80		$6791.42	$109.54	$141.49

Crew No.	Bare Costs Hr.	Daily	Incl. Subs O&P Hr.	Daily	Cost Per Labor-Hour Bare Costs	Incl. O&P
Crew B-5C	Hr.	Daily	Hr.	Daily	Bare Costs	Incl. O&P
3 Laborers	$42.10	$1010.40	$63.25	$1518.00	$48.84	$73.17
1 Equip. Oper. (medium)	56.75	454.00	84.85	678.80		
2 Truck Drivers (heavy)	48.95	783.20	73.35	1173.60		
1 Equip. Oper. (crane)	59.20	473.60	88.50	708.00		
1 Equip. Oper. (oiler)	50.55	404.40	75.55	604.40		
2 Dump Trucks, 12 C.Y., 400 H.P.		1145.80		1260.38		
1 Crawler Loader, 4 C.Y.		1489.00		1637.90		
1 S.P. Crane, 4x4, 25 Ton		1144.00		1258.40	59.04	64.95
64 L.H., Daily Totals		$6904.40		$8839.48	$107.88	$138.12
Crew B-5D	Hr.	Daily	Hr.	Daily	Bare Costs	Incl. O&P
1 Labor Foreman (outside)	$44.10	$352.80	$66.25	$530.00	$46.87	$70.29
4 Laborers	42.10	1347.20	63.25	2024.00		
2 Equip. Oper. (medium)	56.75	908.00	84.85	1357.60		
1 Truck Driver (heavy)	48.95	391.60	73.35	586.80		
1 Air Compressor, 250 cfm		201.50		221.65		
2 Breakers, Pavement, 60 lb.		107.00		117.70		
2 -50' Air Hoses, 1.5"		48.50		53.35		
1 Crawler Loader, 3 C.Y.		1169.00		1285.90		
1 Dump Truck, 12 C.Y., 400 H.P.		572.90		630.19	32.80	36.07
64 L.H., Daily Totals		$5098.50		$6807.19	$79.66	$106.36
Crew B-5E	Hr.	Daily	Hr.	Daily	Bare Costs	Incl. O&P
1 Labor Foreman (outside)	$44.10	$352.80	$66.25	$530.00	$46.87	$70.29
4 Laborers	42.10	1347.20	63.25	2024.00		
2 Equip. Oper. (medium)	56.75	908.00	84.85	1357.60		
1 Truck Driver (heavy)	48.95	391.60	73.35	586.80		
1 Water Tank Trailer, 5000 Gal.		152.25		167.47		
1 High Press. Water Jet, 40 KSI		811.05		892.16		
2 -50' Air Hoses, 1.5"		48.50		53.35		
1 Crawler Loader, 3 C.Y.		1169.00		1285.90		
1 Dump Truck, 12 C.Y., 400 H.P.		572.90		630.19	43.03	47.33
64 L.H., Daily Totals		$5753.30		$7527.47	$89.90	$117.62
Crew B-6	Hr.	Daily	Hr.	Daily	Bare Costs	Incl. O&P
2 Laborers	$42.10	$673.60	$63.25	$1012.00	$45.73	$68.57
1 Equip. Oper. (light)	53.00	424.00	79.20	633.60		
1 Backhoe Loader, 48 H.P.		213.75		235.13	8.91	9.80
24 L.H., Daily Totals		$1311.35		$1880.72	$54.64	$78.36
Crew B-6A	Hr.	Daily	Hr.	Daily	Bare Costs	Incl. O&P
.5 Labor Foreman (outside)	$44.10	$176.40	$66.25	$265.00	$48.36	$72.49
1 Laborer	42.10	336.80	63.25	506.00		
1 Equip. Oper. (medium)	56.75	454.00	84.85	678.80		
1 Vacuum Truck, 5000 Gal.		367.55		404.31	18.38	20.22
20 L.H., Daily Totals		$1334.75		$1854.11	$66.74	$92.71
Crew B-6B	Hr.	Daily	Hr.	Daily	Bare Costs	Incl. O&P
2 Labor Foremen (outside)	$44.10	$705.60	$66.25	$1060.00	$42.77	$64.25
4 Laborers	42.10	1347.20	63.25	2024.00		
1 S.P. Crane, 4x4, 5 Ton		378.10		415.91		
1 Flatbed Truck, Gas, 1.5 Ton		196.15		215.76		
1 Butt Fusion Mach., 4"-12" diam.		426.85		469.54	20.86	22.94
48 L.H., Daily Totals		$3053.90		$4185.21	$63.62	$87.19

For customer support on your **Mechanical Costs** with RSMeans Data, call 800.448.8182.

Crews - Standard

Crew B-6C

Crew B-6C	Hr.	Daily	Hr.	Daily	Bare Costs	Incl. O&P
2 Labor Foremen (outside)	$44.10	$705.60	$66.25	$1060.00	$42.77	$64.25
4 Laborers	42.10	1347.20	63.25	2024.00		
1 S.P. Crane, 4x4, 12 Ton		428.30		471.13		
1 Flatbed Truck, Gas, 3 Ton		820.40		902.44		
1 Butt Fusion Mach., 8"-24" diam.		1076.00		1183.60	48.43	53.27
48 L.H., Daily Totals		$4377.50		$5641.17	$91.20	$117.52

Crew B-6D

Crew B-6D	Hr.	Daily	Hr.	Daily	Bare Costs	Incl. O&P
0.5 Labor Foreman (outside)	$44.10	$176.40	$66.25	$265.00	$48.36	$72.49
1 Laborer	42.10	336.80	63.25	506.00		
1 Equip. Oper. (medium)	56.75	454.00	84.85	678.80		
1 Hydro Excavator, 12 C.Y.		1262.00		1388.20	63.10	69.41
20 L.H., Daily Totals		$2229.20		$2838.00	$111.46	$141.90

Crew B-7

Crew B-7	Hr.	Daily	Hr.	Daily	Bare Costs	Incl. O&P
1 Labor Foreman (outside)	$44.10	$352.80	$66.25	$530.00	$44.88	$67.35
4 Laborers	42.10	1347.20	63.25	2024.00		
1 Equip. Oper. (medium)	56.75	454.00	84.85	678.80		
1 Brush Chipper, 12", 130 H.P.		392.80		432.08		
1 Crawler Loader, 3 C.Y.		1169.00		1285.90		
2 Chain Saws, Gas, 36" Long		82.50		90.75	34.26	37.68
48 L.H., Daily Totals		$3798.30		$5041.53	$79.13	$105.03

Crew B-7A

Crew B-7A	Hr.	Daily	Hr.	Daily	Bare Costs	Incl. O&P
2 Laborers	$42.10	$673.60	$63.25	$1012.00	$45.73	$68.57
1 Equip. Oper. (light)	53.00	424.00	79.20	633.60		
1 Rake w/Tractor		339.45		373.39		
2 Chain Saws, Gas, 18"		104.00		114.40	18.48	20.32
24 L.H., Daily Totals		$1541.05		$2133.40	$64.21	$88.89

Crew B-7B

Crew B-7B	Hr.	Daily	Hr.	Daily	Bare Costs	Incl. O&P
1 Labor Foreman (outside)	$44.10	$352.80	$66.25	$530.00	$45.46	$68.21
4 Laborers	42.10	1347.20	63.25	2024.00		
1 Equip. Oper. (medium)	56.75	454.00	84.85	678.80		
1 Truck Driver (heavy)	48.95	391.60	73.35	586.80		
1 Brush Chipper, 12", 130 H.P.		392.80		432.08		
1 Crawler Loader, 3 C.Y.		1169.00		1285.90		
2 Chain Saws, Gas, 36" Long		82.50		90.75		
1 Dump Truck, 8 C.Y., 220 H.P.		428.05		470.86	37.01	40.71
56 L.H., Daily Totals		$4617.95		$6099.19	$82.46	$108.91

Crew B-7C

Crew B-7C	Hr.	Daily	Hr.	Daily	Bare Costs	Incl. O&P
1 Labor Foreman (outside)	$44.10	$352.80	$66.25	$530.00	$45.46	$68.21
4 Laborers	42.10	1347.20	63.25	2024.00		
1 Equip. Oper. (medium)	56.75	454.00	84.85	678.80		
1 Truck Driver (heavy)	48.95	391.60	73.35	586.80		
1 Brush Chipper, 12", 130 H.P.		392.80		432.08		
1 Crawler Loader, 3 C.Y.		1169.00		1285.90		
2 Chain Saws, Gas, 36" Long		82.50		90.75		
1 Dump Truck, 12 C.Y., 400 H.P.		572.90		630.19	39.59	43.55
56 L.H., Daily Totals		$4762.80		$6258.52	$85.05	$111.76

Crew B-8

Crew B-8	Hr.	Daily	Hr.	Daily	Bare Costs	Incl. O&P
1 Labor Foreman (outside)	$44.10	$352.80	$66.25	$530.00	$48.78	$73.09
2 Laborers	42.10	673.60	63.25	1012.00		
2 Equip. Oper. (medium)	56.75	908.00	84.85	1357.60		
1 Equip. Oper. (oiler)	50.55	404.40	75.55	604.40		
2 Truck Drivers (heavy)	48.95	783.20	73.35	1173.60		
1 Hyd. Crane, 25 Ton		580.85		638.93		
1 Crawler Loader, 3 C.Y.		1169.00		1285.90		
2 Dump Trucks, 12 C.Y., 400 H.P.		1145.80		1260.38	45.24	49.77
64 L.H., Daily Totals		$6017.65		$7862.81	$94.03	$122.86

Crew B-9

Crew B-9	Hr.	Daily	Hr.	Daily	Bare Costs	Incl. O&P
1 Labor Foreman (outside)	$44.10	$352.80	$66.25	$530.00	$42.50	$63.85
4 Laborers	42.10	1347.20	63.25	2024.00		
1 Air Compressor, 250 cfm		201.50		221.65		
2 Breakers, Pavement, 60 lb.		107.00		117.70		
2 -50' Air Hoses, 1.5"		48.50		53.35	8.93	9.82
40 L.H., Daily Totals		$2057.00		$2946.70	$51.42	$73.67

Crew B-9A

Crew B-9A	Hr.	Daily	Hr.	Daily	Bare Costs	Incl. O&P
2 Laborers	$42.10	$673.60	$63.25	$1012.00	$44.38	$66.62
1 Truck Driver (heavy)	48.95	391.60	73.35	586.80		
1 Water Tank Trailer, 5000 Gal.		152.25		167.47		
1 Truck Tractor, 220 H.P.		307.10		337.81		
2 -50' Discharge Hoses, 3"		9.00		9.90	19.51	21.47
24 L.H., Daily Totals		$1533.55		$2113.99	$63.90	$88.08

Crew B-9B

Crew B-9B	Hr.	Daily	Hr.	Daily	Bare Costs	Incl. O&P
2 Laborers	$42.10	$673.60	$63.25	$1012.00	$44.38	$66.62
1 Truck Driver (heavy)	48.95	391.60	73.35	586.80		
2 -50' Discharge Hoses, 3"		9.00		9.90		
1 Water Tank Trailer, 5000 Gal.		152.25		167.47		
1 Truck Tractor, 220 H.P.		307.10		337.81		
1 Pressure Washer		96.95		106.65	23.55	25.91
24 L.H., Daily Totals		$1630.50		$2220.63	$67.94	$92.53

Crew B-9D

Crew B-9D	Hr.	Daily	Hr.	Daily	Bare Costs	Incl. O&P
1 Labor Foreman (outside)	$44.10	$352.80	$66.25	$530.00	$42.50	$63.85
4 Common Laborers	42.10	1347.20	63.25	2024.00		
1 Air Compressor, 250 cfm		201.50		221.65		
2 -50' Air Hoses, 1.5"		48.50		53.35		
2 Air Powered Tampers		67.70		74.47	7.94	8.74
40 L.H., Daily Totals		$2017.70		$2903.47	$50.44	$72.59

Crew B-9E

Crew B-9E	Hr.	Daily	Hr.	Daily	Bare Costs	Incl. O&P
1 Cement Finisher	$49.95	$399.60	$73.45	$587.60	$46.02	$68.35
1 Laborer	42.10	336.80	63.25	506.00		
1 Chip. Hammers, 12 Lb., Elec.		32.40		35.64	2.02	2.23
16 L.H., Daily Totals		$768.80		$1129.24	$48.05	$70.58

Crew B-10

Crew B-10	Hr.	Daily	Hr.	Daily	Bare Costs	Incl. O&P
1 Equip. Oper. (medium)	$56.75	$454.00	$84.85	$678.80	$51.87	$77.65
.5 Laborer	42.10	168.40	63.25	253.00		
12 L.H., Daily Totals		$622.40		$931.80	$51.87	$77.65

Crew B-10A

Crew B-10A	Hr.	Daily	Hr.	Daily	Bare Costs	Incl. O&P
1 Equip. Oper. (medium)	$56.75	$454.00	$84.85	$678.80	$51.87	$77.65
.5 Laborer	42.10	168.40	63.25	253.00		
1 Roller, 2-Drum, W.B., 7.5 H.P.		166.10		182.71	13.84	15.23
12 L.H., Daily Totals		$788.50		$1114.51	$65.71	$92.88

For customer support on your Mechanical Costs with RSMeans Data, call 800.448.8182.

623

Crew No.	Bare Costs		Incl. Subs O&P		Cost Per Labor-Hour	
Crew B-10B	Hr.	Daily	Hr.	Daily	Bare Costs	Incl. O&P
1 Equip. Oper. (medium)	$56.75	$454.00	$84.85	$678.80	$51.87	$77.65
.5 Laborer	42.10	168.40	63.25	253.00		
1 Dozer, 200 H.P.		1504.00		1654.40	125.33	137.87
12 L.H., Daily Totals		$2126.40		$2586.20	$177.20	$215.52
Crew B-10C	Hr.	Daily	Hr.	Daily	Bare Costs	Incl. O&P
1 Equip. Oper. (medium)	$56.75	$454.00	$84.85	$678.80	$51.87	$77.65
.5 Laborer	42.10	168.40	63.25	253.00		
1 Dozer, 200 H.P.		1504.00		1654.40		
1 Vibratory Roller, Towed, 23 Ton		514.75		566.23	168.23	185.05
12 L.H., Daily Totals		$2641.15		$3152.43	$220.10	$262.70
Crew B-10D	Hr.	Daily	Hr.	Daily	Bare Costs	Incl. O&P
1 Equip. Oper. (medium)	$56.75	$454.00	$84.85	$678.80	$51.87	$77.65
.5 Laborer	42.10	168.40	63.25	253.00		
1 Dozer, 200 H.P.		1504.00		1654.40		
1 Sheepsft. Roller, Towed		436.90		480.59	161.74	177.92
12 L.H., Daily Totals		$2563.30		$3066.79	$213.61	$255.57
Crew B-10E	Hr.	Daily	Hr.	Daily	Bare Costs	Incl. O&P
1 Equip. Oper. (medium)	$56.75	$454.00	$84.85	$678.80	$51.87	$77.65
.5 Laborer	42.10	168.40	63.25	253.00		
1 Tandem Roller, 5 Ton		255.65		281.21	21.30	23.43
12 L.H., Daily Totals		$878.05		$1213.02	$73.17	$101.08
Crew B-10F	Hr.	Daily	Hr.	Daily	Bare Costs	Incl. O&P
1 Equip. Oper. (medium)	$56.75	$454.00	$84.85	$678.80	$51.87	$77.65
.5 Laborer	42.10	168.40	63.25	253.00		
1 Tandem Roller, 10 Ton		236.75		260.43	19.73	21.70
12 L.H., Daily Totals		$859.15		$1192.22	$71.60	$99.35
Crew B-10G	Hr.	Daily	Hr.	Daily	Bare Costs	Incl. O&P
1 Equip. Oper. (medium)	$56.75	$454.00	$84.85	$678.80	$51.87	$77.65
.5 Laborer	42.10	168.40	63.25	253.00		
1 Sheepsfoot Roller, 240 H.P.		1348.00		1482.80	112.33	123.57
12 L.H., Daily Totals		$1970.40		$2414.60	$164.20	$201.22
Crew B-10H	Hr.	Daily	Hr.	Daily	Bare Costs	Incl. O&P
1 Equip. Oper. (medium)	$56.75	$454.00	$84.85	$678.80	$51.87	$77.65
.5 Laborer	42.10	168.40	63.25	253.00		
1 Diaphragm Water Pump, 2"		86.80		95.48		
1 -20' Suction Hose, 2"		3.55		3.90		
2 -50' Discharge Hoses, 2"		8.00		8.80	8.20	9.02
12 L.H., Daily Totals		$720.75		$1039.98	$60.06	$86.67
Crew B-10I	Hr.	Daily	Hr.	Daily	Bare Costs	Incl. O&P
1 Equip. Oper. (medium)	$56.75	$454.00	$84.85	$678.80	$51.87	$77.65
.5 Laborer	42.10	168.40	63.25	253.00		
1 Diaphragm Water Pump, 4"		105.25		115.78		
1 -20' Suction Hose, 4"		17.05		18.75		
2 -50' Discharge Hoses, 4"		25.30		27.83	12.30	13.53
12 L.H., Daily Totals		$770.00		$1094.16	$64.17	$91.18
Crew B-10J	Hr.	Daily	Hr.	Daily	Bare Costs	Incl. O&P
1 Equip. Oper. (medium)	$56.75	$454.00	$84.85	$678.80	$51.87	$77.65
.5 Laborer	42.10	168.40	63.25	253.00		
1 Centrifugal Water Pump, 3"		74.45		81.89		
1 -20' Suction Hose, 3"		8.75		9.63		
2 -50' Discharge Hoses, 3"		9.00		9.90	7.68	8.45
12 L.H., Daily Totals		$714.60		$1033.22	$59.55	$86.10

Crew No.	Bare Costs		Incl. Subs O&P		Cost Per Labor-Hour	
Crew B-10K	Hr.	Daily	Hr.	Daily	Bare Costs	Incl. O&P
1 Equip. Oper. (medium)	$56.75	$454.00	$84.85	$678.80	$51.87	$77.65
.5 Laborer	42.10	168.40	63.25	253.00		
1 Centr. Water Pump, 6"		232.90		256.19		
1 -20' Suction Hose, 6"		25.20		27.72		
2 -50' Discharge Hoses, 6"		35.80		39.38	24.49	26.94
12 L.H., Daily Totals		$916.30		$1255.09	$76.36	$104.59
Crew B-10L	Hr.	Daily	Hr.	Daily	Bare Costs	Incl. O&P
1 Equip. Oper. (medium)	$56.75	$454.00	$84.85	$678.80	$51.87	$77.65
.5 Laborer	42.10	168.40	63.25	253.00		
1 Dozer, 80 H.P.		401.40		441.54	33.45	36.80
12 L.H., Daily Totals		$1023.80		$1373.34	$85.32	$114.44
Crew B-10M	Hr.	Daily	Hr.	Daily	Bare Costs	Incl. O&P
1 Equip. Oper. (medium)	$56.75	$454.00	$84.85	$678.80	$51.87	$77.65
.5 Laborer	42.10	168.40	63.25	253.00		
1 Dozer, 300 H.P.		1764.00		1940.40	147.00	161.70
12 L.H., Daily Totals		$2386.40		$2872.20	$198.87	$239.35
Crew B-10N	Hr.	Daily	Hr.	Daily	Bare Costs	Incl. O&P
1 Equip. Oper. (medium)	$56.75	$454.00	$84.85	$678.80	$51.87	$77.65
.5 Laborer	42.10	168.40	63.25	253.00		
1 F.E. Loader, T.M., 1.5 C.Y.		565.80		622.38	47.15	51.87
12 L.H., Daily Totals		$1188.20		$1554.18	$99.02	$129.51
Crew B-10O	Hr.	Daily	Hr.	Daily	Bare Costs	Incl. O&P
1 Equip. Oper. (medium)	$56.75	$454.00	$84.85	$678.80	$51.87	$77.65
.5 Laborer	42.10	168.40	63.25	253.00		
1 F.E. Loader, T.M., 2.25 C.Y.		960.90		1056.99	80.08	88.08
12 L.H., Daily Totals		$1583.30		$1988.79	$131.94	$165.73
Crew B-10P	Hr.	Daily	Hr.	Daily	Bare Costs	Incl. O&P
1 Equip. Oper. (medium)	$56.75	$454.00	$84.85	$678.80	$51.87	$77.65
.5 Laborer	42.10	168.40	63.25	253.00		
1 Crawler Loader, 3 C.Y.		1169.00		1285.90	97.42	107.16
12 L.H., Daily Totals		$1791.40		$2217.70	$149.28	$184.81
Crew B-10Q	Hr.	Daily	Hr.	Daily	Bare Costs	Incl. O&P
1 Equip. Oper. (medium)	$56.75	$454.00	$84.85	$678.80	$51.87	$77.65
.5 Laborer	42.10	168.40	63.25	253.00		
1 Crawler Loader, 4 C.Y.		1489.00		1637.90	124.08	136.49
12 L.H., Daily Totals		$2111.40		$2569.70	$175.95	$214.14
Crew B-10R	Hr.	Daily	Hr.	Daily	Bare Costs	Incl. O&P
1 Equip. Oper. (medium)	$56.75	$454.00	$84.85	$678.80	$51.87	$77.65
.5 Laborer	42.10	168.40	63.25	253.00		
1 F.E. Loader, W.M., 1 C.Y.		302.00		332.20	25.17	27.68
12 L.H., Daily Totals		$924.40		$1264.00	$77.03	$105.33
Crew B-10S	Hr.	Daily	Hr.	Daily	Bare Costs	Incl. O&P
1 Equip. Oper. (medium)	$56.75	$454.00	$84.85	$678.80	$51.87	$77.65
.5 Laborer	42.10	168.40	63.25	253.00		
1 F.E. Loader, W.M., 1.5 C.Y.		423.50		465.85	35.29	38.82
12 L.H., Daily Totals		$1045.90		$1397.65	$87.16	$116.47
Crew B-10T	Hr.	Daily	Hr.	Daily	Bare Costs	Incl. O&P
1 Equip. Oper. (medium)	$56.75	$454.00	$84.85	$678.80	$51.87	$77.65
.5 Laborer	42.10	168.40	63.25	253.00		
1 F.E. Loader, W.M., 2.5 C.Y.		702.10		772.31	58.51	64.36
12 L.H., Daily Totals		$1324.50		$1704.11	$110.38	$142.01

Crew No.	Bare Costs		Incl. Subs O&P		Cost Per Labor-Hour	
Crew B-10U	Hr.	Daily	Hr.	Daily	Bare Costs	Incl. O&P
1 Equip. Oper. (medium)	$56.75	$454.00	$84.85	$678.80	$51.87	$77.65
.5 Laborer	42.10	168.40	63.25	253.00		
1 F.E. Loader, W.M., 5.5 C.Y.		957.50		1053.25	79.79	87.77
12 L.H., Daily Totals		$1579.90		$1985.05	$131.66	$165.42
Crew B-10V	Hr.	Daily	Hr.	Daily	Bare Costs	Incl. O&P
1 Equip. Oper. (medium)	$56.75	$454.00	$84.85	$678.80	$51.87	$77.65
.5 Laborer	42.10	168.40	63.25	253.00		
1 Dozer, 700 H.P.		5120.00		5632.00	426.67	469.33
12 L.H., Daily Totals		$5742.40		$6563.80	$478.53	$546.98
Crew B-10W	Hr.	Daily	Hr.	Daily	Bare Costs	Incl. O&P
1 Equip. Oper. (medium)	$56.75	$454.00	$84.85	$678.80	$51.87	$77.65
.5 Laborer	42.10	168.40	63.25	253.00		
1 Dozer, 105 H.P.		633.85		697.24	52.82	58.10
12 L.H., Daily Totals		$1256.25		$1629.04	$104.69	$135.75
Crew B-10X	Hr.	Daily	Hr.	Daily	Bare Costs	Incl. O&P
1 Equip. Oper. (medium)	$56.75	$454.00	$84.85	$678.80	$51.87	$77.65
.5 Laborer	42.10	168.40	63.25	253.00		
1 Dozer, 410 H.P.		2777.00		3054.70	231.42	254.56
12 L.H., Daily Totals		$3399.40		$3986.50	$283.28	$332.21
Crew B-10Y	Hr.	Daily	Hr.	Daily	Bare Costs	Incl. O&P
1 Equip. Oper. (medium)	$56.75	$454.00	$84.85	$678.80	$51.87	$77.65
.5 Laborer	42.10	168.40	63.25	253.00		
1 Vibr. Roller, Towed, 12 Ton		578.60		636.46	48.22	53.04
12 L.H., Daily Totals		$1201.00		$1568.26	$100.08	$130.69
Crew B-11A	Hr.	Daily	Hr.	Daily	Bare Costs	Incl. O&P
1 Equipment Oper. (med.)	$56.75	$454.00	$84.85	$678.80	$49.42	$74.05
1 Laborer	42.10	336.80	63.25	506.00		
1 Dozer, 200 H.P.		1504.00		1654.40	94.00	103.40
16 L.H., Daily Totals		$2294.80		$2839.20	$143.43	$177.45
Crew B-11B	Hr.	Daily	Hr.	Daily	Bare Costs	Incl. O&P
1 Equipment Oper. (light)	$53.00	$424.00	$79.20	$633.60	$47.55	$71.22
1 Laborer	42.10	336.80	63.25	506.00		
1 Air Powered Tamper		33.85		37.23		
1 Air Compressor, 365 cfm		315.40		346.94		
2 -50' Air Hoses, 1.5"		48.50		53.35	24.86	27.35
16 L.H., Daily Totals		$1158.55		$1577.13	$72.41	$98.57
Crew B-11C	Hr.	Daily	Hr.	Daily	Bare Costs	Incl. O&P
1 Equipment Oper. (med.)	$56.75	$454.00	$84.85	$678.80	$49.42	$74.05
1 Laborer	42.10	336.80	63.25	506.00		
1 Backhoe Loader, 48 H.P.		213.75		235.13	13.36	14.70
16 L.H., Daily Totals		$1004.55		$1419.93	$62.78	$88.75
Crew B-11J	Hr.	Daily	Hr.	Daily	Bare Costs	Incl. O&P
1 Equipment Oper. (med.)	$56.75	$454.00	$84.85	$678.80	$49.42	$74.05
1 Laborer	42.10	336.80	63.25	506.00		
1 Grader, 30,000 Lbs.		1062.00		1168.20		
1 Ripper, Beam & 1 Shank		90.50		99.55	72.03	79.23
16 L.H., Daily Totals		$1943.30		$2452.55	$121.46	$153.28

Crew No.	Bare Costs		Incl. Subs O&P		Cost Per Labor-Hour	
Crew B-11K	Hr.	Daily	Hr.	Daily	Bare Costs	Incl. O&P
1 Equipment Oper. (med.)	$56.75	$454.00	$84.85	$678.80	$49.42	$74.05
1 Laborer	42.10	336.80	63.25	506.00		
1 Trencher, Chain Type, 8' D		1872.00		2059.20	117.00	128.70
16 L.H., Daily Totals		$2662.80		$3244.00	$166.43	$202.75
Crew B-11L	Hr.	Daily	Hr.	Daily	Bare Costs	Incl. O&P
1 Equipment Oper. (med.)	$56.75	$454.00	$84.85	$678.80	$49.42	$74.05
1 Laborer	42.10	336.80	63.25	506.00		
1 Grader, 30,000 Lbs.		1062.00		1168.20	66.38	73.01
16 L.H., Daily Totals		$1852.80		$2353.00	$115.80	$147.06
Crew B-11M	Hr.	Daily	Hr.	Daily	Bare Costs	Incl. O&P
1 Equipment Oper. (med.)	$56.75	$454.00	$84.85	$678.80	$49.42	$74.05
1 Laborer	42.10	336.80	63.25	506.00		
1 Backhoe Loader, 80 H.P.		232.40		255.64	14.53	15.98
16 L.H., Daily Totals		$1023.20		$1440.44	$63.95	$90.03
Crew B-11N	Hr.	Daily	Hr.	Daily	Bare Costs	Incl. O&P
1 Labor Foreman (outside)	$44.10	$352.80	$66.25	$530.00	$50.14	$75.12
2 Equipment Operators (med.)	56.75	908.00	84.85	1357.60		
6 Truck Drivers (heavy)	48.95	2349.60	73.35	3520.80		
1 F.E. Loader, W.M., 5.5 C.Y.		957.50		1053.25		
1 Dozer, 410 H.P.		2777.00		3054.70		
6 Dump Trucks, Off Hwy., 50 Ton		10626.00		11688.60	199.45	219.40
72 L.H., Daily Totals		$17970.90		$21204.95	$249.60	$294.51
Crew B-11Q	Hr.	Daily	Hr.	Daily	Bare Costs	Incl. O&P
1 Equipment Operator (med.)	$56.75	$454.00	$84.85	$678.80	$51.87	$77.65
.5 Laborer	42.10	168.40	63.25	253.00		
1 Dozer, 140 H.P.		762.40		838.64	63.53	69.89
12 L.H., Daily Totals		$1384.80		$1770.44	$115.40	$147.54
Crew B-11R	Hr.	Daily	Hr.	Daily	Bare Costs	Incl. O&P
1 Equipment Operator (med.)	$56.75	$454.00	$84.85	$678.80	$51.87	$77.65
.5 Laborer	42.10	168.40	63.25	253.00		
1 Dozer, 200 H.P.		1504.00		1654.40	125.33	137.87
12 L.H., Daily Totals		$2126.40		$2586.20	$177.20	$215.52
Crew B-11S	Hr.	Daily	Hr.	Daily	Bare Costs	Incl. O&P
1 Equipment Operator (med.)	$56.75	$454.00	$84.85	$678.80	$51.87	$77.65
.5 Laborer	42.10	168.40	63.25	253.00		
1 Dozer, 300 H.P.		1764.00		1940.40		
1 Ripper, Beam & 1 Shank		90.50		99.55	154.54	170.00
12 L.H., Daily Totals		$2476.90		$2971.75	$206.41	$247.65
Crew B-11T	Hr.	Daily	Hr.	Daily	Bare Costs	Incl. O&P
1 Equipment Operator (med.)	$56.75	$454.00	$84.85	$678.80	$51.87	$77.65
.5 Laborer	42.10	168.40	63.25	253.00		
1 Dozer, 410 H.P.		2777.00		3054.70		
1 Ripper, Beam & 2 Shanks		138.75		152.63	242.98	267.28
12 L.H., Daily Totals		$3538.15		$4139.13	$294.85	$344.93
Crew B-11U	Hr.	Daily	Hr.	Daily	Bare Costs	Incl. O&P
1 Equipment Operator (med.)	$56.75	$454.00	$84.85	$678.80	$51.87	$77.65
.5 Laborer	42.10	168.40	63.25	253.00		
1 Dozer, 520 H.P.		3398.00		3737.80	283.17	311.48
12 L.H., Daily Totals		$4020.40		$4669.60	$335.03	$389.13

Crew No.	Bare Costs Hr.	Bare Costs Daily	Incl. Subs O&P Hr.	Incl. Subs O&P Daily	Cost Per Labor-Hour Bare Costs	Cost Per Labor-Hour Incl. O&P
Crew B-11V	Hr.	Daily	Hr.	Daily	Bare Costs	Incl. O&P
3 Laborers	$42.10	$1010.40	$63.25	$1518.00	$42.10	$63.25
1 Roller, 2-Drum, W.B., 7.5 H.P.		166.10		182.71	6.92	7.61
24 L.H., Daily Totals		$1176.50		$1700.71	$49.02	$70.86
Crew B-11W	Hr.	Daily	Hr.	Daily	Bare Costs	Incl. O&P
1 Equipment Operator (med.)	$56.75	$454.00	$84.85	$678.80	$49.03	$73.47
1 Common Laborer	42.10	336.80	63.25	506.00		
10 Truck Drivers (heavy)	48.95	3916.00	73.35	5868.00		
1 Dozer, 200 H.P.		1504.00		1654.40		
1 Vibratory Roller, Towed, 23 Ton		514.75		566.23		
10 Dump Trucks, 8 C.Y., 220 H.P.		4280.50		4708.55	65.62	72.18
96 L.H., Daily Totals		$11006.05		$13981.98	$114.65	$145.65
Crew B-11Y	Hr.	Daily	Hr.	Daily	Bare Costs	Incl. O&P
1 Labor Foreman (outside)	$44.10	$352.80	$66.25	$530.00	$47.21	$70.78
5 Common Laborers	42.10	1684.00	63.25	2530.00		
3 Equipment Operators (med.)	56.75	1362.00	84.85	2036.40		
1 Dozer, 80 H.P.		401.40		441.54		
2 Rollers, 2-Drum, W.B., 7.5 H.P.		332.20		365.42		
4 Vibrating Plates, Gas, 21"		661.40		727.54	19.38	21.31
72 L.H., Daily Totals		$4793.80		$6630.90	$66.58	$92.10
Crew B-12A	Hr.	Daily	Hr.	Daily	Bare Costs	Incl. O&P
1 Equip. Oper. (crane)	$59.20	$473.60	$88.50	$708.00	$50.65	$75.88
1 Laborer	42.10	336.80	63.25	506.00		
1 Hyd. Excavator, 1 C.Y.		795.30		874.83	49.71	54.68
16 L.H., Daily Totals		$1605.70		$2088.83	$100.36	$130.55
Crew B-12B	Hr.	Daily	Hr.	Daily	Bare Costs	Incl. O&P
1 Equip. Oper. (crane)	$59.20	$473.60	$88.50	$708.00	$50.65	$75.88
1 Laborer	42.10	336.80	63.25	506.00		
1 Hyd. Excavator, 1.5 C.Y.		687.55		756.30	42.97	47.27
16 L.H., Daily Totals		$1497.95		$1970.31	$93.62	$123.14
Crew B-12C	Hr.	Daily	Hr.	Daily	Bare Costs	Incl. O&P
1 Equip. Oper. (crane)	$59.20	$473.60	$88.50	$708.00	$50.65	$75.88
1 Laborer	42.10	336.80	63.25	506.00		
1 Hyd. Excavator, 2 C.Y.		951.30		1046.43	59.46	65.40
16 L.H., Daily Totals		$1761.70		$2260.43	$110.11	$141.28
Crew B-12D	Hr.	Daily	Hr.	Daily	Bare Costs	Incl. O&P
1 Equip. Oper. (crane)	$59.20	$473.60	$88.50	$708.00	$50.65	$75.88
1 Laborer	42.10	336.80	63.25	506.00		
1 Hyd. Excavator, 3.5 C.Y.		2158.00		2373.80	134.88	148.36
16 L.H., Daily Totals		$2968.40		$3587.80	$185.53	$224.24
Crew B-12E	Hr.	Daily	Hr.	Daily	Bare Costs	Incl. O&P
1 Equip. Oper. (crane)	$59.20	$473.60	$88.50	$708.00	$50.65	$75.88
1 Laborer	42.10	336.80	63.25	506.00		
1 Hyd. Excavator, .5 C.Y.		452.10		497.31	28.26	31.08
16 L.H., Daily Totals		$1262.50		$1711.31	$78.91	$106.96
Crew B-12F	Hr.	Daily	Hr.	Daily	Bare Costs	Incl. O&P
1 Equip. Oper. (crane)	$59.20	$473.60	$88.50	$708.00	$50.65	$75.88
1 Laborer	42.10	336.80	63.25	506.00		
1 Hyd. Excavator, .75 C.Y.		694.25		763.67	43.39	47.73
16 L.H., Daily Totals		$1504.65		$1977.68	$94.04	$123.60
Crew B-12G	Hr.	Daily	Hr.	Daily	Bare Costs	Incl. O&P
1 Equip. Oper. (crane)	$59.20	$473.60	$88.50	$708.00	$50.65	$75.88
1 Laborer	42.10	336.80	63.25	506.00		
1 Crawler Crane, 15 Ton		792.30		871.53		
1 Clamshell Bucket, .5 C.Y.		67.10		73.81	53.71	59.08
16 L.H., Daily Totals		$1669.80		$2159.34	$104.36	$134.96
Crew B-12H	Hr.	Daily	Hr.	Daily	Bare Costs	Incl. O&P
1 Equip. Oper. (crane)	$59.20	$473.60	$88.50	$708.00	$50.65	$75.88
1 Laborer	42.10	336.80	63.25	506.00		
1 Crawler Crane, 25 Ton		1129.00		1241.90		
1 Clamshell Bucket, 1 C.Y.		68.55		75.41	74.85	82.33
16 L.H., Daily Totals		$2007.95		$2531.30	$125.50	$158.21
Crew B-12I	Hr.	Daily	Hr.	Daily	Bare Costs	Incl. O&P
1 Equip. Oper. (crane)	$59.20	$473.60	$88.50	$708.00	$50.65	$75.88
1 Laborer	42.10	336.80	63.25	506.00		
1 Crawler Crane, 20 Ton		961.35		1057.48		
1 Dragline Bucket, .75 C.Y.		61.25		67.38	63.91	70.30
16 L.H., Daily Totals		$1833.00		$2338.86	$114.56	$146.18
Crew B-12J	Hr.	Daily	Hr.	Daily	Bare Costs	Incl. O&P
1 Equip. Oper. (crane)	$59.20	$473.60	$88.50	$708.00	$50.65	$75.88
1 Laborer	42.10	336.80	63.25	506.00		
1 Gradall, 5/8 C.Y.		846.50		931.15	52.91	58.20
16 L.H., Daily Totals		$1656.90		$2145.15	$103.56	$134.07
Crew B-12K	Hr.	Daily	Hr.	Daily	Bare Costs	Incl. O&P
1 Equip. Oper. (crane)	$59.20	$473.60	$88.50	$708.00	$50.65	$75.88
1 Laborer	42.10	336.80	63.25	506.00		
1 Gradall, 3 Ton, 1 C.Y.		973.80		1071.18	60.86	66.95
16 L.H., Daily Totals		$1784.20		$2285.18	$111.51	$142.82
Crew B-12L	Hr.	Daily	Hr.	Daily	Bare Costs	Incl. O&P
1 Equip. Oper. (crane)	$59.20	$473.60	$88.50	$708.00	$50.65	$75.88
1 Laborer	42.10	336.80	63.25	506.00		
1 Crawler Crane, 15 Ton		792.30		871.53		
1 F.E. Attachment, .5 C.Y.		65.25		71.78	53.60	58.96
16 L.H., Daily Totals		$1667.95		$2157.30	$104.25	$134.83
Crew B-12M	Hr.	Daily	Hr.	Daily	Bare Costs	Incl. O&P
1 Equip. Oper. (crane)	$59.20	$473.60	$88.50	$708.00	$50.65	$75.88
1 Laborer	42.10	336.80	63.25	506.00		
1 Crawler Crane, 20 Ton		961.35		1057.48		
1 F.E. Attachment, .75 C.Y.		70.40		77.44	64.48	70.93
16 L.H., Daily Totals		$1842.15		$2348.93	$115.13	$146.81
Crew B-12N	Hr.	Daily	Hr.	Daily	Bare Costs	Incl. O&P
1 Equip. Oper. (crane)	$59.20	$473.60	$88.50	$708.00	$50.65	$75.88
1 Laborer	42.10	336.80	63.25	506.00		
1 Crawler Crane, 25 Ton		1129.00		1241.90		
1 F.E. Attachment, 1 C.Y.		76.50		84.15	75.34	82.88
16 L.H., Daily Totals		$2015.90		$2540.05	$125.99	$158.75
Crew B-12O	Hr.	Daily	Hr.	Daily	Bare Costs	Incl. O&P
1 Equip. Oper. (crane)	$59.20	$473.60	$88.50	$708.00	$50.65	$75.88
1 Laborer	42.10	336.80	63.25	506.00		
1 Crawler Crane, 40 Ton		1206.00		1326.60		
1 F.E. Attachment, 1.5 C.Y.		87.60		96.36	80.85	88.94
16 L.H., Daily Totals		$2104.00		$2636.96	$131.50	$164.81

For customer support on your Mechanical Costs with RSMeans Data, call 800.448.8182.

Crews - Standard

Crew B-12P

Crew No.	Hr.	Daily	Hr.	Daily	Bare Costs	Incl. O&P
1 Equip. Oper. (crane)	$59.20	$473.60	$88.50	$708.00	$50.65	$75.88
1 Laborer	42.10	336.80	63.25	506.00		
1 Crawler Crane, 40 Ton		1206.00		1326.60		
1 Dragline Bucket, 1.5 C.Y.		65.05		71.56	79.44	87.38
16 L.H., Daily Totals		$2081.45		$2612.16	$130.09	$163.26

Crew B-12Q

Crew No.	Hr.	Daily	Hr.	Daily	Bare Costs	Incl. O&P
1 Equip. Oper. (crane)	$59.20	$473.60	$88.50	$708.00	$50.65	$75.88
1 Laborer	42.10	336.80	63.25	506.00		
1 Hyd. Excavator, 5/8 C.Y.		598.25		658.08	37.39	41.13
16 L.H., Daily Totals		$1408.65		$1872.08	$88.04	$117.00

Crew B-12S

Crew No.	Hr.	Daily	Hr.	Daily	Bare Costs	Incl. O&P
1 Equip. Oper. (crane)	$59.20	$473.60	$88.50	$708.00	$50.65	$75.88
1 Laborer	42.10	336.80	63.25	506.00		
1 Hyd. Excavator, 2.5 C.Y.		1465.00		1611.50	91.56	100.72
16 L.H., Daily Totals		$2275.40		$2825.50	$142.21	$176.59

Crew B-12T

Crew No.	Hr.	Daily	Hr.	Daily	Bare Costs	Incl. O&P
1 Equip. Oper. (crane)	$59.20	$473.60	$88.50	$708.00	$50.65	$75.88
1 Laborer	42.10	336.80	63.25	506.00		
1 Crawler Crane, 75 Ton		2002.00		2202.20		
1 F.E. Attachment, 3 C.Y.		114.15		125.57	132.26	145.49
16 L.H., Daily Totals		$2926.55		$3541.76	$182.91	$221.36

Crew B-12V

Crew No.	Hr.	Daily	Hr.	Daily	Bare Costs	Incl. O&P
1 Equip. Oper. (crane)	$59.20	$473.60	$88.50	$708.00	$50.65	$75.88
1 Laborer	42.10	336.80	63.25	506.00		
1 Crawler Crane, 75 Ton		2002.00		2202.20		
1 Dragline Bucket, 3 C.Y.		71.60		78.76	129.60	142.56
16 L.H., Daily Totals		$2884.00		$3494.96	$180.25	$218.44

Crew B-12Y

Crew No.	Hr.	Daily	Hr.	Daily	Bare Costs	Incl. O&P
1 Equip. Oper. (crane)	$59.20	$473.60	$88.50	$708.00	$47.80	$71.67
2 Laborers	42.10	673.60	63.25	1012.00		
1 Hyd. Excavator, 3.5 C.Y.		2158.00		2373.80	89.92	98.91
24 L.H., Daily Totals		$3305.20		$4093.80	$137.72	$170.57

Crew B-12Z

Crew No.	Hr.	Daily	Hr.	Daily	Bare Costs	Incl. O&P
1 Equip. Oper. (crane)	$59.20	$473.60	$88.50	$708.00	$47.80	$71.67
2 Laborers	42.10	673.60	63.25	1012.00		
1 Hyd. Excavator, 2.5 C.Y.		1465.00		1611.50	61.04	67.15
24 L.H., Daily Totals		$2612.20		$3331.50	$108.84	$138.81

Crew B-13

Crew No.	Hr.	Daily	Hr.	Daily	Bare Costs	Incl. O&P
1 Labor Foreman (outside)	$44.10	$352.80	$66.25	$530.00	$46.04	$69.04
4 Laborers	42.10	1347.20	63.25	2024.00		
1 Equip. Oper. (crane)	59.20	473.60	88.50	708.00		
1 Equip. Oper. (oiler)	50.55	404.40	75.55	604.40		
1 Hyd. Crane, 25 Ton		580.85		638.93	10.37	11.41
56 L.H., Daily Totals		$3158.85		$4505.34	$56.41	$80.45

Crew B-13A

Crew No.	Hr.	Daily	Hr.	Daily	Bare Costs	Incl. O&P
1 Labor Foreman (outside)	$44.10	$352.80	$66.25	$530.00	$48.53	$72.74
2 Laborers	42.10	673.60	63.25	1012.00		
2 Equipment Operators (med.)	56.75	908.00	84.85	1357.60		
2 Truck Drivers (heavy)	48.95	783.20	73.35	1173.60		
1 Crawler Crane, 75 Ton		2002.00		2202.20		
1 Crawler Loader, 4 C.Y.		1489.00		1637.90		
2 Dump Trucks, 8 C.Y., 220 H.P.		856.10		941.71	77.63	85.39
56 L.H., Daily Totals		$7064.70		$8855.01	$126.16	$158.13

Crew B-13B

Crew No.	Hr.	Daily	Hr.	Daily	Bare Costs	Incl. O&P
1 Labor Foreman (outside)	$44.10	$352.80	$66.25	$530.00	$46.04	$69.04
4 Laborers	42.10	1347.20	63.25	2024.00		
1 Equip. Oper. (crane)	59.20	473.60	88.50	708.00		
1 Equip. Oper. (oiler)	50.55	404.40	75.55	604.40		
1 Hyd. Crane, 55 Ton		980.20		1078.22	17.50	19.25
56 L.H., Daily Totals		$3558.20		$4944.62	$63.54	$88.30

Crew B-13C

Crew No.	Hr.	Daily	Hr.	Daily	Bare Costs	Incl. O&P
1 Labor Foreman (outside)	$44.10	$352.80	$66.25	$530.00	$46.04	$69.04
4 Laborers	42.10	1347.20	63.25	2024.00		
1 Equip. Oper. (crane)	59.20	473.60	88.50	708.00		
1 Equip. Oper. (oiler)	50.55	404.40	75.55	604.40		
1 Crawler Crane, 100 Ton		2287.00		2515.70	40.84	44.92
56 L.H., Daily Totals		$4865.00		$6382.10	$86.88	$113.97

Crew B-13D

Crew No.	Hr.	Daily	Hr.	Daily	Bare Costs	Incl. O&P
1 Laborer	$42.10	$336.80	$63.25	$506.00	$50.65	$75.88
1 Equip. Oper. (crane)	59.20	473.60	88.50	708.00		
1 Hyd. Excavator, 1 C.Y.		795.30		874.83		
1 Trench Box		117.75		129.53	57.07	62.77
16 L.H., Daily Totals		$1723.45		$2218.36	$107.72	$138.65

Crew B-13E

Crew No.	Hr.	Daily	Hr.	Daily	Bare Costs	Incl. O&P
1 Laborer	$42.10	$336.80	$63.25	$506.00	$50.65	$75.88
1 Equip. Oper. (crane)	59.20	473.60	88.50	708.00		
1 Hyd. Excavator, 1.5 C.Y.		687.55		756.30		
1 Trench Box		117.75		129.53	50.33	55.36
16 L.H., Daily Totals		$1615.70		$2099.83	$100.98	$131.24

Crew B-13F

Crew No.	Hr.	Daily	Hr.	Daily	Bare Costs	Incl. O&P
1 Laborer	$42.10	$336.80	$63.25	$506.00	$50.65	$75.88
1 Equip. Oper. (crane)	59.20	473.60	88.50	708.00		
1 Hyd. Excavator, 3.5 C.Y.		2158.00		2373.80		
1 Trench Box		117.75		129.53	142.23	156.46
16 L.H., Daily Totals		$3086.15		$3717.32	$192.88	$232.33

Crew B-13G

Crew No.	Hr.	Daily	Hr.	Daily	Bare Costs	Incl. O&P
1 Laborer	$42.10	$336.80	$63.25	$506.00	$50.65	$75.88
1 Equip. Oper. (crane)	59.20	473.60	88.50	708.00		
1 Hyd. Excavator, .75 C.Y.		694.25		763.67		
1 Trench Box		117.75		129.53	50.75	55.83
16 L.H., Daily Totals		$1622.40		$2107.20	$101.40	$131.70

Crew B-13H

Crew No.	Hr.	Daily	Hr.	Daily	Bare Costs	Incl. O&P
1 Laborer	$42.10	$336.80	$63.25	$506.00	$50.65	$75.88
1 Equip. Oper. (crane)	59.20	473.60	88.50	708.00		
1 Gradall, 5/8 C.Y.		846.50		931.15		
1 Trench Box		117.75		129.53	60.27	66.29
16 L.H., Daily Totals		$1774.65		$2274.68	$110.92	$142.17

For customer support on your Mechanical Costs with RSMeans Data, call 800.448.8182.

627

Crews - Standard

Crew B-13I	Bare Costs Hr.	Daily	Incl. Subs O&P Hr.	Daily	Cost Per Labor-Hour Bare Costs	Incl. O&P
1 Laborer	$42.10	$336.80	$63.25	$506.00	$50.65	$75.88
1 Equip. Oper. (crane)	59.20	473.60	88.50	708.00		
1 Gradall, 3 Ton, 1 C.Y.		973.80		1071.18		
1 Trench Box		117.75		129.53	68.22	75.04
16 L.H., Daily Totals		$1901.95		$2414.70	$118.87	$150.92

Crew B-13J	Bare Costs Hr.	Daily	Incl. Subs O&P Hr.	Daily	Cost Per Labor-Hour Bare Costs	Incl. O&P
1 Laborer	$42.10	$336.80	$63.25	$506.00	$50.65	$75.88
1 Equip. Oper. (crane)	59.20	473.60	88.50	708.00		
1 Hyd. Excavator, 2.5 C.Y.		1465.00		1611.50		
1 Trench Box		117.75		129.53	98.92	108.81
16 L.H., Daily Totals		$2393.15		$2955.03	$149.57	$184.69

Crew B-13K	Bare Costs Hr.	Daily	Incl. Subs O&P Hr.	Daily	Cost Per Labor-Hour Bare Costs	Incl. O&P
2 Equip. Opers. (crane)	$59.20	$947.20	$88.50	$1416.00	$59.20	$88.50
1 Hyd. Excavator, .75 C.Y.		694.25		763.67		
1 Hyd. Hammer, 4000 ft-lb		641.50		705.65		
1 Hyd. Excavator, .75 C.Y.		694.25		763.67	126.88	139.56
16 L.H., Daily Totals		$2977.20		$3649.00	$186.07	$228.06

Crew B-13L	Bare Costs Hr.	Daily	Incl. Subs O&P Hr.	Daily	Cost Per Labor-Hour Bare Costs	Incl. O&P
2 Equip. Opers. (crane)	$59.20	$947.20	$88.50	$1416.00	$59.20	$88.50
1 Hyd. Excavator, 1.5 C.Y.		687.55		756.30		
1 Hyd. Hammer, 5000 ft-lb		697.25		766.98		
1 Hyd. Excavator, .75 C.Y.		694.25		763.67	129.94	142.93
16 L.H., Daily Totals		$3026.25		$3702.95	$189.14	$231.43

Crew B-13M	Bare Costs Hr.	Daily	Incl. Subs O&P Hr.	Daily	Cost Per Labor-Hour Bare Costs	Incl. O&P
2 Equip. Opers. (crane)	$59.20	$947.20	$88.50	$1416.00	$59.20	$88.50
1 Hyd. Excavator, 2.5 C.Y.		1465.00		1611.50		
1 Hyd. Hammer, 8000 ft-lb		907.75		998.52		
1 Hyd. Excavator, 1.5 C.Y.		687.55		756.30	191.27	210.40
16 L.H., Daily Totals		$4007.50		$4782.33	$250.47	$298.90

Crew B-13N	Bare Costs Hr.	Daily	Incl. Subs O&P Hr.	Daily	Cost Per Labor-Hour Bare Costs	Incl. O&P
2 Equip. Opers. (crane)	$59.20	$947.20	$88.50	$1416.00	$59.20	$88.50
1 Hyd. Excavator, 3.5 C.Y.		2158.00		2373.80		
1 Hyd. Hammer, 12,000 ft-lb		871.75		958.92		
1 Hyd. Excavator, 1.5 C.Y.		687.55		756.30	232.33	255.56
16 L.H., Daily Totals		$4664.50		$5505.03	$291.53	$344.06

Crew B-14	Bare Costs Hr.	Daily	Incl. Subs O&P Hr.	Daily	Cost Per Labor-Hour Bare Costs	Incl. O&P
1 Labor Foreman (outside)	$44.10	$352.80	$66.25	$530.00	$44.25	$66.41
4 Laborers	42.10	1347.20	63.25	2024.00		
1 Equip. Oper. (light)	53.00	424.00	79.20	633.60		
1 Backhoe Loader, 48 H.P.		213.75		235.13	4.45	4.90
48 L.H., Daily Totals		$2337.75		$3422.72	$48.70	$71.31

Crew B-14A	Bare Costs Hr.	Daily	Incl. Subs O&P Hr.	Daily	Cost Per Labor-Hour Bare Costs	Incl. O&P
1 Equip. Oper. (crane)	$59.20	$473.60	$88.50	$708.00	$53.50	$80.08
.5 Laborer	42.10	168.40	63.25	253.00		
1 Hyd. Excavator, 4.5 C.Y.		3409.00		3749.90	284.08	312.49
12 L.H., Daily Totals		$4051.00		$4710.90	$337.58	$392.57

Crew B-14B	Bare Costs Hr.	Daily	Incl. Subs O&P Hr.	Daily	Cost Per Labor-Hour Bare Costs	Incl. O&P
1 Equip. Oper. (crane)	$59.20	$473.60	$88.50	$708.00	$53.50	$80.08
.5 Laborer	42.10	168.40	63.25	253.00		
1 Hyd. Excavator, 6 C.Y.		3464.00		3810.40	288.67	317.53
12 L.H., Daily Totals		$4106.00		$4771.40	$342.17	$397.62

Crew B-14C	Bare Costs Hr.	Daily	Incl. Subs O&P Hr.	Daily	Cost Per Labor-Hour Bare Costs	Incl. O&P
1 Equip. Oper. (crane)	$59.20	$473.60	$88.50	$708.00	$53.50	$80.08
.5 Laborer	42.10	168.40	63.25	253.00		
1 Hyd. Excavator, 7 C.Y.		3437.00		3780.70	286.42	315.06
12 L.H., Daily Totals		$4079.00		$4741.70	$339.92	$395.14

Crew B-14F	Bare Costs Hr.	Daily	Incl. Subs O&P Hr.	Daily	Cost Per Labor-Hour Bare Costs	Incl. O&P
1 Equip. Oper. (crane)	$59.20	$473.60	$88.50	$708.00	$53.50	$80.08
.5 Laborer	42.10	168.40	63.25	253.00		
1 Hyd. Shovel, 7 C.Y.		4099.00		4508.90	341.58	375.74
12 L.H., Daily Totals		$4741.00		$5469.90	$395.08	$455.82

Crew B-14G	Bare Costs Hr.	Daily	Incl. Subs O&P Hr.	Daily	Cost Per Labor-Hour Bare Costs	Incl. O&P
1 Equip. Oper. (crane)	$59.20	$473.60	$88.50	$708.00	$53.50	$80.08
.5 Laborer	42.10	168.40	63.25	253.00		
1 Hyd. Shovel, 12 C.Y.		5951.00		6546.10	495.92	545.51
12 L.H., Daily Totals		$6593.00		$7507.10	$549.42	$625.59

Crew B-14J	Bare Costs Hr.	Daily	Incl. Subs O&P Hr.	Daily	Cost Per Labor-Hour Bare Costs	Incl. O&P
1 Equip. Oper. (medium)	$56.75	$454.00	$84.85	$678.80	$51.87	$77.65
.5 Laborer	42.10	168.40	63.25	253.00		
1 F.E. Loader, 8 C.Y.		2261.00		2487.10	188.42	207.26
12 L.H., Daily Totals		$2883.40		$3418.90	$240.28	$284.91

Crew B-14K	Bare Costs Hr.	Daily	Incl. Subs O&P Hr.	Daily	Cost Per Labor-Hour Bare Costs	Incl. O&P
1 Equip. Oper. (medium)	$56.75	$454.00	$84.85	$678.80	$51.87	$77.65
.5 Laborer	42.10	168.40	63.25	253.00		
1 F.E. Loader, 10 C.Y.		2674.00		2941.40	222.83	245.12
12 L.H., Daily Totals		$3296.40		$3873.20	$274.70	$322.77

Crew B-15	Bare Costs Hr.	Daily	Incl. Subs O&P Hr.	Daily	Cost Per Labor-Hour Bare Costs	Incl. O&P
1 Equipment Oper. (med.)	$56.75	$454.00	$84.85	$678.80	$50.20	$75.19
.5 Laborer	42.10	168.40	63.25	253.00		
2 Truck Drivers (heavy)	48.95	783.20	73.35	1173.60		
2 Dump Trucks, 12 C.Y., 400 H.P.		1145.80		1260.38		
1 Dozer, 200 H.P.		1504.00		1654.40	94.64	104.10
28 L.H., Daily Totals		$4055.40		$5020.18	$144.84	$179.29

Crew B-16	Bare Costs Hr.	Daily	Incl. Subs O&P Hr.	Daily	Cost Per Labor-Hour Bare Costs	Incl. O&P
1 Labor Foreman (outside)	$44.10	$352.80	$66.25	$530.00	$44.31	$66.53
2 Laborers	42.10	673.60	63.25	1012.00		
1 Truck Driver (heavy)	48.95	391.60	73.35	586.80		
1 Dump Truck, 12 C.Y., 400 H.P.		572.90		630.19	17.90	19.69
32 L.H., Daily Totals		$1990.90		$2758.99	$62.22	$86.22

Crew B-17	Bare Costs Hr.	Daily	Incl. Subs O&P Hr.	Daily	Cost Per Labor-Hour Bare Costs	Incl. O&P
2 Laborers	$42.10	$673.60	$63.25	$1012.00	$46.54	$69.76
1 Equip. Oper. (light)	53.00	424.00	79.20	633.60		
1 Truck Driver (heavy)	48.95	391.60	73.35	586.80		
1 Backhoe Loader, 48 H.P.		213.75		235.13		
1 Dump Truck, 8 C.Y., 220 H.P.		428.05		470.86	20.06	22.06
32 L.H., Daily Totals		$2131.00		$2938.38	$66.59	$91.82

Crew B-17A	Bare Costs Hr.	Daily	Incl. Subs O&P Hr.	Daily	Cost Per Labor-Hour Bare Costs	Incl. O&P
2 Labor Foremen (outside)	$44.10	$705.60	$66.25	$1060.00	$45.25	$68.09
6 Laborers	42.10	2020.80	63.25	3036.00		
1 Skilled Worker Foreman (out)	56.85	454.80	86.00	688.00		
1 Skilled Worker	54.85	438.80	82.95	663.60		
80 L.H., Daily Totals		$3620.00		$5447.60	$45.25	$68.09

For customer support on your Mechanical Costs with RSMeans Data, call 800.448.8182.

Crew No.	Bare Costs		Incl. Subs O&P		Cost Per Labor-Hour	
Crew B-17B	Hr.	Daily	Hr.	Daily	Bare Costs	Incl. O&P
2 Laborers	$42.10	$673.60	$63.25	$1012.00	$46.54	$69.76
1 Equip. Oper. (light)	53.00	424.00	79.20	633.60		
1 Truck Driver (heavy)	48.95	391.60	73.35	586.80		
1 Backhoe Loader, 48 H.P.		213.75		235.13		
1 Dump Truck, 12 C.Y., 400 H.P.		572.90		630.19	24.58	27.04
32 L.H., Daily Totals		$2275.85		$3097.72	$71.12	$96.80
Crew B-18	Hr.	Daily	Hr.	Daily	Bare Costs	Incl. O&P
1 Labor Foreman (outside)	$44.10	$352.80	$66.25	$530.00	$42.77	$64.25
2 Laborers	42.10	673.60	63.25	1012.00		
1 Vibrating Plate, Gas, 21"		165.35		181.88	6.89	7.58
24 L.H., Daily Totals		$1191.75		$1723.89	$49.66	$71.83
Crew B-19	Hr.	Daily	Hr.	Daily	Bare Costs	Incl. O&P
1 Pile Driver Foreman (outside)	$56.20	$449.60	$87.30	$698.40	$55.24	$84.58
4 Pile Drivers	54.20	1734.40	84.20	2694.40		
2 Equip. Oper. (crane)	59.20	947.20	88.50	1416.00		
1 Equip. Oper. (oiler)	50.55	404.40	75.55	604.40		
1 Crawler Crane, 40 Ton		1206.00		1326.60		
1 Lead, 90' High		367.45		404.19		
1 Hammer, Diesel, 22k ft-lb		436.45		480.10	31.40	34.55
64 L.H., Daily Totals		$5545.50		$7624.09	$86.65	$119.13
Crew B-19A	Hr.	Daily	Hr.	Daily	Bare Costs	Incl. O&P
1 Pile Driver Foreman (outside)	$56.20	$449.60	$87.30	$698.40	$55.24	$84.58
4 Pile Drivers	54.20	1734.40	84.20	2694.40		
2 Equip. Oper. (crane)	59.20	947.20	88.50	1416.00		
1 Equip. Oper. (oiler)	50.55	404.40	75.55	604.40		
1 Crawler Crane, 75 Ton		2002.00		2202.20		
1 Lead, 90' High		367.45		404.19		
1 Hammer, Diesel, 41k ft-lb		576.65		634.32	46.03	50.64
64 L.H., Daily Totals		$6481.70		$8653.91	$101.28	$135.22
Crew B-19B	Hr.	Daily	Hr.	Daily	Bare Costs	Incl. O&P
1 Pile Driver Foreman (outside)	$56.20	$449.60	$87.30	$698.40	$55.24	$84.58
4 Pile Drivers	54.20	1734.40	84.20	2694.40		
2 Equip. Oper. (crane)	59.20	947.20	88.50	1416.00		
1 Equip. Oper. (oiler)	50.55	404.40	75.55	604.40		
1 Crawler Crane, 40 Ton		1206.00		1326.60		
1 Lead, 90' High		367.45		404.19		
1 Hammer, Diesel, 22k ft-lb		436.45		480.10		
1 Barge, 400 Ton		858.85		944.74	44.82	49.31
64 L.H., Daily Totals		$6404.35		$8568.83	$100.07	$133.89
Crew B-19C	Hr.	Daily	Hr.	Daily	Bare Costs	Incl. O&P
1 Pile Driver Foreman (outside)	$56.20	$449.60	$87.30	$698.40	$55.24	$84.58
4 Pile Drivers	54.20	1734.40	84.20	2694.40		
2 Equip. Oper. (crane)	59.20	947.20	88.50	1416.00		
1 Equip. Oper. (oiler)	50.55	404.40	75.55	604.40		
1 Crawler Crane, 75 Ton		2002.00		2202.20		
1 Lead, 90' High		367.45		404.19		
1 Hammer, Diesel, 41k ft-lb		576.65		634.32		
1 Barge, 400 Ton		858.85		944.74	59.45	65.40
64 L.H., Daily Totals		$7340.55		$9598.65	$114.70	$149.98
Crew B-20	Hr.	Daily	Hr.	Daily	Bare Costs	Incl. O&P
1 Labor Foreman (outside)	$44.10	$352.80	$66.25	$530.00	$47.02	$70.82
1 Skilled Worker	54.85	438.80	82.95	663.60		
1 Laborer	42.10	336.80	63.25	506.00		
24 L.H., Daily Totals		$1128.40		$1699.60	$47.02	$70.82
Crew B-20A	Hr.	Daily	Hr.	Daily	Bare Costs	Incl. O&P
1 Labor Foreman (outside)	$44.10	$352.80	$66.25	$530.00	$50.55	$75.78
1 Laborer	42.10	336.80	63.25	506.00		
1 Plumber	64.45	515.60	96.45	771.60		
1 Plumber Apprentice	51.55	412.40	77.15	617.20		
32 L.H., Daily Totals		$1617.60		$2424.80	$50.55	$75.78
Crew B-21	Hr.	Daily	Hr.	Daily	Bare Costs	Incl. O&P
1 Labor Foreman (outside)	$44.10	$352.80	$66.25	$530.00	$48.76	$73.34
1 Skilled Worker	54.85	438.80	82.95	663.60		
1 Laborer	42.10	336.80	63.25	506.00		
.5 Equip. Oper. (crane)	59.20	236.80	88.50	354.00		
.5 S.P. Crane, 4x4, 5 Ton		189.05		207.96	6.75	7.43
28 L.H., Daily Totals		$1554.25		$2261.55	$55.51	$80.77
Crew B-21A	Hr.	Daily	Hr.	Daily	Bare Costs	Incl. O&P
1 Labor Foreman (outside)	$44.10	$352.80	$66.25	$530.00	$52.28	$78.32
1 Laborer	42.10	336.80	63.25	506.00		
1 Plumber	64.45	515.60	96.45	771.60		
1 Plumber Apprentice	51.55	412.40	77.15	617.20		
1 Equip. Oper. (crane)	59.20	473.60	88.50	708.00		
1 S.P. Crane, 4x4, 12 Ton		428.30		471.13	10.71	11.78
40 L.H., Daily Totals		$2519.50		$3603.93	$62.99	$90.10
Crew B-21B	Hr.	Daily	Hr.	Daily	Bare Costs	Incl. O&P
1 Labor Foreman (outside)	$44.10	$352.80	$66.25	$530.00	$45.92	$68.90
3 Laborers	42.10	1010.40	63.25	1518.00		
1 Equip. Oper. (crane)	59.20	473.60	88.50	708.00		
1 Hyd. Crane, 12 Ton		471.00		518.10	11.78	12.95
40 L.H., Daily Totals		$2307.80		$3274.10	$57.70	$81.85
Crew B-21C	Hr.	Daily	Hr.	Daily	Bare Costs	Incl. O&P
1 Labor Foreman (outside)	$44.10	$352.80	$66.25	$530.00	$46.04	$69.04
4 Laborers	42.10	1347.20	63.25	2024.00		
1 Equip. Oper. (crane)	59.20	473.60	88.50	708.00		
1 Equip. Oper. (oiler)	50.55	404.40	75.55	604.40		
2 Cutting Torches		25.60		28.16		
2 Sets of Gases		343.10		377.41		
1 Lattice Boom Crane, 90 Ton		1696.00		1865.60	36.87	40.56
56 L.H., Daily Totals		$4642.70		$6137.57	$82.91	$109.60
Crew B-22	Hr.	Daily	Hr.	Daily	Bare Costs	Incl. O&P
1 Labor Foreman (outside)	$44.10	$352.80	$66.25	$530.00	$49.45	$74.35
1 Skilled Worker	54.85	438.80	82.95	663.60		
1 Laborer	42.10	336.80	63.25	506.00		
.75 Equip. Oper. (crane)	59.20	355.20	88.50	531.00		
.75 S.P. Crane, 4x4, 5 Ton		283.57		311.93	9.45	10.40
30 L.H., Daily Totals		$1767.18		$2542.53	$58.91	$84.75
Crew B-22A	Hr.	Daily	Hr.	Daily	Bare Costs	Incl. O&P
1 Labor Foreman (outside)	$44.10	$352.80	$66.25	$530.00	$48.47	$72.84
1 Skilled Worker	54.85	438.80	82.95	663.60		
2 Laborers	42.10	673.60	63.25	1012.00		
1 Equipment Operator, Crane	59.20	473.60	88.50	708.00		
1 S.P. Crane, 4x4, 5 Ton		378.10		415.91		
1 Butt Fusion Mach., 4"-12" diam.		426.85		469.54	20.12	22.14
40 L.H., Daily Totals		$2743.75		$3799.05	$68.59	$94.98

For customer support on your Mechanical Costs with RSMeans Data, call 800.448.8182.

629

Crew No.	Bare Costs		Incl. Subs O&P		Cost Per Labor-Hour	
Crew B-22B	Hr.	Daily	Hr.	Daily	Bare Costs	Incl. O&P
1 Labor Foreman (outside)	$44.10	$352.80	$66.25	$530.00	$48.47	$72.84
1 Skilled Worker	54.85	438.80	82.95	663.60		
2 Laborers	42.10	673.60	63.25	1012.00		
1 Equip. Oper. (crane)	59.20	473.60	88.50	708.00		
1 S.P. Crane, 4x4, 5 Ton		378.10		415.91		
1 Butt Fusion Mach., 8"-24" diam.		1076.00		1183.60	36.35	39.99
40 L.H., Daily Totals		$3392.90		$4513.11	$84.82	$112.83

Crew No.	Bare Costs		Incl. Subs O&P		Cost Per Labor-Hour	
Crew B-22C	Hr.	Daily	Hr.	Daily	Bare Costs	Incl. O&P
1 Skilled Worker	$54.85	$438.80	$82.95	$663.60	$48.48	$73.10
1 Laborer	42.10	336.80	63.25	506.00		
1 Butt Fusion Mach., 2"-8" diam.		156.05		171.66	9.75	10.73
16 L.H., Daily Totals		$931.65		$1341.26	$58.23	$83.83

Crew No.	Bare Costs		Incl. Subs O&P		Cost Per Labor-Hour	
Crew B-23	Hr.	Daily	Hr.	Daily	Bare Costs	Incl. O&P
1 Labor Foreman (outside)	$44.10	$352.80	$66.25	$530.00	$42.50	$63.85
4 Laborers	42.10	1347.20	63.25	2024.00		
1 Drill Rig, Truck-Mounted		771.10		848.21		
1 Flatbed Truck, Gas, 3 Ton		820.40		902.44	39.79	43.77
40 L.H., Daily Totals		$3291.50		$4304.65	$82.29	$107.62

Crew No.	Bare Costs		Incl. Subs O&P		Cost Per Labor-Hour	
Crew B-23A	Hr.	Daily	Hr.	Daily	Bare Costs	Incl. O&P
1 Labor Foreman (outside)	$44.10	$352.80	$66.25	$530.00	$47.65	$71.45
1 Laborer	42.10	336.80	63.25	506.00		
1 Equip. Oper. (medium)	56.75	454.00	84.85	678.80		
1 Drill Rig, Truck-Mounted		771.10		848.21		
1 Pickup Truck, 3/4 Ton		110.85		121.94	36.75	40.42
24 L.H., Daily Totals		$2025.55		$2684.95	$84.40	$111.87

Crew No.	Bare Costs		Incl. Subs O&P		Cost Per Labor-Hour	
Crew B-23B	Hr.	Daily	Hr.	Daily	Bare Costs	Incl. O&P
1 Labor Foreman (outside)	$44.10	$352.80	$66.25	$530.00	$47.65	$71.45
1 Laborer	42.10	336.80	63.25	506.00		
1 Equip. Oper. (medium)	56.75	454.00	84.85	678.80		
1 Drill Rig, Truck-Mounted		771.10		848.21		
1 Pickup Truck, 3/4 Ton		110.85		121.94		
1 Centr. Water Pump, 6"		232.90		256.19	46.45	51.10
24 L.H., Daily Totals		$2258.45		$2941.14	$94.10	$122.55

Crew No.	Bare Costs		Incl. Subs O&P		Cost Per Labor-Hour	
Crew B-24	Hr.	Daily	Hr.	Daily	Bare Costs	Incl. O&P
1 Cement Finisher	$49.95	$399.60	$73.45	$587.60	$48.40	$72.18
1 Laborer	42.10	336.80	63.25	506.00		
1 Carpenter	53.15	425.20	79.85	638.80		
24 L.H., Daily Totals		$1161.60		$1732.40	$48.40	$72.18

Crew No.	Bare Costs		Incl. Subs O&P		Cost Per Labor-Hour	
Crew B-25	Hr.	Daily	Hr.	Daily	Bare Costs	Incl. O&P
1 Labor Foreman (outside)	$44.10	$352.80	$66.25	$530.00	$46.28	$69.41
7 Laborers	42.10	2357.60	63.25	3542.00		
3 Equip. Oper. (medium)	56.75	1362.00	84.85	2036.40		
1 Asphalt Paver, 130 H.P.		2117.00		2328.70		
1 Tandem Roller, 10 Ton		236.75		260.43		
1 Roller, Pneum. Whl., 12 Ton		345.75		380.32	30.68	33.74
88 L.H., Daily Totals		$6771.90		$9077.85	$76.95	$103.16

Crew No.	Bare Costs		Incl. Subs O&P		Cost Per Labor-Hour	
Crew B-25B	Hr.	Daily	Hr.	Daily	Bare Costs	Incl. O&P
1 Labor Foreman (outside)	$44.10	$352.80	$66.25	$530.00	$47.15	$70.70
7 Laborers	42.10	2357.60	63.25	3542.00		
4 Equip. Oper. (medium)	56.75	1816.00	84.85	2715.20		
1 Asphalt Paver, 130 H.P.		2117.00		2328.70		
2 Tandem Rollers, 10 Ton		473.50		520.85		
1 Roller, Pneum. Whl., 12 Ton		345.75		380.32	30.59	33.64
96 L.H., Daily Totals		$7462.65		$10017.08	$77.74	$104.34

Crew No.	Bare Costs		Incl. Subs O&P		Cost Per Labor-Hour	
Crew B-25C	Hr.	Daily	Hr.	Daily	Bare Costs	Incl. O&P
1 Labor Foreman (outside)	$44.10	$352.80	$66.25	$530.00	$47.32	$70.95
3 Laborers	42.10	1010.40	63.25	1518.00		
2 Equip. Oper. (medium)	56.75	908.00	84.85	1357.60		
1 Asphalt Paver, 130 H.P.		2117.00		2328.70		
1 Tandem Roller, 10 Ton		236.75		260.43	49.04	53.94
48 L.H., Daily Totals		$4624.95		$5994.73	$96.35	$124.89

Crew No.	Bare Costs		Incl. Subs O&P		Cost Per Labor-Hour	
Crew B-25D	Hr.	Daily	Hr.	Daily	Bare Costs	Incl. O&P
1 Labor Foreman (outside)	$44.10	$352.80	$66.25	$530.00	$47.54	$71.28
3 Laborers	42.10	1010.40	63.25	1518.00		
2.125 Equip. Oper. (medium)	56.75	964.75	84.85	1442.45		
.125 Truck Driver (heavy)	48.95	48.95	73.35	73.35		
.125 Truck Tractor, 6x4, 380 H.P.		61.66		67.82		
.125 Dist. Tanker, 3000 Gallon		41.27		45.40		
1 Asphalt Paver, 130 H.P.		2117.00		2328.70		
1 Tandem Roller, 10 Ton		236.75		260.43	49.13	54.05
50 L.H., Daily Totals		$4833.58		$6266.14	$96.67	$125.32

Crew No.	Bare Costs		Incl. Subs O&P		Cost Per Labor-Hour	
Crew B-25E	Hr.	Daily	Hr.	Daily	Bare Costs	Incl. O&P
1 Labor Foreman (outside)	$44.10	$352.80	$66.25	$530.00	$47.74	$71.58
3 Laborers	42.10	1010.40	63.25	1518.00		
2.250 Equip. Oper. (medium)	56.75	1021.50	84.85	1527.30		
.25 Truck Driver (heavy)	48.95	97.90	73.35	146.70		
.25 Truck Tractor, 6x4, 380 H.P.		123.31		135.64		
.25 Dist. Tanker, 3000 Gallon		82.54		90.79		
1 Asphalt Paver, 130 H.P.		2117.00		2328.70		
1 Tandem Roller, 10 Ton		236.75		260.43	49.22	54.15
52 L.H., Daily Totals		$5042.20		$6537.56	$96.97	$125.72

Crew No.	Bare Costs		Incl. Subs O&P		Cost Per Labor-Hour	
Crew B-26	Hr.	Daily	Hr.	Daily	Bare Costs	Incl. O&P
1 Labor Foreman (outside)	$44.10	$352.80	$66.25	$530.00	$46.96	$70.35
6 Laborers	42.10	2020.80	63.25	3036.00		
2 Equip. Oper. (medium)	56.75	908.00	84.85	1357.60		
1 Rodman (reinf.)	56.40	451.20	84.90	679.20		
1 Cement Finisher	49.95	399.60	73.45	587.60		
1 Grader, 30,000 Lbs.		1062.00		1168.20		
1 Paving Mach. & Equip.		2474.00		2721.40	40.18	44.20
88 L.H., Daily Totals		$7668.40		$10080.00	$87.14	$114.55

Crew No.	Bare Costs		Incl. Subs O&P		Cost Per Labor-Hour	
Crew B-26A	Hr.	Daily	Hr.	Daily	Bare Costs	Incl. O&P
1 Labor Foreman (outside)	$44.10	$352.80	$66.25	$530.00	$46.96	$70.35
6 Laborers	42.10	2020.80	63.25	3036.00		
2 Equip. Oper. (medium)	56.75	908.00	84.85	1357.60		
1 Rodman (reinf.)	56.40	451.20	84.90	679.20		
1 Cement Finisher	49.95	399.60	73.45	587.60		
1 Grader, 30,000 Lbs.		1062.00		1168.20		
1 Paving Mach. & Equip.		2474.00		2721.40		
1 Concrete Saw		111.50		122.65	41.45	45.59
88 L.H., Daily Totals		$7779.90		$10202.65	$88.41	$115.94

Crew No.	Bare Costs		Incl. Subs O&P		Cost Per Labor-Hour	
Crew B-26B	Hr.	Daily	Hr.	Daily	Bare Costs	Incl. O&P
1 Labor Foreman (outside)	$44.10	$352.80	$66.25	$530.00	$47.77	$71.55
6 Laborers	42.10	2020.80	63.25	3036.00		
3 Equip. Oper. (medium)	56.75	1362.00	84.85	2036.40		
1 Rodman (reinf.)	56.40	451.20	84.90	679.20		
1 Cement Finisher	49.95	399.60	73.45	587.60		
1 Grader, 30,000 Lbs.		1062.00		1168.20		
1 Paving Mach. & Equip.		2474.00		2721.40		
1 Concrete Pump, 110' Boom		470.75		517.83	41.74	45.91
96 L.H., Daily Totals		$8593.15		$11276.63	$89.51	$117.46

Crew B-26C

Crew No.	Bare Costs Hr.	Daily	Incl. Subs O&P Hr.	Daily	Cost Per Labor-Hour Bare Costs	Incl. O&P
1 Labor Foreman (outside)	$44.10	$352.80	$66.25	$530.00	$45.98	$68.89
6 Laborers	42.10	2020.80	63.25	3036.00		
1 Equip. Oper. (medium)	56.75	454.00	84.85	678.80		
1 Rodman (reinf.)	56.40	451.20	84.90	679.20		
1 Cement Finisher	49.95	399.60	73.45	587.60		
1 Paving Mach. & Equip.		2474.00		2721.40		
1 Concrete Saw		111.50		122.65	32.32	35.55
80 L.H., Daily Totals		$6263.90		$8355.65	$78.30	$104.45

Crew B-27

Crew No.	Bare Costs Hr.	Daily	Incl. Subs O&P Hr.	Daily	Cost Per Labor-Hour Bare Costs	Incl. O&P
1 Labor Foreman (outside)	$44.10	$352.80	$66.25	$530.00	$42.60	$64.00
3 Laborers	42.10	1010.40	63.25	1518.00		
1 Berm Machine		271.95		299.14	8.50	9.35
32 L.H., Daily Totals		$1635.15		$2347.15	$51.10	$73.35

Crew B-28

Crew No.	Bare Costs Hr.	Daily	Incl. Subs O&P Hr.	Daily	Cost Per Labor-Hour Bare Costs	Incl. O&P
2 Carpenters	$53.15	$850.40	$79.85	$1277.60	$49.47	$74.32
1 Laborer	42.10	336.80	63.25	506.00		
24 L.H., Daily Totals		$1187.20		$1783.60	$49.47	$74.32

Crew B-29

Crew No.	Bare Costs Hr.	Daily	Incl. Subs O&P Hr.	Daily	Cost Per Labor-Hour Bare Costs	Incl. O&P
1 Labor Foreman (outside)	$44.10	$352.80	$66.25	$530.00	$46.04	$69.04
4 Laborers	42.10	1347.20	63.25	2024.00		
1 Equip. Oper. (crane)	59.20	473.60	88.50	708.00		
1 Equip. Oper. (oiler)	50.55	404.40	75.55	604.40		
1 Gradall, 5/8 C.Y.		846.50		931.15	15.12	16.63
56 L.H., Daily Totals		$3424.50		$4797.55	$61.15	$85.67

Crew B-30

Crew No.	Bare Costs Hr.	Daily	Incl. Subs O&P Hr.	Daily	Cost Per Labor-Hour Bare Costs	Incl. O&P
1 Equip. Oper. (medium)	$56.75	$454.00	$84.85	$678.80	$51.55	$77.18
2 Truck Drivers (heavy)	48.95	783.20	73.35	1173.60		
1 Hyd. Excavator, 1.5 C.Y.		687.55		756.30		
2 Dump Trucks, 12 C.Y., 400 H.P.		1145.80		1260.38	76.39	84.03
24 L.H., Daily Totals		$3070.55		$3869.09	$127.94	$161.21

Crew B-31

Crew No.	Bare Costs Hr.	Daily	Incl. Subs O&P Hr.	Daily	Cost Per Labor-Hour Bare Costs	Incl. O&P
1 Labor Foreman (outside)	$44.10	$352.80	$66.25	$530.00	$44.71	$67.17
3 Laborers	42.10	1010.40	63.25	1518.00		
1 Carpenter	53.15	425.20	79.85	638.80		
1 Air Compressor, 250 cfm		201.50		221.65		
1 Sheeting Driver		7.35		8.09		
2 -50' Air Hoses, 1.5"		48.50		53.35	6.43	7.08
40 L.H., Daily Totals		$2045.75		$2969.89	$51.14	$74.25

Crew B-32

Crew No.	Bare Costs Hr.	Daily	Incl. Subs O&P Hr.	Daily	Cost Per Labor-Hour Bare Costs	Incl. O&P
1 Laborer	$42.10	$336.80	$63.25	$506.00	$53.09	$79.45
3 Equip. Oper. (medium)	56.75	1362.00	84.85	2036.40		
1 Grader, 30,000 Lbs.		1062.00		1168.20		
1 Tandem Roller, 10 Ton		236.75		260.43		
1 Dozer, 200 H.P.		1504.00		1654.40	87.59	96.34
32 L.H., Daily Totals		$4501.55		$5625.43	$140.67	$175.79

Crew B-32A

Crew No.	Bare Costs Hr.	Daily	Incl. Subs O&P Hr.	Daily	Cost Per Labor-Hour Bare Costs	Incl. O&P
1 Laborer	$42.10	$336.80	$63.25	$506.00	$51.87	$77.65
2 Equip. Oper. (medium)	56.75	908.00	84.85	1357.60		
1 Grader, 30,000 Lbs.		1062.00		1168.20		
1 Roller, Vibratory, 25 Ton		664.35		730.78	71.93	79.12
24 L.H., Daily Totals		$2971.15		$3762.59	$123.80	$156.77

Crew B-32B

Crew No.	Bare Costs Hr.	Daily	Incl. Subs O&P Hr.	Daily	Cost Per Labor-Hour Bare Costs	Incl. O&P
1 Laborer	$42.10	$336.80	$63.25	$506.00	$51.87	$77.65
2 Equip. Oper. (medium)	56.75	908.00	84.85	1357.60		
1 Dozer, 200 H.P.		1504.00		1654.40		
1 Roller, Vibratory, 25 Ton		664.35		730.78	90.35	99.38
24 L.H., Daily Totals		$3413.15		$4248.78	$142.21	$177.03

Crew B-32C

Crew No.	Bare Costs Hr.	Daily	Incl. Subs O&P Hr.	Daily	Cost Per Labor-Hour Bare Costs	Incl. O&P
1 Labor Foreman (outside)	$44.10	$352.80	$66.25	$530.00	$49.76	$74.55
2 Laborers	42.10	673.60	63.25	1012.00		
3 Equip. Oper. (medium)	56.75	1362.00	84.85	2036.40		
1 Grader, 30,000 Lbs.		1062.00		1168.20		
1 Tandem Roller, 10 Ton		236.75		260.43		
1 Dozer, 200 H.P.		1504.00		1654.40	58.39	64.23
48 L.H., Daily Totals		$5191.15		$6661.43	$108.15	$138.78

Crew B-33A

Crew No.	Bare Costs Hr.	Daily	Incl. Subs O&P Hr.	Daily	Cost Per Labor-Hour Bare Costs	Incl. O&P
1 Equip. Oper. (medium)	$56.75	$454.00	$84.85	$678.80	$52.56	$78.68
.5 Laborer	42.10	168.40	63.25	253.00		
.25 Equip. Oper. (medium)	56.75	113.50	84.85	169.70		
1 Scraper, Towed, 7 C.Y.		127.80		140.58		
1.25 Dozers, 300 H.P.		2205.00		2425.50	166.63	183.29
14 L.H., Daily Totals		$3068.70		$3667.58	$219.19	$261.97

Crew B-33B

Crew No.	Bare Costs Hr.	Daily	Incl. Subs O&P Hr.	Daily	Cost Per Labor-Hour Bare Costs	Incl. O&P
1 Equip. Oper. (medium)	$56.75	$454.00	$84.85	$678.80	$52.56	$78.68
.5 Laborer	42.10	168.40	63.25	253.00		
.25 Equip. Oper. (medium)	56.75	113.50	84.85	169.70		
1 Scraper, Towed, 10 C.Y.		159.70		175.67		
1.25 Dozers, 300 H.P.		2205.00		2425.50	168.91	185.80
14 L.H., Daily Totals		$3100.60		$3702.67	$221.47	$264.48

Crew B-33C

Crew No.	Bare Costs Hr.	Daily	Incl. Subs O&P Hr.	Daily	Cost Per Labor-Hour Bare Costs	Incl. O&P
1 Equip. Oper. (medium)	$56.75	$454.00	$84.85	$678.80	$52.56	$78.68
.5 Laborer	42.10	168.40	63.25	253.00		
.25 Equip. Oper. (medium)	56.75	113.50	84.85	169.70		
1 Scraper, Towed, 15 C.Y.		176.75		194.43		
1.25 Dozers, 300 H.P.		2205.00		2425.50	170.13	187.14
14 L.H., Daily Totals		$3117.65		$3721.43	$222.69	$265.82

Crew B-33D

Crew No.	Bare Costs Hr.	Daily	Incl. Subs O&P Hr.	Daily	Cost Per Labor-Hour Bare Costs	Incl. O&P
1 Equip. Oper. (medium)	$56.75	$454.00	$84.85	$678.80	$52.56	$78.68
.5 Laborer	42.10	168.40	63.25	253.00		
.25 Equip. Oper. (medium)	56.75	113.50	84.85	169.70		
1 S.P. Scraper, 14 C.Y.		2395.00		2634.50		
.25 Dozer, 300 H.P.		441.00		485.10	202.57	222.83
14 L.H., Daily Totals		$3571.90		$4221.10	$255.14	$301.51

Crew B-33E

Crew No.	Bare Costs Hr.	Daily	Incl. Subs O&P Hr.	Daily	Cost Per Labor-Hour Bare Costs	Incl. O&P
1 Equip. Oper. (medium)	$56.75	$454.00	$84.85	$678.80	$52.56	$78.68
.5 Laborer	42.10	168.40	63.25	253.00		
.25 Equip. Oper. (medium)	56.75	113.50	84.85	169.70		
1 S.P. Scraper, 21 C.Y.		2628.00		2890.80		
.25 Dozer, 300 H.P.		441.00		485.10	219.21	241.14
14 L.H., Daily Totals		$3804.90		$4477.40	$271.78	$319.81

For customer support on your Mechanical Costs with RSMeans Data, call 800.448.8182.

631

Crew B-33F

Crew No.	Bare Costs Hr.	Daily	Incl. Subs O&P Hr.	Daily	Cost Per Labor-Hour Bare Costs	Incl. O&P
1 Equip. Oper. (medium)	$56.75	$454.00	$84.85	$678.80	$52.56	$78.68
.5 Laborer	42.10	168.40	63.25	253.00		
.25 Equip. Oper. (medium)	56.75	113.50	84.85	169.70		
1 Elev. Scraper, 11 C.Y.		1133.00		1246.30		
.25 Dozer, 300 H.P.		441.00		485.10	112.43	123.67
14 L.H., Daily Totals		$2309.90		$2832.90	$164.99	$202.35

Crew B-33G

Crew No.	Bare Costs Hr.	Daily	Incl. Subs O&P Hr.	Daily	Cost Per Labor-Hour Bare Costs	Incl. O&P
1 Equip. Oper. (medium)	$56.75	$454.00	$84.85	$678.80	$52.56	$78.68
.5 Laborer	42.10	168.40	63.25	253.00		
.25 Equip. Oper. (medium)	56.75	113.50	84.85	169.70		
1 Elev. Scraper, 22 C.Y.		1884.00		2072.40		
.25 Dozer, 300 H.P.		441.00		485.10	166.07	182.68
14 L.H., Daily Totals		$3060.90		$3659.00	$218.64	$261.36

Crew B-33H

Crew No.	Bare Costs Hr.	Daily	Incl. Subs O&P Hr.	Daily	Cost Per Labor-Hour Bare Costs	Incl. O&P
.5 Laborer	$42.10	$168.40	$63.25	$253.00	$52.56	$78.68
1 Equipment Operator (med.)	56.75	454.00	84.85	678.80		
.25 Equipment Operator (med.)	56.75	113.50	84.85	169.70		
1 S.P. Scraper, 44 C.Y.		4644.00		5108.40		
.25 Dozer, 410 H.P.		694.25		763.67	381.30	419.43
14 L.H., Daily Totals		$6074.15		$6973.57	$433.87	$498.11

Crew B-33J

Crew No.	Bare Costs Hr.	Daily	Incl. Subs O&P Hr.	Daily	Cost Per Labor-Hour Bare Costs	Incl. O&P
1 Equipment Operator (med.)	$56.75	$454.00	$84.85	$678.80	$56.75	$84.85
1 S.P. Scraper, 14 C.Y.		2395.00		2634.50	299.38	329.31
8 L.H., Daily Totals		$2849.00		$3313.30	$356.13	$414.16

Crew B-33K

Crew No.	Bare Costs Hr.	Daily	Incl. Subs O&P Hr.	Daily	Cost Per Labor-Hour Bare Costs	Incl. O&P
1 Equipment Operator (med.)	$56.75	$454.00	$84.85	$678.80	$52.56	$78.68
.25 Equipment Operator (med.)	56.75	113.50	84.85	169.70		
.5 Laborer	42.10	168.40	63.25	253.00		
1 S.P. Scraper, 31 C.Y.		3667.00		4033.70		
.25 Dozer, 410 H.P.		694.25		763.67	311.52	342.67
14 L.H., Daily Totals		$5097.15		$5898.88	$364.08	$421.35

Crew B-34A

Crew No.	Bare Costs Hr.	Daily	Incl. Subs O&P Hr.	Daily	Cost Per Labor-Hour Bare Costs	Incl. O&P
1 Truck Driver (heavy)	$48.95	$391.60	$73.35	$586.80	$48.95	$73.35
1 Dump Truck, 8 C.Y., 220 H.P.		428.05		470.86	53.51	58.86
8 L.H., Daily Totals		$819.65		$1057.66	$102.46	$132.21

Crew B-34B

Crew No.	Bare Costs Hr.	Daily	Incl. Subs O&P Hr.	Daily	Cost Per Labor-Hour Bare Costs	Incl. O&P
1 Truck Driver (heavy)	$48.95	$391.60	$73.35	$586.80	$48.95	$73.35
1 Dump Truck, 12 C.Y., 400 H.P.		572.90		630.19	71.61	78.77
8 L.H., Daily Totals		$964.50		$1216.99	$120.56	$152.12

Crew B-34C

Crew No.	Bare Costs Hr.	Daily	Incl. Subs O&P Hr.	Daily	Cost Per Labor-Hour Bare Costs	Incl. O&P
1 Truck Driver (heavy)	$48.95	$391.60	$73.35	$586.80	$48.95	$73.35
1 Truck Tractor, 6x4, 380 H.P.		493.25		542.58		
1 Dump Trailer, 16.5 C.Y.		136.70		150.37	78.74	86.62
8 L.H., Daily Totals		$1021.55		$1279.74	$127.69	$159.97

Crew B-34D

Crew No.	Bare Costs Hr.	Daily	Incl. Subs O&P Hr.	Daily	Cost Per Labor-Hour Bare Costs	Incl. O&P
1 Truck Driver (heavy)	$48.95	$391.60	$73.35	$586.80	$48.95	$73.35
1 Truck Tractor, 6x4, 380 H.P.		493.25		542.58		
1 Dump Trailer, 20 C.Y.		151.70		166.87	80.62	88.68
8 L.H., Daily Totals		$1036.55		$1296.24	$129.57	$162.03

Crew B-34E

Crew No.	Bare Costs Hr.	Daily	Incl. Subs O&P Hr.	Daily	Cost Per Labor-Hour Bare Costs	Incl. O&P
1 Truck Driver (heavy)	$48.95	$391.60	$73.35	$586.80	$48.95	$73.35
1 Dump Truck, Off Hwy., 25 Ton		1379.00		1516.90	172.38	189.61
8 L.H., Daily Totals		$1770.60		$2103.70	$221.32	$262.96

Crew B-34F

Crew No.	Bare Costs Hr.	Daily	Incl. Subs O&P Hr.	Daily	Cost Per Labor-Hour Bare Costs	Incl. O&P
1 Truck Driver (heavy)	$48.95	$391.60	$73.35	$586.80	$48.95	$73.35
1 Dump Truck, Off Hwy., 35 Ton		935.20		1028.72	116.90	128.59
8 L.H., Daily Totals		$1326.80		$1615.52	$165.85	$201.94

Crew B-34G

Crew No.	Bare Costs Hr.	Daily	Incl. Subs O&P Hr.	Daily	Cost Per Labor-Hour Bare Costs	Incl. O&P
1 Truck Driver (heavy)	$48.95	$391.60	$73.35	$586.80	$48.95	$73.35
1 Dump Truck, Off Hwy., 50 Ton		1771.00		1948.10	221.38	243.51
8 L.H., Daily Totals		$2162.60		$2534.90	$270.32	$316.86

Crew B-34H

Crew No.	Bare Costs Hr.	Daily	Incl. Subs O&P Hr.	Daily	Cost Per Labor-Hour Bare Costs	Incl. O&P
1 Truck Driver (heavy)	$48.95	$391.60	$73.35	$586.80	$48.95	$73.35
1 Dump Truck, Off Hwy., 65 Ton		1915.00		2106.50	239.38	263.31
8 L.H., Daily Totals		$2306.60		$2693.30	$288.32	$336.66

Crew B-34I

Crew No.	Bare Costs Hr.	Daily	Incl. Subs O&P Hr.	Daily	Cost Per Labor-Hour Bare Costs	Incl. O&P
1 Truck Driver (heavy)	$48.95	$391.60	$73.35	$586.80	$48.95	$73.35
1 Dump Truck, 18 C.Y., 450 H.P.		713.80		785.18	89.22	98.15
8 L.H., Daily Totals		$1105.40		$1371.98	$138.18	$171.50

Crew B-34J

Crew No.	Bare Costs Hr.	Daily	Incl. Subs O&P Hr.	Daily	Cost Per Labor-Hour Bare Costs	Incl. O&P
1 Truck Driver (heavy)	$48.95	$391.60	$73.35	$586.80	$48.95	$73.35
1 Dump Truck, Off Hwy., 100 Ton		2736.00		3009.60	342.00	376.20
8 L.H., Daily Totals		$3127.60		$3596.40	$390.95	$449.55

Crew B-34K

Crew No.	Bare Costs Hr.	Daily	Incl. Subs O&P Hr.	Daily	Cost Per Labor-Hour Bare Costs	Incl. O&P
1 Truck Driver (heavy)	$48.95	$391.60	$73.35	$586.80	$48.95	$73.35
1 Truck Tractor, 6x4, 450 H.P.		601.75		661.92		
1 Lowbed Trailer, 75 Ton		255.05		280.56	107.10	117.81
8 L.H., Daily Totals		$1248.40		$1529.28	$156.05	$191.16

Crew B-34L

Crew No.	Bare Costs Hr.	Daily	Incl. Subs O&P Hr.	Daily	Cost Per Labor-Hour Bare Costs	Incl. O&P
1 Equip. Oper. (light)	$53.00	$424.00	$79.20	$633.60	$53.00	$79.20
1 Flatbed Truck, Gas, 1.5 Ton		196.15		215.76	24.52	26.97
8 L.H., Daily Totals		$620.15		$849.37	$77.52	$106.17

Crew B-34M

Crew No.	Bare Costs Hr.	Daily	Incl. Subs O&P Hr.	Daily	Cost Per Labor-Hour Bare Costs	Incl. O&P
1 Equip. Oper. (light)	$53.00	$424.00	$79.20	$633.60	$53.00	$79.20
1 Flatbed Truck, Gas, 3 Ton		820.40		902.44	102.55	112.81
8 L.H., Daily Totals		$1244.40		$1536.04	$155.55	$192.01

Crew B-34N

Crew No.	Bare Costs Hr.	Daily	Incl. Subs O&P Hr.	Daily	Cost Per Labor-Hour Bare Costs	Incl. O&P
1 Truck Driver (heavy)	$48.95	$391.60	$73.35	$586.80	$52.85	$79.10
1 Equip. Oper. (medium)	56.75	454.00	84.85	678.80		
1 Truck Tractor, 6x4, 380 H.P.		493.25		542.58		
1 Flatbed Trailer, 40 Ton		186.20		204.82	42.47	46.71
16 L.H., Daily Totals		$1525.05		$2012.99	$95.32	$125.81

Crew B-34P

Crew No.	Bare Costs Hr.	Daily	Incl. Subs O&P Hr.	Daily	Cost Per Labor-Hour Bare Costs	Incl. O&P
1 Pipe Fitter	$65.55	$524.40	$98.10	$784.80	$56.52	$84.58
1 Truck Driver (light)	47.25	378.00	70.80	566.40		
1 Equip. Oper. (medium)	56.75	454.00	84.85	678.80		
1 Flatbed Truck, Gas, 3 Ton		820.40		902.44		
1 Backhoe Loader, 48 H.P.		213.75		235.13	43.09	47.40
24 L.H., Daily Totals		$2390.55		$3167.57	$99.61	$131.98

Crew B-34Q

Crew No.	Bare Costs Hr.	Daily	Incl. Subs O&P Hr.	Daily	Cost Per Labor-Hour Bare Costs	Incl. O&P
1 Pipe Fitter	$65.55	$524.40	$98.10	$784.80	$57.33	$85.80
1 Truck Driver (light)	47.25	378.00	70.80	566.40		
1 Equip. Oper. (crane)	59.20	473.60	88.50	708.00		
1 Flatbed Trailer, 25 Ton		135.55		149.10		
1 Dump Truck, 8 C.Y., 220 H.P.		428.05		470.86		
1 Hyd. Crane, 25 Ton		580.85		638.93	47.69	52.45
24 L.H., Daily Totals		$2520.45		$3318.09	$105.02	$138.25

Crew B-34R

Crew No.	Bare Costs Hr.	Daily	Incl. Subs O&P Hr.	Daily	Cost Per Labor-Hour Bare Costs	Incl. O&P
1 Pipe Fitter	$65.55	$524.40	$98.10	$784.80	$57.33	$85.80
1 Truck Driver (light)	47.25	378.00	70.80	566.40		
1 Equip. Oper. (crane)	59.20	473.60	88.50	708.00		
1 Flatbed Trailer, 25 Ton		135.55		149.10		
1 Dump Truck, 8 C.Y., 220 H.P.		428.05		470.86		
1 Hyd. Crane, 25 Ton		580.85		638.93		
1 Hyd. Excavator, 1 C.Y.		795.30		874.83	80.82	88.91
24 L.H., Daily Totals		$3315.75		$4192.93	$138.16	$174.71

Crew B-34S

Crew No.	Bare Costs Hr.	Daily	Incl. Subs O&P Hr.	Daily	Cost Per Labor-Hour Bare Costs	Incl. O&P
2 Pipe Fitters	$65.55	$1048.80	$98.10	$1569.60	$59.81	$89.51
1 Truck Driver (heavy)	48.95	391.60	73.35	586.80		
1 Equip. Oper. (crane)	59.20	473.60	88.50	708.00		
1 Flatbed Trailer, 40 Ton		186.20		204.82		
1 Truck Tractor, 6x4, 380 H.P.		493.25		542.58		
1 Hyd. Crane, 80 Ton		1486.00		1634.60		
1 Hyd. Excavator, 2 C.Y.		951.30		1046.43	97.40	107.14
32 L.H., Daily Totals		$5030.75		$6292.82	$157.21	$196.65

Crew B-34T

Crew No.	Bare Costs Hr.	Daily	Incl. Subs O&P Hr.	Daily	Cost Per Labor-Hour Bare Costs	Incl. O&P
2 Pipe Fitters	$65.55	$1048.80	$98.10	$1569.60	$59.81	$89.51
1 Truck Driver (heavy)	48.95	391.60	73.35	586.80		
1 Equip. Oper. (crane)	59.20	473.60	88.50	708.00		
1 Flatbed Trailer, 40 Ton		186.20		204.82		
1 Truck Tractor, 6x4, 380 H.P.		493.25		542.58		
1 Hyd. Crane, 80 Ton		1486.00		1634.60	67.67	74.44
32 L.H., Daily Totals		$4079.45		$5246.40	$127.48	$163.95

Crew B-34U

Crew No.	Bare Costs Hr.	Daily	Incl. Subs O&P Hr.	Daily	Cost Per Labor-Hour Bare Costs	Incl. O&P
1 Truck Driver (heavy)	$48.95	$391.60	$73.35	$586.80	$50.98	$76.28
1 Equip. Oper. (light)	53.00	424.00	79.20	633.60		
1 Truck Tractor, 220 H.P.		307.10		337.81		
1 Flatbed Trailer, 25 Ton		135.55		149.10	27.67	30.43
16 L.H., Daily Totals		$1258.25		$1707.32	$78.64	$106.71

Crew B-34V

Crew No.	Bare Costs Hr.	Daily	Incl. Subs O&P Hr.	Daily	Cost Per Labor-Hour Bare Costs	Incl. O&P
1 Truck Driver (heavy)	$48.95	$391.60	$73.35	$586.80	$53.72	$80.35
1 Equip. Oper. (crane)	59.20	473.60	88.50	708.00		
1 Equip. Oper. (light)	53.00	424.00	79.20	633.60		
1 Truck Tractor, 6x4, 450 H.P.		601.75		661.92		
1 Equipment Trailer, 50 Ton		205.05		225.56		
1 Pickup Truck, 4x4, 3/4 Ton		175.85		193.44	40.94	45.04
24 L.H., Daily Totals		$2271.85		$3009.32	$94.66	$125.39

Crew B-34W

Crew No.	Bare Costs Hr.	Daily	Incl. Subs O&P Hr.	Daily	Cost Per Labor-Hour Bare Costs	Incl. O&P
5 Truck Drivers (heavy)	$48.95	$1958.00	$73.35	$2934.00	$51.60	$77.27
2 Equip. Opers. (crane)	59.20	947.20	88.50	1416.00		
1 Equip. Oper. (mechanic)	59.15	473.20	88.40	707.20		
1 Laborer	42.10	336.80	63.25	506.00		
4 Truck Tractors, 6x4, 380 H.P.		1973.00		2170.30		
2 Equipment Trailers, 50 Ton		410.10		451.11		
2 Flatbed Trailers, 40 Ton		372.40		409.64		
1 Pickup Truck, 4x4, 3/4 Ton		175.85		193.44		
1 S.P. Crane, 4x4, 20 Ton		582.20		640.42	48.80	53.68
72 L.H., Daily Totals		$7228.75		$9428.10	$100.40	$130.95

Crew B-35

Crew No.	Bare Costs Hr.	Daily	Incl. Subs O&P Hr.	Daily	Cost Per Labor-Hour Bare Costs	Incl. O&P
1 Labor Foreman (outside)	$44.10	$352.80	$66.25	$530.00	$52.54	$78.83
1 Skilled Worker	54.85	438.80	82.95	663.60		
1 Welder (plumber)	64.45	515.60	96.45	771.60		
1 Laborer	42.10	336.80	63.25	506.00		
1 Equip. Oper. (crane)	59.20	473.60	88.50	708.00		
1 Equip. Oper. (oiler)	50.55	404.40	75.55	604.40		
1 Welder, Electric, 300 amp		106.40		117.04		
1 Hyd. Excavator, .75 C.Y.		694.25		763.67	16.68	18.35
48 L.H., Daily Totals		$3322.65		$4664.31	$69.22	$97.17

Crew B-35A

Crew No.	Bare Costs Hr.	Daily	Incl. Subs O&P Hr.	Daily	Cost Per Labor-Hour Bare Costs	Incl. O&P
1 Labor Foreman (outside)	$44.10	$352.80	$66.25	$530.00	$51.05	$76.60
2 Laborers	42.10	673.60	63.25	1012.00		
1 Skilled Worker	54.85	438.80	82.95	663.60		
1 Welder (plumber)	64.45	515.60	96.45	771.60		
1 Equip. Oper. (crane)	59.20	473.60	88.50	708.00		
1 Equip. Oper. (oiler)	50.55	404.40	75.55	604.40		
1 Welder, Gas Engine, 300 amp		147.00		161.70		
1 Crawler Crane, 75 Ton		2002.00		2202.20	38.38	42.21
56 L.H., Daily Totals		$5007.80		$6653.50	$89.42	$118.81

Crew B-36

Crew No.	Bare Costs Hr.	Daily	Incl. Subs O&P Hr.	Daily	Cost Per Labor-Hour Bare Costs	Incl. O&P
1 Labor Foreman (outside)	$44.10	$352.80	$66.25	$530.00	$48.36	$72.49
2 Laborers	42.10	673.60	63.25	1012.00		
2 Equip. Oper. (medium)	56.75	908.00	84.85	1357.60		
1 Dozer, 200 H.P.		1504.00		1654.40		
1 Aggregate Spreader		65.75		72.33		
1 Tandem Roller, 10 Ton		236.75		260.43	45.16	49.68
40 L.H., Daily Totals		$3740.90		$4886.75	$93.52	$122.17

Crew B-36A

Crew No.	Bare Costs Hr.	Daily	Incl. Subs O&P Hr.	Daily	Cost Per Labor-Hour Bare Costs	Incl. O&P
1 Labor Foreman (outside)	$44.10	$352.80	$66.25	$530.00	$50.76	$76.02
2 Laborers	42.10	673.60	63.25	1012.00		
4 Equip. Oper. (medium)	56.75	1816.00	84.85	2715.20		
1 Dozer, 200 H.P.		1504.00		1654.40		
1 Aggregate Spreader		65.75		72.33		
1 Tandem Roller, 10 Ton		236.75		260.43		
1 Roller, Pneum. Whl., 12 Ton		345.75		380.32	38.43	42.28
56 L.H., Daily Totals		$4994.65		$6624.68	$89.19	$118.30

For customer support on your Mechanical Costs with RSMeans Data, call 800.448.8182.

633

Crew No.	Bare Costs		Incl. Subs O&P		Cost Per Labor-Hour	
					Bare Costs	Incl. O&P
Crew B-36B	Hr.	Daily	Hr.	Daily		
1 Labor Foreman (outside)	$44.10	$352.80	$66.25	$530.00	$50.53	$75.69
2 Laborers	42.10	673.60	63.25	1012.00		
4 Equip. Oper. (medium)	56.75	1816.00	84.85	2715.20		
1 Truck Driver (heavy)	48.95	391.60	73.35	586.80		
1 Grader, 30,000 Lbs.		1062.00		1168.20		
1 F.E. Loader, Crl. 1.5 C.Y.		659.75		725.73		
1 Dozer, 300 H.P.		1764.00		1940.40		
1 Roller, Vibratory, 25 Ton		664.35		730.78		
1 Truck Tractor, 6x4, 450 H.P.		601.75		661.92		
1 Water Tank Trailer, 5000 Gal.		152.25		167.47	76.63	84.29
64 L.H., Daily Totals		$8138.10		$10238.51	$127.16	$159.98
Crew B-36C	Hr.	Daily	Hr.	Daily	Bare Costs	Incl. O&P
1 Labor Foreman (outside)	$44.10	$352.80	$66.25	$530.00	$52.66	$78.83
3 Equip. Oper. (medium)	56.75	1362.00	84.85	2036.40		
1 Truck Driver (heavy)	48.95	391.60	73.35	586.80		
1 Grader, 30,000 Lbs.		1062.00		1168.20		
1 Dozer, 300 H.P.		1764.00		1940.40		
1 Roller, Vibratory, 25 Ton		664.35		730.78		
1 Truck Tractor, 6x4, 450 H.P.		601.75		661.92		
1 Water Tank Trailer, 5000 Gal.		152.25		167.47	106.11	116.72
40 L.H., Daily Totals		$6350.75		$7821.98	$158.77	$195.55
Crew B-36D	Hr.	Daily	Hr.	Daily	Bare Costs	Incl. O&P
1 Labor Foreman (outside)	$44.10	$352.80	$66.25	$530.00	$53.59	$80.20
3 Equip. Oper. (medium)	56.75	1362.00	84.85	2036.40		
1 Grader, 30,000 Lbs.		1062.00		1168.20		
1 Dozer, 300 H.P.		1764.00		1940.40		
1 Roller, Vibratory, 25 Ton		664.35		730.78	109.07	119.98
32 L.H., Daily Totals		$5205.15		$6405.78	$162.66	$200.18
Crew B-37	Hr.	Daily	Hr.	Daily	Bare Costs	Incl. O&P
1 Labor Foreman (outside)	$44.10	$352.80	$66.25	$530.00	$44.25	$66.41
4 Laborers	42.10	1347.20	63.25	2024.00		
1 Equip. Oper. (light)	53.00	424.00	79.20	633.60		
1 Tandem Roller, 5 Ton		255.65		281.21	5.33	5.86
48 L.H., Daily Totals		$2379.65		$3468.82	$49.58	$72.27
Crew B-37A	Hr.	Daily	Hr.	Daily	Bare Costs	Incl. O&P
2 Laborers	$42.10	$673.60	$63.25	$1012.00	$43.82	$65.77
1 Truck Driver (light)	47.25	378.00	70.80	566.40		
1 Flatbed Truck, Gas, 1.5 Ton		196.15		215.76		
1 Tar Kettle, T.M.		156.20		171.82	14.68	16.15
24 L.H., Daily Totals		$1403.95		$1965.98	$58.50	$81.92
Crew B-37B	Hr.	Daily	Hr.	Daily	Bare Costs	Incl. O&P
3 Laborers	$42.10	$1010.40	$63.25	$1518.00	$43.39	$65.14
1 Truck Driver (light)	47.25	378.00	70.80	566.40		
1 Flatbed Truck, Gas, 1.5 Ton		196.15		215.76		
1 Tar Kettle, T.M.		156.20		171.82	11.01	12.11
32 L.H., Daily Totals		$1740.75		$2471.99	$54.40	$77.25
Crew B-37C	Hr.	Daily	Hr.	Daily	Bare Costs	Incl. O&P
2 Laborers	$42.10	$673.60	$63.25	$1012.00	$44.67	$67.03
2 Truck Drivers (light)	47.25	756.00	70.80	1132.80		
2 Flatbed Trucks, Gas, 1.5 Ton		392.30		431.53		
1 Tar Kettle, T.M.		156.20		171.82	17.14	18.85
32 L.H., Daily Totals		$1978.10		$2748.15	$61.82	$85.88

Crew No.	Bare Costs		Incl. Subs O&P		Cost Per Labor-Hour	
					Bare Costs	Incl. O&P
Crew B-37D	Hr.	Daily	Hr.	Daily		
1 Laborer	$42.10	$336.80	$63.25	$506.00	$44.67	$67.03
1 Truck Driver (light)	47.25	378.00	70.80	566.40		
1 Pickup Truck, 3/4 Ton		110.85		121.94	6.93	7.62
16 L.H., Daily Totals		$825.65		$1194.34	$51.60	$74.65
Crew B-37E	Hr.	Daily	Hr.	Daily	Bare Costs	Incl. O&P
3 Laborers	$42.10	$1010.40	$63.25	$1518.00	$47.22	$70.77
1 Equip. Oper. (light)	53.00	424.00	79.20	633.60		
1 Equip. Oper. (medium)	56.75	454.00	84.85	678.80		
2 Truck Drivers (light)	47.25	756.00	70.80	1132.80		
4 Barrels w/ Flasher		16.40		18.04		
1 Concrete Saw		111.50		122.65		
1 Rotary Hammer Drill		51.70		56.87		
1 Hammer Drill Bit		25.25		27.77		
1 Loader, Skid Steer, 30 H.P.		177.55		195.31		
1 Conc. Hammer Attach.		118.40		130.24		
1 Vibrating Plate, Gas, 18"		31.55		34.70		
2 Flatbed Trucks, Gas, 1.5 Ton		392.30		431.53	16.51	18.16
56 L.H., Daily Totals		$3569.05		$4980.31	$63.73	$88.93
Crew B-37F	Hr.	Daily	Hr.	Daily	Bare Costs	Incl. O&P
3 Laborers	$42.10	$1010.40	$63.25	$1518.00	$43.39	$65.14
1 Truck Driver (light)	47.25	378.00	70.80	566.40		
4 Barrels w/ Flasher		16.40		18.04		
1 Concrete Mixer, 10 C.F.		140.30		154.33		
1 Air Compressor, 60 cfm		152.95		168.25		
1 -50' Air Hose, 3/4"		7.75		8.53		
1 Spade (Chipper)		8.45		9.29		
1 Flatbed Truck, Gas, 1.5 Ton		196.15		215.76	16.31	17.94
32 L.H., Daily Totals		$1910.40		$2658.60	$59.70	$83.08
Crew B-37G	Hr.	Daily	Hr.	Daily	Bare Costs	Incl. O&P
1 Labor Foreman (outside)	$44.10	$352.80	$66.25	$530.00	$44.25	$66.41
4 Laborers	42.10	1347.20	63.25	2024.00		
1 Equip. Oper. (light)	53.00	424.00	79.20	633.60		
1 Berm Machine		271.95		299.14		
1 Tandem Roller, 5 Ton		255.65		281.21	10.99	12.09
48 L.H., Daily Totals		$2651.60		$3767.96	$55.24	$78.50
Crew B-37H	Hr.	Daily	Hr.	Daily	Bare Costs	Incl. O&P
1 Labor Foreman (outside)	$44.10	$352.80	$66.25	$530.00	$44.25	$66.41
4 Laborers	42.10	1347.20	63.25	2024.00		
1 Equip. Oper. (light)	53.00	424.00	79.20	633.60		
1 Tandem Roller, 5 Ton		255.65		281.21		
1 Flatbed Truck, Gas, 1.5 Ton		196.15		215.76		
1 Tar Kettle, T.M.		156.20		171.82	12.67	13.93
48 L.H., Daily Totals		$2732.00		$3856.40	$56.92	$80.34

Crew B-37I

Crew No.	Bare Costs Hr.	Daily	Incl. Subs O&P Hr.	Daily	Cost Per Labor-Hour Bare Costs	Incl. O&P
3 Laborers	$42.10	$1010.40	$63.25	$1518.00	$47.22	$70.77
1 Equip. Oper. (light)	53.00	424.00	79.20	633.60		
1 Equip. Oper. (medium)	56.75	454.00	84.85	678.80		
2 Truck Drivers (light)	47.25	756.00	70.80	1132.80		
4 Barrels w/ Flasher		16.40		18.04		
1 Concrete Saw		111.50		122.65		
1 Rotary Hammer Drill		51.70		56.87		
1 Hammer Drill Bit		25.25		27.77		
1 Air Compressor, 60 cfm		152.95		168.25		
1 -50' Air Hose, 3/4"		7.75		8.53		
1 Spade (Chipper)		8.45		9.29		
1 Loader, Skid Steer, 30 H.P.		177.55		195.31		
1 Conc. Hammer Attach.		118.40		130.24		
1 Concrete Mixer, 10 C.F.		140.30		154.33		
1 Vibrating Plate, Gas, 18"		31.55		34.70		
2 Flatbed Trucks, Gas, 1.5 Ton		392.30		431.53	22.04	24.24
56 L.H., Daily Totals		$3878.50		$5320.71	$69.26	$95.01

Crew B-37J

Crew No.	Bare Costs Hr.	Daily	Incl. Subs O&P Hr.	Daily	Cost Per Labor-Hour Bare Costs	Incl. O&P
1 Labor Foreman (outside)	$44.10	$352.80	$66.25	$530.00	$44.25	$66.41
4 Laborers	42.10	1347.20	63.25	2024.00		
1 Equip. Oper. (light)	53.00	424.00	79.20	633.60		
1 Air Compressor, 60 cfm		152.95		168.25		
1 -50' Air Hose, 3/4"		7.75		8.53		
2 Concrete Mixers, 10 C.F.		280.60		308.66		
2 Flatbed Trucks, Gas, 1.5 Ton		392.30		431.53		
1 Shot Blaster, 20"		206.20		226.82	21.66	23.83
48 L.H., Daily Totals		$3163.80		$4331.38	$65.91	$90.24

Crew B-37K

Crew No.	Bare Costs Hr.	Daily	Incl. Subs O&P Hr.	Daily	Cost Per Labor-Hour Bare Costs	Incl. O&P
1 Labor Foreman (outside)	$44.10	$352.80	$66.25	$530.00	$44.25	$66.41
4 Laborers	42.10	1347.20	63.25	2024.00		
1 Equip. Oper. (light)	53.00	424.00	79.20	633.60		
1 Air Compressor, 60 cfm		152.95		168.25		
1 -50' Air Hose, 3/4"		7.75		8.53		
2 Flatbed Trucks, Gas, 1.5 Ton		392.30		431.53		
1 Shot Blaster, 20"		206.20		226.82	15.82	17.40
48 L.H., Daily Totals		$2883.20		$4022.72	$60.07	$83.81

Crew B-38

Crew No.	Bare Costs Hr.	Daily	Incl. Subs O&P Hr.	Daily	Cost Per Labor-Hour Bare Costs	Incl. O&P
1 Labor Foreman (outside)	$44.10	$352.80	$66.25	$530.00	$47.61	$71.36
2 Laborers	42.10	673.60	63.25	1012.00		
1 Equip. Oper. (light)	53.00	424.00	79.20	633.60		
1 Equip. Oper. (medium)	56.75	454.00	84.85	678.80		
1 Backhoe Loader, 48 H.P.		213.75		235.13		
1 Hyd. Hammer (1200 lb.)		175.15		192.66		
1 F.E. Loader, W.M., 4 C.Y.		789.35		868.28		
1 Pvmt. Rem. Bucket		63.05		69.36	31.03	34.14
40 L.H., Daily Totals		$3145.70		$4219.83	$78.64	$105.50

Crew B-39

Crew No.	Bare Costs Hr.	Daily	Incl. Subs O&P Hr.	Daily	Cost Per Labor-Hour Bare Costs	Incl. O&P
1 Labor Foreman (outside)	$44.10	$352.80	$66.25	$530.00	$44.25	$66.41
4 Laborers	42.10	1347.20	63.25	2024.00		
1 Equip. Oper. (light)	53.00	424.00	79.20	633.60		
1 Air Compressor, 250 cfm		201.50		221.65		
2 Breakers, Pavement, 60 lb.		107.00		117.70		
2 -50' Air Hoses, 1.5"		48.50		53.35	7.44	8.18
48 L.H., Daily Totals		$2481.00		$3580.30	$51.69	$74.59

Crew B-40

Crew No.	Bare Costs Hr.	Daily	Incl. Subs O&P Hr.	Daily	Cost Per Labor-Hour Bare Costs	Incl. O&P
1 Pile Driver Foreman (outside)	$56.20	$449.60	$87.30	$698.40	$55.24	$84.58
4 Pile Drivers	54.20	1734.40	84.20	2694.40		
2 Equip. Oper. (crane)	59.20	947.20	88.50	1416.00		
1 Equip. Oper. (oiler)	50.55	404.40	75.55	604.40		
1 Crawler Crane, 40 Ton		1206.00		1326.60		
1 Vibratory Hammer & Gen.		2271.00		2498.10	54.33	59.76
64 L.H., Daily Totals		$7012.60		$9237.90	$109.57	$144.34

Crew B-40B

Crew No.	Bare Costs Hr.	Daily	Incl. Subs O&P Hr.	Daily	Cost Per Labor-Hour Bare Costs	Incl. O&P
1 Labor Foreman (outside)	$44.10	$352.80	$66.25	$530.00	$46.69	$70.01
3 Laborers	42.10	1010.40	63.25	1518.00		
1 Equip. Oper. (crane)	59.20	473.60	88.50	708.00		
1 Equip. Oper. (oiler)	50.55	404.40	75.55	604.40		
1 Lattice Boom Crane, 40 Ton		2028.00		2230.80	42.25	46.48
48 L.H., Daily Totals		$4269.20		$5591.20	$88.94	$116.48

Crew B-41

Crew No.	Bare Costs Hr.	Daily	Incl. Subs O&P Hr.	Daily	Cost Per Labor-Hour Bare Costs	Incl. O&P
1 Labor Foreman (outside)	$44.10	$352.80	$66.25	$530.00	$43.63	$65.50
4 Laborers	42.10	1347.20	63.25	2024.00		
.25 Equip. Oper. (crane)	59.20	118.40	88.50	177.00		
.25 Equip. Oper. (oiler)	50.55	101.10	75.55	151.10		
.25 Crawler Crane, 40 Ton		301.50		331.65	6.85	7.54
44 L.H., Daily Totals		$2221.00		$3213.75	$50.48	$73.04

Crew B-42

Crew No.	Bare Costs Hr.	Daily	Incl. Subs O&P Hr.	Daily	Cost Per Labor-Hour Bare Costs	Incl. O&P
1 Labor Foreman (outside)	$44.10	$352.80	$66.25	$530.00	$47.46	$71.67
4 Laborers	42.10	1347.20	63.25	2024.00		
1 Equip. Oper. (crane)	59.20	473.60	88.50	708.00		
1 Equip. Oper. (oiler)	50.55	404.40	75.55	604.40		
1 Welder	57.45	459.60	90.10	720.80		
1 Hyd. Crane, 25 Ton		580.85		638.93		
1 Welder, Gas Engine, 300 amp		147.00		161.70		
1 Horz. Boring Csg. Mch.		314.25		345.68	16.28	17.91
64 L.H., Daily Totals		$4079.70		$5733.51	$63.75	$89.59

Crew B-43

Crew No.	Bare Costs Hr.	Daily	Incl. Subs O&P Hr.	Daily	Cost Per Labor-Hour Bare Costs	Incl. O&P
1 Labor Foreman (outside)	$44.10	$352.80	$66.25	$530.00	$46.69	$70.01
3 Laborers	42.10	1010.40	63.25	1518.00		
1 Equip. Oper. (crane)	59.20	473.60	88.50	708.00		
1 Equip. Oper. (oiler)	50.55	404.40	75.55	604.40		
1 Drill Rig, Truck-Mounted		771.10		848.21	16.06	17.67
48 L.H., Daily Totals		$3012.30		$4208.61	$62.76	$87.68

Crew B-44

Crew No.	Bare Costs Hr.	Daily	Incl. Subs O&P Hr.	Daily	Cost Per Labor-Hour Bare Costs	Incl. O&P
1 Pile Driver Foreman (outside)	$56.20	$449.60	$87.30	$698.40	$54.19	$83.04
4 Pile Drivers	54.20	1734.40	84.20	2694.40		
2 Equip. Oper. (crane)	59.20	947.20	88.50	1416.00		
1 Laborer	42.10	336.80	63.25	506.00		
1 Crawler Crane, 40 Ton		1206.00		1326.60		
1 Lead, 60' High		209.25		230.18		
1 Hammer, Diesel, 15K ft.-lbs.		617.05		678.76	31.75	34.93
64 L.H., Daily Totals		$5500.30		$7550.33	$85.94	$117.97

Crew B-45

Crew No.	Bare Costs Hr.	Daily	Incl. Subs O&P Hr.	Daily	Cost Per Labor-Hour Bare Costs	Incl. O&P
1 Equip. Oper. (medium)	$56.75	$454.00	$84.85	$678.80	$52.85	$79.10
1 Truck Driver (heavy)	48.95	391.60	73.35	586.80		
1 Dist. Tanker, 3000 Gallon		330.15		363.17		
1 Truck Tractor, 6x4, 380 H.P.		493.25		542.58	51.46	56.61
16 L.H., Daily Totals		$1669.00		$2171.34	$104.31	$135.71

For customer support on your Mechanical Costs with RSMeans Data, call 800.448.8182.

635

Crews - Standard

Crew No.	Bare Costs		Incl. Subs O&P		Cost Per Labor-Hour	

Crew B-46

	Hr.	Daily	Hr.	Daily	Bare Costs	Incl. O&P
1 Pile Driver Foreman (outside)	$56.20	$449.60	$87.30	$698.40	$48.48	$74.24
2 Pile Drivers	54.20	867.20	84.20	1347.20		
3 Laborers	42.10	1010.40	63.25	1518.00		
1 Chain Saw, Gas, 36" Long		41.25		45.38	.86	.95
48 L.H., Daily Totals		$2368.45		$3608.97	$49.34	$75.19

Crew B-47

	Hr.	Daily	Hr.	Daily	Bare Costs	Incl. O&P
1 Blast Foreman (outside)	$44.10	$352.80	$66.25	$530.00	$46.40	$69.57
1 Driller	42.10	336.80	63.25	506.00		
1 Equip. Oper. (light)	53.00	424.00	79.20	633.60		
1 Air Track Drill, 4"		1058.00		1163.80		
1 Air Compressor, 600 cfm		421.50		463.65		
2 -50' Air Hoses, 3"		76.70		84.37	64.84	71.33
24 L.H., Daily Totals		$2669.80		$3381.42	$111.24	$140.89

Crew B-47A

	Hr.	Daily	Hr.	Daily	Bare Costs	Incl. O&P
1 Drilling Foreman (outside)	$44.10	$352.80	$66.25	$530.00	$51.28	$76.77
1 Equip. Oper. (heavy)	59.20	473.60	88.50	708.00		
1 Equip. Oper. (oiler)	50.55	404.40	75.55	604.40		
1 Air Track Drill, 5"		1127.00		1239.70	46.96	51.65
24 L.H., Daily Totals		$2357.80		$3082.10	$98.24	$128.42

Crew B-47C

	Hr.	Daily	Hr.	Daily	Bare Costs	Incl. O&P
1 Laborer	$42.10	$336.80	$63.25	$506.00	$47.55	$71.22
1 Equip. Oper. (light)	53.00	424.00	79.20	633.60		
1 Air Compressor, 750 cfm		538.25		592.08		
2 -50' Air Hoses, 3"		76.70		84.37		
1 Air Track Drill, 4"		1058.00		1163.80	104.56	115.02
16 L.H., Daily Totals		$2433.75		$2979.84	$152.11	$186.24

Crew B-47E

	Hr.	Daily	Hr.	Daily	Bare Costs	Incl. O&P
1 Labor Foreman (outside)	$44.10	$352.80	$66.25	$530.00	$42.60	$64.00
3 Laborers	42.10	1010.40	63.25	1518.00		
1 Flatbed Truck, Gas, 3 Ton		820.40		902.44	25.64	28.20
32 L.H., Daily Totals		$2183.60		$2950.44	$68.24	$92.20

Crew B-47G

	Hr.	Daily	Hr.	Daily	Bare Costs	Incl. O&P
1 Labor Foreman (outside)	$44.10	$352.80	$66.25	$530.00	$45.33	$67.99
2 Laborers	42.10	673.60	63.25	1012.00		
1 Equip. Oper. (light)	53.00	424.00	79.20	633.60		
1 Air Track Drill, 4"		1058.00		1163.80		
1 Air Compressor, 600 cfm		421.50		463.65		
2 -50' Air Hoses, 3"		76.70		84.37		
1 Gunite Pump Rig		317.90		349.69	58.57	64.42
32 L.H., Daily Totals		$3324.50		$4237.11	$103.89	$132.41

Crew B-47H

	Hr.	Daily	Hr.	Daily	Bare Costs	Incl. O&P
1 Skilled Worker Foreman (out)	$56.85	$454.80	$86.00	$688.00	$55.35	$83.71
3 Skilled Workers	54.85	1316.40	82.95	1990.80		
1 Flatbed Truck, Gas, 3 Ton		820.40		902.44	25.64	28.20
32 L.H., Daily Totals		$2591.60		$3581.24	$80.99	$111.91

Crew B-48

	Hr.	Daily	Hr.	Daily	Bare Costs	Incl. O&P
1 Labor Foreman (outside)	$44.10	$352.80	$66.25	$530.00	$47.59	$71.32
3 Laborers	42.10	1010.40	63.25	1518.00		
1 Equip. Oper. (crane)	59.20	473.60	88.50	708.00		
1 Equip. Oper. (oiler)	50.55	404.40	75.55	604.40		
1 Equip. Oper. (light)	53.00	424.00	79.20	633.60		
1 Centr. Water Pump, 6"		232.90		256.19		
1 -20' Suction Hose, 6"		25.20		27.72		
1 -50' Discharge Hose, 6"		17.90		19.69		
1 Drill Rig, Truck-Mounted		771.10		848.21	18.70	20.57
56 L.H., Daily Totals		$3712.30		$5145.81	$66.29	$91.89

Crew B-49

	Hr.	Daily	Hr.	Daily	Bare Costs	Incl. O&P
1 Labor Foreman (outside)	$44.10	$352.80	$66.25	$530.00	$50.12	$75.61
3 Laborers	42.10	1010.40	63.25	1518.00		
2 Equip. Oper. (crane)	59.20	947.20	88.50	1416.00		
2 Equip. Oper. (oilers)	50.55	808.80	75.55	1208.80		
1 Equip. Oper. (light)	53.00	424.00	79.20	633.60		
2 Pile Drivers	54.20	867.20	84.20	1347.20		
1 Hyd. Crane, 25 Ton		580.85		638.93		
1 Centr. Water Pump, 6"		232.90		256.19		
1 -20' Suction Hose, 6"		25.20		27.72		
1 -50' Discharge Hose, 6"		17.90		19.69		
1 Drill Rig, Truck-Mounted		771.10		848.21	18.50	20.35
88 L.H., Daily Totals		$6038.35		$8444.34	$68.62	$95.96

Crew B-50

	Hr.	Daily	Hr.	Daily	Bare Costs	Incl. O&P
2 Pile Driver Foremen (outside)	$56.20	$899.20	$87.30	$1396.80	$52.35	$80.15
6 Pile Drivers	54.20	2601.60	84.20	4041.60		
2 Equip. Oper. (crane)	59.20	947.20	88.50	1416.00		
1 Equip. Oper. (oiler)	50.55	404.40	75.55	604.40		
3 Laborers	42.10	1010.40	63.25	1518.00		
1 Crawler Crane, 40 Ton		1206.00		1326.60		
1 Lead, 60' High		209.25		230.18		
1 Hammer, Diesel, 15K ft.-lbs.		617.05		678.76		
1 Air Compressor, 600 cfm		421.50		463.65		
2 -50' Air Hoses, 3"		76.70		84.37		
1 Chain Saw, Gas, 36" Long		41.25		45.38	22.96	25.26
112 L.H., Daily Totals		$8434.55		$11805.73	$75.31	$105.41

Crew B-51

	Hr.	Daily	Hr.	Daily	Bare Costs	Incl. O&P
1 Labor Foreman (outside)	$44.10	$352.80	$66.25	$530.00	$43.29	$65.01
4 Laborers	42.10	1347.20	63.25	2024.00		
1 Truck Driver (light)	47.25	378.00	70.80	566.40		
1 Flatbed Truck, Gas, 1.5 Ton		196.15		215.76	4.09	4.50
48 L.H., Daily Totals		$2274.15		$3336.17	$47.38	$69.50

Crew B-52

	Hr.	Daily	Hr.	Daily	Bare Costs	Incl. O&P
1 Carpenter Foreman (outside)	$55.15	$441.20	$82.85	$662.80	$48.73	$72.97
1 Carpenter	53.15	425.20	79.85	638.80		
3 Laborers	42.10	1010.40	63.25	1518.00		
1 Cement Finisher	49.95	399.60	73.45	587.60		
.5 Rodman (reinf.)	56.40	225.60	84.90	339.60		
.5 Equip. Oper. (medium)	56.75	227.00	84.85	339.40		
.5 Crawler Loader, 3 C.Y.		584.50		642.95	10.44	11.48
56 L.H., Daily Totals		$3313.50		$4729.15	$59.17	$84.45

Crew B-53

	Hr.	Daily	Hr.	Daily	Bare Costs	Incl. O&P
1 Equip. Oper. (light)	$53.00	$424.00	$79.20	$633.60	$53.00	$79.20
1 Trencher, Chain, 12 H.P.		156.95		172.65	19.62	21.58
8 L.H., Daily Totals		$580.95		$806.25	$72.62	$100.78

Crew B-54

	Bare Costs Hr.	Bare Costs Daily	Incl. Subs O&P Hr.	Incl. Subs O&P Daily	Cost Per Labor-Hour Bare Costs	Cost Per Labor-Hour Incl. O&P
1 Equip. Oper. (light)	$53.00	$424.00	$79.20	$633.60	$53.00	$79.20
1 Trencher, Chain, 40 H.P.		401.85		442.04	50.23	55.25
8 L.H., Daily Totals		$825.85		$1075.64	$103.23	$134.45

Crew B-54A

	Bare Costs Hr.	Bare Costs Daily	Incl. Subs O&P Hr.	Incl. Subs O&P Daily	Cost Per Labor-Hour Bare Costs	Cost Per Labor-Hour Incl. O&P
.17 Labor Foreman (outside)	$44.10	$59.98	$66.25	$90.10	$54.91	$82.15
1 Equipment Operator (med.)	56.75	454.00	84.85	678.80		
1 Wheel Trencher, 67 H.P.		1126.00		1238.60	120.30	132.33
9.36 L.H., Daily Totals		$1639.98		$2007.50	$175.21	$214.48

Crew B-54B

	Bare Costs Hr.	Bare Costs Daily	Incl. Subs O&P Hr.	Incl. Subs O&P Daily	Cost Per Labor-Hour Bare Costs	Cost Per Labor-Hour Incl. O&P
.25 Labor Foreman (outside)	$44.10	$88.20	$66.25	$132.50	$54.22	$81.13
1 Equipment Operator (med.)	56.75	454.00	84.85	678.80		
1 Wheel Trencher, 150 H.P.		1324.00		1456.40	132.40	145.64
10 L.H., Daily Totals		$1866.20		$2267.70	$186.62	$226.77

Crew B-54C

	Bare Costs Hr.	Bare Costs Daily	Incl. Subs O&P Hr.	Incl. Subs O&P Daily	Cost Per Labor-Hour Bare Costs	Cost Per Labor-Hour Incl. O&P
1 Laborer	$42.10	$336.80	$63.25	$506.00	$49.42	$74.05
1 Equipment Operator (med.)	56.75	454.00	84.85	678.80		
1 Wheel Trencher, 67 H.P.		1126.00		1238.60	70.38	77.41
16 L.H., Daily Totals		$1916.80		$2423.40	$119.80	$151.46

Crew B-54D

	Bare Costs Hr.	Bare Costs Daily	Incl. Subs O&P Hr.	Incl. Subs O&P Daily	Cost Per Labor-Hour Bare Costs	Cost Per Labor-Hour Incl. O&P
1 Laborer	$42.10	$336.80	$63.25	$506.00	$49.42	$74.05
1 Equipment Operator (med.)	56.75	454.00	84.85	678.80		
1 Rock Trencher, 6" Width		429.85		472.83	26.87	29.55
16 L.H., Daily Totals		$1220.65		$1657.64	$76.29	$103.60

Crew B-54E

	Bare Costs Hr.	Bare Costs Daily	Incl. Subs O&P Hr.	Incl. Subs O&P Daily	Cost Per Labor-Hour Bare Costs	Cost Per Labor-Hour Incl. O&P
1 Laborer	$42.10	$336.80	$63.25	$506.00	$49.42	$74.05
1 Equipment Operator (med.)	56.75	454.00	84.85	678.80		
1 Rock Trencher, 18" Width		1005.00		1105.50	62.81	69.09
16 L.H., Daily Totals		$1795.80		$2290.30	$112.24	$143.14

Crew B-55

	Bare Costs Hr.	Bare Costs Daily	Incl. Subs O&P Hr.	Incl. Subs O&P Daily	Cost Per Labor-Hour Bare Costs	Cost Per Labor-Hour Incl. O&P
2 Laborers	$42.10	$673.60	$63.25	$1012.00	$43.82	$65.77
1 Truck Driver (light)	47.25	378.00	70.80	566.40		
1 Truck-Mounted Earth Auger		390.20		429.22		
1 Flatbed Truck, Gas, 3 Ton		820.40		902.44	50.44	55.49
24 L.H., Daily Totals		$2262.20		$2910.06	$94.26	$121.25

Crew B-56

	Bare Costs Hr.	Bare Costs Daily	Incl. Subs O&P Hr.	Incl. Subs O&P Daily	Cost Per Labor-Hour Bare Costs	Cost Per Labor-Hour Incl. O&P
1 Laborer	$42.10	$336.80	$63.25	$506.00	$47.55	$71.22
1 Equip. Oper. (light)	53.00	424.00	79.20	633.60		
1 Air Track Drill, 4"		1058.00		1163.80		
1 Air Compressor, 600 cfm		421.50		463.65		
1 -50' Air Hose, 3"		38.35		42.19	94.87	104.35
16 L.H., Daily Totals		$2278.65		$2809.24	$142.42	$175.58

Crew B-57

	Bare Costs Hr.	Bare Costs Daily	Incl. Subs O&P Hr.	Incl. Subs O&P Daily	Cost Per Labor-Hour Bare Costs	Cost Per Labor-Hour Incl. O&P
1 Labor Foreman (outside)	$44.10	$352.80	$66.25	$530.00	$48.51	$72.67
2 Laborers	42.10	673.60	63.25	1012.00		
1 Equip. Oper. (crane)	59.20	473.60	88.50	708.00		
1 Equip. Oper. (light)	53.00	424.00	79.20	633.60		
1 Equip. Oper. (oiler)	50.55	404.40	75.55	604.40		
1 Crawler Crane, 25 Ton		1129.00		1241.90		
1 Clamshell Bucket, 1 C.Y.		68.55		75.41		
1 Centr. Water Pump, 6"		232.90		256.19		
1 -20' Suction Hose, 6"		25.20		27.72		
20 -50' Discharge Hoses, 6"		358.00		393.80	37.78	41.56
48 L.H., Daily Totals		$4142.05		$5483.02	$86.29	$114.23

Crew B-58

	Bare Costs Hr.	Bare Costs Daily	Incl. Subs O&P Hr.	Incl. Subs O&P Daily	Cost Per Labor-Hour Bare Costs	Cost Per Labor-Hour Incl. O&P
2 Laborers	$42.10	$673.60	$63.25	$1012.00	$45.73	$68.57
1 Equip. Oper. (light)	53.00	424.00	79.20	633.60		
1 Backhoe Loader, 48 H.P.		213.75		235.13		
1 Small Helicopter, w/ Pilot		2080.00		2288.00	95.57	105.13
24 L.H., Daily Totals		$3391.35		$4168.73	$141.31	$173.70

Crew B-59

	Bare Costs Hr.	Bare Costs Daily	Incl. Subs O&P Hr.	Incl. Subs O&P Daily	Cost Per Labor-Hour Bare Costs	Cost Per Labor-Hour Incl. O&P
1 Truck Driver (heavy)	$48.95	$391.60	$73.35	$586.80	$48.95	$73.35
1 Truck Tractor, 220 H.P.		307.10		337.81		
1 Water Tank Trailer, 5000 Gal.		152.25		167.47	57.42	63.16
8 L.H., Daily Totals		$850.95		$1092.09	$106.37	$136.51

Crew B-59A

	Bare Costs Hr.	Bare Costs Daily	Incl. Subs O&P Hr.	Incl. Subs O&P Daily	Cost Per Labor-Hour Bare Costs	Cost Per Labor-Hour Incl. O&P
2 Laborers	$42.10	$673.60	$63.25	$1012.00	$44.38	$66.62
1 Truck Driver (heavy)	48.95	391.60	73.35	586.80		
1 Water Tank Trailer, 5000 Gal.		152.25		167.47		
1 Truck Tractor, 220 H.P.		307.10		337.81	19.14	21.05
24 L.H., Daily Totals		$1524.55		$2104.09	$63.52	$87.67

Crew B-60

	Bare Costs Hr.	Bare Costs Daily	Incl. Subs O&P Hr.	Incl. Subs O&P Daily	Cost Per Labor-Hour Bare Costs	Cost Per Labor-Hour Incl. O&P
1 Labor Foreman (outside)	$44.10	$352.80	$66.25	$530.00	$49.15	$73.60
2 Laborers	42.10	673.60	63.25	1012.00		
1 Equip. Oper. (crane)	59.20	473.60	88.50	708.00		
2 Equip. Oper. (light)	53.00	848.00	79.20	1267.20		
1 Equip. Oper. (oiler)	50.55	404.40	75.55	604.40		
1 Crawler Crane, 40 Ton		1206.00		1326.60		
1 Lead, 60' High		209.25		230.18		
1 Hammer, Diesel, 15K ft.-lbs.		617.05		678.76		
1 Backhoe Loader, 48 H.P.		213.75		235.13	40.11	44.12
56 L.H., Daily Totals		$4998.45		$6592.26	$89.26	$117.72

Crew B-61

	Bare Costs Hr.	Bare Costs Daily	Incl. Subs O&P Hr.	Incl. Subs O&P Daily	Cost Per Labor-Hour Bare Costs	Cost Per Labor-Hour Incl. O&P
1 Labor Foreman (outside)	$44.10	$352.80	$66.25	$530.00	$44.68	$67.04
3 Laborers	42.10	1010.40	63.25	1518.00		
1 Equip. Oper. (light)	53.00	424.00	79.20	633.60		
1 Cement Mixer, 2 C.Y.		105.70		116.27		
1 Air Compressor, 160 cfm		202.80		223.08	7.71	8.48
40 L.H., Daily Totals		$2095.70		$3020.95	$52.39	$75.52

Crew B-62

	Bare Costs Hr.	Bare Costs Daily	Incl. Subs O&P Hr.	Incl. Subs O&P Daily	Cost Per Labor-Hour Bare Costs	Cost Per Labor-Hour Incl. O&P
2 Laborers	$42.10	$673.60	$63.25	$1012.00	$45.73	$68.57
1 Equip. Oper. (light)	53.00	424.00	79.20	633.60		
1 Loader, Skid Steer, 30 H.P.		177.55		195.31	7.40	8.14
24 L.H., Daily Totals		$1275.15		$1840.91	$53.13	$76.70

For customer support on your Mechanical Costs with RSMeans Data, call 800.448.8182.

637

Crew No.	Bare Costs Hr.	Bare Costs Daily	Incl. Subs O&P Hr.	Incl. Subs O&P Daily	Cost Per Labor-Hour Bare Costs	Cost Per Labor-Hour Incl. O&P
Crew B-62A	Hr.	Daily	Hr.	Daily	Bare Costs	Incl. O&P
2 Laborers	$42.10	$673.60	$63.25	$1012.00	$45.73	$68.57
1 Equip. Oper. (light)	53.00	424.00	79.20	633.60		
1 Loader, Skid Steer, 30 H.P.		177.55		195.31		
1 Trencher Attachment		66.20		72.82	10.16	11.17
24 L.H., Daily Totals		$1341.35		$1913.72	$55.89	$79.74
Crew B-63	Hr.	Daily	Hr.	Daily	Bare Costs	Incl. O&P
4 Laborers	$42.10	$1347.20	$63.25	$2024.00	$44.28	$66.44
1 Equip. Oper. (light)	53.00	424.00	79.20	633.60		
1 Loader, Skid Steer, 30 H.P.		177.55		195.31	4.44	4.88
40 L.H., Daily Totals		$1948.75		$2852.91	$48.72	$71.32
Crew B-63B	Hr.	Daily	Hr.	Daily	Bare Costs	Incl. O&P
1 Labor Foreman (inside)	$42.60	$340.80	$64.00	$512.00	$44.95	$67.42
2 Laborers	42.10	673.60	63.25	1012.00		
1 Equip. Oper. (light)	53.00	424.00	79.20	633.60		
1 Loader, Skid Steer, 78 H.P.		400.10		440.11	12.50	13.75
32 L.H., Daily Totals		$1838.50		$2597.71	$57.45	$81.18
Crew B-64	Hr.	Daily	Hr.	Daily	Bare Costs	Incl. O&P
1 Laborer	$42.10	$336.80	$63.25	$506.00	$44.67	$67.03
1 Truck Driver (light)	47.25	378.00	70.80	566.40		
1 Power Mulcher (small)		221.10		243.21		
1 Flatbed Truck, Gas, 1.5 Ton		196.15		215.76	26.08	28.69
16 L.H., Daily Totals		$1132.05		$1531.38	$70.75	$95.71
Crew B-65	Hr.	Daily	Hr.	Daily	Bare Costs	Incl. O&P
1 Laborer	$42.10	$336.80	$63.25	$506.00	$44.67	$67.03
1 Truck Driver (light)	47.25	378.00	70.80	566.40		
1 Power Mulcher (Large)		326.90		359.59		
1 Flatbed Truck, Gas, 1.5 Ton		196.15		215.76	32.69	35.96
16 L.H., Daily Totals		$1237.85		$1647.76	$77.37	$102.98
Crew B-66	Hr.	Daily	Hr.	Daily	Bare Costs	Incl. O&P
1 Equip. Oper. (light)	$53.00	$424.00	$79.20	$633.60	$53.00	$79.20
1 Loader-Backhoe, 40 H.P.		241.40		265.54	30.18	33.19
8 L.H., Daily Totals		$665.40		$899.14	$83.17	$112.39
Crew B-67	Hr.	Daily	Hr.	Daily	Bare Costs	Incl. O&P
1 Millwright	$55.90	$447.20	$81.00	$648.00	$54.45	$80.10
1 Equip. Oper. (light)	53.00	424.00	79.20	633.60		
1 R.T. Forklift, 5,000 Lb., diesel		269.90		296.89	16.87	18.56
16 L.H., Daily Totals		$1141.10		$1578.49	$71.32	$98.66
Crew B-67B	Hr.	Daily	Hr.	Daily	Bare Costs	Incl. O&P
1 Millwright Foreman (inside)	$56.40	$451.20	$81.70	$653.60	$56.15	$81.35
1 Millwright	55.90	447.20	81.00	648.00		
16 L.H., Daily Totals		$898.40		$1301.60	$56.15	$81.35
Crew B-68	Hr.	Daily	Hr.	Daily	Bare Costs	Incl. O&P
2 Millwrights	$55.90	$894.40	$81.00	$1296.00	$54.93	$80.40
1 Equip. Oper. (light)	53.00	424.00	79.20	633.60		
1 R.T. Forklift, 5,000 Lb., diesel		269.90		296.89	11.25	12.37
24 L.H., Daily Totals		$1588.30		$2226.49	$66.18	$92.77
Crew B-68A	Hr.	Daily	Hr.	Daily	Bare Costs	Incl. O&P
1 Millwright Foreman (inside)	$56.40	$451.20	$81.70	$653.60	$56.07	$81.23
2 Millwrights	55.90	894.40	81.00	1296.00		
1 Forklift, Smooth Floor, 8,000 Lb.		255.90		281.49	10.66	11.73
24 L.H., Daily Totals		$1601.50		$2231.09	$66.73	$92.96
Crew B-68B	Hr.	Daily	Hr.	Daily	Bare Costs	Incl. O&P
1 Millwright Foreman (inside)	$56.40	$451.20	$81.70	$653.60	$59.97	$88.47
2 Millwrights	55.90	894.40	81.00	1296.00		
2 Electricians	61.35	981.60	91.35	1461.60		
2 Plumbers	64.45	1031.20	96.45	1543.20		
1 R.T. Forklift, 5,000 Lb., gas		279.90		307.89	5.00	5.50
56 L.H., Daily Totals		$3638.30		$5262.29	$64.97	$93.97
Crew B-68C	Hr.	Daily	Hr.	Daily	Bare Costs	Incl. O&P
1 Millwright Foreman (inside)	$56.40	$451.20	$81.70	$653.60	$59.52	$87.63
1 Millwright	55.90	447.20	81.00	648.00		
1 Electrician	61.35	490.80	91.35	730.80		
1 Plumber	64.45	515.60	96.45	771.60		
1 R.T. Forklift, 5,000 Lb., gas		279.90		307.89	8.75	9.62
32 L.H., Daily Totals		$2184.70		$3111.89	$68.27	$97.25
Crew B-68D	Hr.	Daily	Hr.	Daily	Bare Costs	Incl. O&P
1 Labor Foreman (inside)	$42.60	$340.80	$64.00	$512.00	$45.90	$68.82
1 Laborer	42.10	336.80	63.25	506.00		
1 Equip. Oper. (light)	53.00	424.00	79.20	633.60		
1 R.T. Forklift, 5,000 Lb., gas		279.90		307.89	11.66	12.83
24 L.H., Daily Totals		$1381.50		$1959.49	$57.56	$81.65
Crew B-68E	Hr.	Daily	Hr.	Daily	Bare Costs	Incl. O&P
1 Struc. Steel Foreman (inside)	$58.15	$465.20	$91.20	$729.60	$57.71	$90.53
3 Struc. Steel Workers	57.65	1383.60	90.45	2170.80		
1 Welder	57.45	459.60	90.10	720.80		
1 Forklift, Smooth Floor, 8,000 Lb.		255.90		281.49	6.40	7.04
40 L.H., Daily Totals		$2564.30		$3902.69	$64.11	$97.57
Crew B-68F	Hr.	Daily	Hr.	Daily	Bare Costs	Incl. O&P
1 Skilled Worker Foreman (out)	$56.85	$454.80	$86.00	$688.00	$55.52	$83.97
2 Skilled Workers	54.85	877.60	82.95	1327.20		
1 R.T. Forklift, 5,000 Lb., gas		279.90		307.89	11.66	12.83
24 L.H., Daily Totals		$1612.30		$2323.09	$67.18	$96.80
Crew B-68G	Hr.	Daily	Hr.	Daily	Bare Costs	Incl. O&P
2 Structural Steel Workers	$57.65	$922.40	$90.45	$1447.20	$57.65	$90.45
1 R.T. Forklift, 5,000 Lb., gas		279.90		307.89	17.49	19.24
16 L.H., Daily Totals		$1202.30		$1755.09	$75.14	$109.69
Crew B-69	Hr.	Daily	Hr.	Daily	Bare Costs	Incl. O&P
1 Labor Foreman (outside)	$44.10	$352.80	$66.25	$530.00	$46.69	$70.01
3 Laborers	42.10	1010.40	63.25	1518.00		
1 Equip. Oper. (crane)	59.20	473.60	88.50	708.00		
1 Equip. Oper. (oiler)	50.55	404.40	75.55	604.40		
1 Hyd. Crane, 80 Ton		1486.00		1634.60	30.96	34.05
48 L.H., Daily Totals		$3727.20		$4995.00	$77.65	$104.06
Crew B-69A	Hr.	Daily	Hr.	Daily	Bare Costs	Incl. O&P
1 Labor Foreman (outside)	$44.10	$352.80	$66.25	$530.00	$46.18	$69.05
3 Laborers	42.10	1010.40	63.25	1518.00		
1 Equip. Oper. (medium)	56.75	454.00	84.85	678.80		
1 Concrete Finisher	49.95	399.60	73.45	587.60		
1 Curb/Gutter Paver, 2-Track		1217.00		1338.70	25.35	27.89
48 L.H., Daily Totals		$3433.80		$4653.10	$71.54	$96.94

For customer support on your Mechanical Costs with RSMeans Data, call 800.448.8182.

Crew No.	Bare Costs		Incl. Subs O&P		Cost Per Labor-Hour	

Crew B-69B	Hr.	Daily	Hr.	Daily	Bare Costs	Incl. O&P
1 Labor Foreman (outside)	$44.10	$352.80	$66.25	$530.00	$46.18	$69.05
3 Laborers	42.10	1010.40	63.25	1518.00		
1 Equip. Oper. (medium)	56.75	454.00	84.85	678.80		
1 Cement Finisher	49.95	399.60	73.45	587.60		
1 Curb/Gutter Paver, 4-Track		791.55		870.71	16.49	18.14
48 L.H., Daily Totals		$3008.35		$4185.10	$62.67	$87.19

Crew B-70	Hr.	Daily	Hr.	Daily	Bare Costs	Incl. O&P
1 Labor Foreman (outside)	$44.10	$352.80	$66.25	$530.00	$48.66	$72.94
3 Laborers	42.10	1010.40	63.25	1518.00		
3 Equip. Oper. (medium)	56.75	1362.00	84.85	2036.40		
1 Grader, 30,000 Lbs.		1062.00		1168.20		
1 Ripper, Beam & 1 Shank		90.50		99.55		
1 Road Sweeper, S.P., 8' wide		715.85		787.43		
1 F.E. Loader, W.M., 1.5 C.Y.		423.50		465.85	40.93	45.02
56 L.H., Daily Totals		$5017.05		$6605.44	$89.59	$117.95

Crew B-70A	Hr.	Daily	Hr.	Daily	Bare Costs	Incl. O&P
1 Laborer	$42.10	$336.80	$63.25	$506.00	$53.82	$80.53
4 Equip. Oper. (medium)	56.75	1816.00	84.85	2715.20		
1 Grader, 40,000 Lbs.		1347.00		1481.70		
1 F.E. Loader, W.M., 2.5 C.Y.		702.10		772.31		
1 Dozer, 80 H.P.		401.40		441.54		
1 Roller, Pneum. Whl., 12 Ton		345.75		380.32	69.91	76.90
40 L.H., Daily Totals		$4949.05		$6297.07	$123.73	$157.43

Crew B-71	Hr.	Daily	Hr.	Daily	Bare Costs	Incl. O&P
1 Labor Foreman (outside)	$44.10	$352.80	$66.25	$530.00	$48.66	$72.94
3 Laborers	42.10	1010.40	63.25	1518.00		
3 Equip. Oper. (medium)	56.75	1362.00	84.85	2036.40		
1 Pvmt. Profiler, 750 H.P.		3448.00		3792.80		
1 Road Sweeper, S.P., 8' wide		715.85		787.43		
1 F.E. Loader, W.M., 1.5 C.Y.		423.50		465.85	81.92	90.11
56 L.H., Daily Totals		$7312.55		$9130.49	$130.58	$163.04

Crew B-72	Hr.	Daily	Hr.	Daily	Bare Costs	Incl. O&P
1 Labor Foreman (outside)	$44.10	$352.80	$66.25	$530.00	$49.67	$74.42
3 Laborers	42.10	1010.40	63.25	1518.00		
4 Equip. Oper. (medium)	56.75	1816.00	84.85	2715.20		
1 Pvmt. Profiler, 750 H.P.		3448.00		3792.80		
1 Hammermill, 250 H.P.		848.80		933.68		
1 Windrow Loader		1427.00		1569.70		
1 Mix Paver, 165 H.P.		2149.00		2363.90		
1 Roller, Pneum. Whl., 12 Ton		345.75		380.32	128.41	141.26
64 L.H., Daily Totals		$11397.75		$13803.61	$178.09	$215.68

Crew B-73	Hr.	Daily	Hr.	Daily	Bare Costs	Incl. O&P
1 Labor Foreman (outside)	$44.10	$352.80	$66.25	$530.00	$51.51	$77.13
2 Laborers	42.10	673.60	63.25	1012.00		
5 Equip. Oper. (medium)	56.75	2270.00	84.85	3394.00		
1 Road Mixer, 310 H.P.		1896.00		2085.60		
1 Tandem Roller, 10 Ton		236.75		260.43		
1 Hammermill, 250 H.P.		848.80		933.68		
1 Grader, 30,000 Lbs.		1062.00		1168.20		
.5 F.E. Loader, W.M., 1.5 C.Y.		211.75		232.93		
.5 Truck Tractor, 220 H.P.		153.55		168.91		
.5 Water Tank Trailer, 5000 Gal.		76.13		83.74	70.08	77.09
64 L.H., Daily Totals		$7781.38		$9869.47	$121.58	$154.21

Crew B-74	Hr.	Daily	Hr.	Daily	Bare Costs	Incl. O&P
1 Labor Foreman (outside)	$44.10	$352.80	$66.25	$530.00	$51.39	$76.95
1 Laborer	42.10	336.80	63.25	506.00		
4 Equip. Oper. (medium)	56.75	1816.00	84.85	2715.20		
2 Truck Drivers (heavy)	48.95	783.20	73.35	1173.60		
1 Grader, 30,000 Lbs.		1062.00		1168.20		
1 Ripper, Beam & 1 Shank		90.50		99.55		
2 Stabilizers, 310 H.P.		2778.00		3055.80		
1 Flatbed Truck, Gas, 3 Ton		820.40		902.44		
1 Chem. Spreader, Towed		83.35		91.69		
1 Roller, Vibratory, 25 Ton		664.35		730.78		
1 Water Tank Trailer, 5000 Gal.		152.25		167.47		
1 Truck Tractor, 220 H.P.		307.10		337.81	93.09	102.40
64 L.H., Daily Totals		$9246.75		$11478.55	$144.48	$179.35

Crew B-75	Hr.	Daily	Hr.	Daily	Bare Costs	Incl. O&P
1 Labor Foreman (outside)	$44.10	$352.80	$66.25	$530.00	$51.74	$77.46
1 Laborer	42.10	336.80	63.25	506.00		
4 Equip. Oper. (medium)	56.75	1816.00	84.85	2715.20		
1 Truck Driver (heavy)	48.95	391.60	73.35	586.80		
1 Grader, 30,000 Lbs.		1062.00		1168.20		
1 Ripper, Beam & 1 Shank		90.50		99.55		
2 Stabilizers, 310 H.P.		2778.00		3055.80		
1 Dist. Tanker, 3000 Gallon		330.15		363.17		
1 Truck Tractor, 6x4, 380 H.P.		493.25		542.58		
1 Roller, Vibratory, 25 Ton		664.35		730.78	96.75	106.43
56 L.H., Daily Totals		$8315.45		$10298.08	$148.49	$183.89

Crew B-76	Hr.	Daily	Hr.	Daily	Bare Costs	Incl. O&P
1 Dock Builder Foreman (outside)	$56.20	$449.60	$87.30	$698.40	$55.13	$84.54
5 Dock Builders	54.20	2168.00	84.20	3368.00		
2 Equip. Oper. (crane)	59.20	947.20	88.50	1416.00		
1 Equip. Oper. (oiler)	50.55	404.40	75.55	604.40		
1 Crawler Crane, 50 Ton		1510.00		1661.00		
1 Barge, 400 Ton		858.85		944.74		
1 Hammer, Diesel, 15K ft.-lbs.		617.05		678.76		
1 Lead, 60' High		209.25		230.18		
1 Air Compressor, 600 cfm		421.50		463.65		
2 -50' Air Hoses, 3"		76.70		84.37	51.30	56.43
72 L.H., Daily Totals		$7662.55		$10149.49	$106.42	$140.97

Crew B-76A	Hr.	Daily	Hr.	Daily	Bare Costs	Incl. O&P
1 Labor Foreman (outside)	$44.10	$352.80	$66.25	$530.00	$45.54	$68.32
5 Laborers	42.10	1684.00	63.25	2530.00		
1 Equip. Oper. (crane)	59.20	473.60	88.50	708.00		
1 Equip. Oper. (oiler)	50.55	404.40	75.55	604.40		
1 Crawler Crane, 50 Ton		1510.00		1661.00		
1 Barge, 400 Ton		858.85		944.74	37.01	40.71
64 L.H., Daily Totals		$5283.65		$6978.14	$82.56	$109.03

Crew B-77	Hr.	Daily	Hr.	Daily	Bare Costs	Incl. O&P
1 Labor Foreman (outside)	$44.10	$352.80	$66.25	$530.00	$43.53	$65.36
3 Laborers	42.10	1010.40	63.25	1518.00		
1 Truck Driver (light)	47.25	378.00	70.80	566.40		
1 Crack Cleaner, 25 H.P.		52.45		57.70		
1 Crack Filler, Trailer Mtd.		154.75		170.22		
1 Flatbed Truck, Gas, 3 Ton		820.40		902.44	25.69	28.26
40 L.H., Daily Totals		$2768.80		$3744.76	$69.22	$93.62

For customer support on your Mechanical Costs with RSMeans Data, call 800.448.8182.

639

Crew No.	Bare Costs		Incl. Subs O&P		Cost Per Labor-Hour	

Crew B-78

	Hr.	Daily	Hr.	Daily	Bare Costs	Incl. O&P
1 Labor Foreman (outside)	$44.10	$352.80	$66.25	$530.00	$43.29	$65.01
4 Laborers	42.10	1347.20	63.25	2024.00		
1 Truck Driver (light)	47.25	378.00	70.80	566.40		
1 Paint Striper, S.P., 40 Gallon		127.10		139.81		
1 Flatbed Truck, Gas, 3 Ton		820.40		902.44		
1 Pickup Truck, 3/4 Ton		110.85		121.94	22.05	24.25
48 L.H., Daily Totals		$3136.35		$4284.59	$65.34	$89.26

Crew B-78A

	Hr.	Daily	Hr.	Daily	Bare Costs	Incl. O&P
1 Equip. Oper. (light)	$53.00	$424.00	$79.20	$633.60	$53.00	$79.20
1 Line Rem. (Metal Balls) 115 H.P.		916.50		1008.15	114.56	126.02
8 L.H., Daily Totals		$1340.50		$1641.75	$167.56	$205.22

Crew B-78B

	Hr.	Daily	Hr.	Daily	Bare Costs	Incl. O&P
2 Laborers	$42.10	$673.60	$63.25	$1012.00	$43.31	$65.02
.25 Equip. Oper. (light)	53.00	106.00	79.20	158.40		
1 Pickup Truck, 3/4 Ton		110.85		121.94		
1 Line Rem.,11 H.P.,Walk Behind		113.45		124.80		
.25 Road Sweeper, S.P., 8' wide		178.96		196.86	22.40	24.64
18 L.H., Daily Totals		$1182.86		$1613.99	$65.71	$89.67

Crew B-78C

	Hr.	Daily	Hr.	Daily	Bare Costs	Incl. O&P
1 Labor Foreman (outside)	$44.10	$352.80	$66.25	$530.00	$43.29	$65.01
4 Laborers	42.10	1347.20	63.25	2024.00		
1 Truck Driver (light)	47.25	378.00	70.80	566.40		
1 Paint Striper, T.M., 120 Gal.		592.95		652.25		
1 Flatbed Truck, Gas, 3 Ton		820.40		902.44		
1 Pickup Truck, 3/4 Ton		110.85		121.94	31.75	34.93
48 L.H., Daily Totals		$3602.20		$4797.02	$75.05	$99.94

Crew B-78D

	Hr.	Daily	Hr.	Daily	Bare Costs	Incl. O&P
2 Labor Foremen (outside)	$44.10	$705.60	$66.25	$1060.00	$43.02	$64.61
7 Laborers	42.10	2357.60	63.25	3542.00		
1 Truck Driver (light)	47.25	378.00	70.80	566.40		
1 Paint Striper, T.M., 120 Gal.		592.95		652.25		
1 Flatbed Truck, Gas, 3 Ton		820.40		902.44		
3 Pickup Trucks, 3/4 Ton		332.55		365.81		
1 Air Compressor, 60 cfm		152.95		168.25		
1 -50' Air Hose, 3/4"		7.75		8.53		
1 Breaker, Pavement, 60 lb.		53.50		58.85	24.50	26.95
80 L.H., Daily Totals		$5401.30		$7324.51	$67.52	$91.56

Crew B-78E

	Hr.	Daily	Hr.	Daily	Bare Costs	Incl. O&P
2 Labor Foremen (outside)	$44.10	$705.60	$66.25	$1060.00	$42.86	$64.38
9 Laborers	42.10	3031.20	63.25	4554.00		
1 Truck Driver (light)	47.25	378.00	70.80	566.40		
1 Paint Striper, T.M., 120 Gal.		592.95		652.25		
1 Flatbed Truck, Gas, 3 Ton		820.40		902.44		
4 Pickup Trucks, 3/4 Ton		443.40		487.74		
2 Air Compressors, 60 cfm		305.90		336.49		
2 -50' Air Hoses, 3/4"		15.50		17.05		
2 Breakers, Pavement, 60 lb.		107.00		117.70	23.80	26.18
96 L.H., Daily Totals		$6399.95		$8694.07	$66.67	$90.56

Crew B-78F

	Hr.	Daily	Hr.	Daily	Bare Costs	Incl. O&P
2 Labor Foremen (outside)	$44.10	$705.60	$66.25	$1060.00	$42.75	$64.22
11 Laborers	42.10	3704.80	63.25	5566.00		
1 Truck Driver (light)	47.25	378.00	70.80	566.40		
1 Paint Striper, T.M., 120 Gal.		592.95		652.25		
1 Flatbed Truck, Gas, 3 Ton		820.40		902.44		
7 Pickup Trucks, 3/4 Ton		775.95		853.54		
3 Air Compressors, 60 cfm		458.85		504.74		
3 -50' Air Hoses, 3/4"		23.25		25.57		
3 Breakers, Pavement, 60 lb.		160.50		176.55	25.28	27.81
112 L.H., Daily Totals		$7620.30		$10307.49	$68.04	$92.03

Crew B-79

	Hr.	Daily	Hr.	Daily	Bare Costs	Incl. O&P
1 Labor Foreman (outside)	$44.10	$352.80	$66.25	$530.00	$43.53	$65.36
3 Laborers	42.10	1010.40	63.25	1518.00		
1 Truck Driver (light)	47.25	378.00	70.80	566.40		
1 Paint Striper, T.M., 120 Gal.		592.95		652.25		
1 Heating Kettle, 115 Gallon		98.65		108.52		
1 Flatbed Truck, Gas, 3 Ton		820.40		902.44		
2 Pickup Trucks, 3/4 Ton		221.70		243.87	43.34	47.68
40 L.H., Daily Totals		$3474.90		$4521.47	$86.87	$113.04

Crew B-79A

	Hr.	Daily	Hr.	Daily	Bare Costs	Incl. O&P
1.5 Equip. Oper. (light)	$53.00	$636.00	$79.20	$950.40	$53.00	$79.20
.5 Line Remov. (Grinder) 115 H.P.		475.57		523.13		
1 Line Rem. (Metal Balls) 115 H.P.		916.50		1008.15	116.01	127.61
12 L.H., Daily Totals		$2028.08		$2481.68	$169.01	$206.81

Crew B-79B

	Hr.	Daily	Hr.	Daily	Bare Costs	Incl. O&P
1 Laborer	$42.10	$336.80	$63.25	$506.00	$42.10	$63.25
1 Set of Gases		171.55		188.71	21.44	23.59
8 L.H., Daily Totals		$508.35		$694.71	$63.54	$86.84

Crew B-79C

	Hr.	Daily	Hr.	Daily	Bare Costs	Incl. O&P
1 Labor Foreman (outside)	$44.10	$352.80	$66.25	$530.00	$43.12	$64.76
5 Laborers	42.10	1684.00	63.25	2530.00		
1 Truck Driver (light)	47.25	378.00	70.80	566.40		
1 Paint Striper, T.M., 120 Gal.		592.95		652.25		
1 Heating Kettle, 115 Gallon		98.65		108.52		
1 Flatbed Truck, Gas, 3 Ton		820.40		902.44		
3 Pickup Trucks, 3/4 Ton		332.55		365.81		
1 Air Compressor, 60 cfm		152.95		168.25		
1 -50' Air Hose, 3/4"		7.75		8.53		
1 Breaker, Pavement, 60 lb.		53.50		58.85	36.76	40.44
56 L.H., Daily Totals		$4473.55		$5891.02	$79.88	$105.20

Crew B-79D

	Hr.	Daily	Hr.	Daily	Bare Costs	Incl. O&P
2 Labor Foremen (outside)	$44.10	$705.60	$66.25	$1060.00	$43.24	$64.94
5 Laborers	42.10	1684.00	63.25	2530.00		
1 Truck Driver (light)	47.25	378.00	70.80	566.40		
1 Paint Striper, T.M., 120 Gal.		592.95		652.25		
1 Heating Kettle, 115 Gallon		98.65		108.52		
1 Flatbed Truck, Gas, 3 Ton		820.40		902.44		
4 Pickup Trucks, 3/4 Ton		443.40		487.74		
1 Air Compressor, 60 cfm		152.95		168.25		
1 -50' Air Hose, 3/4"		7.75		8.53		
1 Breaker, Pavement, 60 lb.		53.50		58.85	33.90	37.29
64 L.H., Daily Totals		$4937.20		$6542.96	$77.14	$102.23

Crew B-79E

Crew No.	Bare Costs Hr.	Bare Costs Daily	Incl. Subs O&P Hr.	Incl. Subs O&P Daily	Cost Per Labor-Hour Bare Costs	Cost Per Labor-Hour Incl. O&P
2 Labor Foremen (outside)	$44.10	$705.60	$66.25	$1060.00	$43.02	$64.61
7 Laborers	42.10	2357.60	63.25	3542.00		
1 Truck Driver (light)	47.25	378.00	70.80	566.40		
1 Paint Striper, T.M., 120 Gal.		592.95		652.25		
1 Heating Kettle, 115 Gallon		98.65		108.52		
1 Flatbed Truck, Gas, 3 Ton		820.40		902.44		
5 Pickup Trucks, 3/4 Ton		554.25		609.67		
2 Air Compressors, 60 cfm		305.90		336.49		
2 -50' Air Hoses, 3/4"		15.50		17.05		
2 Breakers, Pavement, 60 lb.		107.00		117.70	31.18	34.30
80 L.H., Daily Totals		$5935.85		$7912.52	$74.20	$98.91

Crew B-80

Crew No.	Bare Costs Hr.	Bare Costs Daily	Incl. Subs O&P Hr.	Incl. Subs O&P Daily	Cost Per Labor-Hour Bare Costs	Cost Per Labor-Hour Incl. O&P
1 Labor Foreman (outside)	$44.10	$352.80	$66.25	$530.00	$46.61	$69.88
1 Laborer	42.10	336.80	63.25	506.00		
1 Truck Driver (light)	47.25	378.00	70.80	566.40		
1 Equip. Oper. (light)	53.00	424.00	79.20	633.60		
1 Flatbed Truck, Gas, 3 Ton		820.40		902.44		
1 Earth Auger, Truck-Mtd.		200.50		220.55	31.90	35.09
32 L.H., Daily Totals		$2512.50		$3358.99	$78.52	$104.97

Crew B-80A

Crew No.	Bare Costs Hr.	Bare Costs Daily	Incl. Subs O&P Hr.	Incl. Subs O&P Daily	Cost Per Labor-Hour Bare Costs	Cost Per Labor-Hour Incl. O&P
3 Laborers	$42.10	$1010.40	$63.25	$1518.00	$42.10	$63.25
1 Flatbed Truck, Gas, 3 Ton		820.40		902.44	34.18	37.60
24 L.H., Daily Totals		$1830.80		$2420.44	$76.28	$100.85

Crew B-80B

Crew No.	Bare Costs Hr.	Bare Costs Daily	Incl. Subs O&P Hr.	Incl. Subs O&P Daily	Cost Per Labor-Hour Bare Costs	Cost Per Labor-Hour Incl. O&P
3 Laborers	$42.10	$1010.40	$63.25	$1518.00	$44.83	$67.24
1 Equip. Oper. (light)	53.00	424.00	79.20	633.60		
1 Crane, Flatbed Mounted, 3 Ton		236.10		259.71	7.38	8.12
32 L.H., Daily Totals		$1670.50		$2411.31	$52.20	$75.35

Crew B-80C

Crew No.	Bare Costs Hr.	Bare Costs Daily	Incl. Subs O&P Hr.	Incl. Subs O&P Daily	Cost Per Labor-Hour Bare Costs	Cost Per Labor-Hour Incl. O&P
2 Laborers	$42.10	$673.60	$63.25	$1012.00	$43.82	$65.77
1 Truck Driver (light)	47.25	378.00	70.80	566.40		
1 Flatbed Truck, Gas, 1.5 Ton		196.15		215.76		
1 Manual Fence Post Auger, Gas		53.85		59.23	10.42	11.46
24 L.H., Daily Totals		$1301.60		$1853.40	$54.23	$77.22

Crew B-81

Crew No.	Bare Costs Hr.	Bare Costs Daily	Incl. Subs O&P Hr.	Incl. Subs O&P Daily	Cost Per Labor-Hour Bare Costs	Cost Per Labor-Hour Incl. O&P
1 Laborer	$42.10	$336.80	$63.25	$506.00	$49.27	$73.82
1 Equip. Oper. (medium)	56.75	454.00	84.85	678.80		
1 Truck Driver (heavy)	48.95	391.60	73.35	586.80		
1 Hydromulcher, T.M., 3000 Gal.		275.45		303.00		
1 Truck Tractor, 220 H.P.		307.10		337.81	24.27	26.70
24 L.H., Daily Totals		$1764.95		$2412.41	$73.54	$100.52

Crew B-81A

Crew No.	Bare Costs Hr.	Bare Costs Daily	Incl. Subs O&P Hr.	Incl. Subs O&P Daily	Cost Per Labor-Hour Bare Costs	Cost Per Labor-Hour Incl. O&P
1 Laborer	$42.10	$336.80	$63.25	$506.00	$44.67	$67.03
1 Truck Driver (light)	47.25	378.00	70.80	566.40		
1 Hydromulcher, T.M., 600 Gal.		116.95		128.65		
1 Flatbed Truck, Gas, 3 Ton		820.40		902.44	58.58	64.44
16 L.H., Daily Totals		$1652.15		$2103.49	$103.26	$131.47

Crew B-82

Crew No.	Bare Costs Hr.	Bare Costs Daily	Incl. Subs O&P Hr.	Incl. Subs O&P Daily	Cost Per Labor-Hour Bare Costs	Cost Per Labor-Hour Incl. O&P
1 Laborer	$42.10	$336.80	$63.25	$506.00	$47.55	$71.22
1 Equip. Oper. (light)	53.00	424.00	79.20	633.60		
1 Horiz. Borer, 6 H.P.		182.35		200.59	11.40	12.54
16 L.H., Daily Totals		$943.15		$1340.18	$58.95	$83.76

Crew B-82A

Crew No.	Bare Costs Hr.	Bare Costs Daily	Incl. Subs O&P Hr.	Incl. Subs O&P Daily	Cost Per Labor-Hour Bare Costs	Cost Per Labor-Hour Incl. O&P
2 Laborers	$42.10	$673.60	$63.25	$1012.00	$47.55	$71.22
2 Equip. Opers. (light)	53.00	848.00	79.20	1267.20		
2 Dump Truck, 8 C.Y., 220 H.P.		856.10		941.71		
1 Flatbed Trailer, 25 Ton		135.55		149.10		
1 Horiz. Dir. Drill, 20k lb. Thrust		538.65		592.51		
1 Mud Trailer for HDD, 1500 Gal.		309.00		339.90		
1 Pickup Truck, 4x4, 3/4 Ton		175.85		193.44		
1 Flatbed Trailer, 3 Ton		70.30		77.33		
1 Loader, Skid Steer, 78 H.P.		400.10		440.11	77.67	85.44
32 L.H., Daily Totals		$4007.15		$5013.31	$125.22	$156.67

Crew B-82B

Crew No.	Bare Costs Hr.	Bare Costs Daily	Incl. Subs O&P Hr.	Incl. Subs O&P Daily	Cost Per Labor-Hour Bare Costs	Cost Per Labor-Hour Incl. O&P
2 Laborers	$42.10	$673.60	$63.25	$1012.00	$47.55	$71.22
2 Equip. Opers. (light)	53.00	848.00	79.20	1267.20		
2 Dump Truck, 8 C.Y., 220 H.P.		856.10		941.71		
1 Flatbed Trailer, 25 Ton		135.55		149.10		
1 Horiz. Dir. Drill, 30k lb. Thrust		641.20		705.32		
1 Mud Trailer for HDD, 1500 Gal.		309.00		339.90		
1 Pickup Truck, 4x4, 3/4 Ton		175.85		193.44		
1 Flatbed Trailer, 3 Ton		70.30		77.33		
1 Loader, Skid Steer, 78 H.P.		400.10		440.11	80.88	88.97
32 L.H., Daily Totals		$4109.70		$5126.11	$128.43	$160.19

Crew B-82C

Crew No.	Bare Costs Hr.	Bare Costs Daily	Incl. Subs O&P Hr.	Incl. Subs O&P Daily	Cost Per Labor-Hour Bare Costs	Cost Per Labor-Hour Incl. O&P
2 Laborers	$42.10	$673.60	$63.25	$1012.00	$47.55	$71.22
2 Equip. Opers. (light)	53.00	848.00	79.20	1267.20		
2 Dump Truck, 8 C.Y., 220 H.P.		856.10		941.71		
1 Flatbed Trailer, 25 Ton		135.55		149.10		
1 Horiz. Dir. Drill, 50k lb. Thrust		815.80		897.38		
1 Mud Trailer for HDD, 1500 Gal.		309.00		339.90		
1 Pickup Truck, 4x4, 3/4 Ton		175.85		193.44		
1 Flatbed Trailer, 3 Ton		70.30		77.33		
1 Loader, Skid Steer, 78 H.P.		400.10		440.11	86.33	94.97
32 L.H., Daily Totals		$4284.30		$5318.17	$133.88	$166.19

Crew B-82D

Crew No.	Bare Costs Hr.	Bare Costs Daily	Incl. Subs O&P Hr.	Incl. Subs O&P Daily	Cost Per Labor-Hour Bare Costs	Cost Per Labor-Hour Incl. O&P
1 Equip. Oper. (light)	$53.00	$424.00	$79.20	$633.60	$53.00	$79.20
1 Mud Trailer for HDD, 1500 Gal.		309.00		339.90	38.63	42.49
8 L.H., Daily Totals		$733.00		$973.50	$91.63	$121.69

Crew B-83

Crew No.	Bare Costs Hr.	Bare Costs Daily	Incl. Subs O&P Hr.	Incl. Subs O&P Daily	Cost Per Labor-Hour Bare Costs	Cost Per Labor-Hour Incl. O&P
1 Tugboat Captain	$56.75	$454.00	$84.85	$678.80	$49.42	$74.05
1 Tugboat Hand	42.10	336.80	63.25	506.00		
1 Tugboat, 250 H.P.		717.50		789.25	44.84	49.33
16 L.H., Daily Totals		$1508.30		$1974.05	$94.27	$123.38

Crew B-84

Crew No.	Bare Costs Hr.	Bare Costs Daily	Incl. Subs O&P Hr.	Incl. Subs O&P Daily	Cost Per Labor-Hour Bare Costs	Cost Per Labor-Hour Incl. O&P
1 Equip. Oper. (medium)	$56.75	$454.00	$84.85	$678.80	$56.75	$84.85
1 Rotary Mower/Tractor		366.75		403.43	45.84	50.43
8 L.H., Daily Totals		$820.75		$1082.22	$102.59	$135.28

Crew B-85

Crew No.	Bare Costs Hr.	Bare Costs Daily	Incl. Subs O&P Hr.	Incl. Subs O&P Daily	Cost Per Labor-Hour Bare Costs	Cost Per Labor-Hour Incl. O&P
3 Laborers	$42.10	$1010.40	$63.25	$1518.00	$46.40	$69.59
1 Equip. Oper. (medium)	56.75	454.00	84.85	678.80		
1 Truck Driver (heavy)	48.95	391.60	73.35	586.80		
1 Telescoping Boom Lift, to 80'		383.55		421.90		
1 Brush Chipper, 12", 130 H.P.		392.80		432.08		
1 Pruning Saw, Rotary		26.40		29.04	20.07	22.08
40 L.H., Daily Totals		$2658.75		$3666.63	$66.47	$91.67

For customer support on your Mechanical Costs with RSMeans Data, call 800.448.8182.

641

Crews - Standard

Crew No.	Bare Costs Hr.	Daily	Incl. Subs O&P Hr.	Daily	Cost Per Labor-Hour Bare Costs	Incl. O&P
Crew B-86	Hr.	Daily	Hr.	Daily	Bare Costs	Incl. O&P
1 Equip. Oper. (medium)	$56.75	$454.00	$84.85	$678.80	$56.75	$84.85
1 Stump Chipper, S.P.		194.45		213.90	24.31	26.74
8 L.H., Daily Totals		$648.45		$892.70	$81.06	$111.59
Crew B-86A	Hr.	Daily	Hr.	Daily	Bare Costs	Incl. O&P
1 Equip. Oper. (medium)	$56.75	$454.00	$84.85	$678.80	$56.75	$84.85
1 Grader, 30,000 Lbs.		1062.00		1168.20	132.75	146.03
8 L.H., Daily Totals		$1516.00		$1847.00	$189.50	$230.88
Crew B-86B	Hr.	Daily	Hr.	Daily	Bare Costs	Incl. O&P
1 Equip. Oper. (medium)	$56.75	$454.00	$84.85	$678.80	$56.75	$84.85
1 Dozer, 200 H.P.		1504.00		1654.40	188.00	206.80
8 L.H., Daily Totals		$1958.00		$2333.20	$244.75	$291.65
Crew B-87	Hr.	Daily	Hr.	Daily	Bare Costs	Incl. O&P
1 Laborer	$42.10	$336.80	$63.25	$506.00	$53.82	$80.53
4 Equip. Oper. (medium)	56.75	1816.00	84.85	2715.20		
2 Feller Bunchers, 100 H.P.		1177.20		1294.92		
1 Log Chipper, 22" Tree		552.00		607.20		
1 Dozer, 105 H.P.		633.85		697.24		
1 Chain Saw, Gas, 36" Long		41.25		45.38	60.11	66.12
40 L.H., Daily Totals		$4557.10		$5865.93	$113.93	$146.65
Crew B-88	Hr.	Daily	Hr.	Daily	Bare Costs	Incl. O&P
1 Laborer	$42.10	$336.80	$63.25	$506.00	$54.66	$81.76
6 Equip. Oper. (medium)	56.75	2724.00	84.85	4072.80		
2 Feller Bunchers, 100 H.P.		1177.20		1294.92		
1 Log Chipper, 22" Tree		552.00		607.20		
2 Log Skidders, 50 H.P.		1808.40		1989.24		
1 Dozer, 105 H.P.		633.85		697.24		
1 Chain Saw, Gas, 36" Long		41.25		45.38	75.23	82.75
56 L.H., Daily Totals		$7273.50		$9212.77	$129.88	$164.51
Crew B-89	Hr.	Daily	Hr.	Daily	Bare Costs	Incl. O&P
1 Equip. Oper. (light)	$53.00	$424.00	$79.20	$633.60	$50.13	$75.00
1 Truck Driver (light)	47.25	378.00	70.80	566.40		
1 Flatbed Truck, Gas, 3 Ton		820.40		902.44		
1 Concrete Saw		111.50		122.65		
1 Water Tank, 65 Gal.		101.90		112.09	64.61	71.07
16 L.H., Daily Totals		$1835.80		$2337.18	$114.74	$146.07
Crew B-89A	Hr.	Daily	Hr.	Daily	Bare Costs	Incl. O&P
1 Skilled Worker	$54.85	$438.80	$82.95	$663.60	$48.48	$73.10
1 Laborer	42.10	336.80	63.25	506.00		
1 Core Drill (Large)		114.30		125.73	7.14	7.86
16 L.H., Daily Totals		$889.90		$1295.33	$55.62	$80.96
Crew B-89B	Hr.	Daily	Hr.	Daily	Bare Costs	Incl. O&P
1 Equip. Oper. (light)	$53.00	$424.00	$79.20	$633.60	$50.13	$75.00
1 Truck Driver (light)	47.25	378.00	70.80	566.40		
1 Wall Saw, Hydraulic, 10 H.P.		85.50		94.05		
1 Generator, Diesel, 100 kW		496.20		545.82		
1 Water Tank, 65 Gal.		101.90		112.09		
1 Flatbed Truck, Gas, 3 Ton		820.40		902.44	94.00	103.40
16 L.H., Daily Totals		$2306.00		$2854.40	$144.13	$178.40
Crew B-89C	Hr.	Daily	Hr.	Daily	Bare Costs	Incl. O&P
1 Cement Finisher	$49.95	$399.60	$73.45	$587.60	$49.95	$73.45
1 Masonry cut-off saw, gas		61.00		67.10	7.63	8.39
8 L.H., Daily Totals		$460.60		$654.70	$57.58	$81.84

Crew No.	Bare Costs Hr.	Daily	Incl. Subs O&P Hr.	Daily	Cost Per Labor-Hour Bare Costs	Incl. O&P
Crew B-90	Hr.	Daily	Hr.	Daily	Bare Costs	Incl. O&P
1 Labor Foreman (outside)	$44.10	$352.80	$66.25	$530.00	$46.79	$70.14
3 Laborers	42.10	1010.40	63.25	1518.00		
2 Equip. Oper. (light)	53.00	848.00	79.20	1267.20		
2 Truck Drivers (heavy)	48.95	783.20	73.35	1173.60		
1 Road Mixer, 310 H.P.		1896.00		2085.60		
1 Dist. Truck, 2000 Gal.		299.65		329.62	34.31	37.74
64 L.H., Daily Totals		$5190.05		$6904.02	$81.09	$107.88
Crew B-90A	Hr.	Daily	Hr.	Daily	Bare Costs	Incl. O&P
1 Labor Foreman (outside)	$44.10	$352.80	$66.25	$530.00	$50.76	$76.02
2 Laborers	42.10	673.60	63.25	1012.00		
4 Equip. Oper. (medium)	56.75	1816.00	84.85	2715.20		
2 Graders, 30,000 Lbs.		2124.00		2336.40		
1 Tandem Roller, 10 Ton		236.75		260.43		
1 Roller, Pneum. Whl., 12 Ton		345.75		380.32	48.33	53.16
56 L.H., Daily Totals		$5548.90		$7234.35	$99.09	$129.18
Crew B-90B	Hr.	Daily	Hr.	Daily	Bare Costs	Incl. O&P
1 Labor Foreman (outside)	$44.10	$352.80	$66.25	$530.00	$49.76	$74.55
2 Laborers	42.10	673.60	63.25	1012.00		
3 Equip. Oper. (medium)	56.75	1362.00	84.85	2036.40		
1 Roller, Pneum. Whl., 12 Ton		345.75		380.32		
1 Road Mixer, 310 H.P.		1896.00		2085.60	46.70	51.37
48 L.H., Daily Totals		$4630.15		$6044.32	$96.46	$125.92
Crew B-90C	Hr.	Daily	Hr.	Daily	Bare Costs	Incl. O&P
1 Labor Foreman (outside)	$44.10	$352.80	$66.25	$530.00	$48.15	$72.17
4 Laborers	42.10	1347.20	63.25	2024.00		
3 Equip. Oper. (medium)	56.75	1362.00	84.85	2036.40		
3 Truck Drivers (heavy)	48.95	1174.80	73.35	1760.40		
3 Road Mixers, 310 H.P.		5688.00		6256.80	64.64	71.10
88 L.H., Daily Totals		$9924.80		$12607.60	$112.78	$143.27
Crew B-90D	Hr.	Daily	Hr.	Daily	Bare Costs	Incl. O&P
1 Labor Foreman (outside)	$44.10	$352.80	$66.25	$530.00	$47.22	$70.80
6 Laborers	42.10	2020.80	63.25	3036.00		
3 Equip. Oper. (medium)	56.75	1362.00	84.85	2036.40		
3 Truck Drivers (heavy)	48.95	1174.80	73.35	1760.40		
3 Road Mixers, 310 H.P.		5688.00		6256.80	54.69	60.16
104 L.H., Daily Totals		$10598.40		$13619.60	$101.91	$130.96
Crew B-90E	Hr.	Daily	Hr.	Daily	Bare Costs	Incl. O&P
1 Labor Foreman (outside)	$44.10	$352.80	$66.25	$530.00	$47.97	$71.91
4 Laborers	42.10	1347.20	63.25	2024.00		
3 Equip. Oper. (medium)	56.75	1362.00	84.85	2036.40		
1 Truck Driver (heavy)	48.95	391.60	73.35	586.80		
1 Road Mixer, 310 H.P.		1896.00		2085.60	26.33	28.97
72 L.H., Daily Totals		$5349.60		$7262.80	$74.30	$100.87
Crew B-91	Hr.	Daily	Hr.	Daily	Bare Costs	Incl. O&P
1 Labor Foreman (outside)	$44.10	$352.80	$66.25	$530.00	$50.53	$75.69
2 Laborers	42.10	673.60	63.25	1012.00		
4 Equip. Oper. (medium)	56.75	1816.00	84.85	2715.20		
1 Truck Driver (heavy)	48.95	391.60	73.35	586.80		
1 Dist. Tanker, 3000 Gallon		330.15		363.17		
1 Truck Tractor, 6x4, 380 H.P.		493.25		542.58		
1 Aggreg. Spreader, S.P.		848.90		933.79		
1 Roller, Pneum. Whl., 12 Ton		345.75		380.32		
1 Tandem Roller, 10 Ton		236.75		260.43	35.23	38.75
64 L.H., Daily Totals		$5488.80		$7324.28	$85.76	$114.44

Crew B-91B

Crew No.	Bare Costs Hr.	Bare Costs Daily	Incl. Subs O&P Hr.	Incl. Subs O&P Daily	Cost Per Labor-Hour Bare Costs	Cost Per Labor-Hour Incl. O&P
1 Laborer	$42.10	$336.80	$63.25	$506.00	$49.42	$74.05
1 Equipment Oper. (med.)	56.75	454.00	84.85	678.80		
1 Road Sweeper, Vac. Assist.		869.00		955.90	54.31	59.74
16 L.H., Daily Totals		$1659.80		$2140.70	$103.74	$133.79

Crew B-91C

Crew No.	Bare Costs Hr.	Bare Costs Daily	Incl. Subs O&P Hr.	Incl. Subs O&P Daily	Cost Per Labor-Hour Bare Costs	Cost Per Labor-Hour Incl. O&P
1 Laborer	$42.10	$336.80	$63.25	$506.00	$44.67	$67.03
1 Truck Driver (light)	47.25	378.00	70.80	566.40		
1 Catch Basin Cleaning Truck		536.15		589.76	33.51	36.86
16 L.H., Daily Totals		$1250.95		$1662.17	$78.18	$103.89

Crew B-91D

Crew No.	Bare Costs Hr.	Bare Costs Daily	Incl. Subs O&P Hr.	Incl. Subs O&P Daily	Cost Per Labor-Hour Bare Costs	Cost Per Labor-Hour Incl. O&P
1 Labor Foreman (outside)	$44.10	$352.80	$66.25	$530.00	$48.94	$73.34
5 Laborers	42.10	1684.00	63.25	2530.00		
5 Equip. Oper. (medium)	56.75	2270.00	84.85	3394.00		
2 Truck Drivers (heavy)	48.95	783.20	73.35	1173.60		
1 Aggreg. Spreader, S.P.		848.90		933.79		
2 Truck Tractors, 6x4, 380 H.P.		986.50		1085.15		
2 Dist. Tankers, 3000 Gallon		660.30		726.33		
2 Pavement Brushes, Towed		174.60		192.06		
2 Rollers Pneum. Whl., 12 Ton		691.50		760.65	32.33	35.56
104 L.H., Daily Totals		$8451.80		$11325.58	$81.27	$108.90

Crew B-92

Crew No.	Bare Costs Hr.	Bare Costs Daily	Incl. Subs O&P Hr.	Incl. Subs O&P Daily	Cost Per Labor-Hour Bare Costs	Cost Per Labor-Hour Incl. O&P
1 Labor Foreman (outside)	$44.10	$352.80	$66.25	$530.00	$42.60	$64.00
3 Laborers	42.10	1010.40	63.25	1518.00		
1 Crack Cleaner, 25 H.P.		52.45		57.70		
1 Air Compressor, 60 cfm		152.95		168.25		
1 Tar Kettle, T.M.		156.20		171.82		
1 Flatbed Truck, Gas, 3 Ton		820.40		902.44	36.94	40.63
32 L.H., Daily Totals		$2545.20		$3348.20	$79.54	$104.63

Crew B-93

Crew No.	Bare Costs Hr.	Bare Costs Daily	Incl. Subs O&P Hr.	Incl. Subs O&P Daily	Cost Per Labor-Hour Bare Costs	Cost Per Labor-Hour Incl. O&P
1 Equip. Oper. (medium)	$56.75	$454.00	$84.85	$678.80	$56.75	$84.85
1 Feller Buncher, 100 H.P.		588.60		647.46	73.58	80.93
8 L.H., Daily Totals		$1042.60		$1326.26	$130.32	$165.78

Crew B-94A

Crew No.	Bare Costs Hr.	Bare Costs Daily	Incl. Subs O&P Hr.	Incl. Subs O&P Daily	Cost Per Labor-Hour Bare Costs	Cost Per Labor-Hour Incl. O&P
1 Laborer	$42.10	$336.80	$63.25	$506.00	$42.10	$63.25
1 Diaphragm Water Pump, 2"		86.80		95.48		
1 -20' Suction Hose, 2"		3.55		3.90		
2 -50' Discharge Hoses, 2"		8.00		8.80	12.29	13.52
8 L.H., Daily Totals		$435.15		$614.18	$54.39	$76.77

Crew B-94B

Crew No.	Bare Costs Hr.	Bare Costs Daily	Incl. Subs O&P Hr.	Incl. Subs O&P Daily	Cost Per Labor-Hour Bare Costs	Cost Per Labor-Hour Incl. O&P
1 Laborer	$42.10	$336.80	$63.25	$506.00	$42.10	$63.25
1 Diaphragm Water Pump, 4"		105.25		115.78		
1 -20' Suction Hose, 4"		17.05		18.75		
2 -50' Discharge Hoses, 4"		25.30		27.83	18.45	20.30
8 L.H., Daily Totals		$484.40		$668.36	$60.55	$83.55

Crew B-94C

Crew No.	Bare Costs Hr.	Bare Costs Daily	Incl. Subs O&P Hr.	Incl. Subs O&P Daily	Cost Per Labor-Hour Bare Costs	Cost Per Labor-Hour Incl. O&P
1 Laborer	$42.10	$336.80	$63.25	$506.00	$42.10	$63.25
1 Centrifugal Water Pump, 3"		74.45		81.89		
1 -20' Suction Hose, 3"		8.75		9.63		
2 -50' Discharge Hoses, 3"		9.00		9.90	11.53	12.68
8 L.H., Daily Totals		$429.00		$607.42	$53.63	$75.93

Crew B-94D

Crew No.	Bare Costs Hr.	Bare Costs Daily	Incl. Subs O&P Hr.	Incl. Subs O&P Daily	Cost Per Labor-Hour Bare Costs	Cost Per Labor-Hour Incl. O&P
1 Laborer	$42.10	$336.80	$63.25	$506.00	$42.10	$63.25
1 Centr. Water Pump, 6"		232.90		256.19		
1 -20' Suction Hose, 6"		25.20		27.72		
2 -50' Discharge Hoses, 6"		35.80		39.38	36.74	40.41
8 L.H., Daily Totals		$630.70		$829.29	$78.84	$103.66

Crew C-1

Crew No.	Bare Costs Hr.	Bare Costs Daily	Incl. Subs O&P Hr.	Incl. Subs O&P Daily	Cost Per Labor-Hour Bare Costs	Cost Per Labor-Hour Incl. O&P
3 Carpenters	$53.15	$1275.60	$79.85	$1916.40	$50.39	$75.70
1 Laborer	42.10	336.80	63.25	506.00		
32 L.H., Daily Totals		$1612.40		$2422.40	$50.39	$75.70

Crew C-2

Crew No.	Bare Costs Hr.	Bare Costs Daily	Incl. Subs O&P Hr.	Incl. Subs O&P Daily	Cost Per Labor-Hour Bare Costs	Cost Per Labor-Hour Incl. O&P
1 Carpenter Foreman (outside)	$55.15	$441.20	$82.85	$662.80	$51.64	$77.58
4 Carpenters	53.15	1700.80	79.85	2555.20		
1 Laborer	42.10	336.80	63.25	506.00		
48 L.H., Daily Totals		$2478.80		$3724.00	$51.64	$77.58

Crew C-2A

Crew No.	Bare Costs Hr.	Bare Costs Daily	Incl. Subs O&P Hr.	Incl. Subs O&P Daily	Cost Per Labor-Hour Bare Costs	Cost Per Labor-Hour Incl. O&P
1 Carpenter Foreman (outside)	$55.15	$441.20	$82.85	$662.80	$51.11	$76.52
3 Carpenters	53.15	1275.60	79.85	1916.40		
1 Cement Finisher	49.95	399.60	73.45	587.60		
1 Laborer	42.10	336.80	63.25	506.00		
48 L.H., Daily Totals		$2453.20		$3672.80	$51.11	$76.52

Crew C-3

Crew No.	Bare Costs Hr.	Bare Costs Daily	Incl. Subs O&P Hr.	Incl. Subs O&P Daily	Cost Per Labor-Hour Bare Costs	Cost Per Labor-Hour Incl. O&P
1 Rodman Foreman (outside)	$58.40	$467.20	$87.95	$703.60	$52.65	$79.16
4 Rodmen (reinf.)	56.40	1804.80	84.90	2716.80		
1 Equip. Oper. (light)	53.00	424.00	79.20	633.60		
2 Laborers	42.10	673.60	63.25	1012.00		
3 Stressing Equipment		56.70		62.37		
.5 Grouting Equipment		122.08		134.28	2.79	3.07
64 L.H., Daily Totals		$3548.38		$5262.65	$55.44	$82.23

Crew C-4

Crew No.	Bare Costs Hr.	Bare Costs Daily	Incl. Subs O&P Hr.	Incl. Subs O&P Daily	Cost Per Labor-Hour Bare Costs	Cost Per Labor-Hour Incl. O&P
1 Rodman Foreman (outside)	$58.40	$467.20	$87.95	$703.60	$56.90	$85.66
3 Rodmen (reinf.)	56.40	1353.60	84.90	2037.60		
3 Stressing Equipment		56.70		62.37	1.77	1.95
32 L.H., Daily Totals		$1877.50		$2803.57	$58.67	$87.61

Crew C-4A

Crew No.	Bare Costs Hr.	Bare Costs Daily	Incl. Subs O&P Hr.	Incl. Subs O&P Daily	Cost Per Labor-Hour Bare Costs	Cost Per Labor-Hour Incl. O&P
2 Rodmen (reinf.)	$56.40	$902.40	$84.90	$1358.40	$56.40	$84.90
4 Stressing Equipment		75.60		83.16	4.72	5.20
16 L.H., Daily Totals		$978.00		$1441.56	$61.13	$90.10

Crew C-5

Crew No.	Bare Costs Hr.	Bare Costs Daily	Incl. Subs O&P Hr.	Incl. Subs O&P Daily	Cost Per Labor-Hour Bare Costs	Cost Per Labor-Hour Incl. O&P
1 Rodman Foreman (outside)	$58.40	$467.20	$87.95	$703.60	$56.25	$84.51
4 Rodmen (reinf.)	56.40	1804.80	84.90	2716.80		
1 Equip. Oper. (crane)	59.20	473.60	88.50	708.00		
1 Equip. Oper. (oiler)	50.55	404.40	75.55	604.40		
1 Hyd. Crane, 25 Ton		580.85		638.93	10.37	11.41
56 L.H., Daily Totals		$3730.85		$5371.73	$66.62	$95.92

Crew C-6

Crew No.	Bare Costs Hr.	Bare Costs Daily	Incl. Subs O&P Hr.	Incl. Subs O&P Daily	Cost Per Labor-Hour Bare Costs	Cost Per Labor-Hour Incl. O&P
1 Labor Foreman (outside)	$44.10	$352.80	$66.25	$530.00	$43.74	$65.45
4 Laborers	42.10	1347.20	63.25	2024.00		
1 Cement Finisher	49.95	399.60	73.45	587.60		
2 Gas Engine Vibrators		53.70		59.07	1.12	1.23
48 L.H., Daily Totals		$2153.30		$3200.67	$44.86	$66.68

For customer support on your Mechanical Costs with RSMeans Data, call 800.448.8182.

643

Crew No.	Bare Costs		Incl. Subs O&P		Cost Per Labor-Hour	
Crew C-6A	Hr.	Daily	Hr.	Daily	Bare Costs	Incl. O&P
2 Cement Finishers	$49.95	$799.20	$73.45	$1175.20	$49.95	$73.45
1 Concrete Vibrator, Elec, 2 HP		47.45		52.20	2.97	3.26
16 L.H., Daily Totals		$846.65		$1227.40	$52.92	$76.71
Crew C-7	Hr.	Daily	Hr.	Daily	Bare Costs	Incl. O&P
1 Labor Foreman (outside)	$44.10	$352.80	$66.25	$530.00	$45.76	$68.48
5 Laborers	42.10	1684.00	63.25	2530.00		
1 Cement Finisher	49.95	399.60	73.45	587.60		
1 Equip. Oper. (medium)	56.75	454.00	84.85	678.80		
1 Equip. Oper. (oiler)	50.55	404.40	75.55	604.40		
2 Gas Engine Vibrators		53.70		59.07		
1 Concrete Bucket, 1 C.Y.		45.80		50.38		
1 Hyd. Crane, 55 Ton		980.20		1078.22	15.00	16.50
72 L.H., Daily Totals		$4374.50		$6118.47	$60.76	$84.98
Crew C-7A	Hr.	Daily	Hr.	Daily	Bare Costs	Incl. O&P
1 Labor Foreman (outside)	$44.10	$352.80	$66.25	$530.00	$44.06	$66.15
5 Laborers	42.10	1684.00	63.25	2530.00		
2 Truck Drivers (heavy)	48.95	783.20	73.35	1173.60		
2 Conc. Transit Mixers		1170.50		1287.55	18.29	20.12
64 L.H., Daily Totals		$3990.50		$5521.15	$62.35	$86.27
Crew C-7B	Hr.	Daily	Hr.	Daily	Bare Costs	Incl. O&P
1 Labor Foreman (outside)	$44.10	$352.80	$66.25	$530.00	$45.54	$68.32
5 Laborers	42.10	1684.00	63.25	2530.00		
1 Equipment Operator, Crane	59.20	473.60	88.50	708.00		
1 Equipment Oiler	50.55	404.40	75.55	604.40		
1 Conc. Bucket, 2 C.Y.		54.70		60.17		
1 Lattice Boom Crane, 165 Ton		2151.00		2366.10	34.46	37.91
64 L.H., Daily Totals		$5120.50		$6798.67	$80.01	$106.23
Crew C-7C	Hr.	Daily	Hr.	Daily	Bare Costs	Incl. O&P
1 Labor Foreman (outside)	$44.10	$352.80	$66.25	$530.00	$46.01	$69.03
5 Laborers	42.10	1684.00	63.25	2530.00		
2 Equipment Operators (med.)	56.75	908.00	84.85	1357.60		
2 F.E. Loaders, W.M., 4 C.Y.		1578.70		1736.57	24.67	27.13
64 L.H., Daily Totals		$4523.50		$6154.17	$70.68	$96.16
Crew C-7D	Hr.	Daily	Hr.	Daily	Bare Costs	Incl. O&P
1 Labor Foreman (outside)	$44.10	$352.80	$66.25	$530.00	$44.48	$66.76
5 Laborers	42.10	1684.00	63.25	2530.00		
1 Equip. Oper. (medium)	56.75	454.00	84.85	678.80		
1 Concrete Conveyer		204.15		224.57	3.65	4.01
56 L.H., Daily Totals		$2694.95		$3963.36	$48.12	$70.77
Crew C-8	Hr.	Daily	Hr.	Daily	Bare Costs	Incl. O&P
1 Labor Foreman (outside)	$44.10	$352.80	$66.25	$530.00	$46.72	$69.68
3 Laborers	42.10	1010.40	63.25	1518.00		
2 Cement Finishers	49.95	799.20	73.45	1175.20		
1 Equip. Oper. (medium)	56.75	454.00	84.85	678.80		
1 Concrete Pump (Small)		410.35		451.38	7.33	8.06
56 L.H., Daily Totals		$3026.75		$4353.39	$54.05	$77.74
Crew C-8A	Hr.	Daily	Hr.	Daily	Bare Costs	Incl. O&P
1 Labor Foreman (outside)	$44.10	$352.80	$66.25	$530.00	$45.05	$67.15
3 Laborers	42.10	1010.40	63.25	1518.00		
2 Cement Finishers	49.95	799.20	73.45	1175.20		
48 L.H., Daily Totals		$2162.40		$3223.20	$45.05	$67.15

Crew No.	Bare Costs		Incl. Subs O&P		Cost Per Labor-Hour	
Crew C-8B	Hr.	Daily	Hr.	Daily	Bare Costs	Incl. O&P
1 Labor Foreman (outside)	$44.10	$352.80	$66.25	$530.00	$45.43	$68.17
3 Laborers	42.10	1010.40	63.25	1518.00		
1 Equip. Oper. (medium)	56.75	454.00	84.85	678.80		
1 Vibrating Power Screed		75.10		82.61		
1 Roller, Vibratory, 25 Ton		664.35		730.78		
1 Dozer, 200 H.P.		1504.00		1654.40	56.09	61.69
40 L.H., Daily Totals		$4060.65		$5194.60	$101.52	$129.86
Crew C-8C	Hr.	Daily	Hr.	Daily	Bare Costs	Incl. O&P
1 Labor Foreman (outside)	$44.10	$352.80	$66.25	$530.00	$46.18	$69.05
3 Laborers	42.10	1010.40	63.25	1518.00		
1 Cement Finisher	49.95	399.60	73.45	587.60		
1 Equip. Oper. (medium)	56.75	454.00	84.85	678.80		
1 Shotcrete Rig, 12 C.Y./hr		245.40		269.94		
1 Air Compressor, 160 cfm		202.80		223.08		
4 -50' Air Hoses, 1"		32.00		35.20		
4 -50' Air Hoses, 2"		115.60		127.16	12.41	13.65
48 L.H., Daily Totals		$2812.60		$3969.78	$58.60	$82.70
Crew C-8D	Hr.	Daily	Hr.	Daily	Bare Costs	Incl. O&P
1 Labor Foreman (outside)	$44.10	$352.80	$66.25	$530.00	$47.29	$70.54
1 Laborer	42.10	336.80	63.25	506.00		
1 Cement Finisher	49.95	399.60	73.45	587.60		
1 Equipment Oper. (light)	53.00	424.00	79.20	633.60		
1 Air Compressor, 250 cfm		201.50		221.65		
2 -50' Air Hoses, 1"		16.00		17.60	6.80	7.48
32 L.H., Daily Totals		$1730.70		$2496.45	$54.08	$78.01
Crew C-8E	Hr.	Daily	Hr.	Daily	Bare Costs	Incl. O&P
1 Labor Foreman (outside)	$44.10	$352.80	$66.25	$530.00	$45.56	$68.11
3 Laborers	42.10	1010.40	63.25	1518.00		
1 Cement Finisher	49.95	399.60	73.45	587.60		
1 Equipment Oper. (light)	53.00	424.00	79.20	633.60		
1 Shotcrete Rig, 35 C.Y./hr		298.00		327.80		
1 Air Compressor, 250 cfm		201.50		221.65		
4 -50' Air Hoses, 1"		32.00		35.20		
4 -50' Air Hoses, 2"		115.60		127.16	13.48	14.83
48 L.H., Daily Totals		$2833.90		$3981.01	$59.04	$82.94
Crew C-9	Hr.	Daily	Hr.	Daily	Bare Costs	Incl. O&P
1 Cement Finisher	$49.95	$399.60	$73.45	$587.60	$46.79	$69.79
2 Laborers	42.10	673.60	63.25	1012.00		
1 Equipment Oper. (light)	53.00	424.00	79.20	633.60		
1 Grout Pump, 50 C.F./hr.		188.45		207.29		
1 Air Compressor, 160 cfm		202.80		223.08		
2 -50' Air Hoses, 1"		16.00		17.60		
2 -50' Air Hoses, 2"		57.80		63.58	14.53	15.99
32 L.H., Daily Totals		$1962.25		$2744.76	$61.32	$85.77
Crew C-10	Hr.	Daily	Hr.	Daily	Bare Costs	Incl. O&P
1 Laborer	$42.10	$336.80	$63.25	$506.00	$47.33	$70.05
2 Cement Finishers	49.95	799.20	73.45	1175.20		
24 L.H., Daily Totals		$1136.00		$1681.20	$47.33	$70.05
Crew C-10B	Hr.	Daily	Hr.	Daily	Bare Costs	Incl. O&P
3 Laborers	$42.10	$1010.40	$63.25	$1518.00	$45.24	$67.33
2 Cement Finishers	49.95	799.20	73.45	1175.20		
1 Concrete Mixer, 10 C.F.		140.30		154.33		
2 Trowels, 48" Walk-Behind		174.00		191.40	7.86	8.64
40 L.H., Daily Totals		$2123.90		$3038.93	$53.10	$75.97

Crew C-10C

	Bare Costs		Incl. Subs O&P		Cost Per Labor-Hour	
Crew C-10C	Hr.	Daily	Hr.	Daily	Bare Costs	Incl. O&P
1 Laborer	$42.10	$336.80	$63.25	$506.00	$47.33	$70.05
2 Cement Finishers	49.95	799.20	73.45	1175.20		
1 Trowel, 48" Walk-Behind		87.00		95.70	3.63	3.99
24 L.H., Daily Totals		$1223.00		$1776.90	$50.96	$74.04

Crew C-10D

	Bare Costs		Incl. Subs O&P		Cost Per Labor-Hour	
Crew C-10D	Hr.	Daily	Hr.	Daily	Bare Costs	Incl. O&P
1 Laborer	$42.10	$336.80	$63.25	$506.00	$47.33	$70.05
2 Cement Finishers	49.95	799.20	73.45	1175.20		
1 Vibrating Power Screed		75.10		82.61		
1 Trowel, 48" Walk-Behind		87.00		95.70	6.75	7.43
24 L.H., Daily Totals		$1298.10		$1859.51	$54.09	$77.48

Crew C-10E

	Bare Costs		Incl. Subs O&P		Cost Per Labor-Hour	
Crew C-10E	Hr.	Daily	Hr.	Daily	Bare Costs	Incl. O&P
1 Laborer	$42.10	$336.80	$63.25	$506.00	$47.33	$70.05
2 Cement Finishers	49.95	799.20	73.45	1175.20		
1 Vibrating Power Screed		75.10		82.61		
1 Cement Trowel, 96" Ride-On		169.00		185.90	10.17	11.19
24 L.H., Daily Totals		$1380.10		$1949.71	$57.50	$81.24

Crew C-10F

	Bare Costs		Incl. Subs O&P		Cost Per Labor-Hour	
Crew C-10F	Hr.	Daily	Hr.	Daily	Bare Costs	Incl. O&P
1 Laborer	$42.10	$336.80	$63.25	$506.00	$47.33	$70.05
2 Cement Finishers	49.95	799.20	73.45	1175.20		
1 Telescoping Boom Lift, to 60'		310.90		341.99	12.95	14.25
24 L.H., Daily Totals		$1446.90		$2023.19	$60.29	$84.30

Crew C-11

	Bare Costs		Incl. Subs O&P		Cost Per Labor-Hour	
Crew C-11	Hr.	Daily	Hr.	Daily	Bare Costs	Incl. O&P
1 Struc. Steel Foreman (outside)	$59.65	$477.20	$93.55	$748.40	$57.26	$88.92
6 Struc. Steel Workers	57.65	2767.20	90.45	4341.60		
1 Equip. Oper. (crane)	59.20	473.60	88.50	708.00		
1 Equip. Oper. (oiler)	50.55	404.40	75.55	604.40		
1 Lattice Boom Crane, 150 Ton		2269.00		2495.90	31.51	34.67
72 L.H., Daily Totals		$6391.40		$8898.30	$88.77	$123.59

Crew C-12

	Bare Costs		Incl. Subs O&P		Cost Per Labor-Hour	
Crew C-12	Hr.	Daily	Hr.	Daily	Bare Costs	Incl. O&P
1 Carpenter Foreman (outside)	$55.15	$441.20	$82.85	$662.80	$52.65	$79.03
3 Carpenters	53.15	1275.60	79.85	1916.40		
1 Laborer	42.10	336.80	63.25	506.00		
1 Equip. Oper. (crane)	59.20	473.60	88.50	708.00		
1 Hyd. Crane, 12 Ton		471.00		518.10	9.81	10.79
48 L.H., Daily Totals		$2998.20		$4311.30	$62.46	$89.82

Crew C-13

	Bare Costs		Incl. Subs O&P		Cost Per Labor-Hour	
Crew C-13	Hr.	Daily	Hr.	Daily	Bare Costs	Incl. O&P
1 Struc. Steel Worker	$57.65	$461.20	$90.45	$723.60	$56.08	$86.80
1 Welder	57.45	459.60	90.10	720.80		
1 Carpenter	53.15	425.20	79.85	638.80		
1 Welder, Gas Engine, 300 amp		147.00		161.70	6.13	6.74
24 L.H., Daily Totals		$1493.00		$2244.90	$62.21	$93.54

Crew C-14

	Bare Costs		Incl. Subs O&P		Cost Per Labor-Hour	
Crew C-14	Hr.	Daily	Hr.	Daily	Bare Costs	Incl. O&P
1 Carpenter Foreman (outside)	$55.15	$441.20	$82.85	$662.80	$51.36	$76.98
5 Carpenters	53.15	2126.00	79.85	3194.00		
4 Laborers	42.10	1347.20	63.25	2024.00		
4 Rodmen (reinf.)	56.40	1804.80	84.90	2716.80		
2 Cement Finishers	49.95	799.20	73.45	1175.20		
1 Equip. Oper. (crane)	59.20	473.60	88.50	708.00		
1 Equip. Oper. (oiler)	50.55	404.40	75.55	604.40		
1 Hyd. Crane, 80 Ton		1486.00		1634.60	10.32	11.35
144 L.H., Daily Totals		$8882.40		$12719.80	$61.68	$88.33

Crew C-14A

	Bare Costs		Incl. Subs O&P		Cost Per Labor-Hour	
Crew C-14A	Hr.	Daily	Hr.	Daily	Bare Costs	Incl. O&P
1 Carpenter Foreman (outside)	$55.15	$441.20	$82.85	$662.80	$52.88	$79.39
16 Carpenters	53.15	6803.20	79.85	10220.80		
4 Rodmen (reinf.)	56.40	1804.80	84.90	2716.80		
2 Laborers	42.10	673.60	63.25	1012.00		
1 Cement Finisher	49.95	399.60	73.45	587.60		
1 Equip. Oper. (medium)	56.75	454.00	84.85	678.80		
1 Gas Engine Vibrator		26.85		29.54		
1 Concrete Pump (Small)		410.35		451.38	2.19	2.40
200 L.H., Daily Totals		$11013.60		$16359.72	$55.07	$81.80

Crew C-14B

	Bare Costs		Incl. Subs O&P		Cost Per Labor-Hour	
Crew C-14B	Hr.	Daily	Hr.	Daily	Bare Costs	Incl. O&P
1 Carpenter Foreman (outside)	$55.15	$441.20	$82.85	$662.80	$52.77	$79.17
16 Carpenters	53.15	6803.20	79.85	10220.80		
4 Rodmen (reinf.)	56.40	1804.80	84.90	2716.80		
2 Laborers	42.10	673.60	63.25	1012.00		
2 Cement Finishers	49.95	799.20	73.45	1175.20		
1 Equip. Oper. (medium)	56.75	454.00	84.85	678.80		
1 Gas Engine Vibrator		26.85		29.54		
1 Concrete Pump (Small)		410.35		451.38	2.10	2.31
208 L.H., Daily Totals		$11413.20		$16947.32	$54.87	$81.48

Crew C-14C

	Bare Costs		Incl. Subs O&P		Cost Per Labor-Hour	
Crew C-14C	Hr.	Daily	Hr.	Daily	Bare Costs	Incl. O&P
1 Carpenter Foreman (outside)	$55.15	$441.20	$82.85	$662.80	$50.37	$75.59
6 Carpenters	53.15	2551.20	79.85	3832.80		
2 Rodmen (reinf.)	56.40	902.40	84.90	1358.40		
4 Laborers	42.10	1347.20	63.25	2024.00		
1 Cement Finisher	49.95	399.60	73.45	587.60		
1 Gas Engine Vibrator		26.85		29.54	.24	.26
112 L.H., Daily Totals		$5668.45		$8495.14	$50.61	$75.85

Crew C-14D

	Bare Costs		Incl. Subs O&P		Cost Per Labor-Hour	
Crew C-14D	Hr.	Daily	Hr.	Daily	Bare Costs	Incl. O&P
1 Carpenter Foreman (outside)	$55.15	$441.20	$82.85	$662.80	$52.62	$78.99
18 Carpenters	53.15	7653.60	79.85	11498.40		
2 Rodmen (reinf.)	56.40	902.40	84.90	1358.40		
2 Laborers	42.10	673.60	63.25	1012.00		
1 Cement Finisher	49.95	399.60	73.45	587.60		
1 Equip. Oper. (medium)	56.75	454.00	84.85	678.80		
1 Gas Engine Vibrator		26.85		29.54		
1 Concrete Pump (Small)		410.35		451.38	2.19	2.40
200 L.H., Daily Totals		$10961.60		$16278.92	$54.81	$81.39

Crew C-14E

	Bare Costs		Incl. Subs O&P		Cost Per Labor-Hour	
Crew C-14E	Hr.	Daily	Hr.	Daily	Bare Costs	Incl. O&P
1 Carpenter Foreman (outside)	$55.15	$441.20	$82.85	$662.80	$51.21	$76.85
2 Carpenters	53.15	850.40	79.85	1277.60		
4 Rodmen (reinf.)	56.40	1804.80	84.90	2716.80		
3 Laborers	42.10	1010.40	63.25	1518.00		
1 Cement Finisher	49.95	399.60	73.45	587.60		
1 Gas Engine Vibrator		26.85		29.54	.31	.34
88 L.H., Daily Totals		$4533.25		$6792.34	$51.51	$77.19

Crew C-14F

	Bare Costs		Incl. Subs O&P		Cost Per Labor-Hour	
Crew C-14F	Hr.	Daily	Hr.	Daily	Bare Costs	Incl. O&P
1 Labor Foreman (outside)	$44.10	$352.80	$66.25	$530.00	$47.56	$70.38
2 Laborers	42.10	673.60	63.25	1012.00		
6 Cement Finishers	49.95	2397.60	73.45	3525.60		
1 Gas Engine Vibrator		26.85		29.54	.37	.41
72 L.H., Daily Totals		$3450.85		$5097.14	$47.93	$70.79

For customer support on your Mechanical Costs with RSMeans Data, call 800.448.8182.

645

Crew No.	Bare Costs		Incl. Subs O&P		Cost Per Labor-Hour	

Crew C-14G

	Hr.	Daily	Hr.	Daily	Bare Costs	Incl. O&P
1 Labor Foreman (outside)	$44.10	$352.80	$66.25	$530.00	$46.87	$69.51
2 Laborers	42.10	673.60	63.25	1012.00		
4 Cement Finishers	49.95	1598.40	73.45	2350.40		
1 Gas Engine Vibrator		26.85		29.54	.48	.53
56 L.H., Daily Totals		$2651.65		$3921.93	$47.35	$70.03

Crew C-14H

	Hr.	Daily	Hr.	Daily	Bare Costs	Incl. O&P
1 Carpenter Foreman (outside)	$55.15	$441.20	$82.85	$662.80	$51.65	$77.36
2 Carpenters	53.15	850.40	79.85	1277.60		
1 Rodman (reinf.)	56.40	451.20	84.90	679.20		
1 Laborer	42.10	336.80	63.25	506.00		
1 Cement Finisher	49.95	399.60	73.45	587.60		
1 Gas Engine Vibrator		26.85		29.54	.56	.62
48 L.H., Daily Totals		$2506.05		$3742.74	$52.21	$77.97

Crew C-14L

	Hr.	Daily	Hr.	Daily	Bare Costs	Incl. O&P
1 Carpenter Foreman (outside)	$55.15	$441.20	$82.85	$662.80	$49.37	$74.03
6 Carpenters	53.15	2551.20	79.85	3832.80		
4 Laborers	42.10	1347.20	63.25	2024.00		
1 Cement Finisher	49.95	399.60	73.45	587.60		
1 Gas Engine Vibrator		26.85		29.54	.28	.31
96 L.H., Daily Totals		$4766.05		$7136.73	$49.65	$74.34

Crew C-14M

	Hr.	Daily	Hr.	Daily	Bare Costs	Incl. O&P
1 Carpenter Foreman (outside)	$55.15	$441.20	$82.85	$662.80	$51.09	$76.53
2 Carpenters	53.15	850.40	79.85	1277.60		
1 Rodman (reinf.)	56.40	451.20	84.90	679.20		
2 Laborers	42.10	673.60	63.25	1012.00		
1 Cement Finisher	49.95	399.60	73.45	587.60		
1 Equip. Oper. (medium)	56.75	454.00	84.85	678.80		
1 Gas Engine Vibrator		26.85		29.54		
1 Concrete Pump (Small)		410.35		451.38	6.83	7.51
64 L.H., Daily Totals		$3707.20		$5378.92	$57.93	$84.05

Crew C-15

	Hr.	Daily	Hr.	Daily	Bare Costs	Incl. O&P
1 Carpenter Foreman (outside)	$55.15	$441.20	$82.85	$662.80	$49.34	$73.79
2 Carpenters	53.15	850.40	79.85	1277.60		
3 Laborers	42.10	1010.40	63.25	1518.00		
2 Cement Finishers	49.95	799.20	73.45	1175.20		
1 Rodman (reinf.)	56.40	451.20	84.90	679.20		
72 L.H., Daily Totals		$3552.40		$5312.80	$49.34	$73.79

Crew C-16

	Hr.	Daily	Hr.	Daily	Bare Costs	Incl. O&P
1 Labor Foreman (outside)	$44.10	$352.80	$66.25	$530.00	$46.72	$69.68
3 Laborers	42.10	1010.40	63.25	1518.00		
2 Cement Finishers	49.95	799.20	73.45	1175.20		
1 Equip. Oper. (medium)	56.75	454.00	84.85	678.80		
1 Gunite Pump Rig		317.90		349.69		
2 -50' Air Hoses, 3/4"		15.50		17.05		
2 -50' Air Hoses, 2"		57.80		63.58	6.99	7.68
56 L.H., Daily Totals		$3007.60		$4332.32	$53.71	$77.36

Crew C-16A

	Hr.	Daily	Hr.	Daily	Bare Costs	Incl. O&P
1 Laborer	$42.10	$336.80	$63.25	$506.00	$49.69	$73.75
2 Cement Finishers	49.95	799.20	73.45	1175.20		
1 Equip. Oper. (medium)	56.75	454.00	84.85	678.80		
1 Gunite Pump Rig		317.90		349.69		
2 -50' Air Hoses, 3/4"		15.50		17.05		
2 -50' Air Hoses, 2"		57.80		63.58		
1 Telescoping Boom Lift, to 60'		310.90		341.99	21.94	24.13
32 L.H., Daily Totals		$2292.10		$3132.31	$71.63	$97.88

Crew C-17

	Hr.	Daily	Hr.	Daily	Bare Costs	Incl. O&P
2 Skilled Worker Foremen (out)	$56.85	$909.60	$86.00	$1376.00	$55.25	$83.56
8 Skilled Workers	54.85	3510.40	82.95	5308.80		
80 L.H., Daily Totals		$4420.00		$6684.80	$55.25	$83.56

Crew C-17A

	Hr.	Daily	Hr.	Daily	Bare Costs	Incl. O&P
2 Skilled Worker Foremen (out)	$56.85	$909.60	$86.00	$1376.00	$55.30	$83.62
8 Skilled Workers	54.85	3510.40	82.95	5308.80		
.125 Equip. Oper. (crane)	59.20	59.20	88.50	88.50		
.125 Hyd. Crane, 80 Ton		185.75		204.32	2.29	2.52
81 L.H., Daily Totals		$4664.95		$6977.63	$57.59	$86.14

Crew C-17B

	Hr.	Daily	Hr.	Daily	Bare Costs	Incl. O&P
2 Skilled Worker Foremen (out)	$56.85	$909.60	$86.00	$1376.00	$55.35	$83.68
8 Skilled Workers	54.85	3510.40	82.95	5308.80		
.25 Equip. Oper. (crane)	59.20	118.40	88.50	177.00		
.25 Hyd. Crane, 80 Ton		371.50		408.65		
.25 Trowel, 48" Walk-Behind		21.75		23.93	4.80	5.28
82 L.H., Daily Totals		$4931.65		$7294.38	$60.14	$88.96

Crew C-17C

	Hr.	Daily	Hr.	Daily	Bare Costs	Incl. O&P
2 Skilled Worker Foremen (out)	$56.85	$909.60	$86.00	$1376.00	$55.39	$83.74
8 Skilled Workers	54.85	3510.40	82.95	5308.80		
.375 Equip. Oper. (crane)	59.20	177.60	88.50	265.50		
.375 Hyd. Crane, 80 Ton		557.25		612.98	6.71	7.39
83 L.H., Daily Totals		$5154.85		$7563.27	$62.11	$91.12

Crew C-17D

	Hr.	Daily	Hr.	Daily	Bare Costs	Incl. O&P
2 Skilled Worker Foremen (out)	$56.85	$909.60	$86.00	$1376.00	$55.44	$83.80
8 Skilled Workers	54.85	3510.40	82.95	5308.80		
.5 Equip. Oper. (crane)	59.20	236.80	88.50	354.00		
.5 Hyd. Crane, 80 Ton		743.00		817.30	8.85	9.73
84 L.H., Daily Totals		$5399.80		$7856.10	$64.28	$93.53

Crew C-17E

	Hr.	Daily	Hr.	Daily	Bare Costs	Incl. O&P
2 Skilled Worker Foremen (out)	$56.85	$909.60	$86.00	$1376.00	$55.25	$83.56
8 Skilled Workers	54.85	3510.40	82.95	5308.80		
1 Hyd. Jack with Rods		35.90		39.49	.45	.49
80 L.H., Daily Totals		$4455.90		$6724.29	$55.70	$84.05

Crew C-18

	Hr.	Daily	Hr.	Daily	Bare Costs	Incl. O&P
.125 Labor Foreman (outside)	$44.10	$44.10	$66.25	$66.25	$42.32	$63.58
1 Laborer	42.10	336.80	63.25	506.00		
1 Concrete Cart, 10 C.F.		127.85		140.63	14.21	15.63
9 L.H., Daily Totals		$508.75		$712.88	$56.53	$79.21

Crew C-19

	Hr.	Daily	Hr.	Daily	Bare Costs	Incl. O&P
.125 Labor Foreman (outside)	$44.10	$44.10	$66.25	$66.25	$42.32	$63.58
1 Laborer	42.10	336.80	63.25	506.00		
1 Concrete Cart, 18 C.F.		153.45		168.79	17.05	18.75
9 L.H., Daily Totals		$534.35		$741.04	$59.37	$82.34

Crew C-20

	Hr.	Daily	Hr.	Daily	Bare Costs	Incl. O&P
1 Labor Foreman (outside)	$44.10	$352.80	$66.25	$530.00	$45.16	$67.60
5 Laborers	42.10	1684.00	63.25	2530.00		
1 Cement Finisher	49.95	399.60	73.45	587.60		
1 Equip. Oper. (medium)	56.75	454.00	84.85	678.80		
2 Gas Engine Vibrators		53.70		59.07		
1 Concrete Pump (Small)		410.35		451.38	7.25	7.98
64 L.H., Daily Totals		$3354.45		$4836.85	$52.41	$75.58

Crews - Standard

Crew No.	Bare Costs Hr.	Daily	Incl. Subs O&P Hr.	Daily	Cost Per Labor-Hour Bare Costs	Incl. O&P
Crew C-21	Hr.	Daily	Hr.	Daily	Bare Costs	Incl. O&P
1 Labor Foreman (outside)	$44.10	$352.80	$66.25	$530.00	$45.16	$67.60
5 Laborers	42.10	1684.00	63.25	2530.00		
1 Cement Finisher	49.95	399.60	73.45	587.60		
1 Equip. Oper. (medium)	56.75	454.00	84.85	678.80		
2 Gas Engine Vibrators		53.70		59.07		
1 Concrete Conveyer		204.15		224.57	4.03	4.43
64 L.H., Daily Totals		$3148.25		$4610.03	$49.19	$72.03
Crew C-22	Hr.	Daily	Hr.	Daily	Bare Costs	Incl. O&P
1 Rodman Foreman (outside)	$58.40	$467.20	$87.95	$703.60	$56.71	$85.34
4 Rodmen (reinf.)	56.40	1804.80	84.90	2716.80		
.125 Equip. Oper. (crane)	59.20	59.20	88.50	88.50		
.125 Equip. Oper. (oiler)	50.55	50.55	75.55	75.55		
.125 Hyd. Crane, 25 Ton		72.61		79.87	1.73	1.90
42 L.H., Daily Totals		$2454.36		$3664.32	$58.44	$87.25
Crew C-23	Hr.	Daily	Hr.	Daily	Bare Costs	Incl. O&P
2 Skilled Worker Foremen (out)	$56.85	$909.60	$86.00	$1376.00	$55.26	$83.38
6 Skilled Workers	54.85	2632.80	82.95	3981.60		
1 Equip. Oper. (crane)	59.20	473.60	88.50	708.00		
1 Equip. Oper. (oiler)	50.55	404.40	75.55	604.40		
1 Lattice Boom Crane, 90 Ton		1696.00		1865.60	21.20	23.32
80 L.H., Daily Totals		$6116.40		$8535.60	$76.45	$106.69
Crew C-23A	Hr.	Daily	Hr.	Daily	Bare Costs	Incl. O&P
1 Labor Foreman (outside)	$44.10	$352.80	$66.25	$530.00	$47.61	$71.36
2 Laborers	42.10	673.60	63.25	1012.00		
1 Equip. Oper. (crane)	59.20	473.60	88.50	708.00		
1 Equip. Oper. (oiler)	50.55	404.40	75.55	604.40		
1 Crawler Crane, 100 Ton		2287.00		2515.70		
3 Conc. Buckets, 8 C.Y.		323.25		355.57	65.26	71.78
40 L.H., Daily Totals		$4514.65		$5725.68	$112.87	$143.14
Crew C-24	Hr.	Daily	Hr.	Daily	Bare Costs	Incl. O&P
2 Skilled Worker Foremen (out)	$56.85	$909.60	$86.00	$1376.00	$55.26	$83.38
6 Skilled Workers	54.85	2632.80	82.95	3981.60		
1 Equip. Oper. (crane)	59.20	473.60	88.50	708.00		
1 Equip. Oper. (oiler)	50.55	404.40	75.55	604.40		
1 Lattice Boom Crane, 150 Ton		2269.00		2495.90	28.36	31.20
80 L.H., Daily Totals		$6689.40		$9165.90	$83.62	$114.57
Crew C-25	Hr.	Daily	Hr.	Daily	Bare Costs	Incl. O&P
2 Rodmen (reinf.)	$56.40	$902.40	$84.90	$1358.40	$45.50	$71.00
2 Rodmen Helpers	34.60	553.60	57.10	913.60		
32 L.H., Daily Totals		$1456.00		$2272.00	$45.50	$71.00
Crew C-27	Hr.	Daily	Hr.	Daily	Bare Costs	Incl. O&P
2 Cement Finishers	$49.95	$799.20	$73.45	$1175.20	$49.95	$73.45
1 Concrete Saw		111.50		122.65	6.97	7.67
16 L.H., Daily Totals		$910.70		$1297.85	$56.92	$81.12
Crew C-28	Hr.	Daily	Hr.	Daily	Bare Costs	Incl. O&P
1 Cement Finisher	$49.95	$399.60	$73.45	$587.60	$49.95	$73.45
1 Portable Air Compressor, Gas		38.70		42.57	4.84	5.32
8 L.H., Daily Totals		$438.30		$630.17	$54.79	$78.77
Crew C-29	Hr.	Daily	Hr.	Daily	Bare Costs	Incl. O&P
1 Laborer	$42.10	$336.80	$63.25	$506.00	$42.10	$63.25
1 Pressure Washer		96.95		106.65	12.12	13.33
8 L.H., Daily Totals		$433.75		$612.64	$54.22	$76.58
Crew C-30	Hr.	Daily	Hr.	Daily	Bare Costs	Incl. O&P
1 Laborer	$42.10	$336.80	$63.25	$506.00	$42.10	$63.25
1 Concrete Mixer, 10 C.F.		140.30		154.33	17.54	19.29
8 L.H., Daily Totals		$477.10		$660.33	$59.64	$82.54
Crew C-31	Hr.	Daily	Hr.	Daily	Bare Costs	Incl. O&P
1 Cement Finisher	$49.95	$399.60	$73.45	$587.60	$49.95	$73.45
1 Grout Pump		317.90		349.69	39.74	43.71
8 L.H., Daily Totals		$717.50		$937.29	$89.69	$117.16
Crew C-32	Hr.	Daily	Hr.	Daily	Bare Costs	Incl. O&P
1 Cement Finisher	$49.95	$399.60	$73.45	$587.60	$46.02	$68.35
1 Laborer	42.10	336.80	63.25	506.00		
1 Crack Chaser Saw, Gas, 6 H.P.		73.10		80.41		
1 Vacuum Pick-Up System		74.05		81.45	9.20	10.12
16 L.H., Daily Totals		$883.55		$1255.46	$55.22	$78.47
Crew D-1	Hr.	Daily	Hr.	Daily	Bare Costs	Incl. O&P
1 Bricklayer	$52.05	$416.40	$78.90	$631.20	$46.90	$71.10
1 Bricklayer Helper	41.75	334.00	63.30	506.40		
16 L.H., Daily Totals		$750.40		$1137.60	$46.90	$71.10
Crew D-2	Hr.	Daily	Hr.	Daily	Bare Costs	Incl. O&P
3 Bricklayers	$52.05	$1249.20	$78.90	$1893.60	$48.40	$73.31
2 Bricklayer Helpers	41.75	668.00	63.30	1012.80		
.5 Carpenter	53.15	212.60	79.85	319.40		
44 L.H., Daily Totals		$2129.80		$3225.80	$48.40	$73.31
Crew D-3	Hr.	Daily	Hr.	Daily	Bare Costs	Incl. O&P
3 Bricklayers	$52.05	$1249.20	$78.90	$1893.60	$48.18	$73.00
2 Bricklayer Helpers	41.75	668.00	63.30	1012.80		
.25 Carpenter	53.15	106.30	79.85	159.70		
42 L.H., Daily Totals		$2023.50		$3066.10	$48.18	$73.00
Crew D-4	Hr.	Daily	Hr.	Daily	Bare Costs	Incl. O&P
1 Bricklayer	$52.05	$416.40	$78.90	$631.20	$47.14	$71.17
2 Bricklayer Helpers	41.75	668.00	63.30	1012.80		
1 Equip. Oper. (light)	53.00	424.00	79.20	633.60		
1 Grout Pump, 50 C.F./hr.		188.45		207.29	5.89	6.48
32 L.H., Daily Totals		$1696.85		$2484.90	$53.03	$77.65
Crew D-5	Hr.	Daily	Hr.	Daily	Bare Costs	Incl. O&P
1 Bricklayer	52.05	416.40	78.90	631.20	52.05	78.90
8 L.H., Daily Totals		$416.40		$631.20	$52.05	$78.90
Crew D-6	Hr.	Daily	Hr.	Daily	Bare Costs	Incl. O&P
3 Bricklayers	$52.05	$1249.20	$78.90	$1893.60	$47.15	$71.45
3 Bricklayer Helpers	41.75	1002.00	63.30	1519.20		
.25 Carpenter	53.15	106.30	79.85	159.70		
50 L.H., Daily Totals		$2357.50		$3572.50	$47.15	$71.45
Crew D-7	Hr.	Daily	Hr.	Daily	Bare Costs	Incl. O&P
1 Tile Layer	$49.50	$396.00	$72.60	$580.80	$44.15	$64.75
1 Tile Layer Helper	38.80	310.40	56.90	455.20		
16 L.H., Daily Totals		$706.40		$1036.00	$44.15	$64.75
Crew D-8	Hr.	Daily	Hr.	Daily	Bare Costs	Incl. O&P
3 Bricklayers	$52.05	$1249.20	$78.90	$1893.60	$47.93	$72.66
2 Bricklayer Helpers	41.75	668.00	63.30	1012.80		
40 L.H., Daily Totals		$1917.20		$2906.40	$47.93	$72.66

For customer support on your Mechanical Costs with RSMeans Data, call 800.448.8182.

647

Crew No.	Bare Costs Hr.	Bare Costs Daily	Incl. Subs O&P Hr.	Incl. Subs O&P Daily	Cost Per Labor-Hour Bare Costs	Cost Per Labor-Hour Incl. O&P
Crew D-9	Hr.	Daily	Hr.	Daily	Bare Costs	Incl. O&P
3 Bricklayers	$52.05	$1249.20	$78.90	$1893.60	$46.90	$71.10
3 Bricklayer Helpers	41.75	1002.00	63.30	1519.20		
48 L.H., Daily Totals		$2251.20		$3412.80	$46.90	$71.10
Crew D-10	Hr.	Daily	Hr.	Daily	Bare Costs	Incl. O&P
1 Bricklayer Foreman (outside)	$54.05	$432.40	$81.95	$655.60	$51.76	$78.16
1 Bricklayer	52.05	416.40	78.90	631.20		
1 Bricklayer Helper	41.75	334.00	63.30	506.40		
1 Equip. Oper. (crane)	59.20	473.60	88.50	708.00		
1 S.P. Crane, 4x4, 12 Ton		428.30		471.13	13.38	14.72
32 L.H., Daily Totals		$2084.70		$2972.33	$65.15	$92.89
Crew D-11	Hr.	Daily	Hr.	Daily	Bare Costs	Incl. O&P
1 Bricklayer Foreman (outside)	$54.05	$432.40	$81.95	$655.60	$49.28	$74.72
1 Bricklayer	52.05	416.40	78.90	631.20		
1 Bricklayer Helper	41.75	334.00	63.30	506.40		
24 L.H., Daily Totals		$1182.80		$1793.20	$49.28	$74.72
Crew D-12	Hr.	Daily	Hr.	Daily	Bare Costs	Incl. O&P
1 Bricklayer Foreman (outside)	$54.05	$432.40	$81.95	$655.60	$47.40	$71.86
1 Bricklayer	52.05	416.40	78.90	631.20		
2 Bricklayer Helpers	41.75	668.00	63.30	1012.80		
32 L.H., Daily Totals		$1516.80		$2299.60	$47.40	$71.86
Crew D-13	Hr.	Daily	Hr.	Daily	Bare Costs	Incl. O&P
1 Bricklayer Foreman (outside)	$54.05	$432.40	$81.95	$655.60	$50.33	$75.97
1 Bricklayer	52.05	416.40	78.90	631.20		
2 Bricklayer Helpers	41.75	668.00	63.30	1012.80		
1 Carpenter	53.15	425.20	79.85	638.80		
1 Equip. Oper. (crane)	59.20	473.60	88.50	708.00		
1 S.P. Crane, 4x4, 12 Ton		428.30		471.13	8.92	9.82
48 L.H., Daily Totals		$2843.90		$4117.53	$59.25	$85.78
Crew D-14	Hr.	Daily	Hr.	Daily	Bare Costs	Incl. O&P
3 Bricklayers	$52.05	$1249.20	$78.90	$1893.60	$49.48	$75.00
1 Bricklayer Helper	41.75	334.00	63.30	506.40		
32 L.H., Daily Totals		$1583.20		$2400.00	$49.48	$75.00
Crew E-1	Hr.	Daily	Hr.	Daily	Bare Costs	Incl. O&P
1 Welder Foreman (outside)	$59.45	$475.60	$93.25	$746.00	$56.63	$87.52
1 Welder	57.45	459.60	90.10	720.80		
1 Equip. Oper. (light)	53.00	424.00	79.20	633.60		
1 Welder, Gas Engine, 300 amp		147.00		161.70	6.13	6.74
24 L.H., Daily Totals		$1506.20		$2262.10	$62.76	$94.25
Crew E-2	Hr.	Daily	Hr.	Daily	Bare Costs	Incl. O&P
1 Struc. Steel Foreman (outside)	$59.65	$477.20	$93.55	$748.40	$57.14	$88.49
4 Struc. Steel Workers	57.65	1844.80	90.45	2894.40		
1 Equip. Oper. (crane)	59.20	473.60	88.50	708.00		
1 Equip. Oper. (oiler)	50.55	404.40	75.55	604.40		
1 Lattice Boom Crane, 90 Ton		1696.00		1865.60	30.29	33.31
56 L.H., Daily Totals		$4896.00		$6820.80	$87.43	$121.80
Crew E-3	Hr.	Daily	Hr.	Daily	Bare Costs	Incl. O&P
1 Struc. Steel Foreman (outside)	$59.65	$477.20	$93.55	$748.40	$58.25	$91.37
1 Struc. Steel Worker	57.65	461.20	90.45	723.60		
1 Welder	57.45	459.60	90.10	720.80		
1 Welder, Gas Engine, 300 amp		147.00		161.70	6.13	6.74
24 L.H., Daily Totals		$1545.00		$2354.50	$64.38	$98.10

Crew No.	Bare Costs Hr.	Bare Costs Daily	Incl. Subs O&P Hr.	Incl. Subs O&P Daily	Cost Per Labor-Hour Bare Costs	Cost Per Labor-Hour Incl. O&P
Crew E-3A	Hr.	Daily	Hr.	Daily	Bare Costs	Incl. O&P
1 Struc. Steel Foreman (outside)	$59.65	$477.20	$93.55	$748.40	$58.25	$91.37
1 Struc. Steel Worker	57.65	461.20	90.45	723.60		
1 Welder	57.45	459.60	90.10	720.80		
1 Welder, Gas Engine, 300 amp		147.00		161.70		
1 Telescoping Boom Lift, to 40'		278.90		306.79	17.75	19.52
24 L.H., Daily Totals		$1823.90		$2661.29	$76.00	$110.89
Crew E-4	Hr.	Daily	Hr.	Daily	Bare Costs	Incl. O&P
1 Struc. Steel Foreman (outside)	$59.65	$477.20	$93.55	$748.40	$58.15	$91.22
3 Struc. Steel Workers	57.65	1383.60	90.45	2170.80		
1 Welder, Gas Engine, 300 amp		147.00		161.70	4.59	5.05
32 L.H., Daily Totals		$2007.80		$3080.90	$62.74	$96.28
Crew E-5	Hr.	Daily	Hr.	Daily	Bare Costs	Incl. O&P
2 Struc. Steel Foremen (outside)	$59.65	$954.40	$93.55	$1496.80	$57.48	$89.35
5 Struc. Steel Workers	57.65	2306.00	90.45	3618.00		
1 Equip. Oper. (crane)	59.20	473.60	88.50	708.00		
1 Welder	57.45	459.60	90.10	720.80		
1 Equip. Oper. (oiler)	50.55	404.40	75.55	604.40		
1 Lattice Boom Crane, 90 Ton		1696.00		1865.60		
1 Welder, Gas Engine, 300 amp		147.00		161.70	23.04	25.34
80 L.H., Daily Totals		$6441.00		$9175.30	$80.51	$114.69
Crew E-6	Hr.	Daily	Hr.	Daily	Bare Costs	Incl. O&P
3 Struc. Steel Foremen (outside)	$59.65	$1431.60	$93.55	$2245.20	$57.38	$89.25
9 Struc. Steel Workers	57.65	4150.80	90.45	6512.40		
1 Equip. Oper. (crane)	59.20	473.60	88.50	708.00		
1 Welder	57.45	459.60	90.10	720.80		
1 Equip. Oper. (oiler)	50.55	404.40	75.55	604.40		
1 Equip. Oper. (light)	53.00	424.00	79.20	633.60		
1 Lattice Boom Crane, 90 Ton		1696.00		1865.60		
1 Welder, Gas Engine, 300 amp		147.00		161.70		
1 Air Compressor, 160 cfm		202.80		223.08		
2 Impact Wrenches		104.60		115.06	16.80	18.48
128 L.H., Daily Totals		$9494.40		$13789.84	$74.17	$107.73
Crew E-7	Hr.	Daily	Hr.	Daily	Bare Costs	Incl. O&P
1 Struc. Steel Foreman (outside)	$59.65	$477.20	$93.55	$748.40	$57.44	$89.28
4 Struc. Steel Workers	57.65	1844.80	90.45	2894.40		
1 Equip. Oper. (crane)	59.20	473.60	88.50	708.00		
1 Equip. Oper. (oiler)	50.55	404.40	75.55	604.40		
1 Welder Foreman (outside)	59.45	475.60	93.25	746.00		
2 Welders	57.45	919.20	90.10	1441.60		
1 Lattice Boom Crane, 90 Ton		1696.00		1865.60		
2 Welder, Gas Engine, 300 amp		294.00		323.40	24.88	27.36
80 L.H., Daily Totals		$6584.80		$9331.80	$82.31	$116.65
Crew E-8	Hr.	Daily	Hr.	Daily	Bare Costs	Incl. O&P
1 Struc. Steel Foreman (outside)	$59.65	$477.20	$93.55	$748.40	$57.10	$88.63
4 Struc. Steel Workers	57.65	1844.80	90.45	2894.40		
1 Welder Foreman (outside)	59.45	475.60	93.25	746.00		
4 Welders	57.45	1838.40	90.10	2883.20		
1 Equip. Oper. (crane)	59.20	473.60	88.50	708.00		
1 Equip. Oper. (oiler)	50.55	404.40	75.55	604.40		
1 Equip. Oper. (light)	53.00	424.00	79.20	633.60		
1 Lattice Boom Crane, 90 Ton		1696.00		1865.60		
4 Welder, Gas Engine, 300 amp		588.00		646.80	21.96	24.16
104 L.H., Daily Totals		$8222.00		$11730.40	$79.06	$112.79

For customer support on your Mechanical Costs with RSMeans Data, call 800.448.8182.

Crew E-9

	Bare Costs Hr.	Bare Costs Daily	Incl. Subs O&P Hr.	Incl. Subs O&P Daily	Cost Per Labor-Hour Bare Costs	Cost Per Labor-Hour Incl. O&P
2 Struc. Steel Foremen (outside)	$59.65	$954.40	$93.55	$1496.80	$57.31	$89.15
5 Struc. Steel Workers	57.65	2306.00	90.45	3618.00		
1 Welder Foreman (outside)	59.45	475.60	93.25	746.00		
5 Welders	57.45	2298.00	90.10	3604.00		
1 Equip. Oper. (crane)	59.20	473.60	88.50	708.00		
1 Equip. Oper. (oiler)	50.55	404.40	75.55	604.40		
1 Equip. Oper. (light)	53.00	424.00	79.20	633.60		
1 Lattice Boom Crane, 90 Ton		1696.00		1865.60		
5 Welder, Gas Engine, 300 amp		735.00		808.50	18.99	20.89
128 L.H., Daily Totals		$9767.00		$14084.90	$76.30	$110.04

Crew E-10

	Bare Costs Hr.	Bare Costs Daily	Incl. Subs O&P Hr.	Incl. Subs O&P Daily	Cost Per Labor-Hour Bare Costs	Cost Per Labor-Hour Incl. O&P
1 Welder Foreman (outside)	$59.45	$475.60	$93.25	$746.00	$58.45	$91.67
1 Welder	57.45	459.60	90.10	720.80		
1 Welder, Gas Engine, 300 amp		147.00		161.70		
1 Flatbed Truck, Gas, 3 Ton		820.40		902.44	60.46	66.51
16 L.H., Daily Totals		$1902.60		$2530.94	$118.91	$158.18

Crew E-11

	Bare Costs Hr.	Bare Costs Daily	Incl. Subs O&P Hr.	Incl. Subs O&P Daily	Cost Per Labor-Hour Bare Costs	Cost Per Labor-Hour Incl. O&P
2 Painters, Struc. Steel	$45.85	$733.60	$74.45	$1191.20	$46.70	$72.84
1 Building Laborer	42.10	336.80	63.25	506.00		
1 Equip. Oper. (light)	53.00	424.00	79.20	633.60		
1 Air Compressor, 250 cfm		201.50		221.65		
1 Sandblaster, Portable, 3 C.F.		83.80		92.18		
1 Set Sand Blasting Accessories		15.35		16.89	9.40	10.33
32 L.H., Daily Totals		$1795.05		$2661.51	$56.10	$83.17

Crew E-11A

	Bare Costs Hr.	Bare Costs Daily	Incl. Subs O&P Hr.	Incl. Subs O&P Daily	Cost Per Labor-Hour Bare Costs	Cost Per Labor-Hour Incl. O&P
2 Painters, Struc. Steel	$45.85	$733.60	$74.45	$1191.20	$46.70	$72.84
1 Building Laborer	42.10	336.80	63.25	506.00		
1 Equip. Oper. (light)	53.00	424.00	79.20	633.60		
1 Air Compressor, 250 cfm		201.50		221.65		
1 Sandblaster, Portable, 3 C.F.		83.80		92.18		
1 Set Sand Blasting Accessories		15.35		16.89		
1 Telescoping Boom Lift, to 60'		310.90		341.99	19.11	21.02
32 L.H., Daily Totals		$2105.95		$3003.51	$65.81	$93.86

Crew E-11B

	Bare Costs Hr.	Bare Costs Daily	Incl. Subs O&P Hr.	Incl. Subs O&P Daily	Cost Per Labor-Hour Bare Costs	Cost Per Labor-Hour Incl. O&P
2 Painters, Struc. Steel	$45.85	$733.60	$74.45	$1191.20	$44.60	$70.72
1 Building Laborer	42.10	336.80	63.25	506.00		
2 Paint Sprayer, 8 C.F.M.		87.50		96.25		
1 Telescoping Boom Lift, to 60'		310.90		341.99	16.60	18.26
24 L.H., Daily Totals		$1468.80		$2135.44	$61.20	$88.98

Crew E-12

	Bare Costs Hr.	Bare Costs Daily	Incl. Subs O&P Hr.	Incl. Subs O&P Daily	Cost Per Labor-Hour Bare Costs	Cost Per Labor-Hour Incl. O&P
1 Welder Foreman (outside)	$59.45	$475.60	$93.25	$746.00	$56.23	$86.22
1 Equip. Oper. (light)	53.00	424.00	79.20	633.60		
1 Welder, Gas Engine, 300 amp		147.00		161.70	9.19	10.11
16 L.H., Daily Totals		$1046.60		$1541.30	$65.41	$96.33

Crew E-13

	Bare Costs Hr.	Bare Costs Daily	Incl. Subs O&P Hr.	Incl. Subs O&P Daily	Cost Per Labor-Hour Bare Costs	Cost Per Labor-Hour Incl. O&P
1 Welder Foreman (outside)	$59.45	$475.60	$93.25	$746.00	$57.30	$88.57
.5 Equip. Oper. (light)	53.00	212.00	79.20	316.80		
1 Welder, Gas Engine, 300 amp		147.00		161.70	12.25	13.48
12 L.H., Daily Totals		$834.60		$1224.50	$69.55	$102.04

Crew E-14

	Bare Costs Hr.	Bare Costs Daily	Incl. Subs O&P Hr.	Incl. Subs O&P Daily	Cost Per Labor-Hour Bare Costs	Cost Per Labor-Hour Incl. O&P
1 Welder Foreman (outside)	$59.45	$475.60	$93.25	$746.00	$59.45	$93.25
1 Welder, Gas Engine, 300 amp		147.00		161.70	18.38	20.21
8 L.H., Daily Totals		$622.60		$907.70	$77.83	$113.46

Crew E-16

	Bare Costs Hr.	Bare Costs Daily	Incl. Subs O&P Hr.	Incl. Subs O&P Daily	Cost Per Labor-Hour Bare Costs	Cost Per Labor-Hour Incl. O&P
1 Welder Foreman (outside)	$59.45	$475.60	$93.25	$746.00	$58.45	$91.67
1 Welder	57.45	459.60	90.10	720.80		
1 Welder, Gas Engine, 300 amp		147.00		161.70	9.19	10.11
16 L.H., Daily Totals		$1082.20		$1628.50	$67.64	$101.78

Crew E-17

	Bare Costs Hr.	Bare Costs Daily	Incl. Subs O&P Hr.	Incl. Subs O&P Daily	Cost Per Labor-Hour Bare Costs	Cost Per Labor-Hour Incl. O&P
1 Struc. Steel Foreman (outside)	$59.65	$477.20	$93.55	$748.40	$58.65	$92.00
1 Structural Steel Worker	57.65	461.20	90.45	723.60		
16 L.H., Daily Totals		$938.40		$1472.00	$58.65	$92.00

Crew E-18

	Bare Costs Hr.	Bare Costs Daily	Incl. Subs O&P Hr.	Incl. Subs O&P Daily	Cost Per Labor-Hour Bare Costs	Cost Per Labor-Hour Incl. O&P
1 Struc. Steel Foreman (outside)	$59.65	$477.20	$93.55	$748.40	$57.87	$89.95
3 Structural Steel Workers	57.65	1383.60	90.45	2170.80		
1 Equipment Operator (med.)	56.75	454.00	84.85	678.80		
1 Lattice Boom Crane, 20 Ton		1488.00		1636.80	37.20	40.92
40 L.H., Daily Totals		$3802.80		$5234.80	$95.07	$130.87

Crew E-19

	Bare Costs Hr.	Bare Costs Daily	Incl. Subs O&P Hr.	Incl. Subs O&P Daily	Cost Per Labor-Hour Bare Costs	Cost Per Labor-Hour Incl. O&P
1 Struc. Steel Foreman (outside)	$59.65	$477.20	$93.55	$748.40	$56.77	$87.73
1 Structural Steel Worker	57.65	461.20	90.45	723.60		
1 Equip. Oper. (light)	53.00	424.00	79.20	633.60		
1 Lattice Boom Crane, 20 Ton		1488.00		1636.80	62.00	68.20
24 L.H., Daily Totals		$2850.40		$3742.40	$118.77	$155.93

Crew E-20

	Bare Costs Hr.	Bare Costs Daily	Incl. Subs O&P Hr.	Incl. Subs O&P Daily	Cost Per Labor-Hour Bare Costs	Cost Per Labor-Hour Incl. O&P
1 Struc. Steel Foreman (outside)	$59.65	$477.20	$93.55	$748.40	$57.21	$88.73
5 Structural Steel Workers	57.65	2306.00	90.45	3618.00		
1 Equip. Oper. (crane)	59.20	473.60	88.50	708.00		
1 Equip. Oper. (oiler)	50.55	404.40	75.55	604.40		
1 Lattice Boom Crane, 40 Ton		2028.00		2230.80	31.69	34.86
64 L.H., Daily Totals		$5689.20		$7909.60	$88.89	$123.59

Crew E-22

	Bare Costs Hr.	Bare Costs Daily	Incl. Subs O&P Hr.	Incl. Subs O&P Daily	Cost Per Labor-Hour Bare Costs	Cost Per Labor-Hour Incl. O&P
1 Skilled Worker Foreman (out)	$56.85	$454.80	$86.00	$688.00	$55.52	$83.97
2 Skilled Workers	54.85	877.60	82.95	1327.20		
24 L.H., Daily Totals		$1332.40		$2015.20	$55.52	$83.97

Crew E-24

	Bare Costs Hr.	Bare Costs Daily	Incl. Subs O&P Hr.	Incl. Subs O&P Daily	Cost Per Labor-Hour Bare Costs	Cost Per Labor-Hour Incl. O&P
3 Structural Steel Workers	$57.65	$1383.60	$90.45	$2170.80	$57.42	$89.05
1 Equipment Operator (med.)	56.75	454.00	84.85	678.80		
1 Hyd. Crane, 25 Ton		580.85		638.93	18.15	19.97
32 L.H., Daily Totals		$2418.45		$3488.53	$75.58	$109.02

Crew E-25

	Bare Costs Hr.	Bare Costs Daily	Incl. Subs O&P Hr.	Incl. Subs O&P Daily	Cost Per Labor-Hour Bare Costs	Cost Per Labor-Hour Incl. O&P
1 Welder Foreman (outside)	$59.45	$475.60	$93.25	$746.00	$59.45	$93.25
1 Cutting Torch		12.80		14.08	1.60	1.76
8 L.H., Daily Totals		$488.40		$760.08	$61.05	$95.01

Crew E-26

	Bare Costs Hr.	Bare Costs Daily	Incl. Subs O&P Hr.	Incl. Subs O&P Daily	Cost Per Labor-Hour Bare Costs	Cost Per Labor-Hour Incl. O&P
1 Struc. Steel Foreman (outside)	$59.65	$477.20	$93.55	$748.40	$58.91	$91.73
1 Struc. Steel Worker	57.65	461.20	90.45	723.60		
1 Welder	57.45	459.60	90.10	720.80		
.25 Electrician	61.35	122.70	91.35	182.70		
.25 Plumber	64.45	128.90	96.45	192.90		
1 Welder, Gas Engine, 300 amp		147.00		161.70	5.25	5.78
28 L.H., Daily Totals		$1796.60		$2730.10	$64.16	$97.50

For customer support on your Mechanical Costs with RSMeans Data, call 800.448.8182.

649

Crews - Standard

Crew E-27

Crew E-27	Bare Costs Hr.	Daily	Incl. Subs O&P Hr.	Daily	Cost Per Labor-Hour Bare Costs	Incl. O&P
1 Struc. Steel Foreman (outside)	$59.65	$477.20	$93.55	$748.40	$57.21	$88.73
5 Struc. Steel Workers	57.65	2306.00	90.45	3618.00		
1 Equip. Oper. (crane)	59.20	473.60	88.50	708.00		
1 Equip. Oper. (oiler)	50.55	404.40	75.55	604.40		
1 Hyd. Crane, 12 Ton		471.00		518.10		
1 Hyd. Crane, 80 Ton		1486.00		1634.60	30.58	33.64
64 L.H., Daily Totals		$5618.20		$7831.50	$87.78	$122.37

Crew F-3

Crew F-3	Bare Costs Hr.	Daily	Incl. Subs O&P Hr.	Daily	Cost Per Labor-Hour Bare Costs	Incl. O&P
4 Carpenters	$53.15	$1700.80	$79.85	$2555.20	$54.36	$81.58
1 Equip. Oper. (crane)	59.20	473.60	88.50	708.00		
1 Hyd. Crane, 12 Ton		471.00		518.10	11.78	12.95
40 L.H., Daily Totals		$2645.40		$3781.30	$66.14	$94.53

Crew F-4

Crew F-4	Bare Costs Hr.	Daily	Incl. Subs O&P Hr.	Daily	Cost Per Labor-Hour Bare Costs	Incl. O&P
4 Carpenters	$53.15	$1700.80	$79.85	$2555.20	$53.73	$80.58
1 Equip. Oper. (crane)	59.20	473.60	88.50	708.00		
1 Equip. Oper. (oiler)	50.55	404.40	75.55	604.40		
1 Hyd. Crane, 55 Ton		980.20		1078.22	20.42	22.46
48 L.H., Daily Totals		$3559.00		$4945.82	$74.15	$103.04

Crew F-5

Crew F-5	Bare Costs Hr.	Daily	Incl. Subs O&P Hr.	Daily	Cost Per Labor-Hour Bare Costs	Incl. O&P
1 Carpenter Foreman (outside)	$55.15	$441.20	$82.85	$662.80	$53.65	$80.60
3 Carpenters	53.15	1275.60	79.85	1916.40		
32 L.H., Daily Totals		$1716.80		$2579.20	$53.65	$80.60

Crew F-6

Crew F-6	Bare Costs Hr.	Daily	Incl. Subs O&P Hr.	Daily	Cost Per Labor-Hour Bare Costs	Incl. O&P
2 Carpenters	$53.15	$850.40	$79.85	$1277.60	$49.94	$74.94
2 Building Laborers	42.10	673.60	63.25	1012.00		
1 Equip. Oper. (crane)	59.20	473.60	88.50	708.00		
1 Hyd. Crane, 12 Ton		471.00		518.10	11.78	12.95
40 L.H., Daily Totals		$2468.60		$3515.70	$61.72	$87.89

Crew F-7

Crew F-7	Bare Costs Hr.	Daily	Incl. Subs O&P Hr.	Daily	Cost Per Labor-Hour Bare Costs	Incl. O&P
2 Carpenters	$53.15	$850.40	$79.85	$1277.60	$47.63	$71.55
2 Building Laborers	42.10	673.60	63.25	1012.00		
32 L.H., Daily Totals		$1524.00		$2289.60	$47.63	$71.55

Crew G-1

Crew G-1	Bare Costs Hr.	Daily	Incl. Subs O&P Hr.	Daily	Cost Per Labor-Hour Bare Costs	Incl. O&P
1 Roofer Foreman (outside)	$48.20	$385.60	$79.55	$636.40	$43.17	$71.25
4 Roofers Composition	46.20	1478.40	76.25	2440.00		
2 Roofer Helpers	34.60	553.60	57.10	913.60		
1 Application Equipment		192.85		212.13		
1 Tar Kettle/Pot		207.85		228.63		
1 Crew Truck		166.25		182.88	10.12	11.14
56 L.H., Daily Totals		$2984.55		$4613.65	$53.30	$82.39

Crew G-2

Crew G-2	Bare Costs Hr.	Daily	Incl. Subs O&P Hr.	Daily	Cost Per Labor-Hour Bare Costs	Incl. O&P
1 Plasterer	$48.60	$388.80	$72.55	$580.40	$44.37	$66.37
1 Plasterer Helper	42.40	339.20	63.30	506.40		
1 Building Laborer	42.10	336.80	63.25	506.00		
1 Grout Pump, 50 C.F./hr.		188.45		207.29	7.85	8.64
24 L.H., Daily Totals		$1253.25		$1800.10	$52.22	$75.00

Crew G-2A

Crew G-2A	Bare Costs Hr.	Daily	Incl. Subs O&P Hr.	Daily	Cost Per Labor-Hour Bare Costs	Incl. O&P
1 Roofer Composition	$46.20	$369.60	$76.25	$610.00	$40.97	$65.53
1 Roofer Helper	34.60	276.80	57.10	456.80		
1 Building Laborer	42.10	336.80	63.25	506.00		
1 Foam Spray Rig, Trailer-Mtd.		523.85		576.24		
1 Pickup Truck, 3/4 Ton		110.85		121.94	26.45	29.09
24 L.H., Daily Totals		$1617.90		$2270.97	$67.41	$94.62

Crew G-3

Crew G-3	Bare Costs Hr.	Daily	Incl. Subs O&P Hr.	Daily	Cost Per Labor-Hour Bare Costs	Incl. O&P
2 Sheet Metal Workers	$62.30	$996.80	$94.50	$1512.00	$52.20	$78.88
2 Building Laborers	42.10	673.60	63.25	1012.00		
32 L.H., Daily Totals		$1670.40		$2524.00	$52.20	$78.88

Crew G-4

Crew G-4	Bare Costs Hr.	Daily	Incl. Subs O&P Hr.	Daily	Cost Per Labor-Hour Bare Costs	Incl. O&P
1 Labor Foreman (outside)	$44.10	$352.80	$66.25	$530.00	$42.77	$64.25
2 Building Laborers	42.10	673.60	63.25	1012.00		
1 Flatbed Truck, Gas, 1.5 Ton		196.15		215.76		
1 Air Compressor, 160 cfm		202.80		223.08	16.62	18.29
24 L.H., Daily Totals		$1425.35		$1980.85	$59.39	$82.54

Crew G-5

Crew G-5	Bare Costs Hr.	Daily	Incl. Subs O&P Hr.	Daily	Cost Per Labor-Hour Bare Costs	Incl. O&P
1 Roofer Foreman (outside)	$48.20	$385.60	$79.55	$636.40	$41.96	$69.25
2 Roofers Composition	46.20	739.20	76.25	1220.00		
2 Roofer Helpers	34.60	553.60	57.10	913.60		
1 Application Equipment		192.85		212.13	4.82	5.30
40 L.H., Daily Totals		$1871.25		$2982.14	$46.78	$74.55

Crew G-6A

Crew G-6A	Bare Costs Hr.	Daily	Incl. Subs O&P Hr.	Daily	Cost Per Labor-Hour Bare Costs	Incl. O&P
2 Roofers Composition	$46.20	$739.20	$76.25	$1220.00	$46.20	$76.25
1 Small Compressor, Electric		33.80		37.18		
2 Pneumatic Nailers		54.80		60.28	5.54	6.09
16 L.H., Daily Totals		$827.80		$1317.46	$51.74	$82.34

Crew G-7

Crew G-7	Bare Costs Hr.	Daily	Incl. Subs O&P Hr.	Daily	Cost Per Labor-Hour Bare Costs	Incl. O&P
1 Carpenter	$53.15	$425.20	$79.85	$638.80	$53.15	$79.85
1 Small Compressor, Electric		33.80		37.18		
1 Pneumatic Nailer		27.40		30.14	7.65	8.41
8 L.H., Daily Totals		$486.40		$706.12	$60.80	$88.27

Crew H-1

Crew H-1	Bare Costs Hr.	Daily	Incl. Subs O&P Hr.	Daily	Cost Per Labor-Hour Bare Costs	Incl. O&P
2 Glaziers	$51.00	$816.00	$76.40	$1222.40	$54.33	$83.42
2 Struc. Steel Workers	57.65	922.40	90.45	1447.20		
32 L.H., Daily Totals		$1738.40		$2669.60	$54.33	$83.42

Crew H-2

Crew H-2	Bare Costs Hr.	Daily	Incl. Subs O&P Hr.	Daily	Cost Per Labor-Hour Bare Costs	Incl. O&P
2 Glaziers	$51.00	$816.00	$76.40	$1222.40	$48.03	$72.02
1 Building Laborer	42.10	336.80	63.25	506.00		
24 L.H., Daily Totals		$1152.80		$1728.40	$48.03	$72.02

Crew H-3

Crew H-3	Bare Costs Hr.	Daily	Incl. Subs O&P Hr.	Daily	Cost Per Labor-Hour Bare Costs	Incl. O&P
1 Glazier	$51.00	$408.00	$76.40	$611.20	$45.48	$68.67
1 Helper	39.95	319.60	60.95	487.60		
16 L.H., Daily Totals		$727.60		$1098.80	$45.48	$68.67

Crew H-4

Crew H-4	Bare Costs Hr.	Daily	Incl. Subs O&P Hr.	Daily	Cost Per Labor-Hour Bare Costs	Incl. O&P
1 Carpenter	$53.15	$425.20	$79.85	$638.80	$49.51	$74.59
1 Carpenter Helper	39.95	319.60	60.95	487.60		
.5 Electrician	61.35	245.40	91.35	365.40		
20 L.H., Daily Totals		$990.20		$1491.80	$49.51	$74.59

Crew J-1

Crew No.	Bare Costs Hr.	Daily	Incl. Subs O&P Hr.	Daily	Cost Per Labor-Hour Bare Costs	Incl. O&P
3 Plasterers	$48.60	$1166.40	$72.55	$1741.20	$46.12	$68.85
2 Plasterer Helpers	42.40	678.40	63.30	1012.80		
1 Mixing Machine, 6 C.F.		117.35		129.09	2.93	3.23
40 L.H., Daily Totals		$1962.15		$2883.09	$49.05	$72.08

Crew J-2

Crew No.	Bare Costs Hr.	Daily	Incl. Subs O&P Hr.	Daily	Cost Per Labor-Hour Bare Costs	Incl. O&P
3 Plasterers	$48.60	$1166.40	$72.55	$1741.20	$47.11	$70.09
2 Plasterer Helpers	42.40	678.40	63.30	1012.80		
1 Lather	52.05	416.40	76.30	610.40		
1 Mixing Machine, 6 C.F.		117.35		129.09	2.44	2.69
48 L.H., Daily Totals		$2378.55		$3493.49	$49.55	$72.78

Crew J-3

Crew No.	Bare Costs Hr.	Daily	Incl. Subs O&P Hr.	Daily	Cost Per Labor-Hour Bare Costs	Incl. O&P
1 Terrazzo Worker	$49.40	$395.20	$72.45	$579.60	$44.95	$65.92
1 Terrazzo Helper	40.50	324.00	59.40	475.20		
1 Floor Grinder, 22" Path		104.60		115.06		
1 Terrazzo Mixer		161.15		177.26	16.61	18.27
16 L.H., Daily Totals		$984.95		$1347.13	$61.56	$84.20

Crew J-4

Crew No.	Bare Costs Hr.	Daily	Incl. Subs O&P Hr.	Daily	Cost Per Labor-Hour Bare Costs	Incl. O&P
2 Cement Finishers	$49.95	$799.20	$73.45	$1175.20	$47.33	$70.05
1 Laborer	42.10	336.80	63.25	506.00		
1 Floor Grinder, 22" Path		104.60		115.06		
1 Floor Edger, 7" Path		43.95		48.34		
1 Vacuum Pick-Up System		74.05		81.45	9.28	10.20
24 L.H., Daily Totals		$1358.60		$1926.06	$56.61	$80.25

Crew J-4A

Crew No.	Bare Costs Hr.	Daily	Incl. Subs O&P Hr.	Daily	Cost Per Labor-Hour Bare Costs	Incl. O&P
2 Cement Finishers	$49.95	$799.20	$73.45	$1175.20	$46.02	$68.35
2 Laborers	42.10	673.60	63.25	1012.00		
1 Floor Grinder, 22" Path		104.60		115.06		
1 Floor Edger, 7" Path		43.95		48.34		
1 Vacuum Pick-Up System		74.05		81.45		
1 Floor Auto Scrubber		178.40		196.24	12.53	13.78
32 L.H., Daily Totals		$1873.80		$2628.30	$58.56	$82.13

Crew J-4B

Crew No.	Bare Costs Hr.	Daily	Incl. Subs O&P Hr.	Daily	Cost Per Labor-Hour Bare Costs	Incl. O&P
1 Laborer	$42.10	$336.80	$63.25	$506.00	$42.10	$63.25
1 Floor Auto Scrubber		178.40		196.24	22.30	24.53
8 L.H., Daily Totals		$515.20		$702.24	$64.40	$87.78

Crew J-6

Crew No.	Bare Costs Hr.	Daily	Incl. Subs O&P Hr.	Daily	Cost Per Labor-Hour Bare Costs	Incl. O&P
2 Painters	$44.40	$710.40	$66.05	$1056.80	$45.98	$68.64
1 Building Laborer	42.10	336.80	63.25	506.00		
1 Equip. Oper. (light)	53.00	424.00	79.20	633.60		
1 Air Compressor, 250 cfm		201.50		221.65		
1 Sandblaster, Portable, 3 C.F.		83.80		92.18		
1 Set Sand Blasting Accessories		15.35		16.89	9.40	10.33
32 L.H., Daily Totals		$1771.85		$2527.11	$55.37	$78.97

Crew J-7

Crew No.	Bare Costs Hr.	Daily	Incl. Subs O&P Hr.	Daily	Cost Per Labor-Hour Bare Costs	Incl. O&P
2 Painters	$44.40	$710.40	$66.05	$1056.80	$44.40	$66.05
1 Floor Belt Sander		50.15		55.16		
1 Floor Sanding Edger		25.15		27.66	4.71	5.18
16 L.H., Daily Totals		$785.70		$1139.63	$49.11	$71.23

Crew K-1

Crew No.	Bare Costs Hr.	Daily	Incl. Subs O&P Hr.	Daily	Cost Per Labor-Hour Bare Costs	Incl. O&P
1 Carpenter	$53.15	$425.20	$79.85	$638.80	$50.20	$75.33
1 Truck Driver (light)	47.25	378.00	70.80	566.40		
1 Flatbed Truck, Gas, 3 Ton		820.40		902.44	51.27	56.40
16 L.H., Daily Totals		$1623.60		$2107.64	$101.47	$131.73

Crew K-2

Crew No.	Bare Costs Hr.	Daily	Incl. Subs O&P Hr.	Daily	Cost Per Labor-Hour Bare Costs	Incl. O&P
1 Struc. Steel Foreman (outside)	$59.65	$477.20	$93.55	$748.40	$54.85	$84.93
1 Struc. Steel Worker	57.65	461.20	90.45	723.60		
1 Truck Driver (light)	47.25	378.00	70.80	566.40		
1 Flatbed Truck, Gas, 3 Ton		820.40		902.44	34.18	37.60
24 L.H., Daily Totals		$2136.80		$2940.84	$89.03	$122.54

Crew L-1

Crew No.	Bare Costs Hr.	Daily	Incl. Subs O&P Hr.	Daily	Cost Per Labor-Hour Bare Costs	Incl. O&P
1 Electrician	$61.35	$490.80	$91.35	$730.80	$62.90	$93.90
1 Plumber	64.45	515.60	96.45	771.60		
16 L.H., Daily Totals		$1006.40		$1502.40	$62.90	$93.90

Crew L-2

Crew No.	Bare Costs Hr.	Daily	Incl. Subs O&P Hr.	Daily	Cost Per Labor-Hour Bare Costs	Incl. O&P
1 Carpenter	$53.15	$425.20	$79.85	$638.80	$46.55	$70.40
1 Carpenter Helper	39.95	319.60	60.95	487.60		
16 L.H., Daily Totals		$744.80		$1126.40	$46.55	$70.40

Crew L-3

Crew No.	Bare Costs Hr.	Daily	Incl. Subs O&P Hr.	Daily	Cost Per Labor-Hour Bare Costs	Incl. O&P
1 Carpenter	$53.15	$425.20	$79.85	$638.80	$57.49	$86.39
.5 Electrician	61.35	245.40	91.35	365.40		
.5 Sheet Metal Worker	62.30	249.20	94.50	378.00		
16 L.H., Daily Totals		$919.80		$1382.20	$57.49	$86.39

Crew L-3A

Crew No.	Bare Costs Hr.	Daily	Incl. Subs O&P Hr.	Daily	Cost Per Labor-Hour Bare Costs	Incl. O&P
1 Carpenter Foreman (outside)	$55.15	$441.20	$82.85	$662.80	$57.53	$86.73
.5 Sheet Metal Worker	62.30	249.20	94.50	378.00		
12 L.H., Daily Totals		$690.40		$1040.80	$57.53	$86.73

Crew L-4

Crew No.	Bare Costs Hr.	Daily	Incl. Subs O&P Hr.	Daily	Cost Per Labor-Hour Bare Costs	Incl. O&P
2 Skilled Workers	$54.85	$877.60	$82.95	$1327.20	$49.88	$75.62
1 Helper	39.95	319.60	60.95	487.60		
24 L.H., Daily Totals		$1197.20		$1814.80	$49.88	$75.62

Crew L-5

Crew No.	Bare Costs Hr.	Daily	Incl. Subs O&P Hr.	Daily	Cost Per Labor-Hour Bare Costs	Incl. O&P
1 Struc. Steel Foreman (outside)	$59.65	$477.20	$93.55	$748.40	$58.16	$90.61
5 Struc. Steel Workers	57.65	2306.00	90.45	3618.00		
1 Equip. Oper. (crane)	59.20	473.60	88.50	708.00		
1 Hyd. Crane, 25 Ton		580.85		638.93	10.37	11.41
56 L.H., Daily Totals		$3837.65		$5713.34	$68.53	$102.02

Crew L-5A

Crew No.	Bare Costs Hr.	Daily	Incl. Subs O&P Hr.	Daily	Cost Per Labor-Hour Bare Costs	Incl. O&P
1 Struc. Steel Foreman (outside)	$59.65	$477.20	$93.55	$748.40	$58.54	$90.74
2 Structural Steel Workers	57.65	922.40	90.45	1447.20		
1 Equip. Oper. (crane)	59.20	473.60	88.50	708.00		
1 S.P. Crane, 4x4, 25 Ton		1144.00		1258.40	35.75	39.33
32 L.H., Daily Totals		$3017.20		$4162.00	$94.29	$130.06

For customer support on your Mechanical Costs with RSMeans Data, call 800.448.8182.

651

Crews - Standard

Crew No.	Bare Costs		Incl. Subs O&P		Cost Per Labor-Hour	
Crew L-5B	Hr.	Daily	Hr.	Daily	Bare Costs	Incl. O&P
1 Struc. Steel Foreman (outside)	$59.65	$477.20	$93.55	$748.40	$59.83	$90.82
2 Structural Steel Workers	57.65	922.40	90.45	1447.20		
2 Electricians	61.35	981.60	91.35	1461.60		
2 Steamfitters/Pipefitters	65.55	1048.80	98.10	1569.60		
1 Equip. Oper. (crane)	59.20	473.60	88.50	708.00		
1 Equip. Oper. (oiler)	50.55	404.40	75.55	604.40		
1 Hyd. Crane, 80 Ton		1486.00		1634.60	20.64	22.70
72 L.H., Daily Totals		$5794.00		$8173.80	$80.47	$113.53
Crew L-6	Hr.	Daily	Hr.	Daily	Bare Costs	Incl. O&P
1 Plumber	$64.45	$515.60	$96.45	$771.60	$63.42	$94.75
.5 Electrician	61.35	245.40	91.35	365.40		
12 L.H., Daily Totals		$761.00		$1137.00	$63.42	$94.75
Crew L-7	Hr.	Daily	Hr.	Daily	Bare Costs	Incl. O&P
2 Carpenters	$53.15	$850.40	$79.85	$1277.60	$51.16	$76.75
1 Building Laborer	42.10	336.80	63.25	506.00		
.5 Electrician	61.35	245.40	91.35	365.40		
28 L.H., Daily Totals		$1432.60		$2149.00	$51.16	$76.75
Crew L-8	Hr.	Daily	Hr.	Daily	Bare Costs	Incl. O&P
2 Carpenters	$53.15	$850.40	$79.85	$1277.60	$55.41	$83.17
.5 Plumber	64.45	257.80	96.45	385.80		
20 L.H., Daily Totals		$1108.20		$1663.40	$55.41	$83.17
Crew L-9	Hr.	Daily	Hr.	Daily	Bare Costs	Incl. O&P
1 Labor Foreman (inside)	$42.60	$340.80	$64.00	$512.00	$47.81	$72.58
2 Building Laborers	42.10	673.60	63.25	1012.00		
1 Struc. Steel Worker	57.65	461.20	90.45	723.60		
.5 Electrician	61.35	245.40	91.35	365.40		
36 L.H., Daily Totals		$1721.00		$2613.00	$47.81	$72.58
Crew L-10	Hr.	Daily	Hr.	Daily	Bare Costs	Incl. O&P
1 Struc. Steel Foreman (outside)	$59.65	$477.20	$93.55	$748.40	$58.83	$90.83
1 Structural Steel Worker	57.65	461.20	90.45	723.60		
1 Equip. Oper. (crane)	59.20	473.60	88.50	708.00		
1 Hyd. Crane, 12 Ton		471.00		518.10	19.63	21.59
24 L.H., Daily Totals		$1883.00		$2698.10	$78.46	$112.42
Crew L-11	Hr.	Daily	Hr.	Daily	Bare Costs	Incl. O&P
2 Wreckers	$42.10	$673.60	$64.80	$1036.80	$49.10	$74.33
1 Equip. Oper. (crane)	59.20	473.60	88.50	708.00		
1 Equip. Oper. (light)	53.00	424.00	79.20	633.60		
1 Hyd. Excavator, 2.5 C.Y.		1465.00		1611.50		
1 Loader, Skid Steer, 78 H.P.		400.10		440.11	58.28	64.11
32 L.H., Daily Totals		$3436.30		$4430.01	$107.38	$138.44
Crew M-1	Hr.	Daily	Hr.	Daily	Bare Costs	Incl. O&P
3 Elevator Constructors	$85.55	$2053.20	$127.00	$3048.00	$81.28	$120.65
1 Elevator Apprentice	68.45	547.60	101.60	812.80		
5 Hand Tools		50.00		55.00	1.56	1.72
32 L.H., Daily Totals		$2650.80		$3915.80	$82.84	$122.37

Crew No.	Bare Costs		Incl. Subs O&P		Cost Per Labor-Hour	
Crew M-3	Hr.	Daily	Hr.	Daily	Bare Costs	Incl. O&P
1 Electrician Foreman (outside)	$63.35	$506.80	$94.35	$754.80	$64.39	$95.86
1 Common Laborer	42.10	336.80	63.25	506.00		
.25 Equipment Operator (med.)	56.75	113.50	84.85	169.70		
1 Elevator Constructor	85.55	684.40	127.00	1016.00		
1 Elevator Apprentice	68.45	547.60	101.60	812.80		
.25 S.P. Crane, 4x4, 20 Ton		145.55		160.10	4.28	4.71
34 L.H., Daily Totals		$2334.65		$3419.41	$68.67	$100.57
Crew M-4	Hr.	Daily	Hr.	Daily	Bare Costs	Incl. O&P
1 Electrician Foreman (outside)	$63.35	$506.80	$94.35	$754.80	$63.75	$94.94
1 Common Laborer	42.10	336.80	63.25	506.00		
.25 Equipment Operator, Crane	59.20	118.40	88.50	177.00		
.25 Equip. Oper. (oiler)	50.55	101.10	75.55	151.10		
1 Elevator Constructor	85.55	684.40	127.00	1016.00		
1 Elevator Apprentice	68.45	547.60	101.60	812.80		
.25 S.P. Crane, 4x4, 40 Ton		188.55		207.41	5.24	5.76
36 L.H., Daily Totals		$2483.65		$3625.11	$68.99	$100.70
Crew Q-1	Hr.	Daily	Hr.	Daily	Bare Costs	Incl. O&P
1 Plumber	$64.45	$515.60	$96.45	$771.60	$58.00	$86.80
1 Plumber Apprentice	51.55	412.40	77.15	617.20		
16 L.H., Daily Totals		$928.00		$1388.80	$58.00	$86.80
Crew Q-1A	Hr.	Daily	Hr.	Daily	Bare Costs	Incl. O&P
.25 Plumber Foreman (outside)	$66.45	$132.90	$99.45	$198.90	$64.85	$97.05
1 Plumber	64.45	515.60	96.45	771.60		
10 L.H., Daily Totals		$648.50		$970.50	$64.85	$97.05
Crew Q-1C	Hr.	Daily	Hr.	Daily	Bare Costs	Incl. O&P
1 Plumber	$64.45	$515.60	$96.45	$771.60	$57.58	$86.15
1 Plumber Apprentice	51.55	412.40	77.15	617.20		
1 Equip. Oper. (medium)	56.75	454.00	84.85	678.80		
1 Trencher, Chain Type, 8' D		1872.00		2059.20	78.00	85.80
24 L.H., Daily Totals		$3254.00		$4126.80	$135.58	$171.95
Crew Q-2	Hr.	Daily	Hr.	Daily	Bare Costs	Incl. O&P
2 Plumbers	$64.45	$1031.20	$96.45	$1543.20	$60.15	$90.02
1 Plumber Apprentice	51.55	412.40	77.15	617.20		
24 L.H., Daily Totals		$1443.60		$2160.40	$60.15	$90.02
Crew Q-3	Hr.	Daily	Hr.	Daily	Bare Costs	Incl. O&P
1 Plumber Foreman (inside)	$64.95	$519.60	$97.20	$777.60	$61.35	$91.81
2 Plumbers	64.45	1031.20	96.45	1543.20		
1 Plumber Apprentice	51.55	412.40	77.15	617.20		
32 L.H., Daily Totals		$1963.20		$2938.00	$61.35	$91.81
Crew Q-4	Hr.	Daily	Hr.	Daily	Bare Costs	Incl. O&P
1 Plumber Foreman (inside)	$64.95	$519.60	$97.20	$777.60	$61.35	$91.81
1 Plumber	64.45	515.60	96.45	771.60		
1 Welder (plumber)	64.45	515.60	96.45	771.60		
1 Plumber Apprentice	51.55	412.40	77.15	617.20		
1 Welder, Electric, 300 amp		106.40		117.04	3.33	3.66
32 L.H., Daily Totals		$2069.60		$3055.04	$64.67	$95.47
Crew Q-5	Hr.	Daily	Hr.	Daily	Bare Costs	Incl. O&P
1 Steamfitter	$65.55	$524.40	$98.10	$784.80	$59.00	$88.30
1 Steamfitter Apprentice	52.45	419.60	78.50	628.00		
16 L.H., Daily Totals		$944.00		$1412.80	$59.00	$88.30

Crew Q-6

Crew No.	Bare Costs Hr.	Bare Costs Daily	Incl. Subs O&P Hr.	Incl. Subs O&P Daily	Cost Per Labor-Hour Bare Costs	Cost Per Labor-Hour Incl. O&P
2 Steamfitters	$65.55	$1048.80	$98.10	$1569.60	$61.18	$91.57
1 Steamfitter Apprentice	52.45	419.60	78.50	628.00		
24 L.H., Daily Totals		$1468.40		$2197.60	$61.18	$91.57

Crew Q-7

Crew No.	Bare Costs Hr.	Bare Costs Daily	Incl. Subs O&P Hr.	Incl. Subs O&P Daily	Cost Per Labor-Hour Bare Costs	Cost Per Labor-Hour Incl. O&P
1 Steamfitter Foreman (inside)	$66.05	$528.40	$98.85	$790.80	$62.40	$93.39
2 Steamfitters	65.55	1048.80	98.10	1569.60		
1 Steamfitter Apprentice	52.45	419.60	78.50	628.00		
32 L.H., Daily Totals		$1996.80		$2988.40	$62.40	$93.39

Crew Q-8

Crew No.	Bare Costs Hr.	Bare Costs Daily	Incl. Subs O&P Hr.	Incl. Subs O&P Daily	Cost Per Labor-Hour Bare Costs	Cost Per Labor-Hour Incl. O&P
1 Steamfitter Foreman (inside)	$66.05	$528.40	$98.85	$790.80	$62.40	$93.39
1 Steamfitter	65.55	524.40	98.10	784.80		
1 Welder (steamfitter)	65.55	524.40	98.10	784.80		
1 Steamfitter Apprentice	52.45	419.60	78.50	628.00		
1 Welder, Electric, 300 amp		106.40		117.04	3.33	3.66
32 L.H., Daily Totals		$2103.20		$3105.44	$65.72	$97.05

Crew Q-9

Crew No.	Bare Costs Hr.	Bare Costs Daily	Incl. Subs O&P Hr.	Incl. Subs O&P Daily	Cost Per Labor-Hour Bare Costs	Cost Per Labor-Hour Incl. O&P
1 Sheet Metal Worker	$62.30	$498.40	$94.50	$756.00	$56.08	$85.05
1 Sheet Metal Apprentice	49.85	398.80	75.60	604.80		
16 L.H., Daily Totals		$897.20		$1360.80	$56.08	$85.05

Crew Q-10

Crew No.	Bare Costs Hr.	Bare Costs Daily	Incl. Subs O&P Hr.	Incl. Subs O&P Daily	Cost Per Labor-Hour Bare Costs	Cost Per Labor-Hour Incl. O&P
2 Sheet Metal Workers	$62.30	$996.80	$94.50	$1512.00	$58.15	$88.20
1 Sheet Metal Apprentice	49.85	398.80	75.60	604.80		
24 L.H., Daily Totals		$1395.60		$2116.80	$58.15	$88.20

Crew Q-11

Crew No.	Bare Costs Hr.	Bare Costs Daily	Incl. Subs O&P Hr.	Incl. Subs O&P Daily	Cost Per Labor-Hour Bare Costs	Cost Per Labor-Hour Incl. O&P
1 Sheet Metal Foreman (inside)	$62.80	$502.40	$95.25	$762.00	$59.31	$89.96
2 Sheet Metal Workers	62.30	996.80	94.50	1512.00		
1 Sheet Metal Apprentice	49.85	398.80	75.60	604.80		
32 L.H., Daily Totals		$1898.00		$2878.80	$59.31	$89.96

Crew Q-12

Crew No.	Bare Costs Hr.	Bare Costs Daily	Incl. Subs O&P Hr.	Incl. Subs O&P Daily	Cost Per Labor-Hour Bare Costs	Cost Per Labor-Hour Incl. O&P
1 Sprinkler Installer	$63.25	$506.00	$94.80	$758.40	$56.92	$85.33
1 Sprinkler Apprentice	50.60	404.80	75.85	606.80		
16 L.H., Daily Totals		$910.80		$1365.20	$56.92	$85.33

Crew Q-13

Crew No.	Bare Costs Hr.	Bare Costs Daily	Incl. Subs O&P Hr.	Incl. Subs O&P Daily	Cost Per Labor-Hour Bare Costs	Cost Per Labor-Hour Incl. O&P
1 Sprinkler Foreman (inside)	$63.75	$510.00	$95.55	$764.40	$60.21	$90.25
2 Sprinkler Installers	63.25	1012.00	94.80	1516.80		
1 Sprinkler Apprentice	50.60	404.80	75.85	606.80		
32 L.H., Daily Totals		$1926.80		$2888.00	$60.21	$90.25

Crew Q-14

Crew No.	Bare Costs Hr.	Bare Costs Daily	Incl. Subs O&P Hr.	Incl. Subs O&P Daily	Cost Per Labor-Hour Bare Costs	Cost Per Labor-Hour Incl. O&P
1 Asbestos Worker	$58.65	$469.20	$90.05	$720.40	$52.77	$81.03
1 Asbestos Apprentice	46.90	375.20	72.00	576.00		
16 L.H., Daily Totals		$844.40		$1296.40	$52.77	$81.03

Crew Q-15

Crew No.	Bare Costs Hr.	Bare Costs Daily	Incl. Subs O&P Hr.	Incl. Subs O&P Daily	Cost Per Labor-Hour Bare Costs	Cost Per Labor-Hour Incl. O&P
1 Plumber	$64.45	$515.60	$96.45	$771.60	$58.00	$86.80
1 Plumber Apprentice	51.55	412.40	77.15	617.20		
1 Welder, Electric, 300 amp		106.40		117.04	6.65	7.32
16 L.H., Daily Totals		$1034.40		$1505.84	$64.65	$94.11

Crew Q-16

Crew No.	Bare Costs Hr.	Bare Costs Daily	Incl. Subs O&P Hr.	Incl. Subs O&P Daily	Cost Per Labor-Hour Bare Costs	Cost Per Labor-Hour Incl. O&P
2 Plumbers	$64.45	$1031.20	$96.45	$1543.20	$60.15	$90.02
1 Plumber Apprentice	51.55	412.40	77.15	617.20		
1 Welder, Electric, 300 amp		106.40		117.04	4.43	4.88
24 L.H., Daily Totals		$1550.00		$2277.44	$64.58	$94.89

Crew Q-17

Crew No.	Bare Costs Hr.	Bare Costs Daily	Incl. Subs O&P Hr.	Incl. Subs O&P Daily	Cost Per Labor-Hour Bare Costs	Cost Per Labor-Hour Incl. O&P
1 Steamfitter	$65.55	$524.40	$98.10	$784.80	$59.00	$88.30
1 Steamfitter Apprentice	52.45	419.60	78.50	628.00		
1 Welder, Electric, 300 amp		106.40		117.04	6.65	7.32
16 L.H., Daily Totals		$1050.40		$1529.84	$65.65	$95.61

Crew Q-17A

Crew No.	Bare Costs Hr.	Bare Costs Daily	Incl. Subs O&P Hr.	Incl. Subs O&P Daily	Cost Per Labor-Hour Bare Costs	Cost Per Labor-Hour Incl. O&P
1 Steamfitter	$65.55	$524.40	$98.10	$784.80	$59.07	$88.37
1 Steamfitter Apprentice	52.45	419.60	78.50	628.00		
1 Equip. Oper. (crane)	59.20	473.60	88.50	708.00		
1 Hyd. Crane, 12 Ton		471.00		518.10		
1 Welder, Electric, 300 amp		106.40		117.04	24.06	26.46
24 L.H., Daily Totals		$1995.00		$2755.94	$83.13	$114.83

Crew Q-18

Crew No.	Bare Costs Hr.	Bare Costs Daily	Incl. Subs O&P Hr.	Incl. Subs O&P Daily	Cost Per Labor-Hour Bare Costs	Cost Per Labor-Hour Incl. O&P
2 Steamfitters	$65.55	$1048.80	$98.10	$1569.60	$61.18	$91.57
1 Steamfitter Apprentice	52.45	419.60	78.50	628.00		
1 Welder, Electric, 300 amp		106.40		117.04	4.43	4.88
24 L.H., Daily Totals		$1574.80		$2314.64	$65.62	$96.44

Crew Q-19

Crew No.	Bare Costs Hr.	Bare Costs Daily	Incl. Subs O&P Hr.	Incl. Subs O&P Daily	Cost Per Labor-Hour Bare Costs	Cost Per Labor-Hour Incl. O&P
1 Steamfitter	$65.55	$524.40	$98.10	$784.80	$59.78	$89.32
1 Steamfitter Apprentice	52.45	419.60	78.50	628.00		
1 Electrician	61.35	490.80	91.35	730.80		
24 L.H., Daily Totals		$1434.80		$2143.60	$59.78	$89.32

Crew Q-20

Crew No.	Bare Costs Hr.	Bare Costs Daily	Incl. Subs O&P Hr.	Incl. Subs O&P Daily	Cost Per Labor-Hour Bare Costs	Cost Per Labor-Hour Incl. O&P
1 Sheet Metal Worker	$62.30	$498.40	$94.50	$756.00	$57.13	$86.31
1 Sheet Metal Apprentice	49.85	398.80	75.60	604.80		
.5 Electrician	61.35	245.40	91.35	365.40		
20 L.H., Daily Totals		$1142.60		$1726.20	$57.13	$86.31

Crew Q-21

Crew No.	Bare Costs Hr.	Bare Costs Daily	Incl. Subs O&P Hr.	Incl. Subs O&P Daily	Cost Per Labor-Hour Bare Costs	Cost Per Labor-Hour Incl. O&P
2 Steamfitters	$65.55	$1048.80	$98.10	$1569.60	$61.23	$91.51
1 Steamfitter Apprentice	52.45	419.60	78.50	628.00		
1 Electrician	61.35	490.80	91.35	730.80		
32 L.H., Daily Totals		$1959.20		$2928.40	$61.23	$91.51

Crew Q-22

Crew No.	Bare Costs Hr.	Bare Costs Daily	Incl. Subs O&P Hr.	Incl. Subs O&P Daily	Cost Per Labor-Hour Bare Costs	Cost Per Labor-Hour Incl. O&P
1 Plumber	$64.45	$515.60	$96.45	$771.60	$58.00	$86.80
1 Plumber Apprentice	51.55	412.40	77.15	617.20		
1 Hyd. Crane, 12 Ton		471.00		518.10	29.44	32.38
16 L.H., Daily Totals		$1399.00		$1906.90	$87.44	$119.18

Crew Q-22A

Crew No.	Bare Costs Hr.	Bare Costs Daily	Incl. Subs O&P Hr.	Incl. Subs O&P Daily	Cost Per Labor-Hour Bare Costs	Cost Per Labor-Hour Incl. O&P
1 Plumber	$64.45	$515.60	$96.45	$771.60	$54.33	$81.34
1 Plumber Apprentice	51.55	412.40	77.15	617.20		
1 Laborer	42.10	336.80	63.25	506.00		
1 Equip. Oper. (crane)	59.20	473.60	88.50	708.00		
1 Hyd. Crane, 12 Ton		471.00		518.10	14.72	16.19
32 L.H., Daily Totals		$2209.40		$3120.90	$69.04	$97.53

For customer support on your Mechanical Costs with RSMeans Data, call 800.448.8182.

653

Crew No.	Bare Costs		Incl. Subs O&P		Cost Per Labor-Hour	

Crew Q-23	Hr.	Daily	Hr.	Daily	Bare Costs	Incl. O&P
1 Plumber Foreman (outside)	$66.45	$531.60	$99.45	$795.60	$62.55	$93.58
1 Plumber	64.45	515.60	96.45	771.60		
1 Equip. Oper. (medium)	56.75	454.00	84.85	678.80		
1 Lattice Boom Crane, 20 Ton		1488.00		1636.80	62.00	68.20
24 L.H., Daily Totals		$2989.20		$3882.80	$124.55	$161.78

Crew R-1	Hr.	Daily	Hr.	Daily	Bare Costs	Incl. O&P
1 Electrician Foreman	$61.85	$494.80	$92.10	$736.80	$57.35	$85.39
3 Electricians	61.35	1472.40	91.35	2192.40		
2 Electrician Apprentices	49.10	785.60	73.10	1169.60		
48 L.H., Daily Totals		$2752.80		$4098.80	$57.35	$85.39

Crew R-1A	Hr.	Daily	Hr.	Daily	Bare Costs	Incl. O&P
1 Electrician	$61.35	$490.80	$91.35	$730.80	$55.23	$82.22
1 Electrician Apprentice	49.10	392.80	73.10	584.80		
16 L.H., Daily Totals		$883.60		$1315.60	$55.23	$82.22

Crew R-1B	Hr.	Daily	Hr.	Daily	Bare Costs	Incl. O&P
1 Electrician	$61.35	$490.80	$91.35	$730.80	$53.18	$79.18
2 Electrician Apprentices	49.10	785.60	73.10	1169.60		
24 L.H., Daily Totals		$1276.40		$1900.40	$53.18	$79.18

Crew R-1C	Hr.	Daily	Hr.	Daily	Bare Costs	Incl. O&P
2 Electricians	$61.35	$981.60	$91.35	$1461.60	$55.23	$82.22
2 Electrician Apprentices	49.10	785.60	73.10	1169.60		
1 Portable cable puller, 8000 lb.		101.60		111.76	3.17	3.49
32 L.H., Daily Totals		$1868.80		$2742.96	$58.40	$85.72

Crew R-1D	Hr.	Daily	Hr.	Daily	Bare Costs	Incl. O&P
1 Electrician	$61.35	$490.80	$91.35	$730.80	$55.23	$82.23
1 Electrician Apprentice	49.10	392.80	73.10	584.80		
1 Aerial lift		105.00		115.50	6.56	7.22
16 L.H., Daily Totals		$988.60		$1431.10	$61.79	$89.45

Crew R-2	Hr.	Daily	Hr.	Daily	Bare Costs	Incl. O&P
1 Electrician Foreman	$61.85	$494.80	$92.10	$736.80	$57.61	$85.84
3 Electricians	61.35	1472.40	91.35	2192.40		
2 Electrician Apprentices	49.10	785.60	73.10	1169.60		
1 Equip. Oper. (crane)	59.20	473.60	88.50	708.00		
1 S.P. Crane, 4x4, 5 Ton		378.10		415.91	6.75	7.43
56 L.H., Daily Totals		$3604.50		$5222.71	$64.37	$93.26

Crew R-3	Hr.	Daily	Hr.	Daily	Bare Costs	Incl. O&P
1 Electrician Foreman	$61.85	$494.80	$92.10	$736.80	$61.12	$91.08
1 Electrician	61.35	490.80	91.35	730.80		
.5 Equip. Oper. (crane)	59.20	236.80	88.50	354.00		
.5 S.P. Crane, 4x4, 5 Ton		189.05		207.96	9.45	10.40
20 L.H., Daily Totals		$1411.45		$2029.56	$70.57	$101.48

Crew R-4	Hr.	Daily	Hr.	Daily	Bare Costs	Incl. O&P
1 Struc. Steel Foreman (outside)	$59.65	$477.20	$93.55	$748.40	$58.79	$91.25
3 Struc. Steel Workers	57.65	1383.60	90.45	2170.80		
1 Electrician	61.35	490.80	91.35	730.80		
1 Welder, Gas Engine, 300 amp		147.00		161.70	3.67	4.04
40 L.H., Daily Totals		$2498.60		$3811.70	$62.47	$95.29

Crew R-5	Hr.	Daily	Hr.	Daily	Bare Costs	Incl. O&P
1 Electrician Foreman	$61.85	$494.80	$92.10	$736.80	$53.61	$80.36
4 Electrician Linemen	61.35	1963.20	91.35	2923.20		
2 Electrician Operators	61.35	981.60	91.35	1461.60		
4 Electrician Groundmen	39.95	1278.40	60.95	1950.40		
1 Crew Truck		166.25		182.88		
1 Flatbed Truck, 20,000 GVW		201.60		221.76		
1 Pickup Truck, 3/4 Ton		110.85		121.94		
.2 Hyd. Crane, 55 Ton		196.04		215.64		
.2 Hyd. Crane, 12 Ton		94.20		103.62		
.2 Earth Auger, Truck-Mtd.		40.10		44.11		
1 Tractor w/Winch		373.50		410.85	13.44	14.78
88 L.H., Daily Totals		$5900.54		$8372.79	$67.05	$95.15

Crew R-6	Hr.	Daily	Hr.	Daily	Bare Costs	Incl. O&P
1 Electrician Foreman	$61.85	$494.80	$92.10	$736.80	$53.61	$80.36
4 Electrician Linemen	61.35	1963.20	91.35	2923.20		
2 Electrician Operators	61.35	981.60	91.35	1461.60		
4 Electrician Groundmen	39.95	1278.40	60.95	1950.40		
1 Crew Truck		166.25		182.88		
1 Flatbed Truck, 20,000 GVW		201.60		221.76		
1 Pickup Truck, 3/4 Ton		110.85		121.94		
.2 Hyd. Crane, 55 Ton		196.04		215.64		
.2 Hyd. Crane, 12 Ton		94.20		103.62		
.2 Earth Auger, Truck-Mtd.		40.10		44.11		
1 Tractor w/Winch		373.50		410.85		
3 Cable Trailers		192.30		211.53		
.5 Tensioning Rig		55.20		60.72		
.5 Cable Pulling Rig		303.57		333.93	19.70	21.67
88 L.H., Daily Totals		$6451.61		$8978.98	$73.31	$102.03

Crew R-7	Hr.	Daily	Hr.	Daily	Bare Costs	Incl. O&P
1 Electrician Foreman	$61.85	$494.80	$92.10	$736.80	$43.60	$66.14
5 Electrician Groundmen	39.95	1598.00	60.95	2438.00		
1 Crew Truck		166.25		182.88	3.46	3.81
48 L.H., Daily Totals		$2259.05		$3357.68	$47.06	$69.95

Crew R-8	Hr.	Daily	Hr.	Daily	Bare Costs	Incl. O&P
1 Electrician Foreman	$61.85	$494.80	$92.10	$736.80	$54.30	$81.34
3 Electrician Linemen	61.35	1472.40	91.35	2192.40		
2 Electrician Groundmen	39.95	639.20	60.95	975.20		
1 Pickup Truck, 3/4 Ton		110.85		121.94		
1 Crew Truck		166.25		182.88	5.77	6.35
48 L.H., Daily Totals		$2883.50		$4209.21	$60.07	$87.69

Crew R-9	Hr.	Daily	Hr.	Daily	Bare Costs	Incl. O&P
1 Electrician Foreman	$61.85	$494.80	$92.10	$736.80	$50.71	$76.24
1 Electrician Lineman	61.35	490.80	91.35	730.80		
2 Electrician Operators	61.35	981.60	91.35	1461.60		
4 Electrician Groundmen	39.95	1278.40	60.95	1950.40		
1 Pickup Truck, 3/4 Ton		110.85		121.94		
1 Crew Truck		166.25		182.88	4.33	4.76
64 L.H., Daily Totals		$3522.70		$5184.41	$55.04	$81.01

Crew R-10	Hr.	Daily	Hr.	Daily	Bare Costs	Incl. O&P
1 Electrician Foreman	$61.85	$494.80	$92.10	$736.80	$57.87	$86.41
4 Electrician Linemen	61.35	1963.20	91.35	2923.20		
1 Electrician Groundman	39.95	319.60	60.95	487.60		
1 Crew Truck		166.25		182.88		
3 Tram Cars		216.15		237.76	7.97	8.76
48 L.H., Daily Totals		$3160.00		$4568.24	$65.83	$95.17

Crew No.	Bare Costs		Incl. Subs O&P		Cost Per Labor-Hour	
Crew R-11	Hr.	Daily	Hr.	Daily	Bare Costs	Incl. O&P
1 Electrician Foreman	$61.85	$494.80	$92.10	$736.80	$58.36	$87.04
4 Electricians	61.35	1963.20	91.35	2923.20		
1 Equip. Oper. (crane)	59.20	473.60	88.50	708.00		
1 Common Laborer	42.10	336.80	63.25	506.00		
1 Crew Truck		166.25		182.88		
1 Hyd. Crane, 12 Ton		471.00		518.10	11.38	12.52
56 L.H., Daily Totals		$3905.65		$5574.98	$69.74	$99.55
Crew R-12	Hr.	Daily	Hr.	Daily	Bare Costs	Incl. O&P
1 Carpenter Foreman (inside)	$53.65	$429.20	$80.60	$644.80	$49.91	$75.30
4 Carpenters	53.15	1700.80	79.85	2555.20		
4 Common Laborers	42.10	1347.20	63.25	2024.00		
1 Equip. Oper. (medium)	56.75	454.00	84.85	678.80		
1 Steel Worker	57.65	461.20	90.45	723.60		
1 Dozer, 200 H.P.		1504.00		1654.40		
1 Pickup Truck, 3/4 Ton		110.85		121.94	18.35	20.19
88 L.H., Daily Totals		$6007.25		$8402.74	$68.26	$95.49
Crew R-13	Hr.	Daily	Hr.	Daily	Bare Costs	Incl. O&P
1 Electrician Foreman	$61.85	$494.80	$92.10	$736.80	$59.29	$88.35
3 Electricians	61.35	1472.40	91.35	2192.40		
.25 Equip. Oper. (crane)	59.20	118.40	88.50	177.00		
1 Equipment Oiler	50.55	404.40	75.55	604.40		
.25 Hydraulic Crane, 33 Ton		236.34		259.97	5.63	6.19
42 L.H., Daily Totals		$2726.34		$3970.57	$64.91	$94.54
Crew R-15	Hr.	Daily	Hr.	Daily	Bare Costs	Incl. O&P
1 Electrician Foreman	$61.85	$494.80	$92.10	$736.80	$60.04	$89.45
4 Electricians	61.35	1963.20	91.35	2923.20		
1 Equipment Oper. (light)	53.00	424.00	79.20	633.60		
1 Telescoping Boom Lift, to 40'		278.90		306.79	5.81	6.39
48 L.H., Daily Totals		$3160.90		$4600.39	$65.85	$95.84
Crew R-15A	Hr.	Daily	Hr.	Daily	Bare Costs	Incl. O&P
1 Electrician Foreman	$61.85	$494.80	$92.10	$736.80	$53.63	$80.08
2 Electricians	61.35	981.60	91.35	1461.60		
2 Common Laborers	42.10	673.60	63.25	1012.00		
1 Equip. Oper. (light)	53.00	424.00	79.20	633.60		
1 Telescoping Boom Lift, to 40'		278.90		306.79	5.81	6.39
48 L.H., Daily Totals		$2852.90		$4150.79	$59.44	$86.47
Crew R-18	Hr.	Daily	Hr.	Daily	Bare Costs	Incl. O&P
.25 Electrician Foreman	$61.85	$123.70	$92.10	$184.20	$53.85	$80.18
1 Electrician	61.35	490.80	91.35	730.80		
2 Electrician Apprentices	49.10	785.60	73.10	1169.60		
26 L.H., Daily Totals		$1400.10		$2084.60	$53.85	$80.18
Crew R-19	Hr.	Daily	Hr.	Daily	Bare Costs	Incl. O&P
.5 Electrician Foreman	$61.85	$247.40	$92.10	$368.40	$61.45	$91.50
2 Electricians	61.35	981.60	91.35	1461.60		
20 L.H., Daily Totals		$1229.00		$1830.00	$61.45	$91.50
Crew R-21	Hr.	Daily	Hr.	Daily	Bare Costs	Incl. O&P
1 Electrician Foreman	$61.85	$494.80	$92.10	$736.80	$61.36	$91.37
3 Electricians	61.35	1472.40	91.35	2192.40		
.1 Equip. Oper. (medium)	56.75	45.40	84.85	67.88		
.1 S.P. Crane, 4x4, 25 Ton		114.40		125.84	3.49	3.84
32.8 L.H., Daily Totals		$2127.00		$3122.92	$64.85	$95.21

Crew No.	Bare Costs		Incl. Subs O&P		Cost Per Labor-Hour	
Crew R-22	Hr.	Daily	Hr.	Daily	Bare Costs	Incl. O&P
.66 Electrician Foreman	$61.85	$326.57	$92.10	$486.29	$56.16	$83.62
2 Electricians	61.35	981.60	91.35	1461.60		
2 Electrician Apprentices	49.10	785.60	73.10	1169.60		
37.28 L.H., Daily Totals		$2093.77		$3117.49	$56.16	$83.62
Crew R-30	Hr.	Daily	Hr.	Daily	Bare Costs	Incl. O&P
.25 Electrician Foreman (outside)	$63.35	$126.70	$94.35	$188.70	$49.66	$74.29
1 Electrician	61.35	490.80	91.35	730.80		
2 Laborers (Semi-Skilled)	42.10	673.60	63.25	1012.00		
26 L.H., Daily Totals		$1291.10		$1931.50	$49.66	$74.29
Crew R-31	Hr.	Daily	Hr.	Daily	Bare Costs	Incl. O&P
1 Electrician	$61.35	$490.80	$91.35	$730.80	$61.35	$91.35
1 Core Drill, Electric, 2.5 H.P.		62.50		68.75	7.81	8.59
8 L.H., Daily Totals		$553.30		$799.55	$69.16	$99.94
Crew W-41E	Hr.	Daily	Hr.	Daily	Bare Costs	Incl. O&P
.5 Plumber Foreman (outside)	$66.45	$265.80	$99.45	$397.80	$55.91	$83.77
1 Plumber	64.45	515.60	96.45	771.60		
1 Laborer	42.10	336.80	63.25	506.00		
20 L.H., Daily Totals		$1118.20		$1675.40	$55.91	$83.77

Historical Cost Indexes

The table below lists both the RSMeans® historical cost index based on Jan. 1, 1993 = 100 as well as the computed value of an index based on Jan. 1, 2020 costs. Since the Jan. 1, 2020 figure is estimated, space is left to write in the actual index figures as they become available through the quarterly *RSMeans Construction Cost Indexes*.

To compute the actual index based on Jan. 1, 2020 = 100, divide the historical cost index for a particular year by the actual Jan. 1, 2020 construction cost index. Space has been left to advance the index figures as the year progresses.

Year	Historical Cost Index Jan. 1, 1993 = 100		Current Index Based on Jan. 1, 2020 = 100		Year	Historical Cost Index Jan. 1, 1993 = 100	Current Index Based on Jan. 1, 2020 = 100		Year	Historical Cost Index Jan. 1, 1993 = 100	Current Index Based on Jan. 1, 2020 = 100	
	Est.	Actual	Est.	Actual		Actual	Est.	Actual		Actual	Est.	Actual
Oct 2020*					July 2005	151.6	63.4		July 1987	87.7	36.7	
July 2020*					2004	143.7	60.1		1986	84.2	35.2	
Apr 2020*					2003	132.0	55.2		1985	82.6	34.6	
Jan 2020*	239.1		100.0	100.0	2002	128.7	53.8		1984	82.0	34.3	
July 2019		232.2	97.1		2001	125.1	52.3		1983	80.2	33.5	
2018		222.9	93.2		2000	120.9	50.6		1982	76.1	31.8	
2017		213.6	89.3		1999	117.6	49.2		1981	70.0	29.3	
2016		207.3	86.7		1998	115.1	48.1		1980	62.9	26.3	
2015		206.2	86.2		1997	112.8	47.2		1979	57.8	24.2	
2014		204.9	85.7		1996	110.2	46.1		1978	53.5	22.4	
2013		201.2	84.1		1995	107.6	45.0		1977	49.5	20.7	
2012		194.6	81.4		1994	104.4	43.7		1976	46.9	19.6	
2011		191.2	80.0		1993	101.7	42.5		1975	44.8	18.7	
2010		183.5	76.7		1992	99.4	41.6		1974	41.4	17.3	
2009		180.1	75.3		1991	96.8	40.5		1973	37.7	15.8	
2008		180.4	75.4		1990	94.3	39.4		1972	34.8	14.6	
2007		169.4	70.8		1989	92.1	38.5		1971	32.1	13.4	
2006		162.0	67.8		1988	89.9	37.6		1970	28.7	12.0	

Adjustments to Costs

The "Historical Cost Index" can be used to convert national average building costs at a particular time to the approximate building costs for some other time.

Example:

Estimate and compare construction costs for different years in the same city.

To estimate the national average construction cost of a building in 1970, knowing that it cost $900,000 in 2020:

INDEX in 1970 = 28.7

INDEX in 2020 = 239.1

Note: The city cost indexes for Canada can be used to convert U.S. national averages to local costs in Canadian dollars.

Example:

To estimate and compare the cost of a building in Toronto, ON in 2020 with the known cost of $600,000 (US$) in New York, NY in 2020:

INDEX Toronto = 115.6

INDEX New York = 137.1

$$\frac{\text{INDEX Toronto}}{\text{INDEX New York}} \times \text{Cost New York} = \text{Cost Toronto}$$

$$\frac{115.6}{137.1} \times \$600,000 = .843 \times \$600,000 = \$505,908$$

The construction cost of the building in Toronto is $505,908 (CN$).

Time Adjustment Using the Historical Cost Indexes:

$$\frac{\text{Index for Year A}}{\text{Index for Year B}} \times \text{Cost in Year B} = \text{Cost in Year A}$$

$$\frac{\text{INDEX 1970}}{\text{INDEX 2020}} \times \text{Cost 2020} = \text{Cost 1970}$$

$$\frac{28.7}{239.1} \times \$900,000 = .120 \times \$900,000 = \$108,000$$

The construction cost of the building in 1970 was $108,000.

*Historical Cost Index updates and other resources are provided on the following website:
http://info.thegordiangroup.com/RSMeans.html

For customer support on your Mechanical Costs with RSMeans Data, call 800.448.8182.

How to Use the City Cost Indexes

What you should know before you begin

RSMeans City Cost Indexes (CCI) are an extremely useful tool for when you want to compare costs from city to city and region to region.

This publication contains average construction cost indexes for 731 U.S. and Canadian cities covering over 930 three-digit zip code locations, as listed directly under each city.

Keep in mind that a City Cost Index number is a percentage ratio of a specific city's cost to the national average cost of the same item at a stated time period.

In other words, these index figures represent relative construction factors (or, if you prefer, multipliers) for material and installation costs, as well as the weighted average for Total In Place costs for each CSI MasterFormat division. Installation costs include both labor and equipment rental costs. When estimating equipment rental rates only for a specific location, use 01 54 33 EQUIPMENT RENTAL COSTS in the Reference Section.

The 30 City Average Index is the average of 30 major U.S. cities and serves as a national average.

Index figures for both material and installation are based on the 30 major city average of 100 and represent the cost relationship as of July 1, 2019. The index for each division is computed from representative material and labor quantities for that division. The weighted average for each city is a weighted total of the components listed above it. It does not include relative productivity between trades or cities.

As changes occur in local material prices, labor rates, and equipment rental rates (including fuel costs), the impact of these changes should be accurately measured by the change in the City Cost Index for each particular city (as compared to the 30 city average).

Therefore, if you know (or have estimated) building costs in one city today, you can easily convert those costs to expected building costs in another city.

In addition, by using the Historical Cost Index, you can easily convert national average building costs at a particular time to the approximate building costs for some other time. The City Cost Indexes can then be applied to calculate the costs for a particular city.

Quick calculations

Location Adjustment Using the City Cost Indexes:

$$\frac{\text{Index for City A}}{\text{Index for City B}} \times \text{Cost in City B} = \text{Cost in City A}$$

Time Adjustment for the National Average Using the Historical Cost Index:

$$\frac{\text{Index for Year A}}{\text{Index for Year B}} \times \text{Cost in Year B} = \text{Cost in Year A}$$

Adjustment from the National Average:

$$\frac{\text{Index for City A}}{100} \times \text{National Average Cost} = \text{Cost in City A}$$

Since each of the other RSMeans data sets contains many different items, any *one* item multiplied by the particular city index may give incorrect results. However, the larger the number of items compiled, the closer the results should be to actual costs for that particular city.

The City Cost Indexes for Canadian cities are calculated using Canadian material and equipment prices and labor rates in Canadian dollars. Therefore, indexes for Canadian cities can be used to convert U.S. national average prices to local costs in Canadian dollars.

How to use this section

1. Compare costs from city to city.

In using the RSMeans Indexes, remember that an index number is not a fixed number but a ratio: It's a percentage ratio of a building component's cost at any stated time to the national average cost of that same component at the same time period. Put in the form of an equation:

$$\frac{\text{Specific City Cost}}{\text{National Average Cost}} \times 100 = \text{City Index Number}$$

Therefore, when making cost comparisons between cities, do not subtract one city's index number from the index number of another city and read the result as a percentage difference. Instead, divide one city's index number by that of the other city. The resulting number may then be used as a multiplier to calculate cost differences from city to city.

The formula used to find cost differences between cities for the purpose of comparison is as follows:

$$\frac{\text{City A Index}}{\text{City B Index}} \times \text{City B Cost (Known)} = \text{City A Cost (Unknown)}$$

In addition, you can use RSMeans CCI to calculate and compare costs division by division between cities using the same basic formula. (Just be sure that you're comparing similar divisions.)

2. Compare a specific city's construction costs with the national average.

When you're studying construction location feasibility, it's advisable to compare a prospective project's cost index with an index of the national average cost.

For example, divide the weighted average index of construction costs of a specific city by that of the 30 City Average, which = 100.

$$\frac{\text{City Index}}{100} = \% \text{ of National Average}$$

As a result, you get a ratio that indicates the relative cost of construction in that city in comparison with the national average.

3. Convert U.S. national average to actual costs in Canadian City.

$$\frac{\text{Index for Canadian City}}{100} \times \text{National Average Cost} = \text{Cost in Canadian City in \$ CAN}$$

For customer support on your Mechanical Costs with RSMeans Data, call 800.448.8182.

657

4. Adjust construction cost data based on a national average.

When you use a source of construction cost data which is based on a national average (such as RSMeans cost data), it is necessary to adjust those costs to a specific location.

$$\frac{\text{City Index}}{100} \times \frac{\text{Cost Based on}}{\text{National Average Costs}} = \frac{\text{City Cost}}{\text{(Unknown)}}$$

5. When applying the City Cost Indexes to demolition projects, use the appropriate division installation index. For example, for removal of existing doors and windows, use the Division 8 (Openings) index.

What you might like to know about how we developed the Indexes

The information presented in the CCI is organized according to the Construction Specifications Institute (CSI) MasterFormat 2018 classification system.

To create a reliable index, RSMeans researched the building type most often constructed in the United States and Canada. Because it was concluded that no one type of building completely represented the building construction industry, nine different types of buildings were combined to create a composite model.

The exact material, labor, and equipment quantities are based on detailed analyses of these nine building types, and then each quantity is weighted in proportion to expected usage. These various material items, labor hours, and equipment rental rates are thus combined to form a composite building representing as closely as possible the actual usage of materials, labor, and equipment in the North American building construction industry.

The following structures were chosen to make up that composite model:

1. Factory, 1 story
2. Office, 2–4 stories
3. Store, Retail
4. Town Hall, 2–3 stories
5. High School, 2–3 stories
6. Hospital, 4–8 stories
7. Garage, Parking
8. Apartment, 1–3 stories
9. Hotel/Motel, 2–3 stories

For the purposes of ensuring the timeliness of the data, the components of the index for the composite model have been streamlined. They currently consist of:

- specific quantities of 66 commonly used construction materials;
- specific labor-hours for 21 building construction trades; and
- specific days of equipment rental for 6 types of construction equipment (normally used to install the 66 material items by the 21 trades.) Fuel costs and routine maintenance costs are included in the equipment cost.

Material and equipment price quotations are gathered quarterly from cities in the United States and Canada. These prices and the latest negotiated labor wage rates for 21 different building trades are used to compile the quarterly update of the City Cost Index.

The 30 major U.S. cities used to calculate the national average are:

Atlanta, GA	Memphis, TN
Baltimore, MD	Milwaukee, WI
Boston, MA	Minneapolis, MN
Buffalo, NY	Nashville, TN
Chicago, IL	New Orleans, LA
Cincinnati, OH	New York, NY
Cleveland, OH	Philadelphia, PA
Columbus, OH	Phoenix, AZ
Dallas, TX	Pittsburgh, PA
Denver, CO	St. Louis, MO
Detroit, MI	San Antonio, TX
Houston, TX	San Diego, CA
Indianapolis, IN	San Francisco, CA
Kansas City, MO	Seattle, WA
Los Angeles, CA	Washington, DC

What the CCI does not indicate

The weighted average for each city is a total of the divisional components weighted to reflect typical usage. It does not include the productivity variations between trades or cities.

In addition, the CCI does not take into consideration factors such as the following:

- managerial efficiency
- competitive conditions
- automation
- restrictive union practices
- unique local requirements
- regional variations due to specific building codes

United States / Alabama

DIVISION		UNITED STATES 30 CITY AVERAGE			ANNISTON 362			BIRMINGHAM 350 - 352			BUTLER 369			DECATUR 356			DOTHAN 363		
		MAT.	INST.	TOTAL	MAT.	INST.	TOTAL	MAT.	INST.	TOTAL	MAT.	INST.	TOTAL	MAT.	INST.	TOTAL	MAT.	INST.	TOTAL
015433	CONTRACTOR EQUIPMENT		100.0	100.0		101.9	101.9		104.7	104.7		99.5	99.5		101.9	101.9		99.5	99.5
0241, 31 - 34	SITE & INFRASTRUCTURE, DEMOLITION	100.0	100.0	100.0	90.9	89.5	89.9	91.2	94.5	93.5	105.4	84.4	90.9	84.5	88.1	87.0	102.9	84.8	90.4
0310	Concrete Forming & Accessories	100.0	100.0	100.0	85.6	68.7	71.2	90.6	69.7	72.8	82.3	68.8	70.8	90.9	62.8	66.9	90.3	69.1	72.2
0320	Concrete Reinforcing	100.0	100.0	100.0	88.8	71.2	80.3	94.2	71.3	83.1	93.9	70.5	82.6	88.1	67.7	78.2	93.9	71.3	83.0
0330	Cast-in-Place Concrete	100.0	100.0	100.0	82.3	67.7	76.9	103.8	69.3	90.9	80.3	67.8	75.6	101.9	67.2	89.0	80.3	67.7	75.6
03	CONCRETE	100.0	100.0	100.0	89.2	70.4	81.0	92.2	71.4	83.1	90.3	70.4	81.6	92.0	67.2	81.0	89.5	70.7	81.2
04	MASONRY	100.0	100.0	100.0	89.2	63.1	73.2	87.1	64.3	73.1	93.7	63.1	74.9	85.2	62.4	71.2	95.1	62.9	75.3
05	METALS	100.0	100.0	100.0	101.8	94.8	99.7	100.5	94.2	98.7	100.7	94.8	98.9	102.9	93.1	100.0	100.7	95.4	99.1
06	WOOD, PLASTICS & COMPOSITES	100.0	100.0	100.0	82.0	69.9	75.3	91.0	70.0	79.4	77.1	69.9	73.1	95.7	62.1	77.2	88.1	69.9	78.1
07	THERMAL & MOISTURE PROTECTION	100.0	100.0	100.0	96.0	63.0	81.6	94.0	67.6	82.5	96.1	65.9	82.9	93.1	64.5	80.7	96.0	65.7	82.8
08	OPENINGS	100.0	100.0	100.0	94.7	69.2	88.7	102.6	69.8	94.9	94.7	69.5	88.8	106.9	64.6	97.0	94.7	69.7	88.9
0920	Plaster & Gypsum Board	100.0	100.0	100.0	81.8	69.5	73.5	88.9	69.5	75.9	79.3	69.5	72.7	92.7	61.6	71.7	88.7	69.5	75.8
0950, 0980	Ceilings & Acoustic Treatment	100.0	100.0	100.0	77.1	69.5	72.0	84.8	69.5	74.5	77.1	69.5	72.0	83.2	61.6	68.6	77.1	69.5	72.0
0960	Flooring	100.0	100.0	100.0	82.0	69.1	78.3	99.6	69.1	91.0	86.5	69.1	81.6	89.9	69.1	84.1	91.6	69.1	85.3
0970, 0990	Wall Finishes & Painting/Coating	100.0	100.0	100.0	89.1	66.7	75.9	87.0	66.7	75.0	89.1	45.9	63.5	79.4	61.0	68.5	89.1	80.4	84.0
09	FINISHES	100.0	100.0	100.0	80.4	68.8	74.1	91.2	69.2	79.3	83.6	66.6	74.4	85.3	63.5	73.5	86.1	70.3	77.6
COVERS	DIVS. 10 - 14, 25, 28, 41, 43, 44, 46	100.0	100.0	100.0	100.0	74.0	94.2	100.0	83.6	96.3	100.0	74.0	94.2	100.0	72.9	94.0	100.0	74.0	94.2
21, 22, 23	FIRE SUPPRESSION, PLUMBING & HVAC	100.0	100.0	100.0	101.0	51.8	81.7	100.0	66.3	86.8	98.1	67.0	85.9	100.0	64.5	86.1	98.1	65.0	85.1
26, 27, 3370	ELECTRICAL, COMMUNICATIONS & UTIL.	100.0	100.0	100.0	97.7	59.9	79.1	98.4	59.7	79.3	99.8	61.7	81.0	94.5	63.8	79.4	98.6	76.1	87.5
MF2018	WEIGHTED AVERAGE	100.0	100.0	100.0	95.9	67.7	84.0	97.4	71.8	86.6	96.1	70.6	85.3	97.1	69.2	85.3	96.2	72.7	86.3

Alabama

DIVISION		EVERGREEN 364			GADSDEN 359			HUNTSVILLE 357 - 358			JASPER 355			MOBILE 365 - 366			MONTGOMERY 360 - 361		
		MAT.	INST.	TOTAL	MAT.	INST.	TOTAL	MAT.	INST.	TOTAL	MAT.	INST.	TOTAL	MAT.	INST.	TOTAL	MAT.	INST.	TOTAL
015433	CONTRACTOR EQUIPMENT		99.5	99.5		101.9	101.9		101.9	101.9		101.9	101.9		99.5	99.5		102.0	102.0
0241, 31 - 34	SITE & INFRASTRUCTURE, DEMOLITION	105.9	84.4	91.0	89.9	89.5	89.6	84.2	89.4	87.8	89.9	89.5	89.6	97.7	85.4	89.2	95.6	89.9	91.7
0310	Concrete Forming & Accessories	79.4	68.6	70.2	83.9	69.1	71.3	90.9	65.9	69.6	88.5	64.7	68.3	89.4	68.4	71.5	92.8	69.4	72.8
0320	Concrete Reinforcing	94.0	70.4	82.6	93.5	71.3	82.8	88.1	76.3	82.4	88.1	71.3	80.0	91.5	70.5	81.3	99.7	71.4	86.0
0330	Cast-in-Place Concrete	80.3	67.6	75.6	101.9	67.8	89.2	99.2	67.8	87.5	113.0	67.8	96.1	84.2	66.9	77.7	81.5	68.7	76.7
03	CONCRETE	90.7	70.2	81.7	96.5	70.7	85.1	90.8	70.1	81.7	100.1	68.7	86.3	85.5	69.9	78.7	84.1	71.1	78.4
04	MASONRY	93.8	62.9	74.8	83.8	63.1	71.1	86.3	63.3	72.2	81.4	63.1	70.2	92.3	61.4	73.3	89.4	63.2	73.3
05	METALS	100.7	94.6	98.9	100.7	95.2	99.1	102.9	96.7	101.0	100.7	95.1	99.0	102.8	94.9	100.5	101.7	94.3	99.5
06	WOOD, PLASTICS & COMPOSITES	73.8	69.9	71.6	86.7	69.9	77.4	95.7	65.3	79.0	93.0	64.1	77.1	86.7	69.9	77.4	90.3	70.0	79.1
07	THERMAL & MOISTURE PROTECTION	96.0	65.4	82.7	93.3	66.3	81.6	93.0	65.9	81.2	93.3	64.4	80.7	95.6	65.4	82.4	94.2	66.9	82.3
08	OPENINGS	94.7	69.3	88.8	103.4	69.7	95.6	106.6	68.5	97.7	103.4	66.6	94.8	97.3	69.5	90.8	96.2	69.8	90.0
0920	Plaster & Gypsum Board	78.3	69.5	72.4	85.1	69.5	74.6	92.7	64.8	73.9	89.5	63.6	72.1	85.6	69.5	74.8	85.9	69.5	74.9
0950, 0980	Ceilings & Acoustic Treatment	77.1	69.5	72.0	79.1	69.5	72.7	85.0	64.8	71.4	79.1	63.6	68.6	83.0	69.5	73.9	84.6	69.5	74.4
0960	Flooring	84.5	69.1	80.1	86.4	69.1	81.5	89.9	69.1	84.1	88.4	69.1	82.9	91.1	69.1	84.9	91.7	69.1	85.3
0970, 0990	Wall Finishes & Painting/Coating	89.1	45.9	63.5	79.4	57.4	66.4	79.4	63.7	70.1	79.4	66.7	71.9	92.3	45.9	64.8	91.1	66.7	76.7
09	FINISHES	82.9	66.5	74.1	82.9	67.8	74.7	85.7	65.9	75.0	84.0	65.4	73.9	86.2	66.2	75.4	88.3	68.9	77.8
COVERS	DIVS. 10 - 14, 25, 28, 41, 43, 44, 46	100.0	73.9	94.2	100.0	82.9	96.2	100.0	82.5	96.1	100.0	73.4	94.1	100.0	82.5	96.1	100.0	83.3	96.3
21, 22, 23	FIRE SUPPRESSION, PLUMBING & HVAC	98.1	59.8	83.0	102.3	62.3	86.6	100.0	66.9	87.0	102.3	65.7	87.9	100.0	61.4	84.8	100.0	64.3	86.0
26, 27, 3370	ELECTRICAL, COMMUNICATIONS & UTIL.	97.1	56.8	77.2	94.5	59.7	77.4	95.4	65.9	80.8	94.2	59.9	77.3	100.4	56.8	78.9	101.1	76.1	88.8
MF2018	WEIGHTED AVERAGE	95.8	68.3	84.2	97.3	70.2	85.9	97.1	71.7	86.4	97.8	69.8	86.0	96.7	68.7	84.9	96.3	73.1	86.5

Alabama / Alaska

DIVISION		PHENIX CITY 368			SELMA 367			TUSCALOOSA 354			ANCHORAGE 995 - 996			FAIRBANKS 997			JUNEAU 998		
		MAT.	INST.	TOTAL	MAT.	INST.	TOTAL	MAT.	INST.	TOTAL	MAT.	INST.	TOTAL	MAT.	INST.	TOTAL	MAT.	INST.	TOTAL
015433	CONTRACTOR EQUIPMENT		99.5	99.5		99.5	99.5		101.9	101.9		110.5	110.5		112.9	112.9		110.5	110.5
0241, 31 - 34	SITE & INFRASTRUCTURE, DEMOLITION	109.8	85.5	93.0	102.7	85.5	90.8	84.8	89.5	88.0	118.4	121.7	120.7	117.8	124.5	122.5	133.6	121.7	125.4
0310	Concrete Forming & Accessories	85.6	67.6	70.2	85.8	69.1	71.2	90.8	69.2	72.4	120.5	116.2	116.8	124.7	115.0	116.4	125.7	116.2	117.6
0320	Concrete Reinforcing	93.9	66.9	80.9	93.9	71.3	82.8	88.1	71.3	80.0	148.4	120.7	135.0	147.2	120.7	134.4	138.9	120.7	130.1
0330	Cast-in-Place Concrete	80.3	67.9	75.7	80.3	67.8	75.6	103.4	67.9	90.2	105.1	117.5	109.7	109.7	115.6	111.9	116.1	117.5	116.6
03	CONCRETE	93.6	69.3	82.9	89.0	70.7	81.0	92.7	70.8	83.1	107.5	116.7	111.6	101.4	115.5	107.6	114.1	116.7	115.2
04	MASONRY	93.7	63.1	74.9	97.4	63.1	76.3	85.4	63.1	71.7	171.5	119.4	139.5	181.4	117.8	142.3	162.5	119.4	136.0
05	METALS	100.6	93.7	98.6	100.6	95.2	99.0	102.1	95.3	100.1	127.3	104.4	120.5	122.6	105.4	117.5	113.1	104.4	110.6
06	WOOD, PLASTICS & COMPOSITES	81.7	67.7	74.0	78.9	69.9	73.9	95.7	69.9	81.5	112.3	113.5	112.9	126.9	112.7	119.1	121.9	113.5	117.3
07	THERMAL & MOISTURE PROTECTION	96.5	66.1	83.2	95.9	65.9	82.8	93.1	66.3	81.4	176.9	116.9	150.8	186.2	115.8	155.6	188.4	116.9	157.3
08	OPENINGS	94.7	67.4	88.3	94.7	69.7	88.8	106.6	69.7	98.0	130.8	115.8	127.3	132.8	114.7	128.6	131.0	115.8	127.5
0920	Plaster & Gypsum Board	83.0	67.3	72.4	81.1	69.5	73.3	92.7	69.5	77.1	144.1	113.8	123.7	177.5	112.9	134.0	158.2	113.8	128.3
0950, 0980	Ceilings & Acoustic Treatment	77.1	67.3	70.5	77.1	69.5	72.0	85.0	69.5	74.6	131.3	113.8	119.5	123.9	112.9	116.5	136.1	113.8	121.0
0960	Flooring	88.4	69.1	83.0	87.0	69.1	81.9	89.9	69.1	84.1	118.3	119.4	118.6	116.7	119.4	117.4	125.0	119.4	123.5
0970, 0990	Wall Finishes & Painting/Coating	89.1	79.0	83.1	89.1	66.7	75.9	79.4	66.7	71.9	113.1	119.1	116.6	111.6	116.9	114.7	108.7	119.1	114.8
09	FINISHES	85.2	68.9	76.4	83.7	68.8	75.7	85.6	68.8	76.6	127.0	117.1	121.7	128.6	116.0	121.8	128.9	117.1	122.5
COVERS	DIVS. 10 - 14, 25, 28, 41, 43, 44, 46	100.0	82.7	96.1	100.0	70.4	93.4	100.0	82.9	96.2	100.0	114.1	103.2	100.0	113.6	103.0	100.0	114.1	103.2
21, 22, 23	FIRE SUPPRESSION, PLUMBING & HVAC	98.1	64.2	84.8	98.1	65.1	85.1	100.0	67.1	87.1	100.5	107.1	103.1	100.2	107.9	103.2	100.5	107.1	103.1
26, 27, 3370	ELECTRICAL, COMMUNICATIONS & UTIL.	99.2	65.4	82.6	98.3	76.1	87.3	95.0	59.7	77.6	117.2	109.4	113.3	124.2	109.4	116.9	110.2	109.4	109.8
MF2018	WEIGHTED AVERAGE	96.8	70.8	85.8	95.9	72.5	86.0	97.1	71.4	86.3	118.5	113.1	116.2	118.8	113.0	116.3	116.8	113.1	115.2

For customer support on your Mechanical Costs with RSMeans Data, call 800.448.8182.

659

Section 1

DIVISION		ALASKA KETCHIKAN 999			ARIZONA CHAMBERS 865			FLAGSTAFF 860			GLOBE 855			KINGMAN 864			MESA/TEMPE 852		
		MAT.	INST.	TOTAL	MAT.	INST.	TOTAL	MAT.	INST.	TOTAL	MAT.	INST.	TOTAL	MAT.	INST.	TOTAL	MAT.	INST.	TOTAL
015433	CONTRACTOR EQUIPMENT		112.9	112.9		88.2	88.2		88.2	88.2		89.3	89.3		88.2	88.2		89.3	89.3
0241, 31 - 34	SITE & INFRASTRUCTURE, DEMOLITION	169.1	124.6	138.4	71.5	90.1	84.4	91.2	90.1	90.4	109.0	91.0	96.5	71.4	90.1	84.3	97.8	91.0	93.1
0310	Concrete Forming & Accessories	116.3	116.2	116.2	99.2	70.3	74.6	104.6	67.4	72.8	93.5	70.4	73.8	97.4	66.1	70.7	96.5	70.5	74.3
0320	Concrete Reinforcing	110.6	120.7	115.5	105.5	76.9	91.7	105.4	76.9	91.6	110.5	76.9	94.2	105.7	76.9	91.7	111.2	76.9	94.6
0330	Cast-in-Place Concrete	221.8	117.3	182.9	89.8	67.3	81.4	88.6	67.4	81.5	83.3	67.0	77.2	89.5	67.3	81.2	84.0	67.0	77.7
03	CONCRETE	170.9	116.7	147.1	93.4	70.3	83.3	114.1	69.0	94.3	101.9	70.3	88.0	93.1	68.4	82.3	93.1	70.4	83.1
04	MASONRY	188.6	119.4	146.1	96.0	59.4	73.5	96.1	59.4	73.6	101.7	59.3	75.7	96.0	59.4	73.5	101.9	59.3	75.8
05	METALS	122.8	105.4	117.7	103.0	71.6	93.8	103.5	71.7	94.2	105.8	72.5	96.0	103.7	71.6	94.3	106.2	72.6	96.3
06	WOOD, PLASTICS & COMPOSITES	117.1	113.5	115.1	101.8	73.0	85.9	108.2	68.8	86.5	92.7	73.1	81.9	96.4	67.3	80.4	96.7	73.1	83.7
07	THERMAL & MOISTURE PROTECTION	191.6	116.4	158.9	99.9	69.7	86.8	101.8	69.3	87.7	99.1	68.3	85.7	99.9	69.1	86.5	99.0	68.3	85.6
08	OPENINGS	128.6	115.8	125.6	105.8	69.3	97.3	105.9	70.1	97.6	94.9	69.4	88.9	106.0	67.9	97.1	94.9	69.4	89.0
0920	Plaster & Gypsum Board	161.0	113.8	129.2	92.8	72.5	79.1	96.5	68.2	77.5	88.2	72.5	77.6	84.2	66.6	72.4	91.7	72.5	78.7
0950, 0980	Ceilings & Acoustic Treatment	118.4	113.8	115.3	106.8	72.5	83.6	107.6	68.2	81.1	95.5	72.5	80.0	107.6	66.6	80.0	95.5	72.5	80.0
0960	Flooring	116.4	119.4	117.2	89.7	64.5	82.6	92.0	64.5	84.3	98.6	64.5	89.0	88.4	64.5	81.6	100.4	64.5	90.3
0970, 0990	Wall Finishes & Painting/Coating	111.6	119.1	116.0	87.9	58.6	70.5	87.9	58.6	70.5	90.6	58.6	71.6	87.9	58.6	70.5	90.6	58.6	71.6
09	FINISHES	129.2	117.1	122.7	91.3	68.1	78.8	94.5	65.7	78.9	96.6	68.2	81.2	89.9	64.8	76.3	96.4	68.2	81.2
COVERS	DIVS. 10 - 14, 25, 28, 41, 43, 44, 46	100.0	114.2	103.2	100.0	81.8	95.9	100.0	81.3	95.8	100.0	82.1	96.0	100.0	81.2	95.8	100.0	82.1	96.0
21, 22, 23	FIRE SUPPRESSION, PLUMBING & HVAC	98.1	107.2	101.7	97.7	76.8	89.5	100.3	77.0	91.1	96.6	76.9	88.9	97.7	76.8	89.5	100.1	77.0	91.1
26, 27, 3370	ELECTRICAL, COMMUNICATIONS & UTIL.	124.2	109.4	116.9	102.3	66.5	84.7	101.3	60.7	81.3	97.1	62.9	80.2	102.3	62.9	82.9	94.2	62.9	78.7
MF2018	WEIGHTED AVERAGE	128.4	113.4	122.1	98.3	71.8	87.1	102.4	70.5	88.9	99.5	71.4	87.6	98.3	70.4	86.5	98.7	71.5	87.2

Section 2

DIVISION		ARIZONA PHOENIX 850,853			PRESCOTT 863			SHOW LOW 859			TUCSON 856 - 857			ARKANSAS BATESVILLE 725			CAMDEN 717		
		MAT.	INST.	TOTAL	MAT.	INST.	TOTAL	MAT.	INST.	TOTAL	MAT.	INST.	TOTAL	MAT.	INST.	TOTAL	MAT.	INST.	TOTAL
015433	CONTRACTOR EQUIPMENT		92.5	92.5		88.2	88.2		89.3	89.3		89.3	89.3		89.7	89.7		89.7	89.7
0241, 31 - 34	SITE & INFRASTRUCTURE, DEMOLITION	98.3	94.3	95.6	78.4	90.1	86.5	111.4	91.0	97.3	92.9	91.0	91.6	73.0	85.7	81.8	77.2	85.8	83.1
0310	Concrete Forming & Accessories	101.2	70.8	75.3	100.8	70.3	74.8	100.6	70.5	74.9	97.1	67.5	71.8	82.5	61.4	64.5	78.6	61.9	64.4
0320	Concrete Reinforcing	109.1	76.9	93.6	105.4	76.9	91.6	111.2	76.9	94.6	92.3	76.9	84.9	84.4	69.4	77.1	91.0	69.4	80.5
0330	Cast-in-Place Concrete	83.6	67.8	77.7	89.8	67.3	81.4	83.3	67.0	77.3	86.5	67.0	79.2	72.8	74.6	73.5	76.4	74.6	75.7
03	CONCRETE	95.5	70.8	84.6	99.0	70.3	86.4	104.2	70.3	89.3	91.6	69.0	81.7	73.2	68.1	70.9	75.7	68.3	72.4
04	MASONRY	94.8	62.1	74.7	96.1	59.4	73.6	101.8	59.3	75.7	88.9	59.3	70.7	94.1	60.8	73.6	104.7	60.8	77.7
05	METALS	107.8	73.7	97.8	103.5	71.6	94.2	105.6	72.5	95.9	107.0	72.6	96.9	94.7	76.0	89.2	101.1	76.1	91.0
06	WOOD, PLASTICS & COMPOSITES	102.1	72.0	85.5	103.3	73.0	86.6	101.2	73.1	85.7	96.9	69.0	81.6	91.1	63.5	75.9	89.2	64.0	75.4
07	THERMAL & MOISTURE PROTECTION	100.3	69.6	86.9	100.5	69.7	87.1	99.3	68.3	85.8	99.9	67.9	86.0	102.7	61.9	84.9	97.5	61.9	82.0
08	OPENINGS	100.7	71.4	93.9	105.9	69.3	97.4	94.2	71.1	88.8	91.4	70.2	86.5	100.0	61.0	90.9	103.1	61.3	93.4
0920	Plaster & Gypsum Board	99.4	71.2	80.4	93.3	72.5	79.3	93.5	72.5	79.4	96.3	68.2	77.4	73.7	62.9	66.5	84.7	63.5	70.4
0950, 0980	Ceilings & Acoustic Treatment	106.5	71.2	82.7	106.0	72.5	83.4	95.5	72.5	80.0	96.4	68.2	77.4	88.1	62.9	71.1	86.3	63.5	70.9
0960	Flooring	102.4	64.5	91.7	90.6	64.5	83.2	102.2	64.5	91.6	91.8	64.8	84.3	81.8	70.9	78.7	86.1	70.9	81.9
0970, 0990	Wall Finishes & Painting/Coating	95.3	57.2	72.8	87.9	58.6	70.5	90.6	58.6	71.6	91.3	58.6	71.9	92.2	54.5	69.9	88.8	54.5	68.5
09	FINISHES	101.0	68.1	83.2	92.0	68.1	79.1	98.6	68.2	82.2	94.5	65.9	79.0	76.3	62.6	68.9	79.8	62.9	70.7
COVERS	DIVS. 10 - 14, 25, 28, 41, 43, 44, 46	100.0	83.2	96.2	100.0	81.8	95.9	100.0	82.1	96.0	100.0	81.7	95.9	100.0	66.4	92.5	100.0	66.5	92.5
21, 22, 23	FIRE SUPPRESSION, PLUMBING & HVAC	100.0	78.5	91.6	100.3	76.8	91.0	96.6	76.9	88.9	100.1	74.8	90.2	96.5	51.2	78.7	96.4	57.5	81.2
26, 27, 3370	ELECTRICAL, COMMUNICATIONS & UTIL.	99.8	60.7	80.5	101.0	62.9	82.2	94.5	62.9	78.9	96.2	60.6	78.7	95.5	60.5	78.3	95.8	58.1	77.2
MF2018	WEIGHTED AVERAGE	100.6	72.3	88.6	99.9	71.3	87.8	99.7	71.5	87.8	97.7	70.2	86.1	91.6	63.6	79.8	94.0	64.7	81.6

Section 3

DIVISION		ARKANSAS FAYETTEVILLE 727			FORT SMITH 729			HARRISON 726			HOT SPRINGS 719			JONESBORO 724			LITTLE ROCK 720 - 722		
		MAT.	INST.	TOTAL	MAT.	INST.	TOTAL	MAT.	INST.	TOTAL	MAT.	INST.	TOTAL	MAT.	INST.	TOTAL	MAT.	INST.	TOTAL
015433	CONTRACTOR EQUIPMENT		89.7	89.7		89.7	89.7		89.7	89.7		89.7	89.7		112.7	112.7		92.4	92.4
0241, 31 - 34	SITE & INFRASTRUCTURE, DEMOLITION	72.5	85.7	81.6	77.7	85.5	83.1	77.7	85.7	83.3	80.2	85.7	84.0	96.7	102.4	100.7	85.3	90.3	88.8
0310	Concrete Forming & Accessories	78.4	61.8	64.3	95.7	61.6	66.6	86.5	61.8	65.4	76.3	61.8	64.0	85.7	61.9	65.4	93.1	62.4	66.9
0320	Concrete Reinforcing	84.4	62.1	73.6	85.4	66.3	76.2	84.0	69.4	76.9	89.3	69.4	79.7	81.5	65.1	73.6	90.4	69.5	80.3
0330	Cast-in-Place Concrete	72.9	74.6	73.5	83.3	75.5	80.4	80.8	74.5	78.5	78.1	74.6	76.8	79.4	75.8	78.0	80.7	76.9	79.3
03	CONCRETE	72.9	67.0	70.3	80.1	68.0	74.8	79.7	68.2	74.6	78.9	68.2	74.2	77.3	69.0	73.7	81.0	69.3	75.9
04	MASONRY	85.0	60.8	70.1	91.6	60.8	72.7	94.3	60.8	73.7	78.6	60.8	67.7	86.5	60.8	70.7	87.7	62.1	72.0
05	METALS	94.7	73.6	88.5	96.9	74.9	90.4	95.8	76.0	90.0	101.1	76.1	93.7	91.3	89.4	90.8	97.2	76.1	91.0
06	WOOD, PLASTICS & COMPOSITES	87.7	64.0	74.7	107.1	64.0	83.4	96.8	64.0	78.7	86.7	64.0	74.2	95.2	64.4	78.3	98.4	64.2	79.5
07	THERMAL & MOISTURE PROTECTION	103.5	61.9	85.4	103.8	61.7	85.5	103.0	61.9	85.1	97.8	61.9	82.2	108.4	61.9	88.2	98.5	63.1	83.1
08	OPENINGS	100.0	59.6	90.6	102.0	60.1	92.3	100.8	61.3	91.6	103.1	61.3	93.3	105.6	60.5	95.1	95.6	60.7	87.5
0920	Plaster & Gypsum Board	73.0	63.5	66.6	79.7	63.5	68.8	78.8	63.5	68.5	83.4	63.5	70.0	87.3	63.5	71.2	92.0	63.5	72.8
0950, 0980	Ceilings & Acoustic Treatment	88.1	63.5	71.5	89.8	63.5	72.0	89.8	63.5	72.0	86.3	63.5	70.9	92.1	63.5	72.8	87.3	63.5	71.2
0960	Flooring	78.9	70.9	76.7	88.3	70.9	83.4	84.1	70.9	80.4	85.1	70.9	81.1	59.3	70.9	62.6	89.2	81.0	86.9
0970, 0990	Wall Finishes & Painting/Coating	92.2	54.5	69.9	92.2	52.1	68.5	92.2	54.5	69.9	88.8	54.5	68.5	81.7	54.5	65.6	92.8	52.4	68.9
09	FINISHES	75.5	62.9	68.7	79.6	62.6	70.4	78.3	62.9	70.0	79.6	62.9	70.6	74.6	63.2	68.4	85.3	64.9	74.3
COVERS	DIVS. 10 - 14, 25, 28, 41, 43, 44, 46	100.0	66.1	92.4	100.0	77.7	95.0	100.0	68.5	93.0	100.0	66.2	92.5	100.0	67.0	92.6	100.0	78.2	95.1
21, 22, 23	FIRE SUPPRESSION, PLUMBING & HVAC	96.5	60.0	82.2	100.0	48.0	79.6	96.5	49.1	77.8	96.4	51.5	78.8	100.4	51.2	81.1	99.9	49.1	79.9
26, 27, 3370	ELECTRICAL, COMMUNICATIONS & UTIL.	90.0	53.9	72.2	93.1	57.0	75.3	94.2	57.5	76.1	97.7	65.2	81.7	99.0	62.3	81.0	100.1	66.7	83.7
MF2018	WEIGHTED AVERAGE	90.5	64.1	79.3	94.1	62.6	80.8	92.9	62.8	80.2	93.5	64.4	81.2	93.8	66.6	82.3	94.5	65.4	82.2

ARKANSAS / CALIFORNIA

DIVISION		PINE BLUFF 716 MAT.	INST.	TOTAL	RUSSELLVILLE 728 MAT.	INST.	TOTAL	TEXARKANA 718 MAT.	INST.	TOTAL	WEST MEMPHIS 723 MAT.	INST.	TOTAL	ALHAMBRA 917-918 MAT.	INST.	TOTAL	ANAHEIM 928 MAT.	INST.	TOTAL
015433	CONTRACTOR EQUIPMENT		89.7	89.7		89.7	89.7		90.9	90.9		112.7	112.7		95.4	95.4		99.2	99.2
0241, 31 - 34	SITE & INFRASTRUCTURE, DEMOLITION	82.3	85.6	84.6	74.2	85.7	82.2	92.9	87.8	89.3	103.1	102.8	102.9	100.8	105.3	103.9	97.7	105.5	103.1
0310	Concrete Forming & Accessories	76.1	61.9	64.0	83.1	61.7	64.8	81.4	61.5	64.5	90.6	62.2	66.4	114.4	139.5	135.8	102.9	139.7	134.3
0320	Concrete Reinforcing	90.9	69.4	80.5	84.9	69.4	77.4	90.5	69.3	80.2	81.5	63.3	72.7	105.0	133.6	118.9	95.2	133.5	113.8
0330	Cast-in-Place Concrete	78.1	74.6	76.8	76.3	74.5	75.6	85.2	74.4	81.2	83.2	75.9	80.5	83.3	127.8	99.9	86.7	131.0	103.2
03	CONCRETE	79.7	68.3	74.7	76.0	68.1	72.6	77.8	68.1	73.5	83.8	68.8	77.2	94.2	133.1	111.3	96.2	134.3	112.9
04	MASONRY	110.7	60.8	80.0	90.9	60.8	72.4	92.6	60.8	73.0	75.3	60.8	66.4	106.9	139.4	126.9	78.2	138.0	115.0
05	METALS	101.9	76.1	94.3	94.7	75.9	89.1	93.8	75.8	88.5	90.4	89.1	90.0	79.9	113.2	89.7	107.7	114.2	109.6
06	WOOD, PLASTICS & COMPOSITES	86.4	64.0	74.1	92.3	64.0	76.8	93.4	64.0	77.2	100.7	64.4	80.7	99.3	138.0	120.6	97.5	138.3	119.9
07	THERMAL & MOISTURE PROTECTION	97.9	62.0	82.3	103.7	61.9	85.5	98.5	61.9	82.6	108.9	61.9	88.4	104.2	131.8	116.2	108.9	134.9	120.2
08	OPENINGS	104.3	60.8	94.1	100.0	61.3	91.0	109.2	61.3	98.0	103.2	60.1	93.1	87.8	137.4	99.4	103.9	137.6	111.8
0920	Plaster & Gypsum Board	83.1	63.5	69.9	73.7	63.5	66.8	85.7	63.5	70.7	89.2	63.5	71.9	94.5	139.3	124.6	112.7	139.3	130.6
0950, 0980	Ceilings & Acoustic Treatment	86.3	63.5	70.9	88.1	63.5	71.5	89.6	63.5	72.0	90.2	63.5	72.2	111.0	139.3	130.1	109.3	139.3	129.5
0960	Flooring	84.9	70.9	80.9	81.3	70.9	78.4	86.9	43.4	74.7	61.6	70.9	64.2	104.8	121.1	109.4	99.2	121.1	105.3
0970, 0990	Wall Finishes & Painting/Coating	88.8	52.1	67.1	92.2	54.5	69.9	88.8	54.5	68.5	81.7	54.5	65.6	103.9	120.3	113.6	92.6	120.3	109.0
09	FINISHES	79.6	62.6	70.4	76.4	62.9	69.1	81.7	57.2	68.4	75.9	63.2	69.0	101.8	134.1	119.3	99.4	134.2	118.3
COVERS	DIVS. 10 - 14, 25, 28, 41, 43, 44, 46	100.0	77.7	95.0	100.0	66.5	92.5	100.0	66.1	92.5	100.0	67.0	92.6	100.0	119.6	104.4	100.0	120.2	104.5
21, 22, 23	FIRE SUPPRESSION, PLUMBING & HVAC	100.0	51.4	80.9	96.5	50.9	78.6	100.0	54.5	82.1	96.8	64.6	84.2	96.5	130.9	110.0	99.9	130.9	112.1
26, 27, 3370	ELECTRICAL, COMMUNICATIONS & UTIL.	95.9	57.5	77.0	93.1	57.5	75.6	97.6	57.5	77.9	100.5	65.3	83.1	121.0	129.2	125.0	91.5	113.3	102.2
MF2018	WEIGHTED AVERAGE	96.0	63.6	82.3	91.7	63.1	79.6	94.8	63.3	81.5	93.3	69.8	83.3	96.7	128.6	110.2	99.3	126.6	110.9

CALIFORNIA

DIVISION		BAKERSFIELD 932-933 MAT.	INST.	TOTAL	BERKELEY 947 MAT.	INST.	TOTAL	EUREKA 955 MAT.	INST.	TOTAL	FRESNO 936-938 MAT.	INST.	TOTAL	INGLEWOOD 903-905 MAT.	INST.	TOTAL	LONG BEACH 906-908 MAT.	INST.	TOTAL
015433	CONTRACTOR EQUIPMENT		98.7	98.7		100.2	100.2		97.1	97.1		97.3	97.3		96.9	96.9		96.9	96.9
0241, 31 - 34	SITE & INFRASTRUCTURE, DEMOLITION	96.1	106.8	103.5	109.8	106.0	107.2	109.0	102.7	104.7	98.9	103.2	101.8	87.7	102.0	97.6	94.7	102.0	99.7
0310	Concrete Forming & Accessories	105.8	139.2	134.3	113.0	168.1	160.0	111.2	154.1	147.8	102.9	153.3	145.8	107.3	139.9	135.1	102.0	139.9	134.3
0320	Concrete Reinforcing	99.8	133.4	116.0	90.7	135.2	112.2	103.7	136.6	119.6	83.3	134.2	107.9	99.8	133.7	116.2	99.0	133.7	115.8
0330	Cast-in-Place Concrete	88.1	130.2	103.8	110.4	133.5	119.0	94.1	129.6	107.3	95.5	129.2	108.0	81.5	130.2	99.6	92.7	130.2	106.7
03	CONCRETE	91.7	133.7	110.2	105.1	148.2	124.0	106.8	140.8	121.7	95.5	139.9	115.0	90.9	134.2	109.9	100.5	134.2	115.3
04	MASONRY	92.1	137.5	120.0	122.7	154.9	142.5	103.3	153.4	134.1	96.9	145.0	126.5	73.1	139.5	113.9	81.7	139.5	117.2
05	METALS	102.5	112.5	105.4	109.9	116.5	111.9	107.4	116.1	109.9	102.8	115.2	106.4	88.0	114.8	95.9	87.9	114.8	95.8
06	WOOD, PLASTICS & COMPOSITES	96.9	138.3	119.7	104.3	173.4	142.3	111.2	157.3	136.6	101.2	157.3	132.1	99.0	138.4	120.7	92.5	138.4	117.9
07	THERMAL & MOISTURE PROTECTION	108.5	124.3	115.4	112.2	155.7	131.2	112.6	151.2	129.4	98.9	132.9	113.7	107.2	133.4	118.6	107.5	133.4	118.8
08	OPENINGS	92.6	134.3	102.3	91.0	161.7	107.5	103.2	143.2	112.5	94.7	145.1	106.5	87.6	137.6	99.2	87.5	137.6	99.2
0920	Plaster & Gypsum Board	94.4	139.3	124.6	113.4	175.0	154.8	117.4	158.8	145.2	90.9	158.8	136.6	100.3	139.3	126.5	96.2	139.3	125.2
0950, 0980	Ceilings & Acoustic Treatment	100.7	139.3	126.8	101.2	175.0	151.0	113.3	158.8	144.0	98.1	158.8	139.0	114.0	139.3	131.1	114.0	139.3	131.1
0960	Flooring	102.2	121.1	107.5	124.0	143.3	129.4	103.2	143.3	114.4	100.4	121.5	106.3	114.7	121.1	116.5	111.5	121.1	114.2
0970, 0990	Wall Finishes & Painting/Coating	92.5	108.6	102.0	111.6	171.3	146.9	94.1	138.6	120.4	101.3	120.7	112.8	115.1	120.3	118.2	115.1	120.3	118.2
09	FINISHES	96.4	133.0	116.2	110.1	165.4	140.0	104.6	152.0	130.3	94.6	146.3	122.6	108.2	134.3	122.4	107.2	134.3	121.9
COVERS	DIVS. 10 - 14, 25, 28, 41, 43, 44, 46	100.0	117.7	103.9	100.0	134.0	107.6	100.0	130.8	106.9	100.0	130.8	106.9	100.0	120.5	104.6	100.0	120.5	104.6
21, 22, 23	FIRE SUPPRESSION, PLUMBING & HVAC	100.1	128.8	111.4	96.7	169.6	125.3	96.4	130.6	109.9	100.1	130.1	111.9	96.1	131.0	109.8	96.1	131.0	109.8
26, 27, 3370	ELECTRICAL, COMMUNICATIONS & UTIL.	107.6	108.1	107.8	99.9	161.8	130.4	98.4	124.9	111.5	97.0	107.4	102.2	98.5	129.2	113.6	98.3	129.2	113.5
MF2018	WEIGHTED AVERAGE	98.9	124.6	109.8	103.0	151.9	123.7	102.5	134.1	115.9	98.5	129.4	111.5	93.8	128.8	108.6	95.4	128.8	109.5

CALIFORNIA

DIVISION		LOS ANGELES 900-902 MAT.	INST.	TOTAL	MARYSVILLE 959 MAT.	INST.	TOTAL	MODESTO 953 MAT.	INST.	TOTAL	MOJAVE 935 MAT.	INST.	TOTAL	OAKLAND 946 MAT.	INST.	TOTAL	OXNARD 930 MAT.	INST.	TOTAL
015433	CONTRACTOR EQUIPMENT		103.3	103.3		97.1	97.1		97.1	97.1		97.3	97.3		100.2	100.2		96.3	96.3
0241, 31 - 34	SITE & INFRASTRUCTURE, DEMOLITION	93.7	109.2	104.4	105.6	102.6	103.5	100.6	102.6	102.0	91.9	103.4	99.8	115.2	106.0	108.8	99.2	101.6	100.8
0310	Concrete Forming & Accessories	105.4	140.1	135.0	101.6	153.6	145.9	98.1	153.7	145.5	113.6	139.3	135.5	102.3	168.2	158.4	105.5	139.8	134.7
0320	Concrete Reinforcing	100.7	133.7	116.6	103.7	134.1	118.4	107.4	134.1	120.3	100.4	133.4	116.4	92.8	136.9	114.1	98.6	133.5	115.5
0330	Cast-in-Place Concrete	84.1	131.4	101.7	105.2	129.2	114.1	94.2	129.3	107.2	82.6	130.1	100.2	104.8	133.5	115.4	96.1	130.4	108.9
03	CONCRETE	98.3	134.7	114.3	107.1	140.0	121.6	98.2	140.1	116.6	87.4	133.8	107.8	106.3	148.5	124.8	95.5	134.1	112.5
04	MASONRY	88.6	139.5	119.9	104.2	142.9	128.0	101.5	142.9	126.9	94.9	137.4	121.0	131.5	154.9	145.9	97.9	136.8	121.8
05	METALS	94.3	115.0	100.4	106.9	114.7	109.2	103.6	114.8	106.9	100.0	113.4	104.0	105.2	117.2	108.7	98.0	113.9	102.7
06	WOOD, PLASTICS & COMPOSITES	105.6	138.6	123.7	98.3	157.3	130.8	93.9	157.3	128.8	100.8	138.4	121.5	92.8	173.4	137.2	95.6	138.4	119.1
07	THERMAL & MOISTURE PROTECTION	103.4	133.9	116.7	112.0	138.9	123.7	111.6	139.5	123.7	105.6	123.7	113.5	110.8	155.7	130.4	108.6	133.3	119.3
08	OPENINGS	98.1	136.9	107.0	102.5	145.6	112.6	101.3	147.4	112.1	89.8	134.3	100.2	91.1	162.1	107.7	92.4	137.6	102.9
0920	Plaster & Gypsum Board	95.6	139.3	125.0	110.0	158.8	142.8	112.7	158.8	143.7	100.5	139.3	126.6	107.6	175.0	152.9	94.5	139.3	124.7
0950, 0980	Ceilings & Acoustic Treatment	121.4	139.3	133.5	112.5	158.8	143.7	109.3	158.8	142.6	99.4	139.3	126.3	103.8	175.0	151.8	100.8	139.3	126.8
0960	Flooring	112.5	122.2	115.3	98.8	119.5	104.6	99.2	133.0	108.7	101.3	121.1	106.8	117.0	143.3	124.4	93.8	121.1	101.5
0970, 0990	Wall Finishes & Painting/Coating	111.6	120.3	116.7	94.1	134.1	117.8	94.1	138.6	120.4	87.2	109.6	100.5	111.6	171.3	146.9	87.2	114.8	103.6
09	FINISHES	109.4	134.7	123.1	101.7	147.4	126.4	101.0	150.2	127.7	95.0	133.2	115.6	108.1	165.4	139.1	92.5	133.7	114.8
COVERS	DIVS. 10 - 14, 25, 28, 41, 43, 44, 46	100.0	121.0	104.7	100.0	130.8	106.9	100.0	130.8	106.9	100.0	117.9	104.0	100.0	134.0	107.6	100.0	120.4	104.5
21, 22, 23	FIRE SUPPRESSION, PLUMBING & HVAC	99.9	131.0	112.1	96.4	130.0	109.6	99.9	131.7	112.4	96.5	128.8	109.2	100.2	169.6	127.5	100.0	131.0	112.2
26, 27, 3370	ELECTRICAL, COMMUNICATIONS & UTIL.	96.9	131.0	113.7	95.1	120.7	107.7	97.5	109.9	103.6	95.5	108.1	101.7	99.1	161.8	130.0	101.4	117.9	109.5
MF2018	WEIGHTED AVERAGE	98.5	129.8	111.7	101.7	131.3	114.2	100.6	130.6	113.3	95.4	124.4	107.7	103.4	152.0	123.9	98.0	126.7	110.1

For customer support on your Mechanical Costs with RSMeans Data, call 800.448.8182.

661

City Cost Indexes

CALIFORNIA

DIVISION		PALM SPRINGS 922			PALO ALTO 943			PASADENA 910 - 912			REDDING 960			RICHMOND 948			RIVERSIDE 925		
		MAT.	INST.	TOTAL	MAT.	INST.	TOTAL	MAT.	INST.	TOTAL	MAT.	INST.	TOTAL	MAT.	INST.	TOTAL	MAT.	INST.	TOTAL
015433	CONTRACTOR EQUIPMENT		98.1	98.1		100.2	100.2		95.4	95.4		97.1	97.1		100.2	100.2		98.1	98.1
0241, 31 - 34	SITE & INFRASTRUCTURE, DEMOLITION	89.3	103.6	99.2	106.1	106.0	106.0	97.5	105.3	102.9	122.2	102.6	108.7	114.3	106.0	108.5	96.2	103.6	101.3
0310	Concrete Forming & Accessories	99.8	136.1	130.7	100.5	168.2	158.2	103.4	139.6	134.2	105.0	153.6	146.5	115.5	167.8	160.1	103.5	139.7	134.4
0320	Concrete Reinforcing	109.0	133.5	120.9	90.7	136.7	113.0	106.0	133.6	119.3	137.8	134.1	136.0	90.7	136.7	112.9	105.9	133.5	119.3
0330	Cast-in-Place Concrete	82.7	131.0	100.6	93.4	133.5	108.4	79.1	127.8	97.2	107.2	129.3	115.4	107.4	133.3	117.1	89.9	131.0	105.2
03	CONCRETE	90.5	132.7	109.0	95.3	148.5	118.7	89.9	133.1	108.9	114.0	140.1	125.4	108.0	148.2	125.7	96.4	134.3	113.1
04	MASONRY	76.5	137.7	114.1	105.1	151.5	133.6	93.3	139.4	121.6	131.9	142.9	138.7	122.5	151.5	140.3	77.5	137.7	114.5
05	METALS	108.3	114.1	110.0	102.7	116.9	106.9	80.0	113.2	89.7	103.9	114.7	107.1	102.8	116.6	106.8	107.7	114.2	109.6
06	WOOD, PLASTICS & COMPOSITES	92.1	133.5	114.9	90.4	173.4	136.1	85.4	138.0	114.4	107.8	157.3	135.0	107.5	173.4	143.8	97.5	138.3	119.9
07	THERMAL & MOISTURE PROTECTION	108.1	134.3	119.5	110.0	155.0	129.6	103.8	131.8	116.0	128.2	138.9	132.9	110.6	154.8	129.8	109.0	134.8	120.2
08	OPENINGS	99.9	134.9	108.1	91.1	160.2	107.2	87.8	137.4	99.4	115.9	145.6	122.8	91.1	160.2	107.2	102.6	137.6	110.8
0920	Plaster & Gypsum Board	107.2	134.4	125.5	106.0	175.0	152.4	88.8	139.3	122.8	112.6	158.8	143.7	114.5	175.0	155.2	111.9	139.3	130.4
0950, 0980	Ceilings & Acoustic Treatment	106.0	134.4	125.1	102.0	175.0	151.2	111.0	139.3	130.1	148.8	158.8	155.5	102.0	175.0	151.2	114.2	139.3	131.1
0960	Flooring	101.5	121.1	107.0	115.6	143.3	123.4	98.5	121.1	104.9	87.8	128.6	99.3	126.5	143.3	131.3	102.9	121.1	108.0
0970, 0990	Wall Finishes & Painting/Coating	91.1	120.3	108.4	111.6	171.3	146.9	103.9	120.3	113.6	105.2	134.1	122.3	111.6	171.3	146.9	91.1	120.3	108.4
09	FINISHES	98.0	131.4	116.1	106.5	165.4	138.4	99.0	134.1	118.0	107.2	149.0	129.8	111.6	165.4	140.7	101.0	134.2	119.0
COVERS	DIVS. 10 - 14, 25, 28, 41, 43, 44, 46	100.0	119.6	104.4	100.0	134.0	107.6	100.0	119.6	104.4	100.0	130.8	106.9	100.0	133.8	107.5	100.0	119.3	104.3
21, 22, 23	FIRE SUPPRESSION, PLUMBING & HVAC	96.4	130.9	109.9	96.7	170.6	125.7	96.5	130.9	110.0	100.3	130.1	112.0	96.7	164.0	123.1	99.9	130.9	112.1
26, 27, 3370	ELECTRICAL, COMMUNICATIONS & UTIL.	94.6	110.8	102.6	99.0	176.6	137.2	117.5	129.2	123.2	101.3	120.8	110.9	99.6	133.2	116.2	91.3	114.4	102.7
MF2018	WEIGHTED AVERAGE	97.4	125.3	109.2	99.2	153.8	122.3	94.8	128.6	109.1	107.6	131.5	117.7	102.4	146.2	120.9	99.3	126.6	110.8

CALIFORNIA

DIVISION		SACRAMENTO 942, 956 - 958			SALINAS 939			SAN BERNARDINO 923 - 924			SAN DIEGO 919 - 921			SAN FRANCISCO 940 - 941			SAN JOSE 951		
		MAT.	INST.	TOTAL	MAT.	INST.	TOTAL	MAT.	INST.	TOTAL	MAT.	INST.	TOTAL	MAT.	INST.	TOTAL	MAT.	INST.	TOTAL
015433	CONTRACTOR EQUIPMENT		99.0	99.0		97.3	97.3		98.1	98.1		102.2	102.2		109.8	109.8		98.7	98.7
0241, 31 - 34	SITE & INFRASTRUCTURE, DEMOLITION	95.6	111.3	106.4	112.3	103.2	106.0	76.1	103.6	95.1	105.7	107.8	107.2	116.4	113.1	114.1	131.6	97.9	108.3
0310	Concrete Forming & Accessories	100.5	156.0	147.8	108.9	156.6	149.6	107.1	139.7	134.9	103.8	129.1	125.4	106.1	168.3	159.1	103.9	167.9	158.5
0320	Concrete Reinforcing	86.0	134.4	109.4	99.1	134.6	116.3	105.9	133.5	119.3	102.6	133.4	117.5	105.6	132.1	118.4	94.7	135.0	114.2
0330	Cast-in-Place Concrete	88.6	130.2	104.1	95.0	129.6	107.8	62.1	131.0	87.8	90.5	124.5	103.2	116.9	133.3	123.0	109.4	132.4	118.0
03	CONCRETE	97.9	141.1	116.9	105.8	141.6	121.5	71.1	134.3	98.9	100.2	127.3	112.1	117.8	148.1	131.1	103.9	148.1	123.3
04	MASONRY	106.1	145.9	130.6	95.1	148.4	127.9	83.6	137.7	116.8	86.2	133.4	115.2	140.9	154.6	149.3	131.5	151.5	143.8
05	METALS	100.8	109.8	103.4	102.7	115.8	106.5	107.7	114.2	109.6	94.8	114.1	100.5	110.7	122.4	114.2	101.8	121.6	107.6
06	WOOD, PLASTICS & COMPOSITES	87.1	160.2	127.3	100.6	160.2	133.4	101.2	138.3	121.6	95.0	126.1	112.1	95.1	173.4	138.2	106.0	173.2	143.3
07	THERMAL & MOISTURE PROTECTION	122.2	142.5	131.0	106.2	145.4	123.3	107.1	134.8	119.2	108.2	121.5	114.0	116.5	155.7	133.5	108.0	155.1	128.5
08	OPENINGS	103.9	149.0	114.4	93.5	154.4	107.7	100.0	137.6	108.7	99.8	127.8	106.3	99.5	159.0	113.3	92.6	161.5	108.7
0920	Plaster & Gypsum Board	103.4	161.5	142.5	95.4	161.8	140.1	113.3	139.3	130.8	94.4	126.7	116.1	105.8	175.0	152.4	108.3	175.0	153.2
0950, 0980	Ceilings & Acoustic Treatment	102.0	161.5	142.1	99.4	161.8	141.5	109.3	139.3	129.5	121.1	126.7	124.8	111.7	175.0	154.4	109.3	175.0	153.6
0960	Flooring	115.8	128.6	119.4	95.9	143.3	109.2	105.0	121.1	109.5	108.9	121.1	112.3	117.5	143.3	124.7	92.4	143.3	106.7
0970, 0990	Wall Finishes & Painting/Coating	107.9	134.1	123.4	88.1	171.3	137.3	91.1	120.3	108.4	101.5	118.0	111.2	109.6	174.8	148.2	94.4	171.3	139.9
09	FINISHES	105.3	150.7	129.9	94.6	157.3	128.5	99.4	134.2	118.3	106.6	126.1	117.2	109.9	165.8	140.1	99.9	165.2	135.3
COVERS	DIVS. 10 - 14, 25, 28, 41, 43, 44, 46	100.0	131.7	107.1	100.0	131.2	107.0	100.0	117.7	103.9	100.0	115.7	103.5	100.0	128.4	106.3	100.0	133.4	107.5
21, 22, 23	FIRE SUPPRESSION, PLUMBING & HVAC	100.1	131.4	112.4	96.5	137.6	112.7	96.4	130.9	110.0	100.0	129.4	111.5	100.1	182.8	132.6	99.9	170.5	127.7
26, 27, 3370	ELECTRICAL, COMMUNICATIONS & UTIL.	94.8	120.8	107.6	96.5	132.1	114.0	94.6	112.6	103.5	103.8	104.3	104.0	99.2	182.8	140.4	100.7	176.6	138.1
MF2018	WEIGHTED AVERAGE	100.8	133.0	114.4	99.1	137.2	115.2	95.0	126.3	108.2	99.7	121.7	109.0	107.4	158.4	129.0	102.6	153.5	124.2

CALIFORNIA

DIVISION		SAN LUIS OBISPO 934			SAN MATEO 944			SAN RAFAEL 949			SANTA ANA 926 - 927			SANTA BARBARA 931			SANTA CRUZ 950		
		MAT.	INST.	TOTAL	MAT.	INST.	TOTAL	MAT.	INST.	TOTAL	MAT.	INST.	TOTAL	MAT.	INST.	TOTAL	MAT.	INST.	TOTAL
015433	CONTRACTOR EQUIPMENT		97.3	97.3		100.2	100.2		99.4	99.4		98.1	98.1		97.3	97.3		98.7	98.7
0241, 31 - 34	SITE & INFRASTRUCTURE, DEMOLITION	104.4	103.4	103.7	112.1	106.0	107.9	107.6	111.2	110.1	87.8	103.6	98.7	99.1	103.4	102.1	131.3	97.7	108.1
0310	Concrete Forming & Accessories	115.3	139.7	136.1	106.3	168.2	159.0	111.5	167.8	159.5	107.4	139.7	134.9	106.1	139.7	134.8	103.9	156.9	149.0
0320	Concrete Reinforcing	100.4	133.4	116.3	90.7	136.9	113.0	91.3	135.1	112.5	109.6	133.5	121.2	98.6	133.4	115.4	117.1	134.6	125.6
0330	Cast-in-Place Concrete	102.1	130.2	112.6	104.0	133.5	115.0	121.1	132.3	125.2	79.3	131.0	98.6	95.8	132.0	108.6	108.7	131.4	117.1
03	CONCRETE	103.3	134.0	116.8	104.6	148.5	123.9	128.2	147.4	136.6	88.0	134.3	108.3	95.4	134.0	112.4	106.4	142.6	122.3
04	MASONRY	96.5	135.9	120.7	122.2	154.6	142.1	99.1	154.6	133.2	73.4	138.0	113.1	95.2	135.9	120.2	135.3	148.6	143.4
05	METALS	100.7	113.8	104.6	102.6	117.1	106.9	103.7	112.4	106.3	107.8	114.2	109.7	98.5	113.8	103.0	109.3	120.2	112.5
06	WOOD, PLASTICS & COMPOSITES	103.0	138.4	122.4	97.9	173.4	139.5	95.0	173.2	138.0	103.0	138.3	122.4	95.6	138.4	119.1	106.0	160.4	135.9
07	THERMAL & MOISTURE PROTECTION	106.4	132.6	117.8	110.4	156.4	130.4	115.0	155.4	132.5	108.4	134.9	119.9	105.8	132.6	117.5	107.6	147.9	125.1
08	OPENINGS	91.7	134.3	101.6	91.1	162.1	107.6	101.5	159.7	115.1	99.2	137.6	108.2	93.2	137.6	103.6	93.9	154.4	108.0
0920	Plaster & Gypsum Board	101.1	139.3	126.8	111.7	175.0	154.3	113.2	175.0	154.8	114.6	139.3	131.2	94.5	139.3	124.7	116.8	161.8	147.1
0950, 0980	Ceilings & Acoustic Treatment	99.4	139.3	126.3	102.0	175.0	151.2	109.3	175.0	153.6	109.3	139.3	129.5	100.8	139.3	126.8	110.9	161.8	145.2
0960	Flooring	102.0	121.1	107.4	119.5	143.3	126.2	132.6	143.3	135.6	105.6	121.1	110.0	95.0	121.1	102.4	96.6	143.3	109.7
0970, 0990	Wall Finishes & Painting/Coating	87.2	114.8	103.6	111.6	171.3	146.9	107.8	165.5	141.9	91.1	118.0	107.0	87.2	114.8	103.6	94.5	171.3	139.9
09	FINISHES	96.3	133.7	116.5	108.9	165.4	139.5	112.0	164.6	140.5	100.9	134.0	118.8	93.0	133.7	115.0	102.6	157.4	132.3
COVERS	DIVS. 10 - 14, 25, 28, 41, 43, 44, 46	100.0	126.5	105.9	100.0	134.0	107.6	100.0	133.3	107.4	100.0	120.2	104.5	100.0	119.5	104.3	100.0	131.5	107.0
21, 22, 23	FIRE SUPPRESSION, PLUMBING & HVAC	96.5	131.0	110.0	96.7	165.7	123.8	96.7	182.8	130.5	96.4	130.9	110.0	100.0	131.0	112.2	99.9	137.9	114.8
26, 27, 3370	ELECTRICAL, COMMUNICATIONS & UTIL.	95.5	110.8	103.1	99.0	168.0	133.0	96.2	124.9	110.3	94.6	111.9	103.1	94.5	113.7	103.9	99.8	132.1	115.7
MF2018	WEIGHTED AVERAGE	98.2	125.8	109.9	101.5	152.0	122.9	104.5	149.1	123.4	97.0	126.3	109.4	97.2	126.1	109.4	104.6	137.5	118.5

City Cost Indexes

CALIFORNIA / COLORADO

DIVISION		SANTA ROSA 954 MAT.	INST.	TOTAL	STOCKTON 952 MAT.	INST.	TOTAL	SUSANVILLE 961 MAT.	INST.	TOTAL	VALLEJO 945 MAT.	INST.	TOTAL	VAN NUYS 913-916 MAT.	INST.	TOTAL	ALAMOSA 811 MAT.	INST.	TOTAL
015433	CONTRACTOR EQUIPMENT		97.6	97.6		97.1	97.1		97.1	97.1		99.4	99.4		95.4	95.4		90.2	90.2
0241, 31 - 34	SITE & INFRASTRUCTURE, DEMOLITION	101.4	102.6	102.2	100.3	102.6	101.9	129.5	102.6	110.9	96.3	111.1	106.5	115.7	105.3	108.5	141.3	84.4	102.0
0310	Concrete Forming & Accessories	100.7	167.1	157.3	101.9	155.5	147.6	106.2	153.9	146.8	102.3	166.8	157.3	109.8	139.5	135.2	103.8	66.0	71.5
0320	Concrete Reinforcing	104.6	135.1	119.3	107.4	134.1	120.3	137.8	134.1	136.0	92.5	135.0	113.0	106.0	133.6	119.3	114.0	67.9	91.7
0330	Cast-in-Place Concrete	103.3	130.8	113.5	91.7	129.2	105.7	97.5	129.3	109.3	96.4	131.4	109.5	83.4	127.8	99.9	100.1	73.1	90.1
03	CONCRETE	107.1	146.9	124.6	97.3	140.9	116.5	117.0	140.2	127.2	102.4	146.6	121.8	104.4	133.1	117.0	112.5	69.4	93.6
04	MASONRY	101.8	153.5	133.6	101.4	142.9	126.9	130.4	142.9	138.1	77.4	153.5	124.1	106.9	139.4	126.9	130.1	59.5	86.7
05	METALS	108.1	117.7	110.9	103.9	114.7	107.0	102.9	114.7	106.4	103.7	111.7	106.0	79.2	113.2	89.2	105.2	78.5	97.3
06	WOOD, PLASTICS & COMPOSITES	93.6	173.0	137.3	99.5	159.9	132.8	109.6	157.6	136.0	84.8	173.2	133.4	94.1	138.0	118.3	98.5	67.4	81.4
07	THERMAL & MOISTURE PROTECTION	108.9	154.1	128.6	112.1	138.8	123.7	130.2	138.9	134.0	113.1	154.1	130.9	105.0	131.8	116.6	111.9	67.2	92.5
08	OPENINGS	100.8	161.4	114.9	101.3	147.0	112.0	116.8	145.8	123.6	103.2	159.7	116.3	87.7	137.4	99.3	96.3	68.5	89.8
0920	Plaster & Gypsum Board	109.2	175.0	153.5	112.7	161.5	145.5	113.1	159.1	144.1	107.5	175.0	152.9	92.8	139.3	124.1	80.8	66.5	71.2
0950, 0980	Ceilings & Acoustic Treatment	109.3	175.0	153.6	117.5	161.5	147.1	139.9	159.1	152.9	111.1	175.0	154.2	107.8	139.3	129.1	102.7	66.5	78.3
0960	Flooring	102.1	136.5	111.7	99.2	133.0	108.7	88.3	128.6	99.6	126.0	143.3	130.9	101.9	121.1	107.3	107.8	70.1	97.2
0970, 0990	Wall Finishes & Painting/Coating	91.1	165.5	135.2	94.1	134.1	117.8	105.2	134.1	122.3	108.7	165.5	142.3	103.9	120.3	113.6	101.7	76.9	87.0
09	FINISHES	100.0	163.0	134.1	102.6	151.3	129.0	106.7	149.2	129.7	108.4	164.4	138.7	101.2	134.1	119.0	99.8	67.5	82.3
COVERS	DIVS. 10 - 14, 25, 28, 41, 43, 44, 46	100.0	132.4	107.2	100.0	128.0	106.2	100.0	130.9	106.9	100.0	132.9	107.3	100.0	119.6	104.4	100.0	83.0	96.2
21, 22, 23	FIRE SUPPRESSION, PLUMBING & HVAC	96.4	182.2	130.1	99.9	131.7	112.4	96.8	130.1	109.9	100.2	147.2	118.7	96.5	130.9	110.0	96.5	72.4	87.0
26, 27, 3370	ELECTRICAL, COMMUNICATIONS & UTIL.	94.9	124.9	109.7	97.5	114.3	105.8	101.7	120.8	111.1	92.4	127.5	109.7	117.5	129.2	123.2	96.3	63.2	80.0
MF2018	WEIGHTED AVERAGE	101.2	148.4	121.2	100.7	131.3	113.7	107.3	131.6	117.5	100.2	141.6	117.7	97.8	128.6	110.8	103.6	70.3	89.5

COLORADO

DIVISION		BOULDER 803 MAT.	INST.	TOTAL	COLORADO SPRINGS 808-809 MAT.	INST.	TOTAL	DENVER 800-802 MAT.	INST.	TOTAL	DURANGO 813 MAT.	INST.	TOTAL	FORT COLLINS 805 MAT.	INST.	TOTAL	FORT MORGAN 807 MAT.	INST.	TOTAL
015433	CONTRACTOR EQUIPMENT		92.2	92.2		90.2	90.2		99.6	99.6		90.2	90.2		92.2	92.2		92.2	92.2
0241, 31 - 34	SITE & INFRASTRUCTURE, DEMOLITION	97.1	91.4	93.2	99.2	86.7	90.6	104.1	102.5	103.0	134.3	84.4	99.9	109.6	91.2	96.9	99.3	90.9	93.5
0310	Concrete Forming & Accessories	107.0	77.4	81.8	97.3	63.5	68.5	106.1	66.4	72.2	110.9	65.7	72.2	104.5	65.8	71.5	107.6	65.7	71.9
0320	Concrete Reinforcing	108.7	68.1	89.0	107.9	67.9	88.6	107.9	70.1	89.6	114.0	67.9	91.7	108.8	68.0	89.1	108.9	68.0	89.2
0330	Cast-in-Place Concrete	113.8	73.9	98.9	116.8	73.4	100.7	121.8	73.6	103.9	115.1	73.1	99.5	128.5	72.5	107.6	111.6	72.5	97.1
03	CONCRETE	105.2	74.9	91.9	108.7	68.4	91.0	111.4	70.2	93.3	114.2	69.3	94.5	116.3	69.2	95.6	103.7	69.1	88.5
04	MASONRY	101.6	65.4	79.4	103.7	62.3	78.3	104.9	64.1	79.9	117.5	60.6	82.5	118.9	63.1	84.6	116.7	63.1	83.8
05	METALS	94.4	77.9	89.6	97.5	77.7	91.7	99.9	79.0	93.8	105.2	78.4	97.3	95.7	77.7	90.4	94.2	77.8	89.4
06	WOOD, PLASTICS & COMPOSITES	107.1	81.4	93.0	96.6	63.6	78.5	108.4	67.1	85.7	108.4	67.4	85.8	104.6	67.0	83.9	107.1	67.0	85.0
07	THERMAL & MOISTURE PROTECTION	108.7	73.2	93.2	100.9	69.7	92.1	107.1	71.7	91.7	111.9	67.5	92.6	109.0	70.6	92.3	108.6	69.6	91.6
08	OPENINGS	95.4	76.2	90.9	99.5	66.4	91.8	102.5	68.8	94.6	103.1	68.5	95.0	95.3	68.3	89.0	95.3	68.2	89.0
0920	Plaster & Gypsum Board	116.3	81.3	92.8	100.3	62.9	75.2	111.7	66.6	81.4	95.3	66.5	75.9	110.3	66.6	80.9	116.3	66.5	82.8
0950, 0980	Ceilings & Acoustic Treatment	91.9	81.3	84.8	100.0	62.9	75.0	104.4	66.6	78.9	102.7	66.5	78.3	91.9	66.6	74.8	91.9	66.5	74.8
0960	Flooring	111.2	76.7	101.5	102.2	69.0	92.9	108.4	76.7	99.5	113.1	69.0	100.7	107.3	76.7	98.7	111.7	76.7	101.8
0970, 0990	Wall Finishes & Painting/Coating	97.1	76.9	85.1	96.8	76.9	85.0	103.4	76.9	87.7	101.7	76.9	87.0	97.1	76.9	85.1	97.1	76.9	85.1
09	FINISHES	100.6	77.7	88.2	97.4	65.3	80.0	102.5	68.9	84.3	102.4	67.4	83.5	99.2	68.6	82.6	100.7	68.5	83.3
COVERS	DIVS. 10 - 14, 25, 28, 41, 43, 44, 46	100.0	84.5	96.5	100.0	82.2	96.0	100.0	86.3	96.9	100.0	82.9	96.2	100.0	82.1	96.0	100.0	82.1	96.0
21, 22, 23	FIRE SUPPRESSION, PLUMBING & HVAC	96.5	74.8	87.9	100.1	72.2	89.1	99.9	73.8	89.7	96.5	58.7	81.6	100.0	71.7	88.9	96.5	73.5	87.4
26, 27, 3370	ELECTRICAL, COMMUNICATIONS & UTIL.	99.6	80.6	90.3	103.2	73.1	88.4	105.0	80.4	92.9	95.8	50.3	73.4	99.6	80.6	90.2	100.0	78.1	89.2
MF2018	WEIGHTED AVERAGE	98.7	77.0	89.6	101.2	71.5	88.6	103.0	75.4	91.3	103.9	65.6	87.8	102.2	73.5	90.1	99.3	73.5	88.4

COLORADO

DIVISION		GLENWOOD SPRINGS 816 MAT.	INST.	TOTAL	GOLDEN 804 MAT.	INST.	TOTAL	GRAND JUNCTION 815 MAT.	INST.	TOTAL	GREELEY 806 MAT.	INST.	TOTAL	MONTROSE 814 MAT.	INST.	TOTAL	PUEBLO 810 MAT.	INST.	TOTAL
015433	CONTRACTOR EQUIPMENT		93.3	93.3		92.2	92.2		93.3	93.3		92.2	92.2		91.7	91.7		90.2	90.2
0241, 31 - 34	SITE & INFRASTRUCTURE, DEMOLITION	150.9	91.9	110.2	110.4	91.4	97.3	133.7	92.1	105.0	95.9	91.2	92.7	144.0	87.9	105.3	125.2	84.5	97.1
0310	Concrete Forming & Accessories	100.6	65.5	70.7	99.7	66.7	71.5	108.7	66.6	72.8	102.4	65.8	71.2	100.1	65.5	70.6	106.1	63.7	70.0
0320	Concrete Reinforcing	112.8	68.0	91.2	108.9	68.0	89.2	113.2	67.8	91.2	108.7	68.0	89.0	112.7	67.8	91.0	109.0	67.9	89.1
0330	Cast-in-Place Concrete	100.1	72.1	89.6	114.1	73.9	99.5	110.8	73.5	96.9	107.3	72.5	94.3	100.1	72.6	89.8	99.4	73.7	89.8
03	CONCRETE	118.0	68.9	96.4	114.1	70.0	94.7	110.7	69.9	92.8	100.3	69.1	86.6	108.5	69.0	91.1	100.9	68.7	86.7
04	MASONRY	103.1	63.0	78.5	119.7	65.4	86.3	137.8	61.9	91.2	111.9	62.7	81.7	110.0	59.6	79.0	99.5	60.3	75.4
05	METALS	104.8	78.2	97.0	94.3	77.8	89.5	106.5	78.2	98.2	95.7	77.7	90.4	103.9	78.1	96.3	108.3	78.7	99.6
06	WOOD, PLASTICS & COMPOSITES	93.6	67.1	79.0	98.9	67.0	81.3	106.1	67.2	84.7	101.7	67.0	82.6	94.6	67.2	79.6	101.2	63.9	80.7
07	THERMAL & MOISTURE PROTECTION	111.8	68.2	92.8	109.8	71.6	93.2	110.9	69.3	92.8	108.2	70.4	91.8	112.0	67.2	92.5	110.4	67.8	91.8
08	OPENINGS	102.1	68.3	94.2	95.4	68.2	89.0	102.8	68.4	94.7	95.3	68.3	89.0	103.2	68.4	95.1	98.0	66.6	90.7
0920	Plaster & Gypsum Board	121.2	66.5	84.4	107.5	66.5	79.9	135.4	66.6	89.1	108.7	66.6	80.4	79.9	66.5	70.9	84.7	62.9	70.1
0950, 0980	Ceilings & Acoustic Treatment	101.9	66.5	78.0	91.9	66.5	74.8	101.9	66.6	78.1	91.9	66.6	74.8	102.7	66.5	78.3	110.0	62.9	78.3
0960	Flooring	106.9	76.7	98.4	104.9	76.7	97.0	112.4	70.1	100.5	106.2	73.1	96.9	110.4	73.1	100.0	109.1	73.1	99.0
0970, 0990	Wall Finishes & Painting/Coating	101.7	76.9	87.0	97.1	76.9	85.1	101.7	76.9	87.0	97.1	76.9	85.1	101.7	76.9	87.0	101.7	76.9	87.0
09	FINISHES	105.2	68.7	85.4	98.9	69.1	82.8	106.8	67.9	85.8	97.9	67.9	81.6	100.5	68.0	82.9	100.2	66.3	81.9
COVERS	DIVS. 10 - 14, 25, 28, 41, 43, 44, 46	100.0	82.3	96.1	100.0	82.9	96.2	100.0	83.1	96.2	100.0	82.1	96.0	100.0	82.6	96.1	100.0	82.9	96.2
21, 22, 23	FIRE SUPPRESSION, PLUMBING & HVAC	96.5	58.7	81.6	96.5	74.8	87.9	100.0	74.8	90.1	100.0	71.7	88.9	96.5	58.7	81.6	100.0	72.2	89.1
26, 27, 3370	ELECTRICAL, COMMUNICATIONS & UTIL.	93.4	52.3	73.1	100.0	80.6	90.4	95.5	54.6	75.3	99.6	80.6	90.2	95.5	50.3	73.2	96.4	63.3	80.0
MF2018	WEIGHTED AVERAGE	103.9	66.8	88.2	100.9	74.7	89.8	105.7	70.5	90.8	99.3	73.4	88.4	102.6	65.8	87.1	101.8	70.0	88.3

For customer support on your Mechanical Costs with RSMeans Data, call 800.448.8182.

663

COLORADO / CONNECTICUT

DIVISION		SALIDA 812 MAT.	INST.	TOTAL	BRIDGEPORT 066 MAT.	INST.	TOTAL	BRISTOL 060 MAT.	INST.	TOTAL	HARTFORD 061 MAT.	INST.	TOTAL	MERIDEN 064 MAT.	INST.	TOTAL	NEW BRITAIN 060 MAT.	INST.	TOTAL
015433	CONTRACTOR EQUIPMENT		91.7	91.7		95.7	95.7		95.7	95.7		99.0	99.0		96.1	96.1		95.7	95.7
0241, 31-34	SITE & INFRASTRUCTURE, DEMOLITION	134.0	88.2	102.4	106.1	97.8	100.3	105.1	97.7	100.0	101.6	102.4	102.2	102.8	98.5	99.8	105.3	97.7	100.1
0310	Concrete Forming & Accessories	108.6	65.9	72.2	103.4	116.7	114.7	103.4	116.6	114.6	103.5	116.8	114.8	103.1	116.6	114.6	103.9	116.6	114.7
0320	Concrete Reinforcing	112.4	67.9	90.9	116.3	144.5	130.0	116.3	144.5	130.0	111.6	144.6	127.5	116.3	144.5	130.0	116.3	144.5	130.0
0330	Cast-in-Place Concrete	114.7	72.8	99.1	108.5	127.5	115.6	101.6	127.5	111.2	106.5	128.5	114.7	97.8	127.5	108.8	103.3	127.5	112.3
03	CONCRETE	109.1	69.3	91.6	104.1	124.5	113.0	100.9	124.4	111.2	101.1	124.8	111.5	99.1	124.4	110.2	101.7	124.4	111.7
04	MASONRY	138.6	60.6	90.7	109.4	130.9	122.6	100.8	130.9	119.3	100.5	131.0	119.2	100.4	130.9	119.2	102.5	130.9	120.0
05	METALS	103.6	78.6	96.2	98.1	116.6	103.5	98.1	116.5	103.5	103.1	116.0	106.9	95.4	116.5	101.6	94.5	116.5	100.9
06	WOOD, PLASTICS & COMPOSITES	102.4	67.2	83.0	105.0	114.2	110.1	105.0	114.2	110.1	96.2	114.4	106.2	105.0	114.2	110.1	105.0	114.2	110.1
07	THERMAL & MOISTURE PROTECTION	110.8	67.5	92.0	99.7	123.5	110.0	99.8	120.6	108.9	104.4	121.1	111.7	99.8	120.6	108.9	99.8	120.7	108.9
08	OPENINGS	96.3	68.4	89.8	96.4	120.6	102.0	96.4	120.6	102.0	98.1	120.7	103.4	98.5	120.6	103.7	96.4	120.6	102.0
0920	Plaster & Gypsum Board	80.2	66.5	71.0	115.7	114.3	114.8	115.7	114.3	114.8	102.5	114.3	110.5	117.3	114.3	115.3	115.7	114.3	114.8
0950, 0980	Ceilings & Acoustic Treatment	102.7	66.5	78.3	102.0	114.3	110.3	102.0	114.3	110.3	98.8	114.3	109.3	107.5	114.3	112.1	102.0	114.3	110.3
0960	Flooring	115.9	69.9	103.0	92.6	127.0	102.3	92.6	124.5	101.6	96.9	127.0	105.4	92.6	124.5	101.6	92.6	124.5	101.6
0970, 0990	Wall Finishes & Painting/Coating	101.7	76.9	87.0	92.0	125.5	111.8	92.0	129.1	113.9	98.1	129.1	116.4	92.0	129.1	113.9	92.0	129.1	113.9
09	FINISHES	101.1	67.4	82.8	94.7	118.9	107.8	94.7	118.8	107.8	96.2	119.4	108.8	96.0	118.8	108.4	94.7	118.8	107.8
COVERS	DIVS. 10 - 14, 25, 28, 41, 43, 44, 46	100.0	82.7	96.1	100.0	113.9	103.1	100.0	113.9	103.1	100.0	114.4	103.2	100.0	114.0	103.1	100.0	113.9	103.1
21, 22, 23	FIRE SUPPRESSION, PLUMBING & HVAC	96.5	73.6	87.5	100.1	118.6	107.4	100.1	118.6	107.4	100.0	118.6	107.3	96.6	118.6	105.2	100.1	118.6	107.4
26, 27, 3370	ELECTRICAL, COMMUNICATIONS & UTIL.	95.7	63.2	79.7	93.0	105.0	98.9	93.0	103.7	98.3	92.7	109.4	100.9	92.9	105.8	99.3	93.0	103.7	98.3
MF2018	WEIGHTED AVERAGE	103.1	70.9	89.5	99.3	117.0	106.8	98.5	116.7	106.2	99.6	118.0	107.3	97.2	117.0	105.6	98.1	116.7	106.0

CONNECTICUT

DIVISION		NEW HAVEN 065 MAT.	INST.	TOTAL	NEW LONDON 063 MAT.	INST.	TOTAL	NORWALK 068 MAT.	INST.	TOTAL	STAMFORD 069 MAT.	INST.	TOTAL	WATERBURY 067 MAT.	INST.	TOTAL	WILLIMANTIC 062 MAT.	INST.	TOTAL
015433	CONTRACTOR EQUIPMENT		96.1	96.1		96.1	96.1		95.7	95.7		95.7	95.7		95.7	95.7		95.7	95.7
0241, 31-34	SITE & INFRASTRUCTURE, DEMOLITION	105.3	98.4	100.5	97.2	98.1	97.8	105.8	97.6	100.2	106.5	97.7	100.4	105.8	97.8	100.2	105.8	97.7	100.2
0310	Concrete Forming & Accessories	103.2	116.5	114.6	103.1	116.5	114.5	103.4	116.6	114.7	103.4	116.9	114.9	103.4	116.6	114.7	103.4	116.4	114.5
0320	Concrete Reinforcing	116.3	144.5	130.0	91.2	144.5	117.0	116.3	144.5	130.0	116.3	144.5	130.0	116.3	144.5	130.0	116.3	144.5	130.0
0330	Cast-in-Place Concrete	105.1	125.5	112.7	89.5	125.5	102.9	106.7	126.7	114.2	108.5	126.8	115.3	108.5	127.5	115.6	101.3	125.5	110.3
03	CONCRETE	116.3	123.7	119.6	89.2	123.7	104.4	103.3	124.1	112.4	104.1	124.3	113.0	104.1	124.4	113.0	100.8	123.6	110.8
04	MASONRY	101.1	130.9	119.4	99.3	130.9	118.7	100.6	130.1	118.7	101.4	130.1	119.0	101.4	130.9	119.5	100.6	130.9	119.3
05	METALS	94.7	116.4	101.1	94.4	116.4	100.9	98.1	116.5	103.5	98.1	116.8	103.6	98.1	116.5	103.5	97.9	116.3	103.3
06	WOOD, PLASTICS & COMPOSITES	105.0	114.2	110.1	105.0	114.2	110.1	105.0	114.2	110.1	105.0	114.2	110.1	105.0	114.2	110.1	105.0	114.2	110.1
07	THERMAL & MOISTURE PROTECTION	99.9	120.8	109.0	99.7	120.3	108.7	99.9	123.2	110.0	99.8	123.2	110.0	99.8	121.1	109.1	100.0	120.3	108.8
08	OPENINGS	96.4	120.6	102.0	98.7	120.6	103.8	96.4	120.6	102.0	96.4	120.6	102.0	96.4	120.6	102.0	98.8	120.6	103.9
0920	Plaster & Gypsum Board	115.7	114.3	114.8	115.7	114.3	114.8	115.7	114.3	114.8	115.7	114.3	114.8	115.7	114.3	114.8	115.7	114.3	114.8
0950, 0980	Ceilings & Acoustic Treatment	102.0	114.3	110.3	100.2	114.3	109.7	102.0	114.3	110.3	102.0	114.3	110.3	102.0	114.3	110.3	100.2	114.3	109.7
0960	Flooring	92.6	127.0	102.3	92.6	127.0	102.3	92.6	124.5	101.6	92.6	127.0	102.3	92.6	127.0	102.3	92.6	121.1	100.6
0970, 0990	Wall Finishes & Painting/Coating	92.0	125.5	111.8	92.0	129.1	113.9	92.0	125.5	111.8	92.0	125.5	111.8	92.0	129.1	113.9	92.0	129.1	113.9
09	FINISHES	94.7	118.9	107.8	93.9	119.3	107.6	94.7	118.5	107.6	94.8	118.9	107.8	94.6	119.3	107.9	94.5	118.3	107.3
COVERS	DIVS. 10 - 14, 25, 28, 41, 43, 44, 46	100.0	114.0	103.1	100.0	114.0	103.1	100.0	113.9	103.1	100.0	114.1	103.2	100.0	114.0	103.1	100.0	114.0	103.1
21, 22, 23	FIRE SUPPRESSION, PLUMBING & HVAC	100.1	118.6	107.4	96.6	118.6	105.2	100.1	118.6	107.4	100.1	118.6	107.4	100.1	118.6	107.4	100.1	118.4	107.3
26, 27, 3370	ELECTRICAL, COMMUNICATIONS & UTIL.	92.9	106.0	99.4	90.5	107.9	99.1	93.0	104.8	98.8	93.0	150.1	121.1	92.6	107.1	99.7	93.0	108.8	100.8
MF2018	WEIGHTED AVERAGE	99.9	117.0	107.1	95.2	117.2	104.5	98.8	116.7	106.4	99.0	123.2	109.2	98.9	117.2	106.6	98.7	117.1	106.5

D.C. / DELAWARE / FLORIDA

DIVISION		WASHINGTON 200-205 MAT.	INST.	TOTAL	DOVER 199 MAT.	INST.	TOTAL	NEWARK 197 MAT.	INST.	TOTAL	WILMINGTON 198 MAT.	INST.	TOTAL	DAYTONA BEACH 321 MAT.	INST.	TOTAL	FORT LAUDERDALE 333 MAT.	INST.	TOTAL
015433	CONTRACTOR EQUIPMENT		105.3	105.3		118.2	118.2		119.0	119.0		118.4	118.4		99.5	99.5		92.8	92.8
0241, 31-34	SITE & INFRASTRUCTURE, DEMOLITION	103.0	97.8	99.4	106.0	107.5	107.0	104.5	108.7	107.4	102.2	107.8	106.1	117.6	85.0	95.1	95.0	73.2	80.0
0310	Concrete Forming & Accessories	97.7	73.4	77.0	98.0	99.4	99.2	97.0	99.2	98.9	95.9	99.4	98.9	97.1	61.6	66.8	93.4	59.4	64.4
0320	Concrete Reinforcing	103.5	86.8	95.4	101.4	115.1	108.0	97.2	115.1	105.9	101.0	115.1	107.8	96.0	61.0	79.1	94.6	59.9	77.8
0330	Cast-in-Place Concrete	103.6	79.0	94.5	108.1	107.7	107.9	91.2	106.7	97.0	102.6	107.7	104.5	91.8	66.0	82.2	96.3	62.3	83.6
03	CONCRETE	106.0	78.6	94.0	100.4	106.0	102.9	93.4	105.7	98.8	97.7	106.0	101.4	90.5	64.8	79.2	93.4	62.3	79.7
04	MASONRY	97.6	87.8	91.6	99.0	98.5	98.7	101.0	98.5	99.5	94.4	98.5	96.9	86.4	60.9	70.7	87.6	54.4	67.2
05	METALS	103.9	95.6	101.5	102.9	121.9	108.5	104.7	123.1	110.1	103.1	121.9	108.6	101.9	88.9	98.1	96.9	88.2	94.4
06	WOOD, PLASTICS & COMPOSITES	95.4	72.2	82.6	92.3	97.6	95.2	91.0	97.4	94.5	85.7	97.6	92.2	93.2	60.1	75.0	78.3	60.6	68.6
07	THERMAL & MOISTURE PROTECTION	104.3	87.1	96.8	105.2	110.4	107.4	109.2	109.8	109.4	105.1	110.4	107.4	101.4	65.4	85.7	106.3	60.4	86.3
08	OPENINGS	100.7	74.6	94.6	91.2	107.4	95.0	91.2	107.3	94.9	88.6	107.4	93.0	92.9	59.4	85.1	94.2	59.5	86.1
0920	Plaster & Gypsum Board	104.3	71.4	82.2	96.0	97.6	96.9	96.7	97.4	97.2	97.3	97.4	97.4	92.7	59.5	70.4	107.2	60.0	75.5
0950, 0980	Ceilings & Acoustic Treatment	110.9	71.4	84.3	100.2	97.4	98.3	96.0	97.4	96.9	90.8	97.4	95.2	78.3	59.5	65.6	84.8	60.0	68.1
0960	Flooring	95.4	75.5	89.8	96.9	105.4	99.3	91.3	105.4	95.3	96.0	105.4	98.7	99.5	62.7	89.2	97.1	62.7	87.5
0970, 0990	Wall Finishes & Painting/Coating	103.7	73.1	85.6	92.6	119.1	108.3	88.2	119.1	106.4	95.3	119.1	109.4	103.8	61.4	78.7	96.5	58.3	73.9
09	FINISHES	98.5	72.8	84.6	95.9	101.6	99.0	89.7	101.5	96.1	94.2	101.6	98.2	91.5	61.1	75.1	91.3	59.4	74.1
COVERS	DIVS. 10 - 14, 25, 28, 41, 43, 44, 46	100.0	94.5	98.8	100.0	103.7	100.8	100.0	103.4	100.8	100.0	103.7	100.8	100.0	81.4	95.9	100.0	80.7	95.7
21, 22, 23	FIRE SUPPRESSION, PLUMBING & HVAC	100.0	88.4	95.4	100.0	119.4	107.6	100.3	119.3	107.7	100.1	119.4	107.7	99.9	75.6	90.4	100.0	66.7	86.9
26, 27, 3370	ELECTRICAL, COMMUNICATIONS & UTIL.	97.7	100.0	98.8	95.1	109.7	102.3	96.9	109.7	103.2	95.1	109.7	102.3	96.7	58.9	78.1	94.7	67.4	81.3
MF2018	WEIGHTED AVERAGE	101.1	87.5	95.4	99.0	109.6	103.5	98.3	109.7	103.1	97.9	109.7	102.9	97.2	69.5	85.5	96.1	66.4	83.6

664

For customer support on your Mechanical Costs with RSMeans Data, call 800.448.8182.

City Cost Indexes

FLORIDA

DIVISION		FORT MYERS 339, 341			GAINESVILLE 326, 344			JACKSONVILLE 320, 322			LAKELAND 338			MELBOURNE 329			MIAMI 330 - 332, 340		
		MAT.	INST.	TOTAL	MAT.	INST.	TOTAL	MAT.	INST.	TOTAL	MAT.	INST.	TOTAL	MAT.	INST.	TOTAL	MAT.	INST.	TOTAL
015433	CONTRACTOR EQUIPMENT		99.5	99.5		99.5	99.5		99.5	99.5		99.5	99.5		99.5	99.5		94.9	94.9
0241, 31 - 34	SITE & INFRASTRUCTURE, DEMOLITION	106.7	85.2	91.9	127.1	84.8	97.9	117.6	85.2	95.2	108.7	85.2	92.5	125.6	85.1	97.7	96.5	78.9	84.3
0310	Concrete Forming & Accessories	89.6	63.8	67.6	92.5	57.1	62.4	96.9	60.2	65.6	86.4	63.8	67.1	93.7	61.7	66.4	99.3	60.2	66.0
0320	Concrete Reinforcing	95.7	76.1	86.2	101.8	61.5	82.3	96.0	61.3	79.2	98.0	76.4	87.6	97.1	65.6	81.9	101.6	60.0	81.5
0330	Cast-in-Place Concrete	100.4	66.0	87.6	105.4	65.7	90.6	92.7	65.8	82.7	102.7	66.1	89.1	110.7	66.0	94.1	93.1	63.9	82.2
03	CONCRETE	94.0	68.3	82.7	101.7	62.7	84.6	90.9	64.2	79.2	95.8	68.5	83.8	102.1	65.6	86.1	91.7	63.2	79.2
04	MASONRY	82.5	58.8	67.9	99.3	60.8	75.7	86.1	58.7	69.3	96.8	61.0	74.8	84.2	60.9	69.9	88.0	55.5	68.0
05	METALS	99.0	93.6	97.4	100.7	88.4	97.1	100.4	88.6	96.9	98.9	94.7	97.7	110.6	90.6	104.7	97.1	87.6	94.3
06	WOOD, PLASTICS & COMPOSITES	75.4	63.2	68.7	87.2	54.6	69.3	93.2	58.6	74.2	71.0	63.2	66.7	88.9	60.1	73.0	92.9	60.7	75.2
07	THERMAL & MOISTURE PROTECTION	106.1	63.2	87.4	101.8	62.5	84.7	101.7	62.4	84.6	106.0	63.8	87.6	101.9	64.1	85.5	106.1	61.4	86.6
08	OPENINGS	95.5	64.6	88.3	92.5	56.5	84.1	92.9	58.8	84.9	95.5	64.7	88.3	92.2	60.5	84.8	96.6	59.5	88.0
0920	Plaster & Gypsum Board	103.3	62.7	76.0	89.6	53.8	65.6	92.7	58.0	69.3	100.2	62.7	75.0	89.6	59.5	69.4	96.2	60.0	71.9
0950, 0980	Ceilings & Acoustic Treatment	79.7	62.7	68.2	72.9	53.8	60.1	78.3	58.0	64.6	79.7	62.7	68.2	77.5	59.5	65.4	85.6	60.0	68.3
0960	Flooring	94.1	77.2	89.4	97.0	62.7	87.4	99.5	62.7	89.2	92.2	62.7	83.9	97.2	62.7	87.5	98.6	62.7	88.5
0970, 0990	Wall Finishes & Painting/Coating	101.4	63.2	78.8	103.8	63.2	79.8	103.8	63.2	79.8	101.4	63.2	78.8	103.8	80.9	90.2	100.2	58.3	75.4
09	FINISHES	91.1	66.1	77.6	90.4	58.1	72.9	91.6	60.4	74.7	90.2	63.1	75.6	91.0	63.2	76.0	90.9	59.8	74.1
COVERS	DIVS. 10 - 14, 25, 28, 41, 43, 44, 46	100.0	80.0	95.5	100.0	80.8	95.7	100.0	78.0	95.1	100.0	80.0	95.5	100.0	81.4	95.9	100.0	81.5	95.9
21, 22, 23	FIRE SUPPRESSION, PLUMBING & HVAC	98.1	59.3	82.9	98.6	63.1	84.7	99.9	63.1	85.5	98.1	59.9	83.1	99.9	74.2	89.8	100.0	63.8	85.8
26, 27, 3370	ELECTRICAL, COMMUNICATIONS & UTIL.	96.6	60.0	78.6	97.0	58.6	78.0	96.4	61.9	79.4	95.0	58.5	77.0	97.7	62.7	80.5	98.4	77.9	88.3
MF2018	WEIGHTED AVERAGE	96.5	67.7	84.3	98.8	65.8	84.8	96.9	66.6	84.1	97.1	67.6	84.6	100.1	70.3	87.5	96.7	68.0	84.6

FLORIDA

DIVISION		ORLANDO 327 - 328, 347			PANAMA CITY 324			PENSACOLA 325			SARASOTA 342			ST. PETERSBURG 337			TALLAHASSEE 323		
		MAT.	INST.	TOTAL	MAT.	INST.	TOTAL	MAT.	INST.	TOTAL	MAT.	INST.	TOTAL	MAT.	INST.	TOTAL	MAT.	INST.	TOTAL
015433	CONTRACTOR EQUIPMENT		102.0	102.0		99.5	99.5		99.5	99.5		99.5	99.5		99.5	99.5		102.0	102.0
0241, 31 - 34	SITE & INFRASTRUCTURE, DEMOLITION	115.8	89.6	97.7	131.4	85.2	99.5	131.3	85.0	99.3	119.0	85.2	95.7	110.4	85.0	92.9	110.2	89.8	96.1
0310	Concrete Forming & Accessories	101.8	61.5	67.4	96.2	66.2	70.6	94.2	61.6	66.4	93.7	63.7	68.1	92.8	60.8	65.5	99.0	62.2	67.6
0320	Concrete Reinforcing	104.9	65.6	85.9	100.3	67.4	84.4	102.8	67.4	85.7	96.6	76.2	86.7	98.0	76.0	87.4	98.4	61.5	80.5
0330	Cast-in-Place Concrete	111.6	66.7	94.9	97.4	65.9	85.7	120.4	65.3	99.9	105.1	66.1	90.6	103.8	65.8	89.7	92.3	66.7	82.8
03	CONCRETE	100.8	65.7	85.4	100.0	67.9	85.9	109.7	65.6	90.3	95.3	68.4	83.5	97.3	66.9	83.9	90.0	65.3	79.2
04	MASONRY	94.8	60.9	74.0	90.6	60.3	72.0	109.1	59.9	78.9	88.7	61.0	71.6	132.4	58.7	87.1	85.2	60.3	69.9
05	METALS	96.7	89.1	94.5	101.5	90.9	98.4	102.6	90.8	99.1	102.5	93.9	100.0	99.8	93.5	98.0	101.7	87.7	97.6
06	WOOD, PLASTICS & COMPOSITES	91.7	60.2	74.4	92.0	66.2	78.0	90.3	61.1	74.2	92.5	63.2	76.4	79.5	59.3	68.4	95.6	61.2	76.7
07	THERMAL & MOISTURE PROTECTION	109.2	64.0	90.4	102.0	64.1	85.5	101.9	63.2	85.1	100.1	63.8	84.3	106.2	62.2	87.1	96.7	64.0	82.4
08	OPENINGS	97.8	60.6	89.1	91.0	64.5	84.9	91.0	61.4	84.1	97.8	64.2	90.0	94.3	62.5	86.9	97.9	60.2	89.1
0920	Plaster & Gypsum Board	93.3	59.5	70.6	91.9	66.1	74.6	99.6	60.5	73.3	97.2	62.7	74.0	105.4	58.7	74.0	100.0	60.5	73.4
0950, 0980	Ceilings & Acoustic Treatment	89.7	59.5	69.3	77.5	66.1	69.8	77.5	60.5	66.0	83.9	62.7	69.6	81.6	58.7	66.1	86.2	60.5	68.9
0960	Flooring	95.7	62.7	86.4	99.1	72.5	91.6	95.1	62.7	86.0	103.5	54.9	89.9	96.1	60.8	86.2	97.6	61.7	87.5
0970, 0990	Wall Finishes & Painting/Coating	95.6	60.3	74.7	103.8	63.2	79.8	103.8	63.2	79.8	99.0	63.2	77.8	101.4	63.2	78.8	101.4	63.2	78.9
09	FINISHES	93.2	61.1	75.8	92.6	66.8	78.6	92.1	61.6	75.6	96.0	61.8	77.5	92.6	60.5	75.2	94.4	61.8	76.7
COVERS	DIVS. 10 - 14, 25, 28, 41, 43, 44, 46	100.0	81.7	95.9	100.0	80.6	95.7	100.0	79.7	95.5	100.0	80.0	95.5	100.0	79.6	95.4	100.0	81.5	95.9
21, 22, 23	FIRE SUPPRESSION, PLUMBING & HVAC	100.0	56.4	82.9	99.9	63.2	85.5	99.9	62.7	85.3	99.9	58.5	83.6	100.0	59.9	84.2	100.0	65.5	86.5
26, 27, 3370	ELECTRICAL, COMMUNICATIONS & UTIL.	98.4	63.1	81.0	95.5	58.6	77.3	99.0	50.7	75.2	96.9	58.5	78.0	95.0	60.6	78.0	103.7	58.6	81.4
MF2018	WEIGHTED AVERAGE	99.0	66.6	85.3	98.6	68.3	85.8	101.2	65.8	86.2	98.8	67.0	85.4	99.7	66.8	85.8	98.2	67.7	85.3

FLORIDA / GEORGIA

DIVISION		TAMPA 335 - 336, 346 (FLORIDA)			WEST PALM BEACH 334, 349 (FLORIDA)			ALBANY 317, 398 (GEORGIA)			ATHENS 306 (GEORGIA)			ATLANTA 300 - 303, 399 (GEORGIA)			AUGUSTA 308 - 309 (GEORGIA)		
		MAT.	INST.	TOTAL	MAT.	INST.	TOTAL	MAT.	INST.	TOTAL	MAT.	INST.	TOTAL	MAT.	INST.	TOTAL	MAT.	INST.	TOTAL
015433	CONTRACTOR EQUIPMENT		99.5	99.5		92.8	92.8		93.7	93.7		92.6	92.6		96.6	96.6		92.6	92.6
0241, 31 - 34	SITE & INFRASTRUCTURE, DEMOLITION	110.9	87.0	94.4	91.6	73.2	78.9	107.3	76.7	86.1	102.8	90.8	94.5	99.7	96.0	97.1	95.9	91.9	93.2
0310	Concrete Forming & Accessories	95.4	64.1	68.7	96.7	59.1	64.7	90.6	66.5	70.1	90.0	43.6	50.4	94.3	72.3	75.5	90.9	73.0	75.6
0320	Concrete Reinforcing	94.6	76.5	85.8	97.3	57.5	78.1	96.5	71.3	84.3	94.5	63.5	79.5	93.9	71.3	83.0	94.9	70.7	83.2
0330	Cast-in-Place Concrete	101.5	66.2	88.3	91.5	62.2	80.6	94.6	69.1	85.1	109.7	69.5	94.7	113.1	71.1	97.5	103.6	70.0	91.1
03	CONCRETE	95.9	68.7	83.9	90.1	61.8	77.7	86.9	70.1	79.5	99.6	57.4	81.1	101.8	72.4	88.9	92.5	72.2	83.6
04	MASONRY	88.3	61.0	71.5	87.1	52.2	65.7	92.5	68.7	77.9	75.0	77.0	76.2	87.8	68.9	76.2	88.1	68.8	76.2
05	METALS	99.0	95.0	97.8	95.9	87.0	93.3	106.2	97.4	103.6	95.3	79.1	90.6	96.2	83.5	92.5	95.0	82.6	91.4
06	WOOD, PLASTICS & COMPOSITES	83.3	63.2	72.2	83.1	60.6	70.7	80.1	66.4	72.6	92.5	35.8	61.3	97.8	74.0	84.7	93.8	75.3	83.6
07	THERMAL & MOISTURE PROTECTION	106.5	63.8	87.9	106.0	60.6	86.2	100.6	68.0	86.4	95.4	68.3	83.6	96.8	73.8	86.8	95.1	71.4	84.8
08	OPENINGS	95.5	64.7	88.3	93.8	59.0	85.7	85.7	68.5	81.7	91.3	50.2	81.7	99.4	73.5	93.3	91.4	74.0	87.3
0920	Plaster & Gypsum Board	108.2	62.7	77.6	111.9	60.0	77.0	101.1	66.0	77.5	93.0	34.4	53.6	95.0	73.6	80.6	94.0	75.0	81.2
0950, 0980	Ceilings & Acoustic Treatment	84.8	62.7	69.9	79.7	60.0	66.4	77.5	66.0	69.8	96.6	34.4	54.7	89.7	73.6	78.8	97.5	75.0	82.4
0960	Flooring	97.1	62.7	87.5	98.9	54.9	86.6	99.4	70.8	91.4	96.2	81.9	92.2	99.1	70.8	91.1	96.4	70.8	89.2
0970, 0990	Wall Finishes & Painting/Coating	101.4	63.2	78.8	96.5	58.3	73.9	88.7	94.0	91.8	92.4	94.0	93.3	96.1	97.3	96.8	92.4	90.2	91.1
09	FINISHES	93.9	63.1	77.3	91.2	57.8	73.2	91.7	69.7	79.8	95.0	53.9	72.7	95.0	74.5	83.9	94.7	74.5	83.7
COVERS	DIVS. 10 - 14, 25, 28, 41, 43, 44, 46	100.0	80.0	95.6	100.0	80.7	95.7	100.0	85.9	96.8	100.0	82.7	96.1	100.0	87.2	97.2	100.0	87.0	97.1
21, 22, 23	FIRE SUPPRESSION, PLUMBING & HVAC	100.0	59.9	84.3	98.1	58.5	82.6	100.0	68.9	87.8	96.6	66.3	84.7	100.0	69.4	88.0	100.1	69.0	87.9
26, 27, 3370	ELECTRICAL, COMMUNICATIONS & UTIL.	94.7	63.1	79.1	95.7	67.4	81.8	95.5	61.9	79.0	97.8	63.9	81.1	97.6	71.5	84.7	98.3	67.3	83.1
MF2018	WEIGHTED AVERAGE	97.6	68.4	85.3	95.1	64.1	82.0	96.5	71.9	86.1	95.6	66.9	83.5	98.2	75.0	88.4	96.0	73.9	86.6

For customer support on your Mechanical Costs with RSMeans Data, call 800.448.8182.

665

GEORGIA

DIVISION		COLUMBUS 318-319 MAT.	INST.	TOTAL	DALTON 307 MAT.	INST.	TOTAL	GAINESVILLE 305 MAT.	INST.	TOTAL	MACON 310-312 MAT.	INST.	TOTAL	SAVANNAH 313-314 MAT.	INST.	TOTAL	STATESBORO 304 MAT.	INST.	TOTAL
015433	CONTRACTOR EQUIPMENT		93.7	93.7		107.6	107.6		92.6	92.6		103.2	103.2		96.7	96.7		95.5	95.5
0241, 31-34	SITE & INFRASTRUCTURE, DEMOLITION	107.2	76.7	86.2	102.6	95.8	97.9	102.5	90.7	94.3	108.6	91.0	96.4	107.0	82.0	89.7	103.7	78.0	85.9
0310	Concrete Forming & Accessories	90.5	66.6	70.1	83.2	64.5	67.3	93.3	40.7	48.5	90.3	66.6	70.1	93.4	72.0	75.2	78.5	52.2	56.1
0320	Concrete Reinforcing	96.6	71.3	84.4	94.1	60.9	78.0	94.4	63.4	79.4	97.8	71.3	85.0	104.1	70.7	88.0	93.7	71.2	82.8
0330	Cast-in-Place Concrete	94.2	68.8	84.8	106.6	68.0	92.2	115.3	68.7	98.0	93.0	68.3	83.8	98.3	69.5	87.6	109.5	68.6	94.3
03	CONCRETE	86.8	70.0	79.4	98.9	67.0	84.9	101.4	55.8	81.4	86.3	69.8	79.1	88.5	72.5	81.5	99.3	63.4	83.5
04	MASONRY	92.5	68.7	77.9	75.8	75.5	75.6	83.2	76.2	78.9	104.8	68.9	82.7	87.4	68.8	75.9	77.8	77.0	77.3
05	METALS	105.9	97.6	103.4	96.4	94.0	95.7	94.6	78.5	89.9	100.9	97.6	100.0	102.1	96.0	100.3	100.0	98.9	99.7
06	WOOD, PLASTICS & COMPOSITES	80.1	66.4	72.6	75.9	65.2	70.0	96.1	33.2	61.5	85.9	66.5	75.2	87.4	73.9	80.0	69.7	47.2	57.3
07	THERMAL & MOISTURE PROTECTION	100.5	68.9	86.8	97.3	70.1	85.5	95.4	67.7	83.3	99.0	70.6	86.6	98.5	70.4	86.3	95.9	67.1	83.4
08	OPENINGS	85.7	69.3	81.9	92.8	65.8	86.5	91.3	48.8	81.4	85.5	69.3	81.7	95.3	73.2	90.2	93.7	58.5	85.5
0920	Plaster & Gypsum Board	101.1	66.0	77.5	81.9	64.7	70.3	94.9	31.8	52.4	104.7	66.0	78.7	103.3	73.6	83.3	83.6	46.2	58.4
0950, 0980	Ceilings & Acoustic Treatment	77.5	66.0	69.8	110.2	64.7	79.5	96.6	31.8	52.9	72.8	66.0	68.3	87.3	73.6	78.0	105.8	46.2	65.6
0960	Flooring	99.4	70.8	91.4	97.0	81.9	92.8	97.8	81.9	93.3	77.9	70.8	75.9	95.6	70.8	88.6	114.6	81.9	105.4
0970, 0990	Wall Finishes & Painting/Coating	88.7	87.4	87.9	82.8	70.1	75.3	92.4	94.0	93.3	90.7	94.0	92.6	88.0	86.1	86.9	90.8	70.1	78.5
09	FINISHES	91.6	68.9	79.3	104.5	68.2	84.9	95.5	52.1	72.0	80.5	69.7	74.7	92.9	73.2	82.3	107.9	57.9	80.9
COVERS	DIVS. 10-14, 25, 28, 41, 43, 44, 46	100.0	85.9	96.9	100.0	85.3	96.7	100.0	35.5	85.6	100.0	86.0	96.9	100.0	87.0	97.1	100.0	83.8	96.4
21, 22, 23	FIRE SUPPRESSION, PLUMBING & HVAC	100.1	66.8	87.0	96.7	60.8	82.6	96.6	65.6	84.4	100.1	68.2	87.6	100.1	64.8	86.3	97.2	67.3	85.5
26, 27, 3370	ELECTRICAL, COMMUNICATIONS & UTIL.	95.7	67.7	81.9	106.9	62.2	84.9	97.8	70.8	84.5	94.2	62.7	78.6	99.1	63.7	81.7	98.3	63.4	81.1
MF2018	WEIGHTED AVERAGE	96.5	72.2	86.2	97.5	71.4	86.5	96.2	65.6	83.2	95.2	73.1	85.8	97.3	72.7	86.9	97.8	69.7	85.9

GEORGIA — HAWAII — IDAHO

DIVISION		VALDOSTA 316 MAT.	INST.	TOTAL	WAYCROSS 315 MAT.	INST.	TOTAL	HILO 967 MAT.	INST.	TOTAL	HONOLULU 968 MAT.	INST.	TOTAL	STATES & POSS., GUAM 969 MAT.	INST.	TOTAL	BOISE 836-837 MAT.	INST.	TOTAL
015433	CONTRACTOR EQUIPMENT		93.7	93.7		93.7	93.7		97.8	97.8		99.3	99.3		160.5	160.5		94.6	94.6
0241, 31-34	SITE & INFRASTRUCTURE, DEMOLITION	116.9	76.7	89.1	113.7	77.0	88.3	142.5	103.1	115.3	150.0	106.6	120.0	186.3	98.5	125.7	86.8	92.0	90.4
0310	Concrete Forming & Accessories	81.7	41.6	47.5	83.5	64.0	66.8	108.2	123.8	121.5	120.2	123.7	123.2	110.5	51.5	60.2	100.2	81.1	83.9
0320	Concrete Reinforcing	98.7	59.5	79.8	98.7	59.8	79.9	143.4	127.0	135.5	166.8	127.0	147.6	253.7	27.4	144.2	107.6	79.5	94.0
0330	Cast-in-Place Concrete	92.6	68.7	83.7	104.7	68.6	91.2	180.3	123.3	159.1	145.0	123.5	137.0	156.3	95.7	133.7	90.6	96.6	92.8
03	CONCRETE	91.2	56.6	76.0	94.5	66.5	82.2	144.7	123.2	135.3	137.9	123.1	131.4	147.9	63.9	111.0	97.6	86.4	92.7
04	MASONRY	97.8	77.1	85.1	98.7	77.0	85.4	148.6	122.1	132.3	135.5	122.2	127.3	206.7	34.8	101.0	126.3	86.2	101.7
05	METALS	105.3	93.0	101.7	104.4	88.9	99.8	110.8	108.1	110.0	124.1	107.1	119.1	143.1	76.0	123.4	109.6	82.7	101.7
06	WOOD, PLASTICS & COMPOSITES	68.7	33.0	49.0	70.2	63.4	66.5	111.4	124.3	118.5	132.8	124.2	128.1	123.4	53.9	85.1	93.5	79.9	86.0
07	THERMAL & MOISTURE PROTECTION	100.8	66.6	86.0	100.6	69.1	86.9	128.7	119.3	124.6	146.3	120.0	134.9	150.7	58.2	110.4	101.4	86.6	94.9
08	OPENINGS	82.5	47.6	74.4	82.8	61.8	77.9	114.1	123.2	116.2	128.1	123.2	126.9	118.4	43.8	101.0	96.9	75.9	92.0
0920	Plaster & Gypsum Board	93.8	31.6	52.0	93.8	62.9	73.0	115.2	124.8	121.7	160.8	124.8	136.6	236.1	42.5	105.8	92.3	79.4	83.6
0950, 0980	Ceilings & Acoustic Treatment	75.9	31.6	46.0	74.0	62.9	66.5	136.6	124.8	128.6	144.8	124.8	131.3	257.2	42.5	112.4	102.8	79.4	87.0
0960	Flooring	93.0	83.5	90.3	94.3	81.9	90.8	103.8	139.3	113.8	121.3	139.3	126.4	121.2	39.5	98.2	92.9	90.7	92.3
0970, 0990	Wall Finishes & Painting/Coating	88.7	94.0	91.8	88.7	70.1	77.7	99.2	143.8	125.6	110.1	143.8	130.1	105.2	31.5	61.6	92.6	39.5	61.2
09	FINISHES	89.4	52.5	69.4	89.0	67.5	77.4	110.8	129.1	120.7	125.5	129.1	127.4	185.1	48.0	110.9	92.7	78.2	84.9
COVERS	DIVS. 10-14, 25, 28, 41, 43, 44, 46	100.0	82.2	96.0	100.0	85.5	96.6	100.0	113.7	103.1	100.0	113.6	103.0	100.0	66.0	92.4	100.0	88.6	97.4
21, 22, 23	FIRE SUPPRESSION, PLUMBING & HVAC	100.1	68.5	87.7	97.9	63.6	84.4	100.3	112.4	105.0	100.4	112.3	105.1	102.6	33.5	75.5	100.1	74.2	89.9
26, 27, 3370	ELECTRICAL, COMMUNICATIONS & UTIL.	94.1	57.8	76.2	97.7	63.4	80.8	108.1	124.6	116.2	109.8	124.5	117.1	158.1	35.9	97.9	97.2	68.2	82.9
MF2018	WEIGHTED AVERAGE	96.7	66.2	83.8	96.8	70.0	85.5	114.8	118.4	116.3	119.1	118.5	118.9	137.1	51.4	100.9	101.0	79.9	92.1

IDAHO — ILLINOIS

DIVISION		COEUR D'ALENE 838 MAT.	INST.	TOTAL	IDAHO FALLS 834 MAT.	INST.	TOTAL	LEWISTON 835 MAT.	INST.	TOTAL	POCATELLO 832 MAT.	INST.	TOTAL	TWIN FALLS 833 MAT.	INST.	TOTAL	BLOOMINGTON 617 MAT.	INST.	TOTAL
015433	CONTRACTOR EQUIPMENT		89.7	89.7		94.6	94.6		89.7	89.7		94.6	94.6		94.6	94.6		101.9	101.9
0241, 31-34	SITE & INFRASTRUCTURE, DEMOLITION	85.5	86.5	86.2	85.1	92.1	89.9	92.2	87.2	88.7	88.3	92.0	90.9	95.4	91.8	92.9	94.9	96.0	95.7
0310	Concrete Forming & Accessories	108.7	81.4	85.4	94.1	78.3	80.6	113.9	81.1	85.9	100.4	80.5	83.4	101.6	76.5	80.2	84.0	116.0	111.2
0320	Concrete Reinforcing	115.7	96.2	106.3	109.6	77.9	94.3	115.7	96.2	106.3	108.0	79.4	94.2	110.0	77.8	94.4	93.0	102.4	97.5
0330	Cast-in-Place Concrete	98.0	85.2	93.2	86.3	83.4	85.2	101.8	82.0	94.5	93.1	96.4	94.3	95.6	82.3	90.6	98.0	113.7	103.9
03	CONCRETE	103.9	85.3	95.7	89.6	80.2	85.5	107.5	84.1	97.2	96.7	86.0	92.0	104.6	79.0	93.4	91.9	113.6	101.4
04	MASONRY	128.0	87.3	103.0	121.5	87.2	100.4	128.5	84.7	101.6	124.0	86.2	100.8	126.7	82.8	99.7	110.2	121.7	117.3
05	METALS	103.0	88.1	98.6	118.1	81.6	107.4	102.4	88.2	98.3	118.2	82.0	107.6	118.2	81.4	107.4	95.2	123.1	103.4
06	WOOD, PLASTICS & COMPOSITES	96.7	79.8	87.4	87.2	76.3	81.2	102.6	79.8	90.1	93.5	79.9	86.0	94.6	76.3	84.5	81.7	113.0	99.0
07	THERMAL & MOISTURE PROTECTION	158.9	84.3	126.4	100.4	75.3	89.5	159.2	83.1	126.1	101.0	75.0	89.7	101.8	80.5	92.5	95.4	112.1	102.6
08	OPENINGS	113.8	75.9	105.0	99.8	71.3	93.2	106.7	79.6	100.4	97.6	68.8	90.9	100.6	62.4	91.7	91.3	118.7	97.7
0920	Plaster & Gypsum Board	167.0	79.4	108.1	80.3	75.7	77.2	168.9	79.4	108.7	81.9	79.4	80.2	84.0	75.7	78.4	89.1	113.5	105.5
0950, 0980	Ceilings & Acoustic Treatment	136.3	79.4	98.0	104.3	75.7	85.0	136.3	79.4	98.0	110.0	79.4	89.4	106.8	75.7	85.8	83.9	113.5	103.9
0960	Flooring	130.9	77.0	115.8	92.7	77.0	88.3	134.3	77.0	118.2	96.2	77.0	90.8	97.5	77.0	91.7	86.0	120.8	95.8
0970, 0990	Wall Finishes & Painting/Coating	111.2	73.8	89.1	92.6	39.5	61.2	111.2	71.3	87.6	92.5	39.5	61.1	92.6	39.5	61.2	87.6	134.8	115.5
09	FINISHES	158.3	79.5	115.7	90.9	74.0	81.8	159.7	78.9	116.0	93.7	75.9	84.1	94.5	72.9	82.8	86.1	119.0	103.9
COVERS	DIVS. 10-14, 25, 28, 41, 43, 44, 46	100.0	93.0	98.4	100.0	87.4	97.2	100.0	92.6	98.3	100.0	88.5	97.4	100.0	85.9	96.9	100.0	104.5	101.0
21, 22, 23	FIRE SUPPRESSION, PLUMBING & HVAC	99.4	85.5	94.0	101.1	81.8	93.5	100.9	85.1	94.6	100.0	73.6	89.6	100.0	72.2	89.1	96.5	105.4	100.0
26, 27, 3370	ELECTRICAL, COMMUNICATIONS & UTIL.	88.7	81.5	85.2	88.3	70.9	79.7	86.9	78.2	82.6	94.4	66.0	80.4	89.6	70.9	80.4	95.9	90.3	93.1
MF2018	WEIGHTED AVERAGE	108.1	84.4	98.1	100.5	79.9	91.8	108.2	83.5	97.8	102.0	78.4	92.0	103.2	76.8	92.1	95.0	109.5	101.1

ILLINOIS

DIVISION		CARBONDALE 629			CENTRALIA 628			CHAMPAIGN 618 - 619			CHICAGO 606 - 608			DECATUR 625			EAST ST. LOUIS 620 - 622		
		MAT.	INST.	TOTAL	MAT.	INST.	TOTAL	MAT.	INST.	TOTAL	MAT.	INST.	TOTAL	MAT.	INST.	TOTAL	MAT.	INST.	TOTAL
015433	CONTRACTOR EQUIPMENT		110.3	110.3		110.3	110.3		102.8	102.8		100.3	100.3		102.8	102.8		110.3	110.3
0241, 31 - 34	SITE & INFRASTRUCTURE, DEMOLITION	100.9	96.8	98.1	101.3	98.6	99.5	104.0	97.2	99.3	105.4	104.4	104.7	94.8	97.0	96.3	103.5	97.7	99.5
0310	Concrete Forming & Accessories	90.0	108.9	106.1	91.6	114.2	110.9	90.3	116.7	112.8	97.4	159.7	150.5	92.1	117.9	114.1	87.5	114.2	110.2
0320	Concrete Reinforcing	82.2	103.4	92.5	82.2	104.0	92.8	93.0	101.6	97.2	103.8	153.3	127.7	80.6	99.8	89.9	82.2	104.0	92.7
0330	Cast-in-Place Concrete	91.5	102.3	95.7	92.0	119.6	102.3	113.7	110.8	112.6	126.0	155.6	137.0	99.9	114.6	105.4	93.6	118.4	102.8
03	CONCRETE	79.9	107.3	91.9	80.4	115.8	96.0	104.1	112.7	107.9	108.0	156.5	129.3	90.3	114.3	100.8	81.4	115.4	96.3
04	MASONRY	75.0	111.5	97.4	75.0	122.9	104.5	133.1	125.4	128.4	105.3	163.4	141.0	70.9	122.8	102.8	75.3	123.0	104.6
05	METALS	100.9	130.4	109.6	100.9	133.5	110.5	95.2	120.3	102.6	95.9	146.8	110.8	104.9	120.7	109.5	102.1	133.5	111.3
06	WOOD, PLASTICS & COMPOSITES	87.7	105.8	97.7	90.2	111.2	101.7	88.5	114.0	102.5	100.0	158.4	132.1	88.7	116.8	104.2	84.8	111.2	99.3
07	THERMAL & MOISTURE PROTECTION	90.6	101.3	95.3	90.7	111.5	99.8	96.2	115.6	104.6	95.2	150.7	119.4	95.9	113.3	103.5	90.7	110.7	99.4
08	OPENINGS	87.3	116.3	94.1	87.3	119.2	94.8	91.8	116.9	97.7	100.9	169.3	116.9	98.2	118.2	102.9	87.4	118.4	94.6
0920	Plaster & Gypsum Board	93.7	106.2	102.1	95.0	111.6	106.2	91.0	114.5	106.8	101.3	160.0	140.8	96.5	117.4	110.6	92.2	111.6	105.3
0950, 0980	Ceilings & Acoustic Treatment	80.6	106.2	97.8	80.6	111.6	101.5	83.9	114.5	104.5	100.4	160.0	140.6	87.1	117.4	107.5	80.6	111.6	101.5
0960	Flooring	113.7	116.6	114.5	114.6	116.6	115.1	89.2	116.6	96.9	91.3	163.8	111.7	101.8	117.1	106.1	112.6	116.6	113.7
0970, 0990	Wall Finishes & Painting/Coating	103.1	105.0	104.2	103.1	111.6	108.2	87.6	114.9	103.7	96.7	170.3	140.2	95.5	117.8	108.7	103.1	111.6	108.2
09	FINISHES	92.1	110.7	102.1	92.5	114.4	104.4	88.1	116.9	103.7	95.8	162.4	131.8	92.2	118.3	106.3	91.6	114.4	103.9
COVERS	DIVS. 10 - 14, 25, 28, 41, 43, 44, 46	100.0	106.9	101.5	100.0	105.8	101.3	100.0	109.2	102.0	100.0	129.3	106.5	100.0	108.6	101.9	100.0	108.3	101.9
21, 22, 23	FIRE SUPPRESSION, PLUMBING & HVAC	96.3	106.8	100.4	96.3	97.8	96.9	96.5	107.0	100.6	99.9	137.4	114.6	99.9	98.8	99.5	99.9	99.2	99.6
26, 27, 3370	ELECTRICAL, COMMUNICATIONS & UTIL.	91.9	107.8	99.8	93.1	106.3	99.6	99.0	95.4	97.2	97.9	136.2	116.8	95.3	103.7	99.5	92.7	99.7	96.2
MF2018	WEIGHTED AVERAGE	92.5	109.6	99.7	92.8	111.2	100.5	98.5	110.5	103.6	100.1	145.6	119.3	96.6	110.1	102.3	93.8	110.4	100.9

ILLINOIS

DIVISION		EFFINGHAM 624			GALESBURG 614			JOLIET 604			KANKAKEE 609			LA SALLE 613			NORTH SUBURBAN 600 - 603		
		MAT.	INST.	TOTAL	MAT.	INST.	TOTAL	MAT.	INST.	TOTAL	MAT.	INST.	TOTAL	MAT.	INST.	TOTAL	MAT.	INST.	TOTAL
015433	CONTRACTOR EQUIPMENT		102.8	102.8		101.9	101.9		93.7	93.7		93.7	93.7		101.9	101.9		93.7	93.7
0241, 31 - 34	SITE & INFRASTRUCTURE, DEMOLITION	99.3	95.9	97.0	97.3	95.8	96.3	101.2	97.2	98.5	94.7	97.1	96.4	96.7	96.8	96.8	100.5	97.9	98.7
0310	Concrete Forming & Accessories	96.5	114.6	111.9	90.2	117.0	113.0	96.8	156.7	147.8	90.5	145.4	137.3	103.7	125.3	122.1	96.1	158.5	149.3
0320	Concrete Reinforcing	83.3	93.8	88.4	92.5	105.6	98.8	103.8	136.6	119.7	104.6	135.1	119.3	92.7	132.9	112.1	103.8	149.1	125.7
0330	Cast-in-Place Concrete	99.6	109.9	103.4	100.9	109.2	104.0	114.6	143.9	125.5	106.8	135.8	117.6	100.8	121.8	108.6	114.6	152.9	128.9
03	CONCRETE	91.0	109.9	99.3	94.8	113.0	102.8	101.0	147.9	121.6	95.0	139.7	114.7	95.6	126.1	109.0	101.0	154.1	124.3
04	MASONRY	79.1	115.3	101.3	110.4	122.4	117.8	104.5	156.7	136.6	100.6	147.0	129.1	110.4	129.7	122.3	101.2	158.8	136.6
05	METALS	101.9	114.9	105.7	95.2	123.6	103.5	94.1	134.7	106.0	94.1	133.4	105.6	95.3	142.0	109.0	95.1	141.6	108.8
06	WOOD, PLASTICS & COMPOSITES	91.0	114.0	103.6	88.3	115.4	103.2	96.6	157.7	130.2	90.0	144.7	120.1	103.8	123.2	114.5	95.4	158.2	130.0
07	THERMAL & MOISTURE PROTECTION	95.4	109.1	101.4	95.6	110.0	101.9	99.5	144.2	119.0	98.6	139.5	116.4	95.8	122.5	107.4	99.9	147.6	120.7
08	OPENINGS	93.3	114.5	98.2	91.3	115.9	97.0	99.3	161.7	113.9	92.4	156.2	107.3	91.3	133.3	101.1	99.4	167.9	115.4
0920	Plaster & Gypsum Board	96.3	114.5	108.5	91.0	115.9	107.8	93.7	159.5	138.0	92.5	146.1	128.6	98.3	124.0	115.6	97.2	160.0	139.5
0950, 0980	Ceilings & Acoustic Treatment	80.6	114.5	103.4	83.9	115.9	105.5	100.9	159.5	140.4	100.9	146.1	131.4	83.9	124.0	110.9	100.9	160.0	140.8
0960	Flooring	103.0	116.6	106.8	89.1	120.8	98.0	89.4	149.8	106.4	86.3	150.7	104.4	95.7	124.8	103.9	89.9	159.5	109.4
0970, 0990	Wall Finishes & Painting/Coating	95.5	109.3	103.7	87.6	98.5	94.0	89.0	163.0	132.8	89.0	135.7	116.6	87.6	133.0	114.5	91.0	163.9	134.1
09	FINISHES	91.3	115.1	104.2	87.4	116.3	103.1	91.6	157.4	127.2	90.0	145.8	120.2	90.4	124.7	109.0	92.2	160.1	129.0
COVERS	DIVS. 10 - 14, 25, 28, 41, 43, 44, 46	100.0	104.4	101.0	100.0	104.4	101.0	100.0	122.6	105.0	100.0	120.6	104.6	100.0	105.6	101.2	100.0	124.1	105.4
21, 22, 23	FIRE SUPPRESSION, PLUMBING & HVAC	96.4	104.5	99.6	96.5	105.7	100.1	100.0	130.0	111.8	96.5	129.6	109.5	96.5	128.4	109.0	99.9	136.2	114.1
26, 27, 3370	ELECTRICAL, COMMUNICATIONS & UTIL.	93.5	107.8	100.5	96.7	86.5	91.7	96.9	133.3	114.8	92.4	137.0	114.4	94.0	137.0	115.2	96.7	135.5	115.9
MF2018	WEIGHTED AVERAGE	95.1	109.1	101.0	95.7	108.5	101.1	98.3	138.8	115.4	95.0	134.9	111.9	95.9	127.0	109.0	98.3	142.9	117.1

ILLINOIS

DIVISION		PEORIA 615 - 616			QUINCY 623			ROCK ISLAND 612			ROCKFORD 610 - 611			SOUTH SUBURBAN 605			SPRINGFIELD 626 - 627		
		MAT.	INST.	TOTAL	MAT.	INST.	TOTAL	MAT.	INST.	TOTAL	MAT.	INST.	TOTAL	MAT.	INST.	TOTAL	MAT.	INST.	TOTAL
015433	CONTRACTOR EQUIPMENT		101.9	101.9		102.8	102.8		101.9	101.9		101.9	101.9		93.7	93.7		105.1	105.1
0241, 31 - 34	SITE & INFRASTRUCTURE, DEMOLITION	97.9	96.0	96.6	98.1	96.4	96.9	95.5	94.8	95.0	97.4	96.8	97.0	100.5	97.8	98.6	99.8	100.5	100.3
0310	Concrete Forming & Accessories	93.0	115.9	112.5	94.3	113.6	110.8	91.6	99.1	98.0	97.2	127.7	123.2	96.1	158.5	149.3	93.2	115.6	112.3
0320	Concrete Reinforcing	90.1	104.7	97.2	83.0	86.2	84.5	92.5	99.0	95.6	85.3	134.1	108.9	103.8	149.0	125.7	83.2	99.8	91.2
0330	Cast-in-Place Concrete	97.9	115.6	104.5	99.8	106.6	102.3	98.7	98.4	98.6	100.2	126.6	110.0	114.6	152.8	128.8	94.9	109.2	100.2
03	CONCRETE	92.0	114.6	102.0	90.7	107.1	97.9	92.7	100.0	95.9	92.7	129.0	108.7	101.0	154.0	124.3	88.5	111.2	98.5
04	MASONRY	109.7	121.5	116.9	98.9	113.2	107.7	110.2	100.2	104.1	85.6	142.3	120.4	101.2	158.7	136.6	80.7	122.2	106.2
05	METALS	97.9	124.2	105.6	101.9	112.6	105.1	95.2	118.5	102.1	97.9	141.5	110.7	95.1	141.4	108.7	102.5	119.1	107.3
06	WOOD, PLASTICS & COMPOSITES	96.4	113.2	105.6	88.6	114.0	102.6	90.0	97.5	94.1	96.4	123.6	111.3	95.4	158.2	130.0	90.0	114.1	103.3
07	THERMAL & MOISTURE PROTECTION	96.3	112.4	103.3	95.3	106.8	100.3	95.5	98.4	96.8	98.8	129.0	111.9	99.9	147.6	120.7	98.3	114.4	105.3
08	OPENINGS	96.6	119.5	102.0	94.0	112.1	98.2	91.3	103.3	94.1	96.6	136.5	105.9	99.4	167.9	115.4	99.2	116.4	103.2
0920	Plaster & Gypsum Board	95.5	113.7	107.7	95.0	114.5	108.1	91.0	97.6	95.5	95.5	124.4	114.9	97.2	160.0	139.5	99.9	114.5	109.7
0950, 0980	Ceilings & Acoustic Treatment	89.6	113.7	105.8	80.6	114.5	103.4	83.9	97.6	93.2	89.6	124.4	113.1	100.9	160.0	140.8	91.3	114.5	107.0
0960	Flooring	92.6	119.5	100.2	101.8	113.6	105.1	90.2	94.6	91.4	92.6	124.9	101.7	89.9	159.5	109.4	104.8	117.1	108.3
0970, 0990	Wall Finishes & Painting/Coating	87.6	134.8	115.5	95.5	111.5	105.0	87.6	95.5	92.3	87.6	142.7	120.2	91.0	163.9	134.1	98.0	111.5	106.0
09	FINISHES	90.2	118.7	105.6	90.7	114.2	103.4	87.7	97.6	93.0	90.2	128.5	110.9	92.2	160.1	129.0	98.0	115.9	107.7
COVERS	DIVS. 10 - 14, 25, 28, 41, 43, 44, 46	100.0	104.4	101.0	100.0	106.9	101.5	100.0	97.5	99.4	100.0	114.2	103.2	100.0	124.1	105.4	100.0	108.3	101.9
21, 22, 23	FIRE SUPPRESSION, PLUMBING & HVAC	100.0	100.2	100.0	96.4	102.9	98.9	96.5	97.5	96.9	100.1	116.1	106.4	99.9	136.2	114.1	99.9	103.7	101.4
26, 27, 3370	ELECTRICAL, COMMUNICATIONS & UTIL.	97.7	91.5	94.6	91.4	82.2	86.9	89.3	93.1	91.2	98.0	125.6	111.6	96.7	135.5	115.9	98.1	88.7	93.5
MF2018	WEIGHTED AVERAGE	97.6	108.8	102.3	95.7	104.2	99.3	94.6	99.5	96.7	96.6	125.5	108.8	98.3	142.8	117.1	97.5	108.3	102.1

For customer support on your Mechanical Costs with RSMeans Data, call 800.448.8182.

667

INDIANA

DIVISION		ANDERSON 460 MAT.	INST.	TOTAL	BLOOMINGTON 474 MAT.	INST.	TOTAL	COLUMBUS 472 MAT.	INST.	TOTAL	EVANSVILLE 476-477 MAT.	INST.	TOTAL	FORT WAYNE 467-468 MAT.	INST.	TOTAL	GARY 463-464 MAT.	INST.	TOTAL
015433	CONTRACTOR EQUIPMENT		94.5	94.5		81.5	81.5		81.5	81.5		110.3	110.3		94.5	94.5		94.5	94.5
0241, 31 - 34	SITE & INFRASTRUCTURE, DEMOLITION	98.7	89.0	92.0	86.5	87.7	87.3	83.2	87.4	86.1	91.9	115.1	107.9	99.8	88.8	92.2	99.3	92.8	94.8
0310	Concrete Forming & Accessories	94.9	77.4	80.0	100.8	82.1	84.9	95.0	80.2	82.4	94.3	78.6	80.9	92.9	73.7	76.5	95.0	111.1	108.7
0320	Concrete Reinforcing	104.9	83.0	94.3	90.4	86.9	88.7	90.8	86.9	88.9	99.0	81.0	90.3	104.9	77.9	91.9	104.9	115.8	110.2
0330	Cast-in-Place Concrete	106.3	74.9	94.6	101.3	75.7	91.7	100.8	76.5	91.8	96.8	83.2	91.7	113.0	75.2	98.9	111.1	112.2	111.5
03	CONCRETE	95.7	78.1	88.0	99.2	80.3	90.9	98.5	79.6	90.2	99.5	80.8	91.3	98.6	75.7	88.6	97.9	112.1	104.1
04	MASONRY	87.4	74.0	79.2	88.9	74.4	80.0	88.8	74.3	79.9	84.3	77.3	80.0	90.7	71.4	78.8	88.8	110.3	102.0
05	METALS	97.7	88.2	94.9	99.1	75.7	92.2	99.1	74.7	91.9	91.9	83.2	89.3	97.7	86.3	94.3	97.7	107.1	100.5
06	WOOD, PLASTICS & COMPOSITES	93.8	77.6	84.9	109.5	82.9	94.9	104.0	80.5	91.1	90.8	77.7	83.6	93.7	73.8	82.7	91.3	109.1	101.1
07	THERMAL & MOISTURE PROTECTION	109.2	75.0	94.4	95.8	77.9	88.0	95.2	78.7	88.0	100.0	81.7	92.1	108.9	76.8	94.9	107.7	106.2	107.0
08	OPENINGS	93.2	75.8	89.2	97.5	80.0	93.4	93.9	78.6	90.3	91.8	76.0	88.1	93.2	71.2	88.1	93.2	114.6	98.2
0920	Plaster & Gypsum Board	106.4	77.3	86.8	98.6	83.2	88.2	95.4	80.7	85.5	94.2	76.7	82.4	105.8	73.4	84.0	99.5	109.7	106.4
0950, 0980	Ceilings & Acoustic Treatment	89.8	77.3	81.4	77.9	83.2	81.5	77.9	80.7	79.8	81.6	76.7	78.3	89.8	73.4	78.7	89.8	109.7	103.2
0960	Flooring	93.5	75.0	88.3	99.6	83.2	95.0	94.4	83.2	91.2	94.1	71.4	87.7	93.5	71.2	87.2	93.5	110.5	98.3
0970, 0990	Wall Finishes & Painting/Coating	92.8	65.8	76.9	84.4	81.1	82.5	84.4	81.1	82.5	90.3	84.4	86.8	92.8	70.2	79.5	92.8	120.7	109.3
09	FINISHES	91.3	75.7	82.9	90.6	82.4	86.2	88.6	81.0	84.5	89.3	77.7	83.0	91.1	72.9	81.3	90.3	111.8	102.0
COVERS	DIVS. 10 - 14, 25, 28, 41, 43, 44, 46	100.0	87.4	97.2	100.0	88.3	97.4	100.0	88.0	97.3	100.0	91.2	98.0	100.0	87.2	97.1	100.0	107.9	101.8
21, 22, 23	FIRE SUPPRESSION, PLUMBING & HVAC	100.0	77.0	91.0	99.7	79.6	91.8	96.2	79.1	89.5	99.9	78.6	91.5	100.0	72.2	89.1	100.0	106.2	102.4
26, 27, 3370	ELECTRICAL, COMMUNICATIONS & UTIL.	87.7	83.2	85.5	99.9	86.6	93.3	99.1	86.7	93.0	95.7	81.4	88.7	88.4	74.9	81.7	99.5	110.1	104.7
MF2018	WEIGHTED AVERAGE	96.0	79.8	89.1	97.8	81.1	90.7	96.1	80.5	89.5	95.5	82.7	90.1	96.6	76.3	88.0	97.5	108.1	102.0

INDIANA

DIVISION		INDIANAPOLIS 461-462 MAT.	INST.	TOTAL	KOKOMO 469 MAT.	INST.	TOTAL	LAFAYETTE 479 MAT.	INST.	TOTAL	LAWRENCEBURG 470 MAT.	INST.	TOTAL	MUNCIE 473 MAT.	INST.	TOTAL	NEW ALBANY 471 MAT.	INST.	TOTAL
015433	CONTRACTOR EQUIPMENT		86.1	86.1		94.5	94.5		81.5	81.5		100.6	100.6		92.8	92.8		90.6	90.6
0241, 31 - 34	SITE & INFRASTRUCTURE, DEMOLITION	99.8	91.3	93.9	94.9	89.0	90.8	83.9	87.4	86.3	81.2	102.2	95.7	86.6	87.9	87.5	78.5	89.8	86.3
0310	Concrete Forming & Accessories	100.4	85.2	87.4	98.0	78.2	81.1	92.5	78.6	80.7	91.5	78.4	80.3	92.3	76.9	79.2	90.1	76.9	78.9
0320	Concrete Reinforcing	106.1	87.2	97.0	94.9	87.2	91.1	90.4	82.9	86.8	89.7	79.8	84.9	100.0	82.9	91.7	91.0	82.6	86.9
0330	Cast-in-Place Concrete	101.2	84.8	95.1	105.2	81.8	96.5	101.4	77.6	92.5	94.8	74.9	87.4	106.4	73.9	94.3	97.9	73.1	88.7
03	CONCRETE	99.0	84.8	92.8	92.5	81.6	87.7	98.7	78.7	89.9	91.8	78.0	85.7	97.4	77.6	88.7	97.3	76.8	88.3
04	MASONRY	89.8	78.7	83.0	87.1	77.2	81.0	94.0	74.0	81.7	73.8	74.2	74.1	90.5	74.1	80.4	80.2	71.1	74.6
05	METALS	94.7	76.0	89.2	94.2	89.8	92.9	93.9	73.7	90.5	93.9	86.2	91.7	100.8	88.0	97.0	96.0	82.2	91.9
06	WOOD, PLASTICS & COMPOSITES	100.1	86.0	92.4	96.8	77.4	86.1	101.2	79.2	89.1	89.1	78.3	83.2	102.6	77.1	88.6	91.2	77.6	83.7
07	THERMAL & MOISTURE PROTECTION	98.7	82.0	91.4	108.1	77.5	94.8	95.2	77.2	87.4	100.8	76.9	90.4	98.2	76.4	88.7	87.6	72.3	80.9
08	OPENINGS	103.9	81.9	98.8	88.4	76.9	85.7	92.4	76.7	88.8	93.6	75.0	89.3	91.0	75.6	87.4	91.4	76.3	87.9
0920	Plaster & Gypsum Board	96.8	85.8	89.4	111.0	77.1	88.2	93.3	79.4	83.9	71.7	78.4	76.2	94.2	77.3	82.8	92.0	77.5	82.2
0950, 0980	Ceilings & Acoustic Treatment	92.7	85.8	88.0	90.6	77.1	81.5	73.9	79.4	77.6	84.0	78.4	80.2	77.9	77.3	77.5	81.6	77.5	78.8
0960	Flooring	97.2	83.2	93.2	97.3	88.4	94.8	93.3	79.2	89.3	68.9	83.2	72.9	93.7	75.0	88.5	91.6	54.0	81.1
0970, 0990	Wall Finishes & Painting/Coating	96.5	81.1	87.4	92.8	68.8	78.6	84.4	81.4	82.6	85.2	73.0	78.0	84.4	65.8	73.4	90.3	66.3	76.1
09	FINISHES	94.9	84.6	89.4	93.1	79.1	85.5	87.2	79.2	82.8	78.8	78.9	78.8	87.9	75.4	81.1	88.5	71.4	79.2
COVERS	DIVS. 10 - 14, 25, 28, 41, 43, 44, 46	100.0	93.7	98.6	100.0	88.6	97.5	100.0	86.6	97.0	100.0	88.0	97.3	100.0	86.3	97.0	100.0	87.7	97.3
21, 22, 23	FIRE SUPPRESSION, PLUMBING & HVAC	100.1	79.9	92.2	96.5	79.3	89.7	96.2	76.2	88.3	97.0	75.9	88.7	99.7	76.9	90.7	96.4	78.6	89.4
26, 27, 3370	ELECTRICAL, COMMUNICATIONS & UTIL.	101.9	86.7	94.4	92.0	78.3	85.3	98.6	78.6	88.7	93.6	74.1	84.0	91.4	74.0	82.9	94.3	74.6	84.6
MF2018	WEIGHTED AVERAGE	98.7	83.2	92.1	94.2	81.1	88.6	95.8	78.1	88.3	92.5	79.6	87.0	96.2	78.2	88.6	93.9	77.3	86.9

816.2 = 90.7
9 cities

INDIANA / IOWA

DIVISION		SOUTH BEND 465-466 MAT.	INST.	TOTAL	TERRE HAUTE 478 MAT.	INST.	TOTAL	WASHINGTON 475 MAT.	INST.	TOTAL	BURLINGTON 526 MAT.	INST.	TOTAL	CARROLL 514 MAT.	INST.	TOTAL	CEDAR RAPIDS 522-524 MAT.	INST.	TOTAL
015433	CONTRACTOR EQUIPMENT		108.2	108.2		110.3	110.3		110.3	110.3		98.8	98.8		98.8	98.8		95.7	95.7
0241, 31 - 34	SITE & INFRASTRUCTURE, DEMOLITION	97.7	94.2	95.3	93.4	115.4	108.6	93.2	115.7	108.7	98.0	91.8	93.7	87.1	92.6	90.9	99.9	91.4	94.0
0310	Concrete Forming & Accessories	95.1	76.0	78.9	95.0	77.1	79.7	96.0	80.5	82.8	96.5	94.2	94.5	84.1	83.9	83.9	102.5	83.9	86.6
0320	Concrete Reinforcing	104.1	83.6	94.2	99.0	83.0	91.3	91.6	82.6	87.2	94.7	97.8	96.2	95.4	85.8	90.8	95.3	80.2	88.0
0330	Cast-in-Place Concrete	108.2	78.4	97.1	93.7	78.3	88.0	102.0	85.0	95.6	107.3	54.8	87.8	107.3	82.2	97.9	107.6	83.4	98.6
03	CONCRETE	97.7	79.7	89.8	102.5	78.8	92.1	108.3	82.6	97.0	95.8	81.5	89.6	94.7	84.3	90.1	95.9	83.8	90.6
04	MASONRY	91.2	74.0	80.6	92.0	73.3	80.5	84.5	79.6	81.5	99.5	71.9	82.5	101.1	72.4	83.5	105.0	80.0	89.6
05	METALS	101.5	102.4	101.7	92.6	84.4	90.2	87.1	84.7	86.4	87.9	99.1	91.2	87.9	95.4	90.1	90.3	93.1	91.2
06	WOOD, PLASTICS & COMPOSITES	96.6	75.2	84.8	93.0	76.7	84.0	93.2	79.2	85.5	91.6	98.1	95.2	77.8	88.4	83.6	98.9	83.6	90.5
07	THERMAL & MOISTURE PROTECTION	103.2	78.7	92.5	100.1	79.1	91.0	100.0	83.5	92.8	104.4	77.1	92.5	104.7	78.8	93.4	105.5	80.1	94.4
08	OPENINGS	93.1	75.5	89.0	92.3	75.3	88.3	89.3	77.1	86.4	94.5	95.8	94.8	98.8	84.2	95.4	99.3	81.2	95.1
0920	Plaster & Gypsum Board	97.7	74.7	82.2	94.2	75.7	81.8	94.0	78.3	83.4	105.0	98.3	100.5	100.9	88.3	92.4	110.3	83.6	92.3
0950, 0980	Ceilings & Acoustic Treatment	94.2	74.7	81.1	81.6	75.7	77.6	76.7	78.3	77.8	99.0	98.3	98.5	99.0	88.3	91.8	101.4	83.6	89.4
0960	Flooring	90.7	86.2	89.5	94.1	75.7	88.9	95.1	83.2	91.7	93.7	69.1	86.8	88.0	79.5	85.6	107.9	84.3	101.3
0970, 0990	Wall Finishes & Painting/Coating	94.3	82.5	87.3	90.3	79.8	84.1	90.3	84.4	86.8	92.8	84.6	87.9	92.8	84.6	87.9	94.7	70.8	80.6
09	FINISHES	92.5	78.3	84.8	89.3	76.9	82.6	88.9	81.3	84.7	93.9	89.7	91.6	90.1	83.6	86.6	99.3	82.7	90.3
COVERS	DIVS. 10 - 14, 25, 28, 41, 43, 44, 46	100.0	88.7	97.5	100.0	88.7	97.5	100.0	91.8	98.2	100.0	93.4	98.5	100.0	89.1	97.6	100.0	93.4	98.5
21, 22, 23	FIRE SUPPRESSION, PLUMBING & HVAC	99.9	75.0	90.1	99.9	76.5	90.7	96.4	81.0	90.4	96.5	83.8	91.5	96.5	77.8	89.2	100.0	81.2	92.6
26, 27, 3370	ELECTRICAL, COMMUNICATIONS & UTIL.	101.3	83.9	92.7	94.1	83.5	88.8	94.5	83.6	89.2	101.1	72.2	86.9	101.8	76.9	89.5	98.8	80.1	89.5
MF2018	WEIGHTED AVERAGE	98.4	81.8	91.4	96.2	81.7	90.1	94.6	84.8	90.4	95.7	84.3	90.9	95.5	82.4	89.9	98.0	83.7	92.0

City Cost Indexes

IOWA

DIVISION		COUNCIL BLUFFS 515			CRESTON 508			DAVENPORT 527 - 528			DECORAH 521			DES MOINES 500 - 503,509			DUBUQUE 520		
		MAT.	INST.	TOTAL	MAT.	INST.	TOTAL	MAT.	INST.	TOTAL	MAT.	INST.	TOTAL	MAT.	INST.	TOTAL	MAT.	INST.	TOTAL
015433	CONTRACTOR EQUIPMENT		95.0	95.0		98.8	98.8		98.8	98.8		98.8	98.8		102.4	102.4		94.5	94.5
0241, 31 - 34	SITE & INFRASTRUCTURE, DEMOLITION	104.5	88.5	93.4	92.9	93.6	93.4	98.4	94.6	95.8	96.6	91.6	93.2	98.3	99.3	99.0	97.7	88.7	91.5
0310	Concrete Forming & Accessories	83.5	73.6	75.0	79.2	86.7	85.6	102.0	95.9	96.8	94.0	71.8	75.1	95.6	89.5	90.4	84.9	81.5	82.0
0320	Concrete Reinforcing	97.3	79.8	88.8	96.3	85.9	91.3	95.3	99.0	97.1	94.7	85.2	90.1	101.0	101.6	101.3	94.0	79.9	87.2
0330	Cast-in-Place Concrete	111.8	79.1	99.6	114.6	85.7	103.8	103.6	95.2	100.5	104.4	76.0	93.8	97.6	92.1	95.6	105.3	83.0	97.0
03	CONCRETE	98.1	77.6	89.1	97.7	86.8	92.9	94.0	96.7	95.2	93.8	76.6	86.3	91.0	92.8	91.8	92.7	82.6	88.3
04	MASONRY	106.5	76.2	87.8	99.7	79.9	87.5	102.0	92.2	96.0	120.3	69.2	88.9	87.7	89.0	88.5	105.9	68.8	83.1
05	METALS	95.2	92.8	94.5	88.4	96.0	90.6	90.3	105.5	94.8	88.0	94.2	89.9	94.1	96.8	94.9	88.9	92.4	89.9
06	WOOD, PLASTICS & COMPOSITES	76.5	72.1	74.1	69.7	88.4	80.0	98.9	95.3	96.9	88.5	70.9	78.8	87.6	88.5	88.1	78.3	82.2	80.4
07	THERMAL & MOISTURE PROTECTION	104.8	74.8	91.8	106.2	82.2	95.8	104.9	91.6	99.1	104.6	69.9	89.5	98.5	87.6	93.8	105.1	76.5	92.6
08	OPENINGS	98.4	76.8	93.3	108.6	85.9	103.3	99.3	96.7	98.7	97.4	77.9	92.9	100.8	89.6	98.2	98.4	82.4	94.6
0920	Plaster & Gypsum Board	100.9	71.8	81.3	95.8	88.3	90.8	110.3	95.4	100.2	103.7	70.3	81.2	92.8	88.3	89.8	100.9	82.1	88.3
0950, 0980	Ceilings & Acoustic Treatment	99.0	71.8	80.6	90.2	88.3	88.9	101.4	95.4	97.3	99.0	70.3	79.6	93.4	88.3	90.0	99.0	82.1	87.6
0960	Flooring	87.0	84.3	86.2	81.3	69.1	77.8	96.1	89.4	94.2	93.3	69.1	86.5	95.2	94.5	95.0	99.1	69.1	90.7
0970, 0990	Wall Finishes & Painting/Coating	89.6	59.8	72.0	83.6	84.6	84.2	92.8	90.5	91.4	92.8	84.6	87.9	91.6	84.6	87.5	93.9	78.4	84.8
09	FINISHES	91.0	73.7	81.7	85.1	83.4	84.2	95.7	94.1	94.8	93.5	72.4	82.1	93.6	90.0	91.6	94.8	78.9	86.2
COVERS	DIVS. 10 - 14, 25, 28, 41, 43, 44, 46	100.0	91.4	98.1	100.0	91.6	98.1	100.0	96.9	99.3	100.0	88.6	97.5	100.0	95.6	99.0	100.0	92.3	98.3
21, 22, 23	FIRE SUPPRESSION, PLUMBING & HVAC	100.0	72.8	89.4	96.3	79.8	89.8	100.0	93.9	97.6	96.5	74.5	87.9	99.8	86.8	94.7	100.0	75.6	90.4
26, 27, 3370	ELECTRICAL, COMMUNICATIONS & UTIL.	103.9	82.7	93.4	93.5	76.9	85.3	96.9	87.6	92.3	98.8	47.2	73.4	105.1	84.3	94.8	102.5	76.9	89.9
MF2018	WEIGHTED AVERAGE	98.9	79.2	90.6	95.7	84.2	90.8	97.1	94.5	96.0	96.4	73.7	86.8	97.2	90.2	94.3	97.2	79.9	89.9

IOWA

DIVISION		FORT DODGE 505			MASON CITY 504			OTTUMWA 525			SHENANDOAH 516			SIBLEY 512			SIOUX CITY 510 - 511		
		MAT.	INST.	TOTAL	MAT.	INST.	TOTAL	MAT.	INST.	TOTAL	MAT.	INST.	TOTAL	MAT.	INST.	TOTAL	MAT.	INST.	TOTAL
015433	CONTRACTOR EQUIPMENT		98.8	98.8		98.8	98.8		94.5	94.5		95.0	95.0		98.8	98.8		98.8	98.8
0241, 31 - 34	SITE & INFRASTRUCTURE, DEMOLITION	101.4	90.5	93.9	101.5	91.5	94.6	98.0	86.7	90.2	102.6	88.6	92.9	107.7	91.6	96.6	109.6	93.3	98.3
0310	Concrete Forming & Accessories	79.8	78.0	78.3	83.6	71.4	73.2	91.9	87.1	87.8	85.1	77.2	78.3	85.7	37.8	44.9	102.5	75.5	79.4
0320	Concrete Reinforcing	96.3	85.2	90.9	96.2	85.2	90.9	94.7	98.0	96.3	97.3	85.8	91.7	97.3	85.1	91.4	95.3	100.8	98.0
0330	Cast-in-Place Concrete	107.6	41.9	83.1	107.6	71.2	94.0	108.0	63.9	91.6	108.1	82.6	98.6	105.9	55.5	87.2	106.6	89.3	100.2
03	CONCRETE	93.1	67.6	81.9	93.3	74.8	85.2	95.4	81.5	89.3	95.6	81.5	89.4	94.7	54.1	76.8	95.3	85.6	91.0
04	MASONRY	98.6	51.9	69.9	111.6	68.0	84.8	102.3	55.2	73.3	106.0	74.5	86.6	124.6	52.2	80.1	99.2	70.0	81.2
05	METALS	88.5	94.0	90.1	88.5	94.2	90.2	87.9	99.7	91.3	94.2	95.6	94.6	88.1	93.3	89.7	90.3	101.3	93.6
06	WOOD, PLASTICS & COMPOSITES	70.1	88.4	80.2	74.0	70.9	72.3	85.7	97.9	92.4	78.3	78.5	78.4	79.1	34.2	54.4	98.9	74.4	85.4
07	THERMAL & MOISTURE PROTECTION	105.5	65.2	88.0	105.0	70.5	90.0	105.3	70.3	90.1	104.0	73.3	90.7	104.3	55.2	82.9	104.9	73.4	91.2
08	OPENINGS	102.5	75.9	96.3	94.7	77.9	90.8	98.8	92.8	97.4	90.2	76.4	87.0	95.6	46.0	84.0	99.3	81.6	95.1
0920	Plaster & Gypsum Board	95.8	88.3	90.8	95.8	70.3	78.6	101.5	98.3	99.3	100.9	78.3	85.7	100.9	32.5	54.9	110.3	73.8	85.8
0950, 0980	Ceilings & Acoustic Treatment	90.2	88.3	88.9	90.2	70.3	76.8	99.0	98.3	98.5	99.0	78.3	85.0	99.0	32.5	54.2	101.4	73.8	82.8
0960	Flooring	82.6	69.1	78.8	84.6	69.1	80.2	102.2	69.1	92.9	87.7	73.9	83.8	88.9	69.1	83.3	96.1	73.3	89.7
0970, 0990	Wall Finishes & Painting/Coating	83.6	81.3	82.2	83.6	84.6	84.2	93.9	84.6	88.4	89.6	84.6	86.6	92.8	83.6	87.3	92.8	69.1	78.8
09	FINISHES	87.0	78.0	82.1	87.6	72.1	79.2	95.9	84.8	89.9	91.1	77.1	83.5	93.3	46.4	67.9	97.2	73.8	84.5
COVERS	DIVS. 10 - 14, 25, 28, 41, 43, 44, 46	100.0	88.4	96.6	100.0	88.2	97.4	100.0	87.3	97.2	100.0	88.2	97.4	100.0	78.9	95.3	100.0	91.4	98.1
21, 22, 23	FIRE SUPPRESSION, PLUMBING & HVAC	96.3	71.4	86.6	96.3	77.7	89.0	96.5	73.1	87.3	96.5	83.9	91.6	96.5	71.5	86.7	100.0	81.2	92.6
26, 27, 3370	ELECTRICAL, COMMUNICATIONS & UTIL.	99.8	69.4	84.8	98.9	47.2	73.4	100.9	71.0	86.2	98.8	78.4	88.7	98.8	47.2	73.4	98.8	74.0	86.6
MF2018	WEIGHTED AVERAGE	95.5	73.7	86.3	95.3	74.0	86.3	96.3	78.7	88.9	96.1	81.9	90.1	96.8	62.4	82.3	97.7	81.5	90.9

IOWA / KANSAS

DIVISION		IOWA SPENCER 513			IOWA WATERLOO 506 - 507			KANSAS BELLEVILLE 669			KANSAS COLBY 677			KANSAS DODGE CITY 678			KANSAS EMPORIA 668		
		MAT.	INST.	TOTAL	MAT.	INST.	TOTAL	MAT.	INST.	TOTAL	MAT.	INST.	TOTAL	MAT.	INST.	TOTAL	MAT.	INST.	TOTAL
015433	CONTRACTOR EQUIPMENT		98.8	98.8		98.8	98.8		103.7	103.7		103.7	103.7		103.7	103.7		101.9	101.9
0241, 31 - 34	SITE & INFRASTRUCTURE, DEMOLITION	107.8	90.4	95.7	106.9	92.7	97.1	111.3	92.0	98.0	106.7	92.7	97.0	109.4	91.7	97.2	103.1	89.7	93.8
0310	Concrete Forming & Accessories	91.8	37.5	45.5	94.4	68.7	72.5	94.8	54.3	60.3	97.5	61.2	66.6	91.1	61.1	65.5	86.2	70.0	72.4
0320	Concrete Reinforcing	97.3	81.9	91.4	96.9	80.3	88.9	98.9	102.6	100.7	97.1	102.6	99.8	94.7	102.3	98.4	97.5	102.8	100.1
0330	Cast-in-Place Concrete	105.9	65.5	90.9	115.2	84.0	103.6	121.0	83.5	107.1	119.5	87.2	107.5	121.6	86.9	108.7	117.0	87.3	106.0
03	CONCRETE	95.0	57.4	78.5	99.0	77.2	89.4	110.5	74.3	94.6	108.3	78.7	95.3	109.7	78.5	96.0	103.1	82.8	94.2
04	MASONRY	124.6	52.2	80.1	99.4	75.1	84.5	86.4	58.5	69.3	100.5	64.6	78.4	110.7	59.6	79.3	91.8	66.0	76.0
05	METALS	88.1	93.0	89.5	90.7	93.0	91.4	95.6	99.2	96.7	92.8	100.0	94.9	94.2	98.6	95.5	95.3	100.2	96.8
06	WOOD, PLASTICS & COMPOSITES	85.5	34.2	57.3	87.2	64.4	74.6	93.3	51.2	70.1	96.9	57.4	75.2	88.9	57.4	71.5	84.3	68.8	75.8
07	THERMAL & MOISTURE PROTECTION	105.3	55.8	83.8	105.3	77.2	93.0	91.4	62.1	78.7	97.8	65.4	83.7	97.8	66.6	84.2	89.5	75.4	83.3
08	OPENINGS	106.7	46.0	92.5	95.1	72.9	90.0	94.1	62.5	86.7	97.9	65.9	90.4	97.8	65.9	90.4	92.1	74.4	88.0
0920	Plaster & Gypsum Board	101.5	32.5	55.1	103.8	63.6	76.7	88.7	50.0	62.6	96.8	56.3	69.6	90.5	56.3	67.5	85.5	68.1	73.8
0950, 0980	Ceilings & Acoustic Treatment	99.0	32.5	54.2	91.8	63.6	72.8	76.5	50.0	58.6	77.3	56.3	63.2	77.3	56.3	63.2	76.5	68.1	70.8
0960	Flooring	91.6	69.1	85.3	89.6	78.9	86.6	92.9	68.4	86.0	87.6	68.4	82.2	83.8	68.4	79.4	88.3	68.4	82.7
0970, 0990	Wall Finishes & Painting/Coating	92.8	57.2	71.7	83.6	81.3	82.2	89.7	57.9	70.9	96.1	57.9	73.5	96.1	57.9	73.5	89.7	57.9	70.9
09	FINISHES	94.2	42.3	66.1	90.7	70.9	80.0	86.1	55.7	69.6	84.7	60.8	71.8	82.8	60.8	70.9	83.3	67.6	74.8
COVERS	DIVS. 10 - 14, 25, 28, 41, 43, 44, 46	100.0	78.9	95.3	100.0	91.0	98.0	100.0	83.3	96.3	100.0	86.1	96.9	100.0	86.0	96.9	100.0	85.9	96.9
21, 22, 23	FIRE SUPPRESSION, PLUMBING & HVAC	96.5	71.5	86.7	99.9	80.1	92.1	96.4	70.3	86.1	96.5	71.1	86.5	100.0	71.1	88.7	96.4	73.5	87.4
26, 27, 3370	ELECTRICAL, COMMUNICATIONS & UTIL.	100.4	47.2	74.2	95.4	61.6	78.7	104.1	62.7	83.7	95.5	67.9	81.9	92.9	67.9	80.6	101.4	66.9	84.4
MF2018	WEIGHTED AVERAGE	98.2	62.3	83.0	96.8	77.4	88.6	97.9	70.7	86.4	97.3	74.0	87.4	98.6	73.3	87.9	96.1	76.5	87.8

For customer support on your Mechanical Costs with RSMeans Data, call 800.448.8182.

669

KANSAS

DIVISION		FORT SCOTT 667 MAT.	INST.	TOTAL	HAYS 676 MAT.	INST.	TOTAL	HUTCHINSON 675 MAT.	INST.	TOTAL	INDEPENDENCE 673 MAT.	INST.	TOTAL	KANSAS CITY 660-662 MAT.	INST.	TOTAL	LIBERAL 679 MAT.	INST.	TOTAL
015433	CONTRACTOR EQUIPMENT		102.8	102.8		103.7	103.7		103.7	103.7		103.7	103.7		100.4	100.4		103.7	103.7
0241, 31 - 34	SITE & INFRASTRUCTURE, DEMOLITION	99.9	89.7	92.8	111.4	92.2	98.2	90.6	92.7	92.0	110.2	92.7	98.1	94.3	90.2	91.5	111.3	92.3	98.2
0310	Concrete Forming & Accessories	102.8	82.9	85.9	95.2	58.9	64.3	86.0	56.6	61.0	105.8	67.8	73.4	99.4	98.4	98.6	91.5	58.8	63.7
0320	Concrete Reinforcing	96.9	100.8	98.8	94.7	102.6	98.5	94.7	102.6	98.5	94.1	100.8	97.3	94.0	105.3	99.4	96.1	102.3	99.1
0330	Cast-in-Place Concrete	108.5	82.4	98.8	94.2	83.8	90.3	87.3	86.9	87.2	122.1	87.1	109.1	92.9	97.4	94.6	94.2	83.5	90.3
03	CONCRETE	98.4	86.6	93.2	99.8	76.5	89.6	83.0	76.5	80.2	110.8	81.3	97.8	90.7	99.7	94.6	101.7	76.3	90.5
04	MASONRY	92.7	56.1	70.2	109.5	58.6	78.2	100.2	64.5	78.3	97.8	64.5	77.3	93.5	98.1	96.3	108.2	53.6	74.6
05	METALS	95.3	99.0	96.4	92.3	99.8	94.5	92.1	99.0	94.1	92.0	98.8	94.0	103.1	106.2	104.0	92.6	98.7	94.4
06	WOOD, PLASTICS & COMPOSITES	103.8	90.5	96.5	93.9	57.4	73.8	84.0	51.4	66.0	107.3	65.9	84.5	99.6	98.5	99.0	89.6	57.4	71.9
07	THERMAL & MOISTURE PROTECTION	90.5	72.8	82.8	98.1	62.8	82.8	96.6	64.3	82.5	97.8	74.9	87.9	90.3	99.3	94.2	98.3	60.9	82.0
08	OPENINGS	92.1	85.9	90.6	97.8	65.9	90.4	97.8	62.6	89.6	95.9	70.2	89.9	93.2	97.4	94.2	97.9	65.9	90.4
0920	Plaster & Gypsum Board	91.2	90.4	90.7	94.3	56.3	68.7	89.2	50.2	63.0	105.0	65.1	78.1	84.8	98.6	94.1	91.4	56.3	67.8
0950, 0980	Ceilings & Acoustic Treatment	76.5	90.4	85.9	77.3	56.3	63.2	77.3	50.2	59.0	77.3	65.1	69.0	76.5	98.6	91.4	77.3	56.3	63.2
0960	Flooring	102.6	66.9	92.5	86.4	68.4	81.3	80.9	68.4	77.3	92.0	66.9	84.9	82.4	95.3	86.0	84.1	68.4	79.6
0970, 0990	Wall Finishes & Painting/Coating	91.3	77.5	83.1	96.1	57.9	73.5	96.1	57.9	73.5	96.1	57.9	73.5	97.1	102.7	100.4	96.1	57.9	73.5
09	FINISHES	88.5	80.7	84.3	84.5	59.3	70.9	80.3	57.3	67.9	87.3	66.1	75.8	83.6	98.3	91.5	83.7	59.3	70.5
COVERS	DIVS. 10 - 14, 25, 28, 41, 43, 44, 46	100.0	86.8	97.1	100.0	84.0	96.4	100.0	85.4	96.7	100.0	87.0	97.1	100.0	94.5	98.8	100.0	84.0	96.4
21, 22, 23	FIRE SUPPRESSION, PLUMBING & HVAC	96.4	67.4	85.0	96.5	67.4	85.1	96.5	71.1	86.5	96.5	70.4	86.3	99.9	99.1	99.6	96.5	69.3	85.8
26, 27, 3370	ELECTRICAL, COMMUNICATIONS & UTIL.	100.6	67.9	84.5	94.6	67.9	81.4	90.5	61.0	75.9	92.3	72.8	82.7	106.2	98.1	102.2	92.9	67.9	80.6
MF2018	WEIGHTED AVERAGE	96.0	77.1	88.0	96.5	72.0	86.1	92.5	71.9	83.8	97.2	76.0	88.2	97.3	98.6	97.8	96.5	71.7	86.0

KANSAS / KENTUCKY

DIVISION		SALINA 674 MAT.	INST.	TOTAL	TOPEKA 664-666 MAT.	INST.	TOTAL	WICHITA 670-672 MAT.	INST.	TOTAL	ASHLAND 411-412 MAT.	INST.	TOTAL	BOWLING GREEN 421-422 MAT.	INST.	TOTAL	CAMPTON 413-414 MAT.	INST.	TOTAL
015433	CONTRACTOR EQUIPMENT		103.7	103.7		105.0	105.0		106.7	106.7		97.4	97.4		90.6	90.6		96.8	96.8
0241, 31 - 34	SITE & INFRASTRUCTURE, DEMOLITION	99.8	92.2	94.6	98.4	94.4	95.6	96.8	98.5	97.9	114.0	78.8	89.7	78.7	89.4	86.1	87.5	90.7	89.7
0310	Concrete Forming & Accessories	87.8	61.4	65.3	96.7	66.1	70.7	92.9	57.0	62.3	87.2	89.6	89.3	86.5	78.8	79.9	89.3	80.9	82.2
0320	Concrete Reinforcing	94.1	100.5	97.2	93.5	103.8	98.5	92.4	100.4	96.3	92.5	94.2	93.3	89.7	76.8	83.5	90.6	93.0	91.7
0330	Cast-in-Place Concrete	105.6	83.8	97.5	97.6	86.4	93.4	98.7	79.5	91.5	89.1	94.6	91.1	88.4	69.7	81.4	98.6	68.1	87.2
03	CONCRETE	96.8	77.3	88.2	92.3	80.8	87.3	92.4	73.7	84.2	93.1	93.3	93.2	91.6	75.7	84.6	95.0	78.8	87.9
04	MASONRY	125.4	54.9	82.1	87.7	64.3	73.3	97.5	54.1	70.8	90.7	92.0	91.5	92.7	69.2	78.3	89.7	54.7	68.2
05	METALS	94.0	98.9	95.5	99.5	100.0	99.6	96.2	97.5	96.6	95.1	107.7	98.8	96.7	82.8	92.6	96.0	89.8	94.2
06	WOOD, PLASTICS & COMPOSITES	85.4	60.7	71.8	99.9	65.2	80.8	98.7	54.4	74.3	73.4	87.7	81.3	85.7	80.6	82.9	84.2	88.3	86.5
07	THERMAL & MOISTURE PROTECTION	97.2	62.1	82.0	94.0	74.7	85.6	96.2	60.8	80.8	91.2	88.3	90.0	87.6	77.9	83.4	100.2	67.8	86.1
08	OPENINGS	97.8	67.3	90.7	103.5	74.1	96.7	101.3	63.8	92.5	90.6	89.0	90.2	91.4	77.1	88.1	92.6	87.4	91.4
0920	Plaster & Gypsum Board	89.2	59.8	69.4	99.5	64.2	75.7	95.7	53.1	67.0	59.3	87.6	78.4	87.6	80.6	82.9	87.6	87.6	87.6
0950, 0980	Ceilings & Acoustic Treatment	77.3	59.8	65.5	87.2	64.2	71.7	88.0	53.1	64.5	77.0	87.6	84.2	81.6	80.6	80.9	81.6	87.6	85.7
0960	Flooring	82.2	68.4	78.3	98.2	68.4	89.8	94.3	68.4	87.0	73.8	81.3	75.9	89.5	61.9	81.8	91.6	64.3	83.9
0970, 0990	Wall Finishes & Painting/Coating	96.1	57.9	73.5	92.8	67.7	78.0	94.9	57.9	73.0	91.6	90.4	90.9	90.3	67.6	76.9	90.3	53.9	68.8
09	FINISHES	81.1	61.3	70.6	93.9	65.7	78.6	90.3	57.7	72.7	76.1	88.1	82.6	87.1	74.6	80.3	87.9	75.9	81.4
COVERS	DIVS. 10 - 14, 25, 28, 41, 43, 44, 46	100.0	85.3	96.7	100.0	80.3	95.6	100.0	85.1	96.7	100.0	88.2	97.4	100.0	84.4	96.5	100.0	47.8	88.4
21, 22, 23	FIRE SUPPRESSION, PLUMBING & HVAC	100.0	70.5	88.5	100.0	73.2	89.5	99.8	71.0	88.5	96.2	85.0	91.8	99.9	77.0	90.9	96.4	76.4	88.6
26, 27, 3370	ELECTRICAL, COMMUNICATIONS & UTIL.	92.7	73.8	83.4	105.1	71.9	88.7	95.9	73.8	85.0	91.9	88.3	90.1	94.6	74.7	84.8	92.1	88.2	90.2
MF2018	WEIGHTED AVERAGE	97.2	73.5	87.2	98.6	76.5	89.3	97.0	72.6	86.7	93.2	89.7	91.7	94.6	77.2	87.3	94.4	78.0	87.5

KENTUCKY

DIVISION		CORBIN 407-409 MAT.	INST.	TOTAL	COVINGTON 410 MAT.	INST.	TOTAL	ELIZABETHTOWN 427 MAT.	INST.	TOTAL	FRANKFORT 406 MAT.	INST.	TOTAL	HAZARD 417-418 MAT.	INST.	TOTAL	HENDERSON 424 MAT.	INST.	TOTAL
015433	CONTRACTOR EQUIPMENT		96.8	96.8		100.6	100.6		90.6	90.6		100.1	100.1		96.8	96.8		110.3	110.3
0241, 31 - 34	SITE & INFRASTRUCTURE, DEMOLITION	91.6	91.2	91.3	82.7	101.9	95.9	72.9	89.3	84.3	90.0	96.8	94.7	85.3	91.8	89.7	81.7	114.6	104.5
0310	Concrete Forming & Accessories	85.0	76.0	77.3	84.9	70.8	72.8	81.3	72.1	73.5	102.0	78.5	82.0	85.8	81.8	82.4	92.3	77.1	79.3
0320	Concrete Reinforcing	89.6	92.3	90.9	89.3	78.6	84.1	90.2	83.7	87.0	94.0	84.0	89.2	91.0	92.5	91.7	89.8	84.4	87.2
0330	Cast-in-Place Concrete	91.0	72.5	84.1	94.3	79.6	88.9	79.9	67.6	75.3	91.1	77.0	85.8	94.8	69.6	85.4	78.1	84.5	80.5
03	CONCRETE	84.2	78.0	81.5	93.4	76.1	85.8	83.5	73.1	79.0	87.1	79.3	83.6	91.8	79.7	86.5	88.7	81.3	85.5
04	MASONRY	81.9	60.0	68.4	104.0	72.0	84.3	77.4	61.8	67.8	78.6	74.2	75.9	88.6	56.8	69.0	96.0	78.7	85.4
05	METALS	90.9	89.3	90.5	93.9	88.1	92.2	95.9	85.3	92.7	93.1	86.4	91.1	96.0	89.8	94.2	86.8	87.0	86.9
06	WOOD, PLASTICS & COMPOSITES	72.1	77.6	75.1	81.8	68.3	74.3	80.6	73.9	76.9	101.5	77.7	88.4	80.9	88.3	85.0	88.4	75.8	81.5
07	THERMAL & MOISTURE PROTECTION	103.8	69.5	88.9	101.1	72.1	88.5	87.0	68.5	79.0	101.8	75.7	90.5	100.1	69.5	86.8	99.3	82.8	92.1
08	OPENINGS	87.3	75.0	83.4	94.5	71.8	89.2	91.4	73.0	87.1	98.9	78.4	94.1	93.0	87.2	91.6	89.6	78.5	87.0
0920	Plaster & Gypsum Board	88.4	76.6	80.5	68.2	68.0	68.1	86.3	73.7	77.8	91.6	76.6	81.5	86.3	87.6	87.2	90.9	74.8	80.0
0950, 0980	Ceilings & Acoustic Treatment	74.7	76.6	76.0	84.0	68.0	73.2	81.6	73.7	76.2	90.5	76.6	81.1	81.6	87.6	85.7	76.7	74.8	75.4
0960	Flooring	88.4	64.3	81.6	66.3	83.5	71.1	86.9	73.8	83.2	98.9	78.6	93.2	89.8	64.3	82.6	93.2	76.0	88.4
0970, 0990	Wall Finishes & Painting/Coating	87.2	62.0	72.3	85.2	71.1	76.9	90.3	68.1	77.2	94.2	91.7	92.8	90.3	53.9	68.8	90.3	86.7	88.2
09	FINISHES	83.0	72.5	77.3	77.8	72.7	75.0	85.8	71.6	78.1	93.2	80.2	86.2	87.0	76.4	81.3	87.1	77.9	82.1
COVERS	DIVS. 10 - 14, 25, 28, 41, 43, 44, 46	100.0	89.8	97.7	100.0	85.9	96.8	100.0	70.0	93.3	100.0	58.5	90.8	100.0	48.5	88.5	100.0	55.8	90.1
21, 22, 23	FIRE SUPPRESSION, PLUMBING & HVAC	96.5	74.2	87.7	97.0	76.0	88.8	96.6	77.9	89.3	99.9	81.0	92.5	96.4	77.5	89.0	96.6	79.3	89.8
26, 27, 3370	ELECTRICAL, COMMUNICATIONS & UTIL.	90.0	88.2	89.1	95.6	72.4	84.2	91.9	77.9	85.0	100.6	77.9	89.4	92.1	88.2	90.2	94.0	77.9	86.1
MF2018	WEIGHTED AVERAGE	90.8	78.0	85.4	94.3	77.8	87.3	91.3	75.6	84.7	95.4	80.3	89.0	93.8	78.7	87.4	92.2	81.9	87.8

KENTUCKY

		LEXINGTON 403 - 405			LOUISVILLE 400 - 402			OWENSBORO 423			PADUCAH 420			PIKEVILLE 415 - 416			SOMERSET 425 - 426		
DIVISION		MAT.	INST.	TOTAL	MAT.	INST.	TOTAL	MAT.	INST.	TOTAL	MAT.	INST.	TOTAL	MAT.	INST.	TOTAL	MAT.	INST.	TOTAL
015433	CONTRACTOR EQUIPMENT		96.8	96.8		93.7	93.7		110.3	110.3		110.3	110.3		97.4	97.4		96.8	96.8
0241, 31 - 34	SITE & INFRASTRUCTURE, DEMOLITION	93.8	92.8	93.1	87.8	94.4	92.3	91.8	115.3	108.0	84.4	114.8	105.4	125.2	78.0	92.6	77.9	91.2	87.1
0310	Concrete Forming & Accessories	96.5	73.7	77.1	93.5	79.1	81.2	90.8	76.8	78.9	88.8	81.2	82.3	96.0	85.6	87.2	86.8	76.6	78.1
0320	Concrete Reinforcing	98.4	84.0	91.4	94.8	84.2	89.6	89.8	77.3	83.8	90.4	82.5	86.6	93.0	94.1	93.5	90.2	92.5	91.3
0330	Cast-in-Place Concrete	93.1	84.4	89.9	89.3	70.7	82.4	91.1	83.9	88.4	83.2	81.1	82.4	97.9	89.4	94.8	78.1	87.9	81.7
03	CONCRETE	86.9	79.7	83.8	85.8	77.4	82.1	100.7	79.8	91.5	93.3	81.7	88.2	106.8	89.7	99.3	78.6	83.7	80.8
04	MASONRY	80.6	71.1	74.8	78.7	72.0	74.6	88.8	78.6	82.5	91.5	79.2	83.9	88.5	83.2	85.3	83.3	63.2	71.0
05	METALS	93.3	86.7	91.3	95.5	85.5	92.5	88.3	84.7	87.3	85.3	87.5	86.0	95.0	107.5	98.7	95.9	90.0	94.2
06	WOOD, PLASTICS & COMPOSITES	87.3	71.6	78.6	88.8	80.8	84.4	86.3	75.8	80.5	84.0	81.4	82.6	83.3	87.7	85.7	81.5	77.6	79.3
07	THERMAL & MOISTURE PROTECTION	104.1	73.9	90.9	101.0	75.7	90.0	100.1	76.6	89.9	99.4	82.6	92.1	92.1	80.5	87.0	99.3	71.5	87.2
08	OPENINGS	87.5	74.6	84.5	88.8	76.8	86.0	89.6	76.7	86.6	89.0	81.1	87.1	91.2	85.7	89.9	92.1	77.1	88.6
0920	Plaster & Gypsum Board	97.7	70.4	79.4	93.7	80.6	84.9	89.3	74.8	79.5	88.6	80.6	83.2	63.1	87.6	79.6	86.3	76.6	79.8
0950, 0980	Ceilings & Acoustic Treatment	77.9	70.4	72.9	82.8	80.6	81.3	76.7	74.8	75.4	76.7	80.6	79.3	77.0	87.6	84.2	81.6	76.6	78.2
0960	Flooring	93.4	64.3	85.2	95.0	62.9	86.0	92.6	61.9	84.0	91.5	76.0	87.2	78.1	64.3	74.2	90.1	64.3	82.8
0970, 0990	Wall Finishes & Painting/Coating	87.2	79.7	82.8	93.9	68.1	78.7	90.3	86.7	88.2	90.3	73.1	80.2	91.6	91.7	91.7	90.3	68.1	77.2
09	FINISHES	86.4	72.0	78.6	90.8	74.9	82.2	87.3	74.7	80.5	86.5	79.4	82.7	78.7	82.6	80.8	86.4	73.3	79.3
COVERS	DIVS. 10 - 14, 25, 28, 41, 43, 44, 46	100.0	92.1	98.2	100.0	90.5	97.9	100.0	93.0	98.4	100.0	91.1	98.0	100.0	48.2	88.4	100.0	89.2	97.6
21, 22, 23	FIRE SUPPRESSION, PLUMBING & HVAC	100.0	76.4	90.8	99.9	80.0	92.1	99.9	77.1	91.0	96.6	78.2	89.4	96.2	82.3	90.7	96.6	74.0	87.7
26, 27, 3370	ELECTRICAL, COMMUNICATIONS & UTIL.	92.5	74.7	83.7	96.4	77.9	87.3	94.1	73.7	84.0	96.3	77.3	86.9	94.8	88.3	91.6	92.5	88.2	90.4
MF2018	WEIGHTED AVERAGE	93.1	78.1	86.7	93.9	79.6	87.8	94.7	81.0	88.9	92.4	83.2	88.6	95.8	85.4	91.4	91.6	79.5	86.5

LOUISIANA

		ALEXANDRIA 713 - 714			BATON ROUGE 707 - 708			HAMMOND 704			LAFAYETTE 705			LAKE CHARLES 706			MONROE 712		
DIVISION		MAT.	INST.	TOTAL	MAT.	INST.	TOTAL	MAT.	INST.	TOTAL	MAT.	INST.	TOTAL	MAT.	INST.	TOTAL	MAT.	INST.	TOTAL
015433	CONTRACTOR EQUIPMENT		90.9	90.9		91.9	91.9		89.3	89.3		89.3	89.3		88.8	88.8		90.9	90.9
0241, 31 - 34	SITE & INFRASTRUCTURE, DEMOLITION	98.7	87.7	91.1	102.1	91.2	94.6	101.2	85.7	90.5	102.5	87.9	92.4	103.2	87.1	92.0	98.7	87.6	91.1
0310	Concrete Forming & Accessories	77.8	60.0	62.6	99.0	71.8	75.8	79.0	54.9	58.5	95.4	67.8	71.9	96.4	67.8	72.0	77.4	59.4	62.1
0320	Concrete Reinforcing	92.2	54.3	73.9	93.3	54.3	74.5	92.1	54.3	73.8	93.4	54.3	74.5	93.4	54.3	74.5	91.2	54.3	73.3
0330	Cast-in-Place Concrete	89.0	66.2	80.5	88.8	73.1	82.9	87.6	62.6	78.3	87.1	66.1	79.3	91.7	66.1	82.2	89.0	64.1	79.7
03	CONCRETE	82.8	62.0	73.6	85.5	69.8	78.6	85.8	58.3	73.7	86.8	65.4	77.4	89.0	65.4	78.6	82.6	61.0	73.1
04	MASONRY	109.3	64.3	81.6	91.7	66.7	76.3	95.3	60.0	73.6	95.3	66.6	77.7	94.7	66.6	77.5	103.9	63.1	78.8
05	METALS	92.6	70.9	86.2	96.9	74.0	90.1	88.4	69.3	82.8	87.7	69.6	82.4	87.7	69.6	82.4	92.6	70.8	86.2
06	WOOD, PLASTICS & COMPOSITES	89.2	58.8	72.5	103.0	73.2	86.6	83.0	54.3	67.2	103.7	68.0	84.1	101.9	68.0	83.3	88.6	58.8	72.2
07	THERMAL & MOISTURE PROTECTION	99.0	65.6	84.5	97.7	69.2	85.3	96.7	62.9	82.0	97.3	67.4	84.3	97.1	67.2	84.1	99.0	64.8	84.1
08	OPENINGS	110.8	58.6	98.6	97.9	70.4	91.5	96.2	54.9	86.6	99.8	62.2	91.0	99.8	62.2	91.0	110.7	57.1	98.2
0920	Plaster & Gypsum Board	83.6	58.1	66.4	101.0	72.7	82.0	100.0	53.4	68.7	108.6	67.6	81.0	108.6	67.6	81.0	83.3	58.1	66.3
0950, 0980	Ceilings & Acoustic Treatment	87.1	58.1	67.6	92.9	72.7	79.3	96.9	53.4	67.6	94.4	67.6	76.3	95.3	67.6	76.6	87.1	58.1	67.6
0960	Flooring	85.1	68.2	80.4	93.0	68.2	86.0	89.9	68.2	83.8	98.2	68.2	89.8	98.2	68.2	89.8	84.7	68.2	80.1
0970, 0990	Wall Finishes & Painting/Coating	88.8	59.7	71.6	94.7	59.7	74.0	97.7	59.9	75.3	97.7	59.7	75.2	97.7	59.7	75.2	88.8	59.7	71.6
09	FINISHES	80.9	61.2	70.2	92.2	70.3	80.3	90.4	57.5	72.6	93.5	67.2	79.3	93.7	67.2	79.4	80.7	60.9	70.0
COVERS	DIVS. 10 - 14, 25, 28, 41, 43, 44, 46	100.0	79.1	95.3	100.0	84.2	96.5	100.0	80.7	95.7	100.0	83.3	96.3	100.0	83.3	96.3	100.0	78.9	95.3
21, 22, 23	FIRE SUPPRESSION, PLUMBING & HVAC	100.0	63.2	85.6	99.9	64.1	85.9	96.7	61.0	82.7	100.2	64.2	86.0	100.2	64.5	86.2	100.0	61.9	85.0
26, 27, 3370	ELECTRICAL, COMMUNICATIONS & UTIL.	94.3	61.4	78.1	101.7	58.0	80.2	97.0	69.6	83.5	98.1	63.8	81.2	97.7	65.9	82.0	95.9	57.3	76.9
MF2018	WEIGHTED AVERAGE	95.9	65.6	83.1	96.6	69.3	85.1	93.7	64.4	81.3	95.4	67.9	83.8	95.6	68.2	84.1	95.8	64.4	82.5

LOUISIANA / MAINE

		NEW ORLEANS 700 - 701			SHREVEPORT 710 - 711			THIBODAUX 703			AUGUSTA 043			BANGOR 044			BATH 045		
DIVISION		MAT.	INST.	TOTAL	MAT.	INST.	TOTAL	MAT.	INST.	TOTAL	MAT.	INST.	TOTAL	MAT.	INST.	TOTAL	MAT.	INST.	TOTAL
015433	CONTRACTOR EQUIPMENT		88.6	88.6		93.9	93.9		89.3	89.3		98.4	98.4		95.7	95.7		95.7	95.7
0241, 31 - 34	SITE & INFRASTRUCTURE, DEMOLITION	104.3	93.9	97.1	101.1	92.3	95.0	103.6	87.6	92.6	88.2	98.0	95.0	90.6	95.5	94.0	88.3	94.2	92.4
0310	Concrete Forming & Accessories	95.7	68.9	72.8	92.9	60.3	65.1	90.2	64.5	68.2	98.6	78.6	81.5	93.0	76.5	79.0	89.1	78.4	80.0
0320	Concrete Reinforcing	92.9	54.4	74.3	92.5	54.3	74.0	92.1	54.3	73.8	102.6	81.1	92.2	94.1	81.1	87.8	93.1	81.1	87.3
0330	Cast-in-Place Concrete	90.7	69.7	82.9	92.0	65.2	82.0	94.1	64.3	83.0	89.8	115.3	99.3	70.9	114.1	87.0	70.9	114.3	87.1
03	CONCRETE	94.4	66.6	82.2	86.2	61.7	75.4	90.6	63.2	78.5	95.5	92.3	94.1	88.5	91.0	89.6	88.6	91.9	90.1
04	MASONRY	99.8	62.6	77.0	94.5	63.4	75.4	119.9	62.8	84.8	101.0	94.1	96.8	115.8	93.7	102.2	122.5	94.1	105.0
05	METALS	99.1	62.6	88.4	96.7	70.2	88.9	88.4	69.5	82.8	105.4	92.3	101.6	93.7	93.0	93.5	92.1	92.9	92.3
06	WOOD, PLASTICS & COMPOSITES	100.3	71.1	84.2	98.0	59.7	76.9	90.5	65.5	76.8	93.5	75.6	83.6	90.7	73.0	80.9	85.1	75.4	79.8
07	THERMAL & MOISTURE PROTECTION	95.0	60.3	80.3	97.4	65.6	83.6	96.9	65.2	83.1	111.0	100.6	106.5	109.0	100.0	105.1	108.9	100.1	105.1
08	OPENINGS	98.8	64.2	90.7	104.7	56.6	93.5	100.7	56.9	90.5	102.7	78.5	97.1	96.4	77.1	91.9	96.4	78.4	92.2
0920	Plaster & Gypsum Board	96.5	70.5	79.0	91.5	58.8	69.5	101.6	65.0	77.0	109.7	74.5	86.0	116.8	71.9	86.6	112.0	74.5	86.8
0950, 0980	Ceilings & Acoustic Treatment	97.5	70.5	79.3	92.8	58.8	69.9	96.9	65.0	75.4	107.1	74.5	85.1	89.6	71.9	77.7	88.6	74.5	79.1
0960	Flooring	103.6	68.2	93.7	91.2	68.2	84.8	95.8	68.2	88.0	89.9	110.6	95.7	82.5	110.6	90.4	80.9	110.6	89.2
0970, 0990	Wall Finishes & Painting/Coating	103.6	60.6	78.2	87.7	59.7	71.1	98.9	59.9	75.8	96.0	88.9	91.8	90.6	88.9	89.6	90.6	88.9	89.6
09	FINISHES	96.6	68.2	81.2	87.7	61.5	73.5	92.6	64.8	77.6	95.8	84.7	89.8	89.7	83.0	86.1	88.3	84.6	86.3
COVERS	DIVS. 10 - 14, 25, 28, 41, 43, 44, 46	100.0	86.5	97.0	100.0	82.6	96.1	100.0	81.8	95.9	100.0	98.0	99.6	100.0	99.6	99.9	100.0	100.9	100.2
21, 22, 23	FIRE SUPPRESSION, PLUMBING & HVAC	100.1	62.5	85.4	99.9	62.8	85.3	96.7	62.5	83.2	100.0	74.3	89.9	100.3	74.1	90.0	96.7	74.3	87.9
26, 27, 3370	ELECTRICAL, COMMUNICATIONS & UTIL.	102.1	70.6	86.6	101.8	65.5	83.9	95.7	69.6	82.8	101.6	76.8	89.3	100.0	69.4	84.9	98.2	76.8	87.6
MF2018	WEIGHTED AVERAGE	99.0	68.5	86.1	97.1	66.3	84.1	96.0	67.0	83.8	100.4	85.6	94.2	97.1	83.9	91.5	95.9	85.4	91.4

For customer support on your Mechanical Costs with RSMeans Data, call 800.448.8182.

671

MAINE

DIVISION		HOULTON 047 MAT.	INST.	TOTAL	KITTERY 039 MAT.	INST.	TOTAL	LEWISTON 042 MAT.	INST.	TOTAL	MACHIAS 046 MAT.	INST.	TOTAL	PORTLAND 040-041 MAT.	INST.	TOTAL	ROCKLAND 048 MAT.	INST.	TOTAL
015433	CONTRACTOR EQUIPMENT		95.7	95.7		95.7	95.7		95.7	95.7		95.7	95.7		99.2	99.2		95.7	95.7
0241, 31 - 34	SITE & INFRASTRUCTURE, DEMOLITION	90.2	94.2	93.0	79.5	94.3	89.7	88.2	95.5	93.3	89.5	94.2	92.8	87.4	100.6	96.5	86.0	94.2	91.7
0310	Concrete Forming & Accessories	96.7	78.4	81.1	88.8	78.7	80.2	98.2	76.6	79.8	93.8	78.4	80.7	99.8	76.8	80.2	94.9	78.4	80.9
0320	Concrete Reinforcing	94.1	81.1	87.8	89.7	81.1	85.5	115.6	81.1	98.9	94.1	81.1	87.8	106.5	81.1	94.2	94.1	81.1	87.8
0330	Cast-in-Place Concrete	70.9	113.2	86.7	70.7	114.4	87.1	72.4	114.1	87.9	70.9	114.3	87.1	85.8	115.2	96.8	72.4	114.3	86.8
03	CONCRETE	89.5	91.5	90.4	82.1	92.1	86.5	88.1	91.0	89.4	89.0	91.9	90.3	91.5	91.4	91.5	86.0	91.9	88.6
04	MASONRY	97.9	94.1	95.6	106.5	94.1	98.9	98.5	93.7	95.5	97.9	94.1	95.6	104.7	93.7	98.0	92.5	94.1	93.5
05	METALS	92.3	92.8	92.4	86.8	93.1	88.6	97.1	93.0	95.9	92.3	92.8	92.5	103.0	92.5	99.9	92.2	92.9	92.4
06	WOOD, PLASTICS & COMPOSITES	94.5	75.4	84.0	88.2	75.4	81.2	96.3	73.0	83.5	91.5	75.4	82.7	95.8	73.1	83.3	92.4	75.4	83.1
07	THERMAL & MOISTURE PROTECTION	109.1	100.1	105.1	107.8	100.1	104.4	108.7	100.0	104.9	109.0	100.1	105.1	113.6	100.6	107.9	108.6	100.1	104.9
08	OPENINGS	96.5	78.4	92.3	96.1	82.0	92.8	99.3	77.1	94.1	96.5	78.4	92.3	96.8	77.2	92.2	96.4	78.4	92.2
0920	Plaster & Gypsum Board	119.0	74.5	89.0	106.1	74.5	84.8	121.8	71.9	88.2	117.4	74.5	88.5	108.0	71.9	83.7	117.4	74.5	88.5
0950, 0980	Ceilings & Acoustic Treatment	88.6	74.5	79.1	101.8	74.5	83.4	99.3	71.9	80.8	88.6	74.5	79.1	107.4	71.9	83.5	88.6	74.5	79.1
0960	Flooring	83.7	110.6	91.2	86.6	110.6	93.4	85.2	110.6	92.3	82.9	110.6	90.7	89.6	110.6	95.5	83.3	110.6	91.0
0970, 0990	Wall Finishes & Painting/Coating	90.6	88.9	89.6	81.8	102.1	93.8	90.6	88.9	89.6	90.6	88.9	89.6	94.6	88.9	91.2	90.6	88.9	89.6
09	FINISHES	90.2	84.6	87.2	91.6	86.0	88.6	92.6	83.0	87.4	89.7	84.6	86.9	96.2	83.2	89.1	89.4	84.6	86.8
COVERS	DIVS. 10 - 14, 25, 28, 41, 43, 44, 46	100.0	97.6	99.5	100.0	100.9	100.2	100.0	99.7	99.9	100.0	97.6	99.5	100.0	100.1	100.0	100.0	100.9	100.2
21, 22, 23	FIRE SUPPRESSION, PLUMBING & HVAC	96.7	74.3	87.9	96.7	80.4	90.3	100.3	74.1	90.0	96.7	74.3	87.9	100.0	74.1	89.9	96.7	74.3	87.9
26, 27, 3370	ELECTRICAL, COMMUNICATIONS & UTIL.	101.9	76.8	89.5	90.3	76.8	83.6	101.9	76.7	89.5	101.9	76.8	89.5	105.0	76.7	91.0	101.8	76.8	89.5
MF2018	WEIGHTED AVERAGE	95.6	85.2	91.2	92.6	87.0	90.2	97.5	84.9	92.2	95.4	85.3	91.1	99.6	85.4	93.6	94.6	85.4	90.7

MAINE / MARYLAND

DIVISION		WATERVILLE 049 MAT.	INST.	TOTAL	ANNAPOLIS 214 MAT.	INST.	TOTAL	BALTIMORE 210-212 MAT.	INST.	TOTAL	COLLEGE PARK 207-208 MAT.	INST.	TOTAL	CUMBERLAND 215 MAT.	INST.	TOTAL	EASTON 216 MAT.	INST.	TOTAL
015433	CONTRACTOR EQUIPMENT		95.7	95.7		104.5	104.5		104.1	104.1		107.8	107.8		102.4	102.4		102.4	102.4
0241, 31 - 34	SITE & INFRASTRUCTURE, DEMOLITION	90.0	94.2	92.9	100.0	93.5	95.5	100.1	97.7	98.4	100.1	93.6	95.6	92.0	90.3	90.8	98.8	87.7	91.1
0310	Concrete Forming & Accessories	88.6	78.4	79.9	101.0	78.1	81.5	98.4	76.2	79.5	85.1	76.2	77.5	92.3	81.9	83.5	90.4	73.6	76.1
0320	Concrete Reinforcing	94.1	81.1	87.8	108.6	94.9	102.0	101.2	87.3	94.5	103.5	94.0	98.9	93.7	87.4	90.6	92.9	87.1	90.1
0330	Cast-in-Place Concrete	70.9	114.3	87.1	114.1	79.4	101.2	113.8	77.1	100.1	109.4	76.4	97.2	98.0	82.9	92.4	108.8	64.8	92.4
03	CONCRETE	90.2	91.9	90.9	101.4	82.7	93.2	108.2	79.4	95.5	105.2	80.9	94.5	91.4	84.5	88.4	99.7	74.2	88.5
04	MASONRY	109.8	94.1	100.2	94.8	76.6	83.6	104.7	73.1	85.2	111.1	74.6	88.7	99.3	85.4	90.8	113.5	58.6	79.7
05	METALS	92.3	92.8	92.4	105.5	105.2	105.4	103.8	96.6	101.7	90.2	108.3	95.5	100.7	104.2	101.8	101.0	99.7	100.6
06	WOOD, PLASTICS & COMPOSITES	84.6	75.4	79.5	97.1	77.2	86.1	102.5	76.7	88.3	77.3	75.7	76.4	86.7	81.0	83.6	84.5	80.8	82.5
07	THERMAL & MOISTURE PROTECTION	109.1	100.1	105.2	103.5	83.4	94.8	102.7	80.7	93.2	105.3	81.7	95.0	100.9	81.9	92.6	101.0	73.3	89.0
08	OPENINGS	96.5	78.4	92.3	102.3	83.0	97.8	100.5	80.7	95.9	92.7	82.4	90.3	97.6	84.6	94.6	96.0	83.0	92.9
0920	Plaster & Gypsum Board	112.0	74.5	86.8	94.4	76.8	82.5	100.3	76.2	84.0	95.3	75.2	81.8	99.9	80.8	87.0	99.9	80.6	86.9
0950, 0980	Ceilings & Acoustic Treatment	88.6	74.5	79.1	89.1	76.8	80.8	100.5	76.2	84.1	115.1	75.2	88.2	99.0	80.8	86.7	99.0	80.6	86.6
0960	Flooring	80.6	110.6	89.0	90.7	79.5	87.6	92.8	75.7	88.0	87.9	78.4	85.2	85.8	93.1	87.9	85.0	75.7	82.4
0970, 0990	Wall Finishes & Painting/Coating	90.6	88.9	89.6	99.2	72.7	83.5	101.3	75.3	85.9	104.8	72.5	85.7	98.2	84.7	90.1	98.2	72.5	83.0
09	FINISHES	88.4	84.6	86.4	88.5	77.3	82.4	97.0	75.7	85.5	95.6	75.6	84.8	93.3	84.2	88.4	93.5	74.4	83.2
COVERS	DIVS. 10 - 14, 25, 28, 41, 43, 44, 46	100.0	97.5	99.4	100.0	90.3	97.8	100.0	86.8	97.1	100.0	86.7	97.0	100.0	91.3	98.1	100.0	82.9	96.2
21, 22, 23	FIRE SUPPRESSION, PLUMBING & HVAC	96.7	74.3	87.9	100.0	87.0	94.9	100.1	82.0	93.0	96.5	87.4	93.0	96.4	72.0	86.8	96.4	72.8	87.1
26, 27, 3370	ELECTRICAL, COMMUNICATIONS & UTIL.	101.9	76.8	89.5	101.6	88.9	95.3	99.6	86.9	93.3	97.3	101.9	99.5	98.2	80.8	89.6	97.7	62.3	80.3
MF2018	WEIGHTED AVERAGE	96.0	85.2	91.4	100.4	86.3	94.4	101.8	83.2	93.9	97.3	87.7	93.2	96.9	83.8	91.4	98.6	74.8	88.5

MARYLAND / MASSACHUSETTS

DIVISION		ELKTON 219 MAT.	INST.	TOTAL	HAGERSTOWN 217 MAT.	INST.	TOTAL	SALISBURY 218 MAT.	INST.	TOTAL	SILVER SPRING 209 MAT.	INST.	TOTAL	WALDORF 206 MAT.	INST.	TOTAL	BOSTON 020-022, 024 MAT.	INST.	TOTAL
015433	CONTRACTOR EQUIPMENT		102.4	102.4		102.4	102.4		102.4	102.4		99.7	99.7		99.7	99.7		103.0	103.0
0241, 31 - 34	SITE & INFRASTRUCTURE, DEMOLITION	86.3	89.0	88.1	90.6	90.4	90.4	98.8	87.6	91.1	88.9	85.7	86.7	95.1	85.7	88.6	92.7	102.1	99.2
0310	Concrete Forming & Accessories	96.2	92.3	92.9	91.5	78.8	80.7	104.5	50.2	58.2	93.0	75.5	78.1	99.7	75.4	79.0	105.1	139.7	134.6
0320	Concrete Reinforcing	92.9	117.3	104.7	93.7	87.4	90.6	92.9	65.2	79.5	102.3	93.9	98.3	103.0	93.9	98.6	117.7	156.9	136.7
0330	Cast-in-Place Concrete	88.1	72.2	82.2	93.3	82.9	89.5	108.8	62.9	91.7	112.1	76.8	98.9	125.6	76.7	107.4	99.2	141.8	115.1
03	CONCRETE	84.0	90.6	86.9	87.7	83.1	85.7	100.5	59.2	82.4	102.8	80.5	93.0	113.3	80.5	98.9	105.4	142.7	121.8
04	MASONRY	98.6	67.0	79.2	105.1	85.4	93.0	112.9	55.2	77.4	110.6	75.0	88.7	94.7	75.0	82.6	112.0	146.3	133.1
05	METALS	101.0	114.0	104.8	100.9	104.3	101.9	101.0	90.9	98.1	94.8	104.3	97.6	94.8	104.3	97.5	102.8	136.1	112.6
06	WOOD, PLASTICS & COMPOSITES	91.9	101.3	97.0	85.8	76.7	80.8	102.6	51.4	74.4	84.0	75.0	79.0	90.8	75.0	82.1	103.8	140.2	123.8
07	THERMAL & MOISTURE PROTECTION	100.5	79.4	91.3	101.2	83.0	93.3	101.3	68.5	87.0	108.4	86.5	98.9	109.0	86.5	99.2	107.2	137.1	120.2
08	OPENINGS	96.0	102.1	97.4	95.9	81.2	92.5	96.2	60.8	87.9	84.9	82.1	84.2	85.5	81.5	84.5	99.9	147.3	111.0
0920	Plaster & Gypsum Board	102.8	101.7	102.0	99.9	76.4	84.1	109.4	50.4	69.7	101.9	75.2	83.9	104.8	75.2	84.8	109.0	141.0	130.6
0950, 0980	Ceilings & Acoustic Treatment	99.0	101.7	100.8	101.7	76.4	84.6	99.0	50.4	66.2	123.2	75.2	90.8	123.2	75.2	90.8	100.5	141.0	127.8
0960	Flooring	87.2	75.7	84.0	85.4	93.1	87.6	90.7	75.7	86.5	93.8	78.4	89.5	97.2	78.4	92.0	90.5	163.5	111.0
0970, 0990	Wall Finishes & Painting/Coating	98.2	72.5	83.0	98.2	72.5	83.0	98.2	72.5	83.0	112.7	72.5	88.9	112.7	72.5	88.9	94.3	160.4	133.4
09	FINISHES	93.7	88.5	90.9	93.5	80.4	86.4	96.4	56.2	74.6	95.8	75.0	84.5	97.5	75.1	85.4	98.7	147.0	124.8
COVERS	DIVS. 10 - 14, 25, 28, 41, 43, 44, 46	100.0	56.7	90.4	100.0	90.8	98.0	100.0	76.5	94.8	100.0	85.2	96.7	100.0	83.1	96.2	100.0	119.1	104.3
21, 22, 23	FIRE SUPPRESSION, PLUMBING & HVAC	96.4	78.7	89.5	99.9	83.1	93.3	96.4	70.7	86.3	96.5	87.9	93.1	96.5	87.8	93.1	100.1	127.4	110.8
26, 27, 3370	ELECTRICAL, COMMUNICATIONS & UTIL.	99.4	86.9	93.2	98.0	80.8	89.5	96.6	60.1	78.6	94.6	101.9	98.2	92.2	101.9	97.0	102.6	129.5	115.9
MF2018	WEIGHTED AVERAGE	95.9	86.2	91.8	97.4	85.3	92.3	99.0	66.9	85.4	96.5	86.7	92.4	97.3	86.6	92.8	101.9	134.0	115.5

For customer support on your Mechanical Costs with RSMeans Data, call 800.448.8182.

MASSACHUSETTS

DIVISION		BROCKTON 023 MAT.	INST.	TOTAL	BUZZARDS BAY 025 MAT.	INST.	TOTAL	FALL RIVER 027 MAT.	INST.	TOTAL	FITCHBURG 014 MAT.	INST.	TOTAL	FRAMINGHAM 017 MAT.	INST.	TOTAL	GREENFIELD 013 MAT.	INST.	TOTAL
015433	CONTRACTOR EQUIPMENT		98.1	98.1		98.1	98.1		98.9	98.9		95.7	95.7		97.4	97.4		95.7	95.7
0241, 31 - 34	SITE & INFRASTRUCTURE, DEMOLITION	91.1	98.0	95.8	81.2	98.1	92.9	90.3	98.0	95.6	82.7	97.8	93.1	79.5	97.8	92.1	86.4	97.3	93.9
0310	Concrete Forming & Accessories	100.7	121.1	118.1	98.5	120.6	117.4	100.7	120.8	117.8	94.0	118.5	114.9	101.4	121.3	118.4	92.4	117.7	114.0
0320	Concrete Reinforcing	109.7	142.7	125.7	88.0	122.3	104.6	109.7	122.3	115.8	87.9	142.5	114.3	87.9	142.8	114.4	91.5	126.5	108.4
0330	Cast-in-Place Concrete	88.1	136.6	106.2	73.2	136.4	96.7	85.2	136.9	104.4	78.4	136.5	100.0	78.4	136.7	100.1	80.5	122.2	96.6
03	CONCRETE	95.1	129.8	110.3	80.7	126.0	100.6	93.8	126.2	108.0	79.5	128.3	101.0	82.3	130.0	103.2	83.0	120.3	99.4
04	MASONRY	105.4	135.1	123.6	97.9	135.1	120.8	105.9	135.1	123.8	98.2	131.9	118.9	104.7	136.1	124.0	102.7	119.1	112.8
05	METALS	98.8	125.6	106.7	93.6	117.0	100.5	98.8	117.3	104.3	95.3	122.1	103.2	95.3	125.8	104.3	97.8	113.8	102.5
06	WOOD, PLASTICS & COMPOSITES	98.1	119.7	109.9	94.7	119.7	108.4	98.1	119.7	110.1	94.5	116.5	106.6	101.0	119.4	111.2	92.5	119.7	107.4
07	THERMAL & MOISTURE PROTECTION	103.2	128.4	114.2	102.0	125.7	112.3	103.0	124.9	112.5	103.2	122.1	111.4	103.3	128.3	114.2	103.3	109.9	106.2
08	OPENINGS	97.4	128.4	104.6	93.4	118.9	99.3	97.4	118.3	102.2	99.2	126.7	105.6	89.8	128.3	98.7	99.4	120.3	104.3
0920	Plaster & Gypsum Board	92.3	119.9	110.9	87.3	119.9	109.3	92.3	119.9	110.9	111.3	116.7	114.9	114.4	119.9	118.1	112.6	119.9	117.5
0950, 0980	Ceilings & Acoustic Treatment	104.3	119.9	114.9	86.2	119.9	108.9	104.3	119.9	114.9	93.9	116.7	109.3	93.9	119.9	111.5	103.7	119.9	114.6
0960	Flooring	85.0	162.0	106.6	82.6	162.0	104.9	83.8	162.0	105.8	86.7	162.0	107.8	88.2	162.0	108.9	85.9	135.0	99.7
0970, 0990	Wall Finishes & Painting/Coating	85.0	139.2	117.1	85.0	139.2	117.1	85.0	139.2	117.1	86.2	139.2	117.6	87.1	139.2	117.9	86.2	109.3	99.9
09	FINISHES	90.8	130.8	112.5	85.2	130.8	109.9	90.5	131.0	112.4	89.5	128.9	110.9	90.2	130.6	112.1	91.7	120.9	107.5
COVERS	DIVS. 10 - 14, 25, 28, 41, 43, 44, 46	100.0	109.7	102.2	100.0	109.7	102.2	100.0	110.2	102.3	100.0	104.2	100.9	100.0	109.3	102.1	100.0	102.9	100.7
21, 22, 23	FIRE SUPPRESSION, PLUMBING & HVAC	100.4	103.9	101.7	96.8	103.5	99.4	100.4	103.5	101.6	97.0	104.5	99.9	97.0	121.1	106.4	97.0	98.5	97.6
26, 27, 3370	ELECTRICAL, COMMUNICATIONS & UTIL.	101.3	97.7	99.5	98.5	100.1	99.3	101.2	97.7	99.4	100.6	104.1	102.3	97.4	124.4	110.7	100.5	96.1	98.4
MF2018	WEIGHTED AVERAGE	98.5	116.7	106.2	93.2	115.2	102.5	98.3	114.9	105.3	94.6	116.2	103.7	94.0	124.2	106.8	95.9	109.0	101.4

MASSACHUSETTS

DIVISION		HYANNIS 026 MAT.	INST.	TOTAL	LAWRENCE 019 MAT.	INST.	TOTAL	LOWELL 018 MAT.	INST.	TOTAL	NEW BEDFORD 027 MAT.	INST.	TOTAL	PITTSFIELD 012 MAT.	INST.	TOTAL	SPRINGFIELD 010 - 011 MAT.	INST.	TOTAL
015433	CONTRACTOR EQUIPMENT		98.1	98.1		98.1	98.1		95.7	95.7		98.9	98.9		95.7	95.7		95.7	95.7
0241, 31 - 34	SITE & INFRASTRUCTURE, DEMOLITION	87.4	98.1	94.8	91.6	98.0	96.0	90.6	98.1	95.8	88.7	98.0	95.1	91.6	96.7	95.2	91.1	97.1	95.2
0310	Concrete Forming & Accessories	92.7	120.6	116.5	102.5	121.6	118.8	99.0	122.6	119.1	100.7	120.8	117.9	99.0	102.9	102.3	99.3	117.7	115.0
0320	Concrete Reinforcing	88.0	122.3	104.6	108.8	148.3	127.9	109.7	148.1	128.2	109.7	122.4	115.8	90.8	114.1	102.1	109.7	126.5	117.8
0330	Cast-in-Place Concrete	80.4	136.4	101.3	90.7	136.8	107.9	82.4	138.5	103.3	75.1	136.9	98.1	90.0	113.4	98.7	85.8	122.2	99.4
03	CONCRETE	86.7	126.0	104.0	96.2	131.1	111.5	88.0	131.9	107.3	89.1	126.3	105.4	89.0	108.3	97.5	89.5	120.2	103.0
04	MASONRY	104.3	135.1	123.2	110.1	135.9	125.9	97.3	134.9	120.4	104.2	135.7	123.6	98.0	111.5	106.3	97.6	119.1	110.8
05	METALS	95.1	117.0	101.5	98.2	128.4	107.1	98.2	125.0	106.1	98.8	117.4	104.3	98.0	108.2	101.0	100.9	113.7	104.6
06	WOOD, PLASTICS & COMPOSITES	88.0	119.7	105.4	101.5	119.7	111.5	100.8	119.7	111.2	98.1	119.9	110.1	100.8	102.5	101.8	100.8	119.7	111.2
07	THERMAL & MOISTURE PROTECTION	102.5	125.7	112.6	104.0	128.6	114.7	103.7	128.0	114.3	102.9	125.1	112.6	103.8	104.3	104.0	103.7	109.9	106.4
08	OPENINGS	93.9	118.9	99.8	93.4	129.9	101.9	100.4	129.9	107.3	97.4	123.0	103.3	100.4	107.6	102.1	100.4	120.3	105.1
0920	Plaster & Gypsum Board	83.2	119.9	107.9	117.3	119.9	119.1	117.3	119.9	119.1	92.3	119.9	110.9	117.3	102.3	107.2	117.3	119.9	119.1
0950, 0980	Ceilings & Acoustic Treatment	96.2	119.9	112.2	105.5	119.9	115.2	105.5	119.9	115.2	104.3	119.9	114.9	105.5	102.3	103.4	105.5	119.9	115.2
0960	Flooring	80.0	162.0	103.0	88.8	162.0	109.4	88.8	162.0	109.4	83.8	162.0	105.8	89.1	129.8	100.6	88.2	135.0	101.3
0970, 0990	Wall Finishes & Painting/Coating	85.0	139.2	117.1	86.3	139.2	117.6	86.2	139.2	117.6	85.0	139.2	117.1	86.2	109.3	99.9	87.0	109.3	100.2
09	FINISHES	86.3	130.8	110.4	93.8	130.8	113.8	93.7	131.6	114.2	90.4	131.0	112.3	93.8	108.6	101.8	93.9	120.9	108.4
COVERS	DIVS. 10 - 14, 25, 28, 41, 43, 44, 46	100.0	109.7	102.2	100.0	109.8	102.2	100.0	110.8	102.4	100.0	110.2	102.3	100.0	99.5	99.9	100.0	102.9	100.7
21, 22, 23	FIRE SUPPRESSION, PLUMBING & HVAC	100.4	104.2	101.9	100.1	119.7	107.8	100.1	121.2	108.4	100.4	103.4	101.5	100.1	94.2	97.8	100.1	98.5	99.5
26, 27, 3370	ELECTRICAL, COMMUNICATIONS & UTIL.	98.9	97.7	98.3	99.7	122.9	111.1	100.1	119.3	109.6	102.1	97.7	99.9	100.1	96.1	98.2	100.1	93.0	96.6
MF2018	WEIGHTED AVERAGE	95.7	115.0	103.8	98.4	124.2	109.3	97.5	123.8	108.6	97.7	115.2	105.1	97.7	102.6	99.7	98.2	108.5	102.5

MASSACHUSETTS / MICHIGAN

DIVISION		WORCESTER 015 - 016 MAT.	INST.	TOTAL	ANN ARBOR 481 MAT.	INST.	TOTAL	BATTLE CREEK 490 MAT.	INST.	TOTAL	BAY CITY 487 MAT.	INST.	TOTAL	DEARBORN 481 MAT.	INST.	TOTAL	DETROIT 482 MAT.	INST.	TOTAL
015433	CONTRACTOR EQUIPMENT		95.7	95.7		109.1	109.1		96.1	96.1		109.1	109.1		109.1	109.1		97.8	97.8
0241, 31 - 34	SITE & INFRASTRUCTURE, DEMOLITION	91.0	99.7	95.7	81.0	92.2	88.7	93.6	81.7	85.3	72.5	91.7	85.8	80.8	92.3	88.7	98.2	101.2	100.2
0310	Concrete Forming & Accessories	99.6	118.5	115.7	96.3	104.1	103.0	96.0	77.3	80.0	96.4	81.5	83.7	96.2	104.7	103.5	100.4	105.3	104.6
0320	Concrete Reinforcing	109.7	155.5	131.9	101.2	103.5	102.3	98.5	80.9	90.0	101.2	102.7	101.9	101.2	103.6	102.3	101.4	104.8	103.0
0330	Cast-in-Place Concrete	85.3	136.5	104.4	88.3	96.4	91.3	86.1	91.0	87.9	84.5	83.5	84.1	86.4	97.2	90.4	104.5	101.3	103.3
03	CONCRETE	89.3	130.6	107.5	89.8	102.2	95.2	85.7	82.6	84.3	88.0	87.3	87.7	88.9	102.8	95.0	103.1	103.2	103.2
04	MASONRY	97.2	131.9	118.5	99.6	96.4	97.6	99.1	78.4	86.4	99.2	77.7	86.0	99.5	98.2	98.7	102.6	99.8	100.9
05	METALS	100.9	127.4	108.7	102.7	115.7	106.5	103.7	83.7	97.9	103.3	114.0	106.4	102.8	115.8	106.6	104.3	94.3	101.3
06	WOOD, PLASTICS & COMPOSITES	101.2	116.5	109.6	89.2	106.6	98.8	88.4	75.6	81.3	89.2	81.9	85.2	89.2	106.6	98.8	97.1	106.7	102.4
07	THERMAL & MOISTURE PROTECTION	103.8	122.1	111.8	105.3	98.2	102.2	96.1	79.0	88.7	102.7	81.0	93.2	103.6	100.7	102.3	102.6	102.9	102.7
08	OPENINGS	100.4	130.2	107.4	94.2	100.8	95.7	86.5	74.1	83.6	94.2	83.5	91.7	94.2	101.1	95.8	97.2	101.9	98.5
0920	Plaster & Gypsum Board	117.3	116.7	116.9	107.4	106.5	106.8	91.7	71.7	78.2	107.4	81.1	89.7	107.4	106.5	106.8	104.1	106.5	105.7
0950, 0980	Ceilings & Acoustic Treatment	105.5	116.7	113.0	84.8	106.5	99.4	78.7	71.7	74.0	85.7	81.1	82.6	84.8	106.5	99.4	96.9	106.5	103.4
0960	Flooring	88.8	154.9	107.4	91.2	107.2	95.7	91.5	70.4	85.6	91.2	77.0	87.2	90.6	101.6	93.6	95.7	106.7	98.8
0970, 0990	Wall Finishes & Painting/Coating	86.2	139.2	117.6	82.7	95.5	90.3	82.3	76.3	78.7	82.7	79.1	80.6	82.7	93.8	89.3	93.1	100.1	97.3
09	FINISHES	93.7	127.5	112.0	90.8	104.3	98.1	84.7	75.2	79.6	90.6	79.9	84.8	90.6	103.5	97.6	97.7	105.4	101.9
COVERS	DIVS. 10 - 14, 25, 28, 41, 43, 44, 46	100.0	104.2	100.9	100.0	93.7	98.6	100.0	93.7	98.6	100.0	88.0	97.3	100.0	94.1	98.7	100.0	102.4	100.5
21, 22, 23	FIRE SUPPRESSION, PLUMBING & HVAC	100.1	104.7	101.9	100.0	91.3	96.6	100.1	82.6	93.2	100.0	78.6	91.6	100.0	100.0	100.0	99.9	102.7	101.0
26, 27, 3370	ELECTRICAL, COMMUNICATIONS & UTIL.	100.1	104.1	102.1	97.3	103.0	100.1	96.6	77.7	87.3	96.4	82.5	89.5	97.3	96.1	96.7	99.9	103.5	101.6
MF2018	WEIGHTED AVERAGE	98.1	117.0	106.1	97.1	99.8	98.2	95.5	80.4	89.1	96.6	85.3	91.8	96.9	100.9	98.6	100.7	102.0	101.3

		MICHIGAN																
		FLINT			GAYLORD			GRAND RAPIDS			IRON MOUNTAIN			JACKSON			KALAMAZOO		
	DIVISION	484 - 485			497			493, 495			498 - 499			492			491		
		MAT.	INST.	TOTAL	MAT.	INST.	TOTAL	MAT.	INST.	TOTAL	MAT.	INST.	TOTAL	MAT.	INST.	TOTAL	MAT.	INST.	TOTAL
015433	CONTRACTOR EQUIPMENT		109.1	109.1		103.6	103.6		98.3	98.3		91.1	91.1		103.6	103.6		96.1	96.1
0241, 31 - 34	SITE & INFRASTRUCTURE, DEMOLITION	70.2	91.8	85.1	87.4	79.6	82.0	92.9	85.9	88.1	96.4	88.5	90.9	111.0	81.2	90.4	93.9	81.6	85.4
0310	Concrete Forming & Accessories	99.1	81.8	84.4	95.0	73.9	77.0	95.6	75.7	78.6	86.6	78.6	79.8	91.6	82.3	83.7	96.0	77.1	79.9
0320	Concrete Reinforcing	101.2	103.1	102.1	91.9	93.0	92.4	103.4	80.7	92.4	91.6	85.0	88.4	89.3	103.1	96.0	98.5	76.2	87.8
0330	Cast-in-Place Concrete	88.9	85.8	87.8	85.8	78.4	83.1	90.4	89.0	89.9	101.7	67.6	89.0	85.7	89.8	87.2	87.8	91.0	89.0
03	CONCRETE	90.2	88.4	89.4	82.6	80.5	81.6	89.7	81.1	85.9	90.9	76.6	84.6	77.8	89.8	83.1	88.9	81.7	85.7
04	MASONRY	99.7	85.9	91.2	109.7	72.7	87.0	92.0	74.1	81.0	95.4	79.7	85.7	89.2	86.9	87.8	97.6	78.4	85.8
05	METALS	102.8	114.6	106.3	105.5	110.8	107.0	100.8	83.2	95.6	104.8	91.1	100.8	105.7	112.7	107.8	103.7	81.9	97.3
06	WOOD, PLASTICS & COMPOSITES	92.6	80.7	86.1	81.9	73.4	77.2	92.2	73.9	82.2	78.2	78.9	78.6	80.8	79.8	80.2	88.4	75.6	81.3
07	THERMAL & MOISTURE PROTECTION	103.0	84.5	95.0	94.7	74.7	86.0	98.0	71.8	86.6	98.4	74.8	88.1	94.0	88.3	91.5	96.1	79.0	88.7
08	OPENINGS	94.2	82.6	91.5	86.0	78.7	84.3	101.8	73.8	95.2	92.5	69.8	87.2	85.2	85.5	85.3	86.5	73.5	83.4
0920	Plaster & Gypsum Board	109.0	79.8	89.4	91.3	71.8	78.2	99.8	70.2	79.8	51.3	79.0	69.9	90.1	78.3	82.1	91.7	71.7	78.2
0950, 0980	Ceilings & Acoustic Treatment	84.8	79.8	81.5	77.1	71.8	73.5	91.1	70.2	77.0	76.0	79.0	78.0	77.1	78.3	77.9	78.7	71.7	74.0
0960	Flooring	91.2	86.0	89.7	84.9	85.6	85.1	95.7	76.1	90.2	102.0	91.0	98.9	83.4	78.7	82.1	91.5	70.4	85.6
0970, 0990	Wall Finishes & Painting/Coating	82.7	78.3	80.1	78.7	81.6	80.4	89.3	76.9	82.0	94.8	69.6	79.9	78.7	93.8	87.7	82.3	76.3	78.7
09	FINISHES	90.1	81.6	85.5	84.1	76.2	79.8	91.4	75.2	82.6	84.6	80.0	82.1	85.2	81.9	83.4	84.7	75.2	79.6
COVERS	DIVS. 10 - 14, 25, 28, 41, 43, 44, 46	100.0	88.8	97.5	100.0	77.9	95.1	100.0	93.0	98.4	100.0	85.9	96.8	100.0	91.6	98.1	100.0	93.7	98.6
21, 22, 23	FIRE SUPPRESSION, PLUMBING & HVAC	100.0	83.3	93.4	96.8	79.0	89.8	100.0	79.6	92.0	96.7	83.9	91.7	96.8	85.7	92.4	100.1	78.2	91.5
26, 27, 3370	ELECTRICAL, COMMUNICATIONS & UTIL.	97.3	89.1	93.2	94.5	76.3	85.5	102.2	82.9	92.7	100.8	80.7	90.9	98.3	103.0	100.6	96.4	74.1	85.4
MF2018	WEIGHTED AVERAGE	96.8	88.6	93.3	94.5	80.6	88.6	97.9	79.9	90.3	96.5	81.7	90.2	94.0	90.7	92.6	95.8	78.6	88.6

		MICHIGAN															MINNESOTA		
		LANSING			MUSKEGON			ROYAL OAK			SAGINAW			TRAVERSE CITY			BEMIDJI		
	DIVISION	488 - 489			494			480, 483			486			496			566		
		MAT.	INST.	TOTAL	MAT.	INST.	TOTAL	MAT.	INST.	TOTAL	MAT.	INST.	TOTAL	MAT.	INST.	TOTAL	MAT.	INST.	TOTAL
015433	CONTRACTOR EQUIPMENT		111.0	111.0		96.1	96.1		89.8	89.8		109.1	109.1		91.1	91.1		97.7	97.7
0241, 31 - 34	SITE & INFRASTRUCTURE, DEMOLITION	90.7	96.5	94.7	91.5	81.7	84.8	84.8	92.6	90.2	73.5	91.7	86.1	81.8	87.5	85.7	95.1	95.1	95.1
0310	Concrete Forming & Accessories	93.8	74.9	77.7	96.3	79.4	81.9	92.3	105.2	103.2	96.3	79.7	82.1	86.6	72.4	74.5	88.0	82.8	83.6
0320	Concrete Reinforcing	104.8	102.9	103.8	99.3	80.7	90.3	91.7	104.8	98.0	101.2	102.7	101.9	93.1	80.2	86.9	97.4	101.8	99.5
0330	Cast-in-Place Concrete	97.9	85.1	93.1	85.8	90.5	87.5	77.3	97.5	84.8	87.3	83.4	85.8	79.3	76.7	78.3	101.8	104.5	102.8
03	CONCRETE	91.4	84.9	88.5	84.2	83.3	83.8	77.1	101.6	87.9	89.3	86.5	88.1	74.8	76.1	75.4	91.0	95.0	92.8
04	MASONRY	94.5	84.6	88.4	96.3	76.2	83.9	93.2	99.5	97.1	101.2	77.7	86.7	93.5	74.8	82.0	98.9	102.4	101.0
05	METALS	101.6	112.7	104.9	101.5	83.1	96.1	106.3	92.2	102.2	102.8	113.8	106.0	104.8	88.9	100.1	91.2	117.9	99.0
06	WOOD, PLASTICS & COMPOSITES	87.0	71.2	78.3	85.4	78.0	81.3	85.0	106.6	96.9	85.9	79.7	82.5	78.2	71.9	74.8	68.8	77.0	73.3
07	THERMAL & MOISTURE PROTECTION	102.0	83.8	94.1	95.0	73.5	85.7	101.2	100.9	101.1	103.9	80.9	93.9	97.3	74.2	87.3	106.3	90.7	99.5
08	OPENINGS	102.7	77.2	96.8	85.8	76.5	83.6	94.0	101.3	95.7	92.4	82.2	90.0	92.5	65.4	86.2	99.7	101.3	100.1
0920	Plaster & Gypsum Board	100.1	70.2	79.9	71.8	74.2	73.4	104.8	106.5	106.0	107.4	78.8	88.2	51.3	71.8	65.1	104.3	76.8	85.8
0950, 0980	Ceilings & Acoustic Treatment	83.9	70.2	74.6	78.7	74.2	75.6	84.1	106.5	99.2	84.8	78.8	80.8	76.0	71.8	73.2	128.8	76.8	93.7
0960	Flooring	97.2	78.7	92.0	90.2	72.3	85.1	88.2	106.7	93.4	91.2	77.0	87.2	102.0	85.6	97.4	90.7	91.4	90.9
0970, 0990	Wall Finishes & Painting/Coating	91.1	76.9	82.7	80.5	79.7	80.0	84.2	93.8	89.9	82.7	79.1	80.6	94.8	39.4	62.1	89.6	105.8	99.2
09	FINISHES	90.2	74.6	81.7	81.1	77.1	78.9	89.7	104.3	97.6	90.5	78.6	84.0	83.5	71.0	76.7	98.2	86.0	91.6
COVERS	DIVS. 10 - 14, 25, 28, 41, 43, 44, 46	100.0	88.8	97.5	100.0	94.5	98.8	100.0	100.0	100.0	100.0	87.7	97.3	100.0	84.4	96.5	100.0	95.6	99.0
21, 22, 23	FIRE SUPPRESSION, PLUMBING & HVAC	99.9	83.7	93.5	99.9	83.4	93.4	96.7	101.4	98.5	100.0	78.3	91.5	96.7	78.9	89.7	96.7	84.8	92.0
26, 27, 3370	ELECTRICAL, COMMUNICATIONS & UTIL.	99.1	87.6	93.4	96.8	73.0	85.1	99.4	103.1	101.3	95.4	84.5	90.0	96.2	76.2	86.4	105.0	100.7	102.8
MF2018	WEIGHTED AVERAGE	98.0	86.8	93.3	94.3	80.0	88.3	95.1	100.3	97.3	96.5	85.1	91.7	93.4	77.7	86.8	96.7	95.4	96.1

		MINNESOTA																	
		BRAINERD			DETROIT LAKES			DULUTH			MANKATO			MINNEAPOLIS			ROCHESTER		
	DIVISION	564			565			556 - 558			560			553 - 555			559		
		MAT.	INST.	TOTAL	MAT.	INST.	TOTAL	MAT.	INST.	TOTAL	MAT.	INST.	TOTAL	MAT.	INST.	TOTAL	MAT.	INST.	TOTAL
015433	CONTRACTOR EQUIPMENT		100.2	100.2		97.7	97.7		102.9	102.9		100.2	100.2		108.4	108.4		100.7	100.7
0241, 31 - 34	SITE & INFRASTRUCTURE, DEMOLITION	96.0	99.9	98.7	93.3	95.4	94.7	98.2	101.7	100.6	92.6	99.4	97.3	95.0	110.4	104.5	96.0	97.6	97.1
0310	Concrete Forming & Accessories	88.7	83.6	84.4	84.9	82.8	83.1	98.4	94.6	95.2	96.9	96.6	96.6	99.3	114.0	111.8	99.2	94.8	95.4
0320	Concrete Reinforcing	96.2	102.2	99.1	97.4	101.8	99.6	102.6	102.0	102.3	96.0	109.2	102.4	96.1	109.4	102.5	98.9	108.8	103.7
0330	Cast-in-Place Concrete	110.7	109.1	110.1	98.9	107.5	102.1	94.3	97.7	95.6	101.9	99.5	101.0	95.6	115.3	102.9	94.8	95.3	95.0
03	CONCRETE	94.8	97.1	95.8	88.6	96.1	91.9	95.3	98.1	96.5	90.5	100.8	95.0	97.6	114.3	104.9	91.4	98.6	94.6
04	MASONRY	122.5	116.3	118.7	122.6	109.8	114.8	95.3	105.2	101.7	111.3	110.8	111.0	114.5	119.5	117.6	102.9	103.5	103.3
05	METALS	92.2	118.3	99.9	91.1	117.5	98.9	100.2	119.1	105.8	92.1	121.6	100.7	98.7	124.5	106.3	99.3	123.1	106.3
06	WOOD, PLASTICS & COMPOSITES	84.9	74.2	79.1	65.8	74.4	70.6	93.9	91.9	92.8	94.5	95.4	95.0	96.1	110.8	104.2	95.2	92.9	93.9
07	THERMAL & MOISTURE PROTECTION	104.7	102.4	103.7	106.1	99.4	103.2	97.9	102.2	99.8	105.1	93.4	100.1	101.2	115.9	107.6	105.3	90.0	98.6
08	OPENINGS	86.4	99.8	89.5	99.6	99.9	99.7	103.3	104.6	103.6	90.9	112.2	95.9	99.3	121.5	104.5	98.6	112.5	101.8
0920	Plaster & Gypsum Board	91.5	74.1	79.8	103.7	74.1	83.8	94.5	92.1	92.9	95.6	95.9	95.8	101.1	111.2	107.9	103.2	93.2	96.5
0950, 0980	Ceilings & Acoustic Treatment	58.8	74.1	69.1	128.8	74.1	91.9	89.2	92.1	91.1	58.8	95.9	83.8	103.1	111.2	108.6	93.9	93.2	93.4
0960	Flooring	89.6	85.6	88.4	89.5	85.6	88.4	93.2	123.9	101.8	91.4	81.8	88.7	100.6	114.6	104.5	92.3	81.8	89.3
0970, 0990	Wall Finishes & Painting/Coating	84.0	105.8	96.9	89.6	83.4	85.9	86.0	104.4	96.9	95.0	103.7	100.2	101.3	123.8	114.6	85.3	98.2	92.9
09	FINISHES	82.1	85.2	83.8	97.6	82.5	89.4	89.4	100.2	95.3	83.7	94.9	89.7	98.9	115.0	107.6	90.0	93.1	91.7
COVERS	DIVS. 10 - 14, 25, 28, 41, 43, 44, 46	100.0	97.3	99.4	100.0	97.1	99.4	100.0	94.6	98.8	100.0	97.5	99.4	100.0	105.8	101.3	100.0	101.8	100.4
21, 22, 23	FIRE SUPPRESSION, PLUMBING & HVAC	96.0	88.6	93.1	96.7	87.5	93.1	99.8	95.7	97.7	96.0	84.6	91.5	100.0	110.6	104.2	100.0	91.6	96.7
26, 27, 3370	ELECTRICAL, COMMUNICATIONS & UTIL.	102.7	101.7	102.2	104.8	71.1	88.2	100.3	101.7	101.0	109.1	93.1	101.2	107.2	110.3	108.7	103.0	93.1	98.1
MF2018	WEIGHTED AVERAGE	95.6	98.6	96.8	97.4	92.4	95.3	98.6	101.3	99.7	95.7	98.5	96.9	100.7	114.1	106.3	98.3	98.7	98.5

		MINNESOTA																MISSISSIPPI		
	DIVISION	SAINT PAUL			ST. CLOUD			THIEF RIVER FALLS			WILLMAR			WINDOM			BILOXI		
		550 - 551			563			567			562			561			395		
		MAT.	INST.	TOTAL	MAT.	INST.	TOTAL	MAT.	INST.	TOTAL	MAT.	INST.	TOTAL	MAT.	INST.	TOTAL	MAT.	INST.	TOTAL
015433	CONTRACTOR EQUIPMENT		102.9	102.9		100.2	100.2		97.7	97.7		100.2	100.2		100.2	100.2		100.1	100.1
0241, 31 - 34	SITE & INFRASTRUCTURE, DEMOLITION	96.5	102.5	100.7	91.4	100.8	97.9	94.0	94.9	94.7	90.6	99.7	96.9	84.5	98.9	94.4	105.2	85.9	91.9
0310	Concrete Forming & Accessories	98.6	117.0	114.3	86.0	113.5	109.5	88.8	81.9	82.9	85.8	86.8	86.7	90.1	82.2	83.4	92.8	64.1	68.4
0320	Concrete Reinforcing	105.4	109.9	107.6	96.2	109.4	102.6	97.8	101.6	99.6	95.8	109.2	102.3	95.8	108.1	101.8	93.4	48.8	71.8
0330	Cast-in-Place Concrete	99.9	116.3	106.0	97.7	115.0	104.1	100.9	81.3	93.6	99.2	79.5	91.9	85.8	83.3	84.8	114.2	66.3	96.4
03	CONCRETE	95.3	116.1	104.4	86.5	114.0	98.5	89.8	86.5	88.3	86.3	89.5	87.7	77.6	88.5	82.4	94.4	64.0	80.1
04	MASONRY	105.3	127.4	118.9	107.3	121.8	116.2	98.9	102.3	101.0	112.1	114.5	113.6	122.0	87.9	101.0	90.4	62.8	73.4
05	METALS	100.1	125.0	107.4	92.9	123.1	101.8	91.3	116.7	98.7	92.0	121.5	100.7	91.9	119.5	100.0	90.3	84.5	88.6
06	WOOD, PLASTICS & COMPOSITES	95.0	114.2	105.6	82.1	110.3	97.6	70.0	77.0	73.8	81.8	79.9	80.8	86.3	79.9	82.8	93.1	64.5	77.3
07	THERMAL & MOISTURE PROTECTION	102.0	120.5	110.1	104.9	111.7	107.9	107.1	88.1	98.8	104.7	101.0	103.1	104.6	83.0	95.2	100.0	63.1	83.9
08	OPENINGS	98.2	124.3	104.3	91.4	122.1	98.5	99.7	101.3	100.1	88.4	102.8	91.7	91.9	102.8	94.4	96.3	55.0	86.7
0920	Plaster & Gypsum Board	94.5	115.0	108.3	91.5	111.2	104.7	104.3	76.8	85.8	91.5	79.9	83.7	91.5	79.9	83.7	103.8	64.0	77.0
0950, 0980	Ceilings & Acoustic Treatment	91.6	115.0	107.4	58.8	111.2	94.1	128.8	76.8	93.7	58.8	79.9	73.0	58.8	79.9	73.0	86.4	64.0	71.3
0960	Flooring	92.4	122.8	101.0	86.4	120.9	96.1	90.4	85.6	89.0	87.8	85.6	87.2	90.3	85.6	89.0	94.0	65.0	85.8
0970, 0990	Wall Finishes & Painting/Coating	92.1	129.1	114.0	95.0	123.8	112.0	89.6	83.4	85.9	89.6	83.4	85.9	89.6	103.7	98.0	86.0	46.9	62.8
09	FINISHES	91.1	119.1	106.3	81.5	115.9	100.1	98.1	82.6	89.7	81.6	85.3	83.6	81.8	84.8	83.4	89.4	62.8	75.0
COVERS	DIVS. 10 - 14, 25, 28, 41, 43, 44, 46	100.0	108.7	101.9	100.0	103.2	100.7	100.0	95.5	99.0	100.0	97.3	99.4	100.0	93.7	98.6	100.0	71.9	93.7
21, 22, 23	FIRE SUPPRESSION, PLUMBING & HVAC	99.9	116.7	106.5	99.5	110.5	103.8	96.7	84.4	91.9	96.0	101.6	98.2	96.0	81.0	90.1	100.0	55.3	82.4
26, 27, 3370	ELECTRICAL, COMMUNICATIONS & UTIL.	102.9	115.1	108.9	102.7	115.1	108.8	102.1	71.0	86.8	102.7	83.7	93.4	109.1	93.1	101.2	101.0	53.7	77.7
MF2018	WEIGHTED AVERAGE	99.0	117.6	106.9	95.1	114.2	103.2	96.3	89.3	93.3	94.0	98.0	95.7	94.3	91.2	93.0	96.2	63.9	82.6

		MISSISSIPPI																		
	DIVISION	CLARKSDALE			COLUMBUS			GREENVILLE			GREENWOOD			JACKSON			LAUREL		
		386			397			387			389			390 - 392			394		
		MAT.	INST.	TOTAL	MAT.	INST.	TOTAL	MAT.	INST.	TOTAL	MAT.	INST.	TOTAL	MAT.	INST.	TOTAL	MAT.	INST.	TOTAL
015433	CONTRACTOR EQUIPMENT		100.1	100.1		100.1	100.1		100.1	100.1		100.1	100.1		102.6	102.6		100.1	100.1
0241, 31 - 34	SITE & INFRASTRUCTURE, DEMOLITION	101.4	84.4	89.7	103.9	85.2	91.0	107.5	86.1	92.7	104.5	84.1	90.4	101.1	90.6	93.9	109.6	84.5	92.2
0310	Concrete Forming & Accessories	83.8	43.9	49.8	81.9	46.3	51.6	80.7	62.7	65.4	92.5	44.3	51.4	92.0	64.6	68.6	82.0	59.5	62.8
0320	Concrete Reinforcing	104.2	66.5	85.9	99.9	66.6	83.8	104.7	66.7	86.3	104.2	66.5	86.0	104.0	51.7	78.7	100.6	32.4	67.6
0330	Cast-in-Place Concrete	104.1	58.8	87.2	116.5	61.1	95.9	107.2	65.7	91.8	112.0	58.9	92.2	99.6	66.5	87.3	113.9	60.1	93.9
03	CONCRETE	93.8	55.1	76.8	95.8	57.1	78.8	99.3	66.2	84.8	100.1	55.4	80.4	89.2	64.7	78.4	97.9	56.9	79.9
04	MASONRY	89.1	50.8	65.6	115.4	53.6	77.4	132.2	61.4	88.7	89.8	50.7	65.8	95.7	61.5	74.7	111.5	49.4	73.4
05	METALS	92.5	87.9	91.1	87.4	90.4	88.3	93.6	91.2	92.8	92.5	87.8	91.1	96.3	84.6	93.3	87.5	76.0	84.1
06	WOOD, PLASTICS & COMPOSITES	79.4	44.4	60.1	78.8	45.2	60.3	76.2	62.8	68.8	91.8	44.4	65.7	96.3	65.5	79.4	79.9	64.5	71.4
07	THERMAL & MOISTURE PROTECTION	97.7	52.0	77.8	100.0	55.4	80.6	98.2	62.1	82.5	98.1	54.2	79.0	98.3	62.9	82.9	100.2	56.5	81.2
08	OPENINGS	95.1	48.2	84.1	95.9	48.6	84.9	94.8	58.1	86.2	95.1	48.2	84.1	99.5	56.0	89.4	93.2	51.8	83.5
0920	Plaster & Gypsum Board	89.9	43.3	58.6	94.3	44.2	60.6	89.6	62.3	71.2	100.6	43.3	62.1	90.7	64.9	73.3	94.3	64.0	73.9
0950, 0980	Ceilings & Acoustic Treatment	82.3	43.3	56.0	81.5	44.2	56.3	85.0	62.3	69.7	82.3	43.3	56.0	89.7	64.9	73.0	81.5	64.0	69.7
0960	Flooring	97.8	45.0	83.0	87.9	65.0	81.4	96.2	65.0	87.4	103.5	65.0	92.7	92.1	65.0	84.5	86.6	65.0	80.5
0970, 0990	Wall Finishes & Painting/Coating	94.6	46.9	66.4	86.0	46.9	62.8	94.6	56.7	72.1	94.6	46.9	66.4	89.9	56.7	70.3	86.0	46.9	62.8
09	FINISHES	90.2	43.9	65.1	85.4	49.1	65.8	90.8	62.5	75.5	93.7	48.0	68.9	89.9	64.1	75.9	85.5	60.5	72.0
COVERS	DIVS. 10 - 14, 25, 28, 41, 43, 44, 46	100.0	48.0	88.4	100.0	49.0	88.6	100.0	71.3	93.6	100.0	48.0	88.4	100.0	71.9	93.7	100.0	34.2	85.3
21, 22, 23	FIRE SUPPRESSION, PLUMBING & HVAC	98.4	51.3	79.9	98.0	53.1	80.4	100.0	57.4	83.3	98.4	51.7	80.1	100.0	59.1	83.9	98.1	47.0	78.0
26, 27, 3370	ELECTRICAL, COMMUNICATIONS & UTIL.	96.8	41.6	69.6	98.6	55.1	77.2	96.8	55.9	76.6	96.8	39.0	68.3	102.7	55.9	79.6	100.1	57.9	79.3
MF2018	WEIGHTED AVERAGE	95.3	55.2	78.4	95.9	59.1	80.4	98.7	65.5	84.7	96.6	55.5	79.2	97.3	65.6	83.9	96.1	57.8	79.9

		MISSISSIPPI									MISSOURI								
	DIVISION	MCCOMB			MERIDIAN			TUPELO			BOWLING GREEN			CAPE GIRARDEAU			CHILLICOTHE		
		396			393			388			633			637			646		
		MAT.	INST.	TOTAL	MAT.	INST.	TOTAL	MAT.	INST.	TOTAL	MAT.	INST.	TOTAL	MAT.	INST.	TOTAL	MAT.	INST.	TOTAL
015433	CONTRACTOR EQUIPMENT		100.1	100.1		100.1	100.1		100.1	100.1		107.1	107.1		107.1	107.1		101.5	101.5
0241, 31 - 34	SITE & INFRASTRUCTURE, DEMOLITION	96.3	84.3	88.0	100.7	86.1	90.7	98.9	84.3	88.8	87.6	90.0	89.3	89.3	89.9	89.7	101.2	88.4	92.3
0310	Concrete Forming & Accessories	81.9	45.5	50.9	79.4	62.8	65.2	81.2	45.9	51.1	96.0	94.0	94.3	88.9	82.0	83.0	86.2	95.7	94.3
0320	Concrete Reinforcing	101.2	34.3	68.8	99.9	51.6	76.6	101.9	66.5	84.8	93.4	96.5	94.9	94.6	79.6	87.3	100.4	102.6	101.5
0330	Cast-in-Place Concrete	101.1	58.4	85.2	108.2	66.0	92.5	104.1	67.9	90.6	90.7	95.5	92.5	89.8	86.6	88.6	91.7	86.0	89.6
03	CONCRETE	85.9	50.2	70.2	90.1	63.8	78.6	93.4	59.2	78.4	91.2	96.4	93.5	90.4	85.0	88.0	95.0	94.3	94.7
04	MASONRY	116.7	50.3	75.9	90.0	62.1	72.9	121.7	53.3	79.6	109.6	97.9	102.4	105.9	80.5	90.3	99.2	92.2	94.9
05	METALS	87.6	75.6	84.1	88.5	85.6	87.7	92.4	87.9	91.1	91.8	117.7	99.4	92.9	109.4	97.8	85.1	110.6	92.5
06	WOOD, PLASTICS & COMPOSITES	78.8	46.8	61.2	76.2	62.8	68.8	76.7	45.2	59.4	96.7	94.4	95.4	89.2	80.0	84.1	94.0	97.0	95.7
07	THERMAL & MOISTURE PROTECTION	99.4	54.2	79.8	99.6	62.4	83.4	97.6	53.9	78.6	96.4	99.2	97.6	95.9	85.3	91.3	92.7	92.7	92.7
08	OPENINGS	96.0	41.3	83.2	95.6	54.8	86.1	95.0	51.4	84.9	98.7	98.2	98.5	98.6	76.1	93.4	86.0	96.2	88.3
0920	Plaster & Gypsum Board	94.3	45.8	61.7	94.3	62.3	72.8	89.6	44.2	59.0	101.7	94.6	96.9	101.1	79.8	86.8	98.2	96.9	97.3
0950, 0980	Ceilings & Acoustic Treatment	81.5	45.8	57.4	82.4	62.3	68.8	82.3	44.2	56.6	88.9	94.6	92.8	88.9	79.8	82.8	87.4	96.9	93.8
0960	Flooring	87.9	65.0	81.4	86.5	65.0	80.5	96.5	65.0	87.6	95.7	96.9	96.0	92.4	84.5	90.2	94.9	99.7	96.2
0970, 0990	Wall Finishes & Painting/Coating	86.0	46.9	62.8	86.0	46.9	62.8	94.6	45.2	65.4	95.4	106.3	101.8	95.4	67.1	78.6	94.3	103.2	99.6
09	FINISHES	84.8	49.2	65.5	84.8	61.6	72.3	89.7	48.9	67.6	98.2	95.7	96.8	97.0	79.8	87.7	97.1	97.6	97.4
COVERS	DIVS. 10 - 14, 25, 28, 41, 43, 44, 46	100.0	50.9	89.1	100.0	71.5	93.6	100.0	49.0	88.6	100.0	81.4	95.8	100.0	94.3	98.7	100.0	81.3	95.8
21, 22, 23	FIRE SUPPRESSION, PLUMBING & HVAC	98.0	50.7	79.5	100.0	57.6	83.3	98.5	52.7	80.5	96.5	99.5	97.7	100.0	98.7	99.5	96.6	99.9	97.9
26, 27, 3370	ELECTRICAL, COMMUNICATIONS & UTIL.	97.3	56.0	76.9	100.1	55.6	78.2	96.5	55.0	76.1	99.2	76.8	88.2	99.2	98.5	98.9	94.9	75.8	85.5
MF2018	WEIGHTED AVERAGE	94.4	55.8	78.0	94.6	64.5	81.9	96.6	59.0	80.7	96.4	95.5	96.0	97.0	91.2	94.5	93.7	93.8	93.7

For customer support on your Mechanical Costs with RSMeans Data, call 800.448.8182.

675

MISSOURI

DIVISION		COLUMBIA 652			FLAT RIVER 636			HANNIBAL 634			HARRISONVILLE 647			JEFFERSON CITY 650 - 651			JOPLIN 648		
		MAT.	INST.	TOTAL	MAT.	INST.	TOTAL	MAT.	INST.	TOTAL	MAT.	INST.	TOTAL	MAT.	INST.	TOTAL	MAT.	INST.	TOTAL
015433	CONTRACTOR EQUIPMENT		110.3	110.3		107.1	107.1		107.1	107.1		101.5	101.5		113.1	113.1		104.8	104.8
0241, 31 - 34	SITE & INFRASTRUCTURE, DEMOLITION	94.2	93.8	93.9	90.2	89.7	89.8	85.4	89.8	88.4	92.9	89.5	90.6	94.4	98.7	97.4	102.0	92.6	95.5
0310	Concrete Forming & Accessories	84.4	81.0	81.5	102.4	88.3	90.4	94.3	82.5	84.3	83.4	99.0	96.7	96.3	78.6	81.2	97.4	74.2	77.7
0320	Concrete Reinforcing	85.6	93.3	89.3	94.6	102.9	98.6	92.9	96.5	94.6	100.0	110.9	105.3	92.4	93.3	92.9	103.7	84.3	94.3
0330	Cast-in-Place Concrete	88.6	84.8	87.2	93.7	90.6	92.5	85.9	93.5	88.7	93.9	99.7	96.0	93.9	84.1	90.2	99.3	75.5	90.4
03	CONCRETE	80.8	86.3	83.2	94.1	93.3	93.8	87.6	90.5	88.9	91.1	102.0	95.9	87.5	84.8	86.3	93.9	77.7	86.7
04	MASONRY	136.9	86.4	105.9	106.5	76.4	88.0	101.6	94.6	97.3	94.0	99.4	97.3	96.0	86.5	90.1	92.4	80.2	84.9
05	METALS	96.3	115.4	101.9	91.7	119.2	99.8	91.8	117.2	99.3	85.4	115.1	94.1	95.8	114.0	101.1	87.9	98.8	91.1
06	WOOD, PLASTICS & COMPOSITES	80.3	78.6	79.3	105.2	89.6	96.6	94.9	81.1	87.3	90.4	98.7	95.0	97.1	75.6	85.2	106.1	73.5	88.2
07	THERMAL & MOISTURE PROTECTION	90.6	86.1	88.7	96.6	91.0	94.2	96.2	93.4	95.0	91.8	100.8	95.7	97.5	86.1	92.5	91.9	81.2	87.3
08	OPENINGS	94.4	81.7	91.4	98.6	97.4	98.4	98.7	83.7	95.2	85.7	103.2	89.8	94.5	80.0	91.1	86.9	76.3	84.5
0920	Plaster & Gypsum Board	84.7	78.2	80.3	107.7	89.7	95.6	101.4	81.0	87.7	94.1	98.6	97.1	96.2	74.9	81.9	105.4	72.7	83.4
0950, 0980	Ceilings & Acoustic Treatment	88.5	78.2	81.5	88.9	89.7	89.4	88.9	80.1	83.6	87.4	98.6	95.0	93.3	74.9	80.9	88.3	72.7	77.8
0960	Flooring	90.2	98.6	92.5	98.9	84.5	94.9	95.0	96.9	95.5	90.3	100.8	93.2	97.6	71.5	90.3	120.1	72.2	106.6
0970, 0990	Wall Finishes & Painting/Coating	93.1	81.9	86.4	95.4	71.7	81.4	95.4	94.1	94.6	98.6	107.8	104.0	91.2	81.9	85.7	93.9	75.9	83.2
09	FINISHES	84.0	83.5	83.7	100.0	85.1	92.0	97.7	85.3	91.0	94.8	100.2	97.7	91.5	77.1	83.7	103.8	74.4	87.9
COVERS	DIVS. 10 - 14, 25, 28, 41, 43, 44, 46	100.0	93.6	98.6	100.0	94.0	98.7	100.0	78.8	95.3	100.0	82.7	96.2	100.0	93.6	98.6	100.0	79.5	95.4
21, 22, 23	FIRE SUPPRESSION, PLUMBING & HVAC	100.0	96.9	98.8	96.5	96.8	96.6	96.5	97.7	97.0	96.6	100.8	98.2	100.0	96.9	98.8	100.1	71.1	88.7
26, 27, 3370	ELECTRICAL, COMMUNICATIONS & UTIL.	97.8	81.4	89.7	103.8	98.5	101.2	98.0	76.8	87.5	101.5	101.2	101.4	103.2	81.4	92.4	92.9	66.1	79.7
MF2018	WEIGHTED AVERAGE	96.1	90.7	93.8	97.3	94.2	96.0	95.3	91.6	93.7	93.2	100.8	96.4	96.4	89.9	93.7	95.1	77.7	87.7

MISSOURI

DIVISION		KANSAS CITY 640 - 641			KIRKSVILLE 635			POPLAR BLUFF 639			ROLLA 654 - 655			SEDALIA 653			SIKESTON 638		
		MAT.	INST.	TOTAL	MAT.	INST.	TOTAL	MAT.	INST.	TOTAL	MAT.	INST.	TOTAL	MAT.	INST.	TOTAL	MAT.	INST.	TOTAL
015433	CONTRACTOR EQUIPMENT		104.4	104.4		97.4	97.4		99.6	99.6		110.3	110.3		100.4	100.4		99.6	99.6
0241, 31 - 34	SITE & INFRASTRUCTURE, DEMOLITION	94.6	98.6	97.3	88.8	84.9	86.1	76.1	88.5	84.7	92.9	94.1	93.8	92.1	89.3	90.2	79.6	89.1	86.2
0310	Concrete Forming & Accessories	97.2	103.5	102.6	86.9	78.4	79.6	87.4	79.3	80.5	91.6	94.0	93.6	89.6	78.2	79.9	88.2	79.4	80.7
0320	Concrete Reinforcing	98.5	111.1	104.6	93.8	83.3	88.7	96.7	76.0	86.7	86.0	93.4	89.6	84.8	110.3	97.1	96.0	76.0	86.4
0330	Cast-in-Place Concrete	97.3	102.6	99.3	93.6	82.2	89.4	72.0	83.5	76.3	90.6	94.7	92.1	94.7	81.3	89.7	77.0	83.5	79.4
03	CONCRETE	93.8	104.9	98.7	106.9	81.8	95.9	81.5	81.4	81.5	82.5	95.6	88.2	95.7	86.1	91.5	85.2	81.5	83.6
04	MASONRY	99.6	103.3	101.9	113.2	84.8	95.8	104.5	74.1	85.8	110.8	85.3	95.2	117.1	81.6	95.3	104.2	74.1	85.7
05	METALS	94.9	112.8	100.2	91.5	101.0	94.3	92.1	97.8	93.7	95.7	115.9	101.6	94.5	112.8	99.9	92.4	97.9	94.0
06	WOOD, PLASTICS & COMPOSITES	100.1	103.8	102.2	82.2	76.5	79.1	81.3	80.0	80.6	88.0	95.6	92.2	81.6	76.7	78.9	82.9	80.0	81.3
07	THERMAL & MOISTURE PROTECTION	91.8	105.2	97.7	102.2	90.8	97.3	100.5	83.0	92.9	90.9	92.9	91.8	96.1	87.8	92.5	100.7	82.1	92.6
08	OPENINGS	96.5	106.0	98.7	104.1	78.8	98.2	105.0	75.2	98.1	94.4	91.1	93.6	99.4	87.0	96.5	105.0	75.2	98.1
0920	Plaster & Gypsum Board	100.9	104.0	103.0	96.3	76.2	82.8	96.6	79.8	85.3	86.9	95.6	92.8	80.5	76.2	77.6	98.5	79.8	85.9
0950, 0980	Ceilings & Acoustic Treatment	90.5	104.0	99.6	87.3	76.2	79.8	88.9	79.8	82.8	88.5	95.6	93.3	88.5	76.2	80.2	88.9	79.8	82.8
0960	Flooring	95.9	101.8	97.6	73.8	96.4	79.8	87.3	84.5	86.5	93.8	96.4	94.5	72.5	72.9	72.6	87.8	84.5	86.9
0970, 0990	Wall Finishes & Painting/Coating	97.2	107.8	103.5	91.3	78.3	83.6	90.6	67.1	76.7	93.1	90.2	91.4	93.1	103.2	99.1	90.6	67.1	76.7
09	FINISHES	98.1	103.6	101.1	97.4	81.1	88.6	96.9	78.4	86.9	85.4	94.0	90.1	82.9	79.1	80.8	97.5	78.7	87.3
COVERS	DIVS. 10 - 14, 25, 28, 41, 43, 44, 46	100.0	98.4	99.7	100.0	78.3	95.2	100.0	92.2	98.3	100.0	96.7	99.3	100.0	87.9	97.3	100.0	92.2	98.3
21, 22, 23	FIRE SUPPRESSION, PLUMBING & HVAC	100.0	103.4	101.4	96.6	97.1	96.8	96.6	94.9	95.9	96.5	98.6	97.3	96.4	94.9	95.8	96.6	95.0	95.9
26, 27, 3370	ELECTRICAL, COMMUNICATIONS & UTIL.	103.3	101.2	102.2	98.2	76.8	87.6	98.4	98.4	98.4	96.4	79.4	88.0	97.2	101.2	99.2	97.5	98.4	97.9
MF2018	WEIGHTED AVERAGE	97.9	103.8	100.4	98.9	86.5	93.7	95.1	87.7	92.0	94.1	94.3	94.2	96.4	91.5	94.3	95.7	87.8	92.3

DIVISION		MISSOURI SPRINGFIELD 656 - 658			ST. JOSEPH 644 - 645			ST. LOUIS 630 - 631			MONTANA BILLINGS 590 - 591			BUTTE 597			GREAT FALLS 594		
		MAT.	INST.	TOTAL	MAT.	INST.	TOTAL	MAT.	INST.	TOTAL	MAT.	INST.	TOTAL	MAT.	INST.	TOTAL	MAT.	INST.	TOTAL
015433	CONTRACTOR EQUIPMENT		102.8	102.8		101.5	101.5		108.6	108.6		97.9	97.9		97.7	97.7		97.7	97.7
0241, 31 - 34	SITE & INFRASTRUCTURE, DEMOLITION	94.4	91.4	92.4	96.7	87.4	90.3	94.4	97.8	96.7	91.8	93.5	93.0	97.3	93.3	94.5	101.0	93.3	95.7
0310	Concrete Forming & Accessories	97.7	75.4	78.7	96.3	89.1	90.2	99.5	102.5	102.1	99.3	68.3	72.9	86.5	68.2	70.9	99.2	68.2	72.8
0320	Concrete Reinforcing	82.5	92.9	87.5	97.3	110.4	103.6	86.5	105.2	95.5	98.0	81.8	90.2	106.3	81.7	94.4	98.0	81.8	90.2
0330	Cast-in-Place Concrete	96.3	73.7	87.9	92.2	95.5	93.4	101.3	102.5	101.7	118.7	72.7	101.6	130.6	72.5	109.0	138.1	72.5	113.7
03	CONCRETE	92.1	79.0	86.3	90.3	96.0	92.8	97.1	104.0	100.1	98.3	73.1	87.2	101.9	73.0	89.2	106.9	73.0	92.0
04	MASONRY	89.1	79.9	83.5	95.3	88.7	91.2	91.1	107.1	100.9	125.8	83.8	100.0	120.9	82.7	97.4	125.4	83.8	99.8
05	METALS	100.7	101.7	101.0	91.5	113.8	98.0	97.4	119.0	103.8	107.0	89.6	101.9	101.2	89.4	97.7	104.4	89.5	100.0
06	WOOD, PLASTICS & COMPOSITES	89.4	75.0	81.5	106.1	88.7	96.5	97.6	101.0	99.5	90.1	64.6	76.1	76.8	64.6	70.1	91.4	64.6	76.6
07	THERMAL & MOISTURE PROTECTION	94.6	75.5	86.3	92.2	88.7	90.7	94.3	105.5	99.2	108.3	73.7	93.2	107.8	73.3	92.8	108.5	73.7	93.4
08	OPENINGS	101.7	84.3	97.6	89.9	96.7	91.5	100.3	105.8	101.6	98.6	65.9	91.0	96.9	65.9	89.7	99.8	65.9	91.9
0920	Plaster & Gypsum Board	87.8	74.4	78.8	106.9	88.3	94.4	106.5	101.5	103.1	118.3	64.0	81.8	118.8	64.0	81.9	128.7	64.0	85.2
0950, 0980	Ceilings & Acoustic Treatment	88.5	74.4	79.0	94.8	88.3	90.4	90.5	101.5	97.9	96.6	64.0	74.6	104.1	64.0	77.0	105.7	64.0	77.6
0960	Flooring	92.8	71.5	86.8	99.4	99.0	99.3	98.0	99.1	98.3	88.1	80.8	86.0	85.4	90.9	87.0	91.9	80.8	88.8
0970, 0990	Wall Finishes & Painting/Coating	87.6	99.4	94.6	94.3	107.8	102.3	96.4	106.3	102.2	93.8	91.0	92.2	92.3	67.8	77.8	92.3	91.0	91.5
09	FINISHES	87.6	77.6	82.2	100.3	92.8	96.3	100.5	101.9	101.3	92.5	72.3	81.6	93.0	71.8	81.5	96.8	72.3	83.5
COVERS	DIVS. 10 - 14, 25, 28, 41, 43, 44, 46	100.0	90.5	97.9	100.0	94.0	98.7	100.0	102.8	100.6	100.0	91.6	98.1	100.0	91.6	98.1	100.0	91.6	98.1
21, 22, 23	FIRE SUPPRESSION, PLUMBING & HVAC	100.0	69.4	88.0	100.1	87.5	95.2	100.0	105.5	102.2	100.0	76.4	90.7	100.0	71.2	88.7	100.0	71.2	88.7
26, 27, 3370	ELECTRICAL, COMMUNICATIONS & UTIL.	101.3	68.8	85.3	101.5	75.8	88.9	101.8	98.5	100.2	102.3	75.2	88.9	109.2	71.3	90.5	101.7	70.8	86.5
MF2018	WEIGHTED AVERAGE	97.6	79.0	89.7	96.1	90.9	93.9	98.8	104.5	101.2	101.6	78.4	91.8	101.5	76.5	90.9	102.9	76.6	91.8

City Cost Indexes

MONTANA

DIVISION		HAVRE 595			HELENA 596			KALISPELL 599			MILES CITY 593			MISSOULA 598			WOLF POINT 592		
		MAT.	INST.	TOTAL	MAT.	INST.	TOTAL	MAT.	INST.	TOTAL	MAT.	INST.	TOTAL	MAT.	INST.	TOTAL	MAT.	INST.	TOTAL
015433	CONTRACTOR EQUIPMENT		97.7	97.7		99.9	99.9		97.7	97.7		97.7	97.7		97.7	97.7		97.7	97.7
0241, 31 - 34	SITE & INFRASTRUCTURE, DEMOLITION	104.2	93.1	96.6	91.1	96.8	95.0	87.5	93.3	91.5	93.5	93.2	93.3	80.5	93.3	89.3	110.2	93.2	98.4
0310	Concrete Forming & Accessories	79.6	67.2	69.0	100.2	68.3	73.0	89.7	68.3	71.4	97.7	67.3	71.8	89.7	68.4	71.5	90.4	67.3	70.7
0320	Concrete Reinforcing	107.1	78.9	93.5	113.0	81.8	97.9	109.1	85.8	97.8	106.7	78.9	93.3	108.1	85.8	97.3	108.2	78.2	93.3
0330	Cast-in-Place Concrete	140.7	71.2	114.9	102.1	73.3	91.4	113.4	72.5	98.2	124.2	71.3	104.5	96.2	72.6	87.4	139.1	70.2	113.5
03	CONCRETE	109.8	71.6	93.0	94.7	73.3	85.3	91.7	73.7	83.8	98.8	71.7	86.9	80.5	73.8	77.6	113.2	71.2	94.8
04	MASONRY	121.9	80.6	96.5	114.2	82.7	94.8	119.6	82.7	96.9	127.6	80.6	98.7	144.5	82.7	106.5	128.9	80.6	99.2
05	METALS	97.3	88.2	94.6	103.2	88.5	98.9	97.2	90.8	95.3	96.5	88.4	94.1	97.7	91.0	95.7	96.6	88.1	94.1
06	WOOD, PLASTICS & COMPOSITES	68.1	64.6	66.2	93.8	64.7	77.8	80.2	64.6	71.6	88.4	64.6	75.3	80.2	64.6	71.6	79.5	64.6	71.3
07	THERMAL & MOISTURE PROTECTION	108.3	66.2	90.0	103.1	73.9	90.4	107.4	73.6	92.7	107.8	67.7	90.3	106.9	75.6	93.3	108.9	67.5	90.9
08	OPENINGS	97.4	65.2	89.9	96.3	65.9	89.2	97.4	66.8	90.3	96.9	65.2	89.5	96.9	66.8	89.9	96.9	65.1	89.5
0920	Plaster & Gypsum Board	114.1	64.0	80.4	113.5	64.0	80.2	118.8	64.0	81.9	127.8	64.0	84.9	118.8	64.0	81.9	121.8	64.0	82.9
0950, 0980	Ceilings & Acoustic Treatment	104.1	64.0	77.0	107.9	64.0	78.3	104.1	64.0	77.0	101.6	64.0	76.3	104.1	64.0	77.0	101.6	64.0	76.3
0960	Flooring	83.0	90.9	85.3	95.9	90.9	94.5	87.1	90.9	88.2	91.8	90.9	91.6	87.1	90.9	88.2	88.4	90.9	89.1
0970, 0990	Wall Finishes & Painting/Coating	92.3	67.8	77.8	97.6	67.8	79.9	92.3	91.0	91.5	92.3	67.8	77.8	92.3	91.0	91.5	92.3	67.8	77.8
09	FINISHES	92.3	71.3	80.9	100.1	71.8	84.8	92.9	74.3	82.8	95.4	71.3	82.3	92.3	74.3	82.6	95.0	71.3	82.1
COVERS	DIVS. 10 - 14, 25, 28, 41, 43, 44, 46	100.0	88.8	97.5	100.0	91.9	98.2	100.0	89.5	97.7	100.0	88.9	97.5	100.0	92.4	98.3	100.0	88.9	97.5
21, 22, 23	FIRE SUPPRESSION, PLUMBING & HVAC	96.5	68.9	85.6	100.0	71.2	88.7	96.5	69.4	85.8	96.5	75.2	88.1	100.0	71.2	88.7	96.5	75.2	88.1
26, 27, 3370	ELECTRICAL, COMMUNICATIONS & UTIL.	101.7	69.2	85.7	109.0	70.8	90.2	106.0	68.1	87.3	101.7	74.4	88.3	107.0	70.2	88.9	101.7	74.4	88.3
MF2018	WEIGHTED AVERAGE	100.4	74.8	89.6	100.9	76.7	90.7	98.1	76.2	88.9	99.2	77.0	89.8	98.6	77.1	89.5	101.4	76.9	91.1

NEBRASKA

DIVISION		ALLIANCE 693			COLUMBUS 686			GRAND ISLAND 688			HASTINGS 689			LINCOLN 683 - 685			MCCOOK 690		
		MAT.	INST.	TOTAL	MAT.	INST.	TOTAL	MAT.	INST.	TOTAL	MAT.	INST.	TOTAL	MAT.	INST.	TOTAL	MAT.	INST.	TOTAL
015433	CONTRACTOR EQUIPMENT		95.3	95.3		101.9	101.9		101.9	101.9		101.9	101.9		105.0	105.0		101.9	101.9
0241, 31 - 34	SITE & INFRASTRUCTURE, DEMOLITION	98.2	96.2	96.8	101.8	90.6	94.1	106.8	90.7	95.7	105.4	90.6	95.2	94.9	95.7	95.4	101.5	90.6	94.0
0310	Concrete Forming & Accessories	87.0	55.3	60.0	96.5	75.0	78.2	96.1	68.8	72.9	99.3	72.5	76.5	94.4	75.5	78.3	92.3	55.9	61.2
0320	Concrete Reinforcing	107.4	87.4	97.7	100.4	85.9	93.4	99.8	76.1	88.4	99.8	76.1	88.4	97.6	76.4	87.3	100.1	76.3	88.6
0330	Cast-in-Place Concrete	108.1	81.8	98.3	107.3	81.8	97.8	113.6	77.7	100.2	113.6	74.0	98.9	87.2	83.1	85.6	116.9	73.9	100.9
03	CONCRETE	114.3	70.9	95.2	100.2	80.4	91.5	104.9	74.5	91.5	105.1	74.9	91.8	87.5	79.4	83.9	103.7	67.4	87.7
04	MASONRY	106.0	77.4	88.4	111.1	77.4	90.4	104.3	73.8	85.6	112.8	74.0	88.9	93.6	78.2	84.1	101.9	77.4	86.8
05	METALS	99.1	85.2	95.0	91.4	98.0	93.3	93.2	93.7	93.3	94.0	93.7	93.9	95.0	93.3	94.5	93.9	93.8	93.9
06	WOOD, PLASTICS & COMPOSITES	82.6	48.7	64.0	94.3	74.9	83.6	93.4	66.3	78.5	97.2	72.1	83.4	99.0	75.0	85.8	90.7	49.7	68.1
07	THERMAL & MOISTURE PROTECTION	100.7	75.8	89.9	101.6	80.4	92.4	101.8	78.3	91.5	101.8	77.6	91.3	100.1	81.1	91.8	96.0	76.1	87.3
08	OPENINGS	90.8	58.1	83.2	91.3	73.7	87.2	91.3	66.8	85.6	91.3	69.8	86.3	105.4	68.8	96.9	91.2	55.8	83.0
0920	Plaster & Gypsum Board	79.8	47.5	58.1	93.4	74.3	80.6	92.5	65.5	74.3	94.4	71.4	78.9	102.7	74.3	83.6	91.0	48.4	62.4
0950, 0980	Ceilings & Acoustic Treatment	88.6	47.5	60.9	83.4	74.3	77.3	83.4	65.5	71.3	83.4	71.4	75.3	98.3	74.3	82.1	83.9	48.4	60.0
0960	Flooring	91.4	87.3	90.2	84.0	87.3	84.9	83.7	78.9	82.4	85.0	81.3	84.0	96.3	86.6	93.6	90.0	87.3	89.2
0970, 0990	Wall Finishes & Painting/Coating	155.8	51.7	94.2	74.9	60.1	66.1	74.9	62.5	67.6	74.9	60.1	66.1	94.4	79.3	85.5	87.6	45.0	62.4
09	FINISHES	90.8	59.2	73.7	84.9	75.5	79.8	84.9	69.3	76.5	85.6	72.7	78.6	95.6	77.9	86.0	88.4	59.1	72.5
COVERS	DIVS. 10 - 14, 25, 28, 41, 43, 44, 46	100.0	83.9	96.4	100.0	83.5	96.3	100.0	89.7	97.7	100.0	83.1	96.2	100.0	91.0	98.0	100.0	84.1	96.5
21, 22, 23	FIRE SUPPRESSION, PLUMBING & HVAC	96.7	75.0	88.2	96.6	75.3	88.2	100.1	79.5	92.0	96.6	74.6	87.9	100.0	79.6	92.0	96.5	75.2	88.1
26, 27, 3370	ELECTRICAL, COMMUNICATIONS & UTIL.	93.6	65.6	79.8	97.3	81.4	89.5	95.9	65.7	81.0	95.3	79.4	87.5	109.2	65.7	87.8	95.6	65.7	80.9
MF2018	WEIGHTED AVERAGE	98.7	73.3	87.9	96.0	80.7	89.6	97.4	76.8	88.7	97.1	78.2	89.1	98.3	79.7	90.5	96.3	73.1	86.5

NEBRASKA / NEVADA

DIVISION		NORFOLK 687			NORTH PLATTE 691			OMAHA 680 - 681			VALENTINE 692			CARSON CITY 897			ELKO 898		
		MAT.	INST.	TOTAL	MAT.	INST.	TOTAL	MAT.	INST.	TOTAL	MAT.	INST.	TOTAL	MAT.	INST.	TOTAL	MAT.	INST.	TOTAL
015433	CONTRACTOR EQUIPMENT		91.7	91.7		101.9	101.9		94.2	94.2		95.0	95.0		97.9	97.9		94.6	94.6
0241, 31 - 34	SITE & INFRASTRUCTURE, DEMOLITION	84.1	89.7	88.0	102.9	90.4	94.3	89.1	93.9	92.4	86.3	95.2	92.4	83.3	97.4	93.0	67.8	93.8	84.0
0310	Concrete Forming & Accessories	83.2	74.1	75.5	94.8	75.2	78.1	93.1	76.7	79.1	83.5	53.1	57.6	106.1	76.3	80.7	111.4	96.3	98.6
0320	Concrete Reinforcing	100.5	66.5	84.1	99.5	76.2	88.3	100.8	76.5	89.0	100.1	66.3	83.7	115.8	116.1	116.0	124.5	114.5	119.7
0330	Cast-in-Place Concrete	107.9	71.3	94.3	116.9	60.6	96.0	91.4	78.6	86.7	103.1	54.0	84.8	98.0	82.2	92.1	95.0	72.4	86.6
03	CONCRETE	98.8	72.4	87.2	103.8	71.4	89.6	89.8	77.8	84.5	101.2	56.9	81.7	101.0	85.3	94.1	98.5	90.8	95.1
04	MASONRY	117.1	77.3	92.7	90.4	76.9	82.1	93.0	80.0	85.0	101.4	76.8	86.3	118.2	68.8	87.8	125.0	67.7	89.8
05	METALS	94.8	80.4	90.6	93.1	93.2	93.2	95.0	83.9	91.7	105.1	79.4	97.5	111.4	94.1	106.3	115.0	93.6	108.7
06	WOOD, PLASTICS & COMPOSITES	78.4	74.4	76.2	92.7	76.9	84.0	93.4	76.6	84.2	77.5	47.4	61.0	93.6	74.4	83.0	104.7	102.3	103.4
07	THERMAL & MOISTURE PROTECTION	101.2	78.3	91.2	95.9	78.4	88.3	96.7	80.8	89.8	96.5	73.9	86.7	118.1	79.1	101.2	114.2	73.6	96.5
08	OPENINGS	92.7	68.7	87.1	90.6	72.1	86.3	99.8	76.8	94.4	92.7	53.4	83.6	101.1	77.6	95.6	102.5	92.6	100.2
0920	Plaster & Gypsum Board	93.2	74.3	80.5	91.0	76.4	81.2	102.9	76.4	85.1	92.5	46.6	61.6	100.8	73.7	82.6	106.9	102.5	103.9
0950, 0980	Ceilings & Acoustic Treatment	97.6	74.3	81.9	83.9	76.4	78.8	101.0	76.4	84.4	100.4	46.6	64.1	96.0	73.7	80.9	94.5	102.5	99.9
0960	Flooring	104.1	87.3	99.4	91.1	78.9	87.7	100.6	87.3	96.9	116.8	87.3	108.5	99.5	67.1	90.4	103.0	67.1	92.9
0970, 0990	Wall Finishes & Painting/Coating	132.5	60.1	89.7	87.6	58.9	70.6	104.2	61.0	78.6	155.3	61.0	99.5	95.8	79.3	86.1	94.2	79.3	85.4
09	FINISHES	101.8	75.2	87.4	88.7	74.8	81.2	100.4	76.9	87.7	109.1	59.3	82.1	96.5	74.0	84.3	94.0	90.4	92.0
COVERS	DIVS. 10 - 14, 25, 28, 41, 43, 44, 46	100.0	85.1	96.7	100.0	61.1	91.3	100.0	90.2	97.8	100.0	56.1	90.2	100.0	100.2	100.0	100.0	91.0	98.0
21, 22, 23	FIRE SUPPRESSION, PLUMBING & HVAC	96.3	74.3	87.7	100.0	73.7	89.6	100.0	80.3	92.3	96.1	73.5	87.2	100.0	77.0	91.0	98.3	76.9	89.9
26, 27, 3370	ELECTRICAL, COMMUNICATIONS & UTIL.	96.2	81.4	88.9	93.9	65.7	80.0	103.9	83.7	94.0	91.4	65.7	78.7	102.2	88.3	95.3	98.8	88.3	93.6
MF2018	WEIGHTED AVERAGE	97.4	77.5	89.0	96.3	75.6	87.6	97.6	81.5	90.8	98.6	69.2	86.2	102.9	82.5	94.3	102.3	85.2	95.0

For customer support on your Mechanical Costs with RSMeans Data, call 800.448.8182.

677

City Cost Indexes

NEVADA / NEW HAMPSHIRE

DIVISION		ELY 893 MAT.	INST.	TOTAL	LAS VEGAS 889-891 MAT.	INST.	TOTAL	RENO 894-895 MAT.	INST.	TOTAL	CHARLESTON 036 MAT.	INST.	TOTAL	CLAREMONT 037 MAT.	INST.	TOTAL	CONCORD 032-033 MAT.	INST.	TOTAL
015433	CONTRACTOR EQUIPMENT		94.6	94.6		94.6	94.6		94.6	94.6		95.7	95.7		95.7	95.7		98.4	98.4
0241, 31 - 34	SITE & INFRASTRUCTURE, DEMOLITION	73.4	92.5	86.6	76.6	95.5	89.7	73.4	92.6	86.7	80.6	96.1	91.3	74.6	96.1	89.5	88.7	101.2	97.3
0310	Concrete Forming & Accessories	104.3	101.6	102.0	105.2	105.6	105.5	100.6	76.3	79.9	86.6	83.1	83.6	92.1	83.2	84.5	97.2	94.8	95.1
0320	Concrete Reinforcing	123.1	114.7	119.0	113.5	125.9	119.5	116.4	124.2	120.1	89.7	88.6	89.1	89.7	88.6	89.1	100.9	88.7	95.0
0330	Cast-in-Place Concrete	102.0	95.2	99.4	98.7	108.2	102.2	107.9	81.1	97.9	84.7	116.1	96.4	77.8	116.1	92.1	100.9	117.8	107.2
03	CONCRETE	107.5	101.2	104.7	102.4	109.5	105.5	106.7	86.3	98.7	90.4	95.8	92.7	82.8	95.8	88.5	96.8	101.6	98.9
04	MASONRY	130.5	74.8	96.3	117.4	106.2	110.5	124.2	68.7	90.1	89.8	99.5	95.8	89.9	99.5	95.8	97.8	101.4	100.0
05	METALS	115.0	96.5	109.5	124.1	103.9	118.2	116.7	97.4	111.1	94.2	92.1	93.5	94.2	92.1	93.6	101.3	91.8	98.5
06	WOOD, PLASTICS & COMPOSITES	94.8	104.4	100.1	92.4	102.3	97.8	88.4	74.3	80.6	86.6	79.4	82.6	92.4	79.4	85.3	93.3	93.6	93.5
07	THERMAL & MOISTURE PROTECTION	114.7	92.2	104.9	129.4	101.9	117.4	114.2	78.3	98.6	107.3	106.5	107.0	107.1	106.5	106.8	111.3	109.5	110.5
08	OPENINGS	102.4	93.8	100.4	101.4	109.6	103.3	100.2	79.2	95.3	96.1	80.6	92.5	97.2	80.6	93.3	97.0	88.5	95.1
0920	Plaster & Gypsum Board	101.2	104.6	103.5	93.9	102.5	99.7	89.0	73.7	78.7	106.1	78.5	87.5	107.0	78.5	87.9	108.7	93.1	98.2
0950, 0980	Ceilings & Acoustic Treatment	94.5	104.6	101.3	101.0	102.5	102.0	97.8	73.7	81.5	101.8	78.5	86.1	101.8	78.5	86.1	106.1	93.1	97.4
0960	Flooring	100.5	67.1	91.1	92.0	104.1	95.4	97.3	67.1	88.8	85.1	113.1	92.9	87.6	113.1	94.7	96.4	113.1	101.1
0970, 0990	Wall Finishes & Painting/Coating	94.2	112.0	104.7	96.8	119.4	110.2	94.2	79.3	85.4	81.8	90.0	86.6	81.8	90.0	86.6	95.9	90.0	92.4
09	FINISHES	92.9	97.0	95.1	90.8	106.4	99.3	90.8	73.9	81.7	90.2	88.5	89.3	90.6	88.5	89.5	95.9	97.4	96.7
COVERS	DIVS. 10 - 14, 25, 28, 41, 43, 44, 46	100.0	70.7	93.5	100.0	105.1	101.1	100.0	99.8	100.0	100.0	88.5	97.4	100.0	88.5	97.4	100.0	109.5	102.1
21, 22, 23	FIRE SUPPRESSION, PLUMBING & HVAC	98.3	96.2	97.4	100.1	103.0	101.3	100.0	77.1	91.0	96.7	82.2	91.0	96.7	82.2	91.0	100.0	87.2	95.0
26, 27, 3370	ELECTRICAL, COMMUNICATIONS & UTIL.	99.0	93.0	96.0	103.5	107.8	105.6	99.4	88.3	93.9	91.4	75.2	83.4	91.4	75.2	83.4	90.2	75.2	82.8
MF2018	WEIGHTED AVERAGE	103.7	93.2	99.2	105.0	105.1	105.0	103.6	82.6	94.7	94.1	88.4	91.7	93.1	88.4	91.1	98.0	93.2	96.0

NEW HAMPSHIRE / NEW JERSEY

DIVISION		KEENE 034 MAT.	INST.	TOTAL	LITTLETON 035 MAT.	INST.	TOTAL	MANCHESTER 031 MAT.	INST.	TOTAL	NASHUA 030 MAT.	INST.	TOTAL	PORTSMOUTH 038 MAT.	INST.	TOTAL	ATLANTIC CITY 082, 084 MAT.	INST.	TOTAL
015433	CONTRACTOR EQUIPMENT		95.7	95.7		95.7	95.7		99.0	99.0		95.7	95.7		95.7	95.7		93.3	93.3
0241, 31 - 34	SITE & INFRASTRUCTURE, DEMOLITION	88.0	96.2	93.6	74.7	95.2	88.8	87.8	101.3	97.1	89.5	96.3	94.2	83.3	97.0	92.7	88.2	99.7	96.1
0310	Concrete Forming & Accessories	90.7	83.4	84.5	102.4	77.3	81.0	97.5	95.1	95.5	98.5	94.9	95.4	87.9	94.1	93.2	114.8	142.0	138.0
0320	Concrete Reinforcing	89.7	88.6	89.2	90.4	88.6	89.5	108.5	88.7	98.9	112.0	88.7	100.7	89.7	88.7	89.2	81.1	138.0	108.6
0330	Cast-in-Place Concrete	85.1	116.2	96.7	76.3	107.5	87.9	99.2	118.3	106.3	80.5	117.4	94.3	76.3	117.2	91.5	78.5	137.1	100.3
03	CONCRETE	89.9	95.9	92.6	82.2	90.1	85.7	97.0	101.9	99.2	90.0	101.6	95.1	82.4	101.2	90.7	87.7	138.1	109.8
04	MASONRY	92.5	99.5	96.8	100.9	84.1	90.6	94.4	101.4	98.7	94.4	101.4	98.7	90.2	99.8	96.1	104.1	140.0	126.1
05	METALS	94.8	92.5	94.2	94.9	92.2	94.1	102.0	92.2	99.1	100.5	92.7	98.2	96.4	94.3	95.8	98.8	115.0	103.6
06	WOOD, PLASTICS & COMPOSITES	90.9	79.4	84.6	101.9	79.4	89.5	93.6	93.7	93.7	100.2	93.6	96.6	87.8	93.6	91.0	122.8	143.7	134.3
07	THERMAL & MOISTURE PROTECTION	107.9	106.4	107.2	107.2	99.5	103.8	112.5	109.5	111.2	108.3	108.9	108.6	107.8	108.1	107.9	102.1	134.0	116.0
08	OPENINGS	94.8	84.1	92.3	98.1	80.6	94.0	95.7	93.3	95.2	99.0	91.3	97.2	99.6	81.9	95.5	96.4	139.2	106.4
0920	Plaster & Gypsum Board	106.4	78.5	87.6	120.6	78.5	92.3	99.7	93.1	95.3	115.6	93.1	100.5	106.1	93.1	97.4	114.9	144.6	134.9
0950, 0980	Ceilings & Acoustic Treatment	101.8	78.5	86.1	101.8	78.5	86.1	101.2	93.1	95.8	114.1	93.1	100.0	102.7	93.1	96.3	96.4	144.6	128.9
0960	Flooring	87.2	113.1	94.5	97.9	113.1	102.2	93.8	113.1	99.2	91.0	113.1	97.2	85.2	113.1	93.0	97.5	162.5	115.8
0970, 0990	Wall Finishes & Painting/Coating	81.8	103.4	94.6	81.8	90.0	86.6	83.5	119.0	109.3	81.8	102.2	93.9	81.8	102.2	93.9	81.9	141.8	117.4
09	FINISHES	92.1	89.9	90.9	95.5	84.6	89.6	94.8	100.6	97.9	96.8	98.7	97.8	91.1	98.2	95.0	93.5	147.2	122.6
COVERS	DIVS. 10 - 14, 25, 28, 41, 43, 44, 46	100.0	94.7	98.8	100.0	95.6	99.0	100.0	109.7	102.2	100.0	109.4	102.1	100.0	108.7	101.9	100.0	114.9	103.3
21, 22, 23	FIRE SUPPRESSION, PLUMBING & HVAC	96.7	82.3	91.0	96.7	74.1	87.8	100.1	87.3	95.1	100.2	87.3	95.1	100.2	86.3	94.7	99.7	133.1	112.8
26, 27, 3370	ELECTRICAL, COMMUNICATIONS & UTIL.	91.4	75.2	83.4	92.2	49.1	71.0	93.9	79.3	86.7	93.2	79.3	86.4	91.7	75.3	83.6	94.2	143.4	118.5
MF2018	WEIGHTED AVERAGE	94.5	89.0	92.2	94.4	80.2	88.4	98.2	94.5	96.6	97.5	93.7	95.9	94.9	92.4	93.8	96.8	133.2	112.2

NEW JERSEY

DIVISION		CAMDEN 081 MAT.	INST.	TOTAL	DOVER 078 MAT.	INST.	TOTAL	ELIZABETH 072 MAT.	INST.	TOTAL	HACKENSACK 076 MAT.	INST.	TOTAL	JERSEY CITY 073 MAT.	INST.	TOTAL	LONG BRANCH 077 MAT.	INST.	TOTAL
015433	CONTRACTOR EQUIPMENT		93.3	93.3		95.7	95.7		95.7	95.7		95.7	95.7		93.3	93.3		92.9	92.9
0241, 31 - 34	SITE & INFRASTRUCTURE, DEMOLITION	89.3	99.3	96.2	103.8	101.4	102.2	108.9	101.4	103.7	104.8	101.4	102.5	95.2	101.4	99.5	100.0	100.5	100.4
0310	Concrete Forming & Accessories	105.0	137.3	132.5	96.4	148.3	140.6	108.0	148.5	142.5	96.4	148.3	140.7	100.6	148.4	141.3	101.1	139.0	133.4
0320	Concrete Reinforcing	106.5	128.5	117.1	75.3	156.5	114.6	75.3	156.5	114.6	75.3	156.5	114.6	97.7	156.5	126.2	75.3	156.1	114.4
0330	Cast-in-Place Concrete	76.1	133.8	97.6	81.3	131.5	100.0	69.8	140.3	96.1	79.4	140.3	102.1	63.4	131.6	88.8	70.5	133.8	94.0
03	CONCRETE	88.2	133.2	108.0	86.1	142.3	110.8	83.3	145.5	110.6	84.5	145.4	111.2	81.0	142.2	107.9	85.1	138.6	108.6
04	MASONRY	94.2	135.0	119.3	91.8	145.6	124.7	107.2	145.6	130.8	95.3	145.6	126.2	85.3	145.6	122.4	99.6	135.5	121.6
05	METALS	104.7	111.0	106.5	96.5	125.1	104.9	98.0	125.3	106.0	96.5	125.1	104.9	102.4	122.2	108.2	96.6	121.0	103.7
06	WOOD, PLASTICS & COMPOSITES	110.1	137.9	125.4	94.9	148.3	124.3	109.7	148.3	131.0	94.9	148.3	124.3	95.8	148.3	124.7	96.9	139.3	120.2
07	THERMAL & MOISTURE PROTECTION	102.0	133.0	115.5	103.7	142.8	120.7	104.0	144.1	121.4	103.5	136.7	117.9	103.2	142.8	120.4	103.4	129.0	114.5
08	OPENINGS	98.4	132.0	106.3	102.0	145.1	112.1	100.4	145.1	110.8	99.8	145.1	110.4	98.5	145.1	109.4	94.6	138.5	104.8
0920	Plaster & Gypsum Board	110.9	138.6	129.6	111.5	149.4	137.0	119.0	149.4	139.5	111.5	149.4	137.0	114.6	149.4	138.0	113.7	140.2	131.5
0950, 0980	Ceilings & Acoustic Treatment	106.3	138.6	128.1	92.3	149.4	130.8	94.1	149.4	131.4	92.3	149.4	130.8	102.2	149.4	134.0	92.3	140.2	124.6
0960	Flooring	93.7	162.5	113.0	83.7	188.8	113.2	88.7	188.8	116.8	83.7	188.8	113.2	84.7	188.8	114.0	84.9	176.9	110.8
0970, 0990	Wall Finishes & Painting/Coating	81.9	141.8	117.4	83.8	148.6	122.1	83.8	148.6	122.1	83.8	148.6	122.1	83.9	148.6	122.2	83.9	141.8	118.2
09	FINISHES	93.7	143.7	120.8	90.0	156.2	125.8	93.4	156.2	127.4	89.9	156.2	125.8	92.4	156.6	127.1	91.0	147.2	121.4
COVERS	DIVS. 10 - 14, 25, 28, 41, 43, 44, 46	100.0	114.3	103.2	100.0	131.7	107.1	100.0	131.7	107.1	100.0	131.7	107.1	100.0	131.7	107.1	100.0	114.5	103.2
21, 22, 23	FIRE SUPPRESSION, PLUMBING & HVAC	100.0	126.3	110.3	99.7	135.7	113.8	100.0	137.1	114.6	99.9	135.7	113.8	100.0	135.7	114.0	99.7	131.1	112.1
26, 27, 3370	ELECTRICAL, COMMUNICATIONS & UTIL.	98.4	132.1	115.0	96.0	138.5	116.9	96.6	138.5	117.2	96.0	142.3	118.8	100.5	142.3	121.1	95.7	129.9	112.6
MF2018	WEIGHTED AVERAGE	98.0	127.8	110.6	96.4	137.5	113.7	97.5	138.3	114.7	96.1	138.3	114.0	96.5	137.8	114.0	95.8	130.9	110.7

For customer support on your Mechanical Costs with RSMeans Data, call 800.448.8182.

NEW JERSEY

DIVISION		NEW BRUNSWICK 088 - 089			NEWARK 070 - 071			PATERSON 074 - 075			POINT PLEASANT 087			SUMMIT 079			TRENTON 085 - 086		
		MAT.	INST.	TOTAL	MAT.	INST.	TOTAL	MAT.	INST.	TOTAL	MAT.	INST.	TOTAL	MAT.	INST.	TOTAL	MAT.	INST.	TOTAL
015433	CONTRACTOR EQUIPMENT		92.9	92.9		98.3	98.3		95.7	95.7		92.9	92.9		95.7	95.7		97.0	97.0
0241, 31 - 34	SITE & INFRASTRUCTURE, DEMOLITION	101.1	100.9	100.9	110.0	106.2	107.4	106.6	101.4	103.0	102.6	100.5	101.2	106.3	101.4	102.9	88.6	105.4	100.2
0310	Concrete Forming & Accessories	108.5	148.0	142.2	100.7	148.7	141.6	98.3	148.3	140.9	102.4	139.1	133.6	99.0	148.5	141.2	103.5	138.8	133.6
0320	Concrete Reinforcing	82.1	156.1	117.9	96.2	156.6	125.4	97.7	156.5	126.2	82.1	156.1	117.9	75.3	158.1	116.6	107.7	116.2	111.8
0330	Cast-in-Place Concrete	97.0	136.9	111.8	89.8	140.9	108.8	80.8	140.2	102.9	97.0	135.8	111.5	67.5	140.3	94.6	95.6	134.4	110.0
03	CONCRETE	104.2	143.8	121.6	93.1	145.7	116.2	88.8	145.3	113.7	103.9	139.4	119.5	80.6	145.5	109.1	97.4	132.1	112.6
04	MASONRY	102.5	140.6	125.9	94.9	145.7	126.1	91.7	145.6	124.9	91.4	135.5	118.5	93.8	145.6	125.6	95.4	135.5	120.0
05	METALS	98.9	121.4	105.5	104.2	124.7	110.3	97.5	125.1	105.6	98.9	121.0	105.4	96.5	125.3	104.9	104.3	108.8	105.6
06	WOOD, PLASTICS & COMPOSITES	115.7	148.3	133.6	97.8	148.4	125.6	97.5	148.3	125.5	107.3	139.3	124.9	98.7	148.3	126.0	101.6	139.4	122.4
07	THERMAL & MOISTURE PROTECTION	102.5	138.2	118.0	105.2	144.7	122.4	103.8	136.7	118.1	102.5	131.3	115.0	104.1	144.1	121.5	102.9	135.2	117.0
08	OPENINGS	91.6	145.1	104.1	100.5	145.1	110.9	105.0	145.1	114.3	93.4	141.0	104.5	106.1	145.1	115.2	96.9	130.0	104.6
0920	Plaster & Gypsum Board	113.0	149.4	137.5	106.2	149.4	135.3	114.6	149.4	138.0	107.9	140.2	129.6	113.7	149.4	137.7	101.6	140.2	127.5
0950, 0980	Ceilings & Acoustic Treatment	96.4	149.4	132.2	103.0	149.4	134.3	102.2	149.4	134.0	96.4	140.2	125.9	92.3	149.4	130.8	101.2	140.2	127.5
0960	Flooring	95.1	188.8	121.4	96.7	188.8	122.6	84.7	188.8	114.0	92.5	162.5	112.2	84.9	188.8	114.1	100.1	176.9	121.7
0970, 0990	Wall Finishes & Painting/Coating	81.9	148.6	121.4	94.1	148.6	126.3	83.8	148.6	122.1	81.9	141.8	117.4	83.8	148.6	122.1	94.9	141.8	122.7
09	FINISHES	93.8	156.2	127.6	97.2	156.7	129.4	92.5	156.2	127.0	92.3	144.7	120.7	91.0	156.2	126.3	97.0	147.3	124.2
COVERS	DIVS. 10 - 14, 25, 28, 41, 43, 44, 46	100.0	131.5	107.0	100.0	131.8	107.1	100.0	131.7	107.1	100.0	111.6	102.6	100.0	131.7	107.1	100.0	114.7	103.3
21, 22, 23	FIRE SUPPRESSION, PLUMBING & HVAC	99.7	133.8	113.1	100.0	139.2	115.4	100.0	135.5	114.0	99.7	131.1	112.0	99.8	137.1	114.4	100.1	130.8	112.1
26, 27, 3370	ELECTRICAL, COMMUNICATIONS & UTIL.	94.8	133.7	114.0	105.0	142.3	123.4	100.5	138.5	119.2	94.2	129.7	111.8	96.6	138.5	117.2	102.6	128.7	115.4
MF2018	WEIGHTED AVERAGE	98.7	135.6	114.3	100.3	139.7	117.0	98.0	137.7	114.8	98.1	130.8	111.9	96.4	138.3	114.1	99.7	128.9	112.0

DIVISION		NEW JERSEY VINELAND 080, 083			NEW MEXICO ALBUQUERQUE 870 - 872			CARRIZOZO 883			CLOVIS 881			FARMINGTON 874			GALLUP 873		
		MAT.	INST.	TOTAL	MAT.	INST.	TOTAL	MAT.	INST.	TOTAL	MAT.	INST.	TOTAL	MAT.	INST.	TOTAL	MAT.	INST.	TOTAL
015433	CONTRACTOR EQUIPMENT		93.3	93.3		107.5	107.5		107.5	107.5		107.5	107.5		107.5	107.5		107.5	107.5
0241, 31 - 34	SITE & INFRASTRUCTURE, DEMOLITION	92.5	99.4	97.3	91.1	97.5	95.5	109.7	97.5	101.3	96.6	97.5	97.2	97.3	97.5	97.5	107.2	97.5	100.5
0310	Concrete Forming & Accessories	99.2	137.6	131.9	101.0	66.2	71.3	98.6	66.2	71.0	98.6	66.1	70.9	101.1	66.2	71.4	101.1	66.2	71.4
0320	Concrete Reinforcing	81.1	131.2	105.3	99.9	71.5	86.1	116.0	71.5	94.5	117.3	71.4	95.1	109.2	71.5	90.9	104.5	71.5	88.5
0330	Cast-in-Place Concrete	84.6	133.9	103.0	90.5	70.1	82.9	94.8	70.1	85.6	94.7	70.1	85.5	91.3	70.1	83.4	86.0	70.1	80.1
03	CONCRETE	92.2	133.8	110.5	92.9	69.8	82.7	116.7	69.8	96.1	104.3	69.7	89.1	96.4	69.8	84.7	103.3	69.8	88.6
04	MASONRY	92.6	135.5	118.9	106.0	60.5	78.1	108.5	60.5	79.0	108.5	60.5	79.0	113.8	60.5	81.1	116.6	60.5	76.4
05	METALS	98.8	112.6	102.8	108.1	90.6	102.9	106.6	90.6	101.9	106.3	90.5	101.6	105.7	90.6	101.3	104.8	90.6	100.7
06	WOOD, PLASTICS & COMPOSITES	103.7	137.9	122.5	102.1	67.2	82.9	93.5	67.2	79.0	93.5	67.2	79.0	102.2	67.2	82.9	102.2	67.2	82.9
07	THERMAL & MOISTURE PROTECTION	101.9	132.2	115.1	101.0	72.8	88.7	106.3	72.8	91.7	105.0	72.8	91.0	101.3	72.8	88.9	102.5	72.8	89.6
08	OPENINGS	92.9	134.9	102.7	98.5	67.0	91.1	96.5	67.0	89.6	96.6	67.0	89.7	100.8	67.0	92.9	100.8	67.0	93.0
0920	Plaster & Gypsum Board	106.4	138.6	128.1	113.0	66.1	81.5	80.2	66.1	70.7	80.2	66.1	70.7	99.4	66.1	77.0	99.4	66.1	77.0
0950, 0980	Ceilings & Acoustic Treatment	96.4	138.6	124.9	99.8	66.1	77.1	102.7	66.1	78.1	102.7	66.1	78.1	98.6	66.1	76.7	98.6	66.1	76.7
0960	Flooring	91.7	162.5	111.6	88.9	66.8	82.7	97.2	66.8	88.7	97.2	66.8	88.7	90.4	66.8	83.8	90.4	66.8	83.8
0970, 0990	Wall Finishes & Painting/Coating	81.9	141.8	117.4	97.8	52.6	71.1	92.6	52.6	68.9	92.6	52.6	68.9	92.1	52.6	68.7	92.1	52.6	68.7
09	FINISHES	91.1	143.9	119.6	92.3	64.7	77.4	94.2	64.7	78.3	92.9	64.7	77.7	90.7	64.7	76.6	92.1	64.7	77.3
COVERS	DIVS. 10 - 14, 25, 28, 41, 43, 44, 46	100.0	114.5	103.2	100.0	85.1	96.7	100.0	85.1	96.7	100.0	85.1	96.7	100.0	85.1	96.7	100.0	85.1	96.7
21, 22, 23	FIRE SUPPRESSION, PLUMBING & HVAC	99.7	126.6	110.3	100.3	69.0	88.0	97.9	69.0	86.5	97.9	68.6	86.4	100.2	69.0	88.0	98.1	69.0	86.6
26, 27, 3370	ELECTRICAL, COMMUNICATIONS & UTIL.	94.2	143.3	118.5	87.9	69.5	78.8	90.1	69.5	80.0	87.9	69.5	78.8	85.9	69.5	77.8	85.3	69.5	77.5
MF2018	WEIGHTED AVERAGE	96.3	129.8	110.5	98.6	72.5	87.6	101.6	72.5	89.3	99.3	72.4	87.9	99.0	72.5	87.8	99.0	72.5	87.8

NEW MEXICO

DIVISION		LAS CRUCES 880			LAS VEGAS 877			ROSWELL 882			SANTA FE 875			SOCORRO 878			TRUTH/CONSEQUENCES 879		
		MAT.	INST.	TOTAL	MAT.	INST.	TOTAL	MAT.	INST.	TOTAL	MAT.	INST.	TOTAL	MAT.	INST.	TOTAL	MAT.	INST.	TOTAL
015433	CONTRACTOR EQUIPMENT		83.5	83.5		107.5	107.5		107.5	107.5		110.4	110.4		107.5	107.5		83.5	83.5
0241, 31 - 34	SITE & INFRASTRUCTURE, DEMOLITION	97.8	77.5	83.8	96.2	97.5	97.1	99.0	97.5	98.0	99.1	102.8	101.6	92.8	97.5	96.1	112.5	77.7	88.3
0310	Concrete Forming & Accessories	95.4	65.2	69.7	101.1	66.2	71.4	98.6	66.2	71.0	100.0	66.3	71.2	101.1	66.2	71.4	98.8	65.1	70.1
0320	Concrete Reinforcing	113.1	71.3	92.9	106.3	71.5	89.5	117.3	71.5	95.1	99.3	71.5	85.9	108.4	71.5	90.5	101.7	71.3	87.0
0330	Cast-in-Place Concrete	89.5	63.1	79.7	88.8	70.1	81.9	94.7	70.1	85.6	94.3	70.9	85.6	87.0	70.1	80.7	95.4	63.1	83.4
03	CONCRETE	82.1	66.5	75.3	94.0	69.8	83.4	105.3	69.8	89.7	92.9	70.0	82.9	92.9	69.8	82.8	86.1	66.4	77.4
04	MASONRY	104.0	60.2	77.1	101.9	60.5	76.5	119.5	60.5	83.3	95.0	60.6	73.9	101.8	60.5	76.4	98.7	60.2	75.0
05	METALS	105.0	83.0	98.5	104.5	90.6	100.4	107.5	90.6	102.6	101.6	89.6	98.1	104.8	90.6	100.7	104.4	83.0	98.1
06	WOOD, PLASTICS & COMPOSITES	82.9	66.2	73.7	102.2	67.2	82.9	93.5	67.2	79.0	99.0	67.2	81.5	102.2	67.2	82.9	93.5	66.2	78.5
07	THERMAL & MOISTURE PROTECTION	92.2	68.2	81.7	100.8	72.8	88.6	105.2	72.8	91.1	103.5	73.7	90.5	100.8	72.8	88.6	90.1	68.2	80.5
08	OPENINGS	92.0	66.4	86.1	97.3	67.0	90.3	96.4	67.0	89.6	98.8	67.0	91.4	97.2	67.0	90.1	90.6	66.4	85.0
0920	Plaster & Gypsum Board	78.5	66.1	70.2	99.4	66.1	77.0	80.2	66.1	70.7	114.1	66.1	81.8	99.4	66.1	77.0	100.6	66.1	77.4
0950, 0980	Ceilings & Acoustic Treatment	87.3	66.1	73.0	98.6	66.1	76.7	102.7	66.1	78.1	95.2	66.1	75.6	98.6	66.1	76.7	86.2	66.1	72.7
0960	Flooring	128.1	66.8	110.9	90.4	66.8	83.8	97.2	66.8	88.7	100.9	66.8	91.3	90.4	66.8	83.8	119.2	66.8	104.5
0970, 0990	Wall Finishes & Painting/Coating	81.8	52.6	64.5	92.1	52.6	68.7	92.6	52.6	68.9	99.8	52.6	71.9	92.1	52.6	68.7	84.5	52.6	65.6
09	FINISHES	103.0	64.0	81.9	90.5	64.7	76.6	93.0	64.7	77.7	98.2	64.8	80.1	90.4	64.7	76.5	102.4	64.0	81.6
COVERS	DIVS. 10 - 14, 25, 28, 41, 43, 44, 46	100.0	82.7	96.1	100.0	85.1	96.7	100.0	85.1	96.7	100.0	85.2	96.7	100.0	85.1	96.7	100.0	82.6	96.1
21, 22, 23	FIRE SUPPRESSION, PLUMBING & HVAC	100.4	68.7	88.0	98.1	69.0	86.6	100.0	69.0	87.8	100.3	69.8	88.3	98.1	69.0	86.6	98.0	68.7	86.5
26, 27, 3370	ELECTRICAL, COMMUNICATIONS & UTIL.	90.1	83.8	87.0	87.4	69.5	78.6	89.2	69.5	79.5	100.3	69.5	85.2	85.7	69.5	77.7	89.5	69.5	79.6
MF2018	WEIGHTED AVERAGE	96.9	71.4	86.1	97.2	72.5	86.8	100.8	72.5	88.9	99.0	72.9	88.0	96.8	72.5	86.6	96.6	69.4	85.1

For customer support on your Mechanical Costs with RSMeans Data, call 800.448.8182.

679

City Cost Indexes

		NEW MEXICO			NEW YORK														
		TUCUMCARI			ALBANY			BINGHAMTON			BRONX			BROOKLYN			BUFFALO		
DIVISION		884			120 - 122			137 - 139			104			112			140 - 142		
		MAT.	INST.	TOTAL	MAT.	INST.	TOTAL	MAT.	INST.	TOTAL	MAT.	INST.	TOTAL	MAT.	INST.	TOTAL	MAT.	INST.	TOTAL
015433	CONTRACTOR EQUIPMENT		107.5	107.5		115.6	115.6		117.6	117.6		104.4	104.4		109.4	109.4		100.5	100.5
0241, 31 - 34	SITE & INFRASTRUCTURE, DEMOLITION	96.2	97.5	97.1	80.0	105.1	97.4	94.3	89.3	90.8	97.3	109.7	105.9	118.1	120.6	119.8	97.3	102.3	100.7
0310	Concrete Forming & Accessories	98.6	66.1	70.9	97.3	107.3	105.8	99.4	93.4	94.3	95.7	189.1	175.3	105.1	189.1	176.7	101.3	117.1	114.7
0320	Concrete Reinforcing	115.0	71.4	93.9	101.3	115.2	108.0	95.8	106.3	100.9	94.8	182.0	136.9	97.2	242.5	167.5	98.9	116.4	107.3
0330	Cast-in-Place Concrete	94.7	70.1	85.5	81.0	116.4	94.2	106.3	106.9	106.5	85.6	172.0	117.8	108.3	170.5	131.5	109.1	123.5	114.4
03	CONCRETE	103.6	69.7	88.7	87.7	112.8	98.7	94.3	102.6	97.9	88.0	181.1	128.9	106.4	189.9	143.1	104.9	118.7	111.0
04	MASONRY	119.9	60.5	83.4	88.0	117.9	106.4	104.3	104.9	104.7	88.8	188.9	150.3	114.8	188.8	160.3	112.6	122.2	118.5
05	METALS	106.3	90.5	101.6	102.2	125.0	108.9	95.0	133.6	106.3	86.4	174.6	112.3	103.1	175.0	124.2	97.0	107.5	100.1
06	WOOD, PLASTICS & COMPOSITES	93.5	67.2	79.0	94.6	104.0	99.8	101.7	90.2	95.4	97.2	188.6	147.5	103.7	188.3	150.3	98.6	115.8	108.1
07	THERMAL & MOISTURE PROTECTION	105.0	72.8	91.0	105.4	110.4	107.6	109.2	94.8	102.9	102.5	168.8	131.3	109.3	168.2	134.9	102.5	111.7	106.5
08	OPENINGS	96.4	67.0	89.5	96.4	103.8	98.1	90.4	93.7	91.2	92.3	197.7	116.9	87.8	197.5	113.4	99.8	111.4	102.5
0920	Plaster & Gypsum Board	80.2	66.1	70.7	97.8	103.9	101.9	109.2	89.7	96.1	102.8	191.0	162.1	103.3	191.0	162.3	105.8	116.1	112.7
0950, 0980	Ceilings & Acoustic Treatment	102.7	66.1	78.1	97.1	103.9	101.7	96.6	89.7	92.0	89.1	191.0	157.8	91.8	191.0	158.7	104.2	116.1	112.2
0960	Flooring	97.2	66.8	88.7	90.9	110.3	96.4	103.1	102.4	102.9	99.9	182.5	123.1	110.2	182.5	130.5	96.3	116.6	102.0
0970, 0990	Wall Finishes & Painting/Coating	92.6	52.6	68.9	95.9	104.4	100.9	89.3	106.3	99.4	106.2	167.4	142.4	116.0	167.4	146.4	96.9	117.8	109.3
09	FINISHES	92.8	64.7	77.6	90.6	107.1	99.5	93.6	95.9	94.8	98.0	186.2	145.7	107.1	186.0	149.8	102.0	117.6	110.5
COVERS	DIVS. 10 - 14, 25, 28, 41, 43, 44, 46	100.0	85.1	96.7	100.0	103.8	100.9	100.0	96.6	99.2	100.0	141.5	109.3	100.0	140.8	109.1	100.0	106.5	101.4
21, 22, 23	FIRE SUPPRESSION, PLUMBING & HVAC	97.9	68.6	86.4	100.1	110.3	104.1	100.6	97.0	99.2	100.3	177.9	130.7	99.8	177.9	130.5	100.0	102.4	100.9
26, 27, 3370	ELECTRICAL, COMMUNICATIONS & UTIL.	90.1	69.5	80.0	100.0	108.4	104.1	98.7	101.0	99.8	90.6	184.5	136.9	98.5	184.5	140.9	99.5	105.3	102.4
MF2018	WEIGHTED AVERAGE	99.9	72.4	88.3	96.8	111.2	102.8	97.3	101.5	99.1	93.8	175.3	128.3	101.9	177.3	133.8	100.8	110.3	104.8

		NEW YORK																	
		ELMIRA			FAR ROCKAWAY			FLUSHING			GLENS FALLS			HICKSVILLE			JAMAICA		
DIVISION		148 - 149			116			113			128			115, 117, 118			114		
		MAT.	INST.	TOTAL	MAT.	INST.	TOTAL	MAT.	INST.	TOTAL	MAT.	INST.	TOTAL	MAT.	INST.	TOTAL	MAT.	INST.	TOTAL
015433	CONTRACTOR EQUIPMENT		120.0	120.0		109.4	109.4		109.4	109.4		112.6	112.6		109.4	109.4		109.4	109.4
0241, 31 - 34	SITE & INFRASTRUCTURE, DEMOLITION	97.4	89.6	92.0	121.2	120.6	120.8	121.3	120.6	120.8	71.4	99.6	90.9	111.1	119.1	116.6	115.5	120.6	119.0
0310	Concrete Forming & Accessories	85.0	96.3	94.6	92.2	189.1	174.8	95.9	189.1	175.3	81.5	99.9	97.2	88.8	157.2	147.1	95.9	189.1	175.3
0320	Concrete Reinforcing	98.6	105.3	101.8	97.2	242.5	167.5	98.9	242.5	168.4	96.9	112.7	104.5	97.2	174.4	134.6	97.2	242.5	167.5
0330	Cast-in-Place Concrete	98.8	106.4	101.6	117.2	170.5	137.1	117.2	170.5	137.1	78.1	111.6	90.6	99.5	161.5	122.6	108.3	170.5	131.5
03	CONCRETE	90.9	103.7	96.5	112.8	189.9	146.7	113.2	189.9	146.9	81.2	107.4	92.8	98.2	161.8	126.2	105.8	189.9	142.8
04	MASONRY	101.4	106.7	104.6	118.6	188.8	161.8	112.8	188.8	159.5	93.4	112.6	105.2	109.0	174.4	149.2	117.2	188.8	161.2
05	METALS	95.5	135.8	107.3	103.2	175.0	124.3	103.2	175.0	124.3	95.5	124.5	104.0	104.7	172.4	124.6	103.2	175.0	124.3
06	WOOD, PLASTICS & COMPOSITES	84.1	94.4	89.8	87.1	188.3	142.8	91.7	188.3	144.9	82.7	96.1	90.1	83.7	154.0	122.4	91.7	188.3	144.9
07	THERMAL & MOISTURE PROTECTION	108.9	95.2	102.9	109.2	168.3	134.9	109.3	168.3	134.9	98.6	107.1	102.3	108.9	157.8	130.2	109.1	168.3	134.8
08	OPENINGS	96.7	95.7	96.4	86.6	197.5	112.5	86.6	197.5	112.5	90.0	99.1	92.1	87.0	178.6	108.3	86.6	197.5	112.5
0920	Plaster & Gypsum Board	99.5	94.2	96.0	91.7	191.0	158.5	94.2	191.0	159.3	91.0	96.0	94.4	91.4	155.6	134.6	94.2	191.0	159.3
0950, 0980	Ceilings & Acoustic Treatment	106.0	94.2	98.1	81.2	191.0	155.2	81.2	191.0	155.2	86.4	96.0	92.9	80.3	155.6	131.1	81.2	191.0	155.2
0960	Flooring	87.6	102.4	91.8	105.0	182.5	126.8	106.5	182.5	127.9	82.3	107.7	89.5	104.0	179.7	125.2	106.5	182.5	127.9
0970, 0990	Wall Finishes & Painting/Coating	97.2	93.3	94.9	116.0	167.4	146.4	116.0	167.4	146.4	89.3	104.4	98.2	116.0	167.4	146.4	116.0	167.4	146.4
09	FINISHES	94.2	96.9	95.7	102.4	186.0	147.6	103.2	186.0	148.0	84.0	101.1	93.2	101.0	161.6	133.8	102.7	186.0	147.8
COVERS	DIVS. 10 - 14, 25, 28, 41, 43, 44, 46	100.0	97.6	99.5	100.0	140.8	109.1	100.0	140.8	109.1	100.0	96.9	99.3	100.0	132.1	107.1	100.0	140.8	109.1
21, 22, 23	FIRE SUPPRESSION, PLUMBING & HVAC	96.6	95.5	96.2	96.2	177.9	128.3	96.2	177.9	128.3	96.7	109.3	101.6	99.8	161.0	123.8	96.2	177.9	128.3
26, 27, 3370	ELECTRICAL, COMMUNICATIONS & UTIL.	96.4	104.1	100.2	105.3	184.5	144.4	105.3	184.5	144.4	94.3	105.0	99.6	98.0	141.0	119.2	97.0	184.5	140.1
MF2018	WEIGHTED AVERAGE	96.3	102.5	98.9	102.2	177.3	133.9	102.1	177.3	133.9	92.0	107.3	98.5	99.9	157.1	124.1	100.3	177.3	132.8

		NEW YORK																	
		JAMESTOWN			KINGSTON			LONG ISLAND CITY			MONTICELLO			MOUNT VERNON			NEW ROCHELLE		
DIVISION		147			124			111			127			105			108		
		MAT.	INST.	TOTAL	MAT.	INST.	TOTAL	MAT.	INST.	TOTAL	MAT.	INST.	TOTAL	MAT.	INST.	TOTAL	MAT.	INST.	TOTAL
015433	CONTRACTOR EQUIPMENT		90.6	90.6		109.4	109.4		109.4	109.4		109.4	109.4		104.4	104.4		104.4	104.4
0241, 31 - 34	SITE & INFRASTRUCTURE, DEMOLITION	99.8	89.9	92.6	139.6	115.9	123.2	119.2	120.6	120.2	134.9	115.8	121.7	102.7	105.4	104.5	102.2	105.3	104.3
0310	Concrete Forming & Accessories	85.1	90.1	89.3	82.7	130.7	123.7	100.0	189.1	176.0	89.3	130.8	124.7	86.7	138.8	131.1	100.5	135.3	130.1
0320	Concrete Reinforcing	98.9	111.2	104.8	97.3	160.4	127.8	97.2	242.5	167.5	96.5	160.4	127.4	93.5	180.8	135.7	93.6	180.6	135.7
0330	Cast-in-Place Concrete	102.4	105.2	103.4	105.5	146.3	120.7	111.8	170.5	133.7	98.7	146.3	116.5	95.6	150.0	115.8	95.6	149.7	115.7
03	CONCRETE	93.9	99.0	96.2	102.4	140.8	119.3	109.0	189.9	144.5	97.3	140.8	116.4	97.2	150.4	120.6	96.4	148.7	119.4
04	MASONRY	110.2	104.6	106.8	108.1	155.4	137.2	111.7	188.8	159.1	100.8	155.4	134.3	94.6	156.4	132.6	94.6	156.4	132.6
05	METALS	93.0	101.9	95.6	104.0	135.9	113.4	103.1	175.0	124.2	104.0	136.0	113.4	86.2	169.8	110.7	86.4	169.0	110.7
06	WOOD, PLASTICS & COMPOSITES	82.9	86.2	84.7	83.9	124.0	106.0	97.8	188.3	147.6	90.4	124.0	108.9	87.6	132.4	112.3	104.5	128.7	117.8
07	THERMAL & MOISTURE PROTECTION	108.5	95.4	102.8	121.5	145.0	131.7	109.2	168.3	134.9	121.1	145.0	131.5	103.4	147.3	122.5	103.5	144.7	121.4
08	OPENINGS	96.5	93.0	95.7	92.1	142.5	103.9	86.6	197.5	112.5	88.0	142.4	100.7	92.3	168.4	110.1	92.4	166.3	109.6
0920	Plaster & Gypsum Board	89.2	85.8	86.9	91.0	124.9	113.8	99.3	191.0	161.0	91.7	124.8	114.0	98.3	133.2	121.8	111.0	129.3	123.3
0950, 0980	Ceilings & Acoustic Treatment	102.7	85.8	91.3	77.5	124.9	109.4	81.2	191.0	155.2	77.5	124.8	109.4	87.5	133.2	118.3	87.5	129.3	115.7
0960	Flooring	90.3	102.4	93.7	98.6	160.5	116.0	108.2	182.5	129.1	101.0	160.5	117.8	91.2	179.7	116.1	99.2	164.0	117.4
0970, 0990	Wall Finishes & Painting/Coating	98.8	100.6	99.9	116.6	130.9	125.0	116.0	167.4	146.4	116.6	123.4	120.6	104.6	167.4	141.8	104.6	167.4	141.8
09	FINISHES	93.2	92.5	92.9	97.9	135.4	118.2	104.1	186.0	148.4	98.3	134.6	117.9	95.0	148.0	123.7	98.9	142.6	122.5
COVERS	DIVS. 10 - 14, 25, 28, 41, 43, 44, 46	100.0	96.8	99.3	100.0	121.2	104.7	100.0	140.8	109.1	100.0	121.3	104.7	100.0	129.2	106.5	100.0	113.0	102.9
21, 22, 23	FIRE SUPPRESSION, PLUMBING & HVAC	96.5	93.8	95.4	96.4	116.1	104.1	99.8	177.9	130.5	96.6	138.9	113.2	96.8	145.9	116.1	96.8	145.6	116.0
26, 27, 3370	ELECTRICAL, COMMUNICATIONS & UTIL.	95.4	94.9	95.1	95.7	116.3	105.9	97.5	184.5	140.4	95.7	116.3	105.9	88.8	170.6	129.1	88.8	141.7	114.9
MF2018	WEIGHTED AVERAGE	96.4	96.0	96.3	100.6	133.8	114.6	101.6	177.3	133.6	99.1	134.6	114.1	94.1	150.6	118.0	94.4	144.8	115.7

680

For customer support on your Mechanical Costs with RSMeans Data, call 800.448.8182.

NEW YORK

DIVISION		NEW YORK 100 - 102			NIAGARA FALLS 143			PLATTSBURGH 129			POUGHKEEPSIE 125 - 126			QUEENS 110			RIVERHEAD 119		
		MAT.	INST.	TOTAL	MAT.	INST.	TOTAL	MAT.	INST.	TOTAL	MAT.	INST.	TOTAL	MAT.	INST.	TOTAL	MAT.	INST.	TOTAL
015433	CONTRACTOR EQUIPMENT		104.6	104.6		90.6	90.6		94.5	94.5		109.4	109.4		109.4	109.4		109.4	109.4
0241, 31 - 34	SITE & INFRASTRUCTURE, DEMOLITION	106.6	111.9	110.3	100.9	90.8	94.0	109.3	96.7	100.6	135.9	114.8	121.3	114.3	120.6	118.7	112.1	118.6	116.6
0310	Concrete Forming & Accessories	104.3	188.9	176.4	85.0	113.1	108.9	86.8	91.4	90.7	82.7	167.6	155.1	89.0	189.1	174.3	93.2	156.2	146.9
0320	Concrete Reinforcing	100.3	178.3	138.0	97.5	112.0	104.5	101.4	112.0	106.5	97.3	160.0	127.6	98.9	242.5	168.4	99.1	215.6	155.5
0330	Cast-in-Place Concrete	98.9	172.2	126.2	106.0	124.1	112.7	95.5	100.6	97.4	102.2	135.4	114.5	102.9	170.5	128.1	101.2	159.7	123.0
03	CONCRETE	101.6	180.3	136.2	96.2	116.1	105.0	94.3	98.0	95.9	99.7	153.6	123.4	101.5	189.9	140.3	98.7	166.5	128.5
04	MASONRY	102.1	188.8	155.4	117.5	125.1	122.2	88.3	97.4	93.9	100.6	139.5	124.5	106.0	188.8	156.9	114.3	173.3	150.6
05	METALS	97.5	171.0	119.1	95.6	102.4	97.6	99.8	99.5	99.7	104.0	135.5	113.3	103.1	175.0	124.2	105.1	160.6	121.4
06	WOOD, PLASTICS & COMPOSITES	102.0	188.6	149.7	82.8	108.2	96.8	88.8	88.2	88.5	83.9	179.3	136.4	83.8	188.3	141.3	88.7	154.0	124.6
07	THERMAL & MOISTURE PROTECTION	105.2	170.0	133.4	108.6	110.0	109.2	116.0	97.2	107.8	121.4	144.1	131.3	108.8	168.3	134.7	110.0	154.8	129.5
08	OPENINGS	96.0	197.5	119.7	96.5	104.7	98.4	97.4	94.7	96.8	92.1	170.5	110.4	86.6	197.5	112.5	87.0	172.1	106.8
0920	Plaster & Gypsum Board	108.6	191.0	164.0	89.2	108.4	102.1	108.2	87.4	94.2	91.0	181.7	152.0	91.4	191.0	158.4	92.5	155.6	135.0
0950, 0980	Ceilings & Acoustic Treatment	104.5	191.0	162.8	102.7	108.4	106.6	107.3	87.4	93.9	77.5	181.7	147.7	81.2	191.0	155.2	81.1	155.6	131.4
0960	Flooring	101.3	182.5	124.1	90.3	112.1	96.4	104.2	105.1	104.4	98.6	158.6	115.5	104.0	182.5	126.0	105.0	166.9	122.4
0970, 0990	Wall Finishes & Painting/Coating	102.7	167.4	141.0	98.8	111.5	106.3	111.2	95.1	101.7	116.6	123.4	120.6	116.0	167.4	146.4	116.0	167.4	146.4
09	FINISHES	102.9	186.2	148.0	93.3	112.9	103.9	95.0	93.5	94.2	97.7	164.2	133.7	101.3	186.0	147.1	101.5	158.6	132.4
COVERS	DIVS. 10 - 14, 25, 28, 41, 43, 44, 46	100.0	141.5	109.3	100.0	102.7	100.6	100.0	92.7	98.4	100.0	116.5	103.7	100.0	140.8	109.1	100.0	120.1	104.5
21, 22, 23	FIRE SUPPRESSION, PLUMBING & HVAC	100.2	178.3	130.9	96.5	105.1	99.9	96.6	100.3	98.1	96.6	120.3	105.9	99.8	177.9	130.5	100.0	155.9	121.9
26, 27, 3370	ELECTRICAL, COMMUNICATIONS & UTIL.	96.8	184.5	140.0	94.1	98.9	96.4	92.1	91.7	91.9	95.7	121.2	108.3	98.0	184.5	140.6	99.5	134.9	116.9
MF2018	WEIGHTED AVERAGE	99.7	175.2	131.6	97.4	107.5	101.6	97.0	96.6	96.8	99.8	136.8	115.4	100.0	177.3	132.6	100.6	153.5	122.9

NEW YORK

DIVISION		ROCHESTER 144 - 146			SCHENECTADY 123			STATEN ISLAND 103			SUFFERN 109			SYRACUSE 130 - 132			UTICA 133 - 135		
		MAT.	INST.	TOTAL	MAT.	INST.	TOTAL	MAT.	INST.	TOTAL	MAT.	INST.	TOTAL	MAT.	INST.	TOTAL	MAT.	INST.	TOTAL
015433	CONTRACTOR EQUIPMENT		117.6	117.6		112.6	112.6		104.4	104.4		104.4	104.4		112.6	112.6		112.6	112.6
0241, 31 - 34	SITE & INFRASTRUCTURE, DEMOLITION	88.8	105.1	100.0	81.4	100.1	94.3	107.0	109.7	108.9	99.4	103.7	102.4	93.0	98.5	96.8	71.9	98.1	90.0
0310	Concrete Forming & Accessories	105.8	100.3	101.2	96.4	107.1	105.5	86.1	189.3	174.1	94.4	140.8	133.9	98.5	89.1	90.5	99.5	86.3	88.2
0320	Concrete Reinforcing	101.4	105.0	103.1	95.9	115.1	105.2	94.8	216.0	153.4	93.6	152.0	121.9	96.8	106.1	101.3	96.8	100.3	98.5
0330	Cast-in-Place Concrete	97.8	103.3	99.9	94.4	115.4	102.2	95.6	172.1	124.1	92.7	138.9	109.9	98.7	102.3	100.0	90.3	101.0	94.3
03	CONCRETE	99.5	103.3	101.2	94.9	112.4	102.6	99.1	186.4	137.5	93.8	141.3	114.7	97.6	98.1	97.8	95.8	95.4	95.6
04	MASONRY	93.6	103.4	99.7	90.7	117.9	107.4	100.9	188.9	154.9	94.5	142.2	123.8	97.0	101.0	99.5	88.4	99.4	95.2
05	METALS	104.6	119.3	108.9	99.7	126.1	107.4	84.6	174.8	111.1	84.6	131.9	98.5	98.6	119.8	104.9	96.7	117.5	102.9
06	WOOD, PLASTICS & COMPOSITES	108.4	100.2	103.9	100.1	103.8	102.1	86.2	188.6	142.6	96.9	142.5	122.0	98.2	86.0	91.5	98.2	82.7	89.7
07	THERMAL & MOISTURE PROTECTION	116.4	99.8	109.2	100.1	109.8	104.3	102.9	168.8	131.5	103.4	140.9	119.7	104.0	94.1	99.7	92.7	94.0	93.3
08	OPENINGS	100.7	99.3	100.3	95.3	103.7	97.3	92.3	197.7	116.9	92.4	150.7	106.0	92.4	89.5	91.8	95.2	86.3	93.1
0920	Plaster & Gypsum Board	103.8	100.1	101.3	100.3	103.9	102.7	98.4	191.0	160.7	102.1	143.5	130.0	96.8	85.6	89.3	96.8	82.2	87.0
0950, 0980	Ceilings & Acoustic Treatment	100.4	100.1	100.2	92.8	103.9	100.3	89.1	191.0	157.8	87.5	143.5	125.2	96.6	85.6	89.2	96.6	82.2	86.9
0960	Flooring	90.4	105.0	94.5	89.6	110.3	95.4	95.1	182.5	119.6	95.0	171.6	116.5	92.0	92.3	92.1	90.0	92.4	90.7
0970, 0990	Wall Finishes & Painting/Coating	97.7	98.4	98.1	89.3	104.4	98.2	106.2	167.4	142.4	104.6	126.9	117.8	91.9	98.7	95.9	85.6	98.7	93.4
09	FINISHES	95.7	101.4	98.8	89.2	107.0	98.9	96.7	186.2	145.2	96.3	145.8	123.1	92.4	90.0	91.1	90.8	87.9	89.2
COVERS	DIVS. 10 - 14, 25, 28, 41, 43, 44, 46	100.0	98.9	99.7	100.0	103.5	100.8	100.0	141.5	109.3	100.0	114.7	103.3	100.0	95.1	98.9	100.0	89.8	97.7
21, 22, 23	FIRE SUPPRESSION, PLUMBING & HVAC	99.9	88.1	95.3	100.2	106.2	102.6	100.3	177.9	130.8	96.8	123.7	107.4	100.3	94.4	98.0	100.3	92.5	97.2
26, 27, 3370	ELECTRICAL, COMMUNICATIONS & UTIL.	99.5	89.9	94.8	98.7	108.4	103.5	90.6	184.5	136.9	94.8	114.6	104.6	98.6	101.0	99.8	96.6	101.0	98.8
MF2018	WEIGHTED AVERAGE	100.3	99.2	99.8	97.0	109.9	102.4	95.6	176.1	129.6	94.1	130.2	109.4	97.8	98.3	98.0	96.0	96.6	96.2

NEW YORK / NORTH CAROLINA

DIVISION		WATERTOWN 136			WHITE PLAINS 106			YONKERS 107			ASHEVILLE 287 - 288			CHARLOTTE 281 - 282			DURHAM 277		
		MAT.	INST.	TOTAL	MAT.	INST.	TOTAL	MAT.	INST.	TOTAL	MAT.	INST.	TOTAL	MAT.	INST.	TOTAL	MAT.	INST.	TOTAL
015433	CONTRACTOR EQUIPMENT		112.6	112.6		104.4	104.4		104.4	104.4		99.0	99.0		100.4	100.4		104.1	104.1
0241, 31 - 34	SITE & INFRASTRUCTURE, DEMOLITION	79.5	98.6	92.7	97.1	105.4	102.8	104.5	105.4	105.1	96.4	77.8	83.5	98.6	81.9	87.1	98.5	86.3	90.1
0310	Concrete Forming & Accessories	85.3	93.0	91.9	99.2	149.3	141.9	99.4	149.1	141.8	90.5	61.4	65.7	96.6	61.5	66.6	96.2	61.4	66.5
0320	Concrete Reinforcing	97.5	106.1	101.6	93.6	180.8	135.8	97.3	180.8	137.7	95.4	66.0	81.2	99.6	66.8	83.7	104.4	66.8	86.2
0330	Cast-in-Place Concrete	105.1	104.7	105.0	84.9	150.0	109.1	95.0	150.0	115.4	105.9	71.5	93.1	108.0	71.6	94.4	106.3	71.2	93.2
03	CONCRETE	108.8	100.7	105.3	87.6	155.2	117.3	96.6	155.1	122.3	91.1	67.4	80.7	91.2	67.5	80.8	95.2	67.4	83.0
04	MASONRY	89.7	104.8	99.0	93.9	156.4	132.3	97.1	156.4	133.6	84.9	64.4	72.3	88.6	64.4	73.8	84.3	64.4	72.0
05	METALS	96.8	119.8	103.6	86.1	169.8	110.7	94.1	169.9	116.4	101.2	89.9	97.9	102.1	88.9	98.2	117.9	90.2	109.7
06	WOOD, PLASTICS & COMPOSITES	80.3	89.9	85.6	102.4	146.7	126.8	102.3	146.3	126.5	87.6	59.0	71.9	89.9	59.0	72.9	90.6	59.0	73.2
07	THERMAL & MOISTURE PROTECTION	93.0	97.2	94.8	103.2	148.7	123.0	103.5	149.3	123.4	99.1	63.8	83.8	93.7	64.5	81.0	102.8	63.8	85.8
08	OPENINGS	95.2	93.8	94.8	92.4	176.2	112.0	96.0	176.5	114.8	89.0	59.8	82.2	99.0	60.0	89.9	95.6	60.0	87.3
0920	Plaster & Gypsum Board	87.6	89.6	89.0	105.3	147.8	133.9	108.7	147.4	134.7	102.2	57.8	72.3	98.1	57.8	71.0	89.9	57.8	68.3
0950, 0980	Ceilings & Acoustic Treatment	96.6	89.6	91.9	87.5	147.8	128.2	103.7	147.4	133.2	79.4	57.8	64.9	83.5	57.8	66.2	82.8	57.8	66.0
0960	Flooring	83.1	92.4	85.7	97.5	179.7	120.6	96.9	182.5	121.0	94.9	66.8	87.0	95.3	66.8	87.3	102.0	66.8	92.1
0970, 0990	Wall Finishes & Painting/Coating	85.6	93.0	90.0	104.6	167.4	141.8	104.6	167.4	141.8	102.5	57.1	75.7	95.9	57.1	72.9	104.4	57.1	76.4
09	FINISHES	88.2	92.4	90.5	97.1	156.4	129.2	101.0	156.7	131.1	88.8	61.6	74.1	88.3	61.6	73.9	90.4	61.6	74.8
COVERS	DIVS. 10 - 14, 25, 28, 41, 43, 44, 46	100.0	96.4	99.2	100.0	127.5	106.1	100.0	130.7	106.8	100.0	82.5	96.1	100.0	82.5	96.1	100.0	82.5	96.1
21, 22, 23	FIRE SUPPRESSION, PLUMBING & HVAC	100.3	87.8	95.4	100.4	145.9	118.3	100.4	146.0	118.3	100.3	61.3	85.0	99.9	61.3	85.5	100.5	61.3	85.1
26, 27, 3370	ELECTRICAL, COMMUNICATIONS & UTIL.	98.6	90.2	94.4	88.8	170.6	129.1	94.9	170.6	132.2	100.8	57.8	79.6	100.0	60.3	80.4	96.5	57.4	77.2
MF2018	WEIGHTED AVERAGE	97.7	96.8	97.4	93.8	152.8	118.8	97.9	153.0	121.2	96.4	66.6	83.8	97.4	67.6	84.8	100.1	67.2	86.2

For customer support on your Mechanical Costs with RSMeans Data, call 800.448.8182.

681

NORTH CAROLINA

DIVISION		ELIZABETH CITY 279			FAYETTEVILLE 283			GASTONIA 280			GREENSBORO 270, 272 - 274			HICKORY 286			KINSTON 285		
		MAT.	INST.	TOTAL	MAT.	INST.	TOTAL	MAT.	INST.	TOTAL	MAT.	INST.	TOTAL	MAT.	INST.	TOTAL	MAT.	INST.	TOTAL
015433	CONTRACTOR EQUIPMENT		108.0	108.0		104.1	104.1		99.0	99.0		104.1	104.1		104.1	104.1		104.1	104.1
0241, 31 - 34	SITE & INFRASTRUCTURE, DEMOLITION	102.7	87.9	92.4	95.6	86.1	89.0	96.1	78.1	83.7	98.3	86.4	90.1	95.1	85.1	88.2	93.9	84.9	87.7
0310	Concrete Forming & Accessories	82.6	63.6	66.4	90.1	60.1	64.5	96.4	61.4	66.6	96.0	61.3	66.5	87.1	61.2	65.0	83.6	59.9	63.4
0320	Concrete Reinforcing	102.3	69.8	86.6	99.1	66.8	83.4	95.8	66.8	81.7	103.2	66.8	85.6	95.4	66.7	81.5	94.9	66.7	81.3
0330	Cast-in-Place Concrete	106.5	71.7	93.5	111.1	69.3	95.5	103.6	71.2	91.5	105.5	71.2	92.7	105.9	71.2	93.0	102.2	69.2	90.0
03	CONCRETE	95.4	69.1	83.8	93.1	66.2	81.3	89.8	67.5	80.0	94.7	67.4	82.7	90.9	67.3	80.5	88.1	66.1	78.4
04	MASONRY	95.4	61.0	74.2	88.3	61.0	71.5	89.4	64.4	74.0	81.8	64.4	71.1	74.6	64.4	68.3	81.2	61.0	68.8
05	METALS	103.5	92.3	100.2	122.2	90.2	112.8	101.8	90.2	98.4	110.0	90.2	104.1	101.3	90.0	98.0	100.0	90.0	97.1
06	WOOD, PLASTICS & COMPOSITES	75.1	63.4	68.7	86.5	59.0	71.4	95.4	59.0	75.4	90.2	59.0	73.1	82.4	59.0	69.6	79.3	59.0	68.2
07	THERMAL & MOISTURE PROTECTION	102.0	62.5	84.8	98.6	62.3	82.8	99.3	63.8	83.9	102.6	63.8	85.7	99.5	63.8	84.0	99.3	62.3	83.2
08	OPENINGS	92.9	63.1	86.0	89.1	60.0	82.3	92.3	60.0	84.8	95.6	60.0	87.3	89.0	60.0	82.3	89.1	60.0	82.3
0920	Plaster & Gypsum Board	84.4	61.7	69.1	105.8	57.8	73.5	107.9	57.8	74.2	91.1	57.8	68.7	102.2	57.8	72.3	102.2	57.8	72.3
0950, 0980	Ceilings & Acoustic Treatment	82.8	61.7	68.5	81.9	57.8	65.7	83.5	57.8	66.2	82.8	57.8	66.0	79.4	57.8	64.9	83.5	57.8	66.2
0960	Flooring	94.1	66.8	86.4	95.1	66.8	87.1	98.0	66.8	89.2	102.0	66.8	92.1	94.8	66.8	86.9	92.2	66.8	85.0
0970, 0990	Wall Finishes & Painting/Coating	104.4	57.1	76.4	102.5	57.1	75.7	102.5	57.1	75.7	104.4	57.1	76.4	102.5	57.1	75.7	102.5	57.1	75.7
09	FINISHES	87.7	63.4	74.6	89.7	60.8	74.0	91.1	61.6	75.2	90.6	61.6	74.9	88.9	61.6	74.1	88.7	60.8	73.6
COVERS	DIVS. 10 - 14, 25, 28, 41, 43, 44, 46	100.0	85.4	96.8	100.0	81.4	95.8	100.0	82.5	96.1	100.0	79.8	95.5	100.0	82.5	96.1	100.0	81.3	95.8
21, 22, 23	FIRE SUPPRESSION, PLUMBING & HVAC	96.9	58.6	81.9	100.2	59.5	84.2	100.4	60.1	84.6	100.4	61.3	85.0	96.9	60.2	82.5	96.9	58.3	81.7
26, 27, 3370	ELECTRICAL, COMMUNICATIONS & UTIL.	96.2	64.6	80.6	100.6	57.4	79.3	100.3	60.4	80.6	95.7	57.8	77.0	98.6	60.4	79.8	98.4	55.8	77.4
MF2018	WEIGHTED AVERAGE	96.9	68.3	84.8	100.2	66.1	85.8	97.0	66.8	84.2	98.5	67.2	85.3	94.7	67.3	83.1	94.4	65.5	82.2

NORTH CAROLINA / NORTH DAKOTA

DIVISION		MURPHY 289			RALEIGH 275 - 276			ROCKY MOUNT 278			WILMINGTON 284			WINSTON-SALEM 271			BISMARCK 585		
		MAT.	INST.	TOTAL	MAT.	INST.	TOTAL	MAT.	INST.	TOTAL	MAT.	INST.	TOTAL	MAT.	INST.	TOTAL	MAT.	INST.	TOTAL
015433	CONTRACTOR EQUIPMENT		99.0	99.0		106.5	106.5		104.1	104.1		99.0	99.0		104.1	104.1		99.9	99.9
0241, 31 - 34	SITE & INFRASTRUCTURE, DEMOLITION	97.4	76.3	82.8	98.8	90.8	93.3	100.7	86.3	90.8	97.4	77.6	83.7	98.7	86.4	90.2	99.8	97.8	98.4
0310	Concrete Forming & Accessories	97.0	59.8	65.3	96.9	61.0	66.3	88.5	60.9	64.9	91.8	60.1	64.8	97.9	61.4	66.8	110.6	76.6	81.6
0320	Concrete Reinforcing	94.9	64.3	80.1	105.4	66.8	86.7	102.3	66.8	85.1	96.1	66.8	81.9	103.2	66.8	85.6	92.6	98.3	95.3
0330	Cast-in-Place Concrete	109.6	69.2	94.6	109.3	71.3	95.1	104.2	70.5	91.6	105.5	69.3	92.0	108.1	71.2	94.4	106.3	86.6	98.9
03	CONCRETE	93.9	65.6	81.4	93.9	67.2	82.2	95.9	66.9	82.6	93.1	66.2	80.1	96.0	67.4	83.5	93.3	84.4	89.4
04	MASONRY	77.3	61.0	67.3	80.5	63.1	69.8	74.4	63.1	67.4	75.1	61.0	66.4	82.0	64.4	71.2	105.3	83.4	91.8
05	METALS	99.0	89.1	96.1	102.2	89.1	98.3	102.6	90.2	99.0	100.7	90.2	97.6	107.0	90.2	102.1	95.6	94.2	95.2
06	WOOD, PLASTICS & COMPOSITES	96.1	58.9	75.6	90.9	59.2	73.4	82.0	59.0	69.4	89.2	59.0	72.6	90.2	59.0	73.1	106.6	72.7	88.0
07	THERMAL & MOISTURE PROTECTION	99.3	62.3	83.2	97.2	63.8	82.7	102.5	63.2	85.4	99.1	62.3	83.1	102.6	63.8	85.7	109.3	86.0	99.2
08	OPENINGS	89.0	59.4	82.1	97.6	60.1	88.8	92.2	60.0	84.7	89.1	60.0	82.3	95.6	60.0	87.3	103.7	81.9	98.6
0920	Plaster & Gypsum Board	106.9	57.7	73.8	86.5	57.8	67.2	85.8	57.8	67.0	103.7	57.8	72.8	91.1	57.8	68.7	102.4	72.2	82.1
0950, 0980	Ceilings & Acoustic Treatment	79.4	57.7	64.8	83.5	57.8	66.2	80.4	57.8	65.2	81.9	57.8	65.7	82.8	57.8	66.0	108.5	72.2	84.1
0960	Flooring	98.3	66.8	89.4	96.8	66.8	88.3	97.6	66.8	88.9	95.5	66.8	87.3	102.0	66.8	92.1	86.9	53.7	77.6
0970, 0990	Wall Finishes & Painting/Coating	102.5	57.1	75.7	97.8	57.1	73.7	104.4	57.1	76.4	102.5	57.1	75.7	104.4	57.1	76.4	90.9	59.8	72.5
09	FINISHES	90.6	60.7	74.4	90.0	61.4	74.5	88.6	61.3	73.8	89.6	60.8	74.0	90.6	61.6	74.9	96.1	71.1	82.5
COVERS	DIVS. 10 - 14, 25, 28, 41, 43, 44, 46	100.0	81.3	95.8	100.0	82.3	96.1	100.0	82.1	96.0	100.0	81.4	95.8	100.0	82.5	96.1	100.0	94.8	98.8
21, 22, 23	FIRE SUPPRESSION, PLUMBING & HVAC	96.9	58.3	81.7	100.0	60.6	84.5	96.9	59.5	82.2	100.4	59.5	84.3	100.4	61.3	85.0	99.9	77.0	90.9
26, 27, 3370	ELECTRICAL, COMMUNICATIONS & UTIL.	101.7	55.8	79.1	99.1	56.3	78.0	98.0	57.4	78.0	101.1	55.8	78.8	95.7	57.8	77.0	98.7	73.4	86.3
MF2018	WEIGHTED AVERAGE	95.5	64.7	82.5	97.4	67.0	84.5	96.0	66.6	83.6	96.0	65.2	83.0	98.2	67.3	85.2	98.9	81.6	91.6

NORTH DAKOTA

DIVISION		DEVILS LAKE 583			DICKINSON 586			FARGO 580 - 581			GRAND FORKS 582			JAMESTOWN 584			MINOT 587		
		MAT.	INST.	TOTAL	MAT.	INST.	TOTAL	MAT.	INST.	TOTAL	MAT.	INST.	TOTAL	MAT.	INST.	TOTAL	MAT.	INST.	TOTAL
015433	CONTRACTOR EQUIPMENT		97.7	97.7		97.7	97.7		99.9	99.9		97.7	97.7		97.7	97.7		97.7	97.7
0241, 31 - 34	SITE & INFRASTRUCTURE, DEMOLITION	107.5	93.7	98.0	115.7	93.7	100.5	101.3	97.6	98.8	111.5	93.7	99.2	106.5	93.7	97.7	108.9	94.1	98.7
0310	Concrete Forming & Accessories	107.0	71.6	76.8	95.5	71.5	75.1	99.4	72.1	76.1	99.7	71.6	75.7	97.1	71.6	75.4	95.1	71.8	75.2
0320	Concrete Reinforcing	94.8	98.0	96.4	95.7	98.2	96.9	95.4	98.7	97.2	93.4	98.0	95.6	95.4	98.7	97.0	96.6	98.1	97.4
0330	Cast-in-Place Concrete	126.2	81.6	109.6	114.1	81.5	102.0	102.5	85.6	96.2	114.1	81.5	102.0	124.6	81.6	108.6	114.1	81.8	102.1
03	CONCRETE	104.4	80.4	93.8	103.5	80.4	93.3	98.3	82.1	91.2	100.9	80.3	91.8	102.9	80.5	93.1	99.4	80.6	91.1
04	MASONRY	118.0	79.2	94.2	120.1	81.4	96.4	102.7	90.0	94.9	112.1	79.2	91.9	130.9	89.9	105.7	110.1	80.7	92.1
05	METALS	94.7	94.1	94.5	94.6	94.0	94.4	100.2	94.8	98.6	94.6	93.8	94.4	94.6	94.9	94.7	94.9	94.8	94.9
06	WOOD, PLASTICS & COMPOSITES	99.0	67.5	81.7	85.6	67.5	75.7	95.9	67.6	80.4	90.4	67.5	77.8	87.7	67.5	76.6	85.3	67.5	75.5
07	THERMAL & MOISTURE PROTECTION	107.1	82.3	96.3	107.7	83.7	97.2	104.1	87.4	97.0	107.3	83.0	96.7	106.9	85.7	97.7	107.0	83.9	97.0
08	OPENINGS	100.9	79.0	95.8	100.9	79.0	95.8	101.0	79.1	95.9	99.5	79.0	94.7	100.9	79.0	95.8	99.7	79.0	94.8
0920	Plaster & Gypsum Board	118.2	67.0	83.8	109.1	67.0	80.8	101.1	67.0	78.2	110.3	67.0	81.2	110.0	67.0	81.1	109.1	67.0	80.8
0950, 0980	Ceilings & Acoustic Treatment	105.6	67.0	79.6	105.6	67.0	79.6	95.4	67.0	76.3	105.6	67.0	79.6	105.6	67.0	79.6	105.6	67.0	79.6
0960	Flooring	93.8	53.7	82.5	87.4	53.7	77.9	101.2	53.7	87.9	89.2	53.7	79.2	88.0	53.7	78.4	87.1	53.7	77.7
0970, 0990	Wall Finishes & Painting/Coating	86.6	56.4	68.7	86.6	56.4	68.7	93.1	67.4	77.9	86.6	66.0	74.4	86.6	56.4	68.7	86.6	57.5	69.4
09	FINISHES	95.2	66.8	79.8	93.0	66.8	78.8	96.2	68.1	81.0	93.2	67.8	79.5	92.3	66.8	78.5	92.1	66.9	78.5
COVERS	DIVS. 10 - 14, 25, 28, 41, 43, 44, 46	100.0	87.2	97.2	100.0	87.2	97.2	100.0	93.6	98.6	100.0	87.2	97.2	100.0	87.2	97.2	100.0	93.3	98.5
21, 22, 23	FIRE SUPPRESSION, PLUMBING & HVAC	96.6	78.3	89.4	96.6	73.3	87.4	100.0	74.1	89.8	100.1	72.3	89.2	96.6	72.3	87.0	100.1	72.1	89.1
26, 27, 3370	ELECTRICAL, COMMUNICATIONS & UTIL.	94.3	68.3	81.5	101.9	67.9	85.1	98.5	69.4	84.2	97.3	68.3	83.0	94.3	68.3	81.5	100.1	72.6	86.6
MF2018	WEIGHTED AVERAGE	99.2	78.8	90.6	99.9	77.9	90.6	99.7	80.3	91.5	99.4	77.6	90.2	99.2	78.8	90.6	99.3	78.6	90.5

DIVISION		NORTH DAKOTA			OHIO															
		WILLISTON			AKRON			ATHENS			CANTON			CHILLICOTHE			CINCINNATI		
		588			442 - 443			457			446 - 447			456			451 - 452		
		MAT.	INST.	TOTAL	MAT.	INST.	TOTAL	MAT.	INST.	TOTAL	MAT.	INST.	TOTAL	MAT.	INST.	TOTAL	MAT.	INST.	TOTAL
015433	CONTRACTOR EQUIPMENT		97.7	97.7		88.7	88.7		84.9	84.9		88.7	88.7		95.5	95.5		95.9	95.9
0241, 31 - 34	SITE & INFRASTRUCTURE, DEMOLITION	109.2	91.5	97.0	96.2	94.1	94.8	107.3	85.3	92.1	96.3	93.9	94.7	93.8	94.9	94.6	90.0	98.2	95.7
0310	Concrete Forming & Accessories	101.7	71.3	75.8	102.8	82.8	85.8	95.3	78.9	81.4	102.8	74.3	78.5	98.0	81.5	84.0	101.5	80.2	83.3
0320	Concrete Reinforcing	97.5	98.1	97.8	94.7	91.0	92.9	86.9	89.2	88.0	94.7	75.0	85.2	84.0	88.8	86.3	89.1	77.8	83.6
0330	Cast-in-Place Concrete	114.1	81.4	101.9	102.6	89.0	97.6	111.2	95.8	105.5	103.6	87.2	97.5	100.9	92.4	97.8	96.4	77.5	89.3
03	CONCRETE	100.8	80.2	91.7	99.8	85.9	93.7	101.0	86.3	94.6	100.3	78.7	90.8	95.6	86.9	91.8	94.7	79.0	87.8
04	MASONRY	105.1	80.7	90.1	90.9	89.4	90.0	73.6	97.2	88.1	91.5	80.7	84.9	80.2	90.5	86.5	83.0	79.8	81.0
05	METALS	94.8	93.6	94.4	97.5	80.2	92.4	97.1	80.5	92.3	97.5	73.7	90.5	89.4	89.8	89.5	91.5	81.7	88.6
06	WOOD, PLASTICS & COMPOSITES	92.1	67.5	78.6	105.9	81.1	92.2	87.4	74.0	80.0	106.2	71.8	87.3	100.3	78.0	88.0	102.3	80.1	90.1
07	THERMAL & MOISTURE PROTECTION	107.3	83.5	96.9	104.9	91.1	98.9	102.9	91.5	98.0	106.0	87.2	97.8	104.7	88.4	97.6	102.7	80.9	93.2
08	OPENINGS	101.0	79.0	95.8	107.5	82.7	101.7	96.2	75.8	91.5	101.3	69.8	93.9	88.1	78.0	85.7	98.3	76.4	93.2
0920	Plaster & Gypsum Board	110.3	67.0	81.2	98.3	80.5	86.3	93.1	73.2	79.7	99.2	71.0	80.2	96.6	77.8	83.9	96.1	79.9	85.2
0950, 0980	Ceilings & Acoustic Treatment	105.6	67.0	79.6	90.6	80.5	83.8	105.9	73.2	83.8	90.6	71.0	77.4	98.7	77.8	84.6	93.6	79.9	84.3
0960	Flooring	90.1	53.7	79.9	91.7	83.3	89.3	120.7	75.3	108.0	91.8	74.1	86.8	97.5	75.3	91.3	98.7	78.5	93.0
0970, 0990	Wall Finishes & Painting/Coating	86.6	56.4	68.7	98.3	90.9	93.9	100.1	89.1	93.6	98.3	73.9	83.9	97.4	89.1	92.5	97.0	71.7	82.0
09	FINISHES	93.4	66.8	79.0	95.0	83.6	88.8	99.4	78.7	88.2	95.2	73.4	83.4	96.5	80.4	87.8	95.9	79.1	86.8
COVERS	DIVS. 10 - 14, 25, 28, 41, 43, 44, 46	100.0	87.2	97.1	100.0	90.3	97.8	100.0	87.9	97.3	100.0	88.4	97.4	100.0	86.8	97.1	100.0	88.4	97.4
21, 22, 23	FIRE SUPPRESSION, PLUMBING & HVAC	96.6	73.2	87.4	100.0	87.8	95.2	96.5	81.2	90.5	100.0	78.7	91.6	97.0	92.4	95.2	99.9	76.9	90.9
26, 27, 3370	ELECTRICAL, COMMUNICATIONS & UTIL.	97.7	68.7	83.4	98.6	81.8	90.4	97.9	89.7	93.9	97.9	85.0	91.5	97.2	89.7	93.5	96.3	74.2	85.4
MF2018	WEIGHTED AVERAGE	98.4	77.7	89.6	99.4	86.0	93.8	97.1	84.8	91.9	98.9	79.9	90.9	94.3	88.5	91.8	96.1	80.0	89.3

DIVISION		OHIO																	
		CLEVELAND			COLUMBUS			DAYTON			HAMILTON			LIMA			LORAIN		
		441			430 - 432			453 - 454			450			458			440		
		MAT.	INST.	TOTAL	MAT.	INST.	TOTAL	MAT.	INST.	TOTAL	MAT.	INST.	TOTAL	MAT.	INST.	TOTAL	MAT.	INST.	TOTAL
015433	CONTRACTOR EQUIPMENT		92.0	92.0		92.3	92.3		88.9	88.9		95.5	95.5		88.3	88.3		88.7	88.7
0241, 31 - 34	SITE & INFRASTRUCTURE, DEMOLITION	94.9	96.2	95.8	101.8	92.3	95.3	90.0	94.2	92.9	90.0	94.5	93.1	101.1	84.9	89.9	95.5	94.3	94.7
0310	Concrete Forming & Accessories	102.3	90.2	92.0	98.7	78.1	81.1	99.9	77.2	80.6	99.9	77.4	80.7	95.3	76.3	79.1	102.8	74.6	78.7
0320	Concrete Reinforcing	95.2	91.4	93.4	99.9	78.7	89.7	89.1	79.1	84.2	89.1	77.2	83.3	86.9	79.3	83.2	94.7	91.3	93.1
0330	Cast-in-Place Concrete	99.5	97.2	98.6	102.1	80.9	94.2	86.5	80.6	84.3	92.6	80.9	88.3	102.1	89.5	97.4	97.7	91.4	95.4
03	CONCRETE	100.1	92.4	96.7	98.6	79.2	90.0	87.0	78.6	83.3	89.9	79.1	85.1	93.9	81.5	88.5	97.5	83.1	91.2
04	MASONRY	96.9	97.9	97.5	91.4	85.2	87.6	79.1	76.4	77.4	79.5	81.2	80.5	98.8	78.1	86.1	87.8	93.1	91.0
05	METALS	99.1	83.3	94.5	98.1	79.3	92.5	90.9	76.8	86.7	90.9	85.6	89.4	97.2	80.1	92.2	98.1	81.6	93.3
06	WOOD, PLASTICS & COMPOSITES	99.8	88.5	93.6	98.5	77.7	87.0	104.2	77.1	89.3	103.0	77.1	88.7	87.3	75.1	80.6	105.9	70.1	86.2
07	THERMAL & MOISTURE PROTECTION	102.0	96.9	99.8	93.6	83.8	89.3	108.8	79.5	96.1	104.8	79.9	94.0	102.4	83.7	94.3	105.9	91.0	99.4
08	OPENINGS	100.3	86.4	97.0	98.3	74.0	92.7	95.0	74.6	90.3	92.8	74.8	88.6	96.3	73.7	91.0	101.3	76.6	95.5
0920	Plaster & Gypsum Board	97.9	88.1	91.3	94.1	77.0	82.6	98.1	76.9	83.9	98.1	76.9	83.9	93.1	74.3	80.5	98.3	69.2	78.7
0950, 0980	Ceilings & Acoustic Treatment	86.3	88.1	87.5	93.5	77.0	82.4	99.6	76.9	84.3	98.7	76.9	84.0	105.9	74.3	84.6	90.6	69.2	76.2
0960	Flooring	93.1	89.8	92.2	95.1	75.3	89.5	101.2	72.4	93.1	98.5	78.5	92.9	119.8	77.2	107.8	91.8	89.8	91.2
0970, 0990	Wall Finishes & Painting/Coating	101.2	90.9	95.1	99.5	79.6	87.7	97.4	70.7	81.6	97.4	71.3	82.0	100.2	76.5	86.2	98.3	90.9	93.9
09	FINISHES	95.3	90.1	92.5	95.0	77.5	85.5	97.7	75.4	85.6	96.7	76.9	86.0	98.7	76.0	86.4	95.0	78.0	85.8
COVERS	DIVS. 10 - 14, 25, 28, 41, 43, 44, 46	100.0	96.4	99.2	100.0	88.2	97.4	100.0	85.1	96.7	100.0	85.3	96.7	100.0	84.4	96.5	100.0	90.9	98.0
21, 22, 23	FIRE SUPPRESSION, PLUMBING & HVAC	100.0	89.9	96.0	100.0	84.1	93.7	100.7	81.2	93.0	100.5	74.9	90.5	96.5	91.7	94.6	100.0	88.4	95.4
26, 27, 3370	ELECTRICAL, COMMUNICATIONS & UTIL.	98.3	92.7	95.5	99.9	80.2	90.2	94.9	75.7	85.4	95.2	75.5	85.5	98.2	75.7	87.1	98.0	78.2	88.2
MF2018	WEIGHTED AVERAGE	99.1	91.6	95.9	98.4	82.0	91.4	94.8	79.2	88.2	94.8	79.5	88.3	97.2	81.7	90.7	98.4	84.7	92.6

DIVISION		OHIO																	
		MANSFIELD			MARION			SPRINGFIELD			STEUBENVILLE			TOLEDO			YOUNGSTOWN		
		448 - 449			433			455			439			434 - 436			444 - 445		
		MAT.	INST.	TOTAL	MAT.	INST.	TOTAL	MAT.	INST.	TOTAL	MAT.	INST.	TOTAL	MAT.	INST.	TOTAL	MAT.	INST.	TOTAL
015433	CONTRACTOR EQUIPMENT		88.7	88.7		88.7	88.7		88.9	88.9		92.4	92.4		92.4	92.4		88.7	88.7
0241, 31 - 34	SITE & INFRASTRUCTURE, DEMOLITION	91.7	94.0	93.3	95.7	90.6	92.2	90.3	94.2	93.0	141.5	98.2	111.6	99.8	90.9	93.6	96.1	93.9	94.6
0310	Concrete Forming & Accessories	92.1	73.5	76.3	96.6	79.4	81.9	99.9	77.1	80.5	97.6	79.0	81.7	100.3	84.9	87.1	102.8	76.6	80.5
0320	Concrete Reinforcing	86.2	79.0	82.7	92.1	79.0	85.8	89.1	79.1	84.2	89.7	94.9	92.2	99.9	82.9	91.7	94.7	85.2	90.1
0330	Cast-in-Place Concrete	95.1	87.8	92.3	88.5	88.1	88.3	88.8	80.3	85.6	96.1	89.3	93.6	96.9	90.2	94.4	101.7	87.2	96.3
03	CONCRETE	92.1	79.3	86.4	85.4	82.2	84.0	88.1	78.5	83.9	89.4	84.9	87.4	93.2	86.4	90.2	99.4	81.5	91.5
04	MASONRY	89.9	88.9	89.3	92.7	89.9	90.9	79.3	75.9	77.2	80.7	91.1	87.1	98.9	91.8	94.5	91.1	86.6	88.4
05	METALS	98.4	76.0	91.8	97.1	78.9	91.7	90.9	76.8	86.7	93.6	81.8	90.1	97.8	85.3	94.2	97.5	77.9	91.8
06	WOOD, PLASTICS & COMPOSITES	92.8	70.1	80.3	92.8	77.5	84.4	105.6	77.1	89.9	88.1	76.4	81.6	96.9	83.9	89.8	105.9	74.6	88.7
07	THERMAL & MOISTURE PROTECTION	104.1	88.4	97.3	90.5	88.8	89.8	108.7	79.2	95.9	102.0	87.4	95.7	91.0	91.2	91.1	106.1	88.2	98.3
08	OPENINGS	102.3	71.0	95.0	91.2	75.1	87.5	93.2	74.6	88.9	91.3	78.4	88.3	93.8	81.3	90.9	101.3	75.2	95.2
0920	Plaster & Gypsum Board	90.9	69.2	76.3	95.3	77.0	83.0	98.1	76.9	83.9	93.5	75.4	81.3	97.2	83.6	88.0	98.3	73.9	81.9
0950, 0980	Ceilings & Acoustic Treatment	91.4	69.2	76.4	99.5	77.0	84.3	99.6	76.9	84.3	96.7	75.4	82.3	99.5	83.6	88.8	90.6	73.9	79.3
0960	Flooring	86.6	92.7	88.3	94.4	92.7	93.9	101.2	72.4	93.1	122.5	92.7	114.1	94.8	94.4	94.7	91.8	90.0	91.3
0970, 0990	Wall Finishes & Painting/Coating	98.3	77.8	86.2	104.0	77.8	88.5	97.4	70.7	81.6	117.8	87.1	99.6	104.0	87.0	94.0	98.3	80.0	87.5
09	FINISHES	92.4	76.6	83.8	96.3	81.2	88.1	97.7	75.3	85.6	114.5	81.6	96.7	97.0	86.6	91.4	95.1	78.8	86.3
COVERS	DIVS. 10 - 14, 25, 28, 41, 43, 44, 46	100.0	88.3	97.4	100.0	85.9	96.8	100.0	84.9	96.6	100.0	86.2	96.9	100.0	91.3	98.1	100.0	88.5	97.4
21, 22, 23	FIRE SUPPRESSION, PLUMBING & HVAC	96.5	87.5	92.9	96.5	93.1	95.1	100.7	80.9	92.9	96.9	91.3	94.7	100.0	93.6	97.5	100.0	83.9	93.7
26, 27, 3370	ELECTRICAL, COMMUNICATIONS & UTIL.	95.7	90.8	93.3	94.0	90.8	92.4	94.9	80.2	87.7	88.7	106.4	97.4	100.0	103.4	101.7	98.0	74.3	86.3
MF2018	WEIGHTED AVERAGE	96.4	84.2	91.3	94.3	86.7	91.1	94.8	79.7	88.4	96.1	90.0	93.5	97.6	91.2	94.9	98.7	81.9	91.6

For customer support on your Mechanical Costs with RSMeans Data, call 800.448.8182.

683

		OHIO			OKLAHOMA															
	DIVISION	ZANESVILLE			ARDMORE			CLINTON			DURANT			ENID			GUYMON		
		437 - 438			734			736			747			737			739		
		MAT.	INST.	TOTAL	MAT.	INST.	TOTAL	MAT.	INST.	TOTAL	MAT.	INST.	TOTAL	MAT.	INST.	TOTAL	MAT.	INST.	TOTAL
015433	CONTRACTOR EQUIPMENT		88.7	88.7		80.7	80.7		79.9	79.9		79.9	79.9		79.9	79.9		79.9	79.9
0241, 31 - 34	SITE & INFRASTRUCTURE, DEMOLITION	98.9	90.5	93.1	96.9	92.6	94.0	98.1	91.2	93.3	93.4	88.6	90.1	100.0	91.2	93.9	102.4	90.5	94.2
0310	Concrete Forming & Accessories	93.7	78.3	80.6	88.0	55.4	60.2	86.8	55.4	60.1	82.9	55.0	59.1	89.9	55.6	60.7	92.6	55.1	60.7
0320	Concrete Reinforcing	91.6	92.6	92.1	79.5	67.7	73.8	80.0	67.7	74.0	88.6	63.6	76.5	79.5	67.7	73.8	80.0	63.3	71.9
0330	Cast-in-Place Concrete	93.2	86.5	90.7	96.6	70.5	86.9	93.4	70.5	84.9	88.6	70.3	81.8	93.4	70.6	84.9	93.4	70.0	84.7
03	CONCRETE	89.2	83.5	86.7	86.7	62.8	76.2	86.2	62.8	75.9	84.1	61.8	74.3	86.6	62.9	76.2	89.4	61.7	77.2
04	MASONRY	90.7	86.6	88.2	95.2	57.0	71.8	119.3	57.0	81.0	88.6	62.1	72.3	101.3	57.0	74.1	97.5	55.4	71.7
05	METALS	98.5	84.2	94.3	97.3	62.2	87.0	97.4	62.2	87.0	92.7	60.6	83.2	98.8	62.4	88.1	97.8	59.3	86.5
06	WOOD, PLASTICS & COMPOSITES	88.2	77.5	82.3	97.4	54.4	73.7	96.5	54.4	73.4	87.8	54.4	69.5	99.7	54.4	74.8	102.9	54.4	76.2
07	THERMAL & MOISTURE PROTECTION	90.6	84.5	88.0	100.7	64.8	85.1	100.8	64.8	85.1	96.4	64.6	82.6	100.9	64.8	85.2	101.3	62.2	84.3
08	OPENINGS	91.2	79.0	88.4	103.7	55.0	92.3	103.7	55.0	92.3	96.0	54.0	86.2	104.9	55.9	93.5	103.8	54.0	92.2
0920	Plaster & Gypsum Board	91.2	77.0	81.7	90.5	53.7	65.7	90.2	53.7	65.7	78.5	53.7	61.8	91.1	53.7	66.0	91.1	53.7	66.0
0950, 0980	Ceilings & Acoustic Treatment	99.5	77.0	84.3	89.1	53.7	65.3	89.1	53.7	65.3	83.0	53.7	63.3	89.1	53.7	65.3	89.1	53.7	65.3
0960	Flooring	92.6	75.3	87.7	82.7	57.3	75.6	81.7	57.3	74.9	89.7	51.7	79.0	83.3	72.9	80.4	84.8	57.3	77.1
0970, 0990	Wall Finishes & Painting/Coating	104.0	89.1	95.2	86.4	44.8	61.8	86.4	44.8	61.8	92.8	44.8	64.4	86.4	44.8	61.8	86.4	42.2	60.3
09	FINISHES	95.4	78.2	86.1	82.7	53.5	66.9	82.5	53.5	66.8	83.2	52.4	66.5	83.2	56.7	68.8	84.0	54.5	68.0
COVERS	DIVS. 10 - 14, 25, 28, 41, 43, 44, 46	100.0	83.3	96.3	100.0	79.4	95.4	100.0	79.4	95.4	100.0	79.3	95.4	100.0	79.4	95.4	100.0	79.4	95.4
21, 22, 23	FIRE SUPPRESSION, PLUMBING & HVAC	96.5	89.8	93.9	96.5	67.5	85.1	96.5	67.5	85.1	96.6	67.4	85.2	100.0	67.5	87.2	96.5	63.6	83.6
26, 27, 3370	ELECTRICAL, COMMUNICATIONS & UTIL.	94.1	89.7	92.0	95.0	71.0	83.2	95.9	71.0	83.7	97.3	69.2	83.4	95.9	71.0	83.7	97.6	65.6	81.8
MF2018	WEIGHTED AVERAGE	94.9	85.7	91.0	95.2	65.6	82.7	96.4	65.5	83.3	93.1	65.1	81.3	97.0	66.0	83.9	96.3	63.3	82.4

		OKLAHOMA																	
	DIVISION	LAWTON			MCALESTER			MIAMI			MUSKOGEE			OKLAHOMA CITY			PONCA CITY		
		735			745			743			744			730 - 731			746		
		MAT.	INST.	TOTAL	MAT.	INST.	TOTAL	MAT.	INST.	TOTAL	MAT.	INST.	TOTAL	MAT.	INST.	TOTAL	MAT.	INST.	TOTAL
015433	CONTRACTOR EQUIPMENT		80.7	80.7		79.9	79.9		90.9	90.9		90.9	90.9		85.6	85.6		79.9	79.9
0241, 31 - 34	SITE & INFRASTRUCTURE, DEMOLITION	96.2	92.7	93.7	86.8	90.5	89.4	88.1	88.1	88.1	88.7	87.9	88.1	94.2	98.0	96.8	93.9	90.9	91.8
0310	Concrete Forming & Accessories	92.5	55.6	61.1	81.1	41.4	47.3	93.2	55.2	60.8	97.3	54.8	61.1	92.8	63.8	68.0	89.0	55.4	60.3
0320	Concrete Reinforcing	79.7	67.7	73.9	88.3	63.3	76.2	86.9	67.7	77.6	87.7	62.8	75.7	87.9	67.8	78.2	87.7	67.7	78.0
0330	Cast-in-Place Concrete	90.3	70.6	83.0	77.7	70.0	74.8	81.4	71.4	77.7	82.4	71.1	78.2	90.3	73.4	84.0	91.0	70.5	83.4
03	CONCRETE	83.0	62.9	74.2	75.3	55.4	66.6	79.7	64.0	72.8	81.2	62.8	73.1	86.1	67.6	78.0	86.1	62.7	75.8
04	MASONRY	97.4	57.0	72.6	106.1	56.9	75.9	91.3	57.1	70.3	108.1	49.2	71.9	99.8	57.4	73.8	84.1	57.0	67.5
05	METALS	102.5	62.3	90.7	92.6	59.4	82.9	92.6	77.0	88.0	94.1	73.9	88.2	95.7	63.7	86.3	92.6	62.1	83.6
06	WOOD, PLASTICS & COMPOSITES	102.0	54.4	75.8	85.5	36.2	58.4	100.0	54.6	75.0	104.3	54.6	76.9	92.5	65.1	77.4	95.6	54.4	72.9
07	THERMAL & MOISTURE PROTECTION	100.7	64.8	85.1	96.0	61.2	80.9	96.5	64.1	82.4	96.8	61.1	81.2	92.1	66.9	81.1	96.6	65.0	82.9
08	OPENINGS	106.6	55.9	94.8	96.0	43.9	83.9	96.0	55.0	86.5	97.2	54.0	87.1	100.0	61.8	91.1	96.0	55.0	86.4
0920	Plaster & Gypsum Board	92.8	53.7	66.5	77.6	35.0	48.9	84.2	53.7	63.7	86.4	53.7	64.4	94.7	64.6	74.4	82.9	53.7	63.3
0950, 0980	Ceilings & Acoustic Treatment	96.4	53.7	67.6	83.0	35.0	50.6	83.0	53.7	63.3	92.7	53.7	66.4	87.4	64.6	72.0	83.0	53.7	63.3
0960	Flooring	85.0	72.9	81.6	88.6	57.3	79.8	95.7	51.7	83.3	98.2	33.2	79.9	86.4	72.9	82.6	92.8	57.3	82.8
0970, 0990	Wall Finishes & Painting/Coating	86.4	44.8	61.8	92.8	42.2	62.9	92.8	42.2	62.9	92.8	42.2	62.9	89.5	44.8	63.1	92.8	44.8	64.4
09	FINISHES	84.9	56.7	69.6	82.2	42.4	60.7	85.1	52.2	67.3	88.1	48.8	66.9	86.3	63.1	73.7	84.9	54.4	68.4
COVERS	DIVS. 10 - 14, 25, 28, 41, 43, 44, 46	100.0	79.4	95.4	100.0	77.4	95.0	100.0	76.3	94.7	100.0	76.3	94.7	100.0	80.9	95.8	100.0	79.4	95.4
21, 22, 23	FIRE SUPPRESSION, PLUMBING & HVAC	100.0	67.5	87.2	96.6	63.5	83.6	96.6	63.6	83.7	100.1	61.3	84.9	99.9	67.8	87.3	96.6	63.6	83.7
26, 27, 3370	ELECTRICAL, COMMUNICATIONS & UTIL.	97.6	69.2	83.6	95.8	67.0	81.6	97.1	67.0	82.3	95.4	67.2	81.5	103.4	71.1	87.5	95.4	65.6	80.7
MF2018	WEIGHTED AVERAGE	97.3	65.8	84.0	92.4	60.5	78.9	92.7	65.1	81.1	95.0	62.9	81.4	96.4	68.7	84.7	93.1	64.0	80.8

		OKLAHOMA												OREGON					
	DIVISION	POTEAU			SHAWNEE			TULSA			WOODWARD			BEND			EUGENE		
		749			748			740 - 741			738			977			974		
		MAT.	INST.	TOTAL	MAT.	INST.	TOTAL	MAT.	INST.	TOTAL	MAT.	INST.	TOTAL	MAT.	INST.	TOTAL	MAT.	INST.	TOTAL
015433	CONTRACTOR EQUIPMENT		89.7	89.7		79.9	79.9		90.9	90.9		79.9	79.9		97.3	97.3		97.3	97.3
0241, 31 - 34	SITE & INFRASTRUCTURE, DEMOLITION	74.8	83.8	81.0	96.9	90.9	92.7	94.8	87.2	89.5	98.5	91.2	93.5	105.9	98.5	100.8	96.6	98.5	97.9
0310	Concrete Forming & Accessories	87.0	54.9	59.7	82.8	55.3	59.4	97.4	56.0	62.1	86.9	43.7	50.1	107.5	96.6	98.2	104.2	96.6	97.7
0320	Concrete Reinforcing	88.7	67.7	78.5	87.7	65.4	76.9	88.0	67.7	78.2	79.5	67.7	73.8	91.0	112.8	101.5	94.9	112.8	103.5
0330	Cast-in-Place Concrete	81.4	71.3	77.7	93.9	70.5	85.2	89.7	73.2	83.6	93.4	70.5	84.9	117.6	100.6	111.3	113.9	100.6	108.9
03	CONCRETE	81.9	63.8	73.9	87.7	62.3	76.6	86.2	65.0	76.9	86.5	57.5	73.7	110.5	100.3	106.0	101.3	100.3	100.9
04	MASONRY	91.6	57.1	70.4	107.5	57.0	76.5	92.2	57.1	70.6	90.8	57.0	70.0	105.1	100.6	102.3	102.1	100.6	101.1
05	METALS	92.6	76.7	87.9	92.5	61.3	83.3	97.0	77.1	91.2	97.4	62.1	87.0	108.0	96.0	104.5	108.7	95.9	105.0
06	WOOD, PLASTICS & COMPOSITES	92.4	54.6	71.6	87.7	54.4	69.4	103.5	55.6	77.2	96.6	38.7	64.8	99.6	96.1	97.7	95.4	96.1	95.8
07	THERMAL & MOISTURE PROTECTION	96.6	64.1	82.5	96.6	63.9	82.4	96.8	64.3	82.7	100.9	63.3	84.5	118.3	100.2	110.4	117.4	102.7	111.0
08	OPENINGS	96.0	55.0	86.5	96.0	54.4	86.3	98.9	56.2	88.9	103.7	46.3	90.3	96.9	100.0	97.6	97.2	100.0	97.8
0920	Plaster & Gypsum Board	81.3	53.7	62.8	78.5	53.7	61.8	86.4	54.8	65.1	90.2	37.6	54.8	122.7	95.9	104.6	120.8	95.9	104.0
0950, 0980	Ceilings & Acoustic Treatment	83.0	53.7	63.3	83.0	53.7	63.3	92.7	54.8	67.1	89.1	37.6	54.4	85.7	95.9	92.6	86.6	95.9	92.9
0960	Flooring	92.2	51.7	80.8	89.7	57.3	80.6	97.0	61.6	87.1	81.7	54.6	74.1	105.8	105.7	105.7	104.2	105.7	104.6
0970, 0990	Wall Finishes & Painting/Coating	92.8	44.8	64.4	92.8	42.2	62.9	92.8	51.3	68.2	86.4	44.8	61.8	98.6	77.3	86.0	98.6	69.6	81.4
09	FINISHES	82.9	52.5	66.4	83.5	53.2	67.1	88.0	55.8	70.5	82.6	43.7	61.5	100.1	95.9	97.8	98.6	95.0	96.7
COVERS	DIVS. 10 - 14, 25, 28, 41, 43, 44, 46	100.0	79.6	95.5	100.0	79.4	95.4	100.0	76.4	94.7	100.0	77.7	95.0	100.0	102.0	100.4	100.0	102.0	100.4
21, 22, 23	FIRE SUPPRESSION, PLUMBING & HVAC	96.6	63.6	83.7	96.6	67.5	85.2	100.1	63.7	85.8	96.5	67.5	85.1	96.6	105.2	100.0	100.1	98.6	99.5
26, 27, 3370	ELECTRICAL, COMMUNICATIONS & UTIL.	95.5	67.0	81.5	97.4	71.0	84.4	97.3	67.0	82.4	97.4	71.0	84.4	102.0	99.8	100.9	100.7	99.8	100.2
MF2018	WEIGHTED AVERAGE	92.3	64.9	80.7	94.5	65.2	82.2	95.9	65.8	83.2	95.3	62.9	81.6	102.6	100.2	101.6	101.7	98.7	100.5

City Cost Indexes

		OREGON																
	DIVISION	KLAMATH FALLS			MEDFORD			PENDLETON			PORTLAND			SALEM			VALE		
		976			975			978			970 - 972			973			979		
		MAT.	INST.	TOTAL	MAT.	INST.	TOTAL	MAT.	INST.	TOTAL	MAT.	INST.	TOTAL	MAT.	INST.	TOTAL	MAT.	INST.	TOTAL
015433	CONTRACTOR EQUIPMENT		97.3	97.3		97.3	97.3		94.9	94.9		97.3	97.3		99.9	99.9		94.9	94.9
0241, 31 - 34	SITE & INFRASTRUCTURE, DEMOLITION	109.8	98.4	102.0	104.0	98.4	100.2	103.0	91.9	95.4	99.1	98.5	98.7	92.6	102.1	99.2	90.6	91.8	91.5
0310	Concrete Forming & Accessories	100.7	96.3	97.0	99.7	96.3	96.8	101.2	96.6	97.3	105.3	96.8	98.1	105.8	96.8	98.1	107.5	95.3	97.1
0320	Concrete Reinforcing	91.0	112.7	101.5	92.6	112.7	102.3	90.3	112.8	101.2	95.6	112.8	103.9	102.7	112.8	107.6	88.1	112.5	99.9
0330	Cast-in-Place Concrete	117.6	97.0	110.0	117.6	100.4	111.2	118.4	97.5	110.7	117.0	100.6	110.9	107.6	101.7	105.4	93.2	97.8	94.9
03	CONCRETE	113.5	98.9	107.1	107.8	100.1	104.4	93.7	99.3	96.2	103.0	100.4	101.8	98.7	100.7	99.6	77.8	98.8	87.0
04	MASONRY	119.1	100.6	107.7	99.0	100.6	100.0	109.9	100.6	104.2	104.0	100.6	101.9	108.2	100.6	103.5	107.7	100.6	103.4
05	METALS	108.0	95.6	104.4	108.3	95.6	104.6	116.3	96.4	110.4	110.1	96.1	106.0	117.1	95.4	110.8	116.2	95.1	110.0
06	WOOD, PLASTICS & COMPOSITES	90.4	96.1	93.5	89.3	96.1	93.0	92.3	96.2	94.5	96.4	96.1	96.3	90.3	96.3	93.6	100.9	96.2	98.3
07	THERMAL & MOISTURE PROTECTION	118.5	96.8	109.1	118.1	96.1	108.5	110.7	95.4	104.1	117.3	100.2	109.9	114.1	100.9	108.3	110.1	91.4	102.0
08	OPENINGS	96.9	100.0	97.6	99.6	100.0	99.7	93.2	100.0	94.8	95.1	100.0	96.3	102.6	100.0	102.0	93.2	89.0	92.2
0920	Plaster & Gypsum Board	117.1	95.9	102.8	116.5	95.9	102.6	102.4	95.9	98.0	120.3	95.9	103.9	115.4	95.9	102.3	109.0	95.9	100.2
0950, 0980	Ceilings & Acoustic Treatment	93.0	95.9	95.0	98.5	95.9	96.7	61.9	95.9	84.8	88.5	95.9	93.5	96.0	95.9	95.9	61.9	95.9	84.8
0960	Flooring	102.8	105.7	103.6	102.2	105.7	103.2	70.8	105.7	80.6	101.9	105.7	102.9	107.4	105.7	106.9	72.9	105.7	82.1
0970, 0990	Wall Finishes & Painting/Coating	98.6	65.2	78.8	98.6	65.2	78.8	88.3	77.3	81.8	98.4	77.3	85.9	97.3	75.2	84.2	88.3	77.3	81.8
09	FINISHES	100.3	94.6	97.2	100.4	94.6	97.3	70.4	96.0	84.2	98.2	95.9	97.0	99.7	95.8	97.6	71.0	96.0	84.5
COVERS	DIVS. 10 - 14, 25, 28, 41, 43, 44, 46	100.0	101.9	100.4	100.0	101.9	100.4	100.0	96.5	99.2	100.0	102.1	100.5	100.0	102.4	100.5	100.0	102.2	100.5
21, 22, 23	FIRE SUPPRESSION, PLUMBING & HVAC	96.6	105.2	100.0	100.1	105.2	102.1	98.7	112.0	104.0	100.1	111.5	104.6	100.1	106.9	102.8	98.7	71.1	87.9
26, 27, 3370	ELECTRICAL, COMMUNICATIONS & UTIL.	100.7	82.6	91.8	104.1	82.6	93.5	92.8	96.8	94.8	100.9	108.7	104.7	109.3	99.8	104.6	92.8	68.1	80.6
MF2018	WEIGHTED AVERAGE	103.5	97.3	100.9	103.3	97.4	100.8	98.6	100.3	99.3	102.1	102.8	102.4	104.4	100.8	102.9	96.2	87.0	92.3

		PENNSYLVANIA																
	DIVISION	ALLENTOWN			ALTOONA			BEDFORD			BRADFORD			BUTLER			CHAMBERSBURG		
		181			166			155			167			160			172		
		MAT.	INST.	TOTAL	MAT.	INST.	TOTAL	MAT.	INST.	TOTAL	MAT.	INST.	TOTAL	MAT.	INST.	TOTAL	MAT.	INST.	TOTAL
015433	CONTRACTOR EQUIPMENT		112.6	112.6		112.6	112.6		110.7	110.7		112.6	112.6		112.6	112.6		111.8	111.8
0241, 31 - 34	SITE & INFRASTRUCTURE, DEMOLITION	91.7	97.2	95.5	95.1	97.1	96.5	103.6	94.8	97.6	90.5	96.0	94.3	86.3	97.8	94.2	86.3	95.3	92.5
0310	Concrete Forming & Accessories	97.9	108.2	106.7	83.7	82.6	82.8	82.0	81.2	81.3	85.8	96.5	94.9	85.1	94.1	92.8	88.0	76.8	78.5
0320	Concrete Reinforcing	96.8	113.3	104.8	93.8	105.7	99.6	93.0	105.7	99.1	95.8	105.8	100.6	94.4	119.3	106.5	94.5	112.2	103.1
0330	Cast-in-Place Concrete	89.4	99.9	93.3	99.7	86.9	94.9	109.1	85.9	100.5	95.2	90.6	93.5	88.0	96.2	91.0	93.0	93.8	93.3
03	CONCRETE	92.3	107.0	98.8	87.5	89.5	88.4	102.2	88.5	96.2	94.2	90.7	95.5	79.5	100.3	88.7	97.1	90.5	94.2
04	MASONRY	92.1	93.1	92.7	95.7	83.4	88.1	107.1	80.2	90.6	93.1	82.0	86.3	97.6	93.0	94.8	93.8	79.1	84.8
05	METALS	98.9	120.5	105.3	92.9	114.8	99.3	102.2	113.4	105.5	96.7	113.2	101.6	92.6	121.7	101.2	98.3	118.8	104.3
06	WOOD, PLASTICS & COMPOSITES	97.6	111.5	105.3	76.7	81.8	79.5	79.4	81.7	80.7	82.7	101.2	92.9	78.1	94.1	86.9	84.3	75.6	79.5
07	THERMAL & MOISTURE PROTECTION	104.0	107.8	105.7	102.7	90.0	97.2	98.3	87.4	93.6	103.8	88.5	97.2	102.2	94.4	98.8	96.1	83.1	90.4
08	OPENINGS	92.4	107.3	95.9	86.3	85.6	86.1	93.4	85.6	91.6	92.4	95.3	93.1	86.2	99.3	89.3	89.2	81.4	87.4
0920	Plaster & Gypsum Board	94.9	111.8	106.3	86.3	81.2	82.9	97.5	81.2	86.6	87.1	101.2	96.6	86.3	93.9	91.4	108.9	74.9	86.0
0950, 0980	Ceilings & Acoustic Treatment	88.5	111.8	104.2	90.9	81.2	84.4	101.1	81.2	87.7	91.0	101.2	97.9	91.8	93.9	93.2	96.0	74.9	81.8
0960	Flooring	92.0	95.1	92.8	85.3	98.9	89.2	95.0	103.7	97.5	86.2	103.7	91.1	86.3	105.3	91.6	90.5	78.0	87.0
0970, 0990	Wall Finishes & Painting/Coating	91.9	101.8	97.8	87.2	106.2	98.4	97.6	105.7	102.4	91.9	105.7	100.0	87.2	106.2	98.4	92.0	100.7	97.1
09	FINISHES	90.6	105.8	98.8	88.2	87.5	87.8	100.6	87.4	93.5	88.4	99.4	94.3	88.1	97.2	93.0	90.8	78.8	84.3
COVERS	DIVS. 10 - 14, 25, 28, 41, 43, 44, 46	100.0	99.6	99.9	100.0	94.5	98.8	100.0	93.2	98.5	100.0	96.1	99.1	100.0	97.3	99.4	100.0	92.7	98.4
21, 22, 23	FIRE SUPPRESSION, PLUMBING & HVAC	100.3	114.5	105.9	99.8	83.6	93.4	96.6	84.4	91.8	96.8	90.6	94.4	96.3	95.4	96.0	96.7	89.7	94.0
26, 27, 3370	ELECTRICAL, COMMUNICATIONS & UTIL.	98.0	95.8	96.9	88.7	107.7	98.0	94.2	107.7	100.9	92.1	107.7	99.8	89.2	107.7	98.3	92.4	85.8	89.2
MF2018	WEIGHTED AVERAGE	96.7	105.8	100.6	93.3	92.8	93.1	98.8	92.1	96.0	94.9	97.1	95.8	91.3	100.6	95.2	95.1	89.4	92.7

		PENNSYLVANIA																
	DIVISION	DOYLESTOWN			DUBOIS			ERIE			GREENSBURG			HARRISBURG			HAZLETON		
		189			158			164 - 165			156			170 - 171			182		
		MAT.	INST.	TOTAL	MAT.	INST.	TOTAL	MAT.	INST.	TOTAL	MAT.	INST.	TOTAL	MAT.	INST.	TOTAL	MAT.	INST.	TOTAL
015433	CONTRACTOR EQUIPMENT		91.5	91.5		110.7	110.7		112.6	112.6		110.7	110.7		114.7	114.7		112.6	112.6
0241, 31 - 34	SITE & INFRASTRUCTURE, DEMOLITION	104.8	85.9	91.7	108.4	95.2	99.3	92.1	97.4	95.7	99.6	96.7	97.6	86.6	100.3	96.1	85.0	97.1	93.3
0310	Concrete Forming & Accessories	82.9	126.8	120.3	81.6	83.7	83.4	97.2	85.7	87.5	87.8	88.6	88.5	100.5	85.7	87.9	80.6	88.8	87.6
0320	Concrete Reinforcing	93.6	150.9	121.3	92.3	119.4	105.4	95.8	107.3	101.4	92.3	119.0	105.2	103.4	111.0	107.1	94.0	113.1	103.2
0330	Cast-in-Place Concrete	84.5	129.7	101.3	105.2	94.1	101.1	98.0	88.7	94.5	101.3	95.7	99.2	91.7	96.9	93.6	84.5	95.6	88.6
03	CONCRETE	88.0	131.2	107.0	104.4	94.7	100.1	86.4	91.9	88.8	97.2	97.5	97.4	91.9	95.3	93.4	85.0	96.7	90.2
04	MASONRY	95.4	134.9	119.7	107.8	93.4	98.9	85.0	86.8	86.1	117.5	89.1	100.0	88.8	83.9	85.8	104.7	89.5	95.4
05	METALS	96.5	123.9	104.5	102.2	118.8	107.1	93.1	115.6	99.7	102.1	119.7	107.3	105.1	117.4	108.7	98.7	119.7	104.8
06	WOOD, PLASTICS & COMPOSITES	78.5	124.9	104.5	78.3	81.7	80.2	94.0	84.4	88.7	85.3	87.1	86.3	101.8	85.3	92.7	77.1	87.7	83.0
07	THERMAL & MOISTURE PROTECTION	101.1	131.7	114.4	98.7	92.9	96.1	103.2	88.6	96.8	98.2	92.2	95.6	98.9	90.4	99.6	103.3	100.7	102.2
08	OPENINGS	94.8	133.2	103.7	93.4	88.7	92.3	86.4	88.0	86.8	93.4	95.5	93.9	100.9	86.5	97.5	93.0	93.3	93.0
0920	Plaster & Gypsum Board	85.4	126.5	113.0	96.3	81.2	86.2	94.9	83.9	87.5	98.3	86.8	90.5	113.8	84.7	94.2	85.8	86.6	86.4
0950, 0980	Ceilings & Acoustic Treatment	87.7	126.5	113.8	101.1	81.2	87.7	88.5	83.9	85.4	100.3	86.8	91.2	104.0	84.7	91.0	89.4	86.6	87.5
0960	Flooring	76.2	139.3	93.9	94.8	103.7	97.3	92.2	98.9	94.1	98.4	78.0	92.7	94.7	90.4	93.5	83.5	89.3	85.1
0970, 0990	Wall Finishes & Painting/Coating	91.5	143.7	122.4	97.6	106.2	102.7	98.2	92.3	94.7	97.6	106.2	102.7	95.8	84.8	89.3	91.9	104.1	99.1
09	FINISHES	81.9	130.5	108.2	100.9	88.8	94.4	91.6	88.4	89.9	101.3	88.1	94.1	96.0	85.8	90.5	86.6	90.2	88.6
COVERS	DIVS. 10 - 14, 25, 28, 41, 43, 44, 46	100.0	113.0	102.9	100.0	95.2	98.9	100.0	96.0	99.1	100.0	96.4	99.2	100.0	95.4	99.0	100.0	95.5	99.0
21, 22, 23	FIRE SUPPRESSION, PLUMBING & HVAC	96.3	132.6	110.6	96.6	87.5	93.0	99.8	93.6	97.4	96.6	87.3	92.9	100.1	91.7	96.8	96.8	95.5	96.3
26, 27, 3370	ELECTRICAL, COMMUNICATIONS & UTIL.	91.6	132.4	111.7	94.8	107.7	101.1	90.3	93.1	91.7	94.8	107.7	101.2	99.7	85.8	92.9	92.9	90.0	91.5
MF2018	WEIGHTED AVERAGE	94.0	127.1	108.0	99.3	96.0	97.9	93.2	93.9	93.5	98.7	96.3	97.7	98.7	93.0	96.3	94.4	95.9	95.0

For customer support on your Mechanical Costs with RSMeans Data, call 800.448.8182.

685

PENNSYLVANIA

DIVISION		INDIANA 157			JOHNSTOWN 159			KITTANNING 162			LANCASTER 175 - 176			LEHIGH VALLEY 180			MONTROSE 188		
		MAT.	INST.	TOTAL	MAT.	INST.	TOTAL	MAT.	INST.	TOTAL	MAT.	INST.	TOTAL	MAT.	INST.	TOTAL	MAT.	INST.	TOTAL
015433	CONTRACTOR EQUIPMENT		110.7	110.7		110.7	110.7		112.6	112.6		111.8	111.8		112.6	112.6		112.6	112.6
0241, 31 - 34	SITE & INFRASTRUCTURE, DEMOLITION	97.8	95.8	96.4	104.1	96.0	98.5	88.8	97.4	94.7	78.4	95.6	90.3	88.9	97.5	94.8	87.5	97.2	94.2
0310	Concrete Forming & Accessories	82.6	92.9	91.4	81.6	82.5	82.4	85.1	89.7	89.0	90.0	86.4	86.9	91.7	110.3	107.6	81.5	89.3	88.2
0320	Concrete Reinforcing	91.7	119.4	105.1	93.0	118.9	105.5	94.4	119.2	106.4	94.1	111.0	102.3	94.0	108.0	100.8	98.4	116.2	107.0
0330	Cast-in-Place Concrete	99.3	94.4	97.5	110.1	86.7	101.4	91.5	94.7	92.7	79.1	96.7	85.7	91.5	99.2	94.3	89.7	93.0	90.9
03	CONCRETE	94.7	99.1	96.6	103.2	91.6	98.1	82.1	97.7	88.9	85.2	95.6	89.8	91.4	106.9	98.2	90.1	96.6	93.0
04	MASONRY	103.8	93.5	97.5	104.7	83.7	91.8	100.1	88.8	93.1	99.6	86.7	91.7	92.1	96.5	94.8	92.0	91.5	91.7
05	METALS	102.3	119.9	107.4	102.2	117.9	106.8	92.7	120.6	100.9	98.3	118.7	104.3	98.6	119.0	104.6	96.8	120.4	103.7
06	WOOD, PLASTICS & COMPOSITES	80.1	94.0	87.7	78.3	81.7	80.2	78.1	89.3	84.3	87.4	85.1	86.2	89.1	112.6	102.0	77.8	87.0	82.9
07	THERMAL & MOISTURE PROTECTION	98.1	94.2	96.4	98.4	88.5	94.1	102.3	92.4	98.0	95.4	101.0	97.8	103.7	111.9	107.3	103.3	91.0	97.9
08	OPENINGS	93.4	95.4	93.9	93.4	88.7	92.3	86.3	96.7	88.7	89.2	86.4	88.5	92.9	106.6	96.1	89.8	91.9	90.3
0920	Plaster & Gypsum Board	97.8	93.9	95.2	96.1	81.2	86.1	86.3	89.0	88.1	111.8	84.7	93.6	88.0	112.9	104.7	86.2	86.6	86.5
0950, 0980	Ceilings & Acoustic Treatment	101.1	93.9	96.2	100.3	81.2	87.5	91.8	89.0	89.9	96.0	84.7	88.4	89.4	112.9	105.2	91.0	86.6	88.1
0960	Flooring	95.6	103.7	97.8	94.8	78.0	90.1	86.3	103.7	91.2	91.5	95.7	92.7	88.9	94.1	90.4	84.1	103.7	89.6
0970, 0990	Wall Finishes & Painting/Coating	97.6	106.2	102.7	97.6	106.2	102.7	87.2	106.2	98.4	92.0	86.6	88.8	91.9	101.0	97.3	91.9	104.1	99.1
09	FINISHES	100.4	96.1	98.1	100.3	83.9	91.4	88.3	93.4	91.0	90.9	87.3	88.9	88.8	106.5	98.4	87.4	92.7	90.3
COVERS	DIVS. 10 - 14, 25, 28, 41, 43, 44, 46	100.0	96.5	99.2	100.0	94.4	98.8	100.0	96.2	99.1	100.0	95.7	99.0	100.0	102.5	100.6	100.0	96.2	99.2
21, 22, 23	FIRE SUPPRESSION, PLUMBING & HVAC	96.6	85.4	92.2	96.6	78.9	89.6	96.3	90.7	94.1	96.7	92.7	95.1	96.8	117.6	104.9	96.8	94.8	96.0
26, 27, 3370	ELECTRICAL, COMMUNICATIONS & UTIL.	94.8	107.7	101.2	94.8	107.7	101.1	88.7	107.7	98.1	93.7	93.3	93.5	92.9	129.4	110.9	92.1	95.2	93.6
MF2018	WEIGHTED AVERAGE	97.6	97.7	97.7	98.9	91.9	95.9	91.8	97.9	94.4	93.8	94.5	94.1	94.9	111.6	102.0	93.8	96.8	95.1

PENNSYLVANIA

DIVISION		NEW CASTLE 161			NORRISTOWN 194			OIL CITY 163			PHILADELPHIA 190 - 191			PITTSBURGH 150 - 152			POTTSVILLE 179		
		MAT.	INST.	TOTAL	MAT.	INST.	TOTAL	MAT.	INST.	TOTAL	MAT.	INST.	TOTAL	MAT.	INST.	TOTAL	MAT.	INST.	TOTAL
015433	CONTRACTOR EQUIPMENT		112.6	112.6		97.5	97.5		112.6	112.6		99.8	99.8		99.7	99.7		111.8	111.8
0241, 31 - 34	SITE & INFRASTRUCTURE, DEMOLITION	86.7	97.7	94.3	97.3	95.1	95.8	85.2	96.3	92.9	99.8	101.1	100.7	103.7	96.1	98.5	81.3	95.8	91.3
0310	Concrete Forming & Accessories	85.1	93.6	92.3	83.7	125.3	119.1	85.1	92.7	91.6	99.8	142.4	136.2	97.5	97.5	97.5	82.0	87.7	86.8
0320	Concrete Reinforcing	93.3	100.2	96.7	95.9	150.8	122.5	94.4	94.9	94.7	100.4	143.1	121.0	93.5	122.9	107.7	93.4	114.1	103.4
0330	Cast-in-Place Concrete	88.8	94.3	90.8	86.1	127.4	101.5	86.3	94.4	89.3	89.8	135.6	106.8	108.6	101.2	105.8	84.3	96.4	88.8
03	CONCRETE	79.9	96.2	87.0	88.0	129.7	106.3	78.3	95.0	85.6	99.4	139.1	116.8	103.8	103.1	103.5	88.7	96.7	92.2
04	MASONRY	97.3	92.6	94.4	108.3	131.2	122.4	96.5	88.7	91.7	100.2	139.7	124.5	100.4	101.4	101.0	93.3	87.6	89.8
05	METALS	92.7	114.3	99.0	100.3	124.0	107.3	92.7	113.2	98.7	103.8	125.3	110.1	103.7	107.4	104.8	98.6	120.4	105.0
06	WOOD, PLASTICS & COMPOSITES	78.1	94.1	86.9	76.2	125.7	103.4	78.1	94.1	86.9	99.1	143.5	123.6	99.6	96.7	98.0	77.2	85.1	81.6
07	THERMAL & MOISTURE PROTECTION	102.2	92.0	97.8	108.7	130.1	118.0	102.1	90.7	97.1	103.4	139.7	119.2	98.6	99.3	98.9	95.6	99.6	97.3
08	OPENINGS	86.3	91.4	87.5	86.5	133.1	97.4	86.3	94.9	88.3	97.6	144.9	108.6	96.8	102.4	98.1	89.2	93.0	90.1
0920	Plaster & Gypsum Board	86.3	93.9	91.4	84.5	126.5	112.7	86.3	93.9	91.4	103.0	144.6	131.0	97.5	96.4	96.7	106.4	84.7	91.8
0950, 0980	Ceilings & Acoustic Treatment	91.8	93.9	93.2	91.1	126.5	114.9	91.8	93.9	93.2	104.2	144.6	131.5	95.2	96.4	96.0	96.0	84.7	88.4
0960	Flooring	86.3	105.3	91.6	89.4	143.7	104.3	86.3	103.7	91.2	97.8	156.2	114.2	103.1	107.3	104.3	87.5	103.7	92.1
0970, 0990	Wall Finishes & Painting/Coating	87.2	106.2	98.4	89.4	143.7	121.5	87.2	106.2	98.4	97.2	152.6	130.0	100.7	111.3	107.0	92.0	104.1	99.1
09	FINISHES	88.2	96.9	92.9	85.3	129.5	109.2	88.0	96.1	92.4	98.3	146.5	124.4	102.4	100.3	101.2	89.2	91.3	90.3
COVERS	DIVS. 10 - 14, 25, 28, 41, 43, 44, 46	100.0	97.3	99.4	100.0	109.1	102.0	100.0	96.7	99.3	100.0	119.2	104.3	100.0	103.0	100.7	100.0	97.5	99.4
21, 22, 23	FIRE SUPPRESSION, PLUMBING & HVAC	96.3	96.8	96.5	96.6	130.6	110.0	96.3	94.2	95.5	100.1	141.2	116.3	100.0	99.8	99.9	96.7	95.8	96.3
26, 27, 3370	ELECTRICAL, COMMUNICATIONS & UTIL.	89.2	98.2	93.7	91.9	142.0	116.6	90.9	107.7	99.2	98.8	159.0	128.4	97.1	111.3	104.1	92.0	91.3	91.6
MF2018	WEIGHTED AVERAGE	91.4	97.8	94.1	94.7	128.0	108.8	91.3	97.8	94.0	100.2	138.7	116.5	100.7	102.7	101.5	93.7	96.1	94.7

PENNSYLVANIA

DIVISION		READING 195 - 196			SCRANTON 184 - 185			STATE COLLEGE 168			STROUDSBURG 183			SUNBURY 178			UNIONTOWN 154		
		MAT.	INST.	TOTAL	MAT.	INST.	TOTAL	MAT.	INST.	TOTAL	MAT.	INST.	TOTAL	MAT.	INST.	TOTAL	MAT.	INST.	TOTAL
015433	CONTRACTOR EQUIPMENT		119.0	119.0		112.6	112.6		111.8	111.8		112.6	112.6		112.6	112.6		110.7	110.7
0241, 31 - 34	SITE & INFRASTRUCTURE, DEMOLITION	101.8	107.7	105.9	92.2	97.2	95.7	82.7	95.7	91.7	86.6	97.3	94.0	92.6	96.7	95.4	98.3	96.5	97.1
0310	Concrete Forming & Accessories	98.8	86.2	88.0	98.0	86.6	88.3	84.3	82.8	83.0	86.5	90.5	89.9	93.1	84.8	86.0	76.5	93.9	91.4
0320	Concrete Reinforcing	97.2	148.6	122.1	96.8	116.3	106.2	95.1	105.8	100.3	97.1	116.4	106.4	96.0	112.3	103.9	92.3	119.5	105.5
0330	Cast-in-Place Concrete	77.2	97.0	84.6	93.4	93.2	93.3	90.0	87.0	88.9	88.0	94.5	90.4	92.2	94.9	93.2	99.3	95.6	97.9
03	CONCRETE	87.1	102.1	93.7	94.1	95.4	94.7	94.6	89.7	92.5	88.8	97.7	92.7	92.0	94.5	93.1	94.4	100.0	96.9
04	MASONRY	97.5	91.5	93.8	92.5	93.8	93.3	97.8	83.9	89.3	90.3	96.3	94.0	93.6	81.0	85.8	119.3	95.4	104.6
05	METALS	100.6	133.0	110.1	100.9	120.5	106.6	96.6	115.1	102.0	98.7	120.9	105.2	98.3	119.4	104.5	102.0	120.2	107.3
06	WOOD, PLASTICS & COMPOSITES	93.7	82.4	87.5	97.6	83.3	89.8	84.5	81.8	83.0	83.5	87.0	85.5	85.4	85.1	85.2	72.8	94.0	84.4
07	THERMAL & MOISTURE PROTECTION	109.0	103.3	106.6	103.9	91.3	98.4	102.9	97.6	100.6	103.5	91.3	98.2	96.8	92.7	95.0	98.0	95.1	96.7
08	OPENINGS	90.9	99.4	92.9	92.4	89.8	91.8	89.6	85.6	88.7	93.0	94.5	93.3	89.3	89.9	89.4	93.3	99.3	94.7
0920	Plaster & Gypsum Board	94.1	81.9	85.9	96.8	82.8	87.4	87.8	81.2	83.4	86.7	86.6	86.6	105.9	84.7	91.6	94.2	93.9	94.0
0950, 0980	Ceilings & Acoustic Treatment	83.7	81.9	82.5	96.6	82.8	87.3	88.5	81.2	83.6	87.7	86.6	87.0	93.6	84.7	87.6	100.3	93.9	96.0
0960	Flooring	91.3	97.9	93.2	92.0	92.9	92.2	89.4	97.9	91.8	86.8	94.1	88.8	88.3	103.7	92.6	92.0	103.7	95.3
0970, 0990	Wall Finishes & Painting/Coating	88.2	101.8	96.3	91.9	108.3	101.6	91.9	106.2	100.4	91.9	104.1	99.1	92.0	104.1	99.1	97.6	110.0	104.9
09	FINISHES	86.8	88.4	87.7	92.4	89.3	90.7	88.0	87.3	87.6	87.5	91.8	89.8	90.2	89.6	89.9	98.7	97.0	97.8
COVERS	DIVS. 10 - 14, 25, 28, 41, 43, 44, 46	100.0	95.8	99.1	100.0	94.6	98.8	100.0	94.6	98.8	100.0	97.2	99.4	100.0	93.1	98.5	100.0	97.2	99.4
21, 22, 23	FIRE SUPPRESSION, PLUMBING & HVAC	100.3	106.8	102.8	100.3	96.0	98.6	96.8	91.4	94.7	96.8	97.9	97.2	96.7	87.4	93.0	96.6	91.7	94.6
26, 27, 3370	ELECTRICAL, COMMUNICATIONS & UTIL.	97.8	91.3	94.6	98.0	95.2	96.6	91.3	107.7	99.4	92.9	141.8	117.0	92.4	91.9	92.1	91.8	107.7	99.7
MF2018	WEIGHTED AVERAGE	96.5	101.5	98.6	97.4	96.5	97.1	94.5	94.6	94.5	94.3	104.7	98.7	94.6	92.7	93.8	97.8	99.8	98.6

PENNSYLVANIA

DIVISION		WASHINGTON 153			WELLSBORO 169			WESTCHESTER 193			WILKES-BARRE 186 - 187			WILLIAMSPORT 177			YORK 173 - 174		
		MAT.	INST.	TOTAL	MAT.	INST.	TOTAL	MAT.	INST.	TOTAL	MAT.	INST.	TOTAL	MAT.	INST.	TOTAL	MAT.	INST.	TOTAL
015433	CONTRACTOR EQUIPMENT		110.7	110.7		112.6	112.6		97.5	97.5		112.6	112.6		112.6	112.6		111.8	111.8
0241, 31 - 34	SITE & INFRASTRUCTURE, DEMOLITION	98.3	96.8	97.3	94.0	96.6	95.8	103.4	96.1	98.4	84.5	97.2	93.3	83.7	96.8	92.8	82.4	95.6	91.5
0310	Concrete Forming & Accessories	82.8	94.0	92.4	85.2	84.2	84.3	90.0	126.6	121.2	89.1	85.3	85.8	90.1	85.6	86.3	85.1	86.3	86.1
0320	Concrete Reinforcing	92.3	119.5	105.5	95.1	116.1	105.3	95.0	150.8	122.0	95.8	116.3	105.7	95.3	110.9	102.8	96.0	111.0	103.3
0330	Cast-in-Place Concrete	99.3	95.9	98.0	94.4	88.2	92.1	95.4	129.5	108.1	84.5	93.0	87.6	78.0	90.0	82.5	84.7	96.7	89.2
03	CONCRETE	94.8	100.1	97.2	96.9	92.5	95.0	95.8	131.0	111.3	85.8	94.7	89.7	80.2	92.9	85.8	89.8	95.6	92.4
04	MASONRY	103.6	95.5	98.6	98.7	81.4	88.1	102.8	134.9	122.5	105.0	93.5	97.9	85.9	86.1	86.0	94.5	86.7	89.7
05	METALS	101.9	120.6	107.4	96.6	119.7	103.4	100.3	124.0	107.3	96.8	120.4	103.7	98.3	118.1	104.1	100.0	118.6	105.5
06	WOOD, PLASTICS & COMPOSITES	80.2	94.0	87.8	82.1	84.4	83.4	82.8	125.7	106.4	85.9	81.7	83.6	81.9	85.1	83.7	80.7	85.1	83.1
07	THERMAL & MOISTURE PROTECTION	98.1	95.1	96.8	104.1	86.0	96.2	109.1	127.0	116.9	103.3	90.9	97.9	96.1	88.3	92.7	95.6	101.0	98.0
08	OPENINGS	93.3	93.9	94.7	92.4	90.0	91.8	86.5	133.1	97.4	89.8	88.9	89.6	89.3	89.6	89.4	89.2	86.4	88.5
0920	Plaster & Gypsum Board	97.7	93.9	95.1	86.5	83.9	84.8	85.2	126.5	112.9	87.4	81.2	83.2	106.4	84.7	91.8	106.7	84.7	91.9
0950, 0980	Ceilings & Acoustic Treatment	100.3	93.9	96.0	88.5	83.9	85.4	91.1	126.5	114.9	91.0	81.2	84.4	96.0	84.7	88.4	95.1	84.7	88.1
0960	Flooring	95.7	103.7	97.9	85.8	103.7	90.9	89.6	139.3	103.6	87.7	92.9	89.1	87.2	86.8	87.1	88.9	95.7	90.8
0970, 0990	Wall Finishes & Painting/Coating	97.6	110.0	104.9	91.9	104.1	99.1	89.4	143.7	121.5	91.9	101.8	97.8	92.0	101.8	97.8	92.0	84.8	87.7
09	FINISHES	100.3	97.0	98.5	88.1	89.6	88.9	86.7	130.5	110.4	88.4	87.6	88.0	89.7	86.7	88.0	89.5	87.1	88.2
COVERS	DIVS. 10 - 14, 25, 28, 41, 43, 44, 46	100.0	97.2	99.4	100.0	96.2	99.2	100.0	115.3	103.4	100.0	94.3	98.7	100.0	95.5	99.0	100.0	95.6	99.0
21, 22, 23	FIRE SUPPRESSION, PLUMBING & HVAC	96.6	93.2	95.2	96.8	87.4	93.1	96.6	132.6	110.7	96.8	96.6	96.7	96.7	92.2	94.9	100.2	92.7	97.3
26, 27, 3370	ELECTRICAL, COMMUNICATIONS & UTIL.	94.2	107.7	100.9	92.1	95.2	93.6	91.7	132.4	111.8	92.9	90.0	91.5	92.8	83.4	88.2	93.7	82.9	88.4
MF2018	WEIGHTED AVERAGE	97.5	100.2	98.6	95.5	92.9	94.4	95.8	127.9	109.3	94.0	95.5	94.7	92.5	92.3	92.4	95.2	93.0	94.3

PUERTO RICO / RHODE ISLAND / SOUTH CAROLINA

DIVISION		PUERTO RICO — SAN JUAN 009			RHODE ISLAND — NEWPORT 028			RHODE ISLAND — PROVIDENCE 029			SOUTH CAROLINA — AIKEN 298			SOUTH CAROLINA — BEAUFORT 299			SOUTH CAROLINA — CHARLESTON 294		
		MAT.	INST.	TOTAL	MAT.	INST.	TOTAL	MAT.	INST.	TOTAL	MAT.	INST.	TOTAL	MAT.	INST.	TOTAL	MAT.	INST.	TOTAL
015433	CONTRACTOR EQUIPMENT		87.3	87.3		97.9	97.9		100.1	100.1		103.7	103.7		103.7	103.7		103.7	103.7
0241, 31 - 34	SITE & INFRASTRUCTURE, DEMOLITION	81.4	87.9	85.9	86.3	98.0	94.4	89.4	102.2	98.3	128.3	85.2	98.5	123.4	84.7	96.7	108.2	85.2	92.3
0310	Concrete Forming & Accessories	105.9	21.6	34.0	100.7	124.7	121.1	99.9	124.7	121.0	94.0	64.6	68.9	93.1	38.9	46.9	92.3	64.9	69.0
0320	Concrete Reinforcing	114.9	18.7	68.4	109.7	127.1	118.1	103.7	127.1	115.0	94.3	60.6	78.0	93.4	66.7	80.5	93.2	67.3	80.7
0330	Cast-in-Place Concrete	70.9	32.7	56.6	71.7	118.5	89.1	87.5	119.0	99.2	92.9	66.9	83.2	92.9	66.8	83.2	109.1	67.2	93.5
03	CONCRETE	89.8	28.7	61.8	87.5	122.4	102.9	91.8	122.5	105.3	101.0	66.4	85.8	98.4	55.7	79.7	94.9	67.8	81.9
04	MASONRY	110.0	24.2	57.3	99.2	123.1	113.8	103.4	123.1	115.5	79.9	64.8	70.6	94.9	64.8	76.4	96.4	67.5	78.6
05	METALS	116.8	40.6	94.4	98.8	116.7	104.1	103.2	115.9	106.9	102.6	89.1	98.7	102.7	90.7	99.2	104.6	92.1	100.9
06	WOOD, PLASTICS & COMPOSITES	145.4	20.5	76.6	98.0	124.0	112.3	98.4	124.0	112.5	89.7	66.8	77.1	88.4	32.5	57.6	87.2	66.8	76.0
07	THERMAL & MOISTURE PROTECTION	134.1	27.7	87.8	102.6	123.0	111.5	108.3	123.5	114.9	98.1	64.2	83.4	97.8	56.2	79.7	96.8	66.1	83.5
08	OPENINGS	96.8	19.1	78.7	97.4	124.4	103.7	99.0	124.4	104.9	94.0	62.5	86.7	94.1	44.9	82.6	97.6	65.1	90.0
0920	Plaster & Gypsum Board	76.4	25.2	48.0	91.4	124.3	113.6	104.9	124.3	118.0	83.4	65.8	71.6	87.0	30.5	49.0	88.1	65.8	73.1
0950, 0980	Ceilings & Acoustic Treatment	158.8	19.3	64.7	90.3	124.3	113.2	104.1	124.3	117.7	78.8	65.8	70.1	82.1	30.5	47.3	82.1	65.8	71.1
0960	Flooring	117.8	29.7	93.1	83.8	127.3	96.0	86.4	127.3	97.9	97.5	89.8	95.3	99.0	75.8	92.5	98.3	81.9	93.9
0970, 0990	Wall Finishes & Painting/Coating	90.2	24.6	51.4	85.0	118.9	105.0	90.8	118.9	107.4	95.2	66.0	77.9	95.2	56.9	72.5	95.2	70.0	80.3
09	FINISHES	106.0	23.2	61.2	87.7	125.1	107.9	93.7	125.1	110.7	89.0	69.7	78.5	90.1	46.0	66.2	88.2	68.9	77.8
COVERS	DIVS. 10 - 14, 25, 28, 41, 43, 44, 46	100.0	26.6	83.6	100.0	110.2	102.3	100.0	110.3	102.3	100.0	70.3	93.4	100.0	80.1	95.6	100.0	70.3	93.4
21, 22, 23	FIRE SUPPRESSION, PLUMBING & HVAC	88.5	18.8	61.1	100.4	114.8	106.0	100.2	114.8	105.9	96.9	55.2	80.5	96.9	55.6	80.7	100.4	58.4	83.9
26, 27, 3370	ELECTRICAL, COMMUNICATIONS & UTIL.	84.4	25.0	55.2	102.1	95.5	98.9	102.3	95.5	99.0	96.2	61.9	79.3	99.7	30.9	65.8	98.1	58.9	78.8
MF2018	WEIGHTED AVERAGE	98.5	29.8	69.4	97.0	114.7	104.5	99.3	115.0	105.9	97.6	67.1	84.7	98.3	57.5	81.1	98.7	68.2	85.8

SOUTH CAROLINA / SOUTH DAKOTA

DIVISION		SOUTH CAROLINA — COLUMBIA 290 - 292			SOUTH CAROLINA — FLORENCE 295			SOUTH CAROLINA — GREENVILLE 296			SOUTH CAROLINA — ROCK HILL 297			SOUTH CAROLINA — SPARTANBURG 293			SOUTH DAKOTA — ABERDEEN 574		
		MAT.	INST.	TOTAL	MAT.	INST.	TOTAL	MAT.	INST.	TOTAL	MAT.	INST.	TOTAL	MAT.	INST.	TOTAL	MAT.	INST.	TOTAL
015433	CONTRACTOR EQUIPMENT		105.6	105.6		103.7	103.7		103.7	103.7		103.7	103.7		103.7	103.7		97.7	97.7
0241, 31 - 34	SITE & INFRASTRUCTURE, DEMOLITION	105.9	89.5	94.6	117.6	84.9	95.1	112.6	85.4	93.8	110.0	84.3	92.2	112.4	85.4	93.8	99.2	93.4	95.2
0310	Concrete Forming & Accessories	93.2	64.9	69.1	81.1	64.8	67.2	91.8	64.9	68.8	90.1	64.0	67.9	94.4	64.9	69.2	98.9	75.7	79.1
0320	Concrete Reinforcing	98.3	67.3	83.3	92.8	67.2	80.4	92.8	64.4	79.1	93.5	65.8	80.1	92.8	67.3	80.4	101.0	71.4	86.7
0330	Cast-in-Place Concrete	110.3	67.5	94.4	92.9	67.0	83.2	92.9	67.2	83.3	92.8	66.4	83.0	92.9	67.2	83.3	109.5	78.8	98.1
03	CONCRETE	93.2	67.8	82.1	93.3	67.6	82.0	92.0	67.3	81.2	90.1	66.8	79.9	92.2	67.8	81.5	98.7	76.7	89.1
04	MASONRY	89.9	67.5	76.1	79.9	67.4	72.2	77.9	67.5	71.5	101.6	64.0	78.5	79.9	67.5	72.3	109.4	73.3	87.2
05	METALS	101.6	91.0	98.5	103.5	91.4	99.9	103.5	91.7	100.0	102.7	90.5	99.1	103.5	92.1	100.1	93.8	82.1	90.4
06	WOOD, PLASTICS & COMPOSITES	88.6	66.9	76.7	74.8	66.8	70.4	86.8	66.8	75.8	85.4	66.8	75.2	90.6	66.8	77.5	96.9	74.2	84.7
07	THERMAL & MOISTURE PROTECTION	93.3	66.3	81.5	97.2	66.1	83.7	97.1	66.1	83.6	96.9	59.2	80.5	97.1	66.1	83.6	103.1	77.2	91.8
08	OPENINGS	99.7	65.1	91.7	94.1	65.1	87.3	94.0	64.8	87.2	94.1	63.6	87.0	94.1	65.1	87.3	97.2	62.2	89.0
0920	Plaster & Gypsum Board	88.7	65.8	73.3	78.2	65.8	69.9	82.1	65.8	71.2	81.8	65.8	71.1	84.7	65.8	72.0	110.1	73.9	85.7
0950, 0980	Ceilings & Acoustic Treatment	86.2	65.8	72.5	79.6	65.8	70.3	78.8	65.8	70.1	78.8	65.8	70.1	78.8	65.8	70.1	98.8	73.9	82.0
0960	Flooring	90.7	80.4	87.8	89.9	80.4	87.2	96.3	80.4	91.8	95.2	75.8	89.7	97.6	80.4	92.8	91.1	43.8	77.8
0970, 0990	Wall Finishes & Painting/Coating	93.9	70.0	79.7	95.2	70.0	80.3	95.2	70.0	80.3	95.2	66.0	77.9	95.2	70.0	80.3	89.8	33.4	56.4
09	FINISHES	86.5	68.7	76.8	85.0	68.7	76.2	86.9	68.7	77.0	86.3	67.1	75.9	87.6	68.7	77.4	92.4	65.9	78.1
COVERS	DIVS. 10 - 14, 25, 28, 41, 43, 44, 46	100.0	70.5	93.4	100.0	70.3	93.4	100.0	70.3	93.4	100.0	70.0	93.3	100.0	70.0	93.3	100.0	84.8	96.6
21, 22, 23	FIRE SUPPRESSION, PLUMBING & HVAC	100.0	58.3	83.6	100.4	58.2	83.9	100.4	58.2	83.9	96.9	54.8	80.4	100.4	58.2	83.9	100.0	54.2	82.0
26, 27, 3370	ELECTRICAL, COMMUNICATIONS & UTIL.	98.5	62.8	80.9	96.1	62.8	79.7	98.2	59.8	79.3	98.2	59.9	79.3	98.2	59.8	79.3	101.1	65.3	83.4
MF2018	WEIGHTED AVERAGE	97.6	68.9	85.5	96.9	68.6	84.9	97.0	68.2	84.8	96.7	66.4	83.9	97.2	68.3	85.0	98.6	70.1	86.5

For customer support on your Mechanical Costs with RSMeans Data, call 800.448.8182.

687

City Cost Indexes

SOUTH DAKOTA

DIVISION		MITCHELL 573			MOBRIDGE 576			PIERRE 575			RAPID CITY 577			SIOUX FALLS 570-571			WATERTOWN 572		
		MAT.	INST.	TOTAL	MAT.	INST.	TOTAL	MAT.	INST.	TOTAL	MAT.	INST.	TOTAL	MAT.	INST.	TOTAL	MAT.	INST.	TOTAL
015433	CONTRACTOR EQUIPMENT		97.7	97.7		97.7	97.7		99.9	99.9		97.7	97.7		100.9	100.9		97.7	97.7
0241, 31 - 34	SITE & INFRASTRUCTURE, DEMOLITION	95.8	93.2	94.0	95.7	93.2	94.0	97.2	96.3	96.6	97.5	92.8	94.3	92.5	98.7	96.8	95.6	93.2	94.0
0310	Concrete Forming & Accessories	97.9	44.2	52.1	88.2	44.6	51.0	99.7	45.6	53.6	106.9	56.6	64.0	103.3	78.1	81.8	84.6	74.5	76.0
0320	Concrete Reinforcing	100.4	70.5	86.0	103.0	70.5	87.3	97.8	100.2	99.0	94.4	100.4	97.3	106.6	100.6	103.7	97.6	70.5	84.5
0330	Cast-in-Place Concrete	106.4	52.3	86.3	106.4	77.2	95.6	103.1	77.8	93.7	105.6	77.2	95.0	88.3	78.8	84.7	106.4	77.2	95.5
03	CONCRETE	96.5	53.1	77.4	96.2	62.0	81.2	91.8	67.4	81.1	95.9	72.3	85.5	88.6	82.7	86.0	95.3	75.5	86.6
04	MASONRY	97.8	75.5	84.1	106.5	72.8	85.8	106.5	72.6	85.7	106.0	74.0	86.3	91.1	75.6	81.5	132.0	73.7	96.2
05	METALS	92.9	81.6	89.6	92.9	82.0	89.7	96.3	89.5	94.3	95.5	91.2	94.2	95.3	92.6	94.5	92.9	81.9	89.6
06	WOOD, PLASTICS & COMPOSITES	96.6	33.7	62.0	84.4	33.4	56.4	104.2	34.0	65.6	102.2	48.0	72.3	101.6	76.5	87.8	80.5	74.2	77.0
07	THERMAL & MOISTURE PROTECTION	102.8	72.9	89.8	102.9	75.1	90.8	104.1	72.7	90.4	103.6	78.5	92.7	103.2	83.3	94.6	102.6	76.7	91.3
08	OPENINGS	96.3	39.6	83.1	98.9	39.0	84.9	99.8	58.4	90.2	101.0	66.1	92.9	104.2	81.8	99.0	96.3	61.5	88.2
0920	Plaster & Gypsum Board	109.0	32.2	57.3	102.2	32.0	54.9	108.1	32.5	57.2	109.8	46.9	67.5	103.6	76.1	85.1	100.5	73.9	82.6
0950, 0980	Ceilings & Acoustic Treatment	95.5	32.2	52.8	98.8	32.0	53.7	98.6	32.5	54.0	100.4	46.9	64.3	97.7	76.1	83.1	95.5	73.9	80.9
0960	Flooring	90.7	43.8	77.6	86.5	46.4	75.3	98.2	31.2	79.4	90.5	72.7	85.5	93.9	75.0	88.6	85.2	43.8	73.6
0970, 0990	Wall Finishes & Painting/Coating	89.8	36.6	58.3	89.8	37.5	58.8	97.1	95.6	96.2	89.8	95.6	93.2	100.3	95.6	97.5	89.8	33.4	56.4
09	FINISHES	91.4	41.1	64.2	89.9	41.6	63.7	98.4	45.2	69.6	92.5	61.8	75.9	96.4	79.1	87.1	88.6	64.7	75.7
COVERS	DIVS. 10 - 14, 25, 28, 41, 43, 44, 46	100.0	82.0	96.0	100.0	81.9	96.0	100.0	82.3	96.1	100.0	83.5	96.3	100.0	87.0	97.1	100.0	85.2	96.8
21, 22, 23	FIRE SUPPRESSION, PLUMBING & HVAC	96.5	49.0	77.8	96.5	69.0	85.7	100.0	76.8	90.9	100.0	76.8	90.9	99.9	70.2	88.2	96.5	52.1	79.0
26, 27, 3370	ELECTRICAL, COMMUNICATIONS & UTIL.	99.4	63.4	81.7	101.1	39.8	70.9	103.9	47.7	76.2	97.6	47.7	73.0	102.6	63.4	83.3	98.6	63.4	81.2
MF2018	WEIGHTED AVERAGE	96.3	60.9	81.3	96.9	62.9	82.5	99.0	68.5	86.1	98.3	72.0	87.2	97.7	78.4	89.5	97.3	69.0	85.4

TENNESSEE

DIVISION		CHATTANOOGA 373 - 374			COLUMBIA 384			COOKEVILLE 385			JACKSON 383			JOHNSON CITY 376			KNOXVILLE 377 - 379		
		MAT.	INST.	TOTAL	MAT.	INST.	TOTAL	MAT.	INST.	TOTAL	MAT.	INST.	TOTAL	MAT.	INST.	TOTAL	MAT.	INST.	TOTAL
015433	CONTRACTOR EQUIPMENT		104.7	104.7		99.5	99.5		99.5	99.5		105.5	105.5		98.8	98.8		98.8	98.8
0241, 31 - 34	SITE & INFRASTRUCTURE, DEMOLITION	106.6	93.8	97.8	91.5	84.6	86.7	97.3	81.6	86.5	100.9	94.0	96.2	113.6	81.1	91.2	92.6	83.8	86.5
0310	Concrete Forming & Accessories	95.8	57.3	63.0	80.9	62.4	65.1	81.1	31.9	39.2	87.5	41.0	47.8	82.7	57.5	61.2	94.3	61.7	66.5
0320	Concrete Reinforcing	101.6	65.1	83.9	91.4	65.0	78.6	91.4	64.6	78.4	91.3	67.2	79.7	102.2	61.9	82.7	101.6	61.8	82.3
0330	Cast-in-Place Concrete	97.5	64.1	85.1	92.8	63.8	82.0	105.2	56.7	87.2	102.7	66.0	89.0	78.4	58.0	70.8	91.6	64.0	81.3
03	CONCRETE	93.1	62.8	79.8	92.4	65.1	80.4	102.4	48.7	78.8	93.2	56.5	77.0	102.7	60.2	84.0	90.7	64.2	79.0
04	MASONRY	98.6	56.3	72.6	111.4	55.5	77.0	106.7	39.4	65.3	111.7	44.4	70.4	111.0	43.2	69.3	76.3	50.7	60.6
05	METALS	93.0	88.1	91.6	93.5	88.9	92.1	93.5	87.6	91.8	96.0	89.5	94.1	90.3	86.4	89.2	93.4	86.5	91.4
06	WOOD, PLASTICS & COMPOSITES	106.5	56.3	78.9	67.8	63.4	65.4	68.0	29.5	46.8	82.9	39.7	59.1	78.7	62.6	69.8	93.1	62.6	76.4
07	THERMAL & MOISTURE PROTECTION	99.2	61.7	82.9	95.6	62.9	81.4	96.1	52.6	77.2	98.0	55.5	79.5	95.3	55.6	78.0	92.8	60.8	78.9
08	OPENINGS	102.0	57.5	91.6	91.0	53.3	82.2	91.1	35.5	78.1	97.7	44.3	85.2	98.3	59.5	89.3	95.7	55.2	86.2
0920	Plaster & Gypsum Board	80.0	55.6	63.6	87.4	62.9	70.9	87.4	28.0	47.4	90.0	38.5	55.3	101.0	62.1	74.8	109.1	62.1	77.5
0950, 0980	Ceilings & Acoustic Treatment	97.9	55.6	69.4	75.8	62.9	67.1	75.8	28.0	43.6	83.9	38.5	53.3	93.5	62.1	72.3	94.3	62.1	72.6
0960	Flooring	98.3	53.2	86.2	81.7	53.9	73.9	81.7	50.0	72.8	81.0	54.2	73.5	93.3	43.9	79.4	98.4	49.7	84.7
0970, 0990	Wall Finishes & Painting/Coating	98.0	59.5	75.2	85.0	56.5	68.1	85.0	56.5	68.1	87.0	56.5	68.9	94.9	58.7	73.5	94.9	57.6	72.8
09	FINISHES	94.4	56.6	73.9	86.2	60.1	72.1	86.8	36.2	59.4	86.1	43.4	63.0	98.5	55.3	75.1	91.4	58.9	73.8
COVERS	DIVS. 10 - 14, 25, 28, 41, 43, 44, 46	100.0	67.8	92.8	100.0	68.0	92.9	100.0	35.9	85.7	100.0	62.0	91.5	100.0	75.8	94.6	100.0	79.3	95.4
21, 22, 23	FIRE SUPPRESSION, PLUMBING & HVAC	100.1	59.3	84.1	97.8	73.6	88.3	97.8	67.1	85.8	100.1	61.4	84.9	100.0	55.5	82.5	100.0	60.5	84.5
26, 27, 3370	ELECTRICAL, COMMUNICATIONS & UTIL.	100.8	82.9	92.0	94.0	50.6	72.6	95.5	61.3	78.6	100.5	65.7	83.3	91.5	40.8	66.5	97.0	54.9	76.2
MF2018	WEIGHTED AVERAGE	98.0	68.1	85.3	94.8	66.5	82.8	96.2	56.9	79.6	97.6	61.4	82.3	98.1	58.6	81.4	94.8	63.7	81.6

TENNESSEE / TEXAS

DIVISION		TENNESSEE MCKENZIE 382			TENNESSEE MEMPHIS 375, 380 - 381			TENNESSEE NASHVILLE 370 - 372			TEXAS ABILENE 795 - 796			TEXAS AMARILLO 790 - 791			TEXAS AUSTIN 786 - 787		
		MAT.	INST.	TOTAL	MAT.	INST.	TOTAL	MAT.	INST.	TOTAL	MAT.	INST.	TOTAL	MAT.	INST.	TOTAL	MAT.	INST.	TOTAL
015433	CONTRACTOR EQUIPMENT		99.5	99.5		102.0	102.0		105.5	105.5		90.9	90.9		93.9	93.9		92.5	92.5
0241, 31 - 34	SITE & INFRASTRUCTURE, DEMOLITION	96.9	81.8	86.5	91.1	95.4	94.0	102.8	98.1	99.6	91.9	87.4	88.8	91.3	91.3	91.3	95.5	90.1	91.7
0310	Concrete Forming & Accessories	88.4	34.1	42.1	94.6	64.7	69.1	95.9	68.1	72.2	94.0	59.2	64.3	95.6	51.7	58.2	95.3	53.0	59.3
0320	Concrete Reinforcing	91.5	65.5	78.9	110.9	66.0	89.2	107.1	66.8	87.6	88.5	51.7	70.7	95.3	49.6	73.2	86.8	47.2	67.6
0330	Cast-in-Place Concrete	102.9	57.9	86.2	96.5	76.7	89.1	88.2	69.0	81.1	84.6	64.9	77.3	83.2	65.7	76.7	94.0	65.9	83.5
03	CONCRETE	101.0	50.2	78.7	96.5	70.2	84.9	92.9	69.4	82.6	84.5	60.7	74.1	87.1	57.2	74.0	89.4	57.4	75.4
04	MASONRY	110.5	45.1	70.3	92.6	56.3	70.3	87.7	55.9	68.2	99.9	60.6	75.8	97.6	60.2	74.6	91.7	60.7	72.7
05	METALS	93.5	88.0	91.9	87.2	82.5	85.8	101.6	85.1	96.7	107.1	70.4	96.3	101.9	68.7	92.1	100.7	67.4	90.9
06	WOOD, PLASTICS & COMPOSITES	76.6	31.7	51.9	98.9	65.4	80.4	100.1	70.1	83.6	99.2	60.8	78.1	102.6	51.0	74.2	92.9	52.6	70.7
07	THERMAL & MOISTURE PROTECTION	96.1	53.6	77.6	93.0	66.4	81.4	96.9	65.0	83.6	98.3	63.4	83.1	100.1	62.3	83.6	95.1	63.8	81.5
08	OPENINGS	91.1	37.2	78.5	99.5	61.3	90.6	101.0	67.5	93.2	101.8	56.7	91.2	103.2	50.8	91.0	101.1	49.6	89.1
0920	Plaster & Gypsum Board	90.9	30.3	50.1	90.9	64.5	73.1	95.9	69.3	78.0	88.2	60.1	69.3	100.7	50.0	66.6	88.9	51.6	63.8
0950, 0980	Ceilings & Acoustic Treatment	75.8	30.3	45.1	90.8	64.5	73.0	94.1	69.3	77.4	99.2	60.1	72.9	100.7	50.0	66.5	93.3	51.6	65.2
0960	Flooring	84.5	53.9	75.9	97.8	55.2	85.9	97.1	57.5	86.0	93.2	67.6	86.0	93.7	63.8	85.3	94.1	63.8	85.6
0970, 0990	Wall Finishes & Painting/Coating	85.0	44.2	60.8	93.2	59.3	73.1	100.7	68.6	81.7	92.8	51.7	68.5	91.0	51.7	67.7	95.7	44.2	65.2
09	FINISHES	88.0	37.5	60.7	95.9	61.9	77.5	97.8	66.1	80.6	86.1	60.1	72.0	92.6	53.5	71.4	91.1	53.6	70.8
COVERS	DIVS. 10 - 14, 25, 28, 41, 43, 44, 46	100.0	24.0	83.1	100.0	80.0	95.7	100.0	81.1	95.8	100.0	78.9	95.3	100.0	72.7	93.9	100.0	76.5	94.8
21, 22, 23	FIRE SUPPRESSION, PLUMBING & HVAC	97.8	58.6	82.4	100.1	70.3	88.4	99.9	75.1	90.2	100.1	52.1	81.3	99.9	51.1	80.7	99.9	58.7	83.8
26, 27, 3370	ELECTRICAL, COMMUNICATIONS & UTIL.	95.3	55.4	75.6	102.9	63.6	83.5	93.5	63.2	78.6	97.9	54.0	76.3	100.9	58.9	80.2	97.5	57.2	77.6
MF2018	WEIGHTED AVERAGE	96.3	55.1	78.9	96.6	69.8	85.3	98.0	71.8	86.9	97.8	61.4	82.4	98.2	60.0	82.1	97.2	61.4	82.1

TEXAS

DIVISION		BEAUMONT 776-777			BROWNWOOD 768			BRYAN 778			CHILDRESS 792			CORPUS CHRISTI 783-784			DALLAS 752-753		
		MAT.	INST.	TOTAL	MAT.	INST.	TOTAL	MAT.	INST.	TOTAL	MAT.	INST.	TOTAL	MAT.	INST.	TOTAL	MAT.	INST.	TOTAL
015433	CONTRACTOR EQUIPMENT		95.3	95.3		90.9	90.9		95.3	95.3		90.9	90.9		100.6	100.6		106.9	106.9
0241, 31 - 34	SITE & INFRASTRUCTURE, DEMOLITION	90.6	91.8	91.4	101.3	87.3	91.7	81.5	91.9	88.7	101.8	85.8	90.8	144.2	82.8	101.8	106.0	97.6	100.2
0310	Concrete Forming & Accessories	102.3	54.6	61.7	97.6	58.7	64.4	82.1	58.2	61.7	92.6	58.6	63.6	98.4	52.2	59.0	95.0	62.3	67.1
0320	Concrete Reinforcing	85.7	62.0	74.2	81.9	51.1	67.0	87.9	51.2	70.2	88.7	51.0	70.5	79.8	50.9	65.8	91.4	52.3	72.5
0330	Cast-in-Place Concrete	90.7	65.2	81.2	96.8	58.6	82.6	72.6	59.6	67.8	86.7	58.6	76.3	114.1	67.1	96.6	92.6	71.4	84.7
03	CONCRETE	91.0	60.7	77.7	90.3	58.2	76.2	75.8	58.5	68.2	93.4	58.2	77.9	95.2	59.1	79.4	93.3	65.3	81.0
04	MASONRY	102.9	61.7	77.6	122.8	60.6	84.6	143.3	60.6	92.5	103.5	60.2	76.9	81.4	60.7	68.7	99.5	60.2	75.3
05	METALS	99.4	76.2	92.6	102.6	69.8	92.9	99.5	72.1	91.4	104.4	69.6	94.2	97.3	84.4	93.5	101.7	82.6	96.1
06	WOOD, PLASTICS & COMPOSITES	113.6	54.1	80.9	105.8	60.8	81.0	78.6	59.3	68.0	98.6	60.8	77.8	115.6	51.5	80.3	100.8	63.6	80.3
07	THERMAL & MOISTURE PROTECTION	92.8	64.3	80.4	92.0	61.9	78.9	85.2	63.8	75.9	99.0	60.4	82.2	97.7	62.2	82.2	87.3	67.2	78.6
08	OPENINGS	95.0	55.0	85.6	100.4	56.6	90.2	95.7	56.0	86.4	96.1	56.6	86.9	101.3	50.0	89.4	96.5	59.5	87.9
0920	Plaster & Gypsum Board	101.0	53.2	68.9	82.6	60.1	67.5	88.2	58.7	68.3	87.6	60.1	69.1	96.5	50.5	65.5	92.7	62.5	72.4
0950, 0980	Ceilings & Acoustic Treatment	105.3	53.2	70.2	86.6	60.1	68.7	96.7	58.7	71.1	96.8	60.1	72.1	98.8	50.5	66.2	102.4	62.5	75.5
0960	Flooring	116.9	74.6	105.0	75.7	54.6	69.8	86.7	67.0	81.1	91.6	54.6	81.2	107.4	67.7	96.3	94.2	69.4	87.2
0970, 0990	Wall Finishes & Painting/Coating	94.9	53.9	70.6	92.8	51.7	68.4	92.2	57.4	71.6	92.8	51.7	68.5	110.1	44.2	71.1	96.1	54.8	71.6
09	FINISHES	98.0	57.8	76.2	77.6	57.4	66.7	85.4	59.4	71.3	86.2	57.4	70.6	101.9	53.8	75.9	97.9	62.7	78.8
COVERS	DIVS. 10 - 14, 25, 28, 41, 43, 44, 46	100.0	77.8	95.0	100.0	73.4	94.1	100.0	77.0	94.9	100.0	73.4	94.1	100.0	77.8	95.0	100.0	82.5	96.1
21, 22, 23	FIRE SUPPRESSION, PLUMBING & HVAC	100.0	63.0	85.5	96.5	49.5	78.0	96.5	62.4	83.1	96.6	51.7	79.0	100.0	48.7	79.9	99.9	62.3	85.1
26, 27, 3370	ELECTRICAL, COMMUNICATIONS & UTIL.	100.1	63.4	82.0	98.6	46.1	72.7	97.8	63.4	80.8	97.9	55.8	77.1	94.0	63.4	78.9	95.2	63.2	79.4
MF2018	WEIGHTED AVERAGE	97.9	65.5	84.2	97.4	58.7	81.1	95.2	64.9	82.4	97.6	60.3	81.8	98.9	61.4	83.1	98.2	68.0	85.4

TEXAS

DIVISION		DEL RIO 788			DENTON 762			EASTLAND 764			EL PASO 798 - 799, 885			FORT WORTH 760 - 761			GALVESTON 775		
		MAT.	INST.	TOTAL	MAT.	INST.	TOTAL	MAT.	INST.	TOTAL	MAT.	INST.	TOTAL	MAT.	INST.	TOTAL	MAT.	INST.	TOTAL
015433	CONTRACTOR EQUIPMENT		89.9	89.9		100.2	100.2		90.9	90.9		93.9	93.9		93.9	93.9		107.9	107.9
0241, 31 - 34	SITE & INFRASTRUCTURE, DEMOLITION	122.2	85.5	96.9	101.4	79.7	86.4	104.0	85.5	91.2	89.7	92.7	91.8	97.4	92.7	94.2	108.4	90.6	96.1
0310	Concrete Forming & Accessories	95.1	51.0	57.5	105.0	58.8	65.7	98.4	58.6	64.4	95.6	56.7	62.5	98.2	59.5	65.2	90.9	55.8	61.0
0320	Concrete Reinforcing	80.4	47.6	64.5	83.2	51.1	67.7	82.1	51.1	67.1	94.2	51.1	73.4	88.1	50.9	70.1	87.5	59.7	74.0
0330	Cast-in-Place Concrete	122.9	58.4	98.9	74.9	59.7	69.2	102.3	58.5	86.0	73.9	68.0	71.7	89.3	65.9	80.6	96.4	60.6	83.1
03	CONCRETE	115.8	54.0	88.4	70.6	59.7	65.8	94.8	58.1	78.7	82.7	60.5	72.9	86.1	61.0	75.1	92.1	60.3	78.1
04	MASONRY	95.4	60.5	74.0	131.0	59.3	86.9	92.7	60.6	73.0	91.6	62.7	73.8	92.5	59.3	72.1	99.0	60.7	75.5
05	METALS	97.2	67.2	88.4	102.1	84.9	97.1	102.4	69.6	92.7	102.1	69.4	92.5	104.3	69.5	94.1	101.2	91.4	98.3
06	WOOD, PLASTICS & COMPOSITES	96.1	50.4	71.0	117.8	60.9	86.5	113.0	60.8	84.3	90.2	57.1	72.0	99.6	60.9	78.3	95.3	55.9	73.6
07	THERMAL & MOISTURE PROTECTION	94.7	62.0	80.5	90.0	62.4	78.0	92.4	61.9	79.1	91.9	64.4	80.0	89.7	64.4	78.7	84.6	64.4	75.8
08	OPENINGS	96.6	47.8	85.2	118.7	56.2	104.2	70.6	56.6	67.4	96.3	52.8	86.2	101.2	56.6	90.8	99.7	56.0	89.5
0920	Plaster & Gypsum Board	93.3	49.5	63.8	87.3	60.1	69.0	82.6	60.1	67.5	97.3	56.2	69.6	89.8	60.1	69.8	94.8	55.0	68.0
0950, 0980	Ceilings & Acoustic Treatment	95.3	49.5	64.4	90.7	60.1	70.1	86.6	60.1	68.7	94.2	56.2	68.6	92.7	60.1	70.7	100.8	55.0	69.9
0960	Flooring	91.9	54.6	81.4	71.9	54.6	67.1	95.9	54.6	84.3	97.3	69.7	89.5	94.5	64.5	86.0	102.4	67.0	92.7
0970, 0990	Wall Finishes & Painting/Coating	97.8	42.6	65.1	103.3	48.9	71.1	94.2	51.7	69.0	98.6	49.6	69.6	101.2	51.8	72.0	104.6	53.7	74.5
09	FINISHES	94.5	50.3	70.6	76.3	57.2	66.0	84.1	57.4	69.6	92.3	58.1	73.8	89.6	59.5	73.3	95.4	56.9	74.6
COVERS	DIVS. 10 - 14, 25, 28, 41, 43, 44, 46	100.0	73.6	94.1	100.0	79.2	95.4	100.0	71.7	93.7	100.0	76.6	94.8	100.0	79.2	95.4	100.0	78.3	95.2
21, 22, 23	FIRE SUPPRESSION, PLUMBING & HVAC	96.5	59.4	81.9	96.5	55.8	80.5	96.5	49.5	78.0	99.8	65.3	86.3	99.9	55.4	82.4	96.5	62.5	83.2
26, 27, 3370	ELECTRICAL, COMMUNICATIONS & UTIL.	95.9	59.8	78.1	101.0	58.7	80.1	98.5	58.6	78.8	97.5	50.5	74.3	100.2	58.6	79.7	99.6	63.5	81.8
MF2018	WEIGHTED AVERAGE	99.7	60.4	83.1	97.3	62.9	82.7	94.2	60.3	79.8	95.9	63.6	82.3	97.6	62.9	82.9	97.7	66.5	84.5

TEXAS

DIVISION		GIDDINGS 789			GREENVILLE 754			HOUSTON 770 - 772			HUNTSVILLE 773			LAREDO 780			LONGVIEW 756		
		MAT.	INST.	TOTAL	MAT.	INST.	TOTAL	MAT.	INST.	TOTAL	MAT.	INST.	TOTAL	MAT.	INST.	TOTAL	MAT.	INST.	TOTAL
015433	CONTRACTOR EQUIPMENT		89.9	89.9		100.4	100.4		103.3	103.3		95.3	95.3		89.9	89.9		91.8	91.8
0241, 31 - 34	SITE & INFRASTRUCTURE, DEMOLITION	106.8	85.8	92.3	96.4	82.4	86.7	107.7	95.5	99.3	96.8	91.6	93.2	100.2	86.2	90.5	93.7	88.6	90.2
0310	Concrete Forming & Accessories	92.6	51.1	57.3	86.9	57.1	61.5	96.0	58.9	64.4	89.2	54.1	59.2	95.1	52.1	58.5	82.9	58.8	62.3
0320	Concrete Reinforcing	80.9	48.4	65.2	91.9	51.1	72.2	87.3	52.2	70.3	88.2	51.1	70.3	80.4	50.9	66.1	90.9	50.7	71.5
0330	Cast-in-Place Concrete	104.1	58.7	87.2	92.1	59.6	80.0	89.6	68.9	81.9	99.9	59.5	84.9	87.8	65.0	79.3	107.1	58.5	89.0
03	CONCRETE	92.6	54.3	75.8	86.7	58.7	74.4	91.4	62.5	78.7	98.7	56.6	80.2	87.3	57.4	74.2	102.7	58.1	83.1
04	MASONRY	102.2	60.6	76.7	158.6	59.3	97.5	98.0	63.1	76.6	141.3	60.6	91.7	88.9	60.6	71.5	154.2	59.2	95.8
05	METALS	96.7	68.2	88.4	99.2	83.2	94.5	103.9	77.3	96.1	99.4	71.8	91.3	99.6	69.4	90.7	91.8	68.4	84.9
06	WOOD, PLASTICS & COMPOSITES	95.3	50.4	70.6	87.2	58.4	71.4	100.6	59.2	77.8	88.1	54.1	69.4	96.1	51.4	71.5	81.2	60.8	70.0
07	THERMAL & MOISTURE PROTECTION	95.0	62.7	81.0	87.3	61.5	76.1	86.3	68.5	78.5	86.5	63.2	76.4	93.8	63.0	80.4	89.1	60.9	76.8
08	OPENINGS	95.8	48.7	84.8	92.2	55.3	83.6	106.6	56.4	94.9	95.7	52.5	85.6	96.7	49.8	85.8	83.4	56.2	77.0
0920	Plaster & Gypsum Board	92.4	49.5	63.5	82.8	57.5	65.8	94.0	58.0	69.8	92.3	53.2	66.0	94.1	50.5	64.7	81.7	60.1	67.2
0950, 0980	Ceilings & Acoustic Treatment	95.3	49.5	64.4	98.1	57.5	70.7	97.9	58.0	71.0	96.7	53.2	67.4	98.6	50.5	66.1	93.2	60.1	70.9
0960	Flooring	91.7	54.6	81.2	87.7	54.6	78.4	103.5	71.2	94.4	91.3	54.6	81.0	91.8	63.8	83.9	93.7	54.6	82.7
0970, 0990	Wall Finishes & Painting/Coating	97.8	44.2	66.1	90.4	51.7	67.5	102.1	58.6	76.3	92.2	57.4	71.6	97.8	44.2	66.1	82.1	48.9	62.4
09	FINISHES	93.1	50.5	70.0	92.3	56.1	72.7	102.7	60.8	80.1	88.3	54.1	69.8	93.4	52.9	71.5	98.1	57.2	76.0
COVERS	DIVS. 10 - 14, 25, 28, 41, 43, 44, 46	100.0	70.3	93.4	100.0	76.5	94.8	100.0	83.1	96.2	100.0	69.8	93.3	100.0	75.7	94.6	100.0	79.0	95.3
21, 22, 23	FIRE SUPPRESSION, PLUMBING & HVAC	96.5	61.4	82.7	96.5	58.0	81.3	100.0	65.3	86.3	96.5	62.4	83.1	100.0	58.9	83.8	96.5	57.5	81.1
26, 27, 3370	ELECTRICAL, COMMUNICATIONS & UTIL.	93.1	55.6	74.6	92.1	58.7	75.6	101.8	67.8	85.1	97.8	59.1	78.7	96.0	57.1	76.8	92.6	50.9	72.0
MF2018	WEIGHTED AVERAGE	96.1	60.4	81.0	97.4	62.9	82.8	100.4	68.2	86.8	98.7	62.8	83.6	96.4	61.1	81.5	97.5	61.0	82.1

For customer support on your Mechanical Costs with RSMeans Data, call 800.448.8182.

689

City Cost Indexes

TEXAS

DIVISION		LUBBOCK 793 - 794			LUFKIN 759			MCALLEN 785			MCKINNEY 750			MIDLAND 797			ODESSA 797		
		MAT.	INST.	TOTAL	MAT.	INST.	TOTAL	MAT.	INST.	TOTAL	MAT.	INST.	TOTAL	MAT.	INST.	TOTAL	MAT.	INST.	TOTAL
015433	CONTRACTOR EQUIPMENT		102.9	102.9		91.8	91.8		100.7	100.7		100.4	100.4		102.9	102.9		90.9	90.9
0241, 31 - 34	SITE & INFRASTRUCTURE, DEMOLITION	115.0	86.7	95.5	88.6	90.0	89.5	149.0	82.9	103.3	92.8	82.4	85.6	117.7	85.8	95.6	92.1	88.1	89.3
0310	Concrete Forming & Accessories	92.9	53.0	58.9	85.9	54.1	58.8	99.1	51.2	58.3	86.0	57.1	61.4	96.4	59.1	64.6	93.9	59.2	64.3
0320	Concrete Reinforcing	89.7	51.8	71.4	92.6	66.7	80.1	80.0	50.9	65.9	91.9	51.1	72.2	90.7	51.7	71.8	88.5	51.6	70.7
0330	Cast-in-Place Concrete	84.7	68.2	78.6	95.8	58.3	81.8	124.0	59.7	100.1	86.3	59.6	76.4	90.0	65.9	81.1	84.6	64.9	77.3
03	CONCRETE	83.1	60.1	73.0	95.1	58.6	79.0	102.8	56.2	82.3	82.1	58.8	71.9	87.1	62.1	76.1	84.5	60.7	74.1
04	MASONRY	99.2	61.0	75.7	117.3	60.5	82.4	95.1	60.7	74.0	170.8	59.3	102.3	116.2	60.2	81.8	99.9	60.2	75.5
05	METALS	110.9	85.8	103.5	99.4	73.3	91.7	97.1	84.2	93.3	99.2	83.3	94.5	109.2	85.4	102.2	106.4	70.4	95.8
06	WOOD, PLASTICS & COMPOSITES	98.6	52.1	73.0	88.6	54.2	69.6	114.3	50.5	79.2	86.1	58.4	70.9	103.0	60.9	79.8	99.2	60.8	78.1
07	THERMAL & MOISTURE PROTECTION	88.2	63.5	77.5	88.8	60.0	76.3	97.6	62.9	82.5	87.1	61.5	76.0	88.5	63.3	77.5	98.3	62.7	82.8
08	OPENINGS	108.6	52.4	95.5	63.3	56.6	61.7	100.4	48.6	88.3	92.2	55.3	83.6	110.0	56.7	97.6	101.8	56.7	91.2
0920	Plaster & Gypsum Board	88.4	51.1	63.3	80.6	53.2	62.2	97.5	49.5	65.2	82.8	57.5	65.8	90.0	60.1	69.9	88.2	60.1	69.3
0950, 0980	Ceilings & Acoustic Treatment	100.0	51.1	67.0	88.3	53.2	64.7	98.6	49.5	65.5	98.1	57.5	70.7	97.6	60.1	72.3	99.2	60.1	72.9
0960	Flooring	87.5	69.0	82.3	126.2	54.6	106.1	106.9	72.8	97.3	87.2	54.6	78.1	88.6	63.8	81.7	93.2	63.8	84.9
0970, 0990	Wall Finishes & Painting/Coating	103.4	53.7	74.0	82.1	51.7	64.1	110.1	42.6	70.2	90.4	51.7	67.5	103.4	51.7	72.8	92.8	51.7	68.5
09	FINISHES	88.5	55.5	70.6	106.5	53.9	78.0	102.4	54.1	76.3	91.9	56.1	72.5	89.0	59.4	72.9	86.1	59.3	71.6
COVERS	DIVS. 10 - 14, 25, 28, 41, 43, 44, 46	100.0	78.3	95.2	100.0	70.1	93.3	100.0	75.8	94.6	100.0	79.1	95.3	100.0	73.8	94.1	100.0	73.5	94.1
21, 22, 23	FIRE SUPPRESSION, PLUMBING & HVAC	99.7	53.1	81.4	96.4	58.1	81.4	96.5	48.7	77.7	96.5	60.2	82.2	96.1	47.4	77.0	100.1	52.5	81.4
26, 27, 3370	ELECTRICAL, COMMUNICATIONS & UTIL.	96.7	60.1	78.6	93.9	61.4	77.8	93.8	31.6	63.1	92.2	58.7	75.6	96.7	58.9	78.1	98.0	58.9	78.7
MF2018	WEIGHTED AVERAGE	99.2	62.9	83.9	94.7	62.6	81.1	99.7	56.5	81.4	97.2	63.5	82.9	99.7	62.2	83.8	97.7	61.8	82.6

TEXAS

DIVISION		PALESTINE 758			SAN ANGELO 769			SAN ANTONIO 781 - 782			TEMPLE 765			TEXARKANA 755			TYLER 757		
		MAT.	INST.	TOTAL	MAT.	INST.	TOTAL	MAT.	INST.	TOTAL	MAT.	INST.	TOTAL	MAT.	INST.	TOTAL	MAT.	INST.	TOTAL
015433	CONTRACTOR EQUIPMENT		91.8	91.8		90.9	90.9		94.1	94.1		90.9	90.9		91.8	91.8		91.8	91.8
0241, 31 - 34	SITE & INFRASTRUCTURE, DEMOLITION	94.2	88.8	90.5	97.8	87.7	90.8	101.5	93.7	96.1	86.7	87.6	87.3	83.4	91.1	88.7	92.9	88.7	90.0
0310	Concrete Forming & Accessories	77.5	58.8	61.5	97.9	51.4	58.3	94.7	52.6	58.8	101.2	51.2	58.5	93.0	59.4	64.3	87.6	58.9	63.1
0320	Concrete Reinforcing	90.2	51.0	71.3	81.8	51.6	67.2	89.6	47.7	69.3	82.0	50.8	66.9	90.1	50.8	71.1	90.9	50.8	71.5
0330	Cast-in-Place Concrete	87.5	58.5	76.7	91.4	64.9	81.5	85.7	68.3	79.2	74.9	58.6	68.8	88.2	64.9	79.5	105.2	58.6	87.8
03	CONCRETE	97.1	58.1	80.0	86.1	57.2	73.4	88.0	58.0	74.8	73.4	54.7	65.2	88.0	60.6	75.9	102.3	58.2	82.9
04	MASONRY	111.9	59.2	79.5	119.4	60.6	83.3	89.7	60.8	71.9	130.1	60.6	87.4	175.3	59.2	103.9	164.4	59.2	99.7
05	METALS	99.1	68.5	90.1	102.7	70.3	93.2	102.0	65.4	91.3	102.4	69.2	92.7	91.7	69.0	85.0	99.0	68.5	90.1
06	WOOD, PLASTICS & COMPOSITES	79.3	60.8	69.1	106.1	50.4	75.5	98.1	51.7	72.5	115.7	50.4	79.8	93.7	60.8	75.6	90.3	60.8	74.1
07	THERMAL & MOISTURE PROTECTION	89.4	61.8	77.4	91.8	62.6	79.1	86.8	65.6	77.5	91.3	61.7	78.4	88.6	63.0	77.5	89.2	61.8	77.3
08	OPENINGS	63.3	56.6	61.7	100.5	51.0	88.9	102.1	49.5	89.8	67.3	50.8	63.4	83.3	56.5	77.1	63.2	56.5	61.7
0920	Plaster & Gypsum Board	79.0	60.1	66.3	82.6	49.5	60.3	93.3	50.5	64.5	82.6	49.5	60.3	86.8	60.1	68.8	80.6	60.1	66.8
0950, 0980	Ceilings & Acoustic Treatment	88.3	60.1	69.3	86.6	49.5	61.6	95.2	50.5	65.1	86.6	49.5	61.6	93.2	60.1	70.9	88.3	60.1	69.3
0960	Flooring	116.4	54.6	99.0	75.8	54.6	66.9	101.2	67.7	91.8	97.5	54.6	85.4	102.7	63.8	91.7	128.7	54.6	107.8
0970, 0990	Wall Finishes & Painting/Coating	82.1	51.7	64.1	92.8	51.7	68.4	99.4	44.2	66.7	94.2	44.2	64.6	82.1	51.7	64.1	82.1	51.7	64.1
09	FINISHES	104.0	57.5	78.8	77.4	51.3	63.3	104.2	54.0	77.0	83.3	50.5	65.5	100.9	59.3	78.4	107.5	57.5	80.4
COVERS	DIVS. 10 - 14, 25, 28, 41, 43, 44, 46	100.0	73.5	94.1	100.0	76.0	94.7	100.0	78.0	95.1	100.0	72.3	93.8	100.0	79.1	95.3	100.0	79.0	95.3
21, 22, 23	FIRE SUPPRESSION, PLUMBING & HVAC	96.4	56.7	80.8	96.5	52.2	79.1	100.0	59.8	84.2	96.5	53.2	79.5	96.4	55.5	80.3	96.4	58.2	81.4
26, 27, 3370	ELECTRICAL, COMMUNICATIONS & UTIL.	90.1	48.4	69.5	102.8	54.6	79.1	97.5	59.8	78.9	99.8	52.0	76.2	93.7	54.5	74.4	92.6	53.7	73.4
MF2018	WEIGHTED AVERAGE	94.2	60.4	79.9	97.1	59.4	81.2	98.3	62.3	83.1	92.5	58.5	78.2	96.7	62.0	82.1	97.8	61.6	82.5

TEXAS / UTAH

DIVISION		VICTORIA 779			WACO 766 - 767			WAXAHACHIE 751			WHARTON 774			WICHITA FALLS 763			LOGAN 843		
		MAT.	INST.	TOTAL	MAT.	INST.	TOTAL	MAT.	INST.	TOTAL	MAT.	INST.	TOTAL	MAT.	INST.	TOTAL	MAT.	INST.	TOTAL
015433	CONTRACTOR EQUIPMENT		106.6	106.6		90.9	90.9		100.4	100.4		107.9	107.9		90.9	90.9		93.9	93.9
0241, 31 - 34	SITE & INFRASTRUCTURE, DEMOLITION	113.4	88.0	95.8	95.1	88.0	90.2	94.7	82.5	86.3	118.7	90.0	98.9	95.8	88.1	90.5	99.8	89.6	92.7
0310	Concrete Forming & Accessories	90.6	51.3	57.1	99.8	59.2	65.2	86.0	59.0	63.0	85.7	53.1	57.9	99.8	59.2	65.2	103.5	67.2	72.6
0320	Concrete Reinforcing	84.0	50.9	68.0	81.6	47.2	65.0	91.9	51.1	72.2	87.4	50.9	69.7	81.6	51.7	67.1	108.0	84.4	96.6
0330	Cast-in-Place Concrete	108.3	60.5	90.5	81.0	64.9	75.0	91.1	59.8	79.5	111.3	59.5	92.0	86.5	65.0	78.5	86.4	74.3	81.9
03	CONCRETE	99.7	56.7	80.8	79.6	60.0	71.0	85.8	59.7	74.3	103.8	57.1	83.3	82.2	60.8	72.8	105.4	73.2	91.2
04	MASONRY	116.9	60.7	82.4	90.9	60.6	72.3	159.2	59.3	97.8	100.3	60.6	75.9	91.4	60.1	72.2	109.5	60.5	79.4
05	METALS	99.4	87.8	96.0	104.8	68.4	94.1	99.2	83.6	94.6	101.2	87.4	97.1	104.7	70.4	94.6	109.5	82.5	101.5
06	WOOD, PLASTICS & COMPOSITES	98.4	50.5	72.1	113.9	60.8	84.7	86.1	61.0	72.3	88.4	52.7	68.8	113.9	60.8	84.7	85.9	67.1	75.5
07	THERMAL & MOISTURE PROTECTION	88.0	61.1	76.3	92.2	63.7	79.8	87.2	61.6	76.1	85.0	63.7	75.7	92.2	62.9	79.5	103.0	68.8	88.1
08	OPENINGS	99.5	51.1	88.2	78.4	55.6	73.1	92.2	56.7	83.9	99.7	52.3	88.6	78.4	56.7	73.4	92.2	66.1	86.1
0920	Plaster & Gypsum Board	90.9	49.5	63.0	82.7	60.1	67.5	83.2	60.1	67.7	90.4	51.7	64.4	82.7	60.1	67.5	80.6	66.2	70.9
0950, 0980	Ceilings & Acoustic Treatment	101.6	49.5	66.5	87.4	60.1	69.0	99.7	60.1	73.0	100.8	51.7	67.7	87.4	60.1	69.0	104.3	66.2	78.7
0960	Flooring	100.7	54.6	87.7	96.8	63.8	87.5	87.2	54.6	78.1	99.5	66.6	90.2	97.4	71.9	90.2	97.2	60.5	86.9
0970, 0990	Wall Finishes & Painting/Coating	104.8	57.4	76.8	94.2	51.7	69.0	90.4	51.7	67.5	104.6	55.7	75.7	96.1	51.7	69.8	92.6	60.7	73.7
09	FINISHES	92.4	52.0	70.6	83.7	59.3	70.5	92.5	57.6	73.6	94.7	55.2	73.3	84.1	60.9	71.5	93.9	64.9	78.2
COVERS	DIVS. 10 - 14, 25, 28, 41, 43, 44, 46	100.0	69.7	93.2	100.0	78.9	95.3	100.0	79.4	95.4	100.0	70.0	93.3	100.0	73.5	94.1	100.0	84.4	96.5
21, 22, 23	FIRE SUPPRESSION, PLUMBING & HVAC	96.5	52.3	83.1	100.0	58.7	83.8	100.0	59.8	81.4	96.5	61.9	82.9	100.0	52.1	81.2	100.0	69.2	87.9
26, 27, 3370	ELECTRICAL, COMMUNICATIONS & UTIL.	104.9	54.0	79.8	103.0	54.1	78.9	92.1	58.7	75.6	103.9	59.2	81.9	105.0	56.0	80.8	95.2	71.3	83.4
MF2018	WEIGHTED AVERAGE	99.7	62.9	84.2	94.4	62.4	80.9	97.2	63.5	83.0	99.8	64.3	84.8	95.0	61.6	80.9	100.9	71.7	88.5

		UTAH												VERMONT					
	DIVISION	OGDEN			PRICE			PROVO			SALT LAKE CITY			BELLOWS FALLS			BENNINGTON		
		842, 844			845			846 - 847			840 - 841			051			052		
		MAT.	INST.	TOTAL	MAT.	INST.	TOTAL	MAT.	INST.	TOTAL	MAT.	INST.	TOTAL	MAT.	INST.	TOTAL	MAT.	INST.	TOTAL
015433	CONTRACTOR EQUIPMENT		93.9	93.9		92.9	92.9		92.9	92.9		93.8	93.8		95.7	95.7		95.7	95.7
0241, 31 - 34	SITE & INFRASTRUCTURE, DEMOLITION	87.3	89.6	88.9	97.1	87.4	90.4	95.9	88.0	90.5	87.0	89.5	88.7	86.8	96.1	93.3	86.2	96.1	93.1
0310	Concrete Forming & Accessories	103.5	67.2	72.6	106.0	64.2	70.4	105.3	67.2	72.9	105.8	67.2	72.9	96.5	101.7	100.9	94.0	101.6	100.4
0320	Concrete Reinforcing	107.6	84.4	96.4	115.8	84.0	100.4	116.8	84.4	101.1	110.0	84.4	97.6	85.1	86.3	85.7	85.1	86.3	85.6
0330	Cast-in-Place Concrete	87.7	74.3	82.7	86.5	71.6	80.9	86.5	74.3	81.9	95.8	74.3	87.8	89.6	115.9	99.4	89.6	115.8	99.4
03	CONCRETE	94.4	73.2	85.1	106.6	70.8	90.8	104.9	73.2	91.0	114.2	73.2	96.2	88.7	103.7	95.3	88.5	103.6	95.2
04	MASONRY	104.0	60.5	77.3	115.1	65.1	84.4	115.3	60.5	81.6	118.1	60.5	82.7	96.3	86.2	90.1	105.1	86.2	93.5
05	METALS	110.0	82.5	101.9	106.5	82.5	99.5	107.5	82.5	100.1	113.7	82.5	104.5	95.2	91.1	94.0	95.1	90.9	93.9
06	WOOD, PLASTICS & COMPOSITES	85.9	67.1	75.5	89.2	64.0	75.3	87.5	67.1	76.3	87.8	67.1	76.4	102.8	106.8	105.0	100.1	106.8	103.8
07	THERMAL & MOISTURE PROTECTION	101.8	68.8	87.4	104.8	68.3	88.9	104.9	68.8	89.2	110.0	68.8	92.1	100.9	89.1	95.8	100.9	89.1	95.8
08	OPENINGS	92.2	66.1	86.1	96.1	76.3	91.5	96.1	66.1	89.1	94.1	66.1	87.5	100.1	97.5	99.5	100.1	97.5	99.5
0920	Plaster & Gypsum Board	80.6	66.2	70.9	83.4	63.1	69.8	81.2	66.2	71.1	90.9	66.2	74.3	108.9	106.7	107.4	107.9	106.7	107.1
0950, 0980	Ceilings & Acoustic Treatment	104.3	66.2	78.7	104.3	63.1	76.5	104.3	66.2	78.7	97.1	66.2	76.3	97.2	106.7	103.6	97.2	106.7	103.6
0960	Flooring	95.2	60.5	85.4	98.4	56.1	86.5	98.2	60.5	87.6	99.0	60.5	88.1	92.0	98.9	94.0	91.2	98.9	93.4
0970, 0990	Wall Finishes & Painting/Coating	92.6	60.7	73.7	92.6	60.7	73.7	92.6	60.7	73.7	95.7	60.7	75.0	86.4	101.5	95.3	86.4	101.5	95.3
09	FINISHES	92.0	64.9	77.3	95.0	61.7	77.0	94.5	64.9	78.5	93.7	64.9	78.1	90.6	101.8	96.7	90.3	101.8	96.5
COVERS	DIVS. 10 - 14, 25, 28, 41, 43, 44, 46	100.0	84.4	96.5	100.0	81.6	95.9	100.0	84.4	96.5	100.0	84.4	96.5	100.0	91.8	98.2	100.0	91.7	98.2
21, 22, 23	FIRE SUPPRESSION, PLUMBING & HVAC	100.0	69.2	87.9	98.1	63.2	84.4	100.0	69.2	87.9	100.1	69.2	88.0	96.7	87.6	93.1	96.7	87.6	93.1
26, 27, 3370	ELECTRICAL, COMMUNICATIONS & UTIL.	95.5	71.3	83.6	100.2	68.8	84.7	95.7	68.6	82.3	98.1	71.3	84.9	108.3	79.5	94.1	108.3	54.5	81.8
MF2018	WEIGHTED AVERAGE	98.8	71.7	87.4	101.4	70.0	88.1	101.2	71.2	88.5	103.5	71.7	90.1	96.7	92.1	94.8	97.0	88.6	93.4

		VERMONT																	
	DIVISION	BRATTLEBORO			BURLINGTON			GUILDHALL			MONTPELIER			RUTLAND			ST. JOHNSBURY		
		053			054			059			056			057			058		
		MAT.	INST.	TOTAL	MAT.	INST.	TOTAL	MAT.	INST.	TOTAL	MAT.	INST.	TOTAL	MAT.	INST.	TOTAL	MAT.	INST.	TOTAL
015433	CONTRACTOR EQUIPMENT		95.7	95.7		99.0	99.0		95.7	95.7		99.0	99.0		95.7	95.7		95.7	95.7
0241, 31 - 34	SITE & INFRASTRUCTURE, DEMOLITION	87.8	96.1	93.5	92.6	101.0	98.4	85.8	91.9	90.0	89.7	100.8	97.3	90.4	96.1	94.3	85.9	95.2	92.3
0310	Concrete Forming & Accessories	96.7	101.7	101.0	99.4	80.2	83.0	94.6	95.1	95.1	97.3	101.3	100.7	97.0	80.0	82.5	93.5	95.6	95.2
0320	Concrete Reinforcing	84.1	86.3	85.2	107.3	86.2	97.1	85.8	86.1	86.0	101.6	86.2	94.2	106.4	86.1	96.6	84.1	86.2	85.1
0330	Cast-in-Place Concrete	92.5	115.9	101.2	109.6	116.7	112.2	86.9	107.2	94.4	104.5	116.7	109.0	87.8	115.6	98.2	86.9	107.3	94.5
03	CONCRETE	90.7	103.7	96.4	101.7	94.2	98.4	86.2	97.7	91.3	98.5	103.7	100.8	91.4	93.8	92.5	85.9	97.9	91.2
04	MASONRY	104.6	86.2	93.3	103.9	86.4	93.1	104.8	71.5	84.4	101.4	86.4	92.2	87.2	86.3	86.7	131.1	71.5	94.5
05	METALS	95.1	91.1	93.9	102.8	89.8	99.0	95.2	90.0	93.7	101.2	89.9	97.9	101.1	90.3	97.9	95.2	90.4	93.8
06	WOOD, PLASTICS & COMPOSITES	103.1	106.8	105.1	99.2	78.5	87.8	99.0	106.8	103.3	96.0	106.9	102.0	103.2	78.3	89.5	94.1	106.8	101.0
07	THERMAL & MOISTURE PROTECTION	101.0	89.1	95.8	107.4	86.6	98.3	100.7	82.4	92.8	109.4	89.7	100.8	101.1	86.1	94.6	100.6	82.4	92.7
08	OPENINGS	100.1	97.8	99.6	102.6	78.4	96.9	100.1	94.0	98.7	103.1	94.1	101.0	103.2	78.3	97.4	100.1	94.0	98.7
0920	Plaster & Gypsum Board	108.9	106.7	107.4	109.9	77.4	88.1	114.5	106.7	109.2	110.4	106.7	107.9	109.4	77.4	87.9	115.8	106.7	109.6
0950, 0980	Ceilings & Acoustic Treatment	97.2	106.7	103.6	103.9	77.4	86.0	97.2	106.7	103.6	104.9	106.7	106.1	103.2	77.4	85.8	97.2	106.7	103.6
0960	Flooring	92.2	98.9	94.1	101.8	98.9	101.0	95.0	98.9	96.1	98.5	98.9	98.6	92.0	98.9	94.0	98.3	98.9	98.5
0970, 0990	Wall Finishes & Painting/Coating	86.4	101.5	95.3	98.0	88.3	92.3	86.4	88.3	87.5	97.8	88.3	92.2	86.4	88.3	87.5	86.4	88.3	87.5
09	FINISHES	90.8	101.8	96.8	99.1	83.8	90.8	92.2	96.7	94.6	97.9	100.6	99.4	92.0	83.7	87.5	93.3	96.7	95.2
COVERS	DIVS. 10 - 14, 25, 28, 41, 43, 44, 46	100.0	91.8	98.2	100.0	89.1	97.6	100.0	86.7	97.0	100.0	92.2	98.3	100.0	88.6	97.5	100.0	86.8	97.0
21, 22, 23	FIRE SUPPRESSION, PLUMBING & HVAC	96.7	87.6	93.1	100.0	68.3	87.6	96.7	60.5	82.5	96.5	68.3	85.4	100.2	68.2	87.7	96.7	60.5	82.5
26, 27, 3370	ELECTRICAL, COMMUNICATIONS & UTIL.	108.3	79.5	94.1	107.6	53.6	81.0	108.3	54.5	81.8	108.1	53.6	81.3	108.4	53.6	81.4	108.3	54.5	81.8
MF2018	WEIGHTED AVERAGE	97.4	92.1	95.2	101.9	79.8	92.5	96.8	79.0	89.3	100.2	84.4	93.5	98.9	79.4	90.7	98.1	79.3	90.2

		VERMONT			VIRGINIA														
	DIVISION	WHITE RIVER JCT.			ALEXANDRIA			ARLINGTON			BRISTOL			CHARLOTTESVILLE			CULPEPER		
		050			223			222			242			229			227		
		MAT.	INST.	TOTAL	MAT.	INST.	TOTAL	MAT.	INST.	TOTAL	MAT.	INST.	TOTAL	MAT.	INST.	TOTAL	MAT.	INST.	TOTAL
015433	CONTRACTOR EQUIPMENT		95.7	95.7		105.3	105.3		104.1	104.1		104.1	104.1		108.1	108.1		104.1	104.1
0241, 31 - 34	SITE & INFRASTRUCTURE, DEMOLITION	90.3	95.2	93.7	113.7	89.0	96.7	123.9	87.0	98.5	108.2	86.0	92.8	112.7	88.3	95.9	111.4	86.5	94.2
0310	Concrete Forming & Accessories	91.4	96.1	95.4	91.5	71.9	74.8	90.5	71.6	74.4	86.6	64.1	67.5	85.1	61.7	65.2	82.4	71.0	72.7
0320	Concrete Reinforcing	85.1	86.2	85.6	86.6	85.0	85.9	97.8	85.1	91.6	97.7	83.7	91.0	97.1	73.0	85.4	97.8	84.9	91.6
0330	Cast-in-Place Concrete	92.5	108.1	98.3	108.1	77.7	96.8	105.2	77.2	94.8	104.8	44.5	82.3	109.0	75.5	96.5	107.7	74.4	95.3
03	CONCRETE	92.6	98.4	95.2	99.9	77.7	90.1	104.5	77.4	92.6	100.8	62.4	83.9	101.3	70.2	87.7	98.4	76.0	88.6
04	MASONRY	117.2	72.9	90.0	89.8	73.7	79.9	102.3	72.6	84.0	92.3	45.4	63.5	116.9	55.3	79.0	105.1	70.8	84.1
05	METALS	95.2	90.4	93.8	106.0	100.8	104.5	104.6	101.8	103.8	103.4	97.8	101.8	103.7	95.8	101.4	103.7	99.6	102.5
06	WOOD, PLASTICS & COMPOSITES	97.1	106.8	102.4	91.0	70.1	79.5	87.5	70.1	78.0	80.2	69.8	74.5	78.7	59.8	68.3	77.4	70.1	73.4
07	THERMAL & MOISTURE PROTECTION	101.1	83.0	93.2	102.5	81.1	93.2	104.6	80.5	94.1	103.9	58.8	84.3	103.5	61.1	88.8	103.7	78.9	92.9
08	OPENINGS	100.1	94.0	98.7	95.4	74.3	90.5	93.7	74.3	89.2	96.4	67.9	89.7	94.8	65.7	88.0	95.1	73.8	90.1
0920	Plaster & Gypsum Board	106.6	106.7	106.7	98.0	69.2	78.6	94.8	69.2	77.6	91.1	68.9	76.2	91.1	57.9	68.8	91.3	69.2	76.4
0950, 0980	Ceilings & Acoustic Treatment	97.2	106.7	103.6	92.4	69.2	76.8	90.8	69.2	76.2	90.0	68.9	75.8	90.0	57.9	68.4	90.8	69.2	76.2
0960	Flooring	90.2	98.9	92.6	97.1	77.4	91.5	95.6	75.7	90.0	92.3	55.7	82.0	90.8	55.7	80.9	90.8	74.4	86.2
0970, 0990	Wall Finishes & Painting/Coating	86.4	88.3	87.5	115.8	76.0	92.3	115.8	73.1	90.6	102.1	55.3	74.4	102.1	58.6	76.3	115.8	73.1	90.6
09	FINISHES	90.1	97.1	93.9	93.8	72.6	82.3	93.8	71.6	81.8	90.6	62.3	75.3	90.1	59.4	73.5	90.8	71.2	80.2
COVERS	DIVS. 10 - 14, 25, 28, 41, 43, 44, 46	100.0	87.2	97.1	100.0	86.6	97.0	100.0	84.4	96.5	100.0	75.1	94.5	100.0	83.2	96.2	100.0	83.6	96.3
21, 22, 23	FIRE SUPPRESSION, PLUMBING & HVAC	96.7	61.2	82.8	100.4	86.2	94.8	100.4	85.6	94.5	96.8	46.0	76.9	96.8	81.3	90.7	96.8	85.3	92.3
26, 27, 3370	ELECTRICAL, COMMUNICATIONS & UTIL.	108.3	54.3	81.7	96.6	100.2	98.4	94.4	100.2	97.2	96.2	36.3	66.7	96.2	70.4	83.5	98.5	96.9	97.7
MF2018	WEIGHTED AVERAGE	98.2	79.8	90.4	99.6	84.8	93.3	100.4	84.2	93.6	98.2	59.3	81.8	99.4	71.0	87.4	98.7	83.0	92.1

For customer support on your Mechanical Costs with RSMeans Data, call 800.448.8182.

691

VIRGINIA

DIVISION		FAIRFAX 220 - 221			FARMVILLE 239			FREDERICKSBURG 224 - 225			GRUNDY 246			HARRISONBURG 228			LYNCHBURG 245		
		MAT.	INST.	TOTAL	MAT.	INST.	TOTAL	MAT.	INST.	TOTAL	MAT.	INST.	TOTAL	MAT.	INST.	TOTAL	MAT.	INST.	TOTAL
015433	CONTRACTOR EQUIPMENT		104.1	104.1		108.1	108.1		104.1	104.1		104.1	104.1		104.1	104.1		104.1	104.1
0241, 31 - 34	SITE & INFRASTRUCTURE, DEMOLITION	122.4	86.7	97.8	108.5	87.3	93.9	110.9	86.4	93.9	106.0	85.3	91.7	119.3	85.0	95.6	106.8	86.7	92.9
0310	Concrete Forming & Accessories	85.1	63.8	67.0	98.9	60.3	66.0	85.1	65.4	68.3	89.6	38.2	45.8	81.4	56.6	60.2	86.6	70.5	72.9
0320	Concrete Reinforcing	97.8	90.3	94.2	92.9	67.9	80.8	98.5	80.1	89.6	96.4	44.1	71.1	97.8	57.3	78.2	97.1	84.5	91.0
0330	Cast-in-Place Concrete	105.2	78.5	95.2	106.8	83.9	98.3	106.8	74.2	94.7	104.8	50.9	84.7	105.2	52.6	85.6	104.8	74.3	93.4
03	CONCRETE	104.2	75.2	91.5	101.8	71.5	88.5	98.1	72.7	86.9	99.3	45.8	75.8	102.0	57.1	82.3	99.2	75.8	88.9
04	MASONRY	102.2	73.6	84.6	101.8	50.0	69.9	104.0	66.7	81.1	94.9	55.4	70.6	101.4	57.9	74.6	109.3	63.9	81.4
05	METALS	103.8	102.5	103.4	101.3	90.8	98.2	103.8	98.0	102.1	103.4	75.2	95.1	103.7	87.2	98.9	103.6	99.8	102.5
06	WOOD, PLASTICS & COMPOSITES	80.2	59.2	68.6	93.0	59.8	74.7	80.2	64.6	71.6	83.1	30.6	54.2	76.3	59.2	66.9	80.2	71.5	75.4
07	THERMAL & MOISTURE PROTECTION	104.4	79.9	93.7	105.2	69.4	89.6	103.7	76.3	91.8	103.9	57.3	83.6	104.1	69.6	89.1	103.7	70.4	89.2
08	OPENINGS	93.7	69.6	88.0	94.3	54.9	85.1	94.8	68.5	88.7	96.4	30.1	80.9	95.0	57.0	86.2	95.0	68.8	88.9
0920	Plaster & Gypsum Board	91.3	57.9	68.9	103.1	57.9	72.7	91.3	63.6	72.7	91.1	28.5	49.0	91.1	57.9	68.8	91.1	70.7	77.4
0950, 0980	Ceilings & Acoustic Treatment	90.8	57.9	68.6	86.5	57.9	67.2	90.8	63.6	72.4	90.0	28.5	48.6	90.0	57.9	68.4	90.0	70.7	77.0
0960	Flooring	92.5	76.7	88.1	95.6	50.5	82.9	92.5	72.4	86.9	93.5	28.2	75.2	90.6	74.4	86.1	92.3	65.4	84.7
0970, 0990	Wall Finishes & Painting/Coating	115.8	73.6	90.8	99.6	57.2	74.5	115.8	57.0	81.0	102.1	31.2	60.2	115.8	48.3	75.9	102.1	55.3	74.4
09	FINISHES	92.4	65.7	77.9	90.6	57.7	72.8	91.3	65.0	77.1	90.7	33.9	60.0	91.3	58.9	73.7	90.4	67.8	78.2
COVERS	DIVS. 10 - 14, 25, 28, 41, 43, 44, 46	100.0	84.2	96.5	100.0	77.1	94.9	100.0	78.3	95.2	100.0	73.6	94.1	100.0	62.8	91.7	100.0	78.8	95.3
21, 22, 23	FIRE SUPPRESSION, PLUMBING & HVAC	96.8	88.1	93.4	96.9	52.0	79.3	96.8	79.9	90.2	96.8	65.7	84.6	96.8	65.4	84.5	96.8	69.5	86.1
26, 27, 3370	ELECTRICAL, COMMUNICATIONS & UTIL.	97.3	101.1	99.1	91.2	68.3	79.9	94.5	91.8	93.2	96.2	68.3	82.5	96.4	86.4	91.5	97.2	67.8	82.7
MF2018	WEIGHTED AVERAGE	99.5	83.6	92.8	97.9	65.3	84.1	98.3	78.7	90.0	98.1	58.8	81.5	99.0	68.8	86.3	98.7	73.8	88.2

VIRGINIA

DIVISION		NEWPORT NEWS 236			NORFOLK 233 - 235			PETERSBURG 238			PORTSMOUTH 237			PULASKI 243			RICHMOND 230 - 232		
		MAT.	INST.	TOTAL	MAT.	INST.	TOTAL	MAT.	INST.	TOTAL	MAT.	INST.	TOTAL	MAT.	INST.	TOTAL	MAT.	INST.	TOTAL
015433	CONTRACTOR EQUIPMENT		108.2	108.2		110.2	110.2		108.1	108.1		108.0	108.0		104.1	104.1		109.6	109.6
0241, 31 - 34	SITE & INFRASTRUCTURE, DEMOLITION	107.6	88.4	94.3	106.9	93.8	97.9	111.1	88.4	95.4	106.2	88.2	93.8	105.4	85.7	91.8	103.4	92.8	96.1
0310	Concrete Forming & Accessories	98.1	61.1	66.6	103.6	61.2	67.5	91.8	61.8	66.2	87.8	61.3	65.2	89.6	40.4	47.7	94.2	61.9	66.7
0320	Concrete Reinforcing	92.8	66.2	79.9	101.2	66.2	84.3	92.4	73.6	83.3	92.4	66.2	79.7	96.4	83.8	90.3	102.1	73.6	88.3
0330	Cast-in-Place Concrete	103.8	64.5	89.2	111.0	76.1	98.0	110.2	76.6	97.7	102.8	65.6	89.0	104.8	82.4	96.5	95.3	76.6	88.4
03	CONCRETE	98.9	65.0	84.0	102.1	69.0	87.6	104.5	70.8	89.7	97.7	65.5	83.6	99.3	64.7	84.1	94.5	70.7	84.0
04	MASONRY	97.3	55.0	71.3	97.1	55.0	71.3	110.2	64.6	82.2	103.0	55.6	73.9	87.9	54.3	67.3	93.3	64.7	75.7
05	METALS	103.8	93.1	100.6	105.3	91.8	101.3	101.4	96.7	100.0	102.7	93.1	99.9	103.5	96.0	101.3	105.9	95.3	102.8
06	WOOD, PLASTICS & COMPOSITES	91.8	59.8	74.2	99.3	59.7	77.5	83.2	59.5	70.2	79.6	59.8	68.7	83.1	32.0	55.0	93.7	59.7	75.0
07	THERMAL & MOISTURE PROTECTION	105.1	67.7	88.9	104.4	70.0	89.5	105.1	72.3	90.9	105.1	68.0	89.0	103.9	66.6	87.6	101.0	72.9	88.8
08	OPENINGS	94.6	60.1	86.6	95.2	60.1	87.0	94.0	65.6	87.4	94.7	56.0	85.7	96.4	42.9	83.9	101.7	65.7	93.3
0920	Plaster & Gypsum Board	104.3	57.9	73.1	98.5	57.9	71.2	97.8	57.7	70.8	98.6	57.9	71.2	91.1	30.0	50.0	99.5	57.9	71.5
0950, 0980	Ceilings & Acoustic Treatment	91.4	57.9	68.8	88.8	57.9	68.0	88.2	57.7	67.6	91.4	57.9	68.8	90.0	30.0	49.6	89.5	57.9	68.2
0960	Flooring	95.6	51.1	83.1	96.2	51.1	83.5	91.7	71.7	86.1	88.9	51.7	78.4	93.5	55.7	82.9	94.0	71.7	87.8
0970, 0990	Wall Finishes & Painting/Coating	99.6	59.7	76.0	101.2	59.7	76.6	99.6	59.7	76.0	99.6	59.7	76.0	102.1	48.3	70.3	95.4	58.6	73.6
09	FINISHES	91.5	58.4	73.6	90.5	58.3	73.1	89.3	62.5	74.8	88.8	58.6	72.5	90.7	40.9	63.8	91.9	62.5	76.0
COVERS	DIVS. 10 - 14, 25, 28, 41, 43, 44, 46	100.0	83.4	96.3	100.0	83.2	96.3	100.0	83.1	96.2	100.0	83.6	96.3	100.0	73.2	94.0	100.0	83.0	96.2
21, 22, 23	FIRE SUPPRESSION, PLUMBING & HVAC	100.4	66.8	87.2	100.1	66.8	87.0	96.9	69.1	86.0	100.4	67.1	87.4	96.8	66.0	84.7	100.0	69.1	87.9
26, 27, 3370	ELECTRICAL, COMMUNICATIONS & UTIL.	93.6	64.2	79.1	98.0	60.7	79.6	93.8	70.5	82.3	92.1	60.7	76.7	96.2	83.4	89.9	96.8	70.4	83.8
MF2018	WEIGHTED AVERAGE	98.8	68.3	85.9	99.9	68.7	86.7	98.7	72.6	87.7	98.3	67.8	85.4	97.8	67.1	84.8	99.2	72.8	88.1

VIRGINIA / WASHINGTON

DIVISION		VIRGINIA ROANOKE 240 - 241			STAUNTON 244			WINCHESTER 226			WASHINGTON CLARKSTON 994			EVERETT 982			OLYMPIA 985		
		MAT.	INST.	TOTAL	MAT.	INST.	TOTAL	MAT.	INST.	TOTAL	MAT.	INST.	TOTAL	MAT.	INST.	TOTAL	MAT.	INST.	TOTAL
015433	CONTRACTOR EQUIPMENT		104.1	104.1		108.1	108.1		104.1	104.1		90.2	90.2		99.3	99.3		101.6	101.6
0241, 31 - 34	SITE & INFRASTRUCTURE, DEMOLITION	106.0	86.8	92.7	109.3	87.3	94.1	118.1	86.4	96.2	98.1	86.1	89.8	92.1	105.6	101.4	92.8	108.2	103.5
0310	Concrete Forming & Accessories	95.7	71.2	74.8	89.3	61.4	65.5	83.6	66.4	68.9	107.0	64.2	70.5	111.6	104.6	105.7	101.6	104.3	103.9
0320	Concrete Reinforcing	97.5	84.6	91.2	97.1	83.8	90.7	97.1	80.2	88.9	108.4	95.9	102.3	109.2	114.5	111.8	119.5	114.4	117.0
0330	Cast-in-Place Concrete	119.0	86.2	106.8	109.0	85.5	100.2	105.2	63.9	89.8	82.3	82.9	82.5	102.6	109.4	105.1	95.3	109.8	100.7
03	CONCRETE	103.5	80.2	93.3	100.7	75.2	89.5	101.5	69.5	87.4	87.6	76.6	82.8	95.1	107.4	100.5	93.7	107.2	99.7
04	MASONRY	96.3	65.9	77.6	105.0	55.3	74.5	98.7	69.8	80.9	97.8	91.7	94.0	121.0	101.8	109.2	115.3	101.7	107.0
05	METALS	105.8	100.2	104.2	103.6	96.0	101.4	103.8	97.9	102.1	92.8	87.3	91.2	116.8	98.4	111.4	117.8	96.1	111.4
06	WOOD, PLASTICS & COMPOSITES	91.9	71.5	80.7	83.1	59.8	70.3	78.7	64.8	71.1	104.4	58.5	79.2	107.5	104.7	105.9	97.5	104.8	101.5
07	THERMAL & MOISTURE PROTECTION	103.6	76.0	91.6	103.4	67.5	87.8	104.2	75.9	91.9	159.7	82.9	126.3	112.4	105.9	109.6	108.3	103.0	106.0
08	OPENINGS	95.4	68.8	89.2	95.0	58.5	86.5	96.5	68.3	89.9	116.4	65.0	104.4	106.0	107.3	106.3	109.6	104.8	108.5
0920	Plaster & Gypsum Board	98.0	70.7	79.6	91.1	57.9	68.8	91.3	63.8	72.8	154.4	57.3	89.1	112.7	104.9	107.5	106.5	104.9	105.4
0950, 0980	Ceilings & Acoustic Treatment	92.4	70.7	77.8	90.0	57.9	68.4	90.8	63.8	72.6	103.8	57.3	72.4	101.2	104.9	103.7	98.7	104.9	102.9
0960	Flooring	97.1	67.3	88.7	93.1	33.1	76.2	92.0	74.4	87.1	85.4	77.5	83.2	109.9	96.0	106.0	97.6	96.0	97.1
0970, 0990	Wall Finishes & Painting/Coating	102.1	55.3	74.4	102.1	30.7	59.8	115.8	77.6	93.2	84.4	72.1	77.2	97.6	95.2	96.2	99.6	93.1	95.8
09	FINISHES	92.7	68.5	79.6	90.6	52.6	70.0	91.8	68.3	79.1	108.2	65.8	85.2	105.4	101.6	103.4	94.9	101.5	98.5
COVERS	DIVS. 10 - 14, 25, 28, 41, 43, 44, 46	100.0	79.2	95.4	100.0	78.2	95.1	100.0	82.7	96.1	100.0	89.3	97.6	100.0	103.7	100.8	100.0	99.9	100.0
21, 22, 23	FIRE SUPPRESSION, PLUMBING & HVAC	100.4	67.9	87.6	96.8	59.3	82.1	96.8	81.4	90.8	96.7	80.6	90.4	100.0	105.2	102.0	99.9	100.6	100.2
26, 27, 3370	ELECTRICAL, COMMUNICATIONS & UTIL.	96.2	60.8	78.8	95.2	68.4	82.0	94.9	91.9	93.4	89.4	95.6	92.5	107.0	104.1	105.6	105.4	97.7	101.6
MF2018	WEIGHTED AVERAGE	100.1	73.6	88.9	98.6	67.9	85.6	98.9	79.5	90.7	99.1	81.9	91.8	105.1	104.0	104.6	104.0	101.8	103.1

WASHINGTON

	DIVISION	RICHLAND 993			SEATTLE 980 - 981, 987			SPOKANE 990 - 992			TACOMA 983 - 984			VANCOUVER 986			WENATCHEE 988		
		MAT.	INST.	TOTAL	MAT.	INST.	TOTAL	MAT.	INST.	TOTAL	MAT.	INST.	TOTAL	MAT.	INST.	TOTAL	MAT.	INST.	TOTAL
015433	CONTRACTOR EQUIPMENT		90.2	90.2		102.4	102.4		90.2	90.2		99.3	99.3		95.8	95.8		99.3	99.3
0241, 31 - 34	SITE & INFRASTRUCTURE, DEMOLITION	100.3	86.3	90.6	97.8	108.9	105.5	99.7	86.3	90.4	95.2	105.1	102.1	105.2	93.1	96.8	104.0	102.8	103.2
0310	Concrete Forming & Accessories	107.1	80.0	84.0	107.4	109.5	109.2	112.0	79.5	84.3	102.3	104.1	103.9	102.7	95.7	96.7	104.0	76.5	80.5
0320	Concrete Reinforcing	103.9	95.4	99.8	109.2	117.3	113.1	104.6	95.3	100.1	107.8	114.3	110.9	108.8	114.8	111.7	108.8	95.9	102.5
0330	Cast-in-Place Concrete	82.5	84.4	83.2	105.9	113.9	108.9	85.7	84.2	85.2	105.5	108.8	106.7	117.8	99.9	111.2	107.7	90.8	101.4
03	CONCRETE	87.2	84.2	85.9	102.8	111.8	106.7	89.2	83.9	86.8	97.0	106.9	101.4	107.2	100.2	104.1	105.3	84.9	96.3
04	MASONRY	98.8	83.7	89.5	123.5	107.2	113.5	99.4	83.7	89.7	117.6	100.2	106.9	118.8	105.5	110.6	120.9	94.1	104.4
05	METALS	93.2	87.1	91.4	116.1	101.9	111.9	95.4	86.6	92.8	118.7	96.5	112.2	116.0	99.3	111.1	116.0	87.3	107.6
06	WOOD, PLASTICS & COMPOSITES	104.6	77.9	89.9	103.4	108.7	106.3	113.5	77.9	93.9	97.1	104.7	101.2	89.5	94.9	92.5	98.7	73.3	84.8
07	THERMAL & MOISTURE PROTECTION	161.1	84.4	127.7	111.6	110.8	111.3	157.4	84.7	125.8	112.1	102.1	107.8	111.8	101.6	107.4	111.6	88.3	101.5
08	OPENINGS	114.4	74.7	105.2	106.1	110.7	107.2	115.0	74.7	105.6	106.7	104.7	106.2	102.8	101.5	102.5	106.2	73.0	98.5
0920	Plaster & Gypsum Board	154.4	77.2	102.5	111.2	109.1	109.8	144.2	77.2	99.1	110.6	104.9	106.8	109.0	95.2	99.7	113.9	72.7	86.2
0950, 0980	Ceilings & Acoustic Treatment	109.2	77.2	87.6	106.6	109.1	108.3	105.6	77.2	86.4	104.5	104.9	104.8	102.4	95.2	97.5	97.8	72.7	80.9
0960	Flooring	85.7	77.5	83.4	109.5	103.4	107.8	84.8	77.5	82.7	102.2	96.0	100.5	107.4	107.1	107.3	105.3	77.5	97.5
0970, 0990	Wall Finishes & Painting/Coating	84.4	75.1	78.9	107.1	95.8	100.4	84.6	76.4	79.8	97.6	93.1	95.0	100.3	79.9	88.2	97.6	70.9	81.8
09	FINISHES	109.5	78.1	92.5	108.9	106.4	107.6	107.3	78.2	91.6	103.6	101.4	102.4	101.3	95.6	98.3	104.7	74.9	88.5
COVERS	DIVS. 10 - 14, 25, 28, 41, 43, 44, 46	100.0	92.2	98.3	100.0	105.2	101.2	100.0	92.2	98.3	100.0	99.6	99.9	100.0	101.5	100.3	100.0	91.2	98.0
21, 22, 23	FIRE SUPPRESSION, PLUMBING & HVAC	100.3	108.6	103.6	99.9	123.0	109.0	100.3	83.7	93.8	100.0	100.6	100.2	100.0	110.7	104.2	96.5	92.0	94.7
26, 27, 3370	ELECTRICAL, COMMUNICATIONS & UTIL.	87.2	92.8	90.0	106.2	118.9	112.5	85.4	77.1	81.3	106.8	97.7	102.3	113.1	107.3	110.2	107.7	94.5	101.2
MF2018	WEIGHTED AVERAGE	99.8	90.0	95.6	106.3	112.5	108.9	100.1	82.4	92.6	105.4	101.3	103.7	106.6	102.8	105.0	105.6	88.7	98.5

	DIVISION	WASHINGTON YAKIMA 989			WEST VIRGINIA BECKLEY 258 - 259			BLUEFIELD 247 - 248			BUCKHANNON 262			CHARLESTON 250 - 253			CLARKSBURG 263 - 264		
		MAT.	INST.	TOTAL	MAT.	INST.	TOTAL	MAT.	INST.	TOTAL	MAT.	INST.	TOTAL	MAT.	INST.	TOTAL	MAT.	INST.	TOTAL
015433	CONTRACTOR EQUIPMENT		99.3	99.3		104.1	104.1		104.1	104.1		104.1	104.1		106.8	106.8		104.1	104.1
0241, 31 - 34	SITE & INFRASTRUCTURE, DEMOLITION	97.6	103.8	101.9	100.8	87.4	91.6	99.9	87.4	91.3	106.0	87.4	93.1	100.4	93.0	95.3	106.8	87.4	93.4
0310	Concrete Forming & Accessories	102.9	99.2	99.8	85.0	87.1	86.8	86.6	87.1	87.0	86.1	83.7	84.0	96.0	88.6	89.7	83.6	83.8	83.7
0320	Concrete Reinforcing	108.3	95.4	102.1	98.0	87.7	93.1	96.0	82.5	89.5	96.7	82.4	89.8	106.1	87.8	97.2	96.7	95.3	96.0
0330	Cast-in-Place Concrete	112.7	83.0	101.7	99.2	92.5	96.7	102.5	92.5	98.8	102.1	92.0	98.4	95.9	96.6	96.2	111.9	87.6	102.9
03	CONCRETE	101.8	92.4	97.7	91.8	90.2	91.1	95.1	89.3	92.6	98.2	87.5	93.5	90.7	92.2	91.4	102.2	88.2	96.1
04	MASONRY	110.3	81.0	92.3	91.0	89.0	89.8	89.5	89.0	89.2	100.6	85.8	91.5	88.4	90.7	89.8	104.1	85.8	92.9
05	METALS	116.7	87.3	108.0	98.2	103.3	99.7	103.6	101.5	103.0	103.8	101.6	103.2	96.8	102.5	98.5	103.8	106.0	104.5
06	WOOD, PLASTICS & COMPOSITES	97.4	104.7	101.4	80.2	87.4	84.2	82.1	87.4	85.0	81.4	83.0	82.3	89.3	88.2	88.7	78.0	83.0	80.8
07	THERMAL & MOISTURE PROTECTION	112.3	86.1	100.9	105.0	86.5	96.9	103.6	86.5	96.1	104.0	86.6	96.4	101.2	88.1	95.5	103.8	86.3	96.2
08	OPENINGS	106.2	100.5	104.9	94.5	83.0	91.8	96.9	81.8	93.4	96.9	79.4	92.8	94.9	83.4	92.2	96.9	82.1	93.4
0920	Plaster & Gypsum Board	110.4	104.9	106.7	93.5	87.0	89.1	90.2	87.0	88.0	90.8	82.5	85.2	98.7	87.6	91.2	88.6	82.5	84.5
0950, 0980	Ceilings & Acoustic Treatment	99.6	104.9	103.2	78.1	87.0	84.1	87.5	87.0	87.2	90.0	82.5	84.9	89.6	87.6	88.2	90.0	82.5	84.9
0960	Flooring	103.2	77.5	96.0	91.0	98.9	93.2	90.1	98.9	92.6	89.8	95.2	91.3	97.1	98.9	97.6	88.8	95.2	90.6
0970, 0990	Wall Finishes & Painting/Coating	97.6	76.4	85.1	89.3	92.1	91.0	102.1	89.8	94.8	102.1	88.0	93.7	88.6	91.9	90.5	102.1	88.0	93.7
09	FINISHES	103.1	93.3	97.8	86.3	90.2	88.4	88.7	89.9	89.4	89.7	86.1	87.8	92.8	91.0	91.9	89.0	86.1	87.4
COVERS	DIVS. 10 - 14, 25, 28, 41, 43, 44, 46	100.0	96.1	99.1	100.0	90.3	97.8	100.0	90.3	97.8	100.0	90.6	97.9	100.0	92.5	98.3	100.0	90.6	97.9
21, 22, 23	FIRE SUPPRESSION, PLUMBING & HVAC	100.0	107.4	102.9	97.1	80.3	90.5	96.8	84.0	91.8	96.8	89.7	94.0	100.1	81.4	92.8	96.8	89.8	94.1
26, 27, 3370	ELECTRICAL, COMMUNICATIONS & UTIL.	109.9	92.9	101.5	92.9	84.1	88.6	95.2	84.1	89.8	96.6	89.3	93.0	98.0	87.1	92.6	96.6	89.3	93.0
MF2018	WEIGHTED AVERAGE	105.6	95.4	101.3	95.2	87.7	92.0	97.0	88.1	93.2	98.3	88.8	94.3	96.5	89.4	93.5	98.9	89.5	94.9

WEST VIRGINIA

	DIVISION	GASSAWAY 266			HUNTINGTON 255 - 257			LEWISBURG 249			MARTINSBURG 254			MORGANTOWN 265			PARKERSBURG 261		
		MAT.	INST.	TOTAL	MAT.	INST.	TOTAL	MAT.	INST.	TOTAL	MAT.	INST.	TOTAL	MAT.	INST.	TOTAL	MAT.	INST.	TOTAL
015433	CONTRACTOR EQUIPMENT		104.1	104.1		104.1	104.1		104.1	104.1		104.1	104.1		104.1	104.1		104.1	104.1
0241, 31 - 34	SITE & INFRASTRUCTURE, DEMOLITION	103.5	87.4	92.4	105.4	88.5	93.7	115.5	87.4	96.1	104.7	88.0	93.2	100.9	88.2	92.1	109.3	88.4	94.9
0310	Concrete Forming & Accessories	85.5	86.9	86.7	96.0	89.3	90.3	83.9	86.6	86.2	85.0	79.5	80.3	83.9	83.8	83.9	88.0	86.0	86.3
0320	Concrete Reinforcing	96.7	91.6	94.2	99.5	93.1	96.4	96.7	82.4	89.8	98.0	95.1	96.6	96.7	95.3	96.0	96.0	82.3	89.4
0330	Cast-in-Place Concrete	106.9	94.0	102.1	108.2	94.0	102.9	102.5	92.3	98.7	104.0	88.3	98.1	102.1	92.0	98.4	104.4	89.0	98.7
03	CONCRETE	98.6	91.3	95.4	96.8	92.6	95.0	105.1	89.0	98.0	95.3	86.5	91.4	94.9	89.8	92.7	100.3	87.5	94.7
04	MASONRY	105.0	87.9	94.5	91.7	93.4	92.7	94.3	89.0	91.1	93.1	83.9	87.5	122.6	85.8	100.0	80.1	85.2	83.2
05	METALS	103.8	104.8	104.1	100.6	105.7	102.1	103.7	101.0	102.9	98.6	104.4	100.3	103.9	106.0	104.5	104.5	101.4	103.6
06	WOOD, PLASTICS & COMPOSITES	80.4	87.4	84.3	90.8	88.7	89.6	78.3	87.4	83.3	80.2	78.6	79.3	78.3	83.0	80.8	81.5	85.6	83.7
07	THERMAL & MOISTURE PROTECTION	103.6	87.9	96.8	105.2	88.4	97.9	104.6	86.4	96.7	105.3	80.8	94.6	103.7	86.6	96.2	103.9	86.7	96.4
08	OPENINGS	95.2	83.9	92.6	93.7	84.8	91.7	96.9	81.8	93.4	96.2	73.8	91.0	98.0	82.1	94.3	95.8	80.4	92.2
0920	Plaster & Gypsum Board	89.9	87.0	87.9	99.5	88.3	91.9	88.6	87.0	87.5	93.5	77.9	83.0	88.6	82.5	84.5	91.1	85.1	87.1
0950, 0980	Ceilings & Acoustic Treatment	90.0	87.0	88.0	78.1	88.3	85.0	90.0	87.0	88.0	78.1	77.9	78.0	90.0	82.5	84.9	90.0	85.1	86.7
0960	Flooring	89.6	98.9	92.2	98.4	103.3	99.7	88.9	98.9	91.7	91.0	98.9	93.2	88.9	95.2	90.7	92.6	91.9	92.4
0970, 0990	Wall Finishes & Painting/Coating	102.1	91.9	96.0	89.3	89.8	89.6	102.1	64.5	79.8	89.3	83.7	86.0	102.1	88.0	93.7	102.1	83.7	91.2
09	FINISHES	89.2	89.9	89.6	89.3	92.0	90.8	90.1	87.2	88.5	86.6	83.0	84.6	88.6	86.1	87.3	90.6	86.8	88.6
COVERS	DIVS. 10 - 14, 25, 28, 41, 43, 44, 46	100.0	91.1	98.0	100.0	92.3	98.3	100.0	65.4	92.3	100.0	81.4	95.8	100.0	83.5	96.3	100.0	91.3	98.1
21, 22, 23	FIRE SUPPRESSION, PLUMBING & HVAC	96.8	83.4	91.6	100.6	90.6	96.7	96.8	80.2	90.3	97.1	80.9	90.8	96.8	89.8	94.1	100.3	88.3	95.6
26, 27, 3370	ELECTRICAL, COMMUNICATIONS & UTIL.	96.6	87.1	91.9	96.6	89.6	93.1	92.8	84.1	88.5	98.6	76.9	87.9	96.7	89.3	93.1	96.7	88.3	92.5
MF2018	WEIGHTED AVERAGE	98.3	89.0	94.3	97.8	92.2	95.4	98.7	86.1	93.4	96.7	84.1	91.4	98.8	89.6	94.9	98.6	88.6	94.4

For customer support on your Mechanical Costs with RSMeans Data, call 800.448.8182.

693

		WEST VIRGINIA									WISCONSIN								
DIVISION		PETERSBURG 268			ROMNEY 267			WHEELING 260			BELOIT 535			EAU CLAIRE 547			GREEN BAY 541 - 543		
		MAT.	INST.	TOTAL	MAT.	INST.	TOTAL	MAT.	INST.	TOTAL	MAT.	INST.	TOTAL	MAT.	INST.	TOTAL	MAT.	INST.	TOTAL
015433	CONTRACTOR EQUIPMENT		104.1	104.1		104.1	104.1		104.1	104.1		99.0	99.0		99.9	99.9		97.7	97.7
0241, 31 - 34	SITE & INFRASTRUCTURE, DEMOLITION	100.2	88.3	92.0	103.0	88.3	92.9	109.9	88.2	94.9	97.1	101.7	100.3	96.0	99.0	98.1	99.5	95.3	96.6
0310	Concrete Forming & Accessories	87.0	86.4	86.5	83.3	86.8	86.3	89.7	83.8	84.7	99.8	96.2	96.7	98.5	94.7	95.3	108.2	97.5	99.1
0320	Concrete Reinforcing	96.0	91.4	93.8	96.7	95.4	96.1	95.5	95.3	95.4	95.0	137.1	115.4	94.2	114.2	103.9	92.3	109.4	100.5
0330	Cast-in-Place Concrete	102.1	91.8	98.3	106.9	83.5	98.2	104.4	92.0	99.8	105.8	100.2	103.7	99.9	94.8	98.0	103.3	99.9	102.1
03	CONCRETE	95.0	90.2	92.9	98.4	88.2	93.9	100.3	89.8	95.7	96.7	104.7	100.2	93.1	98.4	95.4	96.2	100.6	98.2
04	MASONRY	96.1	85.8	89.8	94.7	87.9	90.5	103.4	85.7	92.5	99.3	100.3	99.9	91.4	96.1	94.3	122.7	97.8	107.4
05	METALS	103.9	104.1	104.0	104.0	105.9	104.5	104.7	106.1	105.1	98.6	112.0	102.5	93.8	105.8	97.3	96.4	104.2	98.7
06	WOOD, PLASTICS & COMPOSITES	82.3	87.4	85.1	77.5	87.4	82.9	83.1	83.0	83.1	96.9	94.1	95.4	100.6	94.2	97.1	107.8	97.2	101.9
07	THERMAL & MOISTURE PROTECTION	103.7	82.5	94.5	103.8	83.8	95.1	104.1	86.7	96.6	105.4	93.4	100.2	104.0	94.6	99.9	106.1	97.4	102.3
08	OPENINGS	98.1	83.6	94.7	98.0	84.5	94.8	96.6	82.1	93.2	98.1	108.3	100.5	103.1	99.3	102.2	99.1	101.2	99.6
0920	Plaster & Gypsum Board	90.8	87.0	88.2	88.3	87.0	87.4	91.1	82.5	85.3	91.6	94.5	93.5	105.5	94.5	98.1	103.3	97.5	99.4
0950, 0980	Ceilings & Acoustic Treatment	90.0	87.0	88.0	90.0	87.0	88.0	90.0	82.5	84.9	84.4	94.5	91.2	90.9	94.5	93.3	83.6	97.5	93.0
0960	Flooring	90.6	95.2	91.9	88.7	98.9	91.6	93.5	95.2	94.0	94.8	118.5	101.4	80.5	106.5	87.8	96.0	114.8	101.3
0970, 0990	Wall Finishes & Painting/Coating	102.1	88.0	93.7	102.1	88.0	93.7	102.1	88.0	93.7	95.0	103.6	100.1	83.3	76.4	79.2	92.0	80.0	84.9
09	FINISHES	89.4	88.7	89.0	88.7	89.5	89.1	90.9	86.1	88.3	92.3	100.7	96.9	87.3	94.9	91.4	91.4	99.1	95.6
COVERS	DIVS. 10 - 14, 25, 28, 41, 43, 44, 46	100.0	55.0	90.0	100.0	91.0	98.0	100.0	84.4	96.5	100.0	98.6	99.7	100.0	94.1	98.7	100.0	95.7	99.0
21, 22, 23	FIRE SUPPRESSION, PLUMBING & HVAC	96.8	83.7	91.7	96.8	83.6	91.6	100.4	89.8	96.2	99.8	95.5	98.1	100.0	88.2	95.4	100.2	83.4	93.6
26, 27, 3370	ELECTRICAL, COMMUNICATIONS & UTIL.	99.7	76.9	88.4	99.0	76.9	88.1	94.1	89.3	91.7	100.6	84.3	92.5	104.3	84.4	94.5	99.0	81.2	90.2
MF2018	WEIGHTED AVERAGE	98.0	85.9	92.9	98.3	87.2	93.6	99.6	89.6	95.4	98.6	98.9	98.7	97.5	94.1	96.1	99.4	93.4	96.9

		WISCONSIN																	
DIVISION		KENOSHA 531			LA CROSSE 546			LANCASTER 538			MADISON 537			MILWAUKEE 530, 532			NEW RICHMOND 540		
		MAT.	INST.	TOTAL	MAT.	INST.	TOTAL	MAT.	INST.	TOTAL	MAT.	INST.	TOTAL	MAT.	INST.	TOTAL	MAT.	INST.	TOTAL
015433	CONTRACTOR EQUIPMENT		97.1	97.1		99.9	99.9		99.0	99.0		101.4	101.4		90.2	90.2		100.2	100.2
0241, 31 - 34	SITE & INFRASTRUCTURE, DEMOLITION	103.0	98.4	99.9	89.8	99.0	96.1	96.0	101.1	99.5	94.7	105.3	102.0	93.6	96.0	95.2	93.3	99.6	97.7
0310	Concrete Forming & Accessories	108.7	106.1	106.5	85.2	94.5	93.2	99.1	95.2	95.7	103.8	95.8	97.0	102.9	113.9	112.3	93.6	92.5	92.6
0320	Concrete Reinforcing	94.8	110.5	102.4	93.9	106.2	99.8	96.3	106.2	101.1	97.8	106.4	101.9	99.9	115.1	107.2	91.3	114.1	102.3
0330	Cast-in-Place Concrete	115.2	102.7	110.5	89.8	96.5	92.3	105.2	99.7	103.2	101.4	100.3	101.0	93.9	112.1	100.7	104.0	102.7	103.5
03	CONCRETE	101.5	105.5	103.3	84.6	97.5	90.3	96.4	98.8	97.4	96.3	99.2	97.6	94.5	112.5	102.4	90.8	100.1	94.9
04	MASONRY	97.1	108.0	103.8	90.5	96.1	93.9	99.4	100.1	99.9	97.4	99.1	98.5	102.3	117.2	111.4	117.9	98.4	105.9
05	METALS	99.6	103.3	100.7	93.7	102.8	96.4	96.0	100.3	97.3	102.8	100.4	102.1	97.5	97.5	97.5	94.1	105.4	97.4
06	WOOD, PLASTICS & COMPOSITES	103.9	105.7	104.9	85.2	94.2	90.2	96.1	94.1	95.0	99.4	94.3	96.6	101.2	112.3	107.3	89.5	90.2	89.9
07	THERMAL & MOISTURE PROTECTION	105.5	104.0	104.9	103.4	93.5	99.1	105.2	93.7	100.2	106.3	100.5	103.7	104.8	112.7	108.2	104.9	100.2	102.9
08	OPENINGS	92.5	105.4	95.5	103.1	92.0	100.5	94.0	91.9	93.5	101.6	98.9	100.9	103.0	111.5	105.0	88.6	90.3	89.0
0920	Plaster & Gypsum Board	80.2	106.4	97.8	100.9	94.5	96.6	90.9	94.5	93.3	101.7	94.5	96.8	96.4	112.6	107.3	91.7	90.5	90.9
0950, 0980	Ceilings & Acoustic Treatment	84.4	106.4	99.2	90.0	94.5	93.0	81.1	94.5	90.1	86.5	94.5	91.9	93.1	112.6	106.3	57.1	90.5	79.6
0960	Flooring	112.8	104.8	110.6	74.7	111.6	85.1	94.4	111.0	99.0	94.3	111.0	99.0	101.7	118.6	106.4	89.7	114.5	96.7
0970, 0990	Wall Finishes & Painting/Coating	106.4	118.7	113.7	83.3	81.7	82.4	95.0	97.7	96.6	94.9	103.6	100.0	104.5	124.3	116.3	95.0	81.7	87.1
09	FINISHES	97.1	109.4	103.7	84.4	96.5	90.9	91.4	98.4	95.2	93.7	99.3	96.7	99.5	116.1	108.5	82.5	95.2	89.4
COVERS	DIVS. 10 - 14, 25, 28, 41, 43, 44, 46	100.0	97.3	99.4	100.0	92.8	98.4	100.0	84.2	96.5	100.0	96.8	99.3	100.0	105.2	101.2	100.0	95.1	98.9
21, 22, 23	FIRE SUPPRESSION, PLUMBING & HVAC	100.0	95.9	98.4	100.0	88.1	95.3	96.3	89.2	93.5	99.8	95.3	98.0	99.8	107.0	102.6	96.0	89.0	93.2
26, 27, 3370	ELECTRICAL, COMMUNICATIONS & UTIL.	100.9	97.1	99.0	104.6	84.4	94.7	100.3	84.9	92.7	101.5	94.0	97.8	100.2	102.4	101.3	102.4	84.9	93.7
MF2018	WEIGHTED AVERAGE	99.4	102.0	100.5	95.9	93.5	94.9	96.7	94.3	95.7	99.7	98.1	99.0	99.2	107.9	102.9	95.4	94.6	95.0

		WISCONSIN																	
DIVISION		OSHKOSH 549			PORTAGE 539			RACINE 534			RHINELANDER 545			SUPERIOR 548			WAUSAU 544		
		MAT.	INST.	TOTAL	MAT.	INST.	TOTAL	MAT.	INST.	TOTAL	MAT.	INST.	TOTAL	MAT.	INST.	TOTAL	MAT.	INST.	TOTAL
015433	CONTRACTOR EQUIPMENT		97.7	97.7		99.0	99.0		99.0	99.0		97.7	97.7		100.2	100.2		97.7	97.7
0241, 31 - 34	SITE & INFRASTRUCTURE, DEMOLITION	91.0	95.3	94.0	86.9	100.9	96.6	96.9	102.1	100.5	103.2	95.3	97.7	90.1	99.3	96.5	86.9	95.8	93.0
0310	Concrete Forming & Accessories	90.7	94.7	94.1	91.3	95.7	95.1	100.1	106.1	105.2	88.3	94.6	93.7	91.6	88.6	89.0	90.1	97.4	96.3
0320	Concrete Reinforcing	92.4	109.3	100.6	96.4	106.4	101.2	95.0	110.5	102.5	92.6	106.4	99.3	91.3	110.2	100.4	92.6	106.2	99.2
0330	Cast-in-Place Concrete	95.8	98.0	96.6	90.3	98.3	93.3	103.8	102.5	103.3	108.6	97.9	104.6	98.0	98.3	98.1	89.3	92.4	90.5
03	CONCRETE	86.2	98.6	91.7	84.2	98.7	90.6	95.8	105.4	100.0	98.0	98.1	98.1	85.5	96.2	90.2	81.4	97.4	88.4
04	MASONRY	104.7	97.8	100.4	98.2	100.2	99.4	99.3	108.0	104.6	121.3	97.6	106.8	117.1	100.4	106.8	104.2	97.6	100.1
05	METALS	94.3	103.5	97.0	96.7	104.3	98.9	100.2	103.3	101.1	94.2	102.5	96.6	95.1	104.6	97.9	94.0	102.9	96.6
06	WOOD, PLASTICS & COMPOSITES	86.7	94.2	90.9	85.8	94.1	90.4	97.2	105.7	101.9	84.2	94.2	89.7	87.9	85.8	86.7	86.2	97.2	92.2
07	THERMAL & MOISTURE PROTECTION	105.0	83.3	95.6	104.5	100.1	102.6	105.5	102.7	104.3	106.0	83.1	96.0	104.5	95.9	100.8	104.8	95.1	100.6
08	OPENINGS	95.6	96.2	95.7	94.1	99.7	95.4	98.1	105.4	99.8	95.6	91.9	94.7	88.0	91.9	88.9	95.8	97.1	96.1
0920	Plaster & Gypsum Board	90.7	94.5	93.2	83.8	94.5	91.0	91.6	106.4	101.5	90.7	94.5	93.2	92.1	86.0	88.0	90.7	97.5	95.3
0950, 0980	Ceilings & Acoustic Treatment	83.6	94.5	90.9	83.6	94.5	90.9	84.4	106.4	99.2	83.6	94.5	90.9	58.8	86.0	77.1	83.6	97.5	93.0
0960	Flooring	87.8	114.8	95.4	90.5	114.5	97.3	94.8	114.8	100.4	87.2	114.5	94.9	90.8	123.1	99.9	87.7	114.5	95.2
0970, 0990	Wall Finishes & Painting/Coating	89.6	104.3	98.3	95.0	97.7	96.6	95.0	115.9	107.4	89.6	80.0	83.9	84.0	105.8	96.9	89.6	83.2	85.8
09	FINISHES	86.5	100.0	93.8	89.2	99.4	94.7	92.3	109.1	101.4	87.4	97.3	92.7	82.1	96.4	89.8	86.2	99.4	93.3
COVERS	DIVS. 10 - 14, 25, 28, 41, 43, 44, 46	100.0	85.6	96.8	100.0	85.6	96.8	100.0	97.3	99.4	100.0	85.9	96.8	100.0	93.9	98.6	100.0	95.7	99.0
21, 22, 23	FIRE SUPPRESSION, PLUMBING & HVAC	96.7	83.1	91.3	96.3	95.8	96.1	99.8	96.6	98.5	96.7	88.6	93.5	96.0	91.1	94.1	96.7	88.8	93.6
26, 27, 3370	ELECTRICAL, COMMUNICATIONS & UTIL.	103.0	79.1	91.2	104.0	94.0	99.1	100.4	98.0	99.2	102.3	78.1	90.4	107.2	98.5	102.9	104.1	78.1	91.3
MF2018	WEIGHTED AVERAGE	95.4	91.9	93.9	95.2	98.0	96.4	98.7	102.3	100.2	97.9	92.2	95.5	95.1	96.5	95.7	94.7	93.4	94.2

For customer support on your Mechanical Costs with RSMeans Data, call 800.448.8182.

WYOMING

DIVISION		CASPER 826			CHEYENNE 820			NEWCASTLE 827			RAWLINS 823			RIVERTON 825			ROCK SPRINGS 829 - 831		
		MAT.	INST.	TOTAL	MAT.	INST.	TOTAL	MAT.	INST.	TOTAL	MAT.	INST.	TOTAL	MAT.	INST.	TOTAL	MAT.	INST.	TOTAL
015433	CONTRACTOR EQUIPMENT		97.3	97.3		94.6	94.6		94.6	94.6		94.6	94.6		94.6	94.6		94.6	94.6
0241, 31 - 34	SITE & INFRASTRUCTURE, DEMOLITION	97.8	94.7	95.7	91.6	90.1	90.6	83.6	89.3	87.6	97.5	89.3	91.9	91.1	89.3	89.8	87.4	89.5	88.9
0310	Concrete Forming & Accessories	101.3	53.6	60.6	102.5	64.6	70.2	93.7	70.5	74.0	97.6	70.8	74.7	92.7	60.3	65.0	99.7	64.7	69.9
0320	Concrete Reinforcing	111.1	83.4	97.7	106.4	83.4	95.3	114.5	83.2	99.3	114.2	83.2	99.2	115.2	83.2	99.7	115.2	81.6	99.0
0330	Cast-in-Place Concrete	103.8	77.8	94.1	97.9	77.4	90.3	99.0	72.8	89.2	99.0	72.9	89.3	99.0	72.8	89.2	99.0	75.0	90.0
03	CONCRETE	103.3	67.9	87.7	100.9	72.8	88.6	101.1	73.8	89.1	116.0	74.0	97.5	110.1	69.2	92.1	101.6	71.7	88.4
04	MASONRY	102.0	64.6	79.0	105.3	65.2	80.7	102.2	59.3	75.8	102.2	59.3	75.8	102.2	59.3	75.8	164.5	61.8	101.4
05	METALS	102.8	80.1	96.1	105.3	80.8	98.1	101.4	79.9	95.1	101.5	80.1	95.2	101.9	79.9	95.2	102.3	79.5	95.6
06	WOOD, PLASTICS & COMPOSITES	95.2	47.8	69.1	91.8	62.6	75.7	82.4	72.7	77.1	86.2	72.7	78.8	81.4	58.7	68.9	91.3	62.6	75.5
07	THERMAL & MOISTURE PROTECTION	112.7	65.9	92.3	107.2	68.0	90.2	108.8	63.9	89.3	110.3	74.9	94.9	109.7	67.7	91.4	111.8	66.5	101.2
08	OPENINGS	109.3	59.4	97.7	107.2	67.7	98.0	111.4	72.3	102.3	111.1	72.3	102.0	111.3	64.6	100.4	111.8	66.5	101.2
0920	Plaster & Gypsum Board	101.7	46.2	64.4	90.6	61.7	71.2	87.6	72.0	77.1	87.9	72.0	77.2	87.6	57.7	67.5	100.2	61.7	74.3
0950, 0980	Ceilings & Acoustic Treatment	105.3	46.2	65.5	98.5	61.7	73.7	100.3	72.0	81.2	100.3	72.0	81.2	100.3	57.7	71.6	100.3	61.7	74.3
0960	Flooring	102.2	73.8	94.2	100.3	73.8	92.8	94.2	44.5	80.2	97.0	44.5	82.3	93.6	60.3	84.2	99.3	59.9	88.3
0970, 0990	Wall Finishes & Painting/Coating	95.7	58.7	73.8	97.9	58.7	74.7	94.6	60.5	74.4	94.6	60.5	74.4	94.6	60.5	74.4	94.6	60.5	74.4
09	FINISHES	99.6	56.3	76.2	96.3	65.1	79.4	91.4	64.6	76.9	93.6	64.6	77.9	92.0	59.5	74.4	94.6	62.7	77.3
COVERS	DIVS. 10 - 14, 25, 28, 41, 43, 44, 46	100.0	92.5	98.3	100.0	87.4	97.2	100.0	91.8	98.2	100.0	91.8	98.2	100.0	77.3	94.9	100.0	86.9	97.1
21, 22, 23	FIRE SUPPRESSION, PLUMBING & HVAC	100.0	70.5	88.4	100.1	73.3	89.6	98.2	71.8	87.8	98.2	71.9	87.8	98.2	71.8	87.8	100.0	73.9	89.8
26, 27, 3370	ELECTRICAL, COMMUNICATIONS & UTIL.	94.1	60.3	77.5	95.2	69.0	82.3	94.0	62.3	78.4	94.0	62.3	78.4	94.0	59.1	76.9	103.4	66.0	79.4
MF2018	WEIGHTED AVERAGE	101.5	68.9	87.7	101.1	72.7	89.1	99.6	71.1	87.5	102.0	71.5	89.1	101.0	68.6	87.3	103.4	71.4	89.9

WYOMING / CANADA

DIVISION		SHERIDAN 828			WHEATLAND 822			WORLAND 824			YELLOWSTONE NAT'L PA 821			BARRIE, ONTARIO			BATHURST, NEW BRUNSWICK		
		MAT.	INST.	TOTAL	MAT.	INST.	TOTAL	MAT.	INST.	TOTAL	MAT.	INST.	TOTAL	MAT.	INST.	TOTAL	MAT.	INST.	TOTAL
015433	CONTRACTOR EQUIPMENT		94.6	94.6		94.6	94.6		94.6	94.6		94.6	94.6		100.1	100.1		99.7	99.7
0241, 31 - 34	SITE & INFRASTRUCTURE, DEMOLITION	91.5	90.1	90.5	88.1	89.3	88.9	85.5	89.3	88.1	85.6	89.3	88.1	117.0	94.7	101.6	105.1	90.6	95.1
0310	Concrete Forming & Accessories	100.5	61.6	67.4	95.4	48.4	55.3	95.5	61.1	66.1	95.5	61.1	66.2	125.6	82.3	88.7	106.1	57.8	64.9
0320	Concrete Reinforcing	115.2	83.2	99.7	114.5	82.6	99.1	115.2	83.2	99.7	117.2	82.3	100.3	175.4	86.9	132.5	140.5	57.5	100.4
0330	Cast-in-Place Concrete	102.3	77.4	93.1	103.3	72.7	91.9	99.0	72.7	89.2	99.0	72.8	89.2	157.7	82.1	129.6	114.0	56.2	92.5
03	CONCRETE	109.9	71.4	93.0	106.3	63.7	87.6	101.3	69.5	87.3	101.6	69.4	87.4	138.9	83.5	114.5	110.3	58.3	87.4
04	MASONRY	102.5	62.0	77.6	102.6	51.2	71.0	102.2	59.3	75.8	102.2	59.3	75.8	165.6	89.3	118.7	159.8	57.4	96.9
05	METALS	105.1	80.5	97.8	101.4	79.1	94.8	101.6	79.8	95.2	102.2	79.1	95.4	111.1	91.5	105.3	114.1	73.6	102.2
06	WOOD, PLASTICS & COMPOSITES	93.3	58.7	74.3	84.2	43.0	61.5	84.3	60.0	70.9	84.3	60.0	70.9	116.5	80.8	96.9	97.5	57.9	75.7
07	THERMAL & MOISTURE PROTECTION	110.1	66.7	91.2	109.1	58.1	86.9	108.9	64.8	89.7	108.3	64.8	89.4	115.3	86.1	102.6	111.8	58.4	88.6
08	OPENINGS	112.0	64.6	101.0	110.0	55.9	97.4	111.6	65.3	100.8	104.5	64.8	95.2	90.5	80.9	88.3	84.0	51.2	76.3
0920	Plaster & Gypsum Board	111.2	57.7	75.2	87.6	41.5	56.6	87.6	58.9	68.3	87.8	58.9	68.4	153.5	80.3	104.3	124.2	56.7	78.8
0950, 0980	Ceilings & Acoustic Treatment	103.9	57.7	72.7	100.3	41.5	60.6	100.3	58.9	72.4	101.1	58.9	72.7	94.3	80.3	84.9	116.8	56.7	76.3
0960	Flooring	97.9	60.3	87.4	95.6	43.8	81.0	95.6	44.5	81.3	95.6	44.5	81.3	119.3	87.3	110.3	99.6	41.5	83.3
0970, 0990	Wall Finishes & Painting/Coating	96.8	58.7	74.2	94.6	58.4	73.2	94.6	60.5	74.4	94.6	60.5	74.4	105.6	83.8	92.7	111.3	47.9	73.8
09	FINISHES	99.2	60.4	78.2	92.1	46.7	67.5	91.8	57.1	73.0	92.0	57.2	73.2	111.3	83.5	96.2	107.7	53.8	78.5
COVERS	DIVS. 10 - 14, 25, 28, 41, 43, 44, 46	100.0	93.4	98.5	100.0	81.4	95.9	100.0	76.1	94.7	100.0	76.2	94.7	139.2	65.0	122.7	131.1	57.9	114.8
21, 22, 23	FIRE SUPPRESSION, PLUMBING & HVAC	98.2	73.3	88.4	98.2	71.8	87.8	98.2	71.8	87.8	98.2	71.8	87.8	103.7	93.5	99.7	103.8	64.8	88.5
26, 27, 3370	ELECTRICAL, COMMUNICATIONS & UTIL.	96.1	59.1	77.9	94.0	76.6	85.4	94.0	76.6	85.4	93.1	86.1	89.7	117.0	82.9	100.2	113.5	55.9	85.1
MF2018	WEIGHTED AVERAGE	102.5	70.2	88.8	100.3	67.1	86.2	99.7	70.7	87.5	99.0	71.9	87.6	116.5	87.1	104.1	110.7	62.3	90.2

CANADA

DIVISION		BRANDON, MANITOBA			BRANTFORD, ONTARIO			BRIDGEWATER, NOVA SCOTIA			CALGARY, ALBERTA			CAP-DE-LA-MADELEINE, QUEBEC			CHARLESBOURG, QUEBEC		
		MAT.	INST.	TOTAL	MAT.	INST.	TOTAL	MAT.	INST.	TOTAL	MAT.	INST.	TOTAL	MAT.	INST.	TOTAL	MAT.	INST.	TOTAL
015433	CONTRACTOR EQUIPMENT		101.5	101.5		99.7	99.7		99.4	99.4		127.1	127.1		100.2	100.2		100.2	100.2
0241, 31 - 34	SITE & INFRASTRUCTURE, DEMOLITION	126.4	92.9	103.3	116.9	94.8	101.6	101.1	92.3	95.0	125.5	118.4	120.6	96.9	93.7	94.7	96.9	93.7	94.7
0310	Concrete Forming & Accessories	145.4	66.1	77.8	127.2	89.1	94.7	99.7	67.9	72.6	125.0	94.9	99.3	132.9	79.2	87.1	132.9	79.2	87.1
0320	Concrete Reinforcing	170.2	54.5	114.3	163.7	85.6	125.9	139.3	47.6	95.0	136.4	82.0	110.1	139.3	72.4	107.0	139.3	72.4	107.0
0330	Cast-in-Place Concrete	109.4	70.2	94.8	130.7	101.8	119.9	134.3	66.9	109.2	136.4	104.3	124.4	105.5	87.5	98.8	105.5	87.5	98.8
03	CONCRETE	118.1	66.6	95.5	121.8	93.2	109.3	120.1	65.1	95.9	126.3	96.8	113.3	107.5	81.5	96.1	107.5	81.5	96.1
04	MASONRY	212.8	60.7	119.3	166.1	93.4	121.4	161.8	65.6	102.6	212.4	89.8	137.1	162.4	76.3	109.5	162.4	76.3	109.5
05	METALS	127.2	78.2	112.8	112.3	92.0	106.3	111.6	76.0	101.1	131.0	102.4	122.6	110.7	85.0	103.1	110.7	85.0	103.1
06	WOOD, PLASTICS & COMPOSITES	150.5	66.8	104.4	120.3	87.7	102.4	89.6	67.4	77.4	102.6	94.4	98.1	131.6	79.0	102.7	131.6	79.0	102.7
07	THERMAL & MOISTURE PROTECTION	128.4	68.7	102.4	121.6	91.4	108.5	116.1	67.6	95.0	132.0	98.0	117.2	114.8	84.1	101.5	114.8	84.1	101.5
08	OPENINGS	100.3	59.9	90.8	87.7	86.6	87.5	82.4	61.3	77.5	82.3	83.7	82.6	89.1	72.3	85.2	89.1	72.3	85.2
0920	Plaster & Gypsum Board	115.4	65.6	81.9	115.8	87.4	96.7	124.3	66.5	85.4	126.4	93.3	104.1	147.6	78.4	101.0	147.6	78.4	101.0
0950, 0980	Ceilings & Acoustic Treatment	123.8	65.6	84.6	103.5	87.4	92.7	103.5	66.5	78.6	149.2	93.3	111.5	103.5	78.4	86.6	103.5	78.4	86.6
0960	Flooring	131.2	61.9	111.7	113.6	87.3	106.2	95.7	58.9	85.4	118.1	84.8	108.7	113.6	86.0	105.8	113.6	86.0	105.8
0970, 0990	Wall Finishes & Painting/Coating	117.9	54.1	80.1	110.3	92.0	99.5	110.3	59.3	80.1	113.7	106.4	109.4	110.3	83.6	94.5	110.3	83.6	94.5
09	FINISHES	122.9	64.3	91.2	108.7	89.3	98.2	103.8	65.5	83.1	123.6	94.8	108.0	112.2	81.2	95.4	112.2	81.2	95.4
COVERS	DIVS. 10 - 14, 25, 28, 41, 43, 44, 46	131.1	60.1	115.3	131.1	66.9	116.8	131.1	60.8	115.5	131.1	92.7	122.6	131.1	75.6	118.8	131.1	75.6	118.8
21, 22, 23	FIRE SUPPRESSION, PLUMBING & HVAC	104.0	78.9	94.1	103.8	96.3	100.9	103.8	79.5	94.3	105.1	90.2	99.2	104.2	84.2	96.4	104.2	84.2	96.4
26, 27, 3370	ELECTRICAL, COMMUNICATIONS & UTIL.	113.9	63.1	88.8	111.5	82.5	97.2	115.8	58.7	87.6	107.1	93.9	100.6	110.6	65.2	88.0	110.6	65.2	88.0
MF2018	WEIGHTED AVERAGE	120.6	70.5	99.4	113.1	90.7	103.6	111.3	70.3	94.0	119.7	95.6	109.5	110.6	80.0	97.7	110.6	80.0	97.7

For customer support on your Mechanical Costs with RSMeans Data, call 800.448.8182.

695

CANADA

DIVISION		CHARLOTTETOWN, PRINCE EDWARD ISLAND			CHICOUTIMI, QUEBEC			CORNER BROOK, NEWFOUNDLAND			CORNWALL, ONTARIO			DALHOUSIE, NEW BRUNSWICK			DARTMOUTH, NOVA SCOTIA		
		MAT.	INST.	TOTAL	MAT.	INST.	TOTAL	MAT.	INST.	TOTAL	MAT.	INST.	TOTAL	MAT.	INST.	TOTAL	MAT.	INST.	TOTAL
015433	CONTRACTOR EQUIPMENT		116.8	116.8		100.3	100.3		100.0	100.0		99.7	99.7		99.7	99.7		98.9	98.9
0241, 31 - 34	SITE & INFRASTRUCTURE, DEMOLITION	134.6	101.5	111.7	102.0	93.6	96.2	130.4	90.5	102.8	115.1	94.3	100.7	101.0	90.6	93.8	117.9	92.0	100.0
0310	Concrete Forming & Accessories	123.7	53.1	63.5	134.9	89.8	96.4	121.9	75.6	82.4	125.0	82.4	88.7	105.9	58.0	65.1	111.8	67.8	74.3
0320	Concrete Reinforcing	156.7	46.6	103.5	106.0	94.5	100.4	156.0	48.8	104.2	163.7	85.3	125.8	142.5	57.6	101.4	163.2	47.6	107.3
0330	Cast-in-Place Concrete	152.1	57.6	116.9	107.4	94.7	102.7	127.3	63.6	103.6	117.5	92.8	108.3	110.5	56.3	90.3	122.9	66.8	102.0
03	CONCRETE	136.8	55.4	101.0	101.6	92.4	97.6	149.1	67.5	113.3	115.7	87.0	103.1	113.1	58.4	89.0	131.9	65.0	102.5
04	MASONRY	187.9	54.9	106.2	161.9	89.8	117.6	208.6	74.3	126.1	165.0	85.6	116.2	163.3	57.4	98.3	222.6	65.6	126.1
05	METALS	136.4	79.6	119.8	113.9	91.9	107.4	127.7	75.2	112.2	112.1	90.7	105.8	105.6	73.7	96.2	128.1	75.7	112.7
06	WOOD, PLASTICS & COMPOSITES	104.3	52.5	75.8	131.8	90.3	108.9	128.1	81.4	102.4	118.5	81.6	98.2	95.8	57.9	74.9	115.9	67.4	89.2
07	THERMAL & MOISTURE PROTECTION	137.8	57.4	102.8	112.3	96.4	105.4	133.0	65.9	103.8	121.4	86.2	106.1	121.0	58.4	93.8	130.1	67.6	102.9
08	OPENINGS	85.7	45.8	76.4	87.8	76.3	85.2	106.3	66.6	97.0	89.1	80.6	87.1	85.5	51.2	77.5	91.3	61.3	84.3
0920	Plaster & Gypsum Board	126.3	50.5	75.3	143.9	89.9	107.6	151.2	80.9	103.9	174.9	81.1	111.8	129.5	56.7	80.5	146.6	66.5	92.7
0950, 0980	Ceilings & Acoustic Treatment	128.9	50.5	76.1	116.0	89.9	98.4	124.6	80.9	95.1	107.6	81.1	89.8	106.6	56.7	72.9	131.1	66.5	87.5
0960	Flooring	115.1	55.3	98.3	115.7	86.0	107.3	112.8	49.5	95.0	113.6	86.0	105.8	101.9	63.4	91.1	107.6	58.9	93.9
0970, 0990	Wall Finishes & Painting/Coating	117.3	39.9	71.5	111.3	104.1	107.1	117.8	56.5	81.5	110.3	85.7	95.7	113.8	47.9	74.8	117.8	59.3	83.2
09	FINISHES	119.8	52.3	83.3	114.6	91.6	102.1	122.6	70.2	94.2	117.3	83.6	99.0	108.4	58.2	81.2	120.3	65.5	90.7
COVERS	DIVS. 10 - 14, 25, 28, 41, 43, 44, 46	131.1	58.4	114.9	131.1	81.5	120.1	131.1	60.6	115.4	131.1	64.6	116.3	131.1	57.8	114.8	131.1	60.8	115.5
21, 22, 23	FIRE SUPPRESSION, PLUMBING & HVAC	104.1	59.0	86.4	103.8	81.0	94.9	104.0	65.9	89.0	104.2	94.3	100.3	103.9	64.8	88.5	104.0	79.5	94.4
26, 27, 3370	ELECTRICAL, COMMUNICATIONS & UTIL.	110.0	47.5	79.2	109.9	84.1	97.2	110.9	51.5	81.6	111.6	83.4	97.7	113.9	52.8	83.8	115.0	58.7	87.2
MF2018	WEIGHTED AVERAGE	121.3	60.2	95.5	110.4	87.6	100.7	124.7	68.4	100.9	113.1	87.3	102.2	110.2	62.5	90.0	121.5	70.2	99.8

CANADA

DIVISION		EDMONTON, ALBERTA			FORT MCMURRAY, ALBERTA			FREDERICTON, NEW BRUNSWICK			GATINEAU, QUEBEC			GRANBY, QUEBEC			HALIFAX, NOVA SCOTIA		
		MAT.	INST.	TOTAL	MAT.	INST.	TOTAL	MAT.	INST.	TOTAL	MAT.	INST.	TOTAL	MAT.	INST.	TOTAL	MAT.	INST.	TOTAL
015433	CONTRACTOR EQUIPMENT		129.1	129.1		101.9	101.9		113.9	113.9		100.2	100.2		100.2	100.2		115.6	115.6
0241, 31 - 34	SITE & INFRASTRUCTURE, DEMOLITION	125.1	121.8	122.8	121.7	95.4	103.5	116.0	100.8	105.5	96.7	93.7	94.6	97.2	93.7	94.8	102.8	103.6	103.4
0310	Concrete Forming & Accessories	128.3	94.9	99.8	125.0	87.5	93.1	126.9	58.6	68.7	132.9	79.1	87.0	132.9	79.0	87.0	124.1	87.7	93.0
0320	Concrete Reinforcing	136.0	82.0	109.9	151.5	81.9	117.8	141.0	57.8	100.8	147.4	72.4	111.1	147.4	72.4	111.1	156.3	77.9	118.3
0330	Cast-in-Place Concrete	145.8	104.3	130.4	173.5	98.2	145.5	112.7	57.5	92.1	104.0	87.5	97.9	107.5	87.4	100.0	99.4	85.1	94.0
03	CONCRETE	130.8	96.8	115.9	139.8	90.6	118.2	116.9	59.7	91.8	107.9	81.4	96.2	109.5	81.4	97.1	111.9	86.1	100.6
04	MASONRY	212.7	89.8	137.2	206.1	86.3	132.5	189.1	58.8	109.0	162.3	76.3	109.5	162.6	76.3	109.6	185.4	86.4	124.6
05	METALS	134.6	102.4	125.1	139.8	90.9	125.5	136.6	83.3	121.0	110.7	84.8	103.1	110.9	84.8	103.2	136.3	99.2	125.4
06	WOOD, PLASTICS & COMPOSITES	101.6	94.4	97.6	114.5	86.7	99.2	106.5	58.2	79.9	131.6	79.0	102.7	131.6	79.0	102.7	103.7	87.8	95.0
07	THERMAL & MOISTURE PROTECTION	143.5	98.0	123.7	129.8	92.6	113.6	137.4	59.4	103.5	114.8	84.1	101.5	114.8	82.6	100.8	137.4	88.7	116.2
08	OPENINGS	80.9	83.7	81.5	89.1	79.5	86.9	88.0	50.3	79.2	89.1	67.9	84.2	89.1	67.9	84.2	88.9	79.2	86.7
0920	Plaster & Gypsum Board	129.8	93.3	105.2	116.6	86.1	96.1	130.3	56.7	80.8	114.4	78.4	90.2	114.4	78.4	90.2	117.8	87.1	97.1
0950, 0980	Ceilings & Acoustic Treatment	144.5	93.3	110.0	112.5	86.1	94.7	137.2	56.7	82.9	103.5	78.4	86.6	103.5	78.4	86.6	139.8	87.1	104.2
0960	Flooring	121.8	84.8	111.4	113.6	84.8	105.5	115.6	66.2	101.7	113.6	86.0	105.8	113.6	86.0	105.8	109.5	81.5	101.6
0970, 0990	Wall Finishes & Painting/Coating	115.6	107.4	110.8	110.4	88.3	97.3	118.8	61.1	84.6	110.3	83.6	94.5	110.3	83.6	94.5	118.4	90.9	102.1
09	FINISHES	123.6	94.9	108.0	111.8	89.4	98.6	125.1	60.4	90.1	107.8	81.2	93.4	107.8	81.2	93.4	117.7	87.5	101.3
COVERS	DIVS. 10 - 14, 25, 28, 41, 43, 44, 46	131.1	93.6	122.8	131.1	90.1	122.0	131.1	58.5	115.0	131.1	75.6	118.8	131.1	75.6	118.8	131.1	68.9	117.3
21, 22, 23	FIRE SUPPRESSION, PLUMBING & HVAC	104.9	90.4	99.2	104.2	93.2	99.9	104.2	73.7	92.2	104.2	84.2	96.3	103.8	84.2	96.1	104.9	82.3	96.0
26, 27, 3370	ELECTRICAL, COMMUNICATIONS & UTIL.	115.0	93.9	104.6	105.4	77.5	91.7	112.2	69.7	91.2	110.2	65.2	88.0	110.9	65.2	88.4	113.3	90.6	102.1
MF2018	WEIGHTED AVERAGE	121.8	95.9	110.9	121.9	88.4	107.7	119.3	69.0	98.0	110.3	79.8	97.4	110.5	79.7	97.5	117.9	88.0	105.2

CANADA

DIVISION		HAMILTON, ONTARIO			HULL, QUEBEC			JOLIETTE, QUEBEC			KAMLOOPS, BRITISH COLUMBIA			KINGSTON, ONTARIO			KITCHENER, ONTARIO		
		MAT.	INST.	TOTAL	MAT.	INST.	TOTAL	MAT.	INST.	TOTAL	MAT.	INST.	TOTAL	MAT.	INST.	TOTAL	MAT.	INST.	TOTAL
015433	CONTRACTOR EQUIPMENT		115.5	115.5		100.2	100.2		100.2	100.2		103.3	103.3		101.9	101.9		101.6	101.6
0241, 31 - 34	SITE & INFRASTRUCTURE, DEMOLITION	105.8	107.0	106.6	96.7	93.7	94.6	97.3	93.7	94.8	120.1	97.1	104.3	115.1	98.0	103.3	94.2	99.8	98.0
0310	Concrete Forming & Accessories	130.0	94.7	99.9	132.9	79.1	87.0	132.9	79.2	87.1	124.3	82.3	88.5	125.1	82.5	88.7	117.3	88.0	92.3
0320	Concrete Reinforcing	136.9	103.3	120.7	147.4	72.4	111.1	139.3	72.4	107.0	109.3	75.8	93.1	163.7	85.3	125.8	96.4	103.1	99.6
0330	Cast-in-Place Concrete	107.6	102.1	105.6	104.0	87.5	97.9	108.4	87.5	100.6	93.8	91.8	93.1	117.5	92.8	108.3	111.7	97.2	106.3
03	CONCRETE	110.2	99.3	105.4	107.9	81.4	96.2	108.8	81.5	96.8	117.2	85.0	103.0	117.5	87.0	104.1	97.1	94.1	95.8
04	MASONRY	173.2	100.5	128.5	162.3	76.3	109.5	162.7	76.3	109.6	169.4	84.0	116.9	171.9	85.7	118.9	142.2	98.6	115.4
05	METALS	134.7	106.4	126.4	110.9	84.8	103.2	110.9	85.0	103.3	112.8	87.1	105.3	113.5	90.7	106.8	124.1	98.0	116.5
06	WOOD, PLASTICS & COMPOSITES	111.4	94.0	101.8	131.6	79.0	102.7	131.6	79.0	102.7	102.2	80.7	90.3	118.5	81.7	98.3	110.3	85.9	96.9
07	THERMAL & MOISTURE PROTECTION	134.0	101.5	119.8	114.8	84.1	101.5	114.8	84.1	101.5	131.2	81.7	109.7	121.4	87.3	106.6	116.4	98.4	108.5
08	OPENINGS	85.3	93.2	87.1	89.1	67.9	84.2	89.1	72.3	85.2	86.1	78.5	84.3	89.1	80.3	87.1	80.0	87.1	81.7
0920	Plaster & Gypsum Board	131.0	93.1	105.5	114.4	78.4	90.2	147.6	78.4	101.0	101.1	79.8	86.8	177.6	81.2	112.8	107.2	85.6	92.6
0950, 0980	Ceilings & Acoustic Treatment	139.9	93.1	108.4	103.5	78.4	86.6	103.5	78.4	86.6	103.5	79.8	87.5	119.0	81.2	93.5	107.9	85.6	92.9
0960	Flooring	115.3	95.0	109.6	113.6	86.0	105.8	113.6	86.0	105.8	112.5	49.2	94.7	113.6	86.0	105.8	101.3	95.1	99.6
0970, 0990	Wall Finishes & Painting/Coating	113.0	104.1	107.8	110.3	83.6	94.5	110.3	83.6	94.5	110.3	76.4	90.3	110.3	79.2	91.9	104.4	93.9	98.2
09	FINISHES	121.0	95.6	107.3	107.8	81.2	93.4	112.2	81.2	95.4	108.3	75.8	90.7	119.9	82.9	99.9	102.1	89.4	95.2
COVERS	DIVS. 10 - 14, 25, 28, 41, 43, 44, 46	131.1	92.5	122.5	131.1	75.6	118.8	131.1	75.6	118.8	131.1	84.4	120.7	131.1	64.6	116.3	131.1	89.6	121.9
21, 22, 23	FIRE SUPPRESSION, PLUMBING & HVAC	105.2	92.1	100.0	103.8	84.2	96.1	103.8	84.2	96.1	103.8	87.5	97.4	104.2	94.4	100.4	104.2	91.3	99.1
26, 27, 3370	ELECTRICAL, COMMUNICATIONS & UTIL.	108.5	103.6	106.1	112.3	65.2	89.1	110.9	65.2	88.4	114.3	74.2	94.5	111.6	82.1	97.1	109.6	99.8	104.8
MF2018	WEIGHTED AVERAGE	116.3	98.8	108.9	110.5	79.8	97.5	110.8	80.0	97.8	113.0	83.5	100.5	114.1	87.4	102.8	108.6	94.6	102.7

CANADA

DIVISION		LAVAL, QUEBEC			LETHBRIDGE, ALBERTA			LLOYDMINSTER, ALBERTA			LONDON, ONTARIO			MEDICINE HAT, ALBERTA			MONCTON, NEW BRUNSWICK		
		MAT.	INST.	TOTAL	MAT.	INST.	TOTAL	MAT.	INST.	TOTAL	MAT.	INST.	TOTAL	MAT.	INST.	TOTAL	MAT.	INST.	TOTAL
015433	CONTRACTOR EQUIPMENT		100.2	100.2		101.9	101.9		101.9	101.9		116.7	116.7		101.9	101.9		99.7	99.7
0241, 31 - 34	SITE & INFRASTRUCTURE, DEMOLITION	97.2	93.9	94.9	114.5	96.0	101.7	114.5	95.4	101.3	103.5	106.8	105.8	113.2	95.5	101.0	104.5	92.5	96.2
0310	Concrete Forming & Accessories	133.2	81.1	88.8	126.4	87.6	93.3	124.6	78.3	85.1	129.1	89.1	95.0	126.4	78.2	85.3	106.1	66.1	72.0
0320	Concrete Reinforcing	147.4	74.4	112.1	151.5	81.9	117.8	151.5	81.8	117.8	126.7	102.1	114.8	151.5	81.8	117.8	140.5	69.8	106.3
0330	Cast-in-Place Concrete	107.5	89.4	100.7	130.1	98.2	118.2	120.7	94.6	111.0	120.2	100.8	113.0	120.7	94.6	111.0	109.8	68.0	94.3
03	CONCRETE	109.5	83.4	98.0	120.0	90.6	107.1	115.5	85.2	102.2	114.6	96.2	106.5	115.7	85.2	102.3	108.4	68.5	90.9
04	MASONRY	162.6	78.4	110.8	180.1	86.3	122.4	161.6	80.1	111.6	183.8	99.5	132.0	161.6	80.1	111.6	159.4	74.6	107.3
05	METALS	110.7	85.9	103.5	134.1	91.0	121.5	112.8	90.8	106.4	134.7	106.8	126.5	113.0	90.8	106.5	114.1	86.8	106.1
06	WOOD, PLASTICS & COMPOSITES	131.7	81.1	103.9	118.0	86.7	100.8	114.5	77.4	94.1	114.8	86.8	99.4	118.0	77.4	95.7	97.5	65.0	79.6
07	THERMAL & MOISTURE PROTECTION	115.4	86.0	102.6	127.1	92.6	112.1	123.5	88.2	108.1	129.9	99.0	116.5	130.5	88.2	112.1	116.6	72.4	97.4
08	OPENINGS	89.1	69.8	84.6	89.1	79.5	86.9	89.1	74.4	85.7	81.6	87.9	83.1	89.1	74.4	85.7	84.0	61.9	78.8
0920	Plaster & Gypsum Board	114.7	80.5	91.7	106.4	86.1	92.7	102.1	76.6	84.9	134.4	85.8	101.7	104.3	76.6	85.6	124.2	64.1	83.7
0950, 0980	Ceilings & Acoustic Treatment	103.5	80.5	88.0	112.5	86.1	94.7	103.5	76.6	85.3	138.5	85.8	102.9	103.5	76.6	85.3	116.8	64.1	81.2
0960	Flooring	113.6	88.4	106.5	113.6	84.8	105.5	113.6	84.8	105.5	109.9	95.1	105.7	113.6	84.8	105.5	99.6	65.6	90.1
0970, 0990	Wall Finishes & Painting/Coating	110.3	85.9	95.8	110.2	96.4	102.1	110.4	75.1	89.5	113.5	100.5	105.8	110.2	75.1	89.4	111.3	83.6	94.9
09	FINISHES	107.8	83.3	94.5	109.6	88.3	98.0	107.4	78.9	92.0	121.5	91.0	105.0	107.5	78.9	92.0	107.7	67.5	86.0
COVERS	DIVS. 10 - 14, 25, 28, 41, 43, 44, 46	131.1	77.5	119.2	131.1	89.2	121.8	131.1	87.1	121.3	131.1	91.6	122.3	131.1	86.1	121.1	131.1	61.0	115.5
21, 22, 23	FIRE SUPPRESSION, PLUMBING & HVAC	104.3	86.5	97.3	104.1	90.1	98.6	104.2	90.0	98.6	105.1	89.2	98.8	103.8	86.8	97.2	103.8	73.7	92.0
26, 27, 3370	ELECTRICAL, COMMUNICATIONS & UTIL.	110.9	67.0	89.3	107.1	77.5	92.5	104.6	77.5	91.2	103.5	102.0	102.8	104.6	77.5	91.2	117.3	88.5	103.1
MF2018	WEIGHTED AVERAGE	110.7	81.6	98.4	117.0	87.9	104.7	111.5	84.7	100.2	116.3	96.5	108.0	111.7	84.0	100.0	110.9	76.0	96.2

CANADA

DIVISION		MONTREAL, QUEBEC			MOOSE JAW, SASKATCHEWAN			NEW GLASGOW, NOVA SCOTIA			NEWCASTLE, NEW BRUNSWICK			NORTH BAY, ONTARIO			OSHAWA, ONTARIO		
		MAT.	INST.	TOTAL	MAT.	INST.	TOTAL	MAT.	INST.	TOTAL	MAT.	INST.	TOTAL	MAT.	INST.	TOTAL	MAT.	INST.	TOTAL
015433	CONTRACTOR EQUIPMENT		117.7	117.7		98.3	98.3		98.9	98.9		99.7	99.7		99.3	99.3		101.6	101.6
0241, 31 - 34	SITE & INFRASTRUCTURE, DEMOLITION	113.0	105.5	107.8	114.4	90.0	97.5	111.9	92.0	98.1	105.1	90.6	95.1	126.8	93.6	103.9	104.6	99.6	101.2
0310	Concrete Forming & Accessories	130.9	90.4	96.4	109.6	54.6	62.7	111.8	67.8	74.3	106.1	58.0	65.1	146.2	80.0	89.8	122.8	91.3	96.0
0320	Concrete Reinforcing	126.8	94.6	111.3	106.9	61.4	84.9	156.0	47.6	103.6	140.5	57.6	100.4	184.7	84.8	136.4	152.7	103.7	129.0
0330	Cast-in-Place Concrete	124.4	97.4	114.3	117.1	63.7	97.2	122.9	66.8	102.0	114.0	56.3	92.6	118.4	80.1	104.1	129.2	106.4	120.7
03	CONCRETE	116.7	94.3	106.8	101.4	60.0	83.2	131.0	65.0	102.0	110.3	58.4	87.5	132.7	81.4	110.2	118.3	98.8	109.8
04	MASONRY	178.8	89.8	124.1	160.3	56.4	96.4	208.2	65.6	120.5	159.8	57.4	96.9	214.4	81.9	133.0	145.1	101.9	118.6
05	METALS	144.3	102.2	131.9	109.6	75.0	99.4	125.6	75.7	111.0	114.1	73.8	102.3	126.5	90.2	115.8	114.2	98.6	109.6
06	WOOD, PLASTICS & COMPOSITES	111.3	90.8	100.0	99.4	53.2	74.0	115.9	67.4	89.2	97.5	57.9	75.7	153.5	80.2	113.2	117.6	89.4	102.1
07	THERMAL & MOISTURE PROTECTION	125.9	97.2	113.4	114.1	60.1	90.6	130.1	67.6	102.9	116.6	58.4	91.3	136.4	82.7	113.0	117.3	104.1	111.6
08	OPENINGS	87.4	78.7	85.4	85.3	51.4	77.4	91.3	61.3	84.3	84.0	51.2	76.3	98.6	78.2	93.8	85.0	90.2	86.2
0920	Plaster & Gypsum Board	126.2	89.9	101.8	98.2	51.9	67.1	145.0	66.5	92.2	124.2	56.7	78.8	140.6	79.8	99.7	110.7	89.2	96.2
0950, 0980	Ceilings & Acoustic Treatment	146.1	89.9	108.2	103.5	51.9	68.7	123.8	66.5	85.2	116.8	56.7	76.3	123.8	79.8	94.1	103.9	89.2	94.0
0960	Flooring	112.4	90.5	106.2	103.6	53.7	89.6	107.6	58.9	93.9	99.6	63.4	89.5	131.2	86.0	118.5	104.2	97.6	102.4
0970, 0990	Wall Finishes & Painting/Coating	115.0	104.1	108.5	110.3	61.0	81.1	117.8	59.3	83.2	111.3	47.9	73.8	117.8	85.0	98.4	104.4	108.1	106.6
09	FINISHES	121.6	92.8	106.0	103.7	54.8	77.2	118.7	65.5	89.9	107.7	58.2	80.9	125.9	81.9	102.1	103.2	93.7	98.1
COVERS	DIVS. 10 - 14, 25, 28, 41, 43, 44, 46	131.1	82.8	120.4	131.1	57.5	114.7	131.1	60.8	115.5	131.1	57.9	114.8	131.1	63.4	116.0	131.1	90.4	122.0
21, 22, 23	FIRE SUPPRESSION, PLUMBING & HVAC	105.1	81.2	95.7	104.2	70.3	90.9	104.0	79.5	94.4	103.8	64.8	88.5	104.0	92.4	99.4	104.2	91.9	99.3
26, 27, 3370	ELECTRICAL, COMMUNICATIONS & UTIL.	107.9	84.1	96.2	113.4	56.3	85.3	111.3	58.7	85.4	112.9	55.9	84.8	112.6	83.3	98.1	110.6	103.8	107.3
MF2018	WEIGHTED AVERAGE	119.1	90.1	106.9	109.0	63.9	89.9	119.6	70.2	98.7	110.7	62.9	90.5	122.5	85.2	106.8	110.8	97.2	105.0

CANADA

DIVISION		OTTAWA, ONTARIO			OWEN SOUND, ONTARIO			PETERBOROUGH, ONTARIO			PORTAGE LA PRAIRIE, MANITOBA			PRINCE ALBERT, SASKATCHEWAN			PRINCE GEORGE, BRITISH COLUMBIA		
		MAT.	INST.	TOTAL	MAT.	INST.	TOTAL	MAT.	INST.	TOTAL	MAT.	INST.	TOTAL	MAT.	INST.	TOTAL	MAT.	INST.	TOTAL
015433	CONTRACTOR EQUIPMENT		117.8	117.8		100.1	100.1		99.7	99.7		101.9	101.9		98.3	98.3		103.3	103.3
0241, 31 - 34	SITE & INFRASTRUCTURE, DEMOLITION	107.0	107.2	107.1	117.0	94.6	101.5	116.9	94.2	101.3	115.4	93.2	100.1	110.2	90.1	96.3	123.4	97.1	105.3
0310	Concrete Forming & Accessories	124.8	88.4	93.7	125.6	78.7	85.6	127.2	81.0	87.8	126.6	65.5	74.7	109.6	54.4	62.5	115.1	77.5	83.0
0320	Concrete Reinforcing	137.0	102.0	120.1	175.2	86.9	132.5	163.7	85.3	125.8	151.5	54.5	104.6	111.7	61.4	87.3	109.3	75.8	93.1
0330	Cast-in-Place Concrete	122.3	103.0	115.1	157.7	76.6	127.5	130.7	81.8	112.5	120.7	69.8	101.8	106.2	63.6	90.3	117.6	91.8	108.0
03	CONCRETE	116.6	96.7	107.9	138.9	79.9	113.0	121.8	82.5	104.6	109.5	66.2	90.5	96.9	59.9	80.7	127.5	82.8	107.9
04	MASONRY	173.2	99.4	127.8	165.6	87.1	117.3	166.1	88.0	118.1	164.8	59.7	100.2	159.5	56.4	96.1	171.4	84.0	117.7
05	METALS	135.5	108.9	127.7	111.1	91.3	105.3	112.3	90.8	106.0	113.0	78.4	102.8	109.7	74.8	99.4	112.8	87.2	105.3
06	WOOD, PLASTICS & COMPOSITES	108.4	85.7	95.9	116.5	77.3	94.9	120.3	79.0	97.6	118.0	66.8	89.8	99.4	53.2	74.0	102.2	74.1	86.7
07	THERMAL & MOISTURE PROTECTION	137.3	99.7	121.0	115.3	83.5	101.5	121.6	88.0	107.0	114.6	68.2	94.4	114.0	59.0	90.1	124.5	81.0	105.6
08	OPENINGS	90.3	87.5	89.7	90.5	77.6	87.5	87.7	79.9	85.9	89.1	59.9	82.3	84.3	51.4	76.6	86.1	74.9	83.5
0920	Plaster & Gypsum Board	129.1	84.6	99.1	153.5	76.7	101.8	115.8	78.5	90.7	104.0	65.6	78.2	98.2	51.9	67.1	101.1	73.0	82.2
0950, 0980	Ceilings & Acoustic Treatment	142.3	84.6	103.4	94.3	76.7	82.4	103.5	78.5	86.7	103.5	65.6	78.0	103.5	51.9	68.7	103.5	73.0	83.0
0960	Flooring	108.3	90.6	103.4	119.3	87.3	110.3	113.6	86.0	105.8	113.6	61.9	99.1	103.6	53.7	89.6	109.0	67.4	97.3
0970, 0990	Wall Finishes & Painting/Coating	113.5	95.9	103.1	105.6	83.8	92.7	110.3	87.1	96.6	110.4	54.1	77.1	110.3	52.0	75.8	107.2	75.0	90.3
09	FINISHES	122.3	89.0	104.3	111.3	80.8	94.8	108.7	82.6	94.5	107.4	64.4	84.0	103.5	53.8	76.7	112.8	76.1	93.0
COVERS	DIVS. 10 - 14, 25, 28, 41, 43, 44, 46	131.1	89.5	121.9	139.2	63.9	122.4	131.1	64.8	116.3	131.1	59.8	115.2	131.1	57.5	114.7	131.1	83.7	120.6
21, 22, 23	FIRE SUPPRESSION, PLUMBING & HVAC	105.1	90.7	99.4	103.7	92.4	99.3	103.8	95.8	100.7	103.8	78.4	93.8	104.2	63.4	88.2	103.8	87.5	97.4
26, 27, 3370	ELECTRICAL, COMMUNICATIONS & UTIL.	104.4	101.9	103.2	118.5	82.1	100.6	111.5	82.9	97.4	113.0	54.4	84.2	113.4	56.3	85.3	111.5	74.2	93.1
MF2018	WEIGHTED AVERAGE	117.5	96.8	108.7	116.7	85.4	103.4	113.1	87.1	102.1	111.6	69.1	93.6	108.2	62.2	88.8	114.0	82.8	100.8

For customer support on your Mechanical Costs with RSMeans Data, call 800.448.8182.

697

City Cost Indexes

CANADA

DIVISION		QUEBEC CITY, QUEBEC			RED DEER, ALBERTA			REGINA, SASKATCHEWAN			RIMOUSKI, QUEBEC			ROUYN-NORANDA, QUEBEC			SAINT HYACINTHE, QUEBEC		
		MAT.	INST.	TOTAL	MAT.	INST.	TOTAL	MAT.	INST.	TOTAL	MAT.	INST.	TOTAL	MAT.	INST.	TOTAL	MAT.	INST.	TOTAL
015433	CONTRACTOR EQUIPMENT		117.4	117.4		101.9	101.9		128.3	128.3		100.2	100.2		100.2	100.2		100.2	100.2
0241, 31 - 34	SITE & INFRASTRUCTURE, DEMOLITION	114.1	104.7	107.6	113.2	95.5	101.0	126.4	119.6	121.7	97.0	93.6	94.6	96.7	93.7	94.6	97.2	93.7	94.8
0310	Concrete Forming & Accessories	127.4	90.5	95.9	139.7	78.2	87.3	131.0	91.5	97.4	132.9	89.8	96.2	132.9	79.1	87.0	132.9	79.1	87.0
0320	Concrete Reinforcing	124.2	94.6	109.9	151.5	81.8	117.8	123.8	88.8	106.9	105.1	94.5	100.0	147.4	72.4	111.1	147.4	72.4	111.1
0330	Cast-in-Place Concrete	131.6	97.0	118.7	120.7	94.6	111.0	153.8	97.7	132.9	109.4	94.7	103.9	104.0	87.5	97.9	107.5	87.5	100.0
03	CONCRETE	119.4	94.3	108.4	116.5	85.2	102.7	131.3	94.0	114.9	105.0	92.4	99.5	107.9	81.4	96.2	109.5	81.4	97.1
04	MASONRY	173.6	89.8	122.1	161.6	80.1	111.6	205.2	86.7	132.4	162.1	89.8	117.6	162.3	76.3	109.5	162.6	76.3	109.6
05	METALS	138.0	103.3	127.8	113.0	90.8	106.5	139.1	101.4	128.0	110.4	91.8	104.9	110.9	84.8	103.2	110.9	84.8	103.2
06	WOOD, PLASTICS & COMPOSITES	114.2	90.7	101.3	118.0	77.4	95.7	108.0	92.2	99.3	131.6	90.3	108.9	131.6	79.0	102.7	131.6	79.0	102.7
07	THERMAL & MOISTURE PROTECTION	125.1	97.2	112.9	141.4	88.2	118.3	143.8	86.6	118.9	114.8	96.4	106.8	114.8	84.1	101.5	115.2	84.1	101.7
08	OPENINGS	88.4	86.0	87.8	89.1	74.4	85.7	87.9	79.8	86.1	88.7	76.3	85.8	89.1	67.9	84.2	89.1	67.9	84.2
0920	Plaster & Gypsum Board	128.5	89.9	102.6	104.3	76.6	85.6	140.8	91.2	107.4	147.4	89.9	108.7	114.2	78.4	90.1	114.2	78.4	90.1
0950, 0980	Ceilings & Acoustic Treatment	131.7	89.9	103.5	103.5	76.6	85.3	154.3	91.2	111.7	102.7	89.9	94.1	102.7	78.4	86.3	102.7	78.4	86.3
0960	Flooring	108.9	90.5	103.8	116.0	84.8	107.2	123.6	95.1	115.6	114.8	86.0	106.7	113.6	86.0	105.8	113.6	86.0	105.8
0970, 0990	Wall Finishes & Painting/Coating	120.8	104.1	111.0	110.2	75.1	89.4	118.0	89.0	100.9	113.4	104.1	107.9	110.3	83.6	94.5	110.3	83.6	94.5
09	FINISHES	117.4	92.7	104.0	108.2	78.9	92.4	130.9	92.9	110.3	112.6	91.6	101.2	107.6	81.2	93.3	107.6	81.2	93.3
COVERS	DIVS. 10 - 14, 25, 28, 41, 43, 44, 46	131.1	82.5	120.3	131.1	86.1	121.1	131.1	70.9	117.7	131.1	81.5	120.1	131.1	75.6	118.8	131.1	75.6	118.8
21, 22, 23	FIRE SUPPRESSION, PLUMBING & HVAC	105.1	81.2	95.7	103.8	86.8	97.2	104.8	86.9	97.8	103.8	81.0	94.9	103.8	84.2	96.1	100.5	84.4	94.1
26, 27, 3370	ELECTRICAL, COMMUNICATIONS & UTIL.	114.7	84.1	99.6	104.6	77.5	91.2	110.8	91.5	101.3	110.9	84.1	97.7	110.9	65.2	88.4	111.5	65.2	88.7
MF2018	WEIGHTED AVERAGE	118.7	90.4	106.7	112.1	84.0	100.2	123.3	92.4	110.2	110.2	87.6	100.6	110.3	79.8	97.4	109.8	79.8	97.1

CANADA

DIVISION		SAINT JOHN, NEW BRUNSWICK			SARNIA, ONTARIO			SASKATOON, SASKATCHEWAN			SAULT STE MARIE, ONTARIO			SHERBROOKE, QUEBEC			SOREL, QUEBEC		
		MAT.	INST.	TOTAL	MAT.	INST.	TOTAL	MAT.	INST.	TOTAL	MAT.	INST.	TOTAL	MAT.	INST.	TOTAL	MAT.	INST.	TOTAL
015433	CONTRACTOR EQUIPMENT		99.7	99.7		99.7	99.7		98.4	98.4		99.7	99.7		100.2	100.2		100.2	100.2
0241, 31 - 34	SITE & INFRASTRUCTURE, DEMOLITION	105.2	92.5	96.4	115.5	94.3	100.9	112.7	92.8	98.9	105.6	93.9	97.5	97.2	93.7	94.8	97.3	93.7	94.8
0310	Concrete Forming & Accessories	126.2	63.4	72.7	125.9	87.6	93.3	109.7	90.7	93.5	115.1	86.2	90.5	132.9	79.1	87.0	132.9	79.2	87.1
0320	Concrete Reinforcing	140.5	69.8	106.3	116.5	86.7	102.1	114.2	88.7	101.9	105.1	85.4	95.6	147.4	72.4	111.1	139.3	72.4	107.0
0330	Cast-in-Place Concrete	112.4	68.0	95.9	120.6	94.2	110.8	113.5	93.1	105.9	108.3	80.5	98.0	107.5	87.5	100.0	108.4	87.5	100.6
03	CONCRETE	110.9	67.3	91.7	111.1	90.0	101.8	101.3	91.2	96.9	97.4	84.7	91.8	109.5	81.4	97.1	108.8	81.5	96.8
04	MASONRY	180.8	74.4	115.4	177.6	90.5	124.1	172.1	86.6	119.5	162.9	90.1	118.2	162.6	76.3	109.6	162.7	76.3	109.6
05	METALS	114.0	86.8	106.0	112.3	91.2	106.1	106.5	88.6	101.2	111.5	95.2	106.7	110.7	84.8	103.1	110.9	85.0	103.3
06	WOOD, PLASTICS & COMPOSITES	120.7	61.4	88.0	119.4	86.7	101.4	96.0	91.3	93.4	106.5	87.9	96.3	131.6	79.0	102.7	131.6	79.0	102.7
07	THERMAL & MOISTURE PROTECTION	116.9	72.2	97.5	121.7	91.5	108.6	117.0	84.7	103.0	120.4	87.3	106.0	114.8	84.1	101.5	114.8	84.1	101.5
08	OPENINGS	83.9	58.9	78.0	90.5	83.5	88.8	85.3	79.4	83.9	82.4	86.1	83.2	89.1	67.9	84.2	89.1	72.3	85.2
0920	Plaster & Gypsum Board	137.9	60.3	85.7	143.2	86.4	105.0	115.6	91.2	99.1	107.1	87.6	94.0	114.2	78.4	90.1	147.4	78.4	101.0
0950, 0980	Ceilings & Acoustic Treatment	121.7	60.3	80.3	109.2	86.4	93.9	124.5	91.2	102.0	103.5	87.6	92.8	102.7	78.4	86.3	102.7	78.4	86.3
0960	Flooring	111.4	65.6	98.5	113.6	94.1	108.1	106.5	95.1	103.3	106.5	92.1	102.4	113.6	86.0	105.8	113.6	86.0	105.8
0970, 0990	Wall Finishes & Painting/Coating	111.3	63.4	82.9	110.3	98.5	103.3	113.8	89.0	99.2	110.3	91.3	99.1	110.3	83.6	94.5	110.3	83.6	94.5
09	FINISHES	114.0	65.4	87.7	113.5	90.2	100.9	112.1	92.3	101.4	104.7	88.2	95.8	107.6	81.2	93.3	112.0	81.2	95.3
COVERS	DIVS. 10 - 14, 25, 28, 41, 43, 44, 46	131.1	61.0	115.5	131.1	66.1	116.6	131.1	69.3	117.4	131.1	87.9	121.5	131.1	75.6	118.8	131.1	75.6	118.8
21, 22, 23	FIRE SUPPRESSION, PLUMBING & HVAC	103.8	75.1	92.6	103.8	101.5	102.9	104.4	86.7	97.5	103.8	90.5	98.6	104.2	84.2	96.3	103.8	84.2	96.1
26, 27, 3370	ELECTRICAL, COMMUNICATIONS & UTIL.	119.9	88.5	104.4	114.7	85.0	100.0	114.6	91.4	103.2	113.2	83.3	98.5	110.9	65.2	88.4	110.9	65.2	88.4
MF2018	WEIGHTED AVERAGE	113.2	75.7	97.3	113.2	91.3	103.9	109.9	88.5	100.9	108.6	88.7	100.2	110.6	79.8	97.6	110.8	80.0	97.8

CANADA

DIVISION		ST. CATHARINES, ONTARIO			ST JEROME, QUEBEC			ST. JOHN'S, NEWFOUNDLAND			SUDBURY, ONTARIO			SUMMERSIDE, PRINCE EDWARD ISLAND			SYDNEY, NOVA SCOTIA		
		MAT.	INST.	TOTAL	MAT.	INST.	TOTAL	MAT.	INST.	TOTAL	MAT.	INST.	TOTAL	MAT.	INST.	TOTAL	MAT.	INST.	TOTAL
015433	CONTRACTOR EQUIPMENT		99.5	99.5		100.2	100.2		123.4	123.4		99.5	99.5		98.8	98.8		98.9	98.9
0241, 31 - 34	SITE & INFRASTRUCTURE, DEMOLITION	94.2	96.1	95.6	96.7	93.7	94.6	115.4	111.8	112.9	94.4	95.7	95.3	121.8	89.4	99.5	108.0	92.0	96.9
0310	Concrete Forming & Accessories	115.3	93.8	97.0	132.9	79.1	87.0	125.3	85.2	91.1	111.0	88.9	92.1	112.0	52.5	61.3	111.8	67.8	74.3
0320	Concrete Reinforcing	97.2	103.2	100.1	147.4	72.4	111.1	172.1	82.4	128.7	98.0	101.4	99.7	154.0	46.5	102.0	156.0	47.6	103.6
0330	Cast-in-Place Concrete	106.7	98.7	103.7	104.0	87.5	97.9	137.1	98.5	122.8	107.6	95.3	103.0	116.7	54.7	93.6	94.8	66.8	84.4
03	CONCRETE	94.8	97.2	95.8	107.9	81.4	96.2	132.7	90.3	114.1	95.0	93.5	94.3	139.1	53.4	101.5	118.1	65.0	94.8
04	MASONRY	141.8	100.8	116.6	162.3	76.3	109.5	200.1	89.9	132.4	141.9	94.9	113.0	207.4	54.8	113.7	205.8	65.6	119.6
05	METALS	114.2	97.9	109.4	110.9	84.8	103.2	137.9	99.1	126.5	113.6	97.1	108.8	125.6	69.2	109.1	125.6	75.7	111.0
06	WOOD, PLASTICS & COMPOSITES	107.8	93.2	99.8	131.6	79.0	102.7	106.9	83.4	94.0	103.5	87.9	94.9	116.4	51.9	80.9	115.9	67.4	89.2
07	THERMAL & MOISTURE PROTECTION	116.4	102.0	110.1	114.8	84.1	101.5	140.2	94.7	120.4	115.7	96.3	107.3	129.4	57.3	98.0	130.1	67.6	102.9
08	OPENINGS	79.5	91.4	82.3	89.1	67.9	84.2	85.4	74.3	82.8	80.2	86.7	81.7	102.5	45.4	89.2	91.3	61.3	84.3
0920	Plaster & Gypsum Board	100.8	93.1	95.6	114.2	78.4	90.1	150.0	82.1	104.3	100.0	87.6	91.7	145.9	50.5	81.8	145.0	66.5	92.2
0950, 0980	Ceilings & Acoustic Treatment	103.9	93.1	96.6	102.7	78.4	86.3	143.7	82.1	102.2	98.2	87.6	91.0	123.8	50.5	74.4	123.8	66.5	85.2
0960	Flooring	100.2	91.9	97.8	113.6	86.0	105.8	116.8	52.2	98.7	98.3	92.1	96.6	107.6	55.3	92.9	107.6	58.9	93.9
0970, 0990	Wall Finishes & Painting/Coating	104.4	104.1	104.2	110.3	83.6	94.5	120.8	98.6	107.7	104.4	95.3	99.0	117.8	39.9	71.7	117.8	59.3	83.2
09	FINISHES	100.1	94.6	97.1	107.6	81.2	93.3	129.2	80.4	102.8	98.3	89.9	93.7	119.8	51.9	83.1	118.7	65.5	89.9
COVERS	DIVS. 10 - 14, 25, 28, 41, 43, 44, 46	131.1	69.7	117.4	131.1	75.6	118.8	131.1	68.6	117.2	131.1	89.6	121.9	131.1	57.0	114.6	131.1	60.8	115.5
21, 22, 23	FIRE SUPPRESSION, PLUMBING & HVAC	104.2	91.1	99.0	103.8	84.2	96.1	104.8	83.0	96.2	103.6	89.4	98.1	104.0	58.9	86.2	104.0	79.5	94.4
26, 27, 3370	ELECTRICAL, COMMUNICATIONS & UTIL.	111.3	101.8	106.6	111.6	65.2	88.7	115.6	81.4	98.7	109.5	102.1	105.9	110.2	47.4	79.2	111.3	58.7	85.4
MF2018	WEIGHTED AVERAGE	106.6	95.6	101.9	110.4	79.8	97.4	122.6	87.4	107.7	106.1	93.6	100.8	121.9	57.9	94.9	117.8	70.2	97.7

698

City Cost Indexes

CANADA

	DIVISION	THUNDER BAY, ONTARIO			TIMMINS, ONTARIO			TORONTO, ONTARIO			TROIS RIVIERES, QUEBEC			TRURO, NOVA SCOTIA			VANCOUVER, BRITISH COLUMBIA		
		MAT.	INST.	TOTAL	MAT.	INST.	TOTAL	MAT.	INST.	TOTAL	MAT.	INST.	TOTAL	MAT.	INST.	TOTAL	MAT.	INST.	TOTAL
015433	CONTRACTOR EQUIPMENT		99.5	99.5		99.7	99.7		117.5	117.5		99.7	99.7		99.4	99.4		135.3	135.3
0241, 31 - 34	SITE & INFRASTRUCTURE, DEMOLITION	98.8	96.0	96.9	116.9	93.9	101.0	108.4	107.7	107.9	108.2	93.4	97.9	101.3	92.3	95.1	117.0	122.8	121.0
0310	Concrete Forming & Accessories	122.8	92.5	97.0	127.2	80.0	87.0	128.7	102.5	106.3	153.7	79.1	90.1	99.7	67.9	72.6	131.0	91.6	97.4
0320	Concrete Reinforcing	86.9	102.6	94.5	163.7	84.8	125.6	136.9	103.9	120.9	156.0	72.4	115.6	139.3	47.6	95.0	132.5	95.2	114.5
0330	Cast-in-Place Concrete	117.5	98.0	110.2	130.7	80.2	111.9	108.4	113.7	110.4	98.2	87.4	94.2	135.7	66.9	110.1	123.5	90.6	111.3
03	CONCRETE	101.9	96.2	99.4	121.8	81.5	104.1	110.5	107.1	109.0	120.3	81.4	103.2	120.8	65.1	96.3	119.7	93.0	108.0
04	MASONRY	142.4	100.5	116.7	166.1	81.9	114.4	173.3	108.7	133.6	209.8	76.3	127.8	161.9	65.6	102.7	165.3	85.6	116.3
05	METALS	114.2	97.0	109.1	112.3	90.4	105.9	135.3	109.4	127.7	124.9	84.7	113.1	111.6	76.0	101.1	133.5	109.8	126.6
06	WOOD, PLASTICS & COMPOSITES	117.6	91.3	103.1	120.3	80.3	98.3	114.8	100.5	107.0	168.2	79.0	119.1	89.6	67.4	77.4	110.3	91.0	99.7
07	THERMAL & MOISTURE PROTECTION	116.6	99.3	109.1	121.6	82.7	104.7	135.6	109.4	124.2	128.6	84.1	109.2	116.1	67.6	95.0	141.7	88.6	118.6
08	OPENINGS	78.9	89.8	81.4	87.7	78.2	85.5	83.9	98.9	87.4	100.3	72.3	93.7	82.4	61.3	77.5	84.5	88.7	85.5
0920	Plaster & Gypsum Board	128.0	91.2	103.2	115.8	79.8	91.6	129.1	99.9	109.5	174.3	78.4	109.8	124.3	66.5	85.4	127.2	89.7	102.0
0950, 0980	Ceilings & Acoustic Treatment	98.2	91.2	93.4	103.5	79.8	87.5	139.9	99.9	112.9	122.2	78.4	92.6	103.5	66.5	78.6	151.0	89.7	109.7
0960	Flooring	104.2	98.5	102.6	113.6	86.0	105.8	109.1	100.8	106.8	131.2	86.0	118.5	95.7	58.9	85.4	124.7	88.5	114.6
0970, 0990	Wall Finishes & Painting/Coating	104.4	96.2	99.5	110.3	85.0	95.3	109.8	108.1	108.8	117.8	83.6	97.6	110.3	59.3	80.1	113.7	100.3	105.8
09	FINISHES	104.1	93.9	98.6	108.7	81.9	94.2	117.8	102.8	109.7	129.3	81.1	103.2	103.8	65.5	83.1	128.0	92.1	108.6
COVERS	DIVS. 10 - 14, 25, 28, 41, 43, 44, 46	131.1	69.7	117.4	131.1	63.4	116.0	131.1	95.4	123.2	131.1	75.6	118.8	131.1	60.8	115.5	131.1	90.8	122.1
21, 22, 23	FIRE SUPPRESSION, PLUMBING & HVAC	104.2	91.2	99.1	103.8	92.4	99.3	105.1	99.6	103.0	104.0	84.2	96.2	103.8	79.5	94.3	105.2	87.4	98.2
26, 27, 3370	ELECTRICAL, COMMUNICATIONS & UTIL.	109.6	101.2	105.4	113.2	83.3	98.5	106.8	103.8	105.3	111.6	65.2	88.8	110.5	58.7	85.0	108.7	78.6	93.9
MF2018	WEIGHTED AVERAGE	107.8	95.0	102.4	113.3	85.3	101.4	116.0	104.2	111.0	120.2	79.9	103.2	110.8	70.3	93.7	117.9	92.4	107.2

CANADA

	DIVISION	VICTORIA, BRITISH COLUMBIA			WHITEHORSE, YUKON			WINDSOR, ONTARIO			WINNIPEG, MANITOBA			YARMOUTH, NOVA SCOTIA			YELLOWKNIFE, NWT		
		MAT.	INST.	TOTAL	MAT.	INST.	TOTAL	MAT.	INST.	TOTAL	MAT.	INST.	TOTAL	MAT.	INST.	TOTAL	MAT.	INST.	TOTAL
015433	CONTRACTOR EQUIPMENT		106.3	106.3		132.8	132.8		99.5	99.5		125.5	125.5		98.9	98.9		130.7	130.7
0241, 31 - 34	SITE & INFRASTRUCTURE, DEMOLITION	124.3	101.3	108.4	135.7	119.6	124.6	90.8	96.0	94.4	113.7	115.3	114.8	111.7	92.0	98.1	146.3	123.6	130.6
0310	Concrete Forming & Accessories	114.9	88.9	92.7	135.0	56.7	68.2	122.8	90.6	95.3	130.4	65.4	75.0	111.8	67.8	74.3	137.3	75.8	84.8
0320	Concrete Reinforcing	111.7	95.0	103.6	167.7	63.3	117.2	95.2	101.9	98.5	127.3	61.0	95.3	156.0	47.6	103.6	138.6	65.8	103.4
0330	Cast-in-Place Concrete	117.6	87.0	106.2	152.4	71.2	122.2	109.3	99.7	105.7	143.6	72.0	116.9	121.6	66.8	101.2	174.9	87.6	142.4
03	CONCRETE	129.4	89.5	111.9	144.0	64.8	109.1	96.2	95.9	96.0	127.0	68.4	101.3	130.4	65.0	101.7	150.7	79.5	119.4
04	MASONRY	175.0	85.5	120.0	246.5	58.1	130.7	141.9	100.1	116.3	194.6	65.3	115.2	208.1	65.6	120.5	234.9	68.9	132.9
05	METALS	107.7	92.2	103.2	145.1	90.5	129.1	114.2	97.3	109.2	141.6	88.0	125.8	125.6	75.7	111.0	140.5	92.7	126.4
06	WOOD, PLASTICS & COMPOSITES	100.0	88.3	93.6	121.3	55.3	85.0	117.6	89.0	101.9	107.7	66.0	84.8	115.9	67.4	89.2	130.1	77.1	100.9
07	THERMAL & MOISTURE PROTECTION	126.9	86.3	109.2	145.8	63.3	109.9	116.4	99.1	108.9	136.1	70.3	107.5	130.1	67.6	102.9	143.7	78.8	115.5
08	OPENINGS	86.8	82.9	85.9	99.5	53.0	88.6	78.7	88.9	81.1	87.5	59.7	81.0	91.3	61.3	84.3	93.9	65.5	87.3
0920	Plaster & Gypsum Board	109.1	87.7	94.7	178.3	53.0	94.0	111.4	88.8	96.2	133.0	64.2	86.7	145.0	66.5	92.2	192.2	75.5	113.7
0950, 0980	Ceilings & Acoustic Treatment	105.8	87.7	93.6	169.3	53.0	90.9	98.2	88.8	91.8	145.3	64.2	90.6	123.8	66.5	85.2	169.0	75.5	105.9
0960	Flooring	112.0	67.4	99.5	122.4	55.2	103.5	104.2	95.9	101.9	115.9	68.5	102.6	107.6	58.9	93.9	121.9	82.9	111.0
0970, 0990	Wall Finishes & Painting/Coating	113.8	100.3	105.8	120.7	53.1	80.7	104.4	97.1	100.1	113.9	53.5	78.1	117.8	59.3	83.2	147.5	75.9	105.1
09	FINISHES	110.7	86.8	97.8	143.7	55.8	96.1	101.6	92.1	96.5	126.9	65.0	93.4	118.7	65.5	89.9	147.5	76.6	109.1
COVERS	DIVS. 10 - 14, 25, 28, 41, 43, 44, 46	131.1	65.8	116.6	131.1	60.6	115.4	131.1	69.2	117.3	131.1	63.8	116.1	131.1	60.8	115.5	131.1	63.9	116.1
21, 22, 23	FIRE SUPPRESSION, PLUMBING & HVAC	103.9	88.7	97.9	104.3	71.6	91.5	104.2	90.7	98.9	105.2	64.6	89.2	104.0	79.5	94.4	104.8	88.5	98.4
26, 27, 3370	ELECTRICAL, COMMUNICATIONS & UTIL.	112.8	82.1	97.7	130.5	57.1	94.3	114.5	100.4	107.6	113.1	63.4	88.6	111.3	58.7	85.4	125.8	77.4	102.0
MF2018	WEIGHTED AVERAGE	114.1	87.6	102.9	132.2	69.2	105.6	107.1	94.4	101.8	122.0	71.2	100.5	119.5	70.2	98.7	131.4	83.2	111.0

For customer support on your Mechanical Costs with RSMeans Data, call 800.448.8182.

699

Location Factors - Commercial

Costs shown in RSMeans cost data publications are based on national averages for materials and installation. To adjust these costs to a specific location, simply multiply the base cost by the factor and divide by 100 for that city. The data is arranged alphabetically by state and postal zip code numbers. For a city not listed, use the factor for a nearby city with similar economic characteristics.

STATE/ZIP	CITY	MAT.	INST.	TOTAL
ALABAMA				
350-352	Birmingham	97.4	71.8	86.6
354	Tuscaloosa	97.1	71.4	86.3
355	Jasper	97.8	69.8	86.0
356	Decatur	97.1	69.2	85.3
357-358	Huntsville	97.1	71.7	86.4
359	Gadsden	97.3	70.2	85.9
360-361	Montgomery	96.3	73.1	86.5
362	Anniston	95.9	67.7	84.0
363	Dothan	96.2	72.7	86.3
364	Evergreen	95.8	68.3	84.2
365-366	Mobile	96.7	68.7	84.9
367	Selma	95.9	72.5	86.0
368	Phenix City	96.8	70.8	85.8
369	Butler	96.1	70.6	85.3
ALASKA				
995-996	Anchorage	118.5	113.1	116.2
997	Fairbanks	118.8	113.0	116.3
998	Juneau	116.8	113.1	115.2
999	Ketchikan	128.4	113.4	122.1
ARIZONA				
850,853	Phoenix	100.6	72.3	88.6
851,852	Mesa/Tempe	98.7	71.5	87.2
855	Globe	99.5	71.4	87.6
856-857	Tucson	97.7	70.2	86.1
859	Show Low	99.7	71.5	87.8
860	Flagstaff	102.4	70.5	88.9
863	Prescott	99.9	71.3	87.8
864	Kingman	98.3	70.4	86.5
865	Chambers	98.3	71.8	87.1
ARKANSAS				
716	Pine Bluff	96.0	63.6	82.3
717	Camden	94.0	64.7	81.6
718	Texarkana	94.8	63.3	81.5
719	Hot Springs	93.5	64.4	81.2
720-722	Little Rock	94.5	65.4	82.2
723	West Memphis	93.3	69.8	83.3
724	Jonesboro	93.8	66.6	82.3
725	Batesville	91.6	63.6	79.8
726	Harrison	92.9	62.8	80.2
727	Fayetteville	90.5	64.1	79.3
728	Russellville	91.7	63.1	79.6
729	Fort Smith	94.1	62.6	80.8
CALIFORNIA				
900-902	Los Angeles	98.5	129.8	111.7
903-905	Inglewood	93.8	128.8	108.6
906-908	Long Beach	95.4	128.8	109.5
910-912	Pasadena	94.8	128.6	109.1
913-916	Van Nuys	97.8	128.6	110.8
917-918	Alhambra	96.7	128.6	110.2
919-921	San Diego	99.7	121.7	109.0
922	Palm Springs	97.4	125.3	109.2
923-924	San Bernardino	95.0	126.3	108.2
925	Riverside	99.3	126.6	110.8
926-927	Santa Ana	97.0	126.3	109.4
928	Anaheim	99.3	126.6	110.9
930	Oxnard	98.0	126.7	110.1
931	Santa Barbara	97.2	126.1	109.4
932-933	Bakersfield	98.9	124.6	109.8
934	San Luis Obispo	98.2	125.8	109.9
935	Mojave	95.4	124.4	107.7
936-938	Fresno	98.5	129.4	111.5
939	Salinas	99.1	137.2	115.2
940-941	San Francisco	107.4	158.4	129.0
942,956-958	Sacramento	100.8	133.0	114.4
943	Palo Alto	99.2	153.8	122.3
944	San Mateo	101.5	152.0	122.9
945	Vallejo	100.2	141.6	117.7
946	Oakland	103.4	152.0	123.9
947	Berkeley	103.0	151.9	123.7
948	Richmond	102.4	146.2	120.9
949	San Rafael	104.5	149.1	123.4
950	Santa Cruz	104.6	137.5	118.5
CALIFORNIA (CONT'D)				
951	San Jose	102.6	153.5	124.2
952	Stockton	100.7	131.3	113.7
953	Modesto	100.6	130.6	113.3
954	Santa Rosa	101.2	148.4	121.2
955	Eureka	102.5	134.1	115.9
959	Marysville	101.7	131.3	114.2
960	Redding	107.6	131.5	117.7
961	Susanville	107.3	131.6	117.5
COLORADO				
800-802	Denver	103.0	75.4	91.3
803	Boulder	98.7	77.0	89.6
804	Golden	100.9	74.7	89.8
805	Fort Collins	102.2	73.5	90.1
806	Greeley	99.3	73.4	88.4
807	Fort Morgan	99.3	73.5	88.4
808-809	Colorado Springs	101.2	71.5	88.6
810	Pueblo	101.8	70.0	88.3
811	Alamosa	103.6	70.3	89.5
812	Salida	103.1	70.9	89.5
813	Durango	103.9	65.6	87.8
814	Montrose	102.6	65.8	87.1
815	Grand Junction	105.7	70.5	90.8
816	Glenwood Springs	103.9	66.8	88.2
CONNECTICUT				
060	New Britain	98.1	116.7	106.0
061	Hartford	99.6	118.0	107.3
062	Willimantic	98.7	117.1	106.5
063	New London	95.2	117.2	104.5
064	Meriden	97.2	117.0	105.6
065	New Haven	99.9	117.0	107.1
066	Bridgeport	99.3	117.0	106.8
067	Waterbury	98.9	117.2	106.6
068	Norwalk	98.8	116.7	106.4
069	Stamford	99.0	123.2	109.2
D.C.				
200-205	Washington	101.1	87.5	95.4
DELAWARE				
197	Newark	98.3	109.7	103.1
198	Wilmington	97.9	109.7	102.9
199	Dover	99.0	109.6	103.5
FLORIDA				
320,322	Jacksonville	96.9	66.6	84.1
321	Daytona Beach	97.2	69.5	85.5
323	Tallahassee	98.2	67.7	85.3
324	Panama City	98.6	68.3	85.8
325	Pensacola	101.2	65.8	86.2
326,344	Gainesville	98.8	65.8	84.8
327-328,347	Orlando	99.0	66.6	85.3
329	Melbourne	100.1	70.3	87.5
330-332,340	Miami	96.7	68.0	84.6
333	Fort Lauderdale	96.1	66.4	83.6
334,349	West Palm Beach	95.1	64.1	82.0
335-336,346	Tampa	97.6	68.4	85.3
337	St. Petersburg	99.7	66.8	85.8
338	Lakeland	97.1	67.6	84.6
339,341	Fort Myers	96.5	67.7	84.3
342	Sarasota	98.8	67.0	85.4
GEORGIA				
300-303,399	Atlanta	98.2	75.0	88.4
304	Statesboro	97.8	69.7	85.9
305	Gainesville	96.2	65.6	83.2
306	Athens	95.6	66.9	83.5
307	Dalton	97.5	71.4	86.5
308-309	Augusta	96.0	73.9	86.6
310-312	Macon	95.2	73.1	85.8
313-314	Savannah	97.3	72.7	86.9
315	Waycross	96.8	70.0	85.5
316	Valdosta	96.7	66.2	83.8
317,398	Albany	96.5	71.9	86.1
318-319	Columbus	96.5	72.2	86.2

For customer support on your Mechanical Costs with RSMeans Data, call 800.448.8182.

Location Factors - Commercial

STATE/ZIP	CITY	MAT.	INST.	TOTAL
HAWAII				
967	Hilo	114.8	118.4	116.3
968	Honolulu	119.1	118.5	118.9
STATES & POSS.				
969	Guam	137.1	51.4	100.9
IDAHO				
832	Pocatello	102.0	78.4	92.0
833	Twin Falls	103.2	76.8	92.1
834	Idaho Falls	100.5	79.9	91.8
835	Lewiston	108.2	83.5	97.8
836-837	Boise	101.0	79.9	92.1
838	Coeur d'Alene	108.1	84.4	98.1
ILLINOIS				
600-603	North Suburban	98.3	142.9	117.1
604	Joliet	98.3	138.8	115.4
605	South Suburban	98.3	142.8	117.1
606-608	Chicago	100.1	145.6	119.3
609	Kankakee	95.0	134.9	111.9
610-611	Rockford	96.6	125.5	108.8
612	Rock Island	94.6	99.5	96.7
613	La Salle	95.9	127.0	109.0
614	Galesburg	95.7	108.5	101.1
615-616	Peoria	97.6	108.8	102.3
617	Bloomington	95.0	109.5	101.1
618-619	Champaign	98.5	110.5	103.6
620-622	East St. Louis	93.8	110.4	100.9
623	Quincy	95.7	104.2	99.3
624	Effingham	95.1	109.1	101.0
625	Decatur	96.6	110.1	102.3
626-627	Springfield	97.5	108.3	102.1
628	Centralia	92.8	111.2	100.5
629	Carbondale	92.5	109.6	99.7
INDIANA				
460	Anderson	96.0	79.8	89.1
461-462	Indianapolis	98.7	83.2	92.1
463-464	Gary	97.5	108.1	102.0
465-466	South Bend	98.4	81.8	91.4
467-468	Fort Wayne	96.6	76.3	88.0
469	Kokomo	94.2	81.1	88.6
470	Lawrenceburg	92.5	79.6	87.0
471	New Albany	93.9	77.3	86.9
472	Columbus	96.1	80.5	89.5
473	Muncie	96.2	78.2	88.6
474	Bloomington	97.8	81.1	90.7
475	Washington	94.6	84.8	90.4
476-477	Evansville	95.5	82.7	90.1
478	Terre Haute	96.2	81.7	90.1
479	Lafayette	95.8	78.1	88.3
IOWA				
500-503,509	Des Moines	97.2	90.2	94.3
504	Mason City	95.3	74.0	86.3
505	Fort Dodge	95.5	73.7	86.3
506-507	Waterloo	96.8	77.4	88.6
508	Creston	95.7	84.2	90.8
510-511	Sioux City	97.7	81.5	90.9
512	Sibley	96.8	62.4	82.3
513	Spencer	98.2	62.3	83.0
514	Carroll	95.5	82.4	89.9
515	Council Bluffs	98.9	79.2	90.6
516	Shenandoah	96.1	81.9	90.1
520	Dubuque	97.2	79.9	89.9
521	Decorah	96.4	73.7	86.8
522-524	Cedar Rapids	98.0	83.7	92.0
525	Ottumwa	96.3	78.7	88.9
526	Burlington	95.7	84.3	90.9
527-528	Davenport	97.1	94.5	96.0
KANSAS				
660-662	Kansas City	97.3	98.6	97.8
664-666	Topeka	98.6	76.5	89.3
667	Fort Scott	96.0	77.1	88.0
668	Emporia	96.1	76.5	87.8
669	Belleville	97.9	70.7	86.4
670-672	Wichita	97.0	72.6	86.7
673	Independence	97.2	76.0	88.2
674	Salina	97.2	73.5	87.2
675	Hutchinson	92.5	71.9	83.8
676	Hays	96.5	72.0	86.1
677	Colby	97.3	74.0	87.4

STATE/ZIP	CITY	MAT.	INST.	TOTAL
KANSAS (CONT'D)				
678	Dodge City	98.6	73.3	87.9
679	Liberal	96.5	71.7	86.0
KENTUCKY				
400-402	Louisville	93.9	79.6	87.8
403-405	Lexington	93.1	78.1	86.7
406	Frankfort	95.4	80.3	89.0
407-409	Corbin	90.8	78.0	85.4
410	Covington	94.3	77.8	87.3
411-412	Ashland	93.2	89.7	91.7
413-414	Campton	94.4	78.0	87.5
415-416	Pikeville	95.8	85.4	91.4
417-418	Hazard	93.8	78.7	87.4
420	Paducah	92.4	83.2	88.6
421-422	Bowling Green	94.6	77.2	87.3
423	Owensboro	94.7	81.0	88.9
424	Henderson	92.2	81.9	87.8
425-426	Somerset	91.6	79.5	86.5
427	Elizabethtown	91.3	75.6	84.7
LOUISIANA				
700-701	New Orleans	99.0	68.5	86.1
703	Thibodaux	96.0	67.0	83.8
704	Hammond	93.7	64.4	81.3
705	Lafayette	95.4	67.9	83.8
706	Lake Charles	95.6	68.2	84.1
707-708	Baton Rouge	96.6	69.3	85.1
710-711	Shreveport	97.1	66.3	84.1
712	Monroe	95.8	64.4	82.5
713-714	Alexandria	95.9	65.6	83.1
MAINE				
039	Kittery	92.6	87.0	90.2
040-041	Portland	99.6	85.4	93.6
042	Lewiston	97.5	84.9	92.2
043	Augusta	100.4	85.6	94.2
044	Bangor	97.1	83.9	91.5
045	Bath	95.9	85.4	91.4
046	Machias	95.4	85.3	91.1
047	Houlton	95.6	85.2	91.2
048	Rockland	94.6	85.4	90.7
049	Waterville	96.0	85.2	91.4
MARYLAND				
206	Waldorf	97.3	86.6	92.8
207-208	College Park	97.3	87.7	93.2
209	Silver Spring	96.5	86.7	92.4
210-212	Baltimore	101.8	83.2	93.9
214	Annapolis	100.4	86.3	94.4
215	Cumberland	96.9	83.8	91.4
216	Easton	98.6	74.8	88.5
217	Hagerstown	97.4	85.3	92.3
218	Salisbury	99.0	66.9	85.4
219	Elkton	95.9	86.2	91.8
MASSACHUSETTS				
010-011	Springfield	98.2	108.5	102.5
012	Pittsfield	97.7	102.6	99.7
013	Greenfield	95.9	109.0	101.4
014	Fitchburg	94.6	116.2	103.7
015-016	Worcester	98.1	117.0	106.1
017	Framingham	94.0	124.2	106.8
018	Lowell	97.5	123.8	108.6
019	Lawrence	98.4	124.2	109.3
020-022, 024	Boston	101.9	134.0	115.5
023	Brockton	98.5	116.7	106.2
025	Buzzards Bay	93.2	115.2	102.5
026	Hyannis	95.7	115.0	103.8
027	New Bedford	97.7	115.2	105.1
MICHIGAN				
480,483	Royal Oak	95.1	100.3	97.3
481	Ann Arbor	97.1	99.8	98.2
482	Detroit	100.7	102.0	101.3
484-485	Flint	96.8	88.6	93.3
486	Saginaw	96.5	85.1	91.7
487	Bay City	96.6	85.3	91.8
488-489	Lansing	98.0	86.8	93.3
490	Battle Creek	95.5	80.4	89.1
491	Kalamazoo	95.8	78.6	88.6
492	Jackson	94.0	90.7	92.6
493,495	Grand Rapids	97.9	79.9	90.3
494	Muskegon	94.3	80.0	88.3

Location Factors - Commercial

STATE/ZIP	CITY	MAT.	INST.	TOTAL
MICHIGAN (CONT'D)				
496	Traverse City	93.4	77.7	86.8
497	Gaylord	94.5	80.6	88.6
498-499	Iron Mountain	96.5	81.7	90.2
MINNESOTA				
550-551	Saint Paul	99.0	117.6	106.9
553-555	Minneapolis	100.7	114.1	106.3
556-558	Duluth	98.6	101.3	99.7
559	Rochester	98.3	98.7	98.5
560	Mankato	95.7	98.5	96.9
561	Windom	94.3	91.2	93.0
562	Willmar	94.0	98.0	95.7
563	St. Cloud	95.1	114.2	103.2
564	Brainerd	95.6	98.6	96.8
565	Detroit Lakes	97.4	92.4	95.3
566	Bemidji	96.7	95.4	96.1
567	Thief River Falls	96.3	89.3	93.3
MISSISSIPPI				
386	Clarksdale	95.3	55.2	78.4
387	Greenville	98.7	65.5	84.7
388	Tupelo	96.6	59.0	80.7
389	Greenwood	96.6	55.5	79.2
390-392	Jackson	97.3	65.6	83.9
393	Meridian	94.6	64.5	81.9
394	Laurel	96.1	57.8	79.9
395	Biloxi	96.2	63.9	82.6
396	McComb	94.4	55.8	78.0
397	Columbus	95.9	59.1	80.4
MISSOURI				
630-631	St. Louis	98.8	104.5	101.2
633	Bowling Green	96.4	95.5	96.0
634	Hannibal	95.3	91.6	93.7
635	Kirksville	98.9	86.5	93.7
636	Flat River	97.3	94.2	96.0
637	Cape Girardeau	97.0	91.2	94.5
638	Sikeston	95.7	87.8	92.3
639	Poplar Bluff	95.1	87.7	92.0
640-641	Kansas City	97.9	103.8	100.4
644-645	St. Joseph	96.1	90.9	93.9
646	Chillicothe	93.7	93.8	93.7
647	Harrisonville	93.2	100.8	96.4
648	Joplin	95.1	77.7	87.7
650-651	Jefferson City	96.4	89.9	93.7
652	Columbia	96.1	90.7	93.8
653	Sedalia	96.4	91.5	94.3
654-655	Rolla	94.1	94.3	94.2
656-658	Springfield	97.6	79.0	89.7
MONTANA				
590-591	Billings	101.6	78.4	91.8
592	Wolf Point	101.4	76.9	91.1
593	Miles City	99.2	77.0	89.8
594	Great Falls	102.9	76.6	91.8
595	Havre	100.4	74.8	89.6
596	Helena	100.9	76.7	90.7
597	Butte	101.5	76.5	90.9
598	Missoula	98.6	77.1	89.5
599	Kalispell	98.1	76.2	88.9
NEBRASKA				
680-681	Omaha	97.6	81.5	90.8
683-685	Lincoln	98.3	79.7	90.5
686	Columbus	96.0	80.7	89.6
687	Norfolk	97.4	77.5	89.0
688	Grand Island	97.4	76.8	88.7
689	Hastings	97.1	78.2	89.1
690	McCook	96.3	73.1	86.5
691	North Platte	96.3	75.6	87.6
692	Valentine	98.6	69.2	86.2
693	Alliance	98.7	73.3	87.9
NEVADA				
889-891	Las Vegas	105.0	105.1	105.0
893	Ely	103.7	93.2	99.2
894-895	Reno	103.6	82.6	94.7
897	Carson City	102.9	82.5	94.3
898	Elko	102.3	85.2	95.0
NEW HAMPSHIRE				
030	Nashua	97.5	93.7	95.9
031	Manchester	98.2	94.5	96.6

STATE/ZIP	CITY	MAT.	INST.	TOTAL
NEW HAMPSHIRE (CONT'D)				
032-033	Concord	98.0	93.2	96.0
034	Keene	94.5	89.0	92.2
035	Littleton	94.4	80.2	88.4
036	Charleston	94.1	88.4	91.7
037	Claremont	93.1	88.4	91.1
038	Portsmouth	94.9	92.4	93.8
NEW JERSEY				
070-071	Newark	100.3	139.7	117.0
072	Elizabeth	97.5	138.3	114.7
073	Jersey City	96.5	137.8	114.0
074-075	Paterson	98.0	137.7	114.8
076	Hackensack	96.1	138.3	114.0
077	Long Branch	95.8	130.9	110.7
078	Dover	96.4	137.5	113.7
079	Summit	96.4	138.3	114.1
080,083	Vineland	96.3	129.8	110.5
081	Camden	98.0	127.8	110.6
082,084	Atlantic City	96.8	133.2	112.2
085-086	Trenton	99.7	128.9	112.0
087	Point Pleasant	98.1	130.8	111.9
088-089	New Brunswick	98.7	135.6	114.3
NEW MEXICO				
870-872	Albuquerque	98.6	72.5	87.6
873	Gallup	99.0	72.5	87.8
874	Farmington	99.0	72.5	87.8
875	Santa Fe	99.0	72.9	88.0
877	Las Vegas	97.2	72.5	86.8
878	Socorro	96.8	72.5	86.6
879	Truth/Consequences	96.6	69.4	85.1
880	Las Cruces	96.9	71.4	86.1
881	Clovis	99.3	72.4	87.9
882	Roswell	100.8	72.5	88.9
883	Carrizozo	101.6	72.5	89.3
884	Tucumcari	99.9	72.4	88.3
NEW YORK				
100-102	New York	99.7	175.2	131.6
103	Staten Island	95.6	176.1	129.6
104	Bronx	93.8	175.3	128.3
105	Mount Vernon	94.1	150.6	118.0
106	White Plains	93.8	152.8	118.8
107	Yonkers	97.9	153.0	121.2
108	New Rochelle	94.4	144.8	115.7
109	Suffern	94.1	130.2	109.4
110	Queens	100.0	177.3	132.6
111	Long Island City	101.6	177.3	133.6
112	Brooklyn	101.9	177.3	133.8
113	Flushing	102.1	177.3	133.9
114	Jamaica	100.3	177.3	132.8
115,117,118	Hicksville	99.9	157.1	124.1
116	Far Rockaway	102.2	177.3	133.9
119	Riverhead	100.6	153.5	122.9
120-122	Albany	96.8	111.2	102.8
123	Schenectady	97.0	109.9	102.4
124	Kingston	100.6	133.8	114.6
125-126	Poughkeepsie	99.8	136.8	115.4
127	Monticello	99.1	134.6	114.1
128	Glens Falls	92.0	107.3	98.5
129	Plattsburgh	97.0	96.6	96.8
130-132	Syracuse	97.8	98.3	98.0
133-135	Utica	96.0	96.6	96.2
136	Watertown	97.7	96.8	97.4
137-139	Binghamton	97.3	101.5	99.1
140-142	Buffalo	100.8	110.3	104.8
143	Niagara Falls	97.4	107.5	101.6
144-146	Rochester	100.3	99.2	99.8
147	Jamestown	96.4	96.0	96.3
148-149	Elmira	96.3	102.5	98.9
NORTH CAROLINA				
270,272-274	Greensboro	98.5	67.2	85.3
271	Winston-Salem	98.2	67.3	85.2
275-276	Raleigh	97.4	67.0	84.5
277	Durham	100.1	67.2	86.2
278	Rocky Mount	96.0	66.6	83.6
279	Elizabeth City	96.9	68.3	84.8
280	Gastonia	97.0	66.8	84.2
281-282	Charlotte	97.4	67.6	84.8
283	Fayetteville	100.2	66.1	85.8
284	Wilmington	96.0	65.2	83.0
285	Kinston	94.4	65.5	82.2

Location Factors - Commercial

STATE/ZIP	CITY	MAT.	INST.	TOTAL	STATE/ZIP	CITY	MAT.	INST.	TOTAL
NORTH CAROLINA (CONT'D)					**PENNSYLVANIA (CONT'D)**				
286	Hickory	94.7	67.3	83.1	177	Williamsport	92.5	92.3	92.4
287-288	Asheville	96.4	66.6	83.8	178	Sunbury	94.6	92.7	93.8
289	Murphy	95.5	64.7	82.5	179	Pottsville	93.7	96.1	94.7
					180	Lehigh Valley	94.9	111.6	102.0
NORTH DAKOTA					181	Allentown	96.7	105.8	100.6
580-581	Fargo	99.7	80.3	91.5	182	Hazleton	94.4	95.9	95.0
582	Grand Forks	99.4	77.6	90.2	183	Stroudsburg	94.3	104.7	98.7
583	Devils Lake	99.2	78.8	90.6	184-185	Scranton	97.4	96.5	97.1
584	Jamestown	99.2	78.8	90.6	186-187	Wilkes-Barre	94.0	95.5	94.7
585	Bismarck	98.9	81.6	91.6	188	Montrose	93.8	96.8	95.1
586	Dickinson	99.9	77.9	90.6	189	Doylestown	94.0	127.1	108.0
587	Minot	99.3	78.6	90.5	190-191	Philadelphia	100.2	138.7	116.5
588	Williston	98.4	77.7	89.6	193	Westchester	95.8	127.9	109.3
					194	Norristown	94.7	128.0	108.8
OHIO					195-196	Reading	96.5	101.5	98.6
430-432	Columbus	98.4	82.0	91.4					
433	Marion	94.3	86.7	91.1	**PUERTO RICO**				
434-436	Toledo	97.6	91.2	94.9	009	San Juan	98.5	29.8	69.4
437-438	Zanesville	94.9	85.7	91.0					
439	Steubenville	96.1	90.0	93.5	**RHODE ISLAND**				
440	Lorain	98.4	84.7	92.6	028	Newport	97.0	114.7	104.5
441	Cleveland	99.1	91.6	95.9	029	Providence	99.3	115.0	105.9
442-443	Akron	99.4	86.0	93.8					
444-445	Youngstown	98.7	81.9	91.6	**SOUTH CAROLINA**				
446-447	Canton	98.9	79.9	90.9	290-292	Columbia	97.6	68.9	85.5
448-449	Mansfield	96.4	84.2	91.3	293	Spartanburg	97.2	68.3	85.0
450	Hamilton	94.8	79.5	88.3	294	Charleston	98.7	68.2	85.8
451-452	Cincinnati	96.1	80.0	89.3	295	Florence	96.9	68.6	84.9
453-454	Dayton	94.8	79.2	88.2	296	Greenville	97.0	68.2	84.8
455	Springfield	94.8	79.7	88.4	297	Rock Hill	96.7	66.4	83.9
456	Chillicothe	94.3	88.5	91.8	298	Aiken	97.6	67.1	84.7
457	Athens	97.1	84.8	91.9	299	Beaufort	98.3	57.5	81.1
458	Lima	97.2	81.7	90.7					
					SOUTH DAKOTA				
OKLAHOMA					570-571	Sioux Falls	97.7	78.4	89.5
730-731	Oklahoma City	96.4	68.7	84.7	572	Watertown	97.3	69.0	85.4
734	Ardmore	95.2	65.6	82.7	573	Mitchell	96.3	60.9	81.3
735	Lawton	97.3	65.8	84.0	574	Aberdeen	98.6	70.1	86.5
736	Clinton	96.4	65.5	83.3	575	Pierre	99.0	68.5	86.1
737	Enid	97.0	66.0	83.9	576	Mobridge	96.9	62.9	82.5
738	Woodward	95.3	62.9	81.6	577	Rapid City	98.3	72.0	87.2
739	Guymon	96.3	63.3	82.4					
740-741	Tulsa	95.9	65.8	83.2	**TENNESSEE**				
743	Miami	92.7	65.1	81.1	370-372	Nashville	98.0	71.8	86.9
744	Muskogee	95.0	62.9	81.4	373-374	Chattanooga	98.0	68.1	85.3
745	McAlester	92.4	60.5	78.9	375,380-381	Memphis	96.6	69.8	85.3
746	Ponca City	93.1	64.0	80.8	376	Johnson City	98.1	58.6	81.4
747	Durant	93.1	65.1	81.3	377-379	Knoxville	94.8	63.7	81.6
748	Shawnee	94.5	65.2	82.2	382	McKenzie	96.3	55.1	78.9
749	Poteau	92.3	64.9	80.7	383	Jackson	97.6	61.4	82.3
					384	Columbia	94.8	66.5	82.8
OREGON					385	Cookeville	96.2	56.9	79.6
970-972	Portland	102.1	102.8	102.4					
973	Salem	104.4	100.8	102.9	**TEXAS**				
974	Eugene	101.7	98.7	100.5	750	McKinney	97.2	63.5	82.9
975	Medford	103.3	97.4	100.8	751	Waxahachie	97.2	63.5	83.0
976	Klamath Falls	103.5	97.3	100.9	752-753	Dallas	98.2	68.0	85.4
977	Bend	102.6	100.2	101.6	754	Greenville	97.4	62.9	82.8
978	Pendleton	98.6	100.3	99.3	755	Texarkana	96.7	62.0	82.1
979	Vale	96.2	87.0	92.3	756	Longview	97.5	61.0	82.1
					757	Tyler	97.8	61.6	82.5
PENNSYLVANIA					758	Palestine	94.2	60.4	79.9
150-152	Pittsburgh	100.7	102.7	101.5	759	Lufkin	94.7	62.6	81.1
153	Washington	97.5	100.2	98.6	760-761	Fort Worth	97.6	62.9	82.9
154	Uniontown	97.8	99.8	98.6	762	Denton	97.3	62.9	82.7
155	Bedford	98.8	92.1	96.0	763	Wichita Falls	95.0	61.6	80.9
156	Greensburg	98.7	96.3	97.7	764	Eastland	94.2	60.3	79.8
157	Indiana	97.6	97.7	97.7	765	Temple	92.5	58.5	78.2
158	Dubois	99.3	96.0	97.9	766-767	Waco	94.4	62.4	80.9
159	Johnstown	98.9	91.9	95.9	768	Brownwood	97.4	58.7	81.1
160	Butler	91.3	100.6	95.2	769	San Angelo	97.1	59.4	81.2
161	New Castle	91.4	97.8	94.1	770-772	Houston	100.4	68.2	86.8
162	Kittanning	91.8	97.9	94.4	773	Huntsville	98.7	62.8	83.6
163	Oil City	91.3	97.8	94.0	774	Wharton	99.8	64.3	84.8
164-165	Erie	93.2	93.9	93.5	775	Galveston	97.7	66.5	84.5
166	Altoona	93.3	92.8	93.1	776-777	Beaumont	97.9	65.5	84.2
167	Bradford	94.9	97.1	95.8	778	Bryan	95.2	64.9	82.4
168	State College	94.5	94.6	94.5	779	Victoria	99.7	62.9	84.2
169	Wellsboro	95.5	92.9	94.4	780	Laredo	96.4	61.1	81.5
170-171	Harrisburg	98.7	93.0	96.3	781-782	San Antonio	98.3	62.3	83.1
172	Chambersburg	95.1	89.4	92.7	783-784	Corpus Christi	98.9	61.4	83.1
173-174	York	95.2	93.0	94.3	785	McAllen	99.7	56.5	81.4
175-176	Lancaster	93.8	94.5	94.1	786-787	Austin	97.2	61.4	82.1

For customer support on your Mechanical Costs with RSMeans Data, call 800.448.8182.

703

STATE/ZIP	CITY	MAT.	INST.	TOTAL
TEXAS (CONT'D)				
788	Del Rio	99.7	60.4	83.1
789	Giddings	96.1	60.4	81.0
790-791	Amarillo	98.2	60.0	82.1
792	Childress	97.6	60.3	81.8
793-794	Lubbock	99.2	62.9	83.9
795-796	Abilene	97.8	61.4	82.4
797	Midland	99.7	62.2	83.8
798-799,885	El Paso	95.9	63.6	82.3
UTAH				
840-841	Salt Lake City	103.5	71.7	90.1
842,844	Ogden	98.8	71.7	87.4
843	Logan	100.9	71.7	88.5
845	Price	101.4	70.0	88.1
846-847	Provo	101.2	71.2	88.5
VERMONT				
050	White River Jct.	98.2	79.8	90.4
051	Bellows Falls	96.7	92.1	94.8
052	Bennington	97.0	88.6	93.4
053	Brattleboro	97.4	92.1	95.2
054	Burlington	101.9	79.8	92.5
056	Montpelier	100.2	84.4	93.5
057	Rutland	98.9	79.4	90.7
058	St. Johnsbury	98.1	79.3	90.2
059	Guildhall	96.8	79.0	89.3
VIRGINIA				
220-221	Fairfax	99.5	83.6	92.8
222	Arlington	100.4	84.2	93.6
223	Alexandria	99.6	84.8	93.3
224-225	Fredericksburg	98.3	78.7	90.0
226	Winchester	98.9	79.5	90.7
227	Culpeper	98.7	83.0	92.1
228	Harrisonburg	99.0	68.8	86.3
229	Charlottesville	99.4	71.0	87.4
230-232	Richmond	99.2	72.8	88.1
233-235	Norfolk	99.9	68.7	86.7
236	Newport News	98.8	68.3	85.9
237	Portsmouth	98.3	67.8	85.4
238	Petersburg	98.7	72.6	87.7
239	Farmville	97.9	65.3	84.1
240-241	Roanoke	100.1	73.6	88.9
242	Bristol	98.2	59.3	81.8
243	Pulaski	97.8	67.1	84.8
244	Staunton	98.6	67.9	85.6
245	Lynchburg	98.7	73.8	88.2
246	Grundy	98.1	58.8	81.5
WASHINGTON				
980-981,987	Seattle	106.3	112.5	108.9
982	Everett	105.1	104.0	104.6
983-984	Tacoma	105.4	101.3	103.7
985	Olympia	104.0	101.8	103.1
986	Vancouver	106.6	102.8	105.0
988	Wenatchee	105.6	88.7	98.5
989	Yakima	105.6	95.4	101.3
990-992	Spokane	100.1	82.4	92.6
993	Richland	99.8	90.0	95.6
994	Clarkston	99.1	81.9	91.8
WEST VIRGINIA				
247-248	Bluefield	97.0	88.1	93.2
249	Lewisburg	98.7	86.1	93.4
250-253	Charleston	96.5	89.4	93.5
254	Martinsburg	96.7	84.1	91.4
255-257	Huntington	97.8	92.2	95.4
258-259	Beckley	95.2	87.7	92.0
260	Wheeling	99.6	89.6	95.4
261	Parkersburg	98.6	88.6	94.4
262	Buckhannon	98.3	88.8	94.3
263-264	Clarksburg	98.9	89.5	94.9
265	Morgantown	98.8	89.6	94.9
266	Gassaway	98.3	89.0	94.3
267	Romney	98.3	87.2	93.6
268	Petersburg	98.0	85.9	92.9
WISCONSIN				
530,532	Milwaukee	99.2	107.9	102.9
531	Kenosha	99.4	102.0	100.5
534	Racine	98.7	102.3	100.2
535	Beloit	98.6	98.9	98.7
537	Madison	99.7	98.1	99.0

STATE/ZIP	CITY	MAT.	INST.	TOTAL
WISCONSIN (CONT'D)				
538	Lancaster	96.7	94.3	95.7
539	Portage	95.2	98.0	96.4
540	New Richmond	95.4	94.6	95.0
541-543	Green Bay	99.4	93.4	96.9
544	Wausau	94.7	93.4	94.2
545	Rhinelander	97.9	92.2	95.5
546	La Crosse	95.9	93.5	94.9
547	Eau Claire	97.5	94.1	96.1
548	Superior	95.1	96.5	95.7
549	Oshkosh	95.4	91.9	93.9
WYOMING				
820	Cheyenne	101.1	72.7	89.1
821	Yellowstone Nat'l Park	99.0	71.9	87.6
822	Wheatland	100.3	67.1	86.2
823	Rawlins	102.0	71.5	89.1
824	Worland	99.7	70.7	87.5
825	Riverton	101.0	68.6	87.3
826	Casper	101.5	68.9	87.7
827	Newcastle	99.6	71.1	87.5
828	Sheridan	102.5	70.2	88.8
829-831	Rock Springs	103.4	71.4	89.9
CANADIAN FACTORS (reflect Canadian currency)				
ALBERTA				
	Calgary	119.7	95.6	109.5
	Edmonton	121.8	95.9	110.9
	Fort McMurray	121.9	88.4	107.7
	Lethbridge	117.0	87.9	104.7
	Lloydminster	111.5	84.7	100.2
	Medicine Hat	111.7	84.0	100.0
	Red Deer	112.1	84.0	100.2
BRITISH COLUMBIA				
	Kamloops	113.0	83.5	100.5
	Prince George	114.0	82.8	100.8
	Vancouver	117.9	92.4	107.2
	Victoria	114.1	87.6	102.9
MANITOBA				
	Brandon	120.6	70.5	99.4
	Portage la Prairie	111.6	69.1	93.6
	Winnipeg	122.0	71.2	100.5
NEW BRUNSWICK				
	Bathurst	110.7	62.3	90.2
	Dalhousie	110.2	62.5	90.0
	Fredericton	119.3	69.0	98.0
	Moncton	110.9	76.0	96.2
	Newcastle	110.7	62.9	90.5
	Saint John	113.2	75.7	97.3
NEWFOUNDLAND				
	Corner Brook	124.7	68.4	100.9
	St. John's	122.6	87.4	107.7
NORTHWEST TERRITORIES				
	Yellowknife	131.4	83.2	111.0
NOVA SCOTIA				
	Bridgewater	111.3	70.3	94.0
	Dartmouth	121.5	70.2	99.8
	Halifax	117.9	88.0	105.2
	New Glasgow	119.6	70.2	98.7
	Sydney	117.8	70.2	97.7
	Truro	110.8	70.3	93.7
	Yarmouth	119.5	70.2	98.7
ONTARIO				
	Barrie	116.5	87.1	104.1
	Brantford	113.1	90.7	103.6
	Cornwall	113.1	87.3	102.2
	Hamilton	116.3	98.8	108.9
	Kingston	114.1	87.4	102.8
	Kitchener	108.6	94.6	102.7
	London	116.3	96.5	108.0
	North Bay	122.5	85.2	106.8
	Oshawa	110.8	97.2	105.0
	Ottawa	117.5	96.8	108.7
	Owen Sound	116.7	85.4	103.4
	Peterborough	113.1	87.1	102.1
	Sarnia	113.2	91.3	103.9

STATE/ZIP	CITY	MAT.	INST.	TOTAL
ONTARIO (CONT'D)				
	Sault Ste. Marie	108.6	88.7	100.2
	St. Catharines	106.6	95.6	101.9
	Sudbury	106.1	93.6	100.8
	Thunder Bay	107.8	95.0	102.4
	Timmins	113.3	85.3	101.4
	Toronto	116.0	104.2	111.0
	Windsor	107.1	94.4	101.8
PRINCE EDWARD ISLAND				
	Charlottetown	121.3	60.2	95.5
	Summerside	121.9	57.9	94.9
QUEBEC				
	Cap-de-la-Madeleine	110.6	80.0	97.7
	Charlesbourg	110.6	80.0	97.7
	Chicoutimi	110.4	87.6	100.7
	Gatineau	110.3	79.8	97.4
	Granby	110.5	79.7	97.5
	Hull	110.5	79.8	97.5
	Joliette	110.8	80.0	97.8
	Laval	110.7	81.6	98.4
	Montreal	119.1	90.1	106.9
	Quebec City	118.7	90.4	106.7
	Rimouski	110.2	87.6	100.6
	Rouyn-Noranda	110.3	79.8	97.4
	Saint-Hyacinthe	109.8	79.8	97.1
	Sherbrooke	110.6	79.8	97.6
	Sorel	110.8	80.0	97.8
	Saint-Jerome	110.4	79.8	97.4
	Trois-Rivieres	120.2	79.9	103.2
SASKATCHEWAN				
	Moose Jaw	109.0	63.9	89.9
	Prince Albert	108.2	62.2	88.8
	Regina	123.1	92.4	110.2
	Saskatoon	109.9	88.5	100.9
YUKON				
	Whitehorse	132.2	69.2	105.6

R011105-05   Tips for Accurate Estimating

1. Use pre-printed or columnar forms for orderly sequence of dimensions and locations and for recording telephone quotations.

2. Use only the front side of each paper or form except for certain pre-printed summary forms.

3. Be consistent in listing dimensions: For example, length x width x height. This helps in rechecking to ensure that, the total length of partitions is appropriate for the building area.

4. Use printed (rather than measured) dimensions where given.

5. Add up multiple printed dimensions for a single entry where possible.

6. Measure all other dimensions carefully.

7. Use each set of dimensions to calculate multiple related quantities.

8. Convert foot and inch measurements to decimal feet when listing. Memorize decimal equivalents to .01 parts of a foot (1/8″ equals approximately .01′).

9. Do not "round off" quantities until the final summary.

10. Mark drawings with different colors as items are taken off.

11. Keep similar items together, different items separate.

12. Identify location and drawing numbers to aid in future checking for completeness.

13. Measure or list everything on the drawings or mentioned in the specifications.

14. It may be necessary to list items not called for to make the job complete.

15. Be alert for: Notes on plans such as N.T.S. (not to scale); changes in scale throughout the drawings; reduced size drawings; discrepancies between the specifications and the drawings.

16. Develop a consistent pattern of performing an estimate. For example:
    a. Start the quantity takeoff at the lower floor and move to the next higher floor.
    b. Proceed from the main section of the building to the wings.
    c. Proceed from south to north or vice versa, clockwise or counterclockwise.
    d. Take off floor plan quantities first, elevations next, then detail drawings.

17. List all gross dimensions that can be either used again for different quantities, or used as a rough check of other quantities for verification (exterior perimeter, gross floor area, individual floor areas, etc.).

18. Utilize design symmetry or repetition (repetitive floors, repetitive wings, symmetrical design around a center line, similar room layouts, etc.). Note: Extreme caution is needed here so as not to omit or duplicate an area.

19. Do not convert units until the final total is obtained. For instance, when estimating concrete work, keep all units to the nearest cubic foot, then summarize and convert to cubic yards.

20. When figuring alternatives, it is best to total all items involved in the basic system, then total all items involved in the alternates. Therefore you work with positive numbers in all cases. When adds and deducts are used, it is often confusing whether to add or subtract a portion of an item; especially on a complicated or involved alternate.

R011105-10   Unit Gross Area Requirements

The figures in the table below indicate typical ranges in square feet as a function of the "occupant" unit. This table is best used in the preliminary design stages to help determine the probable size requirement for the total project.

Building Type	Unit	Gross Area in S.F.		
		1/4	Median	3/4
Apartments	Unit	660	860	1,100
Auditorium & Play Theaters	Seat	18	25	38
Bowling Alleys	Lane		940	
Churches & Synagogues	Seat	20	28	39
Dormitories	Bed	200	230	275
Fraternity & Sorority Houses	Bed	220	315	370
Garages, Parking	Car	325	355	385
Hospitals	Bed	685	850	1,075
Hotels	Rental Unit	475	600	710
Housing for the elderly	Unit	515	635	755
Housing, Public	Unit	700	875	1,030
Ice Skating Rinks	Total	27,000	30,000	36,000
Motels	Rental Unit	360	465	620
Nursing Homes	Bed	290	350	450
Restaurants	Seat	23	29	39
Schools, Elementary	Pupil	65	77	90
Junior High & Middle		85	110	129
Senior High		102	130	145
Vocational		110	135	195
Shooting Ranges	Point		450	
Theaters & Movies	Seat		15	

R011105-20 Floor Area Ratios

Table below lists commonly used gross to net area and net to gross area
ratios expressed in % for various building types.

Building Type	Gross to Net Ratio	Net to Gross Ratio	Building Type	Gross to Net Ratio	Net to Gross Ratio
Apartment	156	64	School Buildings (campus type)		
Bank	140	72	Administrative	150	67
Church	142	70	Auditorium	142	70
Courthouse	162	61	Biology	161	62
Department Store	123	81	Chemistry	170	59
Garage	118	85	Classroom	152	66
Hospital	183	55	Dining Hall	138	72
Hotel	158	63	Dormitory	154	65
Laboratory	171	58	Engineering	164	61
Library	132	76	Fraternity	160	63
Office	135	75	Gymnasium	142	70
Restaurant	141	70	Science	167	60
Warehouse	108	93	Service	120	83
			Student Union	172	59

The gross area of a building is the total floor area based on outside
dimensions.

The net area of a building is the usable floor area for the function intended
and excludes such items as stairways, corridors and mechanical rooms. In
the case of a commercial building, it might be considered as the "leasable
area."

R011105-30   Occupancy Determinations

Function of Space	SF/Person Required
Accessory storage areas, mechanical equipment rooms	300
Agriculture Building	300
Aircraft Hangars	500
Airport Terminal	
Baggage claim	20
Baggage handling	300
Concourse	100
Waiting areas	15
Assembly	
Gaming floors (keno, slots, etc.)	11
Exhibit Gallery and Museum	30
Assembly w/ fixed seats	load determined by seat number
Assembly w/o fixed seats	
Concentrated (chairs only-not fixed)	7
Standing space	5
Unconcentrated (tables and chairs)	15
Bowling centers, allow 5 persons for each lane including 15 feet of runway, and for additional areas	7
Business areas	100
Courtrooms-other than fixed seating areas	40
Day care	35
Dormitories	50
Educational	
Classroom areas	20
Shops and other vocational room areas	50
Exercise rooms	50
Fabrication and Manufacturing areas where hazardous materials are used	200
Industrial areas	100
Institutional areas	
Inpatient treatment areas	240
Outpatient areas	100
Sleeping areas	120
Kitchens commercial	200
Library	
Reading rooms	50
Stack area	100
Mercantile	
Areas on other floors	60
Basement and grade floor areas	30
Storage, stock, shipping areas	300
Parking garages	200
Residential	200
Skating rinks, swimming pools	
Rink and pool	50
Decks	15
Stages and platforms	15
Warehouses	500

Excerpted from the 2012 *International Building Code,* Copyright 2011. Washington, D.C.: International Code Council.
Reproduced with permission. All rights reserved. www.ICCSAFE.org

R011105-50   Metric Conversion Factors

Description: This table is primarily for converting customary U.S. units in the left hand column to SI metric units in the right hand column. In addition, conversion factors for some commonly encountered Canadian and non-SI metric units are included.

If You Know		Multiply By		To Find
Length	Inches	x	25.4[a]	=	Millimeters
	Feet	x	0.3048[a]	=	Meters
	Yards	x	0.9144[a]	=	Meters
	Miles (statute)	x	1.609	=	Kilometers
Area	Square inches	x	645.2	=	Square millimeters
	Square feet	x	0.0929	=	Square meters
	Square yards	x	0.8361	=	Square meters
Volume (Capacity)	Cubic inches	x	16,387	=	Cubic millimeters
	Cubic feet	x	0.02832	=	Cubic meters
	Cubic yards	x	0.7646	=	Cubic meters
	Gallons (U.S. liquids)[b]	x	0.003785	=	Cubic meters[c]
	Gallons (Canadian liquid)[b]	x	0.004546	=	Cubic meters[c]
	Ounces (U.S. liquid)[b]	x	29.57	=	Milliliters[c, d]
	Quarts (U.S. liquid)[b]	x	0.9464	=	Liters[c, d]
	Gallons (U.S. liquid)[b]	x	3.785	=	Liters[c, d]
Force	Kilograms force[d]	x	9.807	=	Newtons
	Pounds force	x	4.448	=	Newtons
	Pounds force	x	0.4536	=	Kilograms force[d]
	Kips	x	4448	=	Newtons
	Kips	x	453.6	=	Kilograms force[d]
Pressure, Stress, Strength (Force per unit area)	Kilograms force per square centimeter[d]	x	0.09807	=	Megapascals
	Pounds force per square inch (psi)	x	0.006895	=	Megapascals
	Kips per square inch	x	6.895	=	Megapascals
	Pounds force per square inch (psi)	x	0.07031	=	Kilograms force per square centimeter[d]
	Pounds force per square foot	x	47.88	=	Pascals
	Pounds force per square foot	x	4.882	=	Kilograms force per square meter[d]
Flow	Cubic feet per minute	x	0.4719	=	Liters per second
	Gallons per minute	x	0.0631	=	Liters per second
	Gallons per hour	x	1.05	=	Milliliters per second
Bending Moment Or Torque	Inch-pounds force	x	0.01152	=	Meter-kilograms force[d]
	Inch-pounds force	x	0.1130	=	Newton-meters
	Foot-pounds force	x	0.1383	=	Meter-kilograms force[d]
	Foot-pounds force	x	1.356	=	Newton-meters
	Meter-kilograms force[d]	x	9.807	=	Newton-meters
Mass	Ounces (avoirdupois)	x	28.35	=	Grams
	Pounds (avoirdupois)	x	0.4536	=	Kilograms
	Tons (metric)	x	1000	=	Kilograms
	Tons, short (2000 pounds)	x	907.2	=	Kilograms
	Tons, short (2000 pounds)	x	0.9072	=	Megagrams[e]
Mass per Unit Volume	Pounds mass per cubic foot	x	16.02	=	Kilograms per cubic meter
	Pounds mass per cubic yard	x	0.5933	=	Kilograms per cubic meter
	Pounds mass per gallon (U.S. liquid)[b]	x	119.8	=	Kilograms per cubic meter
	Pounds mass per gallon (Canadian liquid)[b]	x	99.78	=	Kilograms per cubic meter
Temperature	Degrees Fahrenheit	(F-32)/1.8		=	Degrees Celsius
	Degrees Fahrenheit	(F+459.67)/1.8		=	Degrees Kelvin
	Degrees Celsius	C+273.15		=	Degrees Kelvin

[a]The factor given is exact
[b]One U.S. gallon = 0.8327 Canadian gallon
[c]1 liter = 1000 milliliters = 1000 cubic centimeters
  1 cubic decimeter = 0.001 cubic meter

[d]Metric but not SI unit
[e]Called "tonne" in England and
  "metric ton" in other metric countries

Reference Tables

R011105-60   Weights and Measures

Measures of Length
1 Mile = 1760 Yards = 5280 Feet
1 Yard = 3 Feet = 36 inches
1 Foot = 12 Inches
1 Mil = 0.001 Inch
1 Fathom = 2 Yards = 6 Feet
1 Rod = 5.5 Yards = 16.5 Feet
1 Hand = 4 Inches
1 Span = 9 Inches
1 Micro-inch = One Millionth Inch or 0.000001 Inch
1 Micron = One Millionth Meter + 0.00003937 Inch

Surveyor's Measure
1 Mile = 8 Furlongs = 80 Chains
1 Furlong = 10 Chains = 220 Yards
1 Chain = 4 Rods = 22 Yards = 66 Feet = 100 Links
1 Link = 7.92 Inches

Square Measure
1 Square Mile = 640 Acres = 6400 Square Chains
1 Acre = 10 Square Chains = 4840 Square Yards =
    43,560 Sq. Ft.
1 Square Chain = 16 Square Rods = 484 Square Yards =
    4356 Sq. Ft.
1 Square Rod = 30.25 Square Yards = 272.25 Square Feet = 625 Square
    Lines
1 Square Yard = 9 Square Feet
1 Square Foot = 144 Square Inches
An Acre equals a Square 208.7 Feet per Side

Cubic Measure
1 Cubic Yard = 27 Cubic Feet
1 Cubic Foot = 1728 Cubic Inches
1 Cord of Wood = 4 x 4 x 8 Feet = 128 Cubic Feet
1 Perch of Masonry = 16½ x 1½ x 1 Foot = 24.75 Cubic Feet

Avoirdupois or Commercial Weight
1 Gross or Long Ton = 2240 Pounds
1 Net or Short Ton = 2000 Pounds
1 Pound = 16 Ounces = 7000 Grains
1 Ounce = 16 Drachms = 437.5 Grains
1 Stone = 14 Pounds

Power
1 British Thermal Unit per Hour = 0.2931 Watts
1 Ton (Refrigeration) = 3.517 Kilowatts
1 Horsepower (Boiler) = 9.81 Kilowatts
1 Horsepower (550 ft-lb/s) = 0.746 Kilowatts

Shipping Measure
For Measuring Internal Capacity of a Vessel:
    1 Register Ton = 100 Cubic Feet

For Measurement of Cargo:
    Approximately 40 Cubic Feet of Merchandise is considered a Shipping
    Ton, unless that bulk would weigh more than 2000 Pounds, in which case
    Freight Charge may be based upon weight.

40 Cubic Feet = 32.143 U.S. Bushels = 31.16 Imp. Bushels

Liquid Measure
1 Imperial Gallon = 1.2009 U.S. Gallon = 277.42 Cu. In.
1 Cubic Foot = 7.48 U.S. Gallons

R011110-10   Architectural Fees

Tabulated below are typical percentage fees by project size, for good professional architectural service. Fees may vary from those listed depending upon degree of design difficulty and economic conditions in any particular area.

Rates can be interpolated horizontally and vertically. Various portions of the same project requiring different rates should be adjusted proportionately. For alterations, add 50% to the fee for the first $500,000 of project cost and add 25% to the fee for project cost over $500,000.

Architectural fees tabulated below include Structural, Mechanical and Electrical Engineering Fees. They do not include the fees for special consultants such as kitchen planning, security, acoustical, interior design, etc.

Civil Engineering fees are included in the Architectural fee for project sites requiring minimal design such as city sites. However, separate Civil Engineering fees must be added when utility connections require design, drainage calculations are needed, stepped foundations are required, or provisions are required to protect adjacent wetlands.

Building Types	Total Project Size in Thousands of Dollars						
	100	250	500	1,000	5,000	10,000	50,000
Factories, garages, warehouses, repetitive housing	9.0%	8.0%	7.0%	6.2%	5.3%	4.9%	4.5%
Apartments, banks, schools, libraries, offices, municipal buildings	12.2	12.3	9.2	8.0	7.0	6.6	6.2
Churches, hospitals, homes, laboratories, museums, research	15.0	13.6	12.7	11.9	9.5	8.8	8.0
Memorials, monumental work, decorative furnishings	—	16.0	14.5	13.1	10.0	9.0	8.3

R011110-30   Engineering Fees

Typical **Structural Engineering Fees** based on type of construction and total project size. These fees are included in Architectural Fees.

Type of Construction	Total Project Size (in thousands of dollars)			
	$500	$500-$1,000	$1,000-$5,000	Over $5000
Industrial buildings, factories & warehouses	Technical payroll times 2.0 to 2.5	1.60%	1.25%	1.00%
Hotels, apartments, offices, dormitories, hospitals, public buildings, food stores		2.00%	1.70%	1.20%
Museums, banks, churches and cathedrals		2.00%	1.75%	1.25%
Thin shells, prestressed concrete, earthquake resistive		2.00%	1.75%	1.50%
Parking ramps, auditoriums, stadiums, convention halls, hangars & boiler houses		2.50%	2.00%	1.75%
Special buildings, major alterations, underpinning & future expansion	↓	Add to above 0.5%	Add to above 0.5%	Add to above 0.5%

For complex reinforced concrete or unusually complicated structures, add 20% to 50%.

Typical **Mechanical and Electrical Engineering Fees** are based on the size of the subcontract. The fee structure for both is shown below. These fees are included in Architectural Fees.

Type of Construction	Subcontract Size							
	$25,000	$50,000	$100,000	$225,000	$350,000	$500,000	$750,000	$1,000,000
Simple structures	6.4%	5.7%	4.8%	4.5%	4.4%	4.3%	4.2%	4.1%
Intermediate structures	8.0	7.3	6.5	5.6	5.1	5.0	4.9	4.8
Complex structures	10.1	9.0	9.0	8.0	7.5	7.5	7.0	7.0

For renovations, add 15% to 25% to applicable fee.

For customer support on your Mechanical Costs with RSMeans Data, call 800.448.8182.

711

R012153-10 Repair and Remodeling

Cost figures are based on new construction utilizing the most cost-effective combination of labor, equipment and material with the work scheduled in proper sequence to allow the various trades to accomplish their work in an efficient manner.

The costs for repair and remodeling work must be modified due to the following factors that may be present in any given repair and remodeling project.

1. Equipment usage curtailment due to the physical limitations of the project, with only hand-operated equipment being used.

2. Increased requirement for shoring and bracing to hold up the building while structural changes are being made and to allow for temporary storage of construction materials on above-grade floors.

3. Material handling becomes more costly due to having to move within the confines of an enclosed building. For multi-story construction, low capacity elevators and stairwells may be the only access to the upper floors.

4. Large amount of cutting and patching and attempting to match the existing construction is required. It is often more economical to remove entire walls rather than create many new door and window openings. This sort of trade-off has to be carefully analyzed.

5. Cost of protection of completed work is increased since the usual sequence of construction usually cannot be accomplished.

6. Economies of scale usually associated with new construction may not be present. If small quantities of components must be custom fabricated due to job requirements, unit costs will naturally increase. Also, if only

small work areas are available at a given time, job scheduling between trades becomes difficult and subcontractor quotations may reflect the excessive start-up and shut-down phases of the job.

7. Work may have to be done on other than normal shifts and may have to be done around an existing production facility which has to stay in production during the course of the repair and remodeling.

8. Dust and noise protection of adjoining non-construction areas can involve substantial special protection and alter usual construction methods.

9. Job may be delayed due to unexpected conditions discovered during demolition or removal. These delays ultimately increase construction costs.

10. Piping and ductwork runs may not be as simple as for new construction. Wiring may have to be snaked through walls and floors.

11. Matching "existing construction" may be impossible because materials may no longer be manufactured. Substitutions may be expensive.

12. Weather protection of existing structure requires additional temporary structures to protect building at openings.

13. On small projects, because of local conditions, it may be necessary to pay a tradesman for a minimum of four hours for a task that is completed in one hour.

All of the above areas can contribute to increased costs for a repair and remodeling project. Each of the above factors should be considered in the planning, bidding and construction stage in order to minimize the increased costs associated with repair and remodeling jobs.

R012153-60 Security Factors

Contractors entering, working in, and exiting secure facilities often lose productive time during a normal workday. The recommended allowances in this section are intended to provide for the loss of productivity by increasing labor costs. Note that different costs are associated with searches upon entry only and searches upon entry and exit. Time spent in a queue is unpredictable and not part of these allowances. Contractors should plan ahead for this situation.

Security checkpoints are designed to reflect the level of security required to gain access or egress. An extreme example is when contractors, along with any materials, tools, equipment, and vehicles, must be physically searched and have all materials, tools, equipment, and vehicles inventoried and documented prior to both entry and exit.

Physical searches without going through the documentation process represent the next level and take up less time.

Electronic searches—passing through a detector or x-ray machine with no documentation of materials, tools, equipment, and vehicles—take less time than physical searches.

Visual searches of materials, tools, equipment, and vehicles represent the next level of security.

Finally, access by means of an ID card or displayed sticker takes the least amount of time.

Another consideration is if the searches described above are performed each and every day, or if they are performed only on the first day with access granted by ID card or displayed sticker for the remainder of the project. The figures for this situation have been calculated to represent the initial check-in as described and subsequent entry by ID card or displayed sticker for up to 20 days on site. For the situation described above, where the time period is beyond 20 days, the impact on labor cost is negligible.

There are situations where tradespeople must be accompanied by an escort and observed during the work day. The loss of freedom of movement will slow down productivity for the tradesperson. Costs for the observer have not been included. Those costs are normally born by the owner.

R012909-80   Sales Tax by State

State sales tax on materials is tabulated below (5 states have no sales tax). Many states allow local jurisdictions, such as a county or city, to levy additional sales tax.

Some projects may be sales tax exempt, particularly those constructed with public funds.

State	Tax (%)	State	Tax (%)	State	Tax (%)	State	Tax (%)
Alabama	4	Illinois	6.25	Montana	0	Rhode Island	7
Alaska	0	Indiana	7	Nebraska	5.5	South Carolina	6
Arizona	5.6	Iowa	6	Nevada	6.85	South Dakota	4.5
Arkansas	6.5	Kansas	6.5	New Hampshire	0	Tennessee	7
California	7.25	Kentucky	6	New Jersey	6.625	Texas	6.25
Colorado	2.9	Louisiana	4.45	New Mexico	5.125	Utah	4.85
Connecticut	6.35	Maine	5.5	New York	4	Vermont	6
Delaware	0	Maryland	6	North Carolina	4.75	Virginia	4.3
District of Columbia	6	Massachusetts	6.25	North Dakota	5	Washington	6.5
Florida	6	Michigan	6	Ohio	5.75	West Virginia	6
Georgia	4	Minnesota	6.875	Oklahoma	4.5	Wisconsin	5
Hawaii	4	Mississippi	7	Oregon	0	Wyoming	4
Idaho	6	Missouri	4.225	Pennsylvania	6	Average	5.06 %

Sales Tax by Province (Canada)

GST - a value-added tax, which the government imposes on most goods and services provided in or imported into Canada. PST - a retail sales tax, which five of the provinces impose on the prices of most goods and some services. QST - a value-added tax, similar to the federal GST, which Quebec imposes. HST - Three provinces have combined their retail sales taxes with the federal GST into one harmonized tax.

Province	PST (%)	QST(%)	GST(%)	HST (%)
Alberta	0	0	5	0
British Columbia	7	0	5	0
Manitoba	7	0	5	0
New Brunswick	0	0	0	15
Newfoundland	0	0	0	15
Northwest Territories	0	0	5	0
Nova Scotia	0	0	0	15
Ontario	0	0	0	13
Prince Edward Island	0	0	0	15
Quebec	9.975	0	5	14.975
Saskatchewan	6	0	5	0
Yukon	0	0	5	0

For customer support on your Mechanical Costs with RSMeans Data, call 800.448.8182.

713

R012909-85   Unemployment Taxes and Social Security Taxes

State unemployment tax rates vary not only from state to state, but also with the experience rating of the contractor. The federal unemployment tax rate is 6.0% of the first $7,000 of wages. This is reduced by a credit of up to 5.4% for timely payment to the state. The minimum federal unemployment tax is 0.6% after all credits.

Social security (FICA) for 2020 is estimated at time of publication to be 7.65% of wages up to $137,100.

R012909-86   Unemployment Tax by State

Information is from the U.S. Department of Labor, state unemployment tax rates.

State	Tax (%)	State	Tax (%)	State	Tax (%)	State	Tax (%)
Alabama	6.80	Illinois	6.93	Montana	6.30	Rhode Island	9.49
Alaska	5.4	Indiana	7.4	Nebraska	5.4	South Carolina	5.46
Arizona	12.76	Iowa	7.5	Nevada	5.4	South Dakota	9.35
Arkansas	14.3	Kansas	7.6	New Hampshire	7.5	Tennessee	10.0
California	6.2	Kentucky	9.3	New Jersey	5.8	Texas	6.5
Colorado	8.15	Louisiana	6.2	New Mexico	5.4	Utah	7.1
Connecticut	6.8	Maine	5.46	New York	9.1	Vermont	7.7
Delaware	8.20	Maryland	7.50	North Carolina	5.76	Virginia	6.21
District of Columbia	7	Massachusetts	12.65	North Dakota	10.74	Washington	7.73
Florida	5.4	Michigan	10.3	Ohio	9	West Virginia	8.5
Georgia	5.4	Minnesota	9.1	Oklahoma	5.5	Wisconsin	12.0
Hawaii	5.6	Mississippi	5.6	Oregon	5.4	Wyoming	8.8
Idaho	5.4	Missouri	8.37	Pennsylvania	11.03	Median	7.40%

R012909-90   Overtime

One way to improve the completion date of a project or eliminate negative float from a schedule is to compress activity duration times. This can be achieved by increasing the crew size or working overtime with the proposed crew.

To determine the costs of working overtime to compress activity duration times, consider the following examples. Below is an overtime efficiency and

cost chart based on a five, six, or seven day week with an eight through twelve hour day. Payroll percentage increases for time and one half and double times are shown for the various working days.

Days per Week	Hours per Day	Production Efficiency					Payroll Cost Factors	
		1st Week	2nd Week	3rd Week	4th Week	Average 4 Weeks	@ 1-1/2 Times	@ 2 Times
5	8	100%	100%	100%	100%	100%	1.000	1.000
	9	100	100	95	90	96	1.056	1.111
	10	100	95	90	85	93	1.100	1.200
	11	95	90	75	65	81	1.136	1.273
	12	90	85	70	60	76	1.167	1.333
6	8	100	100	95	90	96	1.083	1.167
	9	100	95	90	85	93	1.130	1.259
	10	95	90	85	80	88	1.167	1.333
	11	95	85	70	65	79	1.197	1.394
	12	90	80	65	60	74	1.222	1.444
7	8	100	95	85	75	89	1.143	1.286
	9	95	90	80	70	84	1.183	1.365
	10	90	85	75	65	79	1.214	1.429
	11	85	80	65	60	73	1.240	1.481
	12	85	75	60	55	69	1.262	1.524

R013113-40 Builder's Risk Insurance

Builder's risk insurance is insurance on a building during construction.
Premiums are paid by the owner or the contractor. Blasting, collapse and
underground insurance would raise total insurance costs.

R013113-50 General Contractor's Overhead

There are two distinct types of overhead on a construction project: Project
overhead and main office overhead. Project overhead includes those
costs at a construction site not directly associated with the installation of
construction materials. Examples of project overhead costs include the
following:

1. Superintendent
2. Construction office and storage trailers
3. Temporary sanitary facilities
4. Temporary utilities
5. Security fencing
6. Photographs
7. Cleanup
8. Performance and payment bonds

The above project overhead items are also referred to as general requirements
and therefore are estimated in Division 1. Division 1 is the first division
listed in the CSI MasterFormat but it is usually the last division estimated. The
sum of the costs in Divisions 1 through 49 is referred to as the sum of
the direct costs.

All construction projects also include indirect costs. The primary components
of indirect costs are the contractor's main office overhead and profit. The
amount of the main office overhead expense varies depending on the
following:

1. Owner's compensation
2. Project managers' and estimators' wages
3. Clerical support wages
4. Office rent and utilities
5. Corporate legal and accounting costs
6. Advertising
7. Automobile expenses
8. Association dues
9. Travel and entertainment expenses

These costs are usually calculated as a percentage of annual sales volume.
This percentage can range from 35% for a small contractor doing less than
$500,000 to 5% for a large contractor with sales in excess of $100 million.

R013113-55 Installing Contractor's Overhead

Installing contractors (subcontractors) also incur costs for general
requirements and main office overhead.

Included within the total incl. overhead and profit costs is a percent mark-up
for overhead that includes:

1. Compensation and benefits for office staff and project managers
2. Office rent, utilities, business equipment, and maintenance
3. Corporate legal and accounting costs

4. Advertising
5. Vehicle expenses (for office staff and project managers)
6. Association dues
7. Travel, entertainment
8. Insurance
9. Small tools and equipment

For customer support on your Mechanical Costs with RSMeans Data, call 800.448.8182.

715

R013113-60 Workers' Compensation Insurance Rates by Trade

The table below tabulates the national averages for workers' compensation insurance rates by trade and type of building. The average "Insurance Rate" is multiplied by the "% of Building Cost" for each trade. This produces the "Workers' Compensation" cost by % of total labor cost, to be added for each trade by building type to determine the weighted average workers' compensation rate for the building types analyzed.

Trade	Insurance Rate (% Labor Cost) Range			Insurance Rate Average	% of Building Cost Office Bldgs.	Schools & Apts.	Mfg.	Workers' Compensation Office Bldgs.	Schools & Apts.	Mfg.
Excavation, Grading, etc.	2.3 %	to	16.9%	9.6%	4.8%	4.9%	4.5%	0.46%	0.47%	0.43%
Piles & Foundations	3.3	to	28.0	15.7	7.1	5.2	8.7	1.11	0.82	1.37
Concrete	3.1	to	25.8	14.4	5.0	14.8	3.7	0.72	2.13	0.53
Masonry	3.3	to	52.1	27.7	6.9	7.5	1.9	1.91	2.08	0.53
Structural Steel	4.4	to	31.8	18.1	10.7	3.9	17.6	1.94	0.71	3.19
Miscellaneous & Ornamental Metals	2.9	to	21.4	12.2	2.8	4.0	3.6	0.34	0.49	0.44
Carpentry & Millwork	3.4	to	29.1	16.2	3.7	4.0	0.5	0.60	0.65	0.08
Metal or Composition Siding	4.8	to	124.6	64.7	2.3	0.3	4.3	1.49	0.19	2.78
Roofing	4.8	to	110.4	57.6	2.3	2.6	3.1	1.32	1.50	1.79
Doors & Hardware	3.1	to	29.1	16.1	0.9	1.4	0.4	0.14	0.23	0.06
Sash & Glazing	3.9	to	21.6	12.8	3.5	4.0	1.0	0.45	0.51	0.13
Lath & Plaster	2.7	to	30.1	16.4	3.3	6.9	0.8	0.54	1.13	0.13
Tile, Marble & Floors	2.0	to	19.0	10.5	2.6	3.0	0.5	0.27	0.32	0.05
Acoustical Ceilings	1.7	to	29.7	15.7	2.4	0.2	0.3	0.38	0.03	0.05
Painting	3.3	to	37.4	20.3	1.5	1.6	1.6	0.30	0.32	0.32
Interior Partitions	3.4	to	29.1	16.2	3.9	4.3	4.4	0.63	0.70	0.71
Miscellaneous Items	1.9	to	97.7	10.3	5.2	3.7	9.7	0.54	0.38	1.00
Elevators	1.3	to	9.0	5.1	2.1	1.1	2.2	0.11	0.06	0.11
Sprinklers	1.8	to	14.6	8.2	0.5	—	2.0	0.04	—	0.16
Plumbing	1.4	to	13.3	7.4	4.9	7.2	5.2	0.36	0.53	0.38
Heat., Vent., Air Conditioning	2.8	to	15.8	9.3	13.5	11.0	12.9	1.26	1.02	1.20
Electrical	1.7	to	10.7	6.2	10.1	8.4	11.1	0.63	0.52	0.69
Total	1.3 %	to	124.6%	—	100.0%	100.0%	100.0%	15.54%	14.79%	16.13%
			Overall Weighted Average	15.49%						

Workers' Compensation Insurance Rates by States

The table below lists the weighted average Workers' Compensation base rate for each state with a factor comparing this with the national average of 9.9%.

State	Weighted Average	Factor	State	Weighted Average	Factor	State	Weighted Average	Factor
Alabama	12.4%	126	Kentucky	10.5%	108	North Dakota	6.4%	65
Alaska	8.6	88	Louisiana	17.5	178	Ohio	5.5	56
Arizona	7.7	79	Maine	8.0	81	Oklahoma	7.4	76
Arkansas	5.1	52	Maryland	9.7	99	Oregon	7.5	76
California	19.6	200	Massachusetts	8.8	90	Pennsylvania	17.2	175
Colorado	6.2	63	Michigan	6.3	64	Rhode Island	10.0	102
Connecticut	12.4	126	Minnesota	13.4	137	South Carolina	16.2	165
Delaware	11.8	120	Mississippi	9.2	94	South Dakota	7.6	77
District of Columbia	7.7	78	Missouri	12.0	122	Tennessee	6.0	61
Florida	8.4	86	Montana	7.1	73	Texas	4.9	50
Georgia	30.0	307	Nebraska	11.7	120	Utah	5.7	58
Hawaii	8.4	86	Nevada	7.6	77	Vermont	8.9	91
Idaho	7.8	79	New Hampshire	8.8	90	Virginia	6.7	69
Illinois	17.5	179	New Jersey	14.2	145	Washington	7.3	75
Indiana	3.1	32	New Mexico	13.0	132	West Virginia	4.2	42
Iowa	9.9	101	New York	15.6	159	Wisconsin	11.2	115
Kansas	5.6	57	North Carolina	11.6	119	Wyoming	4.8	49
			Weighted Average for U.S. is	9.9% of payroll = 100%				

The weighted average skilled worker rate for 35 trades is 9.8%. For bidding purposes, apply the full value of Workers' Compensation directly to total labor costs, or if labor is 38%, materials 42% and overhead and profit 20% of total cost, carry 38/80 x 9.8% = 4.66% of cost (before overhead and profit) into overhead. Rates vary not only from state to state but also with the experience rating of the contractor.

Rates are the most current available at the time of publication.

716

General Requirements | R0131 Project Management & Coordination

R013113-80 Performance Bond

This table shows the cost of a Performance Bond for a construction job scheduled to be completed in 12 months. Add 1% of the premium cost per month for jobs requiring more than 12 months to complete. The rates are "standard" rates offered to contractors that the bonding company considers financially sound and capable of doing the work. Preferred rates are offered by some bonding companies based upon financial strength of the contractor. Actual rates vary from contractor to contractor and from bonding company to bonding company. Contractors should prequalify through a bonding agency before submitting a bid on a contract that requires a bond.

Contract Amount		Building Construction Class B Projects			Highways & Bridges					
					Class A New Construction			Class A-1 Highway Resurfacing		
First	$ 100,000 bid	$25.00 per M			$15.00 per M			$9.40 per M		
Next	400,000 bid	$ 2,500	plus $15.00	per M	$ 1,500	plus $10.00	per M	$ 940	plus $7.20	per M
Next	2,000,000 bid	8,500	plus 10.00	per M	5,500	plus 7.00	per M	3,820	plus 5.00	per M
Next	2,500,000 bid	28,500	plus 7.50	per M	19,500	plus 5.50	per M	15,820	plus 4.50	per M
Next	2,500,000 bid	47,250	plus 7.00	per M	33,250	plus 5.00	per M	28,320	plus 4.50	per M
Over	7,500,000 bid	64,750	plus 6.00	per M	45,750	plus 4.50	per M	39,570	plus 4.00	per M

General Requirements | R0154 Construction Aids

R015423-10 Steel Tubular Scaffolding

On new construction, tubular scaffolding is efficient up to 60' high or five stories. Above this it is usually better to use a hung scaffolding if construction permits. Swing scaffolding operations may interfere with tenants. In this case, the tubular is more practical at all heights.

In repairing or cleaning the front of an existing building the cost of tubular scaffolding per S.F. of building front increases as the height increases above the first tier. The first tier cost is relatively high due to leveling and alignment.

The minimum efficient crew for erecting and dismantling is three workers. They can set up and remove 18 frame sections per day up to 5 stories high. For 6 to 12 stories high, a crew of four is most efficient. Use two or more on top and two on the bottom for handing up or hoisting. They can also set up and remove 18 frame sections per day. At 7' horizontal spacing, this will run about 800 S.F. per day of erecting and dismantling. Time for placing and removing planks must be added to the above. A crew of three can place and remove 72 planks per day up to 5 stories. For over 5 stories, a crew of four can place and remove 80 planks per day.

The table below shows the number of pieces required to erect tubular steel scaffolding for 1000 S.F. of building frontage. This area is made up of a scaffolding system that is 12 frames (11 bays) long by 2 frames high.

For jobs under twenty-five frames, add 50% to rental cost. Rental rates will be lower for jobs over three months duration. Large quantities for long periods can reduce rental rates by 20%.

Description of Component	Number of Pieces for 1000 S.F. of Building Front	Unit
5' Wide Standard Frame, 6'-4" High	24	Ea.
Leveling Jack & Plate	24	
Cross Brace	44	
Side Arm Bracket, 21"	12	
Guardrail Post	12	
Guardrail, 7' section	22	
Stairway Section	2	
Stairway Starter Bar	1	
Stairway Inside Handrail	2	
Stairway Outside Handrail	2	
Walk-Thru Frame Guardrail	2	

Scaffolding is often used as falsework over 15' high during construction of cast-in-place concrete beams and slabs. Two foot wide scaffolding is generally used for heavy beam construction. The span between frames depends upon the load to be carried with a maximum span of 5'.

Heavy duty shoring frames with a capacity of 10,000#/leg can be spaced up to 10' O.C. depending upon form support design and loading.

Scaffolding used as horizontal shoring requires less than half the material required with conventional shoring.

On new construction, erection is done by carpenters.

Rolling towers supporting horizontal shores can reduce labor and speed the job. For maintenance work, catwalks with spans up to 70' can be supported by the rolling towers.

For customer support on your Mechanical Costs with RSMeans Data, call 800.448.8182.

717

Reference Tables

R015433-10 Contractor Equipment

Rental Rates shown elsewhere in the data set pertain to late model high quality machines in excellent working condition, rented from equipment dealers. Rental rates from contractors may be substantially lower than the rental rates from equipment dealers depending upon economic conditions; for older, less productive machines, reduce rates by a maximum of 15%. Any overtime must be added to the base rates. For shift work, rates are lower. Usual rule of thumb is 150% of one shift rate for two shifts; 200% for three shifts.

For periods of less than one week, operated equipment is usually more economical to rent than renting bare equipment and hiring an operator.

Costs to move equipment to a job site (mobilization) or from a job site (demobilization) are not included in rental rates, nor in any Equipment costs on any Unit Price line items or crew listings. These costs can be found elsewhere. If a piece of equipment is already at a job site, it is not appropriate to utilize mob/demob costs in an estimate again.

Rental rates vary throughout the country with larger cities generally having lower rates. Lease plans for new equipment are available for periods in excess of six months with a percentage of payments applying toward purchase.

Rental rates can also be treated as reimbursement costs for contractor-owned equipment. Owned equipment costs include depreciation, loan payments, interest, taxes, insurance, storage, and major repairs.

Monthly rental rates vary from 2% to 5% of the cost of the equipment depending on the anticipated life of the equipment and its wearing parts. Weekly rates are about 1/3 the monthly rates and daily rental rates are about 1/3 the weekly rates.

The hourly operating costs for each piece of equipment include costs to the user such as fuel, oil, lubrication, normal expendables for the equipment, and a percentage of the mechanic's wages chargeable to maintenance. The hourly operating costs listed do not include the operator's wages.

The daily cost for equipment used in the standard crews is figured by dividing the weekly rate by five, then adding eight times the hourly operating cost to give the total daily equipment cost, not including the operator. This figure is in the right hand column of the Equipment listings under Equipment Cost/Day.

Pile Driving rates shown for the pile hammer and extractor do not include leads, cranes, boilers or compressors. Vibratory pile driving requires an added field specialist during set-up and pile driving operation for the electric model. The hydraulic model requires a field specialist for set-up only. Up to 125 reuses of sheet piling are possible using vibratory drivers. For normal conditions, crane capacity for hammer type and size is as follows.

Crane Capacity	Hammer Type and Size		
	Air or Steam	Diesel	Vibratory
25 ton	to 8,750 ft.-lb.		70 H.P.
40 ton	15,000 ft.-lb.	to 32,000 ft.-lb.	170 H.P.
60 ton	25,000 ft.-lb.		300 H.P.
100 ton		112,000 ft.-lb.	

Cranes should be specified for the job by size, building and site characteristics, availability, performance characteristics, and duration of time required.

Backhoes & Shovels rent for about the same as equivalent size cranes but maintenance and operating expenses are higher. The crane operator's rate must be adjusted for high boom heights. Average adjustments: for 150' boom add 2% per hour; over 185', add 4% per hour; over 210', add 6% per hour; over 250', add 8% per hour and over 295', add 12% per hour.

Tower Cranes of the climbing or static type have jibs from 50' to 200' and capacities at maximum reach range from 4,000 to 14,000 pounds. Lifting capacities increase up to maximum load as the hook radius decreases.

Typical rental rates, based on purchase price, are about 2% to 3% per month.

Erection and dismantling run between 500 and 2000 labor hours. Climbing operation takes 10 labor hours per 20' climb. Crane dead time is about 5 hours per 40' climb. If crane is bolted to side of the building add cost of ties and extra mast sections. Climbing cranes have from 80' to 180' of mast while static cranes have 80' to 800' of mast.

Truck Cranes can be converted to tower cranes by using tower attachments. Mast heights over 400' have been used.

A single 100' high material **Hoist and Tower** can be erected and dismantled in about 400 labor hours; a double 100' high hoist and tower in about 600 labor hours. Erection times for additional heights are 3 and 4 labor hours

per vertical foot respectively up to 150', and 4 to 5 labor hours per vertical foot over 150' high. A 40' high portable Buck hoist takes about 160 labor hours to erect and dismantle. Additional heights take 2 labor hours per vertical foot to 80' and 3 labor hours per vertical foot for the next 100'. Most material hoists do not meet local code requirements for carrying personnel.

A 150' high **Personnel Hoist** requires about 500 to 800 labor hours to erect and dismantle. Budget erection time at 5 labor hours per vertical foot for all trades. Local code requirements or labor scarcity requiring overtime can add up to 50% to any of the above erection costs.

Earthmoving Equipment: The selection of earthmoving equipment depends upon the type and quantity of material, moisture content, haul distance, haul road, time available, and equipment available. Short haul cut and fill operations may require dozers only, while another operation may require excavators, a fleet of trucks, and spreading and compaction equipment. Stockpiled material and granular material are easily excavated with front end loaders. Scrapers are most economically used with hauls between 300' and 1-1/2 miles if adequate haul roads can be maintained. Shovels are often used for blasted rock and any material where a vertical face of 8' or more can be excavated. Special conditions may dictate the use of draglines, clamshells, or backhoes. Spreading and compaction equipment must be matched to the soil characteristics, the compaction required and the rate the fill is being supplied.

R015433-15 Heavy Lifting

Hydraulic Climbing Jacks

The use of hydraulic heavy lift systems is an alternative to conventional type crane equipment. The lifting, lowering, pushing, or pulling mechanism is a hydraulic climbing jack moving on a square steel jackrod from 1-5/8" to 4" square, or a steel cable. The jackrod or cable can be vertical or horizontal, stationary or movable, depending on the individual application. When the jackrod is stationary, the climbing jack will climb the rod and push or pull the load along with itself. When the climbing jack is stationary, the jackrod is movable with the load attached to the end and the climbing jack will lift or lower the jackrod with the attached load. The heavy lift system is normally operated by a single control lever located at the hydraulic pump.

The system is flexible in that one or more climbing jacks can be applied wherever a load support point is required, and the rate of lift synchronized.

Economic benefits have been demonstrated on projects such as: erection of ground assembled roofs and floors, complete bridge spans, girders and trusses, towers, chimney liners and steel vessels, storage tanks, and heavy machinery. Other uses are raising and lowering offshore work platforms, caissons, tunnel sections and pipelines.

R015436-50   Mobilization

Costs to move rented construction equipment to a job site from an equipment dealer's or contractor's yard (mobilization) or off the job site (demobilization) are not included in the rental or operating rates, nor in the equipment cost on a unit price line or in a crew listing. These costs can be found consolidated in the Mobilization section of the data and elsewhere in particular site work sections. If a piece of equipment is already on the

job site, it is not appropriate to include mob/demob costs in a new estimate that requires use of that equipment. The following table identifies approximate sizes of rented construction equipment that would be hauled on a towed trailer. Because this listing is not all-encompassing, the user can infer as to what size trailer might be required for a piece of equipment not listed.

3-ton Trailer	20-ton Trailer	40-ton Trailer	50-ton Trailer
20 H.P. Excavator	110 H.P. Excavator	200 H.P. Excavator	270 H.P. Excavator
50 H.P. Skid Steer	165 H.P. Dozer	300 H.P. Dozer	Small Crawler Crane
35 H.P. Roller	150 H.P. Roller	400 H.P. Scraper	500 H.P. Scraper
40 H.P. Trencher	Backhoe	450 H.P. Art. Dump Truck	500 H.P. Art. Dump Truck

Reference Tables

R024119-10 Demolition Defined

Whole Building Demolition - Demolition of the whole building with no concern for any particular building element, component, or material type being demolished. This type of demolition is accomplished with large pieces of construction equipment that break up the structure, load it into trucks and haul it to a disposal site, but disposal or dump fees are not included. Demolition of below-grade foundation elements, such as footings, foundation walls, grade beams, slabs on grade, etc., is not included. Certain mechanical equipment containing flammable liquids or ozone-depleting refrigerants, electric lighting elements, communication equipment components, and other building elements may contain hazardous waste, and must be removed, either selectively or carefully, as hazardous waste before the building can be demolished.

Foundation Demolition - Demolition of below-grade foundation footings, foundation walls, grade beams, and slabs on grade. This type of demolition is accomplished by hand or pneumatic hand tools, and does not include saw cutting, or handling, loading, hauling, or disposal of the debris.

Gutting - Removal of building interior finishes and electrical/mechanical systems down to the load-bearing and sub-floor elements of the rough building frame, with no concern for any particular building element, component, or material type being demolished. This type of demolition is accomplished by hand or pneumatic hand tools, and includes loading into trucks, but not hauling, disposal or dump fees, scaffolding, or shoring. Certain mechanical equipment containing flammable liquids or ozone-depleting refrigerants, electric lighting elements, communication equipment components, and other building elements may contain hazardous waste, and must be removed, either selectively or carefully, as hazardous waste, before the building is gutted.

Selective Demolition - Demolition of a selected building element, component, or finish, with some concern for surrounding or adjacent elements, components, or finishes (see the first Subdivision (s) at the beginning of appropriate Divisions). This type of demolition is accomplished by hand or pneumatic hand tools, and does not include handling, loading,

storing, hauling, or disposal of the debris, scaffolding, or shoring. "Gutting" methods may be used in order to save time, but damage that is caused to surrounding or adjacent elements, components, or finishes may have to be repaired at a later time.

Careful Removal - Removal of a piece of service equipment, building element or component, or material type, with great concern for both the removed item and surrounding or adjacent elements, components or finishes. The purpose of careful removal may be to protect the removed item for later re-use, preserve a higher salvage value of the removed item, or replace an item while taking care to protect surrounding or adjacent elements, components, connections, or finishes from cosmetic and/or structural damage. An approximation of the time required to perform this type of removal is 1/3 to 1/2 the time it would take to install a new item of like kind (see Reference Number R220105-10). This type of removal is accomplished by hand or pneumatic hand tools, and does not include loading, hauling, or storing the removed item, scaffolding, shoring, or lifting equipment.

Cutout Demolition - Demolition of a small quantity of floor, wall, roof, or other assembly, with concern for the appearance and structural integrity of the surrounding materials. This type of demolition is accomplished by hand or pneumatic hand tools, and does not include saw cutting, handling, loading, hauling, or disposal of debris, scaffolding, or shoring.

Rubbish Handling - Work activities that involve handling, loading or hauling of debris. Generally, the cost of rubbish handling must be added to the cost of all types of demolition, with the exception of whole building demolition.

Minor Site Demolition - Demolition of site elements outside the footprint of a building. This type of demolition is accomplished by hand or pneumatic hand tools, or with larger pieces of construction equipment, and may include loading a removed item onto a truck (check the Crew for equipment used). It does not include saw cutting, hauling or disposal of debris, and, sometimes, handling or loading.

R024119-20 Dumpsters

Dumpster rental costs on construction sites are presented in two ways.

The cost per week rental includes the delivery of the dumpster; its pulling or emptying once per week, and its final removal. The assumption is made that the dumpster contractor could choose to empty a dumpster by simply bringing in an empty unit and removing the full one. These costs also include the disposal of the materials in the dumpster.

The Alternate Pricing can be used when actual planned conditions are not approximated by the weekly numbers. For example, these lines can be used when a dumpster is needed for 4 weeks and will need to be emptied 2 or 3 times per week. Conversely the Alternate Pricing lines can be used when a dumpster will be rented for several weeks or months but needs to be emptied only a few times over this period.

R026510-20 Underground Storage Tank Removal

Underground Storage Tank Removal can be divided into two categories: Non-Leaking and Leaking. Prior to removing an underground storage tank, tests should be made, with the proper authorities present, to determine whether a tank has been leaking or the surrounding soil has been contaminated.

To safely remove Liquid Underground Storage Tanks:
1. Excavate to the top of the tank.
2. Disconnect all piping.
3. Open all tank vents and access ports.
4. Remove all liquids and/or sludge.
5. Purge the tank with an inert gas.
6. Provide access to the inside of the tank and clean out the interior using proper personal protective equipment (PPE).
7. Excavate soil surrounding the tank using proper PPE for on-site personnel.
8. Pull and properly dispose of the tank.
9. Clean up the site of all contaminated material.
10. Install new tanks or close the excavation.

Existing Conditions   R0282 Asbestos Remediation

R028213-20   Asbestos Removal Process

Asbestos removal is accomplished by a specialty contractor who understands the federal and state regulations regarding the handling and disposal of the material. The process of asbestos removal is divided into many individual steps. An accurate estimate can be calculated only after all the steps have been priced.

The steps are generally as follows:

1. Obtain an asbestos abatement plan from an industrial hygienist.
2. Monitor the air quality in and around the removal area and along the path of travel between the removal area and transport area. This establishes the background contamination.
3. Construct a two part decontamination chamber at entrance to removal area.
4. Install a HEPA filter to create a negative pressure in the removal area.
5. Install wall, floor and ceiling protection as required by the plan, usually 2 layers of fireproof 6 mil polyethylene.
6. Industrial hygienist visually inspects work area to verify compliance with plan.
7. Provide temporary supports for conduit and piping affected by the removal process.
8. Proceed with asbestos removal and bagging process. Monitor air quality as described in Step #2. Discontinue operations when contaminate levels exceed applicable standards.
9. Document the legal disposal of materials in accordance with EPA standards.
10. Thoroughly clean removal area including all ledges, crevices and surfaces.
11. Post abatement inspection by industrial hygienist to verify plan compliance.
12. Provide a certificate from a licensed industrial hygienist attesting that contaminate levels are within acceptable standards before returning area to regular use.

Thermal & Moist. Protec.   R0784 Firestopping

R078413-30   Firestopping

Firestopping is the sealing of structural, mechanical, electrical and other penetrations through fire-rated assemblies. The basic components of firestop systems are safing insulation and firestop sealant on both sides of wall penetrations and the top side of floor penetrations.

Pipe penetrations are assumed to be through concrete, grout, or joint compound and can be sleeved or unsleeved. Costs for the penetrations and sleeves are not included. An annular space of 1″ is assumed. Escutcheons are not included.

A metallic pipe is assumed to be copper, aluminum, cast iron or similar metallic material. An insulated metallic pipe is assumed to be covered with a thermal insulating jacket of varying thickness and materials.

A non-metallic pipe is assumed to be PVC, CPVC, FR Polypropylene or similar plastic piping material. Intumescent firestop sealants or wrap strips are included. Collars on both sides of wall penetrations and a sheet metal plate on the underside of floor penetrations are included.

Ductwork is assumed to be sheet metal, stainless steel or similar metallic material. Duct penetrations are assumed to be through concrete, grout or joint compound. Costs for penetrations and sleeves are not included. An annular space of 1/2″ is assumed.

Multi-trade openings include costs for sheet metal forms, firestop mortar, wrap strips, collars and sealants as necessary.

Structural penetrations joints are assumed to be 1/2″ or less. CMU walls are assumed to be within 1-1/2″ of the metal deck. Drywall walls are assumed to be tight to the underside of metal decking.

Metal panel, glass or curtain wall systems include a spandrel area of 5′ filled with mineral wool foil-faced insulation. Fasteners and stiffeners are included.

R099100-10   Painting Estimating Techniques

Proper estimating methodology is needed to obtain an accurate painting estimate. There is no known reliable shortcut or square foot method. The following steps should be followed:

- List all surfaces to be painted, with an accurate quantity (area) of each. Items having similar surface condition, finish, application method and accessibility may be grouped together.

- List all the tasks required for each surface to be painted, including surface preparation, masking, and protection of adjacent surfaces. Surface preparation may include minor repairs, washing, sanding and puttying.

- Select the proper Means line for each task. Review and consider all adjustments to labor and materials for type of paint and location of work. Apply the height adjustment carefully. For instance, when applying the adjustment for work over 8' high to a wall that is 12' high, apply the adjustment only to the area between 8' and 12' high, and not to the entire wall.

When applying more than one percent (%) adjustment, apply each to the base cost of the data, rather than applying one percentage adjustment on top of the other.

When estimating the cost of painting walls and ceilings remember to add the brushwork for all cut-ins at inside corners and around windows and doors as a LF measure. One linear foot of cut-in with a brush equals one square foot of painting.

All items for spray painting include the labor for roll-back.

Deduct for openings greater than 100 SF or openings that extend from floor to ceiling and are greater than 5' wide. Do not deduct small openings.

The cost of brushes, rollers, ladders and spray equipment are considered part of a painting contractor's overhead, and should not be added to the estimate. The cost of rented equipment such as scaffolding and swing staging should be added to the estimate.

R220102-20   Labor Adjustment Factors

Labor Adjustment Factors are provided for Divisions 21, 22, and 23 to assist the mechanical estimator account for the various complexities and special conditions of any particular project. While a single percentage has been entered on each line of Division 22 01 02.20, it should be understood that these are just suggested midpoints of ranges of values commonly used by mechanical estimators. They may be increased or decreased depending on the severity of the special conditions.

The group for "existing occupied buildings" has been the subject of requests for explanation. Actually there are two stages to this group: buildings that are existing and "finished" but unoccupied, and those that also are occupied. Buildings that are "finished" may result in higher labor costs due to the

workers having to be more careful not to damage finished walls, ceilings, floors, etc. and may necessitate special protective coverings and barriers. Also corridor bends and doorways may not accommodate long pieces of pipe or larger pieces of equipment. Work above an already hung ceiling can be very time consuming. The addition of occupants may force the work to be done on premium time (nights and/or weekends), eliminate the possible use of some preferred tools such as pneumatic drivers, powder charged drivers, etc. The estimator should evaluate the access to the work area and just how the work is going to be accomplished to arrive at an increase in labor costs over "normal" new construction productivity.

R220105-10   Demolition (Selective vs. Removal for Replacement)

Demolition can be divided into two basic categories.

One type of demolition involves the removal of material with no concern for its replacement. The labor-hours to estimate this work are found under "Selective Demolition" in the Fire Protection, Plumbing and HVAC Divisions. It is selective in that individual items or all the material installed as a system or trade grouping such as plumbing or heating systems are removed. This may be accomplished by the easiest way possible, such as sawing, torch cutting, or sledge hammering as well as simple unbolting.

The second type of demolition is the removal of some items for repair or replacement. This removal may involve careful draining, opening of unions,

disconnecting and tagging electrical connections, capping pipes/ducts to prevent entry of debris or leakage of the material contained as well as transporting the item away from its in-place location to a truck/dumpster. An approximation of the time required to accomplish this type of demolition is to use half of the time indicated as necessary to install a new unit. For example: installation of a new pump might be listed as requiring 6 labor-hours so if we had to estimate the removal of the old pump we would allow an additional 3 hours for a total of 9 hours. That is, the complete replacement of a defective pump with a new pump would be estimated to take 9 labor-hours.

R220523-80   Valve Materials

VALVE MATERIALS

Bronze:
Bronze is one of the oldest materials used to make valves. It is most commonly used in hot and cold water systems and other non-corrosive services. It is often used as a seating surface in larger iron body valves to ensure tight closure.

Carbon Steel:
Carbon steel is a high strength material. Therefore, valves made from this metal are used in higher pressure services, such as steam lines up to 600 psi at 850° F. Many steel valves are available with butt-weld ends for economy and are generally used in high pressure steam service as well as other higher pressure non-corrosive services.

Forged Steel:
Valves from tough carbon steel are used in service up to 2000 psi and temperatures up to 1000° F in Gate, Globe and Check valves.

Iron:
Valves are normally used in medium to large pipe lines to control non-corrosive fluid and gases, where pressures do not exceed 250 psi at 450° F or 500 psi cold water, oil or gas.

Stainless Steel:
Developed steel alloys can be used in over 90% corrosive services.

Plastic PVC:
This is used in a great variety of valves generally in high corrosive service with lower temperatures and pressures.

VALVE SERVICE PRESSURES

Pressure ratings on valves provide an indication of the safe operating pressure for a valve at some elevated temperature. This temperature is dependent upon the materials used and the fabrication of the valve. When specific data is not available, a good "rule-of-thumb" to follow is the temperature of saturated steam at the primary rating indicated on the valve body. Example: The valve has the number 150S printed on the side indicating 150 psi and hence, a maximum operating temperature of 367° F (temperature of saturated steam and 150 psi).

DEFINITIONS
1. "WOG" – Water, oil, gas (cold working pressures).
2. "SWP" – Steam working pressure.
3. 100% area (full port) – means the area through the valve is equal to or greater than the area of standard pipe.
4. "Standard Opening" – means that the area through the valve is less than the area of standard pipe and therefore these valves should be used only where restriction of flow is unimportant.
5. "Round Port" – means the valve has a full round opening through the plug and body, of the same size and area as standard pipe.
6. "Rectangular Port" – valves have rectangular shaped ports through the plug body. The area of the port is either equal to 100% of the area of standard pipe, or restricted (standard opening). In either case it is clearly marked.
7. "ANSI" – American National Standards Institute.

R220523-90   Valve Selection Considerations

INTRODUCTION: In any piping application, valve performance is critical. Valves should be selected to give the best performance at the lowest cost.

The following is a list of performance characteristics generally expected of valves.
1. Stopping flow or starting it.
2. Throttling flow (Modulation).
3. Flow direction changing.
4. Checking backflow (Permitting flow in only one direction).
5. Relieving or regulating pressure.

In order to properly select the right valve, some facts must be determined.

A. What liquid or gas will flow through the valve?

B. Does the fluid contain suspended particles?

C. Does the fluid remain in liquid form at all times?

D. Which metals does fluid corrode?

E. What are the pressure and temperature limits? (As temperature and pressure rise, so will the price of the valve.)

F. Is there constant line pressure?

G. Is the valve merely an on-off valve?

H. Will checking of backflow be required?

I. Will the valve operate frequently or infrequently?

Valves are classified by design type into such classifications as Gate, Globe, Angle, Check, Ball, Butterfly and Plug. They are also classified by end connection, stem, pressure restrictions and material such as bronze, cast iron, etc. Each valve has a specific use. A quality valve used correctly will provide a lifetime of trouble-free service, but a high quality valve installed in the wrong service may require frequent attention.

STEM TYPES
(OS & Y)—Rising Stem-Outside Screw and Yoke

Offers a visual indication of whether the valve is open or closed. Recommended where high temperatures, corrosives, and solids in the line might cause damage to inside-valve stem threads. The stem threads are engaged by the yoke bushing so the stem rises through the hand wheel as it is turned.

(R.S.)—Rising Stem-Inside Screw

Adequate clearance for operation must be provided because both the hand wheel and the stem rise.
The valve wedge position is indicated by the position of the stem and hand wheel.

(N.R.S.)—Non-Rising Stem-Inside Screw

A minimum clearance is required for operating this type of valve. Excessive wear or damage to stem threads inside the valve may be caused by heat, corrosion, and solids. Because the hand wheel and stem do not rise, wedge position cannot be visually determined.

VALVE TYPES
Gate Valves

Provide full flow, minute pressure drop, minimum turbulence and minimum fluid trapped in the line.
They are normally used where operation is infrequent.

Globe Valves

Globe valves are designed for throttling and/or frequent operation with positive shut-off. Particular attention must be paid to the several types of seating materials available to avoid unnecessary wear. The seats must be compatible with the fluid in service and may be composition or metal. The configuration of the Globe valve opening causes turbulence which results in increased resistance. Most bronze Globe valves are rising stem-inside screw, but they are also available on O.S. & Y.

Angle Valves

The fundamental difference between the Angle valve and the Globe valve is the fluid flow through the Angle valve. It makes a 90° turn and offers less resistance to flow than the Globe valve while replacing an elbow. An Angle valve thus reduces the number of joints and installation time.

Check Valves

Check valves are designed to prevent backflow by automatically seating when the direction of fluid is reversed.
Swing Check valves are generally installed with Gate-valves, as they provide comparable full flow. Usually recommended for lines where flow velocities are low and should not be used on lines with pulsating flow. Recommended for horizontal installation, or in vertical lines only where flow is upward.

Lift Check Valves

These are commonly used with Globe and Angle valves since they have similar diaphragm seating arrangements and are recommended for preventing backflow of steam, air, gas and water, and on vapor lines with high flow velocities. For horizontal lines, horizontal lift checks should be used and vertical lift checks for vertical lines.

Ball Valves

Ball valves are light and easily installed, yet because of modern elastomeric seats, provide tight closure. Flow is controlled by rotating up to 90° a drilled ball which fits tightly against resilient seals. This ball seats with flow in either direction, and valve handle indicates the degree of opening. Recommended for frequent operation readily adaptable to automation, ideal for installation where space is limited.

Butterfly Valves

Butterfly valves provide bubble-tight closure with excellent throttling characteristics. They can be used for full-open, closed and for throttling applications.

The Butterfly valve consists of a disc within the valve body which is controlled by a shaft. In its closed position, the valve disc seals against a resilient seat. The disc position throughout the full 90° rotation is visually indicated by the position of the operator.

A Butterfly valve is only a fraction of the weight of a Gate valve and requires no gaskets between flanges in most cases. Recommended for frequent operation and adaptable to automation where space is limited.

Wafer and Lug type bodies when installed between two pipe flanges, can be easily removed from the line. The pressure of the bolted flanges holds the valve in place.
Locating lugs makes installation easier.

Plug Valves

Lubricated plug valves, because of the wide range of service to which they are adapted, may be classified as all purpose valves. They can be safely used at all pressure and vacuums, and at all temperatures up to the limits of available lubricants. They are the most satisfactory valves for the handling of gritty suspensions and many other destructive, erosive, corrosive and chemical solutions.

R221113-50   Pipe Material Considerations

1. Malleable fittings should be used for gas service.
2. Malleable fittings are used where there are stresses/strains due to expansion and vibration.
3. Cast fittings may be broken as an aid to disassembling heating lines frozen by long use, temperature and minerals.
4. A cast iron pipe is extensively used for underground and submerged service.
5. Type M (light wall) copper tubing is available in hard temper only and is used for nonpressure and less severe applications than K and L.
6. Type L (medium wall) copper tubing, available hard or soft for interior service.
7. Type K (heavy wall) copper tubing, available in hard or soft temper for use where conditions are severe. For underground and interior service.
8. Hard drawn tubing requires fewer hangers or supports but should not be bent. Silver brazed fittings are recommended, but soft solder is normally used.
9. Type DMV (very light wall) copper tubing designed for drainage, waste and vent plus other non-critical pressure services.

Domestic/Imported Pipe and Fittings Costs

The prices shown in this publication for steel/cast iron pipe and steel, cast iron, and malleable iron fittings are based on domestic production sold at the normal trade discounts. The above listed items of foreign manufacture may be available at prices 1/3 to 1/2 of those shown. Some imported items after minor machining or finishing operations are being sold as domestic to further complicate the system.

Caution: Most pipe prices in this data set also include a coupling and pipe hangers which for the larger sizes can add significantly to the per foot cost and should be taken into account when comparing "book cost" with the quoted supplier's cost.

For customer support on your Mechanical Costs with RSMeans Data, call 800.448.8182.

R221113-70 Piping to 10' High

When taking off pipe, it is important to identify the different material types and joining procedures, as well as distances between supports and components required for proper support.

During the takeoff, measure through all fittings. Do not subtract the lengths of the fittings, valves, or strainers, etc. This added length plus the final rounding of the totals will compensate for nipples and waste.

When rounding off totals always increase the actual amount to correspond with manufacturer's shipping lengths.

A. Both red brass and yellow brass pipe are normally furnished in 12' lengths, plain end. The Unit Price section includes in the linear foot costs two field threads and one coupling per 10' length. A carbon steel clevis type hanger assembly every 10' is also prorated into the linear foot costs, including both material and labor.

B. Cast iron soil pipe is furnished in either 5' or 10' lengths. For pricing purposes, the Unit Price section features 10' lengths with a joint and a carbon steel clevis hanger assembly every 5' prorated into the per foot costs of both material and labor.

Three methods of joining are considered: lead and oakum poured joints or push-on gasket type joints for the bell and spigot pipe and a joint clamp for the no-hub soil pipe. The labor and material costs for each of these individual joining procedures are also prorated into the linear costs per foot.

C. Copper tubing covers types K, L, M, and DWV which are furnished in 20' lengths. Means pricing data is based on a tubing cut each length and a coupling and two soft soldered joints every 10'. A carbon steel, clevis type hanger assembly every 10' is also prorated into the per foot costs. The prices for refrigeration tubing are for materials only. Labor for full lengths may be based on the type L labor but short cut measures in tight areas can increase the installation labor-hours from 20 to 40%.

D. Corrosion-resistant piping does not lend itself to one particular standard of hanging or support assembly due to its diversity of application and placement. The several varieties of corrosion-resistant piping do not include any material or labor costs for hanger assemblies (See the Unit Price section for appropriate selection).

E. Glass pipe is furnished in standard lengths either 5' or 10' long, beaded on one end. Special orders for diverse lengths beaded on both ends are also available. For pricing purposes, R.S. Means features 10' lengths with a coupling and a carbon steel band hanger assembly every 10' prorated into the per foot linear costs.

Glass pipe is also available with conical ends and standard lengths ranging from 6" through 3' in 6" increments, then up to 10' in 12" increments. Special lengths can be customized for particular installation requirements.

For pricing purposes, Means has based the labor and material pricing on 10' lengths. Included in these costs per linear foot are the prorated costs for a flanged assembly every 10' consisting of two flanges, a gasket, two insertable seals, and the required number of bolts and nuts. A carbon steel band hanger assembly based on 10' center lines has also been prorated into the costs per foot for labor and materials.

F. Plastic pipe of several compositions and joining methods are considered. Fiberglass reinforced pipe (FRP) is priced, based on 10' lengths (20' lengths are also available), with coupling and epoxy joints every 10'. FRP is furnished in both "General Service" and "High Strength." A carbon steel clevis hanger assembly, 3 for every 10', is built into the prorated labor and material costs on a per foot basis.

The PVC and CPVC pipe schedules 40, 80 and 120, plus SDR ratings are all based on 20' lengths with a coupling installed every 10', as well as a carbon steel clevis hanger assembly every 3'. The PVC and

ABS type DWV piping is based on 10' lengths with solvent weld couplings every 10', and with carbon steel clevis hanger assemblies, 3 for every 10'. The rest of the plastic piping in this section is based on flexible 100' coils and does not include any coupling or supports.

This section ends with PVC drain and sewer piping based on 10' lengths with bell and spigot ends and O-ring type, push-on joints.

G. Stainless steel piping includes both weld end and threaded piping, both in the type 304 and 316 specification and in the following schedules, 5, 10, 40, 80, and 160. Although this piping is usually furnished in 20' lengths, this cost grouping has a joint (either heli-arc butt-welded or threads and coupling) every 10'. A carbon steel clevis type hanger assembly is also included at 10' intervals and prorated into the linear foot costs.

H. Carbon steel pipe includes both black and galvanized. This section encompasses schedules 40 (standard) and 80 (extra heavy).

Several common methods of joining steel pipe — such as thread and coupled, butt welded, and flanged (150 lb. weld neck flanges) are also included.

For estimating purposes, it is assumed that the piping is purchased in 20' lengths and that a compatible joint is made up every 10'. These joints are prorated into the labor and material costs per linear foot. The following hanger and support assemblies every 10' are also included: carbon steel clevis for the T & C pipe, and single rod roll type for both the welded and flanged piping. All of these hangers are oversized to accommodate pipe insulation 3/4" thick through 5" pipe size and 1-1/2" thick from 6" through 12" pipe size.

I. Grooved joint steel pipe is priced both black and galvanized, in schedules 10, 40, and 80, furnished in 20' lengths. This section describes two joining methods: cut groove and roll groove. The schedule 10 piping is roll-grooved, while the heavier schedules are cut-grooved. The labor and material costs are prorated into per linear foot prices, including a coupled joint every 10', as well as a carbon steel clevis hanger assembly.

Notes:

The pipe hanger assemblies mentioned in the preceding paragraphs include the described hanger; appropriately sized steel, box-type insert and nut; plus 18" of threaded hanger rod.

C clamps are used when the pipe is to be supported from steel shapes rather than anchored in the slab. C clamps are slightly less costly than inserts. However, to save time in estimating, it is advisable to use the given line number cost, rather than substituting a C clamp for the insert.

Add to piping labor for elevated installation:

10' to 14.5' high	10%	30' to 34.5' high	40%
15' to 19.5' high	20%	35' to 39.5' high	50%
20' to 24.5' high	25%	Over 40' and higher	55%
25' to 29.5' high	35%		

When using the percentage adds for elevated piping installations as shown above, bear in mind that the given heights are for the pipe supports, even though the insert, anchor, or clamp may be several feet higher than the pipe itself.

An allowance has been included in the piping installation time for testing and minor tightening of leaking joints, fittings, stuffing boxes, packing glands, etc. For extraordinary test requirements such as x-rays, prolonged pressure or demonstration tests, a percentage of the piping labor, based on the estimator's experience, must be added to the labor total. A testing service specializing in weld x-rays should be consulted for pricing if it is an estimate requirement. Equipment installation time includes start-up with associated adjustments.

Heating, Ventilating & A.C. R2305 Common Work Results for HVAC

R230500-10 Subcontractors

On the unit cost section of the R.S. Means Cost Data, the last column is entitled "Total Incl. O&P". This is normally the cost of the installing contractor. In the HVAC Division, this is the cost of the mechanical contractor. If the particular work being estimated is to be performed by a sub to the mechanical contractor, the mechanical's profit and handling charge (usually 10%) is added to the total of the last column.

Heating, Ventilating & A.C. R2331 HVAC Ducts & Casings

R233100-30 Duct Fabrication/Installation

The labor cost for fabricated sheet metal duct includes both the cost of fabrication and installation of the duct. The split is approximately 40% for fabrication, 60% for installation. It is for this reason that the percentage add for elevated installation of fabricated duct is less than the percentage add for prefabricated/preformed duct.

Example: assume a piece of duct cost $100 installed (labor only)

Sheet Metal Fabrication = 40% = $40

Installation = 60% = $60

The add for elevated installation is:

$$\frac{\text{Based on total labor}}{\text{(fabrication \& installation)}} = \$100 \times 6\% = \$6.00$$

$$\frac{\text{Based on installation cost only}}{\text{(Material purchased prefabricated)}} = \$60 \times 10\% = \$6.00$$

The $6.00 markup (10′ to 15′ high) is the same.

R233100-40   Steel Sheet Metal Calculator (Weight in Lb./Ft. of Length)

Gauge	26	24	22	20	18	16	Gauge	26	24	22	20	18	16
Wt.-Lb./S.F.	.906	1.156	1.406	1.656	2.156	2.656	Wt.-Lb./S.F.	.906	1.156	1.406	1.656	2.156	2.656
SMACNA Max. Dimension – Long Side	*	30″	54″	84″	85″ Up	*	SMACNA Max. Dimension – Long Side	*	30″	54″	84″	85″ Up	*
Sum-2 sides							Sum-2 Sides						
2	.3	.40	.50	.60	.80	.90	56	9.3	12.0	14.0	16.2	21.3	25.2
3	.5	.65	.80	.90	1.1	1.4	57	9.5	12.3	14.3	16.5	21.7	25.7
4	.7	.85	1.0	1.2	1.5	1.8	58	9.7	12.5	14.5	16.8	22.0	26.1
5	.8	1.1	1.3	1.5	1.9	2.3	59	9.8	12.7	14.8	17.1	22.4	26.6
6	1.0	1.3	1.5	1.7	2.3	2.7	60	10.0	12.9	15.0	17.4	22.8	27.0
7	1.2	1.5	1.8	2.0	2.7	3.2	61	10.2	13.1	15.3	17.7	23.2	27.5
8	1.3	1.7	2.0	2.3	3.0	3.6	62	10.3	13.3	15.5	18.0	23.6	27.9
9	1.5	1.9	2.3	2.6	3.4	4.1	63	10.5	13.5	15.8	18.3	24.0	28.4
10	1.7	2.2	2.5	2.9	3.8	4.5	64	10.7	13.7	16.0	18.6	24.3	28.8
11	1.8	2.4	2.8	3.2	4.2	5.0	65	10.8	13.9	16.3	18.9	24.7	29.3
12	2.0	2.6	3.0	3.5	4.6	5.4	66	11.0	14.1	16.5	19.1	25.1	29.7
13	2.2	2.8	3.3	3.8	4.9	5.9	67	11.2	14.3	16.8	19.4	25.5	30.2
14	2.3	3.0	3.5	4.1	5.3	6.3	68	11.3	14.6	17.0	19.7	25.8	30.6
15	2.5	3.2	3.8	4.4	5.7	6.8	69	11.5	14.8	17.3	20.0	26.2	31.1
16	2.7	3.4	4.0	4.6	6.1	7.2	70	11.7	15.0	17.5	20.3	26.6	31.5
17	2.8	3.7	4.3	4.9	6.5	7.7	71	11.8	15.2	17.8	20.6	27.0	32.0
18	3.0	3.9	4.5	5.2	6.8	8.1	72	12.0	15.4	18.0	20.9	27.4	32.4
19	3.2	4.1	4.8	5.5	7.2	8.6	73	12.2	15.6	18.3	21.2	27.7	32.9
20	3.3	4.3	5.0	5.8	7.6	9.0	74	12.3	15.8	18.5	21.5	28.1	33.3
21	3.5	4.5	5.3	6.1	8.0	9.5	75	12.5	16.1	18.8	21.8	28.5	33.8
22	3.7	4.7	5.5	6.4	8.4	9.9	76	12.7	16.3	19.0	22.0	28.9	34.2
23	3.8	5.0	5.8	6.7	8.7	10.4	77	12.8	16.5	19.3	22.3	29.3	34.7
24	4.0	5.2	6.0	7.0	9.1	10.8	78	13.0	16.7	19.5	22.6	29.6	35.1
25	4.2	5.4	6.3	7.3	9.5	11.3	79	13.2	16.9	19.8	22.9	30.0	35.6
26	4.3	5.6	6.5	7.5	9.9	11.7	80	13.3	17.1	20.0	23.2	30.4	36.0
27	4.5	5.8	6.8	7.8	10.3	12.2	81	13.5	17.3	20.3	23.5	30.8	36.5
28	4.7	6.0	7.0	8.1	10.6	12.6	82	13.7	17.5	20.5	23.8	31.2	36.9
29	4.8	6.2	7.3	8.4	11.0	13.1	83	13.8	17.8	20.8	24.1	31.5	37.4
30	5.0	6.5	7.5	8.7	11.4	13.5	84	14.0	18.0	21.0	24.4	31.9	37.8
31	5.2	6.7	7.8	9.0	11.8	14.0	85	14.2	18.2	21.3	24.7	32.3	38.3
32	5.3	6.9	8.0	9.3	12.2	14.4	86	14.3	18.4	21.5	24.9	32.7	38.7
33	5.5	7.1	8.3	9.6	12.5	14.9	87	14.5	18.6	21.8	25.2	33.1	39.2
34	5.7	7.3	8.5	9.9	12.9	15.3	88	14.7	18.8	22.0	25.5	33.4	39.6
35	5.8	7.5	8.8	10.2	13.3	15.8	89	14.8	19.0	22.3	25.8	33.8	40.1
36	6.0	7.8	9.0	10.4	13.7	16.2	90	15.0	19.3	22.5	26.1	34.2	40.5
37	6.2	8.0	9.3	10.7	14.1	16.7	91	15.2	19.5	22.8	26.4	34.6	41.0
38	6.3	8.2	9.5	11.0	14.4	17.1	92	15.3	19.7	23.0	26.7	35.0	41.4
39	6.5	8.4	9.8	11.3	14.8	17.6	93	15.5	19.9	23.3	27.0	35.3	41.9
40	6.7	8.6	10.0	11.6	15.2	18.0	94	15.7	20.1	23.5	27.3	35.7	42.3
41	6.8	8.8	10.3	11.9	15.6	18.5	95	15.8	20.3	23.8	27.6	36.1	42.8
42	7.0	9.0	10.5	12.2	16.0	18.9	96	16.0	20.5	24.0	27.8	36.5	43.2
43	7.2	9.2	10.8	12.5	16.4	19.4	97	16.2	20.8	24.3	28.1	36.9	43.7
44	7.3	9.5	11.0	12.8	16.7	19.8	98	16.3	21.0	24.5	28.4	37.2	44.1
45	7.5	9.7	11.3	13.1	17.1	20.3	99	16.5	21.2	24.8	28.7	37.6	44.6
46	7.7	9.9	11.5	13.3	17.5	20.7	100	16.7	21.4	25.0	29.0	38.0	45.0
47	7.8	10.1	11.8	13.6	17.9	21.2	101	16.8	21.6	25.3	29.3	38.4	45.5
48	8.0	10.3	12.0	13.9	18.2	21.6	102	17.0	21.8	25.5	29.6	38.8	45.9
49	8.2	10.5	12.3	14.2	18.6	22.1	103	17.2	22.0	25.8	29.9	39.1	46.4
50	8.3	10.7	12.5	14.5	19.0	22.5	104	17.3	22.3	26.0	30.2	39.5	46.8
51	8.5	11.0	12.8	14.8	19.4	23.0	105	17.5	22.5	26.3	30.5	39.9	47.3
52	8.7	11.2	13.0	15.1	19.8	23.4	106	17.7	22.7	26.5	30.7	40.3	47.7
53	8.8	11.4	13.3	15.4	20.1	23.9	107	17.8	22.9	26.8	31.0	40.7	48.2
54	9.0	11.6	13.5	15.7	20.5	24.3	108	18.0	23.1	27.0	31.3	41.0	48.6
55	9.2	11.8	13.8	16.0	20.9	24.8	109	18.2	23.3	27.3	31.6	41.4	49.1
							110	18.3	23.5	27.5	31.9	41.8	49.5

Example: If duct is 34″ x 20″ x 15′ long, 34″ is greater than 30″ maximum, for 24 ga. so must be 22 ga. 34″ + 20″ = 54″ going across from 54″ find 13.5 lb. per foot. 13.5 x 15′ = 202.5 lbs. For S.F. of surface area 202.5 ÷ 1.406 = 144 S.F.

Note: Figures include an allowance for scrap.
*Do **Not** use unless engineer specified. 26GA is very light and sometimes used for toilet exhaust. 16 GA is heavy plate and mostly specified for plenums and hoods.

For customer support on your Mechanical Costs with RSMeans Data, call 800.448.8182.

729

Heating, Ventilating & A.C. R2331 HVAC Ducts & Casings

R233100-50 Ductwork Packages (per Ton of Cooling)

System	Sheet Metal	Insulation	Diffusers	Return Register
Roof Top Unit Single Zone	120 Lbs.	52 S.F.	1	1
Roof Top Unit Multizone	240 Lbs.	104 S.F.	2	1
Self-contained Air or Water Cooled	108 Lbs.	—	2	—
Split System Air Cooled	102 Lbs.	—	2	—

Systems reflect most common usage.
Refer to system graphics for duct layout.

Heating, Ventilating & A.C. R2337 Air Outlets & Inlets

R233700-60 Diffuser Evaluation

CFM = V × An × K where V = Outlet velocity in feet per minute. An = Neck area in square feet and K = Diffuser delivery factor. An undersized diffuser for a desired CFM will produce a high velocity and noise level. When air moves past people at a velocity in excess of 25 FPM, an annoying draft is felt. An oversized diffuser will result in low velocity with poor mixing. Consideration must be given to avoid vertical stratification or horizontal areas of stagnation.

R235000-10 Heating Systems

Heating Systems

The basic function of a heating system is to bring an enclosed volume up to a desired temperature and then maintain that temperature within a reasonable range. To accomplish this, the selected system must have sufficient capacity to offset transmission losses resulting from the temperature difference on the interior and exterior of the enclosing walls in addition to losses due to cold air infiltration through cracks, crevices and around doors and windows. The amount of heat to be furnished is dependent upon the building size, construction, temperature difference, air leakage, use, shape, orientation and exposure. Air circulation is also an important consideration. Circulation will prevent stratification which could result in heat losses through uneven temperatures at various levels. For example, the most

efficient use of unit heaters can usually be achieved by circulating the space volume through the total number of units once every 20 minutes or 3 times an hour. This general rule must, of course, be adapted for special cases such as large buildings with low ratios of heat transmitting surface to cubical volume. The type of occupancy of a building will have considerable bearing on the number of heat transmitting units and the location selected. It is axiomatic, however, that the basis of any successful heating system is to provide the maximum amount of heat at the points of maximum heat loss such as exposed walls, windows, and doors. Large roof areas, wind direction, and wide doorways create problems of excessive heat loss and require special consideration and treatment.

Heat Transmission

Heat transfer is an important parameter to be considered during selection of the exterior wall style, material and window area. A high rate of transfer will permit greater heat loss during the wintertime with the resultant increase in heating energy costs and a greater rate of heat gain in the summer with proportionally greater cooling cost. Several terms are used to describe various aspects of heat transfer. However, for general estimating purposes this data set lists U values for systems of construction materials. U is the "overall heat transfer coefficient." It is defined as the heat flow per hour through one square foot when the temperature difference in the air on either side of the structure wall, roof, ceiling or floor is one degree Fahrenheit. The structural segment may be a single homogeneous material or a composite.

Total heat transfer is found using the following equation:

$Q = AU(T_2 - T_1)$ where
  Q = Heat flow, BTU per hour
  A = Area, square feet
  U = Overall heat transfer coefficient
$(T_2 - T_1)$ = Difference in temperature of air on each side of the construction component. (Also abbreviated TD)

Note that heat can flow through all surfaces of any building and this flow is in addition to heat gain or loss due to ventilation, infiltration and generation (appliances, machinery, people).

R235000-20 Heating Approximations for Quick Estimating

Oil Piping & Boiler Room Piping
Small System . . . . 20 to 30% of Boiler
Complex System
with Pumps, Headers, Etc. . . . . 80 to 110% of Boiler
Breeching With Insulation:
Small . . . . 10 to 15% of Boiler
Large . . . . 15 to 25% of Boiler
Coils: . . . . 15 to 30% of Containing Unit
Balancing (Independent) . . . . 1/2% of H.V.A.C. Estimate

Quality/Complexity Adjustment: For all heating installations add these adjustments to the estimate to more closely allow for the equipment and conditions of the particular job under consideration.
Economy installation, add . . . . 0 to 5% of System
Good quality, medium complexity, add . . . . 5 to 15% of System
Above average quality and complexity, add . . . . 15 to 25% of System

R235000-30 The Basics of a Heating System

The function of a heating system is to achieve and maintain a desired temperature in a room or building by replacing the amount of heat being dissipated. There are four kinds of heating systems: hot-water, steam, warm-air and electric resistance. Each has certain essential and similar elements with the exception of electric resistance heating.

The basic elements of a heating system are:

A. A **combustion chamber** in which fuel is burned and heat transferred to a conveying medium.

B. The **"fluid"** used for conveying the heat (water, steam or air).

C. **Conductors** or pipes for transporting the fluid to specific desired locations.

D. A means of disseminating the heat, sometimes called **terminal units.**

A. The **combustion chamber** in a furnace heats air which is then distributed. This is called a warm-air system.

The combustion chamber in a boiler heats water which is either distributed as hot water or steam and this is termed a hydronic system.

The maximum allowable working pressures are limited by ASME "Code for Heating Boilers" to 15 PSI for steam and 160 PSI for hot water heating boilers, with a maximum temperature limitation of 250° F. Hot water boilers are generally rated for a working pressure of 30 PSI. High pressure boilers are governed by the ASME "Code for Power Boilers" which is used almost universally for boilers operating over 15 PSIG. High pressure boilers used for a combination of heating/process loads are usually designed for 150 PSIG.

Boiler ratings are usually indicated as either Gross or Net Output. The Gross Load is equal to the Net Load plus a piping and pickup allowance. When this allowance cannot be determined, divide the gross output rating by 1.25 for a value equal to or greater than the net heat loss requirement of the building.

B. Of the three **fluids** used, steam carries the greatest amount of heat per unit volume. This is due to the fact that it gives up its latent heat of vaporization at a temperature considerably above room temperature. Another advantage is that the pressure to produce a positive circulation is readily available. Piping conducts the steam to terminal units and returns condensate to the boiler.

The **steam system** is well adapted to large buildings because of its positive circulation, its comparatively economical installation and its ability to deliver large quantities of heat. Nearly all large office buildings, stores, hotels, and industrial buildings are so heated, in addition to many residences.

Hot water, when used as the heat carrying fluid, gives up a portion of its sensible heat and then returns to the boiler or heating apparatus for reheating. As the heat conveyed by each pound of water is about one-fiftieth of the heat conveyed by a pound of steam, it is necessary to circulate about fifty times as much water as steam by weight (although only one-thirtieth as much by volume). The hot water system is usually, although not necessarily, designed to operate at temperatures below that of the ordinary steam system and so the amount of heat transfer surface must be correspondingly greater. A temperature of 190° F to 200° F is normally the maximum. Circulation in small buildings may depend on the difference in density between hot water and the cool water returning to the boiler; circulating pumps are normally used to maintain a desired rate of flow. Pumps permit a greater degree of flexibility and better control.

In **warm-air** furnace systems, cool air is taken from one or more points in the building, passed over the combustion chamber and flue gas passages and then distributed through a duct system. A disadvantage of this system is that the ducts take up much more building volume than steam or hot water pipes. Advantages of this system are the relative ease with which humidification can be accomplished by the evaporation of water as the air circulates through the heater, and the lack of need for expensive disseminating units as the warm air simply becomes part of the interior atmosphere of the building.

C. Conductors (pipes and ducts) have been lightly treated in the discussion of conveying fluids. For more detailed information such as sizing and distribution methods, the reader is referred to technical publications such as the American Society of Heating, Refrigerating and Air-Conditioning Engineers "Handbook of Fundamentals."

D. Terminal units come in an almost infinite variety of sizes and styles, but the basic principles of operation are very limited. As previously mentioned, warm-air systems require only a simple register or diffuser to mix heated air with that present in the room. Special application items such as radiant coils and infrared heaters are available to meet particular conditions but are not usually considered for general heating needs. Most heating is accomplished by having air flow over coils or pipes containing the heat transporting medium (steam, hot-water, electricity). These units, while varied, may be separated into two general types, (1) radiator/convectors and (2) unit heaters.

Radiator/convectors may be cast, fin-tube or pipe assemblies. They may be direct, indirect, exposed, concealed or mounted within a cabinet enclosure, upright or baseboard style. These units are often collectively referred to as "radiatiors" or "radiation" although none gives off heat either entirely by radiation or by convection but rather a combination of both. The air flows over the units as a gravity "current." It is necessary to have one or more heat-emitting units in each room. The most efficient placement is low along an outside wall or under a window to counteract the cold coming into the room and achieve an even distribution.

In contrast to radiator/convectors which operate most effectively against the walls of smaller rooms, **unit heaters** utilize a fan to move air over heating coils and are very effective in locations of relatively large volume. Unit heaters, while usually suspended overhead, may be floor mounted. They also may take in fresh outside air for ventilation. The heat distributed by unit heaters may be from a remote source and conveyed by a fluid or it may be from the combustion of fuel in each individual heater. In the latter case the only piping required would be for fuel, however, a vent for the products of combustion would be necessary.

The following list gives may of the advantages of unit heaters for applications other than office or residential:

a. Large capacity so smaller number of units are required, **b.** Piping system simplified, **c.** Space saved where they are located overhead out of the way, **d.** Rapid heating directed where needed with effective wide distribution, **e.** Difference between floor and ceiling temperature reduced, **f.** Circulation of air obtained, and ventilation with introduction of fresh air possible, **g.** Heat output flexible and easily controlled.

Heating, Ventilating & A.C. | R2356 Solar Energy Heating Equipment

R235616-60 Solar Heating (Space and Hot Water)

Collectors should face as close to due South as possible, but variations of up to 20 degrees on either side of true South are acceptable. Local climate and collector type may influence the choice between east or west deviations. Obviously they should be located so they are not shaded from the sun's rays. Incline collectors at a slope of latitude minus 5 degrees for domestic hot water and latitude plus 15 degrees for space heating.

Flat plate collectors consist of a number of components as follows: Insulation to reduce heat loss through the bottom and sides of the collector. The enclosure which contains all the components in this assembly is usually weatherproof and prevents dust, wind and water from coming in contact with the absorber plate. The cover plate usually consists of one or more layers of a variety of glass or plastic and reduces the reradiation by creating an air space which traps the heat between the cover and the absorber plates.

The absorber plate must have a good thermal bond with the fluid passages. The absorber plate is usually metallic and treated with a surface coating which improves absorptivity. Black or dark paints or selective coatings are used for this purpose, and the design of this passage and plate combination helps determine a solar system's effectiveness.

Heat transfer fluid passage tubes are attached above and below or integral with an absorber plate for the purpose of transferring thermal energy from the absorber plate to a heat transfer medium. The heat exchanger is a device for transferring thermal energy from one fluid to another.

Piping and storage tanks should be well insulated to minimize heat losses.

Size domestic water heating storage tanks to hold 20 gallons of water per user, minimum, plus 10 gallons per dishwasher or washing machine. For domestic water heating an optimum collector size is approximately 3/4 square foot of area per gallon of water storage. For space heating of residences and small commercial applications the collector is commonly sized between 30% and 50% of the internal floor area. For space heating of large commercial applications, collector areas less than 30% of the internal floor area can still provide significant heat reductions.

A supplementary heat source is recommended for Northern states for December through February.

The solar energy transmission per square foot of collector surface varies greatly with the material used. Initial cost, heat transmittance and useful life are obviously interrelated.

Heating, Ventilating & A.C. | R2360 Central Cooling Equipment

R236000-10 Air Conditioning

General: The purpose of air conditioning is to control the environment of a space so that comfort is provided for the occupants and/or conditions are suitable for the processes or equipment contained therein. The several items which should be evaluated to define system objectives are:

Temperature Control
Humidity Control
Cleanliness
Odor, smoke and fumes
Ventilation

Efforts to control the above parameters must also include consideration of the degree or tolerance of variation, the noise level introduced, the velocity of air motion and the energy requirements to accomplish the desired results.

The variation in **temperature** and **humidity** is a function of the sensor and the controller. The controller reacts to a signal from the sensor and produces the appropriate suitable response in either the terminal unit, the conductor of the transporting medium (air, steam, chilled water, etc.), or the source (boiler, evaporating coils, etc.).

The **noise level** is a by-product of the energy supplied to moving components of the system. Those items which usually contribute the most noise are pumps, blowers, fans, compressors and diffusers. The level of noise can be partially controlled through use of vibration pads, isolators, proper sizing, shields, baffles and sound absorbing liners.

Some **air motion** is necessary to prevent stagnation and stratification. The maximum acceptable velocity varies with the degree of heating or cooling

which is taking place. Most people feel air moving past them at velocities in excess of 25 FPM as an annoying draft. However, velocities up to 45 FPM may be acceptable in certain cases. Ventilation, expressed as air changes per hour and percentage of fresh air, is usually an item regulated by local codes.

Selection of the system to be used for a particular application is usually a trade-off. In some cases the building size, style, or room available for mechanical use limits the range of possibilities. Prime factors influencing the decision are first cost and total life (operating, maintenance and replacement costs). The accuracy with which each parameter is determined will be an important measure of the reliability of the decision and subsequent satisfactory operation of the installed system.

Heat delivery may be desired from an air conditioning system. Heating capability usually is added as follows: A gas fired burner or hot water/steam/electric coils may be added to the air handling unit directly and heat all air equally. For limited or localized heat requirements the water/steam/electric coils may be inserted into the duct branch supplying the cold areas. Gas fired duct furnaces are also available.

Note: When water or steam coils are used the cost of the piping and boiler must also be added. For a rough estimate use the cost per square foot of the appropriate sized hydronic system with unit heaters. This will provide a cost for the boiler and piping, and the unit heaters of the system would equate to the approximate cost of the heating coils.

For customer support on your Mechanical Costs with RSMeans Data, call 800.448.8182.

733

R236000-20   Air Conditioning Requirements

BTUs per hour per S.F. of floor area and S.F. per ton of air conditioning.

Type of Building	BTU/Hr per S.F.	S.F. per Ton	Type of Building	BTU/Hr per S.F.	S.F. per Ton	Type of Building	BTU/Hr per S.F.	S.F. per Ton
Apartments, Individual	26	450	Dormitory, Rooms	40	300	Libraries	50	240
Corridors	22	550	Corridors	30	400	Low Rise Office, Exterior	38	320
Auditoriums & Theaters	40	300/18*	Dress Shops	43	280	Interior	33	360
Banks	50	240	Drug Stores	80	150	Medical Centers	28	425
Barber Shops	48	250	Factories	40	300	Motels	28	425
Bars & Taverns	133	90	High Rise Office—Ext. Rms.	46	263	Office (small suite)	43	280
Beauty Parlors	66	180	Interior Rooms	37	325	Post Office, Individual Office	42	285
Bowling Alleys	68	175	Hospitals, Core	43	280	Central Area	46	260
Churches	36	330/20*	Perimeter	46	260	Residences	20	600
Cocktail Lounges	68	175	Hotel, Guest Rooms	44	275	Restaurants	60	200
Computer Rooms	141	85	Corridors	30	400	Schools & Colleges	46	260
Dental Offices	52	230	Public Spaces	55	220	Shoe Stores	55	220
Dept. Stores, Basement	34	350	Industrial Plants, Offices	38	320	Shop'g. Ctrs., Supermarkets	34	350
Main Floor	40	300	General Offices	34	350	Retail Stores	48	250
Upper Floor	30	400	Plant Areas	40	300	Specialty	60	200

*Persons per ton
12,000 BTU = 1 ton of air conditioning

734

R236000-30 Psychrometric Table

Dewpoint or Saturation Temperature (F)

	32	35	40	45	50	55	60	65	70	75	80	85	90	95	100
100	32	35	40	45	50	55	60	65	70	75	80	85	90	95	100
90	30	33	37	42	47	52	57	62	67	72	77	82	87	92	97
80	27	30	34	39	44	49	54	58	64	68	73	78	83	88	93
70	24	27	31	36	40	45	50	55	60	64	69	74	79	84	88
60	20	24	28	32	36	41	46	51	55	60	65	69	74	79	83
50	16	20	24	28	33	36	41	46	50	55	60	64	69	73	78
40	12	15	18	23	27	31	35	40	45	49	53	58	62	67	71
30	8	10	14	18	21	25	29	33	37	42	46	50	54	59	62
20	6	7	8	9	13	16	20	24	28	31	35	40	43	48	52
10	4	4	5	5	6	8	9	10	13	17	20	24	27	30	34
	32	**35**	**40**	**45**	**50**	**55**	**60**	**65**	**70**	**75**	**80**	**85**	**90**	**95**	**100**

Relative humidity (%)

Dry bulb temperature (F)

This table shows the relationship between RELATIVE HUMIDITY, DRY BULB TEMPERATURE AND DEWPOINT.

As an example, assume that the thermometer in a room reads 75° F, and we know that the relative humidity is 50%. The chart shows the dewpoint temperature to be 55° F. That is, any surface colder than 55° F will "sweat" or collect condensing moisture. This surface could be the outside of an uninsulated chilled water pipe in the summertime, or the inside surface of a wall or deck in the wintertime. After determining the extreme ambient parameters, the table at the left is useful in determining which surfaces need insulation or vapor barrier protection.

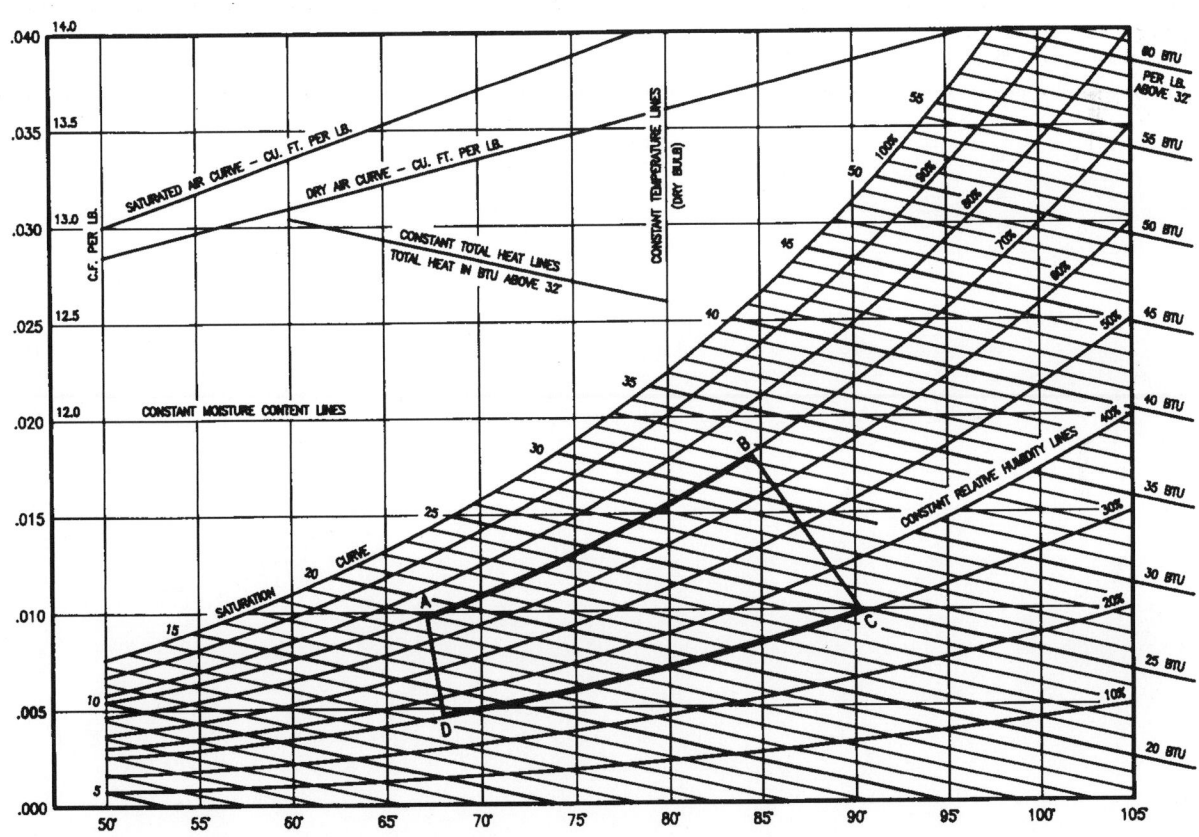

TEMPERATURE DEGREES FAHRENHEIT
TOTAL PRESSURE = 14.696 LB. PER SQ. IN. ABS.

Psychrometric chart showing different variables based on one pound of dry air. Space marked A B C D is temperature-humidity range which is most comfortable for majority of people.

For customer support on your Mechanical Costs with RSMeans Data, call 800.448.8182.

735

R236000-90   Quality/Complexity Adjustment for Air Conditioning Systems

Economy installation, add .... 0 to 5%

Good quality, medium complexity, add .... 5 to 15%

Above average quality and complexity, add .... 15 to 25%

Add the above adjustments to the estimate to more closely allow for the equipment and conditions of the particular job under consideration.

Reference Tables

Fig. R238313-11

R238313-10   Heat Trace Systems

Before you can determine the cost of a HEAT TRACE installation, the method of attachment must be established. There are (4) common methods:

1. Cable is simply attached to the pipe with polyester tape every 12'.
2. Cable is attached with a continuous cover of 2" wide aluminum tape.
3. Cable is attached with factory extruded heat transfer cement and covered with metallic raceway with clips every 10'.
4. Cable is attached between layers of pipe insulation using either clips or polyester tape.

Example: Components for method 3 must include:
  A. Heat trace cable by voltage and watts per linear foot.
  B. Heat transfer cement, 1 gallon per 60 linear feet of cover.
  C. Metallic raceway by size and type.
  D. Raceway clips by size of pipe.

When taking off linear foot lengths of cable add the following for each valve in the system. (E)

In all of the above methods each component of the system must be priced individually.

SCREWED OR WELDED VALVE:			FLANGED VALVE:			BUTTERFLY VALVES:		
1/2"	=	6"	1/2"	=	1' -0"	1/2"	=	0'
3/4"	=	9"	3/4"	=	1' -6"	3/4"	=	0'
1"	=	1' -0"	1"	=	2' -0"	1"	=	1' -0"
1-1/2"	=	1' -6"	1-1/2"	=	2' -6"	1-1/2"	=	1' -6"
2"	=	2'	2"	=	2' -6"	2"	=	2' -0"
2-1/2"	=	2' -6"	2-1/2"	=	3' -0"	2-1/2"	=	2' -6"
3"	=	2' -6"	3"	=	3' -6"	3"	=	2' -6"
4"	=	4' -0"	4"	=	4' -0"	4"	=	3' -0"
6"	=	7' -0"	6"	=	8' -0"	6"	=	3' -6"
8"	=	9' -6"	8"	=	11' -0"	8"	=	4' -0"
10"	=	12' -6"	10"	=	14' -0"	10"	=	4' -0"
12"	=	15' -0"	12"	=	16' -6"	12"	=	5' -0"
14"	=	18' -0"	14"	=	19' -6"	14"	=	5' -6"
16"	=	21' -6"	16"	=	23' -0"	16"	=	6' -0"
18"	=	25' -6"	18"	=	27' -0"	18"	=	6' -6"
20"	=	28' -6"	20"	=	30' -0"	20"	=	7' -0"
24"	=	34' -0"	24"	=	36' -0"	24"	=	8' -0"
30"	=	40' -0"	30"	=	42' -0"	30"	=	10' -0"

For customer support on your Mechanical Costs with RSMeans Data, call 800.448.8182.

737

R238313-10 Heat Trace Systems (cont.)

Add the following quantities of heat transfer cement to linear foot totals for each valve:

Nominal Valve Size	Gallons of Cement per Valve
1/2"	0.14
3/4"	0.21
1"	0.29
1-1/2"	0.36
2"	0.43
2-1/2"	0.70
3"	0.71
4"	1.00
6"	1.43
8"	1.48
10"	1.50
12"	1.60
14"	1.75
16"	2.00
18"	2.25
20"	2.50
24"	3.00
30"	3.75

The following must be added to the list of components to accurately price HEAT TRACE systems:

1. Expediter fitting and clamp fasteners (F)
2. Junction box and nipple connected to expediter fitting (G)
3. Field installed terminal blocks within junction box
4. Ground lugs
5. Piping from power source to expediter fitting
6. Controls
7. Thermostats
8. Branch wiring
9. Cable splices
10. End of cable terminations
11. Branch piping fittings and boxes

Deduct the following percentages from labor if cable lengths in the same area exceed:

150' to 250'	10%	351' to 500'	20%
251' to 350'	15%	Over 500'	25%

Add the following percentages to labor for elevated installations:

15' to 20' high	10%	31' to 35' high	40%
21' to 25' high	20%	36' to 40' high	50%
26' to 30' high	30%	Over 40' high	60%

R238313-20 Spiral-Wrapped Heat Trace Cable (Pitch Table)

In order to increase the amount of heat, occasionally heat trace cable is wrapped in a spiral fashion around a pipe; increasing the number of feet of heater cable per linear foot of pipe.

Engineers first determine the heat loss per foot of pipe (based on the insulating material, its thickness, and the temperature differential across it). A ratio is then calculated by the formula:

$$\text{Feet of Heat Trace per Foot of Pipe} = \frac{\text{Watts/Foot of Heat Loss}}{\text{Watts/Foot of the Cable}}$$

The linear distance between wraps (pitch) is then taken from a chart or table. Generally, the pitch is listed on a drawing leaving the estimator to calculate the total length of heat tape required. An approximation may be taken from this table.

Feet of Heat Trace Per Foot of Pipe

Pitch In Inches	1	1¼	1½	2	2½	3	4	6	8	10	12	14	16	18	20	24
3.5	1.80															
4	1.65															
5	1.46	1.60	1.80													
6	1.34	1.45	1.55	1.75												
7	1.25	1.35	1.43	1.57	1.75											
8	1.20	1.28	1.34	1.45	1.60	1.80										
9	1.16	1.23	1.28	1.37	1.51	1.68										
10	1.13	1.19	1.24	1.32	1.44	1.57	1.82									
15	1.06	1.08	1.10	1.15	1.21	1.29	1.42	1.78								
20	1.04	1.05	1.06	1.08	1.13	1.17	1.25	1.49	1.73							
25		1.04	1.04	1.06	1.08	1.11	1.17	1.33	1.51	1.72						
30				1.04	1.05	1.07	1.12	1.24	1.37	1.54	1.70	1.80				
35					1.06	1.06	1.09	1.17	1.28	1.42	1.54	1.64	1.78			
40						1.05	1.07	1.14	1.22	1.33	1.44	1.52	1.64	1.75		
50							1.05	1.09	1.15	1.22	1.29	1.35	1.44	1.53	1.64	1.83
60								1.06	1.11	1.16	1.21	1.25	1.31	1.39	1.46	1.62
70								1.05	1.08	1.12	1.17	1.19	1.24	1.30	1.35	1.47
80									1.06	1.09	1.13	1.15	1.19	1.24	1.28	1.38
90									1.04	1.06	1.10	1.13	1.16	1.19	1.23	1.32
100										1.05	1.08	1.10	1.13	1.15	1.19	1.23

Note: Common practice would normally limit the lower end of the table to 5% of additional heat and above 80% an engineer would likely opt for two (2) parallel cables.

R260519-90 Wire

Wire quantities are taken off by either measuring each cable run or by extending the conduit and raceway quantities times the number of conductors in the raceway. Ten percent should be added for waste and tie-ins. Keep in mind that the unit of measure of wire is C.L.F. not L.F. as in raceways so the formula would read:

$$\frac{(\text{L.F. Raceway} \times \text{No. of Conductors}) \times 1.10}{100} = \text{C.L.F.}$$

Price per C.L.F. of wire includes:
1. Setting up wire coils or spools on racks
2. Attaching wire to pull in means
3. Measuring and cutting wire
4. Pulling wire into a raceway
5. Identifying and tagging

Price does not include:
1. Connections to breakers, panelboards, or equipment
2. Splices

Job Conditions: Productivity is based on new construction to a height of 15' using rolling staging in an unobstructed area. Material staging is assumed to be within 100' of work being performed.

Economy of Scale: If more than three wires at a time are being pulled, deduct the following percentages from the labor of that grouping:

4-5 wires	25%
6-10 wires	30%
11-15 wires	35%
over 15	40%

If a wire pull is less than 100' in length and is interrupted several times by boxes, lighting outlets, etc., it may be necessary to add the following lengths to each wire being pulled:

Junction box to junction box	2 L.F.
Lighting panel to junction box	6 L.F.
Distribution panel to sub panel	8 L.F.
Switchboard to distribution panel	12 L.F.
Switchboard to motor control center	20 L.F.
Switchboard to cable tray	40 L.F.

Measure of Drops and Riser: It is important when taking off wire quantities to include the wire for drops to electrical equipment. If heights of electrical equipment are not clearly stated, use the following guide:

	Bottom A.F.F.	Top A.F.F.	Inside Cabinet
Safety switch to 100A	5'	6'	2'
Safety switch 400 to 600A	4'	6'	3'
100A panel 12 to 30 circuit	4'	6'	3'
42 circuit panel	3'	6'	4'
Switch box	3'	3'6"	1'
Switchgear	0'	8'	8'
Motor control centers	0'	8'	8'
Transformers - wall mount	4'	8'	2'
Transformers - floor mount	0'	12'	4'

R312319-90 Wellpoints

A single stage wellpoint system is usually limited to dewatering an average 15' depth below normal ground water level. Multi-stage systems are employed for greater depth with the pumping equipment installed only at the lowest header level. Ejectors with unlimited lift capacity can be economical when two or more stages of wellpoints can be replaced or when horizontal clearance is restricted, such as in deep trenches or tunneling projects, and where low water flows are expected. Wellpoints are usually spaced on 2-1/2' to 10' centers along a header pipe. Wellpoint spacing, header size, and pump size are all determined by the expected flow as dictated by soil conditions.

In almost all soils encountered in wellpoint dewatering, the wellpoints may be jetted into place. Cemented soils and stiff clays may require sand wicks about 12" in diameter around each wellpoint to increase efficiency and eliminate weeping into the excavation. These sand wicks require 1/2 to 3 C.Y. of washed filter sand and are installed by using a 12" diameter steel casing and hole puncher jetted into the ground 2' deeper than the wellpoint. Rock may require predrilled holes.

Labor required for the complete installation and removal of a single stage wellpoint system is in the range of 3/4 to 2 labor-hours per linear foot of header, depending upon jetting conditions, wellpoint spacing, etc.

Continuous pumping is necessary except in some free draining soil where temporary flooding is permissible (as in trenches which are backfilled after each day's work). Good practice requires provision of a stand-by pump during the continuous pumping operation.

Systems for continuous trenching below the water table should be installed three to four times the length of expected daily progress to ensure uninterrupted digging, and header pipe size should not be changed during the job.

For pervious free draining soils, deep wells in place of wellpoints may be economical because of lower installation and maintenance costs. Daily production ranges between two to three wells per day, for 25' to 40' depths, to one well per day for depths over 50'.

Detailed analysis and estimating for any dewatering problem is available at no cost from wellpoint manufacturers. Major firms will quote "sufficient equipment" quotes or their affiliates will offer lump sum proposals to cover complete dewatering responsibility.

Description for 200' System with 8" Header		Quantities
Equipment & Material	Wellpoints 25' long, 2" diameter @ 5' O.C.	40 Each
	Header pipe, 8" diameter	200 L.F.
	Discharge pipe, 8" diameter	100 L.F.
	8" valves	3 Each
	Combination jetting & wellpoint pump (standby)	1 Each
	Wellpoint pump, 8" diameter	1 Each
	Transportation to and from site	1 Day
	Fuel for 30 days x 60 gal./day	1800 Gallons
	Lubricants for 30 days x 16 lbs./day	480 Lbs.
	Sand for points	40 C.Y.
Labor	Technician to supervise installation	1 Week
	Labor for installation and removal of system	300 Labor-hours
	4 Operators straight time 40 hrs./wk. for 4.33 wks.	693 Hrs.
	4 Operators overtime 2 hrs./wk. for 4.33 wks.	35 Hrs.

Earthwork   R3141 Shoring

R314116-40   Wood Sheet Piling

Wood sheet piling may be used for depths to 20' where there is no ground water. If moderate ground water is encountered Tongue & Groove sheeting will help to keep it out. When considerable ground water is present, steel sheeting must be used.

For estimating purposes on trench excavation, sizes are as follows:

Depth	Sheeting	Wales	Braces	B.F. per S.F.
To 8'	3 x 12's	6 x 8's, 2 line	6 x 8's, @ 10'	4.0 @ 8'
8' x 12'	3 x 12's	10 x 10's, 2 line	10 x 10's, @ 9'	5.0 average
12' to 20'	3 x 12's	12 x 12's, 3 line	12 x 12's, @ 8'	7.0 average

Sheeting to be toed in at least 2' depending upon soil conditions. A five person crew with an air compressor and sheeting driver can drive and brace 440 SF/day at 8' deep, 360 SF/day at 12' deep, and 320 SF/day at 16' deep.

For normal soils, piling can be pulled in 1/3 the time to install. Pulling difficulty increases with the time in the ground. Production can be increased by high pressure jetting.

R314116-45   Steel Sheet Piling

Limiting weights are 22 to 38#/S.F. of wall surface with 27#/S.F. average for usual types and sizes. (Weights of piles themselves are from 30.7#/L.F. to 57#/L.F. but they are 15" to 21" wide.) Lightweight sections 12" to 28" wide from 3 ga. to 12 ga. thick are also available for shallow excavations. Piles may be driven two at a time with an impact or vibratory hammer (use vibratory to pull) hung from a crane without leads. A reasonable estimate of the life of steel sheet piling is 10 uses with up to 125 uses possible if a vibratory hammer is used. Used piling costs from 50% to 80% of new piling depending on location and market conditions. Sheet piling and H piles can be rented for about 30% of the delivered mill price for the first month and 5% per month thereafter. Allow 1 labor-hour per pile for cleaning and trimming after driving. These costs increase with depth and hydrostatic head. Vibratory drivers are faster in wet granular soils and are excellent for pile extraction. Pulling difficulty increases with the time in the ground and may cost more than driving. It is often economical to abandon the sheet piling, especially if it can be used as the outer wall form. Allow about 1/3 additional length or more for toeing into ground. Add bracing, waler and strut costs. Waler costs can equal the cost per ton of sheeting.

Utilities   R3311 Water Utility Distribution Piping

R331113-80   Piping Designations

There are several systems currently in use to describe pipe and fittings. The following paragraphs will help to identify and clarify classifications of piping systems used for water distribution.

Piping may be classified by schedule. Piping schedules include 5S, 10S, 10, 20, 30, Standard, 40, 60, Extra Strong, 80, 100, 120, 140, 160 and Double Extra Strong. These schedules are dependent upon the pipe wall thickness. The wall thickness of a particular schedule may vary with pipe size.

Ductile iron pipe for water distribution is classified by Pressure Classes such as Class 150, 200, 250, 300 and 350. These classes are actually the rated water working pressure of the pipe in pounds per square inch (psi). The pipe in these pressure classes is designed to withstand the rated water working pressure plus a surge allowance of 100 psi.

The American Water Works Association (AWWA) provides standards for various types of plastic pipe. C-900 is the specification for polyvinyl chloride (PVC) piping used for water distribution in sizes ranging from 4" through 12". C-901 is the specification for polyethylene (PE) pressure pipe, tubing and fittings used for water distribution in sizes ranging from 1/2" through 3". C-905 is the specification for PVC piping sizes 14" and greater.

PVC pressure-rated pipe is identified using the standard dimensional ratio (SDR) method. This method is defined by the American Society for Testing and Materials (ASTM) Standard D 2241. This pipe is available in SDR numbers 64, 41, 32.5, 26, 21, 17, and 13.5. A pipe with an SDR of 64 will have the thinnest wall while a pipe with an SDR of 13.5 will have the thickest wall. When the pressure rating (PR) of a pipe is given in psi, it is based on a line supplying water at 73 degrees F.

The National Sanitation Foundation (NSF) seal of approval is applied to products that can be used with potable water. These products have been tested to ANSI/NSF Standard 14.

Valves and strainers are classified by American National Standards Institute (ANSI) Classes. These Classes are 125, 150, 200, 250, 300, 400, 600, 900, 1500 and 2500. Within each class there is an operating pressure range dependent upon temperature. Design parameters should be compared to the appropriate material dependent, pressure-temperature rating chart for accurate valve selection.

For customer support on your Mechanical Costs with RSMeans Data, call 800.448.8182.

741

Change Orders

Change Order Considerations

A change order is a written document usually prepared by the design professional and signed by the owner, the architect/engineer, and the contractor. A change order states the agreement of the parties to: an addition, deletion, or revision in the work; an adjustment in the contract sum, if any; or an adjustment in the contract time, if any. Change orders, or "extras", in the construction process occur after execution of the construction contract and impact architects/engineers, contractors, and owners.

Change orders that are properly recognized and managed can ensure orderly, professional, and profitable progress for everyone involved in the project. There are many causes for change orders and change order requests. In all cases, change orders or change order requests should be addressed promptly and in a precise and prescribed manner. The following paragraphs include information regarding change order pricing and procedures.

The Causes of Change Orders

Reasons for issuing change orders include:

- Unforeseen field conditions that require a change in the work
- Correction of design discrepancies, errors, or omissions in the contract documents
- Owner-requested changes, either by design criteria, scope of work, or project objectives
- Completion date changes for reasons unrelated to the construction process
- Changes in building code interpretations, or other public authority requirements that require a change in the work
- Changes in availability of existing or new materials and products

Procedures

Properly written contract documents must include the correct change order procedures for all parties—owners, design professionals, and contractors—to follow in order to avoid costly delays and litigation.

Being "in the right" is not always a sufficient or acceptable defense. The contract provisions requiring notification and documentation must be adhered to within a defined or reasonable time frame.

The appropriate method of handling change orders is by a written proposal and acceptance by all parties involved. Prior to starting work on a project, all parties should identify their authorized agents who may sign and accept change orders, as well as any limits placed on their authority.

Time may be a critical factor when the need for a change arises. For such cases, the contractor might be directed to proceed on a "time and materials" basis, rather than wait for all paperwork to be processed—a delay that could impede progress. In this situation, the contractor must still follow the prescribed change order procedures including, but not limited to, notification and documentation.

Lack of documentation can be very costly, especially if legal judgments are to be made, and if certain field personnel are no longer available. For time and material change orders, the contractor should keep accurate daily records of all labor and material allocated to the change.

Owners or awarding authorities who do considerable and continual building construction (such as the federal government) realize the inevitability of change orders for numerous reasons, both predictable and unpredictable. As a result, the federal government, the American Institute of Architects (AIA), the Engineers Joint Contract Documents Committee (EJCDC), and other contractor, legal, and technical organizations have developed standards and procedures to be followed by all parties to achieve contract continuance and timely completion, while being financially fair to all concerned.

Pricing Change Orders

When pricing change orders, regardless of their cause, the most significant factor is when the change occurs. The need for a change may be perceived in the field or requested by the architect/engineer *before* any of the actual installation has begun, or may evolve or appear *during* construction when the item of work in question is partially installed. In the latter cases, the original sequence of construction is disrupted, along with all contiguous and supporting systems. Change orders cause the greatest impact when they occur *after* the installation has been completed and must be uncovered, or even replaced. Post-completion changes may be caused by necessary design changes, product failure, or changes in the owner's requirements that are not discovered until the building or the systems begin to function.

Specified procedures of notification and record keeping must be adhered to and enforced regardless of the stage of construction: *before, during,* or *after* installation. Some bidding documents anticipate change orders by requiring that unit prices including overhead and profit percentages—for additional as well as deductible changes—be listed. Generally these unit prices do not fully take into account the ripple effect, or impact on other trades, and should be used for general guidance only.

When pricing change orders, it is important to classify the time frame in which the change occurs. There are two basic time frames for change orders: *pre-installation change orders,* which occur before the start of construction, and *post-installation change orders,* which involve reworking after the original installation. Change orders that occur between these stages may be priced according to the extent of work completed using a combination of techniques developed for pricing *pre-* and *post-installation* changes.

Factors To Consider When Pricing Change Orders

As an estimator begins to prepare a change order, the following questions should be reviewed to determine their impact on the final price.

General

- *Is the change order work* pre-installation *or* post-installation?

  Change order work costs vary according to how much of the installation has been completed. Once workers have the project scoped in their minds, even though they have not started, it can be difficult to refocus. Consequently they may spend more than the normal amount of time understanding the change. Also, modifications to work in place, such as trimming or refitting, usually take more time than was initially estimated. The greater the amount of work in place, the more reluctant workers are to change it. Psychologically they may resent the change and as a result the rework takes longer than normal. Post-installation change order estimates must include demolition of existing work as required to accomplish the change. If the work is performed at a later time, additional obstacles, such as building finishes, may be present which must be protected. Regardless of whether the change occurs pre-installation or post-installation, attempt to isolate the identifiable factors and price them separately. For example, add shipping costs that may be required pre-installation or any demolition required post-installation. Then analyze the potential impact on productivity of psychological and/or learning curve factors and adjust the output rates accordingly. One approach is to break down the typical workday into segments and quantify the impact on each segment.

Change Order Installation Efficiency

The labor-hours expressed (for new construction) are based on average installation time, using an efficiency level. For change order situations, adjustments to this efficiency level should reflect the daily labor-hour allocation for that particular occurrence.

• *Will the change substantially delay the original completion date?*

A significant change in the project may cause the original completion date to be extended. The extended schedule may subject the contractor to new wage rates dictated by relevant labor contracts. Project supervision and other project overhead must also be extended beyond the original completion date. The schedule extension may also put installation into a new weather season. For example, underground piping scheduled for October installation was delayed until January. As a result, frost penetrated the trench area, thereby changing the degree of difficulty of the task. Changes and delays may have a ripple effect throughout the project. This effect must be analyzed and negotiated with the owner.

• *What is the net effect of a deduct change order?*

In most cases, change orders resulting in a deduction or credit reflect only bare costs. The contractor may retain the overhead and profit based on the original bid.

Materials

• *Will you have to pay more or less for the new material, required by the change order, than you paid for the original purchase?*

The same material prices or discounts will usually apply to materials purchased for change orders as new construction. In some

instances, however, the contractor may forfeit the advantages of competitive pricing for change orders. Consider the following example:

A contractor purchased over $20,000 worth of fan coil units for an installation and obtained the maximum discount. Some time later it was determined the project required an additional matching unit. The contractor has to purchase this unit from the original supplier to ensure a match. The supplier at this time may not discount the unit because of the small quantity, and he is no longer in a competitive situation. The impact of quantity on purchase can add between 0% and 25% to material prices and/or subcontractor quotes.

• *If materials have been ordered or delivered to the job site, will they be subject to a cancellation charge or restocking fee?*

Check with the supplier to determine if ordered materials are subject to a cancellation charge. Delivered materials not used as a result of a change order may be subject to a restocking fee if returned to the supplier. Common restocking charges run between 20% and 40%. Also, delivery charges to return the goods to the supplier must be added.

Labor

• *How efficient is the existing crew at the actual installation?*

Is the same crew that performed the initial work going to do the change order? Possibly the change consists of the installation of a unit identical to one already installed; therefore, the change should take less time. Be sure to consider this potential productivity increase and modify the productivity rates accordingly.

• *If the crew size is increased, what impact will that have on supervision requirements?*

Under most bargaining agreements or management practices, there is a point at which a working foreman is replaced by a nonworking foreman. This replacement increases project overhead by adding a nonproductive worker. If additional workers are added to accelerate the project or to perform changes while maintaining the schedule, be sure to add additional supervision time if warranted. Calculate the

hours involved and the additional cost directly if possible.

• *What are the other impacts of increased crew size?*

The larger the crew, the greater the potential for productivity to decrease. Some of the factors that cause this productivity loss are: overcrowding (producing restrictive conditions in the working space) and possibly a shortage of any special tools and equipment required. Such factors affect not only the crew working on the elements directly involved in the change order, but other crews whose movements may also be hampered. As the crew increases, check its basic composition for changes by the addition or deletion of apprentices or nonworking foreman, and quantify the potential effects of equipment shortages or other logistical factors.

• *As new crews, unfamiliar with the project, are brought onto the site, how long will it take them to become oriented to the project requirements?*

The orientation time for a new crew to become 100% effective varies with the site and type of project. Orientation is easiest at a new construction site and most difficult at existing, very restrictive renovation sites. The type of work also affects orientation time. When all elements of the work are exposed, such as concrete or masonry work, orientation is decreased. When the work is concealed or less visible, such as existing electrical systems, orientation takes longer. Usually orientation can be accomplished in one day or less. Costs for added orientation should be itemized and added to the total estimated cost.

• *How much actual production can be gained by working overtime?*

Short term overtime can be used effectively to accomplish more work in a day. However, as overtime is scheduled to run beyond several weeks, studies have shown marked decreases in output. The following chart shows the effect of long term overtime on worker efficiency. If the anticipated change requires extended overtime to keep the job on schedule, these factors can be used as a guide to predict the impact on time and cost. Add project overhead, particularly supervision, that may also be incurred.

Days per Week	Hours per Day	Production Efficiency					Payroll Cost Factors	
		1st Week	2nd Week	3rd Week	4th Week	Average 4 Weeks	@ 1-1/2 Times	@ 2 Times
5	8	100%	100%	100%	100%	100%	100%	100%
	9	100	100	95	90	96.25	105.6	111.1
	10	100	95	90	85	92.50	110.0	120.0
	11	95	90	75	65	81.25	113.6	127.3
	12	90	85	70	60	76.25	116.7	133.3
6	8	100	100	95	90	96.25	108.3	116.7
	9	100	95	90	85	92.50	113.0	125.9
	10	95	90	85	80	87.50	116.7	133.3
	11	95	85	70	65	78.75	119.7	139.4
	12	90	80	65	60	73.75	122.2	144.4
7	8	100	95	85	75	88.75	114.3	128.6
	9	95	90	80	70	83.75	118.3	136.5
	10	90	85	75	65	78.75	121.4	142.9
	11	85	80	65	60	72.50	124.0	148.1
	12	85	75	60	55	68.75	126.2	152.4

Effects of Overtime

Caution: Under many labor agreements, Sundays and holidays are paid at a higher premium than the normal overtime rate.

The use of long-term overtime is counterproductive on almost any construction job; that is, the longer the period of overtime, the lower the actual production rate. Numerous studies have been conducted, and while they have resulted in slightly different numbers, all reach the same conclusion. The figure above tabulates the effects of overtime work on efficiency.

As illustrated, there can be a difference between the *actual* payroll cost per hour and the *effective* cost per hour for overtime work. This is due to the reduced production efficiency with the increase in weekly hours beyond 40. This difference between actual and effective cost results from overtime work over a prolonged period. Short-term overtime work does not result in as great a reduction in efficiency and, in such cases, effective cost may not vary significantly from the actual payroll cost. As the total hours per week are increased on a regular basis, more time is lost due to fatigue, lowered morale, and an increased accident rate.

As an example, assume a project where workers are working 6 days a week, 10 hours per day. From the figure above (based on productivity studies), the average effective productive hours over a 4-week period are:

$$0.875 \times 60 = 52.5$$

Depending upon the locale and day of week, overtime hours may be paid at time and a half or double time. For time and a half, the overall (average) *actual* payroll cost (including regular and overtime hours) is determined as follows:

$$\frac{40 \text{ reg. hrs.} + (20 \text{ overtime hrs.} \times 1.5)}{60 \text{ hrs.}} = 1.167$$

Based on 60 hours, the payroll cost per hour will be 116.7% of the normal rate at 40 hours per week. However, because the effective production (efficiency) for 60 hours is reduced to the equivalent of 52.5 hours, the effective cost of overtime is calculated as follows:

For time and a half:

$$\frac{40 \text{ reg. hrs.} + (20 \text{ overtime hrs.} \times 1.5)}{52.5 \text{ hrs.}} = 1.33$$

The installed cost will be 133% of the normal rate (for labor).

Thus, when figuring overtime, the actual cost per unit of work will be higher than the apparent overtime payroll dollar increase, due to the reduced productivity of the longer work week. These efficiency calculations are true only for those cost factors determined by hours worked. Costs that are applied weekly or monthly, such as equipment rentals, will not be similarly affected.

Equipment

- *What equipment is required to complete the change order?*

Change orders may require extending the rental period of equipment already on the job site, or the addition of special equipment brought in to accomplish the change work. In either case, the additional rental charges and operator labor charges must be added.

Summary

The preceding considerations and others you deem appropriate should be analyzed and applied to a change order estimate. The impact of each should be quantified and listed on the estimate to form an audit trail.

Change orders that are properly identified, documented, and managed help to ensure the orderly, professional, and profitable progress of the work. They also minimize potential claims or disputes at the end of the project.

Back by customer demand!

You asked and we listened. For customer convenience and estimating ease, we have made the 2020 Project Costs available for download at **RSMeans.com/2020books**. You will also find sample estimates, an RSMeans data overview video, and a book registration form to receive quarterly data updates throughout 2020.

Estimating Tips

- The cost figures available in the download were derived from hundreds of projects contained in the RSMeans database of completed construction projects. They include the contractor's overhead and profit. The figures have been adjusted to January of the current year.

- These projects were located throughout the U.S. and reflect a tremendous variation in square foot (S.F.) costs. This is due to differences, not only in labor and material costs, but also in individual owners' requirements. For instance, a bank in a large city would have different features than one in a rural area. This is true of all the different types of buildings analyzed. Therefore, caution should be exercised when using these Project Costs. For example, for courthouses, costs in the database are local courthouse costs and will not apply to the larger, more elaborate federal courthouses.

- None of the figures "go with" any others. All individual cost items were computed and tabulated separately. Thus, the sum of the median figures for plumbing, HVAC, and electrical will not normally total up to the total mechanical and electrical costs arrived at by separate analysis and tabulation of the projects.

- Each building was analyzed as to total and component costs and percentages. The figures were arranged in ascending order with the results tabulated as shown. The 1/4 column shows that 25% of the projects had lower costs and 75% had higher. The 3/4 column shows that 75% of the projects had lower costs and 25% had higher. The median column shows that 50% of the projects had lower costs and 50% had higher.

- Project Costs are useful in the conceptual stage when no details are available. As soon as details become available in the project design, the square foot approach should be discontinued and the project should be priced as to its particular components. When more precision is required, or for estimating the replacement cost of specific buildings, the current edition of *Square Foot Costs with RSMeans data* should be used.

- In using the figures in this section, it is recommended that the median column be used for preliminary figures if no additional information is available. The median figures, when multiplied by the total city construction cost index figures (see City Cost Indexes) and then multiplied by the project size modifier at the end of this section, should present a fairly accurate base figure, which would then have to be adjusted in view of the estimator's experience, local economic conditions, code requirements, and the owner's particular requirements. There is no need to factor in the percentage figures, as these should remain constant from city to city.

- The editors of this data would greatly appreciate receiving cost figures on one or more of your recent projects, which would then be included in the averages for next year. All cost figures received will be kept confidential, except that they will be averaged with other similar projects to arrive at square foot cost figures for next year.

See the website above for details and the discount available for submitting one or more of your projects.

Same Data. Simplified.

Enjoy the convenience and efficiency of accessing your costs anywhere:

- **Skip the multiplier** by setting your location
- **Quickly search,** edit, favorite and share costs
- **Stay on top of price changes** with automatic updates

Discover more at rsmeans.com/online

50 17 00 | Project Costs		UNIT COSTS			% OF TOTAL			
	UNIT	1/4	MEDIAN	3/4	1/4	MEDIAN	3/4	
01 0000 Auto Sales with Repair	S.F.							01
0100 Architectural		106	119	128	58%	64%	67%	
0200 Plumbing		8.90	9.30	12.40	4.84%	5.20%	6.80%	
0300 Mechanical		11.90	15.95	17.60	6.40%	8.70%	10.15%	
0400 Electrical		18.30	22.50	28.50	9.05%	11.70%	15.90%	
0500 Total Project Costs		178	186	191				
02 0000 Banking Institutions	S.F.							02
0100 Architectural		160	197	239	59%	65%	69%	
0200 Plumbing		6.45	9	12.50	2.12%	3.39%	4.19%	
0300 Mechanical		12.80	17.70	21	4.41%	5.10%	10.75%	
0400 Electrical		31.50	38	58	10.45%	13.05%	15.90%	
0500 Total Project Costs		266	299	370				
03 0000 Court House	S.F.							03
0100 Architectural		84.50	166	166	54.50%	58.50%	58.50%	
0200 Plumbing		3.19	3.19	3.19	2.07%	2.07%	2.07%	
0300 Mechanical		19.95	19.95	19.95	12.95%	12.95%	12.95%	
0400 Electrical		25.50	25.50	25.50	16.60%	16.60%	16.60%	
0500 Total Project Costs		154	284	284				
04 0000 Data Centers	S.F.							04
0100 Architectural		191	191	191	68%	68%	68%	
0200 Plumbing		10.45	10.45	10.45	3.71%	3.71%	3.71%	
0300 Mechanical		26.50	26.50	26.50	9.45%	9.45%	9.45%	
0400 Electrical		25.50	25.50	25.50	9%	9%	9%	
0500 Total Project Costs		281	281	281				
05 0000 Detention Centers	S.F.							05
0100 Architectural		177	187	198	52%	53%	60.50%	
0200 Plumbing		18.65	22.50	27.50	5.15%	7.10%	7.25%	
0300 Mechanical		23.50	34	40.50	7.55%	9.50%	13.80%	
0400 Electrical		39	46	60	10.90%	14.85%	17.95%	
0500 Total Project Costs		299	315	370				
06 0000 Fire Stations	S.F.							06
0100 Architectural		100	129	187	49%	54.50%	61.50%	
0200 Plumbing		10.75	13.95	16.95	4.67%	5.60%	6.30%	
0300 Mechanical		14.95	21	29.50	6.10%	8.45%	10.20%	
0400 Electrical		23	29.50	35.50	10.75%	12.75%	14.95%	
0500 Total Project Costs		210	238	320				
07 0000 Gymnasium	S.F.							07
0100 Architectural		88.50	116	116	57%	64.50%	64.50%	
0200 Plumbing		2.17	7.10	7.10	1.58%	3.48%	3.48%	
0300 Mechanical		3.32	30	30	2.42%	14.65%	14.65%	
0400 Electrical		10.90	21	21	7.95%	10.35%	10.35%	
0500 Total Project Costs		137	204	204				
08 0000 Hospitals	S.F.							08
0100 Architectural		107	176	191	43%	47.50%	48%	
0200 Plumbing		7.85	15.05	32.50	6%	7.45%	7.65%	
0300 Mechanical		52	58.50	76.50	14.20%	17.95%	23.50%	
0400 Electrical		23.50	47.50	61.50	10.95%	13.75%	16.85%	
0500 Total Project Costs		250	375	405				
09 0000 Industrial Buildings	S.F.							09
0100 Architectural		45	71.50	232	46%	54%	56.50%	
0200 Plumbing		1.74	6.55	13.25	2%	3.06%	6.30%	
0300 Mechanical		4.81	9.15	43.50	4.77%	5.55%	14.80%	
0400 Electrical		7.35	8.40	70	7.85%	13.55%	16.20%	
0500 Total Project Costs		80	104	430				
10 0000 Medical Clinics & Offices	S.F.							10
0100 Architectural		89.50	121	161	48.50%	55.50%	62.50%	
0200 Plumbing		8.90	13.15	21	4.47%	6.60%	8.65%	
0300 Mechanical		13.95	22.50	44	7.80%	10.95%	16.10%	
0400 Electrical		20	27	37.50	9.70%	11.65%	14.05%	
0500 Total Project Costs		166	217	292				

50 17 | Project Costs

50 17 00 | Project Costs

			UNIT	UNIT COSTS			% OF TOTAL		
				1/4	MEDIAN	3/4	1/4	MEDIAN	3/4
11	0000	**Mixed Use**	S.F.		*Data*					**11**
	0100	Architectural		91.50	129 *170*	211	48.50%	57%	62.50%
	0200	Plumbing	*8-9%*	6.15	9.35 *11*	11.80	3.23%	3.44%	4.18%
	0300	Mechanical	*19-21%*	16.40	24.50 *36*	47.50	6.10%	13.75%	18.85%
	0400	Electrical		16	26.50	41.50	8.30%	11.40%	14%
	0500	Total Project Costs		189	216 *280*	345			
12	0000	**Multi-Family Housing**	S.F.							**12**
	0100	Architectural		76	114	170	56.50%	62%	66.50%
	0200	Plumbing		6.55	11.55	15.05	5.30%	6.85%	8%
	0300	Mechanical		7	9.45	27	4.92%	6.90%	10.40%
	0400	Electrical		9.90	15.45	19.85	6.20%	7.90%	10.25%
	0500	Total Project Costs		111	208	271			
13	0000	**Nursing Home & Assisted Living**	S.F.							**13**
	0100	Architectural		71.50	94	118	51.50%	55.50%	63.50%
	0200	Plumbing		7.75	11.60	12.80	6.25%	7.40%	8.80%
	0300	Mechanical		6.35	9.35	18.30	4.04%	6.70%	9.55%
	0400	Electrical		10.50	16.55	23	7%	10.75%	13.10%
	0500	Total Project Costs		122	160	192			
14	0000	**Office Buildings**	S.F.							**14**
	0100	Architectural		94.50	129	177	56%	61%	69%
	0200	Plumbing		5.10	8	15.05	2.65%	3.56%	5.85%
	0300	Mechanical		11	17	26	5.60%	8.20%	11.10%
	0400	Electrical		12.75	21.50	34	7.75%	10%	12.70%
	0500	Total Project Costs		161	200	282			
15	0000	**Parking Garage**	S.F.							**15**
	0100	Architectural		31.50	38.50	40.50	70%	79%	88%
	0200	Plumbing		1.04	1.09	2.04	2.05%	2.70%	2.83%
	0300	Mechanical		.81	1.24	4.72	2.11%	3.62%	3.81%
	0400	Electrical		2.78	3.05	6.35	5.30%	6.35%	7.95%
	0500	Total Project Costs		38.50	47	51			
16	0000	**Parking Garage/Mixed Use**	S.F.							**16**
	0100	Architectural		103	112	114	61%	62%	65.50%
	0200	Plumbing		3.30	4.32	6.60	2.47%	2.72%	3.66%
	0300	Mechanical		14.10	15.85	23	7.80%	13.10%	13.60%
	0400	Electrical		14.75	21	22	8.20%	12.65%	18.15%
	0500	Total Project Costs		168	175	181			
17	0000	**Police Stations**	S.F.							**17**
	0100	Architectural		116	130	164	49%	56.50%	61%
	0200	Plumbing		15.30	18.40	18.50	5.05%	5.55%	9.05%
	0300	Mechanical		34.50	48.50	50.50	13%	14.55%	16.55%
	0400	Electrical		26	28.50	30.50	9.15%	12.10%	14%
	0500	Total Project Costs		217	267	305			
18	0000	**Police/Fire**	S.F.							**18**
	0100	Architectural		113	113	345	55.50%	66%	68%
	0200	Plumbing		9.05	9.35	34.50	5.45%	5.50%	5.55%
	0300	Mechanical		13.85	22	79	8.35%	12.70%	12.80%
	0400	Electrical		15.75	20	90.50	9.50%	11.75%	14.55%
	0500	Total Project Costs		166	171	625			
19	0000	**Public Assembly Buildings**	S.F.		*$ Ava*					**19**
	0100	Architectural		115	159 *200*	238	57.50%	61.50%	66%
	0200	Plumbing	*6-7%*	6.10	8.90 *12*	13.40	2.60%	3.36%	4.79%
	0300	Mechanical	*15-16%*	12.85	23 *29*	35.50	6.55%	8.95%	12.45%
	0400	Electrical		19	26	41	8.60%	10.75%	13%
	0500	Total Project Costs		185	253 *314*	375			
20	0000	**Recreational**	S.F.							**20**
	0100	Architectural		106	168	236	53.50%	60%	66%
	0200	Plumbing		7.80	14.75	21.50	3.08%	5%	6.85%
	0300	Mechanical		13.15	20	31.50	5.15%	6.95%	11.70%
	0400	Electrical		15.20	26.50	39.50	7.35%	8.95%	10.75%
	0500	Total Project Costs		168	281	445			

For customer support on your Mechanical Costs with RSMeans Data, call 800.448.8182.

747

		50 17 00 | Project Costs	UNIT COSTS			% OF TOTAL			
			UNIT	1/4	MEDIAN	3/4	1/4	MEDIAN	3/4
21	0000	**Restaurants**	S.F.							**21**
	0100	Architectural		126	193	248	59%	60%	63.50%
	0200	Plumbing		13.80	31.50	40	7.35%	7.75%	8.95%
	0300	Mechanical		14.85	17.55	37	6.50%	8.15%	11.15%
	0400	Electrical		14.80	24	48.50	7.10%	10.30%	11.60%
	0500	Total Project Costs		208	305	415			
22	0000	**Retail**	S.F.							**22**
	0100	Architectural		56	86	127	60%	62%	64.50%
	0200	Plumbing		7	9.25	12.10	5.05%	6.70%	9%
	0300	Mechanical		6.55	9.25	17.05	5.70%	6.20%	10.20%
	0400	Electrical		10.45	18.95	31.50	8.05%	11.25%	12.45%
	0500	Total Project Costs		85.50	114	186			
23	0000	**Schools**	S.F.							**23**
	0100	Architectural		97	125	164	52.50%	56%	61%
	0200	Plumbing		7.75	10.60	16.05	3.85%	4.82%	7.25%
	0300	Mechanical		18.90	26.50	38.50	9.50%	12.35%	15.10%
	0400	Electrical		17.85	25	32.50	9.45%	11.45%	13.30%
	0500	Total Project Costs		169	227	300			
24	0000	**University, College & Private School Classroom & Admin Buildings**	S.F.							**24**
	0100	Architectural		124	153	192	50.50%	55%	59.50%
	0200	Plumbing		7.05	10.95	15.45	2.74%	4.30%	6.35%
	0300	Mechanical		26.50	38.50	46	10.10%	12.15%	14.70%
	0400	Electrical		19.95	28	34	7.65%	9.50%	11.55%
	0500	Total Project Costs		205	284	375			
25	0000	**University, College & Private School Dormitories**	S.F.							**25**
	0100	Architectural		81	142	151	54.50%	65%	68.50%
	0200	Plumbing		10.70	15.10	22.50	6.45%	7.30%	9.15%
	0300	Mechanical		4.79	20.50	32.50	4.13%	9%	12.05%
	0400	Electrical		5.70	19.75	30	4.75%	7.35%	12.30%
	0500	Total Project Costs		119	227	268			
26	0000	**University, College & Private School Science, Eng. & Lab Buildings**	S.F.							**26**
	0100	Architectural		143	164	193	50.50%	56.50%	58%
	0200	Plumbing		9.60	15.15	26.50	3.29%	3.95%	8.40%
	0300	Mechanical		43.50	68.50	70	12.75%	19.40%	23.50%
	0400	Electrical		29	33.50	38.50	9%	11.55%	13.15%
	0500	Total Project Costs		290	315	365			
27	0000	**University, College & Private School Student Union Buildings**	S.F.							**27**
	0100	Architectural		110	289	289	54.50%	54.50%	59.50%
	0200	Plumbing		16.65	16.65	25	3.13%	4.27%	11.45%
	0300	Mechanical		31.50	51	51	9.60%	9.60%	14.55%
	0400	Electrical		27.50	48	48	9.05%	12.80%	13.15%
	0500	Total Project Costs		217	530	530			
28	0000	**Warehouses**	S.F.							**28**
	0100	Architectural		47	73.50	174	61.50%	67.50%	72%
	0200	Plumbing		2.45	5.25	10.10	2.82%	3.72%	5%
	0300	Mechanical		2.91	16.55	26	4.56%	8.20%	10.70%
	0400	Electrical		5.30	19.85	33	7.50%	10.10%	18.30%
	0500	Total Project Costs		70.50	125	243			

Square Foot Project Size Modifier

One factor that affects the S.F. cost of a particular building is the size. In general, for buildings built to the same specifications in the same locality, the larger building will have the lower S.F. cost. This is due mainly to the decreasing contribution of the exterior walls plus the economy of scale usually achievable in larger buildings. The Area Conversion Scale shown below will give a factor to convert costs for the typical size building to an adjusted cost for the particular project.

The Square Foot Base Size lists the median costs, most typical project size in our accumulated data, and the range in size of the projects.

The Size Factor for your project is determined by dividing your project area in S.F. by the typical project size for the particular Building Type. With this factor, enter the Area Conversion Scale at the appropriate Size Factor and determine the appropriate Cost Multiplier for your building size.

Example: Determine the cost per S.F. for a 152,600 S.F. Multi-family housing.

$$\frac{\text{Proposed building area} = 152,600 \text{ S.F.}}{\text{Typical size from below} = 76,300 \text{ S.F.}} = 2.00$$

Enter Area Conversion Scale at 2.0, intersect curve, read horizontally the appropriate cost multiplier of .94. Size adjusted cost becomes .94 x $208.00 = $195.52 based on national average costs.

Note: For Size Factors less than .50, the Cost Multiplier is 1.1
For Size Factors greater than 3.5, the Cost Multiplier is .90

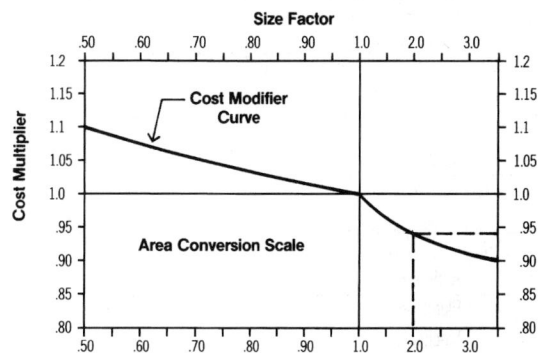

System	Median Cost (Total Project Costs)	Typical Size Gross S.F. (Median of Projects)	Typical Range (Low – High) (Projects)
Auto Sales with Repair	$186.00	24,900	4,700 – 29,300
Banking Institutions	299.00	9,300	3,300 – 38,100
Court House	284.00	47,600	24,700 – 70,500
Data Centers	281.00	14,400	14,400 – 14,400
Detention Centers	315.00	37,800	12,300 – 183,300
Fire Stations	238.00	13,100	6,300 – 49,600
Gymnasium	204.00	52,400	22,800 – 82,000
Hospitals	375.00	87,100	22,400 – 410,300
Industrial Buildings	104.00	22,100	5,100 – 200,600
Medical Clinics & Offices	217.00	20,900	2,300 – 327,000
Mixed Use	216.00	28,500	7,200 – 188,900
Multi-Family Housing	208.00	49,900	2,500 – 1,161,500
Nursing Home & Assisted Living	160.00	38,200	1,500 – 242,600
Office Buildings	200.00	20,500	1,100 – 930,000
Parking Garage	47.00	151,800	99,900 – 287,000
Parking Garage/Mixed Use	175.00	254,200	5,300 – 318,000
Police Stations	267.00	28,500	15,400 – 88,600
Police/Fire	171.00	44,300	8,600 – 50,300
Public Assembly Buildings	253.00	21,000	2,200 – 235,300
Recreational	281.00	28,800	1,000 – 223,800
Restaurants	305.00	6,000	4,000 – 42,000
Retail	114.00	28,700	5,200 – 84,300
Schools	227.00	70,600	1,300 – 410,800
University, College & Private School Classroom & Admin Buildings	284.00	48,300	9,400 – 196,200
University, College & Private School Dormitories	227.00	28,900	1,500 – 126,900
University, College & Private School Science, Eng. & Lab Buildings	315.00	39,800	5,300 – 117,600
University, College & Private School Student Union Buildings	530.00	48,700	42,100 – 50,000
Warehouses	125.00	10,400	600 – 303,800

A	Area Square Feet; Ampere	Brk., brk	Brick
AAFES	Army and Air Force Exchange Service	brkt	Bracket
		Brs.	Brass
ABS	Acrylonitrile Butadiene Stryrene; Asbestos Bonded Steel	Brz.	Bronze
		Bsn.	Basin
A.C., AC	Alternating Current; Air-Conditioning; Asbestos Cement; Plywood Grade A & C	Btr.	Better
		BTU	British Thermal Unit
		BTUH	BTU per Hour
		Bu.	Bushels
ACI	American Concrete Institute	BUR	Built-up Roofing
ACR	Air Conditioning Refrigeration	BX	Interlocked Armored Cable
ADA	Americans with Disabilities Act	°C	Degree Centigrade
AD	Plywood, Grade A & D	c	Conductivity, Copper Sweat
Addit.	Additional	C	Hundred; Centigrade
Adh.	Adhesive	C/C	Center to Center, Cedar on Cedar
Adj.	Adjustable	C-C	Center to Center
af	Audio-frequency	Cab	Cabinet
AFFF	Aqueous Film Forming Foam	Cair.	Air Tool Laborer
AFUE	Annual Fuel Utilization Efficiency	Cal.	Caliper
AGA	American Gas Association	Calc	Calculated
Agg.	Aggregate	Cap.	Capacity
A.H., Ah	Ampere Hours	Carp.	Carpenter
A hr.	Ampere-hour	C.B.	Circuit Breaker
A.H.U., AHU	Air Handling Unit	C.C.A.	Chromate Copper Arsenate
A.I.A.	American Institute of Architects	C.C.F.	Hundred Cubic Feet
AIC	Ampere Interrupting Capacity	cd	Candela
Allow.	Allowance	cd/sf	Candela per Square Foot
alt., alt	Alternate	CD	Grade of Plywood Face & Back
Alum.	Aluminum	CDX	Plywood, Grade C & D, exterior glue
a.m.	Ante Meridiem		
Amp.	Ampere	Cefi.	Cement Finisher
Anod.	Anodized	Cem.	Cement
ANSI	American National Standards Institute	CF	Hundred Feet
		C.F.	Cubic Feet
APA	American Plywood Association	CFM	Cubic Feet per Minute
Approx.	Approximate	CFRP	Carbon Fiber Reinforced Plastic
Apt.	Apartment	c.g.	Center of Gravity
Asb.	Asbestos	CHW	Chilled Water; Commercial Hot Water
A.S.B.C.	American Standard Building Code		
Asbe.	Asbestos Worker	C.I., CI	Cast Iron
ASCE	American Society of Civil Engineers	C.I.P., CIP	Cast in Place
A.S.H.R.A.E.	American Society of Heating, Refrig. & AC Engineers	Circ.	Circuit
		C.L.	Carload Lot
ASME	American Society of Mechanical Engineers	CL	Chain Link
		Clab.	Common Laborer
ASTM	American Society for Testing and Materials	Clam	Common Maintenance Laborer
		C.L.F.	Hundred Linear Feet
Attchmt.	Attachment	CLF	Current Limiting Fuse
Avg., Ave.	Average	CLP	Cross Linked Polyethylene
AWG	American Wire Gauge	cm	Centimeter
AWWA	American Water Works Assoc.	CMP	Corr. Metal Pipe
Bbl.	Barrel	CMU	Concrete Masonry Unit
B&B, BB	Grade B and Better; Balled & Burlapped	CN	Change Notice
		Col.	Column
B&S	Bell and Spigot	CO₂	Carbon Dioxide
B.&W.	Black and White	Comb.	Combination
b.c.c.	Body-centered Cubic	comm.	Commercial, Communication
B.C.Y.	Bank Cubic Yards	Compr.	Compressor
BE	Bevel End	Conc.	Concrete
B.F.	Board Feet	Cont., cont	Continuous; Continued, Container
Bg. cem.	Bag of Cement	Corkbd.	Cork Board
BHP	Boiler Horsepower; Brake Horsepower	Corr.	Corrugated
		Cos	Cosine
B.I.	Black Iron	Cot	Cotangent
bidir.	bidirectional	Cov.	Cover
Bit., Bitum.	Bituminous	C/P	Cedar on Paneling
Bit., Conc.	Bituminous Concrete	CPA	Control Point Adjustment
Bk.	Backed	Cplg.	Coupling
Bkrs.	Breakers	CPM	Critical Path Method
Bldg., bldg	Building	CPVC	Chlorinated Polyvinyl Chloride
Blk.	Block	C.Pr.	Hundred Pair
Bm.	Beam	CRC	Cold Rolled Channel
Boil.	Boilermaker	Creos.	Creosote
bpm	Blows per Minute	Crpt.	Carpet & Linoleum Layer
BR	Bedroom	CRT	Cathode-ray Tube
Brg., brng.	Bearing	CS	Carbon Steel, Constant Shear Bar Joist
Brhe.	Bricklayer Helper		
Bric.	Bricklayer		

Csc	Cosecant
C.S.F.	Hundred Square Feet
CSI	Construction Specifications Institute
CT	Current Transformer
CTS	Copper Tube Size
Cu	Copper, Cubic
Cu. Ft.	Cubic Foot
cw	Continuous Wave
C.W.	Cool White; Cold Water
Cwt.	100 Pounds
C.W.X.	Cool White Deluxe
C.Y.	Cubic Yard (27 cubic feet)
C.Y./Hr.	Cubic Yard per Hour
Cyl.	Cylinder
d	Penny (nail size)
D	Deep; Depth; Discharge
Dis., Disch.	Discharge
Db	Decibel
Dbl.	Double
DC	Direct Current
DDC	Direct Digital Control
Demob.	Demobilization
d.f.t.	Dry Film Thickness
d.f.u.	Drainage Fixture Units
D.H.	Double Hung
DHW	Domestic Hot Water
DI	Ductile Iron
Diag.	Diagonal
Diam., Dia	Diameter
Distrib.	Distribution
Div.	Division
Dk.	Deck
D.L.	Dead Load; Diesel
DLH	Deep Long Span Bar Joist
dlx	Deluxe
Do.	Ditto
DOP	Dioctyl Phthalate Penetration Test (Air Filters)
Dp., dp	Depth
D.P.S.T.	Double Pole, Single Throw
Dr.	Drive
DR	Dimension Ratio
Drink.	Drinking
D.S.	Double Strength
D.S.A.	Double Strength A Grade
D.S.B.	Double Strength B Grade
Dty.	Duty
DWV	Drain Waste Vent
DX	Deluxe White, Direct Expansion
dyn	Dyne
e	Eccentricity
E	Equipment Only; East; Emissivity
Ea.	Each
EB	Encased Burial
Econ.	Economy
E.C.Y	Embankment Cubic Yards
EDP	Electronic Data Processing
EIFS	Exterior Insulation Finish System
E.D.R.	Equiv. Direct Radiation
Eq.	Equation
EL	Elevation
Elec.	Electrician; Electrical
Elev.	Elevator; Elevating
EMT	Electrical Metallic Conduit; Thin Wall Conduit
Eng.	Engine, Engineered
EPDM	Ethylene Propylene Diene Monomer
EPS	Expanded Polystyrene
Eqhv.	Equip. Oper., Heavy
Eqlt.	Equip. Oper., Light
Eqmd.	Equip. Oper., Medium
Eqmm.	Equip. Oper., Master Mechanic
Eqol.	Equip. Oper., Oilers
Equip.	Equipment
ERW	Electric Resistance Welded

E.S.	Energy Saver	H	High Henry	Lath.	Lather
Est.	Estimated	HC	High Capacity	Lav.	Lavatory
esu	Electrostatic Units	H.D., HD	Heavy Duty; High Density	lb.; #	Pound
E.W.	Each Way	H.D.O.	High Density Overlaid	L.B., LB	Load Bearing; L Conduit Body
EWT	Entering Water Temperature	HDPE	High Density Polyethylene Plastic	L. & E.	Labor & Equipment
Excav.	Excavation	Hdr.	Header	lb./hr.	Pounds per Hour
excl	Excluding	Hdwe.	Hardware	lb./L.F.	Pounds per Linear Foot
Exp., exp	Expansion, Exposure	H.I.D., HID	High Intensity Discharge	lbf/sq.in.	Pound-force per Square Inch
Ext., ext	Exterior; Extension	Help.	Helper Average	L.C.L.	Less than Carload Lot
Extru.	Extrusion	HEPA	High Efficiency Particulate Air	L.C.Y.	Loose Cubic Yard
f.	Fiber Stress		Filter	Ld.	Load
F	Fahrenheit; Female; Fill	Hg	Mercury	LE	Lead Equivalent
Fab., fab	Fabricated; Fabric	HIC	High Interrupting Capacity	LED	Light Emitting Diode
FBGS	Fiberglass	HM	Hollow Metal	L.F.	Linear Foot
F.C.	Footcandles	HMWPE	High Molecular Weight	L.F. Hdr	Linear Feet of Header
f.c.c.	Face-centered Cubic		Polyethylene	L.F. Nose	Linear Foot of Stair Nosing
f'c.	Compressive Stress in Concrete;	HO	High Output	L.F. Rsr	Linear Foot of Stair Riser
	Extreme Compressive Stress	Horiz.	Horizontal	Lg.	Long; Length; Large
F.E.	Front End	H.P., HP	Horsepower; High Pressure	L & H	Light and Heat
FEP	Fluorinated Ethylene Propylene	H.P.F.	High Power Factor	LH	Long Span Bar Joist
	(Teflon)	Hr.	Hour	L.H.	Labor Hours
F.G.	Flat Grain	Hrs./Day	Hours per Day	L.L., LL	Live Load
F.H.A.	Federal Housing Administration	HSC	High Short Circuit	L.L.D.	Lamp Lumen Depreciation
Fig.	Figure	Ht.	Height	lm	Lumen
Fin.	Finished	Htg.	Heating	lm/sf	Lumen per Square Foot
FIPS	Female Iron Pipe Size	Htrs.	Heaters	lm/W	Lumen per Watt
Fixt.	Fixture	HVAC	Heating, Ventilation & Air-	LOA	Length Over All
FJP	Finger jointed and primed		Conditioning	log	Logarithm
Fl. Oz.	Fluid Ounces	Hvy.	Heavy	L-O-L	Lateralolet
Flr.	Floor	HW	Hot Water	long.	Longitude
Flrs.	Floors	Hyd.; Hydr.	Hydraulic	L.P., LP	Liquefied Petroleum; Low Pressure
FM	Frequency Modulation;	Hz	Hertz (cycles)	L.P.F.	Low Power Factor
	Factory Mutual	I.	Moment of Inertia	LR	Long Radius
Fmg.	Framing	IBC	International Building Code	L.S.	Lump Sum
FM/UL	Factory Mutual/Underwriters Labs	I.C.	Interrupting Capacity	Lt.	Light
Fdn.	Foundation	ID	Inside Diameter	Lt. Ga.	Light Gauge
FNPT	Female National Pipe Thread	I.D.	Inside Dimension; Identification	L.T.L.	Less than Truckload Lot
Fori.	Foreman, Inside	I.F.	Inside Frosted	Lt. Wt.	Lightweight
Foro.	Foreman, Outside	I.M.C.	Intermediate Metal Conduit	L.V.	Low Voltage
Fount.	Fountain	In.	Inch	M	Thousand; Material; Male;
fpm	Feet per Minute	Incan.	Incandescent		Light Wall Copper Tubing
FPT	Female Pipe Thread	Incl.	Included; Including	M²CA	Meters Squared Contact Area
Fr	Frame	Int.	Interior	m/hr.; M.H.	Man-hour
F.R.	Fire Rating	Inst.	Installation	mA	Milliampere
FRK	Foil Reinforced Kraft	Insul., insul	Insulation/Insulated	Mach.	Machine
FSK	Foil/Scrim/Kraft	I.P.	Iron Pipe	Mag. Str.	Magnetic Starter
FRP	Fiberglass Reinforced Plastic	I.P.S., IPS	Iron Pipe Size	Maint.	Maintenance
FS	Forged Steel	IPT	Iron Pipe Threaded	Marb.	Marble Setter
FSC	Cast Body; Cast Switch Box	I.W.	Indirect Waste	Mat; Mat'l.	Material
Ft., ft	Foot; Feet	J	Joule	Max.	Maximum
Ftng.	Fitting	J.I.C.	Joint Industrial Council	MBF	Thousand Board Feet
Ftg.	Footing	K	Thousand; Thousand Pounds;	MBH	Thousand BTU's per hr.
Ft lb.	Foot Pound		Heavy Wall Copper Tubing, Kelvin	MC	Metal Clad Cable
Furn.	Furniture	K.A.H.	Thousand Amp. Hours	MCC	Motor Control Center
FVNR	Full Voltage Non-Reversing	kcmil	Thousand Circular Mils	M.C.F.	Thousand Cubic Feet
FVR	Full Voltage Reversing	KD	Knock Down	MCFM	Thousand Cubic Feet per Minute
FXM	Female by Male	K.D.A.T.	Kiln Dried After Treatment	M.C.M.	Thousand Circular Mils
Fy.	Minimum Yield Stress of Steel	kg	Kilogram	MCP	Motor Circuit Protector
g	Gram	kG	Kilogauss	MD	Medium Duty
G	Gauss	kgf	Kilogram Force	MDF	Medium-density fibreboard
Ga.	Gauge	kHz	Kilohertz	M.D.O.	Medium Density Overlaid
Gal., gal.	Gallon	Kip	1000 Pounds	Med.	Medium
Galv., galv	Galvanized	KJ	Kilojoule	MF	Thousand Feet
GC/MS	Gas Chromatograph/Mass	K.L.	Effective Length Factor	M.F.B.M.	Thousand Feet Board Measure
	Spectrometer	K.L.F.	Kips per Linear Foot	Mfg.	Manufacturing
Gen.	General	Km	Kilometer	Mfrs.	Manufacturers
GFI	Ground Fault Interrupter	KO	Knock Out	mg	Milligram
GFRC	Glass Fiber Reinforced Concrete	K.S.F.	Kips per Square Foot	MGD	Million Gallons per Day
Glaz.	Glazier	K.S.I.	Kips per Square Inch	MGPH	Million Gallons per Hour
GPD	Gallons per Day	kV	Kilovolt	MH, M.H.	Manhole; Metal Halide; Man-Hour
gpf	Gallon per Flush	kVA	Kilovolt Ampere	MHz	Megahertz
GPH	Gallons per Hour	kVAR	Kilovar (Reactance)	Mi.	Mile
gpm, GPM	Gallons per Minute	KW	Kilowatt	MI	Malleable Iron; Mineral Insulated
GR	Grade	KWh	Kilowatt-hour	MIPS	Male Iron Pipe Size
Gran.	Granular	L	Labor Only; Length; Long;	mj	Mechanical Joint
Grnd.	Ground		Medium Wall Copper Tubing	m	Meter
GVW	Gross Vehicle Weight	Lab.	Labor	mm	Millimeter
GWB	Gypsum Wall Board	lat	Latitude	Mill.	Millwright
				Min., min.	Minimum, Minute

For customer support on your Mechanical Costs with RSMeans Data, call 800.448.8182.

751

Abbreviation	Meaning
Misc.	Miscellaneous
ml	Milliliter, Mainline
M.L.F.	Thousand Linear Feet
Mo.	Month
Mobil.	Mobilization
Mog.	Mogul Base
MPH	Miles per Hour
MPT	Male Pipe Thread
MRGWB	Moisture Resistant Gypsum Wallboard
MRT	Mile Round Trip
ms	Millisecond
M.S.F.	Thousand Square Feet
Mstz.	Mosaic & Terrazzo Worker
M.S.Y.	Thousand Square Yards
Mtd., mtd., mtd	Mounted
Mthe.	Mosaic & Terrazzo Helper
Mtng.	Mounting
Mult.	Multi; Multiply
MUTCD	Manual on Uniform Traffic Control Devices
M.V.A.	Million Volt Amperes
M.V.A.R.	Million Volt Amperes Reactance
MV	Megavolt
MW	Megawatt
MXM	Male by Male
MYD	Thousand Yards
N	Natural; North
nA	Nanoampere
NA	Not Available; Not Applicable
N.B.C.	National Building Code
NC	Normally Closed
NEMA	National Electrical Manufacturers Assoc.
NEHB	Bolted Circuit Breaker to 600V.
NFPA	National Fire Protection Association
NLB	Non-Load-Bearing
NM	Non-Metallic Cable
nm	Nanometer
No.	Number
NO	Normally Open
N.O.C.	Not Otherwise Classified
Nose.	Nosing
NPT	National Pipe Thread
NQOD	Combination Plug-on/Bolt on Circuit Breaker to 240V.
N.R.C., NRC	Noise Reduction Coefficient/ Nuclear Regulator Commission
N.R.S.	Non Rising Stem
ns	Nanosecond
NTP	Notice to Proceed
nW	Nanowatt
OB	Opposing Blade
OC	On Center
OD	Outside Diameter
O.D.	Outside Dimension
ODS	Overhead Distribution System
O.G.	Ogee
O.H.	Overhead
O&P	Overhead and Profit
Oper.	Operator
Opng.	Opening
Orna.	Ornamental
OSB	Oriented Strand Board
OS&Y	Outside Screw and Yoke
OSHA	Occupational Safety and Health Act
Ovhd.	Overhead
OWG	Oil, Water or Gas
Oz.	Ounce
P.	Pole; Applied Load; Projection
p.	Page
Pape.	Paperhanger
P.A.P.R.	Powered Air Purifying Respirator
PAR	Parabolic Reflector
P.B., PB	Push Button
Pc., Pcs.	Piece, Pieces
P.C.	Portland Cement; Power Connector
P.C.F.	Pounds per Cubic Foot
PCM	Phase Contrast Microscopy
PDCA	Painting and Decorating Contractors of America
P.E., PE	Professional Engineer; Porcelain Enamel; Polyethylene; Plain End
P.E.C.I.	Porcelain Enamel on Cast Iron
Perf.	Perforated
PEX	Cross Linked Polyethylene
Ph.	Phase
P.I.	Pressure Injected
Pile.	Pile Driver
Pkg.	Package
Pl.	Plate
Plah.	Plasterer Helper
Plas.	Plasterer
plf	Pounds Per Linear Foot
Pluh.	Plumber Helper
Plum.	Plumber
Ply.	Plywood
p.m.	Post Meridiem
Pntd.	Painted
Pord.	Painter, Ordinary
pp	Pages
PP, PPL	Polypropylene
P.P.M.	Parts per Million
Pr.	Pair
P.E.S.B.	Pre-engineered Steel Building
Prefab.	Prefabricated
Prefin.	Prefinished
Prop.	Propelled
PSF, psf	Pounds per Square Foot
PSI, psi	Pounds per Square Inch
PSIG	Pounds per Square Inch Gauge
PSP	Plastic Sewer Pipe
Pspr.	Painter, Spray
Psst.	Painter, Structural Steel
P.T.	Potential Transformer
P. & T.	Pressure & Temperature
Ptd.	Painted
Ptns.	Partitions
Pu	Ultimate Load
PVC	Polyvinyl Chloride
Pvmt.	Pavement
PRV	Pressure Relief Valve
Pwr.	Power
Q	Quantity Heat Flow
Qt.	Quart
Quan., Qty.	Quantity
Q.C.	Quick Coupling
r	Radius of Gyration
R	Resistance
R.C.P.	Reinforced Concrete Pipe
Rect.	Rectangle
recpt.	Receptacle
Reg.	Regular
Reinf.	Reinforced
Req'd.	Required
Res.	Resistant
Resi.	Residential
RF	Radio Frequency
RFID	Radio-frequency Identification
Rgh.	Rough
RGS	Rigid Galvanized Steel
RHW	Rubber, Heat & Water Resistant; Residential Hot Water
rms	Root Mean Square
Rnd.	Round
Rodm.	Rodman
Rofc.	Roofer, Composition
Rofp.	Roofer, Precast
Rohe.	Roofer Helpers (Composition)
Rots.	Roofer, Tile & Slate
R.O.W.	Right of Way
RPM	Revolutions per Minute
R.S.	Rapid Start
Rsr	Riser
RT	Round Trip
S.	Suction; Single Entrance; South
SBS	Styrene Butadiere Styrene
SC	Screw Cover
SCFM	Standard Cubic Feet per Minute
Scaf.	Scaffold
Sch., Sched.	Schedule
S.C.R.	Modular Brick
S.D.	Sound Deadening
SDR	Standard Dimension Ratio
S.E.	Surfaced Edge
Sel.	Select
SER, SEU	Service Entrance Cable
S.F.	Square Foot
S.F.C.A.	Square Foot Contact Area
S.F. Flr.	Square Foot of Floor
S.F.G.	Square Foot of Ground
S.F. Hor.	Square Foot Horizontal
SFR	Square Feet of Radiation
S.F. Shlf.	Square Foot of Shelf
S4S	Surface 4 Sides
Shee.	Sheet Metal Worker
Sin.	Sine
Skwk.	Skilled Worker
SL	Saran Lined
S.L.	Slimline
Sldr.	Solder
SLH	Super Long Span Bar Joist
S.N.	Solid Neutral
SO	Stranded with oil resistant inside insulation
S-O-L	Socketolet
sp	Standpipe
S.P.	Static Pressure; Single Pole; Self-Propelled
Spri.	Sprinkler Installer
spwg	Static Pressure Water Gauge
S.P.D.T.	Single Pole, Double Throw
SPF	Spruce Pine Fir; Sprayed Polyurethane Foam
S.P.S.T.	Single Pole, Single Throw
SPT	Standard Pipe Thread
Sq.	Square; 100 Square Feet
Sq. Hd.	Square Head
Sq. In.	Square Inch
S.S.	Single Strength; Stainless Steel
S.S.B.	Single Strength B Grade
sst, ss	Stainless Steel
Sswk.	Structural Steel Worker
Sswl.	Structural Steel Welder
St.; Stl.	Steel
STC	Sound Transmission Coefficient
Std.	Standard
Stg.	Staging
STK	Select Tight Knot
STP	Standard Temperature & Pressure
Stpi.	Steamfitter, Pipefitter
Str.	Strength; Starter; Straight
Strd.	Stranded
Struct.	Structural
Sty.	Story
Subj.	Subject
Subs.	Subcontractors
Surf.	Surface
Sw.	Switch
Swbd.	Switchboard
S.Y.	Square Yard
Syn.	Synthetic
S.Y.P.	Southern Yellow Pine
Sys.	System
t.	Thickness
T	Temperature; Ton
Tan	Tangent
T.C.	Terra Cotta
T & C	Threaded and Coupled
T.D.	Temperature Difference
TDD	Telecommunications Device for the Deaf
T.E.M.	Transmission Electron Microscopy
temp	Temperature, Tempered, Temporary
TFFN	Nylon Jacketed Wire

For customer support on your Mechanical Costs with RSMeans Data, call 800.448.8182.

Division Notes

	CREW	DAILY OUTPUT	LABOR-HOURS	UNIT	BARE COSTS				TOTAL INCL O&P
					MAT.	LABOR	EQUIP.	TOTAL	

Division Notes

	CREW	DAILY OUTPUT	LABOR-HOURS	UNIT	BARE COSTS				TOTAL INCL O&P
					MAT.	LABOR	EQUIP.	TOTAL	

Division Notes

	CREW	DAILY OUTPUT	LABOR-HOURS	UNIT	BARE COSTS				TOTAL INCL O&P
					MAT.	LABOR	EQUIP.	TOTAL	

Division Notes

	CREW	DAILY OUTPUT	LABOR-HOURS	UNIT	BARE COSTS				TOTAL INCL O&P
					MAT.	LABOR	EQUIP.	TOTAL	

Division Notes

		CREW	DAILY OUTPUT	LABOR-HOURS	UNIT	BARE COSTS				TOTAL INCL O&P
						MAT.	LABOR	EQUIP.	TOTAL	

Division Notes

	CREW	DAILY OUTPUT	LABOR-HOURS	UNIT	BARE COSTS				TOTAL INCL O&P
					MAT.	LABOR	EQUIP.	TOTAL	

Division Notes

	CREW	DAILY OUTPUT	LABOR-HOURS	UNIT	BARE COSTS				TOTAL INCL O&P
					MAT.	LABOR	EQUIP.	TOTAL	

Other Data & Services
A tradition of excellence in construction cost information and services since 1942

Table of Contents
Cost Data Selection Guide
RSMeans Data Online
Training

For more information visit our website at RSMeans.com

Unit prices according to the latest MasterFormat®

Cost Data Selection Guide

The following table provides definitive information on the content of each cost data publication. The number of lines of data provided in each unit price or assemblies division, as well as the number of crews, is listed for each data set. The presence of other elements such as reference tables, square foot models, equipment rental costs, historical cost indexes, and city cost indexes, is also indicated. You can use the table to help select the RSMeans data set that has the quantity and type of information you most need in your work.

Unit Cost Divisions	Building Construction	Mechanical	Electrical	Commercial Renovation	Square Foot	Site Work Landsc.	Green Building	Interior	Concrete Masonry	Open Shop	Heavy Construction	Light Commercial	Facilities Construction	Plumbing	Residential
1	609	444	465	564	0	533	198	365	495	608	550	310	1078	450	217
2	754	278	87	710	0	970	181	397	219	753	737	479	1197	285	274
3	1745	341	232	1265	0	1537	1043	355	2274	1745	1930	538	2028	317	445
4	960	22	0	920	0	724	180	613	1158	928	614	532	1175	0	446
5	1890	158	155	1094	0	853	1788	1107	729	1890	1026	980	1907	204	746
6	2462	18	18	2121	0	110	589	1544	281	2458	123	2151	2135	22	2671
7	1593	215	128	1633	0	580	761	532	523	1590	26	1326	1693	227	1046
8	2140	80	3	2733	0	255	1138	1813	105	2142	0	2328	2966	0	1552
9	2125	86	45	1943	0	313	464	2216	424	2062	15	1779	2379	54	1544
10	1088	17	10	684	0	232	32	898	136	1088	34	588	1179	237	224
11	1096	199	166	540	0	135	56	924	29	1063	0	230	1116	162	108
12	539	0	2	297	0	219	147	1546	14	506	0	272	1565	23	216
13	740	149	157	252	0	365	124	250	77	716	266	109	756	115	103
14	273	36	0	223	0	0	0	257	0	273	0	12	293	16	6
21	127	0	41	37	0	0	0	293	0	127	0	121	665	685	259
22	1165	7543	160	1226	0	2010	1061	849	20	1154	2109	875	7505	9400	719
23	1170	6906	546	940	0	157	865	775	38	1153	98	887	5143	1919	486
25	0	0	14	14	0	0	0	0	0	0	0	0	0	0	0
26	1513	491	10465	1293	0	860	646	1159	55	1439	649	1361	10246	399	636
27	95	0	448	102	0	0	0	71	0	95	39	67	389	0	56
28	143	79	223	124	0	0	28	97	0	127	0	70	209	57	41
31	1511	733	610	807	0	3263	286	7	1216	1456	3280	607	1568	660	616
32	896	49	8	937	0	4523	408	417	361	867	1941	486	1800	140	533
33	1255	1088	565	260	0	3078	33	0	241	532	3213	135	1726	2101	161
34	107	0	47	4	0	190	0	0	31	62	221	0	136	0	0
35	18	0	0	0	0	327	0	0	0	18	442	0	84	0	0
41	63	0	0	34	0	8	0	22	0	62	31	0	69	14	0
44	75	79	0	0	0	0	0	0	0	0	0	0	75	75	0
46	23	16	0	0	0	274	261	0	0	23	264	0	33	33	0
48	8	0	36	2	0	0	21	0	0	8	15	8	21	0	8
Totals	26183	19027	14631	20759	0	21516	10310	16507	8426	24945	17623	16251	51136	17595	13113

Assem Div	Building Construction	Mechanical	Electrical	Commercial Renovation	Square Foot	Site Work Landscape	Assemblies	Green Building	Interior	Concrete Masonry	Heavy Construction	Light Commercial	Facilities Construction	Plumbing	Asm Div	Residential
A		15	0	188	164	577	598	0	0	536	571	154	24	0	1	378
B		0	0	848	2554	0	5661	56	329	1976	368	2094	174	0	2	211
C		0	0	647	954	0	1334	0	1641	146	0	844	251	0	3	591
D		1057	941	712	1858	72	2538	330	824	0	0	1345	1104	1088	4	851
E		0	0	85	261	0	301	0	5	0	0	258	5	0	5	391
F		0	0	0	114	0	143	0	0	0	0	114	0	0	6	357
G		527	447	318	312	3378	792	0	0	535	1349	205	293	677	7	307
															8	760
															9	80
															10	0
															11	0
															12	0
Totals		1599	1388	2798	6217	4027	11367	386	2799	3193	2288	5014	1851	1765		3926

Reference Section	Building Construction Costs	Mechanical	Electrical	Commercial Renovation	Square Foot	Site Work Landscape	Assem.	Green Building	Interior	Concrete Masonry	Open Shop	Heavy Construction	Light Commercial	Facilities Construction	Plumbing	Resi.
Reference Tables	yes	yes	yes	yes	no	yes	yes	yes	yes	yes	yes	yes	yes	yes	yes	yes
Models					111			25					50			28
Crews	582	582	582	561		582		582	582	582	560	582	560	561	582	560
Equipment Rental Costs	yes	yes	yes	yes		yes		yes	yes	yes	yes	yes	yes	yes	yes	yes
Historical Cost Indexes	yes	yes	yes	yes	yes	yes	yes	yes	yes	yes	yes	yes	yes	yes	yes	no
City Cost Indexes	yes	yes	yes	yes	yes	yes	yes	yes	yes	yes	yes	yes	yes	yes	yes	yes

2020 Seminar Schedule   ☎ 877.620.6245

Note: call for exact dates, locations, and details as some cities are subject to change.

Location	Dates	Location	Dates
Seattle, WA	January and August	San Francisco, CA	June
Dallas/Ft. Worth, TX	January	Bethesda, MD	June
Austin, TX	February	Dallas, TX	September
Jacksonville, FL	February	Raleigh, NC	October
Anchorage, AK	March and September	Baltimore, MD	November
Las Vegas, NV	March	Orlando, FL	November
Washington, D.C.	April and September	San Diego, CA	December
Charleston, SC	April	San Antonio, TX	December
Toronto	May		
Denver, CO	May		

Gordian also offers a suite of online RSMeans data self-paced offerings.
Check our website at RSMeans.com/products/training.aspx for more information.

Facilities Construction Estimating

In this two-day course, professionals working in facilities management can get help with their daily challenges to establish budgets for all phases of a project.

Some of what you'll learn:
• Determining the full scope of a project
• Understanding of Means data and what is included in prices
• Identifying appropriate factors to be included in your estimate
• Creative solutions to estimating issues
• Organizing estimates for presentation and discussion
• Special estimating techniques for repair/remodel and maintenance projects
• Appropriate use of contingency, city cost indexes, and reference notes
• Techniques to get to the correct estimate quickly

Who should attend: facility managers, engineers, contractors, facility tradespeople, planners, and project managers.

Mechanical & Electrical Estimating

This two-day course teaches attendees how to prepare more accurate and complete mechanical/electrical estimates, avoid the pitfalls of omission and double-counting, and understand the composition and rationale within the RSMeans mechanical/electrical database.

Some of what you'll learn:
• The unique way mechanical and electrical systems are interrelated
• M&E estimates—conceptual, planning, budgeting, and bidding stages
• Order of magnitude, square foot, assemblies, and unit price estimating
• Comparative cost analysis of equipment and design alternatives

Who should attend: architects, engineers, facilities managers, mechanical and electrical contractors, and others who need a highly reliable method for developing, understanding, and evaluating mechanical and electrical contracts.

Construction Cost Estimating: Concepts and Practice

This one or two day introductory course to improve estimating skills and effectiveness starts with the details of interpreting bid documents and ends with the summary of the estimate and bid submission.

Some of what you'll learn:
• Using the plans and specifications to create estimates
• The takeoff process—deriving all tasks with correct quantities
• Developing pricing using various sources; how subcontractor pricing fits in
• Summarizing the estimate to arrive at the final number
• Formulas for area and cubic measure, adding waste and adjusting productivity to specific projects
• Evaluating subcontractors' proposals and prices
• Adding insurance and bonds
• Understanding how labor costs are calculated
• Submitting bids and proposals

Who should attend: project managers, architects, engineers, owners' representatives, contractors, and anyone who's responsible for budgeting or estimating construction projects.

RSMeans data Training

Assessing Scope of Work for Facilities Construction Estimating

This two-day practical training program addresses the vital importance of understanding the scope of projects in order to produce accurate cost estimates for facilities repair and remodeling.

Some of what you'll learn:
- Discussions of site visits, plans/specs, record drawings of facilities, and site-specific lists
- Review of CSI divisions, including means, methods, materials, and the challenges of scoping each topic
- Exercises in scope identification and scope writing for accurate estimating of projects
- Hands-on exercises that require scope, take-off, and pricing

Who should attend: corporate and government estimators, planners, facility managers, and others who need to produce accurate project estimates.

Practical Project Management for Construction Professionals

In this two-day course, acquire the essential knowledge and develop the skills to effectively and efficiently execute the day-to-day responsibilities of the construction project manager.

Some of what you'll learn:
- General conditions of the construction contract
- Contract modifications: change orders and construction change directives
- Negotiations with subcontractors and vendors
- Effective writing: notification and communications
- Dispute resolution: claims and liens

Who should attend: architects, engineers, owners' representatives, and project managers.

Maintenance & Repair Estimating for Facilities

This two-day course teaches attendees how to plan, budget, and estimate the cost of ongoing and preventive maintenance and repair for existing buildings and grounds.

Some of what you'll learn:
- The most financially favorable maintenance, repair, and replacement scheduling and estimating
- Auditing and value engineering facilities
- Preventive planning and facilities upgrading
- Determining both in-house and contract-out service costs
- Annual, asset-protecting M&R plan

Who should attend: facility managers, maintenance supervisors, buildings and grounds superintendents, plant managers, planners, estimators, and others involved in facilities planning and budgeting.

Life Cycle Cost Estimating for Facilities Asset Managers

Life Cycle Cost Estimating will take the attendee through choosing the correct RSMeans database to use and then correctly applying RSMeans data to their specific life cycle application. Conceptual estimating through RSMeans' new building models, conceptual estimating of major existing building projects through RSMeans' renovation models, pricing specific renovation elements, estimating repair, replacement and preventive maintenance costs today and forward up to 30 years will be covered.

Some of what you'll learn:
- Cost implications of managing assets
- Planning projects and initial & life cycle costs
- How to use RSMeans data online

Who should attend: facilities owners and managers and anyone involved in the financial side of the decision making process in the planning, design, procurement, and operation of facilities real assets.

Please bring a laptop with ability to access the internet.

Building Systems and the Construction Process

This one-day course was written to assist novices and those outside the industry in obtaining a solid understanding of the construction process - from both a building systems and construction administration approach.

Some of what you'll learn:
- Various systems used and how components come together to create a building
- Start with foundation and end with the physical systems of the structure such as HVAC and Electrical
- Focus on the process from start of design through project closeout

This training session requires you to bring a laptop computer to class.

Who should attend: building professionals or novices to help make the crossover to the construction industry; suited for anyone responsible for providing high level oversight on construction projects.

RSMeans data Training

785

Training for our Online Estimating Solution

Construction estimating is vital to the decision-making process at each state of every project. Our online solution works the way you do. It's systematic, flexible and intuitive. In this one-day class you will see how you can estimate any phase of any project faster and better.

Some of what you'll learn:
- Customizing our online estimating solution
- Making the most of RSMeans "Circle Reference" numbers
- How to integrate your cost data
- Generating reports, exporting estimates to MS Excel, sharing, collaborating and more

Also offered as a self-paced or on-site training program!

Training for our CD Estimating Solution

This one-day course helps users become more familiar with the functionality of the CD. Each menu, icon, screen, and function found in the program is explained in depth. Time is devoted to hands-on estimating exercises.

Some of what you'll learn:
- Searching the database using all navigation methods
- Exporting RSMeans data to your preferred spreadsheet format
- Viewing crews, assembly components, and much more
- Automatically regionalizing the database

This training session requires you to bring a laptop computer to class.

When you register for this course you will receive an outline for your laptop requirements.

Also offered as a self-paced or on-site training program!

Site Work Estimating with RSMeans data

This one-day program focuses directly on site work costs. Accurately scoping, quantifying, and pricing site preparation, underground utility work, and improvements to exterior site elements are often the most difficult estimating tasks on any project. Some of what you'll learn:
- Evaluation of site work and understanding site scope including: site clearing, grading, excavation, disposal and trucking of materials, backfill and compaction, underground utilities, paving, sidewalks, and seeding & planting.
- Unit price site work estimates—Correct use of RSMeans site work cost data to develop a cost estimate.
- Using and modifying assemblies—Save valuable time when estimating site work activities using custom assemblies.

Who should attend: Engineers, contractors, estimators, project managers, owner's representatives, and others who are concerned with the proper preparation and/or evaluation of site work estimates.

Please bring a laptop with ability to access the internet.

Facilities Estimating Using the CD

This two-day class combines hands-on skill-building with best estimating practices and real-life problems. You will learn key concepts, tips, pointers, and guidelines to save time and avoid cost oversights and errors.

Some of what you'll learn:
- Estimating process concepts
- Customizing and adapting RSMeans cost data
- Establishing scope of work to account for all known variables
- Budget estimating: when, why, and how
- Site visits: what to look for and what you can't afford to overlook
- How to estimate repair and remodeling variables

This training session requires you to bring a laptop computer to class.

Who should attend: facility managers, architects, engineers, contractors, facility tradespeople, planners, project managers, and anyone involved with JOC, SABRE, or IDIQ.

Registration Information

Register early to save up to $100!!!

Register 45+ days before the date of a class and save $50 off each class. This savings cannot be combined with any other promotion or discounting of the regular price of classes!

How to register

By Phone
Register by phone at 877.620.6245

Online
Register online at
RSMeans.com/products/seminars.aspx

Note: Purchase Orders or Credits Cards are required to register.

Two-day seminar registration fee - $1,200*.

One-Day Construction Cost Estimating or Building Systems and the Construction Process - $765*.

Government pricing

All federal government employees save off the regular seminar price. Other promotional discounts cannot be combined with the government discount. Call 781.422.5115 for government pricing.

CANCELLATION POLICY:

If you are unable to attend a seminar, substitutions may be made at any time before the session starts by notifying the seminar registrar at 781.422.5115 or your sales representative.
If you cancel twenty-one (21) days or more prior to the seminar, there will be no penalty and your registration fees will be refunded. These cancellations must be received by the seminar registrar or your sales representative and will be confirmed to be eligible for cancellation.
If you cancel fewer than twenty-one (21) days prior to the seminar, you will forfeit the registration fee.
In the unfortunate event of an RSMeans cancellation, RSMeans will work with you to reschedule your attendance in the same seminar at a later date or will fully refund your registration fee. RSMeans cannot be responsible for any non-refundable travel expenses incurred by you or another as a result of your registration, attendance at, or cancellation of an RSMeans seminar.
Any on-demand training modules are not eligible for cancellation, substitution, transfer, return or refund.

AACE approved courses

Many seminars described and offered here have been approved for 14 hours (1.4 recertification credits) of credit by the AACE International Certification Board toward meeting the continuing education requirements for recertification as a Certified Cost Engineer/Certified Cost Consultant.

AIA Continuing Education

We are registered with the AIA Continuing Education System (AIA/CES) and are committed to developing quality learning activities in accordance with the CES criteria. Many seminars meet the AIA/CES criteria for Quality Level 2. AIA members may receive 14 learning units (LUs) for each two-day RSMeans course.

Daily course schedule

The first day of each seminar session begins at 8:30 a.m. and ends at 4:30 p.m. The second day begins at 8:00 a.m. and ends at 4:00 p.m. Participants are urged to bring a hand-held calculator since many actual problems will be worked out in each session.

Continental breakfast

Your registration includes the cost of a continental breakfast and a morning and afternoon refreshment break. These informal segments allow you to discuss topics of mutual interest with other seminar attendees. (You are free to make your own lunch and dinner arrangements.)

Hotel/transportation arrangements

We arrange to hold a block of rooms at most host hotels. To take advantage of special group rates when making your reservation, be sure to mention that you are attending the RSMeans Institute data seminar. You are, of course, free to stay at the lodging place of your choice. (Hotel reservations and transportation arrangements should be made directly by seminar attendees.)

Important

Class sizes are limited, so please register as soon as possible.

*Note: Pricing subject to change.

RSMeans data Training